D1555659

Encyclopedia of Neuroscience

MARC D. BINDER, NOBUTAKA HIROKAWA AND UWE WINDHORST (Eds.)

Encyclopedia of Neuroscience

Volume 4
N–R

With 1625 Figures* and 90 Tables

*For color figures please see our Electronic Reference on www.springerlink.com

Editors:
Marc D. Binder
Department of Physiology & Biophysics
University of Washington School of Medicine
Seattle, Washington, USA
mdbinder@u.washington.edu

Nobutaka Hirokawa
Department of Cell Biology and Anatomy
Graduate School of Medicine
University of Tokyo
Hongo, Bunkyo-ku, Tokyo, Japan
hirokawa@m.u-tokyo.ac.jp

Uwe Windhorst
Göttingen, Germany
siggi.uwe@t-online.de

A C.I.P. Catalog record for this book is available from the Library of Congress

ISBN: 978-3-540-23735-8
This publication is available also as:
Electronic publication under ISBN 978-3-540-29678-2 and
Print and electronic bundle under ISBN 978-3-540-35857-2
Library of Congress Control Number: 2008930846

Springer is part of Springer Science+Business Media

springer.com

Printed on acid-free paper SPIN: 10 84 69 79 2109 — 5 4 3 2 1 0

Preface

Neuroscience is a rapidly expanding endeavor devoted to unraveling the structure and function of the nervous system. It relies on, and keeps close relations to, a number of other disciplines, such as mathematics, physics, chemistry, engineering, computer science, genetics, molecular biology, biochemistry, medicine and philosophy. Indeed, many of its recent successes result from the application of ideas, concepts and methods borrowed from these fields. Thus, neuroscience has become the archetype for interdisciplinary undertakings. This convergence of influences accounts for part of its enormous attractiveness and fascination to students, researchers and lay persons from various walks of life or science. Many of neuroscience's most creative and productive investigators have been lured into the field not only by the excitement inherent in the possibility of uncovering the secrets of the human mind, but by the appeal of venturing into a vast unknown land, requiring the development of new tools for its effective cultivation. Far from simply satisfying our intellectual curiosity, however, neuroscience has become ever more important as a theoretical ground for practical applications in medicine, in particular neurology, and other disciplines.

The explosion of neuroscience has made it virtually impossible for individuals to follow all the ramifications and fast developments in the many corners and branches of this science. This *Encyclopedia* has therefore been designed for a wide variety of readers, from members of the lay public to students, practitioners and researchers in biology, medicine, psychology, sociology, philosophy and their associated auxiliary fields. Moreover, it should also prove useful to advanced researchers of biology and neuroscience who wish to stay abreast of current developments outside their immediate areas of expertise.

In the interest of rapid and convenient access to information, this *Encyclopedia* has adopted a new format. The entire complex of neuroscience has been divided into 38 subject fields organized and surveyed by associated Field Editors. Entries, which are in alphabetical order for rapid localization, are of three type: (1) simple and relatively brief definitions and explanations (glossary entries), (2) structured "essays" of a few pages to provide coherent treatments of particularly important topics, and (3) synopses written by the Field Editors as larger overviews of their fields with links to the essays in their field. Extensive cross-references to definitions and essays serve to lead the reader to additional sources of information.

This *Encyclopedia* is available as a print version (5 volumes, more than 4,500 pages, 6,500 entries and 1,000 illustrations), an eReference version (online version), and as a bundle (print plus online) version.

Thanks are due to a vast number of people who have made this ambitious endeavor possible. First and foremost, we are extremely grateful to our 46 Field Editors who accepted the arduous challenge of organizing their fields, soliciting essays and glossary terms from expert authors, editing the submitted texts, and finally writing their own synopses. Second, many thanks also go to our nearly 1,000 authors who wrote essays and glossary terms. Third, Drs. Thomas Mager, Natasja Sheriff, Michaela Bilic and Jana Simniok of Springer-Verlag investigated much effort, initiative, patience and enthusiasm (at times interrupted by outbursts of frustration) into initiating, administering, pushing ahead and keeping alive this project. Many thanks are due to the numerous unnamed support staff in the background: secretaries, copy editors, computer and graphics specialists at *Springer*.

Marc D. Binder (Seattle)
Nobutaka Hirokawa (Tokyo)
Uwe Windhorst (Göttingen)

Editor-in-Chief

Marc D. Binder
Department of Physiology & Biophysics
University of Washington School of Medicine
Seattle, Washington, USA
mdbinder@u.washington.edu

Nobutaka Hirokawa
Department of Cell Biology and Anatomy
Graduate School of Medicine
University of Tokyo
Hongo, Bunkyo-ku, Tokyo, Japan
hirokawa@m.u-tokyo.ac.jp

Uwe Windhorst
Göttingen, Germany
siggi.uwe@t-online.de

Conceptual Editor

Martin C. Hirsch
iAS interActive Systems,
Marburg, Germany
martin.hirsch@brainmedia.de

Section Editors

Autonomic and Enteric Nervous System

Akio Sato
(deceased)
University of Human Arts and Sciences
Saitama, Japan

Brian Budgell
Departement de Chiropratique
Universite du Quebec a Trois-Rivieres
Quebec, Canada
budgell@uqtr.ca

Sae Uchida
Department of the Autonomic Nervous System
Tokyo Metropolitan Institute of Gerontology
Tokyo, Japan
suchida@tmig.or.jp

Behavior

Hermann Wagner
Institut für Biologie II
RWTH Aachen
Aachen, Germany
wagner@bio2.rwth-aachen.de

Biological Rhythms and Sleep

Martha U. Gillette
Molecular and Integrative Physiology, and Neuroscience Program
Institute for Genomic Biology University of Illinois at Urbana-Champaign
Urbana, IL, USA
mgillett@uiuc.edu

Biomechanics

Walter Herzog
Faculty of Kinesiology
Human Performance Lab University of Calgary
Calgary, AB, Canada
walter@kin.ucalgary.ca

Central Vision

Uwe Windhorst
Göttingen, Germany
siggi.uwe@t-online.de

Andreas K. Engel
Dept. of Neurophysiology and Pathophysiology
University Medical Center
Hamburg-Eppendorf
Hamburg, Germany
ak.engel@uke.de

Cognitive Functions

Fred Mast
Department of Psychology
University of Lausanne
Bâtiment Anthropole
Lausanne, Switzerland
Fred.Mast@unil.ch

Computational Motor Control

Amir Karniel
Department of Biomedical Engineering
Ben-Gurion University of the Negev
Beer Sheva, Israel
akarniel@bgu.ac.il

Development

Fujio Murakami
Laboratory of Neuroscience,
Graduate School of Frontier Biosciences,
Graduate School of Engineering Science
Osaka University
Suita, Osaka, Japan
fujiomurakami@gmail.com

Evolution

Ann B. Butler
Dept. Molecular Neuroscience
Krasnow Institute for Advanced Study
George Mason University
Fairfax, VA, USA
abbutler@gmu.edu

Eye Movements

Adonis K. Moschovakis
Institute of Applied and Computational Mathematics
Foundation for Research and Technology - Hellas
Heraklion, Crete, Greece
moschov@med.uoc.gr

Genetics, Molecular Biology

Sarah McFarlane
Department of Cell Biology and Anatomy
Hotchkiss Brain Institute,
Faculty of Medicine
University of Calgary
Calgary, Alberta, Canada
smcfarla@ucalgary.ca

Hearing

Armin Seidl
Virginia Merrill Bloedel Hearing Research Center

University of Washington
Seattle, WA, USA
armins@u.washington.edu

Edwin W. Rubel
Virginia Merrill Bloedel Hearing Research Center
University of Washington
Seattle, WA, USA
rubel@u.washington.edu

Learning and Memory

Taketoshi Ono
Molecular and Integrative Emotional Neuroscience
Graduate School of Medicine
University of Toyama
Sugitani, Toyama, Japan
onotake@med.u-toyama.ac.jp

Limbic System

Daniel S. Zahm
Department of Pharmacological and Physiolgical Science
Saint Louis University School of Medicine
St. Louis, MO, USA
zahmds@slu.edu

Lennart Heimer
(deceased)
Department of Neurological Surgery
University of Virginia
Charlottesville, VA, USA

Magnetic and Electrical Senses

Wolfgang Wiltschko
Universität Frankfurt
Zoologisches Institut Biologie – Campus der Universität
Frankfurt/Main, Germany
wiltschko@zoology.uni-frankfurt.de

Bernd Kramer
Institut für Zoologie
Animal Behaviour and Behavioural Physiology Research Group,
Universität Regensburg
Regensburg, Germany
bernd.kramer@biologie.uni-regensburg.de

Membrane Biophysics

Peter M. Lalley
Department of Physiology, Medical Sciences Center
University of Wisconsin School of Medicine and Public Health,
Madison, WI, USA
pmlalley@facstaff.wisc.edu

Uwe Windhorst
Göttingen, Germany
siggi.uwe@t-online.de

Muscle

C.J. Heckman
Physiology, Physical Medicine and Rehabilitation
Northwestern University
Feinberg School of Medicine
Chicago, IL, USA
c-heckman@northwestern.edu

Muscle Reflexes

Arthur Prochazka
Professor, Centre for Neuroscience
University of Alberta
Edmonton, AB, USA
arthur.prochazka@ualberta.ca

Neuroanatomy

Farel R. Robinson
University of Washington
Dept. of Biological Structure
Seattle, WA, USA
robinsn@u.washington.edu

Neuroendocrinology

Dick F. Swaab
Netherlands Institute for Neuroscience
Amsterdam, The Netherlands
d.f.swaab@nih.knaw.nl

Paul J. Lucassen
Centre for Neuroscience
Swammerdam Institute of Life Sciences
University of Amsterdam
Amsterdam, The Netherlands
lucassen@science.uva.nl

Neuroimmunology

John J. Haddad
Cellular and Molecular Signaling Research Group
Division of Biological Sciences, Departments of Biology and Biomedical Sciences, Faculty of Arts and Sciences
Lebanese International University (LIU)
Beirut, Lebanon
john.haddad@liu.edu.lb

Neurology

William J. Spain
Department of Neurology
Veterans Affairs Puget Sound Health Care System
University of Washington
Seattle, WA, USA
spain@u.washington.edu

Uwe Windhorst
Göttingen, Germany
siggi.uwe@t-online.de

Neuron Cellular/Molecular
Naweed I. Syed
Dept. Cell Biology and Anatomy
Faculty of Medicine
University of Calgary
Calgary, Alberta, Canada
nisyed@ucalgary.ca

Neuropharmacology
Paul F. Smith
Dept. of Pharmacology and Toxicology
School of Medical Sciences
University of Otago
Medical School
Dunedin, New Zealand
paul.smith@stonebow.otago.ac.nz

Neurophilosophy
Michael Pauen
Institut für Philosophie
Berlin School of Mind and Brain
Humboldt-Universität zu Berlin
Berlin, Germany
michael.pauen@philosophie.hu-berlin.de

Neuropsychiatry
Georg Northoff
Department of Psychiatry
University of Magdeburg
Magdeburg, Germany
Georg.Northoff@med.ovgu.de

Olfaction and Gustation
Pierre-Marie Lledo
Institut Pasteur
Perception and Memory Laboratory
CNRS Unit - Genes, Synapses & Cognition
Paris, Cedex 15, France
pmlledo@pasteur.fr

Pain
Gerald F. Gebhart
Center for Pain Research
University of Pittsburgh
Pittsburgh, PA, USA
gebhartgf@upmc.edu

Posture
Fay B. Horak
Neurological Sciences Institute
Oregon Health and Science University
Portland, OR, USA
fay.horak@gmail.com

Proprioception
Simon Gandevia
Prince of Wales Medical Research Institute
Sydney, Australia
s.gandevia@unsw.edu.au

Regeneration
Chizuka Ide
Institute of Regeneration and Rehabilitation
Department of Occupational Therapy,
Faculty of Nursing and Rehabilitation
Aino University
Ibaraki, Osaka, Japan
c-ide@ot-u.aino.ac.jp

Respiration
Peter M. Lalley
Department of Physiology, Medical Sciences Center
University of Wisconsin School of Medicine and
Public Health
Madison, WI, USA
pmlalley@facstaff.wisc.edu

Retinal Processing
David Vaney
Queensland Brain Institute
University of Queensland
Brisbane, Queensland, Australia
d.vaney@uq.edu.au

W. Rowland Taylor
Neurological Sciences Institute
Oregon Health and Science University
Beaverton, OR, USA
and
Casey Eye Insititute
School of Medicine
Oregon Health and Science University
Portland, OR, USA
taylorw@ohsu.edu

Uwe Windhorst
Göttingen, Germany
siggi.uwe@t-online.de

Rhythmic Movements
Ole Kiehn
Department of Neuroscience
Karolinska Institute Stockholm
Stockholm, Sweden
O.Kiehn@ki.se

Synapse
Masami Takahashi
Department of Biochemistry

Kitasato-University,
School of Medicine
Sagamihara-shi, Kanagawa, Japan
masami@med.kitasato-u.ac.jp

Touch

Yoshiaki Iwamura
Department of Sensory Science
Kawasaki University of Medical Welfare
Okayama, Japan
iwayoshi@mw.kawasaki-m.ac.jp

Vestibular System

Neal H. Barmack
Neurological Sciences Institute
Oregon Health & Science University
Portland, OR, USA
barmackn@ohsu.edu

Vito Enrico Pettorossi
Dipartimento di Medicina Interna
Section of Physiology
Perugia, Italy
vitopett@unipg.it

Voluntary Movements

Martha Flanders
Department of Neuroscience
University of Minnesota,
Minneapolis
MN, USA
fland001@umn.edu

Contributors

Teruo Abe
Department of Cellular Neurobiology
Brain Research Institute
Niigata University
Niigata, Japan
teruoa@bri.niigata-u.ac.jp

Valery V. Abramov
Laboratory of Neuroimmunology
State Research Institute of Clinical Immunology of SB RAMS
Novosibirsk, Russia
valery_abramov@mail.ru

Tatjana Ya. Abramova
Laboratory of Neuroimmunology
State Research Institute of Clinical Immunology of SB RAMS
Novosibirsk, Russia

Monica L. Acosta
Department of Optometry and Vision Science
University of Auckland
Auckland, New Zealand

Masaharu Adachi
Department of Electrical and Electronic Engineering, School of Engineering
Tokyo Denki University
Tokyo, Japan
adachi@d.dendai.ac.jp

Antoine Adamantidis
Department of Psychiatry and Behavioral Sciences
Stanford University
School of Medicine
Palo Alto, CA, USA

Daniel L. Adams
Department of Ophthalmology
Koret Vision Research Laboratory
UCSF, San Francisco, CA, USA

Fabienne Agasse
Center for Neuroscience and Cell Biology
Institute of Biochemistry
Faculty of Medicine,
University of Coimbra
Coimbra, Portugal

Kazuyuki Aihara
Institute of Industrial Science
The University of Tokyo
Tokyo, Japan
aihara@sat.t.u-tokyo.ac.jp

Rachid Ait-Haddou
Human Performance Laboratory
University of Calgary
Calgary, AB, Canada
aihara@sat.t.u-tokyo.ac.jp

Kathryn M. Albers
Department of Medicine
University of Pittsburgh
Pittsburgh, PA, USA
kaa2@pitt.edu

Jessica Albrecht
Department of Neuroradiology
Ludwig Maximilian University Munich
Munich, Germany
Jessica.albrecht@med.uni-muenchen.de

Urs Albrecht
Department of Medicine
Division of Biochemistry
University of Fribourg
Fribourg, Switzerland
urs.albrecht@unifr.ch

Håkan Aldskogius
Uppsala University Biomedical Center
Department of Neuroscience
Uppsala, Sweden
Hakan.Aldskogius@neuro.uu.se

George F. Alheid
Department of Physiology
Feinberg School of Medicine
Northwestern University
Chicago, IL, USA
gfa@northwestern.edu

Douglas W. Allan
Department of Cellular and Physiological Sciences
University of British Columbia
Vancouver, BC, Canada
dwallan@interchange.ubc.ca

Mariana Alonso
Institut Pasteur
Perception and Memory Laboratory
Paris, France
malonso@pasteur.fr

Nadja Althaus
Centre for Brain and Cognitive Development
Birkbeck, University of London
London, UK

Wilko Altrock
Leibniz Institute for Neurobiology
Department of Neurochemistry and Molecular Biology
Magdeburg, Germany

Shimon Amir
Center for Studies in Behavioral Neurobiology,
Department of Psychology
Concordia University
Montreal, QC, Canada
shimon.amir@concordia.ca

Bagrat Amirikian
Department of Neuroscience
University of Minnesota Medical School
Minneapolis, MN, USA
amiri001@tc.umn.edu

Ramón Anadón
University of Santiago de Compostela
Campus Sur Santiago de Compostela
Spain

Holley André
Centre Européen des Sciences du Goût (CESG)
CNRS-INRA-Université de Bourgogne
Dijon, France

Hiroyuki Arakawa
Pacific Biomedical Research Center
University of Hawaii at Manoa
Honolulu, HI, USA

Josephine Arendt
Centre for Chronobiology
Faculty of Health and Medical Sciences
University of Surrey
Guildford, Surrey, UK
arendtjo@aol.com

Lars Arendt-Nielsen
Center for Sensory-Motor Interaction (SMI),
Department of Health Science and Technology
Aalborg University
Aalborg, Denmark
LAN@hst.aau.dk

Paola Arlotta
Center for Regenerative Medicine
Department of Neurosurgery
Massachusetts General Hospital
Harvard Medical
School
Boston, MA, USA

Pamela Arstikaitis
Department of Psychiatry and the Brain Research
Centre
University of British Columbia
Vancouver, BC, Canada
parstika@interchange.ubc.ca

Hideki Asoh
Information Technology Research Institute
National Institute of Advanced Industrial Science and
Technology
Tsukuba, Ibaraki, Japan
h.asoh@aist.go.jp

Claudia Bagni
Department of Experimental Medicine and
Biochemical Sciences, University "Tor Vergata",
Rome, Italy
claudia.bagni@uniroma2.it

Nasser S. Ballani
Biomedical Sciences
Lebanese International University
Beirut, Lebanon
nasser.ballani@liu.edu.lb

Klaus Ballanyi
Department of Physiology
Perinatal Research Centre
Faculty of Medicine & Dentistry
University of Alberta
Edmonton, AB, Canada
klaus.ballanyi@ualberta.ca

James R. Bamburg
Molecular, Cellular and Integrative Neurosciences
Program
Department of Biochemistry and Molecular Biology
Colorado State University
Fort Collins, CO, USA
jbamburg@lamar.colostate.edu

Laura Bamert
Department of Psychology
University of Lausanne
Lausanne, Switzerland

Albert J. Banes
University of North Carolina
North Carolina State University
North Carolina, USA
ajbvault@med.unc.edu

Ai-Min Bao
Netherlands Institute for Neurosciences
Meibergdreef, Amsterdam, The Netherlands
a.bao@nin.knaw.nl

Neal H. Barmack
Neurological Sciences Institute
Oregon Health & Science University
Beaverton, OR, USA
barmackn@ohsu.edu

Scott R. Barnum
Departments of Microbiology and Neurology
University of Alabama at Birmingham (UAB)
Birmingham, AL, USA

Ralf Baron
Division of Neurological Pain Research and Therapy,
Department of Neurology
Universitätsklinikum Schleswig-Holstein
Campus Kiel
Kiel, Germany
r.baron@neurologie.uni-kiel.de

Andres Barria
Department of Physiology and Biophysics
University of Washington
Seattle, WA, USA
barria@u.washington.edu

Donald Bartlett Jr.
Department of Physiology
Dartmouth Medical School
Lebanon, NH, USA
don.bartlett@dartmouth.edu

Alexandra Battaglia-Mayer
Department of Human Physiology and Pharmacology
SAPIENZA, University of Rome
Rome, Italy
alexandra.battagliamayer@uniroma1.it

Philippa R. Bayley
Neurological Sciences Institute
Oregon Health and Science University
Beaverton, OR, USA
bayleyp@ohsu.edu

Christian Beaulé
Department of Biology
Washington University
St. Louis, MO, USA
cbeaule@mac.com

Wolfgang Becker
Sektion Neurophysiologie
Universität Ulm
Ulm, Germany
wolfgang.becker@medizin.uni-ulm.de

Fiona K. Bedford
Department of Anatomy and Cell Biology
McGill University
Montreal, QC, Canada
fiona.bedford@mcgill.ca

Gregory Belenky
Sleep and Performance Research Center
Washington State University
Spokane, WA, USA
Belenky@WSU.edu

Harold J. Bell
Department of Cell Biology and Anatomy
University of Calgary
Calgary, AB, Canada
jbell@math.umbc.edu

Heather Bell
Department of Zoology
University of Oklahoma
Norman, OK, USA
harold.bell@ucalgary.ca

Jonathan Bell
University of Maryland
Baltimore, MD, USA
jbell@math.umbc.edu

Yehezkel Ben-Ari
INMED, INSERM U29
Université de la Méditerranée
Campus de Luminy
Marseille, France
ben-ari@inmed.univ-mrs.fr

Eduardo E. Benarroch
Department of Neurology
Mayo Clinic
Rochester, MN, USA
benarroch@mayo.edu

Karim Benchenane
LPPA
Collèe de France, CNRS
Paris, France

Tim A. Benke
Pediatrics, Neurology, Pharmacology and Neuroscience Program
University of Colorado Health Sciences Center
The Children's Hospital
Aurora, CO, USA

Tomasz Berkowicz
Department of Neurology
Medical University of Lodz
Lodz, Poland
tberkowicz@afazja.am.lodz.pl

Liliana Bernardino
Center for Neuroscience and Cell Biology
Institute of Biochemistry
Faculty of Medicine, University of Coimbra
Coimbra, Portugal

Anthony Berndt
Institute of Maternal and Child Health
Genes and Development Research Group
Department of Cell Biology and Anatomy
Faculty of Medicine
University of Calgary
Calgary, AB, Canada

Sven Bernecker
Department of Philosophy
University of California at Irvine
Irvine, CA, USA
bernecker@Lrz.uni-muenchen.de

Rommy von Bernhardi
Department of Neurology
Faculty of Medicine
Pontificia Universidad Católica de Chile
Santiago, Chile
rvonb@med.puc.cl

Daniel M. Bernstein
Department of Psychology
Kwantlen University College
Surrey, BC, Canada
db6@u.washington.edu

Leslie R. Bernstein
Department of Neuroscience and Department of Surgery
University of Connecticut Health Center
Farmington, CT, USA

Gabriel Bertolesi
The Hotchkiss Brain Institute
Department of Cell Biology and Anatomy
Faculty of Medicine
University of Calgary
Calgary, AB, Canada
gbertole@ucalgary.ca

Martin Biel
Department Pharmazie
Zentrum für Pharmaforschung
Ludwig-Maximilians-Universität München
München, Germany
mbiel@cup.uni-muenchen.de

Klaus Bielefeldt
University of Pittsburgh
Department of Medicine
Pittsburgh, PA, USA
bielefeldtk@dom.pitt.edu

Andrew A. Biewener
Concord Field Station
Department of Organismic and Evolutionary Biology
Harvard University
Cambridge, MA, USA
biewener@fas.harvard.edu

Stuart A. Binder-Macleod
Department of Physical Therapy
University of Delaware
Newark, DE, USA
sbinder@udel.edu

Verner P. Bingman
Department of Psychology and J.P. Scott Center for Neuroscience, Mind and Behavior
Bowling Green State University
Bowling Green, OH, USA
vbingma@bgnet.bgsu.edu

Thomas D. Bird
Department of Neurology
UW Medical Center
Seattle, WA, USA
tomnroz@u.washington.edu

Ingvars Birznieks
Prince of Wales Medical Research Institute
Sydney, NSW, Australia

Eric L. Bittman
Department of Biology
University of Massachusetts
Amherst, MA, USA
elb@bio.umass.edu

D. Caroline Blanchard
Department of Genetics and Molecular Biology
John A. Burns School of Medicine
University of Hawaii at Manoa
Honolulu, HI, USA

Robert J. Blanchard
Department of Psychology
University of Hawaii at Manoa
Honolulu, HI, USA
blanchar@hawaii.edu

Olaf Blanke
Laboratory of Cognitive Neuroscience, Brain-Mind Institute
École Polytechnique Fédérale de Lausanne
Switzerland
Department of Neurology
University Hospital
Geneva, Switzerland
olaf.blanke@epfl.ch

Clark M. Blatteis
University of Tennessee Health Science Center
College of Medicine
Memphis, TN, USA
blatteis@physio1.utmem.edu

Bas Blits
Netherlands Institute for Neurosciences
A Research Institute of the Royal Netherlands
Academy of Arts and Sciences
Amsterdam, The Netherlands

James R. Bloedel
Department of Biomedical Sciences
Iowa State University
Ames, IA, USA
jbloedel@iastate.edu

Joël Bockaert
Université de Montpellier, CNRS UMR5203
Inserm U661, Montpellier, France
Joel.bockaert@igf.cnrs.fr

Ulrich Boehm
Center for Molecular Neurobiology
Institute for Neural Signal Transduction
Hamburg, Germany
Ulrich.Boehm@zmnh.uni-hamburg.de

Heinz Boeker
University Hospital of Psychiatry Zurich
Hospital for Affective Disorders and General
Psychiatry Zurich East
Zurich, Switzerland
boeker@bli.unizh.ch

Karin Boekhoorn
Neurosignalisation Moleculaire et Cellulaire
INSERM U706
Institut du Fer a Moulin
Paris, France

Nikolai Bogduk
University of Newcastle
Newcastle Bone and Joint Institute
Royal Newcastle Centre
Newcastle, NSW, Australia

Bernhard Bogerts
Universitätsklinik für Psychiatrie
Psychotherapie und Psychosomatische Medizin
Otto-von-Guericke
Universität Magdeburg
Magdeburg, Germany
Bernhard.Bogerts@medizin.uni-magdeburg.de

Donald C. Bolser
Department of Physiological Sciences
University of Florida College of Veterinary Medicine
Gainesville, FL, USA

Martha M. Bosma
Department of Biology
University of Washington
Seattle, WA, USA
martibee@u.washington.edu

Annette Bölter
Universitätsklinikum Magdeburg
Abteilung für Psychosomatische Medizin und
Psychotherapie
Magdeburg, Germany

Constantin Bouras
Department of Psychiatry
University Hospitals of Geneva
Geneva, Switzerland
Constantin.Bouras@medecine.unige.ch

Jamie D. Boyd
Department of Biological Sciences
Simon Fraser University
Burnaby, BC, Canada

Vlastislav Bracha
Department of Biomedical Sciences
Iowa State University
Ames, IA, USA
vbracha@iastate.edu

MARK R. BRAFORD JR
Departments of Biology and Neuroscience
Oberlin College, Oberlin
Ohio, OH, USA

CATHERINE BRANDNER
Université de Lausanne
Institut de Psychologie
Lausanne, Switzerland
Catherine.Brandner@unil.ch

ANNA KATHARINA BRAUN
Otto von Guericke University Magdeburg
Institute of Biology
Faculty for Natural Sciences
Leibniz Institute of Neurobiology
Magdeburg, Germany
katharina.braun@nat.uni-magdeburg.de

CHRISTOPHER B. BRAUN
Department of Psychology, Hunter College
Programs in Biopsychology and Behavioral
Neuroscience, and Ecology Evolution and Behavior,
Graduate Center
City University of New York (CUNY)
NY, USA
cbraun@hunter.cuny.edu

HEINZ BREER
University of Hohenheim
Institute of Physiology
Stuttgart, Germany
breer@uni-hohenheim

BJÖRN BREMBS
Freie Universität Berlin
Fachbereich Biologie, Chemie Pharmazie
Institut für Biologie – Neurobiologie
Berlin, Germany
bjoern@brembs.net

PETER BRENNAN
Department of Physiology
University of Bristol
Bristol, UK
p.brennan@bristol.ac.uk

TIMOTHY J. BRENNAN
Departments of Anesthesia and Pharmacology
University of Iowa
Carver College of Medicine
Iowa City, IA, USA
tim-brennan@uiowa.edu

LOÏC BRIAND
Unité Mixte de recherche FLAVIC INRA-ENESAD
Université de Bourgogne
Dijon, France
loic.briand@jouy.inra.fr

DORA BRITES
Centro de Patogénese Molecular - UBMBE iMed.UL,
Faculdade de Farmácia
University of Lisbon
Lisbon, Portugal
dbrites@ff.ul.pt

M. A. BRITO
Centro de Patogénese Molecular - UBMBE iMed.UL,
Faculdade de Farmácia
University of Lisbon
Lisbon, Portugal

ADOLFO M. BRONSTEIN
Division of Neuroscience
Imperial College London
Charing Cross Hospital and the National Hospital for
Neurology and Neurosurgery
London, UK
A.Bronstein@IC.AC.UK

ANNA BROOKS
Laboratory of Perceptual Processing
Southern Cross University
Lismore, NSW, Australia
anna.brooks@scu.edu.au

NILS BROSE
Department of Molecular Neurobiology
Max-Planck-Institute for Experimental Medicine
Göttingen, Germany
brose@em.mpg.de

LAURA L. BRUCE
Department of Biomedical Sciences
School of Medicine
Creighton University
Omaha, NE, USA
lbruce@creighton.edu

PETER C. BRUNJES
Department of Psychology
University of Virginia
Charlottesville, VA, USA
brunjes@virginia.edu

BRIAN BUDGELL
Department de Chiropratique
Universite du Chiropratique
Universite du Quebec a Trois-Rivieres
Quebec, Canada
budgell@uqts.ca

Thomas N. Buell
Department of Neuroscience
University of Connecticut
Health Center
Farmington, CT, USA

Ruud M. Buijs
Instituto de Investigaciones Biomedicas
Department Fisiologia
Universidad Nacional Autonoma de Mexico
Mexico City, Mexico
ruudbuijs@gmail.com

Nathalie Buonviso
Neurosciences Sensorielles, Comportement, Cognition
Université Claude Bernard Lyon 1
CNRS – UMR5020 – INFL LYON
Lyon, France
buonviso@olfac.univ-lyon1.fr

Robert D. Burgoyne
The Physiological Laboratory
School of Biomedical Sciences
University of Liverpool
Liverpool, UK
burgoyne@liv.ac.uk

Robert E. Burke
Laboratory of Neural Control
National Institute of Neurological Disorders and Stroke
National Institutes of Health
Bethesda, MD, USA
reburke@helix.nih.gov

Ralf Busse
Institut für Philosophie
Universität Regensburg
Regensburg, Germany
ralf.busse@psk.uni-regensburg.de

Ann B. Butler
Department of Molecular Neuroscience
Krasnow Institute for Advanced Study
George Mason University
Fairfax, VA, USA
abbutler@gmu.edu

Ronald L. Calabrese
Department of Biology
Emory University
Atlanta, GA, USA
RCalabre@Biology.EMORY.EDU

Mike B. Calford
School of Biomedical Sciences and Hunter Medical
Research Institute
The University of Newcastle
Newcastle, NSW, Australia
mike.calford@newcastle.edu.au

Rose M. Calhoun-Haney
Yerkes National Primate Research Center
Emory University
Atlanta, GA, USA
rcalhou@emory.edu

Gemma A. Calvert
Department of Experimental Psychology
Oxford University
Oxford, UK
g.a.calvert@warwick.ac.uk

Roberto Caminiti
Department of Human Physiology and Pharmacology
SAPIENZA, University of Rome
Rome, Italy

Iain L. Campbell
School of Molecular and Microbial Biosciences
The University of Sydney
Sydney, NSW, Australia
icamp@mmb.usyd.edu.au

Ana R. Campos
Department of Biology
McMaster University
Hamilton, ON, Canada
camposa@univmail.cis.mcmaster.ca

Laura Cancedda
Department of Neuroscience and Brain Technologies
The Italian Institute of Technology
Genoa, Italy
laura.cancedda@iit.it

Ling Cao
Department of Anesthesiology
Dartmouth-Hitchcock Medical Center
Lebanon, NH, USA
LCao@une.edu

Emilio Carbone
Department of Neuroscience
NIS Center of Excellence
CNISM Research Unit
Torino, Italy
emilio.carbone@unito.it

Robert Carlone
Department of Biological Sciences
Brock University
St. Catharines, ON, Canada

José Carlos Dávila
Departamento de Biología Celular, Genética y Fisiología
Facultad de Ciencias
Universidad de Málaga
Málaga, Spain

Bruce A. Carlson
Department of Biology
University of Virginia
Charlottesville, VA, USA
carlson.bruce@gmail.com

Catherine Carr
Department of Biology
University of Maryland
College Park, MD, USA
cecarr@umd.edu

Matt Carter
Department of Psychiatry and Behavioral Sciences
Stanford University
School of Medicine
Palo Alto, CA, USA

Patrizia Casaccia-Bonnefil
Robert Wood Johnson Medical School UMDNJ
Piscataway, NJ, USA

Vivien A. Casagrande
Department of Cell & Developmental Biology
Vanderbilt University Medical School, U3218
Learned Lab
Nashville, TN, USA
vivien.casagrande@vanderbilt.edu

Vivien A. Casagrande
Department of Psychology and Department of Ophthalmology and Visual Sciences
Nashville, TN, USA
vivien.casagrande@vanderbilt.edu

Tristan Cenier
Neurosciences Sensorielles, Comportement, Cognition
Université Claude Bernard Lyon 1
CNRS – UMR5020 – INFL LYON
Lyon, France

Fernando Cervero
Anesthesia Research Unit (Faculty of Medicine), Faculty of Dentistry and McGill Centre for Research on Pain
McGill University
QC, Canada
fernando.cervero@mcgill.ca

Jean Champagnat
UPR CNRS 2216 Neurobiologie Génétique et Intégrative
FRC CNRS Institut de Neurobiologie Alfred Fessard
Gif sur Yvette, France
jean.champagnat@iaf.cnrs-gif.fr

Christine Elaine Chapman
Groupe de Recherche sur le Système Nerveux Central, Département de Physiologie & École de Réadaptation, Faculté de Médecine
Université de Montréal
Montréal, QC, Canada
c.elaine.chapman@umontreal.ca

Melissa J. S. Chee
Department of Pharmacology and Centre for Neuroscience
University of Alberta
Edmonton, AB, Canada
mchee@ualberta.ca

Chien-Fu Chen
Department of Zoology
University of Oklahoma
Norman, OK, USA

Jen-Yung Chen
Dept. of Psychology
Binghamton University
Binghamton, NY, USA
happy_jenyung1@yahoo.com

Paul Cheney
Dept. Molecular and Integrative Physiology
University of Kansas
Medical Center
Kansas City, Kansas, USA
pcheney@kumc.edu

Amanda F. P. Cheung
Department of Physiology, Anatomy and Genetics
University of Oxford
Oxford, UK

Lorenzo Chiari
Dipartimento di Elettronica, Informatica e Sistemistica
Alma Mater Studiorum – Università di Bologna
Bologna, Italia
lchiari@deis.unibo.it

Jakob Christensen-Dalsgaard
Institute of Biology
University of Southern Denmark
Odense, Denmark
jcd@biology.sdu.dk

MacDonald J. Christie
Pain Management Research Institute
Kolling Institute
The University of Sydney at Royal North Shore Hospital
St Leonards, NSW, Australia
macc@med.usyd.edu.au

Paul Cisek
Groupe de Recherche Sur le Système Nerveux Central
Department of Physiology
University of Montreal
Montreal, QC, Canada
cisekp@magellan.unmontreal.ca

Maria Grazia Ciufolini
Department of Infectious, Parasitic and Immune-Mediated Diseases
Istituto Superiore di Sanità
Rome, Italy
ciufolin@iss.it

Julie Chapuis
Laboratoire Systèmes Sensoriels, Comportement et Cognition
UMR 5020 CNRS–UCB Lyon
IFR19 Lyon, France

Andrea L. Clark
Faculty of Kinesiology and Medicine
KNB 304
University of Calgary
Calgary, AB, Canada
jgpa.clark@btinternet.com

Thomas A. Cleland
Department of Psychology
Cornell University
Ithaca, NY, USA
tac29@cornell.edu

Colin W. G. Clifford
School of Psychology
University of Sydney
Sydney, NSW, Australia
colinc@psych.usyd.edu.au

Mayo Clinic
Department of Neurology
Mayo Foundation
Rochester, MN, USA

Jean-François Cloutier
Department of Neurology and Neurosurgery
McGill University
Montreal Neurological Institute
Montréal, QC, Canada
jf.cloutier@mcgill.ca

Daniel A. Cohen
Harvard Medical School, Beth Israel Deaconess Medical Center
Department of Neurology
Brigham and Women's Hospital
Division of Sleep Medicine
Boston, MA, USA
dcohen2@bidmc.harvard.edu

Carol L. Colby
Department of Neuroscience
Center for Neutral Basis Cognition
University of Pittsburgh
Pittsburgh, PA, USA
colby@cnbc.cmu.edu

Jonathan Cole
University of Southampton Clinical Neurological Sciences
Poole Hospital
Dorset, UK
jonathan.cole@poole.nhs.uk

Michael A. Colicos
Department of Physiology and Biophysics
Faculty of Medicine
University of Calgary
Calgary, AB, Canada
mcolicos@ucalgary.ca

Shaun P. Collin
Sensory Neurobiology Group
School of Biomedical Sciences
The University of Queensland
Brisbane, QLD, Australia
s.collin@uq.edu.au

David F. Collins
Faculty of Physical Education and Recreation
Centre for Neuroscience
University of Alberta
Edmonton, AB, Canada
dave.collins@ualberta.ca

William F. Colmers
Department of Pharmacology
University of Alberta
Edmonton, AB, Canada
Centre for Neuroscience
University of Alberta
Edmonton, AB, Canada

Christopher S. Colwell
Department of Psychiatry
University of California
Los Angeles, USA
CColwell@mednet.ucla.edu

Ruth M. Colwill
Department of Psychology
Brown University
Providence, Rhode Island, USA
Ruth_Colwill@Brown.EDU

Charles E. Connor
Krieger Mind/Brain Institute and Department of Neuroscience
Johns Hopkins University
Baltimore, MD, USA
connor@mail.mb.jhu.edu

Jens R. Coorssen
Hotchkiss Brain Institute
Faculty of Medicine
University of Calgary
Calgary, Alberta, Canada
jcoorsse@ucalgary.ca

Gérard Coureaud
Ethology and Sensory Psychobiology Group,
European Center for Taste and Smell
CNRS/University of Burgundy/INRA
Dijon, France
coureaud@cesg.cnrs.fr

Ellen Covey
Department of Psychology
University of Washington
Seattle, WA, USA
ecovey@u.washington.edu

Trinity B. Crapse
Department of Neuroscience
The Center for the Neural
Basis of Cognition, and the Center for Neuroscience
University of Pittsburgh
Pittsburgh, PA, USA

Sarah H. Creem-Regehr
Department of Psychology
University of Utah
Salt Lake City, UT, USA
sarah.creem@psych.utah.edu

Claire Creutzfeld
Harborview Neurology - Comprehensive Stroke Center
University of Washington
School of Medicine
Seattle, WA, USA

Javier Cudeiro
Neurociences and Motor Control Group (Neurocom),
Department of Medicine-INEF-Galicia
University de A Coruña
Campus de Oza, A Coruña, Spain

Kathleen E. Cullen
Department of Physiology
Aerospace Medical Research Unit
McGill University
Montreal, QC, Canada
kathleen.cullen@mcgill.ca

William E. Cullinan
Department of Biomedical Sciences and Integrative Neuroscience Research Center
Marquette University
Milwaukee, WI, USA
william.cullinan@marquette.edu

Yannis Dalezios
Institute of Applied and Computational Mathematics
Crete, Greece
dalezios@med.uoc.gr

Gomez-Merino Danielle
Department of Physiology
Institut de Médecine Aérospatiale du Service de Santé des Armées
Brétigny-sur-Orge, France
dgomez@imassa.fr

Cynthia L. Darlington
Department of Pharmacology and Toxicology
School of Medical Sciences
University of Otago
Medical School
Dunedin, New Zealand
cynthia.darlington@stonebow.otago.ac.nz

Frédérique Datiche
Neurophysiology of Chemoreception Group
European Center for Taste and Smell
CNRS/University of Burgundy/INRA
Dijon, France
datiche@cesg.cnrs.fr

Subimal Datta
Sleep and Cognitive Neuroscience Laboratory,
Department of Psychiatry
Boston University
School of Medicine
Boston, MA, USA
subimal@bu.edu

ANDREA D'AVELLA
Department of Neuromotor Physiology
Santa Lucia Foundation
Rome, Italy

FRANÇOIS DAVID
Neurosciences Sensorielles, Comportement, Cognition
Université Claude Bernard Lyon 1
CNRS – UMR5020 – INFL LYON
Lyon, France

SAMUEL DAVID
Center for Research in Neuroscience
McGill University Health Center
Montreal, Canada
samuel.david@mcgill.ca

STEVE DAVIDSON
Department of Neuroscience
University of Minnesota
Minneapolis, MN, USA
davi1082@umn.edu

JUSTIN R. DAVIS
Brain Research Centre
University of British Columbia
Vancouver, BC, Canada

RANDALL L. DAVIS
Neuroinflammation Research Laboratory
Department of Pharmacology/Physiology
Oklahoma State University
Center for Health Sciences
Tulsa, OK, USA
randadl@chs.okstate.edu

LUIS DE LECEA
Department of Psychiatry and Behavioral Sciences
Stanford University
School of Medicine
Palo Alto, CA, USA
llecea@stanford.edu

PAUL DEAN
Department of Psychology
University of Sheffield
Sheffield, UK

GUSTAVO DECO
Department of Technology, Computational Neuroscience
Universitas Pompeu Fabra
Barcelona, Spain
Gustavo.Deco@upf.edu

CHRISTOPHER A. DEL NEGRO
The Department of Applied Science
The College of William and Mary
Williamsburg, VA, USA
cadeln@wm.edu

JOYCE A. DELEO
Department of Pharmacology
Dartmouth-Hitchcock Medical Center
Lebanon, NH, USA
cadeln@wm.edu

JOSÉ M. DELGADO-GARCÍA
División de Neurociencias
Universidad Pablo de Olavide
Sevilla, Spain
jmdelgar@dex.upo.es

ANNETTE DENZINGER
Tierphysiologie
Zoologisches Institut
Universität Tübingen
Tubingen, Germany

MARI DEZAWA
Department of Anatomy and Neurobiology
Kyoto University Graduate School of Medicine, Yoshidakonoe-cho
Sakyo-ku, Kyoto, Japan
dezawa@anat2.med.kyoto-u.ac.jp

YASIN Y. DHAHER
Department of Biomedical Engineering
McCormick School of Engineering
Department of Physical Medicine and Rehabilitation, Feinberg School of Medicine, Northwestern University, Sensory Motor Performance Program
The Rehabilitation Institute of Chicago
Chicago, IL, USA
y-dhaher@northwestern.edu

HASSAN R. DHAINI
Faculty of Health Sciences
University of Balamand
Aschrafieh, Beirut
hassan.dhaini@balamand.edu.lb

PATRICIA M. DI LORENZO
Dept. of Psychology
Binghamton University
Binghamton, NY, USA
diloren@binghamton.edu

BETTY DIAMOND
Department of Medicine
Columbia University
Medical Center
New York, NY, USA
bd2137@columbia.edu

Sulayman D. Dib-Hajj
Department of Neurology and Center for Neuroscience and Regeneration Research
Yale University
School of Medicine
New Haven, CT, USA
sulayman.dib-hajj@yale.edu

Thomas E. Dick
Division of Pulmonary, Critical Care and Sleep Medicine
Department of Medicine
Case Western Reserve University
Cleveland, OH, USA
thomas.dick@case.edu

Ursula Dicke
Brain Research Institute
University of Bremen
Bremen, Germany

J. David Dickman
Department of Anatomy and Neurobiology
Washington University
St. Louis, MO, USA
ddickman@wustl.edu

Anne Didier
Neurosciences Sensorielles, Comportement, Cognition
Université de Lyon, Université
Claude Bernard Lyon1, Villeurbanne, France
didier@olfac.univ-lyon1.fr

W. Dalton Dietrich
The Miami Project to Cure Paralysis and Neurological Surgery
University of Miami Miller
School of Medicine
Miami, FL, USA
DDietrich@miami.edu

Derk-Jan Dijk
Surrey Sleep Research Centre
University of Surrey
Guildford, Surrey, UK
d.j.dijk@surrey.ac.uk

Christine D. Dijkstra
Department of Molecular Cell Biology and Immunology
VUMC
Amsterdam, The Netherlands
cd.dijkstra@vumc.nl

Paul DiZio
Ashton Graybiel Spatial Orientation Laboratory
Brandeis University
Waltham, MA, USA
dizio@brandeis.edu

Peter R. Dodd
School of Molecular and Microbial Sciences
University of Queensland
Brisbane, Queensland, Australia
p.dodd@uq.edu.au

Rodney J. Douglas
Institute of Neuroinformatics UZH/ETH
University/ETH
Zurich, Switzerland

Megan J. Dowie
Department of Pharmacology and Clinical Pharmacology
University of Auckland
Auckland, New Zealand

Kati E. Draper
Department of Human Nutrition, Foods and Exercise
Virginia Polytechnic Institute and State University
Blacksburg, VA, USA

David Dubayle
CNRS UMR 8119, Neurophysique et Physiologie
Université René Descartes
UFR Biomédicale
Paris, France
dubayle@biomedicale.univ-paris5.fr

Réjean Dubuc
Département de Kinanthropologie
Université du Québec à Montréal
Montréal, QC, Canada
Département de Physiologie
Université de Montréal
Montréal, QC, Canada
rejean.dubuc@gmail.com

Elizabeth Dudkin
Penn State University
Media, PA, USA
ead9@psu.edu

James Duffin
Department of Anaesthesia and Physiology
University of Toronto
Medical Sciences Building
Toronto, ON, Canada
j.duffin@utoronto.ca

Mayank B. Dutia
Centre for Integrative Physiology
Edinburgh University
Hugh Robson Building
Edinburgh, UK
m.b.dutia@ed.ac.uk

Mathias Dutschmann
Department of Neuro and Sensory Physiology
Georg August University of Göttingen
Göttingen, Germany
mdutsch@gwdg.de

Ford Ebner
Department of Psychology
Vanderbilt University
Nashville, TN, USA
ford.ebner@vanderbilt.edu

Victor Reggie Edgerton
Brain Research Institute and Department of Physiological Science
University of California
Los Angeles, CA, USA

Elisabeth Ehler
The Randall Division of Cell and Molecular Biophysics & The Cardiovascular Division
King's College London
London, UK
elisabeth.ehler@kcl.ac.uk

Walter H. Ehrenstein
Leibniz Research Center for Working Environment and Human Factors
University of Dortmund
Dortmund, Germany
ehrenstein@ifado.de

Günter Ehret
Department of Neurobiology
University of Ulm
Ulm, Germany
guenter.ehret@biologie.uni-ulm.de

David D. Eisenstat
Manitoba Institute of Cell Biology,
Departments of Pediatrics and Child Health, Human Anatomy and Cell Science,
Ophthalmology, and Biochemistry and Medical Genetics,
Faculty of Medicine
University of Manitoba
Winnipeg, MB, Canada

Alaa El-Husseini
Department of Psychiatry and the Brain Research Centre
University of British Columbia
Vancouver, BC, Canada

Stacy L. Elliott
Departments of Psychiatry and Urological Sciences and International Collaboration on Repair Discoveries
University of British Columbia
Vancouver, BC, Canada
elliott@icord.org

Bernard T. Engel
Schulich School of Engineering
University of Calgary
Calgary, AB, Canada
Btere@aol.com

Marcelo Epstein
Schulich School of Engineering
University of Calgary
Calgary, AB, Canada
mepstein@ucalgary.ca

Gerhard Ernst
Seminar für Philosophie, Logik und Wissenschaftstheorie
Ludwig-Maximilians-Universität München
München, Germany
Gerhard.Ernst@lrz.uni-muenchen.de

Michael Esfeld
Department of Philosophy
University of Lausanne
Lausanne Switzerland
Michael-Andreas.Esfeld@philo.unil.ch

Greg K. Essick
Department of Prosthodontics and Center for Neurosensory Disorders
School of Dentistry
University of North Carolina
Chapel Hill, NC, USA
essickg@DENTISTRY.UNC.EDU

Naomi Etheridge
School of Molecular and Microbial Sciences
University of Queensland
Brisbane, Queensland, Australia

Thomas Euler
Department of Biomedical Optics
Max-Planck Institute for Medical Research
Heidelberg, Germany
thomas.euler@mpimf-heidelberg.mpg.de

Susan E. Evans
Research Department of Cell and Developmental Biology
UCL, University College London
London, UK
ucgasue@ucl.ac.uk

Manfred Fahle
Bremen University
Department of Human Neurobiology
Bremen, Germany
mfahle@uni-bremen.de

A. S. Falcão
Centro de Patogénese Molecular - UBMBE iMed.UL,
Faculdade de Farmácia
University of Lisbon
Lisbon, Portugal

Jamshid Faraji
Canadian Centre for Behavioural Neuroscience,
Department of Neuroscience
University of Lethbridge
Lethbridge, AB, Canada

David H. Farb
Laboratory of Molecular Neurobiology
Department of Pharmacology and Experimental Therapeutics
Boston, MA, USA
eisensta@cc.umanitoba.ca

Nathan R. Farrar
Department of Biological Sciences
Brock University
St. Catharines, ON, Canada

Paul A. Faure
Department of Psychology, Neuroscience & Behavior,
McMaster University
Hamilton, ON, Canada
paul4@mcmaster.ca

Philippe Faure
Department of Neurobiology
Pasteur Institute and CNRS
Paris, France
phfaure@pasteur.fr

James W. Fawcett
Cambridge University Centre for Brain Repair,
Robinson Way
Cambridge, UK
jf108@cam.ac.uk

Anna Fejtová
Leibniz Institute for Neurobiology
Department of Neurochemistry and Molecular Biology
Magdeburg, Germany

Ursula Felderhoff-Mueser
Department of Neonatology
Charité Universitätsmedizin Berlin,
Campus Virchow
Klinikum
Berlin, Germany

Anatol G. Feldman
Neurological Science Research Center,
Department of Physiology
University of Montreal
Montreal, QC, Canada
feldman@med.umontreal.ca

Daniel J. Felleman
Department of Neurobiology and Anatomy
University of Texas
Medical School-Houston
Houston, TX, USA

Honorary Fellow
Department of Anatomy and Structural Biology
University of Otago
Dunedin, New Zealand
Daniel.Felleman@uth.tmc.edu

Feng Feng
Department of Biology
University of Iowa
Iowa City, IA, USA

Zhong-Ping Feng
Department of Physiology
University of Toronto
Toronto, ON, Canada
Zp.feng@utoronto.ca

André A. Fenton
Department of Physiology and Pharmacology
The Robert F. Furchgott Center for Neural and Behavioral Science,
State University of New York,
Downstate Medical Center
Brooklyn, NY, USA
afenton@biosignalgroup.com

Russell D. Fernald
Department of Biology
Stanford University
Stanford, CA, USA

A. Fernandes
Centro de Patogénese Molecular - UBMBE iMed.UL,
Faculdade de Farmácia
University of Lisbon
Lisbon, Portugal

Joseph R. Fetcho
Department of Neurobiology and Behavior
Cornell University
Ithaca, NY, USA
jfetcho@notes.cc.sunysb.edu

Arie Feuer
Electrical Engineering
Technion, Israel
feuer@ee.technion.ac.il

Howard L. Fields
Ernest Gallo Clinic and Research Center,
University of California
San Francisco, USA
hlf@phy.ucsf.edu

Roger B. Fillingim
Department of Community Dentistry and Behavioral Science,
College of Dentistry
University of Florida and North Florida
South Georgia
Veterans Health System
Gainesville, FL, USA
RFILLINGIM@dental.ufl.ed

Nanna B. Finnerup
Department of Neurology and Danish Pain Research Center
Aarhus University Hospital
Aarhus, Denmark

Maria Fitzgerald
UCL Department of Anatomy and Developmental Biology
University College London
London, UK
m.fitzgerald@ucl.ac.uk

Martha Flanders
Department of Neuroscience
University of Minnesota
Minneapolis, MN, USA
fland001@umn.edu

Tamar Flash
Department of Computer Science and Applied Mathematics
Weizmann Institute of Science
Rehovot, Israel
tamar.flash@weizmann.ac.il

Erica L. Fletcher
Department of Anatomy and Cell Biology
The University of Melbourne
Parkville, VI, Australia

Eric Fliers
Dept of Endocrinology and Metabolism,
Academic Medical Center
University of Amsterdam
Amsterdam, The Netherlands
e.fliers@amc.uva.nl

Kevin C. Flynn
Molecular, Cellular and Integrative Neurosciences Program,
Department of Biochemistry and Molecular Biology
Colorado State University
Fort Collins, CO, USA

Mónica Folgueira
University of A Coruña,
Campus A Zapateira
A Coruña, Spain

Alfredo Fontanini
Department of Neurobiology and Behavior
State University of New York at Stony Brook
Stony Brook, NY, USA
afontanini@brandeis.edu

Robert D. Foreman
Department of Physiology
University of Oklahoma Health Sciences Center
Oklahoma City, OK, USA
Robert-Foreman@ouhsc.edu

Fiona Francis
Département Génétique, Développement et Pathologie Moléculaire
INSERM U567, CNRS UMR 8104, Institut Cochin
Paris, France

Cy Frank
Department of Surgery,
University of Calgary
Calgary, AB, Canada
cfrank@ucalgary.ca

Marcos G. Frank
Department of Neuroscience
University of Pennsylvania
Philadelphia, PA, USA
mgf@mail.med.upenn.edu

Paul Franken
Center for Integrative Genomics
University of Lausanne
Lausanne-Dorigny, Switzerland
paul.franken@unil.ch

Peter Fransson
MR Research Center,
Cognitive Neurophysiology
Department of Clinical Neuroscience,
Karolinska Institute
Stockholm, Sweden
Peter.Fransson@ki.se

Ralph D. Freeman
Vision Science, School of Optometry
Helen Wills Neuroscience Institute,
University of California
Berkeley, CA, USA
freeman@neurovision.berkeley.edu

Curtis R. French
Department of Biological Sciences
University of Alberta
Edmonton, AB, Canada

Bernd Fritzsch
Department of Biomedical Sciences,
Creighton University
Omaha, NE, USA
fritzsch@creighton.edu

Jörg Frommer
Universitätsklinikum Magdeburg,
Abteilung für Psychosomatische Medizin und Psychotherapie
Magdeburg, Germany
joerg.frommer@medizin.uni-magdeburg.de

Elizabeth Fry
Center for Research in Neuroscience
McGill University
Health Center
Montreal, Canada

Toshikatsu Fujii
Department of Behavioral Neurology and Cognitive Neuroscience
Tohoku University
Graduated School of Medicine
Sendai, Japan
fujii@mail.tains.tohoku.ac.jp

Robert S. Fujinami
Department of Pathology
University of Utah School of Medicine
Salt Lake City, Utah, USA
robert.fujinami@hsc.utah.edu

Hidefumi Fukumitsu
Laboratory of Molecular Biology
Gifu Pharmaceutical University
Gifu, Japan
fukumitsu@gifu-pu.ac.jp

Shintaro Funahashi
Kokoro Research Center,
Kyoto University
Kyoto, Japan
h50400@sakura.kudpc.kyoto-u.ac.jp

John B. Furness
Department of Anatomy and Cell Biology and Centre for Neuroscience
University of Melbourne
Parkville, VIC, Australia
j.furness@unimelb.edu.au

Shoei Furukawa
Laboratory of Molecular Biology
Gifu Pharmaceutical University
Gifu, Japan
furukawa@gifu-pu.ac.jp

Volker Gadenne
Department of Philosophy and Theory of Science
Johannes-Kepler-University Linz
Linz, Germany
volker.gadenne@jku.at

Manfred Gahr
Department of Behavioural Neurobiology
Max Planck Institute for Ornithology
Germany
gahr@bio.vu.nl

C. Giovanni Galizia
Universität Konstanz
Faculty of Biology
Konstanz, Germany
galizia@uni-konstanz.de

Milagros Gallo
Department of Experimental Psychology and Physiology of Behavior,
Institute of Neurosciences
University of Granada
Granada, Spain
mgallo@ugr.es

Simon Gandevia
Prince of Wales Medical Research Institute
Sydney, NSW, Australia
s.gandevia@unsw.edu.au

Qian Gao
Section of Comparative Medicine,
Department of Ob/Gyn & Reproductive Sciences and Neurobiology
Yale University School of Medicine
New Haven, CT, USA

P. Gasque
Brain Inflammation and Immunity Group
Department of Medical Biochemistry
Cardiff University
Cardiff, UK

Gary O. Gaufo
Department of Biomedical Sciences
Creighton University
Omaha, NE, USA

Gerard L. Gebber
Department of Pharmacology and Toxicology
Michigan State University
East Lansing, MI, USA
gebber@msu.edu

G. F. Gebhart
Center for Pain Research
University of Pittsburgh
Pittsburgh, PA, USA
gebhartgf@upmc.edu

Karl R. Gegenfurtner
Abteilung Allgemeine Psychologie
Justus-Liebig-Universität
Giessen, Germany
Karl.R.Gegenfurtner@psychol.uni-giessen.de

Yuri Geinisman
Department of Cell and Molecular Biology
Northwestern University's Feinberg School of Medicine and Institute of Neuroscience,
Chicago, IL, USA
yurig@northwestern.edu

Stefanie Geisler
Behavioral Neuroscience Branch,
National Institute on Drug Abuse
Intramural Research Program
Baltimore, MD, USA
geislers@intra.nida.nih.gov

Alan Gelperin
Monell Chemical Senses Center
Philadelphia, PA, USA
agelperin@monell.org

Menno P. Gerkema
Department of Chronobiology
Faculty of Mathematics and Natural Sciences,
University of Groningen
Groningen, The Netherlands
m.p.gerkema@rug.nl

Rémi Gervais
Laboratoire Systèmes Sensoriels, Comportement et Cognition
UMR 5020 CNRS–UCB Lyon IFR19
Lyon, France

Gilles Gheusi
Pasteur Institut,
Laboratory of Perception and Memory
Paris, France
ggheusi@pasteur.fr

Panteleimon Giannakopoulos
Department of Psychiatry
University Hospitals of Geneva
Geneva, Switzerland

Ian L. Gibbins
Department of Anatomy & Histology, and Centre for Neuroscience
Flinders University
Adelaide, SA, Australia
ian.gibbins@flinders.edu.au

Stan C. A. M. Gielen
Department of Biophysics
Radboud University Nijmegen
Nijmegen, The Netherlands
s.giclen@science.ru.nl

Glenn J. Giesler Jr
Department of Neuroscience
University of Minnesota
Minneapolis, MN, USA
giesler@umn.edu

Andrew C. Giles
Brain Research Centre,
University of British Columbia
Vancouver, BC, Canada

Martha U. Gillette
Molecular and Integrative Physiology, and
Neuroscience Program,
Institute for Genomic Biology
University of Illinois at Urbana-Champaign
Urbana, IL, USA
mgillett@uiuc.edu

KEVIN D. GILLIS
Department of Biological Engineering,
Department of Medical Pharmacology and Physiology,
Dalton Cardiovascular Research Center
University of Missouri – Columbia
Columbia, MO, USA
gillisk@missouri.edu

MARTIN GIURFA
Research Center on Animal Cognition
CNRS – University Paul Sabatier
Toulouse, France
giurfa@cict.fr

MICHELLE GLASS
Department of Pharmacology and Clinical Pharmacology
University of Auckland
Auckland, New Zealand
m.glass@auckland.ac.nz

BEN GODDE
Jacobs Center on Lifelong Learning and Institutional Development
Jacobs University Bremen
Bremen, Germany
b.godde@iu-bremen.de

MICHAEL S. GOLD
Department of Medicine, Division of Gastroenterology, Hepatology and Nutrition
University of Pittsburgh School of Medicine
Pittsburgh, PA, USA
GoldM@dom.pitt.edu

TIMOTHY GOMEZ
Department of Anatomy
University of Wisconsin School of Public Health
Madison, WI, USA

IRINA A. GONTOVA
Laboratory of Neuroimmunology
State Research Institute of Clinical Immunology of SB RAMS
Novosibirsk, Russia

AGUSTÍN GONZÁLEZ
Dept. Biología Celular, Fac. Biología
Univ. Complutense
Madrid, Spain

ANTONY W. GOODWIN
Department of Anatomy and Cell Biology
University of Melbourne
Parkville, VIC, Australia
a.goodwin@unimelb.edu.au

GRAHAM C. GOODWIN
ARC Centre for Complex Dynamic Systems and Control,
Department of Electrical and Computer Engineering
The University of Newcastle
Callaghan, NSW, Australia

TESSA GORDON
Centre of Neuroscience,
Division of Physical Medicine and Rehabilitation,
Faculty of Medicine and Dentistry
University of Alberta
Edmonton, AB, Canada

VERENA GOTTSCHLING
Department for Philosophy
York University
Toronto, ON, Canada

DAVID GOZAL
Kosair Children's Hospital Research Institute,
Department of Pediatrics
University of Louisville
Louisville, KY, USA

RICHARD H. GRACELY
Internal Medicine-Rheumatology
University of Michigan
Ann Arbor, MI, USA

WERNER M. GRAF
Department of Physiology & Biophysics
Howard University College of Medicine
Washington, DC, USA

ROBERT W. GRANGE
Department of Human Nutrition, Foods and Exercise
Virginia Polytechnic Institute and State University
Blacksburg, VA, USA
rgrange@vt.edu

ALEXEJ GRANTYN
Laboratoire de Physiologie de la Perception et de l'Action
C.N.R.S – College de France
Paris, France

THOMAS GRAVEN-NIELSEN
Center for Sensory-Motor Interaction (SMI),
Department of Health Science and Technology
Aalborg University
Aalborg, Denmark

JOEL D. GREENSPAN
Department of Biomedical Sciences
University of Maryland Dental School
Baltimore, MD, USA

John J. Greer
Department of Physiology,
Centre for Neuroscience
University of Alberta
Edmonton, AB, Canada

Michael J. Grey
Department of Exercise and Sport Science &
Department of Neuroscience and Pharmacology
The Panum Institute
Copenhagen, Denmark
mgrey@mfi.ku.dk

Kenneth L. Grieve
Faculty of Life Sciences
University of Manchester
Manchester, UK
ken.grieve@manchester.ac.uk

M. Griffiths
Brain Inflammation and Immunity Group
Department of Medical Biochemistry
Cardiff University
Cardiff, UK

Natasha L. Grimsey
Department of Pharmacology and Clinical
Pharmacology
University of Auckland
Auckland, New Zealand

William C. de Groat
Department of Pharmacology
University of Pittsburgh Medical School
Pittsburgh, PA, USA
degroat@server.pharm.pitt.edu

Henk J. Groenewegen
Institute for Clinical and Experimental Neurosciences,
Department of Anatomy and Neurosciences
VU University medical center
Amsterdam, The Netherlands
hj.groenewegen@vumc.nl

Claude Gronfier
Department of Chronobiology
Stem Cell and Brain Research Institute,
Institut National de la Santé et de la Recherche
Médicale
France

Ted S. Gross
Department of Orthopaedics and Sports Medicine
University of Washington
Seattle, WA, USA
tgross@u.washington.edu

Stephen Grossberg
Department of Cognitive and Neural Systems
Center of Excellence for Learning in Education,
Science and Technology, Boston University
Boston, MA, USA
steve@cns.bu.edu

Murray Grossman
Department of Neurology
University of Pennsylvania School of Medicine
Philadelphia, PA, USA
mgrossma@mail.med.upenn.edu

Matthew S. Grubb
MRC Centre for Developmental Neurobiology
King's College London
London, UK
matthew.grubb@kcl.ac.uk

Edward Gruberg
Biology Department
Temple University
Philadelphia, PA, USA
e.gruberg@temple.edu

Ulrike Grünert
The National Vision Research Institute of Australia
Department of Optometry & Vision Sciences
The University of Melbourne
Australia
ugrunert@optometry.unimelb.edu.au

Alain Guillaume
Institut des Sciences du Movement
UMR H6233 CNRS
Université de la Méditerranée
Marseille, France

Gilles J. Guillemin
Centre for Immunology and Department of
Pharmacology, Faculty of Medicine
University of New South Wales, NSW
Australia
g.guillemin@cfi.unsw.edu.au

Salvador Guirado
Departamento de Biología Celular, Genética y
Fisiología, Facultad de Ciencias
Universidad de Málaga
Málaga, Spain
guirado@uma.es

Daniel Guitton
Montreal Neurological Institute
McGill University
Montreal, QC, Canada
dguitt@mni.mcgill.ca

Eckart D. Gundelfinger
Leibniz Institute for Neurobiology
Department of Neurochemistry and Molecular Biology
Magdeburg, Germany
Gundelfinger@ifn-magdeburg.de

Onur Güntürkün
Biopsychology, Institute of Cognitive Neuroscience
Ruhr-University Bochum
Bochum, Germany
onur.guentuerkuen@ruhr-uni-bochum.de

Victor S. Gurfinkel
Neurological Sciences Institute
Oregon Health & Science University
Portland, OR, USA
gurfinkv@ohsu.edu

Serap Gur
Department of Urology
Tulane University Medical Center
New Orleans, LA, USA

Suzanne N. Haber
Department of Pharmacology and Physiology
University of Rochester School of Medicine
Rochester, NY, USA
suzanne_haber@urmc.rochester.edu

Reza Habib
Department of Psychology,
Department of Computer Science, and School of Medicine
Southern Illinois University Carbondale
Carbondale, IL, USA

John J. Haddad
Cellular and Molecular Signaling Research Group,
Division of Biological Sciences,
Departments of Biology and Biomedical Sciences,
Faculty of Arts and Sciences,
Lebanese International University,
Beirut, Lebanon
john.haddad@liu.edu.lb

Thomas Hadjistavropoulos
Centre on Aging and Health and Department of Psychology
University of Regina
Regina, SK, Canada
Thomas.Hadjistavropoulos@uregina.ca

Tatsuya Haga
Institute for Biomolecular Science, Faculty of Science
Gakushuin University
Tokyo, Japan
tatsuya.haga@gakushuin.ac.jp

Pejmun Haghighi
Department of Physiology
McGill University
Montreal, QC, Canada
pejmun.haghighi@mcgill.ca

Akira Haji
Laboratory of Neuropharmacology
School of Pharmacy,
Aichi Gakuin University
Chikusa, Nagoya, Japan
haji@dpc.aichi-gakuin.ac.jp

Mimi Halpern
Facultad de Medicina and Centro Regional de Investigaciones Biomédicas (CRIB)
Universidad de Castilla-La Mancha
Albacete, Spain
mimi.halpern@downstate.edu

Jung-Soo Han
Department of Biological Sciences
Konkuk University
Seoul, South Korea

Herrmann O. Handwerker
Institut für Physiologie und Pathophysiologie
Erlangen, Germany
handwerker@physiologie1.uni-erlangen.de

Thorsten Hansen
Abteilung Allgemeine Psychologie
Justus-Liebig-Universität
Giessen, Germany
Thorsten.Hansen@psychol.uni-giessen.de

Jens Harbecke
Department of Philosophy
University of Witten-Herdecke
Germany

Deborah L. Harrington
Department of Radiology
University of California, San Diego and V.A. San Diego Healthcare Center
San Diego, CA, USA
dharrington@ucsd.edu

Ronald M. Harris-Warrick
Department of Neurobiology and Behavior
Cornell University
Ithaca, NY, USA
rmh4@cornell.edu

Nathan S. Hart
School of Biomedical Sciences
University of Queensland, St. Lucia
Brisbane, QLD, Australia
n.hart@uq.edu.au

Steven E. Harte
Department of Neurology
University of Michigan
VA Ann Arbor Health System
Ann Arbor, MI, USA

Christian Harteneck
Molekulare Pharmakologie & Zellbiologie
Charité Universitätsmedizin Berlin
Berlin, Germany
Christian.Harteneck@charite.de

Yuichi Hashimoto
Institute for Biomolecular Science,
Faculty of Science
Gakushuin University
Tokyo, Japan

Yoshio Hata
Division of Integrative Bioscience
Tottori University Graduate School of Medical Sciences
Yonago, Japan
yhata@grape.med.tottori-u.ac.jp

Yutaka Hata
Department of Medical Biochemistry,
Graduate School of Medicine
Tokyo Medical and Dental University
Tokyo, Japan
yuhammch@med.tmd.ac.jp

Megumi Hatori
The Salk
Institute for Biological Studies
La Jolla, CA, USA

Hanns Hatt
Department of Cell Physiology
Ruhr-University Bochum
Bochum, Germany
Hanns.Hatt@ruhr-uni-bochum.de

Susanne Hausselt
Department of Biomedical Optics
Max-Planck Institute for Medical Research
Heidelberg, Germany

John A. Hawley
Exercise Metabolism Group,
School of Medical Sciences
RMIT University
Bundoora, Victoria, Australia
john.hawley@rmit.edu.au

Michael R. Hayden
Centre for Molecular Medicine and Therapeutics
University of British Columbia
Vancouver, Canada
mrh@cmmt.ubc.ca

John A. Hayes
The Department of Applied Science
The College of William and Mary
Williamsburg, VA, USA

Nicholas P. Hays
Nutrition, Metabolism, and Exercise Laboratory,
Donald W. Reynolds Institute on Aging
University of Arkansas for Medical Sciences
Little Rock, AR, USA
HaysNicholasP@uams.edu

David Hazlerigg
Jackwat Consulting Services Springview
NSW, Australia
jackwat@bigpond.com

C. J. Heckman
Departments of Physiology, Physical Medicine and Rehabilitation, and Biomedical Engineering
Feinberg School of Medicine
Northwestern University
Chicago, IL, USA
c-heckman@nwu.edu

Wayne G. Hellstrom
Department of Pharmacology
Tulane University Medical Center
New Orleans, LA, USA

Thomas Hellwig-Bürgel
Institute of Physiology,
University of Lübeck
Lübeck, Germany
hellwig@physio.uni-luebeck.de

Wayne Hening
Department of Neurology
University of Medicine and Dentistry of New Jersey
New Brunswick, NJ, USA
WAHeningMD@aol.com

Michael Herzog
Laboratory of Psychophysics,
Brain Mind Institute
Ecole Polytechnique Fédérale de Lausanne
Lausanne, Switzerland
michael.herzog@epfl.ch

Walter Herzog
Faculty of Kinesiology
Human Performance Laboratory
University of Calgary
Calgary, AB, Canada
walter@kin.ucalgary.ca

Juan Hidalgo
Institute of Neurosciences and Department of Cellular Biology, Physiology and Immunology,
Animal Physiology Unit,
Faculty of Sciences,
Autonomous University of Barcelona
Bellaterra, Barcelona, Spain

Stephen M. Highstein
Department of Otolaryngology and Department of Anatomy and Neurobiology
Washington University
Medical School
St. Louis, MO, USA
highstes@medicine.wustl.edu

Okihide Hikosaka
Laboratory of Sensorimotor Research
National Eye Institute,
National Institute of Health
Bethesda, MD, USA
oh@lsr.nei.nih.gov

Tina Hinton
Department of Pharmacology,
The University of Sydney
NSW, Australia
tinah@med.usyd.edu.au

Makoto Hirahara
Faculty of Engineering,
Hosei University
Tokyo, Japan
hirahara@hosei.ac.jp

Tomoo Hirano
Department of Biophysics
Graduate School of Science,
Kyoto University
Kyoto, Japan
thirano@neurosci.biophys.kyoto-u.ac.jp

Tatsumi Hirata
Division of Brain Function,
National Institute of Genetics
Graduate University for Advanced Studies (Sokendai)
Mishima, Japan
tathirat@lab.nig.ac.jp

Miyakawa Hiroyoshi
Laboratory of cellular Neurobiology,
School of Life Sciences
Tokyo University of Pharmacy and Life Sciences
Tokyo, Japan
miyakawa@ls.toyaku.ac.jp

Judith A. Hirsch
Department of Biological Sciences
University of Southern California
Los Angeles, CA, USA
jhirsch@usc.edu

Frantisek Hlavacka
Institute of Normal and Pathological Physiology
Bratislava, Slovakia
hlavacka@unpf.savba.sk

William Hodos
Department of Psychology
University of Maryland
College Park, MD, USA
whodos@psyc.umd.edu

Markus J. Hofer
School of Molecular and Microbial Biosciences,
The University of Sydney
Sydney, NSW, Australia

Patrick R. Hof
Department of Neuroscience
Mount Sinai School of Medicine
New York, NY, USA
patrick.hof@mssm.edu

Michel A. Hofman
Netherlands Institute for Neuroscience
Amsterdam, The Netherlands
M.Hofman@nin.knaw.nl

Frank Hofmann
Philosophisches Seminar
Universität Tübingen
Tübingen, Germany
f.hofmann@uni-tuebingen.de

Ralph L. Holloway
Department of Anthropology
Columbia University
USA

Nicholas P. Holmes
Espace et Action
Bron, France
nicholas.p.holmes@wolfson.oxon.org

Ikuo Homma
Department of Physiology
Showa University
School of Medicine
Tokyo, Japan
ihomma@med.showa-u.ac.jp

Fay B. Horak
Neurological Sciences Institute,
Oregon Health and Science University
Portland, OR, USA
horakf@ohsu.edu

Hidenori Horie
Advanced Research Institute for Biological Science
Waseda University
Nishitokyo, Tokyo, Japan

Arata Horii
Department of Otolaryngology
Osaka University School of Medicine
Suita, Osaka, Japan
ahorii@ent.med.osaka-u.ac.jp

David E. Hornung
Dana Professor of Biology
St. Lawrence University
Canton, NY, USA
dhornung@stlawu.edu

Jonathan C. Horton
Department of Ophthalmology
Koret Vision Research Laboratory
UCSF,
San Francisco, CA, USA
hortonj@vision.ucsf.edu

Tamas Horvath
Section of Comparative Medicine,
Department of Ob/Gyn & Reproductive Sciences and Neurobiology
Yale University School of Medicine
New Haven, CT, USA
tamas.horvath@yale.edu

Xintian Hu
Laboratory of Sensory Motor Integration
Kunming Institute of Zoology,
The Chinese Academy of Sciences
Kunming, People's Republic of China

Gema Huesa
Institute of Neurosciences
University Autónoma of Barcelona,
Campus of Bellaterra
Bellaterra, Spain
gema.huesa@uab.es

Peter A. Huijing
Research Instituut Faculteit Bewegingswetenschappen
Vrije Universiteit
Amsterdam, The Netherlands
p_a_j_b_m_huijing@fbw.vu.nl

Inge Huitinga
Netherlands Institute for Neurosciences
Amsterdam, The Netherlands
i.huitinga@nin.knaw.nl

Peter Hunter
Auckland Bioengineering Institute
University of Auckland
Auckland, New Zealand
p.hunter@auckland.ac.nz

Clotilde M. J. I. Huyghues-Despointes
Department of Physiology
Emory University
Atlanta, GA, USA

Masumi Ichikawa
Tokyo Metropolitan Institute for Neuroscience
Tokyo, Japan
mich@tmin.ac.jp

Chizuka Ide
Institute of Regeneration and Rehabilitation
Department of Occupational Terapy
Faculty of Nursing and Rehabilitation
Aino University
Ibaragi, Osaka, Japan
c-ide@ot-u.aino.ac.jp

Junko Iida
Department of Medical Biochemistry,
Graduate School of Medicine
Tokyo Medical and Dental University
Tokyo, Japan

Kazuhiro Ikenaka
National Institute of Physiology Sciences
Division of Neurobiology and Bioinformatics
Myodaiyi, Okazaki, Japan
ikenaka@nips.ac.jp

Kurt R. Illig
Department of Psychology
University of Virginia Charlottesville
VA, USA

Kazuhide Inoue
Department of Molecular and System Pharmacology,
Graduate School of Pharmaceutical Sciences
Kyushu University,
Japan
inoue@nihs.go.JP

Andreas A. Ioannides
Laboratory for Human Brain Dynamics
Brain Science Institute, RIKEN
Saitama, Japan
ioannides@postman.riken.jp

Dexter R. F. Irvine
Department of Psychology
Monash University
VIC, Australia
Irvine@med.monash.edu.au

Yasuyuki Ishikawa
Nara Institute of Science and Technology (NAIST)
Structural Cell Biology
Nara, Japan

Yutaka Itokazu
Department of Anatomy and Neurobiology
Kyoto University Graduate School of Medicine,
Yoshidakonoe-cho
Sakyo-ku, Kyoto, Japan

Alexey M Ivanitsky
Group of Human Higher Nervous Activity
Institute of Higher Nervous Activity and
Neurophysiology
Russian Academy of Sciences
Moscow, Russia
alivanit@aha.ru

George A. Ivanitsky
Laboratory of Human Higher Nervous Activity
Institute of Higher Nervous Activity and
Neurophysiology
Russian Academy of Sciences
Moscow, Russia

Yoshiaki Iwamura
Department of Sensory Science
Kawasaki University of Medical Welfare
Okayama, Japan
iwayoshi@mw.kawasaki-m.ac.jp

Dennis J. McFarland
New York State Department of Health
Laboratory of Nervous System Disorders,
Wadsworth Center
Albany, NY, USA

Mehrnaz Jafarian-Tehrani
Laboratoire de Pharmacologie (UPRES EA2510)
Université Paris Descartes - UFR Pharmacie
Paris, France
mehrnaz.jafarian@univ-paris5.fr

Johannes Jakobsen
Department of Neurology
Aarhus University Hospital
Aarhus, Denmark
jakob@as.aaa.dk

Wilfrid Jänig
Department of Physiology
Physiologisches Institut
Christian-Albrechts-Universität zu Kiel
Kiel, Germany
w.janig@physiologie.uni-kiel.de

Erich D. Jarvis
Department of Neurobiology
Duke University Medical Center
Durham, NC, USA
Jarvis@neuro.duke.edu

John Jeka
Department of Kinesiology
Neuroscience and Cognitive Science Program,
Bioengineering Graduate
Program
University of Maryland
MD, USA
jjeka@umd.edu

Troels S. Jensen
Department of Neurology and Danish Pain Research
Center
Aarhus University Hospital
Aarhus, Denmark
tsjensen@ki.au.dk

Harry J. Jerison
Department of Psychiatry & Biobehavioral Sciences
UCLA Health Sciences Center
Santa Monica, CA, USA
hjerison@ucla.edu

Brett A. Johnson
Department of Neurobiology and Behavior
University of California
Irvine, CA, USA
bajohnso@uci.edu

Bruce R. Johnson
Department of Neurobiology and Behavior
Cornell University
Ithaca, NY, USA
brj1@cornell.edu

GRAHAM A. R. JOHNSTON
Department of Pharmacology
The University of Sydney
NSW, Australia
grahamj@mail.usyd.edu.au

BARBARA E. JONES
Department of Neurology & Neurosurgery
McGill University
Montreal Neurological Institute
Montreal, QC, Canada
barbara.jones@mcgill.ca

LYNETTE A. JONES
Department of Mechanical Engineering
Massachusetts Institute of Technology
Cambridge, MA, USA
ljones@mit.edu

RENÉ J. JORNA
Professor of Knowledge Management & Cognition
Faculty of Economics and Business
University of Groningen
Groningen, The Netherlands
r.j.j.m.jorna@bdk.rug.nl

M. J. JOYNER
Departments of Anesthesiology and Physiology
Mayo Clinic College of Medicine
Rochester, MN, USA
joyner.michael@mayo.edu

JON H. KAAS
Department of Psychology
Vanderbilt University
Nashville, TN, USA
Jon.H.Kaas@vanderbilt.edu

HIDETO KABA
Department of Physiology
Kochi Medical School
Nankoku, Kochi, Japan
kabah@med.kochi-u.ac.jp

TORAH M. KACHUR
Department of Biological Sciences
University of Alberta
Edmonton, AB, Canada

PHILIP J. KADOWITZ
Department of Pharmacology
Tulane University Medical Center
New Orleans, LA, USA
pkadowi@tulane.edu

RYOICHIRO KAGEYAMA
Institute for Virus Research
Kyoto University
Kyoto, Japan
rkageyam@virus.kyoto-u.ac.jp

MICHAEL KALLONIATIS
Department of Optometry and Vision Science
University of Auckland
Auckland, New Zealand
m.kalloniatis@auckland.ac.nz

MIHALY KALMAN
1st Department of Anatomy
Semmelweis University
School of Medicine
Budapest, Hungary
kalman@ana.sote.hu

ANDRIES KALSBEEK
Netherlands Institute for Neuroscience
Amsterdam, The Netherlands

BARBARA KALTSCHMIDT
Department of Cell Biology
Faculty of Biology
University of Bielefeld
Bielefeld, Germany
b.kaltschmidt@uni-wh.de

CHRISTIAN KALTSCHMIDT
Department of Cell Biology
Faculty of Biology
University of Bielefeld
Bielefeld, Germany
C.Kaltschmidt@uni-wh.de

HARUYUKI KAMIYA
Department of Neurobiology
Department of Molecular Neuroanatomy
Hokkaido University School of Medicine
Sapporo, Japan
kamiya@med.hokudai.ac.jp

RYOTA KANAI
Division of Biology
California Institute of Technology
Pasadera, CA, USA
r.kanai@ucl.ac.uk

CHRIS R. S. KANEKO
Department of Physiology and Biophysics
Washington National Primate Research Center,
University of Washington
Seattle, WA, USA
kaneko@u.washington.edu

Hiroshi Kannan
Department of Physiology
Faculty of Medicine
University of Miyazaki
Miyazaki, Japan
kannanh@med.miyazaki-u.ac.jp

Amir Karniel
Department of Biomedical Engineering
Ben-Gurion University of the Negev
Beer Sheva, Israel
akarniel@bgu.ac.il

Haruo Kasai
Laboratory of structured Physiology
Center for Disease
Biology and Integrative Medicine
University of Tokyo
Tokyo, Japan
hkasai@nips.ac.jp

Sabine Kastner
Department of Psychology
Princeton Neuroscience Institute
Center for the Study of Brain, Mind and Behavior
Princeton University
Princeton, NJ, USA
skastner@Princeton.EDU

Yoshifumi Katayama
Department of Autonomic Physiology
Medical Research Institute
Tokyo Medical and Dental University
Tokyo, Japan
kataauto@mri.tmd.ac.jp

Kazuo Kato
Department of Psychology
Kyushu University
Fukuoka, Japan
kaz@hes.kyushu-u.ac.jp

Hiroshi Kawabe
Department of Molecular Neurobiology
Max-Planck-Institute for Experimental Medicine
Göttingen, Germany

Yoshiko Kawai
Department of Physiology
Shinshu University School of Medicine
Matsumoto, Japan

Mitsuo Kawato
ATR Computational Neuroscience Laboratories
Kyoto, Japan
kawato@atr.jp

Leslie M. Kay
Department of Psychology
Institute for Mind and Biology
The University of Chicago
Chicago, IL, USA
LKay@uchicago.edu

Robert W. Keane
Department of Physiology and Biophysics
University of Miami Miller School of Medicine
Miami, FL, USA
rkeane@miami.edu

Geert Keil
Dept. of Philosophy
RWTH Aachen University
Achen, Germany
geert.keil@rwth-aachen.de

Laurie A. Kellaway
Division of Neuroscience
Department of Human Biology
Faculty of Health Sciences
University of Cape Town
Cape Town, South Africa

Bernhard Keller
Department of Neurophysiology
University of Göttingen
Göttingen, Germany
bkeller1@gwdg.de

Edward L. Keller
Smith-Kettlewell Eye Research Institute
San Francisco, CA, USA
elk@ski.org

Daniel Kelly
Department of Philosophy
Purdue University
West Lateyette, IN, USA

Mineko Kengaku
Laboratory for Neural Cell Polarity
RIKEN Brain Science Institute
Wako, Japan
kengaku@brain.riken.jp

Henry Kennedy
INSERM U846, Stem Cell and Brain Research Institute
Department of Integrative Neuroscience
Bron cedex, France
kennedy@lyon.inserm.fr

Robert W. Kentridge
Psychology Department
University of Durham
Durham, UK
robert.kentridge@durham.ac.uk

Roland Kern
Institut für Psychologie II
Münster, Germany

Douglas Steven Kerr
Department of Pharmacology & Toxicology
University of Otago School of Medical Sciences
Dunedin, New Zealand
steve.kerr@stonebow.otago.ac.nz

Dirk Kerzel
Faculté de Psychologie et des Sciences de l'Éducation
Université de Genève
Genève, Switzerland
dirk.kerzel@pse.unige.ch

Emily A. Keshner
Department of Physical Therapy
College of Health
Professions and Department of Electircal and Computer Engineering
College of Engineering
Temple University
Philadelphia, PA, USA
eak@NORTHWESTERN.EDU

Haider Ali Khan
Nuclear Medicine Department
Kuwait Center for Specialized Surgery
Kuwait

Ilya Khaytin
Department of Cell & Developmental Biology
Vanderbilt University
Medical School
U3218 Learned Lab
Nashville, TN, USA

Yoshi Kidokoro
Institute for Molecular and Cellular Regulation
Gunma University
Gunma, Japan
kidokoro@med.gunma-u.ac.jp

Ole Kiehn
Mammalian Locomotor Laboratory
Department of Neuroscience
Karolinska Institute
Stockholm, Sweden
o.kiehn@neuro.ki.se

Tim Kiemel
Department of Kinesiology
University of Maryland
MD, USA
iemel@umd.edu

Joanne S. Kim
Department of Molecular Genetics
University of Toronto
Toronto, ON, Canada

Julia Kim
Laboratory of Molecular Neurobiology
Department of Pharmacology and Experimental Therapeutics
Boston, MA, USA

Hiroshi Kimura
Graduate School of Biostudies
Kyoto University
Kyoto, Japan

Minoru Kimura
Kyoto Prefectural University of Medicine
Kawaramachi-Hirokoji
Kamigyo-ku, Kyoto, Japan
mkimura@koto.kpu-m.ac.jp

Andrew J. King
Department of Physiology, Anatomy and Genetics
Sherrington Building
University of Oxford, Oxford, UK

Jonathan Kipnis
The Weizmann Institute of Science
Rehovot, Israel

Janina Kirsch
Biopsychology, Institute of Cognitive Neuroscience
Ruhr-University Bochum
Bochum, Germany

Stephen J. Kish
Human Neurochemical Pathology Laboratory
Centre for Addiction and Mental Health
Toronto, ON, Canada

Noriyuki Kishi
MGH-HMS Center for Nervous System Repair,
Departments of Neurosurgery and Neurology, and
Program in Neuroscience
Harvard Medical School;
Nayef Al-Rodhan Laboratories
Massachusetts General Hospital
Department of Stem Cell and Regenerative
Biology, and Harvard Stem Cell Institute
Harvard University
Boston, MA, USA
noriyuki_kishi@hms.harvard.edu

Toshihiro Kitama
Center for Life Science Research
University of Yamanashi, Yamanashi
Japan

Christian Klämbt
Institut für Neurobiologie
Münster, Germany

Peter Klaver
Department of Psychology
University of Zurich
Zurich, Switzerland

Elizabeth B. Klerman
Division of Sleep Medicine
Department of Sleep Medicine
Brigham and Women's Hospital
Harvard Medical School
Boston, MA, USA

Keith R. Kluender
Department of Psychology
University of Wisconsin
Madison, WI, USA
krkluend@wisc.edu

Markus Knauff
Center for Cognitive Science
University of Freiburg
Freiburg, Germany

Kenneth Knoblauch
Inserm U846
Bron, France

Mikako Kobayashi-Warren
Department of Pathology
University of Utah
School of Medicine
Salt Lake City, Utah, USA

H. Richard Koerber
Department of Neurobiology
University of Pittsburgh
Pittsburgh, PA, USA

Kiyomi Koizumi
Department of Physiology and Pharmacology
State University of New York
Downstate Medical Center
Brooklyn, New York, USA

Arlette Kolta
Faculté de Médecine Dentaire
Université de Montréal
Succursale Centre-ville
Montréal, QC, Canada

Paul De Koninck
Department of Biochemistry and Microbiology
Laval University
Centre de Recherche Université Laval Robert-Giffard
Quebec, QC, Canada
paul.dekoninck@crulrg.ulaval.ca

Konrad Körding
Institute of Neurology
London, UK
konrad@koerding.com

Helen A. Korneva
Institute for Experimental Medicine
St. Petersburg, Russia

Girish J. Kotwal
Inflamed
Louisville, KY, USA
gjkotw01@yahoo.com

Enikö Kövari
Department of Psychiatry
University Hospitals of Geneva
Geneva, Switzerland

Vladimir A. Kozlov
Laboratory of Neuroimmunology
State Research Institute of Clinical Immunology of SB RAMS
Novosibirsk, Russia

Achim Kramer
Laboratory of Chronobiology
Charité Universitätsmidiain Berlin
Berlin, Germany
achim.kramer@charite.de

Bernd Kramer
Institut für Zoologie
Animal Behaviour and Behavioural Physiology Research Group
Universität Regensburg
Regensburg, Germany
bernd.kramer@biologie.uni-regensburg.de

Nina Kraus
Departments of Communication Sciences, Neurobiology and Physiology, Otolaryngology
Northwestern University
Evanston, IL, USA
nkraus@northwestern.edu

Leah Krubitzer
Department of Psychology
University of California
Davis, CA, USA
lakrubitzer@ucdavis.edu

John L. Kubie
Department of Anatomy and Cell Biology
State University of New York
Downstate Medical Center
Brooklyn, NY, USA

M. Fabiana Kubke
Department of Anatomy with Radiology
Faculty of Medical and Health Sciences
University of Auckland
Auckland, New Zealand
f.kubke@auckland.ac.nz

Yoshihisa Kudo
School of Life Science
Tokyo University of Pharmacy and Life Science
Tokyo, Japan
kudoy@ls.toyaku.ac.jp

Amod P. Kulkarni
Division of Medical Virology
Department of Clinical Laboratory Sciences
Institute of Infectious Disease and Molecular Medicine
amoduct@gmail.com

Shigeru Kuratani
Laboratory for Evolutionary Morphology
Center for Developmental Biology
Riken, Kobe, Japan
saizo@cdb.riken.go.jp

Hiroshi Kuromi
Institute for Behavioral Sciences
Gunma University Graduate School of Medicine
Maebashi, Japan
kuromi@med.gunma-u.ac.jp

Alexander Kusnecov
Rutgers University
Department of Psychology
Piscataway, NJ, USA
kusnecov@rci.rutgers.edu

Frédéric Laberge
Brain Research Institute
University of Bremen
Bremen, Germany

James R. Lackner
Ashton Graybiel Spatial Orientation Laboratory
Brandeis University
Waltham, MA, USA
lackner@brandeis.edu

Bernd Ladwig
Otto-Suhr-Institut für Politik-Wissenschaft
Freie Universität Berlin
Berlin, Germany
berndladwig@hotmail.com

Hugo Lagercrantz
Karolinska Institute
Astrid Lindgren Children's Hospital
Stockholm, Sweden
hugo.lagercrantz@actapaediatrica.se

Peter M. Lalley
Department of Physiology
The University of Wisconsin
School of Medicine and Public Health
Medical Sciences Center
Madison, WI, USA
pmlalley@wisc.edu

Nathalie Lamarche-Vane
Department of Anatomy and Cell Biology
McGill University
Montreal, Quebec, Canada
nathalie.lamarche@mcgill.ca

Michael F. Land
Department of Biology and Environmental Science
University of Sussex
Brighton, UK
m.f.land@sussex.ac.uk

Basile Nicolas Landis
Unité de Rhinologie-Olfactologie
Clinique de Oto-Rhinologie-Laryngologie et de
Chirurgie Cervicofaciale
Hopitaux Universitaires de Genève
Genève, Switzerland
Basile.Landis@hcuge.ch

Gerald Langner
Neuroakustik
Fachbereich Biologie
TU Darmstadt
Darmstadt, Germany
gl@bio.tu-darmstadt.de

Enrique Lanuza
Depts of Biología Cellular i Parasitologia
Facultat de Ciències Biològiques
Universitat de València
Burjassot, Spain
enrique.lanuza@uv.es

Markus Lappe
Psychologisches Institut II
Westf. Wilhelms-Universität
Münster, Germany
mlappe@uni-muenster.de

Mark L. Latash
The Pensylvania State University
University Park
PA, USA
mll11@psu.edu

Françoise Lazarini
Perception and Memory Unit
Neuroscience Department
Pasteur Institute
Paris, France
lazarini@pasteur.fr

Luis de Lecea
Department of Psychiatry and Behavioral Sciences
Stanford University
School of Medicine
Palo Alto, CA, USA
llecea@stanford.edu

Andy K. Lee
Department of Pharmacology
Centre for Neuroscience
University of Alberta
Edmonton, AB, Canada

Samuel C. K. Lee
Research Department
Shriners Hospitals for Children
Philadelphia, PA, USA

Tai Sing Lee
Computer Science Department and Center for Neural Basis of Cognition
Carnegie Mellon University
Pittsburgh, PA, USA
tai@cnbc.cmu.edu

Tzumin Lee
Department of Neurobiology
University of Massachusetts
Medical School
Worcester, MA, USA
Tzumin.Lee@umassmed.edu

Tanya L. Leise
Department of Mathematics
Amherst College
Amherst, MA, USA
tleise@amherst.edu

James C. Leiter
Department of Physiology
Dartmouth Medical School
Lebanon, NH, USA
james.c.leiter@dartmouth.edu

Mado Lemieux
Department of Biochemistry and Microbiology,
Laval University
Centre de Recherche Université Laval Robert-Giffard
Quebec, QC, Canada

Michael Leon
Department of Neurobiology and Behavior
University of California
Irvine, CA, USA
mleon@uci.edu

David A. Leopold
Unit on Cognitive Neurophysiology and Imaging
NIH
Bethesda, MD, USA

Simon Rock Levinson
Department of Physiology and Biophysics
University of Colorado
School of Medicine
Aurora, CO, USA
Rock.Levinson@UCHSC.edu

Joanne M. Lewohl
Genomics Research Centre
School of Medical Sciences
Griffith University
Southport, Queensland, Australia

Alfred J. Lewy
Department of Psychiatry
Oregon Health & Science University
Portland, OR, USA
lewy@ohsu.edu

Jan Lexell
Department of Clinical Science
Lund, Devision of Rehabilitation Medicine,
Lund University
Lund, Sweden
jan.lexell@skane.se

Aihua Li
Department of Physiology
Dartmouth Medical School
Lebanon, NH, USA

David W. Li
Department of Biochemistry and Molecular Biology,
Department of Ophthalmology and Visual Sciences,
College of Medicine
University of Nebraska Medical Center
Omaha, NE, USA
davidli@unmc.edu

Jane E. Libbey
Department of Pathology
University of Utah School of Medicine
Salt Lake City, Utah, USA

David C. Lin
Programs in Bioengineering and Neuroscience
Washington State University
Pullman, WA, USA
davidlin@wsn.edu

Yoav Litvin
Department of Psychology
University of Hawaii at Manoa
Honolulu, HI, USA
litvin@hawaii.edu

Milos Ljubisavljevic
Department of Neurophysiology
Institute for Medical Research
Belgrade, Serbia and Montenegro
milos@uaeu.ac.ae

Pierre-Marie Lledo
Pasteur Institute
Laboratory for Perception and Memory
Paris Cedex, France
pmlledo@pasteur.fr

Anna Lobell
Department of Medical Sciences
Uppsala University
University Hospital
Uppsala, Sweden

Elizabeth F. Loftus
Departments of Psychology and Social Behavior, Criminology, Law and Society, Cognitive Sciences
University of California
Irvine, CA, USA
eloftus@uci.edu

Cairine Logan
Institute of Maternal and Child Health, Genes and Development Research Group
Department of Cell Biology and Anatomy
Faculty of Medicine
University of Calgary
Calgary, AB, Canada
clogan@ucalgary.ca

Catherine M. F. Lohmann
Department of Biology
University of North Carolina
Chapel Hill, NC, USA

Kenneth J. Lohmann
Department of Biology
University of North Carolina
Chapel Hill, NC, USA
klohmann@email.unc.edu

Barbara Lom
Biology Department and Program in Neuroscience
Davidson College
Davidson, NC, USA
balom@davidson.edu

Stephen R. Lord
Prince of Wales Medical Research Institute and
University of New South Wales
Sydney, NSW, Australia
s.lord@unsw.edu.au

Phillip A. Low
Department of Neurology
Mayo Foundation
Rochester, MN, USA
low.phillip@mayo.edu

E. J. Lowe
University of Durham
Durham, United Kingdom
E.J.Lowe@durham.ac.uk

Madeleine Lowery
Rehabilitation Institute of Chicago
Department of Physical Medicine and Rehabilitation
Northwestern University
Chicago, IL, USA
m-lowery@northwestern.edu

Sylvia Lucas
Department of Neurology
University of Washington Medical Center
Seattle, WA, USA
lucass@u.washington.edu

Paul J. Lucassen
Centre for Neuroscience
Swammerdam Institute of Life Sciences
University of Amsterdam
Amsterdam,The Netherlands
P.J.Lucassen@uva.nl

James P. Lund
Faculty of Dentistry
McGill University
Montréal, QC, Canada
james.lund@mcgill.ca

GÖRAN LUNDBORG
Hand Surgery/Department of Clinical Sciences
Malmö University Hospital
Lund University
Lund, Sweden
Goran.Lundborg@hand.mas.lu.se

HUUB MAAS
Departments of Physiology, Physical Medicine and Rehabilitation, and Biomedical Engineering
Northwestern University
Evanston, IL, USA

VAUGHAN G. MACEFIELD
School of Medicine
University of Western Sydney
Sydney, NSW, Australia
v.macefield@uws.edu.au

JEFFREY D. MACKLIS
MGH-HMS Center for Nervous System Repair, Departments of Neurosurgery and Neurology, and Program in Neuroscience
Harvard Medical School;
Nayef Al-Rodhan Laboratories
Massachusetts General Hospital;
Department of Stem Cell and Regenerative Biology, and Harvard Stem Cell Institute
Harvard University
Boston, MA, USA
jeffrey_macklis@hms.harvard.edu

MERRITT MADUKE
Department of Molecular and Cellular Physiology
Stanford University
School of Medicine
Stanford, CA
maduke@stanford.edu

ALEXANDER MAIER
Unit on Cognitive Neurophysiology and Imaging
NIH
Bethesda, MD, USA
maiera@mail.nih.gov

STEVEN F. MAIER
University of New Mexico
Albuquerque, NM, USA

BRIAN E. MAKI
Centre for Studies in Aging, Sunnybrook and Health Sciences Centre
University of Toronto
Toronto, Canada
brian.maki@swri.ca

ADEL MAKLAD
Department of Biology
University of Iowa
Iowa City, IA, USA

JON MALLATT
School of Biological Sciences
Washington State University
Pullman, Washington, WA, USA
jmallat@mail.wsu.edu

MANUEL S. MALMIERCA
Auditory Neurophysiology Unit
Laboratory for the Neurobiology of Hearing
Faculty of Medicine and Insitute for Neuroscience of Castilla y León
Salamanca, Spain
msm@usal.es

JOÃO O. MALVA
Center for Neuroscience and Cell Biology
Institute of Biochemistry
Faculty of Medicine,
University of Coimbra
Coimbra, Portugal
jomalva@fmed.uc.pt

TOSHIYA MANABE
Division of Neuronal Network
Department of Basic Medical Sciences
Institute of Medical Science
University of Tokyo
Tokyo, Japan
tmanabe-tky@umin.ac.jp

NATHALIE MANDAIRON
Neurosciences Sensorielles, Comportement, Cognition
Université de Lyon, Université
Claude Bernard Lyon1, Villeurbanne, France

PAUL R. MANGER
School of Anatomical Sciences
Faculty of Health Sciences
University of the Witwatersrand
Johannesburg, Republic of South Africa
mangerpr@anatomy.wits.ac.za

ABDELJABBAR EL MANIRA
Department of Neuroscience Karolinska Institutet
Stockholm, Sweden
abdel.elmanira@neuro.ki.se

PAUL B. MANIS
Otolaryngology/Head and Neck Surgery and Cell and Molecular Physiology
University of North Carolina at Chapel Hill
NC, USA
pmanis@med.unc.edu

Tadaaki Mano
Gifu University of Medical Science
Seki, Gifu, Japan
tadaaki.mano@nifty.com

Fiona Mansergh
Ocular Genetic Unit, Smurfit Institute of Genetics
Trinity College Dublin
Dublin 2, Ireland
manserghfc@Cardiff.ac.uk

Diego Manzoni
Dipartimento di Fisiologia e Biochimica
Università degli Studi di Pisa
Pisa, Italy
manzoni@dfb.unipi.it

Ying-Wei Mao
Howard Hughes Medical Institute
Massachusetts Institute of Technology
Boston, MA, USA

Uri Maoz
Interdisciplinary Center for Neural Computation
Hebrew University of Jerusalem
Jerusalem, Israel

Denis Mareschal
Centre for Brain and Cognitive Development
Birkbeck, University of London
London, UK
d.mareschal@bbk.ac.uk

Oscar Marín
Instituto de Neurociencias de Alicante
Consejo Superior de Investigaciones Científicas and
Universidad Miguel Hernández
Alicante, Spain
o.marin@umh.es

Lesley Marson
Division of Urology
University of North Carolina at Chapel Hill
Chapel Hill, NC, USA
lmarson@med.unc.edu

Claire Martin
Laboratoire Systèmes Sensoriels, Comportement et Cognition
UMR 5020 CNRS–UCB Lyon
IFR19 Lyon, France

Kevan A. C. Martin
Institute of Neuroinformatics UZH/ETH
Winterthurerstrasse
Zurich, Switzerland
kevan@ini.phys.ethz.ch

Wiesmann Martin
Department of Neuroradiology
Ludwig-Maximilians-University Munich
Germany
Martin.Wiesmann@med.uni-muenchen.de

Paul R. Martin
National Vision Research Institute of Australia
Carlton, VIC, Australia
prmartin@unimelb.edu.au

Luis M. Martinez
Instituto de Neurociencias
CSIC – Universidad Miguel Hernández
Alicante, Spain
l.martinez@umh.es

Gianvito Martino
Neuroimmunology Unit
DIBIT-San Raffaele Scientific Institute
Milano, Italy

Fernando Martínez-García
Biologia Funcional i Antropologia Física
Facultat de Ciències Biològiques
Universitat de València
Burjassot, Spain
fernando.mtnez-garcia@uv.es

Alino Martínez-Marcos
Facultad de Medicina and Centro Regional de Investigaciones Biomédicas (CRIB)
Universidad de Castilla-La Mancha
Albacete, Spain
Alino.Martinez@uclm.es

Shunichi Maruno
Department of Psychology
Kyushu University
Fukuoka, Japan
syunedu@mbox.nc.kyushu-u.ac.jp

Andrew C. Mason
Integrative Behavior and Neuroscience Group,
Department of Biological Sciences
University of Toronto at Scarborough
Scarborough, ON, Canada

Kelby Mason
Department of Philosophy
Rutgers, The State University of New Jersey
New Brunswick, NJ, USA
mason@philosophy.rutgers.edu

Peggy Mason
Department of Neurobiology, Pharmacology and Physiology
University of Chicago
Chicago, IL, USA

Fred W. Mast
Department of Psychology
University of Lausanne
Bâtiment Anthropole, Lausanne, Switzerland
Fred.Mast@unil.ch

Alistair Mathie
Medway School of Pharmacy
Universities of Kent and Greenwich at Medway
Chatham Maritime, Kent, UK
a.a.mathie@kent.ac.uk

Naoya Matsumoto
Department of Anatomy and Neurobiology
Kyoto University Graduate School of Medicine,
Yoshidakonoe-cho, Sakyo-ku, Kyoto, Japan

Carlos Matute
Departamento de Neurociencias
Universidad del País
Vasco, Leioa and
Neurotek-UPV/EHU
Parque Tecnológico de Bizkaia
Zamudio, Spain
carlos.matute@ehu.es

John Robert Matyas
Faculty of Veterinary Medicine
University of Calgary
Calgary, AB, Canada
jmatyas@ucalgary.ca

J. Patrick Mayo
Department of Neuroscience
The Center for the Neural Basis of Cognition, and the Center for Neuroscience
University of Pittsburgh
Pittsburgh, PA, USA

Lawrence Mays
University of North Carolina
Charlotte, NC, USA
lmays@uab.edu

David McAlpine
UCL Ear Institute and Department of Physiology
University College London
London, UK
d.mcalpine@ucl.ac.uk

Catherine A. McCormick
Departments of Biology and Neuroscience
Oberlin College
Oberlin, OH, USA
catherine.mccormick@oberlin.edu

Donald R. Mccrimmon
Department of Physiology
Feinberg School of Medicine
Northwestern University
Chicago, IL, USA
dm@northwestern.edu

Alexander J. McDonald
Department of Pharmacology, Physiology, and Neuroscience
University of South Carolina
School of Medicine
Columbia, SC, USA
MCDONALD@MED.SC.EDU

Sarah McFarlane
The Hotchkiss Brain Institute
Department of Cell Biology and Anatomy
Faculty of Medicine
University of Calgary
Calgary, AB, Canada
smcfarla@ucalgary.ca

Dennis McGinty
Department of Psychology
University of California
Los Angeles, CA, USA
dmcginty@ucla.edu

Patricia A. McGrath
Department of Anaesthesia
Divisional Centre of Pain Management and Research,
The Hospital for Sick Children
Toronto, ON, Canada
patricia.mcgrath@sickkids.ca

Elspeth McLachlan
Prince of Wales Medical Research Institute and the University of New South Wales
Randwick, NSW, Australia
e.mclachlan@unsw.edu.au

Stephanie A. McMains
Department of Psychology
Princeton Neuroscience Institute
Center for the Study of Brain, Mind and Behavior
Princeton University
Princeton, NJ, USA
smcmains@Princeton.EDU

LOWELL T. MCPHAIL
International Collaboration on Repair Discoveries
the University of British Columbia
Vancouver, BC, Canada

LORETA MEDINA
Department of Experimental Medicine
Faculty of Medicine
University of Lleida
Institut of Biomedical Research of Lleida
Lleida, Catalonia, Spain
loreta.medina@cmb.udl.cat

SVYATOSLAV V. MEDVEDEV
Institute of the Human Brain of the Russian Academy of Sciences
Laboratory of the Positron Emission Tomography
St-Petersburg, Russia
Medvedev@ihb.spb.ru

ALFRED R. MELE
Department of Philosophy
Florida State University
Tallahassee, FL, USA
almele@mailer.fsu.edu

D. MENETREY
CNRS UMR 8119, Neurophysique et Physiologie
Université René Descartes
UFR Biomédicale
Paris, France

MANUEL MERCIER
Laboratory of Perceptual Processing
Southern Cross University
Lismore, NSW, Australia
manuel.mercier@epfl.ch

WILLIAM H. MERIGAN
Eye Institute and Center for
Visual Sciences
University of Rochester
Rochester, NY, USA
billm@cvs.rochester.edu

MARTHA MERROW
The University of Groningen
Haren, The Netherlands
m.merrow@rug.nl

GERLINDE A. METZ
Canadian Centre for Behavioural Neuroscience,
Department of Neuroscience
University of Lethbridge
Lethbridge, AB, Canada
Gerlinde.metz@uleth.ca

WALTER METZNER
UCLA Department of Physiological Science
Los Angeles, CA, USA
metzner@ucla.edu

WOLFGANG MEYERHOF
Department of Molecular Genetics
German Institute of Human Nutrition Potsdam-Rehbruecke
Nuthetal, Germany
meyerhof@dife.de

RAJIV MIDHA
Division of Neurosurgery
Department of Clinical Neurosciences
University of Calgary
AB, Canada

AKICHIKA MIKAMI
Primate Research Institute
Kyoto University
Inuyama, Japan
mikami@pri.kyoto-u.ac.jp

ROBERT MILLER
Department of Anatomy and Structural Biology
University of Otago
Dunedin, New Zealand
robert.miller@stonebow.otago.ac.nz

ERIN D. MILLIGAN
Department of Kinesiology and Physical Education,
McGill University
Montreal, QC, Canada
erin.milligan@colorado.edu

THEODORE E. MILNER
School of Kinesiology
Simon Fraser University
Burnaby, British Columbia, Canada
tmilner@sfu.edu

JOSEPH A. MINDELL
Membrane Transport Biophysics Unit
Porter Neuroscience Research Center
National Institute of Neurological Disorders and Stroke
National Institutes of Health
Bethesda, MD
mindellj@ninds.nih.gov

JAN VAN MINNEN
Department Cell Biology & Anatomy
University of Calgary
Calgary, AB, Canada
Jan.van.minnen@falw.vu.nl

Lloyd B. Minor
Department of Otolaryngology – Head and Neck Surgery
The Johns Hopkins University
School of Medicine
Baltimore, MD, USA
lminor@jhmi.edu

Ralph Mistlberger
Department of Psychology
Simon Fraser University
Burnaby, BC, Canada
mistlber@sfu.ca

Takayuki Mitsuhashi
Department of Pediatrics, School of Medicine
Keio University
Tokyo, Japan

Tetsuya Mizuno
Department of Neuroimmunology
Research Institute of Environmental Medicine
Nagoya University
Furo-cho, Chikusa, Nagoya, Japan
tmizuno@riem.nagoya-u.ac.jp

Amalia Molinero
Institute of Neurosciences and Department of Cellular Biology, Physiology and Immunology
Animal Physiology Unit
Faculty of Sciences
Autonomous University of Barcelona
Bellaterra, Barcelona, Spain
Amalia.Molinero@uab.es

Derek C. Molliver
Department of Medicine
University of Pittsburgh
Pittsburgh, PA, USA

Zoltán Molnár
Department of Physiology, Anatomy and Genetics
University of Oxford
Oxford, UK
zoltan.molnar@anat.ox.ac.uk

Bradley J. Molyneaux
MGH-HMS Center for Nervous System Repair, Departments of Neurosurgery and Neurology, and Program in Neuroscience
Harvard Medical School;
Nayef Al-Rodhan Laboratories
Massachusetts General Hospital
Department of Stem Cell and Regenerative Biology, and Harvard Stem Cell Institute
Harvard University
Boston, MA, USA

Barbara Montero
Graduate Center
City University of New York
New York, NY, USA
bmontero@gc.cuny.edu

Jean-Pierre Montmayeur
Centre des Sciences du Goût
Centre National de la Recherche Scientifique
Dijon, France
montmayeur@cesg.cnrs.fr

Robert J. Morecraft
Division of Basic Biomedical Sciences
University of South Dakota
School of Medicine
Vermillion, SD, USA
rmorecra@usd.edu

Nerea Moreno
Dept. Biología Celular
Fac. Biología
Univ. Complutense
Madrid, Spain

Alan Morgan
The Physiological Laboratory
School of Biomedical Sciences
University of Liverpool
Liverpool, UK

Catherine W. Morgans
Neurological Sciences Institute
Oregon Health and Science University
Beaverton, OR, USA
morgansc@ohsu.edu

Katrin E. Morgen
Department of Neurology
Giessen University
Giessen, Germany
Katrin.Morgen@neuro.med.uni-giessen.de

Lawrence P. Morin
Department of Psychiatry
Stony Brook University
Medical Center
Stony Brook, NY, USA
lawrence.morin@stonybrook.edu

Yoshinori Moriyama
Department of Biochemistry
Faculty of Pharmaceutical Sciences
Okayama University
Okayama, Japan
moriyama@pheasant.pharm.okayama-u.ac.jp

Judy L. Morris
Department of Anatomy & Histology, and Centre for Neuroscience
Flinders University
Adelaide, SA, Australia
Judy.Morris@flinders.edu.au

Kendall F. Morris
Department of Molecular Pharmacology and Physiology, Neuroscience
University of South Florida
College of Medicine
Tampa, FL, USA
kmorris@health.usf.edu

John F. B. Morrison
Department of Physiology
Faculty of Medicine and Health Sciences
UAE University
Al Ain, UAE
john.morrison@uaeu.ac.ae

Adonis Moschovakis
Institute of Applied and Computational Mathematics, Foundation of Research and Technology
University of Crete
Heraklion, Crete, Greece
moschov@med.uoc.gr

Aurelie Mouret
Laboratory for Perception and Memory, CNRS URA 2182
Pasteur Institute
Paris, France
amouret@pasteur.fr

Maciej M. Mrugala
Departments of Neurology and Neurosurgery
University of Washington Medical School
Fred Hutchinson Cancer Research Center
Seattle, WA, USA
mmrugala@u.washington.edu

Devin Mueller
Department of Psychiatry
University of Puerto Rico School of Medicine
San Juan, Puerto Rico
devin.mueller@gmail.com

S. Mühlenbrock-Lenter
Brain Research Institute
University of Bremen
Bremen, Germany

Ian Mullaney
School of Pharmacy
Division of Health Sciences
Murdoch University
WA, Australia
I.Mullaney@murdoch.edu.au

Michael Müller
DFG Research Center Molecular Physiology of the Brain
Department of Neurophysiology
Georg-August-Universität Göttingen
Göttingen, Germany

Fujio Murakami
Laboratory of Neuroscience
Graduate School of Frontier Biosciences
Graduate School of Engineering Science
Osaka University
Suita, Osaka, Japan
murakami@fbs.osaka-u.ac.jp

Ikuya Murakami
Department of Life Sciences
University of Tokyo
Tokyo, Japan
ikuya@fechner.c.u-tokyo.ac.jp

Akira Murata
Department of Physiology
Kinki University
School of Medicine
Osaka-sayama, Japan

Subramanyam N. Murthy
Department of Pharmacology
Tulane University Medical Center
New Orleans, LA, USA

Toshitaka Nabeshima
Department of Chemical Pharmacology
Graduate School of Pharmaceutical Sciences
Meijo University
Nagoya, Japan
tnabeshi@med.nagoya-u.ac.jp

Kazunori Nakajima
Department of Anatomy
Keio University School of Medicine
Tokyo, Japan
kazunori@sc.itc.keio.ac.jp

Harukazu Nakamura
Department of Molecular Neurobiology
Graduate School of Life Sciences and Institute of Development,
Aging; Cancer
Tohoku University
Aoba-ku, Sendai, Japan
nakamura@idac.tohoku.ac.jp

Avindra Nath
Department of Neurology
Johns Hopkins University School of Medicine
Baltimore, MD, USA
anath1@jhmi.edu

Eugene Nattie
Department of Physiology
Dartmouth Medical School
Lebanon, NH, USA
Eugene.E.Nattie.Jr@Dartmouth.EDU

J. W. Neal
Department Histopathology
School of Medicine
Cardiff University
Cardiff, UK

Albert Newen
Institut für Philosophie
Ruhr Universitat Bodium
Bodium, Germany
albert.newen@uni-tuebingen.de

Dave Nichols
Department of Biology
University of Texas at San Antonio
Texas, TX, USA

David E. Nichols
Department of Medicinal Chemistry and Molecular Pharmacology
Purdue University
West Lafayette, IN, USA
drdave@pharmacy.purdue.edu

T. Richard Nichols
Department of Physiology
Emory University
Atlanta, GA, USA
trn@physio.emory.edu

Daniel A. Nicholson
Department of Cell and Molecular Biology
Northwestern University's Feinberg School of Medicine and Institute of Neuroscience
Chicago, IL, USA

Trent Nicol
Department of Communication Sciences
Northwestern University
Evanston, IL, USA

Miguel A. L. Nicolelis
Department of Neurobiology
Center of Neuroengineering
Department of Biomedical Engineering
Department of Psychology and Neurosciences
Duke University
Durham, NC, USA

Loredana Nicoletti
Department of Infectious, Parasitic and Immune-Mediated Diseases
Istituto Superiore di Sanità
Rome, Italy

Andreas Nieder
Primate NeuroCognition Laboratory
Department of Cognitive Neurology
Hertie-Institute for Clinical Brain Research
University of Tuebingen
Tuebingen, Germany
andreas.nieder@uni-tuebingen.de

Jens Bo Nielsen
Department of Exercise and Sport Science &
Department of Neuroscience and Pharmacology
The Panum Institute
Copenhagen, Denmark

Tore A. Nielsen
Psychiatry Department
Université de Montréal
Montreal, QC, Canada
tore_a_nielsen@yahoo.com

Andrey R. Nikolaev
Laboratory of Human Higher Nervous Activity
Institute of Higher Nervous Activity and Neurophysiology
Russian Academy of Sciences
Moscow, Russia

Christian Nimtz
Department of Philosophy
University of Bielefeld
Bielefeld, Germany
chn@nimtz.net

Hiroshi Nishimaru
Neuroscience Research Institute
National Institute of Advanced Industrial Science and Technology (AIST)
Tsukuba Center, Higashi
Tsukuba, Ibaraki, Japan
Hiroshi.Nishimaru@neuro.ki.se

Antoine Nissant
Laboratory for Perception and Memory
Pasteur Institute
Paris Cedex,
France
anissant@pasteur.fr

Michael N. Nitabach
Department of Cellular and Molecular Physiology
Yale School of Medicine
New Haven, CT, USA
michael.nitabach@yale.edu

Georg Northoff
Department of Psychiatry
University of Magdeburg
Magdeburg, Germany
Georg.Northoff@Medizin.Uni-Magdeburg.DE

Akinao Nose
Department of Complexity Science and Engineering,
Graduate School of Frontier Sciences
University of Tokyo
Tokyo, Japan
nose@bio.phys.s.u-tokyo.ac.jp

Bobby D. Nossaman
Department of Pharmacology
Tulane University Medical Center
New Orleans, LA, USA
bnossama@tulane.edu

Antonio A. Nunez
Department of Psychology and Neuroscience Program
Michigan State University
East Lansing, MI, USA

Michael P. Nusbaum
Department of Neuroscience
University of Pennsylvania
School of Medicine
Philadelphia, PA, USA
nusbaum@mail.med.upenn.edu

Lars Nyberg
Departments of Radiation Sciences (Radiology) and
Integrative Medical Biology (Physiology)
Umeå University
Umeå, Sweden
Lars.Nyberg@physiol.umu.se

John Philip O'Doherty
Computation and Neural Systems Program
Division of Humanities and Social Sciences
California Institute of Technology
Pasadena, CA, USA
jdoherty@caltech.edu

Hisashi Ogawa
Department of Neurology
Kumamoto Kinoh Hospital;
Department of Sensory and Cognitive Physiology
Kumamoto University
Honjo, Kumamoto, Japan
ogawa@juryo.or.jp

Toshio Ohhashi
Department of Physiology
Shinshu University School of Medicine
Matsumoto, Japan
ohhashi@sch.md.shinshu-u.ac.jp

Frank W. Ohl
BioFuture Research Group
Leibniz Institute for Neurobiology
Magdeburg, Germany
frank.ohl@ifn-magdeburg.de

Takako Ohno-Shosaku
Department of Cellular Neurophysiology
Graduate School of Medical Science
Kanazawa University
Kanazawa, Japan

Ryosuke Ohsawa
Institute for Virus Research
Kyoto University
Kyoto, Japan

Masayoshi Ohta
Department of Anatomy and Neurobiology and
Department of Plastic and Reconstructive Surgery
Kyoto University
Graduate School of Medicine
Kyoto, Japan
ota@kuhp.kyoto-u.ac.jp

Toshiyuki Ohtsuka
Institute for Virus Research
Kyoto University
Kyoto, Japan

Koji Oishi
Department of Anatomy
Keio University
School of Medicine
Tokyo, Japan

Shigeo Okabe
Department of Cellular Neurobiology
Graduate School of Medicine
University of Tokyo
okabe.cbio@tmd.ac.jp

Masato Okada
Graduate School of Frontier Sciences
The University of Tokyo
Kashiwa Chiba, Japan

HIDEYUKI OKANO
Department of Physiology
School of Medicine
Keio University
Tokyo, Japan
hidokano@sc.itc.keio.ac.jp

ALBINO J. OLIVEIRA-MAIA
Instituto de Biologia Molecular e Celular
Faculdade de Medicina
Universidade do Porto
Porto, Portugal

DOUGLAS L. OLIVER
University of Connecticut Health Center
Department of Neuroscience
Farmington, CT, USA
DOliver@neuron.uchc.edu

HIROSHI OMOTE
Department of Biochemistry
Faculty of Pharmaceutical Sciences
Okayama University
Okayama, Japan

HIROSHI ONIMARU
Department of Physiology
Showa University
School of Medicine
Tokyo, Japan
oni@med.showa-u.ac.jp

TAKETOSHI ONO
Molecular and Integrative Emotional Neuroscience,
Graduate School of Medicine
University of Toyama
Sugitani, Toyama, Japan
onotake@med.u-toyama.ac.jp

JOHN ORMOND
Department of Cell & Systems Biology
University of Toronto
Toronto, ON, Canada

INMACULADA ORTEGA-PÉREZ
Laboratory of Perception and Memory
Institute Pasteur
Paris, France
iortega@pasteur.fr

NORIKO OSUMI
Division of Developmental Neuroscience
CTAAR
Tohoku University
School of Medicine
Sendai, Japan

RYLAND W. PACE
The Department of Applied Science
The College of William and Mary
Williamsburg, VA, USA

JEFFREY PADBERG
Center for Neuroscience
University of California
Davis, CA, USA

ANN M. PALKOVICH
Krasnow Institute for Advanced Study
Department of Sociology and Anthropology
George Mason University
Fairfax, VA, USA
apalkovi@gmu.edu

SATCHIDANANDA PANDA
The Salk
Institute for Biological Studies
La Jolla, CA, USA
panda@salk.edu

RAÚL G. PAREDES
Instituto de Neurobiología
Universidad Nacional Autónoma de México
Juriquilla, México
Rparedes@servidor.unam.mx

MAGDA PASSATORE
Department of Neuroscience – Section of Physiology
University of Torino
Torino, Italy
magda.passatore@unito.it

TOSHAL PATEL
Eisai London Research Laboratories
University College London
London, UK

JULIAN F. R. PATON
Department of Physiology
Bristol Heart Institute
School of Medical Sciences,
University of Bristol
Bristol, UK
Julian.F.R.Paton@Bristol.ac.uk

TADD B. PATTON
Department of Psychology
University of South Florida
Tampa, Florida, USA

MICHAEL PAUEN
Institut für Philosophie
Berlin School of Mind and Brain
Humboldt-Universität zu Berlin
Berlin, Germany
m@pauen.com

Keir Pearson
Department of Physiology
University of Alberta
Edmonton, AB, Canada
Keir.pearson@ualberta.ca

John H. Peever
Department of Cell and Systems Biology
University of Toronto
Toronto, ON, Canada
j.duffin@utoronto.ca

Leo Peichl
Max Planck Institute for Brain Research
Frankfurt am Main
Germany
peichl@mpih-frankfurt.mpg.de

Denis Pelisson
Espace et Action Unité 864
INSERM/Université Claude Bernard – Lyon
IFR19 Institut Fédératif des Neurosciences de Lyon
Bron, France
pelisson@lyon151.inserm.fr

Milena Penkowa
Department of Medical Anatomy
The Panum Institute
University of Copenhagen
Copenhagen, Denmark

Keith R. Pennypacker
Departments of Molecular Pharmacology and Physiology
University of South Florida
College of Medicine
Tampa, FL, USA

David J. Perkel
Departments of Biology & Otolaryngology
University of Washington
Seattle, WA
perkel@u.washington.edu

Eric J. Perreault
Department of Biomedical Engineering
Department of Physical Medicine and Rehabilitation
Northwestern University
Chicago, IL, USA
e-perreault@northwestern.edu

Robert J. Peterka
Neurological Sciences Institute
Oregon Health & Science University
Portland, OR, USA
peterkar@ohsu.edu

Robert C. Peters
Department of Functional Neurobiology
Utrecht University
Utrecht, The Netherlands

Ronald S. Petralia
NIDCD/NIH
Bethesda, MD, USA
r.c.peters@bio.uu.nl

Ryan Petrie
Department of Anatomy and Cell Biology
McGill University
Montreal, Quebec, Canada

Vito E. Pettorossi
Department of Internal Medicine
Section of Human Physiology
University of Perugia
Perugia, Italy
vitopett@unipg.it

Axel Petzold
Department of Neuroimmunology,
Institute of Neurology
University College London
London, UK
a.petzold@ion.ucl.ac.uk

Phuc H. Pham
Department of Biology
University of Waterloo
Waterloo, ON, Canada

Anthony G. Phillips
Department of Psychiatry
University of British Columbia
Vancouver, BC, Canada

Stephen J. Piazza
Department of Kinesiology
The Pennsylvania State University
University Park
PA, USA
steve-piazza@psu.edu

Hugh Piggins
Departments of Psychology at Barnard College and at Columbia University, Department of Pathology and Cell Biology, Columbia Health Sciences, New York, NY, USA

Dave B. Pilgrim
Department of Biological Sciences
University of Alberta
Edmonton, AB, Canada
dave.pilgrim@ualberta.ca

Giles W. Plant
Red's Spinal Cord Research Laboratory
School of Anatomy and Human Biology
The University of Western Australia
Crawley, Perth, WA
gplant@anhb.uwa.edu.au

Stefano Pluchino
Neuroimmunology Unit
DIBIT-San Raffaele Scientific Institute
Milano, Italy
pluchino.stefano@hsr.it

Howard Poizner
Institute for Neural Computation
University of California
San Diego, CA, USA
hpoizner@ucsd.edu

Alexander A. Pollen
Department of Physiology, Anatomy and Genetics
University of Oxford
Oxford, UK

Glen Pollock
Pathology and Cell Biology
University of South Florida
College of Medicine
Tampa, FL, USA

Felix Polyakov
Department of Computer Science and Applied Mathematics
Weizmann Institute of Science
Rehovot, Israel

Mu-Ming Poo
Division of Neurobiology
Department of Molecular and Cell Biology
HelenWills Neuroscience Institute
University of California
Berkeley, CA, USA
mpoo@uclink.berkeley.edu

Maria Popescu
Department of Psychology
Vanderbilt University
Nashville, TN, USA
maria.v.popescu@Vanderbilt.Edu

John Porrill
Department of Psychology
University of Sheffield
Sheffield, UK

Alice Schade Powers
Department of Psychology
St. John's University
Jamaica, NY, USA
powersa@stjohns.edu

Claude Prablanc
Espace et Action Unité 864
INSERM/Université Claude Bernard – Lyon, IFR19
Institut Fédératif des Neurosciences de Lyon
Bron, France

Arthur Prochazka
Centre for Neuroscience
University of Alberta
Edmonton, AB, Canada
Arthur.Prochazka@ualberta.ca

Uwe Proske
Department of Physiology
Monash University
Melbourne, VIC, Australia
uwe.proske@med.monash.edu.au

Ignacio Provencio
Department of Biology
University of Virginia
Charlottesville, VA, USA
ip7m@virginia.edu

Dale Purves
Center for Cognitive Neuroscience and Department of Neurobiology
Duke University
Durham, NC, USA
purves@neuro.duke.edu

Martina Pyrski
Department of Physiology
University of Saarland
School of Medicine
Homburg/Saar, Germany
martina.pyrski@uks.eu

Jorge N. Quevedo
Department of Physiology, Biophysics and Neuroscience
Centro de Investigación y de Estudios Avanzados del I. P.N
Mexico City, Mexico
jquevedo@fisio.cinvestav.mx

Albert Quintana
Institute of Neurosciences and Department of Cellular Biology, Physiology and Immunology
Animal Physiology Unit
Faculty of Sciences
Autonomous University of Barcelona
Bellaterra, Barcelona, Spain

Gregory J. Quirk
Department of Psychiatry
University of Puerto Rico
School of Medicine
San Juan, Puerto Rico
gjquirk@yahoo.com

Matt S. Ramer
International Collaboration on Repair Discoveries
the University of British Columbia
Vancouver, BC, Canada
ramer@icord.org

Jan-Marino Ramirez
Department of Organismal Biology and Anatomy
Committees on Neurobiology, Computational
Neuroscience and Molecular Medicine
The University of Chicago
Chicago, IL, USA
jramire@uchicago.edu

Catharine H. Rankin
Brain Research Centre
University of British Columbia
Vancouver, BC, Canada
crankin@psych.ubc.ca

Yong Rao
Centre for Research in Neuroscience
Department of Neurology and Neurosurgery
McGill University Health Centre
Montreal, QC, Canada
yong.rao@mcgill.ca

Theodore Raphan
Department of Computer and Information Science
Institute of Neural and Intelligent Systems, Brooklyn
College of the City University of New York
Brooklyn, NY, USA
raphan@sci.brooklyn.cuny.edu

Daniel L. Rathbun
Center for Neuroscience
University of California
Davis, CA, USA

Nadine Ravel
Laboratoire Systèmes Sensoriels, Comportement et
Cognition
UMR 5020 CNRS–UCB Lyon
IFR19 Lyon, France
nadine.ravel@olfac.univ-lyon1.fr

Adrian Rees
Auditory Group, Institute of Neuroscience
Newcastle University
Newcastle upon Tyne, UK
adrian.rees@ncl.ac.uk

Sandra Rees
Department of Anatomy and Cell Biology
University of Melbourne
Melbourne, Australia
SRees@Unimelb.edu.au

Kathryn M. Refshauge
Faculty of Health Sciences
University of Sydney
Sydney, NSW, Australia
k.refshauge@fhs.usyd.edu.au

Alfonso Represa
INMED, INSERM U29
Université de la Méditerranée
Campus de Luminy, Marseille, France
alfonso.represa@inmed.univ-mrs.fr

Diego Restrepo
University of Colorado at Denver and Health Sciences
Center
Aurora, CO, USA
Diego.Restrepo@uchsc.edu

Mark M. Rich
Department of Neuroscience, Cell Biology and
Physiology
Wright State University
Dayton, OH, USA
mmrich@emory.edu

Robert C. Richardson
Department of Philosophy
University of Cincinnati
Cincinnati, OH, USA

George Richerson
Department of Neurology
School of Medicine
Yale University
New Haven, CT, USA
george.richerson@yale.edu

Diethelm W. Richter
DFG Research Center Molecular Physiology of the
Brain
Department of Neurophysiology
Georg-August-Universität Göttingen
Göttingen, Germany
d.richter@gwdg.de

Justin P. Ridge
School of Molecular and Microbial Sciences
University of Queensland
Brisbane, Queensland, Australia

James K. Rilling
Department of Anthropology
Columbia University
New York, NY, USA

Jürgen A. Ripperger
Department of Medicine
Division of Biochemistry
University of Fribourg
Fribourg, Switzerland

Casto Rivadulla
Neurociences and Motor Control Group (Neurocom),
Department of Medicine-INEF-Galicia
University de A Coruña
Campus de Oza, A Coruña, Spain

Claudio Rivera
Institute of Biotechnology
University of Helsinki
Helsinki, Finland
claudio.rivera@helsinki.fi

Silvestro Roatta
Department of Neuroscience – Physiology Division
University of Torino
Raffaello, Torino, Italy
silvestro.roatta@polito.it

Fernando Rodriguez
Laboratory of Psychobiology
University of Sevilla
Sevilla, Spain

Veronica G. Rodriguez Moncalvo
Department of Biology
McMaster University
Hamilton, ON, Canada

Till Roenneberg
The Univeristy of Munich
Munich, Germany

Malin Rohdin
Karolinska Institute
Astrid Lindgren Children's Hospital
Stockholm, Sweden

Uri Rokni
Center for Brain Science and Swartz Center for
Compentational Neuroscience
Harvard
Cambridge, MA, USA
rokniu@MIT.EDU

Edmund T. Rolls
Department of Experimental Psychology
University of Oxford
Oxford, UK
Edmund.Rolls@psy.ox.ac.uk

Benno Roozendaal
Center for the Neurobiology of Learning and Memory,
Department of Neurobiology and Behavior
University of California
Irvine, CA, USA
broozend@uci.edu

Jonas Rose
Biopsychology
Institute of Cognitive Neuroscience
Ruhr-University Bochum
Bochum, Germany

Andrew M. Rosen
Dept. of Psychology
Binghamton University
Binghamton, NY, USA
andrewri@u.washington.edu

Serge Rossignol
Centre de Recherche en Sciences Neurologiques,
Department of Physiology
Pavillon Paul-G.-Desmarais
Université de Montréal
Montreal, QC, Canada
serge.rossignol@unmontreal.ca

Gerhard Roth
Brain Research Institute
University of Bremen
Bremen, Germany
gerhard.roth@uni-bremen.de

Louise Röska-Hardy
Kulturwissenschaftliches Institut
Essen, Germany
louise.roeska-hardy@kwi-nrw.de

Andre T. Roussin
Dept. of Psychology
Binghamton University
Binghamton, NY, USA
roussinat@yahoo.com

Mark J. Rowe
Department of Physiology and Pharmacology
School of Medical Sciences
The University of New South Wales
Sydney, NSW, Australia
M.Rowe@unsw.edu.au

ARIJIT ROY
Hotchkiss Brain Institute
Department of Medical
Physiology and Biophysics
University of Calgary
Calgary, AB, Canada

ROLAND R. ROY
Brain Research Institute and Department of
Physiological Science
University of California
Los Angeles, CA, USA
rrr@ucla.edu

ARAYA RUANGKITTISAKUL
Department of Physiology
Perinatal Research Centre,
Faculty of Medicine & Dentistry, HMRC
University of Alberta
Edmonton, AB, Canada

SILVIA DE RUBEIS
Department of Biology
University "Tor Vergata"
Rome Department of Molecular and Developmental
Genetics/VIB11,
Catholic University of Leuven
Leuven, Belgium

EDWIN W. RUBEL
Virginia Merrill Bloedel Hearing Research Center
University of Washington
Seattle, WA, USA
rubel@u.washington.edu

MARIO A. RUGGERO
The Hugh Knowles Center and Institute for
Neuroscience
Northwestern University
Evanston, Illinois, USA
mruggero@northwestern.edu

JEFFREY A. RUMBAUGH
Department of Neurology
Johns Hopkins University
School of Medicine
Baltimore, MD, USA

DAVID W. RUSS
School of Physical Therapy
Ohio University
Athens, OH, USA

SHELLEY J. RUSSEK
Laboratory of Molecular Neurobiology
Department of Pharmacology and Experimental
Therapeutics
Boston, MA, USA
srussek@bu.edu

ILYA A. RYBAK
Department of Neurobiology and Anatomy
Drexel University College of Medicine
Philadelphia, PA, USA
rybak@drexel.edu

KLAUS SACHS-HOMBACH
Otto-von-Guericke University Magdeburg
Magdeburg, Germany
klaus.sachs-hombach@gse-w.uni-magdeburg.de

WILLIAM M. SAIDEL
Department of Biology
Rutgers, the State University of New Jersey
Camden, NJ,
saidel@camden.rutgers.edu

TETSUICHIRO SAITO
Graduate School of Medicine
Chiba University
Chiba, Japan
tesaito@faculty.chiba-u.ac.jp

NORIO SAKAI
Department of Developmental Medicine (Pediatrics)
Osaka University
Graduate School of Medicine
Yamadaoka, Suita, Osaka, Japan

HIDEO SAKATA
Department of Administrative Nutrition
Faculty of Health and Nutrition
Tokyo, Japan
sakata@horae.dti.ne.jp

KENJI SAKIMURA
Department of Cellular Neurobiology
Brain Research Institute
Niigata University
Niigata, Japan
npsaki01@bri.niigata-u.ac.jp

YOSHIO SAKURAI
Department of Psychology
Graduate School of Letters
Kyoto University
Sakyo, Kyoto, Japan
ysakurai@bun.kyoto-u.ac.jp

Cosme Salas
Department of Psychology and J.P. Scott Center for Neuroscience, Mind and Behavior
Bowling Green State University
Bowling Green, OH, USA

Ernesto Salcedo
University of Colorado at Denver and Health Sciences Center
Aurora, CO, USA
Ernesto.Salcedo@uchsc.edu

Ali Samii
Department of Neurology
University of Washington
Seattle, WA, USA
asamii@u.washington.edu

Willis K. Samson
Department of Pharmacological and Physiological Science
Saint Louis University
School of Medicine
St. Louis, MO, USA
samsonwk@slu.edu

Thomas Sandercock
Departments of Physiology, Physical Medicine and Rehabilitation, and Biomedical Engineering
Feinberg School of Medicine
Northwestern University
Chicago, IL, USA
t-sandercock@northwestern.edu

Jürgen Sandkühler
Department of Neurophysiology
Center for Brain Research
Medical University of Vienna
Vienna, Austria
juergen.sandkuehler@meduniwien.ac.at

J. C. Sandoz
Research Center for Animal Cognition
CNRS UMR 5169
Paul Sabatier University
Toulouse Cedex, France
sandoz@cict.fr

Vittorio Sanguineti
Dipartimento di Informatica, Sistemistica e Telematica
Università di Genova
Genova, Italy
sangui@dist.unige.it

Robert M. Santer
School of Biosciences
Cardiff University
Cardiff, United Kingdom
santer@cardiff.ac.uk

Samuel Saporta
Pathology and Cell Biology
University of South Florida College of Medicine
Tampa, FL, USA
ssaporta@hsc.usf.edu

Yuka Sasaki
Harvard Medical School
Boston, MA, USA

Marco Sassoè-Pognetto
Department of Anatomy, Pharmacology and Forensic Medicine and National Institute of Neuroscience
University of Torino
Torino, Italy
marco.sassoe@unito.it

Makoto Sato
Division of Cell Biology and Neuroscience, Department of Morphological and Physiological Sciences, Faculty of Medical Sciences
University of Fukui
Matsuoka-Shimoaizuki, Eiheiji, Fukui, Japan
makosato@fmsrsa.fukui-med.ac.jp

Benoist Schaal
Centre Européen des Sciences du Goût, CNRS
Dijon, France
Schaal@cesg.cnrs.fr

Hans-Georg Schaible
Institute of Physiology
Friedrich-Schiller-University of Jena
Jena, Germany
Hans-Georg.schaible@mti.uni-jena.de

Jeffrey D. Schall
Department of Psychology
Center for Integrative and Cognitive Neuroscience,
Vanderbilt Vision Research Center,
Vanderbilt University
Nashville, TN, USA
jeffrey.d.schall@vanderbilt.edu

Henning Scheich
Leibniz-Institute for Neurobiology
Magdeburg, Germany

Marc H. Schieber
Departments of Neurology and of Neurobiology and Anatomy

University of Rochester School of Medicine and Dentistry
Rochester, NY, USA
mhs@cvs.rochester.edu

HANS-ULRICH SCHNITZLER
Tierphysiologie
Zoologisches Institut
Universität Tübingen
Tubingen, Germany
hans-ulrich.schnitzler@uni-tuebingen.de

CHRISTOPH E. SCHREINER
Coleman Memorial Laboratory
W.M. Keck Center for Integrative Neuroscience
Department of Otolaryngology-Head and Neck Surgery
School of Medicine
University of California
San Francisco, CA, USA
chris@phy.ucsf.edu

MARKUS SCHRENK DPHIL
Research Fellow
Department Philosophy
University of Nottingham
University Park
Nottingham, UK
markus.schrenk@ccc.ox.ac.uk

JÜRGEN SCHRÖDER
University of Karlsruhe
Heidelberg, Germany

MICHAEL SCHÜTTE
Institut für Philosophie
Universität Magdeburg
Magdeburg, Germany
Mschuett@philosophie.uni-bielefeld.de

MICHAL SCHWARTZ
The Weizmann Institute of Science
Rehovot, Israel
Charles Scudder
Portland, OR, USA
michal.schwartz@weizmann.ac.il

MICHAEL SEAGAR
INSERM/Université de la Méditerranée, UMR641
Faculté de Médecine Nord
Marseille, France
seagar.m@jean-roche.univ-mrs.fr

ARMIN H. SEIDL
Virginia Merrill Bloedel Hearing Research Center
University of Washington
Seattle, WA, USA
armins@u.washington.edu

KRZYSZTOF SELMAJ
Department of Neurology
Medical University of Lodz
Lodz, Poland

ALICIA SEMAKA
Centre for Molecular Medicine and Therapeutics
University of British Columbia
Vancouver, Canada

ADRIANO SENATORE
B1–173, Department of Biology
University of Waterloo
Waterloo, ON, Canada

SUSAN R. SESACK
Departments of Neuroscience and Psychiatry
University of Pittsburgh
Pittsburgh, PA, USA
sesack@bns.pitt.edu

KAARE SEVERINSEN
Department of Neurology
Aarhus University Hospital
Aarhus, Denmark

ROBERT V. SHANNON
House Ear Institute
Los Angeles, CA, USA
bshannon@hei.org

DAVID SHEINBERG
Department of Neuroscience
Brown University
Providence, RI, USA
David_Sheinberg@brown.edu

SIMING SHEN
Robert Wood Johnson Medical School UMDNJ
Piscataway, NJ, USA

YIRU SHEN
Department of Physiology
Yong Loo Lin School of Medicine
National University of Singapore
and National Neuroscience Institute
Seng, Singapore

GORDON M. SHEPHERD
Department of Neurobiology
Yale University School of Medicine
New Haven, CT, USA
Gordon.shepherd@yale.edu

KIRSTEN SHEPHERD-BARR
Faculty of English
Oxford University
Oxford, UK

Chet C. Sherwood
Department of Anthropology
The George Washington University
Washington, DC, USA

Shigenobu Shibata
Division of Physiology and Pharmacology
School of Science and Engineering
Waseda University
Tokyo, Japan
shibatas@waseda.jp

Kenji Shimamura
Division of Morphogenesis
Institute of Molecular Embryology and Genetics
Kumamoto University
Kumamoto, Japan
imamura@kaiju.medic.kumamoto-u.ac.jp

Toru Shimizu
Department of Psychology
University of South Florida
Tampa, Florida, USA
shimizu@cas.usf.edu

Nahum Shimkin
Department of Electrical Engineering
Technion – Israel Institute of Technology
Haifa, Israel
shimkin@ee.technion.ac.il

René M. Shinal
Department of Community Dentistry and Behavioral Science
College of Dentistry
University of Florida
FL, USA
rshinal@dental.ufl.edu

Sooyoon Shin
Department of Neuroscience
The Center for the Neural Basis of Cognition, and the Center for Neuroscience
University of Pittsburgh
Pittsburgh, PA, USA

Tomomi Shindou
Neurobiology Research Unit
Okinawa Institute of Science and Technology
Uruma, Okinawa, Japan

Sadao Shiosaka
Nara Institute of Science and Technology (NAIST)
Structural Cell Biology
Nara, Japan
sshiosak@bs.aist-nara.ac.jp

Gilles Sicard
Centre Européen des Sciences du Goût
Dijon, France
sicard@cesg.cnrs.fr

Jerome M. Siegel
Center for Sleep Research 151A3
Department of Psychiatry
UCLA School of Medicine
Veterans Administration Greater Los Angeles Health Care System
North Hills, CA, USA

Jonathan Siegel
Department of Communication Sciences and Disorders
Northwestern University
Evanston, IL, USA
Jsiegel@ucla.edu

R. F. M. Silva
Centro de Patogénese Molecular - UBMBE iMed.UL, Faculdade de Farmácia
University of Lisbon
Lisbon, Portugal

Rae Silver
Departments of Psychology at Barnard College and Columbia University, and Department of Pathology and Cell Biology
Columbia Health Sciences Campus
Columbia University
New York, NY, USA
QR@columbia.edu

Narong Simakajornboon
Cincinnati Children's Hospital Medical Center
Cincinnati, OH, USA
Narong.Simakajornboon@cchmc.org

Andrew Simmonds
Department of Cell Biology
Faculty of Medicine and Dentistry
University of Alberta
Edmonton, AB, Canada

Julia Simner
School of Philosophy, Psychology and Language Sciences
University of Edinburgh
Edinburgh, UK
j.simner@ed.ac.uk

Sidney A. Simon
Department of Neurobiology
Duke University
Durham, NC, USA
sas@neuro.duke.edu

Burton Slotnick
Department of Psychology
University of South Florida
Tampa, FL, USA
slotnic@american.edu

Laura Smale
Department of Psychology and Neuroscience Program
Michigan State University
East Lansing, MI, USA
smale@msu.edu

Jeffrey C. Smith
Cellular and Systems Neurobiology Section
National Institute of Neurological Disorders and Stroke
National Institutes of Health
Bethesda, MD, USA
jsmith@helix.nih.gov

Paul F. Smith
Department of Pharmacology and Toxicology
School of Medical Sciences
University of Otago
Medical School
Dunedin, New Zealand
paul.smith@stonebow.otago.ac.nz

Peter G. Smith
Department of Molecular and Integrative Physiology and Kansas Intellectual and Developmental Disabilities Research Center
University of Kansas
Medical Center
Kansas City, KS, USA
psmith@kumc.edu

Philip H. Smith
Department of Anatomy
University of Wisconsin
Medical School – Madison
Madison, WI, USA
Smith@Physiology.wisc.edu

Terence Smith
Eisai London Research Laboratories
University College London
London, UK

Kwok-Fai So
Department of Anatomy
The University of Hong Kong
Hong Kong, People's Republic of China
hrmaskf@hkucc.hku.hk

Beate Sodian
Department of Psychology
Ludwig-Maximillians-Universität München
München, Germany
Beate.Sodian@psy.lmu.de

John F. Soechting
Department of Neuroscience
University of Minnesota
Minneapolis, MN, USA
soech001@umn.edu

U. Shivraj Sohur
MGH-HMS Center for Nervous System Repair, Departments of Neurosurgery and Neurology, and Program in Neuroscience
Harvard Medical School;
Nayef Al-Rodhan Laboratories
Massachusetts General Hospital;
Department of Stem Cell and Regenerative Biology, and Harvard Stem Cell Institute
Harvard University
Boston, MA, USA

Marc A. Sommer
Department of Neuroscience
The Center for the Neural Basis of Cognition, and the Center for Neuroscience
University of Pittsburgh
Pittsburgh, PA, USA
mas@cnbc.cmu.edu

Margaret J. Sonnenfeld
Department of Cellular and Molecular Medicine, Faculty of Medicine
University of Ottawa
Ottawa, ON, Canada
msonnenf@uottawa.ca

Pamela Souza
Department of Speech and Hearing Sciences
University of Washington
Seattle, WA, USA
psouza@u.washington.edu

J. David Spafford
B1–173, Department of Biology
University of Waterloo
Waterloo, ON, Canada
spafford@uwaterloo.ca

William J. Spain
Department of Neurology
University of Washington
Veterans Affairs Puget
Sound Health Care System
Seattle, WA, USA
spain@u.washington.edu

David L. Sparks
Division of Neuroscience
Baylor College of Medicine
Houston, TX, USA
sparks@neusc.bcm.tmc.edu

Charles Spence
Department of Psychology
University of Bath
Bath, UK
charles.spence@psy.ox.ac.uk

Gaynor E. Spencer
Department of Biological Sciences
Brock University
St. Catharines, ON, Canada
gspencer@brocku.ca

Simon G. Sprecher
Department of Biology
Center for Developmental Genetics
New York University
New York, NY, USA
simon.sprecher@gmail.com

James M. Staddon
Eisai London Research Laboratories
University College London
London, UK
James_Staddon@eisai.net

Philip F. Stahel
Department of Orthopaedic Surgery
Denver Health Medical Center
University of Colorado
School of Medicine
Denver, CO, USA
Philip.Stahel@dhha.org

Angela Starkweather
Intercollegiate College of Nursing
Washington State University
Spokane, WA, USA
starkweather@comcast.net

Alexander Staudacher
Institut für Philosophie
Otto-von-Guericke-Universitaet Magdeburg
Magdeburg, Germany
Alexander.Staudacher@gse-w.uni-magdeburg.de

J. van der Steen
Department of Neuroscience
Erasmus MC
Rotterdam, The Netherlands
J.vanderSteen@ErasmusMC.nl

Paul S. G. Stein
Department of Biology
Washington University
St. Louis, MO, USA
stein@biology.wustl.edu

Yossef Steinberg
Electrical Engineering Department
Technion
Haifa, Israel

Maike Stengel
Division of Neurological Pain Research and Therapy,
Department of Neurology
Universitätsklinikum Schleswig-Holstein
Campus Kiel, Kiel, Germany

Achim Stephan
Institute of Cognitive Science
University of Osnabrück
Lower Saxony, Germany
achim.stephan@t-online.de

Susanne J. Sterbing-D'Angelo
Department of Psychology
Vanderbilt University
Nashville, TN, USA
susanne.j.sterbing-dangelo@vanderbilt.edu

Richard J. Stevenson
Department of Psychology
Macquarie University
Sydney, NSW, Australia
Richard.Stevenson@psy.mq.edu.au

Ralf Stoecker
Institut für Philosophie
Universität Potsdam
Potsdam, Germany
Ralf.Stoecker@uni-potsdam.de

Esther T. Stoeckli
Institute of Zoology/Developmental Neuroscience,
University of Zurich
Zurich, Switzerland
esther.stoeckli@zool.uzh.ch

Hans Straka
Laboratoire de Neurobiologie des Réseaux
CNRS UMR 7060 – Université Paris 5
Paris Cedex, France

Volko A. Straub
Department of Cell Physiology & Pharmacology
University of Leicester
Leicester, UK
vs64@le.ac.uk

Arjen M. Strijkstra
Department of Chronobiology
University of Groningen
Groningen, The Netherlands
a.m.strijkstra@rug.nl

Jörg Strotmann
University of Hohenheim
Institute of Physiology
Stuttgart, Germany
strotman@uni-hohenheim.de

Makoto Sugita
Department of Oral Physiology
Graduate School of Biomedical Sciences
Hiroshima University
Hiroshima, Japan
sugisan@hiroshima-u.ac.jp

Kyoungho Suk
Department of Pharmacology
School of Medicine
Kyungpook National University
Daegu, Korea
ksuk@knu.ac.kr

William K. Summers
Alzheimer's Corporation
Albuquerque, NM, USA
acasec@swcp.com

Ying Ying Sung
Department of Physiology
Yong Loo Lin School of Medicine
National University of Singapore
and National Neuroscience Institute
Seng, Singapore

Maki Suzuki
Division of Cyclotron Nuclear Medicine
Cyclotron and Radioisotope Center
Tohoku University
Sendai, Japan

Yoshihisa Suzuki
Department of Anatomy and Neurobiology and
Department of Plastic and Reconstructive Surgery
Kyoto University Graduate School of Medicine
Kyoto, Japan
utsubo@kuhp.kyoto-u.ac.jp

Akio Suzumura
Department of Neuroimmunology
Research Institute of Environmental Medicine
Nagoya University
Nagoya, Japan

Dick F. Swaab
Netherlands Institute for Neuroscience
Meibergdreef, Amsterdam, The Netherlands
d.f.swaab@nih.knaw.nl

Naweed I. Syed
Department of Cell Biology and Anatomy
Faculty of Medicine
University of Calgary
Calgary, AB, Canada
nisyed@ucalgary.ca

John Symons
Department of Philosophy
The University of Texas
El Paso, TX, USA
jsymons@utep.edu

Olga V. Sysoeva
Laboratory of Human Higher Nervous Activity
Institute of Higher Nervous Activity and
Neurophysiology
Russian Academy of Sciences
Moscow, Russia

Gabrielle Szafranski
Department of Psychology
University of South Florida
Tampa, Florida, USA

Ronald Szymusiak
Departments of Medicine and Neurobiology
School of Medicine
University of California
Los Angeles, CA, USA
rszym@ucla.edu

Tamiko Tachibana
Department of Oral Anatomy
Iwate Medical University
School of Dentistry
Uchimaru, Morioka, Japan
tetsu@agr.ehime-u.ac.jp

Masami Takahashi
Department of Biochemistry
Kitasato University
School of Medicine
Sagamihara-shi, Kanagawa, Japan
masami@med.kitasato-u.ac.jp

Takao Takahashi
Department of Pediatrics
School of Medicine
Keio University
Tokyo, Japan
tata@sc.itc.keio.ac.jp

TERRY TAKAHASHI
Institute of Neuroscience
University of Oregon
Eugene, OR, USA

MASAHARU TAKAMORI
Neurological Center
Kanazawa-Nishi Hospital and
Kanazawa University
Kanazawa, Ishikawa, Japan
t-kiyomi@guitar.ocn.ne.jp

SIU LIN TAM
Centre of Neuroscience
Division of Physical Medicine and Rehabilitation
Faculty of Medicine and Dentistry
University of Alberta
Edmonton, AB, Canada

ATSUSHI TAMADA
Laboratory for Neuronal Growth Mechanisms
RIKEN Brain Science Institute
Wako, Saitama, Japan
tamada@brain.riken.go.jp

HIROTAKA TANABE
Department of Neuropsychiatry, Neuroscience
Ehime Graduate School of Medicine
Shitsukawa, Japan
htanabe@m.ehime-u.ac.jp

HIROKAZU TANAKA
ATR Computational Neuroscience Laboratories
Kyoto, Japan
hirokazu@atr.jp

KOHICHI TANAKA
Laboratory of Molecular Neuroscience
School of Biomedical Science and Medical Research Institute
Tokyo Medical and Dental University
Bunkyo-Ku, Tokyo
tanaka.aud@mri.tmd.ac.jp

A. TANDON
Centre for Research in Neurodegenerative Diseases
Toronto, ON, Canada
a.tandon@utoronto.ca

ITARU F. TATSUMI
LD/Dyslexia Centre
Chiba, Japan
itaru_ft@mbp.nifty.com

JANET L. TAYLOR
Prince of Wales Medical Research Institute and
University of New South Wales
Sydney, NSW, Australia
jl.taylor@unsw.edu.au

HANS J. TEN DONKELAAR
Department of Neurology
Radboud University Nijmegen
Medical Centre
Nijmegen, The Netherlands
H.tenDonkelaar@neuro.umcn.nl

HOLM TETENS
Institut für Philosophie
Freie Universität Berlin
Berlin, Germany
Tetens@zedat.fu-berlin.de

CHARLOTTE E. TEUNISSEN
Department of Molecular Cell Biology and
Immunology
VUMC
Amsterdam, The Netherlands

FRÉDÉRIC E. THEUNISSEN
UC Berkeley
Department of Psychology and
Neurosciences Institute
Berkeley, CA, USA
theunissen@berkeley.edu

THIERRY THOMAS-DANGUIN
Flavour Perception Group: Peri-receptor Events and
Sensory Interactions
UMR1129 FLAVIC, ENESAD, INRA
Université de Bourgogne
Dijon Cedex, France
Thierry.Thomas-Danguin@dijon.inra.fr

DAGMAR TIMMANN
Department of Neurology
University of Duisburg-Essen
Essen, Germany
Dagmar.Timmann@uni-essen.de

LENA H. TING
Laboratory for Neuroengineering
The W. H. Coulter
Department of Biomedical Engineering
Emory University and Georgia Institute of Technology
Atlanta, GA, USA
lting@emory.edu

David Tirschwell
Harborview Neurology - Comprehensive Stroke Center
University of Washington
School of Medicine
Seattle, WA, USA

Shelley Tischkau
Department of Pharmacology
Southern Illinois University
School of Medicine
Springfield, IL, USA
stischkau@siumed.edu

Gunnar Tobin
Department of Pharmacology
The Sahlgrenska Academy
Göteborg University
Göteborg, Sweden
gunnar.tobin@pharm.gu.se

Yoav Tock
IBM Haifa Research Laboratory
University Campus
Haifa, Israel
TOCK@il.ibm.com

William H. Tolleson
Division of Biochemical Toxicology
National Center for Toxicological Research
US Food and Drug Administration
Jefferson, AR, USA
william.tolleson@fda.hhs.gov

Gianluca Tosini
Circadian Rhythms and Sleep Disorders Program, Neuroscience Institute
Morehouse School of Medicine
Atlanta, GA, USA
gtosini@msm.edu

Irene Tracey
Department of Clinical Neurology and Nuffield Department of Anaesthetics
Centre for Functional Magnetic Resonance Imaging of the Brain
Oxford University
Oxford, UK
irene@fmrib.ox.ac.uk

Constantine Trahiotis
Department of Neuroscience and
Department of Surgery (Otolaryngology)
University of Connecticut Health Center
Farmington, CT, USA
tino@nso2.uchc.edu

R. Alberto Travagli
Neuroscience, Pennington Biomedical Research Center-LSU System
Baton Rouge, LA, USA
Alberto.Travagli@pbrc.edu

Mats Trulsson
Department of Prosthetic Dentistry
Institute of Odontology
Karolinska Institutet
Sweden

Vinzenz von Tscharner
Human Performance Laboratory
University of Calgary
AB, Canada
tvvon@ucalgary.ca

Amy Tse
Department of Pharmacology and Centre for Neuroscience
University of Alberta
Edmonton, AB, Canada
Amy.tse@ualberta.ca

Frederick W. Tse
Department of Pharmacology & Centre for Neuroscience
University of Alberta
Edmonton, AB, Canada
Fred.tse@ualberta.ca

Minoru Tsukada
Brain Science Institute
Tamagawa University
Tamagawa-gakuen, Machida, Tokyo, Japan
tsukada@eng.tamagawa.ac.jp

Ikuo Tsunoda
Department of Pathology
University of Utah
School of Medicine
Salt Lake City, Utah, USA
Ikuo.Tsunoda@hsc.utah.edu

Ken-Ichiro Tsutsui
Division of Systems Neuroscience
Graduate School of Life Sciences
Tohoku University
Sendai, Japan

A. Russell Tupling
Department of Kinesiology
University of Waterloo
Waterloo, ON, Canada
rtupling@healthy.uwaterloo.ca

Thomas M. Tzschentke
Grünenthal GmbH
Research and Development
Department of Pharmacology
Aachen, Germany
Thomas.Tzschentke@grunenthal.de

Sae Uchida
Department of the Autonomic Nervous System
Tokyo Metropolitan Institute of Gerontology
Tokyo, Japan
suchida@tmig.or.jp

Yasumasa Ueda
Kyoto Prefectural University of Medicine
Kawaramachi-Hirokoji
Kamigyo-ku, Kyoto, Japan

Tadashi Uemura
Graduate School of Biostudies
Kyoto University
Kyoto, Japan
tauemura@lif.kyoto-u.ac.jp

W. Martin Usrey
Center for Neuroscience
University of California
Davis, CA, USA
wmusrey@ucdavis.edu

Ricky van der Zwan
Sleep and Performance Research Center
Washington State University
Spokane, WA, USA
rick.vanderzwan@scu.edu.au

P. A. Van Dongen
Sleep and Performance Research Center
Washington State University
Spokane, WA, USA
hvd@wsu.edu

Rene Vandenboom
Faculty of Applied Health Sciences
Brock University
St. Catharines, ON, Canada
boomerene@hotmail.com

Todd W. Vanderah
College of Medicine
Departments of Pharmacology
and Anesthesiology
University of Arizona
Tucson, AZ, USA
vanderah@email.arizona.edu

Frederique Varoqueaux
Department of Molecular Neurobiology
Max-Planck-Institute for Experimental Medicine
Göttingen, Germany

Michael R. Vasko
Department of Pharmacology and Toxicology
Indiana University
School of Medicine
Indianapolis, IN, USA
vaskom@iupui.edu

Emma L. Veale
Medway School of Pharmacy
Universities of Kent and Greenwich at Medway
Chatham Maritime, Kent, UK

Sigrid C. Veasey
Center for Sleep & Respiratory Neurobiology,
Department of Medicine
School of Medicine
University of Pennsylvania
Philadelphia, PA, USA
veasey@mail.med.upenn.edu

Elly J. F. Vereyken
Department of Molecular Cell Biology and
Immunology
VUMC
Amsterdam, The Netherlands

Joost Verhaagen
Netherlands Institute for Neurosciences
A Research Institute of the Royal Netherlands Academy
of Arts and Sciences
Amsterdam, The Netherlands
j.verhaagen@nih.knaw.nl

Philippe Vernier
CNRS, Institute of Neurobiology A. Fessard
Gif sur Yvette Cedex, France
philippe.vernier@iaf.cnrs-gif.fr

Ronald T. Verrillo
Institute for Sensory Research
Syracuse University
Syracuse, NY, USA
ron_verrillo@isr.syr.edu

Eugene Vlodavsky
Department of Pathology
Rambam Medical Center,
and Faculty of Medicine
Technion-Israel Institute of Technology
Haifa, Israel
e_vlodavsky@rambam.health.gov.il

Brent A. Vogt
Department of Neuroscience and Physiology
State University of New York
Upstate Medical University
Syracuse, NY, USA
vogtb@upstate.edu

Talila Volk
Department of Molecular Genetics
Weizmann Institute
Rehovot, Israel
lgvolk@wicc.weizmann.ac.il

Bruce T. Volpe
Department of Neurology & Neuroscience
The Burke Medical Research Institute
Weill Medical College of Cornell University
White Plains, NY, USA
bvolpe@burke.org

Catherine de Waele
Laboratoire de Neurobiologie des Réseaux
Sensorimoteurs,
CNRS
Paris, France
Catherine.de-waele@univ-paris.fr

Elaine Waddington Lamont
Institute of Neuroscience
Carleton University
Life Sciences Building
Ottawa, ON, Canada
ewlamont@mac.com

Hermann Wagner
Institute for Biology II
RWTH Aachen
Achen, Germany
wagner@bio2.rwth-aachen.de

Tuck Wah Soong
Department of Physiology
Yong Loo Lin School of Medicine
National University of Singapore
and National Neuroscience Institute
Seng, Singapore

David M. Waitzman
University of Connecticut Health Center
Department of Neurology
Farmington, CT, USA
e-waitzman@nso2.uchc.edu

Michelle Wall
Flexcell International Corp.
North Carolina
USA

Sven Walter
Department of Philosophy
University of Saarland
Saarbruecken, Germany
s.walter@philosophy-online.de

Mark M. G. Walton
Division of Neuroscience
Baylor College of Medicine
Houston, TX, USA

Nan Wang
Department of Pharmacology & Centre for
Neuroscience
University of Alberta
Edmonton, AB, Canada

Qiong Wang
Department of Human Nutrition, Foods and Exercise
Virginia Polytechnic Institute and State University
Blacksburg, VA, USA

Meg Waraczynski
Department of Psychology
University of Wisconsin – Whitewater
Whitewater, WI, USA
waraczym@uww.edu

Simon C. Warby
Centre for Molecular Medicine and Therapeutics
University of British Columbia
Vancouver, Canada

Sarah E. Warner
Department of Orthopaedics and Sports Medicine
University of Washington
Seattle, WA, USA

Anne-Kathrin Warzecha
Lehrstuhl Neurobiologie
Fakultät Biologie
Universität Bielefeld
Bielefeld, Germany
ak.warzecha@uni-bielefeld.de

Andrew J. Waskiewicz
Department of Biological Sciences
University of Alberta
Edmonton, AB, Canada
aw@ualberta.ca

Masataka Watanabe
Department of Systems Innovation
Graduate School of Engineering
The University of Tokyo
Tokyo, Japan
watanabe@sk.q.t.u-tokyo.ac.jp

Masumi Watanabe
Department of Speech, Language and Hearing Sciences
Niigata University of Health and Welfare
Niigata, Japan

Shigeru Watanabe
Department of Psychology
Keio University
Mita, Minato-Ku
Tokyo, Japan
swat@flet.keio.ac.jp

Takeo Watanabe
Boston University
Boston, MA, USA
takeo@bu.edu

Linda R. Watkins
University of Colorado at Boulder
Boulder, CO, USA

Stephen G. Waxman
Rehabilitation Research Center
VA Connecticut, USA Healthcare System
West Haven, CT, USA
Stephen.Waxman@yale.edu

Elke Weiler
Department of Neurophysiology
Institute of Physiology
Ruhr-University Bochum
Bochum, Germany
weiler@neurop.ruhr-uni-bochum.de

Peter Wenderoth
Department of Psychology
Macquarie University, North Ryde
Sydney, NSW, Australia
peterw@vision.psy.mq.edu.au

Robert J. Wenthold
NIDCD/NIH
Bethesda, MD, USA
wenthold@nidcd.nih.gov

David A. Westwood
School of Health and Human Performance
Dalhousie University
Halifax, NS, Canada
David.Westwood@Dal.Ca

Heather E. Wheat
Department of Anatomy and Cell Biology
University of Melbourne
Parkville, VIC, Australia

Patrick J. Whelan
Hotchkiss Brain Institute
University of Calgary
Calgary, AB, Canada
whelan@ucalgary.ca

Dennis Whitcomb
Department of Philosophy
Western Washington University
Bellingham, Washington, USA
dporterw@eden.rutgers.edu

Kathleen Whitlock
Department of Molecular Biology & Genetics
Cornell University
Ithaca, NY, USA
kew13@cornell.edu

Jeffery R. Wickens
Neurobiology Research Unit
Okinawa Institute of Science and Technology
Uruma, Okinawa, Japan
ysuzuki@oist.jp

Darius Widera
Department of Cell Biology
Faculty of Biology
University of Bielefeld
Bielefeld, Germany
darius.widera@uni-wh.de

Sidney Wiener
LPPA
Collèe de France, CNRS
Paris, France
sidney.wiener@college-de-france.fr

Walter Wilczynski
Center for Behavioral Neuroscience and Department of Psychology
Georgia State University
Atlanta, GA, USA
wwilczynski@gsu.edu

J. Martin Wild
Department of Anatomy with Radiology
Faculty of Medical and Health Sciences
University of Auckland
Auckland, New Zealand
jm.wild@auckland.ac.nz

Willem C. Wildering
Department of Biological Sciences
Faculty of Science and Hotchkiss Brain Institute,
University of Calgary
Calgary, AB, Canada
wilderin@ucalgary.ca

BRIAN O. WILLIAMS
Department of Human Nutrition, Foods and Exercise
Virginia Polytechnic Institute and State University
Blacksburg, VA, USA

TRICIA WILLIAMS
Neuroscience and Mental Health Program
Research Institute at The Hospital for Sick Children
Department of Anaesthesia
The University of Toronto
Toronto, ON, Canada

DONALD A. WILSON
Department of Zoology
University of Oklahoma
Norman, OK, USA
dwilson@ou.edu

MARTIN WILSON
Department of Neurobiology, Physiology and Behavior
College of Biological Sciences
University of California Davis
Davis, CA, USA
mcwilson@ucdavis.edu

RICHARD J. A. WILSON
Hotchkiss Brain Institute
Department of Medical Physiology and Biophysics
University of Calgary
Calgary, AB, Canada
wilsonr@ucalgary.ca

WOLFGANG WILTSCHKO
Universität Frankfurt
Zoologisches Institut Biologie–Campus der Universität
Frankfurt am Main, Germany
wiltschko@zoology.uni-frankfurt.de

UWE WINDHORST
Göttingen, Germany
siggi.uwe@t-online.de

JEFFERY A. WINER
Division of Neurobiology
Department of Molecular and Cell Biology
University of California at Berkeley
Berkeley, CA, USA

MICHAEL WINKLHOFER
Department für Geo- und Umweltwissenschaften – Sektion Geophysik
Ludwig-Maximilians-Universität Munich
Munich, Germany
michael@geophysik.uni-muenchen.de

PHILIP WINN
School of Psychology
University of St Andrews
Fife, UK
pw@st-andrews.ac.uk

S. WISLET-GENDEBIEN
Centre for Research in Neurodegenerative Diseases
Toronto
ON, Canada
sabine.wislet@utoronto.ca

MARTIN WITT
Smell and Taste Clinic
Department of Otorhinolaryngology
Department of Anatomy
University of Technology Dresden
Medical School
Dresden, Germany
martin.witt@mailbox.tu-dresden.de

MENNO P. WITTER
Kavli Institute for System Neuroscience
Centre for the Biology of Memory
Department of Neuroscience
Norwegian University of Science and Technology (NTNU),
Trondheim, Norway

JOACHIM WITZEL
Central State Forensic Psychiatric Hospital of Saxony-Anhalt
Uchtspringe, Stendal,
Germany
j.witzel@salus-lsa.de

RODGER WOLEDGE
Institute of Human Performance
University College London
London, UK
r.woledge@ucl.ac.uk

LUKE WOLOSZYN
Department of Neuroscience
Brown University
Providence, RI, USA
supawolo@yoshi.neuro.brown.edu

JONATHAN R.WOLPAW
New York State Department of Health
Laboratory of Nervous System Disorders
Wadsworth Center
Albany, NY, USA
wolpaw@wadsworth.org

Stephanie Woo
Department of Biochemistry, Biopyhsics
University of California
San Francisco, CA, USA

Jackie D. Wood
Department of Physiology and Cell Biology and Internal Medicine
College of Medicine
The Ohio State University
Columbus, OH, USA
wood.13@osu.edu

Melanie A. Woodin
Department of Cell & Systems Biology
University of Toronto
Toronto, ON, Canada
m.woodin@utoronto.ca

Marjorie Woollacott
Department of Human Physiology
University of Oregon
Eugene, OR, USA
mwool@uoregon.edu

Michael Wride
Department of Zoology
Trinity College Dublin
Dublin 2, Ireland
wridema@Cardiff.ac.uk

Kenneth P. Wright Jr
Sleep and Chronobiology Laboratory
Department of Integrative Physiology
University of Colorado at Boulder
Boulder, CO, USA
Kenneth.Wright@colorado.edu

Danuta Wrona
Department of Animal Physiology
University of Gdansk
Gdansk, Poland
wronada@biotech.univ.gda.pl

Daw-An Wu
Division of Biology
California Institute of Technology
Pasadera, CA, USA

F. Gregory Wulczyn
Center for Anatomy
Institute for Cell and Neurobiology
Charité University Hospital
Berlin, Germany
gregory.wulczyn@charite.de

Mario F. Wullimann
Ludwig-Maximilians-University
Department of Biology II-Neurobiology
Planegg-Martinsried, Germany
wullimann@zi.biologie.uni-muenchen.de

Robert H. Wurtz
Laboratory of Sensorimotor Research
National Eye Institute
National Institutes of Health
Bethesda, MD, USA
bob@1sr.nei.nih.gov

Douglas R. W. Wylie
University Centre for Neuroscience and Department of Psychology
University of Alberta
Edmonton, AB, Canada

Xiao-Ming Xu
Kentucky Spinal Cord Injury Research Center,
Department of Neurological Surgery
School of Medicine
University of Louisville
Louisville, KY, USA
xmxu0001@louisville.edu

Jayne E. Yack
Department of Biology
Carleton University
Ottawa, ON, Canada

Hiroyuki Yaginuma
Department of Anatomy
School of Medicine
Fukushima Medical University
Fukushima, Japan
h-yaginuma@fmu.ac.jp

Kiyofumi Yamada
Laboratory of Neuropsychopharmacology
Graduate School of Natural Science and Technology
Kanazawa University
Kanazawa, Japan
kyamada@p.kanazawa-u.ac.jp

Masahito Yamagata
Department of Molecular and Cellular Biology
Harvard University
Cambridge, MA, USA
yamagatm@mcb.harvard.edu

Kazuhiko Yamaguchi
Laboratory for Memory and Learning
Brain Science Institute
Riken, Japan
yamaguchi@brain.riken.go.jp

Nobuhiko Yamamoto
Dept Cellular and Molecular Neurobiology
Graduate School of Frontier Biosciences
Osaka University
Osaka, Japan
nibuhiko@fbs.osaka-u.ac.jp

Takashi Yamamoto
Department of Oral Physiology
Graduate School of Dentistry
Osaka University
Osaka, Japan
yamamoto@dent.osaka-u.ac.jp

Lei Yan
Department of Pharmacology & Centre for Neuroscience
University of Alberta
Edmonton, AB, Canada

Qin Yan
Key Laboratory of Protein Chemistry and Developmental Biology of National Education Ministry of China
College of Life Sciences
Hunan Normal University
Changsha, Hunan, China

Xiaohang Yang
Institute of Molecular and Cell Biology
Agency for Science, Technology and Research
Proteos, Singapore
mcbyangn@imcb.a-star.edu.sg

Bill J. Yates
Department of Otolaryngology
University of Pittsburgh
School of Medicine
Eye & Ear Institute Building
Pittsburgh, PA, USA
byates@pitt.edu

Hiromu Yawo
Department of Developmental Biologiy and Neuroscience
Tohoku University
Graduate School of Life Sciences
Sendai, Japan
yawo@mail.tains.tohoku.ac.jp

Valerie Yeung-Yam-Wah
Department of Pharmacology
Centre for Neuroscience
University of Alberta
Edmonton, AB, Canada

Tom C. T. Yin
Department of Physiology and Neuroscience Training Program
University of Wisconsin
Madison, WI, USA
yin@physiology.wisc.edu

Eti Yoles
Proneuron Biotechnologies
Beit-Gamliel, Israel

Elad Yom-Tov
IBM Haifa Research Laboratory
Haifa University Campus
Haifa, Israel
YOMTOV@il.ibm.com

Michio Yoshida
Division of Morphogenesis
Institute of Molecular
Embryology and Genetics
Kumamoto University
Kumamoto, Japan
myoshida@kaiju.medic.kumamoto-u.ac.jp

Yoshihiro Yoshihara
RIKEN Brain Science Institute
Wako, Saitama, Japan
yoshihara@brain.riken.jp

William A. Yost
Speech and Hearing Science
Arizona State University
Tempe, AZ, USA
wyost@luc.edu

Deborah Young
Department of Molecular Medicine & Pathology,
Department of Pharmacology & Clinical Pharmacology
Faculty of Medical & Health Sciences
University of Auckland
Auckland, New Zealand
ds.young@auckland.ac.nz

Hung-Hsiang Yu
Department of Neurobiology
University of Massachusetts
Medical School
Worcester, MA, USA

Li Yu
Centre for Research in Neuroscience
Department of Neurology and Neurosurgery
McGill University Health Centre
Montreal, QC, Canada

Julián Yáñez
University of A Coruña
Campus A Zapateira
A Coruña, Spain

Daniel S. Zahm
Department of Pharmacological and Physiological Science
Saint Louis University
School of Medicine
Saint Louis, MO, USA
zahmds@slu.edu

Lucia Zanotti
Neuroimmunology Unit
DIBIT-San Raffaele Scientific Institute
Milano, Italy

Semir Zeki
Department of Imaging Neuroscience
University College London
London, UK
zeki.pa@ucl.ac.uk

David Zenisek
Yale University School of Medicine
New Haven, CT, USA
shaul.hestrin@stanford.edu

Mei Zhen
Department of Molecular Genetics
University of Toronto
Toronto, ON, Canada
Samuel Lunenfeld Research Institute
Mount Sinai Hospital
Toronto, ON, Canada
zhen@mshri.on.ca

D. W. Zochodne
Department of Clinical Neurosciences
University of Calgary
Calgary, AB, Canada
dzochodn@ucalgary.ca

Stuart M. Zola
Yerkes National Primate Research Center
Emory University
Atlanta, GA, USA
szola@rmy.emory.edu

Frank Zufall
Department of Physiology
University of Saarland School of Medicine
Homburg/Saar, Germany
frank.zufall@uks.eu

Michaël Zugaro
LPPA, Collèe de France
CNRS
Paris, France
michael.zugaro@college-de-france.fr

Na^+ Channels

Definition

►Sodium Channels

Na^+-Ca^{2+} Exchanger (NCX)

Definition

A plasma membrane enzyme that exchanges 3 Na^+ for 1 Ca^{2+}. It can operate in the forward mode (extrusion of Ca^{2+} from the cytosol) or in the reverse mode (uptake of Ca^{2+} into the cytosol).

►Influence of Ca^{2+} Homeostasis on Neurosecretion

Na^+K^+-ATPase

Definition

The Na^+K^+-ATPase transports three Na^+ ions out of and two K^+ ions into the cell, using the energy of ATP hydrolysis (electrogenic transport). It maintains the high sodium and potassium gradient across the cell membrane. The Na^+K^+-ATPase (also called the "cellular sodium pump") is selectively inhibited by cardiac glycosides.

Naked DNA Vaccination

►Neuroinflammation – DNA Vaccination Against Autoimmune Neuroinflammation

Naloxone

Definition

Drug used to antagonize opioid compounds such as morphine. Chemical name = N-allyldihydrohydroxy-normorphinone. Also known by the trade name Narcan.

►Gender/sex Differences in Pain

NANC Transmitters

Definition

NANC transmitters are non-adrenergic, non-cholinergic ransmitters; i.e. other than the classical autonomic ransmitters (noradrenaline and acetylcholine).

►Salivary Secretion Control

Narcolepsy

DANIEL A. COHEN
Harvard Medical School; Department of Neurology, Beth Israel Deaconess Medical Center; Division of Sleep Medicine, Brigham and Women's Hospital, Boston, MA, USA

Synonyms

Gelineau syndrome

Definition

Narcolepsy is a neurological disorder that interferes with the normal regulation of sleep and wakefulness. It has been conceptualized as a condition of behavioral state instability in which there is a low threshold to transition between waking, rapid eye movement (REM)

sleep, and ►non-rapid eye movement (NREM) sleep [1]. This view accounts for the associated symptoms, including daytime sleepiness, disrupted sleep, automatic behavior cataplexy, sleep paralysis, and hypnogogic hallucinations.

Characteristics

Clinical Features

Excessive daytime sleepiness is a prominent complaint and generally the initial reason for seeking medical attention. The feeling of sleepiness is generally higher than normal throughout the day, and there may also be sleep attacks, with an irresistible urge to sleep that arises with very little warning. Semi-purposeful activities may continue during the transition from sleep to wakefulness, a phenomenon known as automatic behavior. Just as sleep may intrude upon wakefulness, nocturnal sleep is often disrupted by excessive awakenings. Increased daytime sleepiness and fragmented nocturnal sleep are symptoms of a variety of sleep disorders and are not specific to narcolepsy. The more specific aspects of narcolepsy relate to the unusual propensity for elements of ►REM sleep to intrude into wakefulness.

Components of REM sleep include: the rich emotional content and perceptual imagery of ►dreams; paralysis of most skeletal muscles; and phasic bursts of muscle activity, including rapid eye movements. These components of REM sleep may be dissociated from each other in narcolepsy and may intrude into wakefulness. Symptoms of REM intrusions during wakefulness include: hypnogogic hallucinations, sleep paralysis, and cataplexy. Hypnogogic hallucinations are hallucinatory experiences that occur at sleep onset. These are vivid perceptual experiences superimposed on the conscious experience of the background environment. There may be a strong emotional context, often a feeling of fear or dread. These symptoms may reflect inappropriate activation of neuronal networks normally activated during REM sleep, including areas that regulate emotional and visual processing [2]. Sleep paralysis reflects the occurrence of REM related muscle atonia that is present upon awakening. This may last seconds to minutes, and it is often a very frightening experience. Patients may try to scream for help, although in the context of muscle atonia there may be little to no actual sound. Tactile stimulation from an outside observer may terminate the episode. Cataplexy is another example of inappropriate REM atonia. It is the most specific clinical symptom of narcolepsy.

Cataplexy is distinguished from sleep paralysis in that the episodes of muscle atonia arise from a background of wakefulness. Strong emotion, particularly laughter, can trigger episodes of cataplexy. The distribution and degree of muscle weakness varies between individuals and may also vary within individual attacks. Episodes may be subtle, such as a change in facial expression or slurred speech, to gross impairment in the ability to retain postural control. Usually there is enough warning to prevent serious falls. The episodes usually last seconds to a couple of minutes, but in rare cases repeated episodes may not allow the return of normal muscle tone for much longer periods, a condition referred to as status cataplecticus. In addition to inappropriate motor inhibition seen in sleep paralysis and cataplexy, patients with narcolepsy may also have excessive motor activity during sleep, including REM sleep without atonia, increased phasic muscle twitches during REM sleep, and increased ►periodic limb movements of sleep [3].

Prevalence

Narcolepsy is generally divided into cases of narcolepsy with cataplexy and narcolepsy without cataplexy. It is unclear whether these entities represent distinct diseases, or rather a spectrum of phenotypic differences with common underlying mechanisms [4]. The prevalence of narcolepsy with cataplexy is estimated to be roughly 50 per every 100,000, with a slightly higher prevalence of cases without cataplexy. The onset of symptoms is often during the second and third decade, but this can vary greatly.

Etiology

Recent insight into the pathophysiology of narcolepsy relates to the discovery of the ►hypocretin (►orexin) neuropeptides, a pair of peptides produced in the lateral hypothalamus that play a role in neurotransmission. Canine narcolepsy has been linked to mutations in the gene encoding the hypocretin (orexin) receptor 2 [5]. A mouse model of narcolepsy was developed by targeted disruption (knock out) of the mouse hypocretin (orexin) gene, confirming the role of this peptide system in the control of sleep-wake behavior [6]. In humans, a low cerebrospinal fluid level of hypocretin 1 is a highly sensitive and specific test for narcolepsy with cataplexy [4]. In addition, hypocretin (orexin) staining in the hypothalamus of human narcoleptic patients is decreased by up to 95%, and recent evidence suggests that this reduction in hypocretin (orexin) staining is the result of targeted cell loss of hypocretin neurons rather than decreased hypocretin production [7].

An autoimmune etiology for human narcolepsy has long been suspected. Focal gliosis, essentially a form of scarring in the brain, has been demonstrated in patients with narcolepsy. Gliosis can occur as the result of an inflammatory process, such as occurs with autoimmune diseases, but this finding is non-specific and can occur with a variety of insults to the brain. Speculation about the autoimmune nature of hypocretin cell loss largely comes from a strong association of narcolepsy with cataplexy and specific alleles of the human leukocyte antigen (HLA) presenting system, particularly the allele DQB1*0602. This allele is present in the majority of narcoleptic patients, but it has a low specificity due to the relatively high frequency of this allele in the general population. Beyond genetic factors, an association of narcolepsy with ►seasonality, with a peak in March births, suggests that early environmental factors may play a role in the pathogenesis of narcolepsy and is consistent with an autoimmune mechanism. Efforts to find specific auto-antibodies in the sera of narcoleptic patients have been mixed, with both positive and negative results. An autoimmune process targeting the destruction of hypocretin (orexin) neurons has been the prevailing hypothesis, but there is recent evidence that circulating antibodies may play a direct functional role in the altered neurotransmission in narcolepsy [8]. The role of the immune system in the pathogenesis of narcolepsy remains to be clarified, and a neurodegenerative mechanism remains a possibility. There are no clear modifiable risk factors for narcolepsy at this point.

Pathophysiology

The regulation of sleep and wakefulness is a complex process, incorporating factors such as: the phase of the endogenous circadian cycle, recent sleep and wake history, food intake, posture, environmental stresses or demands, and emotional state. These inputs act on a relatively discrete set of brain structures that form the basic circuitry for switching between the waking and sleeping states. There has been tremendous progress in the recent years in the development of models to understand the basic mechanisms of these circuits. The role of hypocretin (orexin) neurons in the sleep-wake circuitry has been particularly instrumental in understanding the clinical features of narcolepsy.

A prominent model for the regulation of sleep and wakefulness centers on the concept of a "flip-flop" switch [9]. A flip-flop switch is a circuit in which two opposing sides are mutually inhibitory. By inhibiting the competing side, activity on one side reinforces its own advantage in controlling the state of the system. Transitions between states tend to occur relatively rapidly. The major structures involved in the sleep-wake switch reside in the ►brainstem and hypothalamus. Structures that promote wakefulness include: ascending projections from serotonergic neurons of the dorsal ►raphe nucleus, the noradrenergic neurons of the locus coeruleus, histaminergic neurons of the tuberomammillary nucleus, cholinergic neurons of the pedunculopontine and lateral dorsal tegmental nuclei, cholinergic neurons of the basal forebrain, and dopaminergic neurons of the ventral tegmentum. Neurons that promote sleep are predominantly found in the ►ventrolateral preoptic nucleus (VLPO) of the hypothalamus and contain the inhibitory neurotransmitters GABA and galanin. Projections from the hypocretin (orexin) neurons of the lateral hypothalamus activate the wake promoting structures, which tends to stabilize the sleep-wake switch in the wake state. Individual nuclei within the pons contribute to the regulation and generation of components of REM sleep. There are ►REM-off and ►REM-on regions, both of which are mutually inhibitory and also act together as a flip-flop switch so that all components of REM tend to switch on or off with very little transitional states [10]. Hypocretin (orexin) neurons activate the REM-off cell groups. Therefore, the lack of a functional hypocretin (orexin) system in narcolepsy may destabilize both the sleep-wake switch as well as the REM on-off switch, causing relatively frequent transitions between sleep and waking states and inappropriate activation of components of REM sleep.

Diagnosis

The diagnosis of narcolepsy is generally considered in the evaluation of a patient with excessive sleepiness. Occasionally the diagnosis is considered for symptoms of inappropriate REM fragments during wakefulness, such as isolated episodes of sleep paralysis. However, this can occur in healthy individuals or those who tend to wake during REM sleep such as patients with sleep disordered breathing. Cataplexy is specific to narcolepsy and rarely can be the presenting symptom before the onset of significant daytime sleepiness. Careful clinical history can usually distinguish between the paroxysmal loss of muscle tone in cataplexy and other conditions of transient neurological dysfunction or spells from conditions such as: seizure, syncope, cerebrovascular disease, periodic paralysis, myasthenia, or hyperventilation.

Excessive sleepiness may occur from: processes that disrupt sleep such as ►obstructive sleep apnea, ►periodic limb movement disorder, mood disorders, pain syndromes, and noisy or uncomfortable bed environments; ►chronic insufficient sleep syndrome; medical conditions such as thyroid dysfunction; medication side effects; central nervous system causes such as narcolepsy, ►idiopathic hypersomnia, or ►recurrent hypersomnia. It is not possible to distinguish the etiology based on the degree of sleepiness, since a sudden and irresistible urge to sleep may occur from many causes besides narcolepsy.

A diagnosis of narcolepsy with cataplexy can be made by history and does not necessarily require confirmatory testing. Overnight ►polysomnogram is helpful to search for co-morbid, treatable causes of sleep fragmentation and daytime sleepiness. In order to diagnose narcolepsy without cataplexy, it is necessary to perform the ►multiple sleep latency test (MSLT), which is a daytime napping test. This test is also generally recommended to confirm a diagnosis of narcolepsy even when there is definite cataplexy, but it is less essential when the history of cataplexy is clear. Generally, patients are given five nap opportunities during the daytime, each 20 min in duration and spaced at 2 h intervals. The average time to fall asleep across all naps is roughly 3 min for patients with narcolepsy, whereas it is greater than 10 min for healthy individuals. To minimize both false positive and false negative tests, a cutoff of 8 min or less has been suggested to document pathological sleepiness. In addition to pathological sleepiness, which is not specific to narcolepsy, the presence of REM sleep during at least two of the naps (▶sleep onset REM periods) is required for a diagnosis of narcolepsy. The specificity of this finding for narcolepsy is greater than 90%. The MSLT requires that patients are off REM suppressing medications for at least 15 days for valid interpretation of REM periods during naps, which may be difficult in patients that require stimulant medications to function. Treatment of co-morbid disorders such as sleep disordered breathing is also necessary for accurate interpretation of the MSLT. In patients with questionable cataplexy, particularly those who have the HLA DQB1*0602 allele, CSF testing for hypocretin-1 levels can be done instead of the MSLT. Hypocretin-1 levels that are <110 pg/ml support the diagnosis of narcolepsy. Low CSF hypocretin levels have a specificity of 99% and a sensitivity of 87% in narcolepsy with cataplexy [4]. The sensitivity of this test in narcolepsy without cataplexy is very low.

Treatment

It is conceivable that therapy targeting the deficient hypocretin system will eventually play a fundamental role in the management of narcolepsy, particularly in patients with cataplexy. However, current treatment options are intended to promote daytime alertness as well as suppress the tendency for inappropriate activation of REM fragments during wakefulness. Stimulant medications such as modafinil, methylphenidate, D-amphetamine, or caffeine are generally used to promote daytime alertness. Medications that increase levels of noreprinephrine or serotonin, such as many antidepressants, tend to activate REM-off cells and minimize REM intrusions into wakefulness such as cataplexy [10]. Sodium oxybate (gamma hydroxybutyrate), taken before sleep and a second dose during the sleep period, has recently been demonstrated to minimize both cataplexy and daytime sleepiness in patients with narcolepsy. Non-pharmacological interventions include: scheduled naps, which narcoleptic patients usually find very refreshing; psychosocial support and counseling to minimize the functional disability from this chronic disorder; and treatment of co-morbid sleep disorders such as sleep disordered breathing.

References

1. Mochizuki T et al. (2004) Behavioral State Instability in Orexin Knock-Out Mice. J Neurosci 24(28):6291–6300
2. Braun A et al. (1997) Regional cerebral blood flow throughout the sleep-wake cycle: an H215O PET study. Brain 120:1173–1197
3. Dauvilliers Y et al. (2007) REM sleep characteristics in narcolepsy and REM sleep behavioral disorder. Sleep 30(7):844–849
4. Mignot E et al. (2002) The role of cerebrospinal fluid hypocretin measurement in the diagnosis of narcolepsy and other hypersomnias. Arch Neurol 59:1553–1562
5. Lin L et al. (1999) The sleep disorder canine narcolepsy is caused by a mutation in the hypocretin (orexin) receptor 2 gene. Cell 98:365–376
6. Chemelli RM et al. (1999) Narcolepsy in orexin knockout mice: molecular genetics of sleep regulation. Cell 98:437–451
7. Blouin AM et al. (2005) Narp immunostaining of human hypocretin (orexin) neurons: loss in narcolepsy. Neurology 65:1189–1192
8. Smith AJF et al. (2004) A functional autoantibody in narcolepsy. Lancet 364:2122–2124
9. Saper CB, Scammell TE, Lu J (2005) Hypothalamic regulation of sleep and circadian rhythms. Nature 437 (27):1257–1263
10. Lu J et al. (2006) A putative flip–flop switch for control of REM sleep. Nature 441:589–594

Nasal Airflow

Definition

Nasal airflow is the mechanism by which odorant molecules are delivered to the olfactory receptors although the movement of air through the nose serves important functions beyond odor identification. Nasal airflow allows inspired air to be brought to the temperature and humidity of the lungs and particles ranging from combustion products to airborne bacteria to be trapped in the mucus and removed. In very dry environments nasal airflow also allows water vapor in expired air to be trapped in the mucus and saved.

▶Nasal Passageways

Nasal Passageways

David E. Hornung
Dana Professor of Biology, St. Lawrence University, Canton, NY, USA

Synonyms

Nasal airflow; Inspiration; Breathing cycle

Definition

In mammals, the nasal passageways provide the channel through which odorant molecules are delivered to the airspace above the olfactory receptors thereby allowing the animals to detect and respond to chemical signals in the inspired air [1]. According to some investigators, the structure of the nose itself may even be directly responsible for one of the mechanisms contributing to odor discrimination [2,3,4]. The internal structure of the nose allows air from the outside to be brought to the temperature and humidity of the lungs. It also facilitates a trapping of particles in the inspired air in the nasal mucus thus protecting the lungs from many airborne insults. In very dry environments the structure of the nose allows water vapor in expired air to be trapped in the mucus and saved [5].

Characteristics

From an evolutionary point of view, the nasal passageways begin as simple blind sacs and at their pinnacle become multi-functional, anatomically complex structures. The earliest stage of evolution of nasal passageways is exemplified in elasmobranches (some sharks for example) where the olfactory organs are paired, blind nasal sacs located well anteriorly on the head. The openings of these blind sacs are the external nares which in some sharks are divided into two sections allowing for the inflow and outflow of sea water. Chemoreception through the first cranial nerve (olfaction) in these animals is accomplished as receptors sample the contents of the water trapped in these blind sacs. In animals slightly higher up on the phylogenic tree, the nasal sacs are connected to the mouth through internal nares. This anatomical development makes olfaction more dynamic by permitting water to flow from the external naris, by the olfactory receptors and into the mouth via the internal naris [2,3].

In air breathing animals the internal naris and accompanying connection to the mouth assumes a major role is respiration. For example, as tadpoles undergo the process of metamorphosis into frogs, the connection between the nasal cavity and the mouth is the pathway by which air enters the lungs from outside the animal. In air breathing animals, it is necessary in the mouth to separate the food destined for the stomach from the air heading toward the lungs. As a result olfaction becomes ancillary both to feeding and respiration.

Structure

In air breathing vertebrates the structure of the nasal ►nasal airflow passageways can be as simple as the tube located between the external to internal nares as is observed in some salamanders. In these animals, the incoming air follows a straight path through the nasal air passageways (and by the olfactory receptors) on its way to the lungs. In frogs the nasal airflow patterns are slightly more complex as incoming air is deflected by a baffle plate called the eminentia which is located on the floor of the main nasal cavity [2].

In mammals the surface area of the nasal air passageway is greatly expanded through the elaboration of scrolls of bone (turbinals) from the lateral wall of the nasal chamber. In some species, these scrolls produce a very complex nasal labyrinth as is observed in animals like rodents and canines. In general the greater the complexity of the nasal air passageways the better the animal will be at detecting and identifying low concentrations of airborne odorants. The turbinates are also important in reducing water loss from expired air associated with the high ventilation rate that accompanies the maintenance of a constant internal body temperature [5]. The volume of the nasal cavity has been estimated to be 0.4 cm^3 in rats, 20 cm^3 in beagle dogs and 25 cm^3 in humans.

In most mammals the left and right nasal passageways are anatomically distinct structures separated by a bony plate called the nasal septum. As a result, the inspired air flowing through the left and right nostrils does not mix until it gets to the nasopharynx located in the back of the throat. In rodents, however, there is an incomplete nasal septum which allows some of the air flowing through the nostrils to mix just downstream from the olfactory receptor area.

In primates, the nasal air passageways are much less complicated that what is found in rodents or canines. In humans the internal anatomy of the nose is defined on the lateral side by the inferior, middle and superior turbinates (Figs. 1 and 2) [2]. The respiratory and olfactory epithelial cells lining the inner surface of the nasal cavity are bathed in mucus which is secreted by the goblet cells interspersed in the respiratory epithelium and Bowman's Glands found in the olfactory receptor cell area. The mucus located at the air/epithelium border is watery and it is continually flowing into the back of the throat by the beating of the cilia of the respiratory epithelial cells.

Blood Supply

The nasal mucosa has a very rich blood supply. As a result, the size of the airspace defined by the turbinates can be changed quickly and dramatically by changing the amount of blood flow to the dense capillary beds servicing the nasal epithelium. In this regard the lining of the nose is similar to other better know human erectile tissue. The extensive bleeding sometimes seen following nasal trauma reflects this extensive capillary bed. Additionally, the semi-erectile nature of the nose partially accounts for the nosebleeds that are occasionally observed after a sexual organism.

Nasal Cycle

A pattern has been observed in the nose where congestion on one side of the nose is accompanied by decongestion on the other. This pattern is called the nasal cycle and has been observed in over 50% of the adult human population. The periodicity of the cycle is between

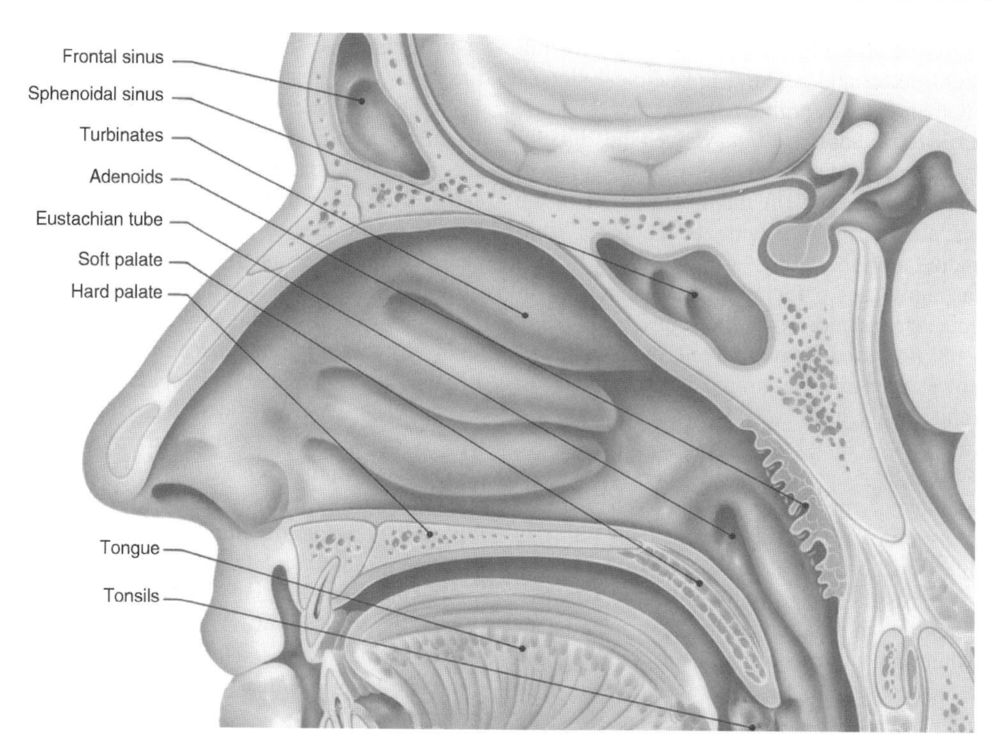

Nasal Passageways. Figure 1 A cut away view of the human head. The inferior, middle and superior turbinates are in order located above the hard palate. The olfactory area is around the superior turbinate (marked "turbinates"; from Merrell Dow Company).

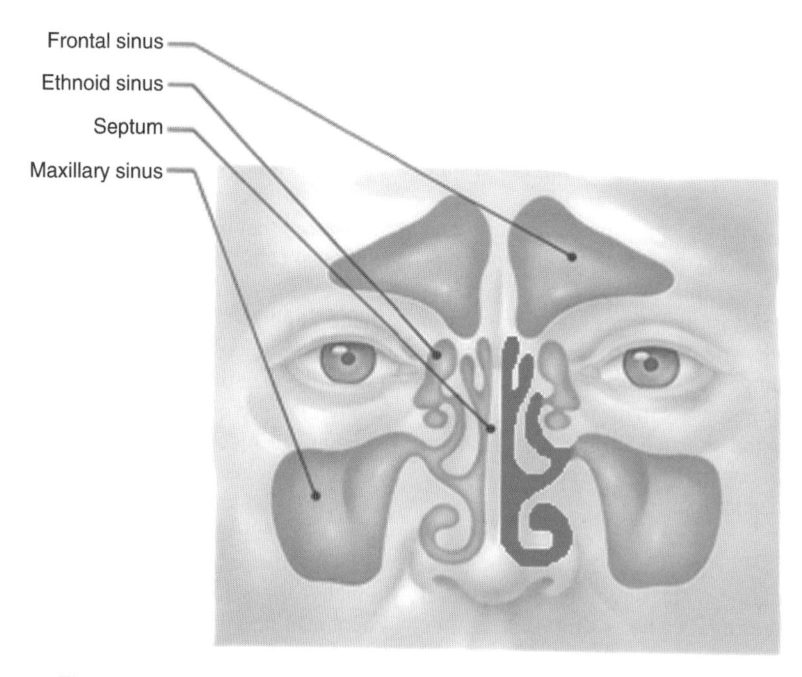

Nasal Passageways. Figure 2 A cross-sectional view of the airspaces in the human nose. The lighter area represents bone and other nasal structures. The nasal air passageways have been darkened on the subject's left side.

50 min and 7 h. There has even been a correlation reported between handedness and the nostril that is open for a larger percent of a 24 h period, such that left handed individuals are more likely to have a more open left nostril and visa versa for right handers. Regardless of the more open nostril, the internal structure of the nose produce convoluted (sometimes turbulent) flow paths for the inspired air. The high airflow seen during sniffing may

increase the amount of turbulence and so may improve the sensitivity to smells [1,2].

Role in Odor Identification

The olfactory receptors in humans are located high in the nose along the septum and on the medial side of the superior turbinate. Because of their location, and because the airspaces defined by the middle and inferior turbinates are relatively much larger that those leading up to the olfactory receptor region, during a ▶sniff only about 10% of the inspired air is directed toward the airspace above the olfactory receptors.

The structure of the nose may play a role in odorant identification as incoming odorant molecules interact with the nasal epithelium as they are directed toward the airspace above the olfactory receptors. The physical and chemical properties of these interactions depend on the natures of the odorants themselves [1,3,4]. For example, for an odorant that is very soluble in the respiratory mucosa, many of the incoming odorant molecules would be expected to be sorbed early in the flow path. As a result, it will take longer for these molecules to get to the receptors as compared to molecules of odorants that were not very mucosa soluble. In addition, there will be a spatial distribution pattern across the receptor sheet itself with most of the mucosal soluble odorant molecules being concentrated early in the flow path as compared to a more even mucosal distribution pattern that would be seen for less mucosa soluble odorants. It has been hypothesized these different arrival times and olfactory mucosal distribution patterns may play some role in odorant identification. These distribution patterns have been called "imposed" patterns. Although imposed patterns may play a role in olfactory function, the fine tuning and spatial distribution of receptors within the olfactory receptor area (inherent patterns) are thought by most investigators to be the primary peripheral mechanism for odorant detection and identification.

Non-Olfactory Functions – Nasal Air Conditioning

The non-olfactory nasal functions of the nasal cavity are together referred to as nasal air conditioning [5]. Some evolutionary biologists suggest the primary selective pressure on nasal structures comes from these non-olfactory functions. During the process of nasal air conditioning, the inspired air is brought to the temperature and humidity of the lungs. In addition, a filtration process occurs in which particles ranging from combustion products to airborne bacteria that are found in the inspired air are trapped in the mucus and removed as the mucus is carried to the back of the throat.

Swell Space

Because of the speed and magnitude of the diameter change that can occur in the area around the inferior turbinate, this area of the nose is sometimes referred to as the "swell space." As cold air enters the nose during inspiration the blood flow to the swell spaces increases dramatically. This increased blood flow causes a swelling of the nasal erectile tissue and so reduces the size of the airspace the incoming air must traverse on its way through the nose. Because the air passageways are now narrower, heat can be more efficiently transferred from the respiratory and olfactory mucosas to the incoming air. As a result, the cold air is effectively warmed before it gets to the back of the throat and the lungs [5].

Humidification

When warmer air (especially if it is also humidified) enters the body the reverse happens and the swell spaces shrink, resulting in a more open nasal air passageways. This temperature and humidity related widening of the nasal cavity explains why, when the nose is blocked because of an upper respiratory infection, breathing warmed, humidified air can sometimes reduce the felling of stuffiness.

Because the nasal mucus has such a high water content, it is able to humidify the incoming air such that when the inspired air has a low humidity, water evaporates from the mucus into the air. As a result of the evaporation from the mucus to the inspired air, the surface of the mucosa is cooled, so that when it is time to discharge the air from the lungs the expired air passes over the cooled surface of the mucosa. Because of the lower surface temperature along the mucosa, some of the water in the expired air condenses and so it is nôt lost. This process is not very efficient in humans since the air passageways are not very convoluted [5]. However, this process can be very efficient in animals with narrower and more convoluted air passageways. For example, this process is so efficient that animals like the Kangaroo Rats of Australia loose no water during breathing even with ambient temperatures in excess of 50°C.

Particle Trapping

In addition to supplying water for humidification purposes, the nasal mucus serves as a trap for particulate matter including smoke, dust particles and airborne bacteria. The respiratory epithelium lining the nasal passageways contain cilia, hair-like protrusions extending into the overlying mucus layer. As these cilia beat, they create a slow wave-like action in the mucus that is responsible for moving the mucus through the nose and to the nasopharynx, where the mucus is then swallowed. The trapped particles flowing with the mucus are more easily dealt with by the alimentary canal than they would be by the blind sacs in the alveoli of the lungs. The mucus likely even contains some white blood cells to deal with trapped airborne bacteria. Cigarette smoke, with contains high levels of carbon dioxide, can temporarily anesthetize the beating of respiratory cilia, thereby slowing the flow of mucus. This would be expected to make it more difficult for the nose to clear trapped particles and may even contribute to the "smoker's cough."

References

1. Hornung DE (2006) Nasal anatomy and the sense of smell. In: Hummel T, Welge-Lussen A (eds) Taste and smell an update. Kerger, Basel, pp 1–22
2. Hornung DE, Mozell MM (1986) Smell – human physiology. In: Rivlin RS, Meiselman RH (eds) Human taste and smell, measurements and uses. Macmillan Publishing, New York, pp 19–38
3. Kent PF, Mozell MM, Murphy SJ, Hornung DE (1996) The interaction of imposed and inherent olfactory mucosal patterns and their composite representation in a mammalian species using voltage-sensitive dyes. J Neurosci 16:345–353
4. Zhao K, Scherer PW, Hajiloo SA, Dalton P (2004) Effect of anatomy on human nasal airflow and odorant transport patterns: implications for olfaction. Chem Senses 29(5): 365–379
5. DeWeese DD, Saunders WH (1968) Textbook of otolaryngology, 3rd edn. Mosby, C. V. St. Louis

Nasopharynx

Passageway from the back of the mouth to the nasal cavity retronasal route: air breathed out passes through the nasopharynx to the nasal cavity.

►The Proust Effect

Natural Hypothermia

Definition

A naturally occurring regulated sub-euthermic body temperature state of endothermic animals. Natural hypothermia is subject to thermoregulation: animals can recover from hypothermia without outside help.

►Hibernation

Natural Kinds

Definition

Natural kinds are categories of things that share some essence (physical, chemical, etc.), e.g. being H2O. This essence can be used in explanations of the perceptible properties of members of the natural kind, e.g. transparency, melting point, etc.

►Information

Natural Stimulus Statistics

Definition

►Sensory Systems

Naturalism

Definition

In the philosophy of mind naturalism is usually thought of as the view that, at a minimum, philosophical theories should be consistent with our scientific picture of the world. A stronger view, which is also sometimes called naturalism, is the view that philosophical investigation really has no place in our understanding of the world, that philosophers, rather than devising theories of their own, should defer to the scientists.

Finally, naturalism is sometimes thought of the view that philosophical investigation should itself be a sort of scientific investigation.

►Physicalism

Naturalized Epistemology

Definition

In its broadest sense a naturalist in epistemology claims that epistemological theorizing is closely tied to theorizing in the natural sciences. There is much disagreement among naturalists as to what exactly the role of the natural sciences within epistemology is or should be.

►Knowledge

Naturally Occurring Cell Death

►Programmed Cell Death

Nausea Syndrome

Definition
Syndrome characterized by epigastric awareness and discomfort, nausea and vomiting.

Nauta Technique

Definition
The Nauta technique is a reduced silver impregnation method for the staining of degenerating axons. It was used to trace the connections of a particular part of the brain by damaging that part and then describing the position and course of degenerating axons. Modifications of the basic Nauta technique have been developed to localize degenerating axons near axon terminals to characterize where the axons of neurons in the damaged area terminate.

Navigation

Definition
Navigation is the process of calculating and executing routes. Animals use navigation to optimize the collection of spatially dispersed resources, to find safety or to interact with individuals. In simple forms of navigation the animal moves towards or away from the goal location which is also to source of beacon signals.

For example, it an animal wants to interact with another, the simplest mechanism to approach the second animal using sensory cues emanating from the second animal.

In many cases navigation is inferential. The animal uses stationary environmental cues – landmarks – to estimate the location of the goal. Landmarks can serve as beacons, where the animal moves in the direction of the landmark to approach the goal, or part of a group, where the animal uses the configuration of landmarks to estimate the location of the goal. When animals compute navigation routes using the relationship among landmarks, the process is considered "mapbased".

Path integration or dead reckoning may be used to return from the current position to the starting position. Path integration requires knowledge of the starting position and direction and the estimation of the current position and direction from the path traveled. It may rely on optic flow and/or internally available information without the use of external landmarks (e.g., in darkness or in visually unstructured environments), which capacity is shown by animals as simple as ants and bees. This in turn requires the continuous monitoring of the intermittent self-generated movements. Location and direction in space are represented by distinct neuronal populations, place cells and head-direction cells.

►Spatial Learning/Memory

NCF

Definition
Nucleus cuneiformis.

Near Response

Definition
Convergence, accommodation of the lens and constriction of the pupil (also called triple response or near reaction).

►Eye Movements Field

Near Response Neurons

Lawrence Mays
University of North Carolina at Charlotte, Charlotte, NC, USA

Synonyms
Vergence neurons; Convergence neurons

Definitions
Ocular convergence (►Convergent eye movement) refers to the nasalward movement (►Ocular adduction) of the eyes required to binocularly view a nearer visual

target. Divergence (►Divergent eye movement) is the opposite movement (►Ocular abduction) of the eyes as gaze is shifted from a near to a far object. These ►vergence (►disparity dependent vergence, ►radial flow dependent vergence) movements are accomplished by the coordinated actions of the medial and ►lateral rectus extraocular muscles. Convergence requires contraction of the ►medial rectus and relaxation of the lateral rectus muscles, while the opposite occurs for divergence. Vergence eye movement may be initiated by the need to reduce binocular disparity, which is the difference between locations of an image of a single target on the two retinas. Convergence of the eyes is accompanied by lens accommodation (►Accommodation of the lens), which is an increase in the refractive power of the crystalline lens needed to focus on the near object.

Characteristics

Higher Order Structures

Single binocular vision requires very precise (≈0.25°) alignment of the two eyes on the object of regard if ►diplopia, or double vision, is to be avoided. Processing of this error signal, which is termed binocular disparity, takes place initially in the primary visual cortex [1]. The mechanism by which binocular disparity is transformed into a motor command signal to converge or diverge the eyes is not known, but such a motor command is seen on midbrain neurons termed near response cells.

Parts of this Structure

The location of midbrain near response cells (including convergence cells) has been identified in macaques, but not yet in humans. They are located in two areas within the midbrain; a peri-oculomotor area, just dorsal and lateral to the oculomotor nucleus and a second area in the ►pretectum [2]. The properties of the neurons in these two zones appear to be similar.

Functions of this Structure

The function of near response cells is to provide the downstream motor elements with appropriate signals to generate vergence eye movements and associated changes in lens accommodation. Considering vergence movements first, the ►extraocular motoneurons encode signals to move the eye by means of a position-rate code and an eye velocity code [3]. The commands for all eye movements include these two elements. Extraocular motoneurons fire at a remarkably (≈5% variation) constant rate for a given eye position, and this rate increases linearly as the eye moves in the motoneuron's on-position. This activity is needed to overcome the elastic restoring forces operating on the oculomotor plant (the mechanical properties of the eye and associated muscles, tendons, and ligaments). In addition, a phasic burst of neural activity by the motoneurons, proportional to eye velocity, overcomes the viscous drag of the oculomotor plant. Medial rectus motoneurons receive inputs related to conjugate eye movements (e.g. ►saccades, smooth pursuit) via the ►medial longitudinal fasciculus (MLF), but this is not the source for vergence commands [4]. Instead, vergence signals are likely due to projections from near response cells in the peri-oculomotor region. Many of these near response cells have tonic firing rates directly proportional to vergence angle and so were termed "convergence cells". A subset has been shown to project to the medial rectus subdivisions of the oculomotor nucleus [5]. Some convergence cells have a position signal and no velocity signal, some show both, and others appear to have a signal related to vergence velocity but not position. Near response cells do not have conjugate eye movement signals. The firing pattern of a convergence cell for convergence is shown in Fig. 1. This activity pattern corresponds to that needed by the medial rectus motoneurons to execute vergence eye movements. In addition to near response cells which increase their activity for convergence, about 25% of near response cells show a linear decrease in activity for convergence, and an increase for divergence. These are termed "divergence cells" [2]. Some of these cells have a divergence velocity signal (divergence burst) alone, or in addition to the divergence position signal. Although no direct projection to the abducens nucleus has been shown, divergence cells have a firing pattern which is appropriate for abducens neurons.

In addition to the signal related to vergence, most ►near response neurons also carry a signal related to lens accommodation. Although it has not been conclusively demonstrated, it is likely that some near response cells provide an input to the Edinger-Westphal nucleus, which in turn provides an input to the ciliary ganglion to effect lens accommodation.

Higher Order Function

Ocular convergence is associated with lens accommodation (see ►Accommodation-vergence interaction) and it is very likely that near response cells are critical elements in this interaction. The model for this interaction is described in the Accommodation-vergence interaction entry, and the role of near response cells in this interaction is described in the following section.

Quantitative Measure for this Structure

With disparity open loop (e.g. one eye occluded), lens accommodation drives convergence, and convergence cells increase their firing rate. Similarly, with accommodation open loop, convergence drives lens accommodation, and convergence cells also increase their firing rate. In order to quantitatively assess the roles of near response cells in the accommodation-vergence interaction, it is necessary to dissociate accommodation

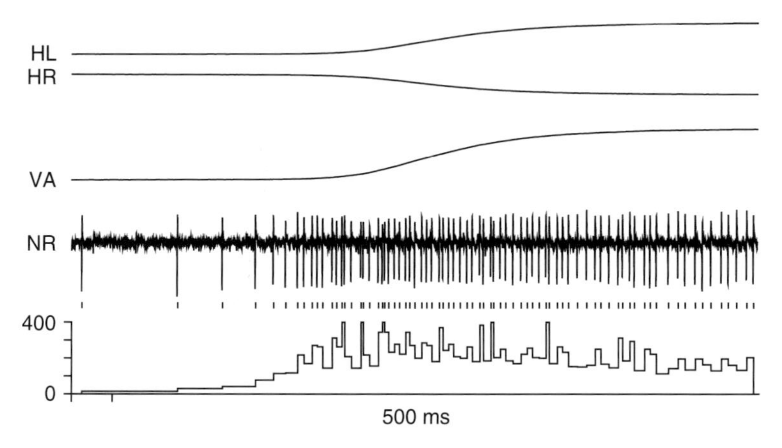

Near Response Neurons. Figure 1 Firing pattern of midbrain near response cell for convergence. Traces are *HL*, horizontal left eye position; *HR*, horizontal right eye position; *VA*, vergence angle (HL-HR); *NR*, extracellular recording of action potentials of near response cell. The histogram at bottom is the firing rate in spikes/s (scale = 400 spikes/s). The time base is 500 ms. This cell shows a linear increase in activity for convergence as well as a small vergence velocity signal.

and vergence in the closed loop condition. This can be accomplished, at least partially, by requiring the subject to maintain clear focus on a target while forcing the eyes to converge, or alternatively, by requiring a constant vergence angle while changing the accommodative response by means of lenses. To the extent to which the accommodative and vergence responses can be dissociated, it is possible to characterize near response cells by the following equation:

$$\Delta FR = R_0 + k_y \times \Delta CR + k_a \times \Delta AR$$

where FR is the firing rate of the cell, R_0 is the firing rate for zero accommodation and convergence, CR is the convergence response, AR is the accommodative response, and k_v and k_a are the coefficients describing the influence of vergence and accommodation, respectively. If a near response cell has a zero k_a value and a non-zero k_v value, then its activity is related exclusively to vergence; if the k_v value is zero and the k_a value is non-zero, then it is related exclusively to accommodation. An analysis of a relatively large number of convergence cells showed that most were not related to accommodation or vergence exclusively, but were related to both (i.e. had a non-zero k_a value and a non-zero k_v value) [5]. Indeed, some near response cells which increased their activity for convergence and accommodation (convergence cells) decreased their activity for convergence when accommodation was held constant (i.e. had a negative k_v value) and *vice versa*. The mechanism for this pattern of results is shown in Fig. 2.

Figure 2 is a representation of the dual-interaction model of accommodation and vergence. Both the accommodative and vergence systems are controlled by negative feedback, and they are cross-linked by the diagonal connections. The near response cells (labeled 1–6) are presumed to be in both systems. Consider the situation in which a subject is required to keep the accommodative response constant while increasing the convergence response. This action will increase the output of the vergence controller (required for more convergence), but it will also increase the cross-link drive to the accommodative system, which is not needed. The accommodation controller will then produce a negative output to counteract this unwanted drive, if accommodation is to remain constant. However, this negative accommodation drive will be sent to the vergence system through the accommodative vergence link, and so will require an increased output from the vergence system to counter it. As long as the cross-link gains are not too high, the system will stabilize at a point at which the increased output of the vergence controller will be great enough to produce the required convergence response and counteract the accommodative vergence cross-link input. In addition, the negative output of the accommodative controller will cancel the input from the convergence accommodation cross-link. Consider near response cell 5, which receives a cross-link input that is equivalent to the average for the accommodative vergence cross-link. For this cell, the negative cross-link input exactly cancels the extra output that the vergence system supplies to deal with the conflicting viewing cues. In this case, the activity of this cell is always associated with the convergence response and is independent of the accommodative response. This cell will have a positive vergence gain (k_v) and a zero accommodative gain (k_a). Consider near response cell 4, which receives a relatively weak cross-link input from the accommodative controller. In the conflict viewing situation described above, this cell's activity will be

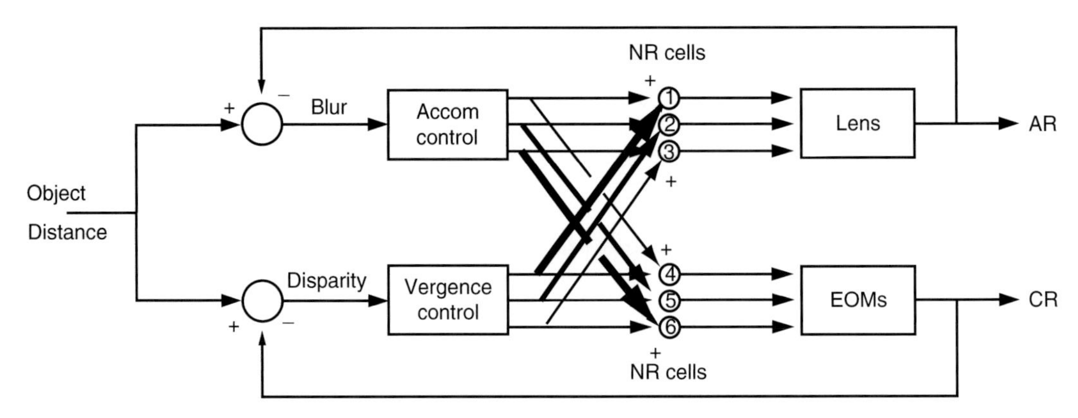

Near Response Neurons. Figure 2 Putative role of near response cells in dual-interaction model of accommodation-vergence interaction. Abbreviations are: *AR*, accommodative response; *CR*, convergence response; *EOMs*, extraocular muscles. The accommodation (top half of figure) and vergence systems (lower half) are controlled by negative feedback mechanisms responding to the distance of the object of regard (object distance). In addition, the accommodative controller provides an input to the vergence system (downward angled cross-link arrows), and the vergence controller also drives the accommodative system (upward angled cross-link arrows). Near response cells (1–6) are thought to be elements in this linkage.

driven primarily by the output of the vergence controller, and thus will fire at a higher rate for a given convergence response when the accommodative response is restrained. Since reinstating the accommodative response would cause this cell to fire at a lower rate, the accommodative gain (k_a) must be characterized as negative, even though the cell's vergence gain (k_v) is positive. Similarly, near response cell 3 would have a positive k_a and a negative k_v, while near response cells 1 and 6 would have positive (non-zero) k_a and k_v values. In this way, small differences in the relative gains of the cross-link and direct inputs to near response cells allow the activity of these cells to be related to both accommodation and vergence. The overall pattern of accommodation and vergence gains of near response cells strongly implies that they are involved in the linkage between accommodation and vergence, but this has not yet been confirmed by more direct physiological or anatomical evidence.

References

1. Freeman RD (2004) Binocular interaction in the visual cortex. In: Chalupa LM, Werner JS (eds) The visual neurosciences. MIT, Cambridge, MA, pp 765–778
2. Mays LE (1984) Neural control of vergence eye movements: convergence and divergence neurons in mid-brain. J Neurophysiol 51:1091–1108
3. Gamlin PDR, Mays LE (1992) Dynamic properties of medial rectus motoneurons during vergence eye movements. J Neurophysiol 67:64–77
4. Gamlin PDR, Gnadt JW, Mays LE (1989) Abducens internuclear neurons carry an inappropriate signal for ocular convergence. J Neurophysiol 62:70–81
5. Zhang Y, Mays LE, Gamlin PDR (1992) Characteristics of near response cells projecting to the oculomotor nucleus. J Neuroplysiol 67:944–960

Nearfield

Definition

The area surrounding an acoustic or hydrodynamic source wherein source energy is dominated by inertial rather than viscous forces. In the nearfield, bulk movements of the medium, (hydrodynamic flow) is of much greater energy than the elastic motions of the medium that make up the propagating pressure wave (i.e., sound in common parlance). The spatial extent of the nearfield depends upon the density of the medium and the frequency of the sound source. The majority of biological relevant sound sources in water have extensive (several meters or more) nearfields, and most fish hearing occurs within the nearfield.

►Evolution of Mechanosensory and Electrosensory Lateral Line Systems

Necessity

Definition

Some statements are not only true but necessarily true (see truth). The fact that they state cannot possibly be otherwise. Consider the necessary "My mother is a woman" versus the non-necessary (i.e. accidental or contingent) "I have a brother." There are, however, different kinds of necessity: a logical necessity, like

"It is raining or it is not raining," has its source in the rules of logic; a conceptual necessity, like "Bachelors are unmarried men," is true in virtue of the meanings of the words of the statement; a metaphysical necessity, like "Water is H2O," is said to be necessarily true because it is the essence of water to be H2O. Nomological necessity, finally, is grounded in the laws of nature:

"Increased heat at constant volume necessitates (or causes) higher pressure" (see also reasoning – a priori).

►Meaning

Necessity, Conceptual

Definition

A statement that cannot turn out to be false, no matter what in fact is said to be necessarily true (opposite: contingently true; see truth). If this necessity has its source in the meaning of the words used in the statement one speaks of conceptual necessity (or analyticity). Examples: "All bachelors are unmarried" or "If an object is red then it has a color." Denials of conceptually necessary truths are baffling if not meaningless: "My car is red but it does not have a color".

►Analyticity
►Meaning
►Necessity, Nomological

Necessity, Nomological

Definition

Scientists and philosophers who do not think that laws of nature are mere summaries or descriptions of the regular events happening in our world claim that the laws bring about or necessitate those events. If it is a law that every C event (or property instantiation) is followed by an E event (or property instantiation), then C nomologically necessitates E. In other words, if C then, according to the law, it must be the case that E. The relation of nomological necessity between events or properties is often thought to be identical to the causation relation (or, at least, a derivative thereof).

Compare nomological to conceptual necessity, which is supposed to be a property of sentences or statements rather than a relation between events or properties.

►Meaning

Nef

Definition

This protein is released by HIV and found to be associated with HIV pathogenesis. It is known to perform an important role in HIV associated neurological disorders and viral replication.

►Central Nervous System Disease – Natural Neuroprotective Agents as Therapeutics

Negative Feedback Control

Definition

A mechanism used to regulate the value or time course of an output variable or signal when the output variable is determined by the value of an input signal or the time course of an input signal. The control mechanism is said to be closed loop when the value of the output variable is sensed or measured, is fed back and compared to some desired reference value and is then used to determine the value of the input signal. The closed loop control mechanism is referred to as negative feedback control when a given change in the output variable is compensated for by changing the input signal to produce an opposite change in the output. Negative feedback control is associated with homeostasis and regulatory processes, but a system regulated by negative feedback control can become unstable.

►Control Theory
►Posture – Sensory Integration

Negative Schizophrenic Symptoms

Definition

Lack of drive and initiative, social withdrawal, depression-like symptoms, anhedonia.

►Schizophrenia

Nematode

►C. elegans Neuroethology

Neocerebellum

Definition

Phylogenetically, a very young part of the cerebellum. The neocerebellum contains the two cerebellar hemispheres and receives afferents mainly from the pons, thus being called also the pontocerebellum.

►Cerebellum

Neocortex

Definition

The evolutionarily most modern part of the vertebrate forebrain that covers with many convolutions most of the visible surface of the human brain. Also called isocortex because of the six-layered structure.

►Isocortex

Neocortical Circuits: Computation in 3-D

Rodney Douglas[1], Henry Kennedy[2,3,4], Kenneth Knoblauch[2,3,4], Kevan Martin[1]
[1]Institute of Neuroinformatics, University/ETH, Zurich, Switzerland
[2]Inserm U846, Bron, France
[3]Stem cell and Brain Research Institute, Bron, France
[4]Université de Lyon, Université Lyon I, Lyon, France

Definition

Connections between neurons within and between cerebro-cortical areas.

Characteristics

Cortical Architecture and Processing

Every thought, every idea, every memory, every decision, and every action we have to make, arise from the activity of neurons in our brains. The results of some of this activity surround us: household objects, books, technology and art. Of all brain structures, the ►neocortex, which forms over 80% of the volume of the human brain is, arguably, the most critical to what makes us human. This is a paradox, because the basic local architecture of the neocortex in all mammals, from mouse to man, appears to be very similar and is determined by the laminar distribution of relatively few types of excitatory and inhibitory neurons organized according to common principles of connectivity. These local circuits are organized in a framework of a six-layered columnar architecture, in which neurons with functional properties in common lie in discrete layers and in vertical slabs or columns [1] (►Striate cortex functions).

The uniformity of its construction suggests that the neocortex provides circuits that are optimized for a class of cortical "algorithm" that can be implemented for the full range of demands of behavior, including perception, cognition, and action. A number of models indicate the forms of general computation that could be carried out in a uniform cortical architecture. Typically these models address a single principle of operation in a small group of neurons; in others a more detailed model is imposed on the columnar architecture of cortex. Experimental results in alert behaving primates and together with theoretical studies, suggest that cognitive operations proceed very rapidly across different cortical areas (►Cerebro-cortical areas; ►Extrastriate visual cortex) by feedforward categorization and feedback modulation, with slower refinement by lateral local interactions. Specification of local and long-distance connections in the cortex will go some way to explaining the implementation of these processes.

Structural Specification of Cortical Connectivity: Integration of Intra- and Inter-Areal Connectivity

What is so special about the circuits of the neocortex? What makes them so efficient and so adaptable to different tasks? A major contribution to our understanding of the structure of the cortical circuit came with the model of a "canonical cortical microcircuit" [1]. This circuit expresses the functional relationships between the excitatory and inhibitory neurons in the different cortical layers and shows how the inputs to a local region of cortex from the sensory periphery via the ►thalamus, or from other cortical areas, are integrated by the cortical circuits. The most critical feature of the canonical circuit is that the neurons are connected in a series of nested positive and negative feedback loops called "recurrent circuits." Because the excitatory and inhibitory neurons are interconnected, excitation and inhibition remain in balance and so the positive feedback does not overexcite the circuit. This organization explains how it is that the relatively tiny numbers of neurons that provide the external inputs to this circuit are nevertheless effective, as they are amplified selectively by recurrent excitatory circuits [1]. Explorations of this model in the visual cortex (►Visual cortex – neurons and local circuits), e.g. [2] have shown how this key notion of recurrent amplification explains the emergence of cortical properties, such as direction sensitivity and velocity sensitivity, orientation selectivity (►Striate cortex functions), masking, and ►contrast adaptation.

The canonical model provides for a richer array of behaviors than the simple feedforward models that preceded it, and is readily applied across the cortex. For example, it is clear that the interlaminar connections have characteristic patterns across cortical areas and across species and thus may perform a generic computation [3]. What has been lacking until very recently is a quantitative model of the vertical (interlaminar) circuits. However, the studies by Binzegger et al. now clearly indicate that, in general, the contribution of the spiny neurons to interlaminar connections exceeds that of their intralaminar connections [4]. Hence, in the infragranular layers (layers 5 and 6), the majority of ►pyramidal cells connect outside their layer of origin. Layer 4 spiny neurons do connect within layer 4 (the "granular" layer), but their major projection is to layer 3. It is only in the supragranular layers (layer 1, 2 and 3) that the pyramidal cells make the majority of their synaptic connections to the same layer. The consequence of this is that the monosynaptic recurrent connectivity of layer 2 and 3 pyramidal cells predominates more than recurrent connectivity in any other layer. The recurrent connectivity of layers 2/3 is intriguing in that the local axons of the pyramidal cells are not uniformly distributed, but form patches or clusters. This pattern of patchy connections, referred to as "lattice connections" by Rockland are embedded within inter-areal feedforward and feedback connections [5]. Because of its appearance when viewed from the surface of the cortex we refer to the local horizontal network formed by a small cluster of pyramidal neurons as a "Daisy" In the neocortex many pyramidal neurons serve a dual function: all of them form the major excitatory neurons in the local cortical circuit, but many of them also project outside their own cortical area to other cortical areas or subcortical structures. Thus many of the same neurons that form a Daisy could also project to other cortical areas.

Inter-Areal Projections

The inter-areal connections come in three flavors: feedforward, feedback, and lateral connections [5,6] (►Visual cortex – neurons and local circuits). Feedforward connections originate principally from the supragranular layers, target layer 4 and connect lower to higher visual areas (►Extrastriate visual cortex) in a sequence tending to show increases in ►receptive field (Visual cortical and subcortical receptive fields) size and response latency. Feedback connections originate from principally infragranular layers, and connect higher to lower visual areas in a sequence suggesting decreases in receptive field size and response latency. It has been suggested that feedforward neurons have a "driving" and feedback neurons a "modulatory" influence. This is why the feedforward and feedback pyramidal cells located in the supragranular layers could also participate in the local Daisy circuits. The feedback neurons located in the infragranular layers likewise may participate in the local Daisy circuit via the local vertical connections with the supragranular layer pyramidal cells [3]. The infragranular feedback neurons probably provide an input to the Daisy, because one of the principal targets of the feedback projections are the supragranular layers (particularly layer 1).

Thus far, most of our knowledge concerning the local horizontal network is derived from studies of the ►primary visual cortex (Visual cortex – neurons and local circuits) of cats and monkeys, where it has been claimed that the horizontal clusters link columns of cortex with representations of like-orientation (►Striate cortex functions). In other cortical areas, including areas of ►prefrontal cortex in the monkey, such as area 46, horizontal clusters are equally apparent, but the representations they link have yet to be defined. At a structural level there are important regularities, whose functionality has yet to be divined. Across all areas and species examined (which include the major divisions of neocortex), there is a linear relationship between the size of the clusters and their spacing [3]. The size of the patches also correlates with the diameter of the lateral spread of the dendrites of pyramidal cells, which increases from ►occipital cortex to prefrontal cortex. It is not known what determines the constancy in the relations of these dimensions.

Inter-Areal Hierarchies

Van Essen and colleagues have gone a long way in exploring the particular hierarchy to be found in the visual system and beyond. They showed that pair-wise comparison of the laminar organization and connections linking cortical areas made it possible to define all inter-areal pathways as either feedforward, feedback or lateral (linking areas on the same hierarchical level) (►Extrastriate visual cortex). While the Felleman and Van Essen model has continued to exert a powerful influence on concepts of neorcortical function and brain organization, it has been questioned by the group of Malcolm Young that showed that there are 150,000 equally plausible solutions to the Felleman and Van Essen model [7].

In order to obtain a determinate model, it is necessary to define the hierarchical distance between stations. Precise quantification of the laminar organization of inter-areal connectivity provides a useful measure of hierarchical distance [5,6]. Injections of retrograde tracers in a mid-level target area show that afferent areas contain both labeled supra- and infragranular layer neurons. Feedforward projections originate predominantly from supragranular layers, and the exact proportion of supragranular neurons labeled relative to all labeled neurons in the same area depends on the hierarchical distance from the target area. Feedforward projections to far-distant areas originate almost exclusively from supragranular layer neurons, and as one approaches the target area, there is a smooth increase in the contribution from the infragranular layers. Likewise in the case of feedback projections, as the hierarchical distance increases there is a steady increase in the proportion of infragranular layers so that far-distant feedback projections are almost uniquely from infragranular layers. This regularity has been encapsulated in a "distance rule" that has the power to define the hierarchical organization of a cortical network from the analysis of the projections to only a small number of key areas [6].

Tracing experiments reveal that around 90% of the projections are local (within 1–2 mm), that is, most of the projections onto a cell are from neurons within the same area. Of the remaining 10%, about two thirds come from neighboring areas and are lateral, so that information flow across the hierarchy is assured by a truly minute proportion of feedforward and feedback neurons. The observation of dense local connections coupled with sparse long-range connections conforms to the idea of a "Small-World" network and goes along with a model of areas as functionally specialized modules, with the long-distance connections serving to communicate the information processed locally within areas rapidly across the cortex.

Physiological Integration of the Daisy Architecture with the Connections between Cortical Areas

How long-range connections influence local circuit functions is an important step in understanding the computational function of the neocortex. One approach is to temporarily inactivate the area by cooling and study what effect the inactivation as on a target area the projecting regions. Cortical areas ►V2 (Cerebro-cortical area V2) and MT (►Cerebro-cortical area MT) have feedback connections to the primary visual cortex (area V1 (►Cerebro-cortical area V1)), and cooling area V2 or area MT reduces the receptive-field center response of area V1 neurons. This suggests that there may be a summing of feedback activity with feedforward input from the thalamic ►lateral geniculate nucleus (LGN), which relays ►retinal activity to area V1. Integration is further suggested by the evidence that feedback projections from extra-striate cortex overlap with clusters of area V1 cortical output neurons [8].

One way to investigate the dynamics of the interaction of inter- and intra-connectivity is to examine the visuo-topic scales of both systems and compare them to the receptive field response of neuronal aggregates in area V1 [8]. In these studies the representation of the ►visual field (►Vision) is determined for the extent of the local Daisy connection as well as for the inter-areal connections. These studies suggest that Daisy connections have the appropriate spatial extent to mediate a restricted portion of the visual response of area V1 neurons, which corresponds to the spatial summation zone within the receptive field. The extent of the Daisy connections was however insufficient in extent to account for the full surround response from beyond the classical receptive field (►Vision). This makes sense because the relatively long delays of the suppressive orientation-selective effects of surround stimulation are similar to those reported for the slow propagation of excitatory activation mediated by horizontal connections.

The visuotopic representation of feedback projections from extra-striate cortex to area VI are commensurate with the full center-surround response of the area V1 neurons (►Visual cortical and subcortical receptive fields). The influence of extra-striate cortex on Daisy connectivity is coherent with the temporal constraints: the timing of the visual responses of higher visual areas largely overlap with area V1 responses, the conduction velocities of the large-caliber fibers projecting from extra-striate cortex to area V1 are considerably faster than those of the horizontal intrinsic fibers, and the inactivation of extra-striate cortex influences the early part of the area V1 neuron visual response. Hence, it would seem that the physiology and the visuo-spatial correspondence between the intra- and inter-areal connection systems provides the basis for the integration of local and global signals in the primary visual cortex [8].

Conclusions

One fundamental question about feedback and feed-forward pathways is whether they constitute distinct functional systems, as implied by the terminology used. Taking the geniculo-cortical pathway (►geniculo-striate pathway) as a model, cortical feedforward pathways supposedly mediate driving influences and feedback mediate modulatory influences. Physiological studies support this general view, e.g. cooling area V1 in the monkey leads to silencing of area V2 neurons, whereas cooling area V2 has only marginal effects on the activity of area V1 neurons. However, if a small driving projection was contained in the feedback pathway that remains dominated by a modulatory function, the driving function might not show up in the cooling experiments. The distance rule suggests that the physiology of feedforward and feedback pathways linking cortical areas is determined by the composition of the parent neurons in terms of supra- and infra-granular layers [6]. The differences in the physiology of feedforward and feedback pathways could be the consequence of (i) differences in the cellular targets and/or (ii) differences in the intrinsic properties of the parent neurons.

The idea that a cortical area is homogenous both in function and structure has been floating for over a century. In a seminal paper, Daniel and Whitteridge [9] showed that while the amount of cortex devoted to a degree of the visual field (the "magnification factor") (►Striate cortex functions) does change across the cortex, there appears to be a constant ratio between the numbers of peripheral receptors and the number of visual degrees represented in the cortex. In the 1970s, Hubel and Wiesel took this a step forward in suggesting that the entire apparatus for representing a point of the retinal image is contained in a small region of cortex a few millimeter in area, which they called a "►hypercolumn." The primary visual cortex thus consists of many such hypercolumns. However, the dynamic properties of neuron response have been shown to change dramatically within a visual cortical area at different eccentricities, and a recent paper shows marked differences in the inputs to the central and peripheral representation of area V1 [10]. The dominance of inter-areal projections by nearby areas could lead to such a specialization because the layout of the visual areas results in different sets of areas being closer to central than peripheral visual field representations. This in turn leads to the prediction that the central representation of early visual areas will be preferentially connected to the ventral processing stream and peripheral to the dorsal processing stream (►Extrastriate visual cortex). Thus while a given cortical area may be formed by multiple copies of a canonical circuit, each region of the area could be modulated independently by its nearest neighbors. Eccentricity-dependent differences in organization would be consistent with the anatomical specializations in the retina (fovea vs. periphery) as well as the behavioral evidence of eccentricity dependence of different tasks (for example, object recognition in central vision vs. global spatial localization in the periphery). Such observations raise questions about how a cortical area should be defined, since it cannot be done by assuming that one area behaves as a single functional entity.

Our approach to investigating cortical hierarchy is based on Graph Theory, in which the distribution of connections between areas is analyzed over the whole network to infer the connectional distances between areas [7]. Such an analysis shows that the distribution of areas bears a close resemblance to their spatial layout in the cortex suggesting organizational principles linking connectivity, adjacency and cortical folding. This study, however, is based only on the presence or absence of a connection between two areas and does not take into account the strength of connections. Because strong connections are very short-range, integrating the strength of neural connections in these models will strongly emphasize the importance of adjacency. Given that the strength of connectivity is eccentricity-dependent, comparing graphs across the cortex will allow us to explore structural features of contextual processing. The challenge will then be to extract the rules allowing integration of Daisy architecture in the contextual process. Significant efforts are now being made to understand how graphical processing can be instantiated in networks of uniform processing elements, how this can be done using asynchronous event-based methods, which are the essence of neuronal computations, and whether graphical processing can be promoted from simple uniform propagation between nodes (whether defined as neurons, clusters of neurons, or cortical areas) to a dynamic "intelligent" selective propagation. The solution to such problems will be an important step to understanding the principles by which biological brains achieve their intelligence.

Acknowledgments

This work was supported by FP6-2005 IST-1583; ANR-05-NEUR-088 (HK, KK); SNF NCCR Neural Plasticity and Repair (KM).

N

References

1. Douglas RJ, Martin KAC, Whitteridge D (1989) A canonical microcircuit for neocortex. Neural comput 1:480–488
2. Douglas RJ, Koch C, Mahowald M, Martin KA, Suarez HH (1995) Recurrent excitation in neocortical circuits. Science 269:981–985
3. Douglas RJ, Martin KA (2004) Neuronal circuits of the neocortex. Annu Rev Neurosci 27:419–451
4. Binzegger T, Douglas RJ, Martin KA (2004) A quantitative map of the circuit of cat primary visual cortex. J Neurosci 24(39):8441–8453
5. Kennedy H, Bullier J (1985) A double-labelling investigation of the afferent connectivity to cortical areas V1 and V2 of the macaque monkey. J Neurosci 5 (10):2815–2830
6. Barone P, Batardiere A, Knoblauch K, Kennedy H (2000) Laminar distribution of neurons in extrastriate areas projecting to visual areas V1 and V4 correlates with the hierarchical rank and indicates the operation of a distance rule. J Neurosci 20(9):3263–3281
7. Vezoli J, Falchier A, Jouve B, Knoblauch K, Young M, Kennedy H (2004) Quantitative analysis of connectivity in the visual cortex: extracting function from structure. Neuroscientist 10(5):476–482

8. Angelucci A, Bullier J (2003) Reaching beyond the classical receptive field of V1 neurons: horizontal or feedback axons? J Physiol Paris 97(2–3):141–154
9. Daniel PM, Whitteridge D (1961) The representation of the visual field on the cerebral cortex in monkeys. J Physiol 159:203–221
10. Falchier A, Clavagnier S, Barone P, Kennedy H (2002) Anatomical evidence of multimodal integration in primate striate cortex. J Neurosci 22(13):5749–5759

Nephron

Definition

The functional unit of the kidney is the nephron. A nephron consists of a glomerulus and a tubule. The glomerulus is a cluster of blood vessels from which the filtrate forms. The tubule is an epithelial structure consisting of many subdivisions designed to convert the blood filtrate into urine.

►Blood Volume Regulation

Nerve Cell Membrane

Definition

A thin membrane comprised of phospholipids, proteins and carbohydrates that encloses the cell cytoplasm, nucleus and other intracellular organelles.

►Membrane Components

Nerve Grafting

Definition

Bridging a gap in nerve continuity by inserting cables of another nerve, transplanted from another part of the body.

►Regeneration: Clinical Aspects

Nerve Growth Factor (NGF)

Definition

Nerve Growth Factor is the first of a series of neurotrophic factors that were found to influence the growth and differentiation of sympathetic and sensory neurons.

►Neural Development
►Neurotrophic Factors
►Regeneration

Nerve Regeneration

Definition

Nerve regeneration is the ability of the nervous system to reestablish functional connection after nerve injury.

The central nervous system is not capable of functional regeneration. The peripheral nervous system has a limited ability to regenerate, although regeneration is often incomplete, misdirected and resulting in neuropathic pain.

►Extrasomal Protein Synthesis in Neurons
►Neuropathic Pain
►Peripheral Nerve Regeneration and Nerve Repair
►Regeneration

Nerve Transfers

Definition

The deliberate direction of a proximal donor nerve to a foreign distal denervated one. Axons from the donor nerve enter into the denervated one and then reinnervate the previously denervated foreign end-organ to allow functional recovery.

►Peripheral Nerve Regeneration and Nerve Repair
►Regeneration: Clinical Aspects

Nervous, Immune and Hemopoietic Systems: Functional Asymmetry

Valery V. Abramov, Irina A. Gontova, Tatjana Ya. Abramova, Vladimir A. Kozlov
Laboratory of Neuroimmunology, State Research Institute of Clinical Immunology of SB RAMS, Novosibirsk, Russia

Synonyms

Asymmetry – absence of symmetry; Contralateral lobes – left and right lobes; Thymocytes – cells of thymus; Splenocytes – cells of spleen; Syngeneic mice – compatible on H-2 antigens mice; Immunization – injection of antigens; Paw preference – preference of paw in taking food; Maturation of cells – differentiation of cells

Definition

The ►asymmetry of the brain hemispheres has been studied for a long time. While studying this phenomenon, numerous facts on structural, functional, and molecular-biological differences of the hemispheres have been accumulated. Based on these data, some studies have been conducted showing that there is asymmetry of not only the brain hemispheres but also of the neuroendocrine system as a whole, including the gonads, adrenal glands, and lobes of the thyroid gland. For example, our experiments discover the existence of functional asymmetry of adrenal glands in ►(CBAxC57Bl/6)F1 hybrid mice.

At the same time, it is known that the hemopoietic and lymphoid organs (bone marrow, thymus, lymph nodes, etc.) as well as the brain hemispheres are presented by two morphologically-divided lobes (contralateral lobes). This enabled us to suppose and prove that not only the neuroendocrine system but also the hemopoietic and immune systems demonstrate the functional asymmetry of bilateral organs.

Characteristics

The role of Functional Asymmetry of a Bone Marrow and of Brain's Hemispheres in Hemopoiesis

Design for Experiment

The recipient mice were prepared by exposure to a whole-body-radiation dose (950 R). Marrow cells were obtained from the femora of mice-donors, suspended in saline and the nucleated cells were counted with a hemacytometer. These suspensions were kept at ice-water temperature until they were injected. The recipient animals then received an intravenous injection with 10^5 of marrow cells (0.5 ml). The effect of functional asymmetry of the bone marrow on hemopoiesis was evaluated by injections of cells from the left or right femoral bone of left- or right-pawed donors to irradiated recipient animals (left- or right-pawed). Eight days after injection of the transplanted cells, the mice were killed and the number of colonies (i.e. in the number of 8-day colony-forming units in spleen – ►CFUs-8) in their spleens was determined by Till and McCulloch method.

Results

Our experiments discovered the existence of functional asymmetry of the bone marrow cells [1–6]. For example, the evaluation of functional asymmetry of the bone marrow showed significant differences in the hemopoiesis (CFUs-8) only in irradiated left-pawed recipients receiving an intravenous injection of marrow cells from the right or left femoral bones of left-pawed donors. No appreciable differences in the formation of CFUs-8 were observed in cases when left and right femur marrow cells were transplanted from right-pawed donors to left-pawed recipients (►left-pawed and right-pawed mice), or marrow cells from either femur of left- and right-pawed donors were injected to right-pawed recipients. These results indicate that the: (i) hemopoietic potential of the bone marrow from the right and left femoral bones is different; and (ii) manifestation of asymmetry of bone marrow hemopoietic functions depends on motor asymmetry of donors and recipients [4–6].

The role of Different Lobes of the Thymus and Different Hemispheres of the Brain in the Development of Humoral Immune Response

Design for Experiment

The mice recipients were thymectomized to evaluate the role of the cells from either left or right thymus lobes in the ►humoral immune response. Five weeks after thymectomy, left and right-pawed recipients animals were intravenously injected with thymocytes from the right or left thymus lobes of left-pawed syngeneic donors (10^7 cells/mouse). These recipient mice were subsequently immunized with sheep red blood cells (►SRBC) 10 days after thymocyte administration. The humoral immune response (antibody-producing cells) in the spleen was counted by the method of Cunningham 4 days after the immunization.

Results

Our experiments showed the existence of functional asymmetry of the thymus and demonstrated that the asymmetry in the nervous and immune systems (hemispheres of the brain and the thymus) plays an important role in the development of humoral immune response to SRBCs. So, the *in vivo* experiments showed that the properties of cells from contralateral lobes of the thymus proved to be a deciding factor that defines the differences at the level of humoral immune response in recipient mice with left-dominant hemispheres. This effect was less pronounced in mice with right-dominant hemispheres. Further analysis showed that left and right-dominant hemisphere mice differ according to the immune response only if mice from both groups received cells from the left but not from the right lobes of the thymus. That is, in the formation of the humoral immune response to SRBCs the functional asymmetry of both the brain and thymus is of great importance [7–9].

The role of Motor Asymmetry of Brain Hemispheres and Functional Asymmetry of Regional Lymph Nodes in the Development of Cellular Immune Response

Design of Experiment

While studying the ►cellular immune responses in the paws, mice were intraperitoneally immunized with 0.5 ml of 5% SRBCs. On the fourth day, the delayed-type hypersensitivity (DTH) reaction was used as an *in vivo* measure of antigen specific T-lymphocyte reactivity. In order to study the cellular immune response, 50% suspension of SRBCs in 0.05 ml of physiologic saline was injected under aponeurosis of the left paw. In addition, 0.05 ml of physiologic saline was injected under aponeurosis of the right paw as a control. The cross-section of a paw at the site of injection was measured 24h later. In order to study the cellular immune response in the right paw, mice were injected with 50% suspension of SRBCs in 0.05 ml of physiologic saline under aponeurosis of the right paw. Further, 0.05 ml of physiologic saline was injected under aponeurosis of the left paw as a control. The index of reaction (DTH) was calculated for each mouse by the formula: $IR = (P_o - P_c)/P_c$, where P_c – paw cross-section in control; P_o – paw cross-section in experiment.

Results

The results obtained showed that the degree of cellular immune response in (CBAxC57Bl/6) F_1 mice depends on the functional asymmetry of regional lymph nodes and paw preference. The thymus functional asymmetry is of insignificant importance in DTH reaction. For example, we found that the intensity of the DTH reaction to SRBCs in the front paws of (CBAxC57Bl/6) F1 mice depends not only on whether the antigen is injected into the left or right paw but also on the motor asymmetry of the hemispheres. While comparing the DTH reaction in the hind left and right paw of mice, we showed that in both right- and left-handed mice it was much more pronounced in the left paw than in the right one [6,10].

Discussion

It is known that autonomic nerves are well presented in the bone marrows, thymus lobes and lymph nodes of mice where, together with specialized cells, they form the neuroendocrine microenvironment that influences the maturation of cells. These data together with our results on differences in the functional properties of cells from contralateral (left or right) bone marrow and lymph nodes, and thymocytes from the thymus lobes allow us to suggest the following: (i) there are differences in the sympathetic and parasympathetic innervation of the contralateral lobes of the organs; and (ii) differences in the functional properties of cells from the contralateral lobes of the organs are caused by differences in the neuroendocrine environment of the lobes mentioned; i.e. preferential influence of catecholamines, acetylcholine and peptides on the cells.

It has been established that sympathetic and parasympathetic activity is preferably regulated by different brain hemispheres. So, sympathetic activity is preferentially regulated by the right hemisphere, whereas parasympathetic activity is regulated by the left hemisphere. In this connection, one can speak about the lateralizing effect of the brain hemispheres on the organs by creating differences in the neuroendocrine environment of the right and left lobes.

At the same time, our data, for example, on the different roles of cells from the contralateral lobes of the thymus in the formation of a humoral immune response confirm the supposition that thymocytes from the lobes mentioned have different functional properties. Moreover, we speculate that the number of T-helper precursors of type 2 might be different in the thymus lobes and/or the precursors mentioned are at different stages of their differentiation. Since the injection of thymocytes from the left thymus to thymectomized recipients is accompanied by the greatest effect on the formation of a humoral immune response, there might be more T-helper precursors in the left lobe and/or they are more mature in comparison with cells from the right lobe. That sympathetic and parasympathetic activity is mainly regulated from different brain hemispheres can help explain their role in the formation of a humoral immune response in thymocyte recipients. If functional differences in cells from the contralateral lobes of the thymus of donors define the preferential influence of catecholamines and acetylcholine, the influence mentioned can also explain the role of the hemispheres in the formation of antibody-forming cells in the recipients who received thymocytes from a given lobe. For example, specific receptors to definite neuromediators might be expressed on the surface of the thymocytes mentioned.

We speculate that the more pronounced DTH reaction on the left than on the right may be connected too with asymmetry of peripheral innervation of contralateral lymph nodes that, in its turn, is controlled by brain hemispheres. That is, peripheral vegetative innervation of regional lymph nodes, regulated by brain hemispheres, might be very important in asymmetrical development of reactions of cellular immunity in front and hind paws of mice. The data testify to an important role of sympathetic innervation in DTH reaction. For

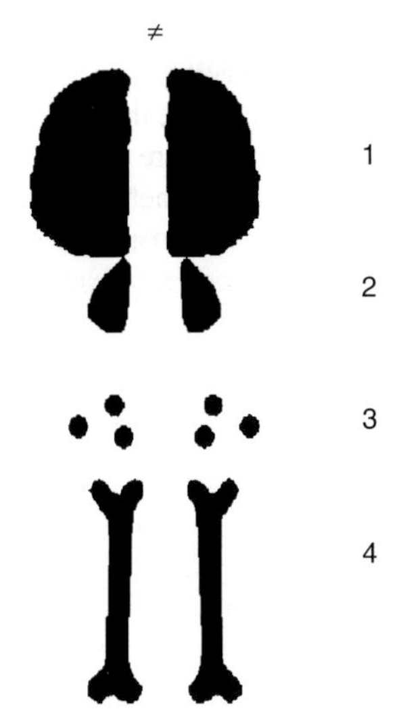

Nervous, Immune and Hemopoietic Systems: Functional Asymmetry. Figure 1 Structural-functional asymmetry of Hemopoietic, Immune, Endocrine and Nervous systems (HIMEN system). ≠ – symbol of asymmetry1 – neuroendocrine system (hemispheres); 2 – thymus; 3 – lymph nodes; 4 – bone marrow.

instance, they showed that stroke lateralized T-cell-mediated cutaneous inflammation. This effect may be mediated by alteration of the cutaneous sympathetic nerve traffic. The authors also demonstrated that there is lateralization of cutaneous inflammatory responses in patients with paresis after poliomyelitis. This lateralization of DTH responses is related to deficiencies in motor and sympathetic innervation of the paretic extremity.

It is possible that sympathetic and parasympathetic activity that is mainly regulated from different brain hemispheres can help to explain their role in the formation of CFUs-8 in mice-recipients.

Conclusion

Thus, our data allow us to speak about the asymmetry of the integrated Hemopoietic, Immune, Endocrine and Nervous systems, i.e. the HIMEN system (Fig. 1) [1,6,9].

References

1. Abramov VV, Abramova TY (1996) Functional asymmetry of hemispheres of brain. In: Komarova LB (ed) Nervous, endocrine, and immune systems asymmetry. Nauka, Novosibirsk, pp 30–43 (in Russian)
2. Grebenshikov AY, Poveshchenko AF, Abramov VV (1999) Expression of IL-1 beta gene in brain after peripheral administration of thymus-dependent and thymus-independent antigens. Dokl Biol Sci 366(1–6): 294–296
3. Abramov VV, Karmatskikh OL, Kozlov VA, Oskina IN (1996) Functional asymmetry of adrenal glands in (CBAxC57Bl/6)F1 mice. Dokl Akad Nauk 347 (6):831–833
4. Abramov VV, Kozlov VA, Karmatskich OL (1990) The asymmetry of exogenous CFUs-12 forming in mice. In: Korneva EA, Polyak AI, Frolov BA (eds) The interactions between the nervous and immune systems. Printing House im. MI Kalinina, Leningrad, Rostov-na-Donu, pp 172
5. Abramov VV, Gontova IA, Kozlov VA (2002) Functional asymmetry of the brain and bone marrow in hemopoiesis in (CBAxC57Bl/6)F_1 mice. Byull Eksp Biol Med 133 (5):468–470
6. Abramov VV, Gontova IA, Kozlov VA (2006) Functional asymmetry in hematopoietic, immune and nervous systems. In: Yegor Malashichev B, Wallace Deckel A (eds) Behavioural and morphological asymmetries in vertebrates. Landes Bioscience – Chapter 12, pp 1–12
7. Abramov VV, Konenkov VI, Gontova IA (1992) Asymmetry of phenotypical and functional characteristics of the cells from lymphoid organs. Dokl Rossiisk Akad Nauk 322(4):802–805
8. Gontova IA, Abramov VV, Kozlov VA (2001) Asymmetry in cerebral hemispheres and thymus lobes during realization of humoral immune response in mice. Byull Eksp Biol Med 131(1):64–66
9. Abramov VV, Gontova IA, Kozlov VA (2001) Functional asymmetry of thymus and the immune response in mice. Neuroimmunomodulation 9(4):218–224
10. Gontova IA, Abramov VV, Kozlov VA (2004) The role of asymmetry of nervous and immune systems in the formation of cellular immunity of (CBAxC57Bl/6)F_1 mice. Neuroimmunomodulation 11(6):385–391

Nervousness

Definition

State of hyperarousal, irritability, hyperactivity, unrest and becoming tired quickly. As this vague term is frequently used in common speech, it is rarely used in medical terminology.

►Personality Disorder

Nervus Terminalis

►Evolution of the Terminal Nerve

Net Torque

Definition

The sum of all muscle torques acting at a joint.

►Impedance Control

Netrins

Definition

Family of bifunctional diffusible guidance molecules with attractive or repulsive effects towards different types of neurons. The role of netrins and their receptors DCC and UNC-5 in guidance of commissural neurons has been extensively studied in vertebrates, worms and flies. DCC receptors mediate attraction and repulsion, while UNC-5 receptors appear to function exclusively in repulsion. These receptors can also mediate cell death.

►Growth Inhibitory Molecules in Nervous System Development and Regeneration

Network

Definition

Neurons that are connected to form a functional unit.

Network Error

Definition

In supervised learning, neural networks learn to approximate a given data set with the help of a teaching signal. During the training process, the network processes input exemplars with its current set of connection weights, resulting in a corresponding output signal. This output is compared to the teaching signal, representing the desired output. The network error is computed as the sum of squared differences between the actual and desired output for each output unit.

►Connectionism
►Neural Networks for Control

Network Interneurons

Definition

Pre-motor neurons that are part of the central pattern generator (CPG).

►Central Pattern Generator (CPG)
►Excitatory CPG Interneurons

Network Oscillations

Definition

Oscillatory activity generated in distinct brain regions by networks of often electrically coupled neurons are fundamental to information processing in the mammalian brain. Many of these rhythms correlate with distinct behavioral states. Electrical synapses appear to play a role in the generation and synchronization of these network oscillations.

►Electrical Synapses

Network Oscillations in Olfactory Bulb

Definition

Oscillatory variations of the field potential can be revealed by extracellular recordings. It presumably results from the synchronized activity of a group of neurons. Several types of oscillations have been described in diverse brain areas. They are classified with respect to their oscillatory frequency. For example, in the main olfactory bulb, there are three types of oscillations. Slow "Θ" oscillations (1–8 Hz) are generated by respiratory cycle. Fast "β" (15–30 Hz) and "γ" (40–80 Hz) oscillations are induced by inhalation of odorant molecules and reflect information processing by the olfactory bulb neuronal network.

►Olfactory Bulb

Network Oscillator

Definition

A neuronal network that produces a rhythmic activity pattern in response to a general excitation as a result of the synaptic connections between the individual neurons. Each neuronal type may discharge during a different phase of the rhythm. In a network oscillator none of the component neurons possess endogenous bursting properties (i.e. the individual neurons on their own would not produce any rhythmic activity). Instead, generation of the rhythmic activity is due to synaptic interactions. The simplest network oscillator configuration is the half-center oscillator that consists of two neurons or groups of neurons, which are connected by mutual inhibitory connections. This configuration can lead to rhythmic switching of activity between the two half-centers producing a biphasic activity pattern.

►Central Pattern Generator
►Half-center
►Respiratory Network
►Rhythmic Movements

Network Reconfiguration

Definition

A process that alters the interactions and output pattern of a neuronal network by changing the number of active neurons, their intrinsic membrane and/or synaptic properties. Network reconfiguration is typically mediated by endogenously released neuromodulators.

The reconfiguration of neuronal networks imbues the nervous system with a high degree of plasticity. It enables, for example, a behaving animal to alter the output of a neuronal network in response to changes in the behavioral, environmental and metabolic conditions.

►Bursting Pacemakers
►Neuromodulator

Neural and Behavioral Responses to Immunologic Stimuli

Alexander Kusnecov
Rutgers University, Department of Psychology, Piscataway, NJ, USA

Introduction: What the Brain Cannot See But Needs to Know

From a behavioral perspective, the nervous system evolved for organisms to approach and avoid stimulus components of their environment. This is predicated on the existence of sensory systems that transduce physical information into the electrochemical events of neural transmission, and which forms the basis for information processing. Each stimulus, no matter how discrete, is an informational package, and can be arranged along a continuum varying in the intensity and magnitude of some specific criterion. For the nervous system, differentiation and identification of objects includes such characteristics as size, texture, and weight, information which is processed and acted on to orchestrate an appropriate response. Thus, pot plants falling off balconies, oncoming motor vehicles, a baseball speeding through the air, all signify to an individual appropriate steps to take in avoiding or engaging a potentially dangerous stimulus. However, this form of signal detection fails in the realm of stimulus encounters with the microbial world of bacteria, viruses and other parasitic entities that represent biological threats against the organism. It is against biological agents at this end of the continuum of environmental stimuli that the nervous system is at a loss to avoid and/or engage with an appropriate response. To address this problem, vertebrate animals evolved a cell-mediated molecular recognition system that recognizes and eliminates microbial organisms. This is the immune system, a heterogeneous collection of cells (collectively referred to as leukocytes) that consists of macrophages and monocytes, which represent the first line of immunologic defense by ingesting and degrading particulate cellular invaders, while T and B lymphocytes (or T and B cells) serve to regulate immune responses, produce antibodies (by B cells) and exert cytotoxic effects by killing virally infected cells. A general review of the immune system is not possible here, although for a general introduction the reader is referred to other sources [1].

The Immune System Communicates with the Brain

For many years it was thought that the immune system functioned independently in exercising this defensive function, and to a large extent in the ►initial phase of encountering and responding to pathogens, it is the main system involved. However, the past several decades have emphasized that the immune system can be influenced by the central nervous system (CNS), and in reciprocal fashion, the activity of the immune system can affect the CNS. A number of publications have compiled reviews of empirical research attesting to this mutually interactive functional relationship [2]. In retrospect, this interaction makes considerable sense, given that the behavior of an organism can serve to control levels of exposure to environments that may contain infectious microorganisms. However, in the absence of foresight, exposure to bacteria or viruses renders too late any evasive and/or protective measures. Therefore, some form of learning and/or knowledge about each infectious episode must be relayed to the CNS to ensure control over future potential encounters. For this reason pathways of communication exist from the immune system to the brain, activating brain mechanisms that essentially perceive the presence of infection in the body. One of the most instructive examples of this is the behavioral phenomenon of conditioned taste aversion (CTA; also referred to as food aversion) a form of avoidant learning in which tastes and foods are rejected if their ingestion had previously been associated with illness. This behavior can be elicited by immune responses and molecular products of the immune system, illustrating the notion that the immune system, as it engages microbial pathogens, simultaneously alerts the brain to initiate adaptive behavioral responses.

The immune system engages microbial antigens by generating *inter alia* a cascade of chemical mediators called cytokines which regulate the cells of the immune system, but in addition, are capable of binding to receptors on afferent nerves (e.g., the vagus) as well as the endothelial cells of the vasculature. The production of cytokines during the immune response signals to the CNS the presence of potential pathogens, leading to the elaboration of various efferent mechanisms (viz., endocrine and autonomic nervous system activity) that result in physiological changes that contribute to the removal of the pathogen. In addition, and perhaps the most important from an adaptational perspective, the CNS response to cytokines initiates behavioral adjustments that result in cognitive processing of all relevant information pertaining to the potential source of the illness and how it can be managed. Returning to the CTA example introduced above, infectious illness following ingestion of a novel food should result in a memory for the contextual circumstances surrounding food intake. Indeed, it is known that memory is enhanced shortly after ingestion of food, and perhaps this phenomenon ensures that in the face of food poisoning there is more effective retrieval of information pertaining to the "eating event." As discussed below, these behaviors are part of a constellation of changes called "sickness behaviors."

The Immune System as a Stressor

Stress is a much used and abused term. The intended (and commonplace) connotation of the term is either as a threat to the organism or a state of the organism that is potentially harmful to health. In order to allow for effective communication of the conditions leading to "stress," investigators use the term "stressor" to denote any stimulus or condition that produces a biological state significantly different from that observed prior to stressor exposure. In recent years, a distinction has been made between two classes of stressors: (i) those that operate through cognitive and exteroceptive sensory stimulation (e.g., pain), and (ii) those occurring within the internal milieu – the molecular signals arising from the cells and tissues of the organism. The former are referred to as *processive* stressors, and involve psychogenic components, while *systemic* stressors involve endogenous changes in the internal milieu which signal the CNS through alternative interoceptive pathways (blood, afferent nerves) [3]. Examples of systemic stressors can include metabolic changes such as insulin elevations and reductions in blood glucose, as well as changes in the chemical composition of the blood as a result of exposure to a pathogen. Therefore, the immune system and the cytokines produced following its stimulation are now viewed as systemic stressors, generating a similar profile of changes in the brain that are produced by processive stressors. These changes include an elevation of the production and release of classical neurotransmitters, including the monoamines (serotonin, dopamine and norepinephrine), in regions of the brain that are known to be involved in the generation of fear, anxiety, reward evaluation, decision-making, and memory. Such regions include the septum, amygdala and associated areas (which process fear and anxiety), hippocampus (involved in memory), prefrontal cortex (decision-making), nucleus-accumbens and other components of the mesolimbic dopaminergic pathway (reward evaluation). All these areas are part of the brain circuitry involved in emotion regulation. Activation of these areas has been determined by measuring the appearance of immediate-early genes (e.g., *c-fos*) that reflect excitation of neurons, as well as through electrophysiological recording of the firing rates of neurons generating action potentials. Many of the changes in these brain regions are observed following exposure of laboratory animals to painful and/or fearful stimuli (i.e. processive stressors). And while the study of these brain regions and their role in cognition and emotion continues to evolve, it is apparent that they are essential to the ability of organisms to adapt in the face of challenges to survival and/or well-being. Consequently, it must be the case that similar engagement of such areas by immunologic factors serves to mediate similarly adaptive behavioral and physiological changes.

The Role of Cytokines

In the early 1970's and 1980's CNS activation by the immune system was demonstrated using sheep red blood cells (SRBC) as the antigen [4], and showed that during the peak of the antibody response there were increases in plasma corticosterone, brain electrophysiological activity, and release of monoamine neurotransmitters in regions of the brain involved in stress and adaptation. This work has been extended to include protein antigens, bacterial toxins, and viruses. Collectively, it is now well established that the CNS is a recipient of "sensory" signals from the activated immune system. Indeed, this was recognized early and led to the concept of the immune system as a "floating brain" or "sensory organ" [4,5].

The search for molecules that might be responsible for the CNS activation observed during the immune response to specific antigens led to important discoveries that opened wide the field of neural-immune research. As stated above, cytokines are the chemical mediators of the immune system, responsible for suppressing, enhancing, and fine-tuning the various cellular mechanisms of the immune system. Cytokines play an important role in the final stage of an immune response which involves an effector component that constitutes the ultimate death blow to an invading pathogen. This effector stage includes T lymphocytes being cytotoxic for bacteria or virally infected cells and B lymphocytes producing antibody that binds and neutralizes antigen. Antibodies also serve as tags that allow macrophages to bind the antibody-bound antigen (e.g., bacterial cell) and summarily eliminate the antigen through cytotoxic means and phagocytosis. The role of cytokines in these processes is to either augment, suppress or maintain control of the effector arm of the immune response. Failure to regulate this process can result in immunopathology, including autoimmune disease and prolonged infection. Numerous cytokines have now been identified, and many more await characterization. However, a number of cytokines have already been designated with neuromodulatory functions. A substantial body of evidence exists on the neural effects of the cytokines interleukin-1 (IL-1), IL-6 and tumor necrosis factor (TNF α). Many other cytokines, including IL-2, interferon α (IFNα) and IFNγ, and transforming growth factor (TGF), have been linked to brain function either through demonstrations of behavioral and neurophysiological changes, or through their involvement in neuropathology. Many more cytokines will continue to be implicated in neuromodulation through research on degenerative and regenerative events in the brain, as for example, in animal models of stroke, dementia and neurological impairment. However, in terms of influences affecting ongoing neural and behavioral functions due to systemic immune responses, there is general acceptance that IL-1, IL-6 and TNFα are involved, given that their administration replicates many of the effects seen in response to many of the experimental antigens shown to activate the brain. Indeed, these cytokines are also produced within the brain by glial cells (viz., astrocytes and microglial cells) which can support neuronal function through regulation of nutrient and neurotransmitter biosynthetic pathways. The production of cytokines by glial cells is an area of intense interest, and clarity of conceptual understanding in terms of precisely how intra-CNS cytokines promote brain function is yet to be realized.

The direct effects of cytokines on CNS function reinforced the principle of immune-derived activation of neural processes. However, demonstrations of the contribution provided by endogenous cytokines, focused largely on the use of lipopolysaccharide (LPS), which has been studied as a model of gram negative bacteremic infection. Exposure of animals to LPS produces significant in vivo production of IL-1, IL-6 and TNFα; and many of the neural and behavioral effects of administered cytokines can be easily induced by LPS. However, LPS activates the cells of the myeloid lineage, such as monocytes and polymorphonuclear phagocytic cells (i.e. macrophages). The study of other antigens that engage T cells, is also needed. These cells are a significant source of cytokines, some of which are exclusive to T cells (e.g., IL-2 and IL-4), while others are produced by both T cells and macrophages (e.g., IL-1, TNF, IL-6, IL-10 and IL-12). A more complete image of the immunological influence on brain function requires the use of antigens that also stimulate T cells.

Bacterial T Cell Superantigens and the CNS

The term "superantigen" (SAg) was coined by Marrack and colleagues [6] to describe the unusually exaggerated proliferative and cytokine response of T cells to staphylococcal enterotoxins derived from gram positive bacteria. The appellation "super" served to contrast these enterotoxins with regularly tested protein antigens (e.g., egg albumin or keyhole limpet hemocyanin), which are highly selective and do not engage as large a T cell population as SAgs, and fail to induce readily observable *in vivo* T cell proliferation and ▶measurable concentrations of cytokines in the blood. A fuller discussion of SAgs and their unique neuroimmunological uses can be found in [7].

The best characterized SAgs are the staphylococcal enterotoxins, which are identified by letter codes, such as A, B, C, and so on (e.g., staphylococcal enterotoxin A, SEA; staphylococcal enterotoxin B, SEB). The impact of SAgs on CNS function has been demonstrated with SEA and SEB, which produce activation of the hypothalamic-pituitary-adrenal (HPA) axis. The HPA axis is exquisitely sensitive to the effects of stress, and its activation is detected by blood plasma or serum measures of glucocorticoids (e.g., corticosterone) released by the adrenal gland, ACTH released by the pituitary gland, and corticotropin releasing hormone (CRH) released by neurons of the hypothalamus in the brain. Collectively, CRH, ACTH and corticosterone represent different hormonal components of the HPA axis, a neuroendocrine pathway essential to the survival of the organism under conditions of stress. Indeed, it is well known that optimal activation of the HPA axis is necessary using models of infection, such as exposure to LPS. Moreover, while SAgs have been shown to activate the HPA axis at all levels, similar activation had also been demonstrated using cytokines, such as IL-1, TNF and IL-6.

Bacterial SAgs also activate stress-related brain pathways that are known to influence the HPA axis. For example, regions previously mentioned, such as the amygdala, hippocampus and septum were activated after injection of SEA and SEB. Activation of these areas likely produces anxiety and/or stress-like states in the animal, supporting much of the research on the effects of LPS and cytokines like IL-1 and TNFα on sickness behavior [8]. Sickness behavior is characterized by lethargy, anorexia and reduced exploration, and is also associated with anhedonia, somnolence, and deficits in learning and memory [9]. The wide range of behavioral changes observed speaks to a general behavioral adjustment in keeping with the protective goals of the immune system. Interestingly, the sickness behavior profile is similar to the symptomatology of depressive illness, which has led many to consider that dysregulation of the immune system may contribute to clinical depression [10,11]. Whether this compelling hypothesis is true remains to be determined.

Evidence exists that the neurobiological effects of bacterial SAgs noted above involve behavioral changes that approximate sickness behavior, but appear to lack signs of malaise that might account for anorexia and lethargy. Administration of SEA and SEB reduces food intake, but only if the food is novel (i.e. unfamiliar). Moreover, there is no evidence of weight loss, although increased body temperature has been noted. Why animals would show reduction of a novel, as opposed to familiar food, has been suggested to reflect an increase in anxiety. It is well known that exposure to novel food causes a neophobic (i.e. fear of novelty) reaction in animals. Therefore, if SAgs activate stress pathways in the brain, this may be altering the threshold level of processing and responding to mildly arousing stimuli. Indeed, the augmented reactivity to a novel food after SAg treatment was also seen in regard to exploration of a novel object, which similarly produces a neophobic reaction (i.e. reduced or more wary exploration). Further discussion of the behavioral effects of SAg treatment is provided elsewhere [7].

Concluding Comments

This brief overview focused primarily on the idea and fundamental findings relating to immunological activation of brain and behavioral functions. The emphasis was in illustrating that defense against challenges to biological integrity are the province of both the nervous system and immune system, and with this common purpose, evolution has ensured that some form of cooperative signaling exists between the two. Ultimately, the organism learns and refines its behavioral repertoire of protection, and in doing so, relies on information relayed by the activities of the immune system. The sensation of illness communicated by elevated cytokines creates a state of arousal that generates a memory for the illness state in and of itself, and through the host of different cognitive mechanisms at the host's disposal, lays down a neural memory that can be used to plan and avoid future exposures to illness. More recent evidence, not covered here, has also demonstrated that cytokines (such as those mentioned in this article) are synthesized and released by glial cells in the brain. The distinction between the immune system and nervous system has become blurred, and it is clear that where defense against infection is concerned, both systems must eventually play a critical role [12].

References

1. Mims C (2000) The war within US: everyman's guide to infection and immunity. Academic Press, New York
2. Ader R (2006) Psychoneuroimmunology, 4th edn. Elsevier, Academic Press, New York
3. Herman JP, Cullinan WE (1997) Neurocircuitry of stress: central control of the hypothalamo-pituitary-adrenocortical axis. Trends Neurosci 20(2):78–84
4. Besedovsky HO, del Rey AE, Sorkin (1985) Immune-neuroendocrine interactions. J Immunol 135 (2 Suppl):750s–754s
5. Blalock JE (1984) The immune system as a sensory organ. J Immunol 132:1067–1070
6. Scherer MT, Ignatowicz L, Winslow GM, Kappler JW, Marrack P (1993) Superantigens: bacterial and viral proteins that manipulate the immune system. Annu Rev Cell Biol 9:101–128
7. Kusnecov AW, Goldfarb Y (2005) Neural and behavioral responses to systemic immunologic stimuli: a consideration of bacterial T cell superantigens. Curr Pharm Des 11 (8):1039–1046
8. Dantzer R, Kelley KW (2006) Twenty years of research on cytokine-induced sickness behavior. Brain Behav Immun 21(2):153–160
9. Anisman H, Merali Z, Poulter MO, Hayley S (2005) Cytokines as a precipitant of depressive illness: animal and human studies. Curr Pharm Des 11(8):963–972
10. Maier SF, Watkins LR (1998) Cytokines for psychologists: implications of bidirectional immune-to-brain communication for understanding behavior, mood, and cognition. Psychol Rev 105(1):83–107
11. Maes M (1999) Major depression and activation of the inflammatory response system. Adv Exp Med Biol 461:25–46
12. Haddad JJ, Harb HL (2005) L-gamma-Glutamyl-L-cysteinyl-glycine (glutathione; GSH) and GSH-related enzymes in the regulation of pro- and anti-inflammatory cytokines: a signaling transcriptional scenario for redox (y) immunologic sensor(s)? Mol Immunol 42:987–1014

Neural Bases of Spatial Learning and Memory

KARIM BENCHENANE, MICHAËL ZUGARO, SIDNEY WIENER
LPPA, Collège de France, CNRS, Paris, France

Definition

For survival many animals have evolved neural mechanisms permitting to localize resources, protection and conspecifics in changing environments. During exploration, information is acquired about cues, cue gradients, landmarks, and appropriate movements for dead reckoning, wayfinding, piloting and other types of spatial navigation. The goal of studying the neural bases of spatial learning and memory is to understand how specific patterns of brain activity underlie these complex spatial capabilities, and in particular how they emerge in novel environments, and how they are dynamically reorganized in changing environments.

Lesion and comparative anatomy studies have pinpointed the ►hippocampal formation as integral for the formation of spatial memory, as well as various types of contextual memories. However, long-term storage of these memories is attributed to other structures, in particular the parahippocampal cortex in humans. *Immediate early gene* imaging and *inactivation* studies by the group of Bontempi have shown that the hippocampus and posterior cingulate cortex are involved in the early phase of spatial memory, but later phases depend on anterior cingulate and prefrontal cortex (it was also found that different layers of the parietal cortex are sequentially involved in spatial memory formation). While the hippocampus is involved in more elaborate navigation capabilities, the striatum is associated with learning spatial habits triggered by simple or more complex cues. For example, in the *Morris water maze*, efficient swimming to a visible platform requires an intact dorsal striatum, while navigating to an immersed, non-visible platform requires an intact hippocampus. Whereas learning spatial layouts is primarily attributed to the hippocampal formation, exploitation of this information for spatial and other contextual based decision-making engages other closely associated structures such as the amygdala, prefrontal cortex and nucleus accumbens. *Brain imagery studies* in humans and studies of brain damaged patients indicate that the right medial temporal lobe, in particular the hippocampal and parahippocampal regions, is important for spatial cognition (see chapter 8 in [1] and chapter 12 in [2]).

Neurophysiological studies in the hippocampal formation and associated structures have demonstrated ►single unit and ►local field potential activity that *computational studies* show to be sufficient to support spatial learning and memory (see chapter 14 in [2]). The primary single unit responses are ►place cells, head direction cells and ►grid cells (Fig. 1) – although spatial responses have also been recorded in other areas (e.g., place selective responses in the subiculum, presubiculum and parasubiculum [8]).

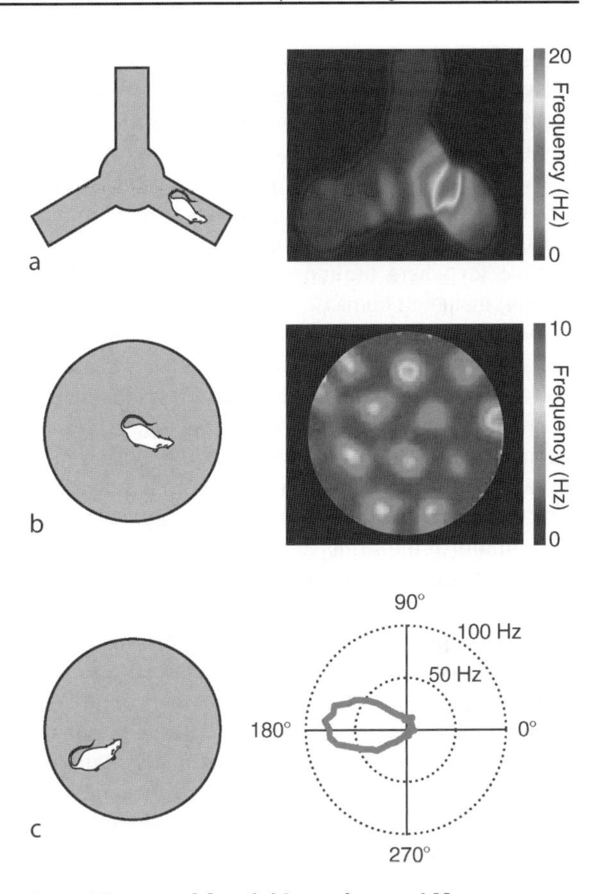

Neural Bases of Spatial Learning and Memory. Figure 1 (a) Place cells selectively discharge when the animal is in a specific location ("place field") in the environment. *Left*. The rat explores a Y-shaped maze (*top view*). *Right*. Color-coded rate map (*blue*, zero firing; *red*, maximal firing rate; *black*, area not visited by the animal). This place cells fires when the animal enters the South-East arm (Benchenane, Larrieu, Wiener and Zugaro, unpublished data). (b) Grid cells fire when the animal is located at one of the nodes of a regular hexagonal lattice. *Left*. The rat explores a circular arena (*top view*). *Right*. Color-coded rate map (adapted from 4). (c) Head direction cells discharge when the head of the animal is oriented in a specific ("preferred") direction of the horizontal plane. *Right*. Tuning curve in polar coordinates. This cell fires maximally when the rat faces 180° (Zugaro, Berthoz and Wiener (2001), J Neurosci 21:RC154).

Characteristics

Place Cells

►Place cells (discovered by O'Keefe and Dostrovsky, in 1971; see Chapter 11 of [2]) are hippocampal pyramidal cells that discharge selectively when the recorded subject occupies a delimited region of its environment referred to as a "firing field" or "place field" [3,10]. Within a given environment, only a fraction of neurons recorded have firing fields. Of these, at any given moment while the animal explores its environment, a different ensemble of place cells are active. Taken together, place cells potentially code for the entire environment: this activity is a supporting pillar of the theory [3] that the hippocampal formation is the neural basis of a ►cognitive map, an internal representation of space which Tolman (1948) postulated in his riposte to

►behaviorism. There is no anatomically topographic arrangement of firing fields (as opposed to, for example, ►somatotopic or ►visuotopic organization in corresponding sensory areas). Place cell activity has been found in the hippocampal formation of various species: rats, mice, birds, bats, and primates (by Ono and colleagues, although Rolls' group observed "view cells" selective for where the monkey looks, rather than its position) including humans.

Place responses in hippocampal neurons demonstrate several properties suggesting that they are key elements for circuits enabling high level cognitive processing [5]. In particular, place cells exhibit *supramodal responses:* although visual cues can dominate the polarization of place responses (e.g., in a cue-controlled environment, if the constellation of background cues is rotated, the positions of the firing fields will rotate in a similar manner), place responses are maintained in darkness and in blind rats, presumably on the basis of idiothetic (vestibular, proprioceptive, and perhaps efferent collaterals of motor commands) and ►somatosensory processing. Thus the same responses can be arrived at independently of sensory modality. Another remarkable property of place cells is their *abstract, stimulus-invariant responses:* during unrestricted movement in an open field, place responses are not directionally modulated - that is, the cell fires when the rat is in the firing field regardless of what direction it is facing and what is in its field of vision. Hippocampal pyramidal cells also have *context-sensitive responses*: they have place responses as well as non-place responses, for example in delayed ►eye-blink conditioning which depend on the current situation. An example in the spatial domain is *remapping*, which refers to the observation that when a rat is transferred from one environment to another, firing fields can shift in relative position, or disappear, while other initially silent neurons may now manifest firing fields. Interestingly, although in markedly different environements the respective layouts of firing fields are uncorrelated, when the rat is exposed to two novel environements with more subtle differences, place cell firing fields are initially similar in both environments, and only gradually shift to more dissimilar positions with repeated exposure.

Recent findings indicate that hippocampal activity reflects more dynamic information than initially envisioned. Many place cells fire maximally or remain virtually silent in the very same location depending on the past or future trajectory. At the population level, hippocampal activity includes information not only about current position, but also about ongoing trajectories. Indeed, it was shown that as the animal walks through the firing field of a place cell, the cell fires at earlier and earlier phases relative to the ongoing theta (7~12Hz) oscillation of the hippocampal local field potential. This property is known as *theta phase precession*. As a consequence, during each theta cycle, cells coding for recently visited locations fire first, then cells coding for current location, and then cells coding for locations further down the trajectory: that is, at each theta cycle, hippocampal discharges encompass information about past, present and future positions.

Strikingly, these patterns of sequential activation observed while the animal explores its environment are later "replayed" during subsequent sleep, in particular during short lasting fast (200 Hz) oscillatory field events called ►ripples, that occur repeatedly during slow wave sleep [7]. This is believed to be a critical mechanism underlying memory consolidation, possibly via ►long term potentiation of ►synapses targeted by the reactivated neurons, consistent with the observation that such *replay* occurs at accelerated rates relative to the original experience.

How are hippocampal place responses generated? Virtually all neocortical inputs to the hippocampus synapse first in the entorhinal cortex (EC). Two principal loops have been highlighted, the *indirect pathway* EC → dentate gyrus → CA3 → CA1 → subiculum + EC, and the *direct pathway* where EC projects directly to CA1. Numerous computational modelling studies have suggested that the recurrent connectivity of area CA3 could endow the hippocampus with ►continuous network attractor properties: this would ensure that at any given moment only a small subset of coactive cells could fire, while the rest remain silent; activity could then be propagated within the network by velocity modulated signals as the animal moves, or be directly imposed onto the network by external signals of sensory origin, the influence of which would be acquired through learning. Surprisingly however, it was shown experimentally that surgical isolation of CA1 from CA3 inputs neither disrupts the ability of CA1 pyramidal cells to acquire place fields nor the capability of the animal to learn a spatial task. This indicates that direct inputs from the EC are sufficient to support the emergence of place cell firing in CA1.

Grid Cells

Grid cells of the dorsolateral part of the medial EC [4] discharge selectively at multiple locations distributed in a regularly spaced (30–70 cm) hexagonal grid - i.e., each node is surrounded by a hexagon of adjacent nodes). The internode spacing and node diameter increase gradually in more ventral zones. At a given depth, nearby neurons have grids laterally displaced relative to one another and can also have different angular polarizations of the grids.

The EC is the principal high-level input to the hippocampal formation. Thus, place cell activity is likely to emerge from a combination of intra-hippocampal dynamics with grid cell and other entorhinal signals.

This can be observed using a protocol known to induce hippocampal remapping (see *Place Cells, above*). There are actually two ways in which hippocampal place cells can remap: either the relative firing rates of the recorded cells change in an unrelated manner, while the spatial position of their firing fields are maintained ("rate remapping"), or the firing fields are relocated in an unpredictable manner ("global remapping"). Interestingly, entorhinal grid cells also demonstrate this dichotomy: in conditions where place cells undergo rate remapping, grid cells maintain stable fields; but when place cells undergo global remapping, grids are shifted and rotated consistently with modifications observed in the hippocampus. Thus, the relative locations of hippocampal firing fields depend on the relative grid layouts of their input grid cells.

In addition, precise spike timing of entorhinal ouputs might also influence spike timing in the hippocampus. Similar to place cells, layer II entorhinal grid cell firings also show phase precession relative to the ongoing theta oscillation. Note that the EC is a principal source of the theta rhythm in hippocampus. It is possible that hippocampal phase precession is not generated purely endogenously, but is at least partly inherited from EC, consistent with previous work where transient perturbation of hippocampal theta rhythm and cell firings did not disrupt phase precession.

Head Direction Cells

Complementary to place cells, head direction (HD) cells (discovered by Ranck, in 1984) discharge selectively when the head is oriented in a particular direction in the horizontal plane, independent of position in space [6]. These responses are not controlled by the geomagnetic field; instead, they are thought to emerge from the interaction between intrinsic neural dynamics and multisensory and possibly motor signals (see chapter 18 of 6). HD cell responses are anchored to background visual cues (likely identified from ►optic field flow cues), but responses are maintained in darkness, presumably engaging ►vestibular and other self-movement cues (see Chapters 6 and 7 of [6]. Different cells are selective for different *"preferred" directions*, and all directions are equally represented. No anatomically topographic organization of directional responses has been reported. HD cells have been found in rats, mice, chinchillas and monkeys.

Unlike remapping in place cells, after environmental changes the preferred directions of simultaneously recorded HD cells rotate coherently, maintaining their angular differences. Thus the HD cell system may provide a more stable spatial reference.

HD cells have been found in over ten different brain areas including parts of Papez' circuit: dorsal tegmental nucleus of Gudden (DTN; which also has cells selective for head velocity) → lateral mammillary nucleus (LMN) → anterior-dorsal thalamic nucleus → post-subiculum + retrosplenial cortex → dorsal part of medial EC → hippocampus (which also has a low incidence of HD cells). Other areas with HD cells include lateral-dorsal thalamic nucleus, dorsomedial striatum and medial precentral cortex (FR2 or AGm).

HD and place responses are both suppressed by inactivation of the vestibular nuclei (sensitive to linear and angular accelerations of the head) which send axonal projections to the DTN. Neurocomputational and neuroanatomical data point to the DTN-LMN circuit as the generator of head direction responses by means of attractor network dynamics. From a computational point of view, this system would effectively compute a mathematical integration of the head angular velocity signal to yield head angular orientation. Progressive drift of this representation due to imperfect integration could be corrected for by sensory inputs (e.g., triggered by familiar visual landmarks).

Space-Related Neuronal Responses and Navigation

A key issue is whether the spatial responses of place cells, grid cells and HD cells do subserve complex navigation capabilities, consistent with theoretical considerations, correlative experimental results and computational models and simulations.

To date, finding a clear link between HD cell responses and behavioral performance in spatial navigation tasks has remained elusive. Given that grid cells have only been discovered in 2005, it should not come as a surprise that no experimental results support their role in spatial navigation yet.

There is however an increasing amount of evidence for a critical role of place cells in spatial learning and memory. Initial support was brought by studies using either ►long term potentiation (LTP) induction protocols, or pharmacological or transgenic approaches. For instance, protein synthesis inhibitors and NMDA antagonists, the two major blockers of LTP, alter both place cell stability and spatial behavior. Similar results have been obtained in mice where the major proteins involved in the molecular cascade of LTP were knocked-out, such as CA1 NMDA knock-out mice, R(AB) mice (with reduced forebrain PKA), GluR2 mutant mice (knock-out for the GluR2 sub-unit of AMPA receptors), αCaMKII-T286A mice (with altered calcium sensitivity of CaMKII) and CREBαΔ- mice (deficient for the α and Δ isoforms of CREB). Remarkably, these studies show relatively good correlations between the level of LTP, place field stability and performance in spatial memory tasks [9].

In addition to these studies, perhaps some of the most convincing evidence to date that place cells do underlie spatial navigation capabilities is that in rats trained to find an unmarked goal in order to receive food rewards, experimental shifts of environmental cues that provoke displacements of place fields relative to the goal location are associated with impaired spatial performance, provided that goal localization requires map-based navigation (see chapter 10 in [1]).

References

1. Jeffrey KJ (ed) (2003) The neurobiology of spatial behavior. Oxford University Press, Oxford, 316 pps
2. Anderson P, Morris R, Amaral D, Bliss T, O'Keefe J (eds) (2007) The hippocampus book. Oxford University Press, New York, 832 pps
3. O'Keefe J, Nadel L (1978) The hippocampus as a cognitive map. Clarendon Press, Oxford, 570 pps
4. Hafting T, Fyhn M, Molden S, Moser MB, Moser EI (2005) Microstructure of a spatial map in the entorhinal cortex. Nature 436:801–806
5. Wiener SI (1996) Spatial, behavioral and sensory correlates of hippocampal CA1 complex spike cell activity: implications for information processing functions. Prog Neurobiol 49:335–361
6. Wiener SI, Taube JS (2005) Head direction cells and the neural mechanisms of spatial orientation. MIT, Cambridge, MA, 480 pps
7. Buzsáki G (2006) Rhythms of the brain. Oxford University Press, Oxford, 464 pps
8. Redish AD (1999) Beyond the cognitive map: from place cells to episodic memory. MIT, Cambridge, MA
9. Tonegawa S, Nakazawa K, Wilson MA (2003) Genetic neuroscience of mammalian learning and memory. Philos Trans R Soc Lond B Biol Sci 358:787–795
10. Wilson MA, McNaughton BL (1993) Dynamics of the hippocampal ensemble code for space. Science 261:1055–1058

Neural Cell Adhesion Molecules (NCAM)

Definition

Belongs to the immunoglobulin superfamily. It mediates cell-cell interactions by homophilic binding, especially during axon fasciculation and nerve sprouting and cell migration.

Neural Coding

Definition

Neurons create, store and convey information. There are two basic codes they may use: rates of action potentials (unitary signals) or timing of action potentials. From a different perspective, neural coding also refers to the question whether specific information is stored in one neurons or in networks of neurons (see Ensemble Coding, Grandmother Neuron).

Neural Coding of Taste

Martin Witt
Smell and Taste Clinic, Department of Otorhinolaryngology; Department of Anatomy, University of Technology Dresden, Medical School, Dresden, Germany

Synonyms

Gustatory neural coding; Taste coding

Definition

Basic gustatory stimuli are conveyed to the brain via activation of neurons connected in series or in parallel. The gustatory information needs several "simple" codes for representation at various neuronal levels. However, "taste" encompasses more than just the activation of the gustatory system by chemical stimuli, but include also touch, vision, olfaction, which all contribute to the perception of foods or drinks. Components of the gustatory pathway are taste receptors, gustatory nerve fibers, as well as relay stations within the brainstem (nucleus of the solitary tract), diencephalon (thalamus, parts of the ventral striatum), and telencephalon (insula/frontal operculum, orbitofrontal cortex, amygdala).

Characteristics

The sense of taste is regarded as the main mechanism protecting the individual from intake of hazardous food. Usually, "taste" comprises at least five relatively distinct sensations such as "salty", "►bitter," "sweet," "acid," and "umami." Other sensations include trigeminally mediated, painful sensations, or the sensations of fattiness or astringency. Identical stimuli may elicit distinct reactions under various behavioral or environmental conditions, implying that various modulatory systems in the central nervous system interact with taste-selective pathways. Most flavors (e.g., cherry or chocolate) are the result from the complex interaction between trigeminal, olfactory and gustatory activations.

These multimodal properties contribute to an understanding how variable taste stimuli may be related to the activity of one or more neurons. In its stricter sense, "taste coding" comprises taste characteristics, which are translated into a specific neuronal firing pattern.

What can be Coded?

- The relative number of stimulus molecules per volume of solution is coded by the rate of impulses in taste-responsive neurons: spike rate.
- Quality of taste stimuli, represented as “breadth of tuning” of ►gustatory neurons. The entropy measure (H) compares how selectively a cell responds to a standard array of chemical stimuli, usually NaCl, quinine hydrochloride, sucrose, and HCl. This method is a mathematical expression with H values ranging from 0.0 representing one of the five basic stimuli, to 1.0 representing equal responses to all five stimuli [1].

Coding of Taste Qualities

There are two competing hypotheses how basic taste qualities may be relayed to the brain:

1. The labeled-line hypothesis: One taste quality is conveyed by one specific neuron type.
2. The across-fiber pattern theory: Taste qualities are coded by the relative activities across the population of responsive neurons.

The Labeled-Line Hypothesis

The old observation that certain regions of the tongue are relatively specialized to convey various taste qualities led to the idea that neurons innervating taste buds in different regions were specialized for these taste sensations. Specifically, certain gustatory neuron types were specific coding channels for taste quality [2]. For example, the strict application of this theory would imply that the perception of the quality “sour” would arise from the activation of only one “sour” neuron type. As an example, Hellekant et al. [3] observed sweet-cluster fibers in the ►Chorda tympani nerve of chimpanzees, which could be specifically blocked by gymnemic acids. However, individual ►gustatory neurons were activated also by different taste stimuli. Consequently, as a variant of the labeled-line theory, taste cells were associated with neuron populations which evoke the greatest number of action potentials after a given gustatory stimulus. This number of neurons reacts as, for example, “sucrose-best” or “NaCl-best” [4]. On the other hand, some animals may behaviorally discriminate between related stimuli, for example, sucrose and maltose, which requires more subtle discrimination abilities than expected for specifically tuned cells. Also, neurophysiological recordings demonstrated that more than one basic taste stimulus was able to activate peripheral gustatory neurons. This strengthens the hypotheses related to an across-fiber pattern taste quality code (see below).

The Across-Fiber Pattern Theory

The across-fiber pattern (ensemble) theory assumes that large ensembles of neurons contribute to the quality of a taste stimulus, because there is no absolute stimulus specificity in the responsiveness of afferent gustatory fibers [2]. On the other hand, reception of complex taste qualities may be understood as a parallel activation of several labeled lines, making the labeled-line theory a special case of the across-fiber pattern theory. The validity of various models of the neural coding of taste quality suffers from a nearly complete reliance on correlative statistical procedures and a lack of available techniques that allow selective manipulation of putative classes of gustatory neurons [5].

Taste Coding in the Gustatory Periphery: Convergence within the Taste Bud vs. Labeled Lines of Gustatory Nerves

Taste stimuli can be mediated either by receptor proteins or apical ion channels. Do taste sensor cells (TSC, Type II cells) respond to a single quality or are they broadly tuned to multiple qualities? TSC are rather neuroepithelial cells than neurons; only a few exhibit synaptic contacts to nerve fibers. Recent studies show that about 80% of TSC (Phospholipase C β2-positive, SNAP-25-negative, i.e., lacking synapses) are narrowly tuned and respond to only one taste quality. In contrast, most presynaptic cells (SNAP-25-positive) are broadly tuned and respond to two or more taste qualities [6]. What is more, TSC may communicate with presynaptic cells by secreting ATP via pannexin-1 hemichannels. Only these broadly-tuned cells form synapses to afferent gustatory nerve fibers [7]. Presumptive transmitters are ATP, serotonin and possibly neuropeptides. Moreover, one afferent nerve fiber branches and synapses to more than one presynaptic taste bud cell, which leads to convergence and breadth of tuning. Chorda tympani fibers of rodent and monkeys have been subdivided into narrowly-tuned S-, N-, and Q-fibers (for sweet, sodium, and bitter), as well as more broadly tuned A- and H-fibers (for acids and electrolytes). Though some authors argue most taste stimuli seem to be transferred in selective populations of neurons in a labeled-line manner [5], others are hesitant to such a conclusion because of difficulties in interpretation of salty stimuli and the fact that neurons are more broadly tuned at higher concentrations [1].

Taste Coding in the Central Nervous System

Central Taste Pathways

Generally, central taste neurons are relatively broadly tuned. Gustatory information is conveyed by afferent fibers of three cranial nerves: [1] Fungiform papillae of the anterior portion of the tongue are innervated by the chorda tympani (CT), a branch of the intermedio-facial nerve complex (CN.VII), which runs with the lingual nerve [2]. Vallate and foliate papillae are innervated by gustatory fibers of the glossopharyngeal nerve (CN.IX), and taste buds on the laryngeal surface of the epiglottis, larynx, and proximal part of the oesophagus are supplied by the superior laryngeal branch of the vagal nerve (CN.X). Their nuclei lie in various peripheral ganglia, but the first central relay stations in the gustatory pathway are in the nucleus of the solitary tract (NST) of the brainstem. In humans, ipsilateral monosynaptic neurons run immediately to the ventral posteromedial nucleus of the thalamus, whereas in rodents an additional relay, the parabrachial nucleus, is interconnected. Dorsal neurons terminate in the "gustatory cortex," the anterior insula/frontal operculum and caudolateral orbitofrontal cortex. In rodents, a ventral affective pathway turns to the lateral hypothalamus, central nucleus of the amygdala, substantia innominata, and bed nucleus of the stria terminalis of the ventral striatum, but the exact pathways are somewhat less clear in primates/humans [8].

There is a Remarkable Intra-Individual Coding Stability, but no Clear Topographical Projection for Basic Taste Qualities in Structures of the CNS

A clear chemo-topographical organization of particular taste qualities is a matter of debate. Recent functional magnetic resonance tomography investigations in humans show inter-individual differences for taste stimuli; taste fields of individual subjects do barely overlap, but there is stable intra-individual activity over a prolonged period of time [9]. In mice, coding stability is maintained over time, even after taste nerve crush and regeneration. Moreover, activity in taste neurons encodes also information about the hedonic value of a gustatory stimulus, regardless if there is a segregation or convergence of taste stimuli in central pathways.

Different neurons of the rodent NST respond best to NaCl, glucose, HCl, and quinine-HCl, but they have relatively broad tuning to different stimuli, as compared to the peripheral gustatory fibers [1]. This is mostly due to convergence of peripheral fibers. Also, many central taste neurons also carry thermal and mechanosensory information. In primates, a similar distribution has been found in neurons of the insular and opercular primary taste cortices [10] showing statistically independent neuron types for the basic taste qualities. Neurons devoted to bitter (quinine) quality comprise approximately 22% of measured insular/opercular cells, and those responding to sweet and salty account for 73%. Only 5% of all measured neurons were activated for the detection of acids [10].

Conclusion

Taste information processing is based on complex interconnections. It has been hypothesized that it exhibits specific cerebral representation fields for the basic gustatory qualities which are nevertheless difficult to identify. Based on molecular, electro-physiological, and behavioral findings, the breadth of tuning at all anatomical levels and multimodal sensitivities of mostly central neurons render it difficult to clearly decide which coding algorithm is "used" for taste processing. Most likely various neuron types define specific across-neuron patterns. New insights have yet to come before clearly understanding syndromes related with taste deregulations such as the ►gourmand syndrome.

References

1. Smith DV, Travers J (1979) A metric for the breadth of tuning of gustatory neurons. Chem Sens Flav 4:215–229
2. Pfaffmann C (1959) The afferent code for sensory quality. Am J Psychol 14:226–232
3. Hellekant G, Ninomiya Y, Danilova V (1998) Taste in chimpanzees. III: Labeled-line coding in sweet taste. Physiol Behav 65(2):191–200
4. Frank M (1973) An analysis of hamster afferent taste nerve response functions. J Gen Physiol 61(5):588–618
5. Spector AC, Travers SP (2005) The representation of taste quality in the mammalian nervous system. Behav Cogn Neurosci Rev 4(3):143–191
6. Tomchik SM, Roberts CD, Pereira E, Stimac R, Roper SD (2007) Responses of taste receptor cells and presynaptic taste cells to taste stimuli. XXIXIth Annual Meeting of the Association for Chemoreception Sciences (AChemS), Sarasota, FL, USA, April 25–30, 2007
7. Huang Y-J, Maruyama Y, Dvoryanchikov G, Pereira E, Chaudhari N, Roper SD (2007) The role of pannexin 1 hemichannels in ATP release and cell-cell communication in mouse taste buds. PNAS 0611280104
8. Small DM (2006) Central gustatory processing in humans. Adv Otorhinolaryngol 63:191–220
9. Schoenfeld MA, Neuer G, Tempelmann C, Schussler K, Noesselt T, Hopf JM, Heinze HJ (2004) Functional magnetic resonance tomography correlates of taste perception in the human primary taste cortex. Neuroscience 127(2):347–353
10. Scott TR, Plata-Salman CR (1999) Taste in the monkey cortex. Physiol Behav 67(4):489–511

Neural Computation

►Connectionism

Neural Control of Eye Movements

Adonis Moschovakis
Institute of Applied and Computational Mathematics, FO.R.T.H., Department of Basic Sciences, University of Crete, Heraklion, Greece

Understanding the neural control of actions is a central goal of the Neurosciences. It is for ►eye movements that we are closest to attaining this goal, at least in the case of fairly complex brains such as those of mammals. Accordingly, ►oculomotor control has become a test bed of theories encapsulating general principles of motor control potentially applicable to additional effectors such as the head or the arm. In part this is due to the simplicity of ►oculomotor dynamics and kinematics. It is also due to the simplicity of the muscular apparatus that controls the eyes (just six ►extraocular muscles per eye). Finally, ►eye orbital mechanics are straightforward and because the eye is a spherical joint, its movements have a small number of degrees of freedom (just three). As with other motor systems, this number is often further reduced in behaving subjects. For example, ocular ►torsion is determined from the horizontal and vertical rotation of the eyes, measured in a coordinate system that moves with it. This fact has been known for more than a century, ever since ►Donders enunciated it in the form of the law that bears his name. It results from a more fundamental one, ►Listing's law [1], according to which for all eye positions, defined as single rotations from primary position, rotation axes lie on a plane perpendicular to the line of sight at primary position (i.e., the position of the eyes when looking approximately straight ahead and the head is upright).

In terms of metrics and kinematics, eye movements can be classified into several distinct types (Fig. 1). Figure 1a provides typical examples of ►saccades (marked with an asterisk). These are the eye movements that rapidly redirect the line of sight in a variety of circumstances, i.e., when exploring the visual world (scanning saccades), toward targets (►pro-saccades) which may appear suddenly in the extrafoveal visual field (►foveating or ►reflexive-saccades) or away from visible targets (►anti-saccades). On the average, the time that elapses between the presentation of a visual target and the onset of a saccade to it (saccadic latency) is equal to about 200 ms. In some human subjects, latency histograms demonstrate a second early peak, made of ►express saccades. Human saccades often undershoot their targets by about 10% and are followed by secondary shorter latency movements, called ►corrective saccades. The goal of other eye movements is to stabilize the projection of visual images on the retina. For example, movements of the head engage the ►vestibulooculomoter system, which evokes compensatory eye movements (slow phases of the ►vestibuloocular reflex – VOR) whose velocity is usually less than or equal, and opposite in sign, to that of the head (Fig. 1d). This ►rotatory VOR is engaged when the head rotates on the neck, and is driven by modulation of afferent signals arising in the semicircular canals. Linear acceleration of the head (e.g., during its forward propulsion on a walking body) evokes the ►translational VOR while keeping the head bent toward the shoulder evokes a static ►ocular counter-rolling response; both of these responses are driven by the otolith organs. Dynamic rotations about off-vertical axes activate both canal and otolith responses. Slow phases can be cancelled when the line of sight fixates a target moving with the head (►VOR suppression) and are interrupted by quick phases (Fig. 1b asterisks) which correspond to saccades. Sequences of slow and quick phases, collectively called ►nystagmus, are also produced after the end of constant velocity rotation (►postrotatory nystagmus), and when the vestibular sensorium is activated by stimuli other than the natural ones, such as after ►caloric stimulation of labyrinths. They are also produced by visual stimuli evoking ►optokinetic eye movements (►optokinetic response – OKR; Fig. 1e), e.g., during rotation of a drum

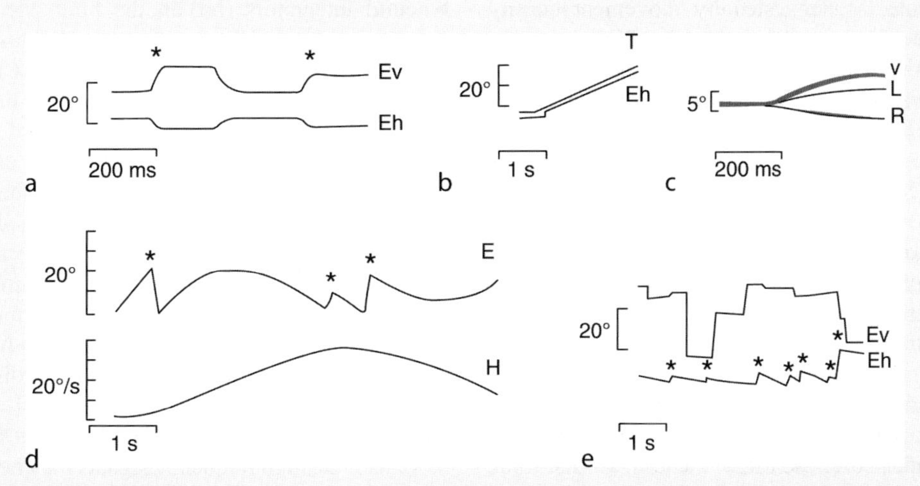

Neural Control of Eye Movements. Figure 1 Distinct types of eye movements readily observed in monkeys. (a) Saccades. (b) Smooth pursuit. (c) Vergence. (d) VOR. (e) OKR. Note the difference in the time scales employed. Upward trace deflection indicates rightward movement. Asterisks in (a), (d) and (e) mark saccades. Abbreviations: dH/dt, head velocity; E, eye position; Eh, horizontal eye position; Ev, vertical eye position; L, eye position – left eye; R, eye position – right eye; T, target position; v, vergence angle.

around the subject at a constant velocity (►optokinetic nystagmus – OKN). Finally, ►ocular following responses (OFR) are the relatively low velocity eye movements that subjects use to follow slowly moving visual scenes. They are further developed in primates which are able to generate ►smooth pursuit eye movements (Fig. 1b) to accurately track small objects moving across a stationary background. Smooth pursuit eye movements can be preceded by saccades (►catch-up saccades; Fig. 1b) and they can be interrupted by saccades. Saccades, smooth pursuit, vestibular and optokinetic eye movements are ►conjugate in the sense that during such movements both eyes move simultaneously in the same direction and by roughly the same amount. Other eye movements (►disjunctive) are not (Fig. 1c). This is the case when the line of sight shifts from near objects to far ones (►divergent eye movements) and vise versa from a far to a near object (►convergent eye movements). The ►near response reflects the intimate relationship of the system that controls the angle between the lines of sight of the two eyes (►vergence angle) and the system that controls the focal point of the lens (►accommodation-vergence interactions).

The ability to accurately record the instantaneous position of the eyes with the ►electrooculogram (EOG) and ►electronystagmographic methods provided the impetus for a large number of studies of the relation between the instantaneous discharge of oculomotor related cells and the parameters of ocular movements. To date, two large families of theories have been proposed to account for the control of movements of body parts. On the one hand, position control theories suggest that neural signals guide effectors to a particular position in space. Alternatively, neural signals could dictate the metrics of the movements that the effectors should execute, as suggested by movement control theories. Recording from the neurons which innervate extraocular muscles, the ►extraocular motor neurons (MNs), allows one to test which of these two families of theories is correct. If a position control theory were implemented in the brain, extraocular motoneurons should be activated in proportion to the position of the eyes in the orbit. Their remarkable conceptual simplicity notwithstanding, the discharge pattern of extraocular motoneurons does not fulfill expectations based on position control theories. During saccades, for example, extraocular MN discharge is characterized by transient bursts accompanying movements and sustained tonic discharges whose intensity is proportional to the position of the eyes in the orbit. This pattern of discharge underscores the use of the term ►burst-tonic (BT) ►neurons to refer to extraocular MNs. In a more quantitative vein, their firing rate (F_R) approximately obeys the expression $F_R = \alpha + \beta.E + \gamma.dE/dt$, where E the instantaneous position of the eyes (vertical or horizontal depending on the cell), dE/dt is the instantaneous velocity of the eyes and α, β, γ. constants of proportionality which differ for different neurons. To obtain an intuitive feeling for these parameters, consider that some neurons discharge at a low rate (small α) and others at high rate (large α) when the eyes point straight ahead, some are deeply (large β) and others little (small β) modulated with eye position while some emit modest (small γ) and others emit considerable (large γ) bursts for saccades in their on direction.

Saccades, smooth pursuit, vestibular, optokinetic and vergence/divergence ocular movements are separately controlled by five, largely distinct neural circuits all of which feed into the final common path embodied by the extraocular MNs. Some of these circuits are anatomically simple. For example, the anatomical substrate of the VOR is a three neuron arc, composed of: (i) primary afferent vestibular fibers, (ii) ►vestibulo-oculomotor connections primarily arising from ►position-vestibular-pause (PVPs) neurons, (iii) extraocular motor neurons. The system that controls disjunctive eye movements is more complex. It comprises ►near response neurons and ►divergence neurons in the brainstem which are in turn driven by ►disparity dependent vergence and ►radial flow dependent vergence cortical signals. At the other extreme of the simplicity-complexity spectrum, saccades rely on more than 30 distinct cell classes found in more than ten widely distributed brain areas [2]. To cope with its structural complexity it is meaningful to divide the saccadic system into smaller sub-circuits, each one of which can be thought of as a central pattern generator responsible for a conceptually distinct operation: the ►brain stem burst generators (BG) in the reticular formation, the horizontal (►Neural integrator - horizontal) and vertical (►Neural integrator - vertical) ►neural integrators (NI) in the brain stem, and the metric computer in the ►superior colliculus. Although largely independent from the point of view of anatomy and physiology, to efficiently control the line of sight, systems responsible for different kinds of eye movements need to interact with each other. For example, the saccadic system needs to interact with the vestibular one (during nystagmus), as well as with the systems controlling smooth pursuit (►pursuit-saccade coordination) and vergence (►saccade-vergence interactions). It also needs to interact with systems controlling the movements of other effectors such as the ►head (►eye-head tracking), the hand (►eye-hand coordination), the eye-lids, etc.

The burst generators (BG) in the brain stem supply MNs with the high frequency signals they need to cause the contraction of the muscles they innervate during a saccade in their pulling direction. They contain cells that exhibit a brief burst of activity before saccades of particular directions (►burst neurons). Depending upon the latency of the burst in relation to the onset of the movement, they are called ►medium lead burst (MLB) neurons or ►long lead burst (LLB) neurons [3], the term ►short lead burst neurons being reserved for the neurons with the shortest burst latencies, i.e., the MNs. The amplitude and direction of saccades are specified in terms of the amplitude of their horizontal and vertical components which are determined relatively independently by distinct groups of premotoneuronal MLB neurons. ►Horizontal (left and right) ►MLB (HMLB) neurons are located in the ►paramedian pontine reticular formation while ►vertical (up and down) ►MLB (VMLB) neurons are located in the ►rostral interstitial nucleus of the medial longitudinal fasciculus (►riMLF). Axons of MLB neurons convey the output of the burst generators directly to extraocular MNs, thus directly providing them with the excitatory drive they need for ipsiversive saccades. One of the most prominent projections of the horizontal ►excitatory burst neurons (►EBNs) is to the ipsilateral abducens nucleus [4] (which contains MNs innervating the ipsilateral lateral rectus muscle, a muscle engaged for ►lateral gaze deviations). Similarly, VMLB neurons ramify extensively within regions of the oculomotor complex housing MNs with vertical pulling directions. For the eyes to move conjugately, which is the case for saccades, synergistic muscles of the two eyes should be activated simultaneously and by roughly the same amount (an expectation known as ►Hering's law of equal innervation [5]). To implement Hering's law, axonal terminations of single VMLB neurons contact motoneurons innervating yoked muscles of both eyes [6]. In the horizontal system, Hering's law is implemented through an additional class of neurons, the abducens ►internuclear ones [7]. The axons of these neurons travel with the ►medial longitudinal fasciculus (MLF) and their interruption causes the syndrome of ►internuclear ophthalmoplegia. In addition to the coactivation of MNs innervating synergist muscles of both eyes, extraocular MNs supplying antagonist muscles are inactivated during saccades (Descartes' principle). For example, activation of abducens MNs in one side of the brain is accompanied by inhibition of the contralateral abducens MNs as well as disfacilitation of ipsilateral medial rectus (MR) MNs (these innervate the MR muscle of the ipsilateral eye). Both are mediated by projections of ►inhibitory burst neurons (IBNs) to the contralateral abducens nucleus [8].

Not all MLB neurons deploy axonal terminations appropriate for influencing extraocular motoneurons. For example, the interstitial nucleus of Cajal (NIC) contains cells whose bursts of discharge are identical to those of downward premotoneuronal MLBs [9]. However, instead of supplying the oculomotor nucleus, their terminal fields overly the riMLF, i.e., a nucleus housing VMLBs and other burst neurons. Such neurons are better suited for conveying corollary discharges through the feedback paths of a BG configured as a ►local feedback loop controller. LLB neurons also vary a lot in terms of projections, discharge pattern and function. Some, such as the ►pontopontine long-lead burst ►(PPLLB) neurons, are likely to embody the front stage of the BG thus receiving presaccadic commands from the superior colliculus (SC) as well as feedback from the MLBs. Other LLB neurons, such as the ►burster-driving neurons (BDNs), are probably involved in the generation of vestibular quick phases. ►"Trigger" neurons, i.e., interneurons mediating higher order commands to initiate saccades, and ►"latch" neurons, i.e., interneurons inverting the sign of MLB discharges to prevent ►omnipause neurons (OPNs) from firing during saccades, may also belong to the LLB class. Not all neurons of the BG burst for saccades. Instead, some pause for saccades in all directions and are for this reason called omnipause neurons (OPNs). Coupling orthogonal (vertical and horizontal) burst generators through OPNs can help coordinate them so that the vertical and horizontal components of oblique saccades have roughly similar duration despite their dissimilar size thus giving rise to oblique saccades with fairly straight trajectories [10].

In addition to the phasic eye displacement signals produced by the BG, extraocular MNs are supplied with tonic signals proportional to eye position. The latter are generated from the former through a process akin to mathematical integration; because it is less than perfect it is referred to as ►leaky integration. The medial vestibular nucleus and the nucleus prepositus hypoglossi are thought to house the ►horizontal neural integrators (HNI) while the interstitial nucleus of Cajal is thought to house the ►vertical neural integrators (VNI). Other oculomotor subsystems, such as those responsible for the VOR, the OKN and the smooth pursuit eye movements, also rely on the neural integrators to function properly [11]. Again contrary to expectations based on position control theories, the discharge of most of the neurons that comprise the vertical and horizontal neural integrators does not reflect the position of the eyes alone. Instead, their discharge resembles that of the extraocular motoneurons they contact and justifies the use of the term burst-tonic (BT) neurons when referring to such cells as well. Intrinsic cellular membrane properties, synaptic mechanisms and network properties could in principle account for the fact that the time constant of the neural integrator (30 s; estimated from the time course of ►ocular drifts in the dark) is more than three orders of magnitude higher than that of single neurons (about 5 ms). A second device with roughly similar impulse responses, known as the "►velocity storage integrator," is also employed by the rotatory VOR to extend the dynamics of its afferent input into the low frequency range.

The ►superior colliculus (SC) is a layered midbrain structure that computes the amplitude and direction of desired saccades and relays appropriate commands to the burst generators. It also plays a crucial role in ►sensorimotor integration in the sense that sensory signals about target location must also be taken into consideration before the requisite motor commands can be issued [12]. The superficial layers of the SC contain a map of visual space while its deeper layers contain additional ►sensory maps as well as a ►motor map of oculomotor space. These are embodied by partially overlapping classes of cells, several of which can be distinguished in the primate SC. The best studied are the ►saccade related burst neurons (in short ►SRBNs), which are characterized by low levels of spontaneous activity and high frequency bursts which they emit shortly before (about 20 ms on the average) saccades of the appropriate amplitude and direction [13]. The SC also contains ►buildup neurons (in short ►BUNs) whose activity increases about 80–100 ms before saccade onset, reaches a peak value preceding saccades by 10–20 ms, and then wanes (discovered by Sparks and his colleagues [14]). Other deeper layer neurons, called ►Visually Triggered Movement cells (VTMs), emit discrete bursts in response to visual stimuli and before saccades to the same stimuli, but not before spontaneous saccades of the same metrics [15]. Each of these neurons discharge only before saccades having a relatively narrow range of sizes and directions, which defines an area known as the cell's ►movement field. Optimal saccades are measured from initial eye position, and not in a frame of reference centered in the head, so the SC works in an ►oculocentric frame of reference. Movement fields are arranged in an orderly topographic map of the SC. Cells discharging for upward saccades are located medially and cells discharging for downward saccades are located laterally while cells preferring small saccades are located more rostrally than cells preferring bigger saccades. It is for this reason that the SC can be thought to use a ►place code to implement the motor map it contains. A particular class of SC presaccadic neurons, the ►tectal long lead burst neurons (►TLLBs), have been characterized both functionally and morphologically [16]. Their axons are responsible, at least in part, for conveying command signals from the SC to the burst generators. To determine saccade size, the strength of these projections to the horizontal and vertical BG increases linearly with the size of the saccades coded by the region they originate from [17]. TLLBs also deploy a rich plexus of recurrent connections which may be instrumental in determining the intensity of discharge and the spread of TLLBs engaged for particular saccades [18]. Another crucial aspect of TLLB physiology is the strong and tonic inhibitory input they receive from the ►basal ganglia, in particular the pars reticulata of the ►substantia nigra. This is relaxed for saccades due to the inhibitory input nigral neurons receive from the ►caudate nucleus.

The input to the saccadic system is retinal error (i.e., the distance of visual targets from the fovea). This is represented in the superficial layers of the SC, retinotopically, in terms of the location of active cells in a neural map of visual space. To account for the subsequent activation of the appropriate presaccadic cells in the deeper SC layers, an early hypothesis of tectal function, the "►foveation hypothesis" [19], invoked the spread of excitation from superficially located visual neurons to underlying presaccadic cells. Because the visual receptive fields of superficial SC cells are aligned with the movement fields of deeper SC cells, flow of information from the superficial to the deeper SC layers would enable the eyes to move accurately and foveate the target. Consistent with the "foveation hypothesis," axons originating from a particular class of superficial SC cells, the L neurons (►interlayer neurons), deploy dense terminal fields in the deeper layers of the primate SC [20]. The ability to make accurate saccades toward targets briefly flashed before the onset of a saccade to a previous target (►double-step saccade paradigm) or elicited in response to the electrical stimulation of the SC demonstrates that the "foveation" hypothesis was too simplistic. An alternative, the "vector subtraction" hypothesis [16], assumes that a neural replica of the eye displacement due to the first saccade is fed back to the SC where it is subtracted from the retinal error vector that signals the location of the second target. The vector that results from this subtraction (represented by a new focus of excitation in the deeper tectal layers) drives the burst generators in a way that produces the correct saccade [18]. Additionally, the "vector subtraction" hypothesis provides an explanation for the response properties of two more classes of neurons. Firstly, it explains the discharge of ►Quasi-visual (Qv) neurons [21] for targets initially presented outside their receptive field and extinguished before saccades such that their location is encompassed by the cell's receptive field. The corollary discharges needed to account for the discharge of Qv neurons are carried by the ►Reticulo-Tectal Long Lead Burst (RTLLB) neurons of the ►central mesencephalic reticular formation [16].

Besides determining saccade metrics, the deeper layers of the SC are also involved in ►eye-head coordination in part due to the ►tectoreticulospinal neurons (TRSNs) they contain [22]. The response properties of TRSNs are more complex than those of primate SRBNs in that they display multisensory convergence, and their bursts are non-obligatorily coupled to saccades. Their projections also target widespread midbrain and rhombencephalic areas. Together with ►reticulospinal long-lead burst (RSLLB) neurons [23] and eye-neck reticulospinal neurons [24] they could cause the fairly widespread facilitation of premotor extracollicular circuits needed to reorient the body, the head and the eyes toward a crudely defined region of space (►orienting reflex). In this manner, body and ►head movements are coordinated with saccades and the VOR to generate shifts of the line of sight (►gaze shifts).

►Precerebellar long-lead burst (PCbLLBs) ►neurons convey to the ►cerebellum saccade related signals originating in the SC and other higher structures [23]. Similarly, mossy fibers arising from several brainstem nuclei (including the dorsolateral pontine, the reticular pontine tegmental and the vestibular ones) supply it with smooth pursuit related signals. Electrical stimulation of the ►cerebellum, in particular lobules VI and VII of the vermis, has been long known to elicit saccades [25]. The saccade related discharges of neurons located in the ►oculomotor vermis influence the contralateral BG in the brain stem through the caudal part of the fastigial nucleus (the ►fastigial oculomotor region) as well as parts of the interpositus and dentate nuclei [26,27]. Both the oculomotor vermis and the fastigial oculomotor region participate in the control of smooth pursuit eye movements, as well [28,29]. At least two more cerebellar regions are also involved in the control of smooth pursuit eye movements: the ►flocculus, which also participates in the modulation of the gain of the VOR (►VOR adaptation) possibly through its projection to ►flocculus target neurons, and the ventral paraflocculus. Both areas contain Purkinje cells discharging for horizontal eye velocity during smooth pursuit and for horizontal head velocity during VOR cancellation (i.e., when the subjects' eyes follow a target moving with the head). Because eye and head velocity signals are roughly equal, and almost cancel each other during VOR in the dark, these neurons are thought to encode the horizontal angular velocity of gaze in space (►horizontal-gaze-velocity Purkinje cells). Purkinje cells with preferred directions other than horizontal can also be found in these regions. Given the variety of oculomotor related signals encoded by cerebellar neurons and the variety of ocular movements evoked in response to the electrical stimulation of several distinct cerebellar regions, it is reasonable to expect that cerebellectomy would cause a variety of oculomotor deficits. These include partial neural integrator failure and impairment of smooth pursuit eye movements [30] as well as ►saccadic dysmetria [31] and disruption of ►saccadic adaptation and VOR adaptation [32].

The neocortex contains several areas participating in the control of ocular movements. One of the best studied is the ►frontal eye field (FEF; [33]), in and near the arcuate sulcus (AS). It contains cells discharging briskly before saccades [34] and its electrical stimulation evokes contraversive saccades [35]. Neighboring subregions of the arcuate sulcus are devoted to the control of eye movements other than saccades. For example, the ►frontal pursuit area is a small region in the depths of the arcuate fissure, sandwiched between the small saccade area of the FEF and the somatic premotor cortex; its electrical stimulation evokes smooth pursuit eye movements and it contains cells discharging for smooth pursuit eye movements. Another subregion, located more rostrally in the prearcuate convexity, contains cells active during the vergent, divergent and accommodative movements accompanying near or far viewing and its electrical stimulation induces convergence and ocular accommodation [36]. Also, a small part of the fundus of the AS near its genu corresponds to a ►fixation zone. The notion that the primate FEF participates in the ►fixation system is supported by the fact that its removal interferes with ►visual fixation and the fact that it contains neurons which discharge tonically while gaze is held stable toward a ►fixation point and pause for saccades (►fixation neurons). Other frontal lobe saccade related areas include the ►supplementary eye field in the dorsal premotor cortex, the principal and periprincipal cortex and the ►anterior cingulate. Finally, regions of the parietal and temporal cortex are also important for the higher order control of ocular movements. The ►parietal eye fields, including the ►lateral intraparietal area (►LIP) in the lateral bank of the intraparietal sulcus, contain saccade related neurons and their lesion causes ►ocular apraxia. Areas in and around the superior temporal sulcus (STS), such as the middle temporal (MT), the medial superior temporal (MST), are also activated for saccades and their lesion impairs smooth pursuit eye movements [37].

References

1. Helmholtz H (1910) Treatise on physiological optics, 1962 Dover Publications, Inc., New York
2. Moschovakis AK, Scudder CA, Highstein SM (1996) The microscopic anatomy and physiology of the mammalian saccadic system. Progr Neurobiol 50:133
3. Luschei ES, Fuchs AF (1972) Activity of brain stem neurons during eye movements of alert monkeys. J Neurophysiol 35:445
4. Strassman A, Highstein SM, McCrea RA (1986) Anatomy and physiology of saccadic burst neurons in the alert squirrel monkey. I. Excitatory burst neurons. J Comp Neurol 249:337
5. Hering E (1868) The theory of binocular vision, 1977 edn., Plenum, New York

N

6. Moschovakis AK, Scudder CA, Highstein SM (1990) A morphological basis for Hering's law: projections to extraocular motoneurons. Science Wash DC 248:1118
7. Baker R, Highstein SM (1975) Physiological identification of interneurons and motoneurons in the abducens nucleus. Brain Res 91:292
8. Strassman A, Highstein SM, McCrea RA (1986) Anatomy and physiology of saccadic burst neurons in the alert squirrel monkey. II. Inhibitory burst neurons. J Comp Neurol 249:358
9. Moschovakis AK, Scudder CA, Highstein SM, Warren JD (1991) Structure of the primate burst generator. II. Medium-lead burst neurons with downward on-directions. J Neurophysiol 65:218
10. Scudder CA (1988) A new local feedback model of the saccadic burst generator. J Neurophysiol 59:1455
11. Moschovakis AK (1997) The neural integrators of the mammalian saccadic system. Front Biosci 2:552
12. Hall WC, Moschovakis AK (eds) (2003) Techniques in neuroscience, the superior colliculus: new approaches for studying sensorimotor integration. CRC Press, Boca Raton
13. Sparks DL (1978) Functional properties of neurons in the monkey superior colliculus: coupling of neuronal activity and saccade onset. Brain Res 156:1
14. Sparks DL, Holland R, Guthrie BL (1976) Size and distribution of movement fields in the monkey superior colliculus. Brain Res 113:21
15. Mohler CW, Wurtz RH (1976) Organization of monkey superior colliculus: intermediate layer cells discharging before eye movements. J Neurophysiol 39:722
16. Moschovakis AK, Karabelas AB, Highstein SM (1988) Structure-function relationships in the primate superior colliculus. II. Morphological identity of presaccadic neurons. J Neurophysiol 60:263
17. Moschovakis AK, Kitama T, Dalezios Y, Petit J, Brandi AM, Grantyn AA (1998) An anatomical substrate for the spatiotemporal transformation. J Neurosci 18:10219
18. Bozis A, Moschovakis AK (1998) Neural network simulations of the primate oculomotor system. III. A one-dimensional one-directional model of the superior colliculus. Biol Cybern 79:215
19. Schiller PH, Koerner F (1971) Discharge characteristics of single units in the superior colliculus of the alert rhesus monkey. J Neurophysiol 34:920
20. Moschovakis AK, Karabelas AB, Highstein SM (1988) Structure-function relationships in the primate superior colliculus. I. Morphological classification of efferent neurons. J Neurophysiol 60:232
21. Mays LE, Sparks DL (1980) Dissociation of visual and saccade-related responses in superior colliculus neurons. J Neurophysiol 43:207
22. Grantyn A, Berthoz A (1985) Burst activity of identified tectoreticulo-spinal neurons in the alert cat. Exp Brain Res 57:417
23. Scudder CA, Moschovakis AK, Karabelas AB, Highstein SM (1996) Anatomy and physiology of saccadic long-lead burst neurons recorded in the alert squirrel monkey. II. Pontine neurons. J Neurophysiol 76:353
24. Grantyn A, Ong-Meang J, Berthoz A (1987) Reticulospinal neurons participating in the control of synergic eye and head movements during orienting in the cat. II. Morphological properties as revealed by intra-axonal injections of horseradish peroxidase. Exp Brain Res 66:355
25. Ron S, Robinson DA (1973) Eye movements evoked by cerebellar stimulation in the alert monkey. J Neurophysiol 36:1004
26. Fuchs AF, Robinson FR, Straube A (1993) Role of the caudal fastigial nucleus in saccade generation. I. Neuronal discharge patterns. J Neurophysiol 70:1723
27. Helmchen C, Büttner U (1995) Saccade-related Purkinje cell activity in the oculomotor vermis during spontaneous eye movements in light and darkness. Exp Brain Res 103:198–208, 103:198
28. Fuchs AF, Robinson FR, Straube A (1994) Participation of the caudal fastigial nucleus in smooth-pursuit eye movements. I. Neuronal activity. J Neurophysiol 72:2714
29. Krauzlis RJ, Miles FA (1998) Role of the oculomotor vermis in generating pursuit and saccades: Effects of microstimulation. J Neurophysiol 80:2046
30. Zee DS, Yamazaki A, Butler PH, Gücer G (1981) Effect of ablation of flocculus and paraflocculus on eye movements in primate. J Neurophysiol 46:878
31. Lhermitte F (1958) Le syndrome cerebelleux: etude anatomo-clinique chez l'adulte. Rev Neurol 98:435
32. Optican LM, Robinson DA (1980) Cerebellar-dependent adaptive control of the primate saccadic system. J Neurophysiol 44:1058
33. Moschovakis AK, Gregoriou GG, Ugolini G, Doldan M, Graf W, Guldin W, Hadjidimitrakis K, Savaki HE (2004) Oculomotor areas of the primate frontal lobes: a transneuronal transfer of rabies virus and [14C]-2-deoxyglucose functional imaging study. J Neurosci 24
34. Bizzi E (1968) Discharge of frontal eye field neurons during saccadic and following eye movements in unanesthetized monkeys. Exp Brain Res 6:69
35. Robinson DA, Fuchs AF (1969) Eye movements evoked by stimulation of frontal eye fields. J Neurophysiol 32:637
36. Gamlin PD, Yoon K (2000) An area for vergence eye movement in primate frontal cortex. Nature 407:1003
37. Dürsteller MR, Wurtz RH, Newsome WT (1987) Directional pursuit deficits following lesions of the foveal representation within the superior temporal sulcus of the macaque monkey. J Neurophysiol 57:1262

Neural Control of the Lower Urinary Tract

►Micturition, Neurogenic Control

Neural Control of Voiding

►Micturition, Neurogenic Control

Neural Correlates of Imprinting

Anna Katharina Braun
Otto von Guericke University Magdeburg, Institute of Biology/Faculty for Natural Sciences, c/o Leibniz Institute of Neurobiology, Magdeburg, Germany

Definition

The term imprinting, derived from Oskar Heinroth and Konrad Lorenz, defines rapid learning events with remarkably stable and long-lasting behavioral outcomes, which occur during specific time windows or sensitive periods in newborn and juvenile vertebrates. The term imprinting is used not only by ►ethologists, but also in more recent literature by psychologists and clinicians in a broader sense, implying a particular ►aetiology of adult behaviors in both animals and human beings. Darwin, who showed great interest in child development, including infancy learning, was one of the first scientists who tried to link child psychology and animal behavior. He thereby prepared the ground for the study of the interplay of instinct and early learning, soon to be taken up by Spalding, and later, in different ways, by both Freud and the ethologists. D. A. Spalding published in 1873 a paper entitled "Instinct, with Original Observations on Young Animals," in which he described the behavior of young domestic chicks. He observed that these animals' ability to recognize the parents is not instinctive, but is in fact learned. Spalding also reported that this learning was confined to a short period soon after hatching, and we now say that during such a critical or sensitive period a chick becomes imprinted to the mother-figure when it learns her characteristics, and forms an emotional bond to her. Heinroth described this type of ►attachment-behavior by the verb "einprägen," corresponding to the English "to stamp in" or "to imprint"; and the word "stamped in" had earlier been used by, among others, Spalding and Thorndike in relation to firmly acquired modes of behavior. Lorenz used the noun Prägung, or imprinting, in his seminal paper in 1935, to refer to the process of rapid bound-formation early in the life of the so-called ►nidifugous birds (e.g., fowl). Accordingly, imprinting occurs when an animal learns to recognize a stimulus that will later release instinctive behavior. For example, a gosling follows the first moving object it sees after hatching. The young goose "recognizes" the moving object as its parent and it later "recognizes" similar objects as members of its own species. Lorenz went further than Spalding in that he claimed that imprinting differed fundamentally from, what he called, ordinary learning, because of its restriction to a brief critical period in the individual's life, its rapid occurrence, and its irreversibility. It was later questioned whether these features would separate imprinting sharply from other forms of learning, and the modern view of imprinting is that it might be continuous with, or be a form of, ►conditioning.

Imprinting can be considered to be an evolutionary old concept of juvenile learning, which is critical for the survival and fitness of the offspring and which occurs in many species including man. Imprinting entails much more plastic mechanisms than were claimed by Lorenz, thus, imprinting is a phenomenon that is of great interest to ontogenetic studies of animal behavior and it has important implications for human developmental psychology and psychopathology. Imprinting is an important aspect of early learning not only in birds but also in ►precocial mammals (e.g., guinea pigs, ►degus, sheep, horses, etc.), and it also plays an important part in the socialization of ►altricial species (e.g., dogs or monkeys) including human and non-human primates. Newborn zebra foals follow any object near to them during the first few days of life, and because of this their mothers show aggressive behavior towards any other zebras that come too close. Many animals, e.g., giraffes, give birth to their young in isolation away from other family members to prevent the newborn from confusing them with its mother, and they only return to the family unit when imprinting has taken place, which usually occurs after three days. Animals can also imprint on humans, and this could be one of the reasons why many captive animals in zoos fail to breed, and similarly, cross fostering in the wild can also lead to fostered animals failing to breed. It appears likely that this failure is due to "faulty" sexual imprinting, i.e., another form of juvenile learning during which the preference of a mate later in life is determined. The sensitive period for sexual imprinting is dependent on the species and lifespan of the animal. Lions, for example, live in prides and form their sexual preference through community bonding over an extended period of time, whereas an animal with a shorter life span, such as insects, need to find a mate quickly and therefore have a shorter sensitive period for sexual imprinting. A third example of imprinting is vocal learning in songbirds (see essay by M. Gahr), and, at least in part, speech learning in humans. The critical period hypothesis of language acquisition claims that a sensitive period exists from birth to puberty, and that during this time window the brain is receptive to language, and is learning rules of grammar quickly. After puberty, language learning becomes more difficult, which is attributed to a drastic change in the way the brain processes language after puberty. The common feature of all forms of imprinting is that "templates" or "concepts" about the species, vocal repertoire/language or behavioral strategies are shaped during early life periods, which are incorporated into adult behaviors.

Characteristics

Higher Level Structure

The brain circuits that are involved in filial imprinting (compare Fig. 1) include the forebrain structures medio-rostral nidopallium/mesopallium MNM (according to the old nomenclature this area was formerly termed medio-rostral neostriatum/hyperstriatum ventrale, MNH), i.e., the homologue of the vocal motor nucleus medialis magnocellularis nidopallii anterioris, MMAN (formerly termed medial nucleus of the anterior neostriatum) and the presumed avian analogue of the mammalian prefrontal cortex/cingulate cortex [1] as well as the intermediate medial mesopallium IMM (which in the old avian nomenclature was termed intermediate hyper-striatum ventrale IMHV), and the nidopallium dorsocau-dale Ndc (which in the old nomenclature was the dorso-caudal neostriatum Ndc), i.e., the presumed avian analogue of the mammalian polysensory association cortex [1].

This brain pathway can be considered to represent the avian analogue of the mammalian ►limbic system, which not only plays a critical role in filial imprinting in the young, but which is also the essential pathway for the regulation of emotional behavior, and in learning and memory formation in the adult.

Lower Level Structure

Whereas the basic wiring diagram of the brain is genetically pre-programmed, its fine tuning throughout different phases of infancy, childhood, and adulthood is highly experience dependent. Normal brain development requires the precise interactions of environmental signals with genes and molecules that drive cellular differentiation and circuit formation. Imprinting, i.e., juvenile learning is "used" to fine tune functional pathways, e.g., sensory, motor as well as limbic circuits, in the brain. This view is supported by experimental studies that have demonstrated that filial imprinting, and most likely also song imprinting, results in the "pre-formatting" of functional brain circuits and thereby determines the behavioral outcome in later life. Whereas during song imprinting a relatively precise

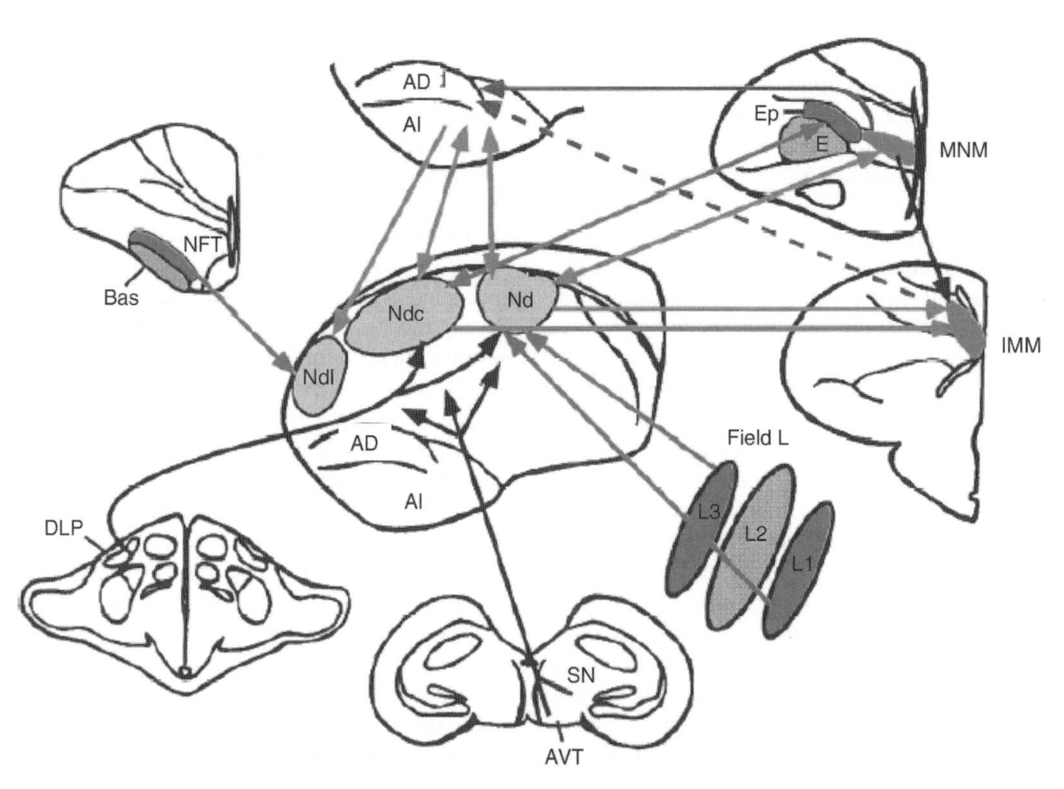

Neural Correlates of Imprinting. Figure 1 Schematic diagram of the brain circuits for filial imprinting in birds including the regions for sensory processing. Modified after Metzger et al 1989 J. Comp. Neurol. 395: 380–404. The belt regions (red) of primary sensory areas (light blue) send projections (red arrows) to distinct areas (light brown) of the dorsocaudal nidopallial complex (dNC) complex. From there, information is conveyed (green arrows) either to the arcopallium intermedium (AI) or to the medio-rostral nidopallium/mesopallium MNM; (green) and the intermediate medial mesopallium IMM; (green). The MNM and IMM project to the AId (blue arrows). Black arrows indicate other major pathways defined in the present study. Note that minor reciprocal connections between the MNM and nidopallium dorsocaudale Ndc are not included here. The dashed blue arrow from the IMM to the archistriatum intermedium, pars dorsale (Aid) indicates a projection defined by Csillag et al. (1994) J Comp Neurol 348:394.

acoustic template is stored, the long-term consequence of filial imprinting, for example, is not so much to form memories for specific details of the imprinting object, the mother, but to establish the "grammar" of social, emotional and cognitive behavioral strategies. On the "hardware" side, imprinting could be viewed as formatting the harddrive, if we compare the brain with a computer, and this will determine its capacity to operate the "software" i.e., behavioral strategies, cognitive and emotional concepts etc. later in life. In fact, there is considerable experimental evidence from a variety of studies on filial imprinting and song learning (cp also essay by M. Gahr) in support of this view, which have shown that such juvenile learning events are accompanied by massive molecular-genetic, neurochemical, physiological and structural changes of neurons and their synaptic circuits.

Structural Regulation

Imprinting is a rapid, efficient and rather stable form of learning that occurs in a more or less immature brain, i.e., in a brain that is still growing. The characteristic features of imprinting (speed and stability) might be explained by the fact that it occurs during developmentally "critical" or sensitive time windows of elevated cellular plasticity. Thus, imprinting takes advantage of the entire spectrum of the molecular, genetic and metabolic cellular cascades, which are highly activated during early postnatal brain development. For visual imprinting, a "time course" of a variety of changes have been described in the chick forebrain region intermediate medial mesopallium, IMM, the earliest changes involve the induction of the ►immediate early gene c-fos, followed by changes in ►phosphorylation of the ►protein kinase C substrate MARCKS, morphological changes in ►axospinous synapses, an increase in ►NMDA receptor number and increases in neural ►cell adhesion molecules [2]. The initial phase of acoustic imprinting is characterized by the activation of the immediate early genes and ►transcription factors zenk and arg3.1 in the chick forebrain regions MNM and the IMM [3]. Quite comparably, during the initial phase of sexual imprinting in male zebra finches, i.e., the first exposure to a female after an isolation period, enhanced zenk expression was measured in a variety of brain areas including lateral nidopallium/mesopallium (formerly termed lateral neostriatum/hyperstriatum LNH), MNM, and the optic tectum. At later stages of acoustic imprinting, an enhanced release of ►glutamate in response to the learned acoustic imprinting stimulus has been observed in the MNM of domestic chicks, which is paralleled by an increased neuronal activation during the presentation of the acoustic imprinting stimulus [4]. Similarly enhanced neuronal responses have been reported in the IMM after visual imprinting [2]. Results from in vivo ►microdialysis experiments in domestic chicks suggest that the emotional bond that develops through auditory imprinting entails an addictive process, which is mediated by the release of ►endorphins and altered ►dopamine and ►serotonin release [5]. Functional imaging revealed elevated metabolic activation in the MNM and Ndc areas of successfully imprinted domestic and guinea chicks in response to the acoustic imprinting stimulus [6]. Quite similarly, young rodents display activations in the anterior cingulate cortex in response to the presentation of the learned acoustic stimulus (maternal vocalizations). Such stimulus-evoked neuronal responses, which occur during filial imprinting, in particular those mediated via the glutamatergic NMDA receptor, appear to trigger the long-term structural changes. There is convincing evidence from several series of experimental studies that acoustic imprinting is linked to a synaptic selection process [7] that serves the fine-tuning of limbic circuits. Successfully imprinted chicks, but not chicks that have been passively, i.e., without the chance to form an association, exposed to the identical acoustic stimulus, display reduced densities of excitatory ►spine synapses in higher associative brain regions (MNM, Ndc) compared to naïve, "Kaspar-Hauser" control animals. Interestingly, during sexual imprinting in zebra finches similar metabolic changes, activations of immediate early genes and reductions of dendritic spines were reported in the MNM and Ndc, and the vocal motor nucleus MMAN of zebra finches undergoes successive pruning of spine synapses during song imprinting [8]. Thus, in contrast to the adult brain, where primarily increased numbers of synaptic connections have been reported after learning [7], imprinting appears to leave its "footprints" within the juvenile brain mainly by using the opposite plasticity mechanism, namely the pruning of synapses. These imprinting-induced synaptic changes may be compared to a sculptor, who creates a statue by removing material from an unshaped marble block, and it may reflect the formatting of the "harddrive." In particular, the long-term structural changes may explain the speed and stability of the memory for a learned imprinting object, stimulus or behavioral strategy.

Function

All animals perform both instinctive actions and learned actions. Instinct almost completely determines the behavior of most insects, spiders, and crustaceans. These animals can learn comparably little, and therefore their survival depends mainly on innate behavioral patterns. Higher animals, including fish, amphibians, reptiles, birds, and mammals can learn more, and they can also modify their instinctive behavior by learning. The primary function of imprinting is to enable the young animal to shape and adapt its behavior, and to optimize the underlying brain circuits (in particular the limbic system) for the behavioral output within its environment.

Pathology

The most fascinating aspect of imprinting is the timely establishment of a particular preference over the course of the animal's development, which is particularly evident in primates, including humans, where "critical periods" develop from neonatal to early childhood, through juvenile to adulthood. For instance, monkeys who are reared separately from all other monkeys will not develop normal social or sexual behavior if, after reaching adulthood, they are placed with other monkeys. As pointed out earlier, brain development requires adequate sensory and also emotional stimulation, provided during filial imprinting, to develop and maintain synaptic connections and to fine tune synaptic circuits. Functional imaging in young rodents revealed that acute separation from the parents induces a massive downregulation of brain metabolism in most limbic regions as well as in sensory and polysensory cortices, whereas the presentation of a learned acoustic imprinting stimulus (the mother's voice) induces metabolic activation in the limbic cortex [9] (Braun and Scheich 1997). Long-term consequences of repeated separation from the parents (causing the disruption of the imprinting process?), chronic social isolation and early weaning are changes of excitatory as well as monoaminergic modulatory synaptic inputs in cortical and subcortical limbic areas. In the rodent anterior cingulate cortex, orbitofrontal cortex, hippocampus, and amygdala region-, cell-, and dendrite-specific changes of synaptic densities [10] have been found in parentally deprived animals. These observations confirm that the prevention or disruption of filial imprinting can affect synaptic development in the prefrontal cortex and other limbic areas. Since the limbic system is critical for a variety of emotional behaviors and associative aspects of learning, such experience-induced morphological changes may lead to altered behavioral and cognitive capacities in later life.

Therapy

Animal studies unveiled the remarkable influence of the parent-infant contact (and most likely also other imprinting-like juvenile learning events) on brain development, in particular on the functional maturation of the limbic pathway. Therefore, learning early in life has always been of great interest on the one hand to all involved in child care and education, but also with indoctrination, and on the other hand to clinical research in child and adult psychiatry, where imprinting also fits into Freud's and other's conception of the aetiology of neurotic symptoms and other mental disorders. The detailed knowledge of the neurobiology of such self-organizing plastic systems may begin to change our conceptual approaches to psychopathology, and open new avenues of therapeutics for the major psychiatric illnesses that are critically dependent on such learning and memory mechanisms. Furthermore, the knowledge of the basic principles of learning- and memory-related neuronal plasticity may in the future be applied to innovative educational concepts for the preschool/elementary school levels.

References

1. Metzger M, Jiang Sh, Wang J, Braun K (1996) Organization of the dopaminergic innervation of forebrain areas relevant to learning: a combined immunohistochemical/retrograde tracing study in the domestic chick. J Comp Neurol 276:1–27
2. Horn G (2004) Pathways of the past: the imprint of memory. Nat Rev Neurosci 5:108–120
3. Bock J, Thode C, Hannemann O, Braun K, Darlison M (2005) Early socioemotional experience induces expression of the immediate-early gene *Arclarg 3.1* in learning-relevant brain regions of newborn chicks. Neurosci 133:625–633
4. Gruss M, Braun K (1996) Stimulus-evoked glutamate in the medio-rostral neostriatum/hyperstriatum ventrale of domestic chick after auditory filial imprinting: an in vivo microdialysis study. J Neurochem 66:1167–1173
5. Baldauf K, Braun K, Gruß M (2005) Opiate modulation of monoamines in the chick forebrain: possible role in emotional regulation? J Neurobiol 62:149–163
6. Maier V, Scheich H (1983) Acoustic imprinting leads to differential 2-deoxy-D-glucose uptake in the chick forebrain. Proc Natl Acad Sci USA 80(12):3860–3864
7. Scheich H, Wallhäußer-Franke E, Braun K (1991) Does synaptic selection explain auditory imprinting? In: Squire LR, Weinberger NM, Lynch G, McGaugh JL (eds) Memory: organization and locus of change. Oxford University Press, New York, pp 114–159
8. Nixdorf-Bergweiler BE, Wallhausser-Franke E, DeVoogd TJ (1995) Regressive development in neuronal structure during song learning in birds. J Neurobiol 27:204–215
9. Poeggel G, Braun K (1996) Early auditory filial learning in degus (Octodon degus): behavioural and autoradiographic studies. Brain Res 743:162–170
10. Poeggel G, Helmeke C, Abraham A, Schwabe T, Friedrich P, Braun K (2003) Juvenile emotional experience alters synaptic composition in the rodent cortex, hippocampus and lateral amygdala. Proc Natl Acad Sci USA 100:16137–16142

Neural Crest

Definition

The neural crest is derived from embryonic ectoderm. It is located in the small space between the dorsolateral part of the neural tube and the overlying ectoderm, extending from the diencephalon to the tail. Cells of the neural crest give rise to glial cells, neurons of sensory, sympathetic, and parasympathetic ganglia, chromaffin cells, and melanocytes.

►Neural Crest and the Craniofacial Development
►Neural Development
►Neural Tube

Neural Crest and the Craniofacial Development

SHIGERU KURATANI
Laboratory for Evolutionary Morphology, Center for Developmental Biology, Riken, Kobe, Japan

Definition

An epithelial neural ridge developing on either side of the neural plate of the early vertebrate embryo. The crest is secondarily de-epithelialized becoming migratory mesenchymal cells called "neural crest cells" (Fig. 1). These move in a stereotypical manner within the embryo following certain specific migratory pathways to be distributed in various locations within the embryo and differentiate into various cell and tissue types, including peripheral neurons, supporting tissues of the peripheral nervous system and some skeletal muscles, endocrinal cells (or many of the cell types belonging to the category of "paraneurons"), pigment cells, smooth muscle cells associated with arteries, cartilage and bones.

Characteristics

Embryonic environments are crucial for the differentiation of the crest cells. Skeletogenic differentiation especially, is specific to the neural crest cells in the region ranging from the head to the neck [2]. Here the cells form an extensive mesenchyme, often called "ectomesenchyme" (meaning the "mesenchyme" derived from the ectoderm; Fig. 1). Thus the neural crest is generally divided into "cephalic" and "trunk" neural crest, depending on the differences in migration pathway, distribution pattern and repertoire of differentiating cell types. The boundary between these two categories, is not clear. Between these two regions, a third population of crest cells, called "vagal crest," has been inferred, corresponding to the region of the neural crest that gives rise to enteric neurons [2]. This part also includes the crest that contributes to the septation of the outflow tract of the amniote heart primordia and is thus alternatively called the "cardiac crest" [3].

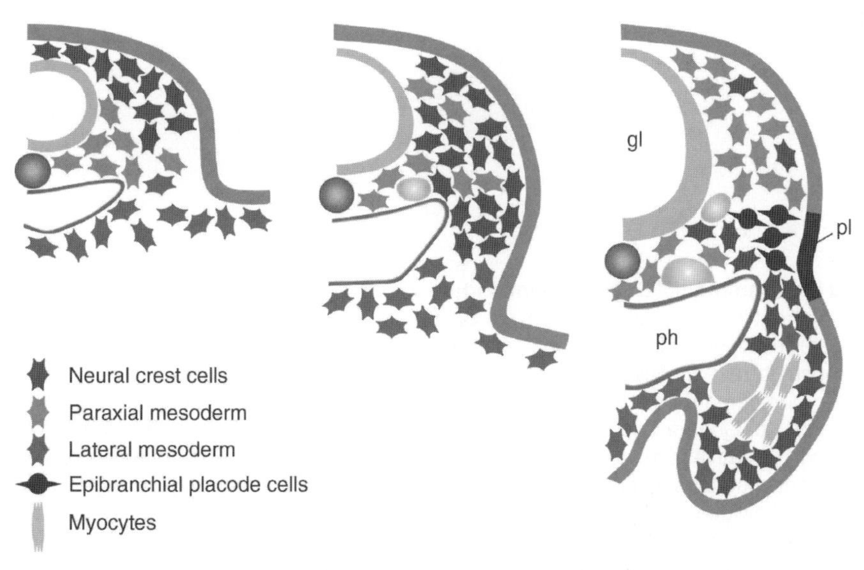

Neural Crest and the Craniofacial Development. Figure 1 Schematic representation of three successive stages of neural crest migration in the head of chicken embryo. The neural crest cells (*blue*) migrate laterally and ventrally from the dorsal aspect of the neural tube to the pharyngeal arch to form an extensive ectomesenchyme. The pharyngeal arch ectomesenchyme will later differentiate into branchial arch skeletons. Note that during crest cell migration, cephalic paraxial mesoderm (*green*) barely changes its position. Also note that there is another group of cells derived from the ectoderm (epibranchial placodes). The placode-derived cells and neural crest cells together form a cranial sensory ganglion. Based on [1].

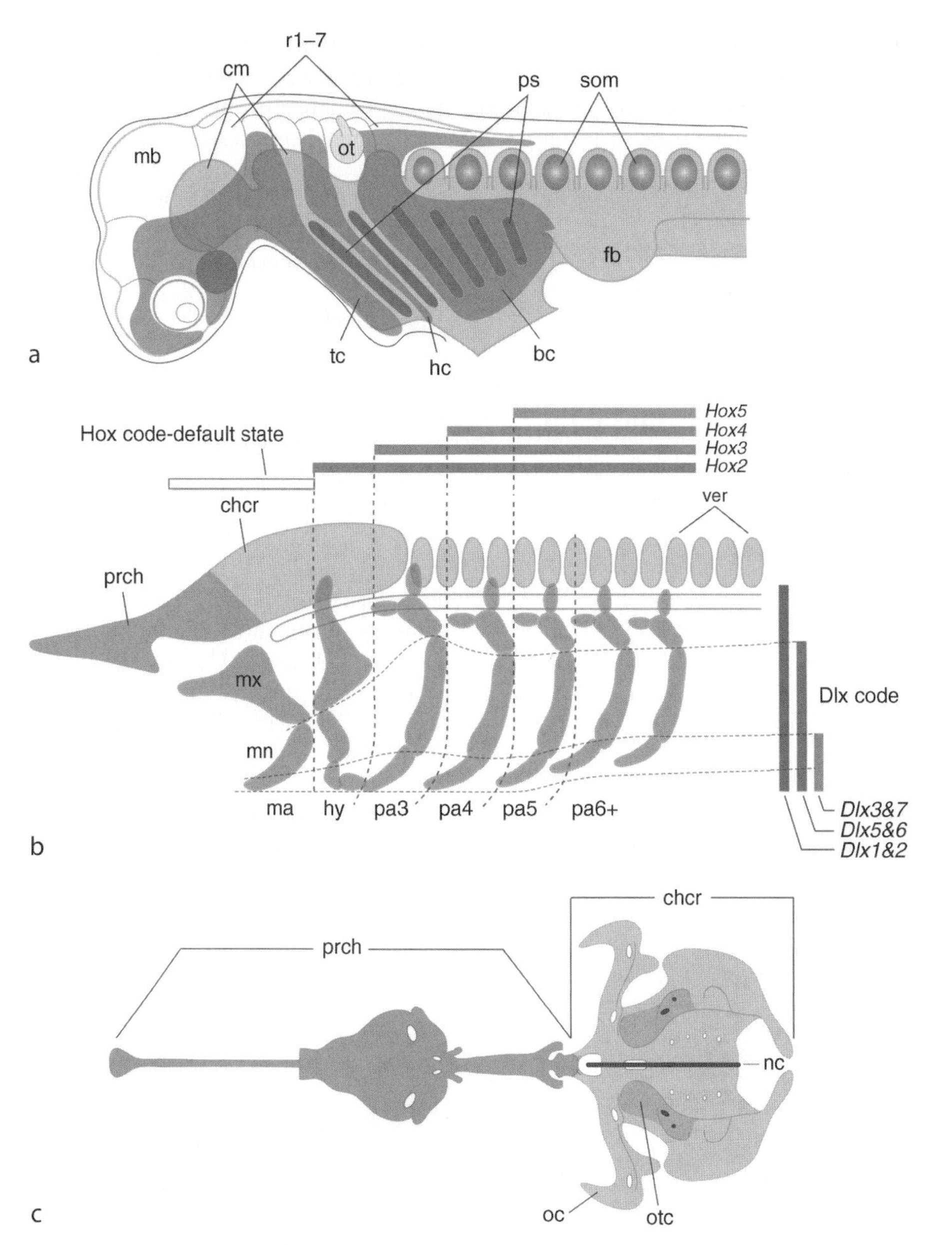

Neural Crest and the Craniofacial Development. Figure 2 Ectomesenchyme in the gnathostome pharyngula, and craniofacial development in the vertebrate head. (a) Schematized model based on a shark embryo. Cephalic crest cells (green) form three distinct cell populations called, from anterior to posterior, the trigeminal crest- (*tc*), hyoid crest- (*hc*) and branchial crest- (*bc*) cells. These cells are located laterally to the cephalic mesoderm (*cm*), proximally attached onto even-numbered rhombomeres (r1–7), separated distally by pharyngeal slits (*ps*) to be distributed in each pharyngeal arch. (b) Generalized vertebrate chondrocranium based on the morphology of a sturgeon larva. Putative mesodermal derivatives are colored *gray*, and crest-derivatives *green*. The chondrocranium consists of dual anatomical components, the dorsal neurocranium and the ventral viscerocranium (pharyngeal arch skeletons). The neurocranium is further divided anteroposteriorly into the crest-derived prechordal (*prch*) and mesodermal chordal (*chcr*) parts. Position-dependent specification of the arch skeleton is based on the Cartesian grid of homeobox gene expression domains, consisting of Hox code along the anteroposterior axis and Dlx code along the dorsoventral axis. Note that the mandibular arch is patterned by the absence of Hox transcripts. (c) Mesenchymal origin of chicken chondrocranium. The scheme *B* is largely based on this mapping performed by Couly et al [6]. Abbreviations: *fb*, fin bud; *mb*, midbrain; *nc*, notochord; *oc*, orbital cartilage; *ot*, otocyst; *otc*, otic capsule; *som*, somites; *ver*, primordial vertebrae.

The pluripotency of the crest cells as well as their highly sophisticated morphogenetic capability are defining features of the neural crest. Although non-vertebrate chordates, including amphioxus and tunicates, share the same basic body plan as vertebrates, for example the notochord and the dorsal neural tube, they do not appear to possess the neural crest or crest-derivatives. In this context, the origin of the neural crest has recently drawn the interest of biologists investigating the evolutionary origin of vertebrates. Vertebrates are characterized by the possession of an overt head with well-developed sensory organs, brain and cranium, associated with vertebrate-specific cell lineages derived from neural crest and ►placodes (both sensory and ganglionic, Fig. 1). The theory of "New Head" by Gans and Northcutt [4] regards the acquisition of placodes and the neural crest as a key innovation that has permitted the development of the vertebrate head. The most recent study, however, has implied the presence of crest-like cells in tunicate larva, although this animal does not develop an extensive ectomesenchyme that characterizes the true vertebrates [5]. Consequently the evolutionary relationship may not be so simple.

Crest-derived cephalic ectomesenchyme is predominantly found in the ventral portion of the head, subdivided into ►pharyngeal arches [1] (Fig. 2a).

This mesenchyme will differentiate into skeletal and connective tissues, that is, the craniofacial and branchial arch skeletal elements, whether they are bones or cartilages, are all of neural crest origin (Fig. 2b). The rostral part of the ►neurocranium is also of crest origin, unlike the more caudal part that is derived from the cephalic mesoderm and somites [6] (Fig. 2c). There still remain controversies as to the origin of the dermal calvarium and there is a possibility that these dermal bones may have different origins of cell lineages regardless of their morphological homology. Nevertheless, the rostral elements such as the amniote frontal and nasal bones are unanimously thought to have a crest origin.

The crest-derived vertebrate cranial skeleton offers a model for the understanding of the morphological specification of the skeletal system, both in development and evolution. Of particular interest is the developmental mechanism that provides positional values to each part of the ectomesenchyme through the spatially organized expression of homeobox genes. One well-known example is the Hox code, the nested pattern of *Hox* gene expression along the anterior–posterior axis of the head. In the pharyngula embryo, no *Hox* genes are expressed in the rostral part of the head including the mandibular arch, *Hox2* genes are expressed in the second (hyoid) and more posterior arches, *Hox3* genes in arch 3 and posteriorly (Fig. 2b). *Hox* genes are tandemly assembled on the DNA as a cluster, with *Hox1* toward the 3′ end and *Hox13* toward the 5′ ends of the cluster. Thus the order of the *Hox* genes roughly coincides with the axial levels of the embryo at which the genes are expressed (colinearity), giving an impression that the genome plays a role as a blueprint for embryonic development [7] (Fig. 2b).

Each *Hox* gene contains a domain called a "homeobox" encoding a DNA-binding protein that recognizes a specific DNA sequence. Thus the Hox proteins can act as a "switch" to regulate specific target genes and the Hox code provides positional values or "identity of pharyngeal arches" to the ectomesenchyme [7]. Consistent with this idea, disruption of *Hoxa-2*, the gene expressed in the hyoid arch, leads to the homeotic transformation of this arch into the identity of the

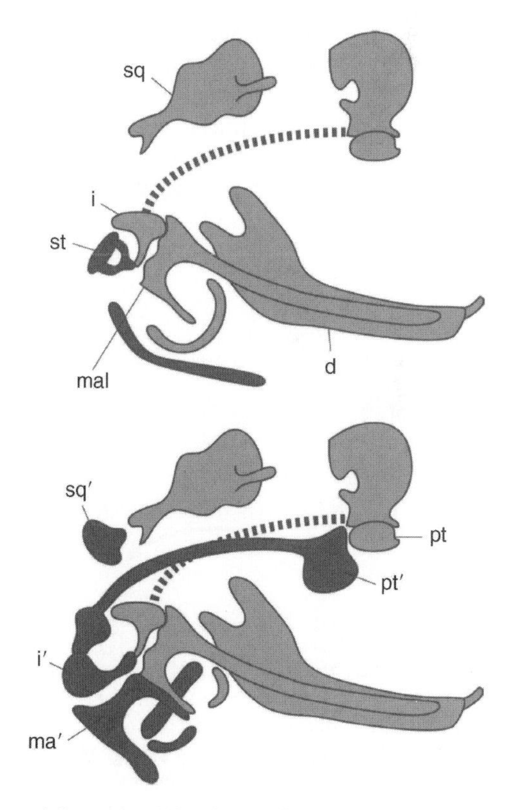

Neural Crest and the Craniofacial Development. Figure 3 Disruption of *Hoxa-2* in the mouse. *Top*. Wild type morphology. Mandibular arch-derivatives are colored *blue* and the hyoid arch-derivatives *red*. In the *Hoxa-2*-mutant mouse, the morphological identities of the hyoid arch derivatives are transformed into those of the mandibular, resulting in a mirror-image duplication of mandibular arch skeletons. Based on [8]. Abbreviations: *d*, dentary; *i*, incus; *mal*, malleus; *pt*, pterygoid; *sq*, squamosal; *st*, stapes.

mandibular, presumed to be a "Hox-code-default state" (Figs. 2b and 3) [8].

The Hox code appears to be conserved in all vertebrates. Even the invention of the jaw seems to have taken place in the mandibular arch based on the same basic code, since the lamprey also seems to have *Hox*-negative mandibular arch [9].

Nested expression of another set of homeobox containing genes, the Dlx code, has been recognized in the head ectomesenchyme of the mouse [10] (Fig. 2b). This code is thought to direct the dorsoventral pattern of each arch by the dorsoventrally nested expression of the *Dlx* genes. Specifically, *Dlx1* and *Dlx2* are expressed in the entire arch ectomesenchyme, *Dlx5* and *Dlx6* in the ventral half and *Dlx3* and *Dlx7* in the ventral tip. Thus the Cartesian grid pattern of homeobox gene expression as a whole is the basis of the skeletal patterning of the vertebrate head (Fig. 2b). To establish such a pattern, epithelial–mesenchymal interaction may play fundamental roles, since information on which skeletal identities the ectomesenchyme is to acquire is not necessarily predetermined in the premigratory crest, but appears to be instructed secondarily by the rostral endoderm of the embryo. The actual shape of the skeletal elements however may rather be obtained through downstream developmental pathways mainly exerted cell-autonomously in the crest cells themselves, since interspecific grafting of cephalic crest results in the host gaining the craniofacial morphology of the donor animal [1] (Fig. 4). The neural crest and its derivatives thus characterize most conspicuously the body plan of vertebrates and its evolution.

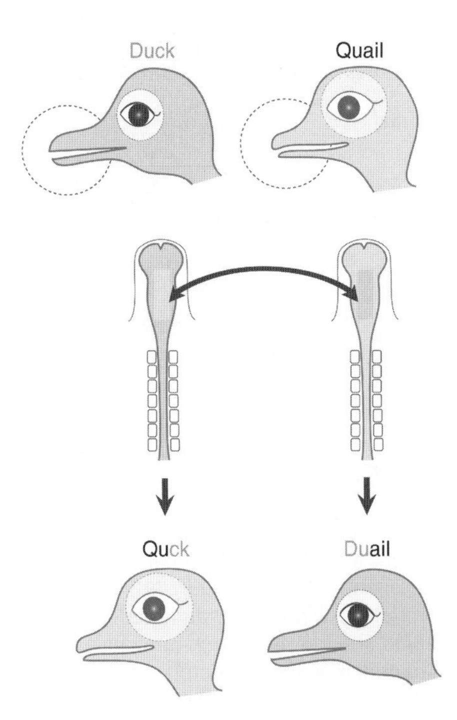

Neural Crest and the Craniofacial Development. Figure 4 Species-specific shape of face? The cephalic neural crest was transplanted between embryos of quail and duck, birds with distinct faces. The cranial shape of the chimera resembles the donor of the crest more than that of the host. Based on [1].

References

1. Kuratani S (2005) Craniofacial development and evolution in vertebrates: the old problems on a new background. Zool Sci 22:1–19
2. Le Douarin NM (1982) The neural crest. Cambridge University Press, Cambridge
3. Kirby ML, Waldo KL (1990) Role of neural crest in congenital heart disease. Circulation 82:332–340
4. Gans C, Northcutt RG (1983) Neural crest and the origin of vertebrates: a new head. Science 220:268–274
5. Jeffery WR, Strickler AG, Yamamoto Y (2004) Migratory neural crest-like cells form body pigmentation in a urochordate embryo. Nature 431:696–699
6. Couly GF, Coltey PM, Le Douarin NM (1993) The triple origin of skull in higher vertebrates: a study in quail-chick chimeras. Development 117:409–429
7. Hunt P, Gulisano M, Cook M, Sham MH, Faiella A, Wilkinson D, Boncinelli E, Krumlauf R (1991) A distinct Hox code for the branchial region of the vertebrate head. Nature 353:861–864
8. Rijli FM, Mark M, Lakkaraju S, Dierich A, Dollé P, Chambon P (1993) Homeotic transformation is generated in the rostral branchial region of the head by disruption of *Hoxa-2*, which acts as a selector gene. Cell 75:1333–1349
9. Takio Y, Pasqualetti M, Kuraku S, Hirano S, Rijli FM, Kuratani S (2004) Lamprey *Hox* genes and the evolution of jaws. Nature 429:1 p following 262. http://www.nature.com/cgi-taf/DynaPage.taf?file=/nature/journal/v429/n6989/full/nature02616_fs.html
10. Depew MJ, Lufkin T, Rubenstein JL (2002) Specification of jaw subdivisions by *Dlx* genes. Science 298:371–373

Neural Crest Cells

Definition

Unique to vertebrates, these cells are derived from the edges of the folds of the neural plate. They migrate to different regions of the body and give rise to a wide variety of tissues, including autonomic and sensory nerves, pigment cells and some cartilage in the head.

►Neural Development

Neural Development

FUJIO MURAKAMI
Laboratory of Neuroscience, Graduate School of Frontier Biosciences, Graduate School of Engineering Science, Osaka University, Suita, Osaka, Japan

Neural Induction

The formation of the nervous system follows ►the body plan, which is composed of the definition of body axes and the allocation of each organ into the body. A portion of dorsal ectoderm is induced to become ►neural ectoderm. This region is called the ►neural plate, which gives rise to the ►neural tube. The tube is formed as the cells surrounding the neural plate. The neural plate cells proliferate, invaginate and pinch off from the surface to form a hollow tube.

The induction of the neural plate occurs owing to signals emanating from the mesoderm of the ►organizer region. In amphibians, the cells that develop in the region of the gray crescent, a light-colored band that appears opposite the point where the sperm enters, migrate into the embryo during gastrulation and form the ►notochord. By utilizing transplantation experiments, Spemann and Mangold demonstrated that this region is capable of inducing a nervous system and called this region as "organizer" [1].

Several molecules that act as antagonists against bone morphogenic protein (BMP) are involved in ►neural induction. These include ►noggin, ►chordin and ►follistatin. These molecules bind to BMP4 suppressing its activity as an inhibitor of neural fate. Fibroblast ►growth factor (Ffg)s also seems to be involved in neural development.

Neural Tube Formation

Regionalization

Regionalization along the Rostrocaudal Axis

►Regionalization of the neural tube is associated with changes in the shape of the neural tube. Early in development, the neural tube differentiates into three distinct regions along the rostocaudal axis: ►prosencephalon, ►mesencephalon and ►rhombencephalon. Caudal to the rhombencephalon, the ►spinal cord differentiates. As development proceeds, the prosencephalon is divided into the ►telencephalon and ►diencephalon, and the rhombencephalon into the ►metencephalon and ►myelencephalon. The telencephalon eventually differentiates into the ►cerebral cortex, ►hippocampus and ►basal ganglia, the diencephalon into the ►thalamus and ►hypothalamus, the metencephalon into the ►cerebellum and the ►pons, and the myelencephalon differentiates into the ►medulla oblongata.

The anteroposterior patterning of the neural tissue is dominated by the head organizer and tail organizer. The former includes FGFs, Wnts and retinoic acids, and the latter includes Cerberus, which inhibits the activity of BMPs, Wnts and nodal-related molecules by strongly binding to these molecules.

Regionalization along the Dorsoventral Axis

The ►differentiation of the neural tube along the dorsoventral axis also takes place concurrent with differentiation along the rostrocaudal axis. In the spinal cord, distinct types of neurons differentiate depending on the position along the dorsoventral axis: Motor neurons, for example, differentiate in the ventral spinal cord and commissural neurons that receive sensory inputs differentiate in the dorsal spinal cord. The differentiation of distinct cell types in the ventral neural tube is dependent on inductive signals derived from axial midline of the notochord and the ►floor plate. The signal is mediated by the secreted protein ►Sonic Hedgehog (Shh) [2]. Neurons generated in progressively more ventral positions are exposed to higher concentration of Shh, which is required for their development.

The differentiation of the dorsal neural tube is also dependent on inductive signals but derived from ectodermal cells; one of the responsible molecules is BMP. Another secreted molecule, Wnt, also appears to function as a dorsalizing signal. These signals and Shh antagonize one another, creating opposing gradients along the dorsoventral axis, which induces expression of different ►transcription factors, ►PAX6 and PAX7 dorsally, and Nkx2.2 ventrally, for example. Expression of a transcription factor or a combination of different transcription factors gives rise to distinct types of neurons along the dorsoventral axis [3].

Secondary Induction from the Midbrain/hindbrain Boundary

The boundary region between the midbrain and hindbrain, the isthmus, has an inducing activity and contributes to the differentiation of the cerebellum and the ►tectum [32,33]. Some brainstem nuclei such as the ►substantia nigra and ►locus coeruleus are also under the influence of molecules released from the ►isthmic organizer. This region expresses several transcription factors such as Pax2, Pax5, Engrailed1 and Engrailed 2, and secreted factors such as FGF8 and Wnt1, which are thought to contribute to the differentiation of the structures adjacent to the isthmus [3]. These molecules are distributed in a graded manner. This, in turn, contributes to the graded expression of guidance molecules in the tectum.

Segmentation

Segmentation in the Hindbrain

The neural tube becomes segmented at some stage of the development. This is an important step for

N

the development of the central nervous system (CNS). The segmentation has been most extensively studied in the hindbrain. Developing hindbrain show 7–8 bulged structures called ►rhombomeres. Rhombomere boundaries are characterized by the expression of a paralogous set of ►Hox genes. Experiments such as the deletion and mis-expression of Hox genes indicate that these genes are involved in specification of rhombomeres. In each rhombomere, a unique set of motor neurons is generated, and these neurons project axons towards unique targets defined by their levels along the rostrocaudal axis. Similarly, sensory axons at different axial levels project to corresponding rhombomeres.

Forebrain

Segmentation of the neural tube can also be seen in the forebrain region. For example, a transcription factor *emx1* is expressed in the anterior half of the forebrain and *emx2* in the posterior half. The expression pattern of additional genes allows to define six rostrocaudally aligned compartments, called ►prosomeres, in the forebrain region (Fig. 1).

Patterning of the Neocortex

The cerebral cortex can be divided into functionally distinct areas such as the ►motor, sensory and visual cortex, which are differentially connected to subcortical structures. Cortical areas appear to be under the control of two transcriptional factors, *pax6* and *emx2*, which are distributed along the anteroposterior axis in a graded manner. The gradient of these genes appears to be regulated, in part, by *fgf8* [4].

Migration and Differentiaion of Neural Crest Cells

A specialized type of cells, called ►neural crest cells emerge from the dorsal-most region of the neural tube at the level of the hindbrain and the spinal cord. These cells migrate ventrally away from this region and differentiate into several types of neuronal and non-neuronal cells. This includes cells of the ►dorsal root ganglia, sensory ganglia of cranial nerves, ►sympathetic chain ganglia, pre-aortic ganglia, ►enteric ganglia, chromaffin cells of the adrenal medulla and melanocytes.

Differentiation and Proliferation of Neurons

Neurogenesis

►Neurogenesis takes place mostly during embryonic stages. Most neurons are generated from progenitors situated in the ventricular zone of the neural tube by symmetric and asymmetric divisions from progenitors, or multipotent neuroepithelial cells or neural stem cells. During the early stage of development, the cells in the

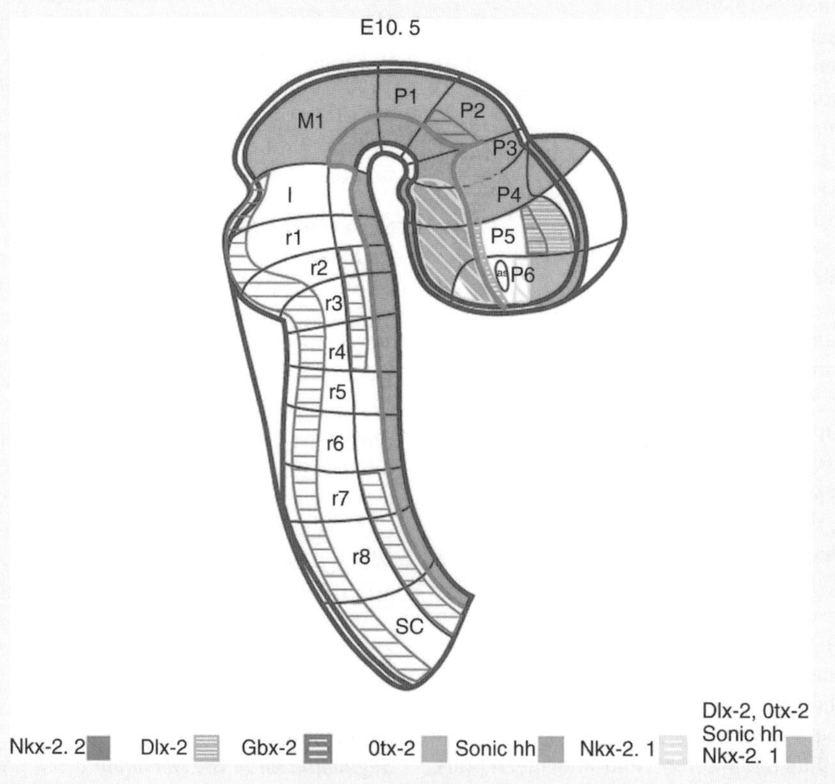

Neural Development. Figure 1 Expression patterns of five transcription factors (Dlx-2, Gbx-2, Nkx-2.1, Nkx-2.2, Otx-2) and one morphogen (sonic hedgehog) in E10.5 mouse neural tube. I, isthmus; M, mesencephalon-midbrain; os, optic stalk; p, prosomere; r, rhombomere; sc, spinal cord (modified from [34]).

ventricular zone extend their process to the pial surface and the nuclei of these cell undergo specialized interkinetic movement, in which the position of the nuclei depends on the phase of ►cell cycle: the nuclei accumulate the ventricular surface during M phase, but in the superficial margin of the neuroepithelium during S phase. Studies of neurogenesis in the neocortex revealed that the subventricular zone is not only the site of ►gliogenesis but also a site of neurogenesis for late-born neurons.

Neuronal Fate

Neural stem cells can produce neural progenitors, neurons and glial cells, but the nervous system is composed of balanced numbers of neuronal and glial cells. There is evidence that secreted factors such as FGF2 and ►Neurotrophin3 promote neural fate, while ►ciliary neurotrophic factor (CNTF) and ►epidermal growth factor (EGF) promote the differentiation into astrocytes. For the generation of neural progenitors and their commitment to the neuronal fate, ►basic helix loop helix (bHLH)-type proneural genes play a pivotal role. These include factors that inhibit the neural differentiation, such as Hes, and neural differentiation promoting factors such as ►Mash1, ►Math1 and ►Neurogenin1. The expression of the latter is regulated by several factors such as ►Notch, which negatively regulates the neurogenesis, and BMP.

Migration of Neurons

Mode of Migration

When neurons become post-mitotic, the neurons must emigrate from the ependymal layer to their future (mature) place/layer/lamina/nucleus. The same is true for the cells from the neural crest, which must find their final (mature) destinations. Such changes in the location of neurons are called ►neuronal migration. Neuronal migration in the neural tube is categorized into two modes; one is radial migration and another is tangential migration: The former occurs in directions of radial glia and perpendicular to the ventricular and pial surfaces and the latter occurs tangential to the pial surface. In the CNS, neurons generated from neuroepithelial cells typically migrate radially (Fig. 2), although some neurons show ventricle-directed radial migration after tangentially traveling over a long distance. Radially migrating neurons are thought to use the process of radial glial cells as substrate and tangentially migrating neurons use axons of other neurons. However, neuronal migration occasionally occurs obliquely to either of these processes, indicating that migration of neurons does not necessarily require physically defined structures.

It is noteworthy that tangential migration of neurons often occurs underneath the pial surface. This would be quite advantageous for migrating neurons, because this region is not crowded by other cells and facilitates migration.

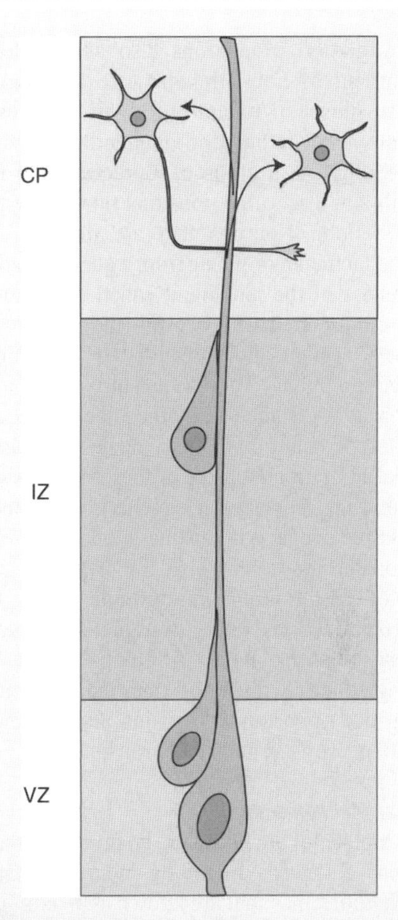

Neural Development. Figure 2 Radial migration of neurons along radial glia and differentiation into postmitotic neuron. CP, cortical plate, IZ, intermediate zone, VZ, ventricular zone.

Significance of Migration

Migration of neurons is required for the establishment of proper laminated structures such as the cerebral cortex, hippocampus and the cerebellum. In the ►pallium of the cerebral cortex, late-generated neurons migrate radially past early-generated neurons, giving rise to inside-out laminated structures of excitatory neurons. Inhibitory cortical neurons also form birthdate-dependent laminated structures, although the route of their migration to their final destinations appears to be more complicated than those of excitatory neurons. These neurons, which originate from ganglionic eminences in the basal forebrain, initially migrate tangentially towards the cortex mainly along the intermediate and subventricular zones, translocate towards the marginal zone, and then descend into the cortical plate to settle in their final destinations. In the marginal zone, these neurons execute multidirectional migration, possibly contributing to dispersion of these neurons [5,6]. A part of these neurons may directly penetrate into the cortical plate without passing through the marginal zone.

The migration of neurons also contributes to the establishment of non-laminated neuronal aggregates, nuclei. A subset of nuclei is formed as a result of a combination of radial and tangential migration. A typical example is a group of ►precerebellar nuclei in the hindbrain, a set of neurons that relay information to the cerebellum. Neurons that are destined to form precerebellar nuclei originate from a germinal zone in the dorsal recess of the hindbrain, called rhombic lip, and migrate circumferentially beneath the pial surface and then they change the direction of migration from tangential to radial. They eventually terminate radial migration at a distance from the pial surface forming aggregates of neurons, namely, nuclei [7]. During the tangential migration, subsets of these neurons cross the ventral midline. Thus, ►nucleogenesis is a consequence of successive events with distinct modes of migration.

Neuronal migration is also important to achieve a balance of excitation and inhibition in mature brain. In the cerebral cortex, for example, excitatory and inhibitory neurons have different origins but are intermingled as a result of neuronal migration [8], creating an opportunity for inhibitory neurons to terminate nearby excitatory neurons.

Molecular Mechanisms

Several molecules are known to be involved in neuronal migration. *In vitro*, migrating neurons respond to diffusible molecules that are known as ►axon guidance molecules. These include Slits, Netrin-1 and ►Semaphorins. Netrin-1, whose role as chemo-attractant for spinal cord and hindbrain commissural axons has been well established, attracts a subset of precerebellar neurons [9], while Slit repels cortical interneurons and inferior olivary neurons. A chemokine, SDF-1, whose expression is prominent in the meningeal tissue, also regulates migration of several types of neuronal cells. SDF-1, for example, attracts GABAergic interneurons *in vitro* and their distribution is disrupted in SDF-1 knock out mice [10]. For radial migration of cortical neurons, a secreted molecule derived from Cajal-Retzius cells, ►Reelin, has been shown to play a pivotal role.

Defects in neuronal migration lead to various ►developmental disorders. This includes human classical ►lissencephaly (smooth brain) and ►double cortex. Genes responsible for these defects are *lissencephaly-1*, encoding Lis1 and *NUDEL*. These molecules form a complex together with ►dynein, which is essential for nucleokinesis, movement of the cell nucleus towards neurites. Doublecortin, loss of which causes human ►subcortical heterotopia, is a microtubule-associated protein and may probably stabilize microtubles. Manipulation of this gene both *in vitro* and *in vivo* affects migration of neurons.

Migrating neurons *in vitro* typically exhibit bipolar morphology headed by a leading process and a trailing process. In many instances, extension of the leading process from migrating neurons *in vitro* is followed by its shrinkage that is associated with forward movement of the cell body. Real-time imaging of cortical neurons revealed that there are at least two modes of migration. One is locomotion and another is translocation [11]. Translocating cortical neurons retain their leading process attached to the pial surface, and their somata move as their leading process shrinks, while locomoting neurons are guided by glial fibers with their processes unrestrained.

Recent progress of imaging technology demonstrated that migrating neurons exhibit a more intricate morphology than has been thought. They have multiple leading processes showing extension and retraction, and the direction of their migration depends on the direction of the longest leading process [5].

Survival of Neurons

The number of neurons is regulated not only by the regulation of its proliferation but also by their naturally occurring death, namely, apoptosis of neurons. The reason for occurrence of the death is not known, but there is evidence suggesting that survival of neurons depends on their synaptic target. The number of neurons of the dorsal root ganglion and motor neurons of the chick, for example, is reduced when the limb bud is removed, whereas more neurons can survive, compared to intact animals, when an extra limb bud is transplanted [12].

The survival of neurons is supported by various growth factors, called ►neurotrophins. The most well known and the first discovered neurotrophin is the ►nerve growth factor (NGF). NGF is secreted from target tissues and supports survival of a subset of neurons.

There is a family of neurotrophins. This includes ►neurotrophin (NT)-3, ►brain-derived neurotrophic factor (BDNF) and ►NT-4/5. Different members of neurotrophins show differential distribution in the nervous system and affect different sets of neurons but by a similar mechanism. These molecules bind to neurons via a family of high affinity receptors, ►TrkA, B or C, with distinct affinities as well as a low-affinity receptor, ►p75. Although these molecules generally support survival (prevent suicide) of neurons, recent studies revealed proNGF, a precursor of NGF can also bind to p75 and negatively regulates cell survival.

Axon Pathfinding

Growth Cone

Postmitotic neurons extend axons, after or during migration. Growing axons are headed by a swelling called ►growth cones, which is a highly motile structure and known to express receptors for various guidance molecules. Growth cones are characterized by thin protrusions called filopodia, which are bounded by sheet-like structures called lamellipodia. Microtubules extending from the axon reach the base of the lamellipodia. The structure of filopodia is supported by a bundle of actin filaments and polymelization of actin molecules causes extension of filopodia. In addition to having motility, growth cones have the capability of detecting environmental cues. They find their growth directions and pathways by responding to such cues (see below).

Long-range Diffusible Cues

A number of guidance molecules are involved in their guidance and are categorized into repulsive and attractive molecules. They can also be categorized into long- and short-range cues.

The most well characterized guidance of axons by a long-rage cue is chemo-attraction of commissural axons by the ventral midline floor plate. *In vitro*, a floor plate explant can attract commissural axons derived from the dorsal spinal cord or the cerebellar primordium in the hindbrain. This is in harmony with in vivo behavior of these axons which course circumferentially through the floor plate. This attraction in vitro can be mimicked by ►laminin-related molecules, Netrin-1 and Netrin-2 [13,14], mRNAs of which are expressed in the floor plate. In mice deficient in Netrin-1 or its receptor, DCC, spinal commissural axons fail to reach the midline [15].

Growth cones that are attracted by a source of chemo-attractant might stall when they arrive at the source. This, however, is not the case. Commissural axons in the hindbrain, for example, continue growing after arriving at the midline. This is because the growth cones lose responsiveness to the midline attractant on their arrival at the midline [16]. In the spinal cord, commissural axons seem to use a similar but somewhat different mechanism: they acquire responsiveness to midline repellent on their arrival at the midline [17].

Floor plate explants also exhibit a repulsive activity to subsets of axons *in vitro*. Candidate molecules for this repulsion are Slits, which typically exert repulsive influence on growth cones via their receptors, Robos.

Long-range attractants expressed by an intermediate target thus contribute to the regulation of crossing and uncrossing axons, but they also serve to guide axons to their final target. Long-range repellents restrict the pathway along which growing axons can advance.

A different kind of diffusible cues contributes to the growth polarity of axons along the rostrocaucal axis. There is evidence that a family of secreted proteins, Wnts, guide post-crossing spinal commissural axons rostrally and corticospinal axons caudally [18,19].

Short-range Cues

Short-range cues can be extracellular matrix molecules such as laminin or membrane-associated proteins such as ►Cadherins and NCAM. It is well established that a number of ►immunoglobulin superfamily cell adhesion molecules are involved in axon guidance. These molecules exert their effect by way of elevating or reducing adhesiveness between growth cones and their environment. In some occasions, these molecules are expressed by pre-existing axonal tracts and later growing axons follow these tracts by utilizing adhesion molecules expressed by them.

Some guidance molecules such as transmembrane ►Semaphorins are associated with cell membranes, and growth cones respond to these molecules by a direct contact with cells expressing such molecules. A member of ►Eph receptors and their ligands, ephrins also play important roles in axon guidance. A notable example of their roles is the establishment of retinotopic organization in the optic tectum [20]. In the tectum, Ephrin A2 and A5 are expressed in a graded manner along the rostrocaudal axis, while one of their ligands, EphA3, is expressed in a graded manner in the retina along the temporal-nasal axis. *In vitro* experiments along with analyses of knockout mice of these molecules indicate that these molecules play pivotal roles in the targeting of retinal axons. Ephrin A/EphA signaling also plays an important role in the regulation of axonal growth polarity along the rostrocaudal axis [21].

Axon Targeting

Targeting of axons is thought to be initially diffuse, but is followed by a sculpturing process, which leads to sharper targeting. For example, corticorubral projections in the cat are initially bilateral but become unilateral as development proceeds [22,23]; rat retinotectal projections, which are initially diffuse become to be topographically organized. These sculpturing processes are associated with elimination of incorrectly projecting axonal branches and proliferation of axon branches in correct targets. Previously, the importance of regressive events such as retraction of axonal branches and death of inappropriately projecting neurons was stressed. However, with development, the proliferation of axonal branches, axon terminals as well as ►dendritic growth takes place, all of which are progressive. Indeed, recent studies have revealed the importance of a progressive process such as the proliferation of axons terminals. In any case, these processes are believed to be regulated by synchronized activities of presynaptic and postsynaptic neurons: synchronized activities of pre- and post-synaptic elements might lead to reinforcement and non-synchronized activity leads to weakening of connections.

Plasticity of Neuronal Connections in the Developing Brain

Developing animals show marked plasticity of neuronal connections compared to adults. Diffuse axonal projections in young animals appear to be related to prominent plasticity of neuronal connections that takes place after early brain damage. Projections of the deep cerebellar nuclei, for example, are bilateral in early postnatal developmental stages in the cat, but become unilateral (crossed projections) as the animal matures [24,25]. On the other hand, removal of the nucleus on one side during the time when bilateral projections are present, causes permanent bilateral projections from the cerebellum. The period during which the lesion of the nucleus is effective in inducing the bilateral connection coincides with the period when bilateral projections are present in normal animals. A similar coincidence of the period of plasticity and the period of the exuberant projection existence can be found in the cortico-rubral system [22,26,27]. Thus, an interesting possibility is that a denervation causes proliferation of pre-existing axonal branches [27].

Synaptogenesis

Structure

The synapse is a key structure for synaptic transmission and the site of transmitter release. The synapse is morphologically defined by electron microscopy: the synaptic differentiation includes the presynaptic density, postsynaptic density, accumulation of synaptic vesicles in the presynaptic membrane and the synaptic cleft, although immature synapses exhibit less pronounced synaptic specialization with few numbers of synaptic vesicles. Typically, synapses are formed on somata and dendrites in mammalian neurons and can be excitatory or inhibitory.

On arrival at their final destinations, growth cones cease growing and form many branches. These branches are associated with swellings both along their length and terminals; such swellings or varicosities contain synaptic vesicles and often form synapses. Synaptogenesis takes place during late developmental stages and is pronounced postnatally.

Process of Synaptogenesis

Synaptogenesis requires the accumulation of synaptic vesicles in the presynaptic site and receptors on the postsynaptic site and adhesion of pre- and postsynaptic membranes. Studies using the ►neuromuscular junction, where acetylcholine serves as a neurotransmitter, indicate that the presynaptic terminal is capable of inducing aggregation of neurotransmitter receptors. Actually, a molecule that has a receptor aggregating activity, ►Agrin, has been identified [28]. Agrin is a proteoglycan synthesized in the motor neuron and released from its terminal. Although Agrin mRNA can be found in the brain, it does not appear to play a role in receptor aggregation in CNS synapses.

Postsynaptic neurons show remarkable dendritic growth during the period of synaptogenesis. In particular, numerous thin filopodial protrusions emerge from dendrites. Although the significance of the presence of these structures remains unknown, these filopodial structures may possibly search for growing axons from presynaptic neurons, and once they encounter growing axons, these structures may retract and form mature synapse [29,30].

Molecules

Several molecules are implicated in synaptogenesis between central synapses. These include Neurexin and its receptor, neuroligin, which are expressed on the pre- and postsynaptic sites, respectively. These molecules can induce synaptic contacts by bidirectional signaling [31].

There is evidence that neuronal activity plays a role in the aggregation of receptor molecules. In this case, the neurotransmitter itself should be involved in maturation of synapses.

During the period of synaptogenesis, other morphological maturation processes proceed. This includes myelination and thickening of axons, both of which contribute to upregulation of conduction velocities.

References

1. Bouwmeester T (2001) The Spemann-Mangold organizer: the control of fate specification and morphogenetic rearrangements during gastrulation in Xenopus. Int J Dev Biol 45:251–258
2. Roelink H, Augsburger A, Heemskerk J, Korzh V, Norlin S, Ruiz i Altaba A, Tanabe Y, Placzek M, Edlund T, Jessell TM (1994) Floor plate and motor neuron induction by vhh-1, a vertebrate homolog of hedgehog expressed by the notochord. Cell 76:761–775
3. Hynes M, Rosenthal A (1999) Specification of dopaminergic and serotonergic neurons in the vertebrate CNS. Curr Opin Neurobiol 9:26–36
4. Fukuchi-Shimogori T, Grove EA (2001) Neocortex patterning by the secreted signaling molecule FGF8. Science 294:1071–1074
5. Tanaka DH, Maekawa K, Yanagawa Y, Obata K, Murakami F (2006) Multidirectional and multizonal tangential migration of GABAergic interneurons in the developing cerebral cortex. Development 133: 2167–2176
6. Tanaka D, Nakaya Y, Yanagawa Y, Obata K, Murakami F (2003) Multimodal tangential migration of neocortical GABAergic neurons independent of GPI-anchored proteins. Development 130:5803–5813
7. Kawauchi D, Taniguchi H, Watanabe H, Saito T, Murakami F (2006) Direct visualization of nucleogenesis by precerebellar neurons: involvement of ventricle-directed, radial fibre-associated migration. Development 133:1113–1123
8. Marin O, Rubenstein JL (2003) Cell migration in the forebrain. Annu Rev Neurosci 26:441–483
9. Taniguchi H, Tamada A, Kennedy TE, Murakami F (2002) Crossing the ventral midline causes neurons to change their response to floor plate and alar plate attractive cues during transmedian migration. Dev Biol 249:321–332
10. Stumm RK, Zhou C, Ara T, Lazarini F, Dubois-Dalcq M, Nagasawa T, Hollt V, Schulz S (2003) CXCR4 regulates interneuron migration in the developing neocortex. J Neurosci 23:5123–5130

11. Nadarajah B, Brunstrom JE, Grutzendler J, Wong RO, Pearlman AL (2001) Two modes of radial migration in early development of the cerebral cortex. Nat Neurosci 4:143–150
12. Hollyday M, Hamburger V (1976) Reduction of the naturally occurring motor neuron loss by enlargement of the periphery. J Comp Neurol 170:311–320
13. Kennedy TE, Serafini T, de la Torre JR, Tessier-Lavigne M (1994) Netrins are diffusible chemotropic factors for commissural axons in the embryonic spinal cord. Cell 78:425–435
14. Serafini T, Kennedy TE, Galko MJ, Mirzayan C, Jessell TM, Tessier-Lavigne M (1994) The netrins define a family of axon outgrowth-promoting proteins homologous to C. elegans UNC-6. Cell 78:409–424
15. Serafini T, Colamarino SA, Leonardo ED, Wang H, Beddington R, Skarnes WC, Tessier-Lavigne M (1996) Netrin-1 is required for commissural axon guidance in the developing vertebrate nervous system. Cell 87:1001–1014
16. Shirasaki R, Katsumata R, Murakami F (1998) Change in chemoattractant responsiveness of developing axons at an intermediate target. Science 279:105–107
17. Stein E, Tessier-Lavigne M (2001) Hierarchical organization of guidance receptors: silencing of netrin attraction by slit through a Robo/DCC receptor complex. Science 291:1928–1938
18. Liu Y, Shi J, Lu CC, Wang ZB, Lyuksyutova AI, Song XJ, Zou Y (2005) Ryk-mediated Wnt repulsion regulates posterior-directed growth of corticospinal tract. Nat Neurosci 8:1151–1159
19. Lyuksyutova AI, Lu CC, Milanesio N, King LA, Guo N, Wang Y, Nathans J, Tessier-Lavigne M, Zou Y (2003) Anterior-posterior guidance of commissural axons by Wnt-frizzled signaling. Science 302: 1984–1988
20. Lemke G, Reber M (2005) Retinotectal mapping: new insights from molecular genetics. Annu Rev Cell Dev Biol 21:551–580
21. Zhu Y, Guthrie S, Murakami F (2006) Ephrin A/EphA controls the rostral turning polarity of a lateral commissural tract in chick hindbrain. Development 133: 3837–3846
22. Higashi S, Yamazaki M, Murakami F (1990) Postnatal development of crossed and uncrossed corticorubral projections in kitten: a PHA-L study. J Comp Neurol 299:312–326
23. Song WJ, Murakami F (1998) Development of functional topography in the corticorubral projection: an in vivo assessment using synaptic potentials recorded from fetal and newborn cats. J Neurosci 18:9354–9364
24. Song WJ, Murakami F (1990) Ipsilateral interpositorubral projection in the kitten and its relation to post-hemicerebellectomy plasticity. Brain Res Dev Brain Res 56:75–85
25. Song WJ, Kobayashi Y, Murakami F (1993) An electrophysiological study of a transient ipsilateral interpositorubral projection in neonatal cats. Exp Brain Res 92:399–406
26. Murakami F, Higashi S, Yamazaki M, Tamada A (1991) Lesion-induced establishment of the crossed corticorubral projections in kittens is associated with axonal proliferation and topographic refinement. Neurosci Res 12:122–139
27. Murakami F, Song WJ, Katsumaru H (1992) Plasticity of neuronal connections in developing brains of mammals. Neurosci Res 15:235–253
28. Nitkin RM, Smith MA, Magill C, Fallon JR, Yao YM, Wallace BG, McMahan UJ (1987) Identification of agrin, a synaptic organizing protein from Torpedo electric organ. J Cell Biol 105:2471–2478
29. Saito Y, Song W-J, Murakami F (1997) Preferential termination of corticorubral axons on spine-like dendritic protrusions in developing cat. J Neurosci 17:8792–8803
30. Ziv NE, Smith SJ (1996) Evidence for a role of dendritic filopodia in synaptogenesis and spine formation. Neuron 17:91–110
31. Dean C, Dresbach T (2006) Neuroligins and neurexins: linking cell adhesion, synapse formation and cognitive function. Trends Neurosci 29:21–29
32. Nakamura H, Katahira T, Matsunaga E, Sato T (2005) Isthmus organizer for midbrain and hindbrain development. Brain Res Brain Res Rev 49:120–126
33. Nakamura H, Watanabe Y (2005) Isthmus organizer and regionalization of the mesencephalon and metencephalon. Int J Dev Biol 49:231–235
34. Rubenstein JL, Martinez S, Shimamura K, Puelles L (1994) The embryonic vertebrate forebrain: the prosomeric model. Science 266:578–580

Neural Filters

Definition

Filters are instruments that let pass only selected information or particles. Tea leaves are filtered out by a tea filter, because they are too large. Likewise, neurons also may filter the incoming information a let pass only a selected part. The mechanisms may be multiple. Thresholding, by which potentials with small amplitudes do not generate an action potential at the output of a neurons, is a powerful means. Other possibilities are coincidence detection or more complex processes.

Neural-Immune Interactions: Implications for Pain Management in Patient with Low-Back Pain and Sciatica

Angela Starkweather
Intercollegiate College of Nursing, Washington State University, Spokane, WA, USA

Synonyms

Neural-immune interactions: Neuroimmunology; Low-back pain: Lumbago; Sciatica: Radiculopathy; Sciatic neuritis

Definition

This essay reviews neuroimmune pathways involved in the initiation and maintenance of low-back pain and sciatica.

Characteristics

Quantitative description: Bidirectional communication between the immune system and the brain and the implications of this communication are emerging concepts in pain research. Although representing a small portion of the disc degeneration syndromes, lumbar herniated discs can cause significant symptoms that may persist even after surgical interventions. Evolving evidence demonstrates that proinflammatory cytokines are key mediators in the process of disc degeneration as well as in the pain experienced by those afflicted with lumbar herniated discs. Activated immune cells release proinflammatory cytokines, which signal the brain through humoral and neural routes. The brain responds by altering neural activity and promoting further production of proinflammatory cytokines within the brain and spinal cord. Increased local cytokine production by disc tissue irritates spinal nerve roots resulting in pain and functional changes in neural activity. This essay explores the importance of cytokine and other related ►neuroimmunology pathways within the context of lumbar disc degeneration and lumbar spine ►pain.

Description of the structure/process/conditions: Disc degeneration is the initial process leading to nontraumatic disc herniation (Fig. 1). This theoretical pathway explains the biomechanical and biochemical events implicated in the process of disc aging, or degeneration, which ultimately leads to the experience of pain phenomena (low-back pain and ►sciatica). Traditionally, it has been believed that the displaced disc tissue was a by-product of the disease and not an interactive element in the disease process itself. However, the discovery of elevated levels of proinflammatory cytokines within injured disc tissue led researchers to conceptualize it as a biologically active tissue. Since that time, connections between the immune system, nervous system, and pain behavior in disc injury have continued to evolve. The current understanding of the mechanisms of disc degeneration and nontraumatic herniation will be reviewed in relationship to the aforementioned Fig. 1.

Regulation of the Structure/Process/Conditions

Mechanical Shear Stress

The function of the intervertebral disc is to provide the spine with mobility while retaining axial stability. While physiological loading helps to maintain metabolism and function in intervertebral discs, excessive mechanical loading appears to be detrimental. Lifestyle, occupational, genetic and biochemical factors influence the effects of mechanical loading on disc degeneration [1].

Neovascularization

Small nonmyelinated nerve fibers grow into the intervertebral disc in areas where there is local production of nerve growth factor (NGF) ("►The role of the NGF family in the regulation of neuroinflammation"). NGF is produced by microvessels, which populate the normally avascular (and aneural) intervertebral disc by extension from adjacent bone. This pattern of nerve growth and receptor expression is implicated in the innervation of painful tissues through NGF-driven axonal growth and maturation. The stimulus that promotes microvessels to release NGF, triggering the process of nerve and vessel ingrowth, remains uncertain. However, IL-1 has the ability to switch chondroctyes from anabolism to catabolism, inducing cartilage breakdown at molecular and morphological levels through stimulating ►matrix metalloproteinases (MMPs). Thus, there is growing evidence of the role of proinflammatory cytokines in matrix degradation (disc degeneration), nerve and vessel ingrowth, and pain. In fact, there is mounting evidence that immune factors are involved not only in the initiation of disc degeneration but also in the progression of disc disease. As the injured disc tissue continues to produce elevated levels of MMPs, thereby losing proteoglycans, the diseased tissue begins to wear down. Annular tears form along the outer wall of the disc, making it more susceptible to splitting, and thus herniating, in the face of exertional forces that raise the intervertebral pressure [2]. Disc-related biomarkers of degeneration are currently being investigated [3].

Direct Mechanical Stimulation and Sensitization of the Dorsal Root Ganglion

Dislocation of intervertebral disc tissue (IVD) by nucleus pulposus (NP) protrusion or extrusion (i.e., herniated disc) is a common source of severe pain. Herniated NP can potentially contact and compress the dorsal root ganglion (DRG) and spinal root that enters the spine at the vertebral level. Acute mechanical compression is sufficient to produce spontaneous activity in the sensory afferents, supporting the classic assumption that mechanical compression is the cause of pain and other neurological symptoms. Mechanical compression has been thought to account for the ischemia, edema, and demyelination that occur in the DRG and the pain that may arise from nerve endings in the outer annulus fibrosus.

Nucleus Pulposus-Induced Nerve Injury

Pathological pain can arise as a consequence of the protrusion of the NP into contact with the DRG and dorsal root. Although pressure per se has classically been considered as a major cause of pain, there is growing evidence that immune-derived substances may be involved as well. Diverse immune cells and equally

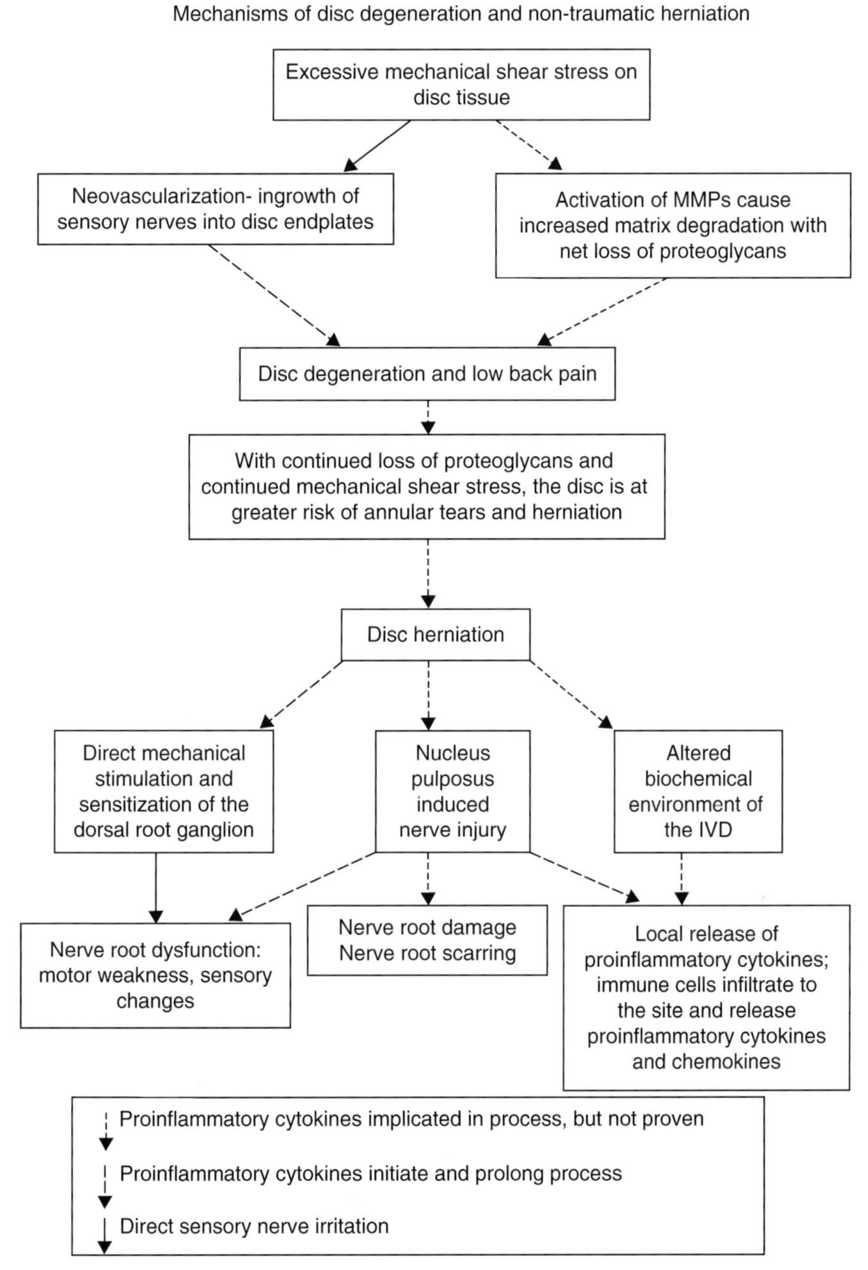

Neural-Immune Interactions: Implications for Pain Management in Patient with Low-Back Pain and Sciatica. Figure 1 Mechanisms of disc degeneration and non-traumatic herniation. MMP = matrix metalloproteinases; IVD = intervertebral disc.

diverse immune cell products are potential mediators. Of these, proinflammatory cytokines have received by far the most attention. Data to date suggest a strong case in support of proinflammatory cytokine involvement in the pain of herniated discs through astrocyte and microglia activation ("►Functions of microglia in immune mechanism in the central nervous system (CNS)") [4]. The cytokines may do this by inducing expression of receptors within DRGs. Also, axonal interactions with proinflammatory cytokines could increase electrical conductivity. Each of these processes could then lead to pain processing.

Biochemical Mediators in Herniated Disc Tissue

Neuropathic pain can occur as a consequence of nerve trauma, with physical damage to nerves altering pain perception and the function of pain transmission pathways ("►Spinal immunology and neuropathic pain"). However, neuropathic pain can also occur in the absence of any detectable physical injury. In these situations, pathological pain appears to be a consequence of immune activation and inflammation, which can also amplify pain as a consequence of physical trauma. The role of immune activation in neuropathic conditions has been firmly established, and a consistent picture has emerged from these models of traumatic and/or inflammatory neuropathic pain. The key cellular mediators are most likely inflammatory cells recruited into the affected area from the general circulation along with locally stimulated cell populations. These cells produce proinflammatory cytokines (tumor necrosis factor [TNF], IL-1, IL-6) within the affected area and create and maintain pathological pain [5].

Alterations in the Biochemical Environment of the Intervertebral Disc (IVD)

Nucleus pulposus-induced effects on adjacent nerve root(s), include alterations in nerve conduction velocity, mechanosensitization, pain behavior, histological degeneration, reduced blood flow, and increased endoneurial fluid pressure. The biochemical changes initiated by exposed NP and the increased production of proinflammatory cytokines create an environment of degradation. This process has been hypothesized to be part of disc resorption, influenced by migrating macrophages. The synergistic effects of nerve compression and the altered chemical environment of the IVD appear to produce the pathophysiologic network leading to the pain experienced by those with herniated discs.

Function

Lifestyle, body weight, aging, and genetics all influence the load environment of the normal IVD. Lifestyle (including occupation) and body weight are capable of accelerating the rate of degeneration and thus further complicates the starting point from which to assess IVD degeneration. The studies to date provide evidence of connective tissue degradation, nerve and vessel ingrowth, and increased production of proinflammatory cytokines that characterize IVD degeneration and herniation. There is considerable need for more investigation into the precise role of cytokines for each of these biological processes. Concurrently, the study of immune involvement in neuropathic pain is in its infancy. Many more immune cells and immune-derived substances may be implicated in the etiology of pathological pain syndromes. Much remains to be learned about the dynamics of immune system modulation of pain and neural function.

The recognition that the immune system may be involved in neuropathic pain has important potential implications. If proinflammatory cytokines contribute to pain and to neuropathological changes in the sensory neurons, it may be possible to devise much-needed alternative approaches for treatment of patients with low-back pain. Surgery for herniated discs is not without cost, and surgical treatment of disc herniation is advised only if nonsurgical treatment fails. Furthermore, resolution of pain is not guaranteed with surgery, as complications and failure rates remain relatively high. Understanding the role of the immune system in disc-related pain may lead to a better appreciation of not only the nature of organic pain but also alternative therapeutic approaches or drug strategies to treat pain and its antecedents. Moreover, the evaluation of immune markers as indices of pain and of immune responsiveness consequent to pain may provide insight into the means by which to fine-tune the therapy provided to individual patients.

A remaining conundrum in clinical practice is how to define disc degeneration. It has been proposed that Modic changes may be an objective marker of discogenic low-back pain. Modic changes are signal intensity changes on plain radiograph X-rays and magnetic resonance imaging that reflect a spectrum of vertebral body marrow changes associated with degenerative disc disease. A correlation between Modic changes on spinal magnetic resonance images and the production of proinflammatory cytokines has been made: Modic 1 changes were more common in patients with discogenic low-back pain, whereas Modic 2 changes occur in patients suffering from sciatica (with increased production of proinflammatory cytokines). Advances in imaging techniques are being applied to disc degeneration which will hopefully lead to a standard method of quantifying disc degeneration in humans [6].

Traditional treatment for low-back pain includes nonsteroidal anti-inflammatory medication, which inhibits prostaglandin synthesis, as first-line therapy. Patients exhibiting ►sciatic symptoms are often prescribed steroids (by mouth or epidurally) to decrease swelling in the affected nerve root. The use of these substances in long-term therapy, however, must be weighed against their side effects. Gabapentin has been added to the armamentarium for treating sciatic pain. Although its mechanism of action is unknown, it is structurally related to the neurotransmitter γ-aminobutyric acid. All of these medications have limited success in relieving symptoms of low-back pain and sciatica, and none prevent progression of degenerative disease.

The recognition of peripheral and central immune cell involvement in neuropathic pain of diverse etiologies may offer a new avenue or approach to pain control. There are multiple situations in which immune-derived proteins (TNF, IL-1, IL-6) have been correlated with and are the likely cause of neuropathic pain conditions. The pervasive and potentially key involvement of the proinflammatory cytokines within an affected body region or within the spinal cord are likely and desirable targets for drug development. The role of gene therapy in disc degeneration is currently being investigated [7–9]. Using genes introduced into target cells, proteins are produced within the degenerate disc, which provide a chemical environment conducive to restoring cell function toward normality. Although some pathologic conditions require immediate decompressive surgery, as in cauda equina syndrome, the role of surgery in disc degeneration syndromes is becoming less clear. As new therapies continue to evolve that are able to target the biochemical factors involved in pain transmission, perhaps the ultimate test will be whether a pathway can be found that reverses the degenerative condition.

References

1. Battié MC (2006) Lumbar disc degeneration: epidemiology and genetics. J Bone Joint Surg 88(A):3–8
2. Benoist M (2002) The natural history of lumbar disc herniation and radiculopathy. Joint Bone Spine 69:155–160
3. Poole AR (2006) Biological markers and disc degeneration. J Bone Joint Surg 88(A):72–75
4. Watkins LR, Maier SF (2002) Beyond neurons: evidence that immune and glial cells contribute to pathological pain states. Physiol Rev 82:981–1011
5. Wieseler-Frank J, Maier SF, Watkins LR (2005) Central proinflammatory cytokines and pain enhancement. Neurosignals 14:166–174
6. Haughton V (2006) Imaging intervertebral disc degeneration. J Bone Joint Surg 88(A):15–20
7. Freemont AJ, Watkins A, Le Maitre C, Jeziorska M, Hoyland JA (2002) Current understanding of cellular and molecular events in intervertebral disc degeneration: implications for therapy. J Pathol 196:374–379
8. Le Maitre CL, Freemont AJ, Hoyland JA (2006) A preliminary in vitro study into the use of IL-1Ra gene therapy for the inhibition of intervertebral disc degeneration. Int J Exp Pathol 87:17–28
9. Le Maitre CL, Richardson SM, Baird P, Freemont AJ, Holyland JA (2005) Expression of receptors for putative anabolic growth factors in human intervertebral disc: implications of repair and regeneration of the disc. J Pathol 207:445–452

Neural Integrator

Definition

A neural network that integrates, in a mathematical sense, its input signals. For example, if a pulse of neural activity was the input to a neural integrating circuit, the circuit's output would be a sustained step of neural activity.

►Neural Integrator – Horizontal
►Neural Integrator – Vertical

Neural Integrator – Vertical

Yannis Dalezios[1], Adonis Moschovakis[2]
[1]Institute of Applied and Computational Mathematics, Crete, Greece
[2]Department of Basic Sciences, Faculty of Medicine, University of Crete, Heraklion, Crete, Greece

Definition

An ensemble of neurons of the vertical oculomotor system that engage in integration in the sense of Newtonian calculus. They receive signals roughly indicative of vertical eye velocity (such as those produced by the vertical burst generators in the rostral interstitial nucleus of the MLF) and generate signals proportional to vertical eye position that they convey to the appropriate ►extraocular motoneurons (those with vertical pulling directions).

Description of the Theory

Although originally thought as a unitary brain device responsible for all eye velocity to position transformations, it is now clear that several cell assemblies are needed to implement neural integration in different planes and for different kinds of eye movements. The nucleus prepositus hypoglossi is crucial for integration in the horizontal plane (►see Neural Integrator – Horizontal) and the same is true for the ►interstitial nucleus of Cajal (NIC) for integration in the vertical plane. For example, unilateral chemical inactivation of the NIC prevents monkeys from holding eccentric vertical eye positions but does not affect their ability to hold eccentric horizontal eye positions. The same is true of human patients suffering from midbrain lesions that encroach on the NIC; they display vertical (but not horizontal) gaze holding failure [1].

The oculomotor subsystem subserving the ►vestibuloocular reflex (VOR) also relies on velocity to position integration. Its input, carried by primary vestibular afferents, encodes (and is in phase with) the angular velocity of the head whereas its output (motoneuron discharge) encodes (and is in phase with) eye position. There is evidence from several species to indicate that the NIC participates in this process, at least in the vertical plane. Its lesions impair the pitch VOR (as documented by gain reduction and phase advancement) while the horizontal VOR remains normal. Because vertical VOR impairments are not as profound as would be expected from the concurrent impairment of vertical gaze holding, it is unlikely that a single neural integrator underlies the generation of both saccadic and VOR signals [2].

To understand how integration is achieved neurally it is important to know the discharge pattern and connections of relevant neurons. Of the several classes of cells contained in the NIC (for reviews see [3,4]), it is the burst-tonic neurons that are of particular interest for the purposes of this essay. The discharge of a typical such unit is illustrated in Figs. 1 and 2.

As shown here, it remains roughly constant in between saccades at a value specified by the vertical position (usually both up and down) of the eyes and its bursts precede saccades with an upward component, whether rightward or leftward. The firing rate (F_R) of such units is described by the expression

$$F_R = F_0 + kE + rdE/dt \quad (1)$$

Where E is the instantaneous vertical position of the eyes. F_0 (the neural discharge at primary position), k (the slope of the rate-position curve) and r (the slope of the rate-velocity curve) are constants that differ for different cells. The average values they obtain have been determined in several species (for a review see [4]). For example, $F_0 = 88$ spikes/s, $k = 3.1$ spikes/s/deg and $r = 0.8$ spikes/s/deg/s (when r was evaluated from smooth pursuit eye movements), in a sample of units of

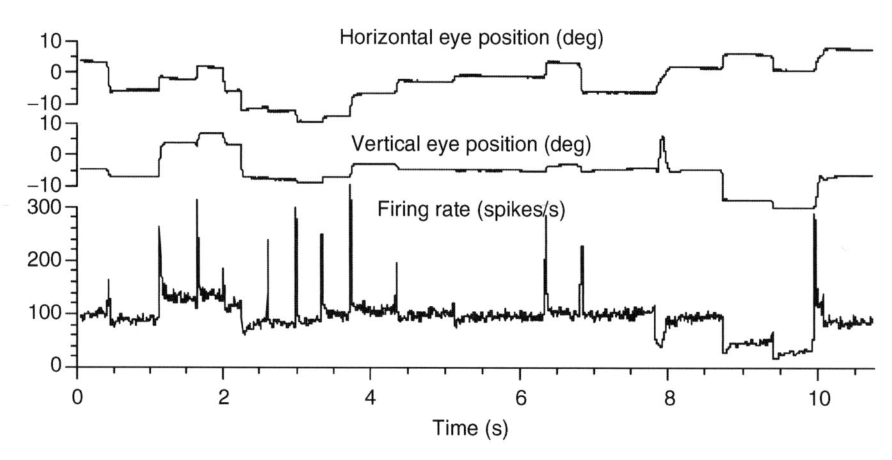

Neural Integrator – Vertical. Figure 1 Oculomotor related discharge pattern of a regular upward efferent fiber of the NIC. (modified from [5], used with permission). Traces from top to bottom illustrate the instantaneous horizontal and vertical eye position and the instantaneous firing rate.

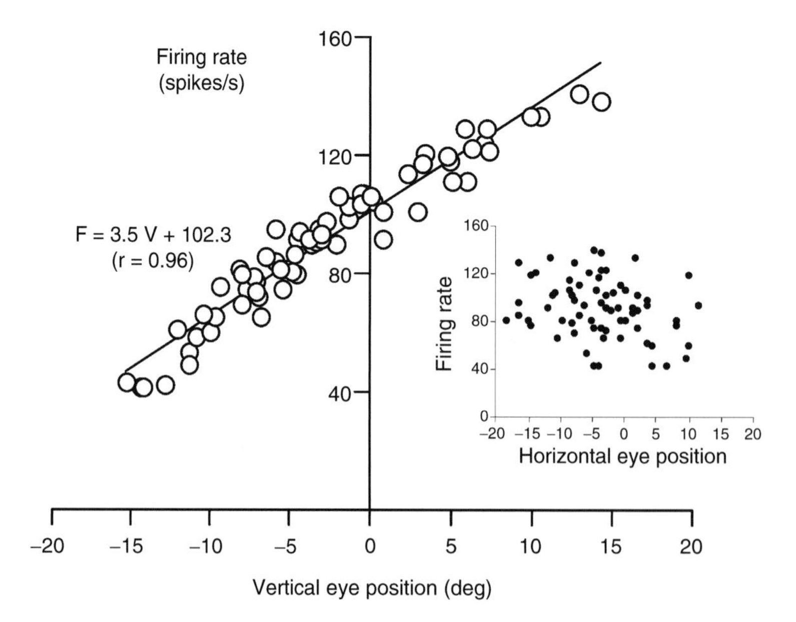

Neural Integrator – Vertical. Figure 2 Quantitative analysis of the relationship (modified from [5], used with permission) between the mean intersaccadic firing rate (ordinate) of the unit shown in Fig. 1 and the mean intersaccadic vertical eye position (abscissa). The solid line is the linear regression line through the data (open circles). The inset is a plot of horizontal eye position (abscissa) versus firing rate (ordinate).

the squirrel monkey, while the same units also emitted about 0.6 spikes per degree of vertical displacement of the eyes during saccades [5]. Differences between NIC burst-tonic units can be appreciable as shown by the fact that 25% of these units modulated their discharge in relation to upward or downward eye position, but not both, while about 25% of them did not emit any bursts for saccades (tonic neurons).

Axons arising from burst tonic units of the NIC have been shown to travel with the posterior commissure (PC; [6]). The trajectory of such PC fibers and the terminal fields they deploy in the contralateral NIC, the oculomotor nucleus and the trochlear nucleus are illustrated in Fig. 3.

Their virtual disappearance, after PC lesions that preceded the injection of tracer in the NIC, indicates that PC fibers are likely to be the conduit of most of the NIC output to these nuclei [7]. Their integrity is necessary for normal velocity to position integration in the vertical plane since inactivation or lesion of the PC disable vertical gaze holding and advance the phase of the vertical VOR in the dark [8].

To gain insight into the operation of this network and understand how the properties of its neurons endow it with the ability to transform velocity signals into position signals it is instructive to consider a model of the vertical neural integrator that captures the crucial biological facts that are known about it such as those described above. To recapitulate, such a model should account for the qualitative (e.g., whether they emit saccade related bursts or not) and quantitative (e.g., the range of values of the parameters F_0, k and r) properties of discharge of its units, the cross-connections between neural integrators located in opposites sides of the brain (via the PC) and the properties of the eye movement deficits which result from lesions of its units (e.g., drifts of appropriately short time constants).

Figure 4 is a block diagrammatic illustration of a lumped version of this model together with typical signals encountered at several of its stages for saccades of two different directions (up, blue; down, red).

The same color code is used to distinguish upward (UNI) from downward (DNI) units while solid lines indicate connections arising from excitatory (UNI_e,

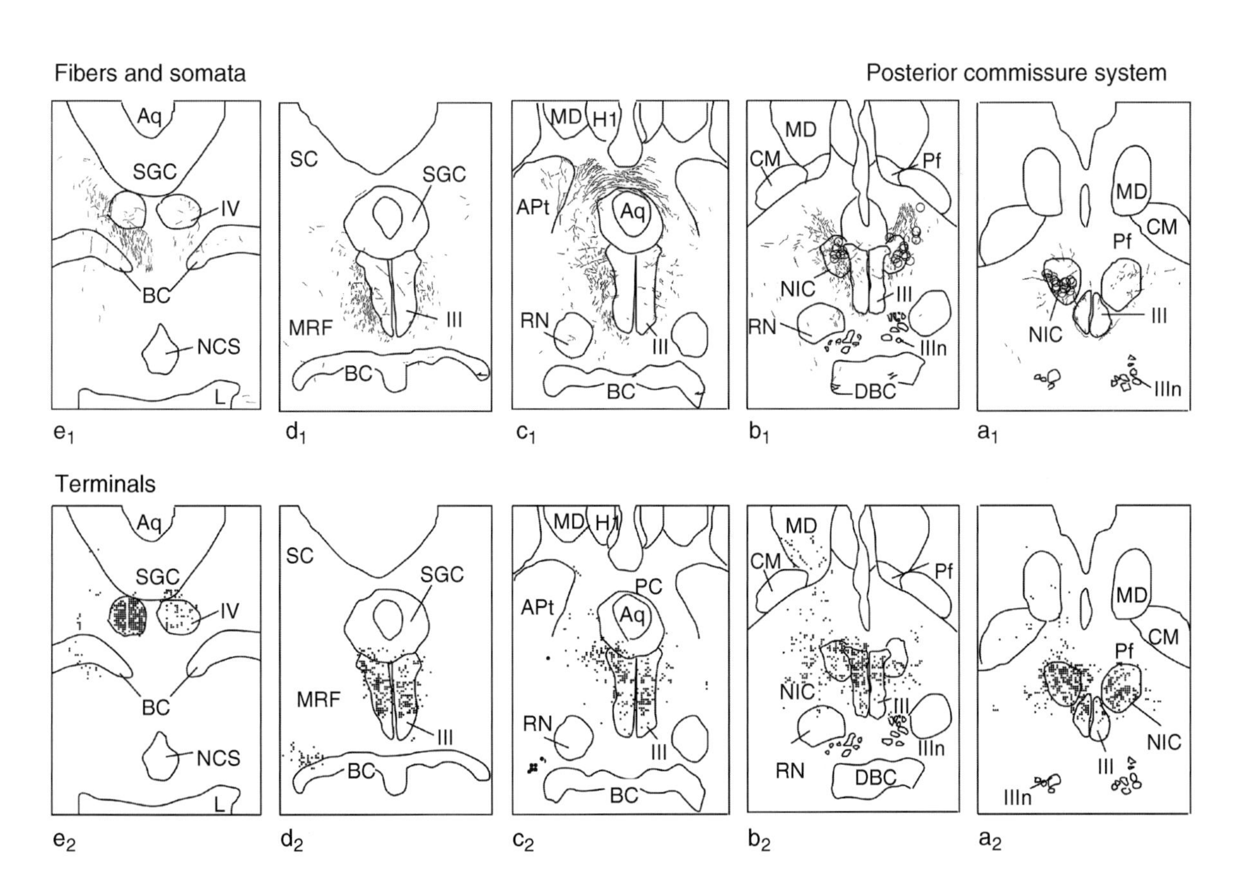

Neural Integrator – Vertical. Figure 3 Location of retrogradely labeled somata (a_1–e_1, open circles), trajectory of PC fibers (a_1–e_1) and terminal fields in and around the oculomotor and trochlear nuclei (a_2–e_2) following biocytin injection in the NIC (reproduced from [7] and used with permission). Sections are shown in the frontal plane and are arranged from rostral (a) to caudal (e).

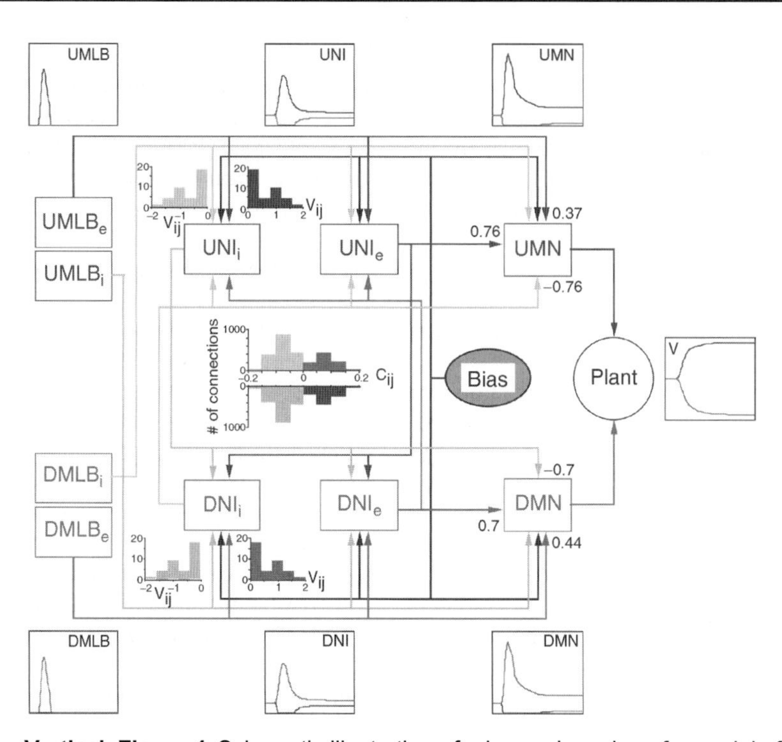

Neural Integrator – Vertical. Figure 4 Schematic illustration of a lumped version of a model of the vertical neural integrator (reproduced from [9], with permission). Units with upward on-direction and the signals carried by all units during upward saccades (waveforms placed inside boxes) are shown in blue. Units with downward on-direction and the signals carried by all units during downward saccades are shown in red. Boxes illustrating discharge patterns measure 200 ms (abscissa) and 800 spikes/s (ordinate). The size of the box containing examples of the eye position output of the system is 200 ms (abscissa) and 20 deg (ordinate). Solid arrows indicate excitatory connections. All other connections are inhibitory. Insets are histograms of connection strengths. Abbreviations: *DMLB*, downward medium-lead burst neuron; *DMN*, downward motoneuron; *DNI*, downward neural integrator; *UMLB*, upward medium-lead burst neuron; *UMN*, upward motoneuron; *UNI*, upward neural integrator. Subscripts indicate the excitatory (e) or inhibitory (i) influence units exert on their targets.

DNI_e) units and stippled lines those arising from inhibitory (UNI_i, DNI_i) units. Each model unit is assumed to establish excitatory (NI_e units) or inhibitory (NI_i units) connections onto NI units with opposite on-direction. They all project to motoneurons as well. The strength of these connections (c_{ij}) and the precise number of excitatory and inhibitory units is less critical. Provided that the number of excitatory neurons is not greater than or equal to the number of inhibitory ones it is always possible to find an average value (of c_{ij}) such that $T = 20$ s. When inhibitory neurons are sufficiently more numerous than excitatory ones (e.g., when $n_i = 2n_e$) T is about equal to 20 s (dashed line), if c is equal to 0.0835. However, there is no need to assume that the weights of all interconnections (c_{ij}) between the NI units of the distributed model obtain this absolute value. Instead, they can vary quite freely around it forming a normal distribution with a coefficient of variation equal to 0.5. This relatively large number implies that the model is quite impervious to the precise values obtained by any of these connection strengths. The histograms near the center of Fig. 4 (solid, excitatory; stippled, inhibitory) illustrate distributions that were obtained with the help of a random number generator after specifying the mean and the standard deviation of the populations and are compatible with a normally operating neural integrator.

The signals carried by the units of such a model (e.g., insets of Fig. 4) are remarkably similar to the discharge pattern of primate NIC neurons. Just like primate neurons, model units generally burst for vertical saccades (up or down) and their tonic discharge in between saccades is proportional to vertical eye position. The similarity is not just a broad, qualitative one. For example, the position sensitivity (k) velocity sensitivity (r) and primary position rate (F_0) of upward model units are statistically indistinguishable from the the position and the velocity sensitivity and the primary position rate of the burst-tonic upward efferent neurons of the NIC [5]. Although most model units (N = 62 of 72) modulated their discharge for saccades and eye position in the same direction (in-phase units) a few behaved quite differently in that they display opposite on-directions for saccades and eye position (anti-phase units). The same is true of NIC neurons found in both cats [3] and monkeys [5]. Finally, as with primate

NIC neurons, a few model units did not emit bursts for saccade which again agrees well with the discharge patterns of neurons encountered in the NIC [5]. Although few in number, such tonic neurons are of particular importance as it has been argued that they carry the output of the neural integrators in contrast to burst-tonic cells that are limited to computational stages closer to its input (see the chapter on ►Neural Integrator – Horizontal). Efforts to test this argument experimentally have had mixed results. Tonic neurons have been shown to project to the abducens nucleus more frequently than burst-tonic cells of the horizontal neural integrator (in the NPH; [10]). In contrast, burst-tonic cells have been shown to comprise the majority of efferent fibers conveying the output of the vertical neural integrators to extraocular motoneurons [5]. All in all, work in the vertical system demonstrates that the neural integrator need not be more than one layer deep and that both its tonic and burst-tonic units send their output to motoneurons.

Given the large variety of their response properties, it is meaningful to ask if the units of the model integrator of Fig. 4 comprise a functional continuum, or alternatively, whether they can be broken up into distinct functional classes and, if so, which. To this end we plotted their position sensitivity (k_V) against their velocity sensitivity ($r_{(s)}$) in Fig. 5.

In this scatter-plot, the units occupy two distinct clouds, to the right and to the left of zero k_V, corresponding to upward and downward units, respectively. No other demarcation point can be found to further subdivide units according to their position sensitivity (high or low).

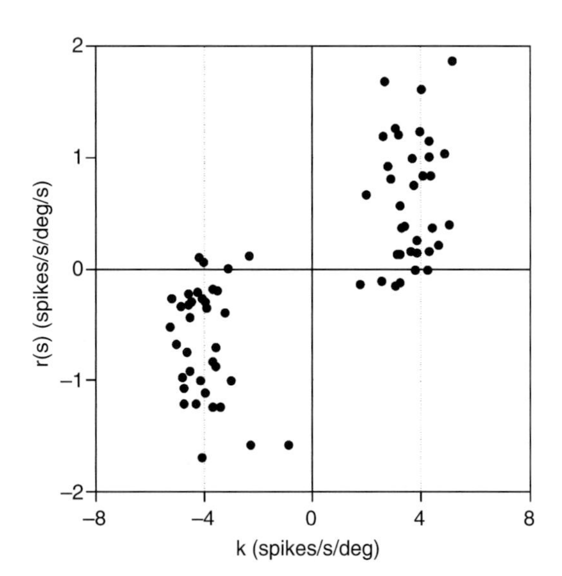

Neural Integrator – Vertical. Figure 5 2-D scatterplot of the slope of the rate position curves (k_V, abscissa) against the slope of the rate-velocity curves ($r_{(s)}$, ordinate: evaluated from saccades) of all 72 units of the model (reproduced from [9], with permission).

Similarly, no demarcation can be found on the orthogonal axis to subdivide units according to their velocity sensitivity. Both up and down units form elongated clouds straddling zero $r_{(s)}$. Most of the points are confined to the upper right (up units) or lower left (down units) quadrants. These correspond to units that are generally sensitive to up or down ocular deviation and emit more or less strong bursts for saccades in the same direction. Both clouds encompass points on or close to the horizontal abscissa of the scatterplot; these correspond to tonic units. Finally, both clouds encompass a few points belonging to the lower right (up units) or the upper left quadrant (down units); these correspond to anti-phase units.

Selective lesions of subpopulations of integrator units could help elucidate how each of these distinct subtypes contributes to neural integration. Such experiments cannot be carried out *in vivo* with presently available techniques. Instead they exemplify one of the strong points of computational approaches. As expected, elimination of the inhibitory units of the model shown in Fig. 4 abolishes integrator function. On the other hand, elimination of excitatory units does not lead to integrator failure provided that the weights of commissural connections are readjusted “postlesionally”. Instead, it abolishes anti-phase activity and more importantly, it raises the slope of the rate position curve dramatically, to between ± 23.5 and ± 33.5 spikes/s/deg. With rate position slopes such as these, and since the eyes reach eccentricities as high as ± 45°, NI neurons would have to sustain very high firing rates (often exceeding 1,000 Hz). In other words, the vertical neural integrator could function normally even if all connections between its neurons were inhibitory ones but not if the dynamic range of the neurons involved must be kept within physiological bounds.

N

References

1. Ranalli PJ, Sharpe JA, Fletcher WA (1988) Palsy of upward and downward saccadic, pursuit, and vestibular movements with a unilateral midbrain lesion: pathophysiologic correlations. Neurology 38:114–122
2. Helmchen C, Rambold H, Fuhry L, Büttner U (1998) Deficits in vertical and torsional eye movements after uni- and bilateral muscimol inactivation of the interstitial nucleus of Cajal of the alert monkey. Exp Brain Res 119:436–452
3. Fukushima K, Kaneko CRS (1995) Vestibular integrators in the oculomotor system. Neurosci Res 22:249–258
4. Moschovakis AK (1997) The neural integrators of the mammalian saccadic system Front. Bioscience 2:552–577
5. Dalezios Y, Scudder CA, Highstein SM, Moschovakis AK (1998) Anatomy and physiology of the primate interstitial nucleus of Cajal. II. Discharge pattern of single efferent fibers. J Neurophysiol 80:3100–3111
6. Moschovakis AK (1995) Are laws that govern behavior embedded in the structure of the C.N.S.? The case of Hering's law. Vision Res 35:3207–3216

7. Kokkoroyannis T, Scudder CA, Highstein SM, Balaban C, Moschovakis AK (1996) The anatomy and physiology of the primate Interstitial Nucleus of Cajal. I. Efferent projections. J Neurophysiol 75:725–739
8. Partsalis A, Highstein SM, Moschovakis AK (1994) Lesions of the posterior commissure disable the vertical neural integrator of the primate oculomotor system. J Neurophysiol 71:2582–2585
9. Sklavos SG, Moschovakis AK (2002) Neural network simulations of the primate oculomotor system. IV. A distributed bilateral stochastic model of the neural integrator of the vertical saccadic system. Biol Cybern 86:97–109
10. Escudero M, de la Cruz RR, Delgado-García JM (1992) A physiological study of vestibular and prepositus hypoglossi neurones projecting to the abducens nucleus in the alert cat. J Physiol 458:539–560

Neural Network in Olfactory Bulb

Definition

Group of interconnected neurons, which are together responsible for a particular neural function, either at the periphery or at the central brain level. The network typically comprises different types of neurons, having different electrical and/or physiological properties. A typical example is the olfactory bulb, primary olfactory center of vertebrates that comprises different neuron types within a number of anatomical and functional units, the glomeruli. Olfactory sensory neurons at the periphery (olfactory mucosa) detect odorants and convey this information to the glomeruli. There, sensory neurons connect onto output neurons (mitral cells/tufted cells), which will further convey olfactory information to higher brain centers (cortex, etc.). Most importantly, two types of local bulb neurons (periglomerular and granule cells) carry out lateral inhibition to neighboring glomeruli. All these neurons constitute the neural network of the olfactory bulb, whose function is to format odor representation, increase signal to noise, and improve odor discrimination ability. Because of the very high number of connections between neurons within neural networks, and the fact that the action of one neuron on the other can be excitatory or inhibitory, neural networks usually induce highly non-linear results at their output. Consequently, their activity should not be studied at the "single-neuron" level but techniques taking into account the connectivity and complexity of the network, like multi-unit electrophysiological recordings or optical imaging techniques should be applied. Furthermore, computational neural networks have proved useful to reproduce the architecture of biological neural networks, in order to understand their processes and outcomes. Such mathematical models have become a whole field in computational neuroscience, and they are now used for developing artificial intelligence applications, for instance for voice or face recognition, automated image analysis, etc.

►Olfactory Bulb

Neural Networks

Kazuyuki Aihara[1], Masato Okada[2], Masaharu Adachi[3], Masataka Watanabe[4]
[1]Institute of Industrial Science, The University of Tokyo, Tokyo, Japan
[2]Graduate School of Frontier Sciences, The University of Tokyo, Kashiwa Chiba, Japan
[3]Department of Electrical and Electronic Engineering, School of Engineering, Tokyo Denki University, Tokyo, Japan
[4]Department of Systems Innovation, Graduate School of Engineering, The University of Tokyo, Tokyo, Japan

Definition

Neural Networks and Architectures

Neural networks, here, mean mathematical models of biological neural networks composed of neurons. There have been many works on neural networks since the seminal paper by McCulloch and Pitts in 1943 [1]. Although neurons are connected through synapses in complicated ways in the living brain, typical architectures of neural network models are classified into ►feedforward networks and ►feedback networks as shown in Fig. 1.

While the former is ►multilayer networks with the input layer, the output layer, and hidden layers between them, the latter is recurrent networks with feedback connections. Typical examples of the feedforward networks are ►perceptron [2] and ►back-propagation learning networks [3]; recently, forward-propagation learning is also proposed [4]. The feedback networks, on the other hand, have been used in many models such as ►associative memory and combinatorial optimization where nonlinear dynamics plays important roles like deterministic ►chaos in ►chaotic neural networks and stochastic synchronization.

Characteristics

Learning Theory in Neural Networks

Learning in models of neural networks is incorporated as rules to update and establish synaptic weights between neurons. The learning rules are classified

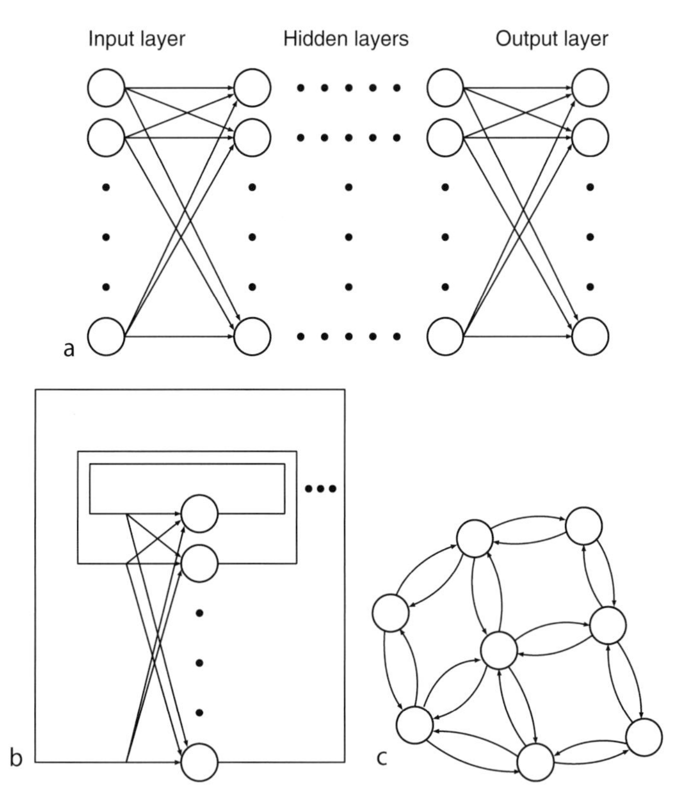

Neural Networks. Figure 1 Network architectures: (a) feedforward networks and (b) feedback networks which are equivalent to (c) networks composed of mutually connected neurons. Each circle represents a neuron.

into supervised and unsupervised learning, respectively, with and without teaching signals. Examples of the former are perceptron [2] and back-propagation learning [3] in feedforward networks, while a typical example of the latter is the Hebbian learning. The beauty of learning in neural networks is its generalization property; acquisition of desired responses to a training data set in feedforward networks can result in correct responses to a new data set. Therefore, unlike conventional software programming, we do not need to figure out the rules and logics behind a given problem, but simply provide examples of correct responses to the networks with respect to input signals.

The learning algorithm in feedforward networks can be roughly categorized into two. One is that the network receives a target vector signal, that is, desired output values of the output layer like the perceptron and the back-propagation learning. Another is reinforcement-learning-based algorithm that assumes only a scalar "reward" signal that tells how much good or bad the present output is, like forward-propagation learning [4].

Perceptron

The perceptron is a multilayer feedforward neural network where changes in synaptic weight take place only at the final output layer [2]. It is a type of supervised learning where a target vector is provided to the final layer as teaching signals.

Let us consider the following two-layer network only with input and output layers for the sake of simplicity:

$$z_i = 1\left(\sum_j w_{ij} a_j\right),$$

where z_i denotes the output of neuron i in the output layer, a_j input j in the input layer with $a_j = 0$ (nonfiring) or 1 (firing), w_{ij} the synaptic weight from input j to neuron i, and function $1(\bullet)$ is the Heaviside output function with $1(u) = 1$ for $u \geq 0$ and $1(u) = 0$ for $u < 0$. We omitted the neuronal threshold, since the threshold can be represented as the synaptic weight from a constantly firing input unit.

The update rule of synaptic weights is given as

$$\Delta w_{ij} = \eta(\delta_i - z_i)a_j,$$

where δ_i denotes the desired output of neuron i and η the learning coefficient. The synaptic weight is changed in the direction to decrease the difference between z_i and δ_i, whenever the outputs do not match the target vector. The perceptron learning has been mathematically proven to realize appropriate values of synaptic weights under a condition that a set of input patterns are linearly separable.

Back-propagation Learning

Back propagation is a learning rule that updates synaptic weights in hidden layers together with the output layer, provided a target vector [3]. Learning ability is much higher than the perceptron, and many practical problems that are nonlinearly separable can be solved with a moderate number of hidden layer neurons.

For simplicity, we assume a three-layer network with a single hidden layer as follows:

$$z_i = f\left(\sum_j s_{ij} x_j\right),$$

$$x_j = f\left(\sum_k w_{jk} a_k\right),$$

whereas x_j denotes the output of neuron j in the hidden layer, f is a continuously differentiable output function like the sigmoid function $f(u) = 1/\{1 + \exp(-u/\varepsilon)\}$ with the steepness parameter ε, s_{ij} the synaptic weight to output neuron i from hidden layer neuron j, and w_{jk} that to hidden neuron j from input k.

To derive the update rules of synaptic weights in each layer [3], we define an error function E as follows:

$$E = 1/2 \sum_i (\delta_i - z_i)^2,$$

which is the sum of squared errors at the output layer.

Here, the synaptic changes in the final layer are implemented as the gradient descent of the error function

$$\Delta s_{ij} = -\eta \frac{\partial E}{\partial s_{ij}}.$$

Defining the learning signal r_i as

$$r_i = (\delta_i - z_i) f'\left(\sum_j s_{ij} x_j\right),$$

the learning rule becomes as follows:

$$\Delta s_{ij} = \eta r_i x_j.$$

Likewise, necessary synaptic change in the hidden layer to decrease the error at the output layer is derived as follows:

$$\Delta w_{jk} = -\eta \frac{\partial E}{\partial w_{jk}}$$

$$= \eta \tilde{r}_j a_k,$$

where

$$\tilde{r}_j = \left(\sum_i r_i s_{ij}\right) f'\left(\sum_m w_{jm} a_m\right).$$

The term "back propagation" originates from the interesting behavior that learning signals in the output layer seem to propagate backward through synaptic weights S_{kj}'S and become learning signals in the hidden layer. Moreover, although the back-propagation learning rule may get stuck in local minima, it has been shown that the learning rules work well in a wide variety of learning tasks.

While the back-propagation learning provides a powerful method to train multilayer neural networks, it suffers from the lack of biological plausibility. Therefore, attempts have been made to train multilayer networks using scalar reinforcement signals like dopaminergic activity [5], which only tell the networks whether the current output is good or bad. The difficulty in this approach is that the reinforcement signal is assumed to be broadcasted to the local neural circuit and hence leads to nonconvergence of learning when directly applied to feedforward network models. Recently, a method to overcome this problem, or forward-propagation learning [4] was introduced, where learning propagates from the input to the output layer under certain neuronal activity conditions that could be observed in the monkey prefrontal cortex. In short, each layer does its best to provide additional dimensions to solve the problem, eventually transforming it linearly separable with a moderate number of hidden neurons.

Associative Memory and Nonlinear Dynamics

Next let us consider a simple feedback neural network model consisting of N mutually connected neurons where output $x_i(t)$ of neuron i at the time step t takes 1 (firing) or 0 (resting) as follows:

$$x_i(t+1) = 1\left(\sum_{j \neq i}^{N} w_{ij} x_j(t)\right),$$

$$w_{ij} = \frac{1}{N} \sum_{\mu=1}^{p} (2\xi_i^{\mu} - 1)\left(2\xi_j^{\mu} - 1\right),$$

where w_{ij} is the synaptic weight from neuron j to neuron i. The neuron i receives inputs from all the other neurons through w_{ij}. Stored patterns $\xi^{\mu} = \left(\xi_1^{\mu}, \ldots, \xi_N^{\mu}\right)$ $(\mu = 1, \ldots, p)$ are embedded as fixed-point attractors of convergent nonlinear dynamics in the recurrent neural network through the Hebbian learning. This model is called the associative memory model. The number of patterns that can be stored as stable fixed points has a certain limit [6]. This limitation is termed storage capacity, which can be theoretically derived by the statistical mechanical methods. The basin of attraction of stored patterns in the convergent dynamics of the feedback networks means error-correcting ability of the associative memory model. It is known that there exist stable fixed points, the number of which increases exponentially with N, besides the stored memory patterns. These are called spurious memory patterns. We can not distinguish the stored patterns and the spurious ones without the information of the memory patterns themselves, since both of them are fixed point attractors of convergent dynamics in the feedback networks of the associative memory.

Chaotic Neural Networks

Neuron models with chaotic dynamics are called chaotic neurons [7]. A discrete-time chaotic neuron model consists of the following terms: the internal states of the external inputs, the feedback inputs, and the relative refractoriness, where the refractoriness means the property of a biological neuron that the firing threshold increases for a certain period after firing. Essential features of the model to exhibit chaotic responses are that all the internal states show exponential decay and the output function of the neuron is continuous like a sigmoid function [7]. Neural network models that are composed of chaotic neurons are called chaotic neural networks. The dynamics of constituent neuron i in the chaotic neural network is represented by the following equations [8]:

$$\xi_i(t+1) = \sum_{j=1}^{M} v_{ij} \sum_{d=0}^{t} k_e^d A_j(t-d),$$

$$\eta_i(t+1) = \sum_{j=1}^{M} w_{ij} \sum_{d=0}^{t} k_f^d x_j(t-d),$$

$$\zeta_i(t+1) = -\alpha \sum_{d=0}^{t} k_r^d x_i(t-d) - \theta_i,$$

$$x_i(t+1) = f(\xi_i(t+1) + \eta_i(t+1) + \varsigma_i(t+1))$$

ξ_i, η_i, and ζ_i are internal state terms, k_e, k_f, and k_r are the decay parameters for external inputs, feedback inputs from the neurons in the network, and refractoriness of neuron i, respectively. x_i and f denote the output of neuron i and the output function that is usually a sigmoid function, respectively. A_j and θ_i are external input j and the threshold of neuron i, respectively. v_{ij} and w_{ij} denote synaptic weight to neuron i from external input j and synaptic weight to neuron i from neuron j, respectively. M and N are the number of the external inputs to neuron i and that of the neurons in the network, respectively.

The model of the chaotic neural networks is applied to associative memory networks and combinatorial optimization networks with chaotic dynamics beyond convergent dynamics of conventional models. An example of recalling behavior of dynamically associative memory in a chaotic neural network [8] is shown in Fig. 2.

In this example, 100 chaotic neurons are interconnected with synaptic weights to store four patterns in Fig. 2a. The filled and open squares in each pattern represent 1 (firing) and 0 (resting) in the neuronal output x_i, respectively. As shown in Fig. 2b, the process of recalling the stored patterns is neither convergent to a fixed point of a stored pattern nor periodic even if the synaptic connection weights among constituent neurons in the network are determined by the Hebbian learning with the stored patterns like the conventional associative memory. Such aperiodic recalling of the stored patterns is caused by the characteristics of the constituent neurons that can exhibit chaotic dynamics as a single neuron.

Synchronization of Neurons and its Modeling with Coupled Phase Oscillators

A visual object is processed in parallel at different visual areas of the brain. Each visual area extracts such specific visual features as orientation, brightness, color, and motion. To recognize the visual object as a whole, these features should be bounded. This is called the binding problem. One hypothesis of the feature binding mechanism is synchronization of neurons in the different visual areas. Some physiological experiments support the synchronization hypothesis. The coupled phase oscillator system is one of the simplest models, which captures essential properties of a general class of nonlinear oscillator systems [9]. Let us consider the following system of two phase oscillators,

$$\frac{d\phi_1}{dt} = \omega_1 + \Gamma(\phi_2 - \phi_1) + \xi_1,$$

$$\frac{d\phi_2}{dt} = \omega_2 + \Gamma(\phi_1 - \phi_2) + \xi_2,$$

where $\phi_i \in [0,2\pi)$ and ω_i denote the phase of a periodic oscillation and its natural frequency for oscillator i, and ξ_i is additive noise. The function of the phase response curve (RPC) $\Gamma(\cdot)$, which is periodic with the period 2π represents the interaction between the two oscillators. Here, $\Gamma(\cdot)$ is assumed to be an odd function. In the case without the additive noise, the dynamics of the phase difference $\Delta\phi$ with $\Delta\phi = \phi_2 - \phi_1$ is described by the following equation,

$$\frac{d\Delta\phi}{dt} = \Delta\omega - 2\Gamma(\Delta\phi),$$

$$\Delta\omega = \omega_2 - \omega_1.$$

The synchronization of the two oscillators occurs under the equilibrium condition that $\Delta\omega = 2\Gamma\ (\Delta\phi)$, while they desynchronize otherwise [9,10]. The stochastic synchronization emerges even in the case that the frequency mismatch $\Delta\omega$ is large enough to be

N

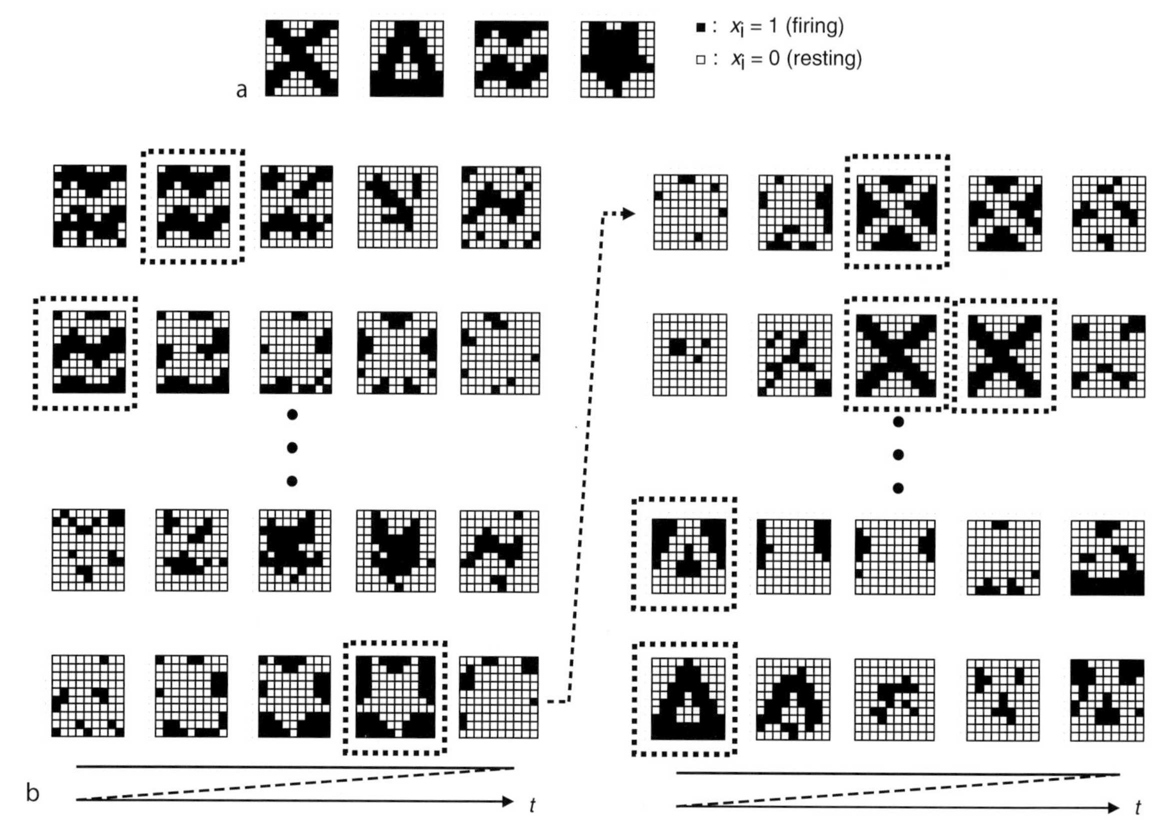

Neural Networks. Figure 2 Associative memory in a chaotic neural network. (a) Stored patterns of the dynamically associative memory [8]. (b) Example of the recalling pattern sequence of the dynamically associative memory in a chaotic neural network.

$\Delta\omega \neq 2\Gamma(\Delta\phi)$ for any $\Delta\phi$ if the noise has appropriate characteristics. Thus, interaction between nonlinear dynamics and noise is an important problem.

References

1. McCulloch WS, Pitts WH (1943) A logical calculus of the ideas immanent in neural nets. Bull Math Biophys 5:115–133
2. Rosenblatt F (1961) Principles of neurodynamics. Spartan, Washington, DC
3. Rumelhart DE, McClelland JL PDP Research Group (1986) Parallel distributed processing, vols 1/2, MIT Press, Cambridge
4. Watanabe M, Masuda T, Aihara K (2003) Forward propagating reinforcement learning: biologically plausible learning method for multi-layer networks. Biosystems 71:213–220
5. Waelti P, Dickinson A, Schultz W (2001) Dopamine responses comply with basic assumptions of formal learning theory. Nature 412:43–48
6. Okada M (1996) Notions of associative memory and sparse coding. Neural Netw 9:1429–1458
7. Aihara K, Takabe T, Toyoda M (1990) Chaotic neural networks. Phys Lett A 144:333–340
8. Adachi M, Aihara K (1997) Associative dynamics in a chaotic neural network. Neural Netw 10:83–98
9. Kuramoto Y (1984) Chemical oscillations, waves, and turbulence. Springer-Verlag, Berlin
10. Pikovsky A, Rosenblum M, Kurths J (2001) Synchronization. Cambridge University Press, Cambridge

Neural Networks for Control

URI ROKNI
Center for Brain Science and Swartz Center for Compentational Neuroscience, Harvard, Cambridge, MA, USA

Definition

In this essay we deal with ►neural networks which interact with the external environment, referred as a ►plant, in order to achieve a specified goal (Fig. 1). The network receives goal related inputs and sensory inputs from the plant, and in response influences the plant by activating ►effectors. At any given time, the state of the plant is fully specified by a set of variables termed the ►state variables. The dynamics of the plant is described

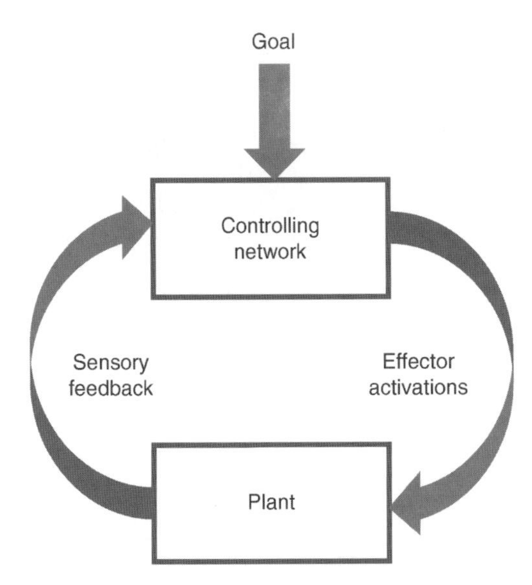

Neural Networks for Control. Figure 1 A controlling network. The network interacts with the external environment, referred as the plant, to achieve a desired goal.

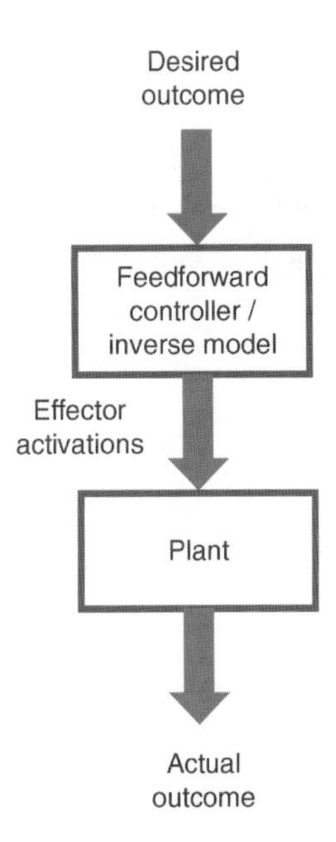

Neural Networks for Control. Figure 2 A feedforward controller. The feedforward controller does not use sensory feedback. It maps the desired outcome to effector activations, inversely to the mapping from effector activations to actual outcome performed by the plant.

by equations of motion which dictate how the state changes as a function of the current state and the current activation of the effectors. The goal may be to reach a desired state, or more generally to minimize a specified ►cost function. Generally, a device which solves such problems is referred as a ►controller. In this essay we are concerned specifically with the implementation of controllers by neural networks in the ►central nervous system (CNS).

Description of the Theory

►**The reaching example**. Consider the problem of arm reaching to a specified target. The plant is the arm and the effectors are the arm muscles. The arm state variables are the joint angles and joint velocities. The equations of motion of the arm specify the joint angle accelerations as a function of the arm state and muscle activations. To reach the target, the controller needs to convert goal related inputs, which represent the target position, and sensory information on the arm state into the appropriate muscle activations.

The Operation of Controllers

Controllers which do not use online sensory feedback from the plant are referred as ►feedforward (open-loop) controllers (Fig. 2). Feedforward controllers are often termed ►inverse models of the plant, because they map the desired outcome of the plant to effector activations, inversely to how the plant maps the effector activations to the actual outcome. For example, the arm plant maps muscle activations into hand acceleration. To properly control the arm, a feedforward controller needs to invert this relation and map the desired hand acceleration to the appropriate muscle activations. Feedforward controllers may be inaccurate because of noise in the plant, or because they do not invert the plant dynamics accurately. In these cases, online sensory feedback to the controller may improve its performance. Controllers which use online sensory feedback are referred as ►feedback (closed-loop) controllers.

Usually, in order to control the plant effectively at any given time, the controller needs to know the plant state, at least implicitly. Therefore, the operation of a controller can be divided into: (i) estimating the plant state, and (ii) mapping the estimated plant state and goal related inputs to effector activations (Fig. 3).

To estimate the plant state at a given time the system may rely on: a previous estimate of the plant state, current sensory feedback, knowledge about the plant dynamics and copies of the current effector activations (referred as ►efferent copies). A module which estimates the plant state from the efferent copies is known as a ►forward model, because it models the plant dynamics in the normal flow from effector activations to plant state. Typically, the different sources of information on the plant state are noisy. For accurate ►state estimation, the system needs to combine these different sources optimally, giving more weight to the more reliable sources of information. An optimal state estimator is known as a ►Bayesian filter. Commonly, engineers approximate the plant dynamics by linear dynamics, and the noise by Gaussian noise. In these

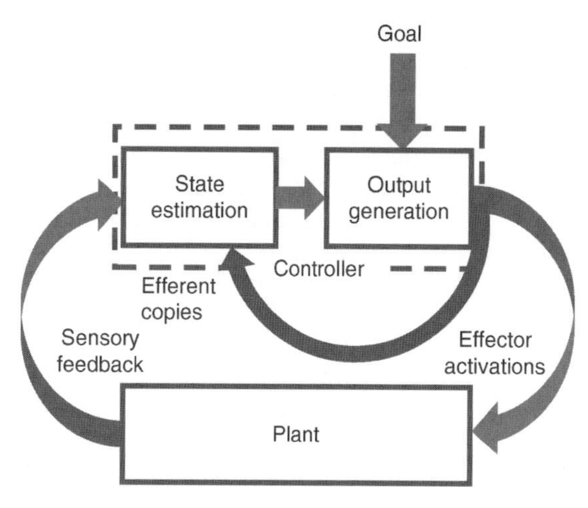

Neural Networks for Control. Figure 3 The control problem is decomposed into two problems. First, estimate the plant state from the sensory feedback and efferent copies of the effector activations. Second, map the estimated plant state and the goal to effector activations.

cases, there is a closed form solution of the Bayesian filter, known as a ►Kalman filter.

The problem of generating the appropriate effector activations given the current plant state may be straightforward when the effector activations can set the plant in its desired state, immediately. But typically, the plant responds gradually and the controller needs to generate the current effector activations that are best *in the long term*. For example, at the beginning of reaching, activating the arm muscles strongly may bring the arm closer to its target at first, but later cause the arm to overshoot. To reach optimally, the controller needs to take into account the future effects of the current effector activations. This problem has been extensively studied in the field of ►optimal control theory.

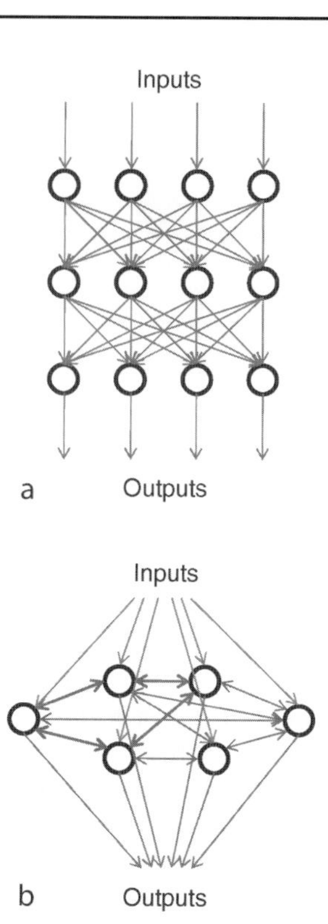

Neural Networks for Control. Figure 4 Feedforward versus recurrent networks. (a) Feedforward networks are arranged in layers without loops in their connections. (b) Recurrent networks contain loops in their connections.

Computing and Learning with Neural Networks

Neurons are believed to be the basic computational elements in the CNS. A neuron integrates action potentials from different cells, performs temporal filtering and transmits the result to other neurons in the form of action potentials. By combining neurons into networks much more complicated computations can be carried out. Networks that are arranged in layers of neurons, each layer projecting to the next, are known as ►feedforward networks (Fig. 4a). Inputs traveling through several layers of neurons may undergo complicated nonlinear transformations. Networks with connections that allow activity to travel in loops are known as ►recurrent networks (Fig. 4b). Inputs traveling through multiple loops in the network may undergo complicated temporal filtering.

To learn a desired input-output relation the strengths of the synaptic connections between neurons need to be tuned. Training data may consist of pairs of inputs and desired outputs. A rule that specifies how synaptic strengths are changed, given the training data, is generally referred as a ►learning rule. A successful learning rule should decrease the output errors of the network (at least on average), so that after repeated applications of the rule the network performance improves. The ►back-propagation learning rule achieves this for feedforward networks by propagating measured output errors back through the network layers to obtain errors of the synaptic strengths. The back-propagation learning rule has been generalized for tuning the synaptic strengths of recurrent networks as well (for a review of neural network theory see [1]).

Control with Neural Networks

The CNS is believed to implement controllers by networks of neurons. In particular, it has been suggested that controllers are implemented by networks that are feedforward, except for recurrent projections from the output layer back to the input layer [2,3, fig. 5]. Static goal related inputs to the network travel through the layers to produce the effector activations. Efferent copies of these muscle activations are fed back to the input layer, thus affecting the effector activations at the

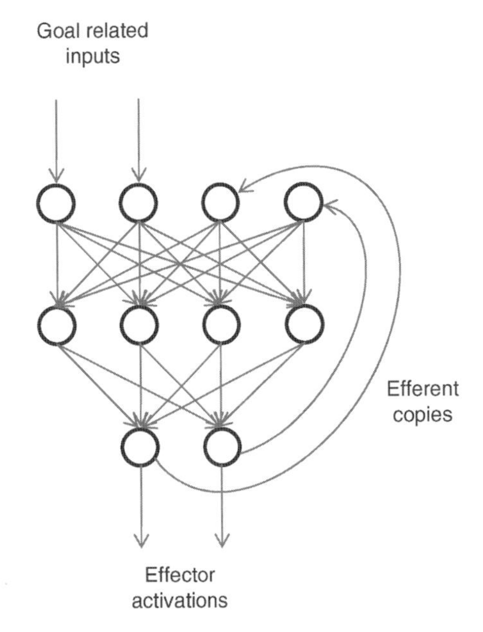

Neural Networks for Control. Figure 5 A sequence generating network. By sending efferent copies back to the input layer the network generates a sequence of effector activations. The network may be used as a feed forword controller.

next time step. In this fashion, the network implements a feedforward controller that maps static goal related inputs to a sequence of effector activations. This approach was generalized to feedback control by adding delayed sensory feedback to the network, and assuming the network is fully recurrent [4].

An alternative model asserts that prior to movement the entire trajectory of the effector activations is represented by static neural activity, where different neuron populations represent the planned effector activations at different times [5]. Recurrent network dynamics are used to find a trajectory that obeys task constraints, e.g. reaching the target, and minimizes a smoothness cost function. To execute the movement, the static representation of the planned trajectory is converted to effector activations that unfold in time.

Another study proposes how the CNS may estimate the plant state [6]. This study implements a Kalman filter for estimating the state of a linear arm model with Gaussian noise, by a recurrent neural network. The activity of the population of neurons in the network represents the current estimate of the arm state. This state estimate is modified by external inputs which convey sensory information on the plant state. The state estimate is also affected by the previous state estimate via the recurrent network connections. Thus, the recurrent connections encode knowledge of the arm dynamics. When simulating the network dynamics its activity estimated the plant state nearly optimally.

Learning Neural Networks for Control

When the CNS encounters an unknown environment (e.g. new loads applied to the arm) it needs to adjust the controlling network. This adjustment is believed to occur by changing synaptic connections between neurons. To implement conventional network learning algorithms, such as back-propagation, the CNS needs to provide the output errors of the network. However, in control problems rather than being provided with the output errors of the network it is provided with the output errors of the plant. Not knowing the plant, which is the problem to begin with, how can the network translate the plant errors into network errors?

The ►direct inverse learning approach solves this problem by obtaining training data from the plant when it is randomly activated [7]. Each input-output pair of the plant is provided to the network *inversely*, as desired output-input, respectively. By training the network with this data it learns the inverse model of the plant. This method suffers from several problems. First, it learns offline. Second, the plant may be non-invertible, meaning that different plant inputs generate the same output. Thus, for a given input (plant output) the network may learn the mean of multiple desired outputs (plant inputs). Generally, this mean will not be either of the correct network outputs (plant inputs). Finally, direct inverse learning minimizes the error in the plant *inputs*, which may not be optimal for minimizing the error in the plant *outputs*.

These problems are solved by an alternative method, known as ►distal supervised learning, which learns the inverse model in tandem with a forward model [8, fig. 6]. During operation of the system, the forward model is trained to imitate the plant by providing the plant input-output pairs as training data. At the same time, the plant errors are back-propagated through the forward model to obtain the output errors of the inverse model. These error signals are used to train the inverse model.

Another approach for learning an inverse model, known as ►feedback error learning, assumes a sensory feedback loop in parallel with the inverse model [9, fig. 7]. The same feedback loop that is used for control, is also used as an error signal for training the inverse model. This approach requires a stable sensory feedback loop to begin with, and leaves open the question of how the feedback loop is learned.

Finally, a different approach, known as ►reinforcement learning, uses noise in the controlling network and a ►scalar performance measure known as the ►reward signal [10]. By correlating the noise with the reward signal, the reinforcement learning rule changes synaptic weights such that outputs that yield greater reward become more likely. Over time the average reward increases to its maximum. Reinforcement learning does

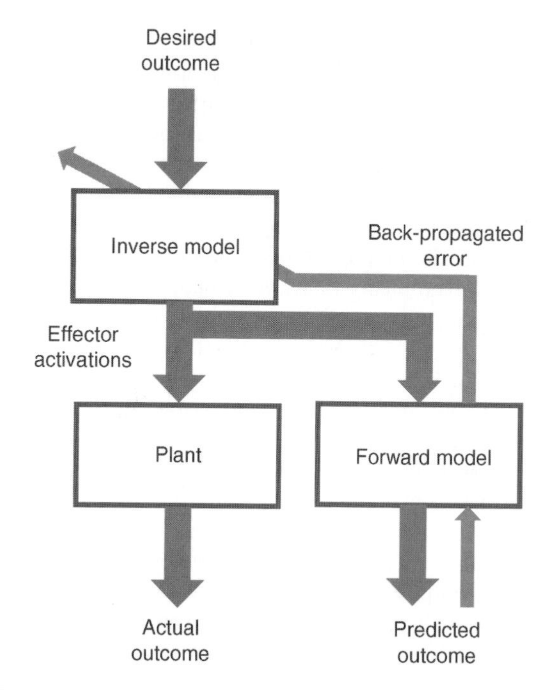

Neural Networks for Control. Figure 6 Distal learning. A network which unitates the plant, referred as a forward model, is learned. At the same time, the errors in the plant output are backpropageted through the forward model network to obtain the errors of the inverse model (green arrow). These errors are used to train the network of the inverse model.

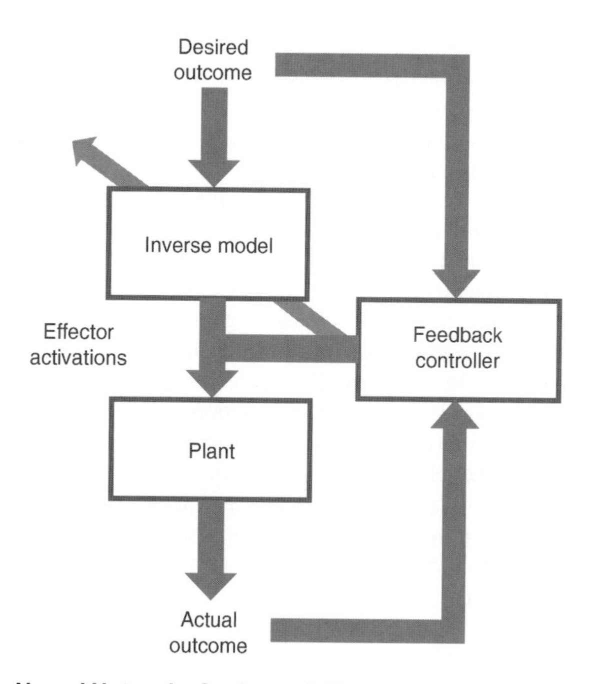

Neural Networks for Control. Figure 7 Error feedback learning. The same sensory feedback that is used for controlling the plant is also used for training the inverse model (green arrow).

not rely on back-propagation of errors. But, it is slower than other methods because it requires averaging over noise and because it uses a scalar performance measure, which is less informative than a ►vector of output errors.

References

1. Hertz J, Krogh A, Palmer G (1991) Introduction to the theory of neural computation. Perseus books, Reading, MA
2. Massone L, Bizzi E (1989) A neural network model for limb trajectory formation. Biol Cybern 61(6):417–425
3. Jordan MI (1986) Attractor dynamics and parallelism in a connectionist sequential machine. In proceedings of the eighth annual conference of the cognitive science society. Erlbaum, Hillsdale, NJ
4. Joshi P, Maass W (2005) Movement generation with circuits of spiking neurons. Neural Comput 17(8): 1715–1738
5. Kawato M, Maeda Y, Uno Y, Suzuki R (1990) Trajectory formation of arm movement by cascade neural network model based on minimum torque-change criterion. Biol Cybern 62(4):275–288
6. Denève S, Duhamel JR, Pouget A (2007) Optimal sensorimotor integration in recurrent cortical networks: a neural implementation of kalman filters. J Neurosci 27 (21):5744–5756
7. Kuperstein M (1988) Neural model of adaptive hand-eye coordination for single postures. Science 239 (4845):1308–1311
8. Jordan MI, Rumelhart DE (1992) Forward models: supervised learning with a distal teacher. Cognitive Science 16:307–354
9. Miyamoto H, Kawato M, Setoyama T, Suzuki R (1988) Feedback-error-learning neural network for trajectory control of a robotic manipulator. Neural Netw 1(3): 251–265
10. Seung HS (2003) Learning in spiking neural networks by reinforcement of stochastic synaptic transmission. Neuron 40(6):1063–1073

Neural Oscillator

►Central Pattern Generator

Neural Pattern Generator

►Central Pattern Generator

Neural Progenitors/Radial Glia

Definition

Generic terms for immature or neural stem cells that can differentiate into a number of neural cell types (different types of neurons, astrocytes, oligodendrocytes).

►Evolution of the Posterior Tuberculum and Preglomerular Nuclear Complex

Neural Regulation of the Pupil

Peter G. Smith
Department of Molecular and Integrative Physiology and Kansas Intellectual and Developmental Disabilities Research Center, University of Kansas Medical Center, Kansas City, KS, USA

Definition

The pupil is an aperture formed by the margins of the iris. Pupillary diameter determines the amount of light traversing the lens to focus on the retina, producing a visual image. The primary factor determining pupil diameter is the concentration of autonomic transmitters at contractile cells of the iris. This is regulated by the ►sympathoadrenal and parasympathetic nervous systems.

Characteristics

Quantitative Description

Pupillary diameter is determined by the relative tone of two opposing contractile elements within the iris. The pupillary sphincter (constrictor) muscle abuts the margin of the pupil. It consists of smooth muscle fibers with their long axes oriented circumferentially with respect to the pupil (Fig. 1).

When these cells contract, the pupil margins are drawn closer together, causing a decrease in pupil diameter (constriction, or ►miosis).

The iris contains a second contractile component, the pupil dilator (radial) muscle. This structure is composed of myoepithelial cells oriented radially, with their long axes directed outward from the pupil center as if spokes in a wheel (Fig. 1). When they contract, the pupil margins are drawn away from the center, increasing pupil diameter (dilation, or ►mydriasis).

Pupil diameter of is thus determined by the relative activity, or tone, of these two opposing muscles.

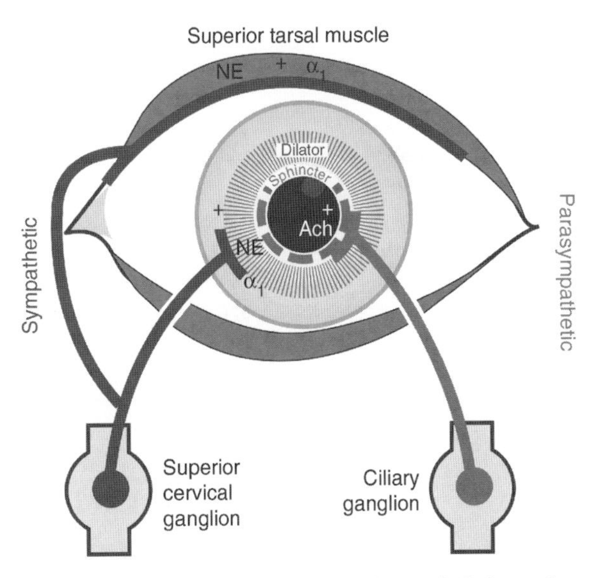

Neural Regulation of the Pupil. Figure 1 Schematic diagram of the innervation of the iris and related structures. The iris sphincter (constrictor) muscle contains smooth muscle cells oriented circumferentially. It is innervated by parasympathetic excitatory (+) nerves from the ciliary ganglion, which release acetylcholine (Ach). The dilator (radial) muscle consists of myoepithelial cells oriented radially. This muscle, as well as the superior and inferior tarsal muscles of the eyelids, is innervated by excitatory (+) sympathetic nerves which release norepinephrine (NE) that act on alpha-1 adrenergic receptors.

Higher Level Structures

Regulation of pupil diameter involves integration of afferent information transmitted by the optic nerve to diencephalic processing structures, as well as emotive pathways incorporating fear, anxiety, pleasure and pain.

Constrictor pathways. Light elicits pupil constriction by way of a subcortical reflex initiated by light passing through the anterior eye and impinging upon photoreceptor cells of the retina. Photoreceptive impulses travel along retinal ganglion cell axons in the optic nerve, which hemi-decussate at the optic chiasm, and travel to the superior colliculus to enter the pretectal nucleus where they synapse on pretectal neurons (Fig. 2).

These neurons send projections to the ipsilateral and contralateral Edinger–Westphal nucleus [1]. This nucleus contains preganglionic parasympathetic neurons, which send axons in cranial nerve III (oculomotor) to postganglionic neurons of the ciliary ganglion, whose axons travel in the short ciliary nerves to innervate the sphincter muscle of the iris (Fig. 2).

Pupillary constriction is also elicited in association with accommodation, which occurs when viewing near objects. Photoreceptive impulses conducted to the visual cortex activate efferent projections to the

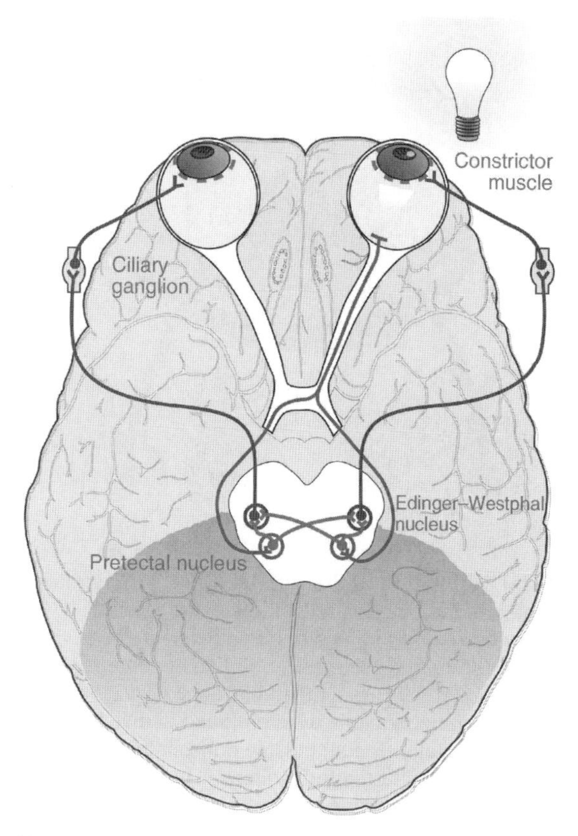

Neural Regulation of the Pupil. Figure 2 Parasympathetic light reflex pathway. When the eye is illuminated, retinal ganglion neurons projecting axons (*green*) to midbrain excite pretectal neurons (red). Pretectal neurons project excitatory synapses to the ipsilateral and contralateral Edinger–Westphal nucleus, which contains parasympathetic preganglionic neurons (*blue*). These neurons provide excitatory innervation to postganglionic neurons in the ciliary ganglion, which innervate the pupillary constrictor (sphincter) muscles in each eye, where they elicit pupil constriction.

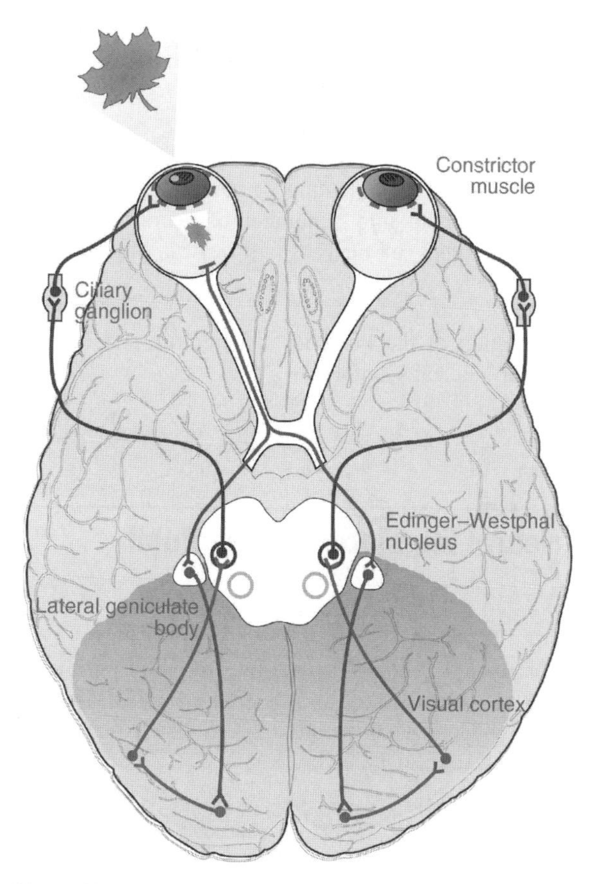

Neural Regulation of the Pupil. Figure 3 Pupillary constriction associated with accommodation. When an object is placed close to the eye for near vision, retinal ganglion axons (green) excite neurons bilaterally in the lateral geniculate body (purple), which project to the visual cortex and associated structures. Descending pathways provide excitatory innervation ipsilaterally to neurons in the Edinger–Westphal nucleus (blue), which excite postganglionic neurons in the ciliary ganglion to elicit constrictor muscle contraction and pupil constriction.

Edinger–Westphal nucleus, which activate oculomotor nerve pathways to the iris constrictor muscle (Fig. 3).

Dilator pathways. Central nervous system pathways mediating pupil dilation are complex and less well defined. Generally, factors leading to sympathoadrenal activation cause pupil dilation. The hypothalamus serves as a central integrative center for pupil dilation [1]. Structures that influence hypothalamic output include inputs from cognitive centers (prefrontal cortex), limbic system including the amygdala, sensory pathways involved in pain perception, and sensory pathways involved in homeostatic regulation such as blood chemical composition (glucose, oxygen, carbon dioxide, pH) and blood pressure. Hypothalamic neurons project to neurons in the lateral medulla, which innervate ipsilateral spinal cord intermediolateral neurons (ciliospinal center of Budge). Axons from these preganglionic neurons exit at C8–T2 and ascend proximate to the carotid artery to the superior cervical ganglion. Postganglionic axons travel in the internal carotid nerve and enter the orbit with the ophthalmic division of the trigeminal nerve. Sympathetic fibers travel in the nasociliary branch of the trigeminal nerve ophthalmic division and long ciliary nerves to innervate the iris dilator muscle (Fig. 4).

Higher Level Processes

Constrictor pathway. The primary stimulus activating parasympathetic pathways is photoreceptive input. The pupillary light reflex is responsive to the intensity of ambient light impinging on retinal rods and cones, as well as melanopsin-expressing retinal ganglion cells that are directly photosensitive [2]. Increasing photointensity produces graded activation of axons projecting to the

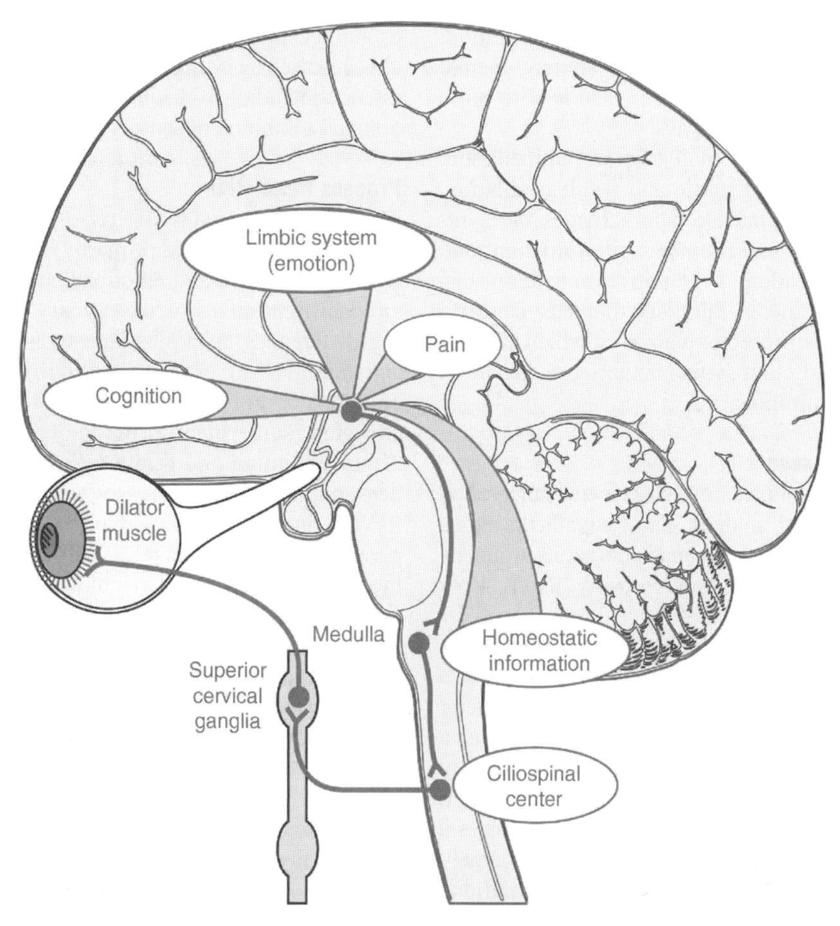

Neural Regulation of the Pupil. Figure 4 Sympathetic pathways mediating pupil dilation. Neurons in the paraventrictular region of the hypothalamus receive input from multiple pathways encoding cognitive information from the cortex, emotion from the limbic system, pain from somatosensory pathways, and homeostatic information concerning blood pressure and composition from ascending pathways. Hypothalamic neurons project to neurons in the medulla, which send axons to the ciliospinal center of the intermediolateral column of the spinal cord. Preganglionic ciliospinal neurons project axons through the ventral roots, which ascend with the paravertebral chain to synapse upon postganglionic sympathetic neurons of the superior cervical ganglion. Postganglionic axons traverse intracranially to enter the orbit and innervate the dilator (radial) muscle of the eye.

pretectal and Edinger–Westphal nuclei, which elicits graded firing of oculomotor neurons bilaterally. The net result is pupillary constriction corresponding roughly to light intensity.

Pupil constriction also occurs when the eye focuses on a near object. A coordinated triad of responses is initiated, involving bilateral ocular convergence, ciliary muscle contraction leading to a more spherical lens with greater refractive index, and pupil constriction. In contrast to the light reflex, accommodation involves higher visual cortical structures associated with near object focality, activating efferent preganglionic motor neurons of the Edinger–Westphal nucleus.

While light intensity and accommodation are major factors eliciting pupil constriction, it is noteworthy that somatic input can also influence this pathway. For example, lingual nerve stimulation can actually inhibit preganglionic activity, thus attenuating pupillary constriction [3].

Dilator pathways. Activation of the pupillary dilator pathway is closely linked to sympathoadrenal tone. The sympathetic nervous system in conjunction with the adrenal medulla is activated when the organism enters into a "fight or flight" response. This occurs when a challenge is perceived – as innocuous as mental arithmetic or as threatening as being confronted by a bear. Visual information recognized as threatening is conveyed via the thalamus to the amygdala and on to the hypothalamus, where descending pathways lead to sympathetic activation and adrenomedullary catecholamine release, leading to a number of responses including pupillary dilator muscle contraction. This response can be further modified by visual information from the sensory and prefrontal cortex, which can alter amygdala output by reinforcement or inhibition. Conditions producing mild or intense anxiety, mental concentration, or sexual arousal can also lead to pupil dilation through similar pathways.

Other types of sensory stimuli promote pupil dilation. Loud noises can lead to startle and result in dilation. In particular, mild to intense pain activates the sympathoadrenal system and promotes pupil dilation, and olfactory inputs leading to emotive activation can influence pupil diameter. Pupillary diameter can also be influenced by baroreceptor pathways, which monitor blood pressure and elicit sympathetic nerve activation when blood pressure falls [4].

Lower Level Processes

The iris dilator and sphincter muscles receive innervation from three ipsilateral sources: parasympathetic axons from the ciliary ganglion, sympathetic axons from the superior cervical ganglion, and sensory axons from the trigeminal ganglion.

Sensory innervation to the iris derives primarily from small diameter trigeminal ganglion neurons giving rise to C fibers, which innervate the dilator and sphincter muscles. These presumptive nociceptor axons display a peptidergic phenotype that includes immunoreactivity for substance P, calcitonin gene-related peptide, galanin and somatostatin. These sensory neurons also express a splice variant of choline acetyltransferase that is largely restricted to peripheral neurons [5]. The function of iridial sensory fibers remains unclear, but they likely provide the CNS with information on noxious stimuli and intraocular pressure, and may subserve efferent functions via postjunctional effects of peptides and possibly acetylcholine on iris smooth muscle, and prejunctional actions on co-projecting autonomic axons.

Postganglionic neurons from the ipsilateral cilliary ganglion provide parasympathetic excitatory innervation to the iris sphincter muscle. Acetylcholine from parasympathetic varicosities elicits contraction via M_3 muscarinic receptors [5].

The primary neural mechanism for pupil dilation is noradrenergic sympathetic innervation of the dilator muscle. Sympathetic axons release norepinephrine within the radial muscle, which acts on excitatory alpha 1A adrenergic receptors [6].

While cholinergic effects dominate in the sphincter muscle and noradrenergic effects dominate in the dilator muscle, it should be noted that sympathetic and parasympathetic nerves co-project to both muscles [5], implying cross-talk between spatially proximate axons. Indeed, acetylcholine from parasympathetic nerves acts on sympathetic prejunctional M_2 muscarinic receptors to inhibit norepinephrine release [7], indicating parasympathetic presynaptic inhibition of sympathetic neurotransmission. Sympathetic neurotransmission can be inhibited prejunctionally by other mechanisms, including α_2 autoreceptors and histaminergic receptors. Noradrenaline released by sympathetic nerves may elicit additional effects, including β-mediated sphincter muscle relaxation and inhibition of sensory neuropeptide release [8].

Process Regulation

The contractile states of iris dilator and constrictor muscles are determined primarily by 3 factors: ambient light, near vision, and emotional state. With moderate ambient light intensity, the primary variable determining differences in pupil diameter is sympathoadrenal activation. If an individual is normotensive and not emotionally aroused, an intermediate pupil diameter is expected (somewhere in the midrange between about 2 mm minimum and 8 mm maximum). In total darkness, pupil diameter increases toward its maximum.

With increasing light, pupil constriction occurs as parasympathetic nerves release acetylcholine to contract the sphincter muscle and to inhibit release of norepinephrine from sympathetic axons innervating the dilator muscle. If only one eye is exposed to light, both ipsilateral and contralateral pupils constrict (►consensual light reflex). This occurs because midbrain pretectal fibers project to the contralateral as well as the ipsilateral Edinger–Westphal nucleus (Fig. 2). A similar mechanism occurs with near-vision in which pupil diameter is constricted directly and consensually, although this does not involve pretectal neurons (Fig. 3).

Function

By constricting the pupil in bright light, light entering the posterior chamber is reduced, preventing excessive bleaching of rods and cones and protecting the eye from possible damage. A reduction in pupil diameter in near vision increases depth of field. As in photography where a smaller aperture provides a greater range of distances in which objects appear in focus, reduced pupil size works similarly, which is important in near vision where greater refractive index of the lens limits the depth of field.

Sympathoadrenal activation dilates the pupil, which has converse effects. In fight or flight, mydriasis serves two functions. First, under threatening conditions, it may be useful to have an enlarged pupil to gather more light and to have a larger peripheral field, making the individual more perceptive to environmental cues. Second, mydriasis occurs in conjunction with alpha adrenoceptor-mediated contraction of the superior and inferior tarsal smooth muscles of the eyelid, which increase palpebral fissure width. Together, these provide important body language cues. The individual may appear "wide-eyed" with rage, signaling a state of emotional activation. On a more subtle note, pupil diameter increases when an individual perceives another to be romantically attractive, thus providing another important cue.

Pathology

Neurological disturbances can occur at several sites, leading to differences in diameters of the two pupils (►anisocoria). Loss of afferent transmission along one optic nerve results in a normal consensual pupillary light reflex when the eye with intact innervation (contralateral to the lesion) is illuminated, but diminished response (apparent dilation) when the light is alternated to the affected eye. This is referred to as the Marcus Gunn pupil.

Degenerative conditions in the central nervous system can affect the light reflex. A fairly common one in advanced syphilis is the ►Argyll–Robertson pupil. The affected eye fails to respond to light but the pupil constricts normally during near vision accommodation. The reason appears to be that pretectal neurons mediating the light reflex degenerate selectively in this disease. Since pretectal neurons are not involved in accommodation, that reflex is preserved.

Damage to efferent postganglionic parasympathetic axons can lead to loss of constriction in response to light and accommodation. This is referred to as Adie's pupil. It is frequently accompanied by an elevated response to cholinomimetic drugs due to parasympathetic denervation supersensitivity.

Interruption of the sympathetic pathway to the pupil results in Horner's syndrome. This consists of unilateral miosis (constrictor predominance in the absence of dilator tone), ptosis (drooping of the upper eyelid and diminished palpebral fissure width due to loss of alpha adrenergic activation of the tarsal smooth muscle), anhydrosis (loss of sympathetically mediated facial sweating), and facial flushing (loss of sympathetic vasoconstrictor tone). These symptoms are all due to interruption of sympathetic innervation to the head. Damage can occur anywhere along the pathway, including higher level first order neurons of the hypothalamus or medulla (frequently due to a stroke or tumor), second order preganglionic neurons leaving the spinal cord and traveling with the carotid artery (tumor, aneurism, trauma) or to the third order postganglionic axons (internal carotid artery dissection, viral infection). Third order lesions can be distinguished from higher order deficits by placing a drop of the hydroxyamphetamine onto the eye. This drug releases norepinephrine from intact sympathetic varicosities. If the lesion involves postganglionic axons, then hydroxamphetamine will not cause mydriasis normally seen in the normal eye because the nerve terminals have degenerated.

Therapy

Adie's pupil often occurs in young females and frequently resolves spontaneously. Because it often accompanies migraines or cluster headaches, it may improve with effective headache management. In Marcus Gunn pupil or Horner's syndrome where a tumor may compress neural pathways, surgical resection or chemotherapy may reverse the deficit. In Horner's syndrome, nerve regeneration and functional restoration usually is more complete if the lesion involves only postganglionic axons. In Argyll–Robertson pupil where a diagnosis of syphilis is confirmed, patients often respond beneficially to antibiotics.

References

1. Pickard GE, Smeraski CA, Tomlinson CC, Banfield BW, Kaufman J, Wilcox CL, Enquist LW, Sollars PJ (2002) Intravitreal injection of the attenuated pseudorabies virus PRV Bartha results in infection of the hamster suprachiasmatic nucleus only by retrograde transsynaptic transport via autonomic circuits. J Neurosci. 22:2701–2710
2. Hattar S, Lucas RJ, Mrosovsky N, Thompson S, Douglas RH, Hankins MW, Lem J, Biel M, Hofmann Foster RG, Yau KW (2003) Melanopsin and rod-cone photoreceptive systems account for all major accessory visual functions in mice. Nature 424:76–81
3. Tanaka T, Kuchiiwa S, Izumi H (2005) Parasympathetic mediated pupillary dilation elicited by lingual nerve stimulation in cats. Invest Ophthalmol Vis Sci 46:4267–4274
4. Heymans C, Neil E (1958) Reflexogenic areas of the cardiovascular system. Little, Brown & Co, Boston
5. Yasuhara O, Aimi Y, Shibano A, Matsuo A, Bellier JP, Park M, Tooyama I, Kimura H (2004) Innervation of rat iris by trigeminal and ciliary neurons expressing pChAT, a novel splice variant of choline acetyltransferase. J Comp Neurol 472:232–245
6. Yu Y, Koss MC (2003) Studies of alpha-adrenoceptor antagonists on sympathetic mydriasis in rabbits. J Ocul Pharmacol Ther 19:255–263
7. Jumblatt JE, Hackmiller RC (1994) M_2-type muscarinic receptors mediate prejunctional inhibition of norepinephrine release in the human iris-ciliary body. Exp Eye Res 58:175–180
8. Fuder H (1994) Functional consequences of prejunctional receptor activation or blockade in the iris. J Ocul Pharmacol 10:109–123

Neural Respiratory Control During Acute Hypoxia

DIETHELM W. RICHTER, MICHAEL MÜLLER
DFG Research Center Molecular Physiology of the Brain, Department of Neurophysiology, Georg-August-Universität Göttingen, Göttingen, Germany

Definition

In mammals, breathing movements are driven by an oscillatory activity generated by a neuronal network located in the lower ►brainstem. Being exposed to

N

severe ►hypoxia this network shows a typical sequence of responses. This essay focuses on the molecular physiology underlying the acute hypoxic responses of this respiratory rhythm generating network (►respiratory network) without considering hypoxia-induced changes in gene regulation.

Characteristics

Structural and Functional Organization of the Respiratory Network

Breathing originates from respiratory movements of the chest and abdomen that are driven by rhythmic discharges of motoneurons innervating inspiratory muscles (diaphragm, external intercostal muscles) and expiratory muscles (internal intercostal muscles and abdominal muscles). The normal (eupneic) pattern of this central respiratory activity is separated into three discrete phases – inspiration, post-inspiration and expiration (Fig. 1).

The phasic activity pattern is generated by a neuronal network composed of distinct sub-populations of pre-motoneurons located within the lower brainstem (medulla oblongata). The kernel of the network is localized bilaterally in the ►pre-Bötzinger complex (►pre-Böztinger neurons and rhythm generation) – a distinct region in the "ventral respiratory group", in the neighboring Bötzinger complex, and – within the "dorsal respiratory group" – in the ventrolateral nucleus of the solitary tract. Another rhythmically active region is localized around the retrotrapezoid nucleus/parafacial respiratory group containing expiratory neurons (►parafacial neurons and respiratory control). In summary, several brainstem regions cooperate to generate a neural respiratory activity that controls effortless quiet breathing movements (►eupnea) (►medullary raphe nuclei and respiratory control; ►pontine control of respiration). The medullary respiratory network communicates synaptically with several other vital neuronal control systems by which it essentially influences amongst others the control of vocalization, swallowing and cardiovascular regulation.

The respiratory neurons receive tonic synaptic excitation from arterial chemoreceptors (►carotid body chemoreceptors and respiratory drive) and the reticular formation that provides a basic activity level on top of which rhythmic activity is generated. An alternating volley of excitatory and inhibitory synaptic inputs then ensures a powerful membrane voltage management and vigorously controls activation or inactivation of ►voltage-gated ion channels ("synaptic voltage clamp"), thereby effectively determining a characteristic rhythmic discharge pattern [1].

Inhibitory synaptic interconnections between the respiratory neurons are vital. ►GABA and ►glycine largely control chloride ion (Cl^-) currents that do not only terminate neuronal discharges, but also precisely define the onset of activity, the activity pattern and the duration of discharge for the different types of respiratory neurons. Furthermore, these currents shape a smooth transition between the individual respiratory phases.

Energy Supply of the Network

Just as other parts of our brain, the respiratory network requires a constant O_2 supply to maintain normal oxidative phosphorylation. "►Cellular respiration" occurs at complex IV of the mitochondrial electron transport chain where O_2 accepts 2 electrons – donated from either NADH or $FADH_2$ – and is reduced to water. In the presence of O_2, the electron transport chain generates and maintains an inwardly directed proton gradient across the inner mitochondrial membrane, thereby supplying the driving force fueling ►ATP synthesis by the mitochondrial ATP synthase (complex V or F_0F_1 ATPase). Yet, when tissue pO_2 drops below 5 mm Hg the proton gradient collapses and ATP synthesis by oxidative phosphorylation ceases. Under such conditions of severe ►hypoxia, only anaerobic glycolysis remains intact, which contributes no more than 1–5% of the normal ATP production. This is, however, not sufficient to ensure undisturbed neuronal function over a longer period of time.

The duration of acute hypoxia that can be tolerated varies among the different brain regions. Neurons are more vulnerable than glial cells, and among the most ►anoxia/►ischemia vulnerable neurons are hippocampal CA1 and cortical pyramidal neurons, cerebellar Purkinje neurons and medium spiny neurons of the striatum. In contrast, most of the brainstem including the respiratory network is far less vulnerable to anoxic/ischemic insults.

In neurons cultured from mouse respiratory center, mitochondria form functionally coupled, dynamically organized aggregates such as "chains" and "clusters." Mitochondrial chains predominate in dendrites and reveal a directed movement that is arrested upon mitochondrial depolarization or blockade of mitochondrial ATP synthesis as it occurs under severe hypoxia [2]. Depolymerization of the cytoskeleton also disrupts mitochondrial chain movements and the mitochondria accumulate in the soma. The consequence of such immobilized mitochondria is that the local energy supply at energetically hot spots like dendritic synapses is no longer assured.

Impaired mitochondrial metabolism is also associated with a change in reactive oxygen species output inducing a strong modulation of the cytosolic redox status. To what degree such redox changes are involved in the hypoxic modulation of the respiratory network is as yet unclear. From the "rhythmic brainstem slice preparation" (►respiratory network analysis: rhythmic slice

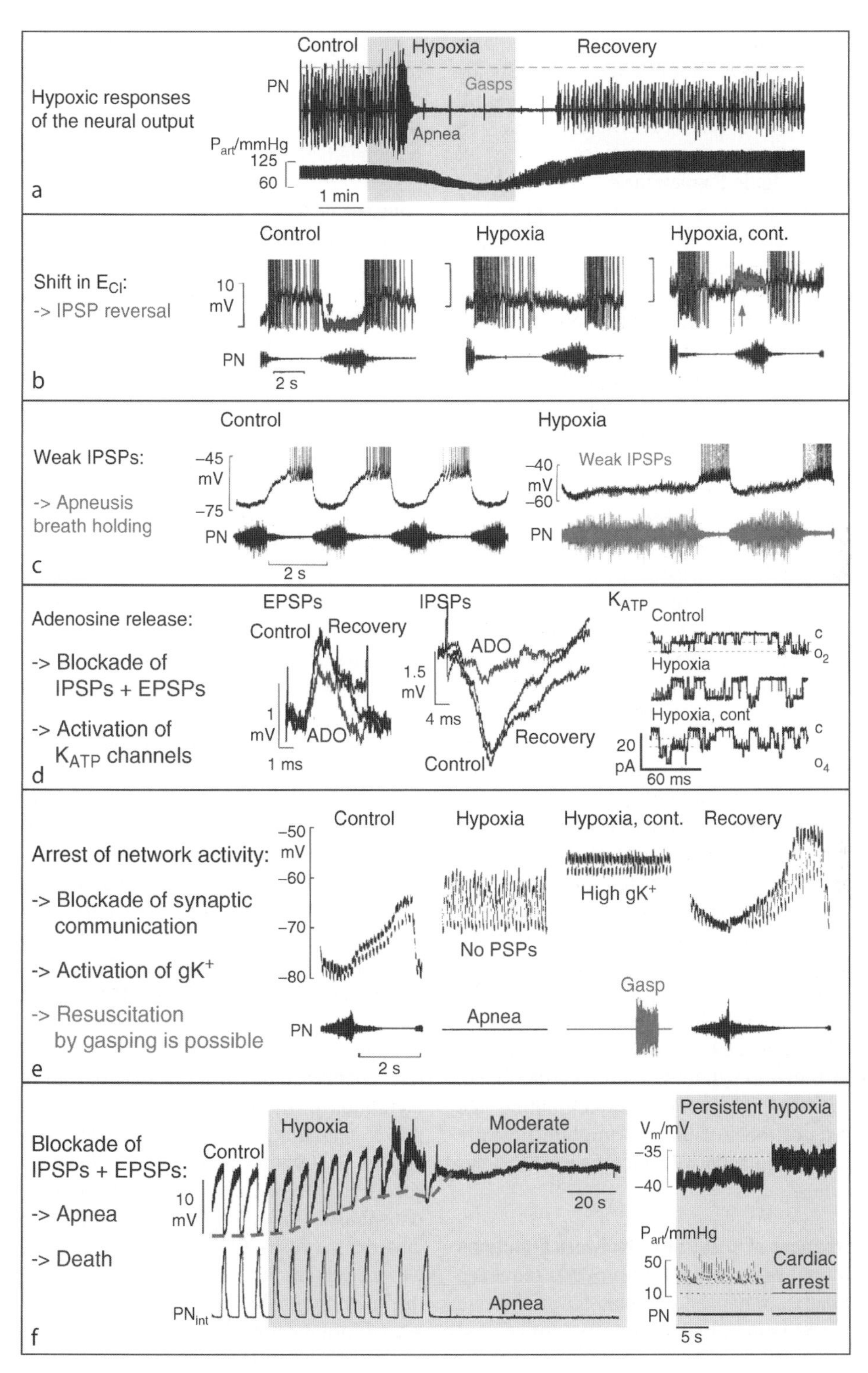

Neural Respiratory Control During Acute Hypoxia. Figure 1 The sequence of hypoxic responses consists of 5 discrete phases. Exposed to severe hypoxia the neural output in the phrenic nerve innervating the diaphragm shows an initial, transient phase of augmentation. This turns into apneustic (►Apneusis) breathing with prolonged inspiratory bursts (the neural equivalent of breath holding), proceeding into respiratory depression and protective ►apnea if hypoxia persists. ►Gasping constitutes the final attempt of auto-resuscitation. If it fails, terminal apnea and irreversible cellular damage will occur.

preparation) of neonatal rats we know, however, that redox changes modulate rhythmic activity. A reducing shift causes an augmentation, while an oxidative shift leads to a depression of inspiratory activity.

Hypoxic Changes in Ionic Homeostasis

Most energy is consumed by ATP–driven pumps that maintain ionic homeostasis. Among these the ubiquitous Na^+/K^+ ATPase is most demanding. Due to the extraordinary number of neurons and glial cells in the brain, the Na^+/K^+ ATPases account for approximately 60% of total ATP consumption. Accordingly, depletion of cellular ATP results in a failure of the Na^+/K^+ ATPase and thus a severe disturbance of ionic homeostasis in and around neurons and glial cells. Na^+ continues to leak into cells and K^+ which is released by the neurons accumulates in the narrow interstitial space, causing a progressive depolarization of neurons and glial cells. Accordingly, irreversible decline of membrane potential and destruction of neural tissue will occur, if O_2 supply is not restored in time. However, such high sensitivity and rapid voltage decline is only seen in hippocampal and cortical brain regions, where hypoxia induces a quick rise in extracellular K^+ to a ceiling level of approximately 9 mM, before it finally continues to rise to levels beyond 40 mM, when the anoxic/terminal depolarization (►ischemic stroke) occurs [3]. In the lower brainstem, however, such massive changes in the extracellular ion levels do not occur as quickly. It is surprising that in the fully intact *in vivo* preparation extracellular K^+ rises only by maximally 1–1.5 mM, even though tissue pO_2 had dropped to zero levels for a duration of several minutes. The robust protection of the respiratory network against hypoxia might in part be referred to a slower decline of ATP, possibly due to an efficient utilization of alternate metabolites and the "loose" tissue architecture of the reticular formation. Together with a "distributed organization" of the respiratory network along the rostrocaudal axis of the medulla this might efficiently protect the network against disturbances in extracellular ion concentrations.

Hypoxic Disturbances of Respiratory Network Functions

A fall of the arterial and tissue pO_2 provokes a sequence of responses that is composed of at least five discrete phases (Fig. 1). As long as the network has not entered the state of terminal apnea, all these changes are reversible and normal network function can be restored by reoxygenation.

Augmentation

In the intact *in vivo* preparation, the respiratory network activity increases very sensitively from the very beginning of hypoxia, because the slightest fall in arterial pO_2 causes an arterial chemoreceptive activation of the entire respiratory network (►carotid body chemoreceptors and respiratory drive). If this reflex activation of breathing movements does not quickly reset normal pO_2 levels and hypoxia persists, a sequence of cellular and systemic trials is initiated, aiming to protect the network.

Apneusis or Breath Holding

During the initial phase of hypoxia, glutamate mediated synaptic excitation is progressively enhanced, but remains stable. Release of increased amounts of glutamate can even activate ►metabotropic receptors of the mGluR1/5 type and ryanodine receptors gating intracellular Ca^{2+} stores. A concomitant activation of Ca(L) channels seems to contribute to a cytosolic Ca^{2+}-rise. This is, however, followed by the activation of hyperpolarizing Ca^{2+}-activated K^+ conductances (►calcium-activated K± channels).

Fairly soon, however, synaptic inhibition starts to be diminished until Cl^- mediated (GABA and glycine induced) IPSPs (inhibitory postsynaptic potentials) finally disappear completely [4] (Fig. 2b, d).

This correlates with the finding that increased release and extracellular accumulation of GABA occurs only initially, is transient and comparably weak and then declines below control levels [5]. Diminished synaptic inhibition disturbs the discharge patterns of the different neuronal subtypes that are no longer phase-locked. As a consequence, inspiration inhibitory processes are fading, resulting in a loss of inspiratory patterning and off-switching and thus pathologic prolongation of inspiratory bursts, which is equivalent to the frequently observed ►breath holding (Fig. 2c). This is a severe disturbance of highest clinical relevance because the dramatic fall in respiratory frequency potentiates hypoxia.

A declining release of GABA indicates an early failure of synaptic inhibition in mature preparations due to a higher anoxia vulnerability of inhibitory interneurons. There is common agreement that the key reason is a disruption of the K^+ and Cl^- gradients. An important player regulating this gradient is the K^+- Cl^- cotransporter KCC2 that is found at inhibitory synapses. It seems that hypoxia induces a rapid loss of tyrosine phosphorylation of KCC2 resulting in a functional disturbance of transport activity and thus intracellular Cl^- accumulation. Consequently the Cl^- gradient across the membrane declines and with a shift of the Cl^- ►equilibrium potential above −70 mV IPSPs that normally hyperpolarize become depolarizing (Fig. 2b). Accordingly, KCC2 knockout mice die immediately after birth due to abolished breathing.

In the rat preparation, IPSPs of respiratory neurons are depolarizing at birth and obviously not yet essential for rhythmogenesis. But interestingly, shortly before delivery, maternal oxytocin release induces a transient reduction in the intracellular Cl^- concentration and an excitatory-to-inhibitory switch of GABA actions in the

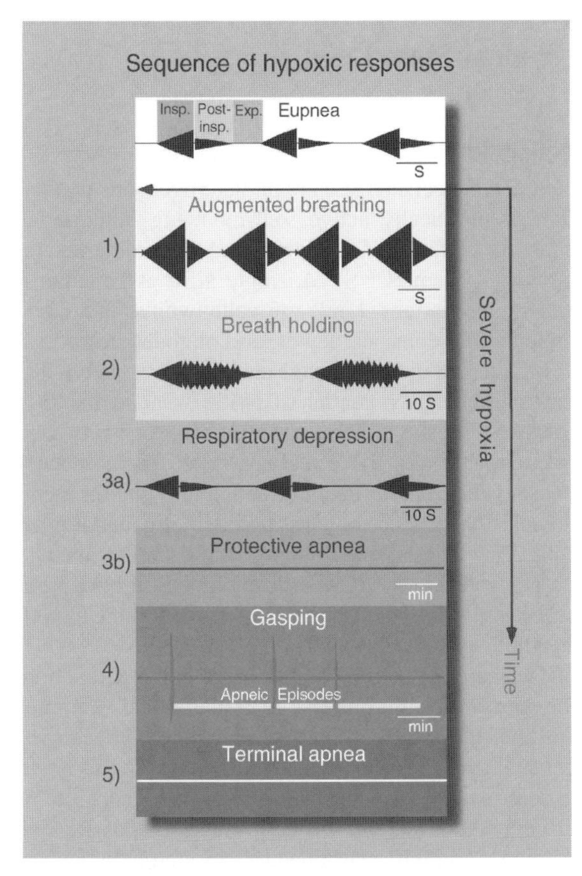

Neural Respiratory Control During Acute Hypoxia. Figure 2 Molecular mechanisms of the hypoxic response. The molecular and cellular mechanisms contributing to the failure of the respiratory network during severe hypoxia originate from a functional loss of synaptic interactions, not the loss of membrane potential or severely disturbed ionic homeostasis. The key processes are firstly reversal of IPSPs, but then attenuation and finally block of synaptic transmission that originates from a strong activation of ATP sensitive pre- and postsynaptic K^+ conductances. Prominent depolarization of the single neurons does not occur even after the network has passed the point of no return and has entered the state of terminal apnea.

fetus. Thus, maternal oxytocin generates hyperpolarizing IPSPs in fetal neurons and protects the fetus against excitotoxic insults during delivery.

Protective Apnea

During progressive periods of hypoxia depression of respiration becomes the principal strategy that is typified by only moderate membrane depolarization of respiratory neurons (Fig. 2f). The release of ►adenosine [5] blocks synaptic transmission by inhibition of Ca^{2+} conductances in axon terminals. Postsynaptically it inhibits the formation of ►cyclic adenosine monophosphate (cAMP and activates ►ATP-sensitive $K^{\pm}$ channels (K_{ATP} channels) (Fig. 2d) [6]. This ATP/ADP/adenosine regulated K_{ATP} conductance keeps the membrane potential of neurons at a relatively negative level of approximately −40 mV to −50 mV and thus protects neurons against massive intracellular Ca^{2+} accumulation that would destroy neurons in such a hypoxic situation. The unfavorable consequence of such a negative membrane potential firmly "clamped" by K^+ currents is that synaptic drive potentials are further reduced and respiratory neurons are completely silenced. Most respiratory neuronal spike discharges cease, although the neurons in fact remain excitable by antidromic stimulation [4].

This "cell-endogenous" protection is getting competent assistance by "cell-external" support through a variety of neurotransmitters and -hormones, such as ►serotonin [5] and endorphins or leucine enkephalin to mention just a few. Together they reinforce protective K^+ effluxes through their action on G protein coupled signaling pathways, inhibiting intracellular cAMP production [7]. Such concerted action of neuromodulatory substances is capable to protect the respiratory neurons for a astonishingly long time that might last up to 20 min as measured in the anesthetized cat [8].

In this respect, it should be mentioned that the time course of hypoxic responses clearly differs between perinatal and adult respiratory networks. In embryonic day E21 rats and in the isolated brainstem of neonatal (1–3 days old) rats, breathing movements even continue relatively unchanged for 25–50 min of hypoxia.

Auto-Recuscitation Through Gasping

After prolonged hypoxia, gasping is the ultimate time point at which re-oxygenation of the blood is still able to completely re-establish brain stem function [9]. Basically, gasping resembles the last operative mechanism and the final attempt to mechanically defeat airway obstruction and anoxia and thus to ensure auto-resuscitation (Fig. 2e). Voltage sensitive dye imaging (►respiratory network analysis: calcium/voltage sensitive dyes imaging) revealed that gasping involves extensive medullary regions covering a much larger region of the rostral ventrolateral medulla and probably the recruitment of additional regions that are normally not active during eupnea [10]. Mechanistically, gasping involves activation of a persistent sodium current (I_{Nap}) that is also discussed to contribute to the endogenous bursting properties of *in vitro* isolated respiratory neurons.

Terminal Apnea

The gasping effort declines steadily in strength and frequency as the respiratory network approaches the "point of no return"and irreversible damage starts to occur. This is the period when the respiratory network proceeds into the terminal and irreversible apnea. Cell swelling occurs, which affects the cytoskeleton and in turn activates the persistent Na^+ current, L-type Ca^{2+} channels (►voltage-gated calcium channels), and Ca^{2+}-sensitive non-selective cationic channels (►calcium-activated non-specific cation current). The resulting accumulation of cytosolic Ca^{2+} is massive and causes irreversible damage of neurons and the collapse of network functions. The cardio-respiratory part of the network also fails and the blood circulatory control is lost as well (Fig. 2f).

In conclusion, the five phases of the hypoxic response, the underlying molecular mechanisms and their time courses impressively demonstrate that the medullary respiratory center is astonishingly resistant to hypoxia. It remains vital even after prolonged periods of severe hypoxia which is due to an efficient protection through metabolic and neuromodulatory adjustment of K^+ conductances. Lack of spontaneous breathing movements, therefore, is not at all a reliable sign for cell destruction (see Fig. 2f) and brainstem death, although most of the structures rostrally of the brainstem including the cerebellum, the striatum and the cortex are definitively irreversibly damaged at that time.

References

1. Richter DW, Ballanyi K, Schwarzacher S (1992) Mechanisms of respiratory rhythm generation. Curr Opin Neurobiol 2:788–93
2. Müller M, Mironov SL, Ivannikov MV, Schmidt J, Richter DW (2005) Mitochondrial organization and motility probed by two-photon microscopy in cultured mouse brainstem neurons. Exp Cell Res 303:114–27
3. Müller M, Somjen GG (2000) Na^+ and K^+ concentrations, extra- and intracellular voltages, and the effect of TTX in hypoxic rat hippocampal slices. J Neurophysiol 83:735–45
4. Richter DW, Bischoff A, Anders K, Bellingham M, Windhorst U (1991) Response of the medullary respiratory network of the cat to hypoxia. J Physiol 443:231–56
5. Richter DW, Schmidt-Garcon P, Pierrefiche O, Bischoff AM, Lalley PM (1999) Neurotransmitters and neuromodulators controlling the hypoxic respiratory response in anaesthetized cats. J Physiol 514:567–78
6. Pierrefiche O, Bischoff AM, Richter DW (1996) ATP-sensitive K^+ channels are functional in expiratory neurones of normoxic cats. J Physiol 494:399–409
7. Lalley PM, Pierrefiche O, Bischoff AM, Richter DW (1997) cAMP-dependent protein kinase modulates expiratory neurons in vivo. J Neurophysiol 77:1119–31
8. Richter DW, Mironov SL, Büsselberg D, Lalley PM, Bischoff AM, Wilken B (2000) Respiratory rhythm generation: Plasticity of a neuronal network. The Neuroscientist 6:181–198
9. Tomori Z, Benacka R, Donic V, Tkacova R (1991) Reversal of apnoea by aspiration reflex in anaesthetized cats. Eur Respir J 4:1117–25
10. Potts JT, Paton JF (2006) Optical imaging of medullary ventral respiratory network during eupnea and gasping in situ. Eur J Neurosci 23:3025–33

Neural Stem Cells

Definition

Neural stem cells (NSC) are an heterogeneous population of mitotically active, self-renewing, multipotent cells of both the developing and the adult central nervous system (CNS). A single NSC is capable of generating various kinds of cells within the CNS, including neurons, astrocytes, and oligodendrocytes.

In vivo and in vitro lineage analyses have shown that the multilineage potential of NSCs is at least partly mediated by the generation of cell-lineage-restricted intermediate progenitor cells, which produce only neurons (neuronal progenitor cells), and glial progenitor cells that produce only astroglial or oligodendroglial cells. Thus, the cellular diversity of the CNS is likely to be generated in a stepwise fashion. NSCs have been successfully isolated from the entire embryonic as well as adult CNS.

The ganglionic eminence(s), in the embryo, and both the subventricular zone (SVZ) of the lateral ventricles and the sub-granular zone (SGZ) of the hippocampus dentate gyrus (DG), in the adult, have been shown to consistently contain stem-like cells capable of driving neuro- and glio-genesis. These regions are then defined as highly specialized CNS germinal niches. Protocols to obtain in vitro large-scale numbers of NSCs are available, thus supporting the concept that these cells might represent a renewable source of uncommitted ready-to-use cells for transplantation purposes.

►Autoimmune Demyelinating Disorders: Stem Cell Therapy
►Neural Development
►Regeneration
►Transplantation of Neural Stem Cells for Spinal Cord Regeneration

Neural Tube

Definition

In vertebrates the presumptive neural ectoderm of the future brain and spinal cord folds dorsally into a tube, the neural tube, from which all neurons and glia in the central nervous system are derived.

►Neural Development

Neuralgia

Definition

Neuralgia denotes a sharp and paroxysmal pain along the course of a nerve, e.g., in ►trigeminal neuralgia)

►Trigeminal Neuralgia (Paroxysmal Facial Pain, Tic Douloureux)

Neurally Controlled Animats

►Computer-Neural Hybrids

Neuraxis

Definition

Central nervous system. Composed of the encephalon located in the skull and the spinal cord running in the vertebral canal.

►General CNS

Neuregulin

Definition

Neuregulin is a family of four structurally related proteins that are part of the epidermal growth factor (EGF) family. They exert their action by activation of ErbB receptors. They have diverse roles in the development of the nervous system.

►Growth Factors
►Synapse Formation: Neuromuscular Junction Versus
►Central Nervous System

Neurexins

Definition

Neurexins are neuronal cell surface proteins and contain single transmembrane region and extracellular domains with repeated sequences. These sequences are similar to sequences in laminin A, slit and agrin, which are proteins shown to be important for axon guidance and synaptogenesis.

►Agrin
►Laminin
►Slits
►Synapse Formation: Neuromuscular Junction Versus Central Nervous System

Neurite

Definition

Often used for nerve cell processes which are in an early stage of growth, or which can not be identified with certainty as an axon or dendrite. The latter is often the case when nerve cells are cultured. Neurons have two kinds of processes, one axon and several dendrites.

Axon is specialized for conduction of action potentials and dendrites for receiving sensory or neural signals.

►Action Potential
►Membrane Components
►Neuronal Changes in Axonal Degeneration and Regeneration

Neurite Extension

►Axon Outgrowth

Neurite Outgrowth

Definition

Neuronal outgrowth begins at the cell body and extends outwards and is strongly influenced by the surrounding cellular environment. Outgrowth is a complex process that requires numerous ultra-structural changes in the growing neurons. These include the insertion of newly synthesized functional membrane, generation of new cytoplasm and the continued expansion and modification of the cytoskeleton as neurites grow and branch.

►Axonal Pathfinding and Network Assembly
►Cytoskeleton

Neuritis

Definition

Neuritis is an inflammation of a nerve associated with continuous pain, paralysis and sensory disturbances.

Neuroactive Steroids

►Neuroendocrinological Drugs

Neuroanatomical Tracer

Definition

Neuroanatomical tracers are molecules that neurons will absorb and move within axons, via intracellular transport mechanisms, either from the neuron cell body to the axon terminals (called orthograde or anterograde transport) or from axon terminals to the cell body (called retrograde transport) or both (called bidirectional transport). Neuroanatomical transporters are visible in tissue either because they are fluorescent, i.e., they emit light at a particular wavelength when illuminated by light of another wavelength, or because they have been reacted with chemical agents that turn them a dark color.

Neuroanatomy

Definition

The branch of biology that explores the structure and organization of tissue that arises as a portion of the nervous system.

Neurocan

Definition

One kind of axon growth inhibitors. It is a secreted molecule and undergoes posttranslational modification in the CNS resulting in a 150 kDa C-terminus and 130 kDa N-terminus fragment.

►Growth Inhibitory Molecules
►Regeneration of Optic Nerve

Neurochemical Remodeling in Retina

Definition

Altered synaptic and/or neurochemical organization secondary to retinal degeneration or retinal damage

►Inherited Retinal Degenerations

Neurochemicals

►Respiratory Neurotransmitters and Neuromodulators

Neurodegeneration and Neuroprotection – Innate Immune Response

Griffiths M.[1], Neal J. W[2], Gasque P.[1,3]
[1]Brain Inflammation and Immunity Group, Department of Medical Biochemistry, Cardiff University, Cardiff, UK
[2]Department Histopathology School of Medicine, Cardiff University, Cardiff, UK
[3]GRII LBGM, Faculty of Sciences and Technology, University of la Reunion, Saint Denis, Reunion, France

Definition

Brain infection, hemorrhage and aging are associated with activation of the innate immune system as expressed by resident glial cells and neurons. The innate immune response (►innate immunity) relies on the detection of "self" and "non-self" (►self, non-self and altered-self) structures in mounting a protective response to promote the clearance of pathogens, toxic cell debris and apoptotic cells accumulating within the brain parenchyma and the cerebrospinal fluid (CSF). Innate immune molecules can also stimulate neurogenesis and contribute to brain tissue repair. However, in some diseases, these protective mechanisms lead to neurodegeneration on the ground that several innate immune molecules can promote neuronal loss. The response is a "double-edged sword," representing a fine balance between protective and detrimental effects. Several key regulatory mechanisms have now been evidenced in the control of the innate immune response and which could be harnessed to explore novel therapeutic avenues.

Characteristics

Quantitative Description: Cellular and Molecular Innate Immune Responses

Classically, innate immune cells are known as neutrophils, natural killer (NK) cells, dendritic cells and macrophages involved in the selective recognition and the clearance of pathogens and toxic cell debris during infection or tissue injury [1]. However, there is little evidence of an immunosurveillance of the brain by these peripheral cells and it is now evident that resident cells, glial cells and neurons, are capable of mounting a robust innate immune response. The local innate immune response is based on the recognition of "non-self" and "altered-self" (self, non-self and altered-self) patterns, also called ►danger signals, by molecules and receptors expressed by microglia, astrocytes, oligodendrocytes and neurons [2]. These molecules and receptors are called pattern-associated recognition receptors (PARRs) and are displayed on the cell membrane or released in soluble forms. The decoding and sampling of the microenvironment for danger signals will contribute to the removal of the harmful intruders. In addition, there is mounting evidence that innate immune molecules (e.g., C3a) can contribute to tissue repair notably by stimulating the mobilization of neural stem cells and with the production of growth factors [3,4]. Critically, several innate immune molecules have also cytotoxic and cytolytic activities and must be controlled to avoid neuronal loss (neurodegeneration) and robust inflammation. Cellular and regulatory mechanisms have recently been described and thought to be at the route of neuroprotective mechanisms.

Description of the Process

Innate Immune Response in Health: The Key Role of Physical Barriers

The brain is isolated from the systemic circulation by a protective blood brain barrier (BBB) composed of endothelial cells linked by tight junctions and surrounded by the end feet of astrocytes (Fig. 1). A further protective barrier composed of specialized ciliated glia, the ependyma, lines the ventricles preventing entry of pathogens from the cerebro spinal fluid (CSF) into the brain. Within the ventricles is the choroid plexus containing Kolmer phagocyte cells. Cells within the choroid plexus and the ependymal layer express receptors that are capable of detecting pathogens in the CSF and regulating the inflammatory response. These innate immune receptors are highly conserved and include Toll-like receptors (TLR), CD14 and ►complement receptors CR3 and CR4. The details regarding the recognition and interaction between glia and pathogens will be discussed below.

Immunoprivileged Status of the Brain by Preventing the Infiltration of Potentially Harmful Systemic Immune Cells

The active destruction of infiltrating T lymphocytes through induction of apoptosis provides the brain with a degree of low immuno-surveillance, preventing the entry of lymphocytes and down regulating inflammation (Fig. 1). Active apoptosis of infiltrating T lymphocytes is induced by neurons and glia utilising the "death signaling pathways" based upon CD95FasL/CD95 (Fas) and the TNF lymphotoxin receptor-TNF receptor 1 (TNFR1) pathway. The initiator of apoptosis, CD95L, is expressed by both astrocytes and oligodendroglia and transmits an apoptotic signal to target T cells. This interaction at the cell surface induces the activation of caspases and subsequent apoptosis of the target cell, with engulfment by microglia and down regulation of their activation. For instance, recovery from experimental EAE is increased through induction of apoptosis in inflammatory T cells by the TNFR signaling pathway and in TNFR knock out mice T cell apoptosis is reduced by 50% in the periphery of demyelinating plaques. In multiple sclerosis, it is possible that this death signaling pathway is functional in reducing the severity of demyelination.

The induction of apoptosis in infiltrating T lymphocytes and virally infected cells will reduce host tissue destruction, but only if they are rapidly cleared by resident cells. Apoptotic cells contain large amounts of toxic enzymes that are able to activate proinflammatory cytokine release. Phagocytosis of apoptotic cells is termed non-phlogistic and is accompanied by a down regulation of proinflammatory cytokines contributing to the limitation of tissue damage and "self" defense [5].

Apoptotic cells are "altered self" and express apoptotic cell associated molecular patterns, ACAMPs. The identity of ACAMPS has not been fully elucidated but includes nucleic acids, sugars, oxidized low density lipoproteins and alteration of membrane electrical charge. The best characterized ACAMP to date is phosphotidylserine (PS). Glia and macrophages express a range of PARRs that recognize ACAMPS including the

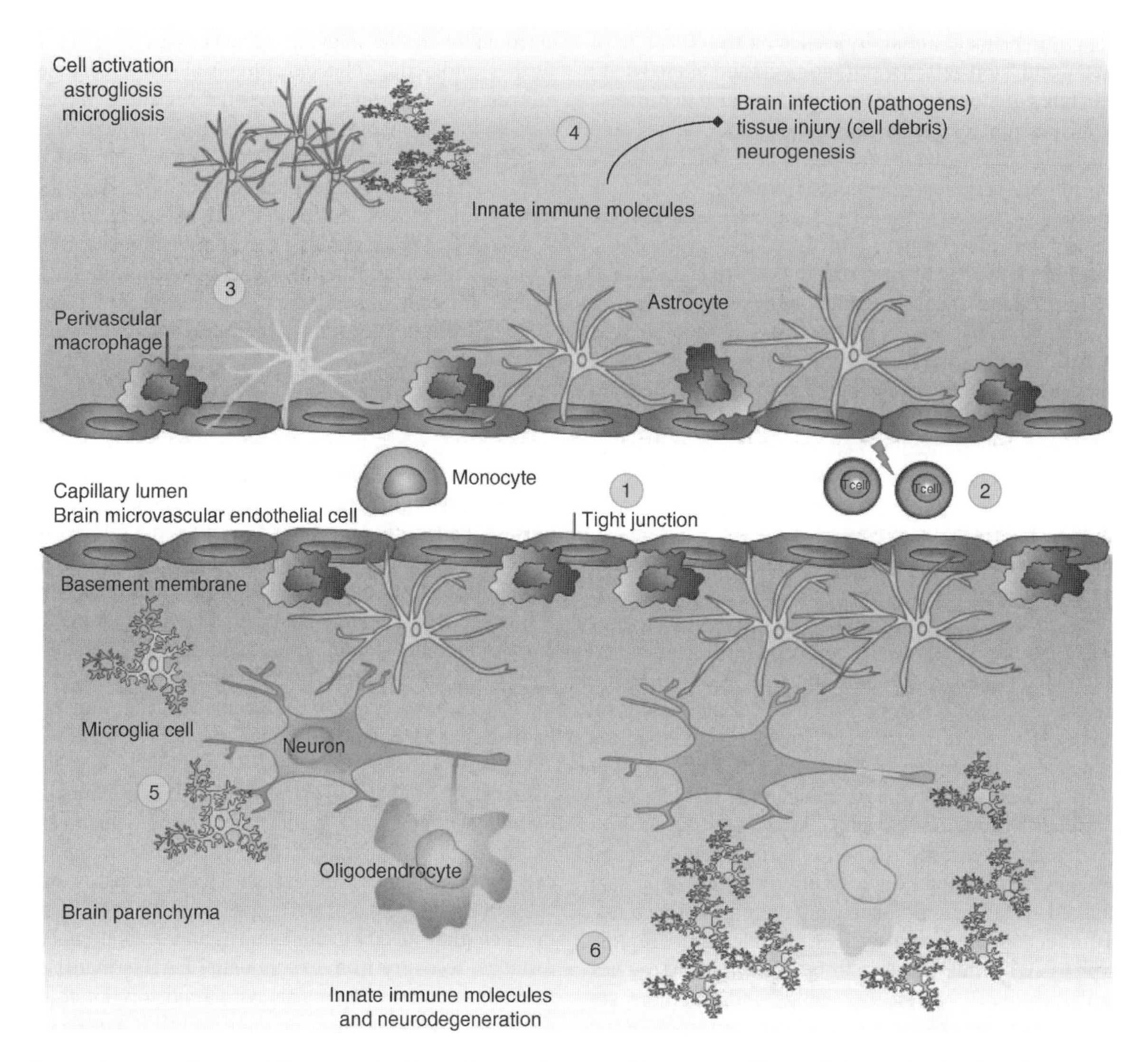

Neurodegeneration and Neuroprotection – Innate Immune Response. Figure 1 Innate (natural) defense mechanisms in health and diseases of the CNS: (i): The physical blood brain barrier composed of endothelial cells linked by tight junctions and surrounded by the end feet of astrocytes prevents the infiltration by pathogens. (ii): Brain cells such as microglia and astrocytes can also induce apoptosis of autoimmune T lymphocytes through the Fas/FasL pathway. (iii): In response to infection or tissue injury, resident cells will be activated, i.e., reactive gliosis with the activation and proliferation of astrocytes and microglia. (iv): Remarkably, innate immune molecules will contribute to the selective recognition and removal of pathogens and toxic cell debris (apoptotic corpses, amyloid fibrils) while preserving self cells. Furthermore, several innate immune molecules will initiate tissue repair (neurogenesis). (v): This response will have to be kept under safe guard through the expression of inhibitory/regulatory molecules for example by neurons to control phagocytosis by microglia. (vi): In sharp contrast, several innate immune molecules have been involved in neuronal loss and oligodendrocyte damage in diseases such as Alzheimer's disease and multiple sclerosis, respectively. For example, host defense complement proteins are known to induce adverse cytotoxic activities against myelin-forming cells and neurons. Hence, a fine balance must exist and which is at the route of future therapeutic strategies.

phosphotidylserine receptor (PSR), CD14, CD36 and milk fat globulin (MFG-EGF 8).

Activation of the classical innate immune complement pathway by virtue of C1q recognition of nucleic acids initiates the generation of ►opsonins C3b and iC3b that bind to apoptotic cells and act as targets for the phagocytic CR3 and CR4 receptors expressed by macrophages and microglia.

Protective Innate Immune Response During Brain Infection and Inflammation: To Promote the Clearance of Pathogens, Apoptotic Cells and Other Cell Debris

After penetrating a damaged BBB, an infiltrating pathogen will encounter the innate immune response delivered by professional (microglia) and non-professional phagocytes (astrocyte, ependyma neurons and oligodendroglia) (Fig. 1). These cells provide the second line of the brain's innate defense against infection, but must also be regulated to prevent destruction of host "self" cells. Astrocytes and microglia express a large array of membrane and soluble PARRs. These include lectins (e.g., phagocytic macrophage mannose receptor (MMR)), scavenger receptors (e.g., SRA), TLRs and associated molecules (TLR2 and 4, CD14) and complement molecules (e.g., C1q, C3, CR1, CR3, CR4). These innate immune receptors are capable of detecting unique arrangements of lipopolysacharides (LPS) and peptidoglycan (PG) molecules termed pathogen associated molecular patterns (PAMPS) within the cell walls of microorganisms. PAMPS are uniquely expressed by pathogens and therefore distinguish "self" (host) from "non-self" (pathogen). Removal of pathogens therefore reduces the severity of an inflammatory response before its destructive effects outweigh the benefits of pathogen clearance. The recognition of pathogens or "not self" also relies upon the defense collagens, a group of PRRs composed of a globular carboxy sequence that recognizes PAMPS together with an amino terminal sequence that binds to specific phagocytic receptors on the surface of phagocytic cells. These include C1q, the first component of the complement classical pathway, mannan-binding lectin (MBL) and surfactant protein A (SPA). C1q and MBL have been shown to be expressed by astrocytes whereas the expression of SPA in the CNS is yet to be defined. The complement receptors (CR), CR3 and CR4 are important for identifying pathogens and apoptotic cells opsonised with complement components C3 fragments (C3b and iC3b).

Innate Immune Responses in Neurodegeneration and Demyelination

Any disruption of the BBB invariably permits entry into the brain of neurotoxic systemic proteins, including thrombin a serine protease vital for blood coagulation and complement proteins. Thrombin, at low concentrations (50–100 pM) is neuroprotective, regulating NGF synthesis and synaptic outgrowth; it is protective against oxygen and glucose deprivation due to its modulation of intracellular calcium. At high levels (500 nM), as found following intra cerebral hemorrhage, thrombin is neurotoxic as it activates NMDA excitotoxic receptors and PAR-1 inhibiting neurite extension and neuronal repair [6].

The complement system is a vital component of the CNS innate immune defense system as it recognizes and clears pathogens and apoptotic cells, but its activation must be closely regulated to prevent excessive tissues destruction. High levels of complement proteins from plasma could infiltrate the brain from a damaged BBB. Moreover, the expression of complement components of both the classical and alternative pathways has been well described in astrocytes, microglia and neurons in vitro and in vivo. The classical complement pathway is activated by C1q interacting with neurons, myelin basic protein (MBP) myelin oligodendrocytic protein (MOG). The alternative pathway is activated independently of C1q and immune complex formation, but through a binding of C3 to activating surfaces that cleaves C5 to initiate the terminal pathway. One component, C3b is abundantly deposited on target cells and functions as an opsonin for microglia and Kolmer cells expressing the PARRs CR1 (CD35), CR3 and CR4 receptors to promote robust phagocytosis. Uncontrolled C3b opsonisation of oligodendrocytes and neurons will lead to demyelination and neuronal loss by phagocytes expressing CR3 and CR4. The terminal pathway provides the membrane attack complex (MAC) composed of C5b-9. This complex is capable of producing cell lysis unless inhibited by complement regulator proteins (regulators of complement activation, RCA), see below. Complement is activated in Alzheimer's Disease as the result of C1q binding to fibrillary beta amyloid, activating complement and increasing C1q, C3 and C5 as part of the protective response promoting clearance of the amyloid plaque. Amyloid plaque formation in complement deficient mice was significantly increased supporting the interpretation that amyloid clearance was related to microglia phagocytosing complement opsonised amyloid. However, in the context of acute inflammation associated with robust expression of complement proteins by microglia and astrocytes, it is plausible that complement activation by myelin/neuronal debris contributes to secondary brain injury. The formation of the MAC and non-specific binding to surrounding cells would cause bystander damage. Complement activation is also present in the tauopathies (including Pick's disease) resulting in localization of complement products on ballooned neurons; the details regarding the identity of the C1q binding molecule in these diseases is not yet known. Interestingly, administration of a C1 inhibitor C1-INH, resulted in neuroprotection after experimental ischaemia, but it protective effect was interpreted as independent of C1q activating the complement pathway. Experimental models of demyelination and multiple sclerosis have demonstrated C9 and MAC deposition in at least half of the MS cases with active myelin destruction. Knock-out mice for CD59, a natural MAC inhibitor, increased the severity of EAE whereas the failure to produce MAC in C6 deficient animals reduced the severity of axon damage in diseased animal compared with the control group.

Regulation of the Process

Innate Immune Regulatory Molecules

To counter the neurotoxic effects of innate immune molecules, both glia and neurons express a range of "self" defense proteins. These include the serine protease inhibitors (serpines) that prevent the synthesis of thrombin. The serpins include the anti thrombin colligin (Hsp47) expressed by microglia and astrocytes, plasminogen activator inhibitor (PAI-1) and protease glial derived nexin-1 (PN-1) both of which are expressed by astrocytes and neurons. Neuroserpin, an inhibitor of plasminogen (tPA) expression is restricted to neurons and astrocytes.

In the systemic organs and brain, the activation of the classical and alternative complement pathways is tightly regulated in order to prevent unrestrained activity resulting in cytolytic destruction of host cells, brain cells being particularly vulnerable to complement attack. The complement regulators can be divided into two groups, the membrane bound and those located within the extracellular fluid, the so called fluid phase regulators. Together they inhibit the activation of the complement pathways (for detailed review see [7]). In brief, the membrane bound regulators, CR1 (CD35) and membrane cofactor protein (MCP, CD46) expressed on nucleated cell membranes bind to C4b or serve as cofactors to increases its cleavage inhibiting this step in the C pathway. Decay accelerating factor (DAF, CD55) inhibits the C3/C5convertase step and CD59 blocks MAC formation in the terminal pathway. The fluid phase regulators are composed of C1 inhibitor (C1-INH) an effective inhibitor of the C1 component of the classical pathway; Factor H and FI (alternative pathway) accelerates C3b/C4b degeneration, whereas S protein and clusterin prevents C5b-7 assisting formation of MAC pathway, inhibiting the final the terminal pathway.

In primary culture, fetal neurons spontaneously activate C classical pathway with formation of MAC resulting in significant lysis because in vitro they express low levels of the membrane regulators (CD59, CD46, C1 INH and FH) and no CD55. By comparison with neurons, astrocytes and microglia express a wider range of inhibitors such as, CD46, CD59 and CD55 together with the fluid phase regulators. Initial observations demonstrated rat oligodendrocytes in culture were susceptible to C attack and did not express CD59. However in human, CD59, CD46 and CD55, together with C1 INH, FH, S protein and clusterin are all expressed and do not spontaneously activate complement. In Alzheimer's disease, AD, and Picks diseases neurons did not express CD35, CD59, CD46 and CD55, whereas in Huntington's disease (HD) neurons expressed high levels of CD46. Overall, the combined data from in vitro and in vivo experiments indicates that astrocytes, microglia and oligodendrocytes are well protected from the effects of direct or bystander complement lysis because they express high levels of CD59, CD46 and CD55. Neurons, particularly lacking CD55, are vulnerable to the detrimental effects of complement attack.

Interestingly, there is a growing body of evidence that neurons may be capable of evading detection by activated microglia and macrophages by expressing the so-called "don't eat me" signals or SAMPs (self associated molecular patterns). SAMPs are markers of "self" preventing recognition of host cells and reducing the severity of any inflammatory response through inhibition of innate immune cells such as microglia and infiltrating macrophages [8].

A number of SAMPS have been identified including CD200 (and its receptor CD200R), the integrin CD47 with its receptor SIRPα, both regulating myeloid cell and lymphocyte activity and, hence, protecting from autoimmunity [9]. On the basis of their regulatory activity in vitro and in vivo CD46, CD55 and FH could also be important don't eat me signals. A further group of SAMPs is sialic acids ubiquitously expressed and interacting with the newly described immunoglobulin (Ig)-like lectins, the siglecs expressed on lymphocytes and microglia.

CD200 is a 41–47 kDa surface molecule and a member of the immunoglobulin Ig supergene (Igsf) family characterized by two IgSF domains [10]. It is a highly conserved molecule found in the invertebrates and vertebrates and many of the glycoproteins containing this arrangement are involved with regulation of the immune system. In the brain OX2 now CD200 is expressed by cerebellar and retinal neurons, together with vascular endothelium. Astrocytes do not express CD200 in contrast to microglia.

The counter receptor to CD200, CD200R, also contains two IgSF domains and is expressed by myeloid cells and brain microglia. In CD200 deficient mice the number of activated microglia and macrophages were more numerous after a lesion, than the wild type animal providing evidence that the CD200/CD200R interaction is related to regulation of microglial activation and local inflammation. This interpretation is supported by experiments in mice inoculated with MOG peptide to induce EAE. Animals that received a blocking monoclonal antibody against CD200R had an increased disease severity as compared with animals without the monoclonal antibody treatment. Furthermore, after EAE in CD200−/− mice, microglia became rapidly activated as compared with type animals. Interestingly in CD200 deficient animals the increased number of phagocytic cells in the retina could also represent the failure to inhibit macrophage entry across the BBB into the eye.

CD47 is constitutively expressed by endothelium, neurons, macrophages and dendritic cells. CD47 has five transmembrane regions with alternatively spliced isoforms of CD47 having a tissue specific expression, form 2 is present in bone marrow, whereas form 4 is highly expressed in brain. The counter receptor for CD47 is signal regulatory protein SIRP alpha (CD172a) a plasma membrane protein with three Ig domains in its extracellular component. CD172 is expressed by myeloid cells and neurons. The interaction between CD47 on a host cell and CD172a recruits tyrosine phosphotases SHP-1and SHP-2 with down regulation of macrophage phagocytosis, complement activation and cytokine synthesis including TGF β all contributing to the reduction of any inflammatory response.

The interaction between CD47 and CD172a has been shown to reduce neutrophil migration across endothelium and blocking CD47 reduced bacterial induced expression of inflammatory cytokines by dendritic cells. Furthermore CD47 is capable of inducing apoptosis in T cells and cells deficient in CD47 are rapidly cleared from the systemic circulation by the spleen. Hence, CD47 represents an important "don't eat me signal" preventing inappropriate phagocytosis of host cells. Whether or not the CD47-CD172a pathway is capable of regulating microglial activity in disease remains to be determined.

Siglecs represent a group of at least 11 recently identified Ig lectins expressed by a wide range of myeloid cells, lymphocytes, macrophages and microglia [11]. Two siglecs, myelin-associated glycoprotein (MAG) and Schwann cell myelinated protein (SMP) are restricted to the CNS and expressed by oligodendrocytes; they are considered important for myelin–axon interaction and control of neuron growth after injury. The non-neural siglecs are characterized by their specific binding to the sialic acids groups, (a group of nine carbon sugars) derivatives of either neuraminic acid or keto deoxynonulosonic acid). Typically these molecules are expressed by "non-self" pathogens and (potentially) apoptotic cells. The interaction between a pathogen expressing the appropriate sialic acid residue and a siglec on microglia will activate phagocytosis and clearance, whereas the absence of the sialic acid residue, for example on a neuron, provides a signal defining "self" and "don't eat me." CD33 (siglec 3) is expressed by myeloid stem cells, monocytes and dendritic cells. CD22 (siglec 2) is expressed by B cells and siglecs 5–7 on macrophages, neutrophils and eosinophils. Microglia express siglec 11 and the tyrosine phosphtases SHP-1 and SHP2, both known to participate in down regulation of phagocytosis.

Moreover, the expression of sialic acids also inhibits the activation of the alternative complement pathway through binding to the complement regulator fH.

Activation of the complement pathway produces C3a and C5a chemoattractant to myeloid cells expressing the anaphylatoxin receptors C3aR and C5aR and contributing to proinflammatory activities. However, C3a also has regulatory properties based on its capacity to block LPS stimulation of macrophage TNF alpha cytokine expression as well as IL 6 and IL1 beta expression by lymphocytes increasing synthesis of IL-10 and NGF. This new "self defense" role for C3a is supported by the presence of C3aR on adrenal and pituitary gland cells, both glands having important roles in the synthesis of corticosteroids to control systemic and central inflammation and infection. More recently, it has been shown that complement anaphylatoxins may be involved in the control of neurogenesis and with anti-apoptotic activities by its capacity to reduce NMDA induced neuronal death.

Innate Immune Regulatory Cells

The presence of large numbers of T lymphocytes inside the brain would normally be expected to have a detrimental effect upon neuronal survival because of increasing the inflammatory response and uncontrolled tissue damage. However, data has accumulated to show that a regulated innate inflammatory response can be beneficial to neuronal survival and enhance axonal repair. Recovery from retinal cell death induced by glutamate was increased in animals with an intact T cell response whereas in those without T cells, recovery was poor. Similarly passive transfer of T cells from injured animals into recovering animals increased the likelihood of recovery, as did immunization against the same antigen/ with glutamate. The vital observation was the protective T cells had to be activated specifically against antigens at the site of axon and neuron injury, this was termed protective autoimmunity. The mechanism responsible for T cell neuroprotection is not clear, although lymphocytes can express neurotrophic growth factors including brain derived neurtrophic factor (BDNF) and ciliary trophic growth factor as do macrophages. These finding have therapeutic implications and studies have shown human T cells in vivo stimulated with glatiramer acetate express BDNF as well as the anti inflammatory cytokine IL 10. Currently a clinical trial involving this compound in MS patients has found some benefit especially reducing the number of new lesions on MRI scanning.

Function

The balance between the protective and harmful effects of the innate immune response mounted against pathogen invasion and brain injury has been termed a "double edged sword." This balance must be critically regulated in order to promote conditions supportive of brain repair but without excessive destruction of "self" or host cells. The CNS innate immune response is regulated by a number of "self defense" pathways. "Self" is distinguished from "non-self" by the detection of surface PAMP and ACAMP molecules and by the expression of SAMPs by host cells. The CNS has a range of defense strategies at its disposal, each capable of regulating the protective components of the innate immune response while at the same time limiting the extent of accompanying brain injury. The therapeutic manipulation of the immuno-regulatory and defense strategies designed to reduce brain injury and promote repair, as described in this review, is now becoming a clinical reality.

References

1. Taylor PR, Martinez-Pomares L, Stacey M, Lin HH, Brown GD, Gordon S (2005) Macrophage receptors and immune recognition. Annu Rev Immunol 23:901–944
2. Medzhitov R, Janeway CA Jr (2002) Decoding the patterns of self and nonself by the innate immune system. Science 296:298–300
3. Gasque P, Neal JW, Singhrao SK, McGreal EP, Dean YD, Van BJ, Morgan BP (2002) Roles of the complement system in human neurodegenerative disorders: pro-inflammatory and tissue remodeling activities. Mol Neurobiol 25:1–17
4. Schwartz M, Moalem G, Leibowitz-Amit R, Cohen IR (1999) Innate and adaptive immune responses can be beneficial for CNS repair. Trends Neurosci 22:295–299
5. Savill J, Dransfield I, Gregory C, Haslett C (2002) A blast from the past: clearance of apoptotic cells regulates immune responses. Nat Rev Immunol 2:965–975
6. Gingrich MB, Traynelis SF (2000) Serine proteases and brain damage - is there a link? Trends Neurosci 23:399–407
7. Morgan BP, Meri S (1994) Membrane proteins that protect against complement lysis. Springer Semin Immunopathol 15:369–396
8. Elward K, Gasque P (2003) "Eat me" and "don't eat me" signals govern the innate immune response and tissue repair in the CNS: emphasis on the critical role of the complement system. Mol Immunol 40:85–94
9. Hoek RMRS, Murphy CA, Wright GJ, Goddard R, Zurawski SM, Blom B, Homola ME, Streit WJ, Brown MH, Barclay AN, Sedgwick JD (2000) Down-regulation of the Macrophage Lineage Through Interaction with OX2 (CD200). Science 290:1768
10. Barclay AN, Wright GJ, Brooke G, Brown MH (2002) CD200 and membrane protein E interactions in the control of myeloid cells. Trends in Immunology 23:285–290
11. Crocker PR (2005) Siglecs in innate immunity. Curr Opin Pharmacol

Neurodegenerative Disease

Definition

A disease that gradually affects the function of the nervous system due to the degeneration of specific populations of neurons in response to various conditions, including inflammation. Acute and chronic immune and inflammatory mechanisms mediated by resident cells such as ►microglia and ►astrocytes contribute to their pathogenesis through the secretion of potent regulatory mediators, including an expanding array of ►cytokines, ►chemokines, proteases, complement proteins and reactive oxygen species.

►Astrocytes
►Chemokines
►Cytokines
►Gene Therapy for Neurological Diseases
►Microglia

Neurodegenerative Diseases – MAPK Signalling Pathways in Neuroinflammation

Rommy von Bernhardi
Department of Neurology, Faculty of Medicine, Pontificia Universidad Católica de Chile, Santiago, Chile

Synonyms

Mitogen-activated protein kinases

Definition

►Neurodegenerative diseases refer to degenerative changes of the nervous system dependent on multiple mechanisms resulting in neurological disease. Acute and chronic immune and inflammatory mechanisms mediated by resident cells such as ►microglia and ►astrocytes contribute to their pathogenesis through the secretion of potent regulatory mediators, including an expanding array of ►cytokines, chemokines, proteases, complement proteins and reactive oxygen species (ROS). Whether inflammatory changes within the brain tissue are a consequence or a primary cause of brain damage still is a matter of debate. However, inflammatory mediators and cellular processes are known to be central to the pathogenesis of many neurodegenerative diseases, such as multiple sclerosis (MS), Alzheimer's disease (AD), stroke, HIV-dementia and others. On the other hand, there is compelling evidence indicating that inflammatory cells and mediators also have beneficial functions in the CNS, assisting in repair and recovery processes. The dual role for glial-mediated ►neuroinflammation suggests that glial cells dysregulation may be involved in the genesis of neurodegenerative diseases. As crucial components of the regulatory machinery underlying inflammation and other phosphorylation/dephosphorylation-dependent mechanisms, ►Mitogen-Activated Protein Kinases (►MAPKs) are good candidates to be involved in the regulation of glial cells and on the progression of neurodegenerative diseases such as AD [2].

Characteristics

Quantitative Description of the Process

Dementia associated with neurodegenerative diseases is accompanied by morphological changes in the brain, including the development of characteristic lesions, but underlying mechanisms ensue before the clinical disease is established. Evidence suggests that the activation of glial cells in response to injury, illness, ageing, or other causes begins a cascade of events leading to a chronic inflammatory process. Through various pathways, inflammatory mediators cause neuronal death, which further activates glial cells that in turn release more inflammatory mediators in a self-amplifying process. Factors secreted by ►microglial cells in response to injury induce activation of astrocytes. Activated astrocytes secrete growth factors promoting microglial growth and activation, but also modulate their cytotoxicity. It has been proposed that microglia-derived factors are responsible for neurotoxicity whereas astrocytes secrete neuroprotective factors. However, astrocytes may also cooperate with microglia enhancing oxidative stress. The association of several pro-inflammatory molecules with neurodegenerative lesions suggests a state of persistent inflammatory activation that could escape endogenous control and become cytotoxic. Genetic research identified susceptibility genes influencing the inflammatory process of neurodegenerative diseases. Polymorphisms in the interleukin genes, IL-1α and IL-1β, are associated with increased risk of early onset AD and the C allele of IL-6, associated to reduced IL-6 activity decreases the risk and delays the onset of sporadic AD. Thus, inherited variations in inflammation mechanisms may influence AD as well as other neurodegenerative diseases pathogenesis.

Astrocytes and microglia are highly reactive to environmental changes and work cooperatively, exerting mutual regulatory activity (Fig. 1). There is constitutive expression of interleukin-1 (IL-1) by astroytes, microglia, neurons and endothelium. Activation of glial cells induces the release of cytokines, such as tumor necrosis factor-α (TNF-α) and IL-1β, which participates at early stages of neuroinflammation [10] and IL-6 at latter stages. IL-1β has been described as a potent activator of host inflammatory responses within and outside the CNS. Microglia appear to be its early source following experimental CNS injury, infection or inflammation. Production by astrocytes usually follows slightly later [10] (Fig. 2).

Besides the central role of cytokines in neuroinflammation, they also influence neuronal and synaptic function via diverse mechanisms contributing to cognitive impairment, including regulation of neurotransmission, neurotransmitters receptors and synaptic efficacy [6]. A cohort study of healthy ageing individuals showed that high levels of IL-6 correlates with lower cognitive functioning in cross-sectional analysis, also predicting subsequent cognitive impairment. Cytokines are also involved in the pathophysiology of neurodegenerative diseases of vascular origin, by

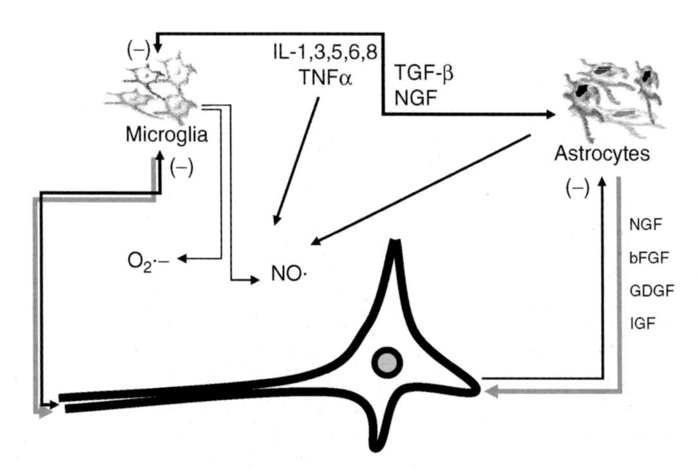

Neurodegenerative Diseases – MAPK Signalling Pathways in Neuroinflammation. Figure 1 Modulation of the CNS: neuron-glia interaction. Homeostasis in the nervous tissue depends on a finely tuned cross talk among glial cells and neurons. There is a reciprocal structural and functional regulation of glial cells and neurons by growth factors and cytokines. Pro-inflammatory cytokines induce the production of NO and ROS, which are deemed as responsible for cell damage. Factors, such as TGF-β, modulate the activation of glial cells. Neurons have an inhibitory regulatory role on glial activation. Neuroinflammation could depend on the inability of microglial cells to respond to modulatory effect or on the failure to establish the modulatory mechanism.

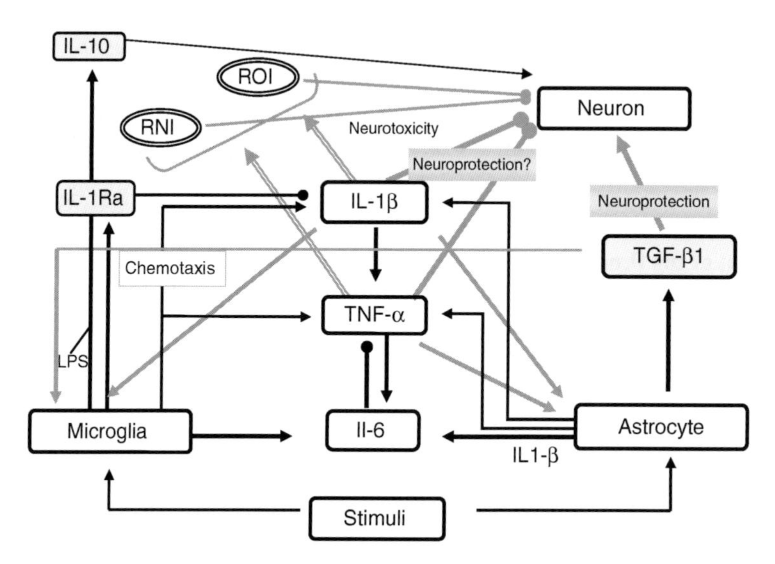

Neurodegenerative Diseases – MAPK Signalling Pathways in Neuroinflammation. Figure 2 Major inflammatory cytokines involved in glial activation. Schematic overview of the principal cytokines produced by glial cells in response to pro-inflammatory stimuli. Stimuli lead to the activation of microglial cells and later of astrocytes. Glia secrete pro-inflammatory cytokines. This, in turn, induces second line cytokines with pro- and anti-inflammatory cytokines, such as IL-6 and TGF-β, IL-10 and IL-1Ra, which also modulate production of pro-inflammatory cytokines, and reactive short life molecules such as oxygen radicals and NO. Pro-inflammatory cytokines and reactive species can promote neurotoxicity. However, pro-inflammatory cytokines also have neuroprotective functions.

influencing the response to ischemia, influencing the coagulation cascade favoring thrombosis and promoting the atherogenic process.

Activated astrocytes and microglia release mediators like ►nitric oxide (►NO) and activated microglia also secrete important amounts of ►reactive oxygen species (►ROS) like superoxide ($O_2^{\bullet -}$) deleterious oxidative molecules capable of inducing hippocampal cell damage [11]. Several neurodegenerative conditions show enhanced oxidative damage. Increased amounts of $O_2^{\bullet -}$ and NO are among the proposed mechanisms for neuronal death and demyelization. Cerebrospinal fluid have increased levels of nitrite and demyelinating lesions show increased levels of nitrotyrosine in multiple sclerosis patients. Increased NO production depends on transcriptional up-regulation of inducible nitric oxide synthase (iNOS) and release of $O_2^{\bullet -}$ by the multicomponent phagocyte NADPH oxidase. Activated microglia also express myeloperoxidase, which generates the potent oxidizing agent hypochlorous acid.

Pro-inflammatory cytokines activate a number of signaling mechanisms (Fig. 3). Activation of IL-1 receptor leads to translocation of NF*k*B to the nucleus and activation of mitogen-activated protein kinases (MAPKs) ERK1/2, P38 and JNK1 [10]. Inhibition of each of the MAPK pathways inhibits inflammation-induced IL-1β production in a concentration-dependent manner. In astrocytes, IL-1β plus TNF-α induce the expression of pro-inflammatory cytokines and iNOS and the production of NO through the activation of ERK1/2 and NF*k*B [7]. Although IL-1β or TNF-α alone is capable of activating NF*k*B, the expression of iNOS appears to require activation of additional transcription factors modulated by MAPKs. IL-1β and IFN-γ induce the activation of the transcription factor AP-1, but the combination of both cytokines markedly inhibits its activation. It has been defined that the activation of AP-1 and expression of iNOS depends on the activation of JNK, an effect observed only during low level of iNOS induction but not during high level of induction by IL-1β and IFN-γ. IL-1 also increases expression and processing of amyloid precursor protein (APP), which could favor production of amyloid β (Aβ) in humans and animal models of AD. Activity of ERK and JNK kinases appear to be required for the induction of APP processing by IL-1.

The main signal pathways induced by IFN-γ are the signal-transducer and activator of transcription-1 (STAT1) and MAPKs (Fig. 3). STAT-1 is the main transcription factor involved in the induction of iNOS by IFN-γ. NF*k*B does not play a key role in the regulation of iNOS expression in response to IFN-γ. However, extracellular signal-regulated kinase (ERK)1/2 MAPK plays an important modulatory role. STAT-1 is activated by a JAK-dependent phosphorylation of a tyrosine residue ($pSTAT1^{tyr}$), whereas its full activation depends on a second phosphorylation at a serine residue

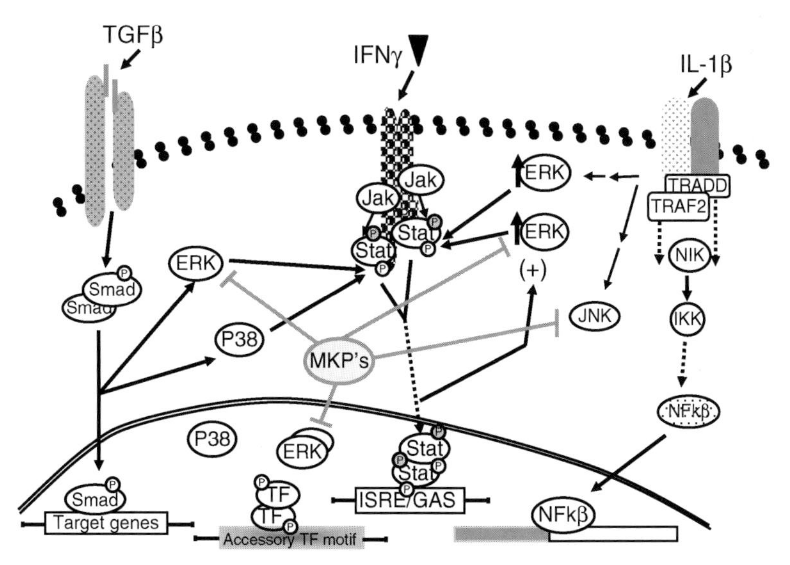

Neurodegenerative Diseases – MAPK Signalling Pathways in Neuroinflammation. Figure 3 Effect of ERK and P38 on IFN-γ signaling. Binding of IFN-γ induces receptor dimerization and their phosphorylation by Jaks. The phosphorylated receptor serves as a docking site for STATs, and adaptors linking to MAPK and PI-3-kinase/Akt. Phosphorylated STATs dimers translocate into the nucleus, regulating the transcription of target genes. TGF-β signaling depends on phosphorylation of Smad proteins and their translocation to the nucleus. Activated Smads regulate diverse effects resulting in cell-state specific modulation of transcription. TGF-β signaling also has Smad-independent pathways, including ERK, JNK and P38 MAPKs. IL-1β binds to IL-1R associated to Nuclear factor-κB (NF-κB) proteins. IL-1β triggers phosphorylation of IκB leading to its degradation, freeing NF-κB. Active NF-κB translocate to the nucleus where, either alone or with other transcription factors including AP-1, Ets and STAT, induce target gene expression. The MAPK/ERK signaling cascade is activated by many receptors. The pathway usually involves small GTP binding proteins (Ras, Rap1), which in turn activate the kinase cascade composed of a MAPKKK (Raf), a MAPKK (MEK1/2) and MAPK (ERK). Activated ERK regulates targets in the cytosol and translocate to the nucleus phosphorylating several transcription factors. Stress-activated protein kinases (SAPK)/Jun N-terminal kinases (JNK) are activated by a variety of signals that are delivered by small GTPases (Rac, Rho, cdc42). SAPK/JNK kinases and P38 MAPKs activation is similar to that of ERKs. P38 is involved in regulation of several transcription factors including ATF-2, Stat1, Max/Myc complex, MEF-2, Elk-1 and indirectly CREB via activation of MSK1. Specific phosphatases for ►MAPK (MKP) end ►MAPK signaling.

($pSTAT1^{ser}$). ERK and P38 signaling are implicated in the expression of iNOS and the generation of NO [5,7]. IFN-γ induce ERK1/2 and P38 phosphorylation while their inhibitors attenuate IFN-γ-induced NO, suggesting that those MAPK signaling pathways are important on the modulation of NO production by glial cells. In contrast, $O_2^{\bullet -}$ production induced by IFN-γ is associated to increased levels of pERK1/2, but not pP38. Inhibition of P38 pathway potentiates LPS-induced ROS production. There is evidence that NADPH oxidase component $p67^{PHOX}$ is phosphorylated by ERK2, suggesting that NADPH oxidase activity is ERK1/2 dependent. Besides regulation of NADPH oxidase, there is a complex cross-talk between ROS and MAPK through multiple mechanisms. MAPKs are implicated in the regulation of pro-inflammatory cytokines, including TNF-α and several ILs, which in turn induce ROS production. On the other hand, ROS signaling modulates not only cytokines transduction pathways but also appears to regulate the MAPK-induced transcription of some pro-inflammatory cytokines [1]. Further more, ROS inactivation of MAPK phosphatases leads to prolonged activation of MAPK pathways.

There is a differential temporal contribution of ERK1/2 and P38 in the full-activation of the STAT-1 pathway in glial cells. pERK1/2 and pP38 levels are increased in glial cultures exposed to IFN-γ. However, whereas phosphorylation of ERK1/2 persisted after 24 h, phosphorylation of P38 rapidly decreased to control levels. Their differential timing suggests that both ERK and P38 modulate STAT-1 at short times, but only ERK1/2 participates after long time stimulation [3]. There are also differences depending on the glial cell type. IFN-γ-induced phosphorylation of P38 only increases in microglial cells but not in mixed glial cell cultures. Similarly, soluble factors secreted by astrocytes decreases IFN-γ-induced phosphorylation of STAT-1 and ERK1/2 in microglial cell cultures [11]. There is evidence that microglial cells are more reactive than mixed glial culture, which is supported by the fact that astrocytes modulate microglial cell reactivity [9].

Lower Level Processes

MAPKs include ERK1/2, ERK3 and ERK5, P38 Hog and JNK/SAPK. Genetic or epi-genetic alterations of MAPKs or of the signaling cascades that regulate them have been implicated in a variety of human diseases including inflammation. Activated MAPKs act in the cytoplasm or translocate into the nucleus phosphorylating other proteins, like transcriptional factors. MAPK signaling is ended by a specific family of phosphatases for MAPK (MKP). ERK1/2 and P38 appear to be key actors in the production of free radicals by glia [7].

Higher Level Processes

Neuroinflammation requires the coordinated activity of a network of intracellular pathways. MAPKs are signaling molecules capable of mediating crosstalk among pathways allowing the generation of complex responses to different combinations of pro-inflammatory stimuli or the presence of other modulatory conditions. It can be achieved through several mechanisms, ranging from direct communication between intracellular pathways to feedback processes depending on autocrine signaling.

The presence of various pro-inflammatory signals, depending on their underlying molecular mechanisms, could be additive (their combined effect is the sum of the effect expected for each of the stimuli), could show a synergy or inhibition as manifestation of a regulatory crosstalk, or saturation if a common upstream signaling component is involved [8]. Factors as age, cell damage, metabolism or oxidative stress are just some of the many input that will influence an inflammatory response involving activation of MAPK, making the response dependent on the context of the environment. MAPK could be involved in the both the magnitude and the temporal pattern of the response. Different outcomes, such as cell proliferation, expression of cytokines or production of ►ROS among others can be expected to be differentially regulated by a certain combination of stimuli.

Regulation of the Process

Feedback mechanisms mediated by crosstalk among brain cells restrain the amplitude and duration of ►glial activation, and restore glial cells to their resting state. Neurons have an inhibitory role on glial activation, regulating the production of many proteins associated with reactive gliosis. Astrocytes attenuate activation of microglial cells and confer protection to hippocampal cells [9]. Hippocampal cells and activated astrocytes secrete transforming growth factor β1 (TGF-β_1). TGF-β_1 modulates ►microglial activation inhibiting production of IL-1 and TNF-α, expression of Class II-MHC and Fas glycoprotein, NOS induction, release of NO [11] and $O_2^{\bullet -}$ production.

There is controversy regarding the negative or beneficial effects of IL-1β, although IL-1β is considered a pro-inflammatory ►cytokine, it can both contribute to and limit neuronal damage serving neuroprotective functions [10]. Beneficial effects are particularly observed when low concentrations are released. IL-1β appears to be capable of inhibiting further activation IL-1β at early stages of inflammation reducing production of NO by microglial cells [11]. In contrast, anti-inflammatory cytokines, like TGF-β1, exerts a delayed effect on the modulation of microglial cell activity. The fact that pro-inflammatory cytokine have inhibitory effects suggests that the timing for the activation of the different pathways is important for the cell response outcome.

TGF-β1 plays a prominent role in homeostasis and tissue repair. Functions exerted mainly through its receptor-activated Smad signaling. It also has neuroprotective effects through the activation of PI3 kinase/Akt and ERK1/2 pathways [12]. The effect of TGF-β1 on glial cell activation is mediated by regulation of ERK1/2 and P38 MAPKs (Fig. 3). Activation of MAPKs in response to TGF-β1 is transient and renders microglial cells refractory to further activation by pro-inflammatory cytokines. TGF-β1 induces a strong expression of MAPK phosphatase type 1 (MKP-1) persistently up-regulated under pro-inflammatory conditions. An increased MKP-1 activity could be the mechanism responsible for the decrease of IFN-γ-induced pERK1/2 in cells exposed to TGF-β1 [3].

It has been proposed the existence of complex regulatory interactions between TGF-β1 and IFN-γ involved in tissue repair. IFN-γ null mice have an increased amount of TGF-β1 and activation of Smad signaling pathway [4]. On the other hand, TGF-β1 null mice have high levels of IFN-γ and over-expression of STAT-1, iNOS and NO production. The over-activation of STAT-1 in TGF-β1 null mice supports the notion that TGF-β1 is an essential immune-regulator for the control of inflammatory events. In glial cells, TGF-β1 treatment results in a reduction of IFN-γ–induced pERK1/2, STAT-1 phosphorylated at serine^{727} (pSTAT-1^{ser}) and tyrosine^{701} (pSTAT-1^{tyr}), and total STAT-1. After long lasting stimulation, IFN-γ decreases TGF-β1-induced pP38 signal transduction. IFN-γ-TGF-β1 crosstalk regulates the production of oxidative molecules through the reduction of STAT-1, ERK1/2 and P38 activation [3]. ERK is involved in the modulation of microglial response by both anti-inflammatory cytokines (TGF-β1) and pro-inflammatory cytokines (IL-1β). TGF-β1 and IL-1β decrease phosphorylation of ERK and the NO production induced by IFN-γ. IL-1β inhibits ERK1/2 phosphorylation after 30 min of activation with IFN-γ and the inhibition is short-lived. In contrast, TGF-β1 inhibits ERK1/2 phosphorylation only after several hours of activation (24 h) and inhibition persists for a long time (Fig. 4). Only certain pro-inflammatory cytokines, like IL-1β has an orchestrating role, TNF-α is unable to inhibit neither production of radical species nor ERK1/2 phosphorylation.

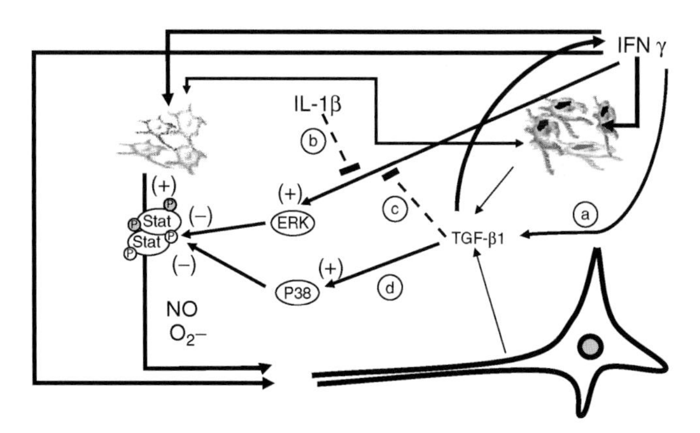

Neurodegenerative Diseases – MAPK Signalling Pathways in Neuroinflammation. Figure 4 Modulation of microglial activation by TGF-β and IL-1β. IFN-γ activates glial cells and neurons. Microglia secrete radical species mediated by the activation of JAK/STAT and MAPK. ERK-dependent phosphorylation of STAT1 in the serine position potentiates its activation. (a) IFN-γ also induce production of cytokines, IL-1β at early times and later TGF-β. IFN-γ and TGF-β show several reciprocal modulatory effects. The increased secretion of TGF-β modulates the production of radical species. (b) IL-1β inhibited IFN-γ-induced activation of ERK. The effect could be part of an early mechanism to limit inflammation when stimuli are mild ending glial cell activation. (c) If the inflammatory stimulus persists, the increased secretion of TGF-β could further inhibit ERK and P38 activation. (d) At later times, activation of P38 will be inhibited by IFN-γ, further reducing the positive modulation on STAT1.

Function

MAPK-regulation in neuroinflammation allows for the integration of specific signaling pathways to yield unique output responses. MAPK can be involved in both convergence and divergence of signaling cascades (see Fig. 3). The combinatorial complexity of signaling may account for the fact that depending on specific background determined by both genetic and environmental factors, pro-inflammatory stimuli can result in limited inflammation and repair or cell damage resulting in a neurodegenerative disease.

Pathology

In a large number of neurodegenerative diseases, including ►multiple sclerosis (MS), Alzheimer's disease, Parkinson's disease, cerebro-vascular diseases and CNS trauma among others, inflammatory response is evident, including complement activation, elevated pro-inflammatory cytokines, chemokines and glial activation. There is increasing evidence that neurotoxicity is mediated by CNS inflammatory processes.

Therapy

The participation of neuroinflammatory mechanisms in the genesis or progression of neurodegenerative diseases has led to extensive investigation of the therapeutic effect of anti-inflammatory and anti-oxidant treatments. There are mixed results depending on the type of neurodegenerative disease and among basic research and clinical trials. Whereas treatment of MS with IFN-β and treatment of CNS trauma or inflammation secondary to radionecrosis with corticosteroids are generally approved, most trials of anti-inflammatory agents report no significant beneficial effect on Alzheimer's patients. Nevertheless, epidemiological studies suggest that the risk of the disease is reduced in patients treated with non-steroidal anti-inflammatory drugs (NSAIDs). Observational studies also suggested that intake of antioxidants, such as vitamin E, could reduce the risk of disease. However randomized controlled clinical trials show at the best only marginal benefits.

Candidate protein kinase inhibitors have been tested in vitro assays and in mouse models for their ability to suppress putative mechanisms of neuroinflammation including iNOS induction and production of IL-1β. The use of MAPKs inhibitors for the treatment of inflammatory disorders has been suggested for infectious diseases, but MAPK cascades have been not proposed as possible therapeutic targets for neurodegenerative diseases. However, due to the variation in responses due to cell type and dose-dependence, the therapeutic use of these inhibitors will demand extensive evaluation. Therapeutic approaches including combination of drugs, each aimed at a different inflammatory target, probably will be more effective than single agents.

Acknowledgements

The author's research was supported by grant FONDECYT 1040831.

References

1. Haddad JJ, Land SC (2002) Redox/ROS regulation of lipopolysaccharide-induced mitogen-activated protein kinase (MAPK) activation and MAPK-mediated TNF-α biosynthesis. Br J Pharmacol 135:520–536

N

2. Haddad JJ (2004) Mitogen-activated protein kinases and the evolution of Alzheimer's: a revolutionary neurogenetic axis for therapeutic intervention? Prog Neurobiol 73:359–377
3. Herrera-Molina R, von Bernhardi R (2007) Modulation of Interferon γ-mediated activation of glial cells by Transforming Growth Factor β1: A role for STAT1 and MAPK Pathways. JBC, submitted
4. Ishida Y, Kondo T, Takayasu T, Iwakura Y, Mukaida N (2004) The essential involvement of cross-talk between IFN-γ and TGF-β in the skin wound-healing process. J Immunol 172:1848–1855
5. Kim SH, Kim J, Sharma RP (2004) Inhibition of p38 and ERK MAP kinase blocks endotoxin-induced nitric oxide production and differentially modulates cytokine expression. Pharmacol Res 49:433–439
6. Li Y, Liu L, Barrer SW, Griffin WS (2003) Interleukin-1 mediates pathological effects of microglia on tau phosphorylation and on synaptophysin synthesis in cortical neurons through a p38-MAPK pathway. J Neurosci 23:1605–1611
7. Marcus J, Karackattu S, Fleegal M, Summers C (2003) Cytokine-stimulated inducible nitric oxide synthase expression in astroglia: Role of Erk Mitogen-activated protein kinase and NF-κB. Glia 41:152–160
8. Natarajan M, Lin KM, Hsueh RC, Sternweis PC, Ranganathan R (2006) A global analysis of cross-talk in a mammalian cellular signaling network. Nature cell biology online publication DOI:10.1038/ncb1418
9. Ramírez G, Toro R, Döbeli H, von Bernhardi R (2005) Protection of rat primary hippocampal cultures form Aβ cytotoxicity by proinflammatory molecules is mediated by astrocytes. Neurobiol Dis 19:243–254
10. Rothwell N, Luheshi G (2000) Interleukin 1 in the brain: Biology, pathology and therapeutic target. Trends Neurosci 23:618–625
11. Saud K, Herrera-Molina R, von Bernhardi R (2005) Pro- and Anti-inflammatory cytokines regulate the ERK pathways: Implication of the timing for the activation of microglial cells. Neurotox Res 8(3,4):277–287
12. Zhu Y, Culmsee C, Klumpp S, Krieglstein J (2004) Neuroprotection by transforming growth factor-β1 involves activation of nuclear factor-κB through phosphatidylinositol-3-OH kinase/Akt and mitogen-activated protein kinase extracellular-signal regulated kinase1,2 signaling pathways. Neurosci 123:897–906

Neurodegenerative Diseases: Tryptophan Metabolism

GILLES J. GUILLEMIN
Centre for Immunology and Department of Pharmacology, Faculty of Medicine, University of New South Wales, NSW, Australia

Synonyms

Kynurenines

Definition

►Tryptophan is an essential amino acid required by all forms of life, for metabolic functions and synthesis of essential proteins including serotonin and melatonin. More than 95% of dietary tryptophan is catabolized through the kynurenine pathway (Fig. 1). The kynurenine pathway leads to the biosynthesis of nicotinamide adenine dinucleotide (NAD), which is an essential cellular cofactor for many cellular reactions, ranging from adenosine tri-phosphate (ATP) synthesis to DNA repair. This metabolic step appears to be highly conserved. In the process, several neuroactive intermediates are produced including kynurenine, kynurenic acid, 3-hydroxykynurenine, 3-hydroxyanthranilic acid, picolinic acid and quinolinic acid [1,2].

Characteristics

Quantitative Description

In physiological conditions, concentrations of most of the kynurenine pathway metabolites are very low. Significant quantitative changes happen in pathological conditions (see below).

Higher-Level Structures

- The cellular location of the kynurenine pathway is only partly understood. It is known to be complete in monocytic lineage cells, including macrophages, microglia and dendritic cells. Whereas in astrocytes, neurons, brain microvascular endothelial cells and oligodendrocytes, the kynurenine pathway is only partly present [3].
- Expression of the enzymes and production of the compounds of the kynurenine pathway vary significantly between species and between cell lines and the respective primary cell type [4].

Process Regulation

The kynurenine pathway is regulated by its first and rate limiting enzyme indoleamine 2,3 dioxygenase (IDO) [5]. IDO is activated during neuroinflammation. Inflammatory molecules such as platelet activating factor, lipopolysaccharides, lentiviral proteins Nef and Tat, amyloid aggregates, and cytokines can induce IDO expression. Among the cytokines, interferon gamma (IFNγ) is the most potent IDO inducer. Interferons beta and alpha, interleukin 1 and tumor necrosis factor α induce IDO to a lesser extent. IDO can also be down regulated by interleukin 4 and nitric oxide.

Functions

The kynurenine pathway has a role in physiological functions.

- The involvement of the kynurenine pathway in psychological functions is mainly due to its ability to divert tryptophan catabolism from the serotonergic

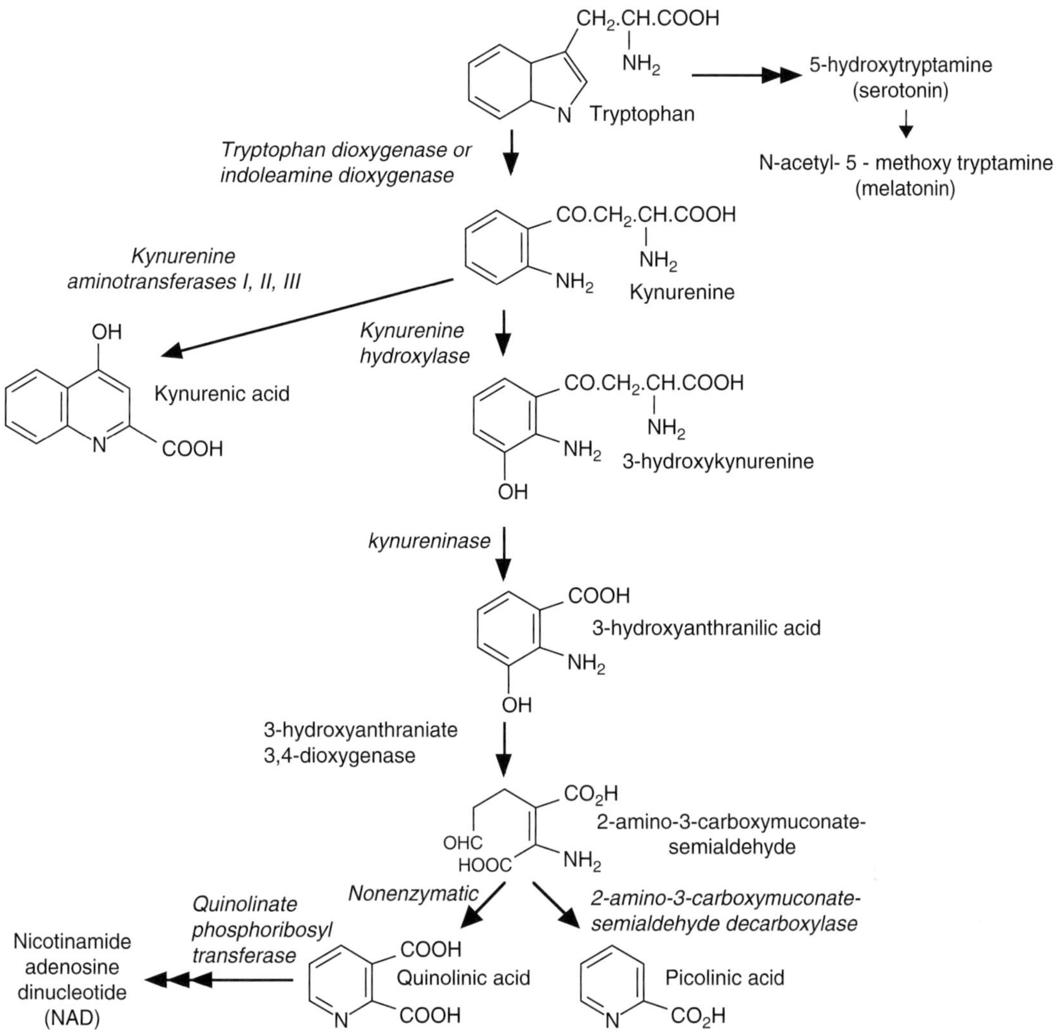
CH2.CH.COOH
NH2
Tryptophan
5-hydroxytryptamine
(serotonin)
N-acetyl- 5 - methoxy tryptamine
(melatonin)
Tryptophan dioxygenase or
indoleamine dioxygenase
CO.CH2.CH.COOH
NH2
NH2
Kynurenine
Kynurenine
aminotransferases I, II, III
OH
Kynurenic acid
COOH
Kynurenine
hydroxylase
CO.CH2.CH.COOH
NH2
NH2
3-hydroxykynurenine
OH
kynureninase
COOH
3-hydroxyanthranilic acid
NH2
OH
3-hydroxyanthraniate
3,4-dioxygenase
CO2H
2-amino-3-carboxymuconate-
semialdehyde
OHC
HOOC
NH2
Nonenzymatic
2-amino-3-carboxymuconate-
semialdehyde decarboxylase
Quinolinate
phosphoribosyl
transferase
Nicotinamide
adenosine
dinucleotide
(NAD)
COOH
Quinolinic acid
COOH
Picolinic acid
CO2H

Neurodegenerative Diseases: Tryptophan Metabolism. Figure 1 Simplified diagram of the kynurenine pathway.

pathway (Fig. 1). Modulation of serotonin (5 hydroxy-tryptophan; ▶5HT) levels is associated with changes in behavior, mood, sleep regulation, and thermo-regulation.

- Recent findings have shown that the kynurenine pathway is one of the major regulatory mechanisms of the immune response [6]. Two theories have been proposed: (i) that tryptophan degradation suppresses T cell proliferation by dramatically depleting the supply of this critical amino acid; (ii) that some downstream kynurenine pathway metabolites act to suppress certain immune cells. Induction of the kynurenine pathway in dendritic cells completely blocks clonal expansion of T cells.
- Tryptophan depletion is associated with IDO and kynurenine pathway activation are implicated in the development of immuno-tolerance associated with pregnancy. IDO activation plays a key role in the protection of the fetus by suppressing T cell-driven local inflammatory responses against the fetal alloantigens [7].

Pathophysiological Conditions

The kynurenine pathway is involved in the anti-bacterial, anti-parasites and anti-viral immune defenses.

- Since tryptophan is an amino acid essential for many metabolic processes, depletion of available tryptophan is an important mechanism for the control of rapid-dividing microbial pathogens including bacteria, parasites and viruses.

Pathology

Two sides have to be considered here: decrease of tryptophan levels and production of kynurenine pathway metabolites (Fig. 2).

- Tryptophan depletion has been associated with mood and psychiatric disorders such as schizophrenia,

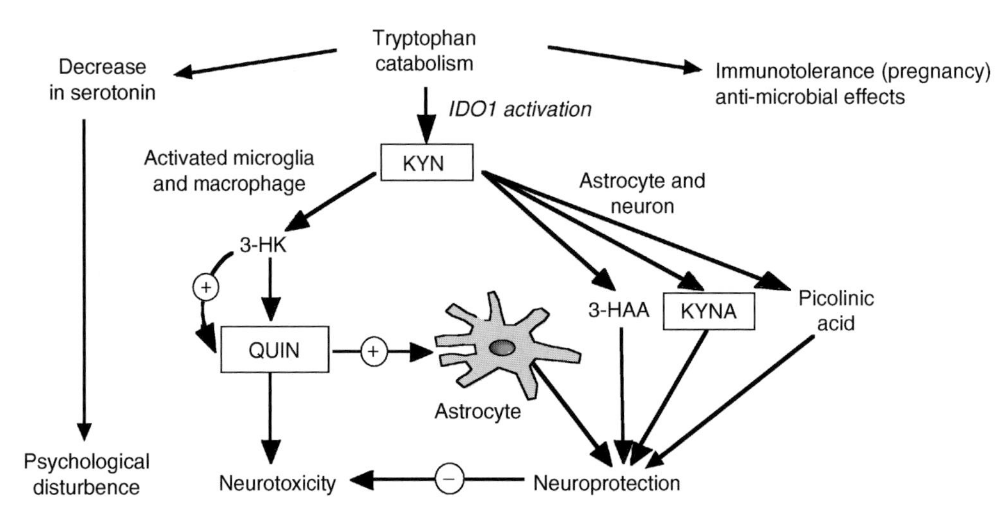

Neurodegenerative Diseases: Tryptophan Metabolism. Figure 2 Summary of the effects of tryptophan depletion through the kynurenine pathway.

depression, panic disorder, seasonal affective disorder and obsessive–compulsive disorder.

- Activation of the kynurenine pathway and production of neuroactive metabolites have been shown to be involved in a number of neurodegenerative disorders, such as Huntington's disease, Parkinson's disease, Alzheimer's disease, amyotrophic lateral sclerosis, multiple sclerosis, AIDS dementia complex, stroke and epilepsy [1,8].

Products derived from the kynurenine pathway can have either neurotoxic and neuroprotective effects. At least four of them are known to have marked effects on neuronal survival: quinolinic acid, 3-hydroxykynurenine, kynurenic acid and picolinic acid.

Quinolinic Acid (QUIN)

Among the kynurenine pathway intermediates, QUIN is the main toxic compound produced (see Fig. 1). In 1981, Stone and Perkins showed for the first time the ability of QUIN to selectively activate neurons expressing NMDA receptors. QUIN neurotoxicity was demonstrated by Schwarcz et al. who showed that intra striatal and intra hippocampal injection of QUIN in the rat brain led to neurodegeneration around the injection site. QUIN leads acutely to human neuronal death and chronically to dysfunction by at least five mechanisms. (i) Activation of the NMDA receptor in pathophysiological concentrations. Neurons within the hippocampus, striatum and neocortex are more sensitive to QUIN, than cerebellar and spinal cord neurons; (ii) QUIN increases glutamate release by neurons and inhibits glutamate uptake by astrocytes leading to excessive microenvironment glutamate concentrations and neurotoxicity; (iii) More recently, it has become clear that a major mechanism of QUIN neurotoxicity is through lipid peroxidation; (iv) QUIN can potentiate its own toxicity and that of other excitotoxins (for example NMDA and glutamate) in the context of energy depletion and (v) Lastly, QUIN leads to astrocyte apoptosis with consequent loss of detoxifying and neurotrophic support of neurons. It is also likely that QUIN toxicity is additive or synergistic with other immune system-derived toxins, which act also via or modulate the NMDA receptor (Fig. 3).

3-Hydroxykynurenine

3-Hydroxykynurenine can produce neuronal damage by the induction of oxidative stress rather than an action on glutamate receptors. 3-Hydroxykynurenine also acts synergistically with QUIN and potentiates its toxicity.

Kynurenic Acid

Kynurenic acid is an antagonist of all ionotropic glutamate receptors including N-methyl-D-aspartate (NMDAR), kainic acid and α-amino-3-hydroxy-5-methyl-4-isoxazole (AMPAR) receptors. Furthermore, kynurenic acid non-competitively inhibits α 7-nicotinic acetylcholine presynaptic receptors (nAChRs). Kynurenic acid is the only known endogenous NMDA receptor inhibitor, which act at the glycine site of the NMDAR and can antagonize some of the effects of quinolinic acid and other excitotoxins. It is noteworthy that in disease states where a large excess quinolinic acid is produced there is insufficient kynurenic acid to block quinolinic acid.

Picolinic Acid

Picolinic acid has been shown to be neuroprotective. Picolinic acid protects against quinolinic acid ands kainic acid-induced neurotoxicity in the brain. However, picolinic acid blocks the neurotoxic but not the excitatory effects of QUIN. The mechanism of its

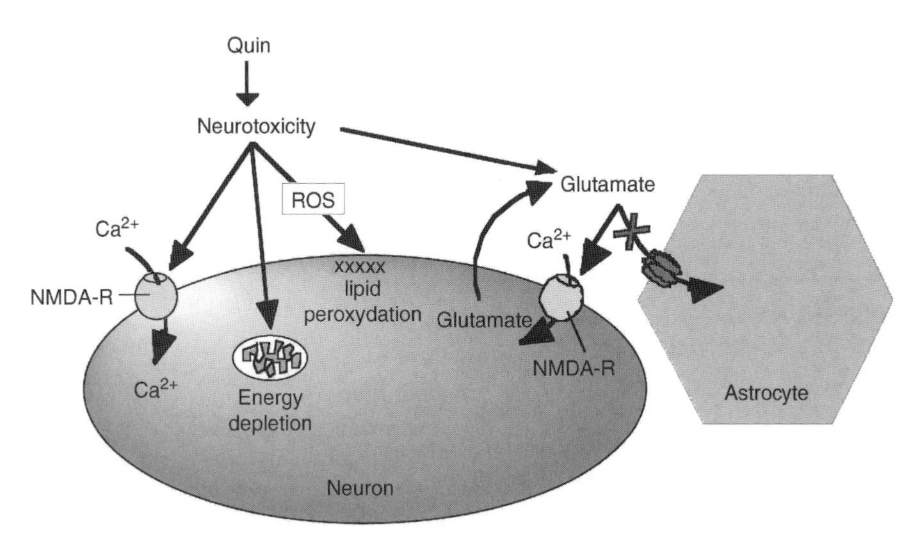

Neurodegenerative Diseases: Tryptophan Metabolism. Figure 3 Mechanisms of quinolinic acid neurotoxicity.

anti-neurotoxic action is unclear but might involved zinc chelation and/or inhibition of nitric oxide synthase. Picolinic acid can influence the immune response and has antifungal, antitumoral and antibacterial activities.

Therapy

Over the last two decades, manipulation of the kynurenine pathway has led to the development of a large number of neuroprotectant and anticonvulsant drugs [9,10].

- Some of these drugs target specific enzymes of the pathway. For example, tryptophan analogs such as 1-methyl tryptophan or 6-chlorotryptophan inhibit IDO activity; another family of compounds such as Ro 61–8,048 and m-nitrobenzoyl-alanine inhibit kynurenine hydroxylase activity.
- Other drugs are developed for their ability to block glutamate receptors and stop excitotoxic mechanisms. Kynurenine and kynurenic acid have been used as the original molecules for several groups of this kind of compounds. For example, 4-chlorokynurenine crosses the blood brain barrier and blocks quinolinic acid toxicity at the glycine site on NMDA receptors. Some kynurenic acid analogues are in or about to enter clinical trials for treatment of epilepsy, stroke and possibly Parkinson's disease.
- More recently, small interfering RNA (siRNA) targeting IDO, a new generation of inhibitors, has shown a great capacity to block the kynurenine pathway.

References

1. Stone TW (2001) Kynurenines in the CNS: from endogenous obscurity to therapeutic importance. Prog Neurobiol 64:185–218
2. Botting NP (1995) Chemistry and neurochemistry of the kynurenine pathway of tryptophan metabolism. Chem Soc Rev 24:401–412
3. Guillemin GJ, Kerr SJ, Smythe GA, Smith DG, Kapoor V, Armati PJ, Croitoru J, Brew BJ (2001) Kynurenine pathway metabolism in human astrocytes: a paradox for neuronal protection. J Neurochem 78:1–13
4. Allegri G, Costa CV, Bertazzo A, Biasiolo M, Ragazzi E (2003) Enzyme activities of tryptophan metabolism along the kynurenine pathway in various species of animals. Farmaco 58:829–836
5. Takikawa O (2005) Biochemical and medical aspects of the indoleamine 2,3-dioxygenase-initiated l-tryptophan metabolism. Biochem Biophys Res Commun 338:12–19
6. Mellor AL, Munn DH (2004) IDO expression by dendritic cells: tolerance and tryptophan catabolism. Nat Rev Immunol 4:762–774
7. Munn DH, Zhou M, Attwood JT, Bondarev I, Conway SJ, Marshall B, Brown C, Mellor AL (1998) Prevention of allogeneic fetal rejection by tryptophan catabolism. Science 281:1191–1193
8. Guillemin GJ, Kerr SJ, Brew BJ (2005) Involvement of quinolinic acid in AIDS dementia complex. Neurotox Res 7:103–123
9. Schwarcz R (2004) The kynurenine pathway of tryptophan degradation as a drug target. Curr Opin Pharmacol 4:12–17
10. Stone TW (2001) Endogenous neurotoxins from tryptophan. Toxicon 39:61–73

Neuroeffector Junction

Definition

A neuroeffector junction is the gap, typically 20–100 nm wide, between a varicosity of an autonomic axon and an effector cell in the target organ. These junctions are wider and not as specialized in structure as synapses in autonomic ganglia or the central nervous system, or motor end plates in skeletal muscle.

►Postganglionic Neurotransmitter

Neuroendocrine Axis

Definition

It provides the structural and functional basis for interactions between brain, hormones, and glands that allow an organism to respond to external stimuli with complex, sometimes long-lasting physiological changes, such as during stress or reproduction.

Typically, it refers to any of these three pathways and their respective feedback mechanisms: the "Hypothalamus-Pituitary-Adrenal (HPA) axis," which extends from the hypothalamus in the brain via the anterior part of the pituitary gland to the adrenal cortex, the "Hypothalamo-Pituitary-Thyroid (HPT) axis," which leads to the thyroid gland, or the "Hypothalamo-Pituitary-Gonadal (HPG) axis," involving the male and/ or female gonads.

►Hypothalamo-Pituitary-Adrenal (HPA) Axis, Stress and Depression
►Hypothalamo-Pituitary-Thyroid (HPT) Axis

Neuroendocrine Regulation

Definition

In the central nervous system several specialized neurons produce peptides and/or proteins that are released as hormones in the circulation. One class of these hormones is the 'releasing factors' secreted in the portal system of the median eminence in order to target especially the adenohypophysis, where they liberate or inhibit the secretion of pituitary hormones. The other class of hormones is directly released into the main circulation either via the neural lobe part of the pituitary or via the pineal. The autonomic nervous system has important influences on the release and the efficiency of all these hormones.

►Hypothalamo-Pituitary-Adrenal (HPA) Axis, Stress and Depression
►Hypothalamo-Neurohypophysial System
►Neuroendocrine Regulation and the Autonomic Nervous System

Neuroendocrine Regulation and the Autonomic Nervous System

Ruud M. Buijs[1,2], Andries Kalsbeek[2]
[1]Instituto de Investigaciones Biomedicas, UNAM, México
[2]Netherlands Institute for Neuroscience, Amsterdam, The Netherlands

Definition

Neuroendocrine Regulation

In the central nervous system several specialized neurons produce peptides and/or proteins that are released as hormones in the circulation. One class of these hormones is the "releasing factors" secreted in the portal system of the median eminence in order to target especially the adenohypophysis, where they liberate or inhibit the secretion of pituitary hormones. The other class of hormones is directly released into the main circulation either via the neural lobe part of the pituitary or via the pineal. The ►autonomic nervous system has important influences on the release and the efficiency of all these hormones.

Autonomic Nervous System

The autonomic nervous system is that part of the central nervous system that operates outside voluntary control. The executing part of the autonomic nervous system is divided into a parasympathetic and a sympathetic branch. These two branches have, in general, antagonistic functions with an anabolic role for the parasympathetic and a catabolic function for the sympathetic branch. In addition information about the functional condition of our organs is transmitted back to the brain via these two branches.

Characteristics

Quantitative Description

Very soon after the discovery of the principle of neurotransmission, Ernst and Bertha Scharrer proposed that certain specialized neurons in the hypothalamus would be able to produce hormones and release them into the blood stream. However, it took the scientific community more than 20 years to get used to this idea and only after Wolfgang Bargmann confirmed and elaborated on these findings was their "Neurosecretion" theory accepted. At about the same time, Frederick Banting and Charles Best made their groundbreaking discovery of insulin as a hormone of the pancreas that could save the life of many diabetic patients. These two findings should have made scientists aware that the areas of endocrinology and neuroscience are intimately integrated. Instead the complexity of the brain and the enormous life saving potential of hormones drove these fields further apart. Now the restoring movement is in progress and more and more it becomes clear that body and mind cannot be separated. Via the autonomic nervous system (ANS), the brain not only affects the production and efficacy of its own hormones but also of the hormones produced by the other organs of the body. Here special attention will be paid to the organization within the central nervous system (CNS) of structures that affect autonomic output and integrate autonomic information in relation with ▶neuroendocrine regulation.

Hypothalamus and the Organization of Autonomic Output and Input

ANS outflow to the periphery is organized in sympathetic and parasympathetic pathways that have, by and large, opposing functions. Autonomic motor neurons responsible for driving this outflow are located in the brain stem and spinal cord respectively. Tracing and stimulation studies demonstrated that mainly the prefrontal cortex, amygdala and hypothalamic nuclei have direct access to autonomic motor centers. Within the ▶hypothalamus, the paraventricular nucleus (PVN) is the single most important structure for neuroendocrine and autonomic integration. The PVN consists of neuroendocrine neurons projecting to the neurohypophysis and median eminence and of pre-autonomic neurons projecting to sympathetic and parasympathetic motor neurons in the intermediolateral column (IML) of the spinal cord and the dorsal motor nucleus of the vagus (DMV) in the (Fig. 1) brain stem respectively.

SCN projections seem to reach the neuroendocrine neurons in the PVN (i.e. CRH, TRH) via an indirect pathway. Interneurons are mainly located in the MPO, subPVN and DMH. As described in the text, this indirect pathway is involved in the control of the daily rhythm of corticosterone release. The neuroendocrine CRH neurons in the PVN project to the median eminence. CRH released in the median eminence will stimulate ACTH secretion from the anterior pituitary, which will finally reach the adrenal cortex via the general circulation and stimulate corticosterone secretion.

The PVN output to non-autonomic sites in the brain in contrast is limited and mainly organized to support its hormonal and autonomic output. For instance, the PVN projects to sites where visceral (sensory) information is transmitted from autonomic sensory fibers to neurons of the CNS (Fig. 2). Via these projections, the PVN is able to modulate incoming information as well. Recent studies have shown that PVN axons projecting to sympathetic and parasympathetic motor neurons have collaterals to the NTS, suggesting that “a copy” of the outgoing signal is sent to the neurons receiving incoming information, as if to inform them what the

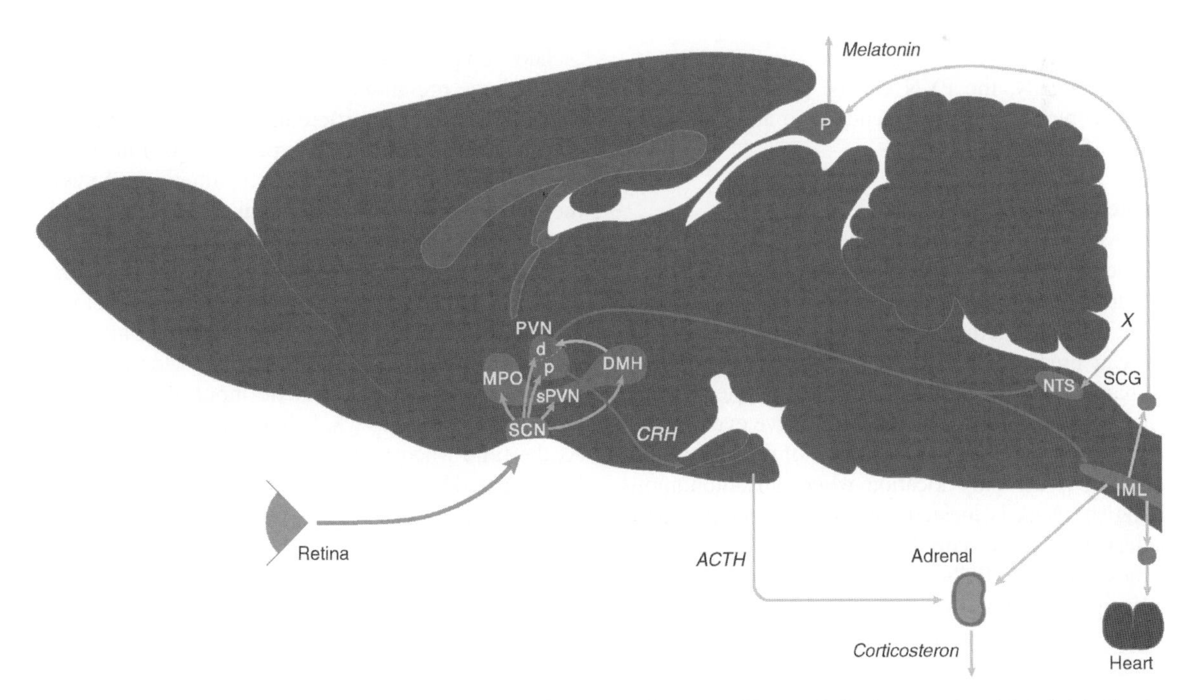

Neuroendocrine Regulation and the Autonomic Nervous System. Figure 1 Pathways via which light affects hormonal and autonomic output by the hypothalamus. Via the retina, light information is transmitted to the SCN and passed on to the pre-autonomic neurons in the PVN via direct SCN–PVN connections. The autonomic neurons of the PVN pass the light information to the sympathetic motor neurons in the IML, with collaterals to the NTS (see Fig. 2) and directly inhibit melatonin secretion from the pineal. Along the same pathway, light will also change the sensitivity of the adrenal cortex to ACTH and thereby affect corticosterone secretion and reduce the heart rate in rats.

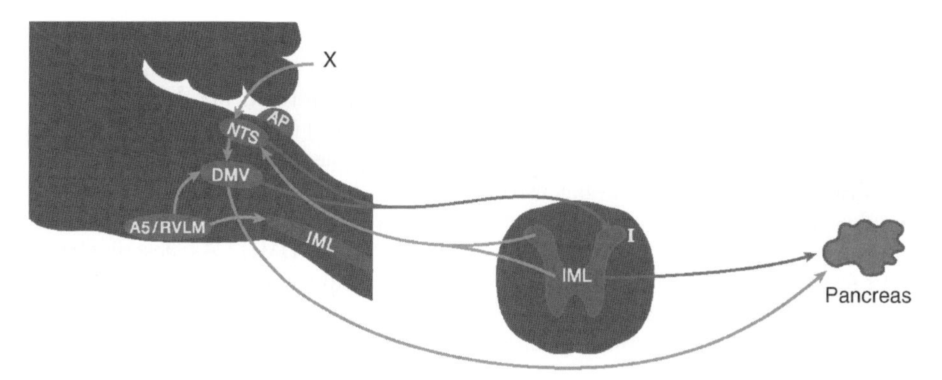

Neuroendocrine Regulation and the Autonomic Nervous System. Figure 2 The autonomic circuit. The parasympathetic-sympathetic interaction illustrates the relationship between the cell groups that may influence the vagal output in *green* or the sympathetic output in *pink*. It is clear that both vagal and sympathetic output influence each other.

action of the brain is. To make this picture of the organization of the autonomic output of the brain complete, it is essential to keep in mind that within the parasympathetic DMV and sympathetic IML, not only motor neurons are present but also interneurons that project to respectively the IML and DMV. In this way the autonomic output and input forms a feedback circuit that can function on its own and forms the basis for the autonomic control that can operate in decerebrate animals. Via the PVN this "autonomic" circuit is open to central (hypothalamic, amygdalar and cortical) modulation (Fig. 2).

Consequently, the PVN can be seen as the major center, controlling the neuroendocrine and autonomic output of the brain.

Balancing Autonomic Output

Our evolutionarily shaped homeostatic systems have "learned" to adapt to the ever-changing light/dark cycle so that the body anticipates coming periods of sleep or activity. In all organisms, mechanisms have developed that can predict when the day ends or starts. These mechanisms are known as the "circadian system." In all cells of the body, clock mechanisms have evolved that in one way or another can keep track of time. However, the brain is the only location where an autonomous biological clock is located, i.e. in the suprachiasmatic nuclei (SCN) of the ventral hypothalamus. By enforcing its message to the PVN, the SCN is able to transmit its daily signal via hormones and the autonomic nervous system to all tissues of the body. Via these signals body tissues are prepared for activity or sleep. The SCN needs to activate body organs and tissues depending on their function at different times of the day, for example muscles work in the active phase when the digestive tract slows down. Thus, during the active phase, an opposite autonomic tone on vasculature will redirect blood away from the abdominal compartment towards the movement compartment. At the same time cerebral blood flow is kept constant. The major advantage of an endogenous oscillator is that it allows the body to anticipate these changes. One of the questions that puzzled investigators for a long time is how SCN neurons, that are mainly active during the daytime, are connected to these two different autonomic systems. Thus in a series of tracing experiments a strict separation between sympathetic and parasympathetic projecting neurons at the level of the PVN was shown to persist up to the SCN (Fig. 3).

The complete segregation of parasympathetic and sympathetic pre-autonomic neurons provides the anatomical basis for a differential control of these two autonomic branches by the hypothalamus. Functionally this separation between parasympathetic and sympathetic pre-autonomic neurons makes perfect sense. For instance, at night the SCN slows down heart function, resulting in a dip in blood pressure. The sympathetic and parasympathetic autonomic nervous systems are proposed to have opposing functions, with alternating activities over the sleep-wake cycle. Moreover, even when we consider sympathetic input alone, it cannot be activated over the whole body at the same moment. At the time when melatonin secretion from the pineal is stimulated by sympathetic terminals, the sympathetic input to the heart needs to be inhibited, at least in humans.

The Autonomic Nervous System Differentiates Between Functionally Different Body Compartments

Recent studies indicate that the body can be divided into different functional autonomic compartments and that at least a thoracic, a movement and a visceral compartment should exist. In this setting, a balanced and flexible autonomic nervous system can oscillate the activities of the organs within the different compartments according to the actual needs of the body. Using

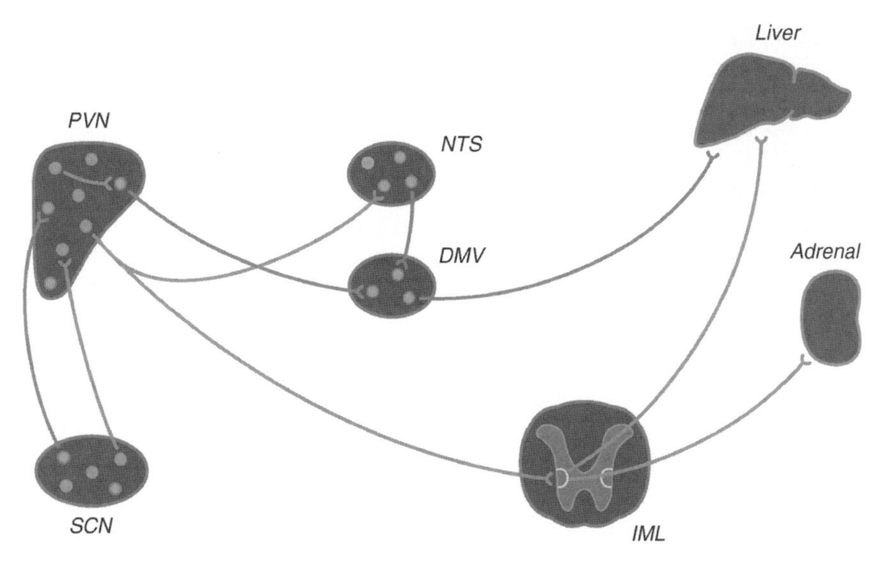

Neuroendocrine Regulation and the Autonomic Nervous System. Figure 3 Sympathetic and parasympathetic differentiation. Scheme of interaction between the hypothalamic suprachiasmatic nucleus (*SCN*) and paraventricular nucleus (*PVN*). Separate sympathetic (*red*) or parasympathetic (*blue*) neurons of the SCN project to pre-autonomic neurons of the PVN, where a similar sympathetic-parasympathetic separation can be observed. Pre-autonomic neurons of the PVN project either to the preganglionic sympathetic neurons in the intermediolateral (*IML*) column of the spinal cord or to the preganglionic neurons of the dorsal motor nucleus of the vagus (*DMV*). The pre-sympathetic PVN neurons have axon collaterals to pre-parasympathetic neurons, either in the PVN itself or via the nucleus tractus solitarius (*NTS*).

viral retrograde tracers to label the motor neurons that innervate specific organs, abdominal organs such as liver, pancreas and intra-abdominal fat were shown to share the same vagal motor neurons. In contrast, distinct sets of vagal motor neurons project to fat tissue that is located in the intra-abdominal or subcutaneous compartments. Again this somatotopic organization was shown to exist up to the hypothalamic biological clock and the amygdalar limbic system. Most probably this astonishing capacity of hypothalamic neurons to specialize also coordinates the "dawn phenomenon," where enhanced glucose production by the liver coincides with enhanced glucose uptake by the target organs at the beginning of the daily activity phase. These two processes will need different actions of the sympathetic neurons that project to the liver (i.e. activity) and to the muscular vasculature (i.e. inactivity). On the other hand, when unbalanced, the shared command of intra-abdominal organs may also result in the simultaneous occurrence of diabetes type 2, dyslipidemia and visceral obesity.

The Hypothalamus Targets the Organs of the Body both by Hormones and by Autonomic Output

The PVN is a hypothalamic center that is able to adapt and coordinate hormonal and autonomic responses, not only by virtue of a highly differentiated hormonal output but also by an equally well-differentiated autonomic output. The enormous variety of different neuroendocrine and autonomic neurons offers the PVN a great potential for integration and harmonization of function. In fact, several studies have demonstrated the presence of an anatomical network for intra-PVN integration and coordination. As an example, the daily rhythm in the activity of the hypothalamo-pituitary-adrenal axis will be discussed in somewhat more detail. The SCN has both direct and indirect connections with the CRH producing neurons in the PVN to control CRH and thus ACTH secretion. The indirect connections run via SCN projections to the interneurons in the DMH that project to the PVN. Next to these neuroendocrine connections the SCN also has direct connections with the pre-autonomic neurons in the PVN projecting to sympathetic motor neurons in the IML that are in direct contact with the adrenal. Probably the SCN also has indirect connections (via the DMH) with the pre-autonomic neurons in the PVN. Using this myriad of connections, the SCN synchronizes its actions so that it affects the autonomic output of the PVN and thus sensitizes the adrenal cortex for ACTH at the same moment that ACTH secretion from the pituitary is stimulated. As a result, the response of the adrenal cortex to ACTH is greatly facilitated.

It is proposed that by the combination of hormones and autonomic output not only the biological clock, but also other brain structures will be able to optimally affect the functioning of organs. Naturally it is also possible that the functionality of organs is affected only by changing their sensitivity and that the hormones are regulated by another mechanism.

References

1. Buijs RM, Kalsbeek A (2001) Hypothalamic integration of central and peripheral clocks. Nat Rev Neurosci 2:521–526
2. Buijs RM, Kreier F (2006) The metabolic syndrome: a brain disease? J Neuroendocrinol 18:715–716
3. Buijs RM, Scheer FA, Kreier F, Yi C, Bos N, Goncharuk VD, Kalsbeek (2006) Organization of circadian functions: interaction with the body. Prog Brain Res 153:341–360
4. Cottrell GT, Ferguson AV (2004) Sensory circumventricular organs: central roles in integrated autonomic regulation. Regul Pept 117:11–23
5. Kreier F, Yilmaz A, Kalsbeek A, Romijn JA, Sauerwein HP, Fliers E, Buijs RM (2003) Hypothesis: shifting the equilibrium from activity to food leads to autonomic unbalance and the metabolic syndrome. Diabetes 52:2652–2656
6. Pavlov VA, Tracey KJ (2004) Neural regulators of innate immune responses and inflammation. Cell Mol Life Sci 61:2322–2331
7. Swaab DF (1997) Neurobiology and neuropathology of the human hypothalamus. Handbook Chem Neuroanatomy 13:39–137
8. Yi CX, Van DV, Dai J, Yin G, Ru L, Buijs RM (2006) Ventromedial arcuate nucleus communicates peripheral metabolic information to the suprachiasmatic nucleus. Endocrinology 147:283–294

Neuroendocrine System

Definition

Hormone producing neurons.

►The Hypothalamo Neurohypophysial System
►Hypothalamo-Pituitary-Adrenal Axis, ►Stress and Depression

Neuroendocrinological Drugs

Cynthia L. Darlington
Department of Pharmacology and Toxicology, School of Health Sciences, University of Otago Medical School, Dunedin, New Zealand

Synonyms

Steroid hormones; Neurosteroids; Neuroactive steroids

Definition

Neuroendocrinological drugs may be naturally occurring steroid hormones, or synthetic hormones that mimic the activity of the naturally occurring steroid hormones. These steroid hormones, referred to as neurosteroids are synthesized in the CNS and in the peripheral nervous system. Neuroactive steroids are steroid hormones that act on neurosteroid receptors, regardless of the origin of the steroid. Examples of neuroactive steroids include ►estrogens, ►progestagens, and progestagen metabolites including allopregnanolone, testosterone and metabolites including 3α-androstanediol. In this entry, "estrogen" is used as a generic term to refer to any neuroactive steroid with estrogenic activity while "progestagen" refers to the naturally occurring ►progesterone or a progestagen, synthetic progesterone. "►Androgens" refers to neuroactive steroids with masculinizing effects. Corticosteroids may also be neuroactive steroids.

Characteristics

Overview

Neuroactive steroids, or their synthetic analogs, exert their actions in the CNS by binding to specific neurosteroid receptors or by binding to neurosteroid binding sites on other types of receptors in the CNS, e.g. the $GABA_A$ receptor.

Neuroactive steroids, like all steroid hormones, are synthesized from cholesterol via discreet enzymatic pathways. Neuroactive steroids are secreted largely under the control of the hypothalamic-pituitary-adrenal (HPA) axis. Specific neurosteroids are secreted within the CNS by several types of cells including oligodendrocytes. The factors controlling neurosteroid secretion are less well understood. In addition, in pregnant females, the neuroactive steroids estrogen and progesterone, are secreted by the placenta and the corpus luteum, respectively.

Neurosteroid receptors are members of the large family of "ligan-dependent nuclear transcription factors" (steroid receptors), with cytosolic or nuclear binding complexes that, in the case of the cytosolic receptors, are translocated to the nucleus, and bind to a promotor region on DNA. The binding of the receptor complex to the promotor region initiates the release of mRNA from the nucleus into the cytoplasm and ultimately, to the production of new protein by the endoplasmic reticulum. However, not all neuroactive steroid activities are modulated by intracellular receptors. It is evident from a number of experimental results that membrane-bound neurosteroid receptors are also active in the CNS. In addition, an allosteric binding site for a ►progesterone metabolite has been identified on the $GABA_A$ receptor and sigma receptors, nicotinic acetylcholine receptors and the NMDA, kainate and AMPA subtypes of the glutamate receptors have been demonstrated to be modulated by neuroactive steroids [1]. As these neurotransmitter receptors are the targets of many non-steroidal neuroactive drugs, the importance of neuroactive steroids as potential therapeutic agents is clear.

One ramification of the discovery and identification of neuroactive steroids has been to identify a mechanism for sex differences in brain function, specifically in brain activity not related to reproductive status. Estrogen, progestagen and androgen receptors are distributed throughout the brains of both females and males. The levels of the hormones, as measured in blood plasma, differ markedly between females and males, depending upon the timing of the measurement. In males, testosterone is released in a pulsatile manner throughout the 24 h cycle, while the low levels of estrogen and progesterone in males remains fairly stable. In females, estrogen and progesterone levels fluctuate on an approximately 28 day cycle, with an estrogen peak on day 14 and a lower estrogen peak again around cycle day 20. Finally, it tapers off by day 28. Progesterone, on the other hand is extremely low until cycle day 17, when it increases rapidly and peaks around day 21, dropping rapidly again by day 28. Changes in mood and behaviour have been associated with these hormonal changes. It is only recently that the implications of the hormonal changes for neurotransmitter function have been recognized.

Historical Perspective on Neurosteroids and Neuroactive Steroids

McEwen and colleagues [2] were responsible for landmark experiments in the 1960s demonstrating the presence of steroid hormones in the brain tissue of rats. Over the next 10 years, extensive research demonstrated the presence of steroid hormone receptors throughout the mammalian brain, and many of these receptors were found outside areas of the CNS associated with reproductive behaviour. This latter result led to the suggestion that steroid hormones might modulate brain activity associated with non-reproductive behaviour. Baulieu [3] coined the term "neurosteroids" in 1981 and in the following years identified many of the neuroactive steroids.

Elucidation of the binding characteristics of neurosteroid receptors has been a major focus of research since the 1980s. While it has been clearly established that neurosteroid receptors have response characteristics in common with most members of the steroid receptor family, some experimental results have not been consistent with intracellular receptor binding. It has been demonstrated that some neuroactive steroids activate membrane bound receptors associated with ion channels and G-proteins, some transporter systems are sensitive to neuroactive steroids, a number of neurotransmitter systems are modulated by neuroactive steroid activity (see above) and both estrogen and progestagen have been reported to modulate the expression of different subunits of the $GABA_A$ receptor [4,5]. It is the interactions with the GABAergic and serotonergic system that have been of particular interest to the development and administration of psychoactive drugs such as benzodiazepines and selective serotonin reuptake inhibitors for the treatment of anxiety and depression [6].

Numerous studies have demonstrated that neuroactive steroids modulate memory and cognition [6]. Estrogen is generally considered an "activator" in the brain, administration of estrogen lowers seizure threshold, prolongs the duration of seizures. The relationship of estrogen to seizure activity is recognized is the term "catamenial epilepsy," a form of epilepsy that is linked to the menstrual cycles in females. Progestagen is known to have sedative effects, administration lowers the seizure threshold, shortens seizure duration, and it has been used as an anesthetic.

Current Therapeutic Strategies

The use of neuroactive steroids as therapeutic agents is relatively recent. Corticosteroids, such as methylprednisolone, have been and continue to be used to prevent and treat edema following head injury. This antiinflammatory effect is, however, independent of the actions of neuroactive steroids that are the focus of this discussion.

Estrogen has been used in conjunction with conventional treatments in females suffering from Parkinson's disease or Alzheimer's disease with limited success [7]. Dementia occurs in some patients with Parkinson's disease and it has been reported that hormone replacement therapy (HRT) decreases the incidence of dementia in older patients. It has also been reported that taking HRT decreases the risk of developing Alzheimer's disease. There are numerous studies showing that estrogen, when taken alone, has a positive effect on mood and cognitive function, but that with the addition of progesterone, the positive effects disappears [6]. This observation on progesterone administration raises one of the potential problems for the use of estrogen as a therapeutic agent. Estrogen has a proliferative effect, enriching the endometrium, causing tissue thickening in preparation for implantation of a fertilized ovum. When fertilization does not occur, the release of progesterone causes the retraction and ultimate sloughing of the excess tissue. If estrogen is given without progestagen administration, either concomitantly or sequentially, the endometrial thickening continues unchecked, increasing the risk for the development of endometrial cancer. In HRT and the contraceptive pill, progestagen is an essential component. An area where the actions of progestagen could be an advantage is in the treatment of catamenial epilepsy. Given the antiepileptic properties of progestagens, the choice of a progestagen only contraceptive might provide antiseizure activity to enhance the action of antiepileptic drugs, while providing the desired contraceptive effect.

A current impediment to the use of neuroactive steroids as therapeutic agents is the side effect profile and adverse events associated with their administration. Deep vein thrombosis, for example, is associated with both progestagen and estrogen administration. Estrogen is contraindicated for administration to individuals with a history of cardiovascular disease, stroke, blood clotting disorders, focal migraine, hypertension and some kinds of cancers. Progestagen, when administered alone as an injectable contraceptive, is associated with a disruption of calcium metabolism and is currently only recommended for an administration period of 2 years.

Neuroendocrinological Drugs. Table 1 Examples of neuroactive steroids

Class	Neuroactive steroid
Estrogen	17β-estradiol
	Estrone
	Estradiol
Progestagen	Progesterone
	Allopregnanolone,
	17α-hydroxyprogesterone
	20α-hydroxyprogesterone
Androgen	Testosterone
	3α-androstanediol
Corticosteroid	3α, 5α-tetrahydrodeoxycortocosterone (THDOC)

Future for Drug Development

The therapeutic uses of neuroactive steroids will depend upon the specificity of the action that can be achieved. There are two known types of estrogen receptors (ER), α and β, with different actions and distribution for each. ERα are found predominantly in the amygdala, septum and hypothalamus, and in smaller numbers in brainstem nuclei including the periaqueductal grey, locus coeruleus and area postrema. ERβ are found in lower concentration than ERα with the highest numbers being found ιv the amygdala, septum, hypothalamus and olfactory cortex. There are two types of progestagen receptors with functions that are yet to be clarified. Androgen receptors are also found in the bran. Undoubtedly, there are additional types and subtypes of these receptors yet to be discovered. There are also neurosteroid binding sites associated with classical neurotransmitter receptors as allosteric binding sites. There are different estrogens (17β-estradiol, estrone and estradiol), progestagens (e.g. progesterone, allopregnanolone, 17α-hydroxyprogesterone, 20α-hydroxyprogesterone) and ►androgens, with slightly differing chemical structures (Table 1). Targeting specific populations of neurosteroid receptors is one obvious direction for drug development. Another direction is the targeting of allosteric binding sites on particular populations of neurons with classical neurotransmitter receptors or transporter proteins such as the dopamine transporter.

The $GABA_A$ receptor is a particularly good candidate for this approach. The $GABA_A$ receptor, in addition to the GABA binding site, has allosteric binding sites for benzodiazepines, barbiturates, progestagens and alcohol. It is the primary inhibitory neurotransmitter receptor found in the brain. The $GABA_A$ receptor is composed of assemblies of subunits arranged around a central chloride channel (see Chapter on Anxiolytics and Hypnotics). The subunits are found in different combinations of subunit type (α, β, γ, δ, ε, π, ρ). The subunit combination of any given receptor determines the pharmacological actions associated with activation of the receptor. Receptors with certain subunit combinations are distributed differently throughout the brain. Binding of ligands to the allosteric binding sites enhances the effect of GABA. It is conceivable that a neurosteroid could be synthesized to target the progestagen binding site, and by modifying the action of benzodiazepines, barbiturates or alcohol overcome some of their adverse side effects.

Epidemiological studies show that females have a higher incidence of anxiety and depressive disorders. Serotonin levels are significantly lower in the female brain than in the male brain. Females are also receive more prescriptions of anxiolytic and antidepressant drugs than males [7]. It may, in the future, be possible to capitalize on the changing levels of neuroactive steroids associated with the menstrual cycle to target and refine drug effects.

The progesterone metabolite allopregnanolone (which may act only on the $GABA_A$ receptor and not progestagen receptors) has been reported to have detrimental effects on mood, memory, cognitive function and has been reported to be associated with premenstrual mood changes. One possibility would be to develop antagonists to allopregnanolone that act specifically to reduce the adverse events associated with its administration. 3β-hydroxypregnane has recently been reported to reduce the learning deficits produced by allopregnanolone administration [6] suggesting that this may be a useful direction for drug development.

References

1. Dubrovsky BO (2004) Steroids, neuroactive steroids and neurosteroids in psychopathology. Prog Neuropsychopharmacol Biol Psychiatry 29:169–192

2. McEwen BS, Weiss JM, Schwartz LS (1968) Selective retention of corticosterone by limbic structures in the rat brain. Nature (London) 220:911–912
3. Baulieu EE (1998) Neurosteroids, a novel function of the brain. Psychoneuroendocrinology 23:963–987
4. Macguire JL, Stell BM, Rafizadeh M, Mody I (2005) Ovarian cycle-linked changes in $GABA_A$ receptors mediating tonic inhibition alter susceptibility and anxiety. Nat Neurosci 8:797–804
5. Pierson RC, Lyons AM, Greenfield LJ Jr (2005) Gonadal steroids regulate $GABA_A$ receptor subunit mRNA expression in NT2-N neurons. Mol Brain Res 138:105–115
6. Birzniece V, Backstrom T, Johansson IM, Lindblad C, Lundgren P, Lofgren M, Olsson T, Ragagnin G, Taube M, Turkmen S, Wahlstrom G, Wang MD, Wihlback AC, Zhu D (2006) Neuroactive steroid effects on cognitive functions with a focus on the serotonin and GABA systems. Brain Res Rev 51:212–239
7. Darlington CL (2002) The female brain. In the series Conceptual advances in brain research. Taylor & Francis, London

Neuroendocrinology

Dick F. Swaab[1], Paul J. Lucassen[2]
[1]Netherlands Institute for Neuroscience, Amsterdam, The Netherlands
[2]Centre for Neuroscience, Swammerdam Institute of Life Sciences, University of Amsterdam, The Netherlands

Definitions

►Neuroendocrinology is the field that studies hormone production by neurons, the sensitivity of neurons for hormones, as well as the dynamic, bidirectional interactions between them. These processes function in concert to allow maintenance of homeostasis for the organism. The central region of interest in this field is the ►hypothalamus, as many endocrine cascades are initiated there and ►feedback regulation of these very same cascades also takes place in this area. In addition to the hypothalamus, most other brain areas are sensitive to hormonal action as well.

The human hypothalamus is a small (4 cm^3) but very complex heterogeneous brain structure. Together with its adjacent areas it is composed of some 20 well-defined (sub)nuclei between 0.25 and 3 mm^3 in size, with very different chemical components and functions (Fig. 1). Their neurons contain one or more of all four types of neuroactive substances; acetylcholine, amines, amino acids, and a multitude of ►neuropeptides.

A special characteristic of the hypothalamus is that some neurons projecting to the ►neurohypophysis or to the portal vessels of the pituitary in the ►median eminence, release their peptide into the blood and act as ►neurohormones. Other neurons project to neurons within and outside the hypothalamus, where they function as ►neurotransmitters or ►neuromodulators, and regulate central functions, including the ►autonomic innervation of all our body organs [1]. In this way, the hypothalamus acts as a center that integrates endocrine, autonomic, and higher brain functions. The fact that neuropeptides are generally very stable after death makes them very suitable as markers of hypothalamic nuclei that enables the identification of structural and functional changes in key hypothalamic nuclei and transmitter systems in postmortem material obtained from patients with e.g. different ►neuroendocrine disorders.

The hypothalamus regulates essential processes throughout all stages of life, and is also the primary structure involved in a number of disorders. A few examples of such hypothalamic disorders will be presented in this synopsis in relation to the different hypothalamic areas.

The Classical Neuroendocrine Neurons in the Supraoptic (SON) and Paraventricular Nucleus (PVN)

The ►supraoptic and paraventricular nucleus (SON and PVN) and their axons running to the neurohypophysis form together the ►hypothalamo-neurohypophysial system (HNS), which represents the classic example of a neuroendocrine system. ►Vasopressin (= ►antidiuretic hormone, ADH) and ►oxytocin are key hormones produced in the large, magnocellular neurons of the SON and PVN, and are e.g. involved in ►antidiuresis, ►labor and ►lactation. A second type of smaller neuroendocrine cells, the parvicellular neurons, is found in the PVN. They release their peptides into the ►portal capillaries that transport them to the ►anterior lobe of the pituitary. Examples are ►corticotrophin releasing hormone (CRH) and ►thyrotropin releasing hormone (TRH). A third type of PVN cells projects to other neurons, where the peptides act as neurotransmitters/neuromodulators. These involve e.g. vaspressin, oxytocin, CRH and TRH neurons that thereby regulate the autonomic nervous system [1]. The entire SON contains some 78,000 neurons on one side from which 88% contains vasopressin and 12% contains oxytocin. The PVN has been estimated to contain about 56,000 neurons of which some 25,000 express oxytocin and 21,000 produce vasopressin. A large number of other, often co-expressing peptides are found in both nuclei [2].

The SON is the main source of circulating vasopressin. Disorders of this vasopressin system involve familial central ►diabetes insipidus with mutations in the vasopressin precursor, autoimmune diabetes insipidus with antibodies against the vasopressin neurons, pregnancy-induced diabetes insipidus due to increased breakdown of plasma vasopressin, ►nephrogenic

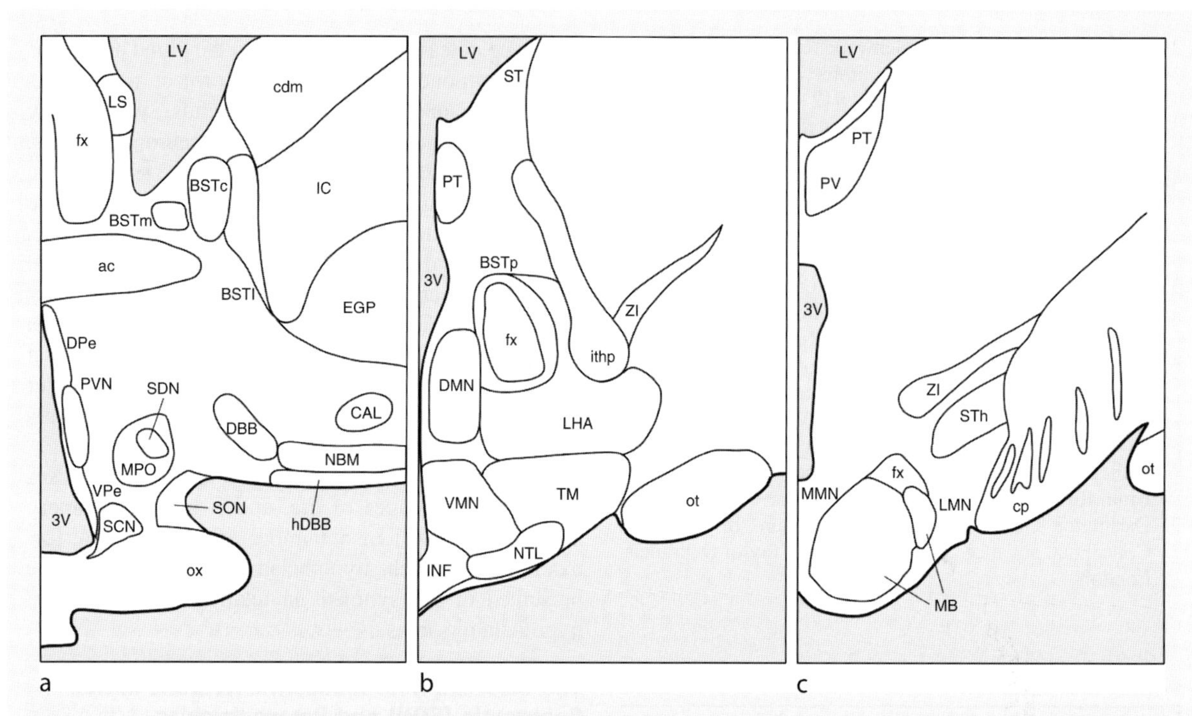

Neuroendocrinology. Figure 1 Schematic representation of the nuclei of the human hypothalamus. Abbreviations: ac: anterior commissure, BST: bed nucleus of the stria terminalis, (c = centralis; m = medialis; l = lateralis; p = posterior); cp: cerebral peduncle, DPe: periventricular nucleus dorsal zone, fx: fornix, hDBB: horizontal limb of the diagonal band of Broca, INF: infundibular nucleus, MB: mamillary body i.e. MMN: medial mamillary nucleus: LMN: lateromamillary nucleus, NBM: nucleus basalis of Meynert, OT: optic tract, Ox: optic chiasma, PVN: paraventricular nucleus, SCN: suprachiasmatic nucleus, SDN: sexually dimorphic nucleus of the preoptic area, SON: supraoptic nucleus, 3V: third ventricle, VMN: ventromedial hypothalamic nucleus, VPe: periventricular nucleus ventral zone (adapted from [2]; Fig. 1.6).

diabetes due to mutations in the gene of the vasopressin receptor-2 or in the gene of the water channel protein aquaporin-2, nocturnal diuresis, ►inappropriate secretion of vasopressin (Schwartz–Bartter syndrome), ►Wolfram's syndrome, and glucocorticoid administration, in which we found a processing disorder of vasopressin, and ►septo-optic dysplasia (De Morsier's syndrome), in which we observed that the SON and PVN were virtually absent [2].

The PVN contains a large number of different neuropeptides [2]. CRH is produced by parvicellular neurons and is a crucial neuropeptide in the regulation of the ►hypothalamo-pituitary-adrenal (HPA)-axis, the final common pathway in the ►stress response. Once a stressor is perceived, rapid rises in CRH in the PVN subsequently stimulate the release of ►adrenocorticotropic hormone (ACTH) from the anterior pituitary into the circulation. The ACTH-releasing activity of CRH is strongly potentiated by vasopressin when co-produced by the same CRH containing parvicellular neurons. ACTH is released into the blood and frees ►cortisol from the adrenal, the main steroid hormone in humans that coordinates various aspects of the stress response throughout the body and CNS. These include changes in behaviour, such as rises in alertness and anxiety, an activation of the autonomic nervous system, causing increases in heart rate and respiration, redistibution of the bloodstream and mobilization of energy sources. At the same time, various processes and bodily systems are temporarily inhibited that are of lesser importance during an acutely threatening situation, including suppression of gastrointestinal function, sleep, sexual activity and growth. In addition, cortisol is one of the most powerful endogenous feedback compounds on the ►pro-inflammatory signal transduction machinery. Hyperactivity of the HPA axis is furthermore a key phenomenon in ►depression and is held responsible for symptoms like decreased ►food intake, decreased sexual activity, disturbed sleep and motor behavior and increased anxiety. The HPA-axis is also activated in ►multiple sclerosis (MS).

After being released from the adrenal, cortisol inhibits its own release through feedback inhibition on the very same brain areas where its release was initially coordinated, through binding to glucocorticoid receptors (GR) in the pituitary and PVN. ►Corticosteroids, at least in rodent, also target the hippocampus that is richly endowed with corticosteroid receptors and is involved in modulation of the stress response. In addition to corticol, also exogenously applied corticosteroids inhibit CRH production in the PVN [3]. In addition to a crucial role in initiation of the stress response, CRH also has central effects, including cardiovascular regulation, respiration, appetite control, stress-related behavior and mood, cerebral blood flow regulation and stress-induced analgesia.

TRH is released in the median eminence as the major hypothalamic hormone stimulating thyroid function, acting on thyroid ►stimulating hormone (TSH) cells of the pituitary, whereas ►somatostatin and dopamine inhibit TSH secretion. The large number of dense TRH fiber terminations in the hypothalamus suggests an important role of this neuropeptide as a neuromodulator [4]. The human PVN is furthermore densely innervated by fibers from the ►infundibular nucleus. In juxtaposition to the TRH neurons in the PVN, NPY, agouti-related peptide (AGRP) and ά-MSH-containing fibers are found, which are involved e.g. in eating behavior. The TRH neurons of the PVN are less active, not only in depression, where decreased amounts of TRH mRNA were found in the PVN [5], but also in ►sick euthyroid syndrome (or ►nonthyroidal illness), a condition often seen in serious illness conditions [6].

The Suprachiasmatic Nucleus (SCN)

The SCN is the biological clock of the brain that regulates the ►circadian and seasonal variations occurring in many endocrine and other functions. The ►circadian pacemaker is localized in the SCN, on top of the optic chiasm, and the clock is entrained to fluctuations in light intensity during the day–night cycle by direct innervation from the ►retinohypothalamic tract. Vasopressin and ►vasoactive intestinal polypeptide (VIP) expressing neurons of the SCN project to the ►dorsomedial nucleus, the subparaventricular zone, and the PVN. Transmeridian flights and shiftwork might lead to disturbances in the functional organization of the biological clock. Moreover, disorders of the circadian timing system are held responsible for advanced and delayed sleepsyndrome, a number of other ►sleep disorders, and nightly restlessness in patients with ►Alzheimer's disease. In addition, day–night fluctuations in the SCN may be the basis of the clear day–night differences found in the course of various diseases and conditions. For example, the moment when death occurs from myocardial infarction, intracerebral hemorrhage, or ischemic stroke, or the moment when complaints about migraine, depressive symptoms, tremor in Parkinson's disease, and sleep disorders in depression are at their most serious, do not occur at random over the day but all have a strong preference for a particular moment of the day [2]. We have shown a decreased amount of vasopressin mRNA in the SCN of depressed patients [7], indicating a diminished activity of the biological clock that might be the neurobiological basis of sleep disturbances and circadian disorders in these patients.

Hypothalamic Nuclei with Structural Sex Differences

►Sex differences in the size of the ►sexually dimorphic nucleus of the preoptic area (SDN-POA) were first described in the rat by [8]. We have found a similar nucleus in the human hypothalamus. Morphometric analysis of the human SDN-POA revealed that its volume is more than twice as large in young adult men as it is in women, and contains about twice as many cells in men [9]. It seems to be homologous to the SDN-POA in the rat in view of its sex difference in size and cell number, localization, cytoarchitecture, and neurotransmitter/neuromodulator content. [10] gave this nucleus another name: ►Interstitial nucleus of the anterior hypothalamus-1 (INAH-1) and also described two other cell groups (►INAH-2 and -3) in the preoptic-anterior hypothalamic area of humans that were larger in the male brain than in the female brain. [11] found a sex difference in INAH-3. Another sex difference was described by [18], in what they called the "►darkly staining posteromedial component of the bed nucleus of the stria terminalis" ►(BNST-dspm). The volume of the BNSTdspm was 2.5 times larger in males than in females. We found a similar sex difference in the ►central nucleus of the bed nucleus of the stria terminalis (BSTc). The BSTc is defined by its dense VIP innervation and by its somatostatin fiber plexus and somatostatin neuron population. The BSTc in men is 40% larger than in women, and men have almost twice as many somatostatin neurons as women [12,13]. We have furthermore found a female-sized central nucleus of the bed nucleus of the BST in male-to-female ►transsexuals [13]. These data were confirmed by neuronal counts of somatostatin cells, the major neuron population in the BSTc [12].

Infundibular (Arcuate) Nucleus

The horseshoe-shaped infundibular (or ►arcuate) nucleus surrounds the lateral and posterior entrance of the ►infundibulum and is situated outside the blood–brain barrier. The infundibular nucleus contains some 520,000 neurons [14] and is involved in reproduction, pain, eating behavior and metabolism, thyroid hormone feedback, growth, and dopamine regulation.

In addition, the infundibular nucleus is continuous with the ►stalk/median eminence region that contains the ►portal capillaries of the adenohypophysis. The ►neuropeptide-Y (NPY) fibers in the median eminence are mainly restricted to the internal zone and only scarcely innervate the neurovascular zone, whereas CRH, LHRH, ►opiomelanocortins, somatostatin, ►GHRH, ►galanin, TRH, and ►substance-P fibers do innervate the stalk/median eminence region. The NPY neurons project to the PVN and NPY together with ghrelin e.g., is one of the most active food intake stimulating peptides.

The infundibular nucleus is chemically characterized by the presence of (pre)proopiomelanocortin neurons, containing e.g. ►ά-melanocyte-stimulating hormone (ά-MSH). The sites of fiber termination of the opiomelanocortin neurons are consistent with the brain sites where pain relief was obtained by deep brain stimulation. Moreover, this nucleus contains peptides that inhibit feeding like ά-MSH that acts by the ►MC-4 receptor and ►cocaine- and amphetamine-regulated transcript (CART). AGRP and NPY stimulate feeding and co-localize in the infundibular nucleus. GHRH neurons are involved in metabolism and are activated during prolonged illness [2,15]. Severe ►obesitas has been reported due to mutations in the genes for preproopiomelanocortin-, MC-4 receptor-, ►leptin, the ►leptin receptor and ►prohormone convertase-1.

The infundibular nucleus also contains ►luteinizing hormone-releasing hormone (LHRH; gonadotropin releasing hormone)-containing cell bodies that, like opioid peptides, play a role in reproduction and sexual behavior, the latter in erections. ►Estrogen and ►androgen receptors and all four thyroid hormone receptors (TR) isoforms are present in the infundibular nucleus.

A subdivision of the infundibular nucleus in ►postmenopausal women and in hypopituitarism was named the ►subventricular nucleus after its location beneath the floor of the third ventricle. The production of neurokinin-B (NKB), substance-P, and estrogen receptor transcripts is strongly increased in postmenopausal women due to the diminished inhibitory action of estrogens. The NKB neurons are presumed to be involved in the initiation of menopausal flushes [16].

Lateral Hypothalamic Area (LHA), Including the Perifornical Area

The LHA is involved in the regulation of food intake and body weight. The classic syndrome following lesions of the lateral hypothalamus involves ►aphagia and ►adipsia. ►Melanin concentrating hormone (MCH) is a neuropeptide produced in the LHA, perifornical and periventricular areas, and in the tuberomammillary and posterior nuclei. MCH increases food intake and lowers plasma glucocorticoid levels in the rat. Quite recently two novel neuropeptides were discovered, designated ►hypocretin 1 and 2 (from hypothalamus and secretin), or ►orexin A and B, that stimulate food consumption in rat. These peptides are localized in the lateral and posterior hypothalamus and accumulate in the perifornical area. The hypocretins are involved not only in food intake, but also in sleep and neuroendocrine control. In ►narcolepsy patients that are ►cataplectic a 85–95% loss of hypocretin neurons was found without gliosis or signs of inflammation, possibly as the result of an autoimmune process [17].

Other Hypothalamic Nuclei

Following destruction of the ►ventromedial hypothalamic nuclei (VMN) e.g. by a tumor or neurochirurgical intervention, a classic tetrad of symptoms has been described, i.e. (i) episodic rage, (ii) emotional lability, (iii) hyperphagia with obesity and (iv) intellectual deterioration, mainly based upon memory loss.

The ►tuberomamillary nucleus (TMN) contains the only ►histaminergic system of the brain and participates in the modulation of the state of arousal, the control of vigilance, sleep and wakefulness, cerebral circulation and brain metabolism, locomotor activity, neuroendocrine and vestibular functions, drinking, sexual behavior, stress, food intake, analgesia, and the regulation of blood pressure and temperature.

The ►nucleus tuberalis lateralis (NTL) can only be recognized as a distinct nucleus in man and higher primates, and contains somatostatin as its main transmitter. The NTL is hypothesized to play a role in feeding behavior and metabolism.

The ►corpora mamillaria, its input from the ►fornix and the efferent ►mamillo-thalamic tract of Vicq d'Azyr are involved in memory processes.In neurodegenerative diseases such as ►Huntington's disease, Alzheimer's disease, and ►Down syndrome, the nucleus tuberalis lateralis, tuberomammillary nucleus, and corpora mamillaria are seriously affected. The latter structure is also destroyed in ►boxer's dementia (punch drunk syndrome).

References

1. Buijs RM, Kalsbeek A (2001). Hypothalamic integration of central and peripheral clocks. Nat Rev Neurosci 2:521–526
2. Swaab DF (2003) The human hypothalamus. Basic and clinical aspects. Part I: Nuclei of the hypothalamus. In: Aminoff MJ, Boller F, Swaab DF (Series eds) Amsterdam Handbook of clinical neurology, Elsevier, 476 pp
3. Erkut ZA, Pool CW, Swaab DF (1998). Glucocorticoids suppress corticotropin-releasing hormone and vasopressin expression in human hypothalamic neurons. J Clin Endocrinol Metab 83:2066–2073
4. Fliers E, Noppen NWAM, Wiersinga WM, Visser TJ, Swaab DF (1994) Distribution of thyrotropin-releasing hormone(TRH)-containing cells and fibers in the human hypothalamus. J Comp Neurol 350:311–323
5. Alkemade A, Unmehopa UA, Brouwer JP, Hoogendijk WJG, Wiersinga WM, Swaab DF, Fliers E (2003) Decreased thyrotropin-releasing hormone gene expression in the hypothalamic paraventricular nucleus (PVN) of patients with major depression. Mol Psychiatr 8:838–839
6. Fliers E, Guldenaar SEF, Wiersinga WM, Swaab DF (1997) Decreased hypothalamic thyrotropin-releasing hormone (TRH) gene expression in patients with nonthyroidal illness. J Clin Endocr Metab 82: 4032–4036

7. Zhou JN, Riemersma RF, Unmehopa UA, Hoogendijk WJ, Van Heerikhuize JJ, Hofman MA, Swaab DF (2001) Alterations in arginine vasopressin neurons in the suprachiasmatic nucleus in depression. Arch Gen Psychiatry 58:655–662
8. Gorski RA, Gordon JH, Shryne JE, Southam AM (1978) Evidence for a morphological sex difference within the medial preoptic area of the rat brain. Brain Res 148:333–346
9. Swaab DF, Fliers E (1985) A sexually dimorphic nucleus in the human brain. Science 228:1112–1115
10. Allen LS, Hines M, Shryne JE, Gorski RA (1989) Two sexually dimorphic cell groups in the human brain. J Neurosci 9:497–506
11. LeVay S (1991) A difference in hypothalamic structure between heterosexual and homosexual men. Science 253:1034–1037
12. Kruijver FPM, Zhou JN, Pool CW, Hofman MA, Gooren LJG, Swaab DF (2000) Male-to-female transsexuals have female neuron numbers in a limbic nucleus. J Clin Endocrinol Metabol 85:2034–2041
13. Zhou JN, Hofman MA, Gooren LJG, Swaab DF (1995) A sex difference in the human brain and its relation to transsexuality. Nature 378:68–70
14. Abel TW, Rance NE (2000) Stereologic study of the hypothalamic infundibular nucleus in young and older women. J Comp Neurol 424:679–688
15. Goldstone AP, Unmehopa UA, Swaab DF (2003) Hypothalamic growth hormone-releasing hormone (GHRH) cell number increased in human illness, but is not reduced in Prader-Willi syndrome obesity. Clin Endocrinol 59:266
16. Rance NE (1992) Hormonal influences on morphology and neuropeptide gene expression in the infundibular nucleus of post-menopausal women. Prog Brain Res 93:221–236
17. Mignot E (2001) A commentary on the neurobiology of the hypocretin/orexin system. Neuropsychopharmacology 25(5 Suppl):S5–S13
18. Allen LS, Gorski RA (1990) Sex difference in the bed nucleus of the stria terminalis of the human brain. PMID: 1707064 [PubMed - indexed for MEDLINE] J Comp Neurol Dec 22; 302(4): 697–706

Neuroendocrinology of Eating Disorders

Dick F. Swaab[1], Paul J. Lucassen[2]
[1]Netherlands Institute for Neuroscience, Meibergdreef, Amsterdam, The Netherlands
[2]SILS Centre for Neuroscience, University of Amsterdam, Amsterdam, The Netherlands

Definition

The crucial role of the human hypothalamus in eating and metabolic disorders may, in the case of a disorder, lead to either increased or decreased body weight. An increasing number of such disorders and mechanisms are distinguished. In disorders of eating and metabolism, frequent hypothalamic co-morbidity occurs in terms of mood, sleep and rhythm disorders that are due to the intense hypothalamic interconnectivity and integrative mechanisms. Moreover, these symptoms often occur in an episodic way and generally have a differential prevalence between the sexes.

Characteristics

Obesity is one of the most pressing health problems in the Western world; anorexia nervosa is a serious psychiatric disorder leading to death in some 10% of the cases. Although for anorexia nervosa the cause is as yet unknown, for many other eating and metabolic disorders, clinical studies have for a long time established a central role for the hypothalamus. In diencephalic syndrome or hypothalamo-optic pathway glioma, emaciation of the entire body is found in infancy and childhood. Lesions in the ventromedial hypothalamus cause increased appetite and obesity, whereas tumors in the lateral hypothalamic area (LHA) can cause anorexia. Seasonal fluctuations in the hypothalamus are the basis for the increases in eating behavior and body weight in fall/winter that reflects the expression of a basic evolutionary process, i.e. ensuring a maximum conservation of energy when food supplies become scarce. Although adaptive in evolution, the same process has become maladaptive in humans when highly palatable, high caloric foods are readily available in present Western society, which may lead for example to the seasonal weight gain in seasonal affective disorder. Narcolepsy is characterized by a tetrad of symptoms, excessive daytime sleepiness, cataplexy, hypnagogic hallucinations and sleep paralysis. There is a substantially (85–95%) reduced number of neurons producing hypocretins/ orexins in the lateral hypothalamus of narcoleptics with cataplexy. An autoimmune cause is considered but not proven.

Genetic Factors

Human obesity certainly has an important inherited component. Yet the genetic factors responsible for obesity in the general population have remained elusive. As far as single gene mutations are concerned, mutations in the gene encoding for leptin or for the leptin receptor have been described in obese subjects. Another genetic defect leads to extreme childhood obesity, abnormal glucose homeostasis, hypogonadotropic hypogonadism, hypocortisolism and elevated plasma proinsulin and (pro-opiomelanocortin) POMC concentrations, but a very low insulin level. This disorder is based upon a mutation in the prohormone processing endopeptidase, prohormone convertase 1 (PC1). Severe early onset obesity, adrenal insufficiency and red hair pigmentation were found to be caused by POMC mutations. Mutations in the MC-4 receptor gene seem

N

to be the cause of monogenic human obesity in up to 4–6% of severely obese humans. A glucocorticoid receptor polymorphism is further associated with obesity and dysregulation of the HPA axis. The brain derived neurotrophic factor (BDNF) Met66 variant is strongly associated with all eating disorders, including bulimia [1]. Even though the inheritability of anorexia nervosa is estimated to be around 70%, it remains difficult to distinguish "psychogenic" and "organic" causes. Psychological disturbances without neurological manifestations may in some cases be due to occult intracranial tumors that can result among other things in anorexia nervosa.

Prader-Willi syndrome (PWS), the most common syndromal form of human obesity, is characterized by grossly diminished fetal activity and hypotonia in infancy, mental retardation (mean IQ of 65) or learning disability and a number of hypothalamic symptoms, i.e. feeding problems in infancy that later develop into insatiable hunger and gross obesity (Fig. 1). The patients usually have a *de novo*, paternally derived, deletion of the chromosome region 15q11–13. Severe fetal hypotonia is often already noticed by the mother during pregnancy; the baby does not seem to move much. Apart from the baby's low level of activity, its position in the uterus at the onset of labor is often abnormal. Timing of the moment of birth is often also abnormal; many children with PWS are born either prematurely or too late. So far, research into the hypothalamus has revealed an intact Neuropeptide-Y (NPY)/Agouti-Related Protein (AGRP) and Growth Hormone Releasing Hormone (GHRH) system in PWS, which is inhibited in a normal way by obesity, but the number of oxytocin expressing neurons in the PVN is clearly diminished. Oxytocin is a satiety peptide and may thus be related to the eating disorder in this syndrome.

Hormones

Cushing syndrome may result from exogenous administration of glucocorticoids or from ectopic ACTH production by a tumor often in the lungs or in the pituitary gland, a supra- or extra-sellar microadenoma. High levels of corticosteroids may lead to central or visceral obesity, hypertension and atypical depression. Metabolic syndrome includes the symptoms insulin resistance, abdominal or visceral obesity, elevated lipids and high blood pressure.

Episodic Disorders

Many eating syndromes are closely related, include a number of hypothalamic symptoms and often have a recurrent nature. Bulimia nervosa for example is characterized by recurrent, often seasonal episodes

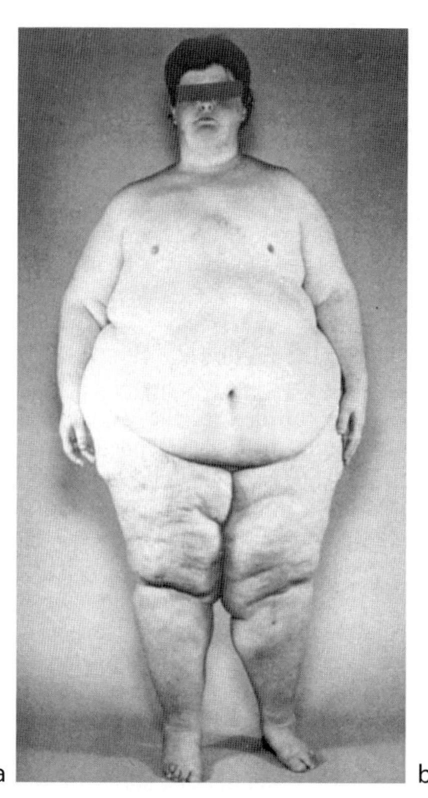

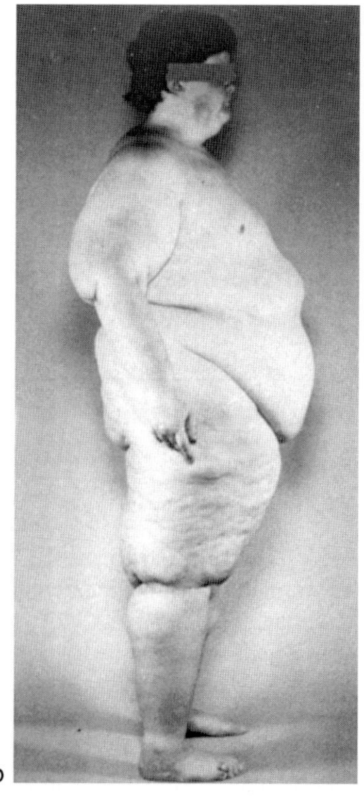

Neuroendocrinology of Eating Disorders. Figure 1 Characteristic pattern of obesity in a patient with Prader–Willi syndrome. (From [2] Fig. 1 with permission.)

of binge eating and a depressed mood. There is also inappropriate compensatory behavior to prevent weight gain such as vomiting or excessive exercise. Light therapy not only improved mood but also the eating disorder in these patients, suggesting an involvement of the hypothalamic circadian timing system. Binge eating disorder differs from bulimia nervosa in that there is little or no behavior related to weight control, such as self-induced vomiting and laxative misuse. A characteristic of this disorder is a mutation in the MC4 receptor. The night eating syndrome is characterized by morning anorexia, evening hyperphagia and insomnia and occurs during periods of stress. In "nighttime eating syndrome" a disconnection is present between the circadian control of eating relative to sleep.

Major Depressive Disorder

Major depressive disorder is not only characterized by a depressed mood but also by a significant weight loss or decrease or increase in appetite. Classically, the melancholic type of depression features anorexia or weight loss. Patients with bipolar disorder may have elevated rates of overweight, obesity and abdominal obesity. The typical patient with seasonal affective disorder (SAD) is a premenopausal woman with marked craving for high carbohydrate/high fat foods and significant weight gain during winter depression. These patients have a high prevalence of the seven repeat allele of the dopamine-4 receptor gene (DRD4). A strong argument for a close relationship between the pathogenetic mechanism of depression and the circadian timing system is the effectiveness of light therapy in patients suffering from SAD.

Fetal programming

Fetal undernutrition has been well documented in a selected group of subjects in follow up studies on the effects of the Dutch Hunger Winter in Amsterdam, 1944–1945 during the German occupation. This condition, which may also occur in the case of placental insufficiency, leads to an adaptive reaction based upon the "fetal expectation" of the presence of a scarcity of food in the environment. This has long-term clinical consequences that are induced by programming effects on fetal hypothalamic function. It leads to an increased risk of obesity, hypertension, hyperphagia, hyperinsulinemia and hyperleptinemia in the offspring. In addition, the offspring have a reduced locomotor behavior and such children are at risk of depression and schizophrenia.

The physiological mechanisms involving the same hypothalamic systems that regulate eating and metabolism in the adult are also proposed to be involved in the fetus for the initiation of parturition. The decreased levels of fetal glucose, increased levels of cortisol and changes in leptin are presumed to activate NPY neurons in the fetal infundibular nucleus, activating the fetal hypothalamo-pituitary- adrenal (HPA) axis and thereby inducing a cascade that will lead to birth. Cortisol not only triggers an increased fetal hypothalamic CRH production, but also a rise in placental CRH. The observation that in anorexia nervosa patients an increased prevalence of obstetric problems is present around the period when they were born indicates that the glucose sensitivity of the hypothalamus of these children was possibly already abnormal at the moment of birth.

References

1. Ribasés M, Gratocoŝ M, Fenández-Aranda F, Bellodi L, Boni C, Anderluh M, Cavallini MC, Cellini E, Di Bella D, Erzegovesi S, Foulon C, Gabrovsek M, Gorwood P, Hebebrand J, Hinney A, Holliday J, Hu X, Karwautz A, Kipman A, Komel R, Nacmias B, Remschmidt H, Ricca V, Sorbi S, Wagner G, Treasure J, Collier DA, Estivill X (2004) Association of BDNF with anorexia, bulimia and age of onset of weight loss in six Europoean populations. Hum Mol Genet 13:1205–1212
2. Kaplan J, Fedrickson PA, Richardson JW (1991) Sleep and breathing in patients with the Prader-Willi Syndrome. Mayo Clin Prog 66:1124–1126
3. Swaab DF, The human hypothalamus in metabolic and episodic disorders. In: Kalsbeck, Fliers, Hofman, Swaab, Van Someren, Buijs (eds) Prog Brain Res Vol 153:3–45

N

Neuroendocrinology of Multiple Sclerosis

INGE HUITINGA
Netherlands Institute for Neurosciences, Amsterdam, The Netherlands

Definition

Multiple sclerosis (MS) is an inflammatory demyelinating disease of the central nervous system that causes severe motor impairment, sensory disturbances and cognitive and memory deficits [1].

Characteristics

MS is the major paralyzing disease amongst young people in the Western World. The disease mostly starts at 25–35 years of age and two thirds of MS patients are female. World wide there are 2.5 million MS patients [1].

Process

In MS, activated macrophages and microglial cells phagocytose myelin, resulting in demyelinated lesions (Fig. 1).

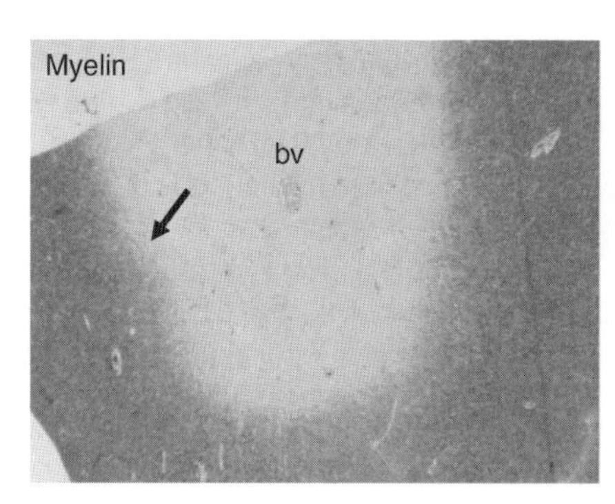

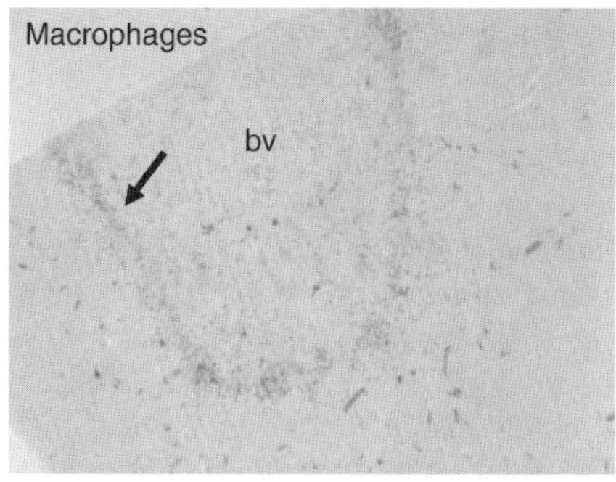

Neuroendocrinology of Multiple Sclerosis. Figure 1 Macrophage mediated demyelination in a chronic active MS lesion.The *left* photograph is a Kluver staining in which myelin is *blue* and demyelination is *white*. The *right* picture is of the same chronic active MS lesion stained for human leukocyte antigen (HLA) DR-DP-DQ. This stains activated demyelinating macrophages *grey-brown* (magnification x 2.5). Note that MS lesions are in many cases surrounding a blood vessel (*bv*). The shelving edge of demyelinating macrophages (*arrow*) moves towards the myelinated area, leaving a gliotic hypocellular plaque behind. Magnification x 2.5. Tissue was obtained from the Netherlands Brain Bank.

Demyelination is confined to the central nervous system (CNS). In the absence of myelin, conduction in the axons is impaired, resulting in neurological deficits. Active demyelinating lesions turn into inactive hypocellular and sclerotic plaques. MS owes its name to the presence of multiple sclerotic plaques. In addition to demyelination, axonal damage also occurs frequently. Axonal damage is caused by the direct effects of inflammatory cells but the harsh environment of the demyelinated sclerotic areas is also thought to induce loss of naked axons. The cause of MS is unknown but involves both genetic and environmental factors. As the cause is unknown, therapeutic approaches are tentative. Most used therapies include anti-inflammatory drugs like synthetic glucocorticoids and immune modulating therapies like beta-interferon and glatimer. MS runs either a relapsing remitting course (RR-MS) or can be primary progressive (PP-MS). In the end stage of relapsing remitting MS, the clinical course often becomes secondary progressive (SP-MS) [1].

Neuroendocrinology of MS

The neuroendocrinology of MS includes the functioning and effects of the hypothalamus-pituitary adrenal (HPA) axis, the hypothalamus-pituitary-gonadal (HPG) axis and the hypothalamus-pituitary-thyroid (HPT) axis in MS.

Quantitative Description

There are several indications that the endocrine status contributes to the start and severity of MS. Sex hormones and glucocorticoids especially have been implicated in both susceptibility to and severity of MS. MS presents itself after adolescence when sex hormone levels rise. Also, pregnancy, especially the third trimester, protects from exacerbations of MS. The period shortly after delivery is highly prone to exacerbations and the first exacerbation of MS often occurs shortly after delivery. The premenstrual period also triggers relapses. In fact, 45% of all exacerbations seem to start in this period. Interestingly, estrogen levels are low just before menstruation, whereas levels comparable to those in the third trimester of pregnancy reduce the number and size of MS lesions as observed by MRI, suggesting a protective role for estrogens [2].

Psychological stress has been considered a contributing factor in MS exacerbation and accumulating evidence is summarized in a recent meta-analysis [3]. How exactly stress influences MS is not known but the stress response systems, the hypothalamus-pituitary-adrenal (HPA) axis and the autonomic nervous system are powerful modulators of immune responses. Indeed the synthetic glucocorticoid methylprednisolone is the major drug used to treat relapses of MS. Interestingly, the activity of the HPA system increases with age as the susceptibility and severity MS decreases [4].

Regulation of the Process

The Hypothalamus-Pituitary-Adrenal (HPA) Axis in MS

Pioneer experiments in the animal model for MS, ►experimental allergic encephalomyelitis (EAE), suggested that decreased activity of the HPA axis may play a role in increased susceptibility and severity of the disease [5]. Glucocorticoids are anti-inflammatory and at physiological concentrations shift antigen specific immune responses from a ►T helper 1 phenotype as is seen in MS, towards a ►T helper 2 phenotype. In turn, inflammatory mediators that are produced in MS lesions such as interleukin-1 (IL-1) activate the HPA axis at the level of the corticotrophin-releasing hormone (CRH) in the hypothalamus and induce a rise in plasma corticosterone levels, the main glucocorticoid in the rat [6]. Indeed, during a clinical episode of EAE, plasma levels of glucocorticoids rise and this rise in corticosterone levels is crucial for recovery from the clinical episode. Rat strains that have a genetically determined low responsive HPA axis are prone to develop EAE. Recently it has been shown that during a chronic model of EAE a first episode of disease was accompanied by high plasma levels of corticosterone. However, during subsequent relapses, plasma glucocorticoid levels were significantly lower. Subsequent experiments showed reduced sensitivity of the HPA system to IL-1 during the course of chronic EAE. Importantly, compensation of plasma corticosterone levels as seen during the first relapse of EAE ameliorated the clinical signs of CR-EAE, indicating that inadequate HPA responses contribute to a more severe course of this chronic model of EAE.

Inspired by the neuroendocrine findings in the animal model for MS, several clinical studies aimed at revealing inadequate cortisol responses related to MS [5]. However, in contrast to the findings in EAE, the HPA system in MS is highly activated. There are indications that this chronic activation of the HPA system relates to both active inflammation in the CNS and neurodegeneration.

Postmortem studies show enlarged adrenals and increased cortisol in the cerebrospinal fluid (CSF) and increased numbers of corticotrophin releasing hormone (CRH) neurons in the hypothalamus, that drive the HPA axis. The increased numbers of CRH neurons concerned solely those CRH neurons that co-localized vasopressin, a sign of chronic activation of the system. Clinical studies showed normal or increased basal cortisol levels. The dexamethasone suppression test showed diminished or normal suppression of cortisol in MS. Activity and responsivity of the HPA system appeared to depend on the type of MS (i.e. PP-MS, RR-MS or SP-MS) and the lesion activity in the CNS. In RR-patients in relapse, it was found by using the ►combined dexamethasone-CRH (Dex-CRH) suppression test that cortisol was elevated. This hyper-responsiveness correlated with inflammatory activity in the brain, as assessed by white cell counts in the CSF and ►gadolinium enhancing ($GD^{\pm}$) lesions on MRI. Interestingly, in a group of both RR-MS and SP-MS patients who were not in relapse, hyper-responsivity in the Dex-CRH test showed no correlation with inflammatory markers in blood and CSF. A negative correlation between hyper-responsivity in the Dex-CRH test with numbers and volume of GD^{+} lesions in this study may furthermore indicate that adequate cortisol responses in MS are needed to control MS lesions.

There are several indications that with progression of the disease and increased axonal damage and disability, the activity of the HPA axis also increases. There is a positive correlation between HPA activity in the Dex-CRH test and atrophy as assessed by MRI. Also, hyperactivity of the HPA system in the progressive stage of the disease (PP-MS and SP-MS) correlates with disability. Recently, in a first longitudinal study, Dex-CRH reactivity significantly predicted disease progression, which implies that the Dex-CRH test might be a prognostic tool.

MS pathology implies the possibility of the occurrence of inflammatory MS lesions within structures controlling the HPA system, i.e. the hippocampus and hypothalamus. Although memory deficits in MS are frequently reported, not much is known about the incidence of lesions in the hippocampus. However, the hypothalamus is prone to the development of MS lesions. Several cases of autonomic disturbances, i.e. temperature and sleep control have been reported in relation to hypothalamic lesions. Systematic post mortem pathological analysis of the hypothalamus in MS showed the occurrence of many MS lesions (Fig. 2).

Although the mean duration of the disease was 20 years in this study, the majority of the lesions in the hypothalamus contained actively demyelinating macrophages. Hypothalamic grey matter was also frequently demyelinated. Apparently the hypothalamus is prone to MS lesion formation. Interestingly, the more active the hypothalamic MS lesions, the less active the CRH neurons were found to be. Whether active MS lesions suppress CRH neurons or MS patients with a genetic predisposition for low CRH develop many MS lesions is currently not clear. The fact is that MS patients that have many MS lesions in the hypothalamus also have a low HPA activity and very severe MS. This favors the idea that low cortisol relates to uncontrolled inflammation and consequently to a severe course of MS.

In the presence of a hyperactive HPA system, the glucocorticoid sensitivity of the immune system is decreased in MS. IL-6 and TNF production by leukocytes is reduced in relapsing remitting MS patients, whereas in progressive MS this reduced sensitivity to glucocorticoids is less pronounced [5]. Mechanisms of glucocorticoid resistance of leukocytes in MS have not yet been elucidated.

In summary, in MS the HPA system is hyperactivated. There is a correlation between activation of the HPA system and inflammation only during a relapse, indicating that inflammatory mediators may activate the HPA system in MS. Hyperactivity of the HPA systems correlates with progressive stages of MS and atrophy and correlates with disease progression. There seems to be a subgroup of MS patients with severe MS and active MS lesions in the hypothalamus that have impaired activation of the HPA system. In particular, younger patients with RR-MS suffer from impaired restraint of inflammation by cortisol due to reduced glucocorticoid sensitivity.

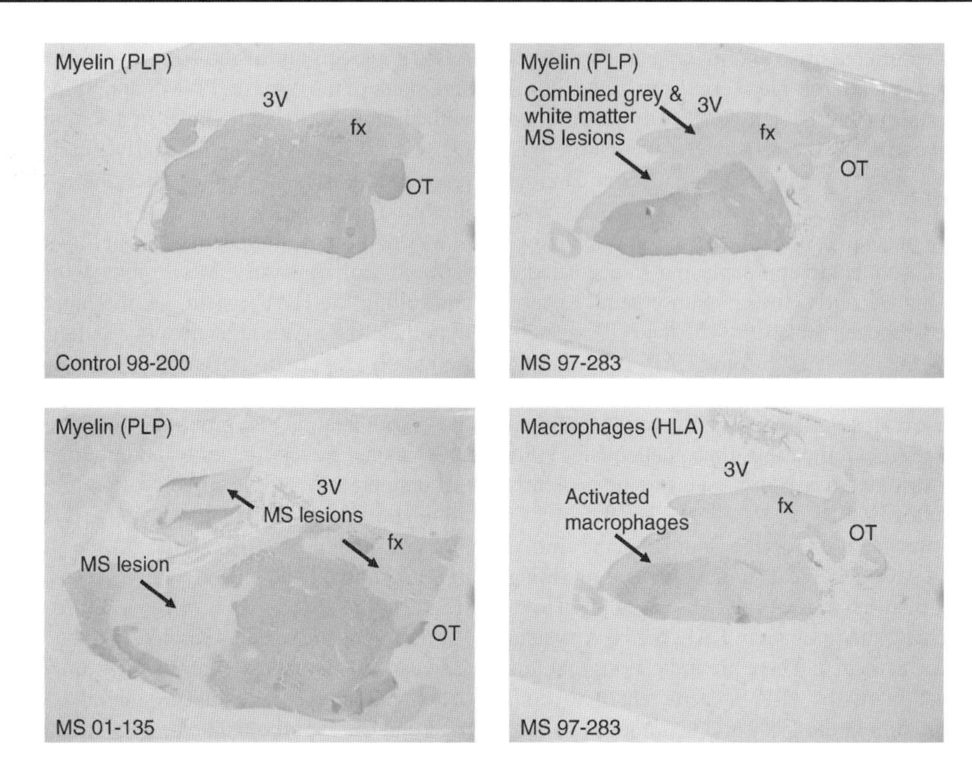

Neuroendocrinology of Multiple Sclerosis. Figure 2 Hypothalamic MS lesions. Immunohistochemical staining of proteolipid protein (PLP) shows myelin (*brown*) on one side of the hypothalamus of control subject NBB # 98–200 (*upper left*) and MS patients NBB #01–135 (*lower left*) and # 97–283 (*upper right*) and HLA DR-DP-DQ to identify activated macrophages in MS patient #97–283 (*lower right panel*). Note the presence of both grey and white matter lesions in the hypothalamus in MS (*arrows*). The lesion in MS #97–283 is active as indicated by the presence of many activated macrophages (*arrows*). *3V* = third ventricle, *fx* = fornix, *OT* = optic tract. Magnification x 1.8. Tissue was obtained from the Netherlands Brain Bank.

The Hypothalamus-Pituitary-Gonadal System in MS

The functioning of the HPG axis is only sporadically investigated in MS. Sex hormones have major effects on immune cells and pregnancy related high levels of estrogens have been related to suppression of relapses during pregnancy, whereas prolactin has been proposed to be responsible for the strongly increased risk of developing relapses of MS after delivery. Increased levels of prolactin, follicle stimulating hormone (FSH) and luteinizing hormone (LH) have been reported in MS, as have decreased levels of estrogens [8]. In several studies, a relationship between altered hormone plasma levels and hypothalamic lesions was observed. Sexual dysfunctions commonly occur in both men and women with MS; some of these cases have been related to decreased sex hormone levels [2,9].

The Hypothalamus-Pituitary-Thyroid System in MS

Like the HPG axis, there have been only a few studies addressing the functioning of the HPT axis in MS. Decreased levels of T3 have been reported in the presence of normal levels of T4 and TSH. Interferon-beta, one of the major drugs for treating MS has been demonstrated to increase the risk of hyperthyroidism slightly. The immune system is known to modulate the activity of the HPT axis at several levels, but much less is known about the effect of TSH, T3 and T4 on the immune system. Interestingly, thyroid hormones are crucial in brain development including myelination and therapeutic possibilities for thyroid-potentiated remyelination in MS have been suggested [10].

References

1. Noseworthy JH, Luchinetti C, Rodriques M, Weinshenker RG (2002) Multiple sclerosis. N Engl J Med 343:983–985
2. El-Etr M, Vukusic S, Gignoux L, Durand-DubiefF, Achiti I, Baulieu EE, Conafreux C (2005) Steroid hormones in multiple sclerosis. J Neurol Sci 233:49–54
3. Mohr DC, Hart SL, Julian L, Cox D, Pelletier D (2004) Association between stressful life events and exacerbation in multiple sclerosis. Brit Med J 328:731–736
4. Swaab DF (2004) Neuroimmunological disorders. In: Aminoff MJ, Boller F, Swaab DF (eds) The human hypothalamus: basic and clinical aspects, pt 2: Neuropathology of the human hypothalamus and adjacent brain structures.Handbook of clinical neurology, Elsevier, Amsterdam, 101–123

5. Gold SM, Mohr DC, Huitinga I, Flachenecker P, Sternberg EM, Heesen C (2005) The role of stress repsonse systems for the pathogenesis and progression of MS. Trends Immunol 26:644–652
6. John CD, Buckingham JC (2003) Cytokines: regulation of the hypothalamo-pituitary-adrenocortical axis. Curr Opin Pharmacol 3:78–84
7. Huitinga I, Erkut ZA, van Beurden D, Swaab DF (2004) Impaired hypothalamus-pituitary-adrenal axis activity and more severe multiple sclerosis with hypothalamic lesions. Ann Neurol 55:37–45
8. Grinsted L, Heltnerg A, Hagen C, Djursing H (1989) Serum sex hormone and gonadotropin concentrations in premenopausal women with multiple sclerosis J Int Med 226:241–244
9. Foster SC, Daniels C, Bourdette DN, Bebo BF Jr (2003) Dysregulation of the hypothalamic-pituitary-gonadal axis in experimental allergic encephalomyelitis and multiple sclerosis. J Neuroimmunol 140:78–87
10. Calza L, Fernanadez M, Giuliani A, D' Intino G, Pirondi S, Sivilia S, Paradisi M, DeSordi N, Giardino L (2005) Thyroid homrone and remyelination in adult nervous system: a lesson from an inflammatory-demyelinating disease. Brain Res Rev 48:644–346

Neuroendocrinology of Psychiatric Disorders

Ai-Min Bao[1], Paul J. Lucassen[2], Dick F. Swaab[1]
[1]Netherlands Institute for Neurosciences, Meibergdreef, Amsterdam, The Netherlands
[2]Swammerdam Institute of Life Sciences, Centre for Neuroscience, University of Amsterdam, Amsterdam, The Netherlands

Definition

Hypothalamic and neuroendocrine alterations in some important psychiatric disorders diagnosed according to DSM-IV will be discussed.

Characteristics

The hypothalamus plays an important role in emotional expression. Tumors, e.g. in the third ventricle region or in the area of the ventromedial hypothalamic nuclei (VMN) may cause overt psychiatric symptoms such as visual hallucinations, violent psychomotor agitation, personality changes and aggression. In addition, changes in hypothalamic nuclei and transmitter/receptor systems are present in different psychiatric disorders and may lead to endocrine alterations that can contribute to the signs and symptoms of these disorders [1].

Depression and Mania

Depression is thought to result from an interaction between environmental stressors and genetic/developmental predispositions that cause a permanent activation of the CRH neurons of the ►HPA axis [2]. The CRH neurons project to the median eminence and co-express vasopressin (AVP) that potentiates the effects of CRH. In addition, the CRH neurons project into the brain. Both centrally released CRH as well as the elevated cortisol levels contribute to the signs and symptoms of depression (see Lucassen and Swaab, HPA-axis). The AVP neurons in the hypothalamic paraventricular nucleus (PVN) and supraoptic nucleus (SON) that project to the neurohypophysis are also activated in depression, which may contribute to the increased release of ACTH from the pituitary. Increased levels of circulating AVP are associated with an increased risk of suicide. The increased activity of oxytocin (OXT) neurons has been implicated in the eating disorder in depression, since OXT acts as a satiety peptide. Moreover, opioid peptides inhibit the HPA-axis, while fewer β-endorphin containing neurons are found in the infundibular nucleus and lower numbers of β-endorphin innervated neurons are present in the PVN of depressed patients. Despite a normal ►body mass index (BMI), lower levels of leptin are found, which may also relate to the changes in appetite, food intake and weight in depressed patients.

In depression nitric oxide synthase containing neurons are found to be reduced in the PVN, but not the SON, which may be related to the increased neuropeptide production of CRH, OXT and AVP in the PVN. ►The hypothalamo-pituitary-thyroid axis also shows a decrease in thyrotropin releasing hormone mRNA in the PVN [3] parallel to alterations in basal thyrotropin (TSH) and thyroxin levels. Consistently with this, thyroid hormone supplements increase the efficacy of antidepressant drugs. Depressed patients showed decreased cerebral spinal fluid (CSF) levels of ►somatostatin in a state related way, while in suicide attempters somatostatin levels are significantly increased.

There is a clear sex difference in depression; the prevalence, incidence and morbidity risk is higher in females than in males, which may be due to both organizing and activating effects of sex hormones on the HPA axis besides social factors. Fluctuations in sex hormone levels, e.g. in the premenstrual period, ante- and post-partum, during the transition phase to the menopause and from the use of oral contraceptives are also involved in the etiology of depression. In mood disorders, the activation of neurons expressing CRH in the PVN is accompanied by increased estrogen receptor (ERα) colocalization in the nucleus of these neurons (Figs. 1 and 2) [4].

Estrogen responsive elements are found in the CRH ►gene promoter region and can stimulate CRH expression. Activation of androgen responsive elements in this region, however, initiates a CRH suppressing effect.

N

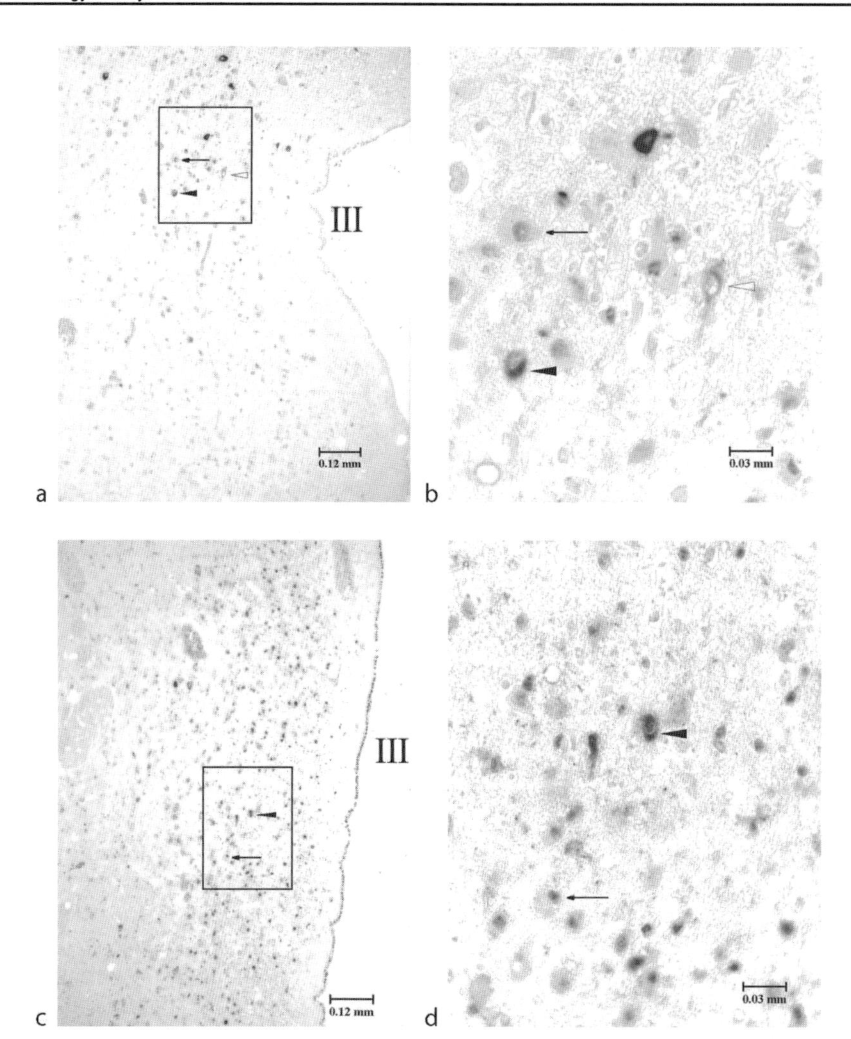

Neuroendocrinology of Psychiatric Disorders. Figure 1 Frontal section of the PVN in a control subject (a, b) and a patient with mood disorder (c, d) stained for CRH (*blue*) and ERα (*red*). b and d represent a 4 x higher magnification of a and c The *arrows, solid and hollow arrowheads* in (a, b) and (c, d) indicate the same place in the preparation to facilitate comparison. Both sections show the central part (mid-level) of the PVN and contain the largest number of stained neurons. It is clear by comparing (a) with (c) and (b) with (d) that the number of stained neurons is markedly increased in the patient with mood disorder. *III:* the third ventricle. The *arrow* points to an ERα nuclear single staining cell; the *solid arrowhead* points to a cytoplasmic CRH-ERα nuclear double staining cell and the *hollow arrowhead* points to a CRH single staining cell.

A decreased activity of the suprachiasmatic nucleus (SCN), the ►hypothalamic clock, is the basis for disturbances of circadian and circannual fluctuations in mood, sleep and other rhythms found in depression [5]. Light therapy for depression activates the SCN, while melatonin improves both sleep and mood.

It should be noted that the interactions between the peptidergic and aminergic networks such as serotonin (5-hydroxytryptamine or 5-HT), noradrenalin, histamine and dopamine contribute to the endocrine changes in depression.

Mixed manic patients, i.e. patients who have both manic and depressive symptoms, have higher HPA axis activity than pure manics. Afternoon plasma cortisol and CSF cortisol levels correlate significantly with a depressed mood, while urinary free cortisol correlates with anxiety indices. Patients with a first episode mania demonstrate significantly larger third ventricular volumes, indicating large hypothalamic changes. In addition, manic patients show hypersecretion of AVP and its ►neurophysin both in CSF and in plasma. Seasonal fluctuations in admissions to psychiatric wards are obvious in mania. Manic-depressive patients generally showed lower levels of melatonin, which again implies the involvement of the circadian and seasonal timing systems.

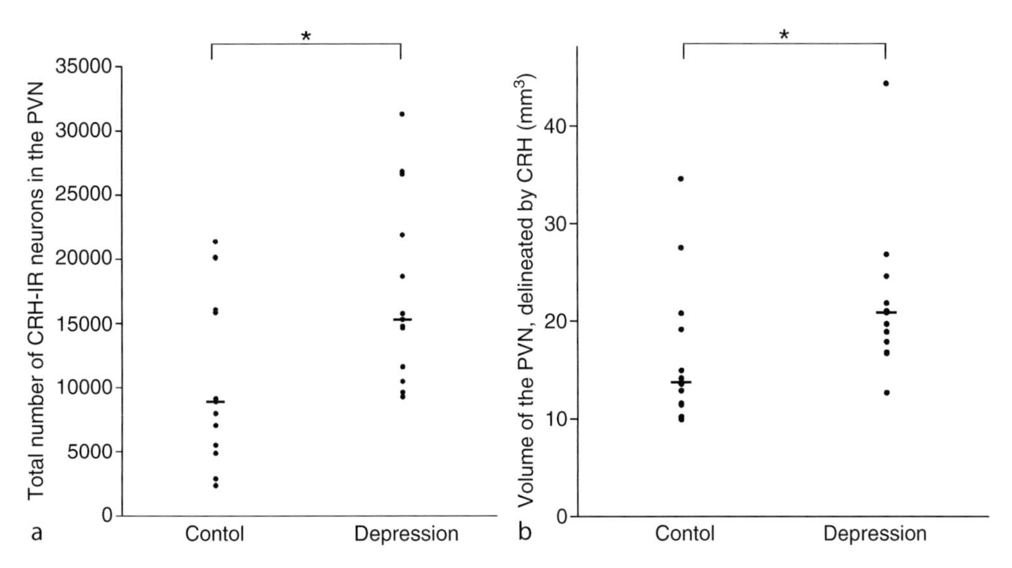

Neuroendocrinology of Psychiatric Disorders. Figure 2 Graph depicting total numbers of CRH-IR neurons in the PVN (a) and the volume of the PVN as delineated by the presence of CRH neurons (b) of control subjects (n = 13) and patients with mood disorders (n = 13). Note that the total number of CRH neurons and the PVN volume in the patients with mood disorders were significantly larger than in controls (a: $z = -2.128$, $p = 0.034$; b: $z = -2.282$, $p = 0.022$). *Horizontal line* indicates the median value.

Anxiety Disorders

Females show a two-fold increased risk for panic disorder as compared to males. Moreover, higher rates have been found in thyroid disease patients. Panic disorder is characterized by hypercortisolemia and increased nocturnal melatonin levels. A hyporesponsive hypothalamic growth hormone system is indicated by blunted growth hormone responses to specific chemical challenges. Patients with a generalized social phobia show significantly higher cortisol response to psychological stressors, although no peripheral HPA axis pathological change has been found, suggesting no basal alterations are present.

Higher rates of ►obsessive compulsive disorder have been found in thyroid disease patients. CSF CRH and somatostatin levels are significantly elevated and the increased secretions of AVP and CRH presumably contribute to persistent behavior. However, it should be noticed that there is no clear relationship between CSF CRH levels and symptom severity.

Posttraumatic Stress Disorder

The clinical symptomatology of post-traumatic stress disorder (PTSD) involves flashbacks, nightmares, sleep problems, emotional numbness or emotional outbursts, anhedonia, inappropriate startle reflexes and problems with memory and concentration. Deficits in short-term verbal memory have been associated with a smaller right side hippocampal volume in these patients. Victims of childhood abuse also have a smaller left side hippocampus. Smaller ►hippocampi may be a risk factor for PTSD rather than the result of this disorder. It is unlikely that hypercortisolism is responsible for the hippocampal atrophy, since PTSD is associated with decreased HPA axis activity and glucocorticoid supersensitivity rather than feedback resistance [6]. Victims of rape or motor vehicle accidents who later developed PTSD appeared to have – by a few hours after the traumatic event – lower cortisol levels than victims who do not subsequently develop a psychiatric disorder or major depression. Pituitary and adrenal hyperactivity to exogenous CRH and ACTH has been demonstrated in these patients. An increased sensitivity or up-regulation of glucocorticoid receptors in PTSD, lowered basal cortisol levels accompanied by an increased sympathetic drive and a pre-existing smaller hippocampal volume thus seems at present the best explanation for all the data.

Aggressive Behavior

Aggression is determined by genetic factors such as ►polymorphisms of enzymes involved in the production and degradation of neurotransmitters, hormones in development and adulthood and specific lesions in the hypothalamus and other brain areas.

Neoplastic or surgical destruction of the VMN may cause the ventromedial hypothalamus syndrome that is characterized by a tetrad of symptoms including episodic rage, emotional liability, hyperphagia with obesity and intellectual deterioration. High rates of aggression are also found in children with ►gelastic seizures due to hypothalamic ►hamartomas or following exposure to androgen based synthetic progestins during gestation. Children and adolescents with conduct disorder are found to have low HPA axis activity correlated with severe and persistent aggression. Adrenal androgen functioning as measured by dehydroepiandrosterone (DHEAS) levels is elevated in patients with conduct disorder. Men are much more aggressive than women, mainly due to the difference in testosterone levels *in utero* and in adulthood. Individuals whose life histories involve numerous antisocial behaviors and personality disorder criminals with multiple offences tend to have higher testosterone levels. Women with bulimia nervosa have increased plasma testosterone levels that correlate with aggression. The use of anabolic androgenic steroids may also lead to aggressive reactions and accompany antisocial personality traits.

The SCN may be involved in rhythmic occurrence of aggressive behavior. Excessive ►cholinergic stimulation can promote serious aggression in man. Central AVP also plays a facilitative role in aggressive behavior.

Schizophrenia

The hypothalamus is atrophied in schizophrenia. Third ventricle enlargement is significantly associated with the persistence of auditory hallucinations and poor response to treatment. The symptoms of schizophrenia can be induced by a tumor in the hypothalamic region [1], which also suggests possible involvement of the hypothalamus in this disease.

Water intoxication, ►polydipsia and ►hyponatremia are serious symptoms of chronic schizophrenia. However, no indication of a corresponding hyperactivity of AVP neurons could be found in schizophrenic patients [7]. ►Hypothalamo-pituitary-gonadal axis abnormalities may also be involved, as evidenced by an irregular menstrual cycle, loss of hair, mid-cycle bleeding and hirsutism. The typical onset of schizophrenia is found during late adolescence and early adulthood when increased levels of sex hormones reach the brain. In women, there is an additional small peak in incidence around the age of 45 when estrogen levels drop. Estrogens may protect against schizophrenia although this is not consistent with the increased incidence of schizophrenia around puberty. Alternatively, androgens may be considered as a risk factor. Abnormal growth hormone responses to TSH and ►luteinizing hormone releasing hormone are present in adolescents but not in adults. There is a high prevalence of thyroid function abnormalities in chronic schizophrenia. In spite of the fact that 36% of schizophrenic patients fulfilled the criteria for major depression, the ►dexamethasone suppression rates were very low, suggesting that depression in schizophrenia may have a different neuroendocrine profile from that in major depressive disorder. A decreased response of the HPA axis to psychological stress or to the stress of lumbar puncture has been found in schizophrenic patients. Reduced numbers of nitric oxide synthase containing neurons in the PVN have also been reported, while plasma leptin levels are decreased in patients with normal BMI.

SCN disorder seems to be present in a subgroup of schizophrenic patients, since circadian rhythm disturbances occur in chronic cases. In addition, in drug free paranoid schizophrenic patients, plasma melatonin circadian rhythm is completely absent, whereas the 24 h profile of plasma cortisol is preserved. A smaller pineal gland has been observed, while high dose melatonin treatment may exacerbate psychosis, both implying a possible relationship between the pineal gland and schizophrenia. The CSF hypocretin levels correlate significantly and positively with sleep latency, which is obviously increased in schizophrenia. ►Hypocretin is produced in the perifornical area of the lateral hypothalamus and involved in ►narcolepsy. The concentrations of α- and γ-►endorphins are elevated, whereas the number of β-endorphin containing neurons in the PVN and the innervation of PVN neurons by β-endorphin containing fibers are reduced in schizophrenic patients. Intravenous injection of β-endorphin resulted in statistically significant but not clinically apparent reduction of symptoms. Increased levels of norepinephrine were found in ►the bed nucleus of the stria terminalis, ventral septum and ►mammillary body in postmortem tissues of patients. The hypothalamic tuberomammillary nucleus (TMN) is proposed to be involved in the pathogenesis of schizophrenia, although both favorable effects of histamine injections, indicating decreased activity of the histaminergic system and elevated CSF histamine metabolite levels, indicating increased activity of the TMN, have been observed in schizophrenia.

Autism

Autism is a developmental disorder characterized by stereotypical repetitive behaviors and disturbed social interactions and communications. The prevalence is four times higher in boys than in girls. AVP and OXT are involved in socialization skills. Male autistic children have lower plasma OXT levels and there is some evidence that OXT infusions significantly reduce repetitive behaviors. A deficiency in AVP was also reported in this disorder. An association of the OXT receptor gene and the AVP receptor-1α gene with autism has recently been described. Some data indicate further HPA axis dysfunction, e.g. abnormal diurnal cortisol rhythm and changes in the ►dexamethasone suppression test. In addition, lower basal levels of TSH, a diminished response of TSH to TRH and abnormalities in dopaminergic and noradrenergic neurotransmission have been found. Increased cell packing density, reduced neuron size and swollen axon terminals (spheroids) are present in different hypothalamic nuclei, suggesting a defect in axonal transport or synaptic transmission.

Chronic Fatigue Syndrome

Noradrenaline, 5-HT and CRH are presumed to be involved in the mechanism of central fatigue. The hypothalamus shows a significant perfusion reduction. HPA axis function is unaffected or reduced, although depression is ubiquitous in this syndrome. Some patients have excessive thirst with a low plasma level of baseline AVP. Lower morning and higher evening cortisol levels further suggest a deficient SCN function.

Fibromyalgic Syndrome

Abnormal function of the HPA-axis has been reported in this disorder. Both the histaminergic TMN system and the SCN have been hypothesized to participate in the daytime somnolence characteristic of fibromyaliga. Some patients show impaired hypothalamic somatotropic reactivity. The nociceptive neurotransmitter substance P is elevated in the CSF. Both growth hormone replacement and DHEAS replacement show beneficial effects.

Postviral Fatigue Syndrome

Most patients with this syndrome have a hypothalamic dysfunction, including changes in body weight and appetite, minor fluctuations in body temperature, excessive sweating, a reversed pattern of sleep or rather excessive sleep, an impaired libido, menstrual irregularities, depression and sometimes fluid retention. Secretion of AVP may be erratic. In addition, an increased sensitivity of hypothalamic 5-HT receptors has been reported (for more detailed references see [1]).

References

1. Swaab DF (2003) The human hypothalamus. Basic and clinical aspects, pt. 2: neuropathology of the hypothalamus and adjacent brain structures. In: Aminoff MJ, Boller F, Swaab DF (eds) Handbook of clinical neurology. Elsevier, Amsterdam
2. Swaab DF, Bao AM, Lucassen PJ (2005) The stress system in the human brain in depression and neurodegeneration. Ageing Res Rev 4:141–194
3. Alkemade A, Unmehopa UA, Brouwer JP, Hoogendijk WJ, Wiersinga WM, Swaab DF et al. (2003) Decreased thyrotropin-releasing hormone gene expression in the hypothalamic paraventricular nucleus of patients with major depression. Mol Psychiatry 8:838–839
4. Bao AM, Hestiantoro A, Van Someren EJ, Swaab DF, Zhou JN (2005) Colocalization of corticotropin-releasing hormone and oestrogen receptor-alpha in the paraventricular nucleus of the hypothalamus in mood disorders. Brain 128:1301–1313
5. Zhou JN, Riemersma RF, Unmehopa UA, Hoogendijk WJ, van Heerikhuize JJ, Hofman MA et al. (2001) Alterations in arginine vasopressin neurons in the suprachiasmatic nucleus in depression. Arch Gen Psychiatry 58:655–662
6. Yehuda R (2001) Biology of posttraumatic stress disorder. J Clin Psychiatry 62:41–46
7. Malidelis YI, Panayotacopoulou MT, van Heerikhuize JJ, Unmehopa UA, Kontostavlaki DP, Swaab DF (2005) Absence of a difference in the neurosecretory activity of supraoptic nucleus vasopressin neurons of neuroleptic-treated schizophrenic patients. Neuroendocrinology 82:63–69

Neuroendocrinology of Tumors

DICK F. SWAAB[1], PAUL J. LUCASSEN[2]
[1]Netherlands Institute for Neuroscience, Meibergdreef, Amsterdam, The Netherlands
[2]SILS Centre for Neuroscience, University of Amsterdam, Amsterdam, The Netherlands

Definition

Primary tumors and metastases when located in, or in the vicinity of the hypothalamus can, depending on their size, location and age induce a wide range of nonspecific symptoms including brain edema, nausea, headaches, vomiting, aphasia, papilledema and even seizures. They can further induce specific autonomic and/or endocrine disturbances that are characteristic for this brain structure.

N

Characteristics

Tumor related symptoms characteristic for the hypothalamus are hyperphagia, obesity, amnesia, diabetes insipidus, dysthermia, circadian rhythm alterations, cachexia, hypogonadism, changes in sexual behavior and precocious puberty. Some of the typical hypothalamic symptoms are in fact nonspecific symptoms of tumors elsewhere. For instance, plasma and cerebrospinal fluid (CSF) levels of vasopressin are increased in those types of brain tumors that are accompanied by brain edema.

Tumors themselves, but also damage inflicted by brain tumor surgery, may affect overall hypothalamic function or induce specific hypothalamic symptoms such as diabetes insipidus (due to destruction of the supraoptic and paraventricular nucleus, SON, PVN), absence of thirst (characteristic of a lesion of the anterior hypothalamus destroying osmoreceptors), hyperphagia (based upon lesion of the PVN or ventromedial nucleus, VMH), diabetes mellitus resulting from hyperphagia, an impairment of temperature regulation (following damage to for example, the preoptic area), or an abnormality of sleep pattern and a reversal of the diurnal-nocturnal sleep rhythms, probably due to a lesion of the suprachiasmatic nucleus. Other symptoms may be present depending on the size of the tumor and its location.

Tumors that affect the posterior region of the hypothalamus, in particular the corpora mammillaria or the pineal region may cause precocious puberty. Tumors of the tuberal and preoptic region of the hypothalamus are often found in hypogonadism. Following a hypothalamic glioma, a patient's sexual orientation has been reported to change from heterosexual to pedophile with impotence (Fig. 1).

Other symptoms of hypothalamic tumors are hyperphagia and obesity, subcutaneous fat depletion, cachexia, autonomic seizures, paroxysm of hypertension, tachycardia and sweating (in diencephalic syndrome) and fits of rage (ventromedial hypothalamus syndrome), amnesia and attacks of laughter or crying (in the case of hamartomas). When tumors cause ventricular obstruction with a rise in intracranial pressure and/or hydrocephalus, a loss of circadian temperature fluctuations and changes in posture and walking pattern or incontinence may occur. Hypothalamic lesions due to craniopharyngioma (Fig. 2) or pilocytic astrocytoma may be accompanied by decreased nocturnal melatonin levels and increased daytime sleepiness.

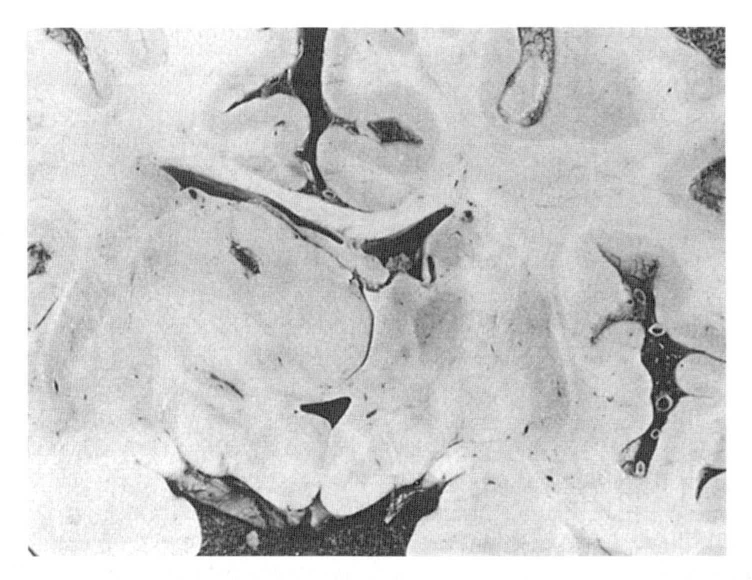

Neuroendocrinology of Tumors. Figure 1 An infiltrating hypothalamic glioma in a patient with a change in sexual orientation from heterosexuality to pedophilia. (From [1] Fig. 3 with permission.)

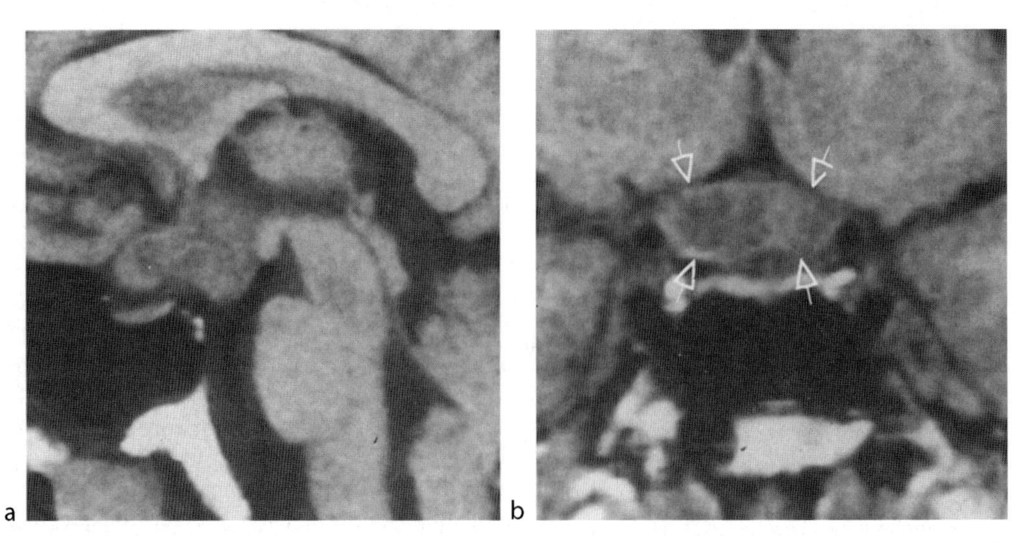

Neuroendocrinology of Tumors. Figure 2 Intrachiasmal craniopharyngioma. Sagittal (a) and coronal (b) T1-weighted MR scans. The tumor has a slightly heterogeneous appearance and has caused marked expansion of the chiasm (*arrows*). (From [2] fig. 20 with permission.)

Other hypothalamic symptoms frequently found in cases with tumors are retarded growth, amenorrhea, panhypopituitarism, dysthermia, bulimia, hydrocephalus, prolonged fever and hyponatremia. Tumors in the region of the optic pathway or infundibulum may cause optic atrophy, visual deficits, visual field defects or visual hallucinations.

Cognitive and psychiatric symptoms are also observed in the case of hypothalamic disorders. Examples are a psychosis and misdiagnosis of schizophrenia, hypersexual behavior, manic excitement, confusional syndromes and hallucinations. Patients with tumors of the region of the third ventricle may exhibit the symptoms of Korsakoff's syndrome i.e. some impoverishment of intellect, changes of personality (usually euphoria or apathy), disorientation, confabulations and memory impairment (for example in the case of tumors that cause bilateral destruction of the fornix). The memory defects caused by hypothalamic tumors may concern both imprinting and retrieval. Akinetic mutism was observed following surgical removal of an epidermoid cyst from the anterior hypothalamus, by a procedure that had probably destroyed the median forebrain bundles that contain the dopaminergic projections. Selective destruction of the hypocretin/orexin system by a tumor in the lateral hypothalamus may cause symptomatic cataplexy.

Hormone production by the tumor itself has also been described. In men, germinomas of the pineal region may cause precocious puberty because these tumors may secrete chorionic gonadotropins (HCG) that stimulate the secretion of testosterone. Hypothalamic neuronal hamartomas are rare malformations that may arise from the mammillary bodies or the tuber cinereum and that occur at the ventral aspect of the posterior hypothalamus. Some of them contain corticotropin releasing hormone (CRH), LHRH, metenkephalin or growth hormone releasing hormone containing neurons. Gelastic seizures, characterized by attacks of laughter, have been noted in 48% of the hamartomas. In addition, visual disturbances, precocious puberty, a great number of psychiatric disorders and cognitive deficits, acromegaly, diabetes insipidus or other endocrinopathies have been reported. Precocious puberty is found in over 74% of the cases and usually small, autonomous LHRH-producing pedunculated hamartomas are present.

Diencephalic syndrome is caused by a hypothalamo-optic glioma or optic pathway glioma (Fig. 3).

This low grade astrocytoma accounts for 5% of all brain tumors. The main clinical features of the diencephalic syndrome include a failure to thrive, extreme cachexia with normal height, hyperkinesis, alert appearance, vomiting, surprisingly happy affect or euphoria, pallor without anemia, hypothermia, excessive sweating, nystagmus and decreased visual acuity. The age of onset ranges from the newborn period to four years.

A low-grade developmental neoplasm, craniopharyngioma, is thought to be derived from Rathke's

a 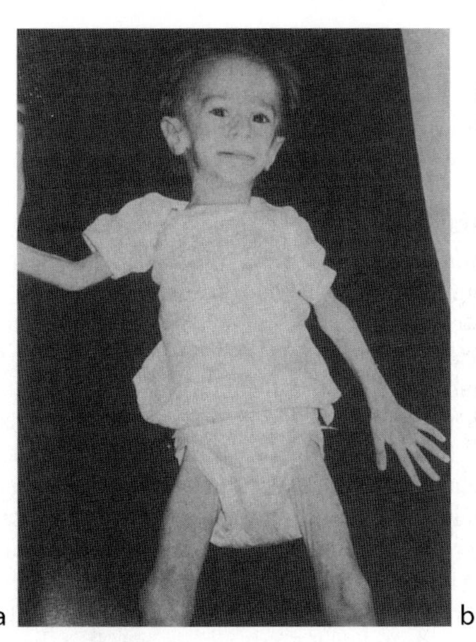

b 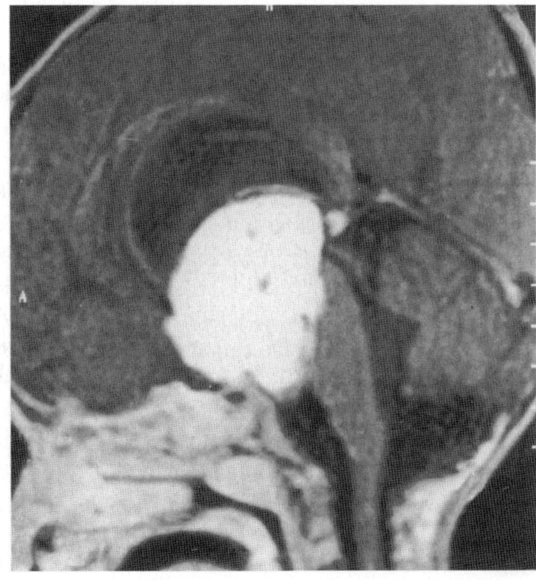

Neuroendocrinology of Tumors. Figure 3 (a) Diencephalic syndrome. Note the severe emaciation of the whole body and the characteristic "pseudohydrocephalic" appearance. (b) MRI of the brain. T1-weighted sagittal images (repetition time/echo time: 570/15) after gadolinium enhancement demonstrate the presence of a large tumor involving the hypothalamic region, distorting the chiasm and brainstem and extending into the third ventricle. Neuropathologically, the tumor proved to be a hypothalamic astrocytoma with pilomyxoid features. (From [3] figs. A, B, with permission.)

pouch, the pituitary anlage and can arise anywhere along the craniopharyngeal canal. This canal is usually obliterated during the 12th week of gestation. In the majority of cases, the craniopharyngioma does not remain confined to the sella and hypopituitarism often ensues. This frequently extends into the third ventricle and stretches the optic chiasm (Fig. 2). The signs and symptoms of a craniopharyngioma are characteristic and the most prominent ones include headache, nausea and vomiting, a failure to grow, increased intracranial pressure and visual loss, depending on the size of the tumor and its location, as well as the age of the patient. Endocrine complaints are infrequently presented, but some typical hypothalamic symptoms may include diabetes insipidus, inappropriate antidiuretic hormone secretion, hyperprolactinemia, deficiencies of LH, FSH, ACTH, TSH or cortisol, panhypopituitarism and hypogonadism.

Brain metastases are common in patients whose systemic cancer is quiescent. Diabetes insipidus due to a tumor in the infundibulum or neurohypophysis is the usual clinical manifestation, especially seen in the terminal stages. Since the posterior lobe and not the anterior lobe of the pituitary is directly supplied by arterial blood from the systemic circulation, the predilection for metastasis in this structure is understandable. The most common sources of metastatic tumors in the pituitary-hypothalamic region are carcinomas of the lungs or breasts and leukemia/lymphoma. Metastatic carcinomas originating from the gastrointestinal tract have also been described.

In the chiasmal and sellar region, approximately 10% of the neoplasms are meningiomas. They may originate from the superior leaf of the diaphragma sellae anterior or posterior of the pituitary stalk or from the inferior leaf of the diaphragma sellae.

A large adenoma of the pituitary may exert upward pressure on the front of the chiasm or between the optic nerves. The first symptom is usually bitemporal hemianoptia and there is optic atrophy. Less often the tumor may impinge on the back of the chiasma. Physical pressure exerted on or in the hypothalamus may lead to fatigue and sleepiness, excessive eating or anorexia, hypothermia, diabetes insipidus, hydrocephalus and hypopituitarism. The patient may also "feel cold" due to subsequent hypothyroidism, may perform less well in daily life activities and suffer from headaches.

References

1. Miller BL, Cummings JL, McIntyre H, Ebers G, Grode M (1986) Hypersexuality or altered sexual preference following brain injury. J Neurol Neurosurg Psychiatry 49:867–873
2. Chong BW, Newton TH (1993) Hypothalamic and pituitary pathology. Radiol Clin North Am 31:1147–1183
3. Zafeiriou DI, Koliouskas D, Vargiami E, Gombakis N (2001) Russell's diencephalic syndrome. Neurology 57:932
3. Choudhury RP (1969) Effects of cholera on the human hypothalamus and hypophysis, pt. 2. J Trop Med Hyg 72:185–192
5. Swaab DF (2003) The human hypothalamus. Basic and clinical aspects, pt. 2: neuropathology of the hypothalamus and adjacent brain structures. In: Aminoff MJ, Boller F, Swaab DF (eds) Handbook of clinical neurology. Elsevier, Amsterdam

Neuroepithelium

Definition

The tissue lining the inside of the neural tube, and containing embryonic neural stem cells. In later stages of mammalian brain development, neuroepithelial cells are sometimes called "radial glial cells" because they have long processes both towards apical (luminal) and basal (pial) sides.

►Neural Tube

Neuroethics

Bernd Ladwig
Otto-Suhr-Institut für Politik-Nissen Schaft,
Freje Universität, Berlin, Germany

Definition

Neuroethics is a field of research concerned with normative and *metaethical* (►Metaethics) problems posed by increasing knowledge of the central nervous system and especially the brain. As the term is not yet established in moral philosophy and neuroscience, other definitions highlight only specific aspects of the field, the *ethical* (►Ethics) implications of treatments for neurological diseases, *moral* principles and rules guiding neuroscience itself or the consequences of neuroscience for the understanding of moral reasoning and moral behavior. Most contributions however, cover a broader range of topics [1–3] for which a more comprehensive definition is appropriate.

Description of the Theory

Research on the brain is research concerning the mind. It is closely connected to the way mankind sees itself as sentient and sapient beings. It also seems to offer new possibilities of reading and controlling mental states and manipulating character traits. Most prominent among the scientific advances in the field are functional neuroimaging, insights into the neurochemistry of thought and a better understanding of the molecular mechanisms of neurotransmitting [4]. For an ethical evaluation it is crucial to weight chances and risks, gains and losses in a principled and nonarbitrary way, aiming at ►norms that are rationally justified for everyone concerned. Normative and metaethical problems associated with new knowledge of the brain can be differentiated as follows:

1. Consequences for ►individual *well-being* and ►*personal autonomy*
2. Consequences for ►*justice*
3. Consequences for the ►*common good* and for objective ►values
4. Consequences for the idea of ►moral agency

Individual Well-being and Personal Autonomy

A central purpose of ►morality is the protection and promotion of fundamental interests that can be subsumed under the terms "individual well-being" and "personal autonomy." Both individual well-being and personal autonomy may be seriously impaired by diseases connected with the central nervous system. Neuroscience can help to identify, cure and prevent shortfalls from normal functioning that cause suffering and reduce the range of individual opportunities. As in bioethics more generally, moral restrictions are justified for the sake of sentient beings – human beings as well as animals – potentially or actually used in research and medical testing. Morality does also matter in the evaluation of serious side effects. It seems reasonable that a significant chance to cure a disease justifies a higher level of risks than a prospect of enhancements beyond the level of normal functioning [4]. This is especially important with respect to young children and other individuals who cannot take responsible decisions by themselves.

Specific problems concerning neuroscience arise whenever interventions in the brain cause personality disorders (e.g. temporarily occurring "religious" experiences induced by magnetic stimulation). Generally speaking, interventions in the brain may affect character traits that are central for an individual's qualitative identity. Forcing a mature person to use pharmacology in order to change her thoughts, feelings or behavior is a clear violation of her autonomy. Taken to the extreme, the subject of the decisions and actions would become blurred and therefore the attribution of autonomy would lose its target.

Autonomy is also diminished if someone voluntarily decides to use mind-enhancing drugs or implants but lacks appropriate information in the light of which he or she would have decided otherwise. Neuroscience might encourage a climate of hope that in turn might make unenlightened decisions more likely. There is some evidence supported by evolution theory that a normally healthy human brain is almost perfect, optimized for purposes of human problem solving. As a consequence, a mentally healthy person might buy a gain in one dimension, e.g. memory, at the price of losses in another dimension, e.g. generalizing [5].

Apart from this assumption, we can ask whether enhancements of brain functions, as distinguished from mere remediation of diseases, would really contribute to individual well-being [6]. Enhancements might consist in the avoidance of challenges that would confront a person with important insights and/or would offer him or her worthwhile opportunities for agency. It is implausible to reduce well-being to pleasant feelings and success without effort. Maybe a lot of problems and even painful feelings are a price that has to be paid for leading full lives as responsible agents.

The more there can be reasonable disagreement on these matters the more important autonomy proves to be. This does also affect duties towards non-autonomous individuals. Persons taking decisions in the name of others should refrain from enhancements, at least from irreversible ones, for which a hypothetical agreement of the persons affected is disputable. Parents projecting and technically imposing their own particular values onto their children neglect the fact that everyone has to lead his or her own life and therefore has a right to an open future [7]. On the other hand, respecting personal autonomy includes respecting decisions made by mature individuals to use mind-enhancing drugs or technologies as long as the enhancement does not result in harming others, violating valid principles of justice or the like. Enhancements as such are a normal objective of human strivings, as the widespread use of caffeine, viagra, and cosmetic surgery shows. It is far from clear whether continuing to do so on a molecular or genetic level would confront mankind with totally new problems.

Functional neuroimaging is another source of threats to autonomy. Although literally reading another person's thoughts is now and for the foreseeable future sheer science fiction, neuroscientists are successful in correlating some psychological states and traits like neuroticism, racial prejudices and intentional deception with distinct patterns of brain activity [8]. This might encourage the use of neuroimaging for the purpose of lie detection, especially in societies obsessed by security issues. Even if neuroimaging cannot really decipher propositional states, its use against the will or without the informed consent of the patient would violate the right to privacy in an especially intimate domain, the privacy of what goes on in the mind.

Justice

The availability of mind-enhancing drugs and techniques as well as neuroimaging can also become a source of discrimination. In order to reduce their risks, insurance companies and employers might expect persons to undergo examinations and pharmacological treatment of the brain. This could lead to different classes of people with the willing on the one hand and the unwilling on the other. Results of those tests could also be used to stigmatize still healthy persons and to look at them as if they were already ill or handicapped. Discrimination of this sort is incompatible with a fundamental principle of justice in modern societies, to treat everybody with equal respect and concern [9]. As far as enhancements are possible, e.g. improving the working memory of older people, they might be expensive and therefore not accessible to all. Without equal access to the new advantages however, equality of opportunity would decline. Preventing some (categories of) people from taking unfair advantage is a well known requirement of justice that is already in tension with respecting the autonomy of parents e.g. in choosing schools for their children. Again, not all that looks new at first sight is new in principle.

The Common Good and Objective Values

Harmful testings, bad side effects, forced interventions, violations of privacy, discrimination and unfairness do not sufficiently explain why many people are disturbed about the new developments in genetics as well as in neuroscience. A fear in the background might be that a society in which those techniques and opportunities would be available and used on a large scale would in important respects be a dehumanizing society. For example, such a society might leave less room for human excellence. There might no longer be reasons to admire other people, e.g. professionals in sports or brilliant thinkers, for what they perform using their natural endowments. Another worry might be that the social relations in such a world would be dominated by reciprocal blaming in domains that are at present up to nature: "Why didn't you enhanced my affiliative behavior and my higher cognitive functions when I was in school?" Taken to the extreme, it might no longer be possible to identify what really belongs to a person and the sense of personal responsibility might become pointless. An even more familiar, yet more dubious ►reason might be that many people take "the natural" to be objectively valuable, independent of its contributions to well-being or goal-attainment. Nature, it seems to them, is a necessary counterpart to human hubris; it is the epitome of what man still cannot and never shall master [10].

Images of a good society and objective values have in common that they do not directly relate to goods that can be possessed by individuals as such. Some goods are essentially shared, e.g. a climate of tolerance and creativity in the public sphere. Other goods might even be totally independent of any interests. They might be good as such, as for many nature seems to be. Such evaluations, however widespread they may be, are much more controversial and even unclear in status, than those concerning fundamental interests and justice. The more disputable they are, the less should they serve as grounds for preventing other people from taking free decisions. Nevertheless, an unrestricted and ongoing public debate about strong evaluations concerning the common good and probably also non-subjective values is an essential part of any reasonable will formation in a political community confronted with developments that might modify the way in which mankind and the world are seen.

The Idea of Moral Agency

Probably the most fundamental change concerns understanding of moral agency and responsibility. Neuroscience and genetics seem to support naturalist positions in the philosophy of mind that in turn seem to undermine the ideas of free will and justified blame. This is not the right place to discuss such positions. If they were true, all that has been said so far about the agenda of neuroethics would prove to be senseless. When engaged in moral deliberations it is necessary to presuppose that it is possible to act out of insights. Every "ought" refers to ►rationality. It is far from self-evident that sense could no longer be made of these presuppositions if the progress in neuroscience were taken seriously. For example, it is in no way metaphysically convincing to see the author of free decisions and actions as a causally independent agent (an *homunculus*) within the person. It is much more convincing to ascribe responsibility to the entire person; given that he or she is able to do what he or she is reasonably convinced he or she has to do. Neuroscience can inform us about the neurological states and processes that enable people to act with reasons – or that prevent people from doing so. It can sharpen the sense of the individual limits of moral responsibility. But that is totally different from undermining the very idea of responsibility.

References

1. Marcus S (2002) Neuroethics: mapping the field. Dana, New York
2. Farah MJ (2002) Emerging ethical issues in neuroscience. Nat Neurosci 5:1123–1129
3. Wolpe PR (2003) Neuroethics on enhancement. Brain Cogn 50:387–395
4. Cardiff Centre for Law Ethics, and Society (2003) Brain research and neuroethics. http://www.ccels.cardiff.ac.uk/issue/caplan.html

5. McClelland JL, McNaughton BL, O'Reilly RC (1995) Why there are complementary learning systems in the hippocampus and neocortex: insights from the successes and failures of connectionist models of learning and memory. Psychol Rev 102:419–457
6. Kass LR (2003) Beyond therapy: biotechnology and the pursuit of human improvement. The president's council on bioethics. http://www.bioethics.gov/background/kasspaper.html
7. Feinberg J (1980) The child's right to an open future. In: Aiken W, LaFollette H (eds) Whose child? Children's rights, parental authority and state power. Rowman and Littlefield, Totowa, NJ
8. Canli T, Amin Z (2002) Neuroimaging of emotion and personality: scientific evidence and ethical considerations. Brain Cogn 50:414–431
9. Dworkin R (1990) Bürgerrechte ernstgenommen. Frankfurt am Main, pp 297–302
10. Sandel M (2002) What's wrong with enhancement? The president's council on bioethics. http://www.bioethics.gov/background/sandelpaper.htm

Neuroethological Aspects of Learning

Frank W. Ohl
BioFuture Research Group, Leibniz Institute for Neurobiology, Magdeburg, Germany

Definition

From an ethological viewpoint, ►learning is the process by which animals and humans can adapt their behaviors and internal states to an ever-changing environment. Simultaneously, from the viewpoint of evolutionary biology, equipping animals with the ability to learn frees natural selection processes from having to genetically fixate a vast variety of different behaviors and signal processing mechanisms to differentially recruit these behaviors in appropriate situations. From the viewpoint of neurophysiology, learning is made possible by a capacity of the nervous system to respond to certain classes of experiences encountered by an animal with changes in some of its functional features, a capacity called ►neuronal learning-induced plasticity.[1]

As the phenomenon of learning will be treated in more generality elsewhere in this volume [1], this essay will focus on some neuroethological aspects of learning. Learning can be defined as the process by which relatively permanent changes occur in behavioral potential as a result of experience [1]. In this definition, the attribute "*relatively permanent*" aims to exclude short-lasting changes in behavior, like fatigue. Experience-induced changes are referred to the "*behavioral potential*", rather than to behavior, because it is known that learning can be "behaviorally silent," i.e., need not express itself in immediate behavior. Finally, the definition clarifies that these changes should be due to *experience*, because other reasons exist for changes in behavioral potential, like the state of arousal, the state of (ontogenetic) development, age, injury, etc.

Central to neuroethological accounts of learning is the question of the relationship between learning and ►neuronal plasticity. This relationship is non-trivial, as the former is an ethological or psychological concept and the latter a physiological concept. Establishing the exact role of neuronal plasticity for learning is therefore predictably difficult, similar to other fields of science where conceptually different levels have to be linked, like for example in the relation between Newtonian mechanics and thermodynamics[2], or in the case of the mind-body problem. Historically, attempts to conceptualize the role of neuronal plasticity for learning have moved from the appreciation of the capacity of the nervous system for rerouting the flow of excitation through a neuronal network in simple learning situations [2], to neural processes that reflect the subjective creation of meaning in cognitively demanding learning situations like category learning and concept formation [3].

This essay will address the following neuroethological aspects of learning: (i) Structure of stimulus relationships that lead to learning, (ii) Motivational aspects of learning, and (iii) Neural aspects of information processing and meaning generation during learning.

Characteristics

Higher Level Processes

Structure of Stimulus Relationships that Lead to Learning

It is now realized that many generalities in learning behavior across species reflect the fact that they are evolutionary designed solutions to similar demands, rather than reflecting similar neuronal mechanisms. Therefore, generalities on the level of learning behavior cannot in general be expected to be supported by generalities on a physiological level [1]. Much research has been conducted to reveal those generalities in environmental or laboratory situations that lead to learning. Studies of the early twentieth century, e.g., by Ivan P. Pavlov, Edward L. Thorndike, Clark L. Hull, Edward C. Tolman, B. F. Skinner, and others, focused on classical conditioning and instrumental conditioning paradigms, and identified the temporal relationship

[1]The addition "learningnduced" distinguishes this form of neuronal plasticity from those observed during ontogeny (►developmental plasticity) or after injury (►compensatory plasticity).

[2]In this example, a conceptual link between both levels has been possible to provide, at least for the equilibrium state of matter, by the framework of .

N

(►contiguity) between stimuli or events as facilitating behavioral changes that could be interpreted as being the result of formed associations between these stimuli or events. Later studies, e.g., by Rescorla and Wagner [4], revealed that the contiguity is neither sufficient nor necessary for the formation of associations. For example, by carefully varying the probabilities of occurrences of events conditional on the occurrence of other events, it was shown that animals can show different levels of association even under constant contiguities. Learning mechanisms rather seem to exploit the probability structure of stimulus relationships, estimated from previous experience, to form associations. Animals can therefore use the information that one stimulus or event carries about the occurrence of another stimulus or event (►contingency).

Motivational Aspects of Learning

For methodological reasons, one of the events referred to above was typically a stimulus that elicits a defined, easily observable, behavioral response. Such stimuli are most often reward stimuli or aversive stimuli that will elicit approach or escape responses, respectively, and are called ►reinforcers. In experiments where reinforcers are applied in dependence of behaviors produced by a subject, reinforcers change the probability with which such behaviors will be emitted in the future. Reinforcers act by at least two discernable mechanisms [5]. The first mechanism, the ►enhancing function [5], works by enhancing the "storage" of information about situations in which they are encountered. This does not require that the animal learns anything about the reinforcer itself. The second mechanism, ►conditioned motivation [5], works by conditioning motivating effects of reinforcers to other brain activity present at a temporal relationship to the experience of the reinforcer. The dopamine system and the striatum have been associated with both types of action [5].

On theoretical grounds, the task of neural coding for reinforcers can be distributed to a number of hypothetical functions [6]. Among those are the *detection* of a reinforcer, i.e., a neural representation of whether a reinforcer is present or not, the *prediction* of a reinforcer, i.e., a neural response to a stimulus which the subject associates with a reinforcer to be followed, and the *expectation* of the reinforcer, i.e., a neural activity which develops temporally between reinforcement prediction and reinforcement detection, for example to support processes of maintaining attention. For all such functions neurons have been found whose firing properties could be associated with all of these functions and their combinations in such structures as the striatum, the prefrontal cortex and orbitofrontal cortex. As learning can be viewed as a process of reducing the discrepancy between a predicted and actually experienced outcome of a learning situation (a magnitude called the ►prediction error), such neurons are likely to have an important function for the neuronal mechanisms underlying learning. For example, in the striatum of monkeys tonically active interneurons have been reported that respond more frequently to unpredicted rewards than to predicted ones [7]. Analogously, microdialysis studies in the rodent medial prefrontal cortex have revealed an increased dopamine efflux in the *very early phase* of the establishment of a behavioral avoidance strategy, but not in later acquisition phases when performance was still increasing or during retrieval sessions. The hypothesis that the dopamine efflux in medial prefrontal cortex correlates with the establishment of *new behavioral strategies* for solving problems in learning situations could be further supported by relearning paradigms [8].

Neural Aspects of Information Processing and Meaning Generation during Learning

For the neuroethological perspective on learning, those learning paradigms of particular importance are those that contain aspects beyond mere associations between stimuli or between stimuli and responses. This is because such paradigms preclude explanation of learning phenomena by a broad class of simple neurophysiological models that are otherwise discussed as elemental for physiological theories of learning. For example, while classical conditioning can in principle be explained by very simple neuronal networks [2], this is not possible for cognitively more demanding learning phenomena like *category learning* (*concept formation*). A few animal models of category learning have been designed to allow the study of the neural basis of such aspects of learning. For example, the Mongolian gerbil (*Meriones unguiculatus*), a rodent with exquisite auditory learning abilities amenable to physiological investigation [e.g., 9], can be trained using frequency-modulated tones to form the categories "rising" and "falling" and sort even novel, previously unheard, frequency-modulated tones into these categories depending on how the pitch of these sounds develops over time [10]. The formation of categories involves development of a new cognitive structure that represents qualities beyond the information given. It allows that a particular meaning (defined by the category) is assigned to even novel stimuli that are processed by a sensory system. The neurophysiological analysis of this learning behavior [3] revealed that during category formation particular spatio-temporal activity states emerge in the auditory cortex. These states have a *metrical structure*, i.e., similarity relations between spatio-temporal patterns correspond to similarities in the perceived category belongingness. This metrical structure is unlike the topographic maps known from various brain structures in which physical stimulus attributes are represented, because it represents the only subjectively valid perceptual scaling experienced by a given individual. These neuronal observables represent

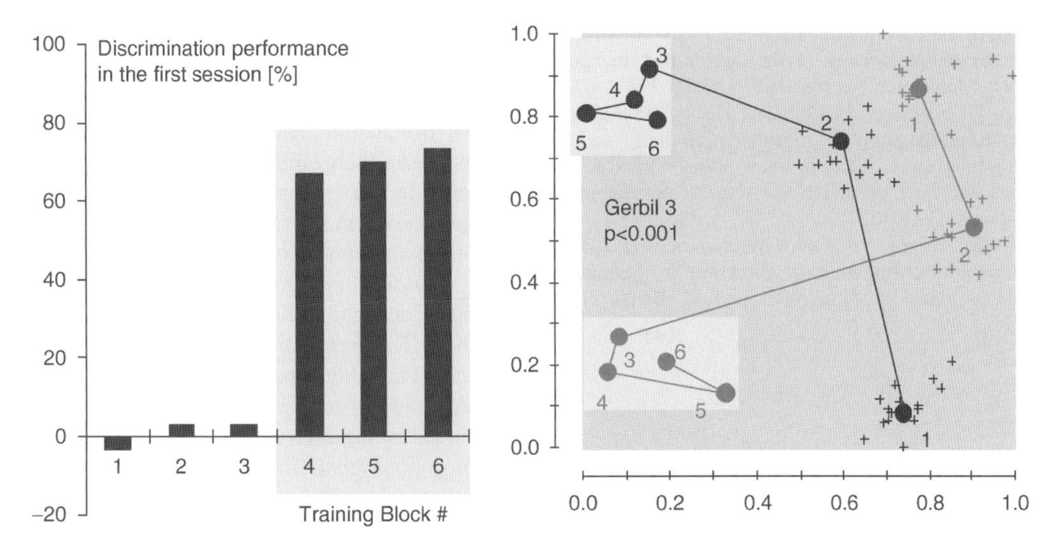

Neuroethological Aspects of Learning. Figure 1 Neurodynamics in auditory cortex during learning of acoustic categories exemplifying the physiological correlate of a learned cognitive structure. Example of behavioral and electrophysiological data from a Mongolian gerbil, which was trained to discriminate rising from falling frequency-modulated tones in a sequence of training blocks. In each training block, a novel pair of a rising and a falling frequency-modulated tone was trained. The left panel displays discrimination performance (quantified by the difference between hit rate and false alarm rate) at the beginning of a training block when the stimuli were novel to the subject. In training blocks 1–3, (*discrimination phase*) initial performance was insignificant but would improve in later sessions of the block (not shown). Starting with training block 4 and for subsequent blocks the subject showed transfer of the learned behaviors to the novel stimuli (*categorization phase*), indicating that a new cognitive structure had been formed in the subject. The transition from discrimination phase to categorization phase occurs abruptly, and for different individuals at different points in time. This behavioral *state transition* is accompanied by a transition in the neurodynamical states of auditory cortex. The right panel is a graphic representation of the similarity and dissimilarity relations between cortical activity patterns during different phases of the learning. Points represent activity states, numbers correspond to the training block, and colors red and blue to the categories “rising” and “falling,” respectively. The spatial distance between any two points in the graph is a measure of the dissimilarity between the corresponding cortical activity states. For blocks 1 and 2 patterns for all trials (+) are shown together with their center of gravity values (•), for subsequent blocks only the center of gravity values are shown to avoid cluttering of the graph. It can be seen that as long as the subject remained in its discrimination phase (blocks 1–3), dissimilarities of cortical activity patterns *within* a category were of the same order of magnitude than *between* categories. After the transition to the categorization phase (blocks 4–6), dissimilarities within a category were abruptly reduced compared to dissimilarities between categories[3]. This indicates the establishment of a new metric in cortical representation of stimuli. This metric does not reflect physical stimulus attributes (as topographic feature maps do), but reflects the subjective perceptual scaling experienced by a given individual. Modified after [3].

an objectively accessible physiological representation of a subjectively existing cognitive structure [3] (Fig. 1).

References

1. Anderson JR (2000) Learning and memory. An integrated approach. Wiley, New York
2. Hawkins RD, Kandel ER (1984) Is there a cell-biological alphabet for simple forms of learning? Psychol Rev 91:375–391

[3] Note that the clustering of the state vectors is due to the fact that in the fourth (and subsequent) training blocks, i.e. when the animal has entered the categorization phase, the state vectors remain in the vicinity of the previously attained vectors.

3. Ohl FW, Scheich H, Freeman WJ (2001) Change in pattern of ongoing cortical activity with auditory category learning. Nature 412:733–736
4. Rescorla RA, Wagner AR (1972) A theory of Pavlovian conditioning: variations in the effectiveness of reinforcement and nonreinforcement. In: Black AH, Prokasy WF (eds) Classical conditioning II: current research and theory, Appleton Century Crofts, NY, pp 64–99
5. White NM, Milner PM (1992) The psychobiology of reinforcers. Annu Rev Psychol 43:443–471
6. Schultz W (2004) Neural coding of basic reward terms of animal learning theory, game theory, microeconomics and behavioural ecology. Curr Opin Neurobiol 14:139–147
7. Apicella P (2002) Tonically active neurons in the primate striatum and their role in the processing of information about motivationally relevant events. Eur J Neurosci 16:2017–2026

N

8. Stark H, Rothe T, Wagner T, Scheich H (2004) Learning a new behavioural strategy in the shuttle-box increases prefrontal dopamine. Neuroscience 126:21–29
9. Schulze H, Neubauer H, Ohl FW, Hess A, Scheich H (2002) Representation of stimulus periodicity and its learning induced plasticity in the auditory cortex: recent findings and new perspectives. Acta Acoustica united Acoustica 88:399–407
10. Wetzel W, Wagner T, Ohl FW, Scheich H (1998) Categorical discrimination of direction in frequency-modulated tones by Mongolian gerbils. Behav Brain Res 91:29–39

Neuroethology

Definition

Part of ethology that deals with the neural basis of behavior.

Neuroethology of Biosonar Systems in Bats

Hans-Ulrich Schnitzler, Annette Denzinger
Tierphysiologie, Zoologisches Institut, Universität Tübingen, Tubingen, Germany

Definition

Evolution has equipped microchiropteran bats for nocturnal life by morphologically, physiologically, and behaviorally adapting their sensory and motor systems. Bats' abilities to echolocate and to fly play a particularly important role in allowing them to exploit resources which are not accessible to other animals. Around 800 species of echolocating bats (suborder Microchiroptera) emit tonal signals and analyze the returning echoes to detect, localize and classify the reflecting targets. All bats use their biosonar systems for spatial orientation and many of them also for food acquisition. Like technical systems bats have a transmitter which produces and radiates the echolocation signals and a receiver which analyzes and evaluates the returning echoes. Comparative neuroethological studies reveal how sound production and hearing in bats have been adapted during evolution to perform habitat-specific echolocation tasks.

Characteristics

The Action-Perception-Loop of Echolocation

Bats' echolocation systems are comparable to other active orientation systems such as radar and sonar systems, and are therefore referred to as biosonar systems. In active systems, the transmitter is connected to the receiver in an action-perception loop (Fig. 1).

Various neuroethological approaches to the study of bats' biosonar systems have been, as demonstrated in Fig. 1, describing the action-perception-loop of echolocation. The ►evolutionary aspect of these systems is examined in studies that develop plausible scenarios for the evolution of echolocation. The ►acoustical aspect is addressed by research that describes the various types of echolocation signals, measures the directionality of signal emission and echo reception, and characterizes the information content of echoes from various targets. The ►behavioral aspect is investigated in studies that define the constraints acting on signal structure, and explain the adaptive value of signal design according to the echolocation tasks and the bats' performance in natural echolocation situations. This aspect is also addressed by psychophysical experiments in the laboratory. The ►structural and functional aspects of echolocation are addressed by studies on the morphology, neuroanatomy, physiology, and pharmacology of the sensory and motor parts of echolocation systems. The ►analytical aspect is explored by those who develop computational theories of echolocation by defining problems related to echolocation tasks, searching for algorithms to solve these problems, and testing these algorithms by implementing them in biomimetic sonar systems. Here, we summarize these different approaches to investigating biosonar systems in bats, by presenting a survey of the reviews that we believe are relevant to understanding the neuroethological basis of echolocation in bats.

The Evolutionary Aspect of Echolocation

The key characteristics that distinguish bats from other mammals are their ability to fly and their use of an active orientation system. Due to the paucity of fossil records, scenarios explaining the evolution of flight and echolocation are speculative and have been developed on the basis of plausibility (Denzinger et al. 2004). The evolution of echolocation presumably took place in several steps (Schnitzler et al. 2004). Echolocation signals may have evolved from high-pitched communication signals, and may have been used by pre-bats to estimate distances before jumping from one branch to the next. With the evolution of gliding and finally of flapping flight, echolocation may have become useful for obstacle avoidance during flight and improved landing control. With increasing maneuverability, the demand for spatial information increased and bats may have used echolocation to identify landmarks and habitat elements, i.e. for spatial orientation within

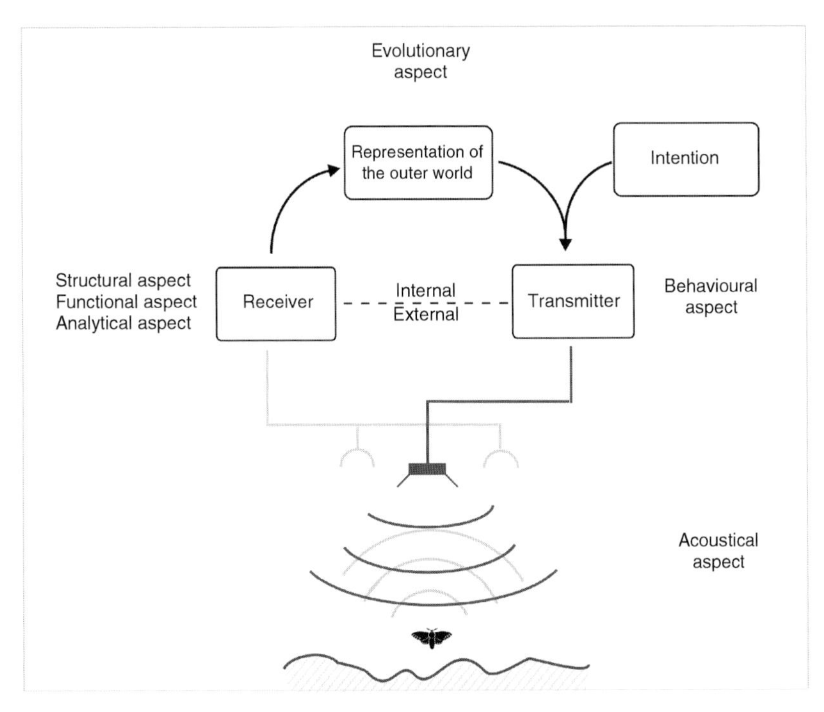

Neuroethology of Biosonar Systems in Bats. Figure 1 The action-perception-loop consists of a transmitter, which generates signals transmitted via a sender antenna, and a receiver, which picks up the returning echoes with receiving antennae and decodes the information contained in the echoes to create a spatial representation of the area covered by the sonar footprint of the emitted signals (Denzinger and Schnitzler 2004). Bats produce their signals in the larynx (phonation), filter them in the vocal tract (articulation), and transmit them either through the open mouth or the nose (transmitter antenna, depicted in black)). The returning echoes are picked up with both ears (receiving antennae, depicted in grey) and echo information is evaluated in the auditory system (receiver). The resulting representation of the outer world and the behavioral intentions determine which signal type of the repertoire will be the next to be generated and transmitted (adapted from Denzinger and Schnitzler 2004).

their home range. Thus, echolocation is likely to have evolved primarily for spatial orientation, and its use for the detection of prey was a later step in evolution. Bats may have encountered situations in which the acoustical cues of an insect colliding with vegetation were preceded by the flight tone of the insect. By reacting with foraging flights towards the flight tones, bats evolved the ability to approach the moving sound source guided by echolocation. Finally, they made the transition to detecting flying insects on the basis of echolocation alone and to hunting for airborne prey on the wing.

Acoustical and Behavioral Aspects of Echolocation

When performing echolocation tasks, microchiropteran bats continuously emit signals that are mostly in the ultrasonic range. These signals differ between species in terms of their duration, pulse interval, frequency, harmonic content, and sound pressure level (SPL). Each species has a specific signal repertoire containing a variety of signal types evolved to perform species-specific echolocation tasks. The tasks performed by bats depend on their behavioral intentions and also on where bats fly and forage, what they eat, and how they acquire their food. Echolocation tasks can be attributed either to spatial orientation or to food acquisition.

All bats use echolocation for spatial orientation. Information derived from returning echoes is used to move in relation to stationary targets along routes and to build up a spatial representation of the environment. Little is known about which part of the information contained in echoes is used to orient in space. Extended targets such as landmarks contain many reflecting facets that generate stochastic echo sequences. Random process parameters distinguish between different vegetation types (Müller and Kuc 2000) and could be used by bats for landmark classification. There is behavioral evidence that bats are able to distinguish between echo trains differing in roughness (Grunwald et al. 2004). In addition, bats emit short, broadband, highly frequency-modulated (FM) signals in classification tasks. Due to the wide range of wavelengths and the strong directionality of high frequencies in such signals, they are well-suited for target classification (Siemers and Schnitzler 2004). Important information might also be encoded in changes within echoes over time.

N

Theoretical studies show that changing echo parameters create time-variant echo features, such as acoustic flow, that might contain information about the position of targets in relation to the bat's motion. Long, constant-frequency (CF) signals are especially well-suited for evaluating acoustic flow information (Müller and Schnitzler 1999, 2000).

Many bats also use echolocation to find their food. Comparative studies reveal that the proximity of prey to background targets is the most relevant ecological constraint on the design of bats' echolocation signals, and can therefore be used to define foraging habitats. Three main habitat types have been described: in the open, or "open space," between and along vegetation, or "edge space," and within vegetation and close to it and the ground, or "narrow space" (Aldridge and Rautenbach 1987; Neuweiler 1990; Fenton 1995; Schnitzler and Kalko 2001; Schnitzler et al. 2003). In open space, bats forage so far off from the background that they do not react to it in their echolocation behavior. In edge space, bats react to the background but the prey echo does not overlap with background echoes. In narrow space, the echoes of prey positioned on substrate or flying very close to the background overlap with background echoes, which may result in masking.

Ecological conditions exert strong selective pressure on signal design, thus favoring species-specific signal types closely connected to habitat type, foraging mode, and prey. This connection can be used to define functional groups based on the preferred habitat type and foraging modes of various bat species. Members of each functional group are confronted by a common set of constraints and must solve similar echolocation tasks. This results in many similarities in signal design within functional groups (Schnitzler et al. 2003). "Open space aerial foragers" catch flying insects and use rather long, narrowband, shallow FM signals of low frequency that are adapted for long-range detection (Fig. 2).

"Edge space aerial/trawling foragers" search for insects flying near vegetation edges, in gaps, near the ground, or drifting on flat water surfaces. They often emit mixed search signals of medium duration. These signals consist of a narrowband, shallow FM component adapted for medium-range detection and a broadband, steep FM component adapted for orienting with respect to background targets. Narrow space bats that either glean their food from the substrate or capture prey close to it face the problem that overlapping clutter echoes may mask the prey echo. Three different strategies have been evolved to cope with this problem. "Narrow space active aerial/gleaning foragers" find their food by echolocation. They either use target-specific echo cues or specific echolocation strategies to separate prey from the background (Denzinger and Schnitzler 2004). "Narrow space passive gleaning foragers" cannot solve the masking problem and use

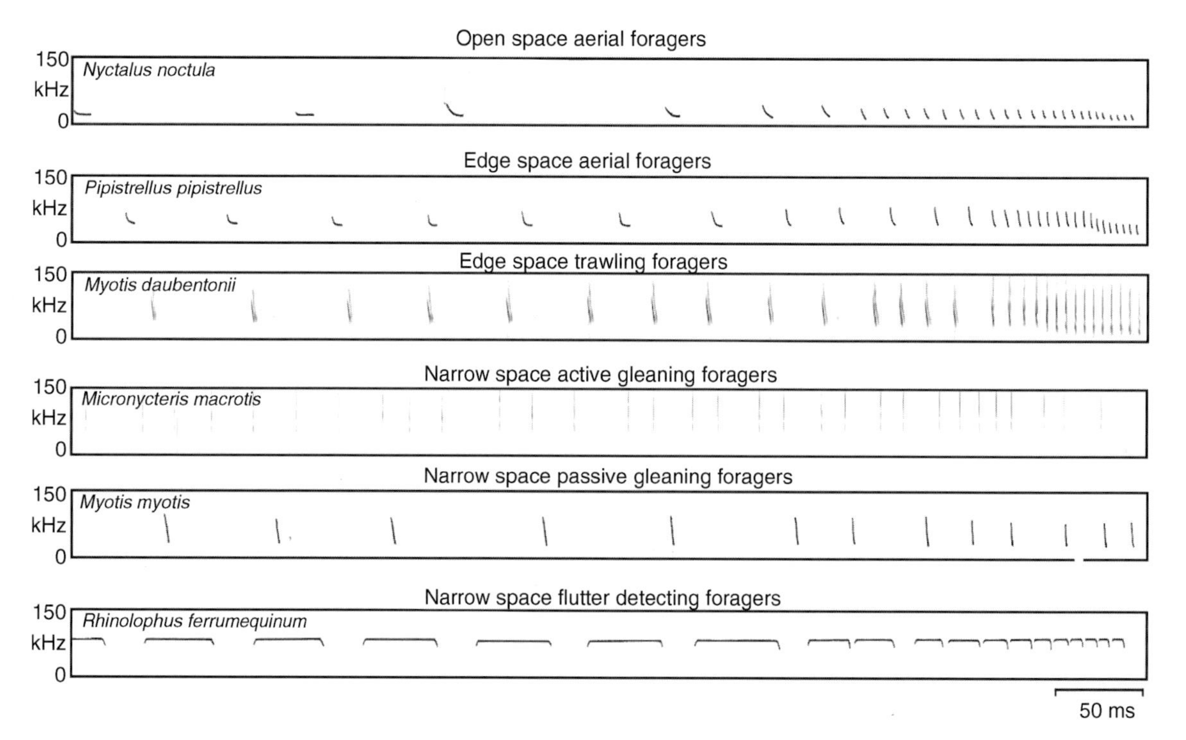

Neuroethology of Biosonar Systems in Bats. Figure 2 Search and approach signals of different functional groups.

prey-generated acoustic or olfactory cues to find their food. Bats in both functional groups emit broadband uni- or multi-harmonic signals of short duration and low sound pressure level. Active gleaners use these signals both for spatial orientation and for food acquisition, whereas passive gleaners use them for spatial orientation only. "Narrow space flutter detecting foragers" have a very specialized echolocation system with signals consisting of a long component of constant frequency followed by a terminal FM component. With Doppler shift compensation and a specialized hearing system, these bats are able to recognize echoes from fluttering prey insects modulated by the rhythm of beating wings in unmodulated background echoes (Schnitzler and Ostwald 1983; Neuweiler 1990; Moss and Schnitzler 1995). Bats have highly variable foraging and echolocation behavior and often forage in more than one habitat (Fenton 1990). According to their behavioral goals and to incoming information, bats choose the most suitable signal types from their repertoire to perform specific echolocation tasks.

A multitude of psychophysical experiments have been conducted to investigate auditory information processing in bats and to clarify which receiver type is implemented in the bat's auditory system. These studies include behavioral audiograms, performance in target echo detection, range estimation and resolution, horizontal and vertical localization, movement discrimination (Moss and Schnitzler 1995; Masters and Harley 2004) and, recently, the bats' performance in stochastic echo parameter evaluation tasks (Grunwald et al. 2004).

Bats hear over a wide range of mainly ultrasonic frequencies, often spanning several octaves. The frequency range of bats' echolocation signals corresponds closely to the range in which they have high auditory sensitivity (Fig. 3). Sensitivity to frequencies below the range of the echolocation signal is crucial for social communication and the detection of prey and predators through passive listening.

For detection, a bat has to decide whether it perceives an echo of its own signal or not. Detection thresholds depend on the bat's signal structure, environmental conditions, and the bat's sonar receiver. Thresholds between 0 and 60 dB SPL have been obtained in various experimental procedures. This variability may have been caused by different masking situations.

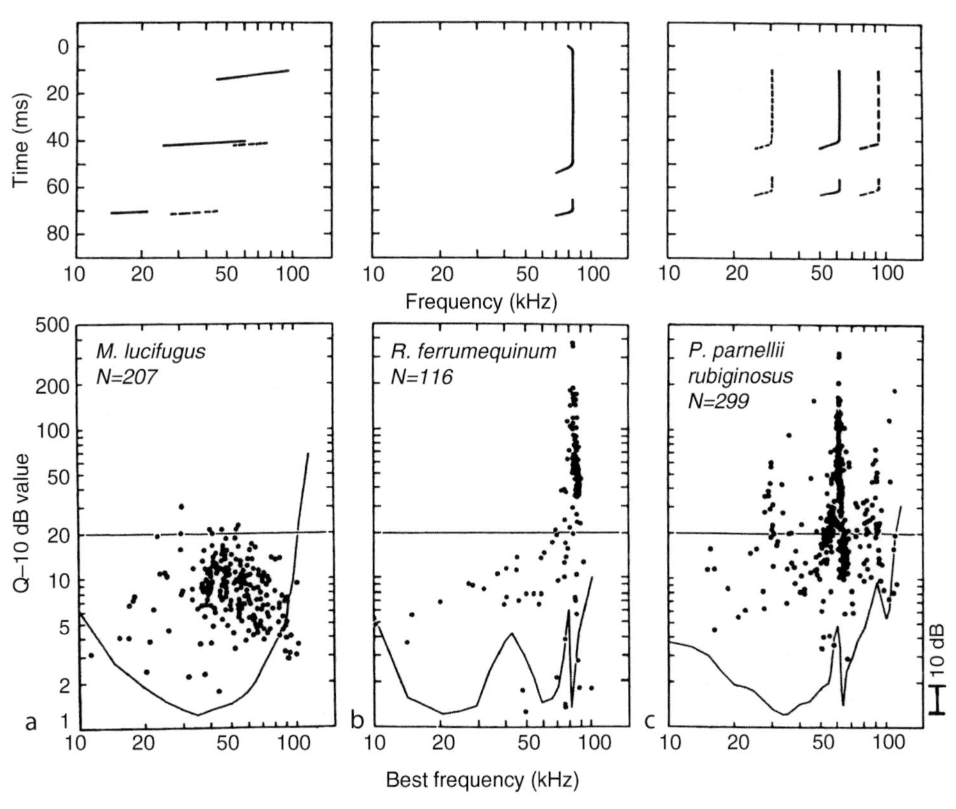

Neuroethology of Biosonar Systems in Bats. Figure 3 Comparison of sonagrams (*upper row*) with audiograms (*solid line*), and Q_{10dB} values of single auditory fibers (*dots*) (*lower row*) of a bat using broadband steep FM signals (*Myotis lucifugus*, A) and of two flutter-detecting foragers (*Rhinolophus ferrumequinum*, B and *Pteronotus parnellii*, C) (adapted from Grinnell 2004). The sonagrams are tilted so that the frequency is displayed in the x-axis as in the audiograms below.

For localization, a bat determines the range and the angle of a target of interest. The range is encoded in the time delay between the emitted signal and the returning echo. The ranging performance of bats has been tested using three different tasks: range difference discrimination, range jitter discrimination, and range resolution. Range difference experiments revealed an accuracy of about 1 cm, which is close to the accuracy estimated from successful prey interception experiments. In range jitter experiments, in which one target jitters while the other is stable, discrimination thresholds down to 10 ns (corresponding to 1.7 μm) have been reported (Simmons et al. 1990). These results have often been questioned because it is hard to imagine how such minimal differences in echo arrival time could be processed in the nervous system. Studies on range resolution tested bats' perception of target depth structure. In range resolution experiments, bats using FM signals were able to discriminate one-wavefront echoes from interfering two-wavefront echoes when the delay offset within the two-wavefront echo was about 12 μs. Range perception is most likely not a single process, but rather a set of perceptual processes in the bat's receiver; the described experimental approaches may be tapping into different processes in the set.

Binaural and monaural echo cues encode the horizontal and vertical angles of targets. Accuracy in determining these angles has only been measured in the FM bat Eptesicus fuscus. Threshold estimates were 1.5° in the horizontal plane and about 3° in the vertical plane.

To classify targets, bats use target-specific spectral and modulation patterns in the echoes or stochastic parameter distributions. Flutter information is used by bats to identify insect echoes and to discriminate them from the stationary background. CF-bats also use flutter information for the classification of insect prey. In behavioral experiments, horseshoe bats were able to sense 8–9% differences in wing beat rate. They also discriminated between different insect species even if they had the same wing beat rate and were presented at novel aspect angles. It has only recently been proven that bats are able to classify echoes according their roughness (Grunwald et al. 2004).

Structural and Functional Aspects of Echolocation

A multitude of publications deal with the question of how echo information is processed in the auditory systems of bats (recently summarized in the books "Hearing by Bats," Popper and Fay (eds.) (1995) and "Echolocation in Bats and Dolphins," Thomas, Moss and Vater (Eds.) (2004)). Most studies were conducted with either vespertilionid bats that use broadband, steep FM signals, or rhinolophids, and the mormoopid bat *Pteronotus parnellii* that uses long, CF-FM signals. The auditory systems of the two groups differ in many ways, reflecting structural and functional differences that are critical to echolocation. Nevertheless, the auditory systems of bats consist of the same basic elements as those of other mammals.

In all bats, the cochlea is specialized for the analysis of high frequencies. Bats that rely on broadband, steep FM signals possess a non-specialized mammal-like frequency representation on the basilar membrane. In flutter-detecting bats, a so-called acoustic fovea is used for the analysis of a narrow band of frequencies around the second harmonic of the CF component of the Doppler-compensated echoes (Grinnell 1995; Kössl and Vater 1995; Vater and Kössl 2004; Vater 2004). Both this difference in frequency representation in the cochlea and in increased sharpness of frequency tuning of neurons connected to the auditory fovea are found throughout the entire ascending auditory pathway (Fig. 3).

In no other mammal is the auditory system proportionally larger and more differentiated than in echolocating bats. Studies on the processing of echo information in the central auditory system of bats focus on the question of how and where the information bearing spectro-temporal attributes of pulse-echo trains are decoded (Covey and Casseday 1995, 1999; Pollak and Park 1995; Wenstrup 1995; O'Neill 1995; Fuzessery et al. 2004).

Between the cochlea and the midbrain, parallel pathways provide multiple transformations of the cochlear signal through the interplay of excitatory or inhibitory outputs, which differ in their temporal discharge patterns and latencies. This results in auditory midbrain neurons that are tuned to parameters relevant for echolocation, such as signal duration, delay between two signals, FM sweep direction, and the rate of periodic frequency and amplitude modulation (Covey and Casseday 1995, 1999). Binaural processing of interaural intensity (and perhaps time differences) in the superior olivary complex provides angular information. However, there is some controversy surrounding the question of whether or not the large and specialized medial superior olive is equivalent to this structure in non-echolocating mammals (Covey and Casseday 1995).

The processing of echolocation information in the colliculo-thalamo-cortical pathway leads to a further decoding of the features that describe echolocation scenes (Suga 1990; Pollak and Park 1995; Wenstrup 1995). Feature extraction is species-specific and reflects the adaptations of the corresponding echolocation systems to species-specific echolocation tasks. Such differences become evident when one compares the analysis and representation of relevant echolocation features in flutter-detecting bats such as *Rhinolophus ferrumequinum* and *Pteronotus parnellii* and bats that use broadband, steep FM signals such as *Myotis lucifugus* (O'Neill 1995; Suga 2004; O'Neill 2004;

Wong 2004). For instance, the functional organization of the auditory cortex of the flutter-detecting forager *Pteronotus parnellii* is characterized by three distinct areas representing different task-relevant features. One region contains neurons that respond solely to certain signal frequencies and amplitudes. There, the area around the second harmonic of the echolocation signals corresponds to the auditory fovea and is greatly enlarged, as it is in all lower nuclei and in the cochlea. A second region contains combinatorial neurons that respond solely to frequency differences between signals. The combinatorial neurons of a third region represent echo delays measured as the time interval between two signals. Within these regions, information-bearing parameters such as echo delay and relative velocity are arranged in maps (Suga 1990, 2004; Pollak et al. 1995; O'Neill 1995, 2004) (Fig. 4).

In rhinolophids, which are also specialized for flutter detection, the functional organization of the cortex is in part similar to that of *Pteronotus parnellii*, but there are also notable differences in the arrangement of feature-encoding cells (Schuller et al. 1991). The cortices of other bats also contain combinatorial neurons that express echolocation features by combining information from two signals that simulate signal echo pairs. The functional organization of the cortex in bats that use broadband, steep FM signals, such as *Myotis lucifugus* and *Eptesicus fuscus*, is very different from that of flutter-detecting foragers. In *Myotis lucifugus*, the cortex consists of two tonotopic regions. One region covers the frequency range of the echolocation signals. It can dynamically change to provide multidimensional feature extraction, which meets the behavioral needs of echolocation in this species (Wong 2004).

Recent studies indicate that cortifugal systems are essential in shaping the response properties of subcortical neurons, thus adjusting and improving signal processing according to auditory experience (Jen et al. 2004; Suga et al. 2004).

Successful echolocation depends on the coordination between auditory and motor systems. Therefore, the control of signal production is tightly coupled to auditory processing of pulse-echo trains (Moss and Shina 2003; Schuller and Moss 2004). Several nuclei at different levels of the brain have anatomical connections and functional responses that indicate their importance for audio-vocal control. The emission of sonar signals is also linked to breathing and other motor activities such as wing beat, pinna movements, and middle ear contraction.

The Analytical Aspect of Echolocation

A model of echolocation based on spectrogram correlation and transformation (the SCAT receiver model) describes the auditory computations necessary to estimate target range (Saillant et al. 1993; Simmons et al. 1996). This model is used to explain the extraordinary range accuracy found by Simmons in his range jitter experiments, results that other authors have cast in doubt (Schnitzler et al. 1985; Pollak 1988, 1993). Another model delivers a biologically-plausible framework for auditory perception in FM bats, by using functional units inspired by what is presently known about the neurobiological elements of the bat's auditory system (Palakal and Wong 2004). In other approaches, properties of bat echolocation systems have been implemented in biomimetic sonar systems. Such systems were able to recognize objects directly from

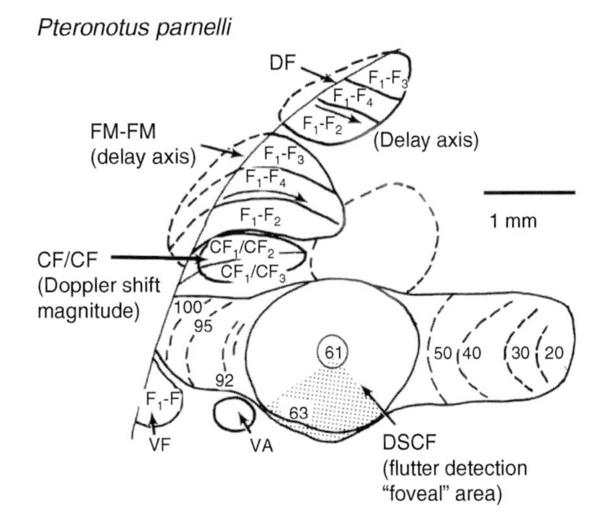

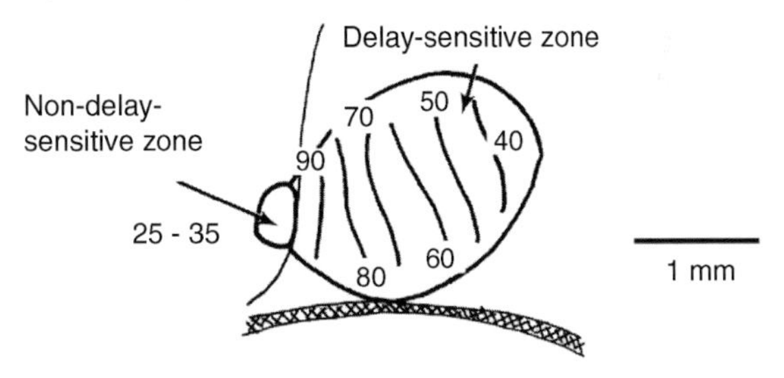

Neuroethology of Biosonar Systems in Bats. Figure 4 Functional organization of the auditory cortex in *Pteronotus parnellii* and *Myotis lucifugus*. In Pteronotus, a tonotopically-organized field with an overrepresentation of foveal neurons (DSCF area) is surrounded by areas containing maps that represent either relative velocity (CF/CF area), as indicated by Doppler shift magnitude, or echo delay (FM-FM and DF area). Neurons in the foveal area are sensitive to flutter information. Myotis has two tonotopically-organized fields. The larger delay-sensitive zone covers the frequency range of sonar pulses and contains delay-sensitive neurons. The tonotopic gradient of the smaller non-delay sensitive zone is reversed and only covers frequencies from 25–35 kHz (adapted from O'Neill 1995).

their echo waveform (Kuc 2004) or from stochastic echo properties (Müller and Kuc 2000), or they were able to determine echo direction using pinna morphology and motion (Walker et al. 2004). A computational theory for the classification of natural biosonar targets was developed on the basis of echoes from real targets recorded with a biomimetic sonar system (Müller 2003).

Outlook

There is still a long way to go before we have attained a complete understanding of echolocation in bats and its neural underpinnings. Most behavioral studies have focused on the detection and localization of single targets, and neurobiological studies have simulated auditory scenes by simply mimicking pulse echo pairs using two succeeding signals. In nature, however, bats are confronted with echolocation scenes that generate complex echo trains from which the relevant information has to be extracted. Further research should therefore focus on the echolocation behavior in natural situations, with a strong emphasis on the adaptive value of signal design and on the adaptive strategies that bats apply to perform complex echolocation tasks. For a better understanding of the neuronal mechanisms underlying echolocation, it will be necessary to study echo processing in vocalizing animals, thereby producing pulse echo trains that are comparable to those occurring in natural scenes. Studies that evaluate the bat's ability to use stochastic echo properties are a promising new approach that may shed new light onto the complex information processing involved in echolocation.

References

1. Aldridge HDJN, Rautenbach IL (1987) Morphology, echolocation and resource partitioning in insectivorous bats. J Anim Ecol 56:763–778
2. Covey E, Casseday JH (1995) The lower brainstem auditory pathways. In: Popper AN, Fay RR (eds) Hearing by Bats. Springer-Verlag, New York, pp 235–295
3. Covey E, Casseday JH (1999) Timing in the auditory system of the bat. Annu Rev Physiol 61:457–476
4. Denzinger A, Schnitzler H-U (2004) Perceptual tasks in echolocating bats. In: Ilg UJ, Bülthoff HH, Mallot HA (eds) Dynamic Perception. Akademische Verlagsgesellschaft Aka GmbH, Berlin, pp 33–38
5. Denzinger A, Kalko EKV, Jones G (2004) Ecological and evolutionary aspects of echolocation in bats. In: Thomas JA, Moss CF, Vater M (eds) Echolocation in Bats and Dolphins. The University of Chicago Press, Chicago, pp 311–326
6. Fenton MB (1990) The foraging behaviour and ecology of animal-eating bats. Can J Zool 68:411–422
7. Fenton MB (1995) Natural history and biosonar signals. In: Popper AN, Fay RR (eds) Hearing by Bats. Springer-Verlag, New York, pp 37–86
8. Fuzessery ZM, Feng AS, Supin A (2004) Central auditory processing of temporal information in bats and dolphins. In: Thomas JA, Moss CF, Vater M (eds) Echolocation in Bats and Dolphins. The University of Chicago Press, Chicago, pp 115–122
9. Grinnell AD (1995) Hearing in bats: An overview. In: Popper AN, Fay RR (eds) Hearing by Bats. Springer-Verlag, New York, pp 1–36
10. Grunwald JE, Schornich S, Wiegrebe L (2004) Classification of natural textures in echolocation. Proc Natl Acad Sci USA 101:5670–5674
11. Jen PHS, Sun X, Chen QC, Zhang J, Xiaoming Z (2004) Cortifugal modulation of midbrain auditory sensitivity in the bat. In: Thomas JA, Moss CF, Vater M (eds) Echolocation in Bats and Dolphins. The University of Chicago Press, Chicago, pp 196–200
12. Kössl M, Vater M (1995) Cochlear structure and function in bats. In: Popper AN, Fay RR (eds) Hearing by Bats. Springer-Verlag, New York, pp 191–234
13. Kuc R (2004) Biomimetic sonar objects from echos. In: Thomas JA, Moss CF, Vater M (eds) Echolocation in Bats and Dolphins. The University of Chicago Press, Chicago, pp 501–506
14. Masters WM, Harley HE (2004) Introduction: Performance and cognition in echolocating mammals. In: Thomas JA, Moss CF, Vater M (eds) Echolocation in Bats and Dolphins. The University of Chicago Press, Chicago, pp 249–259
15. Moss CF, Schnitzler H-U (1995) Behavioral studies of auditory information processing. In: Popper AN, Fay RR (eds) Hearing by Bats. Springer-Verlag, New York, pp 87–145
16. Moss CF, Sinha SR (2003) Neurobiology of echolocation in bats. Curr Opin Neurobiol 13:751–758
17. Müller R, Schnitzler H-U (1999) Acoustic flow perception in cf-bats: Properties of the available cues. J Acoust Soc Am 105:2958–2966
18. Müller R, Schnitzler H-U (2000) Acoustic flow perception in cf-bats: Extraction of parameters. J Acoust Soc Am 108:1298–1307
19. Müller R, Kuc R (2000) Foliage echoes: A probe into the ecological acoustics of bat echolocation. J Acoust Soc Am 108:836–845
20. Müller R (2003) A computational theory for the classification of natural biosonar targets based on a spike code. Network, 14:595–612
21. Neuweiler G (1990) Auditory adaptations for prey capture in echolocating bats. Physiol Rev 70:615–641
22. O'Neill WE (1995) The bat auditory cortex. In: Popper AN, Fay RR (eds) Hearing by Bats. Springer-Verlag, New York, pp 416–480
23. O'Neill WE (2004) Feature extraction in the mustached bats auditory cortex. In: Thomas JA, Moss CF, Vater M (eds) Echolocation in Bats and Dolphins. The University of Chicago Press, Chicago, pp 176–184
24. Palakal MJ, Wong D (2004) A biologically plausible framework for auditory perception in FM bats. In: Thomas JA, Moss CF, Vater M (eds) Echolocation in Bats and Dolphins. The University of Chicago Press, Chicago, pp 459–467
25. Pollak GD (1988) Time is traded for intensity in the bat's auditory system. Hear Res 36:107–124
26. Pollak GD (1993) Some comments on the proposed perception of phase and nanosecond time disparities by echolocating bats. J Comp Physiol A 172:523–531

27. Pollak GD, Winer JA, O'Neill WE (1995) Perspectives on the functional organization of the mammalian auditory system: Why bats are good models. In: Popper AN, Fay RR (eds) Hearing by Bats. Springer-Verlag, New York, pp 481–498
28. Pollak GD, Park TJ (1995) The inferior colliculus. In: Popper AN, Fay RR (eds) Hearing by Bats. Springer-Verlag, New York, pp 296–367
29. Saillant PA, Simmons JA, Dear SP, McMullen TA (1993) A computational model of echo processing and acoustic imaging in frequency-modulated echolocating bats: The spectrogram correlation and transformation receiver. J Acoust Soc Am 94(5):2691–2712
30. Schnitzler H-U, Menne D, Hackbarth H (1985) Range Determination by Measuring Time Delays in Echolocating Bats. In: Michelsen A (eds) Time Resolution in Auditory Systems. Springer-Verlag, New York, pp 180–204
31. Schnitzler H-U, Ostwald J (1983) Adaptation for the detection of fluttering insects by echolocation in horseshoe bats. In: Ewert JP, Capranica RR, Ingle DJ (eds) Advances in Vertebrate Neuroethology. Plenum Press, New York, pp 801–827
32. Schnitzler H-U, Kalko EKV (2001) Echolocation by insect-eating bats. BioScience 51:557–569
33. Schnitzler H-U, Kalko EKV, Denzinger A (2004) Evolution of echolocation and foraging behavior in bats. In: Thomas JA, Moss CF, Vater M (eds) Echolocation in Bats and Dolphins. The University of Chicago Press, Chicago, pp 331–338
34. Schnitzler H-U, Moss CF, Denzinger A (2003) From spatial orientation to food acquisition in echolocating bats. Trends Ecol Evol 18:386–394
35. Schuller G, Moss CF (2004) Vocal control and acoustically guided behavior in bats. In: Thomas JA, Moss CF, Vater M (eds) Echolocation in Bats and Dolphins. The University of Chicago Press, Chicago, pp 3–16
36. Schuller G, O'Neill WE, Radke-Schuller S (1991) Facilitation and delay sensitivity of auditory cortex neurons in CF-FM bats, *Rhinolophus rouxi.* and *Pteronotus p parnellii.* Eur J Neurosci 3:1165–1181
37. Siemers BM, Schnitzler H-U (2004) Echolocation signals reflect niche differentiation in five sympatric congeneric bat species. Nature 429:657–661
38. Simmons JA, Ferragamo M, Moss CF, Stevenson SB, Altes RA (1990) Discrimination of jittered sonar echoes by the echolocating bat, *Eptesicus fuscus*: The shape of target images in echolocation. J Comp Physiol A 167:589–616
39. Simmons JA, Saillant PA, Ferragamo MJ, Haresign T, Dear SP, Fritz J, McMullen TA (1996) Auditory computations for biosonar target imaging in bats. In: Hawkins HL, McMullen TA, Popper AN, Fay RR (eds) Auditory Computation. Springer-Verlag, New York, pp 401–468
40. Suga N (1990) Biosonar and neural computation in bats. Sci Am 6:60–68
41. Suga N (2004) Feature extraction and neural activity: advances and perspectives. In: Thomas JA, Moss CF, Vater M (eds) Echolocation in Bats and Dolphins. The University of Chicago Press, Chicago, pp 173–175
42. Suga N, Zhang Y, Olsen JF, Yan J (2004) Modulation of frequency tuning of thalamic and midbrain neurons and cochlear hair cells by descending auditory system in the mustached bat. In: Thomas JA, Moss CF, Vater M (eds) Echolocation in Bats and Dolphins. The University of Chicago Press, Chicago, pp 214–221
43. Vater M, Kössl M (2004) The ears of whales and bats. In: Thomas JA, Moss CF, Vater M (eds) Echolocation in Bats and Dolphins. The University of Chicago Press, Chicago, pp 89–98
44. Vater M (2004) Cochlear anatomy related to bat echolocation. In: Thomas JA, Moss CF, Vater M (eds) Echolocation in Bats and Dolphins. The University of Chicago Press, Chicago, pp 99–103
45. Walker A, Peremans H, Hallam J (2004) An investigation of active reception mechanisms for echolocators. In: Thomas JA, Moss CF, Vater M (eds) Echolocation in Bats and Dolphins. The University of Chicago Press, Chicago, pp 507–514
46. Wenstrup JJ (1995) The auditory thalamus in bats. In: Popper AN, Fay RR (eds) Hearing by Bats. Springer-Verlag, New York, pp 368–415
47. Wong D (2004) The auditory cortex of the little brown bat, *Myotis lucifugus*. In: Thomas JA, Moss CF, Vater M (eds) Echolocation in Bats and Dolphins, The University of Chicago Press. Chicago, pp 185–189

Neuroethology of Sound Localization in Barn Owls

Terry Takahashi
Institute of Neuroscience, University of Oregon, Eugene, OR, USA

N

Synonyms

Spatial hearing

Definition

Sound localization is the ability to determine the spatial relationships of sound sources in the environment.

Characteristics

Higher Level Structures

The barn owl can hunt in pitch darkness guided by auditory neurons with discrete ►spatial receptive fields (SRFs) [1,2]. These neurons form a topographic map of frontal auditory space in the external nucleus of the inferior colliculus (ICx). The SRFs of these space-specific neurons are based on neuronal sensitivity to interaural differences in the timing and level of sounds (►ITD (interaural time difference) and ►ILD (interaural level difference)), which are also the major cues (►spectral-shape cues) for sound-localization in humans and other mammals [3–5]. Lesions of the space map lead to scotoma-like defects in sound localization, and microstimulation of the optic tectum (OT), which receives a direct, topographic projection from the ICx, and evokes a rapid head turn to that area of space represented at the point of stimulation [6,7]. Finally, the smallest angular separation of sources that the owl can resolve has recently been traced to the granularity of the focal activity evoked on the map [8].

Neurons that are sensitive to the location of sources or the binaural cues associated with source location are found not only in the ICx, but also in Field L, the analog of the mammalian primary auditory cortex, and in the archistriatum, the analog of the mammalian basal ganglia [9–11], although maps of space have not been found [12]. Lesions of the OT lead to inaccurate and long-latency head turns, but do not obliterate these movements, suggesting that forebrain regions may serve as parallel pathways for sound localization [13].

Lower Level Components

Acoustical Cues Barn owls rely primarily on two cues, ITD and ILD, to localize sounds. For the owl, ITD varies with the sound source's azimuth due to the ears' separation along the horizontal axis. For a given azimuth, ITD remains largely constant across the entire range of frequencies (►best frequency) [14] that owls use for sound localization (3–9 kHz) [15]. ILD also varies with the source's azimuth at low frequencies (2–4 kHz), but as frequency increases, the axis along which ILD changes becomes increasingly vertical, allowing for the representation of elevation [14]. This, in turn, is due to the asymmetry in the morphology of the two ears, which causes the right and left ears to be more sensitive to sounds coming from above and below eye-level, respectively.

The manner in which the ears and head alter the magnitude and phase of sounds in a location-specific manner is called the ►head-related transfer function (HRTF). By filtering sounds with the HRTFs for a location in space, we can re-create, over headphones, the sound wave that would have arrived at the eardrums from a sound source at that location. The stimulus is said to have been presented in virtual auditory space (VAS), and the application of VAS techniques allows not only the rapid assessment of neural spatial tuning, but also the analysis of the contribution of the two binaural cues to the neural responses.

Computation of ITD The computation of ITD begins with the encoding of the phase angles of each spectral component by ►phase locking neurons of the nucleus magnocellularis (NM), one of the cochlear nuclei [16]. In the barn owl, phase-locking extends to neurons with best frequencies as high as 9 kHz where strong ILDs are generated. It is for this reason that ITD and ILD can operate over the same frequency range, allowing the owl to localize sounds in two dimensions. NM projects bilaterally to the nucleus laminaris (NL), the avian analog of the mammalian medial superior olive (MSO). In the NL, the ITD of each spectral component is computed by a ►binaural cross-correlation operating over short segments of time [17–22], thus resulting in neurons selective for the ITD.

The NL projects directly to the core of the contralateral ICc that, in turn, projects to the lateral shell of the opposite ICc [23]. As a result of this doubly-crossed pathway, the lateral shell gains a representation of contralateral space. Neurons of the core and lateral shell of the ICc are selective for ITD and frequency and are organized into tonotopic columns. Cells in a column of the ICc-lateral shell project convergently onto a cluster of space-specific neurons, thus endowing them with selectivity for the ITD preserved by the column [24].

Computation of ILD We understand less about the processing of ILD. The sound level in the ipsilateral ear is encoded by cells in the nucleus angularis (NA) [16], which project contralaterally to the nucleus ventralis lemnisci lateral pars posterior (VLVp) [25]. The VLVp of the two sides are interconnected by an inhibitory commissure. The neurons of the VLVp are excited by stimulation of the contralateral ear, via the direct input from NA, and are inhibited by stimulation of the ipsilateral ear, via the commissural input [26,27]. The VLVp projects bilaterally to the lateral-shell of the ICc [28], where ILD and ITD cues are merged. A clear topographical representation of ILD, however, has never been found [29].

The application of VAS techniques has recently shown that if space specific neurons were sensitive to ILD alone, their SRFs would be horizontal swaths of space at the elevation of the cell's normal spatial ►spatial receptive field (SRF). If neurons were sensitive to ITD alone, their SRFs would be a vertical swath at the azimuth of the cell's normal SRF [30,31]. The normal RF thus lies at the intersection of the ITD and ILD-alone RFs where the cell's optimal ITD and ILD-spectra are present and are combined by a multiplication-like process [32].

Higher Level Processes

Multiple sound sources. When there are multiple sources, as is typical in nature, the sound waves from each source will add in the ears, and if the sounds have broad, overlapping spectra, the binaural cues will fluctuate over time in a complex manner [33,34], making it difficult for the space map to image the sources accurately. One key to the space-specific neurons' ability to resolve two simultaneous sources is the difference in spectra [34]. When two sources of identical broadband noises were passed through a neuron's SRF, the neuron discharged maximally when the two speakers flanked the SRF, generating a response function with a single peak. This is to be expected, because the frequency-specific superposition generates binaural cues that are the vector average of those of the two individual sources. When the two sources emitted comb-filtered noises with the energy from each source in alternate frequency bands, the neurons responded when each speaker was in their SRFs, generating bimodal response functions. This too is expected, because superposition of the waves happens within frequency-specific channels and the two sources' energy is contained in alternate bands, resulting in less spectral overlap. Interestingly, neurons were also able to resolve two speakers that emitted ►uncorrelated broadband noises as well as identical broadband noises that were temporally reversed versions of one another. In both cases, the moment-by-moment spectrum differed enough for the neuron to resolve. Thus, the neurons are capable of resolving sounds from two sources as long as their ►short-term spectra differ.

Echoes Early psychophysical studies demonstrated that directional information conveyed by the sound coming directly from the active source (leading source) dominates perception, and that listeners are relatively insensitive to directional information conveyed by reflections (lagging sources). The perceptual dominance of the leading sound and a host of related perceptions are collectively referred to as the "precedence effect." (For reviews, see [33,35,36]).

When two identical sounds are presented within 100 μs of one another from two separate sources, owls and their space-specific neurons respond as though there was a single target located between them [37,38]. This phenomenon, termed "summing localization" can be explained largely by binaural cross-correlation [37]. If the delay between leading and lagging sources is increased to the 1–5 ms range in humans, the single fused phantom target is localized close to the leading source. This is termed "localization dominance" [36,39,40]. At the same time, spatial information regarding the lagging source is degraded, a phenomenon, termed "lag discrimination suppression" [41–46].

Localization dominance has also been reported in studies with non-human subjects, including owls [38,47,48]. At delays between 1 and 10 ms, the owls turned their heads to the side of the leading source. At delays above 10 ms, owls aimed their heads at the lagging source or made double-head turns, first localizing one speaker then turning to the other [38]. The latter observation indicates that the lagging source had become localizable at the longer delays. To our knowledge, the only study of lag discrimination suppression in non-human species is the recent study in the owl [8], which measured the minimal audible angle (MAA), the smallest perceptible change in sound-source location, in the presence of echoes. For a sequence of two noise bursts from locations separated horizontally by 30° and in time by 3 ms, the MAA for lag sources was considerably larger than for lead sources, suggesting that the owl experiences lag discrimination suppression. The MAA for the lead sources was also found to be larger than that for single sources. This ordering of effects, $MAA_{single} < MAA_{lead} < MAA_{lag}$, replicates findings with humans for similar stimuli [49].

There are a growing number of neurophysiological studies of the precedence effect, which generally show that a cell's response to the lag source is weaker than its response to the lead source or to a single source [37,38,50–56]. A recent study in owls has shown that neuronal MAAs, estimated using signal detection theory, have the same ordering found behaviorally; specifically, $MAA_{single} < MAA_{lead} < MAA_{lag}$ [57].

Motion The best evidence for the owl's sensitivity for the direction of object motion is an early observation that whenever an owl struck a mouse in darkness, it arranged its talons so that it grasped the long axis of the mouse's body [1]. When the mouse's trajectory was co-linear with the owl's flight-path, the owl put one set of talons in front of the other to strike, but when the mouse's trajectory was perpendicular to the owl's, the owl arranged its talons side-by-side. Since visual cues were not available in this experiment, the mouse's path was likely determined by acoustical cues.

A number of neurophysiological studies have documented neuronal sensitivity to motion direction using either an array of sequentially-activated speakers simulating saltatory motion [58–62] or continuously varying binaural cues [63,64]. A recent survey demonstrated that neurons in the left and right colliculi prefer, respectively, motion in the clockwise and counter-clockwise directions [62]. The authors suggest that the cells may be involved in orienting movements to the contralateral auditory hemisphere.

Neuronal directional sensitivity has been modeled as a circuit in which the motion-direction sensitive neuron receives an excitatory input from a spatially-selective neuron and a delayed inhibitory input from a second spatially-selective neuron with an SRF some distance away. If a sound source travels from the excitatory neuron to the inhibitory neuron, the excitatory input to the motion sensing neuron arrives first, causing a discharge. If the source travels in the opposite direction, the delay causes the inhibitory input to coincide with the excitatory input (which is stimulated later), nullifying it. This mechanism depends on the delay imposed on the inhibitory input and the distance separating the two spatially tuned inputs, thus allowing for some selectivity for speed. Kautz and Wagner [61] demonstrated that biccuculine reduces sensitivity to motion direction, suggesting GABA-mediated inhibition.

Takahashi and Keller [63] reported auditory motion-unmasking. Using ►binaural beats, which simulates a moving pure-tone source, they showed that space-specific neurons detected the moving tone in noise more easily than static ones. This finding could not be replicated, however, with saltatory-motion [60]. Two psychophysical studies in humans have yielded conflicting results [65,66].

Attention and Working Memory Studies of audiospatial cognition are rare. Knudsen and Knudsen [67] showed that an owl's ability to remember the locations to which it recently oriented is affected by lesions of the archistriatum, suggesting its role in spatial working memory [67].

In a study of attention, the reaction time of owls trained to turn their heads toward a sound source when visual stimuli provided a truthful or a misleading cue to the side from which the sound would appear [68]. As in humans [69], the owl's reaction times were significantly faster when the cues were valid than when they were invalid, suggesting that attention is capable of modulating the owl's orienting behavior. Interestingly, however, even the validly cued trials had reaction times that were considerably longer than when an owl was trained simply to turn its head toward a sound source, without a cue. This may indicate that the cued reaction-time task is difficult for owls.

Process Regulation

Once the binaural cues are computed, these cues must be translated into space. In a baby bird, which has a small head, the maximal ITD, generated by a source to the extreme right or left, may be some 90 μs. A neuron tuned to an ITD of, say, 30 μs represents a location about 30% of the distance to the extreme periphery. The adult, by contrast, has a maximal ITD of about 200 μs [14], so the same neuron represents a more central location.

This process of translation occurs during a critical period, of up to ca. 200 days, during which visual input is crucial [70]. Birds that are blind may form inverted and otherwise distorted maps of auditory space [71,72]. Moreover, if a baby bird matures with a pair of prisms that laterally shifts the visual scene, the now mature bird mis-localizes a sound by the amount that the prisms shifted the visual world [73]. Correspondingly, a space-specific neuron will shift its ►auditory SRF by an amount equal to the prism shift. The plasticity is thought to occur in the ICx, and involves the inputs from the OT, which contributes a visual input, and the ICc, which contributes the spatially selective auditory input [74]. The mechanism of audiovisual calibration in the owl is an example of supervised learning and has been thoroughly investigated and reviewed [75].

Function

Audiospatial Discrimination and Localization Audiospatial discrimination is the ability to detect changes in the position of a sound source. This is to be distinguished from "localization," the ability to point to a sound source in space. In the owl, whose eyes and ears are nearly immobile, sound localization is measured by its head turns. Owls can point their heads to within 2.5° to 5° of a target in azimuth and 3.5° to 8° in elevation [3,76,77]. For both dimensions, accuracy and precision decline slightly with target eccentricity.

Spatial discrimination was recently investigated in the barn owl using a newly developed method based on pupillary dilation response (PDR) [78,79]. In the barn owl, the pupil dilates upon presentation of a sound, and habituates when the same sound is repeated. When the stimulus is changed, for example, by presenting the identical waveform from a different location, the PDR recovers. The degree of recovery is proportional to the magnitude of the difference between the habituating and testing loci, making the PDR similar to a psychophysical ►rating task. The magnitude of the PDR evoked by test stimuli can be expressed in terms of the variance of the PDR evoked by habituating stimuli to derive a ►z-score-like statistic called the "standard separation," D [80]. The MAA, defined as the spatial separation between habituating and test loci at which $D = 0.8$, was found to be about 3° in azimuth and 7° in elevation [81]. This value is considerably finer than that of animals with heads of comparable sizes.

How does owl's MAA compare with the acuity of its space-specific neurons? Such comparisons are often confounded by the lack of a common metric. Signal detection theory, which is based on the premise that the variance in neuronal responses limits behavioral discrimination, provides such a metric [82]. By scaling a neuron's responses to its variance, we can estimate how well an animal might be able to discriminate between two stimuli, if its higher centers were relying strictly upon that neuron's firing. Recent application of this approach by Bala and colleagues [79] to the owl's space-specific neurons predicted its horizontal MAA. It is particularly interesting to note that the SRFs of the space specific neurons are about 2–3 times taller than they are wide. This corresponds roughly to the ratio of the vertical and horizontal MAAs [81].

References

1. Payne RS (1971) Acoustic location of prey by barn owls (Tyto alba). J Exp Biol 54(3):535–73
2. Knudsen EI, Konishi M (1978) Space and frequency are represented separately in auditory midbrain of the owl. J Neurophysiol 41(4):870–84
3. Knudsen EI, Konishi M (1979) Sound localization by the barn owl (Tyto alba) measured with the search coil technique. J. Comp. Physiol. A 133:1–11
4. Moiseff A, Konishi M (1981) Neuronal and behavioral sensitivity to binaural time differences in the owl. J Neurosci 1(1):40–8
5. Pena JL, Konishi M (2001) Auditory spatial receptive fields created by multiplication. Science 292:249–252
6. Wagner H (1993) Sound-localization deficits induced by lesions in the barn owl's auditory space map. J Neurosci 13(1):371–86
7. duLac S, Knudsen EI (1990) Neural maps of head movement vector and speed in the optic tectum of the barn owl. J Neurophysiol 63(1):131–46

8. Spitzer MW, Bala AD, Takahashi TT (2003) Auditory spatial discrimination by barn owls in simulated echoic conditions. J Acoust Soc Am 113(3):1631–45
9. Cohen YE, Knudsen EI (1994) Auditory tuning for spatial cues in the barn owl basal ganglia. J Neurophysiol 72(1):285–98
10. Cohen YE, Knudsen EI (1995) Binaural tuning of auditory units in the forebrain archistriatal gaze fields of the barn owl: local organization but no space map. J Neurosci 15(7 Pt 2): 5152–68
11. Cohen YE, Knudsen EI (1998) Representation of binaural spatial cues in field L of the barn owl forebrain. J Neurophysiol 79(2):879–90
12. Cohen YE, Knudsen EI (1999) Maps versus clusters: different representations of auditory space in the midbrain and forebrain. Trends Neurosci 22(3):128–35
13. Knudsen EI, Knudsen PF, Masino T (1993) Parallel pathways mediating both sound localization and gaze control in the forebrain and midbrain of the barn owl. J Neurosci 13(7):2837–52
14. Keller CH, Hartung K, Takahashi TT (1998) Head-related transfer functions of the barn owl: measurement and neural responses. Hear Res 1181–2:13–34
15. Konishi M (1973) Locatable and nonlocatable acoustic signals for barn owls. Amer Nat 107:775–785
16. Sullivan WE, Konishi M (1984) Segregation of stimulus phase and intensity coding in the cochlear nucleus of the barn owl. J Neurosci 4(7):1787–99
17. Jeffress L (1948) A place theory of sound localization. J Comp Physiol Psychol 41:35–39.
18. Carr CE, Konishi M (1990) A circuit for detection of interaural time differences in the brain stem of the barn owl. J Neurosci 10(10):3227–46
19. Yin TC, Chan JC (1990) Interaural time sensitivity in medial superior olive of cat. J Neurophysiol 64(2):465–88
20. Yin TC, Chan JC, Carney LH (1987) Effects of interaural time delays of noise stimuli on low-frequency cells in the cat's inferior colliculus. III. Evidence for cross- correlation. J Neurophysiol 58(3):562–83
21. Overholt EM, Rubel EW, Hyson RL (1992) A circuit for coding interaural time differences in the chick brainstem. J Neurosci 12(5):1698–708
22. Wagner H (1992) On the ability of neurons in the barn owl's inferior colliculus to sense brief appearances of interaural time difference. J Comp Physiol [A] 170(1):3–11
23. Takahashi TT, Wagner H, Konishi M (1989) Role of commissural projections in the representation of bilateral auditory space in the barn owl's inferior colliculus. J Comp Neurol 281(4):545–54
24. Wagner H, Takahashi T, Konishi M (1987) Representation of interaural time difference in the central nucleus of the barn owl's inferior colliculus. J Neurosci 7(10):3105–3116
25. Takahashi TT, Konishi M (1988) Projections of nucleus angularis and nucleus laminaris to the lateral lemniscal nuclear complex of the barn owl. J Comp Neurol 274 (2):212–38
26. Takahashi TT, Keller CH (1992) Commissural connections mediate inhibition for the computation of interaural level difference in the barn owl. J Comp Physiol [A] 170 (2):161–9
27. Takahashi TT, Barberini CL, Keller CH (1995) An anatomical substrate for the inhibitory gradient in the VLVp of the owl. J Comp Neurol 358(2):294–304
28. Adolphs R (1993) Bilateral inhibition generates neuronal responses tuned to interaural level differences in the auditory brainstem of the barn owl. J Neurosci 13 (9):3647–68
29. Mazer J (1995) Integration of parallel processing streams in the inferior colliculus of the owl., in Division of Biology. California Institute of Technology: Pasadena
30. Euston DR (2000) From spectrum to space: The integration of frequency-specific intensity cues to produce auditory spatial receptive fields in the barn owl inferior colliculus., in Department of Psychology. University of Oregon: Eugene 152
31. Spezio ML, Takahashi TT (2003) Frequency-specific interaural level difference tuning predicts spatial response patterns of space-specific neurons in the barn owl inferior colliculus. J Neurosci 23(11):4677–88
32. Pena J, Konishi M (2002) From postsynaptic potentials to spikes in the genesis of auditory spatial receptive fields. Journal of Neuroscience 22(13):5652–5658
33. Blauert J (1997) Spatial Hearing. The Psychophysics of Human Sound Localization. Cambridge: MIT
34. Takahashi TT, Keller CH (1994) Representation of multiple sound sources in the owl's auditory space map. J Neurosci 14(8):4780–93
35. Zurek PM (1987) Yostand WA Gourevitch Editors. G The precedence effect., in Directional Hearing, Springer Verlag: New York
36. Litovsky RY et al. (1999) The precedence effect. J Acoust Soc Am 106(4 Pt 1):1633–54
37. Keller CH, Takahashi TT (1996) Binaural cross-correlation predicts the responses of neurons in the owl's auditory space map under conditions simulating summing localization. J Neurosci 16(13):4300–9
38. Keller CH, Takahashi TT (1996) Responses to simulated echoes by neurons in the barn owl's auditory space map. J Comp Physiol [A] 178(4):499–512
39. Haas H (1951) Ueber den Einfluss eines Einfachechos auf die Hoersamkeit von Sprache. Acustica 1:49–58
40. Wallach H, Newman E, Rosenzweig M (1949) The precedence effect in sound localization. Amer. J. Psycho 57:315–336
41. Shinn-Cunningham BG, Zurek PM, Durlach NI (1993) Adjustment and discrimination measurements of the precedence effect. J Acoust Soc Am 93(5):2923–32
42. Yost WA, Soderquist DR (1984) The precedence effect: Revisited. J Acoust Soc Am 76:1377–1383
43. Yang L, GD Pollak (1997) Differential response properties to amplitude modulated signals in the dorsal nucleus of the lateral lemniscus of the mustache bat and the roles of GABAergic inhibition. J Neurophysiol 77(1):324–40
44. Yang X, Grantham DW (1997) Cross-spectral and temporal factors in the precedence effect: discrimination suppression of the lag sound in free-field. J Acoust Soc Am 102(5 Pt 1):2973–83
45. Saberi K, Perrott Dr (1990) Lateralization thresholds obtained under conditions in which the precedence effect is assumed to operate. J Acoust Soc Am 87:1732–1737
46. Freyman RL, Clifton RK, Litovsky RY (1991) Dynamic processes in the precedence effect. J Acoust Soc Am 90(2 Pt 1):874–84
47. Cranford JL (1982) Localization of paired sound sources in cats: effects of variable arrival times. J Acoust Soc Am 72(4):1309–11

48. Kelly JB (1974) Localization of paired sound sources in the rat: small time differences. J Acoust Soc Am 55(6):1277–84
49. Litovsky RY (1997) Developmental changes in the precedence effect: estimates of minimum audible angle. J Acoust Soc Am 102(3):1739–45
50. Fitzpatrick DC et al. (1999) Responses of neurons to click-pairs as simulated echoes: Auditory nerve to auditory cortex. J Acoust Soc Am 106:3460–3472
51. Fitzpatrick DC et al. (1995) Neural responses to simple simulated echoes in the auditory brain stem of the unanesthetized rabbit. J Neurophysiol 74(6):2469–86
52. Litovsky RY, Yin TC (1998) Physiological studies of the precedence effect in the inferior colliculus of the cat. II. Neural mechanisms. J Neurophysiol 80(3):1302–16
53. Litovsky RY, Yin TC (1998) Physiological studies of the precedence effect in the inferior colliculus of the cat. I. Correlates of psychophysics. J Neurophysiol 80 (3):1285–301
54. Litovsky RY et al. (1997) Psychophysical and physiological evidence for a precedence effect in the median sagittal plane. J Neurophysiol 77(4):2223–6
55. Litovsky RY, Delgutte B (2002) Neural correlates of the precedence effect in the inferior colliculus: effect of localization cues. J Neurophysiol 87(2): 976–94
56. Yin TC (1994) Physiological correlates of the precedence effect and summing localization in the inferior colliculus of the cat. J Neurosci 14(9):5170–86
57. Spitzer MW Bala AD Takahashi TT (2004) A neuronal correlate of the precedence effect is associated with spatial selectivity in the barn owl's midbrain. J Neurophysiol 92:2051–2070
58. Wagner H, Takahashi T (1990) Neurons in the midbrain of the barn owl are sensitive to the direction of apparent acoustic motion. Naturwissenschaften 77(9):439–42
59. Wagner H, Takahashi T (1992) Influence of temporal cues on acoustic motion-direction sensitivity of auditory neurons in the owl. J Neurophysiol 686:2063–76
60. Wagner H, Trinath T, Kautz D (1994) Influence of stimulus level on acoustic motion-direction sensitivity in barn owl midbrain neurons. J Neurophysiol 71(5):1907–16
61. Kautz D, Wagner H (1998) GABAergic inhibition influences auditory motion-direction sensitivity in barn owls. J Neurophysiol 80(1):172–85
62. Wagner H, Campenhausen M von (2002) Distribution of auditory motion-direction sensitive neurons in the barn owl's midbrain. J Comp Physiol A Neuroethol Sens Neural Behav Physiol 188(9):705–13
63. Takahashi TT, Keller CH (1992) Simulated motion enhances neuronal selectivity for a sound localization cue in background noise. J Neurosci 12(11):4381–90
64. Wang J (1994) Processing of dynamic stimuli by the auditory system of the barn owl (Tyto alba), in Department of Physiology and Neurobiology. University of Connecticut: Storrs. 147
65. Xiao X, Grantham DW (1997) The effect of a free-field auditory target's motion on its detectability in the horizontal plane [letter]. J Acoust Soc Am 102(3):1907–1
66. Deliwala P (1999) Effects of spatial separation and apparent signal movement on masked detection; energetic and informational maskers., in Sargent College of Health and Rehabilitation Sciences. Boston University: Boston. 165
67. Knudsen EI, Knudsen PF (1996) Disruption of auditory spatial working memory by inactivation of the forebrain archistriatum in barn owls. Nature 383(6599):428–31
68. Johnen A, Wagner H, Gaese BH (2001) Spatial attention modulates sound localization in barn owls. J Neurophysiol 85(2):1009–12
69. Posner MI, Nissen MJ, Ogden WC (1978) Pick HL Saltzman. IJ Attended and unattended processing modes: The role of set for spatial location. Modes of perceiving and processing information., Earlbaum: Hillsdale
70. Brainard MS, Knudsen EI (1998) Sensitive periods for visual calibration of the auditory space map in the barn owl optic tectum. J Neurosci 18(10):3929–42
71. Knudsen EI (1988) Early blindness results in a degraded auditory map of space in the optic tectum of the barn owl. Proc Natl Acad Sci USA 85(16):6211–4
72. Knudsen EI, Esterly SD, du Lac S (1991) Stretched and upside-down maps of auditory space in the optic tectum of blind-reared owls; acoustic basis and behavioral correlates. J Neurosci 11(6):1727–47
73. Knudsen EI, Knudsen PF (1989) Vision calibrates sound localization in developing owls. J Neurosci 9:3306–3313
74. Brainard MS, Knudsen EI (1993) Experience-dependent plasticity in the inferior colliculus: a site for visual calibration of the neural representation of auditory space in the barn owl. J Neurosci 13(11):4589–608
75. Knudsen EI (2002) Instructed learning in the auditory localization pathway of the barn owl. Nature 417 (6886):322–8
76. Poganiatz I, Wagner H (2001) Sound-localization experiments with barn owls in virtual space: influence of broadband interaural level different on head-turning behavior. J Comp Physiol [A] 187(3):225–33
77. Whitchurch EA, Takahashi TT (2003) Behavioral and physiological characterization of bimodal interactions in the barn owl. Abst. Soc. Neurosi
78. Bala ADS, Takahashi TT (2000) Pupillary dilation response as an indicator of auditory discrimination in the barn owl. J. Comp. Physiol. A 186:425–434
79. Bala AD, Spitzer MW, Takahashi TT (2003) Prediction of auditory spatial acuity from neural images on the owl's auditory space map. Nature 424(6950):771–4
80. Sakitt B (1973) Indices of discriminability. Nature 241 (5385):133–4
81. Bala AD, takahashi TT (2001) Vertical and horizontal minimal audible angles of the barn owl. Assoc. Res. Otolaryngol. Abs, 264
82. Green DM, Swetts JA (1966) Signal Detection Theory and Psychophysics. New York: Wiley

Neuroethology of Visual Orientation in Flies

Anne-Kathrin Warzecha[1,2], Roland Kern[1]
[1]Lehrstuhl Neurobiologie, Fakultät Biologie, Universität Bielefeld, Bielefeld, Germany
[2]Institut für Psychologie II, Münster, Germany

Definition

The central goal of neuroethological research is to account for the behaviour of an animal in terms of the underlying neuronal mechanisms. The blowfly (Calliphoridae) has proven to be a very fruitful model system for the analysis of how the nervous system controls visually guided orientation behaviour (reviews: [1,2]).

When the blowfly, or any other agent equipped with eyes, moves around, the retinal images of the surroundings are continually displaced. This ►optic flow is characteristic of the animal's path of locomotion and the spatial layout of the environment. Flies exploit optic flow information for controlling various behaviours such as course stabilisation, compensatory head movements, landing, the detection and fixation of objects, and the pursuit of conspecifics in the context of mating behaviour (review: [2,3]). Due to technical advances, it has recently become possible to record the flight behaviour of freely moving flies with great precision. When cruising around, flies usually fly straight and then very quickly change their flight direction by abrupt, saccadic turns of the head and trunk (Fig. 1 review: [1,2]), reminiscent of eye movements in primates. This behaviour constrains the optic flow patterns experienced during cruising flight and may thus facilitate an evaluation of the retinal information.

The computations underlying optic flow processing can partly be understood in terms of the biophysical properties of individual nerve cells and their synaptic interactions.

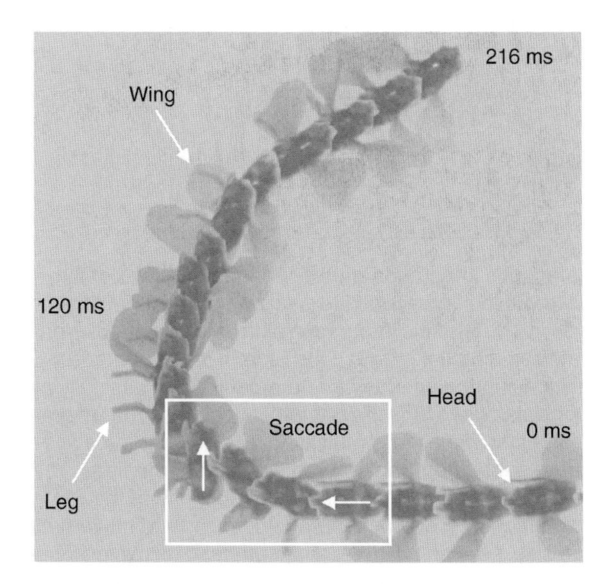

Neuroethology of Visual Orientation in Flies. Figure 1 Flight sequence of blowfly that was videotaped from above with a high-speed camera. Here video images of the fly taken every 12 ms are superimposed. The fly performs a short saccadic turn (indicated by the arrows on the fly's body long axis). Within only 36 ms, it has turned by almost 90°. During such turns the animal can reach turning velocities of more than 3,000° s^{-1} (R. Kern unpublished data).

Characteristics

Motion information is not explicitly represented at the level of the retina. Instead, several processing steps are necessary in order to obtain a neuronal representation of optic flow information (Fig. 2).

Local Information Processing

Optic flow is initially processed by successive layers of retinotopically arranged columnar neurons. The retinal luminance changes are sensed by photoreceptors and are spatio-temporally filtered by them, as well as by their postsynaptic elements, in the lamina. This filtering is thought to remove spatial and temporal redundancies and to play a role in adapting the system to the ambient light level (reviews: [4,5]).

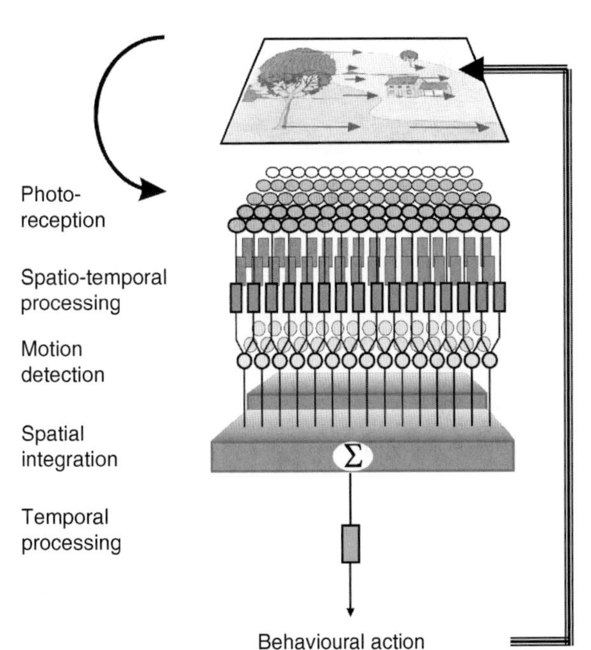

Neuroethology of Visual Orientation in Flies. Figure 2 Scheme of the principle processing steps in visual course control. When moving through a landscape, the images of the surroundings move across the retina of the observer. This optic flow is indicated by the arrows in the uppermost box. The local luminance values of the scene are sensed by a two-dimensional array of photoreceptors. The next layer indicates that this information is spatially and temporally processed in parallel for the entire array of retinotopically organised, local elements. Consecutively, motion is computed on a local basis. Motion information is then spatially pooled over extended parts of the visual field. Again, this process takes place in parallel for different parts of the visual field. In the simplest case, this global motion information is then temporally filtered, and used directly to control behaviour that in turn changes the retinal input. The connections between consecutive elements are only indicated for the foremost row.

Computation of Local Motion Information

Directionally selective responses to motion are computed on a local retinotopic basis in the second visual neuropile, the medulla. The cellular mechanism underlying ►motion detection is still largely unknown because of the difficulties in recording the activity of the extremely small neurons in this brain area (review: [6]). However, the computational properties of this processing step have been revealed in detail. They can be described by a computational model, the so-called correlation-type movement detector (Fig. 3b, reviews: [1,2]). This movement detector is composed of two mirror-symmetrical subunits. In each subunit, the input arising from two neighbouring points in visual space are compared by a multiplication after one of them has been delayed. Subtracting the outputs of the two subunits yields the final detector response.

Dendritic Integration of Local Motion Information

In order to evaluate the global structure of optic flow fields, local motion information needs to be combined from large parts of the visual field. This is accomplished by the large motion-sensitive neurons in the third visual

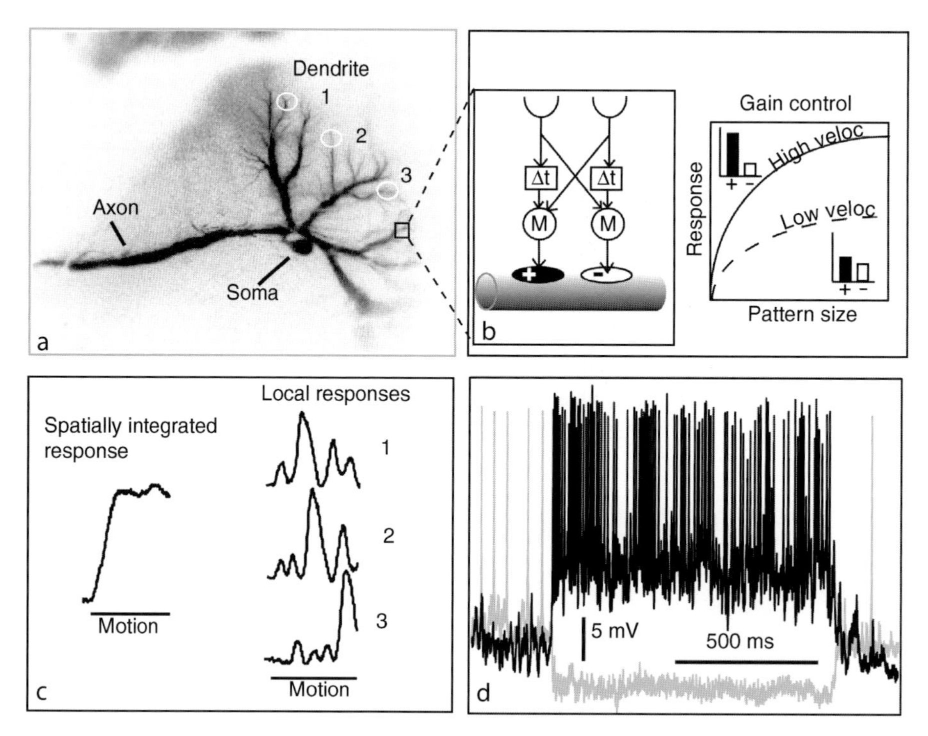

Neuroethology of Visual Orientation in Flies. Figure 3 ►Dendritic integration and response properties of fly tangential cells (TCs). (a) Image of a TC after it has been injected with a fluorescent dye (Warzecha A-K, unpublished fluorescent photography). (b) TCs acquire their direction selectivity by dendritic integration of the signals of two types of input elements with small receptive fields and opposite preferred directions (*left*). One type of input element is excitatory whilst the other is inhibitory. Many properties of the local input elements can be modelled successfully by correlation-type movement detectors. This type of detector consists of two mirror symmetrical subunits that share their input signals. The delayed (Δt) signal of each movement detector input line is multiplied (M) with the un-delayed signal of the other input line. The outputs of the two subunits are subtracted from each other. Due to the fly's panoramic vision, nearly every point in visual space is subserved by such an input pair. Both subunits of the movement detector are activated even during preferred direction motion. Several hundreds of such local movement detectors impinge onto the TC dendrite. With increasing pattern size, i.e. with an increasing number of activated input elements, the postsynaptic potential saturates at a level between the excitatory and the inhibitory reversal potentials (*right diagram*). The exact value of saturation is set by the activation ratio of excitatory and inhibitory input elements, which in turn is a function of stimulus parameters such as velocity or contrast (*right diagram*). (c) Dendritic integration smoothes out pattern dependent fluctuations in the time course of the local inputs. The local responses elicited in three different areas of the dendrite (schematically indicated by *white circles* in a) are plotted underneath each other. In the axon, these time-varying signals are integrated and the fluctuations are largely smoothed out. (d) Single responses of a TC to preferred (*black*) and null (*grey*) direction motion. This TC responds even close to its output terminal with graded changes of the membrane potential that are superimposed by spikes. Electrophysiological data courtesy of Jan Grewe.

neuropile, known as tangential cells (TCs). With their large dendritic trees, they pool motion information from up to about 3,000 retinotopically arranged directionally selective input elements (reviews: [1,2,7,8]). Two types of input elements with opposite preferred direction of motion converge onto the dendrite of TCs, one excitatory and the other inhibitory. Consequently, TCs respond directionally selective to motion in a large part of the visual field (Fig. 3).

The preferred directions of the local retinotopic elements that synapse onto a given TC have been concluded to coincide with the directions of velocity vectors characterising the optic flow induced during particular types of self-motion [8]. Hence, the spatial input organisation of TCs forms a basis for their sensitivity to optic flow.

Approximately 60 different TCs can be individually identified on the basis of their physiological and anatomical characteristics. Some TCs have been suggested to respond specifically to the optic flow resulting from certain types of self-motion (but see below), whereas others detect movements of small objects. Depending on the cell type, TCs respond to visual motion by graded shifts of their membrane potential, by spike-like events superimposed on graded membrane potential shifts (Fig. 3d) or by large amplitude action potentials (reviews: [1,2,7]).

►Dendritic integration of the local motion signals influences the representation of optic flow in several ways (Fig. 3, [1,3]): (i) Representation of velocity: The responses of the retinotopic input elements of the TCs modulate in time even during constant-velocity motion depending on the local luminance changes of the moving pattern (►dendritic integration/dendritic processing). By pooling over many input elements that all experience phase-shifted luminance changes, these response modulations are smoothed out and the integrated signal becomes to some extent proportional to the velocity of the stimulus (Fig. 3c). (ii) Gain control: Saturation nonlinearities of the pooling TCs render their responses largely independent of pattern size while they still depend on other stimulus parameters such as velocity or contrast.

Integration of Global Motion Information from Populations of Motion-Sensitive TCs

Dendritic pooling of local motion information is not the only computation that takes place in the blowfly visual motion pathway to obtain a representation of global optic flow. Instead, motion information from various TCs is pooled at the level of the lobula plate (reviews: [1,10]). This processing step is assumed to render TCs more specific to particular types of self-motion, such as rotation or translation, or to tune them to the detection of a nearby object moving in front of a more distant background.

Pooling motion information from several TCs requires synaptic transfer. Unless the intervening synapses are carefully adjusted to the presynaptic activity levels that occur during sensory stimulation, synaptic transmission may distort the information being transmitted. The properties of the transmission of pre- to postsynaptic signals has been investigated in detail by electrophysiological and optical imaging studies for various pairs of TCs that differed with respect to the response modes of the pre- and postsynaptic neuron (reviews: [2,9]). For one synaptic connection, it could show, for instance, that signal transfer is linear for frequencies up to about 10 Hz (Fig. 4). This synaptic tuning appears to be adaptive, because in this frequency range TCs transmit most information about changes in the velocity of motion stimuli.

Reliability of Encoding Visual Motion Information

In order to understand the performance of motion-sensitive neurons in real time, the analysis of the mechanisms of motion computations need to be complemented by investigating how reliably these neurons represent visual motion information. Due to noise from several sources, neuronal responses to repeated presentation of the same stimulus are variable. Although the spike count variance across trials is relatively small in blowfly TCs (review: [10]), significant neuronal response variability still constrains the timescale on which time varying optic flow characteristic of behavioural situations can be conveyed. The noise sources limiting the performance are still under investigation. However, photon noise could be shown not to limit the reliability of a TC in representing a motion stimulus, at least under photopic conditions (Fig. 5a, review: [2]).

The process of generating spikes from the postsynaptic potential is generally thought not to limit the timescale of motion information processing. Accordingly, spikes in TCs time-lock with a millisecond precision to a motion stimulus, if the stimulus induces fast membrane potential fluctuations that are larger than the membrane potential noise (Fig. 5b, c). If, however, the motion stimulus induces only slow membrane potential fluctuations, the exact timing of spikes is determined by the high frequency components of the membrane potential noise. Then spikes are not precisely time-locked to the stimulus and the spike rate rather than spike timing reflects the stimulus induced changes. Since the computations underlying motion detection require time constants of some tens of milliseconds, motion computation inevitably attenuates the neural representations of high frequency velocity fluctuations. Hence only when the velocity changes in the motion stimulus are very rapid and large, the resulting depolarisations of the TCs are sufficiently pronounced to elicit spikes at a millisecond precision. It is still under debate to what extent rapid and slow velocity

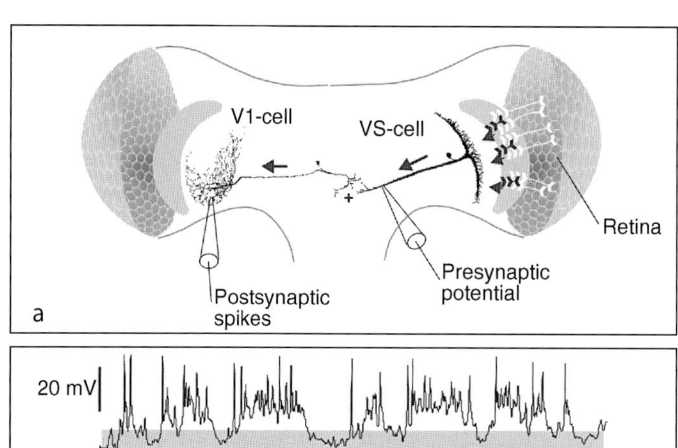

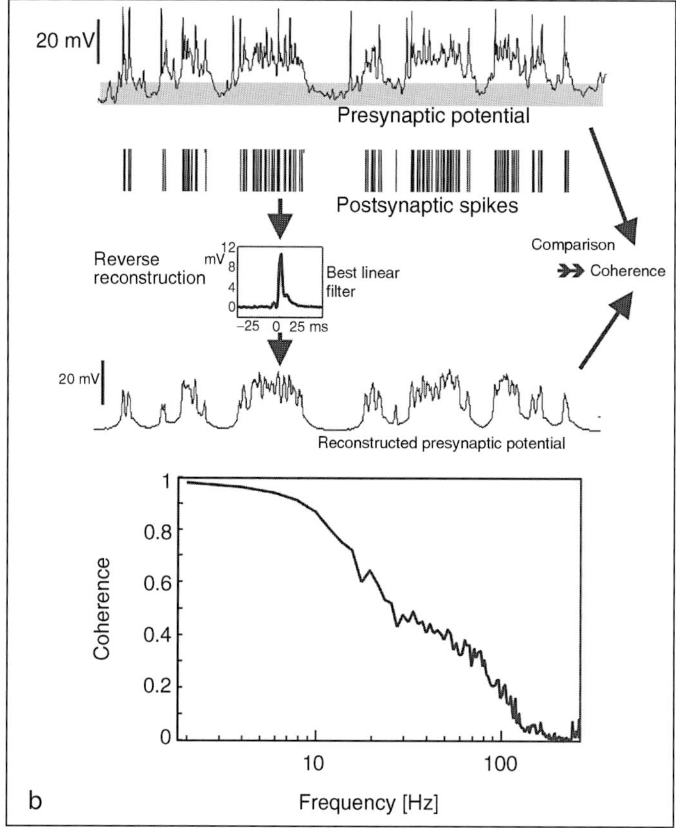

Neuroethology of Visual Orientation in Flies. Figure 4 Synaptic transmission between two motion sensitive tangential cells (TCs). (a) Schematic of the visual motion pathway of the fly illustrating the compound eyes, the local retinotopically organised local elements involved in the local computation of motion, and a particular TC, a so-called VS-cell, that spatially pools the output of the small-field motion sensitive elements with its large dendrite. The VS-cell is synaptically coupled to the V1-cell that transmits optic flow information to the contralateral half of the brain via its extended output arborisation (anatomical reconstructions taken from Hausen K, Egelhaaf M (1989) Neural mechanisms of visual course control in insects. In: Stavenga DG, Hardie RC (eds) Facets of vision. Springer, pp 391–424; and Krapp et al (1998) J Neurophysiol 79:1902–1917, with permissions). (b) Presynaptic membrane potential of a VS-cell and postsynaptic spike train of the V1-cell (occurrence of a spike indicated by a *vertical line*) during presentation of white noise velocity fluctuations. In order to quantitatively relate the time-dependent postsynaptic signal to the presynaptic input, the reverse reconstruction approach was applied as schematically outlined in the figure. For this approach, the linear filter was determined that when convolved with the presynaptic spike train leads to the best estimate of the presynaptic potential. Since hyperpolarisations do not have much effect on the activity of the postsynaptic neuron, the presynaptic potential was rectified at the resting potential (shaded bar underneath the presynaptic potential trace). The coherence function serves as a measure of the similarity of the recorded and the estimated presynaptic membrane potential traces. Coherence values close to 1 for frequencies up to 10 Hz indicate that the system can be regarded as very reliably and approximately linear in this frequency range (modified from Warzecha A-K et al (2003) Neuroscience 119:1103–1112, with permission from Elsevier).

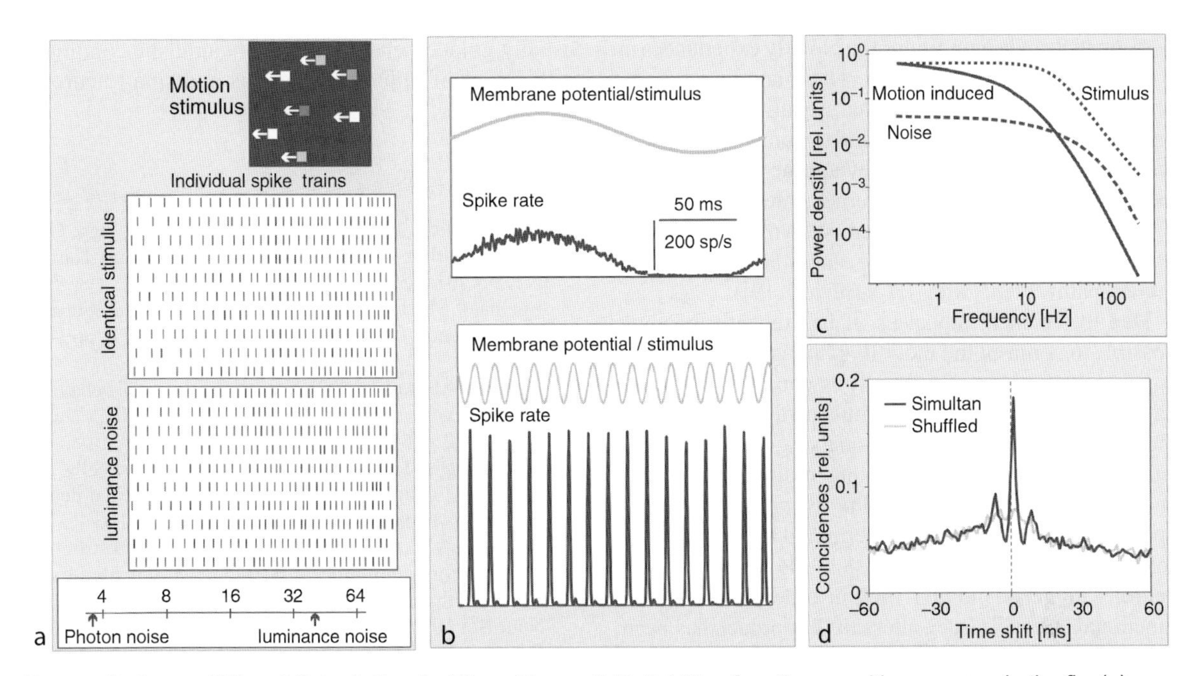

Neuroethology of Visual Orientation in Flies. Figure 5 Reliability of motion sensitive neurons in the fly. (a) Origin of ►neuronal variability. The equivalent noise paradigm was adapted to investigate whether photon noise limits the reliability of the motion detection system. In order to mimic photon noise, the luminance of individual dots was chosen randomly. The motion stimulus consists of a group of coherently moving dots with variable luminance. The two raster plots illustrate a 300 ms section of the responses of a spiking TC. Each vertical bar denotes the occurrence of a spike. Consecutive responses are plotted underneath each other. Either exactly the same luminance values were presented (*upper raster*) or new luminance values were drawn randomly for each trial (*lower raster*). Even if exactly the same stimulus was presented from trial to trial, there is considerable neuronal variability. Analyses derived from signal-detection theory revealed that the response statistics are affected only when luminance noise is much larger than photon noise (*bottom*). Hence, photon noise does not limit the reliability of a motion sensitive neuron under photopic conditions (details in Grewe et al (2003) J Neurosci 23:10776–10783. (b) Time-locking of spikes to sinusoidal stimulus-induced membrane-potential fluctuations (5 and 80 Hz) in a model cell. The model is adjusted to fit the responses of a fly TC to motion stimuli. The stimulus-induced membrane potential fluctuations fed into the model cell were superimposed by stochastic fluctuations mimicking ►neuronal noise. PSTHs illustrate that fast stimulus-induced membrane potential fluctuations are necessary to trigger spikes with a high temporal precision. Slow stimulus-induced fluctuations lead to spike activity with a rate about proportional to the membrane potential (details in Kretzberg et al (2001) J Comput Neurosci 10:79–97, with permission). (c) Dynamic properties of membrane potential fluctuations of a fly TC (HS-cell) elicited by band-limited white-noise velocity fluctuations. Power spectra of the motion stimulus (*dotted*), the motion induced response component (*solid*) and the stochastic membrane potential fluctuations (*dashed*). The motion induced response component was obtained by averaging many individual response components thereby smoothing out stochastic response fluctuations. The stochastic response component results from the difference of each individual response trace to the motion-induced component. The motion-induced component contains most power below 20 Hz, although the stimulus contained higher frequencies. In the low frequency range, the motion-induced response component is larger than the stochastic response component. At higher frequencies, the relationship reverses (details in Warzecha et al (1998) Curr Biol). (d) Cross-correlogram of responses of two TCs (H1 and H2) to band-limited white-noise velocity fluctuations. Either synchronously recorded responses were cross-correlated (*black*) or responses were not recorded synchronously but obtained from the same cell pair in repetitive presentations of the same stimulus (*grey*). AlthoughTCs can generate spikes very precisely (*black*), most spikes time-lock even to dynamical motion stimulation on a much coarser time scale (reprinted from Warzecha et al (1998) Curr Biol 8:359–368, with permission from Elsevier).

changes and thus the exact timing of spikes are functionally significant (review: [10]). This issue can only be resolved if the dynamics of the retinal image displacements as experienced by the blowfly in different behavioural contexts are taken into account.

Encoding of Visual Stimuli Under Naturalistic Conditions

The functional significance of the information that is processed in the visual motion pathway of the blowfly can be assessed only by analysing the neuronal performance under conditions the blowfly experiences in a normal behavioural situation. These conditions comprise of the properties of the visual stimuli with respect to e.g. their dynamics, brightness and textural properties, as well as the context in which they appear with respect to the stimulus history and the internal state of the animal. The stimulus history has been shown to affect the properties of several processing stages along the visual motion pathway including TCs (review: [9]).

Due to technical advances, it has recently become possible to confront the blowfly with visual stimuli that come close to those the blowfly is confronted with in natural situations. Natural stimuli differ largely from those usually employed for an analysis of the mechanisms underlying visual motion processing. This difference in particular pertains to the dynamics of natural optic flow, which is largely determined by the animal's self-motion. Since it is not yet possible to record from neurons in freely moving flies in order to assess the functional role of TCs, an alternative approach has been chosen (review: [3]). The trajectories of freely moving flies were recorded in known surroundings; the optic flow experienced by the flies was reconstructed and then replayed to the animal while recording the activity of a TC in electrophysiological experiments. Using a naturalistic behaviourally generated optic flow led to conclusions that partly contradict those obtained with simple experimenter-designed motion stimuli. The neuronal responses of a TC previously thought to encode rotations of the animal about its vertical body axis astonishingly provide information about the spatial relation of the animal to its surroundings. Thus, one of the great issues of future research is to understand how the neuronal hardware that underlies a given behaviour processes information under natural operating conditions.

Outlook

With their relatively simple and accessible nervous system, blowflies are excellent organisms in which to investigate how visual information is acquired and processed to guide locomotion. Many important aspects of the neuronal computations that underlie visually guided behaviour in blowflies have been elucidated in recent years, ranging from the biophysical properties of individual neurons and their synaptic interactions, to the performance under conditions that come close to what the animal might encounter in normal behaviour. In almost all parts of the above mentioned aspects, models have proved to be an essential element for unravelling the mechanisms underlying visual orientation behaviour in flies (reviews: [1,2]). Also in future research, models will serve as touchstones to challenge our hypotheses about how the blowfly manages to control its virtuosic flight behaviour. It thus appears possible, by studying the blowfly visual system, to understand visually guided behaviour under naturalistic conditions at the level of individual neurons and small neuronal networks.

References

1. Borst A Haag J (2007) Invertebrate Neurobiology. Cold Spring Harbor Laboratory Press, Cold Spring Harbor, NY, pp 101–122
2. Egelhaaf M et al. (2005) Methods in Insect Sensory Neuroscience. CRC Press LLC, Boca Raton, FL, pp 185–212
3. Egelhaaf M et al. (2002) Neural encoding of behaviourally relevant motion information in the fly. Trends Neurosci 25:96–102
4. Laughlin SB (1994) Matching coding, circuits, cells, and molecules to signals: general principles of retinal design in the fly's eye. Prog Retin Eye Res 13:165–196
5. Juusola M, French AS, Uusitalo RO, Weckström M (1996) Information processing by graded-potential transmission through tonically active synapses. Trends Neurosci 19:292–297
6. Douglass JK, Strausfeld NJ (2001) Motion vision: computational, neural, and ecological constraints. Zanker JM, Zeil J (eds) Springer, Berlin Heidelberg New York, pp 67–81
7. Hausen K (1984) Photoreception and vision in invertebrates. Ali MA (ed) Plenum, New York, pp 523–559
8. Krapp HG (2000) Neuronal processing of optic flow. Lappe M (ed) Academic, San Diego, CA, pp 93–120
9. Kurtz R, Egelhaaf M (2003) Natural patterns of neural activity. Mol Neurobiol 27:1–19
10. Warzecha AK, Egelhaaf M (2001) Processing visual motion in the real world: a survey of computational, neural, and ecological constraints. Zanker JM, Zeil J (eds) Springer, Berlin Heidelberg New York, pp 239–277

Neurofibroma

Definition

Neurofibroma is a benign neoplasm composed of the fibrous elements of a nerve.

Neurofibromatosis Types I (NF-I) and Type II (NF-II)

Definition

Autosomal-dominantly inherited neuro-cutaneous disorders presenting with a plethora of manifestations involving the central and peripheral nervous systems, the circulatory system, the skin, and the skeleton.

Lifespan is reduced, frequently due to malignant tumors and hypertension. The common *neurofibromatosis type I (NF-I)* results from an alteration in the long arm of chromosome 17. The diagnosis is clinically based on the presence of two of seven criteria. One criterion is the presence of neurofibroma (a benign peripheral nerve sheath tumor), which is one of the most frequent tumors of neural origin. Others include are central nervous system tumors, skin lesions (café au lait spots), bone malformations and vascular abnormalities. *Neurofibromatosis 2 (NF2)* is clinically and genetically distinct from NF-I by being rarer, exhibiting less skin manifestations and malignant tumors, with the cardinal sign being bilateral vestibular ►schwannomas and ocular abnormalities (cataract).

Neurofilament

Definition

Neurofilaments are intermediate type cytoskeletal proteins which lend structural support to especially axons. They are thought to play a major role in determining the axonal caliber (diameter of the axon). These proteins are intermediate filaments being composed of three different proteins: NF-L (low weight), NF-M (medium weight) and NF-H (high weight).

Developing neurons produce few neurofilaments, however once the growth phase is complete neurofilaments are seen extending from the soma into both axons and dendrites providing structural support to these processes. Neurofilaments can be used to identify and visualize mature neurons using histochemical techniques.

►Axonal Pathfinding and Network Assembly
►Cytoskeleton
►Extrasomal Protein Synthesis in Neurons
►Neuronal Cell Death and Axonal Degeneration:
►Neurofilaments as Biomarkers

Neurofilament Heavy Chain (NfH)

Definition

A 190–210 kDa protein depending on the degree of phosphorylation, encoded on chromosome 22q12.2.

►Neuronal Cell Death and Axonal Degeneration: Neurofilaments as Biomarkers

Neurofilament Light Chain (NfL)

Definition

A 68 kDa protein encoded on chromosome 8p21.

►Neuronal Cell Death and Axonal Degeneration: Neurofilaments as Biomarkers

Neurofilament Medium Chain (NfM)

Definition

A 150 kDa protein encoded on chromosome 8p21.

►Neuronal Cell Death and Axonal Degeneration: Neurofilaments as Biomarkers

Neurogenesis

Koji Oishi, Kazunori Nakajima
Department of Anatomy, Keio University School of Medicine, Tokyo, Japan

Definition

In the narrow sense, “neurogenesis” is the process by which neurons are generated by their progenitor cells. However, the term “neurogenesis” is occasionally used in the broad sense to refer to the entire process of generating functionally mature neurons, including the process of proliferation and neuronal fate specification of progenitor cells and regulation of the cell death of neurons/progenitor cells.

Neurogenesis in the mammalian central nervous system (CNS) used to be thought to occur only during early embryonic development and only recently has it come to be generally accepted that new functional neurons are generated in at least two regions of the adult brain, the ►subventricular zone (SVZ) along the lateral ventricle and the subgranular zone (SGZ) of the dentate gyrus in the hippocampus [1].

Characteristics

Quantitative Description

Approximately 100 billion neurons are thought to be present in the human CNS, and even in mice the number is estimated to be about 40 million. Since each neuronal progenitor cell produces one or two neurons for each neurogenic cell division, a comparable number of progenitor cells is necessary. Moreover, 50–80% (depending on the region) of the neurons generated undergo apoptosis during formation of the neural network in the neonatal CNS, suggesting that much more neurogenesis has to occur to establish the mammalian CNS [2].

It is estimated that about 80,000 and 9,000 new neurons a day are produced in the SVZ and SGZ respectively of the adult rat brain. However, more than 50% of them are eliminated by apoptosis within a week.

Higher Level Structures

Before neurogenesis, the neural plate and neural tube are composed of neuroepithelial cells that form a pseudostratified epithelium. Neuroepithelial cells are bipolar and each of them extends from the apical (ventricular) surface to the basal lamina (radial orientation). Neuroepithelial cells translocate their nucleus from the top to the bottom of the ►neuroepithelium during the cell cycle (referred to as interkinetic nuclear migration or elevator movement) [3]. As the neuroepithelial walls thicken at around E12.5 in the mouse telencephalon, the primitive neuroepithelial cells elongate, while maintaining their orientation and become ►radial glial cells, which are defined as cells with a radial morphology and some glial characteristics (e.g. glycogen granules). The cell bodies of the radial glial cells are in the ►ventricular zone (VZ) and the cells undergo cell divisions at the ventricular surface. One of the most exciting recent discoveries is that radial glial cells, which used to be thought of primarily as migratory scaffolds for young neurons and as glial progenitors, are neurogenic progenitors. A distinct type of progenitor cells, referred to as basal progenitors (also called SVZ, intermediate, or non-surface-dividing progenitors), has been identified as cells that lose contact with the ventricular surface and undergo cell division away from the surface. At later stages basal progenitors are mainly present in the SVZ, a histologically distinct structure adjacent to the VZ. These three cell types, neuroepithelial cells, radial glial cells and basal progenitors, all contribute to the generation of neurons (Fig. 1; see also higher level processes) [4,5].

A primate-specific layer of progenitor cells called the outer SVZ has recently been described and shown to be one of the major sites of cortical neuron production in primates. In contrast to the basal progenitors in rodents, these cells in primates possess radial morphology, strongly suggesting that they are polarized cells [6].

What is the site of neurogenesis? In the embryonic stages, CNS neurons are produced from almost all regions of the neuroepithelium in the neural tube with the exception of a few specialized areas, such as the optic stalk and the floor plate and roof plate of the spinal cord. The onset and period of neurogenesis vary greatly depending on the location along the neuraxis. During the development of the dorsal pallium, for example, neurogenesis, including developmental changes in cell cycle dynamics, maturation of the cortical plate and later events (such as the development of callosal projections or synaptogenesis), progresses along a latero-rostral to medio-caudal gradient.

Neurons in the peripheral nervous system (PNS) are generated from neural crest cells that originate at the border between the neural plate and the prospective epidermis. Around the time of neural tube closure, neural crest cells emigrate from the neural tube, migrate along defined paths in the embryo and differentiate into a wealth of derivatives, including PNS neurons [7].

In most mammals, active neurogenesis occurs throughout life in discrete regions of the intact adult CNS. From rodents to primates, neurons are generated continuously in the SVZ along the lateral ventricle and migrate anteriorly through the rostral migratory stream

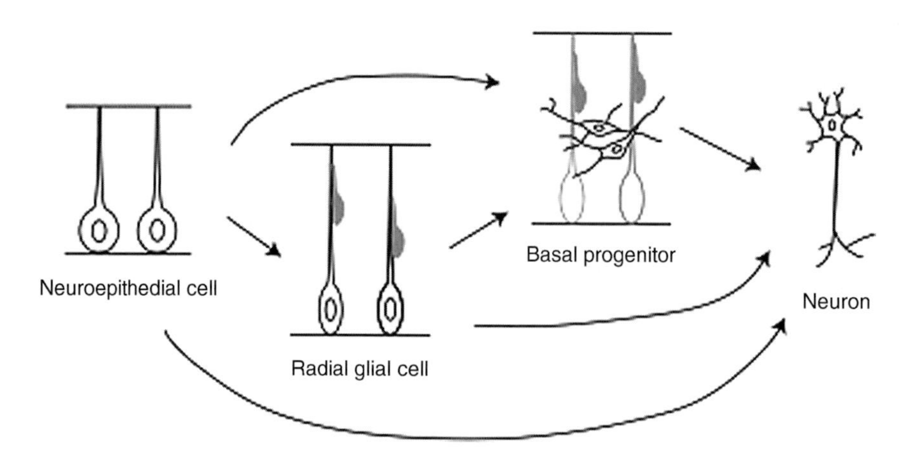

Neurogenesis. Figure 1 Various lineage relationships have been shown to exist or are thought to exist between neuroepithelial cells, radial glial cells, basal progenitor cells and neurons. Three types of progenitor cells contribute to the generation of neurons in the mammalian central nervous system.

into the olfactory bulb to become interneurons. In the SGZ of the dentate gyrus of the hippocampus, new granule neurons have been found to be continuously produced locally in all mammals examined, including humans. Neurogenesis in the intact adult mammalian CNS outside these two regions appears to be extremely limited or nonexistent. After pathological stimulation, such as by a brain insult, adult neurogenesis appears to occur in regions otherwise considered non-neurogenic [1].

Higher Level Processes

Neuroepithelial cells, radial glial cells and basal progenitors have been shown to generate CNS neurons. All three types of these progenitor cells are known to undergo symmetric or ►asymmetric cell divisions during neurogenesis. ►Symmetric cell divisions generate two daughter cells having the same fate while asymmetric cell divisions generate one daughter cell that is identical to the mother cell and a second cell of a different type. Progenitor cells are thought to produce neurons by a variety of processes (Fig. 1) [4,5].

Early in development both neuroepithelial cells and radial glial cells increase in number by undergoing symmetric divisions that lead to the expansion of the progenitor pool and the neural tube wall. There is another type of ►symmetric cell division that produces two neurons. Some neuroepithelial cells and radial glial cells and almost all basal progenitors are thought to divide by this type of cell division. However, the symmetric neurogenic divisions may very well be asymmetric in terms of neuronal subtype, consistent with the observation that some pairs of neurons arising from single progenitors have been found to differ in the expression of certain transcription factors.

Most radial glial cells divide asymmetrically, with some divisions generating another radial glial cell and a neuron, while others generate a radial glial cell and a basal progenitor. This type of cell division enables progenitor cells to generate neurons without losing the characteristics of the original cell.

Neuroepithelial cells and radial glial cells undergo cell division at the ventricular surface, whereas basal progenitors divide away from the surface. Investigation of the relationship between the location of the dividing cell and cell fate has shown that basal divisions only generate neurons, consistent with the expression of neuronal markers in basal progenitors. This finding suggests that basal progenitors are committed to the neuronal lineage. The length of the cell cycle of the basal progenitors is also different. They have a longer G2 phase than progenitors that divide at the ventricular surface [4].

Lower Level Processes

How do the immature neurons generated behave after neuronal differentiation? Various classes of CNS neurons migrate from their birthplace to their final positions [8]. In the developing neocortex, neurons are generated in the VZ and/or SVZ and then move to the developing cortical plate via radial migration. Recent studies by time-lapse imaging in slice cultures have revealed three modes of radial movement by cortical neurons. They are described briefly as ►somal translocation, in which the cell body moves toward the pial surface by shortening its radial process, ►locomotion, i.e. guided migration in which neurons move to a position beneath the pial surface along the basal process of radial glia and ►multipolar migration, migration of neurons residing in the SVZ and intermediate zone in which neurons with multiple processes extending and retracting dynamically move slowly toward the pial surface. In addition, most cortical interneurons in mice are known to derive from the subpallial telencephalon, including the medial and caudal ganglionic eminences and to migrate tangentially into the pallium. While medial ganglionic eminence-derived cells migrate laterally and spread widely throughout the cerebral cortex, caudal ganglionic eminence-derived cells migrate caudally toward the caudal-most end of the telencephalon (named the caudal migratory stream) and spread into the hippocampus and posterior cerebral cortex. There may be additional modes of neuronal migration in distinct areas that lead to the construction of elaborate structures.

Process Regulation

A number of pathways regulating neurogenesis in the CNS have recently been identified [9]. Extrinsic cues, including cell-cell interactions and secreted molecules, are key determinants of progenitor cell fate regulation. Exposure to growth factors, such as fibroblast growth factor (FGF)-2 and epidermal growth factor (EGF), as well as activation of the transmembrane receptor Notch, inhibits neuronal differentiation from progenitor cells, while several extrinsic factors (such as platelet-derived growth factor [PDGF], vascular endothelial growth factor [VEGF], neurotrophic factors, etc.) promote neuronal differentiation.

The molecular machinery governing whether progenitor cells divide symmetrically or asymmetrically has recently begun to be unraveled and centrosomal proteins and heterotrimeric G-proteins have been reported to control these divisions. Nde1, a central component of the centrosome, is expressed in the VZ and Nde1-deficient mice have smaller brains as a result of an increase in asymmetric cell divisions that leads to premature neurogenesis and depletion of the progenitor pool. Other centrosomal proteins, such as ASPM (abnormal spindle-like microcephaly associated), CDK5RAP2 and CENPJ, have been shown to be crucial determinants of cerebral cortex size, suggesting that these proteins also participate in the regulation of symmetric and asymmetric cell divisions. In addition, the Gβγ subunits of heterotrimeric G-proteins have recently been reported to play a key role in regulating spindle orientation and the asymmetric cell fate of progenitor cells [5].

Two transcription factors have been reported to have a role in determining whether progenitor cells undergo symmetric or asymmetric progenitor cell division. Emx2, a homeodomain transcription factor, promotes symmetric cell divisions by progenitor cells, whereas the paired-type homeodomain transcription factor Pax6 promotes asymmetric, neurogenic cell divisions. These transcription factors may therefore regulate cell fate and the appropriate mode of cell division in a coordinated manner.

Pathology

One of the neurodevelopmental disorders, autosomal recessive primary microcephaly (MCPH), has been suggested to be the result of deficient neurogenesis within the neuroepithelium [10]. MCPH is characterized by two principal features, microcephaly (small brain) present at birth and nonprogressive mental retardation. There are at least seven MCPH loci, and four of the genes have been identified, MCPH1, Microcephalin; MCPH3, CDK5RAP2; MCPH5, ASPM and MCPH6, CENPJ. Present evidence suggests that these genes are involved in mitosis by neural progenitor cells and are presumably related to DNA repair and control of mitotic spindles. Autosomal recessive periventricular heterotopia with microcephaly (ARPHM) is another disorder related to deficient neurogenesis and it has recently been reported that mutations in the ARFGEF2 gene cause ARPHM. This finding suggests that vesicle trafficking in neural progenitors is an important regulator of their proliferation and migration during cerebral cortical development.

Genetic studies of human malformations have proved a surprising source for discovery of molecules that regulate neuronal migration [8], including doublecortin and Lis1, mutations of which cause a profound migratory disturbance known as classical lissencephaly (smooth brain). There are other lissencephaly syndromes and mutations in the RELN (reelin) gene have been found in lissencephaly with cerebellar hypoplasia. X-linked lissencephaly with abnormal genitalia is associated with agenesis of the corpus callosum and ambiguous or underdeveloped genitalia. Mutations in the Aristaless-related homeobox transcription factor gene, ARX, have recently been found in these patients.

►Evolution and Embryological Development, of the Cortex in Amniotes

References

1. Ming GL, Song H (2005) Adult neurogenesis in the mammalian central nervous system. Annu Rev Neurosci 28:223–250
2. Oppenheim RW (1991) Cell death during development of the nervous system. Annu Rev Neurosci 14:453–501
3. Fujita S (2003) The discovery of the matrix cell, the identification of the multipotent neural stem cell and the development of the central nervous system. Cell Struc Func 28:205–228
4. Gotz M, Huttner WB (2005) The cell biology of neurogenesis. Nat Rev Mol Cell Biol 6:777–788
5. Huttner WB, Kosodo Y (2005) Symmetric versus asymmetric cell division during neurogenesis in the developing vertebrate central nervous system. Curr Opin Cell Biol 17:648–657
6. Smart IH, Dehay C, Giroud P, Berland M, Kennedy H (2002) Unique morphological features of the proliferative zones and postmitotic compartments of the neural epithelium giving rise to striate and extrastriate cortex in the monkey. Cereb Cortex 12:37–53
7. Nieto MA (2001) The early steps of neural crest development. Mech Dev 105:27–35
8. Kubo K, Nakajima K (2003) Cell and molecular mechanisms that control cortical layer formation in the brain. Keio J Med 52:8–20
9. Temple S (2001) The development of neural stem cells. Nature 414:112–117
10. Woods CG, Bond J, Enard W (2005) Autosomal recessive primary microcephaly (MCPH): a review of clinical, molecular, and evolutionary findings. Am J Human Genet 76:717–728

Neurogenesis and Inflammation

Liliana Bernardino, Fabienne Agasse, João O. Malva
Center for Neuroscience and Cell Biology, Institute of Biochemistry, Faculty of Medicine, University of Coimbra, Coimbra, Portugal

Definition

Inflammation and ►neurogenesis is a research area devoted to dissect the functional interplay between brain immune response and inflammation and production of new neurons and self-repair capacity of the brain.

Characteristics

Introduction

Until recently, the brain was considered as an immune privileged organ, unable to respond with an immune reaction in response to neurodegeneration or to infection, a property mainly attributed to the special protection conferred by the endothelial barrier. However, in the last decade the immune privileged status of the brain has been questioned mostly because in pathological conditions, molecules and cells of the immune system enter the brain and target the brain parenchyma [1]. In healthy conditions, the Central Nervous System (CNS) is routinely surveyed by few immune cells (e.g., T lymphocytes) in search of pathogens. Nevertheless, under pathological conditions, T cells accumulate and can reach relatively high density in inflammatory sites in brain parenchyma. Moreover, it is now well known that the inflammatory response in the CNS (►neuroinflammation) comprises a complex and integrated interplay between different cellular types of the immune system (macrophages, T and B lymphocytes, dendritic cells) and resident cells of the CNS (►microglia, astrocytes, oligodendrocytes, neurons).

Recently, neurogenesis in the CNS has emerged as an important physiological process, especially for plastic structures like the hippocampus and the olfactory bulb. The identification of stem cell niches in the adult brain opens new and fascinating perspectives for future development of strategies for ►brain repair. With this major challenge in mind we must better understand how local cell microenvironment and inflammation (usually accompanying neurodegeneration) affects the final outcome of stem cell proliferation and differentiation.

Inflammation in the Brain

Microglia: The Brain Immune Survey

Microglial cells are the major resident immunocompetent population of cells in the CNS parenchyma. Under physiological conditions, microglia cells display a resting-like phenotype with a ramified morphology and a low or absent expression of immunological molecules and their receptors. Although, in response to an injury, microglial cells became rapidly activated, changing to a round-shaped morphology with thickened and short or retracted processes, proliferate, are recruited to the site of injury and strongly up-regulate the expression and the release of inflammatory mediators such as pro-inflammatory ►cytokines, ►chemokines, reactive oxygen species and nitric oxide [2]. Equipped with receptors and ion channels, microglia can sense their microenvironment and detect the appearance of unusual concentrations of several soluble factors such as neurotransmitters, abnormal endogenous proteins such as beta-amyloid peptide, cell debris and exogenous compounds such as the endotoxin lipopolysaccharide (LPS) and some viruses. LPS, a component of the outer membrane of Gram-negative bacteria, is a well known inducer of cytokines, chemokines and other inflammatory mediators in brain experimental models of neuroinflammation. The most well accepted assumption is that the primary role of activated microglia is to support neuronal function; however, the sustained or excessive activity of these cells may have detrimental consequences and can be associated with the onset and/or exacerbation of neuronal death associated with neurodegenerative disorders [2].

Cytokines

An essential feature of the early immune response to pathogens and cell debris is secretion of pro-inflammatory cytokines by immune cells. Cytokines are low molecular weight proteins responsible for the communication between glial, neuronal and immune cells. Upon binding to their receptors, either soluble or located on the cell membrane, several intracellular pathways are usually triggered, which in turn regulate the activity of transcription factors such as NF-κB, and AP-1. The expression of these molecules and their respective receptors in the CNS are barely detectable under normal conditions, but become stimulated in response to pathological conditions, like ischemia, excitotoxicity and epilepsy. In the brain, cytokines can be originated from two different sources: (i) from infiltrating peripheral immune cells (lymphocytes and macrophages); (ii) from CNS resident cells (neurons, microglia, astrocytes, oligodendrocytes, endothelial cells and perivascular microglia). Functionally, cytokines have been described as being either pro-inflammatory (e.g., Tumor Necrosis Factor (TNF)-α, Interleukin (IL)-1β) or anti-inflammatory (e.g., IL-10, Transforming Growth Factor (TGF)-β, IL-1ra) depending on the final balance of their effects. Although, contributing to the complexity of this classification, the functional role of individual cytokines in CNS inflammation can shift from beneficial to detrimental [3]. Moreover, many cytokine actions are indirect, acting by stimulating the synthesis and function of other cytokines, resulting in a complex "cytokine cascade" triggered by immune and inflammatory responses. Cytokine activity can also be modulated by neurotransmitters, neuropeptides, growth factors and hormones. Therefore, cytokines can act in the CNS as both immunoregulators and neuromodulators in health and disease.

Chemokines

Chemokines are a family of chemo-attractant proteins, structurally related to cytokines with a major role in chemotaxis. In the CNS, chemokines and their receptors are constitutively expressed, at low levels, in astrocytes, microglia, oligodendrocytes, neurons and endothelial cells both in the developing and healthy adult brain, and their expression is induced by inflammatory mediators [4]. They have been reported to be involved in the developmental organization of the brain and in the maintenance of normal brain homeostasis by regulating the migration, proliferation and differentiation of glial and neuronal precursor cells. Moreover, other physiological functions of chemokines have been reported in CNS, such as regulation of synaptic transmission and plasticity, regulation of neurotransmitter release, modulation of ion channel activity, and cell death and survival. Upregulation or dysregulation of chemokines expression plays a crucial role in neurodegeneration and excitotoxicity, being involved in the communication between damaged neurons and surrounding glial cells and in the regulation of neuronal signaling by a diversity of processes. Moreover, the expression of chemokines can be regulated by cytokines, and this has been associated with several acute and chronic inflammatory conditions in the CNS.

Inflammation in Brain Diseases

The efficient regulation of the inflammatory cascade results from a fine-tune equilibrium of events regulating both immune privilege, in health, and effective responses to injury or disease. In general, acute inflammation is beneficial to the organism in limiting the survival and proliferation of invading pathogens and promoting regeneration of the tissue. However, prolonged, excessive inflammation is highly detrimental, leading to the onset and/or to the exacerbation of cell damage in neurodegenerative diseases. Increasing evidence has shown that neurotransmitters released from nerve terminals are involved not only in the communication between neurons, but also between nerve terminals and glial cells, including microglia. Most forms of neuronal injury have been associated with excitotoxicity, i.e., excessive release of excitatory amino acids such as glutamate, and subsequent activation of NMDA and AMPA/Kainate receptors. Besides glutamate, ATP is a co-transmitter released by injured or dying neurons. Thus, glutamate and ATP, play important roles in modulating microglia activation and in determining the fate and extent of neuronal injury. In particular, both IL-1β and TNF-α, released mainly by microglia, can be neuroprotective and/or neurotoxic, contributing directly or indirectly to the initiation, maintenance and outcome of neuronal cell death, including acute insults such as ischemia, trauma and seizures and chronic neurodegenerative disorders such as Parkinson's and Alzheimer's disease.

Neurogenesis in the Brain

Stem/Progenitor Cells and Neurogenesis

New neurons are constantly generated in the adult mammalian brain. This process called neurogenesis occurs mainly in two restricted regions: the subventricular zone (SVZ) and the dentate gyrus of the hippocampus (DG) (Fig. 1). The new neurons generated in the SVZ migrate long distances towards the olfactory bulb where they differentiate into functional interneurons, and contribute to improve odor memory and discrimination [5]. In the hippocampus, new neurons migrate out of the subgranular zone (SGZ) of the dentate gyrus (DG) and differentiate into the granule cell layer (GCL) where they mature and participate in hippocampal functions associated with learning and memory.

In both regions, new neurons arise from a particular population of cells: the ►stem cells/►progenitor cells. In vitro, the stem/progenitor cells constantly proliferate in the presence of growth factors (EGF, FGF-2). Proliferation of a single stem cell leads to the formation of a floating spherical neurosphere formed by a clone of immature cells. Plated on an adherent support, cells differentiate and migrate out of the sphere, where it is possible to identify the presence of neurons, astrocytes (Fig. 2) and oligodendrocytes.

Neurogenesis/Gliogenesis and Self-Repair

Upon brain injury, neurogenesis is modulated in both the SVZ and the DG. Neurogenesis increases in the SVZ following injury, including neurodegenerative diseases, and newly born neurons can migrate towards lesion areas where they can differentiate and replace damaged neurons or oligodendrocytes [6,7]. Neurogenesis also increases in the DG after ischemia or epilepsy. However, in epilepsy, aberrant migration and synaptic

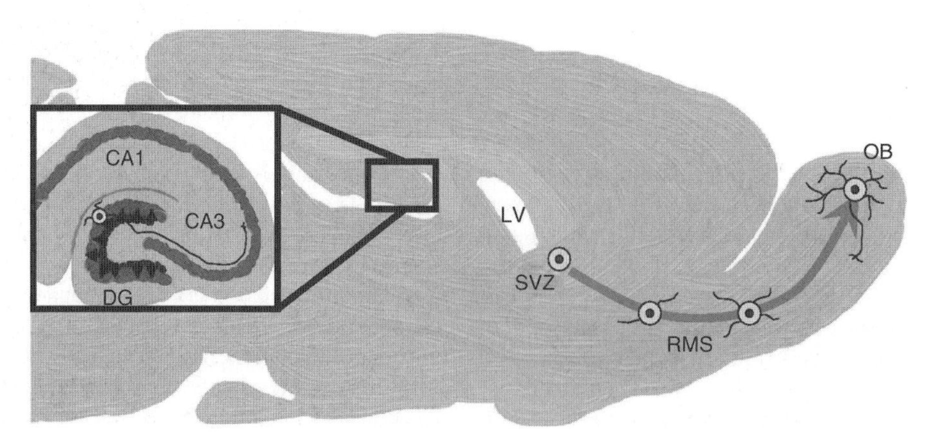

Neurogenesis and Inflammation. Figure 1 Sagittal view of a rodent brain showing the localization of the neurogenic areas: subventricular zone (SVZ) and the dentate gyrus (DG) of the hippocampus. The new neurons generated in the SVZ, adjacent to the lateral ventricles (LV), migrate long distances through the rostral migratory stream (RMS) and integrate in the olfactory bulb (OB) where they differentiate into interneurons. The new neurons generated in the subgranular layer of the hippocampus migrate locally into the granular layer, of the DG, where they differentiate into new granular cells (adapted from Taupin and Gage, 2002, J. Neurosci. Res. 69:745–749).

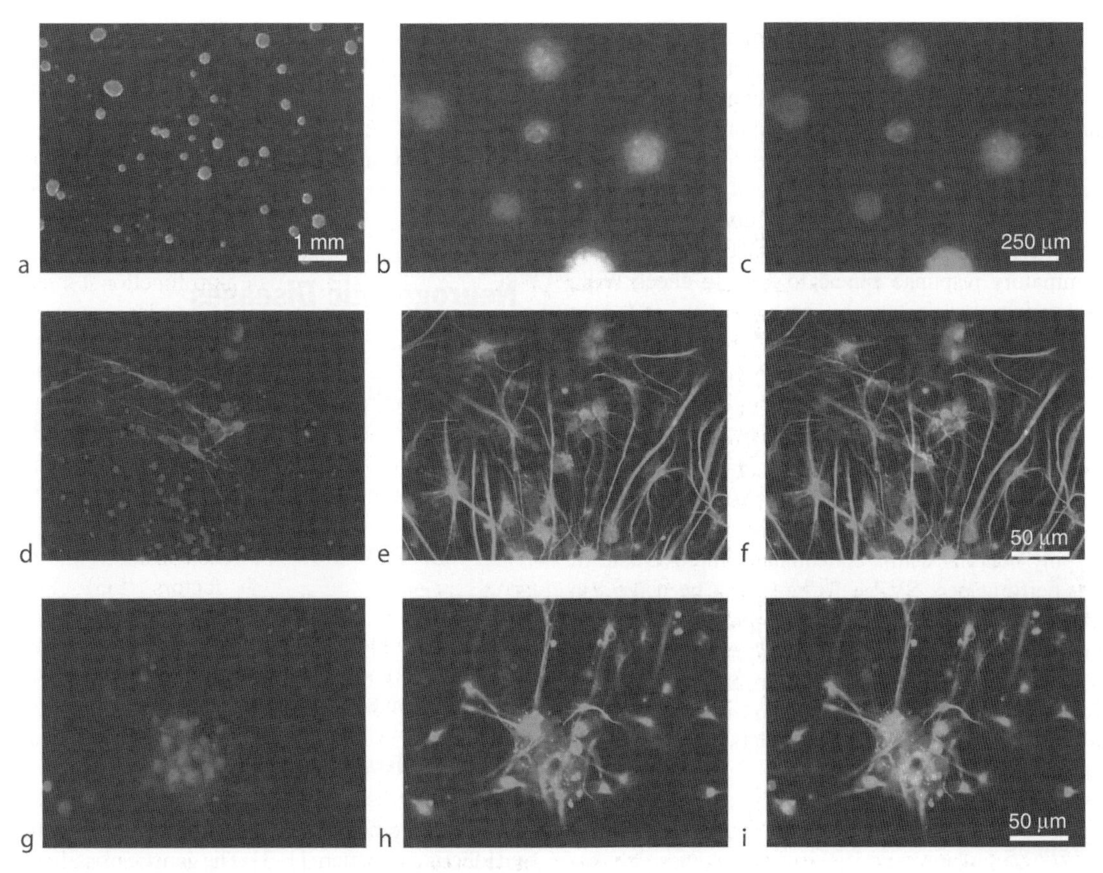

Neurogenesis and Inflammation. Figure 2 Representative photos of SVZ cell cultures from newborn mice in proliferation (a–c) and differentiation conditions (d–i).(a)–(c): in serum free medium with the growth factors EGF and FGF-2, SVZ cells generate clonal aggregates of cells called neurospheres (a–c). SVZ neurospheres obtained from GFP mice represent a useful tool for grafting studies (c).d–i: after plating on poly-D-lysine and withdrawal of the growth factors, cells arise from the edge of the neurosphere and develop into neurons expressing MAP-2 (d) and NeuN (g), and glial cells such as astrocytes expressing GFAP (e, h). (f) and (i) photos are merged of (d), (e) and (g), (h), respectively.

connections of newborn neurons may exacerbate recurrent seizures.

In spite of our growing knowledge about stem/ progenitor cell physiology, efficient replacement of the damaged neuronal circuits is far away from full accomplishment using cell-based strategies of brain repair.

Neurogenesis and Inflammation in the Brain

Inflammation, activation of glial cells and release of inflammatory mediators play ambiguous effects on neurogenesis with stimulatory or inhibitory roles, depending on the specific players and brain disorder.

Inflammation Promotes Neurogenesis/Gliogenesis

Following injury, a great diversity of factors is secreted around the damaged area and these factors can diffuse, act on progenitor cells inducing neurogenesis/gliogenesis. For instance, injury-activated astrocytes and microglial cells produce the growth factors VEGF and FGF-2 and the pro-inflammatory cytokine (TNF-α) able to promote SVZ neurogenesis and proliferation. These processes involve the activation of transcription factors, including NF-κB.

In experimental demyelination, microglial cells secrete IL-1β that induces the secretion of IGF-1 by astrocytes, promoting proliferation of oligodendrocyte progenitor cells and remyelination [8].

Inflammation Inhibits Neurogenesis

On the other edge of the sword, inflammation can inhibit neurogenesis and regeneration. Indeed, activated astrocytes that accumulate in the lesion core forming a glial scar can limit the propagation of the injury. However, the glial scar is also rich in repulsive axon guidance molecules such as proteoglycans, ephrins, semaphorins and Nogo. So, on the one hand the glial scar limits the propagation of the injury, but also constitutes a barrier for efficient regeneration.

In the hippocampus, seizure activity promotes neurogenesis, but subsequent neuronal death and microglial activation inhibits neurogenesis hampering brain repair. In these conditions, systemic administration of minocycline, which inhibits microglia activation, restores neurogenesis in the dentate gyrus. Part of these effects may involve TNF-α, IL-6 and TGF-β since these cytokines are secreted by activated microglial cells and impair neurogenesis in the dentate gyrus [9].

In conclusion, the glial scar and "the dark side" of the inflammatory response appear to hold in check brain repair endeavors.

SVZ Cells as a Tool for Brain Repair

The results of several studies show that SVZ cells are good candidates for cell-based therapy. For instance, in animal models of Parkinson's and Huntington's diseases, SVZ cells grafted in the striatum are able to differentiate respectively into dopaminergic neurons and spiny neurons with a concomitant improvement of motor performance. SVZ cells have also been used in models of multiple sclerosis where chronic inflammation is a major cause of demyelination of the axons. Grafts of SVZ cells into the subcortical white matter of myelin-deficient mice results in the migration of PSA-NCAM progenitors cells, differentiation of oligodendrocytes and successful remyelination [10].

References

1. Nguyen MD, Julien JP, Rivest S (2002) Innate immunity: the missing link in neuroprotection and neurodegeneration? Nat Rev Neurosci 3:216–227
2. Hanisch UK (2002) Microglia as a source and target of cytokines. Glia 40:140–155
3. Bernardino L, Xapelli S, Silva AP, Jakobsen B, Poulsen FR, Oliveira CR, Vezzani A, Malva JO, Zimmer J (2005) Modulator effects of Interleukin-1 beta and Tumor Necrosis Factor–alpha on AMPA-induced excitotoxicity in mouse organotypic hippocampal slice cultures. J Neurosci 25:6734–6744
4. Ubogu EE, Cossoy MB, Ransohoff RM (2006) The expression and function of chemokines involved in CNS inflammation. Trends Pharmacol Sci 27:48–55
5. Rochefort C, Gheusi G, Vincent JD, Lledo PM (2002) Enriched odor exposure increases the number of newborn neurons in the adult olfactory bulb and improves odor memory. J Neurosci 22:2679–2689
6. Picard-Riera N, Decker L, Delarasse C, Goude K, Nait-Oumesmar B, Liblau R, Pham-Dinh D, Evercooren AB (2002) Experimental autoimmune encephalomyelitis mobilizes neural progenitors from the subventricular zone to undergo oligodendrogenesis in adult mice. Proc Natl Acad Sci USA 99:13211–13216
7. Kokaia Z, Lindvall O (2003) Neurogenesis after ischaemic brain insults. Curr Opin Neurobiol 13:127–132
8. Mason JL, Suzuki K, Chaplin DD, Matsushima GK (2001) Interleukin-1beta promotes repair of the CNS. J Neurosci 21:7046–7052
9. Monje ML, Toda H, Palmer TD (2003) Inflammatory blockade restores adult hippocampal neurogenesis. Science 302:1760–1765
10. Cayre M, Bancila M, Virard I, Borges A, Durbec P (2006) Migrating and myelinating potential of subventricular zone neural progenitor cells in white matter tracts of the adult rodent brain. Mol Cell Neurosci 31:748–758

Neurogenetic Diseases

THOMAS D. BIRD
Department of Neurology, UW Medical Center, Seattle, WA, USA

Synonyms

Hereditary neurological disorders

Definition

Neurogenetic Diseases are disorders of the central and peripheral nervous systems caused by molecular defects in heritable material (usually DNA).

Characteristics

Neurogenetic diseases can be classified on the basis of the clinical syndromes or on the basis of the genetic etiology and inheritance pattern [1–3]. The genetic classification is used here with each category containing examples of clinical syndromes.

Chromosomal

Major aberrations of chromosomal material usually produce defects in multiple organ systems and are frequently recognized at birth. They may be caused by a variety of chromosomal aberrations including deletions, duplications, translocations, and ring chromosomes. They are often associated with mental retardation, epilepsy, and dismorphic features. Examples include Down syndrome (trisomy 21), Cri du chat (5p-) and Miller-Dieker Lissencephaly syndrome (17p-).

Mendelian

Diseases in this category are caused by mutations in nuclear DNA, including missense, nonsense, deletions, and duplications of one or more nucleotides. The disease is inherited as an autosomal dominant if the clinical syndrome is present in heterozygous carriers of a single mutation. Examples include Huntington's disease, Myotonic muscular dystrophy, Fascioscapulohumoral muscular dystrophy, tuberous sclerosis, Charcot-Marie-Tooth hereditary neuropathy, ►familial spastic paraplegia, and hereditary ataxias [4]. If clinical manifestations occur only in carriers of two mutations, one on each of the homologous chromosomes (homozygotes or compound heterozygotes), the disease is autosomal recessive. Examples include Friedreich's ataxia, Tay-Sach's disease, Neimann Pick diseases, and phenylketonuria. Mutations on the X chromosome produce x-linked inheritance in which the mutations are usually confined to, or more severe in males and absent or much milder in carrier females. Examples include Duchenne's muscular dystrophy, Pelizaeus-Merzbacher leukodystrophy, and x-linked Charcot-Marie-Tooth hereditary neuropathy.

Many of these neurogenetic syndromes show remarkable genetic heterogeneity. For example, there are more than 20 genetic subtypes of hereditary neuropathy and more than 30 subtypes of familial spastic paraplegia and dominant hereditary ataxias.

Nucleotide repeat expansions: A special category of neurogenetic diseases is caused by abnormally large expansions of normally occurring nucleotide repeats. Many are trinucleotide repeats, often CAG. They are inherited in a mendelian fashion and age of onset is correlated with the size of the repeat expansion (larger expansions having earlier age of onset). Huntington's disease and several autosomal dominant spinocerebellar ataxias (SCA) are caused by CAG repeat expansions. Type 1 myotonic muscular dystrophy is caused by a CTG repeat expansion in the non-coding region of a gene and this expansion interferes with RNA transcription of multiple other genes. This explains the systematic nature of the clinical syndrome affecting many organ systems. X-linked Spinobulbar muscular atrophy is caused by a CTG repeat expansion in the androgen receptor gene on the X chromosome. One form of autosomal recessive myoclonic epilepsy is caused by expansion of a 12 nucleotide repeat (dodeca repeat).

Mitochondrial

►Mitochondrial diseases are the result of primary abnormalities in the respiratory chain electron transport system of mitochondria. Mutations of mitochondrial DNA are inherited from the cytoplasm of the mother's egg. All children of a mother with the mutation will also inherit the mutation. However, males with the mutation do not pass the mutation on to any children. Whether or not the individual expresses clinical signs of the mutation depends on the numbers of mitochondria with the relevant mutation in any given tissue (heteroplasmy). Mitochondrial diseases often have a combination of signs and symptoms frequently including cognitive deficits, seizures, visual loss, hearing loss, peripheral neuropathy, or myopathy. Examples include MELAS (mitochondrial encephalomyopathy with lactic acidosis and stroke), MERRF (myoclonic epilepsy with ragged red fibers), and NARP (neuropathy, ataxia and retinitus pigmentosa). Some mitochondrial disorders are caused by mutations in nuclear genes that control mitochondrial enzymes and these are inherited in a mendelian fashion.

Polygenic/Multifactorial

Individual autosomal, x-linked and mitochondrial neurogenetic diseases are each relatively rare. However, many common neurological disorders also have important genetic contributions to their pathogenesis. It is assumed that they represent the additive affect of several genes (polygenic) interacting with multiple environmental factors (multifactorial) [5]. Examples would be Alzheimer's disease, Parkinson's disease, Epilepsy, Stroke and Multiple Sclerosis. In some instances (e.g. Alzheimer's, Parkinson's and epilepsy) rare forms of the disease are caused by identified mutations in single genes, but the more common form of the disease in the general population is thought to be polygenic/multifactorial. In most instances the multiple genes and environmental factors are unknown. However, genome wide association studies are beginning to identify many candidate genes for these common complex diseases.

References

1. Rosenberg RN, Prusiner SB, DiMauro S, Barchi RL, Nestler EJ (eds) (2008) The Molecular and genetic basis of neurologic and psychiatric disease, 4th edn. Butterworth and Heinemann, PA, USA
2. Pulst SM (ed) (2000) Neurogenetics. Oxford University Press, Oxford, England
3. Lynch DR, Farmer JM (eds) (2006) Neurogenetics: scientific and clinical advances. Taylors and Francis, NY, USA
4. Bird TD, Jayadev S (2008) Genetic diseases of the nervous system. In: Rosenberg RN (eds) Atlas of clinical neurology, 3rd edn. Current Medicine, PA, USA
5. Bird TD, Tapscott SJ (2008) Clinical neurogenetics. In: Bradley WG, Daroff RB, Fenichel GM, Jankovic J (eds) Neurology in clinical practice, 5th edn. Butterworth and Heinemann, PA, USA pp. 781–806

Neurogenic Inflammation

Definition

Precapillary arteriolar vasodilation generated by activity in peptidergic afferent C-fibers and Aδ-fibers and postcapillary plasma extravasation generated by activity in peptidergic afferent C-fibers. The vasodilation is generated by release of calcitonin-gene-related peptide (CGRP) and (to a lesser degree Substance P) and the plasma extravasation by release of Substance P.

►Complex Regional Pain Syndromes: Pathophysiological Mechanisms

Neurogenic Niche

MARIANA ALONSO
Institut Pasteur, Perception and Memory Laboratory, Paris, France

Synonyms

Neuronal regeneration in the nervous system

Definition

A neurogenic niche is a region where neurogenesis takes place. Neurogenesis refers to the entire set of events leading to the production of new neurons from ▶precursor cell in the brain. The degree of neurogenesis depends on the interaction of the microenvironment (niche) with precursor cells that have neurogenic potential.

Characteristics

The adult mammalian central nervous system (CNS) has traditionally been divided into four major cell types: the neurons, the myeling-forming oligodendrocytes, the astrocytes, and the ependymal lining of the central lumen. All of those cell types are generated during development from a common source, the neuroepithelial cells that arise in early embryos in the form of the neural tube. There are three different cell types that contribute to neurogenesis, early neuroepithelial cells for the first neurons, radial glial cells for most neurons in most brain regions, and subventricular zone (SVZ) precursors predominantly at later stages of neurogenesis.

Moreover, it has become clear over the past decades that new neurons are continually generated in the adult brain. This postnatal neuronal production is a conserved biological phenomenon throughout evolution – it has been reported in fishes, amphibians, reptiles, birds, rodents and primates, including humans [1] – but its adaptive functions deserve yet to be explored.

Places were neurogenesis take place are called ▶neurogenic niches. In this neurogenic niche, specialized cells are neural stem cells, capable of self-renewing and generating neurons and glia. In general, stem cells niches are composed of microenvironmental cells that nurture stem cells and enable them to maintain tissue homeostasis. An appropriate spatiotemporal dialog occurs between stem and niche cells in order to fulfill lifelong demands for differentiated cells. The niche concept was introduced in 1978 by Schofield [2]. In ecological terms, an organism's niche refers to where it lives, what it does, and how it interacts with its close environment. Altering an ecosystem (or neurogenic environment) can produce disastrous consequences for an organism (or stem cell). Niche cells provide a sheltering environment that sequesters stem cells from differentiation stimuli, and other stimuli that would challenge stem cell reserves. The niche also safeguards against excessive stem cell production that could lead to cancer. Stem cells must periodically activate to produce ▶progenitor or transit amplifying cells that are committed to produce mature cell lineage. Thus, maintaining a balance of stem cell quiescence and activity is a hallmark of a functional niche. Importantly, stem cells themselves extensively interact with and participate in the niche. Furthermore, niches may in fact be dynamic structures that alter their characteristics over time concomitant with tissue remodeling.

Regulation of the Neurogenic Niches

It is clear now that complex bidirectional interactions between intrinsic programmes and extrinsic cues take place in the neurogenic niches to control its behavior. We can define intrinsic programmes here as the ensembles of factors expressed by stem cells and progenitors that control different neurogenic phases. By contrast, external factors are produced by surrounding tissues to act on stem cells and progenitors. Consequently, whether a cell undergoes self-renewal or differentiation is the result of the spatial and temporal convergence of niche cues and intrinsic state of the cell.

For instance, cell-cell interactions and diffusible signals are key elements allowing feedback control of stem cell activation and differentiation from progeny or the niche support cells (see Fig. 1).

Moreover, an emerging feature of several stem cells niches is the close association with endothelial cells forming blood vessels, which regulate stem cell self-renewal and differentiation. Within a niche, stem cells are frequently anchored to a basal lamina or stromal cells that can provide a substrate for oriented cell division. The basal lamina is also an important regulator of the accessibility of growth factors and other signals, as associated extracellular matrix molecules and glycoproteins can both concentrate and sequester factors in inactive or active forms. In addition, cell anchoring may orient cell division resulting in the segregation of key determinants into one or both daughter cells depending on the plane of division [3]. In this line, given that multipotent progenitor cells can give rise to neuronal and glial cell types in a characteristic order of birth, it is clear that progenitor cell proliferation must be precisely regulated.

Finally, the processes of newborn neuron production, migration, maturation and survival are all subject to modulation by external factors. Neurotransmitters, hormonal status, growth factors and injuries are known to influence proliferation [4] (Fig.1).

Neurogenic Niches in the Adult Mammalian Brain

Two regions, the olfactory bulb (OB) and the dentate gyrus (DG) of the hippocampus, have been shown beyond doubt to receive and integrate constitutively

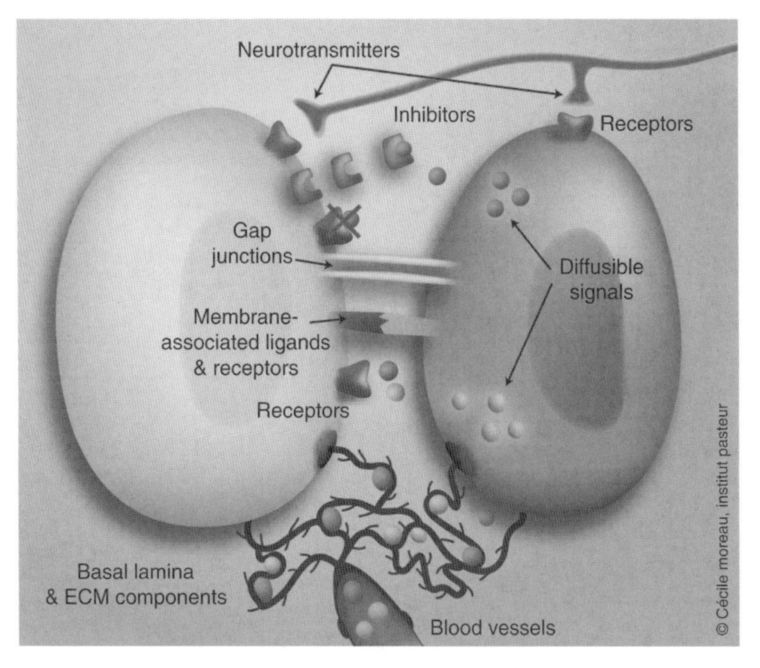

Neurogenic Niche. Figure 1 Membrane-associated receptors and ligands mediate cell-cell contacts, defining cell self-renewal and differentiation. In addition, gap junctions coordinate behavior between couple cells. Diffusible signals can direct stem cells to either self-renew or generated differentiated progeny. The availability of diffusible factors that bind to receptors in turn can be regulated by ligand inhibitors which can sequester these factors and prevent signaling. An extracellular matrix-rich basal lamina, which can be associated with blood vessels, has several functions in stem cell niches, including anchoring cells to the niche, sequestering and presenting diffusible signals, and linking cells and the extracellular matrix. Endothelial cells and the vasculature can regulate stem cells fate decisions. Stem cells can also be regulated by the release of neurotransmitters and other factors from axons.

newborn neurons throughout adult life in the mammalian brain [5] (see Fig. 2a and 2b).

In adult mammals, neurogenic zones where neural stem cells are harbored, exists at least, in two discrete areas of the brain: the subgranular zone (SGZ) of the DG (Fig. 2a) and the SVZ located near the lateral wall of the lateral ventricles (Fig. 2b). The neuroblasts that originate from SVZ precursor cells migrate long distances through the rostral migratory stream (RMS) to populate the OB where they differentiate into granular and periglomerular interneurons and establish contacts with their neuronal targets.

Thousand of young neurons migrate into the OB every day but only a fraction of them survive to complete their differentiation. The newborn cells recruited into the OB and the DG became truly neurons, throughout a unique sequence of events that leads to their functional maturation followed by complete integration.

In the two neurogenic areas of the adult brain, a subset of astrocytes, are the in vivo precursors for adult neurogenesis. In the SVZ, some of these cells contact the ventricle lumen and have a single cilium. Rapidly dividing, transit-amplifying cells derived from stem cells give rise to neuroblast.

As we mentioned before, the neurogenic behavior in both regions appears determined by signals restricted to their niches. Stem cells are in intimate contact with all other cell types, including the rapidly dividing transit amplifying and the neuroblasts (Fig. 2a and 2b).

Therefore, neural stem cells occupy niches formed by both astrocytes and endothelial cells. In general, endothelial cells encourage stem cells to renew themselves, and astrocytes instruct them to become neurons. More surprisingly, mammalian neural stem cells are not passive elements of their microenvironment: they can, under certain conditions, give rise to endothelial cells. This suggests that, if the need arises, stem cells can populate their niche with the features that they need to thrive.

Transplantation studies in mammals support the principle of defining neurogenic and non-neurogenic regions and provide evidence for the role of microenvironment in influencing the potential of neural precursors. If precursor cells are transplanted into neurogenic regions, they can differentiate into neurons in a region-specific manner. SVZ precursor cells generate hippocampal neurons when transplanted into the hippocampus, and SGZ precursor cells generate olfactory interneurons after transplantation into the RMS. When implanted outside the neurogenic regions, both types of precursor cells generate only glia. An inhibitory environment that is refractory to neurogenesis is therefore present throughout

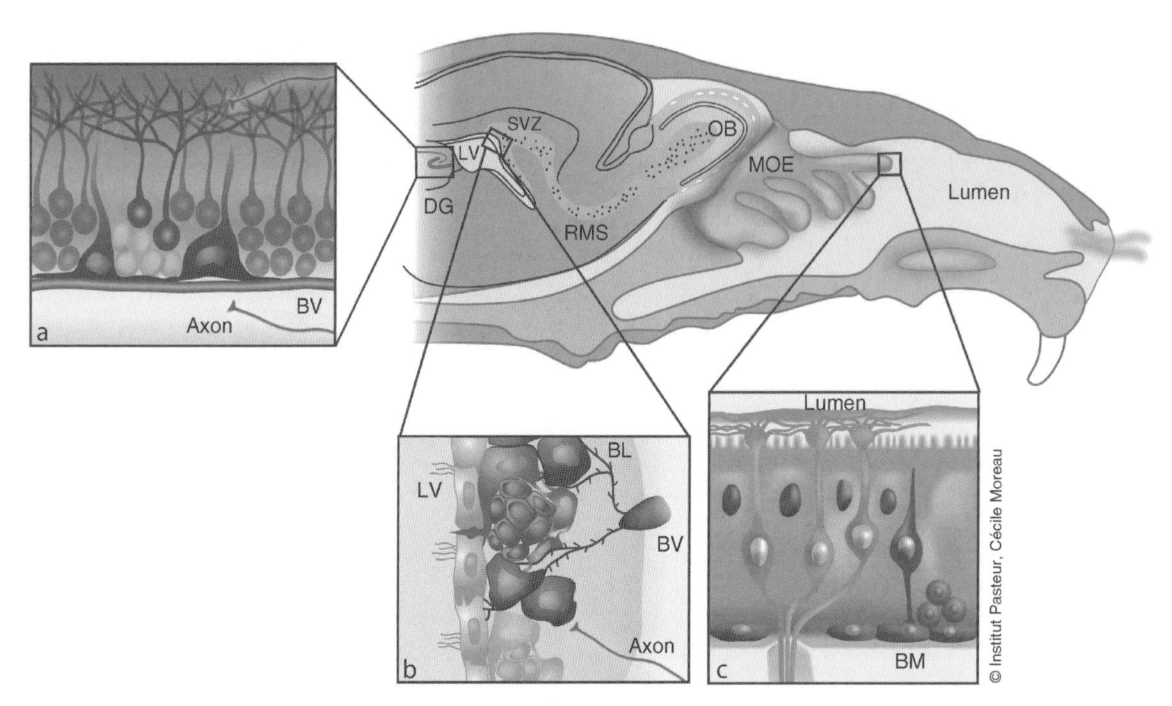

Neurogenic Niche. Figure 2 Neurogenic niches in the adult mammalian brain. Simplified sagital view of rat head showing neurogenic niches. (a) Dentate gyrus (DG) neurogenic niche. Astrocytes (blue) give rise to progenitors (orange), which mature into new granule cells (red). (b) Subventricular zone (SVZ) neurogenic niche. Astrocytes (blue) are the SVZ stem cells and also serve as niche cells. Transit-amplifying cells (green) derived from stem cells give rise to neuroblast (red) that migrate to the olfactory bulb. The basal lamina (BL, brown) extends from the blood vessel (BV) and interdigitates extensively with the SVZ cells. Ciliated ependymal cells (gray) line the lateral ventricule (LV) are shown. (c) Main olfactory epithelium (MOE) neurogenic niche. Horizontal (blue) and globose basal cells (green) are the stem cells. Some globose basal cells are transit-amplifying cells that give rise to olfactory sensory neurons (red). Basement membrane (BM) and supporting cells (gray) are shown. RMS, rostral migratory stream.

most of the adult brain. Thus, adult neurogenic niches have an instructive role in directing neuronal production and stem cell maintenance and shield ongoing neurogenesis from possible external inhibitory influences. Although the components of adult neurogenic niches that mediate these processes are still being elucidated, it is clear that both neural and non-neural cell types are key players. Finally, if it were a static, merely restorative process, adult neurogenesis could not be regarded as a mechanism for adult brain plasticity. However, we know that every aspect of adult-born cell production is tightly regulated and modulated. This strongly suggests that the adult brain can tailor its production of new neurons to match the demands of its environment.

Neurogenesis in the Olfactory Epithelium

In the olfactory system, the main olfactory epithelium (MOE) is also submitted to continual neurogenesis in adults (Fig. 2c). The olfactory sensory neurons (OSNs), which detect odors in the air we breathe, are located within this epithelium and continue to be generated throughout life. This lifelong neurogenesis occurs as a result of continual proliferation and differentiation of progenitors cells located near the base of the epithelium. The olfactory neuroepithelium is in direct contact with the external environment and, as a consequence, has evolved a remarkable ability to replenish sensory neurons lost during natural turnover and in the event of extensive lesions or traumatic injuries. This ongoing adult neurogenesis is essential for maintaining olfactory sensory function.

In the postnatal brain of rodents, and continuing into adulthood, three cell types comprise the OSNs lineage within the olfactory epithelium: the horizontal basal cells (HBCs) which lie directly against the basal lamina, the globose basal cells (GBCs) which are primarily situated immediately apical to the HBCs, and the OSNs (Fig. 2c).

HBCs function as adult olfactory neuroepithelium stem cells and are competent to regenerate both neuronal and non-neuronal lineage in this region (Fig. 2c). HBCs serve as a reservoir of long-lived progenitors that remain largely quiescent during normal neuronal turnover or even after acute, selective loss of mature neurons. Under these conditions, GBCs are largely responsible for tissue maintenance. This characteristic is unique showing a model of adult neurogenesis in which distinct cell population within the same niche mediate normal neuronal turnover and neuronal replacement upon traumatic injury [6]. As was describe previous, different factors that are both made by and found within the microenvironment of the MOE stem and progenitor cells exert crucial growth regulatory effects on these cells. Thus, as with other regenerating tissues, the basis of regeneration in the MOE appears to be a population of stem cells, which resides within a microenvironment (neurogenic niche) consisting of factors crucial for maintenance of its capacity for proliferation and differentiation.

Relevance to Humans

Elucidating the nature of microenvironmental factors released by host tissue in the neurogenic niches that might affect these processes became particularly interesting given that it was demonstrated that adult neurogenesis occurs also in adult human brain. Adult neurogenesis was demonstrated in human hippocampus [7]. As well, a ribbon of SVZ astrocytes lining the lateral ventricles of the human adult brain that proliferate *in vivo* and behave as multipotent progenitor cells *in vitro* was recently describe [8]. In addition, a recent study demonstrates the presence of a human RMS containing migratory progenitor cells some of which become mature neurons in the OB [9]. Moreover, adult neurogenesis was also found in the human MOE [10].

In this context, adult neurogenesis is attracting a lot of attention because of the hope raised by the use of adult neuronal stem cells in regenerating and reconstructing the damaged brain, for instance, in the therapies of neurodegenerative diseases. However, before elaborating strategies aimed at using endogenous progenitors and their relevance to human clinics, it is urgent to precise theirs normal function(s) in a non-pathological brain.

References

1. Alvarez-Buylla A, Garcia-Verdugo JM (2002) Neurogenesis in adult subventricular zone. J Neurosci 22:629–634
2. Schofield R (1978) The relationship between the spleen colony-forming cell and the haemopoietic stem cell. Blood Cells 4:7–25
3. Moore KA, Lemischka IR (2006) Stem cells and their niches. Science 311:1880–1885
4. Abrous DN, Koehl M, Le Moal M (2005) Adult neurogenesis: from precursors to network and physiology. Physiol Rev 85:523–569
5. Lledo PM, Alonso M, Grubb MS (2006) Adult neurogenesis and functional plasticity in neuronal circuits. Nat Rev Neurosci 7:179–193
6. Leung CT, Coulombe PA, Reed RR (2007) Contribution of olfactory neural stem cells to tissue maintenance and regeneration. Nat Neurosci 10:720–726
7. Eriksson PS, Perfilieva E, Bjork-Eriksson T, Alborn AM, Nordborg C, Peterson DA, Gage FH (1998) Neurogenesis in the adult human hippocampus. Nat Med 4:1313–1317
8. Sanai N, Tramontin AD, Quinones-Hinojosa A, Barbaro NM, Gupta N, Kunwar S, Lawton MT, McDermott MW, Parsa AT, Manuel-Garcia Verdugo J, Berger MS, Alvarez-Buylla A (2004) Unique astrocyte ribbon in adult human brain contains neural stem cells but lacks chain migration. Nature 427:740–744
9. Curtis MA, Kam M, Nannmark U, Anderson MF, Axell MZ, Wikkelso C, Holtas S, vanRoon-Mom WM, Bjork-Eriksson T, Nordborg C, Frisen J, Dragunow M, Faull RL, Eriksson PS (2007) Human neuroblasts migrate to the olfactory bulb via a lateral ventricular extension. Science 315:1243–1249
10. Hahn CG, Han LY, Rawson NE, Mirza N, Borgmann-Winter K, Lenox RH, Arnold SE (2005) In vivo and in vitro neurogenesis in human olfactory epithelium. J Comp Neurol 483:154–163

Neurogenic Pain

►Neuropathic Pain

Neurogenic Vasodilatation

Definition

Vasodilatation induced by depolarization of nociceptive nerve terminals in the body periphery. These terminals release calcitonin-gene–related peptide (CGRP), a vasoactive substance. Synonyms: neurogenic inflammation, axon reflex flare.

►Calcitonin Gene Related Peptide (CGRP)
►Nociceptors and Characteristics

Neurogliaform Cells

Definition

Neurogliaform refers to the neuroglial ectodermal cell type. Neurogilaform cells, neuroglial cells, and glial cells are synonymous terms.

Neurogram

Definition

Multi-unit recording from a nerve.

►Extracellular Recording

Neurohormone

Definition

A compound released by a neuron into the circulation, with other neurons as its major target.

Neurohumoral Agent

Definition

Synonym for neurotransmitter. A chemical substance released at a synapse from a presynaptic neuron that binds on a receptor of a postsynaptic neuron and stimulates or inhibits it.

►Neurotransmitter
►Synapse

Neurohypophysis

Definition

Neurohypophysis or the posterior lobe of the pituitary, is the site where the terminals of the Supraoptic and Paraventricular nuclei release neurosecretory granules containing vasopressin or oxytocin.

►Drinking Disorders and Osmoregulation
►Posterior Lobe of the Hypophysis
►Diencephalon

Neuroimaging

Definition

The use of radiographic studies and magnetic resonance imaging to detect structural abnormalities in the central nervous system; visual display of structural or functional patterns of the nervous system as a whole or any of its parts for diagnostic evaluation or visualization of anatomical structures; includes measuring physiologic and metabolic responses to physical and chemical stimuli.

►Magnetic Resonance Imaging (MRI)

Neuroimaging

Modern techniques applied in order to assess alterations of the central nervous system.

►Forensic Neuropsychiatry

Neuroimmune Interactions – Serotonin

David Dubayle, D. Menetrey
CNRS UMR 8119, Neurophysique et Physiologie,
Université René Descartes, UFR Biomédicale,
Paris, France

Synonyms

5-Hydroxytriptamine (5-HT) Neuroimmunomodulation

Definition

The role of ►serotonin (5-hydroxytriptamine or 5-HT) in the bidirectional influences that the nervous and immune systems exert over each other by biochemical and cellular routes.

Characteristics

Characteristics of Serotonin

Dual Location

Serotonin (5-HT) is common to both the nervous and immune systems and, in consequence of its dual location, is involved in the bidirectional interactions these systems generate with each other. By extension, serotonin must be regarded as a fundamental element of the defence system that living organisms have developed against stressors.

Metabolism

Serotonin is formed by tryptophan (5)-hydroxylase from L-tryptophan. L-tryptophan is common to the immune-related ►kynurenine synthesizing pathway (Fig. 1). An increase of kynurenine biosynthesis will lead to a decrease in the availability of L-tryptophan, and reduce 5-HT levels.

Receptors

Serotonin acts through a variety of receptor subtypes. Fourteen subtypes have been recognized to date, with specific location within the central nervous system, and at least four (5-HT_{1A}, 5-$HT_{1B/1D}$, 5-HT_2 and 5-HT_3) are present on immunocompetent cells [2]. 5-$HT_{1B/1D}$ receptor activity is down-regulated by an endogenous tetrapeptide named 5-HT-moduline, likely originating from the adrenal medulla and released by stress.

Functions in Immune Regulation

Four major serotonin actions have been recognized in immune regulation: (i) the activation of T-cells and natural killer-cells, (ii) the production of chemotactic factors, (iii) the modulation of delayed-type hypersensitivity responses, and (iv) the regulation of the natural immunity delivered by macrophages [3].

Responses to Immune Activity

Immune activity affects 5-HT brain levels. Effects depend on circumstances. Acute peripheral immunization raises brain 5-HT turn-over through the local release of specific inflammatory cytokines, mostly interleukin-1, and via sympathetic nervous system activity. On the other hand, chronic immune activation, which is common in normal ageing and accompanies ►neurodegenerative disorders such as Alzheimer's, Huntington's and Parkinson's diseases, leads to 5-HT depletion [1] by promoting L-tryptophan degradation via the kynurenine synthesizing pathway. 5-HT lowering could account for the mood alteration and impaired cognition that accompany ageing and chronic pathological conditions. Light exposure raises 5-HT levels [4] and fights moodiness.

Anatomical Substrate

The anatomical substrate for 5-HT-based neuroimmune interactions is triple including: (i) the central serotonergic system of the brain stem raphé nuclei (►raphé nuclei system), (ii) the "serotonergic/noradrenergic" nerve terminals of the peripheral ►autonomic nervous system (NS) and, (iii) the 5-HT-positive immunocytes that penetrate the nervous system. 5-HT cellular-mediated actions are essential for neuroimmune cross-talk since access of blood-borne 5-HT and

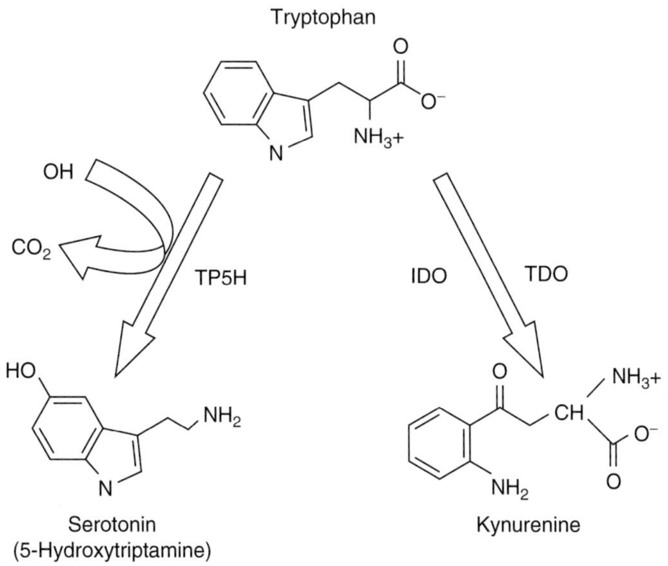

Neuroimmune Interactions – Serotonin. Figure 1 The dual catabolic pathway of tryptophan. On the left, tryptophan (5)-hydroxylase (TP5H) initiates the production of serotonin (5-hydroxytriptamine); on the right, tryptophan (2,3)-dioxygenase (TDO) and indoleamine (2,3)-dioxygenase (IDO) catalyze the formation of kynurenine. Adapted from [1].

L-tryptophan to the brain is highly restricted by the endothelial ►blood-brain barrier (BBB) except for the circumventricular organs where this later is disrupted. The metabolism of L-tryptophan by microvascular endothelial cells is primordial in restricting microbial expansion in the nervous system [5].

Characteristics of the Raphé Serotonergic System

Anatomy

This system, which is the only source of central nervous serotonergic innervation, widely distributes to the brain via arborescent efferents bearing thousands of 5-HT containing varicosities enabling it to interfere with widespread target cells. Both synaptic and non-synaptic contacts have been described suggesting that neuronal communication is relevant of both synaptic – a point-to-point contact – and volume – a passive at distance tissular diffusion – transmission mechanisms. The raphé serotonergic system influences the immune system through two pathways: (i) via the caudal raphé nuclei (nucleus pallidus, obscurus and magnus) brain stem and spinal cord projections that contact the preganglionic motoneurons of both the sympathetic and parasympathetic columns, thus interfering with the autonomic nervous system outflow, (ii) via the rostral raphé nuclei (nucleus median and dorsalis) ►hypothalamic-pituitary gland axis projections, thus interfering with the pituitary neuroendocrine outflow.

Serotonin and Pituitary Hormones

Pituitary hormones and related substances modulate immune functions. Corticosteroids have immunosuppressive effects while prolactin and growth hormone favor immune functions and counter the immunosuppressive effects of the former. 5-HT modulation of pituitary activity would be 5-$HT_{1B/1D}$ receptor dependent [6]. Sumatriptan, a specific agonist for 5-$HT_{1B/1D}$ receptors, lowers levels of plasma prolactin, increases those of growth hormone, but does not affect cortisol concentration. Since 5-$HT_{1B/1D}$ receptors concentrate in the median eminence on non-serotonergic fibers, possibly of hypothalamic origin secreting neuropeptides acting as releasing factors for pituitary neurohormones, 5-HT could interact with the release of these products, finally altering the internal status of the organism [6].

The Peripheral Autonomic Nervous System

The nerve terminals of the peripheral autonomic nervous system are possibly fundamental in neuroimmune interactions as they anatomically ensure the interface between the nervous and immune systems at local peripheral levels. Nerve fibers of the autonomic innervation developing close contacts with lymphocytes, hemopoietic elements, thymocytes, macrophages, ►mast cells and T cells, have been described for a long time. Autonomic fibers comprise noradrenergic terminals that take up and accumulate non-neuronal peripherally released 5-HT in a process that can alter their functions when both transmitters, noradrenaline and 5-HT, are simultaneously released. 5-HT positive immunocytes mast cells are potential providers for this peripheral 5-HT. The 5-HT content of peripheral tissue will evidently increase in inflammatory pathological circumstances. Central nervous autonomic outflow regulation through raphé 5-HT innervation of preganglionic motoneurons also exists, suggesting a central-mediated modulation of local autonomic-immune interactions. Finally, 5-HT could also modulate peripheral nervous inflow since 5-$HT_{1B/1D}$ receptors are present on primary afferent fibers [7]. Primary afferent fibers are one of the ways through which immune changes induced during the course of an infection can generate the central release of cytokines (interleukin 1 and tumor necrosis factor), which in turn acts on 5-HT levels.

The Brain Immunocytes

The CNS of healthy animals is classically viewed as an immune-privileged organ because of the blood-brain barriers (blood-brain, blood-nerve and blood-cerebrospinal fluid barrier) that limit the access of systemic immune cells. This immune privilege is not, however, as strict as it has long been claimed. Both activated T-lymphocytes and mast cells are known to penetrate the brain. Activated T-lymphocytes patrol the brain for immune surveillance; nervous resident mast cells (nsMCs) are present at both ends (primary afferent fibers, diencephalon) of the sensory network and in sympathetic ganglia where they provide a direct cell-to-cell contact opportunity for 5-HT-based neuroimmune cross-talk. NsMCs activities result in significative nervous changes depending on their location.

Primary Afferent Fibers-Associated Mast Cells

These cells lie in close apposition with unmyelinated fibers, contacting both peripheral terminals and ganglionic cells of origin (spinal and nodose ganglia). They promote peripheral fiber elongation and excitability in inflammatory conditions.

Sympathetic Ganglia-Associated Mast Cells

These cells increase ganglionic synaptic transmission when sensitised by immune challenge.

Diencephalon-Associated Mast Cells

These cells comprise several subpopulations that are present in (i) the parenchyma of both the ►thalamus (thalamic mast cells) and main olfactory bulb, as the final central relays for sensory and olfactory inputs before cortex, (ii) the leptomeninges including those surrounding the median eminence, as the key structure for the hypothalamic-pituitary neuroendocrine process, and (iii) the choroid plexus where the cerebrospinal fluid is secreted.

Thalamic Mast Cells

These cells have recently received much attention. Evidence has been given that they (i) have a predominant perivascular location, lying at the vascular interface in the adventia of the arterioles and venules, excluding capillaries, (ii) show some phenotypic specificities, (iii) release substances, including 5-HT, over piecemeal rather than overt degranulation and (iv) constitute a dynamic population that is triggered behaviorally, hormonally, pharmacologically and respond to sensory challenges. Interestingly, they do not express c-kit receptor and cyclooxygenase-2 as their peritoneal and leptomeningeal homogeneic counterparts do, thus precluding classical proliferating and pro-inflammatory properties. Considering properties related to 5-HT dependent mechanisms, evidence have been given that they store 5-HT in proteoglycan-free granules [8] and do not express 5-$HT_{1B/1D}$ receptors although they respond to a systemic injection of sumatriptan [6] (Fig. 2). Recent evidence has been given that thalamic mast cell activation results in neuronal activity variations [9]. Their prevalent location within the thalamic nuclei having cortical projections has been argued for a role in integrative and cognitive sensory processes. It has also been proposed that thalamic mast cells could act on the permeability of the endothelial blood-brain barrier [10], thus making local parenchymal cell populations accessible to humoral signals. Effects at neuronal level could affect the thalamic nuclear receptivity to incoming sensory events; effects at glial level could influence either neuronal or immune surveillance activities.

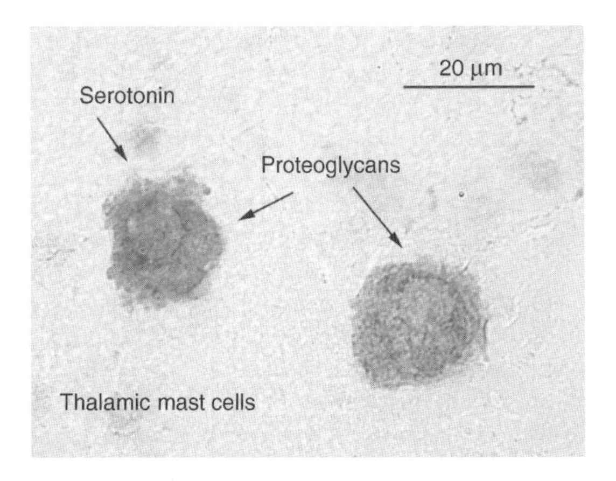

Neuroimmune Interactions – Serotonin. Figure 2 A micrograph (brightfield illumination) of two thalamic mast cells in double-stained material in which purplish-red granules identify proteoglycans (toluidine blue histochemistry) and therefore mast cells, while brilliant green granules identify serotonergic contents (5-HT immunohistochemistry). The fact that granules kept their specific colors, either red or green, suggests that they are mediator-specific. The nuclei remain visible for both cells. Modified from [6]. The staining technique is from [8].

Overview

Given its widespread distribution and the variety of its receptors, 5-HT and its precursor L-tryptophan appear to be involved in the regulation of a number of physiologic functions linked to the afferent and efferent pathways of communication between the nervous and immune systems. 5-HT-based cross-talk between these two systems could be essential for an adapted immune regulation and preventing immune mediated disorders.

References

1. Schröcksnadel K, Wirleitner B, Winkler C, Fuchs D (2006) Monitoring tryptophan metabolism in chronic immune activation. Clin Chim Acta 364:82–90
2. Hoyer D, Martin G (1997) 5-HT receptor classification and nomenclature: towards a harmonization with the human genome. Neuropharmacology 36:419–428
3. Mossner R, Lesch KP (1998) Role of serotonin in the immune system and in neuroimmune interactions. Brain Behav Immun 12:249–271
4. Roberts JE (2000) Light and immunomodulation. Ann NY Acad Sci 917:435–445
5. Adam R, Russing D, Adams O, Ailyati A, Sik Kim K, Schroten H, Daubener W (2005) Role of human brain microvascular endothelial cells during central nervous system infection. Significance of indoleamine 2,3-dioxygenase in antimicrobial defence and immunoregulation. Thromb Haemost 94:341–346
6. Dubayle D, Servière J, Menétrey D (2005) Evidence for serotonin influencing the thalamic infiltration of mast cells in rat. J Neuroimmunol 159:20–30
7. Potrebic S, Ahn AH, Skinner K, Fields HL, Basbaum AI (2003) Peptidergic nociceptors of both trigeminal and dorsal root ganglia express serotonin 1D receptors: implications for the selective antimigraine action of triptans. J Neurosci 23:10988–10997
8. Menétrey D, Dubayle D (2003) A one-step dual-labeling method for antigen detection in mast cells. Histochem Cell Biol 120:435–442
9. Kovacs P, Hernadi I, Wilhelm M (2006) Mast cells modulate maintained neuronal activity in the thalamus in vivo. J Neuroimmunol 171:1–7
10. Zhuang X, Silverman AJ, Silver R (1996) Brain mast cell degranulation regulates blood-brain barrier. J Neurobiol 31:393–403

Neuroimmunology

JOHN J. HADDAD
Cellular and Molecular Signaling Research Group, Division of Biological Sciences, Departments of Biology and Biomedical Sciences, Faculty of Arts and Sciences, Lebanese International University, Beirut, Lebanon; Department of Medical Laboratory Sciences, Faculty of Health Sciences, St Georges Hospital Complex, University of Balamand, Aschrafieh, Beirut, Lebanon

Introduction

The nervous system is the body's master controller. It monitors changes inside and outside the body, integrates sensory input and activates an appropriate response. In conjunction with the ►endocrine system, which is the body's second important regulating system, the nervous system is able to constantly regulate ►homeostasis. The nervous system, along with the endocrine and immune systems, thus forming the nexus of neuroimmune-endocrine interactions, help keep controlled conditions within limits that maintain homeostasis [1]. The nervous system is practically responsible for all of our behaviors, memories and movements. The branch of medical science that deals with the normal functioning and disorders of the nervous system is collectively called neurology and that of neuroimmune interactions neuroimmunology.

Neuroimmunology: Neuroimmune Interactions, Pathways and Mechanisms

Burgeoning research over the past few decades and continuing apace has shown that the immune, nervous and endocrine systems, or the "*trio*," are tightly linked via specialized communication pathways and mechanisms [1]. Interactions between the nervous and immune systems, specifically, provide a physiological (homeostatic) basis for understanding neuroimmune-associated disorders and medical conditions emanating from them. In approximately 200 AD, the Greek author Galen wrote that "melancholic women were more susceptible to breast ►cancer than sanguine women." Since then, a wealth of anecdotal evidence has convinced physicians and researchers of the importance of psychological factors in the prognosis of disease. This belief is now bolstered by substantial evidence that the nervous system output can indeed modulate immune functions and mechanisms of action [2].

Neuroimmune interactions are not by any chance unidirectional. The bidirectional influence emanates from the fact that the immune system can have substantial influence on the nervous system [1]. Anomalies of immune system function or malfunction can certainly cause diseases of, or relating to, the nervous system. It is clear that effective defense mechanisms against infections or immune disorders requires a complex coordination of the activities of the nervous and immune systems, and that abnormalities in the relationships between the two of them can cause disease or pathophysiologic aberrations [3].

Classically, the brain has long been regarded as an "*immunologically*" privileged site [1]. The relative non-immune responsiveness of the brain has been attributed to a lack of lymphatic drainage, the presence of the ►blood-brain barrier (BBB) (as emphasized above), the lack of constitutive expression of the ►major histocompatibility complex (MHC) cluster and the presence of chemical mediators or cofactors purported as capable of inhibiting ►lymphocyte traffic during inflammation [1] (►neuronal cell death and inflammation). This evasion of systemic immunological recognition confers a *privilege* property that is so unique and, in many ways, plays a major role in shaping the grounds for neuroimmune interactions. However, accumulating evidence indicates that immune responses propagate in the nervous system in a manner similar to that in other tissues (non-immune).

The nervous system, in fact, has a number of attributes that influence local immune responses, hence the "*bidirectional*" effect concept [1]. The experimental evidence for neuroimmune interactions can be summarized as follows: (i) alterations or changes in immune responses can be conditioned and regulated; (ii) electrical stimulation or lesions of specific brain sites can alter and modulate immune functions; (iii) ►stress (and the ►hypothalamo-pituitary-adrenal (HPA) axis; see below) alters immune responses and infections in experimental and physiological models; and (iv) activation of the immune system is correlated with altered neurophysiological, neurochemical and neuroendocrine activities of brain tissue [4].

This evidence is elaborated on below, but it is pertinent for now to consider first what the potential links between the nervous and immune systems might be. There is little scope for understanding neuroimmune interactions, but with the benefit of hindsight I can postulate a number of specific neuroimmune mechanisms by which the nervous system might affect immune function. (*This is also evident by the various entitled contributions authors have made to the Neuroimmunology field.*) These interactions include ►glucocorticoids (discussed later) secreted from the ►adrenal cortex, ►catecholamines secreted from ►sympathetic nerve terminals and the ►adrenal medulla, other hormones secreted by the ►pituitary (►hypophysis) and other endocrine organs, and peptides (including endorphins) secreted by the adrenal medulla and autonomic nerve terminals [1]. This network includes not only the ►autonomic nervous system (ANS) and classical neuroendocrine mechanisms, but involves an endocrine function of the immune system. A variety of immune system products (e.g., ►cytokines, peptides and other factors) that function to coordinate the immune response may also provide important signals for the nervous system [2]. Thus, chemical messengers can account for a variety of interactions between both systems. The illustrations in Fig. 1a and b provide a hypothetical schematic model of the most well-known interactions between the nervous system and components of the endocrine and immune systems.

Immunity and the Immune System – An Overview

The immune system is critical to our survival. Examples of what happens when it fails (►acquired immunodeficiency syndrome; AIDS) or when it fails to develop properly (►severe combined immunodeficiency; SCID), just to cite few examples, are abounding [5]. The body is a rich place for microorganism growth. Without our immune system, we, conspicuously, are an excellent propagating growth medium.

What is the Immune System?

Liken it to a colony of ants within us; the immune system is, nonetheless, a restless microenvironment [5]. Instead of separate organisms, however, there are many different cellular components distributed in our organs and tissues and blood stream. These cells are not static but rather move throughout the body, "looking" for

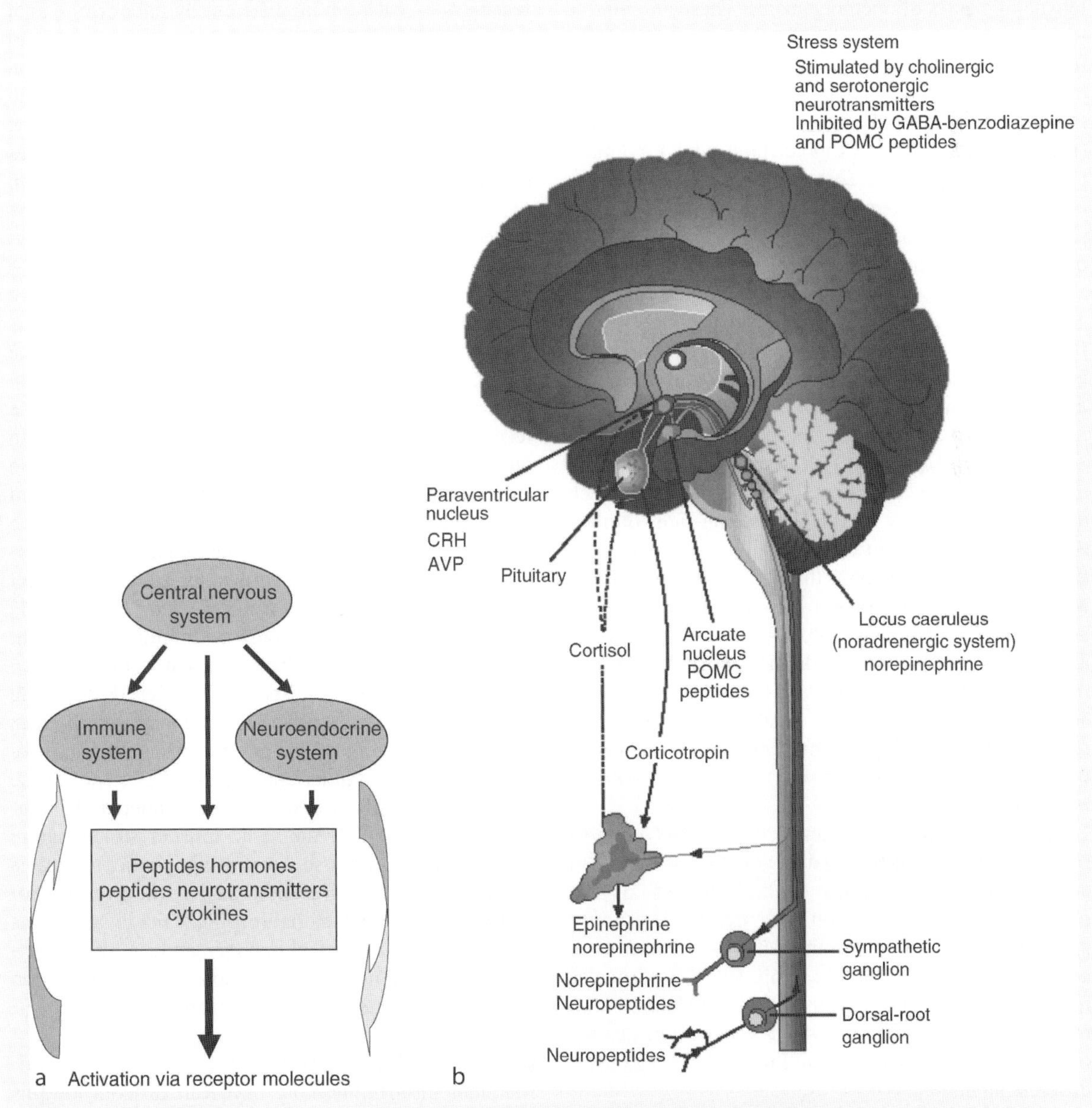

Neuroimmunology. Figure 1 (a) A hypothetical schematic for substantial molecular communication circuits existing between the immune and neuroendocrine systems and involving shared ligands and receptors. (b) Classic components of the CNS systems and the connections with stress and immune system [1].

situations that demand a response. Generally, communication between two or more cells is required before an attack is launched – a system of checks and balances. When we talk about immunity, we must be careful to say just what kind of immunity is meant. For example, there is *innate immunity*. This is a response that is not specific for a particular microorganism or strain of microorganism. It is rather set off by a property that is shared by a whole class of microorganisms (►neurodegeneration and neuroprotection – innate immune response).

Types of Immunity

Innate immunity involves cell surface polymers characteristic of classes of microorganisms. These are referred to as ►pathogen-associated molecular patterns (PAMPs) [6]. They involve major molecular signatures of classes of microorganisms, such as: (i) ►Lipopolysaccharide (►LPS; ►endotoxin) for Gram-negative bacteria (►endotoxic fever); (ii) Lipoteichoic acid for Gram-positive bacteria; and (iii) Lipoproteins for *Mycobacteria, Mycoplasma and Spirochetes*. PAMPs are recognized by ►pattern recognition receptors (PRRs) on the surface of macrophages – phagocytic cells – and also on lymphocytes. The PRRs in mammals resemble Toll, one of a family of receptors in the invertebrate fruit fly, *Drosophila melanogaster* [6]. On binding of infectious organisms to Toll and similar receptors in *Drosophila*, anti-fungal or anti-microbial peptides are released that are appropriate to the infectious organism [7].

►Toll-like receptors (TLRs) share the following molecular properties: (i) an extracellular ►leucine-rich domain (LRD); (ii) a small cysteine-rich domain that differs among different toll-like receptors; and (iii) a cytoplasmic domain that is homologous to the ►interleukin (IL)-1 receptor (IL-1R), a receptor that binds the vertebrate cytokine, IL-1; it is referred to as a ►Toll/IL-1R homology domain (TIR). In mammals, cells interacting with PAMPs through their PRRs appear to release cytokines – small glycoproteins that recruit cells involved in a form of immunity not present in invertebrates like *Drosophila* – *specific acquired immunity* [7].

In specific acquired immunity or *specific adaptive immunity*, the response is against a particular organism and, in fact, usually against multiple aspects of that organism. For example, you may mount a response against a single strain of influenza virus and even against many proteins of that strain. But that particular response will not protect you against a different strain of influenza virus. This is why ►vaccination against one strain does not protect the human body against a different strain [7].

We will learn how the colony of cells that mediate specific acquired immunity is set up and how it operates. In brief, during embryonic development and throughout life, a very large number of cells called lymphocytes are generated. There are several different classes of lymphocytes and millions to billions of cells in each class. Each lymphocyte has a receptor on its cell surface (in fact, many copies of a receptor, but all identical). But each lymphocyte has a different receptor. The job of the receptor on each lymphocyte is to bind to or recognize a potential foreign invader – what is commonly referred to as ►antigen. Antigens, the invading organism or part of organisms against which our immune system must fight, may take many forms [7].

The immune system does not "know" what invaders exist out there in the world. The system has evolved to express such an enormous number of different receptors – each on a different lymphocyte – that at any one time, it contains lymphocytes that could recognize any invader that we encounter. Of course, the particular cell must be able to find and interact with that invader in order to make its protective response. That is the function of the specialized immune tissues (spleen, thymus, lymph nodes) and the circulation of lymphocytes in the blood stream – to bring the protective lymphocytes into contact with the invaders that they must fight [7].

When an invader is encountered by protective lymphocytes and the validity of that encounter is verified, a process is set in motion whereby the lymphocyte is caused to divide multiple times to generate a clone of identical cells expressing the same receptor. These cells set about destroying the invader in one of two major ways: (i) they manufacture and secrete ►antibodies, proteins that bind to the invader and contribute to its demise by one of several means that we will discuss later on. This is called *humoral immunity*; and (ii) the cells destroy the invader directly by direct action of the cells. This is called *cellular immunity* [1] (►stress effects during intense training on cellular immunity, hormones and respiratory infections).

After the invader is effectively beaten down, there now remain an increased number of antibodies, antibody-producing cells, and memory cells than there were before the invader appeared on the scene. These persist in the body and are, in fact, scattered throughout via the blood. If the same invader strikes again, the protective response occurs more quickly and is stronger than it was at first due to the presence of more cells at the outset that recognize the invader. This is called ►immunological memory, and it is why vaccination works (►neuroinflammation – DNA vaccination against autoimmune neuroinflammation).

Developmental Immunity

Immunologists, furthermore, have learned about how this remarkable system is set up during development. In fact, the immune system is perhaps the best-understood developmental system, largely because many of the key cell types are "free floating" (lymphocytes, macrophages), which are much more easily manipulated than solid tissues [1]. This attribute, combined with the ability to study the genetics of immune responses and of the molecules that mediate these responses, has allowed a wealth of information to accumulate – mostly for mice and humans.

Self/Non-self Recognition

How does our immune system distinguish between an invader and our own tissues – our "self?" After all, if we can produce lymphocytes that have receptors that can recognize any invader, surely we produce lymphocytes that recognize molecules in and on our own cells. Why does our own immune system not destroy us? In fact, we do produce lymphocytes that recognize ourselves all the time, but only rarely do they cause ►autoimmune disease (►central nervous system degeneration caused by autoimmune cytotoxic CD8+ T cell clones and hybridomas). We will learn a lot about self-recognition and why it is generally not a serious problem. In the course of addressing this question, we will deal with a subject that I will refer to as the *genetics of the self*. It underlies both our ability to set up a safe and functional immune system, and also the whole area of tissue and organ transplantation.

It revolves around a set of closely linked genes that are lumped together and referred to as the major histocompatibility complex (MHC) of genes. The MHC is the focus for how the immune system avoids attacking our own bodies. The cellular and molecular biology that is involved in MHC-related recognition is one of the most fascinating aspects of how the immune system works. Finally, one should always be considering how we might apply our knowledge of how the immune system works to improve human health. This may involve using vaccines or mediators produced by the immune system itself to enhance immunity when the body's own system is not mounting a strong enough response. In other instances, we may wish to squelch immune activity when it has been misdirected against our own tissues. Also of great interest is the possible use of ►gene therapy to enable the body to manufacture a needed substance that it was unable to make due to an inherited mutation or deletion in an important gene [8–10].

What is the Functionality of the Relationship Between the Nervous and Immune Systems?

There is considerable evidence suggesting that immune system signaling and activation are communicated to the nervous system via specific pathways [1]. This communication essentially occurs through the release of peripheral soluble factors (particularly cytokines, commonly known as "*biologic response modifiers*") by cells of the immune system (lymphoid vs. myeloid) and cells of non-immune origin [11] (►central nervous system inflammation – cytokines and JAK/STAT/SOCS signal transduction). These factors or cofactors function as hormones or modifiers to affect and modulate the responses of the central nervous system (CNS) and peripheral nervous system (PNS). They can affect the CNS directly by crossing (bypassing) the blood-brain barrier (see above) or indirectly by stimulating the ►vagus nerve (see Fig. 1a and b). As a consequence of the diversity of specialized cells and subcellular components in the nervous system, there is a wide range of potential target antigens and clinical syndromes associated with that [1].

Bidirectional Influence: Immune System Effects on the Nervous System – Types of Signaling Molecules and Cofactors

Amongst biological modifiers, cytokines, particularly, can directly influence the electrophysiological function of neurons in the CNS or PNS; this is especially evident during the ensuing inflammation of the brain or PNS, despite the immunologically privileged status [8]. ►Chemokines, on the other hand, resemble a family of proteins associated with the trafficking (emigration) of ►leukocytes in physiological immune surveillance and inflammatory cell recruitment in normal host defense mechanisms [1]. Beside their well-established role in the immune system, evidence indicates that chemokines also play an integral role in the CNS. In fact, they are constitutively expressed by ►microglial cells (or macrophages of the brain) (►microglia – functions in immune mechanisms in the central nervous system), ►astrocytes and neurons, and their expression can be induced with inflammatory mediators, such as cytokines. Chemokines can also modulate neuronal signaling via the regulation of the flow of Ca^{2+} currents [12].

Immunologically active molecules, not necessarily involved with inflammatory reactions, similar to cytokines or chemokines (secreted from cells invading the tissue from the blood stream or secreted centrally by local microglia or astrocytes (►central nervous system inflammation – astroglia and ethanol) as well as bacterial-derived (like lipopolysaccharide (LPS)) or virus-derived molecules can affect voltage-dependent ion currents and ►transmitter receptor-operated ion currents of peripheral or central neurons [1]. Cytokines released by the immune system can influence cognitive processes, for example, and thus modify central neurotransmission and the function of PNS. In addition, cytokines and neuropeptides secreted by peripheral immune cells have dramatic effects on behavior or behavioral aspects of the CNS. Pro-inflammatory cytokines, furthermore, can activate the HPA (discussed below) and thus induce, for example, ►sickness behavior (weakness, malaise, listlessness, inability to concentrate, decreased food and water intake) during the inflammatory ►acute phase response (APR) [1].

The acute phase response has been reported to have specific components (physiological and inflammatory): (i) fever: cytokines such as IL-1, IL-6 and ►tumor-necrosis-factor-α (TNF-α) act at the level of the ►hypothalamus to install and activate thermocenters [1] (►brain inflammation – tumor necrosis factor receptors in mouse brain inflammatory responses); (ii) sickness behavior can be induced by systemic circulating IL-1β and TNF-α; this phenomenon is likely mediated by the vagus nerve since ►vagotomy (severance of the vagus nerve) has been shown to attenuate the behavioral actions of peripheral cytokines; IL-1β, moreover, binds to vagal fibers and thus can increase vagal discharge; (iii) blood-brain barrier: IL-1β can slowly diffuse across the parenchymal blood-brain barrier and activate the basolateral ►amygdala (involved with depressive effects on social behavior) and ►area postrema (this brainstem region activates the HPA axis and gives rise to feelings of nausea) (►bickerstaff's brainstem encephalitis).

Nervous System Components Associated with and/or Affected by Immune Responses

Neurons carry information coded in the form of electrical signals (►Membrane potential, ►Action potential). Evidence indicates that neurons can counter-regulate brain immunity in intact CNS areas. For example, microglia are kept in a quiescent state in the CNS by interactions between the microglia receptor ►cluster of designation *(CD)200* and its ligand, which is inducibly expressed on neurons [13]. Furthermore, neurons can downregulate MHC expression in surrounding glial cells, in particular microglia and astrocytes (►glial and neuronal reactivity to unconjugated bilirubin). Certain chemokines are also associated with the surface of many neurons (e.g., ►fractalkine is bound to the outside of neurons in the dorsal horn of the spinal cord). Microglia, moreover, have been shown to express fractalkine receptors (CX3CR1). When ►spinal cord dorsal horn neurons are activated by nociceptive stimuli, for example, they release fractalkine, which binds to the microglial receptors and stimulates the microglia, causing the release of cytokines [1]. This seems to be a vital mechanism in the generation of ►chronic pain condition [9] (►neural-immune interactions – implications for pain management in low-back pain and sciatica).

Glia (neuroglia) can also form an innate immune system offshoot, within the "immune privileged" CNS, which has the potential to initiate immune responses to exogenous antigens or endogenous degenerative processes [13]. ►Oligodendrocytes (components of the CNS) and ►Schwann cells (components of the PNS), which produce a ►myelin sheath around axons, are sensitive to injurious and/or pharmacologically active agents including antibodies, complements and cytokines (►autoimmune demyelinating disorders – stem cell therapy). Astrocytes, on the other hand, affect neuronal function by the release of ►neurotrophic factors (►neurotrophins), guide neuronal development physiologically, contribute to the metabolism of neurotransmitters and regulate extracellular pH and K^+ concentrations and currents. The astrocyte, specifically, is an immuno-competent cell in the CNS. The reason is that these cells can express MHC class II and co-stimulatory molecules (B7-1 and B7-2 [CD80/CD86] and CD40) that are critical for antigen presentation and subsequent T-cell (lymphocyte) activation [1]. Microglia essentially resemble brain macrophages, the phenotype of which is thought to represent an adaptation to the specialized neural microenvironment. These features include, but are not restricted to, the following: (i) Microglia exhibit a downregulated (less differentiated) phenotype; (ii) serve major homeostatic and reparative functions (can secrete cytokines and neurotrophic factors); (iii) play a role in host defense (can become activated to perform several innate immune functions, such as induction of inflammation; cytotoxicity and regulation of T-cell responses through presentation of antigen); and (iv) are involved in CNS immune surveillance and response [1] (►neuroinflammation – modulating pesticide-induced neurodegeneration).

Nervous System Influence on the Immune System – Unidirectional or Bidirectional?

Converging evidence has demonstrated that the immune system is not regulated in an autonomous fashion, but is influenced by external factors particularly mediated by the nervous system (Fig. 2a and b). What are the major connecting mechanisms mediating neuroimmune interactions? There are at least three putative pathways by which the nervous system can communicate or crosstalk with the immune system: (i) Autonomic nervous system (ANS) route via direct nerve fiber connections; (ii) sensory portion of the nervous system via primary afferent nerve fibers; and (iii) neuroendocrine output via the HPA axis. In particular, accumulating evidence indicates that neural control in immunological phases ranges from induction and activation to effector functions and inactivation zones [1]. Autonomic nervous system influence on the immune system involves the following mechanisms: (i) Sympathetic fibers innervate lymphoid organs; (ii) noradrenergic fibers make synapse-like contacts with systemic lymphocytes (spleen) and release ►noradrenaline (NA; also norephinephrine, NE), ►vasoactive intestinal peptide (VIP) and ►neuropeptide Y (NPY); lymphocytes and macrophages can express α_2 and β_2 adrenoceptors, and lymphocytes also express the Y_1-NPY receptor; (iii) Noradrenaline suppresses immune responses by tonically inhibiting pro-inflammatory cytokine biosynthesis; (iv) sympathetically released NPY inhibits ►natural killer (NK) cell cytotoxic responses and

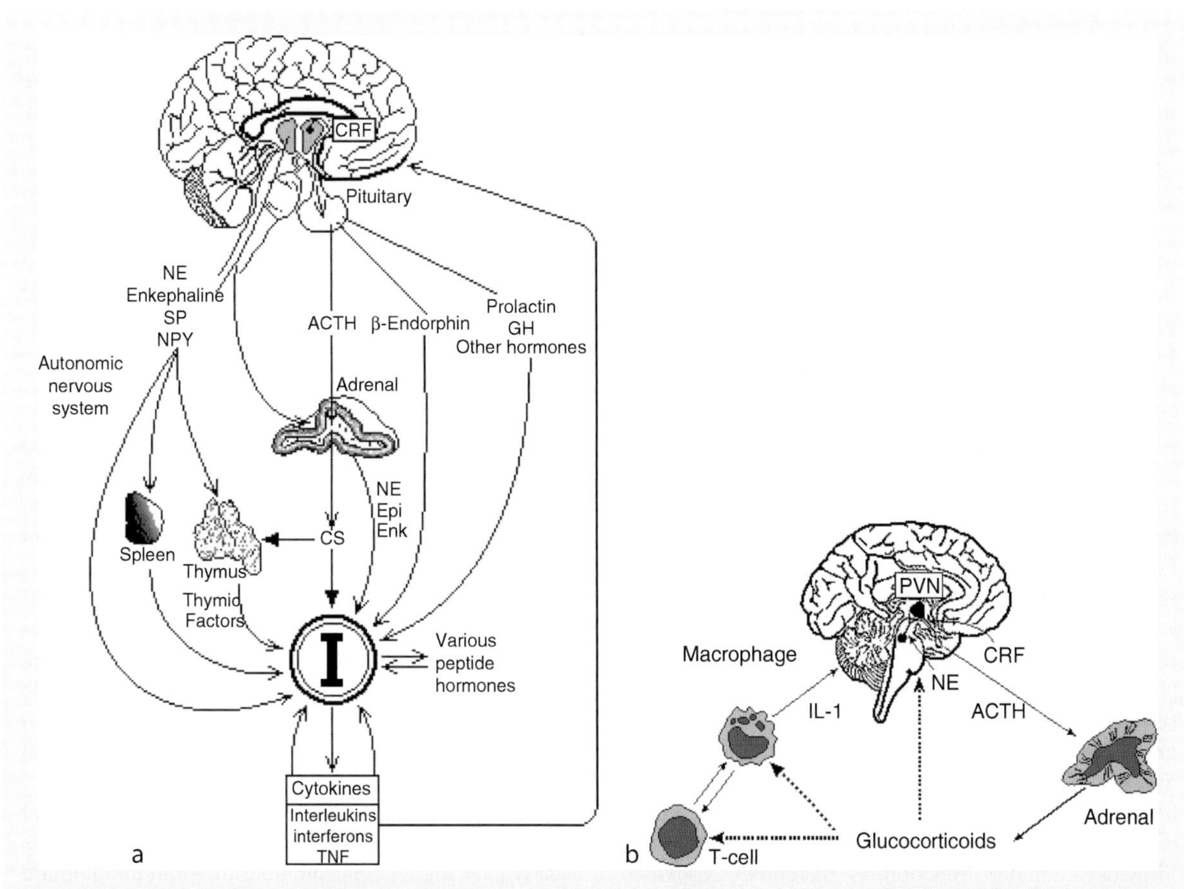

Neuroimmunology. Figure 2 (a) A schematic diagram of the interactions between the brain and components of the endocrine and immune systems. The ability of the brain to alter immune system function by a variety of endocrine pathways and the autonomic nervous system is emphasized, and the effects of peptides and cytokines produced by the immune system on immune cells and the brain is indicated. *CRF*, corticotropin-releasing factor; *CS*, corticosteroids; *Enk*, encephalin; *Epi*, epinephrine; *GH*, growth hormone; *I*, immunocytes; *NE*, norepinephrine; *NPY*, neuropeptide Y; *SP*, substance P; *TNF*, tumor necrosis factor. (b) A schematic diagram of the relationship between the brain, the HPA axis, and immune cells and the role of glucocorticoids. Interleukin-1 (IL-1) produced by lymphocytes during the immune response activates noradrenergic (NE) projections for the brainstem to the hypothalamic paraventricular nucleus (PVN). This input activates the hypothalamic-pituitary-adrenocortical (HPA) axis, stimulating the release of corticotropin-releasing factor (CRF) in the median eminence region of the hypothalamus, which in turn stimulates the secretion of ACTH from the anterior lobe of the pituitary, which then activates the adrenal cortex to synthesize and secrete glucocorticoid hormones. The glucocorticoids in turn provide a negative feedback on cytokine production by lymphocytes [1].

humoral antibody production; and (v) vasoactive intestinal peptide inhibits IL-12 and ►nitric oxide (NO) production by resident macrophages [1].

The major influence of the ►sympathetic nervous system (SNS) on the immune system is to inhibit immune responses. However, there appear to be regional (localized) differences in the effects of NA on immune function. In the thymus, for example, it modulates T-cell proliferation and differentiation; in the spleen and lymph nodes it enhances the primary antibody response. Primary afferents and peptidergic innervations, in particular, exert great influence on the immune system. A number of neuropeptides released from primary afferent nerve fibers have been shown to affect the immune system. For example, ►substance P (SP), a pro-inflammatory undecapeptide, has been localized in the CNS, peripheral sensory neurons, in nerve plexuses of the gut and in the spleen, lymph nodes and thymus (primary and secondary lymphoid organs) [1]. This peptide is found particularly at sites of inflammation and appears to be a regulator of cell-mediated and humoral immune responses. Substance P is involved in (to cite a few examples): (i) early induction of local and systemic host defense responses to inflammation/injury (►traumatic brain injury – rat model of neuroinflammation and expression of matrix metalloproteinases); (ii) causing vasodilatation and increased vascular permeability; (iii) enhancing phagocytosis by neutrophils and/or

N

macrophages; (iv) induction of the release of ►histamine (secondary allergic mediators) and other substances from ►mast cells; and (v) controlling bacterial infections of the gut (►gut-associated lymphoid tissue; GALT); SP blockade has been shown to increase the susceptibility to *Salmonella* infections. Another example is ►calcitonin gene related peptide (CGRP), localized in the bone marrow, lymph node, spleen and thymus where it is released by primary afferent fibers [1]. It has the following properties: (i) it has binding sites on T-cells; (ii) it is a potent vasodilator; and (iii) it inhibits mitogen-stimulated proliferation of T-cells and inhibits T-cell stimulation of ►epidermal Langerhans cells (antigen-presenting cells of dendritic cell origin).

Influence of Hypothalamic/Pituitary Neuroendocrine Hormones on the Immune System

Early studies identified the pituitary gland as an essential component in the regulation of immune system development and activity. Following surgical removal of the pituitary, Gisler and Schenkel-Hulliger [14], for example, observed reduced antibody responses (sera-localized immunoglobulins were diminished); growth-hormone treatment restored antibody production. Since these early observations, numerous studies have implicated neuroendocrine hormones in immune regulation (Fig. 2a). A list of a few major hormones that influence immune function is succinctly provided below (elaborations are expressed in the section entitled "neuroimmune interactions").

Adrenocorticotrophic Hormone (ACTH)

►Adrenocorticotrophic hormone (ACTH) is extracted from the pituitary glands of animals or made synthetically. ACTH stimulates the adrenal glands to release glucocorticoid hormones. These hormones are anti-inflammatory in nature, reducing edema and other aspects of inflammation. Data from the early 1970s indicate that ACTH may reduce the duration of ►multiple sclerosis (MS) exacerbations (►multiple sclerosis – macrophages and axonal loss). In recent years it has been determined that synthetically produced glucocorticoid hormones (e.g., cortisone, prednisone, prednisolone, methylprednisolone, betamethasone, dexamethasone), which can be directly administered without the use of ACTH, are more potent, cause less Na^+ retention and less K^+ loss, and are longer-acting than ACTH. ACTH, in brief, has the following properties: (i) it is derived from ►pro-opiomelanocortin (POMC), an unusual hormone complex manufactured by the anterior lobe of the pituitary (hypophysis). This complex is metabolized into four separate hormones: ACTH, ►melanocyte stimulating hormone (MSH), ►enkephalin and ►β-endorphin; (ii) it is secreted by the anterior pituitary into blood stream; (iii) its release is regulated by ►corticotropin-releasing factor (CRF) (or ►corticotropin-releasing hormone (CRH) or ►corticoliberin), a polypeptide hormone involved in the stress response, stress and hypoglycemia (Fig. 2b); (iv) ACTH initiates the release of adrenal corticosteroids and increases the growth of adrenal cells through actions on ►melanocortin receptors; (v) ACTH receptors are present on both B-cells (lymphocytes) and T-cells; (vi) it reduces antibody responses in vitro and reduces ►IFN-γ (interferon) production; and (vii) systemic injection of lipopolysaccharide (LPS) increases ACTH and corticosterone [1].

Gonadal Steroids

Several lines of evidence implicate sex steroids in immune regulation and in the regulation of neuronal gene expression (transcriptional regulation) [1]. In general, ►androgens (hormones promoting the development and maintenance of male sex characteristics) exert suppressive effects on both humoral and cellular-mediated immune responses and seem to represent natural anti-inflammatory hormones; in contrast, ►estrogens (type of hormones that help develop and maintain female sex characteristics and the growth of long bones) exert immuno-enhancing activities, at least on humoral immune response.

This is based on the following observations: (i) sexual dimorphism exists within the immune system: females usually have higher concentrations of ►immunoglobulin (Ig)G, IgM and IgA than males; (ii) antibody responses to antigens are greater in magnitude and essentially more prolonged in females than males; (iii) females have a higher incidence of autoimmune disease (multiple sclerosis (MS), ►rheumatoid arthritis (RA), systemic lupus erythematosis (SLE),); (iv) manipulation of testosterone or estrogen alters autoimmune disease progression or onset in animal models; (v) sex steroids may influence the immune system, at least in part, via the thymus, where they play a role in development and atrophy; (vi) regulation of the immune system by estrogens is particularly important during pregnancy; in this case the balance between glucocorticoid and estrogen regulation probably plays a role in suppression of the maternal immune system to prevent rejection of the fetus; (vii) estrogen and testosterone can regulate IL-6 expression with loss of IL-6 effect in postmenopausal women and postandropausal men, thus resulting in increased IL-6 being associated with increased occurrence of inflammatory diseases with old age (RA, ►inflammatory bowel disease, osteoporosis); and (viii) in general, females are more sensitive to pain than males, and this is due, in part, to the presence of estrogen, which appears to be pro-nociceptive or of hyperalgesic nature [15].

Adrenocorticotrophic hormone (ACTH), glucocorticoids and gonadal steroids can also directly affect nervous system function. For example, corticosteroid and estrogen receptors are found in discrete locations in the brain and spinal cord (►neuroinflammation – brain and spinal cord injury). Electrical properties of brain neurons are specifically regulated by glucocorticoid-receptor activation. β-Estradiol has been shown to inhibit L-type ►voltage-gated Ca^{2+} channels in brain neurons. Estrogen receptors (a class of proteins found inside the cells of the female reproductive tissue, some other types of tissue, and some cancer cells; estrogen will bind to the receptors inside the cells and may cause the cells to grow) have been found to be coupled to ►metabotrophic glutamate receptors in the ►hippocampus of the brain and thus can affect second messenger systems involved in memory and learning [1].

Immunological Surveillance of the Nervous System by Lymphocytes

Studies on the migration of labeled T cells following intravenous injection have shown that activated T lymphocytes of a rather broad specificity enter the normal CNS parenchyma as early as 3h following administration [1]. Thus, T-cell traffic in the CNS appears to be governed by the same principle as applies to other organs, namely that activated T cells preferentially migrate from the blood into tissues, whereas resting cells exit in lymph nodes via high-endothelial venules (HEVs) (found in lymphoid tissues, excluding the thymus; since endothelial cells are tall and lack tight junctions, this facilitates entry of lymphocytes into lymphoid tissue from the blood.) Low numbers of T cells are consistently demonstrable in normal human and rat brains, indicating that the CNS is continuously patrolled by activated lymphocytes [1].

Conditioning of the Immune Response

Compelling evidence for the influence of the nervous system on the immune system arises from studies that indicate that behavioral conditioning can modify immune responses. A landmark study by Ader and Cohen [16] indicated that after the immunosuppressive drug, cyclophosphamide (a class of drugs known as alkylating agents; it slows or stops the growth of cancer cells), had been paired with the taste of saccharin (the oldest artificial sweetener; in the European Union also known as E954), subsequent ingestion of the saccharin prevented the production of antibodies in response to sheep red blood cell (SRBC) administration (serologic manifestation). This technique has been particularly used to prolong the lives of mice with systemic lupus. There can be little doubt that conditioning can alter immune responses, but the immunological specificity of the effects is not clear, and the mechanisms remain to be unraveled. It is possible that at least some of the immunosuppressive effects are from a conditioning of hormone and neurotransmitter secretion (e.g., glucocorticoids or catecholamines) [1].

Effects of Brain Lesions on Immune Function

Although evidence has indicated that brain lesions may have effects on immunity, consistent coherence is, at best, fragmented, incomplete and complex. Effective lesions are most commonly located in the hypothalamus and are generally inhibitory. Lesions in other ►limbic areas may also be effective, notably in the ►septum, hippocampus and amygdala. Some studies have indicated that cortical lesions can affect immune responses and that the effects depend upon the laterality of the lesion [1]. Renoux and colleagues [17] have reported evidence that lesions of the left cortex, but not the right, produced pronounced immune deficits in spleen cell number, lymphocyte proliferation, and natural killer cell activity (►nervous, immune and hemopoietic systems – functional asymmetry). The lateral specificity indicates that the aforementioned observation cannot be from nonspecific effects of the lesion, and it could account for the greater number of left-handed individuals who exhibit diseases of the immune system. Lesions of the central noradrenergic systems have also been shown to impair various aspects of the immune response (►neurodegenerative diseases – MAPK signaling pathways: cytokine regulation and glial activation).

Effects of Stress on the Immune System

It is established that stress may impair the immune system (see below) [1]. The dogma that stress suppresses immunity is to some extent based on the well-established immunosuppressive effects of glucocorticoids. However, the supra-physiological doses of the steroids used in most of the studies do not allow simple extrapolation to the normal physiological state. In fact, endogenous glucocorticoids at physiological doses are not universally immunosuppressive and actually may enhance immune function. Furthermore, glucocorticoids may not even be the major mechanism by which stress suppresses immune function. Experimental evidence has confirmed the immunosuppressive effect of stress. However, it is important to emphasize that there is considerable evidence to suggest the opposite.

The Role of the Adrenal

Adrenalectomy (surgical removal of one or both adrenal glands) has been shown to prevent the immunosuppressive effects of stress, but other studies have indicated that stress-induced changes in immunity persist in adrenalectomized animals. Adrenalectomy appears to be effective in studies that have examined acute responses to brief stressors (for which the immunosuppressive effects are rapidly reversed), but may be less important for the effects of chronic stress. Adrenalectomy, furthermore, does not permit a distinction among the effects of steroids, catecholamines, or even of neuropeptides secreted by the adrenal gland. More recent studies have suggested an important role for the circulating catecholamines, derived from the sympathetic nervous system and adrenal medulla [1].

The choice of immune parameters measured may also influence the results. Earlier studies relied heavily on mitogen-stimulated proliferation assays, which assess the responsivity (i.e., cell division measured by DNA synthesis) to lectin mitogens [such as concanavalin A (Con A), phytohemagglutinin (PHA), lipopolysaccharide (LPS), or pokeweed mitogen] in vitro (►anti-DNA antibodies against microbial and non-nucleic acid self-antigens). The interpretation of such assays is questionable, because the results are susceptible to a large number of extraneous influences, and the assays are conducted after several days of in vitro incubation separated from normal physiological influences [2]. A measure used more often has recently been that for natural killer cells. There is good evidence that natural killer cells are involved in the rejection of tumors, and therefore at least one of their immunophysiological functions is clear. Stressful treatments have been shown to suppress natural killer cell function. The major effector for the stress-induced effects on natural killer cell function appears to be ►opiates and catecholamines through β- ►adrenergic receptors. Because most of the studies of stress on immune function have used ex vivo procedures, another important factor is whether or not the population of cells sampled may be altered by the in vivo treatment. Cell trafficking, the movement of lymphocytes around the body, is known to be regulated by hormones and other secretions, including those secreted during stress, and it is likely, therefore, that the stressful treatments alter the population of cells harvested for the in vitro analysis [1].

The Role of Glucocorticoids

The best-known mechanism for an influence of the nervous system on the immune system is circulating glucocorticoids secreted by the adrenal cortex (Fig. 2b). Glucocorticoids have long been known to have immunosuppressive effects. The data derive in part from the medical practice of using glucocorticoids postsurgically to decrease tissue inflammation and the rejection of transplanted tissues. However, considerable experimental data suggest that the effects of glucocorticoids are not exclusively immunosuppressive [18] (►neuroinflammation – LPS-induced acute neuroinflammation, rat model). Although it is well established, it is too often forgotten that glucocorticoids are essential for normal immune responses. For example, adrenally compromised individuals are more susceptible to infections. Of particular importance, it was also shown that corticosteroids were essential for normal recovery from infections in adrenalectomized animals [1].

Nevertheless, the extensive evidence for the immunosuppressive effects of glucocorticoids should not be ignored. It should, however, be viewed in the light that most of the data were generated using high doses of synthetic glucocorticoids (e.g., prednisolone, triamcinolone, or dexamethasone, amongst others), which are considerably more potent than the native steroids. The concentrations of these compounds used clinically can cause lysis of immune cells, especially immature ones. The more careful studies have used natural steroids at relatively physiological doses; these have noted stimulatory effects of steroids at lower doses. Inhibitory effects occur at higher (supra-physiological) doses, typically 10^{-6} M, which is close to the maximum concentration of free corticosterone or cortisol found in stressed animals after correcting for that bound by corticosteroid-binding globulins. It is also important that elevations of plasma glucocorticoids following acute stressors are short-lived [1].

Although there are direct effects of glucocorticoids on immune cells in vitro, there may also be indirect ones in vivo. One of the oldest known physiological correlates of stress is the involution of the thymus. This involution, which can decrease thymus weight by more than half (also ageing-related phenomenon), occurs largely because lymphocytes that normally reside there are driven out to the periphery. Stress-induced thymic involution is prevented by adrenalectomy and can be induced by administration of glucocorticoids. Thus glucocorticoids can alter the body's distribution of lymphocytes, which may in itself be an important factor marshalling the immune response to infection. Moreover, as mentioned above, the population of lymphocytes derived by harvesting tissues from animals subjected to experimental treatments may be altered by the redistribution of cells due to glucocorticoid secretion. This should be an important consideration in interpreting the results of ex vivo data [1].

The Role of Catecholamines

Lymphocytes bear both α- and β-adrenergic receptors. Catecholamines appear in the circulation from both the adrenal medulla [noradrenaline (NA) and adrenaline] and from sympathetic terminals (NA). In addition, lymphocytes may be exposed more directly to neuronal secretions while they are resident in the thymus, spleen, and lymph nodes. Anatomical studies have clearly demonstrated a sympathetic innervation of immune structures, such as the bone marrow, thymus, spleen, and lymph nodes. Thus lymphocytes could be exposed to high local concentrations of catecholamines, as well as neuropeptides (►microglial signalling regulation by neuropeptides). A parasympathetic (i.e., cholinergic) innervation of these organs has not been confirmed [19].

In vitro studies have revealed adrenergic effects on lymphocytes. Early studies suggested separate α- and β-adrenergic effects; β-adrenergic receptors were largely inhibitory, whereas α-adrenergic receptors were stimulatory. This generalization has endured to some extent, but the detailed results are very complex. There appear to be separate α- and β-adrenergic stimulatory effects on antibody production in vitro, whereas natural killer cell activity appears to be inhibited by β-adrenergic stimulation. The results of in vivo studies have been of bewildering complexity. Depending on the parameters used, sympathectomy has been shown to impair, enhance, or not change immune responses. In general, sympathectomy in adult animals depresses immune reactivity, but there are also paradoxical effects on lymphocyte proliferation and B-cell differentiation. Among the confounding factors that may contribute to the complexity are compensatory increases in adrenomedullary output, redistribution of lymphocytes, compensatory changes in the number and kind of adrenergic receptors, and the coexistence in sympathetic terminals of peptides, such as NPY [1]. Several studies have suggested that a major mechanism by which natural killer cell activity is regulated in vivo involves catecholamines released by the sympathetic nervous system. For example, the inhibitory effect of intra-cerebroventricular injection of corticotropin-releasing factor (CRF) on natural killer activity is blocked by the ganglionic blocker, chlorisondamine, as is the immunosuppressive effect of IL-1. There is also direct evidence that β-adrenergic receptor blockade can prevent stress-induced effects on natural killer cell activity [19].

The Role of Peptides

Sympathetic nerve terminals contain not only noradrenaline, but also neuropeptides, including endorphins, which may act on the immune system. The presence of NPY, substance P (SP) and vasoactive intestinal peptide (VIP) in the thymus, spleen and lymph nodes, as well as calcitonin gene related peptide (CGRP) in the thymus and lymph nodes, enkephalin and somatostatin in the spleen, tachykinin in the thymus, and peptide histidine isoleucine in lymph nodes have been described [1]. It has been shown that lymphocytes can synthesize and secrete certain peptides. The spectrum of peptides synthesized is large, and includes many of the known peptide hormones, as well as the hypophysiotropic factors. The peptides include ACTH, CRF, growth hormone (GH), thyrotropin (TRH), prolactin, human chorionic gonadotropin, the endorphins, encephalin, SP, somatostatin and VIP. The quantities of the peptides produced are typically very small, and their biochemical characterization has often been perfunctory. Sometimes their existence has been inferred only from the results obtained in the very sensitive assays used to detect their messenger ribonucleic acids (mRNAs), which should not be construed as unequivocal evidence for the presence of the peptides themselves. More careful analyses have not always substantiated the original claims, especially for the endorphins. There is probably considerable variability in the ability of lymphocytes from different sources to produce a specific peptide, but this issue has received no serious attention in the literature [1].

The physiological significance of this production of peptides is not at all clear. Because in many cases lymphocytes display receptors for these same peptides, they may function as chemical messengers within the immune system. However, Blalock has suggested that the peptides may also have systemic functions; for example, ACTH could activate the adrenal cortex. Although there is no good experimental support for this specific example, it is possible that there may be a local bidirectional communication between lymphocytes and other cells. One example of this communication may be in the spleen, where CRF appears to be present in the innervating neurons and CRF-receptors are present on resident macrophages. Another example involves endorphins; β-endorphin produced by lymphocytes in an area of inflammation may exert an analgesic action directly on sensory nerve terminals. Such a mechanism is attractive, because the concentrations of the peptides produced locally may be adequate to exert such effects, and the metabolic lability of peptides would ensure that the effect was localized [1] (see Fig. 2a and b).

Other Hormones of the Hypothalamo-Pituitary-Adrenocortical Axis

Many other hormones are known to affect the immune system. Firstly, there are the hormones of the HPA axis, each of which has been reported to affect immune function: CRF, ACTH, and the endorphins. Corticotropin-releasing factor (CRF) itself has been reported to have a variety of effects. The reported direct effects of CRF on immune cells have generally been stimulatory. For example, CRF has been shown to stimulate B cell proliferation and natural killer activity, as well as IL-1, IL-2, and IL-6 production. Receptors for CRF have been found on immune cells, providing a mechanism for these effects. Although it seems unlikely that CRF in the general circulation ever achieves concentrations high enough to stimulate these receptors, it is possible that local actions may occur, for example, in the spleen. By contrast, CRF injected intra-cerebroventricularly (icv) has largely inhibitory effects on immune function. A major effect of icv CRF is evident on natural killer cell activity and appears to be mediated through the sympathetic nervous system. The footshock-induced reduction of natural killer cell activity appears to be mediated by cerebral CRF, because an antibody to CRF injected icv but not peripherally prevented the shock-related response [1].

Although ACTH has been shown to have some direct effects on immune function, including an inhibition of antibody production and modulation of B cell function, the effects have not been striking. On the other hand, the endorphins have been shown to exert a plethora of effects on immune function. Lymphocytes possess binding sites for opiates, but at least some of these are not sensitive to the opiate antagonist, naloxone. Interestingly, binding sites have been found for N-acetyl-β-endorphin, which is the commonest form of β-endorphin secreted from the anterior pituitary and has no opiate activity. β-Endorphin and other opioid peptides can exert effects on lymphocytes in vitro. By and large, the effects are facilitatory. Such effects have been observed on natural killer cell activity as well as on proliferative responses. Opioid peptides are also chemoattractants for lymphocytes. In contrast to the enhancing effects in vitro, in vivo opiates are largely inhibitory, especially on natural killer cell activity. This apparent contradiction can be explained, because, at least in the case of morphine, the site of opiate action appears to be in the CNS. Moreover, the effects appear to be mediated by the adrenal gland, most probably by catecholamines [1].

Other Hormones

Perhaps the most interesting effect of a pituitary hormone on the immune system is that of prolactin; its effects are largely stimulatory. Reduction of pituitary prolactin secretion (e.g., by dopaminergic agonists or opiate antagonists) impairs immune function and increases susceptibility to infections, such as by *Listeria monocytogenes,* whereas stimulation of prolactin secretion (e.g., by D2 dopaminergic antagonists or opiates) can enhance it. It is postulated that prolactin may be the counter-regulatory hormone to glucocorticoids and thus acts by opposing interactions between these two hormones on immune function as can be demonstrated in vivo. Direct effects of prolactin on lymphocyte function have been difficult to demonstrate, but prolactin antibodies do impair proliferative responses in vitro. Lymphocytes can produce a prolactin-like protein, although its identity with prolactin has not been demonstrated. Thus it appears that prolactin is yet another example of a multifunctional peptide produced by both the pituitary and the lymphocytes [1].

Immune System Signaling of the Brain – Infection as a Stressor

Not that many would technically challenge the notion that sickness is stressful. In his autobiography, Hans Selye (Selye János (1907–1982) was a Canadian endocrinologist of Austrian-Hungarian origin who did much important theoretical work on the non-specific response of the organism to stress.) indicates that it was the common characteristics of sickness regardless of the underlying disease, i.e., "the syndrome of just being sick," that first interested him in stress research and led him to advance his much maligned proposal of the *non-specificity* of stress. That the HPA axis is activated following infections has long been known. During World War I, it was noted that fatalities from infections were associated with striking morphological changes in the adrenal cortex. It was later discovered that endotoxin (lipopolysaccharide, LPS), a potent stimulator of the immune system, stimulated the HPA axis. Subsequently, it was shown that infection of rats with *Escherichia coli* increased the secretion of ACTH [1].

Cytokines, Neuropeptides and the Mechanics of Neuroimmune Interactions

Cytokines are mediators of inter- and intracellular communications [20]. These peptides contribute to a chemical signaling language that regulates development, tissue repair, hemopoiesis, inflammation and the specific and non-specific immune responses [21].

Potent cytokine polypeptides (such as IL-1, IL-6, IL-8 and tumor-necrosis-factor-α (TNF-α)) have pleiotropic (redundant) activities and functional redundancy [1]; in fact, they act in a complex, intermingled network where one cytokine can influence the production of, and response to, many other cytokines. It is also now clear that the pathophysiology of inflammatory ►hyperalgesia, infection and autoimmune and malignant diseases can be explained, at least in part, by the induction of cytokines and the subsequent protracted cellular responses [21]. Of note, cytokines and cytokine antagonists have also exhibited therapeutic potential in a number of chronic and acute diseases [1].

The mechanisms, from both the neural and immunological perspective, involved in stress-induced alteration of immune function are being studied. The immune system is regulated in part by the CNS, acting principally via the HPA axis and the sympathetic nervous system [1]. In recent years, our understanding of the interactions between the HPA axis and immune-mediated inflammatory reactions has expanded enormously. This section outlines the influences that the HPA axis and immune-mediated inflammatory reactions exert on each other and discusses the mechanisms whereby these interactions are mediated. Furthermore, I discuss HPA interactions and oxidative stress evolution within the context of a potential role for the ►transcription factor NF-κB (►NF-κB – activation in the mouse spinal cord following sciatic nerve transection), which regulates a plethora of cellular functions including pro-inflammatory mediated processes [22], and the role of gaseous transmitters (►brain aging and Alzheimer's disease).

Cytokines in the CNS: Neuro-Immune-Endocrine Interactions

The communication between the neuroendocrine and immune systems is bidirectional (see Fig. 1a). The neuroimmune-endocrine interface is mediated by cytokines, such as IL-1 and tumor-necrosis-factor-α

(TNF-α), acting as autocrine/paracrine or endocrine factors regulating pituitary development, cell proliferation, hormone secretion, and feedback control of the HPA axis [1]. Increasing evidence supports the hypothesis that there are bi-directional circuits between the CNS and the immune system. Soluble products that appear to transmit information from the immune compartment to the CNS include ▶thymosins, ▶lymphokines and certain complement proteins.

Opioid peptides, ACTH and ▶thyroid-stimulating hormone (TSH) are additional products of lymphocytes that may function in immunomodulatory neuroendocrine circuits. It was proposed that the term "immunotransmitter" be used to describe molecules that are produced predominantly by cells that comprise the immune system but that transmit specific signals and information to neurons and other cell types [1].

Several cytokines are known to affect the release of anterior pituitary hormones by an action on the hypothalamus and/or the pituitary gland. The major cytokines involved are IL-1, IL-2, IL-6, tumor-necrosis-factor-α (TNF-α) and interferon (IFN) (see Fig. 2a). The predominant effects of these cytokines are to stimulate the HPA axis and to suppress the ▶hypothalamic-pituitary-thyroid (HPT) axis and gonadal axis, and growth hormone (GH) release. However, the relative importance of systemically and locally produced cytokines in achieving these responses and their precise sites of action have not been fully established [1]. There is accumulating evidence that there are significant interactions between the immune and neuroendocrine systems which may explain, at least in part, some of the effects on growth, thyroid, adrenal and reproductive functions which occur in acute and chronic disease (▶central nervous system disease in primary sjögren's syndrome). During stimulation of the immune system (e.g. during infectious diseases), peculiar alterations in hormone secretion occur (hypercortisolism, hyperreninemic hypoaldosteronism, euthyroid sick syndrome, hypogonadism). The role of cytokines is being elucidated [1,23].

IL-1/IL-6/TNF-α and HPA Responses

The bilateral communication between the immune and neuroendocrine systems plays an essential role in modulating the adequate response of the HPA axis to the stimulatory influence of interleukins and stress-related mediators [24] (see Fig. 2a) (▶neuroinflammation – IL-18). It is thus reasonable to assume that inappropriate responses of the HPA axis to interleukins might play a role in modulating the onset of pathological conditions such as infections and related pathologies [1]. Ever since two distinct molecules of IL-1 (IL-1α and IL-1β) were cloned, sequenced and expressed, it has been a matter of investigation whether these two forms of IL-1 possess an identical spectrum of biological activities (▶immunomodulation – brain areas involved).

In situ histochemical techniques were used to investigate the distribution of cells expressing type I IL-1 receptor mRNA in the CNS, pituitary and adrenal gland of the mouse. For instance, hybridization of ^{35}S-labeled antisense cRNA probes derived from a murine T-cell IL-1 receptor cDNA revealed a distinct regional distribution of the type I IL-1 receptor, both in brain and in the pituitary gland [1]. In the brain, an intense signal was observed over the granule cell layer of the ▶dentate gyrus, over the entire midline ▶raphe system, over the choroid plexus and over endothelial cells of postcapillary venules throughout the neuraxis. A weak to moderate signal was observed over the pyramidal cell layer of the hilus and CA3 region of the hippocampus, over the anterodorsal ▶thalamic nucleus, over ▶Purkinje cells of the ▶cerebellar cortex and in scattered clusters over the external-most layer of the ▶median eminence. In the pituitary gland, a dense and homogeneously distributed signal was observed over the entire anterior lobe. Furthermore, no autoradiographic signal above background was observed over the posterior and intermediate lobes of the pituitary, or over the adrenal gland, providing evidence for discrete receptor substrates subserving the central effects of IL-1, thus supporting the notion that IL-1 acts as a neurotransmitter/neuromodulator in the brain. It also supports the fact that IL-1-mediated activation of the HPA axis occurs primarily at the level of the brain and/or pituitary gland [1].

IL-1 and other related pro-inflammatory cytokines are potent activators of the HPA axis [1]. Current studies of IL-1 and its involvement in the HPA axis have indicated that there is a clear-cut differential response to IL-1α and IL-1β. For example, the intravenous injection of human recombinant IL-1β in conscious, freely-moving rats significantly increased the plasma concentrations of ACTH in a dose-related manner, whereas IL-1α did not, suggesting that the two members of the IL-1 family may have a different spectrum of biological actions.

Furthermore, additional investigations clarified the mechanism by which IL-1 activates the HPA axis. For example, the ACTH response to IL-1 was completely abolished by pre-injection of rabbit antiserum generated against corticotrophin-releasing factor (CRF) but not by normal rabbit serum [1]. The IL-1-induced ACTH release did not seem to be caused by a general stress effect of IL-1 because plasma prolactin (PRL) concentrations, another indicator of a stress response, were not altered by IL-1 injection, suggesting that IL-1 acts centrally in the brain to stimulate the secretion of CRF, thereby eliciting ACTH release, and that a direct action of IL-1 on the pituitary gland is unlikely. In addition, it has been reported that intra-peritoneal injection of recombinant IL-1 into mice increased the cerebral concentration of the noradrenaline (NA) catabolite,

3-methoxy-4-hydroxyphenylethyleneglycol (MHPG), probably reflecting increased activity of noradrenergic neurons [1]. This effect was dose-dependent and was largest in the hypothalamus, especially the medial division. Of note, ►tryptophan concentrations were also increased throughout the brain and the increase of MHPG after IL-1 administration paralleled the increase of plasma corticosterone (►neurodegenerative diseases – tryptophan metabolism). In contrast to prior observations [1], both the α- and β-forms of IL-1 were effective, but the activity was lost after heat treatment of the IL-1 [25].

Noradrenergic neurons with terminals in the hypothalamus are known to regulate the secretion of CRF, thus suggesting that IL-1 activates the HPA axis by activating these neurons. Because the initiation of an immune response is known to cause systemic release of IL-1, this cytokine may be an immuno-transmitter communicating the immunologic activation to the brain. The IL-1-induced changes in hypothalamic MHPG may explain the increases of electrophysiological activity, the changes of hypothalamic noradrenaline metabolism and the increases in circulating glucocorticoids reported to be associated with immunologic activation and frequently observed in infected animals [25]. In support of these observations, ACTH secretion by the anterior pituitary has been shown to be stimulated by catecholamines in vivo and in vitro [1] (Fig. 3).

In concert, it has been reported that intra-cerebroventricular injections of IL-1 can cause the release of ACTH. For instance, IL-1β produced an immediate increase in plasma corticosterone and ACTH [1]. Using a potent steroidogenic dose of IL-1β, intra-cerebroventricular injection resulted in the suppression of splenic macrophage IL-1 secretion following stimulation by lipopolysaccharide-endotoxin (LPS) in vitro. Macrophage ►transforming growth factor (TGF)-β secretion, however, was not affected, indicating a differential action of IL-1β on macrophage cytokine production. Following adrenalectomy, the suppressive effect of IL-1β was reversed and resulted in the stimulation of macrophage IL-1 secretion, indicating that the suppression was

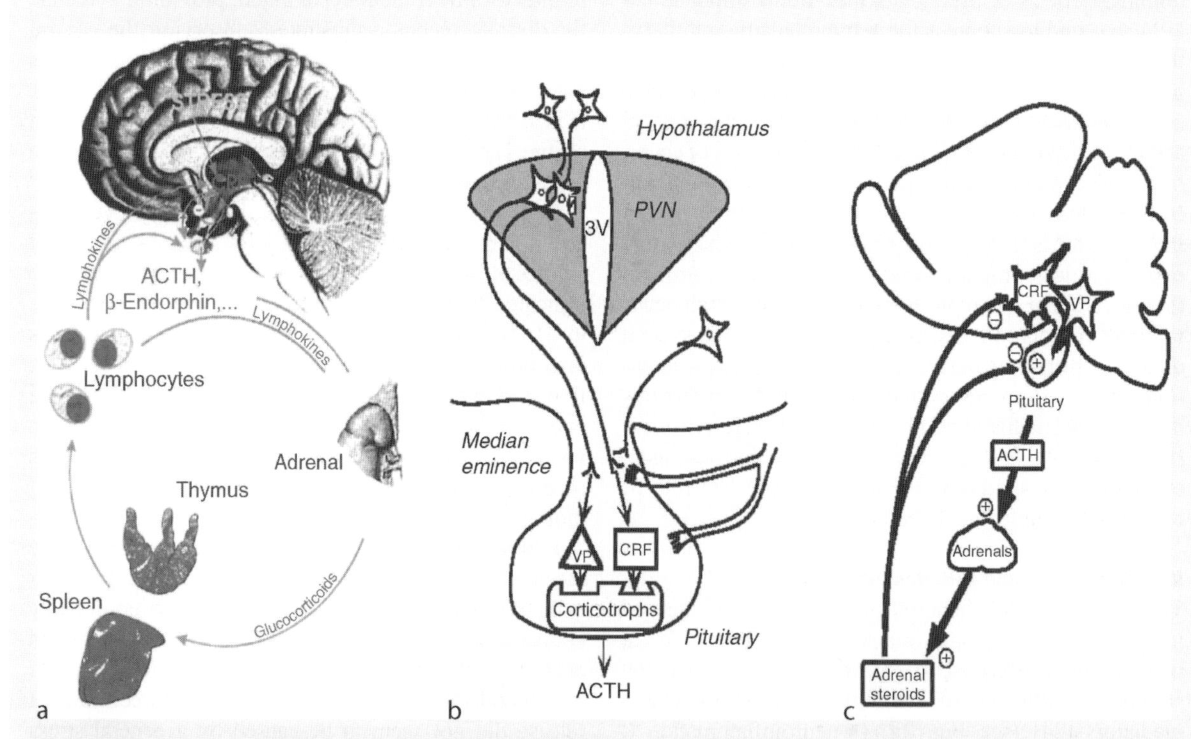

Neuroimmunology. Figure 3 The HPA doctrine. (a) Classic components of the HPA-CNS-immune systems. (b) Neurons of the hypothalamus that synthesize CRF and vasopressin are found in an area called the paraventricular nucleus (PVN). These cell bodies send axons to the median eminence, here peptides are released from the nerve terminals and are transported through vessels of the portal system. When they reach the anterior pituitary, these peptides act on their respective receptors, thereby stimulating ACTH secretion. (c) Following its release into the general circulation, ACTH acts on the cortex of the adrenal glands, which manufacture and secrete glucocorticoids (corticosterone in rodents and cortisol in humans). These glucocorticoids exert a classical negative feedback influence on the pituitary, where they inhibit the effect of CRF and VP, and on the PVN, where they inhibit the synthesis of CRF. Thus after a stimulus stimulates CRF and ACTH release, the production of glucocorticoids will eventually terminate this release, thereby ensuring the maintenance of homeostasis [1].

mediated by adrenocorticol activation. However, surgical interruption of the splenic nerve to eliminate autonomic innervation of the spleen also prevented the macrophage suppressive signal.

Further conflicting reports, however, have been published with regard to a crucial role of catecholamines in IL-1-mediated regulation of the HPA axis [1]. The hypothalamus seems to be an important site of action of IL-1 on the HPA axis, thereby inducing CRF secretion (catecholamines are important modulators of CRF secretion); in turn, IL-1 stimulates catecholamine release from the hypothalamus. In this respect, using an in vitro rat hypothalamic continuous perfusion system, the possible involvement of hypothalamic catecholamines in the effect of IL-1β on hypothalamic CRF secretion was investigated. For instance, neither in vivo pretreatment with an inhibitor of catecholamine synthesis nor in vitro exposure to α- or β-adrenoceptor antagonists (phenoxybenzamine or propranolol, respectively), nor combination of both treatments altered the effect of IL-1 on CRF secretion from superfused hypothalami, indicating that catecholamines are not involved in the in vitro stimulatory action of IL-1 on hypothalamic CRF secretion. In contrast, IL-1-induced corticosterone release was shown to occur by an adrenergic mechanism from rat adrenal gland [1].

An interesting mechanism was reported for IL-1-mediated regulation of the HPA axis. A primary route of peripheral cytokine signaling was proposed through the stimulation of peripheral vagal afferents rather than, or in addition to, a direct cytokine access to the brain. Sub-diaphragmatic, but not hepatic vagotomy, blocked IL-1β-induced hypothalamic norepinephrine depletion and attenuated IL-1β-induced increases in serum corticosterone, suggesting that IL-1 activates the HPA axis via the stimulation of peripheral vagal afferents and further support the hypothesis that peripheral cytokine signaling to the CNS is mediated primarily by stimulation of peripheral afferents [1].

Another major mechanism reported for the action of IL-1 on the HPA axis involves the amygdala. For example, bilateral ibotenic acid lesions of the central amygdala substantially reduced ACTH release and hypothalamic corticotropin-releasing factor and oxytocin cell c-fos expression responses to IL-1 and IL-8, suggesting a facilitatory role for this structure in the generation of HPA axis responses to an immune challenge [1]. Since only a small number of central amygdala cells project directly to the ▸paraventricular nucleus, the authors then examined the effect of central amygdala lesions on the activity of other brain nuclei that might act as relay sites in the control of the HPA axis function. It was found that bilateral central amygdala lesions significantly reduced IL-1β-induced c-fos expression in cells of the ventromedial and ventrolateral subdivisions of the bed nucleus of the stria terminalis and brainstem catecholamine cell groups of the ▸nucleus tractus solitarii (A2 noradrenergic cells) and ventrolateral medulla (A1 noradrenergic and C1 adrenergic cells). These findings, in conjunction with previous evidence of ▸bed nucleus of the stria terminalis and catecholamine cell group involvement in HPA axis regulation, indicated that ventromedial and ventrolateral bed nucleus of the stria terminalis cells and medullary catecholamine cells might mediate the influence of the central amygdala on the HPA axis responses to an immune challenge. Thus these related data established that the central amygdala influences HPA-axis responses to a systemic immune challenge but indicate that it acts primarily by modulating the activity of other control mechanisms [1] (see Fig. 3).

Similarly, an interesting mechanism implicates the vagus nerve. For instance, direct electrical stimulation of the central end of the vagus nerve induced increases in the expression of mRNA and protein levels of IL-1β in the hypothalamus and the hippocampus. Furthermore, expression of CRF mRNA was increased in the hypothalamus after vagal stimulation. In addition, plasma concentrations of ACTH and corticosterone were also increased by this stimulation, indicating that the activation of the afferent vagus nerves can induce production of cytokines in the brain and activate the HPA axis. Therefore, the afferent vagus nerve may play an important role in transmitting peripheral signals to the brain in infection and inflammation. In concert, dorsal and ventral medullary catecholamine cell groups were reported to contribute differentially to systemic IL-1β-induced HPA axis responses. Medial parvocellular paraventricular corticotropin-releasing hormone (mPVN/CRH) cells are critical in generating HPA axis responses to systemic IL-1β. However, although it is understood that catecholamine inputs are important in initiating mPVN/CRH cell responses to IL-1β, the contributions of distinct brainstem catecholamine cell groups are not known [1] (Fig. 4a and b).

The in vivo release of ACTH by IL-1 is reportedly blocked by acute treatment with indomethacin, a non-steroidal anti-inflammatory drug (NSAID), suggesting an involvement of endogenous ▸prostaglandins in the effect of cytokines on the HPA axis [1]. However, indomethacin also increases plasma corticosterone concentrations, raising the possibility that inhibition of ACTH release is due to suppressive effects of hypercorticolemia rather than to blockade of the stimulatory effects of IL-1α. It was observed that the intraventricular administration of indomethacin completely abolished the rise in plasma ACTH levels caused by the peripheral injection of this lymphokine to intact rats. In contrast, implantation of intact rats with indomethacin pellets only partially interfered with IL-1-induced ACTH secretion. To determine whether the effect of indomethacin was due to corticosteroid feedback or represented a modulating action of prostaglandins themselves, a similar series of experiments were carried out in adrenalectomized rats. In the absence of corticoid replacement therapy, acute treatment with indomethacin did not measurably interfere with the stimulatory effect of IL-1α. In contrast, indomethacin blunted, but did not abolish, the effect of IL-1α in adrenalectomized rats pretreated with cortisone or dexamethasone to normalize basal ACTH levels. Thus, the acute ability of indomethacin to totally block IL-1-induced ACTH secretion by intact rats appears to be primarily mediated through corticosteroid feedback. However, results obtained when a similar experiment was carried out in adrenalectomized/corticosteroid-treated rats suggested that the ability of IL-1α to activate the HPA axis might be partially dependent on the release of prostaglandins. In concert, the effects of various cyclo- and lipoxygenase inhibitors on the neuro-chemical and HPA responses to IL-1 indicated a role for prostaglandins in IL-1-mediated activation of the HPA axis (►neuroinflammation – PDE family inhibitors in the regulation of neuroinflammation). For example, pretreatment of mice with the cyclo-oxygenase (COX) inhibitors indomethacin or ibuprofen failed to prevent the elevations of plasma cortisone, or hypothalamic MHPG or tryptophan that followed intraperitoneally administered IL-1 [1].

In contrast, pro-inflammatory cytokines can reduce glucocorticoid receptor translocation and function. Specifically, several studies have found that cytokines induce a decrease in glucocorticoid receptor function, as evidenced by reduced sensitivity to glucocorticoid effects on functional end points [26]. These observations clearly suggested that cytokines produced during an inflammatory response may induce glucocorticoid-receptor resistance in relevant cell types by direct effects on the glucocorticoid receptor, thereby providing an additional pathway by which the immune system can influence the HPA axis [1].

Administration of lipopolysaccharide (LPS) (and other inflammatory mediators) results in the activation of the HPA axis [1] (Fig. 4b). The mechanisms through which LPS stimulates the HPA axis are not well understood, however. In initial studies, the hypothesis that LPS increases plasma ACTH levels by releasing IL-1 was tested. Two experimental tools reported to interfere with the biological activity of IL-1 were used: antibodies directed against IL-1 receptors and α-melanocyte releasing hormone (α-MSH) [27] (►melanin and neuromelanin in the nervous system). The results suggested that LPS activates the HPA axis through a mechanism involving the activation of IL-1 receptors and that the effect of IL-1β, but not IL-1α, on ACTH secretion can be partially blocked by α-MSH. Therefore, LPS acts both at the level of the brain and the gonads to stimulate the HPA axis and inhibits the hypothalamic-pituitary-gonadal (HPG) axis [1].

Exogenously administered IL-1 mimics most of the effects of LPS on pituitary activity. In addition, antibodies against IL-1 receptors can interfere with LPS-induced ACTH secretion, indicating that at least part of the ability of LPS to alter endocrine functions appears to depend upon endogenous IL-1. Of interest, IL-1 and IL-6 share a number of biological functions. Because IL-1 induces IL-6 in vivo, the extent to which IL-6 mediates the effects of IL-1 has come under investigation. The stimulation of the HPA axis by IL-1 and IL-6 is recognized as a critical component of the inflammatory response. In this respect, it was demonstrated that the administration of IL-6 alone did not duplicate the stimulatory effect of IL-1α on ACTH release. On the other hand, sub-optimal amounts of IL-1α and IL-6 synergized to induce an early (30–60 min) ACTH response and produce a later (2–3h) response that was similar to the one observed after IL-1α was administered alone, suggesting that the late response to IL-1 may be dependent on synergy with the endogenous IL-6 it induces systemically and in the CNS (including the hypothalamus and the pituitary gland) [1].

Another mechanism implicates histamine receptors in LPS/IL-1-induced activation of the HPA axis and ACTH release [1]. Lipopolysaccharine (LPS) and LPS-derived cytokines stimulate the release of histamine. Histamine is a known hypothalamic neurotransmitter and activates the HPA axis. To elucidate the role of histamine in LPS- and cytokine-induced ACTH release, Perlstein and colleagues [28] evaluated the effects of several histamine H1 and H2 receptor antagonists on the ACTH response to LPS, IL-1α and histamine in mice. Although all three of the H1 receptor antagonists administered (mepyramine, diphenhydramine or promethazine) were able to block the 10-min ACTH response to histamine, only promethazine (a less selective H1 receptor antagonist than mepyramine) was able to reduce the LPS- or IL-1α-induced ACTH responses. In addition, ranitidine, a powerful and selective H2 receptor antagonist, had little effect on the LPS- and IL-1α-induced ACTH responses, while metiamide, a much less potent first-generation H2 receptor antagonist, substantially diminished ACTH release. It was concluded that the greater effectiveness of promethazine, in contrast to mepyramine or diphenhydramine, probably relates to the ability of phenothiazine derivatives to inhibit non-HA-dependent pathways involved in the stimulation of the HPA axis by cytokines [1].

IL-2 and HPA Responses

The cytokine IL-2 exerts numerous effects within the immune as well as the central nervous system and is thought to serve as a humoral signal in their communication. A major role for IL-2 has been noted in the regulation of the HPA axis responses. Brain-derived or blood-borne

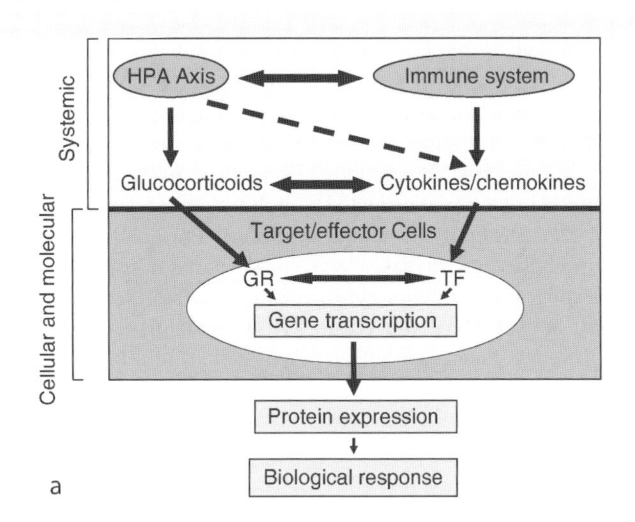

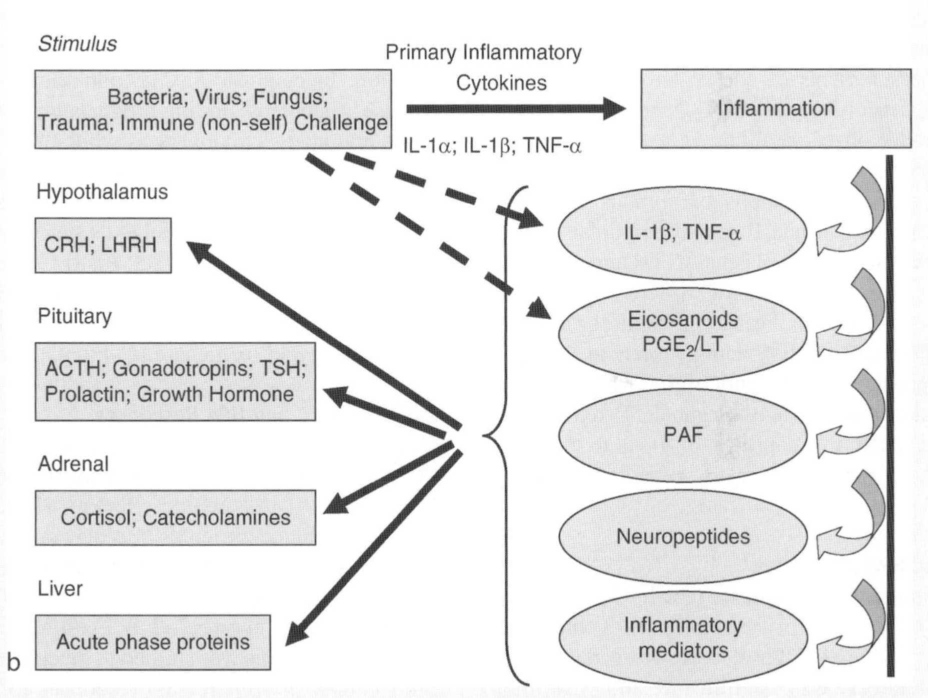

Neuroimmunology. Figure 4 (a) Scheme depicting systemic and cellular/molecular interplay between the HPA axis and the immune system in the regulation of glucocorticoid/cytokine secretion and gene expression. *GR*, glucocorticoid receptor; *TF*, transcription factors. (b) The inflammatory response and the HPA axis. Some of the effects of the inflammatory response on the neuroendocrine system are illustrated. A stimulus such as trauma, stress, immune challenge, or bacterial, viral and fungal toxins acts to provoke the inflammatory process. Inflammatory cells respond by secreting inflammatory mediators such as cytokines. A profound process of inflammation ensues and propagates itself with the auto-induction (autocrine) of inflammatory mediators, including cytokines, eiconsanoids, platelet activating factor, neuropeptides and various other mediators. These agents, particularly inflammatory cytokines, act either directly or indirectly to increase the production of releasing hormones in the hypothalamus (HPA axis), pituitary hormones, cortisol and catecholamines. In addition, the liver participates in this inflammatory-HPA axis by releasing acute-phase proteins. *ACTH*, adrenocorticotropic hormone; *CRH*, corticotropin releasing hormone; *LT, IL*, leukotriene, interleukin; *LHRH*, luteinizing hormone releasing hormone; *PAF*, platelet activating factor; *PGE_2*, prostaglandin; *TSH*, thyroid stimulating hormone; *TNF*, tumor necrosis factor [1].

IL-2 may also control the activity of the HPA axis at various levels of regulation. IL-2, for example, caused a dose-dependent stimulation of secretion of arginine vasopressin (AVP; also called antidiuretic hormone, ADH) from both the intact rat hypothalamus in vitro and hypothalamic cell cultures [1]. IL-2, however, did not increase the secretion of corticotrophin-releasing factor (CRF) in either preparation, nor did it prime the cells to respond to a subsequent dose of IL-2. Both preparations, nevertheless, were able to respond to known CRH secretagogues, such as ►serotonin (5-HT) and K^+. This may provide yet another line of communication between the immune and neuroendocrine systems. In another study [1], it was investigated whether persistently elevated concentrations of central IL-2, which are associated with several diseases or induced during immuno-therapeutic use of this cytokine, could induce long-term activation of the HPA axis. Adult male *Sprague–Dawley* rats received an intra-cerebroventricular infusion of the recombinant cytokine; control animals received heat-inactivated IL-2. IL-2 caused a significant increase in ACTH concentrations during the later portion of the dark phase of the cycle.

Plasma corticosterone concentrations were significantly elevated over almost the whole diurnal cycle. In addition, measurements of corticosterone-binding globulin concentrations revealed IL-2-induced decreases during the dark phase, resulting in a marked increase in free corticosterone. Furthermore, after prolonged chronic infusion, both groups of animals underwent restraint stress. For instance, IL-2-treated animals showed stress-induced increases in plasma ACTH and corticosterone that were not significantly different from those of animals treated with heat-inactivated IL-2. Along with the alteration of HPA activity seen in the IL-2-treated animals, chronic delivery of the cytokine caused periventricular tissue damage and gliosis. Taken together, the data reflected the capacity of IL-2 to modulate neuroendocrine activity over an extended period of treatment [1].

IL-3/IL-6 and HPA Responses

Accumulating evidence indicate that IL-3 can activate the HPA axis [1]. For example, IL-3 and IL-6 equipotently stimulated basal cortisol secretion. The stimulatory effect was significant and maximum cortisol levels were induced later. In contrast to ACTH, which significantly induced cAMP levels in parallel to its steroidogenic effect, IL-3 (or IL-6) had no significant effect on cAMP. Furthermore, the authors showed that specific inhibition of the cyclo-oxygenase (COX) pathway by indomethacin completely blocked the steroidogenic effect of IL-6 while the effect of IL-3 was not affected. In contrast, co-incubation with nordihydroguaiaretic acid (NDGA), a specific inhibitor of the lipoxygenase system, abolished IL-3-stimulated steroidogenesis but had no effect on IL-6-stimulated cortisol secretion, indicating that IL-3 and IL-6 directly stimulate the steroidogenesis at the adrenal level through activation of different, cAMP-independent pathways.

While the stimulatory effect of IL-6 on cortisol secretion from adult human adrenocortical cells seems to be mediated through the cyclo-oxygenase (COX) pathway, the effect of IL-3 on adrenocortical cortisol secretion is dependent on the lipoxygenase pathway. Similarly, the effect of IL-3 and IL-6 on cortisol secretion of bovine adrenocortical cells in primary culture under serum-free conditions was further explored. For instance, both IL-3 and IL-6 stimulated basal cortisol secretion dose-dependently to a similar extent at a similar time course [1]. After incubation with IL-3 or IL-6, a maximum 4.1-fold increase of the cortisol secretion was reached after 12h. Co-incubation of IL-3 and IL-6 revealed, however, no significant synergism. To elucidate a possible involvement of arachidonic acid metabolites in the signal transduction, IL-3 or IL-6 were co-incubated with indomethacin or nordihydroguaiaretic acid. Co-incubation with indomethacin completely abolished the stimulatory effect of IL-6 but had no effect on IL-3-stimulated cortisol secretion. In contrast, specific inhibition of the lipoxygenase system by nordihydroguaiaretic acid blocked IL-3-stimulated steroidogenesis while the effect of IL-6 was not affected. Neither IL-3 nor IL-6 altered cAMP levels significantly, whereas ACTH significantly induced cAMP levels in parallel to its steroidogenic effect. While the stimulatory effect of IL-3 seems to be dependent on the lipoxygenase pathway, the effect of IL-6 on adrenocortical cortisol secretion is mediated through the cyclo-oxygenase (COX) pathway [1].

IL-4/IL-5/IL-10 and HPA Responses

Glucocorticoids are widely used in the therapy of inflammatory, autoimmune, and allergic diseases [1] (Fig. 5a and b). As the end-effectors of the HPA axis, endogenous glucocorticoids also play an important role in suppressing innate and cellular immune responses. The influence of dexamethasone on IL-10 production and the type 1 (T1)/type 2 (T2) T cell balance found in rheumatoid arthritis was studied to determine a possible role for IL-4 in HPA-related responses to rheumatoid arthritis. Peripheral blood mononuclear cells (PBMNC) were isolated from 14 rheumatoid-arthritis patients both before and 7 and 42 days after high-dose dexamethasone pulse therapy. The ex vivo production of IL-10, IFN-γ (T1 cell) and IL-4 (T2 cell) by PBMNC was assessed, along with parameters of disease activity (erythrocyte sedimentation rate, C reactive protein, Visual Analogue Scale, Thompson joint score).

The pro-inflammatory cytokines, IL-1 and tumor-necrosis-factor-α (TNF-α), were among the first to be recognized in this regard. A modulator of these cytokines, IL-10, has been shown to have a wide range of activities in the immune system. IL-10 is produced in pituitary, hypothalamic and neural tissues in addition to lymphocytes [29]. IL-10 enhances CRF and ACTH production in hypothalamic and pituitary tissues, respectively. Further downstream in the HPA axis,

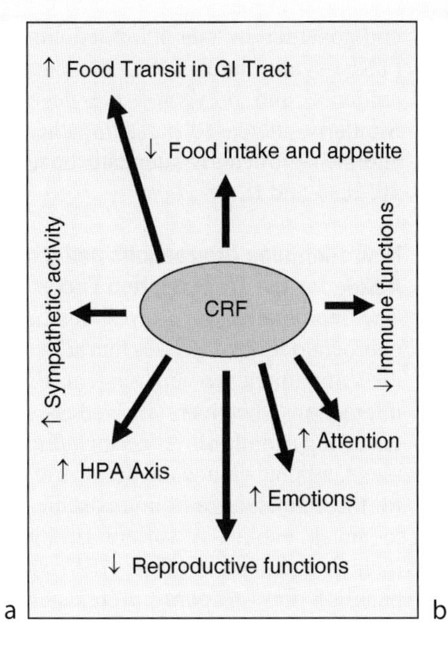

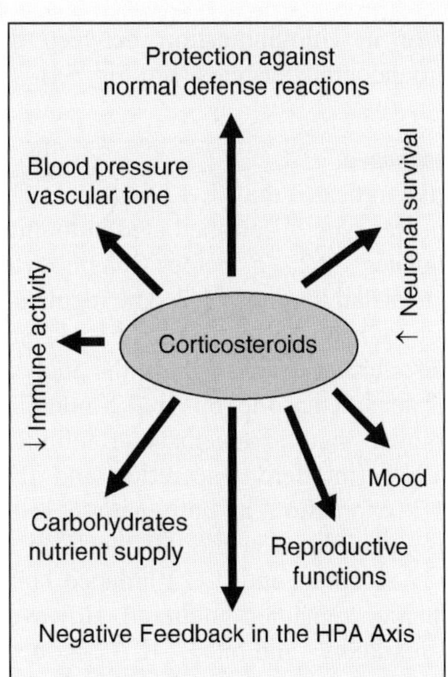

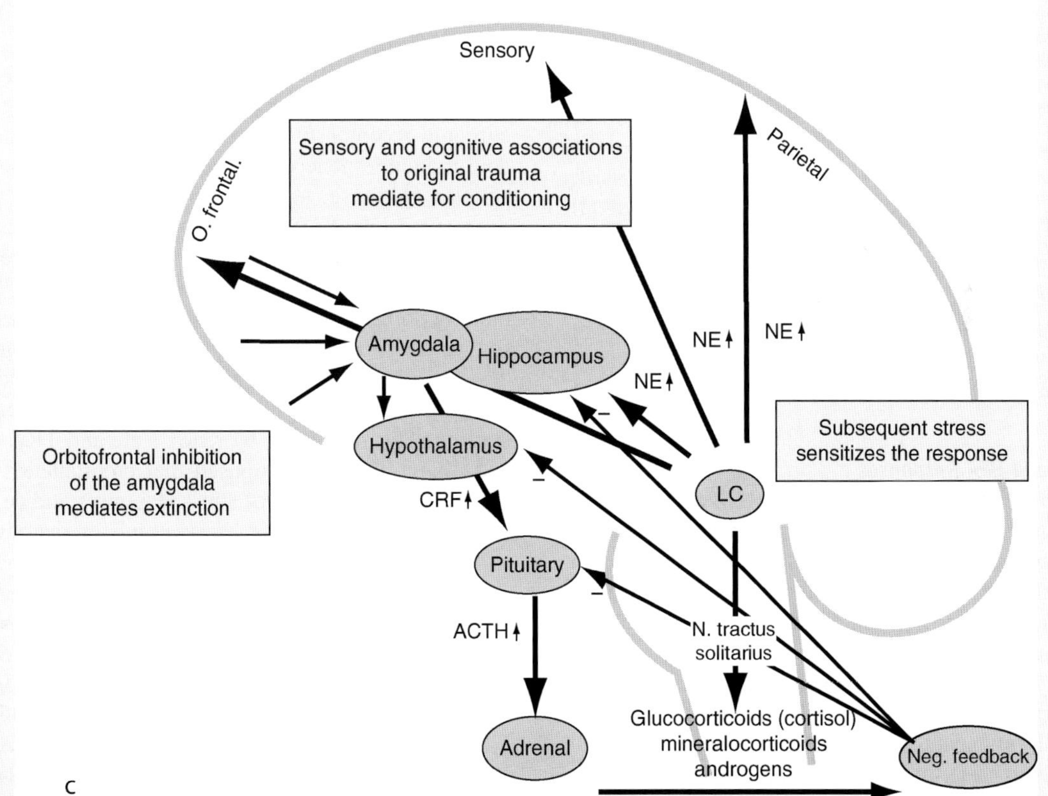

Neuroimmunology. Figure 5 (a, b) The effects of CRF and corticosteroids (glucocorticoids) on a variety of body mechanisms. These wide-ranging effects underscore the significance of HPA axis interactions and the mechanisms involved. (c) Neurochemical-imunologic mechanisms and their sites of action. *ACTH*, adrenocorticotropic hormone; *CRF*, corticotropin-releasing factor; *LC*, locus ceruleus; *NE*, norepinephrine [1].

endogenous IL-10 has the potential to contribute to regulation of glucocorticosteroid production both tonically and following stressors. Evidence indicated that IL-10 might be an important endogenous regulator in HPA axis activity and in CNS pathologies [30]. Thus, in addition to its more widely recognized role in immunity, as anti-inflammatory cytokine, IL-10's neuroendocrine activities point to its role as an important regulator in communication between the immune and neuroendocrine systems (see Fig. 3a).

IL-12 and HPA Responses

Recent studies have indicated that IL-12 promotes Th1 (T helper lymphocytes type 1) cell-mediated immunity, while IL-4 stimulates Th2 (T helper lymphocytes type 2) humoral-mediated immunity [1]. The regulatory effect of glucocorticoids on key elements of IL-12 and IL-4 signaling were further examined. On the analysis of the effect of dexamethasone on IL-12-inducible genes, it was shown that dexamethasone inhibited IL-12-induced IFN-γ (interferon) secretion and IFN regulatory factor-1 expression in both natural killer and T cells. This occurred even though the level of expression of IL-12 receptors and IL-12-induced Janus kinase phosphorylation remained unaltered. However, dexamethasone markedly inhibited IL-12-induced phosphorylation of Stat-4 (a transcription factor) without altering its expression. This was specific, as IL-4-induced Stat-6 phosphorylation was not affected, and mediated by the glucocorticoid receptor, as it was antagonized by the glucocorticoid receptor antagonist RU-486. Moreover, transfection experiments showed that dexamethasone reduced responsiveness to IL-12 through the inhibition of Stat-4-dependent IFN regulatory factor-1 promoter activity. It was concluded that blocking IL-12-induced Stat-4 phosphorylation, without altering IL-4-induced Stat-6 phosphorylation, appears to be a new suppressive action of glucocorticoids on the Th1 cellular immune response and may help explain the glucocorticoid-induced shift toward the Th2 humoral immune response [1] (see Fig. 3b).

IL-18 and HPA Responses

Vertebrates achieve internal homeostasis during infection or injury by balancing the activities of pro-inflammatory and anti-inflammatory pathways (►central nervous system infections – humoral immunity in arboviral infections). The CNS regulates systemic inflammatory responses to lipopolysaccharide (LPS), for instance, through humoral mechanisms (see Fig. 2b). Activation of afferent vagus nerve fibers by LPS or cytokines specifically stimulates HPA anti-inflammatory responses. In this respect, it was described that a previously unrecognized, parasympathetic anti-inflammatory pathway, by which the brain modulates systemic inflammatory responses to LPS, is active at the level of the HPA axis [1,21]. Acetylcholine, the principal vagal neurotransmitter, significantly attenuated the release of pro-inflammatory cytokines, including IL-18, but not the anti-inflammatory cytokine IL-10, in lipopolysaccharide-stimulated human macrophage cultures. Furthermore, direct electrical stimulation of the peripheral vagus nerve in vivo during lethal endotoxemia in rats inhibited tumor necrosis factor (TNF) synthesis in liver, attenuated peak serum TNF amounts, and prevented the development of shock. Similarly, increased parasympathetic tone and acetylcholine significantly attenuate the release of TNF-α, IL-1β, IL-6 and IL-18 [1].

Neuro-Immune Interactions and Oxidative Stress: A Role for the Transcription Factor NF-κB

The mammalian stress response evokes a series of neuroendocrine responses that activate the HPA axis and the sympathetic nervous system (Fig. 5c). Coordinated interactions between the stress response systems, occurring at multiple levels including the brain, pituitary gland, adrenal gland, and peripheral tissues, are required for the maintenance of homeostatic plateau. Adaptation to stress evokes a variety of biological responses, including activation of the HPA axis and synthesis of a panel of stress-response proteins at cellular levels. For example, expression of thioredoxin (TRX), a non-thiol antioxidant, is significantly induced under oxidative conditions. In this regard, it was demonstrated that either antisense TRX expression or cellular treatment with hydrogen peroxide (H_2O_2) negatively modulated glucocorticoid-receptor function and decreased glucocorticoid-inducible gene expression. In addition, impaired cellular response to glucocorticoids is rescued by overexpression of TRX, possibly through the functional replenishment of the glucocorticoid receptor. Moreover, not only the ligand-binding domain but also the DNA binding domain of the glucocorticoid receptor was also suggested to be a direct target of TRX [1]. Together, these observations presented conclusive evidence showing that cellular glucocorticoid responsiveness is coordinately modulated by redox state and TRX level and, thereby, it was proposed that crosstalk between neuroendocrine control of stress responses and cellular antioxidant systems may be essential for mammalian adaptation processes.

Employing primary neurons and clonal cells, it was demonstrated that corticotropin-releasing hormone (CRH; or factor, CRF) has a neuroprotective activity in corticotropin-releasing-hormone-receptor type 1 (CRH-R1)-expressing neurons against oxidative cell death [1]. The protective effect of corticotropin-releasing hormone was blocked by selective and nonselective CRH-R1 antagonists and by protein kinase A (PKA) inhibitors. In addition, overexpression of CRH-R1 in clonal hippocampal cells lacking endogenous CRH-receptors established neuroprotection by corticotropin-releasing hormone (central nervous system disease – natural neuroprotective agents as therapeutics). The activation of CRH-R1 and neuroprotection were accompanied by an increased release of non-amyloidogenic soluble Aβ precursor protein, characteristic of Alzheimer's disease [1] (►brain aging and alzheimer's disease).

At the molecular level, corticotropin-releasing hormone caused the suppression of the DNA-binding activity and transcriptional activity of the oxygen- and reduction-oxidation (redox)-sensitive transcription factor, ►nuclear factor (NF)-κB [31]. Suppression of NF-κB (an inflammatory transcription factor) by overexpression of a super-repressor mutant form of inhibitory-κB (IκB)-α, a specific inhibitor of NF-κB, led to protection of the cells against oxidative stress (►Nf-κB – potential role in adult neural stem cells; NF-κB – activation in the mouse spinal cord following sciatic nerve transection). These observations strongly demonstrated a novel cytoprotective effect of corticotropin-releasing hormone that is mediated by CRH-R1 and downstream by suppression of NF-κB and indicate corticotropin-releasing hormone as an endogenous protective neuropeptide against oxidative cell death in addition to its function in the HPA-system. Moreover, the protective function of corticotropin-releasing hormone proposes a molecular link between oxidative stress-related degenerative events and the CRH-R1 system. Further elaborating on the role of transcriptional regulation in HPA responses, dysregulation of the serotonergic system and abnormalities of the HPA axis function have been implicated in ►neuropsychiatric disorders. Corticosteroid hormones in a variety of animal models suppress serotonin-1A receptors. This effect may play a central role in the pathophysiology of depression. However, little is known about the molecular mechanism underlying this suppressive effect of corticosterone [1].

In this respect, Wissink and colleagues [32] showed by functional analysis of the promoter region of the rat serotonin-1A receptor gene that two NF-κB elements in the promoter contribute to induced transcription of the rat serotonin-1A receptor gene. Furthermore, it was shown that corticosterone represses this NF-κB-mediated induction of transcription. Remarkably, only the glucocorticoid receptor and not the mineralocorticoid receptor was able to mediate this repressive effect of corticosterone, thus arguing that negative cross-talk between the glucocorticoid receptor and NF-κB may provide a basis for the molecular mechanism underlying the negative action of corticosterone on serotonin signaling in the brain [1].

Neuro-Immune Interactions and Oxidative Stress: A Role for Gaseous Transmitters

Recent work has demonstrated that the brain has the capacity to synthesize impressive amounts of the gases nitric oxide (NO) and ►carbon monoxide (CO) [1]. There is growing evidence that these gaseous molecules function as novel neural messengers in the brain. Abundant evidence is presented which suggests that NO has an important role in the control of reproduction due to its ability to control ►gonadotropin-releasing hormone (GnRH) secretion from the hypothalamus. Nitric oxide potently stimulates GnRH secretion and also appears to mediate the action of one of the major transmitters controlling GnRH secretion, glutamate (►glutamate-mediated injury to white matter – mechanisms and clinical relevance). Evidence suggests that NO stimulates GnRH release due to its ability to modulate the heme-containing enzyme, guanylate cyclase, which leads to enhanced production of the second messenger molecule, cGMP. A physiological role for NO in the pre-ovulatory ►luteinizing-hormone (LH) surge was also evidenced by findings that inhibitors and antisense oligonucleotides to ►nitric oxide synthase (NOS) attenuate the steroid-induced and pre-ovulatory luteinizing hormone (LH) surge.

Carbon monoxide (CO) may also play a role in stimulating GnRH secretion as heme molecules stimulate GnRH release in vitro, an effect that requires heme oxygenase activity and is blocked by the gaseous scavenger molecule, hemoglobin. Evidence also suggests that NO acts to restrain the HPA axis, as it inhibits HPA stimulation by various stimulants such as IL-1, vasopressin (VP) and inflammation (►neurogenesis and inflammation). This effect fits a pro-inflammatory role of NO as it leads to suppression of the release of the anti-inflammatory corticosteroids from the adrenals. Although not as intensely studied as NO, CO has been shown to suppress stimulated corticotropin-releasing-hormone release and may also function to restrain the HPA axis [1].

Evidence implicating NO in the control of prolactin (PRL) and growth hormone secretion is plausible, as is the possible role of NO acting directly at the anterior pituitary. Taken as a whole, the current data suggest that the diffusible gases, NO and CO, act as novel transmitters in the neuroendocrine axis and mediate a variety of important neuroendocrine functions. To recapitulate, NO is an unusual chemical messenger. NO mediates blood vessel relaxation when produced by endothelial cells. When produced by macrophages, NO contributes to the cytotoxic function of these immune cells. Nitric oxide also functions as a neurotransmitter and neuromodulator in the central and peripheral nervous systems. The effects on blood vessel tone and neuronal function form the basis for an important role of NO on neuroendocrine function and behavior. NO mediates hypothalamic portal blood flow and, thus, affects oxytocin and vasopressin secretion; furthermore, NO mediates neuroendocrine function in the hypothalamic-pituitary-gonadal (HPG) and hypothalamic-pituitary-adrenal (HPA) axes. Nitric oxide influences several motivated behaviors including sexual, aggressive and ingestive behaviors. Nitric oxide also influences learning and memory (►neuroinflammation – chronic neuroinflammation and memory impairments). Thus, NO is emerging as an important chemical mediator of neuroendocrine function and behavior [1].

Nitric oxide synthetase (NOS), the enzyme responsible for NO formation, is found in hypothalamic neurons containing oxytocin, vasopressin and to a lesser extent corticotropin-releasing factor. Because NO is reported to modulate endocrine activity, the hypothesis that endogenous NO participates in ACTH release by various secretagogues was investigated in vivo. In the adult male rat, the intravenous injection of IL-1β, vasopressin and oxytocin increased plasma ACTH and corticosterone levels [1]. Pretreatment with the L-form, but not the D-form, of *N*-omega nitro-L-arginine-methylester (L-NAME), a specific inhibitor of NOS, markedly augmented the effects of these secretagogues. Blockade of NOS activity also caused significant extensions of the duration of action of IL-1β, vasopressin and oxytocin. In contrast, L-NAME did not significantly alter the stimulatory action of peripherally injected corticotropin-releasing factor, or centrally administered IL-1β. In addition, administration of L-arginine, but not D-arginine, used as a substrate for basal NO synthesis and which did not by itself alter the activity of the HPA axis, blunted IL-1-induced ACTH secretion and reversed the interaction between L-NAME and IL-1β. Then, following prenatal alcohol exposure, immature offspring showed blunted ACTH released in response to the peripheral administration of IL-1β (►prenatal brain injury by chronic endotoxin exposure). Further studies were conducted to investigate the role of changes in corticosteroid feedback (measured by altered adrenal responses to ACTH), corticotropin-releasing-factor content of the median eminence (ME) and the influence of endogenous NO. For instance, the injection of several doses of ACTH failed to indicate measurable differences between the corticosterone responses of offspring born to dams fed *ad libitum* [control (C)], pair-fed (PF), or fed alcohol [ethanol (EtOH)]. Corticotropin-releasing-factor content in the median eminence, taken as an index of the amount of releasable peptide, showed a small, but statistically significant, decrease following prenatal alcohol exposure. A comparable change, however, was also noted in PF rats. As expected, the subcutaneous injection of IL-1β induced smaller increases in plasma ACTH levels of EtOH than C pups. The response of PF animals was intermediate between that of EtOH and C rats. It was also observed that inhibition of NO formation by the administration of the arginine derivative L-NAME augmented ACTH secretion in all three experimental groups and reversed the decreased corticotrophs' response to IL-1β caused by prenatal alcohol [1,33].

Conclusions and Future Prospects

The foregoing indicates that there is now substantial evidence for bidirectional communications between the nervous and immune systems. Communication occurs via chemical messengers, just as it does within the nervous and immune systems. Many of the messengers are already familiar as hormones, neurotransmitters, and cytokines, but presently the messages are poorly understood. Certain messengers from the neuroendocrine system appear to facilitate or inhibit the functions of immune cells, but the specificity remains to be elucidated. Cytokines are clearly potent activators of the HPA axis, but also exert a variety of other physiological effects. In all likelihood, many other messengers remain to be discovered [1,7,8,15,18].

Although our current understanding of the system is limited, it may be important to distinguish local from systemic effects. Whereas circulating concentrations of catecholamines and steroids are probably adequate to exert physiological effects, and this also appears to be true for cytokines, the role of the peptides is less clear. Their systemic concentrations are very low and are unlikely to be sufficient to modulate immune system function in a general way. However, it is possible that peptides secreted by nerve terminals in the thymus, spleen, and lymphoid tissue may achieve local concentrations sufficient to affect immune cells. Such effects may also be possible locally in tissue at sites of inflammation [1,8].

Messengers that can travel more readily and are more stable metabolically may be active systemically, whereas the less stable peptides may be confined to local actions. The chemical nature of the messengers may be suited to their functions. As lipophilic molecules, the glucocorticoids can readily penetrate membrane barriers and affect cells in all bodily tissues, whereas the hydrophilic catecholamines are more labile and their action may be limited to the circulatory systems. Our present knowledge indicates that the glucocorticoids and catecholamines predominantly inhibit immune responses, whereas the peptides are largely facilitatory. When the organism is threatened, the systemic activity of the glucocorticoids to limit immune responses may be important to depress immune activity to prevent undesirable autoimmune actions [1]. By contrast, peptides could facilitate immune responses in small areas close to the site of their release, for example, in an area of inflammation induced by infection or tissue damage. Catecholamines may occupy an intermediate position, existing in sufficient concentrations to have systemic actions but not having broad access to tissues and having relatively short durations of action, except when chronically elevated. Such an arrangement would permit focusing of the activation of immune response in local areas of inflammation, while preventing potentially damaging autoimmune actions that could be triggered by widespread activation (►brain inflammation – biomedical imaging).

The neuroimmune systems communicate bidirectionally (see Fig. 5c). The neuro-immune-endocrine interface is mediated by cytokines acting as auto/paracrine or endocrine factors regulating pituitary development, cell proliferation, hormone secretion and feedback control of the HPA axis. Soluble products that appear to transmit information from the immune compartments to the CNS act as immunotransmitters and function in immunomodulatory neuroendocrine circuits. The relative importance of systemically and locally produced cytokines in achieving these responses and their precise sites of action have been the focus of a burgeoning number of investigations over the past few decades. There is now accumulating evidence that there are important interactions between the immune and neuroendocrine systems, which may explain, in part, some of the effects on growth, thyroid, adrenal and reproductive functions that occur in the pathophysiology of acute and chronic disease [1].

Acknowledgments

The author's work is, in part, supported by the Anonymous Trust (Scotland), the National Institute for Biological Standards and Control (England), the Tenovus Trust (Scotland), the UK Medical Research Council (MRC, London) and the Wellcome Trust (London). Dr. John J. Haddad held the distinguished Georges John Livanos prize (London, UK).

References

1. Haddad JJ, Saade NE, Safieh-Garabedian B (2002) Cytokines and neuro-immune-endocrine interactions: a role for the hypothalamic-pituitary-adrenal revolving axis. J Neuroimmunol 133:1–19
2. Correa SG, Maccioni M, Rivero VE, Iribarren P, Sotomayor CE, Riera CM (2007) Cytokines and the immune-neuroendocrine network: what did we learn from infection and autoimmunity? Cytokine Growth Factor Rev 18:125–134
3. Engelhardt B (2006) Regulation of immune cell entry into the central nervous system. Results Probl Cell Differ 43:259–280
4. Carson MJ, Doose JM, Melchior B, Schmid CD, Ploix CC (2006) CNS immune privilege: hiding in plain sight. Immunol Rev 213:48–65
5. Bukovsky A (2007) Cell commitment by asymmetric division and immune system involvement. Prog Mol Subcell Biol 45:179–204
6. Aderem A, Ulevitch RJ (2000) Toll-like receptors in the induction of the innate immune response. Nature 406:782–787
7. Albiger B, Dahlberg S, Henriques-Normark B, Normark S (2007) Role of the innate immune system in host defence against bacterial infections: focus on the Toll-like receptors. J Intern Med 261:511–528
8. Galea I, Bechmann I, Perry VH (2007) What is immune privilege (not)? Trends Immunol 28:12–18
9. Haddad JJ (2007) Cellular and molecular regulation of inflammatory pain, nociception and hyperalgesia – the role of the transcription factor NF-κB as the lynchpin nocicensor: hyperalgesic or analgesic effect? Curr Immunol Rev 3:117–131
10. Haddad JJ, Harb HL (2005) L-γ-Glutamyl-L-cysteinyl-glycine (glutathione; GSH) and GSH-related enzymes in the regulation of pro- and anti-inflammatory cytokines: a signaling transcriptional scenario for redox(y) immunologic sensor(s)? Mol Immunol 42:987–1014
11. Dinarello CA (2000) Pro-inflammatory cytokines. Chest 118:503–508
12. Biber K, de Jong EK, van Weering HR, Boddeke HW (2006) Chemokines and their receptors in central nervous system disease. Curr Drug Targets 7:29–46
13. Neumann H (2001) Control of glial immune function by neurons. Glia 36:191–199
14. Gisler RH, Schenkel-Hulliger L (1971) Hormonal regulation of the immune response. II. Influence of pituitary and adrenal activity on immune responsiveness in vitro. Cell Immunol 2:646–657
15. McEwen BS (2002) Sex, stress and the hippocampus: allostasis, allostatic load and the aging process. Neurobiol Aging 23:921–939
16. Ader R, Cohen N (1975) Behaviorally conditioned immunosuppression. Psychosom Med 37:333–340
17. Renoux G, Biziere K, Renoux M, Guillaumin JM, Degenne D (1983) A balanced brain asymmetry modulates T cell-mediated events. J Neuroimmunol 5:227–238
18. Tischner D, Reichardt HM (2007) Glucocorticoids in the control of neuroinflammation. Mol Cell Endocrinol (in press)
19. Oberbeck R (2006) Catecholamines: physiological immunomodulators during health and illness. Curr Med Chem 13:1979–1989
20. Oppenheim JJ (2001) Cytokines: past, present and future. Int J Hematol 74:3–8
21. Safieh-Garabedian B, Poole S, Haddad JJ, Massaad CA, Jabbur SJ, Saadé NE (2002) The role of the sympathetic efferents in endotoxin-induced localized inflammatory hyperalgesia and cytokine upregulation. Neuropharmacology 42:864–872
22. Haddad JJ, Olver RE, Land SC (2000) Antioxidant/pro-oxidant equilibrium regulates HIF-1α and NF-κB redox sensitivity: evidence for inhibition by glutathione oxidation in alveolar epithelial cells. J Biol Chem 275:21130–21139
23. Haddad JJ, Lauterbach R, Saadé NE, Safieh-Garabedian B, Land SC (2001) α-Melanocyte-related tripeptide, Lys-D-Pro-Val, ameliorates endotoxin-induced nuclear factor-κB translocation and activation: evidence for involvement of an interleukin-1β$^{193-195}$ receptor antagonism in the alveolar epithelium. Biochem J 355:29–38
24. Spangelo BL, Judd AM, Call GB, Zumwalt J, Gorospe WC (1995) Role of the cytokines in the hypothalamic-pituitary-adrenal and gonadal axes. Neuroimmunomodulation 2:299–312
25. Payne LC, Weigent DA, Blalock JE (1994) Induction of pituitary sensitivity to interleukin-1: a new function for corticotropin-releasing hormone. Biochem Biophys Res Commun 198:480–484
26. Pariante CM, Pearce BD, Pisell TL, Sanchez CI, Po C, Su C, Miller AH (1999) The proinflammatory cytokine, interleukin-1α, reduces glucocorticoid receptor translocation and function. Endocrinology 140: 4359–4366

27. Rivier C, Chizzonite R, Vale W (1989) In the mouse, the activation of the hypothalamic-pituitary-adrenal axis by a lipopolysaccharide (endotoxin) is mediated through interleukin-1. Endocrinology 125:2800–2805
28. Perlstein RS, Mehta NR, Mougey EH, Neta R, Whitnall MH (1994) Systemically administered histamine H1 and H2 receptor antagonists do not block the ACTH response to bacterial lipopolysaccharide and interleukin-1. Neuroendocrinology 60:418–425
29. Smith EM, Cadet P, Stefano GB, Opp MR, Hughes TKJr (1999) IL-10 as a mediator in the HPA axis and brain. J Neuroimmunol 100:140–148
30. Safieh-Garabedian B, Haddad JJ, Saade NE (2004) Cytokines in the central nervous system: targets for therapeutic intervention. Curr Drug Targets CNS Neurol Disord 3:271–280
31. Haddad JJ (2004) On the antioxidant mechanisms of Bcl-2: a retrospective of NF-κB signaling and oxidative stress. Biochem Biophys Res Commun 322:355–363
32. Wissink S, Meijer O, Pearce D, van Der Burg B, van Der Saag PT (2000) Regulation of the rat serotonin-1A receptor gene by corticosteroids. J Biol Chem 275:1321–1326
33. Abreu-Villaca Y, Medeiros AH, Lima CS, Faria FP, Filgueiras CC, Manhaes AC (2007) Combined exposure to nicotine and ethanol in adolescent mice differentially affects memory and learning during exposure and withdrawal. Behav Brain Res 181:136–146
34. Hillhouse EW (1994) Interleukin-2 stimulates the secretion of arginine vasopressin but not corticotropin-releasing hormone from rat hypothalamic cells in vitro. Brain Res 650:323–325
35. Xu Y, Day TA, Buller KM (1999) The central amygdala modulates hypothalamic-pituitary-adrenal axis responses to systemic interleukin-1β administration. Neuroscience 94:175–183

Neuroimmunomodulation

Definition

Neuroimmunomodulation is the modulation of the immune system through the nervous system.

►Neuroimmune Interactions – Serotonin
►Neuroimmunomodulation: The brain areas involved
►Neuroimmunology

Neuroinflammation

Definition

Neuroinflammation is the inflammation of a nerve or of the nervous system.

►Neurogenesis and Inflammation

Neuroinflammation – DNA Vaccination Against Autoimmune Neuroinflammation

ANNA LOBELL
Department of Medical Sciences, Uppsala University, University Hospital, Uppsala, Sweden

Synonyms

DNA immunization; Naked DNA vaccination; Genetic vaccination

Definition

To prevent or treat ►autoimmune neuroinflammation by ►vaccination with DNA encoding one or more autoantigen(s) (►DNA vaccination against autoimmune neuroinflammation). The autoantigen, usually a myelin autoantigen, is transcribed and translated *in vivo*, processed and presented to T cells in the context of major histocompatibility complex after vaccination. Toll-like receptor (TLR) 9 ligand CpG DNA within the plasmid backbone acts as adjuvant to activate the innate immune system. Presence of CpG DNA and the expressed autoantigen are both essential for the protective immune reaction to occur.

Characteristics

Introduction

Failure of immunologic self-tolerance often leads to development of autoimmune disease, which is estimated to afflict up to 5% of the human population. Genetic factors, such as both major histocompatibility complex (MHC) and non-MHC genes, contribute to the risk for most autoimmune diseases. The same non-MHC gene may predispose for several different organ-specific inflammatory diseases such as rheumatoid arthritis and multiple sclerosis (MS). Environmental factors also play an important role in the risk for autoimmune disease. Many infectious agents have been investigated for crossreactivity towards self antigens, potentially leading to an autoimmune disease (molecular mimicry). Dysregulation of inflammatory cells or regulatory T cells has also been suggested as a cause of autoimmunity.

MS is a common disabling disease characterized by inflammation, neurodegeneration and demyelination in the central nervous system (CNS). The major hallmark of MS is the presence of sclerotic plaques in the CNS that are characterized by demyelination, which is associated with an inflammatory reaction that is orchestrated by activated lymphocytes, macrophages and glial cells. The etiology of MS is unknown, necessitating research into the mechanisms underlying pathogenesis. Myelin autoantigens such as myelin basic protein (MBP), proteolipid protein (PLP) and myelin oligodendrocyte glycoprotein (MOG) are candidate autoantigens in MS. Because it is difficult to study CNS autoimmunity in humans, experimental models such as experimental autoimmune encephalomyelitis (EAE) are used. EAE is actively induced by injection of myelin proteins or encephalitogenic myelin peptides together with a proinflammatory adjuvant such as complete Freund's adjuvant (CFA). EAE was previously thought to be a T helper 1 cell (Th1)-mediated autoimmune disease whereas Th2 cells were thought to suppress EAE by inhibiting the Th1 responses. However, interleukin (IL)-23-driven proinflammatory IL-17-producing T cells mediate EAE. The IL-17-producing Th cell (Th17) is a distinct lineage from Th1 and Th2 cells and naïve CD4 T cells surprisingly differentiate into IL-17-producing T cells in the presence of soluble TGF-β or regulatory T cells if IL-6 is present. Th17 cells are then maintained by IL-23. Differentiation of naïve CD4 T cells into Th17 cells is inhibited by Th1, Th2 and regulatory T cells. Thus, the characteristics of the pathogenic T helper cell response in EAE is different from what was previously thought. As a consequence, Th1/Th2 ratios are no longer relevant, as both these T cell lineages suppress pathogenic Th17 cell responses. However, most of the studies described herein were done before the role of Th17 cells in EAE was discovered and Th1/Th2 ratios are therefore discussed.

DNA vaccines can induce antigen-specific CD4 T helper cell and CD8 cytotoxic T cell responses as well as antibody responses. Initially, ►DNA vaccination was thought to prime cytotoxic T cells in a unique manner through intracellular processing of antigen produced in transfected somatic cells. However, ►antigen-presenting cells prime cytotoxic T cells both by direct transfection and by cross-priming after endocytosis of antigen upon DNA vaccination. In fact, antigen-presenting dendritic cells in the spleen process and present the antigen encoded by the DNA vaccine.

DNA vaccines are plasmids produced by *E. coli* and all bacterial DNA contains unmethylated CpG DNA motifs. CpG DNA are pathogen associated molecular patterns (PAMP) which specifically binds the pattern recognition receptor TLR9. The recognition of PAMP by TLR triggers intracellular signaling pathways resulting in the induction of proinflammatory ►cytokines, type I ►interferon (IFN), chemokines and dendritic cell maturation that leads to activation of adaptive immunity. Each TLR recognizes different PAMP: TLR4 recognizes LPS and TLR2 plus TLR1 or TLR6 recognizes bacterial lipopeptides. TLR3 and TLR9 are expressed in endosomes and are involved in the recognition of double-stranded RNA and unmethylated CpG DNA respectively. Because the intracellular pathways of the TLR differ, TLR9 ligation mainly induces type I IFN production, whereas e.g., TLR4 ligation induces IL-12 and IFN-β production. CpG DNA is essential for efficient DNA vaccination and is currently used in therapeutic trials against allergy and as vaccine adjuvant. Furthermore, DNA vaccines coding for antigens can be combined with plasmids encoding various cytokines or ligands important for the ensuing immune response, thus providing methods to specifically modulate the vaccine effect.

In this essay, the effects of DNA vaccination against rodent EAE will be described.

The DNA Vaccine

Mammalian expression vectors, e.g., pCI, pTarget and pcDNA3, are used as plasmid backbones for DNA vaccines. A strong promoter such as the human CMV immediate/early enhancer-promoter controls the transcription of the autoantigen and a Kozak box is introduced around the start codon to enhance transcription. DNA encoding myelin autoantigens or immunodominant myelin protein-derived peptides are ligated into the plasmid backbone. Minigenes encoding myelin protein-derived peptides can be inserted in tandem [1,2]. The DNA insert encoding the myelin autoantigen can be expressed in all mammalian cells, when a strong promoter such as CMV promoter is used. Additionally, presence of CpG DNA within the plasmid backbone of the DNA vaccine is required. Common plasmids such as pCI, pTarget and pcDNA3 all contain at least two immunostimulatory CpG DNA motifs.

N

Immunization Protocols

The DNA vaccine is either injected before or after induction of EAE. In successful preventive trials in rats, the vaccine is administered in PBS as a single injection either intra muscularly or intra dermally three to eleven weeks before induction of EAE [1,3,4], whereas in therapeutic trials the DNA vaccine is given as weekly intra muscular injections after the onset of EAE [5,6]. If the DNA vaccine is injected intra muscularly, the muscles are often pretreated with compounds such as cardiotoxin, to enhance DNA uptake.

Protection from EAE After Vaccination with DNA Encoding Myelin Autoantigens

DNA vaccination prevents EAE [1–4,6–8] with few exceptions [9] (Table 1). Vaccination with DNA encoding myelin peptides or proteins suppresses EAE induced with the corresponding peptide [1,4,7]. Treatment of ongoing EAE has been successful in murine EAE models [5,6]. Despite similar clinical outcomes, the ensuing immune response after DNA vaccination differs in rat and murine EAE models, which suggests that the protective mechanism differ in the two species. Antigen specificity, impact of CpG DNA and the protective mechanism are discussed below.

Neuroinflammation – DNA Vaccination Against Autoimmune Neuroinflammation. Table 1 EAE prevention or treatment using DNA vaccines encoding myelin proteins or peptides

Animal	DNA vaccine	Main findings	Reference
Rat EAE			
Lewis	MBP68-85	Improvement	[1]
Lewis	MBP86-85 and MBP89-101	High antigen specificity	[2]
Lewis	MBP68-85 +/- CpG DNA MBP68-85 + IL-4, IL-10 or TNF-α	CpG DNA is required, cytokine coinjection inhibits the suppressive effect	[3]
Lewis 1Av1, DA, Lewis 1N	MOG91-108 +/- CpG DNA	Improvement, CpG DNA is required	[7]
Lewis 1Av1	MOG91-108	Higher IFN-β expression, lower MHC II expression, lower IL-23p40 expression	[8]
Murine EAE			
SJL/J	PLP139-151	Improvement, lower B7 expression	[4]
SJL/J	MOG	Worsening	[9]
SJL/J	PLP139-151 + IL-4	Improvement	[6]
SJL/J	PLP, MOG, MBP, MAG cocktail + IL-4	Treatment, improvement in relapses	[5]
SJL/J	PLP, MOG, MBP, MAG cocktail + IL-4 + GpG DNA	Treatment, improvement	[10]

Antigen Specificity of DNA Vaccination

Epitope spreading defines the expansion of antigen-specific immune responses beyond those targeted in the initial immunization. EAE and other autoimmune diseases are associated with spreading of autoreactive T and B cell responses. Treatment with a cocktail of DNA vaccines encoding several myelin proteins reduces the number of relapses in EAE and reduces spreading of the autoreactive B cell response, although the treatment does not affect disease severity [5].

Local production of IL-4 or TGF-β is instrumental for a regulatory immune response inhibiting pathogenic T cells of any specificity (bystander suppression) in oral tolerance or altered peptide ligand therapy. Is DNA vaccination acting through bystander suppression, or is the effect antigen specific? A single amino acid exchange in position 79 from serine (non-self) to threonine (self) in MBP68-85 dramatically alters the protective effect of a DNA vaccine encoding MBP68-85 in rats. Furthermore, the DNA vaccine encoding MBP68-85 does not protect from MBP89-101-induced EAE and vise versa [2]. Thus, the protective effect is highly specific and there is no evidence for bystander suppression, nor epitope spreading of tolerance.

The Impact of CpG DNA

DNA vaccines are produced by bacteria and thus contain unmethylated CpG DNA that specifically binds TLR9 and thereby promotes innate immunity and type I IFN production. Treatment with a DNA vaccine containing two CpG DNA motifs suppresses clinical signs of EAE, while a corresponding DNA vaccine without such CpG DNA has no effect [3,7]. Moreover, there are no reports on successful DNA vaccination using plasmid DNA that lacks CpG DNA. Thus, the presence of CpG DNA is decisive for protective DNA vaccination against EAE.

In contrast, co-administration of an antagonist to CpG DNA, GpG DNA, enhances the suppressive effect of DNA vaccination against murine EAE by promoting Th2 immunity [10].

Protective Mechanisms of DNA Vaccination Against EAE

The immune mechanisms involved in protective DNA vaccination are not fully understood, and differ between rat and murine EAE models. Also, the previous work done may have to be re-evaluated in the light of the recent discovery that Th17, but not Th1, responses are responsible for disease. A summary of the main findings is presented in Fig. 1 and Table 1.

The nature of the attenuated T cell function after DNA vaccination against rat EAE has until recently remained elusive since no induction of Th2 or regulatory T cells is observed, and neither the secretion of IFN-γ, nor the proliferative response, can be correlated to the suppressive capability of the DNA vaccine [7]. Instead, coinjection of IL-4 or IL-10 gene with the DNA vaccine inhibits the protective effect, which suggests that Th2 immunity is not involved in the protective mechanism. However, the antigen-specific IL-17 production by T cells is significantly reduced after DNA vaccination and subsequent induction of EAE, compared to controls. Moreover, T cells do not express IL-17 after DNA vaccination – but before induction - of EAE (A. Lobell et al., unpublished). Thus, priming of myelin-specific Th17 cells is indeed impaired after induction of EAE in DNA vaccinated rats which likely contributes to the reduced EAE symptoms (Fig. 1).

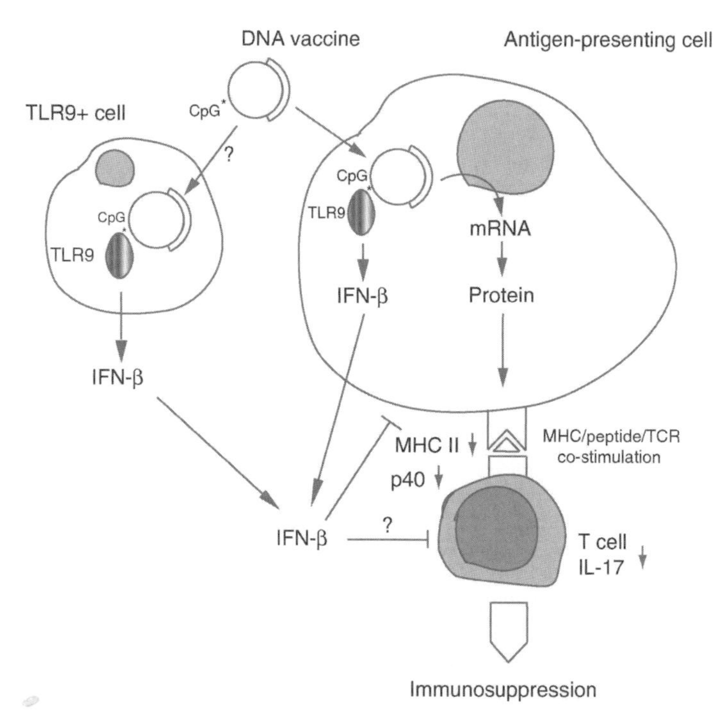

Neuroinflammation – DNA Vaccination Against Autoimmune Neuroinflammation. Figure 1 Illustration of the proposed mechanism for suppressive DNA vaccination against rat EAE. Administration of a DNA vaccine encoding a myelin self peptide results in transcription and translation of the self peptide by antigen-presenting cells. The self peptide is processed and presented on MHC II to Th cells. In either the same antigen-presenting cell or other TLR9-expressing cells, CpG DNA motifs of the DNA vaccine binds TLR9 in the endosome, which induces IFN-β production. Priming of DNA vaccine-induced myelin-specific T cells in the presence of IFN-β results in suppression of subsequently induced EAE and dampened encephalitogenic Th17 responses. IFN-β may act via downregulation of antigen-presenting cell activation and maturation via reduced IL-23 and MHC II expression, and/or directly on encephalitogenic Th17 cells. Lines with bars represent inhibition.

IFN-β expression is upregulated after DNA vaccination, both before and after induction of rat EAE [8]. Silencing of IFN-β *in vivo* reveals a crucial role for the cytokine in suppressive DNA vaccination (A. Lobell et al., unpublished). It is unclear how the CpG DNA-induced IFN-β leads to impaired Th17 responses and protection from EAE. Antigen-presenting cell function may be altered. Indeed, the MHC II and IL-23/IL-12p40 expression are lower in DNA vaccinated rats compared to controls [8], which indicates impaired maturation and activation of antigen-presenting cells.

In contrast, induction of Th2 immunity increases the efficacy of DNA vaccination in mice. Coinjection of IL-4 gene with a DNA vaccine suppresses EAE. The IL-4 gene delivery activates STAT6 locally which shifts the antigen-specific T cells to a protective Th2 phenotype [6]. Th2-promoting CpG DNA antagonist GpG DNA enhances the suppressive effect [10]. *In vitro* exposure of a DNA vaccine to murine splenocytes slightly reduces the expression of costimulatory B7 molecules, which may suggest that anergy is induced by DNA vaccination [4].

DNA Vaccines Targeting Chemokines/Cytokines

Substantial data support the notion that coinjection of cytokine-coding cDNA with DNA vaccines can selectively modulate the ensuing immune response. However, other reports suggest that vaccination with DNA encoding cytokines or chemokines can induce the production of specific antibodies directed against the gene product which in turn suppress EAE.

Nathan Karin and coworkers demonstrate that injection of DNA encoding rat chemokines MIP-1α or MCP-1 suppresses subsequently induced rat EAE, whereas the same strategy for MIP-1β slightly aggravates disease. Likewise, injection of DNA vaccines encoding the proinflammatory cytokine TNF-α confers resistance to EAE and induces antibodies directed against TNF-α. In contrast, production of the respective cytokine seems responsible for the effects in other DNA vaccine trials. Injection of DNA encoding TGF-β1 or an IL-4-IgG1 fusion into rats prior to induction of EAE leads to detectable *in vivo* production of these cytokines and prevention of EAE. An immunization regimen with multiple injections could account for the autoantibody production in the former cytokine/chemokine experiments.

Conclusion

Autoimmune neuroinflammation can be prevented or treated by DNA vaccination. The effect is highly antigen specific and requires CpG DNA within the plasmid backbone of the vaccine.

References

1. Lobell A, Weissert R, Storch MK, Svanholm C, de Graaf KL, Lassmann H, Andersson R, Olsson T, Wigzell H (1998) Vaccination with DNA encoding an immunodominant myelin basic protein peptide targeted to Fc of immunoglobulin G suppresses experimental autoimmune encephalomyelitis. J Exp Med 187:1543–1548
2. Weissert R, Lobell A, de Graaf KL, Eltayeb SY, Andersson R, Olsson T, Wigzell H (2000) Protective DNA vaccination against organ-specific autoimmunity is highly specific and discriminates between single amino acid substitutions in the peptide autoantigen. Proc Natl Acad Sci USA 97:1689–1694
3. Lobell A, Weissert R, Eltayeb S, Svanholm C, Olsson T, Wigzell H (1999) Presence of CpG DNA and the local cytokine milieu determine the efficacy of suppressive DNA vaccination in experimental autoimmune encephalomyelitis. J Immunol 163:4754–4762
4. Ruiz PJ, Garren H, Ruiz IU, Hirschberg DL, Nguyen LV, Karpuj MV, Cooper MT, Mitchell DJ, Fathman CG, Steinman L (1999) Suppressive immunization with DNA encoding a self-peptide prevents autoimmune disease: modulation of T cell costimulation. J Immunol 162:3336–3341
5. Robinson WH, Fontoura P, Lee BJ, de Vegvar HE, Tom J, Pedotti R, DiGennaro CD, Mitchell DJ, Fong D, Ho PP, Ruiz PJ, Maverakis E, Stevens DB, Bernard CC, Martin R, Kuchroo VK, van Noort JM, Genain CP, Amor S, Olsson T, Utz PJ, Garren H, Steinman L (2003) Protein microarrays guide tolerizing DNA vaccine treatment of autoimmune encephalomyelitis. Nat Biotechnol 21:1033–1039
6. Garren H, Ruiz PJ, Watkins TA, Fontoura P, Nguyen LT, Estline ER, Hirschberg DL, Steinman L (2001) Combination of gene delivery and DNA vaccination to protect from and reverse Th1 autoimmune disease via deviation to the Th2 pathway. Immunity 15:15–22
7. Lobell A, Weissert R, Eltayeb S, de Graaf KL, Wefer J, Storch MK, Lassmann H, Wigzell H, Olsson T (2003) Suppressive DNA vaccination in myelin oligodendrocyte glycoprotein peptide-induced experimental autoimmune encephalomyelitis involves a T1-biased immune response. J Immunol 170:1806–1813
8. Wefer J, Harris RA, Lobell A (2004) Protective DNA vaccination against experimental autoimmune encephalomyelitis is associated with induction of IFNbeta. J Neuroimmunol 149:66–76
9. Bourquin C, Iglesias A, Berger T, Wekerle H, Linington C (2000) Myelin oligodendrocyte glycoprotein-DNA vaccination induces antibody-mediated autoaggression in experimental autoimmune encephalomyelitis. Eur J Immunol 30:3663–3671
10. Ho PP, Fontoura P, Platten M, Sobel RA, DeVoss JJ, Lee LY, Kidd BA, Tomooka BH, Capers J, Agrawal A, Gupta R, Zernik J, Yee MK, Lee BJ, Garren H, Robinson WH, Steinman L (2005) A suppressive oligodeoxynucleotide enhances the efficacy of myelin cocktail/IL-4-tolerizing DNA vaccination and treats autoimmune disease. J Immunol 175:6226–6234

Neuroinflammation – IL-18

Philip F. Stahel[1], Ursula Felderhoff-Mueser[2], Scott R. Barnum[3]

[1]Department of Orthopaedic Surgery, Denver Health Medical Center, University of Colorado School of Medicine, Denver, CO, USA

[2]Department of Neonatology, Charité Universitätsmedizin Berlin, Campus Virchow Klinikum, Berlin, Germany

[3]Departments of Microbiology and Neurology, University of Alabama at Birmingham (UAB), Birmingham, AL, USA

Synonyms

Interferon (IFN)-γ inducing factor (IGIF)

Definition

►Interleukin (IL)-18 is a member if the IL-1 family of pro-inflammatory ►cytokines. It was first cloned in the late 1980s and originally designated "interferon (IFN)-γ inducing factor" (IGIF). ►IL-18 is synthesized as an inactive precursor which is cleaved into its active form by ►caspase-1 for functional activity. IL-18 is an important mediator of innate immune responses against infection and has been shown to be critically involved in the pathogenesis of autoimmune diseases [1]. Resident cells of the central nervous system (CNS) express IL-18 and caspase-1 constitutively, thus providing a functional IL-18-dependent immune response in the intracranial compartment. Several studies in recent years have highlighted a crucial role of IL-18 in mediating ►neuroinflammation under a variety of pathological conditions in the CNS, such as bacterial and viral infections, autoimmune demyelinating diseases, hypoxic-ischemic, hyperoxic and traumatic brain injury [5,9].

Characteristics

IL-18 is initially synthesized as an inactive 24 kD precursor protein (pro-IL-18) which is subsequently processed by ►caspase-1 (ICE) (IL-1β converting enzyme) into its mature and biologically active form with a molecular weight of 18 kD. In addition, the secreted pro-form of IL-18 can also be processed into its active form by a variety of extracellular enzymes which are constitutively expressed by leukocytes, such as proteinase-3 (PR-3). The active form of IL-18 can either bind to its heterocomplex IL-18α/β receptor (IL-18Rα/β) or be neutralized by its natural antagonist, IL-18 binding protein (IL-18BP). At least ten IL-1 gene family members have been discovered, which all map to the same region on human chromosome 2, with the notable exception of IL-18, which is encoded for by a single gene located on human chromosome 11q22.2-q22.3.

Biology

As a member of the IL-1 family of cytokines, IL-18 possesses a wide variety of inflammatory and immunoregulatory properties [4]. Human and murine IL-18, which show 65% sequence homology, are secreted by first-line immune cells capable of antigen presentation, such as macrophages, dendritic cells, and Kupffer cells. Other cell-types reported to express IL-18 include T- and B-lymphocytes, vascular endothelial cells, smooth muscle cells, keratinocytes, osteoblasts and synovial fibroblasts. In the rodent brain, IL-18 and its receptor were shown to be constitutively expressed by astrocytes, microglia, neurons and by ependymal cells [2,3]. Different studies have emphasized the important role of IL-18 expression and LPS-dependent upregulation in microglia. IL-18 was also shown to induce signal transduction pathways in microglia. The IL-18 promoter includes binding sites for the transcription factor nuclear factor κB (NFκB). Furthermore, the proteasome inhibitor MG-132 – which regulates NFκB signaling – was found to block LPS- and IFN-γ-mediated IL-18 up-regulation in microglia. These findings imply an important role in IL-18-mediated activation of NFκB in the brain. The IL-18 activating enzyme, caspase-1, was also found to be constitutively expressed in microglia and astrocytes. Thus, maturation and activation of IL-18 can occur in the brain under physiological and inflammatory conditions [7]. The biological effects of IL-18 receptor binding include the induction of other inflammatory mediators, such as IFN-γ, tumor necrosis factor (TNF), IL-1β, IL-8, and of the so called "death proteins" of the extrinsic apoptotic pathway, such as TNF and Fas ligand (FasL). Thus, IL-18 represents a "key" cytokine which controls two distinct immunological regulatory pathways. These include the regulation of monocyte/macrophage function by induction of IFN-γ synthesis by T- and B-lymphocytes, as well as the induction of ►apoptosis through up-regulation of TNF and FasL. Both mechanisms may lead to extensive local inflammation and tissue destruction mediated by IL-18 under pathological conditions, as reported for various autoimmune and inflammatory diseases of the CNS.

Function

IL-18 was first identified as an essential factor promoting IFN-γ production by T cells in the presence of IL-12. This is the most salient biologic property of IL-18 that separates it from IL-1. IL-18 enhances T cell maturation (Th1 > Th2), cytokine production, and cytotoxicity [4]. Other major targets of IL-18 include macrophages, NK cells, B-cells, basophils, and neutrophils. A wealth of data indicate that IL-18 contributes to host defense and inflammation through synergism in a cascade of cytokines associated with innate responses, including IL-12 and IL-15.

The action of IL-18 appears to go beyond immune regulation, as IL-18 (like IL-1) appears to induce sleep in mice, rats and rabbits. IL-18 injected into the brain increases non-rapid eye movement sleep. The sleep effects of IL-18 introduced directly into the brain coincides with increased brain temperature. In contrast, intraperitoneal IL-18 fails to induce fever or sleep. IL-18 mediates its biological functions through binding of a widely expressed heterodimeric receptor consisting of α- and β-chains expressed on many different cell-types including T-lymphocytes, natural killer (NK) cells, monocytes/macrophages, neutrophils, and endothelial cells. Receptor activation by binding of IL-18 leads to the activation of the transcription factor NFκB via a complex intracellular signaling cascade. While the receptor's α–chain (IL-18Rα) is essential for signaling, it binds IL-18 at a relatively low affinity. In contrast, the IL-18Rβ chain, also termed "IL-1 receptor accessory protein-like" (AcPL), binds to the complex formed by IL-18 and the IL-18Rα chain, thus generating a high affinity tricomplex interaction. Both IL-18Rα and IL-18Rβ are structurally related and belong to the extended IL-1 receptor family. The signal transduction pathway subsequent to ligand binding of the IL-18 receptor complex is virtually identical with that of the IL-1 receptor complex. Both IL-18Rα and IL-18Rβ chains are required for signal transduction through the "myeloid differentiation factor 88" (MyD88), the serine-threonine "interleukin-1 receptor-associated kinase" (IRAK), and the "TNFα receptor-associated factor 6" (TRAF6) adapter molecules and involves a series of phosphorylation events that take place during the first few minutes after IL-18R binding. These steps ultimately result in the phosphorylation of the "inhibitor of κB kinase" (IKK) complex, as well as specific "mitogen-activated kinase kinases" (MKKs). The IKKs phosphorylate the NF-κB inhibitor IκB, leading to its ubiquitination and subsequent degradation by the proteasome. This allows NF-κB to translocate to the nucleus and bind to specific promotor sequences. Activated MKKs phosphorylate and activate members of the "c-Jun N-terminal kinase" (JNK) and p38 "mitogen-activated protein kinase" (MAPK) family. These also translocate to the nucleus where they can phosphorylate several transcription factors of the basic leucine zipper family, like c-Jun and c-Fos.

Pathology

CNS Autoimmune Disease

The implication of IL-18 and caspase-1 in contributing to autoimmune neuropathology was investigated in patients with ►multiple sclerosis (MS) and its animal model, ►experimental autoimmune encephalomyelitis (EAE). Caspase-1 was shown to be elevated in EAE brain tissue as well as in peripheral blood mononuclear cells from MS patients, where expression levels correlated with disease activity. In addition, IL-18 levels were found to be slightly elevated in MS patients as compared to healthy controls. In the experimental setting of EAE, the importance of IL-18 on generation of Th1 response has been validated by demonstrating a significant attenuation of disease by administration of neutralizing anti-IL-18 antibodies. Similarly, the neuropathological sequelae of EAE were attenuated by either pharmacological inhibition of caspase-1 or in caspase-1 gene knockout mice. IL-18 and caspase-1 mRNA expression in the CNS was shown during the acute stage of EAE and implicated the IL-18/caspase-1 pathway in the amplification of Th1-mediated immune response in autoimmune CNS disease. This notion was supported by recent findings of resistance to EAE in IL-18 gene-deficient mice, whereas the administration of recombinant IL-18 enhanced the severity of EAE in wild-type mice and restored the ability to generate Th1 immune responses in the IL-18−/− mice.

In human MS patients, IL-18 levels were detected in cerebrospinal fluid taps only in about 3% of all patients studied. However, postmortem brain tissue section analysis from MS patients revealed an increased local expression of IL-18 and IFN-γ in demyelinating cerebral lesions, suggesting that cerebrospinal fluid levels do not accurately reflect the local tissue expression of these mediators in autoimmune CNS disease. This hypothesis is corroborated by a clinical study on patients suffering from the relapsing-remitting form of MS, where individuals with acute exacerbations and active gadolinium-enhancing lesions in MRI had significantly elevated IL-18 levels in serum and cerebrospinal fluid, as compared to MS patients without positive MRI lesions and to control patients without neurological disease. Furthermore, the enhanced expression of IL-18 and its receptor was reported on oligodendrocytes of human tissue samples from patients with active MS, as compared to brain sections from patients with silent MS or from neuropathologically normal subjects. Altogether, these findings imply an important involvement of IL-18 and caspase-1 in the pathogenesis of active stages of autoimmune CNS disease.

Ischemic and Traumatic Brain Injury

IL-18 has been involved in the development of ischemia-induced inflammation in experimental models of middle cerebral artery occlusion (MCAO). As such, focal ischemic brain injury in rats has been shown to induce IL-18 expression in microglia and monocytes/macrophages in the infarcted cortex [6]. In those studies, both mRNA and protein levels increased within 24 hours and reached peaks at 6 days post injury. Interestingly, the expression profile of caspase-1 paralleled the increase of IL-18 levels, but not of IL-1β, suggesting a temporal diversity of expression within cytokines of the IL-1 family and implying a role of the IL-18/caspase-1 pathway in late-stage neuroinflammatory responses to focal cerebral ischemia. In ►stroke patients, elevated IL-18 levels in serum were shown to correlate with the extent of hypodense area volumes in craniocerebral CT scans and with functional disability. Moreover, serum IL-18 levels were shown to be higher in patients with a non-lacunar stroke subtype than in those with lacunar types of stroke. Affirmative data in an experimental model of hypoxic-ischemic brain injury in rats and mice reported the up-regulation of IL-18 and caspase-1 both at the mRNA and protein level within 12 h to 14 days after stroke. While microglia were determined as the major cell-type expressing IL-18 and caspase-1 in the injured hemisphere, IL-18 receptor expression was detected mainly on neurons in the cortex and thalamus. Interestingly, post-injury infarction area and neuropathological scores were significantly decreased in IL-18−/− mice, as opposed to wild-type littermates, suggesting that IL-18 may by functionally involved in the development and exacerbation of hypoxic brain injury.

Similarly to the demonstrated role of IL-18 in contributing to detrimental secondary effects in the injured brain after hypoxic-ischemic injury, IL-18 was shown to represent a "key player" in the pathophysiology of ►traumatic brain injury (TBI) [5]. First evidence of upregulated IL-18 gene and protein expression following optic and sciatic nerve crush injury was reported in rodent models. Interestingly, the constitutive levels of IL-18 mRNA expression were found to be higher in the CNS (optic nerve) than in peripheral nerve tissue (sciatic nerve). After experimental axonal crush injury, IL-18 expression dramatically increased both on injured optic and sciatic nerves. The cellular sources of increased IL-18 levels were determined to be mainly constituted by infiltrating ED1-postive macrophages within two to eight days after axonal injury. In addition, local resident microglia were shown to exhibit enhanced IL-18 expression mainly at sites of myelin degradation, suggesting an involvement of IL-18 mediated microglial neurotoxicity. This notion was confirmed by clinical and experimental data based on studies of severe closed head injury in humans, rats, and mice [8,10]. Significantly elevated IL-18 protein levels were reported in cerebrospinal fluid samples of patients with severe closed head injury for up to 10 days after trauma, as compared to normal controls. Notably, the peaks of intrathecal IL-18 levels in brain-injured patients were almost 200-fold higher than in cerebrospinal fluid from control subjects without neuroinflammatory disease [10].

CNS Infection

Several studies have highlighted a role of IL-18 in mediating the inflammatory response to bacterial, viral and fungal infections of the CNS. In models of pneumococcal and cryptococcal meningits, IL-18 was shown to be up-regulated in the infected brain and to contribute to the neuroinflammatory response. IL-18 −/− mice with pneumococcal meningitis had a prolonged survival and a decreased neuroinflammatory response compared to infected wild-type littermates. These data were supported in a model of fungal infection of the CNS, where mice with cryptococcal meningoencephalitis had increased IL-18 mRNA expression in the infected brain with associated potent neuroinflammatory events. In models of viral CNS infection, IL-18 was shown to play a key role in activating microglial functions by inducing neuronal IFN-γ release in brain parenchyma and thus supporting the viral clearance of infected neurons.

The different roles and pathological effects of IL-18 in the immature and adult brain are outlined in Table 1.

Therapy

Several functional antagonists and inhibitors of IL-18 have been described, which neutralize its pro-inflammatory effects. The most important naturally occurring antagonist is IL-18 binding protein (IL-18BP), a secreted protein which displays high-affinity binding to mature IL-18, but not to the IL-18 precursor. A single copy of the IL-18BP gene exists for humans, mice, and rats. The highest expression of IL-18BP was detected in the spleen

Neuroinflammation – IL-18. Table 1 Role of IL-18 in neuroinflammatory diseases

Neuroinflammatory disease	IL-18 levels in the CNS	Functional role of IL-18
Experimental autoimmune meningo-encephalitis (EAE)	Increased IL-18 gene expression in EAE spinal cord and brain	Neutralizing anti-IL-18 antibodies block the development of EAE
Multiple sclerosis	Elevated IL-18 and receptor protein expression in active MS lesions	n.d.
Experimental dopaminergic neurodegeneration	Increased microglial IL-18 in substantia nigra	Reduced susceptibility to dopaminergic neuronal loss and reduced microglial activation in IL-18 gene-deficient mice
Bacterial meningitis	Increased IL-18 levels in cerebrospinal fluid of bacterial meningitis patients, upregulation of IL-18 protein and gene expression in infected murine brain tissue	Prolonged survival and reduced neuroinflammation in IL-18 gene-deficient mice
Viral encephalitis	Induction of IL-18 and caspase-1 gene expression in murine brains	Protective effect of IL-18 by enhanced clearance of neurovirulent Influenza A infection
Cryptococcal meningoencephalitis	Increased IL-18 gene expression in infected brains	n.d.
Ischemic stroke	Increased intracerebral IL-18 and caspase-1 gene expression in injured brain, elevated IL-18 serum levels	IL-18 levels in serum are predictive of outcome
Axonal injury	Enhanced IL-18 expression on infiltrating macrophages after nerve crush injury	n.d.
Traumatic brain injury (TBI)	Elevated IL-18 protein levels in cerebrospinal fluid and brain tissue	Inhibition by IL-18BP is neuroprotective
Neonatal hypoxia-ischemia	Elevated IL-18 and receptor, IL-1β, caspase-1 protein and gene expression in the injured brain	Reduced infarct volume and neuropathology score in IL-18−/− mice. IL-18 contributes to white matter injury in neonatal brain
Neonatal hyperoxic injury	IL-1β, IL-18 and receptor/caspase-1 protein and gene expression	Inhibition by IL-18BP is neuroprotective

n.d., not determined; CNS, central nervous system; IL-18BP, IL-18 binding protein; MS, multiple sclerosis; EAE, experimental autoimmune encephalomyelitis; TBI, traumatic brain injury.

and intestinal tract which are both immunologically active tissues. Four distinct isotypes of human IL-18BP and two isotypes of murine IL-18BP have been described which are formed by alternative splicing of the respective genes. Two of the four human isotypes and both murine isotypes are biologically functional by neutralizing IL-18. Experimental studies on TBI models revealed that the systemic administration of recombinant IL-18BP after trauma resulted in a significantly improved neurological recovery, both in the adult and immature brain [8, 10]. This neuroprotection was associated with an IL-18BP-dependent downregulation of intracerebral IL-18 levels in mice. Furthermore, hyperoxia-induced neonatal brain injury was largely attenuated by administration of recombinant IL-18BP. Based on these findings, the pharmacological administration of IL-18BP may represent a promising future therapeutic strategy for attenuating the IL-18-mediated neuroinflammation and neurodegeneration in the immature and adult brain.

References

1. Bombardieri M, McInnes IB, Pitzalis C (2007) Interleukin-18 as a potential therapeutic target in chronic autoimmune/inflammatory conditions. Expert Opin Biol Ther 7:31–40
2. Conti B, Park LC, Calingasan NY, Kim Y, Kim H, Bae Y, Gibson GE, Joh TH (1999) Cultures of astrocytes and microglia express interleukin 18. Mol Brain Res 67:46–52
3. Culhane AC, Hall MD, Rothwell NJ, Luheshi GN (1998) Cloning of rat brain interleukin-18 cDNA. Mol Psychiatry 3:362–366
4. Dinarello CA, Fantuzzi G (2003) Interleukin-18 and host defense against infection. J Infect Dis 187(Suppl 2): S370–S384
5. Felderhoff-Mueser U, Schmidt OI, Oberholzer A, Bührer C, Stahel PF (2005) IL-18: a key player in neuroinflammation and neurodegeneration? Trends Neurosci 28:487–493
6. Hedtjärn M, Leverin AL, Eriksson K, Blomgren K, Mallard C, Hagberg H (2002) Interleukin-18 involvement in hypoxic-ischemic brain injury. J Neurosci 22:5910–5959
7. Sekiyama A, Ueda H, Kashiwamura S, Nishida K, Kawai K, Teshima-Kondo S, Rokutan K, Okamura H (2005) IL-18: a cytokine translates a stress into medical science. J Med Invest 52(Suppl.):236–239
8. Sifringer M, Stefovska V, Endesfelder S, Stahel PF, Genz K, Dzietko M, Ikonomidou C, Felderhoff-Mueser U (2007) Activation of caspase-1 dependent interleukins in developmental brain trauma. Neurobiol Dis 25:614–622
9. Stoll G, Jander S, Schroeter M (2000) Cytokines in CNS disorders: neurotoxicity versus neuroprotection. J Neural Transm Suppl 59:81–89
10. Yatsiv Y, Morganti-Kossmann MC, Perez D, Dinarello CA, Novick D, Rubinstein M, Otto VI, Rancan M, Kossmann T, Redaelli CA, Trentz O, Shohami E, Stahel PF (2002) Elevated intracranial IL-18 in humans and mice after traumatic brain injury and evidence of neuroprotective effects of IL-18-binding protein after experimental closed head injury. J Cereb Blood Flow Metab 22:971–978

Neuroinflammation – LPS-induced Acute Neuroinflammation, Rat Model

MEHRNAZ JAFARIAN-TEHRANI
Laboratoire de Pharmacologie (UPRES EA2510), Université Paris Descartes - UFR Pharmacie, Paris, France

Definition

Characterization of a rat model to study acute neuroinflammation induced by ▶lipopolysaccharide (LPS), based on histopathological and biochemical outcomes.

Acute neuroinflammation is a common process accompanying acute brain injuries such as traumatic brain injury (TBI) and cerebral ischemia [1]. Acute inflammatory response due to central nervous system (CNS)-specific glial response within the injured brain exerts detrimental effects by releasing neurotoxic mediators. The model described here mimics some aspects of acute brain injuries with a specific regard to acute neuroinflammation *in vivo*. Lipopolysaccharide (LPS), a known potent immunostimulant, a cell constituent of the cell wall of Gram negative bacteria, is used to induce the inflammatory response within the brain [2]. This model is a powerful tool for mechanistic studies and evaluation of the potential neuroprotective strategies.

Characteristics

Quantitative Description

In the model described here, the LPS used is from *E. Coli* (serotype 0127:B8, Sigma L-3129). The intensity of inflammatory response may differ depending on the dose and source of LPS (*Salmonella versus E. Coli* and the LPS serotype) used, and also the site of the LPS injection. Therefore, caution should be taken in the interpretation and comparison of the data available from the literature.

Description of the Structure

In the model described here, the site of the LPS injection is located in the right hippocampus, precisely in the dentate gyrus at the ▶stereotaxic coordinates [3] relative to the ▶bregma (Fig. 1).

The hippocampus is a part of the brain which is involved in memory and learning. It belongs to the limbic system and its name is due to its seahorse shape. Dentate gyrus is a part of the hippocampal formation containing granule cells, the principal excitatory neurons of the denate gyrus, which project to the pyramidal cells and interneurons of the CA3 subfield of hippocampus.

Description of the Conditions

Rats are anesthetized with chloral hydrate and placed on a stereotaxic frame. During surgery, animals are placed on a heating blanket system, the scalp is incised and a craniotomy is made following the coordinates described above. The injection cannula is implanted unilaterally (Fig. 1), and maintained in the site of injection for 5 min before and after LPS or vehicle (NaCl, 0.9%) infusion. Finally, the scalp is sutured and the animals are returned to their home cage in a room warmed to 26–28°C to recover from the anesthesia.

Description of the Process

Following LPS injection, tissue damage and induced-inflammatory mediators are evaluated by ►cresyl violet staining, ►immunohistochemistry (IHC) and biochemical methods. Brains are removed at different times after LPS injection to establish a time course study. For histological studies, brains are quickly frozen in isopentane at −40°C. Brain sections (20 μm) are prepared at −20°C (Cryostat Jung CM3000) every half-millimeter at six coronal planes, from 2.8 to 5.3 mm posterior to the bregma, to establish a spatial study. Some sections are stained with cresyl violet to assess the tissue damage. Adjacent sections are processed for IHC after being dried and fixed in chilled acetone for 5 min [2].

To assess the inflammatory response, inducible nitric oxide synthase (iNOS) [4] can be used as a marker of inflammation. In fact, iNOS is one of three isoforms of NOS which is inducible under inflammatory conditions and produces a high amount of NO and has a detrimental role especially at the acute phase [5].

Other markers such as Neuronal Nuclei (NeuN), Glial Fibrillary Acidic Protein (GFAP), and OX-42 can be used to visualize the damage to neurones, astrocytes and microglia, respectively. NeuN is a marker of neuronal cell nucleus, GFAP is a marker of astrocytic cytoplasm and OX-42 (or complement receptor 3, CR3) is a marker of microglial cell membrane.

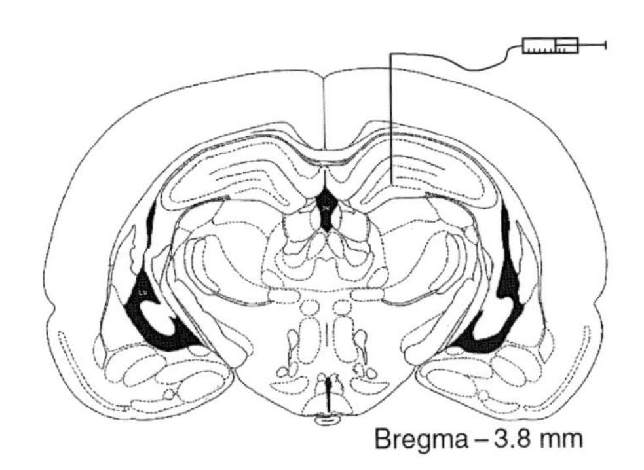

Neuroinflammation – LPS-induced Acute Neuroinflammation, Rat Model. Figure 1 Illustration of LPS injection site in the dentate gyrus at the level of Bregma −3.8 mm [3]. LPS is dissolved in sterilized physiological saline and infused at the dose of 15 μg in a volume of 2 μl, at a rate of 1 μl/min, by using Hamilton syringe (10 μl) and syringe pump.

For biochemical studies, ►ipsilateral and ►contralateral hippocampi are dissected out and quickly frozen to perform one of the following assays. They include NOS activity evaluating iNOS activity [2,6] and also ►myeloperoxidase (MPO) activity which represents an index of monocyte/neutrophil infiltration [7]. The samples can be processed online for tissue NO end products assay, nitrate plus nitrite (NOx) which is an indirect index of NO production [8]. For brain MPO activity assay, caution should be taken to wash out the blood cells from the vasculature by transcardial perfusion through the aorta with NaCl 0.9%. The study of inflammatory mediators can also be extended to pro-inflammatory cytokines such as interleukin-1β (IL-1β), IL-6 and tumor necrosis factor α (TNFα) by using commercially available rat ►ELISA kits.

Regulation of the Structure: Histopathological and Biochemical Outcomes After LPS Infusion

Saline infusion in the hippocampus does not lead to tissue damage and iNOS induction except at the site of injection due to the cannula penetration. In contrast, LPS infusion causes tissue damage, characterized by loss of cresyl violet staining, due to cell necrosis, in the ipsilateral denate gyrus compared to contralateral side (Figs. 2a and 2b). Furthermore, the cellular loss is due in part to neuronal cell loss observed by a marked decrease

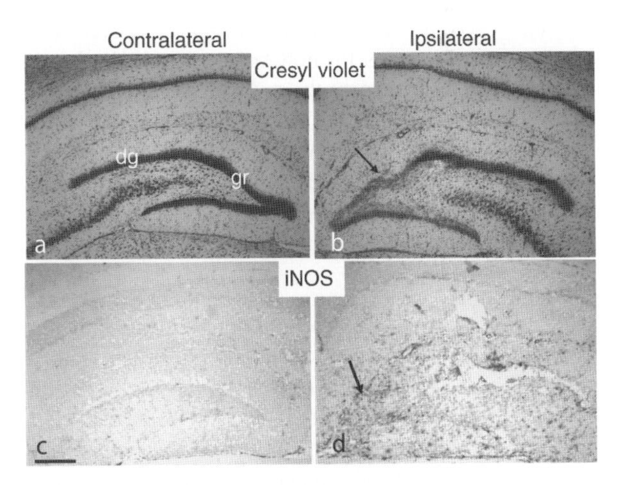

Neuroinflammation – LPS-induced Acute Neuroinflammation, Rat Model. Figure 2 Representative photomicrographs of coronal brain sections for cresyl violet staining (a, b) and iNOS immunoreactivity (c, d) at 0.5 mm posterior to the LPS injection site. Contralateral (a, c) and ipsilateral (b, d) sections to injection site in rats submitted to LPS after 15h. At the site of LPS injection, cell loss was accompagnied by an intense iNOS immunoreactivity. Arrows show the site of brain damage (b) and iNOS immunolabeled cells (d). *dg*, dentate gyrus; *gr*, granular layer. Scale bar = 400 μm.

in NeuN immunolabeling [2]. An intense iNOS immunolabeling is also detected in the lesion area compared to contralateral side (Fig. 2c and 2d). This observation is consistent with intense iNOS activity and brain NOx levels indicating that iNOS is active in the brain parenchyma [2,9]. Following LPS infusion, induction of iNOS is acute and transient, peaked at 24h, with a coronal expansion of 2 mm around the injection site, preceding the peak of cellular loss in the hippocampus at 48–72h [2]. A high level of MPO activity is also observed in the hippocampus showing the infiltration of monocyte/neutrophil in the brain parenchyma following LPS infusion [2]. This model leads to an intense inflammatory response *in vivo* causing neuronal cell loss restricted to the hippocampus. It is a very reproducible model with no mortality over a week post-LPS infusion.

Function Alterations After LPS Infusion

Neuroinflammation induced by LPS in this model is accompanied by some alterations in neurological functions such as impairment in sensorimotor function [2] and spatial memory [9]:

1. Contralateral sensorimotor functions are examined by assessing placing reactions (leg hanging and visual), grasping reflex (left forepaw and left hindpaw), and righting reflex (head tilted; left side and right side) in rats placed on a table [10]. Abnormal postures (thorax twisting and left forelimb flexion) are also examined. The scores for each item are summed and used as a global neurological score. The maximum score is for non-operated rats; the lower the neurological score the more severe the deficit.
2. Spatial memory is a more specific function to examine the function of hippocampus after tissue damage. This function is assessed by both spontaneous alteration behavior in a Y-maze and performance in the Morris water maze task.

Therapy

LPS-induced tissue damage and neuronal dysfunction are limited by reducing the production of inflammatory mediators at its acute phase. One strategy is based on the use of iNOS inhibitors reducing iNOS-derived NO. It is noteworthy that treatment with an iNOS inhibitor prevents LPS-induced spatial memory dysfunction showing the detrimental role of iNOS in LPS-induced brain injury [9]. Since the early and late post-injury inflammatory response may play dual roles, detrimental vs. beneficial, caution should be taken on the timing of anti-inflammatory strategy [1].

References

1. Leker RR, Shohami E (2002) Cerebral ischemia and trauma-different etiologies yet similar mechanisms: neuroprotective opportunities. Brain Res Rev 39:55–73
2. Ambrosini A, Louin G, Croci N, Plotkine M, Jafarian-Tehrani M (2005) Characterization of a rat model to study acute neuroinflammation on histopathological, biochemical and functional outcomes. J Neurosci Meth 144:183–191
3. Paxinos G, Watson C (1982) The rat brain in stereotaxic coordinates. Academic, Sydney
4. Alderton WK, Cooper CE, Knowles RG (2001) Nitric oxide synthases: structure, function and inhibition. Biochem J 357:593–615
5. Louin G, Marchand-Verrecchia C, Palmier B, Plotkine M, Jafarian-Tehrani M (2006) Selective inhibition of inducible nitric oxide synthase reduces neurological deficit but not cerebral edema following traumatic brain injury. Neuropharmacology 50:182–190
6. Louin G, Besson VC, Royo NC, Bonnefont-Rousselot D, Marchand-Verrecchia C, Plotkine M, Jafarian-Tehrani M (2004) Cortical calcium increase following traumatic brain injury represents a pitfall in the evaluation of Ca2+-independent NOS activity. J Neurosci Meth 138:73–79
7. Batteur-Parmentier S, Margaill I, Plotkine M (2000) Modulation by nitric oxide of cerebral neutrophil accumulation after transient focal ischemia in rats. J Cereb Blood Flow Metab 20:812–819
8. Grandati M, Verrecchia C, Revaud ML, Allix M, Boulu RG, Plotkine M (1997) Calcium-independent NO-synthase activity and nitrites/nitrates production in transient focal cerebral ischaemia in mice. Br J Pharmacol 122:625–630
9. Yamada K, Komori Y, Tanaka T, Senzaki K, Nikai T, Sugihara H, Kameyama T, Nabeshima T (1999) Brain dysfunction associated with an induction of nitric oxide synthase following an intracerebral injection of lipopolysaccharide in rats. Neuroscience 88:281–294
10. Wahl F, Renou E, Mary V, Stutzmann JM (1997) Riluzole reduces brain lesions and improves neurological function in rats after a traumatic brain injury. Brain Res 756:247–255

Neuroinflammation – PDE Family Inhibitors in the Regulation of Neuroinflammation

Tetsuya Mizuno

Department of Neuroimmunology, Research Institute of Environmental Medicine, Nagoya University, Furo-cho, Chikusa, Nagoya, Japan

Definition

►Phosphodiesterase (PDE) is a group of enzymes that degrade key second messengers, cyclic AMP (cAMP) and cyclic GMP (cGMP). ►Phosphodiesterase inhibitors (PDEIs) inhibit the function of PDE and elevate cAMP, thereby upregulating PKA/CREB signaling and downregulating NF-κB signaling. Consequently, PDEIs have anti-inflammatory and neuroprotective effects and may ameliorate the neuroinflammation accompanying demyelinating, ►neurodegenerative, and neuroinfectious diseases.

Characteristics

Quantitative Description

Neuroinflammation is involved in the demyelinating diseases, neurodegenerative diseases and neuroinfectious diseases. Activated ►microglia play a key role in the inflammatory processes of these diseases by releasing inflammatory mediators. PDEIs that inhibit the function of phosphodiesterases and elevate cAMP have anti-inflammatory and neuroprotective effects. PDEIs are one of the effective drugs capable of suppressing inflammatory functions of activated microglia.

Involvement of Activated Microglia in Neuroinflammation

Neuroinflammation is a component of demyelinating diseases such as multiple sclerosis (MS) and is observed in an animal model for this disease, experimental allergic encephalomyelitis (EAE). Activated microglia play a key role in inflammatory process by releasing proinflammatory mediators including interleukin (IL)-1β, IL-6, tumor necrosis factor (TNF)-α, nitrite oxide (NO), and reactive oxygen species (ROS). These molecules damage myelinating oligodendrocytes to produce demyelinating lesions.

There is increasing evidence that inflammatory mechanisms are also involved in the pathogenesis of neurodegenerative diseases, such as amyotrophic lateral sclerosis (ALS), Parkinson's disease (PD) and Alzheimer's disease (AD). An immunological mechanism of ALS pathogenesis has been proposed and is supported by the presence of activated microglia within the gray matter of the spinal cord and motor cortex of patients with ALS. In PD, neuroinflammation contributes to the degeneration of neurons in the substantia nigra. Activated microglia and increased levels of inflammatory mediators have been detected in the striatum of PD patients and a large number of animal studies support a role for inflammation in the loss of dopaminergic neurons. In AD, neuroinflammatory mediators are upregulated in affected areas of the brain while fibrillar and oligomeric forms of amyloid β peptide (Aβ) stimulate activation of microglia.

Activated microglia have a key role in chronic neuroinfectious diseases such as human immunodeficiency virus (HIV) encephalitis and prion diseases. Microglia are major targets of infection by HIV-1, and infected microglia are activated to produce inflammatory mediators. The pathogenic isoform of prion protein also stimulates microglia to produce inflammatory mediators. Therefore, inhibiting the release of inflammatory mediators by microglia may be an effective strategy for treating neuroinflammatory, neurodegenerative and neuroinfectious diseases. As PDEIs suppress the inflammatory response of activated microglia, these drugs are good candidates for therapeutic intervention.

Anti-Inflammatory Mechanism of PDEIs

PDEs hydrolyze cyclic nucleotides, and may be specific for cAMP/cGMP or both cAMP and cGMP. There are 11 families of proteins with this enzymatic activity (PDE1-PDE11) and more than 50 isoforms in total [1]. Cyclic nucleotides are second messengers for several G proteins. PDEIs that block one or more PDEs enhance the function of cyclic nucleotides to promote a variety of pharmacologic actions. For example, some PDEIs elevate intracellular cAMP and activate the protein kinase A (PKA) signaling pathway. This suppresses NF-κB-mediated transcription without preventing nuclear translocation of NF-κB complexes [2]. Consequently, PDEIs inhibit the production of inflammatory mediators regulated by NF-κB such as IL-1β, IL-6, TNF-α, and NO (Fig. 1).

PDEIs block one or more phosphodiesterases and elevate intracellular cyclic AMP. Subsequently, they activate PKA signaling pathways that suppress NF-κB-mediated transcription and upregulate CREB. Consequently, PDEIs limit the production of inflammatory mediators and promote LTP that is regulated largely by CREB.

PDE4 and PDE10 are highly expressed in the central nervous system (CNS). The PDE4 inhibitor rolipram has anti-inflammatory, anti-depressant, and memory-enhancing effects. Ibudilast, which has been used in Japan to treat both bronchial asthma and cerebrovascular disorders since 1989, is a broad range PDEI that inhibits PDE3A, PDE4, PDE10 and PDE11(1). Inhibiting of PDE3A and PDE4 may affect tracheal smooth muscle contractility while inhibiting PDE4 and PDE10 has positive effects on neurological conditions.

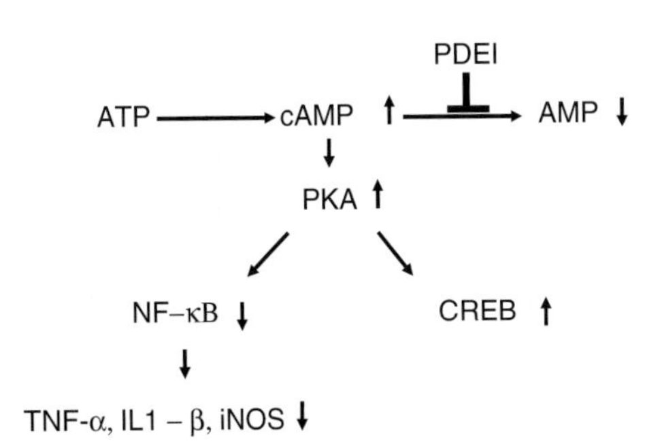

Neuroinflammation – PDE Family Inhibitors in the Regulation of Neuroinflammation. Figure 1 Mechanisms of action of PDEIs.

The Effect of PDEIs on Demyelinating Diseases

Activated microglia perform several functions in the inflammatory process. They are antigen-presenting cells that present myelin-specific antigens to invading T cells in the induction phase of MS and EAE. In addition, they are effector cells that damage oligodendrocytes and neuronal cells by secreting inflammatory cytokines, radicals, and glutamate during the effector phase. However, these cells may also protect neural functions by producing neurotrophic factors. Ibudilast suppresses the production of IL-1β, IL-6, TNF-α, NO, and reactive oxygen species (ROS) by activated microglia. Moreover, it enhances the production of the inhibitory cytokine, IL-10, and neurotrophic factors including nerve growth factor (NGF), glia-derived neurotrophic factor (GDNF), and neurotrophin (NT)-4 by these cells [3].

MS is a T-helper 1 (Th1) lymphocyte-mediated disease. Th1 cells initiate proinflammatory activity while Th2 initiate anti-inflammatory activity. Ibudilast suppresses differentiation of Th1 cells in the CNS, and can shift the cytokine profile such that is dominated by Th2 rather than Th1 cells. Ibudilast significantly suppresses the production of IL-12 by microglia; this cytokine is critical for Th1 differentiation. In addition, ibudilast also suppresses the production of interferon-gamma, but not IL-4 or IL-10, by myelin oligodendrocyte glycoprotein (MOG)-specific T cells reactivated with MOG in the presence of microglia. Thus, PDEIs suppress the activities of activated microglia that contribute to the pathology of MS and EAE.

PDEIs have been examined in clinical studies of MS. The combination of three PDEIs suppresses the frequency of relapse in relapsing remitting MS (RRMS) at the standard therapeutic doses [4]. A randomized, double-blind, placebo-controlled multi-center Phase II clinical trial of ibudilast in patients with RRMS was initiated in Eastern Europe in July, 2005. Enrollment of 297 patients was completed in February, 2006. Interferon (IFN)-β is the first approved therapy for RRMS. However, IFN-β treatment causes several common side effects such as flu-like symptoms (fatigue, chills, and fever) which may be a consequence of elevated levels pro-inflammatory cytokines. PDEIs suppress the upregulation of inflammatory mediators induced by IFN-β. The PDE3 and PDE4B inhibitor pentoxifylline synergistically functions with IFN-β to reduce the production of inflammatory cytokines and upregulate the anti-inflammatory cytokine IL-10 in peripheral blood mononuclear cells from patients with active MS [5]. Ibudilast also suppresses the production of the inflammatory mediators TNF-α, IL-1β, IL-6, and NO concurrent with IFN-β treatment [6] (Fig. 2).

LPS treatment activates microglia to produce TNFα, IL-1β and IL-6. IFN-β enhances the production of these cytokines. Ibudilast significantly suppresses this effect in a dose-dependent fashion. Similarly, LPS and

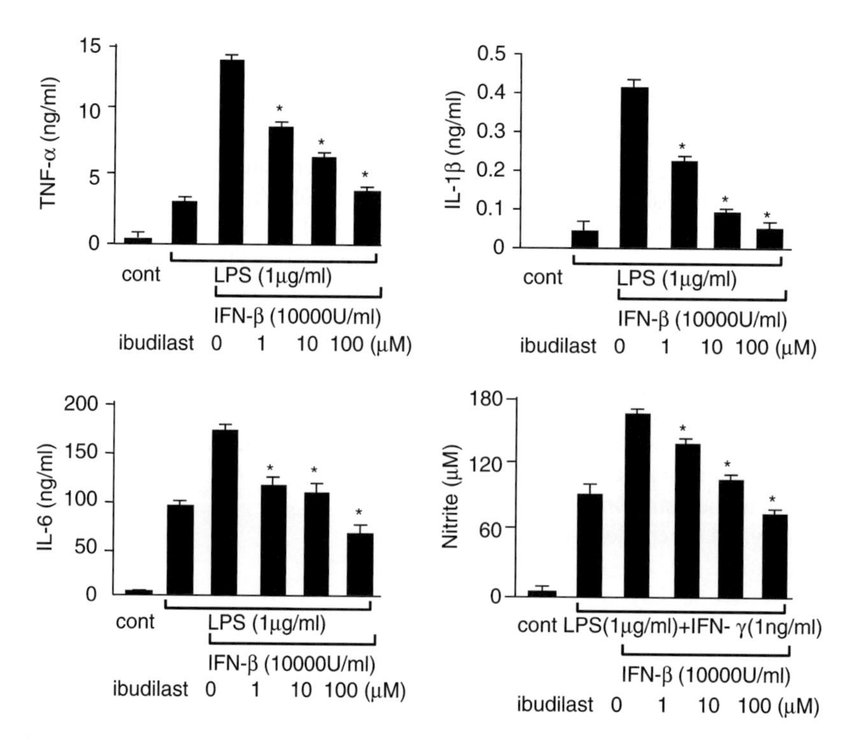

Neuroinflammation – PDE Family Inhibitors in the Regulation of Neuroinflammation. Figure 2 Ibudilast suppresses the enhanced production of inflammatory cytokines and NO induced by IFN-β.

IFN-γ induce the production of NO and its derivative nitrite. Addition of IFN-β enhances the production of these factors as well. Ibudilast significantly inhibits the upregulation of NO in a dose-dependent fashion.

The Effect of PDEIs on Neurodegenerative Diseases

Activated microglia contribute to neuronal degeneration by producing proinflammatory cytokines, glutamate, and peroxynitrite, a product of NO and superoxide. As neuronal degeneration is related to the functional prognosis in MS, suppressing the production of these factors by activated microglia with PDEIs may be an effective strategy for treating the neuronal degeneration associated with MS. For example, ibudilast inhibits the neuronal cell death induced by activated microglia with lipopolysaccharide (LPS) and IFN-γ (Fig. 3).

Microglia activated with LPS (1 μg/ml) and IFN-γ (100 ng/ml) induce neuronal cell death. Addition of ibudilast inhibits this neuronal cell death.

Experimental evidence supports a model for ALS neurodegeneration in which microglia contribute to the cell death of motor neurons. It is generally believed that oxidative stress and glutamate-mediated excitotoxicity are important mechanisms in ALS. NADPH oxidase, the main ROS-producing enzyme during inflammation, is activated in the spinal cord of ALS patients as well as in the spinal cord of animals with a genetic animal model of this disease. Inactivating NADPH oxidase in ALS mice delays neurodegeneration and extends survival [7]. A double-blind, randomized, multicenter, placebo-controlled trial of the PDE4B inhibitor, pentoxifylline, has been conducted with ALS patients. Unfortunately pentoxifylline had a negative effect on survival [8].

Activated microglia play a key role in the initiation and progression of PD. Exposure to a common herbicide, rotenone, induces features of parkinsonism. Rotenone stimulates the release of superoxide from microglia, resulting in the selective destruction of the nigrostriatal dopaminergic system. In animal models, 1-methyl-4-phenyl-1,2,3,6-tetrahydropyridine (MPTP) damages the nigrostriatal dopaminergic pathway to induce parkinsonism. Neuronal damage induced by MPTP is mediated by activated microglia, and postmortem examination of human subjects exposed to MPTP reveals the presence of activated microglia decades after drug exposure. In contrast, PDEIs are reported to stimulate the uptake of dopamine and enhance intracellular dopamine levels in rat mesencephalic neurons.

Neuroinflammation is a characteristic of AD, with activated microglia being the driving force. As nonsteroidal anti-inflammatory drugs (NSAIDs) are the most commonly used of all anti-inflammatory agents, many studies have examined whether NSAIDs might have protective effects on AD. Cyclooxygenase (COX)-1 is upregulated in activated microglia and

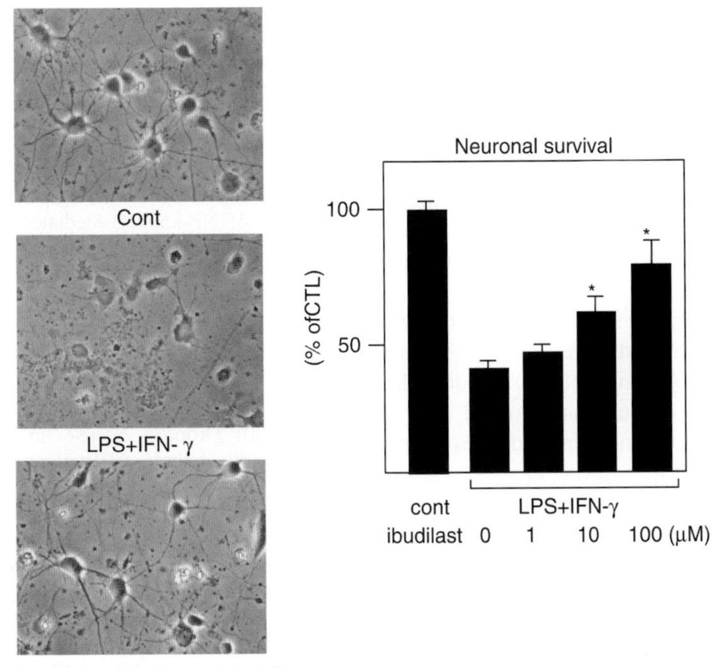

Neuroinflammation – PDE Family Inhibitors in the Regulation of Neuroinflammation. Figure 3 Neuroprotective effects of ibudilast.

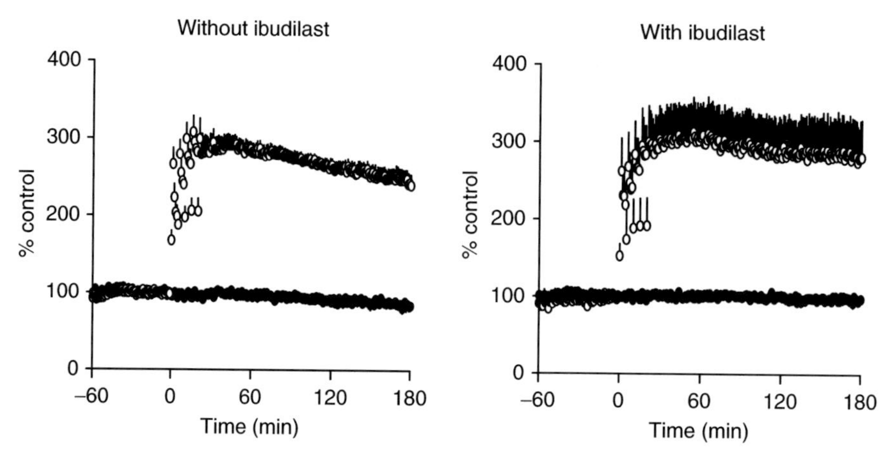

Neuroinflammation – PDE Family Inhibitors in the Regulation of Neuroinflammation. Figure 4 Effect of ibudilast on the suppression of LTP by LPS and IFN-γ.

NSAIDs that block this enzyme could have an ameliorating effect on the disease [9]. Similarly, the effects of PDEIs on activated microglia may also slow the progression of AD.

In addition to its roles in limiting the proinflammatory activity of microglia, PDEIs also affect neuronal activity. PDEIs that activate PKA and the cAMP responsive element-binding protein (CREB) pathway promote ►long-term potentiation (LTP) (Fig. 1). LTP in hippocampal CA1 neurons is essential for memory acquisition. Treating cultured hippocampal neurons with Aβ inactivates the PKA/CREB pathway and inhibits LTP. The PDE inhibitor, rolipram, reverses this inhibition. Moreover, rolipram ameliorates deficits in both LTP and contextual learning in the double-transgenic AD mice [10]. Treating hippocampal slices with LPS and IFN-γ prior to inducing LTP activates microglia and causes the magnitude of LTP decrease gradually. Ibudilast reverses this inhibition (Fig. 4). Thus, PDEIs may have the potential as therapeutics for treating dementia. Recently, novel PDE4 inhibitor, MEM-1414 was developed for treating AD, mild cognitive impairment and depression.

The mean population EPSP slope after tetanic stimulation was gradually attenuated with LPS (1 μg/ml) and IFN-γ (100 ng/ml) (A). Addition of 100 μM ibudilast during LPS and IFN-γ stimulation returned LTP to normal levels (B). The horizontal line indicates control levels at unstimulated sites.

The Effect of PDEIs on Neuroinfectious Diseases

The HIV-1 virus infect macrophages and microglia in the CNS and frequently cause neurocognitive impairment. HIV-1 infected microglia are activated and inflammatory mediators. The HIV-1 envelope glycoprotein 120 (gp120), which is shed from the virus, can cause neuronal cell death. HIV-1 gp120 inhibits LTP, and HIV-1 replication is enhanced by TNF-α. Therefore, PDEI treatment could inhibit cytokine secretion associated with HIV-1 infection and transcriptional regulation of HIV replication. Rolipram is reported to inhibit HIV-1 replication in vitro.

Activated microglia are a predominant feature prion-related encephalopathy. The pathogenic isoform of prion protein causes microglial activation and has a crucial role in neuronal cell death. However, there are few reports examining whether PDEIs are effective in prion diseases.

References

1. Gibson LC, Hastings SF, McPhee I, Clayton RA, Darroch CE, Mackenzie A, Mackenzie FL, Nagasawa M, Stevens PA, Mackenzie SJ (2006) The inhibitory profile of ibudilast against the human phosphodiesterase enzyme family. Eur J Pharmacol 538:39–42
2. Parry GC, Mackman N (1997) Role of cyclic AMP response element-binding protein in cyclic AMP inhibition of NF-kappaB-mediated transcription. J Immunol 159:5450–5456
3. Mizuno T, Kurotani T, Komatsu Y, Kawanokuchi J, Kato H, Mitsuma N, Suzumura A (2004) Neuroprotective role of phosphodiesterase inhibitor ibudilast on neuronal cell death induced by activated microglia. Neuropharmacology 46:404–411
4. Suzumura A, Nakamuro T, Tamaru T, Takayanagi T (2000) Drop in relapse rate of MS by combination therapy of three different phosphodiesterase inhibitors. Mult Scler 6:56–58
5. Weber F, Polak T, Gunther A (1998) Synergistic immunomodulatory effects of interferon-beta1b and the phosphodiesterase inhibitor pentoxifylline in patients with relapsing-remitting multiple sclerosis. Ann Neurol 44:27–34
6. Kawanokuchi J, Mizuno T, Kato H, Mitsuma N, Suzumura A (2004) Effects of interferon-beta on microglial functions as inflammatory and antigen presenting cells in the central nervous system. Neuropharmacology 46:734–742

7. Wu DC, Re DB, Nagai M, Ischiropoulos H, Przedborski S (2006) The inflammatory NADPH oxidase enzyme modulates motor neuron degeneration in amyotrophic lateral sclerosis mice. Proc Natl Acad Sci USA 103:12132–12137
8. Meininger V, Asselain B, Guillet P, Leigh PN, Ludolph A, Lacomblez L, Robberecht W, Pentoxifylline European Group (2006) Pentoxifylline in ALS: a double-blind, randomized, multicenter, placebo-controlled trial. Neurology 66:88–92
9. McGeer PL, McGeer EG (2007) NSAIDs and Alzheimer disease: epidemiological, animal model and clinical studies. Neurobiol Aging 28:639–647
10. Gong B, Vitolo OV, Trinchese F, Liu S, Shelanski M, Arancio O (2004) Persistent improvement in synaptic and cognitive functions in an Alzheimer mouse model after rolipram treatment. J Clin Invest 114:1624–1634

Neuroinflammation: Brain and Spinal Cord Injury

W. Dalton Dietrich[1,2], Robert W. Keane[3]
[1]The Miami Project to Cure Paralysis, University of Miami Miller School of Medicine, Miami, FL, USA
[2]Neurological Surgery, University of Miami Miller School of Medicine, Miami, FL, USA
[3]Department of Physiology and Biophysics, University of Miami Miller School of Medicine, Miami, FL, USA

Synonyms

Cell death after CNS trauma

Definition

Inflammation and Apoptosis after Brain and Spinal Cord Injury

Central Nervous System (CNS) destruction in traumatic brain (TBI) and ►spinal cord injury (SCI) is caused by a complex series of cellular and molecular events. Recent studies have concentrated on signaling by receptors in the interleukin 1 (IL-1) (►interleukin (IL)) and tumor necrosis factor receptor (TNFR) family that mediate diverse biological outcomes. From the basic science research perspective, understanding how receptor signaling mediates these divergent responses is critical in clarifying events underlying irreversible cell injury in clinically relevant models of CNS trauma. From a clinical perspective, this work also provides novel targets for the development of therapeutic agents that have the potential to protect the brain and spinal cord from irreversible damage and promote functional recovery. Here, we discuss how the formation of alternate signaling complexes and receptor membrane localization after TBI and SCI can influence life and death decisions of cells stimulated through IL-1 and TNFR superfamily.

Characteristics

The pathophysiology of acute TBI and SCI is characterized by the shearing of cell membranes and axons, disruption of the blood-spinal cord barrier, cell death, immune cell transmigration, and myelin degradation [1–3]. Deleterious factors such as pro-inflammatory cytokines, proteases up-regulated by immune cells and toxic metabolites, and neurotransmitters released from lysed cells can induce further tissue damage [1–3]. These molecules can also stimulate an inflammatory reaction, with the subsequent release of neurotoxic molecules [1,3]. This subsequent damage, termed the “secondary injury,” causes neuronal cell death and progressive axonal loss over time (days to weeks) laterally and longitudinally to areas undamaged by the initial trauma [2,3]. A primary goal of CNS trauma research has been to prevent or limit secondary cell death that produces further axonal degeneration and creates a significant barrier to the regeneration of descending and ascending fibers [3].

Description of Process

IL-1 family members are known to alter the host response to an inflammatory, infectious, or immunological challenge [1,4]. The best-known members of this family are IL-1α/β, IL-1Ra, and IL-18. IL-1α/β, IL-18 and IL-33 are highly inflammatory cytokines, and dysregulation of their expression can lead to severe pathobiological effects. Accordingly, the expression of these cytokines is highly regulated via soluble receptors (type 2 IL-1 receptor) and natural antagonist proteins (IL-1Ra and IL-18 binding protein), as well as alternatively spliced forms of both ligands and receptors [4]. IL-1 cytokines exert their function through the Toll-like receptor (TLR)-IL-1 superfamily that can be divided into two groups, the TLRs and receptors of the IL-1 family. Currently there are 10 members of the IL-1 receptor family, and IL-1 ligands typically bind to a cellular receptor complex that consists of two members of this family. For example, the receptor complex for IL-1α/β consists of IL-1R and Il-1RAcP, with IL1Ra acting as a natural antagonist of IL-1α/β by trapping IL-1R1 molecules. The hallmark of IL-1 receptor signaling is the activation of the mitogen-activated protein (MAP) kinases p38, JNK and ERK 1/2 and the transcription factor ►nuclear factor kappa beta (NF-κB) [4].

Excessive levels of the proinflammatory cytokines IL-1β and IL-18 are associated with secondary damage following SCI and TBI [3]. Both IL-1β and IL-18 are synthesized as inactive cytoplasmic precursors that are proteolytically processed as biologically active mature forms in response to proinflammatory stimuli by caspase-1, a cysteine protease. IL-33 has also been described as being processed by caspase-1. The processing of pro-IL-1β involves the activation of a multiprotein caspase-1-activating complex termed the ►inflammasome [5,6]. The inflammasome is formed by a member of the NALP protein family, such as NALP1, NALP2 or NALP3, and the adaptor protein ASC that connects the NALPs with caspase-1 [5]. Activation of the inflammasome ultimately results in activation of proinflammatory IL-1β that is secreted by macrophages and triggers another cascade of molecular events that result in inflammation [4]. To date, only neurons have been reported to contain the inflammasome in the CUS [6]. Thus, there is a need to establish the events critical to the assembly and activation of the inflammasome, and to determine if these principles apply to inflammatory processes within the CNS.

Higher Level Structures

CNS inflammatory responses that occur after SCI and TBI are initiated by peripherally-derived immune cells (macrophages, neutrophils, and T-cells), and activated glial cells (astrocytes and microglia) that proliferate or migrate into the lesion site following injury [2,3]. T-cells are essential for activating macrophages and mounting a cellular or immune response. Macrophages and neutrophils have also been proposed to participate in tissue destruction and enlargement of the lesion [2,3]. Macrophages and microglia contribute to the secondary pathological and inflammatory response, in part through the release of cytokines, tumor necrosis factor (►tumor necrosis factor (TNF)), interleukin-1 (IL-1), IL-6, and IL-10, interferon [4], and activation of interleukin receptors (IL-4R and IL-2R) [4]. Cytokines facilitate CNS inflammatory responses by inducing expression of additional cytokines, chemokines, nitric oxide (NO), and reactive oxygen [3]. Since inflammation contributes to both constructive and neurodestructive processes, a more thorough understanding of the autoimmune events that occur following CNS injury may allow us to develop strategies that will harness the beneficial effects of inflammation and, hopefully, help to promote functional recovery [3].

Therapy

Prevention of production of inhibitory proinflammatory molecules by activated mononuclear phagocytes has been demonstrated to be neuroprotective [3]. Various strategies including drug delivery as well as mild hypothermia [3] have been shown to reduce the inflammatory cascade after SCI and provide neuroprotection and improvement in functional outcome. Another strategy has concentrated on targeting selections on the surface of endothelial or inflammatory cells [3]. Interactions of endothelial cell-adhesion molecules with integrins on the white blood cell surface have been shown to promote leukocyte extravasation through the blood-spinal cord barrier and movement into the injured spinal cord.

It is likely that other agents that prevent the synthesis and secretion of IL-1 family members will be effective in CNS trauma. These are the IL-1 Trap, IL-1β-specific monoclonal antibodies, and the caspase-1 inhibitor [4]. It is also possible that agents that target IL-1β secretion may function to limit inflammation after CNS trauma. Moreover, once released, IL-1β must compete for receptor occupancy with the naturally occurring IL-1Ra, the binding and neutralization by the IL-1 type II decoy receptor and the formation of inactive complexes with constitutively secreted soluble IL-1 accessory protein, each of which also limit IL-1β responses. Moreover, we have recently found that therapeutic neutralization of the inflammatory after SCI reduces IL processing, resulting in significant tissue sparing and functional improvement [6]. Thus, continued investigations into the mechanisms underlying the activation of IL-1β inflammatory cascades after SCI and TBI could lead to new strategies to inhibit secondary injury and thus to promote recovery in injured patients.

Regulation of Processes

TBI induces upregulation of TNF-α protein and mRNA in the injured cortex [7,8], and increased levels of TNF-α have been reported in plasma and cerebrospinal fluid of human head injured patients [7]. Gene-targeting studies indicate that the presence of TNF-α in the acute posttraumatic period may be deleterious, whereas this cytokine may play a beneficial role in the chronic period after TBI [7,9]. In a similar fashion, spinal cord trauma leads to increased expression of TNFR1 and TNFR2 receptors and their ligands as well as activation of ►caspases and calpain, but there are conflicting reports as to the role of TNF signaling after SCI that probably reflect the known capacity of TNF to be both pro and anti-apoptotic [7,9]. A solution to this paradox has been proposed in the recent findings that tumor necrosis factor receptor (TNFR) submembrane localization and the formation of alternate signaling complexes can alter the fate of cells stimulated through TNFRs [10].

Mammalian TNF-α signals through two cell surface receptors, TNFR1 (CD120a), and TNFR2 (CD120b). Most cells constitutively express TNFR1 while TNFR2 expression is highly regulated. Activation of TNFR1 leads to the recruitment of the adaptor ►TRADD (TNFR-associated death domain protein) that serves as a platform to recruit additional signaling adaptors [9]. TRADD binds the Ser/Thr kinase receptor-interacting protein (►RIP) and TNF-receptor-associated factors 2 (►TRAF2) and 5 (TRAF5). This TRADD-RIP-TRAF complex causes activation of NF-κB, through an unknown mechanism

[9]. TRAF2 can also recruit secondary adaptors that modulate signaling, i.e. TRAF1 and cellular inhibitor of ►apoptosis protein-1 (►cIAP-1) and -2 (cIAP-2) [9]. cIAP-1 supports ubiquitination and proteasomal degradation of TRAF2 [8], while TRAF2 inhibits signaling through TNFR2 by an unknown process [8]. Additionally, TNFR1 can recruit caspase-8 via TRADD and Fas-associated death domain protein (►FADD) to induce apoptosis [9].

Redistribution of TNFR1 in the plasma membrane is one possible mechanism for regulating efficiency of TNF signaling. Recent *in vitro* evidence suggests that redistribution of TNFR1 into specialized microdomains (►lipid rafts) may account for the outcome of some TNF-α-activated signaling pathways [7,8,9], but TNFR1 localized to nonraft regions of the plasma membrane are capable of initiating different signaling responses [7]. Recently, the role of microdomains in signal transduction emanating from the TNFR family *in vivo* has been addressed [8].

Higher Level Processes

The TNFR superfamily mediates a wide spectrum of important cellular functions ranging from acute inflammation and lymphocyte co-stimulation to apoptosis and other forms of programmed cell death [7,9]. The divergent cellular signaling responses orchestrated by these receptors are dependent on cell-type and environmental factors [9]. In most instances, TNFR1 triggers cellular activation via NF-κB. However, when new protein synthesis is inhibited prior to TNF stimulation, TNFR1 can initiate apoptosis by activation of apical caspases [7,9]. Recent experimental evidence has provided information about how receptor submembrane localization and the formation of alternative signaling complexes by two members of the TNFR family, TNFR1 and Fas, can alter the fates of cells [8,10]. Here, we discuss how programmed cell death after CNS trauma is a tightly regulated process that can be initiated by activation of a specific TNFR family member TNFR1. Deletion of TNFR1 or blocking ligand interactions with different TNFR family members has emerged as a clinically effective therapy for experimental CNS injury.

In cells of the immune system, TNFR1 signaling involves assembly of two molecularly and spatially distinct signaling complexes that sequentially activate NF-κB and caspases [10]. Early after TNF binding to TNFR1, a TNFR1 receptor-associated complex (complex I) forms and contains TRADD, RIP1, TRAF1, TRAF2 and cIAP-1. Complex I transduces signals that lead to NF-κB activation through recruitment of the I-κB kinase "signalsome" high molecular weight complex [10]. TNFR1-mediated apoptosis signaling is induced in a second step in which TRADD and RIP1 associate with FADD and caspase-8 to form a cytoplasmic complex (complex II) that dissociates from TNFR1. However, when complex I triggers sufficient NF-κB signaling, anti-apoptotic gene expression is induced and the activation of initiator caspases in complex II are inhibited. If NF-κB signaling is deficient, complex II transduces an apoptotic signal. Thus, early activation of NF-κB by complex I serves as a checkpoint to regulate whether complex II induces apoptosis at a later time point after TNF binding.

Our recent study has shed new light on how membrane proximal events control fate decisions in signaling by TNFR1 in the CNS after TBI [8]. The results support a model in which a small amount of TNFR1 is constitutively expressed in the lipid raft microdomains. It has been proposed that lipid rafts serve as signaling platforms for variety of receptors including TNFR1 and Fas. TNFR1 signaling complexes in the normal CNS contain adaptor molecules TRADD, RIP, TRAF1, TRAF2 and cIAP-1 (Fig. 1) [8]. Since the TNFR1-TRADD-RIP-TRAF2 complex initiates the pathway leading to survival [9], it is probable that the TNFR1 signaling complex in the normal CNS initiates a survival signal. Moreover, this signaling complex is devoid of FADD, cIAP-2 and caspase-8 [8].

CNS trauma induced rapid translocation of TNFR1 to lipid rafts, altered associations with signaling intermediates, and induced transient activation of NF-κB. RIP and cIAP-1 dissociate from TNFR1, whereas FADD and cIAP-2 increase association with this receptor-signaling complex in lipid rafts. Because the TNFR1-TRADD-FADD complex initiates the pathway leading to apoptosis [9], it is possible that alterations in association of adaptor molecules in the signaling complex are responsible for the switch in the signal transduction pathway from survival in the normal CNS toward apoptosis after trauma (Fig. 1). Dissociation of RIP from the TNFR1 signaling complex induced by trauma may ablate or downregulate the NF-κB pathway and facilitate cell death. Additionally, cIAP-1 and cIAP-2 and TRAF1 have been identified as NF-κB target genes [9]. Trauma-induced interference of the NF-κB pathway may result in altered actions of the caspase-8 inhibitory TNFR1-TRAF-IAP complex to further promote apoptosis [9]. By 30 min after CNS trauma, caspase-8 was present in TNFR1 signaling complexes, supporting the idea that the association of FADD with TRADD initiates the apoptotic program by recruiting caspase-8. Thus, in contrast to TNFR1-mediated signaling in cultured cells, these *in vivo* studies do not reveal an essential role of complex II in the regulation of TNF-α responses after CNS trauma, but rather indicate that in both the normal and traumatized CNS, lipid rafts appear to promote the formation of a receptor-associated signaling complex (complex I) to produce different biological outcomes dictated by these complexes. Moreover, complex I in the traumatized CNS

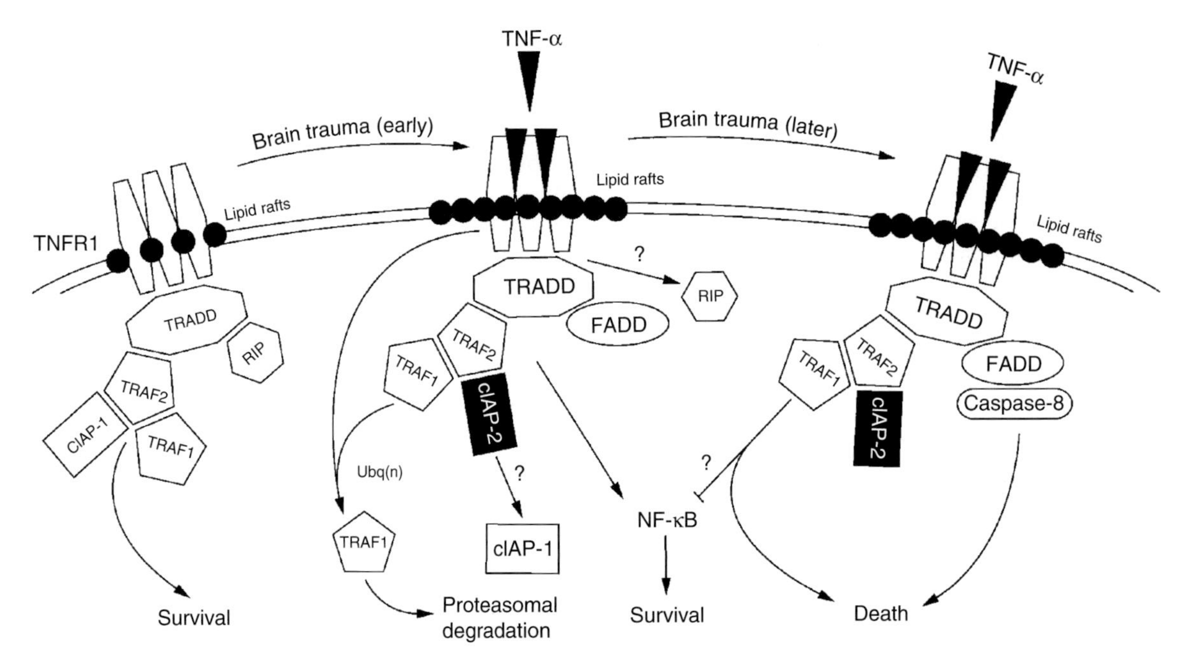

Neuroinflammation: Brain and Spinal Cord Injury. Figure 1 Model of lipid raft mediated TNFR1 signaling after CNS trauma. In a normal rat CNS low levels of TNFR1 are present in lipid rafts and are in complex with TRADD, TRAF1, TRAF2, RIP and cIAP-1 and signals survival. Early after trauma, increased levels of TNFR1 recruit into lipid raft microdomains (•), where they associate with the adaptor protein TRADD, FADD, TRAF2, TRAF1, and cIAP-2. TNFR1 and TRAF1 are polyubiquitinated (Ubq(n)) in lipid rafts after trauma, which leads to degradation via the proteasome pathway. In later stages after injury, RIP and cIAP-1 appear to dissociate from TNFR1 complex by an unknown mechanism, and this complex signals death by activating caspase-8 [8].

harbors activated caspase-8 by 30 min after insult, indicating involvement in downstream signaling cascades. Therefore, the death domain of TRADD may act as a central platform for the recruitment and activation of FADD after CNS trauma, leading to subsequent binding of caspase-8 triggering their activation. These studies support recent evidence that the roles for lipid rafts in Fas and TNFR1 signaling varies between cell types [8,10]. Thus, TNFR signaling is dependent on cell type and subject to influence of other signaling pathways, genetic and environmental factors.

The IL-1 and TNF family of cytokines have been mainly characterized in the immune system and are primarily involved in regulating inflammatory and apoptotic responses. However, these cytokines are detectable in other tissues, for example the normal and traumatized CNS, raising the possibility that these cytokines and their receptors have a role in neurological trauma and disease. There is increasing interest in the role of inflammatory processes in CNS injury, since inactivation neuroprotection in animal models of SCI, stroke and multiple sclerosis [3]. However, a true understanding of how reducing inflammation after CNS injury leads to inhibition of cell death and enhance functional recovery will require more detailed knowledge. For example, the signaling pathways initiated by the IL-1 and TNF receptors in CNS cells have not been delineated. It is not clear if CNS cells exhibit differences in the efficiency of IL-1 or TNF signaling and thus can be categorized as cells in the immune system. The cellular source and target of the ligand in damaged CNS tissues need to be identified, and protocols need to be developed to deliver antibodies to the lesion at later stages to clearly evaluate this therapeutic approach.

Recent experimental evidence has provided information about how receptor submembrane localization and the formation of alternative signaling complexes can alter the fates of cells *in vitro*, but whether these principles apply to signaling mediated by TNFR family members in the normal CNS and after trauma awaits further experimentation. Thus, activation of these signaling pathways might become promising therapeutic targets for the acute treatment of neurological trauma and disease.

Acknowledgments

We would like to thank Dr. George Lotocki for the illustration. The work was supported in part by NIH PO1 NS 38665.

References

1. Benveniste EN (1992) Inflammatory cytokines within the central nervous system: sources, function, and mechanism of action. Am J Physiol 263:C1–C16

2. Bethea JR, Dietrich WD (2002) Targeting the host inflammatory response in traumatic spinal cord injury. Curr Opin Neurol 15:355–360
3. Dietrich WD, Chatzipanteli K, Vitarbo E, Wada K, Kinoshita K (2004) The role of inflammatory processes in the pathophysiology and treatment of brain and spinal cord trauma. Acta Neuochir (Suppl) 89:69–74
4. Dinarello CA (2005) Blocking IL-1 in systemic inflammation. J Exp Med 201:1355–1359
5. Martinon F, Burns K, Tschopp J (2002) The inflammasome: A molecular platform triggering activation of inflammatory caspases and processing of proIL-β. Mol Cell 10:417–426
6. de Rivero Vaccari JP, Lotocki G, Marcillo AE, Dietrich WD, Keane RW (2008) A molecular platform in neurons regulates inflammation after spinal cord injury. J. Neurosci 28:3404–3414
7. Keane RW, Davis AR, Dietrich WD (2006) Inflammatory and Apoptotic signaling after spinal cord injury. J. Neurotrauma 23:335–344
8. Lotocki G, Alonso OF, Dietrich WD, Keane RW (2004) Tumor necrosis factor receptor 1 and its signaling intermediates are recruited to lipid rafts in the traumatized brain. J Neurosci 24:11010–11016
9. Wajant H, Pfizenmaier K, Scheurich P (2003) Tumor necrosis factor signaling. Cell Death Differ 10:45–65
10. Micheau O, Tschopp J (2003) Induction of TNF receptor I-mediated apoptosis via two sequential signaling complexes. Cell 114:181–190

Neuroinflammation: Chronic Neuroinflammation and Memory Impairments

JUNG-SOO HAN
Department of Biological Sciences, Konkuk University, Seoul, South Korea

Synonyms

Neuroinflammation, VSAIDs

Definition

Neuroinflammation and NSAIDs

Neuroinflammatory responses are characteristics of pathologically affected tissue in several neurodegenerative disorders, including Alzheimer's disease (AD). Epidemiological studies have shown that conventional long-term treatments with non-steroidal anti-inflammatory drugs (►NSAIDs) reduce the risk of AD, delay the onset of this disease, ameliorate symptomatic severity, and slow cognitive decline. A transgenic AD Tg2576 mouse has amyloid pathology and activated microglias. Daily intake of ibuprofen, a NSAID, reduces the levels of the inflammatory cytokine, reactive ►astrocytes with glial fibrillary acidic protein (GFAP), β-amyloid deposits, and activated microglia in these AD mice. Rats with intraventricular chronic infusion of β-amyloid or ►lipopolysaccharide (LPS; ►endotoxin) show microglial activation and ►memory impairment. NSAID treatments rescued memory impairment of these rats and lowered inflammatory responses. Thus, these anti-inflammatory agents can significantly delay inflammatory responses characterized by activated glial cells and increased expression of cytokines surrounding amyloid deposits of AD pathology and prevent cognitive decline, as inflammation clearly occurs in the AD brain.

Characteristics

Alzheimer's Disease and Neuroinflammation

AD typically leads to progressive and incapacitating memory loss followed by additional cognitive and behavioral impairments. A neuropathological diagnosis of AD is made upon the detection of amyloid plaques and neurofibrillary ►tangles (NFTs) in the limbic and neocortical areas of the brain. However, AD is now also characterized by neuroinflammatory changes and increased free radicals, as well as classic neuropathological features such as amyloid plaques, neuronal loss, and NFTs [1,2]. *In vivo* measurements of microglial activation using positron emission tomography (PET) and magnetic resonance imaging (MRI) show that inflammation is an early event in the pathogenesis of AD [3]. Further, a chronic inflammatory response characterized by activated microglia, reactive astrocytes, complement factors, and increased inflammatory cytokine expression is associated with amyloid plaques in the AD brain [4].

Chronic inflammatory processes play an important role in the pathogenesis of AD [4]. Clumps of activated microglias and reactive astrocytes appear on ►senile plaques [1]. The levels of inflammatory cytokine interleukin-1-alpha (IL-1α) are increased in the AD brain [5]. The increase in IL-1α might both underlie and be due to widespread astrogliosis in the AD brain. Additionally, IL-1α could induce the expression of the β-amyloid precursor protein (β-APP). Senile plaques contain both β-amyloid and reactive microglial cells that excessively express inflammatory cytokines, including IL-1α and tumor necrosis factor-alpha (TNF-α). Further, activated microglia is a source of free radicals and neurotoxic materials. One potential neurotoxin released by activated microglia is glutamate. Chronic increase of extracellular glutamate impairs the glutamatergic receptor function, leading to the entry of toxic amounts of calcium into neurons and subsequently potentiation of neurotoxicity [3].

There is now overwhelming evidence that a state of chronic inflammation exists in affected regions, although it must still be determined whether inflammation merely occurs to clear the detritus of already existent pathology (plaques/tangles) or inflammatory molecules and mechanisms are uniquely or significantly elevated in the AD brain [6].

Animal AD Model for Neuroinflammation

Well-characterized animal models with important neuropathological features seen in AD have significantly advanced our understanding of the molecular mechanisms of AD and are important in predicting future therapeutic intervention. In the following, I introduce two animal AD models currently used extensively by neuroscientists.

A Transgenic Mice Model for AD

Targeted gene mutation technology represents a powerful new tool for biomedical research. A new gene or an additional copy of an existing gene is added to the genome in the transgenic mice and a gene is missed in the knockout (KO) mice. Transgenic mice expressing mutated human amyloid, human presenilin 1, or both show dramatic parallels to AD. However, none of the models appear to have the full pathological characteristics of human AD.

The most popular Alzheimer transgenic mouse model is the Tg2576 mouse, which carries a human familial AD gene (amyloid peptide protein; β-APP with the "Swedish" double mutation). The model displays age-related neuritic plaque pathology, activated microglias, and reactive astrocytes with increased GFAP (glial fibrillary acid protein) in the hippocampal and neocortical areas. More importantly, these mice show age-related memory deficits linked to defective ►long-term potentiation (LTP).

These AD transgenic models are being used to devise therapeutics strategies for AD. Specifically drugs or procedures that reduce the accumulation of β-amyloid in the mouse models are considered to be potential treatments for AD. One limitation of this approach is that such mice are only partial models of AD. They show abundant β-amyloid deposits, which are comparable to those observed in AD. However, in contrast with AD, the mice do not demonstrate the presence of NFTs. In the transgenic mice, complement staining of the deposits is weak, whereas as in AD it is very strong.

With the emergence of the transgenic/KO mouse models, the need for behavioral studies measuring the cognitive abilities of the mouse has become more urgent. Despite this, few studies to characterize cognitive behaviors with many different mouse strains have been reported. Furthermore, because the behaviors of mice are different from those of the rat, direct comparison of results for mice with those for rats is not fruitful. Therefore, much caution is needed in conducting behavioral experiments measuring memory impairment or enhancement by a given treatment when working with transgenic mice (for example background strains or wild-type littermate controls for a null mutation).

On the contrary, behavioral tasks for measuring cognitive abilities with rats have been well studied and characterized. Conclusive and reliable decisions regarding rat behavioral data can be easily reached through comparison with the results of reported studies. On this basis, an AD rat model has been introduced.

Chronic LPS Infusion Rat Model

β-Amyloid or proinflammagen (such as LPS and interleukin (IL)-2) was chronically infused into the rat ventricle at a very low dose. LPS is a component of the cell wall of gram-negative bacteria and has been used experimentally to stimulate production of endogenous IL-1, β-APP and complement proteins. The chronic infusion of LPS into the brain via the fourth ventricle for 4 weeks reproduces important aspects of the pathology of AD.

Chronic LPS infusions increased the number and density of OX-6-positive reactive microglia, the immune competent cells of the CNS, in the hippocampus of Fischer-344 rat (see Fig. 1). The number and density of astrocyte, observed by GFAP, was also affected. Rats with chronic LPS infusion take longer to find a hidden platform in the Morris water maze, relative to rats with artificial cerebrospinal fluid (aCSF) (see Fig. 2). *Arc* is an immediate-early gene that was cloned from the brain and is a good indicator of neuronal activation by physiological stimuli including LTP induction and seizures. Exploration-induced Arc protein expression within the dentate gyrus (DG) is altered in the hippocampus of rat with cognitive impairment by neuroinflammation. LPS activates microglia to initiate a series of inflammation-induced changes within the hippocampus and entorhinal cortex. The inflammation leads to a reduction in the number of NMDA glutamate receptors within the DG and CA3 hippocampal area without neuronal loss. Furthermore, LPS-induced neuroinflammation impairs the induction of LTP. According to MRI results, the size of the hippocampal formation and the temporal region was decreased. These aspects of the chronic LPS infusion model make it useful for testing potential pharmacotherapies for the prevention of AD [3].

Neuroinflammation of AD and NSAID

Because the inflammatory process is the pathological hallmark associated with AD, it is not surprising that conventional anti-inflammatory therapy using NSAIDs has been shown to slow the progress, or delay the onset, of AD. Untreated elderly demented patients with senile plaques have almost three times more activated microglia than do those patients with senile plaques chronically taking NSAIDs. Neuronal cyclooxygenase (COX)-2 is elevated in the AD brain: long-term

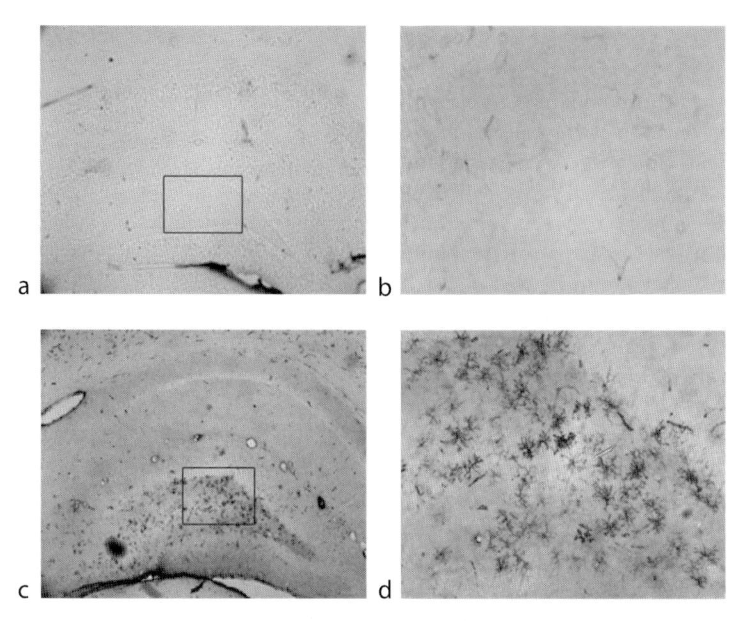

Neuroinflammation: Chronic Neuroinflammation and Memory Impairments. Figure 1 Reactive microglia stained with OX-6 in the hippocampus. Fischer-344 rats infused with aCSF have only a few activated microglias (a, b). Chronic infusion of LPS into the fourth ventricle produces activated microglias distributed throughout the hippocampus of the Fischer-344 rat brain. Activated microglias were expressed highly in dentate gyrus (c, d). (courtesy of Jung-Soo Han).

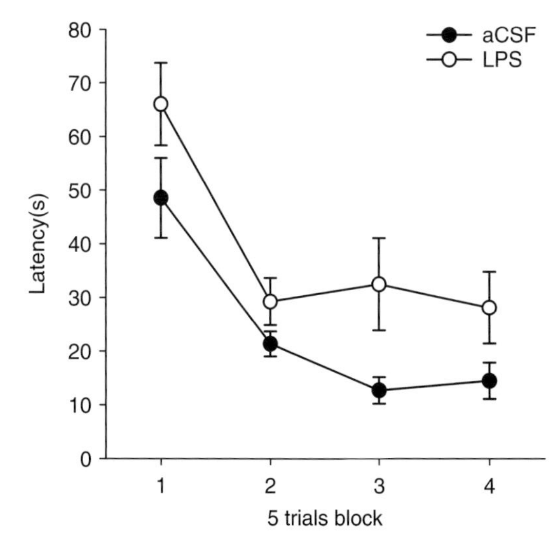

Neuroinflammation: Chronic Neuroinflammation and Memory Impairments. Figure 2 Assessment of spatial learning in rats with LPS infusion and rats with aCSF infusion. Mean latency (±) to reach the escape platform across four blocks of five training trials during the spatial learning task. Fischer-344 rats with LPS or aCSF infusion received three trails/day trainings (1 min intertrial interval, maximum trail duration of 90 s with 30 s on the platform at the end each trial) in the hidden platform training. Rats with LPS infusion perform more poorly in the spatial learning task than rats with aCSF. (courtesy of Jung-Soo Han).

inhibition of this enzyme might underlie the beneficial effects of NSAID therapy in AD [7].

Ibuprofen is a NSAID, and is widely used to reduce pain, fever, and inflammation. The drug inhibits COX enzymes and activates peroxisome proliferators-activated receptors gamma (►PPARγ); both of these actions result in reduced inflammation. In addition, ibuprofen suppresses cerebral plaque formation and inflammation in a mouse model of Alzheimer's disease [8]. However, a major limitation of NSAIDs, such as ibuprofen with respect to the prevention of AD is gastrointestinal and occasional liver and kidney toxicity caused by cyclooxygenase (COX-1) inhibition. These side effects have stimulated a search for alternative anti-inflammatory drugs and, alternatively, attempts to structurally modify existing NSAIDs so as to eliminate their COX-1 inhibition. Many modified NSAID or natural products are tested in animal models or clinically. Two alternatives among them are introduced below.

NO-flurbiprofen, a novel NSAID that lacks gastrointestinal side effects, attenuated the neuroinflammatory reaction and reduced inflammation-induced memory deficit in the chronic LPS infusion rat model [3]. β-Amyloid is also reduced in doubly transgenic (Tg) amyloid precursor protein plus presenilin-1 mice when NO-flurbiprofen is administered between 7 and 12 months of age [9].

Oxidative damage and neuroinflammation are closely associated with the progression of AD and other neurological diseases. In the search for antioxidant and anti-inflammatory agents to reduce ROS and inflammation, the phenolic antioxidant ►curcumin, a yellow curry spice derived from turmeric, has proved to be of interest. This spice is used as a food preservative and herbal medicine in India, where the prevalence of AD in patients between 70 and 79 years of age is 4.4-fold less than that of the United States. In comparison with vitamin E used clinically [10], curcumin is several times more potent as a free radical scavenger. Based on these considerations, curcumin has been evaluated in some animal studies. In an AD transgenic Tg2576 mouse model, curcumin lowers oxidized proteins and IL-1β. The astrocytic marker GFAP and β-amyloid is also reduced by curcumin treatment [11]. Preventive effects of curcumin on the cognitive deficits have also been tested in a rat model with intracerebroventricular infusion of β-amyloid peptides. Dietary intake of curcumin prevents β-amyloid-infusion induced spatial memory deficits in the Morris Water Maze [12].

Summary

Inflammation clearly occurs in the AD brain. In the periphery, degenerating tissue and the deposition of insoluble materials are classical stimulants of inflammation. Likewise, in the AD brain, damaged neurons, neuritis, deposits of insoluble β-amyloid, and neurofibrillary tangles stimulate the inflammation. Direct and bystander damage in AD is cumulated over many years and significantly exacerbates the pathological process. Thus, animal models and clinical studies suggest that inflammation in AD contributes to AD pathogenesis. While anti-inflammatory approaches (for example, NSAID) may not cure AD, it is possible that they will delay onset and slow the progression of this disease as well as slow cognitive decline.

References

1. McGeer EG, McGeer PL (2003) Inflammatory processes in Alzheimer's disease. Prog Neuropsychopharmacol Biol Psychiatry 27:741–749
2. Haddad JJ (2004) Mitogen-activated protein kinases and the evolution of Alzheimer's: a revolutionary neurogenetic axis for therapeutic intervention? Prog Neurobiol 73:359–377
3. Wenk GL, Hauss-Wegrzyniak B (2003) Chronic intercerebral LPS as a model of neuroinflammation. In: Wood PL (ed) Neuroinflammation, 2nd edn. Humana Press, Totowa, NJ, pp 137–150
4. Akiyama H, Barger S, Barnum S, Bradt B, Bauer J, Cole GM, Cooper NR, Eikelenboom P, Emmerling M, Fiebich BL, Finch CE, Frautschy S, Griffin WS, Hampel H, Hull M, Landreth G, Lue L, Mrak R, Mackenzie IR, McGeer PL, O'Banion MK, Pachter J, Pasinetti G, Plata-Salaman C, Rogers J, Rydel R, Shen Y, Streit W, Strohmeyer R, Tooyoma I, Van Muiswinkel FL, Veerhuis R, Walker D, Webster S, Wegrzyniak B, Wenk G, Wyss-Coray T (2000) Inflammation and Alzheimer's disease. Neurobiol Aging 21:383–421
5. Zhu SG, Sheng JG, Jones RA, Brewer MM, Zhou XQ, Mrak RE, Griffin WS (1999) Increased interleukin-1beta converting enzyme expression and activity in Alzheimer disease. J Neuropathol Exp Neurol 58:582–587
6. Rogers J, Webster S, Lue LF, Brachova L, Civin WH, Emmerling M, Shivers B, Walker D, McGeer P (1996) Inflammation and Alzheimer's disease pathogenesis. Neurobiol Aging 17:681–686
7. Ho L, Purohit D, Haroutunian V, Luterman JD, Willis F, Naslund J, Buxbaum JD, Mohs RC, Aisen PS, Pasinetti GM (2001) Neuronal cyclooxygenase 2 expression in the hippocampal formation as a function of the clinical progression of Alzheimer disease. Arch Neurol 58:487–492
8. Lim GP, Yang F, Chu T, Chen P, Beech W, Teter B, Tran T, Ubeda O, Ashe KH, Frautschy SA, Cole GM (2000) Ibuprofen suppresses plaque pathology and inflammation in a mouse model for Alzheimer' s disease. J Neurosci 20:5709–5714
9. Jantzen PT, Connor KE, DiCarlo G, Wenk GL, Wallace JL, Rojiani AM, Coppola D, Morgan D, Gordon MN (2002) Microglial activation and beta-amyloid deposit reduction caused by a nitric oxide-releasing nonsteroidal anti-inflammatory drug in amyloid precursor protein plus presenilin-1 transgenic mice. J Neurosci 22:2246–2254
10. Zandi PP, Anthony JC, Khachaturian AS, Stone SV, Gustafson D, Tschanz JT, Norton MC, Welsh-Bohmer KA, Breitner JC (2004) Reduced risk of Alzheimer disease in users of antioxidant vitamin supplements: the Cache County Study. Arch Neurol 61:82–88
11. Lim GP, Chu T, Yang F, Beech W, Frautschy SA, Cole GM (2001) The curry spice curcumin reduces oxidative damage and amyloid pathology in an Alzheimer transgenic mouse. J Neurosci 21:8370–8377
12. Frautschy SA, Hu W, Kim P, Miller SA, Chu T, Harris-White ME, Cole GM (2001) Phenolic anti-inflammatory antioxidant reversal of Abeta-induced cognitive deficits and neuropathology. Neurobiol Aging 22:993–1005

Neuroinflammation: Modulating Pesticide-induced Neurodegeneration

HASSAN R. DHAINI[1,2]

[1]Faculty of Health Sciences, University of Balamand, Aschrafieh, Beirut

[2]Faculty of Health Sciences, American University of Beirut, Lebanon, Beirut

Synonyms

Biomagnification: high environmental persistence; Reactive oxygen species: superoxide anions; Hydroxyl free radicals; Pesticide: insecticide; Herbicide

Definition

The term ►pesticide describes a chemical capable of being used to control pests to humans, agricultural crops, commercial operations, and households.

Promoting research in this area has resulted in a clear understanding of pesticide toxicological properties. Today, pesticides are considered to be one of the most thoroughly understood chemicals from a toxicological standpoint. Modern strategies in pest control have attempted to design compounds of high selectivity and low ►environmental persistence. Ideally, a pesticide would be selective only to the targeted species and would be non-persistent in the environment. However, these attempts have not been entirely successful. Non-persistent insecticides, such as carbamates and organophosphates, are currently in use but they are not considered as truly selective compounds. Most pesticides currently in use are toxic to humans, and present an environmental and occupational hazard [1].

Many of the used pesticides, whether applied as systemics, aerosols, baits, or fumigants, have been associated with some form of nervous toxicity. Moreover, increasing indications suggest that the pathogenesis of a number of chronic neurodegenerative diseases such as Idiopathic Parkinson's disease (PD), Alzheimer's disease (AD), multiple sclerosis, trauma, and stroke, may be influenced by exposure to infectious agents and pesticides. On the other hand, emerging evidence indicates that the development of these neurological disorders may be mediated by a complex cycle of atypical inflammation steps involving brain immune cells, mainly astrocytes and microglia.

Microglias, considered as the macrophages of the central nervous system (CNS; brain), are normally present in a down-regulated state, and serve the role of immune surveillance. When exposure to an environmental toxicant takes place, microglia change morphology and become active in phagocytosis and in producing inflammatory molecules. In parallel, astrocytes, which normally maintain neuronal homeostasis, also become active and serve in up-regulating the expression of neurotrophic factors and local mediators, and limiting the area of injury. Astrocytes are believed to react whenever the brain is injured by putting down glial scar tissue as part of healing. This whole process, mechanistically distinct from peripheral tissues inflammation, is known as neuroinflammation.

Whether neuroinflammation is harmful or protective to the nervous system, especially in cases of exposure to pesticides, is still a controversial issue. Studies using strategies aimed at both suppressing and inducing the process of neuroinflammation may be successful in identifying new treatments for common neurodegenerative diseases.

Characteristics

Description of Neuroinflammation

Inflammation is the first response of the immune system to infection or irritation. It is characterized by redness (*rubor*), heat (*calor*), swelling (*tumor*), and pain (*dolor*), and constitutes the body's initiation of healing. Although inflammation is considered a defensive mechanism, an exaggerated inflammatory response has been shown to cause additional injury to cells of the host. In the adult CNS, mainly the brain, damage often leads to persistent deficits due to the inability of mature post-mitotic axons to regenerate after injury [2]. This makes host cells in this particular area highly susceptible to injury and inflammatory processes. Inflammation associated with the CNS, known as neuroinflammation, differs from that found in the periphery. It does not involve any pain due to absence of pain fibers in the brain, and does not show any classic signs of inflammation such as redness, swelling, or heat. The process is usually mediated by cytokines following a direct injury to the nervous system, and systemic tissue injury in rare cases. Neuroinflammation involves neural-immune interactions that activate immune cells, glial cells, and neurons in response to injury [3].

Activation of Astrocytes and Neurons

Astrocytes are considered as the most abundant cells in the brain. They are involved in maintaining the functional integrity of neuronal transmission and general activity. Astrocytes become activated in response to brain injuries and produce many pro-inflammatory molecules such as interleukins, prostaglandins, leukotrienes, thromboxanes, coagulation factors, complement proteins, and proteases. In addition, activated astrocytes promote repair of damaged tissues. Certain chemokines released by activated astrocytes attract microglia, which amplifies production of pro-inflammatory molecules. On the other hand, neurons themselves participate in the production of inflammatory molecules, mainly complement proteins.

An increasing amount of data has shown that the production of these molecules by astrocytes and neurons may in fact create an oxidative stress microenvironment that leads to neuronal toxicity and cell death. Studies on neuro-immune-endocrine interactions have shown that the hypothalamic-pituitary-adrenal axis (HPA) plays a key role in protecting cells from oxidative stress through suppression of redox-sensitive transcription factor, nuclear factor (NF)-$_k$B [4]. Several studies have demonstrated that the accumulation of pro-inflammatory and cytotoxic factors by activated glia induces ►neural degeneration. Neurotoxicity has been associated with high levels of nitric oxide (NO), superoxide anions, and other toxic intermediates [5]. Reports on lipid-derived mediators of inflammation, mainly prostaglandins E_2 and I_2, were shown to induce edema, which is deleterious to neuronal function and survival. In addition, several ►reactive oxygen species, such as superoxide anion and hydroxyl radical, were found to be released as byproducts of cyclooxygenase Cox-2 catalytic activity, a key enzyme in inflammatory response, thus leading to brain damage [6]. Emerging experimental evidence demonstrates that the inhibition of the inflammatory response can slow down degeneration of dopamine-containing neurons in PD models. In fact, the use of anti-inflammatory steroids was reported to decrease the production of cytokines and NO, and consequently attenuating degeneration of dopamine-containing neurons in models mimicking PD [7]. Another study conducted in rat animal models has shown that inhibition of neurotoxic factors production, such as tumor necrosis factor alpha (TNF-α), superoxide, and NO, reduces damage to dopamine-containing neurons [8].

Role of Reactive Microglia in Neuroinflammation

The activation of microglia is seen as a major step in brain inflammation. Microglial cells support and protect neurons of the CNS. They are mainly composed of mesodermally-derived macrophages and are able to release a number of inflammatory molecules including pro-inflammatory cytokines, chemokines, superoxide anions, and complement proteins. In addition, microglia have phagocytic and surveillance properties. Studies have shown that microglias are sensitive even to minor disturbances in CNS homeostasis [9]. More importantly, microglia play a key role in cellular responses to pathological lesions such as those of dopaminergic neurons in PD, and such as extracellular deposits of β-amyloid and intracellular neurofibrillary plaques seen in AD. These lesions can recruit and activate microglia around them in the brain. In addition, microglia can express scavenger receptors that facilitate their adhesion to injury sites [10,11].

Reactive microglia exert a protective role through clearance of cellular debris, destruction of foreign particles and release of neurotrophic factors such as the glia-derived neurotrophic factor (GDNF), and the insulin-degrading enzyme (IDE) that destroys damaged protein deposits. However, several studies have demonstrated that the production and buildup of pro-inflammatory and cytotoxic factors may have a negative impact on neurons leading to neurodegeneration. Resulting neurotoxicity is thought to be caused by increased production of NADPH-derived superoxide anions, which leads to additional neuronal damage. In addition, neurotoxicity is aggravated by microglial release of interleukins IL-1, IL-6, IL-8, tumor necrosis factor alpha (TNF-α), and other inflammatory proteins. In addition, increased levels of nitric oxide (NO) have been associated with neurotoxicity [12].

Association between Pesticide Exposure and Neurodegeneration

All chemical pesticides in use today are poisonous to the nervous system of the target species. Pesticides are not highly selective and may affect humans. Exposure to pesticides has long being suspected as a risk factor for a number of neurodegenerative diseases including PD and AD [13]. The identification of chemicals inducing neurodegeneration symptoms, as in the case of 1-methyl-4-phenyl-1,2,3-tetrahydropyridine MPTP-induced Parkinsonian symptoms, is in support of the search for environmental factors at the basis of neurodegenerative diseases. Extensive literature suggests that exposure to pesticides is a risk factor for PD and AD. Many studies have established an association of PD risk with living in rural areas, drinking well water, and farming.

Neurotoxicity of Organochlorines in PD

Organochlorines are a diverse group of agents belonging to three distinct chemical classes that include dichlorodiphenylethane, chlorinated cyclodienes, and chlorinated benzenes. Organochlorines are effective pesticides due to low volatility, chemical stability, lipid solubility, and slow rate of biotransformation and degradation. As a result, these compounds have high persistence and high biomagnification which makes them a serious hazard. Several studies have linked exposure to organochlorines and increased risk of neurodegeneration [14].

One organochlorine, dieldrin, has been detected in postmortem brain samples from PD patients. Studies have shown that dieldrin exhibit selective dopamine-depleting neurotoxicity effects by causing superoxide formation and ►lipid peroxidation. Similarly, dichlorodiphenyl-trichloroethane (DTT), another commonly used organochlorine, has been detected in postmortem brain samples from AD patients. Studies on other members of the organochlorine family showed that production of a direct toxic effect is a function of individual genetic factors, mainly drug-metabolizing enzymes such as cytochromes P-450 (►cytochrome P-450) genetic polymorphism, in addition to frequency of exposure to pesticides [15].

Description of Organophosphates-Induced Neurodegeneration

Currently used organophosphorus ester insecticides (OP) are at least four generations of development away from the early nerve gases. Organophosphates elicit their toxicity through inhibition of acetylcholinesterase AChE, the enzyme responsible for terminating the activity of the neurotransmitter acetylcholine at the level of post-synaptic neurons. Case reports have associated development of Parkinsonism with exposure to organophosphate insecticides.

In addition to general central nervous toxicity symptoms, OPs have been shown to cause a persistent neuropathy with a delayed onset known as organophosphate-induced delayed neuropathy (►OPIDN). This involves slow degeneration of the nervous system seven to fourteen days after exposure to certain OPs at high doses. Causes of OPIDN are still not very well understood; it is hypothesized that OPIDN is caused by the inhibition of Neuropathy Target Esterase (NTE), an enzyme involved in lipid metabolism. In cases of chronic exposure, initial binding of the pesticide to the enzyme is reversible. However, the AChE-OP complex might undergo what is commonly known as "aging." This occurs when the complex dealkylates itself to form an irreversibly inhibited AChE enzyme ultimately leading to neurodegeneration in the axons [16].

Mechanisms of Neurodegeneration by Rotenone

Rotenoids are another class of insecticide whose environmental exposure has been associated with increased neurodegeneration. The naturally occurring rotenone is a widely used rotenoid that inhibits complex I, the first enzyme of the mitochondrial respiratory chain. Complex I dysfunction is a feature of idiopathic PD and is linked to many other neurodegeneration disorders, such as that of retinal ganglion cells in ►Leber's optic neuropathy. Moreover, exposure to rotenone in rats has been shown to produce highly selective neural degeneration similar to that found in PD. In recent years, several studies have demonstrated that continuous exposure to rotenone in rats leads to degeneration of the nigrostriatal dopaminergic system accompanied by movement disorders. In addition, rotenone was shown to exhibit a markedly high toxicity by activating microglia, which releases superoxide free radicals and facilitates degeneration of dopaminergic neurons. Further studies using enzyme inhibitors suggest that rotenone-induced release of superoxide is mediated by microglial NADPH oxidase, a major superoxide generator in immune cells of the nervous system. In addition, rotenone and certain inflammogens have been reported to exert synergistic dopaminergic neurotoxicity [17].

Association between Exposure to Paraquat and PD

Most herbicides are formulated to be toxic to plant biochemical systems that are absent in mammals. Human exposure to paraquat (PQ), a very commonly used weed killer, is known to cause lung fibrosis in addition to liver and kidney damage. Lately, studies on PQ have suggested that this herbicide may be an environmental factor contributing to PD. Paraquat was found to cause damage through generation of highly toxic superoxide anion. Although the biochemical mechanism is not yet fully understood, some evidence suggests that PQ-induced lipid peroxidation and resulting cell death of dopaminergic neurons may underlie the onset of the Parkinsonian syndrome or, to the least, influence the natural course of the disease [18].

The Role of Inflammation in Modulating the Effect of Pesticides

The activation of microglia and astrocytes observed in patients with neurodegenerative disorders, mainly PD and AD, and in animal models suggest an involvement of neuroinflammation in the progression of these diseases. The observed synergism in neurotoxicity between pesticides and inflammogens, in addition to pesticide-pesticide interaction, may support

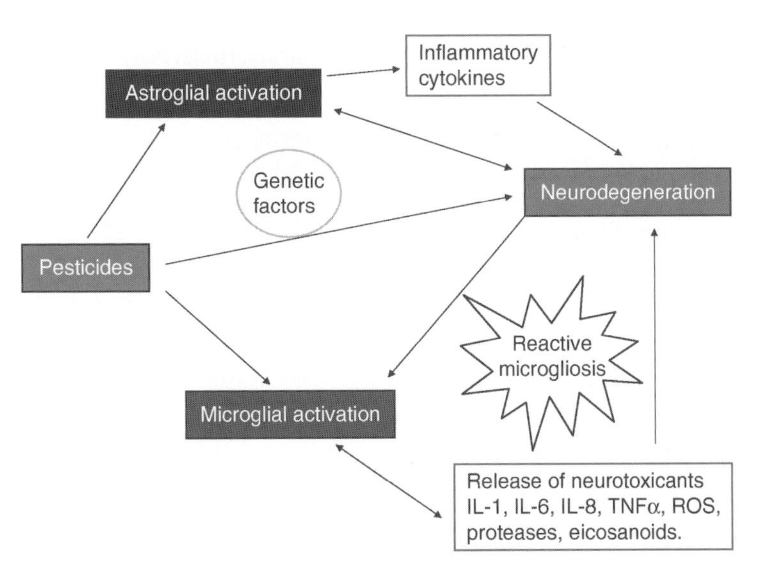

Neuroinflammation: Modulating Pesticide-induced Neurodegeneration. Figure 1 Proposed mechanism of pesticides-induced neurodegeneration and deleterious glial modulation. Exposure to pesticides induces astrocytes and microglia activation while directly causing neuronal injury. Activated glia produce a wide array of pro-inflammatory factors and neurotoxicants, mainly cytokines, Interleukins, reactive oxygen species, proteases, and eicosanoids, which aggravate neuronal damage. This in turn will lead to further microglia activation. A self-propelling cycle is formed: microglial activation cause neurodegeneration while neuronal injury due to direct interaction with pesticides leads to additional glial activation; this further exacerbates neurodegeneration. Abbreviations: *IL*, interleukins; *TNF-α*, tumor necrosis factor-alpha; *ROS*, reactive oxygen species.

N

a multifactorial hypothesis underlying observed neurodegeneration. Experimental evidence showing that inhibition of inflammation correlates with attenuated neuronal damage supports such notion.

Following exposure to pesticides, microglia and astroglia are activated, thus releasing a wide range of neurotoxic endogenous factors, such as superoxide and cytokines. These factors bind directly to their receptors on the targeted neurons to activate an apoptotic pathway. In addition, cytokines lead to release of NO through the induction of NO synthase and Cox-2 within glial cells. Nitric oxide may react with superoxide to form a more potent intermediate: peroxynitrite ($ONOO^-$). Peroxynitrite can cross the cell membrane and cause neuronal injury. In addition, superoxide may convert to hydrogen peroxide (H_2O_2), thus adding to neuronal toxicity directly or by amplifying other neurotoxic factors in microglia [19]. Overall, activated glial cells produce a variety of pro-inflammatory and neurotoxic factors that aggravate neuronal damage. At the same time, direct interaction of pesticides with nerve cells causes neuronal injury. Damaged neurons stimulate an inflammatory response including reactive microgliosis and astrogliosis. These, in turn, will cause further neuronal damage. In fact, regardless of the origin of triggering factors, a vicious cycle is created. Pesticide-induced neuronal damage and neuroinflammation amplify each other in the form of a self-propelling vicious cycle (Fig. 1). Microglial activation leads to neurotoxicity while neuronal injury due to interaction with pesticides leads to additional glial activation; this further exacerbates neurotoxicity leading ultimately to neurodegeneration [20]. The progress of such cycle over a long period of time, especially in cases of chronic or occupational exposure to pesticides, in addition to potential genetically predisposing factors such as drug-metabolizing enzymatic pathway alterations, may lead to synergistic neurodegeneration and the development of symptomatic PD and AD [21].

Much remains to be investigated about the role of pesticides in neurodegeneration, and the potential role of microglial cells in the development of neurodegenerative diseases, mainly PD and AD. However, one thing is certain, the brain's immune system is deeply involved in both diseases, and further studies on neuroinflammation seems promising in contributing significantly to the discovery of new treatments.

References

1. Timbrell J (2002) Introduction to toxicology, 3rd edn. CRC Press, Boca Raton, FL, pp 109–120
2. Hiu G, He Z (2006) Glial inhibition of CNS axon regeneration. Nat rev Neurosci 7(8):617–627
3. Haddad JJ (2007) On the enigma of pain and hyperalgesia: a molecular perspective. Biochem Biophys Res Commun 353(2):217–224
4. Haddad JJ, Saade NE, Safieh-Garabedian B (2002) Cytokines and neuro-immune-endocrine interactions: a role for the hypothalamic-pituitary-adrenal revolving axis. J Neuroimmunol 133(1–2):1–19
5. Gayle DA, Ling Z, Tong C, Launders T, Lipton JW, Carvey PM (2002) Lipopolysaccharide (LPS)-induced dopamine cell loss in culture: roles of tumor necrosis factor-alpha, interleukin-1beta, and nitric oxide. Brain Res Brain Res Rev 133:27–35
6. Basu N, Krady J, Levison S (2004) Interleukin-1: a master regulator of neuroinflammation. J Neurosci Res 78:151–156
7. Kurkowska-Jastrzebska I, Wronska A, Kohutnicka M, Czlonkowski A, Czlonkowska A (1999) The inflammatory reaction following 1-methyl-4-phenyl-1,2,3,6-tetrahydropyridine intoxication in mouse. Exp Neurol 156:50–61
8. Wang MJ, Lin WW, Chen HL, Chang YH, Ou HC, Kuo JS, Hong JS, Jeng KC (2002) Silymarin protects dopaminergic neurons against lipopolysaccharide-induced neurotoxicity by inhibiting microglia activation. Eur J Neurosci 16:2103–2112
9. Tuppo E, Arias H (2004) The role of inflammation in Alzheimer's disease. The International J Biochem Cell Biol 37:289–305
10. Minghetti L (2005) Role of Inflammation in neurodegenerative diseases. Curr Opin Neurol 18:315–321
11. Streit W, Conde J, Fendrick S, Flanary B, Mariani C (2005) Role of microglia in the central nervous system's immune response. Neurol Res 27:685–691
12. Liu B, Hong J (2003) Role of microglia in inflammation-mediated neurodegenerative diseases: mechanisms and strategies for therapeutic intervention. J Pharmacol Exp Ther 304:1–7
13. Kamel F, Hoppin J (2004) Association of pesticide exposure with neurologic dysfunction and disease. Environ Health Perspect 112:950–958
14. Sanchez-Ramos J, Facca A, Basit A, Song S (1998) Toxicity of dieldrin for dopaminergic neurons in mesencephalic cultures. Exp Neurol 150:263–271
15. Kanthasamy A, Kitazawa M, Kanthasamy A, Anantharam V (2005) Dieldrin-induced neurotoxicity: relevance to Parkinson's disease pathogenesis. Neuro Toxicology 26:701–719
16. Ecobichon D (2003) Toxic effects of pesticides. In: Klaassen C, Wakins J III (eds) Casarett & Doulls essentials of toxicology, 2003 edn. McGraw-Hill, New York, pp 333–347
17. Gao H, Hong J, Zhang W, Liu B (2003) Synergistic dopaminergic neurotoxicity of the pesticide rotenone and inflammogen lipopolysaccharide: relevance to the etiology of Parkinson's disease. J Neurosci 23:1228–1236
18. Dinis-Oliveira R, Remiao F, Carmo H, Duarte J, Navarro A, Bastos M, Carbalho F (2006) Paraquat exposure as an etiological factor of Parkinson's disease. Neurotoxicology 27(6):1110–1122
19. Liu G, Gao H, Hong J (2003) Parkinson's disease and exposure to infectious agents and pesticides and the occurrence of brain injuries: Role of neuroinflammation. Environ Health Perspect 111:1065–1073
20. Gao H, Liu B, Zhang W, Hong J (2003) Novel anti-inflammatory therapy for Parkinson's disease. Trends Pharmacol Sci 24:395–401

21. Thiruchelvam M, Richfield EK, Baggs RB, Tank AW, Coryslechta DA (2000) The nigrostriatal dopaminergic system as a preferential target of repeated exposures to combined paraquat and maneb: implications for Parkinson's disease. J Neurosci 20:9207–9214

Neurokinin-1 Receptor (NK1R)

Definition

The tachykinin receptor that is most selective for the endogenous neuropeptide ligand substance P.

►Tachykinins

Neurokinin-2 Receptor (NK2R)

Definition

The tachykinin receptor that is most selective for the endogenous neuropeptide ligand neurokinin A.

►Tachykinins

Neurokinin-3 Receptor (NK3R)

Definition

The tachykinin receptor that is most selective for the endogenous neuropeptide ligand neurokinin B.

►Tachykinins

Neurolab Project

Definition

An international research project on neuroscience carried out in the space shuttle Columbia launched in 1998.

►Autonomic Function in Space

Neurolabyrinthitis

Definition

Inflammation of the neural structures of the labyrinth in the inner ear.

►Disorders of the Vestibuar Periphery
►Peripheral Vestibular Apparatus

Neuroleptic Drugs

Definition

Drugs used to treat psychosis (also called antipsychotic drugs).

►Antipsychotic Drugs

Neuroligins

Definition

Neuroligins are localized to the surface of postsynaptic membranes. They play a role in pre- and postsynaptic differentiation and maintain the functional balance of excitatory and inhibitory synapses.

►Synapse
►Synapse Formation: Neuromuscular Junction Versus Central Nervous System

Neurolipomatosis

Definition

A condition characterized by the formation of subcutaneous multiple fat deposits, with pressure on the nerves resulting in tenderness, pain, and ►paresthesias.

Neurology

William J. Spain
Department of Neurology, University of Washington, Veterans Affairs Puget Sound Health Care System, Seattle, WA, USA

Introduction

Neurology is the main clinical branch of neuroscience. It is a medical discipline practiced by health care professionals. Neurology focuses on the diagnosis and treatment (excluding surgery) of people with afflictions of the central and peripheral nervous system and the skeletal muscles. As more is learned about the biological bases of disorders of thought and behavior, the boundaries between the practice of neurology and psychiatry are becoming less distinct. Medical doctors who practice neurology are called neurologists. To become a neurologist one must first complete a medical degree, followed by a general medical internship. Subsequent resident training normally entails at least 3 years of additional supervised training in providing care for people with neurological disorders. In most countries, there is a governing body that certifies a person as qualified to practice Neurology (boarding). This boarding organization typically sets the training requirements, administers an initial set of exams and periodic recertification exams to insure that those who have gone through the prescribed training have successfully incorporated a broad knowledge base of the elements of neurology and that they are capable of applying that knowledge base to the care of individuals with neurological diseases.

A Brief History of Neurology

The history of neurology is ancient and colorful [1]. A few key milestones are listed here. The earliest known written reference to injury and treatment of the nervous system is contained in the Egyptian Edwin Smith Papyrus that was written around 1700 B.C. It is believed by some to be a copy of work done about a 1000 years earlier by Imhotep, the founder of Egyptian medicine. In ~400 B.C. Hippocrates wrote a treatise entitled *On The Sacred Disease* in which he debunked the idea that epilepsy is a spiritual affliction. He recognized that seizures come from abnormal functioning of the brain. About 200 years later Galen, another Greek physician, documented one of the earliest lesion studies to determine neurologic function. He noted that when the recurrent laryngeal nerve is cut, the "voice of the animal is damaged and its resonance is lost." However, it took about 2000 more years for Neurology to become a well defined and distinct medical discipline. The term neurology was coined by Thomas Willis in his 1664 book on the anatomy of the brain and nerves, *Cerebri anatomi*. This document laid the framework for modern neuroanatomic and neurophysiologic investigations. Neurology became a formal clinical discipline in the 1800's led by the writings and teachings of Jean-Martin Charcot in France, John Hughlings Jackson in England and Silas Weir Mitchell in America. Their observations and methods of clinical evaluation of malfunction of the nervous system formed the basis for the current practice of neurology.

How Neurology is Practiced

A neurologist begins the process of diagnosis by first *talking* to a person who has experienced some set of symptoms, thus obtaining their medical history. The history is obtained in a question and answer format starting with general, open ended questions designed to screen a wide variety of potential problems, followed by more specific questions customized to a person's specific complaints. To paraphrase the great American neurologist H. Houston Merritt of The Neurologic Institute of New York, "the medical history when applied diligently is the most important and revealing tool a doctor has. In the vast majority of cases, a detailed history obtained by a knowledgeable neurologist will usually provide a good idea about the etiology of a person's symptoms. The physical exam and laboratory tests are for the most part confirmatory."

Armed with the information obtained from the history, the neurologist can then form a set of testable hypotheses as to the etiology of the person's complaints. With such a framework, the neurologist makes a set of predictions as to what will be observed on the physical exam and on further laboratory testing. During the physical exam, the neurologist first observes the body structure for asymmetries and abnormalities. Next is a systematic check of mental function, speech, sensory perception and motor functioning obtained by having the patient perform a standard set of simple behaviors or tasks. Because the nervous system is organized through interconnecting circuits, the neurologist (like an electronics repairman) can draw on their detailed knowledge of the nervous system's organization to localize and identify the abnormal structures, cell types and molecules. The logic of diagnosis relies on recognizing common denominators that cause the patients constellation of signs and symptoms from which the neurologist makes a list of the most probable etiologies. This list is called the differential diagnosis.

Sometimes, laboratory testing is needed to narrow the list or confirm the suspected diagnosis. Because there are literally hundreds of tests to choose from, it is essential that the neurologist have a well formed and justified hypotheses in order to keep the workup focused and thereby avoid a lengthy and expensive fishing expedition. The spectrum of tests routinely available include: examining the blood and cerebral spinal fluid for chemical abnormalities, infection and acquired or inherited molecular changes, tissue biopsy for histological examination of cells and molecules, imaging of the internal structures using x-rays, nuclear magnetic resonance signals and scanning techniques for tracer uptake, and physiological testing which can measure electrical activity and the dynamics of metabolism or blood flow in the central nervous system. The results of such testing usually either confirm or confute the initial hypotheses. Throughout the evaluation of the patient, the neurologist draws on prior experience and knowledge of the published experience of others (both anecdotal and results of controlled studies) in order to weigh the historical information, physical findings and results of testing to determine the most probable cause of the disease. Finally, the neurologist chooses the best available course of treatment.

Treatment of Neurological Disease

In the mid 1900s, neurology was primarily a discipline of diagnosis. Knowledge of the chemical and molecular basis of neurologic disease was so limited that only a few effective pharmacologic agents were available for treatment. As a consequence, neurologists were frequently the butt of jokes and sarcasm amongst other physicians; that all they were good for was informing their patients what was wrong with them and what type of suffering and deterioration to expect over what remained of their waning existence. Fortunately, the vast advances in basic and applied neuroscience research in the past 20 years have resulted in numerous effective remedies for diseases of the nervous system. There is, of course, still a long ways to go, as there are no effective means for altering the course of Alzheimer's disease, amyotrophic lateral sclerosis, glioblastoma multiforme and other debilitating diseases of the nervous system.

The treatment of neurological disorders falls into two broad categories; curing the underlying disease and alleviating the symptoms that result in suffering and loss of function. Choosing the best course of treatment is a mixture of applied science, empathy and common sense. The scientific component of treatment is based on the results of controlled studies of the various treatment options in specific situations. The empathy and common sense components of providing care to individuals are sometimes referred to as "the art of medicine." In addition, there is another important aspect to treatment of neurological diseases, namely prevention. Regrettably, patients rarely contact neurologists until they are ill, so that preventative treatment is only a small part of what the typical neurologist does in practice (except for the small number of neurologists involved in public health and education).

Development of Knowledge in Neurology

Until recently much of the knowledge encompassed by the discipline of neurology was derived from the observation of human beings (i.e., uncontrolled experience and "chance experiments of nature"). As such, the descriptions of neurological diseases are often phenomenological. This stands in marked contrast to the basic neurosciences that are founded in the scientific method of asking questions and then drawing conclusions from the results of controlled, hypothesis driven experimentation. Although neurology and basic neuroscience have these fundamental differences, many of the questions asked by basic neuroscientists are motivated by observations that were initially made by neurologists. Further, the comparison of how things work during disease versus during times of health forms a powerful foundation for forming hypothesis. For example, the observation of overactive deep tendon reflexes in patients with motor weakness due to stroke compared to the underactive reflexes in people with weakness due to polio has led to numerous studies of the mechanisms underlying the coupling of somatosensory input to motor output. Similarly, posing the question "why do people who suffer from Alzheimer's disease have problems remembering things?" has motivated many basic studies that have provided insight into how animals learn and remember things. Progress in our understanding of how the brain works has been driven by the combination of innate curiosity and pragmatics.

Classically, the development of knowledge about the normal and abnormal workings of the nervous system followed a method of careful observation of people with neurologic illness for correlative symptoms. These methods are still in practice and complement the modern revolution in laboratory based methods of obtaining knowledge. First, people with similar constellations of signs and symptoms are grouped into syndromes or categories of disease. Then correlations with these abnormal phenotypes are looked for in two main arenas. One involves careful pathological examination of the structures of the nervous system at a gross, cellular and (now with modern techniques) at a molecular level. If differences are found between diseased and normal individuals, they can be correlated with the syndromic grouping and thus, serve to define structures and molecules that are essential to a particular function. So for example, the observation that a lesion in the cerebellum leads to problems with the coordination of motor activity provided insight into the function of the cerebellum. This method of correlation has been advanced by non-invasive methods for high resolution imaging of structures in living patients. The quality of imaging has advanced such that it can provide information formally obtained only from pathological specimens after death (e.g., like Hypocrites did more then 2000 years ago when he correlated the occurrence of epilepsy in individuals with injury to the brain).

Another method of gaining insight from correlation is to find common patterns of behavior and culture within a syndromic grouping (i.e., epidemiology). A notable example of this type of analysis [2] concerned

N

a recent incident when more than one hundred people on Prince Edward Island in Canada became similarly ill with headaches followed by confusion, loss of memory, disorientation, and (in some cases) seizures, coma and death. Their common experience was the recent consumption of cultured blue mussels. This "clue" led to some laboratory-based detective work that identified the accumulation of demoic acid in the mussels as the cause of the illness. Further experimentation revealed that demoic acid is a powerful excitotoxin that activates glutamate receptors. The hippocampus is particularly susceptible to the actions of demoic acid thus explaining the loss of memory and the seizures. (Interestingly, this syndrome which is now known as "amnesic shellfish poisoning" is thought to have caused the 1961 attack of the seaside town of Capitola, California by hundreds of crazed birds – the incident that inspired Alfred Hitchcock's movie *The Birds*.)

Occasionally, a single dramatic case unlocks a mystery of how the nervous system works. Perhaps the most famous is the oft cited case of Phineas P. Gage [3]. In 1848 Gage, while doing railroad work, was the victim of a mishap in which a tamping iron passed through his skull, and the frontal lobes of his brain. This injury changed his personality so much so that his friends said he was "no longer Gage." The scientific reporting of Gage's behavioral changes and associated pathology led to changes in perception about the function of the frontal lobes, particularly with regards to their role in emotion and personality. Before Gage's accident, most scientists thought the frontal lobes had little or no role in behavior. A limitation of the knowledge that comes from such chance "experiments" is that interpretation is confounded by the lack of controls performed in parallel. The case of Phineas Gage was partially responsible for simplistic reasoning behind frontal lobotomies as a cure for unwanted behavior. Lobotomy as a medical treatment has been abandoned, but not before it resulted in many undesirable and irreversible outcomes. Fortunately, neurology has moved beyond the insights it gained as a primarily observational scientific discipline. Now there is a large academic branch of experimental neurology that uses animal-based models of disease and modern methods of analysis of population data.

References

1. McHenry LC Jr (ed) (1969) Garrison's history of neurology. Charles C Thomas Pub Ltd
2. Perl TM, Bard L, Kosatsky T, Hockin JC, Todd E, Remis RS (1990) An outbreak of toxic encephalopathy caused by eating mussels contaminated with domoic acid. New Engl J Med 322:1775–1780
3. Harlow JM (1868) Recovery from a passage of an iron bar through the head. Publ Mass Med Soc 2:327–347

Neuroma

Definition

Tumor in the nervous system. Here referred to the tumor-like structure formed at the end of an injured peripheral nerve, in which some or all of its axons are unable to regenerate to the target tissue. Many lesions formerly called neuromas are now given more specific names such as ganglioneuroma, neurilemmoma, or neurofibroma.

►Neuronal Changes in Axonal Degeneration and
►Regeneration

Neuromalacia

Definition

Necrosis and softening of nerves.

Neuromast

Definition

The sense organ of the mechanosensory lateral line system. Sensitive to minute displacements of an apical cupula.

►Evolution of Mechanosensory and Electrosensory Lateral Line Systems

Neuromast Cell

Definition

Hair cell of the lateral line system.

►Evolution of the Vestibular System

Neuromatosis

Definition

Any disease characterized by multiple ►neuromas.

►Neuroma

Neuromelanin

Definition
Neuromelanin is a brown-black intracellular polymeric pigment derived from dopamine or norepinephrine found within catecholaminergic neurons.

►Melanin and Neuromelanin in the Nervous System

Neuromelanosome

Definition
Neuromelanosome denotes an aggregate of brownblack intracellular pigment granules of varying sizes (0.5–2.5 μm) associated with tightly bound protein and lipid components found within some dopaminergic and noradrenergic neurons.

►Melanin and Neuromelanin in the Nervous System

Neuromere

Definition
Segmental unit of the developing brain. In vertebrates, constrictions seen along the neuraxis. Neuromeres in the hindbrain region are called rhombomeres and have been shown to be lineage restriction units. In insects neuromeres of the thorax and abdomen are largely stereotypical and correspond to the body segments.

Neuromeres of the head are more complex structured.

►Evolution of the Brain: In Fishes
►Evolution of the Brain: Urbilateria
►Evolution of the Telencephalon: In Anamniotes

Neuromeric Model

Definition
Assumes transverse (neuromeres) as well as longitudinal units (roof, alar, basal, floor plates) along the entire anteroposterior neural tube axis, and that their arrangement is guided by selective regulatory gene expression that allows for regionalized developmental processes.

►Evolution of the Brain: In Fishes
►Evolution of the Telencephalon: In Anamniotes

Neuromimes

Definition
Electronic circuits or instruments that mimic the action of neurons or brains.

Neuromodulation

Definition
Actions that change the baseline intrinsic properties of neurons and synapses. Neuromodulators alter the firing properties of neurons (for example, from silent to bursting) and change the strength of synapses. They often act through second-messenger mechanisms such as protein phosphorylation.

Neuromodulation in the Main Olfactory Bulb

Nathalie Mandairon, Anne Didier
Neurosciences Sensorielles, Comportement, Cognition, Université de Lyon, Université Claude Bernard Lyon1, Villeurbanne, France

Synonyms
Centrifugal or feed-back connections to the main olfactory bulb

Definition
Centrifugal connections to the main olfactory bulb refer to the fiber systems originating from central structures, projecting to main olfactory bulb and contributing to olfactory processing.

Characteristics

Noradrenaline

Anatomical Organization

The mammalian main olfactory bulb (MOB) receives a significant noradrenergic input from the ►locus cœruleus. In fact, studies in rat show that approximately 40% of locus cœruleus neurons project to MOB. Noradrenergic fibers are localized in the subglomerular layers where they terminate densely in the internal plexiform and the granule cell layers, and moderately in the external plexiform and mitral cell layers. There are three known classes of noradrenergic receptors: α1, α2 and β. α1 receptors are particularly dense in the external plexiform layer, and moderate in mitral cell layer and granular cell layer. Cellular localization studies demonstrate that mitral/tufted and granule cells express α1, α2 and β receptors.

Cellular Effects

Dendrodendritic reciprocal synapses between mitral and granule cells in the MOB have been recognized as a critical locus where noradrenaline influences the processing of olfactory information.

In the turtle or rat dissociated MOB cultures, noradrenaline disinhibited mitral cells. This effect was attributed to α2 receptor-mediated presynaptic inhibition of granule cell dendrites. Moreover, it has been shown a direct excitatory action on mitral cells by iontophoretic application of noradrenaline in the MOB or by activation of α1 adrenergic receptors in rat [1]. However, field potential studies suggested that noradrenaline, acting at α1 receptors, depolarized granule cells, an effect that would inhibit mitral cells. Finally, in the rat, locus cœruleus activation decreases spontaneous mitral cell discharge but enhances the responses to weak olfactory input.

Locus cœruleus stimulation was also reported to initially decrease and then increase paired-pulse depression of mitral cell-evoked field potentials in the granule cell layer via activation of β receptors in the MOB [2]. It was concluded that noradrenaline release initially decreases then increases mitral cell glutamate release onto granule cells. It should be noted however, that other studies have reported that β receptor agonists have no effect on mitral cells or mitral-to-granule cell transmission. Noradrenaline may thus support opposing actions on the output neurons depending on the type or subtype of activated receptors.

Functional Implications

It now well established that the action of the noradrenaline on the MOB is critical for different kinds of olfactory learning. Olfactory cues trigger rapid increases in noradrenaline levels in the olfactory bulb. Noradrenaline release in the main and accessory OB is critical for the formation and/or recall of specific olfactory memories, pheromonal regulation of pregnancy, postpartum maternal behavior and rapid learning of conditioned odor preferences thought β receptor activation in early postnatal rodents [3]. In addition, noradrenaline levels are increased by sensory deprivation, possibly in order to increase the sensitivity of the mitral cells, in line with their increased response to weak stimuli reported under locus cœruleus stimulation (see above).

Acetylcholine

Anatomical Organization

The cholinergic innervation of the MOB is exclusively extrinsic and originates in the horizontal limb of the ►diagonal band of Broca (Ch3) as demonstrated by the absence of choline acetyl transferase (ChAT)-positive cells revealed by immunohistochemistry or *in situ* hybridization. Using either of these markers or localization of binding to high-affinity uptake sites, cholinergic fibers are found throughout the different layers of the MOB but with great laminar variations: the highest density is found in the glomerular layer and the lowest in the subventricular layer. The cholinergic innervation develops during the first three postnatal weeks in rodents. It is present at birth first in a subset of posterior and medio-dorsal glomeruli, the so called atypical glomeruli, which remain particularly rich in cholinergic fibers in adult and whose function remains unknown [4]. Cholinergic fibers synaptically target dendrites of periglomerular and granular bulbar interneurones. No cholinergic synapses could be identified on mitral cells but rather cholinergic varicosities in close apposition to secondary mitral cell dendrites in the external plexiform layer.

Both nicotinic (ionotropic) and muscarinic (G protein-coupled) receptors are present in the MOB. Nicotinic receptors are pentameres of various subunits (α2–10; β2–4) whose combinations form cationic channels with distinct functional properties. High affinity heteromeric receptors, among which the abundant α4β2 combination, and low affinity α7 homomeric receptors are retrieved in the MOB and show a specific laminar distribution. Quantitative autoradiography indicates that heteromeric receptors are found at high levels in the granular cell layer. In contrast, α7 receptors are concentrated in the glomerular layer and to a lesser amount in the deeper layers of the MOB. Less is known about the cell types expressing the different nicotinic receptors. α2 subunit is expressed by a small group of neurons in the internal plexiform layer and additional rare neurons of the glomerular and external plexiform layers. The β2 subunit strongly labels mitral cells and cells located in the superficial part of the external plexiform layer.

Five subtypes of muscarinic receptors (M1–5) have been cloned in the brain that can be grouped in two families based on their G-protein coupling mechanism and ligand's binding selectivity. The M1 family (M1, 3 and 5) is positively coupled to the activation of phospholipase C and receptors of the M2 family (M2, 4) are negatively coupled to adenylate cyclase and classically act as presynaptic auto- or heteroreceptors. In the MOB, M1-like and M2-like receptors are most abundant in the external plexiform layer compared to the deeper layers while their expression is low in the glomerular layer. Accordingly, M2 receptors have been localized by immunocytochemistry presynaptically on the dendrites of granule cells at synaptic loci in the external plexiform layer and post synaptically on soma of second order bulbar interneurons in the inframitral layers. In the glomerular layer, M2 receptors are expressed by a subset of GABAergic/dopaminergic periglomerular neurons.

Cellular Effects

In line with the heterogeneous distributions of cholinergic fibers and receptors in the MOB, the cellular effect of Ach in the MOB proved to be complex [5]. Through nicotinic receptors, acethylcholine facilitates olfactory information transmission by directly exciting mitral cells in a paracrine manner. Nicotinic receptors activation also induces an increase in periglomerular cells activity which in turn inhibits mitral cells thus supporting an effect opposed to the direct action of Ach onto mitral cells. These two actions are likely mediated by heteromeric high affinity and low affinity α7 containing receptors respectively, in accordance with their laminar distribution (see above).

Through muscarinic receptors, acethylcholine also exerts two distinct actions on two compartments of granule cells. On the soma, it reduces their firing rate, thus producing a disinhibition of mitral cells. Pre-synaptically, through M1 receptors acethylcholine enhances GABA release by granule cells onto mitral cells, thus reinforcing inhibition of the output neurons. Through the several loci at which it influences the bulbar network, acethylcholine actions regulate both the entry and the output signals of the MOB, and is thus a key modulators of olfactory processing.

Functional Implications

Given the well known implications of acethylcholine in memory processes in cortices, most of the studies on the role of acethylcholine in the MOB focused on odor memorization using pharmacological approaches. Systemic administration of scopolamine, a muscarinic antagonist impairs short term olfactory memory in rats and lamb recognition by parturient ewes without affecting olfactory detection.

In addition, a model of cholinergic modulation of the bulbar network suggested that cholinergic inputs may sharpen mitral cell receptive fields, allowing for better discrimination abilities [6]. This assumption was confirmed by the demonstration that intrabulbar administration of a nicotinic antagonist abolished spontaneous discrimination between perceptually close odorants and muscarinic receptors blockade impairs discrimination performances. In both cases, at the doses used in this study, a reward-associated discrimination task was left unaffected [7]. Taken together these data suggest that acethylcholine influences olfactory perception and memorization and thus interdependence between these two aspects of olfactory processing is likely.

Serotonin

Anatomical Organization

The olfactory bulb receives a dense serotonergic input from the ►raphe nuclei. In rats, serotonergic fibers display specific laminar and regional distributions in the MOB. The density of 5-HT fibers in the glomerular layer is 2–3 times greater than in any other layer in MOB. Some serotoninergic fibers were observed in the external plexiform layer, internal plexiform layer, mitral cell layer and granule cell layer. Several types of serotonin receptors are expressed in the MOB, including 5HT1A and 5HT1B and 5HT2C. 5HT2 receptors are mainly distributed in the glomerular, granular cell layers and in the mitral (M) cell layer. By contrast, 5HT1A and 5HT1B receptors are mainly present in the external plexiform and granular cell layers respectively.

Cellular Effects

Serotonin inhibits a subset of mitral cells through an indirect mechanisms involving GABA release by granule or periglomerular cells, an effect that might be modulated by the vigilance states. In contrast, another subset of mitral cells is directly depolarized by serotonin acting at 5HT2A receptors. In addition, serotonin depolarizes some periglomerular cells through 5HT2C receptors [8].

Functional Implications

In the MOB, 5HT has been demonstrated to be involved in olfactory learning. In the literature there is evidence that damage to the MOB or to his serotoninergic innervations may alter olfactory coding and/or memory. For instance, neonates with serotoninergic denervation of the MOB exhibit altered acquisition or expression of an olfaction-based learned behavior. Pharmacological studies indicated that 5HT2 receptors are more likely involved in promoting conditioned olfactory learning in neonatal rats although, it is not clear whether 5HT2A or 5HT2C subtypes are predominantly involved. 5HT2 receptors seem to be required in the acquisition stages but not in the consolidation and retrieval ones [9].

Rats with 5,7-dihydroxytryptamine (5,7-DHT) lesions of serotonergic fibers lose their ability to discriminate odors. It was also shown that damage to serotonergic afferents of the MOB does not induce complete anosmia and does not disrupt the basic mechanisms of olfactory recognition. 5HT depletion caused glomerular layer atrophy. Moreover, the 5HT innervation was hypothesized to act in collaboration with the noradrenergic (NA) innervation in olfactory learning, even if NA alone seems to be able to compensate for the deficit of 5HT in certain learning conditions.

Orexin

Centrifugal orexin-containing fibers originating from the hypothalamic feeding centers (lateral and posterior hypothalamic areas and perifornical area) terminate in the glomerular and mitral cell layers. In addition, a few fibers are seen in the granular layer. Receptors to orexin (G-protein coupled receptors, ORX1–2) are localized to mitral cells principally and to subsets of periglomerular and granular cells. Oxerin indirectly hyperpolarize mitral cells through an increase in GABA release by granule cells while it seems to directly depolarize a small fraction of mitral cells [10]. Orexin infused in the brain stimulate food intake. In the MOB, it is proposed to participate in the regulation of olfactory perception by the feeding status but this remains to be assessed.

Neuromodulators Interfere with Neurogenesis

Bulbar neurogenesis consists of the permanent renewal in the adult of the two populations of interneurons, periglomerular and granular cells. Nicotinic receptors activation increases the death rate of newborn granular neurons, while other studies have shown that Ach promote their survival, possibly then through muscarinic receptors. Chronic administration of a selective 5HT1A and 5HT2C receptor agonists induces an increase in the rate of neurogenesis. Similarly, stimulation of the noradrenergic system by an $\alpha 2$ antagonist promotes adult born cell survival. Neuromodulators are thus able to modulate olfactory processing not only by influencing the existing network but also by regulating the permanent maturation of the MOB circuits.

References

1. Hayar A, Heyward PM, Heinbockel T, Shipley MT, Ennis M (2001) Direct excitation of mitral cells via activation of alpha1-noradrenergic receptors in rat olfactory bulb slices. J Neurophysiol 86:2173–2182
2. Okutani F, Kaba H, Takahashi S, Seto K (1998) The biphasic effects of locus coeruleus noradrenergic activation on dendrodendritic inhibition in the rat olfactory bulb. Brain Res 783:272–279
3. Brennan PA, Keverne EB (1997) Neural mechanisms of mammalian olfactory learning. Prog Neurobiol 51:457–481
4. Durand M, Coronas V, Jourdan F, Quirion R (1998) Developmental and aging aspects of the cholinergic innervation of the olfactory bulb. Int J Dev Neurosci 16:777–785
5. Castillo PE, Carleton A, Vincent JD, Lledo PM (1999) Multiple and opposing roles of cholinergic transmission in the main olfactory bulb. J Neurosci 19:9180–9191
6. Linster C, Cleland TA (2002) Cholinergic modulation of sensory representations in the olfactory bulb. Neural Netw 15:709–717
7. Mandairon N, Ferretti CJ, Stack CM, Rubin DB, Cleland TA, Linster C (2006) Cholinergic modulation in the olfactory bulb influences spontaneous olfactory discrimination in adult rats. Eur J Neurosci 24:3234–3244
8. Hardy A, Palouzier-Paulignan B, Duchamp A, Royet JP, Duchamp-Viret P (2005) 5-Hydroxytryptamine action in the rat olfactory bulb: in vitro electrophysiological patch-clamp recordings of juxtaglomerular and mitral cells Neuroscience 131:717–731
9. McLean JH, Darby-King A, Hodge E (1996) 5-HT 2 receptor involvement in conditioned olfactory learning in the neonate rat pup. Behav Neurosci 110:1426–1434
10. Hardy AB, Aioun J, Baly C, Julliard KA, Caillol M, Salesse R, Duchamp-Viret P (2005) Orexin A modulates mitral cell activity in the rat olfactory bulb: patch-clamp study on slices and immunocytochemical localization of orexin receptors. Endocrinology 146:4042–4053

Neuromodulation of Central Pattern Generators

►Neurotransmitters and Pattern Generation

Neuromodulators

Definition

A neuromodulator is a chemical compound released by particular neurons modulating the activity of targeted cells. In contrast to classical neurotransmitters, a neuromodulator is not reabsorbed by the presynaptic cell. Thus, it can diffuse more widely in the tissue and act on several neurons. Neuromodulators are typically amines or peptides that can phosphorylate ion channels, alter second messenger pathways and intracellular calcium. These modulatory effects change ion channel properties, which in turn lead to the alteration of neuronal discharge patterns.

They can enhance or decrease the activity of neurons or the efficiency of excitatory or inhibitory synapses mediated by classical neurotransmitters as GABA or glutamate.

►Neuropeptides

Neuromodulators in Nociception

Definition

In the transmission from primary nociceptive afferents to central neurons, the combination of glutamate (transmitter) and calcitonin gene related peptide (CGRP) and Substance P (neuromodulators) are most important. Neuromodulators modify and enhance the synaptic transmission.

►Calcitonin Gene Related Peptide (CGRP)
►Nociceptors and Characteristics
►Substance P

Neuromorphic Device

Definition

Electronic chip that emulates biological neural systems by means of analog and/or digital Very Large Scale of Integration (VLSI) technologies. Existing systems have been used to reproduce the functionality of biological sensors (such as silicon retinas and cochleas) and/or implement single neocortical processing modules (such as selective attention modules). Several single-chip neuromorphic systems have been developed, mainly focusing on the sensory periphery (e.g. silicon retinas, silicon cochleas, motion sensors, etc.). More recently, the development of general, standard communication infrastructures also enabled the creation of complex multi-chip systems.

►Computer-Neural Hybrids

Neuromuscular

Definition

Affecting or characteristic of both nerve and muscle.

Neuromuscular Junction

MARK M. RICH
Department of Neuroscience, Cell Biology and Physiology, Wright State University,
Dayton, OH, USA

Synonyms

Endplate; NMJ

Definition

The synapse between a lower motor neuron and a skeletal muscle fiber.

Characteristics

Quantitative Description: The neuromuscular junction in mammals is, on average, 20–40 μm wide and 20–150 μm long.

Higher Level Structures: Neuron, muscle fibers.

Lower Level Components: Synaptic vesicles, sodium channels, calcium channels, acetylcholine receptors, synaptic cleft.

Structural Regulation: The neuromuscular junction consists of a presynaptic motor nerve terminal that contacts the skeletal muscle at a single point along its length. Directly underneath the nerve terminal is a region of specialized postsynaptic muscle membrane in which there is a high concentration of acetylcholine receptors (Fig. 1).

Within the neuromuscular junction there are a number of specialized release sites known as active zones where synaptic vesicles fuse and release acetylcholine (the neurotransmitter at the neuromuscular junction). Between the pre- and postsynaptic cells is extracellular space in which there are a number of structural proteins. This extracellular collection of proteins is known as the basal lamina. The enzyme that breaks down acetylcholine (acetylcholinesterase) is located in the basal lamina and serves to terminate signaling between the pre- and postsynaptic cells.

During development, there is rearrangement of synaptic connections at the neuromuscular junction that is mediated by synaptic activity [2]. This rearrangement results in the mature structure in which only one presynaptic axon innervates each muscle fiber. In the adult, little further rearrangement occurs and the structure of the neuromuscular junction is thought to remain fairly stable. However, the capacity for rearrangement persists at the adult neuromuscular junction and such rearrangement may become important when synaptic function is disrupted in diseases of ►neuromuscular transmission (Fig. 1).

Higher Level Processes: Motor unit function, force generation.

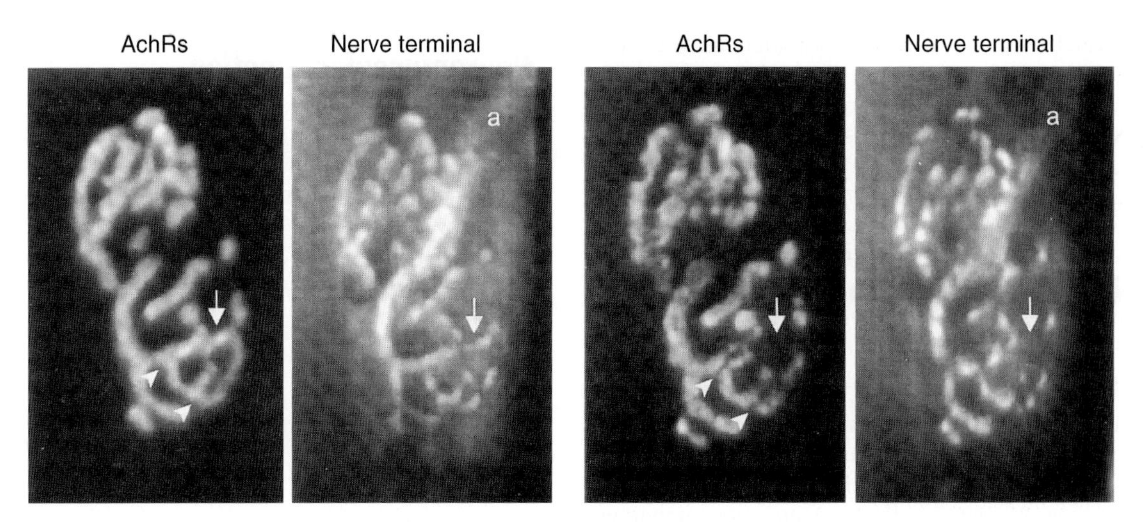

Neuromuscular Junction. Figure 1 Regions of acetylcholine receptors (AChRs) and nerve terminal staining are lost from the neuromuscular junction in autoimmune myasthenia gravis. Shown are images of the AChRs and nerve terminal from an individual mouse endplate before, and 1 week after an immune attack on the postsynaptic AChRs. Prior to immune attack (*on the left*) the structure of the endplate is normal. The postsynaptic AChRs are stained uniformly and precisely align with the nerve terminal. One week after exposure to antibody against AChRs (images of AChRs and nerve terminal on the *right*), regions of AChR staining present in the first view, have been eliminated (*arrows*). Nerve terminal overlying the eliminated regions of AChRs has retracted. In addition, regions of faint AChR staining are present throughout the endplate. Synaptic plasticity of the neuromuscular junction, such as that shown in this figure, may play an important role in various disease states that affect the neuromuscular junction. Each image is 40 μm wide. The presynaptic axon can be seen entering the endplate from the *upper left* in the image of the nerve terminal (a). From [1].

Lower Level Processes: Synaptic vesicle fusion, action potential propagation.

Process Regulation: Synaptic activity regulates function of the neuromuscular junction on both short and long timescales. There is regulation of synaptic function over a millisecond timescale in which the number of synaptic vesicles that fuse following a presynaptic action potential is modulated by previous activity. This category of synaptic plasticity is termed short-term synaptic plasticity and consists of depression, ►facilitation and ►post-tetanic potentiation [3]. In normal calcium there is thought to be a depletion of vesicles that are release-ready by previous pulses, such that fewer vesicles are released during subsequent presynaptic action potentials. This is termed ►synaptic depression and is used clinically to diagnose diseases of neuromuscular transmission such as myasthenia gravis [4]. When extracellular calcium is lowered (or calcium entry is lowered), an opposite response occurs to repetitive stimulation and more vesicles are released following each presynaptic action potential. This is known as facilitation and is a hallmark of diseases of neuromuscular transmission such as Lambert Eaton Myasthenic syndrome in which presynaptic calcium entry is decreased. The cause of facilitation is thought to be summing of the calcium signal from each pulse of the train. A third type of short-term plasticity is known as post-tetanic potentiation and can last for 10s of seconds after a train of pulses given at a high frequency. Post-tetanic potentiation is thought to be due to residual calcium following a train of stimuli.

There is also activity-dependent regulation of neuromuscular function that occurs over the time period of days to weeks. When neuromuscular activity is blocked for several days, nerve terminals at inactive junctions grow new processes that grow over the muscle and form new connections. This process is known as ►sprouting and serves to increase synaptic strength.

Function: The neuromuscular junction's function is to cause the postsynaptic muscle fiber to fire an action potential every time the presynaptic motor neuron spikes. This function differs from that of synapses in the central nervous system. In the central nervous system synapses integrate signals to process information. At the neuromuscular junction there is no information processing. The only time that abnormalities of neuromuscular transmission cause problems is when they become severe enough to cause the muscle fiber to no longer faithfully follow trains of action potentials fired by the presynaptic terminal.

The series of events necessary for neuromuscular transmission occur as follows. An action potential enters the presynaptic nerve terminal. During the action potential there is opening of presynaptic calcium channels, which allows calcium entry. Calcium binds to a calcium sensor (thought to be synaptotagmin) and this leads to fusion of the membrane of synaptic vesicles to the nerve terminal. The precise series of molecular events that underlie vesicle fusion is currently an area of intensive study. The total number of synaptic vesicles that fuse during a presynaptic action potential at mammalian neuromuscular junctions varies between species and ranges from 20 to 100. Following fusion of synaptic vesicle there is release of acetylcholine which diffuses across the synaptic cleft and binds to acetylcholine receptors, causing them to open. Opening of acetylcholine receptors allows for flow of sodium and potassium ions, with the net result being depolarization of the postsynaptic muscle membrane. This depolarization opens muscle sodium channels and triggers a muscle action potential. The total amount of time for this cascade of events is 1 ms.

Pathology: The three diseases most commonly responsible for failure of neuromuscular transmission are myasthenia gravis, botulism, and Lambert-Eaton myasthenic syndrome (LEMS). These diseases present with weakness in the absence of sensory symptoms. Prominent symptoms often include difficulty swallowing as well as double vision. One of the primary diagnostic tests is repetitive nerve stimulation which reveals failure of neuromuscular transmission.

Weakness in myasthenia gravis is most often caused by an autoimmune attack directed at postsynaptic acetylcholine receptors. The result of the attack is that postsynaptic acetylcholine receptor density is reduced (Fig. 1). Thus, when a synaptic vesicle releases acetylcholine, there are fewer acetylcholine receptors available to respond and a smaller postsynaptic current is generated. This reduces the postsynaptic depolarization following a presynaptic action potential. If the postsynaptic depolarization is still large enough, the muscle fiber fires an action potential and there is no weakness. However, during trains of action potentials there is synaptic depression (see above). Depression of acetylcholine release during trains (a normal phenomenon) cause the postsynaptic depolarization caused by opening of acetylcholine receptors to become insufficient to trigger a muscle fiber action potential. Failure to activate muscle fibers causes weakness and a decrement on EMG that is diagnostic of a failure of neuromuscular transmission [4].

In botulism, failure of neuromuscular transmission is caused by cleavage of synaptic proteins that are critical for fusion of synaptic vesicles [5]. The synaptic protein cleaved in botulism depends on the subtype of botulinum toxin. Botulinum A and E cleave SNAP-25, botulinum B and D cleave synaptobrevin (also known as VAMP), and botulinum C cleaves syntaxin. Cleavage of these proteins greatly reduces the number of vesicles that fuse during a presynaptic action potential. This results in reduced release of acetylcholine and failure of neuromuscular transmission.

LEMS is caused by an autoimmune attack on presynaptic calcium channels that reduces calcium entry into the presynaptic terminal. Reduced calcium entry causes a reduction in the number of vesicles that fuse and thus reduces acetylcholine release. During repetitive stimulation of the neuromuscular junction in LEMS, there is dramatic facilitation of the EMG signal due to facilitation of release of synaptic vesicles [4].

Although the three diseases described above are thought to be the primary diseases in which there is failure of neuromuscular transmission, evidence has begun to emerge that neuromuscular dysfunction may be an important contributor to weakness in motor neuron disease as well. The most common form of motor neuron disease is amyotrophic lateral sclerosis (ALS). It has been thought that motor neuron cell death is the sole cause of weakness in ALS. In large part, this is due to the relative ease with which motor neuron cell death can be demonstrated with routine histological examination of autopsy material. Human autopsy results, however, are dominated by disease end stage phenomena, and while there is no doubt that cell death explains the permanent loss of motor units and paralysis of ALS, it remains uncertain whether cell death fully accounts for weakness in earlier stages of the disorder.

Evidence that dysfunction at the neuromuscular junction causes weakness in advance of cell death comes from studies of animal models of motor neuron disease. In both mouse and canine models of motor neuron disease, there is emerging evidence suggesting that loss of neuromuscular innervation (denervation) occurs before motor neuron cell death has begun. In the canine animal model there is further evidence that physiological dysfunction occurs at an even earlier stage of the disease, when no denervation is apparent histologically [6]. The cause of the failure of neuromuscular transmission is a reduction in the number of vesicles that fuse following a presynaptic action potential. If the sequence of events is similar in human motor neuron disease this would suggest that weakness in patients is initially caused by failure of neuromuscular transmission in the absence of a clear structural abnormality. This is then followed by degeneration of the presynaptic terminal, and only at very late stages, when a motor unit is generating almost no force, is there death of the motor neuron. Such a sequence of events would have important treatment implications, since it would mean that treatments aimed at slowing the progression of neuromuscular dysfunction may be important in helping patients with ALS.

Therapy: There are two categories of treatments for diseases of neuromuscular transmission. The first is symptomatic treatment that is aimed at improving neuromuscular transmission and the second is aimed at the underlying disease process. The drug most commonly used to improve neuromuscular transmission is an inhibitor of acetylcholinesterase (usually pyridostigmine). By slowing breakdown of acetylcholine in the synaptic cleft pyridostigmine increases the amplitude and prolongs the time-course of the postsynaptic current. The increase in postsynaptic current allows increased depolarization of the muscle fiber, and thus allows more muscle fibers to reach threshold for action potential initiation. Block of acetylcholinesterase is most appealing on a theoretical basis for disorders such as botulism and LEMS where the underlying problem is a reduction in release of acetylcholine; however, pyridostigmine is most commonly used in treating myasthenia gravis. Another drug, 3,4 diaminopyridine, is used to treat LEMS. This agent inhibits presynaptic potassium channels, and thus prolongs the action potential to allow for increased calcium entry. Treatment for botulism is primarily supportive.

In both myasthenia gravis and LEMS the underlying problem is an autoimmune attack on the neuromuscular junction. Thus, in both of these disorders reversing the underlying disease process by reducing the immune attack on the endplate is the most important part of treatment. LEMS is often a paraneoplastic syndrome in which the antibodies to the neoplasm cross-react with the neuromuscular junction. In these cases, treatment of the neoplasm often results in improvement. Corticosteroid treatment, azathioprine, plasmaphoresis and treatment with intravenous immunoglobulin are all methods of immunosuppression used to treat myasthenia gravis and LEMS. In addition, in myasthenia gravis, thymectomy is often performed to treat more severe cases. For further details on treating these disorders see [7,8].

References

1. Rich MM, Colman H, Lichtman JW (1994) In vivo imaging shows loss of synaptic sites from neuromuscular junctions in a model of myasthenia gravis. Neurology 44 (11):2138–2145
2. Sanes JR, Lichtman JW (1999) Development of the vertebrate neuromuscular junction. Annu Rev Neurosci 22:389–442
3. Zucker RS, Regehr WG (2002) Short-term synaptic plasticity. Annu Rev Physiol 64:355–405
4. Kimura J (1989) Electrodiagnosis in diseases of nerve and muscle: principles and practice. F.A. Davis Company, Philadelphia
5. Huttner WB (1993) Cell biology. Snappy exocytoxins. Nature 365(6442):104–105
6. Pinter MJ et al. (2001) Canine motor neuron disease: a view from the motor unit, in motor neurobiology of the spinal cord. In: Cope TC (ed) CRC Press, New York, pp. 231–250
7. Richman DP, Agius MA (2003) Treatment of autoimmune myasthenia gravis. Neurology 61(12):1652–1661
8. Newsom-Davis J (2003) Therapy in myasthenia gravis and Lambert-Eaton myasthenic syndrome. Semin Neurol 23(2):191–198

Neuromuscular Transmission

Definition

The cascade of events at the neuromuscular junction that causes an action potential in the presynaptic nerve terminal to be propagated to the muscle fiber contacted by that nerve terminal.

►Neuromuscular Junction

Neuromyelitis

Definition

Inflammation of nervous and medullary substance; myelitis associated with neuritis.

►Myelitis
►Neuritis

Neuromyelitis Optica (NMO or Devic's Disease)

Definition

NMO belongs to the group of ►idiopathic inflammatory demyelinating diseases of the central nervous system and has been distinguished from ►multiple sclerosis (MS) by the presence of (usually bilateral, simultaneous, and often severe) ►optic neuritis, spinal cord abnormalities (extending contiguously over three or more vertebral segments), absence of brain abnormalities, and often rapid progression to debility and even death. Pathologically, an antibody-dependent, complement-mediated process is thought to underlie the axonal loss, demyelination and necrosis. A specific serum biomarker, neuromyelitis optica immunoglobulin G (NMO-IgG), which distinguishes neuromyelitis optica from ►multiple sclerosis, targets the blood brain barrier and the water channel aquaporin-4, which is lost in neuromyelitis optica lesions and classifies NMO as an autoimmune ►channelopathy. NMO-IgG is the first specific marker for a central nervous system demyelinating disease. Corticosteroids are used to treat acute attacks and immunosuppressants are the treatment of choice.

►Multiple Sclerosis

Neuromyopathy

Definition

Any disease of both muscles and nerves, especially a muscular disease of nervous origin.

Neuron

Definition

A late nineteenth century Greek term, refers to highly specialized "nerve cells". A neuron exhibits a highly complex repertoire of specialized membranous structures, embedded ion channels, second messengers, genetic and epigenetic elements and unique complements of various proteins such as the receptors.

Neurons are excitable cells (i.e., able to conduct electrical impulses of action potentials), which form elaborate networks through axons and dendrites. This ensemble is responsible for integrating, processing and transmitting information, and forms the basis for e.g., coordinated muscle movements and brain functions, including learning and memory formation.

►Action Potential
►Cell Membrane: Components and Functions

Neuron–Astrocyte Interactions

Yoshihisa Kudo
School of Life Science, Tokyo University of Pharmacy and Life Science, Tokyo, Japan

Synonyms

Tripartite synapse

Definition

Classically the roles of astrocytes in neuron–astrocyte interaction in the central nervous system (CNS) have been as passive and supportive elements, which remove the released neurotransmitters by specific transporters (such as GLT-1, a glutamate transporter), maintain the ionic environment in the ►extracellular space via ion channels and transporters (K^+ channels, Na^+/K^+ exchanger, etc.), supply the energy source through blood vessels to neurons and limit the passage of some toxic substances as a ►blood brain barrier, all optimizing the conditions for neurons and synapses in the CNS (Fig. 1).

However, since the expression of neurotransmitter receptors on astrocytes has been revealed by the dynamic responses in intracellular Ca^{2+} concentration ($[Ca^{2+}]i$) to neurotransmitters, the possibility of cross-talking between neurons and astrocytes has been emphasized as a much more important interaction between those cells. The possibility became much more probable after the discovery of the ability of astrocytes to release neurotransmitters, such as glutamate and ATP. The finding of dynamic interaction between neurons and astrocytes urged the renewal of the concept of astrocytes as possible elements participating in the information processing mechanisms in the brain. Now neuron–astrocyte interaction has been accepted as one of the important systems for establishing higher order brain functions, which had been investigated based

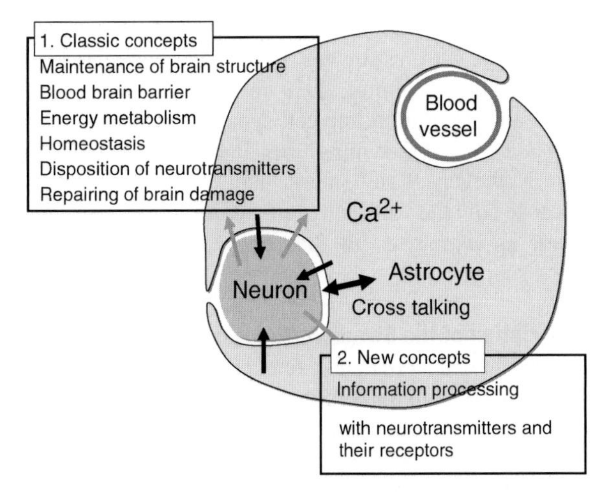

Neuron–Astrocyte Interactions. Figure 1 Classic concepts and new concepts of neuron–astrocyte interaction. According to classic concepts astrocytes are recognized as passive and supporting elements in the brain. Since the discovery of neurotransmitter receptors on astrocytes and the release of transmitters from them, new concepts of neuron–astrocyte interaction as information processing elements have been established.

upon a "neuroncentric" concept until the end of the twentieth century [1,2] (Fig. 1).

Characteristics

Quantitative Description

Astrocytes were named after their stellate shape observed under the light microscope after staining with silver. The cytoskeleton stained by such methods was only 15% of the total volume of the cell. The astrocytes located in the gray matter, called ►protoplasmic astrocytes, have profuse processes that give the cells an appearance that has been referred as to "spongiform." The average volume of a single astrocyte has been calculated to be about 66,000 μm^3. Together with a study that showed that there are about 213 synapses/100 μm^3 in the adult rat hippocampal CA1 subfield, it is estimated that a single astrocyte would be in contact with about 140,000 synapses. A recent 3D high voltage electron microscopic study on protoplasmic astrocytes in the hippocampal CA3 subfield demonstrated the average surface volume ratio as 26.2/μm and taking this value into account, the surface area of a single astrocyte would be about 2,000,000 μm^2. Since the density of astrocytes in the cerebral cortex is high (12,000–30,000/mm^3) and the number of astrocytes in the mammalian brain is estimated to be 1.3–1.4 times larger than the number of neuronal cells, the surface of neuronal cells would be almost completely covered with astrocytes [2].

The percentage of the coverage of astrocyte processes on a single synapse has been estimated by 3D reconstruction studies on electron microscopic images. According to these studies, percentages of all synapses associated with astrocytes have been estimated as 57% in hippocampus, 29% in visual cortex, 69% of parallel fiber–Purkinje cell synapses in the cerebellum and 94% of ascending fiber–Purkinje cell synapses [3]. However, these values may be low estimates because of the difficulty of reproducing the real shape of astrocytes in the preparation for electron microscopy.

Description of the Structure/Process/Conditions

As mentioned above, the fine processes of a single astrocyte envelop numerous synapses in the brain. This structural association strongly suggests possible interaction between neurons and astrocytes, not only in structural and metabolic support but also in information processing. To establish the information processing between them, astrocytes should express some receptors for detecting the molecules released from neighboring neurons and should also release some factors to talk back to neuronal cells. Neurotransmitter receptor expression and dynamic responses were found as early as 1986, in C6Bu1, a clonal astrocyte in culture. The clonal astrocytes responded to serotonin by an increase in intracellular Ca^{2+} concentration ([Ca^{2+}]i), which was measured by a specific Ca^{2+} indicator, fura-2. Since then increases in [Ca^{2+}]i induced by glutamate, GABA, noradrenalin, ATP and acetylcholine have been demonstrated in astrocytes isolated from hippocampus and other brain regions in culture [4]. The majority of receptors expressed on the astrocytes that cause an increase in [Ca^{2+}]i have been classified as Gq-type G-protein coupled receptors, which will cause IP_3 production and thus stimulate the IP_3 receptors on endoplasmic reticulum to release stored Ca^{2+} into the cytosol. However, the Bergman glia cells in the cerebellum express Ca^{2+} -permeable alpha-amino-3-hydroxy-5-methyl-4-isoxazolepropionic acid (AMPA)-type glutamate receptors assembled without the GluR2 subunit [5]. When the increase in [Ca^{2+}]i induced by neurotransmitters was found in astrocytes, researchers expected the role of the [Ca^{2+}]i to be as an activator for transmitter release from the cells. Activation of astrocytes sometimes induces the release of ATP, but the response is not always Ca^{2+} dependent. Although Ca^{2+} dependent release has been found in astrocytes, the processes are shown to be distinct from exocytosis in neuronal terminals. Recently ►gap junction hemichannels have emerged as an additional molecular pathway for transmitter release from astrocytes [6]. The hemichannel of the gap junction has been shown to be activated by lowering the concentration of extracellular divalent cation. The other important pathway is P2X7, a purinergic receptor, which seems to be modulated by intracellular divalent cations (Fig. 2).

Using such multiple machineries, astrocytes can release glutamate, ATP, ►D-serine and some other substances. These substances can send information to neurons by the activation of receptors expressed on their pre- and post-synaptic sites. The structure consisting of astrocyte and pre- and post-synaptic neurons has been emphasized by the coining of the term "►tripartite synapse" [7] (Fig. 3).

High Level Structure/Process/Conditions (Fig. 4)

Astrocyte-Astrocyte Intercommunication

As mentioned above a single protoplasmic astrocyte in gray matter has profuse processes referred to as spongiform, which will make contact at its boundaries with other astrocytes. The processes of adjacent astrocytes do not project into neighboring domains. Thus each astrocyte seems to occupy a separate anatomic domain. Each astrocyte domain makes connections with others using specific gap junction proteins called connexins, which can mediate the passage of current and of rather large molecules (up to 1,000 molecular weight) between astrocytes. This structural feature stabilizes and equalizes membrane potential among groups of astrocytes and ensures common levels of ions and presumably other molecules as well.

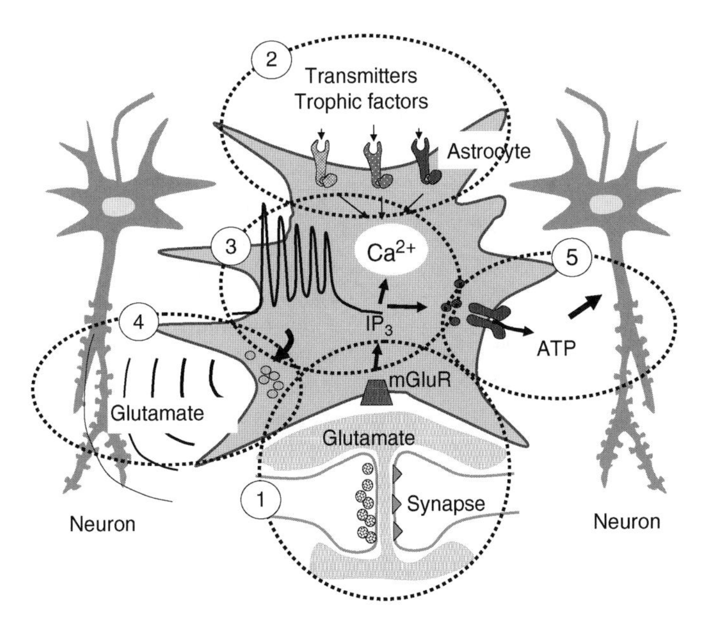

Neuron–Astrocyte Interactions. Figure 2 Neuron–astrocyte interactions. (a) Diffusion of neurotransmitter in extracellular space (spill over). Before disposal of the released neurotransmitters by specific transporters, they diffuse into the extracellular space and activate neurotransmitters expressed on the astrocytes. (b) Expression of many kinds of receptors. Astrocytes have been demonstrated to express receptors for neurotransmitters, such as glutamate, noradrenalin, serotonin, GABA and acetylcholine and also for trophic factors. Activation of these receptors induces the increase in $[Ca^{2+}]$i. (c) Characteristic Ca^{2+} increase. The activation of astrocytes through receptors results in an oscillatory increase in $[Ca^{2+}]$i. (d) Neurotransmitter release. Activation of astrocytes sometimes causes release of neurotransmitters, such as glutamate and ATP. Released glutamate will activate neuronal cells. (e) Inhibitory regulation by ATP released from astrocytes has been demonstrated to depress neuronal activities.

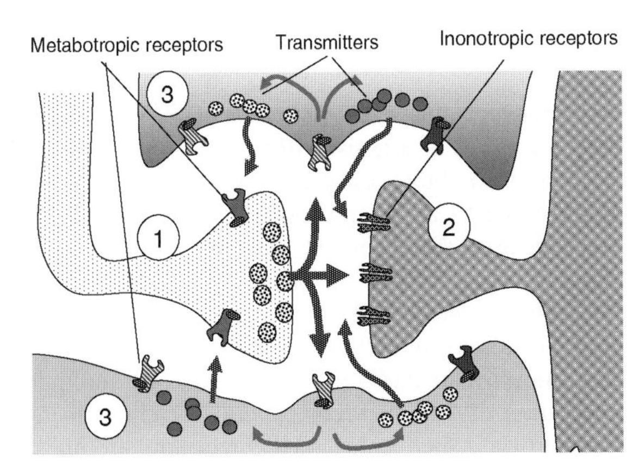

Neuron–Astrocyte Interactions. Figure 3 Tripartite synapse. The structure consisting of (a) pre-, (b) post-synaptic neurons and (c) astrocytes has been emphasized by the coining of the term "tripartite synapse." Information processing between pre- and post-synaptic neurons receives further modulation from astrocytes, which express neurotransmitter receptor and release neurotransmitters.

Astrocyte-Vascular Interaction

New lines of work have shown that receptors and channels essential for the function of astrocytes are densely concentrated in their vascular end-feet. Especially intriguing is the observation that the water channel aquaporin-4 and purine receptors – mediators of astrocytic Ca^{2+} signaling – are expressed primarily at the ►gliovascular interface [2]. The array of these astrocyte-delimited microdomains along the capillary microvasculature allows the formation of higher-order gliovascular units, which serve to match local neural activity and blood flow while regulating neuronal firing thresholds through coordinative glial signaling. By these means, astrocytes might establish the functional as well as the structural architecture of the adult brain.

Neuron-Astrocyte Intercommunication

Activation of astrocytes by neurotransmitters released from the neighboring neuron can evoke the increase in $[Ca^{2+}]$i and the increase can propagate as Ca^{2+} waves for several hundred micrometers [4]. The propagation, however, was found to be limited within a certain area, suggesting the existence of a local circuit. Intercellular communication among astrocytes using Ca^{2+} waves provides astrocyte networks by which astrocytes can signal each other independently from the neuron network and can modulate the activities of neurons over a relatively wide range within the network. The brain structure as an information-processing machine should be recognized as an extraordinarily refined system consisted of the neuron network into

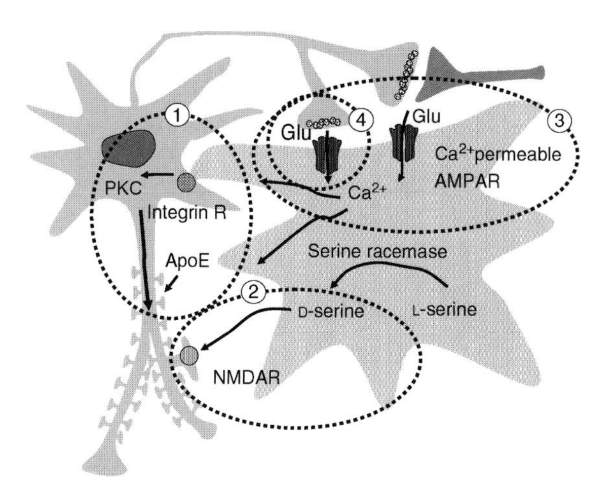

Neuron–Astrocyte Interactions. Figure 4 Interactions among astrocytes, synapses and vasculature. The [Ca^{2+}]i increase in astrocytes induced by neurotransmitter propagates intra- and inter-cellularly. The propagation may be promoted by two mechanisms. One is diffusion of IP3 inside the cell and also through the gap junctions formed between astrocytes. The other is the response mediated by released ATP and its receptor (purinergic receptor). ATP released through the gap junction hemichannel diffuses to adjacent astrocytes and activates their ATP receptors. Functional molecules for regulating the astrocytes, such as purine receptor and aquapolin-4 are expressed mainly on the endfoot of astrocytes, which makes tight contact with a blood vessel. Since synaptic activities give and receive information between astrocytes, the size of the blood vessel may also be regulated depending upon the neuronal activities. The structure consisting of astrocyte, blood vessel and neuron will provide dynamic regulation of information processing in the brain. The gap junction hemichannel will participate in the release not only of ATP but also of some other transmitters and of trophic factors.

which the astrocyte network may be woven tightly and widely. Higher order brain function may be dependent upon this highly sophisticated structure. Although the participation of the "neuron astrocyte network" in the expression of higher order brain functions has not been established yet, these marvelous structures must be taken into account for further understanding the brain function [8].

Regulation of the Structure/Process/Condition

Although the developmental profiles of the regulation of neuron–astrocyte interaction seem to be important in understanding the structure and functions of the brain, only ►synaptogenesis will be discussed in this essay. Some instances of the regulation of the neuron–astrocyte interaction in adult brain will be described (Fig. 5).

Regulation of Synaptogenesis by Astrocytes

The co-culture of purified neurons with astrocytes has been demonstrated to facilitate synaptogenesis. Although the mechanisms of this facilitation have not been elucidated yet, some diffusible factors such as cholesterol complexed with apolipoprotein E-containing lipoproteins have been identified as candidates. Recently astrocytes have been demonstrated to affect neuronal synaptogenesis by the process of adhesion. Local contact with astrocytes via ►integrin receptors elicited protein kinase C activation, which was initially focal but soon spread throughout the entire neuron. This suggests that the propagation of PKC signaling represents an underlying mechanism for synaptogenesis [9].

Regulation of NMDA Receptor Activation by D-Serine Released from Astrocytes

The activation of NMDA receptor by glutamate requires to be co-activated by glycine, which binds a specific binding site. Recently several lines of evidence indicate that D-serine, the stereo-isomer of L-serine, is an endogenous co-activator for NMDA receptor and is three times more potent than glycine in activating its binding site. The D-serine degrading enzyme has been shown to attenuate NMDA mediated transmission. D-serine and serine racemase, a D-serine synthesizing enzyme, have been found to be localized only in astrocytes. Since NMDA receptor is recognized as participating in synaptic plasticity, its regulation by D-serine suggests the important participation of astrocytes in higher order brain functions.

Structural Regulation by the Activation of Ca^{2+} Permeable AMPA Receptors in Bergmann Glia

As mentioned above, Bergman glia cells in the cerebellum express Ca^{2+} -permeable AMPA-type glutamate receptors assembled without the GluR2 subunit [5]. Conversion of these Ca^{2+} permeable receptors into Ca^{2+} impermeable ones by adenoviral mediated delivery of GluR2 results in the retraction of glial cell processes from the spine of Purkinje cells and also the multiple innervation of Purkinje cells by the ascending fibers. The glial Ca^{2+} -permeable AMPA receptors are indispensable for proper structural and functional regulation of Bergmann glia and glutamatergic synapses. Transfer of information from the ascending fiber to the Bergman glia has been shown to depend on the "►ectopic release" of glutamate from the ascending fiber to the receptor expressed on the Bergman cell. This means the existence of active and specific transmission between neuron and astrocyte.

Function

Functions Estimated by In vitro Preparations

Many important findings on neuron–astrocyte interaction have been demonstrated in primary culture or

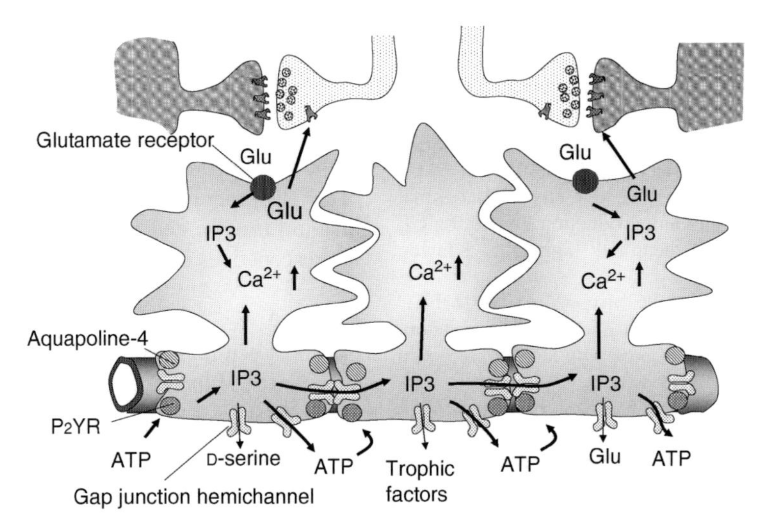

Neuron–Astrocyte Interactions. Figure 5 Regulatory neuron–astrocyte interaction. (a) Synaptogenesis Astrocytes play important roles in synaptogenesis during development. Apoprotein E released from astrocytes has been shown to be a factor facilitating synapse formation. Direct contact of astrocytes and neuronal cells through integrin receptor induces drastic synaptogenesis, which is mediated by protein kinase C (*PKC*) activation. (b) Activation of N-methyl-D-aspartic acid (*NMDA*) receptor by D-serine. D-Serine is produced from L-serine inside astrocytes by a specific enzyme, serine racemase. The amino acid is an effective activator for NMDA receptor, a key receptor for synaptic plasticity. (c) Regulation of neuron–astrocyte interaction by Ca^{2+} -permeable AMPA receptors Ca^{2+} -permeable AMPA-type glutamate receptors expressed on the Bergman glia are indispensable for proper structural and functional regulation of the Bergmann glia and glutamatergic synapses. (d) Ectopic release of glutamate from neuron to astrocyte. The ascending fiber terminal releases glutamate and directly activates Ca^{2+} permeable AMPA receptor expressed on the Bergman cell.

organotypic culture preparations. These studies showed the wide range of inhibitory effects of ATP released from an astrocyte, the facilitatory and inhibitory interactions between astrocytes and neurons due to glutamate released from either cell [10] and facilitation of synaptogenesis [9]. However, since the functions of astrocytes are structure- and environment-dependent as mentioned in this essay, real functional profiles of the neuron–astrocyte interaction in in vivo brain may be difficult measure in such simple in vitro experiments. One possible breakthrough for this difficulty may be provided by in vitro study of retina, which has a closely similar structure to brain. The synapses in the retina are contacted by Müller cells (astrocyte-like radial glia) and make regulatory configurations with neuronal networks similar to those in brain.

Functions Estimated from Pathological Conditions

Recent pathological studies demonstrate that dysfunction of "neuron–astrocyte interaction" is an important causal factor in the development of schizophrenia, depression and some other psychiatric disorders. The activities of a glutamate transporter (GLT-1 type) in astrocytes obtained from schizophrenic patients were significantly higher than in those obtained from normal brain. Furthermore the level of D-serine, a coactivator for NMDA receptor and a product of astrocytes is significantly lower in the schizophrenic brain than in normal brain. These facts suggest that appropriate activities of glutamatergic synapses required for expression of higher order brain functions are regulated by astrocytes.

Many other brain dysfunctions due to abnormality of astrocyte functions, such as epilepsy and dementia, have demonstrated the importance of neuron–astrocyte interactions for the establishment of higher order brain functions.

References

1. Haydon PG (2001) Glia: listening and talking to the synapse. Nat Rev Neurosci 2:185–193
2. Nedergaard M, Ransom B, Goldman SA (2003) New roles for astrocytes: redefining the functional architecture of the brain. Trends Neurosci 26:523–530
3. Reichenbach A, Wolburg H (2005) Astrocytes and ependymal glia. In: Kettenmann H, Ransom BR (eds) Neuroglia, 2nd edn. Oxford University Press, Oxford, pp 19–35
4. Conell-Bell AH, Frankbeiner SM, Cooper MS, Smith SJ (1990) Glutamate-induced calcium waves in cultured astrocytes: long range glial signaling. Science 247:470–473
5. Iino M, Goto K, Kakegawa W, Okado H, Sudo M, Ishiuchi S, Miwa A, Takayasu Y, Saito I, Tsuzuki K, Ozawa S (2001) Glia-synapse interaction through Ca^{2+}-permeable AMPA receptors in Bergmann glia. Science 292:926–929

6. Ye ZC, Wyeth MS, Baltan-Tekkok S, Ransom BR (2003) Functional hemichannels in astrocytes: a novel mechanism of glutamate release. J Neurosci 23:3588–3596
7. Araque A, Papura V, Sanzgiri RP, Haydon PG (1999) Tripartite synapses: glia, unacknowledged partner. Trends Neurosci 22:208–215
8. Baltan-Tekkok S, Ransom BR (2004) The glial-neuronal interaction and signaling: an introduction. In: Hatton GI, Parpura V (eds) Glial neuronal signaling. Kluwer, Amsterdam, pp 1–20
9. Hama H, Hara C, Yamaguchi K, Miyawaki A (2004) PKC signaling mediates global enhancement of excitatory synaptogenesis in neurons triggered by local contact with astrocytes. Neuron 41:405–415
10. Morita M, Higuchi C, Moto T, Kozuka N, Suswuki J, Itofusa R, Yamashita J, Kudo Y (2003) Dual regulation of calcium oscillation in astrocytes by growth factors and pro-inflammatory cytokines via mitogen-activated protein kinase cascade. J Neurosci 23:10944–10952

Neuron/Cell

Definition

The neuron is the functional unit of the nervous system, specialized for the conduction of electrochemical impulses along neuron processes, and the transmission of information from one neuron to another usually by the release of a neurotransmitter from one neuron onto another that expresses receptors for that neurotransmitter.

Neuron/Cellular Doctrine

Definition

The neuron cell doctrine states that the neuron with its processes is a single cell and forms the functional unit of the nervous system. In this doctrine nerve networks (interconnections) and pathways (connections from one collection of neurons to another) are made by synaptic contacts between neurons.

Neuron-Glia-Imaging

ARAYA RUANGKITTISAKUL, KLAUS BALLANYI
Department of Physiology, Perinatal Research Centre, Faculty of Medicine & Dentistry, HMRC, University of Alberta, Edmonton, AB, Canada

Synonyms

Neuronal imaging; *in situ* microscopy

Definition

►Neuron-glia-imaging means videocamera, ►confocal or ►multi-photon microscopy of dynamic changes in the activity and/or morphology of living ►neurons and ►glia at the (sub) cellular level. A rapidly increasing number of cellular processes is being monitored with neuron-glia-imaging *in vitro* and *in vivo* using genetically encoded fluorescent dyes expressed in targeted cells.

Purpose

For most central nervous tissues, the relation between structure and function is largely unknown at the cellular level. Brain structures are monitored *in vivo* with powerful techniques such as ►positron emission tomography (PET). However, analysis of fast activity in central nervous micro-networks is not feasible with such functional brain imaging due to a low temporal and spatial resolution. Rather, video, confocal or multi-photon microscopy (mostly ►two-photon microscopy) is used *in vivo* and *in vitro* for simultaneous imaging of brain cell activity and morphology. Optical techniques are being developed further in conjunction with engineering of genetically encoded fluorescent dyes to allow for neuron-glia-imaging of nervous structures in deeper brain layers of freely moving mammals. This may allow for future endoscopic investigation of neuron-glia networks plus microcirculation in the almost intact mammalian brain and spinal cord, possibly with the attractive option to diagnose and treat nervous diseases such as focal epilepsy.

Principles

Biological Activity Dyes

Biochemical, physiological and morphological features of ►brain cells can be assessed with a variety of (fluorescent) dyes [1–4]. In a morphological study on living ►taste organs of frogs [5], the H^+ - thus, pH-sensitive dye BCECF and the Ca^{2+}-sensitive dye Indo-1 stained glia-like cells, while ►type-II receptor cells were stained with the Cl^--sensitive dye MQAE. In contrast, ►Merkel-like basal cells were only stained with the membrane-labeling dye FM1–43 that is also used for recording synaptic processes such as ►vesicle recycling. The above ion-sensitive dyes as well as the Na^+ -sensitive dye SBFI and the K^+-sensitive dye PBFI are well suited to image nervous activity which is typically associated with notable changes of cellular ions [2–4,6,7]. Neuronal signals in the upper µs-lower ms range are well resolved with high-speed video-microscopy using voltage-sensitive dyes, e.g., Di-8-ANEPPS [1,4] or Ca^{2+} -sensitive dyes [1–4,6–9] (see below). Some dyes stain quite selectively organelles. Amongst these, Rhodamine-123 is used to monitor both mitochondrial structure and membrane potential [1,4]. Dual dye labeling allows for ►ratiometric measurements for a better signal-to-►noise ratio and/or calibration of concentrations of cellular factors [1,3,4]. Ratiometric measurements can also be done with single indicators showing an excitation or emission spectral shift upon ion binding. If ratiometric [Ca^{2+}] measurements are not possible, the change in ►fluorescence can be calculated as *ΔF/F*, i.e., as the background-corrected change in fluorescence (*ΔF*) divided by resting fluorescence (*F*). This calculation allows comparison of fluorescence transients in cellular compartments with different thickness, for example, ►soma or ►spines and/or indicator concentration [1,3,4,7].

Molecular techniques offer a most powerful approach for neuron-glia-imaging [2,4]. For example, cameleons are protein-based, resulting from the fusion of calmodulin with a calmodulin-binding peptide with cyan and yellow mutants of ►green-fluorescent-protein. They utilize ►fluorescence-resonance-energy-transfer for coupling Ca^{2+} binding to changes in fluorescence. In general, this spectroscopic technique allows to monitor changes in both, distance (20–100 nm) and orientation of two ►fluorophores. In addition to Ca^{2+}-sensitive cameleons, specific macromolecule pairs have been designed to use fluorescence-resonance-energy-transfer to record biochemical or physiological signals such as ►membrane potential, ► cyclic-adenosine-monophosphate or protein-protein heterodimerization. Besides, some mutants of green-fluorescent-protein are sensitive to pH and/or Cl^-. Genetically encoded fluorescent probes are being targeted to different tissue and cell types and/or various subcellular structures such as ►endoplasmic reticulum or ►synapses, using ►viral transfection and transgenic techniques [2,4].

Labeling Neurons and Glia

Cultured brain cells form a thin layer and can thus be visualized at reasonable optical resolution with a standard fluorescence microscope attached to a videocamera-based imaging system [1,4]. Neurons and glia in culture can easily be labeled with morphological dyes of the "Alexa Fluor" or "BODIPY" families, or with ►Ca^{2+}-sensitive dyes for ►Ca^{2+} imaging [1,2,4]. Loading with Ca^{2+}-sensitive dye is achieved by adding to the culture medium the membrane-permeant, acetoxymethyl ("AM") form of the dye, which is cleaved into the impermeant, fluorescent form by cellular esterases [1,4]. In contrast to cultures, cells in acute ►brain slices remain in their natural environment *in situ* and thus show often features close to *in vivo* (Fig. 1) [4,6,7]. However, in particular in mature brain structures *in situ*, loading of neurons by addition of the AM form of the Ca^{2+}-sensitive dye to the superfusate may not be successful due to diffusional or uptake problems. In such cases, ►pressure injection of the AM dye can provide adequate loading of glia and neurons in both brain slices and *in vivo* (Fig. 1) [4,6].

In a different *in situ* approach, the membrane-impermeant form of the Ca^{2+}-sensitive dye is injected into a brain cell via the recording ►patch- or ►micro-electrode (Figs. 1, 2) [4,6,7,9]. With this method, morphological features and/or activity-related [Ca^{2+}] rises can be correlated with electrophysiologically recorded biophysical parameters, such as membrane potential, resistance or capacitance. In brain tissue *in situ*, visualization of cellular structures with conventional light microscopy is usually restricted to depths of ~50 μm.

Confocal and Videocamera Microscopy

The spatial resolution of fluorescence imaging is greatly improved by one-photon-excitation ►confocal laser-scanning microscopy, which eliminates out-of-focus and stray light via a pinhole aperture (Figs. 1–4) [4,7,8].

Confocal techniques do not principally improve neuron-glia-imaging in deeper tissue layers, in contrast to one specific subtype of two major video-microscopy techniques [4,8]. Intensified video-microscopy involves imaging a specimen when light levels are too low for standard cameras, while enhanced video-microscopy is used when the specimen is invisible to the eye, either due to lack of contrast or due to its spectral characteristics, i.e., ultraviolet or infrared [4]. As infrared light enters deeper into tissues [4,8,9], infrared ►differential-interference-contrast enhanced video-microscopy produces images of almost three-dimensional quality from cells in tissue depths of 50–100 μm. Besides, the technique allows for dynamic recording of (re)organization of cellular structures [4]. Infrared darkfield video-imaging can be used in brain tissue *in situ* to monitor neuronal activity as cellular ion fluxes and subsequent volume changes affect light scattering. Such imaging of intrinsic optical signals can, e.g., provide cortical activity maps, although not at cellular resolution [1,4].

Two-Photon Microscopy

Two-photon-excitation laser-scanning fluorescence microscopy enables *in situ* visualization of brain cells in tissue depths up to ~2 mm [4,6,9]. For two-photon-excitation, femtosecond pulses from a mode-locked Ti: sapphire infrared laser at megahertz repetition rates are focused to a point in the tissue and scanned over a horizontal optical plane (Figs. 2–4). Within a focus of only about one femtoliter in the tissue, two infrared photons can be simultaneously absorbed by the fluorophore and produce an excited state similar to that from the absorption of a single photon of twice their energy. The fluorescence is collected by a photomultiplier and fed into a computer. Specific software is used to reconstruct a plane of fluorescence intensity reflecting the activity and/or morphology of brain cells within that plane. As for ►confocal microscopy, three-dimensional images can be produced if consecutive image planes are scanned while a stepmotor moves the objective or specimen stage along the z-axis (Figs. 2–4) [4,8,9].

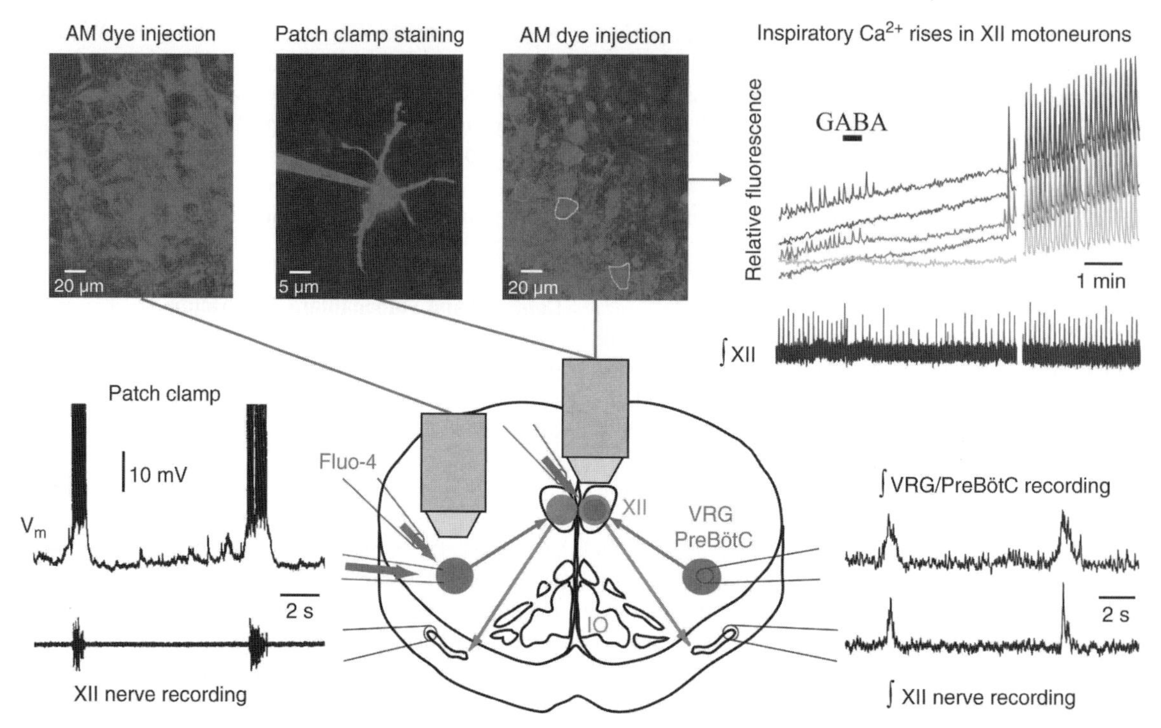

Neuron-Glia-Imaging. Figure 1 Confocal and two-photon ►Ca^{2+} imaging in respiratory neurons. The schematic shows a newborn rat transverse ►brainstem slice containing the neuronal network that initiates and controls inspiratory-related breathing movements. Rhythmogenic ►interneurons are located within the ►pre-Bötzinger Complex (preBötC), a subregion of the bilateral rostrocaudal columns of the ventral respiratory group. preBötC neurons transmit their inspiratory activity to motoneurons of the ►hypoglossal (XII) ►nucleus whose axons project within the same transverse plane to XII ►nerve rootlets. The left lowermost trace shows two bursts of inspiratory XII nerve activity, while the right lowermost trace displays the integrated form of such activity in a different slice. An integrated extracellular signal of preBötC ►neuron population activity is shown above the latter recording, while the trace on top of the XII nerve recording in the left part of the figure shows rhythmic ►membrane potential (V_m) fluctuations of a single VRG-preBötC neuron. Individual respiratory neurons can be labeled via the recording ►patch-electrode with fluorescent dye as exemplified in the central ►confocal image of the upper part of the figure for an inspiratory XII motoneuron stained with the ►Ca^{2+}-sensitive dye Fluo-4. In addition, the activity and morphology of multiple VRG- preBötC neurons (*left*, ►two-photon image) and XII motoneurons (*right*, confocal image) can be visualized with Ca^{2+} imaging. For that purpose, the membrane-permeant form of Fluo-4 (0.5 mM) is ►pressure-injected (0.5–1 psi, 5–15 min) into the VRG- preBötC or XII nucleus. The upper right part of the figure illustrates that two of seven XII motoneurons show Ca^{2+} rises in phase with inspiratory XII nerve activity. Bath-applied (1 mM) ►γ-aminobutyric acid (GABA) slightly depresses respiratory rhythm and abolishes inspiratory Ca^{2+} rises. Upon washout of GABA, Ca^{2+} oscillations of the two cells are potentiated severalfold while pronounced rhythmic activity is induced in further XII motoneurons, probably due to decrease of tonic extracellular GABA levels by stimulated ►GABA-uptake. All recordings from A. Ruangkittisakul and K. Ballanyi, unpublished.

Two-photon microscopy can easily be combined with electrophysiological recording of single brain cell or neuronal population activity, even *in vivo* (Figs. 1, 2), while miniature two-photon-microscopes enable neuron-glia-imaging in freely moving animals [4,6]. Currently, optical fibers and lenses with a diameter <1 mm are being developed for two-photon microscopy. This enables visualization of deep brain structures, such as the ►hippocampus, which is crucial for memory processes while its dysfunction can constitute a focus for ►epilepsy [4,6]. Besides revealing brain cell properties, ►two-photon imaging enables visualization of brain ►microcirculation, which is involved in various nervous diseases such as ►angiopathy or ►stroke [4,6]. The high spatial resolution of two-photon microscopy, but also of other types of neuron-glia-imaging, is further improved by image processing and deconvolution techniques [1,4,8]. Computer power is steadily growing while ultra-fast scanning techniques, such as acousto-optical tuning deflectors, are currently being implemented into modified two-photon microscopy (Fig. 3) [4,8,10]. Thus, four-dimensional imaging of deeper brain structures in real-time is becoming feasible soon. This opens the

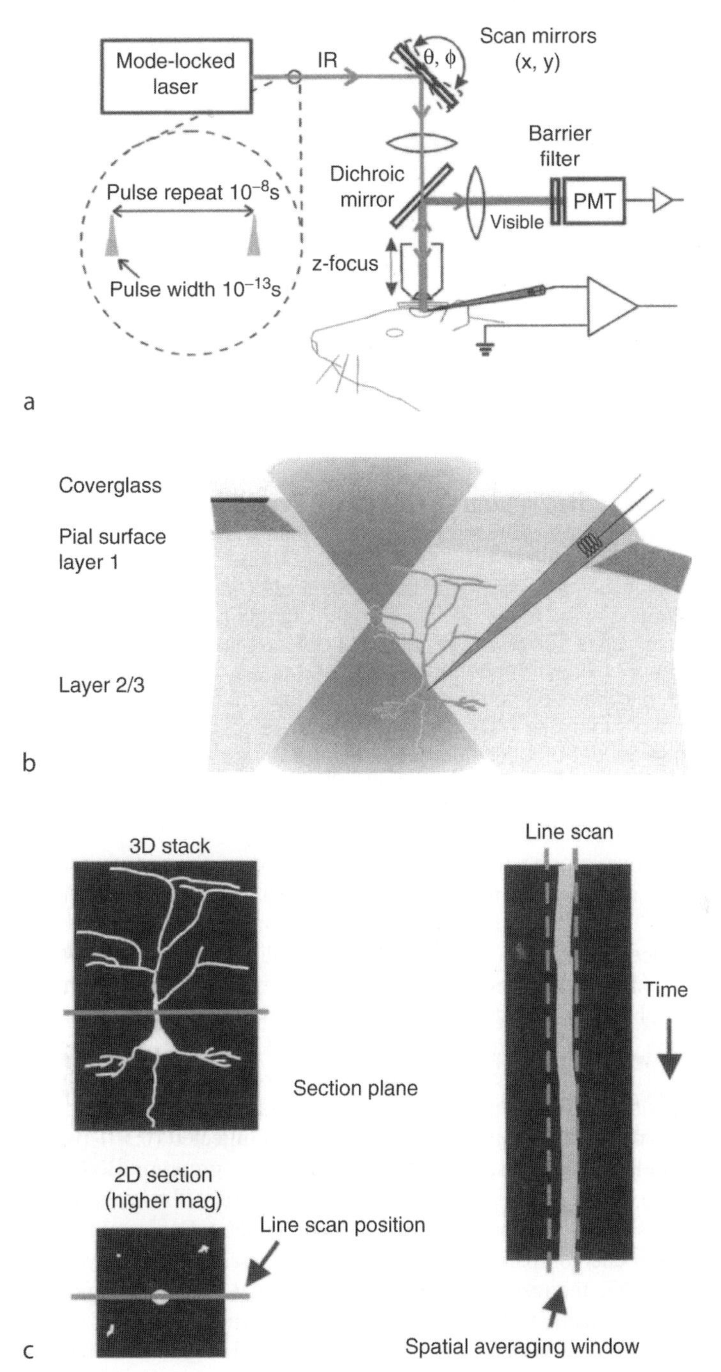

Neuron-Glia-Imaging. Figure 2 *In vivo* two-photon microscopy combined with sharp ►micro-electrode intracellular recording in the ►neocortex of anesthetized rats. A neuron was filled via the micro-electrode with the Ca^{2+}-sensitive dye Calcium-Green-1. This allowed for recording of activity-induced intracellular Ca^{2+} transients in ►dendrites (located in ►cortical layer 1) or ►soma (in layer 2/3) of the neuron simultaneous with electrophysiological recording of membrane potential. (a) Schematic of the microscope. (b) Schematic of the ►craniotomy and recording geometry. (c) Schematic illustrating various imaging modes. Reproduced from Svoboda, Tank, Stepnoski and Denk in [4], with kind permission.

long-term perspective for two-photon microscopy to be used for endoscopy of brain structures to analyze normal and pathophysiologically disturbed neuronal functions [4,6].

Confocal and Two-photon Microscopy of the Isolated Mammalian Respiratory Network

The neuronal network which ultimately initiates and controls muscles mediating inspiratory breathing movement

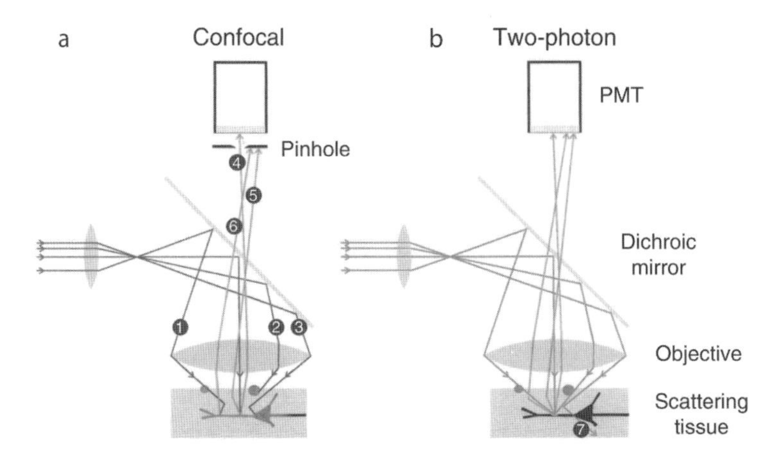

Neuron-Glia-Imaging. Figure 3 Comparison between confocal and two-photon Laser-Scanning microscopy in the intact brain. The fates of typical excitation (*blue* and *red lines*) and fluorescence (*green* lines) photons are shown to illustrate the advantages of two-photon microscopy. The sample consists of a scattering neural tissue permeated by blood vessels (*red ovals*) and containing a labe0000led neuron. In the confocal case (*left*), the short-wavelenth excitation photons have a relatively high chance of being scattered (*1,3*) or absorbed (*2*). Only unscattered fluorescence photons originating at the focus contribute to the signal (*4*). Scattered (*5*) and out-of-focus (*6*) photons are blocked by the pinhole, which is necessary to reject out-of-focus fluorescence. Spurious excitation, and hence photobleaching and phototoxicity, occur throughout a large part of the cell (*green region*). In the two-photon case (right), the long-wavelength excitation photon have a relatively low chance of being scattered or absorbed; scattered photons are too dilute to excite (*7*) and excitation is therefore localized to the tiny focal volume. Since no pinhole is necessary to reject out-of-focus fluorescence, all fluorescence photons entering the objective contribute to the signal. Reproduced from Svoboda, Tank, Stepnoski and Denk in [4], with kind permission.

is currently studied by various groups in transverse brainstem slices of newborn and juvenile rodents (see also essays on "brain slices" and "isolated respiratory centers"). These brain slices contain a kernel of rhythmogenic interneurons within the ►pre-Bötzinger Complex (preBötC), a subregion of the bilateral rostrocaudal columns of the ventral respiratory group (Fig. 1). The preBötC has a rostrocaudal extension of about 200 μm and provides within the slice an inspiratory excitatory drive to ►hypoglossal motoneurons whose axons exit the preparation in the same transversal plane (Fig. 1). This kernel of the ►inspiratory network is devoid of afferent influences from ►sensory systems mediating, e.g., peripheral ►chemosensitivity or the ►lung-stretch reflex. Despite this, the respiratory active brain slices retain central chemosensitivity for CO_2, thus pH, and O_2 and respond to ►neuromodulators in a fashion very similar to that in more intact *in vitro* preparations, and even in (preterm infant) humans. In slices with a rostrocaudal thickness of 500–700 μm, a regular inspiratory rhythm can be recorded at physiological superfusate $[K^+]$ (3 mM) for several hours from hypoglossal ►nerve roots or within the preBötC. Such recording can routinely be combined with ►patch-clamp measurement of oscillatory membrane properties of rhythmogenic preBötC ►interneurons or hypoglossal motoneurons (Fig. 1).

Figure 1 shows that the activity and structure of the cellular elements of this neuronal network can be analyzed with two-photon or confocal microscopy. Individual respiratory neurons are labeled via the recording patch-electrode with fluorescent dye as exemplified for a hypoglossal motoneuron stained with the Ca^{2+}-sensitive dye Fluo-4. Also the activity and morphology of populations of ventral respiratory group/preBötC neurons or hypoglossal motoneurons can be visualized with Ca^{2+} imaging. For that purpose, the membrane-permeant form of Fluo-4 (0.5 mM) is pressure-injected (0.5–1 psi, 5–15 min) into the ventral respiratory group/preBötC or the hypoglossal nucleus. Figure 1 also illustrates that two of seven hypoglossal neurons show Ca^{2+} rises that are in phase with inspiratory hypoglossal nerve activity. Bath-applied ►γ-aminobutyric acid (GABA) slightly depresses the ►respiratory rhythm and abolishes the inspiratory Ca^{2+} rises. Upon washout of GABA, the Ca^{2+} oscillations of the two cells are potentiated severalfold while pronounced rhythmic activity is induced in further hypoglossal motoneurons, probably due to decrease of tonic extracellular GABA levels by stimulated ►GABA-uptake.

Advantages and Disadvantages

Imaging Systems

The choice of hardware for neuron-glia-imaging depends on whether speed or spatial resolution of the acquired image series is more relevant. Digital videocameras with a high pixel number provide a lateral resolution that is indistinguishable from fine-grain photography [1,4].

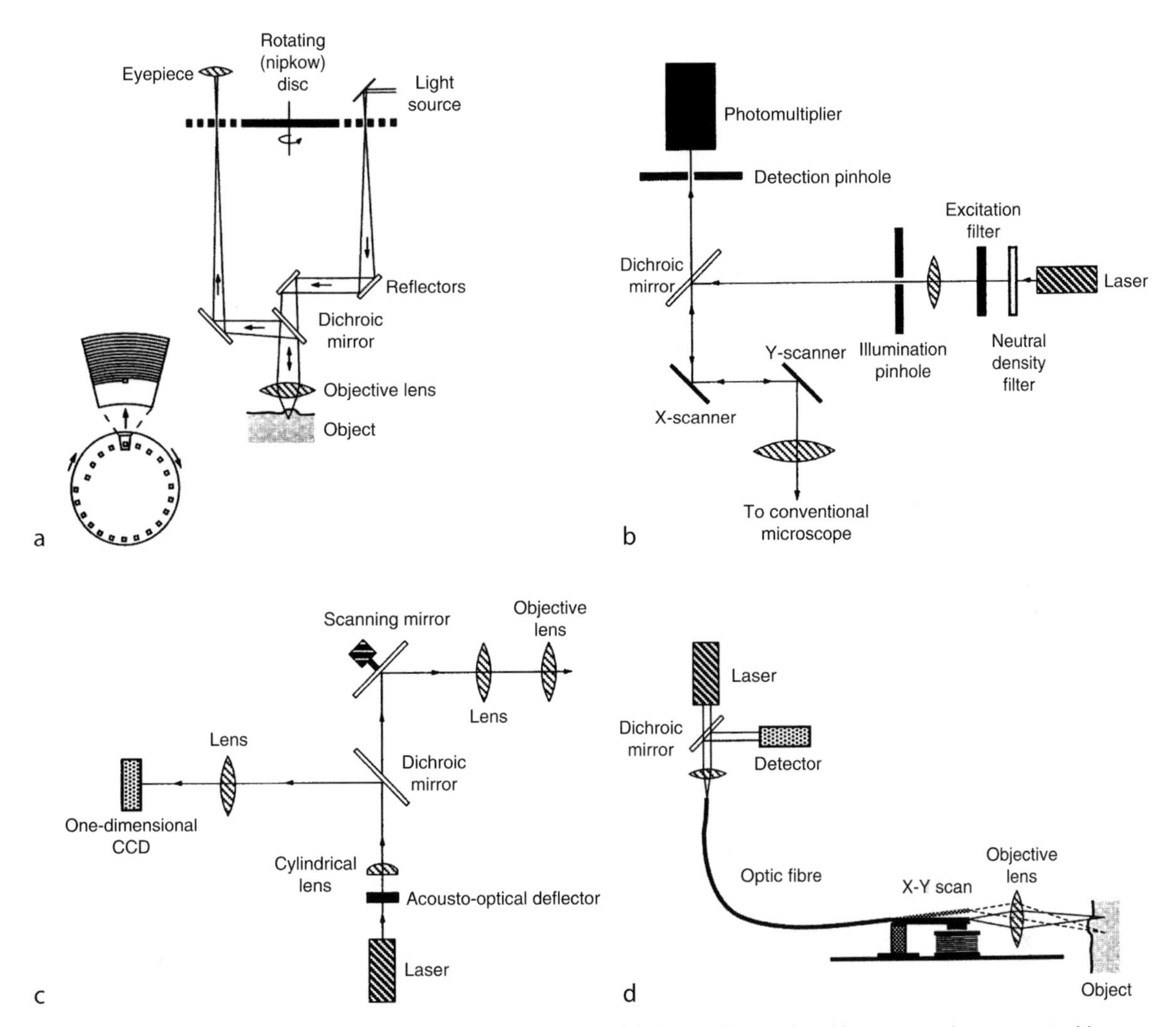

Neuron-Glia-Imaging. Figure 4 Confocal implementation. (a) A two-dimensional image can be generated by spinning a spiral array of illumination apertures (a Nipkow disc) over a window on the specimen, thereby illuminating points on sequential lines (*insert*). A confocal image of the illuminated points is viewed through a similarly moving array in a conjugate image plane. (b) A spot of light from a fixed point source can be scanned across the specimen by mirrors; longer-wavelength fluorescence from the point is reflected back along the same light path through a dichroic reflector to a conjugate fixed-point aperture. The time-varying signal from the detector behind this aperture is converted by computer into a two-dimensional image. (c) Scanning can also be performed by an acousto-optical deflector, but the wavelength dependency of these devices prevents their use for de-scanning the longer-wavelength fluorescence. Instead, a linear detector array can be used, sampling each element in sequence corresponding to the moving point. (d) A simpler configuration uses a fiber optic, vibrating in an image plane, as both point source and detector aperture. Reproduced from A. Fine in [4], with kind permission.

But, the large amount of digital data limits the sampling rate. Conversely, recording with high-speed video-cameras provides full-frame sampling rates >1 kHz at the cost of a low spatial resolution due to digital chips with a low pixel number. A further type of intensified videocameras is optimized to sample fluorescence or luminiscence, at low light intensities, often down to the single-photon level. For example, some voltage-sensitive dyes produce a <0.1% change in fluorescence in response to nervous activity, on top of a high background fluorescence [1,4].

For all its virtues, confocal microscopy has a major flaw in that it excites fluorophores in excess, but detects only a small fraction of the generated fluorescence light [4,8]. Two-photon microscopy reduces problems associated with scattering and absorption of fluorescence compared to one-photon-excitation (confocal) techniques for several reasons. (i) The use of longer excitation wavelengths leads to a better tissue penetration; (ii) The two-photon excitation wavelengths for most useful fluorophores fall into a range (800–900 nm) characterized by low absorption due to water, blood or other intrinsic tissue fluorophores; (iii) Scattered excitation photons are too dilute to excite and, hence, produce negligible background or damage (see below); (iv) In two-photon microscopy, three-dimensional sectioning and resolution are due to the localization of excitation alone. Thus, scattered fluorescence photons constitute useful signals, permitting greatly increased collection efficiencies (Fig. 3) [4,8,9].

In commercial confocal or two-photon microscopes, the sampling rate is limited by the mass of the galvanometric scanning mirrors (Figs. 2–4) [4,8]. It can be increased, at the cost of spatial resolution, to the hundred-millisecond or low millisecond range by acquiring only a small portion of the image plane or a line scan, respectively (Fig. 2) [4]. Considerably higher scan speed can be achieved with acousto-optical tunable filters (deflectors), which use sound to induce refraction waves that behave like a refraction grating (Fig. 4) [4,10]. Imaging of many points at once at rates >100 Hz can also be achieved by coordinated scanning of apertures, e.g., arranged in a spiral pattern of a Nipkow disc (Fig. 4) [4,8]. However, adequate simultaneous illumination of the multiple points is difficult to achieve, thus limiting sensitivity. While two-photon-excitation makes ratiometric measurements with dual-excitation dyes unfeasible, dual-emission rationing is possible. For confocal imaging, the choice of Ca^{2+}-sensitive or other dyes is determined by the availability of lasers operating at the required wavelength [4,8]. This potential limitation for use of dyes can be overcome by Nipkow-disc confocal microscopy (Fig. 4) or videocamera imaging. The latter techniques do not critically depend on dye excitation via a laser and can rather make use of conventional light sources such as a xenon or halogen lamp, often in combination with a monochromator [4,8].

Dye Properties

Due to the limited format of this article, technical considerations will be exemplified only for Ca^{2+}-sensitive dyes (in addition to those properties of such dyes already outlined above) [1,4,7]. Ca^{2+}-sensitive dyes are most powerful tools for neuron-glia-imaging, as they allow simultaneous assessment of (sub)micrometer structures and millisecond electrical activities. Neuronal activity can rapidly increase cytosolic $[Ca^{2+}]$ due to Ca^{2+} influx from the interstitial space via voltage-activated Ca^{2+} channels and/or Ca^{2+}-permeable neurotransmitter receptors, often in conjunction with Ca^{2+} release from intracellular stores. As intracellular Ca^{2+} is a ▶second messenger of ultimate importance, Ca^{2+} imaging is pivotal to the understanding of nervous functions [2–4, 6–9]. Ca^{2+}-sensitive dyes should be chosen according to the primary aim of a study. For assessment of morphological features, e.g., spine formation, it is important that the dye fluoresces brightly already at low nanomolar, thus resting intracellular $[Ca^{2+}]$ levels. For that purpose, Fura-2 can be used with most videocamera imaging systems and two-photon microscopy. Confocal laser-scanning microscopy is often not possible with Fura-2 as most commercial systems cannot use ultraviolet light, while Calcium-Green or Fluo-4 can be used with most imaging systems. Due to its relatively high resting fluorescence, Calcium-Green has a decreased dynamic range for $[Ca^{2+}]$ measurements compared to Fluo-4, while the latter dye fluoresces less brightly at resting $[Ca^{2+}]$ (Fig. 1). Fura-2 has a large dynamic range despite a bright fluorescence at low $[Ca^{2+}]$, as intracellular Ca^{2+} rises decrease its fluorescence intensity [1,3,4,7–9].

The dynamic range of activity-related Ca^{2+} imaging depends greatly also on the dissociation constant (K_d), which describes Ca^{2+} binding to the dye. K_d is affected by many factors, including pH, temperature, protein binding and ions such as Mg^{2+}. Indicators have a detectable response in the concentration range from approximately $0.1 \times K_d$ to $10 \times K_d$. A dye with a low K_d value binds most Ca^{2+} already at low $[Ca^{2+}]$ levels and is saturated at higher values. For example, Fura-2 with a K_d of 145 nM measures effectively $[Ca^{2+}]$ in brain cells at levels between 20–2,000 nM. In contrast, Fura-FF ($K_d = 5{,}500$) is predestined for measurements of micromolar $[Ca^{2+}]$ transients that are observed, e.g., during repetitive or pathological neuronal activity, in particular in small compartments such as dendrites or spines [1,3,4,7,9].

The fact that Ca^{2+} is bound to the dye means that the dye acts as a Ca^{2+} buffer. Accordingly, the magnitude of activity-related $[Ca^{2+}]$ transients is attenuated and the kinetics prolonged. In AM Ca^{2+} dye-loaded cells, the artificial Ca^{2+} buffer adds up to intrinsic Ca^{2+} buffer systems, such as calbindin or parvalbumin. In contrast, in single brain cells loaded with the dye via a recording patch- or micro-electrode, the dye can substitute for endogenous Ca^{2+} buffers that are eventually washed out from the cell. It should be noted that different cell types express substantially different levels of endogenous buffer [1,3,4,7].

The latter considerations show that it is important to choose the right intracellular dye concentration. A high concentration may, on the one hand, provide a better resolution of structures or improve the signal-to-noise ratio for activity measurements. On the other hand, it may distort fast activity related $[Ca^{2+}]$ transients, which often have immediate second messenger function such as activating ▶Ca^{2+}-gated K^+ channels. If neuronal networks in AM Ca^{2+} dye-loaded systems are studied, such as a respiratory brainstem slice (see above and section on brain slices) (Fig. 1), it must be considered that not only the excitability of the imaged cells, but rather major parts of the network, may be affected by high intracellular Ca^{2+} dye concentrations [1,3,4,6,7].

Photobleaching, Phototoxicity, Auto-Fluorescence

A number of natural peptides is auto-fluorescent when excited at wavelengths well into the ultraviolet region. This may cause interference with probes used to measure intracellular ion concentrations requiring excitation in that range of thr spectrum [1,4]. Apart from that, all fluorescent probes will photobleach to a greater or lesser extent when excited with a suitable wavelength, at a rate proportional to the intensity of the incident light. While this may not be a problem for some morphological applications, it does seriously affect any attempt to quantify intracellular concentrations using single-wavelength probes [1,3,4,7–9]. The most obvious practical way to reduce ►photobleaching is to minimize light reaching the probe. Since this reduces the amount of fluorescence, optimum conditions for image analysis must include, for example, optimal dye loading or sampling fluorescence with highly sensitive systems such as cooled CCD-type intensified videocameras or two-photon microscopes. Room light contributes to photobleaching already during the loading procedure and during storing the loaded cells prior to the actual experiment. Also oxygen plays a major role in photobleaching, at least during Ca^{2+} imaging with Fura-2. Genetically encoded fluorescent proteins can have a remarkable resistance to photobleaching in addition to their large extinction ratios and quantum efficiencies [2,4,6].

Two-photon imaging may deteriorate living tissue by production of heat, in particular when a high power of the infrared laser is needed. Accordingly, two-photon microscopy of deeper brain structures *in vivo* may induce a caloric challenge not only to the studied brain region, but indirectly, to the entire animal [4,8,9]. Fluorescent molecules in their excited state react with oxygen to make ►free radicals, which can damage cellular molecules [1,4,7,8]. There are several strategies to reduce such phototoxic cell damage, specifically (i) Use of high numeric aperture lenses allows for collecting more light, thus enabling reduction of excitation light intensity; (ii) Reduction in the number and/or rate of scans. For example, Fluo-4-loaded cells of the ►respiratory network need to be confocally scanned at a rate of 0.3–1 seconds to visualize cytosolic $[Ca^{2+}]$ oscillations. But, it is advisable to record only for several minutes during control and pharmacological treatment and stop scanning for the rest of time; (iii) Using two-photon microscopy, ►phototoxicity is reduced by focal excitation allowing for continuous recording of respiratory oscillations for >1 h at a rate of >2 Hz. For the latter approach, it is advisable to use an external, more sensitive photomultiplier that is located closer to the specimen to enable reduction of excitation light. In some preparations, it may be helpful to reduce potential toxicity chemically by adding antioxidants, such as oxyrase or ascorbic acid to the superfusate [1,4]. The toxicity does probably not solely depend on the intensity of the excitation light. For example, some voltage-sensitive dyes can only be used for a limited time period to monitor nervous activity, before activity of the preparation gets impaired. More recent dyes of that class, such as Di-4-ANEPPS, can be used for up to severely hours without severely impairing nervous functions [1,4,8].

Some of the limitations of conventional dyes can be avoided, or at least attenuated, by fluorescence-resonance-energy-transfer [2,4,8]. This spectroscopic technique is general, non-destructive and easily imaged and has thus proven to be one of the most versatile readouts available to the designer of new optical probes (see above). It is particularly amenable to emission rationing, which is more reliably quantifiable than single-wavelength monitoring and is better suited than excitation rationing for high-speed and laser-excited imaging. Two-photon microscopy can be easily combined with ►fluorescence-lifetime measurements for quantitative fluorescence-resonance-energy-transfer imaging. One major domain of fluorescence-lifetime measurements is the field of ion concentration imaging. This method is insensitive to intensity effects, such as shading, photobleaching, absorption or light source noise. This can be an important advantage, especially in confocal studies of brain cells *in situ*, where absorption effects and photobleaching are important limitations [2,4,8].

N

Acknowledgments

The study was supported by AHFMR, CIHR and CFI-ASRIP. We thank Dr. A. Fine for comments.

References

1. Mason WT (1999) Fluorescent and luminescent probes for biological activity. Academic, London
2. Miyawaki A (2003) Fluorescence imaging of physiological activity in complex systems using GFP-based probes. Curr Opin Neurobiol 13:591–596
3. Takahashi A, Camacho P, Lechleiter JD, Herman B (1999) Measurement of intracellular calcium. Physiol Rev 79:1089–1125
4. Yuste R, Lanni F, Konnerth A (eds) (2000) Imaging neurons. Cold Spring Harbour Laboratory Press, Cold Spring Harbour, NY
5. Li JHY, Lindemann B (2003) Multi-photon microscopy of cell types in the viable taste disk of the frog. Cell Tissue Res 313:11–27
6. Brecht M, Fee MS, Garaschuk O, Helmchen F, Margrie TW, Svoboda K, Osten P (2004) Novel approaches to monitor and manipulate single neurons *in vivo*. J Neurosci 24:9223–9227
7. Eilers J, Schneggenburger R, Konnerth A (1995) Patch clamp and calcium imaging in brain slices. In: Sakmann B, Neher E (eds) Single-channel recording. Plenum, New York, pp 213–229
8. Pawley JB (ed) (1995) Handbook of confocal microscopy. Plenum, New York

9. Svoboda K, Denk W, Kleinfeld D, Tank DW (1997) *In vivo* dendritic calcium dynamic in neocortical pyramidal neurons. Nature 385:161–165
10. Bullen A, Patel SS, Saggau P (1997) High-speed, random excess fluorescence microscopy: I. High-resolution optical recording with voltage-sensitive dyes and ion indicators. Biophys J 73:477–491

Neuron: Structure/Function, Cellular/Molecular

Naweed I. Syed
Department of Cell Biology and Anatomy, Hotchkiss Brain Institute, University of Calgary, Calgary, AB, Canada

Definition

Neuron – a late nineteenth century Greek term, refers to highly specialized "nerve cells" that conduct electrical impulses (►Action potential). This innate propensity to generate and conduct electrical potentials is a unique hallmark of all "excitable cells" of which the neurons are the most specialized. A neuron exhibits a highly complex repertoire of specialized membranous structures, embedded ►ion channels, ►second messengers, genetic and epigenetic elements and unique complements of various proteins such as the ►receptors. A "►synapse," which is the functional building block of all communicating neurons, refers to the juxtaposed point of contact between two excitable cells. As the nerve impulse invades the "►presynaptic terminal," it elicits the release of chemical messenger/s – the ►neurotransmitter into the synaptic cleft. The diffused chemical neurotransmitter substance, such as ►dopamine, ►serotonin or a proteineous ►peptide (►substance P for example) then binds to its respective receptor located on the "postsynaptic" side of the terminal and invokes an electrical response. Synapses are analogous to electrical bulbs that light up when electric current traveling through the nerve ►cables (►Cable theory) is switched on by the neuron. Thus, synapses serve as the functional unit of all neuronal connectivity upon which hinge its marvelous attributes – ranging from the control of simple ►reflexes (Reflexes) to complex motor patterns, ►learning and ►memory, cognition, emotions etc. Perturbations – emanating from either genetic, cellular and molecular malfunction or an injury – disrupt lines of communications between neurons thus rendering the nervous system dysfunctional. Therefore, central to our comprehension of all brain functions and its repair lie an in-depth understanding of the cellular and molecular elements that make up the neuronal architecture.

Characteristics

From Wiring Together to Firing Together: The Marvelous Neuron

The astonishing structural and functional traits of the human brain have eluded many intriguing minds for centuries – and yet our understanding of even the very basic neuronal elements, such as the synapse, remains pedestrian. Notwithstanding tremendous efforts by the neuroscience community over the decades, the sheer numbers of brain cells (tens of billions) and the intricate nature of their connectivity continue to offer formidable challenges. While tools are being developed to visualize and record the activities of functionally active neurons embedded deep within the brain, an alternative paradigm is to understand how the nervous system is put together during ►development in the first instance. A developmental approach to understanding nervous system function and dysfunction is aimed at drawing up the road maps that were originally used to orchestrate the neuronal connectivity patterns. Once a blue print of all such essential, cellular and molecular components used to lay down the original neuronal maps are "decoded," one might be in a much better position to recapitulate these steps in an adult brain to help "rewire" its damaged connectivity.

A variety of animal model systems are being used to define elements that foster neuronal proliferation, migration and differentiation – steps that are central to the normal wiring of the brain. The steps that enable a neuron to get to its final and well-defined destination in the nervous system are highly complex and rely upon a variety of intrinsic cell-cell signaling and extrinsic factors. Having arrived at its final destination, a neuron begins to develop its axonal and dendritic architecture, which is highly ordered and equipped with navigational tools that would enable these newly born processes to reach out to select groups of target cells that are often located at some distance. Such "search and select" tasks are assigned by neurons to highly specialized structures, termed ►growth cones located at the tip of an extending neurite (axon or dendrite). Every growth cone, fueled by specific ►chemotropic molecules and ►growth factors, follows a precise roadmap, rarely deviating from its defined trajectory that is designed for it to seek out its specific target/s. Growth cones are assisted in their navigational tasks by a variety of cell-cell interacting and diffusible molecules comprising the extracellular milieu. A number of molecules, such as netrin, slit etc. and their interacting receptors are eloquently described and discussed in detail by Spencer et al. (►axonal pathfinding and network assembly).

In addition to various growth-permissive molecules described above, a growth cone's navigational ability is also empowered by a number of well-defined growth repulsive factors that are either membrane-bound or diffused along its path, *en route* towards targets. These growth-repulsive molecules such as the Samaphorins, NI35 etc. will, on the one hand, deter growth cone's entry into the wrong territory, and on the other hand, they serve to prevent wiring among functionally "unrelated" neurons. An intriguing aspect of these growth-suppressive or -repulsive molecules is their continued presence in the adult brain – which incidentally offers formidable challenge to brain repair after trauma and injury. Numerous studies in which the activities of these ►growth-inhibitory molecules were neutralized have uncovered an innate regenerative capacity of the adult neurons – thus underscoring their therapeutic importance vis-à-vis functional recovery from stroke and injury. Metz and Faraji have defined some of these ►growth inhibitory molecules in nervous system development and regeneration and have identified their underlying mechanisms. These authors have also offered several therapeutic strategies that might involve perturbation of these growth-inhibitory molecules to ensure functional regeneration and recovery after nerve injury or neuronal degeneration.

In the vicinity of its target tissue, a growth cone slows its advance and makes physical contacts with potential target cells. Cell–cell interactions via a variety of membrane-bound molecules such as neuroligans and neuregulin etc. trigger inductive changes not only in the presynaptic cells but also its postsynaptic partner. On the presynaptic side, the growth cone undergoes dramatic structural changes that begin with the retraction of filopidia while lamellopodia transform into a club shaped structure. Transmitter vesicles and other related synaptic proteins descend into the bulbous ending, which comes to rest at the juxtaposed postsynaptic site. In addition, Ca^{2+} channels (►Calcium channels – an overview) and other elements of the synaptic machinery specifically cluster presynaptically. At the postsynaptic site, neurotransmitter receptors and their respective second messenger molecules cluster – concomitant with the ►postsynaptic density (PSD). Initially, neurons make myriads of synaptic contacts, which are subsequently refined through ►activity-dependent mechanisms. The molecular machinery mediating cell–cell contact coupled with the activity-dependent mechanisms are central to establishing a precise balance between ►inhibitory and ►excitatory synapses and their respective partners. Interplay between various molecules mediating cell–cell interactions and the underlying mechanisms have been described by Arstikaeitis and El-Husseini (►synapse formation: neuromuscular junction vs. central nervous system) and Colicos (►activity-dependent synaptic plasticity). While El-Husseini's lab takes advantage of powerful molecular techniques to unravel various elements of the synaptogenic program, Colicos lab uses novel photoconductive stimulation techniques to decipher how activity-dependent mechanisms either strengthen or weaken certain synapses. Several recent studies from these and other labs have shed significant light on to the mechanisms by which neurons recognize their potential targets and establish synaptic connectivity. Because some developmental aspects of synapse formation are also recapitulated in the adult brain during ►synaptic plasticity that underlies ►learning and memory, many investigators are taking advantage of activity- or plasticity-related changes in the adult brain to understand how synapses may form and subsequently refine during development. The plasticity-related induction of new synapses or the awakening of the ►silent ►synapses has thus provided greater insight into mechanisms that regulate synapse formation during development (activity-dependent synaptic plasticity) This area of research is not only important for our understanding of the mechanisms underlying nervous system development but also synaptic plasticity that forms the basis for learning and memory in the intact animals.

Due in large measure to the complex nature of the neuronal connectivity in the adult brain where cell–cell interactions are often difficult to study at the level of single pre- and postsynaptic neurons, a number of labs have opted to explore various model system approaches to define mechanisms underlying synapse formation. For instance, the ►neuromuscular junction (NMJ) and various invertebrate models have been extensively used to define both the cellular and molecular mechanisms underlying target cell selection, specific synapse formation and synaptic refinement. As a result of these studies as highlighted by Feng in the chapter ►synaptic transmission: model systems we now know a great deal about various steps that determine the specificity of synapse formation both at the NMJ and between central neurons. Molecules such as ►Agrin that are synthesized and secreted by ►Motoneuron (motor neurons) have been shown to bring about specific inductive changes required for the assembly of the postsynaptic machinery at the NMJ. Similarly, postsynaptic cells have been shown to induce clustering of Ca^{2+} channels and other elements of the synaptic machinery at the presynaptic terminal. Newly formed synapses have since been shown to undergo activity-dependent refinement and consolidation.

Among various proteins that are selectively targeted at both the pre- and postsynaptic sites are the ion channels. For instance, Ca^{2+} (►Calcium channels – and overview), Na^{+} (►Sodium channels) and K^{+} channels (►Neuronal potassium channels) are specifically targeted at select synaptic sites, and this targeting is essential not only for normal synapse formation but also the synaptic transmission. In the chapter ►ion channels from development to disease Pham et al. demonstrate how various ion channel sub-types are selectively gated at various synaptic and extrasynaptic sites to serve their well-defined roles in a wide variety of cell types. Perturbation or mutations to various ion channels subtypes either in non-excitable or excitable cells may result in pathologies, such as the neonatal diabetes and ►epilepsy, respectively.

In addition to ion channel targeting to specific synaptic sites, the function of various other synapse-specific and Ca^{2+}-dependent proteins are also highly regulated. Intricate interplays between myriads of ►synaptic vesicle-associated proteins have been an area of intense investigation recently. A combination of biochemical, molecular, imaging and electrophysiological approaches have served to identify how synaptic vesicles might be targeted, primed, docked, released and recycled at the synaptic sites. As outlined by Coorssen in the chapter ►synaptic proteins and regulated exocytosis newly synthesized synaptic vesicles leave the cell body by a series of well-defined pathways. These vesicles are then specifically targeted to select synaptic sites where they get tethered, docked and primed for release. An action potential-induced Ca^{2+} influx through voltage-gated Ca^{2+} channels (VGCCs) (Calcium channels – and overview) is a critical step, which triggers fusion and exocytosis. Following their release at the synapse, the synaptic vesicles undergo endocytosis and are recycled for subsequent re-release. Although the spatio-temporal patterns of the synaptic vesicle behavior have been well characterized, this area of research, however, continues to enjoy its fair share of controversies.

The opening of the VGCCs invokes Ca^{2+} entry into the cytosol. This Ca^{2+} is then rapidly taken up by the fast ►endogenous $Ca^{2\pm}$ buffers, the mitochondria as well as the ►SERCA – sarco-endoplasmic reticulum Ca^{2+}-ATPase pumps (Ion transport). These three steps thus exert a critical regulatory control over the magnitude of the rapid, Ca^{2+}-mediated signaling. In addition to these fast acting steps, the Ca^{2+} ►homeostasis is also maintained by slower endogenous Ca^{2+} buffers, such as the mitochondria, the SERCA pumps, the plasma membrane plasma membrane ►NCX – Na^{+}-Ca^{2+} exchanger and the ►PMCA – plasma membrane Ca^{2+}-ATPase pumps (Ion transport) – all of which curtail subsequent Ca^{2+} signaling. The role/s of these various Ca^{2+}-regulatory steps are not only cell type-specific, but they also vary within a cell from its somal to extrasomal compartments. Recent advances in various imaging and molecular techniques are enabling a greater understanding of the mechanisms by which various regulatory steps maintain Ca^{2+} homeostasis and these are described by Amy Tse et al. (►influence of Ca^{2+} homeostasis on neurosecretion).

In contrast to classical transmitters such a dopamine, serotonin and ►acetylcholine, much less is known about the secretary machinery that regulates the release of dense-cored vesicles containing neuro-►hormones or peptides. Fred Tse (►non-synaptic release) and colleagues have developed reliable carbon fiber ►amperometry approaches to define the kinetics of transmitters (such as ►catecholamines) release at the resolution of single granule cells. The Tse lab and others have also demonstrated the involvement of the ►SNARE complex in the release machinery to provide direct evidence that kinetics of release probability is highly variable from cell to cell and relies, in many important ways, on Ca^{2+} sensitivity of the system. Because the release of polypeptides and peptidergic neurochemical substances occurs at a relatively slower time scale, a great deal is now known about the cellular and molecular mechanisms underlying their mode of release. A variety of peptide messengers have now been shown to regulate important neuronal programs in a number of species.

In their chapter ►neuropeptides in energy balance Chee and Colmers describe how ►neuropeptides modulate hypothalamic circuitry to regulate energy balance. Their work underscores the importance of peptides such as ►melacocortin, ►corticotrophin-releasing hormones (CRH) and CRH-like peptide and ►neuropeptide Y, ►agouti-related peptide (AgRP), ►melanin-concentrating hormone (MCH), ►orexin etc. in regulating food intake and body metabolism. This is an impressive list of candidate molecules that appear to be specifically released to regulate energy balance in various animal models. Deciphering their precise roles is the focus of many laboratories and the studies are deemed important for obesity research.

While it is generally believed that most proteins such as the neuropeptides destined for various extrasomal sites (axons, dendrites and synapses), are synthesized at the soma and then selectively transported to these regions, this dogma has however, been recently challenged. Specifically, several recent studies have provided ample convincing evidence that the extrasomal compartments are able to synthesize a host of synapse- and plasticity-specific proteins *de novo*. Support for this notion stems from earlier studies where a host of mRNA species were identified in dendrites and axons where they were selectively targeted to specific synaptic sites following an activity-dependent mechanism. Subsequent studies using a number of molecular and radio-labeling techniques demonstrated that the targeted mRNA was indeed able to translate specific protein locally. Furthermore, injection of foreign mRNA into the extrasomal compartments was also shown not only to result in the production of encoded proteins but also that these proteins were functional. The impact of this research, which is highlighted by van Minnen in ►extrasomal protein synthesis in neurons are far-reaching and perhaps will be one of the most exciting areas of neuroscience in the years to come.

Once the developmental program has established a complete repertoire of synaptic connectivity, the neuronal networks are put to work through myriad modes of neuronal communication. These range from excitatory to inhibitory to mixed excitatory/inhibitory connections. While the synaptic transmission in general is predominantly chemical, the role of ►electrically coupled networks cannot be underestimated. Specifically, in addition to conventional chemical synapses, many neurons may also connect to each other through ►gap junctions where the membranes of two neurons become contiguous. Current in one cell may pass unabated to another without the need for a ►synaptic delay. While such gap junctions are predominant during development, their presence in the adult nervous system is only beginning to be realized in most vertebrates. In invertebrates, however, electrically coupled networks are quite common where they are often recruited to trigger fast ►escape responses that are critical for their survival and thus cannot afford the synaptic delays which are the hallmark of most chemical synapses. The precise nature of both structural and functional attributes of gap junction/tight junction or electrically coupled cells is wonderfully described by Wildering in the chapter on ►electrical synapses. Blocking gap junctions during early development has been shown to perturb nervous system development; their precise functions in the adult mammalian brain are, however, yet to be fully understood. It is nevertheless generally agreed that one of the hallmarks of gap junctions is to synchronize pattern activity either during a patterned motor program or pathological discharges such as epilepsy.

One of the most fascinating aspects of the neuronal uniqueness is the ability of a network of central pattern-generating neurons to exhibit rhythmical activity in the absence of the peripheral feedback. These networks of neurons, often termed ►central pattern generators (CPG), control a variety of rhythmical behaviors such as ►locomotion, ►respiration, ►feeding, mastication etc. Because CPG neurons can generate fictive, patterned activity underlying a rhythmical behavior, even in an isolated preparation, a great deal is known about intrinsic membrane properties that generate a well-organized motor output. In some instance, neurons are known to possess ►pacemaker potentials, which can generate endogenous bursting patterns; however, the ►rhythmogenesis is always a network phenomenon in both vertebrates and invertebrates. In a serious of chapters written by Bell (►peripheral feedback and rhythm generation), Straub (central pattern generator) and Whelan (►neurotransmitters and pattern generation) we learn a great deal about various intrinsic membrane properties (pacemaker potential, ►endogenous bursters, ►conditional bursters, etc) and synaptic interactions (excitatory/inhibitory, ►half-center model, ►reciprocal inhibition, ►postinhibitory rebound excitation, ramp generators, recurrent inhibition etc.) underlying patterned motor activity. Even though the CPG neurons have been known to generate patterned activity in the absence of any peripheral feedback, Bell (peripheral feedback and rhythm generation) demonstrates how peripheral feedback could be critical for the initiation, modulation and termination of the patterned activity. He specifically focuses on the role of hypoxia-sensitive chemosensory drive from the carotid body chemoreceptors, and how it affects the patterned respiratory discharges. Bell then discusses how these networks of rhythm-generating neurons are similar in both vertebrate and invertebrate animals – assuring us that the fundamental building blocks of CPG neurons are likely conserved throughout the animal kingdom. While Straub (central pattern generator) illuminates various membrane and network properties that are the hallmark of pattern generation, Whelan (neurotransmitters and pattern generation) examines structural, functional and transmitter (serotonin, dopamine etc) organization of the CPG underlying locomotor behavior in mammals. Whereas in some invertebrate models, ►command neurons are thought to be sufficient and necessary to trigger a patterned discharge, it is generally believed that the rhythm generation is a function of polymorphic nature of the network. In this configuration, the network exhibits a highly dynamic repertoire of activity patterns thus allowing greater flexibility within the network. Neuronal networks are thus known not to be hardwired, rather they exhibit great flexibility – allowing a subset of neurons to switch between inter-related networks. A similar reorganization of the network behavior is observed following trauma and injury whereby uninjured neurons either take on additional assignments or switch their roles from one to another.

In contrast to their central counterparts, most peripheral neurons are able to regenerate their axonal projections after injury (►Regeneration). Although this regeneration appears to re-capitulate developmental patterns of growth, the ►reinnervation is often incomplete, mismatched and often accompanied with ►neuropathic pain. Tremendous efforts are therefore being made to improve the outcome of ►peripheral injuries by either manipulating the extracellular environment or the surgical interventions. Zochodne (►axon degeneration and regeneration of peripheral neurons) provides a very comprehensive account for cellular and molecular changes that occur immediately after a peripheral injury (►neurapraxia, ►axonotmesis, neurotemesis and how this signal is conveyed to the cell body to activate the "regenerative program." It is generally believed that an immediate injury response triggers a massive Ca^{2+} influx, which in turn activates a cascade of events that lead to the ►microtubular disorganization and ►neurofilament dissolution. Subsequent SC activation then results in microphage invasion, an upregulation of ►cytokines and ►chemokines – including IL-1β, IL-6, IFN-γ, TNF-α, MCP-1 (monocyte chemoattractant protein-1) and MIP-1α (►macrophage inflammatory protein 1α) followed by a complete breakdown of ►myelin. In the presence of appropriate trophic factors, ►nitric oxide, and various substrate adhesion molecules, a neuron then triggers its regenerative program, which begins with the initiation of new neurites and the re-establishment of synaptic connectivity. In contrast with the above described crush injuries, nerve transsections often result in ►Wallerian degeneration (neurotmesis or Sunderland Type V injury), which involves breakdown of axons and myelin distal to the injury site. It is interesting to note that the regeneration re-activates many but not all elements of the developmental program and as a consequence the functional recovery after nerve injury is often incomplete. Several novel approaches are being developed to enhance the clinical outcomes of nerve injury and are described in detail by Midha (►peripheral nerve regeneration and nerve repair). Specifically, Midha provides extensive overview vis-à-vis the pros and cons of ►nerve grafts, electrical stimulation paradigm and ►nerve conduits that are being used clinically. This chapter also provides a detailed account of various bio-engineering approaches that are being developed to create nerve conduits that may, in the future, play "active" rather than passive roles in promoting nerve regeneration. This approach most certainly holds tremendous potential and is being perused extensively by Zochodne and Midha labs.

Summary

A neuron is considered as the functional unit of the nervous system, whereas a synapse serves as a gatekeeper of all neuronal communication. Over the past 50 years our understanding of both the structural and functional attributes of neurons and synapses has been enhanced tremendously. Specifically, a great deal is now known about the intrinsic membrane properties that contribute to neuronal excitability and shape its unique characteristics. Every unique neuronal trait in turn, makes specific contributions to synaptic properties of the network in which it is embedded. Neurochemical, electrochemical and/or electro-electrical properties empower a network to generate rhythmical patterns, which in turn control important behaviors – ranging from simple reflexes to complex motor patterns and learning and memory. These connectivity patterns are orchestrated early during development and are constantly re-organized and reconfigured throughout life. Perturbation to either the intrinsic membrane or synaptic properties renders the nervous system dysfunctional thus resulting in the permanent loss of neuronal function. Restoration of this connectivity is perhaps one of the greatest challenges facing the neuroscientists – an area that requires extensive efforts not only by the basic scientists, clinical investigators but also the bio-medical engineers and nano-engineers. A multidisciplinary approach is likely to yield bionic hybrids, which can then be interfaced with neurons to resort lost brain function. For instance, bio-compatible and neuron-friendly chips that can be interfaced with networks of brain cells will not only enhance our understanding of brain function but also regain the lost nervous system function. Although challenging – this appears to be the most promising avenue towards regeneration and functional repair of the injured nervous system.

Neuron-to-neuron Communication

►Synapse Formation: Neuromuscular Junction Versus Central Nervous System

Neuronal Cell Death and Axonal Degeneration: Neurofilaments as Biomarkers

AXEL PETZOLD
Axel Petzold, Department of Neuroimmunology, Institute of Neurology, University College London, London, UK

Definitions

The aim of this chapter is to explain why ►neurofilaments (Nf) are a useful ►biomarker for ►axonal degeneration and can be used as a surrogate endpoint in clinical and experimental research (►surrogate outcome).

Neurofilaments: Nf are proteins which are exclusively expressed in neurons and their adjacent axons. Nf are particularly abundant in the axon, where they are key building blocks of the axonal cytoskeleton. The complex protein chemistry of Nf is briefly described.

Biomarker: A biomarker is a characteristic that is objectively measured and evaluated as an indicator of normal biologic processes, pathogenic processes, or pharmacological responses to therapeutic intervention.

Surrogate endpoint: defines a biomarker that is intended to serve as a substitute of a clinically meaningful endpoint and is expected to predict the effect of a therapeutic intervention or the evolution of disease.

Characteristics

Quantitative Description

Nf are obligate heteropolymers (►polymer) that are composed of four subunits: a light (NfL), a medium (NfM), a heavy (NfH) [1] chain and also ►alpha-internexin [2,3]. In some cases, peripherin may be added to the list [2]. These subunits differ not only in their molecular weight, but also in their functional properties, as discussed below.

NfL

The ►neurofilament light chain (NFL) is coded on chromosome 8p21 and consists of 543 amino acids. The molecular mass corresponds to 61 kDa, but due to phosphorylation and glycosylation, migration in sodium dodecyl sulfate (SDS) polyacrylamide gels (PAGE) is slow, and most authors refer to a molecular mass of 68 kDa as determined in SDS-PAGE. NfL forms the back-bone of the Nf heteropolymer and can self-assemble. Mutations in the NfL gene have been associated with Charcot-Marie Tooth disease.

NfM

The ►neurofilament medium chain (NfM) is also coded on chromosome 8p21 and consists of 916 amino acids. The molecular mass is calculated as 102.5 kDa, and runs at 150 kDa in SDS gels. NfM is important for the radial axonal growth. One mutation in the NfM gene has been associated with Parkinsons disease.

NfH

The ►neurofilament heavy chain (NfH) is coded on chromosome 22q12.2 and consists of 1,020 amino acids. The molecular mass of the amino acids corresponds to 111 kDa. Most authors however refer to the molecular mass derived from SDS gels which is also influenced by the charge/weight of bound phosphate and therefore ranges from 190 to 210 kDa for the various phosphoforms. NfH is important for protein-protein interactions which is regulated locally in the axon by phosphorylation. Mutations in the NfH gene have been associated with amyotrophic lateral sclerosis (ALS).

Alpha-Internexin

The 66 kDa alpha-internexin protein is coded on chromosome 10q24.33 and able to form homopolymers. Alpha-internexin has only recently been rediscovered as one of the Nf subunits and the role of alpha-internexin is still poorly understood [3]. Extracellular deposits of alpha-internexin are an important hallmark of a newly discovered neurodegenerative dementia named neurofilament inclusion disease (NFID).

Assembly of the Nf Heteropolymer

Figure 1 illustrated how NfL, NfM and NfH assemble to produce the Nf heteropolymer, which has a diameter of about 10 nm. Because of its size, which is intermediate

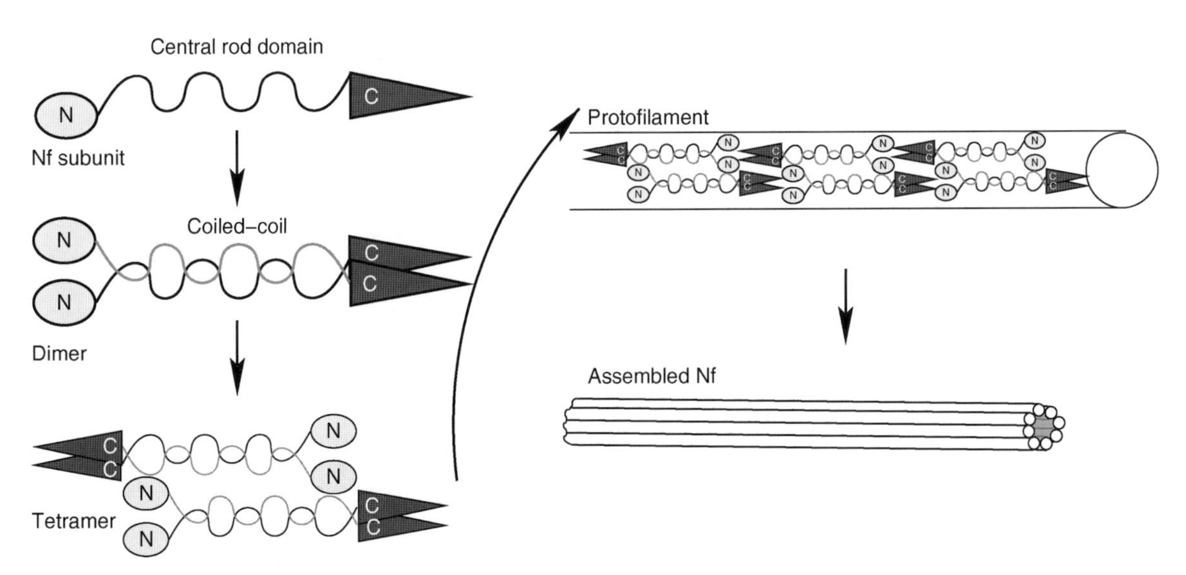

Neuronal Cell Death and Axonal Degeneration: Neurofilaments as Biomarkers. Figure 1 Neurofilament assembly. The central rod domain of the Nf subunits is intertwined in order to form dimers. The dimers are arranged antiparallel to form tetramers. Tetramers combine to form protofilaments, which finally assemble to produce the 10 nm thick Nf (figure reprinted with permission from reference [4]).

between the smaller proteins, e.g. microfilaments (7 nm) and larger proteins such as microtubules (approximately 25 nm), the Nf heteropolymer belongs to the intermediate filaments.

Stoichiometry of the Nf Subunits

The estimated in vitro molar ratio of isolated Nfs from the mouse optic nerve and spinal cord is 4:2:2:1 (NfL:a – internexin:NfM:NfH) [3]. The in vivo stoichiometry of Nfs in body fluids remains unknown.

Classification of Nf

Nf are type IV intermediate filaments (Table 1).

Nf are a Biomarker for Axonal Degeneration

Nf subunits are useful biomarkers for axonal degeneration, as illustrated in Fig. 2. Any insult causing neuronal death or axonal degeneration will inevitably result in disintegration of the axonal membrane. Subsequently the contents of the axonal cytoplasm are released into the ►extracellular fluid (ECF). From the ECF Nfs diffuse into other body fluid compartments such as the ►cerebrospinal fluid (CSF), blood or amniotic fluid. As explained above Nf are a major structural protein component of the axon and the quantification of Nfs from body fluids therefore allows estimation of the degree of axonal degeneration.

The Measurement of Nf Body Fluid Levels

At present, high-throughput quantification of Nfs from body fluids and tissue homogenates is best achieved using enzyme linked immune assays (ELISA). In-house ELISAs have been developed for NfL and NfH [5–8]. These assays are highly robust and have been cross-validated [9,10]. A commercial NfH ELISA kit has recently been made available (Chemicon). Alternatively immunoblots or dot-blot assays have

Neuronal Cell Death and Axonal Degeneration: Neurofilaments as Biomarkers. Table 1 Classification of intermediate filaments and cell-type specificity

Class	Identity	Cell-type specificity
Type I	Acidic keratins	Epithelial
Type II	Neutral & Basic keratins	Epithelial
Type III	GFAP	*Astrocyte*
	Peripherin	Neuronal (peripheral)
	Vimetin	Mesenchymal
	Desmin	Muscle
Type IV	NfL, NfM, NfH	*Neuron and axon*
	Alpha-internexin	*Neuron and axon*
Type V	Laminin A, B, C	Most cells
Type VI	Nestin	CNS stem cells

GFAP = glial fibrillary acidic protein.

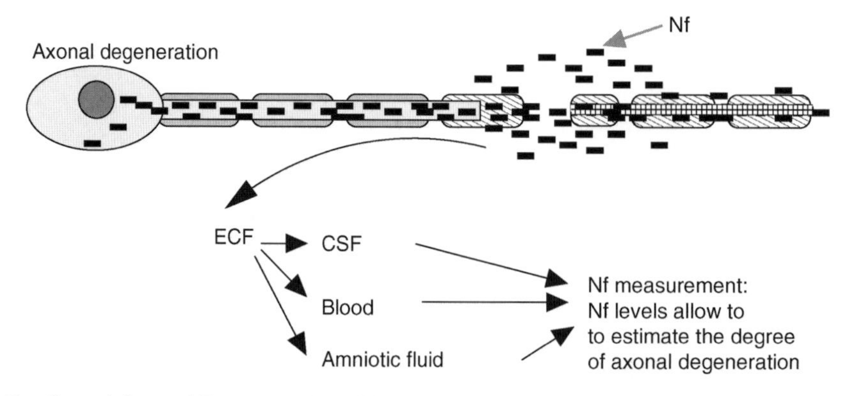

Neuronal Cell Death and Axonal Degeneration: Neurofilaments as Biomarkers. Figure 2 Neurofilaments are released into the extracellular fluid (ECF) following axonal disintegration. From the ECF Nfs equilibrate with the adjacent body fluid compartment. Quantification of Nfs is therefore possible from the cerebrospinal fluid (CSF), blood and amniotic fluid. The degree of axonal degeneration is related to the amount of Nf measured in these body fluids. For this reason body fluid Nf levels permit the estimation of the amount of axonal degeneration. Axonal degeneration is extremely important because the loss of axons is irreversible and may therefore lead to persistent disability.

been used, but generally they are not high-throughput and only semi-quantitative.

The Diseases Associated with High Body Fluid Nf Levels

Neuronal loss and axonal degeneration are a key feature in numerous disorders and frequently represent the endstage of a pathophysiological cascade. Not surprisingly, body fluid levels of Nf subunits have been used to estimate the degree of axonal damage in a number of diseases (Table 2). It is important to remember that body fluid Nf levels are not a diagnostic test for one single disease. In contrast Nf are a biomarker and surrogate enpoint according to the initial definitions.

Conclusion

Neurofilaments are complex proteins composed of four subunits, expressed exclusively in the neuro-axonal compartment. Nf are released into the extracellular fluid from degenerating axons. From the extracellular fluid they diffuse into adjacent body fluid compartments. Using standard ELISA techniques Nf subunits have been quantified from the cerebrospinal fluid, the blood and the amniotic fluid. Because body fluid levels of Nf are related to the amount of neuronal death and axonal loss, they provide valuable prognostic information and correlate with disability in a number of diseases.

Neuronal Cell Death and Axonal Degeneration: Neurofilaments as Biomarkers. Table 2 Diseases in which Nf have been used as a body fluid biomarker for neuronal death and axonal degeneration

Disease	Findings
AD	CSF NfL and NfH levels are elevated in AD. The difference from controls was marginal for CSF NfH levels and more impressive for NfL levels
ALS	CSF NfL and NfH levels are considerably increased in patients with ALS. Rapidly progressing ALS patients had the highest CSF NfH levels
CBD	CSF NfL and NfH levels are elevated in patients with CBD
FTLD	CSF NfL is elevated and CSF NfH marginally elevated in patients with FLTD. The degree of NfH phosphorylation is increased in FTLD compared to AD and controls
GBS	Elevated CSF NfH levels in patients with GBS are a poor prognostic sign, probably due to proximal axonal degeneration. Proximal axonal degeneration at the level of the nerve roots rapidly releases Nfs into the CSF. Proximal axonotmesis requires axonal regrowth over a long distance with the risk of losing chemical and anatomical guidance cues
ICH	CSF NfH levels are high in ICH, probably indicating direct axonal degeneration due to rupture and ischemia
DLB	CSF NfH but not NfL levels are elevated in DLB compared to AD and controls
MMC	Amniotic fluid NfH levels are elevated in mice with MMC and correlated with the size of the lesion
MS	CSF NfH and NfL levels are elevated in MS. CSF NfL levels are highest following a clinical relapse and return to baseline within about 3 months. CSF NfH levels are highest in the secondary progressive phase of the disease when axonal degeneration accumulates. The degree of NfH phosphorylation is increased in patient with more severe disease. High CSF NfH levels are a poor prognostic sign. Both CSF NfL and NfH levels correlate with disability
MSA	CSF NfL and NfH levels are markedly elevated in MSA compared to controls and patients with PD. This may be related to the greater degree and more rapid disease progression in MSA. The highest levels are found in patients with the cerebellar variant of MSA, which may be of help for the differential diagnosis of patients with cerebellar syndromes
NMO	CSF NfH levels are considerably elevated in NMO (synonymous with Devic's disease) suggesting that these patients suffer from substantially more axonal damage than patients with MS or ON
ON	Plasma NfH levels are increased in acute ON. CSF NfH levels are elevated in patients with subacute ON. Plasma and CSF NfH levels correlate with loss of visual function
PD	CSF NfH and NfL levels are increased in PD compared to controls
PSP	CSF NfL and NfH levels are elevated in PSP compared to controls and patients with PD. As with MSA this may be related to the greater degree of axonal loss and more rapid disease progression in PSP patients, who are also very resistant to pharmacological treatment
SAH	CSF NfL and NfH levels are elevated in SAH and correlated with the outcome. Importantly CSF NfH levels showed a secondary increase during the high risk period of vasospasm, probably indicating secondary axonal degeneration following an ischemic insult

AD = Alzheimer's disease, ALS = amyotrophic lateral sclerosis, CBD = cortico-basal degeneration, DLB = Diffuse Lewy body disease, FTLD = fronto-temporal lobar degeneration, GBS = Guillain–Barré syndrome, ICH = intracerebral haemorrhage, MMC = meningo-myelocele, MS = multiple sclerosis, MSA = multiple system atrophy, NMO = neuromyelitis optica, ON = optic neuritis, PD = Parkinson's disease, PSP = progressive supranuclear palsy, SAH = subarachnoid haemorrhage.

Acknowledgement

I apologize to all colleagues whose work has not been cited due to space limitations. The biomarker definitions were adapted from a recent NIH meeting on biomarkers. A more complete list of references can be requested from the author (a.petzold@ion.ucl.ac.uk).

References

1. Lee MK, Cleveland DW (1996) Neuronal intermediate filaments. Ann Rev Neurosci 19:187–217
2. Shaw G (1998) Neurofilaments. Springer, Berlin Heidelberg New York
3. Yuan A, Rao MV, Sasaki T, Chen Y, Kumar A, Veeranna, Liem RK, Eyer J, Peterson AC, Julien JP, Nixon RA (2006) Alpha-internexin is structurally and functionally associated with the neurofilament triplet proteins in the mature CNS. J Neurosci 26:10006–10019
4. Petzold A (2005) Neurofilament phosphoforms: surrogate markers for axonal injury, degeneration & loss. J Neurol Sci 233:183–198
5. Rosengren LE, Karlsson JE, Karlsson JO, Persson LI, Wikkelso C (1996) Patients with amyotrophic lateral sclerosis and other neurodegenerative diseases have increased levels of neurofilament protein in CSF. J Neurochem 67:2013–2018
6. Petzold A, Keir G, Green AJE, Giovannoni G, Thompson EJ (2003) A specific ELISA for measuring neurofilament heavy chain phosphoforms. J Immunol Methods 278:179–190
7. Norgren N, Rosengren L, Stigbrand T (2003) Elevated neurofilament levels in neurological diseases. Brain Res 987(1):25–31
8. Shaw G, Yang C, Ellis R, Anderson K, et al. (2005) Hyperphosphorylated neurofilament NF-H is a serum biomarker for axonal injury. Biochem Biophys Res Comm 336:1268–1277
9. Van Geel WJA, Rosengren LE, Verbeek MM (2005) An enzyme immunoassay to quantify neurofilament light chain in cerebrospinal fluid. J Immunol Methods 296:179–185
10. Petzold A, Shaw G (2007) Comparison of two ELISA methods for measuring levels of the phosphorylated neurofilament heavy chain. J Immunol Methods 319:34–40

Neuronal Cell Death and Inflammation

Jeffrey A. Rumbaugh, Avindra Nath
Department of Neurology, Johns Hopkins University School of Medicine, Baltimore, MD, USA

Synonyms

Excitotoxicity; Apoptosis; Necrosis; Virotexias; Encephalitis; Chemokines; Cytokines; Chemtaxis; Blood brain barrier; Matrix metallo Proteinases

Definition

Exitotoxicity is a mechanism which often leads to Neuronal cell damage or death. Excitotoxicity occurs when glutamate receptors on neurons are overactivated. Neuronophagia occurs when nerve cells are phagocytosed or internalized by macrophages. This often happens to clear debris after the neurons have undergone apoptosis, but in some neuroinflammatory or neurodegenerative conditions, the neurons may be killed as the macrophages digest them. Necrosis is a process of cell death whereby cells swell, rupture, and release their contents, causing ►inflammation and damage to neighboring cells. In the nervous system, this usually happens in response to stroke or trauma. In contrast to necrosis, apoptosis is a form of cell death which does not cause inflammation, but, rather, is frequently the result of inflammatory processes. Therefore, the mechanism of neuronal cell death which usually occurs in the setting of neuroinflammatory and neuro-infectious diseases is apoptosis. Unlike necrosis, apoptosis occurs via a controlled sequence of events, starting with an initial trigger, then proceeding through a specific signaling cascade, resulting in breakdown of the chromatin and shrinkage of the nucleus and cellular contents without rupture of the cell membrane. See Table 1.

Neuronal cell death is the irreversible loss of function of a neuron. Neuroinflammatory diseases such as multiple sclerosis, transverse myelitis, and neurosarcoidosis are characterized by episodic immune activation, which results in nervous system injury. Although an infectious etiology has long been suspected in these diseases, none has been conclusively demonstrated. Other chronic neurodegenerative diseases, such as Alzheimer's disease, also have chronic glial cell activation. It remains unknown if this chronic activation is needed to provide trophic support for injured neurons or if ►cytokines and other host factors released by these cells may be injurious. ►Virotoxins and other infectious agents may trigger immune cascades that can persist long after the infection has been controlled or eradicated, a mechanism of injury that has been termed the "hit and run phenomenon." This persistent immune activation can lead to neuronal injury resulting in neurocognitive impairment. There are many in vitro and in vivo models of neuroinflammatory conditions which allow us to study the effects of immune responses on the nervous system. Although many different conditions produce a neuroinflammatory state, the host repertoire for immune response and the mechanisms for subsequent neuronal death or dysfunction are relatively limited. It is thus hoped that the study of any one of these diseases or model systems will be widely applicable to other autoimmune and neurodegenerative diseases in which immune activation is an important component.

Neuronal cell death is common in patients with both infectious and non-infectious neuroinflammatory

Neuronal Cell Death and Inflammation. Table 1 Features of necrosis versus apoptosis

Feature	Apoptosis	Necrosis
Cell size	Shrunken, small	Swollen, enlarged
Inflammation	Inflammation causes apoptosis	Necrosis causes inflammation
Leakage	Cell contents intact	Cell contents leak out
Membrane	Intact	Disrupted
Nucleus	Fragmented	Condensed
Role	Often physiological	Always pathological

conditions. Even in many infectious conditions, such as HIV infection, neurons themselves are frequently not directly infected. Thus, in all of these conditions, the observed loss of neurons is likely due to indirect effects of inflammatory mediators and/or infectious proteins. A process called apoptosis is the primary mechanism through which neurons die. Apoptosis may also be seen in some other cells of the nervous system, including astrocytes and endothelial cells. Neuronal apoptosis correlates with microglial activation and axonal damage, suggesting that the inflammatory mediators secreted by microglia and other immune cells play a major role in initiating the apoptotic cascades.

Importantly, recent studies have shown that neuronal injury may occur without cell death. The clinical manifestations of neuroinflammatory conditions are likely due to neuronal dysfunction via multiple mechanisms. Although massive neuronal loss may occur in many neuroinflammatory conditions, it often occurs late in disease, and may not be the cause of early clinical manifestations. Some inflammatory pathways, and related dysfunction, may be reversible with strategic neuroprotective and immunomodulatory strategies. Pathological studies have demonstrated injury types which are likely to be reversible, including morphological changes in dendrites and loss of neurites without neuronal cell loss. Such dendritic injury and loss, without neuronal cell body loss, is often called dendritic pruning. Similarly, axonal injury can occur without death of neuronal cell bodies. This raises great hope for the potential use of neurotrophic modes of therapy for neuroinflammatory and neurodegenerative diseases.

Characteristics

The mechanisms by which inflammatory conditions and infections lead to neuronal cell death and dysfunction, as well as associated clinical conditions like ▶encephalitis, ▶vasculitis, or dementia, remain elusive. In many cases, the inflammation generated by the host, in an attempt to combat a presumed or real infection, is itself implicated as a primary factor in causing neuronal dysfunction or degeneration. In this essay, we outline the current state of knowledge regarding the pathophysiology of central nervous system injury in infectious or inflammatory conditions. Understanding these mechanisms should ultimately enable development of immunomodulatory therapies for treating these conditions.

Description of the Process

Inflammation in the nervous system is dependent on two main features: infiltration of monocytes from the peripheral blood into the brain, and activation of microglia, which are the immune cells that are always present in the brain. In viral infections, such as HIV, the virus itself or viral proteins can activate uninfected cells directly. In non-infectious conditions, the inciting agent(s) may be unknown, but these immune cells are likely activated in a similar fashion. Then, in an attempt to eradicate the presumed or real infection in the brain, these immune cells are likely most responsible for the neuroinflammatory and neurotoxic cascades which lead to brain injury. These cells express tumor necrosis factor-α (TNF-α), interleukin-1 (IL-1), interferon-α (INF-α), and nitric oxide synthase (NOS), among other inflammatory mediators [1]. It should be noted, however, that these cells may have neuroprotective properties as well. In fact, immune cells typically serve beneficial functions. This may suggest that the immune response does not become damaging until it becomes dysregulated and chronic.

Cytokines and ▶chemokines are multifunctional proteins which regulate individual cells under physiological or pathological conditions. They are important mediators for communication between nervous tissues and immune cells and are thus very important in the induction and regulation of inflammation in the nervous system, and thus to the progression or inhibition of neurodegeneration [1]. They are expressed by the peripheral immune cells which have entered the brain often across a defective ▶blood–brain barrier, by activated microglia, by astrocytes, and even by certain neurons. They include, but are not limited to, IL-1, IL-8, RANTES, TNF-α, SDF-1, and MCP-1. MCP-1 is a potent chemoattractant for monocytes, drawing them into the brain from the peripheral blood in a process called ▶chemotaxis, so that they, in turn, can produce even more inflammatory cytokines. Many of these cytokines/chemokines have both deleterious and beneficial effects,

N

so the net effect is likely the result of a complex set of interactions and conditions and is very difficult to predict a priori or to determine experimentally.

Neuronal injury can be triggered by various mechanisms. Binding of various cytokines and chemokine receptors either by the cytokines/chemokines themselves (especially TNF-α) or by viral proteins or other neurotoxic agents will lead to increases of intracellular calcium. Similarly, inflammatory conditions may promote excitotoxicity. Neuroinflammatory proteins may either directly stimulate the neuronal glutamate NMDA receptor or may sensitize neurons to the effects of otherwise physiological levels of glutamate. Excessive or chronic stimulation of this receptor results in a long cascade, again leading to increased intracellular calcium. Increased calcium leads to loss of mitochondrial membrane potential, release of cytochrome c, activation of caspases, and apoptosis [2]. Excitotoxicity also leads to production of nitric oxide and free radicals which in turn produce oxidative damage, energy failure, and DNA damage.

Figure 1 depicts the complex interactions between toxic cellular proteins, viral proteins, and the immune response, which lead to neurotoxicity in vitro and to the neurological complications of inflammatory and/or infectious conditions clinically.

Higher Level Processes and Conditions

Reactive astrocytes are common in these neuroinflammatory conditions and they participate in the production of neurotoxic substances, such as TNF-α, and other inflammatory mediators. Activation of astrocytes can alter their function in other ways as well, leading to loss of support for neurons, in turn making the neurons more susceptible to injury or death.

Periventricular white matter pallor is also frequently observed in these conditions. Depending on the specific condition, this can be associated with damage to the oligodendrocytes and/or the myelin sheath which wraps around neuronal axons, or, alternatively, it can be associated with subtle changes of the ►blood–brain barrier (BBB) secondary to inflammation. Inflammation can affect the expression and assembly of ►tight junction proteins, leading to cytoskeletal disruption of endothelial cells and increased endothelial permeability. White matter changes on MRI seem to correlate with perivascular macrophage infiltrates, extravasation of protein, and blood–brain barrier compromise. Endothelial cells and astrocytes functionally form the blood–brain barrier, so injury to either of these cell types can compromise the blood–brain barrier. Endothelial cells may also increase expression of adhesion molecules [3], allowing easier entry of peripheral immune cells into the brain. A compromised blood–brain barrier allows immune cells, inflammatory mediators, and neurotoxins from the peripheral blood to enter the nervous system where they can participate in damaging and killing neurons. Sometimes, however, these white matter changes are reversible, with associated improvement in clinical manifestations.

Various viral infections have been used extensively as model systems for neuroinflammatory conditions,

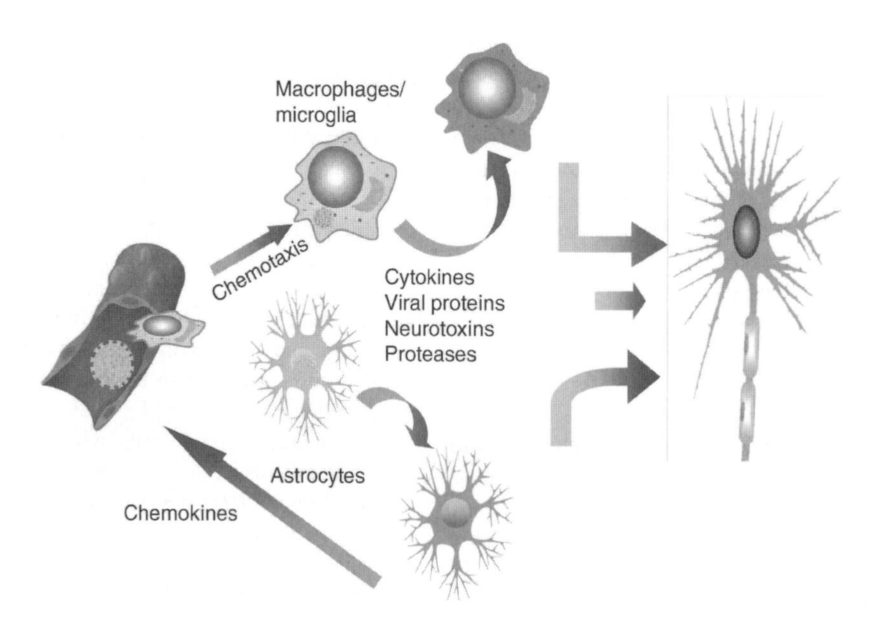

Neuronal Cell Death and Inflammation. Figure 1 Through multiple indirect mechanisms, neurons become dysfunctional and die in the setting of inflammatory and infectious conditions. Inflammatory mediators can be toxic to neurons directly, can alter glial function, and can further activate the immune response, all of which damages the nervous system. The arrows in the figure demonstrate the existence of these complex feed back loops.

and such studies have elucidated mechanisms which are widely applicable. Neurons may die upon interaction with viral proteins, while uninfected microglia, monocytes, and astrocytes are activated upon such interaction. These activated cells release a variety of proinflammatory factors, including cytokines, chemokines, free radicals, matrix metalloproteinases (MMPs), and prostanoids, which may result in secondary neuronal toxicity or further immune cell activation and reactive gliosis. This amplification of the immune cascade after an initial trigger has been termed the "domino effect" [4]. Once the domino effect has been initiated, the inflammatory process may be self-propogating, even if the initial trigger is no longer present, resulting in the "hit and run" phenomenon [5].

MMPs are a family of endopeptidases which enzymatically degrade extracellular matrix proteins and can thus disrupt the blood–brain barrier and neuronal synapses [6]. MMP levels are elevated in the spinal fluid and/or brains of patients with many neuroinflammatory or neurodegenerative conditions, suggesting they may contribute to the neuropathogenesis of these conditions. MMPs can cleave chemokines, such as SDF-1, with the cleavage products subsequently causing neurotoxicity, and they can interact with integrin receptors on neurons, initiating apoptosis. Furthermore, they can become nitrosylated and hyperactive, contributing to neurotoxicity under conditions of oxidative stress [7]. MMP expression in monocytes can facilitate monocyte transmigration through the extracellular matrix. However, it should also be noted that cleavage of the chemokine, MCP-3, by MMP-2 has been shown to decrease the inflammatory response [8], and cleavage of HIV Tat protein by MMP-1 attenuates Tat-induced neurotoxicity [9]. Thus, MMPs may be neuroprotective under certain conditions.

Regulation of the Process

The brains of patients with many of the neuroinflammatory conditions demonstrate up and downregulation of numerous genes compared to control brains. These changes in gene expression profiles are consistent with changes in various neurotransmitter receptor levels and ion currents, which would be expected to alter neuronal excitability. Perhaps these changes in gene expression are in response to the changes in neuronal excitability which are induced by the inflammatory cascades and processes.

Function

Although the mechanisms of inflammation and neuronal cell death and dysfunction may be quite similar in the various neuroinflammatory conditions, the clinical manifestations may be variable. For reasons which are not well understood, certain areas of the brain may be preferentially affected over other areas [10]. For example, as a result of the inflammation caused by HIV infection, the basal ganglia and hippocampus are preferentially damaged, with neuronal losses of up to 50–90% in patients with severe HIV encephalitis. The cortex is also affected with the frontal lobes having 40–60% neuronal loss, but parietal and temporal lobes having only 20% loss. The cerebellum has even less neuronal loss and the occipital lobe has the least. The locations of the predominant changes accounts for the predominant symptoms of HIV dementia, including psychomotor slowing and memory impairment.

It is not surprising that a prolonged, poorly controlled inflammatory reaction in the nervous system would result in neurodegeneration. Neurological outcome from these conditions depends on the interplay of a complicated network consisting of various cell types, pro- and anti-inflammatory factors, viral virulence factors, and host susceptibility factors. A successful therapeutic strategy to treat inflammation-mediated neurodegeneration will likely require a multi-faceted approach, aiming to not only inhibit proinflammatory factors but also to increase neuroprotective and neurotrophic factors.

Acknowledgement

This work was supported by National Institutes of Health grants to JR and AN.

References

1. Merrill JE, Benveniste EN (1996) Cytokines in inflammatory brain lesions: helpful and harmful. Trends Neurosci 19:331–338
2. Kruman I, Nath A, Mattson MP (1998) HIV protein Tat induces apoptosis by a mechanism involving mitochondrial calcium overload and caspase activation. Expt Neurol 154:276–288
3. Briscoe DM, Cotran RS, Pober JS (1992) Effects of tumor necrosis factor, lipopolysaccharide, and IL-4 on the expression of vascular cell adhesion molecule-1 in vivo. Correlation with CD3+ T cell infiltration. J Immunol 149:2954–2960
4. Nath A (1999) Pathobiology of HIV dementia. Sem Neurol 19:113–128
5. Nath A, Conant K, Chen P, Scott C, Major EO (1999) Transient exposure to HIV-1 Tat protein results in cytokine production in macrophages and astrocytes: a hit and run phenomenon. J Biol Chem 274:17098–17102
6. Libby RT, Lavallee CR, Balkema GW, Brunken WJ, Hunter DD (1999) Disruption of laminin beta2 chain production causes alterations in morphology and function in the CNS. J Neurosci 19:9399–9411
7. Gu Z, Kaul M, Yan B, Kridel SJ, Cui J, Strongin A et al. (2002) S-nitrosylation of matrix metalloproteinases: signaling pathway to neuronal cell death. Science 297:1186–1190
8. McQuibban GA, Gong JH, Tam EM, McCulloch CA, Clark-Lewis I, Overall CM (2000) Inflammation dampened by gelatinase A cleavage of monocyte chemoattractant protein-3. Science 289:1202–1206

N

9. Rumbaugh J, Turchan-Cholewo J, Galey D, St Hillaire C, Anderson C, Conant K et al. (2006) Interaction of HIV Tat and matrix metalloproteinase in HIV neuropathogenesis: a new host defense mechanism. FASEB J 20:1736–1738
10. Masliah E, Ge N, Achim CL, Hansen LA, Wiley CA (1996) Selective neuronal vulnerability in HIV encephalitis. J Neuropathol Exp Neurol 51:585–593

Neuronal Changes in Axonal Degeneration and Regeneration

HÅKAN ALDSKOGIUS
Uppsala University Biomedical Center,
Department of Neuroscience, Uppsala, Sweden

Synonyms

Axon reaction; Retrograde neuron reaction; Chromatolysis

Definition

Following injury to the ►axon, the affected nerve cell body (►soma) with its associated processes undergo a sequence of structural and molecular changes, collectively termed the ►axon reaction or ►retrograde neuron reaction. ►Chromatolysis is sometimes used as a synonym, but in a strict sense refers to the marked reduction (dissolution) of basophilic ►Nissl bodies caused by axon injury (Fig. 1).

Neuronal changes to axon injury are fundamental pathophysiological events in any neurological condition, which interrupts axons in the peripheral or central nervous system, or in which the normal peripheral or central target for the neuron is lost.

Characteristics

General Characteristics

The neuronal changes to axon injury include responses to overcome the ►cellular stress imposed by the injury, as well as a reorganization of the overall cell morphology and cellular metabolism from a "transmitting" to a "growing" mode. The latter implies that the expression of genes involved in ►neurotransmitter synthesis and release are typically down-regulated, whereas those promoting ►neurite growth are up-regulated [1,2]. As a result of this phenotypic shift, injured neurons in many ways resemble developing neurons. Under the most favorable circumstances, the injured neuron survives, ►regenerates its axon, and restores functional contact with the ►denervated cells/tissue. A sustained up-regulation of ►growth-associated genes is considered necessary for successful regeneration and target ►reinnervation Neurons with axons in the peripheral nervous system, i.e. ►motoneurons ►sensory ganglion cells and ►autonomic neurons all have this ability, whereas neurons with axons entirely within the central nervous system usually do not.

Axon injury consistently affects ►glial cells in the surroundings of the injured nerve cell body and ►dendrites, as well as neurons with ►synapses on the injured neuron. Under certain circumstances, neurons, which are ►presynaptic or ►postsynaptic to the injured neuron undergo ►trans-synaptic changes [3].

The Phases of Neuronal Changes and Their Regulation

The prototypic neuronal response to injury, as seen following injury to peripheral axons, can be schematically divided in three, overlapping phases. This sequence

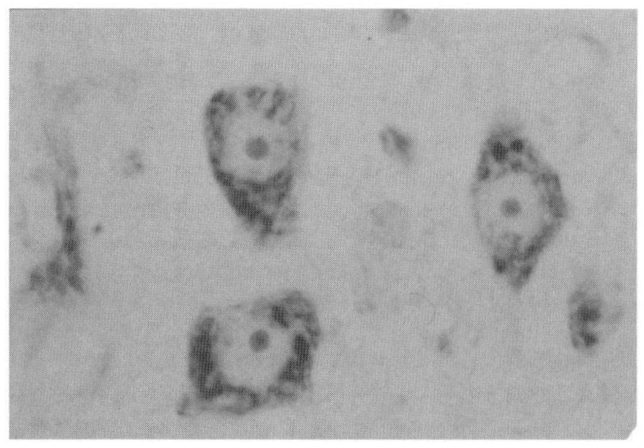

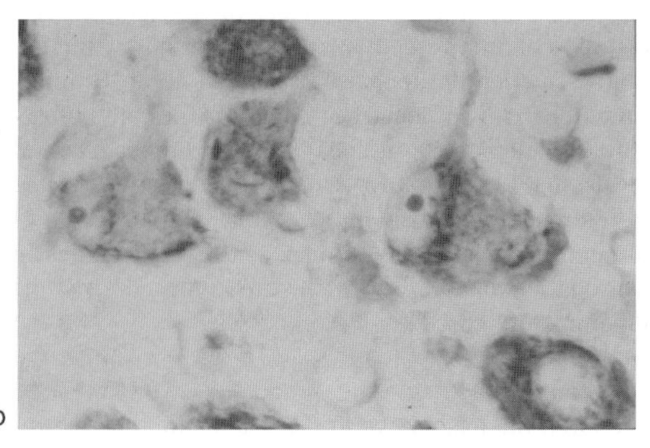

Neuronal Changes in Axonal Degeneration and Regeneration. Figure 1 (a) Normal rat motor neurons showing clumps of dark blue cytoplasmic Nissl bodies stained with a basophilic dye. (b) Rat motor neurons after injury to their axons showing that the Nissl bodies have disappeared, and the cytoplasm acquired a pale color, so-called chromatolysis.

of events applies only in part following injury to axons in the central nervous system (see below).

- The initial phase reflecting the immediate cell stress, and adaptation to loss of target contact; in this phase numerous ►sprouts are formed at the end of the proximal stump of the injured axons
- The phase of axonal elongation; in this phase a subset of the sprouts negotiate their way towards the denervated target; if conditions are sufficiently favorable, target contact is restored
- The phase of maturation of the ►regenerated axon and the return of the neuron to a "transmitting" (normal) mode; in this phase axonal diameter is growing, and ►remyelination of the larger axons is completed

Several factors mediate changes occurring during the initial phase. The injury causes an immediate ►depolarization of the neuron resulting in changes in intracellular ion homeostasis and the release of neurotransmitters and synaptic ►modulators. These early events lead to the induction of ►immediate early genes, which act as ►transcription factors. At a somewhat later point in time, the full transformation of the neuron into a growth state occurs by at least two different processes: (i) molecules from the extracellular environment enter the axon at the injury site and are ►retrogradely transported within the axon to the nerve cell body, and (ii) specific molecules produced by the target cells, so-called ►trophic factors, which normally reach the nerve cell body by retrograde axonal transport, are depleted. Concomitantly, growth promoting molecules, diffusible and associated with the ►extracellular matrix, are up-regulated in cells at the injury site, and in non-neuronal cells in the distal part of the axon. These molecules play a crucial role in the creation of a growth permissive pathway from the injury to the target tissue.

Changes in Neuronal Morphology

The most striking morphological change in nerve cells following axon injury is the loss of Nissl bodies, making the cytoplasm appear paler than normal (chromatolysis). At the same time, the nerve cell bodies often appear round with their nucleus displaced towards the periphery (away from the exit point of the axon). The ultrastructural basis for chromatolysis is loss of granular ►endoplasmic reticulum. The diameter of the axon proximal to the injury gradually becomes thinner, axon collaterals are lost and the dendrites become shorter. Thus, the overall dimension of the injured neuron is reduced. Following target reinnervation, these changes are only partially reversed. The axon diameter and the normal shape of the ►dendritic tree are typically not restored.

Changes in Neuronal Gene Expression

The list of changes in gene expression following axon injury is long. Genes involved in the synthesis of the classical neurotransmitters and with ►postsynaptic receptors are down-regulated, as are also genes contributing to structural stability of the neuron. The latter includes genes for proteins of ►neurofilaments, the major stabilizing component of the axon, and certain classes of ►microtubulus-associated proteins (MAPs), which provide structural support to dendrites. All these changes provide the basis for the morphological changes (see above). Genes belonging to the family of ►stress response (heat shock) protein, and the ►chaperon system are up-regulated, as well as genes supportive of axonal sprouting and extension. The latter includes intracellular molecules, which provide building blocks for the elongating axons, as well as membrane-bound and diffusible molecules, which are necessary for appropriate interactions with the environment. Important intracellular growth-associated molecules include ►actin, ►tubulin and several so-called ►growth-associated proteins (GAPs).

Changes in Glial Cells and Synapses

Striking changes occur in adjacent glial cells as well as in ►synaptic terminals, which cover injured neurons [4–6]. In the central nervous system, non-synaptic neuronal membrane is covered by processes of ►astrocytes. As a result of axon injury, these cells hypertrophy, and increase their coverage of the neuronal membrane in parallel with the disappearance of ►presynaptic terminals. The predominating type of the lost terminals is ►excitatory. ►Microglial cells in the neighborhood of the affected neurons ►proliferate, migrate towards the nerve cell body of the injured neuron, and up-regulate molecules associated with ►immune and inflammatory responses. In autonomic and sensory ganglia of the peripheral nervous system, ►satellite cells, which normally cover nerve cell bodies, proliferate. In addition, monocytes enter from the vascular system, and become transformed to macrophages. In the central nervous system and autonomic ganglia, a large proportion of presynaptic terminals on the cell body and dendrites disappear. The overall result of these changes is that the nerve cell body and dendrites of the injured nerve cell are partially isolated from surrounding influences.

Differences in Neuronal Changes Following Injury to Peripheral or Central Axons

Neurons with there axons confined to the central nervous system initially respond to injury or loss of target in a similar manner as those with axons in the peripheral nervous system [7]. Central neurons produce sprouts, which are capable of making novel synaptic connections within a limited distance [8]. However, central neurons are in general unable to mount the sustained up-regulation of growth supporting gene expression necessary for axon elongation, unless their axons are allowed to grow in a peripheral nervous system environment. The failure to sustain a prompt and long-lasting up-regulation of growth supporting genes is the result of intrinsic neuronal factors, in combination with a powerful inhibitory influence of the environment at the injury site and distal to it.

The Special Case of Sensory Ganglion Cells

Sensory ganglion cells are unique in having one axon projecting to peripheral target tissue, and one centrally, which terminates on postsynaptic neurons in the spinal cord or brainstem. Injury to peripheral sensory axons results in the same sequence of structural and molecular changes in their cell bodies, and proximal axon as described above. In addition, the central process and its terminals are affected. These include a reduction in the diameter of the central process, and changes in the morphology and chemical properties of its central terminals. As a consequence of these changes permanent alterations arise in the transmission of sensory impulses in the spinal cord and brainstem, which may contribute to long-lasting post-injury sensory ►neuropathies, e.g. ►neuropathic pain.

Injury to the central axon is associated with an attenuated neuronal response. The "transmitting" phenotype is essentially intact, and growth supporting gene expression is minimal. The injured axons sprout and elongate, but at a significantly slower rate than after corresponding injury to the peripheral axon. By combining central axon injury with a peripheral one, both axons elongate with the higher rate, and is capable of even moderate growth in the spinal cord itself [9]. The peripheral injury induces a growth state, and act as a ►conditioning lesion that amplifies axon outgrowth from the central process.

Long term Consequences of Axon Injury in the Peripheral Nervous System

There are three principal outcomes of axon injury: (i) neuron survival, axon regeneration, and functional recovery; (ii) neuron survival, failure of axon regeneration, and no functional recovery; (iii) neuron degeneration.

Successful axon regeneration and functional recovery is possible following injury to axons in the peripheral nervous system. This outcome requires that injured axons are able to enter the distal stump unimpeded, and that the target tissue is not too distant. A prime goal in modern research on peripheral axon injury is to eliminate obstacles to axonal elongation, and to increase its rate. With reconnection to a target, the neuron down-regulates genes expressing growth promoting molecules, the dendritic tree expands, and synaptic coverage increase, i.e. the neuron resumes a mature, "transmitting" phenotype. A complete restoration of the pre-injury state is, however, achieved only under the most favorable circumstances.

In case injured axons fail to make functional peripheral connections, the growth state of the neuron will cease, and nerve cell body and its processes enter a state of prolonged, possibly permanent atrophy, which may be severe. Injury to the axon, e.g. by removal of the ►neuroma at the proximal stump, will re-activate the growth state, and a new attempt to regenerate the injured axon [10].

Degeneration and ►death of neurons is a common consequence of axon injury. The risk of this outcome is significantly increased if axon injury (i) affects immature individuals, (ii) the injury leads to complete separation of the proximal and distal parts of the axon, e.g. the nerve is sectioned rather than crushed, (iii) is close to the nerve cell body, (iv) is combined with injury to afferent axons, a common situation in the central nervous system, and (v) affects certain classes of neurons. Neuron degeneration after axon injury commonly occurs by an intrinsic cell death program, ►apoptosis. The relationship between neuron degeneration and regeneration is complex in the sense that injury circumstances that increase the risk of neuron death also leads to the most powerful growth response.

Clinical Aspects

Axons in the peripheral nervous system are injured in trauma and many common disorders, e.g. diabetes. Trauma is also a common cause of axon injury in the central nervous system. More common there are axon injury because of disorders of the cerebral blood vessels (►stroke). Axon injury is at least in part a feature of many chronic disorders of the nervous system, e.g. ►Alzheimer's disease. Nerve cell survival is a prerequisite for functional recovery. Intense research is therefore underway to develop optimal strategies for promoting survival of injured neurons, and allow them to regenerate their axon and restore useful functional synaptic contacts.

References

1. Abankwa D, Küry P, Müller HW (2002) Dynamic changes in gene expression profiles following axotomy of projection fibres in the mammalian CNS. Mol Cell Neurosci 21:421–435
2. Lin H, Hao J, Sung YJ, Walters ET, Ambron RT (2003) Rapid electrical and delayed molecular signals regulate the serum response element after nerve injury: convergence of injury and learning signals. J Neurobiol 57:204–220
3. Ginsberg SD, Martin LJ (2002) Axonal transection in adult rat brain induces transsynaptic apoptosis and persistent atrophy of target neurons. J Neurotrauma 19:99–109

4. Aldskogius H (2001) Microglia in neuroregeneration. Microsc Res Techn 54:40–46
5. Brännstrom T, Kellerth JO (1998) Change in synaptology of adult cat spinal alpha-motoneurons after axotomy. Exp Brain Res 118:1–13
6. Linda H, Shupliakov O, Ornung G, Ottersen OP, Storm-Mathisen J, Risling M, Cullheim S (2000) Ultrastructural evidence for a preferential elimination of glutamate-immunoreactive synaptic terminals from spinal motoneurons after intramedullary axotomy. J Comp Neurol 425:10–23
7. Mason MRJ, Lileberman AR, Anderson PN (2003) Corticospinal neurons up-regulate a range of growth-associated genes following intracortical, but not spinal, axotomy. Eur J Neurosci 18:789–802
8. Chuckowree JA, Dickson TC, Vickers JC (2004) Intrinsic regenerative ability of mature CNS neurons. Neuroscientist 10:280–285
9. Qui J, Cafferty WB, McMahon SB, Thompson SW (2005) Conditioning injury-induced spinal axon regeneration requires signal transducer and activator of transcription 3 activation. J Neurosci 25:1645–1653
10. McPhail LT, Fernandes KJ, Chan CC, Vanderluit JL, Tetzlaff W (2004) Axonal injury reveals the survival and re-expression of regeneration-associated genes in chronically axotomized adult mouse motoneurons. Exp Neurol 188:331–340

Neuronal Determination

►Combinatorial Transcription Factor Codes and Neuron Specification

Neuronal Differentiation

Definition

The process whereby uncommitted neuronal precursor cells gradually accumulate gene products specific to, and required for, the eventual form and function of a specialized neuronal cell type. This process is associated with a cessation of cell division and is usually irreversible.

►Neural Development
►Neural Stem Cells
►Regeneration
►Axonal Pathfinding and Network Assembly
►Combinatorial Transcription Factor Codes and Neuron Specification

Neuronal Ensemble

►Temporal Coding

Neuronal Imaging

►Neuron-Glia-Imaging

Neural Integrator – Horizontal

José M. Delgado-García
Divisiòn de Neurociencias, Universidad Pablo de Olavide, Sevilla, Spain

Definition

A neural network that receives input signals related to horizontal eye velocity and generates a signal proportional to horizontal eye position that it conveys to extraocular motoneurons.

Description of the Theory

The eye moves in the horizontal plane under the action of two antagonist extraocular muscles: the lateral rectus and the medial rectus (see the entry devoted to extraocular motoneurons for a description of their source of innervation). To fully compensate for head movements activating vestibular or optokinetic reflexes, or to maintain a given eye position on a target following a voluntary saccade, motoneurons need to receive an eye position signal [1–5]. Although a common neuronal integrator capable of generating eye position signals for all kinds of eye movements was initially proposed [5], recent experimental data indicates that there are several integrators depending on the neural structure responsible for generating oculomotor commands and on the plane of the movement [6]. It is accepted that horizontal and vertical eye position signals are generated separately in the ►nucleus prepositus hypoglossi (PH) and in the interstitial nucleus of Cajal [2,4,7]. Other brainstem and cerebellar structures, such as the medial vestibular nucleus, the marginal zone between the latter and the PH nucleus, and cerebellar areas including the flocculus and the fastigial nucleus, also contain neurons carrying eye position signals (see [8] for references).

N

How Horizontal Eye Position Signals are Generated

Permanent and transient blockage of the normal function of neuronal integration in the horizontal plane in both cats and monkeys [6,9–11], supports the assumption that it takes place inside the PH nucleus and/or in the functional interactions established by its reciprocal connections with the vestibular nuclei, the contralateral PH, and the cerebellum. However, a still – unanswered question is: how are eye position signals generated by neuronal centers or circuits? Although the intrinsic connectivity of PH neurons is not completely known, in particular regarding the presence of axon collaterals that project in a feedback (or feedforward) fashion onto other oculomotor-related neurons [3,5,6,8], one hypothesis is that the PH neuronal integrator could generate eye position signals from successive synaptic steps in cascade, lateral, or retrograde chain systems [1,2,8]. As illustrated in Fig. 1, the presence of cascade-like, polysynaptic connections could explain the experimental observation of neuronal types with a wide range of eye motor signals (velocity, velocity-position, position-velocity, position, etc.), and the high susceptibility of

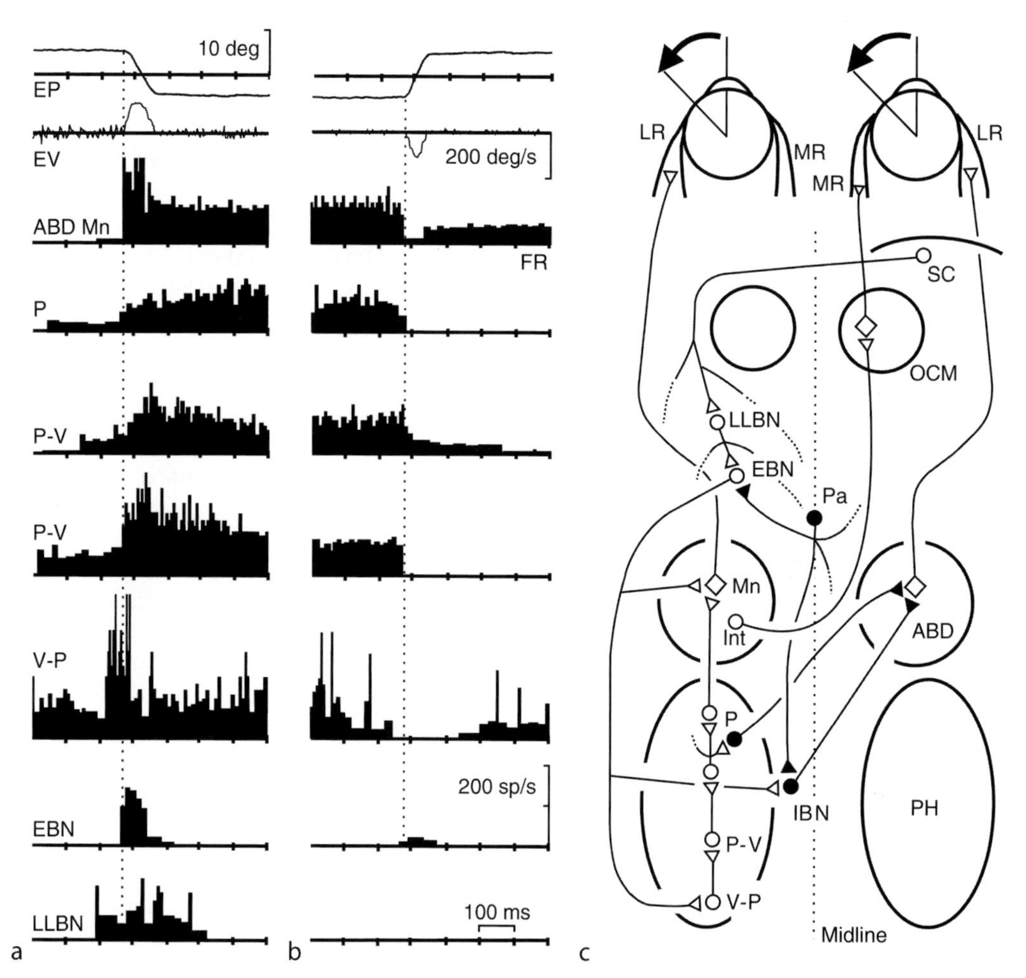

Neural Integrator – Horizontal. Figure 1 Experimental support for a ►cascade model in the generation of eye position signals in the nucleus prepositus hypoglossi (PH). (a,b) Firing rate of seven different types of neuron recorded in alert cats during eye fixations before and after on- and off-directed saccades. From bottom to top are illustrated the firing rates (FR, in spikes/s) of a long-lead burst neuron (LLBN), an excitatory burst neuron (EBN), four PH neurons showing velocity-position (V-P), position-velocity (P-V), or position (P) signals, and an abducens motoneuron (ABD Mn). Abducens interneurons receive the same inputs as ABD Mns, and relay them to medial rectus Mns located in the oculomotor nucleus (OCM). Representatives horizontal eye position (EP, in deg.) and velocity (in deg/s) corresponding to these neuronal activities are illustrated in the two traces at the top. (c) A diagram illustrating the possible pathways generating eye position signals following a saccadic motor command triggered from the superior colliculus (SC). Abbreviations: *LR*, *MR*, lateral and medial rectus muscles; *IBN*, inhibitory burst neuron; *Pa*, omnipause cells. Modified from Escudero et al. [2], and reproduced with permission of the Physiological Society.

the eye position neuronal system to administration of drugs and anesthetics and to the mental state and attentive level [2,7,8,11]. Moreover, these cascade chains could be superimposed upon the shorter, direct pathways carrying eye velocity signals (Fig. 1c). Evidence supporting the participation of neuronal circuits in the generation of the persisting activity that underlies the integration of eye position signals in goldfish has been reported recently [12].

Pure horizontal eye position neurons seem to project monosynaptically from the PH nucleus onto abducens motoneurons [2]. However, other neuronal types carrying mixed vertical position-velocity signals have been reported to project monosynaptically on vertical motoneurons located in the oculomotor complex [3,4]. According to the available information [2,11,13], PH neurons classified as ►principal cells [13] are the ones responsible of the neural integration taking place in this nucleus, and/or of carrying eye position signals to oculomotor nuclei.

Once generated, horizontal (and vertical) eye position signals seem to arrive at extraocular motoneurons, where they are integrated with (i.e. added to) eye velocity signals arriving from specific reticular formation nuclei. The stabilizing role of the intrinsic membrane properties of ocular motoneurons should also be taken into account. The algebraic addition of eye velocity signals arriving preferentially onto motoneuron somata, and of these different sources of eye position signals impinging upon their distal dendrites, can still be further enhanced by the intrinsic active properties of the motoneuron membrane to produce the stable firing rate that these motoneurons display, mainly during eye fixation [1].

Since the seminal contributions of Robinson's group [5], many authors have attempted with the design of more or less realistic mathematical models simulating the generation of eye position signals in mammals. A recent example of implementation for brainstem circuits involved in oculomotor integration processes is illustrated in Fig. 2 [14].

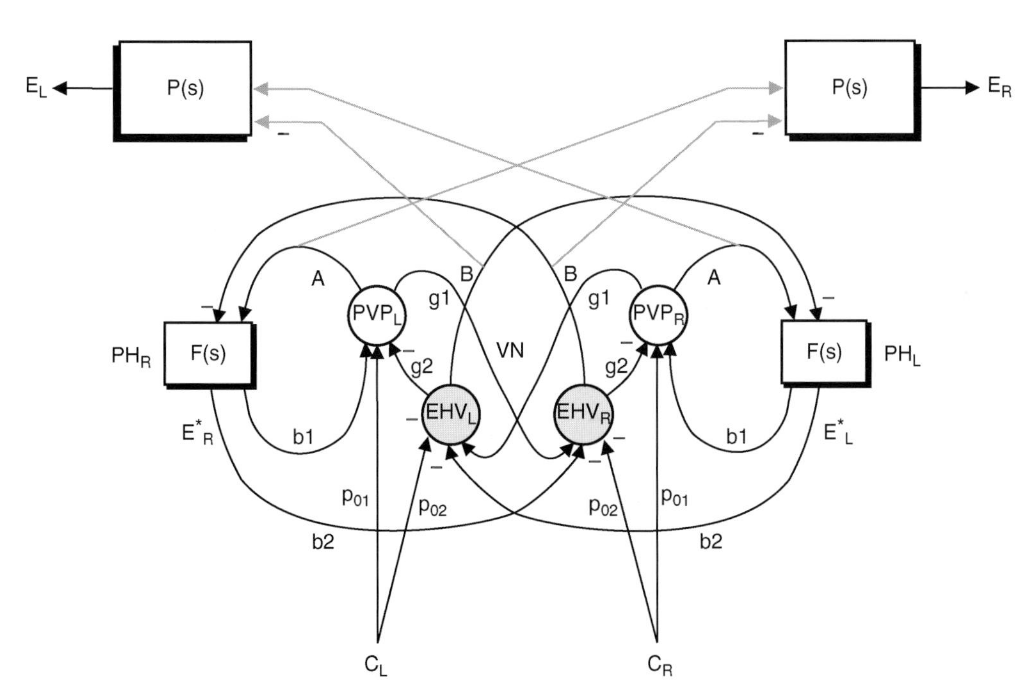

Neural Integrator – Horizontal. Figure 2 A bilateral implementation for brainstem circuits involved in supporting the central ►oculomotor integrator. This model allows merging of both vergence and version integrator functions in the dark. All sensory inputs converge on this circuit, with only semicircular canals ($C_{R,L}$) shown here on both sides. E_R,E_L are monocular eye positions on the right and left; P(s) represent dynamics of each eye plant in Laplace domain (filters); F (s) represent low-pass filters in the PH, which approximately model the true eye-plant dynamics. ►Position-vestibular-pause (PVP) and ►eye-head-velocity (EHV) cells in the vestibular nuclei (VN) are interconnected ipsi- and contralaterally with the PH. All other variables are simply scalar weights on the paths. Each PH receives branches from the same cells eventually driving the contralateral eyeball muscles – as a result, PH cells produce ►monocular efference copies. Due to mirror symmetry, the distributed loops between PH and VN imbed two large time constants, one for vergence and the other for version (conjugate) eye control. Hence, there are actually two integrators available for the two dimensions of horizontal, binocular, eye movements. Modified from A. Green, Visual-Vestibular interaction in a bilateral model of the rotational and translational vestibule-ocular reflexes: An investigation of viewing-context-dependent reflex performance, PhD Thesis, Department of Biomedical Engineering, McGill University, January, 2000."

Role of Neurotransmitters in the Generation of Eye Position Signals

As indicated above, the presence of separate integrators subserving the velocity storage mechanism and eye position signal generator has been proposed [7,13]. In fact, these two integrating mechanisms can be experimentally separated using some pharmacological tools. For example, injections of nitric oxide (NO) synthase inhibitors in the nucleus PH of conscious cats produce alterations of eye velocity, but not of eye position. In contrast, the injection of NO donors in the marginal zone close to the medial vestibular nucleus (i.e. an area rich in NO-sensitive guanylyl cyclase present on GABAergic afferent terminals) seems to affect the generation of eye position signals, a fact confirmed with lesion experiments in monkeys [7,8,14]. Thus, the integrative capabilities of distinct regions of the same PH nucleus subserve different eye-movement subsystems.

Other neurotransmitters (glutamate, acetylcholine) have been proposed to be related to neuronal operations taking place in the PH nucleus during eye fixation. An initial step for the integration of those early findings is represented by the suggestion that a cholinergic synaptically triggered phenomenon participates in the generation of eye position signals, subsequent to glutamatergic velocity signals arriving at the PH nucleus from the paramedian pontine reticular formation, i.e. from the site of excitatory burst neurons (Fig. 1c). It has been reported that the tonic firing present in PH neurons (▶tonic neurons) that follows a velocity motor command are generated, or at least facilitated, by cholinergic inputs acting on post-synaptic muscarinic M1 receptors located on those PH neurons [10]. These findings indicate that eye position signals arriving at the abducens nucleus could be originated in the PH nucleus by the effect of cholinergic inputs, subsequent to the depolarizing effects of glutamatergic excitatory burst neuron inputs, besides the participation of cascade-like [1,2] and/or ipsilateral and contralateral reverberant circuits [12]. Thus, the PH nucleus has more than one control mechanism to transform transient velocity signals into eye position ones [10].

References

1. Delgado-García JM, Vidal PP, Gómez C, Berthoz A (1989) A neurophysiological study of prepositus hypoglossi neurons projecting to oculomotor and preoculomotor nuclei in the alert cat. Neuroscience 29:291–307
2. Escudero M, de la Cruz RR, Delgado-García JM (1992) A physiological study of vestibular and prepositus neurones projecting to the abducens nucleus in the alert cat. J Physiol (Lond) 458:539–560
3. Fukushima K, Kaneko CRS, Fuchs A (1992) The neuronal substrate of integration in the oculomotor system. Prog Neurobiol 39:609–639
4. Moschovakis AK, Scudder CA, Highstein SM (1996) The microscopic anatomy and physiology of the mammalian saccadic system. Prog Neurobiol 50:133–254
5. Robinson DA (1981) The use of control systems analysis in the neurophysiology of eye movements. Ann Rev Neurosci 4:463–503
6. Kaneko CRS (1997) Eye movement deficits after ibotenic acid lesion of the nucleus prepositus hypoglossi in monkeys. I. Saccades and fixation. J Neurophysiol 78:1753–1768
7. López-Barneo J, Darlot C, Berthoz A, Baker R (1982) Neuronal activity in prepositus nucleus correlated with eye movement in the alert cat. J Neurophysiol 47:329–352
8. Delgado-García JM (2000) Why move the eyes if we can move the head? Brain Res Bull 53:475–482
9. Cheron G, Godaux E (1987) Disabling of the oculomotor neuronal integrator by kainic acid injections in the prepositus-vestibular complex of the cat. J Physiol (Lond) 394:267–290
10. Moreno-López B, Escudero M, Delgado-García JM, Estrada C (1996) Nitric oxide production by brain stem neurons is required for normal performance of eye movements in alert animals. Neuron 17:739–745
11. Navarro-López AD, Alvarado JC, Márquez-Ruiz J, Escudero M, Delgado-García JM (2004) A cholinergic synaptically triggered event participates in the generation of persisting activity necessary for eye fixation. J Neurosci 24:5109–5118
12. Aksay E, Gamkrelidze G, Seung HS, Baker R, Tank DW (2001) *In vivo* intracellular recording and perturbation of persistent activity in a neuronal integrator. Nature 4:184–193
13. Galiana HL, Outerbridge JS (1984) A bilateral model for central neural pathways in the vestibuloocular reflex. J Neurophysiol 51:210–241
14. McCrea R, Baker R (1985) Cytology and intrinsic organization of the perihypoglossal nuclei in the cat. J Comp Neurol 237:360–376

Neuronal Migration

Mineko Kengaku
Laboratory for Neural Cell Polarity, RIKEN Brain Science Institute, Wako, Japan

Definition

Cell migration is crucial for a variety of physiological and pathological processes, including leukocyte migration in the inflammatory response and tumor cell metastasis. During development, many cells migrate from their site of origin to their destination and reassemble with other cells for integration into functional tissues. Cell migration is a directional movement distinct from random dispersion and requires some mechanism for guiding cells to their destination. This essay focuses on migration of neurons in the developing brain, which is one of the most significant cell migration events in life.

Characteristics

Quantitative Description

The distance that a cell migrates varies widely from a few to a thousand cell-body diameters. Migration speed depends on the mode of migration. For instance, neurons in the cerebral cortex move at an average speed of 35 μm/h during locomotion and 60 μm/h during somal translocation [1]. Chain migration in the rostral migratory stream is as rapid as 120 μm/h [2].

Description of the Process

Neuronal migration consists of three schematic steps:

1. The ►leading process extends in the direction of travel along the substratum (neuronal or glial processes).
2. The nucleus and other organelles in the cell body move into the leading process.
3. The trailing process at the back detaches from the substratum and retracts to restore the original cell shape.

Harmonious repetition of these steps causes a caterpillar-like movement called locomotion [1]. In another mode of migration called somal translocation, the steps are not typically synchronized so that the cell soma moves within the preformed leading process independently of its extension [1]. In addition, cortical neurons in the intermediate zone migrate irregularly with dynamic extension and retraction of multiple processes; this mode is referred to as multipolar migration [3].

Higher Level Processes

Neuronal migration is classified into two distinct modes by the direction of the travel within developing neural tissue (Fig. 1).

Radial Migration

The vertebrate brain originates from the cylindrical neural tube. Neurons develop in germinative regions on the inner surface of the tube wall (the ventricular zone). Thus, new neurons principally move orthogonally from the ventricular zone toward the outer pial surface. In this mode of migration, which is called radial migration, neurons typically move along the fibers of radial glia traversing the entire depth of the parenchyma. Typical radial migration is thus referred to as "gliophilic migration".

Tangential Migration

In tangential migration, neurons move parallel to the pial surface of the brain, often across segmental or

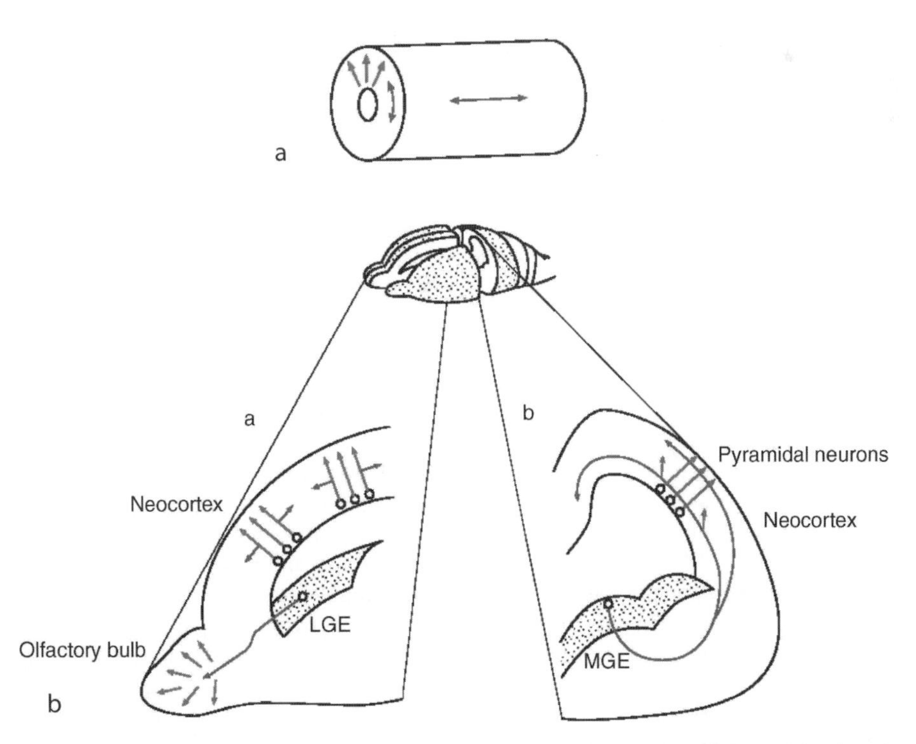

Neuronal Migration. Figure 1 Two modes of migration in the developing brain. (a) Radial (*red*) and tangential (*blue*) migration in the neural tube (b) Neuronal migration in the developing telencephalon. The rostral migratory stream from the subventricular zone of the LGE (*a* sagittal plane), tangential migration of cortical interneurons from the MGE and CGE (*b* coronal plane), and radial migration of cortical pyramidal neurons (*a* and *b*). *LGE, MGE* and *CGE*, lateral, medial and caudal ganglionic eminences.

regional boundaries. Tangentially migrating neurons are often "neurophilic" and extend their leading process along the axons of other neurons.

Higher Level Structure

The following are the relevant brain structures undergoing active neuronal migration during development.

Cerebral Cortex

The mammalian dorsal telencephalon develops into the six-layered cerebral cortex by radial migration of constituent pyramidal neurons. New neurons migrate from the ventricular zone towards the pia and accumulate below the margin of the cerebral wall to form the preplate. Subsequent neurons are deposited within the preplate to form the cortical plate and split the preplate into the superficial marginal zone and the deeper subplate. Cell dating studies have shown that neurons take their position in the cortical plate in an "inside-out" sequence, such that later developed cells migrate past the existing layers of earlier developed neurons and reside at the top of the plate (Fig. 2) [4,5].

Migration of cortical neurons can be classified into several modes with distinct kinetics. Neurons are generated from radial glia, which have a long process reaching the pial surface. Early neurons that inherit the long process from radial glia migrate into the cortical plate by translocation of their nucleus within their own processes independently of other cells. In contrast, later developing neurons often associate with other radial glia and migrate by locomotion in which the leading process moves in harmony with the cell body [1]. In addition, the later neurons have a multipolar shape and migrate in various directions (multipolar migration) in the intermediate zone, before they form a predominant leading process and enter the cortical plate by locomotion (Fig. 2) [3].

Inhibitory interneurons in the neocortex arise in the caudal and medial ►ganglionic eminences (CGE and MGE) of the ventral telencephalon and move to various levels of the developing neocortex in the dorsal telencephalon by tangential migration (Fig. 1) [4]. These neurons form one or two prominent leading processes, which follow the trajectory of axonal plexuses in the marginal zone and intermediate zone.

Rostral Migratory Stream in the Olfactory Bulb

Periglomerular and granule interneurons of the olfactory bulb are generated in the subventricular zone of the lateral ganglionic eminence (LGE) and reach their destination by tangential migration with a rostral orientation (Fig. 1). This migration, known as the "rostral migratory stream," mostly occurs during the early postnatal stage in rodents, but some cells continue to migrate throughout life. Unlike classical neurophilic tangential migration, many of these rostrally migrating cells are mitotic and migrate in chains in a glial tunnel traversing the tissue [2]. Migration of tightly associated strands of neurons is referred to as chain migration.

Cerebellar and Precerebellar Neurons

The developing cerebellum undergoes dynamic morphogenetic movements accompanying active cell migration. Most cell types in the three-layered cerebellar cortex migrate radially from the ventricular zone of the cerebellar anlage in the dorsal aspect of the hindbrain and caudal midbrain. In contrast, precursors of granule cells immediately take a tangential path from their origin in the rhombic lip, the anterior margin of a rhomboid-shaped roof plate lining the edge of the fourth ventricle. Granule cells undergo three successive phases of migration, each perpendicular to the others. First, precursors of granule cells undergo chain migration anteriorly to form the external granule layer over the dorsal surface of the cerebellum. After active proliferation, postmitotic granule cells extend bipolar axons and migrate tangentially following the trajectories of the preformed axons of other granule cells along the long (mediolateral) axis of the cerebellum. The third phase is radial migration along the radial fibers of Bergmann glia to reach the internal granule layer in the deep cerebellar cortex [6] (Fig. 3).

The different populations of rhombic lip cells migrate tangentially to form the precerebellar nuclei in the ventral hindbrain. These include the pontine, lateral reticular and inferior olivary nuclei, which provide the principal input to the cerebellum. The young precerebellar neurons first emit a long leading process circumferentially from the dorsal to the ventral hindbrain and then translocate the nucleus within their own leading process through a tangential migration [7].

Neural Crest

Neurons and glia in the peripheral nervous system arise from the neural crest and migrate dynamically to various regions of the embryo. See accompanying essay in this Encyclopedia.

Regulation of the Process

Leading Process Extension

Leading processes have distinct characteristics depending on the mode of migration. Some tangentially migrating neurons including precerebellar neurons and cerebellar granule cells extend long leading processes tipped by a large growth cone. These tangential leading processes are destined to become axons and their extension is probably controlled by a mechanism similar to that of growth cone steering without cell migration (See accompanying essay in this Encyclopedia). In contrast, gliophilic radial migration is guided by a short, tapering leading process resembling a dendritic tip. Despite some fundamental differences, leading process

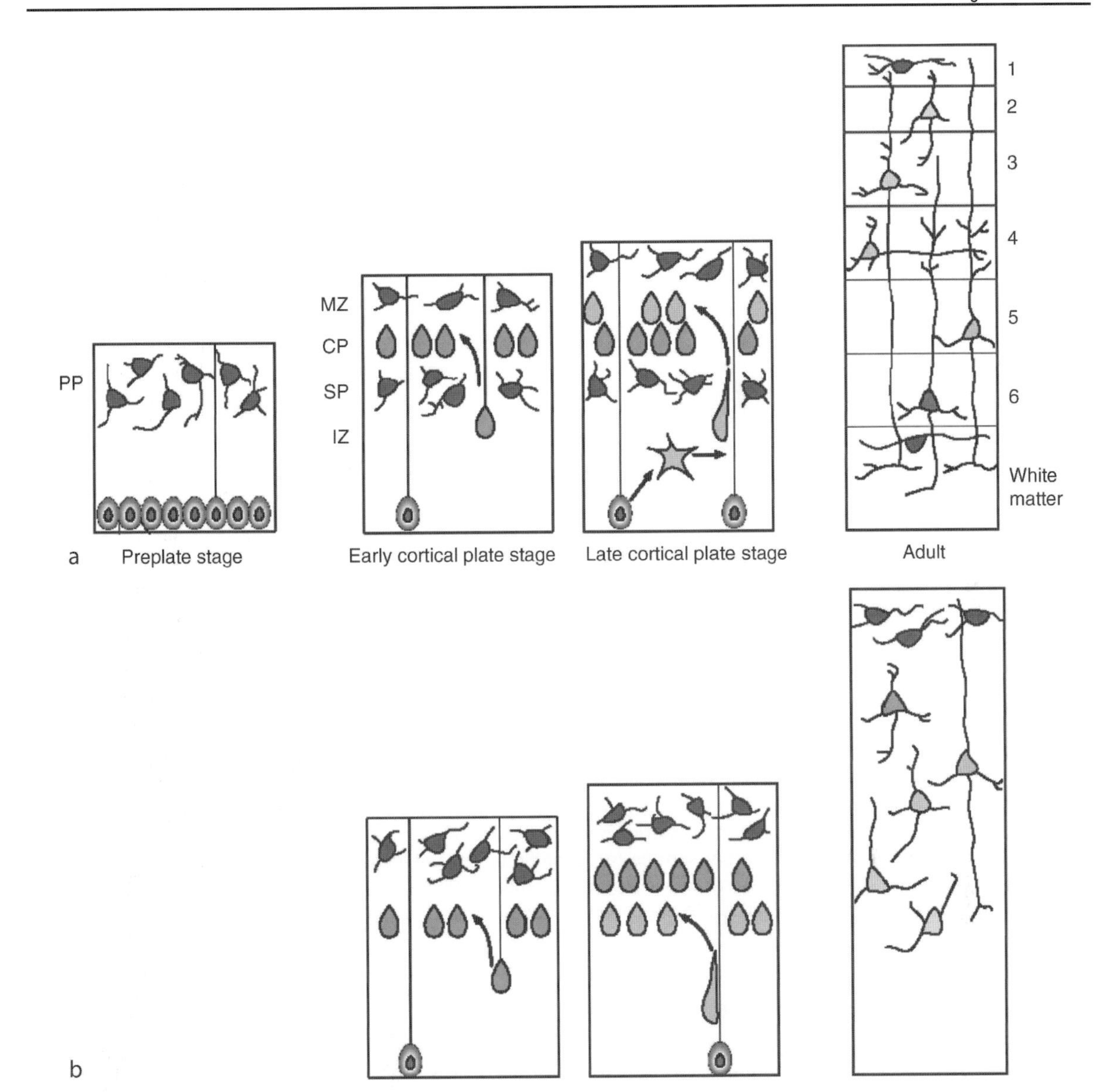

Neuronal Migration. Figure 2 Radial migration and neocortical layer formation a normal cortical development. a The preplate (*PP*) is formed by the first wave of postmitotic cells (*blue*) differentiated from radial glia (*red*) in the ventricular zone (*VZ*). In the early cortical plate (*CP*) stage, new neurons migrate radially from the VZ and split the PP into the marginal zone (*MZ*) and subplate (*SP*). Many neurons adopt somal translocation within the process inherited from radial glia. In the late CP stage, new neurons first move randomly by multipolar migration in the intermediate zone (*IZ*) and then migrate radially toward the CP by locomotion along the fibers of other radial glia. Neurons migrate past their predecessors and expand the CP in an inside-out fashion. The adult stage is marked by the six-layered neocortex. b In *reeler* mice, neurons fail to migrate beyond earlier neurons and pile up underneath the PP. Cortical layering in the adult stage is inverted and disorganized.

extension appears to involve common steps regardless of the mode of migration.

As mentioned in Higher Level Processes above, migrating neurons typically attach to glia or neurons on the path and follow the trajectories of their processes. Cell-substratum attachment is formed by transmembrane adhesion molecules, which recruit actin filaments in the cytoplasm through actin-binding proteins, such as integrin and L1, which immobilize actin filaments at neuron-glia or neuron-neuron contacts. At the cell-substratum attachment, the growing end of the actin filaments orients toward the tip of the leading process and generates a protrusive force for its extension [8]. An actin cross-linking phosphoprotein filamin A has been implicated in the protrusion of leading processes in cortical neurons during the transition from

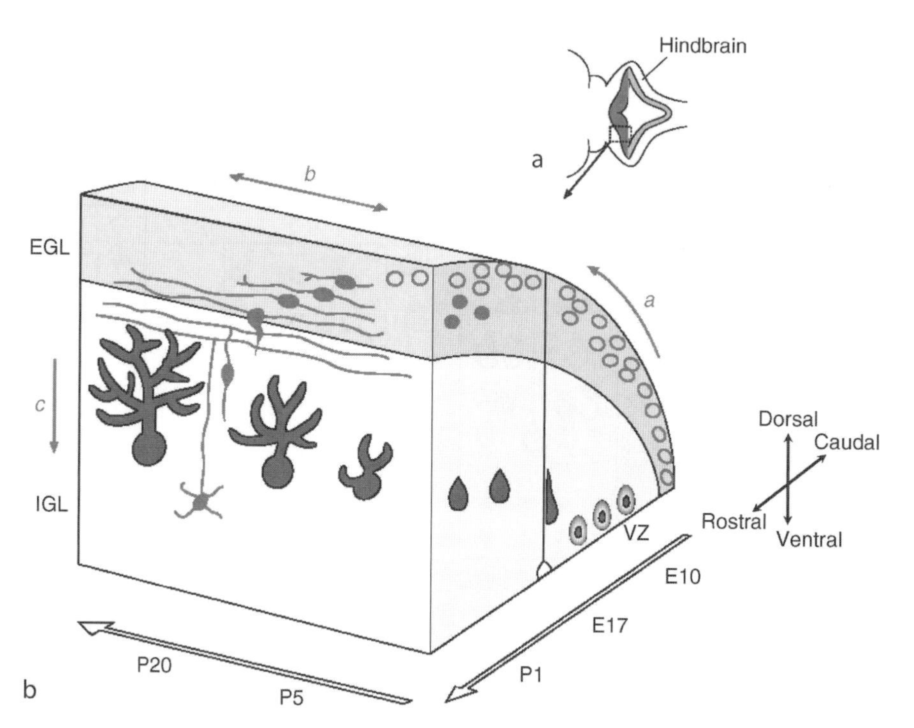

Neuronal Migration. Figure 3 Migration of cerebellar neurons (a) Dorsal view of the developing rhombic lips. The anterior part of the lip (*red*) gives rise to granule cell precursors. Cells from the posterior part of the lip (*green*) migrate ventrally to form the precerebellar nuclei. (b) Embryonic and postnatal development of the cerebellum. Purkinje cells (*blue*) are generated in the ventricular zone (*VZ*) of the cerebellar anlage and reach the cerebellar cortex by radial migration. Granule cells (*red*) undergo three successive phases of migration en route to the internal granule layer (*IGL*) in the cortex: Mitotic precursors (*blank*) migrate rostrally in chains to form the external granule layer (*EGL*) on the dorsal surface (*a*); postmitotic granule cells first extend bipolar axons and tangentially migrate along the mediolateral axis (*b*); cells then make a vertical turn and move ventrally toward the IGL by radial migration (*c*). *Blank arrows* indicate approximate developmental stages in mice.

multipolar migration to locomotion. Actin polymerization can be promoted by guidance cues that include soluble chemicals and adhesive molecules on the path. Receptors for families of such guidance cues can directly or indirectly alter the activity of the Rho family small GTPases, which are key regulators of actin polymerization in various migrating cells (Fig. 4).

Nuclear Migration

Migrations of neurons and other fibroblastic cells are clearly distinguished by their mechanisms of nuclear movement. In fibroblastic cells, nuclear migration is principally served by the actin cytoskeleton in the absence of microtubules; the actin-dependent extension of the leading edge generates a cortical tension that pulls the cell body in all directions. Traction then occurs at the back, because of preferential assembly of the cortex and adhesion to the substrate at the cell front. Finally the nucleus and cytoplasm are dragged forward passively by the traction force [8]. In contrast, some neuronal migration occurs by nuclear-driven cell migration termed "nucleokinesis" in which the nucleus moves within a highly protrusive leading process in a microtubule-dependent manner. In some migrating neurons, microtubules envelop the nucleus and also project into the leading process with the plus-end oriented toward the tip. It is hypothesized that the ►dynein motor complex is anchored to the cell cortex in the leading process and pulls the cell body forward by its minus-end-directed motor activity [9]. Dynein and its regulator LIS1 are colocalized in the cell cortex and ►centrosome and disruption of either of the genes leads to defects in nuclear migration in neurons. The centrosome is typically positioned in front of the nucleus and might mediate the pulling force from the leading process. The dynein complex on microtubules is also localized to the nuclear membrane, which could pull the nucleus toward the centrosome. The two-step nucleokinesis mediated by the centrosome – the movement of the centrosome toward the leading process followed by the movement of the nucleus toward the centrosome – is a favorable model especially for saltatory locomotion (Fig. 4).

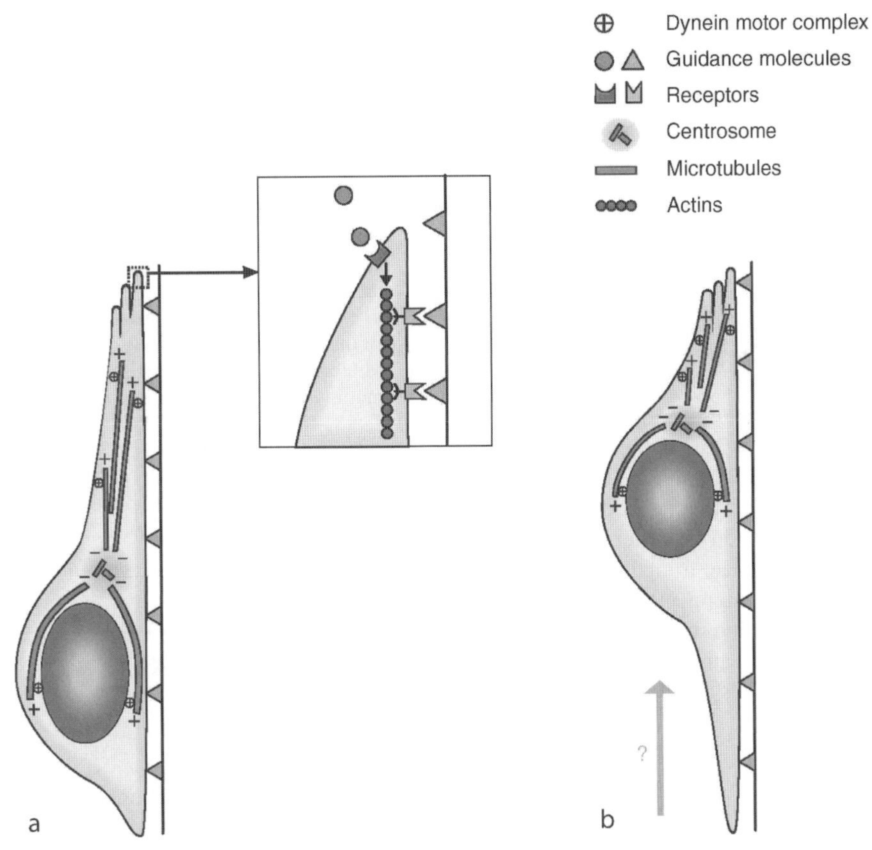

Neuronal Migration. Figure 4 Model for kinetics of neuronal migration. (a) Leading process extension is navigated by guidance molecules, which regulate actin polymerization in the leading edge. (b) Nuclear movement is driven by microtubules radiating from the centrosome. Activation of minus-end-directed motor activity of the dynein complex on the nuclear membrane leads to a displacement of the nucleus toward the centrosome. It is also provable that the dynein motor complex is anchored to the cell cortex in the leading process and pulls the centrosome forward. Additional forces may push the nucleus from the back. The retraction of the trailing process is not well understood.

Traction

As mentioned above, traction of the posterior of the cell is less important for nuclear migration in neurons. Indeed, the trailing processes of some neurons (e.g. cerebellar granule cells) are retained and differentiate into axons after cell migration. The mechanisms of retraction and differentiation of the trailing process are not well understood.

Function

Neuronal migration is required for the formation of defined cell patterns for the development of specific neural circuits.

Radial Migration

Defects in radial migration cause disruption of cortical lamination. Layering of the cerebral cortex is inverted and indistinct in the naturally occurring mouse mutant *reeler*, as radially migrating neurons are able neither to traverse their predecessors nor to assemble into distinct layers (Fig. 2) [5,10]. Reelin, the protein defective in *reeler* mice, is a large extracellular matrix protein secreted by ►Cajal-Retzius cells in the marginal zone. Reelin is a high-affinity ligand for two members of the LDL family of lipoprotein receptors, VLDLR (very-low-density lipoprotein receptor) and LRP8 (low-density lipoprotein receptor-related protein 8, also known as ApoER2), which induce phosphorylation of the tyrosine kinase adaptor Disabled 1 (Dab1) in migrating cortical cells. Mutations in the *Dab1* gene (*scrambler* and *yotari)* or double homozygous null for the genes *VLDLR* and *LRP8*, show identical phenotypes to *reeler*. Reelin signaling might induce events related to the reorganization of microtubules and microfilaments in the cytoskeleton.

Mice deficient in cyclin-dependent kinase 5 (Cdk5), or its activator p35, display similar defects in cortical lamination. Although preplate splitting by early-developed neurons occurs normally, layering of the cerebral cortex is inverted due to the inability of later developed neurons to migrate past their predecessors. The broad substrate range of the Cdk5-p35 complex suggests that it could regulate multiple aspects of migration. Cdk5-p35 is likely to regulate cytoskeletal dynamics by phosphorylating actin- and microtubule-binding proteins. Cdk5 signaling could also regulate neuron-glia attachment. Interestingly, tangential migration of cortical inhibitory interneurons appears relatively unaffected in mice deficient in Cdk5 or p35, suggesting that the Cdk5-p35 pathway is mainly used for radial, gliophilic migration [5,10].

Tangential Migration

Defects in tangential migration also disrupt the formation of specific brain structures and proper neural networks. Most cortical GABAergic interneurons arise in the MGE and CGE and reach the cortex in tangentially migrating streams. In *Dlx1/Dlx2* double-mutant mice, GABAergic interneurons fail to migrate from the MGE and drastically decrease in the neocortex, olfactory bulb and hippocampus [4]. On the other hand, neurons in precerebellar nuclei originate from the rhombic lip in the dorsal hindbrain and migrate tangentially toward the ventral midline secreting netrin 1. The loss-of-function mutant of netrin 1 causes defects in tangential migration of rhombic lip cells and hence the disruption of precerebellar nuclei [7].

Pathology

Disturbances of neuronal migration are implicated in human brain malformations associated with neurological conditions including mental retardation and epilepsy [10]. Mutations in filamin 1 gene lead to X-linked periventricular heterotopia in which a subset of neurons fails to migrate from the ventricular zone. Defects in nucleokinesis by a LIS1 mutation cause type 1 lissencephaly ("smooth brain" without convolutions) characterized by abnormally thickened and incomplete neocortical layers. Another X-chromosome-linked lissencephaly locus has been identified and named doublecortin (DCX). DCX encodes a microtubule-binding protein that is thought to stabilize microtubules. Male patients with mutations in the DCX gene on their single X-chromosome give rise to a phenotype similar to type 1 lissencephaly. Female patients with a heterozygous DCX mutation exhibit double cortex syndrome (subcortical band heterotopia) in which a fraction of neurons expressing a mutant DCX gene halt migration and cluster halfway between the cortex and the ventricle.

Reference

1. Nadarajah B, Parnavelas JG (2002) Modes of neuronal migration in the developing cerebral cortex. Nat Rev Neurosci 3:423–432
2. Wichterle H, Garcia-Verdugo JM, Alvarez-Buylla A (1997) Direct evidence for homotypic, glia-independent neuronal migration. Neuron 18:779–791
3. Tabata H, Nakajima K (2003) Multipolar migration: the third mode of radial neuronal migration in the developing cerebral cortex. J Neurosci 23:9996–10001
4. Marin O, Rubenstein JL (2003) Cell migration in the forebrain. Annu Rev Neurosci 26:441–483
5. Gupta A, Tsai LH, Wynshaw-Boris A (2002) Life is a journey: a genetic look at neocortical development. Nat Rev Genet 3:342–355
6. Hatten ME, Heintz N (1995) Mechanisms of neural patterning and specification in the developing cerebellum. Annu Rev Neurosci 18:385–408
7. Bloch-Gallego E, Causeret F, Ezan F, Backer S, Hidalgo-Sánchez M (2005) Development of precerebellar nuclei: instructive factors and intracellular mediators in neuronal migration, survival and axon pathfinding. Brain Res Rev 49:253–266
8. Mitchison TJ, Cramer LP (1996) Actin-based cell motility and cell locomotion. Cell 84:371–379
9. Tsai L-H, Gleeson JG (2005) Nucleokinesis in neuronal migration. Neuron 46:383–388
10. Gleeson JG, Walsh CA (2000) Neuronal migration disorders: from genetic diseases to developmental mechanisms. Trends Neurosci 23:352–359

Neuronal Network

Definition

A group of neurons that are connected by a specific set of synaptic interactions and fulfil a specific function (e.g. control of locomotion, processing of visual information, etc.). The boundaries of neuronal networks within the nervous system are often not very well defined as individual neurons can be part of a neuronal network under one set of circumstances, but not part of it under different circumstances. Entire networks can be reconfigured by modulatory influences, which enhance or suppress the activity in individual neurons and/or specific synaptic connections. These networks are described as polymorphic neuronal networks. Reconfiguration enables the nervous system to employ the same neuronal elements for the control of related, but different activities (e.g. walking, running, jumping, etc.), which is a more efficient use of resources than the existence of independent neuronal networks for each activity.

►Central Pattern Generator
►Rhythmic Movements

Neuronal Oscillator

Definition

A neuronal circuit or even a single neuron that, owing to the inherent electrical properties of the neuronal membranes and the synaptic connectivity of the component neurons, produces a rhythmic pattern of activity. Central pattern generators (CPGs) include neuronal oscillators, and CPGs for segmentally distributed motor patterns often comprise of neuronal oscillators in each segment of the nervous system participating in the production of the motor pattern; they are then often called segmental oscillators.

▶Central Pattern Generator
▶Intersegmental Coordination
▶Rhythmic Movements

Neuronal Plasticity

Definition

The capacity of neuronal systems (e.g., neurons, parts of neurons, populations of neurons) for change of anatomical and functional features.

▶Activity-Dependent Synaptic Plasticity

Neuronal Polarity

JOANNE SM KIM[1,2], MEI ZHEN[1,2]
[1]Department of Molecular Genetics, University of Toronto, Toronto, ON, Canada
[2]Samuel Lunenfeld Research Institute, Mount Sinai Hospital, Toronto, ON, Canada

Synonyms

Asymmetry in neurons

Definition

Neuronal polarity is the asymmetry in the distribution of cellular components (▶cellular polarity) within ▶neurons. In this essay, we first describe the development and maintenance of neuronal polarity and then demonstrate its function in the nervous system.

Characteristics

Development and Maintenance of Neuronal Polarity

Cells are the basic building blocks of multi-cellular organisms. Some cells appear morphologically and functionally homogeneous. For example, oxygen-carrying erythrocytes and infection-fighting leukocytes in human blood are round and even in shape. Other cells, however, display clear heterogeneity, where specialized regions within a single cell perform different biological functions. Epithelial cells in the vertebrate digestive system are an excellent example of the latter case. Each epithelial cell has an apical side that absorbs outside materials and a basal side that transfers the absorbed materials to the bloodstream. Epithelial cells are therefore termed "polarized."

Neurons are perhaps the most extensively-studied polarized cells. Each neuron performs two distinct functions: receipt of information from signaling cells and transmission of this information to target cells. To accommodate these different functions, neurons form multiple dendrites and in most cases, a single axon (Fig. 1a). Dendrites receive information from other neurons and relay it to the axon of the same cell. The axon then transmits the information to one or more target cells that can be neurons or tissue cells. The process of dendrite and axon differentiation is called neuronal polarization (Fig. 1).

Neuronal polarization has been well-characterized in the cultured hippocampal neurons of rat embryos [1]. This developmental process has been divided into five stages (Fig. 1a). In Stage One, newly-born neurons develop multiple outgrowth extensions called lamellipodia. Lamellipodia extend and elongate into neurites in Stage Two. In Stage Three, one of these neurites begins to grow faster than the others, adopting the axonal fate. Found at the tip of this neurite is a ▶growth cone that leads the axon to its targets. In the fourth stage, while the growing axon matures, the remaining neurites become dendrites. In the fifth and final stage, the axon and dendrites make connections with other cells, forming a neuronal network [1].

These dramatic morphological transformations during neuronal polarization are accompanied by changes in the ▶cytoskeleton structures of the developing neuron. The cytoskeleton supports membrane structure, maintains cell shape, and enables trafficking of organelles and proteins. The cytoskeleton is primarily comprised of polymers of tubulin (microtubules) and actin (actin filaments) (Fig. 1b). During neuronal polarization, microtubules and actin filaments are in highly dynamic states where their subunits assemble onto or disassemble from the filamentous structures, allowing rapid growth or shrinkage of neurites. Bradke and Dotti [2] have provided *in vitro* evidence that the one neurite destined to become the axon is associated with highly dynamic actin filaments. In addition, when a dynamic state was artificially created throughout a developing neuron with an actin-destabilizing drug, this neuron generated multiple axons. Conversely, local application of the drug to a single neurite designated axonal fate to that neurite [2], suggesting that actin dynamics in a neurite is necessary for its axonal fate. Accordingly, cytoskeleton dynamics is thought to be extensively regulated during neuronal polarization.

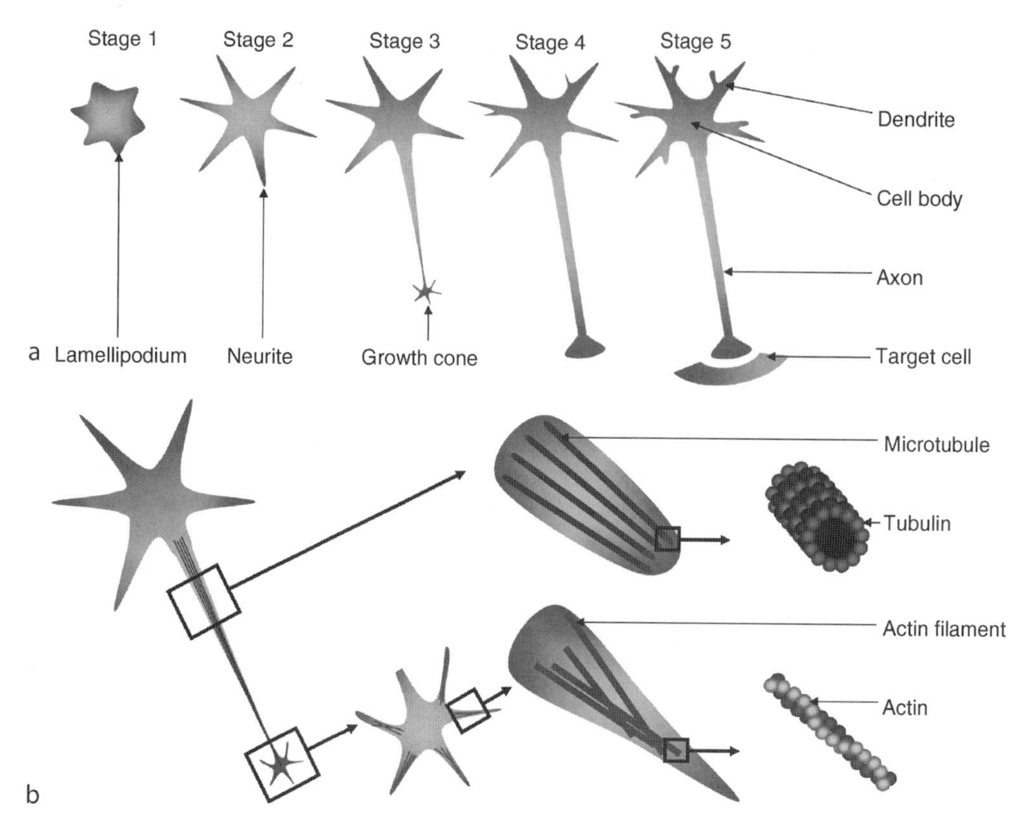

Neuronal Polarity. Figure 1 Neuronal polarization. (a) Five different stages of polarization observed in the cultured rat hippocampal neurons. A mature neuron has multiple short dendrites and a single long axon. Information adapted from Dotti et al. (b) Cytoskeleton structures in the developing neuron. Microtubules are found along the neurite while actin filaments localize to the growth cone. Both microtubules and actin filaments are comprised of protein subunits.

A number of regulators of cytoskeleton dynamics in neuronal polarization have been discovered [3,4]. The first *in vivo* evidence for such regulator was provided by ►Synapses of Amphids Defective 1 (SAD-1) [5]. The SAD-1 protein was identified in ►*Caenorhabditis elegans* (*C. elegans*), a nematode model organism first developed by Nobel laureates Sydney Brenner, Robert Horvitz, and John Sulston and now widely used in the laboratory for its easy handling and amenable genetics. Animals with mutantions in thier SAD-1 gene showed shorter axon lengths in specific neurons. Further, axon-specific proteins were mislocalized to dendrites [5], suggesting that SAD-1 is required for accurate polarization of neurons.

Mammalian ►homologs of *C. elegans* SAD-1 have also been identified [6]. In the mouse, two SAD-1-like proteins were found and named SAD-A and SAD-B. Eliminating both SAD-A and SAD-B causes the cultured hippocampal neurons from embryos to develop multiple projections that are neither dendrites nor axons [6], strongly supporting an evolutionarily conserved role for SAD proteins in neuronal polarity. Another mammalian protein, LKB1, also regulates neuronal polarity [7,8]. Similar to mice lacking the SAD proteins, LKB1 mutant animals show neuronal projections without distinct dendritic or axonal characteristics. Biochemical analyses show that LKB1 regulates neuronal polarity through SAD-A and SAD-B [7,8].

How do the SAD proteins and LKB1 regulate the cytoskeleton dynamics in neurons? The various SAD proteins and LKB1 are ►protein kinases – enzymes that transfer phosphate groups to their target protein. In the mouse, LKB1 first phosphorylates and activates SAD-A and SAD-B [7]. The activated SAD proteins then directly or indirectly (i.e., through another kinase) phosphorylate TAU, a microtubule-associated protein (MAP) [6]. MAPs bind and subsequently stabilize microtubules. Phosphorylated MAPs on the other hand detach from microtubules, resulting in cytoskeleton instability and dynamics. Microtubule dynamics, as mentioned above, is essential for neuronal polarization. Therefore, the SAD proteins and LKB1 regulate the cytoskeleton dynamics by modulating the phosphorylation status of TAU and possibly other MAPs [6–8].

Once developed, neuronal polarity needs to be maintained throughout the animal's lifespan. Presently, little is known about the mechanisms of maintenance. Nevertheless, Hammarlund et al. [9] have identified an essential component in the maintenance of membrane structure in *C. elegans* motor neurons. Supporting and organizing the membrane structure of mature axons is a complex called spectrin [9]. Made up of two protein subunits – alpha and beta spectrins – the spectrin complex maintains membrane integrity by stabilizing actin filaments associated with the membrane. In *C. elegans* animals with mutant beta-spectrin, neurons develop normally. However, mature axons of their motor neurons break in the adult stage [9]. When these animals were restricted from movement, no broken axons were observed, suggesting that breaks occur from the wear and tear of movements. Broken axons continue to regenerate, but fail to form functional axons due to repeated breaks. This study suggests that the membrane structure in mature axons needs to be maintained and that the development and maintenance of neuronal polarity may be governed by separate mechanisms.

Function of Neuronal Polarity

The nervous system is essentially a network of neurons forming connections with each other and other tissues. Communication between neurons is achieved through specialized structures called synapses (Fig. 2a). Each synapse consists of a pre-synaptic terminal in the signaling cell and a post-synaptic terminal in the target cell. Pre-synaptic terminals contain signaling molecules known as ►neurotransmitters in small membrane-bound compartments called vesicles. Upon stimulation, the vesicles release neurotransmitters outside the cell, which then diffuse to the target cells. The target cells contain post-synaptic terminals that are characterized by clusters of receptors for the neurotransmitters, often ion-channels, and other molecules required for signaling (Fig. 2a). Neurotransmitters bind the receptors in the post-synaptic terminals, trigger changes in the membrane potentials and cause subsequent effects in the target cells. In summary, information travels through the neuronal network via synapses formed by individual neurons.

What role does neuronal polarity play in synaptic functions? Dendrites and axons are different not only morphologically but also, more importantly, functionally. The dendrites of a neuron, as well as parts of its cell body, form post-synaptic terminals at the contact sites

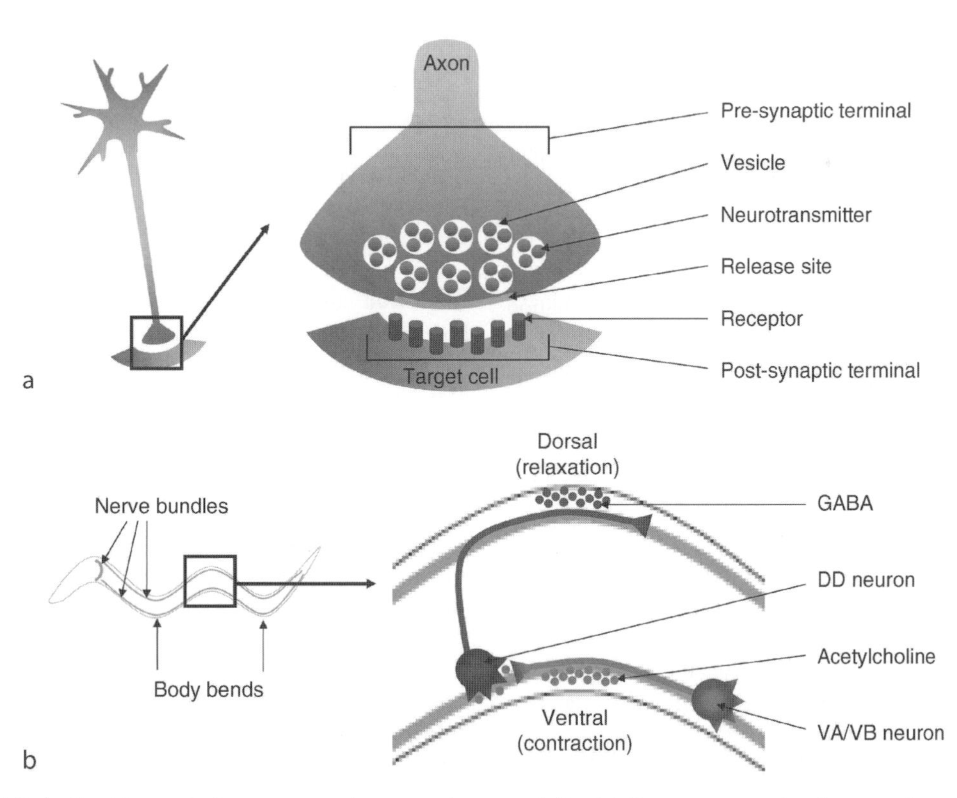

Neuronal Polarity. Figure 2 Synapses and neuronal connectivity. (a) Key components of pre- and post-synaptic specializations. Pre-synaptic terminals contain neurotransmitters in membrane-bound compartments called vesicles. Upon stimulation, neurotransmitters are released and bind the receptors in the post-synaptic terminal of a target cell, triggering changes in that cell. (b) Characteristic body bends generated by neuronal connectivity of motor neurons in *C. elegans*. The VA/VB axon releases acetylcholine which causes contraction of ventral muscles. Acetylcholine also triggers the release of GABA from the DD axon. GABA relaxes dorsal muscles. Information adapted from Schuske et al. [10].

with the axons of signaling cells (Fig. 2a). Axons on the other hand, form pre-synaptic structures that relay signals to target cells upon stimulation (Fig. 2a). Polarity in each neuron, therefore, defines the direction in which a signal travels.

Neuronal polarity and synaptic connections together determine the route by which signals travel throughout the nervous system. A simple example is provided by the *C. elegans* motor circuit [10] (Fig. 2b). The axon of a motor neuron called VA or VB forms synapses with the dendrite of another motor neuron, DD, on the ventral side of the animal. The VA or VB neuron releases neurotransmitters called acetylcholine. Acetycholine stimulates the DD neuron to release its neurotransmitters GABA, from its dorsally located axon. Signals continue to propagate throughout the neuronal network determined by neuronal polarity and synaptic connections.

The motor neurons also form synapses with muscle cells and control motor functions. The VA or VB axon forms synapses with ventral muscles. Acetylcholine, an excitatory neurotransmitter, causes contraction in these muscles (Fig. 2b). The DD axon, on the other hand, forms synapses with dorsal muscles. When these muscles receive GABA, they relax. The result of muscle contraction on the ventral side and relaxation on the opposite dorsal side is a body bend. This pattern of contraction and relaxation on the opposite sides is observed throughout the length of the animal's body, characteristic of roundworms. Alternating the pattern of body bends allows the forward and backward movement of the animal.

Summary
Neuronal polarity refers to the asymmetrical distribution of cellular components within a neuron. In this essay, we described the development of neuronal polarity and its function in the nervous system. During neuronal polarization, a group of molecules work in concert to regulate the dynamics of the cytoskeleton. The cytoskeleton dynamics is essential for neurite extension and axon and dendrite formation. Once established, neuronal polarity needs to be maintained, which likely requires the stabilization of cytoskeleton structures. In the nervous system, neuronal polarity and synaptic connections determine the route by which information travels. The synapses between neurons and other cell types along this route lead to the complex sensory, motor, or cognitive functions of an organism.

References

1. Dotti CG, Sullivan CA, Banker GA (1988) The establishment of polarity by hippocampal neurons in culture. J Neurosci 8:1454–1468
2. Bradke F, Dotti CG (1999) The role of local actin instability in axon formation. Science 283:1931–1934
3. Wiggin G, Fawcett JP, Pawson T (2005) Polarity proteins in axon specification and synaptogenesis. Dev Cell 8:803–816
4. Arimura N, Kaibuchi K (2005) Key regulators in neuronal polarity. Neuron 48:811–884
5. Crump JG, Zhen M, Jin Y, Bargmann CI (2001) The SAD-1 kinase regulates presynaptic vesicle clustering and axon termination. Neuron 29:115–129
6. Kishi M, Pan YA, Crump JG, Sanes JR (2005) Mammalian SAD kinases are required for neuronal polarization. Science 307:929–932
7. Barnes AP, Lilley BN, Pan YA, Plummer LJ, Powell AW, Raines AN, Sanes JR, Polleux F (2007) LKB1 and SAD kinases define a pathway required for the polarization of cortical neurons. Cell 129:549–563
8. Shelly M, Cancedda L, Heilshorn S, Sumbre G, Poo M (2007) LKB1/STRAD promotes axon initiation during neuronal polarization. Cell 129:565–577
9. Hammarlund M, Jorgensen EM, Bastiani MJ (2007) Axons break in animals lacking β-spectrin. J Cell Biol 176:269–275
10. Schuske K, Beg AA, Jorgensen EM (2004) The GABA nervous system in *C. elegans*. Trends Neurosci 27:407–414

Neuronal Potassium Channels

Alistair Mathie, Emma L. Veale
Medway School of Pharmacy, Universities of Kent and Greenwich at Medway, Chatham Maritime, Kent, UK

Definition
Potassium (K^+) channels are proteins which span the membrane of cells and are selectively permeable to K^+ ions.

Characteristics
Potassium (K^+) channels are proteins which span the membrane of cells and which, when open, allow the selective flow of K^+ ions from one side of the membrane to the other (usually from the inside of the cell to the outside). They can be gated by a variety of stimuli including voltage, changes in intracellular Ca^{2+} and certain other physiological mediators. In neurons, they have a number of functional roles related, primarily, to the electrical properties of the membrane. As such, they determine the neuronal action potential frequency, shape the neuronal action potential waveform (►Action potential) and control the strength of synaptic contacts between neurons [see 1]. Additionally, certain K^+ channels regulate the absolute excitability of neurons and set (or contribute to) the neuronal ►resting membrane potential [2] (►Membrane potential – basics). Their physiological importance has been exemplified

by the observations that mutations in K^+ channel sequences in particular individuals leads to such varied clinical disorders as ►epilepsy, episodic ataxia, unregulated insulin secretion and deafness [3].

K^+ Channel Families

K^+ channels are members of the voltage-gated-like ►ion channel superfamily [4]. Since the first molecular cloning of K^+ channel subunits in the 1980s, over eighty different genes have been identified which each encodes for distinct K^+ channel α subunits [see, for example, [5]]. Each K^+ channel consists of a primary pore-forming α subunit often associated with auxiliary regulatory subunits. From the amino acid sequences of K^+ channel α subunits, it is possible to group them into three families (Fig. 1).

These are (i) the six transmembrane (6TM) domain channels which are gated by voltage or in a few cases by both Ca^{2+} and voltage or by Ca^{2+} alone; (ii) the two transmembrane (2TM) or ►inward-rectifier K^+ channels and (iii) the four transmembrane (4TM) or two pore domain (K2P) ►leak K^+ channel(s).

The Six Transmembrane Domain K Channel Family

The 6TM family comprises a number of different subfamilies (Table 1) such as the voltage-gated K_V1–4 family of channels, which underlie functional ►delayed-rectifier channels and ►A-type K^+channel(s). These channels open when the membrane is depolarized. Also in this family are the K_V7 (KCNQ) and K_V10–12 (EAG) channels (including hERG, which are of particular interest due to their role in certain sudden death syndromes). These are low-threshold voltage-gated channels which are often regulated by G protein coupled receptors such as ►muscarinic acetylcholine receptors and have a non-inactivating component open close to the resting membrane potential of the cell. Finally, in this family are the Ca^{2+} and Na^+ activated K^+ channels (K_{Ca}1–5). These K^+ channels can be divided into three broad groups. These are the large conductance maxi-K^+ channels (or ►BK channels) corresponding to the *slo* family of K^+ channels, the small-conductance or ►SK channels (corresponding to SK1-SK3) and the intermediate conductance K^+ channels, corresponding to SK4 channels. Within the 6TM family, the pore-forming

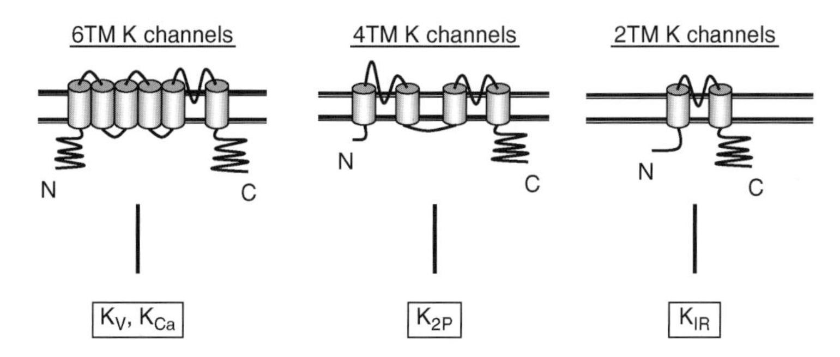

Neuronal Potassium Channels. Figure 1 Schematic representation of the structure of the α subunit of the three primary K^+ channel families, 6TM, 4TM and 2TM.

Neuronal Potassium Channels. Table 1 The 6TM K channel family

Subfamily group	Subtypes	Functional characteristics	Associated subunits
K_V1.x	K_V1.1–1.8	K_V (1.1–1.3, 1.5–1.8) K_A (1.4)	$K_V\beta_1$, $K_V\beta_2$
K_V2.x	K_V2.1–2.2	K_V (2.1)	K_V5.1, K_V6.1–6.3 K_V8.1, K_V9.1–9.3
K_V3.x	K_V3.1–3.4	K_V (3.1, 3.2) K_A (3.3, 3.4)	MiRP2 (K_V3.4)
K_V4.x	K_V4.1–4.3	K_A	KChIP, KChAP
K_V7.x	K_V.7.1–7.5 (KCNQ1–5)	K_V7.1 – cardiac IK_S K_V7.2/7.3–M current	minK, MiRP2 (K_V.7.1)
K_V10.x	K_V10.1–10.2 (eag1 - 2)	–	–
K_V11.x	K_V11.1–11.3 ((h)erg1 - 3)	K_V11.1 – cardiac IK_R	minK, MiRP1 (K_V11.1)
K_V12.x	K_V12.1–12.3 (elk1 - 3)	–	–
K_{Ca}1.x K_{Ca}4.x K_{Ca}5.x	K_{Ca}1.1 K_{Ca}4.1–4.2 K_{Ca}5.1	BK_{Ca} K_{Na}	KCNMB1–4 (K_{Ca}1.1)
K_{Ca}2.x K_{Ca}3.x	K_{Ca}2.1–2.3 K_{Ca}3.1	SK_{Ca} (K_{Ca}2.1–2.3) IK_{Ca} (K_{Ca}3.1)	–

N

α subunits form tetramers. Heteromeric channels may be formed within subfamilies (e.g. $K_V1.1$ with $K_V1.2$; $K_V7.2$ with $K_V7.3$).

The Two Transmembrane Domain K Channel Family

The 2TM domain family of K^+ channels is also known as the inward-rectifier K^+ channel family. Current flows through them more easily in an inward direction because they are blocked by intracellular polyamines and/or Mg^{2+} ions at depolarized voltages. This family includes the strong inward-rectifier K^+ channels ($K_{IR}2$), the G-protein-activated inward-rectifier K^+ channels ($K_{IR}3$) and the ►ATP-sensitive K^+ channels ($K_{IR}6$, which combine with sulphonylurea receptors (SUR)) (see Table 2). Like the 6TM family, the pore-forming a subunits of the 2TM family form tetramers. Heteromeric channels may be formed within subfamilies (e.g. $K_{IR}3.2$ with $K_{IR}3.3$).

The Four Transmembrane Domain K⁺ Channel Family

The 4TM family of K^+ channels (TWIK, TREK, TASK, TALK, THIK and TRESK channels, Table 3) are the most recently identified K^+ channel family and underlie ►leak currents open at all voltages and expressed heterologously throughout the nervous system. They are regulated by a wide array of ►neurotransmitters and biochemical mediators. The primary pore-forming α subunit contains two pore domains (hence K2P) and so it is envisaged that they form functional dimers rather than the usual K^+ channel tetramers. There is some evidence that they can form heterodimers within subfamilies (e.g. $K_{2P}3.1$ with $K_{2P}9.1$).

Structural Features of K⁺ Channels

Whilst the structural properties of the different K^+ channel families vary considerably, one region, the pore (P) region (and within this region the selectivity filter particularly), is highly conserved between K^+ channels. The P region forms a hydrophobic hairpin loop in the membrane, within which is located the selectivity filter of the channel that confers K^+ selectivity. The selectivity filter has a highly conserved sequence, usually TYGYG. This region allows K^+ channels to be both extremely selective in which ions they allow to pass, yet still allow extremely fast transport rates, close to the aqueous diffusion limits (see Fig. 2).

The unique structure of the selectivity filter, in particular the arrangement of these five conserved residues which have their carbonyl oxygen atoms aligned towards the center of the selectivity filter pore, forms "customized oxygen cages" which mimic the arrangement of water molecules around K^+ ions. The K^+ ions can then enter the selectivity filter easily by diffusion. Sodium (Na^+) ions (although smaller) have a different arrangement of water molecules around them in solution and so do not enter the selectivity filter so easily. The selectivity filter also allows multiple ion occupancy, i.e. more than one K^+ ion (from two to two and a half on average) can sit in the selectivity filter at one time. The positively charged ions repel each other and ions are pushed through the

Neuronal Potassium Channels. Table 2 The 2TM K channel family

Subfamily group	Subtypes	Functional characteristics	Associated subunits
$K_{IR}1.x$	$K_{IR}1.1$ (ROMK1)	Inward-rectifier current	–
$K_{IR}2.x$	$K_{IR}2.1$–2.4 (IRK1–4)	IK_1 in heart, "strong" inward–rectifier current	–
$K_{IR}3.x$	$K_{IR}3.1$–3.4 (GIRK1–4)	G-protein-activated inward-rectifier current	–
$K_{IR}4.x$	$K_{IR}4.1$–4.2	Inward-rectifier current	–
$K_{IR}5.x$	$K_{IR}5.1$	Inward-rectifier current	–
$K_{IR}6.x$	$K_{IR}6.1$–6.2 (K_{ATP})	ATP-sensitive, inward-rectifier current	SUR1, SUR2A, SUR2B
$K_{IR}7.x$	$K_{IR}7.1$	Inward-rectifier current	–

Neuronal Potassium Channels. Table 3 The 4TM K channel family

Subfamily group	Subtypes	Functional characteristics
TWIK	$K_{2P}1.1$ (TWIK1) $K_{2P}6.1$ (TWIK2) $K_{2P}7.1$ (KNCK7)	Leak current
TREK	$K_{2P}2.1$ (TREK1) $K_{2P}10.1$ (TREK2) $K_{2P}4.1$ (TRAAK)	Leak current
TASK	$K_{2P}3.1$ (TASK1) $K_{2P}9.1$ (TASK3) $K_{2P}15.1$ (TASK5)	Leak current
TALK	$K_{2P}16.1$ (TALK1) $K_{2P}5.1$ (TASK2) $K_{2P}17.1$ (TASK4)	Leak current
THIK	$K_{2P}13.1$ (THIK1) $K_{2P}12.1$ (THIK2)	Leak current
TRESK	$K_{2P}18.1$ (TRESK1)	Leak current

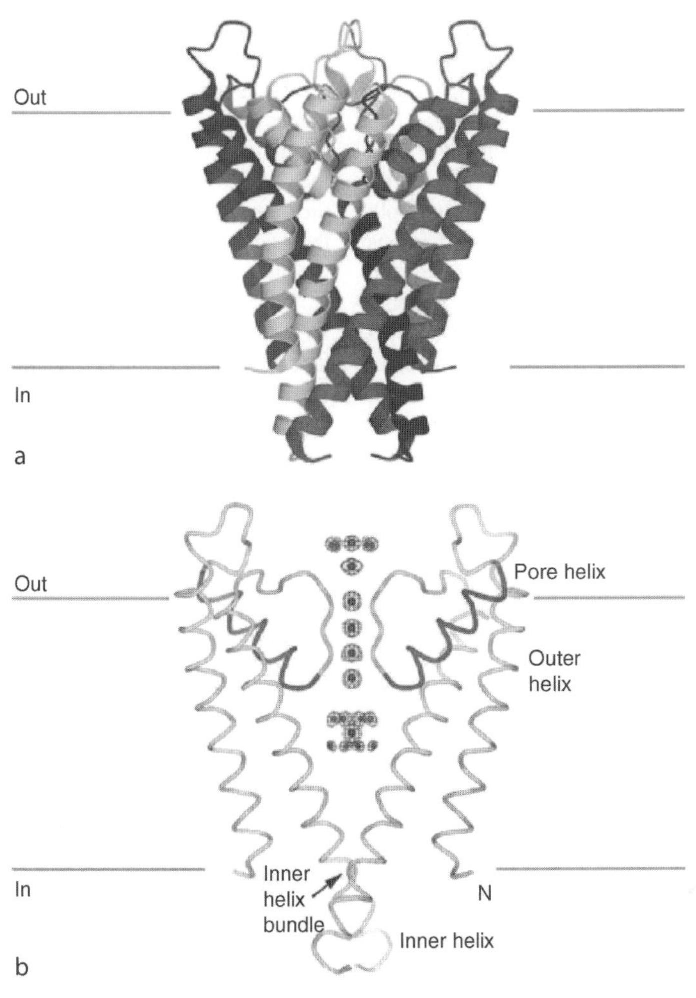

Neuronal Potassium Channels. Figure 2 (a) A ribbon representation of the KcsA K^+ channels with its four subunits colored differently. (b) The same channel with front and back subunits removed. The electron density along the ion pathway is shown as a blue mesh whilst the selectivity filter is shown in yellow. From [6] with kind permission from Springer Science and Business Media.

channel, usually down their concentration gradient so that K^+ ions flow from inside the cell to the outside [6,7].

Other regions of K^+ channels vary more widely from family to family and gene to gene. These include the regions of the channel that sense stimuli which act to alter the activity of K^+ channels (such as voltage, Ca^{2+}, phosphorylation, activated G protein subunits, etc), regions of the protein concerned with gating and regions which form protein/protein interactions with other proteins. For example, the fourth transmembrane domain (S4) of voltage-gated K_V channels contains many positively charged amino acids and this is thought to be the region that senses changes in membrane voltage. However, exactly how this region then responds to the voltage change it senses is an area of intense debate [see 8]. For a subgroup of K_V channels the intracellular N terminus region is mobile and interacts with the channel pore when the channel is open leading to current ►inactivation (termed N-type inactivation). Functionally, this is seen as fast inactivation characteristic of ►A-type K^+ currents (►I_A) (see below).

K^+ Channel Auxiliary Subunits

Most primary α subunits of K^+ channels interact with auxiliary subunits which can act to alter both channel function and channel expression levels at the cell membrane. For example, K_V β1 and β2 subunits accelerate inactivation when co-expressed with certain K_V1 channel subunits. Furthermore K_V5, K_V6, K_V8 and K_V9 subunits do not form functional channels when expressed alone but act as auxiliary subunits to modify the function of K_V2 channel subunits when co-expressed with these (see Table 1). In most cases, the detailed role and importance of auxiliary K^+ channel subunits has still to be established. Perhaps the most well known of K^+ channel auxiliary subunits described to date are the sulphonylurea receptors which form multimeric complexes with $K_{IR}6$ channels to give functional ATP-sensitive K^+ channels.

Functional Properties of Neuronal K⁺ Channels

The large number of K^+ channel genes, the possibility of heteromeric combinations and the existence of both auxiliary subunits and post-translational modifications such as phosphorylation, when taken together, suggest that the potential number of functional K^+ channel types in neurons is extremely large. Despite this, only a comparatively small number of distinct K^+ channel currents have been characterized to date, suggesting that many channel combinations have properties that differ from each other only subtly. This often makes it extremely difficult to be certain which subunit combinations underlie which functional currents seen in particular neurons. Nevertheless a few functional profiles are seen in many neuron populations. These include delayed-rectifier K^+ currents (K_V), ►A type currents (K_A), ►M currents, Ca-activated K currents and leak K currents.

Delayed rectifier, K_V, currents (e.g. $K_V1.1$, $K_V3.1$, $K_V3.2$ homomers) are the main K^+ current in many excitable cells (Fig. 3). Once a threshold voltage is reached, their conductance increases upon membrane depolarization rising sigmoidally and they inactivate slowly. From one cell to another (or one protein to another) they have diverse kinetics, pharmacology and voltage-dependence but they are usually sensitive to relatively low concentrations of ►tetraethylammonium ions (►TEA). Their activity controls action potential depolarization and hence the duration of the action potential (Action potential).

►*Transient K^+ currents* such as ►K_A currents or ►K_D currents (e.g. $K_V1.2$, $K_V1.4$, $K_V3.4$, $K_V4.2$ homomers) are usually found in cells in addition to delayed rectifier currents. Their conductance also increases with depolarization but they characteristically inactivate quickly, albeit with varying timescales when comparing one to another. They are often selectively localized within a neuron and have a role in regulating the interspike interval in a train of action potentials, thus their activity helps to determine the latency to the first action potential spike.

The *M current* is encoded by members of the K_V7 or KCNQ subfamily (usually seen as $K_V7.2$ and $K_V7.3$ heteromers in neurons). The current gets its name because it is inhibited following activation of muscarinic acetylcholine receptors. The M current is a sustained current which activates slowly at subthreshold voltages and normally keeps the neuronal membrane hyperpolarized. However, suppression of the current following muscarinic receptor activation leads to membrane depolarization. The current controls spike frequency accommodation; thus typically on depolarization, one might see a few spikes before the M current activates fully to act as a break to further firing.

Ca^{2+}-activated K^+ currents are activated following a rise in intracellular Ca^{2+}. BK currents have a role in action potential depolarization whilst SK currents contribute to after hyperpolarizations which control the action potential firing rate.

Leak K^+ currents are open at all voltages (i.e. they are not voltage gated) and contribute to the cell membrane potential and neuronal excitability. K2P channels often underlie leak K^+ currents and these channels are highly regulated by agents such as neurotransmitters and other physiological mediators, thereby constantly tuning neuronal excitability.

So, why are there so many different K^+ channel subunits and different K^+ current functional profiles?

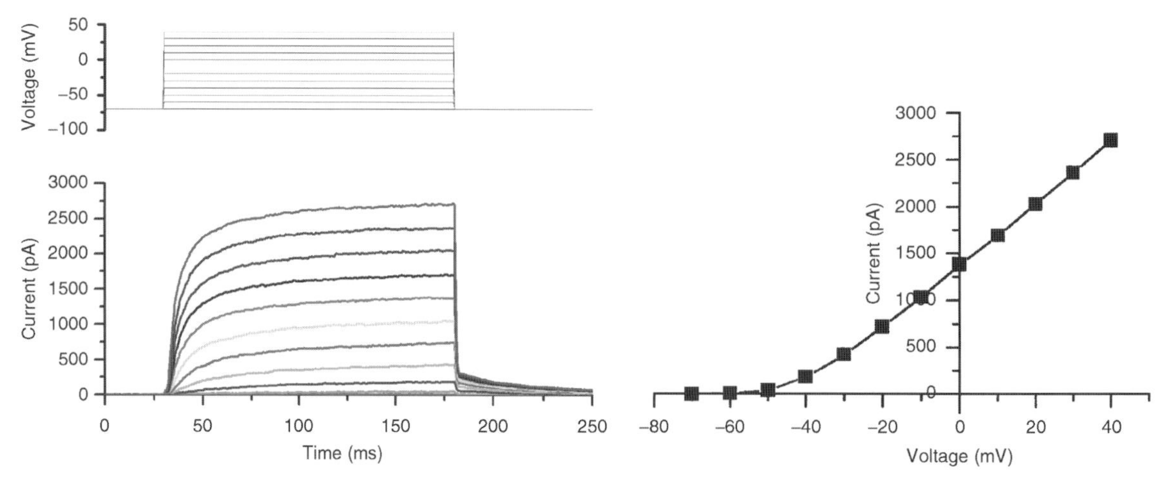

Neuronal Potassium Channels. Figure 3 Left hand side – a family of K^+ currents through $K_V1.1$ potassium channel homo-tetramers expressed in HEK cells, evoked by step depolarizations in membrane potential. Right hand side – the peak current is plotted against the voltage for each step. Note that current is activated at voltages positive to −50 mV.

The most likely explanation is that different neurons express different subsets of K^+ channels in order to uniquely tailor their responses to particular synaptic inputs. For example, neurons that are required to fire at extremely high frequencies (such as neurons in the auditory brainstem nuclei [9] are often found to have high expression levels of K_V3 channels. These particular channels activate and deactivate extremely quickly which allows the membrane potential to depolarize quickly following an action potential, ready to depolarize again very rapidly as required for high frequency firing. Studies into the expression levels of K^+ channels are mapping their distinct distribution throughout the central nervous system [e.g. 10]. This differential distribution of K^+ channels underlies the differential neuronal responses seen throughout the central nervous system to synaptic inputs.

References

1. Bean BP (2007) The action potential in mammalian central neurons. Nat Rev Neurosci 8:451–465
2. Hille B (2001) Ion channels of excitable membranes. Sinauer Associates, Sunderland MA
3. Ashcroft FM (2000) Ion channels and disease. Academic Press, London
4. Yu FH, Yarov-Yarovoy V, Gutman GA, Catterall WA (2005) Overview of molecular relationships in the voltage-gated ion channel superfamily. Pharmacol Rev 57:387–395
5. Coetzee WA, Amarillo Y, Chiu J, Chow A, Lau D, McCormack T, Moreno H, Nadal MS, Ozaita A, Pountney D, Saganich M, Vega-Saenz de Miera E, Rudy B (1999) Molecular diversity of K channels. Ann NY Acad Sci 868:233–285
6. MacKinnon R (2004) Nobel Lecture: Potassium channels and the atomic basis of selective ion conduction. Biosci Rep 24:75–100
7. Yellen G (2002) The voltage-gated potassium channels and their relatives. Nature 419:35–42
8. Tombola F, Pathak MM, Isacoff RY (2006) How does voltage open an ion channel? Annu Rev Cell Dev Biol 22:23–52
9. Song P, Yang Y, Barnes-Davies M, Bhattacharjee A, Hamann M, Forsythe ID, Oliver DL, Kacmarek LK (2005) Acoustic environment determines phosphorylation state of the Kv3.1 potassium channel in auditory neurons. Nat Neurosci 8:1335–1342
10. Trimmer JS, Rhodes KJ (2004) Localisation of voltage-gated ion channels in mammalian brain. Annu Rev Physiol 66:477–519

Neuronal Proliferation

Definition

Cell division of neuronal progenitor cells in the ventricular layer of the vertebrate neural tube. Cell cycle genes control the number of neurons generated from progenitor cells in the central nervous system. Cells must withdraw from the cell cycle prior to migration and differentiation.

►Neural Development
►Neural Tube

Neuronal Tropism

Definition

The preference of a viral vector for infecting neurons.

►Gene Therapy for Neurological Diseases

Neuronitis

Definition

Inflammation of one or more neurons (former name for acute idiopathic polyneuritis).

Neuronopathy

Definition

Polyneuropathy involving destruction of the cell bodies of neurons.

Neuropathic Pain

MAIKE STENGEL, RALF BARON
Division of Neurological Pain Research and Therapy, Department of Neurology Universitätsklinikum Schleswig-Holstein, Campus Kiel, Kiel, Germany

Synonyms

Neurogenic pain

Definition

Neuropathic pain is defined as a “Pain arising as a direct consequence of a lesion or disease affecting the

N

somatosensory system" [1]. The term "disease" refers to identifiable disease processes such as inflammatory, autoimmune conditions or channelopathies, while lesions refer to macro- or microscopically identifiable damage. The restriction to the somatosensory system is necessary, because diseases and lesions of other parts of the nervous system may cause nociceptive pain. For example, lesions or diseases of the motor system may lead to spasticity or rigidity, and thus may indirectly cause muscle pain. These pain conditions are now explicitly excluded from the condition neuropathic pain.

Characteristics

Epidemiology of Neuropathic Pain

Chronic neuropathic pain is common in clinical practice, greatly impairs the quality of life of patients and is a major economical health problem. Estimates of point prevalence for neuropathic pain in the general population are as high as 5%, a quarter of them suffering from severe pain. Moreover, a recent prospective cross-sectional survey in 13,000 chronic pain patients with both nociceptive and neuropathic pain types who were referred to pain specialists in Germany revealed that 13% of these patients suffer from the two classical neuropathic disorders, ▶post-herpetic neuralgia (PHN) and painful diabetic neuropathy, and 40% of all have at least a neuropathic component to their discomfort (especially patients with chronic back pain and ▶radiculopathy) [2]. Comorbidities such as poor sleep, depressed mood and anxiety are common in neuropathic pain and have a significant impact on the global pain experience.

Classification

Disease/Anatomy-Based Classification

It is common clinical practice to classify neuropathic pain according to the underlying etiology of the disorder and the anatomical location of the specific lesion. The majority of patients fall into four broad classes (Table 1): painful peripheral neuropathies (focal, multifocal or generalized, e.g. traumatic, ischemic, inflammatory, toxic, metabolic, hereditary), central pain syndromes (e.g. stroke, multiple sclerosis, spinal cord injury), complex painful neuropathic disorders (complex regional pain syndromes, CRPS) and mixed pain syndromes (combination of nociceptive and neuropathic pain, e.g., chronic low back pain with radiculopathy).

Painful Peripheral (Focal, Multifocal, Generalized) Neuropathies

The anatomical distribution pattern of the affected nerves provides valuable differential diagnostic clues as to possible underlying causes. It is therefore common clinical practice to group painful neuropathies into symmetrical generalized polyneuropathies, affecting

Neuropathic Pain. Table 1 Disease/anatomy-based classification of painful peripheral neuropathies Painful peripheral neuropathies

Painful peripheral neuropathies
Focal, multifocal
Phantom pain, stump pain, nerve transection pain (partial or complete)
Neuroma (post-traumatic or postoperative)
Posttraumatic neuralgia
Entrapment syndromes
Mastectomy
Post thoracotomy
Morton's neuralgia
Painful scars
Herpes zoster and postherpetic neuralgia
Diabetic mononeuropathy, diabetic amyotrophy
Ischemic neuropathy
Borreliosis
Connective tissue disease (vasculitis)
Neuralgic amyotrophy
Peripheral nerve tumors
Radiation plexopathy
Plexus neuritis (idiopathic or hereditary)
Trigeminal or glossopharyngeal neuralgia
Vascular compression syndromes
Generalized (polyneuropathies)
Metabolic or nutritional
Diabetic, often "Burning feet syndrome"
Alcoholic
Amyloid
Hypothyroidism
Beri beri, Pellagra
Drugs
Antiretrovirals, Cisplatin, Oxaliplatin, Disulfiram, Ethambutol, Isoniazid, Nitrofurantoin, Thalidomid, Thiouracil
Vincristine, Chloramphenicol, Metronidazole, Taxoids, Gold
Toxins
Acrylamide, Arsenic, Clioquinol, Dinitrophenol, Ethylene oxide, Pentachlorophenol, Thallium
Hereditary
Amyloid neuropathy
Fabry's disease
Charcot-Marie-Tooth disease type 5, type 2B
Hereditary sensory and autonomic neuropathy (HSAN) type 1, type 1B
Malignant
Carcinomatous (paraneoplastic)
Myeloma
Infective or post-infective, immune
Acute or inflammatory polyradiculoneuropathy (Guillain-Barré syndrome)

Neuropathic Pain. Table 1 Disease/anatomy-based classification of painful peripheral neuropathies Painful peripheral neuropathies (Continued)

Borreliosis
HIV
Other polyneuropathies
Erythromelalgia
Idiopathic small-fiber neuropathy
Central pain syndromes
Vascular lesions in the brain (especially brainstem and thalamus) and spinal cord:
Infarct
Hemorrhage
Vascular malformation
Multiple sclerosis
Traumatic spinal cord injury including iatrogenic cordotomy
Traumatic brain injury
Syringomyelia and syringobulbia
Tumors
Abscesses
Inflammatory diseases other than multiple sclerosis; myelitis caused by viruses, syphilis
Epilepsy
Parkinson's disease
Complex painful neuropathic disorders
Complex regional pain syndromes type I and II (Reflex sympathetic dystrophy, causalgia)
Mixed-pain syndromes
Chronic low back pain with radiculaopathy
Cancer pain with malignant plexus invasion
Complex regional pain syndromes

many nerves simultaneously, and into asymmetrical neuropathies with a focal- or multifocal distribution or processes affecting the brachial or lumbosacral plexuses. One important subgroup of polyneuropathies is characterized by a predominant, or in some cases even isolated, involvement of small afferent fibers (i.e. unmyelinated C-fibers and small myelinated Aδ-fibers). In many cases, autonomic efferent small fiber systems are also affected. Different etiologies may lead to small fiber polyneuropathies, but up to 20% of cases, however, are of unknown cause. It is important to realize that conventional electrophysiological techniques like NCS (nerve conduction study), SEP (somatosensory evoked potential), etc. only assess the function of myelinated peripheral axonal systems and the contribution of small fibers will be missed. Therefore, especially in small fiber neuropathies, alternative diagnostic procedures have to be used, like ►quantitative sensory testing (QST).

Central Pain Syndromes

Central pain is defined as chronic pain following a lesion or disease of the central nervous system. The cause of pain is a primary process within the CNS (central nervous system). The highest incidence is observed after spinal cord injury, lesions in the lower brainstem and thalamus. An involvement of spino-thalamo-cortical pathways seems to be crucial for the development of central pain, whereas isolated lesions of the lemniscal system are never associated with pain. Many kinds of lesions can induce central pain. The most common are cerebrovascular lesions, multiple sclerosis (MS) and traumatic spinal cord injuries (SCI). Central pain often develops with a latency of weeks or months after the inciting event.

Complex Painful Neuropathic Disorders

In addition to the classical neuropathic syndromes like painful diabetic neuropathy, postherpetic neuralgia or phantom limb pain are certain chronic painful conditions that share many clinical characteristics. These syndromes were formerly called reflex sympathetic dystrophy, M. Sudeck or causalgia and are now classified under the umbrella term complex regional pain syndromes (CRPS). CRPS are painful disorders that may develop as a disproportionate consequence of trauma typically affecting the limbs. CRPS type I usually develops after minor trauma with no obvious nerve lesion at an extremity (e.g. bone fracture, sprains, bruises or skin lesions, surgeries). CRPS type II develops after trauma that typically involves a large nerve lesion.

Mixed Pain Syndromes

Both nociceptive and neuropathic processes contribute to many chronic pain syndromes and these different mechanisms may explain the qualitatively different symptoms and signs that patients experience. In particular, patients with chronic low-back pain, cancer pain and CRPS seem to fit into this theoretical construct.

Mechanism-Based Classification

In neuropathic pain a disease/anatomy-based classification is often insufficient. Despite obvious differences in etiology, many of these diseases share common clinical phenomena; for example, touch-evoked pain in post-herpetic neuralgia and painful diabetic neuropathy. Conversely, different signs and symptoms can be present in the same disease; for example, pain paroxysms and stimulus-evoked abnormalities in post-herpetic neuralgia. Classification on the basis of location also has its shortcomings, as neuroplastic changes following nervous system lesions often give rise to sensory and pain distributions that do not respect nerve, root, segmental or cortical territories. These observations have raised the question whether an entirely different strategy, in which pain is analyzed on the basis of underlying mechanisms [3], could provide an alternative approach for examining and classifying patients, with the ultimate aim of obtaining a better treatment outcome [4,5].

Signs and Symptoms in Neuropathic Pain

Patients with neuropathic pain demonstrate a variety of distinct sensory symptoms that can coexist in combinations. Bedside sensory examination should include touch, pinprick, pressure, cold, heat, vibration, and temporal summation ([6], definitions in Table 2).

Responses can be graded as normal, decreased or increased to determine whether negative or positive sensory phenomena are involved. Stimulus-evoked (positive) pain is classified as dysesthetic, hyperalgesic or allodynic, and according to the dynamic or static character of the stimulus. Touch can be assessed by gently applying cotton wool to the skin, pinprick sensation by the response to sharp pinprick stimuli, deep pain by gentle pressure on muscle and joints, cold and heat sensation by measuring the response to a thermal stimulus, for example by thermo-rollers kept at 20 or 45°C. Cold sensation can also be assessed by the response to acetone spray. Vibration can be assessed by a tuning fork placed at strategic points (interphalangeal joints, etc).

Pathophysiological Mechanisms in Patients

Peripheral and Central Sensitization of Nociceptive Neurons

Abnormal nociceptor sensitization and abnormal spontaneous afferent activity has been demonstrated in many peripheral nerve injury models. Partial nerve lesion is associated with dramatic changes in the regulation of receptors and channels in damaged as well as undamaged primary afferent neurons. These neurons develop spontaneous activity (ectopic discharge) and an increased sensitivity to chemical, thermal and mechanical stimuli. Ectopic impulse generation following nerve injury is associated with enhanced expression and changes in the distribution of certain voltage gated

Neuropathic Pain. Table 2 Definition and assessment of negative and positive sensory symptoms or signs in neuropathic pain

	Symptom/Sign	Definition	Assessment bedside exam	Expected pathological response
Negative signs and symptoms	Hypoesthesia	Reduced sensation to non painful stimuli	Touch skin with painters brush, cotton swab or gauze	Reduced perception, numbness
	Pall-hypoesthesia	Reduced sensation to vibration	Apply tuning fork on bone or joint	Reduced perception threshold
	Hypoalgesia	Reduced sensation to painful stimuli	Prick skin with single pin stimulus	Reduced perception, numbness
	Therm-hypoesthesia	Reduced sensation to cold/warm stimuli	Contact skin with objects of 10°C (metal roller, glass with water, coolants like acetone)	Reduced perception
			Contact skin with objects of 45°C (metal roller, glass with water)	
Spontaneous sensations/pain	Paraesthesia	Non-painful ongoing sensation (ant crawling)	Grade intensity (0–10)	–
			Area in cm^2	
	Paroxysmal pain	Shooting electrical attacks for seconds	Number per time	–
			Grade intensity (0–10) Threshold for evocation	
	Superficial pain	Painful ongoing sensation often of burning quality	Grade intensity (0–10)	–
			Area in cm^2	

Neuropathic Pain. Table 2 Definition and assessment of negative and positive sensory symptoms or signs in neuropathic pain (Continued)

	Symptom/Sign	Definition	Assessment bedside exam	Expected pathological response
Evoked pain	Mechanical dynamic allodynia	Normally non painful light moving stimuli on skin evoke pain	Stroking skin with painters brush, cotton swab or gauze	Sharp burning superficial pain
				Present in the primary affected zone but spread beyond into unaffected skin areas (secondary zone)
	Mechanical static allodynia	Normally non painful gentle static pressure stimuli at skin evoke pain	Manual gentle mechanical pressure at the skin	Dull pain
				Present in the area of affected (damaged or sensitized) primary afferent nerve endings (primary zone)
	Mechanical punctate hyperalgesia	Normally stinging but not painful stimuli evoke pain	Manual pricking the skin with a safety pin, sharp stick or stiff von Frey hair	Sharp superficial pain
				Present in the primary affected zone but spread beyond into unaffected skin areas (secondary zone)
	Temporal summation	Repetitive application of identical single noxious stimuli is perceived as increasing pain sensation (Wind-up like pain)	Pricking skin with safety pin at interval <3 s for 30 s	Sharp superficial pain of increasing intensity
	Cold allodynia (hyperalgesia)	Normally non (slightly) painful cold stimuli evoke pain	Contact skin with objects of 20°C (metal roller, glass with water, coolants like acetone)	Painful often burning temperature sensation
			Control: contact skin with objects of skin temperature	Present in the area of affected (damaged or sensitized) primary afferent nerve endings (primary zone)
	Heat allodynia (hyperalgesia)	Normally non (slightly) painful heat stimuli evoke pain	Contact skin with objects of 40°C (metal roller, glass with water)	Painful burning temperature sensation
			Control: contact skin with objects of skin temperature	Present in the area of affected (damaged or sensitized) primary afferent nerve endings (primary zone)
	Mechanical deep somatic hyperalgesia	Normally non painful pressure on deep somatic tissues evoke pain	Manual light pressure at joints or muscle	Deep pain at joints or muscles

sodium channels in primary afferent neurons, leading to a lowering of the action potential threshold [7].

As a consequence of peripheral nociceptor hyperactivity, dramatic secondary changes in the spinal cord dorsal horn also occur. Partial peripheral nerve injury leads to an increase in the general excitability of spinal cord neurons. This so-called central sensitization is probably due to activity in pathologically sensitized C-fibers, which sensitize spinal cord dorsal horn neurons by releasing glutamate and the neuropeptide substance P. Neuronal voltage-gated calcium channels that are located presynaptically at spinal nociceptive terminals are up-regulated after peripheral nerve injury and play an important role in the process of central sensitization by mediating the release of glutamate and substance P. If central sensitization is established, normally innocuous tactile stimuli become capable of activating spinal cord pain signaling neurons via Aß-low threshold mechanoreceptors [8] [see ►Hyperalgesia and Allodynia]. By this mechanism, light touching of the skin induces pain (i.e. mechanical allodynia).

Changes in the Brain

Most animal experiments have concentrated on the dorsal horn as the location of central sensitization. However, in rodents, sensitized neurons are also found in the thalamus and primary somatosensory cortex after partial peripheral nerve injury. Furthermore, MEG, PET and functional MRI studies demonstrate fundamental changes in somatosensory cortical representation and excitability in patients with phantom limb pain, CRPS and central pain syndromes as well as experimental pain models. Interestingly, these changes correlate with the intensity of the perceived pain and disappear after successful treatment of the pain.

Deafferentation: Hyperactivity of Central Pain Transmission Neurons

Although the above convincingly supports a role for peripheral and central sensitization in the generation of neuropathic pain, in some patients there is a profound cutaneous deafferentation of the painful area with no significant ►allodynia. Assuming that the ►dorsal root ganglion cells and the central afferent connections are lost in such patients, their pain must be the result of intrinsic CNS changes. In animal studies, following complete primary afferent loss of a spinal segment, many dorsal horn cells begin to fire spontaneously at high frequencies. There is some evidence that a similar process may underlie the pain that follows extensive denervating injuries in humans. Recordings of spinal neuron activity in a pain patient whose dorsal roots were injured by trauma to the ►cauda equina revealed high frequency regular and paroxysmal bursting discharges. The patient complained of spontaneous burning pain in a skin region that was anesthetic by the lesion (anesthesia dolorosa).

Inflammation in Neuropathic Pain

After nerve lesion, activated macrophages infiltrate from endoneural blood vessels into the nerve and dorsal root ganglia, releasing pro-inflammatory cytokines, in particular tumor necrosis factor alpha (TNF-α) [see ►Immune System and Pain]. These mediators induce ectopic activity in both injured and adjacent uninjured primary afferent nociceptors at the lesion site.

In patients with inflammatory neuropathies, such as vasculitic neuropathies or HIV neuropathy, deep proximal aching and paroxysmal pain are characteristic phenomena. COX2 (cyclooxygenase 2) and proinflammatory cytokines were found to be up-regulated in nerve biopsy specimens of these patients. In the affected extremities of CRPS patients, the fluid of artificially produced skin blisters contain significantly higher levels of IL-6 and TNF-α as compared with the uninvolved extremity.

Quantitative Sensory Testing as a Diagnostic Tool in Neuropathic Pain

The modern theoretical concept of a mechanism-based therapy assumes that a specific symptom predicts a specific underlying mechanism [3]. However, this approach carries certain important caveats. Clinical experimental studies indicate that a specific symptom might be generated by several entirely different underlying pathophysiological mechanisms. Therefore, a specific symptom profile rather than a single symptom might be required to predict the underlying mechanism. To translate these ideas into the clinical framework, it is important to characterize the somatosensory phenotype of a patient as precisely as possible. A standardized quantitative sensory testing protocol (QST) was introduced by the German Research Network on Neuropathic Pain, including 13 parameters and encompassing thermal as well as mechanical testing procedures for the analysis of the somatosensory phenotype. To judge plus or minus symptoms in patients, an age- and gender-matched data-base for absolute and relative QST reference data for several body regions in healthy human subjects was established. The precise phenotypic mapping with QST is an important step to establish a future mechanism-based drug therapy of neuropathic pain. If the symptoms are closely related to mechanisms, clinical assessment of the symptoms may give an idea of the concert of the distinct mechanisms that operate in one individual patient. This knowledge may lead in the future to an optimal poly-pragmatic therapy with drugs that address the specific combination of mechanisms in each patient.

Present Medical Treatment

The number of trials for peripheral neuropathic pain has expanded greatly in the last few years [9]. In summary, the medical management of neuropathic pain consists of five main classes of oral medication (serotonin/norepinephrine modulating antidepressants, sodium-blocker-anticonvulsants, calcium-modulator-anticonvulsants, tramadol and opioids) and two categories of topical medications mainly for patients with cutaneous ►allodynia and ►hyperalgesia (capsaicin and local anaesthetics). For central neuropathic pain, there are limited data. Since more than one mechanism is at work in most patients, a combination of two or more analgesic strategies to address multiple mechanisms will generally produce greater pain relief. Therefore, in most patients a stepwise process using successive monotherapies is not appropriate. Early combinations of two or three drugs form different classes is the general practical approach.

►Voltage-gated Sodium Channels: Multiple Roles in the Pathophysiology of Pain
►Development of Nociception

References

1. Treede RD, Jensen TS, Campbell JN, Cruccu G, Dastrovsky JO, Griffin JW, Hansson P, Hughes R, Nurmikka T, Serra J (2007) Neuropathic pain. Redefinition and a grading system for clinical and research purposes. Neurology Nov 14; [EPub ahead of Print].
2. Freynhagen R, Baron R, Gockel U, Tolle TR (2006) PainDETECT: a new screening questionnaire to identify neuropathic components in patients with back pain. Curr Med Res Opin 22(10):1991–1920
3. Jensen, TS, Baron, R (2003) Translation of symptoms and signs into mechanisms in neuropathic pain. Pain 102:1–8
4. Woolf CJ, Bennett GJ, Doherty M, Dubner R, Kidd B, Koltzenburg M, Lipton R, Loeser JD, Payne R, Torebjork E (1998) Towards a mechanism-based classification of pain? Pain 77(3):227–229
5. Baron R (2006) Mechanisms of disease: neuropathic pain – a clinical perspective. Nat Clin Pract Neurol 2(2):95–106
6. Cruccu G, Anand P, Attal N, Garcia-Larrea L, Haanpaa M, Jorum E, Serra J, Jensen TS (2004) EFNS guidelines on neuropathic pain assessment. Eur J Neurol 11:153–162
7. Binder A, Baron R (2007) Sodium channels in neuropathic pain – friend or foe? Nat Clin Pract Neurol 3(4):179
8. Tal M, Bennett GJ (1994) Extra-territorial pain in rats with a peripheral mononeuropathy: mechano-hyperalgesia and mechano-allodynia in the territory of an uninjured nerve. Pain 57:375–382
9. Finnerup NB, Otto M, McQuay HJ, Jensen TS, Sindrup SH (2005) Algorithm for neuropathic pain treatment: an evidence based proposal. Pain 118(3):289–305

Neuropathy

Definition

Any peripheral nerve disease, which usually causes weakness and numbness (see ►peripheral neuropathy).

Neuropeptide S (NPS)

Definition

A twenty-residue peptide expressed in a few discrete nuclei in the brainstem (pericoerulear area and parabrachial nucleus). The name arises because its N-terminal residue is a Serine (S), which is conserved across vertebrate species. NPS receptors (formerly, GPRA receptors) are distributed widely throughout the brain. Infusion of NPS in the brain has anxiolytic-like properties and induces wakefulness.

►Hypocretin/Orexin
►Neuropeptides
►Ventrolateral Preoptic Nucleus (VLPO)

Neuropeptide Y (NPY)

Definition

A 36 amino acid neuromodulator well known as a contributor to the regulation of feeding. In the context of circadian rhythm regulation, it is important because it is synthesized in neurons of the intergeniculate leaflet (IGL) that project to the suprachiasmatic nucleus (SCN) via the geniculohypothalamic tract (GHT). NPY is released from GHT terminals in the SCN and acts as a neuromodulator mediating the ability of certain nonphotic environmental stimuli to modulate circadian rhythm phase.

►Circadian Rhythm
►Intergeniculate Leaflet
►Neuromodulators
►Neuropeptides
►Suprachiasmatic Nucleus (SCN)

Neuropeptides

Willis K. Samson
Department of Pharmacological and Physiological Science, Saint Louis University School of Medicine, St. Louis, MO, USA

Synonyms

Neuromodulators; Neurotransmitters

Definition

►Neuropeptides are small proteins produced in and released from neurons. They exert potent membrane effects on both adjacent neurons (i.e., synaptic) and glia (i.e., paracrine) and can be released from neurons to gain access to the hypothalamo-pituitary-portal vessels (i.e., neuroendocrine) or general circulation (i.e., endocrine). Within the limbic system, these small chains of amino acids exert potent behavioral effects that can be divided into several categories, motivation, appetite, fear and anxiety, arousal, memory, addiction and aggression to name but a few.

Characteristics

Quantitative Description

No one rule is agreed upon by the scientific community. Neuropeptides are considered to be proteins consisting of fewer than 100 amino acids, although some investigators would assign such status to peptides comprised of fewer than 50 amino acids. Clearly these biologically active factors are produced as part of a larger pre-prohormone that, following translation of the encoding mRNA, undergoes numerous catalytic modifications resulting in the assembly of the bioactive components in secretory granules. Along the way, biochemical adjustments to the amino acid backbone that are essential for expression of full biologic activity are completed, including N- and C-terminal modifications and in some cases unique changes to single amino acid structure such as bromination and surprisingly in the case of the peptide ghrelin the addition of an octanoyl group [1]. In general, neuropeptides are thought to be extremely labile and the termination of their biological activity is a combination of uptake (of bound peptide to cognate receptor) or enzymatic processing by cell surface proteases abundant in the neuropil. Many but not all the identified neuropeptide receptors are members of the G-protein classes of biologically active receptors; however, some neuropeptides bind directly to ion channels, membrane spanning guanylyl cyclases or tyrosine kinase-like growth factor receptors. While direct actions of neuropeptides can be demonstrated on the more classically described neurotransmitters (e.g., acetylcholine, norepinephrine, dopamine, glutamate, GABA), important interactions of neuropeptides with other neuropeptide producing neurons are clearly the basis for the pharmacological and in some cases the demonstrated physiologically relevant actions of the peptides. Since neuropeptides are unique products of single genes, molecular engineering technologies [2,3] have allowed for the development of informative animal models in which the peptide itself or its receptor are selectively ablated in either the embryonic stage (and thus throughout life, gene knockouts) or at any stage during postnatal development (selective gene ablation, Cre-LOX strategies) using a technique that can be initiated without the deleterious effects of loss of the neuropeptide during fetal development. These technologies also allow for continuous or regulated over-expression of the neuropeptide by insertion of a transgene that is constitutively expressed or induced chemically for either short or long periods of time. These technologies have clearly established the importance of these "non-classical" neurotransmitters in a wide variety of neural events and provided insight into therapeutic strategies for the treatment of CNS disorders.

Higher Level Structures

It is impossible in the space allotted to catalog all the currently identified neuropeptides, their localization in limbic structures or their multiple and diverse biologic actions. One peptide will be used as an example for each class of characteristic biologic effects exerted by neuropeptides within limbic structures.

Motivation is a behavior intimately linked to reward. The endogenous opioid systems including the neuropeptides beta-endorphin and the enkephalins are important neural factors that organize learned responses and provide signals that impact motivation [4]. Aversive stimuli are also transmitted by neuropeptides, including CCK-8 and oxytocin. Included in motivation is a component of memory and the role of vasopressin in this behavior has been demonstrated in experimental animals and humans. This neuropeptide surprisingly affects both the consolidation and recovery of memory and contributes, via actions in the medial septal nuclei, to the emotional components of memory including aggression.

By far the most comprehensively cataloged behavioral response to neuropeptides within limbic structures is appetite, although it is difficult to separate appetitive behaviors from the additional limbic actions of many of these neuropeptides such as arousal state, anxiety and even addiction. Currently there are more than 50 neuropeptides that have been characterized for their effects on appetite (Table 1).

These can be divided into those that elicit hunger and those that contribute to ►satiety. While the actions of these "feeding peptides" are exerted in many distinct limbic structures, it is in the medial-basal hypothalamus where the primary interactions take place, in particular in the arcuate nuclei, the ventromedial hypothalamic nuclei and the lateral hypothalamic nuclei. Cells in these areas integrate ascending neuronal input from brainstem structures such as the nucleus tractus solitarius (in particularly the vagal afferent relaying information relevant to ongoing and metabolic state), from temporal lobe structures including hippocampus and amygdala and from circumventricular organs such as the subfornical organ, area postrema and organum vasculosum lamina terminalis.

Circulating hormones derived from the gut and adipocytes activate cells in blood brain barrier free sites and in some cases directly affect the activity of neurons in these hypothalamic nuclei, informing these integrative centers of the fed/fasted state of the individual.

Neuropeptides. Table 1 Neuropeptides acting in limbic structures to alter feeding behaviors

Stimulators	Inhibitors
Agouti-Related Peptide (AGRP)	Adrenomedullin
Dynorphin	Amylin
Beta-Endorphin	Anorectin
Galanin	Bombesin
Ghrelin	Brain-derived neurotrophic factor (BDNF)
Growth Hormone Releasing Hormone	Calcitonin Gene Related Peptide (CGRP)
Hypocretins/Orexins	CART
Melanin Concentrating Hormone (MCH)	Cholecystokinin (CCK)
Neuropeptide Y (NPY)	Ciliary Neurotrophic Factor (CNTF)
RF-amide peptides	Corticotropin Releasing Hormone (CRH)
VGF	Enterostatin
	Galanin-Like Peptide (GALP)
	Glucagon-like Peptide 1 (GLP-1)
	Insulin-like Growth Factors
	Insulin
	Interleukin-1
	Leptin
	Motilin
	Nesfatin
	Neuromedin B
	Neuromedin U
	Neuropeptides B (NPB)
	Neuropeptide K (NPK)
	Neuropeptide W
	Neurotensin (NT)
	Obestatin
	Oxytocin
	Peptide YY (3–36)
	Prolactin Releasing Peptide (PrRP)
	POMC
	Thyrotropin Releasing Hormone (TRH)
	Urocortin

Metabolic factors themselves such as plasma glucose and free fatty acids are also sensed by neurons in these "feeding centers" and it is neuropeptides that then transmit the appropriate signals to motivational centers in brain organizing the appropriate response to ambient nutrient availability [5,6]. Central to this organization are neurons in the arcuate nucleus, which provide the critical organization of these inputs. Neuropeptide Y (NPY) and agouti-related peptide (AGRP) are produced in the same arcuate neurons and these two neuropeptides are central to the stimulatory drive for feeding [2]. They are inhibited by circulating leptin, a peptide derived from adipocytes and by insulin levels. High circulating levels of free fatty acids also exert inhibitory effects on NPY/AGRP neurons [6]. Peptide YY_{3-36} is a gut hormone released in response to feeding that potently inhibits NPY/AGRP neurons [7]. On the other hand ghrelin, an octanoylated peptide produced in the gastric mucosa, is released during the fasted state and exerts direct, stimulatory actions on these neurons [1]. Thus the peripheral tissues can directly regulate motivation and appetite via metabolic and hormonal factors.

Also located in the arcuate nucleus are neurons that express the pro-opiomelanocortin (POMC) gene and, after post-translational processing of the encoded mRNA package for synaptic release, the neuropeptide alpha-melanocyte stimulating hormone (alpha-MSH). This peptide binds to a family of melanocortin receptors expressed on neurons in the feeding center and exerts potent anorexigenic actions, thus reducing the motivation for food seeking behaviors. These POMC neurons are stimulated by insulin and leptin [2,6]. They are inhibited by NPY produced in adjacent neurons and binding of alpha-MSH to its melanocortin receptors is antagonized by AGRP. The integrative nature of these neuronal systems is further demonstrated by the ability of alpha-MSH to inhibit the activity of the NPY/AGRP neurons. Thus, even simply within the small population of arcuate neurons, a push-pull mechanism exists for the integration of feeding signals from the periphery. Neurons in the adjacent lateral hypothalamic area (LHA) contribute to the integration. These neurons produce the neuropeptides orexin A and orexin B (products of the same gene, also called hypocretin 1 and hypocretin 2). They are activated by NPY and inhibited by POMC. Important efferent projections from these LHA neurons innervate reward centers and brain stem satiety centers [8]. These neurons are also responsible for the behavioral arousal (locomotor activity, autonomic nervous system activation) that accompanies food seeking and eating [3].

The orexin neurons in LHA appear to play a central role in the reward aspects of food [9] and they are influenced by recognized interactions of the ascending dopaminergic pathways from the ventral tegmental area (VTA) to the nucleus accumbens (NAc). GABA-ergic neurons from the NAc project to the orexin neurons in LHA. Thus the neuropeptide systems of the "feeding centers" in medio-basal hypothalamus are intimately associated with brain reward circuitry in limbic structures.

Lower Level Components

As mentioned above, orexin neurons in the LHA also control behavioral arousal and are essential for normal sleep-wakefulness. Administration of exogenous orexin into the brains of experimental animals results in increased cardiovascular function (increased sympathetic outflow resulting in increased mean arterial pressure and heart rate) and in increases in spontaneous locomotor activity, including ambulatory and grooming behaviors. The behavioral arousal observed in response to pharmacological application of orexin predicted the result of loss of orexin neurons during the development of ▶narcolepsy/cataplexy in animal models and humans [3].

Anxiety and fear behaviors may also play a role in the limbic responses to changes in the endocrine and metabolic state. This is best illustrated by the actions of another hypothalamic neuropeptide, corticotropin-releasing hormone (CRH) on food intake and avoidance behaviors. The main function of CRH is expressed in anterior pituitary gland where it exerts primary control of the production and release of adrenocorticotropin (ACTH). ACTH in turn stimulates glucocorticoid (cortisol in humans, corticosterone in lower mammals) production in the adrenal gland. In addition to their mixed catabolic and anabolic effects in peripheral tissues, glucocorticoids feedback into the hypothalamus to control CRH release. In the absence of glucocorticoid negative feedback, the activity of CRH neurons increases and appetite is suppressed (the anorexigenic action of CRH). Behaviors characterized by fear and anxiety increase as well, due to the limbic actions of CRH. Alternatively, when glucocorticoid levels are high, such as in multiple models of stress and anxiety, CRH neurons are suppressed and appetite increases.

Social recognition and maternal behaviors are also regulated by the actions of neuropeptides in limbic structures. In an extensive series of studies, Thomas Insel and colleagues demonstrated the importance of the neuropeptide oxytocin in the organization and maintenance of affiliative behaviors in rodents [10]. Additional studies by Insel and others established the importance of oxytocin in the organization and imprinting of maternal behaviors.

Neuropeptides also play important roles in the organization of survival responses to changes in an animal's external environment including physical and emotional stressors. Neuropeptides that exert important metabolic actions, particularly those that stimulate feeding (e.g., NPY, orexin) are also potent activators of the sympathetic response to stress. Oxytocin, in addition to acting on social behaviors, is one of the most important neurotransmitters controlling sodium balance and therefore fluid and electrolyte homeostasis, which in itself is essential for the appropriate responses to stress.

Structural Regulation

As detailed above, neuropeptides are pivotal factors in the assembly and integration of interoceptive and exteroceptive information. Neuropeptide producing neurons are targets of ascending and descending neural networks and furthermore are the detectors of hormonal and metabolic information. A highly organized system of communication is being discovered in which neuropeptides interact with other neurotransmitters systems in brain to organize the appropriate responses to those changes in the internal and external milieus. Not only do these neuropeptide systems control behavior and autonomic responses in normal physiology, they serve as potential targets for therapeutic interventions in disease [7].

Pathology

Loss of the expression of individual neuropeptides or their receptors results in distinct pathologies, reflective of the physiological role played by those peptidergic systems of communication within limbic structures. Failure of the gonadotropin releasing hormone (GnRH) neurons to migrate into the hypothalamus during development results not only in hypogonadism (loss of GnRH action in pituitary gland), but also the absence of characteristic gender based behaviors. Obesity caused by failure of satiety factors to inhibit neurons in hypothalamic feeding centers occurs when leptin receptors are mutated or absent [6]. Behavioral arousal is severely compromised when orexin neurons in the lateral hypothalamic area begin to disappear during the onset of narcolepsy.

Therapy

Neuropeptides are relatively easy to mass-produce by classic synthetic chemistry or recombinant technologies. Subtle modifications in peptide structure can be made, resulting in increased or decreased biological activity and thus neuropeptides are attractive compounds for the development of super-active agonists or potent antagonists. While the problem of access into the neuropil must be overcome when designing peptide analogs for therapeutic actions within limbic structures, some synthetic neuropeptides readily access the brain or at least act on blood brain barrier free sites to affect limbic function. New strategies are being developed to facilitate delivery of synthetic neuropeptides into the brain and thus take advantage of their inherent activities to treat various pathologies. Particularly in the behaviors associated with feeding, these studies show strong promise. Indeed, PYY_{3-36} has been shown to be an effective anorexic agent in human trials because of its ability to cross the blood brain barrier and activate the neuropeptide systems in limbic centers that inhibit feeding while inhibiting those neurons that produce ►orexigenic neuropeptides [7]. The potential to reverse the catastrophic consequences of loss of orexin neurons during the development of narcolepsy/cataplexy [3] by treatment with a blood brain barrier permeant analog of the neuropeptide has stimulated intense interest in not only the physiologic actions of the neuropeptide itself, but also the convergence of feeding signals and arousal state in general.

References

1. Hosoda H, Kojima M, Mizushima T, Shimizu S, Kangawa K (2003) Structural divergence of human ghrelin. Identification of multiple ghrelin-derived molecules produced by post-translational processing. J Biol Chem 278:64–70
2. Sindelar DK, Palmiter RD, Woods SC, Schwartz MW (2005) Attenuated feeding responses to circadian and palatability cues in mice lacking neuropeptide Y. Peptides 26:2597–2602
3. Willie JT, Chemelli RM, Sinton CM, Tokita S, Williams SC, Kisanuki YY, Marcus JN, Lee C, Elmquist JK, Kohlmeier KA, Leonard CS, Richardson JA, Hammer RE, Yanagisawa M (2003) Distinct narcolepsy syndromes in orexin receptor-2 and orexin null mice: molecular genetic dissection of non-REM and REM sleep regulatory processes. Neuron 38:715–730
4. Garzon M, Pickel VM (2004) Ultrastructural localization of Leu5-enkephalin immunoreactivity in mesocortical neurons and their input terminals in rat ventral tegmental area. Synapse 52:38–52
5. Cowley MA, Pronchuk N, Fan W, Dinulescu DM, Colmers WF, Cone RD (1999) Integration of NPY, AGRP, and melanocortin signals in the hypothalamic paraventricular nucleus: evidence of a cellular basis for the adipostat. Neuron 24:155–163
6. Morton GJ, Cummings DE, Baskin DG, Barsh GS, Schwartz MW (2006) Central nervous system control of food intake and body weight. Nature 443:289–295
7. Batterham RL, Cowley MA, Small CJ, Herzog H, Cohen MA, Dakin CL, Wren AM, Byrnes AE, Low MJ, Ghatei MA, Cone RD, Bloom SR (2002) Gut homone PYY(3–36) physiologically inhibits food intake. Nature 418:650–654
8. Thorpe AJ, Mullett MA, Wang C, Kotz CM (2003) Peptides that regulate food intake: regional, metabolic, and circadian specificity of lateral hypothalamic orexin A feeding stimulation. Am J Physiol 284:R1409–R1417
9. Harris GC, Wimmer M, Aston-Jones G (2005) A role for lateral hypothalamic orexin neurons in reward seeking. Nature 437:556–559
10. Ferguson JN, Young LJ, Insel TR (2002) The neuroendocrine basis of social recognition. Front Neuroendocrinol 23:200–224

Neuropeptides in Energy Balance

Melissa J.S. Chee[1,2], William F. Colmers[1,2]
[1]Department of Pharmacology, University of Alberta, Edmonton, AB, Canada
[2]Centre for Neuroscience, University of Alberta, Edmonton, AB, Canada

Synonyms

Polypeptide; Energy homeostasis; Energy state; Appetite regulation; Food intake; Hypothalamus

Definition

►Energy balance refers to the homeostasis between positive and negative energy states in an organism and is controlled by the regulation of food intake and energy expenditure. In higher organisms, these processes are regulated by the brain, where the hypothalamus assumes an important role. The hypothalamic circuitry involved includes ►peptide chemical messengers. These neuropeptides can be orexigenic, stimulating increased energy intake and tending to decrease energy expenditure, or anorexigenic, reducing energy intake and promoting increased energy expenditure. Here, we discuss the main hypothalamic neuropeptides that modulate the central circuitry involved in energy homeostasis, and their integration of peripheral peptide and protein hormone signals of energy balance.

Characteristics

Neuropeptide Modulation of Hypothalamic Circuitry

Early experiments investigating the role of the hypothalamus in energy balance focused on the ventromedial nucleus (VMH) and the lateral hypothalamic area (LHA). Lesions to the VMH in rats resulted in ►hyperphagia and increased weight gain, while lesions to the LHA caused anorexia and weight loss. The “dual centre hypothesis” thus proposed that the VMH and LHA were the hypothalamic ►satiety centre and feeding centre, respectively [1]. However, this hypothesis is now believed to be overly simplistic. Lesions to the VMH and LHA may have interfered with intra- and extra-hypothalamic neural pathways passing through these regions. Furthermore, lesions at other hypothalamic regions can also influence energy balance.

The regulation of energy balance is now thought to involve a number of hypothalamic nuclei and extra-hypothalamic brain regions. These include the area postrema and subfornical organ, which are ►circumventricular organs (CVO) that mediate communication between the brain and peripheral appetite signals [2]; the nucleus tractus solitarius (NTS) of the brainstem, which receives inputs from vagal sensory afferents, and is capable of mediating some feeding responses [3]; the amygdala, which contributes an emotional component to the acquisition, storage, and recall of experiences with food; the nucleus accumbens and ventral tegmental area, which processes motivational and reward-mediated behaviors [4]; and the neocortex, which integrates taste, olfactory, visual, memory, and social factors that influence food intake [5].

The information regarding the nutrient and energy levels in the body is communicated to the brain, via neural and endocrine signals. The arcuate nucleus (ARC) of the hypothalamus plays an important role in integrating these signals. The ARC projects within the hypothalamus to the VMH, LHA, paraventricular nucleus (PVN), and dorsomedial nucleus (DMH) [6]. Communication within the hypothalamus is mediated by several classes of chemical messengers, including the amino acid and biogenic amine transmitters, cytokines, cannabinoids, and most notably, neuropeptides. These peptide neurotransmitters can act as either orexigenic (Table 1) or anorexigenic (Table 2) signals to stimulate or inhibit food intake, respectively.
A brief discussion of some individual peptides involved in energy balance regulation follows.

Melanocortin Peptides

Activation of the central melanocortin system reduces food intake and increases energy expenditure. It includes α-, β-, and γ-melanocyte stimulating hormone (MSH), all of which are derived from the proopiomelanocortin (POMC) peptide precursor. POMC expression in the CNS is restricted to the neurons of the NTS and ARC. POMC expression in the ARC is reduced in fasted animals. α-MSH is thought to represent the main melanocortin signal in this circuit. The ARC POMC neurons project to the PVN, DMH, and LH, where the melanocortins are released and act via MC3 and MC4 receptors. α-MSH and β-MSH are the main ligands of the MC4 receptor while γ-MSH is the main ligand at the MC3 receptor. Acute central administration of α-MSH reduces food intake and chronic administration reduces body weight. Genetic modulation of the melanocortin system, unlike most genetic models targeting specific hypothalamic circuits

Neuropeptides in Energy Balance. Table 1 Peptides with orexigenic action(s) in the hypothalamus

Peptide		Receptor	Site of synthesis	Site of action
			Hypothalamic	*Hypothalamic*
AgRP		MC3, MC4	ARC	PVN, DMH
Galanin		Gal1, Gal2	PVN, ARC, DMH	ARC
GALP		Gal1, Gal2	ARC, ME, DMH	ARC, PVN, LH
MCH		MCH1, MCH2	LH	PVN, VMH, DMH
NPY		Y1, Y2, Y4, Y5	ARC, DMH	ARC, PVN, DMH, LHA, VMH, PFA
Opioids	Dynorphin	κ-OR	ARC, PVN	PVN, DMH, LHA, VMH, ARC
	β-endorphin	μ-OR	ARC, VMH, DMH, PVN	ARC, VMH, DMH, PVN
	Enkephalin	δ-OR	ARC	VMH, DMH, PVN
Orexins	Orexin A	OX_1, OX_2	LH, DMH, PFA	ARC, PVN, LH, PFA, VMH
	Orexin B			
			Periphery	*Hypothalamic*
Ghrelin		GHSR1a	Stomach	ARC, PVN, VMH

Neuropeptides in Energy Balance. Table 2 Peptides with anorexigenic action(s) in the hypothalamus

Peptide		Receptor	Site of synthesis	Site of action
			Hypothalamic	*Hypothalamic*
CART		Unknown	ARC, PVN, DMH, LHA	PVN, DMH, LHA
CRH		CRH1, CRH2	PVN	PVN
Melanocortin	α-MSH	MC4	ARC	PVN, VMH, DMH, ARC
	β-MSH	MC4		VMH, DMH, ARC
	γ-MSH	MC3		ARC
UCN I–IIII		CRH2, CRH1	LHA, PVN	PVN, ARC, VMH
			Periphery	*Hypothalamic*
Amylin		CGRP relative	Pancreas	LH
Bombesin		BB1, BB2	GI tract	PVN
CCK		CCK1, CCK2	Small intestine	PVN, DMH
Glucagon-like peptides	GLP1	GLP1	Small intestine	ARC
	GLP2	GLP2		PVN, ARC
	OXM	GLP1		DMH
Insulin		IR-A, IR-B	Pancreatic islet	ARC, DMH, PVN
Leptin		OB-Rb	Adipose tissue	ARC, VMH, PVN, DMH
PP		Y4	Pancreas	PVN, ARC
PYY		Y2	GI tract	ARC

that regulate appetite, produces significant changes in food intake and body weight. ►Knockout of the MC4 or MC3 receptor results in hyperphagia and obesity in mice, and humans with MC4 receptor or POMC mutations are obese.

Corticotrophin-Releasing Hormone (CRH) and CRH-like Peptides

CRH plays a central role in mediating stress by action on the hypothalamic-pituitary-adrenal axis (HPA). Central administration of CRH also reduces appetite in rodents. However, the anorexigenic effects of the CRH-like peptides, including urocortin (UCN) I–III, are more potent than that of CRH. CRH is expressed in the PVN and UCN is expressed in both the PVN and the LHA. Interestingly, CRH can play a role in initiating food-seeking behavior via the HPA axis.

Cocaine- and Amphetamine-Regulated Transcript (CART)

CART is expressed in several hypothalamic nuclei and codes for a peptide that reduces food intake following ►intracerebroventricular (ICV) injection to the third cerebral ventricle. However, some orexigenic actions of the CART peptide have also been demonstrated. It has been suggested that CART can activate inhibitory autoreceptors that downregulate the anorexigenic CART signaling.

Neuropeptide Y (NPY)

NPY, a member of the pancreatic polypeptide (PP) family, is one of the most potent known orexigenic neuropeptides. It is mainly produced by neurons in the ARC and brainstem and acts at several sites in the hypothalamus, including the PVN, VMH, DMH, LHA, and the perifornical area (PFA), a region of the LHA surrounding the fornix. Central administration of NPY stimulates food intake, which is thought to be mediated via Y1 and Y5 receptors. The greatest effect of NPY on food intake has been observed when administered into the PFA. The orexigenic actions of NPY peak during the appetitive, "food-seeking" phase of feeding. NPY release is enhanced immediately prior to the onset of feeding and gradually decreased as food intake continues. Despite the powerful acute effects of NPY on feeding, neither the overexpression nor knockout of NPY or its receptors causes profound changes in body weight. There appears to be compensatory mechanisms that protect orexigenic signaling in the hypothalamus when NPY signaling is compromised. This is observed in germline mutations of NPY signaling, which take effect throughout the development and growth of the animal. However, this is not true for adult animals, as the targeted disruption of NPY neurons in the ARC of adult mice results in lethal ►hypophagia [7].

Agouti-Related Peptide (AgRP)

AgRP has potent orexigenic effects by its action as an endogenous antagonist of the melanocortin receptors. This suggests that melanocortin signaling is tonic and modulated by AgRP, which is ►co-localized exclusively with NPY in ARC neurons. ARC AgRP expression is greatly increased in fasted animals.

Central injection of AgRP potently stimulates food intake in rodents and can increase food intake for up to one week.

Melanin-Concentrating Hormone (MCH)

MCH is produced mainly by two groups of LH neurons: Type A MCH neurons co-express CART and send descending projections to the brainstem while Type B neurons do not express CART and send ascending projections to the forebrain. The orexigenic effects of MCH are mediated by MCH1 receptors. Mice overexpressing the MCH gene are more susceptible to high-fat diet-induced obesity. In contrast, animals lacking the MCH or MCH1 receptor genes are lean and are resistant to obesity. Such conditions of MCH deficiency are accompanied by a marked suppression of ARC POMC expression, perhaps responding to counterbalance the decrease in satiety and body weight.

Orexin

The orexins play an important role in maintaining the arousal system, which has been associated with changes in energy homeostasis. The orexins – Orexins A and B – are also known as the hypocretins and are produced primarily in the LHA but also in the PFA and DMH. The orexigenic effects of Orexin A are more potent than that of Orexin B. Orexin A has equal affinity for the OX_1 and OX_2 receptor. The OX_1 receptor is expressed in the VMH and LHA. The OX_2 receptor is expressed in neurons of the PVN, DMH, and ARC, which receives the highest density of LHA orexin fibers. Orexin A delays the onset of satiety, thereby increasing the duration of a meal to prolong feeding behavior. Disruption of orexin signaling by the ablation of orexin neurons or knockout of the prepro-orexin gene produces a hypophagic phenotype, which is partly attributed to the development of narcolepsy in these animals.

Galanin and Galanin-like Peptide (GALP)

Galanin is expressed in the PVN, DMH, and ARC. GALP expression is restricted to the ARC. The orexigenic effects of galanin and GALP are fat-sensitive. Galanin acts via the Gal1 receptor and the positive feedback circuitry between galanin and dietary fat contributes to large meal sizes. The orexigenic effects of GALP, which are mediated by the GalR2 receptor, are enhanced following a high-fat diet when preceded by a period of hypophagia.

Opioid Peptides

The opioid peptides (β-endorphin, enkephalin, and dynorphin) are all expressed in the ARC. Dynorphin is also expressed in the PVN and enkephalin in the PVN, VMH, and DMH. The orexigenic actions of the opioids are most commonly associated with β-endorphin, though all opioids stimulate food intake. The opioids preferentially increase fat ingestion over that of proteins or carbohydrates. Central blockade of opioid receptors prevents orexigenic actions of other peptides such as NPY, AgRP, orexin, and galanin.

Contribution of Peripheral Peptides and Hormones

The central feeding circuitry is influenced by peripheral signals that indicate nutritional state and adiposity level. Peptides and other hormones that are produced and released from the gastrointestinal (GI) tract, pancreas, and adipose tissue play an important role in the short- and long-term regulation of energy balance. These peripheral signals can have central effects by:

1. Stimulation of vagal sensory afferents in the GI tract to signal brainstem nuclei, such as the NTS [8]
2. Stimulation of neurons in CVOs that project to the hypothalamus [2]
3. Transport across the ►blood-brain barrier (BBB) to act directly within the CNS [9]

Ghrelin

Ghrelin is secreted by the stomach and is the only peripheral factor known to signal to the CNS to increase food intake. It plays a role in meal initiation; circulating ghrelin levels rise immediately before a meal and remain elevated during periods of food deprivation, and rapidly decline following food intake. Ghrelin can have central effects by crossing the BBB, acting via the vagal afferent pathway, and acting at CVOs. In addition, ghrelin is locally produced in the hypothalamus. In the hypothalamus, ghrelin acts at the ARC, LH, and PVN. The effects of ghrelin are mediated by NPY and AgRP release and are most potent in the PVN. In the ARC, ghrelin activates NPY neurons and inhibits POMC neurons.

Leptin

Leptin is the protein product of the obese (*OB*) gene and plays a significant role in reducing food intake. Leptin is produced in white adipose tissue and levels of circulating leptin are directly correlated with body fat, glucose uptake, and food intake. Humans with congenital leptin deficiency, animals with defects in leptin (*ob/ob* mouse) or the leptin receptor (*db/db* mouse, fatty Zucker rat) are grossly obese. However, this is not a major cause of obesity, since most obese humans are leptin resistant and have elevated leptin levels. Leptin is transported into the brain via truncated forms of the leptin receptor (Ob-Ra, c, or d) and leptin signaling occurs via the long form of the receptor (Ob-Rb). In the ARC, leptin activates anorexigenic POMC neurons and inhibits orexigenic NPY/AgRP neurons. In the LHA, leptin inhibits the orexigenic MCH- and orexin-expressing neurons. Thus, when circulating leptin levels are low during periods of food-restriction or fasting, orexigenic neurons are activated and expression of orexigenic neuropeptides is increased while anorexigenic neuronal activity and neuropeptide expression is inhibited. Conversely, high plasma leptin levels inhibit orexigenic pathways and activate anorexigenic pathways. Leptin clearly participates in long-term energy balance regulation. The absence of leptin signaling produces drastic changes in the energy state while leptin overexpression produces only mild phenotypic changes; this suggests that leptin is an essential signal indicating the adequacy of energy stores, which is required for the maintenance of vital activities such as reproduction [10].

Insulin

Insulin is produced by the pancreas and, in addition to its well characterized role in the regulation of blood glucose levels, can act as a circulating adiposity signal. Plasma insulin levels change directly with changes in adiposity so that positive energy states stimulate, and negative energy states decrease, insulin secretion. Little or no insulin is produced in the brain but it can have significant central effects. Insulin moves by transporter-mediated entry into the brain and acts on insulin receptors in the ARC, PVN, and DMH. The actions of insulin in the CNS are anorexigenic. The orexigenic NPY and anorexigenic melanocortin systems are both downstream mediators of insulin signaling. Thus, insulin can increase POMC expression and prevent fasting-induced increases in NPY expression. Insulin is also implicated in the long-term regulation of energy balance as the specific deletion of neurons expressing the central insulin receptor leads to obesity.

Glucagon-like Peptides (GLP)

The GLP, including GLP-1, GLP-2, and oxyntomodulin (OXM), are anorexigenic peptides released by the gut following food intake. GLP-1 and OXM, but not GLP-2 can affect food intake when administered peripherally. However, all three GLPs inhibit food intake when administered centrally; this is mediated by the GLP-1 and GLP-2 receptors. In addition, central or peripheral administration of GLP-1 potently stimulates insulin release. The anorectic effects of these peptides are mediated by the brainstem, which is reciprocally connected with the hypothalamus.

Cholecystokinin (CCK)

CCK is secreted by the small intestine following food intake and is the first gut peptide to have been shown to affect food intake. The CCK1 and CCK2 receptors mediating its actions have been cloned and characterized. Its role appears to be restricted to that of a short-term satiety signal contributing to meal termination.

Pancreatic Polypeptide (PP)

PP, a member of the PP fold peptide family, is released from the pancreas following the ingestion of food and reduces appetite and food intake via the Y4 receptor. However, counter-intuitively, the global deletion of the Y4 receptor produces hypophagic mice with reduced body weight gain. PP can enter the CNS via the area postrema, a CVO rich in the expression of Y4 receptors. However, the central effects of PP are orexigenic, possibly as a result of pharmacological activation of Y5 receptors.

Peptide YY (PYY)

PYY is released into the circulation from enteroendocrine cells of the gut after a meal. In the circulation, PYY is rapidly converted to the C-terminal PYY_{3-36} fragment, the major circulating form. Peripheral administration of PYY_{3-36} reduces food intake in rodents and humans by activation of ARC neurons. Direct administration of PYY_{3-36} to the ARC also suppresses food intake. This is mediated by the activation of postsynaptic Y2 receptors that directly inhibits anorexigenic POMC ARC neurons and activation of Y2 autoreceptors on NPY/AgRP ARC neurons that suppresses orexigenic tone. However, ICV injection of PYY_{3-36} increases food intake in rodents, presumably due to the pharmacological activation of Y1 and/or Y5 receptors.

Experimental Methods Employed

Several experimental approaches have been used to demonstrate orexigenic or anorexigenic effects of a peptide:

1. Acute administration of orexigens such as by intracerebroventricular injection or direct microinjection into the specific hypothalamic nuclei stimulates feeding behavior in satiated animals, while anorexigens can prevent feeding following periods of food-deprivation.
2. Chronic administration of orexigens induce hyperphagia and increase body weight gain while chronic administration of anorexigens results in hypophagia and reduced body weight gain.
3. Under conditions of food-deprivation, orexigenic and anorexigenic peptide gene expression is increased and decreased, respectively; while opposite patterns emerge in animals on a high-fat diet.
4. Genetic manipulation at embryogenesis, such as by transgenic peptide overexpression, or a generation of a peptide- or receptor-knockout mouse, rarely results in animals with profound changes in feeding behavior. This implies that the regulation of energy balance is protected by a redundant system that can compensate for compromised orexigenic or anorexigenic signaling systems, during development. However, current evidence suggests that such redundancies are not effective in adult animals.

Thus the homeostatic regulation of energy intake and expenditure is remarkably complex. This is unsurprising given the importance of energy balance to the survival of the organism. The inbuilt redundancies in the hypothalamic networks that play a key role in this regulation may explain the resistance of obesity to pharmacological interventions based on targeting single chemical signals.

References

1. Brobeck JR (1946) Mechanism of the development of obesity in animals with hypothalamic lesions. Physiol Rev 26:541–559
2. Fry M, Hoyda TD, Ferguson AV (2007) Making sense of it: roles of the sensory circumventricular organs in feeding and regulation of energy homeostasis. Exp Biol Med (Maywood) 232:14–26
3. Grill HJ (2006) Distributed neural control of energy balance: contributions from hindbrain and hypothalamus. Obesity (Silver Spring) 14:216S–221S
4. Berthoud HR (2004) Mind versus metabolism in the control of food intake and energy balance. Physiol Behav 81:781–793
5. Berthoud HR (2006) Homeostatic and non-homeostatic pathways involved in the control of food intake and energy balance. Obesity (Silver Spring) 14:197S–200S
6. Schwartz MW, Woods SC, Porte D Jr, Seeley RJ, Baskin DG (2000) Central nervous system control of food intake. Nature 404:661–671
7. Luquet S, Perez FA, Hnasko TS, Palmiter RD (2005) NPY/AgRP neurons are essential for feeding in adult mice but can be ablated in neonates. Science 310:683–685
8. Schwartz GJ (2006) Integrative capacity of the caudal brainstem in the control of food intake. Philos Trans R Soc Lond Ser B Biol Sci 361:1275–1280
9. Banks WA (2006) Blood-brain barrier and energy balance. Obesity (Silver Spring) 14:234S–237S
10. Margetic S, Gazzola C, Pegg GG, Hill RA (2002) Leptin: a review of its peripheral actions and interactions. Int J Obesity Relat Metab Disord 26:1407–1433

Neuropharmacology

Paul F. Smith
Department of Pharmacology and Toxicology, School of Medical Sciences, University of Otago Medical School, Dunedin, New Zealand

Definition

"Neuropharmacology" is the subdiscipline of pharmacology devoted to the study of the action of drugs on the nervous system. This includes the effects of therapeutic drugs as well as recreational drugs and toxins. Neuropharmacology is distinct from related subjects, such as neurochemistry, neurophysiology and neuroanatomy, in that the emphasis is on *drug action*. Whereas the neurochemist and the neurophysiologist study the same cellular machinery of the nervous system as the neuropharmacologist, the latter does so with the chemical structure of drugs and how it affects that machinery, in mind. In this regard, neuropharmacology interacts with medicinal chemistry in order to understand how various molecular structures affect receptors and other drug targets within the cells of the nervous system. No matter how diverse different branches of pharmacology may be, what pharmacologists, including neuropharmacologists, have in common is the desire to understand how exogenous chemical structures manipulate living cells, and the discovery of drugs is of course one of the major routes for the clinical application of neuroscientific knowledge to the management of clinical neurological disorders (see chapters on ►Analgesics, ►Anticonvulsants, ►Antipsychotics for examples).

A Changing Field...

Despite the neuropharmacologist's emphasis on drug action, more than ever before, neuropharmacology interacts with neurochemistry, neurophysiology and neuroanatomy, as the level of detail available on neural functions becomes ever greater. One obvious example of this is the striking increase in the understanding of the complexity of the biochemical pathways that are affected by drugs. For example, the interactions of ►G protein-coupled receptors (GPCRs) with other proteins and the subunit-specific actions of ►$GABA_A$ receptor agonists, are understood to a degree not possible previously (see chapters on ►Pharmacodynamics and ►Anxiolytics and hypnotics). Hence, for the last decade and a half, there has been a strong molecular influence in neuropharmacology.

New Waves of Influence...

Neuropharmacology is influenced by progress in every area of biology and medicine. The increasing interest in the effects of growth factors and ►cytokines in the nervous system has led to the investigation of whether some of these endogenous molecules, or their synthetic derivatives, might be of use in the treatment of neurological disease (see chapter on ►Growth factors). At the same time, gene therapy, which traditionally has not been considered to be a form of drug treatment (but which must be according to Goodman and Gillman's definition of a drug as any chemical that affects living processes...) is now being investigated, in some cases instead of, and in other cases, in addition to, conventional drug therapy (see chapter on ►Gene therapy).

Another important theme in recent neuropharmacology has been the recognition of the differences in drug action in the male and female nervous system. While historically, most drugs have been tested mainly in males, even including those such as ►benzodiazepines that were preferentially prescribed to females, it is now recognized that there is an urgent need to understand the differences between males and females in the action of neurological drugs, and also how drugs affect females differently through the menstrual cycle (see chapter on ►Neuroendocrinological drugs).

Since 1988 when the first cannabinoid receptor was reported, cannabinoid pharmacology has undergone a revolution. The endocannabinoid system has emerged as one of major importance in human biology and in the brain it has been shown to interact with almost every neurotransmitter system. The chapter on ►Cannabinoids reviews the latest developments in this fast moving area of neuropharmacology, including the new therapeutic drugs that are developing from it.

Neuropharmacology includes not only therapeutic drugs, but also recreational and lifestyle-enhancing drugs, and an important topic in this respect is the recreational use of stimulants, such as ►methamphetamine and ►methylenedioxymethamphetamine (MDMA). These drugs are used and abused around the world, and the investigation of how they affect the nervous system is an active area of research in neuropharmacology (see chapter on ►Stimulants).

Finally, some topics are common to all or at least most drugs affecting the nervous system, and the analysis of drug tolerance and dependence is one such topic. Because of the fact that the nervous system adapts to any repeated stimulus, its response to a drug is rarely the same twice, and therefore understanding how its reaction changes over time and how this may be related in some cases, to dependence or addiction, is a critical issue for both therapeutic and non-therapeutic drugs (see chapter on ►Tolerance and dependence).

Neuropharmacology in the twenty-first century is an exciting, compelling discipline that has all of the attractions of other areas of neurobiology, but with a strong focus on clinical applications and drug development for the treatment of the many neurological disorders that continue to afflict society.

Neurophilosophy

Michael Pauen
Institut für Philosophie, Humboldt-Universität Zu Berlin, Unter den Linden 6, Berlin

Neurophilosophy is an emerging field at the interface between neuroscience and philosophy. It encompasses, first, methodological, conceptual, anthropological, and ethical questions that are of relevance to neuro- and cognitive science. Second, neurophilosophy uses results from these sciences in order to shed light on related philosophical problems.

The Term "Neurophilosophy"

The term became popular after Patricia Churchland [1] used it as a title for a book and has since then received increasing attention, both in scientific literature and in academic teaching [2]. Still, "neurophilosophy" is not yet a received term for an established branch in academic philosophy like "philosophy of science" or "philosophy of history" with their own journals, academic programs, scientific societies, etc.

Nevertheless, the close connection to empirical science sets neurophilosophy apart from its predecessors, particularly from traditional philosophy of mind, although there is a significant overlap between both fields. Even proponents of ►materialism in the second half of the twentieth century used only placeholders for neural states like the legendary "C-fiber firings" rather than real neurobiological data. Functionalism, a widely accepted view in the 1970's and 1980's, held that a detailed understanding of the brain could not help us to get a significantly better understanding of conscious states. However, with the success of neuroscience, it became increasingly evident that neuroscientific findings *are* relevant for philosophical theories. Likewise, it turned out that neuroscience raises conceptual, methodological, and ethical problems that are of philosophical relevance. Neurophilosophy tries to account for these problems and it makes use of neuroscientific insights in order to solve them.

What is the Rationale of Neuro-Philosophical Cooperation?

From the philosophical perspective, brain research is of special interest because the brain is the material substrate of those distinctive human abilities and ►properties like consciousness, ►free will, the ►self, cognition, memory, and emotion that have always taken center stage in philosophical thinking. Although even most ►materialist philosophers would insist that there is an important difference between ►*knowledge about* the brain and *knowledge about* certain cognitive abilities, the former knowledge can be of central importance if we want to know more about the details of our cognitive abilities. Given that our current views of these abilities are based on pre-scientific assumptions, it would follow that new empirical findings might lead to a profound revision of these views. Since these abilities, in turn, are substantial for human self-understanding, it is not difficult to see why neuroscience is of particular relevance from a philosophical point of view.

Conversely, there are at least two reasons why neuroscience can take advantage of philosophical work.

First, with the advent of new experimental techniques, particularly with the availability of non-invasive imaging, questions of fundamental relevance for human self-understanding like consciousness, the self, or the free will problem, became subject to neuroscientific research. One might feel tempted to conclude that this enables neuroscience to solve old philosophical problems. On reflection, however, it turns out that, in order to come up with experimental designs or to derive conclusions that are relevant for our self-understanding, we first need clear cut concepts of what it takes to be conscious, self-conscious, or to act freely. This is true, in particular, because we have strong but usually fuzzy and incoherent pre-scientific intuitions concerning these ideas, and philosophers have considerable expertise in transforming such intuitions into coherent concepts. Of course, there may be irresolvable conflicts between competing intuitions; but even then, the ensuing debates about the correct understanding are useful because they clarify the available conceptual and theoretical choices. The second reason why neuroscience can take advantage of philosophical considerations results from the complexity of the brain itself and of the related research programs. As a consequence, any theory that tries to come close to a comprehensive picture of the mind/brain raises fundamental methodological and ►epistemological questions, partly because it has to account for findings from a vast number of disciplines, ranging from molecular biology up to cognitive psychology. Philosophers have discussed these questions for quite some time, thus exposing the advantages and disadvantages of the options at hand in order to make a reasonable choice possible.

Neurophilosophical Tools

Given that philosophy tries to explore the premises and foundations of empirical knowledge rather than providing such knowledge itself, it should not be surprising that there are neither uncontroversial philosophical results nor established methodologies. If a question becomes tractable by such a methodology or if we can come up with uncontroversial answers, then the question looses its philosophical interest. This is true for neurophilosophy, too. Still, there are certain universally accepted "tools" that are useful in this endeavor. ►*Logic* is, of course, one of them, especially if we try to expose the consequences and implications of a particular view. *Conceptual analysis* is another important tool. In this case, philosophers try to characterize or even define a pre-theoretic concept. In an ideal case, such a definition provides necessary and sufficient conditions for the application of a concept. Since definitions in a strict sense are often hard to come by, we must be content with *characterizations* of the typical way of using a concept. In many cases, intuitions play an important role in our pre-scientific understanding. In order to explore these intuitions, philosophers use thought experiments as "intuition-pumps" [3]. Many of these thought experiments are based on extreme scenarios in order to enable clear distinctions in difficult cases. So if philosophers come up with their notorious Zombies, Zimbos and the like, they do not maintain the empirical possibility of suchlike beings in some distant future; rather, they try to find out how we would classify these strange beings, and knowing that can be quite useful if we want to make conceptual distinctions in less extreme cases.

The Mind-Body Problem

In what follows, some of the most important neurophilosophical issues will be discussed. The most basic problem, of course, is the notorious mind-body problem which can also serve as an example for the "division of labor" between philosophers and scientists: While scientists try to find out facts about the relation between mind and brain, philosophers have to clarify the conceptual distinctions between the different available options, to expose the theoretical advantages and disadvantages of each of them, and to outline possible empirical evidence that would support or disprove each of these options. In doing so, neurophilosophy clarifies the criteria for the assessment of empirical evidence and limits the number of possible options for the interpretation of this evidence, thus enabling us to make a reasonable choice between these options in light of the available evidence.

Trivially, there are two fundamental alternatives: Either the mental is some kind of physical process or it is not. Monists hold that it is, Dualists hold that it is not. Due to its initial plausibility, psychophysical dualism can be found in many mythological and religious writings including the Bible, but it is also part and parcel of our commonsense beliefs about the mind-body relation. Depending on the ►causal relation between the mental and the physical, philosophers have distinguished different varieties of dualism. Psychophysical parallelism in its original form, as it was defended by Leibniz, holds that there is no psychophysical interaction at all. Mind and body run in parallel like two clocks that remain synchronized without any causal interaction between them, once they have been started together. ►Epiphenomenalism, a theory that was brought forward by T. H. Huxley [34], Jackson [4], Robinson [5], and, in principle, also by Chalmers , states that there is only a one-way causation: Neural events cause mental events, but mental events, in turn, do not cause anything; they are causally inefficacious by-products. Certainly the most widely held dualist view is psychophysical interactionism as it was defended by Descartes. More recently, interactionist views were held by philosophers like Popper but also by neuroscientists like Eccles [32] or [35]. Interactionism holds that mental events like volitional acts cause physical events like brain processes, and physical events, in turn, cause mental events, like ►perceptions. Philosophers have exposed at some length the implications and problems of each of the available alternatives [7]. In the case of interactionist dualism, psychophysical causation is one of the notorious difficulties because, on this view, mental processes like acts of will would have to interrupt the physical causal chain. This would violate almost uncontroversial doctrines like the principle of the causal closure of the physical realm or the law of the conservation of energy. Dualists have made various suggestions how to solve these problems, but none of them has been universally accepted. Other varieties of dualism face severe problems, too: Parallelism has difficulties to explain the synchronicity between the mental and the physical realm, given that the existence of God is not an acceptable scientific hypothesis. Epiphenomenalism is counterintuitive because mental states, due to their causal inefficacy, cannot be among the causes, why we talk about these very states, react upon them, or even design epiphenomenalist theories about them [8].

However, seen from a neurophilosophical perspective, there is still another objection against dualism as a scientific hypothesis. Presumed that natural science tries to explain empirical facts with reference to natural laws and physical entities, the postulate of a non-physical mind doesn't appear as a sensible scientific hypothesis because it implies that mental facts are not amenable to a naturalistic ►explanation. While this does not rule out that dualism is in fact true, it seems unreasonable for natural science, including neuroscience, to start with such an assumption that would seriously limit its own explanatory scope. Natural science should extend the realm of naturalistic explanations as far as possible rather than beginning with the premise that there are facts beyond explanation in principle.

It would follow that monism is a much more plausible position to take from the viewpoint of natural science. Monists assume that there is only one type of entities. According to current versions of monism there are only physical entities; thus monism is in fact a sort of ►physicalism. In the last five or so decades, philosophers have explored a vast number of varieties of monism that differ with respect to the status of mental states in such a physical world. While identity-theorists think that we can keep our pre-scientific beliefs about causally efficacious mental states as long as we accept that these states are physical states, more radical materialists like eliminativists and ►logical behaviorists think that mental states and the related mentalistic idiom have no place in a scientific picture of the world.

According to eliminative materialism, the existence of mental states like beliefs and desires is more than questionable [1,9,10]. Rather than being subject to immediate access from the first person perspective, these states are postulates of a vernacular theory, usually called "►folk psychology". Folk psychology has originally been introduced by our remote ancestors in order to explain and predict human behavior. In the meantime, we became so much used to this postulate that it now appears to us as if we had immediate access to the postulated mental states.

However, neuroscience will eventually provide a better theory of human behavior that gets along without any reference to beliefs and desires. This theory will replace its primitive precursor, thus eliminating not only folk psychology itself but also mental states as its theoretical postulates. Our successors will substitute the precise terminology of neuroscience for the unclear mentalistic idiom, and in the long run even the alleged "direct access" to mental states from the first person perspective will disappear. Mental states will be "eliminated" because the underlying theory has to be given up, just like phlogiston or caloric were eliminated when their theoretical basis vanished. Eliminative materialism is a substantial part of Patricia Churchland's original version of neurophilosophy [1]. If eliminativism is true, then natural science should be able to solve or better: *dissolve* the mind-body problem. All there is that needs to be explained is the brain, and it would seem that neuroscience is able, in principle, to explain neural processes.

Many philosophers think however that this solution is all too simple. Apart from the fact that it is counterintuitive to think that mental states are only postulates of a theory, eliminativism seems to undermine its own theoretical basis: Eliminativists, after all, have to believe in their own theory, but if their theory is true, then there are no beliefs.

This is one of the reasons why eliminative materialism has lost some support in the recent past. Its strongest monistic competitor is the identity theory which was proposed by Feigl [11], and Place [12] in the 1950s. Identity theorists have no doubt that mental states exist; according to their theory, one could dispense with mental states only at the cost of giving up the corresponding physical states, too. Identity theorists can even insist that mental states are perfectly adequate subjects of scientific inquiry, say in cognitive or volitional psychology. Another advantage of this theory is that it provides a simple solution to the problem of mental causation: If mental states are identical with certain physical states, then they *are* physical states and should have causal powers just like other physical states.

According to the original version of the identity theory, there is a one-to-one relation between types of mental states (say pain) and types of physical states (say some neural type *N*). Thus, each particular token of a certain mental type (pain) will also be a token of the corresponding physical type (*N*), and vice versa. Talking about a pain state and talking about a neural state of type *N* is talking about one and the same type of states. This is the reason why this variety of the identity theory has been called "type identity."

The most severe objection that appeared almost detrimental to the identity theory in its original form refers to the so called "multiple realizability" of mental states. Organisms and even artificial systems whose physical makeup differs considerably from the human neuroanatomy, so the objection goes, might be able to feel pain. It would follow that, contrary to what type identity theorists would have us to believe, not all pain states are neural states of type *N*, or in general: Mental states of a certain type can exist in the absence of the related physical type.

Solutions to this problem have been proposed in the meantime [13], but it was originally thought that this objection forces us to reject the type identity theory. As a consequence, another variety of the identity theory emerged, namely the so-called "token identity" theory [14]. According to token identity theorists, every token of a certain mental state is identical with a token of *some* physical state. Thus, token-identity theorists postulate a one-to-*many* relation between mental and physical types: Tokens of *one* mental type (pain) can be realized by the tokens of *many* different physical types (neural states *N, O, P,* state *Q* in a silicon chip, etc.). Typically, token identity-theorists subscribe to functionalism [15]. According to functionalism, mental states have a distinctive functional role that can be captured in an extended ►behavioral terminology. Accordingly, being in a pain state is to have certain functional or behavioral properties, e.g. the tendency to say "ouch" under certain conditions, to take painkillers, to think about other ways to get rid of pain, etc. A complete list of these features should be able to capture what we mean if we talk about pain. Consequently, every physical state that performs this functional role would count as a "realizer" of pain because it meets the relevant conceptual criteria.

While it seems that functionalism was successful in providing a solution to the problem of multiple realization, its fundamental premise, namely that mental states can be captured by a distinctive functional role, has been challenged. It seems that, say, our concept of color-experiences cannot be captured by any distinctive functional role, given that it seems perfectly possible to imagine cognitive systems with identical functional roles that have different color-experiences or no color-experiences at all [16]. This assumption has played a major role in the discussion of the so called explanatory gap problem (see below).

One might conclude that all these ►arguments, objections, and counterarguments show that neurophilosophy has failed to provide a solution to its most basic problem. But this would be a misunderstanding of the objectives of philosophy. Of course: Philosophers have to clarify the conceptual criteria for each of the available options, they can even rule out options with severe theoretical disadvantages, but they cannot provide the empirical data themselves. More conceptual clarity and a better understanding of the implications of each of the available options can help us to make a reasonable choice between these options, once the relevant data are available. But since it is one of the distinctive characteristics of neurophilosophy that it relies on empirical data in order to make progress on philosophical problems, it cannot provide a solution before the relevant data are in.

Particular Questions

The "Explanatory Gap Problem"

But how could we make progress towards a solution? One obvious strategy is to look for the "neural correlates of consciousness," or better, for strict and specific correlations between distinctive and well-defined mental states on the one hand and distinctive and well-defined neural activities on the other. If there is no evidence for psychophysical interaction and if it is true that eliminative materialism, psychophysical parallelism, and epiphenomenalism are subject to severe theoretical objections, we might be justified to believe that mental states *are* physical states and some version of the psychophysical identity claim is true.

But even then, one might feel somewhat worried. It seems difficult to understand that the different qualia or qualitative properties of mental states, the "way it feels" to be in a pain state or to have a red-experience should be identical with the uniform activity of simple neurons. The worry has already been felt by Locke and Leibniz and it has received increased attention in the recent past. Many philosophers have argued that this worry results from an "explanatory gap" that cannot be closed in principle [4,17,18].

In trying to come to terms with this worry, one should note, first, that the problem does not concern the factual relation between neural activities and mental processes. Asking why neurons bring about consciousness would obviously imply a distinction between the mental and the physical that is incompatible with the claim that mental processes *just are* neural activities. Rather, the question concerns the relation between our *knowledge about* neural activities and our *knowledge about* mental processes, that would allow us to account for problems on the mental level in neurobiological terms.

Normally, we use ►reductive explanations in order to understand problems concerning higher level properties of a complex entity (e.g. heat) in terms of knowledge about the lower level constituents of this entity (e.g. mean kinetic energy of molecules), say the laws that apply to these lower level constituents. Reductive explanations make it intelligible why an entity with certain lower level constituents has specific higher level features. A higher level property that cannot be reductively explained, in principle, is an ►emergent property.

Reductive explanations have two basic requirements: First an explanation why the constituents have a certain lower level feature (a specific kinetic energy), second a bridge between the higher and the lower level that makes it intelligible why having the lower level feature *really is* having the higher level feature in question (i.e. a certain temperature). Given these requirements, our lower level explanation should help us to understand the presence of the higher level property in question.

Provided that scientific explanations pertain to observable properties rather than to subjective features, any bridge between mental and physical properties would require a determination of subjective mental features in terms of objective observable properties which guarantees that the organism in question really has the mental features that should be explained. Many philosophers think that this is impossible, in principle. In their view, subjective, qualitative properties of mental states cannot be captured by objective properties that are observable from the third person perspective. Two organisms that are identical on the molecular level may have completely different qualitative mental properties. Consequently, a reductive explanation would be impossible, in principle. While some philosophers conclude that some sort of dualism must be true, others hold that the whole question is misguided [19] or that additional knowledge concerning mental states might eventually improve our abilities to operationalize even qualitative mental properties [20] (PS Churchland 1996). If these latter authors are right, scientific progress might eventually lead to a solution of the explanatory gap problem.

Mental Representation

A quite similar puzzle pertains to the problem of mental representation. Basically, a theory of mental representation has to explain how individual states of a cognitive system can refer to external entities, how they can be "about" something. For a physicalist approach, the theory as to make it intelligible how this is possible in a physical system. According to the "computational theory of mind" which was popular in the 1970's and 1980's, the problem can be solved on the basis of an analogy between the brain and a traditional Von Neumann Computer. Following Fodor [21], mental representations are discrete symbols of an innate "►language of thought." Cognitive operations are formal manipulation-processes of these symbols according to certain syntactic rules (the "program"), and because the semantic content of mental representation is mirrored by their syntactic properties of the symbols, the manipulation processes are sensitive to semantic content.

Many philosophers have criticized that it is implausible to assume the existence of an innate and therefore unchanging language of thought; in addition, it has been doubted whether Fodor can explain in a naturalistically acceptable way how the states of a physical system get their meaning (►naturalization of intentionality). Moreover, it seems that there are ►mental images or "iconic" representations [22] that cannot be accounted for in this theory, in principle, and scientists have pointed out that the architecture of the brain differs in several crucial respects from formal symbol manipulation devices like traditional computers [23].

Connectionism is an alternative that tries to account for these observations. According to this approach, the brain doesn't work like a traditional computer but, rather, it is a layered neural network, and mental representations are not discrete symbols but rather neural activities that are distributed over different areas of the brain [24]. Although is has been questioned whether this theory can account for higher level cognitive processes like logical reasoning, many philosophers have come to believe that it provides a much better basis for an account of mental representation, given that this theory is not amenable to the objections against the computational theory of mind and it better accounts for the neuroscientific facts.

The ►mental model theory is another alternative [25]. While the symbols of a language of thought are only arbitrarily related to their objects, mental models preserve the relations between the objects they represent: If there is an order between certain parameters of an object in the outside world (e.g. hue, temperature, size), this order should be preserved by the model of the object in question. In addition, representations from different aspects of an object or a scene (e.g. size, weight, sound, smell) can be combined in the respective model.

Self

Philosophers have also addressed several more detailed questions, among them the problem of the self or the problem of free will. Again, such considerations are not intended to replace empirical research; rather they try to specify the criteria that a person has to meet in order to count as self-conscious or free. While these criteria have to capture our pre-scientific intuitions they should also do justice to new scientific findings that may prompt for a revision of our concepts, just like science has prompted us to revise the concept of an atom.

It may be tempting to conceive of the self as a monolithic entity, a kind of "central-observer" that is realized by a single neural process. Since there are good reasons to believe that no such monolithic entity exists, neither on the neural nor on the psychological level, one might conclude that the self is only a fiction [3].

Many philosophers, however, maintain the self cannot be conceived of as a monolithic entity. As David Hume has noted already in the eighteenth century, we don't experience a "self" when we direct our attention inwards. This need not lead to skepticism concerning the self but it should be considered when the conceptual criteria for the ascription of a self or of self-consciousness are discussed. It seems evident that, in order to count as self-conscious, a person needs access, at least in principle, not only to her *present* feelings, experiences, and beliefs, but also to the related states in the past. The second, almost uncontroversial requirement is that a self-conscious person has to recognize these feelings, experiences, and beliefs *as her own* feelings, experiences, and beliefs. And third, she needs some kind of *self-concept,* that is, a more or less stable and coherent idea of those features and abilities that are characteristic for herself.

In addition, it would seem that being a self requires also being an agent [26]. Other issues are more controversial, e.g. whether it is really possible to understand how self-consciousness emerges, how the term "I" refers, and whether first-person access to our own mental states is privileged or even immune to error. Again, empirical findings may call for a more or less fundamental revision of our pre-scientific concepts.

In any case, even if we could determine the criteria for self-consciousness, it would be still another, empirical question, whether or not an actual person, or maybe human beings in general, meet these criteria. This raises questions concerning our autobiographical knowledge, that is, concerning our ability to remember events and experiences in our own past, and to integrate these memories as well as our present experiences into an adequate and coherent picture of our self or into a "self-model" [25,27]. Another empirical issue concerns the cognitive mechanism that is required for recognizing one's own experiences *as* one's own experiences. One promising candidate is the ability for perspective-taking as it has been identified in theory-of-mind research (►theory-theory). It seems obvious that a confirmation of this hypothesis would alter our concept of self-consciousness, demonstrating how empirical findings might lead to conceptual revisions.

Free Will

One of the most substantial implications of our pre-scientific self-understanding is that we can be held responsible for what we do, at least in principle. Provided that responsibility requires freedom and freedom, in turn, requires the ability to do otherwise, it would seem obvious that there is no free will and thus no responsibility in a determined world. In such a world, everything including our ►actions is determined by natural laws, so only those events *could* have happened that actually *did* happen, and only those actions *could* be performed that actually *were* performed. Conversely, freedom and responsibility would require the absence of determination.

Again, all these statements are based on certain standards for free action, and these standards need justification. Probably the most important question is whether or not freedom and determination are *compatible.* ►Compatibilists think they are, incompatibilists think they are not. Incompatibilists typically argue that freedom requires the ability to do otherwise under identical conditions which is obviously impossible in a determined world. Compatibilists, by contrast, may argue that freedom requires the ability to act according to one's will, and this seems possible even in a determined world. Unfortunately, there are many cases in which we would not say that a person who acts according to her will is free because the person's will lacks freedom. An addict may have the wish to take his drug and act accordingly, but we wouldn't say that he is free because his will is determined by his addiction.

Compatibilists have tried to respond to such objections. According to Harry Frankfurt [28], freedom requires not only that one performs the action that one wishes to perform, but also that one has the wish one really wants to have. Thus, an addict who has the desire to take his drug but despises this desire because he wants to get rid of this addiction, would *not* count as free; conversely, a person who does what she wants and approves her own wishes would be free. Compatibilists have also argued that getting rid of determination does not enhance freedom because an action that is not determined, cannot be determined by the agent either, it's just a random event.

Some philosophers even maintain that the criteria for freedom are such that it is impossible in principle that human actions are free. But even if one does not accept this view, skepticism concerning free will seems to be supported by empirical results from e.g. [29]. According to these experiments, conscious will is controlled by subconscious brain processes that are beyond a person's control. However, these experiments are still subject to a fierce debate. Apart from problems concerning the timing of the relevant mental and neural processes one might doubt that Libet has really investigated genuine decisions given that his subjects had no choice between different alternatives.

In any case, it is possible that future empirical results will show that human agents lack the abilities that are required for free actions. Philosophers have already discussed possible consequences, e.g. for our common-sense understanding of responsibility and for the justification of our legal system. Apart from that, one might suspect that neuroscientific findings may have an impact on the related concepts: If there is evidence that human actions in general fail to meet the traditional criteria for freedom, there may still be other interesting differences that call for a revision of our current concept of freedom.

Another important issue concerns the role of emotions in action and decision. Experiments and case studies by Antonio Damasio [30] and his group show that emotions play an important role even in what seem to be purely rational decisions. This raises the question whether the traditional distinction between rational and irrational decisions has to be revised.

Neuroethics

It seems obvious that neuroscience may have severe ethical implications. As a consequence, neuroethics as a specialized discipline has emerged in recent years. One of the reasons for this development is that neuroscience raises a number of specific issues that require a distinctive expertise. By and large, these issues fall into two categories. On the one hand, a "neuroscience of ethics" addresses neuroscientific findings that are relevant for ethics. Those findings concern the neural basis of the human ability to act according to moral standards, but also pathological alterations of the relevant structures that might affect and even destroy these abilities. On the other hand, an "ethics of neuroscience" addresses ethical standards for neuroscientific research and its applications like neuroprosthetics or psychotropic drugs. One example for these drugs is Ritalin which has well established short-term benefits in the case of ADHD (attention deficit/hyperactivity disorder). On the other hand, Ritalin seems to have long-term effects on the brain-structure, and is subject to abuse by healthy persons who seek to enhance their cognitive performance [31]. A clear distinction between treatment and enhancement might be helpful but has yet to be established; in addition, one might ask whether improving cognitive performance by a drug rather than by education or psychotherapy does not alter the way we look at humans in a fundamental way.

Neuroprosthetics, i.e. electronic devices like cochlea implants or electrodes for deep brain stimulation in the case of Parkinson's disease, raise ethical questions because they may have psychological side effects that involve the patient's personality. Unlike conventional prosthetics, neuroprosthetics may be regarded as a part of the person's self, thus changing the very person whose prosthetics they are.

It would seem, then, that there are quite some fields in which neurophilosophy can help to promote scientific progress and to assess the consequences of this development. This requires close cooperation between neuroscience and philosophy but it requires also a clear distinction between those issues that can be treated empirically and those that require genuine philosophical research. Traditional armchair reasoning, by contrast, doesn't appear as a very promising strategy if philosophy wants to play its role in the development of the neuro- and cognitive sciences.

Caveat!

References

1. Churchland PS (1986) Neurophilosophy. Cambridge MA
2. Bickle J, Mandik P (2002) The philosophy of neuroscience. The Stanford Encyclopedia of Philosophy (Winter 2002 edn) URL = <http://plato.stanford.edu/archives/win2002/entries/neuroscience/>
3. Dennett DC (1991) Consciousness explained. Boston
4. Jackson F (1982) Epiphenomenal qualia. Philos Q 32:127–136
5. Robinson WS (1982) Causation, sensations and knowledge. Mind 41:524–540
6. Chalmers DJ (1996) The conscious mind. Oxford, New York
7. Ryle G (1949) The concept of mind. London
8. Pauen M (2000) Painless pain. Am Philos Q 37:51–64
9. Rorty R (1970/71) In defense of eliminative materialism. Rev Metaphys 24:112–121
10. Churchland PM (1979) Scientific realism and the plasticity of mind. Cambridge
11. Feigl H (1958) The 'Mental' and the 'Physical'. Minnesota Studies Philos Sci 2:370–497
12. Place UT (1956) Is consciousness a brain process? Br J Psychol 47:44–50
13. Pauen M (2003) Is type identity incompatible with multiple realization? Grazer Philosophische Studien 65
14. Davidson D (1970) Mental events. In: Davidson D (ed) Essays on actions and events. Oxford, pp 207–224
15. Lewis D (1980) Psychophysical and theoretical identifications. In: Block N (ed) Readings in philosophy of psychology, vol. I. London, pp 207–215
16. Block N (1980) Troubles with functionalism. In: Block N (ed) Readings in philosophy of psychology, vol. I. London, pp 268–305
17. Nagel T (1974) What Is It Like to Be a Bat? Philos Rev 83:435–450
18. Levine J (1983) Materialism and qualia. Pacific Philos Q 64:354–361
19. Dennett DC (1993) Quining qualia. In: Goldman AI (ed) Readings in philosophy and cognitive science. Cambridge MA, pp 381–414
20. Hardin CL (1990) Color and illusion. In: Lycan WG (ed) Mind and cognition. Oxford, pp 555–566
21. Fodor JA (1975) The language of thought. New York
22. Tye M (1991) The imagery debate. Cambridge MA
23. Edelman GM (1992) Bright air, Brilliant fire. New York
24. Rumelhart DE, McClelland JL (1986) Parallel distributed processing. Cambridge MA
25. Johnson-Laird, PN (1983) Mental models. Cambridge MA
26. Strawson G (1997) The self. JCS 4:405–428
27. Metzinger T (2003) Being no one. Cambridge MA
28. Frankfurt HG (1971) Freedom of the will and the concept of a person. J Philos 68:5–20
29. Libet B (1985) Unconscious cerebral initiative and the role of conscious will in voluntary action. Behav Brain Sci 8:529–539
30. Damasio AR (1994) Descartes' error. New York

31. Butcher J (2003) Cognitive enhancement raises ethical concerns. Lancet 362:132–133
32. Popper KR, Eccles JC (1977) The self and its brain. New York
33. Strawson G (1989) Consciousness, free will, and the unimportance of determinism. Inquiry 32:3–27
34. Thomas H. Huxley (1904) On the Hypothesis that Animals are Automata. In: ders., Collected Essays Bd. I. London: Macmillan, p. 199–250
35. Benjamin Libet (1994) A testable field theory of mind-brain interaction. Journal of consciousness studies 1:119–126
36. Churchland, Patricia smith (1996) The Hornswoggle Problem. Journal of Consciousness Studies 3:S402–405

Neurophysins

Definition

Part of the precursors of the neurohormones vasopressin and oxytocin that are generated in the hypothalamus and released from the posterior pituitary.

►Neuroendocrinology of Psychiatric Disorders

Neurophysiology of Sexual Spinal Reflexes

Stacy L. Elliott
Departments of Psychiatry and Urological Sciences, University of British Columbia, Vancouver, BC, Canada
International Collaboration on Repair Discoveries, University of British Columbia, Vancouver, BC, Canada
BC Center for Sexual Medicine, Vancouver Hospital, Vancouver, BC, Canada

Synonyms

Sexual neurophysiology

Definition

The ►sexual spinal reflexes are spinal cord reflexes (consisting of afferent and efferent components) which instigate the genital vasocongestion and neuromuscular tension responsible for sexual arousal (erection in men and vaginal lubrication and elongation in women), the triggering of ejaculation in men, and possibly orgasm in both sexes.

Characteristics

Sexual Arousal in Men and Women

Both men and women have erectile tissue in their genitalia. Nerve mediated vasocongestion to the pelvis allows for genital tissue engorgement. In men, penile erection occurs when the sinusoidal spaces (the trabeculae) of the corpora cavernosal bodies fill with blood, expanding them to their anatomical limit set by a surrounding, stocking-like, elastic tunica albuginea. Arterial penile filling, or "tumescence," is further assisted by the cessation of venous outflow from compression of the emissary veins which pierce the tunica. This veno-occlusive mechanism, in conjunction with pelvic floor musculature contraction, increases the intracavernosal pressure making the penis rigid. In women, a similar tumescence of the cavernosal bodies of the clitoris occurs with sexual stimulation along with extensive pelvic vasocongestion (uterine, vaginal, urethral, labial and pelvic ligaments), resulting in swelling of the external genitalia, lengthening of the vagina and the production of vaginal lubrication (a transudate extruded through the vaginal epithelial cells). Physiologically, arousal is primarily a parasympathetic event. The sympathetic nervous system dominates during male ejaculation and probably also at orgasm.

Ejaculation in Men

Ejaculation is the process of delivery of semen through the penis to the distal urethral opening. ►Seminal emission, the first phase of ejaculation, involves transport of spermatozoa into the prostatic urethra via the ejaculatory ducts in the prostate. The accumulation of fluid immediately proceeds propulsatile ejaculation and contributes to the sensation of *ejaculatory inevitability*. Expulsion, or propulsatile ejaculation, the second phase of ejaculation, propels the seminal bolus distally out the urethral meatus by rhythmic contractions of the bulbocavernosus and ischiocavernosus muscles. Emission is under voluntary control, whereas expulsion is not [1].

Orgasm in Men and Women

Orgasm is the pleasant physical and mental sensation of sexual climax. While it most often occurs in men with ejaculation, it is not synonymous with it. Orgasm can occur in men and women during sleep, during hypnotic suggestion or even after loss of external genitalia or CNS injury. However, it can still be elusive to some persons with normal physiology despite their best efforts (orgasmic dysfunction). In men, orgasm has been variously described as the result of cerebral processing of ►pudendal nerve sensory stimuli from increased pressure in the posterior urethra and prostate, and from sensory stimuli arising from the veramontanum and contraction of the urethral bulb and accessory sexual organs [1]. Women's orgasm can be described as a pleasurable peak sensation accompanied by involuntary rhythmic contraction of the pelvic floor, often with concomitant uterine and anal contractions. Most men have post-ejaculatory refractory periods, but women do not [2], allowing for the potential of serial orgasms.

Characteristics

Experimental evidence demonstrates that the spinal cord contains all the neural circuitry involved in the generation of genital arousal [3]. However, sexual functioning is extremely complicated beyond the spinal cord: all body senses, emotions and social awareness will determine whether an individual person will orient towards sexual activity (the situation is safe, erotic and likely to be sexually rewarding) or lose interest (the situation is unrewarding, unsafe or nonsexual). The brain is the ultimate controller. For example, in healthy men, heightened sexual arousal sends descending signals to the spinal cord allowing for the natural unfolding of the sexual spinal reflexes, with the resultant erection acting as a positive reinforcement. This "supraspinal control" is primarily inhibitory in nature, and needs to be "removed" so excitatory signals can pass through to the spinal cord. In male rats, the anatomical site for the descending inhibitory action has been identified in the rostral pole of the paragigantocellular nucleus bilaterally located in the oblongata [4]. A human infant will have rapid reflex erection or lubrication to touch of the genitalia: when myelination of the spinal nerve tracts is complete the brain can then impose inhibition or facilitation of this reflex at a cortical level. Only men and women with complete spinal cord injury (SCI), i.e. whose cortical control is interrupted by the spinal cord injury itself, can provide data for the practical understanding of the neurology of the spinal reflexes. However, their reaction to their sexual experience will ultimately affect their sexual neurophysiology and function. In humans, recognizing this complex, moment-to-moment mind-body interaction linked to reward and expectation is what distinguishes the sexual reflexes from other more automatic physiological reflexes that are also under excitatory and inhibitory control from the brain.

Autonomic spinal nuclei are activated by excitatory descending projections from the brain. While the traditional concept of sexual responses follows a "hard wired" model with neural connections extending from the brain to the peripheral nerve receptors at the sexual end organs, there are neuroendocrine and other biological factors, as well as other non-CNS pathways that are possibly significant. For example, there is growing evidence of a nociceptive vagal-solitary tract pathway from the vaginocervix region in women with complete spinal cord injuries (so far not identified in men with SCI), consistent with pathways already described in basic rat studies and confirmed by fMRI in the human female [5].

Organization of the Sexual Reflexes

Central Control

Brain initiated excitatory descending signals are the result of positive "interpretation" of sexual imagery and fantasy, of visual, auditory and olfactory inputs (including the little known role of pheromones in humans), and of cerebral evaluation of ascending signals from somatic (usually tactile) stimulation from the body (particularly the genitalia, nipples, etc). The control of genital arousal is located in the limbic system (linked to processes of motivation and reward), and the hypothalamus (the coordinating center for complex autonomic responses), and other midbrain structures [3]. This has been supported by various functional imaging modalities done in humans undergoing sexual arousal including positron-emission tomography (PET) and functional magnetic resonance (fMRI) showing activation of the higher cortex, the limbic and paralimbic cortex and other subcortical regions: these studies show that sexual responses require brain areas that integrate cognitive, motivational and autonomic components [6]. Seminal emission and ejaculation are controlled by the paraventricular nucleus (PVN) of the anterior hypothalamus and the medical preoptic area (MPOA).

Ascending and Descending Pathways

Both the lateral spinothalamic pathway (which terminates in the thalamus) and spinoreticular pathway (which crosses the cord to the opposite side and travels in the lateral spinal columns, terminating in the reticular formation) relay sexual sensory information to the brain. Descending signals from the brain travel in the dorsal and dorsolateral white matter and enter into the spinal gray matter. The spinal interneurons connect the afferent and efferent pathways and coordinate various components of the sexual response [2].

Spinal Cord and Peripheral Innervation

The spinal cord is the integration site for afferent signals from the periphery and descending modulation (excitatory and inhibitory) of interneurons from supraspinal areas. The spinal cord contains neurons projecting to every anatomic genital element participating in the sexual response [2]. S*pinal interneurons* located in and around the intermediolateral cell column and in the medial gray form a column of neurons through segments T12–S1 [2]: these interneurons relay afferent input from the genitalia to efferent somatic and autonomic spinal neurons en route to structures involved in the sexual response. They receive projections from the supraspinal structures and are involved in modulation/coordination of the sexual response.

At the effector organ, locally released nitric oxide (NO) is the primary molecule responsible for genital arousal. Other facilitory neurotransmitters such as vasoactive intestinal polypeptide (VIP) supplying the arterioles of the corpora results in smooth muscle relaxation and erection in males and tumescence of clitoral tissue in women, whereas noradrenaline and neuropeptide Y (NPY) are the primary inhibitors of genital arousal response. Acetylcholine (Ach) likely acts synergistically with other vasodilators released by nerves or contained within vascular structures in both men and women: sympathetic pathways may produce erection via a cholinergic mechanism [2].

Spinal Cord Reflex Pathways

Genital Arousal

Recognition of genital arousal either externally or internally is a combination of both somatic cutaneous sensitivity (pressure, touch, temperature, vibration and pain) and visceral sensitivity of the internal organs (movement of the uterus and ligaments during intercourse, bladder pressure, etc) [2]. Afferent messages travel to the sacral 2, 3, 4 segments along various pathways, depending on the anatomical structure, but in general, touch and temperature signals from the clitoris and glans penis travel along the dorsal nerve which merges into the *pudendal* nerve (S1–S3), touch and vibration (especially from the vagina and cervix) travel via the ►*pelvic nerve* (S1–S3) and potentially via the ►vagus nerve in women (which may be related to vaginocervical stimulation only), whereas deep pressure and visceroreceptive stimuli likely travel along the *pelvic* (S1–S3) and ►*hypogastric* (►T12–L1) *nerves* [2]. These afferent signals reach the brain through the ascending pathways.

Descending from the brain, two areas in the spinal cord provide the main transmission of efferent signals, (i) T10–12 and L1–3 segments (sympathetic and other fibers) known to be responsible for relaying the messages responsible for psychogenic or mental arousal respectively, and (ii) sacral 2, 3, 4 (parasympathetic and motor fibers) responsible for reflexogenic arousal (genital vasocongestion and muscular pelvic floor activation).

Clinical Correlate: Spinal Cord Injury

Depending on the level and completeness of injury, isolated SCI can disrupt either psychogenic or reflex genital arousal. Reflex arousal results from stimulation of the dorsal nerve, which propagates signals to the sacral spinal cord, synapsing with the parasympathetic efferent neurons whose axons travel back to the corpora cavernosa and pelvic viscera. If the SCI is above this sacral level, the reflex erection is rarely disrupted. Due to loss of tonic inhibitory control that reduces the sensory threshold and onset of erectile responses [7], men with SCI at the cervical level will have preserved, if not enhanced, reflex erection, especially if the lesion is complete [8]. The ability to have reflex erections or vaginal lubrication is lost if the sacral spinal cord is injured or if the pudendal nerve or pelvic nerve is destroyed [8]. The pathways of the sympathetic nervous system can compensate for parasympathetic deficits and preserve erectile function. However, the intense mental concentration needed to maintain psychogenic erection via the sympathetic chain can also provoke seminal emission, leading to unwanted detumescence.

Ejaculatory Reflexes

Afferent sensory information from the genitalia (primarily the glans penis) travels in the pudendal nerve (within the dorsal nerve of the penis) to the S4 level of the spinal cord. Afferent autonomic fibers within the hypogastric plexus transmit information from the sympathetic ganglia located alongside the spinal cord [1]. The efferent component involves sequential contraction of internal accessory sexual organs, which can be associated with pleasurable sensations. Spinal cord segments extending T10–S4 are involved in the ejaculatory process.

Seminal emission, consisting of sperm transport and seminal fluid formation, is under sympathetic control (T10–12), and closure of the bladder neck (internal urinary sphincter) to prevent retrograde ejaculation is controlled via the sympathetic fibers emerging at L1, 2. The actual process of expulsion (propulsatile ejaculation) of seminal fluid occurs through the intermittent relaxation of the external urinary sphincter. Spasmodic contractions of the seminal vesicles, prostate and urethra (parasympathetic fibers of S2, 3, 4) and rhythmic contractions of the bulbocavernosus, ischiocavernosus, levator ani and related muscles (somatic/motor signals of S2, 3, 4 via the pudendal nerve), propel the seminal bolus distally.

Two other local ejaculatory reflexes have been identified: stimulation of the glans penis brings semen to the posterior urethra (glans–vasal reflex) and a urethromuscular reflex is responsible for propulsion of the semen to the exterior (urethral meatus) [9].

Clinical Correlate: Spinal Cord Injury

It is possible to trigger the ►ejaculation reflex through intense penile stimulation with high intensity vibrators (*vibrostimulation*) in about 60–90% of men with SCI lesions above T10 neurological level. No longer under cortical control, the T10–L2 levels can be triggered as a coordinated reflex as long as the afferent and/or efferent signals are intact, or vibrostimulation can activate enough remaining afferent fibers to complete the reflex. If the spinal cord lesion is below T10 with resultant interference of the lumbosacral reflex, then *electroejaculation*, which electrically stimulates (via the periprostatic nerves) the efferent pathways of seminal emission, will invariably work.

Orgasm

The neurology of orgasm is not fully understood, but likely entails somatic and autonomic components [7].

Orgasm, or components of orgasm, while most often generated from external genital stimulation, are not wholly dependant on such. Orgasm triggered by non-genital stimulation (erogenous zones), or from the brain alone (as in fantasy or during sleep) may be a form of autonomic excitation and release. Furthermore, in spinal cord injured women, the vagus nerve has been identified as a non-spinal cord pathway for orgasm [5]. However, orgasm is primarily triggered by stimulation of the clitoris (and/or anterior vaginal wall) or glans penis and appears to have a reflex component that can be reinforced with practice [2,10]. An experimental model, developed in anesthetized spinalized female rats, mimics the human orgasmic response: assuming the removal of descending inhibitory inputs, the *urogenital reflex* consists of stimulation of the urethra resulting in rhythmic contractions of the vagina, uterus and anal sphincter [7].

Regarding the subjective pleasure component of orgasm (especially in women), there are two arguments: one suggests orgasm is a simple reflex stimulus-response reaction, and the other that sensations identified as orgasm are closely attached to contextual meaning at the time [7]. Confusion between orgasmic etiologies makes the reflex component of orgasm difficult to define. The distinction between genital and non-genital orgasm may help resolve these differences. For example, those men and women with SCI who have intact pain and temperature sensations from the genitalia through the lateral spinothalamic tract, and intact descending corticospinal tracts as demonstrated by voluntary anal contraction, are potentially able to experience *genital* orgasm, whereas loss of these specific ascending and descending reflex pathways do not rule out the potential for *non-genitally* induced (i.e. cerebral or other non-genital erotic zone generated) orgasm [10]. After SCI, orgasm is less likely to be experienced if there is complete disruption of the sacral reflex arc (such as occurs with conus medularis injuries), or if T10–L1 sensation is not intact [2]. Likewise, orgasm can occur despite damage to the sympathetic ganglia, but it is rarely possible after injury to the pudendal nerve.

References

1. McMahon CG (2004) The ejaculatory response. In: Seftel AD (ed) Male and female sexual dysfunction. Mosby, London, pp 43–57
2. Giuliano F, Julia-Guilloteau V (2006) Neurophysiology of female genital response. In: Goldstein I, Meston CM, Davis SR, Traish AM (eds) Women's sexual function and dysfunction. Taylor & Francis, New York, pp 168–173
3. Allard J, Giuliano F (2004) Central neurophysiology of penile erection. In: Seftel AD (ed) Male and female sexual dysfunction. Mosby, London, pp 3–18
4. Marson L, McKenna KE (1990) The identification of a brainstem site controlling the spinal sexual reflexes in male rats. Brain Res 515:303–308
5. Komisarik BR, Whipple B, Crawford A, Liu WC, Kalnin A, Mosier K (2004) Brain activation during vaginocervical self-stimulation and orgasm in women with complete spinal cord injury: fMRI evidence of mediation by the vagus nerves. Brain Res 1024 (1–2):77–88
6. Rees PM, Fowler CJ, Maas CP (2007) Sexual Function in men and women with neurological disorders. Lancet 369:512–525
7. McKenna KE, Chung SK, McVary KT (1991) A model for the study of sexual function in anesthetised male and female rats. Am J Physiol 261 (5 Pt 2):R1276–R1285
8. Chang AT, Steers WD (1999) Neurophysiology of penile erection. In: Carson C, Kirby R, Goldstein I (eds) Textbook of erection dysfunction. ISIS Medical Media, Oxford, UK, pp 59–72
9. Shafik A (1998) The mechanism of ejaculation: the glans–vasal and urethromuscular reflexes. Arch Androl 41(2):71–78
10. Elliott S (2002) Ejaculation and orgasm: sexuality in men with SCI. Top Spinal Cord Inj Rehabil 8(1):1–15

Neuropile

Definition

The integration center within the central nervous system (CNS) surrounding the neuronal cell bodies and composed of a network of axonal and dendritic fibers interconnected by synapses.

Neuroprostheses

►Computer-Neural Hybrids

Neuroprotection

Definition

Any therapeutic intervention that prevents or slows down secondary neurodegeneration. After an acute central nervous system (CNS) injury, and in most chronic neurodegenerative conditions, three types of neurons operate: those damaged by the primary insult that will inevitably die, healthy neurons, and marginally damaged neurons. The last two types, unless protected, are susceptible to secondary degeneration and death. "Neuroprotection" also refers to therapeutic interventions that reduce the rate of secondary degeneration.

►Autoimmune Demyelinating Disorders: Stem Cell Therapy
►Central Nervous System Disease – Natural Neuroprotective Agents as Therapeutics

Neuroprotective Agents

Definition

Agents that protect neurons against various injuries such as excitotoxicity, ischemia, and hypoxia.

Neuropsychiatry – Historical Development and Current Concepts

GEORG NORTHOFF
Department of Psychiatry, University of Magdeburg, Leipsiger Sharse 44, 39120 Magdeburg, Germany,

Introduction: The Concept of "Neuropsychiatry"

A clinician taking care of a patient suffering from Parkinson's disease does not have to treat motor symptoms alone (►tremor, ►akinesia, ►rigor), which belong to the field of neurology, but has to diagnose and treat effective (e.g., depressions) or cognitive symptoms (e.g., ►bradyphrenia, retardation of thinking), which belong to the field of psychiatry. However, it is quite unlikely that he will meet all requirements of both fields because he is either a neurologist or a psychiatrist, and he therefore has learned different ways of thinking and different methods. This reflects the paradox situation in contemporary relationship between neurology and psychiatry: On the one hand neurology and psychiatry are regarded as two different disciplines with different content concerning methodology, diseases, diagnostics, and therapy. On the other hand the boundaries of separation between both disciplines melt away by applying neurological methodology and diagnostics in psychiatric diseases (►CT, ►Spect, ►PET) as well as by the interest of neurology in complex mental functions and psychiatric symptoms in neurological diseases. Particularly in the Anglo-American region this resulted in the foundation of the discipline of "neuropsychiatry": "Although half a century ago neurology and psychiatry seemed to be diverging from a common purpose and sanding as two stools apart, these recent advances have not only seen the stools bridged by a plank, but the whole structure has come gradually to resemble a bench, which for some seems quite comfortable. The last decade has seen not only an exponential growth of knowledge in the field of the neurosciences, but also a resolution of some interdisciplinary rivalry, and laid the foundations of neuropsychiatry for at least the rest of this century [1]." Neuropsychiatry in this sense can best be understood by considering its historical development, which is discussed briefly in the following. One crucial feature of Neuropsychiatry in this sense is that it considers subjective experience of mental states in First-Person Perspective, as distinguished from mere scientific observation of neuronal states in Third-Person Perspective, as crucial.

Neurology, Psychiatry, and Neuropsychiatry

Historic Development of the Separation of Neurology and Psychiatry

Neurology as an independent discipline has developed at the beginning of the nineteenth century when Parkinson's disease and ►Multiple sclerosis were defined as neural diseases [2] (see also [3]). Following advances in anatomy and pathology in the nineteenth century correlations between the clinical appearance and the pathological–anatomical substrate became possible to an increasing degree; this has been successful in numerous neurological diseases and symptoms (e.g. ►aphasia) and consequently established the field of neurology as an independent discipline of medicine. On the contrary clinical–pathological correlations in psychiatry did not show significant success, which strengthened the separation of medicine/neurology on the one hand and psychiatry on the other. The previously apparent connection between psychiatry and philosophy/humanities [2], the letter dealing with mental states as implicated in psychiatric diseases as mental disorders, additionally supported the separation between neurology and psychiatry.

The application of scientific principles in psychiatry has later been established by Griesinger, Kahlbaum, Kraepelin, and Maudsely who created a consistent nosology and regarded psychiatric diseases as diseases of the brain. Nevertheless correlations between the clinical appearance and the pathological–anatomical substrate still remained unsuccessful. Although in the field of neurology such correlations showed rising success since neurological diseases could be localized anatomically and morphologically in the brain implying that they were regarded as "structural" diseases. The failure of structural localization of psychiatric diseases, in contrast, resulted in their acceptance as "functional" diseases with "functional" meaning "psychological" and thus mental origin as distinguished from "structural" describing the "organic" and thus neuronal origin [2]. This opposition of "structure versus function" and "localization versus nonlocalization" accounted for the separation of neurology and psychiatry in a decisive manner; the neurologist dealt with structural diseases of the brain, the psychiatrist focused on functional diseases of the mind. This leads to the development of neurology and psychiatry as independent disciplines in the Anglo-American region whereas in the German-speaking regions both disciplines have been kept together for a long time in the common discipline of "Nervenheilkunde."

Changes in the Relationship between Neurology and Psychiatry

Different developments in the last decades question the comparison of neurology and psychiatry. These developments mounted from both the field of neurology and psychiatry, which is described briefly in the following. By the discovery of variability of the brain's neuronal structures, i.e., the plasticity, and the introduction of new imaging techniques (►MRI, PET; see below) neurology changed its appearance: Solely static-structural observations have been replaced by a dynamic-functional anatomy [3]. This "functional neuroanatomy" not only examines different static structures but tries to point out the connections between these structures, to show plasticity of connections and structures as well as the influence of function on structure [1]. The boundaries between anatomy and physiology melt away because of this "functional neuroanatomy," where the contrast of anatomical-static structure and physiological-dynamic function is dismantled and rather regarded as a relationship of mutual complementation: "The boundaries between anatomy and physiology, between form and function, break down at an ultrastructural level. Anatomy is not static. Pharmacotherapy may alter structure as well as function." [3]. The new imaging techniques on the one hand allow to image anatomic structures more exactly and in detail. On the other hand they make it possible to draw a connection between functional changes and anatomic structure – they are a kind of "window for the brain function" [4]. These developments make it possible to look at psychiatric diseases too. The physiologic-functional examination, which is still getting off the ground, could shed new light on the pathophysiology of psychiatric diseases.

The development of psychopharmacology supported the hypothesis already set by Griesinger that mental diseases are diseases of the brain. The question of the mechanisms of the pharmacological effects in the brain by psychopharmacological drugs resulted in an intensive exploration of ►neurotransmitters, ►synapses, and receptors in the brain, which resulted in a better and extended understanding of physiological brain functions. Out of this the discipline of "biological psychiatry" emerged, which tries to correlate psychopathological phenomena with functional and structural changes in the brain [5]. Initially "Biological psychiatry" mainly dealt with synapses, neurotransmitters, and receptors; by examination of the mechanisms of effectiveness one tries to gain insight into the structural and functional events in the brain in psychiatric symptoms and diseases.

Not only in psychiatry changes have happened, but also in neurology. In neurology a rising interest in complex phenomena and behavior, which could not clearly be traced back to reflexes, developed [6]. Here the insufficiency of classic neurology became clear [7] – because of this, mainly in America, the discipline of "behavioral neurology" developed. It deals with complex phenomena like aphasia and ►amnesia and tries to localize these phenomena neurologically and neuroanatomically, respectively [8]. The old method of clinical–pathological correlations (see above) is applied here in a new field of interest, the field of higher-order cognitive phenomena and complex behavior – in doing so the focus is still on structure and the aim of localization [3].

The discipline of "neuropsychology," constantly developing in the last years, too, examines the "correlations between brain function and psychological processes" [9] closely following classical neuropathology [9,10]. The separation between neuropsychology on the one hand and "behavioral neurology" on the other hand is mainly pursued in the USA. In "behavioral neurology" mainly aphasia and amnesia are investigated. The main topic of neuropsychology is to provide objective methods for the examination of psychological function, which then can be applied on the clinical problems as dealt with in "behavioral neurology" and set into relationship to brain structure and function [11]. The "behavioral neurologist" primarily looks at the structures of the brain and secondarily their relation to complex psychological functions. The neuropsychologist, on the contrary, regards primarily the psychological functions and secondarily their relation to the structures of the brain [8]. The neuropsychologist captures affective and cognitive alterations in the Parkinson's disease using standardized tests in an objective manner – the correlation of these results with structures and functions of the brain is left to the neurologist to a large extent. The quantitative, operationalized measurement of psychological functions, the so-called psychometrics [10], is increasingly used in psychiatry too, where psychopathological phenomena are captured quantitatively and objectively by operationalization of the psychopathological symptoms. Thus the psychopathology which previously has often been called nonscientific becomes affiliated with an "empiric-scientific methodology" [12] thereby gaining scientific status.

The above described developments of different disciplines in the border area between neurology and psychiatry aim to bridge the contrasts between both fields from different directions (biological *psychiatry*, behavioral *neurology*, neuro*psychology*) [3]. All of these three disciplines would explain the above described example of Parkinson's disease differently: The biological psychiatrist localizes the affective and cognitive symptoms in the microstructure of the transmitters, synapses, and receptors; the "behavioral neurologist" localizes the same symptoms in the macrostructures of the brain; the neuropsychologist objectivizes and standardizes psychological functions. All of them place motor/neurological and psychiatric symptoms next to each other and explain them more or less independently from each other – the internal connection of motor and psychological alterations in the Parkinson's disease as experienced by the patient gets lost. Because of this in the following we want to show that neuropsychiatry could be able to demonstrate these internal connections between psyche and motor activity.

Neuropsychiatry: Characterization and Definition

The discipline of neuropsychiatry tries to bridge the gap between neurology on the one hand and psychiatry on the other – in doing so, psychological functions and neurological structures are not only to be observed at the same time but are to be connected internally coherently: "This new orientation of which Jellife spoke, and of which he himself was a notable exemplar, did not involve merely combining neurological and psychiatric knowledge (as every neurologist and psychiatrist does to some extent), but conjoining them seeing them as inseparable, seeing how psychiatric phenomena might emerge from the physiological, or how, conversely, they might be transformed into it – ..." [3]. The static-structural, strictly localizing observation of the classic neurology and the "behavioral neurology" will be contrasted by a dynamic-functional approach of the neuropsychiatry. Psychological and motor alterations in the Parkinson's disease are no longer explained separately and independently but are regarded as two expressions of a uniform dynamic-functional structure – the alteration of this structure, and not of the two different symptom complexes as two different phenomena, have to be explained. Thus the "neurologilization" of psychiatric functions [3] is impossible – behavior can not only be observed with neurological methodology, but has to be assessed by integration of neurological and psychiatric knowledge [3].

Biological psychiatry cannot be identified with and reduced to neuropsychology, because it does not exclusively deal with complex macro phenomena of behavior and psychological functions, but with micro phenomena of the synapses, transmitters, and receptors – psychological functions mainly remain beyond observation [13]. While biological psychiatry uses functional neuroanatomy, restricting it to microstructures (synapses), behavioral neurology focuses on macrostructures of the brain though considering them solely in a static, anatomic-structural sense. Neuropsychiatry should aim to combine the dynamic-functional approach with the observation of macrostructures in the phenomena of behavior and higher-order psychological functions. In other terms, neuropsychiatry in this sense would take a middle position between a static-structural, localizing neurology on the one hand and a dynamic-functional, holistic/anti-localizing psychiatry on the other. As a bridge between neuroanatomy and psychopathology, neuropsychiatry has to examine functional and dynamic processes being positioned in between strictly localizable neurological functions and strictly holistic psychological processes [14]. This middle level of neuropsychiatry undermines the traditional opposition of structure versus function and localization versus nonlocalization.

Function and Neuronal Integration in Neuropsychiatry

Function and Localization

As already mentioned the contrast of structure versus function and charactzerized the latter as crucial for the discipline of neuropsychiatry. What does the term "function" mean? "Functional" can be understood in two senses [15]: First, "functional" means just nonorganic and thus psychological, as it has been understood in psychiatry. Second, "functional" can be understood in a physiological sense in contrast to anatomic – here "functional" describes dynamic, plastic, and variable physiological processes, being in contrast to the static, anatomic structure; this second meaning of the word "functional" has been used in the characterization of functional neuroanatomy (see above). We want to follow the second and original meaning of the word "functional" [5] and regard the physiologic-functional description as middle level between static-structural pathology and dynamic-functional psychiatry as the specific neuropsychiatric level [14]. Physiologic functions may not correlate with a specific static anatomic structure any more, but they develop in so-called functional systems [7,16]. These systems produce distinct functions by a dynamic constellation of changes between different parts of the brain reflecting what may be called neuronal integration (see below). They are plastic, show a systemic (and not a concrete) structure, and operate by dynamic autoregulation [16]. The realization of function depends then on dynamic systems that include different brain regions, which Luria characterized as "functional systems":

"According to this view a function is, in fact, a functional system (...) directed towards the performance of a particular biological task and consisted of a group of interconnected acts that produce the corresponding biological effect. The most significant feature of a functional system is that, as a rule, it is based on a complex dynamic "constellation" of connections, situated at different levels of the nervous system, that in the performance of the adaptive task, may be changed with the task itself may be unchanged" [16].

The brain here is regarded as a network consisting of different, overlapping functional systems with an internal dynamic, a so-called "neurodynamic" [3], which has already been demonstrated in the form of "resonant oscillator circuits" for the brain [17]. For our example of Parkinson's disease this would mean that the functional interaction of the functional systems of motor action, affect/emotion and cognition and their "interfunctional relation" [18] are altered. The motor action cannot be observed separated from affect/emotion and cognition because of the mutually overlapping systems – between the different functional systems there are so-called "functional knots" [7], enabling the motor action to influence affect/emotion and cognition directly, and reverse.

What does this neuropsychiatric network with its interconnections between the different functional systems imply for the problem of localization versus nonlocalization and consecutively for the separation between neurology and psychiatry? Historically and currently, the discussion of mental processes is often characterized by the opposition of localizationists, who claim for exact localizability of mental processes in structures of the brain, and holists or aquipotentionalists, who believed that all structures of the brain are necessary for mental processes [19]. Both approaches can be considered "psycho morphological attempts" [7], which give priority to either the structural-functional differentiation/specialization of the brain (localizationists) or to the plasticity of the brain (holists) as visible in functional restitution following structural lesions. A dynamic-functional neuropsychiatry regards both positions as different aspects of the organization of the neuronal network without either aspect prevailing or dominating. Neuropsychiatry aims to combine both positions in a concept of "systemic-dynamic localization" [7]. In this concept, a function cannot be localized in a distinct anatomic structure but in a functional system with its functional interconnections. On the other hand no anatomic structure of the brain can be assigned to only one function, but it is always involved in different functional systems simultaneously or successively – Luria calls this "functional pluripotentialism" [7].

This "functional pluripotentialism" the functional overlaps the "functional knots," of the functional systems of affection, cognition, and motor action as well as their connection in the case of the Parkinson's disease. This can be localized neither in a distinct anatomic structure (localizationists) nor in the whole brain (holists) – a distinct kind of alteration of the functional interaction of the functional systems of motor action, affect/emotion, and cognition is represented by the dynamic-functional localization in the Parkinson's disease. This makes clear that traditional contrasts of structure versus function and localization versus nonlocalization, which resulted in the separation of neurology and psychiatry (see above), can no longer be maintained in the present form and could be bridged by a dynamic-functional neuropsychiatry. The middle level between structural localization and functional holism of mental states may be characterized dynamic-functional, which may be realized by what can be called neuronal integration.

Neuronal Integration

Neuronal integration describes the coordination and adjustment of neuronal activity across multiple brain regions. The interaction between distant and remote brain areas is considered necessary for complex functions to occur, such as emotion or cognition [20]. Neuronal integration focusing on the interaction between two or more brain regions must be distinguished from neuronal segregation [20]. Here a particular cognitive or emotional function or processing capacity is ascribed to neural activity in a single area that is both necessary and sufficient; one can subsequently speak of neuronal specialization and localization. We assume that higher psychological functions as complex emotional–cognitive interactions cannot be localized in specialized or segregated brain regions. Instead, we assume that higher psychological functions require interaction between different brain regions and thus neuronal integration.

For neuronal integration to be possible, distant and remote brain regions have to be linked together, which is provided by connectivity. Connectivity describes the relation between neural activity in different brain areas. There is anatomical connectivity for which we will use the term connections in order to clearly distinguish it from functional connectivity. In addition, Friston and Price [21] distinguish between functional and effective connectivity: Functional connectivity describes the "correlation between remote neurophysiological events," which might be due to either direct interaction between the events or other factors mediating both events. A correlation can either indicate a direct influence of one brain area on another or their indirect linkage via other factors. In the first case the correlation is due to the interaction itself whereas in the second the correlation might be due to other rather indirect factors like for example, stimuli based on common inputs. In contrast, effective connectivity describes the direct interaction between brain areas, it "refers explicitly to the (direct) influence that one neural system exerts over another, either at a synaptic or population level" [21]. Here, effective connectivity is considered on the population level because this corresponds best to the level of different brain regions investigated here. For example, the ▶prefrontal cortex might modulate its effective connectivity with subcortical regions thereby influencing specific functions like for example interoceptive processing. Based upon connectivity, neural activity between distant and remote brain regions has to be adjusted, coordinated, and harmonized. Coordination and adjustment of neural activity might not be arbitrarly but guided by certain principles of neuronal integration [22]. These principles describe functional mechanisms according to which the neural activity between remote and distant brain regions is organized and coordinated as for instance in top-down modulation (see also [23]).

Top-down Modulation

Top-down modulation might be considered a typical example of specific mechanisms of neuronal integration; it can be described as modulation of hierachically lower regions by those being higher in the hierarchy. Often top-down modulation concerns modulation of neural activity in subcortical regions by cortical regions. For example, premotor/motor cortical regions might modulate neural activity in subcortical ▶basal ganglia like the ▶caudate and ▶striatum [23,24]. Yet another example is top-down modulation of ▶primary visual cortex by prefrontal cortical regions, which has been shown to be essential in visual processing [25]. Top-down modulation might be related to the concepts of "re-entrant circuitry" [26] and feedback modulation. These concepts allow for circuiting of information and readjustment of neural activity in one area according to another rather distant area. This provides the possibility of adjusting, filtering, and tuning neural activity in the lower area according to the one in the higher area. For example, top-down modulation allows for attentional modulation of visual input, which makes selective visual perception possible.

We want to focus on the medial prefrontal cortex. Neural activity in both medial prefrontal cortex and ▶amygdala has been shown to be involved in emotional processing [27]. Their functional relationship is supposed to be characterized by top-down modulation of the amygdala by the medial prefrontal cortex [28], [29]. Medial prefrontal cortical regions seem to exert also top-down control of neural activity in the ▶insula [30] that is densely and reciprocally connected with subcortical medial regions like the ▶hypothalamus, the ▶periaqueductal grey (PAG), the ▶substantia nigra, and various brain stem nuclei such as the ▶raphe nuclei and the ▶locus coeruleus [31]. Both the amygdala and the subcortical medial regions are involved in regulating internal bodily functions whereas medial prefrontal cortical regions have been associated with emotional processing [27,32]. The three regions, medial prefrontal cortex, amygdala, and subcortical medial regions, show dense and reciprocal connections [33,34]. Therefore, one might assume modulation between all of them. This might not only include top-down modulation, as illustrated, but also the reverse kind of modulation, bottom-up modulation. In the case of bottom-up modulation a hierachically lower area modulates activity in an area being higher in the hierarchy. For example, subcortical midline regions might modulate neural activity in medial prefrontal cortex via the insula thus concerning the same regions as top-down modulation. Accordingly, bottom-up and top-down modulation might co-occur across the same regions.

Functionally, this co-occurrence of bottom-up and top-down modulation might allow for reciprocal adjustment between emotional and internal bodily processing. Internal bodily processing concerns only stimuli from the own body, so-called internal self-related stimuli. These include, for example, stimuli from autonomic-vegetative or other humoral functions. Whereas emotional processing concerns both internal self-related and thus internal bodily stimuli and external self-related stimuli from the environment. For example, emotional processing might be induced by specific events within the environment which in turn might induce internal bodily stimuli. This is well compatible with the co-occurrence of bodily and emotional symptoms in ▶Posttraumatic Stress Disorder (PTSD) where such top-down modulation between amygdala and medial prefrontal cortex is assumed to be altered [35]. The elucidation of specific mechanisms of neuronal integration for higher-order functions and mental states, their control by genetic and social factors, and their changes in psychiatric disorders will be the central task for future neuropsychiatry.

References

1. Beaumont JG (1987) Einführung in die Neuropsychologie. Psychologie Verlags- Union, München, Weinheim
2. Berrios GE, Markova IS (2002) The concept of neuropsychiatry: a historical overview. J Psychosom Res; 53(2):629–638
3. Caine CD, Joynt RJ (1986) Neuropsychiatry again. Arch Neurol 43:325–327
4. Churchland PA (1986) Neurophilosophy. Toward a unified science of the mind/brain. MIT Press, Bradford, Cambridge
5. Cummings JL (1985) Clinical Neuropsychiatry. Grund & Stratton, New York
6. Davidson RJ (2002) Anxiety and affective style: role of prefrontal cortex and amygdala. Biol Psychiatry 51(1): 68–80
7. Filskow SB, Boll TJ (1986) Handbook of clinical neuropsychology, vol 2. Wiley, New York, pp 45–81
8. Friston KJ, Price CJ (2001) Dynamic representations and generative models of brain function. Brain Res Bull 54(3): 275–285

9. Harrington A (1989) Psychiatrie und die Geschichte der Lokalisation geistiger Funktion. Nervenarzt 60:603–611
10. Kolb B, Whishaw JQ (1985) Fundamentals of human neuropsychology, 2nd edn. W. Freeman Company, New York
11. Lamme VA (2004) Separate neural definitions of visual consciousness and visual attention; a case for phenomenal awareness. Neural Netw 17(5–6):861–872
12. Luria AR (1966) Higher cortical functions in man (transl: Haigh B). Basic Books, New York
13. Luria AR (1973) The working brain. An introduction to neuropsychology (transl: Haigh B). Basic Books, New York
14. Masterman DL, Cummings JL (1997) Frontal-subcortical circuits: the anatomic basis of executive, social and motivated behaviors. J Psychopharmacol 11(2):107–114
15. Mueller J (ed) (1989) Neurology and psychiatry. A meeting of minds. Karger, New York
16. Mundt Ch (1989) Psychopathologie heute. In: Kister KP et al. Brennpunkte der Psychiatrie. Psychiatrie der Gegenwart 9, 3. Aufl. Springer, Heidelberg, p 147–185
17. Nagai Y, Critchley HD et al. (2004) Activity in ventromedial prefrontal cortex covaries with sympathetic skin conductance level: a physiological account of a "default mode" of brain function. Neuroimage 22(1): 243–251
18. Northoff G (2002) What catatonia can tell us about "top-down modulation": a neuropsychiatric hypothesis. Behav Brain Sci 25(5):555–577; discussion 578–604
19. Northoff G, Bermpohl F (2004) Cortical midline structures and the self. Trends Cogn Sci 8(3):102–107
20. Northoff G, Heinzel A et al. (2004) Reciprocal modulation and attenuation in the prefrontal cortex: an fMRI study on emotional-cognitive interaction. Hum Brain Mapp 21(3):202–212
21. Oepen G (ed) (1988) Psychiatrie des rechten und linken Gehirns. Neuropsychologische Ansätze zum Verständnis von 'Persönlichkeit,' 'Depression' und 'Schizophrenie.' Deutscher Ärzte Verlag, Köln
22. Ongur D, Price JL (2000) The organization of networks within the orbital and medial prefrontal cortex of rats, monkeys and humans. Cereb Cortex 10(3):206–219
23. Panksepp J (1998) Affective Neuroscience: the foundations of human and animal emotions. Oxford University Press, New York
24. Pessoa L, Ungerleider LG (2004) Neuroimaging studies of attention and the processing of emotion-laden stimuli. Prog Brain Res 144:171–182
25. Phan KL, Wager T et al. (2002) Functional neuroanatomy of emotion: a meta- analysis of emotion activation studies in PET and fMRI. Neuroimage 16(2):331–348
26. Poeck K (1982) Klinische Neuropsychologie. Thieme Verlag, Stuttgart, New York
27. Price CJ, Friston KJ (2002) Degeneracy and cognitive anatomy. Trends Cogn Sci 6(10):416–421
28. Rauch SL, Shin LM, Phelps EA (2006) Neurocircuitry models of posttraumatic stress disorder and extinction: human neuroimaging research–past, present, and future. Biol Psychiatry 60(4):376–382
29. Reynolds EH, Trimble MR (1989) The bridge between neurology and psychiatry. Churchill Livingstone, Edinborough, London, New York
30. Rogers D (1987) Neuropsychiatry. Br J Psychiatry 150:425–427
31. Singer W (1989) Search for coherence: a basic principle of cortical self- organization. Concepts Neurosci 1(1):1–21
32. Tononi G, Edelman GM (2000) Schizophrenia and the mechanisms of conscious integration. Brain Res Brain Res Rev 31(2–3):391–400
33. Trimble MR (1981) Neuropsychiatry. Wiley, Clichester
34. Trimble MR (1988) Biological psychiatry. Wiley, Clichester
35. Vygotsky LS (1978) Mind in society. The development of higher psychological processes. Harvard University Press, Cambridge, London

Neuropsychopharmacology

►Behavioral Neuropharmacology

Neuroregeneration

Definition

Growth of new neural tissue to replace tissues that were injured or lost.

Unlike neuroprotection, neuroregeneration implies replacement of degenerating axons with newly formed fibers.

►Neuroprotection
►Regeneration

Neurosarcoidosis

Definition

Complication of sarcoidosis involving inflammation and abnormal deposits in the tissues of the nervous system.

Neurosecretosome

►Synaptosome

N

Neurosemiotics

RENÉ J. JORNA
Professor of Knowledge Management & Cognition
Faculty of Economics and Business, University of Groningen, Groningen, The Netherlands

Definition

Neurosemiotics is about the sign or signal related aspects of synapses, neurons and neural nets [20,29]. Because neurons are the physiological tissue within which the brain or mind resides, neurosemiotics is also about cognition or about humans as information processing and connectionist systems [30].

Description of the Theory

To understand the complex relationships between signs, signals and semiosis as aspects of semiotics on the one hand and brain and cognition on the other hand, an aid for understanding the distinction in various (ontological) levels of description [5,26,23] is used. The explanation of this conceptual help is followed (Neuro-psychological consequences) by a description of its neuropsychological implications and after that (Neurosemiotics) by an analysis of various categorizations of signs. The article ends (The role of Neurosemiotics) with a determination of the role of neurosemiotics.

Levels of Description

The intriguing intangible phenomenon of the human mind seems to require various perspectives at the same time. The perspectives depend on certain levels of aggregation, also called levels of description. "Levels are clearly abstractions, being alternative ways of describing the same system, each level ignoring some of what is specified at the level beneath it." [23]. The idea of levels of description for a (cognitive) system has been elaborated most elegantly by Dennett [5,6,7]. He distinguishes three independent levels: (i) the physical (or physiological) stance, (ii) the functional (or design) stance, and (iii) the intentional stance. Other authors [22,23,26] give similar accounts in which, however, the number of levels varies. Newell (1990;26, p.46) says: "Often we describe the same system in multiple ways [...]. The choice of what description to use is a pragmatic one, depending on our purpose and our own knowledge of the system and its character." In the following we use Dennett's explanation of levels [5,6].

The physical (or physiological) stance explains behavior in terms of physical properties of the states and the behavior of the system under concern. For its proper functioning the human organism requires a complex interaction of its parts and with the external world. The central nervous system in al its subdivisions and the endocrine system are there to transmit data (signals, information) that reveal the state of one part of the system to other parts. We can also mention the transmission of currents in the synaptive system of neurons. Within the study of brain and cognition, the physiological stance is the endpoint of ontological reduction.

The second level concerns the functional design of a system. At a functional level it is important to know the components of a system, how they are defined and how the components and sub-components are connected. Stated differently, if the input and output of each component are known, then with a certain input at the beginning of the system the resulting behavior (output) can be predicted based on the characteristics of the components. The behavior of a system is conceived of as the result of the interaction between several functional components or processes. The physical structure of the system is not explicitly taken into account, although it surely imposes constraints on the behavior of the system at the higher level. The capacity limitations of human memory, for instance, impose boundary conditions on the complexity of decision making at the higher intentional level.

Thirdly, Dennett distinguishes the intentional stance. Complex behavior that is adapted to the prevailing circumstances, according to some criterion of optimality is said to be rational or intelligent. A behaving system to which we can successfully attribute rationality or intelligence qualifies as an intentional system. It is not necessary for a behaving system to "really" possess rationality or intelligence, as long as the assumption allows us to correctly predict the behavior of the system based on our knowledge of the circumstances in which the system is operating.

One may deal with this level's distinction in two different ways: instrumentalistic and ontological. Dennett [5] has taken the levels distinction in a strictly instrumentalistic way, claiming only pragmatic validity. Summarizing his position twenty years later, he wrote: "As I have put it, physical stance predictions trump design [or functional] stance predictions which trump intentional stance predictions - but one pays for the power with a loss of portability and a (usually unbearable) computational cost." [8]. In contrast to Dennett, authors such as [11,23,26] assign an ontological meaning to each level. The higher levels introduce emergent qualities into human behavior that make no sense if we maintain only an instrumentalistic point of view.

(Neuro-)Psychological Consequences

Neurons, synapses, neuronal nets are all located at the physiological level. The most important question resulting from the levels distinction is the art of the relationship between the levels. Is the relationship reductionist? Or is ultimate reduction not possible. If the intentional level can be reduced without loss of information to the functional level and from there to the physiological level we have a reductionist position [3]. If one accepts that levels do exist in a parallel way in which lower levels set constraints for the higher levels also implying that a higher level can be incorporated in various concrete designs at the lower levels (the so called "under-determination" issue) one is non-reductionist, Sometimes this position is called the "emergence" or "supervenient" position. In that case properties and characteristics of levels exist on their own (level) [21].

From a neurological and psychological point of view, the (non-)reductionist position and the level's distinction determine how one sees and therefore studies human beings and how one deals with neurosemiotics. If human beings in all their complexity of behavior and cognition can be reduced to neurons, the study of humans in the long-run equals neuroscience. In that case humans as biological, psychological and social entities will be explained in an analytic and scientific way in terms of the well elaborated lowest domain of neuroscience. Neurosemiotics then is a matter of extremely complex signal analysis (see section 3). However, if behavior and cognition cannot be ultimately reduced and can only be studied at their appropriate level of description, more holistic and emergent descriptions of humans have value on their own. This implies that signs and symbols requiring an autonomous interpretation mechanism are proper object of study, even, or especially, within neurosemiotics.

One consequence of this distinction relates to the possibility of "real" artificial intelligence. If one follows the reductionist line of reasoning resulting in theories, predictions and explanations at the neuroscientific level, it is tempting to argue that the only way intelligence can be mimicked is by doing neuroscience. A neurological basis of "real" intelligence is essential. In contrast with this position one can follow Good Old Fashioned Artificial Intelligence GOFAI [12] which says that intelligent (functional or intentional) systems can "run" on different kinds of "hardware": neurons or chips. In the latter case, the program (Deep Blue 2, see [15] that beat the world champion in chess is an intelligent system, comparable to a human, but wired with different material. However, in the former case the software program is a nice piece of programming, but is not intelligent in the real sense, because it lacks synapses, neurons and similar tissues.

Neurosemiotics

The discussion of various levels and (non-)reductionism determines the view on neurosemiotics in two respects. First, semiotics in relation to a domain of study can never exist without a precise demarcation of that domain [28]. Second, humans whether they are conceived of as information processing system or as neurological systems are signals, signs and symbols using systems themselves [16]. Therefore, neurosemiotics, largely, is also about our selves. I will first give a definition of semiotics and illustrate its relevance for neurons, then I will introduce and categorize various kinds of "signs" and finally I will discuss semiosis or sign interpretation.

Extending the definition of semiotics (Posner, 1979), a demarcation of semiotics is as follows: "Semiotics is the study of all sorts of sign or symbol processes in the communication and the exchange of knowledge, in the sense of data, between and inside information processing systems, such as humans, other organisms and machines" [17]. Other definitions can be found in [1,9,25,24], to name but a few. In this demarcation a sign is defined as "something that stands for something else" (aliquid stat pro aliquo), which is the same definition as for the notion of representation [18]. In the definition of semiotics three aspects in relation to neurosemiotics are important. First, signals are not mentioned and signs and symbols are equivalent. Second, data, information and knowledge seem to be interchangeable, whereas the usual distinction is that knowledge is interpreted information, which is interpreted data. Third, behind communication and exchange an interpretation mechanism is presupposed.

Within semiotics, various categorizations of semiotic elements are used. For neurosemiotics the following distinctions are relevant: (i) signal and sign, (ii) index, icon and symbol and (iii) sign or symbol sets (see for overviews: [10,14,31].

First, we have the classical distinction in signals and in signs. Signals are positioned at the physical/chemical level of description and their modus operandi is causal. In this respect signals do not fall into the definition of "something standing for something else." Part of the firing of a neuron is the chemical substances that are emitted: signals. A signal is part of the functioning of a cell, an organelle or a neuron as elements of the material world. Signs, however, are semantic in their functioning. For something to be a sign implies an interpretation mechanism. Smoke is part of a fire and therefore can be called a signal, but as soon as an interpreting mechanism, an animal, a human or an electronic sensor, takes action, the signal is a sign. It stands for something and it is interpreted. It is an open question whether the firing of a neuron or synapse activity is a signal or a sign. The choice has consequences for the kinds of research questions: mainly cognitive in the case of signs and merely physiological in the case of signals (see also [2], who accepts the functional level, but abandons the notion of central processors).

Second, a distinction can be made in various kinds of signs: index, icon or symbol [25]. An index is a sign that is not arbitrarily, but directly connected to what it represents. It resembles very much the signal, but a signal is not interpreted, an index is. In case of an icon, the sign resembles, mimics or imitates what it stands for, without incorporating all characteristics. An image of a nerve with MRI resembles the nerve, without consisting of proteins or potassium (K). A symbol is a sign that has an arbitrary or conventional relationship with what it stands for. The word "apple" has an arbitrary relationship with apple (at least in English) and words of the language or codes have to be learned.

Third, [13,14] distinguishes various kinds of sign sets (Goodman speaks of symbol sets) in order to define notational systems. A notional system, such as mathematics, logic or musical scores, fulfills syntactic and semantic requirements and realizes communication without ambiguity. If the elements of a sign set met none of the requirements it is just a collection of markers, strokes or noises. If the elements fulfill the syntactic requirements of disjointness and finite differentiation it is a notational scheme, such as language or the Arabic number system. If the elements also fulfill the semantic requirements, that is to say they are unambiguous, semantically disjoint and semantically finitely differentiated, the sign set is a notational system.

In the above categorizations neurons and synapses can be seen as signals, and if one accepts that firing of neurons involves information transmission it is a sign, that is to say an index and not an icon or a symbol, and finally in terms of Goodman's distinction it is a set of markers, without syntactic and semantic characteristics.

The distinctive element in the question whether we are dealing with signals or signs and the comprehensive effect of neurosemiotics is the place and role of an interpretation mechanism, a signal interpreter, which activity some would call semiosis (sign understanding). An interpretation mechanism implies that meaning, sense or signification is attached to signals. If that takes place we have a sign, in whatever variety. From a neurosemiotic point of view the question is whether we can demarcate such a mechanism at the physiological or neuronal level. If the answer to this question is no, the question moves to the next functional level and it is generally agreed that cognition - in a better defined sense - does the work here, in the form of processing units [19]. Here we speak of information processing resulting in knowledge. However, if the answer to the question at the physiological level is affirmative, it is the assignment of neuroscience and neuropsychology to describe and explain an "interpreter" at this physiological level without jumping to the next higher level. Despite the vast research in this area this has not been accomplished yet.

The role of Neurosemiotics

Neurosemiotics is definitely not the first domain where neuroscientists seek their primary inspiration. Nevertheless, looking at neuroscience from a semiotic perspective has three advantages. It raises questions of data or information transmission in unconventional terms, in unusual conceptualizations. Because of its higher level and multi-disciplinary perspective it brings to neuroscience results from domains such as logic, systems engineering or aesthetics, where issues of sign exchange and understanding are omni-present. Third, it brings modesty to the field of neuroscience, where - with exaggeration - the dominant view at the moment is that it should and can replace psychology, physiology, cognitive science and information science. Semiotics had that dream in the past.

References

1. Bouissac P (1998) Encyclopedia of Semiotics. Oxford University Press, Oxford
2. Brooks RA (1998) Cambrian intelligence: the early history of the new AI. MIT, Cambridge, MA
3. Churchland PS (1986) Neurophilosophy: toward a unified science of the mind-brain. MIT, Cambridge, MA
4. Deely J (1990) Basics of Semiotics. Indiana University Press, Bloomington
5. Dennett DC (1978) Brainstorms: Philosophical essays on mind and psychology. Harvester Press, Hassocks, Sussex
6. Dennett DC (1987) The intentional stance. MIT, Cambridge, MA
7. Dennett DC (1991) Consciousness explained. Penguin, London
8. Dennett DC (1998) Brainchildren: essays on designing minds. MIT, Cambridge, MA
9. Eco U (1976) A theory of semiotics. Indiana University Press, Bloomington
10. Eco U (1984) Semiotics and the philosophy of language. Indiana University Press, Bloomington
11. Fodor JA (1975) The language of thought. Thomas Y. Crowell, New York
12. Fodor JA (1983) The modularity of mind: an essay on faculty psychology. MIT, Cambridge, MA
13. Goodman N (1968/1981) Languages of art, 2nd edn. The Harvester Press, Brighton, Sussex
14. Goodman N (1986) Of mind and other matters. Harvard University Press, Cambridge MA
15. Hsu, Feng-hsiung (2002) Behind deep blue: building the computer that defeated the world chess champion. Princeton University Press, Princeton
16. Jorna RJ, Smythe WE (1998) Editors special issue: semiotics and psychology: the signs we live by. Theory Psychol 8(6):723–730
17. Jorna RJ (1990a) Wissenrepräsentation in künstlichen Intelligenzen. Zeichentheorie und Kognitionsforschung. Zeitschrift für Semiotik 12:9–23
18. Jorna RJ (1990b) Knowledge representation and symbols in the mind. Probleme der Semiotik. Tübingen. Stauffenburg Verlag
19. Michon JA, Jackson JL, Jorna RJ (2003) Semiotic Aspects of Psychology. In: Posner R, Robbering K, Sebeok TA (eds) Semiotics: a handbook on the sign-theoretic foundations of nature and culture. de Gruyter, Berlin

20. Müller A, Wolff JR (2003) Semiotische Aspekte der Neurophysiologie: Neurosemiotik. In: Posner R, Robbering K, Sebeok TA (eds) Semiotics: a handbook on the sign-theoretic foundations of nature and culture. de Gruyter, Berlin
21. Nagel E (1961) The structure of science. Routledge & Kegan Paul, London
22. Newell A (1982) The knowledge level. Artif Intell 18:81–132
23. Newell A (1990) Unified theories of cognition. Harvard University Press, Cambridge MA
24. Nöth W (2000) Handbuch der Semiotik (2e Auflage). Metzler Verlag, Stuttgart
25. Peirce CS (1931–1958) The collected papers of Charles Sanders Peirce. Harvard University Press, Cambridge, MA
26. Pylyshyn ZW (1984) Computation and cognition: towards a foundation for cognitive science. MIT, Cambridge, MA
27. Posner R (1979) Editorial opening. Zeitschrift für Semiotik 1
28. Posner R, Robbering K, Sebeok TA (eds) (2003) Semiotics: A Handbook on the Sign-Theoretic Foundations of Nature and Culture. de Gruyter, Berlin
29. Roepstorff A (2003) Cellular neurosemiotics: outlines of an interpretive framework. University of Aarhus, Aarhus
30. Rumelhart DE, McClelland JL (1986) Parallel distributed processing. Explorations in the microstructure of cognition. MIT, Cambridge, MA
31. Sebeok TA (1994) An Introduction to Semiotics. Pinter Publishers. Groningen 22-01-2005, London

Neurosis

Definition

Large group of non-psychotic disorders. As defined by S. Freud, a disease that has psychological causes, due to conflicts and traumata in early life history. The symptoms are direct consequence and symbolic expression of an unconscious psychical conflict. This conflict has its origin in the early childhood. The symptoms are a compromise between an instinctual desire and its psychical defense.

►Personality Disorder

Neurosphere Culture

Definition

Neurosphere culture enables the selective expansion of NSCs in floating culture within a serum-defined medium containing growth factors, such as EGF and/or FGF2. A neurosphere derived from a single cell has been shown by the differentiation assay to be capable of generating all the major three cell lineages of the CNS, i.e., neurons, astrocytes and oligodendrocytes, indicating the multi-potency of the neurosphere-initiating cell.

When a neurosphere is dissociated into single cells, each cell starts to form secondary neurospheres at a high frequency. For human NSCs, however, some different protocols are being used.

►Transplantation of Neural Stem Cells for Spinal Cord ►Regeneration

Neurosteroids

►Neuroendocrinological Drugs

Neurosyphilis

Definition

Chronic infectious disease caused by Treponema pallidum, occurring in various forms: ►syphilitic meningitis, ►meningovascular syphilis, ►tabes dorsalis, ►dementia paralytica.

►Dementia Paralytica
►Meningovascular Syphilis
►Syphilitic Meningitis
►Tabes Dorsalis

Neuroticism

Definition

As defined by H.J. Eysenck, hereditable poor emotional stability, that predisposes the person to develop neurotic symptoms during excessive stress. Using questionnaires neuroticism can be tested objectively.

►Personality Disorder

N

Neurotransmitter

KAZUHIDE INOUE
Department of Molecular and System Pharmacology, Graduate School of Pharmaceutical Sciences, Kyushu University, Japan

Synonyms

Chemical transmitter

Definition

Neurotransmitter is a chemical that achieves chemically mediated transmission, a major mode of interneuronal and neuron-effector communication. The chemical should meet several criteria to be recognized as a neurotransmitter. The criteria are: (i) A neurotransmitter must be synthesized in a neuron and released from a presynaptic terminal, (ii) A neurotransmitter should reproduce the specific responses that are evoked by the stimulation of presynaptic neurons at the postsynaptic neuron or effector cells, (iii) The effect of the chemical should be blocked by antagonists in a dose-dependent manner, and (iv) A neurotransmitter must be reabsorbed into the presynaptic neuron or glia, or metabolized into an inactive form by enzymes to terminate the stimulation.

Characteristics

Function

Neurotransmitters transmit information from presynaptic neurons to postsynaptic neurons or peripheral effector cells through the activation of a receptor/channel. The function of a neurotransmitter depends on the receptor/channel. A neurotransmitter that stimulates an ionotropic receptor/channel causes an excitatory or inhibitory current in a target cell resulting in depolarization or hyperpolarization, respectively. A neurotransmitter that stimulates GTP binding of a protein-coupled metabotropic receptor causes second messenger signal transduction in a target cell, resulting in cell response.

Further reading

1. Deutch AY, Roth RH (2002) Neurotransmitter. In: Squire LR, Bloom FE, McConnel SK, Roberts JL, Spitzer NC, Zigmond MJ (eds) Fundamental neuroscience. Academic, San Diego, pp 163–196
2. Cooper JR, Bloom FE, Roth RH (2003) Cellular and molecular foundations of neuropharmacology. In: Cooper JR, Bloom FE, Roth RH (eds) The biochemical basis of neuropharmacology. Oxford University Press, New York, pp 7–64
3. Kuba K (2002) Transmission of Excitation. In: Toyoda J, Kumada K, Ozawa S, Fukuda K, Honma K (eds) Standard textbook, Igakushoin, Tokyo, pp 112–136

Neurotransmitter Receptor Trafficking

►Receptor Trafficking

Neurotransmitter Release: Priming at Presynaptic Active Zones

HIROSHI KAWABE, FREDERIQUE VAROQUEAUX, NILS BROSE
Department of Molecular Neurobiology, Max-Planck-Institute for Experimental Medicine, Göttingen, Germany

Synonyms

Fusion competence of secretory vesicles; Readily releasable secretory vesicles

Definition

The Vesicle Priming Step in the Neurotransmitter Release Process

Ca^{2+}-triggered fusion of ►synaptic vesicles is the key step in neurotransmitter release and synaptic signalling. Before vesicles can fuse with the plasma membrane in response to an arriving action potential and concomitant Ca^{2+} influx, they have to be translocated and tethered/docked to the plasma membrane. A subset of these tethered vesicles are then primed to fusion competence. Only these docked and primed vesicles, also referred to as readily releasable vesicles, can fuse with the plasma membrane in a Ca^{2+}-dependent manner. In the absence of vesicle priming, transmitter release is completely blocked.

In synapses, vesicle docking and priming are spatially restricted to and temporally coordinated at active zones, which represent a highly specialized presynaptic subcompartment (Figs. 1a and b) [1].

Vesicle translocation, tethering (►Translocation and tethering of synaptic vesicles), and priming are rather slow processes. However, the fact that these processes are spatially restricted to and temporally coordinated at active zones, leading to the generation of a readily releasable vesicle pool at synaptic release sites, guarantees that synaptic transmitter release is spatially accurate and very fast. In fact, excitation–secretion coupling at many synapses occurs in a sub-ms time frame. This, in turn, allows neuronal networks to propagate information at very high density and with very high reliability.

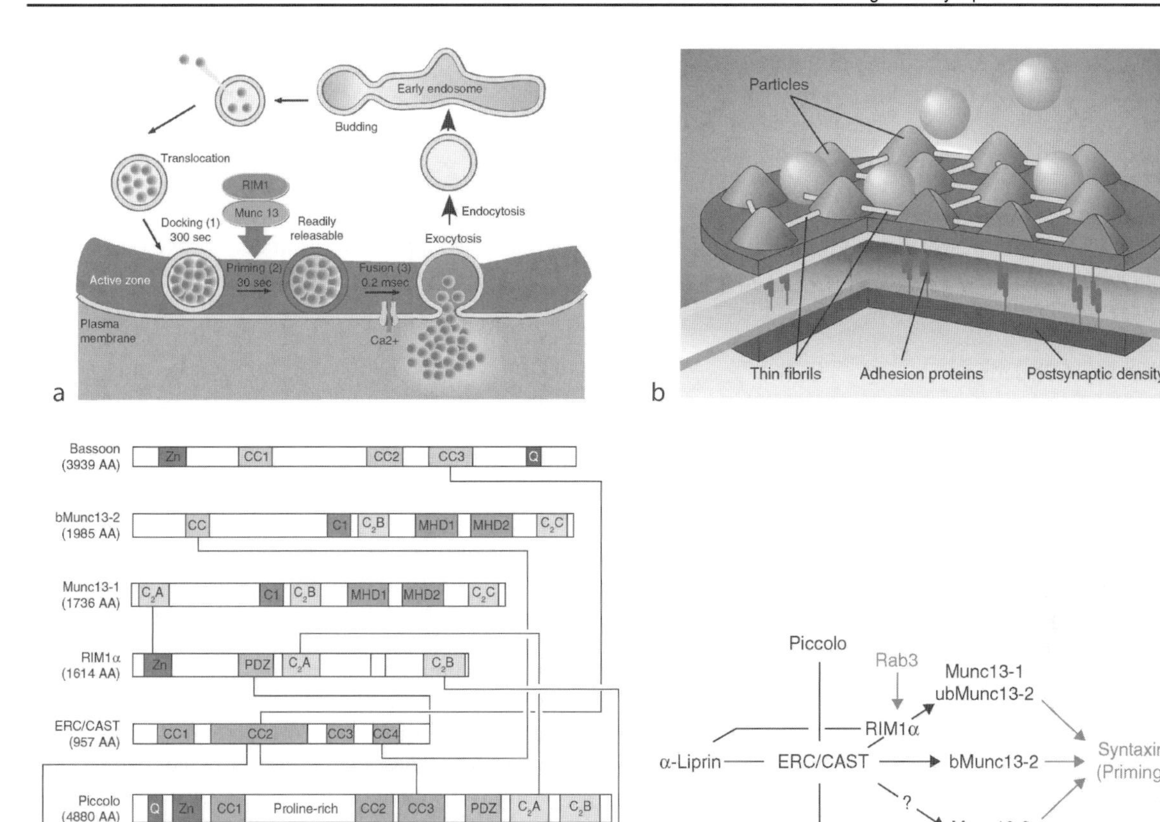

Neurotransmitter Release: Priming at Presynaptic Active Zones. Figure 1 (a) Schematic representation of neurotransmitter release at the presynaptic active zone. Shown is a segment of a presynaptic terminal, and the trafficking and maturation steps that synaptic vesicles pass through. Synaptic vesicles filled with neurotransmitters translocate from the presynaptic cytoplasmic pool to the plasma membrane, where they dock (1) and undergo a priming step (2) to become fusion competent. Fusion competent vesicle exocytose in response to an arriving action potential and concomitant Ca^{2+} influx (3) Note that all these steps are spatially restricted at the active zone. (b) The structure of an active zone. Pyramid-like particles are connected to each other with thin fibrils forming a particle web. Synaptic vesicles can access the presynaptic plasma membrane through a grid surrounded by fibrils. Figure adapted from the review by Dr. Hugo Bellen. (c) Structures of active zone components. Any given active zone specific protein has a multi-domain structure and interacts with several other active zone proteins. Direct binding between proteins is indicated by black lines. (d) Regulation of Munc13 proteins by ERC/CAST. ERC/CAST can bind to Munc13 proteins either directly or indirectly through RIM1. This interaction likely controls the Munc13 priming efficiency. In addition, ERC/CAST can directly interact with all other active zone components, and it is possible that this protein is at the core of an active zone protein network that forms "particles" of the active zone.

Characteristics

Quantitative Description

Active zones are defined morphologically on the basis of three criteria: (i) They constitute an electron dense area at the presynaptic plasma membrane that is located opposite of the ▶postsynaptic density, (ii) they harbour morphologically docked synaptic vesicles, and (iii) they provide the tethering site for synaptic vesicles on electron-dense protrusions, which emanate from the active zone and form a presynaptic proteinaceous cytomatrix (Fig. 1b). The electron dense structures of active zones contain a network of at least six types of active zone specific proteins, whose functional characteristics and interactions constitute the molecular basis of active zone morphology and function (Figs. 1c and d).

The number of active zones at each synaptic terminal can vary between synapse types and can range from one to several hundred. In cultured primary hippocampal neurons, for example, 70% of presynaptic boutons contain only a single active zone [2]. On the other hand, at the calyx of Held in the medial nucleus of the trapezoid body, one synaptic terminal contains some 550 active zones [3].

At most active zones in the central nervous system, synaptic vesicles are recruited and docked at the presynaptic plasma membrane within about 300 s, and docked vesicles are then primed within 30s or even faster. After arrival of an action potential and the subsequent influx of Ca^{2+} through voltage-gated Ca^{2+} channels, the local Ca^{2+} concentration rises beyond 10–20 μM and fusion events occur within 0.1–1 ms (Fig. 1a). Priming appears to involve multiple reversible maturation steps of synaptic vesicles, and primed vesicles are stabilized before the influx of Ca^{2+} [1].

Higher Level Structures

Chemical synapses are asymmetric intercellular junctions that typically form between the axon of a presynaptic neuron and the dendrite, cell body, or, in some cases, the axon of a postsynaptic neuron. These junctions are maintained by adhesion and scaffolding proteins (Fig. 1b). At chemical synapses, signal propagation is achieved by transducing the axonal electric signal into the exocytosis of neurotransmitters at the presynaptic active zone. Released transmitter diffuses to the postsynaptic neuron where it binds to surface receptors. Their activation, in turn, changes the physiological state of the receiving neuron.

Lower Level Components

Active zones vary in appearance from one synapse to another, especially in size and shape of electron dense particles [4]. The three-dimensional structure of the cytoskeletal matrix of the active zone (CAZ) at the frog neuromuscular junction has been studied in detail. It is 1–2 μm long, 75 nm wide and extends 50–75 nm from the presynaptic membrane into the cytoplasm. It consists of three structures: beams, ribs, and pegs. The presynaptic membrane underneath the CAZ curves outwards forming a ridge and the beams run parallel to the ridge's long axis, whereas the pegs connect the CAZ and the presynaptic membrane. Most interestingly, the ribs extend orthogonally to the ridge's long axis and form 7–12 connections to docked vesicles located on each flank of the ridge. The actual protein constituents of these beams, ribs and pegs are unknown.

In view of these and other data, proteins forming the active zone should not only be restricted in their localization to the CAZ, but they should also be of large molecular size and contain multiple protein interaction sites. In the mammalian central nervous system, active zones form "particle web" structures (Fig. 1b), in which ~50 nm pyramid-like "particles" are spaced uniformly and connected to each other with thin fibrils forming a "web" structure. This particle web likely contains all of the known active zone components: Munc13s, Bassoon, Piccolo, RIM1, α-Liprin, and ERC/CAST (Fig. 1c). Each one of these proteins binds to several others, thus forming a large multimolecular complex (Fig. 1c and d). While it is not known if these CAZ proteins depend on each other for localization, mutual functional regulation has been demonstrated for some combinations (e.g. RIM1 and Munc13-1).

Structural Regulation

Synaptogenesis and Active Zone Formation

Functional synapses are formed in two steps [5]: (i) Initial contact formation between immature axons and dendrites – in this step, several classes of cell adhesion molecules, including ►Cadherins and Protocadherins, Neurexins and Neuroligins, and ►SynCAMs/Nectins, are implicated; and (ii) recruitment of pre- and postsynaptic transmembrane and cytoplasmic proteins to the sites of initial contacts.

Trans-synaptic interactions of cell adhesion molecules are thought to trigger the recruitment of presynaptic active zone components, and postsynaptic scaffolding molecules and receptors to the forming synapse. Interestingly, many postsynaptic components are recruited to the initial synaptic contact in a gradual manner, while presynaptic active zone components are recruited in a stepwise manner. A fascinating hypothesis has emerged recently to explain this difference. According to this hypothesis, active zones are formed by the delivery of precursor active zones in the form of transport vesicles. These contain active zone specific proteins such as Piccolo and Bassoon, and are thus named Piccolo/Bassoon transport vesicles or PTVs, in addition to RIMs, Munc13s, Syntaxins, and SNAP-25. Upon fusion of such vesicles with the plasma membrane of a nascent synapse, active zone proteins are deposited and localized. In cultured hippocampal neurons, one functional release site appears to be composed of 2–3 PTVs. In mature neurons, after the establishment of functional synapses, PTVs are extremely rare. Thus, it is possible that mature neurons employ alternative transport mechanisms for active zone components, and alternative mechanisms to form active zones and synapses.

Lower Level Processes

The SNARE Complex and Munc13 Proteins

The synaptic SNARE (soluble NSF attachment protein receptors) complex is a trimeric complex composed of one synaptic vesicle protein, Synaptobrevin 2, and two synaptic plasma membrane proteins, Syntaxin and SNAP-25 (Fig. 2).

This complex catalyzes synaptic vesicle fusion by bringing the fusing vesicle and plasma membranes into close proximity, thus facilitating lipid bilayer mixing and subsequent membrane fusion.

Structural studies on Syntaxins showed that members of this protein family contain an autonomously folded N-terminal domain in addition to the C-terminal SNARE domain. These two domains can interact in an intramolecular fashion to form a "closed" conformation, which is unable to form a SNARE complex with

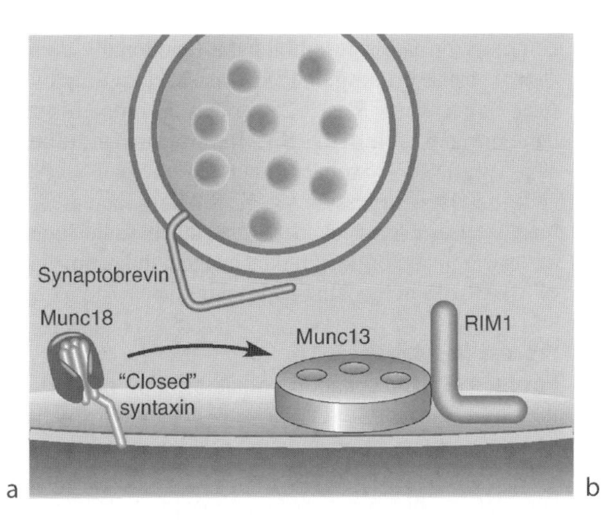

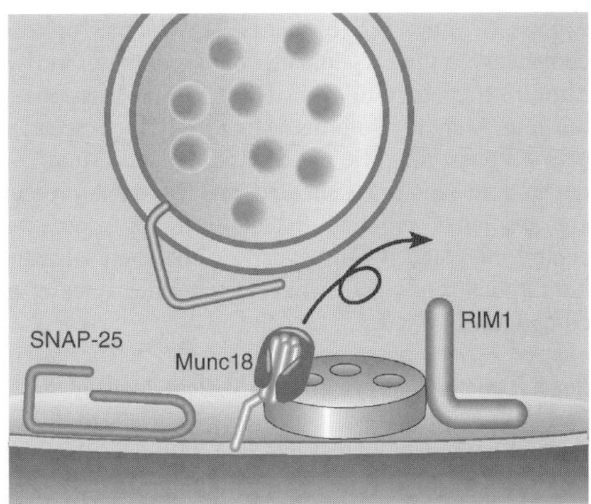

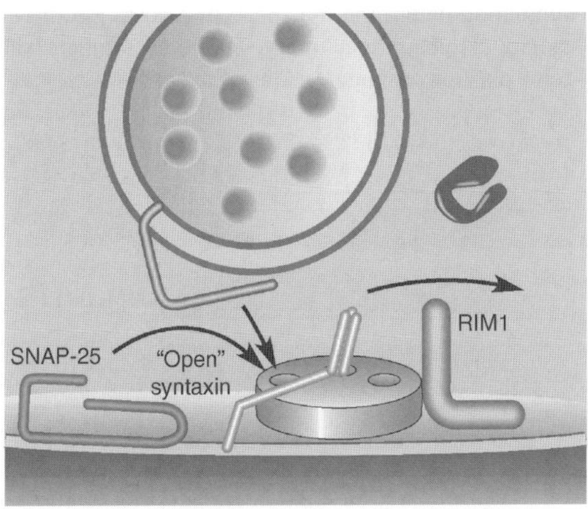

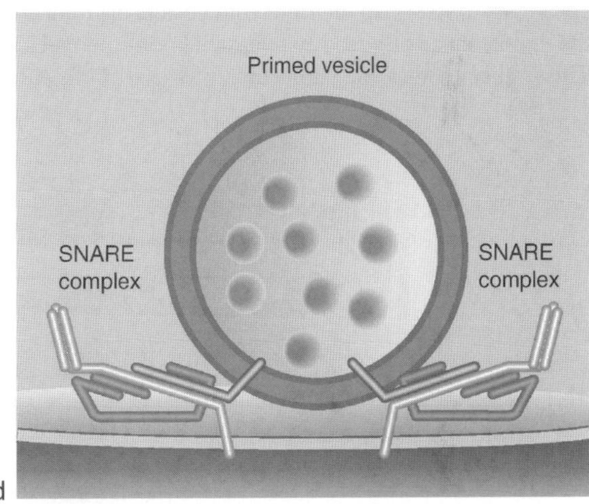

Neurotransmitter Release: Priming at Presynaptic Active Zones. Figure 2 Hypothetical regulation of SNARE complex formation during the priming step. Syntaxin in its "closed" conformation forms a stable complex with Munc18 (a). Munc13 displaces Munc18 from the Syntaxin-Munc18 complex and facilitates the structural switch of Syntaxin from the "closed" to the "open" conformation (b and c), thus promoting the assembly of the SNARE complex (d). Figure adapted from images by Dr. Erik Jorgensen.

Synaptobrevin 2 and SNAP-25. Only "open" Syntaxins are thought to be able to enter SNARE complexes.

SNARE complex formation is controlled by several types of regulatory proteins. Two of these, Munc18-1 and Munc13s, are absolutely essential for synaptic vesicle fusion. The molecular mechanism of Munc18-1 function is still poorly understood. Based on the structure of the Syntaxin 1/Munc18-1 complex, it was postulated that Munc18-1 binds to the "closed" Syntaxin conformation and blocks its SNARE motif from participating in SNARE complex formation [6]. Thus, Munc18-1 has to dissociate from Syntaxins in order to allow SNARE complex formation.

Interestingly, Munc18-1, Syntaxin 1/2, and SNAP-25 are localized to the presynaptic terminal, but their distribution is not restricted to active zones. Specific proteins localized to the active zone plasma membrane or cytomatrix are likely to coordinate the active zone processes of vesicle tethering, priming, Ca^{2+}-influx, and vesicle fusion, and to spatially restrict the synaptic vesicle exocytosis process to active zones. Indeed, neurons appear to employ a very simple molecular mechanism to limit transmitter release to active zones: They restrict the localization of essential components of the release machinery, Munc13s, to this subcellular compartment. Munc13-1, the best characterized Munc13 protein, binds to the N-terminal domain of Syntaxin, and likely promotes the conformational switch in Syntaxin from the "closed" to the "open" form, thus permitting the assembly of the synaptic SNARE complex (Fig. 2).

Rodents and humans express five Munc13 genes, Munc13-1, -2, -3, and -4, and Bap3. Two splice variants of Munc13-2 are known, the ubiquitously-expressed Munc13-2 (ubMunc13-2) and the brain-specifically expressed Munc13-2 (bMunc13-2). Munc13 proteins contain two or three C_2 domains, one C_1 domain, and two Munc13-homology domains (Fig. 3a). At least one of the four Munc13 isoforms, Munc13-1, is enriched at active zones. Studies on Munc13-1 and Munc13-2 deletion mutant mice showed that Munc13 proteins are absolutely required for the priming process in central nervous system synapses. In their absence, both spontaneous and evoked transmitter release are completely blocked.

Regulation of Readily Releasable Vesicle Pools by SNARE Proteins

Several perturbations that interfere with SNARE complex assembly or stability (e.g. antibodies to SNAP-25, deletion of SNARE regulators, or Clostridial toxins that cleave SNAP-25) also interfere with vesicle priming and the maintenance of a readily releasable pool of vesicles. This indicates that the molecular process of SNARE complex formation resembles the functionally defined step of vesicle priming. Indeed, a recent study employing over-expression of SNAP-25 variants in chromaffin cells demonstrated a role of this SNARE protein in the priming step. Two SNAP-25 splice variants are known, SNAP-25a and -b. Over-expression of SNAP-25b increases the readily releasable vesicle pool size but leaves the kinetics of vesicle fusion or docking unchanged, indicating a role of SNAP-25b in blocking depriming.

Process Regulation

Functionally, the synaptic vesicle priming rate defines the size of the readily releasable vesicle pool, which is a key determinant of presynaptic release probability and synaptic efficacy. During periods of high frequency stimulation, the basal priming rate and its dynamic regulation determine the speed of recovery of the presynaptic release machinery, and thus regulate short-term plasticity characteristics of synapses.

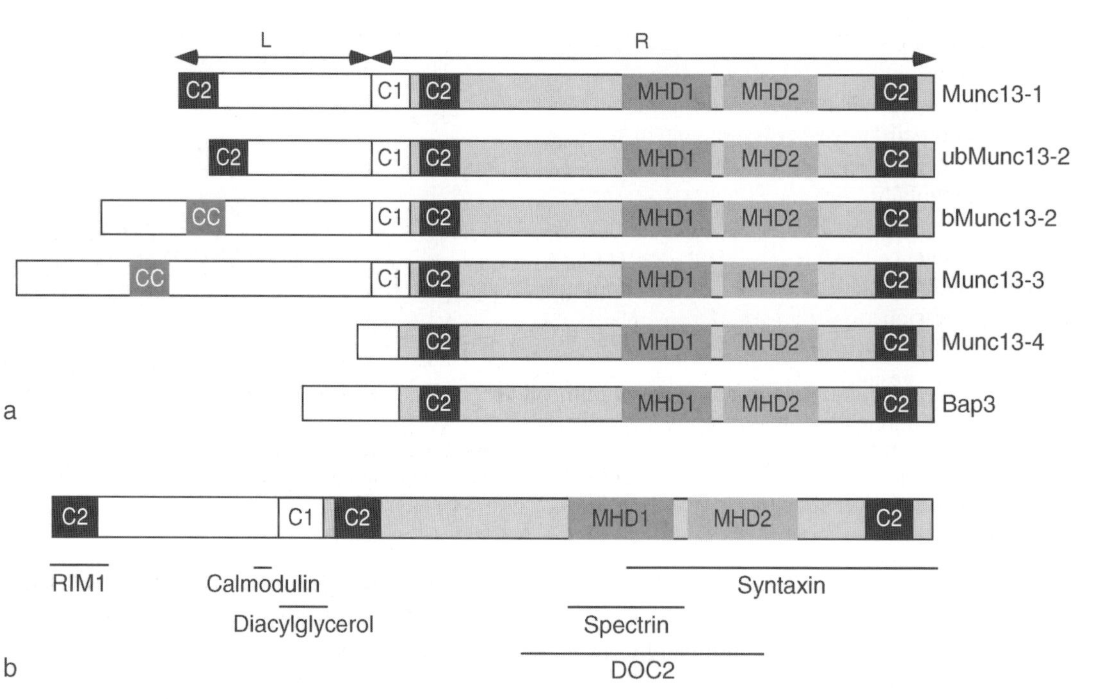

Neurotransmitter Release: Priming at Presynaptic Active Zones. Figure 3 (a) Domain structures of Munc13 proteins. The Munc13 protein family contains five isoforms, Munc13-1, -2, -3, and -4, and Bap3. Two Munc13-2 splice variants are known, ubMunc13-2 (ubiquitously expressed), and bMunc13-2 (brain-specific). All Munc13 proteins share a common, highly homologous C-terminal region (R region) with C_1, C_2, and Munc13 homology domains, but have unrelated N-termini (L region). This modular structure and differential evolution indicate that the conserved R-regions play essential roles, and unrelated L-regions play regulatory roles in Munc13 function. (b) Binding partners of Munc13-1. The N-terminus of Munc13-1 binds to another component of the active zone, RIM1, and to second messengers, Ca^{2+}/Calmodulin and diacylglycerol. These interactions are not essential for priming of synaptic vesicles, but modify the efficacy of priming and thus regulate short-term plasticity. The C-terminus of Munc13-1, -2, and -3 binds to the SNARE protein Syntaxin. This interaction is thought to be essential for the actual priming step and to involve the regulation of the Syntaxin conformation. In addition, the C-terminus of Munc13-1 binds to DOC2 (Double C2 protein), β-Spectrin, and Msec7, but the physiological significance of these interactions is not known.

Munc13 proteins are regulated by multiple proteins and second messengers. As a consequence, synaptic vesicle priming is a highly dynamic and tightly regulated presynaptic process (Fig. 3b) [5]. Synapses driven by Munc13-1 exhibit ►synaptic depression during high frequency stimulation, whereas synapses using ub-Munc13-2 as the main priming protein show ►synaptic facilitation during, and ►augmentation following periods of high frequency stimulation. Munc13 mediated synaptic depression, facilitation, and augmentation are regulated by two second messenger dependent processes: (i) Diacylglycerol activates Munc13s by binding to their C_1 domain, resulting in Munc13 activation and frequency facilitation/augmentation (in the case of Munc13-2) or at least compensation of synaptic depression (in the case of Munc13-1); (ii) Ca^{2+}-dependent binding of Calmodulin to Munc13 proteins resulting in an increase of the priming activity and functional changes in short-term plasticity that are similar to the effects of diacylglycerol.

In addition to second messengers, several proteins bind to and regulate Munc13 proteins (Fig. 3b). The most striking example is RIM1. RIM1 is a component of the active zone and was originally identified as a target of Rab3 small GTPases. RIM1 binds to the N-terminal C_2 domain of Munc13-1 and ubMunc13-2. Several lines of evidence support the notion that RIM1 is a functional regulator of Munc13 proteins but not an actual mediator of the priming step: (i) Over-expression of Munc13-1 mutants lacking RIM1 binding activity leads to a 50% reduction in priming activity as compared to the over-expression of wild type Munc13-1; (ii) in the absence of RIM1α, the dominant RIM isoform, the readily releasable vesicle pool size of hippocampal neurons is reduced by only 50%, while deletion of Munc13s eliminates primed vesicles entirely; (iii) RIM1 deficient mice show a 50% reduction in Munc13-1 expression levels; (iv) application of diacylglycerol to RIM1α deficient neurons can still increase Munc13 mediated priming and transmitter release.

Pathology

Vesicle Priming and Memory Formation

Abolishing RIM1-expression in mice results in compromised ►long-term potentiation in hippocampal and cerebellar mossy fibre terminals [7,8], and dramatic deficits in associative learning and locomotor responses to novelty [9]. As one main function of RIM1 is the regulation of Munc13 priming activity, it is likely that the dynamic regulation of synaptic vesicle priming is a key process in mossy fibre long-term potentiation. These considerations, and the finding that Munc13 mRNAs are affected in fragile-X type mental retardation, indicate that the priming step regulated by the Munc13-RIM1 complex might be important for the pathology of learning disorders and mental retardation.

Cytotoxic Granule Exocytosis Deficiency in Patients With Mutant Munc13-4 Expression

An immunoproliferative syndrome, familial hemophagocytic lymphohistiocytosis or FHL, is an often fatal childhood disorder characterized by infiltration of multiple organs by activated T cells and macrophages. One of three forms of FHL, FHL3, is caused by mutations of the gene encoding Munc13-4, a ubiquitously expressed distantly related Munc13 isoform [10]. In Munc13-4 deficient cytotoxic T cells, lytic granules are associated with the plasma membrane, indicating that Munc13-4 is dispensable for vesicle trafficking and docking in these cells, but is necessary for a step subsequent to docking. Although it is not clear whether a priming step exists in the exocytosis of lytic granules, Munc13-4 appears to play a similar role in a post-docking step of this type of exocytosis as other Munc13 proteins do in synapses.

References

1. Sorensen JB (2004) Formation, stabilisation and fusion of the readily releasable pool of secretory vesicles. Pflügers Arch 448:347–362
2. Schikorski T, Stevens CF (1997) Quantitative ultrastructural analysis of hippocampal excitatory synapses. J Neurosci 17:5858–5867
3. Satzler K, Sohl LF, Bollmann JH, Borst JG, Frotscher M, Sakmann B, Lubke JH (2002) Three-dimensional reconstruction of a calyx of Held and its postsynaptic principal neuron in the medial nucleus of the trapezoid body. J Neurosci 22:10567–10579
4. Zhai RG, Bellen HJ (2004) The architecture of the active zone in the presynaptic nerve terminal. Physiology (Bethesda) 19:262–270
5. Rosenmund C, Rettig J, Brose N (2003) Molecular mechanisms of active zone function. Curr Opin Neurobiol 13:509–519
6. Jahn R, Lang T, Südhof TC (2003) Membrane fusion. Cell 112:519–533
7. Castillo PE, Schoch S, Schmitz F, Südhof TC, Malenka RC (2002) RIM1alpha is required for presynaptic long-term potentiation. Nature 415:327–330
8. Lonart G, Schoch S, Kaeser PS, Larkin CJ, Südhof TC, Linden DJ (2003) Phosphorylation of RIM1alpha by PKA triggers presynaptic long-term potentiation at cerebellar parallel fiber synapses. Cell 115:49–60
9. Powell CM, Schoch S, Monteggia L, Barrot M, Matos MF, Feldmann N, Südhof TC, Nestler EJ (2004) The presynaptic active zone protein RIM1alpha is critical for normal learning and memory. Neuron 42:143–153
10. Feldmann J, Callebaut I, Raposo G, Certain S, Bacq D, Dumont C, Lambert N, Ouachee-Chardin M, Chedeville G, Tamary H, Minard-Colin V, Vilmer E, Blanche S, Le Deist F, Fischer A, de Saint Basile G (2004) Munc13-4 is essential for cytolytic granules fusion and is mutated in a form of familial hemophagocytic lymphohistiocytosis (FHL3). Cell 115:461–473

Neurotransmitter Transporter

KOHICHI TANAKA
Laboratory of Molecular Neuroscience, School of Biomedical Science and Medical Research Institute, Tokyo Medical and Dental University, Bunkyo-Ku, Tokyo

Synonyms

Uptake carrier

Definition

►Neurotransmitter transporters are uptake carriers in the plasma membrane of neurons and glial cells, which pump neurotransmitters from the extracellular space into the cell (Fig. 1) [1].

Characteristics

Quantitative Description

The size of neurotransmitter transporters ranges between 523 and 798 amino acids (Table 1).

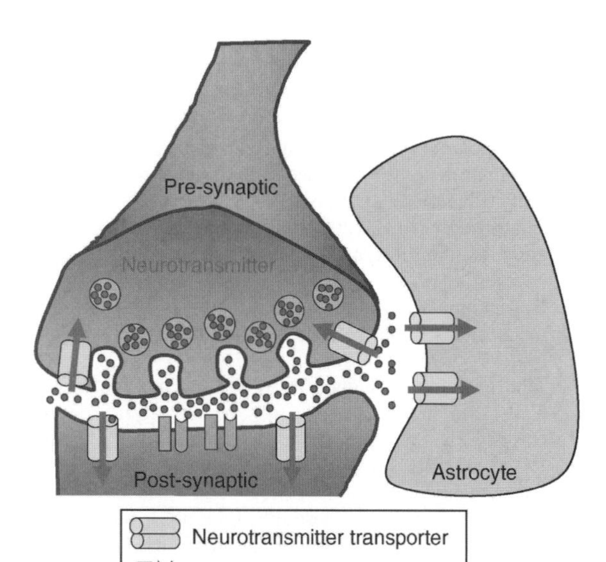

Neurotransmitter Transporter. Figure 1 Schematic Representation of the Main Neurotransmission Steps at a Synapse. The neurotransmitter is synthesized in the presynaptic neuron, stored in synaptic vesicles and released into the synaptic cleft by exocytosis to act on neurotransmitter receptors. To terminate the neurotransmission, the released neurotransmitter has to be removed promptly from the synaptic cleft. High-affinity neurotransmitter transporters present at pre-synaptic or post-synaptic neurons and/or astrocytes are responsible for rapid removal of neurotransmitter [1].

Chemical and stereological quantification shows that the number of glial glutamate transporter GLAST (EAAT1) is 2,300 μm^{-2} of membrane in the hippocampus [2].

Higher Level Structures

Since the physiological action of neurotransmitter is terminated by its ►removal by high-affinity neurotransmitter transporters, they are concentrated in the plasma membrane of neurons and glial cells surrounding the synaptic clefts [1].

Higher Level Processes

Synaptic transmission is a fundamental process in neuronal communication in the brain (Fig. 1). Neurotransmitter transporters influence many aspects of synaptic transmission, including the timecourse of the postsynaptic response and the peak postsynaptic receptor occupancy, by modulating the duration and the intensity of neurotransmitters action at the synapse. Therefore, they are involved in the fine tuning of information processing in the brain. They are also responsible for the replenishment of neurotransmitter pools within nerve endings [3,4].

Lower Level Processes

The neuronal neurotransmitter transporters shuttle transmitters from the extracellular fluid and concentrate them up to 10,000 times higher within the cytosol of the presynaptic terminal. They use the transmembrane ion gradients of Na^+, which are ultimately set up by Na^+/K^+ pump, as an energy source for moving transmitter molecules up to steep concentration gradients. They also display absolute requirements for other ions and can be divided into two families depending on their ionic dependence: (i) the Na^+/Cl^--dependent transporters, and (ii) the Na^+/K^+-dependent transporters (Fig. 2). [1]. Glutamate transporters co-transport 3 Na^+ and 1 H^+ ions, and countertransport 1 K^+ ion along with one glutamate molecule [5]. In contrast, the other family of transporters for most of the amino acid (except glutamate) and amine neurotransmitters, co-transport 2–3 Na^+ ions and Cl^- ion [1].

Process Regulation

Activation of kinases can modulate the activity of neurotransmitter transporters. Protein kinase C (PKC) activation inhibits the activity of gamma-amino butyric acid (GABA), serotonin, and glycine transporters. Glutamate transporters are also modulated by phosphorylation by PKC. Furthermore, several proteins have been shown to interact with neurotransmitter transporters and have effects on their activity. The most commonly observed way of dynamically regulating transport activity is thought to be the removal and recycling of the protein from the cell surface [6–8].

Neurotransmitter Transporter. Table 1 Plasma membrane neurotransmitter transporters

Name	Substrate	Coupling ions	Size (amino acids)
GLAST	L-Glutamate	Na^+, H^+ and K^+	543
GLT1	L-Glutamate	Na^+, H^+ and K^+	572
EAAC1	L-Glutamate	Na^+, H^+ and K^+	523
EAAT4	L-Glutamate	Na^+, H^+ and K^+	561
EAAT5	L-Glutamate	Na^+, H^+ and K^+	559
GAT1	GABA	Na^+ and Cl^-	598
GAT2	GABA	Na^+ and Cl^-	614
GAT3	GABA	Na^+ and Cl^-	602
GAT4	GABA	Na^+ and Cl^-	627
GLYT1a	Glycine	Na^+ and Cl^-	633
GLYT1b	Glycine	Na^+ and Cl^-	637
GLYT2	Glycine	Na^+ and Cl^-	798
DAT	Dopamine	Na^+ and Cl^-	619
SERT	Serotonin	Na^+ and Cl^-	630
NET	Norepinephrine	Na^+ and Cl^-	617

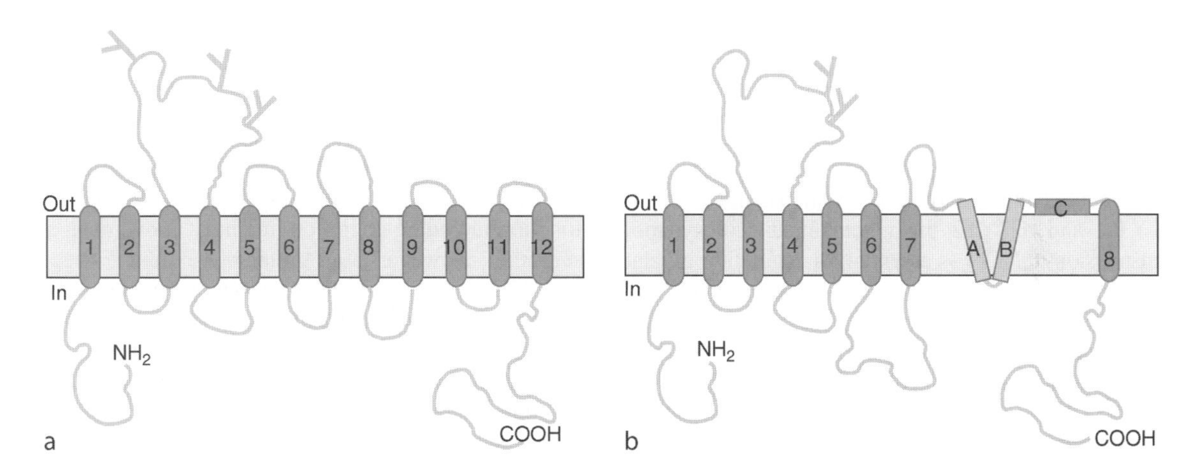

Neurotransmitter Transporter. Figure 2 Schematic Structural Organization of Neurotransmitter Transporters. (a) Schematic topology of Na^+/Cl^--dependent transporters depicting 12 transmembrane domains connected by intracellular (*in*) and extracellular (*out*) loops. Y, potentially *N*-glycosylated asparagines residue. (b) Schematic topology of Na^+/K^+-dependent glutamate transporters showing the transmembrane domains (1–8), the "re-entrant hairpin loops" (A and B) similar to the ion-permeating pore of ion channels, and a "loop" that is predicted to extend partially into the "translocation pore" between transmembrane domains 7 and 8.

Function

Glutamate Transporters [2,5,9]

Glutamate transporters are well positioned to regulate extracellular glutamate concentrations, but their contributions to shaping excitation at glutamatergic synapses vary between different types of synapses. Glutamate transporters do not determine the decay rate of the synaptic currents in the hippocampus, whereas glutamate transporters (GLAST and EAAT4) are the dominant factors that determine the kinetics of excitatory postsynaptic currents (EPSCs) at the synapses from cerebellar parallel fibers and climbing fibers to Purkinje cells. Glutamate transporters clear glutamate not only inside the synaptic cleft, but also from the extracellular space outside the synapse. Thus, glutamate transporters play a critical role in the specificity of synaptic communication in the brain by determining the amount of glutamate efflux from the synaptic cleft and the distance it diffuses. Furthermore, glutamate transport by glial transporters triggers astrocytic glycolysis and release of lactate, which in turn nourish neurons and sustains neuronal activity.

GABA Transporters [7]

GABA transporters regulate the extracellular levels of GABA, the main inhibitory neurotransmitter in the mammalian brain. Four distinct genes encoding GABA transporters have been identified (Table 1). A fraction of GABA transporters is located in the vicinity of symmetric synapses and are responsible for GABA uptake at inhibitory synapses, thus contributing to terminating GABA's action and to shaping inhibitory postsynaptic responses. A study of GABAergic synaptic transmission in the hippocampus of GAT-1 knockout mice supports this view, and suggests that GAT-1 deficiency results in elevated GABA levels, thus inducing post- and presynaptic changes in GABAergic synapses.

Glycine Transporters [6,7]

Glycine exerts multiple functions in the central nervous system (CNS), as one of the major inhibitory neurotransmitters and as a positive modulator on glutamatergic neurotransmission through *N*-methyl-D-aspartate receptors. The synaptic action of glycine ends by active recapture through specific high-affinity glycine transporters located in neuronal and glial plasma membranes. Two genes encoding glycine transporters, GLYT1 and GLYT2, have been cloned (Table 1). GLYT1 is widely expressed in astroglial cells throughout the mammalian CNS, whereas GLTT2 is localized in the axon terminals of glycinergic neurons. Knockout mice deficient in glycine transporters revealed distinct roles of GLYT1 and GLYT2 in glycine-mediated synaptic transmission. GLYT1 is essential for regulating glycine concentrations at glycine receptors, whereas GLYT2 plays a critical role in replenishing the cytoplasmic pool of glycine that is needed for transmitter loading of synaptic vesicles in glycinergic nerve terminals.

Dopamine Transporter [4,8]

Dopamine is a mediator of many functions, such as movement, emotion and cognition. The dopamine transporter (DAT) is present exclusively in neurons and is the primary mechanism for clearance of dopamine from the extracellular space. Deletion of DA in mice results in disrupted clearance of dopamine, an elevated extracellular concentration of dopamine and dramatically decreased intraneuronal storage of dopamine.

Serotonin Transporter [4,8]

The serotonin transporter (SERT) plays a critical role in the maintenance of normal neurotransmission by serotonin, and is the primary target for several antidepressants and psychostimulants. SERT can be used as a marker of serotoninergic neurons because it is present exclusively on dendrites, perikarya, axons, and nerve endings of the serotoninergic neurons. The SERT knockout mice showed a six-fold elevation in the extracellular concentration of serotonin and a marked reduction (60–80%) in intracellular concentration.

Norepinephrine Transporter [4,8]

Norepinephrine is an important neurotransmitter in the CNS. It regulates affective states, learning and memory, endocrime and autonomic functions. The norepinephrine transporter (NET) in the plasma membrane acts to terminate noradrenergic transmission by uptaking released norepinephrine back into the cell for cyclic use, and is a direct target of various antidepressants and psychostimulants. NET is a specific marker of noradrenergic neurons in the CNS. Similar to the findings obtained with DAT knockout mice and SERT knockout mice, the prolonged synaptic lifetime of noradrenaline in NET knockout mice results in elevation of the extracellular level of noradrenaline and depletion of intraneuronal stores.

The profound neurochemical alternations that were observed in mice lacking DAT, NET or SERT show that the plasma membrane monoamine transporters are crucial not only in terminating neurotransmission, but also in replenishing transmitter stores.

Pathology

Glutamate Transporters [2,5,9]

L-glutamate is the major excitatory neurotransmitter in the brain, and its interactions with specific receptors are responsible for most aspects of normal brain function including cognition, memory, movement, and sensation. However, high glutamate exposure triggers neuronal death, a process known as excitotoxicity. Excitotoxicity has been implicated as the mechanism of neuronal injury resulting from acute insults such as ischemia, epilepsy and trauma as well as chronic neurodegenerative diseases such as amyotrophic lateral sclerosis (ALS), Huntington's disease (HD), Alzheimer's disease (AD), multiple sclerosis (MS) and HIV-1-associated dementia. The linkage between impaired glutamate transporter function and a rise in extracellular levels of neurotoxic glutamate suggests that transporter malfunction is a plausible mechanism of these neurologic diseases.

Increased extracellular glutamate during ischemia triggers the death of neurons. During ischemia, the reversal of glutamate transporter (particularly GLT1), due to the depletion of energy and the rundown of ionic gradients, leads to the release of glutamate into the extracellular space, exacerbating excitotoxicity in the ischemic region.

In ALS, a decrease in the glutamate transporter activity due to the reduction of the GLT1 in affected areas of the CNS, or the expression of aberrantly spliced transcripts from the GLT1 gene, are associated with impaired glutamate uptake and increased extracellular levels of glutamate, resulting in excitotoxic damage to motor neurons.

GABA Transporters [7,10]

Electrophysiological studies of human temporal-lobe epilepsy suggest that a loss of hippocampal GABA-mediated inhibition may underlie the neuronal hyperexcitability. Temporal-lobe epilepsy is characterized in part by a loss of reversal of GABA transport that is secondary to a reduction in the number of GABA transporters [10].

Glycine Transporters [3,6]

GLTY1 knockout mice show severe motor deficits accompanied by lethargy, hypotonia and hyporesponsivity. This overall reduction of motorsensory functions is similar to the symptoms associated with glycine encephalopathy. GLYT2 knockout mice display a severe neuromotor disorder characterized by spasticity, muscular rigidity, tremor and impaired righting response. These symptoms are similar to those associated with human hyperekplexia. Although neither of the human GLTY genes has been linked to either disease, mutations in glycine receptors and enzymes responsible for degrading glycine have been implicated.

Dopamine Transporter [4,8]

DAT is the major target for psychostimulants such as cocaine methylphenidate and amphetamine. The large increase in extracellular dopamine levels that are produced by these drugs result in continuous stimulation of target neurons, a key event leading to the rewarding action of cocaine and thus to addiction.

Mice lacking DAT display symptoms found in attention-deficit/hyperactive disorder (ADHD), and their increased locomotor activity is inhibited by the psychostimulants that are commonly used to treat ADHD. Furthermore, genetic studies indicate that the polymorphisms in the DAT gene is associated with ADHD.

DAT densities are affected in several brain disorders, including Parkinson's disease, Wilson's disease, Lesh-Nyhan disease, Tourette's syndrome, and major depression.

Serotonin Transporter [4,8]

SERT is implicated in the etiology of various neurological or psychiatric syndromes. Previous studies have shown that midbrain SERT levels were reduced in patients with impulsive aggressive behavior, alcoholism or depression. Multiple polymorphisms are found in the 5′-flanking promoter region and in the second intron of the SERT gene. The two variants in the 5′ region are associated with different rates of SERT expression, and the one leading to the lower transcriptional efficacy seems to be more frequent in subjects with anxiety-related personality traits and in alcoholics with suicidal behavior. Moreover, the polymorphisms at the second intron seem to be associated with bipolar and unipolar disorders.

Norepinephrine Transporter [4,8]

NET is a major target of tricyclic antidepressants and several psychostimulants. NET levels are reduced in the locus coeruleus in people with major depression. A single mutation in the coding sequence of NET has been linked to Orthostatic Intolerance.

References

1. Masson J, Sagne C, Hamon M, Mestikawy SEL (1999) Neurotransmitter transporters in the central nervous system. Pharmacol Rev 51:439–464
2. Danbolt NC (2001) Glutamate uptake. Prog Neurobiol 65:1–105
3. Brasnjo G, Otis TS (2003) Glycine transporters not only take out the garbage, they recycle. Neuron 40:667–669
4. Gainetdinov RR, Sotnikova TD, Caron MG (2002) Monoamine transporter pharmacology and mutant mice. Trends Pharmacol Sci 23:367–373
5. Kanai Y, Hediger MA (2004) The glutamate/neutral amino acid transporter family SLC1: molecular, physiological and pharmacological aspects. Eur J Physiol 447:469–479
6. Aragon C, Lopez-Corcuera B (2003) Structure, function and regulation of glycine neurotransporters. Eur J Pharmacol 479:249–262
7. Conti F, Minelli A, Melone M (2004) GABA transporters in the mammalian cerebral cortex: localization, development and pathological implications. Brain Res Rev 45:196–212
8. Torres GE, Gainetdinov RR, Caron MG (2003) Plasma membrane monoamine transporters: structure, regulation and function. Nat Rev Neurosci 4:13–25
9. During MJ, Ryder KM, Spencer DD (1995) Hippocampal GABA transporter function in temporal-lobe epilepsy. Nature 376:174–177
10. Tanaka K (2000) Functions of glutamate transporters in the brain. Neurosci Res 37:15–19

Neurotransmitters and Pattern Generation

PATRICK J. WHELAN
Hotchkiss Brain Institute, University of Calgary, Calgary, AB, Canada

Synonyms

Neuromodulation of central pattern generators

Definition

Many motor behaviors fundamental to life such as chewing, breathing, and stepping are produced by networks of interconnected neurons termed Central Pattern Generators (CPGs). CPGs are defined as networks of neurons that can generate rhythmic behaviors in the absence of phasic input. The motor outputs of CPGs are determined by a complex interaction between synaptic, cellular and network properties. This system ensures that CPGs can produce a rich ensemble of behaviors. The purpose of this essay is to describe the components of CPGs and briefly highlight how new technology is providing tools to unravel network architecture.

Characteristics

Introduction to CPGs

Perhaps the first experiments suggesting the existence of an intrinsic network of neurons that could produce rhythmic patterns were published in 1911 by Thomas Graham Brown [1]. These seminal experiments demonstrated that rhythmic activity of the hindlimbs of cats could be evoked for a short period after the spinal cord was fully transected. Since that time his observations have been confirmed in a number of species and for different behaviors. An important advance in the field was the development of ►in vitro techniques where tissue containing CPGs could be removed and maintained outside the body. Mainly as a result of these types of experiments it was found that CPGs are formed from interconnected groups of neurons whose output is a result of a complex interaction between cellular, synaptic and network properties [2]. The rhythmic behaviors they produce must be robust and yet be capable of being modified to meet the needs of the animal. Within motor control there have been a number of classic preparations used to examine motor rhythms such as the digestive centers of crustaceans, the lamprey and xenopus swimming centers, the heart beat system of the leech and the rodent locomotor network. Here we will discuss how synaptic, intrinsic and cellular properties contribute to network function (Fig. 1).

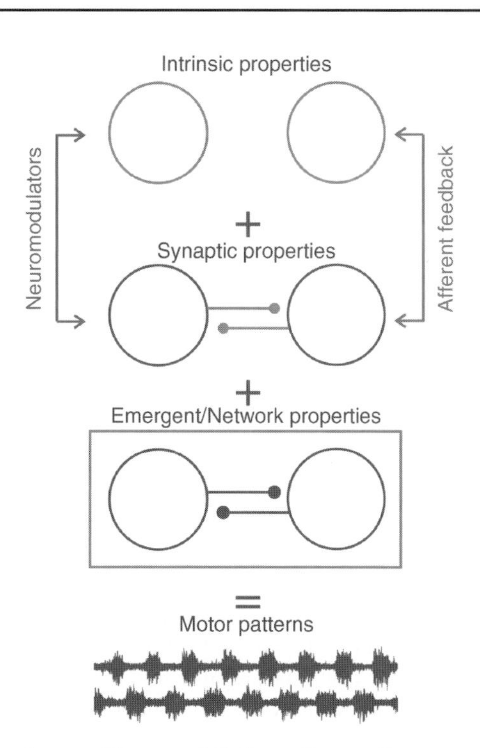

Neurotransmitters and Pattern Generation. Figure 1 Schematic illustrating the components of Central Pattern Generators that interact to produce rhythmic motor output.

1. *Motor rhythms are often generated by neurons that have rhythmogenic capabilities (so called endogenous and ►conditional bursters).* Pacemaker cells have voltage-dependent ion channels that when expressed allow a neuron to oscillate and produce bursts of action potentials. Pacemaker cells come in two flavors; ►endogenous bursters can produce oscillations in the absence of synaptic input, while conditional bursters require synaptic input to produce oscillations. For example, the mollusk *Clione* moves through the water by alternating movements of its dorsal and ventral fins. Pacemaker interneurons critical to this behavior can generate a rhythmic membrane oscillation without external inputs. Ablation of these interneurons has established that these pacemaker neurons are necessary to produce fictive swimming. More commonly, pacemaker cells found in CPGs are conditional bursters. For example, in protovertebrates such as the lamprey, excitatory interneurons have been identified that are part of the swim CPG [3]. These excitatory neurons show oscillatory membrane properties that are N-methyl-D-Aspartate (NMDA) receptor dependent. The voltage-dependent conductances that produce endogenous and conditional oscillations can be controlled by external neuromodulators. In this way, the oscillation frequency, burst duration, and burst spike rate can all be modulated. More dramatically, cells with pacemaker properties can become non-oscillatory in a state-dependent manner. In invertebrate networks, the consequence of changes in these state-dependent properties can often be correlated with the output of the network. However, in mammalian networks that contain many thousands of cells it is difficult to ascribe a function to pacemaker cells. This is because of the redundancy that is inherent in complex interconnected networks. However, with the advent of genetic tools (see Future Directions) it may be possible to circumvent this issue by selectively silencing candidate interneuronal populations.

2. *Intrinsic cellular properties can promote and stabilize motor rhythms.* Neurons are endowed with a rich set of voltage-dependent conductances that can sculpt the firing characteristics of a neuron to a given synaptic input [4]. Therefore, two cells with similar synaptic inputs can produce very different outputs. ►Intrinsic properties can have profound effects on the behavior of networks. For example, in many CPGs, alternating activity between populations of cells are mediated by mutual inhibitory connections. When one centre is active the other is inhibited. After inhibition is removed, the inhibited side "rebounds" with a burst of activity, which then inhibits the other population of cells, leading to rhythmic bursting behavior. Cellular properties can promote oscillatory behavior. One of these properties is spike frequency adaptation [3]. Progressive increases in ►intracellular Ca^{2+} during a burst can lead to a longer after hyperpolarization thereby increasing the interval between spikes. This slowing in the spike rate over time reduces the inhibition of the mutually inhibited population of cells. Another intrinsic property that promotes oscillatory behavior is postinhibitory rebound. When cells that exhibit this property are inhibited, a long-lasting voltage-dependent mixed cationic conductance is often activated (I_h current). When inhibition is removed this excitatory conductance is still active and pushes the membrane potential above rest making the cell more excitable. At a network level this boost in cellular excitability promotes activity on the formally inhibited side.
3. *Synaptic connectivity (inhibition/excitation) within the CPG is central for establishing patterns of output, such as alternation or synchronicity.* ►Synaptic properties are important for establishing the strength of connections between neurons in the network. For example, in many vertebrates, inhibitory commissural neurons couple together populations of cells that are rhythmically active and promote alternating activity between the two active populations. Interestingly, early in development intracellular chloride concentrations are high and as a result ►GABA and glycinergic (►glycine) effects are functionally excitatory. Therefore although the inhibitory commissural neurons still couple the two populations together, the bursting pattern is synchronized across the populations [5]. Thus changes in the sign of synaptic properties can fundamentally alter the motor pattern produced by CPGs. While chemical transmission is obviously critical, electrical synapses also contribute to network function. For example, in the pyloric rhythm of crustaceans, mixed synapses consisting of both chemical and electrical synapses can result in an activity-dependent reversal of sign. In vertebrate networks electrical synapses are common early in development, where they can lead to synchronization of cellular firing. One possible function of electrical synapses is to enhance network connectivity when the level of chemical transmission is low during embryonic stages of development [5]. It is now recognized that electrical synapses are also present in mammalian networks of adults, but their role is not well understood.
4. *CPGs are not static hard-wired networks.* A general principle of CPGs is that they can be reorganized to function in more than one motor task [6]. For example, the pyloric rhythm network in the stomatogastric ganglia of crustaceans consists of a network of 14 neurons connected by electrical and chemical connections. The pyloric rhythm activates striated muscles in the stomach, which helps push food through the digestive system, and is one of four networks within the stomatogastric ganglia. Neurons are able to switch from one network to another within the stomatogastric ganglia following administration of neuromodulators. Likely of importance to vertebrate systems, neuromodulators cause separate CPGs to blend together and even form entirely new networks. An intriguing possibility is that neuromodulators reconfigure vertebrate CPG circuits, similar to what occurs in invertebrate systems. Work on a variety of vertebrate preparations has demonstrated that 5-HT, noradrenaline and ►dopamine can modulate CPGs leading to the production of distinct patterns [5]. For example application of dopamine to the spinal cord can evoke a locomotor pattern that resembles walking whereas application of ►serotonin evokes a pattern closer to that of swimming. Of interest is how neuromodulators alter intrinsic and synaptic properties of interneurons that are components of the CPG. The lamprey system is arguably the best-described vertebrate system where 5-HT and dopamine's effects have been correlated to changes in cellular properties at the network level [3]. Both 5-HT and dopamine increase the duration of burst discharge thereby slowing the rhythm. The increase in burst discharge is due to actions on excitatory interneurons. Both neuromodulators act to decrease N, P/Q Ca^{2+} conductances, thereby indirectly reducing the amplitude of apamin sensitive K_{Ca} conductances. This leads to an increase in spike frequency and also results in longer lasting bursts of spikes. The cycle period increases because the increase in the population burst duration on one side ensures that inhibitory commissural interneurons inhibit activity on the opposite side. As a result of a longer period of inhibitory commissural activity it takes a longer time for the other side to "rebound" and the overall cycle period of the rhythm increases.
5. ►*Afferent feedback can affect the timing and pattern of CPG output.* Once networks are activated,

N

multiple changes in reflex sign and gain occur to allow afferent feedback to influence ongoing behavior. As a consequence, afferent feedback can stabilize the operation of CPGs and affect the timing and patterning of their output [7,9,10]. These types of observations have been made in many species performing diverse behaviors such as cockroach and cat walking, chewing in crustacea, feeding in the snail, scratching in the turtle and breathing in mammals. For example, under quiescent conditions, input from Golgi tendon organs that encode contractile force in extensor muscles has an inhibitory effect on motoneurons. However, when spinal CPGs are activated, input from extensor GTOs results in excitation of extensor motoneurons which delays the onset of the next flexor burst. Once CPGs are activated these effects are thought to occur by inhibition of the normal inhibitory pathway and an opening of a long-latency excitatory pathway. Afferent feedback can also select different patterns of CPG output. For example different scratch patterns in the turtle can be selected by afferent input from different regions. Similar types of results have been obtained when afferent input onto the gastric mill (crustacean chewing rhythm centre) is manipulated.

6. *Many of the patterns produced by CPGs are non-intuitive based on anatomical connectivity.* At first blush it would seem that once you understand what connects to what in a network you would be close to understanding its output. However, the output of a network also depends on the weighting of the connections, and the intrinsic properties of the neurons themselves. Importantly, these network, synaptic or intrinsic properties are interdependent and it is because of this interaction that many networks display ►emergent properties. An important tool for examining the dynamic performance of networks is to construct computational models [7]. For many reasons it is not realistic to model every parameter present in biological networks and all models make assumptions. Nevertheless, modeling a network offers several advantages for probing the operation of networks. For example, the effect of altering specific voltage-dependent conductances in a population of cells can be easily accomplished using a model, but is currently difficult to do experimentally. Modeling allows the exploration of parameter space that would be impossible to do experimentally which can lead to important insights into network function. Having said that, models should be able to offer predictions regarding network dynamics that can be tested experimentally. This process is important in order to validate the model as well as suggesting new avenues for exploration. An elegant example of the power of modeling is a study that took advantage of the stomatogastric ganglia system and demonstrated that multiple combinations of synaptic weights and intrinsic properties could produce similar network outputs in the pyloric network [7]. This type of insight would have been impossible to gain using conventional experimental techniques.

Future Directions

One of the attractive features of some invertebrate networks is that the number of cells in the CPG is often manageable and the cells can be identified. This allows experiments to be performed where the output of these identified cells can be manipulated by injecting hyperpolarizing or depolarizing current. On the other hand vertebrate CPGs present several challenges for experimentalists. The underlying networks are large compared to invertebrates and furthermore different classes of inhibitory and excitatory cells are often not clustered into discrete regions. Examination of mammalian spinal networks provides a perspective on these issues. The interneurons involved in spinal CPGs are not localized to specific nuclei and thus have eluded ready identification [8]. Recently, genetic approaches have provided new tools that allow classes of interneurons within the spinal cord to be identified. These interneurons can be silenced chronically or acutely allowing experimentalists to test whether the output of the network is disrupted. For example, the excitability of a class of inhibitory interneurons was recently manipulated using an allatostatin based system to silence the cells acutely [9]. Allatostatin receptors are found in insects but are not normally expressed in mammalian systems. Allatostatin receptors activate GIRK channels which hyperpolarize cells. This means that when allatostatin is applied to the spinal cord only the classes of neurons in which allatostatin receptors were artificially introduced are hyperpolarized. At the moment these are among the best tools we have for determining whether certain classes of interneurons are important for regulating network function in mammals. Having said that, approaches that examine connectivity between oscillators will be easier to interpret compared to deletion of classes that are putatively part of the oscillator itself. One of the major issues here is addressing necessity and sufficiency. If deletion of a class of interneurons blocks activity, it is difficult to conclude that they are essential components of the network. This is because deletion of the ceus may transiently reduce the excitability of the network. However homeostatic mechanisms could compensate for this reduction in excitability. Likewise, if the perturbation produces no effect we cannot easily conclude that these cells are not part of the network. Therefore, we will need to develop techniques that rigorously examine the effectiveness of manipulation techniques at a population rather than a cell-by-cell level. To do this we will need new tools to understand network function. Critically, we need to target candidate CPG interneurons and examine their connectivity within the network. It is clear that sampling the output of each neuron using intracellular recording techniques will not be sufficient to make conclusions regarding connectivity within the population of interneurons in mammalian networks. What is needed is perhaps the use of voltage sensitive dyes combined with genetic approaches that would double label populations of interest. In the future we will likely see the merging of these two approaches to examine network connectivity. Another critical test will be the ability to quickly activate and inactivate discrete populations of cells during network activity, like those performed in other preparations such as the STG network of crustacea. Once classes of interneurons that contribute to CPG function are identified, it will be necessary to test their ►in vivo functionality. The zebrafish embryo provides an example of what is possible. The zebrafish embryo is transparent and cells in the escape network can be visualized in the behaving animal [10]. This has allowed experimenters to ablate cells in the circuit and examine changes in behavior. While similar experiments in mammals will obviously be much more difficult, the use of multiphoton approaches in combination with genetic silencing approaches should allow new insights into the operation of mammalian CPG circuits.

References

1. Graham-Brown T (1911) The intrinsic factors in the act of progression in the mammal. Proc R Soc Lond 84:308–319
2. Getting PA (1989) Emerging principles governing the operation of neural networks. Annu Rev Neurosci 12:185–204
3. Grillner S (2003) The motor infrastructure: from ion channels to neuronal networks. Nat Rev Neurosci 4 (7):573–586
4. Harris-Warrick RM (2002) Voltage-sensitive ion channels in rhythmic motor systems. Curr Opin Neurobiol 12 (6):646–651
5. Whelan PJ (2003) Developmental aspects of spinal locomotor function: insights from using the in vitro mouse spinal cord preparation. J Physiol 553 (pt 3):695–706
6. Marder E, Thirumalai V (2002) Cellular, synaptic and network effects of neuromodulation. Neural Netw 15 (4–6):479–493
7. Prinz AA (2006) Insights from models of rhythmic motor systems. Curr Opin Neurobiol 16(6):615–620
8. Kiehn O (2006) Locomotor circuits in the mammalian spinal cord. Annu Rev Neurosci 29:279–306
9. Gosgnach S et al. (2006) V1 spinal neurons regulate the speed of vertebrate locomotor outputs. Nature 440 (7081):215–219
10. Fetcho JR, Higashijima SI, McLean DL (2008) Zebrafish and motor control over the last decade. Brain Res Rev 57(1):86–93

Neurotransmitters in the Auditory System

Ronald S. Petralia, Robert J. Wenthold
NIDCD/NIH, Bethesda, MD, USA

Synonyms

Excitatory and inhibitory neurotransmission at synapses in the central auditory system and cochlea

Definition

In the auditory (hearing) system, as in other parts of the nervous system, neurotransmitters are chemicals that cross a synapse and mediate nerve impulse transmission from one neuron to the next neuron, or from sensory cells (hair cells in the cochlea of the inner ear) to neurons, or from neurons to effector cells (in this case, the same sensory hair cells of the cochlea). Neurotransmitters typically excite or inhibit the neuron to enhance or reduce nerve impulse transmission, but they can also mediate long-term changes in the neuron, including maturation and learning. All of these functions involve interplay of different neurotransmitters, with some kinds modulating the effects of other kinds, to shape the response. For the auditory system, this results in accurate sound identification and localization, and the integration of audition with other sensory modalities and with locomotor responses.

Characteristics

The auditory system is a complex system of interconnected neural centers extending from the sensory hair cells of the cochlea to the auditory cortex, and employs a wide variety of neurotransmitters and their receptors. As in other neural systems in the brain, neurotransmission in the auditory system largely involves a balanced release of excitatory and inhibitory neurotransmitters. The ascending excitatory neurotransmission from the cochlea to the cortex relays the transduced auditory information; along the way, the information is filtered and refined utilizing inhibitory neurotransmission. Excitatory neurotransmission is mediated primarily by glutamate, while GABA and glycine mediate inhibitory neurotransmission. In general, discussion here of excitatory connections within the brain will refer to glutamatergic-type connections, although the glutamatergic nature of the connection may not have been established definitively in all cases. In addition, a number of other neurotransmitter compounds (described below) can modulate neural responses in a variety of ways.

The auditory system serves to identify both the nature of a sound and its localization in space and time. In particular, sound localization requires unusually rapid neurotransmission. Thus, several components of the auditory system are designed for this very fast transmission and include specialized ▶glutamate receptors (GluRs) for fast excitatory connections, and a preference for glycinergic inhibitory neurotransmission over the somewhat slower GABAergic neurotransmission, as discussed below.

Basic Patterns of Excitatory and Inhibitory Innervation in the Auditory System

The component structures making up the auditory system are numerous and their interconnections are complex [1–3]. They are discussed in detail in other essays in this encyclopedia and can only be summarized here (Fig. 1).

Sound elicits vibrations in the cochlea; these vibrations are increased and sharpened by outer hair cells (OHCs), which act as a cochlear amplifier. These vibrations are then transduced into nerve impulses by the inner hair cells (IHCs), which have a glutamatergic connection with afferent dendritic nerve endings from spiral ganglion cells. The latter send axons into the brain (auditory nerve) and these axons form glutamatergic connections with the ventral (VCN) and dorsal (DCN) cochlear nuclei. In addition to these IHC connections, OHCs form synapses with a small number of ganglion cells and these also send input to the cochlear nuclei, but their function is not well known. Axons from various neuron types in the cochlear nuclei form excitatory connections with the nuclei of the superior olivary complex (SOC) and lateral lemniscus (LL) and the inferior colliculus (IC). In the SOC, the best-studied nuclei include the lateral and medial superior olive (LSO, MSO) and the medial nucleus of the trapezoid

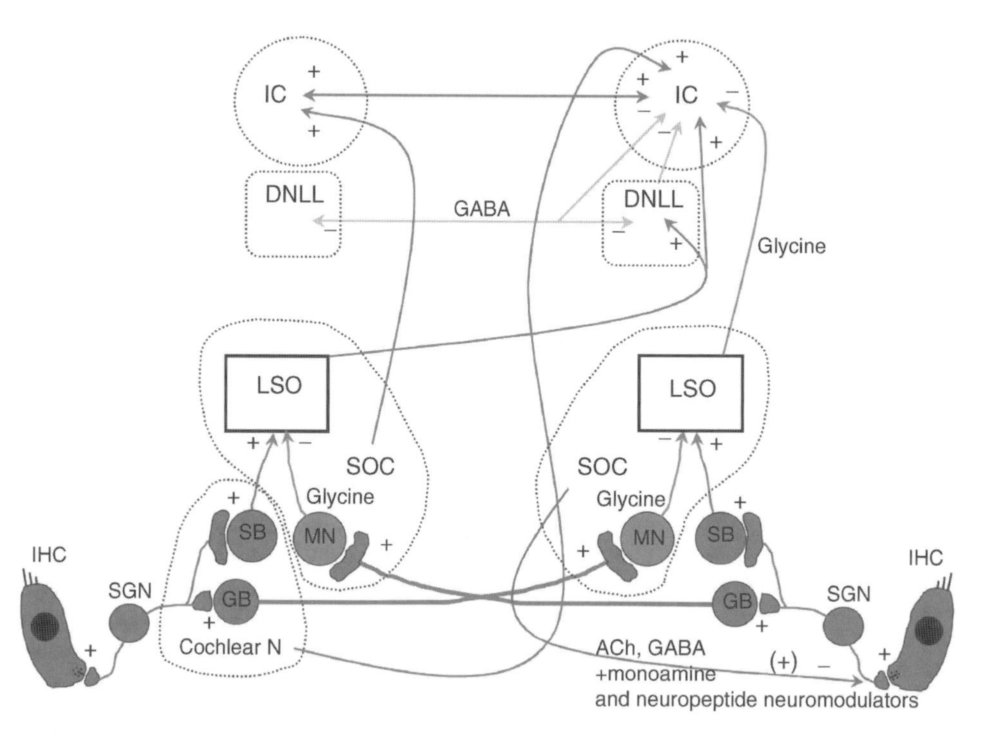

Neurotransmitters in the Auditory System. Figure 1 *Simplified diagram of a sample of neurotransmission in the Auditory System.* Only a few of the main cell types, nuclei, and connections are shown. In the cochlea, sounds induce vibrations, which are amplified by OHCs (not shown), and these vibrations are transduced into nerve impulses by IHCs (IHCs). The first few connections are mainly rapid relays. IHCs form excitatory synapses (*red color*, +) with spiral ganglion neurons (SGN), and these form excitatory synapses with neurons in the cochlear nuclei. Within the cochlear nuclei, spherical bushy cells (SB) send excitatory input to LSO in the SOC. Beginning at this level, some more complicated connections allow neurons to compare sound input from both ears (i.e., for sound localization). LSO neurons receive ipsilateral excitatory input from SB and contralateral input from globular bushy cells (GB) via an excitatory synapse with neurons of the MNTB (MN). MNTB neurons make glycinergic inhibitory (*green color*, –) synapses with LSO neurons. In turn, cochlear nuclei and LSO and other SOC nuclei make excitatory and inhibitory connections with higher centers. Combination of excitatory (probably mainly glutamatergic) and inhibitory (GABA {*orange color*} or glycine {*green color*}) connections allows binaural integration of sound information at these higher centers (DNLL; IC). The latter brain regions then send connections to MGB, which connects to auditory cortex (not shown). In addition, sound transduction in the cochlea is modulated by the olivocochlear efferent input from the SOC to both IHCs and OHCs.

body (MNTB). The major excitatory connections from the cochlear nuclei are with the ipsilateral LSO, bilateral MSO, and contralateral MNTB. This contralateral MNTB then makes a glycinergic inhibitory connection with its ipsilateral LSO. Thus, the SOC is the lowest level where the ascending auditory inputs from the two ears converge. Next, the three nuclei of the LL (NLL) receive contralateral (i.e., monaural only) excitatory input from the cochlear nucleus. Convergence of auditory input is seen again in the IC, which receives contralateral excitatory input from the cochlear nuclei, LSO and other IC, ipsilateral excitatory input from the MSO, and ipsilateral glycinergic inhibitory input from the LSO. The IC also receives ipsilateral GABA, glycine and excitatory input from the three NLL, as well as GABA inhibitory input from the contralateral dorsal nucleus of the NLL (DNLL; i.e., 1 of the 3 NLL) and IC. The IC then sends glutamatergic excitatory and GABA inhibitory connections to the ipsilateral medial geniculate body (MGB) in the thalamus, and the MGB sends excitatory connections to the auditory cortex.

Major Examples of Excitatory Inhibitory Control of Auditory Signal Processing

Many of the complex ipsilateral and contralateral excitatory and inhibitory connections discussed in the previous paragraph serve to localize sounds based on interaural intensity disparities (i.e., between the two ears). This is for high frequency sounds; for low frequency sounds, sound localization depends on interaural timing/phase differences. Basically, neurons that measure interaural intensity differences are excited by stimulation of one ear and are progressively inhibited by increasing stimulus intensity at the other ear, and are called EI neurons [1]. Neurons that act in this way are found in the LSO, DNLL and IC (Fig. 1). Although the combined roles of these three structures is not fully understood, the LSO may be responsible for the basic sound localization; it has an ipsilateral excitatory input, and an ipsilateral glycinergic MNTB input from neurons that receive a contralateral excitatory (cochlear nuclear) input. In contrast, the IC combines inputs from the LSO (excitatory contralateral and glycinergic ipsilateral) and DNLL (GABAergic bilateral) with other direct inputs, including those from the cochlear nuclei; the IC may utilize this input to distinguish the first sound from the typically abundant echoes that follow it.

While the previous examples look at the convergence of excitatory and inhibitory connections between different brain regions, most parts of the brain have local inhibitory interneurons within the brain structure, and these impinge on the principal excitatory (usually glutamatergic) neurons of that structure. This local inhibitory circuitry helps to refine the output of the excitatory principal neurons, making the response more specific. The best example of this is the auditory cortex [2], which has pyramidal neurons that send glutamatergic, descending efferent fibers to lower auditory structures. These excitatory neurons receive their major excitatory input from the MGB, and receive GABA-mediated inhibition from a variety of neuron types within the cortex. Another good example is the DCN [2]. Its (pyramidal) neurons project excitatory fibers to the contralateral IC (and MGB) and receive glutamatergic input from the auditory nerve onto their basal dendrites; the auditory nerve also provides glutamatergic input to small GABA and glycinergic inhibitory neurons that then form synapses on these basal dendrites. In contrast, the apical dendrites of these fusiform cells receive glutamatergic input from small, local ►granule cells. Granule cells also form glutamatergic synapses on local glycinergic (+GABA) cartwheel cells and GABAergic stellate cells; both of these inhibitory neurons form synapses on the apical dendrites of the fusiform cells [4]. The major input to granule cells appears to come from somatosensory inputs from the external ear (pinna). Thus, basal dendrites receive auditory nerve input and associated inhibition and apical dendrites receive mainly somatosensory input and associated inhibition. This arrangement may allow the DCN to coordinate pinna orientation and sound localization [5].

Two notable exceptions to this pattern of principal excitatory neurons modified by local inhibitory neurons are found in the rat auditory system – almost all DNLL neurons are GABAergic and almost all MGB neurons are excitatory. The MGB may show an evolutionary trend, with almost no GABAergic neurons in rats and bats, while about 25% of the neurons in the MGB in monkeys and cats are GABAergic; this may reflect more complex auditory communication in the latter animals [2,6]. For rats and bats, refinement of the MGB output via inhibitory input depends on GABAergic fibers from the IC and the thalamic reticular nucleus; the latter inhibitory input may affect synchronization of MGB processing with general arousal.

Specific Neurotransmitters of the Auditory System

Ideally, identification of a specific type of neurotransmission will include studies of neurotransmitters and their receptors, transporters and vesicular transporters using biochemistry, in situ hybridization and light and electron microscope immunocytochemical localization [7].

Excitatory Neurotransmitters

As noted above, most excitatory neurotransmission in the auditory system is mediated by glutamate or related compounds; we also consider acetylcholine (ACh) neurotransmission in this section, because they classically are considered excitatory.

As noted above, glutamate or a related compound probably acts as the excitatory neurotransmitter for the IHCs, spiral ganglion neurons (auditory nerve), and principal projection neurons in most auditory brain structures. However, for many of these excitatory projections, the glutamatergic nature of the neurotransmitter has not been determined definitively. GluRs are common at auditory nerve terminals, granule cell terminals, and other kinds of terminals in neurons throughout the cochlear nuclei [8]. Both ipsilateral and contralateral glutamatergic input was confirmed for the LSO, MSO, and another SOC nucleus [9]. Cochlear nuclear input that forms giant terminals, the calyces of Held, on neurons of the MNTB is also probably glutamatergic (Fig. 2). Other studies have supported the glutamatergic nature of inputs to the DNLL, various inputs to the IC, IC commissural connections, inputs from the central nucleus of the IC to the MGB, input from the IC and auditory cortex to the ventral division of the MGB, and in general the output of the auditory cortex, including that from the layer V pyramidal cells to the IC [2].

GluRs include ionotropic receptors, which form a sodium or calcium channel made up of four or five subunits, and metabotropic receptors, which link to G proteins to elicit long-term changes in neurons [7,8,10]. The major ionotropic GluRs that handle fast neurotransmission are AMPA receptors. AMPA receptors vary in function and properties depending on their subunit makeup (GluR1–4); the most common type contains GluR2 and passes sodium but is impermeable to calcium. While the latter type is common in many parts of the auditory system, some auditory synapses use a special kind of ►AMPA receptor that contains primarily GluR3

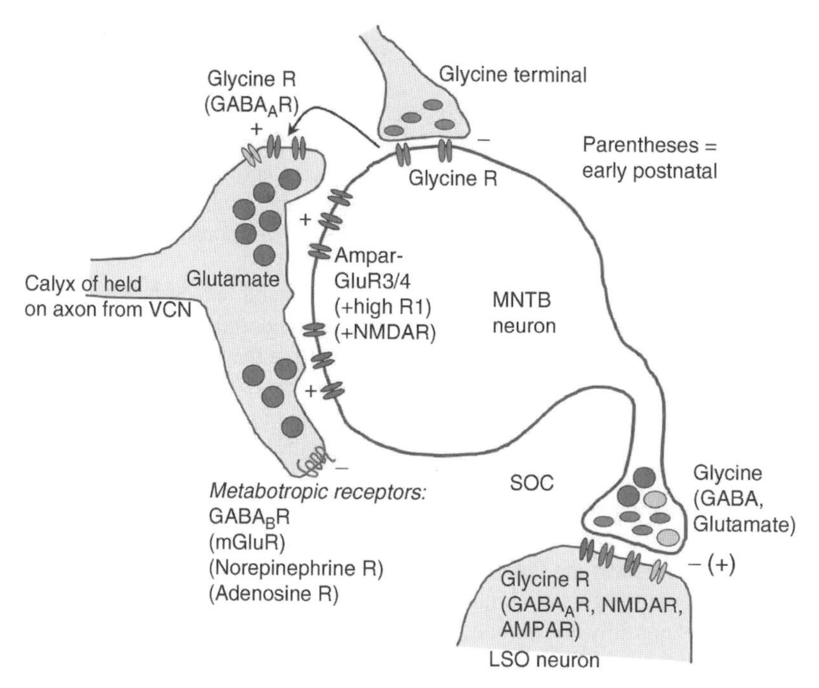

Neurotransmitters in the Auditory System. Figure 2 Example of how various neurotransmitters and their receptors are integrated in part of the auditory system, and how this changes during postnatal development. *Red, green, and orange colors* represent excitatory, glycinergic, and GABAergic neurotransmission, respectively, as in Fig. 1. During early postnatal development, some neurotransmitters and their receptors are present transiently (shown in parentheses in the diagram) and mediate synaptic plasticity, e.g., establishment of final synaptic connections and maturation of their physiological properties. The basic adult circuit shown here includes a thick axon from a VCN-globular bushy cell that forms a calyx of Held-glutamatergic synapse on a neuron of the MNTB. The postsynaptic AMPA receptors contain the GluR3 and GluR4 subunits; these mediate a fast relay of the signal. In addition, presynaptic receptors (mainly glycine and GABA) help to modulate the signal. The MNTB neuron then relays the signal via a glycinergic inhibitory connection with neurons in LSO in the SOC. However, during early postnatal development, neurotransmission is slower, being mediated through postsynaptic AMPA receptors that are high in GluR1 subunit, and through NMDA receptors, as well as being modulated by several neurotransmitters impinging on presynaptic receptors. These include $GABA_A$ ionotropic receptors (later replaced by glycine receptors) and $GABA_B$, mGluR (glutamatergic), norepinephrine, and adenosine metabotropic receptors (only $GABA_B$ remains common in adult). This early arrangement is required for maturation of this synapse. An even more radical change is seen during development of MNTB neuron synapses with LSO neurons. During early postnatal ages, these terminals secrete both excitatory and inhibitory neurotransmitters, although the adult synapse is just inhibitory. The early presence of glutamate and its associated receptors at an inhibitory synapse indicates the fundamental role that GluRs play in maturation of synapses.

and GluR4 (especially the "flop" subtype of GluR4), is calcium permeable, and is unusually fast – this facilitates the very rapid neurotransmission needed to effect the rapid relay of auditory information, such as needed for sound localization. For example, this type is found in the giant specialized endbulb-type glutamatergic synapses between the auditory nerve and the spherical ►bushy cells of the VCN, and in the ►calyx of Held between axons from globular bushy cells of the VCN and neurons of the MNTB (Fig. 2). Interestingly, neurotransmission at this synapse is relatively slow in early postnatal development, probably due to the prevalence of GluR1 in the AMPA receptors; neurotransmission becomes faster coincident with an increase in the prevalence of GluR3/4-containing AMPA receptors [11]. A similar reduction in GluR1 with development may occur in some kinds of neurons of the cochlear nuclei [7].

Another important ionotropic GluR is the ►NMDA receptor, which passes calcium and is implicated in neuronal plasticity during development and learning; it is most famous for its role in the establishment of long-term potentiation of synaptic current. Typical NMDA receptors contain the main subunit, NR1, combined with one or more NR2 subunits. NMDA receptor levels vary widely in the auditory system [2,8]) and they probably participate in modification of synaptic responses where they are prevalent; this may be particularly important for the development of many synapses (Fig. 2). In addition to typical NMDA receptors, there may be other, less studied forms containing subunits of NR3. NR3A is found in a number of auditory brain regions [12] and may form a special kind of excitatory glycine receptor when combined with the most common subunit, NR1.

Delta (δ) ionotropic GluRs are poorly understood and do not normally form a functional channel. Interestingly, the δ1 subunit is highly expressed in IHCs and ganglion cells of the cochlea [13]. Expression of δs is generally low in the adult brain, except in the cerebellum (δ2) and outer DCN [8]. In the DCN, δ labeling is high in granule cell-parallel fiber synapses of cartwheel cell neurons and those of apical dendrites of fusiform cells; δ is relatively uncommon in basal dendrites of fusiform cells. In contrast, in these fusiform cells, the AMPA receptor subunit, GluR4, and the metabotropic GluR, mGluR1α, are found only in synapses of basal dendrites. This differential distribution of GluRs is related to co-processing of functionally different sensory inputs on fusiform cells, as discussed above. Thus, fast auditory nerve transmission to basal dendrites probably requires high GluR4, while granule cell input to apical dendrites is a type that is not modulated by metabotropic GluRs. The need for high δ in synapses between granule cells and outer DCN neurons is not understood; probably the function of these synapses corresponds to that of similar δ2-containing, granule cell-parallel fiber synapses in the cerebellum and as in the latter, δ probably plays a modulatory role.

Cholinergic neurotransmission has been studied best in the olivocochlear system. This consists of special medial and lateral groups of neurons in the SOC that send efferent axons to outer and inner cochlear hair cells, respectively [2]. Those going to OHCs probably modify OHC function to enhance transduction or signal detection, while those associated with IHCs mainly form synapses on the afferent fibers beneath the IHCs and probably have a modulatory role. Interestingly, cholinergic neurotransmission that is mediated through α9/α10 nicotinic ACh receptors has an indirect hyperpolarizing effect on hair cells, and thus inhibits hair cell neurotransmitter release [14]. In addition to this efferent system to the cochlea, some cholinergic connections are seen throughout the auditory system. Both groups of olivocochlear efferents provide some collateral input to the VCN. There is also evidence for a substantial number of cholinergic neurons in the external cortex of the IC. Cholinergic input probably exerts regulatory influences on basic neural circuitry. ACh receptors in the auditory system include both nicotinic ionotropic and muscarinic metabotropic types. In particular, α7 nicotinic ACh receptors may be important during postnatal development throughout the brain auditory system [15,16]. Presynaptic α7 nicotinic ACh receptors at immature glutamatergic synapses, activated via diffusion of ACh from nearby cholinergic terminals, may facilitate glutamate release at these terminals and consequently help mediate maturation of these synapses [16]. The auditory system is also involved with the cholinergic system of the brain in an indirect way. Auditory stimuli activate cholinergic neurons in the reticular activating system in the brainstem; these neurons project ascending and descending tracts extending from the cortex to the spinal cord and provide an arousal mechanism that precedes locomotor response [17].

Inhibitory Neurotransmitters

As noted above, both GABAergic and glycinergic inhibitory neurotransmission are prevalent throughout the auditory system [2,3,7]. Generally, inhibitory synapses may be distinguished from excitatory synapses by the shape of the vesicles (oval to flat in inhibitory and round in excitatory) and the proportion of pre/postsynaptic densities (thicker postsynaptic density in excitatory synapses). ►GABA receptors include both ionotropic ($GABA_A$) and metabotropic ($GABA_B$) types. A portion of the lateral group of neurons of the olivocochlear system, associated with IHCs, is GABAergic. In fact, in the mouse, GABA may co-localize with ACh in olivocochlear efferent terminals going to both IHCs and OHCs [18]. Auditory brain regions typically have mixtures of GABAergic and glycinergic neurons in different proportions. In the cochlear nuclei, inhibitory neurons and both GABA and ►glycine receptors are more common in

DCN than VCN. Other auditory brain regions send out particularly important groups of inhibitory fibers (as noted above) – GABAergic especially for those fibers from DNLL to IC and from IC to MGB, and glycinergic especially for those from MNTB to LSO, LSO to IC, and VNLL to IC. GABA and glycine can also occur in the same neuron populations. Glycinergic inputs tend to mediate faster neurotransmission than GABAergic. Not surprisingly, glycinergic neurotransmission may replace GABAergic neurotransmission in some neurons during postnatal development, as some auditory connections speed up (Fig. 2; as described above for a similar developmental change in AMPA-type GluRs). Thus, in the calyx of Held (glutamatergic) synapses on MNTB neurons, presynaptic glycine receptors replace $GABA_A$ receptors during postnatal development, coincident with development of inhibitory glycinergic input on postsynaptic MNTB neurons [19,20]. The presynaptic glycine receptors probably respond to spillover of glycine released from the nearby glycinergic synapses; activation of these presynaptic receptors enhances glutamate release from the calyx terminals. Another interesting association between excitatory and inhibitory neurotransmission at the same synapse is found in MNTB-neuron terminals in LSO (Fig. 2). During early postnatal development, these terminals release GABA, glycine and glutamate, and the postsynaptic membrane bears receptors for all three neurotransmitters [21]. Glutamate activation of NMDA receptors at these synapses probably mediates activity-dependent refinement of this inhibitory circuit. Finally, as this synapse matures, it becomes predominantly glycinergic.

Monoamine Neurotransmitters/Neuromodulators

The monoamines (biogenic amines) include the catecholamines (dopamine, norepinephrine, and epinephrine), serotonin (=5HT), and histamine. They act as neuromodulators (►Monoamine Neurotransmitters/Neuromodulators) that activate second messenger systems in the affected neuron, thus modifying synaptic responses or having other broad effects on neurons. The auditory regions of the brain are widely innervated by different monoamines, especially norepinephrine (=noradrenaline; from locus coeruleus and other brainstem cell groups) and serotonin (from midline raphe nuclei of the brainstem reticular formation), although their function in the auditory system remains mainly speculative (e.g., [22–24]). Norepinephrine helps regulate glutamate release from the calyx of Held synapses on MNTB neurons, via inhibition mediated by presynaptic norepinephrine receptors (Fig. 2; [25]).

Both norepinephrine and serotonin are involved in arousal mechanisms, so they probably affect selective attention in the auditory system [17,22]. In the cochlear nuclei, norepinephrine may have effects on detection of sounds in noise and on timing mechanisms [23]. The cochlea contains norepinephrine, dopamine, and serotonin; changes in turnover of the former two occur in response to white noise [26]. Serotonergic fibers are closely associated with olivocochlear efferent fibers near the bases of IHCs and OHCs.

Neuropeptides and Other Substances as Auditory Neuromodulators

A wide variety of ►neuropeptides are found throughout the auditory system and they modulate neuron excitability and other functions, usually via metabotropic receptors [27–29]. In the cochlea, opioid peptides and calcitonin gene-related peptide (CGRP) are found in the olivocochlear efferents; these innervate hair cells and afferents to IHCs (Fig. 2). Two opioid peptides, endomorphin-1 and dynorphin B, inhibit α 9/α10 nicotinic ACh receptor-mediated inhibition of hair cells (see above; [14]. Opioids are probably released from cholinergic efferent terminals when the firing frequency passes a certain threshold (e.g., after noise exposure), which is sufficient to induce exocytosis of opioid-containing large dense-core vesicles. Opioid secretion has been implicated in hyperacusis and peripheral tinnitus [14,28]. In auditory brain regions, neuropeptide input includes that from other auditory regions. For example, somatostatin is found in some neuron somata in VCN, some cells near LSO, and some cells of LL and IC [27]. The greatest convergence of neuropeptide innervation may be in the ventral nucleus of the trapezoid body (VNTB) and the small cell cap plus granule cell regions of the cochlear nuclei; these include substance P, CGRP, enkephalins, dynorphins, cholecystokinin, and somatostatin [28]. These peptides exert different effects substance P strongly excites VNTB neurons, leu-enkephalin strongly inhibits VNTB neurons, and cholecystokinin can show either excitatory or inhibitory effects on VNTB neurons, while somatostatin modulates the release of other neurotransmitters.

There are probably numerous other substances that can act as neurotransmitters or neuromodulators. Nitric oxide (NO) can act as a neuromodulator in the brain and may be involved in NMDA GluR-mediated synaptic plasticity; NO has been described in IC [2] and cochlear spiral ganglion neurons [30]. The purine ATP can act as a neuromodulator in the cochlea in the spiral ganglion neurons, and specifically at their afferent synapses with hair cells. ATP activates ionotropic ►purine receptors (P2X) on spiral ganglion neuron membranes. This initiates a calcium influx that induces NO production in the neuron. Purines can also act on metabotropic receptors, including ATP-sensitive P2Y and adenosine-sensitive P1 receptors. Adenosine, converted from ATP that is probably secreted with glutamate from immature calyx of Held synapses, binds to presynaptic adenosine receptors (Fig. 2; see above), inhibiting glutamate release to help regulate high frequency neurotransmission at this synapse [31].

References

1. Pollak GD, Burger RM, Klug A (2003) Dissecting the circuitry of the auditory system. Trends Neurosci 26:33–39
2. Malmierca MS, Merchán MA (2004) Auditory system. In: Paxinos G (ed) The rat nervous system, 3rd edn. Elsevier Academic, New York, pp 997–1082
3. Wenthold RJ, Hunter C (1990) Immunocytochemistry of glycine and glycine receptors in the central auditory system. In: Ottersen OP, Storm-Mathisen J (eds) Glycine Neurotransmission. Wiley, New York, pp 391–416
4. Rubio ME (2004) Differential distribution of synaptic endings containing glutamate, glycine, and GABA in the rat dorsal cochlear nucleus. J Comp Neurol 477:253–272
5. Kanold PO, Young ED (2001) Proprioceptive information from the pinna provides somatosensory input to cat dorsal cochlear nucleus. J Neurosci 21:7848–7858
6. Winer JA, Larue DT (1996) Evolution of GABAergic circuitry in the mammalian medial geniculate body. Proc Natl Acad Sci USA 93:3083–3087
7. Wenthold RJ, Hunter C, Petralia RS, Niedzielski AS, Wang Y-X, Safieddine S, Zhao H-M, Rubio ME (1997) Receptors in the auditory pathways. In: Berlin CI (ed) Neurotransmission and hearing loss: basic science, diagnosis, and management. Singular, San Diego, pp 1–23
8. Petralia RS, Rubio ME, Wang Y-X, Wenthold RJ (2000) Differential distribution of glutamate receptors in the cochlear nuclei. Hear Res 147:59–69
9. Srinivasan G, Friauf E, Löhrke S (2004) Functional glutamatergic and glycinergic inputs to several superior olivary nuclei of the rat revealed by optical imaging. Neuroscience 128:617–634
10. Wenthold RJ, Hunter C, Petralia RS (1993) Excitatory amino acid receptors in the rat cochlear nucleus. In: Merchan MA, Juiz JM, Godfrey DA, Mugnaini E (eds) The mammalian cochlear nuclei: organization and function. Plenum, New York, pp 179–194
11. Joshi I, Shokralla S, Titis P, Wang L-Y (2004) The role of AMPA receptor gating in the development of high-fidelity neurotransmission at the calyx of Held synapse. J Neurosci 24:183–196
12. Wong H-K, Liu X-B, Matos MF, Chan SF, Pérez-Otaño I, Boysen M, Cui J, Nakanishi N, Trimmer JS, Jones EG, Lipton SA, Sucher NJ (2002) Temporal and regional expression of NMDA receptor subunit NR3A in the mammalian brain. J Comp Neurol 450:303–317
13. Safieddine S, Wenthold RJ (1997) The glutamate receptor subunit δ1 is highly expressed in hair cells of the auditory and vestibular systems. J Neurosci 17:7523–7531
14. Lioudyno MI, Verbitsky M, Glowatzki E, Holt JC, Boulter J, Zadina JE, Elgoyhen AB, Guth PS (2002) The α9/α10-containing nicotinic ACh receptor is directly modulated by opioid peptides, endomorphin-1, and dynorphin B, proposed efferent cotransmitters in the inner ear. Mol Cell Neurosci 20:695–711
15. Happe HK, Morley BJ (2004) Distribution and postnatal development of α7 nicotinic acetylcholine receptors in the rodent lower auditory brainstem. Dev Brain Res 153:29–37
16. Metherate R, Hsieh CY (2004) Synaptic mechanisms and cholinergic regulation in auditory cortex. Prog Brain Res 145:143–156
17. Skinner RD, Homma Y, Garcia-Rill E (2004) Arousal mechanisms related to posture and locomotion: 2. Ascending modulation. Prog Brain Res 143:291–298
18. Maison SF, Adams JC, Liberman MC (2003) Olivocochlear innervation in the mouse: immunocytochemical maps, crossed versus uncrossed contributions, and transmitter colocalization. J Comp Neurol 455:406–416
19. Turecek R, Trussell LO (2002) Reciprocal developmental regulation of presynaptic ionotropic receptors. Proc Natl Acad Sci USA 99:13884–13889
20. Awatramani GB, Turecek R, Trussell LO (2005) Staggered development of GABAergic and glycinergic transmission in the MNTB. J Neurophysiol 93:819–828
21. Gillespie DC, Kim G, Kandler K (2005) Inhibitory synapses in the developing auditory system are glutamatergic. Nat Neurosci 8:332–338
22. Behrens EG, Schofield BR, Thompson AM (2002) Aminergic projections to cochlear nucleus via descending auditory pathways. Brain Res 955:34–44
23. Thompson AM (2003) A medullary source of norepinephrine in cat cochlear nuclear complex. Exp Brain Res 153:486–490
24. Thompson AM, Hurley LM (2004) Dense Serotonergic innervation of principal nuclei of the superior olivary complex in mouse. Neurosci Lett 356:179–182
25. von Gersdorff H, Borst JGG (2001) Short-term plasticity at the calyx of Held. Nat Rev Neurosci 3:53–64
26. Gil-Loyzaga P, Bartolomé V, Vicente-Torres A, Carricondo F (2000) Serotonergic innervation of the organ of Corti. Acta Otolaryngol 120:128–132
27. Wynne B, Robertson D (1997) Somatostatin and substance P-like immunoreactivity in the auditory brainstem of the adult rat. J Chem Neuroanat 12:259–266
28. Robertson D, Mulders WHAM (2000) Distribution and possible functional roles of some neuroactive peptides in the mammalian superior olivary complex. Microsc Res Tech 51:307–317
29. Martín F, Coveñas R, Narváez JA, Tramu G (2003) An immunocytochemical mapping of somatostatin in the cat auditory cortex. Arch Ital Biol 141:157–170
30. Yukawa H, Shen J, Harada N, Cho-Tamaoka H, Yamashita T (2005) Acute effects of glucocorticoids on ATP-induced $Ca2^+$ mobilization and nitric oxide production in cochlear spiral ganglion neurons. Neuroscience 130:485–496
31. Kimura M, Saitoh N, Takahashi T (2003) Adenosine A_1 receptor-mediated inhibition at the calyx of Held of immature rats. J Physiol 553.2:415–426

Neurotransmitters in the Gut

Definition

Enteric neurons synaptically communicate with each other using neurotransmitters. Acetylcholine (ACh) is the main transmitter mediating fast excitatory postsynatptic potentials (fast EPSPs, nicotinic in nature).

Many other neuroactive substances have also been identified as putative neurotransmitters for fast EPSPs

(ATP, glutamate) and slow EPSPs (serotonin (5-HT), substance P, vasoactive intestinal peptide (VIP), and other peptides) and inhibitory PSPs (noradrenalin, GABA, somatostatin) within the enteric nervous system. Furthermore, neurotransmitters released from enteric efferent neurons to intestinal effectors include: motor excitatory (substance P); motor inhibitory (ATP, nitric oxide, VIP, PACAP); and secretary (VIP). Many substances, including amines, amino acids and peptides may act as neuromodulators in the gut. So called braingut peptides are a group of peptides present both in the enteric and central nervous systems.

►Acetylcholine (ACh)
►ATP
►Bowel Disorders
►GABA
►Gutamate
►Noradrenalin
►PACAP
►Serotonin

Neurotrophic Factors

Definition

Neurotrophic factors are a family of proteins responsible for the growth and survival of neurons during development and for the maintenance of adult neurons.

They are also capable of promoting damaged axons to regenerate after various peripheral and central nervous system injuries.

►Neural Development
►Neurotrophic Factors in Nerve Regeneration
►Regeneration
►Growth Factors

Neurotrophic Factors in Nerve Regeneration

Matt S. Ramer, Lowell T. McPhail
International Collaboration on Repair Discoveries, the University of British Columbia, Vancouver, BC, Canada

Definition

Neurotrophic factors (NTFs) are naturally-occurring multifunctional secreted proteins, expressed throughout development and into adulthood, and in part serve to promote neuronal survival, to support axonal outgrowth and target innervation, and in some cases to modulate synaptic transmission. This essay covers only their involvement in axonal regeneration. Three families of NTFs (the neurotrophins, the glial cell line-derived NTF family, and the ►interleukin-6 (IL-6) family), and their effects on regeneration of two classes of peripheral nerve axons (primary sensory axons, and the axons of lower motoneurons) are used to illustrate the principle that regeneration is often, but not always, accelerated or otherwise improved by endogenous or exogenous NTFs. Where known, mechanisms of action are also described.

Characteristics

Quantitative Description

The Model Systems

Vertebrate primary sensory neurons are pseudounipolar and clustered in bilaterally-symmetrical ganglia (dorsal root ganglia, DRG) adjacent to the spinal cord (Fig. 1).

Their single axon bifurcates within the DRG: one branch travels to the periphery where it innervates sense organs in skin, muscles or viscera; the other projects via the dorsal root to the spinal cord where it innervates spinal and/or supraspinal neurons. Most (~70%) DRG neurons detect temperature or noxious (painful) stimuli. These are small-to-medium in diameter, with thin, unmyelinated and slowly-conducting axons which innervate superficial spinal laminae centrally. Large diameter DRG neurons subserve proprioception and mechanoreception and terminate in deeper spinal laminae. Many of these axons also bifurcate central to their entry point within the cord and send a long projection via the dorsal columns to nuclei in the brainstem (Fig. 1). While injury to sensory axons can occur in any of these regions (peripheral nerve, dorsal roots and spinal cord), the differing environments in which the axons are injured dictate vastly differing outcomes: peripheral nerve injury can be followed by successful regeneration and target reinnervation; axons injured within dorsal roots regenerate up to the PNS: CNS interface, but do not penetrate the spinal cord; axons injured within dorsal columns do not regenerate at all, and may retract from the injury site for several hundred microns.

Lower motoneurons are situated in the ventral spinal cord and brainstem. Their axons exit the CNS via ventral roots or cranial motor nerves. Spinal motor axons join mixed spinal nerves just distal to the DRG, and innervate intra- and extrafusal muscle fibers in the periphery (Fig. 1). Motor axons can be injured anywhere along their length, but as with sensory axons, regenerative success depends on the site of injury: if axons are severed between their exit point in the spinal cord and their peripheral targets, they may regenerate and

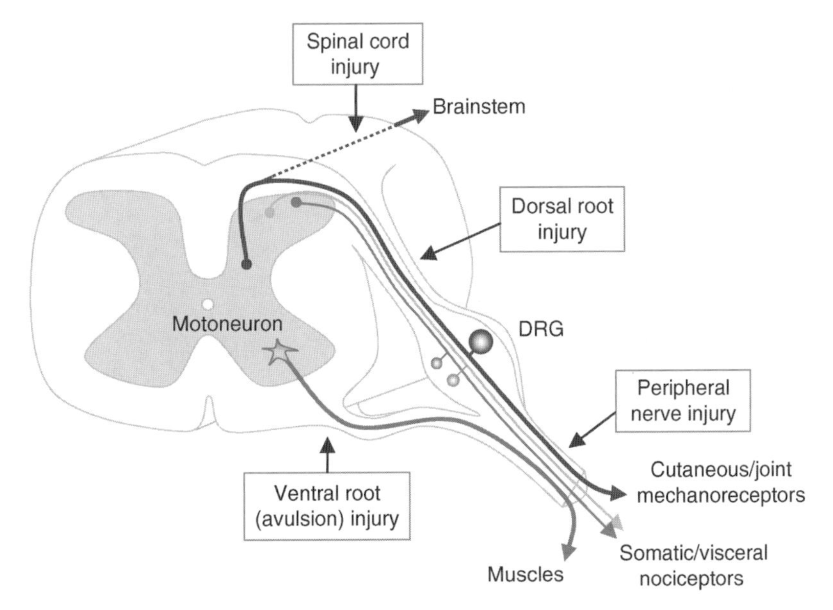

Neurotrophic Factors in Nerve Regeneration. Figure 1 The axons of sensory neurons located in the ►dorsal root ganglia (DRG) bifurcate, one branch travels to peripheral sensory organs; the other projects via dorsal roots to the spinal cord where it innervates spinal and/or supraspinal neurons. Small DRG neurons detect temperature or noxious (painful) stimuli. These are small-to-medium in diameter, with unmyelinated, slowly-conducting axons that innervate superficial laminae in the spinal cord. Large DRG neurons subserve proprioception and mechanoreception and terminate in deeper laminae. Many of the DRG axons also bifurcate central to their entry point within the cord and send a long projection via the dorsal columns to nuclei in the brainstem. Lower motoneurons are situated in the ventral spinal cord and brainstem. Their axons exit the CNS via ventral roots or cranial motor nerves. Spinal motor axons join mixed spinal nerves just distal to the DRG, and innervate muscle fibres in the periphery.

restore function; if they are avulsed (torn from the cord such that the injury site lies deep to the pial surface), regeneration fails.

Neurotrophic Factors

The first NTF to be discovered, based on its ability to promote neuronal survival and neurite outgrowth, was nerve growth factor (NGF). NGF is the prototypical member of a family which includes brain-derived NTF (BDNF), neurotrophin 4/5 (NT-4/5) and neurotrophin-3 (NT-3) (Fig. 2).

These molecules act through specific receptors: NGF binds to tropomyosin related kinase A (TrkA), BDNF and NT-4/5 bind to TrkB, and NT-3 binds to TrkC. All of these receptors are expressed in dorsal root ganglia, and (with the exception of TrkA) motoneurons. All neurotrophins also bind a receptor, ►$p75^{NTR}$, which is co-expressed with the Trks but whose role in regeneration remains enigmatic. The neurotrophin sensitivity of DRG neurons is subtype-specific and has been repeatedly demonstrated *in vitro* and *in vivo*: half of all nociceptors/thermoreceptors express TrkA (the other half do not express any Trk but do express GDNF receptor components), whereas the majority of mechano/proprioceptors express TrkC, and a minority express TrkB. Motoneurons express both TrkB and TrkC.

The glial cell line-derived NTF (GDNF) family includes GDNF, neurturin, ►artemin and ►persephin (Fig. 2). GDNF was isolated based on its ability to promote survival of midbrain dopaminergic neurons, and the rest were identified based on sequence homology. All GDNF family members have been shown to augment neurite outgrowth *in vitro*, but only GDNF and neurturin have proved to enhance regeneration *in vivo*. GDNF family members signal through receptor complexes which involve a common signaling component (►Ret) and ligand-specific binding components (GFRα1–4). GFRα1–3 are expressed in DRG neurons, mainly among thermo/nociceptors, while GFRα1, 2 and 4 are expressed by motoneurons.

Three members of the interleukin-6 (IL-6) family of neurotrophic ►cytokines will be considered here: ciliary NTF (CNTF), leukemia inhibitory factor (LIF) and IL-6 (Fig. 2). Peripheral nerve injury increases the exposure of severed axons to all three factors, which are synthesized in nonneuronal cells. IL-6 upregulation also occurs following axotomy in large-diameter DRG neurons and in motoneurons. These molecules share a common receptor component, gp130, but activate it through receptor complexes: IL-6 and CNTF first bind non-signaling receptor components [IL-6Rα and ►CNTFRα (Cilary Neurotrophic Factor Receptor)] and then to a gp130 homodimer (in the case of IL-6) or a

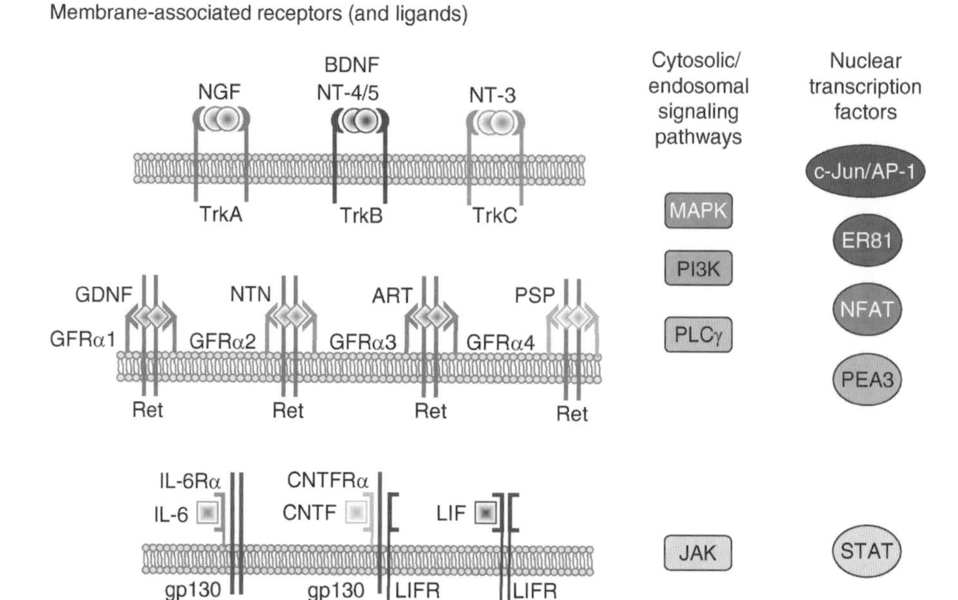

Neurotrophic Factors in Nerve Regeneration. Figure 2 The neurotrophin family of NTFs consists of nerve growth factor (NGF), brain-derived NTF (BDNF), neurotrophin 4/5 (NT-4/5) and neurotrophin-3 (NT-3). NGF binds tropomyosin related kinase A (TrkA), BDNF and NT-4/5 bind to TrkB, and NT-3 binds to TrkC. The glial cell line-derived NTF (GDNF) family includes GDNF, ►neurturin (NTN), ►artemin (ART) and persephin (PSP). GDNF members signal through receptor complexes involving (Ret) and specific ligand binding components ►Glial cell line – Derived neurotrophic factor receptors (GFRα1–4). Three examples of the interleukin-6 (IL-6) family of neurotrophic ►cytokines are, ciliary NTF (CNTF), ►leukemia inhibitory factor (LIF) and IL-6. IL-6 members share the receptor component, ►gp130, but activate it through specific receptors: IL-6 with ►Interleukin-6 Receptor IL-6Rα, CNTF with ►CNTFRα and LIF with ►LIFR. Cytosolic/endosomal signaling pathways following NTF ligand receptor binding include; Phosphatidyl inositol-3-kinase (PI3K), mitogen-activated protein kinase (MAPK), Phospholipase C (►PLC γ) and Janus kinase (JAK). Transcription factors induced by NTF activated pathways include; ►c-Jun (AP-1), ►ER81, ►NFAT, PEAT and STAT.

heterodimer consisting of gp130 and ►leukemia inhibitory factor receptor (LIFR, in the case of CNTF). LIF signals through LIFR: gp130 heterodimers. All DRG neurons express gp130, IL-6Rα, and CNTFRα, while thermoceptors and nociceptors express LIFR. While LIFR, CNTFRα and gp130 are also found in motoneurons, data on motoneuronal IL-6R are lacking.

Higher Level Processes

Positive and Negative Signaling Associated With Axotomy

Peripheral nerve injury results in a switch in the neurons from a transmissive to a regenerative state. The change of state of the neuron is accomplished predominately by an alteration in activation and/or expression of transcription factors, resulting in a decrease in molecules involved synaptic transmission and a concomitant increase in regeneration-associated structural and cytoskeletal proteins. Injury-induced changes are mediated by three broad classes of signals. The first is the immediate entry of ions such as sodium and calcium into the open end of the proximal axon stump, resulting in depolarization. The two other signals are those related to the retrograde transport of molecules in the axon. There are those signals which appear as a result of injury (positive signals): factors released from cells at the injury site which have direct actions on the severed axons or are retrogradely transported to the cell body. There are also signals that disappear (negative signals) as a result of the interruption of retrogradely transported molecules from axonal targets. Positive signals include NTFs derived from Schwann cells and other non-neuronal cells at the site of injury. The loss of target-derived NTFs contributes to negative signaling. It has been proposed (although in many cases evidence is still lacking) that NTFs involved in both types of signaling promote regeneration: positive signals at the injury site may help begin the regenerative process, while target-derived signals may consolidate functional reinnervation. This reasoning has underpinned the application of exogenous NTFs following nerve injury *in vivo*.

Effects of NTFs at the Axonal Growth Cone

NTFs can promote regeneration through local signaling at the growth cone, where they act to increase the motility of filopodial actin and microtubules, the principal cytoskeletal elements governing growth cone dynamics. This has been most clearly demonstrated for the Trk receptors, and for TrkA in particular [1] (Fig. 3).

TrkA activation results in the activation of ►phosphatidyl inositol-3-kinase (PI3K) which leads to filopodial elongation via its effectors Rac and ►Cdc42, Rho family ►GTPases whose activation is generally associated with increased axonal outgrowth. Trk activation also leads, via increases in intracellular cyclic adenosine monophosphate (cAMP), to protein kinase A (PKA) activation, which has a number of positive effects on filopodial extension including activation of an actin anti-capping protein, ►Ena/VASP, the effect of which is to allow for profilin-mediated increases in filopodial length. Microtubule stability is another prerequisite for regenerative growth that is positively regulated by Trk signaling. Trk-mediated activation of a pathway including the kinases Ras, ►Raf and ►Erk2 (a mitogen-activated protein kinase, MAPK) results in phosphorylation of microtubule associated proteins (MAPs) leading to increased stability. More recently it has been shown that TrkA activation allows microtubule plus-end capping by adenomatous polyposis coli (APC), through inhibition of glycogen synthase kinase 3β (GSK-3β). In the absence of NGF signaling, GSK-3β phosphorylates APC, preventing microtubule plus-end capping, reducing stability and decreasing motility [2] (Fig. 3). GDNF family members also activate MAPK and PI3K pathways, and have similar effects on growth cone motility. IL-6-related NTFs are not known to have direct effects on cytoskeletal dynamics at the growth cone.

Effects of NTFs at the Neuronal Cell Body

Receptor-bound NTFs, including NGF, BDNF, NT-3, and NT-4/5, CNTF and LIF, are transported from the growth cone and axon to the cell body where they can effect transcription of ►regeneration-associated genes (RAGs) (see essay in this Encyclopedia by H. Aldskogius). Such transport may be mediated by dynein-microtubule interactions, and is thought to involve a "signaling endosome", which includes, in addition to the neurotrophin receptor and its ligand, MAPK, PI3K and ►PLCγ pathway-associated signaling molecules [3] (Fig. 3). Therefore, the retrograde transport of the internalized neurotrophins and their receptor complex as well as the retrograde propagation of signaling pathways may work in concert to support survival and growth of the injured axon. Once at the cell body, retrogradely transported NTFs effect regeneration via both activation and *de novo* synthesis of appropriate transcription factors (TFs) (Fig. 2): the TFs CREB and ►NFAT (Nuclear factor of activated T cells) are both activated by retrogradely transported neurotrophins; NT-3 signaling results in the upregulation of ER81 while GDNF upregulates ►PEA3 – both are required for

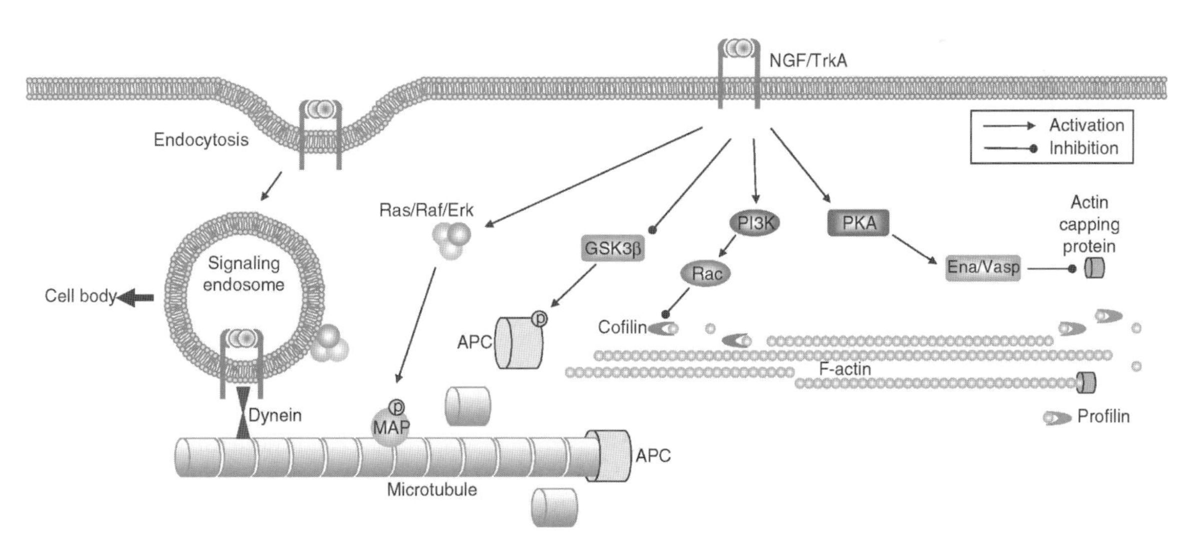

Neurotrophic Factors in Nerve Regeneration. Figure 3 NTFs promote regeneration through signaling at the growth cone or via signaling endosomes directed towards the cell body. TrkA binding results in the activation of phosphatidyl inositol-3-kinase (PI3K) and filopodial elongation/axonal growth via the ►GTPase Rac. Trk binding activates protein kinase A (PKA) producing filopodial extension and activation of an actin anti-capping protein, ►Ena/VASP allowing a profilin-mediated increased filopodial length. Trk-mediated activation of the kinases Ras, ►Raf and ►Erk2 (►Extracellular stress regulated kinase) results in phosphorylation of ►microtubule associated proteins (►MAPs) and increased microtubule stability. TrkA activation also allows microtubule plus-end capping by ►adenomatous polyposis coli (APC), through inhibition of ►glycogen synthase kinase 3β (GSK-3β). GDNF NTFs activate MAPK and PI3K pathways, and have similar effects on growth cone motility. IL-6-related NTFs have no known direct effects on growth cone motility. Following endocytosis, the ligand-receptor complex forms a signaling endosome and is retrogradely transported to the cell body via the microtubule associated motor dynein.

NTF-mediated outgrowth. The immediate-early gene product ►c-Jun, which associates with the AP-1 TF complex, is perhaps the best characterized downstream effector of NTF signaling. Not only is c-Jun upregulated following peripheral nerve injury in sensory and motoneurons, but also it is activated by ►Jun kinases (►JNK), recently shown to be downstream effectors of neurotrophin-dependent MAPK signaling [4]. Finally, retrograde signaling by IL-6 family members results in the activation of Janus kinases (JAKs) and the subsequent nuclear translocation of activated ►signal transducers and activator of transcription 3 (STAT3), effecting RAG expression (Fig. 2).

Regulation of the Process

NTFs and Regeneration of Sensory Axons

The majority of sensory neurons respond to either NGF, NT-3 or GDNF. Surprisingly little is known about the efficacy of these factors in enhancing regeneration of sensory axons following peripheral nerve injury, possibly because peripheral nerve regeneration is often successful. However, there appears to be little requirement for endogenous NTFs in naturally-occurring regeneration: axotomized TrkA- and TrkC- expressing nociceptive and mechanosensitive axons, for example, reinnervate their peripheral targets normally in the presence of antibodies which block NGF and NT-3 [5]. The effects of exogenous NTFs have been more clearly demonstrated following dorsal root lesions [6]: while sensory axons normally fail to regenerate beyond the PNS:CNS interface, NGF, NT-3 and GDNF treatment resulted in the regrowth of appropriate populations of axons into the spinal cord. Furthermore, regrowth was accompanied by functional recovery, which was demonstrated both electrophysiologically and behaviorally. Mechano/proprioceptive axons injured within dorsal columns are also able to regenerate when treated with NT-3, but not with BDNF or GDNF [7].

The neurotrophic cytokines LIF and IL-6 do not induce neurite outgrowth form ►DRG neurons directly, but collaborate with other growth-promoting factors or manipulations: LIF enhances DRG neurite elongation in NGF-treated cultures, and IL-6 has a similar effect in cultures treated with either NGF or NT-3 [8]. *In vivo*, peripheral nerve regeneration is impaired in LIF knockout mice [8]. Additionally, the growth-enhancing effect on axotomized dorsal column sensory axons of a prior peripheral nerve "conditioning" injury was abolished in the absence of IL-6 [8]. No data are available on the effect of CNTF on sensory axon regeneration.

NTFs and Regeneration of Motor Axons

While direct actions of NTFs on sensory axons are evident from their differential effects on distinct subpopulations of adult DRG neurons, the same cannot be said for adult motoneurons, which are more homogenous and, by virtue of their poor survival in isolation, difficult to study *in vitro*. Like DRG neurons, motoneurons do not appear to require endogenous NTFs for successful axonal elongation, since function-blocking antibodies against BDNF, CNTF and GDNF (the most potent motoneuronal NTFs) do not prevent axonal elongation following injury to the facial nerve (a purely motor nerve) [9]. On the other hand, exogenous NTFs, including BDNF, NT-3, GDNF, CNTF and IL-6, have been shown to improve functional recovery, and in several cases increase muscle innervation, following peripheral nerve injury or ventral root avulsion. Despite these findings, it is as yet unclear how exogenous NTFs promote functional recovery. Motoneuron survival following axotomy (particularly following ventral root avulsion injury), collateral sprouting of spared axons, and even trophic or migration-enhancing effects on nonneuronal cells are all able to explain increased target reinnervation and improved motor performance [10]. While all of these factors have well-described survival promoting effects on motoneurons, there is as yet little evidence supporting direct effects of NTFs on the regeneration of motor axons *in vivo*.

References

1. Markus A, Patel TD, Snider WD (2002) Neurotrophic factors and axonal growth. Curr Opin Neurobiol 12:523–531
2. Zhou FQ, Zhou J, Dedhar S, Wu YH, Snider WD (2004) NGF-induced axon growth is mediated by localized inactivation of GSK-3beta and functions of the microtubule plus end binding protein APC. Neuron 42:897–912
3. Zweifel LS, Kuruvilla R, Ginty DD (2005) Functions and mechanisms of retrograde neurotrophin signalling. Nat Rev Neurosci 6:615–625
4. Waetzig V, Herdegen T (2005) MEKK1 controls neurite regrowth after experimental injury by balancing ERK1/2 and JNK2 signaling. Mol Cell Neurosci 30:67–78
5. Diamond J, Foerster A, Holmes M, Coughlin M (1992) Sensory nerves in adult rats regenerate and restore sensory function to the skin independently of endogenous NGF. J Neurosci 12:1467–1476
6. Ramer MS, Priestley JV, McMahon SB (2000) Functional regeneration of sensory axons into the adult spinal cord. Nature 403:312–316
7. Bradbury EJ, McMahon SB, Ramer MS (2000) Keeping in touch: sensory neurone regeneration in the CNS. Trends Pharmacol Sci 21:389–394
8. Cafferty WB, Gardiner NJ, Das P, Qiu J, McMahon SB, Thompson SW (2004) Conditioning injury-induced spinal axon regeneration fails in interleukin-6 knock-out mice. J Neurosci 24:4432–4443
9. Streppel M, Azzolin N, Dohm S, Guntinas-Lichius O, Haas C, Grothe C, Wevers A, Neiss WF, Angelov DN (2002) Focal application of neutralizing antibodies to soluble neurotrophic factors reduces collateral axonal branching after peripheral nerve lesion. Eur J Neurosci 15:1327–1342

10. Boyd JG, Gordon T (2003) Neurotrophic factors and their receptors in axonal regeneration and functional recovery after peripheral nerve injury. Mol Neurobiol 27:277–324

Neurotrophin 4/5 (NT-4/5)

Definition

Neurotrophin 4/5 is a neurotrophic factor that has been shown to have a protective effect on the survival of retinal ganglion cells.

►Regeneration of Optic Nerve
►Retinal Ganglion Cells

Neurotrophins

Definition

Neurotrophins are molecules important for the health and maintenance of neurons. They are structurally and functionally related to nerve growth factor, the first identified neurotrophin, and signal through high affinity tyrosine kinase receptors (Trk A-C) and a low affinity p75 neurotrophin receptor. Neurotrophins can stimulate neurite outgrowth and act as a long-range diffusible guidance cue. In addition, neurotrophins can stimulate gene expression and promote neuron survival.

►Neural Development
►Regeneration

Neuroturin

Definition

A member of the glial cell line-derived neurotrophic factor (GDNF) family of neurotrophic factors that also includes artemin and persephin. GDNF family members use a receptor complex that consists of the common receptor tyrosine kinase signaling component Ret and one of the GPI-linked receptors (GFRα1 to 4) that regulate ligand binding specificity. GFRα2 is the preferred receptor for neuroturin.

►Glia Cell Line-derived Neurotrophic Factor
►Neurotrophic Factors
►Neurotrophic Factors in Nerve Regeneration

Neurovascular Tract

Definition

The extramuscular collagen fiber-reinforced sheet of connective tissue in which nerves and blood vessels are embedded. This tract is connected to other extramuscular elements of the muscular compartment.

Neurovegetative Function in Outer Space

►Autonomic Function in Space

Neutralizing Antibody

Definition

Neutralizing antibody (NA) is an antibody which inhibits the infectivity of a blood-borne virus or bacterium.

►Central Nervous System Infections: Humoral Immunity in Arboviral Infections

Neutrophils

Definition

Neutrophils are often called polymorphonuclear cells due to their multilobulated nucleus. As the cells that respond first, normal circulating neutrophils enter into local tissue where infection or injury occurs. Neutrophils exert their protection mainly by phagocytosing damaged tissue, infected cells and invading pathogens.

New Developments in G Protein-Coupled Receptor Theory

IAN MULLANEY
School of Pharmacy, Division of Health Sciences, Murdoch University, WA, Australia

Synonyms

Cell signaling; Molecular pharmacology; Signal transduction

Definition

Approximately 1% of the genome of higher organisms encodes the family of G protein-coupled receptors (GPCRs). These receptors are characterized by containing seven transmembrane domains, being linked to guanine nucleotide-binding proteins (GTP-binding proteins) and ubiquitously distributed on the plasma membrane of all cell types. The first wave of research into these molecules identified family members, assigned coupling to individual GTP-binding proteins, characterized signaling pathways and elucidated the molecular apparatus by which GPCRs are switched off and recycled back to the membrane. A new wave of GPCR research is underway, looking at the more complex relationship that these molecules have with a variety of processes both cellular and pharmacologic.

Characteristics

Dimerization/Oligomerization

Although protein-protein interactions have been shown to be instrumental in the organizational structure and function of many cell signaling processes, it has been the belief that G protein-coupled receptors occur and function as monomeric, non-interacting species. However, it has become clear that a number of GPCR species can form dimers (both ►GPCR homodimers and ►GPCR heterodimers) and/or larger oligomeric (►GPCR oligomer) complexes. Although the precise cellular function of GPCR homodimerization is unclear, a number of roles have been proposed for this process. These revolve around the participation of dimerization at the level of synthesis or during protein maturation in the Golgi apparatus and result in the successful delivery of the protein to the cell surface. Crucial to the process is the involvement of specific dimerization motifs that occur in various structural domains within the protein. Disruption of such a motif in TMD VI (transmembrane domain VI) of the β2 adrenoceptor resulted in loss of homodimerization and subsequent insertion of this receptor into the plasma membrane. With over 400 genes encoding non-sensory GPCRs now having been identified, it is no surprise that co-expression and subsequent dimerization between different receptor species has been observed. Heterodimerization has been proposed to promote the formation of receptors with unique pharmacological properties, contributing to the pharmacological diversity of GPCRs. Jordan and Devi [1] first provided biochemical and pharmacological evidence for a fully functional heterodimer comprising kappa and delta opioid receptors. This new formation resulted in a novel receptor that exhibited ligand binding and functional properties distinct from those of either receptor. Furthermore, the kappa-delta heterodimer synergistically binds highly selective agonists and potentiates signal transduction. More recently, other heterodimer receptor pairs including the orexin-1 receptor (►Orexin receptor) and cannabinoid CB1 receptor dimer and the angiotensin AT1 receptor and bradykinin B2 receptor dimer have been identified. These have been demonstrated to display changes in GTP-binding protein-coupling specificity as well as altered receptor-mediated endocytosis.

GPCR – Protein Complexes in Living Cells

In addition to binding to protein and peptide ligands, GTP-binding proteins and forming complexes amongst themselves, GPCRs also appear to interact with a large and diverse group of proteins that have a role in a number of receptor functions including trafficking, signal transduction, desensitization (►Receptor desensitization) and down-regulation and receptor recycling. Candidate molecules for interaction include ►GPCR kinases (GRKs), arrestins, protein kinase A, protein kinase C, molecular chaperones and receptor activity-modifying proteins (RAMPs). The use of novel biophysical techniques such as fluorescence resonance energy transfer (FRET) and ►bioluminescence resonance energy transfer (BRET), which utilize energy transferred from a fluorescent donor to an acceptor molecule, has allowed observation of protein-protein interactions to be analyzed in real time in living cells.

The receptor-activity-modifying proteins (RAMPs) are single transmembrane proteins that heterodimerize with GPCRs. To date 3 RAMPs (RAMP1,2,3) have been molecularly identified. Calcitonin-gene-related peptide (CGRP) and adrenomedullin are related peptides with distinct pharmacological profiles and signaling pathways. Each of these ligands function through the calcitonin-receptor-like receptor (CRLR). It is association with the various RAMPs that determines ligand specificity. RAMP1 presents the receptor at the cell surface as a mature glycoprotein and a CGRP receptor whilst RAMP2-transported receptors are core-glycosylated and are adrenomedullin receptors. Adrenomedullin and CGRP receptors are potential therapeutic targets for several diseases including migraine, hypertension, pulmonary hypertension and sepsis. Thus, understanding how ligand binding to the receptor complex is regulated by RAMPs is crucial for the development of pharmaceutical agents. Recent studies showing association of RAMPs with other GPCR families has broadened the importance of this class of accessory proteins [2].

Inverse Agonists/Constitutive Activation

GPCRs are able to achieve and maintain a spontaneously active conformation that results in a constitutively active receptor, a process termed ►negative efficacy.

Certain ligands (termed inverse agonists) have been demonstrated to decrease this ►constitutive activation. Indeed, many drugs that were originally classified as neutral antagonists, including prazosin, pirenzipine, trihexylphenidyl and losartan, can now be reclassified as inverse agonists (►Inverse agonism). In 1996, five GPCR subtypes (δ opioid receptor, β2 adrenoceptor, 5-HT_{2C} receptor, bradykinin B2 receptor, M1 muscarinic acetylcholine receptor) were noted to display negative efficacy. At present several dozen constitutively active GPCR subtypes have been identified [3]. Although original studies used systems that overexpressed GPCRs, native GPCRs expressed at normal levels demonstrate constitutive activity. In addition, this activity has a physiological role. For example, in rat brain synaptosomes, inverse agonists, but not competitive antagonists, acting at the H_3 histamine receptor suppress K^+-induced release of histamine. Thus a degree of constitutive H_3 receptor activity in presynaptic terminals *in vivo* limits histamine release [4].

Furthermore, the existence of this phenomenon has allowed for the consideration of inverse agonists to be used as pharmacotherapeutic agents. In diseases where mutation in receptors results in increased constitutive signaling via the receptor (such as male precocious puberty where there is a mutation in the LH receptor in the testes that results in early onset of puberty), use of an appropriate inverse agonist could eliminate the constitutive activity of the receptor [3]. A number of pharmacological agents originally determined to be neutral antagonists but now reclassified as inverse agonists, are in use within a clinical setting. Included in this category are a series of dopamine D2 receptor ligands including haloperidol, clozapine and olanzapine, that are currently used in the treatment of schizophrenia [5]. Similar results were obtained demonstrating that the putative antipsychotic agents L-745,870 and U-101958, thought to be dopamine D4-specific antagonists, display inverse agonist effects at this receptor [6].

Free Fatty Acids as GPCR Ligands

Although originally thought primarily as direct energy sources, free fatty acids also play key roles in regulating a variety of physiological effects. Although these responses were thought to occur as a consequence of their metabolism, it is now clear that free fatty acids including α-linolenic acid and docosahexanoic acid, function directly as agonists at GPCRs. The breakthrough came when a series of orphan GPCRs (►Orphan receptors) were identified from a variety of human tissues. Although most were involved with glucose metabolism in pancreatic β-cells, one, ►GPR40, was highly expressed in brain. This receptor couples to the Gq/G11 group of GTP-binding proteins and when activated by the FFA palmitate, results in substantial elevation in intracellular calcium. Although the physiological role of these receptors in brain is unclear, it has been speculated that they may play a fundamental role in the neurological disorders of aging, Alzheimer's disease, Parkinson's disease, and stroke [7].

Addiction, GPCR Desensitization and Tolerance to Drugs

It has been well characterized that desensitization of agonist-occupied GPCR occurs first through phosphorylation of the receptor by GPCR kinases (GRKs) then subsequent binding of the phosphorylated receptor to arrestins that uncouple it from GTP-binding proteins. This in turn targets the receptors for internalization via clathrin-coated vesicles.

The frequent administration of certain drugs leads to the development of tolerance to the effects of the drug. This is best characterized by opiates such as the alkaloid morphine where the analgesic, rewarding and respiratory effects of the drug are diminished upon repeated administration, limiting its clinical usefulness (see Chapter on ►Tolerance and Dependence). Morphine causes its behavioral effects by activating μ opioid receptors, an effect verified by the loss of drug effects in μ opioid receptor knockout mice [8]. This receptor subtype couples to Gi/Go activating inwardly rectifying potassium channels, inhibiting voltage-activated calcium channels and decreasing intracellular cAMP levels through inhibition of adenylyl cyclase. As with other GPCRs, the μ opioid receptor is rapidly desensitized upon activation with agonist. However, the route of signal attenuation is dependent on the type of agonist activating the μ opioid receptor. DAMGO, a high-affinity peptide agonist, induces desensitization through GRK whereas morphine, which is a partial agonist at these receptors, desensitizes the GPCR through activation of protein kinase C. Although tolerance to morphine is a multi-faceted process, it is recognized that adaptive changes at the level of the μ opioid receptor is an important factor in tolerance in the intact animal. Indeed, there are numerous reports that implicate protein kinase C in both acute and chronic tolerance to the analgesic properties of opioids. Such studies employ the use of protein kinase C inhibitors such as H7 and bisindolylmaleimide to reverse tolerance in animals even when they are administered some days after morphine infusion [9]. Thus the development of inhibitor drugs, to be used in negating tolerance to morphine, increasing the clinical efficiency of opiate drugs and reducing the social problems that accompany recreational opioid abuse, is of prime importance.

The 5-HT_{2C} receptor is used as a target for drug addiction. Agonists at this receptor have been demonstrated to moderate addiction-related behaviors such as hyperactivity, place preference and compulsive behavior connected with common recreational drugs including cocaine, alcohol, nicotine and cannabinoids. However the use of such agonists is associated with a series of unwanted adverse effects including anxiety, hypophagia, hypolocomotion and disturbances of motor function. Activation of these receptors with agonists results in receptor phosphorylation then dephosphorylation, anevent demonstrated to mediate many of the adverse effects associated with addiction. One new strategy in the treatment of addiction has been to prevent the dephosphorylation of the receptor by inhibiting ►PTEN, the phosphatase responsible, with specific peptides that blocks its interaction with the receptor. This mimics the agonist activation process but suppresses the adverse effects associated with dephosphorylation [10]. This approach could lead to a more effective strategy for the treatment of drug addiction.

References

1. Jordan BA, Devi LA (1999) G protein-coupled receptor heterodimerization modulates receptor function. Nature 399:697–700
2. Parameswaran N, Spielman WS (2006) RAMPs: the past, present and future. Trends Biochem Sci 31:631–638
3. Bond RA, Ijzerman AP (2006) Recent developments in constitutive receptor activity and inverse agonism, and their potential for GPCR drug discovery. Trends Pharmacol Sci 27:92–96
4. Morisset S, Rouleau A, Ligneau X et al. (2000) High affinity constitutive activity of native H3 receptors regulates histamine neurons in brain. Nature 408:860–864
5. Hall DA, Strange PG (1997) Evidence that antipsychotic drugs are inverse agonists at D2 dopamine receptors. Br J Pharmacol 121:731–736
6. Gazi L, Bobirnac I, Danzeisen M et al. (1998) The agonist activities of the putative antipsychotic agents, L-745,870 and U-101958 in HEK293 cells expressing the human dopamine D4.4 receptor. Br J Pharmacol 124(5):889–896
7. Brown AJ, Jupe S, Briscoe CP (2005) A family of fatty acid binding receptors. DNA Cell Biol 24:54–61
8. Matthes HW, Maldonado R, Simonin F et al. (1996) Loss of morphine-induced analgesia, reward effect and withdrawal symptoms in mice lacking the μ opioid receptor gene. Nature 383:819–823
9. Bailey CP, Smith FL, Kelly E et al. (2006) How important is protein kinase C in μ-opioid receptor desensitization and morphine tolerance. Trends Pharmacol Sci 27:558–565
10. Ji SP, Zhang Y, Van Cleemput J et al. (2006) Disruption of PTEN coupling with 5-TH2C receptors suppresses behavioural responses induced by drugs of abuse. Nat Med 12:324–329

Newtonian Mechanics

Definition

A formulation of the laws of mechanics based on Newton's approach, whereby forces are primary quantities to be related to the motion of the system by certain vectorial laws. In the case of particle mechanics, these laws are essentially those formulated by Newton himself. In continuum mechanics, this approach is generalized by (i) the introduction of appropriate balance laws in terms of densities, (ii) the independent postulation of the balance of angular momentum and (iii) the inclusion of the first and second laws of thermodynamics for continuous systems.

►Mechanics

NF-κB

Definition

NF-κB (nuclear factor-kappa B) is a transcription factor protein complex involved in many important biological processes, including the immune response to infection, learning and memory.

►Chromatin Immunoprecipitation and the Study of Gene Regulation in the Nervous System
►NF-κB – Potential Role in Adult Neural Stem Cells

NF-κB – Potential Role in Adult Neural Stem Cells

Darius Widera, Christian Kaltschmidt, Barbara Kaltschmidt
Department of Cell Biology, Faculty of Biology, University of Bielefeld, Bielefeld, Germany

Synonyms

NF-κB: (NF-kappaB; NF-κB); DNA-binding subunits of NF- κB; p65 (RelA); p50 (NFKB1); p52 (NFKB2);

c-Rel (REL); RelB (RELB); Inhibitory subunits of NF-κB: IκB-α (NFKBIA); IκB-β (NFKBIB); IκB-ε (NFKBIE); IκB-ζ (NFKBIZ)

Definition

Neural Stem Cells

Stem cells are defined as undifferentiated cells with the ability to (i) proliferate, (ii) exhibit self-maintenance, (iii) generate a large number of progeny, including the principal phenotypes of the tissue, (iv) retain their multilineage potential over time, and (v) generate new cells in response to injury or disease.

Neural stem cells can be found within their complex niche in the mammalian brain. Neural stem cells could be maintained in culture via propagation of floating cell clusters called "neurospheres." Neurospheres contain committed progenitors, differentiated astrocytes, neurons and neural stem cells. A progenitor is defined as mitotic cell with a fast cell-devision cycle that retains the ability to proliferate and give rise to terminally differentiated cells but that is not capable of indefinite self-renewal.

Nuclear Factor-κB

Nuclear factor kappa B (NF-κB) is a ►transcription factor (TF) composed of homo- or heterodimeric DNA-binding subunits (e.g., p50 and p65). Inducible forms of NF-κB reside in the cytoplasm due to an interaction with one inhibitory subunit (e.g., IκB-α). Upon activation by growth factors, etc. (see Fig. 1), the IκB Kinase complex (IKK) catalyzes phosphorylation of the inhibitory subunit IκB, which leads to proteasomal degradation. This exposes the nuclear localization signals (see green dots in Fig. 1). Nuclear import of NF-κB via importins activates transcriptional target genes driving ►proliferation, such as Cyclin D1 (see Fig. 1). NF-κB is involved in many biological processes, such as inflammation and innate immunity, development, apoptosis and anti-apoptosis [1]. In the nervous system NF-κB plays a crucial role in neuronal plasticity, learning, neuroprotection and neurodegeneration. In addition, recent data suggest a crucial role of NF-κB on proliferation, migration and ►differentiation of neural stem cells.

In the G_1 phase, NF-κB activates cyclin D1 expression by direct binding to multiple sites in the cyclin D1 promotor. This promotes G_1-to-S progression. In contrast, NF-κB action in the M-phase leads to differentiation or migration induction.

In this essay we suggest a model explaining the multiple action of NF-κB within neural stem cells.

CNS

The Central Nervous System (CNS, systema nervosum centrale), consisting of the brain and spinal cord is one of the two major parts of the nervous system. CNS integrates all nervous activities. The second part is the peripheral nervous system (PNS) which is outside the brain and spinal cord and regulates e.g., the heart muscle, the muscles in blood vessel walls or glands.

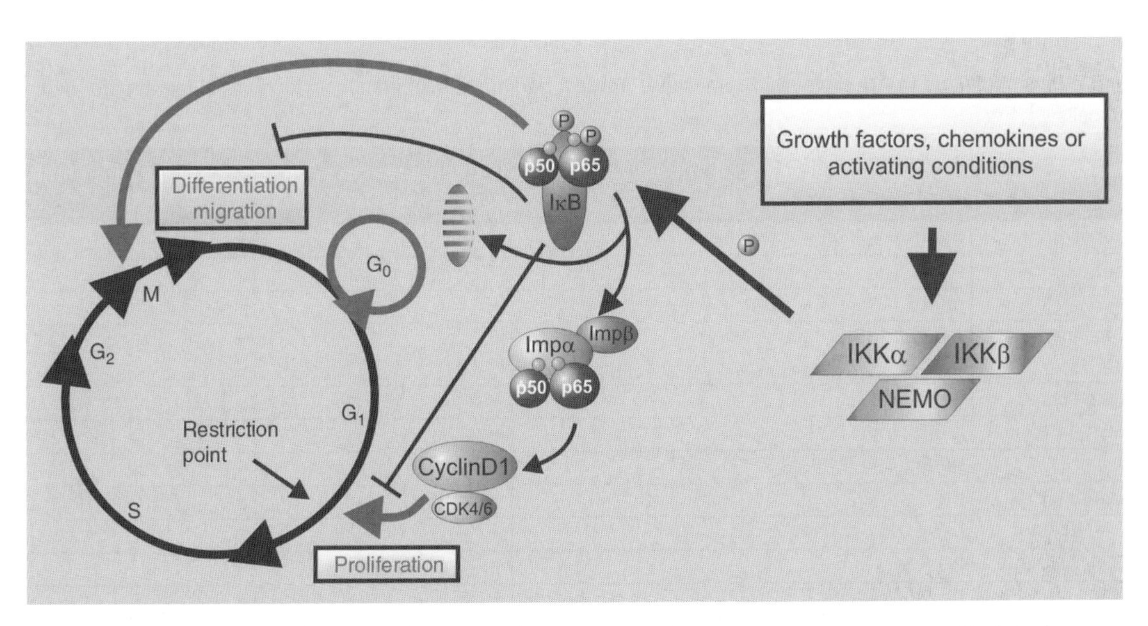

NF-κB – Potential Role in Adult Neural Stem Cells. Figure 1 Model for the involvement of NF-κB in the molecular machinery of the cell cycle essential for proliferation, differentiation and migration of neural stem cells. Activation of IKK complex and subsequent NF-κB activation via growth factors, chemokines or activating conditions leads to ubiquitination and proteasomal degradation of IκB. After nuclear translocation NF-κB binds to specific promoter regions of the target genes and activates their transcription. Depending on the cell cycle point these targets are genes regulating proliferation or migration and differentiation.

On the cellular level, CNS consists of a network of nerve cells, glial cells (e.g., astrocytes and microglia) and neural stem cells. Neurons, the primary cells of the CNS, are responsible for information processing and storage. Glial tissue surrounds and supports neurons and is important for response against infection and tissue repair (e.g., microglia). Novel data define astrocytes and radial glia as a potential stem cell pools.

Characteristics

Adult Neural Stem Cells

Until recently, the dogma existed that stem cells are not present in the adult CNS. Currently, there are many reports clearly demonstrating neurogenesis in different regions of the adult brain [2,3]. Stem cells within the adult brain were found within the subgranular zone of the hippocampus and in the subventricular zone (SVZ). The immunocytochemical markers expressed by NSCs include inter alia the intermediate filament Nestin, the transcription factors Sox1 and Sox2, the RNA binding protein Musashi and the transmembrane protein prominin-1(CD133) (see Table 1).

Isolated and cultured NSCs may have the ability to replace lost cells within the central nervous system, - an important issue for future therapy of neurodegenerative diseases as Parkinson's and Alzheimer's disease. Moreover NSCs also offer hope for fighting against cancer by the delivery of chemotherapy agents directly to tumor cells.

NF-κB in the Nervous System

In the nervous system, the most frequent form of NF-κB is a heterodimer composed of p50 and p65. Activating stimuli like ►Tumor Necrosis Factor (TNF) (see Fig. 2) or Erythropoietin (EPO) activate a kinase complex composed of two IκB-specific kinases (IKKα and IKKβ) and a modulatory subunit (IKKγ/NEMO). The IKK-α/β complex phosphorylates the inhibitory IκB, which is then ubiquitinilated and degraded via the proteasome. This degradation triggers the translocation of NF-κB into the nucleus followed by initiation of transcription. For a detailed discussion on the action of NF-κB in the CNS see also [4,5 and 6].

Apart from the inducible NF-κB activity, there are reports on constitutively active NF-κB in several cell types, such as hippocampal neurons, or numerous brain-related cancer types such as gliobastomas.

NF-κB and Neural Stem Cell Proliferation

Most of the culture protocols for NSCs use bFGF (FGF-2) and EGF for keeping the cells in undifferentiated and proliferating state [2–3]. Over the years many additional molecules and cultivation conditions were identified to influence the NSC proliferation (see Table 2).

Here we summarize several evidences for a crucial involvement of NF-κB in proliferation control.

An enhanced proliferation of NSCs in vitro and in vivo after Erythropietin (EPO) treatment has been reported. Demonstrably, the authors provide evidence

NF-κB – Potential Role in Adult Neural Stem Cells. Table 1 Examples for immunocytochemical markers for neural stem cells

Marker	Detected in species: m: mouse; r: rat; h: human	Expression in adult NSCs	Expression in fetal NSCs
Nestin	m/r/h	+	+
Sox1	m/r/h	+	+
Sox2	m/r/h	+	+
prominin-1 (CD133)	m/h(CD133)	+	+
Musashi	m/r/h (MSI)	+	+
SSEA-1/LeX	m/r/h	+	+
L1	m/r/h	+	+
ABCG2 (Bcrp1)	r/h	+	+
PSA-NCAM	m/r/h	+	+
CD24	h	+	+
CD44	h	+	+
CD81	h	+	+
CD90	h	+	+
CD184	h	+	+
Dnmt3a	m	–	+
Vimentin	m/r/h	+	+

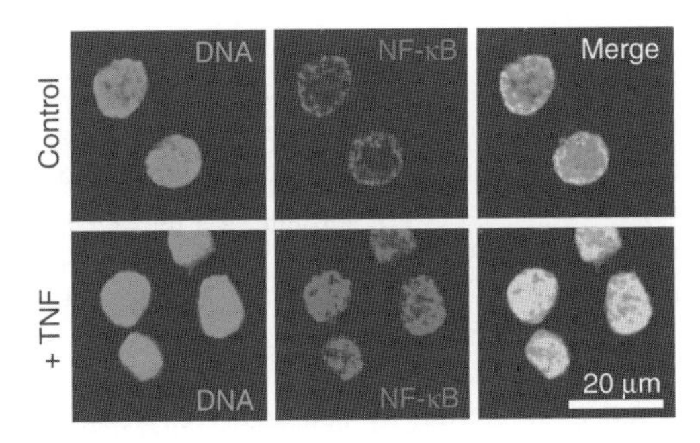

NF-κB – Potential Role in Adult Neural Stem Cells. Figure 2 TNF-induced nuclear localization of the transactivating NF-κB subunit p65 in neural stem cells. NSCs were fixed and stained with an antibody against the p65 subunit of NF-κB. Nuclei (DNA) were stained with SYTOX (*green*). The activation of the NF-κB pathway is shown as nuclear translocation of NF-κB visualised using an antibody against the p65 subunit. The nuclear translocation of NF-κB is followed by the transcription of target genes responsible for proliferation, migration and differentiation of neural stem cells.

that EPO is a homeostatic autocrine-paracrine signaling molecule with actions mainly mediated by NF-κB [7].

EPO and EPO receptors are upregulated in the CNS after hypoxia. Similarly, hypoxia activates NF-κB in neonatal rat hippocampus and cortex. Under hypoxic conditions, hypoxia-inducible-factor1α (HIF-1α), an important transcription factor for regulation of the oxygen response, translocates into the nucleus and binds to promoter region of the *epo* gene leading to upregulated expression (for review see [8]). Induction of proliferation is also conceivable for culture density. The level of reactive oxygen species (ROS) is significantly elevated under low density conditions, leading to increased proliferation. Bonello et al. recently reported that ROS activates HIF-1α itself via a functional NF-κB binding site in pulmonary artery smooth muscle cells.

We and others demonstrated that TNF-α triggers the proliferation of NSCs. NF-κB has been identified as the main driving force of TNF-mediated proliferation [9].

NF-κB – Potential Role in Adult Neural Stem Cells. Table 2 Examples for molecules and/or conditions inducing proliferation or migration of NSCs

Proliferation	
Molecule or condition	**Influence on proliferation + positive/- negative**
EGF	+
bFGF	+
TNF	+
EPO	+
Hypoxia	+
L1	–
GM-CSF	+
ROS/Density	+
Neurofibromin	+
mAChR-stimulation	+
Cerebral infarction	+
Soluble amyloid precursor protein	+
Abeta	–
Sphingosine-1-phosphate	+
NO	–
Traumatic brain injury	+
Glutamate	+
Migration	
MCP-1	
SCF	
SDF-1α	
PDGF	
Cerebral cortex injury	
Microglia culture supernatants	
ischemia stroke	
Seizure	

In the nervous system glutamate is described as a potent activator of NF-κB. In respect of the influence on NSCs, also glutamate enhances survival and triggers the proliferation of SVZ derived NSC.

As another molecule which increases the proliferation of NSCs, Sphingosine-1-phosphate (S1P) was described. Studies investigating endothelial cells showed that S1P induces the activation of NF-κB-mediated transcriptional activity.

Amyloid beta-peptide (Aβ), a self aggregating peptide and responsible for Alzheimer's disease, significantly decreases the proliferation of neural stem cells in vitro and in vivo. This result correlates with the fact that high amounts of Aβ acts as repressor of NF-κB. In contrast – the soluble secreted form of amyloid precursor protein, a well known NF-κB target gene, increases the proliferation of neural stem cells. In this context it is of importance that secreted beta-amyloid precursor protein counteracts the proapoptotic action of mutant presenilin-1 by activation of NF-κB.

Neurofibromin is able to increase the proliferation of NSCs. Neurofibromin, a product of the neurofibromatosis 1 (*nf1*) gene is one of the key regulators of the RAS oncogene. Noteworthy, expression of activated RAS stimulates NF-κB.

Cyclin dependent kinase 4 and 6 (CDK4/6) signaling is essential in ►cell cycle regulation in NSCs. In addition, the formation of the complex of CDKs 4 and 6 with Cyclin D1 is necessary for the cell cycle progression. Demonstrably, NF-κB controls growth and differentiation through transcriptional regulation of Cyclin D1 (see Fig. 1).

NO is a physiological inhibitor of neurogenesis. In addition, nitric oxide synthesis inhibition increases proliferation of neural precursors. It is noteworthy that NO is a well known repressor of NF-κB in neurons providing a link between NO dependent increase of progenitor proliferation and decreased NF-κB activity.

TGF-β1 is one of the well known inhibitors of NF-κB. According with our theory of proliferation control by NF-κB, TGF-β1 has been identified as a potent inhibitor of neurogenesis inducing a cell cycle arrest in the G_0/G_1 phase.

Taken together there are numerous evidences for a crucial role of NF-κB in control of neural stem cell proliferation.

NF-κB and Migration of Neural Stem Cells

In spite of many important proceedings on the field of neural stem cell biology, the factors that orchestrate homing of NSCs are largely unknown. There are only few reports identifying factors inducing migration of NSCs (see Table 2).

The expression of several chemokine receptors by NSCs such as CCR2, CXCR4 and c-kit (Stem Cell Factor Receptor) is well described.

MCP-1 is a very potent chemotactic factor for neural precursors [10]. Interestingly, MCP-1 expression can be strongly induced by TNF. In addition, the *mcp-1* gene contains a functional NF-κB binding site in its promoter region which is necessary for response to TNF. In the hematopoietic system, binding of MCP-1 to its receptor strongly activates NF-κB, providing a further hint for NF-κB regulation of migration.

In another approach Sun et al. demonstrated potent induction of migration by neuronally expressed stem cell factor (c-kit Ligand, SCF) in vitro and in vivo. Analogous to *mcp-1*, also the *scf* gene contains a NF-κB binding site. Stromal derived factor 1α (SDF-1α), a well known ligand of the CXCR4, is described as a further chemokine, inducing migration of NSCs. Here, directed migration of neural stem cells induced by the SDF-1α secreted by astrocytes and endothelium was demonstrated. Furthermore, the over-expression of IκB results in loss of SDF-1α mediated migration of breast cancer cells in vivo.

All those results let us hypothesize that NF-κB is not only crucially involved in NSC proliferation, but also in control of migration.

NF-κB and Neural Stem Cell Differentiation

The IL-6 family of cytokines, including the leukemia inhibitory factor (LIF) and ciliary neurotrophic factor (CNTF) promote astrocytic differentiation by activating transcription factors, such as STAT 3, AP-1 and NF-κB. Both CNTF and LIF triggers the recruitment of glycoprotein 130 (gp 130) to their specific receptors leading to activation of the RAS-MAP kinase pathway. Downstream of RAS MAP kinases and PKC transduce the signals to their substrates activating nuclear transcription factors (NF-κB, AP-1 and NF-IL6). This cross-talk of those transcription factors and co-activators induce astrocytic fate specification in NSCs.

Recent reports demonstrated that NF-κB is required for neuronal differentiation of neuroblastoma cells. Cells induced to differentiate with retinoic acid, show nuclear NF-κB localization. In contrast, over-expression of NF-κB super-repressor suppressed neuronal differentiation.

NF-κB activity, induced via activation of the Rho family of small GTPases, regulates neurite outgrowth and dendritic spine formation in neuroblastoma cells.

Some studies demonstrated that bone morphogenetic proteins (BMPs) promote astroglial lineage commitment by mammalian subventricular zone progenitor cells. In contrast other approaches suggest that BMPs promote neuronal differentiation of NSCs in SVZ. These controversial findings can be explained by dose-dependent action and complex signaling via several cooperating transcription factors. Interestingly, it has been suggested, that NF-κB may positively regulate BMP-2 gene transcription and that overexpression of a NF-κB superrepressor may lead to changes in downstream signals including BMP-4.

All these results suggest a very complex control mechanism and clearly indicate an involvement of NF-κB in differentiation regulation. Further studies should investigate the involved mechanisms in detail.

References

1. Bonizzi G, Karin M (2004) The two NF-kappaB activation pathways and their role in innate and adaptive immunity. Trends Immunol 25(6):280–288
2. McKay R (1997) Stem cells in the central nervous system. Science 276(5309):66–71
3. Gage FH (2000) Mammalian neural stem cells. Science 287(5457):1433–1438
4. Kaltschmidt B, Widera D, Kaltschmidt C (2005) Signaling via NF-kappaB in the nervous system. Biochim Biophys Acta 1745(3):287–299
5. Widera D, Mikenberg I, Kaltschmidt B, Kaltschmidt C (2006) Potential role of NF-kappaB in adult neural stem cells: the underrated steersman? Int J Dev Neurosci 24(2–3):91–102
6. Mattson MP, Meffert MK (2006) Roles for NF-kappaB in nerve cell survival, plasticity, and disease. Cell Death Differ 13(5):852–860
7. Shingo T, Sorokan ST, Shimazaki T, Weiss S (2001) Erythropoietin regulates the in vitro and in vivo production of neuronal progenitors by mammalian forebrain neural stem cells. J Neurosci 21 (24):9733–9743
8. Haddad JJ (2003) Science review: redox and oxygen-sensitive transcription factors in the regulation of oxidant-mediated lung injury: role for hypoxia-inducible factor-1alpha. Crit Care 7(1):47–54
9. Widera D, Mikenberg I, Elvers M, Kaltschmidt C, Kaltschmidt B (2006) Tumor necrosis factor alpha triggers proliferation of adult neural stem cells via IKK/NF-kappaB signaling. BMC Neurosci 7:64
10. Widera D, Holtkamp W, Entschladen F, Niggemann B, Zanker K, Kaltschmidt B, Kaltschmidt C (2004) MCP-1 induces migration of adult neural stem cells. Eur J Cell Biol 83(8):381–387

NF-κB: Activation in the Mouse Spinal Cord Following Sciatic Nerve Transection

Keith R. Pennypacker[1], Glen Pollock[2], Samuel Saporta[2]

[1]Departments of Molecular Pharmacology and Physiology, University of South Florida College of Medicine, Tampa, FL, USA

[2]Pathology and Cell Biology, University of South Florida College of Medicine, Tampa, FL, USA

Definition

NF-κB is a ubiquitous nuclear ►transcription factor (TF) found in the cytoplasm that regulates a number of physiological processes, such as inflammation, apoptosis, and cellular growth (reviewed in [1]). NF-κB is upregulated after injury in the central nervous system (CNS), and there have been suggestions that neurons, microglia and astrocytes upregulate NF-κB in response to injury. In this essay, we present data that NF-κB is upregulated in neurons, but not microglia or astrocytes, following peripheral nerve transection, and likely plays a role in the immediate survival of these neurons.

Characteristics

Description of the Structure/Process

NF-κB is a family of proteins that includes p50, p52, p65/RelA, RelB, and c-Rel which form homo- and heterodimers. The inactive form of NF-κB consists of dimers that are complexed with an inhibitory factor, IκB, in the cytoplasm. Diverse extracellular signals such as TNF, IL-1, and multiple growth factors activate alternate transduction pathways that converge to activate this transcription factor, which regulates more genes than any other DNA regulatory protein. Phosphorylation of IκB by protein kinases, such as inhibitory factor kappa B kinase alpha (IKKα), leads to its ubiquination and ultimate degradation. Once IκB is degraded, NF-κB is released and translocated to the nucleus, where it interacts with a specific DNA sequence in the promoter region stimulating the transcription of a wide array of genes. In the CNS, neurons contain constitutively active NF-κB, which is upregulated after injury [2–3]. Both microglia and astrocytes have been shown to induce NF-κB activity in response to injury [4].

NF-κB activity has been implicated in opposing apoptosis [5] and enhancing neuronal survival following CNS injury [7]. For example, activated NF-κB increases the expression of genes that block the activation of caspase-8 and caspase-3 apoptotic factors. NF-κB enhances the transcription of cytokines, such as IL-6, which are reported to be neuroprotective. The activation of NF-κB in response to noxious stimuli may indicate a survival mechanism that blocks apoptosis. The increased activity of NF-κB in response to Bcl-2, which is also an NF-κB target gene and a known anti-apoptotic factor, supports its role in the inhibition of apoptosis and neuronal rescue. However, some evidence has been presented suggesting that NF-κB may enhance apoptosis in cerebral ischemia [8]. The role of NF-κB in neuronal survival has recently generated a great deal of interest as a possible site of future therapeutic intervention following brain or spinal cord injury.

Role of NF-κB in Spinal Cord Injury

Studies have supported the hypothesis that NF-κB plays a role in the survival of neurons within the hippocampus [6]. NF-κB activation enhances neurosurvival by increasing the expression of genes that block apoptotic signaling [5].

N

Complete sciatic nerve transection leads to increased NF-κB activation in neurons on the side of the spinal cord ipsilateral to the side of axotomy in transgenic mice [9]. This increased activation is statistically significant at 3 and 5 d, and returns to baseline 10 d after transection. The number of neurons containing NF-κB activation was greater in the dorsal horn than the ventral horn at all time points. Neurons that activated NF-κB were located bilaterally in the dorsal horn of the spinal cord in un-operated animals as well as in the experimental animals, which is consistent with previous reports of constitutive NF-κB expression in the spinal cord [2].

Based on the results of immunofluorescent double-labeling, we have shown that the constitutive and induced activation of NF-κB, which was reported by production of β-galactosidase, occurs in neurons and not in astrocytes or microglia/macrophages (Fig. 1). Previous studies that measured p65 (RelA) levels in the spinal cord of rats that had undergone spinal cord injury (SCI) by contusion showed an increase in NF-κB activity in microglia and endothelial cells as well as neurons following crush SCI. However, crush SCI violates the blood–brain barrier and subjects the spinal cord to the presence and effects of blood borne cells and their inflammatory mediators. Pollock et al. also reported a more rapid increase in NF-κB activity, which was increased at 24 h and remained elevated for at least 72 h, than the time course in the present study. This earlier activation is most likely the result of the direct influence of immune cells releasing inflammatory cytokines.

NFκB activity is also increased in experimental autoimmune encephalomyelitis primarily in microglia, macrophages and T lymphocytes. In this study, myelin basic protein specific T lymphocytes were injected into the experimental animals causing an increase in NF-κB activity that peaked at 6 d and returned to normal by day 14. This follows a similar timeline of NF-κB activation to the present study, though Kaltschmidt and colleagues [10] did not examine NF-κB in neurons. Their results dealt primarily with the immune cell activation and the presence of perivascular infiltrates.

The time course of NF-κB activity is consistent with a study that evaluated dorsal root ganglion cells following sciatic nerve injury. This study demonstrated no increase in NF-κB activity 14 d after complete sciatic nerve transection, but increased activity after partial sciatic nerve ligation and chronic constriction injury at this time point. Our results suggest that the upregulation of NF-κB activity would have returned to baseline in the dorsal horn by this time point following complete sciatic nerve transection. This time course does not correspond to degenerative changes that occur within the spinal cord following axotomy. For example,

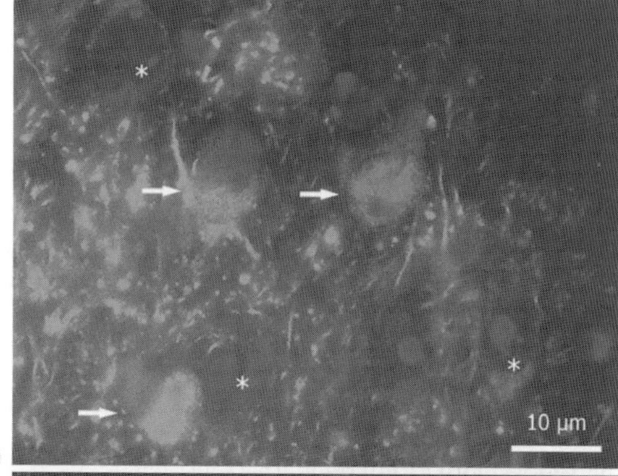

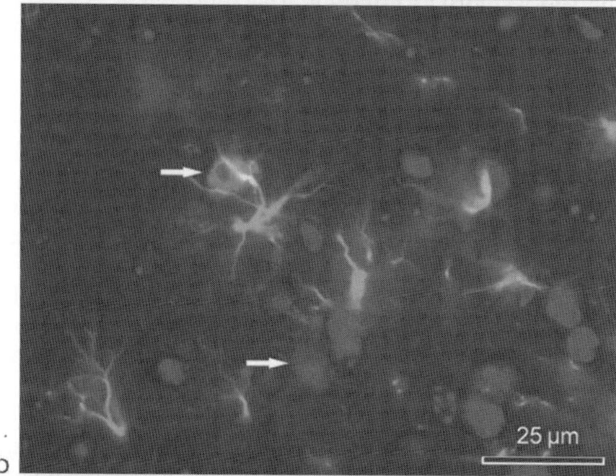

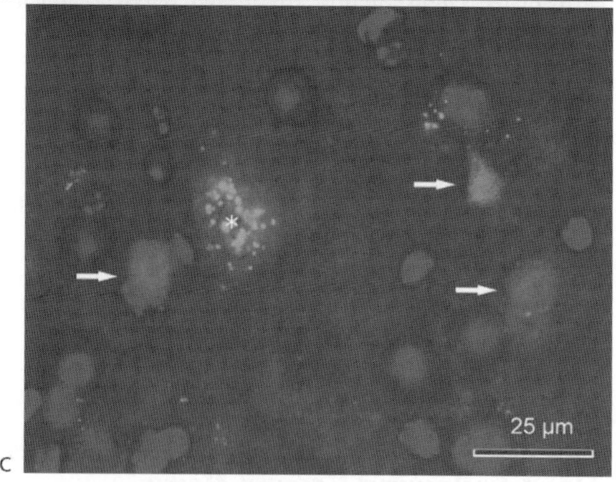

NF-κB: Activation in the Mouse Spinal Cord Following Sciatic Nerve Transection. Figure 1 Fluorescent photomicrographs of β-gal-positive cells (*red*) in the spinal dorsal horn double labeled with a green fluorescent chromagen for (a) MAP-2 (neurons), (b) GFAP (astrocytes) and (c) cd11b (microglia). β-gal-positive cells double label with MAP-2 (*arrows*), but not GFAP or cd11b, suggesting that they are neurons, rather than astrocytes or microglia.

degeneration of neurons in the dorsal horn in adult rats after sciatic nerve transection does not occur until 30 d post-axotomy. Reports have appeared of positive

TUNEL staining in the dorsal horn 14 d following a constricting sciatic nerve injury but not after complete transection, which showed no TUNEL positive cells at 14 d post-axotomy. Motor neurons in adult rats undergo atrophy, but not apoptosis, following sciatic nerve transection, unlike neonatal rats which show apoptosis in both the dorsal and ventral horns. Pollock and coworkers [9] showed very little NF-κB activity in the ventral horns and none in the alpha motor neurons at any time examined following complete sciatic nerve transection. These previous studies demonstrate a difference between sciatic nerve transection in neonatal rodents and adult rodents with significant changes in the ventral horn of young animals but minimal changes in the ventral horns of the adult animals. Our results are consistent with these findings as there is very little NF-κB activity in the ventral horn of any of the experimental animals.

Many studies have shown that sciatic nerve transection in the neonatal mouse or rat leads to a marked cell loss in the ventral horn. There is ample evidence that in neonatal rodents, axotomy leads to motor neuron loss via apoptosis that is usually evident by 21 d and continues to increase out to 30 and 60 d. However, there is a significant difference between adult and neonatal animals in this response. Pollin and coworkers [6] showed that age is a major determinant of the response of motor neurons to axotomy in the mouse. While neonatal mice show massive apoptosis in the ventral horn in response to axotomy, adult mice show very little apoptosis of motor neurons in response to peripheral nerve transection. These data are consistent with the lack of changes seen during this time period in the ventral horn of the adult mice used in this experiment. While actual cell loss is seen at later time periods, the changes in transcription factor activity and apoptotic pathways are seen during the first few weeks following neuronal injury.

A trend in the total number of NF-κB-positive cells appears to show a decrease below baseline for the animals sacrificed at the 1 d time period and then a rise above baseline for the 3 and 5 d time periods, followed by a return to baseline at the 10 d time period. An initial decrease in NF-κB activation could be consistent with a decrease in p65 levels which occur as a result of a decrease in the retrograde transport of trophic factors, however, the decrease in NF-κB activation was not statistically significant at the 1 d time point.

Function of NF-κB in Glial Cells and Neurons in Spinal Cord Injury

Glial cells have been shown to proliferate in response to sciatic nerve transection in both neonatal and adult rats. There was no evidence of NF-κB activity in any cells other than neurons in the work by Pollock et al. [9]. It has been shown that glial cells interact with neurons to regulate NF-κB activity in the CNS [2]. The sciatic nerve transection model does not violate the blood–brain barrier in the spinal cord and this may account for the lack of NF-κB activity seen in cells other than neurons. This may have important implications as to the NF-κB activity response to different types of spinal cord insults as the response appears to be only in neurons in this model but is also evident in microglia and endothelial cells in the contusion model. This may present an opportunity to target different cells in peripheral nerve injury versus direct spinal cord injury.

NF-κB plays a central role in promoting neuronal survival within the injured spinal cord and thus is an important target for research and future interventions. The extensive variety and availability of transgenic and knockout strains of mice provide an advantage over the rat for investigative opportunities in spinal cord injury.

References

1. Memet S (2006) NF-kappa B functions in the nervous system: from development to disease. Biochem Pharmacol 72:1180–1195
2. Kaltschmidt B, Kaltschmidt C (2000) Constitutive NF-kappa B activity is modulated via neuron-astroglia interaction. Exp Brain Res 130:100–104
3. Mattson MP (2005) NF-kappa B in the survival and plasticity of neurons. Neurochem Res 30:883–893
4. Nonaka M, Chen XH, Pierce JE, Leoni MJ, McIntosh TK, Wolf JA, Smith DH (1999) Prolonged activation of NF-kappa B following traumatic brain injury in rats. J Neurotrauma 16:1023–1034
5. Choi JS, Kim JA, Kim DH, Chun MH, Gwag BJ, Yoon SK, Joo CK (2000) Failure to activate NF-kappa B promotes apoptosis of retinal ganglion cells following optic nerve transection. Brain Res 883:60–68
6. Pennypacker KR, Kassed CA, Eidizadeh S, Saporta S, Sanberg PR, Willing AE (2001) NF-kappa B p50 is increased in neurons surviving hippocampal injury. Exp Neurol 172:307–319
7. Kassed CA, Butler TL, Patton GW, Demesquita DD, Navidomskis MT, Memet S, Israel A, Pennypacker KR (2004) Injury-induced NF-kappa B activation in the hippocampus: implications for neuronal survival. Faseb J 18:723–724
8. Schwaninger M, Inta I, Herrmann O (2006) NF-kappa B signalling in cerebral ischaemia. Biochem Soc Trans 34:1291–1294
9. Pollock G, Pennypacker KR, Memet S, Israel A, Saporta S (2005) Activation of NF-kappa B in the mouse spinal cord following sciatic nerve transection. Exp Brain Res 165:470–477
10. Kaltschmidt C, Kaltschmidt B, Lannes-Vieira J, Kreutzberg GW, Wekerle H, Baeuerle PA, Gehrmann J (1994) Transcription factor NF-kappa B is activated in microglia during experimental autoimmune encephalomyelitis. J Neuroimmunol 55:99–106

NG2

Definition
One kind of chondroitin sulfate proteoglycans that is expressed by a distinct mature glial population in the adult brain and spinal cord.

NGF

Definition
►Nerve Growth Factor (NGF)

NI-35/250

Definition
Inhibitory fractions of CNS myelin with molecular weights of 35 and 250 kD. The neurite growth inhibitory (NI)-35/250 proteins are of highly conserved structure and were found to be associated with oligodendrocytes and myelin sheaths in mammals, including human tissue. NI-250 likely corresponds to Nogo-A. In vitro assays revealed that the application of NI-35/250 inhibits spreading of fibroblasts and causes collapse of growth cones thus arresting neurite growth.

►Growth Inhibitory Molecules in Nervous System Development and Regeneration

Niche in Ecology

Definition
An evolutionary or ecological concept that describes the precise role or place an organism holds in an ecosystem.

►The Phylogeny and Evolution of Amniotes

Niche in Neurogenesis

Definition
Region where the degree of neurogenesis is related to the interactions of the precursor cell with its microenvironment.

►Adult Neurogenesis

Nidifugous

Definition
Animals (birds) leaving the nest shortly after birth/hatching

►Neural Correlates of Imprinting

Niemann-Pick Disease

Definition
Niemann-Pick disease is subdivided into two classes. In types A and B, the genes for acid sphingomyelinase are deficient; in types C and D, the genes for the NPC-1 protein are deficient. Niemann-Pick disease type C (NPC) is a fatal autosomal-recessive, neuro-visceral lipid storage disorder resulting from mutations in either the NPC1 (95% of families) or NPC2 gene and presenting with a plethora of symptoms related to hepatic and pulmonary diseases, and various neuropsychiatric disorders. The clinical spectrum ranges from a neonatal rapidly fatal disorder to an adult-onset chronic neurodegenerative disease. In the latter, symptoms include ►cerebellar ataxia, movement disorders, vertical supranuclear ►ophthalmoplegia, ►dysarthria, ►dysphagia, psychiatric and cognitive impairments as well as, less frequently, ►epilepsy and ►cataplexy.

Night Terrors

Definition
Night terrors, also known as Sleep Terrors or Pavor Nocturnus, consist of sudden arousals from slow-wave

(delta) Non-REM sleep accompanied by a cry or piercing scream and behavioral manifestations of intense fear. There is pronounced autonomic discharge, with tachycardia, tachypnea, flushing of the skin, diaphoresis, mydriasis, and increased muscle tone.

Amnesia for the episode generally occurs, but some adults recall vivid dreaming that accompanied their sleep terrors.

►Non-REM Sleep

Nightmare

►Dreaming

Nigrostriatal Fibers

Definition

Somatotopic, dopaminergic projection of the substantia nigra, pars compacta, to the corpus striatum.

►Pathways

Nissl Bodies

Definition

Collections of granular endoplasmic reticulum and ribosomes in nerve cells, and which show strong staining in histological sections of nerve cells impregnated with basophilic dyes.

►Chromatolysis
►Neuronal Changes in Axonal Degeneration and Regeneration

Nissl Stain

Definition

A method of staining brain tissue due to interactions between a basic stain (e.g. cresyl violet) and acidic groups, e.g. in DNA, RNA ("Nissl substance"). Since most negatively charged groups (DNA, RNA) are in the soma, this technique stains somata of neurons, but not their distal dendrites and axons. The technique is named after its discoverer Franz Nissl (1860–1919).

Nissl Technique

Definition

The Nissl (turn of nineteenth century German Neurologist) technique is a neuron staining method that uses basic dyes to stain the chromatin material (ribose nucleic acid) in the cell body.

Nitric Oxide

Definition

Nitric Oxide (NO) has been proposed to play a role in intercellular communication and is considered to be an unconventional transmitter. Unlike classical neurotransmitters it is not stored in vesicles, does not bind to specific target receptors on the membrane surface, and does not have an active process to terminate its action. The gas nitric oxide is synthesized from arginine a reaction that depends on the enzyme nitric oxide synthase (NOS).

NLL

►Nuclei of the Lateral Lemniscus

NMDA

Definition

N-methyl-D-aspartatic acid is an amino acid derivative that binds to the NMDA-type glutamate receptor.

N

NMDA Receptors

Definition

A variety of ionotropic receptors for the common excitatory neurotransmitter glutamate. They are ligand-gated ion channels that are gated open when glutamate binds to them but only when the membrane potential is relatively depolarized; thus they have voltage-dependent properties. Voltage dependence is conferred on these channels by the binding of Mg^{2+} ions in the channel pore at relatively hyperpolarized membrane potentials. NMDA (N-methyl-D-aspartate) is a specific agonist for these receptors, i.e., it is a drug that selectively activates NMDA receptors and not other glutamate receptor types.

Receptor activation induces the influx of sodium and calcium ions, and the efflux of potassium ions. The movement of ions causes depolarization of the neuronal membrane and increases the probability that an action potential will be generated. NMDA receptors consist of NR1 (GluRζ1) and some of NR2A (GluRε1), NR2B (GluRε2), NR2C (GluRε3) and NR2D (GluRε4) subunits and are localized at the postsynaptic site on the dendritic spine.

The NR1 subunit is an essential component of NMDA receptors and the composition of the NR2 subunits determines the properties of NMDA receptor channels.

►Glutamate Receptor Channels
►Memory, Molecular Mechanisms
►Associative Long-Term Potentiation (LTP)
►Long-Term Potentiation (LTP)

NMDA-LTP

Definition

LTP that requires activation of the NMDA receptor, one type of glutamate receptors, for its induction. NMDA receptors are usually activated by high-frequency activity of afferent fibers that causes strong depolarization of postsynaptic cells. A typical example is LTP in the hippocampal CA1 region.

►Memory, Molecular Mechanisms
►Associative Long-Term Potentiation (LTP)
►Long-Term Potentiation (LTP)

NMJ

►Neuromuscular Junction

NMR

Definition

►Nuclear Magnetic Resonance

Nociception

Definition

Nociception is the process of detecting and transmitting signals in the presence of a noxious stimulus.

Nociception and Growth Factors

►Growth Factors and Pain

Nociceptive Modulation

►Descending Modulation of Nociception

Nociceptive Neurons

Definition

According to the International Association for the Study of Pain, a nociceptor is a receptor that is preferentially sensitive to a noxious stimulus or to a stimulus which would become noxious if prolonged.

Nociceptors respond selectively to damaging stimuli and transmit information via Aδ and C afferent neurons. They are broadly classified as mechanical nociceptors that respond to strong mechanical stimulation, heat nociceptors that respond to temperatures above 45°C, and polymodal receptors that respond more generally to noxious stimuli (including mechanical, heat, and chemical). Although the primary nociceptive responses are generated by nociceptors in the periphery, neurons throughout the central nervous system that respond to noxious stimuli are also nociceptive, including those in the midline and intralaminar thalamus and parts of cingulate cortex.

►Cutaneous Pain, Nociceptors and Adequate Stimuli
►Joint Pain, Nociceptors and Adequate Stimuli
►Nociceptors and Characteristics
►Visceral Pain, Nociceptors and Adequate Stimuli

Nociceptive Pathways

Definition

A nociceptive pathway is the pathway that the information from the nocipteror follows within the nervous system.

►Ascending Nociceptive Pathways

Nociceptive Reflexes

Definition

Nociceptive reflexes are motor responses to noxious stimuli.

►Integration of Spinal Reflexes
►Viscero-somatic Reflex

Nociceptors: Adequate Stimulus

H. Richard Koerber
Department of Neurobiology, University of Pittsburgh, Pittsburgh, PA, USA

Definition

The ►adequate stimulus for a given sensory neuron is a sensory event that is of sufficient magnitude to elicit action potentials, typically resulting in a sensory experience. A ►nociceptor is defined by the International Association for the Study of Pain (IASP) as a receptor preferentially sensitive to a noxious stimulus (one which is damaging to normal tissues [see ►Visceral pain for exception re. visceral nociceptors]), or to a stimulus which would become noxious if prolonged. Thus the adequate stimulus for a nociceptor would be one that has the capability to cause pain or discomfort. While most sensory neurons respond best to stimuli of a single ►sensory modality (e.g., mechanical or thermal), nociceptors can often be activated by stimuli of different modalities (i.e., are polymodal).

Characteristics

Historical Perspective

The concept of nociception was first introduced by Sherrington [1] at the turn of the twentieth century. He proposed that the sensation of pain was mediated by specific sensory organs that were responsive to painful or noxious stimuli. However, it was several decades before the existence of specific pain sensing fibers or nociceptors was firmly established [2]. During the intervening period of time, many different theories about how peripheral sensory neurons could code for the sensation of pain were hypothesized. A recent eloquent historical review of the subject by E. R. Perl [3] provides a detailed timeline of these events. In brief, the differing theories on the roles of peripheral receptors can be generalized as specificity versus intensity. In the specificity theory of pain, only those sensory neurons that respond to stimulus intensities in the noxious range would be responsible for signaling pain, whereas in the intensity theory most, if not all, sensory neurons would be responsive to innocuous stimuli, but would be more responsive (generate more action potentials over time) to stimuli of noxious intensity. This increased number of inputs would signal the fact that the intensity of the stimulus had reached a noxious level. Here, I will briefly describe what has been reported about primary sensory neurons in general, and more specifically putative nociceptors over the past several decades and then discuss how these findings fit with these two contrasting theories.

Primary Sensory Neurons

Primary sensory neurons provide constant feedback on the external environment as well as the internal state of the body. These neurons are located in sensory ganglia that lie outside of the central nervous system. While some of these ganglia are associated with cranial nerves such as the trigeminal ganglion, the majority of these neurons are to be found in dorsal root ganglia associated with each spinal nerve (see ►Visceral pain for role of sensory neurons in the nodose ganglion). Their axons bifurcate within the ganglion and give rise to a peripheral branch that innervates various tissue types in the trunk, limbs and viscera and a central branch that travels through the appropriate dorsal root to enter the spinal cord, where it forms synapses with second-order neurons.

There are several different ways of classifying primary sensory neurons. The most common means of classification is based on the conduction velocity of their peripheral axons, which is directly related to the axon diameter and to whether the axon is myelinated. Based on the distribution of these peripheral nerve conduction velocities, primary sensory neurons are routinely divided into different groups: Aα/β, Aδ, and C. While the actual conduction velocity of these groupings varies across different mammalian species, the Aα/β group consists of large myelinated axons that have the fastest peripheral conduction velocities and approach 100 m/s in some species. The Aδ group contains smaller caliber fibers that are thinly myelinated and conduct at an intermediate velocity, and the C-fiber group is comprised of the smallest, unmyelinated and most slowly conducting fibers, usually <2 m/s. Another convenient means of classification of primary sensory neurons is the peripheral tissue they innervate. Fibers innervating skin are described as cutaneous sensory neurons. Similarly, afferent fibers innervating thoracic, abdominal or pelvic viscera are termed visceral afferents.

Primary sensory fibers are responsive to a variety of sensory modalities, including mechanical, thermal and chemical stimuli. Fibers responsive to a specific sensory modality are further classified according to the intensity of the adequate stimulus. Those primary sensory fibers that respond to gentle mechanical forces or innocuous thermal stimuli, and fail to encode a broad range of stimulus intensities, are classified as low threshold mechanoreceptors or innocuous cooling or warming fibers, respectively. As defined above, sensory fibers that respond only to stimulus intensities that would be considered tissue threatening or have the potential to be damaging are termed nociceptors. The majority of these nociceptive fibers is responsive to two or more stimulus modalities, and for that reason are referred to as polymodal nociceptors.

Sensory neurons responding to different intensities of peripheral stimuli are distributed across the different conduction velocity groups. Most sensory neurons with fibers conducting in the Aα/β range respond to innocuous mechanical stimuli, and are classified as low threshold mechanoreceptors. A subset of these fibers can respond to relatively innocuous mechanical stimuli, but also encode stimulus intensities into the noxious range and in some cases respond to noxious heating of the skin. This trend reverses with decreasing conduction velocity as a majority of Aδ-fibers and most C-fibers are classified as nociceptors. The relative numbers of functional types in specific conduction velocity groups varies between species and the areas of the body the fibers innervate. However, it is important to point out that both nociceptors and non-nociceptors exist in all three conduction velocity groups.

Primary sensory neurons also exhibit diversity in many other properties, including cell membrane properties, laminar location of central projections and neurochemical content [4]. Myelinated fibers that respond to innocuous mechanical stimulation of the skin have narrow somal action potentials without breaks in the rising or falling phase of the spike. Unmyelinated and myelinated nociceptive fibers have broader action potentials that most often have a distinct inflection on the falling phase of the spike [4]. While this relationship is quite constant for myelinated fibers, all unmyelinated fibers have broad inflected somal action potentials regardless of their peripheral response properties.

Spinal Projections

The central branches of myelinated low threshold mechanoreceptive fibers enter the spinal cord and bifurcate into main ascending and descending branches that travel in the dorsal columns. Additional collateral branches turn ventrally and pass through the dorsal horn before terminating in lamina III-V. These laminae are largely involved in processing inputs elicited by innocuous mechanical stimuli. The central projections of myelinated nociceptors were first described by Light and Perl [5], who focused on fibers conducting in the intermediate Aδ range. They found that upon entry into the spinal cord, the main branches were laterally located in the dorsal column often in or near ►Lissauer's tract. Terminal arbors from these afferents were centered primarily on laminae I and IIo, spinal laminae largely involved in the processing of inputs elicited by noxious peripheral stimuli, with some passing ventrally to terminate in lamina V. More recently, Woodbury and Koerber [6] have demonstrated an additional central projection pattern for myelinated nociceptive fibers. These fibers have projections very similar to those of low threshold mechanoreceptors in laminae III-V, however, their projections also extend dorsally through laminae I and II.

Unmyelinated C-fibers, most of which are considered to be nociceptors, enter the spinal cord and usually bifurcate and run rostrally and caudally along the surface of the dorsal horn in Lissauer's tract. The primary collaterals send off several additional branches that penetrate ventrally into the superficial dorsal horn laminae I and II where they end in dense terminal fields. While individual C-fibers have projections that are focused more or less in different parts of the superficial dorsal horn, most are found in both laminae I and II with the primary focus usually in lamina II [7,8].

In summary, the sensory neurons that have peripheral response characteristics of nociceptors (i.e., code stimulus intensity in the noxious range), consistently have direct projections to the superficial dorsal horn laminae known to be involved in nociceptive processing. However, it is also important to note that not all sensory neurons projecting to these spinal laminae are nociceptors.

Neurochemical Properties

As a group, nociceptors have been the main target of neurochemical analysis in recent years, and they have been shown to exhibit the most pronounced phenotypic diversity among sensory neurons. Unmyelinated nociceptive sensory neurons contain a large number of neuroactive compounds and express receptors and have been divided into two major groups based on the combination of neurochemical phenotype and sensitivity for different ►neurotrophins. The first group of nociceptive neurons is sensitive to nerve growth factor and expresses its high-affinity receptor tyrosine kinase receptor A (trkA); they also usually contain ►neuropeptides such as calcitonin gene-related peptide (CGRP), substance P and galanin. The second group is responsive to members of the glial cell line-derived neurotrophic factor family, and expresses the cognate ►receptor tyrosine kinase. This group usually does not contain peptides, have binding sites for the isolectin IB4 (obtained from *Bandeiraea simplicifolia*), and contains the ►purinergic receptor $P2X_3$, suggesting sensitivity to ATP [7].

Less is known about the different neurochemical phenotypes of myelinated nociceptive fibers. However, they can be divided into two groups based on whether they express the trkA receptor and contain peptides such as (CGRP), or as, it has recently been shown, contain the neurtrophic factor 3 receptor trkC and lack peptides [9].

Nociceptors: Intensity vs. Specificity

As discussed above, it has been shown that there is a population of sensory neurons that only respond to stimulus intensities in the noxious range, corresponding to the classical definition of a nociceptor. In addition, most individual nociceptors can respond to several stimulus modalities. These sensory fibers are distinctly different from those that are responsive to innocuous stimuli and do not encode stimulus intensity.

However, not all sensory neurons fit neatly into these two different categories. It has long been known from animal studies that some fiber types are not easily classified as associated with nociceptors or non-nociceptors. For example, Burgess and Perl [2] reported a group of myelinated fibers that were distinct from low threshold or innocuous myelinated fibers, but had thresholds below those considered to be nociceptors. They referred to these as moderate pressure units. In addition, more recent experiments in humans have demonstrated that many putative nociceptive fibers, including polymodal fibers, have mechanical and/or thermal thresholds at intensities that are well below those necessary to evoke pain. However, it was also shown that there was a clear and significant relationship between the numbers of action potentials elicited by suprathreshold stimuli in these fibers and the perception and intensity of pain in these subjects [10]. Taken together, these findings suggest that while it is clear that some sensory neurons clearly fit the original concept of nociceptors put forth by Sherrington, others that have ►absolute response thresholds that would suggest a non-nociceptive function may also function as nociceptors (i.e., they contribute to the sensation of pain). These two groups of fibers share many other properties, including broad somal action potentials, neurochemical content and central projections in the superficial laminae of the spinal dorsal horn.

In summary, nociceptors, the necessary substrate for the specificity theory, are clearly present. However, it is also clear that sensory neurons that do not fit the Sherringtonian definition of a nociceptor do code for the intensity of a stimulus into the noxious range and thus have the potential to contribute to the sensation of pain.

N

References

1. Sherrington CS (1906) The integrative action of the nervous system. Cambridge University Press, Cambridge
2. Burgess PR, Perl ER (1967) Myelinated afferent fibres responding specifically to noxious stimulation of the skin. J Physiol 190:541–562
3. Perl ER (2007) Ideas about pain, a historical view. Nat Rev Neurosci 8:71–80
4. Koerber HR, Mendell LM (1992) Functional heterogeneity of dorsal root ganglion cells. In: Sensory neurons: diversity, development and plasticity. Oxford University Press, New York
5. Light AR, Perl ER (1979) Spinal termination of functionally identified primary afferent neurons with slowly conducting myelinated fibers. J Comp Neurol 186:133–150
6. Woodbury CJ, Koerber HR (2003) Widespread projections from myelinated nociceptors throughout the substantia gelatinosa provide novel insights into neonatal hypersensitivity. J Neurosci 23:601–610
7. Snider WD, McMahon SB (1998) Tackling pain at the source: new ideas about nociceptors. Neuron 20:629–632

8. Sugiura Y, Lee CL, Perl ER (1986) Central projections of identified, unmyelinated (C) afferent fibers innervating mammalian skin. Science 234:358–361
9. McIlwrath SL, Lawson JJ, Anderson CE, Albers KM, Koerber HR (2007) Overexpression of neurotrophin-3 enhances the mechanical response properties of slowly adapting type 1 afferents and myelinated nociceptors. Eur J Neurosci 26:1801–1812
10. Torebjork HE, LaMotte RH, Robinson CJ (1984) Peripheral neural correlates of magnitude of cutaneous pain and hyperalgesia: simultaneous recordings in humans of sensory judgments of pain and evoked responses in nociceptors with C-fibers. J Neurophysiol 51:325–539

Nociceptors and Characteristics

HERRMANN O. HANDWERKER
Institut für Physiologie und Pathophysiologie, Erlangen, Germany

Synonyms

Nociceptors (human)

Definition

The term "nociceptor" has been derived from the Latin "nocere," which means to harm or to damage. Nociceptors are characterized by two distinctive features: (i) they are responsive preferentially to tissue threatening stimuli and encode their intensity, (ii) they mediate nocifensive motor and vegetative reactions by their central connections.

Characteristics

General

The nociceptor concept was introduced by the famous physiologist Sir Charles Sherrington in a monograph published in 1906. Sherrington used it to describe primary afferent neurons which were processing information on tissue-threatening stimuli in animal experiments. He observed that noxious stimuli induced withdrawal reflexes and defined nociceptors as a type of afferent nerve inducing withdrawal (nocifensive) reflexes, which became the surrogate of human pain experiences in animal experiments. In addition, nocifensive motor reflexes and the respective reactions of the autonomic nervous system (i.e., rise in blood pressure, increase in heart rate, etc.) can also be observed in human subjects, even in states of unconsciousness, when pain experiences are precluded.

For obvious reasons, a proof of nociceptor functions cannot be based solely on responses to damaging stimuli as the Latin word stem would suggest. Many neurons may be excited by damaging stimuli at least temporarily in the process of their decay. Therefore, the concept of the nociceptor requires at least one additional element: central nervous system connections.

One may note that nociceptors are not just "pain receptors." Pain is related to functions of the conscious central nervous system. Excitation of nociceptors is neither a necessary nor a sufficient condition for pain experiences. Nevertheless, it is true that in conscious healthy human beings a more or less linear relationship has been found (e.g., between the activation of heat sensitive nociceptors in the skin and the intensity of burning pain). In addition, the ►hyperalgesia of inflamed tissue is reflected in the increased activity of nociceptors [1], provided that no other factors interfere, such as placebo suggestions, analgesic medications, etc.

On the other hand, the term nociceptor can not be restricted to the receptor and channel molecules in the nerve terminals that are activated by noxious physical or chemical stimuli. One such a membrane receptor, which is often erroneously labeled as "the nociceptor," is the TRPV1 (transient receptor potential vanilloid) receptor channel which is opened by binding of the pungent substance capsaicin, by heating, and by application of acids. This molecule is also expressed in non-sensory cells and in neurons with arguable nociceptive functions.

Figure 1 shows a systematic diagram of a nociceptor neuron. The cell bodies of these neurons are in the dorsal root ganglia and in the trigeminal ganglia, respectively. The peripheral terminals may be found in most tissues of the body. The central terminals form synapses with secondary neurons in the superficial dorsal horn of the spinal cord and in the nucleus caudalis of the trigeminal complex. An "ideal" nociceptor should provide the central nervous system exclusively with nociceptive information, and hence encode only stimulus intensities in or near the noxious range. In the CNS, nociceptor input must contribute to the information processing within the central nociceptive neuronal network and hence to the subjective experience of pain. Visceral afferents provide a particular challenge for the nociceptor concept. For example, in the wall of the urinary bladder and of intestines, numerous slowly conducting afferent nerve fibers have been found which may serve two purposes (see ►Visceral Pain). In the case of the urinary bladder, they control micturition and mediate sensations of increasing urgency with filling of the bladder. At higher tension of the bladder wall, these sensations become painful while the same neuronal population is more vigorously excited. Thus it seems as if these afferent nerves contribute to two functions: nociceptive and non-nociceptive. The complex issue of peripheral coding of visceral pain has been controversially discussed [3].

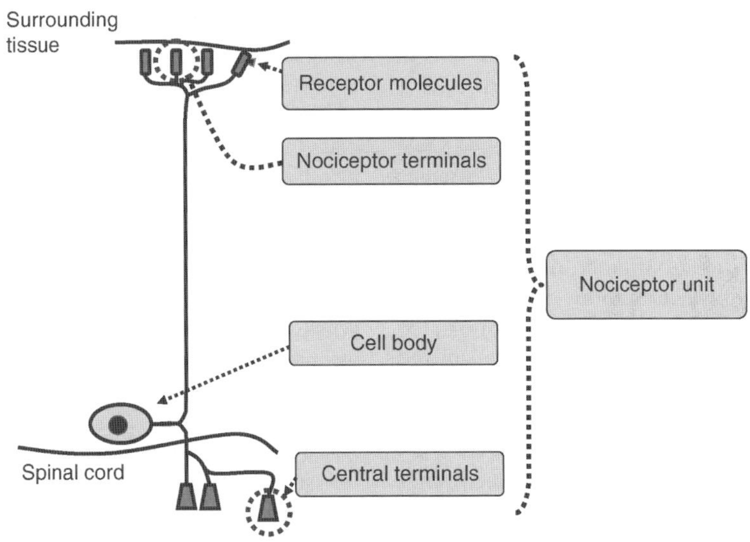

Nociceptors and Characteristics. Figure 1 Schematic diagram of the relevant parts of a nociceptive neuron. The cell bodies (pericaria) are situated in the dorsal root ganglia and in the ganglia trigemini, respectively. Peripheral terminals can be found in most tissues of the body, central terminal make synapses with secondary nociceptive neurons in the CNS. (Reprinted from [2], by courtesy of Elsevier Pb).

The Trp (Transient Receptor Potential) Receptor Family and Other Molecules in Nociceptor Terminals

The fast progress of the molecular analysis of receptor and channel structures in nerve terminals has lead to a deeper understanding of the molecular mechanisms of stimulus transduction in nociceptor terminals [4]. This topic can only marginally be covered by this essay. A large group of nociceptor neurons express the ►trk-A receptor, the high affinity receptor for nerve growth factor (NGF). This applies also to humans, since a genetic defect of the trk-A receptor leads to congenital insensitivity to pain (CIPA-syndrome) [5]. Recently it has been discovered that a genetic defect of the voltage dependent NaV1.7 channel also leads to congenital pain intensity. This finding points to a major role of this molecule in safeguarding spike conduction in nociceptive nerve endings [6]. Most of the peripheral neurons expressing trk-A receptors are also peptidergic (their cell bodies synthesize the ►neuropeptides CGRP [calcitonin gene related peptide] and to a minor extent also substance P). These neuropeptides may be released from central nerve terminals as ►neuromodulators of synaptic transmission in the dorsal horn of the spinal cord. They can also be released from peripheral nerve terminals, leading to ►neurogenic vasodilatation and plasma-extravasation (see ►Inflammatory Pain). This is the basis of the axon reflex reaction following noxious stimulation of the skin (neurogenic flare reaction). However, from animal experiments it can be extrapolated that another group of nociceptors exists which are not dependent on NGF, but on other nerve growth factors, in particular BDNF (brain-derived neurotrophic factor). Immuno-cytochemically, many of these neurons are characterized by staining with a plant isolectin, IB4.

Many human nociceptors are equipped with receptor channels of the trp family [7]. Most important is the above mentioned TRPV1. This molecule forms an nonspecific cation channel which is operated by binding of the plant derived molecule capsaicin. TRPV1 receptors are common in trkA expressing peptidergic primary afferent units, but it is not expressed in all nociceptors. As mentioned above, the TRPV1 receptor is also operated by noxious heating and acids. From experiments with genetically modified knockout mice, it is known that this receptor is partly, but not exclusively responsible for the nocifensive responses to acids and noxious heat. There are other ►trp receptor molecules such as TRPA1 which may also explain part of the sensitivity of nociceptor terminals to algogenic chemical substances (e.g., mustard oil). Another group of receptor molecules important for the functioning of nociceptors are G-protein coupled receptors for bradykinin (B1, B2) and for prostaglandins of the E-group (EP1,2,3). Activation of these G-protein coupled receptors by endogenous algogenic substances released in the course of inflammatory tissue reactions leads to sensitization of nociceptors to mechanical, heat and chemical stimuli. Intracellularly, this process is mediated by second messenger cascades involving protein kinases, which leads to a characteristic lowering of thresholds and greater magnitude of suprathreshold responses to mechanical and thermal stimuli, and forms the basis of hyperalgesia due to inflammation.

Microneurography of Human Nociceptors

The functional properties of nociceptors in man are mainly studied with the methods of ►microneurography, which is the extracellular recording of spike potentials from individual afferent nerve fibers in peripheral nerves of awake healthy volunteers and patients [8]. These studies have been performed by about a dozen laboratories worldwide during the last 30 years. Studies on muscle nociceptors are difficult and rare [9] and even in cutaneous nerves, with a few exceptions, only unmyelinted C-fibers have been studied because it is difficult to obtain stable recordings from small myelinated nociceptive nerve fibers [10]. For this fiber class, one has to resort to data from monkeys which have a similar afferent nerve fiber spectrum. Apparently, myelinated nociceptors (A-δ) are heterogeneous. Some are high threshold mechanosensors with little chemical or thermal responsiveness. These units may, however, play a role in motor withdrawal reflexes. Another group of small myelinated nociceptors are similarly equipped with molecular receptor molecules in their terminals, as is the case with C-fiber nociceptors.

As indicated above, the nociceptor nerve terminal expresses a mosaic of receptor molecules. However, this mosaic is differentially distributed in different nociceptor populations. Cutaneous nociceptors have been sorted into subclasses according to their thresholds for certain types of stimuli and the coding of suprathreshold noxious stimuli. The most common group of C-nociceptors respond to mechanical stimulation with von Frey bristles (thresholds in most cases 30–150 mN) and to heating (thresholds in most cases 40–44°C). These units have been named C-MH units because they are responsive to *m*echanical stimulation and to *h*eating. They are comparable to the "polymodal nociceptors" originally described in the hairy skin of the cat. Many of those units are also responsive to capsaicin and to vasoneuroactive endogenous agents such as bradykinin. However, on intracutaneous injection of capsaicin, responses are usually short lived and followed by desensitization of nociceptor terminals. Another group of C-units in human skin are insensitive to noxious mechanical stimulation, including pricking the skin with a hypoderminc needle. Due to their insensitivity to mechanical stimuli, they have often been called "►silent nociceptors." However, they gain mechanical sensitivity comparable to that of C-MH units in the course of inflammation; therefore, they have also been poetically named "sleeping nociceptors." These units are sensitive to capsaicin and indeed show a greater response to capsaicin than C-MH units. Because capsaicin treatment leads to sensitization to mechanical stimuli, the mechanoinsensitive nociceptors are the source of primary mechanical and heat hyperalgesia following intracutaneous capsaicin injection. We have named this nociceptor class C-M_iH or C-H units because in intact skin they are also sensitive to heating, having slightly higher thresholds than C-MH units (around 46°C). In addition, nociceptor terminals that express G-protein coupled receptor molecules seem to be different in the C-MH and C-M_iH classes. A subgroup of C-M_iH units is particularly sensitive to histamine and express the histamine H_1 receptor, and it has been shown that responses of this group of nociceptors parallels closely the itch sensation induced by intracutaneous application of histamine [9]. Discovery of histamine sensitive "itch-nociceptors" does not solve the puzzle of the peripheral mechanism of itching entirely, however. There are forms of itching which are not mediated by this group of primary afferents.

In human skin, CM_iH units and not the more common C-MH units mediate sustained flare responses by releasing CGRP upon noxious stimulation [8]. All these features indicate that the class of mechanoinsensitive C-nociceptors (i.e., C-M_iH nociceptors) is more important for inflammatory pain and probably also for other types of pathological pain than the more common C-MH units.

References

1. Torebjörk HE, LaMotte RH, Robinson CJ (1984) Peripheral neural correlates of magnitude of cutaneous pain and hyperalgesia: simultaneous recordings in humans of sensory judgements of pain and evoked responses in nociceptors with C-fibers. J Neurophysiol 51:325–339
2. Handwerker HO (2006) Nociceptors: neurogenic inflammation. In Cerverp F, Jensen TS (eds) Handbook of Clinical Neurology, vol 81 (3rd series) pain. Elsevier, Edinburgh, pp 23–33
3. Cervero F, Jänig W (1992) Visceral nociceptors: a new world order? Trends Neurosci 15:374–378; Cervero F, Jänig W (1993) Reply. Trends Neurosci 16:139
4. Lumpkin EA, Caterina MJ (2007) Mechanisms of sensory transduction in the skin. Nature 445:858–865
5. Indo Y, Tsuruta M, Hayashida Y, Karim MA, Ohta K, Kawano T, Misubuchi H, Tonnoki H, Awaya Y, Matsuda I (1996) Mutations in the TRKA/NGF receptor gene in patients with congenital insensitivity to pain with anhidrosis. Nat Genet 13:485–488
6. Cox JJ, Reiman F, Nicholas AK, Thronton G, Roberts E, Springell K, Karbani G, Jafri H, Mannan J, Raashid Y, Al-Gazali L, Hamamy H, Valente EM, Gorman S, Williams R, McHale DP, Wood JN, Gribble FM, Woods CG (2006) An SCN9A channelopathy causes congenital inability to experience pain. Nature 444:894–898
7. Levine JD, Allessandri-Haber N (2007) TRP channels: targets for the relief of pain. Biochem Biophys Acta [Jan. Epub ahead of print]

8. Torebjörk HE, Schmelz M, Handwerker HO (1996) Functional properties of human cutaneous nociceptors and their role in pain and hyperalgesia. In: Belmonte C, Cervero F (eds) Neurobiology of Nociceptors. Oxford University Press, Oxford, pp 349–369
9. Simone DA, Marchettini P, Caputi G, Ochoa JL (1994) Identification of muscle afferents subserving sensation of deept pain in humans. J Neurophysiol 72:883–889
10. Adriaensen H, Gybels J, Handwerker HO, VanHees J (1983) Response properties of thin myelinated (A-delta) fibers in human skin nerves. J Neurophysiol 49:111–122

Nocturnal Myoclonus

Definition

Periodic stereotyped leg twitches during sleep.

Nocturnal/Diurnal

LAURA SMALE, ANTONIO A. NUNEZ
Psychology Department, Michigan State University, East Lansing, MI, USA

Definition

A diurnal animal is one that is most active during the day, and a nocturnal animal is one that is most active at night.

Characteristics

General Considerations

Virtually all organisms undergo rhythms associated with the day-night cycle, but the patterns of these rhythms can vary considerably from one species to the next. The most striking differences have to do with the coupling between the rhythms and the day-night cycle. Specifically, in some animals activity is elevated during the day, while in others activity is highest during the night. The former are referred to as "diurnal" organisms and the latter as "nocturnal" ones. In both cases, an internal "circadian" ►clock generates endogenous rhythms that have a period of approximately 24 h and that are synchronized, or "►entrained," to a 24 h day by environmental cues such as the ►light-dark cycle.

One of the key adaptations that made the evolution of mammals possible was a change in this circadian time-keeping system from one that promoted a diurnal pattern in reptiles to one that supported a nocturnal pattern of adaptation in mammals. Although nocturnality remains most common, diurnality resurfaced in a variety of independent mammalian lineages, including our own. At each of these transitions changes occurred in the biological timekeeping mechanisms that coordinate a wide range of behavioral and physiological processes. More is known about how circadian systems operate in nocturnal than diurnal species because the animals traditionally used in biomedical research are nocturnal (e.g., mice, rats and hamsters). However, research on diurnal mammals has accelerated in recent years. Below, some of the ways in which rhythms in diurnal and nocturnal mammals are similar and different are first described briefly, and then the neural mechanisms that might cause the differences are outlined.

Patterns of Rhythmicity in Diurnal and Nocturnal Mammals

Day- and night-active animals differ with respect not only to their patterns of general activity, but also their rhythms in virtually every aspect of behavior and physiology [1]. Peaks in rhythms in secretion of most hormones, body temperature, digestive function, heart rate, mating behavior and feeding all occur approximately 180° out of phase in diurnal and nocturnal mammals. Some examples are highlighted below.

The circadian timekeeping system produces rhythms in secretion of many hormones, almost all of which are quite different in nocturnal and diurnal species [2]. The major period of prolactin secretion occurs at night in humans, while in rats this hormone does not rise until the second half of the day. The rhythms in adrenal secretion of corticosteroids are also inverted in nocturnal and diurnal species, with the rise always occurring in this case at the beginning of the active period of the day. Pituitary hormones regulating the gonads undergo rhythms that can differ between the sexes and can change across development and with reproductive states such as pregnancy and lactation.

However, in each case these patterns are not the same in diurnal and nocturnal species. For example, as young men go through puberty, their plasma levels of gonadotropins change from being arrhythmic to rhythmic, with the highest levels during the night, and then they revert back to an essentially arrhythmic state when adulthood is reached. Such changes are not apparent in nocturnal species. Just before ovulation, the pituitary gland produces a surge in luteinizing hormone release that is precisely timed in many rodents to precede the onset of the daily active period by three to four hours. This surge thus occurs 180° out of phase in day- and night-active animals. The rhythm in secretion of the hormone ►melatonin is notable in that it is actually the same across species, regardless of an animals' activity pattern. The ►pineal gland releases this hormone into the blood stream at night, when diurnal animals sleep and nocturnal animals are active. The duration of the period during which melatonin is secreted varies with daylength and can consequently serve as a signal of the changing seasons, regardless of an animal's activity pattern. There is some evidence that melatonin has a soporiphic effect in humans, and that its administration to people whose endogenous release is low, such as the elderly, can help in the consolidation of sleep at night. As one might expect, nocturnal mammals do not exhibit this response.

Body temperature (Tb) is influenced by a variety of factors that govern its pattern of change across the day. While ►homeostatic mechanisms can elevate Tb when ambient temperatures drop in both diurnal and nocturnal animals, circadian systems have opposite effects on them, promoting a rise during the day in diurnal animals and at night in nocturnal ones. Although rhythms in Tb persist when we are completely inactive, their amplitude is ordinarily increased by activity. This consequently increases the magnitude of the difference between nocturnal and diurnal animals. Many other factors, such as pregnancy and lactation, can alter the pattern of daily rhythms in Tb, but the circadian drive behind the rhythms is always quite different in diurnal and nocturnal species.

Rhythms in feeding behavior also are typically inverted in day- and night-active animals, but they are quite unusual with respect to their plasticity [3]. As one might expect, when food is freely available, diurnal animals eat most during the day, and nocturnal ones at night. However, restricting the period of food availability to short intervals during daylight hours eventually causes activity of nocturnal rats to begin prior to the presentation of the food, effectively entraining the animals' circadian rhythms in activity and feeding. The result is that nocturnal rats become considerably more diurnal. Such studies have established the existence of two interacting oscillators that together shape the degree to which an animal is more, or less, active during the day than night. One of these is referred to as a "light entrainable oscillator" and the other as a "►food entrainable oscillator." Although their interactions are poorly understood in diurnal species, they may provide some insight into mechanisms that govern when the active phase of an animal will be.

The Circadian Oscillator and its Synchronization to Environmental Cycles

At the simplest level, one way in which the nervous system might generate different circadian patterns in nocturnal and diurnal animals is that the central clock and mechanisms responsible for its synchronization with a light-dark cycle could differ, while the coupling of that clock to the physiological and behavioral functions that it regulates could be the same. However, a sizeable body of data has accumulated to support the idea that mechanisms coupling the clock to the environment are actually the same.

This issue has been examined via behavioral studies in a variety of mammals, including humans. One line of such work has involved studies looking directly at mechanisms that synchronize the clock with a light-dark cycle. This entrainment process involves a rhythm in how the clock itself can be shifted by light. In both nocturnal and diurnal animals, light has relatively little effect during the day, but causes delays early in the night and advances late at night. As this rhythm is responsible for the pattern of entrainment of the clock itself to light, it suggests that the coupling between the clock and photic inputs to it are the same in nocturnal and diurnal species. This model is further supported by more direct studies of the brain mechanisms involved in the process.

At the center of this system is a small collection of neurons within the ►hypothalamus referred to as the ►suprachiasmatic nucleus ►(SCN) [4]. Cells within it receive input from the retina that is responsible for the entrainment of that clock to the light-dark cycle, and that emit signals that broadcast temporal information directly and indirectly to a variety of other regions of the brain [4]. Several indices of overall SCN activity fluctuate on a daily basis and have patterns that are very similar from one species to another, regardless of whether they are nocturnal or diurnal [5]. For example, rhythms in metabolism within the SCN, as measured by glucose uptake, peak during the day in representatives of widely varying taxa exhibiting a range of activity patterns (e.g., lab rats, Turkish hamsters, golden hamsters, mice, opossums, squirrel monkeys, sheep, house sparrows and cats). Another indication that the SCN functions similarly in species with different behavioral rhythms is that firing rates of neurons peak during the day in diurnal chipmunks as well as in several nocturnal rodents. The most direct evidence that the key differences between nocturnal and diurnal species lie downstream of the primary circadian ►oscillator has been obtained recently through direct examination of molecular elements intrinsic to it. *Per1* and *Per2* are ►clock genes representing important components of that clock and their expression patterns have now been examined in several diurnal and nocturnal species. In all of these, rhythms in messenger RNA (mRNA) for *Per1* and *Per2* peak during the day. Taken together, the considerations above suggest that entrainment of the circadian oscillator to the light-dark cycle are the same in nocturnal and diurnal mammals and that the coupling of the clock to the functions that it controls are therefore likely to differ.

Clock-Driven Output Signals

Differences in the coupling of the circadian oscillator to the systems that it controls could theoretically reside in the SCN or in tissues that receive direct or indirect signals from it. In nocturnal mammals the SCN uses a combination of classical axonal outputs and an as yet unidentified factor that is released into, and diffuses through, the ►ventricular system. This diffusible factor has not been identified in any species, and there are no data bearing on whether one exists in diurnal animals or not. The discussion here therefore focuses on axonal output systems. The SCN projects to a relatively restricted set of targets located primarily within the hypothalamus in nocturnal rodents. Information on this issue is more limited in diurnal mammals, but recent studies have revealed that the SCN projects to the same regions in a diurnal rodent, *Arvicanthis niloticus*, (also referred to as a grass rat), as in nocturnal ones [6].

Another feature of SCN outflow that has been examined is the nature of the molecules the SCN uses to transmit temporal information to other regions of the brain. Three such molecules have been identified in both nocturnal and diurnal species: vasopressin (VP), vasoactive intestinal polypeptide (VIP) and ►prokineticin 2 (PK2). VP has been seen in cells within the SCN of almost every species that has been examined and the projections of these cells represent a major output of the SCN. Evidence that VP release is the same in diurnal and nocturnal species was obtained originally through its measurement in ►cerebral spinal fluid in many mammals and these rhythms peaked during the day in all of them. *VP* mRNA in the SCN fluctuates on a daily basis in two diurnal species, (*Arvicanthis ansorgei* and *Arvicanthis niloticus*), in a pattern that is the same as in several nocturnal rodents. Thus, both transcription of its gene and the daily rhythm in its secretion suggest that the VP signal is the same in nocturnal and diurnal species.

This is likely to be the case for the VIP output as well. Neurons containing VIP exist in the SCN of all species examined to date, and their axons form a major output system. Rhythms in *Vip* mRNA are very different in male and female lab rats, with the rise occurring approximately ten hours earlier in females than in males. The same patterns are seen in the SCN of female and male *A. niloticus*, respectively.

Recent evidence has implicated prokineticin 2 (PK2) as a third output signal, contributing in this case to rhythms in locomotor activity. The gene for PK2 in *A. niloticus* is expressed according to the same rhythmic patterns as in mice.

Although there are likely to be other molecular outputs of the SCN, and these might vary across species, all of those that have now been examined are the same in the nocturnal and diurnal animals that have been looked at. The current data thus support the conclusion that the primary differences between nocturnal and diurnal species emerge from differences in responsiveness of SCN targets to SCN signals, or downstream of these targets, rather than from temporal or spatial characteristics of output systems within the SCN. Studies of SCN targets in diurnal mammals, discussed next, are more scarce and have focused primarily on the diurnal grass rat, *A. niloticus*.

The Lower Subparaventricular Zone (sPVZ)

It has been suggested that one source of the differences between diurnal and nocturnal mammals may be a group of cells located just dorsal to the SCN [6]. The primary projection of the SCN extends dorsally into a region known as the ►subparaventricular zone (sPVZ). The ventral portion of the sPVZ, referred to as the lower sub-paraventricular zone (LSPV) shows significant functional differences when diurnal grass rats are compared to nocturnal lab rats. This was first seen is studies of what is referred to as an ►immediate early gene called cFos which regulates the expression of a host of other genes. Rhythms in cFos are apparent in the LSPV of both lab rats and grass rats, but in grass rats the rising phase of the rhythm begins four to five hours after lights go off in a light-dark cycle with 12 h of light and 12 h of darkness, whereas in lab rats this rise does not occur until eight to nine hours later, when the lights come on. Another important difference is that, whereas in grass rats the cFos rhythm in this cell population persists in constant darkness, it goes away in lab rats held in the same conditions. This region also exhibits a rhythm in a calcium-binding protein, calbindin, in grass rats but not lab rats.

Further support for a role of the LSPV in the circadian regulation of diurnality has come from work on activity rhythms of grass rats with axon-sparing lesions in this region [6]. Animals with damage to the LSPV exhibited severe disturbances in both free-running and entrained activity rhythms. Furthermore, an increase in the proportion of activity that occurs at night was positively correlated with the extent of damage produced by these lesions. It therefore seems likely that although the LSPV is not essential for the maintenance of diurnality, it contributes to the organization of the normal stable diurnal activity rhythms in these animals.

In grass rats and lab rats the SCN and LSPV project to virtually all of the same target regions. Taken together, the data suggest that cells in the LSPV could function as oscillators that modulate the ways in which target cells respond to other signals originating in the SCN. Given the species differences in the rhythms in the LSPV, its influence on rhythms in other SCN targets is likely to differ considerably in lab rats and grass rats. It has been proposed that the LSPV could thus contribute to the maintenance diurnality through effects on rhythms within a host of other cells that receive direct input from the SCN [6]. These include cells with highly specialized functions as well as ones with more widespread projections and more general functions.

Reproductive Rhythms

Several specialized groups of neuroendocrine cells receive direct input from the SCN, including those associated with estrus-related events. As noted above, many behavioral and neuroendocrine events associated with reproduction undergo rhythms that peak at very different times of day in diurnal and nocturnal animals [7]. This is the case for sexual behavior and for the secretion of hormones associated with it. Neuroendocrine functions are particularly amenable to analysis of SCN target systems because direct projections from the SCN to at least some neuroendocrine cells have been identified. This is the case for some neurons that contain estrogen receptors (ERs) and others containing gonadotropin-releasing hormone (GnRH). Below, the roles of the latter population of cells in regulation of the timing of the ovulatory surge in luteinizing hormone (LH), and how it differs in nocturnal and diurnal animals, are discussed.

Ovulation is triggered by a surge in LH that occurs at the end of the follicular phase of the female reproductive cycle. In female lab rats and hamsters it occurs three to four hours prior to the onset of estrus behavior, which begins at the time of lights-off. In these species, both the behavior and the LH surge are promoted by rising levels of ovarian hormones in conjunction with a signal originating in the SCN. The SCN signal reaches both GnRH neurons and estrogen receptor-containing cells in the ►preoptic area (POA), a region critical for the surge. The system is therefore one in which the circadian regulation of specific behavioral and physiological events depends upon well-defined populations of cells to which the SCN projects directly in nocturnal rodents. These pathways have now been examined in diurnal grass rats in efforts to identify where and how the system might differ in nocturnal and diurnal species.

The LH surge associated with estrus occurs 12 h apart in grass rats and lab rats, as does a rise in cFos within GnRH neurons associated with it [7]. Activation of these cells occurs 12 h apart in ►ovariectomized grass rats and lab rats that are kept in constant darkness, providing evidence that endogenous circadian mechanisms regulate the timing of GnRH neuron responses to steroid hormones. This could occur via direct projections from the SCN to these cells [6].

One important signal released by SCN cells projecting to GnRH neurons in rats is VIP. Evidence for such a projection comes from the demonstration of VIP fibers forming synapses, most of which are eliminated by SCN lesions, on GnRH cells. A role for this projection in the timing of the surge in nocturnal rodents is further suggested by evidence that it is delayed and attenuated by interference with VIP. Another way in which the SCN appears to promote the surge is through the release of VP, which, when administered into the POA generates a surge in SCN-lesioned lab rats. In diurnal grass rats, appositions between fibers containing either VIP or VP and GnRH cells are numerous in regions critical for the generation of the LH surge [7]. As rhythms in mRNA for these two peptides are very similar in lab rats and grass rats, it is likely that VP and VIP rise and fall at the same times of day in these two species.

Taken together, the data from lab rats and grass rats suggest that the SCN releases the same rhythmic signals onto a population of cells that is very different with respect to their circadian pattern of function in diurnal and nocturnal rodents. The question of diurnality in this system is therefore a question of what might make GnRH neurons respond differently to VIP and VP in grass rats and lab rats (Fig. 1). Theoretically, the answer could be that GnRH neurons are intrinsically different with respect to the ways in which they respond to these two signals. Another possibility is that GnRH neurons are fundamentally the same in the two species, and that differences are produced by inputs to them from other regions. The LSPV could represent such an area.

Sleep and Arousal

The neural control of the ►sleep-wake cycle involves numerous neuronal populations in the forebrain, midbrain and hindbrain [8]. Two mutually antagonistic components of this complex network are found in the in the hypothalamus-preoptic area. One of these is a small cluster of cells in the ►ventrolateral preoptic nucleus (VLPO) that contains ►galanin and provides inhibitory inputs to a series of ascending arousal systems originating in the hypothalamus and ►brainstem. In nocturnal animals these cells become active during the light phase of the cycle when levels of sleep are highest (Sherin et al., 1996). The primary ►hypothalamic system that antagonizes the sleep promoting influence of the VLPO is made up of ►orexin/hypocretin-positive cells located in the lateral hypothalamus. These neurons provide excitatory inputs to most of the arousal systems that are inhibited by the VLPO, and in lab rats these neurons are most active during the dark phase of the day-night cycle when these animals are active. The diurnal grass rat shows a very similar distribution of galanin-positive neurons in the VLPO and orexin-positive cells in the lateral hypothalamus, and rhythms in both of these cell groups are 180° out of phase in diurnal grass rats and nocturnal lab rats [5].

The question of what is responsible for this complete reversal in the rhythms in VLPO and orexin/hypocretin neurons is of considerable interest. In the case of the arousal system, fibers originating from both the LSPV and the SCN of grass rats appear to project to cells in the lateral hypothalamus that contain orexin/hypocretin. Thus, interactions between these two sources of circadian signals may influence the phase of the rhythm in activity of orexin/hypocretin cells and consequently the arousal systems to which they project. As rhythms in

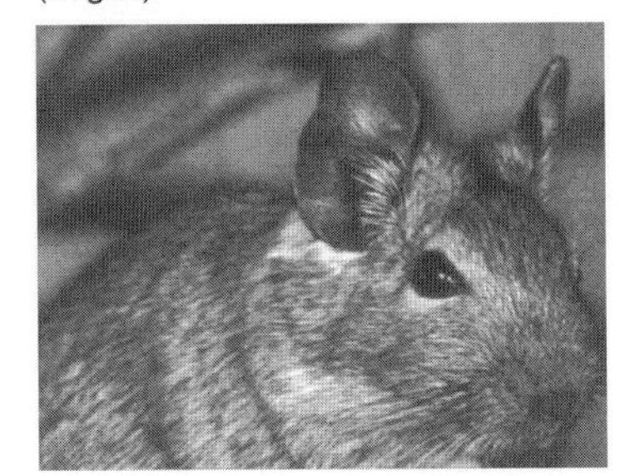

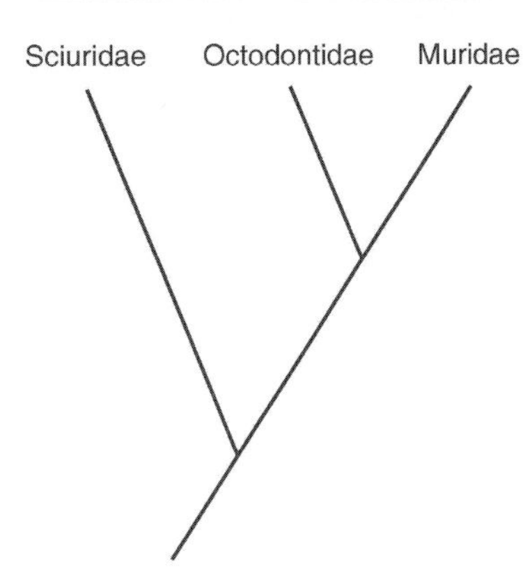

Nocturnal/Diurnal. Figure 1 Phylogenetic relationships of three diurnal rodents that are currently being studied. Neural mechanisms associated with circadian timkeeping systems are similar in some, but not all, ways in these species. This may reflect the fact that diurnality evolved independently in the three lineages.

the LSPV are very different in grass rats and lab rats, these interactions are likely to have very different effects.

The circadian influence on the sleep-promoting system originating in the VLPO may be achieved in a more indirect fashion, as projections to this region from the LSPV and the SCN are sparse in both grass rats and lab rats. One route through which circadian signals may reach the VLPO includes SCN cells that project to the ▶dorsomedial nucleus of the hypothalamus, which, in turn, contains cells that project to the VLPO. It has been suggested that the integration of circadian signals that influence the phase of rhythms of neural activity in the VLPO may take place in the ▶dorsomedial nucleus [9]. Differences within these regions of the hypothalamus could therefore contribute to differences in the daily organization of sleep of nocturnal and diurnal animals.

Summary and Significance for Humans

The system through which the SCN regulates specific rhythms comprises multiple parallel but intersecting output pathways. Rhythms within the SCN appear to be the same in nocturnal and diurnal species, but direct and indirect components of this output system differ between them in a variety of ways. Some SCN targets have widespread projections and play regulatory roles in a variety of different functions, some of which are quite specialized (e.g., VLPO) and others that play more general roles (e.g., LSPV). Rhythms in these components of the system exhibit an array of patterns of rhythmicity with varying phase relationships in nocturnal and diurnal species. Other SCN targets, such as neuroendocrine GnRH cells, receive inputs that converge on them from a variety of regions but have very restricted output pathways, are more specialized and are linked more directly to the endpoints whose functions they regulate. In the case of GnRH cells, the rhythms are inverted by 180° in nocturnal laboratory rats and diurnal grass rats.

Taken together, the overall pattern of results emerging from recent studies on the issue suggests that there is not one simple all-or-none "switch" that determines whether an animal is diurnal or nocturnal. The ultimate differences between mammals expressing these different patterns are more likely to emerge through a variety of interrelated mechanisms operating at varying points between the SCN and the behavioral and physiological systems that it regulates. However, these data supporting this general model come from comparisons between only one nocturnal species, lab rats, and one diurnal one, grass rats. Although it seems likely that this general principle will apply to other diurnal species, including humans, the specifics of where the differences are, and the details of the rhythmic patterns in such cell populations, may vary.

References

1. Moore-Ede M, Sulzman F, Fuller C (1983) The clocks that time us. Harvard University Press, Cambridge, MA
2. Turrek FW, van Cauter E (1994) Rhythms in reproduction. In: Knobil E, Neill JD (eds) The physiology of reproduction, Raven Press, New York, pp 487–540
3. Stephan FK (2002) The "other" circadian system: food as a zeitgeber, J Biol Rhythms 17:284–292
4. Klein D, Moore RY, Reppert SM (1991) Suprachiasmatic Nucleus: the mind's clock. Oxford University Press, New York
5. Smale L, Lee TM, Nunez AA (2003) Mammalian diurnality: Some facts and gaps. J Biol Rhythms 18:356–366
6. Schwartz MD (2006) Neural substrates of diurnality in the Nile grass rat, Arvicanthis niloticus. PhD thesis, Michigan State University, East Lansing, MI
7. Mahoney M (2003) Sex, surges and circadian rhythms: The timing of reproductive events in a diurnal rodent. PhD thesis, Michigan State University, East Lansing, MI
8. Fuller PM, Gooley JJ, Saper CB (2006) Neurobiology of the sleep-wake cycle: sleep architecture, circadian regulation, and regulatory feedback. J Biol Rhythms 21:482–493
9. Saper CB, Lu J, Chou TC, Gooley J (2005) The hypothalamic integrator for circadian rhythms. Trends Neurosci 28:152–157

Node

Definition

Primary organizer present in all chordates, called node in chick and mouse, Spemann's organizer in frog, and the shield in fish. It produces signals that are involved in neural induction.

►Evolution and Embryological Development of the Forebrain

Node of Ranvier

Definition

The node of Ranvier is a small stretch of bare axonal plasma membrane (axolemma) that separates myelin segments along individual myelinated nerve fibers.

In the peripheral nervous system, individual Schwann cells form each myelin sheath segment, while in the central nervous system, oligodendrocyte processes form myelin sheath segments. Easily identified in longitudinal sections, nodes separate the terminal paranodal loops of adjacent myelin internodes. Nodal length is related to the diameter of the axon and can vary from less than 1 μm in the small fibers of the optic nerve to more than 5 μm in the large fibers of the spinal cord.

The node is the site of Na^+ channels clusters to allow for saltatory conduction of action potentials, as well as the site for axonal sprout formation in peripheral nerve regeneration. (Louis Antoine Ranvier, 1835–1922).

►Action Potential Propagation
►Oligodendrocyte
►Peripheral Nerve Regeneration and Nerve Repair
►Myelin Sheath
►Schwann Cell

Nodes in Acoustics

Definition

A point, line, or surface of a standing wave in which the net energy flux is zero at all points.

►Acoustics

Nodulus

Synonyms

Nodule

Definition

The vermis segment nodulus and the hemisphere segment flocculus together form the flocculonodular lobe.

Phylogenetically it is very old and is thus called the archicerebellum. Since its afferents come mainly from the vestibular nuclei (vestibulocerebellar tract), the "vestibulocerebellum" is another synonym.

►Cerebellum

Nogo

Definition

Member of the reticulon protein family with potent inhibitory effects for neurite outgrowth. The membrane- bound Nogo is associated with CNS myelin and exists as three splice variants (Nogo-A, Nogo-B and Nogo-C). Neutralization of Nogo in vivo results in enhanced regeneration of severed axons, increased plasticity of uninjured fibers, and functional recovery in animal models of CNS injury.

►Growth Inhibitory Molecules in Nervous System Development and Regeneration
►Regeneration
►Regeneration of Optic Nerve

Nogo-Neutralizing Antibody IN-1

Definition

An antibody generated using the rat CNS myelin N1250 protein. It is raised in hybridoma cells and has been shown to promote axonal regeneration in the damaged central nervous system (CNS).

Noise

Definition

A signal, which is superimposed onto the signal of interest. Noise can be random or deterministic.

►Signals and Systems

Noise, Colored

Definition

A random noise whose spectrum is not flat, i.e., its amplitude at some frequencies is different from that at other frequencies.

►Signals and Systems

Noise, White

Definition

Random noise, whose spectrum is flat, i.e., its amplitude is the same at all frequencies. White noise at different time intervals is uncorrelated, unless the interval is zero.

►Signals and Systems

Noise-induced Transport

Definition

A phenomenon in which the fluctuations inherent in the system create or enhance directed motion.

►Brownian Ratchet

Nominalism

Definition

Nominalism is the doctrine that there are no universals, i.e. no abstract, general entities like e.g. Platonic forms.

Only particular things are thought to exist. The functions of universals are attributed to the signs of human language, i.e., as predicates or common nouns that can be applied to many particular things.

►Information
►Possible World
►Property

NOMPC

Definition

NOMPC forms a Ca^{2+}-permeable channel in Drosophila mediating sensory hair cell mechanotransduction.

►TRP Channels

Non-associative Learning

Definition

Non-associative learning can be defined as a change in the behavioral response that occurs over time in response to a single type of stimulus. Habituation and sensitization are typical examples.

▶Learning

Non-conceptual Knowledge

Definition

Non-conceptual knowledge is knowledge which cannot be communicated (fully and satisfactorily) by the use of concept words, for example pictorial knowledge or knowledge how to do something.

▶Knowledge

Nondeclarative (Implicit) Memory

Definition

Nondeclarative memory refers to a heterogeneous collection of nonconscious memory abilities such as skills (e.g., riding a bicycle), conditioned responses, priming, skills, habits and other learnings, which are displayed through performance and not conscious recollection. These components are preserved in amnesic patients.

▶Amnesia
▶Long-Term Memory

Non-dystrophic Myotonias

Definition

Heterogenous group of rare hereditary diseases characterized by ▶myotonia or electrical myotonia and resulting from mutations in several ▶ion channel genes (Cl^-, Na^+, Ca^{2+}, K^+ channels). There are three main groups. Sodium channel dysfunctions underlie paramyotonia congenita (which is cold-sensitive), K^+-aggravated myotonia, and ▶hyperkalemia periodic paralysis with myotonia. Chloride channel dysfunctions underlie the myotonia congenita disorders (ClC-1 mutations cause 'pure' myotonia congenitalis), which are not sensitive to temperature. Channel myotonia comes in a recessive (Becker type) form and a dominant (Thomsen type) form. In contrast to ▶myotonic dystrophy, paramyotonia congenita is a non-dystrophic, rather benign disorder, in which the skeletal muscle is not dystrophic, but rather shows hypertrophy secondary to 'exercise' by prolonged contractions.

Non-holonomic Constraint

Definition

A non-integrable constraint involving the velocities of the particles of a system.

▶Mechanics

Non-invasive Imaging Techniques

Definition

Non-invasive imaging techniques like electroencephalography (EEG), magnetoencephalography (MEG) and functional magnetic resonance imaging (fMRI) allow the study of the human brain in vivo during active processing. Whereas EEG and MEG measure the electrical neuronal activity directly with high temporal (ms) but less spatial resolution (cm), fMRI is based on the hemodynamic changes following electrical brain activity. Since the hemodynamic response is slow (several seconds), fMRI has a poor temporal resolution.

However, spatial resolution is in the range of millimeters.

▶Electroencephalography
▶Functional Magnetic Resonance Imaging (fMRI)
▶Magnetoencephalography

Nonlinear Control Systems

Nahum Shimkin
Department of Electrical Engineering, Technion – Israel Institute of Technology, Haifa, Israel

Definition

▶Nonlinear control systems are those ▶control systems where nonlinearity plays a significant role, either in the controlled process (plant) or in the controller itself.

Nonlinear plants arise naturally in numerous engineering and natural systems, including mechanical and biological systems, aerospace and automotive control, industrial process control, and many others.

Nonlinear control theory is concerned with the analysis and design of nonlinear control systems. It is closely related to nonlinear systems (▶System – nonlinear) theory in general, which provides its basic analysis tools.

Characteristics

Numerous methods and approaches exist for the analysis and design of nonlinear control systems. A brief and informal description of some prominent ones is given next. Full details may be found in the textbooks [1–6], and in the Control Handbook [7].

Most of the theory and practice focus on feedback control. A typical layout of a feedback control system is shown in Fig. 1.

A basic (finite dimensional, time invariant) nonlinear system in continuous time may be specified by the standard state-space model:

$$\begin{aligned} \tfrac{d}{dt}x(t) &= f(x(t),u(t));\ x(0)=x_0 \\ y(t) &= h(x(t),u(t)) \end{aligned} \tag{1}$$

or, more succinctly, as $\dot{x}=f(x,u)$ (the state equations) and $y=h(x,u)$ (the output equation). Here $x(t)\in R^n$ is the state vector, $u(t)\in R^m$ is the vector of input signals, and $y(t)\in R^q$ is the output vector. This model may apply to the plant (see Fig. 1), as well as to the controller (with appropriately modified inputs and outputs). The state of the overall feedback system is then the combined state of the plant and the controller. A specific class of systems that has been studies in depth is linear-in-control systems, where $f(x,u)=f_0(x)+\sum_{i=1}^{m} f_i(x)u_i$. We limit the discussion here to continuous-time systems, although similar theory exists for the discrete-time case.

Nonlinear models may be classified into smooth and non-smooth ones. The latter are often associated with parasitic effects such as dry friction and actuator saturation. When significant, these effects may enter as constraints in the design, or even require specific compensation techniques. Our discussion below pertains mainly to smooth nonlinearities.

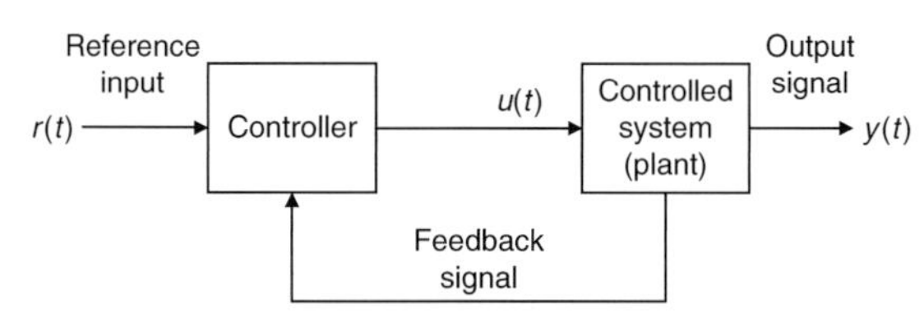

Nonlinear Control Systems. Figure 1 Basic feedback control system.

Basic Concepts from Systems Theory

The following notions from systems theory are of particular importance and relevance to nonlinear control, and are dealt with in depth in the cited texts.

a. *Equilibrium points:* For the nonlinear system $\dot{x}=f(x)$, a point x_e in the state space is an equilibrium point if $f(x_e)=0$. Similarly, for the controlled system $\dot{x}=f(x,u)$, the pair is an equilibrium point if $f(x_e,u_e)=0$.

b. *Lyapunov stability:* This is the basic notion of stability that deals with the asymptotic behavior of trajectories that start off an equilibrium point. An equilibrium point x_e of the system $\dot{x}=f(x)$ is (weakly) *stable* if all solutions *x(t)* that start near x_e stay near it forever. It is *asymptotically stable* if, in addition, *x(t)* converges to x_e whenever started near enough to it. If this convergence occurs for any initial state then x_e is *globally* asymptotically stable. Exponential stability requires an exponential rate of convergence to the equilibrium. Note that a nonlinear system may have several ▶equilibrium points, each with different stability properties. For input-driven state equations with unspecified input, namely $\dot{x}=f(x,u)$, these stability notions are generalized by the concept of *input-to-state stability*, which requires the state vector to be close to equilibrium whenever both the initial state and the control input *u(t)* are close to their equilibrium values.

c. *Lyapunov's direct method:* The most general approach to date for stability analysis of nonlinear systems is Lyapunov's method, which relies on the concept of a ▶Lyapunov function or generalized energy function. Essentially, a Lyapunov function for an equilibrium point x_e of the system $\dot{x}=f(x)$ is a differentiable function *V(x)* which has a strict minimum at x_e, and so that its derivative $\dot{V}(x)\triangleq\frac{\partial V(x)}{\partial x}\cdot f(x)$ along the system trajectories is negative in some neighborhood of the equilibrium. Existence of such *V(x)* implies stability of x_e, and further implies a symptotic stability if *V(x)* is stricty negative for $x\pm x_e$. Many extensions and refinement of this result exist, covering various stability properties such as exponential stability, global stability, and estimates on the domain of attraction of the equilibrium.

We note that various converse theorems establish the existence of a Lyapunov function whenever the equilibrium point is stable (in the appropriate sense); however no general procedure exists for finding such a function.

d. *Linearization:* The small-signal behavior of the nonlinear system (1) around an equilibrium point (x_e, u_e) may be captured through a linear state equation of the form: $\dot{\tilde{x}}=A\tilde{x}+B\tilde{u}$, where $\tilde{x}(t)=x(t)-x_e$, $\tilde{u}(t)=u(t)-u_e$, and the matrices (*A,B*) are computed as corresponding gradients of the system function *f(x,u)* at (x_e,u_e). A similar relation holds for the output equation. A basic stability result (also known as

N

Lyapunov's indirect method) is that the Hurwitz-stability of the matrix A (namely, all eigenvalues have strictly negative real part) implies the asymptotic stability of the respective equilibrium point.

e. *Input-output stability and gain:* The dynamic system (1) is said to input-output stable with respect to a signal norm $\|\cdot\|$, if $\| y(\cdot) \| \leq \gamma \| u(\cdot) \| + \beta$ for some constants $\gamma \geq 0$ and β (and every input $u(\cdot)$ in the input space). The name BIBO stable is also used when the norm is the max norm, namely $\| y(\cdot) \| = \sup_t \| y(t) \|$. The system gain is the smallest number γ that satisfied the above bound. For state-space models, various results relate input-output stability to corresponding stability properties of the state.

f. *Passivity:* The system-theoretic notion of passivity essentially captures the physical notion of a system that does not generate energy. As such, many mechanical and other systems satisfy this property. Passivity provides a useful analysis tool, and a basis for design methods. Notably, passivity implies stability, and the feedback connection of passive systems is passive.

g. *Controllability and Reachability:* These two closely-related concepts that apply to the state equation $\dot{x} = f(x, u)$ concern the possibility of reaching a given state from any other state (controllability) or reaching any other state from a given state (reachability) by choosing appropriate controls. Local versions focus on small neighborhoods of any given point. These properties have been studied in depth, especially for the class of linear-in-control systems, using tools from differential geometry.

h. *Observability:* Observability concerns the ability to distinguish between two (initial) states based on proper choice of input and observation of the system output. This concept, roughly, indicates whether a feedback controller that uses only the output y can fully control the state dynamics.

Analysis of Feedback Systems

Alongside general tools and methods from system theory, a number of results and analysis methods apply specifically to feedback systems, and some of these are described next. The basic feedback connection of two subsystems is shown in Fig. 2.

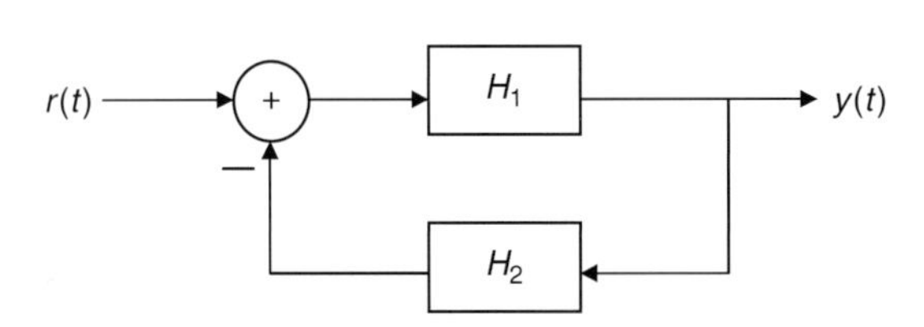

Nonlinear Control Systems. Figure 2 Negative feedback connection.

a. *Limit Cycles and ▶Describing Function Analysis:* Limit cycles, or sustained oscillations, are common in nonlinear feedback systems, and are usually not desired in control systems. The describing function method checks for the possibility of oscillations by (i) approximating the response of nonlinear elements to sinusoidal inputs of given amplitude and frequency by their first harmonics only (ii) checking for the possibility of a loop gain of 1 with phase shift of 180° (the so called harmonic balance equation, for a negative feedback system). In case of a positive answer the analysis yields estimates for the frequency and amplitude of oscillations. While the method is essentially heuristic it is often useful for initial analysis.

b. *Small Gain Theorem:* The small gain theorem allows to establish the input-output stability of a feedback system from properties of its subsystems. Assume that H_1 and H_2 are both input-output stable, with respective upper-bounds γ_1 and γ_2 on their gains. If $\gamma_1\gamma_2 < 1$, then the feedback system in Fig. 2 is input-output stable, and its gain is bounded by $\gamma_1/(1 - \gamma_1$ that $\gamma_2)$.

c. *Circle Criterion:* Consider the special case H_1 is a linear time-invariant system, and H_2 is a static nonlinearity $h_2(\cdot)$ that satisfies a $[k_1, k_2]$ sector condition; in the scalar case this means that $h_2(x)/x \in [k_1, k_2]$. The circle criterion (and the related Popov criterion) provides frequency-domain conditions on the transfer function of H_1 that imply the stability of the feedback system.

Design Methods

Control system design in general aims to satisfy certain performance objectives, such as stability, accurate input tracking, disturbance rejection, and robustness or insensitivity to parameter uncertainty (see the Control section for further details). The diverse nature of nonlinear systems necessarily calls for a variety of design approaches of different nature, and some of the more notable ones are briefly described below.

One design viewpoint is to consider the controlled system as an approximately linear one, or linearize the system by appropriate transformation, to which well-established linear control techniques may be applied.

a. *PID Control:* The PID (Proportional-Integral-Derivative) regulator is a simple linear controller, which is often cited as the most prevalent feedback controller. In particular, it finds use in many non-linear applications, from industrial process control to robotic manipulators. On-site tuning of the ▶PID controller parameters is often used, especially in the process control industry, and numerous manual and auto-tuning procedures exist based on direct measurement of some characteristics of the system response. Analytical (model-based) design is of course also used, often building on one of the ▶linearization methods below.

b. Local linearization and Gain Scheduling: The simplest analytical approach to controller design for a nonlinear system relies on fitting an approximate linear model to the controlled system, usually through local linearization around a typical working point (see above), and then designing a linear controller for this model. Evidently this method may fail when non-linear effects are significant. ▶Gain Scheduling takes this approach a step further: Linear controllers are designed for a range of possible operating points and conditions, and the appropriate controller is put into play according to the current system state.

c. Feedback Linearization: ▶Feedback linearization or ▶global linearization uses input and state variable transformations to arrive to an equivalent linear system. As a simple example, the scalar system $\dot{x} = u^3 + f(x)$ is readily transformed to $\dot{x} = v$ by defining an auxiliary input $v = u^3 + f(x)$. A control law to determine v can now be designed for the linear system, and the actual control u may then be computed using the inverse relation $u = (v - f(y))^{1/3}$. The latter equation, which is in the form of state feedback, gives the method its name. One can distinguish between full state linearization, where the state equation is fully linearized, and input-output linearization, where the input-output map is linearized. In either case, measurement of the entire state vector is required to implement the transformation. The theory provides conditions under which feedback linearization is possible, and procedures to compute the required transformations.

The following methods approach the design problem directly using non-linear tools, notably ▶Lyapunov stability and Lyapunov functions, and are notable examples of *robust* nonlinear control.

d. ▶*Lyapunov Design and Redesign:* In Lyapunov-based design, a stable system is synthesized by first choosing a candidate Lyapunov function V, and then selecting a state-feedback control law that renders the derivative of V negative. The Lyapunov redesign method provides the system with robustness to (bounded) uncertainly in the system dynamics. It starts with a stabilizing control law and Lyapunov function for the nominal system, and adds certain (non-smooth) terms to the control that ensure stability in the face of all admissible uncertainties. While Lyapunov redesign is restricted to systems that satisfy a matching condition, so that the uncertainty terms enter the state equations at the same point as the control input, the basic approach has been extended to more general situations using recursive or *backstepping* methods.

e. ▶*Sliding Mode Control:* In this robust design approach, also known as Variable Structure Control, an appropriate manifold (often a linear surface) in the state space is first located on which the system dynamics takes a simple and stable form. This manifold is called the sliding surface or the switching surface. The control law is designed to force trajectories to reach that manifold in finite time, and stay there thereafter. As the basic control law is discontinuous by design around the switching surface, unwanted chattering around that may result and often require some smoothing of the control law.

Many other techniques from control engineering are applicable to the design of nonlinear systems. Among these we mention:

- Optimal Control: Here the control objective is to minimize a pre-determined cost function. The basic solution tools are Dynamic Programming and variational methods (Calculus of Variations and Pontryagin's maximum principle). The available solutions for nonlinear problems are mostly numeric.
- Model Predictive Control: An approximation approach to optimal control, where the control objective is optimized on-line for a finite time horizon. Due to computational feasibility this method has recently found wide applicability, mainly in industrial process control.
- Adaptive Control: A general approach to handle uncertainty and possible time variation of the controlled system model. Here the controller parameters are tuned on-line as part of the controller operation, using various estimation and learning techniques.
- Neural Network Control: A particular class of adaptive control systems, where the controller is in the form of an Artificial Neural Network.
- Fuzzy Logic Control: Here the controller implements an (often heuristic) set of logical (or discrete) rules for synthesizing the control signal based on the observed outputs. Defuzzification and fuzzification procedures are used to obtain a smooth control law from discrete rules.

A detailed description of these are related approaches, which are often considered as separate fields of control engineering, may be found in [7].

References

1. Isidori A (1995) Nonlinear control systems, 3rd edn. Springer, London
2. Khalil HK (2002) Nonlinear systems, 3rd edn. Prentice-Hall, Englewood Cliffs, NJ
3. Nijmeijer H, van der Schaft AJ (2006) Nonlinear dynamical control systems, 3rd edn. Springer, New York, NY
4. Sastry S (2004) Nonlinear systems: analysis, stability and control, Springer, New York, NY
5. Slotine J-J, Li W (1991) Applied nonlinear control, Prentice-Hall, Englewood Cliffs, NJ
6. Vidyasagar M (2002) Nonlinear systems analysis, 2nd edn. SIAM Classics in Applied Mathematics, Philadelphia, PA
7. Levine WS (ed) (1996) The Control Handbook, CRC Press, Boka Raton, FA

Nonlinear System

Definition

A system whose output spectrum contains frequency components in addition to those in the input to the system.

Nonmonotone Dynamics

Definition

Equations specifying state transition of a recurrent neural network consisting of units with nonmonotonic output functions.

Non-NMDA-LTP

Definition

LTP that does not require activation of the NMDA receptor for its induction. In most cases, voltage-gated Ca^{2+} channels at presynaptic terminals play a role in its induction. A typical example is LTP in the hippocampal mossy fiber-CA3 synapse and it is believed that activation of presynaptic protein kinase A is involved in its expression.

►Memory, Molecular Mechanisms
►Associative Long-Term Potentiation (LTP)
►Long-Term Potentiation (LTP)

Non-peripheral Vestibular Disorders

►Central Vestibular Disorders

Non-photic

Definition

Circadian rhythm phase is primarily under the influence of light acting through photoreceptors connected directly or indirectly to the circadian clock. However, certain phase shifts are induced by non-photic stimuli that influence the clock via pathways not directly involving retinal projections. Such shifts can be as great in magnitude as light-induced phase shifts. The term, non-photic stimulus, has had a non-specific use in reference to any non-light input to the circadian clock, including such endogenous neurotransmitters as serotonin agonists that are internally administered to test animals. A more specific application of the term is in reference to externally applied stimuli that exert phase control over the circadian clock. These non-photic (Stimuli, Shifts, Resetting) stimuli have a phase response curve (PRC) similar to that obtained by infusing neuropeptide Y (NPY) directly into the suprachiasmatic nucleus (SCN). This "NPY-type PRC" differs from the "light-type PRC" in several respects: (i) maximum phase advances occur during the mid- to late subjective day; (ii) phase delays occur during the subjective night; and (iii) there is no dead zone. Such stimuli apparently require an intact geniculohypothalamic tract (GHT) and activation of NPY neurons in the intergeniculate leaflet (IGL) that results in NPY release in the SCN.

►Circadian Rhythm
►Intergeniculate Leaflet
►Neuropeptide Y (NPY)
►Phase Response Curve
►Suprachiasmatic Nucleus (SCN)

Non-rectifying Gap Junctions

Definition

Non-rectifying gap junctions conduct ionic current equally well in both directions. In contrast to electrical synapses comprised of rectifying gap junctions, electrical synapses comprised of non-rectifying gap junctions transmit electrical signals in a bidirectional fashion.

►Electrical Synapses

Non-REM (NREM) Cells

Definition

Also Called Sleep-Active Neurons; Neurons that exhibit increases in extracellularly recorded discharge rate during NREM sleep compared to waking and REM sleep. This type of cell is mostly located in the anterior hypothalamus and preoptic area (POA) and some are also located in the basal forebrain area, solitary tract nucleus, and in the dorsal raphe nucleus. The neurotransmitter identity of these neurons is not definitive but they are most likely to contain the neurotransmitter GABA.

►Rapid Eye Movement (REM) Sleep

Non-REM Sleep

Definition

In mammals there are two types of sleep – rapid eye movement (REM) and non-REM (NREM). Normally, when we first enter the sleep state it is via quiet NREM sleep.We lie passively, breathing slowly. Our eyes drift slowly back and forth and every once in a while we shift our sleep position. During non-REM sleep there are decreases in blood pressure, heart rate, and respiratory rate. One could therefore characterize NREM sleep as an exceedingly dormant behavioral state. There are many important events that occur during this state, however, such as increases in pulsatile release of growth and sex hormones from the pituitary, antibody production, and elimination of unwanted and excess mental traces. REM and NREM stages are defined in terms of electro-physiological signs that are detected with a combination of electroencephalography (EEG), electrooculography (EOG) and electromyography (EMG), the measurement of which in humans is collectively termed polysomnography. In a human, NREM sleep is divided into four stages, each corresponding to an increasing depth of sleep. As the depth of sleep increases, the EEG recordings are progressively dominated by high-voltage, low-frequency wave activity. Stage I NREM sleep is characterized by relatively low voltage (<50 μV), mixed frequency activity (4–7 Hz: theta frequency range) and vertex sharp waves in the EEG. Stage II NREM sleep is characterized by slow (<1 Hz) oscillations with distinctive sleep spindles (waxing and waning of 12–14 Hz waves lasting between 0.5 and 1.0 s; peak amplitudes of 100 μV) and K-complex waveforms (a negative sharp wave followed immediately by a slower positive component). Stage III NREM sleep is demarcated by the addition to the spindling pattern of high voltage (>100 μV) slow waves (1–4 Hz; delta frequency waves), with no more than 50% of the record occupied by the latter. In Stage IV, the record is dominated by high-voltage (150–250 μV) slow waves (1–3 Hz). Collectively, stages III and IV NREM sleep are also called slow-wave sleep (SWS). Distinctions between stages of NREM sleep in animal models (mouse, rat, cat, and non-human primates) differ slightly from that of humans. In these animals, NREM sleep is normally divided into two stages (SWS I and II). SWS-I is identified by the presence of sleep spindles in the cortical EEG. SWS-II is considered deep sleep, also termed delta sleep, and is identified by the presence of high amplitude, low-frequency waves (0.1–4.0 Hz) in the cortical EEG. NREM sleep can also be confidently identified in most reptiles and birds but is not seen with convincing clarity in either fish or amphibians.

►EEG in Sleep States
►Electroencephalography
►Electromyography
►Electrooculogram (EOG)
►Rapid Eye Movement (REM) Sleep
►Sleep States

Nonsense Mutation

Definition

A nucleotide substitution that causes premature termination of protein translation, which invariably results in nonfunctional proteins.

Non-spanning Muscle Fiber

Definition

A muscle fiber that ends in the middle of the muscle belly and is attached with a tapering end to the stroma of the muscle. It does not span the distance between two aponeuroses (tendon plates) of a muscle, but is attached to one of them by a myotendinous junction.

►Intramuscular Myofascial Force Transmission

Nonspecific Control Parameter

Definition

A nonspecific control parameter does not prescribe or contain the code for an emerging pattern. Rather, the nonspecific control parameter leads a dynamical system through a variety of patterns or states. When the control parameter passes through a critical point, a qualitative change (bifurcation) occurs in the dynamical system (e.g., coupled oscillators) leading to a new output pattern.

Non-steroidal Anti-inflammatory Drugs (NSAID)

Definition

Non-steroidal anti-inflammatory drugs reduce pain, fever, and inflammation. The term "nonsteroidal" is used to distinguish these drugs from steroids, which have a similar anti-inflammatory action.

►Analgesia
►Neuroinflammation: Chronic Neuroinflammation and Memory Impairments

Non-Synaptic Release

FREDERICK W. TSE, NAN WANG, LEI YAN, AMY TSE
Department of Pharmacology & Centre for Neuroscience, University of Alberta, Edmonton, AB, Canada

Synonyms

Exocytosis; Vesicle fusion

Definition

Vesicular release of neurotransmitters, hormones or mediators for cellular communications at sites other than the synaptic cleft.

Characteristics

Diversity of the Molecular Machinery for Exocytosis

The process of exocytosis mediates the cellular secretion of chemicals (including neurotransmitters, hormones and mediators) that are stored in intracellular vesicles or granules. In addition, exocytosis also regulates the delivery of transmembrane proteins to the plasma membrane. In the last 10 years, over a dozen proteins (each typically with multiple isoforms [1]) in the molecular machinery that mediate, or regulate, exocytosis have been identified. It is increasingly clear that this machinery is not identical in different cell types, or even among different types of granules in the same cell. Here, we shall discuss how changes in the molecular machinery of exocytosis may contribute to some of the variability in the quantal release and Ca^{2+}-sensitivity among the different types of granules.

Detection of Quantal Release from Different Types of Granules

In most neurons, neurotransmitters (e.g. dopamine) are stored in small synaptic vesicles (SVs; average diameter of ~50 nm) and peptides (e.g. arginine vasopressin) are stored in large dense core granules (LDCGs; average diameter of ~200 nm). In electron micrographs, the lumen of a SV is electron lucent but that of a LDCG is electron opaque. The electron dense material (dense core) in the LDCG is largely contributed by the presence of either a gel-like matrix, or the dense packaging of crystalline cargo. In the neuroendocrine system, secretory cells such as adrenal chromaffin cells store catecholamine hormones (e.g. adrenaline) in LDCGs. In some interneurons (e.g. in sympathetic ganglia) and chemosensory cells (e.g. carotid glomus cells), neurotransmitters are stored in small dense core granules (SDCGs; average diameter of ~100 nm).

The development of carbon fiber ►amperometry has enabled researchers to study in detail the release kinetics of easily oxidizable transmitters (e.g. catecholamines) from a single granule. Each molecule of catecholamine released from the cell is oxidized by the carbon fiber electrode (placed on the cell surface) and the resultant electrical signal is measured as an amperometric current. Figure 1 shows an example of a large amperometric signal recorded from a rat chromaffin cell during the exocytosis of a LDCG. Upon the fusion of a LDCG with the plasma membrane, the lumen of the granule forms a transient connection (a ►fusion pore) with the exterior of the cell.

The fusion pore first opens to a semi-stable state that lasts for up to tens of milliseconds, and results in a leakage of catecholamines (detected as the "foot" signal; Fig. 1). Occasionally some fusion pores flicker and close at this stage and give rise to the "stand-alone" foot signals. For the vast majority of LDCGs, the semi-stable fusion pore suddenly starts to dilate very rapidly and the rapid release of catecholamine gives rise to the "spike" phase of the amperometric signal. The rapid dilation of the fusion pore is significantly driven by the rapid decondensation (and probably rapid expansion) of the gel matrix. The decondensation of the gel matrix is in turn regulated by the change in vesicular pH and the concentrations of other ions (e.g. Ca^{2+}). The amount of catecholamines released from a single granule (i.e. quantal size, Q) can be estimated from the time integral of the amperometric signal. The

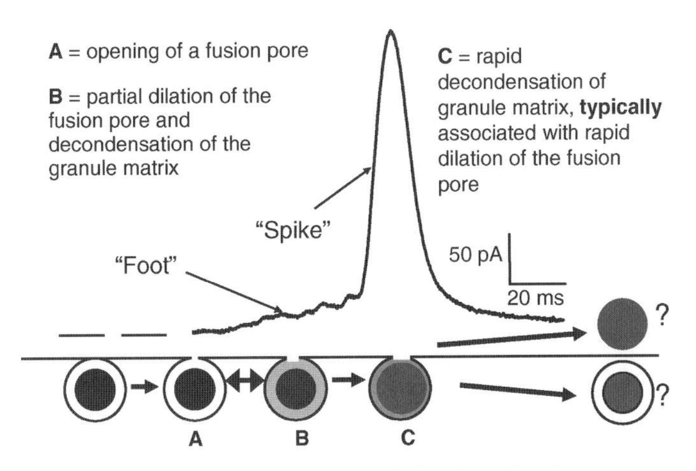

Non-Synaptic Release. Figure 1 An amperometric signal showing the complex kinetics of catecholamine release from a large dense core granule (of a rat chromaffin cell). The overall release kinetics was determined by the complex interactions between the fusion pore and the gel matrix.

kinetics of main spike provides information on the rapid phase of release from a single granule, and the kinetics of the semi-stable fusion pore can be inferred from the foot signal.

In general, granules with larger vesicular diameters have been associated with larger values of Q as well as slower quantal release kinetics. For example, in leech Retzius cells, the LDCGs have larger Q values and the main spike of the amperometric signal from LDCGs also exhibits slower kinetics than those from SVs [2]. However, the slowing in the kinetics of quantal release does not increase monotonically with granule size. Our recent study in chromaffin cells [3] shows that in some LDCGs with very large Qs, their most rapid phase of release is comparable or even faster than LDCGs with smaller Qs. While the semi-stable fusion pore of a LDCG with a larger Q tends to persist for a longer duration, it also tends to reach a larger size before the onset of rapid dilation. Moreover, once the fusion pore of a LDCG with very large Q starts to dilate, the spike portion of the amperometic signal typically has very rapid kinetics [3]. Thus, other than the differences that are directly caused by the vesicular size, differences in fusion pore structures can also contribute to the variability in release kinetics among the different types of granules.

Different Forms of Exocytosis

The fusion of a chromaffin LDCG shown in Fig. 1 reflects the "►full fusion" type of exocytosis in which a single granule fuses with the plasma membrane and releases its entire content (Fig. 2a). However, other forms of exocytosis have also been reported (Fig. 2).

Amperometric studies have shown that SVs in midbrain dopamine neurons [4] as well as chromaffin LDCGs [5] can undergo "►kiss and run" exocytosis (Fig. 2b). During this mode of release, the fusion pore can either close before the complete discharge of if its vesicular content, or undergo rapid flickering such that only a fraction of the vesicular content is released during each flicker [4]. This mode of release allows a vesicle to be reused with minimal reorganization of the molecular machinery that is involved in exocytosis. A recent study with two photon fluorescent microscopy [6] has shown that chromaffin LDCGs can also undergo "►serial compound" exocytosis (Fig. 2c). During this mode of exocytosis, a granule that is deeper in the cytoplasm fuses with another granule which has already fused the plasma membrane. Because of the increased distance for the diffusion of transmitters to the cell's surface, the release kinetics of granules deeper in the cytoplasm is expected to be slower. This mode of exocytosis is found to occur predominantly in the intercellular space in a cluster of chromaffin cells and only during high intensity stimulation [6]. Other forms of exocytosis, which have been described in white blood cells, include "►classical compound exocytosis" and "►piecemeal degranulation." "Classical compound exocytosis" involves the pre-fusion of granules before they finally fuse with the plasma membrane (Fig. 2d) and "piecemeal degranulation" involves the exocytosis of small vesicles that are probably formed by a fission process from certain type of LDCGs (Fig. 2e). What determines the prevalent form of exocytosis in each type of granule in a certain cell type is unclear, but both the selective expression of specific molecular machinery of exocytosis in different types of granules (see next section), as well as their selective regulation by specific intracellular messengers such as Ca^{2+} (for example, see [5] have been implicated.

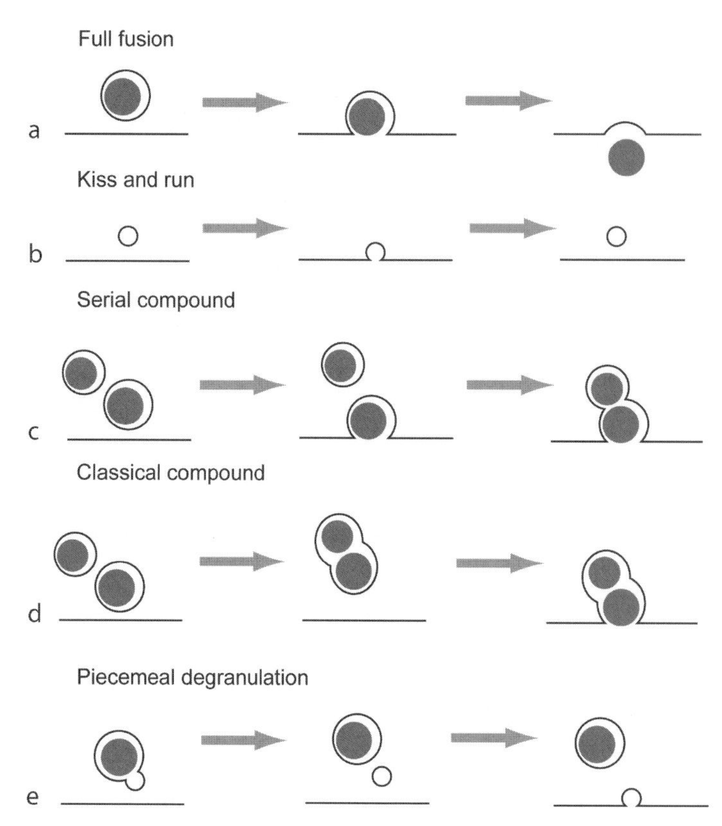

Non-Synaptic Release. Figure 2 Different forms of exocytosis. The dense core in individual dense core granules is depicted in *blue*.

Difference in the Molecular Machinery of Fusion may Contribute to the Variability in Release Kinetics and Ca^{2+}-Sensitivity Among Granule Subtypes

As described earlier, one factor which may contribute to the differences in the release kinetics between SVs and LDCGs is the structure of the fusion pore. This raises the question whether the molecular machinery for fusion is different among the various types of granules. It is generally accepted that a protein complex, called the soluble ►N-ethylmaleimide-sensitive factor (NSF) attachment protein receptor (SNARE, typically concentrated at lipid rafts), can act as a physical link between a "docked" vesicle (or granule) and the plasma membrane (Fig. 3a), and is part of the essential machinery for regulating exocytosis of SVs and LDCGs.

In the presynaptic terminal, the ►soluble NSF attachment protein receptor (SNARE) complex comprises three proteins, syntaxin, ►synaptosome-associated protein of 25 kDa (SNAP-25) and synaptobrevin (also called *v*esicle *a*ssociated *m*embrane *p*rotein, abbreviated as VAMP). In the trans-configuration of the SNARE (Fig. 3b), syntaxin and SNAP-25 are predominantly on the plasma membrane, and synaptobrevin is on the vesicular membrane. Each of the three proteins in the SNARE complex has multiple isoforms, and different cell types express different isoforms [1].

It turns out that cell types which undergo the more unusual forms of exocytosis (such as piecemeal degranulation in some white blood cells, see Fig. 2), indeed have some unusual isoforms of SNAREs expressed on their granules. Moreover, the proportion or molecular conformation of the dominant isoform(s) in an individual cell type can also be differentially regulated (e.g. by cAMP, diacylglycerol or protein kinase C). In view of these complexities, it is likely that variations in the molecular structure of the SNARE complex contribute to the diversity among the fusion pore structure and kinetics in different types of granules.

In addition to the more rapid kinetics for quantal release, the exocytosis of SVs has been suggested to have a higher Ca^{2+} requirement than LDCGs. Moreover, even among the SVs, both synchronous (i.e. those triggered with a minimal delay after a rise in the concentrations of intracellular Ca^{2+} ($[Ca^{2+}]_i$) and asynchronous (i.e. those triggered with an obviously longer delay) releases in synapses appear to have different Ca^{2+} requirements. However, a recent study suggests that the same Ca^{2+}-sensing mechanism (involving five Ca^{2+}-binding sites interacting allosterically) may mediate both synchronous, as well as asynchronous release, and interference with the SNARE-complex may affect the Ca^{2+}-sensitivity of SVs [7]. Multiple proteins

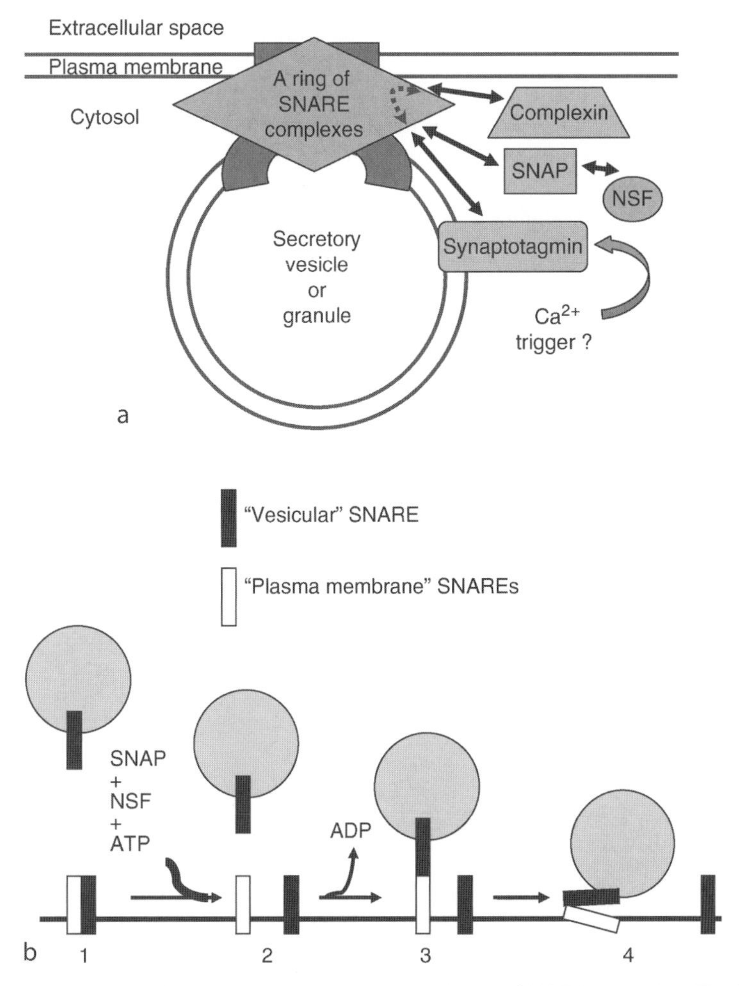

Non-Synaptic Release. Figure 3 (a) Multiple proteins interact with the SNARE complex. The lipid raft is depicted in *red*. The *arrow in blue* represents the interactions among complexin, SNAP and synaptotagmin at the SNARE complex. (b) Cis- and trans-SNARE complexes. (1) A cis-SNARE complex (from a previous cycle of exocytosis) on the plasma membrane. (2) The cis-SNARE complex is dissociated by the recruitment of NSF (ATPase) to the complex in the presence of SNAP. (3) The formation of a trans-SNARE complex in a "loose" conformation. (4) The transformation of a trans-SNARE complex into a "tight" conformation.

are known to interact with the SNARE complex (Fig. 3a). Among these proteins is synaptotagmin, which is a key Ca^{2+} sensor for triggering exocytosis. Synaptotagmin has at least a dozen isoforms, some with vastly different Ca^{2+}-sensitivities, and the dominant expression of specific isoform(s) varies even among different types of synapses. Therefore, the absence of synaptotagmin, or the presence of specific synaptotagmin isoform(s) in different types of granules, may already contribute significantly to the variations in their Ca^{2+}-sensitivity for exocytosis.

The site on the SNARE complex that interacts with synaptotagmin also interacts with at least two other proteins: (►soluble NSF attachment protein (SNAP), not related to SNAP-25) and complexin (Fig. 3a). The three main known isoforms of SNAP (α-, β-, γ-) all recruit the ATPase, ►N-ethylmaleimide-sensitive factor (NSF), to the SNARE complex. The ATPase activity of NSF is essential for breaking the cis-SNARE complex (in which all SNARE components are topologically on the same membrane, see Fig. 3b) which allows the subsequent formation of the trans-SNARE complex (the one depicted in Fig. 3a). Other than this important function of priming granules for fusion, α-SNAP may also affect the Ca^{2+} sensitivity in some LDCGs. In chromaffin cells, an oversupply of exogenous α-SNAP selectively enhances a component of exocytosis that is only prominent at submicromolar $[Ca^{2+}]_i$ [8]. Since α-SNAP can displace synaptotagmin from the SNARE complex, one possible interpretation is that oversupply of α-SNAP allows the formation of a larger proportion of trans-SNARE complex on some LDCGs either with no synaptotagmin, or with a certain isoform of synaptotagmin that has a higher-sensitivity for Ca^{2+}. On the other hand, binding of complexin to SNARE is suggested to activate SVs into a "superprimed metastable state." Subsequently, the binding of Ca^{2+} to synaptotagmin 1 results in the displacement of complexin from the SNARE complex, which triggers fast synchronous synaptic release [9]. Thus, the Ca^{2+}-requirement for exocytosis in the different types of granules may be dependent on the complex interactions among synaptotagmin, SNAP and complexin at the SNARE complex, and a major challenge in the study of exocytosis is to understand how these complex sequential interactions are regulated *in vivo*.

Do All Fusion Pores for Exocytosis have the Same Macroscopic Structure?

Currently there are two main models of the initial fusion pore for exocytosis [10]. The first model postulates that the initial fusion pore is similar to a gap junction, which is a ring of protein complexes that connects the vesicular lumen to the extracellular space via an opening whose wall is entirely proteinaceous (Fig. 4I).

The second model postulates that a ring of protein complexes first causes hemi-fusion of the lipid monolayer of cytosolic leaflets; then when the other monolayer (on the extracellular leaflet of the plasma membrane and the luminal monolayer of the granule) also fuse to open the fusion pore, the wall of the initial fusion pore is essentially lined by lipid molecules (Fig. 4III). For both models, the existence of an initial ring of protein complexes arises mainly from the assumption that a fusion pore has radial symmetry, and each protein complex is probably a SNARE complex with some of its associated proteins. Published

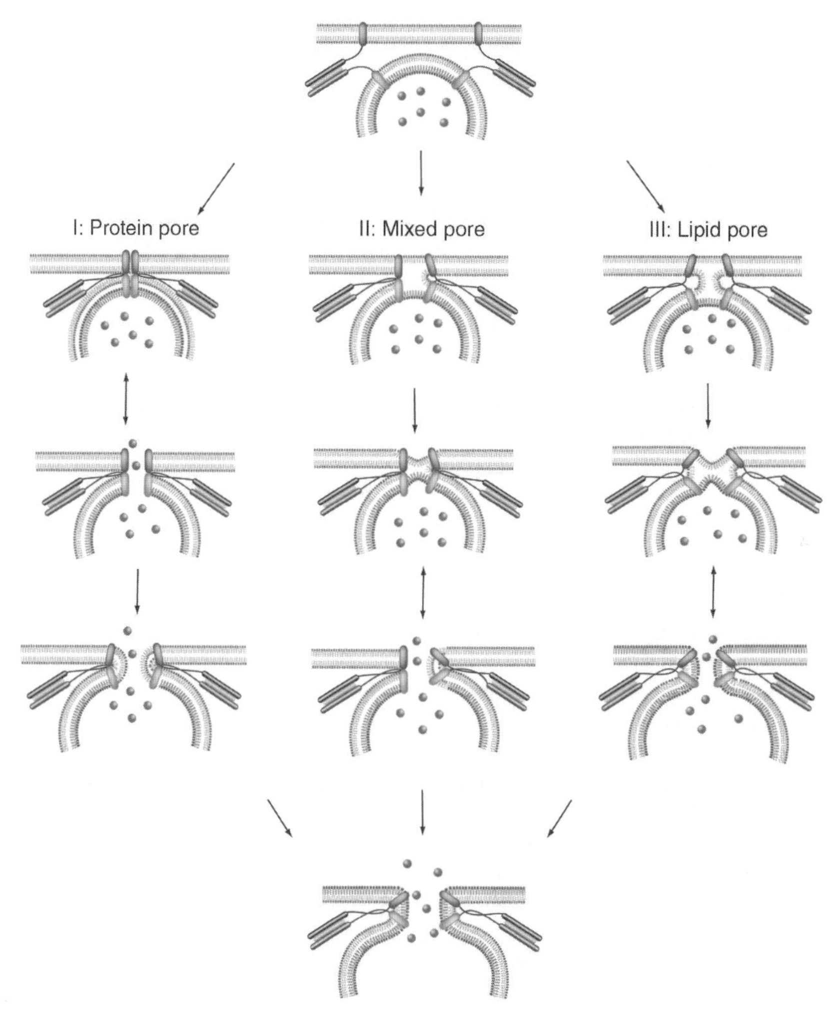

Non-Synaptic Release. Figure 4 Three models for the macroscopic structure of a fusion pore. Note that in (I) and (III), all subunits in the ring of proteins that initially surrounded the fusion pore underwent identical conformational changes simultaneously. In contrast, the subunits in (II) underwent similar changes, but not synchronously. The opening of the semi-stable fusion pore in each model is depicted as reversible.

versions of both models typically depict that every complex in the ring simultaneously undergoes the same macroscopic structural changes to open the fusion pore. Moreover for both models, if the fusion pore indeed proceeds to further rapid dilation, there must be significant influx of lipids between at least some pairs of adjacent protein complexes.

Building on the scenario that the specific isoforms of individual SNAREs, as well as the SNARE-interacting proteins, may not be identical in every SNARE complex that form the initial ring around the fusion pore, we propose a third model in which the initial fusion pore for exocytosis can be lined with both lipids and proteins (Fig. 4II). If we consider that some SNARE complexes in the ring are more sensitive to the trigger Ca^{2+}, then with an intermediate elevation of $[Ca^{2+}]_i$, it is unlikely that all complexes in the ring respond synchronously. If a certain protein complex (e.g. the one on the right hand side of the fusion pore shown in Fig. 4II) can be triggered more readily to fuse the vesicular and plasma membranes near it, the resultant fusion pore can be initially lined by lipid on the right side, but lined by proteins in the left side. According to this model, the initial fusion pore can also dilate or close in ways that are very similar to the two other models (depicted in Fig. 4I & Fig. 4III). Also similar to the other two models, our model suggests that the initial size of the fusion pore is determined by the number of SNARE complexes that surrounds it. However, in our model, the influx of lipid adjacent to each triggerd SNARE complex and the total number of triggered SNARE complex in each fusion pore can increase over a period of time with an intermediate elevation of $[Ca^{2+}]_i$ (that does not trigger essentially all SNARE complexes in each fusion pore synchronously). Therefore the initial dilation of individual fusion pores up to the triggering of all SNARE complexes can vary considerably even if fusion pores with the same number of SNARE complexes are compared. Furthermore, the influx of lipid adjacent to each activated SNARE complex also predicts that the size of the dilating fusion pore is unlikely to increase in precisely "quantized" steps as each additional SNARE complex is triggered.

Summary

The diversities among the proteins in, and associated with, the SNARE complex can contribute significantly to the variability in kinetics of release as well as the Ca^{2+}-sensitivity of exocytosis among the different types of secretory vesicles and granules. These diversities may be part of the molecular adaptation to allow different patterns of exocytosis to regulate diverse physiological functions in time scales that range from submillisecond (e.g. in synapses), to seconds (e.g. in endocrine cells) to days (e.g. in cell growth).

References

1. Brunger AT (2005) Structure and function of SNARE and SNARE-interacting proteins. Q Rev Biophys 38:1–47
2. Bruns D, Jahn R (1995) Real-time measurement of transmitter release from single synaptic vesicles. Nature 377:62–65
3. Tang KS, Wang N, Tse A, Tse FW (2007) Influence of quantal size and cAMP on the kinetics of quantal catecholamine release from rat chromaffin cells. Biophys J 92:2735–2746
4. Staal RG, Mosharov EV, Sulzer D (2004) Dopamine neurons release transmitter via a flickering fusion pore. Nat Neurosci 7:341–346
5. Elhamdani A, Azizi F, Artalejo CR (2006) Double patch clamp reveals that transient fusion (kiss-and-run) is a major mechanism of secretion in calf adrenal chromaffin cells: high calcium shifts the mechanism from kiss-and-run to complete fusion. J Neurosci 26:3030–3036
6. Kishimoto T, Kimura R, Liu TT, Nemoto T, Takahashi N, Kasai H (2006) Vacuolar sequential exocytosis of large dense-core vesicles in adrenal medulla. EMBO J 25:673–682
7. Schneggenburger R, Forsythe ID (2006) The calyx of Held. Cell Tissue Res 326:311–337
8. Xu J, Xu YM, Ellis-Davies GCR, Augustine GJ, Tse FW (2002) Differential regulation of exocytosis by alpha- and beta-SNAPs. J Neurosci 22:53–61
9. Tang J, Maximov A, Shin OH, Dai H, Rizo J, Sudhof TC (2006) A complexin/synaptotagmin 1 switch controls fast synaptic vesicle exocytosis. Cell 126:1175–1187
10. Jackson MB, Chapman ER (2006) Fusion pores and fusion machines in Ca^{2+}-triggered exocytosis. Annu Rev Biophys Biomol Struct 35:135–160

Non-verbal Communication

Definition

Communication without words. General form of communication in the animal kingdom.

Nootropic Drugs

PAUL F. SMITH
University of Otago Medical School, School of Medical Sciences, Department of Pharmacology and Toxicology, Dunedin, New Zealand

Synonyms

Cognitive-enhancing drugs; "Smart" drugs

Definition

The word "►nootropic" is derived from the Greek word for "mind", which is "noos." ►Nootropic drugs are intended to enhance cognitive function, either in individuals with neurological and psychiatric disorders that include symptoms such as memory loss, or in healthy individuals who seek to increase their cognitive function beyond the normal level. Nootropic drugs comprise a heterogeneous, often controversial, collection of drugs that work with varying degrees of success.

Characteristics

Cognitive Enhancement: What does it mean?

The term "cognitive enhancement" is often used to refer to an increase in memory, but it can also refer to an increase in other aspects of cognitive processing, such as attention and even elements of a behavioral response that are not easily distinguished from memory itself. There is a basic distinction between enhancing cognition in a patient who has a neurological deficit of some sort, for example, a brain lesion, and enhancing cognition in someone with normal cognitive function. In the former case, the drug therapy is intended to replace a brain chemical that is missing or reverse a pathological change; in the latter case, the drug is intended to enhance normal neurochemical function in some way. While drug treatment for memory deficits in diseases such as ►Alzheimer's disease have had limited success, it is debatable whether any nootropic drug has clearly been shown to enhance normal cognition in humans. In the case of drug treatment to restore memory in a neurological disorder, there is a clear objective, which is to improve a patient's performance on a memory test to within normal limits. In the case of drug treatment to enhance normal cognition, the objective is less clear. The major obstacle for the development of all nootropic drugs is that the precise mechanisms of cognition, including memory, have not yet been identified. Consequently, it is difficult to design drugs to manipulate a neural system that is not fully understood. However, progress is being made in understanding the neural basis of memory, using models of memory such as ►long-term potentiation (LTP), and this is gradually leading to the development of novel nootropic drugs [1,2].

Types of Nootropic Drugs

Most nootropic drugs that are currently used to treat cognitive deficits in neurological disorders manipulate the brain neurotransmitter, ►acetylcholine. Acetylcholine is diminished in Alzheimer's disease [3]; therefore, these drugs are intended to replace the missing neurotransmitter. Nootropic drugs in this category include inhibitors of the cholinesterase enzymes, which metabolize acetylcholine in the synaptic cleft, in addition to drugs that activate the acetylcholine receptors (►acetylcholine receptor agonists). ►Cholinesterase inhibitors prolong the action of acetylcholine in the synapse. Acetylcholine receptor agonists increase the level of activation of acetylcholine receptors. ►Tacrine was the first cholinesterase inhibitor used to treat Alzheimer's disease; however, ►donepezil, ►rivastigmine and ►galantamine are now the first line treatments [4–8]. All three drugs inhibit acetylcholinesterase, but rivastigmine also inhibits butyrylcholinestase, and galantamine acts as an agonist at nicotinic acetylcholine receptors [4]. Unfortunately, their beneficial effects are limited, they work more effectively in some patients than others [5], and they can cause adverse side effects such as nausea, vomiting and liver toxicity [4].

Many other drugs have been investigated for possible efficacy in the treatment of cognitive deficits, including ►memantine, selegiline, modafanil, vitamin E, *Ginkgo biloba* extracts and a variety of other herbal extracts, and ►α-amino-3-hydroxy-5-methylisoxazole-4-propionic acid (AMPA) receptor potentiators. Of these, only memantine and AMPA receptor potentiators have been shown to have any clinical effect [4]. Memantine blocks the calcium ion channel associated with the ►N-methyl-D-aspartate (NMDA) subtype of glutamate receptor (an uncompetitive NMDA antagonist). Since the NMDA receptor is involved in increases in synaptic efficacy associated with LTP, it may at first seem paradoxical that memantine could benefit cognition. However, over-stimulation of the NMDA receptor causes neurotoxicity and therefore memantine may improve cognition in diseases such as Alzheimer's disease by limiting this form of neural damage [3,4,6,7]. ►Piracetam is one of a number of AMPA receptor potentiators that has been actively researched but is not yet prescribed for Alzheimer's disease. Along with aniracetam, it is a pyrrolidone that increases the response of the AMPA subtype of glutamate receptor [9]. Other AMPA receptor potentiators, sometimes referred to as "►ampakines," include benzylpiperidines such as CX-516 and CX-546, which are in clinical trials [1,9].

Most of the drugs that are used clinically in the treatment of cognitive disorders do not enhance cognition in people without a neurological deficit. However, many drugs that were first investigated for their general nootropic effects, were then tested in patients with Alzheimer's disease and other neurological disorders to determine whether they would have any beneficial clinical effect. Some herbal drugs such as *Ginkgo biloba* extracts have been claimed to improve memory in both Alzheimer's disease patients and in neurologically intact individuals; however, there is no convincing evidence for a consistent effect in either case [4,10].

Summary of Nootropic Sites of Action

Where in the brain do nootropic drugs act to produce their effects? Many areas of the brain are involved in the encoding, consolidation and retrieval of memories. Areas such as the hippocampus and other areas of the medial temporal lobe are believed to be important for encoding new memories; the neocortex is thought to be important for long-term storage of memories [1,2].

Molecular Mechanisms of Action of Nootropic Drugs

The precise mechanisms of the formation and retrieval of memories are not understood and therefore the precise mechanisms of action of nootropic drugs are not known. However, it is well established that acetylcholine is important for the formation new memories. Activation of acetylcholine receptors causes intracellular changes in neurons that result in the activation of proteins such as the cyclic adenosine monophospshate (cAMP) response element binding protein (►CREB), which many researchers regard as a form of "molecular switch" that converts short-term memories into long-term ones [2]. The neural model of memory, LTP, has been used to better understand the increases in synaptic efficacy that are likely to underlie the formation of memories. In LTP, the activation of post-synaptic NMDA receptors is thought to be a critical step in producing the biochemical changes that lead to enhanced synaptic efficacy. While excessive activation of NMDA receptors results in excessive calcium influx and neurotoxicity, a smaller elevation of NMDA receptor activation may enhance memory. For example, the ampakines have been developed so that they elevate NMDA receptor activation indirectly, by potentiating AMPA receptor activity, which then leads to increased depolarization, thus lowering the threshold for NMDA receptor activation [1,9]. The downstream effects of AMPA receptor modulation include an increase in growth factors such as brain-derived neurotrophic factor (BDNF), which is known to be important in synaptic plasticity [9]. While still in clinical trials, these drugs may be one class of nootropic drug that is used to treat Alzheimer's disease and other related neurological disorders in the future. Other possibilities include drugs that modulate dopaminergic and serotonergic function. Many researchers believe that no one drug with a single mechanism of action is likely to provide successful therapy for cognitive disorders.

Will the same nootropic drugs that are used to treat cognitive disorders be useful for enhancing cognition in healthy individuals? It is possible but more likely that there are natural limits to the extent that the normal neurochemical machinery of memory can be enhanced before adverse side effects develop.

References

1. Lynch G (2002) Memory enhancement: the search for mechanism-based drugs. Nat Neurosci 5 (suppl):1035–1038
2. Barco A, Pittenger C, Kandel ER (2003) CREB, memory enhancement and the treatment of memory disorders: promises, pitfalls and prospects. Expert Opin Ther Targets 7:101–114
3. Roberson ED, Mucke L (2006) 100 years and counting: prospects for defeating Alzheimer's disease. Science 314:781–784
4. Evans JG, Wilcock G, Birks J (2004) Evidence-based pharmacotherapy of Alzheimer's disease. Int J Neuropsychopharmacol 7:351–369
5. Rockwood K, Fay S, Jarrett P et al. (2007) Effect of galantamine on verbal repetition in AD. Neurology 68:1116–1121
6. Beier MT (2007) Treatment strategies for the behavioural symptoms of Alzheimer's disease: focus on early pharmacologic intervention. Pharmacotherapy 27:399–411
7. Feldman HH, Schmitt FA, Olim JT (2006) Activities of daily living in moderate-to-severe Alzheimer's disease: an analysis of the treatment effects of memantine in patients receiving stable donepezil. Alzheimer Dis Assoc Disord 20:263–268
8. Brodaty H, Woodward M, Boundy K et al. (2006) A naturalistic study of galantamine for Alzheimer's disease. CNS Drugs 20:935–943
9. O'Neill MJ, Bleakman D, Zimmerman DM et al. (2004) AMPA receptor potentiators for the treatment of CNS disorders. CNS Neurol Disord Drug Targets 3:181–194
10. Maclennan K, Darlington CL, Smith PF (2002) The CNS effects of *Ginkgo biloba* extracts and ginkgolide B. Prog Neurobiol 67:236–258

Noradrenaline or Norepinephrine

Definition

Noradrenaline (also called norepinephrine) is a biogenic amine neurotransmitter that is widely distributed throughout the brain and is also present in sympathetic adrenergic neurones and adrenal gland. Central noradrenergic cells are involved in the ascending arousal system and attention. Noradrenaline released from sympathetic nerve endings can cause increased heart rate and blood pressure.

Noradrenergic Neuron/Cell

Definition

A noradrenergic neuron uses noradrenaline (norepinephrine) as its neurotransmitter. Groups A1-A7 are examples of noradrenergic neurons located in the medulla.

►A1-A7 cell groups Cellulae noradrenergicae/A1 – A7)

Norepinepherine

►Noradrenaline

Normosmia/Hyposmia/Anosmia

Definition

Normosmia is the subjectively perceived normal olfactory function, usually defined as the ability to detect the great majority of tested odors in a given olfactory test. Hyposmia means the decrease of this olfactory function and anosmia the total loss of any olfactory function. Beside total anosmia, specific anosmias have been described, where only certain odors are not perceived and most odors are smelt normally.

►Smell Disorders

Northern Blot

Definition

A molecular assay that identifies specific messenger RNA components by using a radioactively complementary probe. mRNA from a cell is isolated and separated based on size by polyacrylamide gel electrophoresis.

The antisense probe is then used to identify the mRNA species in the gel.

Notochord

Definition

Axial mesoderm that lies beneath (ventral to) the neural palte/tube and extends from head to tail in chordates. It produces signals that are involved in the specification of ventral parts of the neural tube.

►Evolution and Embryological Development of the Forebrain

Noxious (Algogenic) Chemical Stimulation of the Heart

Definition

Cardiac afferent fibers are activated by injecting noxious chemicals via a catheter that is placed inside the pericardial sac of an anesthetized animal. The noxious chemicals are substances such as bradykinin, capsaicin or an algogenic cocktail composed of serotonin, bradykinin, prostaglandin E2, histamine and adenosine. An algogen is a chemical mediator that is usually generated within diseased or damaged tissue and produces pain behavior.

►Viscero-Somatic Reflex

Noxious Stimuli

Definition

Stimuli that are intense enough to potentially or actually damage body tissue. For example, thermal stimulation of the skin up to 46°C evokes a sense of warmth, while that over 48°C is noxious.

Noxious Stimulus-evoked Vocalizations

Definition

These are animal vocalizations that occur during noxious stimulation. These vocalizations are mediated by brainstem mechanisms and exhibit different spectrographic characteristics than vocalizations that occur after a noxious stimulus (vocalization afterdischarges, VADs). VADs are considered to be a direct index of the unpleasantness associated with the sensations evoked by noxious stimulation. VADs are mediated by brain structures involved in mediating pain unpleasantness in humans (medial thalamus, anterior cingulate cortex, amygdala) and suppressed by drug treatments that reduce pain unpleasantness in humans.

►Emotional/Affective Aspects of Pain

NRSF

Definition

Neuron-restrictive silencing factor (NRSF), also known as repressor element-1 silencing transcription factor (REST), silences neuronal gene expression in neural progenitors as well as non-neuronal cell types.

►Chromatin Immunoprecipitation and the Study of Gene Regulation in the Nervous System

NRTP

Definition

Nucleus reticularis tegmenti pontis.

NST

►Nucleus Tractus Solitarius

NT-4/5

Definition

►Neurotrophin 4/5

NTS

►Nucleus Tractus Solitarii

Nuclear Factor Kappa Beta (NF-κB)

Definition

Family of transcription factors called Rel involved with cytokine-induced activation of gene expression by binding to the enhancer; essential for HIV expression; involved in neuroinflammation, peripheral and systemic inflammations.

►Cytokines
►Human Immunodeficiency Virus (HIV)
►NF-κB: Potential Role in Adult Neural Stem Cells

Nuclear Factor of Activated T Cells (NFAT)

Definition

A transcription factor that is activated as a result of axonal injury. Following binding of a neurotrophin to its receptor, both serine/threonine phosphotase and nuclear GSK3β regulate NFAT transcriptional activity and its translocation to the nucleus. NFAT interacts with members of the AP-1 complex, Fos and Jun.

►Neurotrophic Factors in Nerve Regeneration

Nuclear Magnetic Resonance

Definition

Nuclear Magnetic Resonance is used to describe the phenomenon where an atomic nucleus positioned in a strong magnetic field absorbs energy in the form of electromagnetic radiation at a specific frequency (Lamor frequency). A NMR signal is thereafter recorded as an induced current in a receiver coil near the sample as the atomic nuclei return to their thermal equilibrium state. The NMR technique is used to study molecular structure.

►Magnetic Resonance Imaging

Nuclear Matrix

Definition

The nuclear matrix is defined as the non-chromatin structure resistant to detergent extraction and nuclease digestion followed by high salt treatment.

Nuclei of Posterior Column

Synonyms

Nuclei columnaepost.

Definition

In the cuneate nucleus and gracile nucleus terminate the epicritic afferents of the posterior column – funiculus dorsalis – (cuneate fasciculus and gracile fasciculus), which is the reason why they are also called posterior column nuclei.

- Gracile nucleus: afferents from the trunk and lower extremities.
- Cuneate nucleus: afferents from the upper extremities and neck (medial cuneate nucleus) and vestibular organ (lateral cuneate nucleus). The efferents of both nuclei cross to the contralateral side in the medulla as the internal arcuate fibers and join the trigeminal efferents (epicritic sensibility of the face) to form the medial lemniscus, before passing to the thalamus (ventral posterolateral thalamic nucleus), from where they project into the somatosensory cortex (postcentral gyrus).

Nuclei, Telencephalic, Deep

Definition

Nuclei that form within the basal parts of the walls of the paired telencephalic vesicles, including the amygdaloid and septal nuclei, caudate nucleus, putamen, accumbens, globus pallidus, ventral pallidum, bed nuclei of the stria terminalis and nuclei of the diagonal band.

► Striatopallidum

Nuclei, Ventral Tier, Thalamic

Definition

Principal or "relay" nuclei comprising anterior motor and posterior sensory groups. Anteriorly, cerebellar inputs spread widely in the ventrolateral nucleus, whereas basal ganglia inputs are more restricted in the ventromedial nucleus and anterior, medial parts of the ventrolateral (ventral anterior nucleus, in primates). The ventral posterolateral and posteromedial nuclei relay sensory information from the dorsal column system of the spinal cord and trigeminal nerve, respectively.

► Striatopallidum

Nuclei of the Lateral Lemniscus

Ellen Covey
Department of Psychology, University of Washington, Seattle, WA, USA

Synonyms

Cell groups of the lateral lemniscus; Lateral lemniscal nuclei; NLL

Definition

The nuclei of the lateral lemniscus (NLL) comprise several groups of neuron cell bodies in the mammalian brainstem embedded within or lying near the fiber tract known as the lateral lemniscus. Neurons in the nuclei of the lateral lemniscus constitute a major component of the ascending auditory pathway and include both monaural and binaural cell groups. They are a major source of input to the inferior colliculus.

Characteristics

Quantitative Description

Figure 1 shows a schematic drawing of the nuclei of the ►lateral lemniscus (LL) as seen in a frontal section through the lower brainstem of a cat and an echolocating bat.

The NLL of the cat is typical of most mammals. In the bat, the nuclei of the lateral lemniscus are hypertrophied and much more clearly differentiated than they are in other mammals. For this reason, the bat has proven to be an ideal species in which to study the properties of specific cell types in the NLL. In all mammals the NLL can be divided into two major subdivisions, the dorsal nucleus of the lateral lemniscus (DNLL) which receives input from ►binaural structures, and a ventral complex of nuclei comprising the intermediate nucleus of the lateral lemniscus (INLL) and the ventral nucleus of the lateral lemniscus (VNLL). The INLL and VNLL receive their primary input from one ear only and collectively make up the ►monaural pathway (for review, see [1]).

Monaural Pathways

The INLL and VNLL receive projections from the contralateral ear, mainly via the ventral ►cochlear

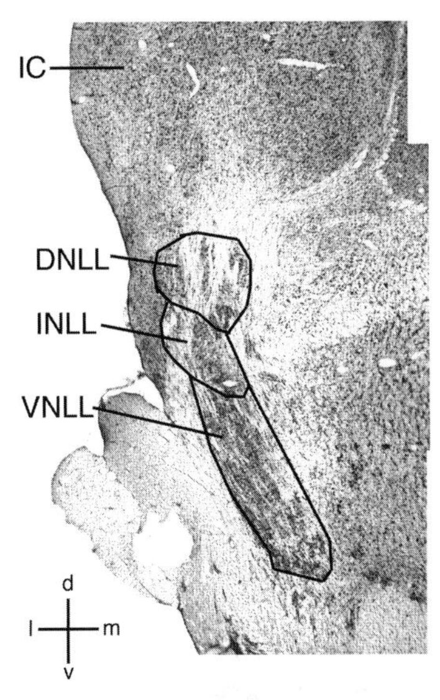

Nuclei of the Lateral Lemniscus. Figure 1 Frontal section through the brainstem of a cat showing the nuclei of the lateral lemniscus. DNLL, dorsal nucleus of the lateral lemniscus; IC, inferior colliculus; INLL, intermediate nucleus of the lateral lemniscus; VNLL, ventral nucleus of the lateral lemniscus; d, dorsal; l, lateral; m, medial; v ventral.

nucleus, medial nucleus of the trapezoid body, and ►periolivary nuclei. Because their input originates mainly from one side, they are sometimes referred to as the monaural nuclei of the lateral lemniscus. Both INLL and VNLL project to the central nucleus of the ►inferior colliculus and constitute the largest single source of projections to the auditory midbrain.

Figure 2 summarizes the principal cell types that are found in the INLL and VNLL. In all species, the INLL contains predominantly elongate cells whose dendrites are oriented orthogonal to the ascending fibers of the lateral lemniscus.

The VNLL contains a population of small round neurons with a single highly branched dendrite. These neurons resemble ►spherical bushy cells in the cochlear nucleus, and receive ►calyx-like terminals on their cell body. The calyces originate from cells in the ventral cochlear nucleus. In most mammals the bushy cells are intermingled with other cell types throughout the VNLL, but in echolocating bats they are segregated into a highly organized and homogeneous structure termed the ►columnar nucleus of VNLL (VNLLc) [2,3]. Neurons in the VNLLc are ►glycinergic, so are thought to provide inhibitory input to their target cells. The other main cell type in VNLL is mutipolar neurons, which in bats are segregated into a subdivision referred to as the ►multipolar cell region of VNLL (VNLLm). Figure 2a is a block diagram summarizing the ascending monaural pathways via the nuclei of the lateral lemniscus.

Binaural Pathways

The DNLL receives most of its input from the binaural structures of the superior olivary complex, including projections from the ipsilateral ►medial superior olive (MSO) and projections from the ►lateral superior olive (LSO) of both sides. The right and left DNLLs are reciprocally connected with one another via a fiber tract, the ►commissure of Probst. The DNLL provides bilateral projections to the inferior colliculus. Most neurons in the DNLL are ►GABAergic and are therefore thought to be inhibitory to their postsynaptic neurons. Figure 2b is a block diagram summarizing the binaural components of the ascending pathway via the DNLL.

Function

The nuclei of the lateral lemniscus are a major component of the complex system of parallel pathways in the mammalian auditory brainstem. However, their role in the neural processing underlying sound perception is not well understood. What we do know is that most neurons in the INLL and VNLL respond to sounds at the contralateral ear and are unaffected by sounds at the ipsilateral ear while most neurons in the DNLL are sensitive to differences in the intensity or timing of sounds at the two ears. Thus, it has been suggested that the DNLL plays a role in localizing sound sources while the monaural nuclei play a role analyzing temporal patterns of sound (for review, see [1,4]).

Monaural Pathways: INLL and VNLL

Based on the limited evidence available, the structure, connectivity, and functional properties of INLL and VNLL seem to be similar in all mammals. However, because these structures are unusually large, clearly organized, and accessible in echolocating bats, much of what we know about their physiology comes from these animals. It is thought that the different neuron types of INLL and VNLL transform inputs from the cochlear nucleus in various ways depending on their ►intrinsic properties, and integrate inputs from multiple sources to provide a system of ►delay lines that are important for creating temporal ►feature detector neurons in the inferior colliculus. Response latencies of neurons in the ventral cochlear nucleus range from about 1–6 ms, but in the INLL and VNLL, this range lengthens to include latencies from about 2–20 ms [5]. Delayed input from some neurons converging with rapid input from others provides a mechanism through which it is possible to compare sound events that occur at different times (see [6], for review).

Neurons in the INLL and VNLL respond to sounds with a variety of discharge patterns, which are probably shaped through each neuron's intrinsic properties and integration of multiple synaptic inputs. Sustained responses provide a real-time representation of a sound's duration, since the neuron fires for as long as the sound is

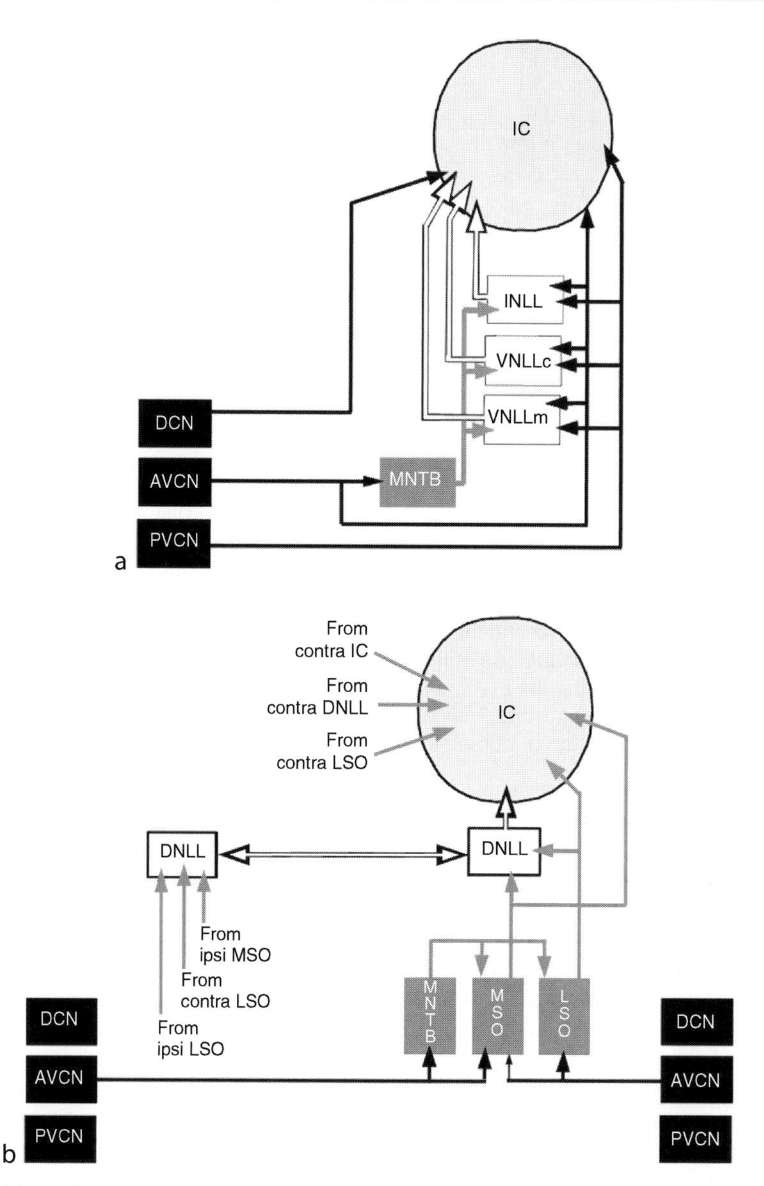

Nuclei of the Lateral Lemniscus. Figure 2 Block diagram of monaural and binaural pathways to the midbrain via the nuclei of the lateral lemniscus. (a) Monaural pathways via INLL and VNLL. (b) Binaural pathways via DNLL. Abbreviations same as in Fig. 1; AVCN, anteroventral cochlear nucleus; DCN, dorsal cochlear nucleus; LSO, lateral superior olive; MNTB, medial nucleus of the trapezoid body; MSO, medial superior olive; PVCN, posteroventral cochlear nucleus; VNLLc, columnar nucleus of VNLL; VNLLm, multipolar nucleus of VNLL.

present. Onset responses provide a precise marker for the time when a sound begins, or when a changing sound enters the neuron's area of sensitivity. Echolocating bats have a specialized group of cells in VNLL, the columnar nucleus, which are extraordinary in having timing precision on the order of several tens of microseconds, across a wide range of conditions. It is possible that these neurons have evolved to provide precise timing markers for when the bat emits an echolocation call and when the echo of that call returns, allowing the bat's neural circuitry to calculate the distance to the object from which the echo was reflected. Although this cell type is segregated in bats, it is present in other mammals as well, intermingled among other cell types. This suggests that the precise timing information about sound onset provided by the VNLL is important for all animals that hear, albeit on a different level of time resolution (for reviews see [1,4]).

The Binaural Pathway: DNLL

The DNLL represents an intermediate binaural processing stage between the superior olivary complex (SOC) and inferior colliculus. It receives inhibitory and excitatory input from the SOC on the same side, excitatory input from the SOC on the opposite side, and inhibitory input from the opposite DNLL (see Fig. 2b). As a result, responses of DNLL neurons are influenced by sound at both ears, in complex ways, being binaurally facilitated under some conditions and inhibited in under other conditions. It has been suggested that DNLL may play a role in perception of sounds originating from multiple sources [7], but at present its function is not well understood.

References

1. Covey E (1993) The monaural nuclei of the lateral lemniscus: parallel pathways from cochlear nucleus to midbrain. In: Merchan MA, Juiz JM, Godfrey DA (eds) The mammalian cochlear nuclei: organization and function. Plenum, New York, pp 321–334
2. Covey E, Casseday JH (1986) Connectional basis for frequency representation in the nuclei of the lateral lemniscus of the bat, *Eptesicus fuscus*. J Neurosci 6:2926–2940
3. Covey E, Casseday JH (1991) The ventral lateral lemniscus in an echolocating bat: parallel pathways for analyzing temporal features of sound. J Neurosci 11:3456–3470
4. Oertel D (1999) The role of timing in the brainstem auditory nuclei of vertebrates. Annu Rev Physiol 61:497–520
5. Haplea S, Covey E, Casseday JH (1994) Frequency tuning and response latencies at three levels in the brainstem of the echolocating bat, *Eptesicus fuscus*. J Comp Physiol A 174:671–683
6. Covey E, Casseday JH (1999) Timing in the auditory system of the bat. Annu Rev Physiol 61:457–476
7. Burger RM, Pollak GD (2001) Reversible inactivation of the dorsal nucleus of the lateral lemniscus reveals its role in the processing of multiple sound sources in the inferior colliculus of bats. J Neurosci 21:4830–4843
8. Covey E (1993) Response properties of single units in the dorsal nucleus of the lateral lemniscus and paralemniscal zone of an echolocating bat. J Neurophysiol 69:842–859
9. Covey E, Casseday JH (1995) The lower brainstem auditory pathways. In: Popper AN, Fay RR (eds) Handbook of auditory research, vol 5: hearing and echolocation in bats. Springer, New York, pp 235–295

Nucleus, Bed, of the Stria Terminalis

Definition

The bed nucleus of the stria terminalis is a deep telencephalic nucleus comprising a number of variably distinct subnuclei, of which all resemble, in terms of intrinsic composition and extrinsic connections, the centromedial part of the amygdaloid complex.

The bed nucleus of the stria terminalis and the centromedial part of the amygdala are densely interconnected by long associational connections of which one part traverses the stria terminalis and another part traverses the sublenticular part of the basal forebrain. Hence, the bed nucleus is regarded as an integral part of the extended amygdala.

►Striatopallidum

Nucleus, Caudate

Definition

The caudate nucleus is one of the deep telencephalic nuclei consisting of a large globular head that occupies the rostral part of the hemispheric wall, an elongated body that arches upward and backward over the internal capsule and a long slender tail that arches downward and forward into the temporal lobe in relation to the hippocampus and amygdaloid complex. The caudate nucleus receives massive cortical inputs largely from the frontal lobe and a number of subcortical inputs, particularly from intralaminar thalamic nuclei and dopaminergic neurons in the substantia nigra pars compacta. It projects strongly to the globus pallidus and substantia nigra.

►Basal Ganglia
►Striatopallidum

Nucleus, Lentiform (Lenticular)

Definition

The lentiform nucleus is a large cone-shaped gray mass forming the central core of the hemisphere with a convex lateralward-facing base, comprising the putamen, and medial-ward pointing apex, comprising the globus pallidus. Grouping of the globus pallidus and putamen as the lentiform nucleus reflects an artificial association, in so far as the globus pallidus is structurally and functionally dissimilar to the putamen, receiving dense projections from the striatum (caudate nucleus and putamen) and projecting to the subthalamic nucleus, substantia nigra, brainstem reticular formation, and via a relay in the thalamus, to motor staging areas of the cortex.

►Basal Ganglia
►Striatopallidum

Nucleus, Mediodorsal, Thalamic

Definition

The mediodorsal (MD) thalamic nucleus is situated within the internal medullary lamina at the dorsomedial extremity of the thalamus and surrounded by nuclei of the midline-intralaminar group.

The mediodorsal nucleus (MD) is always included with the thalamic association nuclei, sometimes as a subcategory of the principal or "relay" nuclei. Comprising medial, central and lateral segments involved with, respectively, emotional expression, olfaction and visual attention, the MD is reciprocally connected with frontal lobe association cortex including the orbitomedial and agranular insular (medial and central segments) and the frontal eye field (lateral segment). In addition, the MD receives projections from basal ganglia structures including ventral striatopallidum (medial and central segments) and the substantia nigra reticulata (lateral segment). The cortical projection field of the MD has been said to define the extent of the "prefrontal" cortex.

►Thalamus
►Striatopallidum

Nucleus Accumbens

Definition

A large gray mass located in the medial part of the basal forebrain that, over the years, has been variably associated with the septal nuclei and striatum. Previously called the "nucleus accumbens septi", it is interposed without clear boundaries between the ventral parts of the caudate nucleus and putamen and the olfactory tubercle or its homologue. Contemporarily regarded as an integral part of the striatal complex, it is no longer considered to be a nucleus in its own right. It receives cortical inputs from the basal amygdala, hippocampus and medial prefrontal cortex and projects to the ventral pallidum, lateral hypothalamus and ventral mesencephalon. Experimental histochemical and connectional studies have revealed that the accumbens comprises sub-territories, including a core, shell and arguably a rostral pole.

►Striatopallidum

Nucleus Ambiguus

Definition

Like the dorsal nucleus of the vagus nerve and the nucleus of the hypoglossal nerve, the ambiguus features a cellular column of at least 2 cm in length and also runs parallel to these nuclei. This is no surprise as it is the origin of somatomotor (actually special visceromotor) fibers of glossopharyngeal nerve (IX) and vagus nerve (X), which are responsible for innervation of the pharynx and larynx muscles.

►Myelencephalon

Nucleus Basalis Magnocellularis

Definition

Group of cholinergic cells in the basal forebrain, term used in nonprimates to refer to the area equivalent to the nucleus basalis of Meynert in primates

►Evolution of Subpallial Cholinergic Cell Groups

Nucleus Basalis (NB) of Meynert

Definition

The NB is a group of neurons in the basal forebrain that receive inputs from limbic and paralimbic regions and send cholinergic excitatory projections to the entire brain, particularly the neocortex. Projections to the sensory cortical areas are strictly topographically ordered. The NB seems to play a key role in the control of selective attention.

►Basal Forebrain

Nucleus Intermediolateralis (IML)

Definition

The IML is present in the thoraco-lumbar (upper lumbar) and sacral spinal cord and contains the highest density of spinal preganglionic neurons. It is part of the intermediate zone.

►Autonomic Reflexes

Nucleus Isthmi

Definition

It is found at the dorsocaudal end of the dorsal tegmentum; and is homologous to the mammalian parabigeminal nucleus. Indirect visual input to the nucleus isthmi originates from the ipsilateral tectum.

An ipsilateral isthmotectal projection to several retinorecipient layers and a contralateral isthmotectal projection to only the superficial layer of retinal afferents is found in amphibians and all other vertebrate taxa. The isthmotectal projection is topographically organized and in register with the retinal maps. Recordings from isthmic neurons and lesion experiments of the nucleusisthmi reveal that the representation of the visual space differs from that of the tectal representation of the visual space. The isthmic nucleus is essentially involved in object localization and selection.

►Evolution of the Visual System: Amphibians

Nucleus of the Optic Tract

Definition

As the optic tract curves around the brain stem toward the lateral geniculate nucleus there are a number of terminations in a nucleus within the pretectal area, just dorsal to the superior colliculus. The subcortical neuronal substrate of the optokinetic reflex has been investigated in many mammals. In all of these animals, the pretectal nucleus of the optic tract and the dorsal terminal nucleus of the accessory optic tract (NOTDTN), with its strongly direction-selective neurons, links the visual information from the retina and visual cortex, via projections to the inferior olive, the nucleus praepositus hypoglossi, the nucleus reticularis tegmenti pontis and the dorsolateral pontine nucleus, with the cerebellum and the oculomotor structures. In most mammals, the direct retinal input to the NOT-DTN comes almost exclusively from the contralateral eye, whereas in the monkey the retinal input is strongly bilateral. Data show that the motion-sensitive areas in the superior temporal sulcus (STS) provide the main input to the NOT-DTN.

►The Central Mesencephalic Reticular Formation – Role in Eye Movements

Nucleus of the Solitary Tract

►Nucleus Tractus Solitarii

Nucleus Prepositus Hypoglossi

Definition

The nucleus prepositus hypoglossi is located on the surface of the fourth ventricle between the abducens and the hypoglossal motor nuclei. Main oculomotor-related inputs to prepositus neurons arrive from the ipsilateral paramedian pontine reticular formation, from both vestibular nuclei, and from the contralateral prepositus nucleus. Prepositus neurons project to the abducens nucleus, superior colliculus, cerebellum, and other brainstem centers related to eye movements.

►Neural Integrator – Horizontal
►Vestibular Secondary Afferent Pathways

Nucleus Proprius

Synonyms

Nucl proprius

Definition

Nucleus in the middle of the posterior horn of the spinal cord. Present in all spinal cord segments, it is a synaptic center for proprioceptive afferents from the locomotor apparatus. The impulses pass on to the cerebellum via the anterior spinocerebellar tract where they are compared with setpoint values.

►Medulla Spinalis

Nucleus Reticularis Gigantocellularis (NRG)

Definition

The rostral and medial portion of the medullary reticular formation named for the giant cells that it contains. It is a major source of reticulospinal neurons.

►Reticulospinal Long-Lead Burst Neurons

Nucleus Reticularis Pontis Caudalis (NRPc)

Definition

Roughly the caudal half of the pontine reticular formation. Its borders are based on cytoarchitecture but are poorly delineated. NRPc gives way to nucleus reticularis gigantocellularis roughly at the caudal border of the abducens nucleus.

Nucleus Reticularis Tegmenti Pontis (NRTP)

Definition

A large reticular nucleus at the bottom of the pontine reticular formation overlying the pontine nuclei. NRTP receives input from cortical and higher-level subcortical structures and conveys (presumably integrated) information to the cerebellum. NRTP also receives substantial feedback from the cerebellum by way of efferents from the deep cerebellar nuclei.

Nucleus Tractus Solitarii (NTS)

Definition

►Nucleus of the Solitary Tract

Nucleus Tractus Solitarii

R. Alberto Travagli
Neuroscience, Pennington Biomedical Research Center-LSU System, Baton Rouge, LA, USA

Synonyms

Nucleus of the tractus solitarius; NTS; Nucleus of the solitary tract; NST

Definition

The nucleus tractus solitarius (NTS) is the principal visceral sensory nucleus in the brain and comprises neurochemically and biophysically distinct neurons located in the dorsomedial medulla oblongata. The NTS conveys information from the gustatory, cardiorespiratory, esophageal and the subdiaphragmatic gastrointestinal viscera that is subsequently assimilated with homeostatic signals arriving from other integrative centers of the pons, diencephalon and forebrain.

An exceptionally large diversity of neurochemical phenotypes and receptor proteins coupled with an impressive network of connections to and from the NTS and a loose blood brain barrier characterizes this brainstem nucleus as a vital controller of homeostatic functions.

The old concept of the NTS (and the closely related dorsal motor nucleus of the vagus and area postrema) as a simple relay center is overcome by an increasingly large body of work showing that the NTS has segregated lines of specificity in controlling sensory and pre-motor visceral information.

Characteristics

Anatomical and Neurochemical Description

The NTS is a Y shaped nucleus whose caudal pole is at the level of the pyramidal decussation and rostral pole is at the lower limit of the facial nucleus (VII cranial nerve). The lateral limits of the NTS are adjacent to the substantia reticulata, its medial portions are fused caudally to form the subnucleus commissuralis. Moving rostrally, the NTS abuts the ►area postrema (AP) medially and, rostral to AP, the walls of the fourth ventricle. The ►dorsal motor nucleus of the vagus (DMV) is located ventral to the NTS while the nucleus of the hypoglossus (XII cranial nerve) is located medially to NTS but only in regions rostral to the AP. Dorsally the NTS is separated from the nucleus gracilis and, rostral to AP, from the nucleus cuneatus by their respective fasciculi.

Separate subnuclei can be distinguished within the NTS of several animal species, but these subnuclei are more useful in defining the general regions of the NTS rather than its neuronal subclasses, neurochemical characteristics, visceral representations or projection targets. In general, the visceral projections sites within the NTS reveal a rostro-caudal pattern reflecting the head-to-toe location of the viscera; the gustatory (i.e., tongue) NTS, for example, is uppermost, while the pharyngeal, cardiorespiratory and esophageal areas are somewhat overlapping at levels spanning both rostrally and caudally to the AP. The subdiaphragmatic viscera are represented in the more caudal portions of the NTS, where the distal gastrointestinal projections are denser. A loose viscerotopic organization is also recognizable along the medio-lateral extent of the NTS; moving from lateral toward medial NTS, the afferent receptive fields of the tongue, soft palate and pharynx, lungs, esophagus, baro- and chemoreceptor nerves, stomach and distal intestines, respectively, are located. Reviewed in [1–3].

The NTS integrates the inputs received from viscera with the inputs it receives from the spinal cord, the spinal trigeminal nucleus (V cranial nerve), ventrolateral medulla, raphe nuclei, several components of the reticular formation, A5 and A6 areas, parabrachial and Kolliker Fuse nuclei, dorsal tegmental regions including the periaqueductal gray, and bed nucleus of the stria terminalis. From the forebrain, the NTS receives descending input from a large and interconnected complex of neurons in the paraventricular and lateral hypothalamic areas and the central nucleus of the amygdala. Significant cortical projections also come from the medial prefrontal and insular regions. These regions projecting to the NTS are richly interconnected and the NTS itself maintains reciprocal connections with practically all of the abovementioned areas.

Neurons in the NTS are of medium size (10–15 μm diameter) and can be distinguished morphologically into multipolar (or stellate) neurons with 3-four dendrites exiting the soma or bipolar neurons with two dendrites only exiting the soma at opposite poles. These neuronal dendrites can extend outside the boundaries of the NTS itself to make contact with adjacent nuclei or the ependymal layer of the fourth ventricle. No apparent morphological or biophysical characteristic can be correlated with certainty to a specific physiological role of any given neuron, although recent works by MC Andresen and RA Travagli's groups has started to investigate this possibility.

The NTS contains a large array of membrane receptors and an even larger variety of neurochemical phenotypes, from "classical" neurotransmitters such as glutamate, GABA, glycine, serotonin, nitric oxide and catecholamines to an assorted content of neuropeptides or neuromodulators such as neuropeptide Y, enkephalins, cholecystokinin, somatostatin, glucagon-like peptide-1, etc. Peculiar to NTS, compared to other brain regions, is the apparently fundamental role in the modulation of neuronal NTS activity played by co-transmitters, in particular peptides such as substance P and CGRP, released from vagal afferent fibers. Similar to previous description, no specific physiologic role can be attributed to any given cell containing a particular neurochemical phenotype, although the localization of neuronal groups with a similar phenotype is, in some instances, very localized. Catecholaminergic neurons, for example, are confined to the A2 area whereas cholecystokinin-containing neurons are located in the caudal portions of the NTS.

Physiological Description

Fibers from the trigeminal (V cranial nerve), facial (VII cranial nerve), glossopharyngeal (IX cranial nerve) and vagus (X cranial nerve) form the solitary tract (or tractus solitarius) and their terminals overlap along the rostro-caudal extent of the NTS. Afferent fibers belong to the myelinated A- and unmyelinated C-groups. Regardless of their origin or type, though, afferent fibers enter the brainstem in the tractus solitarius and activate NTS neurons via the release of excitatory neurotransmitters, mainly glutamate.

The majority of NTS neurons do not possess pacemaker activity, i.e., they do not have spontaneously occurring action potentials; thus, to convey the sensory information to other areas, NTS neuronal projections must be driven and are modulated by synaptic activity, either from the afferent fibers of the tractus solitarius, from other neuronal areas, or via circulating hormones. In fact, we have to consider that the NTS is a circumventricular organ with a leaky blood brain barrier, fenestrated capillaries and enlarged perivascular space that allows the passage of large molecules, including circulating hormones and neurotransmitters. Additionally, the adjacent area postrema, which lies entirely outside the blood brain barrier, has a series of short, communicating vessels that potentially send postremal venous drainage to the NTS. These morphological characteristics allow NTS neuronal activity to be open to modulation by a wide variety of transmitter and hormonal peptides, including, among many others, glutamate itself, cholecystokinin, leptin, and ATP.

By being subject to a vast array of modulatory activity, the synaptic connections between NTS and the motor neurons controlling cardiorespiratory, pharyngo-esophageal and gastrointestinal functions, by implication, play a major role in shaping the efferent output. Further, recent studies suggested a potential role of NTS neurons in the integration of nociceptive signals and cardiorespiratory afferent fibers.

Role of NTS in

►Swallowing: NTS neurons behave as pattern generators in the initiation and organization of the sequential or rhythmic motor pattern of swallowing.

Application of excitatory aminoacids (glutamate) or antagonism of inhibitory amino acids (GABA) in NTS induces sequential and rhythmic motor pattern of swallowing. Swallowing can also be evoked by cortical stimulation, however, this is abolished by lesions of the NTS indicating that the NTS is the main central system responsible for swallowing. Since batches of NTS neurons are activated sequentially during swallowing, it is likely that distinct neuronal subgroups control different regions of the swallowing canal. Integrated swallowing information from the NTS is transmitted mainly to the nucleus ambiguus, whose motoneurons innervate the esophagus, pharynx, and intrinsic laryngeal muscles [4].

►Gustatory: fibers from the facial (VII cranial nerve) and glossopharyngeal (IX cranial nerve) carry gustatory afferent fibers to the rostral portions of the NTS. Although morphologically distinct, the gustatory-related NTS neuronal subtypes all receive synaptic inputs that use excitatory (glutamate) or inhibitory (GABA) amino acids as the main neurotransmitters. Gustatory-related NTS neurons project rostrally to the pontine gustatory relay area, to the reticular formation and to motoneurons of cranial nerves V (trigeminal), VII (facial), IX (glossopharyngeal), X (vagus) and XII (hypoglossus) [5].

►Cardiorespiratory: most cardiovascular afferent fibers and practically all fibers from the aortic depressor nerve and the carotid sinus nerve converge on the dorsomedial portion of the NTS. Electrical or pharmacological excitation of NTS neurons mimics baroreceptor activation and induces a decrease in heart rate, blood pressure and sympathetic nerve activity. Baroreflexes are blocked by pressure application of glutamate antagonists in the NTS, while GABA antagonists increase blood pressure, indicating glutamate and GABA as major players in the reflexive cardiovascular control from the NTS. Cardiovascular-related neurons of the NTS provide the premotor integration to neurons of the caudal ventrolateral medulla and neurons of the nucleus ambiguus, which control sympathetic and parasympathetic output, respectively [6].

Fibers innervating the slowly- and rapidly-adapting stretch receptors and bronchopulmonary C-fibers innervating the lungs and airways terminate in the caudal lateral NTS where they target different neuronal subtypes, suggesting that each afferent fiber may contribute to several different components of the respiratory phases. Excitatory (glutamate) and inhibitory (GABA and glycine) amino acids are the main neurotransmitters used by fiber terminals onto NTS neurons. Respiratory-related NTS neurons project to areas involved in rhythm- and pattern-generating neurons of the pre-Botzinger and Botzinger regions, to cranial and bulbo-spinal premotor neurons involved in respiratory pattern formation and to other neuronal areas devoted to a further integration with ingestive, cardiovascular, orofacial, airway protective and pain pathways [7].

►Gastrointestinal: mechanical or chemical manipulations of the proximal gastrointestinal tract activates NTS neurons, mainly via a vagal-mediated release of excitatory aminoacids such as glutamate. Localized pressure applications of glutamate directly into the medial NTS can evoke rapid, large, and vagally mediated gastric relaxations similar to those evoked by stimulation of vagal afferent fibers. The inhibition of gastric motor function occurs probably via activation of GABAergic or catecholaminergic NTS neurons projecting to the efferent motoneurons of the DMV. Antagonism of inhibitory amino acids (GABA) in NTS induces a dramatic increase in gastric motor functions indicating the presence of a robust GABAergic tone that dampens vagal gastrointestinal motor output [3].

►Feeding: signals that relate to ingestive behavior are generated by the interaction of food with chemo- and mechano-sensors in the alimentary canal before and during absorption. Signals related to satiation (i.e., short term feedback signals) and meal size are detected by sensory afferent neurons terminating in the NTS, which orchestrates the basic features of reflexive ingestion and gastrointestinal needs. Since afferent fiber terminals use excitatory amino acids (glutamate) as their main neurotransmitter, blockade of ionotropic glutamate receptors delays satiation and increases meal size. Circulating hormones and robust hypothalamic projections contribute to the NTS control of the integration of food intake [8].

In summary, neurons of the NTS provide the framework for the hardware responsible for coordinating vital homeostatic responses; emerging evidence shows that NTS neurons are functionally distinct and comprise an important station in segregated lines of specificity controlling sensory and pre-motor visceral information. In its static form, one may presume that activation of any given neuronal circuit within the NTS elicits the same hard-wired efferent response. Recent studies have, however, revealed a high degree of plasticity in available responses, such as, for example, in the control over the stomach or airway circuits [9,10]. One agonist signal may "gate" another, and the tonic effects of afferent input may "gate" agonist responses that are related to the "►state of activation" of a particular NTS circuit.

References

1. Blessing WW (1997) The lower brainstem and bodily homeostasis. Oxford University Press, Oxford
2. Barraco R, El-Ridi M, Parizon M, Bradley D (1992) An atlas of the rat subpostremal nucleus tractus solitarius. Brain Res Bull 29:703–765

3. Travagli RA, Hermann GE, Browning KN, Rogers RC (2006) Brainstem circuits regulating gastric function. Annu Rev Physiol 68:279–305
4. Jean A (2001) Brainstem control of swallowing: neuronal network and cellular mechanisms. Physiol Rev 81:929–969
5. Bradley RM, King MS, Wang L, Shu W (1996) Neurotransmitter and neuromodulator activity in the gustatory zone of the nucleus tractus solitarius. Chem Senses 21:377–385
6. Andresen MC, Kunze DL (1994) Nucleus tractus solitarius – gateway to neural circulatory control. Annu Rev Physiol 56:93–116
7. Kubin L, Alheid GF, Zuperku EJ, McCrimmon DR (2006) Central pathways of pulmonary and lower airway vagal afferents. J Appl Physiol 101:618–627
8. Berthoud HR, Sutton GM, Townsend RL, Patterson LM, Zheng H (2006) Brainstem mechanisms integrating gut-derived satiety signals and descending forebrain information in the control of meal size. Physiol Behav 89:517–524
9. Bonham AC, Chen CY, Sekizawa SI, Joad JP (2006) Plasticity in the nucleus tractus solitarius (NTS) and its influence on lung and airway reflexes. J Appl Physiol 101:322–327
10. Browning KN, Travagli RA (2006) Short-term receptor trafficking in the dorsal vagal complex: an overview. Auton Neurosci 126–127:2–8

Null Mutation

Definition

Mutation to induce no expression of a certain gene.

Equivalent to knockout of gene.

Nyquist Sampling Theorem

Definition

The sampling rate needed in order to be able to reconstruct a sampled signal back to its original (continuous) form, without loss of information, should be sampled at least twice as fast as the highest frequency component in the signal.

►Signals and Systems

Nystagmus

Definition

The rhythmic, alternating movement of the eyes, usually consisting of a rapid or saccadic movement in one direction followed by a slower, smooth movement of the eyes in the opposite direction. Nystagmus is designated based on the direction of the rapid eye movement.

►Cerebellar Functions
►Saccade

Nystagmus – Optokinetic (OKN)

Definition

Nystagmus produced by movement of the visual surround relative to a subject. In primates, the slow phase velocity rises sharply and maintains a steady state value throughout the rotation.

►Velocity Storage

Nystagmus – Pathological

Definition

Pathological nystagmus occurs at rest and can be caused by disorders in the brainstem, cerebellum or vestibular system, and is associated with vertigo and dizziness.

►Disorders of the Vestibular Periphery
►Ischemic Stroke

Nystagmus – Per-Rotatory Vestibular

Definition

Nystagmus produced during prolonged rotation of the head in space. For rapid initiation of rotation, the slow phase velocity rises sharply at the start of rotation and declines to zero as the rotation continues at a constant velocity.

► Velocity Storage

Nystagmus – Post-Rotatory Vestibular

Definition

Nystagmus produced upon stopping after long term rotation (>10 s) of the head in space. Its characteristics are the same as those of per-rotatory vestibular nystagmus.

► Velocity Storage

Nystagmus – Vestibular

Definition

Vestibular nystagmus results from an asymmetry in the resting activity between the two labyrinths. The nystagmus that occurs in cases of unilateral vestibular hypofunction has fast and slow components. The slow components are directed toward the labyrinth that has diminished activity (hypofunction).

► Disorders of the Vestibular Periphery

Obesity

Definition

Increased body weight due to excessive amounts of fat; partly due to genetic disposition.

Object Perception

Definition

►Form Perception

Object-based Attention

Definition

Object-based attention refers to mechanisms by which an entire object is selected for further processing. Object-based attention will spread to all features of the selected object. For instance, if subjects perform a task on the shape of an object, the processing of the color of the object will also be enhanced. In addition, if subjects are attending to only a portion of the object, attention will spread over the entire spatial extent of the object (see Visual attention).

►Visual Attention

OBPs

►Odorant Receptor

Observability

Definition

Observability represents the ability to detect (observe) physical system behavior by means of the sensors connected to it. In terms of system states, it is the ability to infer state values using sensor outputs.

►Control

Observational Learning

Definition

Learning that occurs as a function of observing, retaining and replicating behavior observed in others. Observation[al] learning is a looser concept than imitation learning. Social learning is a type of observation[al] learning. Motivation, attention and/or simple paring of novel stimuli may contribute to this process. Thus even invertebrates such as octopus can perform observation[al] learning (Florito and Scotto, 1992).

►Imitation Learning

Obsessive-compulsive Disorder [OCD]

Definition

A neurotical disease (in DSM-IV: anxiety disorder) in which the mind is flooded with persistent and uncontrollable thoughts or the individual is compelled to repeat certain acts again and again, causing significant distress and interference with everyday functioning.

►Personality Disorder

Obstructive Sleep Apnea

SIGRID C. VEASEY
Center for Sleep & Respiratory Neurobiology, Department of Medicine, School of Medicine, University of Pennsylvania, Philadelphia, PA, USA

Synonyms

Sleep apnea; Obstructive sleep apnea hypopnea syndrome; Obstructive sleep-disordered breathing

Definition

►Apnea: a cessation in ventilatory airflow lasting for 10 s or longer.

►Sleep apnea: A diverse group of disorders with a common feature of repeated sleep-dependent cessations in airflow.

►Obstructive sleep apnea: A syndrome in which repeated cessations in airflow occur as a direct consequence of sleep-dependent collapse of the upper airway. The syndrome is characterized by sleepiness, fatigue and snoring.

Characteristics

Overview

Obstructive sleep apnea is a rapidly evolving syndrome. For decades, snoring was believed to be more of an inconvenience for the bed partner than a sign of significant disease. Recent studies, however, have clearly established that snoring in association with unrefreshed sleep is a warning sign of obstructive sleep apnea, and obstructive sleep apnea is now widely recognized as an independent risk factor for significant cardiovascular and neurological morbidities. Therefore, obstructive sleep apnea is a disorder for which high clinical suspicion, early diagnosis, and effective intervention are of utmost importance.

Pathogenesis

A unique feature of disorders of sleep apnea is that the brief cessations in ventilation occur exclusively in sleep. In wakefulness, neural mechanisms ensure continued respiration. In sleep, both ►non-rapid-eye-movement (NREM) and/or ►rapid-eye-movement (REM) sleep, ventilatory drive may fall sufficiently to allow apneas to develop. The obstructive nature of events occurs in part because ►NREM and ►REM sleep are associated with reductions in muscle activity, with a greater decline in ►pharyngeal muscle activity than reduction in pump muscle activity [1]. This occurs as a normal physiological response to sleep in all individuals. However, individuals with obstructive sleep apnea rely upon specific muscles surrounding the pharynx to stent open the airway for ventilation. Because the ►oropharynx is a highly collapsible tube, and one that may be collapsed from any direction, a number of pharyngeal muscles must act in concert to stent open the pharynx. These pharyngeal muscles include the tongue (►genioglossus), ►soft palate muscles, and muscles that stiffen the posterior wall or extend the lateral walls of the pharynx. In sleep, reductions in upper airway dilator muscle tone result in collapse of the upper airway [1]. Collapse of the upper airway is most likely to occur during inspiration, when negative pressures are generated in the lumen of the oropharynx [1]. Occlusion of the upper airway results in cessation of airflow, or apnea, that in turn results in ►hypercapnia, hypoxia and stimulation of upper airway afferents, resulting in arousal and resumption of the necessary upper airway muscle activity to reopen the upper airway. This process can be repeated up to 100 times/h of NREM and REM sleep. In many individuals, the slee p-dependent events involve a reduction in flow, rather than complete cessation of flow. A sleep-related reduction in flow, associated with a drop in oxygen saturation and an arousal, is termed a ►hypopnea. These events disrupt sleep and oxygenation and are considered as clinically significant as ►apneas. One of these events is illustrated in Fig. 1, showing a reduction in airflow just at sleep onset, result in a drop in oxygen saturation and arousal.

The intermittent hypoxia and frequent arousals from both apneas and hypopneas can increase sympathetic activity and induce inflammatory and smooth muscle changes in vessels, exacerbating hypertension and ►atherosclerosis. In addition, recent studies suggest the intermittent hypoxia can result in irreversible cognitive impairments and neural injury.

Epidemiology

Symptomatic obstructive sleep apnea is present in 4–7% of adult males and 2–3% of adult females in North America, Europe and in select regions of Asia [2]. Despite the 2:1 male:female predominance in adults, there are no apparent gender differences in ►OSA prevalence in pre-pubertal children. Prevalence increases with age and is estimated to approach 40% in elderly individuals. While prevalence does not vary with race, the severity of OSA is greater in age-matched African American individuals than in Caucasians, and it appears that OSA develops at an earlier age in African Americans than in Caucasians. Familial aggregation has been established for OSA. Approximately 40% of the variance in the apnea hypopnea frequency may be explained by familial factors, and a positive family history increases the relative risk by 2- to 4-fold [2]. How much of this variance is simply ►obesity remains to be determined [2].

►Craniofacial anomalies that compromise upper airway space and stability provide additional risk

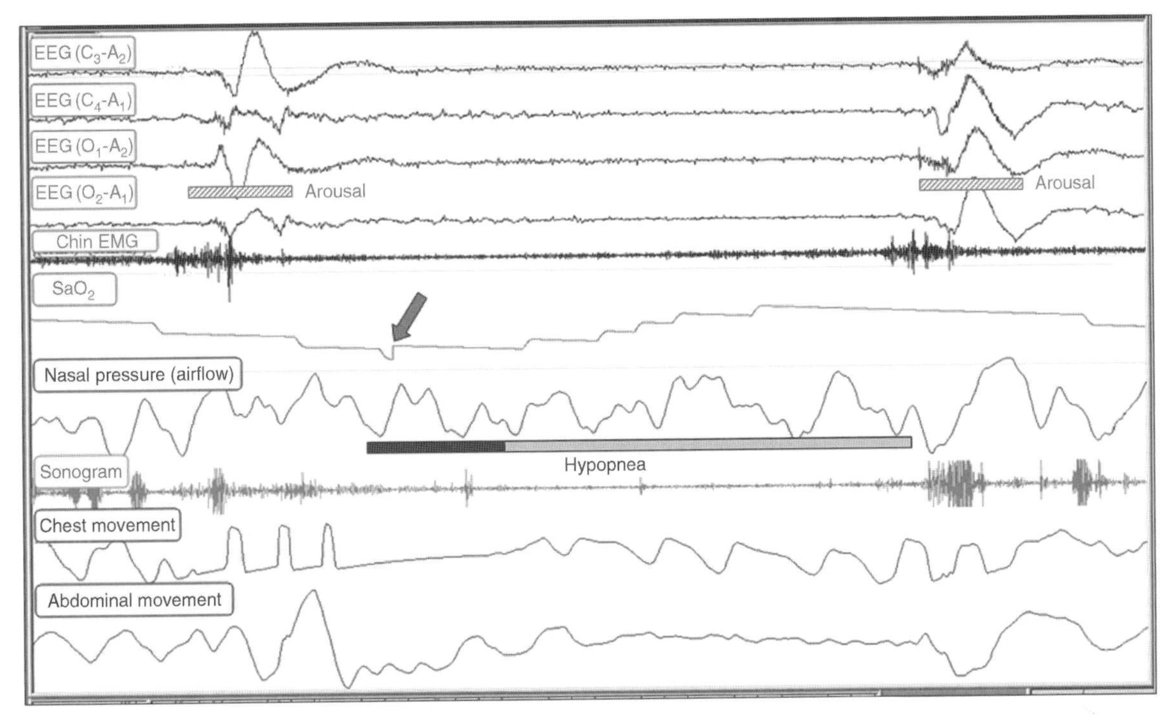

Obstructive Sleep Apnea. Figure 1 Polysomnography for the diagnosis of obstructive sleep apnea. The electroencephalographic activity across the frontal, parietal and occipital cortices, is present in the top four channels. Channel 5 shows the chin elctromyogram. The arterial saturation (SaO_2) is presented in channel 6, and airflow measured with a nasal pressure transducer, the snore signal and chest and abdominal movements are shown below. An arousal from one hypopnea, underlined by the first red bar, is rapidly followed by sleep onset an another hypopnea. The hypopnea terminates with the second arousal, underlined by a red line. Notice the SaO_2 appears to fall just after the arousal. This is attributed to a delay in circulation time required for detection of the peripheral signal. Notice the snoring is quiet across the hypopnea when little airflow is exchanged. Polysomnography typically records four additional channels for leg movements, electrocardiogram and eye movements. These signals were removed to highlight the respiratory events.

factors for OSA. For example, ►micrognathia in ►Treacher-Collins syndrome and ►Pierre-Robin syndrome and ►maxillary insufficiency in ►Down and ►Apert syndromes predispose to collapsible upper airways. ►Hypothyroidism and ►acromegaly increase tongue soft tissue that impinges upon the oropharynx. The most common risk factor for OSA, however, is obesity, defined in adults as a body mass index >30 kg/m^2. In children, the major risk factor for OSA has been enlarged ►adenoid and ►tonsillar tissue. With the increasing prevalence of childhood obesity, obesity is becoming the major risk factor for OSA in children as well as in adults. A minority of patients diagnosed with OSA is non-obese. In these cases, chronic nasal obstruction from allergies, ►polyps or ►septal deviation, or craniofacial variances, e.g., ►retrognathia or ►macroglossia, may contribute to the increased collapsibility of the upper airway and OSA. In summary, the most important risk factor for OSA is obesity. Nonetheless, it is important to recognize that OSA occurs in diverse groups of patients, and in light of the significant neurobehavioral and cardiovascular morbidities, it is important to explore the possibility of OSA in all individuals presenting with snoring and unrefreshed sleep or poor sleep quality.

Presentation and Diagnosis

The presence of snoring and excessive sleepiness or fatigue despite adequate time allowed for sleep (7–9 h) in adults should prompt evaluation for OSA. It is important to recognize that in females and in children the presentation may be less straightforward. In both women and children, the snoring may be subtle, and rather than sleepiness, adult females may complain of fatigue or insomnia. Children are far more likely to exhibit increased motor activity, poor performance in school, and/or impulsive behavior than to have daytime sleepiness. Snoring alone does not make a good screening tool. In the United States, habitual snoring is present in 40% of adult males, 20% of adult females and 10% of children. Thus, it is essential to identify associated neurobehavioral complaints in snorers, e.g., unrefreshed sleep, fatigue, insomnia, restless sleep or irritability before proceeding with diagnostic testing.

The physical exam in many individuals in which a clinical suspicion for OSA is raised will suggest upper airway compromise. A neck circumference (>17 in. in adult males or >15 in. in adult females, a low lying soft palate, a small space behind the soft palate, large tonsils, a small mandible, and thickened lateral walls of the oropharynx all suggest increased upper airway collapsibility. Figure 2 shows a typical oropharynx in a person with mild OSA. There is no obvious obstruction.

The tonsils and tonsillar pillars are somewhat medial and may result in lateral wall collapse in sleep, but it is entirely possible that the point of initial collapse is lower in the airway or caused by retrograde placement of the tongue in sleep. Because airway examination occurs with the patient awake, a clear obstruction is unlikely to be identified.

The gold standard diagnostic tool for OSA is ►polysomnography. Polysomnography refers to the recording of multiple physiological signals during sleep [3]. Electroencephalographic and electromyographic signals are used to score specific sleep stages, as described elsewhere in this text. Airflow is measured indirectly with use of a thermistor, or with a nasal pressure transducer, and chest and abdominal movements are measured using piezosensors or strain gauges. Arterial oxygenation is recorded with ►pulse oximetry. An example of 30 s polysomnographic recording is presented in Fig. 1.

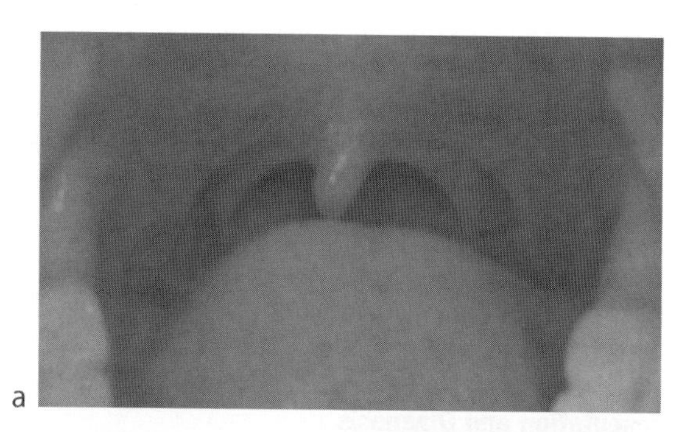
a

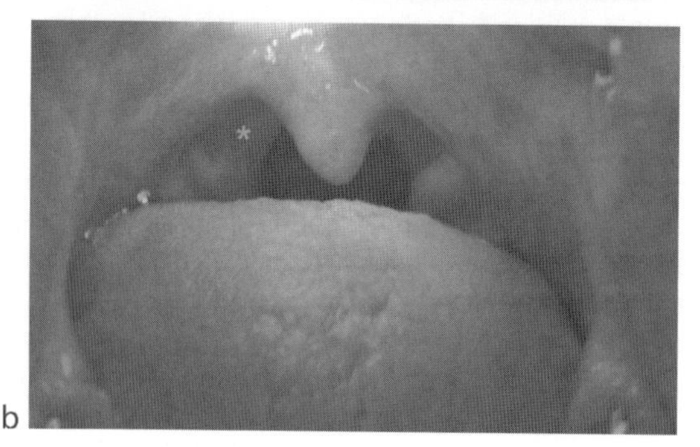

b

Obstructive Sleep Apnea. Figure 2 Upper airway physical findings. (a) top panel shows a normal wide oropharynx. (b) In this individual with mild OSA, the tonsils are only mildly enlarged and the soft palate (uvula) is readily visible with some lateral wall narrowing at the tonsillar pillars (*).

Complimentary channels include channels to detect leg movements or snoring and the electrocardiogram. The diagnosis of OSA in adults requires >5 apneas or hypopneas/h, on average, across sleep with symptoms, as above [3]. In children, neurobehavioral symptoms and an ►apnea index >1 is sufficient for the diagnosis. Presently, the majority of polysomnographies are performed in clinical sleep laboratories; however, because obesity is on the rise and the clinical suspicion for OSA is heightened, it is anticipated that there will be a shift in the near future towards the implementation of simpler, more cost-effective screening tools for OSA.

Treatment

The primary goal of therapy for OSA is to prevent collapse of the upper airway. The mainstay therapy for OSA is a remarkably effective mechanical therapy; ►positive airway pressure (PAP) titrated to an optimal pressure in each individual can fully prevent collapse of the upper airway in all stages of sleep in almost all patients with OSA [4]. Each individual with OSA will require a unique pressure to stent open her upper airway across all of NREM and REM sleep. The pressure needed will vary with sleep stage (NREM sleep vs. REM sleep) and with position and with nasal obstruction and sleeping position [4]. All of these factors must be taken into consideration when identifying the optimal pressure to alleviate OSA. Thus, a properly performed titration must confirm that apneas and hypopneas are alleviated in all sleep stages, all sleeping positions and that sleep is less fragmented. The latter ensures that subtle events have also been prevented. Prescribed pressures typically vary between 5 and 15 cm H_2O. Although remarkably effective for OSA, PAP therapy is cumbersome, requiring a tightly fitted mask over the nose and/or mouth. Figure 3 shows one of the newer PAP interfaces that allows an individual improved visibility for reading prior to sleep onset.

Despite advancements in mask comfort and PAP delivery, less than half of the individuals prescribed PAP regularly use this therapy. Nonetheless, every effort should be made in individuals to encourage use of PAP regularly, as this is the only therapy for OSA shown to lessen cardiovascular and neurobehavioral morbidity. For patients with claustrophobia and other mask difficulties, behavioral therapy to adjust to mask use has been shown highly effective. Recent developments in PAP therapy include machines that can self-adjust the level of PAP based on airflow patterns, and these, too, may increase usage in select groups of patients [4].

Alternative therapies for OSA should be considered in individuals with mild sleep apnea and in individuals unable to acclimate to PAP use. These alternative

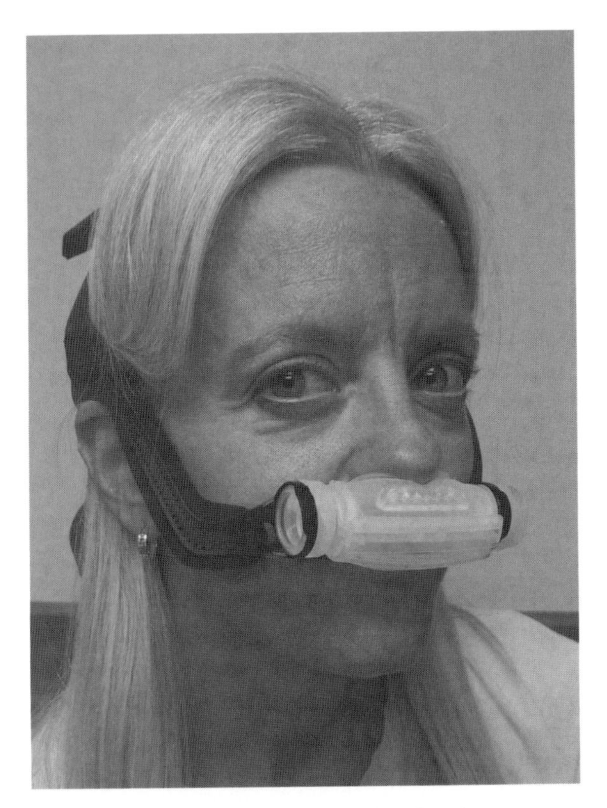

Obstructive Sleep Apnea. Figure 3 Continuous positive airway pressure interface. This system is designed to deliver positive airway pressure to the nares and allow improved visibility. Flexible tubing connects the nasal mask to a small air pump to deliver positive pressure. Newer machines have the capability of detecting snoring, apneas and hypopneas, hours of usage and mask leaks.

therapies include surgical procedures to shorten the soft palate and reduce collapsibility of the pharynx (►uvulopalatoplasty), or to reduce the tongue volume (►genioglossectomy) or to advance the genioglossus forward (genioglossus advancement hyoid myotomy). These therapies in select groups of patients are expected to improve OSA in 50% of patients [5]. In patients with persistent symptomatic OSA, a second phase of surgery may be necessary to increase pharyngeal space (maxillary advancement or maxillary and ►mandibular osteotomy). Laser-assisted uvulopalatoplasty and temperature-controlled radio frequency are most effective for benign snoring. Some patients who do not tolerate PAP or in whom OSA is mild may benefit from oral appliances that advance the mandible. As with surgical therapies, the oral appliances are most likely to work in individuals with mild disease.

Weight loss should be recommended in all obese individuals with OSA. Dietary counseling should be the first step taken, and all patients should understand that reduced caloric intake is the critical factor for successful weight loss. Behavioral modification programs enhance sucess of weight loss. Exercise may help maintain weight, but in most non-athletic individuals, healthy caloric restriction should be the primary strategy for weight loss. ►Bariatric surgery should be reserved for individuals with morbid obesity who have failed dietary weight loss programs. The majority of individuals who have substantial weight loss after bariatric surgery will experience marked reductions in OSA, if not lasting reversal of the disease [6]. Treatment of OSA in persons with hypothyroidism or acromegaly should begin with PAP therapy, but across the treatment of the underlying endocrine disorder the PAP settings may need adjusting, as the soft tissues remodel. Several medical therapies for OSA may be considered as second line therapies for mild OSA. There may be subsets of individuals who respond to supplemental oxygen, positional therapy and rarely to pharmacotherapies such as selective serotonin reuptake drugs in individuals with mild ►REM sleep-predominant apnea [7]. Because these adjunctive therapies are rarely fully effective, treatment success should be determined with repeated polysomnography.

Stimulant therapy to reduce residual sleepiness in treated OSA has been recently examined [8]. The effect size for objective sleepiness is small, and it should be understood that individuals with residual sleepiness remain at high risk for motor vehicle accidents.

Associated Morbidities

One of the most important advances in OSA has been the substantiation of OSA as an independent risk factor for cardiovascular, endocrine and neurological morbidities. OSA is now widely accepted as an independent risk factor for several cardiovascular diseases, including hypertension, congestive heart failure, and stroke [9]. Importantly, the relative risk for hypertension increases even at levels of mild OSA (5–15 events/h), and use of PAP therapy can reduce this risk. The rates of significant cardiovascular events across 10 years in a large prospective European trial were found to be fourfold larger in untreated vs. treated untreated OSA [10]. The risk of cardiovascular death is also reduced with PAP therapy in persons with severe OSA [10]. Children with OSA show left ventricular dysfunction and increased levels of circulating inflammatory markers associated with atherosclerosis [9]. The mechanisms are poorly understood, but contributing factors include increased sympathetic activity, endothelial inflammation and ►oxidative stress [9]. In light of the seriousness of morbidities and the disease interactions associated with OSA, even in children, every effort to treat OSA effectively should be made. There have been several recent reports suggesting that OSA is an independent risk factor for insulin resistance. This risk persists after controlling for obesity, and several studies have demonstrated improvement in glucose control and insulin sensitivity with successful use of PAP therapy. Whether long-term PAP therapy reduces the occurrence of complications of diabetes remains to be studied. Several recent reports suggest that OSA may impair liver function and might contribute to non-alcoholic fatty liver disease, a major risk for liver failure in developed countries. OSA is an independent risk factor for motor vehicle crashes, raising the relative risk by 2.5-fold, and a direct link between OSA and motor vehicle accidents is supported by the reduction in car crash risk with successful treatment of sleep apnea.

Future Directions

►Obstructive sleep apnea is now widely accepted as a serious disorder, associated with significant morbidity. The importance of recognition and treatment of obesity in children and young adults is critical for reducing the prevalence of this disorder. For the millions of individuals with undiagnosed OSA, there is a readily appreciable need to improve screening methodologies to dramatically increase availability. PAP is a remarkably effective therapy and efforts to improve its acceptance must continue, while we await the development of effective pharmacotherapies.

References

1. Remmers JE, deGroot WJ, Sauerland EK, Anch AM (1978) Pathogenesis of upper airway occlusion during sleep. J Appl Physiol 44(6):931–938
2. Palmer LJ, Redline S (2003) Genomic approaches to understanding obstructive sleep apnea. Respir Physiol Neurobiol 135(2–3):187–205
3. Redline S, Budhiraja R, Kapur V, Marcus CL, Mateika JH, Mehra R, Parthasarthy S, Somers VK, Strohl KP, Sulit LG, Gozal D, Wise MS, Quan SF (2007) The scoring of respiratory events in sleep: reliability and validity. J Clin Sleep Med 3(2):169–200
4. Basner RC (2007) Continuous positive airway pressure for obstructive sleep apnea. N Engl J Med 356(17):1751–1758
5. Elshaug AG, Moss JR, Southcott AM, Hiller JE (2007) Redefining success in airway surgery for obstructive sleep apnea: a meta analysis and synthesis of the evidence. Sleep 30(4):461–467
6. Fritscher LG, Canani S, Mottin CC, Fritscher CC, Berleze D, Chapman K, Chatkin JM (2007) Bariatric surgery in the treatment of obstructive sleep apnea in morbidly obese patients. Respiration 74(6):647–652
7. Veasey SC, Guilleminault C, Strohl KP, Sanders MH, Ballard RD, Magalang UJ (2006) Medical therapy for obstructive sleep apnea: a review by the Medical Therapy for Obstructive Sleep Apnea Task Force of the Standards of Practice Committee of the American Academy of Sleep Medicine. Sleep 29(8):1036–1034
8. Santamaria J, Iranzo A, Ma Montserrat J, de Pablo J (2007) Persistent sleepiness in CPAP treated obstructive sleep apnea patients: evaluation and treatment. Sleep Med Rev 11(3):195–207
9. McNicholas WT, Bonsigore MR (2007) Sleep apnoea as an independent risk factor for cardiovascular disease: current evidence, basic mechanisms and research priorities. Eur Respir J 29(1):156–178
10. Marin JM, Carrizo SJ, Vicente E, Agusti AG (2005) Long-term cardiovascular outcomes in men with obstructive sleep apnoea-hypopnoea with or without treatment with continuous positive airway pressure: an observational study. Lancet 365(9464):1046–1053

Obstructive Sleep-disordered Breathing

►Obstructive Sleep Apnea

Occipital Cortex

Definition

The posterior part of the cerebral cortex.

Occipital Lobe

Synonyms

Lobus occipitalis

Definition

Extends from the occipital pole to the parietooccipital sulcus.

►Telencephalon

Occlusal Table

Definition

The space between the upper and lower teeth.

►Tactile Sensation in Oral Region

Occlusion

Definition
Artificial increase in low-frequency level produced by blocking the ear canal.

►Hearing Aids

Occlusion in Audition

Definition
Artificial increase in low-frequency level produced by blocking the ear canal.

►Hearing Aids

Octaval Nuclei

Definition
Primary hindbrain recipient targets for inner ear afferents. This complex of nuclei may be homologous (in whole or in part) with the mammalian cochlear nuclei complex.

►Evolution of Mechanosensory and Electrosensory Lateral Line Systems

Octave

Definition
The ratio between two sound frequencies of two.

►Acoustics

Octavolateralis System

Definition
A set of sensory organs, both mechanosensitive and electrosensitive, in aquatic vertebrates that are innervated by the eighth cranial nerve and by the lateral line nerves. More specifically: the sense of hearing, the sense of equilibrium, the sense of rotation, the mechanosensitive lateral line system, and the electric sense.

►Electroreceptor Organs
►Evolution of the Mechanosensory and Electrosensory Lateral Line Systems

Octopus Cells

Definition
Typical neuron of the posteroventral cochlear nucleus (PVCN) that receive small auditory nerve terminals on their dendrites and project to the ventral nucleus of the lateral lemniscus.

►Cochlear Nucleus

Ocular Abduction

Definition
Horizontal movement of the eye away from the nose.

Ocular Counter-rolling Response

Definition
Conter-rotation of the eyes about the optic axis, i.e., torsion, during an imposed head or body tilt to the right or to the left about the naso-occipital axis (see also "VOR-tilt VOR").

►Vestibulo-Oculomotor Connections
►Vestibulo-Oculomotor System: Functional Aspects

Ocular Dominance

Definition
The degree to which one eye dominates a given neuron in the visual pathway or the perception of a scene.

►Binocular Vision

Ocular Drift Movements

Definition

Involuntary, smooth, and mostly slow, eye movements that do not correspond to a target movement. Some types of drift occur predictably in certain behavioral contexts such as: glissades in the aftermath of saccades, anticipatory drift in the direction of an imminent target movement, centripetal drift during the attempt to maintain an eccentric eye position in darkness. Others are predominantly random such as the miniature drifts during fixation with velocities of the order of 0.1°/ s which can cause deviations from the intended fixation point of up to 0.2°, or the slow wanderings of the eyes during drowsiness which result in considerably larger excursions.

►Oculomotor Control
►Saccade, Saccadic Eye Movements

Ocular Following Responses (OFR)

Definition

Smooth eye movement elicited by optic flow from relative motion between observer and visual scene (or parts thereof) in a highly automatic manner (unconscious reaction, no instruction required) and at short latency (70–80 ms). It often is initiated by a series of brief acceleration peaks creating a mean acceleration of up to $100°/s^2$; considerably larger values are achieved in the aftermath of saccades, though. OFR is considered to be part of the early or direct component of the optokinetic reflex.

►Oculomotor Control

Ocular Micromovements

Definition

Involuntary movements occurring during fixation consisting of (i) tremor, (ii) slow drifts and (iii) microsaccades. Tremor and drifts are uncorrelated in the two eyes whereas microsaccades have the same direction and similar – though not identical -amplitudes in both eyes. As a result of these micromovements, the line of sight describes an erratic, two-dimensional path about the intended fixation point.

►Oculomotor Control
►Saccade, Saccadic Eye Movements

Ocular Motoneurons

Motoneurons that innervate the ocular muscles.

►Evolution of Oculomotor System

Ocular Muscles

Muscles that move the eye in the orbit.

►Evolution of Oculomotor System

Ocular Tremor

Definition

Involuntary ocular micromovement occurring during fixation and consisting of waxing and waning irregular oscillations with frequencies between 70 and 90 Hz and mean amplitudes of about 0.002°.

►Oculomotor Control

Oculocentric Frame of Reference

Definition

Also, "Retinotopic frame of reference." A frame of reference centered on the eyes and moving with them.

►Eye Movements Field

Oculo-manual Synergy

►Eye-Hand Coordination

Oculomotor

►Evolution of the Vestibular System

Oculomotor Cerebellum

Definition

Usually refers to the medial parts of the cerebellum that regulate the generation of saccadic and smooth-pursuit eye movements.

►Cerebellum, Role in Eye Movements
►Saccade, Saccadic Eye Movements
►Smooth Pursuit Eye Movements

Oculomotor Control (Theory)

Wolfgang Becker
Sektion Neurophysiologie, Universität Ulm, Albert-Einstein-Allee, Ulm, Germany

Definition

The theory of oculomotor control aims at metaphorically understanding which types of innervation patterns are required to generate the various types of eye movements (►saccades, ►reflexive saccades, ►micro-saccades, ►express saccades, ►corrective saccades, ►pro-saccades, anti-saccades, ►catch-up saccades, smooth pursuit, vergence, fixation), and how afferent (mostly visual and vestibular) and efferent information is processed to shape these patterns (sensori-motor transformation). The metaphors it uses mostly draw on control systems theory and are referred to as models; typically, the modeling approach disregards the intricacies and variety of the neural substrates, lumping many of them into a small number of processing stages with either mathematically or empirically defined transfer characteristics between input and output. Processing stages interact by way of signals which can represent a flow of neural activity along axons or physical parameters such as position or velocity. Formerly, models had to be simple to be amenable to mathematical analysis, whereas nowadays the behavior of very complex structures can be rapidly determined by simulation software.

Characteristics

Description of the Theory

Common Characteristics of Visual Eye Movement Control

A prototypical scheme of visual eye movement control using a high level of abstraction is shown in Fig. 1. In particular, this and the following schemes do not explicitly show the bilateral symmetry of the oculomotor system and the elaborate push-pull interactions of its constituent elements; rather, they represent the net effect of these interactions.

Visually controlled eye movements aim at bringing the retinal image of visual target objects into the foveal area and at stabilizing them there. The basic information available to this end is *R,* the target's retinal eccentricity with respect to the fovea. *R* reflects the difference between target (*T*) and eye (*E*) position, and represents the current error in eye position; because of the "built-in" retroaction of *E* on *R*, that is, on the very signal it is reacting to, visual eye movement control constitutes a *negative feedback* system and is said to be *closed loop*. *R* is processed by a number of parallel, semi-independent pathways that perform the visuo-motor transformations required for the various types of visually controlled eye movements. In Fig. 1, these pathways are symbolized by dashed signal paths, while the typical structure of one of them is shown in more detail as a reference for the further description.

R first must be detected and processed by the visual system to obtain the information (e.g., error velocity) based on which the *controller* can generate an error-correcting motor command. Detection and processing of R, but also the operation of the controller and of other stages, require considerable time. These delays can be lumped into a single delay time (d) that represents the latency of the eye's response to a change of target position or velocity.

Interestingly, for all types of eye movements, including those controlled by non-visual signals (e.g., vestibular), the primordial motor commands issued by their respective controllers appear to specify *eye velocity* rather than eye position. In Fig. 1, these commands are shown to converge at a summing junction whose output represents a compound eye

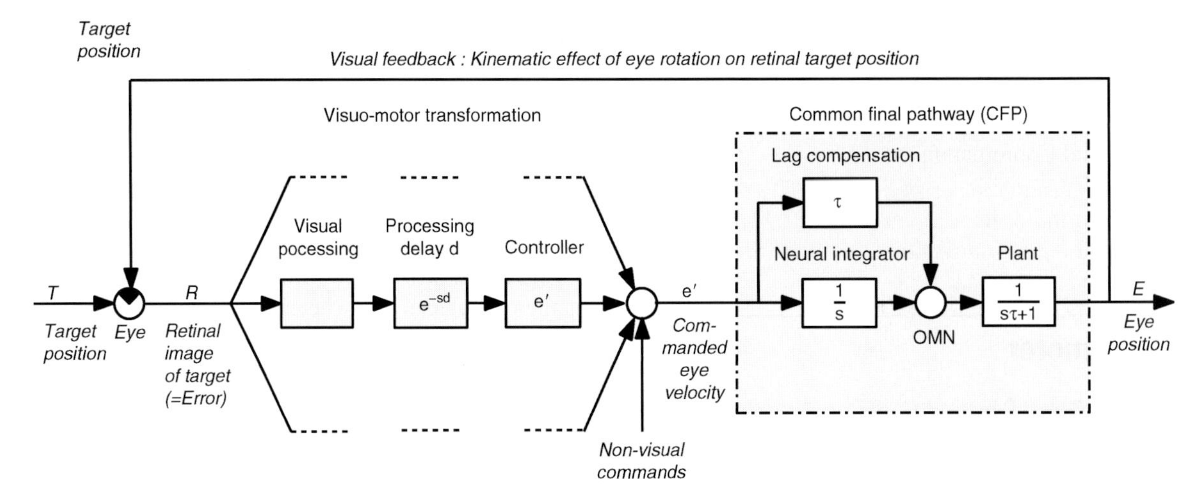

Oculomotor Control (Theory). Figure 1 Basic structure of visual eye movement control. Italicized text and symbols denote *signals*; symbols beginning with lower case denote neural activity; upper case refers to physical quantities. Normal print describes functions and identifies the various elements of the scheme; symbols inside boxes describe global transfer characteristics of these elements by Laplace transforms (s, complex frequency; τ, time constant of plant).

velocity command (e'). This signal is converted into an eye position command by a stage that calculates its time integral. The substrate of this so-called *neural integrator* (NI) has been located to the medial vestibular nucleus and the nucleus prepositus hypoglossi and their reciprocal connections with the vestibulo-cerebellum. NI is also being referred to as *hold integrator* because it is responsible for holding the eyes at whatever position they have been brought to by a preceding, but now gone, e'-signal. NI has been the target of interesting attempts to explain integration in terms of a network of neurones that excite themselves via reciprocal connections to neighboring neurones [1].

The output of NI reaches the oculomotor nuclei (OMN); in the case of horizontal eye movements the abducens nucleus (nVI) in the first place, from whence it is forwarded to the rectus medialis complex of the contralateral oculomotor nucleus (nIII). OMN, in turn, send the position command to the extraocular eye muscles. The mechanical compound consisting of these muscles, the eye ball, and its connective tissue is collectively referred to as *plant*. The dynamics of the plant is dominated by visco-elastic forces, while the mass of the globe plays a minor role. As a first order approximation it can be described by a first order lag system with time constant $\tau = 150$–200 ms. Thus, a step increase of OMN activity causes the eye to exponentially approach the position coded by this step, with fairly sluggish creeping in the final phase. To overcome this sluggishness, there is a direct projection of e' to OMN, which adds a velocity component to the position command obtained from NI. This combination of position and velocity components becomes particularly clear during saccades, where a *pulse-step pattern* of innervation is observed in OMN.

Theoretically, if the gain of the direct projection assumes the numerical value of τ, the compound labeled "Common final pathway" in Fig. 1 behaves like an ideal integrator which accurately converts e' into eye position E, and for many purposes such a simplification is an acceptable approximation. The term *common final pathway* (CFP) for the aggregate consisting of NI, lag compensation, OMN, and plant reflects the belief that all velocity commands – visual and non-visual, saccadic and smooth – are processed by the same integrator and that their direct projections all converge at the same pool of motoneurones. This notion is a useful approximation for many purposes but should not be overvalued. Already at the level of the extraocular eye muscles, the occurrence of different types of muscle fibres raises the suspicion of a functional division according to, for example, fast and slow eye movements. Also, as yet there is no agreement as to how far ►vergence movements share the integrator for ►conjugate eye movements or use separate pathways.

Controllers

The oculomotor system's closed loop character combined with its considerable delay time (d = 100–200 ms) causes a major complication: If its response to a change of target position or velocity is to be accurate and fast (i.e., not much longer than d), the gain of its controller – essentially the ratio E/R – must be large. On the other hand, long delays combined with a large gain cause instability (oscillations) in a closed loop system. Different strategies have been developed by the saccadic and smooth pursuit systems, to arrive at a viable compromise between response velocity and accuracy on the one side and stability on the other side.

Saccades

Basically, the control of saccades [2,3] (Fig. 2) can be likened to the operation of a sample-and-hold system: The error $R(t_0)$ existing at time t_0 is measured by the visual system and transferred to a memory and a decision stage. After a processing delay (d), the decision stage triggers a neural *pulse generator* which emits a high-amplitude, short-duration pulse of neural activity, whose mathematical integral approximately equals the value of $R(t_0)$ held in memory. Fed into CFP, this pulse acts as the saccadic velocity command e' and produces a fast, ramp-like movement of duration d_S – the saccade. At the end of this saccade, the sequence of events repeats: the now existing error $R(t_0 + d + d_S)$ is again visually measured and, if non-zero, corrected by a further saccade. In this way the execution proper of saccades is not visually controlled, but *open-loop* with respect to visual feedback. It is, however, thought to be under *local feedback* control: a copy of the e'-command sent to CFP would be fed into a neural replica of CFP – basically an integrator – whose output therefore images current eye displacement (cE) without incurring visual delays. Subtraction of cE from the *desired eye displacement* held in memory (dE) yields the current *motor error* (mR) which indicates how far the eye still has to go. If switch OPN is closed, mR drives a pool of burst neurones (BN, located in the vicinity of nIV and in the mesencephalon) which emit the e'-pulse. The pulse, therefore, would last until $mR = 0$ (implying $\int e'dt = dE$, as desired). BN activity would vary linearly with small mR but level off with large mR, thus determining the well-known non-linear relationship between saccade velocity and amplitude. Switch OPN provides for the discrete nature of saccades; to close it, a trigger impulse representing an "explicit" decision for a saccade is required, and in order to remain closed BN activity must entertain a "latch" signal. Therefore, as mR approaches zero and silences BN, the switch opens making the loop refractory for new motor errors until a new decision is issued. In the mean time, both the replica of CFP and the memory would be reset to zero (the former is often referred to as "*resettable integrator*"). A likely substrate of switch OPN are the omnipause neurones in the pontine raphe, which inhibit BN during fixation and are silenced before BN activity starts.

Closer experimental analysis of the saccadic system has revealed important features that are not covered by the basic scheme in Fig. 2 (i) The decision for a saccade and the desired amplitude (dE) put into memory are not necessarily based on the same sample $R(t_0)$; rather, dE reflects some kind of average of $R(t)$ from the interval $t_0 < t <$ d-80 ms. (ii) Processing of successive saccades does not always occur in a strictly serial manner but can considerably overlap in time. (iii) Saccades are not completely open loop but can to some degree be modulated by concurrent visual events. (iv) A local (i.e., non-visual) feedback signal also reaches the visual processing stage, where it anticipates the visual consequences of ongoing (and perhaps even of as yet only decided-upon) saccades before the visual afferents can signal the resulting change of R.

Two other aspects not dealt with by the scheme in Fig. 2 are (i) The co-ordination between the eyes and the

O

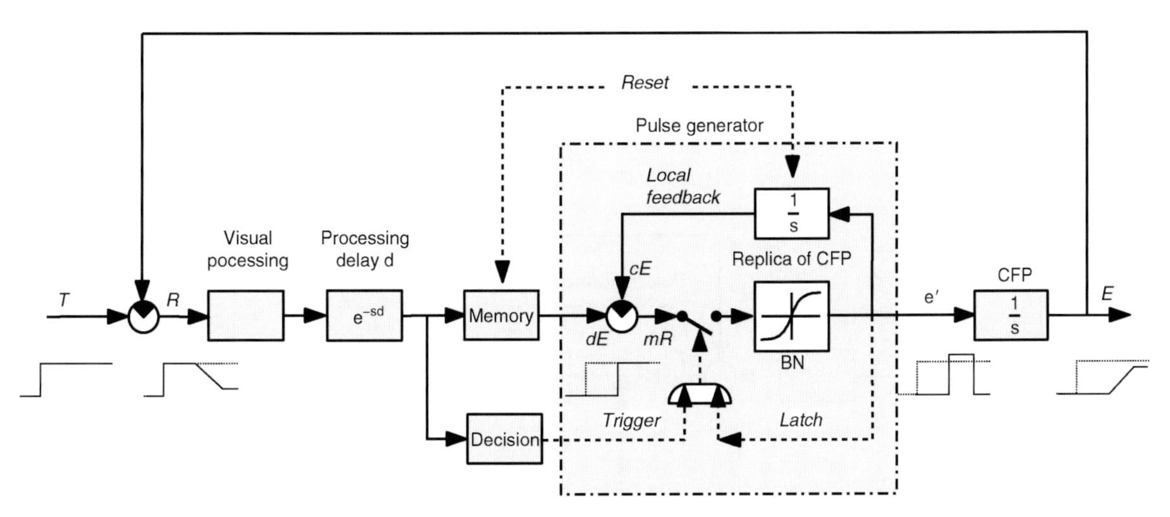

Oculomotor Control (Theory). Figure 2 Structure of saccadic eye movement control. *dE*, desired eye displacement; *cE*, copy of current eye displacement; *mR*, motor error; BN, burst neurones (icon sketches non-linear relation between motor error and magnitude of *e'*); OPN, switch disabling burst neurones during fixation; other symbols and conventions as in Fig. 1. *Insets* show time course of signals *T, R, dE, e'* and *E* in relation to a step of *T* (*dashed*).

head during natural gaze shifts, which are generally executed with the head moving in support of the eyes; several expansions and variations of the basic structure have been proposed in which vestibular mechanisms play a crucial role for this co-ordination [4]. (ii) Whereas small displacements in 2D-space can be essentially accounted for by two orthogonal systems of the type sketched in Fig. 2, the laws of spherical geometry require that not only *R* but also *E* be taken into account when creating the *e′*-signal for large displacements between arbitrary positions.

Smooth Pursuit Eye Movements (SPEM)

Basically, two alternative control structures are being discussed to account for the experimentally observed characteristics of SPEM [5] (Fig. 3). The *error-driven model* (also called *image motion model* or *closed-loop model*; Fig. 3a) posits that the motor output is driven exclusively by the current error in the way implied by Fig. 1. In accordance with SPEM's function of stabilizing the image of moving targets on the retina, first the current *error velocity r′* (*retinal slip* of target image) is extracted from *R*. A second differentiating stage also calculates error acceleration *r″*. Signals *r′* and *r″* are then combined by weighted summation to obtain a signal representing the desired eye acceleration ($dE'' = g_v \cdot r' + g_a \cdot r''$) which in turn is converted, by integration in the controller proper, into the eye velocity command *e′* sent to CFP. The lumped effect of this processing is that $E'(t + d) = g_a \cdot R'(t) + g_v \cdot \int R'(t)dt$; therefore, it can be likened to that of a PI-controller with delay time d. Dependent on the relative weights of the proportional and the integrating contributions, such a system can oscillate at frequencies from 0.5/d ($g_v = 0$) to 0.25/d ($g_a = 0$); given d = 0.1 s, this corresponds to the range 2.5–5 Hz. The SPEM responses of man to sudden target movements indeed exhibit damped oscillations of 3.8 Hz; yet, there is no combination of g_v and g_a that would account at the same time for this frequency and other essential features of SPEM responses (e.g., rise time, steady state accuracy, and dependence on target velocity). For a satisfactory explanation of all relevant SPEM characteristics, several non-linear gain elements (saturating with increasing input) have to be inserted into the pathways preceding the I-controller of Fig. 3a.

The alternative approach, the *perceived velocity* (or *open- loop*) model (Fig. 3b), tries to reconcile the various characteristics of SPEM by assuming that SPEM oscillations are caused by an "inner," local feedback loop rather than by the "outer," visual loop.

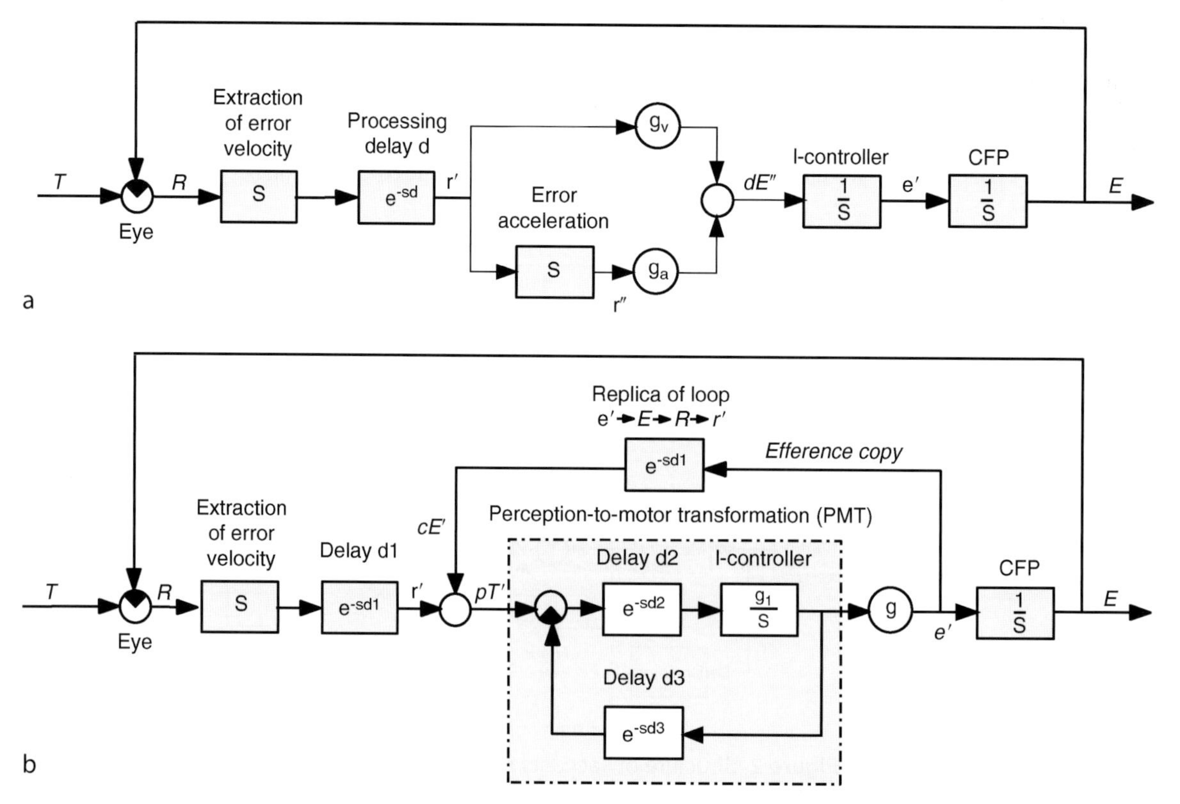

Oculomotor Control (Theory). Figure 3 Smooth pursuit control: (a) error-driven model; (b), perceived velocity model. Conventions as in Fig. 1. *r′* (*r″*) error velocity (acceleration); *dE″*, desired eye acceleration; *cE′*, delayed copy of eye velocity; *pT′*, perceived target velocity; *g*, gain coefficients. Other symbols as in Fig. 1.

The model posits that the target's velocity, as it existed one visual delay time (d1) earlier, i.e., T'(t-d1), is *reconstructed* using delayed neural representations of (i) the retinal slip: $r' = R'$(t-d1), and of (ii) eye velocity: $cE' = E'$(t-d1); their sum, $pT' = T'$(t-d1), represents the "perception" of T' by the SPEM-system and may also determine conscious perception of target velocity. cE' would be obtained from an efference copy of the velocity command e', passed through a neural replica of the pathway mediating the retroaction of e' upon r'; with the simplifying assumptions of Fig. 3, this replica reduces to delay d1 (since $1/s \cdot s = 1$). If perceived target velocity pT', is then translated one-to-one into the velocity command e', SPEM velocity will faithfully follow T' except for a delay. During steady state operation, the *perception-to-motor transformation* (PMT) stage, when envisioned as a local feedback loop with integrating controller, has indeed a gain of one. The fact that in most people tracking a target of constant speed SPEM is slightly slower than the target, can be accounted for by gain element g (e.g., $g \approx 0.9$); a value $g < 1$ also insures stable operation of the positive (or "regenerative") feedback loop through which the efference copy of e' entertains the perception of T'.

By adding cE' to r' a positive, non-visual loop is created which offsets the subtraction of E from T and, hence, functionally neutralizes the negative visual feedback around the outer loop. Thus, SPEM becomes a virtually open-loop, feed-forward response to target movement. Therefore, oscillations cannot arise from the system architecture as a whole, but only in its constituents; specifically, it has been suggested that the experimentally observed damped oscillations arise in the inner (PMT-) loop, with frequency determined by delay times d2 and d3, and amplitude by integrator gain g_I. However, to render all relevant characteristics of SPEM, the perceived velocity model also requires the addition of non-linearities. Furthermore, proponents of the closed loop model point out that it has difficulties in rendering the effects observed during artificial prolongations of the delay time.

Both models apply only to the pursuit of targets moving at constant speed. With periodically moving targets, very effective predictive mechanisms dominate behavior which can virtually eliminate the delay between target and eye and, therefore, require more sophisticated models.

Vergence Movements

The control of vergence movements (Fig. 4) differs from that of saccades and SPEM in several aspects: It must move the two eyes in opposite directions and it is not only driven by errors in eye position or velocity (here: by retinal disparity) but also by an input unrelated

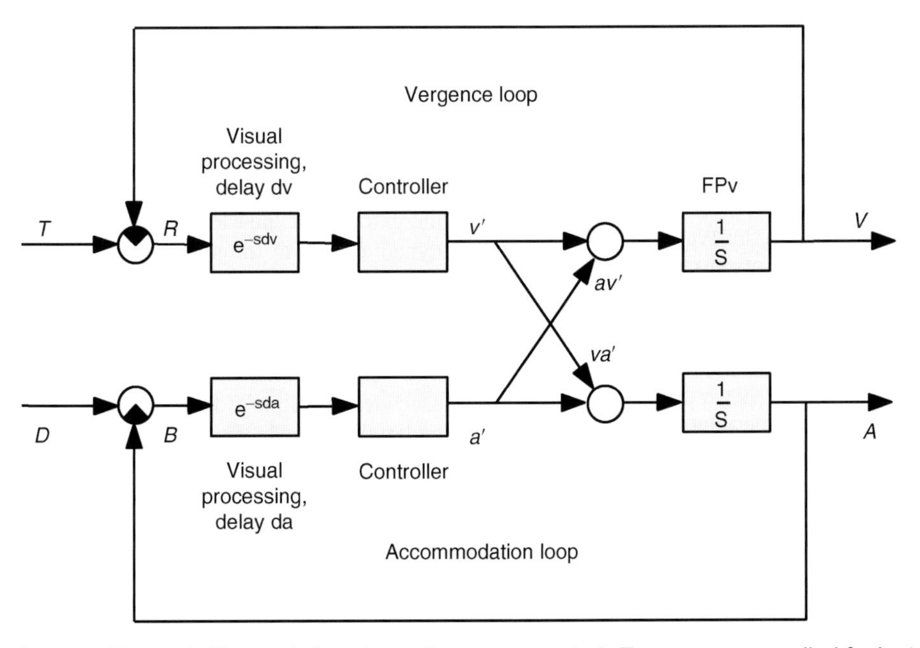

Oculomotor Control (Theory). Figure 4 Structure of vergence control. *T*, convergence called for by target; *R*, retinal disparity; *v'*, commanded velocity of vergence; *V*, vergence angle of eyes; *D*, target distance^{-1} (diopters); *B*, error in accommodation (blur); *a'*, commanded rate of change of accommodation; *A*, accommodation; *av'* and *va'A*, contributions of accommodation to vergence and vice versa (mostly denoted AC/A and CA/C in the literature); FPv, final pathway of vergence system (partially overlapping with CFP). Conventions as in Fig. 1. (For a broad synopsis of the use of models in oculomotor physiology see Carpenter RHS (1988) Movements of the eyes (2nd edition). Pion, London For examples of how models benefit the analysis of neuro-ophthalomological problems see Leigh RJ, Zee DS (1999) The Neurology of Eye Movements (3rd edition). Oxford University Press, New York, Oxford).

to eye position, namely the error in accommodation (retinal blur); the response to this input is known as *accommodative vergence*. As accommodation, in turn, is not only driven by blur but also by retinal disparity (*convergence accommodation*), two mutually coupled feedback circuits result with fairly similar constituent elements, except for a significantly larger delay (da) in the accommodative loop as compared to the vergence loop (dv). The situation is further complicated by the possibility that, much as with conjugate movements of the two eyes, there might be separate systems for pursuit vergence (tracking a target moving slowly in depth) and saccade-like vergence (called for by sudden changes of fusional demands) [6]; therefore, no details of the controller are specified in Fig. 4. As with other oculomotor subsystems, the controller signal reaches the plant both via a direct (velocity coding) and an integrating (position coding) pathway which, when lumped with the plant, could be roughly equated to an integrator (box FPv). Due to the disconjugate character of vergence, the integrating pathway cannot be identical to the neural integrator of the common final pathway (CFP) of conjugate movements, although it may partially overlap with it [7]; hence, the notion of a CFP is not applicable here in a strict sense. Finally, it is not clear whether the cross-coupling between vergence and accommodation occurs before the integration of the commanded vergence (v') and accommodation (a') velocities (as shown in Fig. 4), or thereafter [8].

References

1. Arnold DB, Robinson DA (1997) The oculomotor integrator: Testing of a neural network model. Exp Brain Res 113:57–74
2. Becker W (1989) Metrics. In: Wurtz RH, Goldberg ME (eds) The Neurobiology of Saccadic Eye Movements. Elsevier, Amsterdam, New York, Oxford, pp 13–67
3. Scudder CA, Kaneko CRS, Fuchs AF (2002) The brainstem burst generator for saccadic eye movements. A modern synthesis. Exp Brain Res 142: 439–462
4. Galiana HL, Guitton D (1992) Central organization and modeling of eye-head coordination during gaze shifts. Ann NY Acad Sci 656: 452–471
5. Churchland MM, Lisberger SG (2001) Experimental and computational analysis of monkey smooth pursuit eye movements. J Neurophysiol 86:741–759
6. Zee DS, Fitzgibbon EJ, Optican LM (1992) Saccade-vergence interactions in humans. J Neurophysiol 65:1624–1641
7. McConville K, Tomlinson RD, King WM, Paige G, Na EQ (1994) Eye position signals in the vestibular nuclei: Consequences for models of integrator function. J Vestib Res 4:391–400
8. Schor CM (1992) A dynamic model of cross-coupling between accommodation and convergence: Simulations of step and frequency responses. Optom Vis Sci 69:258–269

Oculomotor Dynamics

Charles Scudder
Portland, OR, USA

Synonyms

Oculomotor plant; Orbital dynamics

Definition

►Oculomotor dynamics are the properties of the oculomotor system that determine the time-course of the rotation of the eye in response to the discharges of ocular motoneurons. These properties are a product of the inertia of the eye, the viscoelastic properties of the tissue surrounding the eye, and the dynamic properties of the extraocular muscles that control its movements. These properties are usually described mathematically using differential equations or their equivalent (e.g. computer models).

Characteristics

Measurement of Oculomotor Dynamics

The time course of an eye movement is not a replica of the aggregate discharge rate of the ocular motoneurons, but is modified by oculomotor dynamics. The difference between the two can be quantified and used to describe oculomotor dynamics. This measurement is a composite of the three factors listed above. To interpret this measurement, it is also important to directly measure inertia, tissue viscoelasticity, or muscle properties in isolation using mechanical methods, as described below. Force transducers placed in series with the extraocular muscles have also helped to isolate the dynamics due to the muscles and the dynamics due to the eye and orbit [1,2].

The force produced during the generation of a saccade is illustrated in Fig. 1. The time course is divided into three components; the "pulse" that occurs during the saccade, the "slide" (decay in force) occurring after the end of the saccade, and the "step" (long-term force) that keeps the eye in a static position until the next eye movement.

Viscoelastic Properties of the Orbital Tissue

Rotation of the eye causes a displacement of the orbital tissue, such as the conjunctivum, Tenon's capsule, fat in the orbit, and the connective tissue of the extraocular muscles. These tissues resist rotation, their resistance displaying both an elastic and a viscous component. The elastic component provides a static restoring force that depends only on the angle of rotation away from straight ahead. This force increases with angular deviation nearly linearly over a range of angles, but

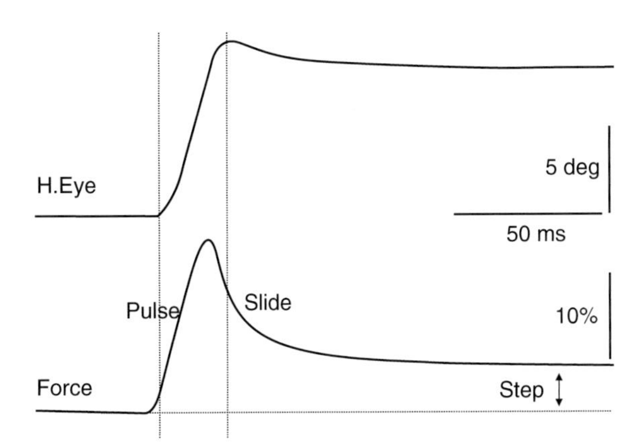

Oculomotor Dynamics. Figure 1 Recording of the horizontal eye position (H. Eye) and muscle tension (Force) in the lateral rectus muscle recorded during an abducting horizontal saccade. Tension increases slightly before saccade onset and peaks somewhat before saccade termination. This phase is frequently called the "pulse" of force because of the waveshape of the associated motoneuron discharge, and is responsible for producing the rapid velocity of saccades. This is followed by an initially rapid and then slow decline in force commonly called the "slide." Normally the eye would be stationary during this time, but the force transducer has caused a minor abnormality. Force never declines to its initial value, but rather, there is a persistent force (the "step") that holds the eye in its final abducted position. Force is expressed as a percentage of the maximum force developed by the muscle during any saccade, probably 50–60 g-force. Dotted lines mark saccade onset and termination. Figure modified from Miller & Robins, Fig. 9 [2].

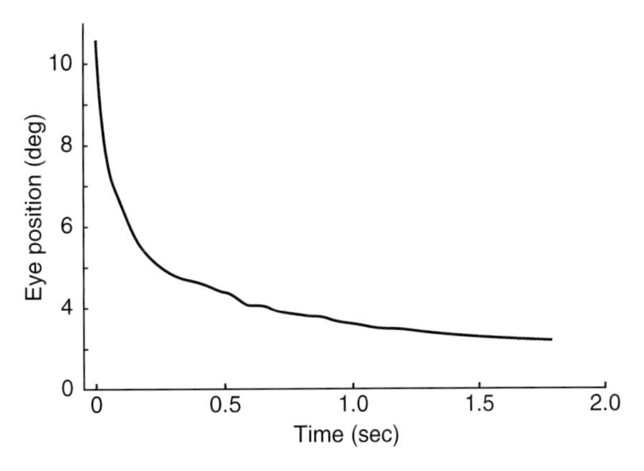

Oculomotor Dynamics. Figure 2 The return of eye position towards straight-ahead gaze after being released from an abducted position. Movement is initially very rapid, then slows down, and finally creeps toward a final position over several seconds (not shown). Data is replotted from Sklavos et al. [4].

increases more rapidly after about halfway to the maximum of natural eye movements (see ►Orbital mechanics). The viscous component resists an ongoing rotation of the eye with a force that is proportional to the velocity of rotation.

The mechanical method of measuring the viscoelastic forces is to pull on the eye with a constant tangential force, and measure the time course of the change in angular position. The eye rotates rapidly during the first few milliseconds, but progressively slows down and continues to move increasingly slowly over succeeding seconds. The process is characteristic of most tissue in the body and is known as "tissue creep" [3]. An equivalent experiment is illustrated in Fig. 2, where the eye is held in a static position and then released (isotonic force = 0). The eye rotates back towards straight ahead as described above. Technically, the change in position is described by the sum of an infinite number of exponentials with different ►time constants [3], but practically, a very good fit to the data can be obtained with a small number of exponentials. For the data in Fig. 1, one exponential accounts for 85% of the variance, two account for 98.5%, and four account for 99.9% [4].

For a quantitative description of the viscoelastic properties of the orbit, each exponential is modeled as the product of one "Voigt element," which is a spring (the elastic component with spring constant K) in parallel with a dashpot (the viscous element with viscosity R) as in Fig. 3a. Two exponentials are modeled as two Voigt elements in series, as in Fig. 3b.

Using a single Voigt element to describe oculomotor dynamics [5] has intuitive appeal because it requires a neural controller for eye movements having only two components. One is a velocity command, such as the burst of saccadic burst neurons (see ►MLBNs) or the discharge of vestibular afferents during the ►vestibuloocular reflex (►VOR) that is needed to overcome the viscosity of the orbital tissue. The second is a position command, thought to be obtained by integrating the velocity command (see ►Neural integrator), that is needed to overcome the elasticity of the tissue. However, this model cannot explain the presence of the slide (Fig. 1) or the frequency-dependent characteristics of motoneuron firing-rate modulation during sinusoidal ►smooth pursuit [6], and predicts an unrealistically high force to move the eye during a saccade [6]. Using two Voigt elements (Fig. 3b) greatly reduces all three problems, and is quite adequate for the didactic purposes of this article.

Muscle Dynamics

Force in the extraocular muscles varies with the number of motoneurons recruited and the firing rate of each motoneuron. Three factors contribute to the dynamics of force buildup (or decline) in the muscles; the "twitch

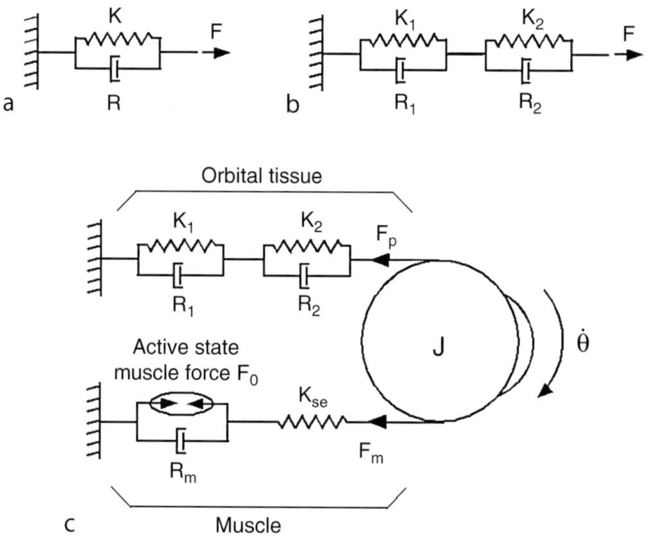

Oculomotor Dynamics. Figure 3 Components used in modeling oculomotor dynamics, including a nearly complete model. A single Voigt element (a) is composed of a spring with spring constant K (representing elastic restoring forces in the orbit) and a dashpot with viscosity R (representing the viscous, velocity dependent, properties of orbital tissue). A single Voigt element responds to a step change in force with a change in length fit by a single exponential having a time constant of R/K. The viscoelastic properties of the orbit are more accurately modeled by two (b) or more (not shown) Voigt elements in series. This model of the orbital tissue is shown attached to the eyeball (top of (c)), with a model of the muscle attached at the bottom. The parallel elastic component of the muscle has been lumped with the other orbital tissue. The active-state force generator (muscle crossbridges) is in parallel with a dashpot with viscosity R_m, which models the reduction in force F_m according to the ▶force-velocity relationship. The series elastic component (spring constant K_{se}) lengthens during the initial buildup of force at the onset of a saccade. For all practical purposes, the inertia of the eye (moment J) can be ignored, meaning that the magnitude of $F_p \approx F_m$.

time" of the muscle, the distributed recruitment of motoneurons over time, and the firing rate of the active motoneurons. These factors make little difference during slow eye movements (smooth pursuit, VOR), but a major one during rapid saccadic eye movements [6].

An action potential in a motoneuron and the muscle fibers it innervates produces a rapid buildup to a peak of force and then a gradual decline. The time to peak is called the "twitch time" [7], and is 5–7 ms in monkeys [8]. The finite twitch time is due to the fact that connective tissue and muscle proteins are springy (series elastic component [7]) in combination with the fact that rapidly shortening muscle develops less force (▶Force velocity relationship [7]; see ▶Muscle twitch).

During the repetitive firing of a motoneuron, the force of each twitch adds to the force that remains from the preceding twitches. At the start of repetitive firing, this superposition produces a cumulative force that builds up and saturates with a roughly exponential envelope whose ▶time constant decreases as the firing rate increases [9]. During an actual saccade, there is a complex interplay between the ▶extraocular motoneuron firing rates and the force-velocity and length-tension properties of the shortening muscle, which makes exact modeling difficult. In practice, it has proven satisfactory to approximate all these dynamics as a single spring and dashpot in series and parallel, respectively, with the active (force-producing) components of the muscle (F_o in Fig. 3c).

The bursts of repetitive firing in extraocular motoneurons occurring just prior to saccades do not start simultaneously, but their onsets are spread over 6–8 ms in monkeys [5,9] and could be longer in humans, who have slower saccades. The effect of this distributed recruitment is to slow the initial acceleration of the eye during saccades.

Inertia of the Eye

Calculations, modeling, and measurement all show that the force required to overcome the inertia of the eye is negligible during slow eye movements and is very small during saccades [6,10]. For all practical purposes, inertia can be ignored in models of oculomotor dynamics. However, the inclusion of an unrealistically high moment of inertia has been used in some modeling studies [11] to account for the discrepancy between the slow acceleration of the eye relative to the almost instant buildup of firing rate in single saccadic burst neurons (see ▶Burst cells – medium lead). This discrepancy, however, is the product of finite twitch times in the extraocular muscles and the spread of burst-neuron and motoneuron recruitment times in relation to saccade onset, with a minimal contribution from the inertia of the eye.

Cumulative Orbital Dynamics

A model of oculomotor dynamics is illustrated in Fig. 3c. The viscoelastic properties of the orbital tissue are illustrated at the top of the eye, and the muscle is illustrated at the bottom. The agonist and antagonist muscles, which are reciprocally innervated and have mirror-image force profiles [2], have been lumped together into one muscle. The "parallel elastic component" of the muscle, which is sometimes modeled as a separate element, has been lumped into the orbital tissue. This is consistent with the fact that the passive muscle has viscous as well as elastic properties that are measured with the other orbital tissues in the release experiments described above [4,10]. The moment of inertia is denoted as J. Treating J as negligible, the differential equation describing eye acceleration as a function of muscle force (F_m), rate of change of force, and the viscoelastic impedance is below:

$$\ddot{\theta} = \frac{1}{\mu}\left(F_m + T_s\dot{F}_m - K_o\theta - R_o\dot{\theta}\right)$$

Ks, Rs, and Ts are composite spring, rate, and time constants [6,10]. Muscle force (F_m) is the active-state force (F_o) reduced by the rate of change of force and eye velocity:

$$F_m = F_o - T_m\dot{F}_m - R_m\dot{\theta}$$

Equations that include inertia can be found in Robinson [10]. Values for all parameters can be found in references [4,6,10].

The interaction of all the dynamic elements will be illustrated for a saccade. To begin the saccade, motoneurons begin their bursts over a range of times leading to a gradual buildup in active-state tension. The buildup of force delivered to the eyeball (F_m) is further slowed by the dynamic properties of the muscle, as discussed above. This buildup is illustrated in Fig. 1 during the so-called "pulse" phase. Shortly after the onset of force, the eye begins to rotate. As the inertia of the eye is small, F_p is almost equal to F_m, reflecting that the primary impedance to motion is provided by the viscoelastic properties of the orbital tissue. At the end of the pulse, motoneuron firing rate drops rapidly at first, and then more gradually with a "slide" similar to that illustrated in the force trace of Fig. 1. In a normal eye (without a force transducer), the eye would stop moving at this point. The decline in muscle force compensates for the decline in the reactive force in the orbital tissues as they "creep" to a new steady state. In terms of the model in Fig. 3, the Voigt element with the faster time constant was initially stretched disproportionately, and relaxes during the slide as the Voigt element with the slower time constant stretches. At the end of the slide, which can take several seconds in the actual eye or a model with more than two Voigt elements, there is a residual force (the "step" in Fig. 1) that is needed to maintain the eye at its new position against the just stretched elastic components of the orbital tissue.

References

1. Collins CC, O'Meara D, Scott AB (1975) Muscle tension during unrestricted human eye movements. J Physiol (London) 245:351–369
2. Miller JM, Robins D (1992) Exraocular muscle forces in alert monkey. Vis Res 32:1099–1113
3. Fung YC (1993) Biomechanics; mechanical properties of living tissues. Springer-Verlag, New York
4. Sklavos S, Porrill J, Kaneko CRS, Dean P (2005) Evidence for wide range of time scales in oculomotor plant dynamics: implications for models of eye-movement control. Vis Res 45:1525–1542
5. Robinson DA (1970) Oculomotor unit behavior in the monkey. J Neurophysiol 33:393–404
6. Fuchs AF, Scudder CA, Kaneko CRS (1988) Discharge patterns and recruitment order of identified motoneurons and internuclear neurons in the monkey abducens nucleus. J Neurophysiol 60:1874–1895
7. Aidley DJ (1998) The physiology of excitable cells. Cambridge University Press, Cambridge
8. Fuchs AF, Luschei ES (1971) Development of isometric tension in simian extraocular muscle. J Physiol (London) 219:155–166
9. Fuchs AF, Luschei ES (1970) Firing patterns of abducens neurons of alert monkeys in relationship to horizontal eye movement. J Neurophysiol 33:382–392
10. Robinson DA (1964) The mechanics of human saccadic eye movement. J Physiol (London) 174:245–264
11. van Gisbergen JAM, Robinson DA, Gielen S (1981) A quantitative analysis of generation of saccadic eye movements by burst neurons. J Neurophysiol 45:417–442

Oculomotor Nerve (III)

Synonyms

N. oculomotorius (N.III)

Definition

The oculomotor nerve is a motor cranial nerve endowed with both somato- and visceromotor components, for which one complex is responsible in each case. Together with the trochlear nerve (IV) and abducens nerve (VI) it controls eye movements.

It is involved in the lateral and medial eyeball movements (lateral rectus muscle and superior oblique muscle), raising of the palpebra as well as accommodation (ciliary muscle) and adaptation (sphincter muscle of pupil). Skull: superior orbital fissure.

►Nerves

Oculomotor Nucleus

Definition

A nucleus which contains both motoneurons and interneurons. The motoneurons send direct projections to all extraocular muscles except for the superior oblique muscle and the lateral rectus muscle.

Oculomotor Plant

►Eye Orbital Mechanics
►Oculomotor Dynamics

Oculomotor Systems

►Evolution of Oculomotor System

Oculomotor Vermis

Definition

The circumscribed portion of the cerebellar vermis (lobules VIc and VII) that appears to be integral to the control of saccadic and smooth-pursuit eye movements. The time course of changes in eye position or the firing rate of neurons can sometimes be described mathematically by an exponential, $X = X_0e^{-(t/T)}$, where X is the position or firing rate, X_0 is the initial value of X, t is time, and T is the "time constant" of the exponential. The time constant is equivalent to the amount of time required for X to decay to 36% (1/e) of X_0.

►Cerebellum, Role in Eye Movements
►Saccade, Saccadic Eye Movements
►Smooth Pursuit Eye Movements

Odor

Martina Pyrski, Frank Zufall
Department of Physiology, University of Saarland School of Medicine, Homburg/Saar, Germany

Synonyms

Odor; Odorant; Olfactory cue; Smell; Scent, Aroma

Definition

"Odor" refers to an emanation composed of multiple different odor molecules termed odorants, whose individual chemical properties are perceived by the sense of smell. In humans, this term is frequently used to describe a sensation as a result of odor perception, for example the pleasure resulting from the floral smell of roses (good odor) or the disgust following the smell of spoiled food (bad odor).

Characteristics

In contrast to the senses of vision, hearing and touch, the chemical senses - smell (and taste) - are challenged by an enormous number of molecularly distinct stimuli. Natural odors derived from food and plants and social stimuli, such as those present in urine, sweat and saliva, represent complex mixtures that contain a multitude of chemically diverse compounds. The information contained in these molecules is detected and processed by the sense of smell, a sensory modality that emerged very early in the evolution of living forms. Detection of olfactory cues is initiated by interaction of odor molecules with specific receptors located in the cellular membrane of olfactory sensory neurons in the nasal epithelium. The initial chemical odor information is then translated into neuronal activity patterns and subsequently converted into perceived odor quality and behavioral responses as a result of pattern recognition and evaluation by the brain.

Odor Detection in Mammals Occurs Through Multiple Olfactory Subsystems

In vertebrates, the cellular, molecular and genetic mechanisms underlying odor detection and the sense of smell are probably best understood in the mouse olfactory system. Odor detection begins in the olfactory sensory neurons (OSNs) located in the main olfactory epithelium (mOE) of the nasal cavity. Volatile odor molecules enter the nasal cavity with each breath and dissolve in the mucus covering the epithelial surface, a process that may be facilitated by small carrier molecules or odor binding proteins. The next step is a direct contact of odor molecules with the olfactory cilia which emanate from the dendritic knob of each OSN (Fig. 1).

These cilia contain all the necessary components for odor detection and subsequent chemo-electrical signal transduction. The electrical output signal produced by each OSN travels along a single axonal projection toward the main olfactory bulb (mOB) in the forebrain, the first relay station of odor processing in the brain. The axons from several millions of OSNs coalesce to form the olfactory nerve, also known as 1st cranial nerve.

In addition to a main olfactory system, most mammals have evolved an accessory olfactory (or vomeronasal) system (Fig. 1), which is anatomically and functionally distinct from the main system. Odor detection in the accessory olfactory system begins in the paired vomeronasal organ (VNO), located ventrally at the base of the nasal septum and rostral to the mOE. Odor stimuli are actively transported into the lumen of the VNO by a vascular pumping mechanism. The sensory epithelium of the VNO covers the inner medial side of each tube and, in analogy to the mOE, contains vomeronasal sensory

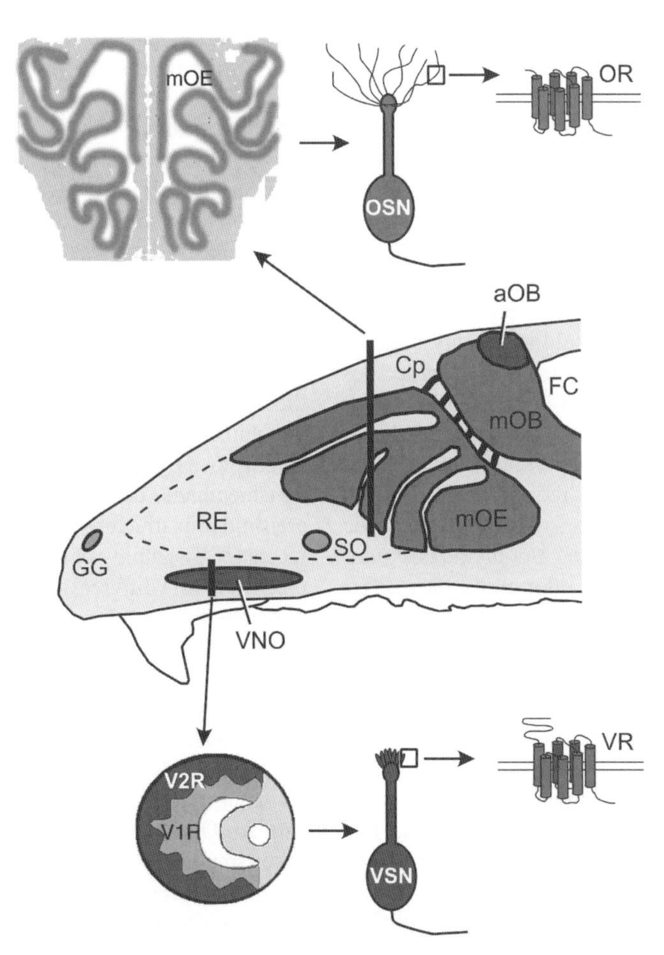

Odor. Figure 1 Schematic of a hemisected head of a mouse (sagittal view) illustrating the anatomical location of different olfactory subsystems and key structures. *Cp* cibriform plate; *FC* frontal cortex; *GG* Grüneberg ganglion; *mOB* main olfactory bulb; *RE* respiratory epithelium; *SO* septal organ of Masera; The black bar in the main olfactory epithelium (mOE, red) refers to the coronal section at the top left (arrow) that depicts the bilateral symmetry of the mOE. Olfactory sensory neurons (OSNs, red) contain numerous cilia that carry odor receptors (OR, red) of the GPCR type. The black bar in the vomeronasal organ (VNO, blue) refers to the coronal section shown at the bottom left with V1Rs and V2Rs expressed in the apical (light blue) and basal (dark blue) halves of the vomeronasal sensory epithelium, respectively. Vomeronasal neurons (VSN, blue) carry numerous microvilli that express vomeronasal receptors (VR, blue) of the GPCR type.

neurons (VSNs). These extend microvilli instead of cilia towards the lumen of the VNO. VSN axons project to the accessory olfactory bulb (aOB) located posterior and dorsal to the mOB (Fig. 1).

The traditional distinction that the mammalian main olfactory system recognizes general odor molecules and the vomeronasal system detects pheromones is no longer valid. The emerging picture is that both systems have considerable overlap in terms of the chemosignals they detect and the effects that they mediate [1]. Other, functionally less well characterized olfactory subsystems in rodents comprise of the septal organ of Masera and the Gruneberg ganglion (Fig. 1). Finally, some odor molecules such as menthol and phenylethyl alcohol can be detected by free nerve endings of the 5th cranial nerve which are part the somatosensory system. These nerves are often sensitive to pain as well as temperature stimuli and terminate in the nasal cavity.

Odor Molecules

The olfactory environment is estimated to comprise hundreds of thousands of structurally distinct compounds that potentially can be detected and discriminated by the olfactory system. These odor molecules are classified by several means, most commonly by the presence of specific physical and chemical properties or encoded odor quality, but also by the characteristics of the corresponding receptors, resulting activity patterns in the brain, and function. Typical odor molecules of air-breathing species are small hydrophobic chemicals of organic origin with a molecular weight of less than 300 Da, i.e., they are volatile at ambient temperature. In aquatic animals, requirements for odor molecules are different, with non-volatile, hydrophilic compounds like amino acids being among the best odor ligands identified. Chemically, odor molecules differ by many parameters including size, functional groups, 3D-structure, and flexibility. They encompass the whole array of aliphatic acids, alcohols, aldehydes, ketones, and esters. To the human nose, changes of functional groups can cause pronounced differences in perceived odor quality, e.g., octanoic acid has the smell of sweat whereas the structurally related aldehyde octanal (Fig. 2) has the smell of oranges. The presence of functional groups is not always a prerequisite for odor. Alkenes such as 2,4,4-trimethylpentane and cyclooctane both have pronounced camphor quality as a consequence of molecular shape.

Further chemical features that are subject to olfactory discrimination include differences in carbon bond branching and saturation, as well as substitutions by aromatic, alicyclic, polycyclic, and heterocyclic ring structures or halogens in numerous possible positions. For some substances, substitutions can be exchanged without altering odor quality, e.g., exchanging the aldehyde group in benzaldehyde with other groups of similar size and charge does not affect its bitter almond quality. Most intriguingly, humans are capable to distinguish between the enantiomers of chiral odor molecules, such as (+)-carvone (caraway) and (−)-carvone (spearmint), which is likely mediated by stereo-selective receptors (Fig. 2). Enantio-selectivity is also exemplified by the pheromonal compound androstenone that induces mating stance in female pigs. (+)-Androstenone (Fig. 2) has an unpleasant (sweat, urine) odor quality to some humans and a pleasant (floral, sweet) odor quality to others, while (−)-androstenone is generally perceived as odorless. In contrast to mice, enantio-selectivity in humans is less pronounced and restricted to few odor molecules, while most enantiomers encode identical odor quality. Furthermore, carvone and androstenone are typical examples for which specific anosmias - the inability to detect particular odor molecules - have been identified in a certain percentage of humans.

Odor Receptors

How is the neural recognition of this almost infinite number of structurally diverse odor molecules achieved? Early on it has been noted that for a molecule to have an odor it needs to possess a molecular configuration that is complementary to specific sites of its receptor system [2]. This stereospecific theory has been validated by the discovery of a multi-gene family encoding odor receptors (ORs) [3], a finding that has set a milestone in the molecular understanding of odor detection (http://nobelprize.org/medicine/laureates/2004/press.html). ORs belong to the superfamily of G-protein coupled seven-transmembrane domain receptors (GPCR) (Fig. 1), and are similar in structure to the rhodopsin and β-adrenergic receptors. The ability of the olfactory system to recognize thousands of different odor molecules derives from the large size and diversity of the OR family. Based on genome sequencing projects (http://www.ncbi.nlm.nih.gov/Genbank), more than 1,000 potentially functional OR genes have been identified in mouse, while humans are left with about 400 potentially functional OR genes. Phylogenetically, ORs are preserved from fish to mammals and divide into two major classes. Class-1 or fish-like ORs are encoded by aquatic animals detecting water-soluble molecules, but are also present in ~10% of the mouse gene repertoire. Class-2 ORs are unique to terrestrial vertebrates detecting volatile odors.

ORs are highly divergent, especially in transmembrane domains 3–5. As a result of multiple OR sequence alignments across species and the developing of computational prediction models, odor binding is envisioned to occur in a binding pocket formed by the OR. Specific amino acid residues in key positions, predominantly located in the highly variable transmembrane domains are thought to interact with different parts of the odor molecules. However, exactly which parts of

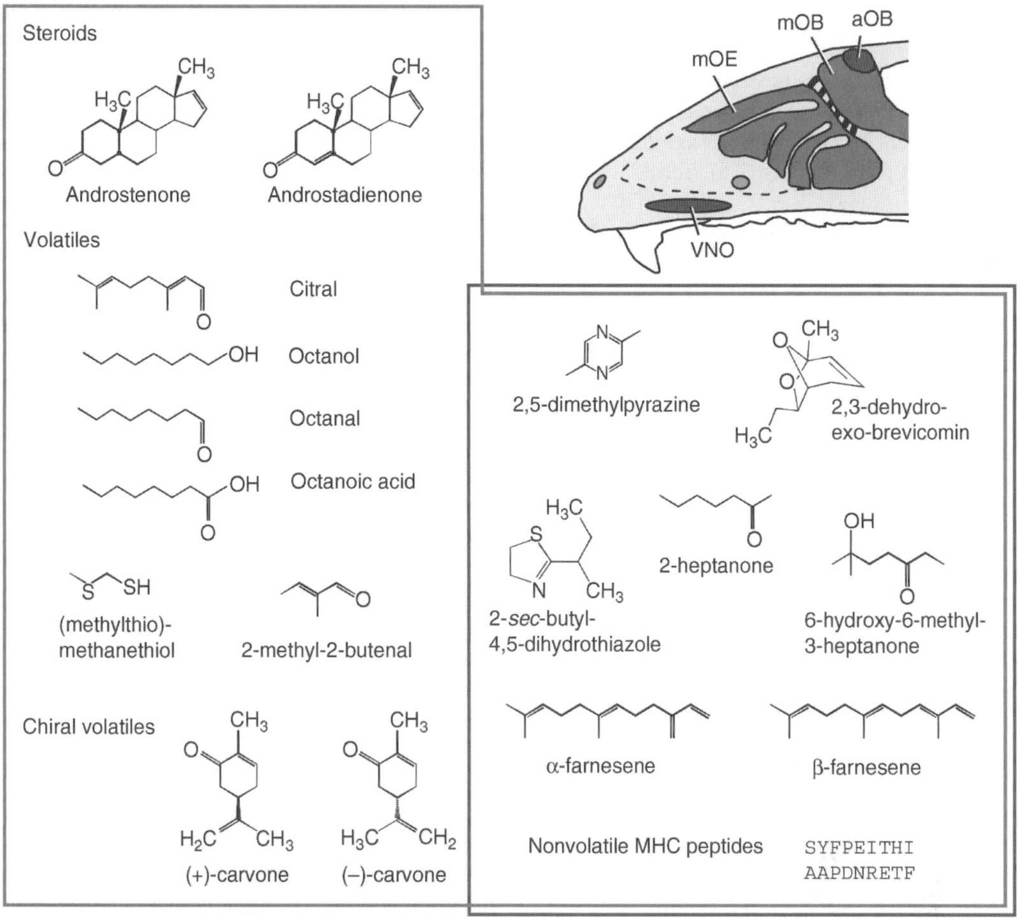

Odor. Figure 2 Chemical structure of odor molecules detected by the main olfactory epithelium (mOE) and the vomeronasal organ (VNO, blue). Steroids, volatiles including chiral volatiles and nonvolatile MHC peptides are detected by the mOE (red box). Overlapping odor cues that are detected by the mOE and the VNO encompass volatiles as well as nonvolatile MHC peptides (overlayed red and blue boxes). The main olfactory bulb (mOB, red) receives odor information from the mOE and the accessory olfactory bulb (aOB, blue) from the VNO.

the odor molecules are recognized by the ORs is still subject to intense investigation.

In situ hybridization studies show that expression of ORs in the rodent mOE is organized in a zonal pattern and that each individual OSN expresses only one OR. OSNs that express the same OR are confined to one out of four rostro-caudal zones and axonal projections of homologous OSNs coalesce into two glomeruli (one lateral and one medial) in each mOB.

Odor Coding: Molecular Level

Despite the large size and diversity of the OR family, the question arises how a limited number of $\sim$1,000 different ORs is capable of detecting an exceedingly larger variety of environmental olfactory cues? Identification of the first functional OR–odor ligand pairs [4,5], a process known as the "deorphanizing" of an OR, has solved this apparent discrepancy. Functional recordings of physiological odor reponses and polymerase chain reaction analyses of single OSNs have revealed that the discriminatory power of the olfactory system depends on combinatorial receptor activation as a result of an unusually broad ligand-tuning of individual ORs. Given that single OSNs express only a single OR-type, different odor molecules activate specific, partially overlapping sets of OSNs with distinct sensitivities. In other words, a single OSN has a receptive field composed of different odor ligands that bind its OR with distinct affinity. The fact that ORs detecting the same ligands can be both highly homologous or extremly divergent suggests that these ORs recognize identical or different odotopes (i.e., functional groups of an odor molecule), respectively. The resulting neural activity patterns are thus concentration-dependent: OSNs expressing ORs with the lowest threshold for a given odor are activated first, and the less sensitive ones are recruited at higher concentrations. This concentration dependence may explain the psychophysical phenomenon that some odor molecules are perceived differently at different concentrations. For example, with increasing concentration the perception of indole by humans ranges from "flowery" to "fecal."

Chemical Properties of Odor Molecules

Despite the relatively small number of ORs that have been deorphanized thus far, several features underlying odor recognition have emerged. The receptive field of a given OR appears to be determined by the functional groups, structure, and flexibility of an odor ligand. Some ORs accept 2–3 functional groups such as aldehydes, alcohols, and aliphatic acids [5] in combination with 3–4 consecutive carbons, while other ORs appear to be restricted to single functional groups. For example, the rat I7 OR is activated by straight-chained aldehydes ranging from C7-C10, with octanal (Fig. 2) representing the best ligand identified thus far [4]. Unsaturated C-double bonds that confer molecular rigidity or carbon backbone branches are, depending on position, tolerated, but structurally related molecules with different functional groups, such as octanal and octanoic acid (Fig. 2), yield no receptor activation. Aldehydes are potent ligands with low detection thresholds, and more than 30 different octanal-responsive, yet unidentified rat ORs, have been estimated from octanal evoked activity patterns in the mOE. However, not all ORs exhibit such broad tuning and some receptors appear to be specialists for a single or very few odor molecules.

Ligand binding does not always induce receptor activation. Several studies show that odor molecules exhibit dual functions and are agonists for some ORs, but antagonists for others. Citral for example strongly reduces the response of OR-I7 to octanal (Fig. 2). Antagonistic effects of odor molecules add another level of complexity to olfactory coding and may coincide with the psychophysical observation that both perceived quality and intensity of odor is not necessarily the sum of its single components, and that single substances are perceived differently than the same substances in a mix. Thus, at the molecular level, odor coding is a function of OSN activity patterns emerging from the combinatorial activation (and inhibition) of subsets of ORs both of which depend on concentration and chemical features of the odor molecules.

Odor Sensing by the Vomeronasal Organ

The VNO expresses a different set of chemosensory receptors, termed vomeronasal receptors (VRs), that also belong to the superfamily of GPCRs but are otherwise distinct from ORs [6]. VRs consist of two unrelated families, V1Rs and V2Rs, that are expressed in the apical and basal layers of the VNO sensory epithelium, respectively. Recent years have shown that vomeronasal sensory neurons detect a number of pheromones (see [7] for historic definition of the term pheromone) that mediate species-specific behavioral repertoires [1]. However, the VNO also detects some general odors without known pheromonal actions.

Compared to ORs, little is known about the chemical features or binding characteristics of VRs. From a chemical perspective, some of the molecules that stimulate the apical, V1R-expressing VSNs represent typical volatiles that for the human nose, would encode a specific odor quality. In some cases, the compounds are not specific for the VNO, but are detected by both mOE and VNO [1]. For example, 2, 5 dimethylpyrazine, a candidate key-food odorant for humans (with a smell of roasted beef), is also present in mouse urine and is known to delay puberty in mice (Fig. 2). The volatile 2-heptanone that has a fruity odor quality, is a male urinary compound that conveys pheromonal action by extending estrus in female mice (Fig. 2). For 2-heptanone two distinct mouse receptors have been identified, the vomeronasal receptor V1R2b and the olfactory receptor OR912–93 both of which are activated at nanomolar concentrations.

The basal, V2R-expressing layer of the VNO appears to be involved in the detection of nonvolatile ligand families, consisting of peptides and proteins, which requires direct physical with the stimulus source. One such family consists of antigenic peptides – the major hitocompatibility complex (MHC) class 1 peptides – that are crucial in the context of immune surveillance and carry information about the genetic make-up of an individual [1]. Interestingly, such MHC peptides are also detected in the mOE, which gives further support to a model involving parallel processing of the same social odor cues by the two olfactory subsystems (Fig. 2). Convergent information derived from the two olfactory systems is likely integrated by higher brain centers.

Odor Processing by Higher Brain Centers

How is odor information represented in the brain? The olfactory glomeruli of the main olfactory bulb (mOB) form the first relay station in the brain where axonal projections of OSNs synapse onto second order neurons, the mitral and tufted cells. It is well-established that odor stimulation evokes spatially and temporally distinct glomerular activation patterns in the mOB that result from the differential activation of specific sets of ORs in the mOE [8] (e.g., see http://leonserver.bio.uci.edu). The brain then needs to extract the features of these bulbar activity patterns. These depend to some extent on chemical odor properties, mainly functional group and structure. For example, molecules with identical functional group, but different C-chain length activate in part overlapping glomeruli that are not activated by structurally related compounds with different functional groups; single molecules with two different functional groups activate glomeruli that are distinct from those responding to binary mixtures.

Attempts to correlate molecular and functional results on odor coding in rodents with those derived from human psychophysics show that the relation between odor structure and perceived odor quality is still poorly understood. The fact that chemically closely related molecules can confer different odor qualities, whereas molecules that smell alike do not necessarily share chemical similarity suggests that molecular properties and their translation into neuronal activity patterns and spatial odor images in the mOB are not the only determinants in defining odor quality. Further processing of olfactory information by higher brain centers that eventually produce an olfactory percept is only beginning to be understood. Many olfactory-associated brain functions derive from psychophysical studies on humans with discrete brain lesions. Functional imaging of brain activity in humans provides a promising technique to decipher the neural basi of odor perception.

Mitral and tufted cells in the mOB transmit their output signals to the olfactory cortex (Fig. 3), a broadly defined area that consists of the anterior olfactory nucleus, the olfactory tubercle, the piriform cortex, the entorhinal cortex, and the cortical amygdaloid nuclei. The amygdala, which is part of the limbic system and associated with emotional state, participates in formation and storage of olfactory memory.

The entorhinal cortex that projects to the hippocampus plays a role in associative learning and olfactory memory. The orbitofrontal cortex receives afferents from parts of the olfactory cortex through the thalamus and is involved in the conscious perception and discrimination of odor. Recent studies connect the piriform cortex with mechanisms in odor identification as well as olfactory memory and learning. Its anterior region, the principal target of mOB output signals, has been suggested to synthetize information about odor structure into a quality percept.

Mitral cells of the aOB project to the medial amygdala, which regulates social behaviors such as mating and recognition of conspecifics. Odor information of the mOB and the aOB is possibly integrated by hypothalamic gonadotropin-releasing hormone (GnRH) neurons resulting in changes in endocrine status and social/sexual behavioral outputs [9]. Furthermore, odor information undergoes additional refinement by higher cortical centers that integrate olfactory input with previous odor experience, afferents from other sensory systems, and in the case of humans, information obtained through language.

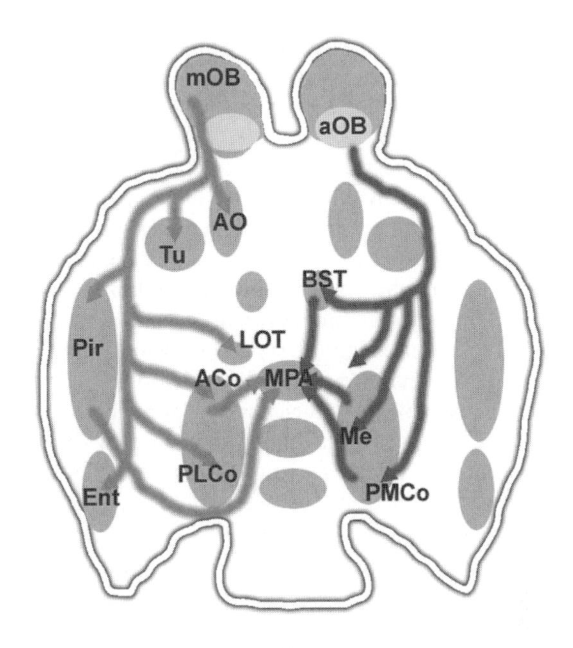

Odor. Figure 3 Brain pathways for odor processing emerging from the main olfactory bulb (mOB, red), the accessory olfactory bulb (aOB, blue), and their predicted targets in the mouse brain. *Aco* anterior cortical amygdaloid nucleus; *AO* anterior olfactory nucleus; *BST* bed nucleus of the stria terminalis; *Ent* entorhriunal cortex; *LOT* nucleus of the lateral olfactory tract; *Me* medial amygdala; *MPA* medial preoptic area; *Pir* piriform cortex; *PLCo* posterolateral cortical amygdaloid nucleus; *PMCo* posteromedial cortical amygdaloid nucleus; *Tu* olfactory tubercle.

Odor Function

Odor cues play an important role in the perception of the environment and in the overall survival of a species. During breathing, air-composition is constantly and involuntarily evaluated. In addition to locating potential food sources, detection (and secretion) of odor has multiple functions in inter- and intraspecies chemical communication, i.e., in the identification of prey, predators, mates, and in the adjustment of social and reproductive behavior. Social behaviors are mediated by both the main and accessory olfactory systems. Common to many mammals is the marking of landscape by depositing individual odors. These complex odor messages carry information about gender, sexual and social status, territoriality, mood, and fitness. In chemical communication, scent marks often serve to deter rivals and attract potential mates. Across many species, scent marks elicited by predators are interpreted as warning signs causing escape behavior. Dogs and wolves produce scent marks through urination and defecation, whereas foxes have developed a specialized supracaudal gland that constantly secretes a mixture of volatile terpenes.

A particularly well-established, odor-induced social behavior is the suckling behavior of rabbit pubs. The milk of female rabbits contains 2-methyl-2-butenal (Fig. 2), a volatile pheromone that guides pups towards their mother's nipples and triggers immediate suckling. Another well-known odor-mediated behavioral change depends on the steroid androstenone (Fig. 2), which induces mating stance in female pigs during heat.

In humans, the smell of androstenone is described as both unpleasant (sweat, urine) or pleasant (floral, sweet). Although present in human axillary sweat and urine, it is not yet clear whether androstenone represents a human pheromone. However, androstadienone (Fig. 2), a related compound in male human sweat, is known to affect endocrine status by maintaining high levels of the hormone cortisol in exposed women. Another example of odor-induced endocrine change in humans derives from odor stimulation of females with armpit or vagina secretions from donor females. Estrus cycles of acceptor females synchronize with that of the donor female ("McClintock" effect) by either advancing or retarding menstruation.

References

1. Brennan PA, Zufall F (2006) Pheromonal communication in vertebrates. Nature 444:308–315
2. Amoore JE (1963) Stereochemical theory of olfaction. Nature 198:271–272
3. Buck L, Axel R (1991) A novel multigene family may encode odorant receptors: a molecular basis for odor recognition. Cell 65:175–187
4. Zhao H, Ivic L, Otaki JM, Hashimoto M, Mikoshiba K, Firestein S (1998) Functional expression of a mammalian odorant receptor. Science 279:237–242
5. Malnic B, Hirono J, Sato T, Buck LB (1999) Combinatorial receptor codes for odors. Cell 96:713–723
6. Dulac C, Axel R (1995) A novel family of genes encoding putative pheromone receptors in mammals. Cell 83(2):195–206
7. Karlson P, Lüscheri M (1959) Pheromones: a new term for a class of biologically active substances. Nature 183:55–56
8. Johnson BA, Leon M (2007) Chemotropic odorant coding in a mammalian olfactory system. J Comp Neurol 503:1–34
9. Boehm U, Zou Z, Buck LB (2005) Feedback loops link odor and pheromone signaling with reproduction. Cell 123(4):683–695

Odor Memory

Peter Brennan
Department of Physiology, University of Bristol, Bristol, UK

Synonyms

Memory-odor; Odor learning; Olfactory learning

Definition

►Odor memory is the store of information about an odor that enables an animal to recognize an odor along with its associations and meaning and link it to an appropriate behavioral response. Olfactory learning is process by which the nervous system forms such odor memories.

Characteristics

The ability to learn to recognize a particular odor and associate it with a meaning or a predictive value can be demonstrated in a wide variety of vertebrates and invertebrates, and has been studied intensively in terrestrial mollusks, fruit flies, honeybees, rodents and primates. The odor stimuli used in such experiments can consist of individual ►odorants, odorant mixtures or complex, naturally occurring odors containing hundreds of constituents. Olfactometers can be used to carefully control the concentration of odorants and the composition of odor mixtures in the sampled air. Odorant mixtures are generally perceived and learnt as unitary sensory objects, rather than being analyzed in terms of their individual components [1].

One of the simplest form of olfactory learning is a "Go, No-Go" successive odor discrimination, in which an animal is rewarded for making a behavioral response to the rewarded odor (CS+), with no reward delivered in response to the unrewarded odor (CS−). This type of learning is comparatively rapid, occurring within tens of conditioning trails and the memory can last for months (Fig. 1).

Furthermore, the memory for the correct responses to a pair of odors is robust to subsequent learning of other odor pairs, and the learning of subsequent odor pairs is more rapid as the animal learns a win-stay, lose-shift strategy. This results in the ability of rats to learn the correct responses to a new pair of odors after only few trials, which is comparable to the ability of primates to learn visual discriminations. Rodents can also learn to discriminate odors presented simultaneously, either at separate odor ports or in air flowing down separate arms of a Y maze. These discriminations can be made extremely rapidly and it has been estimated that the time to make the discrimination is as small as 220 ms, less than the time taken for a single sniff [2].

Short-term memory (►memory, short-term) for odors can be tested using a delayed non-matching to sample procedure. In this procedure, subjects are presented with a sample odor, which is removed and then, after a variable delay, the subjects are presented with the simultaneous choice of the same odor and a different odor. Responses to the odor that is different from the sample odor are rewarded. Using this task, rats can be shown to have short-term memory for odors of at least 60 s. Moreover, rats can also be trained to learn odor sequences where the correct odor choice depends on the sequence of preceding

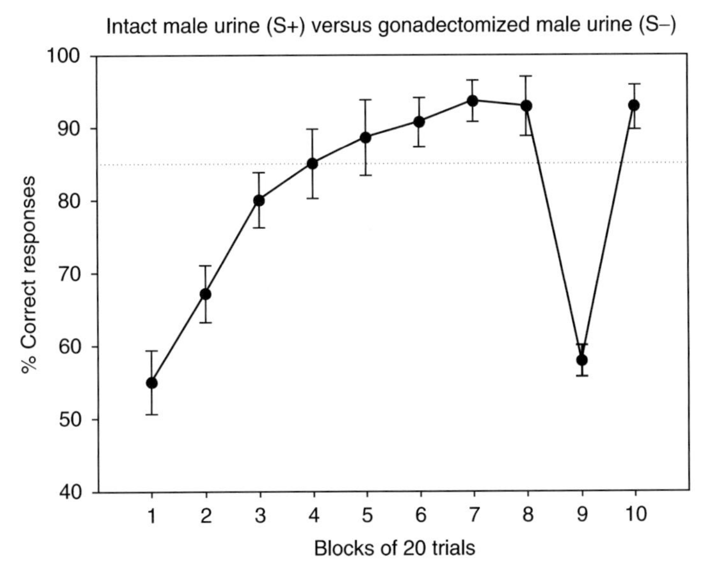

Odor Memory. Figure 1 Typical learning curve for a "go, no-go" odor discrimination task for a group of seven inbred mice (Keller and Bakker unpublished data). The learning criterion of 85% correct responses was achieved by the fourth daily block of 20 trials. In block nine the same (S+) stimulus was used for S+ and S− trials. The drop in performance to chance levels demonstrates that the mice were using odor cues to perform the task rather than any extraneous sensory cues associated with the training procedure.

odors, which has been proposed as a test of episodic memory [3].

Rewarding and aversive training stimuli result in learning to approach or avoid the conditioned odor, respectively. However, animals can also learn about the familiarity of odors that have not been paired with any overt training stimulus. If an animal is presented with a novel odor it will initially spend time investigating it. Subsequent presentations of the same odor elicit reduced investigation, as the response habituates, whereas presentation of a novel odor elicits intense investigation. This forms the basis of the ►habituation/dishabituation test of olfactory discrimination, which in some ways is a more natural test of olfactory behavior, but requires the animal to be motivated to investigate the stimulus in the first place. Many innately attractive odors, such as urine odors in the case of rodents, may contain pheromonal components that can act as rewarding stimuli for the associative learning of non-pheromonal odors [4].

In many ways, odor learning has been most extensively studied in humans, who have the advantage of not requiring explicit reinforcement for learning to occur, as they can give verbal responses. However, the very fact that humans can name odors poses problems, in that it is often difficult to dissociate the odor memory from the memory for the verbal label. The association of a verbal label with an odor is a separate process from the recognition of an odor [1]. This is demonstrated by the "tip-of-the-nose" phenomenon in which a person reports that they recognize an odor, and its name is on the tip of the tongue, but they can't quite recall it.

Neural Changes Underlying Odor Learning

A vast number of individual odorants are able to stimulate ►olfactory receptor proteins on olfactory sensory neurons (OSNs). If present at a sufficient concentration then the neural activity that is evoked by an odorant can lead to the perception an odor. However, most odors in nature are not the result of single odorants, but arise from complex mixtures of many odorants. The neural activity evoked by a mixture of odorants that come from a single source are associated to synthesize a unitary neural representation of the odor, known as an odor object. This allows the odor to be discriminated from similar odors and used to recognize and locate the source of the odor. The neural representation of the odor is also associated with neural representations of the object derived from other sensory systems, as well as the context in which it is perceived, and ultimately its meaning for the animal. The ability to subsequently recall these associations in terms of recognizing the odor and its meaning constitute the odor memory. This is not a trivial task. For instance, over 500 individual odorants contribute to the odor of fresh coffee. A few major components will be common between different varieties of coffee. These are the main contributors to coffee odor and will lead to different varieties being classified as coffee. It is the differences in the numerous minor components that give rise to the fine distinctions between the different coffee varieties, and the ability to make such fine discriminations is enhanced by prior experience with the odors and the importance of making the discrimination [1].

In mammals, the process of associating odorant features into an odor object that can be readily discriminated from similar odors is primarily a function of the ►main olfactory bulb (MOB), anterior olfactory nucleus and anterior piriform cortex, at the initial stages of olfactory processing. OSNs in the main olfactory epithelium express a single odorant receptor type and respond to a small range of odorants with certain shared structural and functional attributes. ►Mitral cells in the MOB receive input from OSNs expressing a single receptor type. However, the responses of mitral cells in the MOB do not simply depend on the input that they get from the sensory neurons. They are also influenced by the arousal and motivational state of the animal, such as whether it is hungry, or the possible presence of a predator. In addition, mitral cell activity is likely to be influenced by a centrally generated expectation of an odor arising from reciprocal connections with higher-level olfactory processes, and activated in situations such as a predator searching for a particular type of prey.

Significantly, there is accumulating evidence that the responses of mitral cells in the MOB depend not only on information provided by OSNs, but also on the meaning of the odor. Hence, the odor-evoked activity of mitral cells in the MOB has been found to change following learning a new reward association for an odor [5]. Such learning-dependent changes in the odor-evoked pattern of neural activity in the MOB are likely to be at least partially the result of changes in gain of lateral and recurrent inhibition from granule cell interneurons. The change in spatiotemporal pattern of mitral cell activity following learning has been hypothesized to "pull apart" the representation of the learned odor from those of similar odors generated by the MOB. This could increase the probability that they could be discriminated reliably and linked to different behavioral responses. There is evidence for this type of decorrelation of odor-evoked patterns of activity in the honeybee antennal lobe (the insect equivalent of the mammalian MOB) following appetitive odor conditioning [6].

Information from individual mitral cells in the MOB is distributed to a large number of pyramidal cells by virtue of their highly divergent projections to the anterior piriform cortex. Conversely, each anterior piriform cortex pyramidal neuron samples information from a large number of mitral cell neurons across a large extent of the MOB. It is thought that these pyramidal cells can therefore act as coincidence detectors. According to this hypothesis, when the pyramidal cells in the anterior piriform cortex receive synchronized input from a sufficient number of mitral cells, the strength of those inputs is enhanced, increasing the probability that the same combination of inputs will cause that pyramidal cell to fire in the future. This is supported by evidence arising from ►cross-adaptation to odorant mixtures. This suggests that whereas mitral cells at the level of the olfactory bulb respond to individual odorants, pyramidal cells in the anterior piriform cortex respond to specific combinations of odorants that form an odor object [1]. Moreover, the pattern of interconnectivity of the anterior olfactory nucleus and anterior piriform cortex is thought to confer pattern completion properties on the network, in which a degraded pattern of input is able to trigger activity in the complete network of pyramidal neurons that respond to the odor object [7]. This might underlie the ability of the olfactory system to cope with the naturally occurring variability in odorant mixtures that are generalized to a particular odor memory.

Higher-Level Brain Areas Involved in Odor Learning

The network of cells in the anterior piriform cortex that represent an odor object communicate with cells in the posterior piriform cortex, which also receive direct input from the olfactory bulb and have widespread reciprocal connections with other brain regions. The interconnections of the posterior piriform cortex suggest that it is likely to function at a similar level to association cortex in other sensory systems and may be involved in forming multimodal representations of stimuli [7]. For instance, the posterior piriform cortex is likely to be involved in associating the sight, sound and smell of a predator into a single representation that can be recalled by input from any one ►modality. Perhaps the most important multimodal representations of odors are in relation to the taste, smell and texture of food, which combine to a representation of flavor, which appear to be stronger than those formed between odors and other sensory modalities. Neurons with multimodal responses to both taste and smell have been found in the orbitofrontal cortex and insular cortex of primates.

Finally these odor representations have to drive an appropriate response. This can be an innate response – especially in the case of ►pheromones. However for the majority of odors, the appropriate response is learned as a result of experience. The amygdala is particularly involved in eliciting learned emotional responses to odors, whereas neurons in the orbitofrontal cortex have been shown to respond to the meaning of an odor and the context in which it occurs. However, it should be remembered that there are extensive reciprocal connections among these areas, and the neural changes that underlie odor memory are distributed throughout all levels of the olfactory system (Fig. 2).

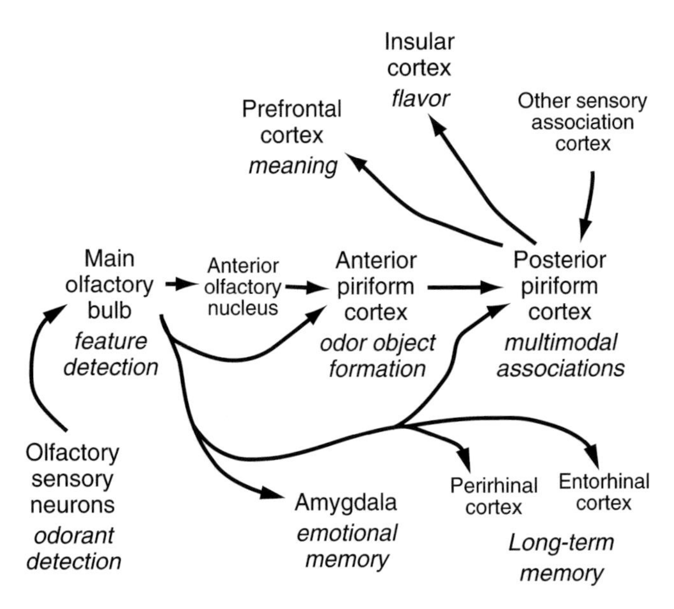

Odor Memory. Figure 2 Major brain areas involved in distinct aspects of odor memory. Extensive reciprocal connections and interconnections between brain areas have been omitted for clarity.

Importance of Odor Learning in Mammals

Odor memory is vital for the recognition of significant elements of the environment, such as food, predators and prey, as well as social cues that enable individual and kin recognition, and odor cues used for navigation. The sense of smell plays a particularly important part in mother-offspring interactions, which are vital for the reproductive success of most mammals. For example, ewes rapidly learn to recognize their own lamb by its odor, within a few hours of giving birth. This odor memory enables the ewe to discriminate between its own lamb, to which it shows acceptance behavior, and alien lambs, which it rejects. Formation of the memory for own lamb odors occurs during a period of a few hours, triggered by the vaginocervical stimulation of birth, and involves dramatic changes in the responsiveness of mitral cells in the MOB to lamb odors [8].

Odor learning is also important for neonatal mammals, especially those that are altricial, in which hearing and sight are poorly developed at birth. For instance, the rabbit mammary pheromone 2-methyl-but-2-enal not only acts as a pheromone to elicit nipple search behavior, but also acts as an unconditioned stimulus to induce memory formation to the maternal odors, or artificial odors that have been applied to the mother [4]. These conditioned odors are then able to elicit full nipple search behavior and therefore reinforce the innate response to the pheromone.

Adult rats can readily be trained to avoid an odor, which has previously been associated an aversive stimulus, such as a mild electric shock. This conditioned fear response is dependent on the amygdala, and is adaptive in helping the rat to avoid potentially dangerous environmental situations. However, if rat pups are exposed to an odor that has been paired with the aversive stimulus of a mild electric shock before postnatal day 10, they will learn to approach the odor [9]. Again this odor memory is adaptive, as at this age the rat pups are normally confined to the nest and are dependent on their mother for their survival. No matter how rough the nest environment or their mother is towards them, the pups remain attracted to the maternal odor. Therefore the function of odor memory can alter to adapt to the changing behavioral priorities of an animal during the course of postnatal and adult life. Moreover, the long-term memory (▸memory, long-term) of neonates for odors learned in the nest environment, or even *in utero*, can have lasting effects on their behavior as adults, such as post-weaning food preferences or their choice of mate.

References

1. Wilson DA, Stevenson RJ (2006) Learning to smell. The John Hopkins University Press, Baltimore, MD
2. Uchida N, Mainen ZF (2003) Speed and accuracy of olfactory discrimination in the rat. Nat Neurosci 6:1224–1229
3. Fortin NJ, Agster KL, Eichenbaum HB (2002) Critical role of the hippocampus in memory for sequences of events. Nat Neurosci 5:458–462
4. Coureaud G, Moncomble A-S, Montigny D, Dewas M, Perrier G, Schaal B (2006) A pheromone that rapidly promotes learning in the newborn. Curr Biol 16:1956–1961
5. Kay LM, Laurent G (1999) Odor- and context-dependent modulation of mitral cell activity in behaving rats. Nat Neurosci 2:1003–1009

O

6. Faber T, Joerges J, Menzel R (1999) Associative learning modifies neural representations of odors in the insect brain. Nat Neurosci 2:74–78
7. Haberly LB (2001) Parallel-distributed processing in olfactory cortex: new insights from morphological and physiological analysis of neuronal circuitry. Chem Senses 26:551–576
8. Brennan PA, Kendrick KM (2006) Mammalian social odors: attraction and individual recognition. Philos Trans R Soc Lond B Biol Sci 361:2061–2078
9. Sullivan RM, Landers M, Yeaman B, Wilson DA Good memories of bad events in infancy. Nature 407:38–39

Odor-binding Proteins

►Odorant-Binding Proteins

Odor Cells in Hippocampus

Definition

In a series of studies aimed at exploring the role of hippocampal function in memory using the model system of olfactory-hippocampal pathways and odor learning in rats, it has been demonstrated that hippocampus itself is not essential to memory for single odors, but is critical for forming the representations of relations among odor memories, and for the expression of odor memory representations in novel situations. The studies that exploit the exceptional qualities of olfactory learning are helping to clarify the nature of higher order memory processes in all mammals, and extending to declarative memory in humans.

►Olfaction
►The Hippocampus: Organization

Odor Code

►Odor Coding
►Olfactory Information

Odor Coding

Martin Giurfa
Research Center on Animal Cognition,
CNRS – University Paul Sabatier, Toulouse, France

Synonyms

Odor code; Olfactory code; Olfactory coding; Odor representation

Definition

The processes by which essential features of odor molecules are translated into patterns of neural activity in the olfactory circuit.

Characteristics

A code is a set of rules allowing the translation of information from one form or dimension into a different one, in such a way that essential features of the original message are preserved and made available for further unambiguous reading and information extraction. In the case of odors, the nervous system translates the information pertaining to chemical stimuli into patterns of neural activity at the first stages of processing in the brain. We focus here on odor encoding within the vertebrate olfactory bulb (OB) and the analogous circuit in insects, the antennal lobe (AL), excluding specialized pheromonal centers and higher-order centers of the olfactory circuit.

Odor molecule determinants such as chain length (number of carbon atoms), functional group (aldehyde, alcohol, ketone, etc.) and concentration, among others, seem to be the sensory primitives that are processed by the olfactory pathways. They are transduced from the chemical world into the neural domain by differential activation of olfactory receptor proteins on the surface of olfactory sensory neurons, on the insect antenna or in the nose of vertebrates.

The olfactory message is first processed at the level of the primary olfactory centers in the brain (the AL in the case of insects and the OB in the case of vertebrates). Both the AL and the OB are organized according to similar anatomical principles. They are constituted by glomeruli, which are the anatomical and functional units involved in the first steps of odor processing in the nervous system. Olfactory receptor neurons expressing the same receptor type converge to one or a few glomeruli [1] so that the response of a glomerulus is an amplified version of the responses of the receptor type under consideration. There are up to several hundred glomeruli in an insect antennal lobe and several thousand in a vertebrate olfactory bulb. Glomeruli are not simple convergence sites of olfactory receptor axons; they are interconnected by different sets of local inhibitory neurons, which release the inhibitory neurotransmitter GABA (γ-aminobutyric acid), thus producing complex patterns of firing activity in response to an odor. In insects, local inhibitory and excitatory interneurons may connect laterally few or multiple glomeruli. In vertebrates, lateral inhibitory connections are provided by periglomerular cells whose dendrites are restricted to one glomerulus and by short axon cells which have dendrites and axons extending throughout several glomeruli. In addition, a second level of powerful inhibitory connections is provided by the interaction between granular cells and output cells to the OB.

The processed signal is further conveyed to higher-order centers by such output neurons, the projection neurons in insects and the mitral/tufted cells in vertebrates. Thus, once odors activate groups of receptor neurons, the information does not simply flow through the AL/OB to downstream areas via projection neurons or mitral cells. Instead, the presence of inhibitory neurons within the neural network of the AL/OB determines a global reformatting of odor representations, in the form of a stimulus-dependent, spatio-temporal redistribution of activity across the AL/OB [2].

The olfactory code is a spatio-temporal code in that it contains two complementary components, the spatial and the temporal dimensions. Each of these two dimensions has been studied using different techniques, mainly imaging for the spatial code, and electrophysiology for the temporal code. The impression that such different analyses correspond to separate, unconnected properties of the olfactory code should be avoided. Spatial and temporal properties of the olfactory code represent, in fact, different sides of the same coin.

Spatial Coding of Odors

Odors may be encoded at the level of the AL/OB in terms of a specific spatial pattern of glomerular activation (Fig. 1).

Such activity pattern constitutes an odor map, which is proper to each odor, symmetric between hemispheres and conserved between individuals [3]. Spatially distributed activity patterns relate to certain structural features of the odor molecules as molecules with similar structural properties are encoded in terms of partially overlapping activity patterns. Neural similarity, measured in terms of the amount of overlap of glomerular activation patterns, correlates directly with perceptual similarity, measured in terms of behavioral odor choices [4], i.e. odors judged as similar correspond to partially coincident odor maps.

Odor concentration affects the odor map as generally, the number of activated glomeruli increases with increasing concentrations of the stimulating odor. A critical question would be, therefore, how ►concentration invariance is achieved given the changing nature of this odor representation. A possible answer comes from the fact that, as mentioned above, spatial coding is not the unique form of translating chemical stimulus features into patterns of neural activity (see below "Temporal coding").

Quantifying glomerular activity requires identifying individual glomeruli across preparations in the same or different individuals. To this end, atlases of the primary olfactory center have been established in the case of the antennal lobe of some insects (honeybees, moths, flies) where such an approach is accessible due to a lower number of constitutive glomeruli.

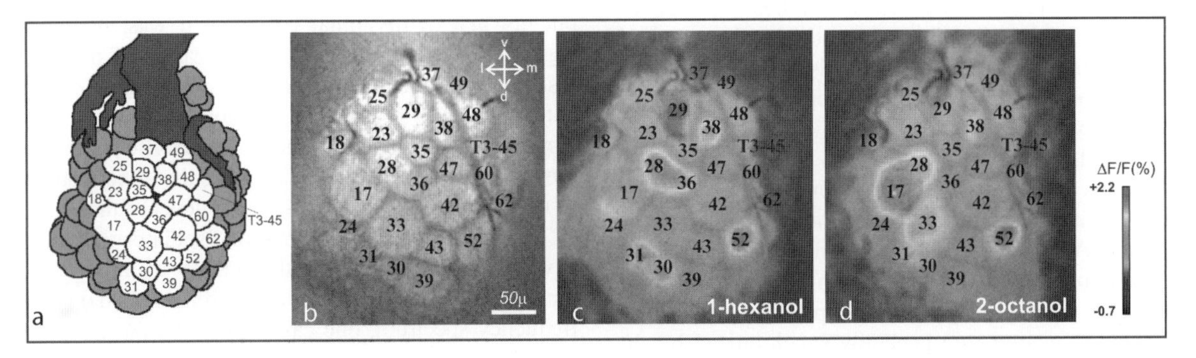

Odor Coding. Figure 1 Spatial coding of odors at the level of the antennal lobe of the honeybee *Apis mellifera*. (a) Atlas of the honeybee antennal lobe showing 24 glomeruli individually identified. (b) Example of an anatomical staining of the frontal part of a left antennal lobe with the 24 identified glomeruli (*d* dorsal; *l* lateral; *m* medial; *v* ventral). (c) Calcium-imaging recordings of neural activity *in vivo* upon odor stimulation of a honeybee. Superimposed activity map in response to the odor 1-hexanol, showing which glomeruli were activated. The colors (see scale on the right) represent activity levels in terms of fluorescence variation (ΔF/F %) with respect to a basal level (no olfactory stimulation). (d) Superimposed activity map in response to the odor 2-octanol. Each odorant is encoded by a specific spatial pattern of glomerular activation.

Olfactory maps can be visualized using different kinds of techniques allowing measurements of neural activity upon olfactory stimulation. Markers of neural activity vary from radiolabels ([^{14}C] 2-deoxyglucose) and antibodies (*c-fos*) to fluorescent dyes (voltage-sensitive or calcium reporters), or intrinsic optical properties of the tissue. Using some of these and other techniques it is possible to disentangle the contributions of olfactory receptors conveying the olfactory message to the brain from that of local interneurons and projection neurons conveying such a message to higher-order brain centers. In this way, the role of the different neural subpopulations in the elaboration of the odor map can be understood. Assessing the respective contributions of pre- and postsynaptic elements is crucial for understanding the computations carried out at the level of the AL/OB.

Activity maps in the AL/OB are not static but dynamic odor representations. Such a dynamics mostly reflects interglomerular interactions within the AL/OB. However, at the level of sensory afferences to the glomeruli of the OB, diverse, glomerulus- and odorant-dependent temporal dynamics are already present, thus showing that glomerular maps of primary sensory input to the OB are temporally dynamic, even before further processing within the bulb. These dynamics may contribute to the representation of odorant information and affect information processing in the central olfactory system.

Temporal Coding of Odors

Comprehensive studies on the temporal coding of odors have been performed in several species but studies on locusts have been crucial to understand the principles governing this coding [2]. Such studies have shown that both monomolecular and complex odors are encoded combinatorially by dynamical assemblies of projection neurons. Information about odor identity is contained in the timing of action potentials in an oscillatory population response, rather than on the mere spiking frequency of the response.

Indeed, each projection neuron in an odor coding assembly responds with an odor-specific temporal firing pattern consisting of periods of activity and silence. Any two projection neurons responding to the same odor are usually co-active only during a fraction of the population response. The spikes of coactivated projection neurons are generally synchronized by the distributed action of local interneurons in the AL, which release GABA. Because projection neurons convey the olfactory information to higher-order structures, the mushroom bodies, the coherence of projection neuron activity can be measured in this target area in terms of local field potential (LFP) oscillations [2]. LFP oscillations have a frequency of 20–35 Hz. Each successive cycle of the odor-evoked oscillatory LFP can therefore be characterized by a co-active subset of projection neurons. As a consequence, each odor is encoded by a specific succession of synchronized assemblies [2]. The action potentials produced by a projection neuron during its odor-specific phases of activity are not necessarily all phase-locked to the LFP. For each odor–projection neuron combination, however, precise and consistent epochs of phase-locked or non-phase-locked activity can be identified (Fig. 2).

Increased odor concentration leads to changes in the firing patterns of projection neurons, similar to those caused by changes in odor identity, potentially confounding representations for identity and concentration. However, concentration-specific response patterns cluster by identity, resolving the apparent confound. Thus, odor encoding comprises three main aspects: the identity of the odor-activated neurons, the temporal evolution of the ensemble, and oscillatory synchronization.

Besides oscillatory synchronization, the odor-evoked responses of local interneurons and projection neurons also contain prolonged and successive periods of increased and decreased activity (slow response patterns), which are cell and odor specific and are stable from trial to trial. Hence, oscillatory synchronization and slow patterning together shape a complex, distributed representation in which odor-specific information appears both in the identity and in the time of recruitment and phase-locking of projection neurons.

Experiments on honeybees [5] showed that oscillatory synchronization between projection neurons is selectively abolished by picrotoxin, an antagonist of the $GABA_A$ receptor acting on GABA-ergic local interneurons of the antennal lobe, and that such a picrotoxin-induced desynchronization impairs the behavioral discrimination of molecularly similar odorants, but not that of dissimilar odorants. It was, therefore, suggested that oscillatory synchronization of neuronal assemblies is functionally relevant, and essential for fine, but not coarse olfactory discrimination. Interestingly, picrotoxin has no effect on the slow response patterns of projection neurons [2], thus showing that other sources of neural inhibition are at play at the level of the AL.

In vertebrates, three types of oscillatory rhythms have been distinguished in the activity of mitral cells, the pendant of insect projection neurons. Based on their frequency spectrum, one can distinguish three oscillation types:

1. θ oscillations (1–8 Hz) are generated by the respiratory rhythm and are correlated with increased and decreased stimulation of olfactory afferences upon inspiration and expiration, respectively. Different mitral cells may exhibit different response latencies to the same odorant and odor coding

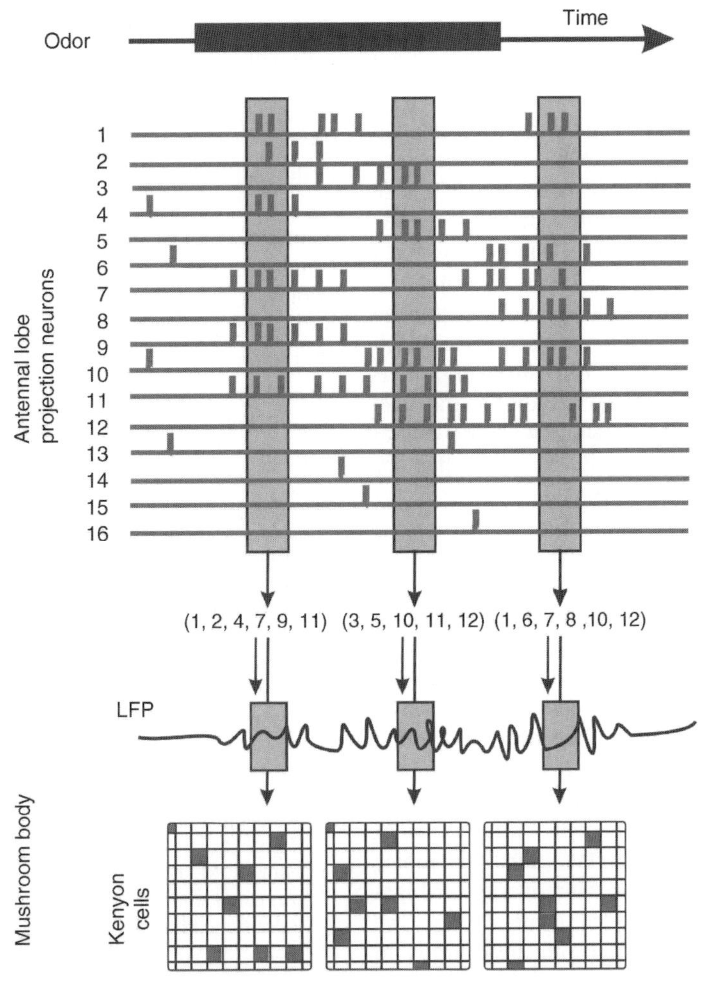

Odor Coding. Figure 2 Temporal coding of odors in the locust olfactory system.The presumed odor representation is combinatorial, spatially distributed and relies on synchronized and evolving neural assemblies. An odor stimulus elicits spiking activity in several projection neurons (1–16), which constitute the output to the antennal lobe. For each odor–projection neuron combination, however, precise and consistent epochs of phase-locked or non-phase-locked activity can be identified. The coherence of projection neuron activity can be measured at the level of the mushroom bodies in terms of local field potential (LFP) oscillations. Only few Kenyon cells, the constitutive cells of the mushroom bodies, are activated by projection neuron input (sparse coding) (adapted from Laurent G, Trends in Neurosci 19:489–496, 1996).

models have been proposed based on the phase relationship between action potentials of mitral cells and the phase of a θ cycle [6].

2. β oscillations (15–30 Hz) are induced by the inhalation of odor molecules and their origin is a matter of debate. While some theories posit that β oscillations originate not in the OB itself but in downstream structures (e.g., olfactory cortex) that feedback on it, other theories postulate that rhythmic input on granular cells induce these oscillations. The function of β oscillations is still unclear but it has been shown that olfactory learning and habituation can enhance the prevalence of β rhythm over γ rhythm in an odor-specific manner [7].
3. γ oscillations (40–80 Hz) are present in the olfactory system of several vertebrate species and can be related to those evinced in the olfactory system of insects (see above). These oscillations are generated in the olfactory bulb upon inhalation of odor molecules. Mitral cell activity is synchronized with γ oscillations and such synchronization arises from the interaction between mitral and granular cells. Glutamate released from mitral cell dendrites excites the dendrites of granule cells, which in turn mediate GABA-ergic inhibition back onto mitral cells [8]. Granular cells do not synchronize with γ oscillations; it has been proposed that they release GABA in a rhythmic manner and in absence of action

potentials [8]. Such a rhythmic inhibitory activity seems to play a fundamental role in the modulation of the oscillatory frequency.

Importantly, not all mitral cells are synchronized with γ oscillations during the response to an olfactory stimulus. In fact, the two neural populations, those exhibiting and those not exhibiting synchrony, encode different properties of the odor: non-synchronized action potentials allow encoding the fine identity of an odor while synchronized action potentials encode the category (ensemble of similar odor molecules) to which the perceived odor belongs [9]. Thus, different properties of an odor can be encoded by the same mitral cells depending on their synchronization with the neural population. The role of γ oscillations in olfactory perception in rodents has been demonstrated by experiments on transgenic mice presenting alterations of inhibitory activity in the OB. Such alterations result in significant changes in olfactory discrimination.

The picture emerging from studies on the temporal coding of odors in the AL/OB suggests that the transfer of odor-evoked signals from receptors to the AL/OB circuits is accompanied by a reshaping of odor representations so that stimulus-dependent, temporal redistribution of activity arises across these circuits. Such a reshaping exploits time as a coding dimension and results from the internal connectivity of the AL/OB circuits and from the global dynamics that these connections produce. Moreover, centrifugal connections from higher order centers (e.g. the mushroom bodies in insects, the olfactory cortex in vertebrates) to the AL/OB may also play an important role in reshaping of odor representations. This top-down process is specifically involved during learning condition in which neutral odorants are transformed into ►aversive or ►attractive ones.

Conclusions

All in all, the antennal lobe of insects and the olfactory bulb of vertebrates act similarly upon olfactory stimulation: to prevent ►adaptation, they format and reshape odor representations, increase the signal-to-noise ratio and improve odor discrimination. It appears that spatial and temporal dimensions are complementary aspects of odor coding in the AL/OB circuits and that their separated analysis responds to the use of different recording techniques that have put the emphasis on one aspect or the other. As we have detailed above, temporal variations of the spatial code are observed in imaging experiments, and in the temporal code, odor-specific information appears also in the identity of the active projection neurons, i.e. a spatial-related property. For instance, synchronization of output neuron activity at specific sites within the odor map is crucial as shown by studies in the moth where odors elicit high synchrony of action potentials in paired cells connected to the same glomerulus but low synchrony in cells connected to different glomeruli [10]. Such studies revealed a strong relationship between recording positions, temporal correlations, and similarity of odor response profiles, thus supporting the notion that the olfactory system uses both spatial and temporal coordination of firing to encode chemosensory signals [10]. As shown by this example, future neurophysiological studies should bring the spatial and the temporal dimensions of the odor code together, using recording methods that allow both good spatial and temporal resolution. In this way, characterizing the fine relationship between the temporal and the spatial dimension of olfactory coding at the level of the AL/OB will be possible.

References

1. Mombaerts P, Wang F, Dulac C, Chao SK, Nemes A, Mendelsohn M, Edmondson J, Axel R (1996) Visualizing an olfactory sensory map. Cell 87:675–686
2. Laurent G, Stopfer M, Friedrich RW, Rabinovich MI, Volkovskii A, Arbanel H (2001) Odor encoding as an active, dynamical process: experiments, computation, and theory. Annu Rev Neurosci 24:263–297
3. Kauer JS, White J (2001) Imaging and coding in the olfactory system. Annu Rev Neurosci 24:963–979
4. Guerrieri F, Schubert M, Sandoz JC, Giurfa M (2005) Perceptual and neural olfactory similarity in honeybees. PLoS Biol 3(4):e60
5. Stopfer M, Bhagavan S, Smith BH, Laurent G (1997) Impaired odour discrimination on desynchronization of odour-encoding neural assemblies. Nature 390:70–74
6. Schaefer AT, Margrie TW (2007) Spatiotemporal representations in the olfactory system. Trends Neurosci 30:92–100
7. Martin C, Gervais R, Chabaud P, Messaoudi B, Ravel N (2004) Learning-induced modulation of oscillatory activities in the mammalian olfactory system: the role of the centrifugal fibres. J Physiol Paris 98:467–478
8. Lagier S, Carleton A, Lledo PM (2004) Interplay between local GABAergic interneurons and relay neurons generates gamma oscillations in the rat olfactory bulb. J Neurosci 24:4382–4392
9. Friedrich R, Laurent G (2001) Dynamic optimization of odor representation by slow temporal patterning of mitral cell activity. Science 291:889–894
10. Lei H, Christensen TA, Hildebrand JG (2004) Spatial and temporal organization of ensemble representations for different odor classes in the moth antennal lobe. J Neurosci 24:11108–11119

Odor Detection

Definition

The sensory process by which an external odorant stimulus elicits an odor sensation, without necessarily identifying the exact quality of the detected stimulus. Odor detection is released when a stimulus reaches the detection threshold (or absolute threshold), that is the lowest stimulus capable of producing a sensation that something has changed in reference to a control stimulus. Odor detection can be conscious or unconscious. Conscious odor detection is ordinarily revealed by behavioral or verbal responses. Unconscious odor detections can be revealed by recording the alteration in the reactivity of the autonomous nervous system. Odor detection is compromised in many conditions, especially in Parkinson disease and the later stages of Alzheimer's disease and following damage to the olfactory mucosa or bulb.

►Alzheimer's Disease
►Olfactory Hallucinations
►Parkinson Disease
►Smell Disorders

Odor Discrimination

Definition

This is the ability to detect differences between odors. This is measured in several ways, all of which involve presenting two different smells and having the participant judge whether they are the same or different. Odor discrimination allows to extract an olfactory signal from a background and to make a distinction between different odorant molecules. Whilst compromised odor detection will always affect identification and discrimination, impaired discrimination (or identification) can occur independently of detection.

►Odor
►Olfactory Hallucinations

Odor Expertise

►Olfactory Perceptual Learning

Odor-exposure Learning

►Olfactory Plasticity

Odor Familiarity

►Olfactory Perceptual Learning

Odor Identification

Definition

This is the ability to correctly provide a name for an odor, when no other cue to its identity is present. This may be measured by simply asking a person to generate a name, or by providing a list of names from which the person has to choose. The most well established test of olfactory functioning, the Smell Identification Test (SIT), utilizes the latter method.

►Odor
►Olfactory Hallucinations

Odor Image

►Odor Maps

Odor Learning

►Odor – Memory

Odor Maps

Aurelie Mouret
Laboratory for Perception and Memory,
CNRS URA 2182, Pasteur Institute, Paris, France

Synonyms

Odor image

Definition

Extraction of information from an odor stimulus is a multi-level task for the brain, involving levels of neuronal processing from the odorant receptors up to the olfactory cortex. Sensory modality at each level is represented by activity patterns in two-dimensional neural space. Various sensory signals activate topographically distinct subsets of neurons. Such patterns represent odor ►maps.

Characteristics

Olfactory Epithelium

Information processing begins with the mapping of an odorant to the subset of receptors that it activates. More than 400,000 compounds are thought to be odorous to the human nose; mammals have developed nearly 1,000 types of odorant receptors to cope with this huge variety of odorants. Odorant receptors are expressed on the cilia of olfactory sensory neurons situated in the nasal olfactory epithelium. Each receptor presumably detects particular ►molecular features of odorants and thus binds to a specific range of odorants sharing common features [1]. However, each odorant can bind to multiple, but specific, odorant receptors. Thus, an odorant or a mixture of odorants will activate a specific combination of odorant receptors located within the olfactory epithelium. At this level, the odor ►map may be considered a map of receptor space, providing practically unlimited coding capacity for the olfactory system. Olfactory sensory neurons usually produce only one odorant receptor type, so the odor map of receptor space directly translates to a map of activated olfactory sensory neurons. A spatial dimension is also present, at least in mammals in which olfactory epithelium is divided into four zones [1]. A given odorant receptor is only produced by sensory neurons distributed throughout one of the four zones (Fig. 1). Domains of receptor production have also been identified in fish and insects.

Olfactory Bulb

In rodents, with few exceptions, olfactory sensory neurons producing the same odorant receptor converge onto two topographically fixed glomeruli, one in the lateral and the other in the medial part of the main olfactory bulb, arising from sensory neurons in the lateral and medial epithelium, respectively. In mice, there are approximately 2,000 glomeruli and their localization is roughly conserved among individuals. Each glomerulus represents a single odorant receptor; thus, the glomerular sheet of the olfactory bulb forms a map of odorant receptors [1]. Furthermore, neurons that are segregated in the epithelium extend to distinct regions of the bulb, such that the spatial topography of the nasal epithelium is preserved in the glomerular sheet, following the principle of a "zone-to-domain" projection (Fig. 1). Thus, two symmetrical sensory maps are generated; one is in the rostrolateral hemisphere and the other is in the caudomedial hemisphere. However, a zone-specific expression pattern in the olfactory epithelium is not present in a small group of odorant receptors. An individual odorant receptor of

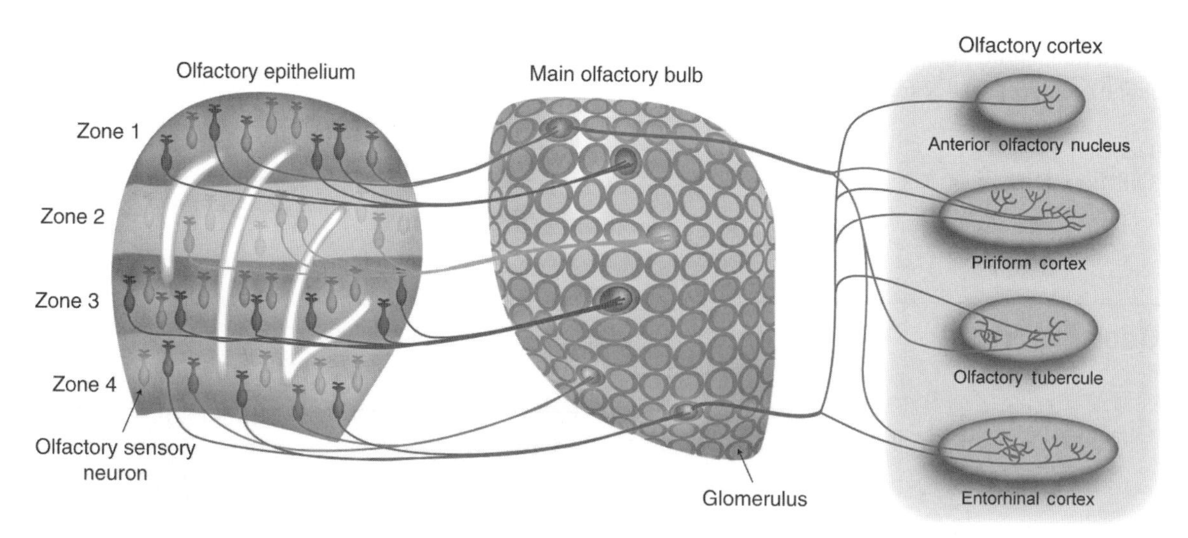

Odor Maps. Figure 1 *From odorant receptors to the olfactory cortex:* Olfactory sensory neurons expressing a given odorant receptor are distributed widely in one of the four zones and converge their axons onto a few topographically fixed glomeruli in the olfactory bulb. Each glomerulus represents a single odorant receptor. Mitral cells from the glomeruli form synapses with clusters of neurons in multiple olfactory cortical areas. Inputs from different odorant receptors overlap spatially. (*Olfactory bulb outputs from two glomeruli only are displayed for more clarity*).

this group is typically represented by a single glomerulus located at the most ventral portion of the bulb [2]. Thus, the non-zonal odorant receptors generate a small map at the most ventral part of each main olfactory bulb. Convergence is less strict in the accessory olfactory system (which processes some pheromones) and similar olfactory sensory neurons can converge onto multiple neighboring glomeruli. Nevertheless, odorant quality is represented through spatial patterns of glomerular activation in both cases, reflecting differential activation of olfactory sensory neurons (Fig. 1). This principle of odor mapping is widely observed in various vertebrate species and in invertebrates, including honeybees, moths and flies. Moreover, the odorant-specific spatial positions of activated glomeruli are conserved in animals of the same species.

Individual glomeruli in the olfactory bulb function as molecular-feature detecting units: they respond to a range of odorants sharing specific combinations of molecular features. Furthermore, glomeruli with similar response properties are located in close proximity and form molecular-feature clusters [1]. This is consistent with evidence that sensory neurons expressing homologous odorant receptor genes project their axons to neighboring glomeruli. A precise chemotopic organization is sometimes present within glomerular clusters. For instance, a chemotopic progression with increasing odorant carbon number has been detected in multiple response clusters [2]. So, the glomerular sheet of the bulb topographically represents the characteristic molecular features in a systematic, gradual and multidimensional fashion.

Olfactory Cortex

Each glomerulus is a spherical neuropil containing the axons of several thousand olfactory sensory neurons that establish synapses with dendrites of approximately 50 mitral and tufted cells (the olfactory bulb projection neurons) and local interneurons. Axons of mitral cells carrying input from a given olfactory receptor synapse with multiple specific clusters of pyramidal neurons in the olfactory cortex, generating a stereotyped map of olfactory receptor inputs that is different from that in the olfactory bulb. The projections to the olfactory cortex are diffuse and have characteristics of a combinatorial array, with extensive overlap of afferent inputs and widespread intracortical association connections. Thus, inputs from different odorant receptors are mapped onto partially overlapping clusters of pyramidal neurons [3] (Fig. 1). It appears that individual neurons receive signals from various odorant receptors. Thus, although inputs from various odorant receptors are segregated in the olfactory epithelium and olfactory bulb, single neurons in the olfactory cortex seem to combine multiple inputs. The olfactory cortex is thought to be important for integrating signals from various molecular-feature-detecting units of the bulb.

Mapping Methods

Several mapping methods have been used to identify odor-specific spatial activation patterns in the olfactory bulb in mammals [1]. These methods can be classified into two complementary groups, each with their advantages and disadvantages. The first group has the advantage that the responses are mapped over the entire bulb and includes methods involving functional MRI (fMRI) and assessment of 2-deoxyglucose uptake, expression of immediate early genes (e.g. *c-fos, c-jun, Arc* and *zif268*) and production of phosphorylated ERK. These methods allow investigation of how individual odorants are represented within the entire glomerular sheet of the bulb. The disadvantage of this group is that, with the exception of fMRI, these methods map the response to only one odorant in each animal.

On the contrary, the second group of methods facilitate mapping of the responses to many odorants in the same bulb of an animal. This group includes optical imaging of intrinsic signals, imaging with calcium-sensitive or voltage-sensitive dyes, imaging using pHluorin, electrophysiological recording of single neuron activity and fMRI. With these methods, it is possible to determine the range of odorants that activate an individual glomerulus. However, again except for the fMRI method, these methods allow us to map only the exposed surface of the bulb. Only the dorsal and posterolateral surfaces have been successfully mapped thus far.

Functional Relevance of the Spatial Arrangement of Glomeruli in the Olfactory Bulb

If the spatial map is important to olfactory behavior, then disrupting the map should impair one or more olfactory functions. Slotnick and colleagues tested this hypothesis in a series of behavioral experiments and showed that ablations of large portions of the olfactory bulb and other destruction of olfactory inputs did not significantly impair odor discrimination and detection. Furthermore, animals trained before such manipulations can often still recognize the same odors after ablation. Even rats with no bulb can carry out olfactory discriminations, supported by olfactory nerve inputs that reinnervate areas of the olfactory cortex [4]. Rather than concluding that spatial maps have only a minor function in the olfactory bulb, it may be argued that discrimination and detection of odors are not the computations facilitated by these mechanisms.

Moreover, the chemotopic arrangement of glomeruli in the bulb seems to have a functional relevance. Even though the relationship between the molecular structures of odorants and their subjectively perceived odors is not entirely clear, odorants with similar combinations of molecular features tend to have similar odor qualities, at least for the human nose. Thus, it is possible that molecular-feature clusters of glomeruli are part of the representation of basic odor quality [1]. There are various lines of evidence that favor this hypothesis. Measurements of spontaneous responses show that rats generalize between odor pairs with very similar glomerular activity maps, but not between odor pairs with different glomerular maps. However, rodents can be trained by differential reinforcement to discriminate between all odor pairs tested to date with high accuracy. Nevertheless, although discrimination performance in rats is always very good, there is still a significant correlation between glomerular map dissimilarity and discrimination accuracy. Lateral inhibition among neighboring glomeruli may allow mitral cells to respond to a narrower range of stimuli than their associated sensory neurons. This possibly permits a smaller overlap in the number of highly activated mitral cells responding to two similar odorants, thus facilitating their discrimination. Lateral projections of interneurons that are distributed more densely between neighboring than distant glomeruli confirm this hypothesis. Therefore, the spatial clustering of glomerular responses may coordinate the principle responses of bulbar projection neurons by way of center-surround functionality implicating inhibitory interneuronal networks.

Despite the accepted correlation between odor maps and odorant structural commonalities, this relationship breaks down if odorant concentration is included as a variable. If odorant concentrations are increased, more glomeruli respond and odor maps broaden and intensify [5]. The recruited glomeruli are located near the originally activated glomeruli due to chemotopic clustering of glomeruli with similar odorant specificities. Higher odorant concentrations recruit additional sensory neuron populations with progressively lower affinities for the presented agonist. However, the qualitative perception of odors is usually not affected by variability in concentration, suggesting that various neural normalization mechanisms can preserve concentration-independent odor quality information. Regardless of concentration, relative levels of glomeruli activation in the bulb are stable and the representation of odor quality may rely on these activity patterns [5]. The impact of stimulus concentration is not as high in mitral cells and increasing odorant concentrations do not monotonically increase their spiking rates. The mechanisms for normalization of olfactory representations are not precisely known, but it is possible that they do not rely on center-surround inhibition, as global normalization has to be carried out for the entire bulb.

Development of the Glomerular Map in the Olfactory Bulb

Creation of the map begins prenatally when axons of olfactory sensory neurons navigate toward the bulb, resort in a receptor-specific manner and terminate in a broad area of the bulb surface, interdigitated with other axon populations. Only postnatally, the axons segregate into completely separate glomerular structures. This maturation process requires various amounts of time, ranging from a few days to about one month, depending on the glomerulus [6]. Very precise axonal targeting is achieved, even for populations expressing highly related odorant receptors and innervating neighboring glomeruli.

The complex processes of axon navigation, fiber sorting and cell recognition are governed by a hierarchical system of recognition and adhesion molecules. Attractive or repulsive interactions apparently drive the growing axons towards or away from regions of the bulb. However, the diversity of the guidance molecules that have been identified is not sufficient to explain the precise topographical glomerular map observed in the bulb. The odorant receptor protein is itself involved in axon guidance and may control the production of guidance molecules and adhesion molecules [7]. Whereas the initial (prenatal) process of glomerulization mainly requires molecular determinants, postnatal activity-dependent processes refine glomerular organization. Whether genetic or activity-dependent mechanisms are dominant in this process of map formation, it is clear that the cues organizing these connections must be present throughout the life of the animal and not only during the initial phases of olfactory development. The olfactory epithelium is continuously self-renewed and olfactory sensory neurons are continuously replaced by newborn neurons that can re-establish good glomerular connections. Thus, the glomerular map does not change throughout adulthood.

Dynamics of the Odor Maps in the Olfactory Bulb

Odorant responses are often considered static spatial entities. However, various factors may influence the primary sensory input to the olfactory bulb and give rise to differences in the timing of glomerular responses to odorants. First, the nature of the airflow in the naris of rodents causes stimuli to arrive at receptors in various expression zones within the olfactory epithelium at different times [2]. There is a chromatographic effect in the nasal cavity and various odorants are chemically converted in the nasal mucosa before linking to their receptors. This may explain how various odorants can activate the same glomeruli with different kinetics. Furthermore, individual sensory neurons expressing the same odorant receptor may have identical odorant response profiles, but different activation thresholds and their axon terminals may be modulated presynaptically. This widens the range of population terminals converging into a single glomerulus. The dynamics of glomerular activation also depends on the breathing cycle and changes within a respiration cycle and from one cycle to the next [8]. Thus, whereas some glomeruli respond less strongly during the second breathing cycle, suggesting that adaptation occurs, others respond more strongly, indicating that other processes also contribute to the dynamics observed. The active inhalation pattern of the animals also controls adaptive filtering to detect changes in odor landscape. Thus, neural representations of the same odorant sampled during low-frequency passive respiration and high-frequency sniffing differ [9]. Consequently, glomerular odorant responses differ in amplitude, latency and rise time in an odorant-specific manner and is also dependent on sniffing behavior for a particular odorant. Conjointly, mitral and tufted cell activities also demonstrate stimulus-specific temporal structure. Thus, a temporal code for odorant quality may be embedded in these temporal bulbar activation differences. Spatial distribution and the temporal structure of neuronal activity should therefore not be studied in isolation, but considered as a single entity of the same coding process. Currently, although there is increasing evidence for the importance of temporal structure in bulb odorant-evoked output, little is known about how this temporal patterning is translated within cortical neural ensembles.

Most studies on odor maps have been done with naive animals and have confirmed that they are conserved from one individual to another within the same species. Depending on the mapping method, these maps are not entirely similar because they require animals that are either awake or anesthetized. Anesthesia may itself modify odor processing. In animals that are awake, the output of the olfactory bulb represents the integration of odor stimuli and behavioral variables relevant to odor expectation, discrimination, context and predictive associations. Thus, a certain degree of map flexibility is expected, depending on the behavioral context and on the physiological state of the animal. The fact that the spatio-temporal output of the bulb is affected by learning is consistent with this theory. Training can modify the odor map [10], challenging the findings of studies that put in parallel behavioral performances of trained animals and odor maps of naive animals. Odor maps are dynamic and various changes, particularly those induced by training, may be long-lasting. The centrifugal fibers that richly innervate the bulb can modulate odorant perception and may affect spatial and temporal patterning of glomerular activation. Another factor of bulbar functional plasticity is the continuous neurogenesis occurring in the bulb. Learning induces changes in neurogenesis in the bulb, which may support long-lasting changes in odor maps.

References

1. Mori K, Takahashi YK, Igarashi KM, Yamaguchi M (2006) Maps of odorant molecular features in the Mammalian olfactory bulb. Physiol Rev 86:409–433
2. Johnson BA, Leon M (2007) Chemotopic odorant coding in a mammalian olfactory system. J Comp Neurol 503:1–34
3. Zou Z, Li F, Buck LB (2005) Odor maps in the olfactory cortex. Proc Natl Acad Sci USA 102:7724–7729
4. Slotnick B, Cockerham R, Pickett E (2004) Olfaction in olfactory bulbectomized rats. J Neurosci 24:9195–9200
5. Cleland TA, Johnson BA, Leon M, Linster C (2007) Relational representation in the olfactory system. Proc Natl Acad Sci USA 104:1953–1958
6. Strotmann J, Breer H (2006) Formation of glomerular maps in the olfactory system. Semin Cell Dev Biol 17:402–410
7. Serizawa S, Miyamichi K, Takeuchi H, Yamagishi Y, Suzuki M, Sakano H (2006) A neuronal identity code for the odorant receptor-specific and activity-dependent axon sorting. Cell 127:1057–1069
8. Spors H, Wachowiak M, Cohen LB, Friedrich RW (2006) Temporal dynamics and latency patterns of receptor neuron input to the olfactory bulb. J Neurosci 26:1247–1259
9. Verhagen JV, Wesson DW, Netoff TI, White JA, Wachowiak M (2007) Sniffing controls an adaptive filter of sensory input to the olfactory bulb. Nat Neurosci 10:631–639
10. Salcedo E, Zhang C, Kronberg E, Restrepo D (2005) Analysis of training-induced changes in ethyl acetate odor maps using a new computational tool to map the glomerular layer of the olfactory bulb. Chem Senses 30:615–626

Odor Memory

►Olfactory Perceptual Learning

Odor Perception

Definition

The ability to detect and recognize an odor

►Olfactory Perception
►Olfactory Sense

Odor Receptor

►Odorant Receptor

Odor Recognition

Definition

The perceptual process by which an odor sensation is cognitively related to its source or, in humans, by which an odor sensation evokes a verbal label that designates its source. In theory, odor recognition occurs at the recognition threshold, that is when a odor stimulus reaches the quantitative level at which it can be qualitatively recognized.

►Olfactory Perception

Odor Representation

►Odor Coding

Odor Sampling

Definition

Active exploration of an odor including acceleration of respiratory rhythm called sniffing behavior.

►Odor-sampling Behavior

Odor-Sampling Behavior

GÉRARD COUREAUD[1], FRÉDÉRIQUE DATICHE[2]
[1]Ethology and Sensory Psychobiology Group, European Center for Taste and Smell, CNRS/University of Burgundy/INRA, Dijon, France
[2]Neurophysiology of Chemoreception Group, European Center for Taste and Smell, CNRS/University of Burgundy/INRA, Dijon, France

Synonyms

Sniffing behavior (mammals); Wing fanning (insects); Flicking behavior (crustaceans); Coughing (fishes)

Definition

Odor-sampling designates a behavior by which animals actively collect air-borne or water-borne odor stimulus carrying information from the surroundings, in order to localize and/or identify the source of the emitted odor, and to respond in an adaptive manner (e.g. approach, avoidance) to the stimulation. To collect the odor stimulus, the organism may sniff, flick, fan, cough or bubble (according to the species and the environment), behaviors that consist in the active drive of air or water across or into the olfactory organ (sniffing, fanning, nasal sac compressing - coughing -, bubbling), or in the moving of the organ through the fluid carrying the stimulus (flicking).

Characteristics

Environment as a World of Odors

In the animal kingdom, odors are important vectors of information likely to elicit behavioral decisions supporting adaptive responses to social and feeding needs. Thus, from early to late development, olfaction is involved in detection and localization of, and communication with, conspecifics, detection of competitors and predators, selection of habitats, localization of preys and more generally of food.

However, animals are only intermittently exposed to odor stimuli. Indeed: (i) the olfactory organ (e.g. nose, antenna) is a structure which anatomically protects the substructures carrying the olfactory receptors, and therefore limits or blocks the continuous access of odor molecules to the receptors; (ii) informative odor cues (signals) are often sporadically emitted from odor sources spatially dispersed; (iii) odors are transported in the environment by wind or water currents submitted to physical turbulences (i.e. odors generally consist of plumes, patches or filaments in aerial and marine environments). In other words, odors in the ambient air or water are fluctuating both temporally and spatially. This creates the necessity for animals to sample their olfactory environment, i.e. to extract and gain access to the odor cues. Odor-sampling behavior responds to this necessity, in allowing a voluntary (intermittent) exposure to specific and ephemeral olfactory information emanating from the surroundings. In addition, odor-sampling behavior, coupled to the olfactory organ morphology, may form the first level of signal filtering, before its processing at the receptor then neural levels.

Odor-Sampling Behavior in Terrestrial Environment

In humans, and mammals in general, odor-sampling that follows the detection of an odor is supported by a so-called "sniffing" behavior. During a sniff, air enters through the nostrils (anterior nares), and continues through the nasal cavity, then out the posterior nares to the top of the throat. Part of the airflow reaches the olfactory epithelium, which lines the roof of the nasal cavity (below the cribiform plate). Usually, a single human sniff approximately has a duration of 1.6 s, an average inhalation velocity of 30 l/min (twice that of a normal inspiration), and a volume of 500 cm^3. However, humans generally take several successive sniffs to sample odors, thus displaying sniffing episodes rather than single sniffs. During an episode, each sniff has a reduced duration and volume as compared to a single sniff, but the average inhalation velocity remains the same. Multiple sniffs are quite surprising knowing that odor presence and intensity can be determined, in laboratory conditions, in a single sniff. But sniffing episodes are certainly necessary in natural conditions, where the localization, identification and discrimination of odors constitute difficult tasks due to air turbulences and exposure to complex mixtures (emanating from biological sources) [1]. Human odor-sampling may for instance impact scent-tracking abilities, and is correlated with food neophobia.

In rodents, nostrils act as flow diverters during sniffing, permitting to inspire air from the immediate front of the snout and to expire it backward. Such aerodynamics makes sense, allowing extracting odor cues from the environment while reducing the disturbance of the olfactory sample. The sniffing behavior by itself consists in a relatively stereotyped sequence divided in two successive phases. During the first phase, the animal fixes the head, protracts the vibrissae, inhales briefly, and retracts the tip of the nose. Then, during the second phase, it retracts the vibrissae, exhales and protracts the nose. Generally, the entire sequence is repeated, after repositioning of the head, at around 4–12 Hz and occurs in bouts lasting 1–10 s. Sniffing behavior is considered to be synchronized with whisking, head bobbing and heartbeat. Recently, it was suggested to be constituted, in rats, by two successive modes: type-I sniffing, displayed with a respiration frequency of 6–9 Hz, allowing the acquisition of odor information; then type-II sniffing (9–12 Hz), preparing the animal to display the behavioral response accompanying its final decision [2].

In insects, olfactory receptors are borne by chemosensory sensilla carried by the antennae. Usually, the sensilla form a dense boundary layer between the whole antennae and the receptors. To sample odors from the surroundings, animals display particular wing motions that induce pulses of air flowing to the body, from front to rear. The consequence is an increase in the interception of chemical signals on the olfactory sensilla, due to a decrease in the depth of the boundary layer. Typical wing motions allowing such sampling happen during flight (these motions differ in angle and amplitude from those typically used to fly), or during walking in flying and non-flying insects. Wing motions displayed by walking insects to sample odors are named "wing fanning". This latter behavior severely increases the air penetration and rate of interception of odorant molecules both into the antennae and the sensilla: in silkworm moth (*Bombyx mori*), the airflow produced is 15 times faster at the level of the antennae, and 560 times faster at the level of the sensilla [3], as compared to walking.

Whatever the species, and in addition to the increase in the capture rate of odorants, sniffing and wing fanning may also have a second function: to replace the fluid volume being sampled, i.e. the fluid volume adjacent to the surface of the chemosensory structures. Both functions may occur with a single increase in velocity of airflow, or with periodic fluctuations in velocity (thus minimizing ▶habituation and ▶familiarization processes) [3].

Odor-Sampling Behavior in Aquatic Environment

Among arthropods, crustaceans present adaptations illustrating odor-sampling behavior. Crustaceans have different chemosensory organs, among which the lateral flagella of the first antennae (lateral antennules) constitute olfactory organs. In the American lobster (*Homarus americanus*) and Spiny lobster (*Panulirus argus*), for instance, olfactory sensilla (called aesthetascs) form a dense "toothbrush" on the distal half of the antennules. The brush forms, as in insects, a boundary layer which shields the receptors from odor access. When they perceive a chemical signal, lobsters generally wave their antennae and increase the rate of "antennule flicking" (the right and left antennules may flick independently). This behavior allows water to be driven at high velocity through the brush, the boundary layer to be decreased, and then stimulus access to the chemoreceptors (carried by the antennules) to be increased. In other words, antennular flicking is a form of "sniffing" in this taxon, and allows odor perception. It constitutes a behavioral expression which can be easily quantified, and which is therefore used to determine the biological relevance of stimuli. Antennular flicking is critical for efficient orientation behavior [4].

In fishes, odor-sampling behavior has often been thought to be relatively involuntary. In teleostean fishes, olfaction occurs when the water flow is sufficient to bring odor molecules in contact with the receptors embedded in the ciliated olfactory epithelium. The epithelium is located in two nasal sacs (situated in the dorso-anterior part of the head) opened by one or two nares. "Passive" increase of the water flow is induced by ciliary action of cells from the epithelium and by the increase in swimming speed (isosmate fishes), or by continuous pumping in the nasal chambers related to respiration (cyclosmates). However, voluntary sniffing behavior, named "coughing", has also been suggested. In pleuronectid flounders (e.g. *Lepidopsetta bilineata*, *Platichthys stellatus*; cyclosmates), coughing consists in the rapid protrusion of the jaw, coupled with an expulsion of water from the mouth and an entrance of water in the nasal chambers through the nares. Then, the mouth closes, and water is rapidly expulsed from the nares. This behavior is usually displayed into a stereotyped behavioral sequence including the lift of the head off the substratum, and the orientation to the odor source. Coughing is, for instance, strongly displayed in response to food odorants. It is suggested to support voluntary and frequent sampling of small odorant patches, allowing to gain access to specific odor cues more efficiently than through the continuous circulation of water tied to respiration. Coughing may also have another function: the ejection of foreign material from the olfactory chambers or gills [5].

Finally, it is generally considered that mammals cannot sniff and smell in aquatic environment (except fetuses in the womb) since they are not able to inspire air. However, a recent study brings evidence that in semi-aquatic mammals, a particular mechanism may allow to sample odor underwater: the star-nosed mole (*Condylura cristata*) and the water shrew (*Sorex palustris*) are indeed able to exhale air bubbles onto objects or scent trails before re-inspiring these bubbles. The re-inspiration brings back into the nose the smell of the environmental targets contacted through the bubbles. Interestingly, the volume of air corresponding to these bubbles, the rate of airflow and the frequency characterizing this behavior appear similar to that related to sniffing in small rodents living above water. Such underwater sampling behavior can therefore be considered equivalent to sniffing in the air [6].

Functional Aspects of Odor-Sampling

Odor-sampling behavior is not only dedicated to the transport of odorants from the environment to the olfactory receptors. It is a dynamic process which directly participates in the temporal and spatial coding of odor stimuli. More generally, it constitutes a main component of olfactory processing and influences olfactory percept. For instance, in humans, functional magnetic resonance imaging (fMRI) demonstrates that odor-sampling (sniffing) induces activity in the primary olfactory cortex, and that this activation reflects the encoding of air flow as a factor contributing to the computation of odor intensity and identity [7].

The changes in air flow induced by sniffing through the nasal cavity (in mammals) could influence the mechanical component of the odor perception: in the olfactory epithelium, olfactory neurons detect the chemical but also mechanical stimulation caused by odorant molecules. Regarding the olfactory perception per se, variations in air flow result first in distinct retention of the odorants carried by the flow, and therefore in distinct perception. Thus, high or low velocities respectively optimize perception of odorants presenting higher, or lower, sorption rate. Second, the air flow related to sniffing also influences the distribution of odorant molecules over the epithelium. By the way of this active mechanism, distinct odorants are spatially directed to distinct regions of the nasal cavity and to different populations of olfactory receptors, a process called "zonation" [8]. Subsequently, sniffing impacts the spatial representation of an odor at the level of the olfactory bulb, and influences the detection, identification, and discrimination abilities of animals.

Moreover, sniffing behavior carries temporal information about volatile cues throughout the olfactory system, from the olfactory bulb to higher cerebral structures. This impact is important knowing that temporal properties of an odor cue contribute to its representation. From this point of view, electrophysiological recordings reveal that odor-related activity in the olfactory bulb is strongly modulated by respiration and that the phase of spiking relative to the sniff cycle might encode information regarding odorant intensity and quality. Interestingly, the slow theta rhythm (4–12Hz in rats) generally recorded during sniffing in the mitral cell layer of the bulb, is also observed in the hippocampus, a structure involved in memory and orientation behavior. Such coherence in frequency between distinct brain areas might illustrate the cooperation of sensory, motor and cognitive cerebral regions expressed when the animal is engaged in an adaptive task. For instance, theta oscillations are both displayed in the olfactory bulb and dorsal hippocampus of rats that are sniffing during the initial stages of a reversal odor learning [9].

Finally, in complex natural scenes, sniffing plays a role in odor perception through successive sniffing cycles (even if a single sniff supports odor detection and identification). Successive samplings participate in progressive change of olfactory network dynamics which may then lead to a might converge, by the repetition of sniffing actions, in a more precise odor representation. From this point of view, multiple sniffs compose a synthetic memory-based system forming "perceptual gestalts" [10], which might be determinant for analysis of complex olfactory mixtures, identification of relevant odor cues and scent-tracking.

References

1. Laing D (1983) Natural sniffing gives optimum odor perception for humans. Perception 12:99–117
2. Kepecs A, Uchida N, Mainen ZF (2007) Rapid and precise control of sniffing during olfactory discrimination in rats. J Neurophysiol 98:205–213
3. Loudon C, Koehl MAR (2000) Sniffing by a silkworm moth: wing fanning enhances air penetration through and pheromone interception by antennae. J Exp Biol 203:2977–2990
4. Atema J (1995) Chemical signals in the marine environment: Dispersal, detection, and temporal signal analysis. Proc Natl Acad Sci USA 92:62–66

5. Nevitt GA (1991) Do fish sniff? A new mechanism of olfactory sampling in pleuronectid flounders. J Exp Biol 157:1–18
6. Catania KC (2006) Underwater "sniffing" by semi-aquatic mammals. Nature 444:1024–1025
7. Mainland J, Sobel N (2006) The sniff is part of the olfactory percept. Chem Senses 31:181–196
8. Schoenfeld TA, Cleland TA (2006) Anatomical contributions to odorant sampling and representation in rodents: zoning in on sniffing behavior. Chem Senses 31:131–144
9. Macrides F, Eichenbaum HB, Forbes WB (1982) Temporal relationship between sniffing and the limbic theta rhythm during odor discrimination reversal learning. J Neurosci 2:1705–1717
10. Wilson DA (2001) Receptive fields in the rat piriform cortex. Chem Senses 26:577–584

Odor Selectivity

Definition

Property of neuron responses (firing rate or other measure of odor response) that varies dependent on the odorant stimulus.

►Olfactory Information

Odor Tracking (Localization)

Definition

The chain of motor actions by which animals search and efficiently orient to a source of odor cues over short or long distances. The recipient organism displays general body movements (as in male moth approaching a female) or local head movements (as in mammalian newborns locating the mother's nipple) to create sensory asymmetry in the plumes released by an odor source in order to stimulate chemosensors located in or on bilateral organs (antennae, nasal fossae).

►Social Chemosignal

Odorant

Definition

An odorant is a volatile chemical molecule that that naturally exists as a component of an odor.and is sensed by the olfactory system. Odorants stimulate sensory neurons of the olfactory system in the nasal cavity by binding to odorant (olfactory) receptor proteins on the cell membrane, triggering an electrical response that can be transmitted to the brain. The ability of an odorant to bind to and activate an olfactory receptor protein depends on molecular features such as the size, shape and presence of functional groups. Naturally occurring odors may be composed of hundreds of odorants.

►Glomerular Map
►Memory – Odor
►Odorant Receptor Protein
►Odor
►Olfactory Perceptual Learning

Odorant-Binding Proteins

Loïc Briand
Unité Mixte de recherche FLAVIC INRA-ENESAD-Université de Bourgogne, Dijon, France

Synonyms

Odor-binding proteins; Olfactory binding proteins

Definition

Odorant-binding proteins (OBPs) are abundant small soluble proteins secreted in the ►nasal mucus of a variety of species, from insects to vertebrates including human beings. OBPs reversibly bind odorants with dissociation constants in the micromolar range and are good candidates for carrying airborne odorants, which are commonly hydrophobic molecules, through the aqueous nasal ►mucus towards olfactory receptors. Although the physiological function of vertebrate OBPs is not yet clearly established, their essential role in eliciting the behavioral response and odor coding have been demonstrated in the fruit fly [1].

Characteristics

General Properties of Vertebrate Obps

OBPs are secreted by the olfactory epithelium in the nasal ►mucus at high concentration (~10 mM). They reversibly bind odorants with dissociation constants in the micromolar range [2]. OBPs have been identified in a variety of vertebrates including cow, pig, rabbit, mouse, rat, xenopus, elephant and human beings [2–4]. Different OBP subtypes have been reported to occur simultaneously in the same animal species, two in pig, four in mouse, three in rabbit and at least eight in porcupine. In rat, three OBPs have been cloned with quite different sequences and binding properties [5]. Molecular weights of OBPs fall within a narrow range (around 18 kDa). They are highly soluble proteins belonging to the lipocalin superfamily. As regards their quaternary structure, some OBPs were observed as monomers, such as porcine, rat OBP-3 or human OBP, while some others are found as dimers, such as bovine OBP, rat OBP-1 and OBP-2. OBP heterodimers have also observed in mouse. The typical isoelectric point of OBPs is in the acidic range, between 4 and 5. However, some rare OBPs exhibit a neutral or slightly basic isoelectric point, such as rat OBP-2 and human hOBP-2A. As sites of production, OBPs are synthesized within the nasal cavity, but in different glands and areas. Some OBPs have been clearly shown to be expressed in the olfactory area by the Bowman's glands.

OBPs are also found in the sensillary lymph of insect antennae. Although insect OBPs seem to play a similar role in olfaction, they do not share any amino acid sequences or structural similarities with vertebrate OBPs [5].

Human OBPs

Two putative human OBP genes (named $hOBP_{IIa}$ and $hOBP_{IIb}$) localized on chromosome 9q34 were first described before evidence of human OBP expression in the mucus covering the olfactory cleft [4]. The $hOBP_{IIa}$ gene codes for a protein, called hOBP-2A, which is 45.5% homologous to rat OBP-2. This gene is transcribed in the nasal cavity, in contrast to$hOBP_{IIb}$, which is transcribed in the genitals and codes a protein that is 43% identical to the human tear lipocalin-1. The presence of human OBP expression appears limited to the uppermost region of the ►nasal passage where odorant molecules are detected by olfactory receptor neurons.

Ligand Binding Properties of OBPs

OBPs bind with high efficiency a large number of odorants belonging to different chemical classes (Fig. 1).

Although no preferential binding was observed with the porcine and bovine OBPs, a broad specificity was revealed by the study of the 3 rat OBPs, which are specially tuned towards distinct chemical classes of odorants. Rat OBP-1 preferentially binds heterocyclic compounds such as pyrazine derivatives and OBP-2 appears to be more specific for long-chain aliphatic aldehydes and carboxylic acids, whereas OBP-3 was described to interact strongly with odorants composed of saturated or unsaturated ring structure [6].

Human OBP-2A was observed to bind many diverse odorants with dissociation constants in the micromolar range, as found in all known vertebrate OBPs [4]. However, specificity of hOBP-2A is more restricted than those of porcine and rat OBP-1 and 3. A chemical specificity of this OBP for aldehydes, either aliphatic or aromatic, enhanced by the size of the odorant molecule, ir clear comparing odorant chemical series. Note that hOBP-2A can also be characterized by its low affinity for a very potent odorant, 2-isobutyl-3-methoxy pyrazine, and a very high affinity for large aliphatic acids.

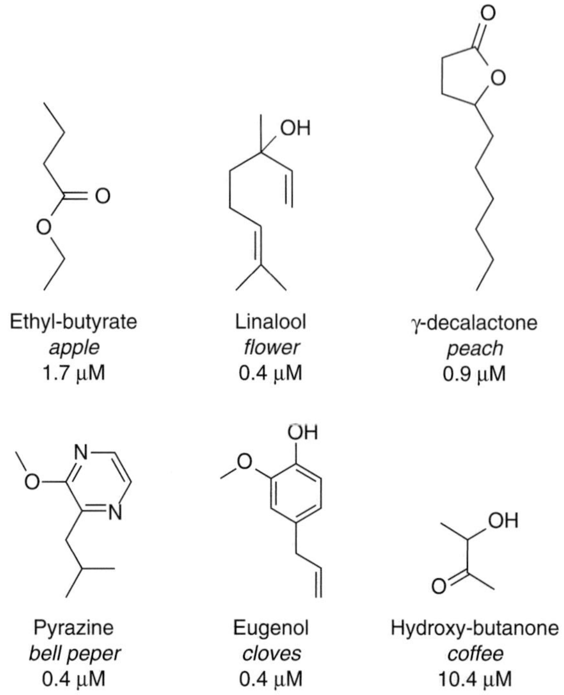

Odorant-Binding Proteins. Figure 1 Examples of odorants presenting different odors, which bind tightly or weakly to rat OBP-1. The dissociation constants of these compounds for rat OBP-1 are indicated in italics.

Consensus Sequence, Homology and Disulfide Bond

All known vertebrate OBPs belong to the lipocalin superfamily. All members of this family have low sequence identity, but few characteristic signatures allow their identification: a GxW motif at about 15–20 residues from the N-terminus, two cysteines in the middle and a glycine at the C-terminal end (Fig. 2).

One of the conserved cysteine residues, located on the fourth strand of the first β-sheet, forms a disulfide bridge tightening the α-helix C-terminal domain and the β-barrel. When comparing OBP sequences, note that the percentage of identity among OBPs is low (21–26% on average) with the bovine and porcine OBP showing a maximal identity (42%), whilst rat OBP-2 exhibits the lowest identity (12–19%) when compared to all other OBPs. Consequently, tissue expression (i.e. in the olfactory epithelium) and ligand binding properties should be systematically taken into account in order to classify OBPs.

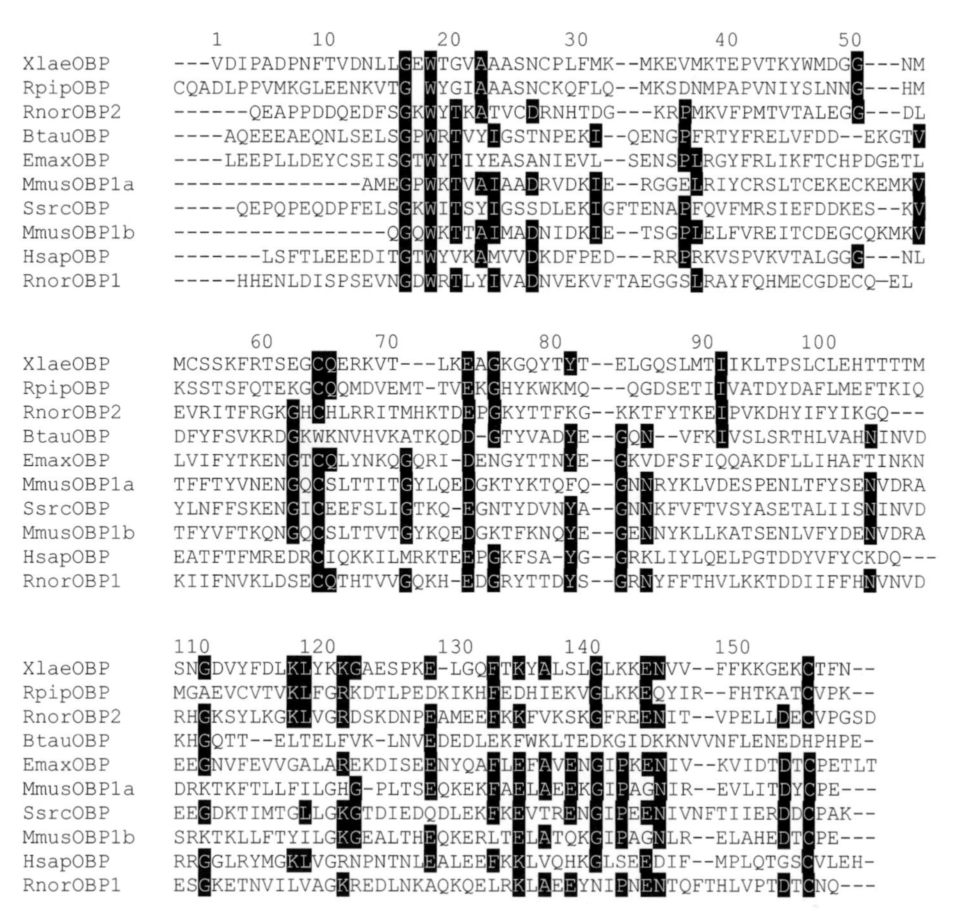

Odorant-Binding Proteins. Figure 2 Sequence alignments of vertebrate OBPs. Conserved amino acid residues are shown white on black background. OBPs are: XlaeOBP (*Xenopus laevis* OBP), RpipOBP (*Rana pipiens* OBP), RnorOBP1 (Rat OBP-1), RnorOBP2 (Rat OBP-2), MmusOBP1a (Mouse OBP subunit IA), MmusOBP1b (Mouse OBP-1B), BtauOBP (Bovine OBP), EmaxOBP (Elephant OBP), SsrcOBP (Porcine OBP) and HsapOBP (Human OBP-2A).

Structural Properties of OBPs

Vertebrate OBPs like other members of the lipocalin superfamily display low sequence similarities, but share a conserved folding pattern made of an 8-stranded anti-parallel β-barrel linked together by seven loops, and connected to an α-helix (Fig. 3). The β-barrel defines a central apolar cavity, called the calyx, whose role is to bind and transport hydrophobic molecules such as odorants [3].

Bovine OBP, which forms a dimer with an elongated shape, was the first OBP whose structure was deciphered through X-ray crystallography [7] and was therefore considered as the prototype of OBP, in spite of the absence of the second disulfide bridge. However, the molecule is not a classical lipocalin, since it exhibited a structural feature called domain swapping. The β-barrel of each monomer comprises its own strands 1–8, but the eighth strand originates from the other monomer. By this mechanism, the C-terminal part of one of the homodimers rotates and takes the place of that of the other. In addition to the buried cavity in the middle of the β-barrel, as in monomeric OBPs, a central pocket, composed of residues belonging to the β-barrel domains and to the C-terminal ends, is located at the dimer interface in communication with the solvent.

Porcine OBP is a monomer whose 3D-structure is typical of a lipocalin. Two cysteine residues form a disulfide bridge between the C-terminal and the loop joining strands 3 and 4 of the β-barrel [8] and a single cavity is observed inside the β-barrel, which does not communicate directly with the external solvent. A few amino acid side chains, which block the access to the solvent, would therefore have to move to make the binding of odorants possible. The cavity is mainly covered with hydrophobic and aromatic side chains.

Structure of the Odorant-Binding Pocket

Up to now, only a few odorant-OBP complexes have been submitted to structural analysis. It has been observed that two odorant molecules could occupy

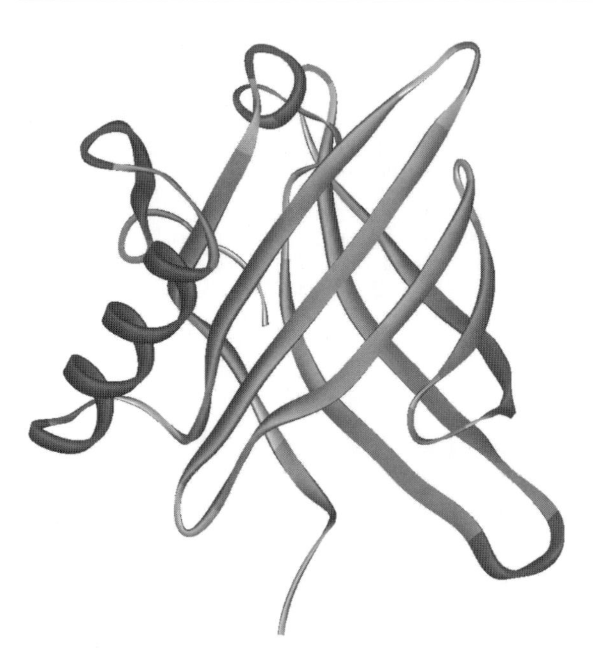

Odorant-Binding Proteins. Figure 3 Ribbon representation of porcine OBP-1 forming a typical lipocalin eight-strand β-barrel, flanked by a single α-helix. The color coding is according to the secondary structure; helices, red; L-strands, cyan; other motifs, green.

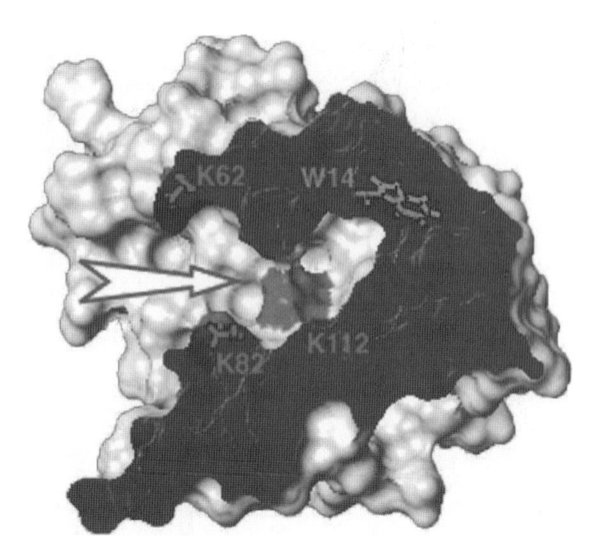

Odorant-Binding Proteins. Figure 4 Slabbed view through the molecular surface and binding-pocket of the predicted 3D-structure of human OBP-2A. In the binding-pocket (*arrow*), lysine side chains and surfaces are colored in red, tryptophan in violet.

the β-barrel cavity of bovine OBP [8]. On the basis of porcine OBP data [9], the most likely binding site is inside the β-barrel, since this may be general for all OBPs. The size of the β-barrel pocket was found to be 780 Å^3 [8] for the bovine protein and about 500–550 Å^3 for the porcine OBP [9]. Using porcine OBP, a limited number of odorants, with relatively good affinity (affinity constants > 10^6 M^{-1}) and different chemical groups (aromatic ring, aliphatic chain or polar group) were co-crystalized with porcine OBP [9]). In the crystalline complexes, the odorant orientation inside the cavity have been proved to be opportunistic with no specific target patches for aromatic or charged group. Interactions between the different odorants and the β-barrel involve most of the residues in the cavity. Except for the two asparagines, which display a polar interaction between the amino acid side chain and the keto oxygen of benzophenone, all interactions are hydrophobic. The number of these interactions appears to be roughly related to the size of the odorant, but without any correlation with affinity measured in solution. Although the odorant-binding pocket is shielded from the solvent, openings have been observed using molecular simulations and it has been proposed that, tyrosine residue Y82 constitutes the door of the cavity. As regard human OBP-2A, its three-dimensional structure have not been yet described but a model has been proposed (Fig. 4). It has been shown using site-directed mutagenesis that affinity enhancement of OBP-2A for aldehydes compared to the corresponding aliphatic acids, could result from an interaction between aldehyde function and lateral chain of a lysyl residue K112, stabilizing odorant docking [10].

Hypothetical Physiological Functions

In mammals and in insects, olfactory receptors are separated from air by a protective layer of hydrophilic secretion, the nasal mucus and sensillar lymph, respectively. Hydrophobic airborne odorants have to cross this aqueous barrier to reach their neuron receptors. OBPs, which have been hypothesized to play such a transporter role, likely appeared during the adaptation to terrestrial life. This carrier role is also supported by their relatively low affinity constant for odorants associated with their high concentration in the olfactory fluids. Their involvement in olfactory discrimination has also been proposed, because of the presence in the mucus of rat of three different OBP subtypes, specifically tuned toward distinct chemical classes of odorants [6]. In addition to the solubilization of odorants, various hypotheses have been proposed for other OBP functions [2]. They could either, (i) filter and buffer odorants in the mucus, then narrow the wide range of odorant intensities, (ii) eliminate odorants after olfactory receptor binding, or (iii) directly interact with olfactory receptors. The essential role of OBPs in eliciting the behavioral response and coding of odor has only been demonstrated in insects. It has been demonstrated that drosophila OBP LUSH is mandatory for the activation of pheromone-sensitive chemosensory neurons [1]. In mammals, it is stimm a matter of debate whether there might be involved ►specific anosmia or ►parosmia.

References

1. Xu P, Atkinson R, Jones DN, Smith DP (2005) *Drosophila* OBP LUSH is required for activity of pheromone-sensitive neurons. Neuron 45:193–200
2. Pelosi P (2001) The role of perireceptor events in vertebrate olfaction. Cell Mol Life Sci 58:503–509
3. Tegoni M, Pelosi P, Vincent F, Spinelli S, Campanacci V, Grolli S, Ramoni R, Cambillau C (2000) Mammalian odorant binding proteins. Biochimica et Biophysica Acta 1482:229–240
4. Briand L, Eloit C, Nespoulous C, Bezirard V, Huet J-C, Henry C, Blon F, Trotier D, Pernollet J-C (2002) Evidence of an odorant-binding protein in the human olfactory mucus: location, structural characterization, and odorant-binding properties. Biochemistry 41:7241–7252
5. Tegoni M, Campanacci V, Cambillau C (2004) Structural aspects of sexual attraction and chemical communication in insects. Trends Biochem Sci 29:257–264
6. Löbel D, Jacob M, Volkner M, Breer H (2002) Odorants of different chemical classes interact with distinct odorant binding protein subtypes. Chem Senses 27:39–44
7. Tegoni M, Ramoni R, Bignetti E, Spinelli S, Cambillau C (1996) Domain swapping creates a third putative combining site in bovine odorant binding protein dimer. Nat Struct Biol 3:863–867
8. Spinelli S, Ramoni R, Grolli S, Bonicel J, Cambillau C, Tegoni M (1998) The structure of the monomeric porcine odorant binding protein sheds light on the domain swapping mechanism. Biochemistry 37:7913–7918
9. Vincent F, Spinelli S, Ramoni R, Grolli S, Pelosi P, Cambillau C, Tegoni M (2000) Complexes of porcine olfactory-binding protein with odorant molecules belonging to different chemical classes. J Mol Biol 300:127–139
10. Tcatchoff L, Nespoulous C, Pernollet JC, Briand L (2006) A single lysyl residue defines the binding specificity of a human odorant-binding protein for aldehydes. FEBS Lett 580:2102–2108

Odorant Receptor

FRANÇOISE LAZARINI
Perception and Memory Unit, Neuroscience Department, Pasteur Institute, Paris, France

Synonyms

Olfactory receptor; Odor receptor; Olfactory receptor protein; OBPs

Definition

►Odorant receptor proteins are G protein-coupled seven transmembrane proteins, which number more than 1,000 in some mammalian species, and mediate the detection of thousands of volatile odorants. They are expressed, in mammals, in the cilia of the olfactory sensory neurons residing in the olfactory neuro-epithelium in the nasal cavity. They are located, in adult insects, on either the antennae or maxillary palp. They are expressed by sperm cells, and are thought to trigger ►chemotaxis toward the oocyte. A second class of odorant receptor proteins was described in 2001 for volatile amines, and called "trace amine-associated receptors" (TAAR). Most odorant receptors recognize multiple related odors and most odorants are recognized by several receptors.

Characteristics

Quantitative Description

In 1991, Buck and Axel discovered the odorant receptor gene family in rat [1]. In 2004, Linda Buck and Richard Axel won the Nobel Prize in Physiology or Medicine for this major discovery. Odorant receptors are seven-transmembrane-domain proteins encoded by large gene families. *Drosophila* has a highly diverse family of 60 odorant receptor genes [2]. In mammals, the odorant receptor family of genes, comprising some 1,100 functional genes in the mouse, 347 in the human, respectively, is the largest family of G protein-coupled receptors in the genome, which may make up as much as 3% of the genome. Only a small part of odorant receptor genes form functional ►odor receptors. In the mouse, 1,296 odorant receptor genes (including 20% pseudogenes) were found, which can be classified into 2,228 families [3]. Mouse odorant receptor genes are distributed in 27 clusters on all mouse chromosomes except 12 and Y. The distribution was not uniform, with more than half of the genes contained in a few large, compact clusters on chromosomes 7, 11 and 9. Class I odorant receptors correspond to fish-like receptors that bind water-soluble odorants, and separate clearly in the phyllogenetic tree from the classical, mammalian-specific class II odorant receptors. There are 147 Class I odorant receptors in the mouse odorant receptor subgenome, 120 of them potentially functional. All the class I odorant receptor genes are located in a single large cluster on chromosome 7 (cluster 7–3). Class I odorant receptors are prevalent in the mammalian genome and may be centrally involved in mammalian olfaction. In the mouse, they are expressed in the most dorsal zone of the olfactory epithelium. Conversely, Class II receptors have been found in all four zones.

Humans have lost nearly two-third of the odorant receptor genes as compared to mice, providing a possible explanation for the reduced sense of smells of humans compared to rodents. The human odorant receptor genome repertoire is organized similarly to the mouse one [4]. Human odorant genes are dispersed in more than 50 chromosomal locations and organized mostly in clusters. Most subfamilies are encoded by a single locus and most loci encode a single or very few subfamilies. Odorant receptors of a single locus recognize structurally related odorants, suggesting that different parts of the genome are involved in the detection of different odorant type.

A second class of odorant receptors was described in 2001 for volatile amines, metabolic derivatives of classical biogenic amines, and called 'trace amine-associated receptors' (TAAR). Encoding TAAR are present in human, mouse and fish olfactory neurons [5]. They show sequence similarities to the receptors for the neurotransmitters serotomin and dopamine. TAAR1 is thought to be a receptor for thyronamines, decarboxylated and deiodinated metabolites of the thyroid hormones, while the mouse mTAAR2- mTAAR9 receptors are most probably olfactory receptors for volatile amines.

Description of the Structure and Pharmacology

Odorant receptors are in every species heptahelical G-protein-coupled receptors. In mammals, odor receptors belong to class A of the G protein-coupled receptors that are characterized by a long second extracellular loop, containing an extra pair of conserved cysteins, and specific short sequences [6]. Odor receptors share a similarity from 40 to 90% identity. They also have a region of hypervariability, which is the binding site for ligands. This region consists in the third, fourth and fifth alpha – helical transmembrane regions, thought to face each other and form a pocket into the membrane. Mammalian odor receptors are related phyllogenetically to other chemosensory receptors (taste receptors, vameronsal receptors and gustatory receptors). Invertebrate odor receptors bear no homology to vertebrate odorant receptors. *Drosophila* odorant receptors have a mildly conversed region in the seventh transmembrane domain [2].

Odorant receptors bind to structures on odor molecules. They are generally able to recognize multiple related but not identical molecules. They are able to discriminate between thousands of low molecular mass, aliphatic and aromatic molecules with varied carbon backbones and diverse functional groups, including aldehydes, esters, ketones, alcohols, alkenes, carboxylic acides, amines, imines, thiols, halides, nitriles, sulphides and ethers. For many odors, the dose-response curves in single cells have relatively elevated EC50 values, or midpoint, ranging from 10 to 100 μM. They can be activated by multiple odors, and conversely most odors are able to activate more than one type of receptor.

Members of the TAAR family are activated by the trace amines found in the central nervous system (beta-phenylethylamine, tyramine, tryptamine and octopamine). Individual TAAR are specific for different amine structures; three of them are activated by volatile amines found in urine (a source of pheromonal cues of a variety of chemical compositions), some of which have been involved in regulating reproductive behavior [5].

Olfactory Signal Transduction

Odorant receptors are specialized to detect certain odorants and to convert external stimuli into intracellular signals [7]. Once the odorant has bound to the odorant receptor, the receptor undergoes structural changes and sequentially activates the specific olfactory-type G protein (Golf) and the lyase – adenylate cyclase type III (ACIII)- which converts ATP into cyclic AMP (cAMP), a molecule that has numerous signaling roles in cells. See Fig. 1.

The cAMP opens specific cyclic nucleotide-gated (CNG) channels, which allow calcium and sodium ions to enter into the cell, depolarizing the ►olfactory sensory neuron and triggering action potentials which then carry odor information to the olfactory bulb in the brain. The second-messenger cascade of enzymes provides amplification and integration of odor-binding events. The binding of one odor molecule to an odorant receptor activates tens of Golf proteins, each of which will activate an adenylyl cyclase III molecule able to produce about a 1,000 molecules of cAMP per second. Three cAMP molecules are sufficient to open a CNG channel, which can allow the crossing of hundred of thousands of cations, depolarizing the cell and inducing an action potential. The calcium ions entering through the CNG channels are capable of activating and thus opening channels permeable to negatively charged chloride ion (Cl−). When the Cl− channels open, the Cl− efflux further depolarizes the olfactory sensory neuron, thus adding to the excitatory response magnitude. On the other hand, calcium ions entering through the CNG channels act on these channels, probably with calmodulin, to decrease their sensitivity to cAMP, thus requiring a stronger odor stimulus to produce sufficient cAMP to activate the channels. This negative feedback (inhibitory) pathway constitutes a crucial adaptation response allowing olfactory sensory neurons to adjust their sensitivity to odor stimuli. In invertebrates, both excitatory and inhibitory responses to odors have been described, suggesting the existence of multiple transduction pathways.

Expression and Function

In insects, olfaction is a critical sensory modality for controlling behaviors such as mate selection, food choice and navigation toward suitable oviposition sites. Odorant receptors are located, in adult insects, in small subsets of olfactory receptor neurons in either the antenna or maxillary palps, which constitute the olfactory sensory organs. In mammals, the sense of smell is triggered by odorant receptors, which are expressed in the cilia of the olfactory sensory neurons of the olfactory neuroepithelium lining the nasal cavity. In mice, odorant receptors are also involved in mating and other social

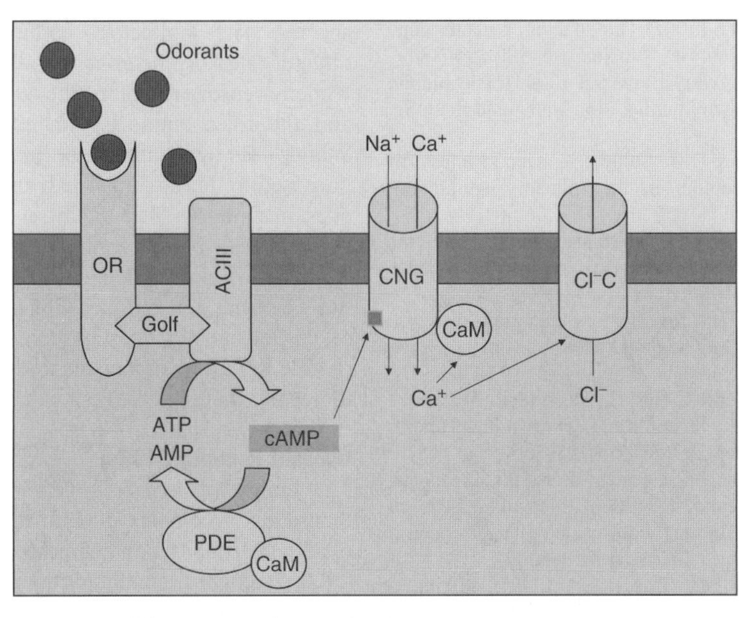

Odorant Receptor. Figure 1 ►Olfactory transduction. Within the olfactory sensory neuron, a cascade of enzymatic activity transduces the binding of an odorant molecule to an odorant receptor into an action potential that can be transmitted to the central nervous system. OR, odorant receptor; ACIII, adenylyl cyclase III; CNG, cyclic nucleotide-gated channel; Cl−C, negatively charged clhoride ion channel; CaM, calmodulin; PDE, phosphodiesterase.

behaviors. Moreover, in mammals, a subset of odorant receptors is specifically expressed in the testis and odorant receptors have been identified in spermatids and mature spermatozoa [8]. These odorant receptors may play a role in chemotaxis of spermatozoa toward the oocyte.

In *drosophila*, each sensory neuron express only a single odorant receptor, and all sensory neurons expressing the same receptor contact a single restricted target, named ►glomerulus in a relay station called the antennal lobe of the brain, analogous to the vertebrate olfactory bulb. *Drosophila* has about 50 types of olfactory receptor neurons, corresponding to about 50 identified glomeruli in the antennal lobe [2].

In mammals, with a few exceptions, each olfactory sensory neuron expresses only one of the 1,000 odorant receptor genes [9]. All cells expressing the same receptor converge onto one or a few glomeruli, in the olfactory bulb [6]. Glomeruli (nearly 2,000 in the rat) are spherical conglomerate of neuropil (diameter of 50–100 μ) that consists of the incoming axons of the olfactory sensory neurons and the dendrites of the main projection cells (mitral cells) in the olfactory bulb. Mitral axons of olfactory sensory neurons leaving the olfactory bulb project to higher brain structures including the piriform cortex, the olfactory cortex, hippocampus and amygdala, allowing for both the conscious perception of odors and their emotional and motivational effects. Lateral processing of the message occurs though two populations of inhibitory GABAergic interneurons in the olfactory bulb: periglomerular cells and granule cells. Each glomerular unit presents a receptive field that is thought to be defined by the molecular range, or pharmacological profile of each odorant receptor.

The mammalian olfactory system uses a combinational receptor coding scheme to encode odor identity and to discriminate odors [10]. A given odor activates a set of odorant receptors, and then a set of olfactory sensory neurons, and then a set of glomeruli in the olfactory bulb, forming a spatial map of sensory information. Different odors activate overlapping but non-identical patterns of receptors and thus glomeruli. Slight changes in the structure of an odorant or changes in its concentration results in changes in the combination of receptors that recognize the odorant. Receptors that recognize similar odors (such as ►enantiomers) generally map in the same area in the olfactory bulb. Individual TAAR are sparsely expressed in discrete subdomains of the neuroepithelium, and are co-expressed with neither other TAAR, nor probably the odorant receptors. In mice, TAAR may mediate behavioral and physiological responses to amine-based social cues present in urine, as urine from sexually mature male mice, but not from females or sexually immature mices, could stimulate mTAAR5, a receptor activated by trimethylamine [5].

O

References

1. Buck L, Axel R (1991) A novel multigene family may encode odorant receptors: a molecular basis for odor recognition. Cell 65(1):175–187
2. Vosshall LB, Wong AM, Axel R (2000) An olfactory sensory map in the fly brain. Cell 102(2):147–159

3. Zhang X, Firestein S (2002) The olfactory receptor gene superfamily of the mouse. Nat Neurosci 5(2):124–133
4. Malnic B, Godfrey PA, Buck LB (2004) The human olfactory receptor gene family. Proc Natl Acad Sci USA 101(8):2584–2589
5. Liberles SD, Buck LB (2006) A second class of chemosensory receptors in the olfactory epithelium. Nature 442(7103):645–650
6. Mombaerts P (1999) Seven-transmembrane proteins as odorant and chemosensory receptors. Science 286(5440):707–711
7. Firestein S (2001) How the olfactory system makes sense of scents. Nature 413(6852):211–218
8. Spehr M, Gisselmann G, Poplawski A, Riffell JA, Wetzel CH, Zimmer RK, Hatt H (2003) Identification of a testicular odorant receptor mediating human sperm chemotaxis. Science 299(5615):2054–2058
9. Serizawa S, Miyamichi K, Nakatani H, Suzuki M, Saito M, Yoshihara Y, Sakano H (2003) Negative feedback regulation ensures the one receptor-one olfactory neuron rule in mouse. Science 302(5653):2088–2094
10. Malnic B, Hirono J, Sato T, Buck LB (1999) Combinatorial receptor codes for odors. Cell 1999 96(5):713–723

Odorant Receptor: Genomics

Jean-François Cloutier
Department of Neurology and Neurosurgery, McGill University, Montreal Neurological Institute, Montréal, QC, Canada

Definition

►Odorant receptor genomics refers to the study of the structure and function of genes encoding receptors involved in the sense of smell. It includes defining the number and chromosomal arrangements of odorant receptor genes present in various genomes, as well as the molecular mechanisms that regulate their expression in an organism.

Characteristics

The survival and well being of most terrestrial vertebrates is dependent on their ability to detect ►odors in their environment and to respond to social cues. Neurons located in sensory epithelia of the nasal cavity detect volatile and water-soluble molecules and transmit the information gathered to the brain where it is further processed to generate odor perception and behavioral outputs. While the detection of volatile odorant molecules plays an important role in the modulation of acquired behavior such as food foraging, detection of ►pheromones is thought to control innate responses such as male-to-male aggression in many vertebrate species. The ►olfactory epithelium (OE) contains olfactory sensory neurons that express two classes of ►chemosensory receptors: the odorant receptors (ORs) and the trace amine-associated receptors (TAARs). While ORs recognize odor molecules, TAARs are proposed to detect compounds that can provide social cues. In contrast to the OE, the sensory epithelium of the vomeronasal organ contains sensory neurons that express two classes of putative pheromone receptors, the V1R and V2R families. Together, these families of seven-transmembrane G protein-coupled receptors (►GPCRs) allow organisms to detect a large range of molecules that regulate their behavior (Fig. 1).

Odorant Receptors (ORs)

The ability of terrestrial vertebrates to discriminate thousands of complex odors in the environment relies on the detection of odorant molecules by ORs. A single OR can recognize a multitude of odorant molecules and a specific odorant can bind to several ORs perhaps eliciting different levels of neuronal activity. The combination of ORs activated by odorant molecules present in a complex odor leads to the propagation of signals that ultimately renders a representation of the odor in the central nervous system. In light of the complexity of this ►combinatorial code, it is not surprising that ORs represent one of the largest mammalian gene families.

In some terrestrial vertebrates that rely heavily on their sense of smell for survival, such as the mouse, larger OR gene repertoires have been described than in humans whose sense of smell is considered to be more aesthetic. The mouse genome contains ~1400 genes that are organized in clusters located on almost all chromosomes [1]. While the majority of these genes encode functional ORs, ~15% of them are ►pseudogenes. The coding region of OR genes consists of a single ►exon preceded by an ►intron that separates it from non-coding exons in the 5′ region. The coding exon gives rise to OR proteins that are 300–350 amino acids in size. ORs contain structural features that are common to most GPCRs such as the seven hydrophobic stretches that form the transmembrane domains and specific conserved cysteines that form potential disulfide bonds. In addition, ORs contain sequences that distinguish them from other GPCRs including a long second extracellular loop, as well as conserved amino acid motifs in an intracellular loop and in some of the transmembrane domains. The presence of these conserved features in ORs are usually enough to classify a gene as belonging to the large family of ORs. Nonetheless, aside from these conserved features, there is on average an overall low amino acid similarity (37%) between ORs. This may allow the OR repertoire to recognize a large number of structurally diverse odorants.

The OR superfamily is subdivided into two classes of receptors. Class I ORs were originally identified in fish

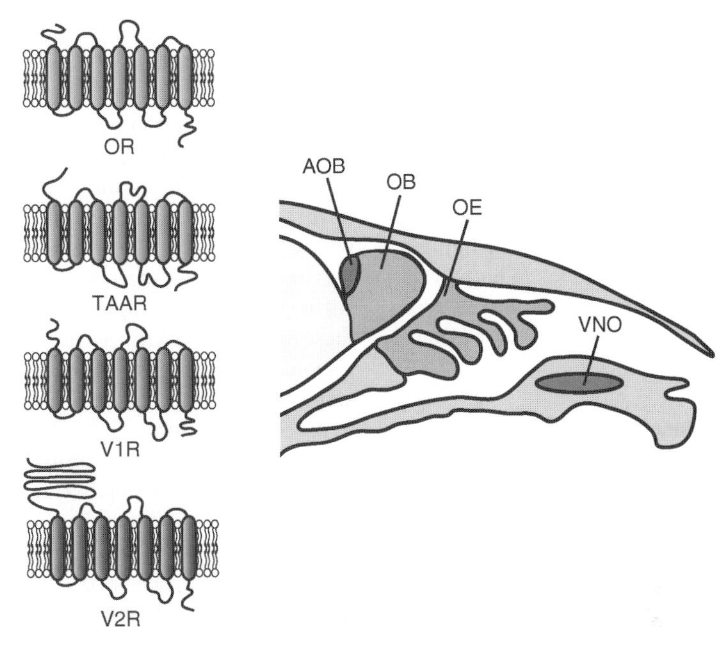

Odorant Receptor: Genomics. Figure 1 Anatomy of the olfactory systems and structure of olfactory receptors. Olfactory sensory neurons located in the olfactory epithelium (OE) project axons that connect with second-order neurons in the olfactory bulb (OB). In contrast, vomeronasal neurons located in the vomeronasal organ (VNO) project their axons to the accessory olfactory bulb (AOB) where they form synapses with second-order neurons. The information processed by second-order neurons is relayed to various regions of the brain where an odor representation is generated. Olfactory receptors belong to the large family of G-protein coupled seven-transmembrane receptors. While odorant receptors (OR) and trace amine-associated receptors (TAAR) are expressed in the OE, two families of vomeronasal receptors (VR), V1R and V2R, are expressed in the VNO.

but later shown to represent approximately 10% of the mouse OR gene repertoire. In contrast, class II genes have so far been identified only in terrestrial vertebrates and represent the majority of the OR gene repertoire in mouse. While all Class I OR genes are segregated in a single cluster on chromosome 7, class II OR genes are located in clusters on all chromosomes except 12 and Y. The functional relevance of the sequence divergence observed between these two classes of receptors is still unclear. However, it has been proposed that Class I and II receptors bind volatile odorants that have low and high levels of hydrophobicity, respectively.

The gene structure and chromosomal arrangements of OR gene clusters observed in mouse is conserved in humans with an OR gene repertoire consisting of approximately 950 ORs [2]. While this total number may not seem that different from the number of OR genes present in the mouse genome, it is estimated that ~60–70% of these genes could be pseudogenes. Hence, humans may express approximately 300–350 functional ORs, three times less than are expressed in mice. The pseudogenization of the OR repertoire appears to parallel the evolution tree. The highest percentages of OR pseudogenes are observed in the human (~63%) and old-world monkey (~30%) genomes, while New World monkeys have a similar fraction of pseudogenes as found in the mouse genome (~20%). The increase in pseudogenes observed in humans, as well as in old-world primates, is likely the result of decreased selective pressure for olfactory function throughout evolution.

Vomeronasal Receptors (VRs)

The ►accessory olfactory system plays a critical role in the detection of and responsiveness to pheromones. Vomeronasal sensory neurons located in the ►vomeronasal organ express members of the Vomeronasal Receptor (VRs) superfamily that are putative pheromone receptors. These receptors are seven transmembrane GPCRs that are distinct from the OR superfamily. Two large families of VRs have been identified, V1R and V2R. In mouse, the V1R and V2R families are respectively comprised of ~200 and ~60 putative functional genes that are dispersed across several chromosomes [1,3]. While V1Rs, as ORs, are encoded by a single exon, the V2R gene structure is more complex and contains several coding exons. This difference in gene structure is also reflected in the overall V2R protein structure. In addition to features common to ORs and V1Rs, such as the seven transmembrane domains, V2Rs contain a large extracellular N-terminal domain that binds ligands.

In humans, the majority (95%) of V1R sequences identified are pseudogenes. Five V1R genes that are predicted to encode functional receptors have been described, with at least one of them observed at the mRNA level in human olfactory mucosa [4]. Moreover, no intact V2R genes have been reported in humans. The high occurrence of VR pseudogenes in humans, as well as in primates, suggests that pheromone detection in these species is either not prevalent or mediated through other families of receptors.

Trace Amine-Associated Receptors (TAARs)

In addition to ORs, a second class of chemosensory receptors has been identified in the OE of mice. TAARs can recognize volatile amines and at least one of them is activated by urine from sexually mature male mice [5]. These observations suggest that TAARs may be implicated in the detection of social cues in mice. The mouse genome contains 16 TAAR genes, including 1 pseudogene, that are all located in a compact region of chromosome 10 and that share high sequence identities [6]. Of these 16 genes, 8 have so far been shown to be expressed in the OE. The coding region of TAAR genes consists of a single exon, which gives rise to proteins of approximately 350 amino acids that contain seven hydrophobic stretches of amino acids and conserved extracellular cysteine residues. In humans, 9 TAAR genes have been identified, including 3 pseudogenes [6]. It remains to be determined whether they are expressed in the human olfactory mucosa.

Regulation of odorant receptor gene expression

The development of a functional olfactory system is dependent on the tight regulation of OR gene expression in olfactory sensory neurons. Each OR is expressed in a small subset of neurons that are distributed in one of four defined but partially overlapping expression domains within the OE. Within each of these domains, neurons expressing the same OR are randomly distributed and each neuron expresses a single OR gene from the large repertoire available. Furthermore, a functional OR is expressed from only one of two gene ►alleles in a process termed ►monoallelic exclusion. The expression of a single OR per neuron is critical to define the profile of odorants recognized by this neuron. In addition, expression of the OR has also been shown to play a role in the accurate elaboration of ►topographic connections in the ►olfactory bulb. Mechanisms must therefore exist to first determine which subgroup of ORs will be expressed in a neuron based on its location in the OE. This is followed by the stochastic expression of a single receptor and by inhibition of expression of other OR genes in the same neuron. The mechanisms underlying these two levels of regulation of OR gene expression are beginning to be unraveled.

The spatial regulation of OR genes in neurons of the OE may be achieved through the combinatorial expression of various families of transcription factors in different regions of the OE. For some class II OR genes, the presence of short sequences upstream of the transcriptional start sites have been shown to be sufficient to induce appropriate spatial expression of these ORs in the OE. These short sequences contain regions recognized by homeodomain-containing transcription factors and by Olf1/EBF (O/E) family transcription factors. The LIM-homeodomain protein, Lhx2, can bind to the promoter region of at least one OR gene and is required for expression of class II OR genes [7]. Three members of the O/E family, O/E-1 to 3, are expressed in developing olfactory sensory neurons and the presence of O/E binding sequences in several OR gene promoter regions suggests they may also control OR gene expression [8]. However, the overlapping expression of these three family members in olfactory sensory neurons has made it difficult to establish their requirement for OR gene expression using gene-targeting approaches in mice.

The stochastic selection of expression of a single OR in a neuron is first dependent on the positive activation of gene expression through a *cis* or *trans*-acting mechanism (Fig. 2). It has been proposed that a region of homology upstream of each OR gene cluster, termed H, can act as a ►locus control region (LCR) to regulate expression of these genes in *cis* [9]. A similar mechanism is used to regulate the expression of photopigment genes in the visual system. This LCR would recruit proteins to form an activation complex that can randomly promote transcription of a single gene within the locus following chromatin rearrangements. Such a regulatory sequence has been identified far upstream of the mouse MOR28 gene cluster. Alternatively, a single H region could also regulate expression of OR genes in *trans* through interchromosomal interactions. In support of this hypothesis, the H region found upstream of the MOR28 gene cluster has been shown to interact with the promoter of several OR genes located on different chromosomes [10]. However, while deletion of H from the mouse genome affects expression of OR genes proximal to the location of H, the expression of OR genes outside of this gene cluster is unaffected [11,12].

Since only one allele of an OR gene is expressed in a single neuron, a mechanism must also exist to prevent transcription of the other allele as well as to prevent expression of other OR genes in the neuron. This may be achieved through a negative feedback mechanism in OSNs [9]. Expression of a full-length mRNA giving rise to a functional OR protein

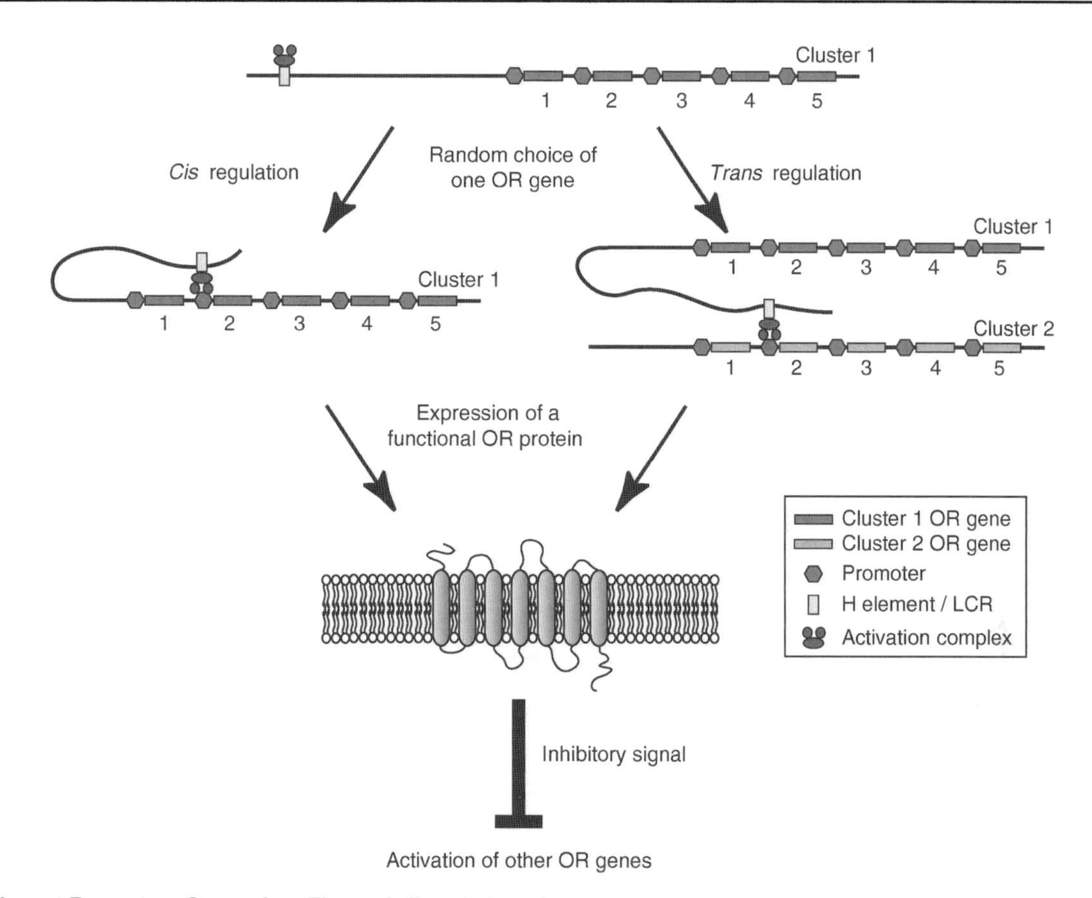

Odorant Receptor: Genomics. Figure 2 Regulation of odorant receptor gene expression. Odorant receptor (OR) genes are arranged in clusters located on almost all chromosomes. A single OR gene is expressed per neuron through positive and negative mechanisms of regulation. An activation complex is recruited to a locus control region (LCR), termed H, located upstream of an OR gene cluster. Through chromosomal remodeling, this activation complex interacts in either *cis* or *trans* with a single OR gene promoter within a gene cluster to induce gene expression. This stochastic expression of a single OR protein leads to the generation of an unidentified signal that inhibits activation of other OR genes in the neuron by a negative feedback mechanism.

prevents the secondary activation of other OR genes. In contrast, expression of a full-length mRNA containing a premature stop codon from a pseudogene does not prevent activation of another OR gene. These observations suggest that expression of a functional OR protein leads to an as yet unidentified inhibitory signal that negatively regulates expression of other OR genes. In addition, the OR coding region contains regulatory elements importent to suppress expression of additional receptors [13]. Taken together, these mechanisms do not only prevent expression... Such a mechanism does not only prevent expression of two types of receptors in a single neuron but also serves to avoid the generation of receptorless neurons. The control of VR, and possibly TAAR, gene expression also ensures that a single receptor is expressed per neuron generated. Whether regulation of these families of genes is under the control of similar mechanisms to the ones identified for OR genes remains to be determined.

References

1. Zhang X, Zhang X, Firestein S (2007) Comparative genomics of odorant and pheromone receptor genes in rodents. Genomics 89:441–450
2. Glusman G, Yanai I, Rubin I, Lancet D (2001) The complete human olfactory subgenome. Genome Res 11:685–702
3. Yang H, Shi P, Zhang Y, Zhang J (2005) Composition and evolution of the V2r vomeronasal receptor gene repertoire in mice and rats. Genomics 86:306–315
4. Rodriguez I, Mombaerts P (2002) Novel human vomeronasal receptor-like genes reveal species-specific families. Curr Biol 12:R409–R411
5. Liberles SD, Buck LB (2006) A second class of chemosensory receptors in the olfactory epithelium. Nature 442:645–650
6. Lindemann L, Ebeling M, Kratochwil NA, Bunzow JR, Grandy DK, Hoener MC (2005) Trace amine-associated receptors form structurally and functionally distinct subfamilies of novel G protein-coupled receptors. Genomics 85:372–385

7. Hirota J, Omura M, Mombaerts P (2007) Differential impact of Lhx2 deficiency on expression of class I and class II odorant receptor genes in mouse. Mol Cell Neurosci 34:679–688
8. Michaloski JS, Galante PAF, Malnic B (2006) Identification of potential regulatory motifs in odorant receptor genes by analysis of promoter sequences. Genome Res 16:1091–1098
9. Serizawa S, Miyamichi K, Sakano H (2004) One neuron-one receptor rule in the mouse olfactory system. Trends Genet 20:648–653
10. Lomvardas S, Barnea G, Pisapia DJ, Mendelsohn M, Kirkland J, Axel R (2006) Interchromosomal interactions and olfactory receptor choice. Cell 126:403–413
11. Fuss SH, Omura M, Mombaerts P (2007) Local and as effect of the H element on expression of odorant receptor genes in mouse Cell 130:373–384
12. Nishizumi H, Kamasaka K, Inoue N, Nakashima A, Sakano H (2007) Deletion of the core-H region in mice abolishes the expression of three proximal odorant receptor genes in as, Proc Natl Acad Sci USA 104:20067–20072
13. Nguyen MQ, Zhou Z, Marks CA, Ryba NJP, Belluscio L, (2007) Prominent roles for odorant receptor coding sequences in allelic exclusion. Cell 131:1009–1017

Odorant Receptor Protein

Definition

►Odorant Binding Proteins.

►Odorant Receptor
►Odorant Receptor: Genomics

Odorants

►Olfactory Information

Odors

►Olfactory Information

Odotopic Representation

Definition

Odotopic representations involve a unique spatial pattern of activity in the olfactory system (e.g. a unique pattern of activated olfactory glomeruli) for odorant stimuli that evoke unique odor perceptions.

►Glomerular Map

Off Center Cells

Definition

►Visual Cortical and Subcortical Receptive Fields

Ohm's Law

Definition

The electrical current (I, in Amperes) that flows through an electrical resistor equals the potential difference (voltage, V, in Volts) across the resistor divided by the resistor's electrical resistance (Ohm, in Ω): $I = V/\Omega$.

►Action Potential
►Membrane Potential: Basics

Old/new Recognition

►Recognition Memory

Olfaction

Definition

The sense of smell. The process whereby odorant molecules bind to receptors in the olfactory epithelium and leading to the generation and propagation of neural signals responsible for odor perception.

►Odor
►Odorant
►Odorant Receptor Neuron
►Odor Perception
►Olfactory Epithelium
►Olfactory Sense

Olfaction and Gustation Aging

Nicholas P. Hays
Nutrition, Metabolism, and Exercise Laboratory, Donald W. Reynolds Institute on Aging, University of Arkansas for Medical Sciences, Little Rock, AR, USA

Synonyms

Senescence; Gerontology; Elderliness

Definition

Elderly adults often have an impaired ability to detect and recognize ►tastes and ►odors. Olfactory and gustatory impairment can be particularly harmful in aged individuals, given the likely contribution of such dysfunction to poor appetite, lower dietary energy and nutrient intakes, and the consumption of inappropriate food choices such as spoiled food. These phenomena may in turn influence body composition, nutritional stores, immune function, and disease status. Olfactory dysfunction can also be dangerous as it may prevent the detection of smoke or natural gas odors during household emergencies. Although the precise mechanisms underlying age-related changes in taste and smell remain uncertain, physiological changes associated with the aging process itself, diseases, medication usage, trauma, and environmental factors are all possible contributors. Flavor enhancement, increased dietary variety, and other interventions have been identified that can improve food intake and enhance eating enjoyment. Given the projected increases in the size and longevity of the elderly population in the U.S. and worldwide, additional effective interventions that can maintain or improve chemosensory function in this vulnerable population are needed.

Characteristics

Introduction

It is generally accepted that all sensory modalities, including ►gustation, olfaction [see ►olfactory senses], vision [see ►binocular vision], audition [see ►auditory system], and somatosensation [see ►somatic sense] commonly decline with increasing age. Vision and auditory losses are perhaps most typically associated with the aging process, and these impairments are indeed highly prevalent among elderly adults, with approximately 34% of adults aged 65 years and older reporting vision and/or hearing impairment [1]. Taste and smell dysfunctions are also recognized as a common characteristic of old age, but frequently receive less attention, perhaps because their impact on mortality, morbidity, and functional status is less direct. Figure 1 illustrates the nearly exponential increase in self-reported taste and/or smell dysfunction with increasing age in a representative cohort of U.S. adults [2]. These data indicate that individuals aged 65+ years account for almost half (~41%) of the total number of individuals reporting chronic chemosensory problems. The prevalence of impairment is likely even higher when considering that self-reported data may underestimate the level of actual impairment as measured by objective testing.

The basic anatomy and neurobiology of the gustatory and olfactory systems have been fully described elsewhere [see taste, odor]. Briefly, taste signals are received by receptor cells located in ►taste buds in the ►gustatory papillae of the tongue and other structures of the oral cavity. Taste information is then transmitted to the ►gustatory cortex, orbitofrontal cortex [see ►cortex – orbitofrontal], ►amygdala, and lateral ►hypothalamus of the ►brain. Smell sensations are received by a small area of ►olfactory epithelial tissue located on the dorsal surface of the nasal cavity [see ►nasal passages], where odorants bind to receptors in olfactory neurons, which then transmit information about the identity and concentration of the chemical signal to the ►olfactory cortex in the brain.

Aging can influence different aspects of gustatory and olfactory sensory perception and sensitivity. Older individuals often require higher concentrations of an odorant or tastant to be present before detection and recognition of the chemical stimulus can be achieved. In other words, the detection and recognition ►thresholds for various tastes and smells are higher in older adults compared to younger. The magnitude of these changes can also vary across specific sensory qualities; salt taste [see ►taste – salt] thresholds appear to increase more during the aging process than sweet [see ►taste – sweet] thresholds. In addition, older individuals may have alterations in the ►suprathreshold perception of tastes and smells, such that more concentrated chemical stimuli are not perceived as more intense. Odor identification is also frequently poor among the elderly, although this may be due to both sensory impairment as well as cognitive and memory dysfunction resulting in difficulty with odor-naming tasks. In general, olfactory dysfunction is more common

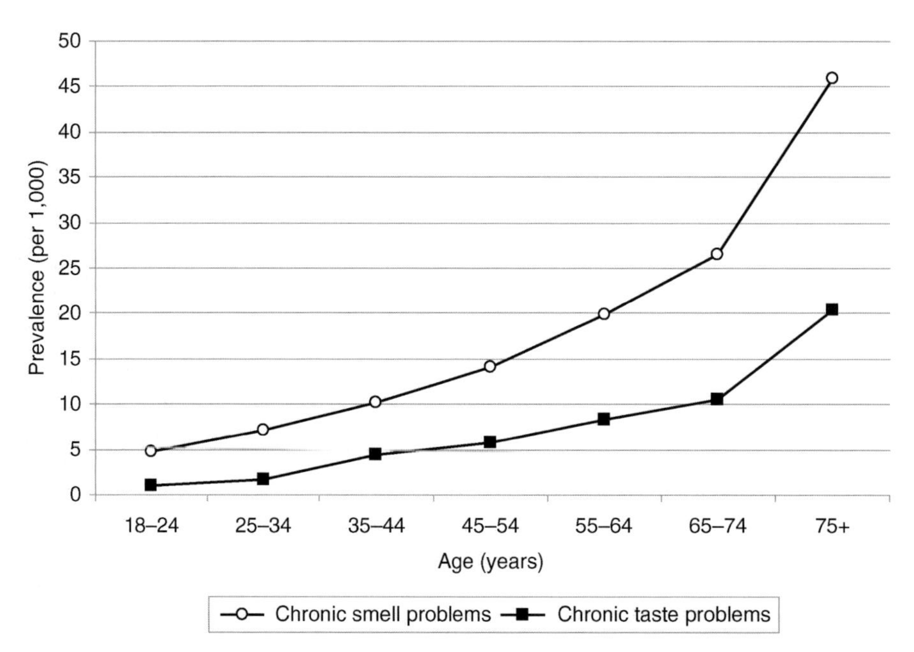

Olfaction and Gustation Aging. Figure 1 Age specific prevalence rates (per 1,000) of self-reported chronic (≥3 months duration) chemosensory problems among individuals living in 42,000 randomly selected U.S. households (1994 Disability Supplement to the National Health Interview Survey). Adapted from Hoffman et al. [2] with permission © 1998 New York Academy of Sciences.

than taste dysfunction among the elderly population and individuals who describe problems with their sense of "taste" typically exhibit olfactory and not gustatory dysfunction, since it is difficult to distinguish true taste from ▶retronasal olfaction [3]; strictly defined changes in taste alone are rare. Olfactory perception, however, declines with increasing age even in generally healthy men and women. As shown below, smell and taste changes associated with aging can manifest along a continuum of sensitivity, and can range from the total absence of sensation (e.g. ▶ageusia) to a diminished or distorted sensation (e.g. ▶hypogeusia, ▶dysgeusia).

Terminology

Gustation	
Normogeusic	Normal taste sensory function
Hypogeusia	Diminished sensitivity of taste
Dysgeusia	Distortion of normal taste
Ageusia	Absence of taste
Olfaction	
Normosmic	Normal smell sensory function
Hyposmia	Diminished sensitivity of smell
Dysosmia	Distortion of normal smell
Parosmia	Distortion of odor perceptions when odor is present
Phantosmia	Odor sensations in absence of odor stimulus (i.e. olfactory hallucination)
Anosmia	Absence of smell
Cacosmia	Feeling ill in response to odors

Etiology

The causes of taste and smell dysfunction among elderly individuals are not completely understood. The olfactory epithelium is particularly vulnerable to age-associated dysfunction because of its anatomical location and proximity to environmental trauma, as well as a greater susceptibility to decreased ▶neurogenesis secondary to its relatively small size (1–2 cm^2) and thinness. Declines in taste sensitivity were thought historically to result from a loss of functional taste buds over time, but more recent work indicates that taste bud numbers do not decrease with age and thus declines may be due to changes in taste cell membrane ion channels and receptors. The etiology of age-associated chemosensory dysfunction is further complicated by the varied environmental and medical factors that can also influence these systems and which frequently impact the elderly. Several possible causal factors are briefly described below:

Normal aging. One hypothesis for the decline in taste sensitivity with age is reduced taste receptor cell turnover rate, resulting in alterations in taste bud structure and subsequent dysfunction in older subjects. In addition, the olfactory mucosa may be gradually replaced by respiratory epithelium during the normal aging process, reducing smell perception and sensitivity. Animal data suggests that menopause may be associated with changes in olfactory perception, potentially contributing to further alterations in olfactory function among older women.

Diseases/infection. Acute or chronic nasal and sinus problems can lead to olfactory dysfunction by obstruction of the nasal passage, by viral-mediated damage to the olfactory receptors, and by altering the amount or composition of the mucus layer that odorants must traverse to reach the olfactory epithelial surface [3]. Neurodegenerative diseases such as ►Alzheimer's and ►Parkinson's disease have been associated with olfactory deficits. Recent work indicates that difficulty in odor identification predicts the transition from normal to mildly impaired cognition [4], and from mildly impaired cognition to Alzheimer's disease, suggesting that tests of olfactory perception may be useful in identifying apparently healthy and cognitively intact individuals who are at increased risk of developing severe cognitive impairment. Other representative diseases associated with impaired olfaction and gustation are listed below.

Medical conditions associated with taste or smell dysfunction

Neurological
- Alzheimer's disease
- Bell's palsy
- Damage to the chorda tympani
- Down's syndrome
- Epilepsy
- Familial dysautonomia
- Guillain-Barré syndrome
- Head trauma
- Korsakoff's syndrome
- Multiple sclerosis
- Parkinson's disease
- Raeder's paratrigeminal syndrome
- Tumors and lesions

Nutritional
- Cancer
- Chronic renal failure
- Liver disease including cirrhosis
- Niacin deficiency
- Thermal burn
- Vitamin B_{12} deficiency
- Zinc deficiency

Endocrine
- Adrenal cortical insufficiency
- Congenital adrenal hyperplasia
- Cretinism
- Cushing's syndrome
- Diabetes mellitus
- Hypothyroidism
- Kallmann's syndrome
- Panhypopituitarism
- Pseudohypoparathyroidism
- Turner's syndrome (gonadal dysgenesis)

Local
- Allergic rhinitis, atopy, and bronchial asthma
- Glossitis and other oral disorders
- Leprosy
- Oral aspects of Crohn's disease
- Radiation therapy
- Sinusitis and polyposis
- Xerostomic conditions including Sjögren's syndrome

Viral infections
- Acute viral hepatitis
- HIV infections
- Influenza-like infections

Other
- Amyloidosis and sarcoidosis
- Cystic fibrosis
- High altitude
- Hypertension
- Laryngectomy
- Psychiatric disorders

Adapted from Schiffman et al. [5] with permission © 2004 Humana Press Inc.

Medication usage. Taste alterations can be a common side effect of many medications. Medications typically do not produce total taste losses, but may produce metallic or bitter dysgeusias. Certain medications can be absorbed and then excreted in the saliva, where they can stimulate an adverse taste sensation or alter normal taste signal transduction. Other medications can diminish salivary output, decreasing the ability of tastant molecules to be dissolved and carried to the taste buds, or alter the composition of the olfactory mucus layer, modifying the absorption of odorants [6]. More than 250 medications are thought to interfere with smell and taste acuity, with selected medications listed below.

Medications associated with taste or smell dysfunction

Antianxiety agents
- Alprazolam (Xanax)
- Buspirone (BuSpar)

Antibiotics
- Ampicillin
- Azithromycin (Zithromax)
- Ciprofloxacin (Cipro)
- Clarithromycin (Biaxin)
- Enalapril (Vaseretic)
- Griseofulvin (Grisactin)
- Metronidazole (Flagyl)
- Ofloxacin (Floxin)
- Terbinafine (Lamisil)
- Tetracycline
- Ticarcillin (Timentin)

►Anticonvulsants
- Carbamazepine (Tegretol)
- Phenytoin (Dilantin)

►Antidepressants
- Amitriptyline (Elavil)

Clomipramine (Anafranil)
Desipramine (Norpramin)
Doxepin (Sinequan)
Imipramine (Tofranil)
Nortriptyline (Pamelor)
Antihistamines and decongestants
Chlorpheniramine
Loratadine (Claritin)
Pseudoephedrine
Antihypertensives and cardiac medications
Acetazolamide (Diamox)
Amiloride (Midamor)
Amiodarone (Pacerone, Cordarone)
Betaxolol (Betoptic)
Captopril (Capoten)
Diltiazem (Cardizem)
Enalapril (Lexxel, Vasotec, Vaseretic)
Hydrochlorothiazide (Esidrix)
Nifedipine (Procardia)
Nitroglycerin
Propafenone (Rythmol)
Propranolol (Inderal)
Spironolactone (Aldactone)
Tocainide (Tonocard)
Anti-inflammatory agents
Auranofin (Ridaura)
Beclomethasone (Beclovent, Beconase)
Budesonide (Rhinocort)
Colchicine
Dexamethasone (Decadron)
Flunisolide (Nasalide, AeroBid)
Fluticasone (Flonase)
Gold (Myochrysine)
Hydrocortisone
Penicillamine (Cuprimine)
Antimanic drugs
Lithium
Antimigraine agents
Dihydroergotamine (Migranal)
Naratriptan (Amerge)
Rizatriptan (Maxalt)
Sumatriptan (Imitrex)
Antineoplastics
Cisplatin (Platinol)
Doxorubicin (Adriamycin)
Levamisole (Ergamisol)
Methotrexate (Rheumatrex)
Vincristine (Oncovin)
Antiparkinsonian agents
Levodopa (Larodopa; with carbidopa: Sinemet)
►Antipsychotics
Clozapine (Clozaril)
Trifluoperazine (Stelazine)
Antithyroid agents
Methimazole (Tapazole)
Propylthiouracil
Antiviral agents
Ganciclovir (Cytovene)
Interferon (Roferon-A)
Zalcitabine (HIVID)
Bronchodilators
Bitolterol (Tornalate)
Pirbuterol (Maxair)
Lipid-lowering agents
Atorvastatin (Lipitor)
Fluvastatin (Lescol)
Lovastatin (Mevacor)
Pravastatin (Pravachol)
Muscle relaxants
Baclofen (Lioresal)
Dantrolene (Dantrium)
Pancreatic enzyme preparations
Pancrelipase (Cotazym)
Smoking cessation aids
Nicotine (Nicotrol)
Adapted from Doty and Bromley [7] with permission © 2004 Elsevier Inc.

Trauma/surgical interventions. Olfactory sensory information is transmitted by a single nerve (►cranial nerve I) which can be severed by a sharp upward blow to the nose (e.g. during an automobile accident or severe fall) proximal to the location where the nerve passes through the ethmoid bone. Gustatory sensation is transmitted via three cranial nerves (VII, IX, X) and thus is more resistant to trauma-induced dysfunction. In fact, even if one taste nerve is damaged or severed during surgery of the middle-ear region, the remaining nerves appear to compensate for the resultant loss of taste in that area of the mouth, thereby preserving overall taste perception [3].

Environmental factors. Olfactory neurons are the receptors for odorant chemical signals and therefore are directly exposed to potential airborne environmental toxins; taste receptors are specialized cells and thus the taste neurons are protected from this type of direct exposure. As a result, the olfactory system is vulnerable to damage from chemical fumes or metallurgical dust from occupational, industrial, household, or ambient sources. Tobacco smoke-induced hyposmia has also been documented.

Oral health and hygiene. Poorly fitting dentures or other dentition problems that impair chewing and mouth movements during eating can negatively impact retronasal olfaction by reducing the volatilization and movement of odor molecules from the oral cavity to the olfactory epithelium. Dentures may also cover the taste buds located in the soft palate in the roof of the mouth.

Consequences

Age-related losses of taste and smell perception can result in poor appetite, reduced energy and nutrient intakes, and diminished eating enjoyment and motivation to eat. Consequently, chemosensory losses can lead to impaired nutritional status, reduced immune function, protein-energy malnutrition, involuntary weight loss, increased disease susceptibility or exacerbation of existing disease states, and overall decreased quality of life [6]. Poor taste and smell perception may lead to consumption of spoiled food and subsequently increased likelihood of food-borne illness. Taste and smell signals are important factors in meal initiation (via cephalic-phase stimulation of salivary, gastric, and pancreatic secretions; see ▶food anticipatory behavior), continuation of food intake during a meal, and meal termination (via sensory-specific satiety). Taste and smell enhance enjoyment of meals and are the primary reinforcements of eating; maximal chemosensory acuity is thus especially important in elderly individuals for whom other sources of personal gratification may be infrequent.

The evidence for alterations of food intake as a result of olfactory or gustatory dysfunction alone is limited, however. Although chemosensory disturbances likely play an important role, other factors may also contribute to food intake dysregulation among older adults. Additional physiological factors, such as delayed gastric emptying and altered digestion-related hormone secretion and hormonal responsiveness, often act concurrently with chemosensory losses as well as with social, psychological, and medical factors to reduce food intake and promote weight loss in elderly adults.

Other consequences of taste and smell dysfunction include a decreased ability to detect natural gas leaks, volatile chemical fumes, and fires, which can result in increased risk for serious injury and death among elderly adults, their family members, and the general public. Elderly adults can have a heightened concern with personal hygiene and may overuse perfumes and colognes as a result of a lack of ability to detect offensive bodily or breath odors.

Therapeutic Strategies

While specific medical or pharmacological causes of olfactory and/or gustatory dysfunction can be resolved via appropriate treatment or pharmacotherapeutic modifications, chemosensory dysfunction that results from more intractable causes such as increasing age or environmental damage may be more resistant to improvement. In these cases, therapeutic strategies have been developed that improve food palatability and food intake, but do not alter impaired chemosensory pathways directly.

One intervention that is commonly employed is the use of flavor enhancements. Naturally-derived or chemically-synthesized concentrated odorants and flavorings can be added to individual foods to amplify or supplement the sensory signals provided by these foods. Flavor enhancement has been shown to increase the appeal of certain foods, attenuate decreases in energy intake, and improve immune status among elderly individuals [e.g. 8]. Many of these studies are limited by small sample sizes, short duration, and a lack of data regarding total dietary energy intake or nutritional status, and thus additional research is warranted. Olfactory declines tend to result in the predomination of ▶bitter tastes, but this bitterness can be masked with salt, sweet, or flavored (e.g. coffee, chocolate) extracts. Other flavor enhancements such as spices, herbs, salt, or other compounds (e.g. monosodium glutamate, concentrated meat flavor, etc.) can improve food palatability and increase dietary intake. Recent media reports examining the increasing availability and marketing of spicy and highly flavored foods in U.S. groceries and restaurants attribute this national trend to an aging population and a resultant demand for spicier foods to overcome age-related sensory declines.

Another commonly employed strategy is to alter patterns of dietary variety in order to decrease sensory specific satiety and increase food intake. A recent study examined potential associations between low dietary variety and low body mass index (BMI) and dietary energy intake in older adults. In contrast to some but not all previous reports suggesting that dietary variety typically decreases with age, adults 61 years of age or older were shown to consume a greater total food variety compared with adults 60 years or younger [9]. However, older adults with low BMIs (<22 kg/m^2) consumed a lower variety of energy-dense foods and had a lower overall energy intake compared to older adults with higher BMIs [9]. Thus the results of this study suggest that consumption of a diet containing a high variety of energy dense foods may be associated with higher energy intake and greater body weight in older adults. Presentation of a variety of palatable foods with different textures, temperatures, and appearances can also promote increased intake even though the chemosensory characteristics of the foods are not altered. The order of foods eaten can also be rotated to stimulate intake and reduce sensory-specific satiety.

Zinc supplementation and hormone replacement therapy have been suggested as additional therapeutic methods for improving taste and smell function, respectively. Hormone replacement in healthy postmenopausal women does not appear to improve performance on olfactory detection, discrimination, or recognition tasks. A recent randomized, double-blind, placebo-controlled study examining zinc supplementation in older Europeans aged 70–87 years demonstrated that supplementation with 30 mg zinc per day resulted in increased salt taste acuity, but not sweet, sour, or bitter taste acuity, among subjects recruited in one of two geographical regions [10], suggesting the efficacy of this approach may be limited to individuals with poor baseline zinc status or another trait common to those subjects examined in this region.

Audible and visual gas detection systems exist that will notify an individual of a natural gas leak; these systems are especially important for elderly individuals with olfactory losses who may be unable to detect the "rotten-egg" smell of mercaptan which is added to natural gas as a warning agent. Novel mechanisms for visually indicating the presence of food-borne pathogens, via sensors or temperature logs integrated within food packaging materials, are in development and may ultimately help older adults who cannot discriminate spoiled from wholesome food using taste or smell cues.

Conclusion

An awareness of changes in taste and smell in association with increased age has existed for thousands of years – the Roman statesman Cicero (106–43 BC) stated that "I am grateful to old age because it has made me less interested in good food and more interested in good conversation." Projected increases in both the size and average lifespan of the elderly population in the U.S. and worldwide will lead to an increased prevalence of chemosensory dysfunction, with a concomitant increase in the negative health consequences of these dysfunctions. Additional effective interventions that can maintain or improve chemosensory function in this vulnerable population are needed.

References

1. Lam BL, Lee DJ, Gómez-Marín O, Zheng DD, Caban AJ (2006) Concurrent visual and hearing impairment and risk of mortality. Arch Ophthalmol 124:95–101
2. Hoffman HJ, Ishii EK, Macturk RH (1998) Age-related changes in the prevalence of smell/taste problems among the United States adult population. Results of the 1994 Disability Supplement to the National Health Interview Survey (NHIS). Ann N Y Acad Sci 855:716–722
3. Duffy VB, Chapo AK (2006) Smell, taste, and somatosensation in the elderly. In: Chernoff R (ed) Geriatric nutrition: the health professional's handbook. Jones and Bartlett Publishers, Sudbury, MA, pp 115–162
4. Wilson RS, Schneider JA, Arnold SE, Tang Y, Boyle PA, Bennett DA (2007) Olfactory identification and incidence of mild cognitive impairment in older age. Arch Gen Psychiatry 64:802–808
5. Schiffman SS, Rogers MO, Zervakis J (2004) Loss of taste, smell, and other senses with age: effects of medication. In: Bales CW, Ritchie CS (eds) Handbook of clinical nutrition and aging. Humana Press Inc., Totowa, NJ, pp 211–289
6. Seiberling KA, Conley DB (2004) Aging and olfactory and taste function. Otolaryngol Clin N Am 37:1209–1228
7. Doty RL, Bromley SM (2004) Effects of drugs on olfaction and taste. Otolaryngol Clin N Am 37:1229–1254
8. Schiffman SS, Warwick ZS (1993) Effect of flavor enhancement of foods for the elderly on nutritional status: food intake, biochemical indices, and anthropometric measures. Physiol Behav 53:395–402
9. Roberts SB, Hajduk CL, Howarth NC, Russell R, McCrory MA (2005) Dietary variety predicts low body mass index and inadequate macronutrient and micronutrient intakes in community-dwelling older adults. J Gerontol A Biol Sci Med Sci 60A:613–621
10. Stewart-Knox BJ, Simpson EEA, Parr H et al (2008) Taste acuity in response to zinc supplementation in older Europeans. Br J Nutr Jul 99:129–136

Olfaction/Gustation Sensing Chemical Stimuli

PIERRE-MARIE LLEDO
Pasteur Institute, Laboratory for Perception and Memory, Paris Cedex, France

Introduction

The ►sensory systems are the devices with which we perceive the external world, while sensory perception amounts to the deconstruction of this external world for subsequent reconstruction of the internal representation. Animals indeed discriminate and recognize numbers of physical and chemical signals in their environment, which profoundly influence their behavior and provide them with essential information for survival [1].

A number of sophisticated sensory ►modalities available for that purpose all rely on a specific ►coding, that is a set of rules by which information is transposed from one form to another. For the ►chemical senses, this transposition concerns the ways by which chemical information give rise to specific neuronal responses in a dedicated sensory organ [2]. ►Olfaction is applied to chemosensory systems that detect chemicals emanating from a distant source. In contrast, when chemical senses require physical contact with the source for detection, they are called ►gustatory.

The origin of chemical detection (also called ►chemosensation) dates back to prokaryotes and has evolved into four distinct modalities in most vertebrates [3]. As we shall see below, the ►main olfactory system, the ►accessory olfactory system, the gustatory system and the so-called ►common chemical sense mostly carried by ►trigeminal sensory ►neurons, all differ with respect to receptor molecules, ►receptor cells and wiring of the receptor cells with the central nervous system (CNS).

Unlike most animals, humans primarily rely on ►vision and ►audition. The relevance of these two senses for human life have driven intense research into the elucidation of visual and auditory perception, leaving the understanding of the more primitive chemical senses behind. Nevertheless, during the last two decades, modern neuroscience has made considerable progress in understanding how the brain perceives, discriminates, and recognizes ►odorant molecules. This growing knowledge took over when the chemical senses were no longer considered only as a matter for poetry or the perfumes industry [4]. Over the last decades, chemical senses captured the attention of scientists who started to investigate the different stages of chemosensory systems. Distinct fields such as genetic, biochemistry, cellular biology, neurophysiology and ethology have contributed to provide a picture of how chemical information is processed in the olfactory and gustatory systems as it moves from the periphery to higher areas of the brain. So far, the combination of these approaches has been most effective at the cellular level but there are already signs, and even greater hope, that the same is gradually happening at the systems level. ►Taste and olfaction researches caught up with the advance in other sensory systems through dramatic developments achieved in a recent past. All these advances started with the discovery of the genes encoding the chemosensory receptors of the olfactory [5] and taste [6] systems. Then, further achievements were performed following the development of new experimental tools brought into play by geneticists and molecular biologists and subsequently used by the physiologists.

Although far from being complete, to date we have a fairly comprehensive view about how chemicals interact with their cognate receptors to initiate signal ►transduction in the sensory receptor cells [7]. We know now how the sensory information is first transduced in the olfactory and gustatory systems by specialized ►receptor neurons located in dedicated sensory organs. Among the different relays along the olfactory and gustatory pathways, local circuits in the second- and third-order brain areas then process the simple mono-phasic sensory signal conveyed by the sensory neurons [8] to convert it into a multi-dimensional code, including among others a ►combinatorial coding. We are now about understanding how chemical information is encoded and processed, but it is the challenge for the next decade to uncover how sensory information triggers specific behavioral outputs.

Two Distinct Chemical Stimuli

According to the phylogenetic position of the species, a number of very different but sophisticated ways, based on distinct sensory channels, have risen in order to process information from the external world in subsequently reformatted internal states [8,9]. The following synopsis describes the contributions of our understanding of chemical sensory systems that encompasses two intermingled senses: olfaction and taste. For the sake of clarity, these two modalities are presented separately, although they act, most often, in a concerted manner that gives rise to the so-called ►flavor [10,11]. Although chemical perception of a food or a flower arises from the central integration of multiple sensory inputs, it is possible to distinguish the different modalities contributing to it, especially when attention is drawn to particular sensory characteristics. Nevertheless, our experiences of the flavor or a fragrance are simultaneously of an overall unitary perception [11]. Research aimed at understanding the mechanisms behind this integrated chemical perception is, for the most part, relatively recent. However, ►psychophysical, neuroimaging and neurophysiological studies on cross-modal sensory interactions involved in olfaction and taste perception have started to provide an understanding of the integrated activity of sensory systems that generate such unitary perceptions, and hence the mechanisms by which these signals are functionally united when anatomically separated. Below I present the emerging picture that originates from the recent researches on ►odor and taste. The current model of chemosensory information processing supposes a particular combination of sensory inputs, ►temporal and ►spatial concurrence, and ►memory functions [12].

The Sense of Smell

Mammalian olfactory system regulates a wide range of multiple and integrative functions such as physiological regulation, emotional responses (e.g., anxiety, fear, pleasure), reproductive functions (e.g., sexual and maternal behaviors) and social behaviors (►social chemosignals are involved in the recognition of conspecifics, family, clan or outsiders, for examples) [13]. To achieve this large variety of functions, two anatomically and functionally separate sensory organs are required. First, the ►vomeronasal organ is specialized to sense chemical compounds (e.g., ►pheromones), specific regarding the origin of the source. By transferring information through the ►accessory olfactory bulb, this sensory organ provides information about the social and sexual status of other individuals within the species. However, recent evidence also suggests some cross-talk between the main and accessory systems. Recent molecular and neurophysiological approaches have offered new insights into the mechanisms of pheromone detection in rodent and into the sensory coding of pheromone signals that lead to the gender discrimination or aggressive behavior, for example. They show that the vomeronasal organ does not have an exclusive function with regard to pheromone recognition but it responds also to molecules other than pheromones, at least in rodents. Thus, it is highly debated today, to what extent only the vomeronasal organ can detect pheromones, and also to what extent it can only detect pheromones [14].

In mammals, the second sensory organ is represented by the ►olfactory epithelium, which recognizes more than a thousand airborne volatile molecules called odorant compounds (or odorants) [15]. This neuroepithelium is connected to the next central station for processing ►olfactory information: the main olfactory bulb (referred to below as the olfactory bulb). While advances in understanding olfactory transduction were taking place, interest in the olfactory bulb, was also intensified [15]. This growing interest has been spurred on by discovering the way the sensory organ connects to the olfactory bulb. Finally, several observations indicate that descending forebrain axons from various areas can selectively modulate olfactory bulb odorant-evoked responses. These data clearly show, at the very least, that olfaction processing does not involve simple feed-forward pathways. Rather, in real world situations where information has to be continually updated, olfactory responses that originate from the periphery are modulated by forebrain circuits and their projections to the olfactory bulb circuit.

Evolutionary Dimensions of Olfaction

The main olfactory system detects only volatile odorants, whereas the accessory system picks up less volatile or even water-soluble odorants. It is generally thought that the accessory system specializes in pheromone detection, whereas the main system detects common odorants [2]. In terrestrial environments, chemical signals can be either volatile or non-volatile. Accordingly, terrestrial vertebrates have two functionally and anatomically distinct olfactory systems: one detecting volatile cues (the main olfactory system) and another thought to process mostly non-volatile signals (the ►vomeronasal system). Such a dichotomy has been brought into play to support the long-standing hypothesis according to which the vomeronasal system evolved as an adaptation to terrestrial life. Today, accumulated evidence rather contests this assumption. The evolution of a vomeronasal system in aquatic species might rather provide a selective advantage for terrestrial life, and consequently it could have been retained in many species of terrestrial vertebrates. In spite of this, anatomical studies, and most recently molecular studies indicate that the selective pressure to retain vomeronasal chemosensory input has been lost in higher primates. As a result, Old World primates, apes and humans might not have retained a functional vomeronasal system. Alternatively, species without a distinct vomeronasal system may still have an accessory olfactory system intermingled within the main system. Thus, it is yet possible that the accessory system did not "arise" at some point of the vertebrate evolution, but rather it just became anatomically separated from the main system [16].

As our knowledge about the neurobiology of olfaction is growing, it is becoming incredibly evident that the main olfactory systems of animals in disparate phyla have many striking features in common. For instance, vertebrate and insect olfactory systems display common organizational and functional characteristics [17]. Further recent works that were undertaken to broaden this scope to include nematodes, mollusks and crustaceans have only strengthened this assumption. The initial common event, shared by all odorant detection systems, requires the specific interaction of odorant molecules with specific receptors expressed on the cilia of sensory olfactory neurons before conveying information to central structures [18]. Basically, four features are shared by all olfactory systems. They include: (i) the presence of ►odorant binding proteins [19] in the fluid overlying the receptor cell dendrite; (ii) the requirement of ►G-protein-coupled receptors (GPCRs) [20] as ►odorant receptors ([5,21]; even though some sensory neurons may use transmembrane guanylate cyclase receptors such as in *C. elegans* and mammals); (iii) the use of a two-step signaling cascade in odorant transduction; and (*iv*) the presence of functional structures at the first central target in the ►olfactory pathway. All these characteristics may represent adaptations that have evolved independently, and therefore might provide us with valuable information about the way the nervous system processes odorant stimuli. Alternatively, these shared properties may instead reflect underlying homology, or could have arisen independently due to similar constraints.

Similarly, the perception of odorant molecules arises from invariant series of information-processing steps that occur in anatomically distinct structures. In mammals, the olfactory epithelium contains several thousands of bipolar olfactory sensory neurons, each projecting to one of several modules in the olfactory bulb. These discrete and spherical structures, called ►olfactory glomeruli, are both morphological and functional units made of distinctive bundles of neuropil. This term reflects both the homogeneity of the sensory inputs conveyed by the ►olfactory nerve, and the degree to which the neurons in the same glomerular unit are interconnected. In different species, each glomerular structure results from the convergence in the olfactory bulb of 5–40,000 axon terminals of sensory neurons that express the same odorant receptor. As each group of glomerulus-specific output neurons is odorant receptor-specific, they form a morphologically defined network somewhat analogous to ►ocular-dominance columns in ►visual cortex or to ►barrels in the ►somatosensory cortex [22]. It is also worth noting that a number of mechanisms have evolved to ensure that only a single odorant receptor is expressed per sensory cell. In rodents, tight transcriptional control results in the choice of one among a possible thousand odorant receptor genes. This extremely large repertoire of odorant receptors is undergoing rapid evolution, with at least 20% of the genes lost to frame-shift mutations, deletions and point mutations that are the hallmarks of ►pseudogenes [3,4]. Facing a changing environment, this characteristic may reflect the pressure made on a gene family to diversify and generate large numbers of new receptors that might confer new selective advantages. Interestingly, approximately 50% of human odorant receptor genes carry one or more coding region disruptions and are therefore considered pseudogenes. This massive pseudogenization of the odorant receptors repertoire in humans and Old World primates is preceded by a moderately high level of pseudogenes (approximately 30%). Thus, there has been a decrease in the size of the intact odorant receptor repertoire in apes relative to other mammals, with a further deterioration of this repertoire in humans. Since such decline occurred concomitant with the evolution of full ►trichromatic vision in two separate primate lineages, it is possible that the weakening of olfaction results from the evolution of full ►color vision in our primate ancestors [23]. However, several overlooked human features such as the structure of the nasal cavity, retronasal smell, olfactory brain areas, and language call for reassessing the status of the sense of smell in human beings [24].

From Odorant Molecules to Cortical Centers

The olfactory system is responsible for correctly coding sensory information from thousands of odorous stimuli. To accomplish this, odor information has to be processed throughout distinct levels. At each one, a modified representation of the odor stimulus is generated. To understand the logic of olfactory information processing, one has first to appreciate the coding rules generated at each level, from the odorant receptors up to the level of the ►olfactory cortex [25,26]. In mammals, the initial event of odor detection takes place at a peripheral olfactory system, the olfactory epithelium of the nasal cavity. There, olfactory transduction starts with the activation of some of the thousand different types of odorant receptors located on the cilia of sensory neurons that comprise the olfactory neuroepithelium. The sensory neurons project to a small number of olfactory glomeruli paired on both the medial and lateral aspects of the olfactory bulb. About 20–50 second-order neurons emanate for each glomerulus and project to a number of higher centers, including the olfactory cortex. Using a trans-synaptic tracer expressed in olfactory receptor neurons under the control of two specific olfactory receptor promoters, it was possible to demonstrate that the projection of bulbar output neurons receiving sensory inputs from homologous glomeruli, form reliable discrete clusters in different regions of the olfactory cortex. Such clusters can be partly overlapping, but clearly distinct between odorants (a process called ►odor maps). A certain overlap between more diffuse projections to higher olfactory centers may constitute the anatomical basis for crosstalk between information strands emanating from different odorant receptors. This characteristic is probably helpful to integrate multiple modules of olfactory information into a composite ►gestalt, specific for a particular scent made of numerous chemical compounds.

From the External World of Odorants to Internal States

Even in humans, during the first hours of life in the open air, the newborn child behaves like a macrosmatic animal. Meanwhile, the human being is totally dominated by ►affect. During the rest of the development period and all of adult life, olfaction will remain the sense that opens the most direct route to the affective sphere [27].

To achieve this privileged relationship between olfaction and affect, the two olfactory systems connect different areas. The vomeronasal system mainly projects to the ►hypothalamus and ►amygdala that are known to control innate endocrine or behavioral responses. In contrast, in the main olfactory system, information is processed in cortical areas, which may give rise to the conscious representation of odorant molecules. In primates, the projections from the olfactory bulb reach medial olfactory areas including the ►piriform (►primary olfactory) ►cortex, ►entorhinal cortex, cortico-medial nucleus of the amygdala, and ►olfactory tubercle. From the ►piriform cortex (Primary olfactory cortex), projections reach ►area 13, a part of the caudal ►orbitofrontal cortex, and from there on to different orbitofrontal areas.

Odors are important in emotional processing; yet relatively little is known about the representation of the affective qualities of odors in the human brain. Recent results suggest that there is a ►hedonic map of the sense of smell in brain regions such as the orbitofrontal cortex [26,27]. These results have implications for understanding the psychiatric and related problems that follow damage to these brain areas. It is remarkable that amongst all the senses, olfaction possesses a particular link with the ►limbic system that was taken to be the "nose-brain" (the actual meaning of ►rhinencephalon). Today, it is clear that the primary olfactory cortex projects to the entorhinal area, which in turn projects to the ►hippocampus. Thus, we see reintroduced, after years of fervent affirmation followed by years of fervent denial, the idea that the hippocampus receives olfactory inputs. The pathway that links olfaction to the limbic system seems to be privileged. The path from the olfactory epithelium is more direct than the path from sensory surfaces such as the skin. Moreover, the primary olfactory cortex projects to the amygdala, in large part onto a particular cell group, the lateral nucleus of the amygdala, by bypassing the neocortex. However, while it is clear that the olfactory bulb projects to the amygdala in rodents, one wonders whether such a connection is still present in humans. For instance, the vomeronasal organ and the corresponding region of the accessory ►olfactory bulb are thought to form an apparatus dedicated to the processing of sexually significant odors, but in the fully formed human body none of these structures has been identified.

Non-invasive functional imaging studies of the human olfactory system revealed that the sense of smell is organized similarly to other sensory modalities, and that the specific psychological characteristics of olfaction should be attributed to an early involvement of the limbic system rather than a conceptually different mode of processing. Taking into account the high connectivity of limbic structures and the fact that activation of the amygdala immediately induces ►emotions and facilitates the coding of memories, one should not be so surprised to uncover the special relationship that links olfaction with emotions and memory.

In sum, as a result of unprecedented developments in methods for examining the structure, function, and neurochemistry of olfactory system circuits, research in olfaction has progressed dramatically in recent years. Applying new technologies, including those of neurophysiology and functional imaging should help to unravel the mysteries of how chemical perception gives rise to unique olfactory experiences such as those triggered by the exquisite fragrance of jasmine.

The Sense of Taste

The ►gustatory sense enables animals to detect and discriminate among foods, to select nutritious diets, and to initiate, sustain and terminate ingestion for the purpose of maintaining energy balance [11]. For most mammals, the decision to ingest a particular food depends not only on its taste but also on its appearance, familiarity, odor, texture, temperature and, importantly, its post-ingestive effects (for example, the ability to reduce hunger). For humans, such factors also include cultural acceptance as well as the social, emotional and cognitive contexts under which a given food is eaten. Revealing the logic of the neural mechanism of gustation is currently a major topic in modern neurobiology, given the efforts made so far towards the understanding of how complex feeding behaviors can become dysfunctional (as in the case of anorexia or obesity) [28].

In marked contrast to the olfactory system, the gustatory system has little discriminative power. Sapid stimuli come as five basic tastes, sweet, umami, bitter, salty and sour while the olfactory sensory organ recognizes about 10,000 airborne volatile molecules, in human beings. Taste stimuli are detected by assemblies of about 100 cells that form well-known specialized morphological structures, the ►taste buds, which are located in the chemosensory papillae on the tongue. However, we know astonishingly little about the precise function of these small chemosensory organs. Their characterization has largely relied on cytological and ultrastructural data [29].

The Peripheral Gustatory System

Although the sense of taste is generally associated solely with the activation of taste buds, placing food or drinks in the mouth automatically elicit responses from several distinct systems that monitor the temperature, the sound when chewing, and texture of the food. In this regard, gustation is inherently multisensory [30]. Every gourmet worth his/her salt is aware that the list of the five basic tastes should also include further perceptual categories such as astringent, fatty, tartness, water, metallic, starchy, cooling, tingling and pungent. The subjective sensations associated with these non-primary tastes result from the co-activation of taste and specialized somatosensory neurons located in the oral cavity. These specialized neurons surround taste buds, and include different classes of mechano- and chemoreceptors that transmit information on the food's texture, weight and temperature to the brain mainly via the ►trigeminal system.

Transduction Pathways for Primary Tastes

In the oral chemosensory epithelia, taste buds contain about 50–100 ►taste receptor cells (TRCs) of various types. These TRCs are embedded in stratified epithelia and are distributed throughout the tongue, palate, epiglottis and esophagus. On their apical end, taste cells make contact with the oral cavity through a small opening in the epithelium called the taste pore, which is filled with microvilli. The plasma membranes of these microvilli contain many of the receptors responsible for detecting the presence of various tastants. Small clusters of TRCs are electrically and chemically coupled by ►gap junctions allowing their synchronous activation.

On the palate and the anterior tongue, TRCs are innervated by the ►chorda tympani nerve and greater superior petrosal branches of the ►facial nerve, respectively. These nerves transmit information about the identity and quantity of the chemical nature of the tastants. On the epiglottis, esophagus and posterior tongue, TRCs are innervated by the lingual branch of the ►glossopharyngeal nerve and the superior laryngeal branch of the ►vagus nerve. These nerves are responsive to tastants and participate primarily in the brainstem-based arch reflexes that mediate swallowing (ingestion) and gagging (rejection). TRCs transmit information to the peripheral nerves by releasing ►ATP to ►P2X purinergic receptors located on the postsynaptic membrane of primary afferents. Other transmitters such as ►serotonin, ►glutamate and ►acetylcholine might also be released.

The key to understanding how TRCs transduce chemical stimuli lies in determining the identification and operation of different types of taste receptor and their downstream signaling pathways. As for olfaction, proteins belonging to the G-protein-coupled receptor superfamily have been established as the receptors for sweet tastants (taste receptor, type 1, member 2 (T1R2)/T1R3), amino acids (T1R1/T1R3) and bitter (T2Rs) tastants. The sensations associated with the other two primary tastants, sour and salt (NaCl), are mediated by ion channels of the ►transient receptor potential (TRP) and ►epithelium sodium channel (ENaC) superfamilies, respectively.

Taste Pathway

Receptor cell depolarization leads to the release of neurotransmitter, which generates first post-synaptic potentials and then action potentials in the associated nerve endings. The axons, whose cell bodies lie in the sensory ganglia of the cranial nerve, enter the medulla and synapse in the region of the ►nucleus of the solitary tract (N. tractus solitarii). The nerve cells in this part of the medulla are important in mediating salivation and other gastrointestinal reflexes. Their axons also cross over, and relay via the contralateral ►medial lemniscus, to the ►thalamus, and thence to the post-central gyrus in the region of the ►insula.

Coding in the Periphery

Two schemes have been proposed to explain how taste processing is achieved through the interaction of TRCs with their associated afferent nerve fibers: the ►"labeled line" model and the "►across-fiber pattern" (or "distributed") model [11]. The assessment of experimental data supporting either of these hypotheses constitutes an important source of debate in the field of gustatory physiology. The ►labeled line hypothesis (model) implies that sensory information is processed through segregated and feed-forward circuitry that connects peripheral sensory receptors to higher-order structures in the CNS. By contrast, across-fiber pattern models propose that sensory fibers (or neurons) are broadly tuned, in such a way that stimulus identity and intensity are specified by a unique combinatorial pattern of activity distributed across populations of neurons.

At the peripheral level one can find experimental support for both labeled line and across-fiber pattern models, but recent data from genetic studies strongly favor the existence of labeled lines. The validity of either model at the periphery should not necessarily be generalized to CNS circuits. In contrast to the periphery, the CNS possesses the anatomical structure required for ►multisensory integration and this ability might determine a difference in coding strategies between the CNS and peripheral nervous system (PNS). In fact, much of the current neurophysiological data describe gustatory processing as multisensory and distributed across several brain regions [31].

In contrast to the highly specialized information transfers performed by TRCs and peripheral fibers, central gustatory processing seems to be distributed, probably as a result of its capacity for ►multimodal (multisensory) integration. Approaching the encoding of a gustatory stimulus in this manner will provide new insights into how information is encoded, beyond the theories that have been historically proposed to model the mechanisms by which taste quality is coded in the periphery. Indeed, how these sensory modalities are synthesized into a single percept, which allows animals to rapidly decide whether to ingest or reject a particular food, is the greatest challenge for the near future in gustatory physiology.

The growth of our knowledge in gustation has not yet reached the level of olfaction. Many fundamental problems in the emerging field of taste are still to be resolved. For example, what is the coding logic for multisensory integration? How is a taste percept generated from activation of labeled- or distributed-lines? Answers to these basic questions might help us understand how the brain makes sense of chemical compounds that we daily place in the mouth.

Concluding Remarks

Smell and taste problems can have a big impact on our lives. Because these senses contribute substantially to our enjoyment of life, our desire to eat, and be social, smell and taste disorders can be serious. When smell and taste are impaired, we eat poorly, socialize less, and as a result, feel worse. Many older people experience this problem. But not only chemical signaling make us happier, smell and taste also warn us about dangers, such as fire, poisonous fumes, and spoiled food. Certain jobs require that these senses be accurate – chefs and firemen rely on taste and smell. Loss or reduction of the sense of smell (►anosmia or ►hyposmia) may be due to damage to the olfactory mucosa (e.g., in smoking) or to the olfactory bulbs or tracts. CNS disorders (e.g., some types of ►epileptic seizures) can cause ►parosmia (disturbed sense of smell). Like olfaction, the sense of taste is important in regulating appetite and to some degree, dietary intake. Loss or reduction of the ability to taste is termed ►ageusia or ►hypogeusia and is a widely distributed feature of ageing as more than 200,000 people visit a doctor with smell and taste disorders every years in the United States.

Strikingly, olfactory and gustatory systems are endowed with rejuvenating properties throughout life. Olfactory and taste cells are one of the few cell types of the nervous system to be continuously replaced when the sensory organs become old or damaged. Scientists are examining this phenomenon, called adult ►neurogenesis [32], while studying ways to use this potential to replace other damaged nerve cells of the CNS.

Smell and taste have here been presented separately but one should keep in mind that about 75% of what we perceive as taste actually comes from smell. It is the odor molecules from food that give us most of our taste sensation as taste buds allow us to perceive only five flavors. Of all our senses, smell is our most primal. Animals need the sense of smell to survive. Although a blind rat might survive, a rat without its sense of smell can't mate or find food. For humans, the sense of smell communicates many of the pleasures in life–the aroma of a pot roast in the oven, fresh-cut hay, a rose garden. Although our sense of smell is our most primal, it is also very complex. To identify the smell of a rose, the brain analyzes simultaneously over 300 odor molecules. The aroma of a baking apple pie sends one message when someone is hungry and quite another when that person has just finished a six-course meal!

Although recent discoveries in the field of molecular biology raise the hope of a future understanding of the transduction and peripheral coding of odors and tastes, it seems that they imply a risk: to make us forget that in the other extreme of knowledge, that of maximal complexity, the evolution of cognitive sciences allows an epistemologically fruitful reformulation of information-processing problems. In the future, we have to try to find out to what extent higher-order processes interact with the sensory level in order to produce sufficiently reliable representations, as compared with what we know about vision and audition, for instance. After all, we should not forget what Sigmund Freud, addressing the members of the Vienna Psycho-analytical Society, said about olfaction: "*the organic sublimation of the sense of smell is a factor of civilization*."

References

1. Finger T, Simon SA (2002) The Cell Biology of Lingual Epithelia. In: Finger T, Silver WL, Restrepo D (eds) The neurobiology of taste and smell. Vol 2. Wiley-Liss, New York, 12: 287–314
2. Ache BW (1991) Phylogeny of smell and taste. In: Getchell TV, Bartoshuk LM, Doty RL, Snow JB (eds) Smell and taste in health and disease. Raven, New York pp 3–18
3. Mombaerts P (2004) Genes and ligands for odorant, vomeronasal and taste receptors. Nat Rev Neurosci 5:263–278
4. Mombaerts P (2004) Love at first smell – the 2004 Nobel Prize in physiology or medicine. N Engl J Med 351:2579–2580
5. Buck L, Axel R (1991) A novel multigene family may encode odorant receptors: a molecular basis for odor recognition. Cell 65:175–187
6. Margolskee RF (2005) Sensory systems: taste perception. Sci STKE 290:tr20
7. Ache BW (1994) Towards a common strategy for transducing olfactory information. Semin Cell Biol 5:55–63
8. Dalton P, Doolittle N, Nagata H, Breslin PAS (2000) The merging of the senses: integration of subthreshold taste and smell. Nat Neurosci 3:431–432
9. Olender T, Feldmesser E, Atarot T, Eisenstein M, Lancet D (2004) The olfactory receptor universe – from whole genome analysis to structure and evolution. Genet Mol Res 3:545–553
10. Smith DV, St. John SJ (1999) Neural coding of gustatory information. Curr Opin Neurobiol 9:427–435
11. Simon SA, de Araujo IE, Gutierrez R, Nicolelis MAL (2006) The neural mechanisms of gustation: a distributed processing code. Nat Rev Neurosci 7:890–901
12. Algom D, Cain WS (1991) Chemosensory representation in perception and memory. In: Bolanowski SJ, Gescheider GA (eds) Ratio scaling of psychological magnitude. Hillsdale. Lawrence Erlbaum Associates, N.J., pp 183–198
13. Dulac C, Torello AT (2003) Molecular detection of pheromone signals in mammals: from genes to behaviour. Nat Rev Neurosci 4:551–562
14. Wysocki CJ, Preti G (2004) Facts, fallacies, fears, and frustrations with human pheromones. Anat Rec 281:1200–1210
15. Lledo PM, Gheusi G, Vincent JD (2005) Information processing in the mammalian olfactory system. Physiol Rev 85:281–317
16. Keverne EB (2004) Brain evolution, chemosensory processing, and behavior. Nutr Rev 62:S218–S223
17. Hildebrand JG, Shepherd GM (1997) Mechanisms of olfactory discrimination: converging evidence for common principles across phyla. Annu Rev Neurosci 20:595–631
18. Ache BW, Young JM (2005) Olfaction: diverse species, conserved principles. Neuron 48:417–430
19. Tegoni M, Pelosi P, Vincent F, Spinelli S, Campanacci V, Grolli S, Ramoni R, Cambillau C (2000) Mammalian odorant binding proteins. Biochim Biophys Acta 1482:229–240
20. Ronnett GV, Moon C (2002) G proteins and olfactory signal transduction. Annu Rev Physiol 64:189–222
21. Buck LB (2004) Olfactory receptors and odor coding in mammals. Nutr Rev 62:S184–S188
22. Johnson BA, Woo CC, Leon M (1998) Spatial coding of odorant features in the glomerular layer of the rat olfactory bulb. J Comp Neurol 393:457–471
23. Gilad Y, Wiebe V, Przeworski M, Lancet D, Pääbo S (2004) Loss of olfactory receptor genes coincides with the acquisition of full trichromatic vision in primates. PLoS Biol 2:120–125
24. Shepherd GM (2004) The human sense of smell: are we better than we think? PLoS Biol 2:572–575
25. Malnic B, Hirono J, Sato T, Buck LB (1999) Combinatorial receptor codes for odors. Cell 96:713–723
26. Rolls ET (2001) The rules of formation of the olfactory representations found in the orbitofrontal cortex olfactory areas in primates. Chem Senses 26:595–604
27. Bensafi M, Rouby C, Farget V, Bertrand B, Vigouroux M, Holley A (2003) Perceptual, affective, and cognitive judgments of odors: pleasantness and handedness effects. Brain Cogn 51:270–275
28. Scott K (2005) Taste recognition: food for thought. Neuron 48:455–464

29. Breslin PA, Huang L (2006) Human taste: peripheral anatomy, taste transduction, and coding. Adv Otorhinolaryngol 63:152–190
30. Rolls ET (2004) Convergence of sensory systems in the orbitofrontal cortex in primates and brain design for emotion. Anat Rec 281:1211–1224
31. Cytowic RE (2002) Synesthesia: a union of the senses. MIT, Cambridge
32. Lledo P-M, Grubb M, Alonso M (2006) Adult neurogenesis and functional plasticity in neuronal circuits. Nat Neurosci Rev 7:179–193

Olfactometer

Definition

A device for delivering odorant stimuli with controlled odorant concentrations and durations.

► Brain States and Olfaction

Olfactory Acuity

► Olfactory Perception

Olfactory Adaptation

Definition

The decrease over time in the neural response to a continuous odorant presentation is known as olfactory adaptation.

► Glomerular Map
► Odorant

Olfactory Amygdala

Definition

Group of nuclei of the amygdaloid complex that receives olfactory information directly from the main olfactory bulbs or indirectly from other amygdaloid nuclei.

► Evolution of the Amygdala: Tetrapods
► Olfactory Bulb

Olfactory Aura

► Olfactory Hallucinations

Olfactory Awareness

► Olfactory Perception

Olfactory Binding Proteins

► Odorant-Binding Proteins

Olfactory Bulb

Marco Sassoè-Pognetto
Department of Anatomy, Pharmacology and Forensic Medicine and National Institute of Neuroscience, University of Torino, Torino, Italy

Synonyms

Main olfactory bulb, as opposed to "accessory olfactory bulb"

Definition

The olfactory bulb is the first relay station in the olfactory pathway, situated at the rostral end of the brain. It receives sensory input from olfactory receptor neurons located in the nasal cavity and sends output fibers to a group of hemispheric regions collectively termed the olfactory cortex.

Characteristics

The olfactory bulbs develop from the ventral surface of the cerebral hemispheres and in the large majority of vertebrates they represent the most rostral extension of the neural axis. In apes and humans, however, the olfactory bulbs lie on the ventral surface of the frontal lobes, just above the nasal cavities, from which they are separated by the cribriform plate of the ethmoid bone (Fig. 1).

The unmyelinated axons of the olfactory nerve pass through this bone and reach the olfactory bulb, where they terminate in spheroidal regions of neuropil called glomeruli. Head trauma can lesion the olfactory nerve fascicles as they traverse the cribriform plate, resulting in anosmia. The olfactory tract leaves the posterior pole of the olfactory bulb. It contains the output projections of the bulb as well as afferent fibers originating from a variety of brain regions.

In most vertebrate species, the olfactory bulb is organized according to the same basic plan and, as in other cortical regions, it shows a laminated structure, consisting of seven concentric layers (Fig. 2).

The principal (output) neurons of the bulb are the mitral and tufted cells (M/T cells), which can be divided into multiple subtypes based on position, dendritic morphology and axonal projection patterns. Mitral and tufted cells receive excitatory sensory inputs in the glomerular layer and send their axons to different regions of the olfactory cortex. Apart from this straight-through pathway, there are other neurons that act locally and make predominantly inhibitory connections with the principal cells. Synaptic inhibition plays a crucial role in the processing of olfactory information before it is transmitted to the olfactory cortex (see below).

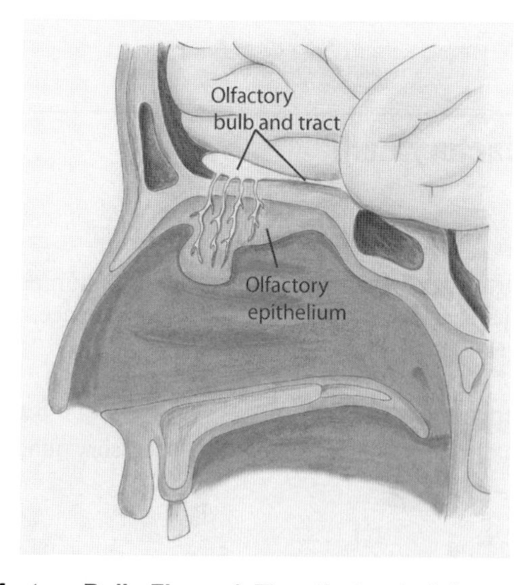

Olfactory Bulb. Figure 1 The olfactory bulb is a small ovoid structure that lies on the cribriform plate of the ethmoid bone. It receives input from olfactory sensory neurons located in the nasal cavity and projects to the olfactory cortex through the olfactory tract. (Courtesy of Dr. Alessandro Ciccarelli.)

It should be noted that neuronal excitation in the olfactory bulb is mediated primarily by glutamate, whereas GABA is the principal inhibitory neurotransmitter. In addition, the olfactory bulb is rich in several other neuroactive substances, which likely exert a neuromodulatory function.

The circuit organization of the olfactory bulb can be conveniently separated into two distinct levels [2]. Sensory inputs from the ►olfactory sensory neurons are first processed within the glomeruli, where they are subject to amplification and attenuation. The second level of processing involves reciprocal interactions between the principal neurons and local interneurons, which provide GABAergic inhibition through dendro-dendritic synapses.

Input Processing

Within the glomeruli, sensory axons make excitatory synaptic connections with the apical dendrites of the output neurons and a heterogeneous class of intrinsic neurons called periglomerular cells (►periglomerular cell in olfactory bulb) (PG) (Fig. 3).

In addition, PG cells are interconnected with M/T cells through dendro-dendritic synapses. Several types of dendro-dendritic synapses have been described, including excitatory synapses from M/T cells to PG cells, inhibitory synapses from PG cells to M/T cells, and synapses between distinct subtypes of PG cells. While there is no evidence for dendro-axonic synapses onto olfactory nerve axons, it has been shown that GABA and dopamine inhibit glutamate release from these axons by activating presynaptic receptors ($GABA_B$ receptors and dopamine D_2 receptors, respectively). It is likely that the paucity of glial barriers within the glomerular neuropil facilitates the diffusion of neurotransmitter and the occurrence of nonsynaptic interactions. The current model suggests that intraglomerular microcircuits contribute to regulate transmission from the olfactory nerve to M/T cells and thus serve as a gain control mechanism of incoming sensory signals.

A remarkable specificity exists in the projection pattern of olfactory receptor neurons to the glomeruli. First, each glomerulus is the site in which thousands of sensory axons converge on the dendrites of just ~20–50 relay neurons. A notable feature is that the axon of each sensory neuron terminates in only one glomerulus and, similarly, the apical dendrite of each principal neuron (mitral or tufted cell) arborizes into a single glomerulus. Second, all of the olfactory axons terminating in a glomerulus express the same odorant receptors (out of a repertoire of ~1,000 genes in rodents). Third, olfactory sensory neurons expressing a specific type of odorant receptor usually innervate two distinct glomeruli, which

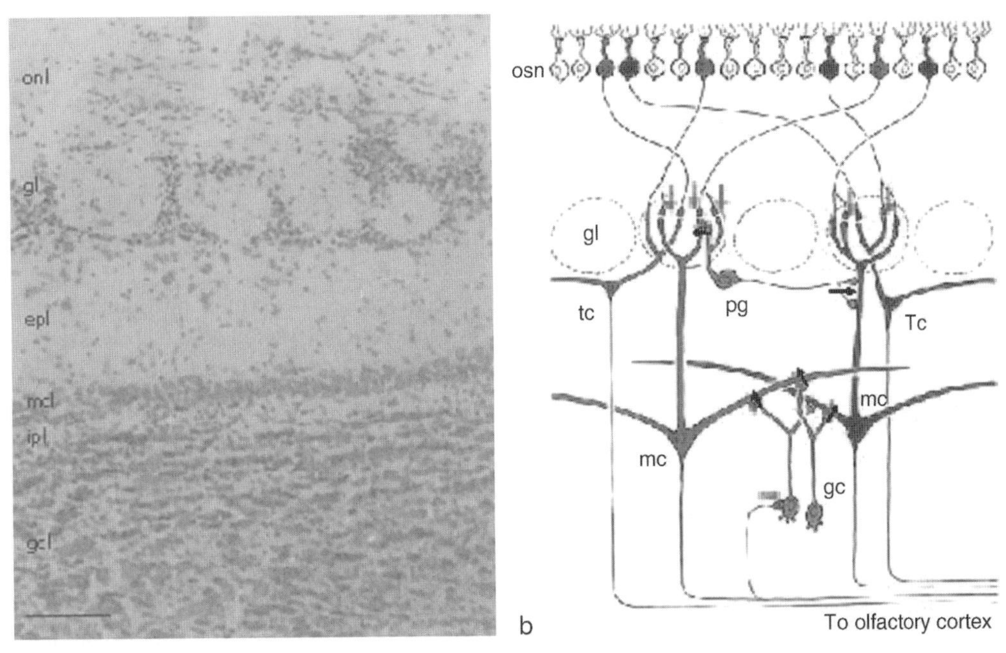

Olfactory Bulb. Figure 2 (a) Coronal section of the rat olfactory bulb illustrating the laminar organization. The most superficial layer is the olfactory nerve layer (onl), which contains the axons of olfactory sensory neurons. Deep to the granule cells is a periventricular or subependymal layer (not visible in this micrograph), which contains migrating neuroblasts. gl: glomerular layer; epl: external plexiform layer; mcl: mitral cell layer; ipl: internal plexiform layer; gcl: granule cell layer. Scale bar: 100 μm. (b) Circuit diagram summarizing the basic synaptic organization of the olfactory bulb. Two glomerular units are shown, each receiving input from olfactory sensory neurons (osn) expressing a given type of odorant receptor and connecting to a subset of mitral cells (mc) and tufted cells (tc). Periglomerular cells (pg) and granule cells (gc) mediate feedback and lateral inhibition of principal neurons through axo-dendritic and dendro-dendritic synapses. *Red arrows* indicate excitatory synapses, and *black arrows* indicate inhibitory synapses. (Adapted from [1]; courtesy of Dr. Alessandro Ciccarelli.)

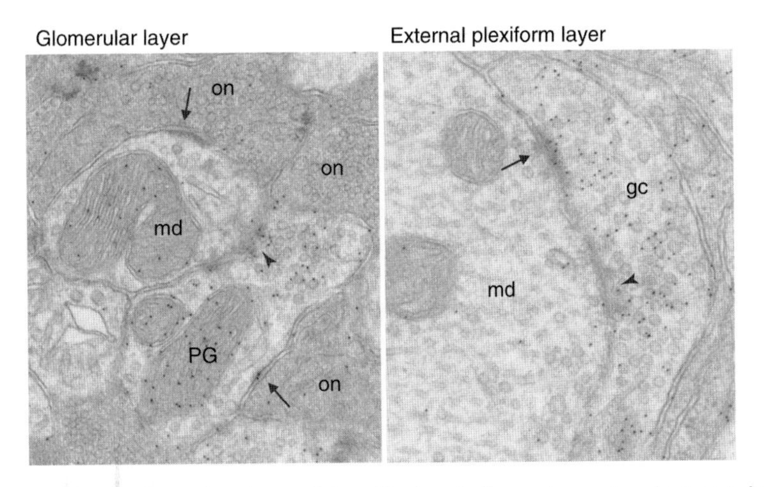

Olfactory Bulb. Figure 3 Synaptic connections of the olfactory bulb as shown by electron microscopy after postembedding immunogold labeling with an antiserum against GABA (courtesy of Dr. Patrizia Panzanelli). Gold particles of 10 nm identify GABA-immunopositive structures. In the glomerular layer, olfactory nerve axons (on) make asymmetrical synapses (*arrows*) with two dendritic profiles. One dendrite is GABA-positive and therefore belongs to a periglomerular cell (pg). The other dendrite (md) likely belongs to a mitral/tufted cell. Note that the pg dendrite also makes a dendro-dendritic synapse (*arrowhead*) with the md profile. External plexiform layer. A reciprocal dendro-dendritic synapse between a mitral cell dendrite (md) and a granule cell spine (gc) is shown. Note that the granule cell spine is GABA-positive. The mitral-to-granule synapse (*arrow*) is asymmetrical, whereas the granule-to-mitral synapse (*arrowhead*) is symmetrical.

are bilaterally symmetrical and similarly located in the olfactory bulbs of different animals [3]. Therefore, a glomerulus can be defined as a convergence center for inputs originating from a given type of odorant receptor. This specificity implies that glomeruli represent basic functional units, analogous to cortical columns, and that different odors are represented by different patterns of spatial activity in such glomerular units [1] (see glomerular map).

Inhibitory Control of Mitral/Tufted Cells

The second level of information processing in the olfactory bulb is based on reciprocal synapses between the principal neurons and the granule cells (Fig. 3). Granule cells are axonless neurons, whose cell bodies give rise to an apical dendrite that extends radially in the external plexiform layer [4]. The dendrites of granule cells are characterized by the presence of large spines (also called gemmules), that establish reciprocal synapses with the dendrites of M/T cells [5]. In these reciprocal connections, both sides of the synapse are dendrites capable of releasing neurotransmitter. The dendrites of M/T cells release glutamate and excite the spines of granule cells, which in turn release the inhibitory neurotransmitter GABA back onto the principal neurons (Fig. 2). As a result, activated M/T cells can inhibit themselves (feedback inhibition), as well as their neighbors (lateral inhibition).

There is compelling evidence that lateral inhibition is crucial in refining olfactory information, as it enhances the contrast between the activity of M/T cells connected to different glomerular units, and thus sharpens the tuning specificity of the output neurons to different odor molecules [1]. In other words, activation of M/T cells associated with one glomerulus results in inhibition of other glomerular units through the reciprocal dendrodendritic interactions. Therefore, the lateral inhibition mediated by granule cells enhances the contrast between strongly activated and faintly activated glomerular units and increases the specificity of individual M/T cells to odor molecules. Given that the basal dendrites of mitral cells have a projection field with a radius of about 1 mm, they potentially can influence the activity of glomerular units over long distances. This is consistent with experimental evidence that odor maps are widely distributed in the glomerular layer [6].

Lateral inhibition is also important for synchronizing the output responses of M/T cells connected to functionally related glomeruli. It has been known for a long time that stimulation with odor molecules elicits γ-frequency (30–80 Hz) oscillations of local field potentials, reflecting synchronized spike discharges of the principal neurons. This synchronization likely serves as a mechanism for temporal summation of signals from different glomerular units, and may play an important role in odor discrimination [7]. Of particular interest is the possibility that plastic changes in the strength of dendro-dendritic synapses may represent one mechanism underlying olfactory learning.

Other Neuronal Populations

In addition to PG cells and granule cells, there is a relatively small population of short-axon cells, which are distributed in the glomerular and granule cell layers. Recent studies suggest that interglomerular interactions mediated by short-axon cells represent a mechanism by which activated glomeruli can influence the activity of other glomerular units and contribute to enhance the spatial responses to odors [8]. In addition, one type of short-axon cell located in the granule cell layer provides GABAergic inhibition onto granule cells and therefore can control the strength of feedback and lateral inhibition onto the principal neurons [9].

Centrifugal Afferents

The olfactory bulb receives a prominent innervation by centrifugal fibers from a variety of sites in the brain. The best characterized are cholinergic fibers arising from the basal forebrain and noradrenergic and serotoninergic fibers arising, respectively, from the *locus coeruleus* and the mesencephalic raphe nucleus. These centrifugal afferents mediate a considerable degree of control over olfactory processing, which seems to be important for adapting olfactory function to different behavioral states. Of particular interest is the action of noradrenaline, which suppresses granule cell inhibition of M/T cells. Noradrenergic modulation of dendro-dendritic inhibition has been involved in some forms of olfactory learning.

Parallel Processing of Olfactory Stimuli

As in other sensory systems, the olfactory bulb contains several parallel pathways for processing olfactory information. An obvious case is the accessory olfactory bulb, a structure present in most terrestrial vertebrates that receives sensory inputs from the vomeronasal organ. Within the main olfactory bulb, there is evidence for specialized glomerular units that process certain types of olfactory stimuli. For instance, the so called "modified glomerular complex" has been implicated in suckling behavior in neonatal animals. Mitral and tufted cells also appear give rise to parallel output pathways from the olfactory bulb. These neurons interact with different subpopulations of granule cells and project their axons to different cortical regions [10]. However, our understanding of how mitral and tufted cells process distinct aspects of olfactory information is still preliminary.

Plasticity

The olfactory bulb is one of the few brain regions in which neurogenesis in maintained throughout life.

Bulbar interneurons are continuously replaced from a population of stem cells located in the subventricular zone of the lateral ventricle. Neuroblasts generated in this area migrate along the rostral migratory stream to the olfactory bulb, where they complete their differentiation into GABAergic neurons. Similarly, olfactory sensory neurons undergo continuous turnover during adult life. Remarkably, these neurons can reestablish functional synaptic connections with their target cells in the olfactory bulb. This degree of plasticity is unmatched in the brain and makes the olfactory bulb a unique model for studying the mechanisms of neural development and cell replacement.

References

1. Mori K, Nagao H, Yoshihara Y (1999) The olfactory bulb: coding and processing of odor molecule information. Science 286:711–715
2. Shepherd GM (2004) Olfactory bulb. In: Shepherd GM (ed) The synaptic organization of the brain, 5th edn. Oxford University Press, New York, pp 165–216
3. Vassar R, Chao SK, Sitcheran R, Nuñez JM, Vosshall LB, Axel R (1994) Topographic organization of sensory projections to the olfactory bulb. Cell 79:981–991
4. Shepherd GM, Chen WR, Willhite D, Migliore M, Greer CA (2007) The olfactory granule cell: from classical enigma to central role in olfactory processing. Brain Res Rev 55:373–382
5. Price JL, Powell TPS (1970) The synaptology of the granule cells of the olfactory bulb. J Cell Sci 7:125–155
6. Leon M, Johnson BA (2003) Olfactory coding in the mammalian olfactory bulb. Brain Res Rev 42:23–32
7. Laurent G (2002) Olfactory network dynamics and the coding of multidimensional signals. Nat Rev Neurosci 3:884–895
8. Aungst JL, Heyward PM, Puche AC, Karnup SV, Hayar A, Szabo G, Shipley MT (2003) Center-surround inhibition among olfactory glomeruli. Nature 426:623–629
9. Pressler R, Strowbridge B (2006) Blanes cells mediate persistent feedforward inhibition onto granule cells in the olfactory bulb. Neuron 49:889–904
10. Zou Z, Horowitz LF, Montmayeur JP, Snapper S, Buck LB (2001) Genetic tracing reveals a stereotyped sensory map in the olfactory cortex. Nature 414:173–179

Olfactory Bulb Glomerulus

Definition

An olfactory glomerulus is a compartmentalized mass of neuropil in the glomerular layer of the olfactory bulb that contains synapses between olfactory sensory neuron axon terminals and dendrites of both projection neurons (mitral and tufted cell apical dendrites) and local periglomerular cell inhibitory interneurons. Glomeruli also contain numerous dendrodendritic synapses between mitral or tufted cells and both periglomerular and so-called short-axon cells. A typical rodent olfactory glomerulus receives convergent projections only from sensory neurons expressing the same odorant receptor gene. At the neuronal circuit level, an individual glomerulus in the olfactory bulb may function as a molecular-feature detecting unit.

►Flavor
►Glomerular Map
►Olfactory Bulb
►Olfactory Bulb Mitral Cells
►Olfactory Sensory Neuron
►Periglomerular Cells in Olfactory Bulb

Olfactory Bulb Granule Cells

Definition

These are a large population of small GABAergic interneurons in the vertebrate olfactory bulb that do not receive sensory input directly. They form dendrodendritic reciprocal synapses with mitral cell lateral dendrites and also receive axodendritic synapses from mitral cell axon collaterals and centrifugal fibers. Most of their inputs are glutamatergic, but they also receive GABAergic inputs. Most olfactory bulb centrifugal inputs target the granule cells.

►Olfactory Bulb
►Olfactory Bulb Mitral Cells

Olfactory Bulb Mitral Cells

Definition

Glutamatergic projection neurons lying in the mitral cell layer of the olfactory bulb. They receive direct input from olfactory sensory neuron terminals in olfactory bulb glomeruli, and project directly to olfactory cortex. They also have multiple, complex interactions with olfactory bulb interneurons, both periglomerular cells and granule cells, through conventional and dendrodendritic synapses.

►Olfactory Bulb
►Olfactory Cortex
►Olfactory Sensory Neuron

Olfactory Code

►Odor Coding

Olfactory Coding

►Odor Coding

Olfactory Cortex

GILLES SICARD
Centre Européen des Sciences du Goût, Dijon, France

Synonyms

Downstream neural structure of the olfactory bulb

Definition

Referring to multiple structures receiving olfactory information and presenting the classical cyto-architecture of nervous cortex, "olfactory cortices" is a more correct definition of the topic of this article.

Stock of knowledge. Details of the ►primary cortical projections of the olfactory system indicate the diversity of the structures that are directly connected to the olfactory bulb neurons (Fig. 1).

Characteristics

The graph is soon a divergence from the canonical hierarchical organization of a sensory pathway. The bulbar output is conveyed by the lateral olfactory tract. On the functional point of view, we do assume that a topographical representation of the olfactory stimulus based on the chemical features takes place in the glomerular layer of the olfactory bulb.

From the receptor level, the primary ►olfactory cortex is reached through two synapses only. The receptor neurons are connected to mitral and tufted cells in the olfactory bulb. These relay neurons feed the pyramidal cells of the cortex, a three-layered paleocortex. The primary olfactory projections are annexed to the ►limbic system, an associative area. This system plays a role in social and emotional processing and supports some of the mechanisms of the memory.

Two neurons, two synapses: It is noticeable that this short pathway bypasses the thalamus before displaying cortical representations of the stimulus. This peculiar arrangement differing from those observed in other sensorial modalities can be explain by the fact that olfactory modality got ahead the emergence of the thalamic structures in the phylogenesis. On a functional point of view, this also means that probably in the olfactory system the processes fulfilled by the thalamus are implemented in other structure(s), and logically, in the downstream structures (thus the olfactory bulb) and/or the structures described the present chapter. It looks likely as both olfactory bulb and olfactory cortices receive modulating influences from diverse centers, including for instance the arousal or the satiety control systems.

Focusing on the functional properties of the olfactory cortices, they are considered both as the targets of relay neurons from the olfactory bulb and as the origin of neurons contacting neocortical associative territories such as the orbito-frontal cortex, the neocortical temporal cortex … and even parts of the thalamus!

The ►primary olfactory cortex includes contiguous or dispersed structures in the medial aspect of the temporal lobe of the brain which homologous equivalents are not easy to identify among different species. In order to describe the functions of the olfactory cortex, we get information from different animal models, rat, mouse, rabbit or frog, including man. In the view of this complexity, the terminology itself can be misleading: For instance the anterior olfactory nucleus which is funded by the fibers of the lateral olfactory tract, is a true cortex, characterized by the presence of pyramidal neurons. For those reasons, we have limited the description to the main structures: the anterior olfactory nucleus, the piriform cortex, the olfactory part of the amygdala and the entorhinal cortex. A ►secondary olfactory cortical area taking information from ►primary olfactory cortices, the orbito-frontal cortex will be also envisaged in the article.

With the olfactory cortex processing, important integrations of the olfactory signal follows the first sharpening of the information captured from the chemical environment by the receptor organ. If in the olfactory bulb the chemical nature of the stimulus is decomposed (de-constructed representation of odorants), in the olfactory cortex several tasks of reconstruction (re-constructed representation) take place, add memorized information and finally these levels of processing tend to confer a "meaning" to the actual olfactory message.

Primary Cortical Projections

Anterior olfactory nucleus. Natural odorants are mixtures of chemicals. To imagine the integrative processes of the neurons, one can test how the neurons are responding to chemical mixtures and to their isolated components. While bulbar neurons show a

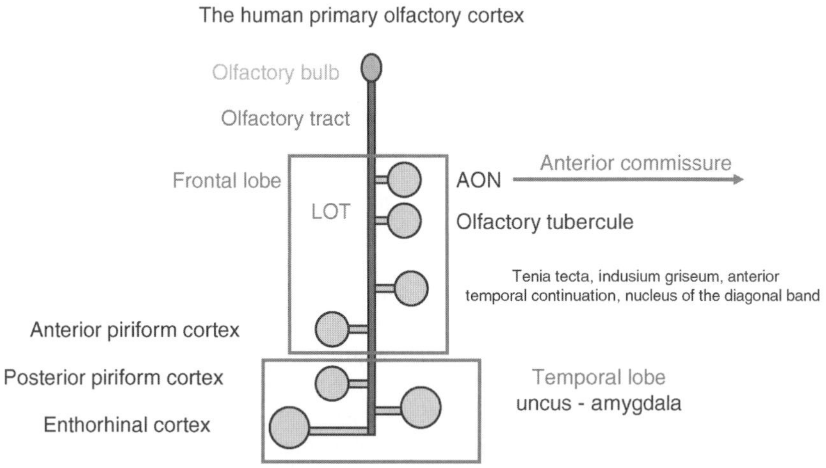

Olfactory Cortex. Figure 1 The human primary olfactory cortex: The hierarchical representation of the olfactory pathway describes a multi-unit network: The olfactory tract is constituted of the mitral and tufted-neuron axons, the relay neurons directly connected to the receptor neurons from the olfactory receptor organ in the olfactory bulb. By this pathway a number of cortical structures receive direct sensory inputs. They are distributed in different parts – frontal temporal – of the cerebral cortex. These primary cortical elements are largely interconnected and receive modulations from different higher centers (arousal, satiety…). The first part, anterior olfactory nucleus and piriform cortex are concerned by sharpening of the sensory message. The other units are involved in control of emotional responses, behavioral responses to odorant stimulations and in olfactory learning.

sparse responsiveness they show high selectivity and they often respond to only one of the components in a mixture. In the anterior olfactory nucleus, the majority of neurons can respond to mixture of dissimilar chemicals and to their isolated components. In addition, the responses to the mixture exceed the simple sum of the responses to each of its components [1]. These properties point out a first kind of integrative process that the neuronal populations of ►primary cortical olfactory level are able to realize: This is a simple sharpening of the sensory input message.

Piriform cortex. Extensively connected with higher-order cortical areas, the piriform cortex received also direct afferences from the olfactory bulb. The receptor fields of its neurons have been characterized by neuronal tracing, giving a spatial idea of the coding of the odorant stimuli [2] while electrophysiological recording of the neuronal responses to odorants gave a functional view [3]. As a ►primary cortex, it could take part to the extraction of specific features from the olfactory message, thus used the combinatorial analytic representation of the sensory signal provided by the olfactory bulb. At this level, some neurons require particular combinations of chemicals to respond, thus suggest a combination of signals from distinct samples of bulbar neurons [4]: The cortex plays a role in discrimination of odorant signals. Nevertheless, at this early level, some modulations of the neuronal responses in behaving rats by non olfactory information (reward, expectation) were found, adding associative functions to its competences. In that sense, this "►primary" cortex differs from primary cortices (►primary, secondary cortices) of the other sensorial modalities, which are rather dedicated to sharpen the input message. The olfactory piriform cortex must be regarded as a piece of olfactory learning and memory [5] It is of great importance to note that the receptive fields of the neurons in this cortex change with the experience: This property is indicative of upper-stream associative areas in the other sensorial modalities.

According to neuronal tracing, the partial overlapping of the projections from different receptor channels in the piriform cortex suggests that the cortex is able to merge different elements of the peripheral signal. In the anterior olfactory nucleus or in anterior piriform cortex, the selectivity of neurons to diverse chemical or perceptual categories appears to be broader than that of the bulbar relay neurons. This is true assuming that, due to the functional convergence of receptor neurons on bulbar glomeruli, the output neurons, mitral and tufted cells, have relatively narrow ►selectivity profiles. (We must notice some discrepancy about the chemical selectivity of the bulbar neurons reported in different studies). Nevertheless, the complex selectivity profiles of the cortical output neurons means that these neurons integrate several odorant features.

Odorant quality coding in the olfactory piriform cortex. Tracing the projections area of the output bulbar neurons, it is possible to discriminate an anterior part and an posterior part of the piriform cortex [2]. This is confirmed by functional observations of the spatial organization of the responses to hedonic contrasted chemical stimuli in human [6]. While the anterior part seems to encode the chemical features, the posterior part could discriminate stimuli along a qualitative dimension, i.e. their odor. Following the partial functional convergence shown by the bulbo-cortical relationships, the anterior piriform Cortex could reconstruct the complex environmental stimuli that have been decomposed by the topographic arrangements of the bulbar projections of the hundreds of specific olfactory receptors. The mechanisms or the rules of this reconstruction are not known. In this debate, the representation suggested by the topographical combinatorial theory plays a central role. However, one must notice that several studies on electrophysiological reports confirms that the selectivity of the bulbar neurons is scarce, but indicate that these neurons convey activations elicited by very different chemical structures [3,4]. The integration of this information on the discrimination processing by the cortical neuronal population must be further examined.

Contributions of olfactory cortices to behavioral controls. Several other olfactory cortices, recruiting even a larger amount of influences, are also directly connected to the olfactory bulb.

The amygdala is in fact a series of nucleus, receiving inputs from multiple ascending sensory pathways, including olfactory, gustatory, visual, auditory and visceral information, more or less directly from the sensory organ, thus after more or less stages of treatment. Here again the afferent olfactory pathway is the shortest. Extending influences on the hypothalamus, the medulla or the spinal chord, the amygdala is implicated in the modulation of the visceral functions in relation with emotional status. By its connections with the nearest olfactory structures in the rostral temporal lobe, it modulates their activity according to the mood or the emotional life of the animal.

Different implications of the olfactory amygdala on animal behavior have been investigated. In fact the different cortical olfactory structures, including amygdala, entorhinal cortex, perirhinal cortex are interconnected: Consequently, the exerted controls supported by olfactory cues are the effects of a network of specialized structures.

Fear olfactory conditioning or olfactory conditioned food or beverage aversion are examples that can give an idea of the functions of these networks.

Differential implication of theses area has been shown in their contribution to olfactory and contextual fear conditioning. The amygdala participates in the acquisition and the expression of fear conditioned to both an olfactory conditioned stimulus and to the training context. The perirhinal cortex participates to olfactory, but not contextual, fear conditioning. In addition, the perirhinal cortex seems to play a prominent role in recognition of the conditioned stimuli [7].

Another behavioral register is intensively explore: the ►odor conditioned aversion. Several parts of the primary olfactory cortex are implicated in its mechanisms. For instance, the effects of lesions of the entorhinal cortex are coherent with a role of this cortex in conditioned odor-aversion learning. A subdivision of this cortex as indicated by the heterogeneity of its connections is confirmed by functional arguments. The lateral part only is involved in the control of the olfactory memory trace during the conditioned olfactory aversion process. In addition the data are consistent with the idea that the lateral part represent the input of the structure while the medial part represent the output to hippocampus [8]. Here again, an olfactory cortex network is implicated. Interestingly, it has been shown that electrophysiological stimulations of the lateral entorhinal cortex is able to inhibit the olfactory input from the amygdala.

Integration. The primary olfactory cortex is of course inserted in a larger cerebral network and is a target for numerous modulating impacts. For instance, in the rat, 800 neurons from the anterior hypothalamus are secreting the peptide ►GnRH. Influence of these neurons on primary cortical structures of the olfactory system: Some neurons of the anterior olfactory nucleus, anterior and posterior piriform cortex, anterior cortical amygdaloidal nucleus and the lateral entorhinal cortex as it is shown by anterograde barley lectin labeling receive projection of the hypothalamic GnRH neurons [9]. This particular pathway illustrates one of the nervous supports of the integration of the olfactory sensitivity in the physiology and behavior. Odors signals or pheromones could have effects on the neuroendocrine status but in return, mediated by cerebral feed-back loops under the influence of sexual or reproductive hormones, other parts of the brain could modulate the olfactory abilities.

Secondary Cortical Areas

Axons of neurons from the primary cortical areas reached a number of others brain structures. Focusing on the olfactory sense, the orbito-frontal cortex and temporo-lateral neocortical structures are the most extensively studied. At this level, it is a neocortex that receives and processes the olfactory information.

As a main property, the *olfactory orbito-frontal cortex* receives afferent axons from several other sensorial sources. Among the other important influences, the cortex receives information from the gustatory pathway and had been explored as a centre related to feeding behavior and food choice. As seen using brain imagery, the orbito-frontal cortex is consistently activated by olfactory stimuli [10] and is sensitive to context. These are functional characteristics of a secondary cortex (primary, secondary cortices). Moreover, in this cortex, we find converging fibers from multiple sensory areas, i.e. the primary somatosensory cortex, the primary taste cortex (frontal operculum), the inferior temporal visual cortex, the striatum, the amygdala and the olfactory piriform cortex. Additional fibers from ▶hunger neurons confer to this area a central role in the food-related evaluation of odor and taste. Some neurons of this cortex are responsive to odor and taste for instance. Some of them decrease their response to food eaten to satiety.

This last remark illustrates an important view of the sensory physiology: The multimodality appears as an ultimate refinement of the environment representation. In the orbito-frontal cortex, representations of taste and other mouth feels, smell sight are converging. This is why the representation of food stimuli, and finally appetite, are modulated by sensory-specific controls, involving olfactory cues.

References

1. Lei H, Mooney R, Katz LC (2006) Synaptic integration of olfactory information in mouse anterior olfactory nucleus. J Neurosci 26:12023–12032
2. Zou Z, Li F, Buck LB (2005) Odor maps in the olfactory cortex. Proc Natl Acad Sci 102:7724–7729
3. Davidson IG, Katz LC (2007) Sparse and selective odor coding by mitral/tufted neurons in the main olfactory bulb. J Neurosci 27:2091–2101
4. Zou Z, Buck LB (2006) Combinatorial effects of odorant mixes in olfactory cortex. Science 331:1477–1481
5. Ross RS, Eichenbaum H (2006) Dynamics of hippocampal and cortical activation during consolidation of a nonspatial memory. J Neurosci 26:4852–4859
6. Gotfried JA, Winston JS, Dolan RJ (2006) Dissociable codes of odor quality and odorant structure in human piriform cortex. Neuron 49:467–479
7. Otto T, Cousens G, Herzog C (2000) Behavioral and neuropsychological foundations of olfactory fear conditioning. Behav Brain Res 110:119–128
8. Ferry B, Ferreira G, Traissard N, Majchzak M (2006) Selective involvement of the lateral entorhinal cortex in the control of the olfactory memory trace during conditioned odor aversion in the rat. Behav Neurosci 120:1180–1186
9. Boehm U, Zou Z, Buck LB (2005) Feedback loops link odor and pheromone signaling with reproduction. Cell 123:683–695
10. Zatorre RJ, Jones-Gotman M, Evans AC, Meyer E (1992) Functional localization and lateralization of human olfactory cortex. Nature 360:339–340

Olfactory Cortex – Piriform Cortex

Alfredo Fontanini
Department of Neurobiology and Behavior, State University of New York at Stony Brook, Stony Brook, NY, USA

Synonyms

Piriform cortex; Pyriform cortex; Prepyriform cortex

Definition

At a very general level the term "olfactory cortex" can be used for all those areas in the rostro-ventral portion of the forebrain which receive direct projections from the olfactory bulb. These areas are: the anterior olfactory nucleus (also called anterior olfactory cortex), the olfactory tubercle, the ▶piriform cortex, the entorhinal cortex, the insular cortex and the amygdala [1]. More specifically, however, the term has been – and will be, in the context of this entry – used in reference to the piriform cortex, by far the largest cortical area primarily involved in perception and learning of olfactory stimuli.

Characteristics

Introduction

The piriform cortex, also referred to as paleocortex for its old phylogeny, has an evolutionarily well-conserved cellular and synaptic organization [2]. Differently from the neocortex, which appeared more recently in evolution and has a complex multilayered architecture [3], the olfactory cortex is organized in a simpler and experimentally more tractable three layered architecture. Despite this different organization, however, the olfactory cortex and neocortical sensory areas share many functional properties [4]. The study of the olfactory cortex offers, therefore, the unique opportunity to understand how general properties of cortical organization and functioning can be produced by simpler and phylogenetically older structures. As such, a deep understanding of the olfactory cortex will not only help us in the study of olfaction, but also likely advance our knowledge of the general functional organization of the cerebral cortex [4].

Cytoarchitecture

The olfactory cortex is vertically organized into three layers, each characterized by a different composition of cell types and axonal fibers which spread horizontally [1]. Layer I, the most superficial, is a low cell-density layer composed, in its most superficial part (Ia), by afferent sensory fibers horizontally organized and coming from the olfactory bulb through the lateral olfactory tract (LOT), and in its deeper portion (Ib) by cortico-cortical (associative) horizontal axons coming from other parts of the olfactory cortex and other olfactory areas. Afferent and associative fibers contact the apical dendrites of excitatory neurons located in layer II and III and dendrites of inhibitory interneurons. The next layer, layer II, is composed by densely packed somata of excitatory (pyramidal and semilunar) and inhibitory (stellate and bipolar) cells. Finally, layer III shows a gradual decline in cell density with increasing distance from layer II, and contains somata and dendrites of deep pyramidal neurons, multipolar interneurons, basal dendrites of layer II pyramidal cells and cortico-cortical associative fibers.

As in the case of neorcortex, the circuit of the olfactory cortex is organized around principal excitatory neurons. The different subtypes of excitatory neurons, which are characterized by distinct functional properties, are all embedded in the same, apparently stereotyped, circuit (Fig. 1a) [1]: pyramidal and semilunar neurons receive feed-forward excitatory input from mitral cells in the olfactory bulb and recurrent associative excitatory inputs from other principal neurons within the olfactory cortex, in turn, they send their outputs within the olfactory cortex itself and to other cortical areas (entorhinal and perirhinal cortices, hippocampus, amygdala and orbitorfrontal cortex among them [1]).

Principal neurons are also embedded into two inhibitory circuits (Fig. 1b) [1]: one of which is based on a feedforward input from inhibitory cells in layer I directly activated by afferents from the bulb, the second is a feedback inhibitory loop carried by inhibitory interneurons in layer Ib and III which are activated by associative recurrent fibers from pyramidal cells. Bipolar interneurons, which receive both afferent and associative inputs can take part to both circuits. This

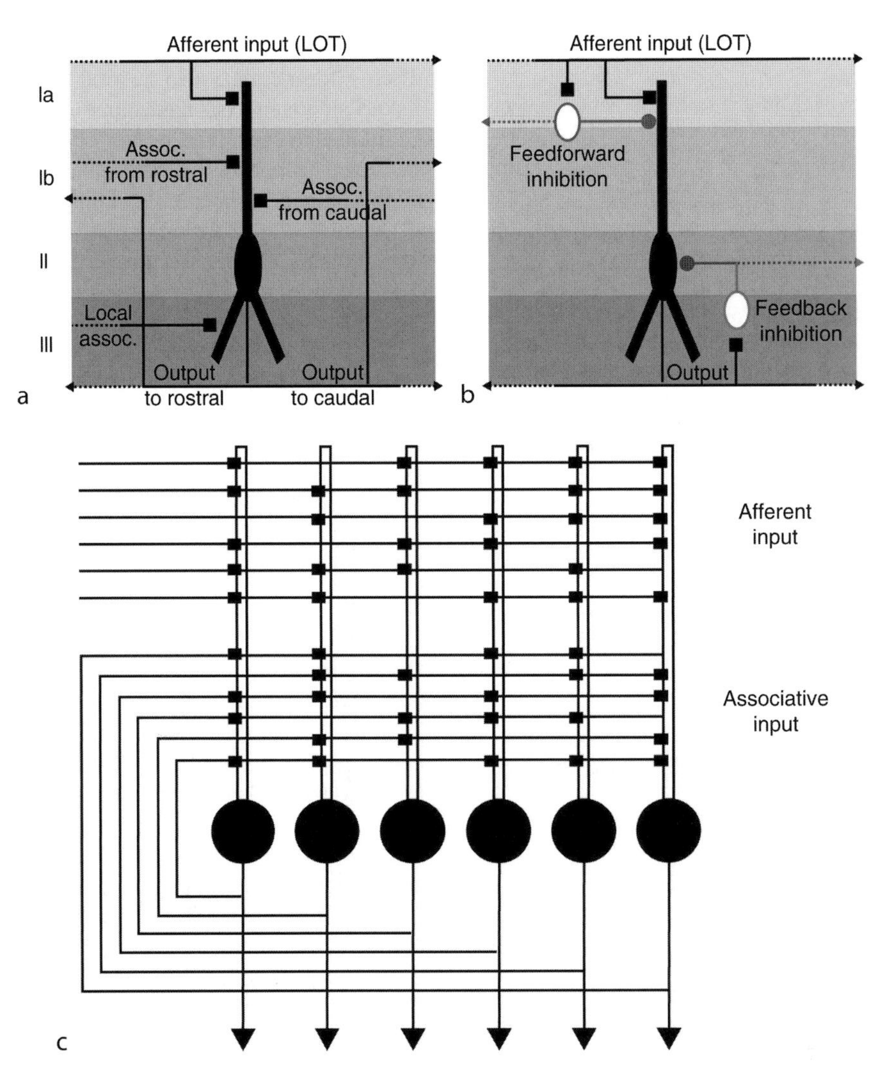

Olfactory Cortex – Piriform Cortex. Figure 1 Architecture of the olfactory cortex and of an autoassociative network. (a) Excitatory inputs and outputs of a pyramidal neuron. All fibers are organized vertically and segregated in different layers. Afferent inputs come from the LOT, associative inputs can come from distant or be local. Black squares represent excitatory synaptic contacts, lines with arrows represent the outputs of the circuit or, in case of the LOT, the signal propagating caudally. (b) Simplified inhibitory circuit impinging on pyramidal neurons. White circles are feedforward and feedback inhibitory interneurons; grey circles represent inhibitory contacts. (c) Schematic of an autoassociative network: synaptic contacts from afferent and associative inputs are represented as black squares contacting the neural units. (a) and (b) modified from [1].

structure, which is the foundation of the olfactory cortex basic electrophysiological behavior, is however far from rigid and immutable. Previous patterns of activity, sensory experience, as well as neuromodulators play a major role in inducing synaptic plasticity at different sites, shaping this architecture and resulting in different functional configurations [5].

Functional Organization

The characteristic extension of the associative system, and the suggestion that afferent inputs might be diffuse and without major topographical organization lead to the formulation of the most influential functional view of the olfactory cortex to date [5,6]. According to this view, the olfactory cortex can be seen as a biological analogue of a typical autoassociative artificial neural network. These types of artificial neural networks, characterized by neural units (or nodes) receiving sparse external inputs and also recurrent autoassociative inputs coming from the nodes themselves (Fig. 1c), are ideally suited for performing tasks analogous to those thought to be performed by the olfactory cortex: they can detect and discriminate complex mixtures of odors, reconstruct known mixtures on the basis of some of its components and dynamically switch between processing, storing and recalling of inputs and memories. Learning and dynamics are ensured, in this artificial network as well as in the olfactory cortex, by plasticity and neuromodulation of afferent and associative synapses [5].

This functional view of the olfactory system has been recently challenged by new results coming from genetic tracing and showing that the organization of the cortex is not as homogeneous as previously believed, but rather individual odors are processed by spatially organized quasi-specific subsets of neurons (Fig. 2) [7].

These results have given strength to a different view of the cortex, according to which odors are represented by the feedforward activation of specific sets of partially overlapping neural populations (labeled lines) and that complex mixtures are coded – and learned – by patterns of coactivation of the subset of neurons receiving convergent inputs.

In reality these two views, the distributed/associative versus the labeled-line/feedforward, can be integrated in several ways. The olfactory cortex is divided into an anterior part and a posterior part [1]: the anterior olfactory cortex is principally driven by afferent bulbar inputs which are functionally organized into large (and to some degree also overlapping) patches; the posterior part, on the other hand, is less driven by afferent inputs and they are organized in a more distributed fashion. Taking this evidence into account it is possible to imagine that the organization of each of the two subdivisions could be biased toward one or the other coding scheme. Additionally, and more importantly, while genetic tracing shows that specific cells code for a specific odor, the degree of convergence seen in the cortex for inputs carrying information for different odors is remarkable and compatible with the model of an autoassociative network. Therefore some of the properties of the autoassociative framework, like the importance of associative fibers, the complex temporal evolution of processing due to cortico-cortical associative connections and the ability to dynamically switch between different network configurations, can be incorporated in the feedforward theory to add complexity, flexibility and ecological realism. Recent work employing simultaneous recordings from multiple neurons in the olfactory cortex has shown that odors activate spatially scattered populations of neurons, which are only partially non-overlapping, and that the patterns of activity become more complex and overlapping as the time course of the response evolves [8]. These results provide support to the fact that the simple labeled line feedforward processing scheme needs to be integrated into a more complex distributed coding paradigm.

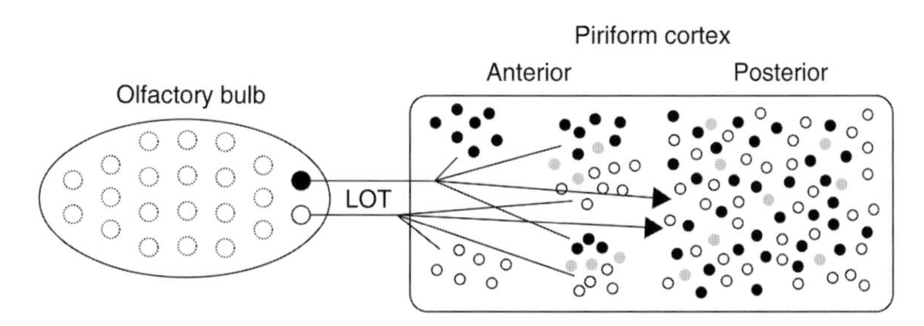

Olfactory Cortex – Piriform Cortex. Figure 2 Topographical organization of bulbar inputs to the olfactory cortex. Outputs from different glomeruli project to partially overlapping but overall spatially distinct patches of neurons in the anterior olfactory cortex. Projections to the posterior cortex are more distributed. Black and white circles in the piriform cortex represent cells activated by distinct glomeruli, grey circles are cells receiving convergent inputs. Modified from [7].

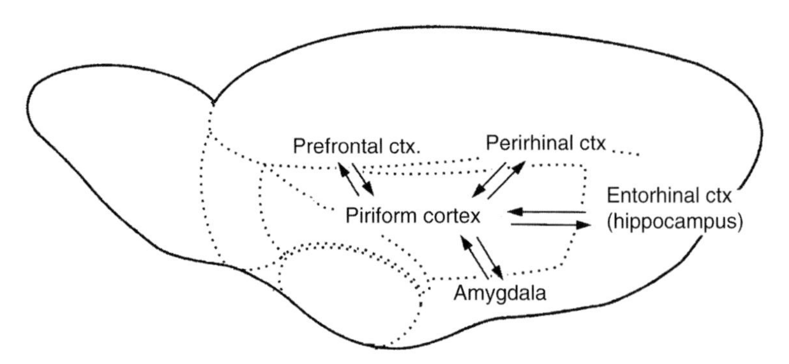

Olfactory Cortex – Piriform Cortex. Figure 3 Bidirectional connections between the piriform cortex and other high order cortical areas. Modified from [1].

Macroscopic Dynamics

Regardless of the coding scheme, electrophysiological recordings from the olfactory cortex of animals engaged in purposeful behaviors have revealed an even more complex picture: odor processing is inherently dependent on the behavioral and environmental context. Pioneering work from Walter Freeman [see for a review 9], for instance, has shown that the olfactory cortex produces different patterns of activity depending on the physiological and cognitive state of the animal: odors presented to hungry or thirsty cats, for instance, produce oscillatory activity larger than the one evoked by the same stimuli presented to satiated animals. These and other more recent observations imply that olfactory coding mechanisms are constantly modulated by dynamic activity from other brain areas involved in different cognitive states [10]. The anatomy is consistent with this view, as the olfactory cortex receives direct or indirect inputs from high order brain areas, such as the hippocampus, entorhinal cortex, orbitofrontal cortex, amygdala and hypothalamus; additionally several brainstem neuromodulatory nuclei provide noradrendergic, cholinergic, serotoninergic and dopaminergic modulation [1]. These projections are the anatomical substrate through which emotional states (sustained by amygdala), memories (hippocampus), expectations (amygdala and orbitofrontal cortex), hunger and thirst (hypothalamus) and arousal levels (neuromodulatory nuclei) could influence patterns of spontaneous and odor-evoked olfactory cortex activity (Fig. 3).

Summary

The olfactory cortex is the largest area devoted to processing of olfactory information. It shares many functional properties with other sensory areas, but it has the advantage of a relatively simpler organization. The enhanced experimental and conceptual tractability deriving from this simpler organization has favored the use of the piriform cortex as a study model for complex issues such as sensory coding and behavioral modulation of sensory responses. Future studies of the olfactory cortex will therefore help us understand not only olfaction, but also fundamental functional properties of sensory systems in general.

References

1. Neville KR, Haberly LB (2004) Olfactory cortex, In: Shepherd GM (eds), The synaptic organization of the brain. Oxford University Press, New York, pp 415–454
2. Haberly LB (1990) Comparative aspects of olfactory cortex, In: Jones EG, Peters A (eds) Comparative structure and evolution of cerebral cortex. Cerebral cortex. Plenum, New York, pp 137–166
3. Aboitiz F, Morales D, Montiel J (2003) The evolutionary origin of the mammalian isocortex: towards an integrated developmental and functional approach. Behav Brain Sci 26:535–552; discussion 552–585
4. Fontanini A, Bower JM (2006) Slow-waves in the olfactory system: an olfactory perspective on cortical rhythms. Trends Neurosci 29:429–437
5. Linster C, Hasselmo ME (2001) Neuromodulation and the functional dynamics of piriform cortex. Chem Senses 26:585–594
6. Haberly LB, Bower JM (1989) Olfactory cortex: model circuit for study of associative memory? Trends Neurosci 12:258–264
7. Zou Z et al (2001) Genetic tracing reveals a stereotyped sensory map in the olfactory cortex. Nature 414:173–179
8. Rennaker RL et al (2007) Spatial and temporal distribution of odorant-evoked activity in the piriform cortex. J Neurosci 27:1534–1542
9. Freeman WJ (2001) Neurodynamics: an exploration in mesoscopic brain dynamics. Springer-Verlag, New York
10. Kay LM, Freeman WJ (1998) Bidirectional processing in the olfactory-limbic axis during olfactory behavior. Behav Neurosci 112:541–553

Olfactory Cue

►Odor

Olfactory Discernment

►Olfactory Perception

Olfactory Disorders

►Smell Disorders

Olfactory Ensheathing Cells

Definition

Glial cells unique to the olfactory system, which ensheath the axons of the olfactory receptor neurons, without providing full myelination. The primary olfactory system is an unusual tissue in that it can support neurogenesis throughout life. This unique regenerative property depends, in part, on the presence of olfactory ensheathing cells, and has recently been shown to have a remarkable ability to repair spinal cord injury.

►Myelin
►Regeneration

Olfactory Epithelium

Definition

The olfactory epithelium is a specialized chemosensory portion of the nasal epithelial tissue that contains the olfactory sensory neurons. In humans, it occupies an area of about 5 cm^2 covering the posterior part of the roof of each nasal cavity and the superior nasal concha. The olfactory epithelium is composed of three types of cells: the olfactory sensory neurons, which transduce odorants into electrical signals, the supporting, glia-like cells and the basal cells, which are stem cells capable of replacing the olfactory cell population. Because of this regenerative capacity, damage to the olfactory epithelium may results in only temporary anosmia.

►Anosmia
►Evolution of Olfactory and Vomeronasal Systems
►Odorant
►Olfactory Sensory Neuron

Olfactory Glomerular Module

Definition

Also known as a glomerular domain, an olfactory glomerular module is a spatial cluster of olfactory glomeruli responding to chemically similar odorant stimuli. Spatial clustering of glomeruli with similar response profiles into glomerular modules may facilitate the use of local center-surround lateral inhibitory networks to restrict the molecular receptive range of mitral cell projection neurons to a more narrow range of stimuli. Thus, odorants that stimulate strongly overlapping sets of receptors may be represented by a smaller set of mitral cells.

►Glomerular Map
►Odorant
►Olfactory Bulb Mitral Cell
►Olfactory Glomerulus

Olfactory Glomerulus

Definition

►Olfactory Bulb Glomerulus

Olfactory-guided Behavior Studies

►Behavioral Methods in Olfactory Research

Olfactory Hallucinations

Richard J. Stevenson
Department of Psychology, Macquarie University, Sydney, NSW, Australia

Synonyms

Olfactory aura; Phantosmia

Definition

An olfactory hallucination is a subjective experience of smell, which occurs in the absence of an appropriate stimulus.

Characteristics

Olfactory hallucinations (OHs) can occur in normal participants, as an unaccompanied primary symptom (phantosmia), and as a secondary symptom in a range of medical and psychiatric disorders [1]. Whilst the term simple or complex has been used to classify hallucinations in the auditory and visual domains (e.g., spots of light vs. an elephant) this distinction does not readily transfer to OHs. Most OHs appear to be complex, in that the person perceives a fully formed odor object (e.g., the smell of cooked chicken) rather than an unformed olfactory event. However, something akin to the simple versus complex distinction may be reflected in the integration of the OH with other concurrent events (real or hallucinated). For example, a Charles Bonnet syndrome patient reported hallucinating both a visual image of a girl *and* the smell of her perfume.

There are several other characteristic features of OHs. First, they show the same range of odor qualities (what it smells like) as real odors and they vary in intensity and hedonics, with most OHs reported as unpleasant. Second, when an OH is first experienced, they may be accompanied by highly odor-appropriate behavior, such as searching for a "gas leak." This is the only objective evidence we have for the presence of an OH. Third, where OHs occur repeatedly, the person may gain insight into the nature of these experiences, although this may depend upon whether there is an underlying psychopathology (e.g., insight appears more common in epileptic than in schizophrenic OHs).

There are two other features that warrant comment. The first is the perceived locus of the OH. This can be in the nose or mouth, on the surface of the body or in the external environment. A defining feature (more below) of some forms of OH is their location, notably in olfactory reference syndrome, in which a person is convinced that their own body emanates a foul smell. In these cases, the person may not in fact be hallucinating a smell, rather the person infers the presence of a smell from other peoples reaction to them.

Presentation

Healthy Adults

Olfactory hallucinations (OHs) are widely reported in healthy adults. A large study of the frequency of all types of hallucination, conducted in Western Europe, revealed that 8.6% of the sample had experienced an OH, and that 0.9% of the sample experienced these several times a week [2]. OHs were the commonest reported daytime hallucination across all modalities. Other studies of non-clinical populations have found that OHs occur more frequently in individuals scoring higher on measures of psychosis-proneness.

Primary Symptom

OHs can occur as a sole presenting symptom in the condition termed phantosmia. The prevalence of phantosmia is unknown, but according to Leopold [3] it occurs more frequently in women, and is a progressively worsening, relapsing and remitting condition, with lifelong duration. Whilst OHs may be brief, phantosmia may be considerably more persistent, in some cases the hallucination may last hours or days, nonetheless even with this different time-course, it still fulfils the general definition of an OH.

Secondary Symptom

Schizophrenia: With the exception of epilepsy (more below), OHs have been studied most extensively in schizophrenia. An early view was that the presence of OHs was indicative of a poor prognosis, but there does not appear to be any substantial support for this notion. Rather, OHs appear to co-occur with tactile hallucinations and other positive symptoms of the disease. Phenomenologically, schizophrenic OHs are qualitatively varied, but may occasionally include descriptions, which suggest a delusion rather than an OH (e.g., smell of aliens, devils breath and angels). In most cases the OHs are reported as unpleasant or disgusting. Prevalence estimates vary between 2–35%. Most OHs are attributed to an external source (with some notable exceptions – see [4], for an excellent and representative set of examples), are of a similar time-course to real olfactory experiences and can result in behaviors consistent with the OH (e.g., escaping a building smelling of smoke).

Epilepsy: OHs can occur in the hours or days before a seizure (prodromal) or immediately, within minutes, preceding a seizure. These experiences are usually termed auras and estimates vary as to their prevalence (1–30%; [5]). Phenomenologically, these OHs cover all odor qualities, are brief, localized to the environment, and are predominantly unpleasant. An interesting feature is that they may be repetitive, in that the same person always experiences the same OH.

Migraine: OHs can occur prior to a migraine (again described as auras), with the same time course and features (immediate vs. prodromal) as in Epilepsy.

Post-traumatic stress disorder (PTSD): Several papers have documented OHs in PTSD under circumstances where the person is re-exposed (or imagines) to contextual cues associated with the event (e.g., smelling smoke/gasoline whilst traveling in a car following a traumatic motor vehicle accident). In all cases, the OH appears specific and appropriate to the traumatic event.

Brain injury: Both traumatic brain injury, stroke and aneurysm can result in OHs. In some cases these more closely resemble phantosmia (and may share similar causation via damage to peripheral olfactory structures) whilst in others, especially aneurysm and stroke, the OHs may be complex (integrated) and hedonically varied.

Drug abuse: OHs have been reported in both chronic cocaine and alcohol users, but studies are few and so prevalence cannot be estimated. These reports indicate a presentation akin to that observed in Epilepsy – predominantly negative, brief and qualitatively varied OHs.

Miscellaneous: OHs have also been described, albeit rarely, in Parkinson's disease, Charles Bonnet syndrome, Depression and Alzheimer's disease.

Cause

Whilst there has been fairly long history of theoretical and empirical work on visual and auditory hallucinations, especially in schizophrenia, relatively little work has been undertaken in respect to olfactory hallucinations (OH). This section starts by examining the association between the olfactory system and the two clinical conditions in which OHs are most well documented (epilepsy and schizophrenia), and then outlines theories that may account for OHs in these conditions. The second part of this section examines phantosmia, and the final part OHs in normal participants.

Epilepsy and Schizophrenia

The neural basis of epilepsy and schizophrenia can overlap with brain areas known to be involved in olfactory function. In epilepsy, olfactory abnormalities tend only to accompany the disorder when the focus for the seizure is in the temporal lobe. Here the seizure may start or propagate to the amygdala and uncus and then into primary olfactory processing areas located on the boundary of the frontal and temporal lobes. Not surprisingly then, OHs (auras) tend to be associated with temporal lobe epilepsy. In schizophrenia, abnormalities have been detected in the orbito frontal cortex (OFC), amygdala and medio-dorsal nucleus of the thalamus (MDNT). Respectively, the OFC is secondary olfactory cortex, the amygdala is involved in processing the hedonic valence of odors and the MDNT is one of the routes by which information flows from primary olfactory cortex to secondary olfactory cortex, and may be instrumental in attributing the source of sensory stimulation ("that's a smell").

There are several contemporary theories of hallucinations [6], including cortical irritation, cortical release, intrusion of imagery or dreams, and attentional/sensory impairment theories. How well do these models account for OHs in epilepsy and schizophrenia? Cortical irritation is the oldest hypothesis and suggests that excess neural activity at a particular brain loci results in the activation of memory traces that are then experienced as real events. Whilst this was heavily based on electrical brain stimulation (EBS) studies, it turns out that EBS results in *very few* olfactory-related experiences. This conclusion is based upon a large number of reported studies, stimulating many regions in the temporal/frontal regions. The rarity of these events suggests that focal irritation in brain areas known to be abnormal, especially in temporal lobe epilepsy, is an unlikely explanation.

A second class of explanation (of varying form) is that hallucinations arise as a result of abnormal – typically reduced – sensory input. This results in cortical release or hyperexcitability, causing memories of prior sensory experience to be re-experienced as real. There is one major problem with this account for OHs in epilepsy and schizophrenia. This is that patients who experience OHs may not have reduced sensory input. Three studies have examined schizophrenic participants with OHs. They find no consistent deficit in ►odor detection, no abnormal changes to olfactory mucosa, and no history of disease states that might affect olfactory function. With epilepsy, the picture is less clear, with no systematic studies as yet. However, olfactory deficits in temporal lobe epilepsy are usually indicative of central (i.e., ►odor identification and ►odor discrimination) rather than peripheral pathology (i.e., detection is typically intact). Thus there is likely to be no reduction of sensory input that this class of theory would require.

The third class of explanation suggests that hallucinations result from the intrusion of dreams into the waking state or the misattribution of imagery to the external environment, rather than correctly to oneself. Whilst both of these types of explanation have been extensively explored, especially in respect to auditory hallucinations of people conversing, they have significant obstacles to overcome as an account of OHs. Whilst olfactory dreams and images certainly do occur, the former are rare and the latter are hard to generate [7]. Indeed, some argue that we may have no capacity to consciously experience odor images at all. In this case, misattribution accounts may not have much utility in explaining OHs.

A further, and more recent class of model suggests that hallucinations arise from a combination of attentional deficits and impaired sensory functioning. As noted above, impaired sensory functioning (detection) does not appear to be a salient feature of either epilepsy or schizophrenia.

Finally, there are a number of other possible causes of OHs in epilepsy and schizophrenia that have not been widely canvassed. First, impaired odor identification might lead to what *appears* to be an OH (e.g., misidentifying the smell of table polish for smoke). Second, the likely presence of amygdala abnormalities in schizophrenia and epilepsy, the predominantly unpleasant nature of OHs and the amygdala's role in mediating aversive reactions to odors, might suggest this as a possible neural locus for these events. In summary, there is at present no well-defined model of OHs and there is a need to test the various theoretical accounts described above more directly.

Phantosmia

Whilst epileptic and schizophrenic OHs likely involve a dominant central cause, phantosmia almost certainly derives from a combination of both peripheral and central causes [3]. Evidence favoring a peripheral basis for phantosmia is that it typically disappears if the olfactory mucosa is treated with a local anesthetic and that examination of excised mucosal tissue from phantosmia patients reveals disordered axon growth and an abnormal ratio of mature to immature neurons. Evidence favoring a central locus comes from the finding that many phantosmia patients have no detectable abnormality in odor detection and that such patients typically have no history of upper respiratory tract infection or head injury prior to onset. Interestingly, magnetic resonance spectroscopy imagining has revealed significantly lowered GABA levels in several central sites, including the amygdala [8]. Given the overwhelming predominance of unpleasant OHs in phantosmia, this again suggests possible amygdala pathology as a common feature of OHs.

Normal Participants

Several studies suggest that OHs are more common in healthy individuals who score higher on measures of schizotpy or psychosis-like dimensions, although it is not currently possible to estimate the proportion of variance accounted for by this variable [9]. What it does suggest, however, is that normal variation in schizotypy may reflect proneness to OHs, implying a similar causal explanation to those described above for schizophrenia. In addition, a proportion of OH-like experiences may also be accounted for by more mundane failures to identify an odor source, misperceptions (which may be more common in olfaction than in other senses), illicit drug use, alcohol, anxiety and depression, and lack of sleep [2].

Conclusion

The study of hallucinations can offer important insights into clinical conditions such as schizophrenia, as well as revealing much about routine perceptual processing. The study of OHs is not well advanced, empirically or theoretically, but it will be important in testing the generality of current theories of hallucinations.

References

1. Greenberg MS (1992) Olfactory hallucinations. In: Serby MJ, Chobor KL (eds) Science of olfaction. Springer, New York, pp 467–499
2. Ohayon MM (2000) Prevalence of hallucinations and their pathological associations in the general population. Psychiatry Res 97:153–164
3. Leopold D (2002) Distortion of olfactory perception: diagnosis and treatment. Chem Senses 27:611–615
4. Bromberg W, Schilder P (1934) Olfactory imagination and olfactory hallucinations. Arch Neurol Psychiatry 32:467–492
5. West SE, Doty RL (1995) Influence of epilepsy and temporal lobe resection on olfactory function. Epilepsia 36:531–542
6. Collerton D, Perry E, McKeith I (2005) Why people see things that are not there: a novel perception and attention deficit model for recurrent complex visual hallucinations. Behav Brain Sci 28:737–794
7. Stevenson RJ, Case TI (2005) Olfactory imagery: a review. Psychon Bull Rev 12:244–264
8. Levy LM, Henkin RI (2004) Brain GABA levels are decreased in patients with phantageusia and phantosmia demonstrated by magnetic resonance spectroscopy. J Comput Assist Tomogr 28:721–727
9. Bell V, Halligan PW, Ellis HD (2006) The Cardiff Anomalous Perceptions Scale (CAPS): a new validated measure of anomalous perceptual experience. Schizophrenia Bulletin 32:366–377

Olfactory Information

Leslie M. Kay
Department of Psychology, Institute for Mind and Biology, The University of Chicago, Chicago, IL, USA

Synonyms

Odors; Odorants; Odor code; Olfactory system dynamics

Definition

Olfactory information can refer to the chemical stimuli (odorants), the perceptual effect of the stimuli (odors), the individual neural responses which receive this input (odor code), and the dynamical interaction of the many brain subsystems which comprise the central olfactory pathways (olfactory system dynamics).

Characteristics

►Odor signals can be viewed as stereotyped activation maps defined by neuronal ►receptive fields and also as perceptual objects in which the odorant stimuli are associated with meaning, behavior and experience. The anatomy and physiology of the mammalian olfactory system has been studied from both perspectives. The ►olfactory bulb receives direct input from ►olfactory receptor neurons in the olfactory epithelium. These neurons project in a "receptor-topic" arrangement, such that an individual ►glomerulus receives input from only one type of receptor (Fig. 1).

Because an individual odorant can activate multiple ►olfactory receptors, glomerular input maps are fragmented and highly distributed representations in the form of glomerular activation patterns. The patterns also have a dynamic structure which can be seen using ►Ca^{++} imaging.

The mammalian ►olfactory system is also characterized by dense bidirectional connectivity among its many structures. The ►olfactory bulb may receive more synaptic input from the brain than it does from the olfactory receptor sheet, similar to a comparison of retinal and V1 projections to the thalamic ►lateral geniculate nucleus. Olfactory bulb structure has been likened at different times to the ►retina, primary visual cortex and more recently the sensory ►thalamus [2]. This essay concentrates on the mammalian system, but some references to the analogous insect systems are made [1]. The peripheral input structure, glomerular architecture and ►centrifugal input all have perceptual and physiological consequences.

Odor Psychophysics in Animals

Psychophysical studies examining odor similarity use generalization methods, in which animals are trained to recognize one odorant, and similarities are judged by generalization of a behavioral response to other odors. Taking the ►glomerular input maps produced by various imaging methods and ►mitral cell responses corresponding to these areas as a guide, many compounds have been shown to exhibit similarity gradients along changes in molecular features, such as carbon chain length [3].

Thus, there are similarities in chemical composition, receptor activation and input patterns, which then correspond to similarities in odor quality. On the other hand, most animals are also very good at distinguishing even very similar odorants, and ►reinforcement learning can help an individual to discern even very small differences in glomerular activation patterns (Fig. 2).

Psychophysical responses to monomolecular odorants are relatively stable over a range of concentrations, due in part to mechanisms within the input layer. A subpopulation of ►GABAergic periglomerular cells

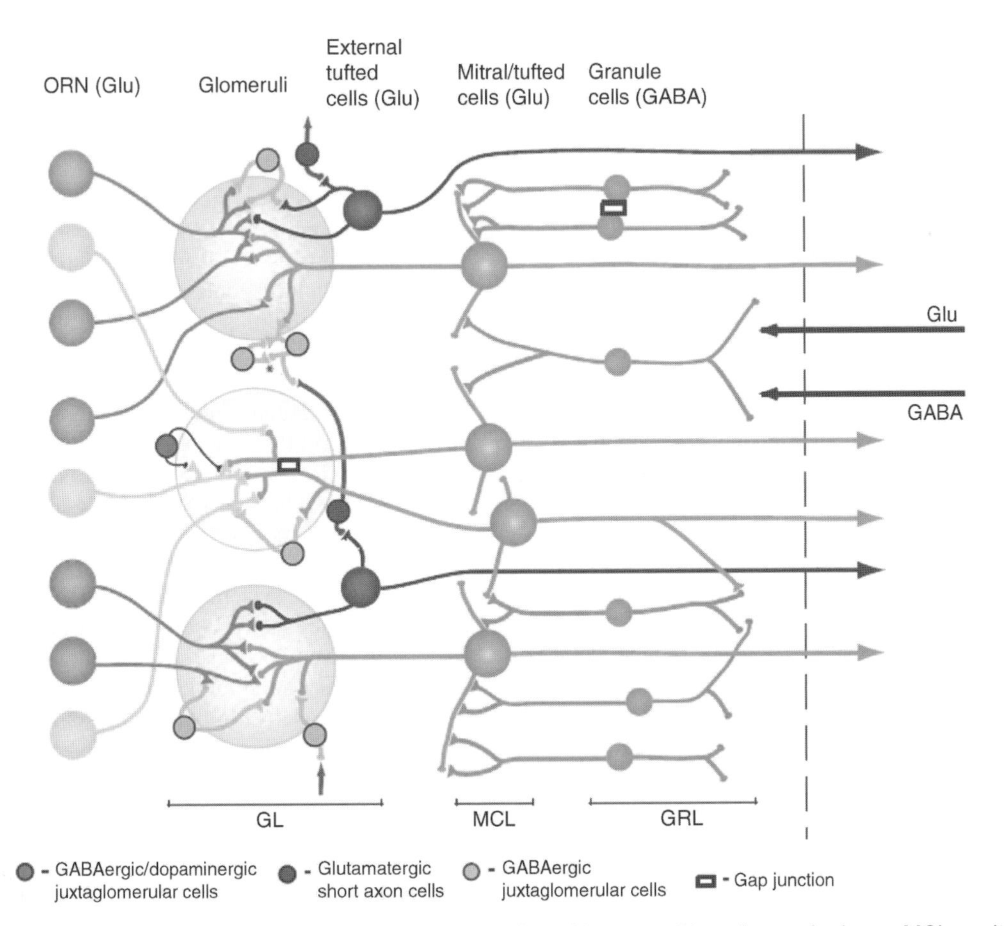

Olfactory Information. Figure 1 Schematic of olfactory bulb architecture. GL – glomerular layer; MCL – mitral cell layer; GRL – granule cell layer. Pial surface on the *left*, centrifugal inputs on the *right* (Reprinted from [1] with permission; Elsevier).

O

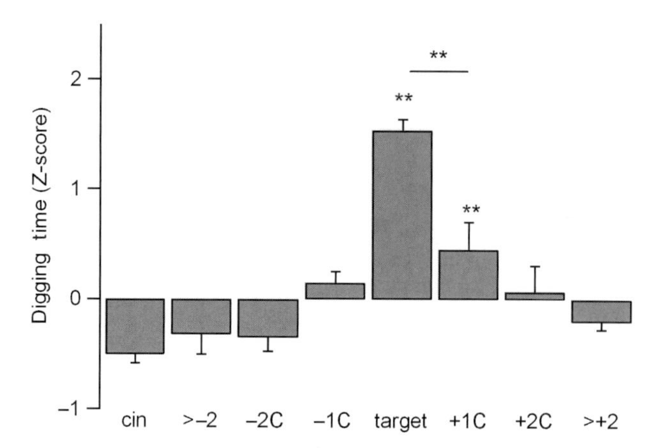

Olfactory Information. Figure 2 Example of carbon chain length generalization pattern for a series of aldehydes. Mice trained to dig in cage bedding scented with a single aliphatic aldehyde (target) generalize the response to a nearby aldehyde (1 carbon difference in chain length). Digging times are compared to other aldehydes of chain lengths longer and shorter than the target and a control odor (cin – cineole).

receives direct input from the ►olfactory nerve and mediates feedforward inhibition onto the ►mitral cell apical dendrites. Release of ►dopamine by ►juxtaglomerular cells and ►acetylcholine by the horizontal nucleus of the diagonal band of Broca in the ►cholinergic basal forebrain also modulate the incoming ►afferent activity. ►Excitatory and ►inhibitory connections among glomeruli and lateral inhibition between mitral and granule cells have been proposed as mechanisms for ►contrast enhancement and ►gain control.

Odor mixtures present a more complex picture, and two behavioral methods have been used to investigate their perceptual properties in animals. The first looks at mixture quality, in which ►associative learning or ►habituation is used to train an animal to recognize a given mixture, and components are then tested in a ►generalization paradigm. These studies suggest a general theoretical principle: odors that smell alike or activate significantly overlapping receptor or glomerular populations produce a ►synthetic or ►configural (►Configural/Configurational) quality. Mixtures of dissimilar or nonoverlapping odors produce ►elemental qualities. However, there is growing evidence that mixture perception may not be so simple, as compounds with similar structures can produce elemental responses, and those with very different structures can produce synthetic responses in binary mixtures. Furthermore, as the number of compounds in a mixture grows, humans experience more synthetic effects. Concentration and pungency also significantly affect mixture perception.

The second method of assessing mixture perception addresses animals' ability to recognize the ratio of various odor components, in which they choose a response associated with the component represented at higher concentration [4]. Responses in this case follow a ►psychometric curve (►Psychometric Curve/Psychometric Function). This method does not specifically address odor mixture quality, but it can be used to manipulate odor discrimination difficulty. What this method has been able to show is that rodents can identify some odors in 1–2 sniffs, but as discrimination becomes more difficult this brief sampling time results in poorer performance. Training rats to sniff longer results in greater performance levels in more difficult discriminations; this suggests a ►speed-accuracy tradeoff in odor sampling.

The mechanisms for learning differences between odors in a behavioral context involve areas of the brain beyond the ►glomerular maps and are addressed at the physiological level.

Physiology of Olfactory Information

Ease of access to the olfactory bulb and the importance of olfactory information for rodents drove this research to very deep levels even before single unit recordings in waking and mobile animals became technically feasible or practical. Thus, this field proceeded from its beginning at the systems level, only more recently addressing issues such as ►odor coding and ►receptive fields. However, because of the high-dimensional nature of olfactory stimuli, we still know relatively little about the relative importance of salient molecular features, concentration, ►pungency or even the existence of odor ►categories. (Much of the anatomical, physiological and computational background is reviewed in a few sources [1,5,6].)

Individual Neuron Responses

►Mitral cells in the ►olfactory bulb typically respond in a ►burst-like manner around the peak of inhalation. They receive input from a single ►glomerulus, and those with dendrites in the same glomerulus can excite each other. In anesthetized mammals, mitral and ►tufted cells in the ►olfactory bulb and ►pyramidal cells in the ►piriform cortex can respond with an increase or decrease in firing rate upon presentation of odorants in front of an animal's nose. In this situation mitral cells show relatively stable odor responses that correspond roughly to the ordered representations suggested by mapping studies. However, there are exceptions to this simple ordering, since many mitral cells respond to many different odor classes, and in any given place in the olfactory bulb, one can often find cells that respond to an odor class.

Mitral and tufted cells in the olfactory bulb respond in a graded fashion to similar odorants, reminiscent of classical ►receptive fields with broad ►tuning curves that can be shifted by prolonged exposure to non-optimal odorants within a cell's ►receptive field. This plasticity is similar to that in other sensory systems, such as receptive field ►learning-induced plasticity in ►auditory cortex. Mitral cells show significant cross-habituation to odors within their receptive fields. Odor responses of pyramidal cells in piriform cortex of anesthetized rats are somewhat different. While these cells exhibit tuning curve properties similar to mitral cells, the responses of single neurons to related odorants do not cross-habituate, suggesting that odor responses within the piriform cortex are more selective overall than those within the olfactory bulb.

Waking mammals present a somewhat different picture. ►Odor selectivity has been recorded in a handful of studies, limited by the difficulty of recording isolated mitral cells in waking mammals. The phase of the respiratory cycle in which a mitral cell fires during periods of slow breathing (< 5 Hz in rats) represents the identity of a relatively long (5 s) odor stimulus associated with reinforcement. However, when rats perform odor discriminations with a briefer sampling time (1–2 s), they sniff at high rates (6–12 Hz), and mitral cells uncouple from the respiratory cycle. Firing rate responses in waking rats predict behavior most strongly, and only a small part of a cell's response varies with odor. When the behavioral association (positive or negative reinforcement) of an odor is changed, a cell's odor selectivity also changes. Studies of single neuron firing patterns in the ►piriform cortex of waking mammals are scarce, but odor responses there are also modified by changes in behavioral associations.

Population Activity

Population physiology presents a window into system-level dynamics. The ►local field potential has been very useful for understanding how the various parts of the olfactory and limbic systems interact with and control each other. Many early studies described the parameters which govern oscillatory responses in many parts of the olfactory system and those which relate local field potentials to single neuron activity [7,8]. The olfactory bulb exhibits two major classes of oscillations, slow (< 12 Hz) and fast (>12 Hz) (Fig. 3).

Slow Temporal Structure

Slow oscillations are in the ►theta frequency range (2–12 Hz in the olfactory bulb) for rodents and are generally correlated in phase and frequency with the respiratory cycle and with mitral cell burst firing. They are supported by afferent input and by intrinsically bursting cells like the ►external tufted cells in the ►glomerular layer. The burst behavior of mitral cells leads to a loose temporal structure within the olfactory bulb, in which within a 100–150 ms time window many cells are activated, and in the exhalation phase and prior to the next inhalation fewer cells are activated. Thus, the ►theta oscillation in the olfactory bulb represents these high and low firing states. At low respiratory rates, this leads to a sampling of the olfactory environment in the nose approximately every 300 ms in a ►saccade-like fashion. However, respiration does not completely describe these rhythms or mitral cells' firing patterns even in anesthetized animals, and there is evidence that ►centrifugal inputs can modulate both. During fast sniffing, mitral cells tend to fire ►tonically and the theta rhythm no longer represents high and low firing rates in the mitral cell population. Also during fast sniffing coupling between the hippocampal theta rhythm and sniffing or olfactory bulb oscillations in the high theta range (>5 Hz) have been associated with learning and performance of odor discriminations. Otherwise, these two rhythms are uncorrelated. This low frequency coupling may aid information transfer between the olfactory and hippocampal systems.

Fast Temporal Structure: Circuit Properties

Within the respiratory cycle there is structure at a finer timescale. At the end of inhalation the ►gamma oscillation (~40–100 Hz) is initiated. This odor-evoked oscillation was first described by Adrian [9]. The gamma burst lasts for 60–100 ms at low respiratory rates (~6–8 cycles per burst; Fig. 3). These fast odor-evoked oscillations have been well-studied at the physiological and computational levels in this system and in the analogous insect system [5]. Most researchers agree that olfactory bulb gamma oscillations arise from the reciprocal dendrodendritic (►Reciprocal Dendro-dendritic Synapse) interaction between mitral and granule cells in the ►external plexiform layer in a ►negative feedback circuit. Olfactory bulb mitral cells' firing times are probabilistically related to the population-level gamma oscillation (Fig. 4).

While this oscillation is often referred to as a source of ►synchrony between individual neurons, it more

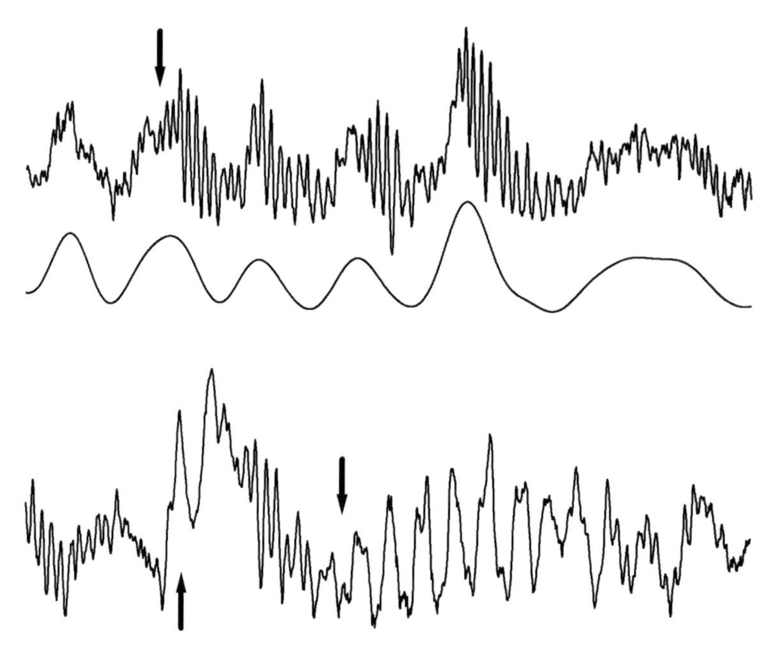

Olfactory Information. Figure 3 Olfactory bulb oscillations (local field potential; each trace is 1 s long). *Top* trace shows gamma oscillations initiated at the peak of inhalation (*downward arrow*). The ►theta band part of the signal is shown just below, with each cycle representing a sniff. *Bottom* trace shows an odor-evoked beta oscillation. *Upward arrow* is the ►sensory evoked potential, and *downward arrow* shows the onset of the ►beta oscillation.

precisely represents the level of synchrony between individual neurons and the ►emergent local field potential. In this case, an increase in gamma oscillation power and a decrease in spectral width are associated with mitral cells firing in more restricted time windows, rather than precise temporal synchrony between neurons. This suggests increased precision in the temporal structure of the olfactory information.

Odor-evoked oscillations also occur in the insect ►antennal lobe, which is an analogue of the olfactory bulb, with very similar circuit properties. While insect oscillations are ~20 Hz, they are similar to mammalian gamma oscillations in the relationship of the principal neurons' firing patterns to the oscillatory local field potential and the dependence of the oscillations on the interaction between excitatory ►projection neurons and the ►GABAergic local neurons. In the insect system it has been shown that a group of projection neurons fires in an odor-specific temporal pattern across cycles of the fast oscillations [1], which has led some to conclude that the mammalian system may use a similar mechanism during periods of high amplitude gamma oscillations.

In the mammalian system, sources of ►desynchronization of the local field potential associated with this system lie in the centrifugal and intrabulbar sources of drive to the ►granule cell layer, both ►GABAergic and ►glutamatergic (Fig. 4). Desynchronization is seen as a source of stability and flexibility in this system, and may be important for understanding the functional differences between the mammalian systems and the simpler insect system.

In waking rats and mice, the gamma band has been further subdivided into two bands that are distinct in their behavioral associations but sometimes overlap in frequency. Gamma 1 (~70 Hz in waking rats and mice) is used to refer to the classical odor-evoked gamma described above. Gamma 2 (~55 Hz) is used to refer to the somewhat lower frequency oscillation that occurs between breaths during periods of alert immobility and low breathing rates. The source of gamma 2 oscillations is different from that of gamma 1, likely arising from ►GABAergic drive to the granule cells. The functional association of these oscillations is unknown, but may be related to attentional processes or dynamic stability.

Fast Temporal Structure: Perceptual Properties

Activity in the ►gamma frequency band has been associated with odor discrimination circuitry in many species. Walter J. Freeman and colleagues showed that over the surface of the olfactory bulb there is a common ►gamma band waveform of the ►EEG [10]. The spatial patterns of amplitude of this waveform were the best indicator of an odor, and the patterns were produced reliably only when meaning (positive or negative reinforcement) was associated with an odor.

Gamma band (and gamma-like) oscillatory population synchrony is one specific mechanism associated with more difficult or highly overlapping odor discriminations

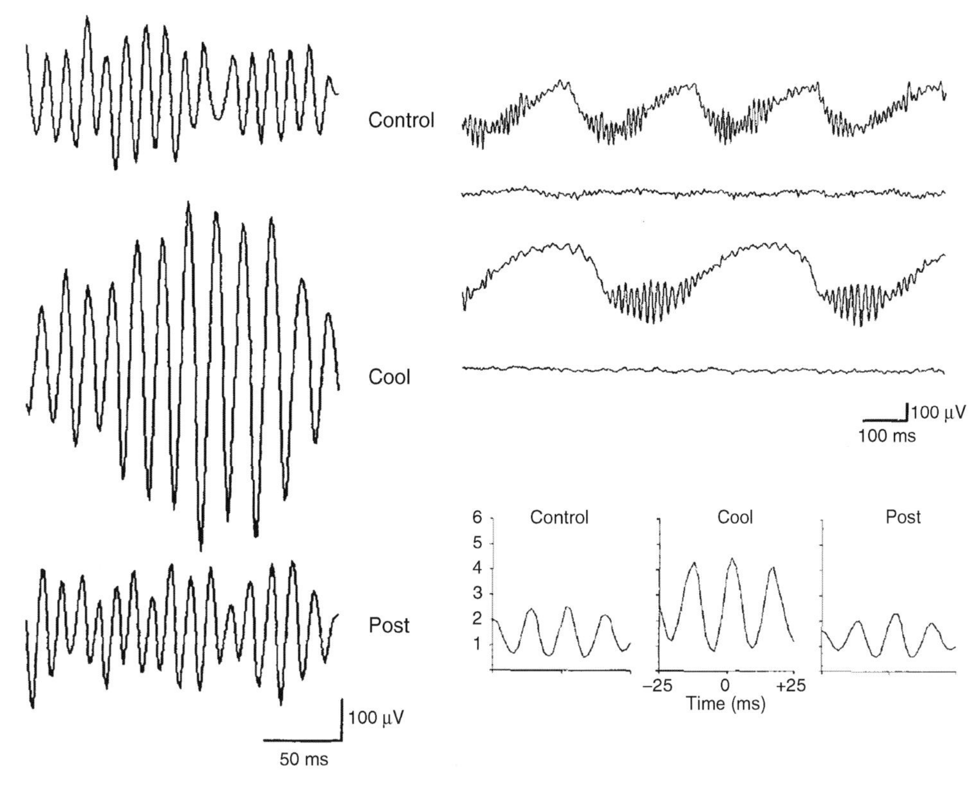

Olfactory Information. Figure 4 Centrifugal input to the olfactory bulb causes desynchronization of the local field potential. Cooling the rear portion of the olfactory bulb effectively blocks input from the rest of the brain and produces a large increase in ►gamma oscillation power. Pulse probability density (*bottom traces*) shows that single mitral cells are more strongly coupled with the local field potential gamma oscillation without centrifugal input (Compiled and reprinted with permission from Springer, Gray and Skinner, *Exp Brain Res* 1988. 69(2):378–386.).

in rodents and insects. Disruption of these oscillations in honeybees leads to a selective decrease in discriminating highly overlapping odorants (fine discrimination). Increased olfactory bulb gamma power in β3 knockout mice leads to a selective increase in fine odor discrimination. In both studies, coarse discrimination was unaffected. Unmanipulated rats dramatically increase the power of gamma oscillations when performing fine odor discrimination, relative to coarse discrimination in a two-alternative choice task, suggesting that temporal precision in mitral cell firing patterns is enhanced.

Odor-associated beta band oscillations (15–30 Hz) are also seen in waking rats, where they predict the onset of correct performance in Go/No-Go odor discrimination tasks. Beta oscillations occur concurrently in the ►olfactory bulb, ►piriform cortex, ►entorhinal cortex, and dorsal and ventral ►hippocampus. Similar oscillations occur in the olfactory bulb, piriform cortex, entorhinal cortex and hippocampus during repeated passive odor stimulation in a ►sensitization-like fashion (Fig. 3). Beta oscillations differ significantly from gamma oscillations in that they require a complete bidirectional loop between the olfactory bulb and the rest of the olfactory system, suggesting temporal structure distributed across many brain areas. In anesthetized rats, beta oscillations occur at the end of exhalation, and this period has been associated with enhanced firing in the granule cell layer.

Summary

The combination of ordered but highly complex input maps combines with centrifugal input to the olfactory bulb and oscillatory dynamical states to produce odor perception. Input pattern overlap predicts odor similarity and discrimination difficulty, and animals can adjust their sniffing behavior along with changes in the olfactory system to interpret and respond to odors. Fast oscillations represent cell assemblies that process odors within and between olfactory areas, and slow oscillations at the respiratory frequency can serve momentary system wide coupling possibly to facilitate information transfer.

References

1. Kay LM, Stopfer M (2006) Information processing in the olfactory systems of insects and vertebrates. Semin Cell Dev Biol 17(4):433–442
2. Kay LM, Sherman SM (2007) An argument for an olfactory thalamus. Trends Neurosci 30(2):47–53

0

3. Cleland TA et al (2002) Behavioral models of odor similarity. Behav Neurosci 116(2):222–231
4. Rinberg D, Koulakov A, Gelperin A (2006) Speed-accuracy tradeoff in olfaction. Neuron 51(3):351–358
5. Cleland TA, Linster C (2005) Computation in the olfactory system. Chem Senses, 30(9):801–813
6. Shipley MT, McLean JH, Ennis M (1995) Olfactory system. In: Paxinos G (ed) The rat nervous system. Academic Press, San Diego
7. Freeman WJ (1975) Mass action in the nervous system. Academic Press, New York, p 489
8. Rall W, Shepherd GM (1968) Theoretical reconstruction of field potentials and dendrodendritic synaptic interactions in olfactory bulb. J Neurophysiol, 31(6):884–915
9. Adrian ED (1942) Olfactory reactions in the brain of the hedgehog. J Physiol, 100:459–473
10. Freeman WJ, Schneider W (1982) Changes in spatial patterns of rabbit olfactory EEG with conditioning to odors. Psychophysiology 19(1):44–56

Olfactory Learning

►Odor – Memory
►Olfactory Plasticity

Olfactory Marker Protein

Definition

A cytoplasmic protein expressed at high levels ubiquitously and exclusively throughout the soma, cilia, and axon of olfactory sensory neurons. Its function remains obscure.

►Olfactory Sensory Neuron

Olfactory Nerve

Matthew S. Grubb
MRC Centre for Developmental Neurobiology, King's College London, London, UK

Synonyms

First cranial nerve; Olfactory sensory inputs

Definition

The olfactory nerve consists of the axonal projections of olfactory sensory neurons, which extend from the olfactory epithelium in the nose through the cribriform plate of the skull to contact postsynaptic targets in the glomeruli of the olfactory bulb. Uniquely among pathways in the central nervous system, the entire nerve is continuously regenerated throughout adult life and has a remarkable capacity for recovery from injury.

Characteristics

Anatomy, Morphology, and Molecular Characteristics

The olfactory nerve is the shortest of the cranial nerves, and is one of only two – along with the optic nerve – which do not project to the brainstem. It is composed primarily of the axons of olfactory sensory neurons (OSNs), which sit in the olfactory epithelium (OE) of the nasal cavity and whose job is to transduce information in airborne odorant molecules into electrical signals that are sent to the brain's olfactory bulb (OB). OSN axons are small (~0.2μm diameter) and unmyelinated, and extend from the OE into the underlying lamina propria of the olfactory mucosa, where they coalesce into small-sized bundles. These bundles increase in size as they exit the lamina propria, and form branches of the olfactory nerve that cross through perforations of the skull's cribriform plate before entering the outer nerve layer (ONL) of the OB (Fig. 1).

Having crossed the boundary between the peripheral and central nervous systems, OSN axons then exit the

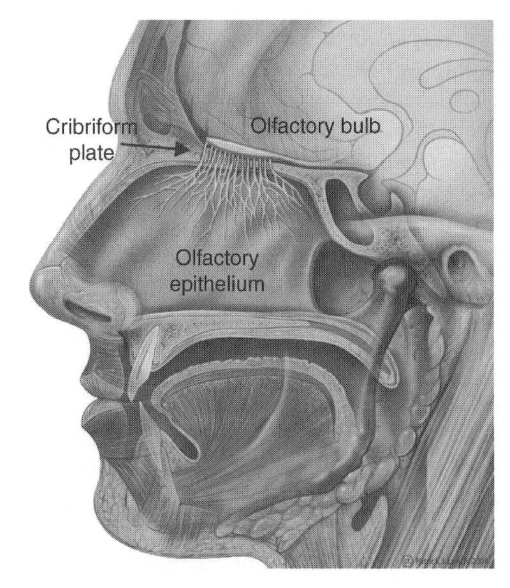

Olfactory Nerve. Figure 1 Olfactory nerve. The axons of olfactory sensory neurons in the olfactory epithelium, shown here in yellow, project through the cribriform plate of the skull to the olfactory bulb, also shown in yellow. Illustration © PJ Lynch and CC Jaffe.

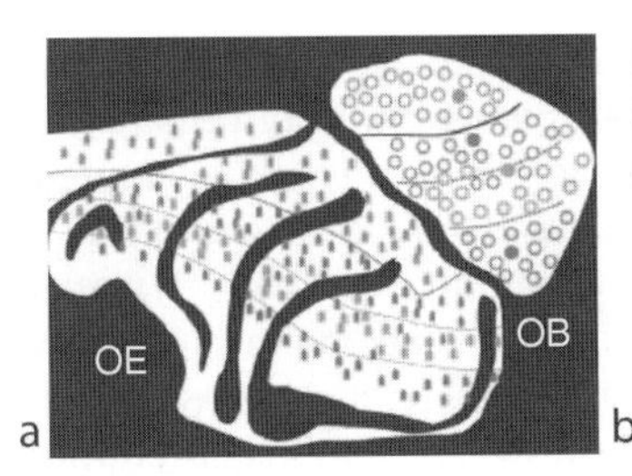

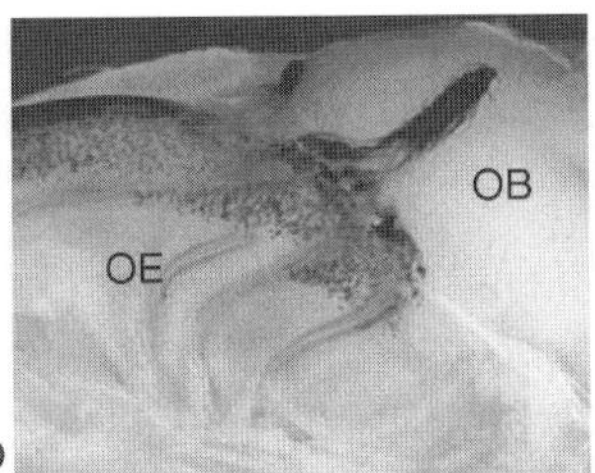

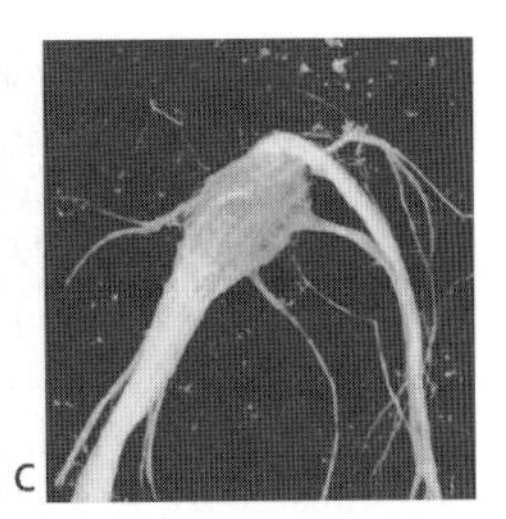

Olfactory Nerve. Figure 2 Organization of olfactory nerve inputs to the olfactory bulb. (a) Zones in the olfactory epithelium (OE) project to particular regions of the olfactory bulb (OB). (b) Axons from olfactory sensory neurons that express a single type of olfactory receptor, labeled here in blue, project onto a single glomerulus in the medial OB. (c) Inputs to a single glomerulus are untidy, with axons entering the structure from all angles. (a) and (b) reprinted with permission from [1], (c) reprinted with permission from [2].

ONL to terminate in OB glomeruli (▶Olfactory bulb glomeruli), specialized and highly complex arrangements of axons and ▶dendrites that host the very first steps in odor information processing.

The organization of axons within the olfactory nerve is based on the olfactory receptor (OR) molecules expressed by OSNs. Each OSN expresses a single OR, and OSNs that express a particular OR lay scattered randomly within one of four OE zones. Each zone provides OSN axons that project to a particular region of the OB, although while the dorsal zone of the OE projects exclusively to the anterior dorsal bulb, the projections from other OE zones overlap somewhat [1] (Fig. 2a).

More striking is the astonishingly precise projection of OSN axons onto individual glomeruli: all of the OSNs expressing a given OR project onto only 2, mirror-symmetric glomeruli per bulb, and each glomerulus receives input only from axons expressing a single OR [1,3] (Fig. 2b). This huge OR-specific convergence only begins when the olfactory nerve reaches the ONL. Up to this point, axons from OSNs expressing different ORs are all completely intermingled, but on entry to the OB they begin a process of ▶homotypic fasciculation whereby axons from OSNs with the same OR run together in bundles. These bundles then converge onto individual glomeruli, a highly specific process which is nonetheless surprisingly untidy [2] (Fig. 2c).

Once in the correct glomerulus, OSN axons make glutamatergic, excitatory synaptic connections with the dendrites of three main types of OB neuron (Fig. 3). Mitral cells (▶Olfactory bulb mitral cells) and ▶tufted cells are glumatergic projection neurons that receive olfactory nerve input and project directly to olfactory cortex. ▶Periglomerular cells, in contrast, constitute a heterogeneous population of local interneurons that receive olfactory nerve input and make modulatory connections within and between glomeruli.

Along their route from OE to OB, OSN axons are surrounded and supported by the processes of

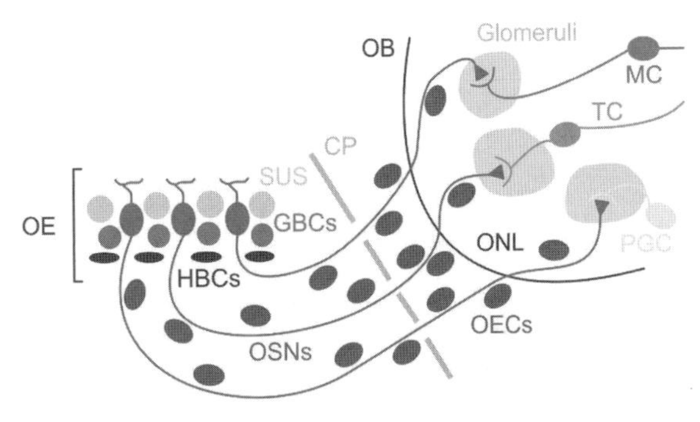

Olfactory Nerve. Figure 3 Cell types associated with the olfactory nerve. *OE* olfactory epithelium; *OB* olfactory bulb; *ONL* outer nerve layer; *CP* cribriform plate; *SUS* sustentacular cells; *GBCs* globose basal cells; *HBCs* horizontal basal cells; *OSNs* olfactory sensory neurons; *OECs* olfactory ensheathing cells; *MC* mitral cell; *TC* tufted cell; *PGC* periglomerular cell.

▶olfactory ensheathing cells (OECs), glia that are unique to the olfactory nerve and which possess characteristics of both Schwann cells and astrocytes [4] (Fig. 3). OECs do not provide proper Schwann cell-style myelination, but instead extend thin processes which each wrap up to 200 OSN axons, providing them with mechanical and metabolic support. In addition, it appears that OECs are essential for the growth-permitting environment of the olfactory nerve, expressing guidance cues and neurotrophic factors which allow new OSN axons to make their way to the OB. Indeed, OECs have been used successfully to promote axon outgrowth and repair in models of CNS injury [4].

As well as possessing unique glia, the olfactory nerve also contains unique axons. Adult OSN axonal compartments contain molecules that are not found in most other axons of the mature CNS. These include mRNA, which appears to be transported along OSN axons rather than locally translated, transcription factors, and cytoskeletal proteins such as MAP5 and vimentin which are more commonly found in developing neuronal processes [5]. OSN axons also contain ►olfactory marker protein (OMP), a molecule expressed strongly, ubiquitously, and uniquely throughout the olfactory nerve, but whose function remains obscure. Along with the permissive environment created by OECs, these unique axonal features of the olfactory nerve may underlie, or at least reflect, its regenerative capacity (see Adult Neurogenesis below).

Development

The olfactory nerve is initially established in rather early prenatal development. In mice, the first OSN axons arrive in the brain around embryonic day (E) 12, having extended from the OE through a "migratory mass" that includes OEC progenitors and guidepost mesenchyme cells [1]. Just before entering the presumptive OB, growing OSN axons wait for a short time before entering the ONL and fasciculating with other axons expressing the same OR. Fasciculated bundles are then directed to the region of their appropriate target glomerulus by molecular guidance cues including semaphorins, ephrins, and surface carbohydrates [1], with axons reaching specific domains in the presumptive glomerular layer as early as E15.5. The precise direction of OSN axons to their appropriate glomeruli depends at least in part on the particular ORs they express: aberrant glomerular targeting results when OR expression is genetically altered in a subset of OSNs [1]. Spontaneous, but not sensory activity in OSNs also appears necessary for the correct initial formation of OB glomeruli [6,7].

By postnatal day (P) 0, glomeruli in the rostral OB are clearly formed, while it takes a further 2–3 days for those in the caudal OB to catch up. However, the development of the olfactory nerve does not end there: at this stage, many OSNs axons expressing a particular OR terminate in two or more glomeruli. Over the next month or so of postnatal maturation, these diffuse projections are pruned to produce a tight, single glomerular target structure in each half-bulb (Fig. 4), a process that is highly dependent upon olfactory sensory experience [7].

Adult Neurogenesis

The olfactory nerve is unique among CNS axon tracts in that its generative capacity extends past postnatal development and continues throughout adult life. Unlike other CNS neurons, the nature of OSNs' function as detectors of airborne odorants means they are directly exposed to the external environment, and thus to the accompanying risk of damage by toxins and pathogens. In order to maintain normal olfactory nerve function in the face of this threat, OSNs keep fresh by a process of continual turnover – after a lifetime of around 3 months, those that have not been killed already undergo programmed cell death and are replaced by new OSNs born from stem cells residing in the basal layer of the OE [8]. These cells migrate up to more superficial layers of the OE and extend an axon towards the OB, taking approximately 1 week post-mitosis to express mature markers such as OMP and to form functional glomerular synapses [8]. This normal replacement occurs with very high accuracy – there is no sign of degradation in the glomerular map with routine ageing. In addition, if a subpopulation of OSNs expressing the same OR is specifically removed, the replacement population extends axons to the OB and forms a glomerulus in precisely the right location. This entire process of OSN regeneration, and particularly the regrowth of olfactory nerve axons, probably involves many of the guidance factors and activity-dependent processes that orchestrate the initial formation of the olfactory nerve during brain development. In particular, OECs appear crucial to the growth-permissive status of the olfactory nerve environment throughout adult life.

The continual turnover of OSNs, and the presence of stem cells in the OE mean that the olfactory nerve is unique in the CNS in being able to recover from injury. After even drastic interventions such as section of the olfactory nerve or chemical lesion of the entire OE, recovery is possible – new OSN axons can extend and find the correct target zone of the OB after ~2–3 weeks [8]. There, recovery is not perfect: there are substantial targeting errors in an en-masse regenerating ON, producing multiple glomerular foci and incorrect terminal locations. However, although we currently know nothing about how the olfactory nerve functions following recovery from injury, we do know that olfactory behavior recovers extremely well. Whilst not anatomically perfect, then, the recovery capability of the olfactory nerve is easily good enough to restore useful olfactory function. Unsurprisingly, this unique ability has been the spur for many studies looking to use elements of the olfactory nerve niche to promote recovery in other models of CNS injury. Indeed, promising results have so far come from approaches involving ectopic transplantation of OECs.

Physiology and Function

The fundamental function of the olfactory nerve is to transmit olfactory information from its site of transduction in the OE to the site of its first processing in the glomeruli of the OB. This information is carried solely in the form of sodium-based action potentials, which are propagated along unmyelinated OSN axons at a speed of ~0.5m/s. Whether or not an action potential occurs in a given OSN axon depends on the particular OR expressed by the cell, and the presence of particular odorants in the olfactory environment. Individual OSNs are actually rather broadly-tuned

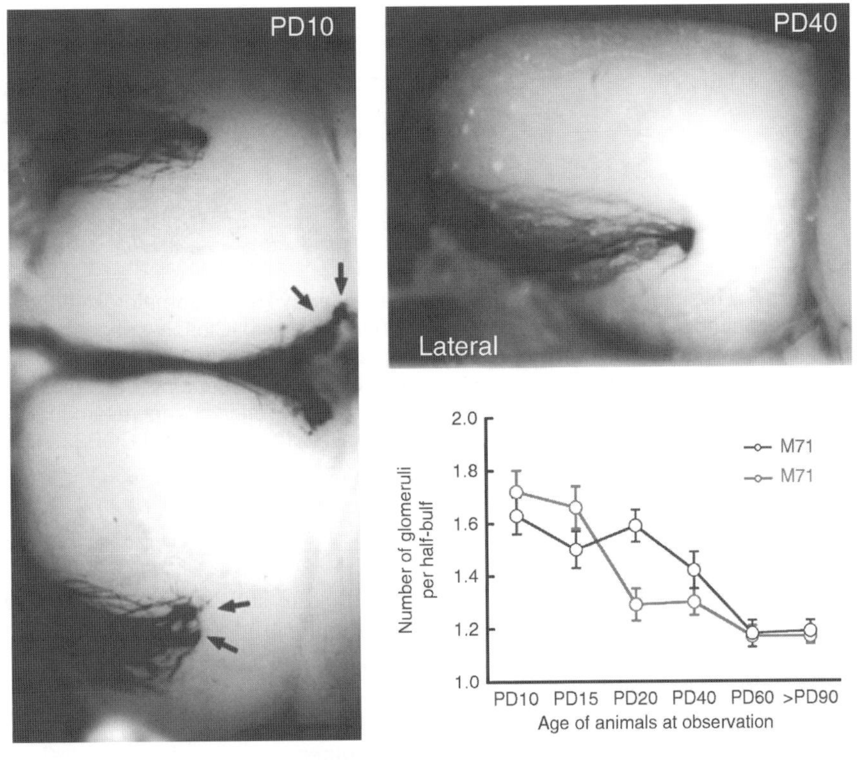

Olfactory Nerve. Figure 4 Postnatal refinement of olfactory nerve projections in mice. At postnatal day (PD) 10 (*left*), axons from olfactory sensory neurons expressing a single olfactory receptor type converge onto multiple glomeruli in the medial and lateral olfactory bulbs (arrows). By PD40 (*right*, *top*), axons converge onto a single glomerulus in the lateral bulb. The plot at bottom right shows the refinement of glomeruli with postnatal development for two distinct olfactory receptor types. Reprinted with permission from [7].

to odorants, since ORs can bind a relatively large number of different odorant molecules. Furthermore, even within a subgroup of OSNs that all express the same OR, variations in transduction processes mean that odorant responses can be markedly different. This means that the information carried by any one olfactory nerve axon actually says very little about which odorants are present or absent in the environment. Only a combinatorial code for odors, embedded in the activity of the ensemble of fibers constituting the olfactory nerve, can allow olfactory detection and discrimination to take place.

As well as the type of odorant stimulus present, the information carried by the olfactory nerve also depends on the strength of the activating odorants. As in all sensory systems, increasing the intensity of the stimulus produces an increase in firing frequency in olfactory nerve fibers. But this may not be the only temporal code present in the pathway, since different odorant concentrations are also known to evoke different firing *patterns* in olfactory nerve fibers. In addition, recordings of calcium activity in olfactory nerve axon terminals have revealed glomerulus-specific dynamics – some glomeruli are quicker, or longer-lasting than others. These differences in temporal dynamics are consistent for the same glomeruli across individual animals, and are only weakly correlated with odorant strength, suggesting they might represent another way, as well as firing frequency, that olfactory information is coded in the axons of individual OSNs.

Finally, coding in the axons of the olfactory nerve may be influenced by a rather unique process in the brain – ►ephaptic interactions between fibers. In most major axon tracts, firing in component axons is kept independent by myelination. The olfactory nerve, however, consists of bundles of hundreds of small axons loosely held together by the processes of OECs, meaning that the insulation of individual axons may not be very good. In these conditions, action potentials in one OSN axon could spread passively to activate other neighboring OSN axons. Indeed, mathematical models of the olfactory nerve suggest that such ephaptic interactions are possible, and even likely. Since OSN axons are not sorted by OR types until they reach the OB, these ephaptic effects could only act to disrupt OR-specific activity in particular fibers. If ephaptic interactions do occur in the real olfactory nerve, then, they may render the transmission of olfactory information from the nose to the brain far less than perfect.

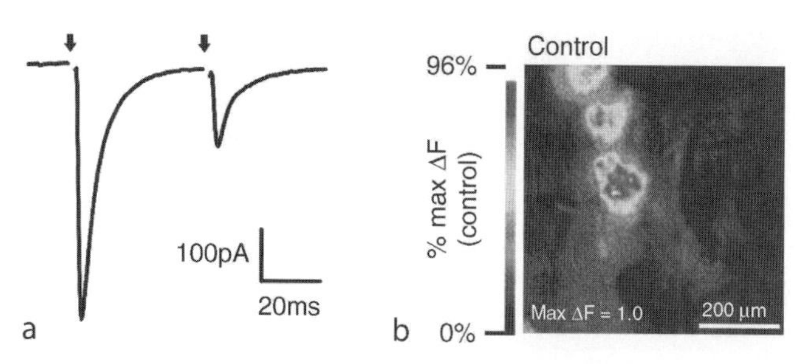

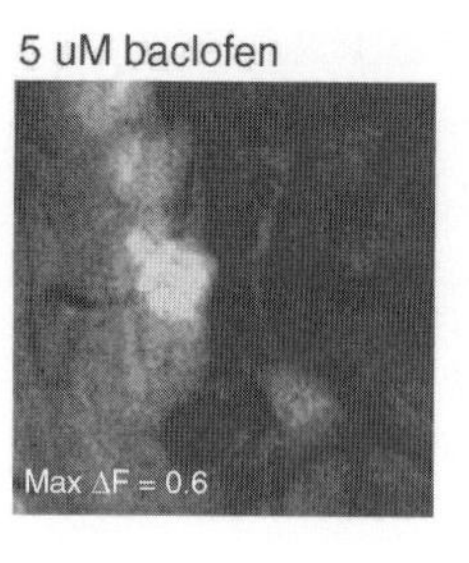

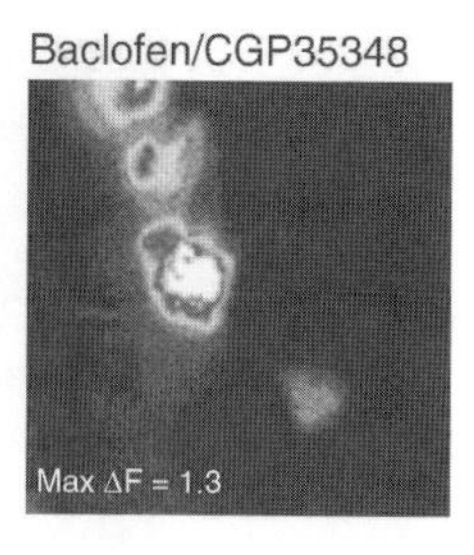

Olfactory Nerve. Figure 5 Physiology of olfactory nerve terminals. (a) Evoked glutamatergic responses recorded in a periglomerular cell after paired stimulation of olfactory nerve inputs (arrows). Closely-spaced stimulation produces a depression of the second response, a feature characteristic of high release probability at olfactory nerve synapses. (b) Modulation of glutamate release at olfactory nerve synapses by $GABA_B$ receptors. Each blob shows release levels in an entire glomerulus in response to odorant stimulation. Release at olfactory nerve terminals is decreased by the $GABA_B$ receptor agonist baclofen, and increased by the $GABA_B$ receptor antagonist CGP35348. (a) recorded by the author, (b) reprinted with permission from [10].

In other sensory systems, primary sensory neurons transfer freshly-transduced electrical information about the world to their postsynaptic target cells via very reliable and morphologically specialized synaptic connections. In contrast, the connections of the olfactory nerve with its postsynaptic targets in OB glomeruli appear, structurally, to be rather normal glutamatergic synapses. However, functional experiments in OB slices have shown that these connections too are extremely reliable. Unusually for the brain, olfactory nerve terminals have very high release probability – ~0.8 or more [9] (Fig. 5a) which should ensure the highly reliable transfer of olfactory information from the OE to the brain. The underlying mechanisms subserving such high release probability are not known, although it is not due to multivesicular release, and the relationship between calcium entry and glutamate release appears to be nearly linear [9].

Whilst they transmit presynaptic activity with high fidelity, olfactory nerve terminals are unique among primary sensory afferents in being sites of extensive modulation. Although ultrastructural experiments have found no synapses *onto* olfactory nerve terminals in the OB, electrophysiological experiments have revealed strong modulation of release probability by GABA acting through $GABA_B$ receptors (Fig. 5b), by dopamine acting through D_2 receptors, and by cyclic nucleotides acting through terminally-expressed cyclic nucleotide-gated channels. This modulation is almost all intraglomerular, meaning that the immediate periglomerular cell postsynaptic targets of olfactory nerve terminals can release either GABA or dopamine, or both, to influence both their own inputs and others in the vicinity [10]. Such feedback modulation may ensure that, despite the high release probability at olfactory nerve synapses, the dynamic range of the terminals is maintained. In other words, the modulation ensures that the OB can still respond to a range of odorant concentrations, even after repeated or prolonged presentation of a strong stimulus.

References

1. Strotmann J, Breer H (2006) Formation of glomerular maps in the olfactory system. Semin Cell Dev Biol 17:402–410
2. Potter SM, Zheng C, Koos DS, Feinstein P, Fraser SE, Mombaerts P (2001) Structure and emergence of specific olfactory glomeruli in the mouse. J Neurosci 21:9713–9723
3. Treloar HB, Feinstein P, Mombaerts P, Greer CA (2002) Specificity of glomerular targeting by olfactory sensory axons. J Neurosci 22:2469–2477
4. Fairless R, Barnett SC (2005) Olfactory ensheathing cells: their role in central nervous system repair. Int J Biochem Cell Biol 37:693–699
5. Nedelec S, Dubacq C, Trembleau A (2005) Morphological and molecular features of the mammalian olfactory sensory neuron axons: what makes these axons so special? J Neurocytol 34:49–64
6. Yu CR, Power J, Barnea G, O' Donnell S, Brown HE, Osborne J, Axel R, Gogos JA (2004) Spontaneous neural activity is required for the establishment and maintenance of the olfactory sensory map. Neuron 42:553–566
7. Zou DJ, Feinstein P, Rivers AL, Mathews GA, Kim A, Greer CA, Mombaerts P, Firestein S (2004) Postnatal refinement of peripheral olfactory projections. Science 304:1976–1979
8. Schwob JE (2002) Neural regeneration and the peripheral olfactory system. Anat Rec 269:33–49
9. Murphy GJ, Glickfield LL, Balsen Z, Isaacson JS (2004) Sensory neuron signaling to the brain: properties of transmitter release from olfactory nerve terminals. J Neurosci 24:3023–3030
10. McGann JP, Pírez N, Gainey MA, Muratore C, Elias AS, Wachowiak M (2005) Odorant representations are modulated by intra- but not interglomerular presynaptic inhibition of olfactory sensory neurons. Neuron 48:1039–1053

Olfactory Pathways

ALBRECHT J, WIESMANN M
Department of Neuroradiology, Ludwig-Maximilians-University Munich, Germany

Synonyms

Olfactory structures; Olfactory cortical areas; Olfactory cortex

Definition

The perception of a smell is an integration of various sensations (olfactory, trigeminal, tactile, thermal, as well as gustatory sensations). This article is engaged with the olfactory pathways in particular. The human olfactory pathways can be divided into three parts [1,2] (Fig. 1):

(1) The olfactory receptors are located in the mucosa of the nasal cavities. From there olfactory nerves run to the olfactory bulb which is located inside the bony skull beneath the orbital forebrain. From an evolutionary point of view the olfactory bulb is not a ganglion but a part of the telencephalon, one of the oldest portions of the brain. Following this it is postulated that the olfactory bulb constitutes the genuine primary olfactory cortex [3], which is contradictory to the common literature.

(2) The olfactory tract connects the olfactory bulb to secondary olfactory cortex consisting of the anterior olfactory nucleus, the ►olfactory tubercle, the piriform cortex, parts of the amygdala (►periamygdaloid cortex, anterior and posterior cortical nuclei, nucleus of the lateral olfactory tract) and a small anteriomedial part of the entorhinal cortex. Since the recognition of the olfactory bulb as a cortical structure these areas are called secondary olfactory cortex [3].

(3) Regions known to receive projections from the secondary olfactory cortex include the orbitofrontal cortex, agranular insular cortex, additional subnuclei of the amygdala, medial and lateral hypothalamus, medial thalamus, basal ganglia, and hippocampus. These regions are termed tertiary olfactory regions.

Although the current understanding of the organization of the olfactory pathways depends basically on observations made in rodents and non-human primates, it is generally assumed that the human olfactory system owns the same basic organization.

Characteristics

Olfactory nerves/Primary Olfactory Cortex (POC)

►Olfactory receptors (OR): Olfactory receptor neurons are located in the olfactory epithelium, on the roof of the nasal cavity, above or below the anterior middle turbinate insertion, and are covered by a layer of olfactory mucosa. In humans, several million olfactory receptor neurons are found in both nasal cavities constituting the first-order neurons of the olfactory system. Olfactory receptor neurons are the only sensory neurons in the human body that are directly exposed to the external environment and can therefore be damaged by external harmful substances. Thus the average lifetime of the neurons is only a few months. Afterwards they are replaced through differentiation of neuronal stem cells [4]. It is known that cAMP or cGMP gated ion-channels activated by G_{olf}-protein coupled receptor proteins are responsible for odor induced activity of olfactory receptor cells. Between 350 and 400 different types of olfactory receptors are found in the human nasal mucosa. Every olfactory receptor cell expresses only one or maybe two of odorant receptor types. In addition, all neurons expressing the same receptor protein send their axons to the same two glomeruli in each olfactory bulb. In vertebrates, an olfactory stimulus, e.g., the odor of roses, does not activate one specific OR only. Instead, a large number of receptors are activated, although the intensity of activation differs between all of them. A different olfactory stimulus will activate a different set of ORs, of which some may have been activated by the first stimulus as well, while others may not. Again, however, there is a characteristic intensity pattern of the activated receptors. Hence, quality coding seems to be related to neuronal analysis of the topographical distribution of activated receptor proteins [5].

►Olfactory nerves: The axons from the olfactory receptor neurons group into small bundles to form the olfactory nerves, or Fila olfactoria. On average, 12–16 branches of olfactory nerves run along the nasal septum on each side medially and additionally 12–20 branches course along the lateral wall of each nasal cavity [2].

►Olfactory bulb: The olfactory nerves run upwards through the foramina of the cribriform plate of the ethmoid, entering the anterior cranial fossa. On the way from epithelium to olfactory bulb the axons regroup to form more homogeneous bundles. The olfactory nerves terminate at the ipsilateral olfactory bulb. The two olfactory bulbs, one on each hemisphere, lie in a bony groove formed by the cribriform plate. In the olfactory bulb, the axons of the olfactory receptor neurons synapse with dendrites of second-order neurons in the olfactory system (mitral and tufted cells) forming discrete glomeruli.

Secondary Olfactory Cortex

The olfactory bulbs are connected to the secondary olfactory cortex via the ►olfactory peduncles. The olfactory peduncles consist of the olfactory tracts as well as a thin layer of grey matter which belongs to

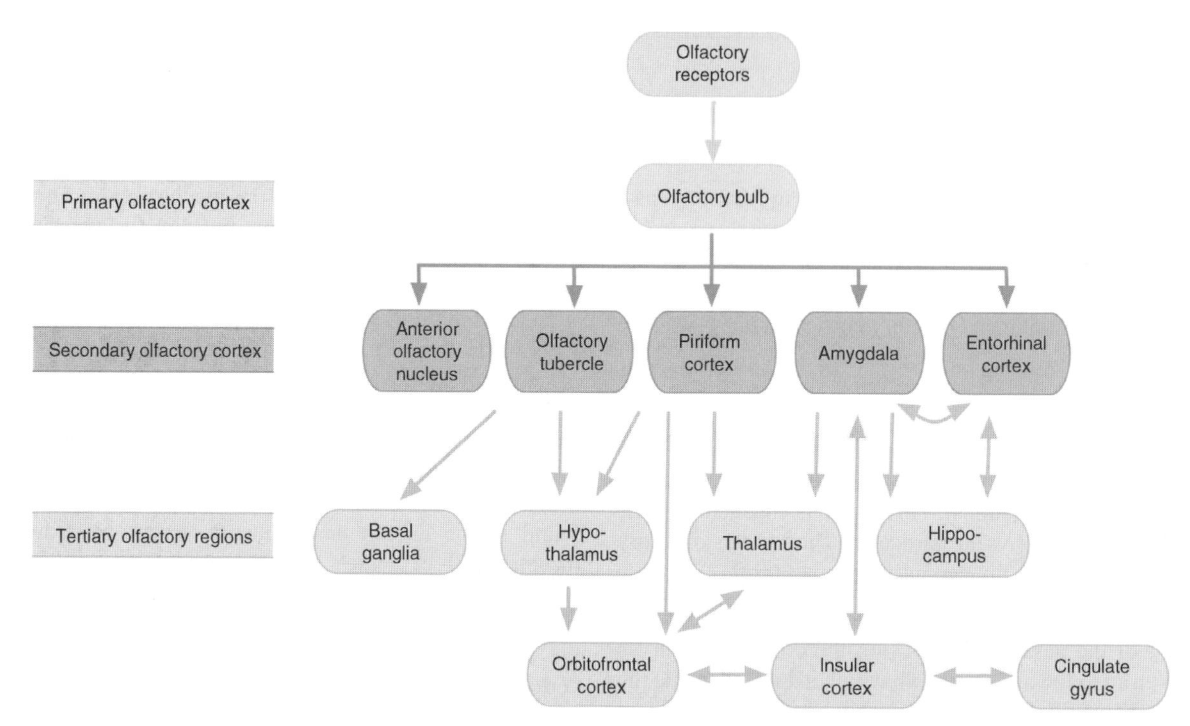

Olfactory Pathways. Figure 1 Schematic illustration of the major central nervous projections of the olfactory receptor neurons. Shown are the three parts of the olfactory pathways (olfactory receptors/primary olfactory cortex, secondary olfactory cortex, and tertiary olfactory regions) and their connections.

the anterior olfactory nucleus. The postsynaptic axons of the mitral and tufted cells leave the olfactory bulb forming the lateral olfactory tract, one on each hemisphere. The lateral olfactory tract is situated in the ►olfactory sulcus of the orbital surface of the frontal lobe, lateral to the gyrus rectus. It transfers olfactory information to a number of ipsilateral brain areas within the posterior orbital surface of the frontal lobe and the dorsomedial surface of the temporal lobe [5]. Unlike in several non-mammalian species, there is no medial olfactory tract in mammals, including primates [4]. The lateral olfactory tract runs along the olfactory sulcus until it reaches the rostral part of the ►anterior perforated substance, where it divides into three roots, or striae. This area is called the ►olfactory trigone. The medial olfactory stria curves upwards to the ►septal region. The lateral olfactory stria curves laterally and leads to the medial surface of the temporal lobe. Delineated by the medial and lateral striae is the anterior perforated substance. The posterior border of the anterior perforated substance is delimited by a band of fibers that passes from the amygdala to the ►septum pellucidum. This band is called the diagonal band of Broca. The intermediate olfactory stria continues onto the anterior perforated substance, ending at the olfactory tubercle. Although well documented in animals, the intermediate and medial striae are extremely rudimentary in humans. Thus the lateral olfactory stria provides the only source of bulbar afferents to the brain. All areas receiving a direct projection from the lateral olfactory stria constitute the secondary olfactory cortex, consisting of the anterior olfactory nucleus, the olfactory tubercle, the piriform cortex, parts of the amygdala (periamygdaloid cortex, anterior and posterior cortical nuclei, nucleus of the lateral olfactory tract) and a small anteriomedial part of the entorhinal cortex.

►Connections within the secondary olfactory cortex: In rodents and carnivores, it has been shown that there is an extensive system of associational connections within the areas of the secondary olfactory cortex [4]. These fibers originate in all of the olfactory areas except the olfactory tubercle. Many of the associational fibers also extend into cortical regions beyond the areas that receive fibers from the olfactory bulb, including portions of the entorhinal, perirhinal, and insular cortex, and the medial amygdaloid nucleus.

►Contralateral connections: The projection of the olfactory bulb itself is entirely unilateral. However, fiber bundles from the olfactory peduncle cross in the ►anterior commissure to reach the contralateral olfactory bulb and cortex, providing the major route of interhemispheric olfactory information transfer. Although these fibers run with the olfactory tract, they do not originate from mitral or tufted cells of the olfactory bulbs. Instead, they originate from those cells of the anterior olfactory nucleus, which are located in the olfactory bulb. Similar commissural fibers also originate more caudally, in the anterior part of the piriform cortex [4]. In humans all contralateral olfactory projections exert inhibitory effects only.

►Centrifugal projections to the olfactory bulb: Many of the olfactory cortical areas, including the anterior olfactory nucleus, piriform cortex, and periamygdaloid cortex send fibers back to the olfactory bulb. The projection of the anterior olfactory nucleus is bilateral. There is also a substantial projection from the nucleus of the horizontal limb of the diagonal band to the superficial layers of the olfactory bulb.

So far, a clear transformation of the highly ordered topographic map of the bulb onto the olfactory cortex has not been demonstrated. Small areas of the olfactory bulb project to virtually the entire olfactory cortex, and small areas of the cortex receive afferents from virtually the entire olfactory bulb [6]. However the results of a recent genetic tracer study in rodents indicate that a given olfactory receptor subtype projects to discrete neuronal clusters within the olfactory cortex, suggesting a topographical organization in olfactory cortex which is similar to the bulbar organization [7].

►Piriform cortex: The piriform cortex is the largest olfactory cortical area in humans as well as in most mammals. It is situated along the lateral olfactory tract on the caudolateral part of the orbital cortex, near the junction of the frontal and temporal lobes, and continues onto the dorsomedial aspect of the temporal lobe. Due to this it is defining two subdivisions: the anterior (frontal) piriform (or "prepiriform") cortex and the posterior (temporal) piriform cortex. Both parts of the piriform cortex are histologically identical. However it has been suggested that human frontal and temporal piriform cortex are functionally distinct [5]. The piriform cortex is activated by olfactory stimuli but habituates rapidly to repetitive stimulation. It has been shown "that sniffing, whether an odorant is present or absent, induces activation primarily in the piriform cortex" [8] leading to the assumption that the sniff primes the piriform cortex for an optimal perception of an odor [5]. It is suggested that the temporal part of the piriform cortex mediates basic odor perception independent of odor valence while the frontal part of the piriform cortex is receptive to hedonic value of the odor. Additionally the piriform cortex is involved in olfactory learning and memory [5].

►Amygdala: Projections from the olfactory bulb terminate in several discrete portions of the amygdala (periamygdaloid region, anterior and posterior cortical nuclei, nucleus of the lateral olfactory tract). The cytoarchitectonic transition from the amygdala to the temporal piriform cortex is poorly demarcated. The olfactory areas of the amygdala send projections back to the bulb as well as provide direct input to lateral, basolateral, central amygdaloid nuclei and to basal ganglia, thalamus, hypothalamus, and prefrontal cortex [5]. It is suggested that the amygdala is highly responsive to odor stimulation. The amygdala is proposed to play an important role in affective responses in general, and in olfactory hedonics in particular. The amygdala is responsible for the interaction between valence and intensity of an odorant, as well as for olfactory memory. Of all the senses, olfaction possesses the most intimate relation with the amygdala.

Tertiary Olfactory Regions

From secondary olfactory cortex, information is transmitted to several other parts of the brain, including orbitofrontal cortex, agranular insular cortex, additional subnuclei of the amygdala, medial and lateral hypothalamus, medial thalamus, basal ganglia, and hippocampus. These areas have been referred to as tertiary olfactory regions. Projections to and among these areas are complex and cannot be discussed here in detail. Most of these areas are not specific for processing of olfactory stimuli and show activation by other sensory inputs as well. This complex network of brain areas provides the basis for odor-guided regulation of behavior, feeding, emotion, autonomic states, and memory [5].

►Orbitofrontal cortex (OFC): The OFC is situated at the basal surface of the frontal lobes. It receives input from all secondary olfactory regions (except the olfactory tubercle) in the absence of an obligatory thalamic intermediary and in turn provides feedback connections to each of these regions. The OFC represents the main neocortical projection site of the olfactory cortex and is responsible for initial processing of olfactory information [5]. There is converging evidence that specialized areas within the OFC are engaged depending on the specific task of olfactory processing and that there is some functional lateralization. The posterior OFC is known to be associated with low-level aspects of olfactory processing, such as passive smelling and odor detection whereas the anterior OFC is engaged with higher-order olfactory processing, including associative learning, working memory, and odor recognition memory. Additionally there is evidence for different brain activation associated with odorants of different pleasantness. Whereas pleasant odors evoke activity in medial OFC, unpleasant odors lead to an activation in lateral OFC. Furthermore the OFC receives input from other sensory areas, especially from gustatory, visual, and visceral centers, providing the basis for multisensory integration, resulting in feeding-related and odor-guided behaviors [5]. A functional imaging study demonstrated that regions of the OFC are related to olfactory sensory-specific satiety [9]. The activation of some regions within the OFC produced by the odor of a food eaten to satiety decreased, whereas there was no similar decrease for the odor of food which was not eaten in the meal.

Other Brain Areas Involved in Olfactory Processing

►Cingulate: Although there are connections between the cingulate gyrus and frontal areas involved in olfaction, the cingulate gyrus has not typically been considered as a part of the olfactory system. The cingulate gyrus is involved in processing of information of various kinds. More specifically, the anterior cingulate is frequently involved in tasks requiring attention to sensory features in the environment. In olfactory studies, activations have been reported in anterior as well as in posterior parts of the cingulum. Interestingly, the cingulate gyrus has also been reported to be of critical importance in the processing of painful sensations. Thus, one might speculate that emotions induced by either odors or pain relate to a similar pattern of brain activation in the cingulate gyrus [2].

►Cerebellum: In several studies, cerebellar activation following olfactory stimulation has been reported. Yet, the functional significance of these findings remains unclear. In a functional imaging study the effects of smelling versus sniffing an odor on cerebellar activation were compared and it was hypothesized that the cerebellum maintains a feedback mechanism that regulates sniff volume in relation to odor concentration [10].

In conclusion, functional imaging data support a model of hierarchical organization of olfactory processing. From the ORs and olfactory bulbs (primary olfactory cortex), information are projected to the secondary olfactory cortex. The piriform cortex is the most prominent part of the secondary olfactory cortex in man. Neuroimaging as well as neuroanatomical data suggest that this area is at least minimally engaged during all olfactory tasks. A variety of tertiary regions have been shown to receive projections from the secondary olfactory cortex, among which the OFC seems to be engaged in most tasks of olfactory processing. Thus, core areas within the olfactory system may play a mandatory initial role. However, the involvement of tertiary regions seems to vary with specific task demands, e.g., whether odor processing is related to recognition or emotional response. Finally, another level of organization appears to involve brain areas that fall outside of the typically defined olfactory system, which become engaged during specific types of processing. Examples of these areas are the activation of the cingulate cortex as a multimodal sensory processing area or involvement of the cerebellum which is involved in the adjustment of sniff volume in regard to odor concentration. The ipsilateral nature of olfactory projections, the absence of the thalamic relay during information transmission to the cortex, and the overlap with limbic brain areas are properties of the olfactory system, which sharply distinguish olfaction from other sensory modalities. In summary, odor processing seems to comprise a serial processing of information from primary to secondary and tertiary regions, and also a parallel, distributed processing engaging a complex and distributed network of brain regions whose pattern of activation varies depending on the specific requirements of the task.

References

1. Weismann M, Yousry I, Heuberger E, Nolte A, Ilmberger J, Kobal G, Yousry TA, Kettenmann B, Naidich TP (2001) Functional magnetic resonance imaging of human olfaction. Neuroimaging Clin N Am 11(2):237–250
2. Wiesmann M, Kettenmann B, Kobal G (2004) Functional magnetic resonance imaging of human olfaction. In: Taylor AJ, Roberts DD (eds) Flavor perception. Blackwell, Oxford, pp 203–227
3. Cleland TA, Linster C (2003) Central olfactory structures. In: Doty RL (eds) Handbook of olfaction and gustation. Marcel Dekker, New York, pp 165–180
4. Price JL (2004) Olfactory system. In: Paxinos G, Mai JK (eds) The human nervous system. Elsevier, Amsterdam, pp 1197–1211
5. Gottfried JA (2006) Smell: central nervous processing. Adv Otol Rhinol Laryngol 63:44–69
6. Haberly LB, Price JL (1977) The axonal projection patterns of the mitral and tufted cells of the olfactory bulb in the rat. Brain Res 129(1):152–157
7. Zou Z, Horowitz LF, Montmayeur JP, Snapper S, Buck LB (2001) Genetic tracing reveals a stereotyped sensory map in the olfactory cortex. Nature 414(6860):173–179
8. Sobel N, Prabhakaran V, Desmond JE, Glover GH, Goode RL, Sullivan EV, Gabrieli JD (1998) Sniffing and smelling: separate subsystems in the human olfactory cortex. Nature 392(6673):282–286
9. O'Doherty J, Rolls ET, Francis S, Bowtell R, McGlone F, Kobal G, Renner B, Ahne G (2000) Sensory-specific satiety-related olfactory activation of the human orbitofrontal cortex. Neuroreport 11(4):893–897
10. Sobel N, Prabhakaran V, Hartley CA, Desmond JE, Zhao Z, Glover GH, Gabrieli JD, Sullivan EV (1998) Odorant-induced and sniff-induced activation in the cerebellum of the human. J Neurosci 18(21):8990–9001

Olfactory Peduncle

Definition

The olfactory peduncle runs bilaterally from the olfactory bulb to the anterior perforated substance. The olfactory peduncle contains the olfactory tract as well as thin layers of grey matter which are part of the anterior olfactory nucleus.

►Olfactory Bulb
►Olfactory Pathways
►Olfactory Tract

Olfactory Perception

BURTON SLOTNICK[1], ELKE WEILER[2]
[1]Department of Psychology, University of South Florida, Tampa, FL, USA
[2]Department of Neurophysiology, Institute of Physiology, Ruhr-University Bochum, Bochum, Germany

Synonyms

Olfactory awareness; Olfactory sensitivity; Olfactory discernment; Olfactory acuity

Definition

Olfactory ►perception is a process that starts in the nose with the stimulation of olfactory sensory neurons and terminates in higher cerebral centers which, when activated, make us consciously aware of an odor. In humans this awareness is generally confirmed by verbal reports while in animal studies some sort of odor detection or discrimination task is used. In mammals, olfactory stimuli are received and processed by multiple systems (the main olfactory system, vomeronasal, and the ►septal organ system). Activation (particularly by irritants) of trigeminal, ►vagal and glossopharyngeal receptors in the respiratory tract may contribute to the perceptual experience. However, most research has concentrated on the main olfactory system which also appears to be the only functional olfactory system in humans.

Among the more remarkable aspects of olfactory perception are a seemingly infinite number of odors and odor combinations that can be discriminated, that for humans, most odors generate an emotional response that can range from extreme disgust to extreme pleasantness, and that, in many species, odor exposure can exert profound influence on social, including reproductive, behavior. The neuroscience of olfactory perception has been driven largely by these and related behavioral outcomes and may be viewed as attempts to understand their neurobiological basis.

Characteristics

The Biological Basis of Odor Perception

Molecular biological studies identifying the large family of odorant receptor genes have revealed principles in the organization of sensory neurons and their pattern of projection to the olfactory bulb. Each sensory neuron expresses one of a large number of receptor proteins (about 1,000 in rodents) and the axons of neurons that express the same receptor converge to terminate in the same glomerular areas in the olfactory bulb. While receptor–ligand interactions define which odorant molecules will activate a sensory neuron, the stimulus spectrum (range of sensitivity) of any one sensory neuron appears broadly rather than narrowly tuned. Consequently, each class of neurons may respond to a wide variety of odorants, more strongly to some, more weakly to others (depending on structural interactions of ligand–receptor binding). As a result, many and perhaps hundreds, of different classes of sensory neurons may respond more or less strongly to even simple (monomolecular) odorants [1,2].

The inputs to the bulb from sensory neurons are relayed to more central brain areas by second order (mitral and tufted) cells whose axons converge to form the lateral olfactory tract, the primary projection pathway from the bulb to the brain. Although the olfactory cortex (piriform and lateral entorhinal cortices) is the primary termination for these outputs, there are fairly direct projections to four other target areas: prefrontal orbital cortex (via the dorsal medial thalamic nucleus), hippocampus (via the lateral entorhinal cortex), the corticomedial division of the amygdala, and the hypothalamus [3]. As described below, it is tempting to associate each of these projection targets with different known olfactory functions: analysis of complex olfactory signals (primary olfactory cortex), acquisition of cognitive based olfactory tasks and, perhaps, conscious awareness of an odor (the medial dorsal thalamic-orbital frontal cortex system), excellent olfactory memory (the entorhinal–hippocampal system), emotional component of odors (amygdala, limlaic system), and olfactory influenced neuroendocrine changes (projections to hypothalamus).

Odor Quality Perception

Perhaps the most active area of research relevant to olfactory perception concerns the neural mechanisms that code for odor discrimination and odor quality. Work here has concentrated largely on the olfactory bulb because the functional organization of its inputs is now well understood and because bulbar activity in response to odor stimulation can be visualized using a variety of methods including functional magnetic resonance imaging (fMRI), optical imaging of intrinsic signals, indexing increases in metabolic activity using 2-deoxyglucose (2-DG) and expression of molecular activity markers such as c-FOS [2].

The so-called “combinatorial” view of odor coding is the most widely accepted explanation for the physiological basis of odor quality perception. The convergence of inputs from sensory neurons expressing the same type of receptor plus the many different types of sensory neurons that respond to any one odor results in activation of multiple discrete regions in the olfactory bulb upon odor stimulation. Although structurally similar odors may activate similar or overlapping areas in the olfactory bulb, in all cases examined, each odor produced a unique pattern of glomerular activation.

This "odotopy" or odotopic map representation of different odors at the level of the olfactory bulb provides the primary evidence for the generally accepted "combinatorial" view of odor coding.

While the details of this scheme are topics in other chapters of this volume, its potential significance for understanding odor perception is clear: according to this view, the pattern of inputs from the sensory epithelium to the olfactory bulb provides the neural basis for odor discrimination and, hence, largely determines the perceived quality of an odor. In general, this combinatorial hypothesis has considerable face validity; it provides a reasonably parsimonious account of odor coding, and is solidly grounded in both the molecular biological studies on the organization of inputs to the olfactory bulb and the results of numerous mapping studies. Nevertheless, this view has been challenged by results obtained using fast imaging methods, by studies using awake, behaving animals, and by recent work suggesting that the organization of olfactory cortex may be more suitable for coding complex odor signals.

Temporal Parameters and Early Events in the Olfactory Bulb

The minimum time required to identify a stimulus helps define the temporal period during which neural coding occurs. Both human and animal subjects can identify an odor after only a few hundred milliseconds of exposure (i.e., after one or two sniffs). What neural events occur during this brief period? Fast imaging methods demonstrate that, within the first few hundred milliseconds of odor exposure, activity across the glomerular layer of the olfactory bulb evolves, is temporally complex and that responses to different odors vary in many parameters including latency of onset, rise time, amplitude, modulation by respiration cycle, temporal dynamics of activation, sniff rate, and the extent to which rise time and amplitude are correlated [4]. The important point is that within the brief time needed to identify an odor, numerous neural events are potential candidates for odor coding. Because odotopic maps of the olfactory bulb are based on averaging activity over many seconds or minutes of odor exposure, it remains unclear whether such maps represent the temporally dynamic changes that occur during the first few sniffs of an odor [2].

Disruption of Bulbar Inputs

One method for examining the functional significance of odor maps is to assess odor detection and discrimination after surgical or toxicant destruction of bulbar sites activated by a target odor. Surprisingly, even extensive disruption in the patterns of bulbar inputs in rats fails to produce a specific anosmia or hyposmia, or to significantly disrupt ability to discriminate between odors [5]. In related behavioral studies only mixed results have been obtained in attempts to assess other predictions based on the proposed odotopic view of odor coding (e.g., that similarity in patterns of bulbar activation should predict perceived similarity or difficulty in discriminating between odors).

Perception of Complex Odors and the Olfactory Cortex

The question of whether we experience the individual components of odorant mixtures (i.e., analytic perception) or as a single odor (i.e., synthetically) is complex because, in mixtures, odorants having different vapor pressures and solubilities may produce complex outcomes, and the resulting molecules probably compete for sites on olfactory sensory neurons. Nevertheless, except in the laboratory, most odors encountered represent complex mixtures of vapors. Behaviorally, the issue has been largely resolved by a variety of studies in which human subjects are asked to identify the number of or components of different odors in mixtures. Even with training, subjects are rarely able to identify individual components or accurately identify the number of components in mixtures of three or more odorants.

The evidence from these and related studies strongly supports the view that olfaction is synthetic and that complex mixtures, such as the many volatile molecules that contribute to the odor of urine or coffee, are perceived as single odor "objects." It follows then that analytic or feature detection functions that occur at the level of the olfactory bulb may be early events in further signal processing that result in mixtures being perceived as a single identifiable odor. Where might such synthesis occur? The organization of inputs from olfactory epithelium to the olfactory bulb effects a relatively simple transformation in which signals from sensory neurons expressing the same membrane receptor are represented in spatially discrete areas of the bulb. In contrast, bulbar output neurons are subject to numerous synaptic interactions within the bulb as well as feedback from ►centrifugal projections originating in deeper brain structures and have extensive connections within olfactory cortex. These provide the opportunity for more complex modification in the representation in olfactory cortex of the initial sensory signals. For example, whereas mitral/tufted cells in the olfactory bulb receive input from just a single type of odor receptor, each neuron in the olfactory cortex appears to receive information from multiple bulbar output neurons and some neurons are activated only if two different odor receptor signals are received. Further, responses in olfactory cortex may have considerable plasticity: unit responses to components of odor mixtures are readily modified by exposure to the mixture and, in trained animals, modified as a function of whether the odor was associated with a reward [6].

In brief, our understanding of the biological basis of odor quality perception is incomplete. The results of behavioral studies with rodents, the enumeration of neural events during odor sampling and initial studies on olfactory cortex provide important data but not, as yet, an alternative scheme of odor coding.

Perceptual Subqualities

Can odors be classified into types or subqualities? For other modalities stimulation produces only a limited number of qualitative differences or subqualities such as the basic types of tastes, skin sensation, colors or tonal frequencies that can be discriminated. For olfaction, literally thousands of monomolecular odorants may each produce a qualitatively different perception, and combinations of odorants may produce additional unique qualitative experiences. A number of odor classificatory schemes have been proposed, some of which are based on multivariate analyses of odor judgments by a panel of subjects sampling a wide variety of odors. None, however, are able to accommodate the full range of perceptual experience generated by monomolecular odors or have strong predictive value for how a novel odorant or a mixture of odorants would be judged. Nevertheless, there appears to be reasonably broad agreement for a limited number of descriptors (such as camphor, musk, floral, peppermint, ether, pungent and putrid, the seven primary odors suggested by Amoore) and such schemes have heuristic value. However, odorants within any such class often have diverse physiochemical properties and, with few exceptions, it has not proven possible to reliably predict odor quality from the molecular structure of an odorant.

Affective Responses to Odors

Few olfactory stimuli are judged as hedonically neutral; most elicit a clear like or dislike reaction on the part of the perceiver. The ubiquitous use of odorants in cosmetics and foods attests to the fact that many odors are pleasing and can influence mood and appetite. In humans, the hedonic valence of an odor is largely learned and the experience associated with an odor probably determines its hedonic valence. There are obvious cultural differences in odor preference: for example the odor of the durian fruit is judged generally as fetid by Westerners but is described as heavenly by natives in South East Asia.

▶Trigeminal, ▶glossopharyngeal and vagus nerves in the respiratory tract respond to airborne irritants and their activation together with olfactory sensory neurons may contribute to perceived intensity and unpleasantness of some odors. Except for fear or aversive responses shown by some animals to the odor of predators, it has proven difficult to assess odor preferences in laboratory animals. Human fMRI studies demonstrate arousal of the amygdala by both pleasant and unpleasant odors but, interestingly, not by more neutral odors. These outcomes are in agreement with the more general findings that the amygdala plays an important role in emotional arousal.

Odor Memory

The "Proust effect" provides a popular example of long-term odor memory, and déjà vu phenomena are often triggered by odors. Clearly, odors, particularly those associated with an emotion arousing event, are remembered for years if not the lifetime of an individual. Studies with rodents demonstrate near perfect retention of odor discrimination tasks even after a brief exposure to the conditioning odor or after manipulations specifically designed to maximize proactive and retroactive interference with odor memory [7]. Where such long-term memories are stored is uncertain; in rats, neither surgical disruption of the olfactory thalamic-orbital prefrontal cortex or projections to the amygdala disrupt odor memory. In humans, fMRI studies reveal activation of many brain areas during the encoding of odor stimuli but more restricted areas and especially olfactory cortex and orbital prefrontal areas in recall or identification of familiar odors.

Cognitive Function

In humans, olfaction is generally not viewed as an essential sensory modality and does not appear to play an important role in cognitive or higher mental processes (i.e., we don't "think with our noses"). In contrast, rats, whose behavior is largely guided by and dependent on odors, become quite competent in performing complex, cognitive based tasks when odors are provided as discriminative cues. Thus, rats quickly acquire strategies for nearly errorless solutions for a series of simple discrimination tasks and more difficult matching to sample problems (i.e., they acquire a "learning set"), demonstrate paired associate learning, and even solve problems requiring a form of transitive inference. It is unlikely that other sensory cues could support such learning and, indeed, rats perform more poorly or fail when trained on learning set or matching to sample tasks if visual or auditory cues are used. These cognitive abilities appear to be dependent on thalamic-orbital frontal cortical projections: lesions of this system, including those confined largely to the olfactory component of the medial dorsal thalamic nucleus, have little or no effect on simple odor discrimination problems but disrupt acquisition of complex olfactory tasks [7].

Odors, Reproduction and Unconscious Perception

Olfaction is a critically important sensory modality for most mammals and is used in a variety of behaviors from homing to identifying sources of food and the social status of conspecifics. The demonstration that exposure of gravid female mice to the odor of males from a different strain can disrupt pregnancy (the

"Bruce Effect") led to studies demonstrating clearly the influence of conspecific odors on neuroendocrine changes involved in sexual maturity, mate selection and other aspects of reproduction and social interactions in rodents and, to some extent, in primates. Whether odors play a similar role in humans remains a continuing topic of interest. In humans, exposure to steroidal and other odors from exocrine glands appears to have subtle and gender-specific effects on a number of physiological indices and may alter mood [8,9]. Of particular interest is the evidence that such changes may occur without the subject's conscious awareness of the odor stimulus. It is unclear whether this "unconscious perception" is mediated by neural pathways that bypass olfactory cortex. Such pathways exist in mammals with well developed vomeronasal/accessory olfactory bulb structures but there is scant evidence for the existence of a similar accessory olfactory system in humans.

Olfaction, Schizophrenia and Neurodegenerative Disease

Deficits in odor identification together with signs of degeneration in central olfactory structures are a pervasive concomitant of schizophrenia, Wilson's, Parkinson's disease (PD) and Alzheimer-type dementia (AD). Patients diagnosed as schizophrenic perform poorly on an odor identification task despite having reasonably normal odor detection thresholds. Indeed, olfactory dysfunction may be near universal in neurodegenerative diseases and occurs even in those with cerebellar ataxia; its onset may predate the first clinical signs of the disease and, thus, be diagnostic, particularly for patients at risk for psychosis or AD (e.g., those with an ApoE4 allele).

Other sensory systems do not exhibit the extensive degenerative changes that occur in the olfactory system in PD and AD and it is unclear why a progressive loss of smell function should be characteristic of and even predate movement and cognitive disorders [10].

In brain imaging studies, identification deficits appear to be more closely associated with changes in olfactory cortex or the temporal lobe than with the frontal lobe. Interestingly, olfactory auras often precede the onset of temporal lobe psychomotor epilepsy. The temporal lobe may also be involved in other olfactory disorders including olfactory hallucinations (phantosmia) and altered or distorted perception of odors (parosmia).

References

1. Buck LB (1996) Information coding in the vertebrate olfactory system. Annu Rev Neurosci 19:517–544
2. Wachowiak M, Shipley MT (2006) Coding and synaptic processing of sensory information in the glomerular layer of the olfactory bulb. Semin Cell Dev Biol 17:411–423
3. Carmichael ST, Clugnet MC, Price JL (1994) Central olfactory connections in the macaque monkey. J Comp Neurol 346:403–434
4. Spors H, Wachowiak M, Cohen LB, Friedrich RW (2006) Temporal dynamics and latency patterns of receptor neuron input to the olfactory bulb. J Neurosci 26:1247–1259
5. Slotnick B, Bodyak N (2002) Odor discrimination and odor quality perception in rats with disruption of connections between the olfactory epithelium and olfactory bulbs. J Neurosci 22:4205–4216
6. Wilson DA, Kadohisa M, Fletcher ML (2006) Cortical contributions to olfaction: plasticity and perception. Semin Cell Dev Biol 17:462–470
7. Slotnick B (2001) Animal cognition and the rat olfactory system. Trends Cogn Sci 5:216–222
8. Snowdon CT, Ziegler TE, Schultz-Darken NJ, Ferris CF (2006) Social odours, sexual arousal and pairbonding in primates. Philos Trans R Soc Lond B Biol Sci 361:2079–2089
9. Jacob S, Hayreh DJ, McClintock MK (2001) Context-dependent effects of steroid chemosignals on human physiology and mood. Physiol Behav 74:15–27
10. Albers MW, Tabert MH, Devanand DP (2006) Olfactory dysfunction as a predictor of neurodegenerative disease. Curr Neurol Neurosci Rep 6:379–386

Olfactory Perceptual Learning

Donald A. Wilson, Heather Bell,
Chien-Fu Chen
Department of Zoology, University of Oklahoma,
Norman, OK, USA

Synonyms

Odor memory; Odor familiarity; Odor expertise

Definition

Perceptual learning is an improvement through experience in the ability or potential ability to detect and/or discriminate sensory stimuli. Perceptual learning can be demonstrated in nearly all sensory systems, for example through the enhanced ability of musicians to identify or discriminate musical notes, or of visual artists to identify similar colors. In the sense of smell, most ►odors experienced in nature or everyday life are complex mixtures of many different ►odorant molecules. Being able to discriminate these different mixtures from each other is one of the main functions of the olfactory system. In mammals, recognition and discrimination of such odors appears to involve an initial analysis of the inhaled stimulus into its component molecular and submolecular features, and a subsequent merging of those features into a unitary odor object, such as "coffee" or "rose." As odors become more familiar, both the encoding of the features and their synthesis into objects are enhanced, leading to improvements in fine sensory discrimination. Experience-dependent changes within the nervous system underlying this olfactory perceptual learning occur throughout the olfactory sensory pathway.

Characteristics

Sensory discrimination - the ability to determine whether two stimuli are the same or not - can improve with experience. Slight differences between two stimuli that originally went undetectable, can become detectable with experience and training. This experience-dependent improvement is called perceptual learning, and generally regarded as a form of ►implicit learning, not requiring conscious awareness. Typical examples of perceptual learning include improvements in visual vernier acuity, where the ability to determine whether two vertical lines are either exactly in line or slightly horizontally displaced from each other can be improved through training. Similar examples have been described for auditory pitch perception and haptic (sense of touch) texture discrimination. One common characteristic of perceptual learning is that the effect is largely limited to the familiar stimulus set. Thus, improvements in vernier acuity for vertical lines does not transfer to acuity for horizontal lines.

The improvement in sensory discrimination with experience implies a change in the underlying sensory system which encodes the stimuli. Sensory systems generally encode stimuli in the external world by having populations of neurons tuned to slightly different aspects of those stimuli. Thus, peripheral receptors, transducing sensory input into neural activity, may only respond to a narrow range of energy – a certain wavelength or location of light in vision, a certain frequency of sound in audition, or a certain molecular shape or charge in olfaction. Through the cooperative action of large ensembles of such neurons, information about the identity of the original stimulus emerges, which can then guide perception and behavioral responses.

This basic sensory system function leads to several potential mechanisms through which perceptual learning may arise. Experience with a specific range of sensory inputs could lead to changes in peripheral receptor number or relative tuning distribution, tuning of neurons within the central nervous system, and/or local circuit interactions within the large ensembles. There is evidence for all of these experience-dependent changes occurring in the olfactory system associated with perceptual learning.

Behavioral Evidence of Olfactory Perceptual Learning

In humans, experience with specific odors enhances subsequent discrimination and identification of those odors. Thus, familiar odorants are more easily discriminated than unfamiliar odorants [1]. This experience-dependent improvement can be induced either through specific exposure or training, or emerge over a lifetime of experience. This latter process may contribute to strong cultural differences in perception and categorization of odors.

In animal models, as in humans, odor experience enhances discriminability of familiar odors [2,3]. Naïve rodents, for example, fail to respond differentially to many monomolecular odorants differing by a single hydrocarbon in their molecular structure. This can be tested in a habituation/cross-habituation paradigm, where one odorant is repeatedly presented until some behavioral response habituates. Then, a second odorant is presented. If the animal discriminates between the odorants, the new odorant evokes a behavioral response. If the animal does not discriminate between the odorants, the response to the new odorant is comparable to the habituated odorant. Using such a paradigm, naïve animals that were habituated to, for example the four carbon odorant molecule ethyl butyrate, showed cross-habituation to the five carbon odorant molecule ethyl valerate, suggesting they cannot discriminate between these odorants (i.e., the odors are similar). However, if given prior experience with these odorants, they subsequently do show differential responses to the two odorants. These experience-induced changes appear selective to the familiar odorants and do not create a general enhancement for discrimination of all odorants.

In addition to experience-induced enhancement of odorant discrimination, perceptual learning can also improve identification of components within odorant mixtures [4]. With simple mixtures of pure odorants, the intensity of individual components plays a major role in the ability to identify those components. Thus, as might be expected, as one component within a binary mixture becomes more intense (higher relative concentration) than the other, that component comes to dominate the perception of the mixture. However, familiarity of the components produces a similar effect. Familiar components are more easily identified within a mixture than unfamiliar components. This consequence of perceptual learning may underlie the ability of professional flavorists and perfumers to identify components with mixtures, although human psychophysical data suggest even professionals have only a limited ability to analyze complex mixtures that include greater than 3–4 components into their constituent parts.

Finally, in addition to experience-induced enhancement of discriminability, odorant exposure may also enhance detectability of odorants [5]. Perhaps the best example of this is perception of the odorant androstenone, though other odorants show similar effects. Androstenone is a component of human sweat, and is more concentrated in males than females. Many individuals appear to have very high thresholds for detecting androstenone, or are even ►anosmic to it. However, repeated exposure over multiple days can significantly improve detection in these individuals, dramatically lowering detection thresholds. There is some evidence that females may acquire this experience-dependent sensitivity faster than males.

Neurobiology of Olfactory Perceptual Learning

At the neurobiological level, memory for odors and their associations is distributed throughout the sensory pathway, with evidence for changes from the receptor sheet all the way to the primary olfactory cortex [2,6]. The olfactory systems of all vertebrates and many invertebrates share several basic structural features. Peripheral olfactory receptor neurons express one or a few olfactory receptor genes which code for proteins that bind to odorant molecules sharing a particular structure. These receptor neurons then project to the second order neurons within a central nervous system structure called the olfactory bulb (vertebrates) or antennal lobe (invertebrates). The connections between receptor and second order neurons occurs within structures called glomeruli, which receive input from receptor neurons all expressing the same olfactory receptor genes. Thus, stimulation with a particular odorant activates a unique combination of glomeruli based on which receptors that odorant molecule binds. The response of second order neurons reflects the homogeneous receptor input, as well as local circuit interactions. The second order neurons then project to the olfactory cortex (mammals) or mushroom bodies (invertebrates), where convergence of the different molecular features extracted by the periphery occurs on individual third order neurons. In different behavioral paradigms and different species, olfactory experience has been found to change the response patterns of receptor neurons and glomeruli, and both single cell and ensemble activity of second and third order neurons.

Experience-induced responsiveness to odorants, such as androstenone, may involve both peripheral and central changes. Evidence in humans and rodents suggests that repeated or prolonged exposure to an odorant such as androstenone produces enhanced olfactory receptor sheet responses as measured with electro-olfactogram [7]. The electro-olfactogram is a measurement of summed receptor sheet activity, much as the electroencephalogram measures summed cortical activity. The specific mechanism of enhanced receptor sheet responsiveness to exposed odors in currently unknown. In addition to these peripheral changes, there is some evidence for central sensitization [5]. Humans exposed unilaterally to androstenone will become able to smell it through either nostril, despite the lack of a direct connection between the two receptor sheets. This suggest that central neurons, that receive convergent information from the two airways, may partially mediate the exposure-induced sensitization.

Experience-induced enhancements in discrimination appear to rely on changes within the central nervous system. Exposure to an odor for as little a few minutes can produce a long-lasting shift in the tuning of second order neurons, such as olfactory bulb mitral cells [3]. These shifts enhance the number of second order neurons encoding familiar odorant features. These changes in individual neuron activity are accompanied by large scale neural ensemble changes, as evidenced by changes in odorant-evoked local field potentials within the olfactory bulb. At least two mechanisms may contribute to these changes in stimulus-evoked activity. First, connectivity between existing neurons may be altered during perceptual learning through synaptic plasticity. Plasticity of synapses within glomerular and/or between second order neurons and local interneurons could affect feedback, feedforward and lateral inhibition. These changes in inhibition could influence both responses of single neurons to familiar stimuli and timing of evoked activity. A change in odorant-evoked spike timing, for example increased synchrony, is hypothesized to enhance the salience of familiar stimulus features to downstream neurons, thus facilitating their identification and discrimination.

A second mechanism of perceptual learning associated change within the olfactory bulb is anatomical restructuring of local circuits. A major class of local interneurons in the mammalian olfactory bulb, granule cells, undergo continual neurogenesis throughout life in many animals. Survival and incorporation of granule cells into local circuits is dependent on odor experience. Given the precise projections of olfactory receptor neurons to olfactory bulb glomeruli, different stimuli evoke different spatial patterns of activity across the olfactory bulb, with activation of a given glomerulus associated with activity of a local, spatially defined column of second order neurons and interneurons such as granule cells. Repeated stimulation of a given glomerulus over several weeks by exposure to a particular odorant, enhances survival of granule cells near that glomerular column, while sensory deprivation reduces granule cell survival [8]. Granule cells not only control excitability of second order neurons, but are also the target of cortical feedback to the olfactory bulb. Thus, they may play an important role in familiarity induced effects on olfactory bulb odor encoding.

Finally, olfactory perceptual learning is associated with changes within mushroom bodies of invertebrates and olfactory cortex of mammals [2,9]. As noted above, the olfactory cortex is hypothesized to synthesize disparate, co-occurring odorant features into perceptual wholes, or odor objects. As this synthesis occurs, a template is formed in cortical circuits, allowing a rapid match of subsequent input to that stored template and enhanced discrimination and recognition. This cortical learning may also contribute to perceptual stability of complex odors, even in the face of slight alterations in intensity or presence of some components [6]. Olfactory perceptual learning may involve changes in both the anterior and posterior piriform cortices, as well as the orbitofrontal cortex. In both humans [9] and rodents

[10], the anterior piriform cortex appears to encode stimulus identity, with experience creating a unique encoding of a mixture stimulus distinct from that of its components. In contrast, the posterior piriform cortex appears to encode information about odor quality (e.g., fruitiness) or categorical information, a process again enhanced by experience and odor familiarity.

The types of modifications in neural coding and perception described here associated with olfactory perceptual learning most likely occur in all cases when odors become familiar or are actively learned. The result is that our perception of familiar odors is different than our perception of novel odors, allowing enhanced discrimination and identification of the familiar.

References

1. Rabin MD (1988) Experience facilitates olfactory quality discrimination. Percept Psychophys 44:532–540
2. Davis RL (2004) Olfactory learning. Neuron 44:31–48
3. Fletcher ML, Wilson DA (2002) Experience modifies olfactory acuity: acetylcholine-dependent learning decreases behavioral generalization between similar odorants. J Neurosci 22:RC201
4. Livermore A, Laing DG (1996) Influence of training and experience on the perception of multicomponent odor mixtures. J Exp Psychol Hum Percept Perform 22:267–277
5. Mainland JD, Bremner EA, Young N, Johnson BN, Khan RM, Bensafi M, Sobel N (2002) Olfactory plasticity: one nostril knows what the other learns. Nature 419:802
6. Wilson DA, Stevenson RJ (2006) Learning to smell: olfactory perception from neurobiology to behavior, Johns Hopkins University Press, Baltimore, p 309
7. Wang L, Chen L, Jacob TJ (2003) Evidence for peripheral plasticity in human odour response. J Physiol 554:236–244
8. Lledo PM, Gheusi G (2003) Olfactory processing in a changing brain. Neuroreport 14:1655–1663
9. Li W, Luxenberg E, Parrish T, Gottfried JA (2006) Learning to smell the roses: experience-dependent neural plasticity in human piriform and orbitofrontal cortices. Neuron 52:1097–1108
10. Kadohisa M, Wilson DA (2006) Separate encoding of identity and similarity of complex familiar odors in piriform cortex, Proc Natl Acad Sci USA 103:15206–15211

Olfactory Plasticity

J. C. Sandoz
Research Center for Animal Cognition, CNRS UMR 5169, Paul Sabatier University, Toulouse Cedex, France

Synonyms

Olfactory learning; Odor-exposure learning; Olfactory priming

Definition

Olfactory plasticity is a general term referring to all types of changes in odor-evoked responses resulting from individual experience. These changes, usually monitored behaviorally, rely on short- or longer-lived structural and/or functional modifications at different levels of the olfactory circuits. Olfactory plasticity is therefore a form of ►neural plasticity.

Characteristics

A general rule of sensory systems is that they constantly adapt to environmental conditions, inducing modifications of the way they process sensory stimuli. Such modifications can be very short (in the range of seconds, for instance receptor desensitization) or very long (in the range of years, plasticity of central representations), depending on the type of experience and of the species considered. Due to obvious differences in lifespan, modifications that are considered to correspond to a medium-term range in a given species may be assigned to the long-term range in a different species.

Olfactory plasticity is found in a wide range of species, from nematodes (*C. elegans*) to humans, with prominent examples in insects (fruit flies, bees, etc.) and mammals (rabbits, rats, mice, humans, etc.). These changes can affect all levels of the olfactory circuits, from the most peripheral (olfactory receptors) to the most central ones (cortical representation). Olfactory plasticity can be demonstrated in behavioral experiments, and its neural basis is usually the subject of neurophysiological and/or neuroanatomical experiments. We will first provide a brief description of a generalized olfactory system (for details, see essays on ►olfactory perception, or ►odor coding). We will then detail the types of sensory/associative experiences that induce olfactory plasticity at the behavioral level. To finish, we will present the current view of the neural basis of olfactory plasticity.

The Olfactory System

The anatomical organization of the olfactory system of vertebrates and of invertebrates, like insects, shows many fundamental similarities. Odors are detected at the periphery (olfactory mucosa within the nose *or* antenna) by olfactory sensory neurons (OSN), which each express a given type of olfactory ►G-protein-coupled receptor. These neurons relay odor information to a first olfactory centre, the olfactory bulb (OB) in vertebrates or its equivalent in insects, the antennal lobe (AL). Both structures are organized in a similar modular way: each of their subunits, the glomeruli, receives input from OSNs expressing the same olfactory receptor type. Glomeruli are sites of intensive synaptic contacts between several neuron types, in particular inhibitory neurons providing local inter-glomerular computation (periglomerular cells/local interneurons), and second-order neurons (mitral cells/projection neurons) that relay processed information to higher brain centers. Between mitral cells, granule cells provide additional lateral inhibition in vertebrates. The complex ►neural network of the AL/OB is considered to be a major site for olfactory plasticity. It performs computations which are thought to mediate better discrimination between similar olfactory inputs, allowing more segregated spatio-temporal odor representations to be conveyed to higher brain centers such as the piriform cortex, the entorhinal cortex and the peri-amygdaloid cortex in mammals, or the mushroom bodies and the lateral protocerebral lobe in insects (see essay on ►odor coding). These structures are thought to be involved in higher-order processing of odor information, like providing the synthetic part of mixture representation, but also in associative learning and memory of odors, and, at least in mammals, providing emotional and hedonic values to odors. As we will see, olfactory plasticity can take place at all levels of the olfactory system.

Sensory and Associative Experience Inducing Olfactory Plasticity

Experimentally, olfactory plasticity is often demonstrated by the result of behavioral experiments, during which a particular olfactory experience induces changes in the way animals or subjects respond to odors. We will review these types of experiences from simple olfactory exposures to much more complex forms of associative learning between particular odors and different outcomes.

Olfactory Exposure

Simple odor exposure, even a very short one, can have consequences on the way the olfactory system will respond to subsequent odor presentations. The most peripheral of these phenomena is called ►olfactory adaptation [1], during which exposure to an odor (from very short pulses to stimulations of a few seconds) decreases reversibly the sensitivity of olfactory receptor neurons (usually in the range of seconds to a few minutes). Functionally, this is believed to allow an animal to constantly adapt its olfactory system to environmental odors, avoiding saturation of the cellular transduction machinery and thereby keeping the ability for the animal to detect more relevant short-lasting odors. Different forms of odor adaptation have been described, depending on the length of the odor stimulation inducing it (short or long puff) and the length of the adaptation (short or long-lived). These different forms are thought to depend on slightly different but interconnected cellular feedback loops within olfactory sensory neurons. Odor adaptation is considered to be reversible. Experimentally, it has provided previous researchers an interesting way of testing whether two odorants are detected by different or overlapping sets of olfactory receptor neurons, in so-called cross-adaptation experiments: animals are first exposed to a mono-molecular odorant A until they adapt to it. Then a second odorant B is presented. If response to B is affected by the former presentation of A, it suggests that detection of B depends on receptors used for the detection of A.

Simple odor exposures do not only affect the periphery, and the changes that they induce at the central level are then considered forms of ►perceptual learning . On a quantitative level, repeated presentations of an odor can have two kinds of effects. On the one hand, the probability of a behavioral response provided by an animal to the presentation of the odor (for instance, a startle or a sniffing response) will tend to decrease through repeated presentations of this odor. This effect is termed ►odor habituation . In some cases, even if a decrease of an odor-evoked response is observed, this effect can be more related to a reduction of the animal's attention or of its overall responsiveness than to a decrease of odor detection ability or changes in odor processing. In fact, repeated experience with an odor can have the opposite effect, reducing the olfactory detection threshold (odors are detected at lower concentration) and can even allow odor detection by seemingly anosmic subjects. On a more qualitative level, repeated experience with a range of different odors can greatly improve the discrimination ability of subjects among these, but also novel, odorants. Furthermore, experience with an olfactory mixture can strongly modify the way the individual components of the mixture are perceived. For instance, a given odor presented to a subject together with a "smoky" odor will tend to be perceived afterwards as smoky, while the same odor would smell cherry-like after being presented together with a "cherry" odor. Such effects are usually interpreted as forms of ►implicit memory and are thought to rely on neural plasticity at different levels of central areas, from the OB where it would modify the receptive range of mitral/tufted cells to the piriform cortex and the orbito-frontal cortex where the synthetic representation of odors may change. These olfactory forms of perceptual learning can take place rapidly, but are usually long-lasting.

Associative Learning

The most prominent forms of olfactory plasticity relate to associative conditioning, during which animals learn to associate odors with particular outcomes or behaviors, which have a positive or negative significance for the animal. It is generally accepted that most of our hedonic relationship to odors is not innate, but rather acquired throughout our lifetime by associations between these odors and particular events or contexts. Odor learning starts even before birth from the mother's amniotic fluid, as the olfactory system is already functional in utero by 12 weeks of gestation. For instance, children of mothers who consume particular odors (garlic, cumin, etc.) and were therefore exposed to these odors during gestation and/or breast-feeding show specific preferences for these odors afterwards. Throughout young age, children learn to associate particular scents or tastes with edibility and/or positive and negative events, and it is generally accepted that by the age of 8, most of our adult olfactory preferences are acquired, although adult experience certainly continues to shape olfactory preference [2]; see also learning during a sensitive period, below].

Experimental psychology distinguishes two main forms of associative learning which both are very prominent in the olfactory domain:

1. In ►classical (Pavlovian) ►conditioning , an animal learns to associate an originally neutral, ►conditioned stimulus (CS – here an odor) with a biologically relevant, ►unconditioned stimulus (US). For instance, honeybees learn to associate odors with sucrose solution in the paradigm of the proboscis extension response (PER) conditioning. In a hungry bee, sucrose solution triggers the reflex extension of the mouthparts (the PER), allowing the insect to drink. Prior to conditioning, odors are ineffective. However, after a single CS/US association, the odor can now elicit the PER and after a few such associations an odor-sucrose memory is formed that can last for the bee's lifespan.
2. In ►operant (instrumental) ►conditioning , the animal learns to associate a behavioral action to a ►reinforcement , and a ►discriminative stimulus (e.g., an odor) can function as a signal for producing the learned behavior. For instance, an odor may act as the signal for a rat to poke its nose in a particular box in order to receive a food reward. Although conditioning creates an association between nose poking and the food reward, odor-food and odor-poking associations are also built and will drive the rat's choice.

In both learning paradigms, odor-outcome (US or reinforcement) associations are established, which can be either ►appetitive or ►aversive.

More complex olfactory learning tasks can be conceived, either in a classical or an operant framework, establishing multiple associations between different odors and multiple outcomes. A simple example of such tasks is differential conditioning (A+, B–), in which an odor A is associated with a US/►reinforcer and another odor B is left without consequence. Experimentally, such conditioning has often been used in the study of neural olfactory plasticity [3–4], because it provides the experimenter with a within-animal control as the same animal has to learn to respond to odor A but not to odor B: usually, specific changes in neural responses are found for A but not for B. In some cases, learning can induce a decorrelation of the neural representations of A and B, making them more discernible for the olfactory system. More complex forms involve ambiguities between odors and outcomes, and give a special meaning to the concomitant presentation of two or more odors: for instance, in biconditional discrimination (AB+, CD+, AC–, BD–), each odor is as often reinforced as not, and the right behavioral response can only be found after linking different odor representations (here the animal should respond to odor A when it is presented together with B but not when A is presented together with C). All these different forms of olfactory learning are based on increasingly complex associations, and pose each different constraints to the olfactory system. In this case, one expects a decorrelation of the representations of odor combinations with different outcome, irrespective of the common presence of a given odor (e.g., AB+ vs. BD–).

Olfactory Plasticity During a Sensitive Period

The olfactory plasticity phenomena detailed above can take place at any moment in an animal's life. There are, however, instances of olfactory plasticity that can only happen during sensitive periods such as after mating or short after birth. Thus, newly-mated female mice learn the specific odor of the mating male, and any encounter with a different male will provoke pregnancy failure [5]. Another prominent example is neonatal learning in rabbit pups, which learn extremely fast – during the first three days after birth, odors that are present on the doe's belly. Recently, a mammary pheromone was found, which alone triggers stereotyped orocephalic movements of nipple search in young rabbit pups. Normally, odors do not elicit this response. However, a single simultaneous presentation of an odor together with the pheromone dramatically changes the pups' behavior, such that it will now respond to the odor presented alone [6]. This form of classical olfactory conditioning is particular, not only for the existence of a strict sensitive period, but also for the fact that an odor, the mammary pheromone, acts as a reinforcer.

Neural Basis of Olfactory Plasticity

Changes in odor-evoked behavioral responses can rely on neural plasticity at all levels of the olfactory circuits, from the most peripheral during olfactory adaptation to the more central, OB/AL and/or higher brain centers for perceptual and associative conditioning. Olfactory plasticity is manifested at the neuron level through both structural and functional neuronal changes.

On a structural level, the number and/or repartition of synaptic contacts between olfactory neuronal populations can be modified. For instance, differential olfactory conditioning is accompanied in ►pyramidal neurons of the piriform cortex by an increased density of ►dendritic spines linked to intra-cortex connections, but also to pruning (reduction) of spines linked to afferent input from the olfactory bulb, suggesting intense rearrangements of olfactory connectivity through learning [7]. Such structural changes can sometimes be correlated with a change in the volume or shape of neuronal structures like the glomeruli. In the particular case of the olfactory bulb, neural olfactory plasticity can take the form of the genesis and preferential survival of novel neurons that will integrate the neural network, specifically as inhibitory interneurons (periglomerular and granule cells). It could be shown that this process is increased after differential olfactory learning [3] and that the novel production (or the loss) of such interneurons has important consequences for OB activity. Structural plasticity is usually related to long-term forms of olfactory plasticity, as they need time to take place.

On a functional level, the strength and efficacy of synaptic transmission can be modified. This can imply many changes at the level of neurotransmitter release, receptor equipment, intra-cellular cascades, second messengers and for long-term forms, it relies on novel protein synthesis. Most functional work on olfactory plasticity has concentrated on describing changes observed in odor-evoked responses within olfactory structures. Depending on the recording technique, modifications are observed on the amplitude, frequency or synchronisation of electrophysiological responses [8,9], on the intensity or repartition of optically-monitored activity [4], on the pattern of production of synaptic proteins etc. In some experiments, plasticity is assessed on whole brain structures with awake and behaving animals. For instance, olfactory bulb field potential activity can be monitored from freely-behaving rats. Odor stimuli usually produce a frequency change in the ►field potential, with a power decrease in the γ frequency range (60–90 Hz) associated with a power increase in the β range (15–40 Hz). This pattern of response was found to be strongly amplified in animals trained in an olfactory learning task, precisely at the moment when they started mastering the task [9]. In such cases, it is difficult to determine precisely the location of this plasticity, which can reveal synaptic efficacy changes over a whole olfactory network. In other approaches, particular neuron populations can be monitored, usually in fixed animals. Thus, in rats, it could be shown in electrophysiological recordings of mitral cell activity that an olfactory exposure can modify their receptive range, so that they would respond to a wider range of odors after olfactory exposure than before [8]. In fruitflies, associative aversive conditioning based on odor-electric shock associations can be applied on a fixed fly under the microscope. Optical imaging experiments coupled to the genetic expression of a reporter of synaptic activity (synapto-PHluorin) in particular antennal lobe populations showed that projection neurons that were initially not activated by an odor prior to conditioning could be recruited shortly after differential aversive conditioning. Recordings from sensory neuron or inhibitory local interneuron populations did not show any change, demonstrating that plasticity took place at the level of second order neurons [10].

Until now, most available data on neural plasticity underlying olfactory learning was obtained from primary olfactory centers (OB/AL) that are easier to access. However, research on higher-order structures is growing. Thus, experiments have already shown that electrophysiological responses of neurons in the olfactory cortex are strongly influenced by previous odor stimulations, and are certainly involved in perceptual learning. But the kind of activity changes appearing at this level, in contrast to those found within the primary centers, correspond to higher-order computations allowing, for instance, the discrimination between a mixture and its components, a task deemed as one of the most critical for odor perception. Moreover, such higher-order structures are good candidates for harboring associative olfactory memories. In fruitflies, Kenyon cells (third order neurons) within the mushroom bodies displayed dramatic increases of calcium responses to the learned odor several hours after a differential aversive conditioning task, with even a localization of changes within specific branches of these neurons [10]. In fact, as only a few third order neurons are activated by a given odor, as opposed to many second-order neurons, neurons in more central areas constitute an ideal substrate for the associative memory trace, giving a particular odor a particular meaning.

Conclusion: Odor Processing Plasticity or Odor-Reinforcement Memory?

As detailed above, many electrophysiological, functional imaging or neuroanatomical studies find strong neural plasticity within olfactory circuits, especially after associative conditioning. However, it is often difficult to relate such neural plasticity to its exact function. Are the observed changes related to modifications of odor processing, modulating for instance the neural representation of the learned odors so that it can be better distinguished from environmental background? Or are they related to an olfactory ►engram, revealing the storage of odor-reinforcement associations in the brain? The picture emerging from the studies carried out so far suggests that primary olfactory centers (OB/antennal lobe) may be responsible for the former, and higher olfactory centers for the latter, but considerable work is still needed to confirm this hypothesis. Future neurobiological studies of olfactory plasticity will have to answer these questions, using a combination of approaches, asking in particular whether the observed cells (and their plasticity) are necessary and sufficient for the expression of olfactory plasticity at the behavioral level.

References

1. Zufall F, Leinders-Zufall T (2000) The Cellular and molecular basis of odor adaptation. Chemical Senses 25:473–481
2. Herz RS (2002) Influences of odors on mood and affective cognition. In: Rouby C et al (eds) Olfaction, Taste and Cognition Cambridge University Press, UK, pp 160–177
3. Alonso M, Viollet C, Gabellec MM, Meas-Yedid V, Olivo-Marin JC, Lledo PM (2006) Olfactory discrimination learning increases the survival of adult-born neurons in the olfactory bulb J Neurosci 26:10508–10513
4. Faber T, Joerges J, Menzel R (1999) Associative learning modifies neural representations of odors in the insect brain. Nat Neurosci 2:74–78
5. Brennan PA, Keverne EB (1997) Neural mechanisms of mammalian olfactory learning Prog Neurobiol 51:457–481
6. Coureaud G, Moncomble AS, Montigny D, Dewas M, Perrier G, Schaal B (2006) A pheromone that rapidly promotes learning in the newborn. Curr Biol 16:1956–1961
7. Knafo F, Libersat F, Barkai E (2005) Dynamics of learning-induced spine redistribution along dendrites of pyramidal neurons in rats. Eur J Neurosci 21:927–935
8. Fletcher ML, Wilson DA (2003) Olfactory bulb mitral-tufted cell plasticity: odorant-specific tuning reflects previous odorant exposure. J Neurosci 23:6946–6955
9. Martin C, Gervais R, Hugues E, Messaoudi B, Ravel N (2004) Learning modulation of odor-induced oscillatory responses in the rat olfactory bulb: a correlate of odor recognition? J Neurosci 24:389–397
10. Berry J, Krause WC, Davis RL (2008) Chapter 18 Olfactory memory traces in Drosophila. Prog Brain Res 169:293–304

Olfactory Priming

►Olfactory Plasticity

Olfactory Receptor

Definition

Olfactory receptors are members of the seven transmembrane domain G-protein coupled family of receptor proteins. The binding of an odorant molecule to an olfactory receptor initiates a conformational change that activates the G-protein and leads to an electrical response in the olfactory sensory neuron that can be transmitted to the brain. Around 1,000 genes encoding functional olfactory receptor proteins have been identified in the mouse genome, with around 350 functional olfactory receptors identified in the human genome. Individual receptor types are typically activated by small and partially overlapping ranges of odorants. The identity of an odorant is therefore conveyed by the pattern of different odorant receptor types that it activates, i.e. an across-fiber pattern code.

►G-protein Coupled Receptors (GPCRs): Key Players in the Detection of Sensory and Cell-Cell Communication Messages
►Odorant
►Odorant Receptor
►Odor Coding
►Olfactory Sensory Neuron

Olfactory Receptor Neuron (ORN)

Definition

Olfactory receptor neurons are cells in the olfactory epithelium in the nasal cavity. They are bipolar neurons with an apical dendrite with cilia facing the interior space of the nasal cavity and a basal axon that via the first cranial (olfactory) nerve passes through the cribriform plate and enters the olfactory bulb. Each olfactory receptor neuron probably expresses a single type of olfactory receptor protein, and neurons with the same receptors are scattered through one of four zones in the epithelium. Olfactory sensory neurons are also sometimes called "olfactory receptors," although this term can be confused with the odorant receptor proteins themselves. It should be noted that the olfactory epithelium is also innervated by the trigeminal nerve, which is responsible for mechanical sensations (touch and pressure), as well as pain and temperature. Trigeminal fibers also respond to chemicals found in onions, mustard and chile powder.

►Odorant Receptor Protein
►Olfactory Bulb
►Olfactory Epithelium
►Olfactory Nerve

Olfactory Receptor Protein

►Odorant Receptor

Olfactory-recipient

Definition

Parts of the basal telencephalon receiving inputs from the main olfactory bulb. Olfactory-recipient areas include the olfactory amygdala and, depending on the species, the ventral telencephalon, lateral pallium, lateral cortex or olfactory cortex.

►Evolution of Olfactory and Vomeronasal Systems
►Olfactory Amygdala
►Olfactory Bulb
►Olfactory Cortex

Olfactory Recognition

Definition

Olfactory recognition, first and foremost, refers to the process by which an odor molecule is sensed and detected by an olfactory receptor. This process is not yet well understood, mainly because of the fact that the protein structure and function of many G-protein coupled receptors (GPCRs) is still under investigation. In contrast to most other GPCRs that recognize their ligands through ionic or hydrogen bond interactions, it appears that olfactory receptors recognize odorants primarily by weak hydrophobic and van der Waals interactions, which allow the observed broad but selective odor ligand binding of olfactory receptors.

►G-protein Coupled Receptors (GPCRs): Key Players in the Detection of Sensory and Cell-Cell Communication Messages
►Odor
►Odorant
►Odorant Receptor Protein
►Olfactory Receptor
►Olfactory Receptor Neuron

Olfactory Sense

Hanns Hatt
Department of Cell Physiology, Ruhr-University Bochum, Bochum, Germany

Synonyms

Sense of smell; Chemosensation; Odor perception

Definition

The olfactory system enables most animals to continuously monitor their chemical environment. The sensitivity and range of olfactory systems is remarkable, enabling organisms to detect and discriminate between thousands of low molecular mass, mostly organic compounds which we commonly call odors. The task is accomplished by specialized olfactory sensory neurons which encode the strength, duration and quality of odorant stimuli into distinct patterns of afferent neuronal signals. Thus, the molecular structure of an odorant molecule is converted into a pattern of electrical activity, which intern is processed in the olfactory bulb and higher brain centres and ultimately perceived as a characteristic odor quality. Odor perception is a result of complex biochemical and electrophysiological reaction mechanisms.

Characteristics

Measurable characteristics of olfaction are:

1. Anatomical organization
2. Signal transduction pathway (molecular basis of sensitivity and specificity)
3. Odorant information processing
4. Olfactory receptors outside the nose

The human nose is often considered something of a luxury. However, even if we have lost faith in our noses, we are still strongly influenced by smells even if only subconsciously. Smells can evoke memories and emotions, influence our mood and are important for our enjoyment when eating. All the delicate nuances of an excellent cuisine or of a noble glass of wine are, in the final analysis, savored through our sense of smell. In addition, before the spirit and beauty of a person can fascinate us, our nose must become infatuated. The olfactory systems have developed, the main olfactory system (described here in detail) and the accessory system, known as the ►vomeronasal system (►Vomeronasal Organ (system)), which is specialized for chemical communication between one another (see glossary). An indication of the importance of the olfactory system in humans is the significant proportion – more than one percent – of the genome is devoted to encoding the proteins of smell. Let us follow the odor trail from molecule to perception.

Atomical Organization

A flower or any odorous subject has to release molecules according to their vapor pressure into the air. During inhalation they can reach our nasal cavity. There is a series of conchal formations, called turbinates. In the most upper one the olfactory epithelium is located, which consists of three mature cell types: bipolar primary sensory olfactory neurons, supporting (sustentacular) cells and basal cells (adult stem cells) which generate olfactory receptor neurons and sustentacular cells throughout our whole life (Fig. 1a). The turnover of the about 20 million olfactory neurons in less than one month. At the apical pole of the cell body of an olfactory sensory neuron (OSN) is a single dendrite that reaches up to the surface of the tissue and ends in a knob like swelling from which project some 20–25 very fine cilia. These cilia, which actually lie in the thin layer of mucus covering the tissue, contain all the molecular components necessary to convert the chemical odor stimulus into an electrical cell signal [1]. On the proximal pole the cell body of OSN narrows into an axon that joins with other axons to form small nerve bundles that then project into a region of the brain, known as the olfactory bulb. Molecular genetic studies have shown that all the neurons, expressing a particular olfactory receptor protein terminate within a single target in the olfactory bulb, called glomerulus: Spherical conglomerates of neuropil some 50–100 μm in diameter that consist of the incoming axons of OSN and the dendrites of the main projection cells in the bulb, the mitral cells. In human, as in other vertebrates, the number of glomeruli correlates with the number of different types of OSN (about 350).

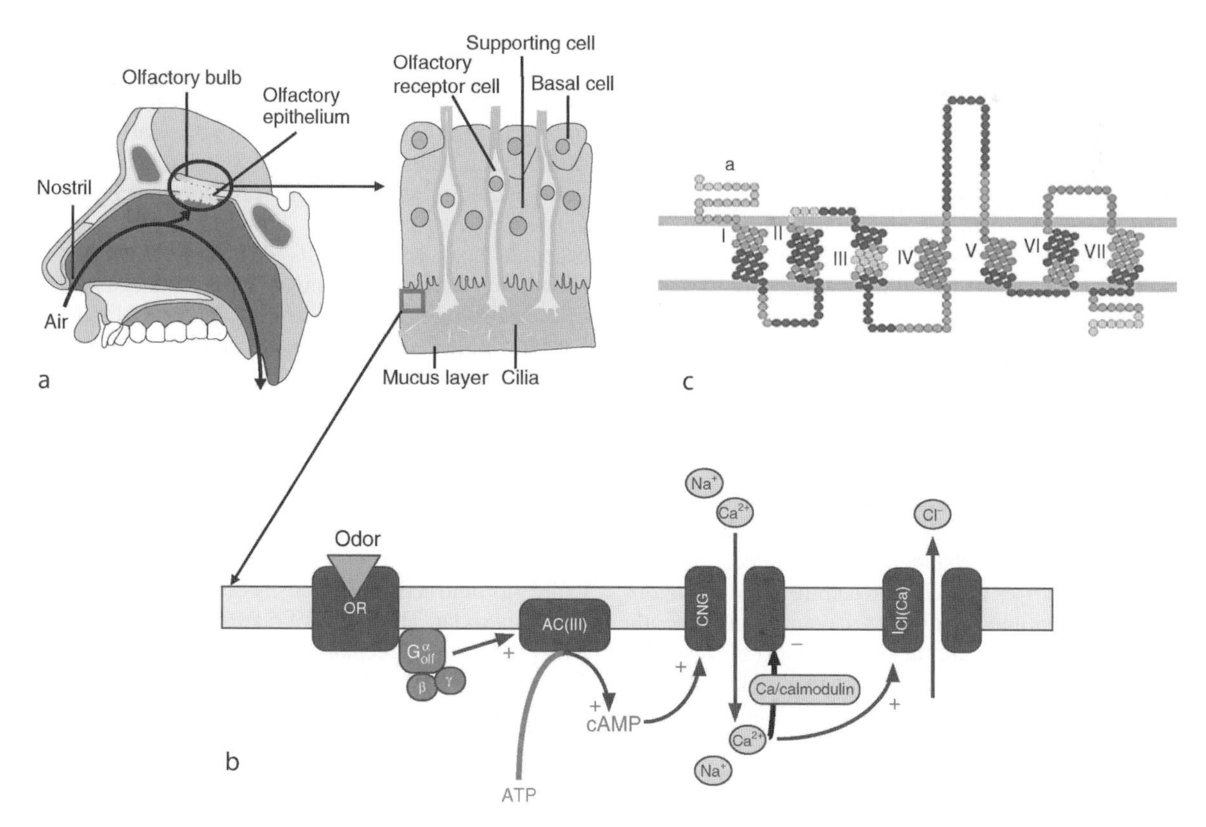

Olfactory Sense. Figure 1 (a) General layout of the nasal chemoreceptive area (*left side*) and the olfactory epithelium (*right side*). (b) Molecular processes during transduction of odor stimuli in an electrical cell response. (c) Molecular structure of a human olfactory receptor protein. The amino acid chain passes through the cell membrane seven times.

Signal Transduction Pathway

Recent advances of electrophysiological and molecular biological methods have provided new insights into the mechanisms of chemosensory signal transduction. The transduction process begins when odorants are dissolved in the mucus. Here the discovery of small, water soluble proteins in the mucus fluid, which are produced by glands of the nasal cavity, has led to the concept that these so-called odorant binding proteins (OBP) may accommodate hydrophobic odor molecules in an aqueous environment and enhance their access to the receptor sides. Several distinct OBP-subtypes have been identified and each subtype appears to have an unique ligand binding profile suggesting a more specific role of these proteins [1]. Meanwhile, it is generally accepted that the interaction of odor molecules with the receptor protein leads to the activation of a so-called ►G_{olf}-protein as mediator to activate the enzyme adenylate cyclase which produces large amounts of cyclic adenosine monophosphate (cAMP) as second messenger. The cAMP molecules now act directly within the cell membrane to change the structure (conformation) of a channel protein (cyclic nucleotide gate channel, CNG) in its open state (Fig. 1b), enabling it to conduct specific cations (Na^+, Ca^{2+}) from the nasal mucosa into the cell [2]. As a result, the negative membrane potential (about −70 mV at rest) is shifted to more positive values, called depolarization or cell excitation. Above a certain threshold (−50 mV) this analog sensor potential is converted into a digital action potential frequency near the axon hill of the soma of the OSN. The action potentials are conducted along the neurites into the olfactory bulb. This signal transduction cascade provides amplification and integration of odorant binding events. One olfactory receptor protein activated by an odor molecule can produce about a thousand molecules of a second messenger (cAMP) per second. The calcium ions entering through the CNG channel have a double function. First, they are able to activate another ion channel that is permeable to the negatively charged chloride ion [2]. Because OSN maintain an unusual high intracellular chloride concentration such that there is a chloride efflux when these channels are activated. Thus, it further depolarizes the cells and adding to the excitatory response magnitude. However, calcium ions entering the CNG-channels are also important in response adaptation through a negative feedback pathway. Calcium acts probably via a ►calmodulin dependent mechanism to decrease the affinity of the channel for cAMP and therefore making the channel after a longer period of opening more and more insensitive. This is one of several mechanisms for adaptation. Others include phosphorylation of olfactory receptor proteins sending them into ►internalization, and of fast sodium channels leading to inhibit of action potentials.

The initial step in the recognition of an odorant is its binding to the olfactory receptor protein. The discovery of a large family of genes which encode heptahelical transmembrane proteins (Fig. 1c) and are expressed exclusively in the olfactory epithelium by Linda Buck and Richard Axel (1991) was the ground-breaking work which opened new avenues of research for better understanding of odorant recognition [3]. The odorant receptor proteins are classical G-protein coupled receptors and the about 320 amino acids are highly homologous and Southern blots of genomic libraries suggested that the gene family consists in mice of at least 1,300 putative members. In the human genome about 900 olfactory receptor genes were identified, but two third of these turned out to be non-functional or "pseudogenes" which have lost their function during evolution. A total of 347 putative functional olfactory receptor genes in man was determined [4]. It is still the largest gene family in the human genome. The high proportion of pseudogenes indicate a variable repertoire of functional olfactory receptor genes in the human population. Many specific anosmia, e.g., the inability to smell particular odors, could be due to hereditary defects of OR genes. Interestingly, out of the 347 functional OR genes, each olfactory sensory cell expresses only one type which implies a sophisticated mechanism of olfactory gene choice. The members of the olfactory receptor gene family are distributed on nearly every human chromosome except 20 and Y, often found in large clusters. Chromosome 11 is particularly notable in that it contains nearly half of all olfactory receptor genes including the two largest olfactory receptor gene clusters [4].

In 1998, six years after its identification, it could be shown by functional expression and characterization of olfactory receptor genes that they encode for odorant receptors [5]. One year later the first human olfactory receptor was deorphanized. The receptor hOR17–40 reacts specifically to Helional and structurally related substances [6]. The functionality of the protein was demonstrated by a recombinant expression of the receptor in HEK 293 cells and calcium imaging measurement to demonstrate the cell response after odor application. Unfortunately, it has not been possible to get a functional expression and activation of many of the human olfactory receptors so far. The ability of olfactory sensory neurons to express cloned receptors while other cells could not is further evidence for the involvement of some olfactory specific ►chaperone or cofactor necessary for functional receptor expression. So only a few other human olfactory receptors have been successfully expressed and characterized. Most data existing from the receptor hOR17–4 which is activated by odorants like Bourgeonal, Cyclamal and Lilial (smelling like Lilly of the Valley). A detailed molecular receptive field (Fig. 2a) could be described [7]. From these data it was suggested that the receptor

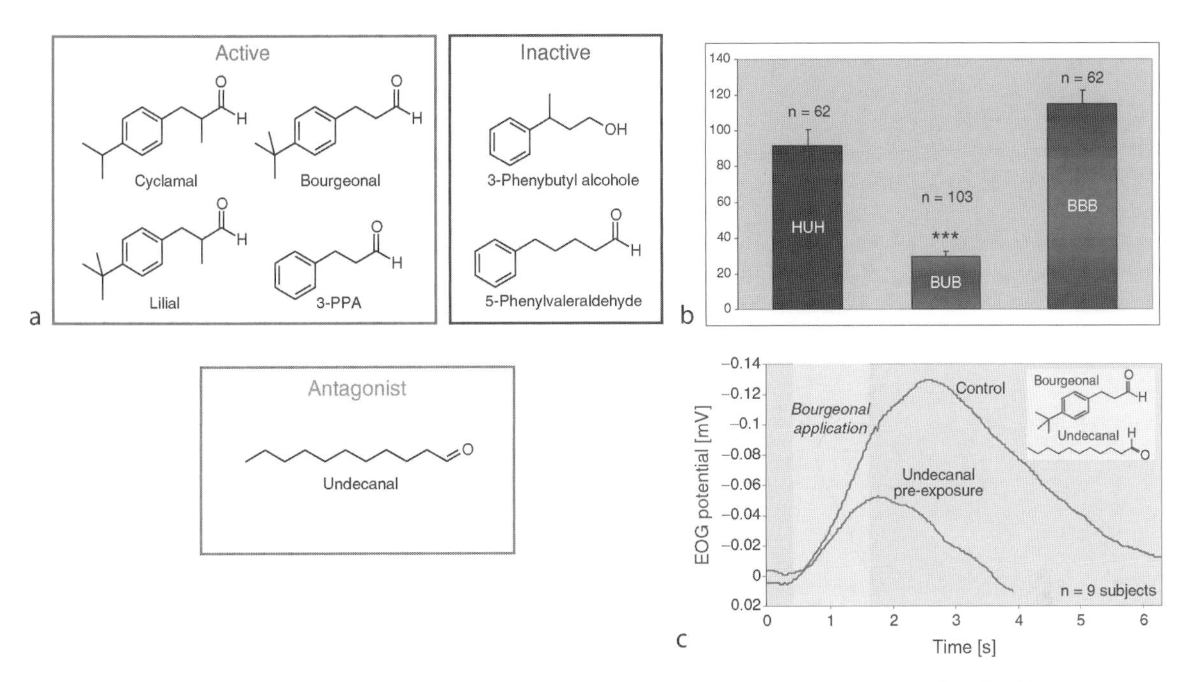

Olfactory Sense. Figure 2 (a) Effective versus ineffective agonists and antagonists towards hOR17–4. (b, c) In psychometric measurements and electro-olfactogram recordings Undecanal was identified as ►competitive antagonist for hOR17–4.

recognizes a particular feature of different ligands, in analogy to a ►pharmacophore in medical chemistry. In addition another analogy to pharmacology, the existence and the effectiveness of antagonists, could be shown. It was speculated for many years that it should be possible to construct antagonists for olfactory receptors in a similar way as in the case of the medically used blockers of adrenergic or dopaminergic receptors. Interestingly, under the many substances tested, Undecanal showed a clear competitive antagonistic effect highly specific for the receptor hOR17–4 [8]. Variations of agonist/antagonist concentrations ratios indicate competition of both compounds for the receptors ligand binding pocket (Fig. 2b, c). Most odor molecules are recognized by more than one receptor and most receptors recognize several odor molecules, related by chemical properties. Thus, the recognition of an odorant molecule depends on which receptors are activated and to what extent. For each odorant there are best receptors, but also others that are able to recognize the odorant only in a higher concentration and will participate in the discrimination of that compound. Thus, all data indicate that the nose uses a combinatory coding scheme to discriminate the waist number of different smells [4].

Odorant Information Process

To inform the brain, olfactory sensory neurons extend axons from the olfactory epithelium to the olfactory bulb. There is a considerable amount of data demonstrating that all neurons expressing the same receptor type convert their axons into the same glomerulus: usually two glomeruli which are located on the lateral and medial hemisphere of the bulb, respectively [4]. These findings indicate that an individual glomerulus is dedicated to receiving input from a single receptor type and so serves as a functional unit in the coding of olfactory information. The wiring process is still largely unknown. The basic olfactory map is probably established by a developmental hardwired strategy. The convergence of signals from thousands of neurons expressing the same olfactory receptor protein onto a few glomeruli by optimize the sensitivity to low concentrations of odorants by allowing the integration of weak signals from many olfactory epithelium neurons. The invariant pattern of inputs might have a different advantage, ensuring that the neuronal representation (code) from odorant remains constant over time, even though olfactory epithelium neurons are short lived cells that are continuously replaced. Many natural odors such as flowers, scents and perfumes consist of hundreds of individual chemical compounds. When such a complex mixture reaching our nasal cavity, out of the about 350 different types of olfactory sensory cells, only those are activated which bearing receptors for one of the chemicals in the mixture. Having in mind that all the sensory cells have the same receptor proteins, wherever they may be located in the olfactory epithelium (Fig. 3), all send their neuronal processes to one and the same glomerulus in the olfactory bulb, thus producing a

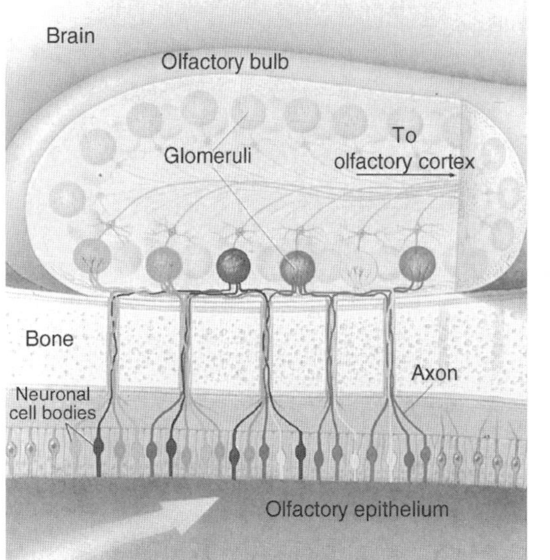

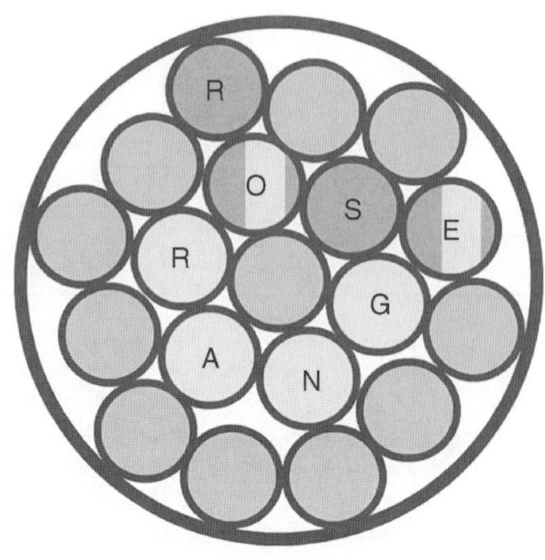

Olfactory Sense. Figure 3 (a) Olfactory receptor neurons expressing the same olfactory receptor protein project to a specific glomerulus in the olfactory bulb. (b) Schematic activation model of the glomeruli after stimulation with the scent of rose or orange.

constant activation pattern. For instance, when we smell the odor of a rose, the complex odorant mixture in a rose essential oil activates about hundred different receptor types and a similar number of glomeruli. The result is a reproducible, but complex pattern of glomerular activation, from which it is possible to interfere by reverse logic which odor mixture has been smelt [9]. The rose scent activation pattern is clearly distinct from e.g., an orange-scent pattern (Fig. 3). Although individual chemical components are present in both odor mixtures, the patterns in activated glomeruli can overlap but are clearly discriminable. In psychology, this representation by a particular shape could be described with the terms "Odor Gestalt" or "Gestalt Recognition." Once we have learned an odor, we can recognize it again, even though some of the information it normally contains may be missing. Many artificial rose or orange scents that are industrially produced take advantage of this knowledge.

Olfactory Receptors Outside the Nose

Recently it could be shown that olfactory receptors also exist and play an important functional role outside the olfactory epithelium: in human sperm cells. The latter possess olfactory receptor proteins as well as all the other members of the second messenger cascade, the G-protein, adenylate cyclase (Type III) and cyclic nucleotide gated channels [7,10]. Oversimplifying one could say that a sperm cell is nothing more than an olfactory neuron with a tail. Using molecular biological techniques (►Polymerase Chain Reaction (PCR)), biochemical methods (antibodies) and proteome analysis, it was clearly demonstrated that the receptor hOR17–4 is functionally expressed in human spermatozoa. By calcium imaging experiments it was shown that sperm cells indeed get activated by odorants like Bourgeonal or Cyclamal in a concentration dependent manner (Fig. 4). The threshold was in the micromolar range. Sperm react exactly to the same profile of active and inactive substances of the hOR17–4 as the recombinantly expressed receptor. Interestingly, the activation of hOR17–4 is completely inhibited by simultaneous presentation of the competitive inhibitor Undecanal [7]. These studies on the pharmacology of the sperm odorant receptor were then extended to the physiology of spermatozoa: Human sperm cells showed a concentration dependent positive chemotactic behavior to stimulating odorants (Bourgeonal, Cyclamal) and doubled their speed in presence of the odor. When the antagonist was applied, the effects of Bourgeonal on sperm navigation and swim speed were strongly inhibited. These data suggest that hOR17–4 signaling potentially governs chemical communication between sperm and egg cell. Additional studies made the important finding that this sperm receptor is in fact also expressed in human olfactory receptor neurons. Careful analysis of human tissue revealed bonafide expression of hOR17–4 in nasal epithelium [8]. The nose smells what sperm attracts. These data could potentially be used to manipulate fertilization with important consequences for contraception and procreation, but also to develop sniffing tests for identification of patients with fertilization problems based on functional olfactory receptors.

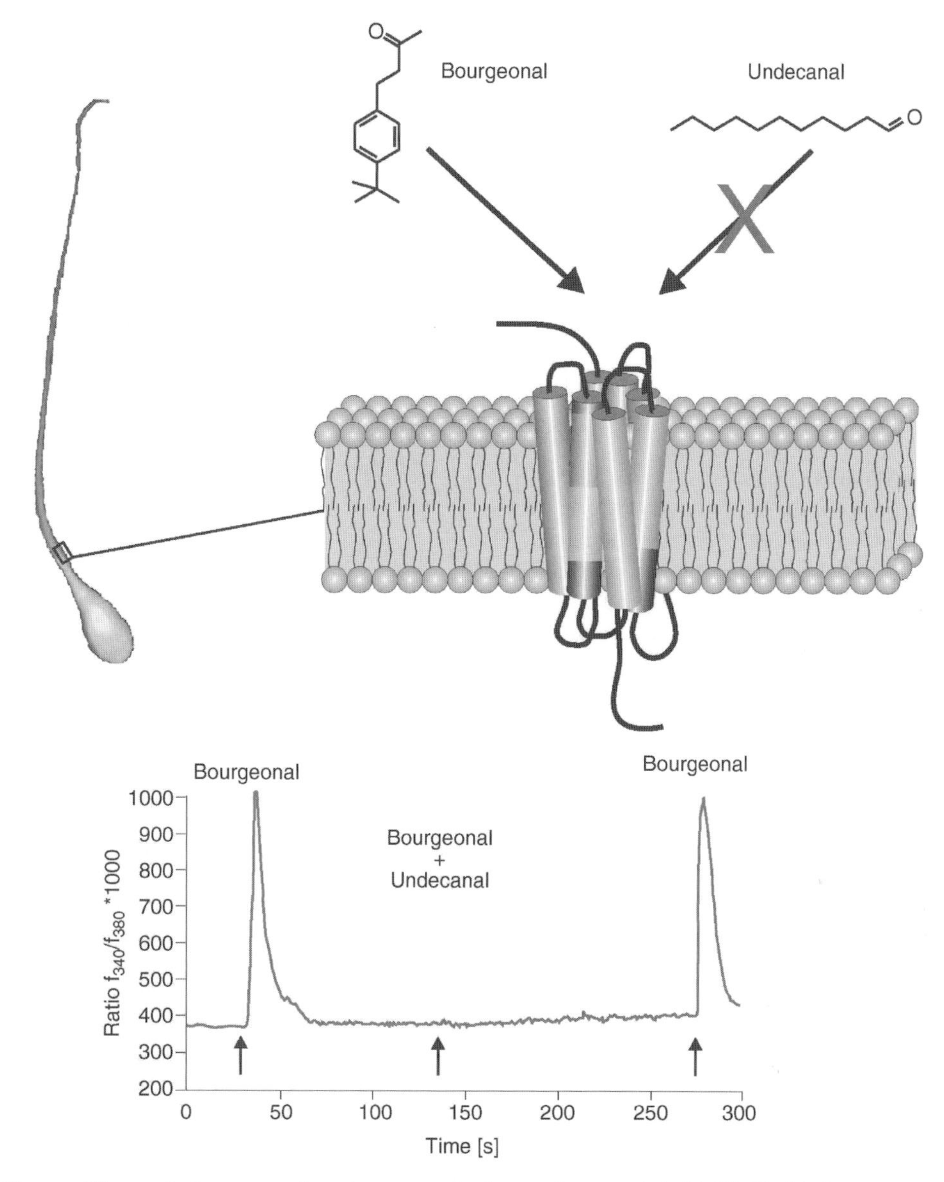

Olfactory Sense. Figure 4 Bourgeonal works as a potent receptor agonist of hOR17–4 in human spermatozoa, whereas Undecanal inhibits this effect.

References

1. Breer H (2003) Olfactory receptors: molecular basis for recognition and discrimination of odors. Anal Bioanal Chem 377:427–433
2. Frings S (2001) Chemoelectrical signal transduction in olfactory sensory neurons of air-breathing vertebrates. CMLS Cell Mol Life Sci 58:510–519
3. Buck L, Axel R (1991) A novel multigene family may encode odorant receptors: a molecular basis for odor recognition. Cell 65:175–187
4. Malnic B, Godfrey PA, Buck LB (2004) The human olfactory receptor gene family. Proc Natl Acad Sci USA 101:2584–2589
5. Firestein S (2001) How the olfactory system makes sense of scents. Nature 413:211–218
6. Wetzel ChH, Oles M, Wellerdieck Ch, Kuczkowiak M, Gisselmann G, Hatt H (1999) Specificity and sensitivity of a human olfactory receptor functionally expressed in human embryonic kidney 293 cells and *Xenopus laevis* oocytes. J Neurosci 19:7426–7433
7. Spehr M, Gisselmann G, Poplawski A, Riffell JA, Wetzel CH, Zimmer RK, Hatt H (2003) Identification of a testicular odorant receptor mediating human sperm chemotaxis. Science 299:2054–2058
8. Spehr M, Schwane K, Heilmann S, Gisselmann G, Hummel H, Hatt H (2004) Dual capacity of a human olfactory receptor. Curr Biol 14(19):832–833
9. Shepherd GM (2006) Smell images and the flavour system in the human brain. Nature 444:316–321
10. Weyand I, Godde M, Frings S, Weiner J, Müller F, Altenhofen W, Hatt H, Kaupp UB (1994) Cloning and functional expression of a cyclic-nucleotide-gated channel from mammalian sperm. Nature 368:859–863

Olfactory Sensitivity

►Olfactory Perception

Olfactory Sensory Neuron

Definition

►Olfactory Receptor Neuron

Olfactory Sulcus

Definition

The olfactory sulcus runs bilaterally along the orbital surface of the forebrain. It divides gyrus rectus from medial orbital gyrus. In the olfactory sulcus the olfactory peduncle runs from the olfactory bulb to the anterior perforated substance.

►Olfactory Bulb
►Olfactory Peduncle
►Olfactory Pathways

Olfactory System

Definition

Main chemosensory system in vertebrates. It is composed of an olfactory epithelium, located in the postero-dorsal nasal cavity, a main olfactory bulb and olfactory-recipient areas of the telencephalon. It is able to detect numerous odorants, mainly volatiles, present in the environment.

►Chemical Senses
►Evolution of Olfactory and Vomeronasal Systems
►Odorant
►Olfactory Bulb
►Olfactory Epithelium

Olfactory System Dynamics

►Olfactory Information

Olfactory Tract

Definition

Nerve fibers connecting the olfactory bulb to the olfactory cortex.

►Olfactory Bulb
►Olfactory Cortex
►Olfactory Pathways

Olfactory Transduction

Definition

Intracellular cascade of enzymes induced by the binding of odorants to odorant receptors. The interaction between an odorant and its cognate receptor induces a transduction pathway, involving the activation of specific Golf proteins, adenylate cyclase III, cyclic nucleotide-gated (CNG) and negatively charged chloride ion channels, providing amplification and integration of odor-binding events. This olfactory transduction ultimately transmits an electric signal to the central nervous system that results in a sensation of smell.

►Odorant
►Odorant Receptor

Olfactory Trigone

Definition

The olfactory trigone is a small portion of the olfactory peduncle. The olfactory peduncle runs from the olfactory bulb to the anterior perforated substance.

There, its diameter increases before it divides into three roots, or striae. This portion is termed olfactory trigone.

►Olfactory Bulb
►Olfactory Pathways
►Olfactory Peduncle

Olfactory Tubercle

Definition

From the olfactory trigone, the intermediate olfactory stria continues onto the anterior perforated substance. On top of the anterior perforated substance, there is a layer of gray matter, which is called the olfactory tubercle. In most mammals, the olfactory tubercle is a prominent bulge on the ventral surface of the frontal lobe situated caudally to the olfactory peduncle and medially to the lateral olfactory tract of mammals. It receives afferent input from the lateral olfactory tract. The olfactory tubercle differs from the piriform cortex in that it does not send output projections to the olfactory bulb or to any other secondary olfactory structure. The outputs of the olfactory tubercle are directed towards the thalamus, ventral pallidum, nucleus accumbens and, in monkeys, the orbitofrontal cortex. The inputs and projections to and from olfactory tubercle can vary substantially among species. The olfactory tubercle resembles the underlying corpus striatum and thus is often combined with the nucleus accumbens to the ventral striatum. In humans the olfactory tubercle is poorly developed resulting in a difficult visualization using functional imaging techniques.

►Olfactory Pathways
►Olfactory Tract
►Olfactory Trigone

Oligoclonal Bands (OCBs)

Definition

OCBs are distinct bands of IgG seen in electrophoretic analysis of CSF in MS patients. A few antibodyproducing plasma cell clones produce the IgG within the CNS. This pattern is not normally seen since most IgG in CSF is derived from serum and appears as diffuse broad bands in CSF as well as in serum. In MS, two or more bands must be seen in CSF and be absent in serum indicating intrathecal synthesis of IgG. Though approximately 90% of CDMS patients have OCBs, they may also be found in patients with other CNS inflammatory or infectious diseases.

►Multiple Sclerosis

Oligodendrocyte

Definition

Oligodendrocytes are a type of glial cell in the CNS. The cytoplasmic extensions of these cells form myelin, which wraps around large axons. One oligodendrocyte can myelinate up to 30 axons. Oligodendrocytes are found predominantly in the white matter of the CNS. Diseases of oligodendrocytes include demyelinating diseases such as multiple sclerosis, leukodystrophies and tumors named as oligodendrogliomas.

►Inhibitory Molecules in Regeneration
►Multiple Sclerosis
►Myelin
►Regeneration

Oligodendrocyte-Myelin Glycoprotein (OMgp)

Definition

►Regeneration

Olivary Pretectal Nucleus

Definition

The olivary pretectal nucleus (OPN) is a midbrain structure that is part of the circuit mediating the pupillary light reflex. It receives direct retinal input, including inputs from melanopsin expressing retinal ganglion cells. The firing rate of OPN neurons is directly related to the intensity of light stimulation on the retina and correspondingly to the degree of pupillary constriction.

►Neural Regulation of the Pupil
►Pupillary Light Reflex
►Retinal Ganglion Cells

Olive

Synonyms

►Oliva

Definition

- Inferior olive is the actual "olive" and is located directly beneath the pons, in the myelencephalon. This large nucleus plays a major role in movement coordination.
- Superior olive: nuclear conglomeration in the ►Mesencephalon, is a component of the auditory tract.

Olivocerebellothalamic Circuit

Definition

Neuronal circuit between the thalamus, the dentate nucleus of the cerebellum, and the inferior olivary nucleus.

►Essential Tremor

OMgp

Definition

OMgp stands for oligodendrocyte myelin glycoprotein. It is a glycosylphosphatidylinositol-anchored protein expressed mainly by oligodendrocytes in the central nervous system (CNS). It is found concentrated at nodes of Ranvier and plays a part in the control of myelination. Through its interaction with the Nogo receptor OMgp can cause growth cone collapse and inhibition of neurite outgrowth.

►Glial Scar
►Node of Ranvier
►Oligodendrocyte

Omnipause Neuron Area

Definition

A small region on the midline of the brainstem near the boundary of the pons and medulla. Neurons in this structure discharge at high tonic rates whenever an animal is fixating, but then turn off sharply and completely for saccades in all directions. These cells function as an inhibitory brake on other saccade-related cells in the saccadic system during fixation and help to prevent unwanted saccades from occurring.

►Omnipause Neuron
►Saccade, Saccadic Eye Movements

Omnipause Neurons

Chris R. S. Kaneko
Department of Physiology and Biophysics, Washington National Primate Research Center, University of Washington, Seattle, WA, USA

Synonyms

Pause neurons (pns); OPNs

Definition

Omnipause neurons (OPNs) are the neurons that control saccadic eye movements by inhibiting the activity of all burst neurons. Burst neurons, in turn, directly drive the saccadic burst in motoneurons that produces the saccade. These neurons are located in the medial pons between the rootlets of the abducens nerves as they leave the brainstem ([4], Fig. 1). They are normally tonically active and discharge at a constant high rate (up to 200 ►spikes/s in ►rhesus monkey) that is unrelated to eye position (rasters and histogram, Fig. 2 bottom two traces in each panel). They cease firing (pause) before and during all saccades (Fig. 2). Their

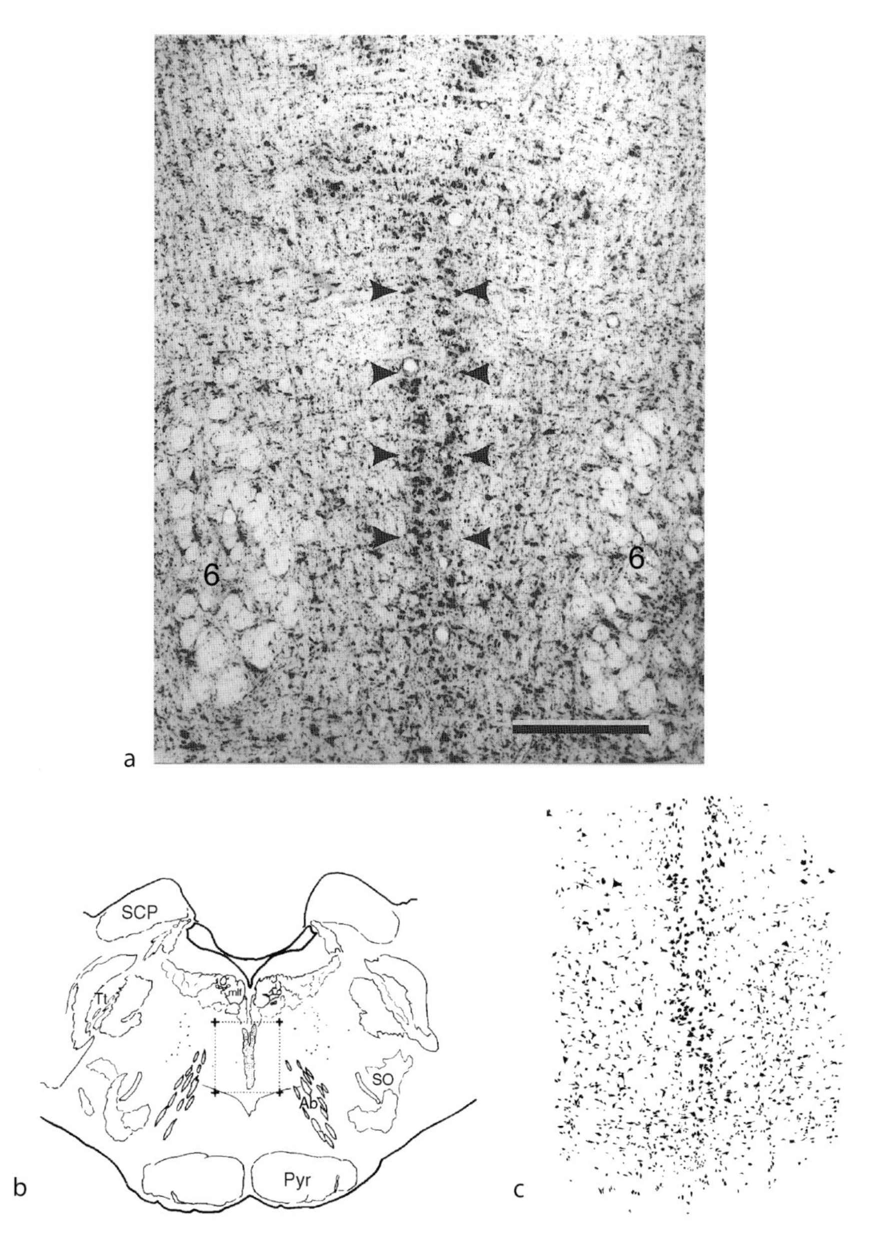

Omnipause Neurons. Figure 1 OPN Anatomy. (a) Photomicrograph of the nucleus raphe interpositus (rip; arrowheads). OPNs are co-extensive with the rip and form a bilaminate columnar nucleus that straddles the midline between the rootlets of the abducens nerve [1]. (b) Tracing of the frontal section for orientation. The photomicrograph in (a) was taken from this section (dotted box). (c) Drawing of every neuron in (a) to show the distinct appearance of the OPNs is not due to poorly stained neurons, that the OPNs are larger than neighboring neurons, and are further distinguished by their isolated position between the longitudinal fiber tracks. Abbreviations: *6*, abducens nerve rootlets; *pyr*, pyramidal track; *SCP*, superior cerebellar peduncle. Calibration is 1 mm in (a).

pause begins just before (~15 ms) the onset (Fig. 2, thin vertical lines) of the movement and a few ms before the burst in medium lead burst neurons (mlbns). The duration of the pause is highly linearly correlated with saccade duration (Fig. 2b, middle). Anatomical studies have shown that OPNs project directly to both horizontal and vertical burst neuron regions. Physiological studies confirm that OPNs monosynaptically inhibit burst neurons. Based on their discharge and their connections, there is no doubt that OPNs control saccades by gating the activity of burst neurons.

OPNs are perhaps the most studied of the neurons that comprise the saccadic ►burst generator, and saccades may be the best characterized motor system, so we know quite a bit about OPN anatomy and physiology. In 1972, Luschei and Fuchs [1] and Cohen

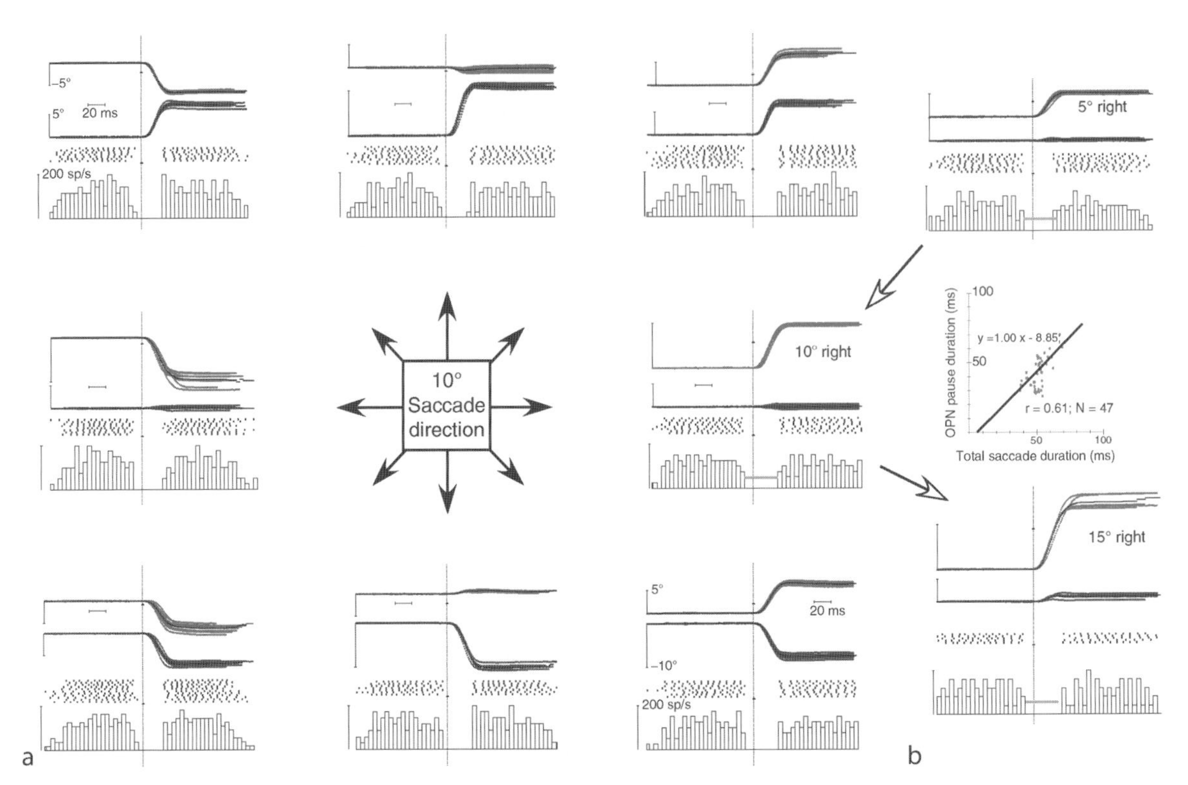

Omnipause Neurons. Figure 2 Discharge of OPNs. (a) In each panel, the traces are (top to bottom) horizontal eye position (red); vertical eye position (blue); rasters; histogram variability at ends is due to variable duration of each trace. 7–11 saccades in the direction indicated by the center plot are overlayed and the average (black line) shows the consistency of the movements. Note the pause lead. (b) Pause duration for different size saccades. 5° (upper) and 15° (lower) saccades show comparison of pause duration with 10° rightward saccades (indicated by green horizontal bar). Note the bar (redrawn) overlaps the pause for 5° and is shorter than that for 15° saccades. Traces as in (a). Middle, scatter plot of pause duration as a function of saccade duration for rightward saccades. Linear regression is least-squares fit to plot showing slope of one, i.e. pause duration equals saccade duration.

and Henn [2] reported recording eye movement related neurons in the pontine and medullary ►reticular formation of alert monkeys. One class, the OPNs, discharged at a high tonic rate (Fig. 2, rasters and histograms) but ceased firing in association with saccades or ►quick phases in any direction (Fig. 2a). Shortly thereafter, Keller [3] showed that electrical microstimulation of OPNs prohibited saccades and quick phases of ►nystagmus. This result immediately suggested their function was to control saccades by discharging at a high tonic rate in order to tonically inhibit burst neurons. These early results led to Robinson's model of saccadic control [4] that posited the role of OPNs was to prevent the discharge of the high gain, burst neurons that might otherwise cause instabilities in the system and thus, unwanted eye movements. He further suggested that saccades were initiated by a trigger signal of unknown origin mediated by an inhibitory interneuron and originating from more central structures like the superior colliculus. A final element was that the OPNs were modeled as being actively inhibited during the saccade to prevent unwanted interruptions of the saccade by means of a latch circuit comprised of burst neuron feedback to OPNs via another inhibitory interneuron.

While the basic circuit has been confirmed thoroughly in both cats and monkeys, some of the other details of the Robinson model [4] have garnered only rudimentary support. Anatomical tracing studies (e.g. [5]) showed that OPNs projected to each of the areas that contained saccadic burst neurons (see ►HMLBs (horizontal medium lead burst neurons), ►PPLLBs (ponto-pontine LLBs), ►PCbLLBs (precerebellar LLBs), and ►RSLLBs (reticulospinal LLBs)). Later, intracellular staining and modern tracing studies using transneuronal labeling have unequivocally demonstrated the projection from OPNs to burst neurons. Electrophysiological studies in cats have shown that this monosynaptic connection is inhibitory in all cases. Recent immuno-labeling suggests that OPNs use glycine to inhibit burst neurons. Recordings from alert cats and monkeys and anatomical studies of their afferents has shown that the high rate of tonic discharge is probably due to a multiplicity of afferent input from all sensory modalities

[6]. This surmise is corroborated by the fact that OPNs are silent when animals go to sleep, and that a burst of OPN activity can be recorded if an afferent volley is synchronized by, for example, a click of sound. On the other hand, a neural basis for the trigger and the latch is yet to be established. Intracellular recording from identified cat OPNs has shown that they are inhibited during saccades. The inhibitory ►postsynaptic potentials (ipsps) are characterized by an initial abrupt hyperpolarization that decays back to resting, with a time course that is well correlated with saccadic eye velocity, consistent with them receiving both trigger and latch inputs. The ipsp decay is expected if it is caused by burst neuron input whose discharge is also highly correlated with eye velocity. Electrical microstimulation amongst long-lead burst neurons (LLBs) suggests some of them may be appropriate inhibitory interneurons to provide some of that input, but the juxtaposition of these elements has made it difficult technically to affect each element independently and thereby produce more substantive proof.

OPNs receive their major saccadic input from the contralateral superior colliculus. The input appears to be heavier from the caudal than the rostral portions of the colliculus and more concentrated from the lateral portions than the medial. The input is both monosynaptic (excitatory) and disynaptic (inhibitory), and it is assumed that the inhibitory input is relayed via an inhibitory interneuron and acts as a trigger for saccade generation. As mentioned, their high tonic rate is maintained by multiple afferent sources that use gamma-aminobutyric acid, glycine and glutamate but not monoamines as transmitters [7].

OPNs have been identified in man by immunohistochemistry and damage to OPNs has been invoked to explain a variety of eye movement pathologies, like square wave jerks, that result in oscillopsia. However, either transient or permanent inactivation of OPNs in monkeys leads to slower saccades (longer durations and lower peak velocities), possibly due to inactivation of ►post-inhibitory rebound in the EBNs that they innervate.

Characteristics

Higher Order Structures

There are three higher order structures that influence OPNs directly. OPNs receive input from the contralateral superior colliculus. Whether they also receive an ipsilateral input remains controversial. There also may be inputs from the frontal eye fields. One that projects directly to the pons, but this is still uncertain, and another that is indirect via the superior colliculus. Based on anatomical evidence, OPNs may also receive direct input from the caudal fastigial nucleus of the cerebellum that is presumably excitatory. The fastigial input may play a role in adaptive plasticity of saccade amplitude and/or saccadic error correction during on-going saccades by allowing fastigial output to terminate the saccade.

Parts of This Structure

OPNs have been studied extensively in cat and monkey and there are a number of differences between the species. The somata of the majority of OPNs (Fig. 1) are located in the nucleus raphe interpositus (rip) [8] in the monkey and in the nucleus raphe pontis in the cat [5]. In both species, occasional OPNs can be found in the surrounding reticular formation; specifically the caudal nucleus reticularis tegmenti pontis in the monkey [5] and the superior central nucleus in the cat. In the monkey, it appears that virtually all neurons in the rip are OPNs [5]. They are medium-sized (~35 μm diameter), multipolar neurons in monkey (Fig. 1a). In cat, their shape ranges from spindle shaped to spheroid and they are slightly larger (~46 μm, [9]). In monkey, OPNs send long horizontal dendrites in both directions, and the contralateral branches extend across the midline and into the longitudinal fiber tracts that traverse this portion of the pons in the ventral portion of, and below the medial longitudinal fasciculus. In contrast, cat OPN dendritic fields are ellipsoidal and only a minority have dendrites that cross the midline. Axons arise from the soma and bifurcate either ipsilaterally, or more usually, contralaterally after crossing the midline. In the cat, the stem axons are about 4 μm in diameter and the branch axons are about 3 μm in diameter. In the cases, from cat, where axons could be traced to terminal boutons, all were found in burst neuron regions and were either en passage or terminaux endings. The former were 2.6 μm in diameter and the latter 2.8 μm. Detailed intracellular fills are not available for monkey OPNs.

Function of This Structure

OPNs provide tonic inhibition to the saccadic burst neurons to prohibit saccades except when they are silenced. In addition, clinical and inactivation evidence suggests that the inhibition, when interrupted, contributes to activation of a post-inhibitory rebound in burst neurons that potentiates the very high-frequency discharge of burst neurons. Besides this permissive role in saccades, the OPNs also serve to coordinate various types of eye movements. The horizontal and vertical components of oblique saccades are mediated via separate horizontal and vertical burst neuron groups and are coordinated via OPN disinhibition. Thus, the OPNs serve to cross couple the burst neurons and control oblique saccade duration. This function is featured prominently in models of saccade generation that include both horizontal and vertical burst generators. The coordinating function seems to extend to other types of eye movements because OPNs are silenced during combined eye and head movements, combined ►vergence and ►version movements, as well as during blinks. They may also have a role in slow pursuit eye movements, but the exact nature of that role is uncertain. Thus, OPNs seem to assist in the timing of coordinated eye movements in general.

Higher Order Function

OPNS are low-level premotor neurons with no higher order (e.g., cognitive) functions yet indicated. The function of the potential direct, cortical inputs is not clear, but all of the OPN inputs seem to share at least a portion of the responsibility for triggering saccades. Although still somewhat controversial, there don't appear to be any OPNs that are specialized either for head or coordinated eye and head movements, even though some LLBs are so specialized. As mentioned, their connectivity mediates the co-ordination of oblique saccades (►Hering's Law of equal innervation), and the push-pull organization of EBNs and IBNs results in relaxation of antagonist during agonist activation (►Sherrington's Law of reciprocal innervation). There is also emerging evidence that OPNs may play a role in the coordination of smooth pursuit and saccadic eye movements in both cats and monkeys, but the nature of that role has not yet been elucidated.

Quantitative Measure for This Structure

Just as for other elements of the saccadic burst generator, the number of OPNs is not clear because of technical limitations in marking all of them so that they may be counted. Perhaps transneuronal retrograde labeling techniques will allow an estimate in the near future. Likewise, virtually nothing is known about the unitary ipsps OPN output to burst neurons or the membrane biophysics of OPNs.

References

1. Luschei ES, Fuchs AF (1972) Activity of brain stem neurons during eye movements of alert monkeys. J Neurophysiol 35:445–461
2. Cohen B, Henn V (1972) Unit activity in the pontine reticular formation associated with eye movements. Brain Res 46:403–410
3. Keller EL (1974) Participation of medial pontine reticular formation in eye movement generation in monkey. J Neurophysiol 37:316–332
4. Robinson DA (1975) Oculomotor control signals. In: Lennerstrand G, Bach-y-Rita P (eds) Basic mechanisms of ocular motility and their clinical implication. Pergamon, Oxford, pp 337–374
5. Langer TP, Kaneko CRS (1990) Brainstem afferents to the oculomotor omnipause neurons in monkey. J Comp Neurol 295:413–427
6. Evinger C, Kaneko CRS, Fuchs AF (1982) Activity of omnipause neurons in alert cats during saccadic eye movements and visual stimuli. J Neurophysiol 47:827–844
7. Horn AKE, Büttner-Ennever JA, Wahle P, Reichenberger I (1994) Neurotransmitter profile of saccadic omnipause neurons in nucleus raphe interpositus. J Neurosci 14:2032–2046
8. Büttner-Ennever JA, Cohen B, Pause M, Fries W (1988) Raphe nucleus of the pons containing omnipause neurons of the oculomotor system in the monkey and its homologue in man. J Comp Neurol 267:307–332
9. Ohgaki T, Curthoys IS, Markham CH (1987) Anatomy of physiologically identified eye-movement-related pause neurons in the cat: Pontomedullary region J Comp Neurol 266:56–72

On Center Cells

Definition

►Visual Cortical and Subcortical Receptive Fields

Ongoing Neurogenesis

►Adult Neurogenesis

Oniric Mentation

►Dreaming

Ontogenetic

Definition

Pertaining to the biological development of an individual.

Ontological Status

Definition

Something's ontological status can be determined by answering the question whether it exists. Bill Clinton and Sherlock Holmes, although both human beings, thus currently differ in ontological status.

►Logical

Ontology

Definition

Ontology is the study of being or of what there is. Typically, ontologies of philosophers might comprise concrete objects like chairs or electrons, abstract objects like numbers or ►propositions, properties like the property of being a chair, facts like the fact that Paris is west of Warsaw, or events like the 2004 World Series.

►Epiphenomenalism

Opacity

Definition

Primarily a feature of certain sentences, e.g., of many ascriptions of propositional attitudes. The truth of such ascriptions does not systematically depend on the truth or falsity of the proposition involved. Consider the following two belief-ascriptions: "Mary believes that 1 + 1 = 2" and "Mary believes that 2756 + 488 = 3244." Even though both propositions ("1 + 1 = 2" and "2756 + 488 = 3244") are true, the two beliefascriptions can differ in truth-value. Whether it is true that Mary believes that 2756 + 488 = 3244 therefore does not systematically depend on the truth of "2756 + 488 = 3244."

►Representation (Mental)

Open Loop Behavior

Definition

Behavior that is executed without feedback control. This may, in nature, be due to completing a task before feedback is possible.

Open Reading Frame

Definition

The region of the gene between the start and stop codon that encodes for the protein.

Operant

Definition

Control by the consequences, i.e. by positive or negative reinforcement (=punishment) that is the result of a particular behavior and that shapes the future expression of that behavior.

Operant Conditioning

Björn Brembs
Freie Universität Berlin Fachbereich Biologie, Chemie, Pharmazie, Institut für Biologie – Neurobiologie, Berlin, Germany

Synonyms

Instrumental conditioning

Definition

Operant conditioning describes a class of experiments in which an animal (including humans) learns about the consequences of its behavior and uses this knowledge to control its environment.

Characteristics

Our life consists of a series of experiences in which we learn about our environment and how to handle it. Learning about the environment ("the plate is hot") and learning the skills to control it ("riding a bike") have been experimentally conceptualized as classical and operant conditioning, respectively. The two are so intertwined that a treatment of operant conditioning is impossible without reference to classical conditioning.

Operant Conditioning

Operant (instrumental) conditioning [1] is the process by which we learn about the consequences of our actions, e.g., not to touch a hot plate. The most famous operant conditioning experiment involves the "Skinner-Box" in which the psychologist B.F. Skinner trained rats to press a lever for a food reward. The animals were placed in the box and after some exploring would also press the lever, which would lead to food pellets being dispensed into the box. The animals quickly learned that they could control food delivery by pressing the lever. However, operant conditioning is not as simple as it first seems. For instance, when we touch a hot plate (or the rat the lever), we learn more about the hot plate than about our touch: we avoid contact of any body part with the plate, not only the hand that initially touched it. Obviously, we learned that the hot plate burns us. It is not only confusing that this type of environmental learning is usually called classical conditioning, we cannot even be sure that it is the only process taking place during conditioning.

Classical Conditioning

Classical (Pavlovian) conditioning [2] is the process by which we learn the relationship between events in our environment, e.g., that lightning always precedes thunder. The most famous classical conditioning experiment involves "Pavlov's dog": The physiologist I.P. Pavlov trained dogs to salivate in anticipation of food by repeatedly ringing a bell (conditioned stimulus, CS) before giving the animals food (unconditioned stimulus, US). Dogs naturally salivate to food. After a number of such presentations, the animals would salivate to the tone alone, indicating that they were expecting the food. The dog learns that the bell means food much as we learn that the plate is hot in the operant example above. Therefore, it is legitimate to ask if operant conditioning is in essence a classical process. Both operant and classical conditioning serve to be able to predict the occurrence of important events (such as food or danger). However, one of a number of important differences in particular suggests that completely different brain functions underlie the two processes. In classical conditioning, external stimuli control the behavior by triggering certain responses. In operant conditioning, the behavior controls the external events.

The Relationship Between Operant and Classical Conditioning

Ever since operant and classical conditioning were distinguished in 1928, their relationship has been under intense debate. The discussion has shifted among singular stimulus-response concepts, multiprocess views, and a variety of unified theories. Today, modern neuroscience distinguishes between procedural memories (skills and habits) and declarative memories (facts or events). The intensity and duration of the debate can in part be explained by the fact that most learning situations comprise operant and classical components to some extent: one or more initially neutral stimuli (CS), the animal's behavior (BH), and the ►reinforcer (US). The example above of learning to avoid touching a hot plate is very instructive. Extending the hand (BH) toward the round hotplate (CS) leads to the painful burn (US). In principle, our brain may store the situation as memory of the pain associated both with the hotplate (classical conditioning, CS-US) and with the extension of the hand (operant conditioning, BH-US) to predict the consequences of touching the plate at future encounters.

Habit Formation

A phenomenon called habit formation [3] confirms the tight interaction between operant and classical components in operant conditioning. In the early stages of an operant conditioning experiment (e.g., a rat pressing a lever for food in a Skinner box), the animal performs the lever presses spontaneously with the aim of obtaining the food (goal-directed actions). This can be shown by feeding the animals to satiety after training: they now press the lever less often when they are placed back in the box, because they are not hungry anymore. However, the same treatment fails to reduce lever pressing after the animals have been trained for an extended period. The behavior has now become habitual or compulsive; whenever the animals are placed now in the box, they frantically press the lever even if they are not hungry (or even if the food will make them sick). Although in the early stage of operant conditioning the behavior controls the environment (lever pressing to obtain food), habit formation effectively reverses the situation such that now the environment (box, lever) controls the behavior (lever pressing). One could say that overtraining an operant situation leads to a situation very similar to a classical one. Thus, operant conditioning consists not only of two components (operant and classical) but also of two phases (goal-directed and habitual behavior), with the relationship of the components changing with the progression from one phase to the next. Despite many decades of research filling bookshelves with psychological literature, our neurobiological understanding of the mechanisms underlying these processes is rather vague. What little is known comes from a number of different vertebrate and invertebrate model systems on various levels of operant conditioning. This essay is an attempt to integrate the neuroscience gained from many such disparate sources.

Neuroscientific Principles in Operant Conditioning

If there is a consensus for a critical early-stage process in operant conditioning, it is that of reafference. To detect the consequences of behavior, the brain has to compare its behavioral output with the incoming sensory stream and search for coincidences. The neurobiological concept behind this process is that of corollary discharges (or efference copies). These efference copies are "copies" of the motor command sent to sensory processing stages for comparison. Thus, neurobiologically, any convergence site of operant behavior and the US is very interesting with regard to potential plasticity mechanisms in operant conditioning. The efference copies serve to distinguish incoming sensory signals into self-caused (reafferent) and other, ex-afferent signals [4]. Modern theories of operant conditioning incorporate and expand this reafference principle into two modules: one is concerned with generating variable behavior and another predicts and evaluates the consequences of this behavior and feeds back onto the initiation stage [5]. Some evidence exists that the circuits mediating these functions are contained within the dorsal and ventral striatum of the vertebrate brain. We have only very poor mechanistic knowledge about the first module. Behavioral variability could be generated actively by dedicated circuits in the brain or simply arise as a by-product of accumulated errors in an imperfectly wired brain (neural noise). Despite recent evidence supporting the neural control of behavioral variability, the question remains controversial. Only little more is known about the neurobiology of the second module. Promising potential mechanisms have been reported recently from humans, rats, crickets, and the marine snail *Aplysia*. These studies describe conceptually similar neural pathways for reafferent evaluation of behavioral output (via efference copies) and potential cellular mechanisms for the storage of the results of such evaluations at the convergence site of operant behavior and US. However, to this date, a general unifying principle such as that of synaptic plasticity in classical conditioning is still lacking.

From a larger perspective, there is evidence suggesting that the traditional distinction of entire learning experiments into either operant or classical conditioning needs to be reconsidered. Rather, it appears that an experimental separation of classical and operant components is essential for the study of associative learning. As outlined above, most associative learning situations comprise components of both behavioral (operant) and sensory (classical) predictors. Vertebrate research had already shown that operant and classical processes are probably mediated by different brain areas. Research primarily from the fruit fly *Drosophila* and *Aplysia* has succeeded in eliminating much if not all of the classical components in "pure" operant conditioning experiments, a feat which has so far proven difficult to accomplish in any modern vertebrate preparation. This type of operant conditioning appears more akin to habit formation and lacks an extended goal-directed phase. These paradigms successfully reduce the complexity of operant conditioning by isolating its components and as such are vital for the progress in this research area. The new invertebrate studies revealed that pure operant conditioning differs from classical conditioning not only on the neural, but also on the molecular level. Apparently, the acquisition of skills and habits, such as writing, driving a car, tying laces, or our going to bed rituals is not only processed by different brain structures than our explicit memories, but also the neurons use different biochemical processes to store these memories.

The realization that most learning situations consist of separable skill-learning and fact-learning components opens the possibility to observe the interactions between them during operant conditioning. For instance, the early, goal-directed phase is dominated by fact learning, which is facilitated by allowing a behavior to control the stimuli about which the animal learns. Skill learning in this phase is suppressed by the fact-learning mechanism. This insight supports early hypotheses about dominant classical components in operant conditioning [6], but only for the early, goal-directed phase. If training is extended, this suppression can be overcome and a habit can be formed. Organizing these processes in such a hierarchical way safeguards the organism against premature stereotypization of its behavioral repertoire and allows such behavioral stereotypes only if they provide a significant advantage. These results have drastic implications for all learning experiments: as soon as the behavior of the experimental subject has an effect on its subsequent stimulus situation, different processes seem to be at work than in experiments where the animal's behavior has no such consequences, even if the subject in both cases is required to learn only about external stimuli. Conversely, apparently similar procedural tasks that differ only in the degree of predictive stimuli present may actually rely on completely different molecular pathways. The hierarchical organization of classical and operant processes also explains why we sometimes have to train so hard to master certain skills and why it sometimes helps to shut out dominant visual stimuli by closing our eyes when we learn them.

References

1. Skinner BF (1938) The behavior of organisms. Appleton, New York
2. Pavlov IP (1927) Conditioned reflexes. Oxford University Press, Oxford
3. Yin HH, Knowlton BJ (2006) The role of the basal ganglia in habit formation. Nat Rev Neurosci 7:464–476
4. von Holst E, Mittelstaedt H (1950) Das Reafferenzprinzip. Wechselwirkungen zwischen Zentralnervensystem und Peripherie. Naturwissenschaften 37:464–476
5. Dayan P, Balleine BW (2002) Reward, motivation, and reinforcement learning. Neuron 36:285–298
6. Rescorla RA (1987) A Pavlovian analysis of goal-directed behavior. Am Psychol 42:119–129

Operant Conditioning

Definition

A Definition of operant conditioning, also called instrumental conditioning, requires a distinction between elicited and emitted behavior. Elicited behavior is a response that is associated with a biologically relevant stimulus. Pavlovian or classical conditioning is an example of elicited behavior since there is always a formal, temporal relationship between the conditional stimulus (for example, a bell) and the unconditional stimulus (for example, meat powder to the tongue which elicits salivation). After a number of pairings, the conditional signal is seen to elicit a response that is similar to that elicited by the unconditional stimulus. Emitted behavior is behavior, which is produced by the subject in order to obtain a desirable outcome (commonly called a reinforcer): such behavior is said to operate upon the environment to produce reinforcement. In typical studies of operant conditioning, the availability of the reinforcer is signaled by a cue of some sort. Thus, the relationship between elicited and emitted behavior is complex. However, any discussion of this issue goes well beyond the subject matter of this essay.

Operational Closure

Definition

Operational (or organizational) closure means that certain relations and processes define a system as a unity, in determining the dynamics of interaction and transformations that the system may undergo as such a unity (Maturana/Varela). Operationally closed systems are not causally closed, i.e. they may interact causally with the environment.

Operculum

Definition

Part of the posterior portion of the inferior frontal gyrus of the frontal lobe in the brain.

Ophiid (Type)

Definition

"Snake-like," "snake-type."

►Evolution of the Brain: At the Reptile-Bird Transition

Opioid

Definition

Any compound or substance that binds to the opioid receptor resulting in the activation of the receptor.

►Analgesia

Opioid Peptides

Definition

Opioid peptides are short sequences of amino acids which mimic the effect of opiates in the brain. Endogenous opioid peptides are derived from three gene families, β-endorphins, enkephalins and dynorphins. Three types of opioid receptors, μ, δ and κ receptors, are pharmacologically identified.

Opisthotonus

Definition

Arched back produced by tonic contractions of the back muscles, for example in ►tetanus.

►Tetanus (Pathological)

OPN4

►Melanopsin

OPNs

►Omnipause Neurons

Opsin Evolution

►Evolution of Eyes

Opsonin

Definition

A terminology derived from the Greek and meaning, sauce or seasoning, in other words making the target cells such as pathogen more palatable to the phagocyte and more easily eaten. For example, C3b is an opsonin bound to target cells following complement activation and promoting phagocytosis by macrophages expressing C3 receptors.

►Neurodegeneration and Neuroprotection – Innate Immune Response

Optic Ataxia

Definition

Specific impairment of the visual control of limb movements observed in patients with lesion of the posterior parietal cortex. This deficit is expressed as errors both in final limb position in reaching/pointing tasks and in the shaping of hand aperture in grasping tasks. These deficits are exacerbated when the movements are programmed and executed under peripheral vision by asking the patient to keep gaze on a fixation point. Pure forms of optic ataxia, without sensory or motor deficits, indicate a role of the posterior parietal cortex in visuo-motor transformations for limb movement control.

►Eye-Hand Coordination
►Visual Neurosychology
►Visual Space Representation for Reaching

Optic Axis

Definition

Where we look, i.e., roughly coincidental with the line of sight.

Optic Chiasm

Definition

The optic chiasm is a landmark between the optic nerve and optic tract in the pathway between the retina and lateral geniculate nucleus of the thalamus. It contains the crossing of fibers of the so-called optic nerve to form its continuation, the optic tract of the opposite side. The fibers arise from ganglion cells in the retina. The crossing fibers in the optic chiasm contain information from the temporal visual fields (retinal nasal fields) of both eyes. Uncrossed fibers in the optic chiasm contain information from the nasal visual fields (temporal retinal fields) of both eyes. The chiasm is located on the ventral surface of the brain at the level of the anterior hypothalamus.

O

Optic Flow

Markus Lappe
Psychologisches Institut II, Westf. Wilhelms-Universität, Fliednerstrasse, Münster, Germany

Synonyms

Optical flow; (optic) Flow field; Retinal flow

Definition

Optic flow is the pattern of motion induced on the retina of a moving observer.

Characteristics

Mathematical Properties

Optic flow arises from the movement of an observer through a static visual scene. The movement of the observer creates relative movement between the visual objects in the scene and the eye of the observer. The projection of the relative movement of the scene objects onto the ▶visual field of the observer creates ▶visual motion. The collection of all the visual motions from throughout the visual field forms the optic flow. Since the motion in the visual field is first sensed by its projection on the retina, retinal flow is the collection of all image motion on the retina that arises from observer movement.

The retinal projection of the relative movement of a point in the scene can be described as a motion *vector*, i.e., by noting the motion direction and speed on the retina. The direction depends on the particular self-motion that the observer performs. When the observer moves to the left, all image motion is directed to the right. When the observer moves straight forward, all image motion is directed radially away from a point in the movement direction of the observer. This point is known as the ▶focus of expansion. The speed of a particular motion vector in the optic flow depends on the distance of the point from the eye of the observer. Points near to the observer move faster in the retinal projection than points further away. The difference in the speeds of two points in the same visual direction but in different distances from the observer is known as ▶motion parallax.

Optic flow not only arises from linear translations of the observer, such as sideward or forward movement, but also from rotations. Such rotations can occur either from moving along a curve or from eye movements of the observer. For example, when the observer performs an eye movement from right to left then rightward visual motion is induced on the retina. However, unlike in the case of leftward linear translation, the speeds of the motion vectors induced by eye rotation do not depend on the distance of the respective scene points from the observer. All points move with the same speed which is exactly opposite to the speed of the eye movement.

Thus, a single optic flow vector θ of a point R in the scene is mathematically a function of the translation T and rotation Ω of the eye of the observer and the distance Z of the point from the eye: $\theta = f(T, \Omega, Z)$. The precise equation is derived from perspective geometry [1]. Important for many aspect of flow analysis is the fact that in this equation the observer speed T and the depth Z are coupled such that the flow depends only on the quotient T/Z, not on Z directly.

The simplest optic flow is that of the radial outward movement obtained from linear forward movement. However, this is only a special case and the combination of translation, rotation, and scene distances can give rise to very different optic flow patterns. Since observer movement naturally triggers gaze stabilization reflexes such as the ▶vestibulo-ocular reflex or the ▶optokinetic reflex the optic flow observed under natural conditions will often result from a combination of translation and eye rotation.

Figure 1 shows a few characteristic examples. The observer moves across a ground plane. Heading is marked by a cross, gaze direction by a circle. Panels c to f show cases where the same heading is combined with different gaze directions. These gaze directions are shown in panel b. Panel c shows the retinal flow when gaze coincides with heading, i.e., when the observer looks straight into the direction of movement. In this case a focus of expansion is centered on the retina. In panel d, the observer looks off to the side. Gazing at some fixed point on the horizon allows him to keep his eyes stationary, i.e., no eye movements occur. A focus of expansion identifies heading, but now it is displaced from the center of the visual field. In panel e, the observer's gaze is directed at some element of the ground located in front of him and to the right. Because gaze is directed downward the horizon is in the upper visual field. Moreover, since the observer now looks at a point that is moving relative to himself, an eye movement is induced to stabilize gaze on this point. The resulting retinal flow field, a combination of translational and rotational flow, resembles a distorted spiraling motion around the fovea. There is no focus of expansion in the direction of heading (+). In panel f the observer looks at the same point as in panel d, but now he tracks an object that moves leftward along the horizon (for instance a car). This leftward pursuit induces rightward retinal image motion. The combination with the forward movement results in a motion pattern that resembles a curved movement and does not contain a focus of expansion.

Behavioral Aspects

From its conception by Gibson in the 1950s [3] optic flow has been assumed to play a role in the control of self-motion. Since then, experimental studies have shown that optic flow is involved in many behavioral tasks:

Control of Stance. Direction and speed of the optic flow are used as feedback signals for postural stability. When standing observers are exposed to a large flow field that periodically expands and contracts they sway in phase with the flow field [4]. The coupling between optic flow and posture maintenance is particularly strong in children and decreases in strength with age as the influence of ▶vestibular and somatosensory contributions to postural stability increases.

Control of Speed. Walking observers use the speed of the optic flow as a control signal for walking speed. Normally, a particular forward movement leads to a particular optic flow speed. If the flow speed is artificially increased, as has been done for observers walking on a treadmill in front of a projection screen on which a flow pattern was presented, walking speed increased proportionally [5]. Similar effects are seen for bicycling and car driving. When a mismatch between flow speed and walking speed is maintained for a

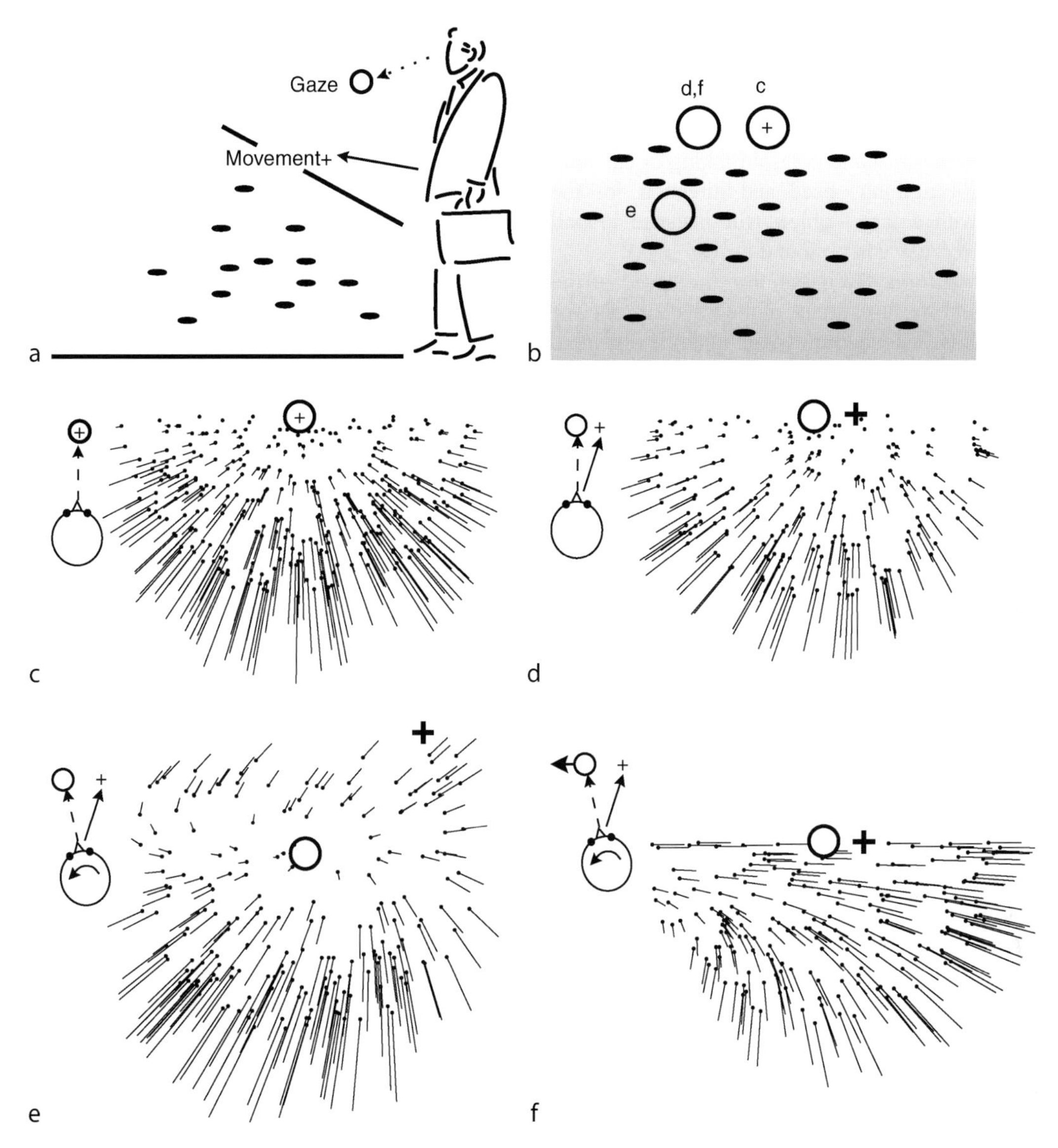

Optic Flow. Figure 1 Examples of optic flow fields induced by combinations of forward movement and eye movement. Taken with modifications from [2]. See text for detailed explanation. (a) Observer moves towards the cross while lookins at the circle. (b) different directions of gaze used in panels c to f. (c) Optic flow for straight translation int the direction of gaze. (d) Optic flow when directon of motion differs from direction of gaze. (e) Optic flow when direction of motion differs from direction of gaze and gaze stabilizing eye movement reflexes are taken into account. (f) Optic flow when direction of motion differs from direction of gaze and the observer tracks a moving object.

several minutes, for example when the flow speed is constantly lower than normal for a runner on a treadmill, an after effect is observed in which the walker inadvertently advances when attempting to run in place on solid ground with eyes closed.

3D Scene Perception. Because of motion parallax the optic flow contains information about the distances of the points of the scene. This information can be extracted to estimate the relative distances between objects in the scene and to recover surface layout [1]. Absolute distances cannot be retrieved from the optic flow because flow magnitude depends on the quotient of observer speed (T) and point distance (Z). For example, in an airplane flying high above the ground optic flow speed is very low even for very high forward speed of the plane. Thus, distance can only be calculated when the observer speed is known, which is usually not the case.

►*Time-to-Contact*. Information in the optic flow allows to estimate the time-to-contact or the time-to-passage with an obstacle during forward motion. By itself, the speed of an optic flow vector of a particular object is insufficient for the estimation of distance to the object (because it depends on T/Z) but a combination of speed with object size or of speed with the object's visual angle allows a direct calculation of time-to-contact. This information may be used to control braking or catching and to control running speed and direction for the intersection with a target object (for instance in ball sports). An overview can be found in [6].

►*Path Integration*. By integrating the speed of the optic flow over time an estimate of the travel distance or path length of an extended movement can be obtained. This estimate is subject to a scale factor since the speed of the flow depends on both the speed of the self-movement and the distance to the objects in the environment, but in many natural circumstances the height of the observer above the ground can provide the required scale. The estimation of travel distance from optic flow is based on an the integration of an estimate of observer velocity that is derived from the optic flow [7].

Heading. Heading refers to the direction of the movement of the observer. Gibson's original proposal for the use of optic flow was the identification of the heading (for example when landing an aircraft) by locating the focus of expansion in the flow field. Most optic flow research since then has centered on heading perception (overviews in [8] and [9]). Indeed, human observers are quite accurate in finding the focus of expansion in an expanding flow field. However, the situation is much more complicated because in most natural situations the optic flow on the retina is influenced by rotations and the flow field does not contain a focus of expansion (cf. Fig. 1). Yet, geometric calculations prove that the optic flow in these cases also contains sufficient information to separate the translational and rotational contributions if several flow vectors are available [1]. Many computational algorithms have been developed for this task, among them a few that are formulated as biologically plausible neurocomputational models (overview in [2]). Human observers can indeed estimate heading from flow fields of translation and rotation with reasonable accuracy (a few degrees of visual angle). An important finding was that heading estimation can be performed solely from the information in the flow field, i.e., from the direction and speed of the flow vectors, without any other sensory signals necessary. However, in natural situations eye movements that influence the structure of the retinal flow are accompanied by extra-retinal eye movement signals such as the ►efference copy signal or eye muscle ►proprioception. These signals are also used in optic flow analysis and increase the accuracy of the heading estimate. Rotational contributions to the retinal flow may also arise from movements on a curved path, in addition to, or instead of eye movements. Therefore, a separation of translational and rotational contributions may only provide the momentary or instantaneous heading but not the full information about the future path of the observer, since the rotational contributions are ambiguous. Estimations of path curvature, which are required for steering for instance, can be derived from successive independent heading estimates or from a combination of optic flow and extraretinal eye movement signals. Alternatively, specialized behavioral strategies, such as fixating a specific point in the flow field, may allow the estimation of steering-relevant information directly from the retinal velocities.

Although the above descriptions refer to human observers, optic flow is used for such behavioral tasks throughout the animal kingdom (see [2] for several examples). The use of optic flow for the control of speed, distance, time-to-contact, and course control has been shown in insects, birds, and mammals, exemplifying the ecological importance of optic flow. Moreover, the above descriptions show that optic flow is often part of multi-modal mechanisms for behavioral control, interacting with ►proprioceptive, vestibular, and internal feedback signals. Exposure to optic flow is also known to induce ►vection, the subjective feeling of self-movement in a physically static observer.

Neurophysiological Processing

In the visual system of primates visual motion information is routed via V1 and V2 to the ►middle temporal (MT) and subsequently to the ►medial superior temporal area (area MST) and other visual areas in the parietal lobe. Most clearly related to optic flow is area MST (detailed reviews in [2]). Many neurons in area MST respond selectively to entire optic flow patterns and not just to an individual motion vector in a particular flow field. A neuron might respond selectively to a particular flow pattern, such as an expansion as in Fig. 1c, but when tested with small stimuli the selectivities in subfields of the ►receptive field do not match one-to-one the pattern of the preferred large flow field. Thus, MST neurons are genuinely selective for optic flow. Their selectivity arises from complex interactions between selectivities in local subfields. Functional ►brain imaging in humans has confirmed an area selective for optic flow which is part of the human ►MT± complex.

When tested with multiple different flow patterns such as visual expansions, rotations and translations, MST reveals a continuum of response selectivities. Some neurons respond to several different patterns or to flow fields that combine translational and rotational contributions. Instead of classifying the selectivity of MST neurons by the preferred pattern of flow it is also possible to describe their selectivity in terms of heading. Indeed, it is possible to calculate heading from the firing rates of the neuronal population in MST. Next to visual motion signals, area MST also receives extra-retinal eye movement information. This information is used to counteract the effects of eye movements on the retinal flow and maintain selectivity for heading in the presence of eye movements. There are also interactions with vestibular signals during self-motion.

Other areas of the parietal lobe, the ►ventral intraparietal area (VIP) and area 7A, as well as the ►fundus of the superior temporal sulcus (FST) also respond to optic flow. Neurons in area MT, the major input to area MST, respond to optic flow but their responses can be explained by their selectivity to local image motion within their receptive field. However, some global properties of the visual field map in MT seem related to optic flow analysis. Preferred speeds increase with eccentricity similar to the increase of speed with eccentricity in typical flow fields. The distribution of preferred directions for neurons with peripheral receptive fields is biased towards centrifugal motion similar to the radial motion directions in a typical optic flow. The increase of the receptive field sizes with eccentricity is well adapted to the size of image patches over which neighboring flow signals are uniform. These patches are small in the center of the visual field, where optic flow vectors point in different directions, and large in the peripheral visual field where neighboring flow vectors are usually very similar. Computational modeling shows that this adaptation of receptive field sizes leads to significant noise reduction in the optic flow representation in area MT.

As mentioned above, optic flow is used by many animals. A brief description of the neuronal pathways of optic flow analysis in birds can be found in the essay on *visual-vestibular interactions*. In flies, optic flow is analyzed by a small number of neurons of the horizontal (HS) and the vertical (VS) system in the lobula plate (Krapp in [2]). Unlike neurons of primate MST, which show no simple correlation between local motion selectivities and flow patterns selectivity, the flow selectivity of these neurons in the fly is matched by the sensitivity to local motion in subfields of their very large receptive fields. These neurons seem to form matched filters for particular flow patterns. Like in primate MST, information about the translation and rotation of the animal can be decoded from the population activity.

References

1. Longuet-Higgins HC, Prazdny K (1980) The interpretation of a moving retinal image. Proc Roy Soc Lond B 208:385–397
2. Lappe M (ed) (2002) Neuronal processing of optic flow. International Review of Neurobiology, vol 44. Academic Press, New York
3. Gibson JJ (1950) The perception of the visual world. Houghton Mifflin, Boston
4. Lee DN, Aronson E (1974) Visual proprioceptive control of standing in human infants. Percept Psychophys 15:529–532
5. Prokop T, Schubert M, Berger W (1997) Visual influence on human locomotion – modulation to changes in optic flow. Exp Brain Res 114:63–70
6. Hecht H, Savelsbergh GJP (eds) (2004) Time-to-contact. Advances in Psychology, vol 135. Elsevier, Amsterdam
7. Frenz H, Bremmer F, Lappe M (2003) Discrimination of travel distances from 'situated' optic flow. Vision Res 43:2173–2183
8. Warren WH Jr (1998) Visually controlled locomotion: 40 years later. Ecol Psychol 10:177–219
9. Lappe M, Bremmer F, van den Berg AV (1999) Perception of self-motion from visual flow. Trends Cogn Sci 3:329–336

Optic Flow Dependent OFR

Definition

►Ocular Following Responses (OFR).

►Oculomotor Control
►Optic Flow

Optic Nerve

Definition

The optic nerve is the portion of the visual pathway between the retina and lateral geniculate nucleus of the thalamus that lays rostral to the optic chiasm. The continuation of the path caudally is the optic tract. The cell body of origin for this pathway is the ganglion cell in the retina.

Optic Neuritis

Definition

Sudden inflammation of the ►optic nerve occurring most often between 20 and 40 years of age, and may be a ►demyelinating disease of unknown origin or a co-manifestation of ►multiple sclerosis. The inflammation may occasionally be the result of a viral infection.

►Multiple Sclerosis

Optic Radiation

Synonyms

Radiatio optica

Definition

The visual radiation is the term used to designate the ray-shaped fiber bundles that leave the lateral geniculate body and at the lateral wall of the lateral ventricle pass on to the area 17 (striate cortex) at the occipital pole. They conduct the visual raw material after being processed by the LGB. Also called geniculocalcarine tract.

►Geniculo-striate Pathway
►Lateral Geniculate Nucleus (LGN)
►Primary Visual Cortex
►Striate Cortex Functions

Optic Tract

Definition

The optic tract is the portion of the visual pathway between the retina and lateral geniculate nucleus of the thalamus that lies caudal to the optic chiasm. The portion that is rostral to the optic chiasm is the optic tract. The cell body of origin for this pathway is the ganglion cell in the retina.

Optic Tract Nucleus

Synonyms

►Nucl. tractus optici; ►Nucleus of optic tract

Definition

The optic tract nucleus lies in the Myelencephalon near the superior colliculus. The nucleus is fused with the dorsal terminal nucleus and is an important center of the subcortical pathway which mediates horizontal optokinetic nystag

►Diencephalon

Optical Coherence Tomography (OCT)

Definition

OCT is an emerging ocular imaging technique to measure optic structures with micrometer resolution. It is useful in the measurement of retinal nerve fiber layer (RNFL) thickness and total macular volume corresponding to the ganglion cell body layer. The thickness of these unmyelinated nerve fiber layers may reflect axonal integrity, and function (vision) may be directly correlated with structure. Though RNFL thickness may be significantly decreased in multiple scerosis (MS) patients with optic neuritis compared to healthy controls, even in MS patients with no history of optic neuritis, RNFL may still be decreased in thickness consistent with a neurodegenerative disease model of MS.

►Inherited Retinal Degenerations
►Multiple Sclerosis
►Optic Neuritis
►Retinal Ganglion Cells

Optical Flow

►Optic Flow

Optical Illusions

►Visual Illusions

Optimal Control

Definition

Optimal Control is a particular control technique in which the controller is designed to minimize a certain performance index. For example, in human postural control, the performance index may be a combination of center of mass variance and mean squared ankle torque.

►Adaptive Control
►Modeling of Human Postural Control
►Motor Control Models

Optimal Control Theory

Definition

The mathematical theory of how controllers should be designed to achieve optimal performance.

►Neural Networks for Control

Optimal Muscle Length

Definition

The optimal length of a muscle is defined as the length at which a muscle can exert its maximal isometric steady-state force.

►Force Depression/Enhancement in Skeletal Muscles
►Length-tension

Optimization

Definition

An algorithm to achieve a particular goal while minimizing one, or a set of criteria. Mathematical optimization is defined by minimizing, maximizing, or optimizing a specific function (typically called the objective or cost function) while simultaneously satisfying any equality and/or inequality constraints. Mathematical Optimization has been the preferred approach to solve the distribution problem in biomechanics.

►Distribution Problem in Biomechanics
►Motor Control Models

Optimization Model for Motor Control and Learning

Definition

Computational models based on the idea that motor control and learning are planned and executed so as to achieve a behavioral goal, namely a tradeoff between task performance, body stability, and energy consumption. These models explain invariant movement features as a result of optimality and motor learning as a relaxation process toward a global minimum of a behavioral goal. Voluntary arm reaching, for example, has been modeled as smoothness or accuracy maximization, and locomotion as gait optimization in such a way as to maximize traveling distance using minimal muscle work.

►Theories on Motor Learning

Optocollic Reflex

Definition

A reflexive compensatory head movement elicited in response to motion of the entire visual world.

►Visual-Vestibular Interaction

Optogenetic

Definition

A method to manipulate the activity of genetically identified neurons using light-sensitive ion channels.

►Hypocretin/Orexin

Optokinetic After-Nystagmus (OKAN)

Definition

When subjects are placed in darkness following optokinetic nystagmus, the nystagmus continues and the slow phase velocity has characteristics similar to Per- and Post-Rotatory Nystagmus. The presence of OKAN can be directly related to activation of velocity storage.

►Optokinetic Nystagmus
►Per-rotatory Vestbular Nystagmus
►Velocity Storage

Optokinetic Nystagmus (OKN)

Definition

A physiological nystagmus that occurs when a large part of the image moves uniformly over the retina, such as when viewing objects from a moving train or turning around. It consists of two components of eye movements: slow phase, which moves the eyes to follow the visual scene motion (called optokinetic response), and quick phase, which rapidly reset the eye position deviation by slow phase.

►Nystagmus
►Optokinetic Response

Optokinetic Reflex

Definition

►Optokinetic Nystagmus (OKN)

Optokinetic Response

Definition

Compensatory head, eye and body movements in response to motion of the entire visual world. They function to control gaze, posture and locomotion (alternatively known as optomotor responses).

►Visual-Vestibular Interaction

Optokinetic Response Adaptation

Charles A. Scudder
Portland, OR, USA

Definition

Optokinetic response (OKR) adaptation is a behavioral change and underlying neural process that increases the ability of the optokinetic system to move the eyes and track moving large-field visual stimuli (see ►Optokinetic nystagmus). The adaptation is stimulated by motion of the visual image across the retina (►Retinal slip), and is prominent in species where the performance of the optokinetic system is normally low, such as rodents and fish. The increased efficacy of the OKR acts to reduce image motion across the retina and thereby improve visual acuity.

Methods to Produce and Measure Adaptation

Methods to produce and measure OKR adaptation are an extension of those used to produce and measure OKR itself. Subjects typically sit at the center of a large cylindrical drum with a visual pattern on the inside that takes up most of the subject's visual field (Fig. 1).

Oscillation of the "optokinetic drum," usually in a horizontal plane, evokes eye-movements that tend to track the motion of the drum (see Optokinetic eye movements). The ability of the eye-movements to track drum motion is often measured as gain, which is the ratio of eye angular velocity to drum angular velocity. Perfect tracking would produce equal eye and drum velocities and a gain of 1.0. Actual gains are always less than one, and cannot exceed one.

Whereas measurement of OKR gain requires only a few minutes, continued drum motion is used to produce adaptation. Adaptation takes place anytime there is retinal slip, but a measurable change in OKR gain requires an hour or so of drum oscillation. Figure 2 illustrates adaptation in a rabbit. OKR gain at the start of adaptation is about 0.5 (eye movement only compensates half of the drum motion). After an hour, OKR efficacy has increased to 0.74, and an additional two hours of adaptation increases gain only slightly more to 0.78.

Characteristics

Species Dependencies

OKR adaptation has mainly been observed in rodents [1–3] and goldfish [4] where the gain of the OKR is typically well below 1.0 except at very low drum velocities. This is in part because gains less than one allow adequate retinal slip to produce adaptation in a suitable paradigm, and there is sufficient room below the maximum gain of one for the increased OKR efficacy to

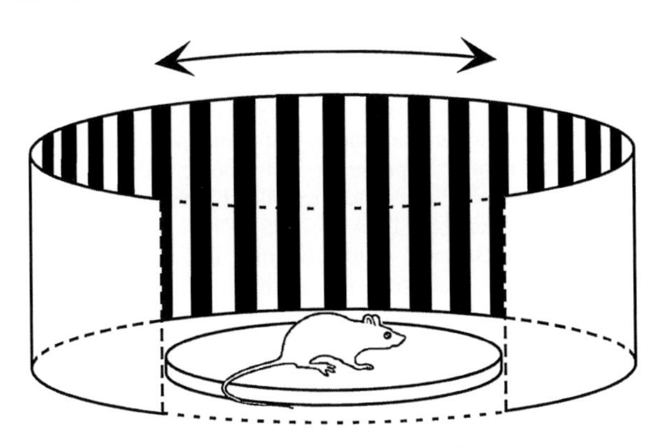

Optokinetic Response Adaptation. Figure 1 Illustration of the apparatus used to generate and adapt the optokinetic response. A mouse sits on a stationary platform surrounded by an optokinetic drum which oscillates back and forth about a vertical axis. The mouse would be restrained in an actual experiment. The drum is illustrated as being lined on the inside with vertical black and white stripes, but other high contrast patterns have been used. Vestibular responses can be produced by rotating the platform on which the mouse sits. To induce the vestibuloocular reflex (VOR), rotation takes place in the dark, but various VOR adaptation paradigms combine rotation of the mouse with motion of the optokinetic drum in the light.

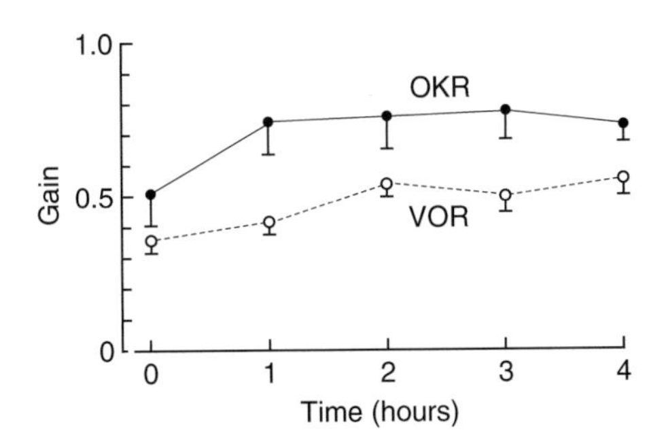

Optokinetic Response Adaptation. Figure 2 Plot of the gain of the optokinetic response (OKR – solid line) as a function of the duration of OKR adaptation. OKR gain increases rapidly in the first hour and then plateaus. The paradigm for increasing OKR gain also has the effect of increasing the gain of the vestibuloocular reflex (VOR – dotted line) in rodents. Figure adapted from Nagao et al. [6].

be observed. In one case where OKR gain was close to one at low drum velocities, the effect of OKR adaptation could still be observed at higher drum velocities where OKR gain normally drops well below one [2].

OKR adaptation has not been reported in primates, but neither has it been systematically tested. The excellent tracking of optokinetic targets at velocities less than 60°/s leaves little opportunity for adaptation to occur or to be observed. Moreover, differences between primate and rodent physiology argue that OKR adaptation is less likely in primates. Primates lack the directionally selective retinal ganglion cells that participate in rodent OKR, they have a fovea instead of a visual streak, and ►smooth pursuit rather than OKR dominates primate responses to motion in the visual field.

Velocity Dependence

OKR adaptation has been produced using optokinetic drum velocities past the limit at which the eyes reliably track the drum. In rodents, this is at low stimulus frequencies (0.1–0.4 Hz) and at peak drum velocities of 3°/s–10°/s. Retinal slip velocities are then between 2°/s and 8°/s. It has been reported that low retinal-slip velocities (<1°/s) do not produce adaptation [3], but this has not been extensively tested.

OKR Adaptation and Head Movement

The vestibulo-ocular reflex (VOR), which is not a visual-following reflex, acts to counter-rotate the eyes in the head whenever the head moves with the goal of stabilizing the visual scene on the retina (see ►VOR). Adaptation of the VOR occurs when this ocular compensation is imperfect, or in other words, when movement of the head produces retinal slip (see ►VOR adaptation). This retinal slip might be expected to produce OKR adaptation as a byproduct, and indeed it does [1,2]. This is best demonstrated by using different combinations of forced head motion and drum motion in order to create retinal slip velocities that are either in the same or opposite direction as eye motion. For instance, when the drum motion is in the opposite direction as head motion (a paradigm that increases VOR gain), slip velocity is in the same direction as eye velocity. However, when drum motion is in the same direction as the head motion (a paradigm that decreases VOR gain), slip velocity is in the opposite direction as eye velocity. In rodents, both paradigms produce OKR adaptation and the effect of both is to increase OKR gain [1,2]. Apparently retinal slip of any kind augments OKR gain in these animals.

However in monkeys and cats, the situation is different [5]. Paradigms that increase VOR gain also increase OKR gain, and those that decrease VOR gain also decrease OKR gain. In each case, the result is to improve the VOR in the sense that there is less retinal slip during head rotation at the end of the particular paradigm, but the concomitant reduction of OKR gain is maladaptive. This has been interpreted to mean that the primary function of the adaptive mechanisms is to adjust VOR gain, but that the VOR and OKR pathways share a common structure that changes the gain of both systems simultaneously.

Finally, there is the possibility that the OKR-adaptation paradigm (drum motion with no head motion) could alter the VOR because of the retinal slip. In rodents and goldfish, this paradigm does increase the gain of the VOR [1,4,6] (Fig. 2). However in monkeys, the effect is negligible [7]. The above differences between rodents and primates reinforce the idea that primates are not the same as rodents and goldfish regarding the existence of OKR adaptation.

Upstream Conditions

As noted above, the existence of retinal slip of adequate velocity is required for OKR adaptation to occur.

Involved Structures

Adaptation of the OKR presumably involves plasticity at synapses that are part of the normal OKR pathway (see ►Optokinetic eye movements). In most species, this involves indirect projections of from the accessory optic system to the floccular lobe of the cerebellum and then to the vestibular nuclei. Rodents may have an additional pathway that has not been found in primates from the pretectum directly to the vestibular nuclei.

Experimental interventions that diminish or abolish OKR adaptation precisely parallel those that diminish or abolish adaptation of the VOR (see VOR adaptation and ►Flocculus hypothesis). Among these interventions are those known to disrupt long-term depression (LTD) at the cerebellar parallel-fiber to Purkinje-cell synapse. They include destruction or inactivation of the flocculus [6], destruction of the climbing-fiber afferent pathway to the cerebellum [3], disruption of metabotropic glutamate receptors either by direct blockage or by elimination in mutant mice [8], blockage of nitric oxide synthase [9], and disruption of phosphokinase C. The first two appear to implicate the flocculus in OKR adaptation, but the latter four are not necessarily specific. Measurements of Purkinje-cell activity during adaptation show that changes do occur within the flocculus, and that the changes probably produce the changes in OKR gain rather than reflect feedback from the altered eye velocity [6]. In different experiments, physiological changes have also been observed in synapses the vestibular nuclei [10]. Shutoh et al. [10] argue that short-term plastic changes (about a day) reside in the flocculus while long term plastic changes reside in the vestibular nuclei. As has been strongly indicated for the VOR, it seems likely that plastic changes of some sort occur in both the flocculus and the vestibular nuclei.

References

1. Collewijn H, Grootendorst AF (1979) Adaptation of optokinetic and vestibulo-ocular reflexes to modified visual input in the rabbit. Prog Brain Res 50:771–781
2. Nagao S (1983) Effects of vestibulocerebellar lesions upon dynamic characteristics and adaptation of vestibulo-ocular and optokinetic responses in pigmented rabbits. Exp Brain Res 53:36–46
3. Katoh A, Kitazawa H, Itohara S, Nagao S (1998) Dynamic characteristics and adaptability of mouse vestibulo-ocular and optokinetic response eye movements and the role of the flocculo-olivary system revealed by chemical lesions. Proc Natl Acad Sci USA 95:7705–7710
4. Marsh E, Baker R (1997) Normal and adapted visuo-oculomotor reflexes in goldfish. J Neurophysiol 77:1099–1118
5. Lisberger SG, Miles FA, Optican LM, Eighmy BB (1981) The optokinetic response in monkey: underlying mechanisms and their sensitivity to long term adaptive changes in V.O.R. J Neurophysiol 45:869–890
6. Nagao S (1989) Role of cerebellar flocculus in adaptive interaction between optokinetic eye-movement response and vestibulo-ocular reflex in pigmented rabbits. Exp Brain Res 77:541–551
7. Lisberger SG, Miles FA, Zee DS (1984) Signals used to compute errors in monkey vestibuloocular reflex: possible role of flocculus. J Neurophysiol 52:1140–1153
8. Shutoh F, Katoh A, Kitazawa H, Aiba A, Itohara S, Nagao S (2002) Loss of adaptability of horizontal optokinetic response eye movements in mGluR1 knockout mice. Neurosci Res 42:141–145
9. Katoh A, Kitazawa H, Itohara S, Nagao S (2000) Inhibition of nitric oxide synthesis and gene knockout of neuronal nitric oxide synthase impaired adaptation of mouse optokinetic response eye movements. Learn Mem 7:220–226
10. Shutoh F, Ohki M, Kitazawa H, Itohara S, Nagao S (2006) Memory trace of motor learning shifts transsynaptically from cerebellar cortex to nuclei for consolidation. Neuroscience 139:767–777

Optomotor Response

Definition

In a broad sense the motor response to a visual stimulus. In narrower sense, the response of an animal to wide-field, visual stimulation (synonym: optokinetic response).

Oral Mucosa

Definition

The epithelium lining the inside of the mouth, the tongue and the palate.

►Tactile Sensation in Oral Region

Oral-facial Dyskinesias

Definition

Repetitive, rhythmic, bizarre movements in the face region.

Orbital Dynamics

►Oculomotor Dynamics

Orbital Pulleys

Definition

When the eyes move from the primary position, the eye muscles do not slide freely within the orbital tissue. Instead their paths are restricted, possibly by rings of connective tissue and smooth muscle that have been termed orbital pulleys.

►Eye Orbital Mechanics

Orbital Tissues Definition

Orbital tissues are the fat and connective tissues that surround the eyeball in the bony orbit.

►Eye Orbital Mechanics

Orbitofrontal Cortex

Definition

The orbitofrontal cortex is a region situated at the ventral surface of the frontal part of the brain. It is the subpart of the prefrontal cortex that receives projections from the magnocellular medial nucleus of the mediodorsal thalamus. The orbitofrontal cortex is an important brain region for the processing of rewards and punishments. The medial orbitofrontal cortex activity is related to monitoring the reward value of many different reinforcers, whereas lateral orbitofrontal cortex activity is related to the evaluation of punishers, which may lead to a change in ongoing behavior. The subjective hedonic experience is mediated by mid-anterior orbitofrontal cortex.

Orexigenic

Definition

Orexigenic means possessing activity that stimulates food intake. [Anorexigenic: opposite of orexigenic.]

►Neuropeptides

Orexin/Hypocretin

Definition

Orexins (OxA and OxB) are two neuroexcitatory peptides derived from the same precursor produced in a few thousand neurons restricted to the perifornical area of the hypothalamus. The orexins bind to two receptors (Ox1 and Ox2). Orexin is a synonym of hypocretin, and was given its name (orexi, appetite in Greek) because of initial studies showing increase in food intake following infusion of pharmacological doses of the peptides in the brain. The orexin/hypocretin system stabilizes wakefulness and sets the arousal threshold, enhances catabolism and is a gate to drug reinstatement. Dysfunctional orexin may be associated with the sleep disorder narcolepsy.

►Brain States and Olfaction
►Hypocretin/Orexin
►Memory and Sleep
►Narcolepsy
►Nocturnal/Diurnal
►Sleep – Motor Changes
►Sleep – Sensory Changes
►Ventrolateral Preoptic Nucleus (VLPO)

Organ Discharge

►Electric Organ Discharge

Organ of Corti

Definition

The mammalian organ of hearing proper, lying between the basilar membrane and the tectorial membrane of the cochlea. It contains the inner hair cells, the outer hair cells and the peripheral synapses of the afferent and efferent neurons of the auditory nerve.

►Cochlea

Organizational Hormonal Effects

Definition

Hormone-induced alterations occurring during the early development of an organism that give rise to chronic changes in structure and/or function of particular anatomic systems. For example, manipulating the gonadal hormonal milieu of neonate rodents can produce durable effects on the developing nervous system resulting in lifelong changes in nociception and antinociception.

►Gender/sex Differences in Pain

Organizer

Definition

Area, tissue or cell group of an embryo able to produce signals (or signaling proteins) that have an effect at a distance on the fate of adjacent tissue, in a concentration dependent manner (this requires the expression of specific receptors in the tissue). Examples of organizers are the node and the notochord, which produce signals that have an effect either on the ectoderm (node signals related to neural induction) or on the ventral neural plate/tube (notochord signals related to dorsoventral patterning). These are cases of organizers acting early in development and are many times referred to as "primary organizers." Later in development, there are "local organizers" inside the neural tube having an effect on patterning and specification of adjacent areas (for example, the isthmic organizer or the zona limitans intrathalamica). These local organizers of the neural tube that appear later in development are called "secondary organizers."

►Evolution and Embryological Development of the Forebrain
►Node
►Notochord

Organizing Centers and Patterning

Definition

Restricted regions of the embryo that secrete specific signalling molecules, responsible for specifying distinct domains (molecularly, anatomically, functionally distinct) in competent neighbouring tissues. This process is called patterning.

►Evolution of the Brain: In Fishes
►Evolution of the Telencephalon: In Anamniotes

Orientation Behavior

Definition

Ability to move in space either with respect to an external reference system (passive) or by actively generating spatial information (like in echo location).

Orientation Selectivity in Vision

Definition

Neurons in the retina and lateral geniculate nucleus of the thalamus are sensitive to local changes in light levels, much like sensors in a digital camera. But these cells are not able to resolve higher order features of the visual scene. By contrast, cortical cells respond best to elongated contours, or edges, formed by extended boundaries between relatively dark and bright regions of the image – contrast borders. Importantly, almost all cortical neurons are orientation selective: individual cells are strongly excited by contours that share a common spatial orientation but respond weakly if at all to stimuli tilted perpendicular to the optimal angle. Different neurons prefer different stimulus orientations. Also, some neurons are tuned to a narrow range of stimulus angles while others are less selective. Orientation selectivity is the most widely studied aspect of visual cortical function; its origin in different species and its role in visual processing remain a subject of great interest.

►Visual Cortical and Subcortical Receptive Fields

Orientation Sensitivity in Cutaneous Mechanosensation

Definition

Subjects can discriminate a 10% angular difference in the orientation of a cylinder indented into the fingertip. Discriminating the orientation of a grating (usually vertical vs. horizontal) is also used to assess spatial resolution. Orthogonal gratings can be discriminated for groove widths around 1 mm at the fingertips and around 4 mm at the more proximal regions of the fingerpad.

►Processing of Tactile Stimuli

Orienting Linear Vestibulo-ocular Reflex (lVOR)

Definition

The reflex that responds to low frequency linear accelerations of the head in space to produce eye movements that tend to align the coordinate frame of the eyes with the net direction of the linear or equivalent linear acceleration of the head. This has also been referred to as the tilt response.

►Vestibuloocular Reflexes

Orienting Movement

Definition

►Orienting Reflex

Orienting Reflex

Definition

Also known as orienting response(s). In a general sense, it is the complex behavioral pattern aimed at optimizing the perception of biologically significant events in the environment and to make rapid and efficient choice of an appropriate motor response. Orienting is truly "reflexive" toward particularly intense or previously unexperienced, novel sensory stimuli. Accordingly, the Pavlovian school used the term "what happens?-reflex." Its earliest manifestation is the generalized alerting. Sensory events signaling a potential danger or a positive reinforcement, such as prey for a predator or food delivery for an operantly conditioned animal, are also highly efficient to induce orienting. Motor responses to such stimuli are, respectively, either avoidance or approach. To make the choice between these strategies, the source of the stimulus must be rapidly identified. Alignment of the line of sight on the stimulus (gaze shifting) is the most important motor component of orienting reflex in animals whose behavior is dominated by vision and, in particular, in those having a small central region of the retina specialized for fine-grain visual discrimination (e.g., fovea in primates, area centralis in felines).

►Operant Conditioning
►SC-Tectoreticulospinal neurons (TRSNs)
►Vision

Orienting Responses

Definition

Movements that direct the line of sight and/or the ears towards sensory stimuli.

Orthodromic Action Potential Propagation

Definition

Propagation of action potentials in the naturally occurring direction (from "orthos", Greek for straight, correct; "dromos", Greek for run).

►Action Potential Propagation

Orthostatic Intolerance

Definition

Orthostatic intolerance is difficulty in maintaining standing posture due to orthostatic hypotension. Astronauts returning on Earth after spaceflights often complain of this symptom. Similar orthostatic problems occur after long-term bed rest.

►Autonomic Function in Space

Oscillations and Plasticity in the Olfactory System

Nadine Ravel, Rémi Gervais,
Julie Chapuis, Claire Martin
Laboratoire Systèmes Sensoriels, comportement et Cognition, UMR 5020 CNRS–UCB Lyon, IFR19 Lyon, France

Definition

Learning induces neural assemblies formation detectable in the network through modulation of oscillatory activities.

Characteristics

The Concept of Neural Assemblies

Current theories put forward that information storage in the brain relies on changes in functional interactions within widely distributed neural areas. This concept of distributed memory suggests in turn the idea that stimuli representations could be achieved through assemblies of simultaneously active neurons. Such assemblies could be found within a given structure or between distinct neural areas. As a consequence, memory should be considered as a dynamical process involving spatio-temporal patterns of reactivation of previously reinforced neural ensembles within and across different brain areas. These assemblies involve both sensory and limbic areas.

If we accept this concept of distributed representations one have to face the problem, commonly addressed as the "binding problem" of how such distributed activities could be put back together to elicit stable and unambiguous representations of objects in the brain. Indeed, according to this theory a given neuron would be able at different time to take part to different stimuli representations. As a consequence, neuronal elements belonging to the same assembly must be identifiable and differentiated from members of other assemblies (see Fig. 1). Twenty years ago, von der Malsburg proposed that neurons joining into an assembly should establish temporal synchronization on a millisecond time scale. This temporal tagging has two major advantages: Synchronization of neuronal activities facilitates signal transmission to target structures because temporal coincidence of action potential volleys on post-synaptic higher areas increases probability of eliciting action potentials. In addition, this coincidence is very important in voltage-dependent processes like NMDA-receptor-gated conductance which are of prime importance in induction of synaptic plasticity.

Synchronous activities in assemblies often occur in a repetitive way and give rise to well-known brain rhythms also called oscillations. They can be recorded with macro electrodes either directly from the scalp (electroencephalogram, EEG) or from intracerebral inserted electrodes (►local field potentials, LFPs). They have been observed in many different brain areas especially those showing a laminar organization like cortices. These oscillations of LFPs exhibit a large variety of frequencies from 1 to 100 Hz depending on the vigilance state (arousal, attentiveness, sleep, etc.) or the presence of a sensory stimulation or the necessity to control a motor behavior. The origin of oscillations is still a matter of debate but one major hypothesis is that they could be an emergent property of a given network resulting from inhibitory interneurons and reciprocal connections. In relay neurons of any cortical area, these oscillations likely reflect current source generated by neuronal synaptic input in the dendritic tree and action potentials generated at the cell body level. LFP activities are a good indicator of how and when a large set of neurons synchronize and desynchronize during information processing. This review will illustrate how the study of the mammalian olfactory system brings information on the functional significance of neural oscillations in sensory processing and memory.

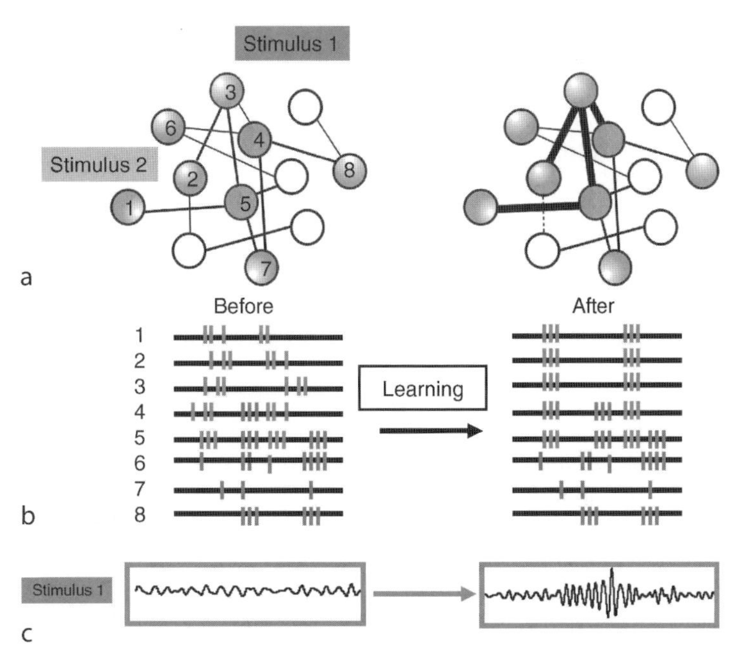

Oscillations and Plasticity in the Olfactory System. Figure 1 Illustration of the concept of neural assemblies. (a). As symbolized by the colour code, each stimulus co activates a specific ensemble of neurons. However, some neural elements (4 and 5) could be co activated by both stimuli. Learning of stimulus 1 is associated with reinforcement of synaptic contacts between neurons previously co activated by this stimulus (*thick lines*). (b). Temporal organization of the discharge of each neuron clearly differentiates two neural assemblies (in green for stimulus 1 and red for stimulus 2). Neurons 4 and 5 could take part to both representations depending on their discharge timing. Before learning, units taking part to the same assembly are simply co-activated. Repeated presentation of stimulus 1 refines of neural discharge synchronization. As a consequence, the amplitude of ►local field potential oscillatory activity is increased and the dominant frequency corresponds to the periodicity of the synchronization.

O

Oscillatory Activities in the Olfactory System

In the mammalian olfactory system, Adrian in the 50's initially described prominent oscillations in field potential activities. In ►awake animals, in the absence of any olfactory stimulation, the signal derived from the first relay of olfactory processing, the olfactory bulb, exhibits a well structured activity as shown on Fig. 2.

Slow modulations of LFP associated with inhalation are easy to observe. These high amplitude oscillations are in the theta range (2–12 Hz). They have been shown to follow the respiratory activity and hence might vary in frequency. Moreover, during period of exploration associated with active sniffing, the respiratory modulation has a frequency range which overlaps with the theta activity typically observed in limbic areas such as the hippocampus (4–12 Hz).

Recordings also show regular spindle bursts of oscillations during each inspiration phase of the respiratory cycle. This second type of oscillatory activity is in the gamma range (30–90 Hz). Interestingly, in a given animal, even in the absence of any olfactory stimulus, the distribution of amplitude of gamma bursts forms a stable map at the surface of the OB. Presentation of an odor in a specific experimental context modifies this distribution. However, this new map is more related to the behavioral meaning of the stimulus than to its chemical quality. Indeed, if the same odor is presented in another context, a different map is obtained [1]. Recently, Kay [2] proposed to distinguish two types of gamma activity, type 1 (65–90 Hz) corresponding to the bursts associated to the peak of inhalation and type 2 (35–65 Hz), lower in frequency. These rhythms seem to be associated with different behavioral features and are likely to be produced by different synaptic interactions within the olfactory bulb.

Whereas gamma and theta activities have been studied for a long time, at first, little attention was paid to an intermediate type of periodic activity in the beta range (15–35 Hz).

This activity has now been reported by several authors to be selectively associated with ►odor sampling not only in the olfactory bulb, but also at higher level of olfactory processing like the piriform cortex and lateral entorhinal cortex. These studies pointed out to a more or less prominent increase in the amplitude of this oscillatory activity in response to behaviorally relevant odors [3,4] or odors experimentally associated with a reward [5–8].

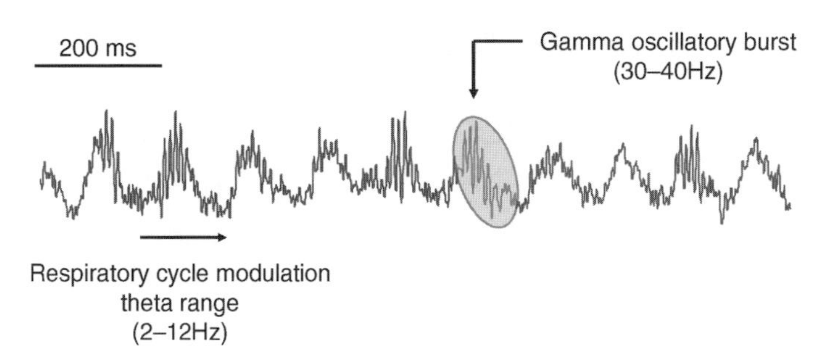

Oscillations and Plasticity in the Olfactory System. Figure 2 Spontaneous activity recorded in the olfactory bulb in ▶awake animal.

Thus, both gamma and beta oscillatory activities were associated to perception and cognitive processing of olfactory stimulus. In awake animals, gamma oscillatory activity is prominent in the absence of odor and seems more related to attention toward an expected stimulus or a given experimental context. Beta oscillatory activity has never been reported in the absence of odor. This activity emerges during odor sampling and is modulated both by the chemical nature of the odor and its behavioral significance.

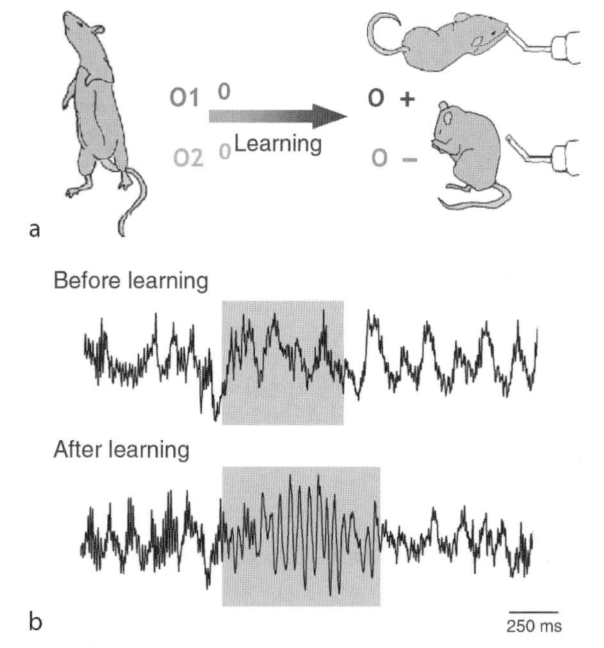

Oscillations and Plasticity in the Olfactory System. Figure 3 Learning-induced modulation of beta oscillatory activity. (a). Experimental protocol. Two odors without any a priori behavioral signification (O1 and O2) are assigned with two different values by pairing their presentation either with a sweet (O+) or a bitter (O−) solution. (b) Comparison of odor-induced activity in the olfactory bulb before and after O1 has acquired a positive value for the rat. The shaded zone corresponds to the odor sampling period. After learning, an oscillatory burst in the beta range (around 27 Hz) is clearly observed.

Oscillations and Construction of Odor Representations

According to the concept of neural assemblies proposed above, synchrony in a given neural network favors both signal transmission and synaptic plasticity. Hence, if this view is correct one could predict that olfactory learning should induce reinforcement of excitatory transmission between cells responding to the odor to be learned. As a consequence, learning should modify oscillatory regimes associated with the processing of learned odors. A first step toward the experimental demonstration of this hypothesis has been made by a few studies in which multisite neural recordings were performed in animals engaged in two different olfactory discrimination learning paradigms [7–9].

In the first paradigm (see Fig. 3), two odors without any a priori ▶behavioral signification (▶odor with behavioral signification) were assigned with two different values by pairing their presentation either with a sweet (O+) or a bitter (O−) solution. At the beginning of the experiment, the two odors induced the same behavioral response but after a few experimental sessions, thirsty rats exhibited a differential response to each odorant. Indeed, they learnt to run promptly to drink when O+ was delivered and avoid drinking when O− was presented. In parallel to this behavioral response modification, a clear oscillatory activity in the beta band (near 27 Hz) emerged in the olfactory bulb in response to odors used in the learning paradigm. In respect to a potential role in olfactory coding, we found that this activity exhibited different characteristics in amplitude and latency according to the recorded region in the olfactory bulb (anterodorsal vs. posteroventral) and the chemical nature of the odorants. More interestingly, the large beta oscillatory activity emerged a few trials before the animal reached the criterion level. As a whole, results stressed out the possible role of the beta oscillatory activity in both odor representation and olfactory recognition. The same type of activity was also found in other structures involved in odor stimulus processing like the piriform cortex. Moreover, a pharmacological inactivation of feedback connections from piriform cortex to olfactory bulb prevented in both structures the emergence of beta activity in response to learned odors suggesting that this oscillatory activity could be the signature of a neural network set up through learning and involving well-known reciprocal excitatory cortico-cortical connections between the olfactory bulb and the piriform cortex.

Recently, using a two-alternative choice odor discrimination, Beshel and colleagues [9] also showed a functional link between gamma range oscillatory activities in the OB and plasticity. In this paradigm, task demand was manipulated using either dissimilar or similar odorants ("coarse" vs. "fine" discrimination). Gamma oscillatory power progressively increased over the course of fine discrimination learning in contrast to coarse discrimination. This modulation was specific to gamma frequency range (65–85 Hz) and independent of changes in the theta or beta frequency range. It was also restricted to the OB despite gamma activity was also reported during spontaneous activity in the piriform cortex. This experimental result is in favor of a functional role of gamma oscillatory activity in pattern disambiguation. However, in mammals, data establishing a direct link between oscillatory activity disruption and behavioral performance alteration are still lacking.

Until now, the only demonstration that oscillatory synchronization might play a determinant role in fine stimulus encoding and odor recognition was brought by a work on honeybees [10]. In this animal model, odors evoke oscillatory synchronizations of groups of neurons in the antennal lobe, a structure functionally equivalent to the vertebrate olfactory bulb. These oscillations, in the beta range (around 30 Hz) could be selectively disrupted with picrotoxin, a pharmacological antagonist of $GABA_A$ receptors without affecting neural response and selectivity to odors. Behavioral experiments combining pharmacological disruption of odor-evoked oscillatory activity and evaluation of olfactory discrimination performance showed that picrotoxin-treated animals failed to discriminate between similar odorants although they were unimpaired for coarse discrimination. These observations were the first real argument for a role of neural synchronization in separation of spatially overlapping neural networks. It is of course tempting to speculate that neural oscillatory synchronization might play a similar role in other animal models.

In conclusion, one can point out that the detailed investigation of neural rhythms through LFPs recordings in behaving animals brings important insight on neural correlates of sensory discrimination and recognition. One of the main advantages of this approach is the relative ease with which one can obtain signal from several recording sites simultaneously and over the course of training (several days). This allows investigation of some neural correlates which sustain learning and memory in a time scale which characterized many forms of knowledge acquisition.

References

1. Freeman WJ, Schneider W (1982) Changes in spatial patterns of rabbit olfactory EEG with conditioning to odors. Psychophysiology 19:44–56
2. Kay LM (2003) Two species of gamma oscillations in the olfactory bulb: Dependence on behavioural state and synaptic interactions. J Integr Neurosci 2(1):31–44
3. Zibrowski EM, Vanderwolf CH (1997) Oscillatory fast wave activity in the rat pyriform cortex: relations to olfaction and behaviour. Brain Res 766:39–49
4. Chabaud P, Ravel N, Wilson DA, Mouly AM, Vigouroux M, Farget V, Gervais R (2000) Exposure to behaviourally relevant reveals differential characteristics in rat central olfactory pathways as studied through oscillatory activities. Chem. Senses 25:561–573
5. Boeijinga PH, Lopes da Silva F (1989) Modulations of EEG activity in the entorhinal cortex and forebrain olfactory areas during odour sampling. Brain Res. 478:257–268
6. Ravel N, Chabaud P, Martin C, Gaveau V, Hugues E, Tallon-Baudry C, Bertrand, Rémi Gervais (2003) Olfactory learning modifies the expression of odour-induced oscillatory responses in the gamma (60–90 Hz) and beta (15–40 Hz) bands in the rat olfactory bulb. Eur J Neurosci 17:350–358
7. Martin C, Gervais R, Hugues E, Messaoudi B, Ravel N (2004) Learning modulation of odor-induced oscillatory responses in the rat olfactory bulb: a correlate of odor recognition? J Neurosci 24(2):389–397
8. Martin C, Gervais R, Messaoudi B, Ravel N (2006) Learning-induced oscillatory activities correlated to odour recognition: a network activity. Eur J Neurosci 23:1801–1810
9. Beshel J, Kopell N, Kay LM (2007) Olfactory bulb gamma oscillations are enhanced with task demands. J Neurosci 27(31):8358–8365
10. Stopfer M, Bhagavan S, Smith BH, Laurent G (1997) Impaired odour discrimination on desynchronization of odour-encoding neural assemblies. Nature 390:70–74

Oscillations in the Brain

Definition

Oscillation is the variation, typically in time, between two boundary values of some measure. In the brain, oscillatory activities have been widely observed. At the level of the neurons and networks of neurons, it has been shown that intrinsic (mainly due to ion channel) and networks properties (connectivity, inhibition and excitation), endowed the neuron and the network with dynamical properties, including abilities to oscillate at multiple frequencies. Oscillations are groups into category that depend on their frequency and their relation with particular behaviors. Among other one can distinguished oscillations in the gamma band (30–90 Hz) that have been involved in perception, problem solving, fear and other higher brain function.

►Brain Rhythms
►Network Oscillations
►Network Oscillations in Olfactory Bulb

Oscillator

Definition

A device that generates a periodic signal.

►Signals and Systems

Oscillator for Circadian Rhythm

Definition

A system that produces rhythmic output or whose state varies in a periodic fashion in the absence of external stimuli. A circadian oscillator produces a rhythm whose period is approximately 24 h when the organism is maintained in constant conditions. This may be detected in the cycle of activity and rest, in gene expression as reflected by mRNA abundance or protein concentration, etc. Negative feedback loops involving regulation of transcription by translational products have been found to generate such circadian oscillations in a number of organisms. Circadian oscillators are typically entrainable by environmental cues within a range of periods close to 24 h, and often vary little in period over a range of temperatures.

Oscillator Versus Hourglass Timers

Definition

Time-measurement can be achieved by different types of mechanisms that change state in a predictable way before returning to the starting point. Once started, an oscillator may continue to generate cycles (have an endogenous rhythm) indefinitely, or it may damp out. In contrast, an hourglass measures a fixed interval and then must be restarted (by some external stimulus) in order to measure a second interval. Whether a biological system acts as an oscillator or as an hourglass can be a function of its environment; changes in parameters may alter the behavior of an oscillator so that it damps rapidly and thus functions as an hourglass.

Osmolality

Definition

Osmolality refers to the total concentration of all particles that are free in a solution. Thus, glucose contributes one particle, whereas fully dissociated NaCl contributes two. In all body fluid compartments, humans have an osmolality – expressed as the number of osmotically active particles per kilogram of water – of approximately 290 mOsmoles/kg water (290 mOsm).

Osmotic Energy

Definition

The energy associated with a concentration gradient.

►Energy/Energetics

Osseoperception

Definition

The tactile sense relayed through dental impacts placed in the jaws to serve as replacements for lost teeth.

►Tactile Sensation in Oral Region

Osteoarthritis

Definition

Osteoarthritis is a joint degenerative disease characterized by the breakdown of articular cartilage, osteophyte formation, joint swelling, stiffness and pain. The disease progresses from an initial hypertrophy of the articular cartilage to degeneration of the cartilage and underlying bone. Osteophytes also grow throughout the affected joint.

►Articular Cartilage
►Measurement Techniques (Pressure)

Osteoblast

Definition

Cell of fibroblast lineage responsible for secreting unmineralized bone matrix.

►Bone

Osteoclast

Definition

Cell of macrophage lineage responsible for resorbing bone.

►Bone

Osteocyte

Definition

Former osteoblasts, which are entombed within mineralizing matrix, reside within the bone in caverns termed lacunae, and appear to play an integral role in maintaining bone vitality and the tissue's ability to respond to altered loading states.

►Bone

Osteoporosis

Definition

A systemic disease in which bone mass and morphology have degraded sufficiently to elevate the risk of fracture.

►Bone

Osteostracans

Definition

A group of early jawless craniates that lived around 425–415 Ma BP and resembled gnathostomes in having pectoral fins, but not pelvic ones.

►The Phylogeny and Evolution of Amniotes

Other Minds Problem

Definition

The other minds problem is the problem of how we know (or are justified in believing) that other human beings exemplify mental properties similar to the ones we exemplify, given that their conscious mental life is not accessible from our third person point of view. The existence of other minds is typically justified by an argument from analogy, stated in its classic form by John Stuart Mill and Bertrand Russell, according to which one's own body and outward behavior are observably similar to the body and the behavior of others, so that one is justified by analogy in believing that they also exemplify similar mental properties.

►Epiphenomenalism

Otic Placode

Definition

Thickening of the ectoderm and precursor of the otocyst.

►Evolution of the Vestibular System

Otoconia

Definition

Dense calcium carbonate particles ("ear stones") that are attached to the gelatinous otolith membrane over the utricular and saccular maculae. Otoconia serve as inertial sensors of linear acceleration.

►Evolution of the Vestibular System
►Peripheral Vestibular Apparatus
►Sacculus
►Utriculus

Otocyst

Definition

Invagination of the otic placode forming a cycst at first that later subdivides and gives rise to the complex adult three-dimensional structure of the labyrinth.

►Evolution of the Vestibular System
►Otic Placode

Otoencephalitis

Definition

Inflammation of the brain due to an extension from an inflamed middle ear.

Otolith

Definition

The vestibular receptor organ that responds to linear accelerations of the head. Otoliths contain receptor cells in a small patch of neuroepithelium termed the macula. Above the macula is a gelatinous membrane into which the stereocilia of the hair cells project. Otoconia lie embedded in and attached to the top of the membrane.

►Otoconia
►Peripheral Vestibular Apparatus

Otolith Organs

Definition

The parts of the vestibular labyrinth composed of the utricles and saccules that sense linear acceleration or equivalent linear acceleration of the head.

►Peripheral Vestibular Apparatus
►Sacculus
►Utriculus

Otx

Definition

Member of a gene family (orthodenticle)

►Evolution of the Vestibular System

Otx1, Hominids

►Evolution of the Vestibular System

Outer Hair Cells

Definition

The hair cells of the mammalian cochlea responsible for amplifying the vibrations of the basilar membrane and the hair cell stereocilia.

►Cochlea

Outer Plexiform Layer

Definition

Synaptic layer in the outer (distal) retina where photoreceptors make synapses with horizontal and bipolar cells.

►Inherited Retinal Degenerations

Output Unit

Definition

A model network neuron that provides the network response to activity propagated through hidden units due to signals received by input units.

►Neural Networks

Ovariectomize

Definition

Surgical removal of the ovaries.

Overdetermined System

Definition

A mathematical system is called overdetermined if it has more system equations than unknowns. Overdetermined systems typically do not have a solution.

►Distribution Problem in Biomechanics

Overfitting

Definition

Overfitting refers to a problem that can arise during the training of artificial neural networks (or other statistical learning systems). During training the network learns a mapping from the input domain to the desired output. The target of this process is to capture the underlying regularities in the data that are to be modelled. However, since there are, in general, limited amounts of training data, the network may learn to approximate these correctly, while failing to process new data appropriately. This is referred to as overfitting: the network has learned a function that is too complex, modelling not only the regularities of the dataset, but also its noise.

►Connectionism

Overhang (DNA)

Definition

When a restriction cleaves DNA asymmetrically a stretch of single stranded nucleotides is left. If the single stranded bases end in a 3' hydroxyl a 3' overhang remains. Similarly, a 5' overhang remains when the single stranded bases end in a 5' phosphate. Overhangs are often generated in molecular biology by use of DNA endonucleases. Larger overhangs of several nucleotides, such as those created by restriction endonucleases, are often called "sticky-ends" since a DNA molecule with complimentary sequence in the overhang region can anneal to each other. This phenomenon is used in molecular biology to piece together DNA molecules from different sources which are then covalently linked with DNA ligase.

►Serial Analysis of Gene Expression

Overlap Zone in Skeletal Muscle

Definition

The overlap zone in skeletal muscle designates the area of overlap between the contractile proteins actin and myosin. At short muscle length, the overlap zone is big, and for increasing muscle length, the overlap zone becomes smaller. When there is no overlap between actin and myosin (i.e. the overlap zone has vanished), active force production is not possible anymore.

►Actin
►Force Depression/Enhancement in Skeletal Muscles
►Myosin
►Sarcomere Structural Proteins

Overshadowing

Definition

The ability of a conditioned stimulus (CS) to elicit a conditioned response (CR) is reduced when its pairings with the unconditioned stimulus (US) take place in the presence of another neutral stimulus. Assessment of the magnitude of overshadowing is made through comparison with a control group that receives only pairings of the CS and the US. This is one of several examples of cue competition or stimulus selection effects that prompted development of predictive-driven learning models.

►Theory on Classical Conditioning

Overshoot (of Action Potential)

Definition

Reversal of membrane potential during the action potential peak.

►Action Potential

Overtraining Syndrome (OTS)

Definition

When prolonged, excessive training stress are applied concurrent with inadequate recovery, many of the positive physiological changes associated with physical training are reversed with overtraining. Chronic physiological maladaptations and performance decrements occur. Throughout the twentieth century, many names have been given to this chronic maladaptive state (e.g., underperformance syndrome, sports fatigue syndrome), but presently the term overtraining syndrome (OTS) is used. A large number of symptoms associated with overtraining have been reported in the literature and categorized according to physiological performance, psychological/information processing, immunological, and biochemical parameters. It is also probable that other signs/symptoms typically associated with overtraining are evident before a deterioration in performance. These might include generalized fatigue, depression, muscle and joint pain, and loss of appetite. However, it is the decline in performance frequently associated with an increased volume or load of training that captures the attention of the athlete and coach.

►Stress Effects During Intense Training on Cellular Immunity

Owl

Definition

Mostly night active bird species of the order Strigiformes. The barn owl, especially, is a model system for investigating mechanisms of sound localization (see essay on "Sound localization in the barn owl"), depth vision and plasticity of the nervous system.

Oxidative Potential

Definition

Motoneurons, like the muscle fibers that they supply, derive their energy from metabolism that either requires oxygen or does not. The metabolism that does require oxygen to generate adenosine triphosphate for energy is referred to as oxidative energy, the cells having oxidative potential.

►Axonal Sprouting in Health and Disease
►Motoneuron

Oxidative Stress

Definition

Oxidative stress is a medical term for damage to animal or plant cells caused by reactive oxygen (ROS) and nitrogen (RNS) species, which include superoxide radical, singlet oxygen, peroxynitrite or hydrogen peroxide. It is defined as an imbalance between prooxidants and antioxidants, with the former prevailing.

► Alzheimer's Disease – Oxidative Injury and Cytokines

Oxytocin

Neuropeptide secreted as hormone by the neurohypophysis and involved in labor, lactation and reproduction.

► The Hypothalamo Neurohypophysial System
► Hypothalamo-Pituitary-Adrenal Axis
► Stress and Depression

Oxytocinergic Central Pathways

Definition

Oxytocinergic central pathways are involved in reproduction, cognition, tolerance, adaptation and the regulation of cardiovascular and respiratory functions. Centrally released oxytocin would also give rise to sedation.

► The Hypothalamo Neurohypophysial System

P0

Definition

A major protein component of the insulating myelin sheath around axons of the peripheral nervous system.

►Protein Zero

P2

Definition

Event-related brain potential approximately 150–250 ms after a stimulus during wakefulness. The P2 is linked to the N1 component, is reported to be elicited by attended and non-attended stimuli, and its amplitude is reported to be associated with the intensity of the eliciting stimulus, yet the amplitude is reported to be reduced in response to attended stimuli.

►Sleep – Motor Changes
►Sleep – Sensory Changes

p53

Definition

This is a transcription factor important for cell cycle regulation. p53 was one of the first tumour suppressor genes identified. p53 gene mutations are frequently identified in a variety of cancers.

►Chromatin Immunoprecipitation and the Study of Gene Regulation in the Nervous System

p75

►Brain Inflammation: Tumor Necrosis Factor Receptors in Mouse Brain Inflammatory Responses

P300

Definition

Event-related brain potential approximately 300 ms after a stimulus during wakefulness when stimuli are both detected and attended.

►Sleep – Motor Changes
►Sleep – Sensory Changes

P Element

Definition

P element – transposable elements are a heterogeneous class of genetic elements that can move and insert at new locations on a chromosome. P elements contain terminal inverted repeat sequences and encode the protein transposes to allow for integration into the genome.

►GAL4/UAS

p38 MAPK

Definition

p38 MAPK is a member of mitogen-activated protein kinase (MAPK) family. The name MAPK was given, because another member of the family extracellular signal-regulated kinase (ERK) was first recovered as a kinase activity in the cytosol of EGFtreated cells. Later, MAPK has been found to operate in three pathways: ERK, p38, and c-Jun N-terminal kinase/stress-activated protein kinase (JNK/SAPK). p38 MAPK is a kinase of 38 kDa that responds to stress condition such as tumor necrosis factor-alpha (TNFα), UV, and H_2O_2.

►Microglial Signaling Regulation by Neuropeptides
►Mitogen Activated Protein Kinase (MAPK)

p50 (NFKB1)

►Nf-κB – Potential Role in Adult Neural Stem Cells

p65 (RelA)

►Nf-κB – Potential Role in Adult Neural Stem Cells

PACAP

Definition

Pituitary adenylyl cyclase activating peptide; a neuropeptide expressed in nervous system where it functions to regulate cellular communication. PACAP has emerged as a likely retinal messenger to the suprachiasmatic nucleus (SCN), acting in concert with glutamate to communicate photic information to the circadian system. PACAP-like immunoreactivity is found in terminals of retinal ganglion cells (RGCs) innervating the SCN and two of the receptors sensitive to PACAP (PAC1 and VPAC2) are expressed in the SCN. To date, all of the available evidence indicates that the PAC1 receptor is responsible for mediating the effects of PACAP on SCN neurons.

Mechanistically, PACAP pre-synaptically enhances the release of glutamate onto SCN neurons and postsynaptically enhances the magnitude of the response to glutamate within the SCN. PACAP can also increase calcium in SCN neurons by causing a release from intracellular stores as well as an enhancement of voltage-dependent calcium currents. Increasing calcium has the consequence of activating the mitogen-activated protein kinase (MAPK) signaling cascade and increasing transcription. At a systems level, application of PACAP can shift the phase, or alter the magnitude of glutamate-induced phase shifts, in the circadian rhythm of SCN neuronal firing in a brain slice preparation. Similarly, microinjections of PACAP into the SCN region in vivo can cause phase shifts. Administration of a PACAP receptor antagonist or an antibody against PACAP attenuates light-induced phase delays. The circadian system of mice deficient in PACAP or the PAC1 receptor exhibit altered behavioral responses to light. Overall, these studies point to a role for PACAP in increasing the functional coupling between the melanopsin-containing RGCs and the retino-recipient subpopulation of SCN neurons.

►Circadian Rhythm
►Melanopsin
►p38 MAPK
►Phase Response Curve (PRC)
►Retinal Ganglion Cells
►Suprachiasmatic Nucleus (SCN)

Pacemaker

Definition

In a rhythmically active neuronal network, the neuron or neurons that have the major role in generating the rhythm. These may exhibit spontaneous oscillations of their membrane potential leading to action potentials that are very nearly equally spaced (endogenous bursting neurons), or a mutually excitatory set of neurons. An oscillator in a multi-oscillatory system entrains the other oscillators in the system, and so sets their phase relative to the pacemaker's phase, as well as forcing each oscillator's period to equal that of the pacemaker. An entrained oscillator's phase may lead or lag (occur either earlier or later than) that of the pacemaker, depending on the type of coupling (inhibitory or excitatory) and the relative periods of the oscillators in the system. Experimental proof that an oscillator can behave as a pacemaker in a physiological system is provided when transplantation confers the period or phase of the donor to the reconstituted system.

►Stomatogastric Ganglion
►Tonic Activity of Sympathetic Nerves

Pacemaker Neurons

►Bursting Pacemakers

Pacemaker Potential

Definition

An intrinsically generated rhythmic membrane fluctuation that is caused by voltage-dependent and voltage-independent ion fluxes. A pacemaker potential that gives rise to action potentials is called a pacemaker "burst." Synonym for pacemaker potential: "drive potential."

►Bursting Pacemakers

Pacinian Corpuscle

Definition

Pacinian corpuscles are the largest cutaneous mechanoreceptors, but are also found in deep structures (e.g., abdominal mesentery). They are exquisitely sensitive to light mechanical stimuli, including high frequency vibration (>60 Hz).

►Vibration Sense
►Pacinian Corpuscle Regeneration

Pacinian Corpuscle Regeneration

Chizuka Ide
[1]Department of Anatomy and Neurobiology, Kyoto University Graduate School of Medicine, Yoshidakonoe-cho, Sakyo-ku, Kyoto, Japan;
[2]Department of Occupational Therapy, Aino University Faculty of Nursing and Rehabilitation, Shigashiohta 4-5-4, Ibaragi City, Osaka, Japan

Synonyms

Vater-Pacini corpuscle; Vater-Pacini's corpuscle; Pacini corpuscle; Pacini's corpuscle; Pacinian corpuscle; Corpusculum lamellosum

Definition

The Pacinian corpuscle is the largest ellipsoidal sensory corpuscle functioning as a very rapidly adapting mechanoreceptor. It consists of one straight axon terminal extending at the center of the corpuscle along its long axis, the inner core surrounding the axon terminal, and the outer core occupying the outermost part of the corpuscle.

Characteristics

Structure

The Pacinian corpuscle is ellipsoidal in shape, measuring 0.2–1 mm at the long axis, and can be seen by the naked eye in dissection. It is found in connective tissues of the subcutaneous layer, joint capsule and periosteum. The corpuscle is innervated by a single thick myelinated axon that ends as a straight axon terminal in the center. The axon terminal is sandwiched by hemicircular inner cores, so that the axon terminal is oval in cross section [1]. The inner core consists of stacks of numerous thin cytoplasmic processes of specialized Schwann cells. The outer core is composed of loose lamellae of modified perineurial cells. The inner core cells express p75 and TrkB, while the outer core cells express only p75 [2]. Small axoplasmic protrusions are formed at the oval edges as well as at the extreme end of the axon terminal. These protrusions are considered to be the site of mechano-electric transduction [3].

Regeneration

Following denervation, the axon terminal disappears, and the inner core becomes somewhat atrophic but remains with the outer core for an extended period. Following re-innervation, an axon enters the corpuscle and the inner core lamellae become "active," as in the case of the Meissner corpuscle. After regeneration, most Pacinian corpuscles are innervated by a single axon, but there are a few that receive two axons, or remain non-innervated. Some corpuscles have multi-terminals associated with inner cores, resulting partly from the branching of regenerating axon terminals [4]. Aberrant regenerating nerves other than sensory axons can enter Pacinian corpuscles [5].

Pacinian corpuscles can be regenerated even in the non-cellular environment; the connective tissue scaffolds of the Pacinian corpuscle remain after the cellular components have been degraded by local freeze-treatment. A regenerating axon enters such acellular scaffolds, accompanied by Schwann cells migrating from the proximal stump (Fig 1). Schwann cells develop into inner core cells associated with axon terminals. Perineurial cells develop into outer core cells within the scaffold of the original outer core region. Although atypical in its organization, the newly regenerated corpuscle possesses the three basic components including axon terminals, inner and outer cores [6]. This indicates

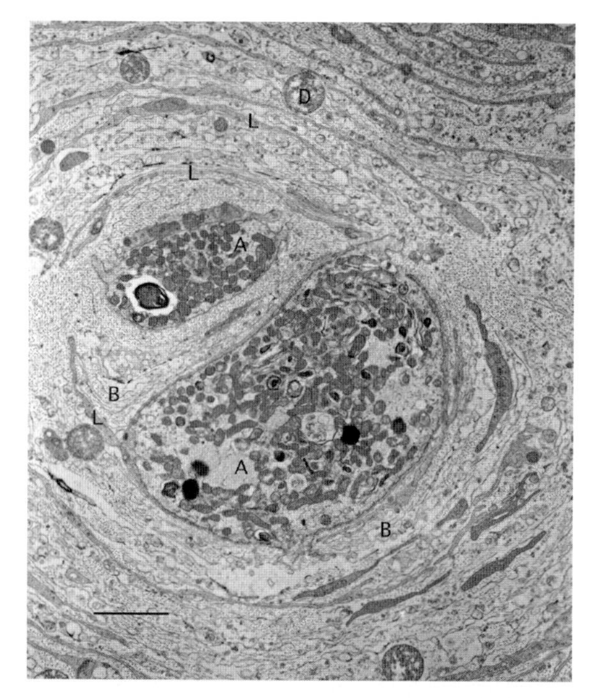

Pacinian Corpuscle Regeneration. Figure 1 The periosteum of the tibia, in which numerous Pacinian corpuscles are located, was freeze-treated to kill cellular components of the corpuscle in the rat. The acellular matrix including basal laminae (B) of the corpuscle remained after the cellular components had been degraded. Regenerating axons accompanied by immature Schwann cells enter the matrix; the axon terminals (A) are situated at the center of the matrix and Schwann cells extend thin cytoplasmic lamellae (L) along the basal lamina scaffolds around axon terminals with a pattern similar to that of the normal corpuscle. New Pacinian corpuscles, although atypical in overall cellular structure, can develop in the acellular matrix of the old corpuscle. (D) cell debris. Scale bar: 2 μm.

that the acellular matrix of the Pacinian corpuscle has the ability to induce the innervating axons, Schwann cells and perineurial cells to develop into axon terminals, inner core cells, and outer core cells, respectively.

References

1. Munger BL, Yoshida Y, Hayashi S, Osawa T, Ide C (1988) A re-evaluation of the cytology of cat Pacinian corpuscle. I. The inner core and clefts. Cell Tissue Res 253:83–93
2. Stark B, Risling M, Carlstedt T (2001) Distribution of the neurotrophin receptors p75 and trkB in peripheral mechanoreceptors; observations on changes after injury. Exp Brain Res 136:101–107
3. Ide C, Hayashi S (1987) Specialization of plasma membrane in Pacinian corpuscles: Implications for mechanoelectric transduction. J Neurocytol 16:759–773
4. Zelena J, Zacharova G (1997) Reinnervation of cat Pacinian corpuscles after nerve crush. Acta Neuropathol 93:285–293
5. Zelena J, Jirmanova I, Lieberman AR (1990) Reinnervation of transplanted Pacinian corpuscles by ventral root axons: Ultrastructure of the regenerated nerve terminals. J Neurocytol 19:962–969
6. Ide C (1987) Role of extracellular matrix in the regeneration of a Pacinian corpuscle. Brain Res 413:155–169

Paciniform Endings

Definition

Small mechanoreceptors resembling Pacinian corpuscles, located in the deeper regions of the dermis of glabrous skin.

►Pacinian Corpuscle
►Vibration Sense

Paedomorphosis

Definition

The brains of amphibians appear to be much simpler than those of other vertebrates. Lissamphibians, i.e., living amphibians, have undergone secondary simplification, which arises from paedomorphosis, a form of heterochronic evolution. This process has affected the three amphibian orders differently: anurans appear to be least and salamanders most paedomorphic, while caecilians exhibit an intermediate degree of paedomorphosis.

It commonly involves different degrees of retardation, reduction or absence of traits in otherwise fully developed organisms when compared with phylogenetic outgroups. Thus, a mosaic of fully adult traits, weakly expressed traits, and missing characters appears in terminal ontogenetic stages. Accordingly, amphibian brains are expected to have fewer cells, a lower degree of morphological differentiation of cells, and reduced migration, but retain the plesiomorphic structural, functional and developmental organization found among other vertebrates.

►Evolution of the Visual System: Amphibians

PAG

Definition

►Periaqueductal Gray Matter (PAG)
►Pain Imaging

Pain

G. F. Gebhart
Center for Pain Research, University of Pittsburgh, Pittsburgh, PA, USA

Introduction

Pain is appreciated to be a complex sensory experience, characterized by both discriminative and emotional/cognitive dimensions. Moreover, the peripheral and central nervous system components that comprise the pain "network" are highly plastic, meaning that neural and non-neural constituents undergo changes in behavior and excitability in painful conditions. Thus, tissue insult commonly leads to changes in either or both the quality and intensity of perceived stimuli, typically beginning at peripheral sites and including central components in persistent pain states.

Stimuli that evoke pain are termed noxious and the peripheral sensory receptors/transduction sites acted upon by ►noxious stimuli are termed ►nociceptors. Input from nociceptors is widely distributed throughout the central nervous system and can evoke simple ►nociceptive reflexes that are organized at the level of the spinal cord (e.g., ►nociceptive withdrawal reflexes), engage ►autonomic centers in the brainstem that increase heart rate and blood pressure, or lead to expression of emotional-affective responses that can be influenced by gender, age, previous experience, ►stress and mood (among other factors) (►Emotional/affective aspects of pain; ►Pain in older adults; ►Pain in children; ►Gender/sex difference in pain).

More than 100 years ago, Sherrington [1] advanced the operational definition of a noxious stimulus and anticipated by decades the discovery of sensory receptors (nociceptors) in skin that responded only to noxious intensities of stimulation. Sherrington's experimental work established that mechanical, thermal or chemical stimuli that damaged or threatened damage to skin were *adequate* for activation of nociceptors to cause pain. We know now that adequate noxious stimuli differ for skin, muscle, joints and internal organs, and also that some nociceptors can be activated by low-threshold stimuli, revealing the importance of encoding of stimulus intensity by peripheral sensory receptors (►Cutaneous pain, nociceptor and adequate stimuli). For example, cutting, crushing or burning stimuli, which reliably produce pain when applied to skin, are not reliable noxious stimuli when applied to internal organs. Pain arising from internal organs is more commonly produced by over-distension, traction on the mesenteries, ischemia or inflammation (e.g., appendicitis). Similarly, adequate noxious stimuli for muscle and joints, which also are not exposed to the external environment, include chemicals typically associated with inflammatory processes. In further distinction from skin, deep pain such as arises from muscle and viscera are relatively poorly localized and commonly referred to other sites, including overlying skin and muscle. The clinical presentation and characteristics of muscle pain (►Muscle pain including fibromyalgia), ►joint pain (►Joint pain, nociceptors and adequate stimuli) and ►visceral pain are discussed in detail in essays in this section of the Encyclopedia (see also ►incisional/post-op pain; ►Low back/spine pain).

As indicated above, tissue insult commonly alters either the quality and/or intensity of applied stimuli. Hyperalgesia (an increased response to a stimulus which is normally painful) and allodynia (pain due to a stimulus which does not normally provoke pain [►Hyperalgesia and allodynia]) represent increases in the excitability of nociceptors (hyperalgesia) and activation of low-threshold mechanoreceptors (allodynia – by mechanisms not fully understood), respectively. Increases in excitability of nociceptors have been documented to arise from changes in ►voltage-gated ion channels (e.g., ►Ca^{2+} channels, ►K^+ channels and ►Na^+ channels [►Voltage-gated ion channels and pain]) and ►ligand-gated channels and receptors (e.g., transient receptor potential [TRP] channels (►TRP channels), purinergic receptors, including both P2X and P2Y, etc. [►G-protein coupled receptors (GPCRs) in sensory neuron function and pain]). Such changes can be initiated by a variety of peripheral mediators, including those associated with inflammation (e.g., ►prostaglandins, protons, etc. [►Inflammatory pain]), ►growth factors (►Growth factors and pain) and the ►immune system (►Immune system and pain). Typically, increased excitability of nociceptors (for example produced by inflammatory mediators) is relatively short-lived and reversible. However, these changes can persist, such as occurs in ►autoimmune diseases characterized by dysregulation of an immune response or following tissue insult early in life. ►Rheumatoid arthritis, ►multiple sclerosis and some viral infections can produce ►chronic pain that is difficult to manage.

After nociceptor neurogenesis and maturation (►Development of nociception) [2], tissue damage in the neonatal period can lead to increased pain sensitivity [3] and exaggerated responses to noxious stimuli in adult non-human animals well after the early insult has fully recovered. For example, organ insult in neonatal animals [4,5], skin incisions/surgery as well as stressful events (e.g., maternal separation [6]; all have been shown to lead to increased sensitivity to noxious stimuli in adult life. This increased sensitivity is not always apparent when acute noxious stimuli are tested, but is clearly evident when tissue is re-inflamed or injured.

These peripheral events, whether introduced early in life or produced in adults, have far-reaching central consequences. ►Sensitization (increased excitability of nociceptors [7]; leads to changes in excitability of neurons in the spinal cord as well as sites rostral in the brain. By analogy to the periphery, increases in the excitability of spinal and supraspinal neurons is referred to as "central sensitization" [8]. The initial impetus for the increase in excitability of central neurons arises from the increased input from peripheral nociceptors, including ►silent nociceptors (►Nociceptors and characteristics) [9], and increased release of ►neurotransmitters from their central terminals. Central sensitization, either at the level of the spinal cord or at supraspinal sites, represents plasticity of central neurons, which apparently can be sustained well beyond recovery from the peripheral insult.

At the level of the spinal cord, because of similarities in neurotransmitters released from nociceptor terminals and characteristics of neuron response properties, central sensitization shares characteristics with learning (►Synaptic long-term potentiation (LTP) in pain pathway). At supraspinal sites, nociceptive input is widely distributed to sites important for the discriminative aspects of pain (e.g., location, duration and intensity) (►Ascending nociceptive pathways) and also to brain areas associated with emotion and cognition (►Pain imaging; Emotional/affective aspects of pain). This nociceptive input also influences sites in the brainstem important to descending control of spinal input (►Descending modulation of nociception). Descending influences from the midbrain and medulla were initially believed to be principally, if not exclusively, inhibitory in nature and selective for nociceptive input. We now know that descending influences can contribute to chronic pain states, either by reduced descending inhibition or active facilitation of spinal input and, moreover, and are not selective for spinal nociceptive input, but also modulate non-noxious (innocuous) spinal input. Indeed, it is now considered that chronic disorders such as functional gastrointestinal diseases, ►fibromyalgia, etc. may be contributed to by disordered descending modulation of spinal input.

It addition to the nociceptive component of the experience of pain (e.g., activation of nociceptors, spinal pathways, mechanisms of peripheral and central sensitization, etc.), nociceptive input at supraspinal sites also engages emotional, affective and cognitive dimensions of pain (Emotional/affective aspects of pain). That supraspinal influences can be potent is attested to by the ►placebo analgesic response, in which expectancy has been identified as the most relevant psychological mediator. Expectancy can be manipulated by verbal instruction or by conditioning. The discriminative dimension of pain is associated with ►thalamic and ►somatosensory cortex and ►motor cortex whereas the emotional/affective dimension of pain activates the amygdala, cingulate gyrus and ►prefrontal cortex (Pain imaging). Brain imaging has thus confirmed and extended our anatomical understanding of the discriminative and emotional/cognitive anatomical dimensions of the experience of pain. Interestingly, both the discriminative and emotional/cognitive dimensions of pain may be sexually dimorphic. A growing literature reveals differences between males and females with respect to sensitivity to noxious stimuli as well as to the ability to tolerate pain (►Gender/sex differences in pain). Women access physicians for pain-related problems far more frequently than do men, and while there may be many reasons why this is so, one contributing factor certainly relates to distinct differences in responses to noxious stimuli, likely contributed to by hormonal influences.

Despite our significantly increased knowledge about pain and pain mechanisms, there remain significant challenges in several areas. ►*Neuropathic pain*, including ►*central pain*, arises from damage to the nervous system, either in the periphery or centrally. Unlike *inflammatory pain*, damage to the nervous system results in pain commonly produced by normally innocuous stimuli (i.e., touch-evoked pain, or allodynia), which is difficult to manage. Pain in neonates (Development of nociception) and children (Pain in children), as well as pain in older adults, including those with ►dementia (Pain in older adults), similarly present challenges in terms of both pain assessment/measurement (►Pain psychophysics) and pain management. It was incorrectly assumed until relatively recently that the nociceptive system was undeveloped or underdeveloped in neonates and, accordingly, that they did not feel pain (Development of nociception; Pain in children) or require analgesia or anesthesia. This flawed thinking has fortunately been corrected and we now know that untreated, unattended pain in neonates and young children can lead to long-term changes in responses to noxious stimuli. Because young children and adults with dementia cannot effectively communicate their pain, they too tend to be under-treated.

In the United States, pain accounts for 20% of patient visits to physicians and 10% of prescription drug sales [10], figures likely comparable to those in many other countries (but not including those where governments restrict access to analgesic drugs, and unrelieved pain is commonplace). Pain management can be daunting for health care providers, even with access to all available drugs and management strategies. As recounted in the essays in this section of the Encyclopedia, significant progress has been achieved in understanding mechanisms underlying painful disease conditions, with consequent improvement in pain management.

References

1. Sherrington CS (1906) The integrative action of the nervous system. Charles Scribner's Sons, New York
2. Woolf CJ, Ma Q (2007) Nociceptors – noxious stimulus detectors. Neuron 55:353–364
3. Hermann C, Hohmeister J, Demirakca S, Zohsel K, Flor H (2006) Long-term alteration of pain sensitivity in school-aged children with early pain experiences. Pain 125:278–285
4. Al-Chaer ED, Kawasaki M, Pasricha PJ (2000) A new model of chronic visceral hypersensitivity in adult rats induced by colon irritation during postnatal development. Gastroenterol 119:1276–1285
5. Randich A, Uzzell T, DeBerry JJ, Ness TJ (2006) Neonatal urinary bladder inflammation produces adult bladder hypersensitivity. J Pain 7:469–479
6. Coutinho SV, Plotsky PM, Sablad M, Miller JC, Zhou H, Bayati AI, McRoberts JA, Mayer EA (2002) Neonatal maternal separation alters stress-induced responses to viscerosomatic nociceptive stimuli in rat. Am J Physiol 282:G307–G316
7. Bessou P, Perl ER (1969) Response of cutaneous sensory units with unmyelinated fibers to noxious stimuli. J Neurophysiol 32:1025–1043
8. Ji R-R, Woolf CJ (2001) Neuronal plasticity and signal transduction in nociceptive neurons: implications for the initiation and maintenance of pathological pain. Neurobiol Dis 8:1–10
9. Schaible H-G, Schmidt RF (1984) Effects of an experimental arthritis on the sensory properties of fine articular afferent units. J Neurophysiol 54:1109–1122
10. Max MB (2003) How to move pain and symptom research from the margin to the mainstream. J Pain 4:355–360

Pain, Neuropathic

Definition

►Neuropathic Pain

Pain among Seniors

►Pain in Older Adults (Including Older Adults with Dementia)

Pain and Growth Factors

►Growth Factors and Pain

Pain and Immune System

►Immune System and Pain

Pain and Ligand-gated Channels/Receptors

►G-Protein Coupled Receptors in Sensory Neuron Function and Pain

Pain and Voltage-gated Ion Channels

►Voltage-Gated Ion Channels and Pain

Pain Distress

►Emotional/Affective Aspects of Pain

Pain Emotion

Definition

Or pain emotional component. The emotional reactions and feeling states associated with thoughts (cognitions) about pain, such as anxiety, depression, fear, and despair. These feeling states involve thoughts about the present, past and future, and are distinct from the immediate feelings of pain unpleasantness that motivate

P

or inhibit behaviors and that are similar to other immediate feelings such as hyperthermia or the urge to breathe.

►Emotional/Affective Aspects of Pain

Pain Hypersensitivity

►Hyperalgesia and Allodynia

Pain Imaging

Irene Tracey
Department of Clinical Neurology and Nuffield Department of Anaesthetics, Centre for Functional Magnetic Resonance Imaging of the Brain, Oxford University, Oxford, UK

Definition

Pain imaging is the capacity to identify functionally relevant neuronal activity within the central nervous system (brain and spinal cord) correlated with the subjective experience of pain. Imaging relates principally to studies in humans, but not exclusively. Different technologies provide this capability to image pain with varying degrees of invasiveness, spatial and temporal resolution.

Characteristics

Why Image Pain?

Most neuroimaging methods provide a non-invasive, systems-level understanding of the central mechanisms involved in pain processing. To date, the focus has been to dissect the physiological, psychological and cognitive factors that influence nociceptive inputs to alter pain perception in healthy subjects and patients suffering from chronic pain. Obtaining reliable objective information related to the individual's subjective pain experience provides a powerful means of understanding not only the central mechanisms contributing to the chronicity of pain states, but also potential diagnostic information. Identifying non-invasively where plasticity, sensitization and other amplification processes might occur along the pain neuraxis for an individual, and relating this to their specific pain experience or measure of pain relief, is of considerable interest to the clinical pain community and pharmaceutical industry. This is why imaging pain is useful.

Imaging the Brain – Methods Available

Figure 1 illustrates the main imaging modalities in use today and what physiological correlate of brain activity they measure. There is a "cost" or balance between the spatial and temporal information achievable and how "invasive" you have to be if you want high resolution in both domains. Therefore, when choosing your imaging modality, pros and cons must be considered, dependent upon your hypothesis and goal. Other methods provide different sorts of information about the brain (i.e., structural or metabolic rather than functional), and these newer ways of examining the human brain are providing exciting and highly novel information about pain processing.

Pain as a Perception

Pain is a conscious experience, an interpretation of nociceptive inputs influenced by memories, emotional,

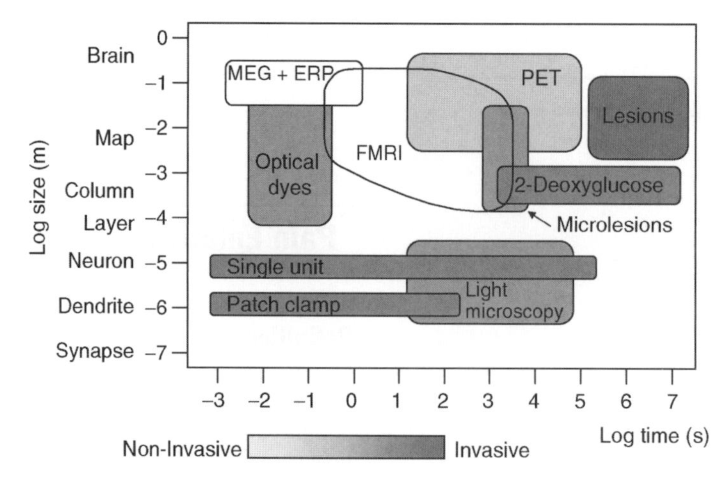

Pain Imaging. Figure 1 A schematic displaying the relationship between the spatial and temporal resolution, as they relate to non-invasiveness, for the main current imaging tools.

pathological, genetic and cognitive factors. Resultant pain is therefore not always related linearly to nociceptive drive or input, neither is it solely for vital protective functions; this is especially true in, chronic pain states. Furthermore, the behavioral response of a subject to a painful event is modified according to what is appropriate or possible in any particular situation. Pain is, therefore, a highly subjective experience. Figure 2 illustrates the mixture of physical, cognitive and emotional factors that influence nociceptive inputs to amplify, attenuate and color the pain experience.

Clearly, the majority of factors influencing pain percepts are centrally mediated and our ability to unravel and dissect their contribution has only been feasible since neuroimaging allowed us non-invasive access to the human CNS. Determining the balance between peripheral versus central influences, and ascertaining which are due to pathological versus emotional or cognitive influences, will clearly aid decisions regarding the targeting of treatments (i.e., pharmacological, surgical, cognitive behavioral or physical rehabilitation). This is perhaps where imaging might provide its greatest contribution in the field of pain.

The "Cerebral Signature" for Pain Perception

Because pain is a complex, multifactorial subjective experience, a large distributed brain network is accessed during nociceptive processing; this is often called the "pain matrix" and simplistically can be thought of as having lateral components (sensory–discriminatory, involving areas such as primary and secondary somatosensory cortices, thalamus and posterior parts of insula) and medial components [affective–cognitive-evaluative, involving areas like the anterior parts of insula, anterior cingulate cortex (ACC) and prefrontal cortices]. However, because different brain regions play a more or less active role depending upon the precise interplay of the factors involved in influencing pain perception (e.g., cognition, mood, injury, and so forth), the "pain matrix" is not a defined entity [1]. A recent meta-analysis of human data from different imaging studies provides clarity regarding the commonest regions found active during an acute pain experience as measured by PET and ►FMRI [2] (See Fig. 3).

These areas include: primary and secondary somatosensory, insular, anterior cingulate, and prefrontal cortices as well as the thalamus. This is not to say these areas are the fundamental core network of human nociceptive processing (and if ablated would cure all pain), although studies investigating acute pharmacologically induced analgesia do show predominant effects on this core network, suggesting their overall importance on influencing pain perception. Other regions such as basal ganglia, cerebellum, amygdala, hippocampus, and areas within the parietal and temporal cortices can also be active dependent upon the particular set of circumstances for that individual (see Fig. 2). A "cerebral signature" for pain is perhaps how we should define the network that is necessarily unique for each individual.

To understand how nociceptive inputs are processed and altered to subsequently influence changes in the pain experienced, it is useful to separately examine the main factors listed in Fig. 2 that alter pain perception.

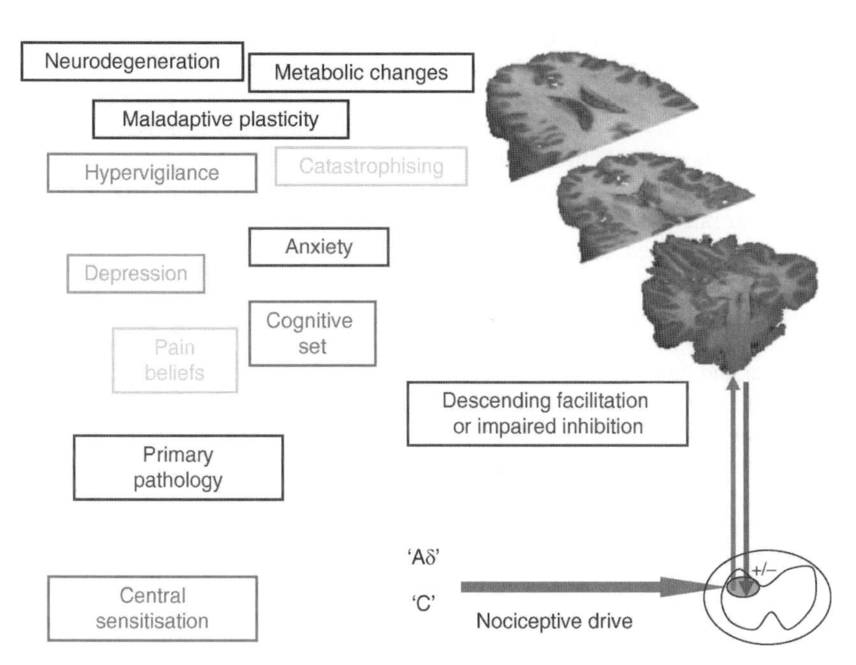

Pain Imaging. Figure 2 Schematic illustrating the main factors that influence nociceptive inputs to alter pain perception.

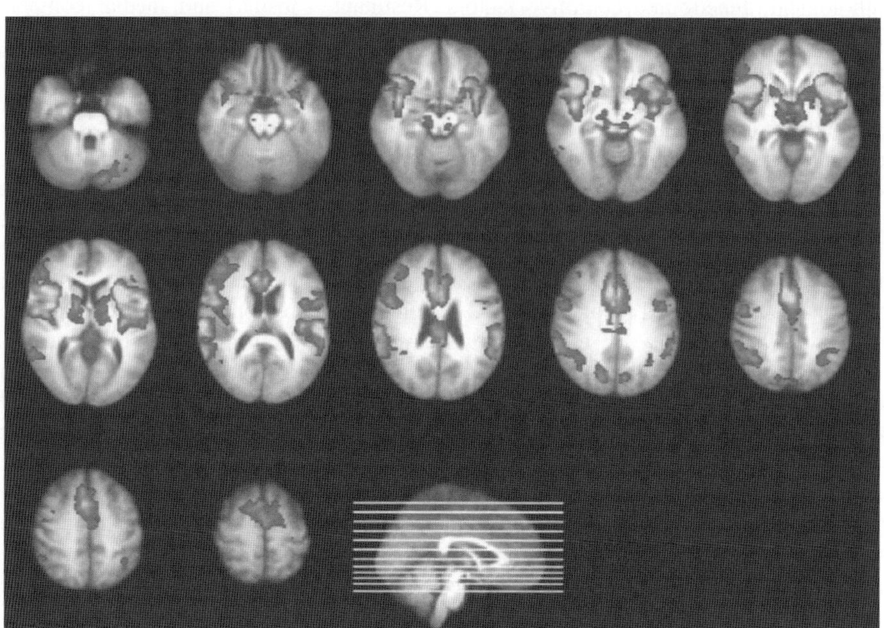

Pain Imaging. Figure 3 Neuroanatomy of pain processing. Main brain regions that activate during a painful experience are highlighted as bilaterally active but with more dominant activation on the contralateral hemisphere (more yellow).

Genetics: We cannot ignore the possibility that our genes influence both how nociceptive stimuli are processed and how the brain reacts to peripheral injury and increased nociceptive inputs. Similarly, we cannot ignore the central role that our life experiences have on both these processes. Imaging studies have investigated whether individuals claiming to be more "sensitive" to pain, compared with others, activate more brain regions involved in pain perception. Early work suggests that subjects who rated the pain highest exhibit more robust pain-induced activation of ►S1, ACC, and ►PFC compared with those who rated pain lowest. The key question is whether this increased pain report and correlated objective readout is nature or nurture driven. Studies are beginning to link genetic influences on human nociceptive processing with physical processes within the brain. Zubieta and colleagues examined the influence of a common functional genetic polymorphism affecting the metabolism of catecholamines on the modulation of responses to sustained pain in humans using psychophysical assessment and ►PET [3]. Individuals homozygous for the met158 allele of the catechol-*O*-methyltransferase (COMT) polymorphism (val158met) showed diminished regional μ-opioid system responses to pain (measured using PET) and higher sensory and affective ratings of pain compared with those homozygous for the valine polymorphism. This study and others are providing good evidence regarding how our genes influence nociceptive processing within the brain and consequently our pain experience.

Attention

We know from experience that attention is very effective in modulating the sensory and affective aspects of pain. FMRI and neurophysiological studies showed attention- and distraction-related modulations of pain-evoked activations in many parts of the pain "matrix." From these studies, regions that appear critical during the attentional modulation of pain include the descending pain modulatory system as well as key elements of the pain "matrix."

The descending pain modulatory system

This is a well characterized anatomical network that enables us to regulate nociceptive processing (largely within the dorsal horn) in various circumstances to produce either facilitation (pro-nociception) or inhibition (anti-nociception) [4] (see also ►Descending modulation of nociception). The pain-inhibiting circuitry, of which the periaqueductal grey (►PAG) is a part, is best known and contributes to environmental (e.g., during the fight or flight response) and opioid-mediated analgesia. There are descending pathways that facilitate pain transmission, however, and it is thought that sustained activation of these circuits may underlie some states of chronic pain (see below). Recently, researchers have investigated whether alteration in people's attention influences brainstem activity and, therefore, nociceptive processing via cortical–brainstem influences. In an early study using high-resolution imaging of the human brainstem, we showed significantly increased activity within the PAG in subjects who were distracted compared to when they paid attention to their pain, with concomitant changes in pain ratings. Indeed, the change in pain rating between attending and distracting conditions correlated with the change in PAG activity across the group, suggesting a varying capacity to engage the descending inhibitory system in normal individuals. Further work by others has extended these observations and shown that the cingulo-frontal cortex exerts top–down influences on the PAG and posterior thalamus to gate pain modulation during distraction. These studies, and others, provide clear evidence for the involvement of brainstem structures in the attentional modulation of pain perception, and recent work using diffusion tractrography confirms that anatomical connections exist between cortical and brainstem regions in the human brain, thereby enabling such top-down influences.

Placebo

Recent work in humans has helped provide a framework by which the placebo effect and subsequent analgesia is mediated (see ►Placebo analgesic response). Again, the brainstem is critically involved in mediating placebo analgesia. Descending influences from the diencephalon, hypothalamus, amygdala, ►ACC) insula and prefrontal cortex that elicit inhibition or facilitation of nociceptive transmission via brainstem structures are thought to occur during placebo analgesia. Using PET, it has been confirmed that both opioid and placebo analgesia are associated with increased activity in the rostral ACC, and that a covariation between the activity in the rostral ACC and the brainstem during both opioid and placebo analgesia, but not during pain alone, exists. Wager and colleagues extended these early observations to examine placebo expectation effects [5]. Using a conditioning design, they found that placebo analgesia was related to decreased brain activity in classic pain processing brain regions (e.g., thalamus, insula, and ACC), but was additionally associated with increased activity during anticipation of pain in the prefrontal cortex (PFC), an area involved in maintaining and updating internal representations of expectations. Stronger PFC activation during anticipation of pain was found to correlate with greater placebo-induced pain relief and reductions in neural activity within pain regions. Furthermore, placebo-increased activation of the PAG was found during anticipation, the activity within which correlated significantly with dorsolateral PFC (DLPFC) activity. This is consistent with the concept that prefrontal mechanisms can trigger opioid release within the brainstem and, thereby, influence the descending pain modulatory system to modulate pain perception during the placebo effect.

Mood

For both chronic and acute pain sufferers, one's mood and emotional state has a significant impact on resultant pain perception and ability to cope. For example, it is a common clinical and experimental observation that anticipating and being anxious about pain can exacerbate the pain experienced. Anticipating pain is highly adaptive, but for the chronic pain patient it becomes maladaptive and can lead to fear of movement, avoidance, anxiety, and so forth. Studies aimed at understanding how anticipation and anxiety cause a heightened pain experience have been performed using imaging methods [6]. Critical regions involved in amplifying or exacerbating the pain experience include the entorhinal complex, amygdala, anterior insula and prefrontal cortices.

With regard to mood, depressive disorders often accompany persistent pain. Although the exact relationship between depression and pain is unknown, with debate regarding whether one condition leads to the other or if an underlying diathesis exists, studies have attempted to isolate brain regions that may mediate their interaction. Early studies indicated that activation in the amygdala and anterior insula appears to differentiate fibromyalgia patients with and without major depression. Another fibromyalgia study found that pain catastrophizing (defined as a set of negative emotional and cognitive processes), independent of the influence of depression, was significantly associated with increased activity in brain areas related to anticipation of pain (medial frontal cortex, cerebellum), attention to pain (dorsal ACC, dorsolateral PFC), emotional aspects of pain (claustrum, closely connected to amygdala) and motor control [7]. The construct of catastrophizing incorporates magnification of pain-related symptoms, rumination about pain, feelings of helplessness, and pessimism about pain-related outcomes. The results by Gracely and colleagues support the notion that catastrophizing influences pain perception through altering attention and anticipation, as well as heightening emotional responses to pain (see ►Emotional/affective aspects of pain).

The prefrontal cortex and pain

It is clear from these few studies described above and others in the literature that pronounced PFC activation is consistently found across clinical pain conditions, irrespective of underlying pathology. We are only beginning to unravel the roles of specific PFC regions in pain perception; it is thought they reflect emotional, cognitive and interoceptive components of pain conditions, as well as perhaps processing of negative emotions, response conflict and detection of unfavorable outcomes in relation to self. Interestingly, imaging studies attempting to capture the neural signature of the ongoing, spontaneous pain that patients commonly experience are finding increased medial PFC, including rostral ACC, activity during episodes of sustained high ongoing pain. These early data suggest a very different neural "signature" for the patient's ongoing pain, compared to the acute nociceptive network found active in response to provoked stimulation, as described above in most FMRI studies. A specific role for the lateral PFC as a "pain control center" has been put forward in a study of experimentally induced allodynia in healthy subjects [8]. In this study, increased lateral PFC activation was related to decreased pain affect, supposedly by inhibiting the functional connectivity between medial thalamus and midbrain, thereby driving endogenous pain-inhibitory mechanisms.

It is important to also note that the PFC (specifically the dorsolateral PFC) is one site of potential major neurodegeneration and cell death in chronic pain patients. These latest findings suggest that severe chronic pain could be considered a neurodegenerative disorder that especially affects this region. However, determining what the possible causal factors are that produce such neurodegeneration is difficult. Candidates include the chronic pain condition itself, the pharmacological agents prescribed for pain management or perhaps the physical lifestyle change subsequent to becoming a chronic pain patient. Carefully controlled longitudinal studies are needed.

Pain without a nociceptive input

Recent imaging data display activity of the near entire "pain matrix" without any nociceptive input during empathy and hypnosis manipulations, suggesting it is time to reconsider how we define central pain processing with respect to the origin of the input and resultant perception and meaning. This is not to say that pain experienced without a nociceptive input (sometimes referred to as psychogenic pain) is any less real than "physically" defined pain; indeed, neuroimaging studies have highlighted the physiological reality of such experiences due to the extensive neural activation that occurs.

Injury

Recently, changes within the descending pain modulatory network have been implicated in chronic pain (central sensitization) and in functional pain disorders [9] (see also ►Descending modulation of nociception). Changes are defined in terms of patients having either a dysfunctional descending inhibitory system or an activated and enhanced descending facilitatory system. There has been convincing evidence advanced regarding the differential involvement of the PAG, rostroventromedial medulla (►RVM), parabrachial nucleus (►PB), dorsal reticular nucleus and nucleus cuneiformis (►NCF) in the generation and maintenance of central sensitization states and hyperalgesia in both animal models and, for the first time, in a human model of secondary hyperalgesia [10]. Changes within the descending pain-modulatory network in chronic pain, in terms of patients having either a dysfunctional descending inhibitory system or an activated and enhanced descending facilitatory system, are clearly implicated in these and increasingly in other clinical studies.

Understanding which CNS areas are involved in engaging or disengaging this descending modulatory system has significant potential to not only further our understanding of how pain is perceived, but in developing mechanism-based therapies for treating different types of acute and chronic pain.

Spinal cord imaging

Clearly, to determine the extent of changes present within the CNS we must develop methods that allow noninvasive access to the changes within the human spinal cord, and these are currently being successfully developed.

Altered opioidergic and dopaminergic pathways

The availability of PET ligands for opioid and dopamine receptors has allowed the study of these receptor systems in several clinical pain states. Early opioid receptor ligand studies showed decreased binding in patients with chronic pain that normalized after reduction of their pain symptoms. Regional differences in ligand binding have recently been found in neuropathic pain studies with decreased binding in several key areas involved in pain perception. The dopaminergic pathways have also been implicated in pain processing in animal and patient studies. Early studies in fibromyalgia patients indicate reduced presynaptic dopaminergic activity in several brain regions in which dopamine plays a critical role in modulating nociceptive processes. Similar to the endogenous opioid system, the issue of cause and effect between a "functional hypodopaminergic state" and pain has yet to be resolved, making this an exciting area of current research.

Conclusion

Knowledge regarding how pain is perceived at a central level in humans is growing. An extensive network is recruited that is highly modifiable depending upon genetics, the environment, mood and the particular injury sustained. Combined, these produce a unique cerebral signature that produces an individualized pain experience.

References

1. Tracey I, Mantyh PW (2007) The cerebral signature for pain perception and its modulation. Neuron 55:377–391
2. Apkarian AV, Bushnell MC, Treede RD, Zubieta JK (2005) Human brain mechanisms of pain perception and regulation in health and disease. Eur J Pain 9:463–484
3. Zubieta JK, Heitzeg MM, Smith YR, Bueller JA, Xu K, Xu Y, Koeppe RA, Stohler CS, Goldman D (2003) COMT val158met genotype affects mu-opioid neurotransmitter responses to a pain stressor. Science 299:1240–1243
4. Fields H (2005) Central nervous system mechanisms of pain modulation. In: Wall PM, R, (ed) Textbook of pain. Churchill Livingstone, London,125–142
5. Wager TD, Rilling JK, Smith EE, Sokolik A, Casey KL, Davidson RJ, Kosslyn SM, Rose RM, Cohen JD (2004) Placebo-induced changes in FMRI in the anticipation and experience of pain. Science 303:1162–1167
6. Ploghaus A, Narain C, Beckmann CF, Clare S, Bantick S, Wise R, Matthews PM, Rawlins JN, Tracey I (2001) Exacerbation of pain by anxiety is associated with activity in a hippocampal network. J Neurosci 21:9896–9903
7. Gracely RH, Geisser ME, Giesecke T, Grant MA, Petzke F, Williams DA, Clauw DJ (2004) Pain catastrophizing and neural responses to pain among persons with fibromyalgia. Brain 127:835–843
8. Lorenz J, Minoshima S, Casey KL (2003) Keeping pain out of mind: the role of the dorsolateral prefrontal cortex in pain modulation. Brain 126:1079–1091
9. Suzuki R, Rygh LJ, Dickenson AH (2004) Bad news from the brain: descending 5-HT pathways that control spinal pain processing. Trends Pharmacol Sci 25:613–617
10. Zambreanu L, Wise RG, Brooks JC, Iannetti GD, Tracey I (2005) A role for the brainstem in central sensitisation in humans. Evidence from functional magnetic resonance imaging. Pain 114:397–407

Pain in Children

PATRICIA A. MCGRATH[1,2], TRICIA WILLIAMS[1]

[1]Department of Anaesthesia, Divisional Centre of Pain Management and Research, The Hospital for Sick Children, Toronto, ON, Canada

[2]Neuroscience and Mental Health Program, Research Institute at The Hospital for Sick Children; Department of Anaesthesia, The University of Toronto, Toronto, ON, Canada

Synonyms

Pediatric pain; Adolescent pain

Definition

The unique aspects of nociceptive processing and pain perception associated with a developing pain system and a maturing child, in contrast to those of a mature adult.

Children are not "little adults" with respect to nociceptive processing and pain perception. The ▶developing nociceptive system responds differently to injury (i.e., increased excitability and sensitization) when compared to the mature adult system [1,2]. Moreover, a child's pain appears to have a greater degree of ▶plasticity when compared to that of adults – more influenced by cognitive, behavioral, and emotional factors [3].

Characteristics

Children's Pain Problems

Like adults, children can experience many different types of pain throughout their lives – acute pain due to disease or trauma, recurrent episodes of headache, stomach ache, or limb pain unrelated to disease, and chronic pain due to injury, disease, psychological factors, or of unknown etiology. However, the prevalence of certain types of pain is different for adults and children. For example, chronic back pain is a major problem for adults but not for children. Recurrent pain syndromes (i.e., abdominal, headache, limb pains or "growing pains") are more common pain problems for children.

Pain prevalence increases with age and certain pain conditions vary with sex and age. For example, clinical referrals indicate that Complex Regional Pain Syndrome-Type 1 affects girls more than boys with a ratio of ~6–9:1 and affects children primarily in their pre- and early teen years. Complex idiopathic pain conditions and somatization disorders seem to predominantly affect older adolescents.

Although we lack precise data on the incidence and prevalence of many childhood pain conditions, an increasing number of epidemiological studies are focused on obtaining such data, identifying individual risk and prognostic factors and documenting the long term impact for children and their families.

Developmental Considerations

Considerable neuronal plasticity is evident throughout the developing system from the periphery to the brain (for review, [1,2]) (see ▶Development of nociception). Although basic nociceptive connections are formed before birth, these systems are immature and exhibit increased responsivity in comparison to the adult animal. The conduction velocity of afferent fibers, action potential shape, receptor transduction, firing frequencies and receptive field properties change substantially over the postnatal period. High threshold Aδ mechanoreceptors (which respond maximally to noxious mechanical stimuli) and low threshold Aβ mechanoreceptors (which respond maximally to innocuous stimuli) respond with lower firing frequencies than those in the adult animal. The receptive fields of dorsal horn cells are larger in the newborn. The larger receptive fields and dominant A-fiber input increases the likelihood of central cells being excited by peripheral sensory stimulation and acts to increase the sensitivity of infant sensory reflexes. Some inhibitory mechanisms (►Inhibitory mechanisms in developing system) in the dorsal horn are immature at birth and descending inhibition is delayed. The lack of descending inhibition in the neonatal dorsal horn means that an important endogenous analgesic system that should attenuate noxious input as it enters the spinal cord is lacking, and thus the effects of the input may be more profound than in the adult [2].

Most studies in developmental neurobiology have been conducted on rat pups because they have comparable developmental timetables with respect to the anatomy, chemistry, and physiology of maturing human pain pathways. To study neural function in human infants, investigators have monitored behavioral and neurophysiologic responses, and revealed comparable findings of plasticity and increased excitability in the developing nervous system (for review, [1,4]). In comparison to adults, young infants have exaggerated reflex responses (i.e., lower thresholds and longer lasting muscle contractions) in response to certain types of trauma, such as needle insertion. Repeated mechanical stimulation at strong (but not pain-producing) intensities can cause sensitization in very young infants, while repeated painful procedures such as those required during intensive care can profoundly affect sensory processing in infants. Infants after surgery can develop a striking hypersensitivity to touch, as well as to pain.

While we do not know specifically how such injuries may affect the mature human pain system or influence adult pain perception, increasing attention is focused on the possible consequences of untreated pain, particularly in infants [5]. For example, circumcised newborn infants display a stronger pain response to subsequent routine immunizations at 4 and 6 months than uncircumcised infants, but application of lidocaine-prilocaine anesthetic cream at circumcision attenuates the pain response to the subsequent immunizations [6]. Studies of former premature infants who required intensive care have shown behavioral differences related to early pain experiences. The results of behavioral studies in infants, like those from neurobiological studies in animals, indicate increased responsivity to pain.

Factors that Modify Children's Pain

A child's pain perception can be regarded as plastic from a psychological, as well as biological perspective. Tissue damage initiates a sequence of neural events that may lead to pain, but many developmental, social, and psychological factors can intervene to alter the sequence of nociceptive transmission and thereby modify a child's pain. Child characteristics, such as cognitive level, sex, gender, temperament, previous pain experience, family, and cultural background shape generally how children interpret and cope with pain (for review [7–9]).

In contrast, ►situational factors vary dynamically, depending on the specific circumstances in which a child experiences pain. For example, a child receiving treatment for cancer may have repeated injections, central venous port access and lumbar punctures – all of which can cause pain (depending on the analgesics, anesthetics, or sedatives used). Even though the tissue damage from these procedures is the same each time, the particular set of situational factors for each treatment is unique for a child. The expectations, behaviors and emotional state of the child, parent and health care provider all play a critical role. "What children and parents understand, what they (and health care staff) do, and how children and parents feel" can profoundly impact a child's pain experience. Certain situational factors can intensify pain and distress, while others can eventually trigger pain episodes, prolong pain related disability, or maintain the cycle of repeated pain episodes in recurrent pain syndrome [3]. Parents and health care providers can dramatically improve a child's pain experience and minimize their disability by modifying children's understanding of a situation, their focus of attention, perceived control, expectations for obtaining eventual recovery and pain relief, and the meaning or relevance of the pain.

Situational factors may affect children even more than adults. Adults typically have experienced a wide variety of pains (i.e., diverse etiology, intensity and quality), providing them with a broad base of knowledge and coping behaviors. When adults encounter new pains, they evaluate them primarily from the context of their cumulative life experience. In contrast, children with more limited pain experience must evaluate new pains primarily from the context of the immediate circumstances. Children's understanding of pain, pain coping strategies, and the impact of pain increase with age, but many questions remain about the interplay of maturation, cognitive development, and experience in mediating a child's pain.

Pain Measures for Infants and Children

Pain assessment is an intrinsic component of pain management in infants and children. Clinicians need an objective measure of pain intensity and an understanding of the factors that cause or exacerbate pain for an individual child. More than 60 pain measures are now available for infants, children, and adolescents (for review, [10]). While no single pain measure is appropriate for all children and for all situations in which they experience pain, we should be able to evaluate pain for almost every child.

►Physiological parameters including heart rate, respiration rate, blood pressure, palmar sweating, blood cortisol and cortisone content, O_2 levels, vagal tone and endorphin concentrations have been studied as potential pain measures. However, they reflect a complex and generalized stress response, rather than correlate with a particular pain level. As such, they may have more relevance as distress indices within a broader behavioral pain scale. Behavioral scales record the type and amount of pain-related behaviors children exhibit. Since a child's specific pain behaviors depend on the type of pain experienced, different scales are usually required for acute and persistent pain. Clinicians monitor children for a specified time period and then complete a checklist noting which distress behaviors (e.g., crying, grimacing, guarding) occur. Behavioral scales must be used for infants and children who are unable to communicate verbally. Recently, investigators are validating pain scales for children who are ►developmentally disabled. However, the resulting pain scores are indirect estimates of pain and do not always correlate with children's own pain ratings. Even though clinicians may use diaries rather than formal scales, prospective evaluation of a child's behavior is an essential component of pain management, providing information about medication use, compliance with treatment recommendations, and the extent of pain-related disability (i.e., school attendance, physical activities, and social activities with peers).

Psychological or self-report measures include a broad spectrum of projective techniques, interviews, questionnaires, qualitative descriptive scales, and quantitative rating scales designed to capture the subjective experience of a child's pain [11]. By the age of five, most children can differentiate a wide range of pain intensities, and many can use simple ratio and interval pain scales (e.g., visual analog scales, numerical scales, faces, verbal descriptor scales) to rate their pain intensity. Many scales have excellent psychometric properties, are convenient to administer, easy for children to understand, adaptable to many clinical situations, and help parents to monitor their child's pain at home. Interviews, usually conducted independently with a child and parents, are the cornerstone of assessment for children with persistent pain, enabling clinicians to identify relevant child, family, and situational factors that contribute to children's pain and disability problems.

Child-centered Pain Management

Pain control is not merely "drug versus nondrug therapy," but rather an integrated approach to reduce or block nociceptive activity by attenuating responses in peripheral afferents and central pathways, activating endogenous pain inhibitory systems, and modifying situational factors that exacerbate pain. Adequate analgesic prescriptions, administered at regular dosing intervals, must be complemented by a practical cognitive-behavioral approach to ensure optimal pain relief. Pain control is achieved practically by adjusting both drug and nondrug therapies in a rational child-oriented manner based on the assessment process [12]. Analgesics include acetaminophen, non-steroidal anti-inflammatory drugs, and opioids. ►Adjuvant analgesics include a variety of drugs with analgesic properties, such as anticonvulsants and antidepressants that were initially developed to treat other health problems, but whose therapeutic uses have been expanded. The use of adjuvant analgesics has become a cornerstone of pain control for children with chronic pain, especially when pain has a neuropathic component. Children with severe pain may require progressively greater and more frequent opioid doses due to drug tolerance and should receive the doses they need to relieve their pain. The fear of opioid addiction in children has been greatly exaggerated. Neonates and infants require the same three categories of analgesic drugs as older children. However, premature and term newborns show reduced clearance of most opioids. The differences in pharmacokinetics and pharmacodynamics among neonates, preterm infants, and full-term infants, warrant special dosing considerations for infants and close monitoring when they receive opioids.

An extensive array of nondrug therapies is available to treat a child's pain including physical, psychological and complementary and alternative approaches. Counseling, attention and distraction, guided imagery, hypnosis, relaxation training, biofeedback, and behavioral management are used routinely to treat a child's procedural pain and chronic pain. Children seem more adept than adults at using psychological therapies, presumably because they are generally less biased than adults about their potential efficacy. Strong and consistent scientific evidence supports the efficacy of many psychological therapies for relieving children's procedural pain and for relieving childhood headache, but few rigorous evaluations have been conducted on their efficacy for relieving other types of chronic pain – even though they are considered an essential component of many treatment programs.

Clinical and Research Challenges

As a result of extensive research, we have gained better insights about how the developing nociceptive system responds to tissue injury, how children perceive pain, how to assess pain in infants and children, and which drug and nondrug therapies will alleviate their pain. The emphasis has shifted gradually from an almost exclusive disease-centered focus – detecting and treating the putative source of tissue damage – to a more child-centered perspective, assessing the child with pain, identifying contributing psychological and contextual factors, and then targeting interventions accordingly. However, serious challenges remain from both research and clinical perspectives [13].

We have discovered much about the plasticity of the developing nociceptive system, but still have much to learn about how signals from painful stimuli are processed, especially at higher levels (see ►Pain imaging). Although we need further developmental research in neurobiology, neurophysiology, and pharmacology, we now know that infants seem particularly vulnerable because of their heightened responsivity to tissue injury and we must devote particular attention to their pain management.

We need to apply the existing knowledge about pain assessment and pain management more consistently within our clinical practice. Regrettably, many hospitals still do not require consistent documentation of children's pain, preventing us from ensuring that children's pain is adequately controlled. Hospital administrators or accreditation organizations should establish children's pain control as a priority. In spite of established analgesic dosing guidelines for infants and children, the undertreatment of postoperative and chronic pain is a continuing problem in many centers.

Moreover, increasing responsibility for evidence-based practice dictates that health care providers adopt clear guidelines for determining when treatments are effective and for identifying children for whom they are most effective. We lack data from well-designed cohort studies and randomized controlled trials to validate the efficacy of many interventions (both drug and nondrug therapies) used extensively in clinical practice. Although cognitive-behavioral interventions are critical components of pain management programs for chronic pain, most of the data supporting their efficacy is derived from studies of childhood headache [14].

We critically need data on child-centered treatment efficacy – that is, when interventions are selected for the individual child with pain, based on an assessment of the specific cognitive, behavioral, and emotional factors contributing to their pain and disability. We need longitudinal studies to identify key risk factors that influence a child's vulnerability to chronic pain, in particular the apparent increased vulnerability in females. Future studies should use brain imaging technology and psychophysical measurement to evaluate the neural mechanisms underlying chronic pain and cognitive function in children. Our ultimate and continuing challenges are to better understand the experience of children's pain and to improve clinical practice, so that health care providers use the existing "state of the art" pain scales, interpret children's pain scores to guide therapeutic decisions, and document treatment effectiveness.

References

1. Andrews Campbell K (2007) Infant Pain Mechanisms. In: Schmidt RF, Willis WD (eds) Encyclopedic reference of pain. Springer-Verlag, New York, pp 976–981
2. Fitzgerald M, Howard RF (2003) The neurobiologic basis of pediatric pain. In: Schechter NL, Berde CB, Yaster M (eds) Pain in infants, children, and adolescents, 2nd edn. Lippincott Williams and Wilkins, Baltimore, pp 19–42
3. McGrath PA, Dade LA (2004) Effective strategies to decrease pain and minimize disability. In: Price DD, Bushnell MC (eds) Psychological modulation of pain: integrating basic science and clinical perspectives. IASP Press, Seattle, pp 73–96
4. Johnston C, Stevens B, Boyer K et al. (2003) Development of psychologic responses to pain and assessment of pain in infants and toddlers. In: Schechter NL, Berde CB, Yaster M (eds) Pain in infants, children, and adolescents, 2nd edn. Lippincott Williams and Wilkins, Baltimore, pp 105–127
5. Grunau RE (2000) Long-term consequences of pain in human neonates. In: Anand KLS, Stevens BJ, McGrath PJ (eds) Pain in neonates, 2nd edn. Elsevier, Amsterdam, pp 55–76
6. Taddio A, Katz J, Ilersich AL et al. (1997) Effect of neonatal circumcision on pain response during subsequent routine vaccination. Lancet 349:599–603
7. Schechter NL, Berde CB, Yaster M (eds) (2003) Pain in infants, children, and adolescents. Lippincott Williams and Wilkins, Baltimore
8. McGrath PJ, Finley GA (eds) (2003) Pediatric pain: biological and social context. IASP Press, Seattle
9. Unruh AM, Campbell MA (1999) Gender variations in children's pain experiences. In: Finley GA, McGrath PJ (eds) Chronic and recurrent pain in children and adolescents. IASP Press, Seattle, pp 199–241
10. McGrath PA (2007) Pain assessment: children. In: Schmidt RF, Willis WD (eds) Encyclopedic reference of pain. Springer-Verlag, New York, pp 1644–1648
11. Champion GD, Goodenough B, von Baeyer CL et al. (1998) Measurement of pain by self-report. In: Finley GA, McGrath PA (eds) Measurement of pain in infants and children. IASP Press, Seattle, pp 123–160
12. Brown SC (2007) Analgesic guidelines for infants and children. In: Schmidt RF, Willis WD (eds) Encyclopedic reference of pain. Springer-Verlag, New York, pp 78–83
13. McGrath PA (2007) Pain in children. In: Schmidt RF, Willis WD (eds) Encyclopedic reference of pain. Springer-Verlag, New York, pp 1665–1669
14. Eccleston C, Morley S, Williams A et al. (2002) Systematic review of randomised controlled trials of psychological therapy for chronic pain in children and adolescents, with a subset meta-analysis of pain relief. Pain 99:157–165

Pain in Older Adults (Including Older Adults with Dementia)

Thomas Hadjistavropoulos
Centre on Aging and Health and Department of Psychology, University of Regina, Regina, SK, Canada

Synonyms

Pain in the elderly; Pain among seniors; Pain in patients with dementia

Definition

Most investigations focusing on pain among older adults (elderly persons) involve participants who are at least 60 or 65 years of age.

Characteristics

Prevalence

Although the prevalence of acute pain remains steady across the lifespan, there is an increased prevalence of chronic pain at least until the seventh decade of life [1]. Limited evidence also suggests a plateau or even a slight reduction in the frequency of pain complaints after age 80 [1].

Pain is a very common problem among older adults. Chronic pain affects at least 50% of seniors living in the community and approximately 80% of residents of long-term care facilities. Moreover, in a large scale Canadian investigation of nursing home residents, it was shown that conditions likely to cause pain occur with equal frequency in residents with and without dementia [2]. Despite the increasing prevalence of most pain problems with age, the study of pain among older adults had not received much literature attention until recently [1].

Pain Perception, Thresholds and Tolerance

Age-related changes in peripheral, spinal and central nervous system ►nociceptive pathways would be expected to alter pain sensitivity and therefore the perception of noxious stimulation [3]. Indeed, evidence suggests that age can have an impact on the function of nociceptive pathways and mechanisms, including alterations in afferent transmission and descending modulation [1]. More specifically, for example, research on the perceptual experience that tends to accompany activations of nociceptive fibers has suggested the presence of a selective age-related impairment in A-fiber function and a greater reliance on C-fiber information in older adults. Considering that A-fibers subserve the epicritic, first warning aspects of pain, while C-fiber information is more diffuse, dull and prolonged, it might be reasonable to expect some changes in pain intensity and quality among elderly persons [3].

Research has also revealed evidence that temporal summation (i.e., the enhancement of pain sensation that is associated with repeated stimulation) is altered in older adults. Temporal summation is the result of transient, repetitive activation of dorsal horn neurons in the spinal cord and is believed to play a central role in the development of ►hyperalgesia and post-injury tenderness [3]. Based on the findings that are available in the literature, it is likely that post-injury tenderness and hyperalgesia may take longer to resolve among older adults [3]. An additional age-related change has been demonstrated by Washington, Gibson and Helme [4], who have shown that endogenous inhibitory pain control mechanisms that descend from the cortex and midbrain onto spinal cord neurons decline with advancing age. Such a decline could be expected to reduce the ability of older adults to cope with persistent pain states [3].

With respect to neurochemical and morphological age-related changes in the central nervous system, Gibson, Gorman and Helme [5] used the pain-related encephalographic response to index the central nervous system processing of noxious stimulation. These researchers found that that older adults tended to display a significant reduction in peak amplitude and an increase in response latency. They concluded that these findings were suggestive of a reduced cortical activation and slowing in the cognitive processing of noxious information. Nonetheless, it is important to remember that despite such limited laboratory evidence of reduced sensitivity to pain with advancing age, there is no evidence to suggest that seniors who report pain suffer any less than their younger counterparts [1].

Research has also examined the possible impact of dementia on pain responses. In general, the findings have shown no difference in the ►pain threshold of those with mild to moderate Alheimer's disease, despite an increase in ►pain tolerance when compared to age matched controls [6,7]. Nonetheless, Benedetti et al. [6] have demonstrated that whereas the sensory-discriminative components of pain are preserved even in advanced stages of Alzheimer's disease, the cognitive and affective functions, which are related to both anticipation and autonomic reactivity, are severely affected. This sensory-affective dissociation is well correlated with neuropathological findings in Alzheimer's disease. Moreover, Benedetti et al. [6,7] showed that pain tolerance among older adults with Alzheimer's disease is tightly related to the severity of the disease. That is, more severe cognitive impairment and more significant electroencephalogram (EEG) changes were associated with higher pain tolerance. Thus, despite the preservation of pain thresholds in the presence of dementia, there is an increase in pain tolerance with increased severity of the disease. It is noted, however, that clinical research has shown that the reflexive reactions that dementia patients show to painful stimulation (e.g., pain due to discomforting physiotherapy exercises) are comparable or more intense than the reactions of cognitively intact patients [8]. Such clinical findings underscore the importance of managing pain effectively regardless of patient cognitive status.

Age Differences in Psychosocial Aspects of Pain

There is evidence of psychosocial differences in the mediators and context of pain. Although not perfectly consistent across studies, the evidence shows age-related differences in beliefs and attributions about pain as well as coping strategies [9]. There is, for example, evidence of increased stoicism among older adults when it comes to the reporting of symptoms. This stoicism could lead to an underreporting of pain among older adults [9]. Moreover, the social context and stressors that affect seniors with pain differ from those of younger persons. For example, younger adults with chronic pain conditions are often concerned about issues relating to return to work whereas older persons are often retired, may be widowed and may be more concerned about loneliness and possible social isolation.

The Assessment of Pain in the Older Adult

Given age-related differences in the social context and co-morbidities of chronic pain, research has focused increasingly on the validation of pain assessment tools among older adults [10]. Instruments that are specialized in the assessment of the older adult have also been developed [10].

The accurate assessment of pain in the older adult is especially challenging when it comes to persons with severe dementia who have limited ability to communicate. Because pain is a subjective phenomenon, clinicians tend to rely on self-report. Recently, there have been worthwhile efforts to develop and validate observational measures of pain that rely on the recording of pain-related behaviors such as facial expressions and paralinguistic vocalizations (such assessment tools have been reviewed elsewhere [8]).

The Treatment of Pain in the Older Adult

Although the prevalence of chronic pain increases with age, pain is always the result of pathology and is never a natural part of being old. As such, it is always important to manage chronic pain. Recommended doses of drugs used for pain management in older adults are often lower than doses used in younger persons because of age-related physiological changes (e.g., age-related changes in fat to muscle ratio, slowing of metabolic rates, lower protein levels in the blood). Although numerous drugs have been shown to be effective in treating pain in older adults, more research concerning the ►pharmacokinetics, ►pharmacodynamics, efficacy and safety of medications in older persons is needed [9]. In addition, many of the painful conditions that elderly persons tend to suffer are responsive to physiotherapy, although special adaptations may be required for frail seniors [9]. Finally, initial evidence suggests that ►cognitive behavior therapy can be helpful in assisting older adults with pain management [9].

References

1. Charlton JE (ed) (2005) Core curriculum for professional education in pain, 3rd edn. IASP Press, Seattle
2. Proctor W, Hirdes J (2001) Pain and cognitive status among nursing home residents in Canada. Pain Res Manag 6:191–205
3. Gibson SJ, Chambers C (2004) Pain over the life span, in Pain: Psychological perspectives. In: Hadjistavropoulos T, Craig KD (eds) Lawrence Erlbaum Associates, Mahwah, NJ
4. Washington LL, Gibson SJ, Helme RD (2000) Age-related differences in endogenous analgesic response to repeated cold water immersion in human volunteers. Pain 89:89–96
5. Gibson SJ, Gorman MM, Helme RD (1990) Assessment of pain in the elderly using event-related cerebral potentials. In: Bond MR, Charlton JE, Woolf C (eds) Proceedings of the VIth World Congress on Pain, Elsevier, Amsterdam
6. Benedetti F et al. (2004) Pain reactivity in Alzheimer patients with different degrees of cognitive impairment and brain electrical activity deterioration. Pain 111 (1–2):22–29
7. Benedetti F et al. (1999) Pain threshold and tolerance in Alzheimer's disease. Pain 80(1–2):377–382
8. Hadjistavropoulos T (2005) Assessing pain in older persons with severe limitations in ability to communicate. In: Gibson SJ, Weiner DK (eds) Pain in older persons. IASP Press, Seattle
9. Gibson SJ, Weiner DK (eds) (2005) Pain in older persons. IASP Press, Seattle
10. Herr KA (2005) Pain assessment in the older adult with verbal communication skills. In: Gibson SJ, Weiner DK (eds) Pain in older persons. IASP Press, Seattle, pp 111–133

Pain in Patients with Dementia

►Pain in Older Adults (Including Older Adults with Dementia)

Pain in the Elderly

►Pain in Older Adults (Including Older Adults with Dementia)

Pain in the Head

►Headache

Pain Modulation

▶Descending Modulation of Nociception

Pain Psychophysics

Joel D. Greenspan
Department of Biomedical Sciences, University of Maryland Dental School, Baltimore, MD, USA

Synonyms

Quantitative sensory testing QST

Definition

The systematic evaluation of the quantitative relationship between physical stimuli and the pain they evoke.

Characteristics

The discipline of ▶psychophysics was developed in the German experimental psychology laboratories of the early nineteenth Century. It has been applied to every sensory system, including pain. The earliest published work in pain psychophysics recognized today is that of Ernst Weber [1] While most of Weber's work in psychophysics addresses tactile perception, a portion encompasses pain. The first major opus in pain psychophysics of the twentieth Century was that of James Hardy and colleagues at Cornell University. Over 200 papers from this group were distilled for the book "Pain sensations and reactions"[2]. This body of work was a principal reference for pain psychophysics for decades, despite a reliance on a narrow range of approaches that were not always found to generalize in subsequent studies.

The two essential components for psychophysical evaluation of any sensory system are (i) controlled stimuli and (ii) a valid method of quantifying sensory experience. For pain psychophysics, the development of reliable pain-evoking stimuli was complicated by the need to avoid tissue damage. The most commonly used stimulation techniques are cutaneous heating and mechanical pressure applied to cutaneous and/or deeper tissues. However, several other forms of stimulation are used for pain psychophysics, including cold, electrical, chemical, laser, and visceral distention [3].

Pain Threshold

One principal psychophysical measure is ▶threshold. Simply stated, threshold is the minimal level of stimulation needed to evoke a sensation. By extension pain threshold is the minimal level of stimulation needed to evoke a sensation of pain. Accordingly, thresholds are reported in terms of stimulus values, such as temperature (in °C) or mechanical forces (kg equivalent weight, or Newtons). An alternative measure is the time it takes a constant, sustained stimulus to be perceived as painful, thus measuring pain threshold in terms of seconds.

The assessment of pain threshold is more complicated than other sensory thresholds because of the nature of pain and people's concept of it. In other sensory modalities, threshold is recognized by the "step" between no sensation and sensation. Thus, the subject is attempting to distinguish between sensing nothing and something. For pain thresholds, the subject is instead distinguishing between two types of sensation – one considered painful and one non-painful. Thus, a critical element in pain threshold determination is the particular sensory experience an individual considers painful. This factor will be influenced by, among other things, the subject's pain experience history, and the instructions given by the experimenter. For instance, thresholds are likely to be different if a subject is instructed to indicate when he perceives "pain" versus when he perceives "a sharp or burning sensation" or "an uncomfortable sensation." It is also possible that a subject's criterion for judging what sensation is painful changes in the course of an evaluation session. Furthermore, the likelihood and extent of such changes can vary depending upon the range and number of stimuli applied [4]. Several other factors can influence pain threshold values, including features of the psychophysical protocol (e.g., the specific design, stimulus parameters, the range of stimuli, the threshold calculation procedure), which makes it questionable to compare threshold values across studies that vary with respect to these and any other protocol features.

Another important fact to recognize is that threshold is a statistical entity. While we are accustomed to representing thresholds as very precise values (i.e., heat pain thresholds expressed as temperature at a 0.1°C level of precision), it is not the case that weaker stimuli are necessarily painless, and stronger ones are always painful. Instead, there is a range of stimuli for which the lower values are less frequently painful and the higher values are more frequently painful. The threshold value, in principle, is the midpoint of that range (Fig. 1).

This concept applies whether one is considering data derived from a single person tested repeatedly, or from a group of people.

Heat pain threshold has been found to be fairly consistent across many body sites. However, heat pain thresholds are significantly higher on ▶glabrous skin than on hairy skin of the extremities. This relative consistency across the body allows one to assess regional pain threshold abnormalities by comparing

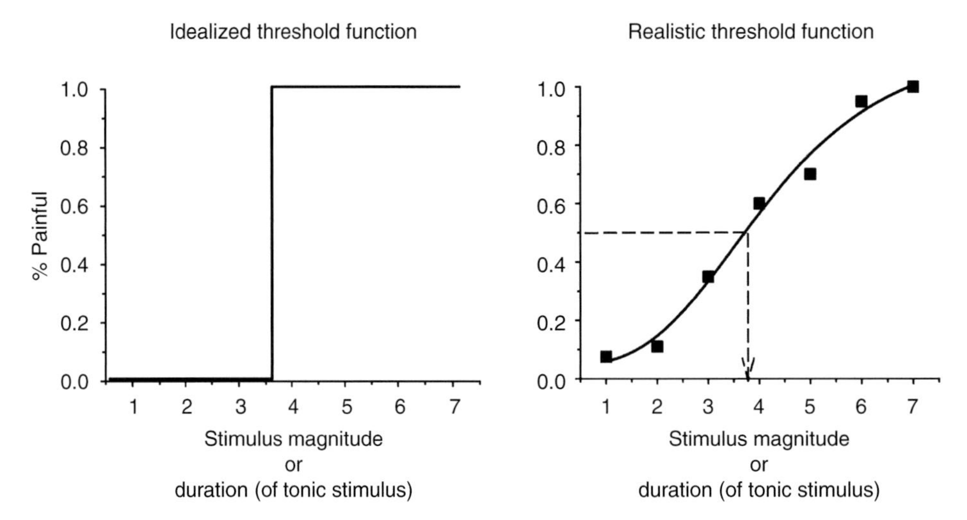

Pain Psychophysics. Figure 1 Ideally, pain threshold can be envisioned as the stimulus intensity that divides non-painful from painful intensities of stimulation (*left*). But, in reality, there is a range of stimulus intensities that are sometimes perceived as painful and sometimes as non-painful. Thus, pain threshold is typically regarded as the stimulus intensity that is painful 50% of the time (*right*).

thresholds between two body sites, when one of them is accepted as a reference. This approach is often done by comparing thresholds between homolateral body sites in the cases of unilateral sensory abnormalities [5]. Despite this intra-personal consistency, there is considerable inter-personal variability in pain threshold. This feature has been demonstrated in studies that have evaluated a large number of subjects in attempts to develop a normative database of pain thresholds [6].

While the concept of psychophysical threshold has been used for almost two centuries, it has been criticized as a strict measure of sensory perception because it can be influenced by psychological states such as expectancy and anticipated rewards. In an attempt to account for these and other "non-sensory" factors that could influence threshold determination, the approach of ►signal detection theory (SDT) was adapted for psychophysics [7]. This approach allowed for the distinction between stimulus discriminability (d') and response bias (β), in which the former was the "bias-free" measure of sensory detection. Threshold measures cannot make this distinction. Approaches based on SDT have been applied to pain psychophysics; however, problems particular to pain psychophysics have been noted. Despite the advantage of SDT approaches in distinguishing stimulus discriminability from response bias in psychophysical assessments, the major drawback of this approach is the need for many more stimulus trials than are needed for most threshold protocols. In addition, SDT assumes a perceptual stability over the course of testing, which is not necessarily the case for pain. Another concern is that knowing how discriminable two (or more) stimuli are from one another is not the same information as how painful those stimuli are. Thus, the kind of information derived from STD-derived protocols are supplemental to, rather than replacements for, the type of information gathered using other psychophysical approaches to pain perception.

Suprathreshold Pain Scaling

Another major psychophysical endpoint is evaluation of perception above threshold (suprathreshold perception). The principle is to have the subject represent the sensory experience on a quantitative continuum – often referred to as "scaling" perception. There are many ways to accomplish this, and protocols are based on either direct or indirect methods. Indirect scaling methods require the subject to use another continuum to match the perception under investigation. An example of this is to have the subject adjust the volume of a sound so that the loudness matches the intensity of another sensory dimension, such as pain. In this way, the pain intensity can be measured in terms of decibels of sound. Another approach is to have the subject draw a line length to represent the intensity of a sensation, allowing the sensation intensity to be measured in millimeters.

The direct scaling methods require the subject to choose a number that reflects perceptual intensity, and thus do not require an intermediate modality such as another sensory dimension or a motor task. These direct scaling methods have been more frequently employed for pain psychophysics over the last few decades. In most pain studies, subjects are provided a number

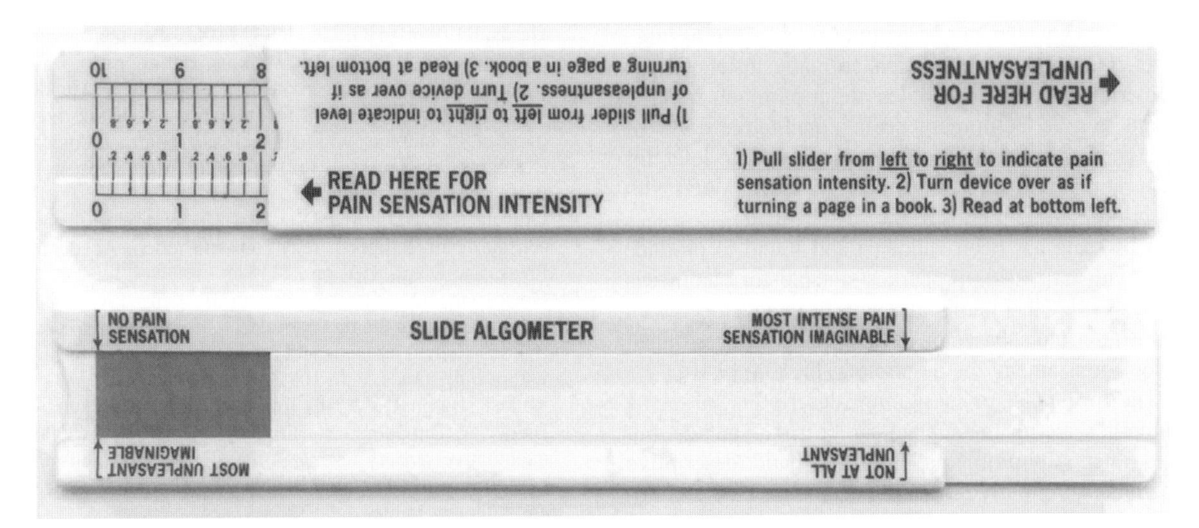

Pain Psychophysics. Figure 2 A mechanical visual analog scale for pain rating developed by Dr. Donald D. Price and colleagues. *Bottom*: A sliding plastic piece moves to reveal a red bar. The subject adjusts the length of the red bar to match the perceived pain intensity (in the orientation shown), or the perceived unpleasantness (when rotated 180°). *Top*: The number at the edge of the adjusted plastic piece is the numerical value assigned either pain intensity or unpleasantness. Figure courtesy of Dr. Price.

Pain intensity scale		Unpleasantness scale	
20		20	
19		19	
18	Extremely intense	18	
17	Very intense	17	Very intolerable
16	Intense	16	
15		15	Intolerable
14	Strong	14	
13	Slightly intense	13	Very distressing Slightly intolerable
12	Barely strong	12	Very annoying
11	Moderate	11	Distressing
10		10	Very unpleasant
9		9	Slightly distressing
8	Mild	8	Annoying
7		7	Unpleasant
6	Very mild	6	Slightly annoying
5	Weak	5	Slightly unpleasant
4	Very weak	4	
3		3	
2		2	
1	Faint	1	
0	No pain sensation	0	Neutral
	Gracely box SL		Gracely box SL

Pain Psychophysics. Figure 3 Pain rating scales with descriptors developed by Dr. Richard H. Gracely and colleagues. The subject reports the perceived pain intensity (a) or unpleasantness (b) by choosing a number between 0 and 20. Placement of descriptors along the length of the numeric scale, which serve to provide connotative meaning to the numbers, was based on psychometric procedures described in [9]. Figure courtesy of Dr. Gracely.

P

scale to use, which can be as limited as 0–5, or as large as 0–100, or even unbounded. In many instances, the numeric scale also includes descriptors at specific points to give the numeric scale a qualitative frame of reference. One of the most commonly used scales for pain ratings is a 0–100 visual analog scale (VAS) with descriptors at both ends of the scale (Fig. 2).

This scale has been validated and found to produce data with ratio scale properties suitable for parametric analysis [8]. Other pain scales have been developed with more descriptors, based on psychometrically determined associations among the descriptors (Fig. 3a) [9,10].

The principle of assigning descriptors along a numeric scale for the subject can be an advantage, but also potentially problematic. On one hand, descriptive anchors serve to give the numbers a consistent connotative meaning to the subjects, and thereby help to standardize the scale. On the other hand, if different people interpret the same term differently (Is your concept of "extremely intense" the same as mine?), the presence of these terms can introduce an idiosyncratic bias, rather than standardizing the numeric scale. Despite this possibility, and the inherent uncertainty of measuring a subjective phenomenon, these types of scales have been successfully used for many psychophysical studies of pain over the last few decades.

While the aforementioned scales are designed to measure pain intensity, a similar set of scales have been developed to measure pain affect or unpleasantness (Figs. 2 and 3b). In principle, any perceptual dimension could be measured with a similarly constructed scale.

Pain Tolerance

Pain tolerance is less frequently used than pain threshold in scientific studies, and it has some significant disadvantages: (i) For some forms of stimulation, pain tolerance cannot be reached without risking tissue injury; (ii) Pain tolerance generally shows greater variability than threshold, both within and across subjects; (iii) It is more widely altered by subject bias or past experience than threshold. However, pain tolerance measures a qualitatively different aspect of the pain experience than does pain threshold. Arguably, pain tolerance is a measure more reflective of the affective and motivational aspects of the pain experience, while threshold is a measure of the discriminative aspect. One common form of this test is the "►cold pressor test," which involves submersion of a body part in ice water. Another test involves ischemic pain, produced by applying a pressure cuff on the subject's arm.

References

1. Ross HE, Murray DJ (eds) (1978) The sense of touch (transl. Weber EH). Academic Press, New York
2. Hardy JD, Wolff HG, Goodell H (1952) Pain sensations and reactions. Williams and Wilkins, Baltimore
3. Graven-Nielsen T, Sergerdahl M, Svensson P, Arendt-Nielsen L (2001) Methods for induction and assessment of pain in humans with clinical and pharmacological examples. In: Kruger L (ed) Methods in pain research. CRC Press, Boca Raton, FL, pp 263–304
4. Gracely RH, Naliboff BD (1996) Measurement of pain sensation. In: Kruger L (ed) Pain and touch. Academic Press, San Diego, CA, pp 243–313
5. Greenspan JD, Ohara S, Sarlani E, Lenz FA (2004) Allodynia in patients with post-stroke central pain (CPSP) studied by statistical quantitative sensory testing within individuals. Pain 109:357–366
6. Rolke R, Baron R, Maier C, Tolle TR, Treede RD, Beyer A, Binder A, Birbaumer N, Birklein F, Botefur IC (2006) quantitative sensory testing in the German research network on neuropathic pain (DFNS): standardized protocol and reference values. Pain 123:231–243
7. Green DM, Swets JA (1966) Signal detection theory and psychophysics. Robert E. Krieger, Huntington, NY
8. Price DD, McGrath PA, Rafii A, Buckingham B (1983) The validation of visual analogue scales as ratio scales measures for chronic and experimental pain. Pain 17:45–56
9. Gracely RH, McGrath P, Dubner R (1978) Ratio scales of sensory and affective verbal pain descriptors. Pain 5:5–18
10. Greenspan JD, Roy EA, Caldwell PA, Farooq N (2003) Thermosensory intensity and affect throughout the perceptible range. Somatosens Mot Res 20:19–26

Pain System

►Ascending Nociceptive Pathways

Pain Threshold

Definition

The International Association for the Study of Pain defines pain threshold as the least amount of pain that a person can recognize.

►Pain in Older Adults (Including Older Adults with Dementia)

Pain Tolerance Level

Definition

The International Association for the Study of Pain defines pain tolerance level as the greatest amount of pain that a person can tolerate.

►Pain in Older Adults (Including Older Adults with Dementia)

Pain Unpleasantness

Definition

The immediate, disagreeable aspect of pain, similar to feelings of thermal distress (too hot or cold), thirst, or hunger, that motivates behaviors to reduce this feeling state. This immediate state is in contrast to the emotional reactions and feeling states associated with thoughts about pain (pain emotion), such as anxiety, depression, fear, and despair. In the case of injury, the unpleasantness of pain usually motivates movements to escape or to minimize the injury (pain evoked movement). After the injury, during the healing phase, the unpleasantness of pain may inhibit movement to protect the injured area and promote healing (movement evoked pain).

►Emotional/Affective Aspects of Pain

Painful Neuropathies

►Voltage-gated Sodium Channels: Multiple Roles in the Pathophysiology of Pain

Palaeocerebellum

Synonyms

►Paleocerebellum

Definition

Phylogenetically, a very old part of the cerebellum. Corresponds to the vermis cerebelli with its surrounding intermediate part (paravermal part). The afferents of this region come from the spinal cord, hence this part is also called the spinocerebellum.

►Cerebellum

Palaeomagnetism

Definition

The study of remanent magnetization of rocks and sediments to unravel information of the ancient magnetic field.

►Geomagnetic Field

Paleocortex

Definition

The paleocortex (Greek for old cortex) is a phylogentically older type of cortex with less than the six layers seen in the neocortex, but more than the three layers seen in the archicortex (hippocampal formation). The parahippocampal gyrus has cortex of this type.

Paleoencephalon

Definition

The paleoencephalon describes phylogentically older parts of the cerebral hemisphere that evolved along with the olfactory system. Sometimes the term rhinencephalon is used as a synonym. In the strict olfactory sense, the paleoencephalon would include the olfactory bulb, anterior olfactory nucleus, olfactory tubercle, and portions of the amygdala and nearby piriform cortex.

Paleoneurology

Definition
The study of the endocasts of fossil animals.

►Evolution of the Brain in Humans – Paleoneurology

Paleopallium

Definition
The palopallium refers to the cortex of the paleoencephalon, i.e., paleocortex (see above).

Paleopallium and Archipallium

►Evolution of the Pallium: in Amphibians

Pallia Dorsale and Piriforme

►Evolution of the Pallium: in Amphibians

Pallial Amygdala

Definition
Portion of the amygdaloid complex derived from pallial regions. It posses layered cortical and nuclear components in amniotes, whereas only a nuclear portion is present in anamniotes (anurans amphibians).

►Evolution of the Amygdala: Tetrapods

Pallial Primordia

►Evolution of the Pallium: in Amphibians

Palliative

►Placebo Analgesic Response

Pallidum

►Globus pallidus
►Diencephalon

Pallidum, Ventral

Definition
A rostroventral extension of the globus pallidus that protrudes into the basal forebrain and olfactory tubercle beneath the anterior commissure. The ventral pallidum receives projections from the accumbens and medium cell (striatal) districts of the olfactory tubercle and projects to the lateral hypothalamus, medial extremity of the subthalamic nucleus, ventral tegmental area and adjacent medial part of the substantia nigra and ventrolateral part of the periaqueductal gray.

►Hypothalamus
►Hypothalamus, Lateral
►Striatopallidum

Pallium

Definition
The roof of the forebrain (telencephalon) which includes the cerebral cortex, hippocampus, olfactory cortex, claustrum, and some amygdalar groups – pallial is the adjective.

►Evolution of the Brain in Reptiles
►Evolution of the Wulst

Pallium (Medial, Dorsal)

Definition

The dorsal portion of the telencephalon with a cytoarchitectural organization that is primarily cortical (suggestively layered).

►Evolution of Hippocampal Formation

Palmitoylation

Definition

Addition of palmitic acid, a saturated fatty acid containing 16 carbon molecules via an enzymatic reaction involving a Palmitoyl-acyl transferase enzyme.

Palmitic acid is covalently attached to proteins via thioester bonds at cytosolic cysteine residues and it is reversible reaction.

►Receptor Trafficking

PAN/PVC Tube

Definition

Semipermeable polyacrylonitrile/polyvinylcholoride polymer guidance tube for placing cells within and transplantation to the spinal cord.

►Transplantation of Olfactory Ensheathing Cells

Papez Neuronal Circuit

Definition

The mammillothalamic fasciculus, Vicq d'Azyr bundle conducts efferents of the mammillary body to the thalamus (anterior thalamic nucleus). This in turn projects via the cingulum to the hipppocampus, while the latter projects back via the fornix to the mammillary body and anterior thalamic nucleus.

This creates a neuronal feedback circuit, which is called the Papez neuronal circuit and plays a role in memory formation. Being a vital component of the Papez neuronal circuit, the hippocampus is involved in memory formation. Lesions result in loss of the ability transfer the contents from short-term memory to long-term memory (anterograde amnesia).

►General CNS

Par Protein

Definition

Partitioning defective (Par) proteins include par-3, par-6, cdc42, and atypical protein kinase-C (aPKC). These proteins form a complex that exhibits a polarized distribution in the cell and is involved in establishing cellular polarity.

Parabolic Flight

Definition

A flight trajectory of parabolic climbs and dives in which the aircraft and its contents are in free fall during the pushover periods, simulating a 0 g environment in the sense that objects in the aircraft are weightless. The length of the weightless phases depends on the air speed of the aircraft.

►Autonomic Function in Space
►Proprioception Effect of Gravity

Parabrachial Area

Synonyms

►Nuclei parabrachiales; ►Parabrachial nuclei

Definition

The parabrachial area comprises three nuclear areas:

- Lateral parabrachial nucleus
- Medial parabrachial nucleus
- Kolliker-Fuse nucleus

Brainstem second relay station both for taste and visceral sensory pathways located in the pons. It is formed by several nuclei. The medial parabrachial area receives gustatory afferents from the nucleus of the tractus solitarius while the lateral parabrachial receives visceral afferents both vagal and from the area postrema. It is considered a primary site for taste– visceral integration relevant for conditioned taste aversion acquisition in rodents.

►Conditioned Taste Aversion
►Diencephalon
►Parabrachial Nuclei

Parabrachial Complex

Definition

A compact cluster of relay nuclei located rostrally in the dorsolateral pons, surrounding the middle cerebral peduncle (brachium conjunctivum). Individual nuclei within the parabrachial complex receive various ascending axonal inputs that provide information about viscerosensory function, metabolic status, and pain (arriving from nuclei in the spinal cord, nucleus of the solitary tract, and other brainstem sites). This ascending information is integrated with substantial descending inputs from the hypothalamus, amygdala, bed nucleus of the stria terminalis, and other brain sites. Different nuclei within the parabrachial complex deliver this integrated information to subcortical regions of the forebrain (primarily to subnuclei within the amygdala, hypothalamus, thalamus, and basal forebrain), and to nuclei in the midbrain and brainstem, thus influencing processes that include ingestive behavior, arousal, emotion, and autonomic function.

Parabrachial Nuclei

Definition

Latin: Nuclei Parabrachiales; Nuclear complex that is located in the dorsolateral tegmentum of the pons and serves as a major relay center for converging visceral, nociceptive, and thermoreceptive information to the forebrain. The parabrachial complex includes several subnuclei involved in taste sensation and control of gastrointestinal, cardiovascular activity, and respiratory functions.

►Central Regulation of Autonomic Function
►Parabrachial Area

Paradoxical Embolism

Definition

Cardiac embolism that contains material from the venous system and reached the arterial system through a cardiac shunt.

►Ischemic Stroke
►Stroke

Parahippocampal Gyrus

Synonyms

►Gyrus parahippocampalis

Definition

The gyrus marks the transition from hippocampus with its allocortex to the isocortical structure of the temporal lobe. A cross-section shows four discrete cortical regions: presubiculum and parasubiculum on the hippocampal sulcus, entorhinal area and the perirhinal cortex deep in the calcarine sulcus.

►Telencephalon

Parallel Arrangement

Definition

A combination of two rheological elements, such that the elongation is common to both and the forces are to be added to obtain the force of the combined element.

►Mechanics

Parallel Processing

Definition

In the brain information is processed not only in one stream as in a typical computer but in many parallel and independent streams. Good examples are the different sensory pathways.

Parallel Visual Processing Streams

▶Visual Processing Streams in Primates

Parallelism

Definition

Mental and physical events run parallel to each other without any causal relations obtaining between mental and physical events.

▶Causality

Paralysis

Definition

Severe loss of motor strength resulting from damage to ▶motor units or to descending tracts impinging on them. Lower motoneuron paralysis presents with possible involvement of individual muscles, severe atrophy, flaccidity and ▶hypotonia with absent ▶tendon reflexes, possible ▶fasciculations and ▶fibrillations. Upper motoneuron paralysis usually presents with diffuse distribution of affected muscles, little atrophy, ▶spasticity, ▶Babinski sign, ▶fasciculations.

▶Babinski Reflex
▶Fasciculations
▶Fibrillations
▶Motoneuron
▶Motor Unit
▶Spasticity
▶Tendon Reflex

Paralysis Agitans

Definition

▶Parkinson Disease

Paralytic Ileus

▶Bowel Disorders

Paramedian Pontine Reticular Formation (PPRF)

Definition

Anatomically, the PPRF is just the medial portion of the pontine reticular formation (<2 mm from the midline in macaques), but in the oculomotor literature, the PPRF is a functional unit that contains many of the neuronal populations and much of the neuronal circuitry involved in the generation of (mainly horizontal) eye movements. It first came to prominence when it was shown that lesions of the PPRF eliminated or drastically impaired most horizontal eye movements. The PPRF includes most elements of the brainstem burst generator involved in the generation of horizontal saccades, namely excitatory burst neurons, long-lead burst neurons, omnipause neurons, and arguably, inhibitory burst neurons which are on the ponto-medullary border.

Intermixed with these neurons are saccade-related neurons that project to the cerebellum, and reticulospinal neurons that mediate head movements and eye-head coordination. Embedded in the PPRF are other nuclei that are usually considered to be distinct from the reticular formation. The most important is the abducens nucleus, which contains lateral rectus motoneurons and internuclear neurons that project to medial rectus motoneurons. There are also circumscribed precerebellar relay nuclei that convey oculomotor signals to the oculomotor vermis, fastigial nucleus, and the floccular lobe, namely raphe pontis, the intrafascicular nucleus, the rostral pole of the abducens nucleus, and medial nucleus reticularis tegmenti pontis. Finally, the PPRF contains the fibers connecting these and other oculomotor structures (e.g., the superior colliculus and the vestibular nuclei), so the drastic effects of lesions are a product of destroying both the neuronal populations and the inputs to these populations.

►Brainstem Burst Generator
►Cerebellum – Role in Eye Movements
►Long-Lead Burst Neurons (LLBNs)
►Omnipause Neurons
►Saccade, Saccadic Eye Movement
►Superior Colliculus
►Vestibular Nuclei

Parameters

Definition

Constants or variables that are not conditioned by natural laws but define essential characteristics of the system's behavior under the action of the laws.

►Equilibrium Point Control

Parametric Control

Definition

►Equilibrium Point Control

Paramyotonia Congenita

Definition

►Non-dystrophic myotonias

Paraphasia

Definition

Incorrect use of words occurring in conduction aphasia.

►Aphasias

Paraplegia

Definition

Bilateral paralysis of the lower body including the two legs, most commonly resulting from damage to the spinal cord (complete transection), spinal nerve roots or peripheral nerves.

Parasomnias

Definition

Undesired physical events that occur during the entry into sleep, within sleep or during arousals from sleep. They include sleep walking and sleep terrors.

►Sleep-Wake Cycle

Parasthesia

Definition

Altered sensation, usually ascribed to pins and needles or tingling but which can also be ascribed to burning or pricking.

►Proprioception: Effect of Neurological Disease

Parasympathetic

►Central Integration of Cardiovascular and Respiratory Activity Studied In Situ

Parasympathetic Ganglia

►Autonomic Ganglia

Parasympathetic Nervous System

Definition

Parasympathetic refers to the branch of the autonomic nervous system that arises from specific cranial nerve nuclei and from the sacral spinal segments. This system supplies visceral organs with specific functions, such as the sphincter pupillae and the ciliary body in the eye, secretory glands producing fluid including acid in the stomach, enhancing motility of stomach and distal colon, bladder, etc.

►Ageing of Autonomic/Enteric Function
►Autonomic Ganglia
►Autonomic Reflexes
►Parasympathetic Pathways

Parasympathetic Pathways

Definition

Beside the sympathetic pathways, the parasympathetic pathways are the second major component of the autonomic nervous system (there are also enteric pathways and an afferent or sensory component). Some parasympathetic pathways involve brain stem and cranial ganglia, and some are centered on the sacral region of the spinal cord and the pelvic ganglia.

In the brain stem the main nuclei of the parasympathetic pathways (the cranial parasympathetic outflow) are the Edinger-Westphal nucleus, the superior and inferior salivatory nuclei, dorsal vagal nucleus and nucleus ambiguus. The neurons of these nuclei issue axons then enter the oculomotor nerve, the facial, the glosso-pharyngeal nerve and vagus nerve, respectively.

The oculomotor branch projects to the ciliary ganglion (its neurons innervate the iris and the ciliary muscle, hence providing accommodation and pupil constriction).

The facial nerve branch project to the pterygopalatine and the submandibular ganglia (their neurons innervate the lachrymal, the submandibular and the sublingual glands), and the glosso-pharyngeal branch project to the otic ganglion (its neurons innervate the parotid gland). The fibers from the nucleus ambiguus and dorsal vagal nucleus enter the vagus nerve, of which they represent the motor component, and extend a long distance in the neck, thorax and upper part of the abdomen. The fibers terminate synapsing on neurons in small ganglia close to the tracheal and bronchial muscles and the esophagus and in some enteric ganglia of stomach and intestine. From these minute intramural ganglia, post-ganglionic fibers emerge that innervate glands and smooth musculature of airways, esophagus and gastro-intestinal tract. In the spinal cord, preganglionic parasympathetic neurons are assembled into columns in the second, third and fourth sacral segments. Their preganglionic fibers, predominantly cholinergic, project onto pelvic ganglia, whose post-ganglionic fibers innervate mainly the urogenital organs. In some organs, typically the heart or the pupil, which receive both sympathetic and parasympathetic fibers, these exert antagonistic effect. Many organs, however, are controlled predominantly by one or the other of the two pathways.

►Sympathetic Pathways

Paraventricular Nucleus (PVN)

Definition

A subdivision of the hypothalamus located adjacent to the third ventricle. The PVN contains many distinct neurosecretory cells that regulate important physiological functions in the neuroendocrine system. These cells can be anatomically and functionally divided into a magnocellular division and a parvocellular division.

Magnocellular neurons produce the hormones oxytocin and vasopressin and project to the posterior pituitary gland. Parvocellular neurons that synthesize corticotrophin releasing hormone project to the median eminence at the ventral surface of the brain where they release peptides into the blood vessels of the hypothalamo- pituitary portal system, vasculature that carries peptides to the anterior pituitary gland. Some neurons, projecting to autonomic nuclei of the brainstem and spinal cord, are activated, in a stimulus-specific fashion, by hypoglycemia, hypovolemia, cytokines, pain, and environmental stressors. The PVN receives afferent projections from multiple brain areas, such as the brainstem and other hypothalamic regions, that provide information about the homeostatic state of the organism.

►Central Regulation of Autonomic Function
►Hypothalamo-neurohypophysial System
►Hypothalamo-pituitary-adrenal Axis, Stress and Depression
►Hypothalamus
►Pituitary Gland
►Ventrolateral Preoptic Nucleus (VLPO)

Paravertebral Ganglia

Definition

The paravertebral ganglia are interconnected autonomic ganglia that lie close to the spinal nerves and the vertebrae, from the lower cervical/upper thoracic level to the sacral level of the spinal cord. The chains of paravertebral ganglia are paired, and lie just lateral to the bodies of the vertebrae.

►Autonomic Ganglia
►Sympathetic Nervous System
►Sympathetic Pathways

Paresis

Definition

A weakening of a muscle due to pathology of the muscle, the motoneurons that innervate it, or the efferent nerve that carries the latter's axons.

Paresthesia

Definition

Abnormal sensory experiences such as numbness, pins-and-needles sensations and tingling, occurring spontaneously without external sensory stimulation and with some sensory ►peripheral neuropathies.

►Peripheral Neuropathies

Parietal Lobe

Synonyms

►Lobus parietalis

Definition

Extends from the central sulcus to the parietooccipital sulcus.

►Telencephalon

Parietal Organ

Definition

A heritage of ancient, sea bottom-dweller vertebrates, an "eye" on the top of the head, the pineal body evolved in correlation with it.

►Evolution of the Brain: At the Reptile-Bird Transition

Parinaud Syndrome

Definition

This syndrome results from dorsal midbrain lesions and is characterized by impaired vertical eye movements (especially upwards) and absence of the pupillary light reflex.

►Pupillary Light Reflexes

Parkinson Disease

Ali Samii
Department of Neurology, University of Washington Medical Center, Seattle, WA, USA

Synonyms

Idiopathic Parkinson's disease; Idiopathic Parkinsonism

Definition

A progressive neurological disease named after James Parkinson.

Characteristics

Clinical Features, Epidemiology, and Etiology

The four cardinal features of Parkinson disease are tremor, bradykinesia, rigidity, and ►postural instability. ►Parkinsonism is a non-specific term used to describe a constellation of signs on physical examination similar to those seen in Parkinson disease. Parkinson disease is defined as asymmetric Parkinsonism with no known cause, characterized by most of the four cardinal features, and responsive to anti-Parkinson medications (usually ►dopaminergic drugs) [1]. The diagnostic criteria for Parkinson disease have become more rigorous with gradations of diagnostic certainty. Any one of resting tremor, rigidity, or bradykinesia would suggest clinically possible Parkinson disease. Any two of the four cardinal signs (especially if asymmetric) would suggest clinically probable Parkinson disease. Clinically probable Parkinson disease with significant improvement in motor signs with dopaminergic drugs would suggest clinically definite Parkinson disease. The non-motor features of Parkinson disease include loss of sense of smell, depression, anxiety, autonomic dysfunction (constipation, urinary urgency, sexual dysfunction, orthostatic hypotension), sleep disturbance, including ►rapid-eye-movement (REM) sleep behavior disorder, and cognitive impairment. The non-motor symptoms of Parkinson disease are gaining much more attention than before since they contribute greatly to disability as the disease progresses and they do not respond to dopaminergic drugs.

The prevalence of Parkinson disease in industrialized countries is estimated at 0.3% of the general population and approximately 1% of the population aged over 60 [2]. The prevalence is slightly higher in men than in women. The mean age of onset is about 60, but approximately 5% of patients have young onset Parkinson disease, with motor symptoms appearing before age 40. *The pathology underlying the motor symptoms of PD is injury to the dopaminergic projections from the subtantia nigra pars compacta to the striatum. Lewy Bodies are the pathological hallmarks of PD, but they are not confined to the substantia nigra. Pathology is widespread in PD and involves the amygdala, the olfactory bulb, dorsal motor nucleus of the vagus nerve, locus ceruleus, pedunculopontine nucleus, raphe nuclei, the cortex, and the peripheral autonomic nervous system. This extensive pathology may account for the non motor symptoms of PD.* The etiology of PD is unknown, but aging, environmental factors, and genetic predisposition probably all play a role in causing Parkinson disease [3]. The recent discovery of several genetic loci related to familial Parkinson disease has led to the hypothesis that failure of the ►ubiquitin-proteasome system and protein misfolding are the final common pathways in the pathogenesis of Parkinson disease [4]. Mutations of the leucine-rich repeat kinase 2 (LRRK2) gene are the most common identifiable cause of Parkinson disease, in that 1% of patients without a family history and more than 5% of patients with a first degree relative with Parkinson disease have a LRRK2 mutation [5].

Medical Treatment of Motor Symptoms

There is no definitive agent known to slow down disease progression at the cellular level in Parkinson disease [6]. Therefore, treatment remains symptomatic with mostly dopaminergic drugs. A pilot study suggested that high dose ►coenzyme Q10 may slow symptom progression in early Parkinson disease [7]. These results have yet to be confirmed in larger studies with longer follow-up periods. Treatment is typically initiated when motor symptoms cause disability. The treatment of early Parkinson disease is with either a monoamine oxidase-B (MAO-B) inhibitor (selegiline or rasagiline), a non-ergot-derived dopamine agonist (pramipexole, ropinirole, transdermally absorbed rotigotine) or levodopa [8]. MAO-B inhibitors are generally safe and well tolerated, although there are warnings about certain drug interactions and tyramine containing foods when using MAO-B inhibitors. Side effects of dopamine agonists include nausea, hypotension, leg edema, vivid dreams, hallucinations (especially in the older population with cognitive deficits), somnolence (even sleep attacks), and disinhibited behavior (such as gambling). Dopamine agonists have more antiparkinson efficacy than MAO-B inhibitors, but they are less effective than levodopa. Treatment with an MAO-B inhibitor combined with a dopamine agonist may control motor symptoms for the first 2–5 years, but the likelihood of requiring levodopa after that increases significantly.

Levodopa remains the most potent anti-parkinson drug and is the backbone of therapy throughout much of the course of the disease. It is the preferred initial drug in the older population and those with cognitive deficits or serious co-morbid conditions. Levodopa is combined with carbidopa or benserazide to prevent peripheral conversion to dopamine by dopa-decarboxylase. Side effects of levodopa are similar to those of dopamine agonists, except that somnolence, hallucinations, and leg edema are less common. Complications of long-term levodopa therapy include ►motor fluctuations, including "►end-of-dose wearing off", and dyskinesia [9]. Dividing protein intake throughout the day may help reduce motor fluctuations. Controlled-release forms of levodopa may provide a longer duration of benefit, but their absorption is more unpredictable than immediate-release levodopa. ►Catechol-O-methyl transferase (COMT) inhibitors, entacapone or tolcapone, prolong the half-life of circulating levodopa and improve end-of dose wearing off. Dyskinesia can be reduced by decreasing levodopa dosage, but at the expense of worsening motor symptoms. In a patient with motor fluctuations and dyskinesia, adding a dopamine agonist to levodopa may help reduce motor fluctuations. It may also allow for levodopa reduction, which in turn alleviates dyskinesia. The subcutaneously injectable dopamine agonist, apomorphine, is useful for rapid treatment of "off" periods in Parkinson disease. However, given the severity of apomorphine-induced nausea, pre-medication with domperidone or trimethobenzamide is needed. Amantadine may help suppress dyskinesia. MAO-B inhibitors may also be added to levodopa to help alleviate motor fluctuations.

Medical Treatment of Non-Motor Symptoms

Non-motor symptoms in Parkinson disease may occur as part of the disease or as complications of treatment [10]. These include depression, cognitive impairment, psychosis, constipation, sleep disturbance, orthostatic hypotension, drooling, and urinary symptoms. Depression in Parkinson disease is usually treated with a selective serotonin reuptake inhibitor (SSRI). There are no controlled head-to-head studies to suggest one SSRI is superior to another in Parkinson disease. Constipation should be treated aggressively using multiple modalities such as stool softeners, increased fiber intake, and suppositories. Disorders of sleep in Parkinson disease include daytime somnolence, sleep attacks, night-time awakenings due to over night bradykinesia, rapid eye movement (REM) behavior disorder, and restless legs/periodic limb movements. Daytime somnolence and sleep attacks may be associated with dopamine agonists and the agonist may have to be stopped. Overnight bradykinesia and restless legs may be alleviated with a bedtime dose of long acting levodopa sometimes with entacapone, or a dopamine agonist. Clonazepam is effective in treating REM behavior disorder. Psychosis in Parkinson disease is thought to be mostly drug-induced, and it occurs more frequently in demented patients. Dopamine agonists are more likely to cause hallucinations than levodopa. First, the agonist and/or anticholinergic agent should be stopped, and the lowest dose of levodopa should be used. Adding an ►atypical neuroleptic may be necessary. Quetiapine is the more popular atypical neuroleptic in Parkinson disease. It has fewer extrapyramidal adverse effects than risperidone and olanzapine and there is no need for weekly or bi-weekly blood count measurements that would be required with clozapine. ►Centrally acting cholinesterase inhibitors (rivastigmine, donepezil, and galantamine) are somewhat effective in treating the dementia associated with Parkinson disease. Rivastigmine is more commonly used because the data to support its efficacy in Parkinson disease are more robust. Treatment options for hypotension include reducing the dosage of antiparkinson medications, increased salt and fluid intake, and adding fludrocortisone or midodrine. Drooling may be reduced by the peripheral anticholinergic agent, glycopyrrolate, but this drug may worsen constipation. Injection of botulinum toxin into salivary glands improves drooling. Urinary urgency may be treated with peripheral anticholinergic agents (oxybutynin and tolterodine) or alpha-adrenergic blocking agents (prazosin and terazosin). Unfortunately, the former worsen constipation and the latter exacerbate hypotension.

Surgical Treatment of Motor Symptoms

Deep brain stimulation of "hyperactive" nuclei relieve motor symptoms in patients who have severe motor fluctuations and dyskinesia [9]. High frequency stimulation of deep brain targets presumably reduces neural activity in tissue surrounding the electrode contact. The "suppression" of the target induced by deep brain stimulation can be sculpted by adjustments of the electrode configuration, stimulation intensity, pulse width, and frequency. Bilateral stimulation of the globus pallidus internus (GPi) or the subthalamic nucleus (STN) is effective in relieving motor symptoms. Some claim that bilateral STN stimulation is superior to bilateral GPi stimulation because it allows a reduction in antiparkinson medications, but there is continued debate about the optimal stimulation target for Parkinson disease. A large randomized multi-center study comparing bilateral STN to bilateral GPi stimulation in 300 patients with Parkinson disease is currently under way in the United States, the results of which should be available by 2009. Adverse effects of deep brain stimulation surgery include brain hemorrhage, infarct, seizures, and death. Other complications include lead breakage, other hardware failure, pulse generator malfunction, and hardware infection. Side effects from the stimulation itself include worsening dyskinesia, paresthesias, and cognitive, mood, speech, and gait disturbances. The stimulation-related side effects may be reversible by adjusting stimulation parameters. The key to successful outcome is appropriate patient selection. The surgical patient must have clinically definite Parkinson disease with documented motor improvement on levodopa. There should be no dementia, untreated psychiatric condition, or serious medical illness. ►Fetal mesencephalic tissue transplantation has been studied in Parkinson disease. Although there was marginal improvement in some patients, disabling dyskinesia occurred in many patients. Therefore, fetal mesencephalic tissue transplantation is not a treatment option for Parkinson disease at present.

References

1. Samii A, Nutt JG, Ransom BR (2004) Parkinson's disease. Lancet 363(9423):1783–1793
2. de Rijk MC, Launer LJ, Berger K, Breteler MM, Dartigues JF, Baldereschi M, Fratiglioni L, Lobo A, Martinez-Lage J, Trenkwalder C, Hofman A (2000) Prevalence of Parkinson's disease in Europe: a collaborative study of population-based cohorts. Neurologic Diseases in the Elderly Research Group. Neurology 54 (11 Suppl 5):S21–S23
3. Litvan I, Chesselet MF, Gasser T, Di Monte DA, Parker D Jr, Hagg T, Hardy J, Jenner P, Myers RH, Price D, Hallett M, Langston WJ, Lang AE, Halliday G, Rocca W, Duyckaerts C, Dickson DW, Ben-Shlomo Y, Goetz CG, Melamed E (2007) The etiopathogenesis of Parkinson disease and suggestions for future research. Part II. J Neuropathol Exp Neurol 66(5):329–336
4. Olanow CW, McNaught KS (2006) Ubiquitin-proteasome system and Parkinson's disease. Mov Disord 21(11):1806–1823
5. Schapira AH (2006) The importance of LRRK2 mutations in Parkinson disease. Arch Neurol 63(9):1225–1228
6. Biglan KM, Ravina B (2007) Neuroprotection in Parkinson's disease: an elusive goal. Semin Neurol 27(2):106–112
7. Shults CW, Oakes D, Kieburtz K, Beal MF, Haas R, Plumb S, Juncos JL, Nutt J, Shoulson I, Carter J, Kompoliti K, Perlmutter JS, Reich S, Stern M, Watts RL, Kurlan R, Molho E, Harrison M, Lew M (2002) Parkinson study group. Effects of coenzyme Q10 in early Parkinson disease: evidence of slowing of the functional decline. Arch Neurol 59(10):1541–1550
8. Horstink M, Tolosa E, Bonuccelli U, Deuschl G, Friedman A, Kanovsky P, Larsen JP, Lees A, Oertel W, Poewe W, Rascol O, Sampaio C (2006) European federation of neurological societies; movement disorder society-European section. Review of the therapeutic management of Parkinson's disease. Report of a joint task force of the European Federation of Neurological Societies and the Movement Disorder Society-European Section. Part I: early (uncomplicated) Parkinson's disease. Eur J Neurol 13(11):1170–1185
9. Horstink M, Tolosa E, Bonuccelli U, Deuschl G, Friedman A, Kanovsky P, Larsen JP, Lees A, Oertel W, Poewe W, Rascol O, Sampaio C (2006) European federation of neurological societies; movement disorder society-European section. Review of the therapeutic management of Parkinson's disease. Report of a joint task force of the European Federation of Neurological Societies (EFNS) and the Movement Disorder Society-European Section (MDS-ES). Part II: late (complicated) Parkinson's disease. Eur J Neurol 13(11):1186–1202
10. Miyasaki JM, Shannon K, Voon V, Ravina B, Kleiner-Fisman G, Anderson K, Shulman LM, Gronseth G, Weiner WJ (2006) Quality standards subcommittee of the American academy of neurology. Practice parameter: evaluation and treatment of depression, psychosis, and dementia in Parkinson disease (an evidence-based review): report of the quality standards subcommittee of the American academy of neurology. Neurology 66(7):996–1002

Parkinsonism

Definition

Syndrome characterized by rigidity, tremor, and bradykinesia, of which Parkinson disease is the most common cause.

►Parkinson Disease

Parosmia

Definition

Distorted perception of smells in the presence of an odor source. Many patients not only suffer from quantitative olfactory dysfunction (anosmia, hyposmia), but also experience qualitative olfactory dysfunctions, classified under terms such as dysosmia or olfactory distortion.

These distortions can be roughly divided into parosmias (also called troposmia) and phantosmias with the major difference that distorted olfactory sensations are experienced in the presence (parosmia) or absence of an odor (phantosmia), respectively. Patients suffering from parosmia have distorted sensations of smell elicited by an odorant, therefore it is also called stimulated olfactory distiortion. Parosmia is described as a qualitatively "wrong" perception of odors. For example, a patient may perceive the smell of rotten eggs whenever he takes a smell at roses. In most cases, the "wrong" smell is considered unpleasant. Parosmia is typically, but not always associated with quantitative olfactory loss (hyposmia). Parosmia mainly occurs in combination with post-traumatic or post-infectious olfactory loss. Rare causes of parosmia such as brain tumors, side-effects of drugs, paraneoplastic syndromes, endocrine disorders, neurologic disorders, psychiatric disorders or intracerebral haemorrhage have been reported. Although the exact site of the generation of parosmia remains unknown, most parosmias are likely to be generated at the level of the olfactory epithelium and/or the olfactory bulb. On the other hand, parosmia may also be a problem of the central nervous system. Important clinically is the observation that most parosmic impressions tend to diminish over months and finally disappear after years.

►Olfactory Pathways
►Smell Disorders

Paroxysmal Extreme Pain Disorder (PEPD)

Definition

Previously referred to as familial rectal pain. The severe pain in PEPD patients along with redness in the lower body can start in infancy (and possibly in utero), and is induced by defecation or probing of the perianal areas, and is accompanied sometimes by tonic non-epileptic seizures, syncopes, bradycardia and occasionally asystole.

Pain progresses with age to ocular and maxillary/mandibular areas and is triggered by cold, eating or emotional state. Pain episodes can last seconds to minutes (and hours in extreme cases), and gradually subside over minutes.

►Voltage-gated Sodium Channels: Multiple Roles in the Pathophysiology of Pain

Paroxysmal Hemicrania (PH)

Definition

Rare headache belonging to the group of ►trigeminal autonomic cephalalgias and is characterized by severe, unilateral pain attacks localized to orbital, supraorbital, and temporal areas, lasting 2 to 30 minutes, and accompanied by ipsilateral ►autonomic features.

►Headache

Pars Tubualis

Definition

An area of the pituitary stalk rich in melatonin receptors.

►Melatonin

Partial Seizures (Focal Seizures)

Definition

These seizures are characterized by focal motor or sensory symptoms indicating the location of brain lesions. For instance, aversive seizures characterized by deviation of eyes or head to one side indicate a focus in the opposite ►prefrontal cortex. Focal (Jacksonian) motor seizures may result from localized lesions (injury or tumor) to the contralateral ►primary motor cortex (M1) and may start as local rapid (clonic) contractions, often at a finger, great toe or mouth corner, and then spreading over the body with loss of consciousness (Jacksonian march).

►Jacksonian Motor Seizures
►Prefrontal cortex
►Primary Motor Cortex (M1)

Parvocellular Cells

Definition

Small cells located in two to four layers of the LGN of primates that have been proposed to be part of a pathway (the P pathway) from the retina to visual cortex concerned with detail and color vision.

►Evolution of the Visual System: Mammals – Color Vision and the Function of Parallel Visual Pathways in Primates

Passband

Definition

The frequency region is which the magnitudes of sound are not attenuated by a filter.

►Acoustics

Passive Avoidance Learning

Definition

Passive avoidance is a task in which animals avoid an aversive stimulus by inhibiting a previously punished response (compare with Active avoidance learning).

Behaviors that are more compatible with natural defensive responses to aversive stimuli (see SSDR in glossary) are more easily learned.

►Active Avoidance Learning
►Aversive Learning

Patch Clamp

Definition

Erwin Neher and Bert Sakmann developed the patch clamp in the late 1970s and early 1980s. They received the Nobel Prize in Physiology or Medicine in 1991 for this work. This electrophysiological method allows to record ion channels' activity in an individual cell using a glass electrode with an open tip diameter of about 1 μm. There are three major configurations of this method. The "cell attached" mode (the glass pipette is tightly sealed on the cell membrane), the "excised configuration" (the patch of membrane isolated by the pipette is removed from the cell) and the "whole cell" configuration (the membrane under the pipette is broken, providing an access to the intracellular space of the cell). The "cell attached" and the "excised" configurations allow recording the activity of individual ion channels embedded in the patch of membrane isolated by the glass pipette. The "whole cell" configuration allows to record global electrical activity passing through the entire cell membrane. This technique can be used in the "voltage clamp" mode, keeping the voltage constant in order to see changes in the current. Conversely, it can be used in the "current clamp" mode, in order to see changes in the voltage.

►Intracellular Recording

Path Integration

Definition

A process of estimating one's own position in space by integrating (vectorially summing) the distances and directions covered by previous self-movement (idiothetic inputs). This is also known as dead reckoning. The composite vector points from the start position to the current position. There are two forms of path integration: path integration with a map and path integration without a map. When path integration is used with a map it can be used to update the traveler's location on the map. Therefore, it is an important contributor to mapping. When path integration is used without a map it can only be used for returning to a start location, or "homing". This is done by the simple process computing the outbound vector from a start (home) location, inverting the composite vector and following the inverted vector. Homing by means of path integration has been shown in insects (ants) and rodents.

►Optic Flow
►Spatial Learning/Memory

Pathology

Definition

Pathology literally is the study of pathos (disease, suffering). Traditionally, pathology is a morphologic study of histological abnormalities detected by microscopic examination. More recently, pathology uses molecular, microbiological, and immunological techniques.

Pathology also means a condition produced by disease.

Pause Neurons (pns)

►Omnipause Neurons

Paw Preference

Definition

Preference of paw in taking food.

►Nervous, Immune and Hemopoietic Systems: Functional Asymmetry

Pax Genes

NORIKO OSUMI
Division of Developmental Neuroscience, CTAAR, Tohoku University School of Medicine, Sendai, Japan

Definition

Pax genes are originally identified as homologs structurally related to the *Drosophila* pair-rule gene, *paired*, which encodes a ►transcription factor [1]. There are nine members in the *Pax* gene family, which are categorized into four subclasses (Fig. 1) [2,3].

All members have a DNA-binding paired domain (PD) together with or without an octapeptide (OP), and the members except Pax1 and Pax9 have another DNA-binding homeodomain (HD). Group 1 (Pax1, Pax9) have only PD and OP, Group 2 (Pax2, Pax5, Pax8) have PD, OP, and incomplete HD, Group 3 (Pax3, Pax7) have PD, OP, and complete HD, and Group 4 (Pax4, Pax6) have PD and complete HD. PD consists of 6 alpha-helices, and is divided into two subdomains (N-terminal PAI domain and C-terminal RED domain), each recognizing distinct half-sites of the bipartite binding site in adjacent major grooves of the DNA helix [4].

Characteristics

Expression patterns of each *Pax* gene are highly region-specific, and observed in ►CNS, ►PNS, and various ectodermal, mesodermal, and endodermal tissues. The importance of the Pax family in organogenesis can be assumed from various congenital diseases and cancers related to mutations of *Pax* genes and down- or up-regulation of *Pax* gene expressions [5]. Interestingly, there are spontaneous mouse mutants for *Pax1* (*undulated*), *Pax3* (*Splotch*), and *Pax6* (*Small eye*), which is quite in contrast with other transcriptional factors that are crucial in organogenesis. Since *Pax1* and *Pax9* are predominantly expressed in mesodermal tissues, and *Pax4* is more important in pancreatic development, the other members of the *Pax* family are taken into consideration here.

Pax2/5/8

Pax2, Pax5, and Pax8 are structurally close and work similarly. Pax5 is originally identified as BSAP (B-cell specific activator protein) that is essential for development of B-lymphocytes. Pax2/5/8 expressed in the boundary region between the midbrain and hindbrain (midbrain/hindbrain boundary: MHB; (Fig. 2a), and important in establishment of MHB that works as an organizer for brain patterning [6].

In ovo mis-expression of Pax2 and Pax5 by electroporation in chick embryos can change the fate of presumptive diencephalons to the tectum. Expression of Pax2, Pax5, En, Wnt1, and Fgf8 consists of a feedback loop at the MHB in formation of the optic tectum in the midbrain and of the cerebellum in the hindbrain. Similar machineries are also work in the formation of MHB in mammalian embryos, which is shown by the phenotype

Group	Pax genes	Basic structure (PD OP HD)	Localization: Mouse	Localization: Human	Mouse mutant: Natural	Mouse mutant: Targeted	Human syndrome
Group 1	*Pax1*	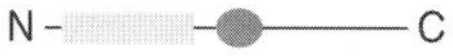 N-PD-OP-C	2	20p11	*Undulated*		Spina bifida (?)
	Pax9		12	14q12-q13			
Group 2	*Pax2*	 N-PD-OP-HD-C	19	10q25	*1Neu Pax2*	Yes	Renal coloboma
	Pax5		2	2q12-q14			
	Pax8		4	9p13		Yes	
Group 3	*Pax3*	 N-PD-OP-HD-C	1	2q35	*Splotch*		Waadenburg syndrome I, III
	Pax7		4	1p36.2		Yes	
Group 4	*Pax4*	 N-PD-HD-C	6	7q32		Yes	
	Pax6		2	11p13	*Small eye*	Yes	Aniridia Peters anomaly

Pax Genes. Figure 1 Structures and genomic positions of *Pax* genes. *Pax1-9* genes are categorized into four groups from their molecular structures. Natural mutations of *Pax* genes in the mouse and human diseases are also shown. PD paired domain; OP octapeptide; HD homeodomain.

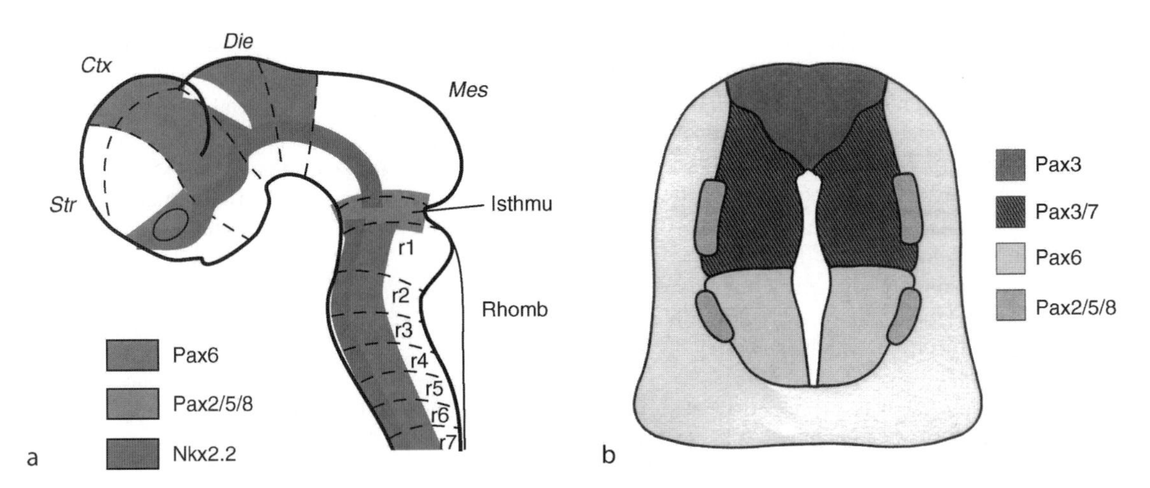

Pax Genes. Figure 2 Expression patterns of *Pax* genes in the early neural tube. Schematic illustration of the lateral (a) and transverse (b) views showing region specific expression of *Pax* genes. (b) is taken from ref. [1]. *Ctx* cerebral cortex; *Str* striatum; *Die* diencephalon; *Mes* mesencephalon; *r1-7* rhomobomere 1–7.

of knockout mice of the genes. Pax2 and Pax5 are also expressed in the specific dorsoventral region of the developing neural tube (Fig. 2b) [7], and may be involved in specification of inverneurons. Pax2 also works to establish the forebrain/midbrain boundary through interaction with Pax6 (see below). Mutation in *Pax2* is reported to be related to a kidney disease, renal coloboma.

Pax3/7

Pax3 and Pax7 are structurally close and work similarly. Pax3/7 is expressed in the dorsal region of the developing neural tube (Fig. 2b) [1] and the dermomyotome, an embryonic primordium of the muscle and dermis. Electroporated Pax3/7 forces the midbrain to differentiate into the tectum in chick embryos. Spontaneous mutant *Splotch* (*Sp*) mice have mutations in *Pax3* and show abnormalities in development of tissues originated from the neural crest that is formed at the most dorsal part of the neural tube. Heterozygous *Sp/+* mice show white spots in the abdomen, legs, and tail, which is due to abnormal development of melanocytes derived from the neural crest. Homozygous *Sp/Sp* mice show spina bifida (separated spinal bones), exencephaly (open brain), reduced or loss of dorsal root ganglia, and hypoplastic leg muscles [1–3]. Pax7 is expressed in neuronal cells of the optic tectum/ superior colliculus. Neurons expressing Pax7 migrate towards the pia and concentrate in the stratum griseum superficiale, the target site for retinal axons.

In humans, patients of Waardenburg syndrome type 1 show mutation at 2q37, and those of type 3 show deletion in 2q35-37 covering *PAX3* gene [5]; both sets of patients suffer from hearing loss and abnormal skin color. The hearing loss of Waardenburg patients is thought to be caused by abnormal migration of neural crest cells into the otic primordium from the similar phenotype seen in *Sp/+* mice. Involvement of *Pax3* and *Pax7* in cancer is also reported [5]. For example, alveolar rhabdomyosalcoma is caused by translocation of chromosomes including *PAX3* or *PAX7*.

Pax6

PAX6/Pax6 gene is first isolated as a responsible gene for human congenital aniridia (lack of the iris in the eye) and mouse *Small eye* mutant [8], and more intensively studied than any other *Pax* genes. *Pax6* gene is highly conserved throughout the phylogeny, and related to development of the sense organs, eyes, brain, pancreas, and pituitary gland [1–3]. In the early eye primordium, Pax6 is strongly expressed in the head ectoderm that will form the lens and cornea, and in the optic cup that will form the neural and pigment retina. Later in development, Pax6 is expressed in prospective ganglion-cells and amacrine-cells, but not in photoreceptors. Pax6 also influences eye development through a non-cell autonomous manner by regulating migration of neural crest cells.

In the vertebrate CNS, Pax6 is expressed in the forebrain, hindbrain, and spinal cord from the earliest stage of brain development. *Small eye* mice and rats are spontaneous *Pax6* mutant strains, homozygotes which lack the eyes and nasal structures. They also exhibit severe malformation in various brain regions where *Pax6* is expressed, showing the importance of the gene in brain patterning, neuronal migration, and axon extension [9,10]. For example, mutual repression of Pax6 and Pax2 defines the boundary between the forebrain and midbrain, and that of Pax6 and Gsh2 in the telencephalon establishes the boundary between the cerebral cortex and the striatum. In the dorsal telencephalon, Pax6 and Emx1 show gradient expression patterns in opposite directions, thereby patterning the telencephalon along the anterior-posterior axis. In the hindbrain and spinal cord, Pax6 is essential in specification of the ventral neurons: *Pax6* homozygous mutant mouse and rat embryos lack the hypoglossal nerve (a somatic motor nerve of the XIIth cranial nerves) and En1-expressing interneurons. Pax6 is involved in navigation of certain neurons such as mitral cells in the olfactory bulb and olfactory cortex neurons, which is done by non-cell autonomous manners. *Pax6* homozygous mutant rats show defects in thalamocortical projection, which may include Neuregulin1 signaling pathways.

In the adult brain, Pax6 is expressed in neurons of the olfactory bulb, amygdala, thalamus, and cerebellum [7]. Since homozygous *Pax6* mutants die at birth and therefore cannot survive into the postnatal stage, the function of Pax6 in the adult brain remains largely unsolved. It has recently been shown that Pax6 is important in postnatal neurogenesis by maintaining neural stem/progenitor cells; Pax6 heterozygous rats show decreased cell proliferation in the hippocampus and subventricular zone of the lateral ventricle. Pax6 can also promote neuronal differentiation, so it is likely that Pax6 works multifunctionally in highly context-dependent manners.

Known target/downstream molecules for Pax6 transcription factor in CNS are a bHLH transcription factor Neurogenin2, cell adhesion molecules L1 and R-cadherin, a secreted factor Wnt7b, a fucosyltransferase FucT9, and a fatty acid binding protein Fabp7 (BLBP). It is of interest that FucT is involved in the synthesis of LewisX carbohydrate epitopes that are used as markers in blastcystes, hematopoietic stem cells, and neural stem cells, and that Fabp7/BLBP is a well-known marker of radial glia and works to maintain embryonic stem cells. For patterning the brain, mutual repression of Pax6 and other transcription factors such as Pax2 (forebrain/midbrain regionalization), Gsh2 (cortex/striatum regionalization), and Nkx2.2 (somatic motor precursor domain formation) delineates distinct brain regions.

Clinically, expression of *PAX6* is reported to be significantly reduced in glioblastoma, the most common primary malignant brain tumors, molecular mechanisms of which remain unsolved.

References

1. Gruss P, Walther C (1992) Pax in development. Cell 69(5):719–722
2. Mansouri A, Stoykova A, Gross P (1994) *Pax* genes in development. J Cell Sci Suppl 18:35–42
3. Dahl E, Koseki H, Balling R (1997) *Pax* genes and organogenesis. Bioessays 19:755–765
4. Czerny T, Shaffner G, Busslinger M (1999) DNA sequence recognition by Pax proteins: bipartite structure of the paired domain and its binding site. Genes Dev 7(19):2048–2061
5. Chin N, Epstein JA (2002) Getting your Pax straight: Pax proteins in development and disease. Trends Genet 18(1):41–47
6. Nakamura H (2000) Regionalization of the optic tectum: combinations of gene expression that define the tectum. Trends Neurosci 24:32–39
7. Stoykova A, Gruss P (1994) Roles of *Pax*-genes in developing and adult brain as suggested by expression patterns. J Neurosci 14:1395–1412
8. Hanson I, Van Heyningen V (1995) Pax6: more than meets the eye. Trends Genet 11(7):268–272
9. Osumi N (2000) The role of Pax6 in brain patterning. Tohoku J Exp Med 193(3):163–174
10. Manuel M, Price DJ (2005) Role of Pax6 in forebrain regionalization. Brain Res Bull 66(4–6):387–393

PC Afferents

Definition

Rapidly adapting mechanoreceptive afferents (also called fast adapting type II) with large receptive fields, ending in relation to Pacinian corpuscles (PC).

►Pacinian Corpuscle
►Vibration Sense

PCD

►Programmed Cell Death

PDZ Domain

Definition

PDZ domain is a structural domain of 80–90 aminoacids found in signaling proteins, which provides structural integrity to protein signaling complexes and helps anchor transmembrane proteins to the actin cytoskeleton. There are approximately 200 PDZ containing proteins in the human genome.

PEA3

Definition

A member of the ETS class of DNA-binding transcription factors. EPEA3 is phosphorylated and activated by Ras via MAP kinase signaling pathways and regulates the expression of several genes in a variety of cell types.

►Neurotrophic Factors in Nerve Regeneration

Pediatric Pain

►Pain in Children

Pedophilic Perpetrators

Definition

Men who abuse children sexually for different reasons.

►Forensic Neuropsychiatry

Pedunculopontine Tegmental Nucleus

Synonyms

►Nucl tegmentalis pedunculopontinus (Ch.5)

Definition

An important nucleus from the cholinergic cell group of the lateral reticular formation. It has two parts:

- Pedunculopontine tegmental nucleus, compact part
- Pedunculopontine tegmental nucleus, diffus part

Pedunculopontine Tegmental Nucleus, Compact Part

Synonyms

►Nucl tegmentalis pedunculopontinus; Pars compacta

Definition

This densely packed part of the pedunculopontine tegmental nucleus lies in the caudo-lateral Mesencephalon and has reciprocal connections with the motor centers and the limbic system. Efferents go to the spinal cord. Electrical stimulation of this area causes coordinated locomotion ("mesencephalic locomotor region") in decerebrated animals.

►Mesencephalon

Pedunculopontine Tegmental Nucleus, Diffuse Part

Synonyms

►Nucl tegmentalis pedunculopontinus; Pars dissipata; ►Pedunculopontine tegmental nucleus; Dissipated part

Definition

In addition to the pedunculopontine tegmental nucleus, compact part, the pedunculopontine tegmental nucleus also contains a cholinergic region with loosely arranged cell bodies. But their function is not clear unlike that of the locomotor tasks of the compact part.

►Mesencephalon

Pelvic Afferents

►Visceral Afferents

Pelvic Floor

Definition

The pelvic organs are supported by striated muscles (levator ani and coccygeus) and connective tissue that close the caudal end of the abdomen spanning the space between the pubic bones anteriorly, the ischial spines laterally and the sacrum posteriorly.

►Micturition

Pelvic Nerve

Definition

The pelvic nerve connects the pelvic viscera (urinary bladder, urethra, distal bowel, vagina, uterine cervix) with the sacral S2-S4 (and in some species the lower lumbar) segments of the spinal cord. The nerve contains efferent autonomic neurons (mainly parasympathetic, but not entirely so), and also afferent nerves that convey sensory information to the central nervous system. It is responsible for the neural control of the hindgut, bladder and reproductive organs.

►Micturition
►Neurogenic Control
►Neurophysiology of Sexual Spinal Reflexes
►Visceral Afferents

Pendular Nystagmus

Definition

Nystagmus with a quasi-sinusoidal waveform (as opposed to "saw tooth" or "jerk" nystagmus in which there is alternation of slow and fast phases of nystagmus.

►Central Vestibular Disorders
►Nystagmus

Penetrance in Inheritance

Definition

In dominantly inherited disorders, penetrance refers to the proportion of persons with a mutation who show clinical symptoms. Complete penetrance refers to a situation where symptoms are present in everyone who has the mutation, and incomplete penetrance refers to a situation where symptoms are not always present in those with the mutation.

Penumbra

Definition

The marginally perfused area in the brain that surrounds the most deeply infarcted area during an ischemic stroke. This tissue is still viable if adequate perfusion can be maintained or restored.

►Ischemic Stroke
►Stroke

PEPD

Definition

►Paroxysmal Extreme Pain Disorder

Percept

Definition

The conscious experience of a sensory stimulus. The percept reflects stimulation of the sensory system (e.g. eye, ear, skin), but is also determined by higher-level cognitive processes (e.g. attention, memory).

►Perception

Perception in Vision

Dirk Kerzel
Faculté de Psychologie et des Sciences de l'Éducation, Université de Genève, Genève, Switzerland

Definition

Perception is the conscious experience of sensory stimulation. The perceptual process begins with the transformation of the external stimulus energy into the firing of neurons. The sensory signals originating in the sense organs are analyzed into different perceptual attributes such as pitch, color, form, or motion.

What is perceived depends not only on the raw sensory signal, but also on higher-level cognitive processes such as attention and memory. These processes are not always conscious, and they organize and interpret the information coming from the eyes (vision), ears (audition), nose (olfaction), tongue (taste), skin (tactile sense), and inner ears (vestibular senses). As a result, we perceive meaningful objects and events that are defined in both space and time. The perceptual process is very different from a mere image taken by a camera because objects and events are recognized and thereby linked to previously acquired knowledge stored in memory.

Characteristics

Perception is our mind's window on the world. It enables us to create mental representations of objects (flowers, cars, etc.) or events (walks in a park, accidents, etc.), and enables us to interact with objects in the world. Vision is by far the most important sensory modality and this contribution is restricted to this modality. Nonetheless, most of the concepts presented here can be applied to other sensory modalities (audition, olfaction, taste, tactile and vestibular senses). The importance of vision is evident in the space allotted to it in the human neocortex: the primary visual cortex occupies about 15% of the neocortex, and more than 30 visual areas have been identified. Altogether, 60% of the human neocortex is involved in the processing of visual stimuli.

Intuitively, visual perception is a passive process; as soon as we open our eyes, we see the world around us. Another intuition about visual perception is that the basic units of visual perception are objects because we typically talk about objects (cars, flowers, etc.) when we report what we see. Both intuitions are wrong. Research in psychology and neurophysiology has shown that perception is a highly active process and that the attributes of an object such as its color or its movement are processed independently. The characteristic division of labor starts in the retina, and continues as visual information is transmitted to the corpus geniculatum laterale (CGL) in the thalamus, and from there to the primary visual cortex. We will examine each of these stages in turn and show how complementary systems guarantee reliable perception under different conditions and for different purposes.

Light enters the eye through a small aperture, the pupil, and passes through the cornea and the lens before it reaches the retina (see Fig. 1).

Cornea and lens bring the image that is projected onto the retina into focus [1]. While the cornea has a larger focusing power (42 diopters), only the lens' focusing power (about 18 diopters) can be adjusted to bring objects into focus. To this end, the ciliar muscles contract and the curvature of the lens increases, a process that is called accommodation. A high curvature of the lens is associated with high focusing power that is necessary to project a clear image of nearby objects on the fovea. Across the lifetime, the lens loses elasticity and its maximal curvature decreases. As a consequence, a clear image would be projected on a plane behind the retina, but what is captured by the retina is blurred. This condition is known as presbyopia ("old eye"). Similar blurred retinal images result when the eye ball is too long or too short such that the focusing power of the lens is too weak or too strong, respectively. All these impairments may be corrected by glasses that either focus and thereby increase the focusing power of the optical system or disperse the light and thereby decrease its focusing power.

To be treated in the nervous system, the stimulus energy has to be transformed into electrical signals of neurons. In the visual modality, the process of transduction is achieved by two types of receptors: rods and cones. They are hidden behind a transparent layer containing amacrine, horizontal, bipolar and ganglion cells. The ganglion cells transmit the neuronal signals originating in the receptors to subcortical centers. Their axons leave the eye through an aperture in the retina

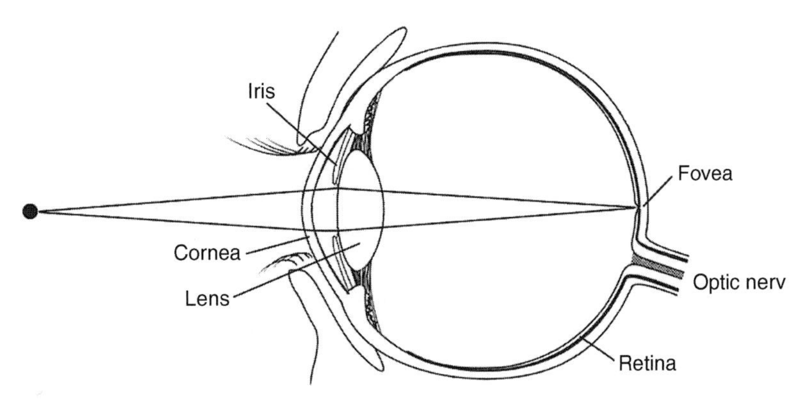

Perception in Vision. Figure 1 A cross section of the human eye. The light passes through a small aperture (pupil) and is focused by the cornea and lens. The rays of light reach receptors in the retina that are hidden behind a translucent layer of cells (not shown).

devoid of receptors, the blind spot. Although no visual information is received in this region, we do not perceive a "hole" in our visual field when we close one eye. The visual system fills the void rather actively by extrapolating visual information from the neighboring retina.

Light entering the receptor engenders a biochemical cascade when hitting a photopigment contained in the receptor. The electrical potential of the receptor changes as a result of the cascade. This process is extremely sensitive: A single photon may change the potential of the receptor and light is perceived when only seven receptors are stimulated simultaneously. However, receptors respond only to electromagnetic energy at wavelengths between 400 and 700 nm. Small wavelengths evoke the color blue, medium wavelengths the colors green and yellow, and long wavelengths the color red. Wavelengths shorter (e.g. X-rays) than 400 or longer than 700 nm (e.g. radio waves) are not absorbed by the receptors.

The two receptor classes, rods and cones, have very different characteristics and involve different neural circuits. Rods are larger than cones and contain a photopigment that responds best to light at a wavelength of 500 nm. That is, light of this wavelength evokes a response more easily than light with higher or lower wavelengths. In contrast, cones may contain one of three different photopigments that differ in their preferred frequency: the short wave pigment (419 nm), the medium-wavelength pigment (531 nm) and the long wavelength pigment (558 nm). Taken together, the three cone types respond best to a wavelength of 560 nm. The difference in the preferred wavelength of rods and cones is evident in the Purkinje phenomenon: The colors in the lower part of the spectrum seem brighter (e.g. green objects appear more salient) when seen under conditions where rods are active compared to conditions where cones are active.

Rods enable us to see at low light intensities, but do not contribute to our visual experience in the daylight. The rod's high sensibility is achieved by summing up signals across a large number of neighboring rods. While this strategy makes the system more sensitive, it produces a loss in spatial resolution. The activity of neighboring receptors cannot be discriminated and we only see blurred outlines in the dark. Also, the rod system cannot discriminate between different wavelengths and is therefore color-blind.

Cones are active during the daylight and do not contribute to nocturnal vision. They show far less summation across space than rods and thereby allow for the discrimination of spatial detail. Because the three types of cone receptors have different spectral sensitivities, they are differently activated by the incoming light. For instance, a monochrome yellow light of 575 nm will activate the long wavelength cone more than the medium and short-wavelength cones. The pattern of activation of the three receptors is the basis for color vision and any physical stimulus that produces the same pattern of activity in the receptors will be perceived as equal (a so-called metamer color). For instance, a blend of medium and long wavelength is also perceived as yellow. If one of the cone types is missing due to a genetic deficiency, color vision is abnormal. Dichromats are predominantly male because the deficient gene is on the X-chromosome. They cannot discriminate between red and green and their color perception is limited to a continuum from blue to yellow or from blue to red, depending on the missing cone type.

The distribution of rods and cones across the retina is not uniform. While most of the retina contains both rods and cones, there is a small area located on the line of sight, referred to as fovea, which contains only cones. Few of the foveal cones (as few as one) converge on one ganglion cell and the density of cones and ganglion cells is higher than in the periphery. Therefore, the fovea is the retinal area with the highest spatial resolution (acuity). Most of what we consciously perceive is being projected on the fovea, in part because of the disproportionately large cortical representation of the fovea with respect to the rest of the retina. The cortical magnification of the fovea is due to the high density of ganglion cells in the fovea and to a larger cortical representation of ganglion cells projecting from the fovea.

The projections from the fovea to the cortex are highly ordered [2]. Adjacent locations on the retina will also be represented in adjacent locations in V1, a principle that is referred to as retinotopy. Cells in V1 combine the circular receptive fields of ganglion cells to form elongated receptive fields. The receptive field of a neuron refers to the area on the receptor surface from where the neuron receives information. While ganglion cells are maximally stimulated by small points of light, cells in V1 respond best to lines or bars. Cells in V1 have been classified according to their preferred stimulus. Simple cells in V1 respond best to bars of a certain orientation. Complex cells respond to oriented bars moving in a certain direction. End-stopped cells are selective to oriented bars of a certain length moving in a certain direction. Another characteristic of V1 is its organization into columns that share common processing characteristics. Location columns comprise neurons that respond to one particular location of the retina. Within such a location column, each eye is represented by two ocular dominance columns with neurons responding preferably to stimulation of the left or right eye. Finally, each ocular dominance column is composed of a complete set of orientation columns covering vertical to horizontal orientations. Neurons in an orientation column respond best to lines at a certain orientation.

While we do not directly perceive the representation of the stimulus in V1, V1 is necessary for conscious perception. Lesions of V1 lead to blindness in the affected region. While this deficit, referred to as scotoma, eliminates conscious perception of stimuli presented in the respective visual area, it may not completely eliminate perceptual processing of those stimuli [3]. If patients who deny conscious perception of stimuli presented in the scotoma are forced to respond to these stimuli, their responses may show some residual sensitivity. For instance, they may be able to point to a stimulus presented in the scotoma with above chance (but far from perfect) performance. Even if cortical stimulus processing is precluded in this condition, retinal projections to a subcortical center in the midbrain are still intact. About 10% of the projections from the retina go to the superior colliculus, a structure that is also implied in the control of eye movements. Subcortical processing via this route may enable us to localize an object while circumventing consciousness.

From V1, visual processing in the cortex continues along two major pathways: A ventral pathway from V1 to the inferotemporal cortex and a dorsal pathway from V1 to the posterior parietal cortex (see Fig. 2).

Even if there is considerable crosstalk between the two pathways, they show important functional specialization. The dorsal pathway is responsible for determining an object's location ("where" pathway), while the ventral pathway is responsible for determining an object's identity ("what" pathway). More recently, the dorsal pathway has been characterized as action pathway ("how" pathway) because it determines how a motor action is being carried out [4]. Obviously, information about where an object is located in space is crucial for successful motor interaction with it. A case study provided neuropsychological evidence for the distinction of "what" and "how." Patient DF suffered from damage to her ventral pathway after carbon monoxide poisoning. She was unable to identify simple geometrical forms or

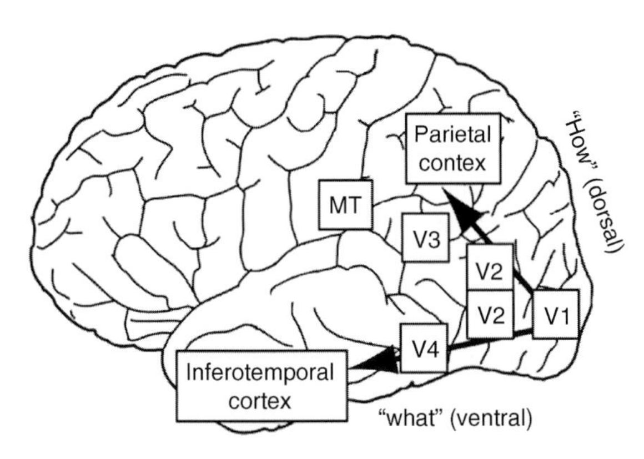

Perception in Vision. Figure 2 Pathways from V1, in the occipital lobes, to the temporal and parietal cortices. The dorsal pathway from V1 to the posterior parietal cortex is important for object localization and action control. The ventral pathway from V1 to the inferotemporal cortex is important for object recognition.

to name objects, a condition known as visual form agnosia. For instance, she could not identify a screwdriver or describe the orientation of a slot. However, her actions toward these objects were unimpaired and she could place a card in the slot which cannot be done without information about its orientation. Presumably, information about the slot's orientation was available in the dorsal pathway that guided her actions, but not in the ventral pathway where conscious recognition of the object would usually take place. Conversely, patients with lesions in the posterior parietal area often show impairments in the visual guidance of actions, while object recognition is unimpaired.

The broad distinction between two visual pathways was further refined by the identification of cortical modules. Modules are cortical areas specialized in the processing of a particular perceptual dimension, such as form (see above), color, or motion in the visual modality. Damage to such a module results in an inability to perceive the respective dimension appropriately; a condition referred to as agnosia [5]. For instance, lesions of V4 make the perception of color difficult and patients perceive the world in shades of gray even if their cone system (see above) is intact. Lesions of V5 make the perception of motion impossible. In one such patient, moving objects appeared as static images in separate positions with no transitions between these positions. Thus, the patient was unable to pour a liquid into a glass because the liquid seemed to be frozen. Also, her interaction with other people was disturbed because she could not follow the movements of her interlocutor's mouth.

After demonstrating the modular, analytic processing of visual information, the intuitive unity of objects mentioned at the beginning requires further explanation. How can we perceive objects as basic units given that the brain analyzes objects into their component attributes? According to feature integration theory, the binding of attributes belonging to one and the same object requires attention [6]. That is, visual focal attention "glues" together the various properties of an object that are processed in a distributed manner in various modules. The role of attention has been studied using a paradigm referred to as visual search. In a typical trial, the observer is asked to look for a target, for instance a blue circle (features "blue" and "round"), and to indicate its absence or presence by a key press. The target is accompanied by a number of distractors. If the target can be discriminated from the distractors on the basis of a single attribute, the search is effortless and the target is readily detected even in a large set of distractors. For instance, the blue circle is easily seen among red circles. In this case, attention is not necessary to tie together the form and shape of the objects because the work of the color module is sufficient to signal the presence of the target. In this condition, the blue target is "salient" and will "pop out" from the red distractors. If, however, a conjunction of attributes has to be detected, the search is more effortful. For example, it is more difficult to detect a blue circle among blue squares and red circles. According to the feature integration theory, the observer has to scan one object after another and focus attention on each individual object to determine whether the required conjunction is present. The work of a single module is not sufficient (color and form is important), such that serial, attention-demanding integration of attributes is necessary. Consequently, reaction times in a conjunction search increase with the number of distractors.

There is general consensus that attention is necessary for conscious perception to occur. In a phenomenon called "inattentional blindness," observers fail to detect large changes in a picture if their attention is diverted by a secondary task [7]. For instance, observers may fail to notice a black gorilla walking through a scene if they are asked to count the passes of a team dressed in white playing against a team dressed in black. Even if the gorilla was clearly processed at the sensory level, it failed to be consciously perceived because the observer was focusing on white elements in the scene. Thus, both high-level cognitive and low-level sensory processes determine our visual world.

References

1. Gregory RL (1997) Eye and brain: The psychology of seeing. Oxford University Press, Oxford, UK
2. Hubel DH, Wiesel TN (2005) Brain and visual perception: the story of a 25-year collaboration. Oxford University Press, Oxford, UK
3. Weiskrantz L (2004) Roots of blindsight. Progr Brain Res 144:229–241
4. Goodale MA, Milner D (2004) Sight unseen. Oxford University Press, Oxford, UK
5. Farah MJ (2004) Visual agnosia. MIT Press, Cambridge, MA
6. Treisman AM (1998) Feature binding, attention and object perception. Philos Trans R Soc Lond B Biol sci 353:1295–1306
7. Rensink RA (2002) Change detection. Annu Rev Psychol 53:245–277

Perception, Philosophy

Alexander Staudacher
Otto-von-Guericke-Universitaet Magdeburg, Institut für Philosophie, Magdeburg, Germany

Definition

In current philosophical debates on perception the term "perception" refers almost exclusively to sense perception, although it can be used (like the French "perception" or the German "Wahrnehmung" as well) for the acquisition of knowledge in general. Commonly there are taken to be five senses: sight, hearing, smell, touch and taste. Discussions concentrate for the most part on the sense of sight, however. Although it was at one time argued that perceptual verbs like "to see," "to hear" etc. don't refer to a kind of mental state, episode or event, because they qualify certain observations as successful in the same vein as "to win" does not describe an event or episode but reports an achievement, it is common now to acknowledge the existence of perceptual states and to take perceptual verbs as referring to those states.

Description of the Theory

One may distinguish in the philosophy of perception basically two kinds of different issues: the first issue concerns the question of how perception as a particular mental phenomenon is to be construed. The second issue concerns epistemological questions, that is, questions dealing with the role of perception in the acquisition of factual knowledge. Making such a distinction is not to deny that both issues are connected in many ways. Indeed, it seems obvious that satisfying answers to the epistemological questions presuppose that at least some questions concerning the first issue have been settled. The main focus of this article will lie, however, on the first issue. The following questions are of particular importance: How is the representational content (henceforth simply: content) of perceptual states to be understood? Mental states in general represent the world to be a certain way; they tell us how the world is in certain respects. In this sense my perception of the tree outside the window represents the world to be a certain way. Correspondingly, there can arise two cases of misrepresentation here, illusion and hallucination. Illusions are cases where something in perception appears to be different from the way it actually is (e.g. a dummy of a tree appearing as a tree). In hallucinations objects appear to be there where no such object is present at all (e.g. the rats hallucinated by a delirious alcoholic). Accordingly, the philosophy of perception not only has to specify the nature of perceptual content (including the question as to whether or not it is similar to the content of other mental states like belief or thought) but has to give also an adequate account of illusion and hallucination. Mental states and their contents are related in various ways to one another and to our actions: we entertain beliefs with certain contents (e.g. that it will rain soon) because we have other contentful mental states (e.g. we perceive a sky full of grey and heavy clouds, we believe this to be a reliable sign of coming rain etc.) and we will therefore act in a certain way (we take our umbrella with us) and so on. This leads to the question of how perceptual content is related to other mental content and to the further question of what kind of structure it has to possess in order to be able to stand in these relations.

A further question concerning the nature of perceptual states is whether they can be exhaustingly characterized by reference to their content. Perceptual states are often taken to be phenomenally conscious, that is, it is somehow for the perceiving subject to have them. They have a certain distinctive "feel" to them which distinguishes them from mere beliefs and thoughts. While seeing a tree I experience the colors of its leaves in a vividly conscious way which is lacking when I am merely thinking of these colors.

At a most general level one can divide theories of perception into two different classes which give different answers to the question what we have to take as the immediate objects of perceptual awareness. According to the one class we are immediately aware of the physical objects of perception, such as tables, trees, clouds and people etc. These objects are public in the sense that they can be perceived by more than one subject and they exist independently of whether someone preceives them. Furthermore, they change their properties neither when perceived under different perceptual conditions (e.g. varying perspectives or lighting-conditions) nor in the light of varying mental conditions of the perceiver which can influence her perceptions (e.g. expectations, drugs etc.). Theories belonging to the first class are generally called ►direct realism.

According to the other class we are immediately aware of "►sense data." An example of a sense datum is the more or less round, red, bulgy expanse you are immediately aware of when you are seeing a ripe tomato in normal daylight. A sense datum isn't public in the sense that two subjects can be aware of it. Furthermore, it has been assumed that sense data cannot exist independently of our awareness of them and that they regularly change when the perceptual conditions or our mental preconditions change. If we look, for example, at the tomato under different lighting-conditions what we will be immediately aware of will be a sense datum of a different color than before. Furthermore, you will also be directly aware of sense data in cases where no physical object is present at all (hallucination) or where this object is different from the way it appears to you in perception (illusion) (see sense data).

Sense data theories have claimed that an adequate picture of our perceptual awareness in cases of hallucinations and illusions requires the assumption that we are in these cases immediately aware of sense data which instantiate the properties no physical object actually present has. Furthermore, they have tried to show that the immediate objects of our perceptual awareness in cases of veridical perception have to be sense data as well. The vividly experienced colors (felt temperatures etc.) making up the phenomenal aspect of conscious perception can be seen according to these theories as properties of the sense data we are aware of. Concerning the question of how perceptions are related to other mental states sense data theories have traditionally assumed that perceptual states can be a secure fundament for all our empirical knowledge because we are immune against error as far as our sense data are concerned. These data will be as they appear to us. More recent defenses of sense data have put more emphasis on the role of these data in an adequate account of perceptual awareness; the epistemological presuppositions which gave rise to the search for a secure fundament of empirical knowledge have come into discredit in the last decades. There are basically two rival theories within this class, ►indirect or representative realism and a certain version of ►phenomenalism. According to ►indirect realism we have to distinguish between sense data and the mind-independent physical objects of the world surrounding us. Whereas we are in perception directly or immediately aware of sense data, our perceptual awareness of physical objects is only indirect in the following way: in the case of veridical perception the physical object is the cause of the sense datum; it is the first member of the causal chain that leads via the stimulation of our senses and the ensuing processes in the brain to the occurrence of the latter. Sometimes this causal account is supplemented with the claim that sense data function as natural signs for the physical objects they are caused by. In the case of hallucinations the causal chain won't start with the physical object; in the case of illusions the physical object will play a causal role but the chain will be influenced by further factors. Historically, this account has also been strongly motivated by the claim that physical objects don't possess the kind of properties they seem to possess according to our perceptions of them; meaning they are, e.g. neither colored in the way they appear to our eyes nor warm or cold in the way they feel to our hands, but possess only those properties physics must postulate in order to explain their causal interaction (including the interaction with our bodies). A theory of this kind can be traced back to the writings of René Descartes (1596–1650) and John Locke (1623–1704); for more recent defenses see [2].

Given that our perception of physical objects is indirect in that indicated way it seems that our knowledge of them is also indirect. We may either conclude that physical objects are the causes of our sense data or we just believe it to be that way, but are such beliefs or conclusions justified at all? According to one of the major objections to indirect realism the answer has to be negative and consequently indirect realism can't be justified either because one of its central claims is that physical objects are often among the causes of our sense data. According to this objection indirect realism cannot justify this claim because information about causal relationships has to be established empirically by the observation of cause and effect. But if indirect realism is right, the only available empirical information is our immediate perceptual awareness of our sense data. Therefore, the only information we can get will be about the effect but not about the cause. Consequently, indirect realism seems to undermine itself and to lead to skepticism concerning the existence of the physical world.

A typical indirect realist answer to this challenge is to use a strategy which is familiar in the philosophy of science when it comes to justifying the existence of unobservable scientific theoretical entities like molecules, atoms and so on. It is admitted that we have no empirical access to physical objects in the strict sense but it is held that we are nevertheless justified in claiming their existence, because this claim gives us the best explanation for the fact that the sequence of our sense data possesses the order and structure it actually does [2].

Phenomenalism can be traced back to central claims of the philosophy of George Berkeley (1685–1753); for a classical defense in the twentieth century see [1]. Roughly put, phenomenalism holds that physical objects can be identified with complex sequences of actual and possible sense data, and concludes that direct awareness of these comes down to direct awareness of physical objects. A tomato is then nothing but the complex sequence of sense data I have when I am actually looking at it or I would have if I took a different perspective of it or touched it and so on. If physical objects are nothing but sequences of actual or possible sense data a claim like "There exists a rock in the desert nobody has seen so far" has to be interpreted as the claim "If you go to a certain place in the desert you will enjoy sense data of a rock."

It has been objected to phenomenalism that our talk about existing physical objects can't be replaced in that way by talk about actual and possible sense data. To be complete, the required interpretations would not only have to be of a kind of complexity no one has been able to arrive at so far (note, that these analyses would also have to comprise phrases like "you go to a certain place in the desert"), they seem to be at odds with the causal roles we ascribe to physical objects that no one perceives (think of the roots supporting the trunk of a tree); to be the cause of something requires that it actually exists. However, unperceived objects are according to phenomenalism only sequences of possible sense data a perceiver would have if he were in the right situation. More recent statements of phenomenalism have therefore tried to defend versions which don't require such interpretations [4].

Direct realism has been traced back to Aristotle (384–322 B.C.), Thomas Aquinas (c. 1224–1274), and the Scottish common sense philosopher Thomas Reid (1710–1796), although these historical ascriptions are a matter of controversy. Direct realism comes in a variety of different forms, differing are mainly as to how the content of perceptual states is best to be construed.

According to the ▶belief-theory of perception, perceptions are a certain kinds of beliefs we acquire with the help of our sense organs (for more qualifications as to what special kind of beliefs perceptions are see [9]). Beliefs have propositional content, that is, content which can be specified with the help of a that-clause in which something is classified in a certain respect with the help of a concept (e.g. "Peter believes that grass is green"). The possession of concepts ("grass," "green") implies among other things at least the ability of the believer to recognize things as being able to under that concept. This proposal accords well with the fact that we often express the content of perceptions with the help of that-clauses and imply the possession of the relevant concepts by the perceiver when we say things like "Sarah sees that there is milk left in the fridge." It can also explain how perceptions serve to justify other beliefs (Sarah's belief that she doesn't have to go the super-market now), because justification proceeds by inferences and inferential relations can be explained best by reference to a content with propositional structure.

Nevertheless, this account faces serious difficulties: Sometimes the content of our perceptions can't be equated with the contents of our beliefs: we may perceive the arrows of the well known Müller-Lyer figure to be of a different length, although we know quite well that they are in fact of the same length. Although we modify our beliefs in the light of new beliefs, our perceptions often aren't accessible to this kind of revision. Perception in contrast to beliefs seems to be, as it is called, "modular" (for more on modularity see [6]). With their equation of perception to belief and thought belief-theories have notorious difficulties in doing justice to the fact that perceptions are phenomenally conscious. In order to avoid the introduction of sense data at this point the so-called ▶adverbial theory of experience has held that the phenomenal aspect of our sense experiences can be analyzed in the following way: a vivid conscious experience of something red (be it veridical or not) can be understood as a certain way of perceiving; that is, we are not conscious of something instantiating the property which is responsible for the character of our conscious experience (a sense datum), but our perceptual state itself is characterized by that property, in the case of an experience as of something red we "sense in a redly manner", as it has been put [7]. It has been claimed, however, that this account cannot deal adequately with situations where we have an awareness of a manifold of different items with different colors and forms because this requires being aware of different items instantiating these properties [2]. An alternative account that may be able to deal with the problems of the belief-theory is ▶representationalism (not to be confounded with representative or indirect realism) [8]. Representationalism seems especially apt to avoid a third problem of the belief-theory, the fact that we can also perceive things without disposing of the relevant concepts; or, as it is often put, that perception can have nonconceptual content. We not only render the content of perceptions with the help of that-clauses, but by saying also things like "He sees the computer" or "He heard the Tristan-accord." With statements like these we often don't imply that the perceiving subject has to have the concept of a computer or of the Tristan-accord in order to see or hear these things (think of a baby or a dog as the perceiving subjects). According to representationalism we have non-conceptual perceptual representations of the items in question. The admission of the existence nonconceptual content leads to the question of how this kind of content is related to conceptual content and how it can play a role in the justification of our empirical beliefs. Here it has been argued that perceptual content can fulfill its role in the justification of empirical beliefs only if it is taken to be conceptual, which means that the representationalist claim concerning the nonconceptual character of our perceptual experience finally can't be right [9]. Whether this objection is sound is a matter of ongoing discussions.

Representationalism also uses the idea that perceptions have nonconceptual content in order to explain the fact that perceptions are phenomenally conscious. To put it very roughly, phenomenally conscious states are conceived as nonconceptual representations which deliver information to the perceiver he which can use to form his beliefs and desires. These states acquire their nonconceptual content by being dependent in the right way on the features they represent (e.g. they are caused by them under optimal conditions; therefore the presence of red objects leads in general to perceptions of red). Because the typical phenomenal aspects of our sense experiences are experienced by the perceiver as features of the represented surroundings (e.g. in a conscious perception of the blue sky the blue is experienced as a feature of the sky), representationalism claims that perceptual states acquire their phenomenal character by being related in the right way to certain features of our surroundings: we represent the blue of the sky in a nonconceptual way. Whether such a conception of the nonconceptual content of perception can be used to explain the phenomenally conscious aspect of perception is, however, also a matter of controversy [1].

All positions considered so far share the presupposition that veridical perceptions and hallucinations have something in common. In both cases the perceiver will either have sense data, or perceptual beliefs, or nonconceptual representations or states that can be characterized by an adverbial modification. There is, however with the so-called "▶disjunctive account" of sense experience [10] a further variety of direct realism which puts this presupposition in question: according to this position a perceptual experience is either a perception or a hallucination (a disjunction of these two types of states), but these two types of states need not share a common element: perception implies that we are aware of physical objects and the content is just constituted by the perceived aspect of the physical object itself, hallucinations however don't imply this because there is no such object present. In the case of hallucinations we simply take ourselves to perceive something although it isn't the case.

One consequence of this position which might be hard to swallow is that it demands that we are in the case of hallucinations not only in error about the physical world but also about the conscious mental states we are in; we assume that we are perceiving while we aren't. A second problem is that the disjunctive account has difficulties with the fact that hallucinations and veridical perceptions are the causal consequences of the same type of brain states; one may evoke a hallucination by a stimulation of those parts of the brain which are also activated in the case of veridical perception. However according to a widely held principle on causation the same kinds of causes lead to the same kind of effects. It seems, therefore, that the adherent of the disjunctive account has to give up this principle, because according to his account there is nothing in common between hallucinations and veridical perceptions eventhough they result from the same kind of brain stimulation (for further discussion see [4]).

P

References

1. Maund B (2003) Perception. McGill-Queen's University Press, Montreal
2. Jackson F (1977) Perception. A representative theory. Cambridge University Press, Cambridge
3. Ayer AJ (1940) The Foundations of empirical knowledge. Macmillan and Co., London
4. Robinson H (1994) Perception. Routledge, London
5. Pitcher G (1971) A theory of perception. Princeton University Press, Princeton
6. Fodor J (1983) The modularity of the mind. MIT Press, Cambridge/Mass
7. Chisholm R (1957) Perceiving: a philosophical study. Cornell University Press, Ithaca/NY
8. Tye M (1995) Ten problems of consciousness. MIT Press, Cambridge/Mass
9. McDowell J (1994) Mind and world. Harvard University Press, Cambridge/Mass
10. Haddock A, Macpherson F (ed.) (2007) Disjunctivism: Perception, Action, Knowledge. Oxford University Press, Oxford

Perceptive Field

Definition

Psychophysical counterpart of receptive field organization.

Term first introduced by Jung and Spillmann (1970) to link perceptual properties to neuronal function. Although psychophysical data represent the final stage of integration of the activity of numerous neurons, they typically resemble those obtained with single cell recordings (single neuron doctrine by Barlow 1972). Perceptive fields may thus be regarded as psychophysical equivalents of receptive fields with their various forms of center-surround antagonism and selective spatio-temporal sampling characteristics that allows for non-additive, Gestalt-like integration of the stimulus input.

►Psychophysics

Perceptron

Definition

The Perceptron was the first neurocomputer (artificial neural network) to be developed. The original Perceptron consisted of two layers of units (input and output layer) and used a simple learning rule for weight adjustment according to the difference of the network output and the target vector. It is capable of solving linearly separable problems. It can be proven that the network converges to a solution (i.e., reaches its error minimum) within a finite number of steps for any linearly separable problem. Multilayer Perceptrons with hidden (intermediate) layers of units are more powerful and can learn complex non-linearly separable problems.

►Connectionism
►Neural Networks

Perceptual Completion

►Perceptual Filling-In

Perceptual Constancy (in Vision)

Definition

The tendency of perceiving objects and scenes as being invariant in size, shape, lightness, color, etc., despite variation of sensory inputs originating in the same objects and scenes observed from varying viewing distance, at varying viewing angle, in varying illumination, etc.

►Color Constancy

Perceptual Correlates

►Visual Neuropsychology

Perceptual Discrimination

Definition

Ability to distinguish between different levels of a perceptual dimension such as color, pitch, or pressure.

The perceptual dimension may map directly onto a physical dimension (e.g. the perceived length of a bar may also be measured) or may be created by the perceptual system (e.g. we perceive colors but the rays of light are not colored).

►Perceptual Impairment
►Perception

Perceptual Filling-In

Ikuya Murakami
Department of Life Sciences, University of Tokyo, Tokyo, Japan

Synonyms

Perceptual completion

Definition

Perceptual filling-in refers to the visual phenomenon in which a certain part of the ►visual field appears to be overwhelmed or filled with the visual attributes, such as brightness, color, and texture, of the surrounding area. Filling-in connotes planar interpolation of surface properties; linear interpolation of contour and bar is often referred to as perceptual completion. Though filling-in at the ▶blind spot has been studied most extensively, mechanisms of perceptual filling-in are considered to be rather ubiquitous in the normal visual field as well.

Characteristics

Phenomenal Variations

Since vision is a process of ecologically valid estimation of our outer world, our visual system has evolved so as to infer object properties from impoverished visual inputs on the ▶retina. As a result, in a great variety of situations the brain assumes a surface to be filled with color and texture from outside, especially when retinal inputs to that surface part are unavailable or poor.

One of the most striking examples is filling-in at the blind spot, an oval-shaped area (approximately 5 degrees in diameter) defined for each eye, located at approximately 15 degrees horizontally from the center of the visual field. This area is insensitive to light stimulation because ▶photoreceptors are totally absent at the corresponding retinal region (optic disc). Why does this visual deficit go unnoticed? As the blind spot of one eye is a normal field of view for the fellow eye, a complete visual scene is of course available with both eyes open. However, the visual world also seems as complete with only one eye open. Clearly, perceptual filling-in is constantly at work there. One can easily convince oneself of this phenomenon by looking at an annular shape fully covering the border of the blind spot (Fig. 1): the annulus appears to be a large solid disk uniformly filled with the color and texture of the figure.

FP

+

Perceptual Filling-In. Figure 1 Perceptual filling-in at the blind spot. As the annulus covers the border of the blind spot of the right eye, the yellow inner disk perceptually disappears; the inside appears to be filled with the same color and texture as in the annulus. The reader can experience filling-in by looking straight at the fixation point labeled “FP” with only the right eye open, from an appropriate viewing distance.

Perceptual filling-in can also occur at a small deficit (▶scotoma) of the visual field that arises for a few people from an acquired local damage to the retina (or other visual pathways). This might explain why patients with local ▶retinal degenerations are sometimes unaware of their scotoma before they take ophthalmologic testing across the visual field. The monkey has also been shown to see perceptual filling-in both at the natural blind spot and at a scotoma caused by retinal damage [1].

Whereas perceptual filling-in at the blind spot and scotoma occurs in the region without visual inputs, filling-in also occurs in the region without visual-input *changes*. When an image is artificially stabilized on just the same part of the retina by using a special experimental device, this image is initially visible but soon fades away from one’s ▶perception, filled-in with its background color. Normally, incessant ▶fixational eye movements produce tiny wobbling of retinal images, preventing such perceptual fading of stationary things in everyday life. Nevertheless, by looking sideways at a faint or blurry stationary figure on a uniform background for a long time, one may experience similar image fading and filling-in with background color (▶Troxler effect). Likewise, when one looks sideways at a stationary surface stimulus on a background of randomly twinkling texture, this surface gradually disappears from perception, filled-in with similar random twinkles of texture (▶artificial scotoma) [2].

Other phenomena of perceptual filling-in in normal observations include the ▶Craik-O’Brien-Cornsweet effect (Fig. 2a) and the ▶watercolor illusion (Fig. 2b).

In these illusions, different surface parts physically share identical luminance and chromaticity, but they are perceived to have different colors when there are sharp ▶contrasts of luminance and chromaticity at the very boundary [3]. In these cases, clear physical edges delineate one area to fill-in with one color and another area with another color. Interestingly, however, perceptual filling-in of surface can also be bounded by contours that are perceptually completed themselves (Fig. 2c). In a figure designed to induce an ▶illusory contour, the color of certain parts of the figure perceptually fills-in the inside of the subjective square of the illusory contour but not beyond (▶neon-color spreading effect). Altogether, these phenomena may be viewed as examples demonstrating that the visual system infers surface properties of featureless areas by using conspicuous cues nearby.

Models

How are such ▶percepts of filling-in accomplished? To the extent that the phenomenal variety is diverse, there may exist as many theories for the mechanisms underlying them. Of these, three influential ideas have been proposed for perceptual filling-in at the blind spot and other poor regions [4].

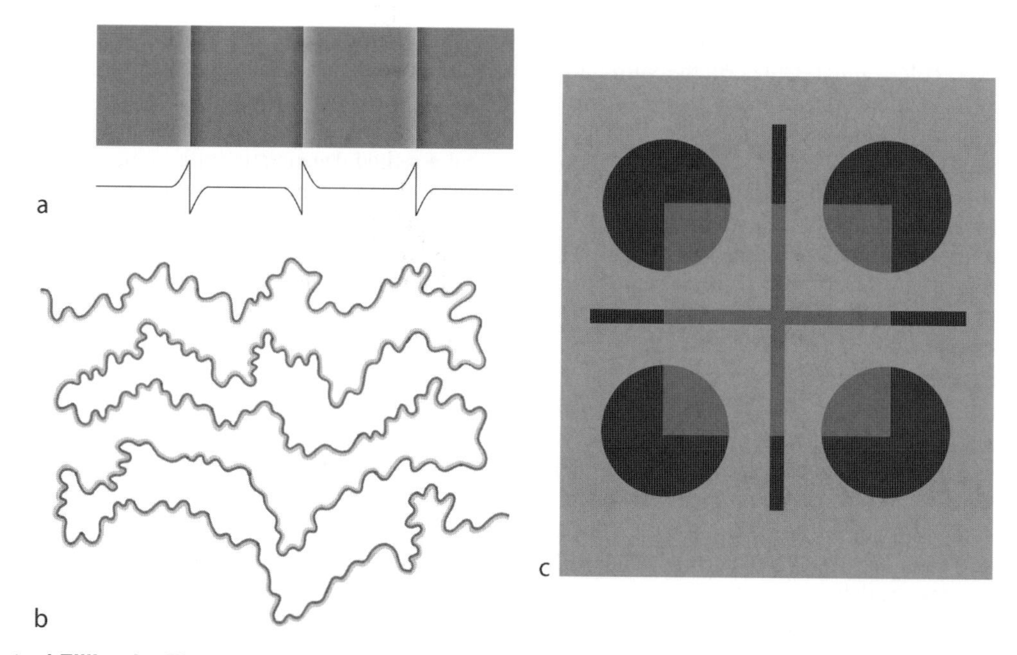

Perceptual Filling-In. Figure 2 Various filling-in phenomena. (a) The Craik-O'Brien-Cornsweet effect. The inset indicates luminance as a function of horizontal position. The area just to the left of the central edge appears uniformly darker than adjacent areas, although the luminance changes are confined within the borders between areas. (b) Watercolor illusion. The two areas delineated by the wiggly borders appear to be filled uniformly with faint yellowish and bluish colors, although the color changes are confined within the borders, with the remaining areas left achromatic, in reality, Adapted from [3], with kind permission from the author. (c), Neon-color spreading effect. A square-shaped illusory contour is formed by the black inducers, and a bluish translucent color appears to be spreading inside the subjective square seen in front.

1. *Isomorphism.* When perceptual filling-in occurs, a two-dimensional neural map (or ►visual topography) representing that location of the visual field is actually activated as such, so that individual neurons within that map are actually excited in some fashion parallel with perceived surface properties.
2. *Symbolic Labeling.* Attributes inside a poor region can be cognitively designated by symbolic or logical operations using reliable information nearby, without invoking isomorphic activation – the brain is not "filling in but finding out" the missing surface properties.
3. *Ignorance.* One does not see a black hole of the blind spot simply because one is only aware of visual events that are abundant in the surround and "ignores the absence."

Some recent findings have been more or less in favor of isomorphism, as we will see below, although labeling and ignorance may operate at crucial processing stages for one's ►conscious perception and ►awareness.

Neurophysiological Studies

In the brain, visual signals from the retina are registered in visual topography, or a two-dimensional map of neurons topographically representing the visual field. Are the neural representations of the blind spot and scotoma on this topography really activated when perceptual filling-in occurs? Physiologists are still on their way to solving this important question.

Perhaps the simplest way of neural activation would be to distort the map itself (often called the "►sewn-up" model). The representation of the blind spot could "see" its surroundings if it received direct innervations from retinal portions surrounding the blind spot. However, current studies of functional anatomy have indicated quite orderly mapping from retina to ►striate cortex without distortion at the blind spot, and at the scotoma as well [5]. Nonetheless, some ►single-unit recording studies have suggested re-mapping of visual topography of a scotoma [6]. This scotoma was induced in an animal by producing small local lesions in the retinae. After the damage, cortical neurons that had originally coded positions inside the scotoma expanded and shifted their ►receptive fields towards the outside – as if they abandoned the damaged part and started to look out. Thus, some ►cortical plasticity might be initiated in certain conditions for adaptive compensation for retinal damage.

Besides the feedforward mapping, there are at least a couple of hypothetical schemes to make neural filling-in possible in early visual cortex. The first is lateral propagation of visual signals into the cortical representation of the blind-spot region via horizontal connections. The second is feedback from ►extrastriate visual cortex back to early visual cortex. Yet another scenario would be that the neural correlate of filling-in resides not in early visual cortex containing neurons with tiny receptive fields, but in higher cortical areas where neurons have larger receptive fields covering the blind spot and beyond. More research is needed to determine which is the case.

A few attempts have been made to seek activities of individual neurons that parallel the filling-in phenomenon. One of the most recent findings is about the neuronal activity in striate cortex (►primary visual cortex, ►V1) of a behaving monkey (Fig. 3) [7].

These neurons had receptive fields largely overlapping the blind spot of one eye, i.e., with the other eye open, the neurons were excited by visual stimuli presented in these fields. The blind-spot eye only was open. A small bar stimulus (labeled *b*) on one side of the blind spot fell onto the tip of the receptive field, and thus caused mild firing responses as expected. Interestingly, responses increased when another bar stimulus (labeled *c*) was added on the other side, so as

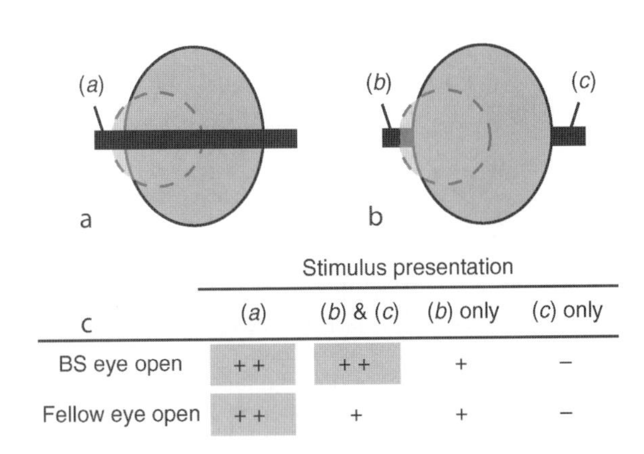

	Stimulus presentation			
	(a)	(b) & (c)	(b) only	(c) only
BS eye open	+ +	+ +	+	–
Fellow eye open	+ +	+	+	–

Perceptual Filling-In. Figure 3 Perceptual completion at the blind spot and neuronal responses [7]. (a) Schematic illustration of the stimulus. The receptive field (*broken circle*) of the recorded neuron partially overlapped the blind spot (*solid oval*). A horizontal bar (*a*) was presented such that it should penetrate both the receptive field and the blind spot. (b) In other cases, the retinal input was confined within a small sub-region (*b* and/or *c*). (c) Schematic diagram illustrating responses of a few V1 neurons exhibiting increased activities during perceptual filling-in. The neuronal responses were recorded during four types of stimulus presentations while either the blind-spot eye (BS eye) only or the fellow eye only was opened. The symbols "++," "+," and "–" indicate strong, weak, and no responses, respectively. The *yellow boxes* indicate the viewing conditions in which the monkey would perceive a long bar across the blind spot.

to induce perceptual completion of a single bar across. The second bar stimulus per se did not cover the receptive field, but the perceptually completed bar did. In contrast, such a response increase was not observed with the fellow eye only open. Therefore, these neurons fired vigorously when the filled-in bar coincided with their receptive fields inside the blind spot.

The neural correlate of the "artificial scotoma" (see above) has also been investigated in the awake monkey [8]. A static, gray square was located to cover the receptive field of the recorded neuron, and the background was filled with twinkling random noise (Fig. 4a). Over prolonged observation, the static square was perceptually replaced with twinkling noise at some time. The firing activities of neurons in a higher visual area, ►V3, were initially weak but became stronger just at the time this "artificial scotoma" would be perceptually filled-in with noise (Fig. 4b).

These findings seem consistent with isomorphic activations in the two-dimensional cortical map of the visual field during perceptual filling-in. Meanwhile, other studies indicate mixed results among different types of neurons, cortical areas, methodologies, and types of filling-in phenomena. At present, many important questions yet remain to be answered: what kinds of isomorphic activations of visual neurons are necessary and sufficient, how these could be implemented in a biologically feasible manner, and whether only one kind of architecture is enough for all kinds of filling-in to occur.

Perceptual Studies

Other lines of evidence for explicit representations corresponding to perceptual filling-in come from ►psychophysics of ►visual illusions. In one such study, a phenomenon of ►motion aftereffect was used (Fig. 5) [9].

After prolonged viewing of a moving stimulus, a static stimulus presented at the adapted location appears to move in the opposite direction, and this effect is known to transfer between eyes if different eyes are used in adaptation and subsequent test. The observer was adapted to a moving stimulus covering the blind spot of one eye and thus perceptually filling-in the inside of it. Later, the stimulus was abruptly switched to a smaller static stimulus well inside the blind spot of the adapted eye, but as it was now given to the fellow eye, the observer could see it. It was perceived to move in the opposite direction to the adapting stimulus, thus the motion aftereffect was elicited after adaptation to motion filling-in inside the blind spot without direct stimulation. This finding suggests that filled-in motion and actual motion share a common pathway.

Another observation of brightness filling-in suggests topographic representation (Fig. 6) [10]. Presented on a white background was a uniformly lit disk-shaped figure, whose luminance decreased continuously from white

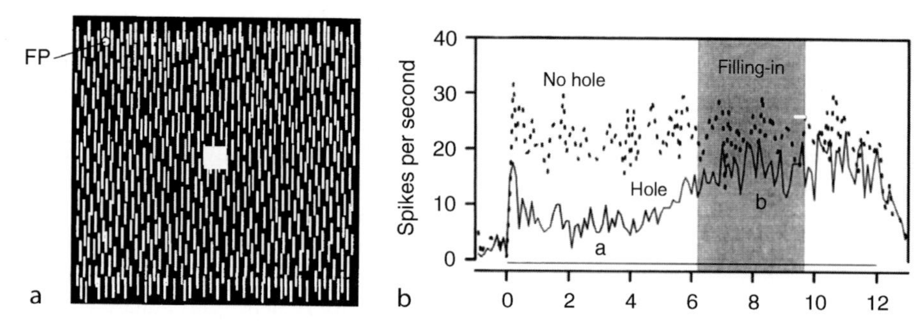

Perceptual Filling-In. Figure 4 "Artificial scotoma" and neuronal responses [8]. (a) Schematic illustration of the stimulus, which consisted of randomly positioned bars and a gray square covering the receptive field of the recorded neuron. Three such random-bar patterns alternated at 20 Hz. The monkey was rewarded for maintaining fixation at the fixation point (FP). (b) Typical responses from a single neuron in area V3. The firing rate is plotted against time after stimulus onset. The neuron initially exhibited a transient response to the stimulus onset followed by a low firing rate (labeled "Hole"). With prolonged observation, however, the neuron's responses gradually increased, reaching a plateau (labeled "Filling-in") at the time observers would experience perceptual filling-in of the square with random bars. The plateau was comparable to the firing rate of the same neuron to the same stimulus without the gray square (labeled "No-hole"). Adapted from [8], with kind permission from Nature Publishing Group.

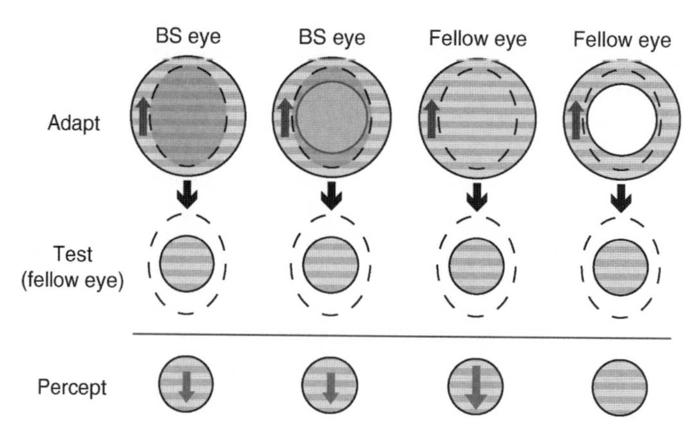

Perceptual Filling-In. Figure 5 Perceptual filling-in of motion and a subsequent motion aftereffect inside the blind spot [9]. The adapting stimulus was a horizontal grating drifting vertically (its direction is indicated by the *blue arrows*). The stimulus region was either a disk covering the entire blind spot (*broken oval*) or an annulus covering the border of the blind spot. The observer was adapted with only the blind-spot eye (BS eye) or only the fellow eye open. Note that the disk and annulus were equivalent for the BS eye. The test stimulus was a stationary horizontal grating well inside the blind spot, delivered to the fellow eye. A motion aftereffect occurred (its direction indicated by the *red arrows*) where motion had been perceptually filled-in during adaptation.

to black in half a second. The observer saw it darkening continuously, but its brightness was perceived as inhomogeneous: the brightness change towards the center of the disk appeared to lag behind the change at its edge. In other words, the center appeared brighter than the edge at each time slice, and darkness "swept" inward. This phenomenon could be seen in a disk placed in a normal visual field and in the same disk covering the blind spot just as well, either with the blind-spot eye or the fellow eye open. This effect appears to corroborate the propagation of brightness signals from edge to interior on the cortical map, suggesting that neural filling-in operations obey the same principle all around the visual field.

Functional Significance

The retinal image is a two-dimensional array of "raw" light intensity. From this image, the visual system has to estimate lots of things, such as object contours, surface assignment to objects, their volumetric structures, and their material properties, just to list a few. On the other hand, retinal inputs never provide the brain with sufficient information for solving these problems. For example, the brain has to determine what surface is lit by what light source, but the retinal image is only a result of their interactions. Thus, the same surface can project very different images onto the retina depending on viewing conditions. Also, the image of the same surface under the same illumination may be registered

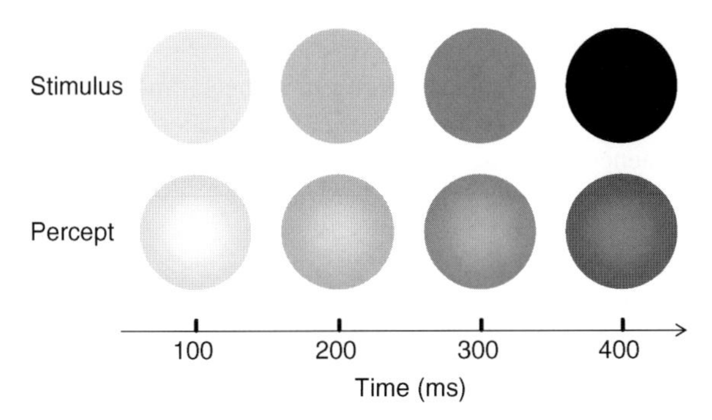

Perceptual Filling-In. Figure 6 Brightness filling-in in a normal visual field [10]. On a computer monitor, a small spot of light (1 degree in diameter) gradually changed its luminance from white to black over time. Its perceived brightness was different between the edge and inside: the brightness at the edge appears to fill inward. The same phenomenon occurs at the blind spot as well, with a larger spot of light (8–10 degrees in diameter).

as having different values of chromaticity, because of differences in our spectral sensitivity across space and time. Furthermore, the same surface may even be partially unregistered (blind spot and scotoma). Therefore, filling-in has a clear functional significance of maintaining our ►perceptual constancy in spite of imperfection of visual information from the eye. As perceptual filling-in is viewed this generic way, a great number of puzzles remain to be solved in future research.

References

1. Komatsu H (2006) The neural mechanisms of perceptual filling-in. Nat Rev Neurosci 7:220–231
2. Ramachandran VS, Gregory RL (1991) Perceptual filling in of artificially induced scotomas in human vision. Nature 350:699–702
3. Pinna B, Brelstaff G, Spillmann L (2001) Surface color from boundaries: a new 'watercolor' illusion. Vision Res 41:2669–2676
4. Pessoa L, Thompson E, Noë A (1998) Finding out about filling in: a guide to perceptual completion for visual science and the philosophy of perception. Behav Brain Sci 21:723–748
5. Smirnakis SM, Brewer AA, Schmid MC, Tolias AS, Schüz A, Augath M, Inhoffen W, Wandell BA, Logothetis NK (2005) Lack of long-term cortical reorganization after macaque retinal lesions. Nature 435:300–306
6. Gilbert CD, Wiesel TN (1992) Receptive field dynamics in adult primary visual cortex. Nature 356:150–152
7. Matsumoto M, Komatsu H (2005) Neural responses in the macaque V1 to bar stimuli with various lengths presented on the blind spot. J Neurophysiol 93:2374–2387
8. De Weerd P, Gattass R, Desimone R, Ungerleider LG (1995) Responses of cells in monkey visual cortex during perceptual filling-in of an artificial scotoma. Nature 377:731–734
9. Murakami I (1995) Motion aftereffect after monocular adaptation to filled-in motion at the blind spot. Vision Res 35:1041–1045
10. Paradiso MA, Hahn S (1996) Filling-in percepts produced by luminance modulation. Vision Res 36:2657–2663

Perceptual Grouping

Definition

►Form Perception

Perceptual Impairment

Definition

Inability to distinguish between different levels of a perceptual dimension such as color, pitch, or pressure.

The impairment may originate at a peripheral level (e.g. absence of a particular cone type in the retina leads to color blindness) or at a central level (e.g. lesions of a cortical region). The impairment may regard particular regions of space, particular perceptual dimensions (e.g. color, movement, etc.) or the recognition of what is perceived (e.g. inability to recognize familiar faces).

►Perceptual Discrimination
►Perception

Perceptual Processing

►Visual Neuropsychology

Perceptual Saliency (in Vision)

Definition

Degree to which a target stimulus "pops out" in a set of stimuli. If the target stimulus differs by a single attribute (e.g. its color) from the other objects, it is highly salient.

If the target stimulus differs by a combination of attributes (e.g. a combination of color and form) from the others, it is less salient.

► Perception

Perceptual Task

Definition

Detection or classification of a stimulus or discrimination between different stimuli.

► Sensory Plasticity and Perceptual Learning

Perforated Patch

Definition

A form of patch-clamp recording in which an antibiotic such as nystatin, amphotericin, or gramicidin are included in the patch solution to create ionophores in the plasma membrane that permit intracellular recording without modifying the cytosolic contents of the neuron.

► Patch Clamp

Perforated Synapse

Daniel A. Nicholson, Yuri Geinisman
Department of Cell and Molecular Biology, Northwestern University's Feinberg School of Medicine and Institute of Neuroscience, East Chicago Avenue, Chicago, IL, USA

Synonyms

Synapse with a perforated or discontinuous postsynaptic density (PSD); Synapse with subsynaptic plate perforations; Synapse with a fenestrated, Horseshoe-shaped or segmented PSD; Synapse with an annulate, Horseshoe-shaped or multifocal presynaptic vesicular grid

Definition

►Perforated synapses belong to a special morphological variety of synaptic junctions, characterized by the presence of aligned discontinuities (gaps) in their postsynaptic and presynaptic densities.

Characteristics

Quantitative Description

Ultrastructural features of perforated synapses, their quantitative characteristics and changes under various experimental conditions were the subject of numerous investigations. The results of these studies, which were previously reviewed in detail [1–5], are summarized below.

Ultrastructure

The hallmark of a synapse at the electron microscopic level is the ►postsynaptic density (PSD). The PSD is a plate(s) of electron-dense material on the cytoplasmic face of the postsynaptic membrane. PSDs are most noticeable in ultra thin sections of tissue conventionally prepared (i.e., osmicated) for electron microscopy (Fig. 1).

Synapses can be subdivided based on the thickness of their PSD, relative to that of the ►presynaptic density, into two main types: asymmetric synaptic junctions with a thicker PSD are considered to be excitatory; and symmetric ones with a PSD nearly as thin as the presynaptic density are considered to be inhibitory. Electron microscopic analyses of asymmetric synapses have demonstrated that they can be further subdivided, based on the configuration of their PSDs into either perforated or nonperforated (or macular) ones. When sections are made perpendicular or at an angle to a perforated PSD, a discontinuity or perforation is usually observed in a subset of its sectional profiles (Fig. 1a–c). Perforated PSD shapes can also be visualized in sections passing parallel to PSDs. Planar reconstructions of perforated PSDs from consecutive serial sections show that they assume three basic shapes: a fenestrated PSD exhibiting a hole(s) in its single plate; a horseshoe-shaped PSD having a single plate; and a segmented PSD consisting of several (2–4) separate plates. ►Nonperforated synapses, on the contrary, are characterized by a continuous PSD plate of a relatively simple shape, which may be approximated by that of a circular or elliptical disk. Sectioning of such a plate at different angles produces PSD profiles that lack perforations (Fig. 1e–h). Although dimensions of nonperforated PSDs are generally smaller than those of perforated ones, this is not an invariant feature of nonperforated PSDs because some of them overlap in size with perforated PSDs.

Distribution

Perforated synapses are widely distributed throughout the mammalian brain, including various neocortical areas, the hippocampal formation, striatum, putamen,

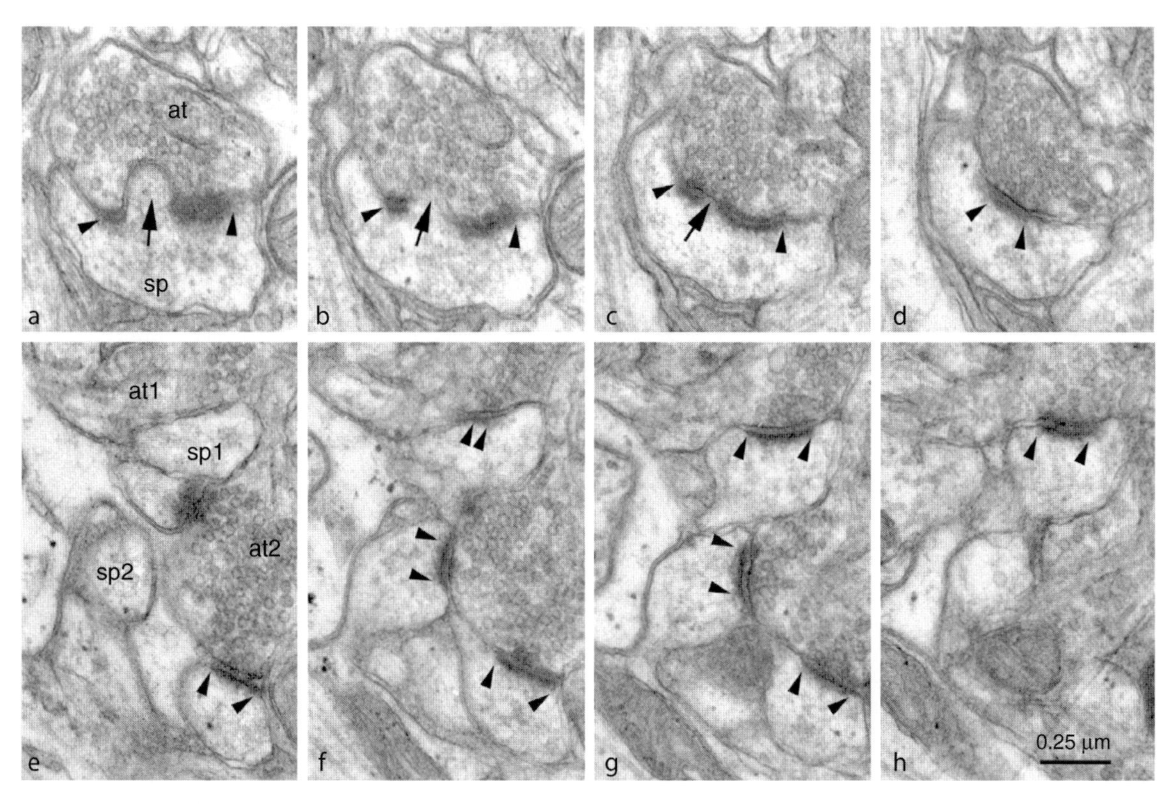

Perforated Synapse. Figure 1 Perforated (a–d) and nonperforated (e–h) axospinous synapses, seen in electron micrographs of consecutive serial sections through the middle molecular layer of rat dentate gyrus. (a–d) The perforated synapse (its presynaptic axon terminal and postsynaptic spine head are labeled in (a) by "at" and "sp," respectively) exhibits PSD profiles (*arrowheads*) that have a discontinuity (arrowheads) in some sections (a–c). A synaptic spinule (a, b) is a finger-like protrusion of the spine head, which invaginates through a discontinuity in the PSD (*arrows* in a and b) into the axon terminal. (e–h), The nonperforated synapses formed between axon terminals (labeled in (e) by "at1" and "at2") and spine heads (labeled in e by "sp1," "sp2" and "sp3") display continuous PSD profiles (*arrowheads*) in all sections.

deep cerebellar nuclei, and hypothalamus. Both perforated and nonperforated ►asymmetrical synapses are found mainly on dendritic spines and occasionally on dendrites. Axospinous perforated synapses have been extensively studied, and the present essay is focused on these synaptic junctions, which will be referred to as perforated synapses.

Frequency

Perforated synapses constitute 10–25% of the entire axospinous synaptic population. Nevertheless, the number of perforated synapses per neuron is substantial. For example, the total number of pyramidal neurons in rat CA1 hippocampal region is ~400,000, whereas the total number of perforated synapses in one of its layers (stratum radiatum) reaches ~800,000,000. Therefore, each CA1 pyramidal neuron receives ~2,000 perforated synapses in the stratum radiatum alone.

Lower Level Components

A ►synaptic spinule or spine partition is a postsynaptic component of some perforated synapses. In single sections, it is seen as a small outgrowth that arises from the spine head at a PSD discontinuity and protrudes into the presynaptic axon terminal (Fig. 1a, b). Three-dimensional reconstructions of perforated synapses demonstrate that the extent of spinules differs depending on the shape of perforated PSDs [2]. In fenestrated synapses, they assume the form of a focal partition, the base of which is limited to a hole in the PSD plate (Fig. 2b).

Horseshoe-shaped synapses exhibit a spinule whose base is restricted to the interval between the two arms of the PSD horseshoe (Fig. 2c). In segmented synapses, however, a complete spine partition divides the spine head cavity into separate compartments, each one containing a discrete transmission zone between the axon terminal and a PSD segment (Fig. 2d). Each perforated synaptic subtype can have or lack spine partitions (Fig. 2e–g). In the molecular layer of rat dentate gyrus, the proportion of synaptic junctions with partitions in total samples of fenestrated, horseshoe-shaped and segmented synapses amounts to 38, 60 and 92%, respectively.

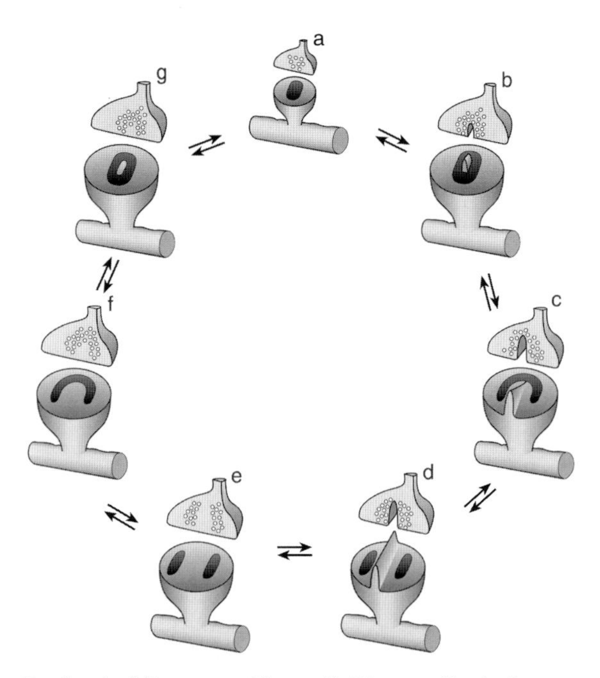

Perforated Synapse. Figure 2 Diagram illustrating a hypothetical synapse restructuring that may underlie activity-dependent alterations in synaptic strength. The schematic shows the following axospinous synaptic subtypes: nonperforated synapse (a) perforated synapses that have a fenestrated PSD and focal spine partition (b) a horseshoe-shaped PSD and sectional partition (c) a segmented PSD and complete partition (d) or that lack spine partitions but exhibit a segmented (e) horseshoe-shaped (f) and fenestrated (g) PSD. The sequence of synapse remodeling from a through to b, and c to d is postulated to be a rapid process that supports an initial maximal synaptic enhancement. The sequence from d through to e, f and g to a may lead to the return of elevated synaptic responses to the control level.

►AMPA receptors and NMDA receptors (AMPARs and NMDARs, respectively) are types of ionotropic glutamate receptors associated with the PSD of excitatory synapses. These PSDs also contain cytoskeletal scaffolding and adaptor proteins as well as signaling molecules, which together form a cluster of synaptic signal transduction machinery located at or near the site of synaptic transmission. Compelling electrophysiological evidence indicates that the number of postsynaptic AMPARs and NMDARs is the major determinant of synaptic strength. ►Postembedding immunogold electron microscopy reveals striking differences in the expression of AMPARs and NMDARs between perforated and nonperforated axospinous synapses, which may indicate corresponding differences in the strength of synaptic transmission at these synaptic subtypes [6,7]. Interestingly, some nonperforated synapses exhibit NMDAR but not AMPAR immunoreactivity, whereas perforated synapses are invariably immunostained for both receptor types. This finding may mean that perforated synapses are never postsynaptically "silent" at resting membrane potentials. Moreover, perforated synapses from rat CA1 stratum radiatum express significantly more AMPARs (by 660%) and NMDARs (by 80%) than their immunopositive nonperforated counterparts. The total area of the PSD plate(s) is generally larger in perforated synapses as compared to nonperforated ones, and postsynaptic AMPAR content positively correlates with PSD size. The difference in the expression of glutamate receptors between the two synaptic subtypes remains significant, however, when their PSD areas are equalized, indicating that this difference is related to both PSD configuration and PSD size.

Function

Numerous studies report activity-dependent increases in the number or proportion of perforated synapses. Such changes occur in cortical regions of the rodent brain during postnatal development and under various experimental conditions including housing of animals in complex environments, hippocampal kindling, NMDA-dependent hippocampal LTP and behavioral learning. Based on these observations, perforated synapses have been widely implicated in synaptic plasticity. It is necessary, however, to note here that many of the reported results should be considered with caution. The majority of studies published so far employed inappropriate procedures for quantification of perforated synapses. For example, in studies that analyzed synapses on single (rather than consecutive serial) sections, perforated synapses might not have been reliably recognized and quantified because they frequently exhibit nonperforated PSD profiles (Fig. 1b).

Such methodological drawbacks notwithstanding, there is a growing body of evidence in favor of the longstanding notion that an addition of perforated synapses is required to support an enhancement of synaptic strength. Especially demonstrative in this respect are the findings of LTP experiments [8,9], showing that perforated synapse number is markedly increased early (15–60 min) after LTP induction, when synaptic responses are maximally enhanced, and returns to control levels afterwards. The observed increase in the proportion of perforated synapses reflects a selective change in the number of segmented, completely partitioned synaptic junctions [8,9].

The existence of distinct morphological subtypes of axospinous synapses suggests a hypothetical model [2,4] of structural synaptic modifications associated with activity-induced alterations of synaptic strength (Fig. 2). According to this model, LTP induction triggers an enlargement of nonperforated PSDs, which is followed by the consecutive formation of perforated synapses having initially a focal spine partition with a fenestrated PSD (Fig. 2b), then a sectional partition with a horseshoe-shaped PSD (Fig. 2c), and finally a complete partition with a segmented PSD (Fig. 2d). The latter perforated subtype has an exceptionally high level of AMPAR immunoreactivity [7], which likely translates into an unusually strong synaptic conductance relative to that at other axospinous synapses. A high number of AMPARs at segmented synapses may be necessary for mediating the maximal level of synaptic responses characteristic of certain forms of synaptic plasticity, such as the early LTP phase. The subsequent enduring retention of a relatively lower level of synaptic enhancement during the late LTP phase, and its return to control levels over time, may involve the conversion of segmented, completely partitioned synapses into nonpartitioned perforated subtypes (Fig. 2e–g) and eventually back into non-perforated synaptic junctions (Fig. 2a). The proposed model helps to explain the large degree of morphological heterogeneity among perforated synapses, and indicates a likely association between activity-dependent synaptic restructuring and enhancements in synaptic strength. It has also been postulated that the perforated synaptic subtypes may be intermediates in the process of synapse splitting and division, but there are no data demonstrating the existence of this process.

Pathology

Recent observations indicate that learning and memory disturbances during aging are associated with structural changes in perforated synapses that are likely to have a deleterious effect on their function. It has long been known that aging-related impairments of cognitive functions do not affect all aged individuals: some of them have preserved learning and memory capacities even at advanced chronological age. The reason for such pronounced individual differences in mnemonic functions remains unknown. In a study exploring this problem [10], aged rats were behaviorally tested on a hippocampus-dependent water maze task and separated into groups with impaired or unimpaired spatial learning as compared with young adults. PSD area was estimated in perforated and nonperforated synapses from the stratum radiatum of hippocampal field CA1. The results show that aged learning-impaired rats exhibit a marked (~30%) reduction in perforated PSD area, whereas learning-unimpaired rats do not. This change is highly selective because it does not involve nonperforated PSDs. Given the strong positive correlation between PSD size and AMPAR content among perforated synapses [7], the removal of postsynaptic AMPARs from the PSD may underlie the reduction in PSD area and weaken the efficacy of synaptic transmission. Such a mechanism is also a prime candidate for contributing to the cognitive decline in the subset of aged learning-impaired animals.

References

1. Calverley RKS, Jones DG (1990) Contributions of dendritic spines and perforated synapses to synaptic plasticity. Brain Res Rev 15:215–249
2. Geinisman Y (1993) Perforated axospinous synapses with multiple, completely partitioned transmission zones: probably structural intermediates in synaptic plasticity. Hippocampus 3:417–434
3. Jones DG, Harris RJ (1995) An analysis of contemporary morphological concepts of synaptic remodeling in the CNS: perforated synapses revisited. Rev Neurosci 6:177–219
4. Geinisman Y (2000) Structural modifications associated with hippocampal LTP and behavioral learning. Cereb Cortex 10:952–962
5. Nikonenko I, Jourdain P, Alberi S, Toni N, Muller D (2002) Activity-induced changes of spine morphology. Hippocampus 12:585–591
6. Ganeshina O, Berry RW, Petralia RS, Nicholson DA, Geinisman Y (2004) Differences in the expression of AMPA and NMDA receptors between axospinous perforated and nonperforated synapses are related to configuration and size of postsynaptic densities. J Comp Neurol 468:86–95
7. Ganeshina O, Berry RW, Petralia RS, Nicholson DA, Geinisman Y (2004) Synapses with a segmented, completely partitioned postsynaptic density express more AMPA receptors than other axospinous synaptic junctions. Neuroscience 125:615–623
8. Geinisman Y, deToledo-Morrell L, Morrell F, Heller RE, Rossi M, Parshall RF (1993) Structural synaptic correlate of long-term potentiation: formation of axospinous synapses with multiple, completely partitioned transmission zones. Hippocampus 3:435–446
9. Toni N, Buchs P-A, Nikonenko I, Povilaitite P, Parisi L, Muller D (2001) Remodeling of synaptic membranes after induction of long-term potentiation. J Neurosci 21:6245–6251
10. Nicholson DA, Yoshida R, Berry RW, Gallagher M, Geinisman Y (2004) Reduction in size of perforated postsynaptic densities in hippocampal axospinous synapses and age-related spatial learning impairments. J Neurosci 24:7648–7653

Periamygdaloid Cortex

Definition

The periamygdaloid cortex is a paleocortical brain region on the medial surface of the amygdala. It belongs to the uncus of the temporal lobe, and has been described as anatomically heterogeneous.

►Olfactory Pathways

Periaqueductal Gray Matter (PAG)

Definition

The periaqueductal gray matter is the midbrain area surrounding the cerebral aqueduct (of Sylvius) and consisting of different longitudinal columns that initiate stimulus-specific autonomic and antinociceptive responses to external stressors. The lateral column of the periaqueductal gray initiates flight-or-flight sympathoexcitatory responses associated with opioidindependent analgesia. The ventrolateral column elicits hypotension, bradycardia, immobility, and hyporeactivity to the environment, associated with opioid-dependent analgesia.

►Central Regulation of Autonomic Function

Periaxin

Definition

Periaxin is a Schwann cell-specific protein of 147 kDa. It is expressed exclusively by myelinating Schwann cells and is predominately localized to their abaxonal surface. Periaxin is a myelin-related protein that undergoes dynamic changes in its localization during ensheathment and myelination.

►Myelin
►Schwann Cell
►Schwann Cells in Nerve Regeneration

Perifornical Area/Lateral Hypothalamus

Definition

Brain area of the lateral hypothalamus surrounding the fornix, a bundle of fibers that connects the hippocampus with the septal area and tuberomammillary nucleus.

►Hypothalamus
►Hypothalamus, Lateral

Periglomerular Cell in Olfactory Bulb

Definition

Periglomerular cells constitute one type of intrinsic neurons of the olfactory bulb. Their cell bodies are among the smallest in the brain and surround olfactory glomeruli. They usually give rise to a single dendrite that arborizes within a glomerule, and an axon, that extends laterally in the periglomerular region, as far as five glomeruli. Therefore, periglomerular cells participate in both intraglomerular and interglomerular circuits.

Periglomerular cells are not homogeneous, but occur in several distinct subtypes that differ in connectivity, neurochemical content (expressing GABA, dopamine, calbindin, or calretinin, or combinations of these) and electrophysiological properties. Based on connectivity, they are currently classified as type 1 cells, which receive excitatory input from the olfactory nerve, and type 2 cells, which establish few or no synapses with olfactory nerve axons.

Periglomerular cells are one of the few types of neurons that undergo continuous neurogenesis throughout life.

►Dopamine
►GABA
►Olfactory Bulb
►Olfactory Bulb Glomerulus
►Olfactory Nerve

Perilymph

Definition

Fluid similar to cerebrospinal fluid (CSF) inside the bony labyrinth that surrounds the delicate membranous labyrinth. Perilymph has a high sodium content and low potassium content.

►Peripheral Vestibular Apparatus

Perimysium

Definition

The fascia that covers the full perimeter of each muscle fascicle, with the exception of its ends. It forms the wall of a "tunnel" in which the muscle fascicle operates. It is continuous with the endomysial stroma within the fascicle.

►Intramuscular Myofascial Force Transmission
►Skeletal Muscle Architecture

Period

Definition

1. The time it takes a periodic function to repeat itself once, the time for the completion of a cycle.
2. Name of a clock gene whose mutation alters period length of a circadian cycle, identified originally in the fruit fly, which together with the Cryptochrome proteins suppress transcriptional activation during the dark phase.

►Acoustics
►Clock Genes
►Clock-Controlled Genes

Period (Tau, τ) of Circadian Rhythm

Definition

The time that it takes for the biological oscillator to complete one cycle, to go from start to finish. In the case of circadian oscillators, the period is close to, but not equal to, 24 h. The measurement of period should be made over the course of a number of cycles when the organism is held under conditions of constant dark and fixed temperature. The periods of circadian oscillators are also temperature compensated; that is, the period length does not change in response to alterations in the external temperature. For diurnal organisms, the period is typically longer than 24 h; for nocturnal organisms, however, the period is typically less than 24 h (a finding referred to as Aschoff's rule). The molecular genetic factors responsible for the generation of the circadian period have been the subject of much research.

Although the period length of circadian oscillations is largely determined by these genetic factors, there is also evidence for modest history-dependent regulation of period. The persistence of circadian oscillations under constant conditions and the maintenance of period in the face of temperature changes are fundamental features of circadian oscillators.

►Circadian Rhythm
►Phase Response Curve (PRC)

Periodic Behavior

Definition

A rhythmic behavior which can be observed to repeat, and is separated by a regular interval of time between adjacent events.

►Peripheral Feedback and Rhythm Generation

Periodic Brain Activation

Definition

Periodic brain activation is associated with the REM state and plays a role in localized recuperative processes and in emotional regulation during sleep.

►Memory and Sleep

Periodic Limb Movements of Sleep

Definition

A sleep related movement disorder consisting of repetitive, involuntary limb movements during sleep.

Electromyographic sensors on the limbs reveal trains of muscle contractions with characteristic shape, amplitude, frequency, and distribution. The limb movements may be temporally associated with arousals from sleep.

The high association with restless leg syndrome and the same targets of therapy suggest that the two disorders share common mechanisms.

►Sleep – Developmental Changes

Periodicity Pitch

Definition

The pitch that is nearly equal to the frequency of a sinusoidal amplitude modulation of a tone or a complex sound.

► Acoustics
► Tonotopic Organization (Maps)

Periodontal Ligament

Definition

The collagenous connective tissue that attaches a tooth to the surrounding alveolar bone of the lower (mandibular) or upper (maxillary) jaw.

► Tactile Sensation in Oral Region

Periodontal Mechanoreceptors

Definition

Receptors that innervate the periodontal ligament and signal information about loads applied to the teeth.

► Periodontal Ligament
► Tactile Sensation in Oral Region

Periodontal Pressoreceptors

Definition

Rapidly conducting trigeminal sensory afferent neurons that innervate specialized receptors in the periodontal ligament that anchors the roots of the teeth in the jawbones. They respond to pressure applied to the crowns of the teeth.

► Mastication
► Periodontal Ligament

Peri-Personal Space

Definition

► Visual Space Representation for Reaching

Peripheral Autonomic (Parasympathetic, Sympathetic) Pathway

Definition

Pathway consisting of a population of preganglionic neurons and a population of postganglionic neurons transmitting impulses in the autonomic ganglia and to the effector cells. Each pathway is specified by the target cells it innervates, i.e. as muscle vasoconstrictor, cutaneous vasoconstrictor, sudomotor, cardiomotor etc pathway.

► Autonomic Reflexes

Peripheral Chemoreception in Respiration

Definition

The major peripheral chemoreceptors are the carotid bodies and the aortic bodies, which convert the hypoxic signals into an increased neural activity to produce reflex responses in the respiratory system. Hypoxic signals from the arterial chemoreceptors are conveyed through the carotid sinus and vagal afferents, which terminate almost exclusively in the nucleus tractus solitarii (NTS) area.The neurons in the NTS project excitatory inputs to the ventral respiratory group (VRG) neurons. Hypoxia initially increases and then slowly decreases ventilation. The initial increase is attributed to stimulation of arterial chemoreceptors and the ensuing slow decline of ventilation is due to the resulting hypocapnia and the central effects of hypoxia.

► Carotid Body Chemoreceptors and Respiratory Drive
► Neural Respiratory Control during Acute Hypoxia
► Nucleus of the Solitary Tract
► Respiratory Network Responses to Hypoxia

Peripheral Chemoreceptors

►Respiratory Reflexes

Peripheral Clock

►Peripheral Oscillator

Peripheral Feedback and Rhythm Generation

HAROLD J. BELL
Department of Cell Biology and Anatomy, University of Calgary, Calgary, AB, Canada

Synonyms

Sensory modulation of central pattern generators; CPGs; Afferent input to rhythm generating networks

Definition

Rhythmic motor behaviors such as walking, wing beating, chewing and breathing are all governed by networks of one or more neurons, named central pattern generators (CPGs). By definition, the CPG is able to generate and maintain a basic patterned neuronal activity that is absolutely fundamental to producing the behavior which it governs. However, an animal's behavior must take into account conditions or changes in the environment such that the behavior remains relevant and appropriate. Behavior is kept relevant and appropriate by integrating sensory feedback into the rhythm generation process. Sensory feedback is a general term which refers to neural signals transduced through a wide variety of sensory modalities (chemoreception, mechanoreception (►Mechanoreceptor), etc.). Peripheral feedback specifically refers to sensory feedback that is transduced outside of the central nervous system, and provides the CPG network with neural signals encoding information relevant to behavior. Thus, peripheral feedback is intimately related to and is an important component of the process of rhythm generation.

Characteristics

Rhythm Generation and Behavior

Repetitive motor acts are involved in a seemingly endless number of behaviors across animal species: walking, crawling, flying, swimming, feeding and breathing are all excellent examples. Because of the complex and repetitive nature of the motor acts involved in coordinating these behaviors, networks of neurons have evolved that are extremely efficient at generating the basic timing and pattern of motor neuron discharge that underlie them, in a highly reproducible fashion. These networks of neurons are almost exclusively located in ganglia or the nuclei of brains within intricate central nervous systems, and therefore are commonly called central pattern generators (CPGs).

By definition, a CPG network is capable of spontaneously generating a rhythmic activity that is sufficient for eliciting the motor behavior that it governs, even in the absence of other synaptic inputs.

CPGs have been studied in considerable depth for a number of behaviors observed across numerous species [1]. Given the diversity of behaviors and therefore the CPGs that govern them, it is difficult to suggest that any one specific behavior or CPG is typical of all others. Nevertheless, this diversity also means it is advantageous to focus upon a specific behavior in order to discuss how CPGs can be modulated by peripheral feedback. For this reason we shall focus our discussion upon breathing behavior, and the respiratory rhythm generator that drives this behavior.

Respiratory Rhythm

Breathing is perhaps the quintessential example of a rhythmic behavior, as it is essential to the survival of countless animal species from insects, to fish, to mammals. Breathing is a homeostatic motor behavior, which allows the animal to obtain oxygen from, and eliminate carbon dioxide into the external environment. Across *Animalia* many breathing patterns, from regular periodic rhythms (►Periodic behavior), to episodic patterns (►Episodic behavior) can be found, and even within some species these patterns are labile. As a motor behavior, breathing requires the coordinated activation of respiratory pump muscles. The CPGs controlling breathing in most animals are very complicated, and so in most cases very little is known about the detailed synaptic connectivity and neurons involved in respiratory rhythm generation. Indeed, if one considers the extent of behaviors and therefore related CPGs present in the animal kingdom, relatively few examples are available of fully described networks.

Invertebrate model systems have been fundamental in advancing our understanding of nearly every aspect of modern neuroscience, and in no field has this been more so than in the study of central rhythm generation.

Presently, fully described CPG networks are limited to those identified in invertebrate species. This is because the central neurons of invertebrates exist within ganglia, and many are readily identifiable based upon location and morphology, making it possible to study an individual cell involved in a particular behavior across multiple animals of the same species. Therefore invertebrates are especially amenable to the study of rhythm generation since networks can be mapped from identified cells. One model system wherein the essential central neuronal components of the respiratory CPG have been identified is the pulmonate freshwater mollusk *Lymnaea stagnalis* [2]. *Lymnaea* are aquatic air breathers, and their breathing is a hypoxia-dependent behavior. *Lymnaea* breathe through coordinated motor output to the muscles of the mantle cavity and pneumostome. Contraction of muscles in the mantle cavity and pneumostome opening muscles allow stale air to be expelled from and fresh air to subsequently be drawn into the animal's rudimentary lung. The core of the CPG controlling this behavior consists of three identified cells which have been studied both *in vivo* and *in vitro* (see Fig. 1).

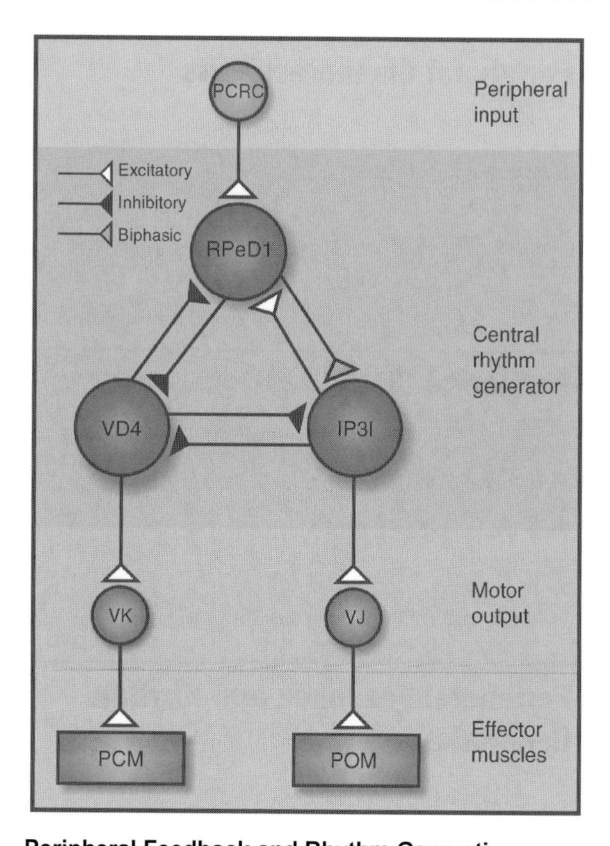

Peripheral Feedback and Rhythm Generation. Figure 1 A schematic representation of the only fully described respiratory CPG; that identified in the pulmonate freshwater snail, *Lymnaea stagnalis*. This respiratory CPG is composed of three interneurons, which have been identified and named as follows: right pedal dorsal 1 (RPeD1), visceral dorsal 4 (VD4), and the "input 3" interneuron IP3I. The opening and closing of the respiratory orifice, the pneumostome, is accomplished through patterned activity of motor neurons controlling pneumostome opening muscles (POM) and pneumostome closing muscles (PCM). Note that the central rhythm generating neuron RPeD1 receives an excitatory chemical synaptic input from oxygen-sensitive peripheral ►chemoreceptor cells (PCRCs) located in the periphery, which act to initiate and subsequently regulate the activity of this rhythm generator network. As such, peripheral inputs to the CPG are an important component of the rhythm generation process.

By contrast, in mammals the respiratory central pattern generator is comparatively complex, and for good reason; respiratory control also needs to accommodate other related behaviors aside from breathing such as feeding, sniffing, licking, and vocalization. Not surprisingly, the intricacy of the mammalian central nervous system has made it rather difficult to characterize the respiratory CPG. Impressive progress has been made however, and central respiratory-related neuronal populations are known to be distributed bilaterally throughout a number of nuclei in the medulla and pons (see Fig. 2). Species differences aside, the mammalian CPG controlling breathing behavior is the topic of great controversy and active research [3]. Regardless of this complexity, breathing remains a motor behavior involving the rhythmic activation of respiratory pump muscles enabling lung ventilation.

Inspiratory muscles, mainly the diaphragm and external intercostal muscles, act to increase thoracic volume and draw atmospheric air into the lungs. Expiration is normally a passive process of elastic recoil during resting breathing, however expiratory muscles are required for forceful expiration at higher levels of ventilation, and these muscles are mainly the abdominal muscles and the internal intercostal muscles. During resting breathing, the most important respiratory muscle by far is the diaphragm, which contracts during inspiration and is controlled by motor innervation via the phrenic nerve. Characterization of respiratory neurons in the medulla has therefore been accomplished by describing their activity in relation to phrenic nerve activity. If the rhythmic activity of a central neuron has the same period as phrenic activity, then the cell can be designated a respiratory neuron. Further characterization takes into account the phase relationship between the cell and phrenic activity (see Fig. 3). For example a respiratory neuron that fires bursts of action potentials in sync with phrenic activity is called an inspiratory neuron. Conversely, a cell that fires during phrenic quiescence, it is called an expiratory neuron. More specific classification takes into account more specific phases of the respiratory cycle (e.g. pre-inspiratory, late expiratory), and the pattern of discharge of the neuron during its period of activity (e.g. augmenting, decrementing, or constant), and

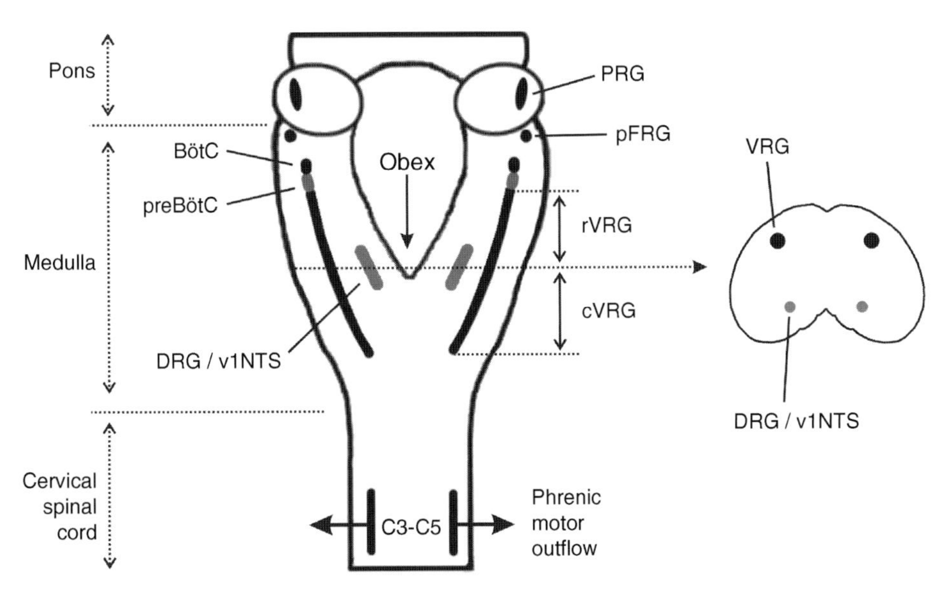

Peripheral Feedback and Rhythm Generation. Figure 2 A schematic representation of those brainstem regions known to contain respiratory neurons in mammals, along with their approximate anatomical locations. Shown, is a schematic view of the brainstem from the dorsal coronal perspective, with a transverse section at the level of the obex. In contrast to the respiratory CPG of *Lymnaea* (shown in Fig. 1), the rhythmic motor output to respiratory muscles in mammals is generated by a complex network of many neurons bi-laterally distributed throughout several regions of the pons and medulla. Relevant neuronal populations are represented as follows: *PRG*, pontine respiratory group; *pFRG*, parafacial respiratory group; *BötC*, Bötzinger complex; *preBötC*, pre Bötzinger complex; *DRG*, dorsal respiratory group; *cVRG*, caudal ventral respiratory group; *rVRG*, rostral ventral respiratory group; *vlNTS*, ventrolateral nucleus tractus solitarii.

these characteristics are used to functionally differentiate respiratory neurons from one another [4].

Of the populations of neurons in the medulla that have been classified as respiratory neurons, it has long been debated as to which of these were actually involved in generating the base rhythm, and could therefore be designated, "the CPG." Two groups of neurons in the brainstem have been identified that are presently believed to be essential to the generation of a basic respiratory rhythm: (i) A region of the ventrolateral medulla named the preBötzinger complex (PreBÖT), and (ii) a region of the medulla ventrolateral to the facial nucleus near the ventral surface, named the parafacial respiratory group (pFRG). While the precise role of these two neuronal populations in respiratory rhythm generation remains the focus of much ongoing research, it has recently been proposed they may form a coupled network oscillator, with the former responsible for generating inspiration, and the latter generating active expiration phases of the respiratory cycle.

The Role of Peripheral Feedback in Respiratory Rhythm

It has been widely accepted since the early 1960s that while a CPG can independently function to generate a basic rhythmic motor behavior, sensory feedback from the periphery is necessary for modulating the timing and amplitude of events so that this behavior remains relevant to the environmental demands on the animal [5]. In this regard, the respiratory CPG is certainly no different, and several forms of afferent information provide a potent modulating influence over rhythmic output. Input from airway receptors can elicit cough or sneeze airway defense mechanisms. Inputs from ►carotid body chemoreceptors provide a potent drive to ventilation when arterial blood becomes hypoxic, hypercapnic, or acidic. Group III and IV afferents in the skeletal muscle are able to stimulate breathing, and are believed to be important in coupling the cardiovascular and respiratory systems to regulate gas exchange. Stretch receptors in the lungs are important in shaping the respiratory cycle and timing the inspiratory and expiratory phases of the breathing cycle. Cutaneous thermoreceptors can elicit a powerful gasp reflex which has been documented during sudden cold water immersion. In aquatic air breathing mollusk *Lymnaea stagnalis*, mechanosensory input to the respiratory CPG is required to gate the episodic breathing rhythm so that the respiratory orifice only opens at the water surface, and a similar mechanism is likely to exist in other aquatic air breathers.

Clearly there are many different modalities of peripheral ►sensory input which modulate the respiratory CPG, but there are also a range of influences that sensory input can have on the characteristics of the respiratory rhythm. Some inputs typically affect rhythm only very briefly over one respiratory cycle (i.e. cough or sneeze reflexes), while some involve

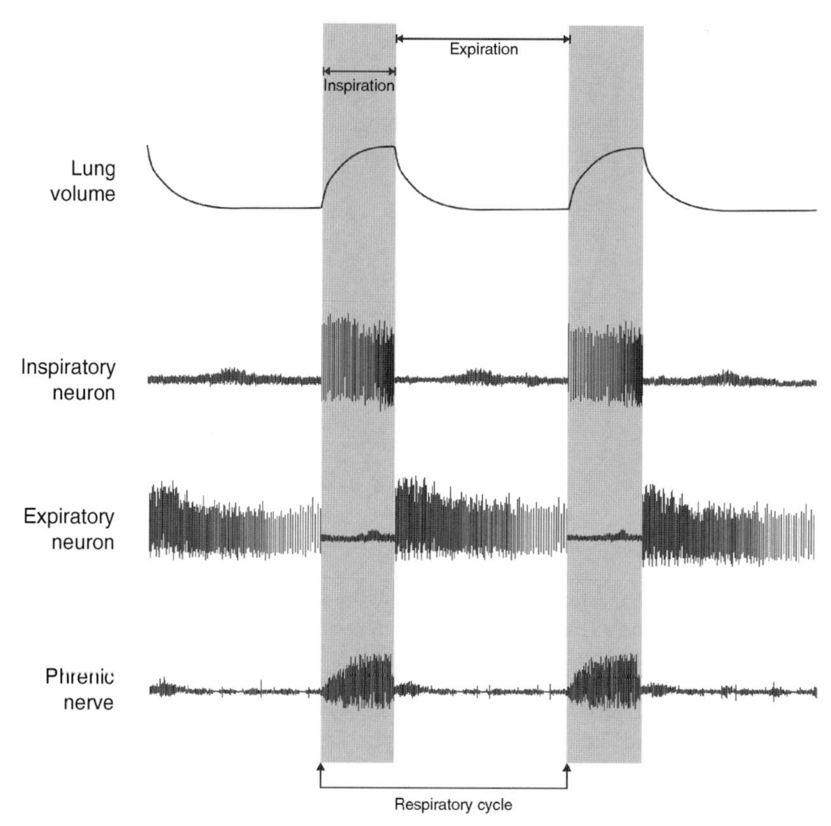

Peripheral Feedback and Rhythm Generation. Figure 3 Phase relationships between rhythmic breathing behavior, phrenic nerve activity, and inspiratory and expiratory respiratory neurons in the mammalian brainstem. A neuron which fires during the interval of phrenic activity can be described as an inspiratory neuron, since this neuron is active during the inspiratory phase of rhythmic breathing behavior (i.e. when lung volume is increasing). Conversely, a neuron which selectively fires during phrenic quiescence can be designated a expiratory neuron since it is active during the expiratory phase of rhythmic breathing behavior (i.e. when lung volume decreases). Further characterization takes into account the firing pattern of the respiratory neuron during its period of activity (e.g. augmenting, decrementing or constant firing frequency), and the specific phase of the inspiratory or expiratory cycle in which it is active (e.g. early or late).

shaping the rhythm during each breath cycle (i.e. lung stretch receptors) and others affect the entire respiratory cycle over the course of many breaths (carotid body chemoreceptors).

Mechanisms of Peripheral Modulation of Respiratory Rhythm

As we have seen, peripheral feedback is involved in modulating respiratory rhythm such that breathing can be matched to the needs of the organism. Whether this is a sustained respiratory drive as occurs during exercise, or a transient effect as occurs during protective airway reflexes, the respiratory rhythm is open to modulation from afferent input.

The main characteristics of respiratory rhythm that can be modulated are breathing frequency, or the number or breathing cycles per unit time, and the amplitude of the motor act, or breath size. These features of the breath cycle are shown in more detail in Fig. 4. Since the pathways and putative mechanisms of modulation of breathing rhythm depend greatly upon the input signal, we will examine one select example at greater depth.

Carotid Body Chemoreceptor Inputs

The mammalian carotid body chemoreceptors are arterial chemosensory organs that increase breathing in conditions of ►hypoxia, hypercapnia, or acidosis. Increased chemoreceptor stimulation causes an increase in afferent output that can be measured in the carotid sinus nerve which innervates the organ. Carotid body afferent inputs to the brainstem are first conducted via neurons located mainly within the petrosal ganglion, subsequently via afferent fibers in the glossopharyngeal (IXth cranial) nerve. The ►carotid body chemoreceptor inputs (►Chemosensory input) affect respiratory rhythm as follows: Afferent input resulting from either carotid body stimulation or direct stimulation of the carotid sinus nerve elicits a decrease in inspiratory duration leading to an increase in breathing frequency, and an increase in the amplitude of integrated phrenic

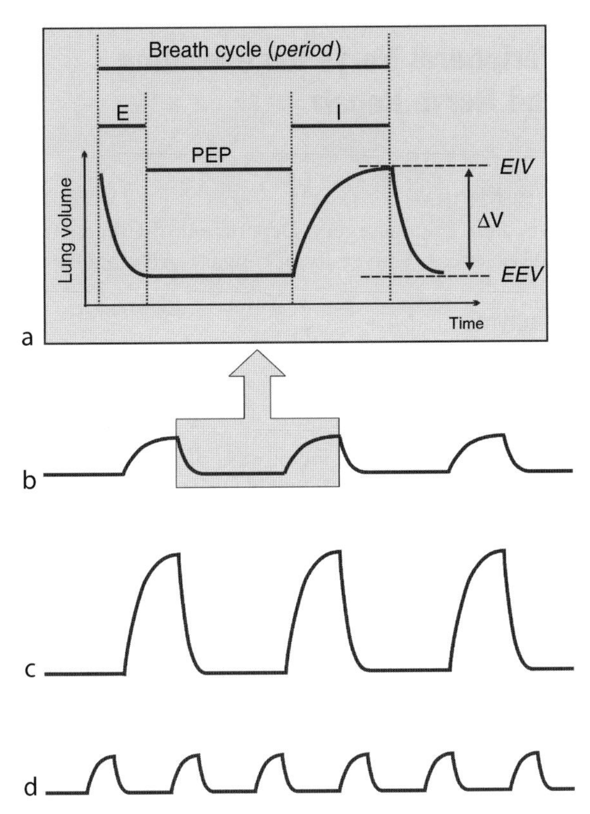

Peripheral Feedback and Rhythm Generation. Figure 4 A diagrammatic representation of those aspects of the breathing cycle which are dictated by final output of the respiratory rhythm generator. (a) an expanded view of one cycle of rhythmic breathing behavior, as shown in (b). The change in lung volume during the breath cycle (ΔV) can be altered by changing the end-inspiratory volume (EIV) or the end-expiratory volume (EEV). In addition, the durations of inspiration (I), expiration (E), and the post-expiratory pause (PEP) can be altered so as to change the period of the breath cycle. Lung ventilation can be increased by augmenting ΔV (as shown in (c) or by decreasing the period of the breath cycle (as shown in (d), or any combination of the two. Peripheral inputs modulate respiratory rhythm and therefore aspects of the breath cycle, as shown in (a), such that rhythmic breathing behavior remains relevant to the needs of the animal.

whole nerve activity. Simply stated, carotid input to the respiratory CPG leads to an increase in ventilation through increasing breathing rate and breath size.

The central respiratory pathways involved in this response have been studied in some further detail. Within the brainstem, most of the sensory afferents from the carotid sinus nerve show arborizations within the nucleus of the solitary tract, or nucleus tractus solitarii (NTS), specifically an area of the commissural subnucleus of the NTS including what has been named the dorsal respiratory group (DRG) [6]. The synaptic connectivity and therefore the mechanisms of influence of carotid sensory afferents are not well described beyond this level of the medullary respiratory network. However, carotid sensory afferents are known to release glutamate in the NTS which is believed to be primarily responsible for the increase in breathing via its action on NMDA receptors at this site. The involvement of multiple neurotransmitters and neuromodulators is likely, though specific details remain a topic of ongoing study.

The overt effects of activation of carotid afferents terminating in the NTS upon rhythm generation have been further documented at the level of respiratory neuronal populations in the medulla including those neurons of the Pre-BÖT complex believed to be an important component of the respiratory CPG. Unilateral activation of carotid afferents terminating in the NTS elicits bi-lateral changes in the rhythmic activity of inspiratory driver neurons in the rostroventrolateral medulla; these inspiratory neurons are inhibited and therefore their firing duration is decreased. Afferent stimulation also has the effect of concurrently exciting pre-motor inspiratory neurons in the caudal ventral respiratory group and in the DRG, increasing the drive to phrenic motoneurons [7].

Kinetics of Neuromodulation of Respiratory Rhythm

The activation of carotid afferents is able to elicit changes in respiratory rhythm, which take only a couple of respiratory cycles to achieve their maximal effect ($\tau \sim 10$ s in cats). The removal of carotid input results in a return to baseline rhythmic activity that has a slightly longer time course ($\tau \sim 45$ s, again in cats). This apparent "inertia" in the respiratory rhythm generator, which has been called "short term potentiation" (STP) [8] has been well documented, yet the mechanisms involved are not well understood. Slower onset, longer-lasting alterations in respiratory rhythm have also been documented, that can persist long after the input stimulus has returned to baseline levels. This phenomenon, called long term facilitation (LTF), results in augmented respiratory rhythm lasting for more than one hour after removal of peripheral input to the CPG. One particular stimulus that is capable of eliciting LTF is ►intermittent hypoxia, as sensed by the carotid body chemoreceptors. After repeated hypoxic episodes (for example 3×5 min episodes, separated by 5 min of normoxia), both the frequency and amplitude of phrenic nerve discharge has been documented to remain augmented for hours post-stimulus [9]. This phenomenon has now been documented to various extents in multiple animals in different degrees of reduction, including rats, cats and rabbits. While the phenomenon has proven less robust in some cases than others, LTF in respiratory rhythm is nevertheless a fascinating example of how afferent input can modulate rhythm generation via a CPG, even long after the afferent signal has been removed.

Summary

As we have seen, breathing is an excellent example of a rhythmic behavior that is governed by a central network of neurons. This network in mammals resides in the ponto-medullary brainstem, and receives peripheral inputs from many modalities that are capable of modulating the depth and rate of breathing. While the pathways and mechanisms by which the respiratory CPG is influenced by sensory afferents are specific to this system, it nevertheless illustrates the importance of peripheral feedback in the process of rhythm generation. This is a general neurophysiological principle that applies equally to most rhythmic behaviors across *Animalia*: afferent feedback is essential for ensuring that rhythm generation via CPG networks remains appropriate to the environmental demands on the organism.

References

1. Marder E, Calabrese RL (1996) Principles of rhythmic motor pattern generation. Physiol Rev 76(3):687–717
2. Syed NI, Bulloch AG, Lukowiak K (1990) In vitro reconstruction of the respiratory central pattern generator of the mollusk Lymnaea. Science 250(4978):282–285
3. Onimaru H, Homma I (2006) Point:Counterpoint: the parafacial respiratory group (pFRG)/pre-Botzinger complex (preBotC) is the primary site of respiratory rhythm generation in the mammal. J Appl Physiol 100(6): 2094–2095
4. Duffin J, Ezure K, Lipski J (1995) Breathing rhythm generation: focus on the rostral ventrolateral medulla. News Physiol Sci 10:133–140
5. Wilson D (1961) The central nervous control of locust flight. J Exp Biol 38:471–490
6. Finley JC, Katz DM (1992) The central organization of carotid body afferent projections to the brainstem of the rat. Brain Res 572(1–2):108–116
7. Morris KF, Arata A, Shannon R, Lindsey BG (1996) Inspiratory drive and phase duration during carotid chemoreceptor stimulation in the cat: medullary neurone correlations. J Physiol 491(Pt 1):241–259
8. Wagner PG, Eldridge FL (1991) Development of short-term potentiation of respiration. Respir Physiol 83(1):129–139
9. Bach KB, Mitchell GS (1996) Hypoxia-induced long-term facilitation of respiratory activity is serotonin dependent. Respir Physiol 104(2–3):251–260

Peripheral Glial Cell

Definition

Glial cells wrapping peripheral neurons in Drosophila.

►Alternative Splicing and Glial Maturation
►Glia Cells

Peripheral Nerve Regeneration and Nerve Repair

Rajiv Midha
Division of Neurosurgery, Department of Clinical Neurosciences, University of Calgary, AB, Canada

Synonyms

Axonal regeneration

Definition

►Nerve regeneration is the process of axonal regrowth within peripheral nerve following injury. Basic science and clinical studies underscore both the possibilities and failure of spontaneous axonal regeneration after peripheral nerve injuries. Historical and current methods of nerve repair at the tissue level take advantage of the natural tendency of peripheral axons to grow when provided with a suitable (micro)environment. New and emerging nerve repair strategies, building on key discoveries at the cellular and molecular, have the potential to further significantly improve the outcome from the repair of nerve injuries.

Characteristics

Nerve Regeneration Outcomes and Challenges

Peripheral nerves have the potential to regenerate axons and reinnervate end-organs, with resulting good functional recovery. Indeed, this is the case with all minor nerve injuries, such as neuropraxia, where the axon remains intact, and most purely axontemetic injuries, where axons are interrupted but the degree of internal damage is minimal. In the latter circumstance, regenerating axons use their existing ►endoneurial pathways to specifically reinnervate their own precise target end-organs, as confirmed in recent experiments using bioengineered fluorescent mice [1].

Outcome following more severe peripheral nerve injury (PNI) however remains variable and often very poor. At least three important factors contribute to the relatively poor outcome. The first challenge is the pathology at the tissue level. The majority of clinical PNI exhibit both a loss of axon continuity and a significant disruption in the internal connective tissue structures. The resulting scarring within the nerve or a frank gap (with lacerating injuries) presents a formidable barrier to regenerating axons, preventing them from effectively innervating the distal nerve stump. These are currently managed with a repair of the divided nerve or, for the usual scenario of longer gaps or scar segments that need to be resected, placement of interposed nerve grafts. Direct nerve or nerve graft repair unfortunately do not obviate the misdirection of axons. This introduces the second biological challenge, which is at the cellular level. Even with the most meticulous repair, the endoneurial tubes can never be reapproximated exactly and this results in mismatching of regenerating axons at the site of suture, or within the graft, leading to inappropriate (non-specific) reinnervation and subsequent poor recovery in function. The third challenge may be considered as a quantitative one, which is associated with chronic denervation (►Chronic nerve denervation) of the distal nerve. Chronic denervation is common because of the often extensive injury zone that prevents any axonal outgrowth or (even if outgrowth occurs) the relatively slow rate of regeneration. As a consequence, the distal nerve segment remains chronically devoid of regrowing axons. The resulting prolonged denervation of Schwann cells (SCs) appears to a critical factor which makes them unreceptive for axonal regeneration, and is the single most important quantitative contributor to poor end-organ reinnervation [2].

Spontaneous Axonal Regeneration After Nerve Injuries

The details of the basic processes involved in nerve degeneration and early regeneration are outlined in detail elsewhere in this Encyclopedic Series (see the contribution by Doug Zochodne). Herein, I will summarize some key concepts, with a more detailed focus on the critical role played by the denervated nerve, distal to the nerve injury.

The injury to an axon is associated with a plethora of changes in the neuronal cell body that shift its phenotype from a neuron involved in maintenance/ neurotransmission, to a regenerating one, with a corresponding up-regulation of ►regeneration associated genes (RAGs). Regeneration from neurons requires severed axons to redeem their original axoplasmic volume by extending their processes distally. New axons sprout from one or two internodes proximal to an injury zone. These spontaneously sprouting daughter axons, supported by SCs, and surrounded by a common basal lamina tube are termed a "regenerating unit." Therefore in regeneration, the number of neurons does not increase, but rather surviving neurons regenerate their cellular processes in concert with their surrounding microenvironment. Key components of the microenvironment involve a very close and intimate relationship between regrowing axons, and the supporting glial SCs, which must both proliferate and migrate [3]. Indeed, the initial stages of axonal regeneration are highly dependent on SCs which lead the regrowing axons across the nerve injury site.

Severance of a peripheral nerve results in ►Wallerian degeneration in all axons distal to the injury site, evidenced by the disintegration of axoplasmic microtubules and neurofilaments. The majority of axons along the distal stumps of transected nerves are reduced to granular and amorphous debris within 24 h; by 48 h, the myelin sheath has begun to transform into short segments that then form into ovoids. Activated macrophages invade the degenerating distal nerve stump and phagocytose the disintegrating nerve fibers and myelin. There is an accompanying proliferation of the acutely denervated SCs, which now alter their gene expression, and change phenotypically from myelinating to growth supportive [4].

As axons from the proximal nerve stump arrive in the nerve distal to original injury, SCs and basal lamina tubes are reinnervated and thereby supported, which in turn, provides an excellent trophic and cell-adhesive environment for further axonal regeneration [5]. Regenerating axons, when contacting SCs, release neuregulins from their growth cones which bind to erbB receptors on the SCs to mediate a second phase of SC proliferation. SCs secrete chemoattractive factors, including cytokines such as interleukin-1β, leukemia inhibitory factor and monocyte chemoattractant protein-1, that recruit macrophages into the denervated nerve stumps [5]. Cytokines, derived from both the SCs as well as the macrophages that enter the nerve, drive the expression of the non-myelinating, dedifferentiated "denervated" phenotype of the SCs which proliferate, form the ►bands of Bungner (SC lined basal lamina tubes) and guide regenerating axons within the distal nerve stump [4]. The switch in SC phenotype is associated with up-regulation of several growth associated genes including several neurotrophic factors, p75 NTR, glial fibrillary acidic protein, GAP-43, netrin-1 and the transcription factor, Krox-24, which all support axonal regeneration [5]. In summary, the capacity of the denervated distal nerve to support axonal regeneration depends on proliferating acutely denervated SCs within the basal lamina tube which are essential for guiding regenerating axons to the end-organs.

Repairing Nerve Injury Gaps: Nerve Grafts, Electrical Stimulation and Nerve Conduits

The majority of clinical PNI leave the nerve grossly intact, but nevertheless exhibit loss of axon continuity and a severe disruption in the internal connective tissue structures. Scarring within the nerve presents a barrier to regenerating axons. Nerve repair requires resection of the scarred segment and repair of the resulting lengthy nerve gap, usually with a nerve graft, typically procured from another area of the patient's own body (nerve autograft). The nerve graft contains surviving SCs and basal lamina endoneurial tubes, which provide neurotrophic support, as well as favorable cell and endoneurial tube surface adhesion molecules to regenerating axons [5]. Nerve grafts therefore provide a seemingly optimal tissue bridge that regenerating axons from the proximal nerve stump exploit to innervate the distal stump. With the advent of the operating microscope and microsurgical techniques, Millesi improved clinical results and popularized the use of nerve autografts. The outcomes, associated with using nerve grafts, are best for nerves which are primarily innervating one or a few discrete motor or sensory targets, and especially where the repair is close to the target end-organ. For the more proximal injuries, ones requiring lengthy grafts and ones involving nerve elements which innervate a variety of motor and sensory targets, the outcomes remain poor. Moreover, results of microsurgical repair, which is essentially at the tissue level, have reached a plateau over the last few decades.

A major shortcoming with the nerve graft technique is the biological constraint, which cannot be overcome by further progress in microsurgical techniques. Even with the most meticulous repair, regenerating axons at the site of suture, or within the graft, get misdirected or lost, leading to inappropriate (non-specific) and incomplete reinnervation, respectively. Innovative recent investigations with nanoscale engineered devices suggest that some day surgery at the cellular level to splice and repair individual axons may be feasible. Recent insights into the nuances of axonal regeneration however provide potential for some new therapies that are even readily accessible soon. We now understand that axonal outgrowth is not synchronous, but rather staggered, so that some pioneering axons grow out early, while others lag far behind. In fact, many of the lagging axons get delayed at suture repair sites, which in the case of a nerve graft are compounded, involving two separate repair locations. Emerging evidence also suggests that pioneering motor axons, for example, may prefer to associate with corresponding "motor" SCs, based on their surface-specific basement membrane molecules. These pioneering axons may start the process, and they together with their migrating SCs facilitate similar axons to follow by offering these cues as required. Studies by Brushart and colleagues show that motor axons particularly may be biased in their selection of distal targets, preferring phenotypically appropriate rather than inappropriate endoneurial pathways in the distal nerve, so that ultimately some ►preferential motor reinnervation (PMR) is exhibited [6]. Exciting recent work by Gordon and Brushart suggests that epochs of electrical stimulation as short as one hour have a significant influence not only on synchronizing the initial re-growth of motor axons, but also on possibly enhancing PMR. Clinical trials to assess the utility of short duration electrical stimulation on improving the outcome of nerve repair are therefore warranted.

In an attempt to provide a more suitable environment for regenerating axons to sample and respond to appropriate *endogenous* directional cues, many investigators have proposed using an artificial (non-nerve) conduit interposed between the proximal and distal nerve stumps. Moreover, a bioengineered graft may allow the introduction of *exogenous* therapies that build on our rapidly expanding knowledge of axonal guidance. Axons are guided to their targets by growth cones, specialized structures at the tip of axons, that respond to the coordinated action of many contact-mediated cues provided directly or indirectly by SCs and diffusible cues, which are either attractive or repulsive. A short list of candidate molecules to consider include the classical neurotrophins, neuropoietic cytokines (including CNTF, IL-6, oncostatin, and LIF), other neurotrophic factors (including insulin, IGF-1 and GDNF), cell adhesion molecules (including NCAM, L1 and N-cadherin), and extracellular matrix proteins (including laminin, tenascin C, fibronectin and heparan sulfate proteoglycan) [5]. By exogenously providing the most appropriate or suitable cues, we may be able to profoundly influence axonal regeneration within the nerve conduit. The clinical utility of such a strategy is apparent when considering the example of repair of nerves to the hand, where the median nerve is paramount for sensation whereas the ulnar nerve is critical for discrete motor function of the digits. In these instances, the conduit repair can be endowed with specific growth factors or other molecules that will bias the regeneration towards a population of axons (motor vs. sensory) to achieve improved specificity of reinnervation. Artificial nerve conduits or tubes are already proven and in clinical use for short injury gaps of 3 cm or less in humans. Advances in bioengineering, coupled with our understanding of how to effectively deliver growth factors, cell adhesion molecules and other therapies within the artificial nerve graft, should lead to major advances in improving both the quantity and specificity of axonal regeneration through the nerve conduit [7].

Deleterious Changes Associated with Chronic Nerve Denervation

So far, the therapies considered in this essay have focused on the repair of the nerve injury site or nerve injury gap. We now turn our attention to the critical distal nerve environment. Denervated SCs initially up-regulate neural cell adhesion molecule and basement membrane components (including laminin), which are used for attachment and growth of the regenerating axons. A growth supportive environment of the distal nerve is unfortunately not maintained unless axonal contact is re-established. Progressive regression of the capacity of the denervated SCs to sustain their growth permissive phenotype [8] and the progressive decline in numbers of SCs in the chronically denervated distal nerve stumps underlie the loss of capacity to support axonal regeneration [9]. Deleterious changes that affect axonal growth become increasingly pronounced over time, and are coupled with a progressive decline in the number of reinnervated motor units after chronic denervation [2]. In recent experiments, we have demonstrated that immediate innervation of the distal nerve, as compared to chronic denervation, greatly improves subsequent re-innervation, confirming that the failure in regeneration is related to the profound changes in the distal stump from chronic denervation. A critical component of this unreceptive environment in the distal nerve is the chronically denervated SC whose growth support properties appears to be "turned off" [5].

The above biological issue is very clinically relevant as delayed nerve repairs occur frequently in clinical practice. The reason this occurs is because the majority of nerve injuries leave the nerve in physical continuity with an often unknown propensity to spontaneously recover, a process which unfolds over several weeks to months. Hence, most patients undergo appropriate surgical exploration several months after injury and receive a delayed nerve repair. Even patients who have immediate nerve repair are subject to distal nerve denervation for considerable periods as the rate of regeneration is relatively slow (~1 mm/day in humans), and, given that nerve injuries are far from their end-organs, the distances that regenerating nerve fiber need to grow very long. Hence, almost all severe human nerve injuries creates a situation in which the SCs of the distal nerve are chronically denervated [5].

Therapies to Counteract Chronic Distal Nerve Denervation

Fortunately, the clinical paradigm of ►nerve transfers has emerged a powerful means of bypassing the long zone of chronically denervated nerve. Nerve transfers essentially involve the repair of a distal denervated nerve element using a proximal foreign nerve as the donor of neurons and their axons which will reinnervate the distal targets. The concept initially arose to sacrifice the function of a donor (lesser valued) nerve/muscle to revive function in the recipient nerve and muscle that will undergo reinnervation. Nerve transfers have become increasingly utilized for the repair of brachial plexus injuries, especially where the proximal motor source of the denervated element is absent because of avulsion from the spinal cord. Increasingly advocated are the use of transfers in situations where the proximal motor source is available, but the regeneration distance is so long that the outcome would be poor. A nerve transfer into the denervated distal nerve stump, but deliberately chosen so as to be very close to the motor end-organ, would then restore greater function as compared to a very proximal nerve graft repair.

Biological means of protecting the distal denervated nerve are emerging. One strategy is the exogenous use of cytokines (such as TGF-β), normally produced during Wallerian degeneration, to counteract the deleterious effects of chronic denervation and to maintain the growth-promoting denervated SC. In a recent experiment, chronically denervated SCs, exposed to TGF-β and forskolin, when infused into a nerve regeneration chamber, dramatically increased the number, size and myelination of regenerating axons, as compared to SCs without pre-treatment. Other research laboratories are exploring gene therapy approaches, using lentiviruses or other vectors to augment or resurrect the repertoire of regeneration associated molecules expressed by the denervated SC.

Since chronically denervated SCs appear to become effete, another logical approach is to support the distal denervated nerve environment by replacing lost cells with those derived exogenously. SC infusion has been successful in promoting regeneration and remyelination of the injured peripheral nerve. However, human SCs must be derived from invasive nerve biopsies and have a limited, lengthy expansion *in vitro*. It is thus desirable to identify a more accessible source of SCs for transplant therapies. Bone marrow stromal cells, adipose tissue derived stem cells and skin derived precursor stem cells have all been shown to generate functional SCs that can be used for transplantation in nerves [10]. When stem cells derived from skin were transplanted into artificial nerve guidance tubes bridging a 16-mm gap in rodent sciatic nerve, there was promising improvement in behavioral, electrophysiological, and morphometric parameters measured over vehicle control. It is therefore conceivable that skin derived SCs may be very useful in models of chronic nerve injury, either by infusing them into a chronically denervated distal nerve or perhaps immediately after injury as a protective treatment in an attempt to avoid distal changes associated with chronic injury. If experimental studies prove benefit, we anticipate that nerve repair in the future in patients will be augmented by the use of their own (autologous and cultured) skin-derived SCs, easily procured, via relatively non-invasive skin graft harvesting.

Conclusion

Our basic understanding of nerve and axonal regeneration has increased tremendously over the last century, with corresponding gains in our ability to repair clinical nerve injuries. We can already overcome the challenge associated with lacerating nerve injuries and severe nerve injuries in continuity with precise repairs at a tissue level using microsurgical techniques, nerve grafts and short length artificial nerve conduits. We are on the verge of the next (cellular) generation of nerve repair using modalities such as specific molecular therapy, electrical stimulation, and bioengineered and micro fabricated devices. These approaches promise to allow improved synchronization of axonal regeneration and more exact specificity of reinnervation, While distal targeted nerve transfers have already had a clinical impact, in the near future gene and (stem) cell therapy approaches to augment and resurrect the distal denervated nerve environment are anticipated.

References

1. Nguyen QT, Sanes JR, Lichtman JW (2002) Pre-existing pathways promote precise projection patterns. Nat Neurosci 5:861–867
2. Fu SY, Gordon T (1995) Contributing factors to poor functional recovery after delayed nerve repair: prolonged denervation. J Neurosci 15:3886–3895
3. Chen YY, McDonald D, Cheng C, Magnowski B, Durand J, Zochodne DW (2005) Axon and Schwann cell partnership during nerve regrowth. J Neuropathol Exp Neurol 64:613–622
4. Scherer SS, Salzer JL (1996) Axon-Schwann cell interactions during peripheral nerve degeneration and regeneration. In: Jessen KR, Richardson WD (eds) Glial cell development. Bios Scientific, Oxford, UK, pp 169–196
5. Fu SY, Gordon T (1997) The cellular and molecular basis of peripheral nerve regeneration. Mol Neurobiol 14:67–116
6. Brushart TM (1993) Motor axons preferentially reinnervate motor pathways. J Neurosci 13:2730–2738
7. Belkas JS, Shoichet MS, Midha R (2004) Peripheral nerve regeneration through guidance tubes. Neurol Res 26:151–160
8. Li H, Terenchi G, Hall SM (1997) Effects of delayed re-innervation on the expression of c-erb receptors by chronically denervated rat Schwann cells in vivo. Glia 20:333–347
9. Dedkov EI, Kostrominova TY, Borisov AB, Carlson BM (2002) Survival of Schwann cells in chronically denervated skeletal muscles. Acta Neuropathol (Berl) 103:565–574
10. McKenzie IA, Biernaskie J, Toma JG, Midha R, Miller FD (2006) Skin-derived precursors generate myelinating Schwann cells for the injured and dysmyelinated nervous system. J Neurosci 26:6651–6660

Peripheral Nervous System (PNS)

Definition

Peripheral Nervous System (PNS) is part of the vertebrate nervous system comprised of the nerves outside of the central nervous system and including cranial, spinal nerves and sympathetic and parasympathetic systems.

Peripheral Neurons

Definition

Motor, sensory and autonomic neurons with axons that connect the body to the central nervous system (brain, spinal cord).

►Peripheral Nervous System

Peripheral Neuropathies

Definition

Peripheral neuropathies are diseases of peripheral nerves and can occur in acute and chronic forms. Often motor and sensory fibers are afflicted together. Motor symptoms include muscle weakness and depression or loss of tendon reflexes. Sensory symptoms vary and may include ►paresthesias such as numbness, pins-and-needles sensations, tingling, impaired cutaneous pain and temperature sensations (with risk of injuries), varied impairment of cutaneous mechanical sensation and ►proprioception. Often, peripheral body parts are involved most strongly, leading to the so-called glove-and-stocking pattern. There is a specific form of ►large-fiber sensory neuropathy.

►Tendon reflex

Peripheral Oscillator

Achim Kramer
Laboratory of Chronobiology, Charité
Universitätsmediain Berlin, Berlin, Germany

Synonyms

Peripheral clock

Definition

The term "peripheral oscillator" refers to a circadian oscillator, which is located in cells, tissues or organs outside of the suprachiasmatic nucleus (the site of the master oscillator in mammals).

Characteristics

Properties

In mammals, the circadian system is organized in a hierarchical manner. In addition to the ►master clock in the ►suprachiasmatic nucleus (►SCN), peripheral tissues (such as liver, kidney, heart, skeletal muscle etc.), some extra-SCN neuronal tissues (such as the olfactory bulb) as well as even some immortalized cell lines (e.g. rat-1 and NIH3T3 fibroblasts [1]) exhibit circadian gene expression. The molecular mechanism of these ►oscillations is similar to that in the SCN [2].

Peripheral oscillators are cell-autonomous and ►self-sustained, but – in contrast to ►cellular clocks in SCN neurons – probably not synchronized within a tissue. Therefore, explants of peripheral tissues display damped oscillations on the tissue-level *ex vivo* with self-sustained, but gradually desynchronizing individual ▶cellular oscillators [3].

In vivo, peripheral oscillations ▶phase-lag the rhythms in the SCN by several hours and are dominated by the circadian ▶pacemaker in the SCN (master-slave oscillators). For example, embryonic fibroblasts from ▶*Period1*-deficient mice (which have an ▶endogenous period of 20 h) exhibit wild-type period oscillations when implanted in wild-type mice [4].

Function

Peripheral oscillators are thought to regulate ▶circadian rhythms in local physiology. Genome-wide transcriptional profiling of various peripheral tissues revealed that about 5–10% of all mRNAs show a circadian expression pattern. Interestingly, while rhythmic transcripts expressed in many or all peripheral tissues are rare (they include components of the core oscillator machinery), the majority of oscillating transcripts are either only expressed or only rhythmic in one particular tissue. In the liver, for example, some of these rhythmic genes code for enzymes involved in rate-limiting steps of rhythmic hepatocytic processes suggesting that these rhythms are generated by a local liver oscillator.

Formally, however, it is difficult to decide whether a tissue-specific rhythm is regulated by a peripheral oscillator or driven by rhythmic systemic factors or by a combination thereof. To investigate these possibilities, tissue-specific clock knockout mice have been investigated. In the liver, for example, the majority of rhythmic transcripts are only rhythmic when the liver clock is functioning. A small fraction of transcripts (including *Period2*), however, continue to oscillate without a functional liver clock suggesting that these rhythms are driven by systemic circadian cues [5]. In the ▶retina, the local clock seems to be required for a normal physiology of vision. It regulates both the rhythmic expression of genes in a ▶light-dark cycle that are not rhythmic in ▶constant conditions as well as inner retinal electrical responses to light [6]. In the heart, disruption of the local clock leads to a severely attenuated induction of myocardial fatty acid-responsive genes during fasting [7]. Together, these results indicate that peripheral clocks significantly contribute to important physiological functions *in vivo*.

The circadian expression of genes within a given tissue is often widely distributed among circadian phases. This ensures that within a single cell biochemically incompatible processes are separated in time. Molecularly, different phases of expression can be regulated in various ways: the expression of so-called first-order ▶clock-controlled genes is directly regulated by components of the core oscillator. These rhythmic gene products themselves can regulate rhythmic processes further downstream, which may even be tissue-specific. Thereby, a complex hierarchy of rhythmic expression patterns with varying phases may emerge regulating the rhythmic physiology specific for a given tissue.

Entrainment

While the SCN clock is synchronized to the geophysical time (▶entrainment) by light-dark cycles, peripheral cells of mammals are not light-sensitive. Hence, daily ▶non-photic resetting cues are required for a correct phase relation among SCN and peripheral tissues. Up to now, the identity of these cues is unknown, although it is likely that many factors contribute to the entrainment of peripheral oscillators. Humoral as well as neuronal signals emanating from the SCN have been suggested to entrain peripheral oscillators. There are neural outputs from the SCN to peripheral organs via the autonomic nervous system, indicating direct (multisynaptic) neural control. Glucocorticoid hormones are prominent candidate factors, since (i) they are able to strongly phase-shift peripheral oscillators both *in vitro* and *in vivo* and (ii) the SCN regulates the rhythmic expression of glucocorticoid hormones *via* the hypothalamic-pituitary-adrenal axis. In addition, several more indirect routes are discussed to be involved in the daily resetting of peripheral oscillators. (i) The SCN directly regulates daily activity-rest cycles and thus also rhythmic food consumption. While feeding time has almost no effect on the SCN clock, peripheral oscillators are strongly influenced by restricted feeding schedules [8]. If food is available only in the inactive phase, the molecular circadian clock in the periphery is completely uncoupled from the SCN and synchronizes to the restricted feeding rhythm (see also: ▶food entrainable oscillator). It has therefore been speculated that peripheral clocks sense the metabolic state, which would impinge on the molecular properties of the cellular oscillator and thereby phase-reset the molecular clock [9]. (ii) The circadian clock in the SCN regulates rhythmic fluctuations in core body temperature. These temperature rhythms are sufficient to entrain cultured fibroblasts [10]. Thus, it is conceivable that body temperature rhythms contribute to the entrainment of peripheral oscillators. Together, it seems likely that a variety of pathways are involved in the daily entrainment of peripheral oscillators, and different tissues may require different resetting cues.

References

1. Balsalobre A, Damiola F, Schibler U (1998) A serum shock induces circadian gene expression in mammalian tissue culture cells. Cell 93:929–937
2. Yagita K, Tamanini F, van Der Horst GT, Okamura H (2001) Molecular mechanisms of the biological clock in cultured fibroblasts. Science 292:278–281

P

3. Nagoshi E, Saini C, Bauer C, Laroche T, Naef F, Schibler U (2004) Circadian gene expression in individual fibroblasts: cell-autonomous and self-sustained oscillators pass time to daughter cells. Cell 119:693–705
4. Pando MP, Morse D, Cermakian N, Sassone-Corsi P (2002) Phenotypic rescue of a peripheral clock genetic defect via SCN hierarchical dominance. Cell 110:107–117
5. Kornmann B, Schaad O, Bujard H, Takahashi JS, Schibler U (2007) System-driven and oscillator-dependent circadian transcription in mice with a conditionally active liver clock. PLoS Biol 5:e34
6. Storch KF, Paz C, Signorovitch J, Raviola E, Pawlyk B, Li T, Weitz CJ (2007) Intrinsic circadian clock of the Mammalian retina: importance for retinal processing of visual information. Cell 130:730–741
7. Durgan DJ, Trexler NA, Egbejimi O, McElfresh TA, Suk HY, Petterson LE, Shaw CA, Hardin PE, Bray MS, Chandler MP, Chow CW, Young ME (2006) The circadian clock within the cardiomyocyte is essential for responsiveness of the heart to fatty acids. J Biol Chem 281:24254–24269
8. Damiola F, Le Minh N, Preitner N, Kornmann B, Fleury-Olela F, Schibler U (2000) Restricted feeding uncouples circadian oscillators in peripheral tissues from the central pacemaker in the suprachiasmatic nucleus. Genes Dev 14:2950–2961
9. Rutter J, Reick M, Wu LC, McKnight SL (2001) Regulation of clock and NPAS2 DNA binding by the redox state of NAD cofactors. Science 293:510–514
10. Brown SA, Zumbrunn G, Fleury-Olela F, Preitne N, Schibler U (2002) Rhythms of mammalian body temperature can sustain peripheral circadian clocks. Curr Biol 12:1574–1583

Peripheral Proteins

Definition

Protein molecules that are anchored to either the cytoplasmic or extracellular side, such as intracellular signaling components, immunoglobulins and structural proteins.

►Membrane Components

Peripheral Receptor

Definition

The peripheral receptor is the junction site of peripheral autonomic nerve terminals on target organs. Receptors of sympathetic nerves contain noradrenaline (norepinephrine), except sweat glands, as neurotransmitter. These include adrenergic α and β receptors. Receptors of sweat glands contain acetylcholine as a neurotransmitter. Parasympathetic nerve receptors contain acetylcholine as a transmitter.

►Acetylcholine
►Noradrenaline
►Parasympathetic Pathways
►Sympathetic Pathways

Peripheral Rhythms

Definition

Circadian rhythms of molecular, physiological or behavioural parameters, which are generated in cells, tissues or organs outside of the suprachiasmatic nucleus (the site of the master oscillator in mammals).

►Circadian Rhythm
►Clock Coupling Factors
►Suprachiasmatic Nucleus

Peripheral Sensitization

Definition

Sensitization of nociceptors for mechanical, thermal and chemical stimuli.

►Hyperalgesia and Allodynia
►Joint Pain

Peripheral Synapse

Definition

A synapse that is located outside of the central nervous system. An example of a peripheral synapse is the neuromuscular junction where axons of motoneurons synapse with muscle fibers. Another example is the myenteric plexus of the enteric nervous system.

►Neuromuscular Junction (NMJ)

Peripheral Vestibular Apparatus

J. David Dickman
Department of Anatomy and Neurobiology,
Washington University, St. Louis, MO, USA

Definition

The *peripheral vestibular apparatus* detects head motion and position of the head in space relative to gravity. Motion detection and our sense of position in space are determined by receptors of the vestibular system that lie in the inner ear. Although not considered to be a cognitive sense, the vestibular system nonetheless contributes to the fine control of visual gaze, posture, spatial orientation and navigation. Here the peripheral receptor apparatus and its role in motion detection and spatial orientation is discussed.

Characteristics

Structure of the Peripheral Vestibular Sensors

In every day life, two types of motion, rotational and linear are experienced. Orientation relative to gravity is constantly updated. Rotational motion *(angular acceleration)* is experienced during head turns, while *linear acceleration* occurs during walking, falling, leaning and during vehicular travel. *Linear accelerations* are also are experienced during head tilts relative to gravity. Detection of motion and spatial position begins with the vestibular receptors lying in the inner ear. These receptors then send this information to the brain, where it is integrated into a uniform signal regarding direction and speed of motion, as well as the position of the head in space. In the brain, signals from vestibular receptors combine with information from other systems detecting motion such as the muscle proprioceptors and visual receptors. Central processing of these multimodal signals occurs very rapidly to ensure adequate coordination of visual gaze and postural responses (balance), autonomic responses and awareness of spatial orientation.

Vestibular Labyrinth

The vestibular labyrinth of the inner ear is located in the temporal bone, lateral and posterior to the cochlea. It consists of two parts. The *bony labyrinth* houses and protects the more fragile sensory structures contained inside the *membranous labyrinth*. Five separate receptor structures are represented in the vestibular portion of the membranous labyrinth. These include *three* ►semicircular canals and *two* ►otolith organs. The five vestibular receptor organs on each side of the head complement each other in function. The three semicircular canals, including the horizontal, anterior and posterior canals, lie in three different head planes and respond to rotational head movements [1]. The two otolith organs, including the ►utricle and ►saccule, perceive linear motions of the head (*linear accelerations*) and the orientation of the head relative to gravity [2]. Each of the semicircular canals and otolith organs are spatially aligned so as to be maximally sensitive to movements in specific directions. For example, the horizontal semicircular canal and the utricle both lie in a plane roughly equivalent to that of the head held during normal walking posture. In humans, that plane lies about 30° elevated from the naso-occipital axis. In contrast, the vertical canals and the saccule lie in vertical head planes, nearly orthogonal to the horizontal semicircular canal. Each of the canals on one side of the head works in opposite fashion to their counterparts in the contralateral ear. Together, receptors inside the semicircular canals and otolith organs can respond to head motion in any spatial direction. The membranous labyrinth consists of a series of fluid-filled tubes and sacs where the mechanics of motion detection and transduction occur. Surrounding the membranous labyrinth is a fluid called ►perilymph. It is similar to cerebral spinal fluid and has a high concentration of sodium. Inside the membranous labyrinth, where the vestibular receptors lie is a very different fluid called ►endolymph, which is similar to intracellular fluid, with a high concentration of potassium. Endolymph is important, because it is the high potassium concentration that drives transduction of the motion detection mechanoreceptors. The receptor cells of the vestibular organs are innervated by primary afferent neurons that make up part of the *vestibulocochlear* or *VIIIth cranial nerve*. The somas of these bipolar afferents lie in the vestibular ganglion (Scarpa's ganglion) nestled in the internal *acoustic meatus*, a small shelf-like opening through which axons from the ganglion pass into the ipsilateral brainstem, cerebellum and reticular formation.

Hair Cells

Motion detection begins with mechanoreceptor cells in the vestibular system called ►hair cells due to the many ►stereocilia that project from the apical portion of the receptor cell. Each hair cell contains 50–100 stereocilia and a single longer ►kinocilium. The stereocilia are oriented in a number of rows of ascending height, where the tallest stereocilia lie next to the kinocilium. There are two types of hair cells that differ in their morphology, afferent terminations and channel currents. Type I hair cells look like chalices (Fig. 1). They are completely surrounded by a unique calyx-shaped afferent terminal [3]. These type I hair cells are characterized by an inward-rectifying potassium current. Type II hair cells are cylindrically shaped and are innervated by simple synaptic boutons from the vestibular afferent fibers. Both types of hair cells exhibit excitatory synapses upon VIIIth nerve afferents, with glutamate or aspartate as the neurotransmitter. Both types of hair

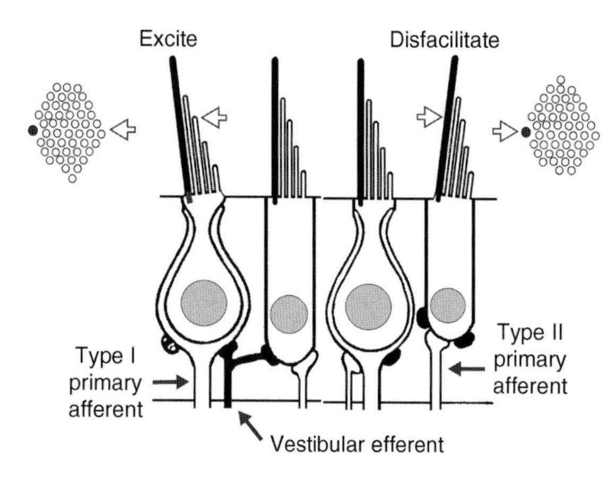

Peripheral Vestibular Apparatus. Figure 1 Schematic of hair cells. Type I hair cells have a calyx-like form that is encapsulated by a primary afferent. Type II hair cells have a cylindrical form. The hair cell bundles contain a single kinocilium (*black*) and multiple stereocilia (*clear*). The *open arrows* indicate a deflecting force applied to hair cell bundles. To the *left*, the Type I hair cell bundle is deflected in towards the kinocilium, exciting the hair cell, increasing release of transmitter (aspartate or glutamate) and increasing the discharge of the vestibular primary afferent. To the *right*, the Type II hair cell bundle is deflected away from the kinocilium, disfacilitating the hair cell, decreasing release of transmitter and decreasing the discharge of the vestibular primary afferent. Vestibular efferents (*illustrated in black*) synapse on the primary vestibular afferents of Type I hair cells and directly on Type II hair cells.

cells also receive vestibular efferent input (acetylcholine, calcitonin gene-related peptide) from the brainstem, which is believed to be involved in controlling receptor sensitivity.

For each of the semicircular canals, the receptor hair cells lie in a specialized patch of neuroepithelium termed the *crista*. The crista is contained in an enlarged region of the membranous duct termed the ▶ampulla, in the approximate center of the canal. A gelatinous structure, termed the ▶cupula, completely covers the crista and forms a fluid-tight partition across the ampulla. The stereocilia of the hair cells are embedded in the gelatinous cupula. During rotational head movements, endolymph is displaced (lags behind) inside the membranous ducts due to inertia and pushes the cupular partition in a direction opposite to the head turn. In each of the three semicircular canals, the hair cells have a uniform anatomical orientation. In each hair cell the kinocilium and stereocilia have the same spatial polarization. Cupular movement causes the stereocilia and kinocilium to flex towards or away from each other. When the stereocilia bend towards the kinocilium, an excitatory generator potential is produced (see below). When they bend away from the kinocilium, hair cells become less polarized. The anatomical polarization of hair cells within a particular semicircular canal gives rise to the directional selectivity of each vestibular organ.

Linear accelerations are detected by receptor hair cells in the otolith organs, which lie in a specialized neuroepithelium termed the *macula*. The stereocilia of the otolith hair cells extend into a gelatinous coating above the macula that is covered by thousands of calcium carbonate crystals, termed ▶otoconia (Greek for ear stones). The otoconia, being much more dense than the surrounding endolymph are not displaced by normal endolymph movements, but instead are only moved during linear motion or changes in head position relative to gravity (linear accelerations) due to their inertia. These otoconia displacements produce bending of the underlying hair cell stereocilia.

Mechanoelectric Transduction

Motion detection begins with the receptor hair cells, which are directionally selective to stereocilia displacement. With movements of the stereocilia towards the kinocilium, hair cell membranes are depolarized and the innervating vestibular afferent fibers increase their firing rate. However, if the stereocilia are deflected away from the kinocilium, the hair cell is hyperpolarized and the afferent fibers decrease their firing rate. This works through *mechanoelectric transduction* of specific potassium (K^+) channels in the apical portion of the kinocilium [4]. When the stereocilia are deflected toward the kinocilium, small actin filaments open the potassium channels, allowing K^+ to enter the hair cell (due to concentration gradients from the endolymph) and depolarize the cell membrane. Through a cascade of events, depolarization leads to synaptic vesicle and neurotransmitter (aspartate or glutamate) release. The transmitter binds with excitatory receptors in the post-synaptic terminal of the vestibular afferent, depolarizing the afferent and increasing its action potential firing rate. When the kinocilium and stereocilia are returned to their normal positions, voltage sensitive potassium channels at the base of the hair cell are allowed to open and release K^+, thereby repolarizing the membrane to its resting potential. When the stereocilia are deflected away from the kinocilium, more potassium channels close and further release of K^+ through the basolateral potassium channels occurs, resulting in cell hyperpolarization. When the head is stationary, vestibular primary afferent fibers have a high spontaneous firing rate (approximately 90 impulses/s). The high rate allows for bidirectional response of the afferents so that silencing (rectification) of the neural response to most natural head motions does not occur.

Morphological Polarization of Hair Cells

Since the hair cells are directionally selective, their orientation on the cristae and maculae are important for signaling direction of movement. For example, all hair cells in the horizontal semicircular canal cristae are arranged similarly, with their kinocilium lying closest to (i.e. pointing toward) the utricle. Horizontal head rotations that produce endolymph movement toward the utricle causes deflections of the stereocilia towards the kinocilium and all of the hair cells in the horizontal semicircular canal are depolarized. Since all hair cells in each of the semicircular canals have the same anatomical orientation, the geometry of the canals determines directional selectivity (Fig. 2a).

The kinocilia of utricular hair cells are oriented towards an imaginary line that runs through the center of the utricular macula. This imaginary line corresponds to a physical structure termed the ▶striola, a dense pile of small otoconia on the surface membrane of the macula. (Fig. 2c). On either side of the striola the kinocilium-stereocilia axes are oppositely polarized. The hair cells in the saccular macula have a similar polarization on opposite sides of the striola (Fig. 2b).

Hair cells in the utricular macula encode linear acceleration in the horizontal plane. Hair cells in the saccular macula encode linear acceleration in the vertical plane. However, since both the saccular and utricular maculae have curved surfaces and since the striola is also curved, linear motions along any direction in three-dimensional space will excite a subpopulation of hair cells.

Semicircular Canal Function

The membranous semicircular duct consists of a fluid-filled tube with a partition (the cupula) in the middle. When the head is stationary, no rotational acceleration is imparted to the semicircular canals and no endolymph flow occurs. The afferents from the complementary canals on both sides of the head have nearly equivalent firing rates. When the head turns, say about the vertical axis, as in shaking the head to indicate "no," the horizontal semicircular canals turn with it, leaving the endolymph fluid behind (lagging the duct) due to inertial forces and viscous drag between the fluid and the walls of the canals. The relative endolymph movement on one side of the head pushes the cupula partition and produces stereocilia deflection towards the kinocilium, while the endolymph movement on the opposite side causes stereocilia deflection away from the kinocilium. Depending upon the direction (toward/away from the kinocilium) stereocilial deflection elicits either depolarization or hyperpolarization of the hair cell membrane. With a leftward head turn, the stereocilia in the left horizontal semicircular canal will be deflected toward the kinocilium resulting in increased discharge of the left VIIIth nerve afferents. Conversely, the right horizontal canal hair cells will be hyperpolarized, since the stereocilia are deflected away from the kinocilium, producing decreased firing rates in right VIIIth nerve fibers. With a rightward head turn, the opposite activation pattern in the hair cells and afferents will be produced.

This functional coupling of the two horizontal semicircular canals also applies to pairs of vertical semicircular canals. For example, the left posterior semicircular canal lies in roughly the same plane as the right anterior semicircular canal. When the head is moved in an angle of about 45° off an imaginary sagittal plane through the head, the discharge from afferents of one of these vertical canals increases and the discharge from the other decreases.

Neural information carried on vestibular afferent fibers from both the left and right semicircular canals is transmitted into the vestibular nuclei. Many neurons in the vestibular nuclei receive information from receptors on both sides of the head, so that coding of rotation direction can be very precise. Due to the mechanics of the semicircular canal system, as well as the transduction properties of pre- and post-synaptic processing, semicircular canal afferents encode head velocity and

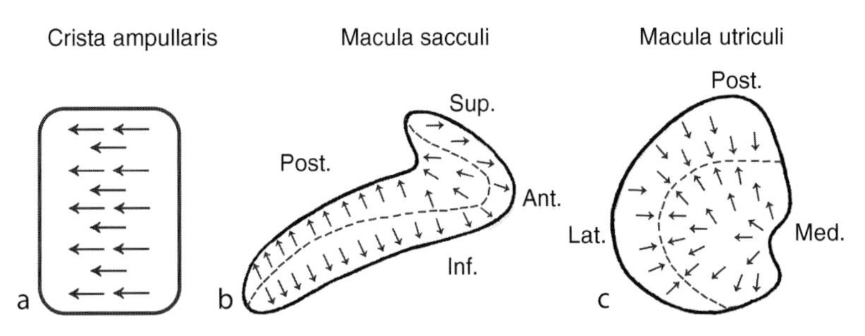

Peripheral Vestibular Apparatus. Figure 2 Hair cell polarization a semicircular canal crista and in saccular and utricular maculae. Hair cell polarization is indicated by the orientation of the multiple stereocilia with respect to the single kinocilium. (a) In a semicircular canal crista ampullaris the polarization of hair cells is uniform as indicated by the alignment of the *arrows*. (b) The polarization of hair cells in the saccular macula is opposite on either side of the dividing line that corresponds to the striola, indicated by the *dashed line*. Note that the saccular macula is aligned with the sagittal plane. (c) The polarization of hair cells in the utricular macula is also opposite on either side of the striola. The predominant polarization of hair cells in the utricular macula is mediolateral.

are band limited in their response to motion. Very slow head rotations (e.g. less than 3–4°/s) produce little response. Instead, afferents are most responsive to rotation speeds between 10 and 150°/s [5], which is in the most often produced range of human head movements.

Otolith Function

Linear motion or changes of head position with respect to gravity (rolled or pitched) causes otoconial displacement due to inertia and consequent deflection of stereocilia in otolithic hair cells. Similar to semicircular canal receptors, otolith hair cells are either depolarized or hyperpolarized by stereocilial deflection toward or away from the kinocilium. Thus, a topographic coding of directional space or movement is represented by the activation of hair cells in particular regions of the maculae. The innervating VIIIth nerve fibers maintain the directional signal, since each afferent only innervates hair cells from a small region on the macular neuroepithelium. Unlike semicircular canal afferents, otolith afferents respond to static head positions, slow head movements, fast head movements and even very fast vibrations with high fidelity.

Vestibular Efferent System

Hair cells and vestibular primary afferents receive information from the brain conveyed by vestibular efferents. The vestibular efferent system (VES) consists of a group of 100–500 cells bordering the genu of the facial nerve in the dorsal brainstem whose axons terminate on vestibular primary afferents and hair cells in the labyrinth [6,7]. Neurons of the VES are cholinergic, identified by acetylcholinesterase histochemistry [7] and choline acetyltransferase immunohistochemistry [8] as well as by the retrograde transport of [^{3}H]-choline.

The functional importance of the VES has remained an enigma because of the technical difficulty of physiologically identifying putative neurons of the VES and analyzing the response characteristics of these neurons *in vivo*. In fish, vestibular efferents can be identified physiologically and intracellularly labeled [9]. In the chinchilla, indirect evidence suggests that vestibular efferents receive vestibular inputs from both the ipsilateral and contralateral labyrinths.

In the cat and the monkey, it is possible to study the action of the VES indirectly by electrically stimulating the region of the brain stem where the cell bodies of vestibular efferents are located and simultaneously recording the effect on the activity of single primary afferents [6]. Primary vestibular afferents in both the fish and the monkey are excited by electrical stimulation of vestibular efferents [6,9].

Two similar ideas have been proposed to explain the function of the VES. (i) Vestibular efferents are used to establish an operating point on the sensitivity curve for primary vestibular afferents so that in different vestibular environments these afferents can be modulated within a linear range [6] and (ii) vestibular efferents detect behavioral arousal and thereby lower the threshold of vestibular primary afferents, thereby facilitating guidance of escape behaviors [9]. Implicit in these two ideas is the necessity of a central correlative efferent signal, so that changes in primary afferent activity induced by vestibular efferent signals can be distinguished from changes in primary afferent activity caused by peripheral vestibular stimulation. Such modifications could take tens of millisecsonds or tens of minutes.

Regeneration of Hair Cells

Remarkably, in many vertebrates including amphibians, reptiles and birds, hair cells and their innervating afferents spontaneously regenerate [10]. Spontaneous *regeneration* of mammalian hair cells does not occur. However, recent developments in promoting mammalian regeneration provide renewed hope for restoring hearing and balance following hair cell loss.

References

1. Fernandez C, Goldberg JM (1971) Physiology of peripheral neurons innervating semicircular canals of the squirrel monkey. II. Response to sinusoidal stimulation and dynamics of peripheral vestibular system. J Neurophysiol 34:661–675
2. Fernandez C, Goldberg JM (1976) Physiology of peripheral neurons innervating otolith organs of the squirrel monkey. I. Response to static tilts and to long-duration centrifugal force. J Neurophysiol 39:970–984
3. Lindeman HH (1973) Anatomy of the otolith organs. Adv Otorhinolaryngol 20:405–433
4. Hudspeth AJ (2005) How the ear's works work: mechanoelectrical transduction and amplification by hair cells. C R Biol 328:155–162
5. Hullar TE, Della Santina CC, Hirvonen T, Lasker DM, CareyJP, Minor LB (2005) Responses of irregularly discharging chinchilla semicircular canal vestibular-nerve afferents during high-frequency head rotations. J Neurophysiol 93:2777–2786
6. Goldberg JM, Fernandez C (1980) Efferent vestibular system in the squirrel monkey: anatomical location and influence on afferent activity. J Neurophysiol 43:986–1025
7. Warr WB (1975) Olivocochlear and vestibular efferent neurons of the feline brain stem: Their location, morphology and number determined by retrograde axonal transport and acetylcholinesterase histochemistry. J Comp Neurol 161:159–181
8. Schwarz DWF, Satoh K, Schwarz IE, Hu K, Fibiger HC (1986) Cholinergic innervation of the rat's labyrinth. Exp Brain Res 64:19–26
9. Highstein SM, Baker R (1985) Action of the efferent vestibular system on primary afferents in the toadfish, Opsanus tau. J Neurophysiol 54:370–384
10. Zakir M, Dickman JD (2006) Regeneration of vestibular otolith afferents after ototoxic damage. J Neurosci 26:2881–2893

Peripheral Vision

Definition

Vision excluding the 5°–10° most central area of the visual field (foveal and peri-foveal vision). Peripheral vision is specialized for low spatial frequency and high temporal frequency stimuli, is not sensitive to wavelength (achromatic) and has a low sensitivity/rapid adaptation to light intensity (scotopic).

►Visual Field

Peristalsis

Definition

Peristalsis is the propagating contraction of the circular muscle layer of the bowel. The contractions develop into peristaltic waves moving along the gut and provide the forces for propelling bowel contents usually in an aboral direction. The peristaltic waves occur intrinsically at a frequency of 3 per/min, but they usually travel only a short distance.

Peristalsis in the Small Bowel

Definition

Peristalsis is the propagating contraction of the circular muscle layer. The contractions develop into peristaltic waves moving along the gut and provide the forces for propelling bowel contents usually in an aboral direction.

The peristaltic waves occur intrinsically at the frequency of 3/min, but they usually travel only a short distance.

►Bowel Disorders

Peristimulus Time Histogram (PSTH)

Definition

Cumulative histogram of the number of spikes occurring within discrete time bins in and around the time of a repeated presentation of an auditory stimulus.

Perisynaptic Schwann Cells

Definition

Schwann cells normally encase axons in peripheral nerves to form the insulating myelin sheath. At the neuromuscular junction between the motor nerve and the endplate region of each skeletal muscle cell, there are Schwann cells that do not form myelin but normally respond to the chemical neurotransmitter substance, acetylcholine, that s released from active nerve terminals to excite the skeletal muscle fibers and in turn, lead to muscle contraction. These Schwann cells are termed perisynaptic Schwann cells because they are at the synaptic region of each muscle fiber.

►Acetylcholine
►Axonal Sprouting in Health and Disease
►Myelin
►Schwann Cell

Periventricular Zone

Definition

The periventricular region of the thalamus has poorly-defined cell groups that are sometimes homologized with the midline thalamic groups seen in non-primates such as nucleus reunions, rhomboid nucleus, and median central nucleus. It is likely that the region is involved in visceral activities in that it has connections to the hypothalamus, amygdala, and cingulate cortex.

Permeabilized Skeletal Muscle Fibers

Definition

Experimental model using single muscle fibers after membrane destruction via detergents or other means. Membrane destruction renders the contractile proteins actin and myosin directly accessible to the external media.

►Force Potentiation in Skeletal Muscle

Per-Rotatory Vestibular Nystagmus

Definition
Nystagmus produced during prolonged rotation of the head in space. For rapid initiation of rotation, the slow phase velocity rises sharply at the start of rotation and declines to zero as the rotation continues at a constant velocity.

►Nystagmus

Persephin

Definition
A member of the glial cell line-derived neurotrophic factor (GDNF) family of neurotrophic factors that also includes artemin and neuroturin. GDNF family members use a receptor complex that consist of the common receptor tyrosine kinase signaling component Ret and one of the GPI-linked receptors (GFRα1 to 4) that regulate ligand binding specificity. GFRα4 is the preferred receptor for persephin.

►Neurotrophic Factors in Nerve Regeneration

Persistent Na$^+$ Currents

Definition
Persistent Na$^+$ currents are TTX-sensitive, voltage-dependent Na$^+$ currents flowing at voltages between −65 and −40 mV, and thus significantly influencing sub-threshold membrane potential changes and the firing rate and pattern of discharge.

►Action Potential
►Sodium Channels
►Tetrodotoxin

Persistent Vegetative State

Definition
This state results from damage of cortical neurons (particularly ►pyramidal cells) in ►hippocampus and ►cerebral cortex after prolonged hypoxia or ischemia. One to two weeks of coma may be followed by a state similar to that of ►hydrencephalic children, in which patients appear awake, may smile, cry, fixate objects, or eat food put in their mouth, but have no cognitive relation to their environment.

Personal Knowledge

Definition
Personal knowledge is knowledge possessed by a person (not for example by a scientific community or culture etc.).

►Knowledge

Personality Disorder

Annette Bölter, Jörg Frommer
Universitätsklinikum Magdeburg, Abteilung für Psychosomatische Medizin und Psychotherapie, Magdeburg, Germany

Synonyms
Character neurosis; Characterogenic neurosis; Autopsychic neurosis; Psychopathic personality

Definition
A heterogeneous group of disorders, regarded as long-standing, pervasive, inflexible and maladaptive personality traits, patterns of behavior and thinking that impair social and occupational functioning.

Characteristics
Personality disorders are coded on Axis II of DSM. They are often comorbid with an Axis I disorder and can serve as a context for Axis I problems, for example depression or anxiety disorder, shaping them in different ways. The symptoms come close to describing characteristics that all people possess from time to time and in varying degrees, however the diagnoses of a personality disorder is defined by the extremes of several traits and is not used unless the patterns of behavior are enduring, pervasive and dysfunctional.

Beside personality disorders DSM-IV distinguishes personality change, for example after a major medical condition. Five subtypes of personality change are listed: labile, disinhibited, aggressive, apathetic and paranoid.

Catergories and Clusters of Personality Disorders

DSM-IV distinguishes ten subgroups: (i) Paranoid, (ii) Schizoid, (iii) Schizotypal, (iv) Borderline, (v) Histrionic, (iv) Narcissistic, (vii) Antisocial, (viii) Dependent, (ix) Avoidant, (x) Obsessive-compulsive (►Obsessive-Compulsive Disorder [OCD]). In DSM-IV these subtypes are also grouped into three clusters: Cluster A (paranoid, schizoid, schizotypal) is defined by characteristics of being odd or eccentric. Because the symtpoms of these disorders are similar to those of the prodromal or residual phases of schizophrenia, they are considered by some researchers to be less severe variants of schizophrenia. Individuals in cluster B (antisocial, borderline, histrionic and narcissistic) appear dramatic, emotional or erratic. Those in cluster C (avoidant, dependent and obsessive-compulsive) seem anxious or fearful.

Diagnosis

Although there do exist some diagnostic questionnaires, personality disorders are preferably assessed by structured interviews. In recent years these diagnoses have become reliable in some degree. But it is still usual for a disordered person to meet diagnostic criteria for more than one personality disorder. The categorical diagnostic system of DSM-IV therefore may not be ideal for classifying personality disorders. A dimensional approach may be more appropriate. During recent years different contributions on personality disorders emerged from several psychological paradigms: understanding on the basis of modern psychoanalysis and attachment theory, approaches based upon the five-factor model of personality, cognitive theories and Linehan`s behavioristic approach, interpersonal concepts as shown by Benjamin, neurobiological models, and a self-developed approach focusing on private theories of the patients themselves.

Therapy

A therapist who is working with patients that have a personality disorder is typically also concerned with Axis I problems because most patients even enter treatment because of Axis I disorder. For example, a person with avoidant personality disorder may be seen for social phobia. Psychodynamic therapy of personality disorders has a long tradition. It aims to remove repressions and to correct the problems underlying the personality disorder. Transference-Focuses Psychotherapy as developed by Kernberg is a psychodynamic treatment designed especially for patients with severe personality disorders, e.g. ►borderline personality disorders. The focus of treatment is on a deep psychological make up – a mind structured around a fundamental split that determines the patient's way of experiencing self and others and the environment. Since this internal split determines the nature of the patient's perceptions, it leads to the chaotic interpersonal relations, impulsive self-destructive behaviors, and other symptoms. The core task in Transference-Focused Psychotherapy is to identify the patient's moment-to-moment experience of the therapist because it is believed that the patient lives out his/her predominant object relation patterns in the patient-therapist-relationship. Another successful approach is Dialectical Behavior Therapy as developed by Linehan. During individual and group therapy, therapist and client work towards improving skill use. These skills are broken down into four modules: mindfulness (derived from Zen tradition), interpersonal effectiveness, distress tolerance, and emotion regulation. In addition common treatments of personality disorder are psychopharmacological drugs. The choice is determined by the associated Axis I disorder.

References

1. American Psychiatric Association (1994) Diagnostic and statistical manual of mental disorders, DSM-IV. American Psychiatric Association, Washington
2. Costa PT, Widiger TA (eds) (1993) Personality disorders and the five-factor model of personality. American Psychological Association, Washington
3. Frommer J, Reissner V, Tress W, Langenbach M (1996) Subjective theories of illness in patients with personality disorders Psychother Res 6:56–69
4. Kernberg OF (1984) Severe personality disorders: psychotherapeutic strategies. Yale University Press, New Haven
5. Lenzenweger MF, Clarkin JE (2005) Major theories of personality disorders, 2nd edn. Guilford Press, New York
6. Linehan MM (1993) Cognitive-behavioral treatment of borderline personality disorder. Guilford Press, New York
7. Millon T (1996) Disorders of personality: DSM-IV and beyond, 2nd edn. Wiley, New York
8. Yeomans FE, Clarkin JF, Kernberg OF (2002) A primer of transference-focussed psychotherapy for the borderline patient. Jason Aronson, Northvale, NJ

Pesticide: Insecticide

►Neuroinflammation: Modulating Pesticide-Induced Neurodegeneration

P

PET

Definition
►Positron Emission Tomography

Petit Mal Seizures

Definition
►Absence Epilepsy

PFC

Definition
Prefrontal cortex.

►Prefrontal Cortex

PGO Spikes

Definition
Ponto-geniculo-occipital; high amplitude sharp waves, which originate in the pons, are transmitted to the lateral geniculate and from there up to the occipital cortex, where they are recorded in cats during rapid eye movement (REM) sleep. PGO spikes reflect bursts of discharge by neurons in the pontine reticular formation which are transmitted rostrally into the visual and also other relay nuclei of the thalamus by which they are transmitted to cortical areas. This phasic activity is also transmitted to brainstem motor nuclei, as reflected in rapid eye movements, and spinal motor nuclei, as reflected in twitches of the distal extremities. Such twitches can be observed as manifestation of the phasic excitation through the brain during REM sleep which occurs in all mammals. As proposed by some, PGO spiking might signify a programming of the brain and organism during development in species-specific motor activities and behaviors. It might also entail enhancement of learning new information and behaviors in the adult during REM sleep.

►Rapid Eye Movement (REM) Sleep
►Sleep-Wake Autonomic Regulation

P23H Rat

Definition
Transgenic rat model of human Proline-23-Histidine rhodopsin mutation leading to Retinitis Pigmentosa.

►Inherited Retinal Degenerations

Phagocytic

Definition
Activity of cells called phagocytes which engulfs and absorbs waste material, harmful microorganisms, or other foreign bodies in the bloodstream and tissues.

Phantom Limb Sensation and Pain

Mike B. Calford
School of Biomedical Sciences and Hunter Medical Research Institute, The University of Newcastle, Newcastle, NSW, Australia

Definition
When appropriately studied, it is usually reported that all adults who have had a limb amputated either by trauma or surgery have at some stage experienced the continued presence of the missing limb within their perceptual body image. The full range of somatosensory percepts (cutaneous touch, deep touch, vibration, itch, tickle, joint movement etc) are variously ascribed to the phantom. Attribution of pain to a phantom limb is a distinct phenomenon with widely varying reports of prevalence. Recent studies have reported higher proportions of cases with phantom limb

pain (60–80%) and there is a possibility that earlier reports were affected by patient reluctance to describe their experience.

Those with spinal cord injury (paraplegia, quadriplegia) rarely localize non-painful somatosensory percepts to their affected limbs. However they usually retain these limbs within their body image and sometimes attribute full function to them – a form of *anosognosia* also found in some cases of stroke-induced hemiplegia. There is no consensus view of the prevalence of phantom pain after spinal cord injury. This may be attributable to varying pathology and to the extent of the lesion. Irrespective of this, there are many well-described cases of phantom limb pain after complete spinal transection.

Supernumerary limb is a term that encompasses various presentations of the appearance of a third (or more) arm or leg in the perceptual body image. A similar misperception can occur temporarily with brachial plexus nerve block, but it is a rare chronic condition usually following transient middle cerebral artery ischemia or mild neurotrauma. *Alien hand* syndrome describes the converse condition in which, after mild neurotrauma, an intact limb is lost to the perceptual body image and its presence denied or disowned.

Characteristics

Phantom limb experiences manifest immediately or within a few days of an amputation. In a minority of cases, phantom sensations persist relatively unchanged. In other cases, the phantom sensations and the perceptual body image alters. With upper limb amputation, a loss of the image of the arm but retention of the phantom hand is classically described – but there are many variations. The phantom sensations disappear after a few months to years in around 50% of cases. However, in laboratory conditions with stimulation of the amputation stump (or other body parts) or with imagery, elements of a phantom limb experience can be elicited in most amputees even years after they have reported loss of the phantom sensations. Cortical activation achieved through caloric ear stimulation is reported to recall lost phantom sensations and also to alleviate temporarily painful phantom sensation by replacement with a nonpainful phantom [1].

Phantom limb pain shares many characteristics with neuropathic pain and is equally difficult to account for. Pain sensations are rarely continuous but chronic presentation of repetitive bouts is common. Phantom pain can mimic the presentation of a pre-amputation chronic pain. It is becoming widespread practice to provide limb analgesia prior to a surgical amputation. This practice has followed a number of case reports indicating that such post-amputation mimicry is thereby avoided. Nevertheless, clear evidence of efficacy for pre-surgery limb analgesia is not apparent [2].

The highly plastic nature of the somatosensory perception of the limbs can be easily demonstrated. When operating a hammer, pointer, golf club, stilts or sword, we readily telescope our body image to include the extension. In such an operation, sensory inputs from receptors in the hand or arm are combined and interpreted to provide a "percept" from the tip of the extension. While training improves performance, the extension of percept to encompass tool use by either upper or lower limb is largely automatic. In complex multisensorimotor tasks with machines (such as driving a car), we are capable of integrating a variety of somatosensory, visual and auditory cues to expand our body image to the limits of the machine or task. Within this context, it is then somewhat surprising that essentially 100% of limb amputees experience a phantom experience of their missing part, rather than adapt their body image to encompass the amputation. Nevertheless, after amputation the plastic nature of the body image is not lost, for in most cases the phantom image locks onto a prosthetic and adapts to tool use.

Melzack [3] reasoned that the perceptual experience of a limb does not derive from activation of a single brain area but from the combined output of a distributed network, termed the neuromatrix. To account for the persistent nature of phantom limb perception, in the light of considerable physiological and psychophysical evidence of plasticity, it was concluded that, when activated, elements of the overall network retain the identity of the former limb. On the basis of reports of phantom limb experiences in some cases of congenitally missing limbs, it was further suggested that the identity of a given neuromatrix is genetically predetermined. This latter suggestion remains contentious, but the broader neuromatrix concept is well accepted and is consistent with recent work.

Ramachandran and others [4,5] have described many cases in which passive cutaneous stimulation of either the amputation stump or a distant body area (e.g. face for upper limb amputation) produces a dual sensation with one being localized to the phantom. In some cases, a clear topographic map of the phantom can be plotted. These demonstrations reveal both plasticity and rigidity in the somatosensory representation of the limb consistent with the concept that some elements of the neuromatrix retain their original perceptual attribution.

Extensive plasticity of the primary somatosensory representation has been demonstrated in monkeys and other mammals. With upper limb amputation or deafferentation the former arm and hand representation is taken over by a responsiveness to adjacent body areas, the back of the head and the face [6]. There is a considerable immediate unmasking effect [7] such that some areas immediately switch to show new responsiveness, but the total filling-in of the former representation develops over a longer period. The somatosensory representation in the cortex is distributed across at least seven fields in central and parietal regions. Each is organized as a topographic map of the contralateral body and there is a partial specialization of function. Thus, for example, muscle spindles are the predominant input to one field (area 3a) while cutaneous receptors dominate others. There is, however, no cortical field devoted to or dominated by pain perception. Rather the nociceptive pathway, although separate in the spinal cord and brainstem, feeds into the major (lemniscal) somatosensory thalamocortical pathway. This may be an important aspect in any explanation of phantom limb pain, as those with persistent phantom pain could be shown with neuroimaging to have a reorganization within the primary somatosensory cortex, whereas those with an innocuous phantom did not [8]. Consistent with the reorganization reported in the monkey studies, this reorganization involved expansion of the cutaneous representation of the chin and lips into the former arm and hand area. The extent of the reorganization was highly correlated with the degree of phantom pain suggesting that activation of the cutaneous receptors of other body areas provides inappropriate stimulation to the cortical representation of the former pain signaling neurons of the missing limb. It needs to be noted that such reasoning is over simplistic if applied to any single representation since pain perception is likely to involve activation of multiple fields.

A consistent approach to managing or treating phantom pain has not developed to date. The linking of phantom pain to cortical reorganization phenomena and the knowledge that cortical representations are inherently plastic has led to investigations of potential therapies based around manipulating the reorganization, using pharmacological or physical therapy approaches [2]. Use of mirrors to aid in mental imagery to "move" the phantom limb has been reported to be successful in some cases [9]. Whereas treatments (e.g. stump analgesia) aimed at silencing a supposed overly active peripheral nerve are not generally useful, there were persistent early reports that a subgroup are helped by reducing sympathetic activity and sympathetic disturbance is often found. A plausible explanation for sympathetic activity directly affecting sensory nerves has come from animal work showing sprouting of noradrenergic sympathetic axons into the dorsal root ganglion following a peripheral nerve injury. Nevertheless, overall there is no supporting evidence for using sympathectomy as a treatment for neuropathic pain and hence for phantom pain [10].

References

1. Andre JM, Martinet N, Paysant J, Beis JM, Le Chapelain L (2001) Temporary phantom limbs evoked by vestibular caloric stimulation in amputees. Neuropsychiatry Neuropsychol Behav Neurol 14:190–196
2. Flor H (2002) Phantom-limb pain: characteristics, causes, and treatment. Lancet Neurol 1:182–189
3. Melzack R (1990) Phantom limbs and the concept of a neuromatrix. Trends in Neurosci 13:88–92
4. Halligan PW, Marshall JC, Wade DT, Davey J, Morrison D (1993) Thumb in cheek? Sensory reorganization and perceptual plasticity after limb amputation. Neuroreport 4:233–236
5. Ramachandran VS, Stewart M, Rogers-Ramachandran DC (1992) Perceptual correlates of massive cortical reorganization. Neuroreport 3:583–586
6. Pons TP, Garraghty PE, Ommaya AK, Kaas JH, Taub E, Mishkin M (1991) Massive cortical reorganization after sensory deafferentation in adult macaques. Science 252:1857–1860
7. Calford MB, Tweedale R (1990) Interhemispheric transfer of plasticity in the cerebral cortex. Science 249:805–807
8. Flor H, Elbert T, Knecht S, Wienbruch C, Pantev C, Birbaumer N, Larbig W, Taub E (1995) Phantom-limb pain as a perceptual correlate of cortical reorganization following arm amputation. Nature 375:482–484
9. Ramachandran VS, Hirstein W (1998) The perception of phantom limbs. The D. O. Hebb lecture. Brain 121:1603–1630
10. Mailis A, Furlan A (2003) Sympathectomy for neuropathic pain. Cochrane Database Syst Rev CD002918

Phantosmia

Definition

Describes the distorted perception of smells in the absence of an odor source. Most often, phantosmias occur after trauma or URTI and consist of unpleasant odors occurring without being elicited through environmental odor sources. Phantosmias also have a tendency to disappear over the course of years.

►Smell Disorders
►Olfactory Hallucinations

Pharmacodynamics

Definition

Pharmacodynamics refers to the biochemical and physiological effects of drugs on the body.

Pharmacogenomics

Definition

The science of understanding how an organism's genetic inheritance (i.e. genotype) affects the body's response to drugs.

Pharmacokinetics

Definition

A branch of pharmacology that concerns itself with the process in which drug is absorbed, distributed, metabolized and eliminated by the body.

Pharmacophore

Definition

Pharmacore is defined as a set of structural features in a molecule that is recognized at a receptor site and is responsible for that molecule's biological activity.

Phase

Definition

Any point on a cycle; the instantaneous state of a periodic process.

Phase Advance

Definition

A shift that accelerates (advances) the arrival of an oscillator at a particular event marker (phase). For example, flying from New York to Paris results in a 6 h phase advance of the light-dark cycle, since dawn in Paris is 6 h ahead of dawn in New York. When people are getting up in New York, people in Paris are eating lunch – they are already at a later phase of their daily routine. (Lunch typically tastes better in Paris.) In order for an oscillator to entrain to signals from a pacemaker that has a shorter period, it must execute phase advances.

►Phase Advance Curve
►Oscillator
►Pacemaker

Phase Angle

Definition

The displacement, in units of time or angular degrees, between phases of two coupled oscillators, e.g., a circadian oscillator and a light-dark cycle.

►Circadian Rhythm
►Phase Advance Curve

Phase Delay

Definition

A shift that sets back the time of arrival of an oscillator at a particular event marker phase. For example, flying from Paris to New York results in a 6 h phase delay of the light-dark cycle, since dusk arrives 6 h later in New York than in Paris. When people are eating supper in Paris, people in New York are eating lunch – they are at an earlier phase of their daily routine. In order for an oscillator to entrain to signals from a pacemaker that has a longer period, it must execute phase delays.

►Oscillator
►Pacemaker
►Phase Advance Curve

Phase Locking

Definition

Phase locking is a term that describes the auditory nerve's ability to discharge in synchrony with the acoustic stimulus. Auditory neurons no not fire on every cycle of a sinusoidal acoustic sound, but rather fire stochastically on a specific phase of the stimulus.

►Cochlear Implants

Phase Locking in Auditory System

Definition

Phase locking describes the ability of a neuron to fire action potentials that are time locked to a stimulus event. In auditory neurons, phase locking is used in the context of pure tones, consisting of a single sine wave. Auditory neurons in barn owls phase lock to as high as 10 kHz, in mammals phase locking does not occur above, roughly, 4 kHz.

Phase locking requires temporal precision in action potential timing, with standard deviations of fewer than 100 μs at some frequencies.

►Intrinsic Properties of Auditory Neurons
►Neuroethology of Sound Localization in Barn Owls

Phase Relations

Definition

Phase relations refer to the temporal relationship between coupled elements expressed as a fraction (degrees or radians) of the cycle time.

Phase Response Curve

Christopher S. Colwell
Department of Psychiatry, University of California, Los Angeles, USA

Synonyms

PRC

Definition

Phase response curve (PRC) in a graph that plots the magnitude of phase shifts resulting from perturbations from discrete stimuli.

Characteristics

The discrete stimuli often come in the form of light pulses, which induce ►phase delays and ►phase advances during early and late ►subjective night, respectively, but produce negligible phase shifts during subjective day. The phases in which light does not produce a ►phase shift (ΔΦ) is referred to as the "dead zone" of the PRC. A variety of non-photic stimuli can also generate phase-dependent phase shifts and could potentially be used to entrain the circadian system [1]. In general, these non-photic PRCs are characterized by phase shifts during the subjective day and a dead zone during the subjective night. With both photic and non-photic stimuli, the amplitude of the PRC can vary with the strength of the stimulus.

The most straightforward way to generate a PRC is to have a population of animals in constant conditions and to measure their ►free-running rhythms. Based on these rhythms, individual organisms can then be exposed to stimuli at discrete phases of the daily cycle and the resulting phase shifts (see definition) measured by the next cycle. By convention, phase advances are plotted as positive values while phase delays are plotted as negative values. After some perturbations (especially those that cause phase advances), there can be a few days of transients before the full magnitude of the phase shift can be determined. Under the constant conditions required for these measurements, the ►period (Tau, τ) of the rhythm will not be equal to 24-h and the phase will need to be normalized by the endogenous period. Typically, the phases of the endogenous cycle are designated as circadian time (CT) 0–24 with the number of minutes in each hour of CT being equal to Tau/24 multiplied by 60. By definition, CT 0 is the time of activity onset for a diurnal organism while CT 12 is the time of activity onset for a nocturnal one. The phases of the endogenous cycle that coincide with the prior daytime are called "subjective day" while the phases that coincide with the prior nighttime are called "subjective night".

PRCs have been used to explain how 24-h ►entrainment is maintained between an endogenous circadian oscillator (which has a period that is not equal to 24-h) and environmental cues. While a description of this model is well beyond the scope of this article, a simple example may be useful. If the PRC is known, then the phase relationship between environment and the endogenous ►oscillator can be predicted based on the periods of the biological oscillation (Tau) and the ►zeitgeber cycle (T) [2]. For example, if the circadian oscillator has a period of 25 h and is entrained to a light pulse every 24 h. Stable entrainment can only be reached when the light causes a phase shift that corrects the difference between Tau and T which in this case would be a phase delay of 1-h per cycle (Tau – T = ΔΦ). Only when the light falls on the portion of the PRC in the early night causing the 1-h phase delay will the biological oscillator be entrained to the physical cycle.

References

1. Johnson CH, Elliott JA, Foster R (2003) Entrainment of Cicardian Progemas. Chronobiology International. Neuron 20(5):741–774
2. Roenneberg T, Daan S, Merroa M (2003) The Art of Entrainment. J Biological Rythms 18(3):183–194

Phase Shift ($\Delta\Phi$)

Definition

A shift in the phase of the biological oscillation. Phase (Φ) is one of the most important parameters describing any oscillation as it refers to the time points within the cycle. To measure the phase of the rhythm, a reliable reference point must be chosen. In the case of circadian oscillations, the onset of activity or the peak expression of a biochemical parameter are commonly used. These biological markers are typically expressed relative to the time of lights-on or lights-off when the organism is in a light cycle. For circadian oscillators, exposure to light is the most physiologically relevant and widely studied perturbation that results in phase shifts. Many treatments that cause phase shifts of the circadian system produce different effects depending on the time of day during which the treatment was applied.

►Circadian Rhythm
►Phase Response Curve

Phase Spectrum

Definition

A description of the relationship between starting phase and frequency of the sinusoidal components of a complex sound wave.

►Acoustics

Phasic

Definition

Transiently occurring at the onset or offset of an event.

Phasic responses indicate that the adaptation of the nervous system under consideration is rapid. Opposite term is tonic.

Phenethylamine

Definition

A simple molecule that serves as the framework for many biologically-active molecules, including dopamine, noradrenaline (norepinephrine), and certain hallucinogens. It is essentially a phenyl ring separated by a two carbon chain from an amino group.

►Dopamine
►Noradrenaline
►Hallucinogens

Phenomenal Character

Definition

The basic feature of many mental states to feel a certain way. It is somehow for an organism to be in pain, and it is also somehow for an organism to have a sensation of warmth. But these two mental states differ in phenomenal character in that it feels different for an organism to be in pain and to have a sensation of warmth.

►Behaviorism
►Logical

Phenomenal Concepts

Definition

Phenomenal concepts are those concepts that are immediately related to perceptions and sensations, in such a way that one is immediately inclined to apply these concepts whenever one undergoes the appropriate experience and introspects on one's experience. Having a red experience, for example, makes one inclined to apply the phenomenal concept of a red experience.

►The Knowledge Argument

Phenomenology

Definition

"Phenomenology" means doctrine of appearances. In general it means the reflective inquiry into one's own [→] consciousness. It also names Edmund Husserl's philosophical project of studying how the world appears to us in [→] intentional consciousness. He held that intentional consciousness is intrinsically "directed" to objects. In some intentional episodes objects are merely "meant," while in others (intuitions) objects are (partially) "given." The "fulfillment" of the former episodes by intuitions is the basis of knowledge.

►Argument
►Logic

Phenophysics

►Psychophysics

Phenotype

Definition

Two specimens of a diploid organism, such as a flowering plant, may look alike (same phenotype), even though one is homozygous for, say, red petal color (identical alleles for the gene concerned on the two homologous chromosomes) and one is heterozygous (one allele coding for white, the other one for red which is dominant in certain plants; different genotypes).

►Electric Fish

Pheomelanin

Definition

A polymeric pigment varying in color from yellow to red and produced from 1,4-benzothiazinylalanine derived from L-tyrosine and L-cysteine by melanocytic cells.

►Melanin and Neuromelanin in the Nervous System

Pheomelanosomes

Definition

Spherical organelles 0.7 μm diameter found within melanocytes that compartmentalize pheomelanin synthesis and storage.

►Melanin and Neuromelanin in the Nervous System

Pheromone

Definition

Pheromones were originally defined in relation to insects as "substances secreted to the outside of an individual and received by a second individual of the same species in which they release a specific reaction, for example, a definite behavior or developmental process." This led to the distinction between two categories of pheromonal effect: releaser pheromones that elicit immediate and relatively stereotyped behavioral responses, such as sexual attraction; and primer pheromones that elicit a longer-term change in hormonal or developmental state, such as acceleration of puberty. Many in the field now regard the original definition as over-restrictive and would expand the term to include signaler pheromones that convey information such as individual identity, and modulator pheromones that have an effect on mood or emotion.

►Accessory Olfactory System
►Chemical Senses
►Evolution of Olfactory and Vomeronasal Systems

Philosophy of Action

►Action, Action-Theory

Phobic Neurosis

Definition

An anxiety disorder in which there is intense fear and avoidance of specific objects or situations, most frequently fear of wide places (agoraphobia), closed spaces (claustrophobia) or animals. The fear is recognized as irrational by the individual.

►Personality Disorder

Phoneme

Definition

The minimal contrastive unit in the sound system of a language; substituting one phoneme for another changes the meaning of a word (e.g., /t/ vs. /n/ because/bat/ differs in meaning to /ban/).

Phonotaxis

Definition

Sound-induced, directional movement of an organism relative to the sound source. It is usually either directed towards the source (positive phonotaxis) or away from it (negative phonotaxis). It is most commonly found in acoustic communication behaviors in insects.

►Auditory-Motor Interactions

Phosphacan

Definition

One kind of axon growth inhibitors. It is expressed in the CNS as a secreted splice variant of the gene encoding the extracellular domain.

►Regeneration of Optic Nerve

Phosphagen

Definition

The term given to both high-energy phosphate compounds, 5'-adenosine triphosphate and phosphocreatine.

►Energy Sensing and Signal Transduction in Skeletal Muscle

Phosphatidyl Inositol-3-kinase (PI 3-kinase or PI3K)

Definition

Part of a family of related enzymes that phosphorylate the 3 position hydroxyl group of the inositol ring of phosphatidylinositol. The phosphorylated phosphoinositides produced by PI 3-kinases function via the phosphoinositide-binding domains, which are recruited to cellular membranes for various signaling funtions.

PI3K and its downstream effector the actin-associating Protein kinase (Akt) pathway are involved in the regulation of neuronal soma size and axon caliber.

►Neurotrophic Factors in Nerve Regeneration

Phosphene

Definition

A phosphene is a consciously perceived visual experience, typically a localized flash of light, which, rather

P

than being elicited by a visual stimulus, is induced by non-visual stimulation of nerves within the visual system. Phosphenes may, for example, be produced by pressure in the eyeball inducing neural activity in the retina or by magnetic or electrical stimulation of visual areas in the cerebral cortex.

►Blindsight

Phosphodiesterase Inhibitors (PDEIs)

Definition

Drugs that inhibit the function of phosphodiesterases (PDEs). These molecules elevate cAMP, resulting in upregulation of PKA and CREB signaling and downregulation of NF-κB. Consequently, they are neuroprotective and anti-inflammatory agents.

►CREB
►Cyclic AMP
►Neuroinflammation – PDE Family Inhibitors in the Regulation of Neuroinflammation
►NF-κB

Phosphodiesterase (PDE)

Definition

Denotes a class of enzymes that hydrolyze the cyclic nucleotides cAMP and cGMP (second messengers).

PDEs have an important role as regulators of signal transduction mediated by these second messengers.

There are 11 families of proteins with this enzymatic activity (PDE1-PDE11) and more than 50 isoforms in total.

►Phosphodiesterase: A family of inhibitors in the regulation of neuroinflammation
►Neuroinflammation – PDE Family Inhibitors in the Regulation of Neuroinflammation

Phospholipids

Definition

The major lipid molecule of the cell membrane, composed of two fatty acids linked through glycerol phosphate to a polar group.

►Membrane Components
►Plasma Membrane - Structure and Functions

Phosphorylation

Definition

A reaction involving the addition of a phosphate group to a molecule. Many enzymes are activated by the covalent bonding of a phosphate group. The oxidative phosphorylation of ADP forms ATP.

►Energy Sensing and Signal Transduction in Skeletal Muscle

Photocycles

Definition

Cycles, or rhythms, in environmental lighting condition.

As with any rhythm, these can vary with respect to several parameters such as their period, phase and amplitude. Photocycles also vary with respect to the relative duration of their different phases and their waveforms. Some, for example, are sinusoidal and others have abrupt transitions between their light and dark phases. Animals use changing photocycles to anticipate and prepare for daily and seasonal changes in the environment, and their responses to these changes can be influenced by all of the parameters noted above.

Photo-Inactivation Technique

Definition

A technique to selectively lesion individual neurons within the nervous system. Fluorescent dyes are intracellularly injected into cells that die upon illumination by a laser beam or ultraviolet light source. Other nearby neurons are not affected. This technique has been successfully used to isolate putative pacemaker neurons in functional neuronal networks.

►Bursting Pacemakers
►Stomatogastric Ganglion

Photon Catch

Definition

The number of photons absorbed over a given duration by an individual photoreceptor, or class of photoreceptors. The probability of photon absorption depends on photon wavelength, but all subsequent steps of phototransduction are wavelength-independent.

►Photoreceptors
►Phototransduction
►Retinal Color Vision in Primates

Photoperiod

Definition

The term "photoperiod" refers to the duration of the light phase of a light-dark cycle. In biology, the term is most commonly used in the context of changes in daylength that many temperate zone organisms use to anticipate and prepare for seasonal changes in their environments.

►Photocycle

Photoperiodic Time Measurement

Definition

The measurement by organisms of the duration of the light relative to the dark phase of a light-dark cycle, often abbreviated as PTM. Photoperiodic time measurement enables temperate zone organisms to anticipate and prepare for seasonal changes in their environments.

In vertebrates, the pineal gland plays an important role in the system via its seasonally changing patterns of secretion of the hormone melatonin.

►Photocycle
►Pineal Gland

Photoperiodism

Definition

The use of daylength by organisms to prepare for seasonal changes in their environments. Animals may use either the rising or the falling phase of the annual rhythm in photoperiod as a signal to anticipate the arrival of spring or the coming of winter conditions, respectively. In mammals, photoperiodism can promote changes in a variety of parameters including, for example, pelage, body weight, reproductive function, activity levels, and aggressive behavior. Some animals use declining photoperiods to prepare for winter conditions and others use the rising phase to prepare for the arrival of spring. In mammals, photoperiodism involves the pineal gland, which signals the changing seasons via its changing pattern of secretion of the hormone melatonin.

►Hibernation
►Seasonality

Photopic

Definition

Daylight conditions or vision.

Photopigments

Nathan S. Hart
School of Biomedical Sciences, University of Queensland, St. Lucia, Brisbane, QLD, Australia

Synonyms

Visual pigments; Rhodopsins; Porphyropsins; Visual purple

Definition

Photopigments are ►G-protein-coupled transmembrane proteins contained within the ►photoreceptors. Their function is to absorb the incident light and trigger a biochemical cascade that alters the electrical properties of the photoreceptors and, ultimately, modulates the rate of ►glutamate release (see ►Phototransduction).

Characteristics

Description of the Structure

Photopigments are light-sensitive single-chain polypeptides, belonging to the family of ►G-protein-coupled receptors (GPCRs), capable of activating heterotrimeric G-proteins such as *transducin* (see ►Phototransduction). The photopigments of all animals, and even some archaebacteria, share a common structural conformation that consists of seven α-helical transmembrane segments and an extracellular (or, in the case of vertebrate rod photoreceptors (►Photoreceptors), intradiscal) amino (N) terminus [1]. These similarities reflect the ancient evolutionary origin of photopigment molecules. Photopigments differ from other GPCRs in that light, rather than another molecule, is the ►ligand that stimulates the receptor. This photosensitivity arises from the physical behavior of a ►chromophore molecule embedded in a pocket formed by the protein's transmembrane domain [2].

Opsins

The proteinaceous component of the photopigment is called an ►opsin, and the amino acid sequence of its polypeptide chain determines both its receptor properties and, in combination with the chromophore, its spectral absorbance. Opsins contain 350–500 amino acids, and comparisons of sequence homology reveal that all vertebrate ►visual pigments belong to one of five distinct classes: SWS1, SWS2, RH1, RH2 and M/LWS [3]. Moreover, the gene duplication events that led to the separation of these different opsin classes predate the divergence of the jawed and jawless vertebrate lineages [4].

In functional terms, photopigments are classified by their wavelength of maximum absorbance (λ_{max}). In most instances, photopigments belonging to the same opsin gene class have a λ_{max} value in a similar region of the visible spectrum, although there is considerable overlap between classes. SWS1 opsin genes code for ultraviolet- or violet-sensitive cone photopigments with λ_{max} values between 355–440 nm. SWS2 opsin genes produce short-wavelength-sensitive ("blue") photopigments with λ_{max} values between 410 and 475 nm. RH1 and RH2 opsin genes generate medium-wavelength-sensitive ("green") photopigments with λ_{max} values between 460–540 nm in rod and cone photoreceptors, respectively. M/LWS opsin genes code for long-wavelength-sensitive ("red") cone photopigments with λ_{max} values between 505–630 nm.

Birds, turtles and some fish have retained and express all five opsin genes. These species usually possess at least four spectrally distinct cone types and have tetrachromatic color vision (►Color processing). However, throughout vertebrate evolution there has been a tendency for certain taxa to lose opsin classes that were present in their ancestors. For example, some reptiles (e.g. lizards) have lost the RH1 rod opsin as a consequence of the loss of rod photoreceptors from their ►retina. Most placental mammals lack both the SWS2 and RH2 cone opsin genes and are dichromats. Many marine mammals have even lost the SWS1 gene and with it any chance of color vision. However, some primates (including humans) have re-evolved a third cone opsin via a duplication of their M/LWS opsin gene and have trichromatic color vision.

In most species, only one type of opsin is expressed in a given photoreceptor type. However, there are instances where two or more opsins are co-expressed, giving rise to a mixture of photopigments with different λ_{max} values in a single *outer segment* (►Photoreceptors). In mice and rabbits, SWS1 and M/LWS opsins are co-expressed within the same cone photoreceptor and the relative proportion of the two photopigments varies with retinal location. Co-expression of two or more opsins may also occur in some fish while they undergo an ontogenetic shift in habitat light environment.

Chromophores

Vertebrate photopigments employ two different chromophores, 11-*cis* retinal and 11-*cis*-3,4-didehydroretinal, which are the aldehydes of vitamin A_1 and A_2, respectively. Photopigments based on 11-*cis* retinal (►rhodopsins), are generally found in mammals, birds and marine fish. Photopigments based on 11-*cis*-3,4-didehydroretinal (►porphyropsins) are characteristic of freshwater fish and turtles. Different chromophores can be used interchangeably with the same opsin. However, a porphyropsin photopigment will have a λ_{max} value shifted towards longer wavelengths compared to a rhodopsin photopigment utilizing the same opsin, a phenomenon called the "chromophore shift".

Many aquatic species, including some lampreys, teleost fish, elasmobranchs and amphibians, are capable of producing both rhodopsin and porphyropsin photopigments. In most cases, the chromophore type changes from 11-*cis* retinal to 11-*cis*-3,4-didehydroretinal in response to an ontogenetic shift in habitat type from fresh to salt water (or vice versa). The long-wavelength spectral shift induced by porphyropsin use is considered to be an adaptation to the relatively red-shifted illumination in most freshwater or estuarine habitats compared to terrestrial or marine light environments. Intriguingly, some terrestrial reptiles, such as the true chameleons (Chamaeleondiae), maintain ►rhodopsin/►porphyropsin mixtures of the same opsin type within individual cone photoreceptors, although in the absence of ontogenetic shifts in habitat type the functional significance is unclear.

Spectral Absorbance

The absorbance spectrum of an unbleached photopigment is characterized by four distinct peaks (Fig. 1). The first and most visually relevant of these is the so called α-peak. When classifying the spectral absorbance of photopigments, the λ_{max} value quoted usually refers to the wavelength position of the α-peak, which represents the main absorbance band of the bound chromophore. The α-peak varies in λ_{max} from about 350 to 620 nm, depending on both chromophore type and its physicochemical interactions with the opsin apoprotein (see section entitled Spectral Tuning and the Opsin Shift). A second, smaller peak (β-peak) occurs at shorter wavelengths and is due to the *cis*-band of the chromophore. The wavelength position of the β-peak is positively correlated to that of the α-peak but varies over a much smaller wavelength range (310–390 nm). The third (γ-peak; 280 nm) and fourth (δ-peak; 231 nm) absorbance peaks are caused by tyrosine and tryptophan residues in the opsin and a variety of organic bonds in the photopigment, respectively [5].

Generally speaking, we are only concerned with the absorbance of the α-peak because wavelengths to which the other absorbance peaks are sensitive are prevented from reaching the photoreceptors by absorption of short wavelengths by the ►ocular media or other spectral filters in the retina. For example, the human lens and macula contain carotenoid pigments (lutein and zeaxanthin) that prevent light of wavelengths shorter than about 400 nm from reaching the photoreceptors. However, in several other vertebrate (and invertebrate) species, ultraviolet light (UV, 300–400 nm) is allowed to enter the eye to stimulate specialized UV-sensitive photopigments and the β-peak of photopigments with longer λ_{max} values may absorb some of this light.

Regulation of the Structure

Chromophore Photosensitivity and the Schiff's Base Linkage

All opsins contain a lysine residue at a particular site on the interior surface of the chromophore-binding pocket that couples to the chromophore via a Schiff's base bond (aldimine linkage). In vertebrate photopigments, the Schiff's base is usually ►protonated, and its positive charge is balanced by a negatively charged residue

P

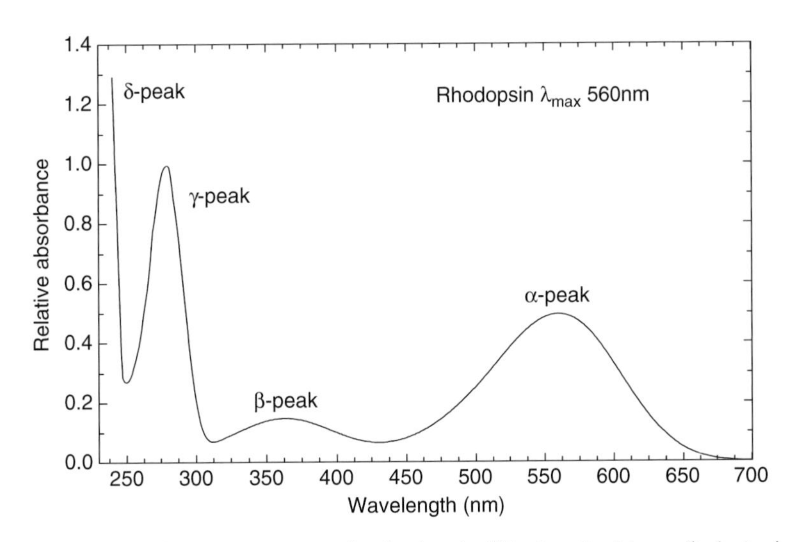

Photopigments. Figure 1 Absorbance spectrum of a rhodopsin (11-*cis* retinal-based) photopigment. See text for details.

(almost invariably glutamate) that acts as a ►counterion to stabilize the linkage electrostatically. This charge distribution is fundamental to the photosensitivity and wavelength specificity of the photopigment. When the chromophore absorbs a photon of light, it undergoes a conformational change and flips from the less stable 11-*cis* configuration to the more stable all-*trans* ►isomer (see ►Phototransduction). This causes a major structural rearrangement that displaces the Schiff's base from its interaction with the glutamate counterion. The subsequent loss of electrostatic stability results in further structural rearrangement of the opsin, deprotonation of the Schiff's base and generation of the active form of the photopigment (metarhodopsin II) [6].

Spectral Tuning and the Opsin Shift

Free chromophore (retinaldehyde) has a peak absorbance at 375 nm. When combined with a simple amino-group ($-NH_2$) containing-compound (n-butylamine), the protonated retinaldehyde-Schiff's base complex formed has a λ_{max} at 440 nm [7]. The difference in λ_{max} between the protonated retinaldehyde-Schiff's base complex and the α-peak of a given opsin-retinaldehyde photopigment is called the "opsin shift."

The magnitude and direction of the opsin shift depend on a variety of molecular interactions, all of which are directly attributable to the amino acid sequence of the opsin. In particular, the interactions of charged, polar or polarisable amino acid residues with the chromophore may affect the strength of the interaction between the protonated Schiff's base and the glutamate counterion, and/or alter the delocalization of charge along the length of the chromophore. A reduction in charge delocalisation along the chromophore, or a strengthening of the Schiff's base-counterion interaction (which helps to prevent charge delocalisation) increases the stability of the chromophore and short-wavelength-shifts the photopigment λ_{max} as a result of the higher energy required for photoisomerization. In contrast, an increase in charge delocalisation, or a weakening of the Schiff's base-counterion interaction, decreases the energy required for photoisomerization and results in a long-wavelength-shifted photopigment [8].

Thus, by altering the types of amino acid present within the opsin polypeptide, at specific locations that are close enough to interact electrostatically with the chromophore, the λ_{max} of a rhodopsin photopigment can be varied almost continuously from 350 to 575 nm (as far as 630 nm in the case of porphyropsin photopigments). In addition to a λ_{max} 500nm RH1 rod photopigment, the human retina contains three cone photopigments with λ_{max} values at 430 (SWS1), 530 and 560 nm (M/LWS). The mechanisms of spectral tuning of vertebrate (especially mammalian) M/LWS photopigments are probably the best understood of all opsin genes and, in most species, their λ_{max} values are determined by various combinations of amino acid substitutions at just five locations in the opsin apoprotein, the so called "five sites rule" [3].

Anion Sensitivity and Spectral Tuning

In addition, most vertebrate M/LWS photopigments are thought to be spectrally tuned in part by the binding of anions, more specifically chloride ions, at locations on the opsin close to the Schiff's base linkage. For example, chicken M/LWS photopigment has a native λ_{max} value of 565 nm that can shift to 520 nm in the absence of chloride ions. It is thought that the Schiff's base counterion is complex, and the inclusion of an exchangeable chloride ion helps to maintain a relatively delocalized distribution of charge on the chromophore and, consequently, red shifts the λ_{max} of the photopigment [9].

Function

Photopigments absorb photons of light and trigger the enzymatic activity of the photoreceptor G-protein transducin. This leads to a biochemical cascade that ultimately results in a change in the rate of release of ►neurotransmitter by the photoreceptor and a detectable neural signal. Although the primary function of a photopigment is to detect light of any wavelength, the divergence of the ancestral photopigment gene into multiple spectral types very early in evolution suggests that there is considerable selection pressure on the spectral tuning of photopigments. Although generalisations are difficult to make, the number of different photopigment types and their λ_{max} values are almost certainly determined by the spectral distribution of the available light, the need to find food and potential mates, and avoid predators.

Pathology

►Retinitis pigmentosa (RP) (►Inherited retinal degenerations) is a collection of genetically inherited diseases that result in the degeneration of photoreceptors and the retinal pigment epithelium. Symptoms include night blindness and the loss of peripheral vision. Over 100 different point (single amino acid) mutations in the gene encoding rod opsin, located on autosomal chromosome 8, are known to cause RP, possibly as a result of the failure of the mutant opsin to fold correctly, bind retinal or activate transducin [6].

The genes coding for green- and red-sensitive cone pigments (M/LWS opsins) in humans are carried on the X chromosome. Mutations in these genes are responsible for the different forms of red-green ►color blindness and, because of their location on the X chromosome, are more apparent in males, which have only one copy of the gene.

These include amino acid substitutions that alter the λ_{max} of the green- (deuteranomaly) or red-sensitive (protanomaly) photopigments and result in anomalous color vision, or cause the complete loss of green- (deuteranopia) or red-sensitive (protanopia) cones [10].

The SWS1 gene encoding the human blue-sensitive cone opsin is located on chromosome 7. Mutations in the SWS1 gene that cause a spectral shift in the expressed pigment (tritanomaly) or a loss of functional blue cones (tritanopia) result in blue-yellow color blindness. As the SWS1 gene is carried on a pair of autosomal chromosomes, mutations in both copies are required to produce tritanopia. Consequently, blue-yellow color blindness occurs less frequently than red-green color blindness.

References

1. Stenkamp RE, Filipek S, Driessen CA, Teller DC, Palczewski K (2002) Crystal structure of rhodopsin: a template for cone visual pigments and other G protein-coupled receptors. Biochimica Biophys Acta 1565:168–182
2. Sakmar TP, Menon ST, Marin EP, Awad ES (2002) Rhodopsin: insights from recent structural studies. Annu Rev Biophys Biomol Struct 31:443–484
3. Yokoyama S (2000) Molecular evolution of vertebrate visual pigments. Prog Retin Eye Res 19:385–419
4. Collin SP, Knight MA, Davies WL, Potter IC, Hunt DM, Trezise AE (2003) Ancient colour vision: multiple opsin genes in the ancestral vertebrates. Curr Biol 13: R864–R865
5. Rodieck RW (1973) The vertebrate retina. WH Freeman and Company, San Francisco
6. Pepe IM (2001) Recent advances in our understanding of rhodopsin and phototransduction. Prog Retin Eye Res 20:733–759
7. Nakanishi K (1991) 11-cis-retinal, a molecule uniquely suited for vision. Pure Appl Chem 63:161–170
8. Hunt DM, Wilkie SE, Bowmaker JK, Poopalasundaram S (2001) Vision in the ultraviolet. Cell Mol Life Sci 58:1583–1598
9. Kleinschmidt J, Hárosi FI (1992) Anion sensitivity and spectral tuning of cone visual pigments *in situ*. Proc Natl Acad Sci USA 89:9181–9185
10. Rodieck RW (1998) The first steps in seeing. Sinauer Associates, Sunderland, MA

Photoreception

Definition

The sensory process by which light energy is converted into a biologically relevant signal.

Photoreceptor, Variety and Occurence

Megumi Hatori, Satchidananda Panda
The Salk Institute for Biological Studies, La Jolla, CA, USA

Definition

Photoreceptors regulate light-dependent physiologies; image formation and non-image forming adaptation such as regulation of circadian entrainment, seasonal reproduction and body color change. The vertebrate retina contains three types of photoreceptors: rods, cones and ipRGCs. In non-mammalian vertebrates, the pineal complex, deep brain, skin, parapineal and parietal eye are also known to contain photoreceptors. Each vertebrate photoreceptor contains a photopigment consisting of a protein called opsin and vitamin-A-based light-absorbing molecule (chromophore), 11-*cis* retinal.

Characteristics

Introduction

Dynamic adaptation to ambient light is central to survival of most animals. Light adaptation is achieved by two basic mechanisms: image formation for rapid adaptation to the physical environment and non-image forming adaptation of physiology and behavior to ambient light quality. The image-forming function is exclusively mediated by the eye – a specialized optical structure with a lens that projects an image to the ►retina. ►Cone and ►rod photoreceptors of the retina capture the image and send the information to the visual cortex for image reconstruction [1]. The non-image forming (NIF) photoresponses vary widely, from rapid adjustment of pupil diameter to progressively slow responses, such as skin color adaptation in amphibians, adaptation of the circadian clock to the daily day–night cycle, and seasonal reproductive behavior. Accordingly, the underlying ►photopigments vary widely in (i) the cell types of expression, (ii) anatomical location and (iii) target cells or effecter process.

The photopigment for most of the above responses comprises an opsin family of G-protein coupled receptor and a light-sensitive vitamin-A based chromophore. The amino acid sequence of the opsin scaffold determines (i) the spectral properties, (ii) specificity of the downstream signaling pathway and (iii) interaction with other regulatory molecules. Although flavin-based photopigments, such as ►cryptochromes, have been implicated in the circadian photoresponses of insects, no such light-dependent function of cryptochrome molecules in vertebrates has been conclusively established.

In mammalian vertebrates all of the photoresponses originate from photopigments of the retina. In non-mammalian vertebrates, however, additional anatomical

P

structures such as the parietal eye, ►pineal-, parapineal-complex, deep brain, and skin are also known to contain functional photoreceptors which primarily mediate NIF photoresponses.

Ocular Photoreceptors in the Vertebrates

The rod and cone cells are the predominant photoreceptors of the retina. They exhibit exquisite subcellular specialization such that the different cellular functions are stratified along the length of the cells. The nuclei mark a virtual functional boundary, with the distal segment functioning in photoreception, while the segment proximal to the lens makes synaptic connections to other cell types of the retina. The photopigment in these cells is tightly packed into specialized membranes that are organized into rod or cone like structures in the outer segment of rod or cone cells, respectively (Fig. 1).

The outer segment is connected to the rest of the cells by a narrow structure, the ciliary stalk. Between the ciliary stalk and the nuclei, various organelles such as mitochondria, the Golgi apparatus and the endoplasmic reticulum are concentrated in stratified layers. The inner segment of the rod/cone cells contain the synaptic structures for signal transduction to other cell types of the retina, which ultimately connect to the ►retinal ganglion cells (RGCs) – the only retinal cell type that makes synaptic connections to the brain. No other cell types of the mammalian vertebrate retina were known to be directly light sensitive until the discovery of ►intrinsically photosensitive RGCs (see section ►*ipRGCs*).

In non-mammalian vertebrates, however, opsin photopigments are expressed in additional retina cell types. The horizontal and amacrine cells in the retina of teleost fish express vertebrate ancient (VA)-opsin. In birds and amphibians melanopsin mRNA is extensively expressed in both inner and outer nuclear layers of the retina. However, in all animals the rod and cone cells are the most characterized photoreceptors of the retina.

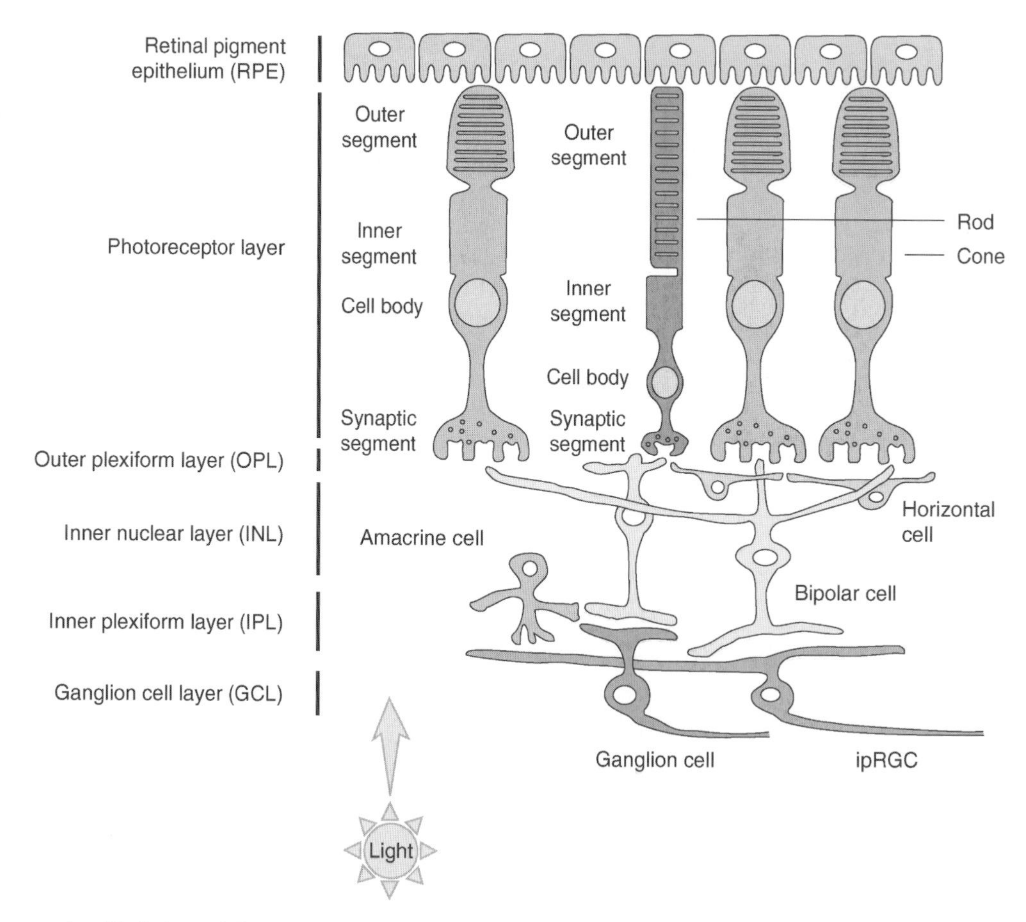

Photoreceptor, Variety and Occurence. Figure 1 Schematic structure of the vertebrate retina. Vertebrate retina is composed of multiple cell layers with specialized functions. The rods and cones of the outer nuclear layer and a few RGCs of the ganglion cell layers are bona fide photoreceptors or intrinsically photosensitive. Other cell types such as horizontal, bipolar, amacrine and ganglion cells participate in light signal processing and signal transduction to the brain.

Rods

Typically, a rod cell is sensitive enough to respond to a single photon of light and therefore is responsible for scotopic or dim-light vision in night (Table 1) [1,2]. With increasing light intensity at dawn, the rods become saturated or "bleached", thus under daylight the less sensitive cones mediate visual function. Rods respond slowly to light and have longer integration time than cones do.

Generally, each rod or cone cell expresses only one type of opsin photopigment, thus the photoresponses of the cell reflect that of the photopigment. Most vertebrate retina contain a single type of rod expressing rhodopsin with peak sensitivity ~500 nm (Fig. 2), although several amphibians like toads and salamanders are known to have two types of rods, called the red-rods and the green-rods, which have distinct absorbance spectra.

The photopigment in rod cells is densely packed into specialized disc membranes that are stacked to form the rod outer segment. The outer segment of both rods and cones are continuously renewed. In rods, new rhodopsin molecules are synthesized just before dawn and the distal discs are shed and phagocytosed by the retinal pigment epithelium (RPE) cells in the morning.

Photoreceptor, Variety and Occurence. Table 1 Comparison of rods, cones and ipRGCs. Modified from [3]

	Rods	Cones	ipRGCs
Major function	Scotopic/night vision	Photopic/daytime vision	Non-image forming photoresponses
Location	Outer nuclear layer	Outer nuclear layer	Ganglion cell layer
Photopigment	Rod opsin (rhodopsin)	Cone opsins	Melanopsin
Localization of opsins	Outer segment	Outer segment	Soma, dendrites, and axon.
Sensitivity	High	Moderate (less sensitive than rods)	Low
Integration time	Longer integration time than cones (tens of milliseconds)	Respond faster than rods	Seconds
Light response of membrane potential	Fast hyperpolarizing	Fast hyperpolarizing	Slow depolarizing
Action potentials	No	No	Yes
Role of RPE for photopigment regeneration	Essential	Essential	Apparently unnecessary
Receptive field	Very small	Very small	Very large (tens of microns)
Number of cells in human eye	~ 120 million	~ 6 million	<5,000
Spatial distribution in human eye	Peripheral	Fovea	Uniform

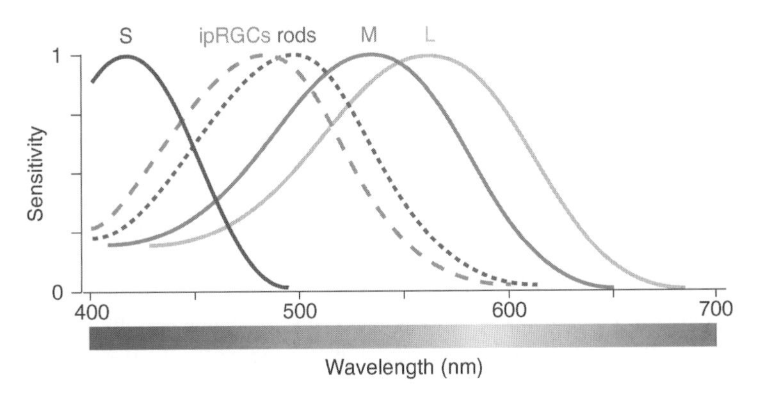

Photoreceptor, Variety and Occurence. Figure 2 Spectral sensitivity of mammalian ocular photoreceptors. Spectral sensitivities of cones (S-cones, M-cones and L-cones), rods and ipRGCs.

P

Such temporal regeneration of rods coincides with their primary role in scotopic vision.

Cones

Cones are cone-shaped photoreceptors in the retina that can function under bright light and are responsible for photopic or bright-light vision (Table 1) [1,2]. Cones are less sensitive to light intensity than the rods. On the other hand, response times to stimuli are faster than those of rods, and they are also able to perceive color, finer detail and more rapid changes in images. Unlike the disc membranes of the rod, the outer segment of the cones is formed by invagination of the plasma membranes. The photopigments of the cones are primarily synthesized prior to dusk, and membranes from the tips of the cones are shed in the evening. Based on peak spectral sensitivity, most animals contain two types of cones, one at >500 nm (long wavelength cone or L-cone) and one at <500 nm (short wavelength cone or S-Cone). Some animals including teleost fish, Old World monkeys and humans have one additional cone-type absorbing in the medium wave (M-cone). There are several exceptions; for example, several nocturnal mammals like owl monkeys, a New World monkey, have only one type of cone. Sensitivities of human S-cones, M-cones and L-cones are to ~420 nm (blue), ~530 nm (green) and ~560 nm (yellowish–green) wavelengths, respectively (Fig. 2). S-cones are smaller in number than M- and L-cones and are sparse in the fovea region. The difference in the signals received from the three cone types allows the brain to distinguish millions of colors. As cones are functional under bright light, they require rapid regeneration of the *all-trans* retinal. There is some evidence that unlike the rods which depend on the RPE cells, the cones depend on a different type of retinal cycle involving cones and the Müller cells of the retina.

ipRGCs (Intrinsically Photosensitive Retinal Ganglion Cells)

The ipRGCs are intrinsically photosensitive and constitute only 1–2% of the retinal ganglion cells of the adult mammalian retina. They do not play any major role in image-forming vision but are important photoreceptors for NIF photoresponses, which include light entrainment of the master circadian oscillator resident in the hypothalamic ►suprachiasmatic nucleus (SCN), light suppression of pineal melatonin synthesis and release, light modulation of activity and pupillary light reflex (Table 1) [3,4].

These adaptive photoresponses are almost intact in several animal models with complete rod/cone degeneration and exhibit an action spectrum characteristic of the absorption spectrum of the opsin class of photopigments with a peak around 480 nm. ►Melanopsin, a novel photopigment expressed in these cells, mediates photosensitivity (Fig. 2). Unlike the rod/cone photoreceptors, ipRGCs exhibit no polarized expression of the photopigment; melanopsin immunoreactivity is localized to membranes of the dendrites, somas and possibly axons of these cells. Cell bodies of the ipRGCs are mostly resident in the RGC sublayer, while the dendrites heavily arborize within the inner plexiform layer spreading to as many as tens-of-microns in diameter of receptive fields (Fig. 1). The dendrites make synaptic contacts with both rod and cone bipolar cells and receive rod/cone inputs, and hence are unique photoreceptor cells of the retina with two almost independent major functions: (i) to initiate photochemical reaction by melanopsin as well as (ii) to transmit rod/cone initiated photoresponse. Unlike other RGCs, axons of these ipRGCs primarily project to the SCN, where they are almost equally distributed between contra- and ipsi-lateral SCN. They also send additional projections to other brain regions implicated in NIF photoresponses.

Isolated ipRGCs depolarize in response to light and generate action potentials in a melanopsin-dependent manner. They exhibit long latency of light response and are depolarized as long as the lights are on and have long deactivation kinetics. The native molecules and channels underlying the photoresponses of the ipRGCs have not conclusively been established. However, the persistence of the non-image forming adaptive photoresponses in animal models deficient in several signaling components and channels mediating rod/cone photoresponses imply the ipRGCs recruit a distinct set of signaling molecules.

Phototransduction Mechanism

In all opsin-based photopigments, the first step in photoresponse begins with the photoisomerization of a vitamin-A based *cis*-isomer of retinal to its *trans*-isomer. The resultant conformational change of the protein triggers activation and release of a Gαi/Gαo (most vertebrate opsins) or Gαq/Gα11 (melanopsin) class of G-protein. The Gαi/Gαo class of effector G-protein of rod/cones, also known as transducin (Gt), activates a phosphodiesterase, which in turn rapidly degrades cGMP and causes closing of the cyclic nucleotide gated channels. Hence, photoactivation of rod/cone photoreceptors leads to membrane hyperpolarization and a pause in neurotransmitter glutamate release. On the other hand, it is presumed that photoactivated melanopsin activates the Gαq/Gα11 class of G-proteins, which in turn activates phospholipase C (PLC). Activated PLC can signal through several mechanisms, including intracellular calcium release, increase in IP3 and DAG, and ultimately opening of a channel leading to membrane depolarization, and release of neurotransmitters, such as glutamate and adenylate cyclase- activating peptide 1. Signal termination occurs at various steps including receptor, G-protein and downstream components. Finally, regeneration of active photopigment occurs, enabling another round of photoresponses to be initiated. The all-*trans* retinal from rod/cone cells is transported to the RPE cells where an elaborate multistep enzymatic process ensures sufficient retinal regeneration for use by the rod/cone photoreceptors. Melanopsin, on the other hand is presumed to isomerize the resultant all-*trans* retinal to the active 11-*cis* isomer by an intrinsic photoisomerase activity.

Extraocular Photoreceptors in Nonmammalian Vertebrates

While mammals use ocular photopigments for all types of photoresponses, several non-mammalian vertebrates express functional photopigments outside their eyes, including the parietal eyes, pineal, parapineal glands, dermal cells, and in brain. Organisms use them primarily for physiological adaptation to the ambient light quality. Some of these tissues such as pineal and parietal eyes develop embryologically from the diencephalons, which also gives rise to the eye. Accordingly the photoreceptor cells of these organs show some morphological similarities with those of the retina, and they also express multiple photopigments in distinct or in the same cell types.

Pineal Photoreceptor

In lower vertebrates such as lampreys, fishes, amphibians, lizards and birds, the pinealocytes have both intrinsic photosensitivity as well as neuroendocrine function, such that under cultured conditions of isolated pineal gland, they are light-sensitive and have the ability to produce ►melatonin in a circadian and light-dependent manners.

The action spectrum of the photosensitivity of the isolated chicken pineal resembles the absorption spectrum of rhodopsin. Consistent with this prediction, an opsin-like protein was identified in the chicken pineal gland and named pinopsin (=pineal opsin), the first functional opsin to be discovered outside the retina [5]. Pinopsin upon reconstitution with 11-*cis* retinal forms a functional photopigment with peak sensitivity at 468 nm and can couple to Gt. Besides pinopsin, a chicken red-sensitive cone pigment, called iodopsin, and melanopsin are also expressed in the chicken pineal gland. Pinopsin is not detected in fish and mammals. Alternatively, the pineal gland of zebrafish expresses exo-rhodopsin (=extra-ocular ►rhodopsin). Exo-rhodopsin is expressed in the majority of pineal cells, but not in retinal cells. The pineal photopigments might function to modulate melatonin secretion in a cell autonomous manner. Additionally, they also make synaptic connections with neurons that project to various thalamic and hypothalamic regions. Mammals have lost the intrinsic photosensitivity of pinealocytes, but they still retain circadian regulation of melatonin secretion, which is also light regulated by ocular photopigments via a multisynaptic pathway.

Parietal Eye Photoreceptors

Several lizards harbor an extra eye-like structure called the parietal eye that does not appear to have any role in image-forming vision, but rather helps adapt to light quality. Like the eye, the parietal eyes also contain various cell types stratified into distinct layers. However, unlike the image-forming photoreceptors, which hyperpolarize in response to light, the parietal eye photoreceptors either depolarize or hyperpolarize in response to specific wavelengths of light. These photoreceptor cells express two different types of opsins – pinopsin and parietopsin in the same cell type [6]. Distinct absorption spectra of these two opsins and functional coupling to distinct signaling mechanisms may produce such unique membrane potential properties of the parietal eye photoreceptor cells.

Photoreceptors in the Deep Brain

Most photoreceptor cells described above are located outside or immediately beneath the skull. However, in some vertebrates there is evidence for photoreceptors in the deep brain which largely mediate long term seasonal photo-adaptations. The photoperiodic change in gonad function persists in pinealectomized and enucleated Japanese quails and exhibits peak sensitivity around 500 nm. Actually, ~500 nm wave-length light can transmit through the scalp and skull and reach to the deep brain.

Several opsins of the eye and pineal glands are shown to be expressed in deep brain, but their functions are still almost unknown. Rhodopsin was cloned from lateral septum of pigeons, pinopsin-expressing cells exist in the anterior preoptic nucleus of toads, VAL (vertebrate ancient-long) opsin is localized in a small portion of cells surrounding the diencephalic ventricle of central thalamus in zebrafish, and melanopsin is expressed in hypothalamic sites of *Xenopus laevis*.

References

1. Rodieck RW (1998) The first steps in seeing. Sinauer Associates, Sunderland, MA
2. Solomon SG, Lennie P (2007) The machinery of color vision. Nat Rev Neurosci 8:276–286
3. Berson DM (2003) Strange vision: ganglion cells as circadian photoreceptors. Trends Neurosci 26:314–320
4. Nayak SK, Jegla T, Panda S (2007) Role of a novel photopigment, melanopsin, in behavioral adaptation to light. Cell Mol Life Sci 64:144–154
5. Okano T, Yoshizawa T, Fukada Y (1994) Pinopsin is a chicken pineal photoreceptive molecule. Nature 372:94–97
6. Su CY, Luo DG, Terakita A, Shichida Y, Liao HW, Kazmi MA, Sakmar TP, Yau KW (2006) Parietal-eye phototransduction components and their potential evolutionary implications. Science 311:1617–1621

P

Phototransduction

NATHAN S. HART
School of Biomedical Sciences, University of Queensland, St. Lucia, Brisbane, QLD, Australia

Synonyms

Visual transduction cascade

Definition

Phototransduction is the process by which light energy that is absorbed by ►photopigments contained within ►retinal photoreceptors (►Photoreceptors) is converted into a biochemical signal that leads to a ►hyperpolarization of the photoreceptors.

Characteristics

Description of the Process

Photoisomerization

Phototransduction begins when a photon of light is absorbed by a photopigment molecule. The energy imparted by the photon induces a conformational change in the ►chromophore (see ►Photopigments), in the case of 11-*cis* retinal causing it to flip to the all-*trans* isomer. This transition is called ►photoisomerization, and is the only light-dependent step in the phototransduction process; it is also extremely rapid, taking less than 200 fs [1]. Photoreceptor ►outer segments are densely packed with photopigment molecules to capture as much of the incident light as possible. Nevertheless, only two out of every three photons absorbed by the photopigment succeed in isomerising the chromophore; photopigments have a ►quantum efficiency of about 0.67, regardless of photon wavelength [2]. Energy from photons that are absorbed but do not induce chromophore isomerization is dissipated as heat.

The probability that a photon of light will be absorbed by a photopigment of given spectral sensitivity is dependent on the photon's wavelength. However, an individual photoreceptor is incapable of discriminating between photoisomerization events caused by photons of different wavelength and can only signal the rate of photon capture, a concept known as the "principle of univariance." Wavelength discrimination requires photoreceptors with differing photopigment spectral sensitivities and the ancillary neural mechanisms to compare their output signals, i.e. color vision (►Color processing).

Photopigment Activation

The change in chromophore confirmation during photoisomerization alters the local structure of the chromophore-binding pocket of the photopigment apoprotein (►opsin). These structural rearrangements propagate throughout the tertiary structure of the photopigment, which progresses step-wise through a series of physically and spectroscopically distinct photobleaching intermediates until, through deprotonation of the Schiff's base linkage that binds the chromophore to the opsin, the biochemically active form of the photopigment (R*) is created [3]. In the case of ►rhodopsin (11-*cis* retinal based) photopigments, the activated form of the molecule is called metarhodopsin II and appears approximately 2 ms after photoisomerization [4].

Transducin Activation

The next stage in the phototransduction process occurs when R* triggers a biochemical cascade that amplifies the visual signal to a detectable threshold in a stereotyped fashion (Fig. 1). As each R* is free to diffuse rapidly within the plane of the outer segment disk membrane, and does so randomly as a result of ►Brownian motion, it encounters and activates several hundred ►transducin molecules within about 100 ms [5]. Transducin is a membrane-bound heterotrimeric G protein that consists of three subunits, T_α, T_β and T_γ. Rods and cones contain different isoforms of the three subunits, but the proteins perform the same function in the cascade [6].

Amino acids on the cytoplasmic loops of R* bind briefly (<0.1 ms) to ►epitopes on the transducin complex, and induce a conformational change in the structure of the T_α subunit that results in the release of a molecule of bound guanosine diphosphate (GDP). When the GDP released from the T_α nucleotide-binding pocket is subsequently replaced with a molecule of guanosine triphosphate (GTP), a further conformational change activates the transducin complex, causing both the detachment of R* and the dissociation of T_α-GTP from the heterotrimer [1]. At this stage, R* is then free to move away and activate other transducin molecules before it is eventually deactivated through phosphorylation and binding to ►arrestin (see below).

Phosphodiesterase Activation

Activated GTP-bound transducin T_α subunits have a high affinity for guanosine 3′, 5′ cyclic monophosphate (cGMP)-specific phosphodiesterase (PDE). PDE is a membrane-bound protein consisting of two catalytic subunits (PDE_α and PDE_β) and two identical inhibitory subunits (PDE_γ). In the dark-adapted state, the catalytic activity of the PDE_α and PDE_β subunits is blocked by their respective PDE_γ subunits [7]. Following illumination and activation of transducin, T_α-GTP binds to PDE and displaces either or both of the PDE_γ subunits, thereby exposing its catalytic sites.

In this second stage of signal amplification, activated PDE_α and PDE_β subunits catalyze the conversion of

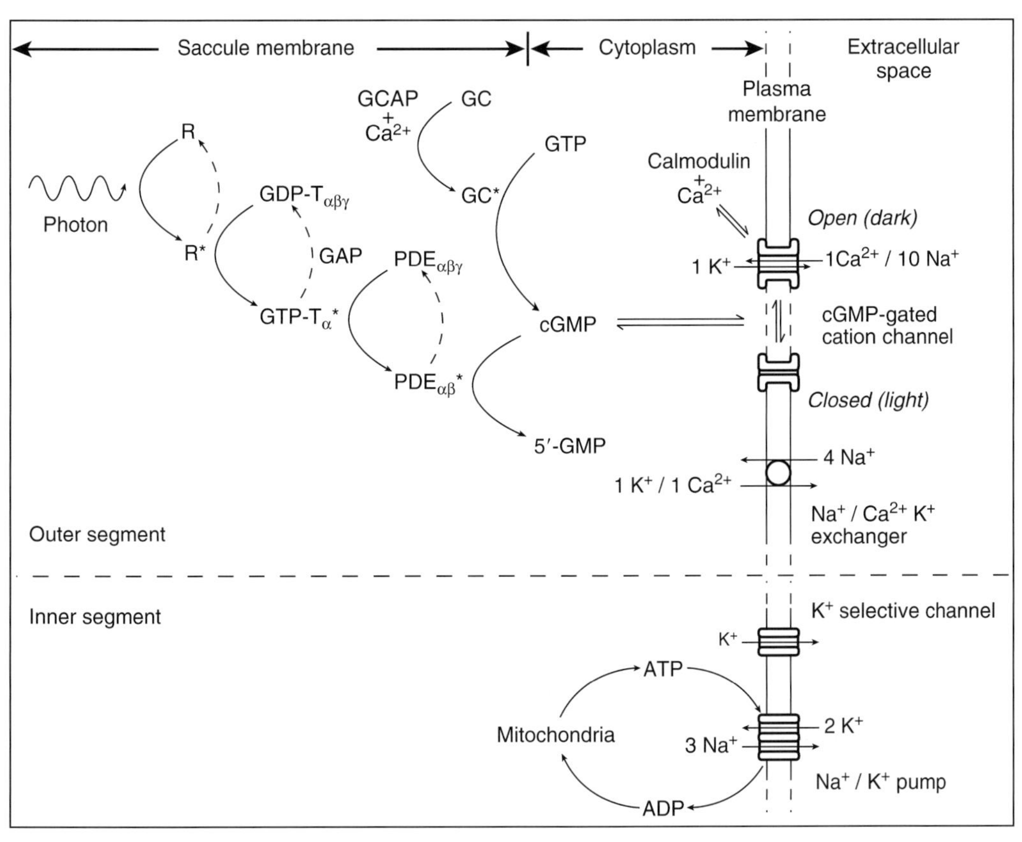

Phototransduction. Figure 1 Simplified schematic diagram of the phototransduction cascade in a mammalian rod. Relevant membrane channels are also depicted. Abbreviations: *R*, rhodopsin; *GDP*, guanosine diphosphate; *GTP*, guanosine triphosphate; *T*, transducin; *PDE*, phosphodiesterase; *cGMP*, cyclic guanosine monophosphate; *GMP*, guanosine monophosphate; *GAP*, GTP-ase activating proteins; *GC*, guanylate cyclase, *GCAP*, guanylate cyclase activating proteins; *ATP*, adenosine triphosphate; *ADP*, adenosine diphosphate. An asterisk signifies the activated form of the enzyme. Subscripts indicate the enzymatic subunit(s) present. Dashed arrows represent regeneration steps.

many molecules of cytoplasmic cGMP to 5′ guanosine monophosphate (GMP) and rapidly reduce the concentration of cGMP molecules present in the outer segment. CGMP is a ►second messenger in the phototransduction process, and the fall in its concentration is detected by cGMP-gated cation channels in the outer segment plasma membrane, which close. The closure of ionic channels reduces the influx of Na^+ and Ca^{2+} ions and, because K^+ ions continue to be removed from the cell via K^+-selective channels in the ►inner segment, causes a local hyperpolarization of the photoreceptor's transmembrane potential [4].

Photoreceptor Dark Current, Photocurrents and Hyperpolarization

Phototransduction works by changing the electrical potential across the photoreceptor plasma membrane. Like all living cells, the interior of a photoreceptor is charged negatively with respect to the extracellular space (►Membrane potential – basics); the resting transmembrane potential of a rod is about −37 mV, whereas that of a cone is around −46 mV [8]. In the dark, this transmembrane potential is maintained by a balanced flow of cations (predominantly Na^+) into and out of the photoreceptor, known as the dark current (►Photoreceptor dark current), which has a magnitude of up to −34 pA in rods and −30 pA in cones. In the outer segment, voltage-insensitive cGMP-gated ionic channels allow the entry of ten Na^+ and one Ca^{2+} for every K^+ they expel, and $Na^+/Ca^{2+}K^+$ exchangers (►Ion transporters) expel one Ca^{2+} and one K^+ while allowing four Na^+ to enter. The influx of 14 Na^+ to the outer segment is balanced by the activity of ATP-driven Na^+/K^+ pumps (Ion transporters) in the inner segment, which admit 2 K^+ for every 3 Na^+ they expel. Several cycles of the Na^+/K^+ pump are required, and the additional K^+ that enter the cell reciprocally pass out of the cell down their concentration gradient via K^+-selective channels [1,4].

Each cGMP-gated cation channel consists of four subunits, and each subunit is capable of binding one molecule of cGMP at an intracellular domain close to the C terminus of the polypeptide chain. The channel pore only allows the influx of cations when at least three molecules of cGMP are bound and, to a first approximation, the number of open channels varies as the cube of cytoplasmic cGMP concentration [4]. When the concentration of cGMP falls as a result of light-induced PDE catalytic activity, the number of open channels falls dramatically and the influx of cations to the outer segment is restricted. However, because Na^+ and K^+ ions are still pumped out of the inner segment there is a net loss of positive charge within the photoreceptor. The change in cation flow induced by the absorption of photons is called the ►photocurrent. In the case of rod photoreceptors, a single photoisomerization event can hyperpolarize the transmembrane potential with a photovoltage of 1.2 mV [8].

Synaptic Deactivation

Initially, the transmembrane hyperpolarization induced by cGMP-gated cation channel closure is localized to the cytoplasm, near the activated segment of disk membrane (rods) or plasma membrane infolding (cones). However, this charge displacement is gradually redistributed along the entire surface area of the plasma membrane, and the hyperpolarization spreads electronically (►Electrotonic spread) to the inner segment and ►synaptic terminal.

The effect of hyperpolarization on the synaptic terminal is to close inward voltage-gated Ca^{2+} channels (►Calcium channels – an overview) in the plasma membrane. However, free Ca^{2+} ions continue to be extruded from the cell by $Na^+/Ca^{2+}K^+$ exchangers, and so the concentration of Ca^{2+} in the synaptic terminal falls [1]. As the rate of ►synaptic vesicle fusion with the presynaptic membrane is proportional to the concentration of cytoplasmic Ca^{2+}, the hyperpolarization-induced closure of voltage-gated Ca^{2+} channels results in a graded reduction in the rate of ►glutamate release into the synapse.

Regulation of the Process

Deactivation of R and the Phototransduction Cascade*

During its random diffusion across the membrane surface, R* inevitably encounters and binds to a molecule of ►rhodopsin kinase, which rapidly phosphorylates serine and threonine residues on the photopigment C-terminus [6]. Phosphorylation of R* dramatically increases its affinity for another cytoplasmic protein, arrestin, which replaces the kinase and deactivates R*. Arrestin remains bound to the deactivated photopigment until the chromophore is reduced to all-*trans* retinol by retinal dehydrogenase, and dissociates from the opsin protein, approximately 1 s after photoisomerization.

The other components of the biochemical cascade must also be deactivated to ensure a reliable stereotyped response to each photoisomerization event and allow recovery of the photoreceptor. Once T_α-GTP has bound to and activated PDE, the complex is recognized by other membrane proteins, known as ►GTPase activating proteins (GAPs). Assisted by the inhibitory γ-subunit of PDE, GAP stimulates the intrinsic GTPase activity of T_α, which hydrolyses the bound nucleotide to GDP [9]. Phosphorylated T_α-GDP subsequently detaches from the PDE_γ subunit, thereby inactivating the catalytic activity of PDE, and reassociates with the $T_{\beta,\gamma}$ dimer.

Recovery and Light Adaptation

The fall in concentration of cytoplasmic Ca^{2+} associated with cGMP-gated cation channel closure and transmembrane hyperpolarization provides a mechanism for negative feedback, which controls both the recovery of the photoreceptor and its response to increasing levels of illumination. Decreasing Ca^{2+} levels stimulate, indirectly, the activity of a guanylate cyclase (GC) enzyme that synthesizes cGMP from GTP. GC is activated by guanylate cyclase activating proteins, or GCAPs, that are also Ca^{2+}-binding proteins [6]. The GCAPs in question are unable to activate GC when Ca^{2+} ions occupy two of their specific Ca^{2+}-binding sites. However, when the Ca^{2+} concentration drops, some of the GCAPs are disinhibited and stimulate GC to increase the rate of conversion of GTP to cGMP by a factor of 5–10 [4]. The subsequent increase in cGMP production counteracts the decrease caused by PDE activity and allows some of the cGMP-gated cation channels to reopen, thereby helping to restore the dark current.

Additionally, Ca^{2+} levels modulate the behavior of the cGMP-gated cation channels through the actions of another Ca^{2+}-binding protein, ►calmodulin. Calmodulin is a small (17 kD) protein that becomes activated when at least three of its four Ca^{2+}-binding sites are occupied by Ca^{2+} ions. The tetrameric cGMP-gated channel protein consists of two types of subunits, α and β. The β-subunit is larger (155 kD) than the α-subunit (80 kD), and has a calmodulin binding domain on an intracellular portion of the polypeptide chain near the N-terminus [4]. Activated calmodulin can bind to the β-subunit and reduce the affinity of the channel proteins for cGMP, leading to fewer open channels. However, when the cytoplasmic Ca^{2+} concentration is reduced during phototransduction, the subsequent decrease in activated calmodulin levels results in an increased affinity of the channel proteins for cGMP [5]. This leads to a greater number of bound cGMP molecules and the opening of more cation channels.

Both of these Ca^{2+}-modulated negative feedback mechanisms subserve recovery, and potentially ►light adaptation, by preventing saturation of the light response, and allow the photoreceptors to respond to incremental changes in light intensity over a wide range of background illumination levels [10]. The importance of these feedback mechanisms is evident if one considers the responses of photoreceptors to dim flashes of light. At very low light levels, the magnitude of the photocurrent scales linearly with light intensity. If the dark current of a rod is −34 pA and each photoisomerization event causes a photocurrent of 0.7 pA, the rod response would saturate with a flash intensity that caused less than 50 photoisomerizations. However, at higher flash intensities, typically those generating photocurrents of more than one third of the maximal dark current, the increase in photocurrent with increasing light intensity becomes increasingly smaller in an exponential fashion. In this way, rods are able to signal incremental changes in brightness, up to flash intensities that cause around 400–500 photoisomerizations before saturating [4].

Function

Light is a visible form of electromagnetic radiation and, as such, exhibits both wave and particle properties. One property of the particle nature of light is that light energy travels in discrete packets (quanta) called photons. This has two important consequences for vision. Firstly, the visual system must detect quantal events that involve the transfer of miniscule amounts of energy. Secondly, the arrival of photons at the retina is a stochastic Poisson process. Photoreceptor neurons are exquisitely sensitive and can respond to the absorption of a single photon by a photopigment molecule. Phototransduction is the process by which this singular event is sufficiently amplified to create an electrical signal that can reliably be detected by the rest of the visual system.

Pathology

A number of inherited diseases affecting the mammalian retina are caused by mutations in genes that encode components of the phototransduction cascade [1]. ►Retinitis pigmentosa (►Inherited retinal degenerations) – a general term for a number of inherited conditions that cause degeneration of the photoreceptors and retinal pigmented epithelium – can result from mutations in both guanylate cyclase and PDE. In addition, various forms of ►night blindness are caused by mutations in either the α-subunit of transducin, which reduce the endogenous GTPase activity that mediates PDE inactivation (autosomal dominant night blindness), or in arrestin or rhodopsin kinase (►Oguchi disease).

►Evolution of Eyes

References

1. Pepe IM (2001) Recent advances in our understanding of rhodopsin and phototransduction. Prog Retin Eye Res 20:733–759
2. Rodieck RW (1973) The vertebrate retina. WH Freeman, San Francisco
3. Sakmar TP, Menon ST, Marin EP, Awad ES (2002) Rhodopsin: insights from recent structural studies. Annu Rev Biophys Biomol Struct 31:443–484
4. Rodieck RW (1998) The first steps in seeing. Sinauer Associates, Sunderland, MA
5. McNaughton PA (1995) Rods, cones and calcium. Cell Calcium 18:275–284
6. Ebrey T, Koutalos Y (2001) Vertebrate photoreceptors. Prog Retin Eye Res 20:49–94
7. Granovsky AE, Natochin M, Artemyev NO (1997) The gamma subunit of rod cGMP-phosphodiesterase blocks the enzyme catalytic site. J Biol Chem 272:11686–11689
8. Schneeweis DM, Schnapf JL (1995) Photovoltage of rods and cones in the macaque retina. Science 268:1053–1056
9. He W, Cowan CW, Wensel TG (1998) RGS9, a GTPase accelerator for phototransduction. Neuron 20:95–102
10. Rebrik TI, Korenbrot JI (2004) In intact mammalian photoreceptors, Ca^{2+}-dependent modulation of cGMP-gated ion channels is detectable in cones but not in rods. J Gen Physiol 123:63–75

Phrenic Nerve

Definition

Nerve innervating the diaphragm originating from C3 to C5 cervical cord.

►Hering-Breuer Reflex

Phrenology

Definition

Phrenology is an outdated field of study that proposed that mental functions could be related to the position of protuberances on the skull.

Phyletic Method

Definition

Uses cladistic methodology (cladograms, outgroup comparison) for establishing evolutionary polarity (i.e. ancestrality vs. derivedness) of characters.

►Evolution of the Brain: In Fishes
►Evolution of the Telencephalon: In Anamniotes

P

Phylogenetic

Definition

Relating to or based on evolutionary development or history.

Phylogenetic Scale

►Evolution and the Scala Naturae

Phylogenomic Studies

Definition

Comparisons of whole genomes, or of large numbers of genes within genomes, to construct relationships among organisms.

Phylogeny

►Evolution and Phylogeny: Chordates
►Evolution and Phylogeny of Vertebrates

Phylogeny and Evolution of Chordates

►Evolution and Phylogeny: Chordates

Physical Exercise

►Stress Effects During Intense Training on Cellular Immunity, Hormones and Respiratory Infections

Physicalism

Barbara Montero
Graduate Center, City University of New York, New York, NY, USA

The Basic Formulation

The predominant philosophical theory about the world and our place in it is physicalism, the view, simply put, that everything is physical. In many circles, physicalism is not so much taken as the subject of debate but is rather assumed as the starting point around which other debates evolve. For example, a central problem in philosophy of mind is how to understand the mental given the truth of physicalism. But what, exactly, is the theory of physicalism? The simple formulation of physicalism as the view that everything is physical admits a number of interpretations, indeed, each term – "everything," "is," and "physical" – can be understood in different ways. To understand physicalism, then, let us look at these three aspects of the physicalist doctrine.

Scope

For some, physicalism is a theory about everything whatsoever, where "everything" is interpreted in its broadest sense to include, for example, concrete objects such as rocks and trees, abstract objects such as numbers and sets, properties such as the property of being conscious, events such as the event of my thinking about philosophy, even God, if She exists.[1]Others, however, think that arguments for physicalism have to be restricted to a certain scope? go only so far and restrict its scope accordingly. For example, some take physicalism as a theory only about the concrete world, that is, roughly about phenomena[2] in space or time.[3] Others take it to be a theory about the empirical world, that is, about the phenomena that we come to know via our senses, or to put it more carefully, about the phenomena, the knowledge of which is justified via our sense experience. Still others take it to

[1] Or at least this is one way to state it. As a matter of terminology some physicalists reserve the term "physical" for the fundamental entities and properties of physics. Physicalism, on their view, can be true even if not everything is physical (in the narrow sense specified above), as long as everything nonundamental is related to the physical (again, in the narrow sense) in the right way.

[2] I use "phenomena" in the broadest sense to include whatever exists.

[3] The idea that the concrete also includes anything that is either spatial or temporal is important since if it were to include only the spatio-emporal than a disembodied nonspatial soul would not count as a counterexample to physicalism.

be a theory about or the contingent world, or the causal world, or the contingent and/or causal world.

How one restricts the scope of physicalism depends on one's purposes. And since the main physicalist target is typically the mental, it is not unusual to simply focus on the question of whether the mental is physical, or in other words, focus on physicalism with respect to the mental. Indeed, it is not unusual to see the theory that the mental is physical simply referred to as "physicalism." While "physicalism" in this sense refers just to the theory that the mental is physical, a more encompassing type of physicalism is sometimes evoked to justify physicalism with respect to the mental. One finds this when physicalists argue, that the mental is very likely to be physical because everything else is physical. Here, again, one wants to know, "just what is the scope of everything else?"

In considering physicalism about the mental, one sometimes finds another sort of restriction. While typically, philosophers speak of physicalism with respect to the mental as the view that all mental properties, such as the property of feeling pain, or seeing red, are physical properties, occasionally physicalism is taken to be a theory only about things and not about properties. For example, some will claim that the view that human beings are physical things with nonphysical properties is consistent with physicalism. Most, however, would take this to be an anti-physicalist view; if physicalism is true, all properties must be physical.

The view that all properties are physical is taken to imply the view that all entities are physical. The idea being that if all of your properties are physical, then you are physical.

Let us say, then, that physicalism in the broad sense is the view that everything, or some substantial subset of everything, is physical; in the narrow sense it is the view that all mental properties are physical.

The Dependence Relation

When the physicalist claims everything is physical or that all mental properties are physical properties, what is being said of everything or of all mental properties? Typically something is taken to be physical if its existence depends in the right way on basic or fundamental physical properties, where the fundamental physical properties are taken to be the microphysical properties countenanced by physics, such as the property of having a charge, of being a quark, and so forth. But what exactly is the ►relation between the fundamental physical properties and the higher level properties, such as mental properties, that suffices to make the higher level properties count as physical. In other words, when we say that everything is physical, just what is meant by "is?"

Some hold that the relation between higher level properties and fundamental physical properties is that of explanation. On this view it is thought that physicalism is true only if everything is either a fundamental physical property or law, or it can be explained in terms of such properties and laws. As such, physicalism is an epistemic thesis that is, a thesis about what we know about the world rather than a thesis about what the world is really like. Still, it may have ontological implications since typically we think that a good indication of whether the fundamental nature of r is p is that we can explain r in terms of p.

Most philosophers, however, see physicalism primarily as an ontological thesis, a thesis that tells us about what the world is like. For this reason dependence relations are typically not formulated in terms of explanation, but rather in terms of supervenience, determination, realization, or constitution. Let us look at these relations.

Supervenience-physicalists hold that physicalism is true if everything either is a fundamental physical phenomenon or supervenes on fundamental physical phenomena. In terms of physicalism with respect to the mental, the view is simply that the mental supervenes on the fundamentally physical.[4] What is the relation of supervenience? There are actually many supervenience relations and a large literature discussing them. The basic idea, however, is this: to say that *A* properties supervene on *B* properties, means that there cannot be a change in A properties without a change in B properties. So, for example, mental properties are said to supervene on fundamental physical properties if and only if there cannot be a change in mental properties without a change in fundamental physical properties. If your mood changes from being happy to being sad, for example, supervenience implies that there must be a change in your fundamental physical properties as well. Sometimes, the idea is explained in terms of mind and brain: when you change from feeling happy to sad, supervenience physicalism says there must be a change in your neural structure. But since a neural change, must ultimately involve a change the fundamental level, the idea amounts to the same thing.

Is supervenience physicalism really physicalism? Some think not since it seems that one kind of property could supervene, in this sense, on another kind of property yet be utterly different in kind from that property. For example, epiphenomenalism, a type of antiphysicalism, states that pain is a nonphysical property, which nonetheless supervenes on properties of the brain. Pain, on the epiphenomenalist view, is caused by the brain yet is utterly distinct from it and has no causal influence on it. Because of concerns such as these, supervenience is often thought of as merely a necessary condition for physicalism.

Philosophers who are looking for a sufficient condition for physicalism often turn to the relations of determination, constitution, and realization. While differing in details, these relations are all supposed to capture the idea that while the higher level properties,

[4] What if there is no fundamental level? Physicalism needs to be reformulated. I address this in the next section.

P

such as mental properties, are not identical to certain lower level properties, such as neural properties, they are still not entirely distinct from these properties. The determination relation between the mental and the physical is something like the relation between, say, Mt. Everest and the little pebbles that constitute it. Mt. Everest is arguably entirely determined by its parts (once you have all the little pebbles, you have the mountain) and in this sense it is not distinct from its parts. Yet nonetheless, it might be argued, it is not identical to its parts since you could easily substitute one pebble from another mountain for a pebble from Mt. Everest without turning Mt. Everest into a different mountain. As we say that Mt. Everest is constituted or determined by its parts, the physicalist says that the mind is entirely constituted or determined by its physical parts.

Often the physicalist's claim is spelled out in terms of a thought experiment about the creation of the world. Imagine that in the beginning God created all the fundamental particles of physics and set them in motion according to a plan. If after creating the quarks and leptons and so forth and making sure that their patterns of motion were in accord with certain laws, would God have more work to do? For example, would she also need to create human minds? Physicalists think not since they think that the world we are familiar with, which includes rocks and trees and people with thoughts, is entirely constituted by complex arrangements of microphysical particles. This is not to say that physicalists must hold that there is nothing but complex arrangements of microphysical particles. Rather, most physicalists hold that higher level features of the world, such as rocks and trees and minds, do exist. It is just that the higher level features of the world are created in creating all of the microphysical aspects of the world.[5]

The Physical

Now we must address the question of what is the physical? When we say, for example, that everything is determined by fundamental physical phenomena what are these fundamental physical phenomena? As I said above, most define the fundamental physical in terms of the entities and properties and perhaps laws posited by microphysics: the fundamental physical phenomena are those entities and properties mentioned in the theories of microphysics. But what is meant by microphysics? Some take microphysics to refer to current microphysical theory. This gives us a relatively clear position: physicalism becomes the view that everything that exists either is or is determined by the entities, properties, and laws of current microphysics. Unfortunately, this is a theory that is rather difficult to accept since we know that current microphysics is most likely neither entirely true nor complete and thus we know now that it is not true that everything is determined by such phenomena.

Because formulating the microphysical in terms of current microphysics leads to a theory we currently know is false, most physicalists formulate physicalism in terms of a true and complete microphysics. As such, physicalism is the view that everything will be accounted for by the entities and properties of a true and complete microphysics. But what is a true and complete microphysics? It would seem to be a theory that tells us about the fundamental nature of everything. But if so, physicalism turns out to be true by definition since the fundamental nature of everything, of course, will be accounted for by a theory that accounts for the fundamental nature of everything. While there is nothing necessarily wrong with being true by definition, most philosophers working on the question of physicalism do not think that at this point in the debate the truth of physicalism simply follows from its definition. Rather, most think that physicalism requires argument and empirical support, support which they speculate is forthcoming.

If we can neither formulate physicalism in terms of current physics nor in terms of a true and complete physics, how are we to formulate physicalism? It is not clear that we can formulate physicalism so as to assign pride of place to physics, however, we can reformulate physicalism with respect to the mind so that it captures at least most of what physicalists and antiphysicalists think is at issue. This is done by turning physicalism into a thesis not about the fundamental physical aspects of the world, but rather about the fundamental non-mental aspects of the world. Physicalism with respect to the mind then amounts to the view that all mental properties are determined by nonmental properties. One way to think about this view is that the mind is physical if all mental properties are determined by neural properties (couldn't a physicalist concede that some mental properties may be determined by other physical properties, say those of silicon chips?) and neural properties are themselves ultimately determined by non-mental properties.[6]

[5] I have not discussed questions of the modality of the determination relation. Most physicalists think that physicalism could have been false, that God could have created nonphysical minds along with the creation of the fundamental physical particles, but some think of it as a necessary truth, that it is not possible that there could have been anything nonphysical. Moreover, while most think that physicalism could have been false, they also presumably think that it is not just by chance that it is true in our world; that is, they would deny that our world is such that if, say, two kinds of substances, which had never been brought together, were brought together, they would produce a nonphysical substance. In this sense, the creation thought experiment does not quite capture the content of physicalism.

[6] This formulation would not be acceptable to an externalist, that is to someone who thinks that some mental content depends on features of the world outside of the brain. Externalists, then must rely on the more general formulation of physicalism as the view that all mental properties are determined by nonmental properties.

This way of formulating physicalism captures a central point of contention between physicalists and antiphysicalist: whether mentality is a fundamental feature of the world. But what if there is no fundamental level of reality? For example, if mental properties are determined by non-mental properties, which are themselves determined by mental properties, which are determined by non-mental properties and so on *ad infinitum*, or if it is mental "all the way down," then, in either case, all fundamental properties are non-mental properties, vacuously, since there are no fundamental properties, yet physicalists would probably want to reject these world views. To address this issue, we can formulate physicalism with respect to the mental as the view that *all mental properties are eventually decomposable into non-mental properties such that all further decompositions of these properties are non-mental.*

As a general theory of physicalism, that is, as theory about the nature of everything, this formulation is inadequate. This is because there may be phenomena that for certain purposes would be unacceptable yet are nonmental. For example, a world with fundamental moral properties or abstract entities would seem to be inconsistent with physicalism, yet such things are not mental. That there is no general theory of physicalism, however, does not mean that much interesting work cannot proceed on specific topics, including questions about the ultimate nature of the mind.

▶Causality
▶The Knowledge Argument

Physiological Cell Death

▶Programmed Cell Death

Physiological Pain Parameters

Definition

Physiological pain measures include heart rate, respiration rate, blood pressure, palmar sweating, cortisol and cortisone levels, O_2 levels, vagal tone and endorphin concentrations. They are indirect pain measures and reflect a complex and generalized stress response, rather than correlate with a particular pain level.

▶Pain in Children

Physiology of Body Fluid Balance

▶Blood Volume Regulation

Pia Mater

Synonyms

▶Meninges; ▶Cisterns

Definition

Together with the arachnoid, the pia mater forms the leptomeninx.

Whereas the arachnoid follows the course of the dura mater and hence of the calvaria, the pia mater rests on the surface of the brain, pursuing a joint course along the sulci. Lying between the pia mater and arachnoid is the subarachnoid space that is filled with CSF.

Pick's Disease

Definition

Pick's disease is a severe neuro-degenerative disease in the neocortex of the ▶frontal lobe and anterior ▶temporal lobe, and at times of neurons in the ▶striatum.

Pickwickian Syndrome

Definition

Named after a Charles Dickens tale involving a character with clinical signs of the syndrome of obstructive sleep apnea and obesity hypoventilation.

The combination of upper airway obstruction and reduced lung volumes from the restrictive ventilatory effect of obesity causes low blood oxygen and elevated carbon dioxide, with their associated physiological consequences.

▶Sleep – Developmental Changes

PID Control

Definition

A popular closed-loop control method that uses a simple (Proportional-Integral-Derivative) linear controller with easily tunable parameters.

▶Nonlinear Control Systems

Piloerection or Pilomotor Reflex

Definition

Piloerection or pilomotor reflex, also called horripilation, consists of involuntary hair erection induced by contraction of arrectores pilorum muscles, i.e., the tiny muscles located at the origin of each body hair. It is a reaction to cold temperature or strong emotions producing the so-called cutis anserina or goose bumps/pimples. Arrectores pilorum muscles receive the contractile command by the sympathetic nervous system.

▶Sympathetic Nervous System

Pineal Body

Synonyms

▶Glandula pinealis; ▶Pineal gland

Definition

This nuclear region, shaped like a pine cone, above the quadrigeminal lamina forms, together with the habenular nuclei, the so-called epithalamus. The function of the pineal body, also called epiphysis or pineal gland, is to produce the hormone melatonin, which plays a role in light-controlled circadian rhythms. Afferents come from the suprachiasmatic nucleus, pre-geniculate nucleus and posterior commissural nucleus.

▶Diencephalon

Pineal Gland

Definition

The pineal gland (or the epiphysis cerebri, or epiphysis) is an endocrine gland located in the brain. The primary role of the pineal gland is the synthesis and release of the hormone melatonin. This hormone plays an important role in the regulation of many neurobiological aspects (e.g., circadian rhythms, sleep and reproduction).

The pineal gland receives a sympathetic innervation from the superior cervical ganglion. In nonmammalian vertebrates the pineal gland is directly photosensitive and usually contains circadian oscillators.

▶Avian Pineal Gland as "Third Eye"
▶Pineal Oscillators

Pineal Hormone

▶Melatonin

Pineal Oscillators

Definition

In many animals the pineal gland contains circadian pacemakers that directly control the synthesis of melatonin. A single pineal cell (pinealocyte) is capable of driving the circadian rhythms in melatonin release.

Therefore, it is composed of many "clocks" that are capable of producing a synchronized output (melatonin).

Indeed, in birds and in many non-mammalian vertebrates, the pineal gland acts as a circadian pacemaker that regulates the entire circadian system.

In mammals the pineal gland does not contain circadian oscillators. However, recent studies have reported that the mammalian pineal gland may also contain a circadian clock, although this circadian oscillator does not regulate melatonin synthesis and, therefore, its functional role is unclear.

▶Circadian Rhythm

Pinealectomy

Definition

Surgical removal of the pineal gland.

►Seasonality

Pioneer Axons

Definition

Pioneer axons are the first axons to extend into novel regions of the developing embryo. Growth cones of pioneer axons generally have a broad hand-like appearance with filopodia, consistent with making guidance choices based on cues in the environment.

Later growing axons will often fasciculate with these pioneer axons and use them as tracks on which to grow to their target region. In most of the embryo, motor neurons of the CNS send out the pioneer axons traveling to peripheral structures and sensory axons later grow along these tracks. However, in the developing head, sensory ganglion neurons send out the pioneer axons and motor axons later travel along these tracks.

►Growth Cones

Piriform Cortex

Definition

The allocortical piriform cortex (also termed olfactory or pyriform or prepiriform cortex) is the largest primary area involved in the perception and learning of olfactory stimuli. It is located in the rostral part of the forebrain, ventrally to the rhinal sulcus, and it is reciprocally connected with other olfactory regions and with high order brain areas. It is commonly divided in two regions (an anterior and a posterior part) which are believed to have different functional roles.

►Olfactory Cortex

Pit Organ

Definition

Receptor of infrared radiation, pits near the nostrils, actually primitive "eyes," which register not only the heat, but also the direction of its source, e.g. a small rodent.

►Evolution of the Brain: At the Reptile-Bird Transition

Pituitary Adenylate Cyclase-Activating Polypeptide1 (PACAP)

Definition

Neuropeptide involved in the phase shift response of the circadian clock in response to nocturnal light and also stimulates the release of melatonin from the adult pineal gland.

►Clock-Controlled Genes

Pituitary Gland

Definition

Also called hypophysis. The pituitary gland is an appendix to the hypothalamus at the base of the forebrain. The pituitary consists of an anterior adenohypophysis and a posterior neurohypophysis.

The hypothalamus emits a number of releasing hormones and release-inhibiting hormones, which regulate the release of hormones from the adenohypophysis.

►Hypophysis
►Hypothalamo-neurohypophysial System
►Hypothalamo-pituitary-adrenal Axis, Stress and Depression
►Hypothalamo-pituitary-thyroid Axis
►Hypothalamus
►Releasing-Hormone and Release-Inhibiting Hormone

P

PKA

Definition

►Protein Kinase A

PKC

Definition

►Protein Kinase C

Place Cells

Definition

Neurons discharging selectively when the subject is in a delimited region of its environment referred to as a "firing field" or "place field." First discovered in the hippocampus by O'Keefe and Dostrovsky (1971).

Found in rats, mice, birds, bats, and primates including humans. Although most observers agree that the hippocampus has a critical role in learning and memory, there remains considerable debate about the precise functional contribution of the hippocampus to these processes.

Two of the most influential accounts hold that the primary function of the hippocampus is to generate cognitive maps and to mediate episodic memory processes. The well-documented spatial firing patterns (place fields) of hippocampal neurons in rodents, along with the spatial learning impairments observed with hippocampal damage support the cognitive mapping hypothesis. The amnesia for personally experienced events seen in humans with hippocampal damage and the data of animal models, which show severe memory deficits associated with hippocampal lesions, support the episodic memory account.

►The Hippocampus: Organization
►Neural Bases of Spatial Learning and Memory
►Hippocampus: Organization, Maturation, and Operation in Cognition and Pathological Conditions

Place-versus-Response Controversy

Definition

This controversy was a well-known debate among groups of psychologists in the early twentieth century.

The response group (behaviorists) felt that all behavior, including navigation, must be described in simple stimulus-response terms. The place group (cognitivists) felt that complex behavior could only be explained by positing that the rat brain contained a central representation – a map of the places in the environment.

►Spatial Learning/Memory

Placebo Analgesic Response

Howard L. Fields
Ernest Gallo Clinic and Research Center, University of California, San Francisco, USA

Synonyms

Expectancy effect; Palliative; Dummy treatment; Control treatment

Definition

The ►placebo analgesic response is a reduction in ►pain resulting from the recipient's expectation that a treatment received has analgesic efficacy. The term placebo is usually used when the treatment is intentionally given deceptively and has no intrinsic analgesic efficacy.

Characteristics

Some individuals experience a reduction of their pain when they are given a completely inert substance with the understanding that it is an effective treatment. If the person giving it knows that the treatment is inert (i.e., it is given with the intention to deceive the recipient), the treatment is a placebo. However, even if the individual receiving the placebo experiences a reduction in their pain following placebo administration, it is not necessarily the case that the placebo caused the reduction. The reason for the uncertainty is that in clinical syndromes it is common to observe frequent fluctuations in pain over time in the absence of any treatment. This fluctuation is called the ►natural history of the painful condition. Consequently, one cannot conclude that a given instance of a reduction in pain is due to the preceding treatment manipulation, whether placebo or active medication. Because of this uncertainty, well designed clinical trials of analgesic medications generally require a comparison group receiving placebo treatment. The efficacy of an analgesic medication is established by showing that it produces a reduction in pain superior to placebo.

This uncertainty about the cause of a fluctuation in pain also presents challenges for research into the mechanisms underlying the placebo analgesic response [1]. One obvious difficulty is that if the individual improves, it is uncertain whether it is because of a placebo analgesic response or would have occurred in the absence of treatment. Thus, placebo analgesic responses in an individual are usually inferred, not observed. On the other hand, placebo analgesic effects are quite robust when studied comparing two *groups* of subjects, one of which receives a placebo and the other no treatment. The ►placebo effect in that situation is the difference between those two groups. The difficulty in identifying individual, as opposed to group placebo analgesic responses has made it very difficult to determine whether there are specific characteristics of individuals that make them more or less likely to respond to a placebo manipulation. On the other hand, comparing groups given placebo versus no-treatment has increased our understanding of the psychological and neural mechanisms of the placebo analgesic response.

In addition to its usefulness in controlled clinical trials, responses to placebo likely play a significant role in clinical practice. Placebo administration can provide effective relief for severe pain conditions, including those following major surgical procedures. It is also important to point out that even when effective analgesic agents are given to patients, part of the pain relief obtained may be due to a ►placebo response. For example, when a moderate dose of morphine is given to a pain patient by hidden infusion (i.e., from a preprogrammed remote pump at an unknown time), it is much less effective than when it is given openly. When patients with postoperative pain are told that they may receive morphine, but instead are given an open administration of saline through an intravenous catheter, they obtain pain relief equivalent to a moderate dose of morphine given by hidden infusion [2]. This suggests that in many cases the relief obtained when individuals take an analgesic agent is a sum of a placebo response plus the effect of the active agent.

Studies directed at the psychological determinants of the placebo response have identified ►expectancy as the most relevant psychological mediator. Expectancy can be manipulated verbally for example by simply informing the subject that the treatment they are about to receive is a powerful analgesic [3]. Expectancy can also be manipulated by explicit training, which leads to ►conditioned memory. One form of conditioning is to treat subjects in pain with a powerful analgesic agent and subsequently administering a similar appearing placebo treatment to them [4]. Another approach is through surreptitiously lowering the intensity of an experimental painful stimulus in the presence of a placebo treatment [5] and then re-administering the placebo treatment. For either approach, the conditioning markedly increases the analgesic effectiveness of a placebo treatment. Furthermore, there is a clear relationship between the subject's expectations and the degree of relief experienced [5].

Progress has also been made in understanding the neural mechanisms underlying the placebo analgesic effect. One of the first mechanistic proposals was that expectation of reward somehow led to the release of endogenous ►opioid peptides which acted at ►opioid receptors in the central nervous system to produce analgesia. This idea was supported by the finding that placebo relief of dental postoperative pain could be blocked by the opioid receptor antagonist ►naloxone [6]. Naloxone was later shown to block placebo analgesia in experimental pain models in normal volunteers [7]. Furthermore, there is a central nervous system pathway that modulates pain transmission (see ►Descending modulation of nociception). This pathway is organized in a top-down fashion. It includes the ►anterior cingulate cortex and other regions of the ►prefrontal cortex, the ►hypothalamus, ►amygdala and brainstem and it terminates in the ►spinal cord [8]. The anterior cingulate cortex and other prefrontal areas have been implicated in expectancy effects including placebo analgesia. Furthermore, endogenous opioid peptides are released throughout this pathway and contribute to its analgesic effects. Human ►functional imaging studies, including both ►positron emission tomography and ►functional magnetic resonance imaging, have provided evidence that supports a role for this ►pain modulating pathway in placebo analgesia. Placebo analgesic effects correlate with activation of prefrontal cortex and brainstem regions areas that overlap with the endogenous opioid mediated ►pain modulatory pathway [9,10].

In summary, the expectation of pain relief, whether induced by verbal instruction or conditioning, can bestow robust analgesic potency on a treatment that would otherwise be ineffective. The expectancy effect is mediated by a pain modulating pathway with endogenous opioid links. Placebo analgesic responses may also contribute to the efficacy of active analgesic agents through summation with direct drug effects.

References

1. Amanzio M, Benedetti F (1999) Neuropharmacological dissection of placebo analgesia: expectation-activated opioid systems versus conditioning-activated specific subsystems. J Neurosci 19:484–494
2. Benedetti F, Amanzio M (1997) The neurobiology of placebo analgesia: from endogenous opioids to cholecystokinin. Prog Neurobiol 52:109–125

3. Bingel U, Lorenz J, Schoell E, Weiller C, Buchel C (2006) Mechanisms of placebo analgesia: rACC recruitment of a subcortical antinociceptive network. Pain 120:8–15
4. Fields HL, Basbaum AI, Heinricher MM (2006) Central nervous system mechanisms of pain modulation. In: McMahon SB, Koltzenburg M (eds) Wall and Melzack's Textbook of pain, 5th edn. Elsevier Churchill Livingstone, Edinburgh, pp 125–142
5. Hoffman GA, Harrington A, Fields HL (2005) Pain and the placebo: what we have learned. Perspect Biol Med 48:248–265
6. Levine JD, Gordon NC, Fields HL (1978) The mechanism of placebo analgesia. Lancet 2:654–657
7. Levine JD, Gordon NC, Smith R, Fields HL (1981) Analgesic responses to morphine and placebo in individuals with postoperative pain. Pain 10:379–389
8. Pollo A, Amanzio M, Arslanian A, Casadio C, Maggi G, Benedetti F (2001) Response expectancies in placebo analgesia and their clinical relevance. Pain 93:77–84
9. Price DD, Milling LS, Kirsch I, Duff A, Montgomery GH, Nicholls SS (1999) An analysis of factors that contribute to the magnitude of placebo analgesia in an experimental paradigm. Pain 83:147–156
10. Wager TD, Rilling JK, Smith EE, Sokolik A, Casey KL, Davidson RJ, Kosslyn SM, Rose RM, Cohen JD (2004) Placebo-induced changes in FMRI in the anticipation and experience of pain. Science 303:1162–1167

Planes of Section

Definition

Planes of section are usually expressed in relation to the upright midline axis of the body (or spinal column). The plane of section that passes through or parallel to the midline and extends in the dorsal-ventral direction is the sagittal plane. Parasagittal planes pass to the left or to the right of the midline sagittal plane that is marked by the sagittal suture of the skull. The horizontal plane is at a right angle to the sagittal plane and parallel with the ground. Horizontal planes progress inferiorly or superiorly. The frontal (coronal) plane is orthogonal to the other two. It is in the plane of the coronal suture of the skull. Frontal planes progresses anteriorly and posteriorly.

Plant

Definition

The external environment that needs to be controlled, e.g. an arm.

►Neural Networks for Control

Plasma Membrane

Definition

►Cell Membrane Components and Functions

Plasma Membrane Ca^{2+}-ATPase (PMCA)

Definition

A pump on the plasma membrane that couples ATP hydrolysis to the extrusion of Ca^{2+} from the cytosol to the extracellular space.

►Influence of Ca^{2+} Homeostasis on Neurosecretion

Plasmalemma

Definition

The term denotes the cell membrane separating the interior from the exterior of a cell except in muscle cells, where the cell membrane is called "sarcolemma."

►Cell Membrane Components and Functions

Plasmid Vector

Definition

Small circular molecules of double stranded DNA derived from natural plasmids found in bacteria. They themselves are distinct from the chromosomal genome of bacteria. An exogenous piece of DNA can be easily inserted into a plasmid if both the plasmid and the exogenous DNA source have recognition sites for the same restriction endonuclease.

►Serial Analysis of Gene Expression

Plasticity

Definition

The ability of the brain to change its functional organization and neural representations (e.g. motor and sensory maps) as a result of damage or experience and in response to altered peripheral conditions or behavioral demands.

▶Motor Cortex – Hand Movements and Plasticity
▶Somatosensory Cortex, Plasticity

Plasticity in Central Auditory System

DEXTER R. F. IRVINE
Department of Psychology, Monash University, VIC, Australia

Definition

Changes in the structural and functional characteristics of neurons in the central auditory system (CAS), which occur in response to altered patterns of input (i.e., not as a direct consequence of aging or of other changes in the organism's state), and are not explicable as passive consequences of the altered input, but involve some form of dynamic change in neural properties [1].

Characteristics

Higher Level Structures

The CAS comprises the various ▶auditory subcortical nuclei and the multiple cortical fields that comprise the auditory cortex (▶auditory cortical areas), together with the (ascending and descending) pathways that link these structures.

Lower Level Components

The lower level components are the individual neurons and their processes that make up the various nuclei and fiber tracts comprising the CAS, the synaptic connections between these neurons, and the networks they comprise.

Higher Level Processes

CAS plasticity has been demonstrated in a variety of experimental conditions, and is assumed to underlie a number of forms of auditory behavioral plasticity; however, it is by no means clear that the same mechanisms are involved in all cases. A broad distinction can be made between forms of plasticity that occur only during development (in many cases within restricted ▶critical periods), and those that are observed in both young and adult animals. In the following sections, the major paradigms that have been used to study developmental and adult plasticity will be briefly reviewed.

Developmental Plasticity: Neonatal Cochlear Ablation

Neonatal ablation of one ▶cochlea in mammals, which eliminates afferent input from that ear to ▶binaural neurons in the CAS, results in structural and functional changes in the input from the intact ear to these neurons. Axons from the ventral division of the ▶cochlear nucleus (CN) on the side of the intact ear, which normally terminate in restricted regions of the major nuclei of the ▶superior olivary complex, project to, and make synapses in, regions of these nuclei normally innervated by the CN on the side of the ablated cochlea. Similarly, the terminal fields in the central nucleus of the ipsilateral ▶inferior colliculus (IC) of axons from the CN on the side of the intact ear are much larger than those in normal animals [2]. Both of these results indicate that, following unilateral cochlear ablation, axons from the CN on the intact side sprout to innervate additional territory. These structural changes are associated with increased excitatory responses to stimulation of the intact ear in the ipsilateral IC and ▶primary auditory cortex (AI) [2].

Developmental Plasticity: Space Map Plasticity

In barn owls, and in at least some mammals, the deep layers of the ▶superior colliculus (SC) contain maps of auditory and visual space that are in register (i.e., bimodal neurons, or auditory and visual neurons at the same locus, have corresponding auditory and visual ▶spatial receptive fields (RFs)). The auditory ▶space map is derived via orderly projections from the ▶external nucleus of the (ICX), and is based on neural sensitivity to ▶interaural time and level differences (ITDs and ILDs, respectively) and, in mammals, on sensitivity to ▶spectral cues produced by the effects of the head and ▶pinnae on the sound field [2]. In both barn owls [3] and ferrets [2] reared from infancy with one ear plugged (thus altering the values of the interaural disparities associated with any given spatial location), the visual and auditory maps are in register. In the barn owl, this reflects the fact that the tuning to ITDs and ILDs of neurons in ICX and SC has changed to maintain alignment of the auditory and visual RFs. This change in neural tuning is associated with the appearance of novel projections from the central nucleus of IC (ICC) to ICX: axons sprout into, and form synapses in, regions outside their normal projection zone [3]. Immediately after ear plugging, young owls mislocalize sounds, but in parallel with the map changes, they recover accurate localization ability. Ear plugging in adult animals does not result in adaptive changes in interaural

P

disparity tuning, and the auditory and visual space maps therefore remain out of register while the plug is in position. Analogous developmental plasticity occurs when owls are raised wearing prismatic spectacles that displace the visual field in the horizontal plane: the tuning of SC neurons to ITDs (which serve as the cue for azimuthal location in barn owls) shifts to maintain the correspondence between auditory and visual RFs [3]. The plasticity demonstrated in these studies using experimental manipulations of sensory input occurs in all animals during development as the size of the head increases, and the values of ITDs and ILDs associated with particular spatial locations consequently change.

Adult Plasticity: Injury-Induced

Lesions of a restricted region of the cochlea (which result in a partial hearing loss) result in a reorganization of the ►frequency map in AI. The nature of this reorganization is that the region deprived of its normal input by the cochlear lesion is occupied by an expanded representation of the frequencies represented at the edge(s) of the cochlear lesion. The thresholds, latencies, and sharpness of ►frequency tuning of neurons in this expanded representation at their new ►characteristic frequency (CF) indicate that the changed frequency organization is not a passive consequence of the peripheral lesion, but reflects a dynamic process of reorganization. Such reorganization is observed after mechanical cochlear lesions, and after lesions produced by ►ototoxic drugs or by ►noise trauma [1]. Analogous reorganization is seen in visual and somatosensory cortices after restricted retinal lesions and digit amputation, respectively [4]. After mechanical cochlear lesions, similar reorganization is seen in the major auditory thalamic nucleus (the ventral division of the ►medial geniculate nucleus), but reorganization is patchy in the ICC and is not observed in the dorsal nucleus of the CN, suggesting that this form of CAS plasticity is a characteristic of ►thalamo-cortical circuitry.

Adult Plasticity: Learning-Induced

Changes in the stimulus selectivity of neurons in the higher levels of the auditory pathway, as a consequence of behavioral conditioning procedures in which an acoustic stimulus serves as the ►conditioned stimulus or discriminative (henceforth "training") stimulus, were the first demonstrations of CAS plasticity in adult animals, and this remains the most active area of research on this topic. The most common finding in these studies is that the response of cortical neurons at the training frequency increases, while the response at other frequencies decreases, such that the training frequency becomes the neurons' ►best frequency. Such effects have been described in AI and in secondary auditory cortical fields, and in the auditory thalamus [1,5,6]. These studies have commonly used ►fear conditioning procedures, but appropriate controls, and the specificity of the effects to the training frequency, establish that the observed effects are manifestations of plasticity rather than consequences of changes in state variables [6]. A substantial body of evidence indicates that the ►cholinergic basal forebrain plays an important modulatory role in this form of CAS plasticity [1].

Another form of auditory learning that might involve CAS plasticity is ►perceptual learning, the improvement in discriminative capacity with training that is a common observation in all sensory modalities both in psychophysical experiments and in everyday life [1,7,8]. The specificity of many forms of visual perceptual learning to particular features of the training stimuli, or to the region of the receptor to which the stimuli are presented, has led to the proposal that the learning might involve changes in neural tuning in primary sensory cortex [7,8]. In the auditory system, the evidence for changes in AI associated with perceptual learning is equivocal [7], and it is possible that the learning involves changes in higher-order decision making processes. Indirect evidence for auditory cortical changes is provided by ►functional imaging evidence of larger auditory cortical responses to musical tones in trained musicians, although in this case it is not clear whether the larger cortical responses are a consequence of training, or whether people with this innate characteristic are more likely to undertake such training [1].

Two forms of auditory perceptual learning that are of great practical significance relate to speech processing. The first is the effect of language experience on the perception of speech sounds, and thus on language acquisition, during a critical period of development [9]. The second is the improvement in speech discrimination shown by people with ►cochlear implants over the months and years following implantation. Whether these forms of learning reflect plasticity in the CAS itself or in regions involved in cognitive processes is not clear.

Adult Plasticity: Microstimulation-Induced

A series of studies in adult bats and gerbils have indicated that focal electrical stimulation of a region of AI can change the frequency selectivity of neurons around the stimulation site in AI, and in the tonotopically corresponding area of the central nucleus of the IC [5]. ►Centrifugal fibers from AI to IC have been shown to play a role in this plasticity.

Lower-Level Processes

The cellular mechanisms of CAS plasticity (and those of plasticity in other sensory systems) are incompletely understood. Different forms of plasticity have different time courses, as do different phases of particular forms of plasticity. The first stage of injury-induced plasticity in sensory cortices appears to be an expansion of RFs, reflecting an unmasking of normally-inhibited excitatory inputs from outside the classically defined RF [10]. A similar unmasking is involved in space map plasticity in the barn owl [3]. The subsequent establishment of smaller RFs in injury-induced plasticity presumably involves changes in the efficacy of both excitatory and inhibitory synapses, and there is evidence for the strengthening of excitatory synapses via ►NMDA-receptor mediated ►long-term potentiation in developmental plasticity in the barn owl [3] and in adult plasticity in the visual and somatosensory systems [10]. Longer-term changes in the barn owl involve axonal sprouting and ►synaptogenesis [3]; axonal sprouting is also involved in the changes observed after unilateral cochlear ablation in neonatal animals [2]. Axonal sprouting of horizontal fibers in the superficial layers of the cortex has also been shown to be involved in visual cortical plasticity after retinal lesions in adults [8]. Structural changes of this sort have not yet been shown in injury-induced auditory plasticity in adults. Finally, as mentioned previously, corticofugal projections have been shown to be involved in some forms of plasticity [5,8].

Function

Auditory developmental plasticity and learning-induced plasticity are undoubtedly adaptive, in that they enhance the organism's ability to adjust to altered patterns of input and, in the case of human language acquisition and adaptation to a cochlear implant, make important auditory discriminations. It is less-clear that adult injury-induced plasticity is adaptive, as the organism remains deaf in the frequency range affected by the cochlear lesion, and there is little evidence that the CAS reorganization consequent on the lesion in any way compensates for this deafness. It is likely that this form of plasticity is an extreme manifestation of the processes that underlie other forms of plasticity in response to altered input, but it is also possible that similar processes are involved in recovery of function after central damage such as that produced by stroke.

Pathology

There is little evidence that these forms of plasticity have pathological consequences, although it has been suggested that tinnitus might be a consequence of cortical reorganization consequent on a peripheral lesion [1].

Therapy

The possibility that some forms of learning impairment might involve deficiencies in aspects of auditory processing that exhibit plasticity, such that these deficiencies can be modified by training, has resulted in the recent development of a number of auditory training software packages. As discussed previously, the success of cochlear implants undoubtedly rests in part on the plasticity of CAS structures, and plasticity therefore contributes to the therapeutic effects of these devices.

References

1. Irvine DRF, Wright BA (2005) Plasticity in spectral processing. In: Malmierca M, Irvine DRF (eds) Auditory spectral processing. Elsevier, San Diego pp 435–472
2. Moore DR, King AJ (2004) Plasticity of binaural systems. In: Parks TN, Rubel EW, Popper AN, Fay RR (eds) Plasticity of the auditory system. Springer, New York, pp 96–172
3. Knudsen EI (2002) Instructed learning in the auditory localization pathway of the barn owl. Nature 417:322–328
4. Kaas JH, Florence SL (2001) Reorganization of sensory and motor systems in adult mammals after injury. In: Kaas JH (ed) The mutable brain. Harwood Academic Publishers, New York, pp 165–242
5. Suga N, Ma X (2003) Multiparametric corticofugal modulation and plasticity in the auditory system. Nat Rev Neurosci 4:783–794
6. Weinberger NM (2004) Experience-dependent response plasticity in the auditory cortex: issues, characteristics, mechanisms, and functions. In: Parks TN, Rubel EW, Popper AN, Fay RR (eds) Plasticity of the auditory system. Springer, New York, pp 173–227
7. Irvine D, Brown M, Martin R, Park V (2005) Auditory perceptual learning and cortical plasticity. In: Koenig R, Heil P, Budinger E, Scheich H (eds) The auditory cortex: a synthesis of human and animal research. Lawrence Erlbaum, Mahwah, NJ, pp 409–428
8. Tsodyks M, Gilbert C (2004) Neural networks and perceptual learning. Nature 431:775–781
9. Kuhl PK (2000) A new view of language acquisition. Proc Natl Acad Sci USA 97:11850–11857
10. Calford MB (2002) Dynamic representational plasticity in sensory cortex. Neuroscience 111:709–738

Plasticity in Nociception

Definition

Plasticity with respect to nociception refers to modulation of nociceptive afferent signals. Plasticity with respect to pain refers to the modulation of pain perception. The biological response to tissue damage, and the subsequent perception of pain, can vary depending on activation of other ascending and descending sensory systems.

►Spinal Dorsal Horn Plasticity

Plateau Potential

Definition

A membrane potential depolarization that is sustained by intrinsic properties even after the stimulus that triggered it has been terminated. In contrast to a pacemaker potential, the plateau potential is not rhythmically activated and terminated by the neuron.

Instead, onset and termination of plateau potentials are typically triggered by excitatory and inhibitory synaptic inputs. Most commonly plateau potentials are caused by the action of persistent inward currents (e.g. persistent Na^+ current, low-voltage activated Ca^{2+} currents, etc.). Once these currents have been activated by a synaptic input or brief depolarization, the resulting inward current is sufficient to sustain the depolarization and maintain their activation. The depolarization can be terminated by inhibitory synaptic inputs that hyperpolarize the membrane potential sufficiently to remove the activation of the persistent inward current. The plateau potential can also be terminated by intrinsic mechanisms, e.g. persistent Ca^{2+} influx can activate Ca^{2+}-dependent K^+ channels that cause an outward current which will repolarize the membrane potential and terminate the plateau potential.

►Calcium Channels – an Overview
►Central Pattern Generator
►Neuronal Poassium Channels
►Sodium Channels

Play

►Learning and Motivation

PLC γ

Definition

Is a member of the Phospholipase C family that consists of PLC-Δ, -β, -γ and –ε which all require calcium for their catalytic activity. PLCγ is activated by transmembrane receptors with tyrosine kinase activity and is a key enzyme involved in the activation of protein kinase C (PKC) leading to Akt activation and increased survival following axonal injury.

►Neurotrophic Factors in Nerve Regeneration

Plegia

Definition

Severe loss of motor strength by motor paralysis of entire limbs or parts thereof.

Plexus

Definition

A plexus (Latin for braid) is a region of exchange of nerve fibers of different spinal cord levels to form specific peripheral nerves.

PMBSF

►Barrel Cortex

Pneumotaxic Center

►Pontine Control of Respiration

PNS

Definition

Peripheral nervous system including cranial and spinal nerves.

p75NTR

Definition
p75NTR (p75 neurotrophin receptor) is a 75 kDa member of the Tumour Necrosis Factor (TNF) receptor family whose members share a cysteine-rich common extracellular binding domain. It was originally identified as a low-affinity receptor for NGF (nerve growth factor). It has since been shown that p75NTR binds all known members of the neurotrophin family (NGF, BDNF, NT-3, NT-4/5), and subsequent intracellular activation is upstream of both the PKC and JNK pathways resulting either survival or death. Denervated Schwann cells express high levels of p75NTR as a consequence of the loss of axonal contact. The function of p75NTR in Schwann cells is currently not known.

▶Neurotrophic Factors in Nerve Regeneration
▶Schwann Cells in Nerve Regeneration
▶Tumor Necrosis Factor-α

Point Contacts

▶Integrin-dependent Adhesion Contacts

Point-to-Point Movements

Definition
Point-to-point movements are a sequence of discrete movements, where each movement starts from one position aiming to a next position.

▶Motor Control Models

Polar Decomposition Theorem

Definition
A theorem in linear algebra stating that every nonsingular square matrix is uniquely decomposable into the product of an orthogonal matrix and a positivedefinite symmetric matrix.

▶Mechanics

Polarity

Definition
Polarity in the case of neurons, polarity refers to the fact that neurons are polarized structures with an axon and dendrites.

Polarity of Neurons

Definition
In the case of neurons, polarity refers to the fact that neurons are polarized structures with an axon and dendrites.

Poles

Definition
The roots of the denominator of a transfer function.

▶Signals and Systems
▶Transfer Function

Polioencephalitis (Anterior Poliomyelitis)

Definition
Acute infectious viral disease affecting several parts of the central nervous system, classically involving spinal ▶motoneurons with final degeneration and paralysis.

P

Poliomyelitis

Definition

Acute infectious disease of humans, particularly children, caused by any of three serotypes of the human poliovirus; infection is usually limited to the gastrointestinal tract and nasopharynx, and is often asymptomatic. The central nervous system, primarily the spinal cord, may be affected, leading to rapidly progressive ►paralysis, coarse fasciculation and ►hyporeflexia. ►Motoneurons are primarily affected, and ►encephalitis may also occur.

Poly(A) Tail

Definition

A polyadenosine tail is the product following polyadenylation of pre-mRNA. Most messenger RNA molecules end with a poly-A stretch at their 3′ ends in eukaryotic organisms. The poly(A) tail protects mRNA from exonucleases and plays a role in transcription termination. Typically 50–200 adenosines are added to pre-mRNA.

►Serial Analysis of Gene Expression

Polyadenylation

Definition

Polyadenylation is a covalently-linked tail of a long stretch of adenosines added to the 3′ end of the mRNA, and is required to generate mature mRNA.

Polydipsia

Definition

A symptom in which the patient ingests abnormally large amounts of fluids.

►Neuroendocrinology of Psychiatric Disorders

Polyhydramnios

Definition

Increased amounts of amniotic fluid.

►Endocrine Disorders of Development and Growth

Polymer

Definition

A compound consisting of many repeated linked units.

Polymerase Chain Reaction (PCR)

Definition

Polymerase chain reaction (PCR) is used to isolate and amplify in an exponential fashion a DNA sequence of interest. mRNA is isolated from a cellular or tissue source, a cDNA copy is made by reverse transcriptase, and the DNA found between a 3′ and a 5′ nucleotide single strand DNA primer (complementary DNA sequence) is amplified by approximately 20–30 cycles that include denaturation of double stranded DNA, annealing of the primers to the cDNA, and DNA synthesis.

Polymodal

►Multimodal Integration

Polymodal Receptor

Definition

Polymodal receptor denotes a sensory receptor (e.g. nociceptor) responsive to more than one modality or sub-modality (quality), e.g. to pressure, temperature and/or certain chemical substances. Polymodality is also prevalent in central neurons.

►Nociceptors and Characteristics
►Sensory Systems

Polymorphic Network

Definition

A neuronal network that can change its functional connectivity in response to modulatory influences by higher order interneurons. Modulatory interneurons can enhance or suppress the function of individual synapses and neurons within a network. This can result in the functional removal or addition of neurons to a neuronal network, which leads to the reconfiguration of the network.

►Central Pattern Generator

Polymorphism

Definition

Sequence variants that exist naturally in the population and that do not cause disease. Polymorphisms account for all interpersonal variation (e.g., height, eye, skin and hair color, etc.)

►Bioinformatics

Polymyositis Syndrome

Definition

Inflammatory myopathy prevailing in proximal muscles and resulting in weakness.

Polyneuropathy

Definition

Disease affecting many nerves.

Polypeptide

►Neuropeptides in Energy Balance

Polypeptide Synthesis

►RNA Translation

Polyradiculopathy

Definition

Disease of many spinal nerve roots.

Polyribosomes

Definition

Polyribosomes are multiple ribosomes attached to a single messenger RNA (mRNA) molecule. The ribosomes translate the information encoded on the mRNA into a protein.

►Extrasomal Protein Synthesis in Neurons

Polysomnogram

Definition

A sleep monitoring technique combining electrophysiological technologies to determine sleep stages and other sleep phenomena. Minimally, polysomnography (PSG) includes the electroencephalogram (EEG) to record electrical activity in the brain; the electrooculogram (EOG) to record eye movements; and the electromyogram (EMG) to record muscle tone. More extensive montages including electrocardiography (ECG) and measurements of airflow, breathing effort, body position, snoring sounds, and blood oxygen saturation are used for the diagnosis and treatment of disorders of sleep.

►Alertness Level
►Brain States and Olfaction

►Electroencephalography
►Electromyography
►Electrooculography

Polysynaptic Accessory Olfactory System

►Accessory Olfactory System

Pons (Varolius)

Definition

The pons (Latin for bridge, describing the large fiber bundle running across the ventral surface of the brainstem, perpendicular to its long axis) is the portion of the brainstem between the medulla and midbrain. The cerebellum sits dorsal to the pons and is attached to it by the large fiber bundles of the middle cerebellar peduncle on each side. The pons consists of two parts: base of pons and tegmentum. The typical protruding base of pons accommodates the pyramidal tracts. Interspersed here are the pontine nuclei, where corticopontine fibers synapse. The tegmentum area contains cranial nerve nuclei (V,VI, VII, VIII), trapezoid body, medial lemniscus, parts of the reticular formation and the medial longitudinal fasciculus.

Pontine Control of Respiration

Thomas E. Dick[1], Mathias Dutschmann[2], Kendall F. Morris[3]
[1]Division of Pulmonary, Critical Care and Sleep Medicine/Department of Medicine, Case Western Reserve University, Cleveland, OH, USA
[2]Department of Neuro and Sensory Physiology, Georg August University of Göttingen, Göttingen, Germany
[3]Department of Molecular Pharmacology and Physiology Neuroscience, University of South Florida College of Medicine, Tampa, FL, USA

Synonyms

Pontine respiratory group; Pneumotaxic center; Ponto-medullary respiratory pattern generator

Definition

Pontine control of respiration involves the modulation of the timing and amplitude of the muscle activities that execute breathing, and airflow in and out of the lungs. Breathing is a vital function. Together with the cardiovascular system, breathing supplies oxygen to maintain general metabolism and is thus essential for bodily function. Importantly, breathing is also integrated with other mammalian behaviors like vocalizing, swallowing, coughing, and sniffing. Thus, breathing interacts with the environment not only in terms of inhaling and exhaling gas but also in terms of communication, food consumption, airway protection, and odor detection. The pons modulates breathing not only for gas exchange but also in these different behaviors.

Characteristics

Pontine Influences on the Respiratory Motor Pattern

Breathing in mammals is controlled by a distributed network of neurons located bilaterally in columns of neurons that extend from the ventrolateral medulla to the parabrachial/Kölliker-Fuse complex in the rostral dorsolateral pons [1]. The medulla generates the primary respiratory rhythm and motor pattern, while the pontine nuclei exert strong modulatory influences on the respiratory frequency and shape of the respiratory motor pattern.

The respiratory motor pattern is defined by a sequence of bursts of different motor activities that are commonly divided into three major phases: (i) inspiration, (ii) postinspiration (early expiration or passive expiration), and (iii) late expiration (active expiration) (Fig. 1).

During the inspiratory and late expiratory phases, spinal motoneurons receive excitatory drive to contract thoracic and abdominal striated muscles including the diaphragm, to pump air in and out of the lungs (Fig. 1). Complementary motor activities transmitted in the cranial nerves (glossopharyngeal (IX), vagal (X), and hypoglossal (XII) nerves) target muscles that modulate airway resistance. During the inspiratory phase, posterior cricoarytenoid muscles and laryngeal abductors decrease upper airway resistance by pulling the vocal chords apart, whereas during the postinspiratory phase, the thyroarytenoid muscles that act as adductors increase resistance by drawing the vocal chords together. Laryngeal adductor activity depends on the pons (Fig. 2). Even in resting breathing, activity of laryngeal adductors (as well as inactivity of the airway dilators) limits or "brakes" expiratory airflow to prevent ►atelectasis especially in mammals with highly compliant chest walls. Thus, even though we separate pontine modulation of rhythm or timing from that of motor activity, these two variables are not independent. Pumping and resistance motor activities are associated with the phases of the respiratory cycle and

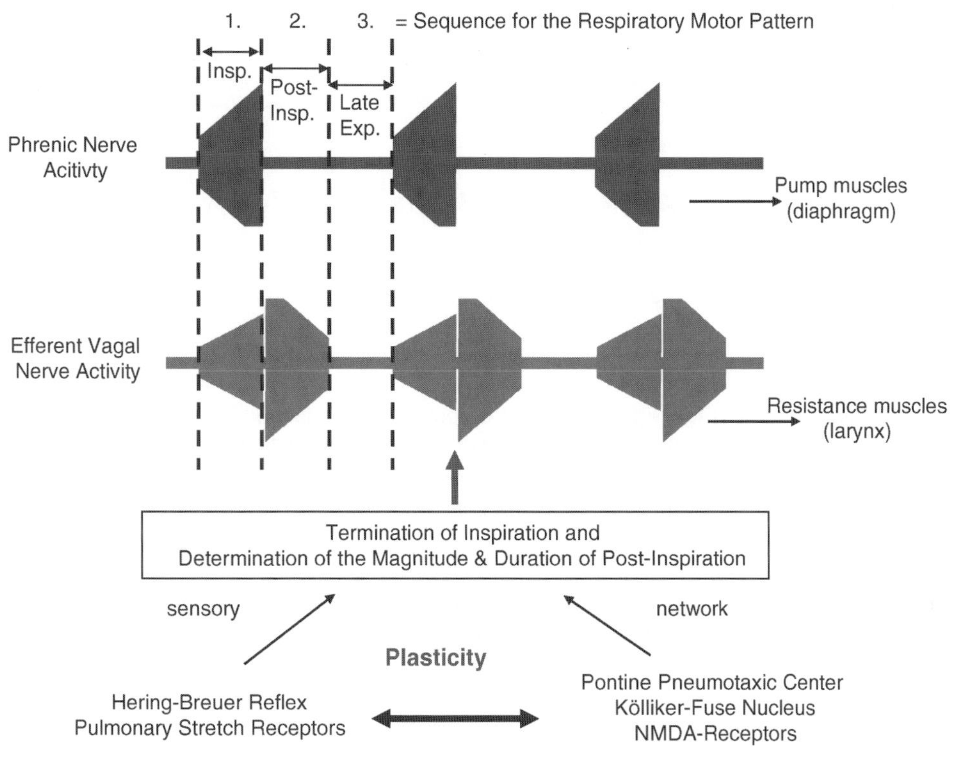

Pontine Control of Respiration. Figure 1 Schematic drawing illustrating the coordinated pattern of two representative motor activities critical in defining the breathing pattern – phrenic nerve activity (top) and efferent vagal nerve activity (bottom). In these recordings, the sequential phases of the respiratory cycle ((i) inspiration, (ii) postinspiration, (iii) late expiration) can be identified. Two convergent pathways, an afferent pathway mediating the Hering-Breuer reflex and a central pathway that requires the dl pons, control the termination of inspiration and the duration of the postinspiratory phase of the breathing cycle. Plasticity in the expression of the breathing pattern depends on the pons and results from interaction between afferent and network pathways.

are coordinated by synaptic interactions within circuits of the pontomedullary respiratory network.

Pontine and Vagal Afferent Interaction Influences Timing of the Pattern

The importance of the pons in the neuronal control of breathing was established by Marckwald in 1888 [1]. He investigated breathing in rabbits and noted that after transecting both vagal nerves, respiratory rhythm became apneustic (►apneusis) after mid-pontine transection (compare phrenic nerve activity pattern before and after pontomedullary transection in Fig. 2). These studies complement those of Breuer, conducted in Hering's laboratory in 1868 [2]. They showed that lung inflation evokes a reflex that shortens inspiration and lengthens expiration and that this reflex, namely, the "►Hering-Breuer Reflex" was vagally dependent and mediates the afferent input from pulmonary stretch receptors. An interpretation of Marckwald's work was that the pons acted like an "internal vagus" acting on the same medullary neurons that mediate the Hering Breuer reflex. Subsequent studies have supported this interpretation [3]. Stimulating the lateral pons evokes phase switching, and recording neuronal activity in the dl pons shows that the magnitude and strength of its respiratory modulation increases in the absence of vagal input [4,5].

Interestingly, the Hering-Breuer reflex is of minor importance in the regulation of breathing in humans. However, breathing is not simply a stereotyped rhythmic activity that is controlled reflexly from the pulmonary stretch receptors of the lungs and chemoreceptors of the vasculature but rather a behavior whose function and control is incorporated into the context of other behaviors like vocalizing. Thus, the importance of the pons in respiratory control may be in determining breathing pattern in the context of behavior.

Expression of Plasticity in the Breathing Pattern Depends on the Pons

Activity-dependent plasticity (See Encyclopedia Article Respiratory Neuroplasticity by Morris and Bolser) of the breathing pattern is evident after brief (seconds to minutes) activation of afferent pathways whether they are mechanosensory (vagal afferent) [6] or chemosensory [7] pathways. Following stimulation, the breathing

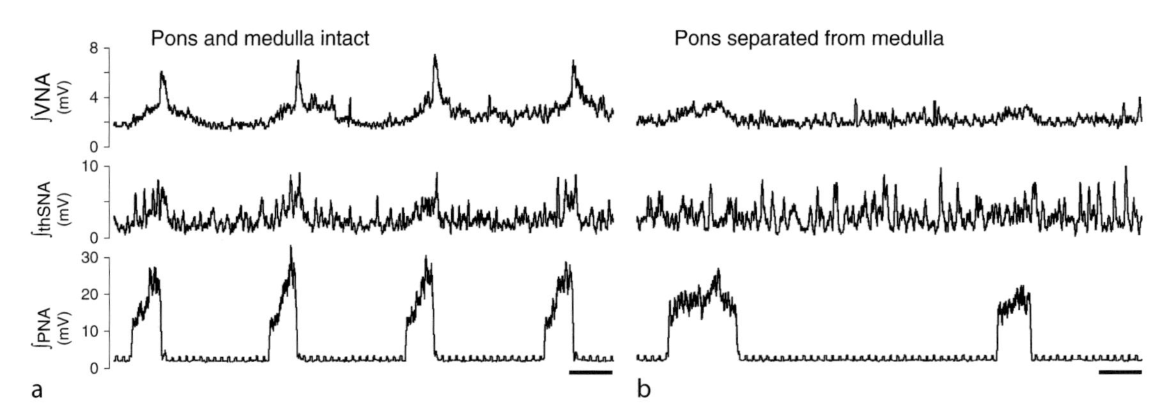

Pontine Control of Respiration. Figure 2 Recordings of respiratory-modulated motor activities recorded from a vagus nerve (VNA, upper trace), phrenic nerve (PNA, lower trace), and thoracic sympathetic chain (thSNA), before (left) and after (right) pontomedullary transection. (a) With the pons connected to the medulla (recording from working heart brainstem preparation, see Encyclopedia Article Central Integration of Cardiovascular and Respiratory Activity Studied *In Situ* by Paton), breathing is regular as depicted by uniform pattern of activity from cycle-to-cycle. Specifically, VNA, which has some inspiratory activity increases sharply in the postinspiratory phase immediately after inspiration ceases when PNA burst ends. Sympathorespiratory coupling is evident in the inspiratory modulation of thSNA. (b) After pontomedullary transection, the respiratory rhythm as well as motor activity is irregular and disrupted, in particular, nonuniform durations of respiratory phases and no postinspiratory activity in the vagal motor nerve (Bar 1s).

pattern does not return immediately to baseline even though the controlled physiological variables such as blood gases have returned to baseline. Instead, the pattern gradually returns to baseline and this depends on the lateral pons because if input from the lateral pons is blocked, then the breathing pattern returns to baseline immediately. The pons may be acting directly on the pattern generator, particularly postinspiratory medullary neurons and may be acting through its ability to gate or regulate the incorporation of afferent information by the medulla [6]. The regulation of sensory input is a common way that behaviors override reflex control.

Anatomical studies indicate that the lateral pons and nuclei of the solitary tract (nuclei of the medulla (NTS) that receive mechano- and chemo-sensory input) are connected reciprocally (Fig. 1 – horizontal two-headed arrow at the bottom). Sensory information is transmitted to the Kölliker-Fuse nucleus and, in turn, the Kölliker-Fuse nucleus "gates" or modulates the efficacy of afferent input in the NTS; in particular, activity of the Kölliker-Fuse nucleus can suppress sensory input in NTS [8]. However, the precise synaptic interactions between NTS and Kölliker-Fuse nucleus still need to be elucidated.

The Role of the Pons in Respiratory Pattern Dysfunction

The respiratory pattern is highly variable in a mouse model for the Rett syndrome, which is a neurodevelopmental disease with severe respiratory disorders and lack of vocalization. This respiratory disturbance is associated with impairment of pontine and vagal modulation of postinspiratory activity [9]. These data suggest that disturbance of the postinspiratory gating mechanisms causes severe respiratory disorders and that the pons has influence over variability, especially for other behaviors like vocalization.

Future Directions

Presently, while we know much about reflex control of the breathing and blood pressure both of which are controlled by pontomedullary networks, we know very little about the factors that integrate their control. This is especially true of those intrinsic to the brainstem. For instance, sympathorespiratory coupling is at least partially dependent on the pons (Fig. 2). Thus, the pons acts to coordinate the cardiorespiratory system to maintain homeostasis and oxygenation of vital organs including the CNS.

Oscillations (0.1 Hz) that are expressed in the blood pressure as ►Mayer Waves can influence the breathing pattern. Mayer Waves are thought to be related to baroreceptor reflex control of blood pressure. Preliminary analysis of pontine neuronal spike trains recorded continuously before and after vagal nerve transection suggests that in addition to the respiratory rhythm, a slow rhythm (~0.1 Hz) is also expressed by respiratory-modulated pontine activity. In some cases, this rhythm, which is normally masked, is strong enough that the respiratory rhythm itself becomes synchronized with it after vagotomy. The 0.1-Hz rhythm may be synchronized with Mayer Waves. In experiments in dogs, Mayer waves and the respiratory modulation of blood pressure, ►Traube-Hering waves, were found to synchronize after vagotomy [10]. Additional preliminary analysis suggests that many pontine neurons that have respiratory modulation are also modulated with the blood pressure pulse of the cardiac cycle. These lines of evidence may indicate that individual pontine neurons are arranged in networks that help coordinate the intertwined control of breathing and perfusion. However, a great deal of work remains to elucidate the mechanisms of that coordination.

References

1. Alheid GF, Milsom WK, McCrimmon DR (2004) Pontine influences on breathing: an overview. Respir Physiol Neurobiol 143:105–114
2. Ullmann E (1970) About Hering and Breuer. In: Porter R (ed) Breathing: Hering-Breuer centenary symposium. London, Churchill, pp 3–15
3. Cohen MI, Wang SC (1959) Respiratory neuronal activity in pons of cat. J Neurophysiol 22:33–50
4. Feldman JL, Gautier H (1976) Interaction of pulmonary afferents and pneumotaxic center in control of respiratory pattern in cats. J Neurophysiol 39:31–44
5. St John WM (1987) Influence of pulmonary inflations on discharge of pontile respiratory neurons. J Appl Physiol 63:2231–2239
6. Siniaia MS, Young DL, Poon CS (2000) Habituation and desensitization of the Hering-Breuer reflex in rat. J Physiol London 523:479–491
7. Dick TE, Coles SK (2000) Ventrolateral pons mediates short-term depression of respiratory frequency after brief hypoxia. Respir Physiol 121:87–100
8. Dutschmann M, Morschel M, Kron M, Herbert H (2004) Development of adaptive behaviour of the respiratory network: implications for the pontine Kölliker-Fuse nucleus. Respir Physiol Neurobiol 143:155–165
9. Stettner GM, Huppke P, Brendel C, Richter DW, Gartner J, Dutschmann M (2007) Breathing dysfunctions associated with impaired control of postinspiratory activity in Mecp2-/y knockout mice. J Physiol (London) 579: 863–876
10. Cherniack NS, Edelman NH, Fishman AP (1969) Pattern of discharge of respiratory neurons during systemic vasomotor waves. Am J Physiol 217:1375–1383

Pontine-Geniculate-Occipital (PGO) Waves

Definition

Bursts of excitation that arise in the brainstem and are subsequently detected in the visual thalamus and visual cortex.

►Sleep States

Pontine Micturition Center (PMC)

Definition

The PMC mediates spino-bulbo-spinal reflexes activated by sacral afferent neurons from the urinary bladder that are involved in micturition and continence of the urinary bladder. It consists of the medial PMC (Barrington's nucleus) which triggers micturition (contraction of the detrusor vesicae and relaxation of the external vesical sphincter) and the lateral PMC which maintains continence (inhibition of mechanisms leading to micturition, activation of the external vesical sphincter).

►Autonomic Reflexes
►Micturition

Pontine Nuclei

Definition

A group of neuronal populations located in the pons region of the hindbrain. They are the major source of the mossy fiber input to the cerebellar cortex in birds and mammals. They receive their input from the spinal cord, the striatum of the forebrain, and the tectum of the midbrain.

►Evolution of the Cerebellum

Pontine Respiratory Group

►Pontine Control of Respiration

Pontine Wave (P-Wave)

Definition

REM sleep-associated phasic field potential recorded in the pons. Lasting for 75–150 ms, the P-wave appears during REM sleep as clusters containing a variable number of waves (3–5 waves/burst) or a singlet with amplitudes from 100 to 150 μV and a frequency range of 30–60 spikes/min. P-wave is the pontine component of ponto-geniculo-occipital (PGO) wave. The P-wave is generated by the phasic activation of a group of glutamatergic cells in the pons. The P-wave is critically involved in the reactivation of both the hippocampus and amygdala to reprocess cognitive information and to form memory traces in the cortex.

►Sleep

Pontocerebellum

Definition

The hemispheres belong to the phylogenetic young neocerebellum and receive their afferences via the moss fibers of the pontocerebellar tract from the pontine nuclei. Therefore, one also likes to summarize all hemispheric sections to the so-called pontocerebellum.

►Cerebellum

Pontomedullary Respiratory Pattern Generator

►Pontine Control of Respiration

Ponto-Pontine Long-Lead Burst Neurons

Charles Scudder
Portland, OR, USA

Definition

Ponto-pontine neurons, in general, are neurons that have both their somata and terminal fields in the pontine reticular formation. Ponto-pontine ►long-lead burst neurons (►burst cells – long lead (LLBNs)) (PP-LLBNs) are ponto-pontine neurons that exhibit long-lead burst discharges during ►saccades. Although some PP-LLBNs have been experimentally identified, those having the well defined functions expected from theory remain hypothetical at this point.

Characteristics

Higher Order Processes

Based on experimental studies of the saccadic system and on models of the saccadic system that incorporate these experimental results, PP-LLBNs are needed to relay signals from higher saccadic command centers to the ►saccadic burst generator. These command centers consist of the deep and intermediate layers of the ►superior colliculus with weaker projections from ►frontal eye fields (FEF). The superior colliculus, in turn, coalesces saccade-related information from the frontal eye fields, the ►supplementary eye fields (SEF), the ►lateral intraparietal area (LIP), and the ►substantia nigra and issues the principal saccadic command. Neurons in the FEF and superior colliculus both have long-lead discharge patterns, and provide the long-lead signal to the PP-LLBNs.

Lower Level Processes

PP-LLBNs are thought to convey the saccadic command to the ►excitatory burst neurons (EBNs), ►inhibitory burst neurons (IBNs), and ►omnipause neurons (OPNs) of the saccadic burst generator. There are at least two, and possibly three, populations of PP-LLBNs that project to these neurons.

Parts of the Ponto-Pontine LLBN System

Excitatory Relay PP-LLBNs

The generation of saccades requires a powerful excitatory input to the EBNs and IBNs. This excitation could be provided by a direct input from the superior colliculus, or by way of PP-LLBNs intercalated between the superior colliculus and the burst neurons [1]. Experimental data supports both possibilities. In the monkey, the superior colliculus does not project directly to EBNs, but does project to long-lead burst neurons (LLBNs) located in the pontine reticular formation [2]. PP-LLBNs do project to the region containing EBNs (see below), and must connect with EBNs, or else EBNs would have no suitable input. In the cat, a direct connection from the superior colliculus to EBNs has been demonstrated, but a parallel indirect pathway exists that most likely involves PP-LLBNs [3–5]. Similarly, cat IBNs receive direct and indirect input from the superior colliculus [3–6]. This set of connections is illustrated in Fig. 1a.

Inhibitory Trigger Neurons

Another population of PP-LLBNs, called ►trigger neurons, is needed to initiate the saccade. Activity in the superior colliculus activates the trigger neurons, which inhibit the tonically active OPNs. The pause

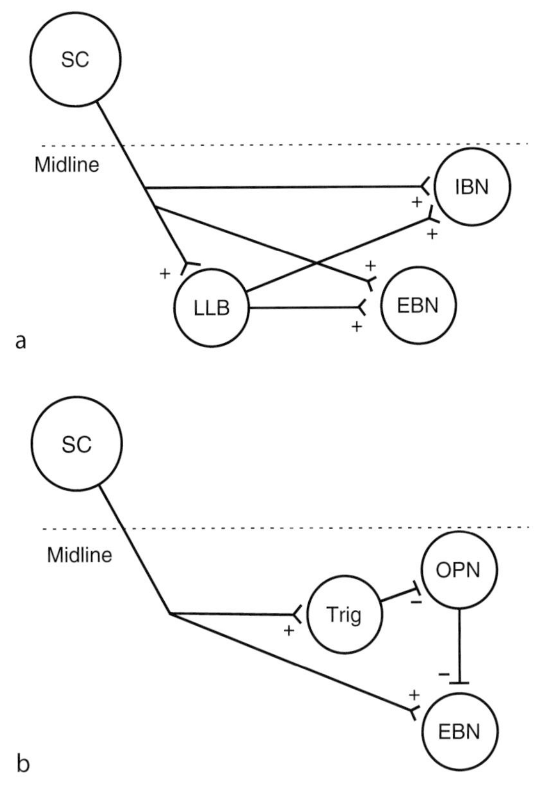

Ponto-Pontine Long-Lead Burst Neurons.
Figure 1 A portion of the wiring of the saccadic burst generator showing the connections of two pools of PP-LLBNs. (A) Relay PP-LLBNs (LLB) convey the output of the superior colliculus (SC) to the premotor burst neurons; excitatory burst neurons (EBN) and inhibitory burst neurons (IBN). The connections are all excitatory (+). The SC also has direct connections to the EBNs and IBNs in the cat. (B) Inhibitory PP-LLBNs (Trig) convey the output of the SC to omnipause neurons (OPN) in order to silence them and trigger the start of the saccade. Inhibition is signified by the "−" sign. A more complete diagram showing the relation of these components to the whole burst generator is presented in the brainstem burst generator section.

in OPN activity disinhibits the burst neurons (see ▶Brainstem burst generator) and allows them to discharge. The inhibitory trigger neurons intercalated between the superior colliculus and OPNs (Fig. 1b) are needed to accomplish the inhibition of the OPNs, because the efferents of the superior colliculus are excitatory. In agreement, microstimulation of the superior colliculus leads to inhibitory potentials in OPNs at disynaptic latencies, and intracellular recordings from OPNs in alert cats reveals a pre-saccadic phase of inhibition that is possibly mediated by trigger LLBNs [6–8]. Microstimulation in the ▶paramedian pontine reticular formation (PPRF) where LLBN somata are prevalent produces inhibition in OPNs [6].

Identified PP-LLBNs

Definitive anatomical and physiological evidence for the existence of PP-LLBNs consists of four PP-LLBNs revealed using the intraaxonal labeling technique [9]. The small sample size probably does not reflect a paucity of PP-LLBNs, but rather the difficulty and biases of the intraaxonal technique. All four neurons are anatomically different, but nonetheless innervate many of the same parts of the reticular formation. The diversity of these neurons, together with various fragments of incompletely labeled neurons, shows that there may be a large variety of PP-LLBNs.

Somata of three neurons were located in the rostral PPRF (NRPo) caudal and ventral to the trochlear nucleus. The soma of the fourth was located at the mid PPRF just rostral to the EBN area. Axons of all four descended in the PPRF, and all four axons gave off one or two branches that ascended to, and sometimes terminated in, the ipsilateral mesencephalic reticular formation (cf. Fig. 2). Most (3/4) had branches that terminated in ipsilateral NRPo, and all had branches that terminated in the part of ipsilateral caudal PPRF (NRPc) where EBNs are located. They also all had branches that innervated ▶raphe interpositus (the locus of OPNs) or the immediately adjacent region containing OPN dendrites. Less frequent targets included ▶NRTP, raphe pontis, the contralateral PPRF, the IBN area, and ▶nucleus reticularis gigantocellularis (one origin of reticulospinal pathways). One axon could be followed well into the medulla and may have been headed for the spinal cord.

As with other LLBNs, PP-LLBNs were mostly silent during fixation, and gradually began firing on average 44–107 ms preceding the start of ▶ipsiversive saccades. Firing rate peaked just before the saccade start, and ended just before saccade end. The spatial properties of their discharges were as varied as their anatomy. Two were ordinary ▶vectorial burst neurons (discharging only for saccades to a small circumscribed region of visual space – the ▶movement field), one was a vectorial burst neuron with an unusually wide movement field, and the fourth was a ▶directional burst neuron that discharged for all saccades into the ipsiversive visual space.

Pathology

PP-LLBNs and their axonal and terminal processes are intermixed in the PPRF with a multitude of eye-movement related and other neurons. Experimental and natural lesions of the region that contains PP-LLBNs necessarily destroy these other neurons as well, so deficits cannot be attributed to one population alone. Lesions of the PPRF can impair most types of horizontal eye movements; ▶VOR, ▶smooth pursuit, ▶optokinetic nystagmus, and saccades [10]. Many of these deficits can be traced to damage of fibers that traverse the PPRF, but

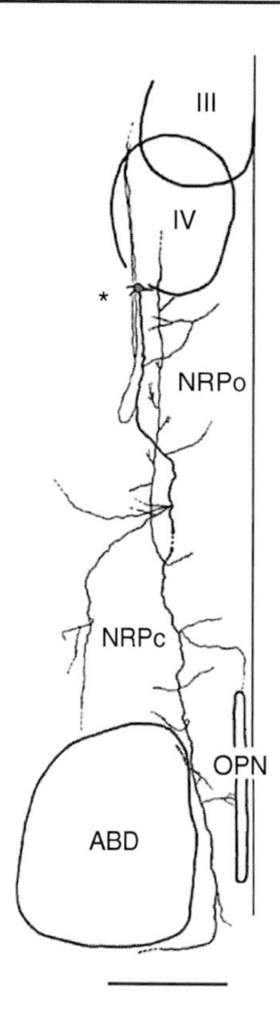

Ponto-Pontine Long-Lead Burst Neurons. Figure 2 Camera lucida drawing of a PP-LLBN in the left side of the pontine reticular formation reconstructed in a horizontal plane. The soma (*) is ventral and immediately caudal to the trochlear nucleus (IV), and the axon descends through the nucleus reticularis pontis oralis (NRPo) and ►nucleus reticularis pontis caudalis (NRPc). Terminal arborizations innervate these structures, including the part of NRPc containing excitatory burst neurons. Other branches terminate just lateral to the omnipause neurons (OPN). ABD = abducens nucleus, III = oculomotor nucleus. Dashed lines indicate fading of the label. Calibration bar is 1 mm.

deficits in saccades are surely the result of destruction of the saccade-related neurons in the PPRF. Small unilateral lesions cause a slowing and shortening of ipsiversive saccades. Larger unilateral lesions eliminate all ipsiversive saccades, and large bilateral lesions eliminate all horizontal saccades. Large caudal lesions that involve the OPNs also affect vertical saccades.

References

1. Hepp K, Henn V (1983) Spatio-temporal recoding of rapid eye movement signals in the monkey paramedian pontine reticular formation (P.P.R.F.). Exp Brain Res 52:105–120
2. Keller EL, McPeek RM, Salz T (2000) Evidence against direct connections to PPRF EBNs from SC in the monkey. J Neurophysiol 84:1303–1313
3. Chimoto S, Iwamoto Y, Shimazu H, Yoshida K (1996) Monosynaptic activation of medium-lead burst neurons from the superior colliculus in the alert cat. J Neurophysiol 75:2658–2661
4. Grantyn AA, Grantyn R (1976) Synaptic actions of tectofugal pathways on abducens motorneurons in the cat. Brain Res 105:269–285
5. Izawa Y, Sugiuchi Y, Shinoda Y (1999) Neural organization from the superior colliculus to motoneurons in the horizontal oculomotor system of the cat. J Neurophysiol 81:2597–2611
6. Kamogawa H, Ohki Y, Shimazu H, Suzuki I, Yamashita M (1996) Inhibitory input to pause neurons from pontine burst neuron area in the cat. Neurosci Lett 203:163–166
7. Yoshida K, Iwamoto Y, Chimoto S, Shimazu H (1999) Saccade-related inhibitory input to pontine omnipause neurons: An intracellular study in alert cats. J Neurophysiol 82:1198–1208
8. Yoshida K, Iwamoto Y, Chimoto S, Shimazu H (2001) Disynaptic inhibition of omnipause neurons following electrical stimulation of the superior colliculus in alert cats. J Neurophysiol 85:2639–2642
9. Scudder CA, Moschovakis AK, Karabelas AB, Highstein SM (1996) Anatomy and physiology of saccadic long-lead burst neurons recorded in the alert squirrel monkey. II. Pontine neurons. J Neurophysiol 76:353–370
10. Leigh RJ, Zee DS (1999) The neurobiology of eye movements. Oxford University Press, New York

Population Code

Synonyms

Also Ensemble Code

Definition

Population code (also ensemble code) denotes a code by which neural information is encoded in the spatiotemporal activity patterns of many neurons.

►Sensory Systems

Population Vector

Definition

An algorithm for estimating a vectorial quantity encoded by the spiking activity of a neuronal population given the response characteristics of each individual neuron to different values of that vectorial quantity (tuning curve). Originally proposed for decoding the hand movement direction from the firing rate of a population of directionally tuned neurons in the motor cortex of macaque monkeys. To compute the population vector in a specific condition (e.g. a specific movement direction), given the tuning curve of each cell (determined from the cell's response in several conditions) and the response vector associated to its maximum (e.g. the cell's preferred direction), all the response vectors, weighted by the cell firing rates in that specific condition, are summed together.

►Reaching Movements

Pore Loop

Definition

The pore loop represents a short amino acid sequence that forms the ion permeation pathway of tetrameric cation channels. In voltage-gated cation channels this domain is localized between the S5 and S6 segment.

The X-ray structure of the pore loop has been resolved in the bacterial KcsA channel. The domain consists of an α helical portion (the pore helix) and an uncoiled strand of 4–5 amino acid residues (the selectivity filter) forming the narrowest part of the pore.

►Cyclic Nucleotide-Regulated Cation Channels

Pore-loop Channels

Definition

Pore-loop channels all bear an extracellular, re-entrant loop, which provides a highly selective aqueous pore for particular ions. All pore-loop channels are structural derivatives of inward rectifying potassium (K^+) channels, and are the largest class of channels within the ion channel family. Pore loop channels include the voltage-gated channels, including the potassium, calcium (Ca^{2+}) and sodium (Na^+)-selective channels, the inward rectifying and two pore potassium channels and the glutamate receptors.

►Calcium Channels – an Overview
►Ion Channels from Development to Disease
►Glutamate Receptors
►Neuronal Potassium Channels
►Sodium Channels

Porphyropsins

►Photopigments

Position Sense

Definition

The ability to detect the position of joints of the body. It is tested clinically as part of assessments of proprioception.

►Joint position sense
►Proprioception and Orthopedics
►Proprioception Role of Joint Receptors

Positional Alcohol Nystagmus

Definition

Nystagmus resulting from alcohol intoxication. The nystagmus is due to the passage of alcohol into the cupula of each of the semicircular canals, which renders them lighter than the surrounding endolymph. The semicircular canals then respond to changes in head position resulting in positional nystagmus. A second phase is noted when alcohol diffuses from the cupula and into the endolymph. An oppositely directed positional nystagmus is then noted.

►Disorders of the Vestibular Periphery
►Semicircular Canals

Positional Cloning

Definition

Positional cloning is a technique used to identify genes, either associated with disease or with a mutant in an animal model, based on their location on a chromosome.

Position-Vestibular-Pause Neurons

KATHLEEN E. CULLEN
Department of Physiology, Aerospace Medical Research Unit, McGill University, Montreal, QC, Canada

Definition

Traditionally, the vestibulo-ocular reflex (VOR) is considered to be a stereotyped reflex that effectively stabilizes gaze by moving the eye in the opposite direction to concurrent head motion. The three neuron arc responsible for mediating the VOR was first described by Lorente de No' in 1933. This pathway consists of projections from vestibular afferents to interneurons in the vestibular nuclei, which in turn project to extraocular motoneurons. The simplicity of this three neuron arc is reflected in the fast response time of the VOR; compensatory eye movements lag head movements by only 5–6 ms in the primate [1]. Horizontal and vertical position-vestibular-pause (PVP) neurons are thought to constitute most of the intermediate leg of the direct pathway that mediate the VORs, which are evoked by yaw and pitch rotations, respectively. As the response of horizontal PVP neurons have been more extensively investigated in alert animals, they are the focus of this essay.

The results of recent investigations have changed our view of the VOR. In particular, neurophysiological recordings from PVP neurons provide firm evidence that the VOR is not a hard wired reflex, as had been commonly assumed. This essay first describes the responses of PVP neurons during field standard tests including passive whole body rotations and eye-movements in head-restrained monkeys. Next, the results of studies that have characterized PVP neurons during (i) redirections of gaze that are produced by coordinated eye-head movements, and (ii) gaze stabilization while viewing near versus far targets during head movements, are summarized. These findings are considered in relation to the sensory-motor transformations needed to guide behavior in a manner consistent with current gaze strategies.

Characteristics

Response During Standard Head-Restrained Rotations and Eye Movements

Type I Neurons

►Type I position-vestibular-pause (PVP) neuron constitute most of the intermediate leg of the direct pathway that produces the horizontal VOR (reviewed in [2]). The majority of type I PVP neurons are located in the rostral medial vestibular nucleus, and send an excitatory projection to the motoneurons of the (i) contralateral ►abducens nucleus, or (ii) ipsilateral medial rectus subdivision of the ►oculomotor nucleus (Fig. 1a). A minority send inhibitory projections to the motoneurons of the ipsilateral abducens nucleus. Within the abducens nucleus, motoneurons project directly to the ipsilateral lateral rectus muscle (LR), while a separate class of neurons (internuclear neurons) project to the contralateral oculomotor nucleus, which projects to the medial rectus muscle (MR; Fig. 1b). During horizontal eye movements, the effective force moving the eye is generated by the sum of the forces of the LR and MR.

Type I PVP neurons derive their name from the signals they carry during head-restrained rotation and eye movement paradigms. Their firing rates increase with contralaterally directed eye position; they are sensitive to

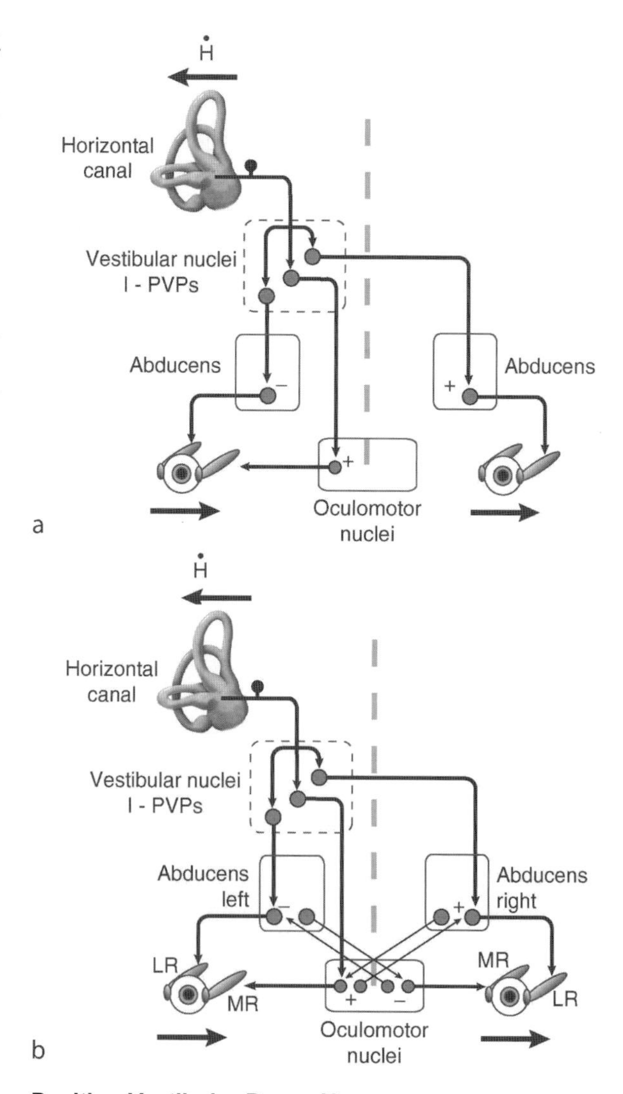

Position-Vestibular-Pause Neurons. Figure 1 (a) Direct VOR pathway; rotation of the head to the left generates right eye movements. (b) Connections between the abducens and oculomotor nuclei.

ipsilaterally directed head velocity during vestibular stimulation (i.e. a type I response); and their discharges cease (pause) for ipsilaterally directed saccades and vestibular quick phases. In addition, these neurons show modulation for contralaterally directed eye movements during ▶smooth pursuit tracking. The activity of a typical PVP neuron is illustrated in Figs. 2a–c. The neuron's mean firing rate is linearly related to eye position during periods of steady ocular fixation (Fig. 2a). During sinusoidal passive whole-body rotation in the dark at 0.5 Hz, neuronal modulation leads ipsilateral head rotation by about 10–20° (Fig. 2b), and the neuron pauses or stops

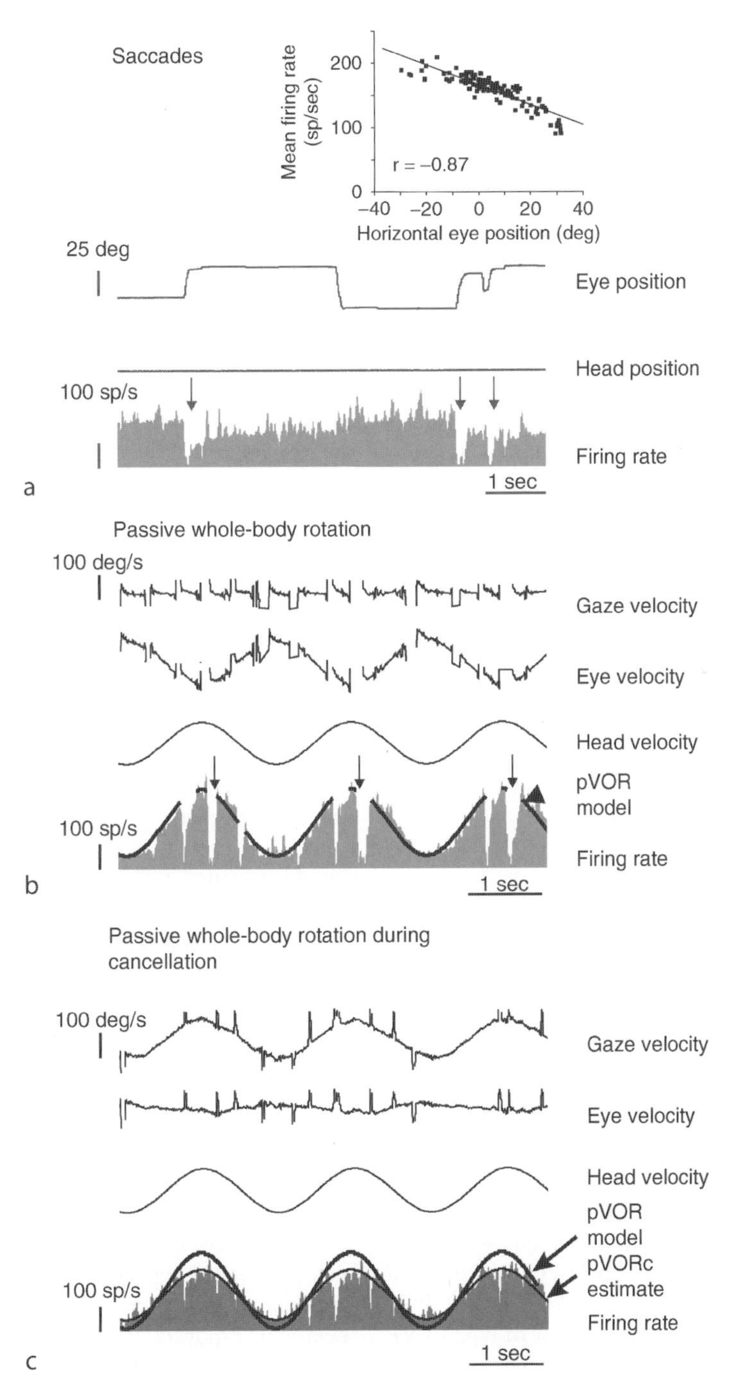

Position-Vestibular-Pause Neurons. Figure 2 Activity of an example type I PVP neuron in the head-restrained condition. (a) Responses are correlated with horizontal eye position during periods of steady fixation. (b, c) Responses to passive whole-body rotation during (b) the VOR in the dark (pVOR), and (c) cancellation of the VOR by fixation of a target that moves with head (pVORC).

firing during ipsilaterally directed vestibular quick phases (Fig. 2b, downward arrows). The behavior of PVP neurons during the compensatory slow phase VOR evoked by passive rotation can be well described by the equation:

$$fr = a + kE + gH' + cH''$$

where fr = neuronal firing rate, a is the resting discharge, E is eye position, k is the eye position sensitivity of the neuron during ocular fixation, H′ and H″ are head velocity and acceleration, respectively, and g and c are neuronal sensitivities to head velocity and acceleration, respectively.

In order to dissociate the vestibular-related modulation of PVP neurons from their eye-movement related responses, vestibular physiologists have traditionally utilized a paradigm in which the monkey "cancels" its VOR by tracking a target that moves with the head. The resulting vestibular stimulation does not lead to eye motion in the opposite direction to the head motion, since trained subjects can accurately follow the target at frequencies <1.5 Hz. Type I PVP neurons respond robustly to ipsilaterally directed head velocity in this condition, but show a 30% decrease in modulation as compared to VOR in the dark (Fig. 2c [3,4]). This reduction in head-velocity sensitivity occurs at latencies that are too short to be mediated by smooth pursuit pathways (see Fig. [3]), and supports the idea that there is a parametric adjustment of the gain of the direct VOR pathway while the VOR is voluntarily suppressed. Signals carried to the extraocular motorneurons by other premotor inputs help offset the residual modulation of PVP neurons, so that the eye remains immobile (see essay on VOR suppression).

Type II Neurons

During head-restrained rotations and eye movements, the head and eye movement sensitivities of type II PVP neurons are opposite to those of type I PVP neurons; firing rates increase in response to contralaterally-directed head rotation and ipsilaterally directed eye position and velocity. Otherwise, the firing pattern of these two types of neurons is very similar. The projections of type II PVP neurons are not known, however it is thought that they support inhibitory commissural pathways between vestibular nuclei [2] and that they contribute to the weak three neuron arc that, in part, mediates the translational VOR (see section on "Translational VOR" below). In addition, it is likely that the inhibitory inputs from type II PVP neurons contribute to the pause-behavior of type I PVP neurons during ipsilaterally directed saccades, vestibular quick phases, and ►gaze shifts. This proposal is consistent with known interconnections between the brainstem saccade generator and the vestibular nuclei [5].

PVP Neuron Modulation: Voluntary Head Movements

Recent work has shown that PVP neurons process vestibular information in a manner that depends principally on the subject's current gaze strategy, rather than whether the head movement was actively generated or passively applied. As described below, the head velocity signal carried by the direct VOR pathway is reduced when the behavioral goal is to redirect the visual axis of gaze. In contrast, PVP neurons robustly encode head velocity signals when the behavioral goal is to stabilize the visual axis of gaze relative to space, regardless of whether head movement is actively or passively generated.

Gaze Redirection

There is much accumulated evidence from studies in head-restrained monkeys to indicate that both type I and II PVP neurons differentially encode head-velocity during gaze redirection versus gaze stabilization. First, as described above, while PVP neurons encode head velocity during the compensatory slow phase component of the VOR evoked by passive whole-body rotation, they pause or significantly decrease their firing during vestibular quick phases where gaze is redirected. In addition, PVP neuron responses are significantly attenuated, as compared to passive rotation in the dark, when the VOR is suppressed during passive whole-body rotation by tracking a target that moves with the head. In this latter condition, the goal is to redirect the axis of gaze relative to space rather than stabile gaze [3,4].

It is useful to suppress PVP transmission in each of the above circumstances, since eye movement commands generated by the direct VOR pathways would function to drive the eye in the opposite direction to the intended change in gaze. An analogous argument can be made for situations in which the axis of gaze is voluntarily redirected, a combination of eye and head movements. In order to rapidly redirect the visual axis towards a target of interest, primates commonly generate coordinated eye-head movements, termed gaze shifts. Similarly, coordinated smooth head and eye movements (i.e. ►gaze pursuit) are frequently generated in order to track moving targets. Attenuating the modulation of the direct VOR pathways (i.e. type I PVP neurons) during either gaze shifts or gaze pursuit would also be behaviorally advantageous.

Indeed, during rapid orienting gaze shifts, the head-velocity related signals carried by type I PVP neurons are dramatically reduced [2,5,6]. As shown in Fig. 3a, neuronal responses to head velocity during ipsilateral gaze shifts are consistently attenuated relative to passive whole-body rotation in the dark (i.e. during the VOR, Fig. 2c). As a result, a model based on a neuron's response during passive rotation (Fig. 3a; heavy line: pVOR model) will systematically over-predict

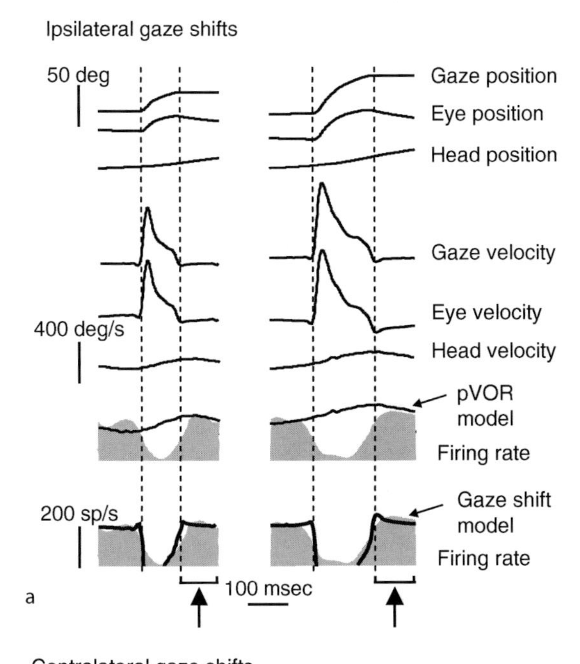

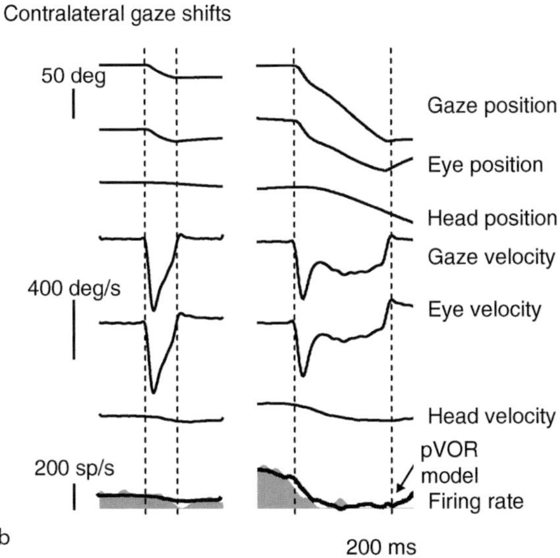

Position-Vestibular-Pause Neurons. Figure 3 The activity of a type I PVP neuron during and following ipsilaterally (a) and contralaterally (b) directed gaze shifts. Arrows indicate the post gaze shift intervals in (a).

its discharge during gaze shifts. Moreover, neuronal modulation is increasingly attenuated for larger amplitude gaze shifts reaching 70% attenuation for gaze shifts >60°, [2,5]. Type I PVP neurons also contribute little to the generation of the VOR during large contralaterally directed gaze shifts, since they are typically driven into inhibitory cut-off (Fig. 3b). Overall, the amplitude-dependent attenuation of the discharge of these neurons is consistent with the results of behavioral studies demonstrating that the VOR is more strongly suppressed during large than during small gaze shifts (see essay on "VOR suppression"). Similar results have been obtained from characterizations of type II PVP neurons [2].

The head-velocity related modulation of PVP neurons is also reduced when gaze is redirected to follow a moving target by means of coordinated eye-head pursuit. Responses to the voluntary head movements that are generated during eye-head pursuit are attenuated by approximately 30% compared to responses to passive whole-body rotation. Thus, the attenuation observed during eye-head pursuit is comparable to that observed when the VOR is suppressed by fixating a target that moves with the head (Fig. 2c), as well as that observed during small rapid gaze shifts (<25°). The histogram in Fig. 4 summarizes the head velocity-related signals that are carried by PVP neurons across different voluntary behaviors, and emphasizes the fact that transmission through the VOR pathways is attenuated during all behaviors that involve gaze redirections.

Gaze Stabilization

As described above, an intact VOR is not beneficial when the behavioral goal is to redirect gaze to a new target. In contrast, a fully functional VOR is clearly critical when the behavioral goal is to maintain stable gaze. Moreover, there have been reports that VOR performance is enhanced in response to active head rotations compared to passive whole-body rotations [1]. These findings have led to the suggestion that neck proprioceptive signals and/or a copy of the command to the neck motoneurons might function to augment transmission in the VOR pathways (i.e. PVP neurons) during active head movements.

Single unit recording experiments, however, indicate that neither neck proprioceptive nor neck motor efference copy signals augment the modulation of PVP neurons during active head-on-body movements. First, in rhesus monkeys, PVP neurons encode head movement in the same manner during passive head-on-body rotations and passive whole-body rotations (Fig. 5a; [1,2]). In addition, PVP neurons are not modulated in response to passive rotation of the body under a stationary head (Fig. 5b [2]). Thus, the passive activation of neck proprioceptors does not significantly alter the sensorimotor transformations carried out at the level of the direct VOR pathways. Second, neuronal responses are comparable during passive whole-body rotations and active head movements that are produced during periods of stable gaze [2,5,6]. For example, as shown in Fig. 3a, immediately following a rapid orienting gaze shift, the head continues to move even after gaze has stabilized relative to space (intervals denoted by the arrows). During this interval, a neuron's activity can be predicted based on its response to

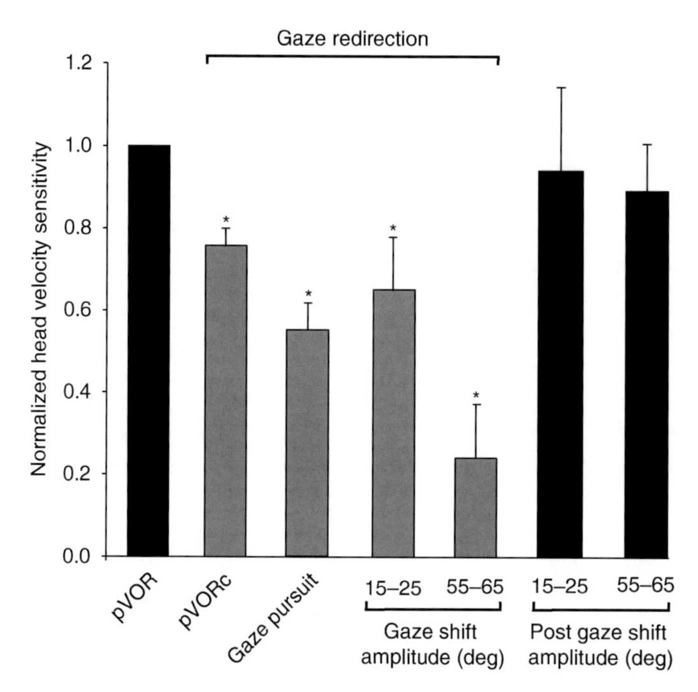

Position-Vestibular-Pause Neurons. Figure 4 Summary of type I PVP neuron discharge activity during passive and voluntary head motion. The (*) symbol denotes significant attenuation as compared to pVOR.

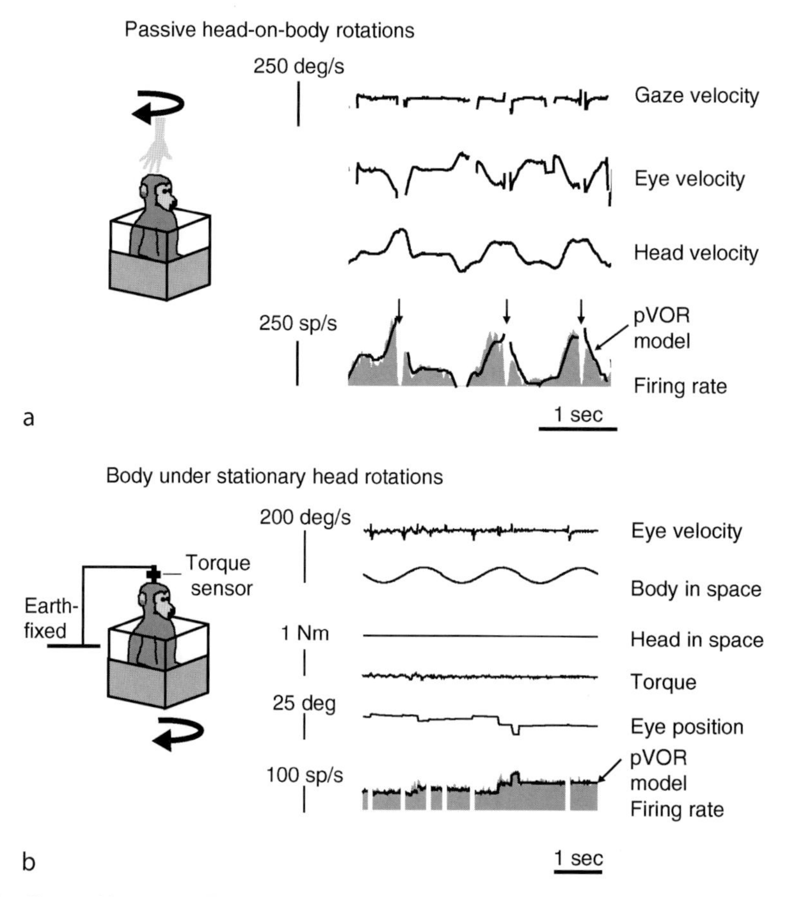

Position-Vestibular-Pause Neurons. Figure 5 (a) The head was passively rotated on a stationary body. (b) Neuronal responses are not modulated by passive stretching of neck proprioceptors produced by passively rotating the body under a stationary head.

passive whole body rotation (heavy line). Taken together, these results are consistent with accumulating evidence from behavioral studies showing that VOR performance is generally comparable during passive and active rotations of the head-on-body in primates [1].

Summary

Recent single unit experiments show that the head velocity signals carried by the direct VOR pathways are modulated in a manner that is consistent with the current behavioral goal. PVP neurons demonstrate robust head-velocity related modulation in response to self-generated and passively applied head rotations when gaze is stable. In contrast, when the behavioral goal is to redirect gaze relative to space, the head-velocity signals carried by PVP neurons are significantly reduced.

PVP Neuron Modulation: Near Versus Far Viewing During Rotations and Translations

Angular VOR

A second situation where the VOR shows behaviorally-dependent modulation is during the fixation of near versus far earth-fixed targets. During head rotations, the eyes translate as well as rotate relative to space, since they cannot both be perfectly aligned with the axis of rotation. Consequently, for the same amplitude of head rotation, a larger VOR gain is necessary to stabilize a near than a far earth-fixed target due to the differences in the translation of the target relative to the eyes (Fig. 6a). Differences in the responses of the type I PVP neurons that mediate the direct VOR pathways are consistent with these distance-related changes in VOR gain (Fig. 6b [7]).

Translational VOR

Recently, several investigations have specifically focused on the premotor pathways that generate the VOR in response to stimulation of the otoliths during translations (i.e. the translational VOR (TVOR); [8]). The latency of the TVOR is somewhat longer than that of the angular VOR; compensatory eye movements generally lag head movements by >10 ms in the primate. Although a direct disynaptic pathway has been shown to exist between the otoliths and the abducens nucleus, the longer latency of the TVOR suggests that it is primarily mediated by more complex polysynaptic pathways. During translation along the interaural axis, the gain of the horizontal TVOR response depends on

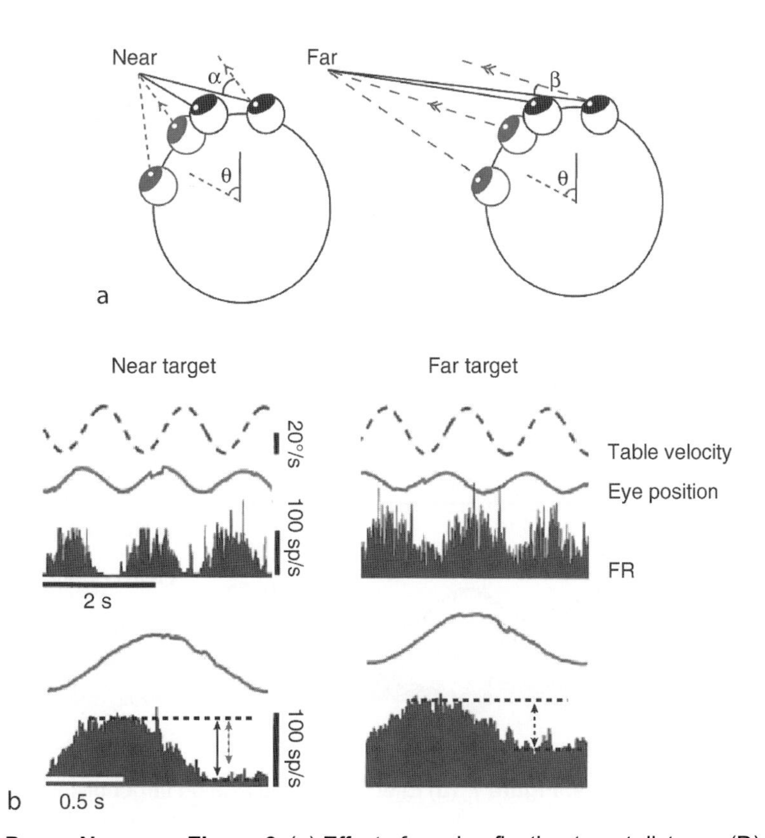

Position-Vestibular-Pause Neurons. Figure 6 (a) Effect of varying fixation target distance (D) on VOR gain for a fixed axis of rotation. (b) Responses of a type I PVP neuron when passive whole body rotation was applied while viewing a near (*left panel*) versus far (*right panel*) target.

the distance of the target being viewed (Fig. 7a). Moreover for translations along the nasal-occipital axis, the amplitude and sign of the eye movements evoked by the TVOR depend on gaze angle as well as viewing distance. The latter finding is of particular interest, since it demonstrates that direction as well as the amplitude of the TVOR response is modified in a behaviorally-dependent manner.

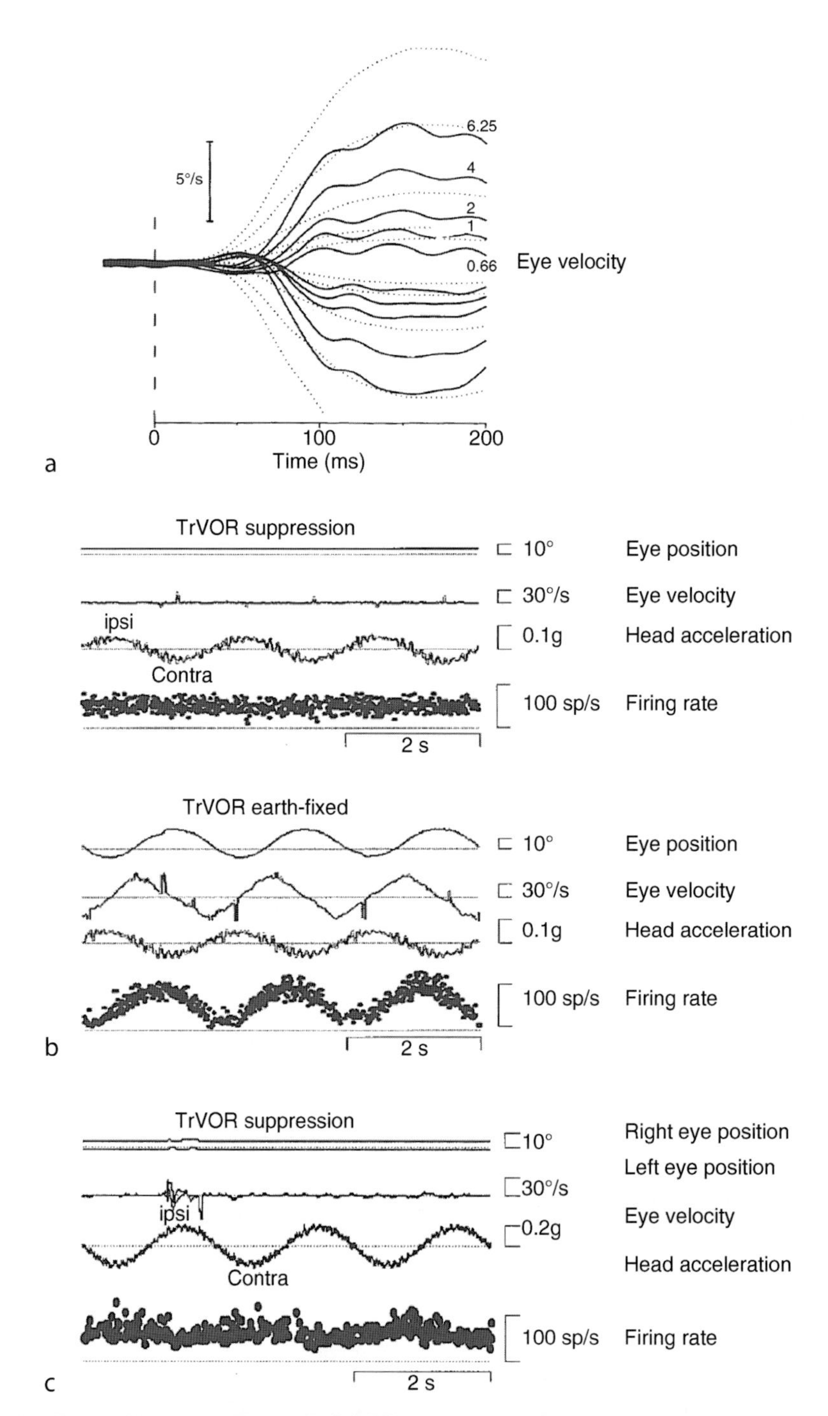

Position-Vestibular-Pause Neurons. Figure 7 (a) Effect of varying fixation target distance on VOR gain during translation along the interaural axis. Continuous lines represent average eye velocity, while dotted lines indicate the ideal response that would be required for perfect gaze stability. (b, c) Responses of an example type I (b) and type II (c) PVP neuron during lateral translation.

Type I PVP neurons, which constitute the main interneuron in the direct angular VOR pathway, do not receive direct otolith inputs and thus do not contribute to the three neuron pathway that mediates the direct TVOR pathway. This is shown in Fig. 7b, where a typical neuron shows negligible response modulation during translation when the subject fixates a head-fixed target (TrVOR suppression; [8,9]). Comparison with Fig. 2c highlights the striking difference in response to type I PVP neurons during suppression of the TVOR and suppression of the angular VOR. However, when a subject fixates an earth-fixed target during translation (Fig. 7b, TrVOR earth fixed), neurons modulate in a manner that is consistent with their oculomotor-related response during smooth pursuit [9]. These results are in general agreement with previous studies that have compared responses during on and off-centered rotations [7,10]. Given that the type I PVP neurons show robust modulation during the TVOR, and that they project directly to the extraocular motoneurons (Fig. 1), it can be concluded that they contribute to the generation of the reflex via inputs from polysynaptic (but not direct) pathways.

In contrast, type II PVP neurons show slight response modulation during suppression of the VOR during translation (Fig. 7c), suggesting that these neurons contribute to the relatively weak ipsilaterally projecting three neuron arc that mediates the direct TVOR pathway. However, these neurons do not change their response amplitude for near versus far viewing during TVOR suppression [9]. Additional inputs, most likely transmitted via cerebellar/floccular pathways, appear to be necessary to modulate the TVOR as a function of gaze angle and viewing distance [7,8,9].

References

1. Cullen KE, Roy JE (2004) Signal processing in the vestibular system during active versus passive head movements. J Neurophysiol 91:1919–1933
2. Roy JE, Cullen KE (2002) Vestibuloocular reflex signal modulation during voluntary and passive head movements. J Neurophysiol 87:2337–2357
3. Cullen KE, McCrea RA (1993) Firing behavior of brain stem neurons during voluntary cancellation of the horizontal vestibuloocular reflex. I. Secondary vestibular neurons. J Neurophysiol 70:828–843
4. Scudder CA, Fuchs AF (1992) Physiological and behavioural identification of vestibular nucleus neurons mediating the horizontal vestibuloocular reflex in trained rhesus monkeys. J Neurophysiol 68:244–264
5. Roy JE, Cullen KE (1998) A neural correlate for vestibulo-ocular reflex suppression during voluntary eye-head gaze shifts. Nat Neurosci 1:404–410
6. McCrea RA, Gdowski GT (2003) Firing behaviour of squirrel monkey eye movement-related vestibular nucleus neurons during gaze saccades. J Physiol 546:207–224
7. Chen-Huang C, McCrea RA (1999) Effects of viewing distance on the responses of vestibular neurons to combined angular and linear vestibular stimulation. J Neurophysiol 81:2538–2557
8. Angelaki DE, Green AM, Dickman JD (2001) Differential sensorimotor processing of vestibulo-ocular signals during rotation and translation. J Neurosci 21(11):3968–3985
9. Meng H, Green AM, Dickman JD, Angelaki DE (2005) Pursuit – vestibular interactions in brain stem neurons during rotation and translation. J Neurophysiol 93(6):3418–3433
10. McConville KM, Tomlinson RD, Na EQ (1996) Behavior of eye-movement-related cells in the vestibular nuclei during combined rotational and translational stimuli. J Neurophysiol 76:3136–3148

Positive Feedback Control

Definition

A mechanism used to regulate the value or time course of an output variable or signal when the output variable is determined by the value of an input signal or the time course of an input signal. The control mechanism is said to be closed loop when the value of the output variable is sensed or measured, is fed back and compared to some desired reference value and is then used to determine the value of the input signal. The closed loop control mechanism is referred to as positive feedback control when a given change in the output variable produces a change in the input signal that causes a further change in the output in the same direction.

Positive feedback control is often unstable, but when appropriately configured systems can remain stable while using positive feedback control.

►Posture – Sensory Integration

Positive Schizophrenic Symptoms

Definition

Symptoms associated with reality distortion; most frequently auditory hallucinations (hearing voices), feeling of being observed or persecuted.

►Schizophrenia

Positron Emission Tomography

SVYATOSLAV V. MEDVEDEV
Institute of the Human Brain of the Russian Academy of Sciences, Laboratory of the Positron Emission Tomography, St-Petersburg, Russia

Definition

Positron emission tomography (PET) is a nuclear imaging technique that allows quantitative evaluation of biochemical and physiological processes in vivo, by using ►radiopharmaceuticals (RPs) labeled with short-lived positron-emitting radionuclides, which are detected by their annihilation radiation with electronic coincidence detector systems.

Purpose

PET can be used in both research and clinical purposes for quantitative mapping of various physiological and biochemical processes. A number of these processes depend on the ►radiotracers available.

Clinical Purpose

Oncology: The most common application of PET is to determine the presence, severity and staging of cancer, its recurrences and responses to treatment. In neuro-oncology, PET has been found useful for the differentiation between radiation necrosis and ►glioma recurrence with 2–^{18}F-2-deoxy-D-glucose (►FDG) (less with labeled amino acids), glioma grade determination (FDG), and guidance for biopsy [1].

Whole-body FDG-PET enables the identification of primary lesions as well as local recurrence, lymph node involvement and distant metastases; so far, it has been quite a useful technique for tumor staging. PET provides more benefits for patients with non-small cell lung cancer (NSCLC), malignant melanoma, breast cancer, lymphoma, and colorectal cancer. PET also proved to be useful in radiation-treatment planning as well as in monitoring treatment responses (radio- or chemotherapy) [2,3].

Neurology: PET has significant implications in making a precise diagnosis of various ►neuropsychiatry and ►movement disorders. Clinical application includes lateralization of epileptic foci in ►temporal lobe epilepsy prior to surgery (FDG or rarely ►FMZ – flumazenil); differentiation between various types of ►dementia (FDG); assessment of dopaminergic neuron degeneration in ►Parkinsonism (6-FDOPA); and recognition of ►depression syndrome (WAY100635) [4].

Cardiology: PET is a modern tool for quantitative measurements of myocardial blood flow (^{13}N-ammonia; ^{82}Rb) and assessment of myocardial viability (FDG), and provides important criteria for a patient's selection for a revascularization. Imaging of the ►sympathetic nervous system with PET is possible but not very common in clinical cardiology [5].

Research

The most common research [4,6] application of PET involves ►activation studies using high-sensitivity imaging of regional cerebral blood flow (rCBF) from multiple [^{15}O]water injections. The alteration in neuronal activity in particular brain regions correlates with rCBF changes and the under performance of particular activities. Thus, PET is used to examine the spatial brain organization of the maintenance of cognitive, sensory, motor, emotional and other processes and responses to drug application.

Due to its high sensitivity and high specific activity (SA, the amount of radioactivity per mole) of the radiotracers, PET allows one to obtain quantitative information about the distribution of the target receptors throughout the brain, their affinity and density. The results are employed in fundamental studies of the pathogenesis of various neuropsychiatric disorders.

PET is a modern tool for the quantitative evaluation of drug-binding sites in living humans. By conducting PET studies with a suitable ►radioligand before and after treatment by a drug, the fraction of the total number of binding sites that are occupied by the drug can be quantified ("drug occupancy study)." Thus, the mechanism of drug action and optimal dosage can be evaluated in a very safe manner using the so-called "PET micro-dosing concept." With this approach, the number of patients to be studied in phase II clinical trials can be minimized from thousands to tens, resulting in a reduction in the time and costs of the studies.

Principles

Basic Principles in PET Imaging

When a PET [4,7] ►radionuclide is introduced into the human body, at any given time, part of the nucleus will decay emitting a positron (positively charged electron) and a neutrino. The neutrino leaves the body without interaction. The positron, after a series of scatterings, annihilates with an electron. As a result, two photons (gamma quantum) of equal energy, i.e., 511 keV, are emitted in opposite directions at almost 180° (Fig. 1a) and can be detected by an external detector system.

If arrays of gamma detectors are gathered around the body, a coincident event – simultaneous signal from a pair of detectors - means that annihilation took place somewhere on the straight line (line of response) connecting this pair (Fig. 1b).

The next annihilation induces coincident events, usually in another pair of detectors, and gives another line of response. The density of all intersections of such lines yields a spatial distribution corresponding to the

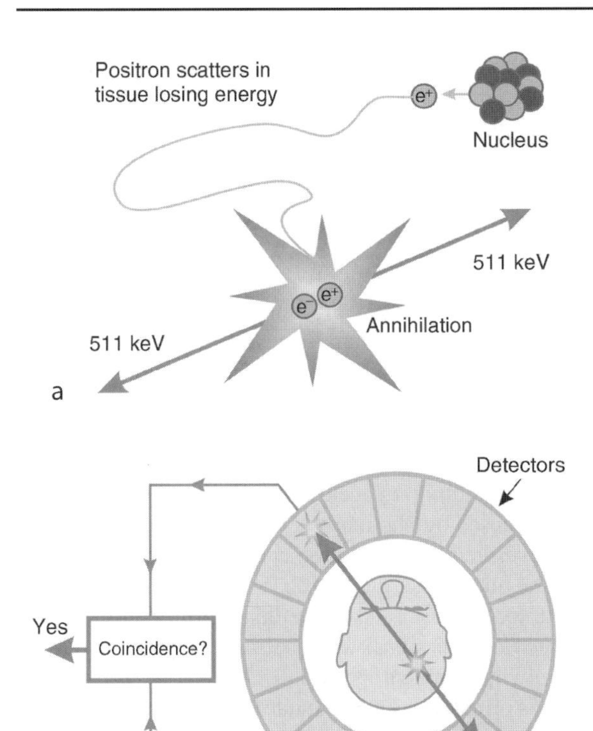

Positron Emission Tomography. Figure 1 (a) When a PET radionuclide is introduced into the human body, at any given time part of the nucleus will decay emitting a positron. After a series of scattering, the positron annihilates with an electron. As a result, two photons (gamma quantum) of equal energy, i.e., 511 keV are emitted in opposite directions very close to 180° apart. (b). If arrays of gamma detectors are gathered around the body, a coincidence event – simultaneous signal from a pair of detectors – means that annihilation took place somewhere on the straight line (line of response) connecting this pair.

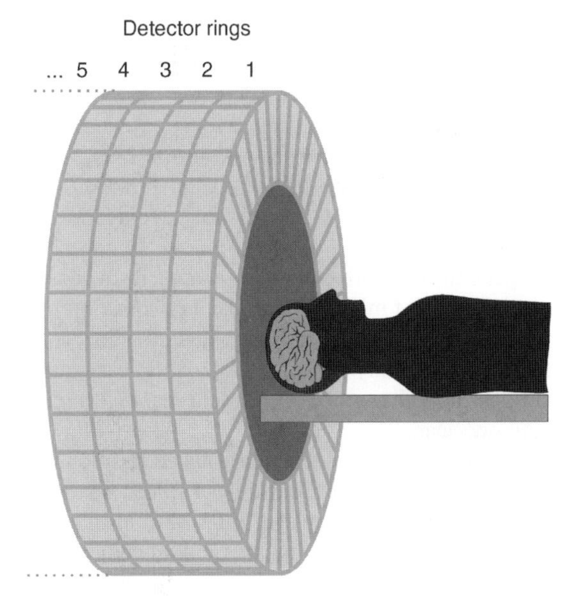

Positron Emission Tomography. Figure 2 A PET scanner consists of up to 30,000 detectors (scintillation crystals with photomultipliers), arranged in rings, formed in a cylinder around the body.

isotope concentration map. This method of coincidence registrations is the essence of PET and provides for its high efficiency. Information is contained in the direction of the photon movements. In other systems such as gamma cameras, this information is provided by using special metal tubes – collimators, which allow only the minority of all photons to enter the collimator and thus reach the detectors, while the majority is absorbed in the metal. Thus, in this system, each detector detects only photons coming from one given direction. The coincidence mode of registration and the use of short-lived radionuclides (2–110 min half-life) mostly contribute to the high sensitivity of PET, which allows the detection of pikomole amounts of substances. As a result, studies of drug abuse (cocaine, amphetamine) and toxic compounds become possible at concentration levels that do not cause any pharmacological effect [8].

A PET scanner consists of up to approximately 30,000 detectors (scintillation crystals with photomultipliers) arranged in rings formed in a cylinder around the body (Fig. 2).

These detectors involve a coincidence registration circuit, which collects information about coincidence events – counts. Using conventional algorithms for image reconstruction including different types of corrections, a volume isotope concentration map can be obtained. The unique feature of the PET coincidence technique is that corrections for radiation losses can be performed within the body (attenuation correction). This procedure is called “transmission scan.”

Theoretically, the spatial resolution of PET scanners is limited by physical characteristics of positron flight within the tissues. When a positron is emitted, it travels a short distance from the nucleus, typically about 2–8 mm maximum range. It loses its kinetic energy during this flight and then annihilates with an electron. It is the distance between the decaying nucleus and the point of annihilation, and the fact that the annihilation photons are not emitted at exactly 180° apart (deviation from 180° is up to 0.25°), which ultimately limit the spatial resolution of PET brain scanners to 4–5 mm (on average). However, in general, PET offers higher resolution than compared with 7–15 mm for single-photon emission computer tomography (SPECT).

PET allows registration of the counts per pixel, which are later transformed to counts per minute per ml of tissue. However, in the counts acquisition process, the allowable radiation dose (regulated by the authorities) and limited acquisition time (depends on the instrumentation and study protocol) have to be considered. The result may be a poor ▶signal-to-noise ratio, which can be overcome by the use of spatial filtration. The latter is an additional source for degradation in resolution.

In general, the collection time varies from tens of seconds to tens of minutes. It is like an exposition time in photography and creates a strong limitation to the study protocol. Only steady states or at least quasi-steady or periodic processes should be investigated during one scan.

In modern PET scanners for data acquisition, so-called 2D and 3D modes are widely accepted. In the 2D mode, special septa between rings are installed to reduce the effect of scattering and random coincidence.

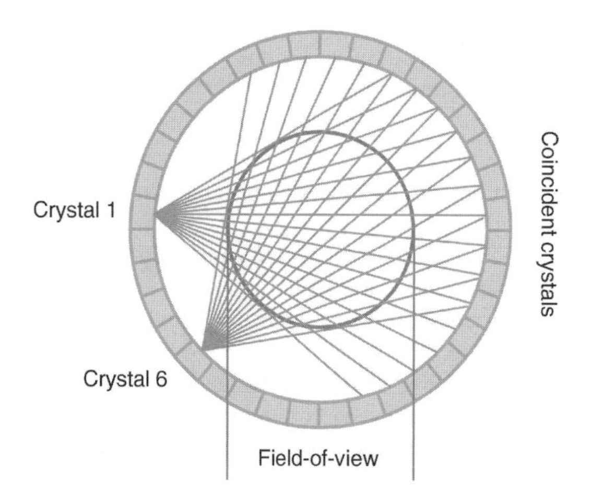

Positron Emission Tomography. Figure 3 Each detector of the ring forms a fan with a number of opposite detectors. Overlapping of all fans forms the field of view.

As a result, each detector forms three fans between itself and an array of other detectors of the ring (Fig. 3) as well as the two nearest rings.

As a result, there is a reconstruction of three slices: one in the ring and two so-called cross slices. It means that 15 rings give 29 slices. This procedure gives better resolution and accuracy, but is rather wasteful because many lines of response between other rings are not collected. If the septa are retracted, coincidences are admitted from large axial acceptance angles. The 3D mode allows increasing the number of counts by five times. This results in increases of necessary computing resources and, what is more important, in increases of mis-positioned events caused by scattered photons and in the registration of accidental coincidences, including some caused by photons outside the field of view. This, however, reduces resolution and quantitative accuracy. On the other hand, the 3D mode allows the reduction of the injected dose, and/or the duration of study, and/or improving the signal-to-noise ratio. The 3D mode is mainly used in neurophysiology, where the duration of scan is important. The 2D mode is more often applied in oncology (Fig. 4).

Finally, PET provides a map of spatial distribution of a positron-emitting isotope density in the field of view of the rings. When a compound labeled by this isotope is introduced into humans (usually by intravenous injection), it distributes within the body via the blood circulation in accordance with the delivery, uptake, metabolism and excretion of the particular tracer. To translate the measured radioactivity distribution into functional or physiological parameters, compartment models are used for the radiotracers with known metabolism. One of the most important characteristics of cell functioning is energy consumption. It can be compared to "gasoline consumption" in an engine, so for the living cells, the gasoline is glucose. However, it is like a gas that an engine absorbs and excretes.

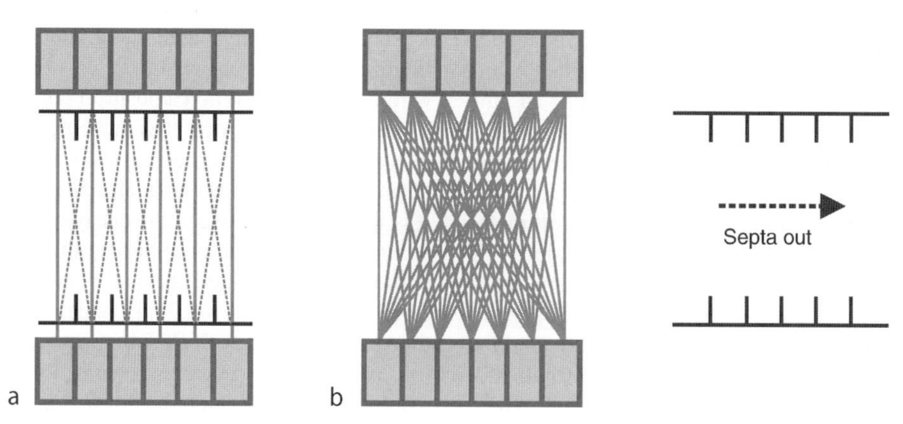

Positron Emission Tomography. Figure 4 *Left*: 2D mode. Septa allow each detector to form lines of responses between itself and an array of other detectors of the same ring and the two nearest rings only. *Right*: 3D mode. Lines of responses between all rings are allowed.

2–^{18}F-2-deoxy-D-glucose (FDG), a glucose analog, has a similar rate of consumption as normal glucose, but a different way for excretion. It is accumulated in a cell at a rate proportional to the energy consumption (metabolism) of this cell. Therefore, PET with FDG allows direct assessment of the level of glucose consumption, which is one of the most important processes responsible for vital functions [6]. As glucose metabolism is greatly enhanced in malignancies, FDG is the most important and popular radiotracer for PET oncology. In fact, this tracer has many other applications in neurology and cardiology and is considered the "working horse" of PET (like ^{99}mTc in conventional nuclear medicine).

In addition to the high sensitivity (see above), the use of short-lived radioisotopes allows injection of a relatively high activity of the tracer (185 MBq for FDG brain scan), leaving the total radiation dose within acceptable limit. As PET radionuclides belong to the major elements of life, unlimited numbers of radiotracers can be prepared to track various physiological processes. In practice, although more than 2000 RPs have been evaluated in PET, no more than 10–15 specimens have been introduced into clinical routines. The reasons are the difficult synthesis and the necessity for automation in operating high levels of radioactivity, high running costs, and very strict regulations.

The logic of a PET study is as follows. First, a field of interest has to be specified (i.e., neurophysiology, oncology, and cardiology). Within this field (let's say oncology), the process of interest has to be identified (glycolysis or amino acid transport) and an appropriate tracer considered (FDG or ^{11}C-metionine). The next steps are radiotracer synthesis and PET study design using an appropriate pharmacokinetic model, after this the PET study itself. The final stage is data processing and assignment of the image to the pathology under study.

It should be emphasized that PET is a functional imaging technique, giving an isotope distribution map, which reflects a particular biochemical process, not the anatomy. The introduction of a hybrid system (PET-CT: PET-computer tomography) has greatly enhanced the performance and accuracy of PET imaging. The CT component is used to relate the signal of radiotracer to anatomical landmarks and to correct for non-uniform attenuation (instead of traditional transmission scans) [3].

PET Radionuclides

PET employs radiotracers (radiopharmaceuticals, RPs) labeled with short-lived positron-emitting radionuclides [9]. The four conventional radionuclides are: ^{15}O (half-life $T_{1/2} = 2$ min); ^{13}N ($T_{1/2} = 10$ min); ^{11}C ($T_{1/2} = 20.4$ min) and ^{18}F ($T_{1/2} = 110$ min). Carbon, oxygen, and nitrogen are elements of life and the building blocks of nearly every molecule of biological importance. A fluorine-18 is often used to replace a hydrogen atom or hydroxyl group in a molecule.

Due to this short half-life, the PET radionuclides have to be produced in the vicinity, normally with a small dedicated cyclotron. PET cyclotrons accelerate charged particles (protons, deuterons) at a fixed energy (10–18 MeV for protons). Modern PET cyclotrons are negative-ions machines, which are characterized by an easy extraction process and dual beam option. PET radionuclides are produced in cyclotron targets via various nuclear reactions and delivered into a shielded hot cell by either gas flow (15O, 11C) or extra-pressure of helium (^{13}N, ^{18}F).

PET Radiopharmaceuticals

For PET applications, the radionuclides have to be tagged to specific pharmaceuticals, referred to as "radiopharmaceuticals" [10]. PET radionuclides are produced in a simple chemical form. They have to be transferred into tracers of interest via a complex synthesis using special automated modules.

The 2-min half-life of ^{15}O is very short; therefore, only very simple radiotracers like water-^{15}O and ^{15}O-butanol are produced as rCBF agents. Quite rarely, ^{15}O-labelled gases are used in inhalation studies. ^{13}N-ammonia ($T_{1/2} = 10$ min) is available from a cyclotron target to fulfill the needs of heart-perfusion tracers.

The chemistry of carbon-11 ($T_{1/2} = 20$ min) is extensively developed. Depending on the target gas, ^{11}C is taken from the target in the form of $^{11}CH_4$ or $^{11}CO_2$. The latter is a versatile agent for labeling of carboxylic acids, such as 1–^{11}C-acetate, a tracer for oxidative myocardial metabolism. Most of the ^{11}C-preparations are based on ^{11}C-methylations, including L-^{11}C-methyl-methionine, a second important tumor-seeking agent after FDG. Receptor radioligands such as ^{11}C-SCH23390 (D_1), ^{11}C-raclopride (D_2), ^{11}C-PE2I (dopamine transporter), ^{11}C-MADAM (serotonin transporter), ^{11}C-flumazenil (central BZ), ^{11}C-PK1195 (peripheral BZ), and ^{11}C-OH-BTA1 (β-amyloids) are obtained by this method.

The longest-living ^{18}F (110 min), allows several doses of the RPs to be obtained in one batch and to be delivered to other hospitals without access to a cyclotron. Irradiation of ^{18}O-enriched water by protons is most commonly used for generating high amounts of ^{18}F (35–70 GBq). Radionuclide is used in nucleophilic fluorination reactions to produce FDG. Although FDG is used in more than 80% of the routine PET studies, it is not a specific tracer for tumors as it enters other glucose-utilizing cells. New radiotracers for accurate characterization of tumors include *O*-(2′-^{18}F-fluoroethyl)-L-tyrosine (FET) or 2-[^{18}F]fluoro-L-tyrosine (2-FTYR). Due to the low accumulation in gray matter, these amino acids provide higher contrast

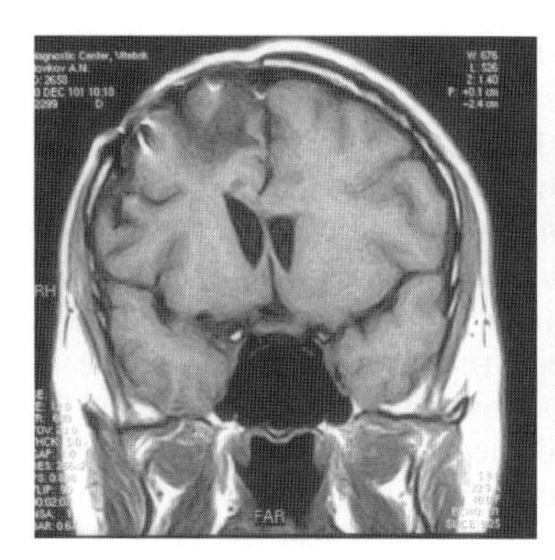

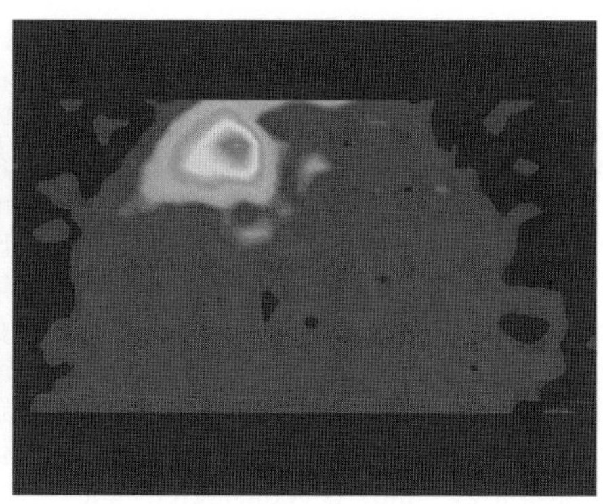

Positron Emission Tomography. Figure 5 Patient after tumor removal and radiotherapy. *Left*: MRI diagnosis splits between radiation necrosis and recurrence of tumor. *Right*: PET with 11C-metionine proved recurrence.

images of brain tumors. FLT, a labeled thymidine analog, was introduced for prognostic assessment and evaluation of responses to anti-proliferative therapy in colorectal, lung and other cancers. Assessment of tumor hypoxia using hypoxia markers (^{18}F-FMISO, ^{18}F-FAZA) allows one to select patients for treatments specifically designed to attack poorly oxygenated (hypoxic) tumor cells.

Data Processing and Analysis

Modern software for PET data processing and analysis usually includes means for: (i) preliminary data processing (smoothing, filtering, co-registration of images from different modalities, spatial normalization, i.e., image deformation to match the standard one); (ii) data 2D and 3D visualization; and (iii) statistical analysis. The most widely used software for research environment is the Statistical Parametric Mapping (SPM) software package (http://www.fil.ion.ucl.ac.uk/spm/).

For some study purposes (activation study, some studies in oncology), it is often enough just to compare the numbers of counts from different body areas without calculating the real concentration of the radiotracer.

Advantages and Disadvantages

At present, single-photon emission computer tomography (SPECT) makes up the majority of nuclear medicine procedures, mostly due to lower costs and availability of radiotracers from commercial sources (^{123}I, ^{111}In) or isotopic generators (^{99m}Tc). Due to the higher sensitivity of PET, the detectable amounts of molecules are lower than with SPECT. This is extremely important for receptor and drug development studies with very low amounts (pikomoles) of the substances involved. Unlike SPECT, PET allows a quantitative evaluation of the results using tracer kinetic modeling. In clinical studies, PET is usually used after ►magnetic resonance imaging (MRI) studies, and sometimes PET results can radically change a diagnosis (Fig. 5). In whole-body PET studies, PET highlights peculiarities, which exist but are unrecognizable on MRI images. Due to the unlimited number of natural substrates, substrate analogs and drugs that can be labeled, PET allows the study of practically all varieties of physiological processes.

The major limitations of PET are the complexity of PET studies, high capital investments and running costs for the production of RPs requiring an on-site cyclotron. Recently, FDG has been delivered over distances corresponding to 2 h flight. Many stand-alone PET scanners are installed and served from one central "cyclotron/radiochemistry factory." PET has proved to be cost-effective in staging and managing of certain malignancies such as NSCLC, by reducing the overall health care reimbursement. Due to the higher diagnostic accuracies of PET procedures, patients always benefit, even though the costs may be higher than with CT or MRI, which basically rely on morphological changes for tumor detection.

With respect to activation studies, functional MRI (fMRI), using the ►blood oxygenation level-dependent (BOLD) contrast method with echo-planar imaging, competes with PET. However, high levels of noise and ►claustrophobia effects result in several limitations in cognitive function assessments using this technique.

References

1. Wiebe LI (2004) PET radiopharmaceuticals for metabolic imaging in oncology. Int Congr Ser 1264:53–76
2. Bergstrom M, Grahnen A, Langstrom B (2003) Positron emission tomography microdosing: a new concept with application in tracer and early clinical drug development. Eur J Clin Pharmacol 59:357–366
3. McQuade P, Rowland DJ, Lewis JS, Welch MJ (2005) Positron-emitting isotopes produced on biomedical cyclotrons. Curr Med Chem 12(7): 807–818
4. Mazziotta JC, Toga AW, Frackowiak JRS (2000) Brain mapping. The Disorders. Academic Press, New York
5. Dobrucki LW, Sinusas AJ (2005) Cardiovascular molecular imaging. Semin Nucl Med 35:73–81

6. Frackowiak RSJ, Friston KJ, Frith CD, Dolan RJ, Mazziotta JC (1997) Human brain function. Academic Press, New York
7. Zanzonico P (2004) Positron Emission Tomography: a review of basic principles, scanners design and performance, and current systems. Semin Nucl Med 34:87–111
8. Fowler JS, Ido T (2002) Initial and subsequent approach for the synthesis of 18FDG. Semin Nucl Med 32:6–12
9. McQuade P, Rowland DJ, Lewis JS, Welch MJ (2005) Positron-emitting isotopes produced on biomedical cyclotrons. Curr Med Chem 12(7):807–818
10. Welch MJ, Redvanly CS (2003) Handbook of radiopharmaceuticals. Radiochemistry and application. Wiley, London, p 848

Posner Paradigm

Definition

►Visual Attention

Possibilism

Definition

The view that possible worlds exist in addition to the actual world and have the same ontological status.

►Possible World

Possible World

Volker Gadenne
Department of Philosophy and Theory of Science, Johannes-Kepler-University Linz, Linz, Germany

Definition

A possible world is a complete way things might be. Possible worlds are alternative worlds one of which is the actual world. Philosophers use the notion of a possible world to define and discuss ideas such as possibility or necessity.

Description of the Theory

Leibniz's Idea of a Possible World

Consider the actual world, that is, the whole of what is the case, not only here and now but also in the past and in the future throughout all time. Thinking how things are in the actual world, lots of other worlds can easily be imagined, simply by changing one or more features. In one possible world Schubert composed one more symphony. In another world, the dinosaurs did not die out, with all consequences and so on. Even the slightest difference, say one more atom in this table, makes a world different from the actual one.

Note that changes of states of affairs may have consequences. Some things that are possible in separate worlds are not possible in combination. There is a possible world in which George visits a conference in 2004 and there is another possible world in which George dies in a car accident in 2003. But there is (probably) no possible world that contains both these states of affairs.

Leibniz used the idea of possible worlds in his philosophy of creation. God had in his mind infinitely many worlds he could have created. He chose the best of these possible worlds and made it actual. On the basis of this theory, Leibniz attempted to provide a ►theodicy, a "justification of God" in the face of all the evils in the actual world. God had perfect reason to bring into existence the actual world despite all pain and suffering it contains, since it is the best of all possible worlds. Leibniz was convinced that even god could not create anything. He could have made other laws of nature, but not worlds that are logically impossible. [*Vielleicht genauer: Leibniz' Gott kann logisch Unmögliches nicht schaffen, wohl aber nomologisch unmögliches].

Modal Logic and Possible World Semantics

The concept of possible worlds plays a central role in ►modal logic. Kripke [1,2] was most influential on the development of ►possible world semantics and its application to metaphysical problems (for an introduction, see [3]). Modal claims are fundamental to the ways the world is talked about. Consider, for example, the sentence: "There might be ten planets." This is a sentence in the mode of *possibility*. Another modal notion is *necessity*, as in the sentence: "Bachelors are unmarried." [*Sollte man hier den modalen Charakter nicht explizit machen so wie in dem vorangegangenen Beispiel?] What makes such statements true or false? According to the traditional view, modal statements are made true or false by relations of ideas or by linguistic conventions. In this example, the meaning of "unmarried" is part of the meaning of "bachelor." However, some philosophers find it hard to see how all ►propositions thought to be necessarily true should be true by convention. How could the way that a thing is talked about make it true, e.g. that John is a human being or that infinitely many primes exist? Here the idea of possible worlds has its part. With its help the proposition may be expressed as follows. "There is at least one possible world in which the sun has exactly ten planets." Necessity can be defined as truth in all possible worlds. "In all possible worlds bachelors are unmarried." "In all possible worlds there are infinitely many primes."

Generally speaking, to say that a proposition is true is just to say that it is true in the actual world. But to say that a proposition is necessary or necessarily true is to say that it is true in every possible world. And to say that a proposition is possible or possibly true is to say that it is true in some possible world. According to this theory, the modal notions of necessity and possibility are explained in terms of quantification over worlds. In order to speak of a proposition p as necessarily true, a *universal quantifier* over worlds is needed. "For all possible worlds W, p is true in W." To speak of a proposition p as possibly true, an *existential quantifier* over worlds has to be used. "There is at least one possible world W, such that p is true in W."

Logicians found that on this neo-Leibnizian account they could give clear sense to the modal notions as they function in the various modal logics. Further concepts such as "validity," "soundness" and "completeness" can be defined in terms of models constructed from sets of alternative worlds. Important results have been obtained by these methods (which can however not be presented here since they require a lot of technical details).

Modal notions are central to many of the traditional areas of philosophy, e.g. the nature of causation or free will. These notions have traditionally been challenged by empiricists. The most prominent critic was Quine [4]. In the 1950s, many philosophers became convinced that the ideas of necessity and possibility could have no place in philosophy. The development of possible world semantics gave many of them new reason to believe that the empiricist challenge can be met.

The Existence of Possible Worlds: Possibilism and Actualism

One of the most difficult problems of possible world theories concerns the ontological status of such worlds. The quantification over worlds seems to require that all these worlds exist. There are two main views dealing with this question, ►possibilism and ►actualism. For *possibilism*, as held by Lewis [5], there really is a plurality of possible universes of the same kind as this one. Each of them is conceived as a very comprehensive *concrete object*, having as its parts less comprehensive concrete objects such as stones, trees and persons. All the concrete objects that inhabit the various possible worlds are fully real. They are supposed to be really out there. Lewis denies that this world, which is called the "actual" one, has a special ontological status. The actual world is just a part of total reality, it is the part spatially and temporally related to this world. There is however no causal interaction between different possible worlds, because each of them is spatiotemporally closed.

Can the same individuals exist in different worlds? Lewis denies this. There are no "transworld individuals." Each object exists in just one possible world. But how can he then account for the idea that there are different ways the same things could have gone? Instead of relating different possible worlds by strict numerical identity of some objects he ties them by what he calls the "counterpart relation." For example, a particular person is in the actual world and no other, but has "counterparts" in several other worlds. The counterpart is not really the person, but it resembles the person closely in important respects.

Lewis argues for his theory by emphasizing its fruitfulness. Starting with the concept of a possible world as a primitive and with the means of set theory, he could not only define necessity and possibility, but also concepts such as "property" or "proposition." His theory is committed to the program of an austere ►nominalism. Properties and propositions are reduced to sets of concrete objects. However, many philosophers find it hard to accept that all those possible objects should be regarded as fully real. The strict nominalistic account of modal notions was also criticized. It leads to some unsatisfactory consequences.

Another view about the existence of possible worlds is ►actualism. Actualists too, start with the assumption that the actual world is not the only possible world. But unlike possibilism, actualism gives the actual world a special ontological status. Only what actually exists, exists at all. This seems to imply that possible worlds do not exist. However, not all actualists draw this consequence. The leading advocates of actualism, like Plantinga [6] and Stalnaker [7], rather think that possible worlds can be identified with something that belongs to the actual world. To express this idea, they do not restrict themselves to the resources of nominalism, but refer to abstract entities (►abstract entity), especially, states of affairs. Every possible ►state of affairs is supposed to *exist in the actual world*. However, not all states of affairs *obtain*. Not everything that might be the case (and therefore exists as a state of affairs) is really the case (obtains). For example, the state of affairs that Aristotle became Plato's successor at the academy exists, but failed to obtain.

Possible worlds are regarded by Plantinga as "maximally comprehensive possible states of affairs." This is a possible state of affairs, W, so comprehensive that, for any state of affairs S, W either includes S or precludes S (and thus encompasses a whole world). As a consequence, all the possible worlds *exist*. But only one of them *obtains* – the *actual* world. Possible worlds are *abstract entities*, not concrete objects (►concrete entity). The same individuals can exist in different possible worlds. Actualism therefore needs no counterpart relation. Possible world theories raise a lot of problems, technical ones as well as metaphysical difficulties, which are still unsolved. Nevertheless many philosophers are convinced of the fruitfulness of this program.

Applications to Other Fields

The notion of a possible world has been applied to several areas of philosophy to formulate and discuss special problems, e.g. in the philosophy of mind. Kripke [2] himself used it to analyze *mind-brain identity* and he argued against identity theory. Roughly, the structure of his argument is identity theory implies that pain is identical with a certain type of brain state. Such an identity statement would have to be necessarily true, i.e. true in all possible worlds, if it was true at all. But the tie between pain and a certain type of brain state is plainly contingent. Therefore, they cannot be identical.

A central idea of contemporary philosophy of mind is *supervenience*. Kim [8] analyses mind-brain supervenience in terms of possible worlds. The assumption that two persons with the same brain states must necessarily have the same mental states can be formulated as follows. "Mental properties supervene on physical properties in that if any x (in any possible world) and y (in any possible world) have the same physical properties (in their respective worlds), then x and y have the same mental properties (in those worlds)."

References

1. Kripke SA (1963) Semantical considerations on modal logic. Acta Philos Fenn 16:83–94
2. Kripke SA (1980) Naming and necessity. Harvard University Press, Cambridge
3. Loux MJ (1998) Metaphysics, Chap 5. Routledge, London
4. Quine WVO (1960) Word and object. MIT, Cambridge
5. Lewis DK (1986) On the plurality of worlds. Blackwell, Oxford
6. Plantinga A (1974) The nature of necessity. Oxford University Press, Oxford
7. Stalnaker R (1984) Inquiry. MIT, Cambridge
8. Kim J (1996) Philosophy of mind. Westview, Boulder

Possible World Semantics

Definition

A type of formal semantics that uses the notion of a possible world as a central concept; formal semantics is the study of the interpretations of formal languages.

►Possible World
►Property

Postactivation Potentiation (PAP)

►Force Potentiation in Skeletal Muscle

Postcentral Gyrus

Synonyms

►Gyrus postcentralis

Definition

= primary somatosensory cortex = SI
= area 3 + 1 + 2

The postcentral gyrus lies in the parietal lobe directly behind the central sulcus. Observing strict somatotopic arrangement, the somatosensory tracts of the contralateral body half terminate here.

Conscious localization and differentiation of quality and intensity of a tactile stimulus are effected in cooperation with the postcentral gyrus. Lesions of the postcentral gyrus reduces the response to tactile, thermal and noci stimuli from the contralateral body half.

►Telencephalon

Posterior Cerebellar Lobe

Synonyms

►Lobus cerebelli post.; ►Posterior lobe of cerebellum

Definition

The posterior lobe is the part of the cerebellum caudal to the primary fissure, and is composed of vermis portions (declive, folium, tuber, pyramid and uvula) as well as hemisphere portions (simple lobule, semilunar, gracile and biventer lobules as well as tonsil). Functionally this subdivision has practically no significance, since the cerebellum evidences a functional arrangement in a vertical direction (vermis, intermediate part, lateral part).

►Cerebellum

Posterior Colliculus

►Inferior Colliculus

Posterior Column

Synonyms

►Funiculus post; ►Posterior funiculus

Definition

The cuneate fasciculus and gracile fasciculus together form the posterior column and are the main axes of epicritic sensibility: – gracile fasciculus: it collects the epicritic fibers from the sacral, lumbar as well as lower thoracic cord and terminates in the gracile nucleus. – cuneate fasciculus: contains the fibers from the upper thoracic cord as well as from the cervical cord and terminates in the cuneate nucleus.

►Pathways

Posterior Commissure

Synonyms

►Commissura post

Definition

Here cross the fibers that are vital for controlling vertical eye movement and consensual light reaction of the pupils, including fibers from the superior colliculus, pretectal region as well as tegmentum of Mesencephalon.

►Telencephalon

Posterior Cortical Atrophy

Definition

Degenerative disorder of the posterior part of the brain beginning with visual symptoms and then proceeding into more general ►dementia. Initially, elementary visual functions are lost, but then more complex syndromes show up, including visual agnosia, topographical problems, ►optic ataxia, simultanagnosia, ocular apraxia (►Balint's syndrome), right-left confusion, ►alexia, ►acalculia, ►agraphia (►Gerstmann's syndrome).

►Balint's Syndrome
►Gerstmann's Syndrome
►Optic Ataxia

Posterior Horn

Synonyms

►Cornu post

Definition

He majority of primary afferents entering through the posterior horn terminate in the posterior horn of the spinal cord. Three zones can be distinguished:

- Marginal cells
- Substantia gelatinosa
- Nucleus proprius

►Medulla spinalis

Posterior Lobe of the Hypophysis

Synonyms

►Neurohypophysis

Definition

The posterior lobe of the hypophysis is also called the neurohypophysis since it is composed of hypothalamic nervous tissue. Its proximal segment is formed by the tuber cinerum and infundibulum, and its distal segment is the posterior lobe of the hypophysis. Via the infundibular nucleus, axons of the paraventricular nucleus and of the supraoptic nucleus pass to the blood vessels in the posterior lobe, where they release the hormones ADH and oxytocin.

►Diencephalon

Posterior Nuclei

Definition

The thalamic nuclei that project to the parietal somatosensory cortex, relaying nociceptive inputs from the periphery.

►Somatosensory Cortex I

Posterior Parietal Cortex (PPC)

Definition

Cerebral cortex posterior to the postcentral gyrus.

►Visual Space Representation for Reaching

Posterior Spinocerebellar Tract

Synonyms

►Tractus spinocerebellars post

Definition

The posterior spinocerebellar tract carries primary afferents from the spinal cord to the cerebellum.

It has its origin in Clarke's column in the thoracic cord and conducts proprio- and exteroceptive impulses (skin receptors, muscle spindles, tendon spindles) from the posterior limbs to the cerebellum.

►Cerebellum

Posterior Tuberculum

Definition

A caudal part of the diencephalon present in cartilaginous and bony fishes that contains some dopaminergic neurons as well as several laterally migrated nuclei of the preglomerular nuclear complex that are involved in the relay of ascending sensory pathways, particularly for the gustatory and lateral line systems.

►Evolution of the Somatosensory System: In Non-mammalian Vertebrates

Posterolateral Column

Synonyms

►Tractus postervlat. (Lissauer); ►Posterolateral tract (Lissauer)

Definition

The white matter between the ventral root and dorsal root gives rise to the lateral column, containing:

1. anterolateral column with
   - anterolateral fasciculus
   - parts of the anterior spinocerebellar tract.
2. posterolateral column with
   - posterior spinocerebellar tract
   - parts of the anterior spinocerebellar tract
   - lateral pyramidal tract.

►Medulla Spinalis

Posteromedial Barrel Subfield

►Barrel Cortex

Postganglionic Fiber (Neuron)

Definition

Ganglion neurons of autonomic ganglia all issue an axon (which turns into a nerve fiber), and virtually all these fibers exit the ganglion directed toward a peripheral target organ, along different paths depending on the ganglion of origin. Some form discrete nerves (splanchnic, pelvic, urinary), others reach somatic nerves and the mixed nerves of autonomic and somatic fibers thus formed reach the periphery (the limbs in particular and all the skin). The post-ganglionic fibers can be very long and are usually unmyelinated, hence of slow conduction velocity. Upon reaching the target organ they branch extensively and make contact with muscle elements (mainly smooth muscle cells) and with secretory elements (glands). Post-ganglionic fibers exert their effect on muscle and glands by releasing neurotransmitters that stimulate (and sometimes inhibit) contraction and secretion. The neurotransmitters are released from "terminals" that are not only the expansions at the anatomical end of each nerve branch but also at bulbous expansion (varicosities) scattered along a substantial part of the terminal portion of the axons. Each ganglion neuron issues one fiber traveling to the periphery, which has many branches within the terminal organ, which have many thousands of varicosities along their terminal branches.

In the bladder muscle, for example, a few thousand ganglion neurons directly innervate millions of muscle cells.The exact relationship between varicosities (nerve endings) and muscle cells (or gland cells) varies from a close contact (a neuro-muscular junction) in some tissues (the bladder, for example) to a loose relationship with a wide gap (some blood vessels, for example).

►Autonomic Ganglia
►Parasympathetic Pathways
►Sympathetic Pathways

Postganglionic Nerves

Definition

Autonomic nerves going to the end organ, e.g., cavernous nerve and the dorsal nerve of the penis and clitoris.

►Sexual reflexes

Postganglionic Neurotransmitter

Judy L. Morris, Ian L. Gibbins
Department of Anatomy & Histology, and Centre for Neuroscience, Flinders University, Adelaide, SA, Australia

Synonyms

Autonomic neurotransmitter

Definition

Postganglionic autonomic neurons have their cell body in an ►autonomic ganglion and an axon that extends out to a target organ. These neurons regulate activity of most organs of the body by releasing combinations of neurotransmitters. Postganglionic neurotransmitters are released from multiple swellings along the axons, or ►varicosities, separated from the target cell membrane by gaps of 20–100 nm to form ►neuroeffector junctions (Fig. 1).

Each axon has thousands of varicosities that can release neurotransmitter from the postganglionic neuron. It is now clear that the earliest identified neurotransmitters, acetylcholine and noradrenaline (adrenaline in some non-mammalian vertebrates), do not mediate all actions of postganglionic autonomic neurons. Nearly all neurons releasing acetylcholine or noradrenaline also synthesize and release various combinations of other neurotransmitter molecules including adenosine triphosphate (ATP), nitric oxide (NO) and one or more neuropeptides such as vasoactive intestinal peptide (VIP), neuropeptide Y (NPY) or opioid peptides [1,2,3,4,5]. Furthermore, many neurons intrinsic to the gastrointestinal tract or airways use nitric oxide, ATP and one or more neuropeptides but do not synthesize acetylcholine or noradrenaline (Table 1).

The release of more than one transmitter from the same postganglionic neuron, termed ►co-transmission, is now accepted as the rule rather than the exception. As well as regulating activity of the target organs (►postjunctional actions), transmitters released from postganglionic neurons can act back on terminals of the same or other nearby nerve terminals to alter further transmitter release (►prejunctional actions). Some transmitters have both pre-junctional and postjunctional actions, while others act at only one site (Fig. 1).

Characteristics

Quantitative Description

Twenty or more different molecules have been identified as neurotransmitters in postganglionic autonomic neurons including enteric neurons (Table 1). Some individual neurons contain five or more co-transmitters [1].

Higher Level Structures

Neurotransmitters are stored in membrane-bound vesicles within axon terminals. Synaptic vesicles can vary in size from small (40–60 nm diameter) to large (80–120 nm diameter). Nerve terminals contain both small and large vesicles in varying ratios. In varicose terminals of most postganglionic autonomic neurons, small vesicles are more abundant than large vesicles. Sometimes small vesicles tend to be clustered towards the cell membrane adjacent to the neuroeffector junction, but this is not always apparent. In contrast, large vesicles are not concentrated near the neuroeffector junction. Small vesicles

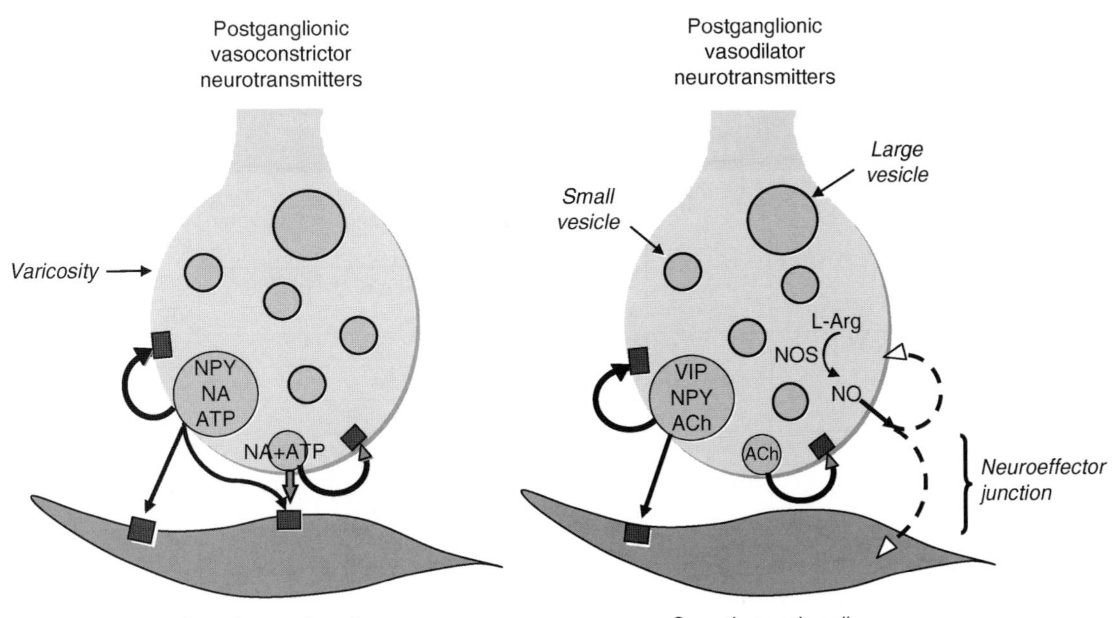

Postganglionic Neurotransmitter. Figure 1 Example of two types of postganglionic neurons releasing multiple neurotransmitters in a single blood vessel. Co-transmitters stored in and released from varicosities of postganglionic vasoconstrictor neurons and postganglionic vasodilator neurons into neuroeffector junctions with smooth muscle cells in the uterine artery (see [1]). Large vesicles in vasoconstrictor neurons contain noradrenaline (NA), adenosine triphosphate (ATP) and neuropeptide Y (NPY), while small vesicles contain only NA and ATP. Transmitters released from small and large vesicles can act on the postjunctional receptors on the smooth muscle cells to produce vasoconstriction, as well as prejunctional receptors on the membrane of the varicosity where they usually inhibit transmitter release. Nearby varicosities of vasodilator neurons contain large vesicles with acetylcholine (ACh), vasoactive intestinal peptide (VIP), NPY and several other peptides [6]. Small vesicles contain ACh alone. Nitric oxide is synthesized in the cytoplasm by nitric oxide synthase conversion of L-arginine, before diffusing across the neuroeffector junction and into smooth muscle cells to activate cyclic GMP and relax the smooth muscle. VIP has a potent vasorelaxant action, while ACh only has a prejunctional effect to inhibit neurotransmitter release. The function of NPY released from the vasodilator neurons is not known, but it may produce relaxation of an already constricted vessel and is likely to have a prejunctional inhibitory effect on transmitter release. It is likely that postganglionic neurotransmitters also can affect neurotransmission from adjacent varicosities.

in postganglionic neurons contain noradrenaline, ATP or acetylcholine. Large vesicles often contain neuropeptides in addition to non-peptide transmitters. Nitric oxide is an unusual neurotransmitter – it is not stored in synaptic vesicles but synthesized on demand in the nerve terminals (Fig. 1).

Lower Level Structures

The chemical structure of postganglionic neurotransmitters encompasses low molecular weight gases such as nitric oxide (and possibly carbon monoxide; [2]), purine nucleotides like ATP, catecholamines and acetylcholine, and neuropeptides ranging in size from less than a dozen amino acids (e.g., enkephalin, substance P) to more than thirty amino acids (VIP, NPY, calcitonin gene-related peptide (CGRP; Table 1). Almost all of these substances also are neurotransmitters in the central nervous system. The nature and sequence of neurotransmitters has been remarkably conserved through evolution, so that only very small if any chemical differences occur between postganglionic neurotransmitters in different vertebrate classes.

Higher Level Processes

As in all other neurons using chemical neurotransmission, synaptic vesicles in postganglionic neurons release their neurotransmitters by exocytosis, and vesicle membranes are recycled at the nerve terminal. Large vesicles are formed in the cell body where they are packaged with a variety of proteins (such as neurotransmitter synthesizing enzymes and transporters) and neuropeptides, then are transported down to the nerve terminal. In some neurons post-translational processing of peptides such as dynorphin can occur within large vesicles as they are transported down the axon. The non-peptide transmitters noradrenaline and acetylcholine are synthesized or taken up into vesicles in the cell body, axon and terminals. In contrast, neuropeptides cannot be taken up and repackaged into vesicles at the nerve terminal. This differential processing of peptide and non-peptide transmitters may

Postganglionic Neurotransmitter. Table 1 Substances localized in postganglionic autonomic neurons of most vertebrates

Neurotransmitter	Molecular weight	Postganglionic neurons with neurotransmitter	Common co-transmitters
Acetylcholine (ACh)	146	Parasympathetic neurons.	NO, VIP, Som
		Enteric motor neurons, Enteric secretomotor neurons.	SP, NKA
			NPY, Som, Gal, CCK, CGRP
		Subpopulation of sympathetic nerves e.g., sudomotor neurons	VIP
Adenosine triphosphate (ATP)	507	Sympathetic neurons.	NA, NPY
		Pelvic nerves to the bladder.	Ach
		Enteric inhibitory neurons.	NO, VIP
Adrenaline (Ad, epinephrine)	183	Sympathetic neurons in amphibians, fish.	NPY
Noradrenaline (NA, norepinephrine)	169	Most sympathetic neurons in mammals, birds, reptiles.	ATP, NPY
Calcitonin gene-related peptide (CGRP)	3,807	Parasympathetic neurons.	ACh, NO, VIP
		Enteric secretomotor neurons.	ACh, NPY, Som, Gal, CCK, CGRP
Cholecystokinin (CCK8)	1,142	Enteric secretomotor neurons	ACh, NPY, Som, Gal, CGRP
Dynorphin (DynA1–17)	2,148	Sympathetic neurons.	NA, +/- NPY
		Enteric neurons.	VIP, NO, GRP
Enkephalin (Enk)		Some sympathetic neurons.	NA, +/- NPY
Met-Enk, Leu-Enk	574, 556	Enteric neurons.	Dyn, VIP, NO
Galanin (Gal)	3,211	Sympathetic neurons.	NA, +/- NPY
		Intrinsic cardiac neurons.	ACh, Som
		Enteric neurons.	
Gastrin releasing peptide (GRP)	2,806	Enteric neurons.	VIP, Dyn, Gal, NO
5-Hydroxytryptamine (5-HT, serotonin)	1,76	Taken up into and released from some sympathetic and enteric neurons	NA, NPY
			Ach
Neurokinin A (NKA)	1,133	Enteric motor neurons.	SP, Ach
Neuropeptide Y (NPY)	4,254	Many sympathetic neurons.	NA, ATP
		Some parasympathetic neurons.	ACh, +/- VIP
		Enteric secretomotor neurons.	ACh, Som, CGRP, CCK
Nitric oxide (NO)	30	Parasympathetic neurons.	ACh, VIP
		Enteric inhibitory neurons.	VIP, ATP
Peptide histidine isoleucine (PHI)	2,996	Most parasympathetic neurons.	ACh, VIP, PACAP
		Enteric inhibitory neurons.	VIP, NO, ATP
Pituitary adenylate cyclase activating peptide (PACAP) 38	4,538	Enteric neurons	VIP, GRP
		Parasympathetic neurons.	ACh, VIP, PHI
Somatostatin (Som)	1,638	Intrinsic cardiac neurons.	ACh, +/- Gal
		Enteric neurons.	ACh, NPY, CGRP, CCK
Substance P (SP)	1,348	Cranial parasympathetic neurons.	Ach
		Enteric motor neurons.	ACh, NKA
Vasoactive intestinal peptide (VIP)	3,326	Most parasympathetic neurons.	ACh, PHI, PACAP
		Enteric inhibitory neurons.	PHI, PACAP, NO, ATP

Details of co-transmitters derived mostly from animal studies, concentratial on guinea-pegs (see [1,4]). A definitive neurotransmitter role has been established for all substances in all locations listed. Many entric neurons do not receive direct inputs from the central nervous system but form part of intrinsic neural circuits.

contribute to selective depletion of co-transmitters after intense activation of autonomic neurons. Nitric oxide is synthesized in postganglionic nerve terminals by the calcium-dependent enzyme, nitric oxide synthase (NOS), located in the cytoplasm and not in vesicles (Fig. 1).

The action potential-dependent exocytosis of neurotransmitters from small vesicles in all neurons, including postganglionic autonomic neurons, occurs through specific interactions between proteins located on the vesicle membrane and proteins attached to the inner surface of the nerve terminal membrane, the ►SNARE proteins (soluble NSF attachment protein receptor proteins). Exocytosis is a multi-step process that is calcium-dependent. First, the vesicles are released from a framework of actin filaments and move close to the terminal membrane where they become docked. This is followed by fusion of the vesicle membrane with the nerve terminal membrane, allowing release of vesicle contents into the extracellular space of the neuroeffector junction. Exocytosis of transmitters from small vesicles can be inhibited by botulinum neurotoxins that act intracellularly to cleave the SNARE proteins. However, it is not clear whether exocytosis of neuropeptides from large vesicles uses the same SNARE proteins that mediate small vesicle exocytosis. Release of neuropeptides from postganglionic autonomic neurons certainly is less sensitive to blockade by botulinum toxin than release of non-peptides from small vesicles [6]. Nitric oxide release from postganglionic autonomic neurons is completely resistant to botulinum toxin, confirming that this transmitter is not associated with vesicular storage and exocytosis.

Lower Level Processes

After release from the terminals of postganglionic neurons, noradrenaline, ATP and acetylcholine are rapidly removed from the neuroeffector junction by degradation or uptake by the nerve terminal and target tissue. These mechanisms limit the time course and distance over which the transmitters can act on prejunctional or postjunctional receptors. Nitric oxide acts on target tissues after diffusion across the membrane of both the nerve terminal and the postjunctional cell. This molecule diffuses rapidly from the neuroeffector junction and potentially can act outside the neuroeffector junction. However, superoxide radicals can rapidly inactivate nitric oxide, so diffusion of the transmitter away from the nerve terminal may be quite limited. Neuropeptides released from postganglionic neurons are not rapidly taken up across the pre- or postjunctional membrane and largely remain in the extracellular space until they are enzymatically degraded. This can result in neurally released neuropeptides interacting with receptors at considerable distances from the neuroeffector junction, a phenomenon called volume transmission. Nevertheless, neuropeptides bound to postjunctional receptors can be taken up into the target cell via ►endocytosis, thus contributing to ►receptor desensitisation.

Process Regulation

The actions of postganglionic neurotransmitters can be regulated by altering synthesis, transport, release, breakdown or reuptake of the transmitter itself, by altering expression or availability of neurotransmitter receptors, or by altering intracellular messengers and ion channels mediating neurotransmitter actions in the target cell. The major regulator of postganglionic neurotransmitter release is the frequency and pattern of action potentials travelling down the postganglionic axon. This is determined primarily by the pattern of impulses leaving the central nervous system via preganglionic neurons. However, this pattern can be modulated by local and circulating hormones changing the excitability of postganglionic neurons in autonomic ganglia. The excitability of postganglionic neurons also can be changed by ongoing activation of sensory nerves passing through autonomic ganglia, such as happens in inflammation. The sensory neurotransmitter substance P, and the hormone angiotensin both increase the excitability of many postganglionic neurons so that they fire more often in response to the same pattern of preganglionic nerve activity. Many substances also can modulate the expression of neurotransmitters or their synthetic enzymes by altering gene transcription in the postganglionic nerve cell body, or affect the release of transmitter from the varicosities of postganglionic neurons.

Ultimately, the pattern of firing of postganglionic neurons determines which co-transmitters are released from the varicosities. With a low frequency of impulses, <2 pulses per second (Hz), the small vesicles preferentially release their transmitters, so catecholamines, ATP and acetylcholine are released without neuropeptides. NOS also can be activated by low impulse frequency, releasing nitric oxide. Generally, large vesicles containing neuropeptides are not released until impulse frequencies reach at least 5, and up to 20, per second. An irregular pattern of high frequency activation is more effective in releasing neuropeptides from large vesicles than is a continuous high frequency firing. This frequency-dependent release of co-transmitters allows postganglionic neurons to produce a wide range of actions on their target tissues.

Function

Neurotransmitters released from postganglionic autonomic neurons have a wide range of functions. They activate or inhibit target cells such as smooth muscle cells of blood vessels, viscera, airways and skin, cardiac muscle, secretory cells in many glands, and other neurons in autonomic ganglia including enteric neurons. These actions regulate vital processes such as heart rate and arterial blood pressure, control of regional blood flow, the gastrointestinal system, respiration, reproduction and thermoregulation.

Many neurotransmitters can act both on the target tissue (postjunctional action), and back on the nerve terminal that released them (prejunctional action) to regulate further release of neurotransmitter (Fig. 1). Postganglionic neurotransmitters typically act via receptors on the target cell membrane that in turn can switch on or off a large array of second messenger pathways that influence intracellular calcium levels and usually affect membrane potential. Some transmitters, for example ATP, also can act via receptors that are themselves ion channels, called ionotropic receptors. In contrast, nitric oxide does not use receptors on the target cell membrane but diffuses freely across the membrane of both the nerve terminal and target cell, where it stimulates production of cyclic GMP. The detailed actions of multiple co-transmitters released from the same nerve terminal are not fully known. However, it is clear that some co-transmitters may have only a postjunctional action and some only a prejunctional action [1,6]. Further research also is required to clarify the roles of co-transmitters that potentially can have opposite effects on the target cells (Fig. 1). Nevertheless, the very different molecular sizes of co-transmitters, their different post-junctional signalling systems together with their different methods of inactivation, results in wide variations in the time course of neurotransmitter action. While noradrenaline, ATP, acetylcholine and nitric oxide typically have post-junctional actions lasting seconds to minutes, neuropeptide effects are slow in onset and can last up to tens of minutes, if not hours. Thus, release of neuropeptide transmitters by higher levels of impulse activity provides an efficient way to produce long-lasting functional changes in the target tissues.

Pathology

Pathological conditions involving dysfunction of the autonomic nervous system include those affecting post-ganglionic transmitter synthesis, release or post-junctional actions. Congenital deficiency in ▶dopamine-ß-hydroxylase (DßH) has been demonstrated in a small number of patients with autonomic dysfunction restricted to sympathetic pathways. These patients fail to synthesize adequate noradrenaline, and noradrenaline and adrenaline are undetectable in the plasma while plasma dopamine is elevated. The most obvious symptom is severe postural hypotension from an early age [7]. Other symptoms include ptosis and retrograde ejaculation. In contrast, some forms of hypertension involve hyperactivity of cardiovascular sympathetic nerves that results in increased release of noradrenaline from postganglionic neurons. This increased sympathetic activity is thought to be involved in both the development and maintenance of arterial hypertension [7]. Autoimmune neuropathies affecting autonomic nerve function also can occur. Autoantibodies leading to autonomic dysfunction most often are directed at nicotinic receptors in autonomic ganglia. However, autoantibodies directed at muscarinic receptors on peripheral target tissues have been reported in Sjögren's syndrome and scleroderma. These antibodies prevent acetylcholine released from parasympathetic nerve terminals from producing secretion in the lacrimal and salivary glands via post-junctional muscarinic receptors, resulting in the characteristic sicca symptoms of the disease [8].

Therapy

Therapeutic interventions for autonomic dysfunction include use of a wide variety of agents affecting post-ganglionic transmitters. DOPS (dihydroxyphenylserine) is useful in overcoming DßH deficiency, as it is converted directly to noradrenaline by dopa-decarboxylase, thus alleviating the requirement for DßH ([7]. Antagonists for post-junctional ß-adrenoceptors (beta blockers) are used widely to treat hypertension by reducing sympathetic cardio-excitation. In Sjögren's syndrome, agonists of M3 muscarinic receptors improve sicca symptoms, and use of antiidiotypic antibodies to neutralize autoantibodies has been proposed [8]. In conditions involving hyperactivity of postganglionic ▶cholinergic nerves, such as hyperhidrosis, anticholinergic agents are sometimes used although surgical sympathectomy has long been the treatment of choice. Recently, botulinum neurotoxin treatment has been used to block exocytosis of acetylcholine from the sudomotor neurons. Botulinum toxin A (Botox) has been popular, but it has been suggested that botulinum toxin B (Neurobloc) might produce more prolonged therapy due to its decreased immunogenic nature [9]. Botox also is used increasingly to treat autonomic dysfunctions such as neurogenic urinary incontinence or oesophageal achalasia. Sildenafil (Viagra), an inhibitor of phosphodiesterase-5, is used for treatment of erectile dysfunction in males, although its benefit in females is more controversial. This agent enhances the action of nitric oxide released from pelvic autonomic nerves, by reducing breakdown of cyclic GMP in smooth muscle [10]. Thus, sildenafil is only beneficial if the pelvic nerve pathways are intact.

References

1. Gibbins IL, Morris JL (2000) Pathway specific expression of neuropeptides and autonomic control of the vasculature. Regul Pept 93:93–107
2. Baranano DE, Snyder SH (2001) Neural roles for heme oxygenase: contrast to nitric oxide. Proc Natl Acad Sci. 98:10996–11002
3. Burnstock G (2004) Cotransmission. Curr Opin Pharmacol 4:47–52
4. Furness JB, Costa M (1987) The enteric nervous system. Churchill Livingston, Edinburgh, p 290

5. Lundberg JM, Hökfelt T (1986) Multiple coexistence of peptides and classical transmitters in peripheral autonomic and sensory neurons – functional and pharmacological implications. Prog Brain Res 68:241–262
6. Morris JL, Jobling P, Gibbins IL (2001) Differential inhibition by botulinum neurotoxin A of cotransmitters released from autonomic vasodilator neurons. Am J Physiol Heart Circ Physiol 281:H2124–H2132
7. Bannister R, Mathias CJ (eds) (1993) Autonomic failure. A textbook of clinical disorders of the autonomic nervous system, 3rd edn. Oxford University Press, Oxford, p 953
8. Cavill D, Waterman SA, Gordon TP (2003) Antiidiotypic antibodies neutralize autoantibodies that inhibit cholinergic neurotransmission. Arthritis Rheum 12:3597–3602
9. Nelson L, Bachoo P, Holmes J (2005) Botulinum toxin type B: a new therapy for axillary hyperhidrosis. Br J Plast Surg 58:228–232
10. Gibson A (2001) Phosphodiesterase 5 inhibitors and nitrergic transmission – from zaprinast to sildenafil. Eur J Pharmacol 41:1–10

Postherpetic Neuralgia (PHN)

Definition

Postherpetic neuralgia (PHN) is a neuropathic pain syndrome and is the most common complication of herpes zoster (HZ, shingles). PHN occurs mainly in the 60 years plus age group and in individuals suffering more severe acute pain and rash with HZ. Herpes zoster (shingles, HZ) results from reactivation of varicella zoster virus (VZV), one of a family of human herpes viruses, which has remained latent in sensory ganglia following primary infection with varicella (chicken pox). The incidence of HZ increases with age reflecting an age related decline in cell mediated immunity (CMI).

►Neuropathic Pain

Post-inhibitory Rebound

Definition

The ability of a neuron to respond to a hyperpolarization with a depolarization of the membrane potential above the level of the normal resting membrane potential. This can either be due to the activation of a hyperpolarization- activated inward current, Ih, or the action of low-voltage activated inward current, ICa(T), that is inactivated at the normal resting membrane potential. In this case, hyperpolarization of the membrane potential removes the inactivation of the current, so that it can be activated when the membrane potential returns to its resting level at the end of the hyperpolarization. This activation causes a transient depolarization of the membrane potential, the post-inhibitory rebound, that can trigger action potentials.

►Calcium Channels – an Overview
►Central Pattern Generator
►Omnipause Neurons
►Stomatogastric Ganglion

Post-junctional

Definition

Post-junctional refers to the target cell distal to a neuroeffector junction, responding to neurotransmitters.

Post-saccadic Drift

Definition

In an ideal saccade, the eye rapidly accelerates and then abruptly stops so that gaze remains stationary during the subsequent period of fixation. In reality, many saccades are followed by continued eye movements that have sub-saccadic velocities (e.g., 2–30°/sec in primates) but sufficient durations to appreciably change the direction of gaze. These post-saccadic drifts may be onwards or backwards relative to the saccade, and have several origins. After undershooting saccades in cats, the superior colliculus may continue to fire at low rates and gaze may drift onward toward the target. It has been suggested that is one mechanism the saccadic system uses to correct the undershoot. Less pronounced drifts seen in humans may serve the same purpose. Short duration onward or backward post-saccadic drifts, often called glissades, are thought to be due to a mismatch between the neural signal that moves the eyes during a saccade (the saccadic burst) and that which holds the eyes in position after the saccade (the tonic signal). Small disconjugate glissades occur after saccades in normal subjects, but more prominent glissades can result from central (e.g., cerebellar or cortical) damage and from damage to the ocular motor nerves, muscles,

P

or orbital tissue. Finally, the recurring post-saccadic centripetal drift and inability to hold eccentric gaze (gaze-evoked nystagmus) that results from damage to the brainstem neural integrator is also a form of post-saccadic drift.

►Brainstem Burst Generator
►Oculomotor Dynamics
►Saccade, Saccadic Eye Movement
►Superior Colliculus

Postsurgical Pain

►Incisional/Postoperative Pain

Postsynaptic Currents (EPSCs and IPSCs) or Potentials (EPSPs and IPSPs)

Definition

By the patch clamp method, it is possible to record the electrical events occurring in a postsynaptic cell as a result of neurotransmitters' release by the presynaptic terminal and the consecutive opening of ionotropic receptors. In the "voltage clamp" mode, the voltage is kept constant, so it is possible to record the current passing through the open ion channels, called "postsynaptic current" (PSC). In the "current clamp" mode, it is possible to record the changes in membrane potential induced by the opening of ion channels, called "postsynaptic potentials" (PSP).

For an excitatory synapse, the binding of neurotrasmitters induces the opening of cationic channels, which is depolarizing the cell. The induced electrical events are called "excitatory postsynaptic currents" (EPSCs) and "excitatory postsynaptic potentials" (EPSPs). For an inhibitory synapse, the binding of neurotrasmitters induces the opening of chloride channels, which is hyperpolarizing the cell. The induced electrical events are called "inhibitory postsynaptic currents" (IPSCs), and "inhibitory postsynaptic potentials" (IPSPs).

►Patch Clamp
►Postsynaptic Potential

Postsynaptic Potential

Miyakawa, Hiroyoshi
Laboratory of Cellular Neurobiology, School of Life Sciences, Tokyo University of Pharmacy and Life Sciences, Tokyo, Japan

Definition

At chemical synapses, transmitter molecules are released from presynaptic terminals to the synaptic cleft or extracellular space as the mediator of transmission, and bind to receptors located on the membrane surface of postsynaptic cells. Binding of transmitters often give rise to transient change in the membrane potential of the postsynaptic cells, which is referred to as postsynaptic potentials (Fig. 1) [1–5]. In the case of electrical synapses [6], electric current flows directly from one cell to the other as the mediator of transmission, and causes a change in membrane potential. Although this can also be referred to as postsynaptic potential, the term "postsynaptic" is conditional, because the electrical synapses are bidirectional. Hence, this term is used mostly for chemical synapses.

Characteristics

Mechanisms for Generating Postsynaptic Potentials

Some types of receptors are directly coupled with ion channels while others are either indirectly coupled to ion channels or not coupled at all. Activation of the receptors, either directly or indirectly coupled to ion channels, gives rise to changes in the open probabilities of ion channels, which can be detected as the change in the ion conductance of the membrane. Depending upon the ion selectivity, ion conductance, composition of the ion species, and the membrane potential, electric current flows through ion channels and charges the membrane capacitance to generate postsynaptic potentials. The direction of the postsynaptic currents depends on the ion-selectivity of the ion-channels coupled to the receptors, the composition of the ion species that permeate through the ion-channels, and the membrane potential. The direction of the current reverses at a certain potential, when the membrane potential of the postsynaptic cell is varied (Fig. 2). This potential is called the reversal potential. For instance, the reversal potential of the nicotinic acetylcholine receptor is around zero mV in physiological conditions, and the reversal potential of $GABA_A$ receptor is near the resting potential.

When the direction of the induced current is inward, and hence gives rise to depolarizing potential change, the current is referred to as excitatory postsynaptic current (EPSC), and the potential as excitatory postsynaptic potential (EPSP). When the direction of the

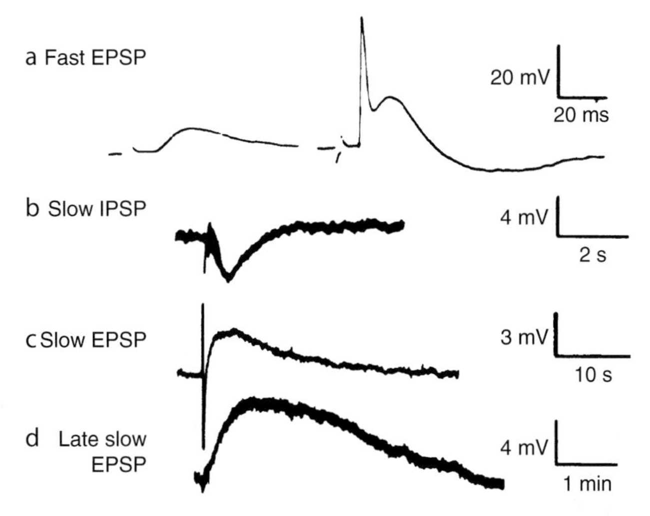

Postsynaptic Potential. Figure 1 Various postsynaptic potentials generated in frog sympathetic ganglion neurons. a: fast EPSP induced by stimulating preganglionic fibers. This potential is mediated by nicotinic acetylcholine receptors, which is coupled directly with ion channels. A stronger stimulation induces larger depolarization, which triggers action potential (right). b,c: slow IPSP and slow EPSP induced by repetitive stimulations. The slow IPSP and the EPSP are mediated by muscarinic acetylcholine receptors. d: A late slow EPSP evoked by repetitively stimulating spinal nerves is mediated by LHRH-like peptide. (Adapted from reference [7]).

current is outward and gives rise to hyperpolarizing potential change, the current is referred to as inhibitory postsynaptic current (IPSC), and the potential as inhibitory postsynaptic potential (IPSP).

Factors that Determine the Time Course of Postsynaptic Potentials

The time course of postsynaptic potentials range from a few milliseconds to a few minutes (Fig. 1). In general, fast postsynaptic potentials are mediated by ion-channel-coupled receptors (ionotropic receptors) for small molecule transmitters, and slow postsynaptic potentials are mediated by GTP-binding-protein-coupled receptors (metabotropic receptors). In the case of ion-channel-coupled receptors, channels open upon binding of transmitters, postsynaptic current flows through the channel, and the current charges the membrane to cause a change in the membrane potential (Fig. 3). Activation of GTP-binding-protein-coupled receptors usually generates second messengers in the cytoplasm, and the second messengers directly or indirectly regulate ion channels. Indirect regulation, in many cases, is by phosphorylation or dephosphorylation of ion channels.

In excitable cells, action potentials can be triggered when the summation of postsynaptic potential provides enough depolarization. The amplitude and the time course of postsynaptic potentials are important in triggering action potentials. Many factors are involved in determining the amplitude and the time course of postsynaptic potentials. Some of the factors are the amount and the time course of transmitter release, the structure of synapses, the lifetime of transmitters in the synaptic cleft, the properties of postsynaptic receptors, and the electrical properties of postsynaptic cells. The lifetime of transmitters in the synaptic cleft is determined by uptake mechanisms or degrading mechanisms, and estimated to be very short in the case of small molecule transmitters, such as glutamate and acetylcholine. The opening of channels indirectly linked to the receptors usually takes longer. The activation of ionotropic receptors is usually fast; hence, the rise time of postsynaptic currents is fast. The decay of postsynaptic currents also depends not only on the lifetime of transmitters in the cleft, but on the properties of receptors, channels, and the membrane as well. Some of the ligand-gated channels for small molecule transmitters show desensitization, and may be involved in determining the time course of postsynaptic potential. In the case of fast transmission, such as transmission at a neuromuscular junction, the decay time of postsynaptic potential is close to the time constant of the postsynaptic membrane, and is usually longer than that of the postsynaptic current (Fig. 3).

Dual whole cell recordings from neurons with long dendrites, such as neocortical pyramidal neurons, show that the postsynaptic potential measured from the cell body is slower compared to the potential measured near the input sites. This kind of distortion depends on the properties of the dendrites, and is important in determining cell response.

Higher Level Processes

The process of postsynaptic potential generation is part of ►**synaptic transmission**. Processes involved in synaptic transmission are classified as ►**presynaptic processes**, ►**postsynaptic processes** and the rest. The

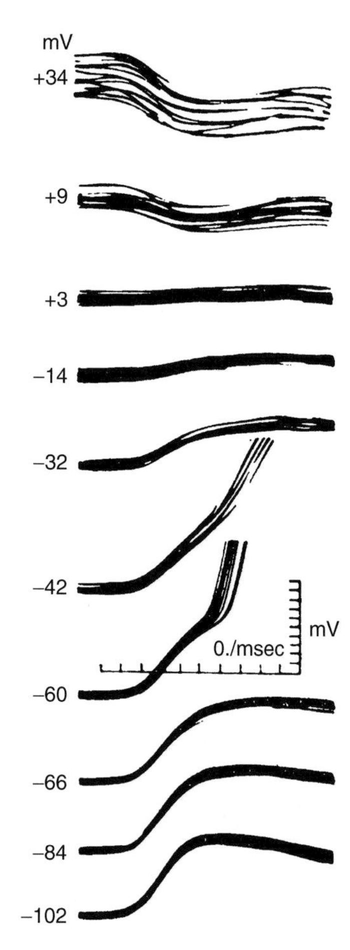

Postsynaptic Potential. Figure 2 Reversal of Postsynaptic Potential. EPSP was measured from a cat motoneuron. The membrane potential was varied by injecting current through an intracellular electrode. The postsynaptic potential is depolarizing at potentials below 3 mV, and hyperpolarizing above 7 mV. Triggered action potentials are shown in the traces at −60 and −42 mV. (Adapted from reference [8]).

presynaptic processes are the processes concerning ►**transmitter release**, and the postsynaptic processes concerning ►**reception** of transmitters including the generation of postsynaptic potential. Upon reception of transmitters, the postsynaptic cells do not necessarily generate postsynaptic potential, but may activate intracellular signaling processes that include protein phosphorylation-dephosphorylation and protein synthesis through gene expression.

Lower Level Processes

- Diffusion of transmitters.
  - Activation and desensitization of postsynaptic receptors.
  - Charging and discharging of membrane capacitance by postsynaptic currents.
  - Intracellular signaling.
  - Clearance of transmitters from synaptic cleft.

Process Regulation

Generation of postsynaptic potential is regulated by various ligands including biogenic amines, neuropeptides, and hormones. This kind of regulation is referred to as neuromodulation. The important mechanisms for neuromodulation are phosphorylation-dephosphorylation of proteins and gene expression of proteins, which take part in the generation of postsynaptic potential. The activities of the cells themselves, which take part in generating postsynaptic potentials, also regulate the process of postsynaptic potential generation. This kind of regulation is referred to as ►**neuronal plasticity**. Neuronal plasticity shares common mechanisms with neuromodulations, and is described elsewhere.

Function

To convey information for significant distance without decay, the neurons generate action potentials that propagate along the axonal processes. When the action potentials arrive at presynaptic terminals of chemical synapses, the information is converted into the form of chemical substances called transmitters. Upon reception of transmitters by the postsynaptic cells, the information is converted back to the form of electrical signal. The electrical signal can easily spread within the postsynaptic cell and can be summated and integrated to generate the output of the postsynaptic cells.

In the case of neuromuscular junctions, muscles receive a single presynaptic fiber. However, the amplitude of a postsynaptic potential (end-plate potential), induced by a single action potential in the presynaptic fiber, is large enough to trigger an action potential at the junction, which propagates along the entire length of the muscle fiber, and also into the transverse tubules, to induce a transient Ca rise and thereby generate a twitch response. The postsynaptic current generated by the activation of a postsynaptic receptor is greater than the current generated by the presynaptic action potential, and provide a current large enough to significantly depolarize the potential of the muscle membrane, the area of which is larger than that of the presynaptic membrane. Since the end-plate potential has large amplitude and a fast rate of rise, the muscle membrane can generate action potential. In contrast, most neurons in the central nervous system receive many thousands of inputs, and single synaptic inputs do not usually trigger action potentials in the postsynaptic cells because the amplitude of postsynaptic potentials is not large enough. Before an action potential is generated in the axons, many synaptic inputs need to be integrated. As synaptic inputs are converted into the form of an electrical signal, and electrical signals can quickly spread along the somato-dendritic axis, inputs

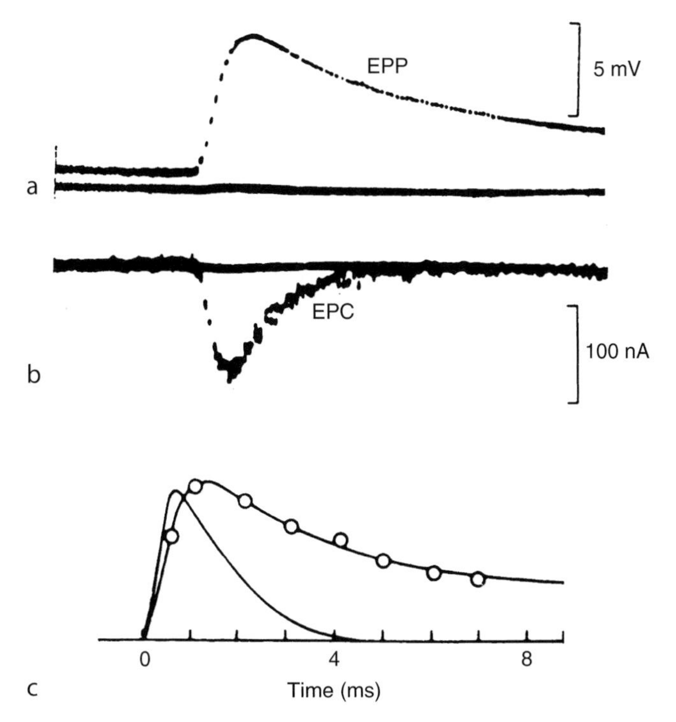

Postsynaptic Potential. Figure 3 Postsynaptic potential and current. a: End-plate potential (EPP) in a curarized frog muscle fiber. b: End-plate current (EPC) measured from the same fiber voltage-clamped at the resting potential. c: The continuous lines show the actual EPP and EPC. The circles show the EPP calculated from the EPC, assuming the time constant of the membrane to be 25 ms (adapted from reference [9]).

that arrive at various locations at various times can easily be summated and integrated. Thus, the postsynaptic potentials serve as mediators of information integration, which can lead to generation of action potentials in the postsynaptic cells that travel down the axon, and eventually induce transient Ca rise in the presynaptic terminals to trigger transmitter release.

There are cases in which the generation of action potential is skipped and postsynaptic potentials directly trigger transmitter release. In those cases, the extent of transmitter release depends on the amplitude of the potentials, and the synaptic transmission can be graded.

References

1. Cowan WM, Sudhof TG, Stevens CF (2001) Synapses. The Johns Hopkins University Press
2. Kuno M (1995) The synapse: function, plasticity, and neurotrophism. Oxford University Press, Oxford
3. Pappas GD, Purpura DP (ed) (1972) Structure and function of synapses. Raven, New York
4. Eccles JC (1964) The physiology of synapses," Springer-Verlag
5. Katz B (1996) "Nerve, Muscle, and Synapse," McGraw-Hill
6. Bennett MVL (2000) "Electrical synapses, a personal perspective (or history)." Brain Research Review 32:16–28
7. Kuffler SW (1980) Slow synaptic responses in the autonomic ganglia and the pursuit of a peptidergic transmitter. Journal of Experimental Biology 89:257–286
8. Coombs JS, Eccles JC, Fatt P (1955) Excitatory synaptic action in motoneurones. Journal of Physiology 130:374–395
9. Takeuchi A, Takeuchi N (1960) On the permeability of end-plate membrane during the action of transmitter. Journal of Physiology 154:52–67

Postsynaptic Receptor Trafficking

►Receptor Trafficking

Postsynaptic Receptors

Definition

Receptors that are expressed at the postsynaptic membrane and are responsible for mediating changes in the excitability of the postsynaptic cell.

►Synaptic Transmission: Model Systems

P

Posttetanic Potentiation

Definition

An increase in postsynaptic response to presynaptic release of neurotransmitter following a single stimulus that is applied at various times after a train of stimuli.

The cause is thought to be increased release of neurotransmitter from the presynaptic terminal.

►Force Potentiation in Skeletal Muscle
►Neuromuscular Junction

Posttraumatic Pain

►Incisional/Postoperative Pain

Posttraumatic Stress Disorder

Definition

A disorder that develops as a consequence of exposure to highly traumatic experiences, characterized by inappropriate fear responses to stimuli associated with those experiences.

►Learning and Extinction
►Stress

Postural Control

FAY B. HORAK
Neurological Sciences Institute, Oregon Health and Science University, Portland, OR, USA

Introduction

To inhabit the world, in all of its unpredictable, variable environments and situations, requires a powerful, yet flexible, system of postural control. For example, the ability to move from sitting to standing; to take a step; to respond to a slip or trip; to predict and avoid obstacles; to carry a glass of wine without spilling it, even when walking across a rolling boat; and to orient your body to a speeding soccer ball, all require excellent postural control. Although neural control of postural orientation and equilibrium involves most of the nervous system and all body segments, the postural system is often forgotten because it usually operates at an automatic, non-voluntary level. Only after an injury to the nervous system or musculo-skeletal system, when we have to really "think about" our balance and postural alignment or battle dizziness and spatial disorientation, do we begin to appreciate the complex systems involved in postural control.

Biomechanical Goals of Postural Control

Postural control involves neural control of ►postural equilibrium and ►postural orientation [1]. Postural equilibrium involves coordination of sensory and motor ►strategies to maintain balance, that is, to stabilize the body's ►center of mass over its ►base of support. An important goal of postural equilibrium control is to prevent falls during both self-initiated and externally-triggered disturbances of stability. The postural equilibrium system controls stability during stance posture as well as during locomotion and performance of voluntary tasks. Postural orientation involves the positioning of body alignment with respect to gravity, the support surface, visual environment and other ►sensory reference frames. The goals of postural equilibrium and postural orientation are independently controlled and sometimes subjects give up one goal for another. For example, an athlete may give up the goal of postural equilibrium in order to achieve their goal to orient their body appropriately to a ball.

Stance Posture

Although the musculoskeletal system affords some passive stability, humans and most animals require active postural muscle activation to maintain stance posture against gravity and to orient their body segments appropriately to their environment. To oppose the destabilizing effects of gravity, standing humnas are continuously making small correction to upright body position, called ►postural sway. ►Postural muscle tone provides antigravity support and flexibly adjusts to changes in support, alignment, and environmental conditions [2]. Besides postural tone, control of postural sway requires integration of sensory information to detect body motion with respect to the environment and the activation of muscles to maintain equilibrium and alignment of segments. Postural sway during stance can be measured with ►stabilometry; quantification of forces under the feet as continuous displacement of the ►center of pressure [3]. Displacement of the center of pressure represents the combination of motion of the center of body mass as well as the

▶ground reaction forces used to control the body ▶center of mass over the base of foot support.

Several different types of ▶control theories have been used to describe how the nervous system maintains consistent reference values for posture. Because posture is so adaptable and flexible, depending on the situation, models of human posture control include ▶optimal control and ▶adaptive control, such as ▶Kalman filers, of more than one variable, such as position of center of body mass, orientation of the trunk and head in space, energy efficiency, etc. Postural sway during human stance is often modeled as an ▶inverted pendulum biomechanical system in which the center of mass of the body is situated at the upper end of a rigid link that pivots about a joint at the base (i.e., the ankle), although actual body sway includes control of multiple segments.

Automatic Postural Responses

▶Automatic postural responses counteract unexpected disturbances to equilibrium. In humans, postural responses are triggered at 100 ms in response to external perturbations. This latency of automatic postural responses is faster than the fastest voluntary postural reactions but slower than the fastest ▶stretch reflexes. Stretch reflexes are triggered by muscle spindles and result in activation of the stretched muscles but these reflexes contribute little functional torque to correct postural equilibrium. Automatic postural responses include responses in muscles that are shortened, as well as stretched, as well as muscles far from the site of perturbation that can exert torque against surfaces to correct posture [4]. The recruitment of muscles in a postural response depends on the goal of maintaining equilibrium and not on stereotyped reflexes.

Automatic postural responses depend on ▶central set so that they are specific to the conditions of support and adapt to prior experience. Central set is the readiness of the central nervous system for an upcoming event based on initial conditions, prior experience and expectations. For example, leg muscles are activated in response to surface perturbations during free stance but arm muscles are activated and leg muscles suppressed in response to surface perturbations when holding onto a stable support [5]. In addition, muscles on the back of the legs are activated in response to forward body sway while standing but muscles on the front of the legs and in the arms are active when supported on the hands and feet [6]. Postural responses change even in the first trial after a change in body configuration but continue to adapt with repeated trials to continue to optimize the response for the particular conditions. For example, a gradual adaptation of the postural response can be observed during repeated trials of surface rotation. In response to the first rotation, a destabilizing response may be seen in the stretched ankle extensor muscle but with repeated rotations this activation of the extensor is suppressed and activation of the stabilizing ankle flexor gradually increases.

Subjects can also influence which postural response is selected and the magnitude of their response based on experience, expectations, and intention [7]. For example, the stretched ankle extensor muscle responses are inhibited and the shortened tibialis muscles are triggered when subjects are instructed to step in response to a forward body perturbation [8]. Poor coordination of automatic postural responses can result in failure to return to equilibrium in response to external perturbations. Automatic postural responses can be defined by their ▶postural strategies and ▶postural synergies.

Postural Strategies and Postural Synergies

Postural strategies can be defined by their functional goals and described based either on body kinematics (relationship of body segmental motion) or body kinetics (relationship of body segmental forces). Two main types of ▶postural movement strategies can be used to return the human body to equilibrium when perturbed while standing: strategies that return the center of mass back over the ▶base of foot support and strategies that change the base of support under the falling center of mass by stepping or reaching. The ▶fixed-support strategies, that return the body center of mass over the base of foot support, form a continuum from the ankle strategy to the hip strategy. The ▶ankle strategy, in which the body moves as a flexible inverted pendulum, is appropriate for small amounts of sway when standing on a firm surface [9]. ▶The hip strategy, in which the body exerts torque at the hips to quickly move the body center of mass, is used when standing on surfaces not allowing adequate ankle torque or when the body center of mass must be moved more quickly such as for a faster, larger disturbance [9]. When subjects suddenly change from standing on a wide to a narrow surface, or vice versa, there is a gradual adaptation from an ankle to a hip strategy and vice versa with repeated perturbations. This gradual change in postural strategies suggest that they not only depend upon sensory feedback and current sensory conditions but also upon prior conditions based on central set.

▶Change-in-support strategies of stepping and/or reaching to recover equilibrium in response to perturbations are also common, especially during gait and when it is not important to keep the feet in place [10]. However, even when subjects step in response to an external perturbation, they first attempt to return the body center of mass to the initial position by exerting angle torque. If a railing or other stable surface is available, subjects forced to extend their base of support by external displacement will also use a ▶reach-to-grasp strategy [11]. Reaching reactions are initiated even faster than stepping reactions. Change-in-support strategies are often used even under conditions in which it is biomechanically possible for subjects to return to equilibrium using a fixed-support strategy. Figure 1 illustrates fixed-support and change-in-support strategies to correct forward and lateral postural displacements.

►Postural synergies are groups of muscles activated together by the nervous system to maintain equilibrium [4]. By eliminating the need to control each muscle independently, postural synergies are thought to simplify the neural control task of selecting and coordinating multiple muscles across the body. Postural synergies define the muscle activation patterns that are used by the nervous system to implement various postural strategies. For example, Fig. 2 shows several muscles activated in the ankle, hip and mixed ankle-hip postural muscle synergies in response to forward sway perturbations.

Anticipatory Postural Adjustments

Voluntary movements are accompanied by ►anticipatory postural adjustments that act to counter, in a predictive manner, postural destabilization associated with a forthcoming movement [12]. Anticipatory postural adjustments are activated as ►feedforward postural control,

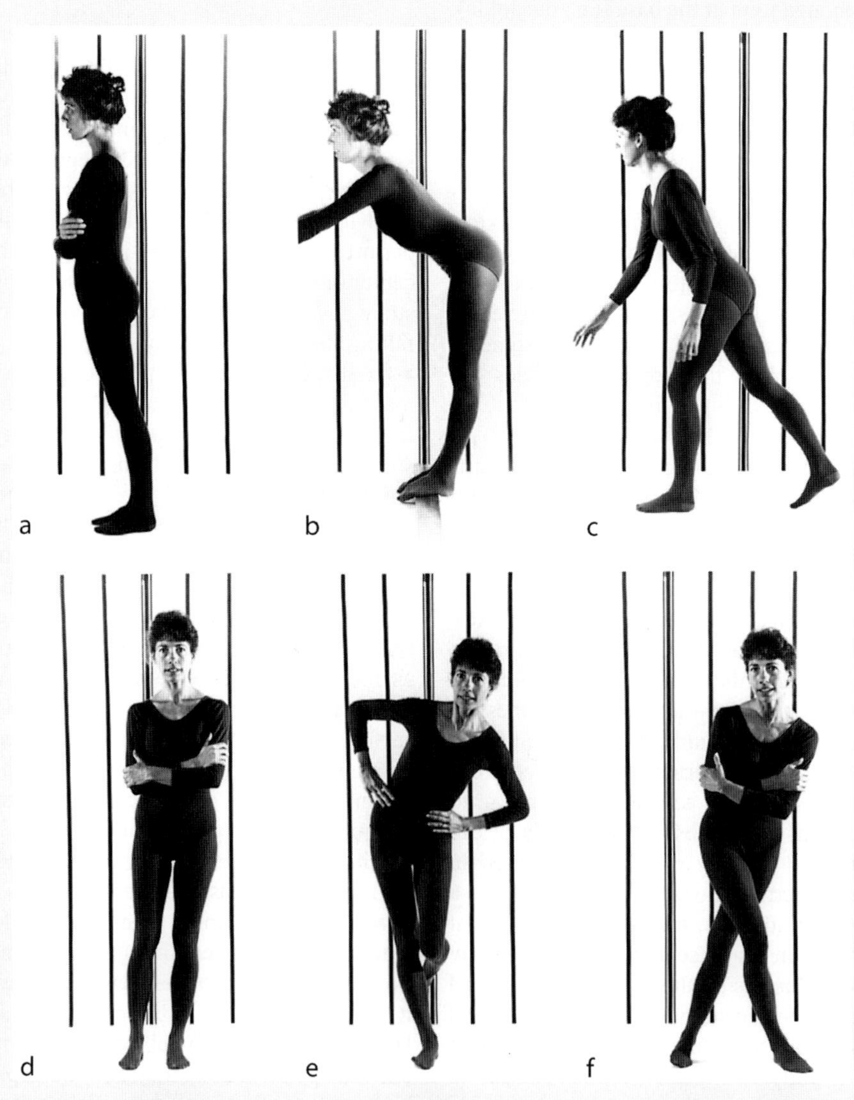

Postural Control. Figure 1 Shows examples of feet-in-place and stepping strategies to correct forward and lateral postural displacements. In response to small CoM displacements, humans use a strategy that maintains upright trunk orientation. In response to more forceful displacements, humans add rapid trunk and hip movements to move the CoM over the base of foot support. Stepping and reaching strategies can also be used to recover equilibrium by moving the base of support under the falling CoM. Lateral stepping includes both a cross-over strategy, as shown, and a step by the loaded leg to widen the stance width.

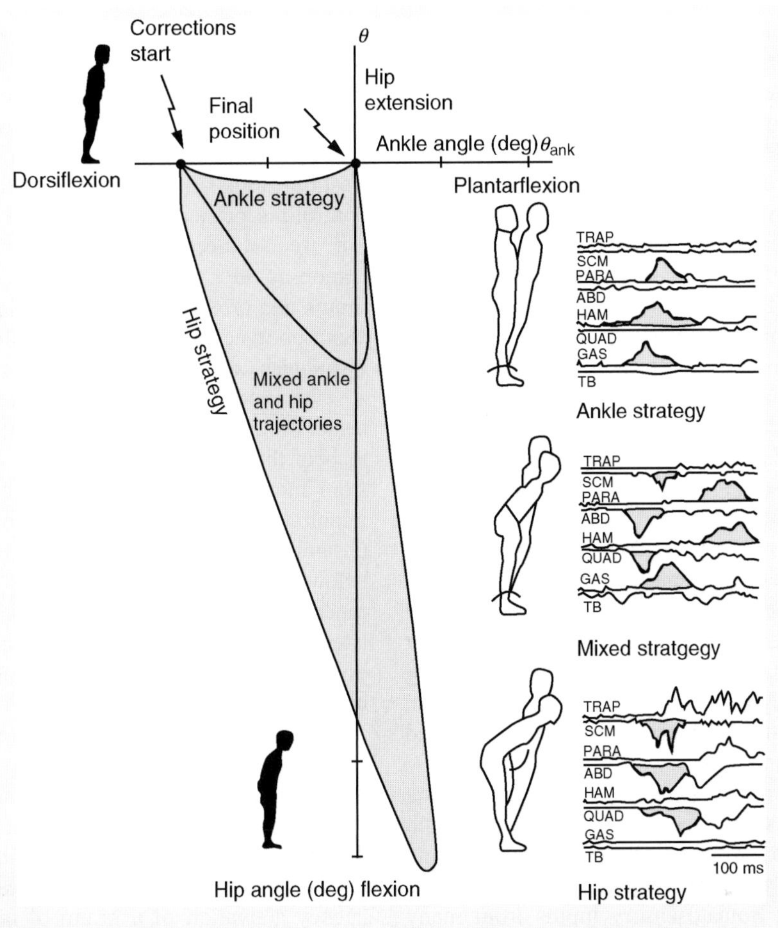

Postural Control. Figure 2 Plots the change in ankle and hip angles using the ankle and hip strategies and the continuum of mixed ankle-hip strategies used to return the body to upright stance equilibrium after a forward sway external perturbation.

prior to any sensory feedback indicating postural instability. For example, prior to taking a step, anticipatory postural adjustments move the body forward and onto the stance leg prior to lifting the stepping leg. In addition, when a standing subject rapidly moves their arms, leg and trunk muscles are activated more than 50 ms in advance of the prime mover arm muscles [13]. Anticipatory postural adjustments are specific to the biomechanical requirements of each specific movement and adapt when the biomechanical requirements change. For example, anticipatory postural adjustments in the legs associated with arm movements are reduced or disappear when subjects are supported at the trunk and no longer need the anticipatory postural muscle activity in the legs for stability [5]. These studies suggest that there is a preselection of an anticipatory postural muscle synergy associated with every voluntary movement requiring postural stability. This pre-selection or preparation of the sensorimotor nervous system in advance of movement has been called central set [14].

During locomotion, both anticipatory postural adjustments, via feedforward control, and automatic postural responses, via feedback control, contribute to postural stability. Unperturbed walking or running in healthy individuals consists of placing the feet under a falling center of body mass so the nervous system must anticipate where the feet need to be to maintain equilibrium during walking [15]. During bipedal locomotion, the trunk segment and thus, the body center of mass, is inherently unstable in the lateral direction and thus requires frequent corrections of lateral trunk orientation and/or lateral foot placement. When an individual slips or trips or makes voluntary movements while walking or running, the same automatic postural strategies observed during stance (See *Postural Strategies*) are added to the locomotor pattern [16]. Somatosensory feedback is also used to modify joint stiffness and quick responses to accommodate unanticipated changes in surface configuration.

Sensory Integration

Sensory information from the ►somatosensory, visual and vestibular systems must be integrated in order to interpret complex sensory environments because sensory information from a single sensory channel can be ambiguous and misleading. Postural control depends on the central neural interpretation of convergent sensory information from somatosensory, vestibular, visual systems. Thus, the nervous system controls posture via estimates of position and motion of the body and the environment by combining sensory inputs from several modalities. In addition, ►kinematic and ►kinetic body information must be integrated for control of posture. Sensory systems that signal kinematic position and motion of the body provide ►negative feedback control to minimize postural motion whereas sensory systems that signal kinetic force input provide ►positive feedback control to maximize joint torque when tilting [17]. Interpretation of sensory information by integrating sensory information across modalities is also thought to involve internal models of the body's sensory and motor dynamics, also called the body schema, as well as internal models of the environment. These internal models are based on expected sensory inputs from prior experience and provide the basis for central set. Errors between expected and actual sensory information is thought to be the basis for disorientation, dizziness, and motion sickness in both pathology and challenging environments.

Somatosensory inputs for posture include pressure information from skin in contact with surfaces, limb segment orientation from muscle proprioceptors and joint receptors, as well as muscle length, velocity and force information. Somatosensory inputs from many different types of peripheral sensory receptors converge onto neurons in the spinal cord to encode intersegmental and limb orientation in space [18]. Somatosensory inputs are important for triggering the earliest automatic postural responses in response to external perturbations. Thus, people with neuropathies that slow conduction of somatosensory inputs such as from diabetes or multiple sclerosis have longer than normal latencies of automatic postural responses. Somatosensory inputs are also important for providing information about the direction of perturbation and about the texture and stability of the support surface so that appropriate postural strategies can be selected. Somatosensory inputs can provide confusing, ambiguous information about body center of mass motion because they cannot distinguish between body motion over a stable surface and surface motion under a stable body, such as when standing on a moving boat or pier.

Vestibular inputs for posture are important for orientation of the trunk and head to gravity, especially when the surface is unstable. The vestibular system consists of two types of structures located in the inner ear, the ►labyrinths that encode head rotational acceleration and the ►otoliths that encode head linear acceleration, including gravity. The labyrinth consists of three, fluid-filled, semicircular canals that each sense a different direction of head rotation via motion of hair cells imbedded in the ►cristae, the sensory tissue. Within the otoliths, the ►utricle senses horizontal linear acceleration such as during walking and the ►saccule senses vertical acceleration such as during falling. Vestibulospinal inputs are particularly important for controlling orientation of the head and trunk in space but are not necessary to trigger automatic postural responses to external perturbations [19]. Vestibular inputs can provide confusing, ambiguous information about body center of mass motion because they cannot distinguish, on their own, between head motion over a stable body and head motion accompanying body center of mass motion. Vestibular information is thought to help the somatosensory system distinguish a stable from an unstable surface and then become increasingly important for controlling postural orientation the more unstable is the surface (see ►sensory re-weighting, below). Thus, patients who have lost all vestibular function can still stand and walk and show normal latencies of automatic postural responses to a slip or trip although they will orient to moving surfaces and become unstable when vision is not available [20].

Vestibular inputs must be interpreted via somatosensory inputs for the nervous system to control posture. For example, ►galvanic vestibular stimulation from direct current behind the ears can activate or inhibit the vestibular nerve and result in ►vestibulospinal responses. Vestibulospinal responses consist of medium latency activation of a group of muscles that tilt the body toward the side of the inhibited vestibular nerve when standing. The direction of body tilt depends on the direction the head is facing with respect to the base of foot support [21]. The muscles activated depend on which muscles can exert forces against the surface such that leg muscles are activated in free stance but arm muscles are activated with holding onto a stable surface [22]. Vestibular control of head orientation in space also depends on the close interaction between the vestibular and somatosensory systems via the ►vestibulocollic and ►cervico-collic reflexes.

Visual information provides knowledge of body sway and orientation in the environment and provides advanced information about potentially destabilizing situations. Vision can provide information about the direction and speed of body sway. For example, forward body sway is signaled by the visual system as backward visual flow across the peripheral retina and looming across the central retina. Visual information also allows perception and body orientation with respect to the vertical and horizontal visual environment (see perception of visual vertical, below). Thus, standing subjects exposed to slowly moving visual surrounds will sway with reference to the visual motion, even when unaware of it. Visual inputs can provide confusing, ambiguous information about body center of mass motion because they, alone, cannot distinguish between body motion with respect to a stable visual surround and visual surround motion with respect to a stable body. For example, when stationary subjects view large moving scenes, especially in their peripheral vision, they often momentarily perceive self-motion in the opposite direction. When actually moving the body through space, vision also provides advanced, or ▶feedforward, information to position body parts to avoid obstacles, navigate complex terrain, and plan motor strategies. For example, subjects tend to view obstacles in order to plan foot placement and clearance about 3 steps before they reach the obstacles [23].

The ability to orient the body with respect to gravity, the support surface, visual surround and internal references and to automatically alter how the body is oriented in space, depending on the context and task requirements is an important attribute of postural control. For example, a subject may automatically orient their body perpendicular to the support surface unless the support surface becomes unstable, when they will orient themselves to gravity or to their visual surround.

▶Sensory re-weighting is an important mechanism for changing the relative contributions made by different sensory systems for postural control. Figure 3 shows a model of sensory integration for postural control in which somatosensory, vestibular and visual inputs can change weighting depending on changes in the environment. Using this model, studies have shown that in a well-lit environment with a firm base of support, healthy subjects refy on somatosensory 70%, vision 10% and vestibular 20% [20]. However, when healthy subjects stand on an unstable surface, they increase sensory weighting to vestibular and vision as they decrease dependence on surface somatosensory inputs for postural orientation [24]. Ability to reweight sensory information depending on the sensory context is important for maintaining stability when moving from one sensory context to another, such as from a moving boat to firm ground. Individuals with loss of somatosensory, vestibular or visual input from pathology are limited in their ability to reweight postural sensory dependence and thus, are at risk of falls in particular sensory contexts. In addition, some central nervous system disorders may impair the ability to quickly reweight sensory dependence, even when the peripheral sensory systems are intact. Subjects can use ▶sensory substitution to replace one sensory modality for another to help control posture. For example, light touch on a cane can be used to substitute haptic sensory cues for missing vestibular or somatosensory inputs dues to pathology and thereby reduce postural sway in stance [25,26]. Biofeedback systems that provide visual, auditory or somatosensory inputs to the nervous system correlated with body sway have also been shown to provide effective sensory substitution to improve postural stability in patients with loss of sensory information.

Healthy individuals also have a conscious perception of vertical spatial orientation. ▶Perception of verticality, or upright, may have multiple neural representations [27]. In fact, ▶perception of visual vertical, or ability to align a line in the dark with gravity, is independent of ▶perception of postural (or proprioceptive) vertical, or ability to align the body in space without vision [28]. For example, the internal representation of visual, but not postural, vertical is tilted in subjects with unilateral vestibular loss, whereas the internal representation of

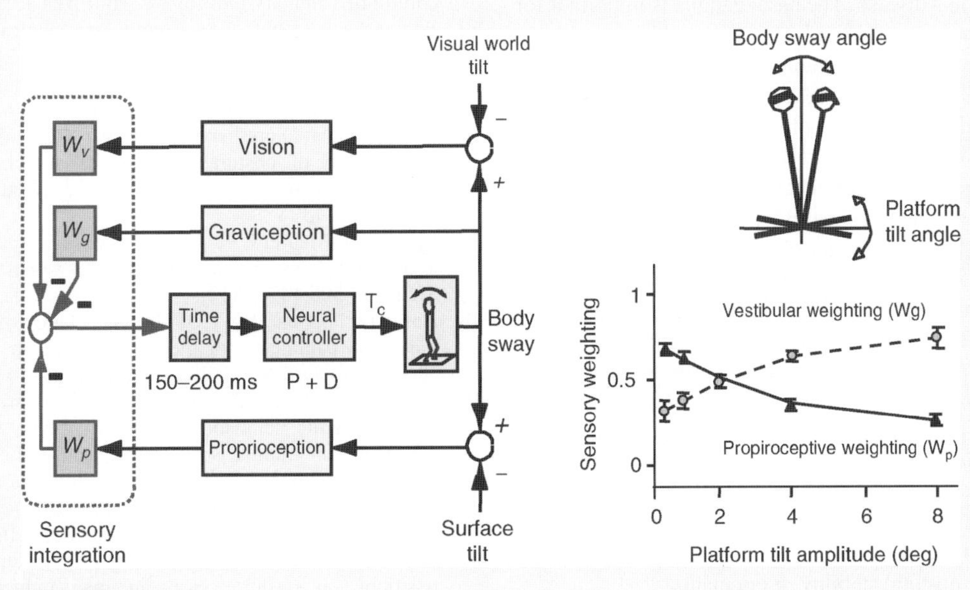

Postural Control. Figure 3

postural, but not visual vertical is tilted in some subjects with stroke. A tilted or inaccurate internal representation of vertical will result in automatic postural alignment that is not aligned with gravity.

Neuroanatomy of Posture Control and Clinical Implications

Control of posture is distributed in the nervous system and the musculoskeletal system such that pathology almost anywhere in the nervous system or musculoskeletal system can impair postural equilibrium and/or postural orientation. The spinal cord is sufficient for maintaining antigravity support and locomotor patterns but not for maintaining balance [29]. Sensory pathways in the spinal cord carry somatosensory information about limb orientation as well as motor pathways such as the medially located vestibulospinal and reticulospinal pathways for activating postural muscle synergies. In the brainstem, the vestibular nuclei are important for integrating sensory information across modalities for postural orientation and the reticular formation is likely involved in organizing postural synergies. The important ►role of the cerebellum in posture can be seen by the severe problems with postural stability and postural orientation in patients with damage to the cerebellum. Damage to the spinocerebellum, specifically, impairs postural stability by causing larger than normal automatic and anticipatory postural adjustments and by impairing the ability to optimize postural strategies based on prior experience [30]. In contrast, damage to the vestibulocerebellum results in difficulty using vestibular or visual information to orient the body with reference to gravity or visual references. The basal ganglia's importance to postural control can be seen by the frequent falls in patients with pathology involving the basal ganglia, such as Parkinson's disease. The basal ganglia is important for quickly changing postural strategies when conditions change, for regulating postural muscle tone, for generating forceful anticipatory and reactive postural responses and for perception of postural orientation [31]. The cerebral cortex is involved in postural control in as many complex ways as voluntary movement [32]. The cortex is involved in changing postural responses with alterations in cognitive state, initial sensory-motor conditions, prior experience, and prior warning of a perturbation, all representing changes in central set. In addition, the supplementary motor cortex is involved in generating anticipatory postural adjustments and the primary motor cortex participates in longer latency postural responses to perturbations. Parietal and temporal association cortical areas are involved in perception of spatial orientation and in formulating the internal models of the body and the environment so important to postural control. Thus, damage to almost any part of the cortex from a cerebral vascular accident can impact postural stability or orientation.

References

1. Horak FB, Macpherson JM (1996) Postural orientation and equilibrium. In: Rowell LB, Shepherd JT (eds) Handbook of physiology. Exercise: regulation and integration of multiple systems. Oxford University Press, New York, 255–292
2. Gurfinkel V, Cacciatore TW, Cordo P, Horak F, Nutt J, Skoss R (2006) Postural muscle tone in the body axis of healthy humans. J Neurophysiol 96(5):2678–87
3. Chiari L, Rocchi L et al. (2002) Stabilometric parameters are affected by anthropometry and foot placement. Clin Biomech 17(9–10):666–77
4. Ting LH, Macpherson JM (2005) A limited set of muscle synergies for force control during a postural task. J Neurophysiol 93:609–613
5. Cordo PJ, Nashner LM (1982) Properties of postural adjustments associated with rapid arm movements. J Neurophysiol 47:287–302
6. Macpherson JM, Horak FB et al. (1989) Stance dependence of automatic postural adjustments in humans. Exp Brain Res 78:557–566
7. Horak F, Kuo A (eds) (2000) Postural adaptation for alterered environments, tasks, and intentions. Biomechanics and neural control of posture and movement. Springer, New York
8. Burleigh AL, Horak FB (1994) Influence of prior experience and instruction on postural organization of perturbed step initiation. Soc Neurosci Abstr 20:794
9. Horak FB, Nashner LM (1986) Central programming of postural movements: adaptation to altered support-surface configurations. J Neurophysiol 55(6):1369–1381
10. Maki BE, McIlroy WE (1997) The role of limb movements in maintaining upright stance: the "change-in-support" strategy. Phys Therapy 77:488–507
11. McIlroy WE, Maki BE (1995) Early activation of arm muscles follows external perturbation of upright stance. Neurosci Lett 184:1–4
12. Massion J (1992) Movement, posture and equilibrium: interaction and coordination. Prog Neurobiol 38:35–56
13. Belenkii VY, Gurfinkel VS et al. (1967) Elements of control of voluntary movements. Biofizika 12:135–141
14. Prochazka A (1989) Sensorimotor gain control: a basic strategy of motor systems? Prog Neurobiol 33:281–307
15. Winter DA, Patla AE et al. (1990) Assessment of balance control in humans. Med Prog Technol 16:31–51
16. Nashner LM, Forssberg H (1986) Phase-dependent organization of postural adjustments associated with arm movements while walking. J Neurophysiol 55(6):1382–1394
17. Mergner T, Maurer C et al. (2002) Sensory contributions to the control of stance: a posture control model. Adv Exp Med Biol 508:147–152
18. Bosco G, Poppele RE (1997) Representation of multiple kinematic parameters of the cat hindlimb in spinocerebellar activity. J Neurophysiol 78(3):1421–1432
19. Mergner T (2002) The matryoshka dolls principle in human dyamic behavior in space: a theory of linked references for multisensoryu perception and control of action. Curr Psychol Cogn 21(2–3):129–212
20. Peterka RJ, Benolken (2002) Sensorimotor integration in human postural control. J Neurophysiol 88(3):1097–1118
21. Lund S, Broberg C (1983) Effects of different head positions on postural sway in man induced by a

reproducible vestibular error signal. Acta Physiologica Scandinavica 117:307–309
22. Storper IS, Honrubia V (1992) Is human galvanically induced triceps surae electromyogram a vestibulospinal reflex response? Otolaryngol Head Neck Surg 107:527–535
23. Patla AE (1992) The neural control of locomotion. 1–36
24. Peterka RJ, Loughlin PJ (2004) Dynamic regulation of sensorimotor integration in human postural control. J Neurophysiol 91(1):410–423
25. Creath R, Kiemel T et al. (2002) Limited control strategies with the loss of vestibular function. Exp Brain Res 145(3):323–333
26. Dickstein R, Peterka R et al. (2003) Effects of light fingertip touch on postural responses in subjects with diabetic neuropathy. J Neurol Neurosurg Psychiatry 74:620–626
27. Karnath HO, Fetter M et al. (1998) Disentangling gravitational, environmental, and egocentric reference frames in spatial neglect. J Cogn Neurosci 10(6):680–690
28. Anastasopoulos D, Haslwanter T et al. (1997) Dissociation between the perception of body verticality and the visual vertical in acute peripheral vestibular disorders in humans. Neurosci Lett 233(2–3):151–153
29. Macpherson JM, Fung J (1999) Weight support and balance stance in the chronic spinal cat. J Neurophysiol 82(6):3066–3081
30. Horak FB, Diener HC (1994) Cerebellar control of postural scaling and central set in stance. J Neurophysiol 72(2):479–493
31. Horak FB, Frank J et al. (1996) Effects of dopamine on postural control in parkinsonian subjects: scaling, set and tone. J Neurophysiol 75(6):2380–2396
32. Jacobs JV, Horak FB (2007) Cortical control of postural responses. J Neural Transm 114(10):1339–1348

Postural Equilibrium

Definition

A state in which the body is either at rest, moving at constant velocity or executing a repeatable (periodic) pattern of motion. A stable system is one that returns to a state of equilibrium after it has been perturbed.

►Postural Strategies

Postural Instability

Definition

Impairment of balance when standing, walking, or turning.

Postural Muscle Tone

VICTOR S. GURFINKEL
Neurological Sciences Institute, Oregon Health & Science University, Portland, OR, USA

Definition

Postural tone is the steady contraction of muscles that are necessary to hold different parts of the skeleton in proper relation to the various and constantly changing attitudes and postures of the body.

Description of the Theory

Decerebrate Posture

Because postural muscle tone is completely suppressed by narcosis, postural tone has mainly been studied using an experimental model called the decerebrate animal [1]. In the decerebration of mammals, the cerebral cortex and thalamus are surgically inactivated by intercollicular cross-section of the brain stem under general anesthesia. Once the effects of the anesthesia have dissipated, the condition known as "decerebrate rigidity" can be seen. This condition is characterized by a strong extended neck, trunk, tail and limbs, which resist attempts to flex them. In decerebrate rigidity, there is no sensation of pain. Because of this rigidity, when the decerebrate cat is placed on its four limbs, tension in the limb muscles is enough to maintain its body posture. This muscle tone has been named "postural tone" [1]. In decerebrate animals, the neuronal structures of the brain stem and spinal cord are in an active condition. Therefore, this model is useful for studying many questions of neurophysiology. For example, on a background of high muscle tone, it is possible to study not only influences of excitation, leading to the enhancement of muscle tone, but also to study inhibition, which results in the suppression of muscle tone.

In the past, many researchers have been devoted to studying the nature of decerebrate rigidity. It was found that deafferentation (i.e., sectioning appropriate dorsal roots) abolishes decerebrate rigidity of limb muscles [1]. In addition, it was found that the tonus of the extensor muscles is autogenous, in that each muscle is dependent on afferent nerve fibers from the muscle itself ("myotatic component" in decerebrate rigidity). These findings were reproduced many times [2]. This showed that the origin of decerebrate rigidity cannot be completely explained by the myotatic component. The actual situation is more complex. In studies of decerebrate cats, it was shown that in addition to proprioception, there are other sources of postural tonic activity that are connected to the position of the head in space and the position of the head relative to the trunk [3]. These neck and vestibular tonic reflexes strongly influence the level of decerebrate rigidity and cause a redistribution of muscle tension.

Another type of experiment elucidated mechanisms of decerebrate rigidity. This experiment involved, as described above, the cutting of the dorsal roots to the forelimbs in a decerebrate preparation to make forelimbs flaccid. When the spinal cord was transected below the level of origin of the brachial nerves (postbrachial transection), the forelegs became rigid, although still deafferented. This effect can be explained by the fact that post-brachial transection of the spinal cord cuts off the flow of inhibitory pulses to the motoneurons that ascend from tonically active propriospinal neurons located in L2–L3 segments (the "Schiff-Sherrington inhibition").

It is known that the cerebellum is involved in the control of muscle tone and that damage to the cerebellum is accompanied by a reduction in tone. However, it has been shown that the removal of the cerebellum in a cat that has undergone intercollicular brain stem transection increases decerebrate rigidity. This same effect can be caused by bilateral destruction of the fastigial nuclei. Here again, rigidity returns to the deafferented forelimbs of the decerebrate preparation.

Another model of decerebrate rigidity involves properties that are different from the intercollicular preparation and has extended our knowledge about the mechanisms involved in decerebrate rigidity. It is the so-called anemic decerebration (high ligation of the basilar artery and of both carotids) that functionally inactivates all the nervous structures that are supplied by blood vessels arising cephalad to the basilar ligature, including the anterior lobe of cerebellum. After anemic decerebration, strong tension in all extensor muscles also develops. However, decerebrate rigidity, after anemic decerebration, is not eliminated by transection of the dorsal roots. In the anemic preparation, rigidity is caused by the nonmyotatic component of muscle tension, represented mainly by the tonic labyrinthine influences on the spinal motoneurons. The nonmyotatic component of extensor rigidity is released by anemic damage to the anterior lobe of the cerebellum.

One cogent interpretation of the evolution of the cerebellar regulation of postural tone is that postural extensor mechanisms are tonically inhibited through bulbospinal relays by the paleocerebellum, whereas a tonic facilitating influence is exercised by the neocerebellum on the cerebral cortex [2]. After efferent innervation (γ innervation) of muscle spindles was established, intercollicular decerebrate rigidity was called "γ rigidity," and anemic rigidity was called "α rigidity" [4]. A comparison of these two kinds of decerebrate rigidity showed that they change differently under the influence of various factors, suggesting that both muscle and labyrinthine receptors tonically support extensor rigidity. The disappearance of decerebrate rigidity following deafferentation indicated that in the intercollicular animal, the vestibular component is tonically inhibited by cerebellar and spinal mechanisms (Schiff-Sherrington). However, in anemic decerebrate animals, the importance of the myotatic component of decerebrate rigidity is reduced by γ paralysis, while that of the vestibular component is increased through a release mechanism. It is possible that other brain structures also participate in the formation of decerebrate rigidity. The posture of the decerebrate animal may be the consequence of a disturbance in balance between different sources of tonic excitatory and inhibitory influences.

Postural Adaptation

As mentioned above, a decerebrate cat can be stood on its legs and will maintain this position; a position that is a caricature not only because the cat's legs are hyper-extended but also because the body is absolutely motionless due to constant muscle tension. Such immobility can also be observed naturally in intact animals (e.g., a hunting dog becomes motionless when pointing to detected game, rabbits experience immobilization catatonia when faced with danger and many animals undergo hypnosis). However, immobilization, that is muscles in a state of constant tension, is the exception not the rule. Postural tone of muscles ranges from very high tension at exaggerated decerebrate rigidity, up to complete atonia in the vertebra prominens reflex. Although the decerebrate animal exhibits rigidity, it still has the capacity to adapt to experimental conditions. In 1909, it was shown for the first time that when a decerebrate animal's limb was flexed, the limb did not move back to its initial position; rather it adapted and maintained the new position [5]. This condition was explained as: "The forced stretch causes a relaxation of the tonic extensor, and this condition of relaxed tonus persists after the forced stretch itself has ceased. This reaction, may for brevity, be termed the 'lengthening reaction'" [5]. Similarly, when a decerebrate animal's limb is moved, the antagonist muscles shorten and adapt to their shorter length. In this shortening, the muscle insertion and origin are brought closer together, which appears to induce heightened tonus in the muscle—a reaction that has been termed the "shortening reaction" [5]. In other words, the limb that is brought by movement into a new posture remains in that new posture when released. This property is especially present in skeletal muscles when the nerve centers are operating to maintain posture [5]. Such behavior of tonically active muscles has been termed "plasticity." Skeletal muscles in this form of reflex contraction quite readily adapt themselves to different lengths while counteracting one and the same load [6].

The shortening reaction can be observed in humans. In some neurological diseases that are accompanied by tone abnormalities, passive movement of a joint results in the contraction of shortened muscles. This contraction is tonic, that is, it persists after the movement terminates.

Another form of postural tone adaptation in decerebrate animals deals with neck and vestibular tonic reflexes [3]. In decerebrate animals, neck reflexes result in a redistribution of tonic muscle tension that is dependent on the relative position of the head and trunk. Dorsiflexion of the neck produces extension of forelimbs and flexion of hind limbs, ventriflexion of the neck produces flexion of the forelimbs and extension of the hind limbs, lateral flexion of the neck (ear toward the shoulder) produces extension of the fore and hind limbs on the side toward which the head is turned and rotation of the head on the neck produces extension of both the fore and hind limbs on the side toward which the chin is turned. The changed distribution of muscle tonus in the limbs continues as long as the head retains a specific relationship to the trunk. Changes in head position in space when neck reflexes are inactivated, also result in a redistribution of postural tone of neck, trunk and legs muscles. There is one position of the head in space in which the extensor tone of the limbs is maximal and one position of the head in space in which it is minimal. The maximal and minimal positions differ from each other by 180°. When the position of the head in space is changed when the body is in different positions, tonic neck and labyrinth reflexes combine to modify postural tone in various ways.

Another factor that influences the distribution of postural tone is how the body or limbs contact the support surface. When a limb contacts the support surface, dorsiflexion occurs in the distal part of the limb (fingers and hands, toes and feet), which results in significant enhancement (strengthening) of tone in the extensor muscles of all joints in the limb (termed "the positive reaction of support"). The limb, in this enhanced tonic state, is capable of maintaining body weight. This reaction has been observed in animals without a cerebellum [7]. In decerebrate animals, this reaction is more difficult to observe because the limbs are in an initial state of high tonic tension. This type of reaction has also been observed in humans with brain pathology. In the examples discussed above, the adaptive changes of postural tone were tonic in character, but in natural conditions, adaptive changes are also phasic in character. In natural conditions, such reactions are mainly directed to preserving balance and maintaining the stability of body parts during movement. Almost all movements of the hands and legs in natural conditions inevitably require a redistribution of postural tone. For example, when a human moves the arms to catch a falling object, contact with the object is preceded by the contraction of appropriate hand muscles. Similarly, when a human moves a leg when taking a step, this movement is preceded by a change in muscle tone distribution in the trunk and in the opposite leg. Adaptation of muscle tonic activity maintains harmony between mobility and stability.

References

1. Sherrington CS (1898) Decerebrate rigidity and reflex coordination of movements. J Physiol 22:319–332
2. Bremer F (1932) Le tonus musculaire. Ergebn d Physiol 34:678–740
3. Magnus R, de Kleijn A (1912) Die Abhängigkeit des Tonus der Extremitätenmuskeln von der Kopfstellung. Arch f d ges Physiol 145:455–548
4. Granit R (1955) Receptors and sensory perception. Yale University Press, New Haven
5. Sherrington CS (1909) On plastic tonus and proprioceptive reflexes. Quart J Exper Physiol 2:109–156
6. Sherrington CS (1915) Postural activity of muscle and nerve. Brain 38:191–234
7. Rademaker GGJ (1931) Das Stehen: Statische Reaktionen, Gleichgewichtreaktionen und Muskeltonus unter besenderer Berücksichtigung ihres Verhaltens bei kleihirlosen Tieren. Springer, Berlin Heidelberg New York

Postural Orientation

Definition

Postural orientation refers to the positioning of the body or body segments with respect to some reference frame.

The selection of the reference frame is arbitrary and there are many choices. A commonly used and often implicit reference is a global or earth fixed reference frame. Other common reference frames include clinical joint based frames where joint motions can be described, for example in terms of extension or flexion and local reference frames which might move with the body or be affixed to a particular body segment.

►Posture – Sensory Integration

Postural Reactions

►Postural Strategies

Postural Strategies

BRIAN E. MAKI
Centre for Studies in Aging, Sunnybrook and Health Sciences Centre, University of Toronto, Toronto, Canada

Synonyms

Balance-recovery reactions; Postural reactions; Postural synergies

Definition

Postural strategies are specific patterns of muscle activation, joint torque, joint rotation and/or limb movement that are evoked by balance perturbation. These reactions serve to prevent the body from falling and act to re-establish a state of ►postural equilibrium. Triggered by multiple sensory inputs, they involve polysynaptic spinal and supraspinal neural pathways and are highly adaptable to meet functional demands. Strategy selection and modulation are dependent on: (i) the features of the perturbation (timing, direction, magnitude, predictability), (ii) the "►central set" of the individual (affect, arousal, attention, expectations, prior experience), (iii) ongoing activity (cognitive or motor) and (iv) environmental constraints (on reaction force generation and limb movement).

Description of the Theory

Biomechanical Requirements for Postural Equilibrium

The mechanics of the upright human body can be modeled as a multi-link ►inverted pendulum, with each link corresponding to a body segment (e.g., foot, shank, thigh, trunk) [1]. Static postural equilibrium requires the ►center of mass (COM) of this linkage to be positioned over the ►base of support (BOS); however, the linkage is inherently unstable, due to the force of gravity. Additional destabilizing forces arise due to movement of the body and interaction with the environment. The BOS is usually defined by the position of the feet, but may include the arms when grasping or touching an object for support. In the absence of arm support, static equilibrium requires the COM to be positioned over the feet and the perimeter of this foot area can be considered to represent the static stability limits associated with the BOS. Dynamic equilibrium takes into account the additional requirement of controlling the momentum associated with movement of the COM [2]. If the COM is moving with sufficient horizontal velocity, it is possible for the body to be dynamically unstable, even when the COM is positioned over the BOS. Conversely, it is possible for the body to be dynamically stable even though the COM is located outside the static stability limits of the BOS, provided that the COM is moving toward the BOS with sufficient velocity that it can eventually be repositioned over the BOS.

Postural Strategies for Responding to Perturbation

Passive muscle stiffness could, in theory, be sufficient to maintain a stable upright posture under static conditions; however, the reality is that coordinated muscle activation is required to keep the body upright in the activities of daily life. To maintain upright stance, the central nervous system (CNS) must actively regulate the static and dynamic relationship between the COM and the BOS [2]. Responses to perturbations must involve deceleration of the COM, but can also involve changes in the BOS. Accordingly there are two distinct classes of postural strategies: (i) "►fixed-support" ("feet-in-place") strategies, in which the BOS is not altered and (ii) "►change-in-support" reactions, where the BOS is altered via rapid ►stepping or reaching movements of the limbs [3]. See Fig. 1.

Although the focus here is on reactions to perturbation of stance, it should be noted that fixed support and change-in-support reactions are also used to respond to perturbations experienced during gait. However, the gait responses may show some differences, e.g., bilateral asymmetries in muscle activation, phase dependent gating of sensory inflow and triggering of additional strategies (e.g., elevation of the swing foot in response to a trip perturbation). In addition, stepping reactions during gait may involve modulation of an ongoing step, while arm reactions may be affected by gait-related arm swing.

Postural reactions can be controlled, to some extent, in a predictive manner provided that the characteristics of the destabilization are known in advance (e.g., the "►anticipatory postural adjustments" that normally precede preplanned volitional movement or the reactions evoked by a periodic sinusoidal perturbation). In general, however, sensory information about the body orientation and motion is also required, particularly when balance is disturbed unexpectedly by a sudden perturbation (e.g., due to a force applied to the body or motion of the support surface). This sensory information is used to detect instability and to generate appropriate stabilizing responses, either by triggering and scaling preprogrammed "feedforward" reactions or by continuously updating ongoing "feedback" corrections. The initial phases of the reactions to sudden, transient perturbation are thought to be triggered feedforward corrections, whereas the later phases may also involve ongoing feedback control.

Postural strategies involve multiple sensory inputs (somatosensory, vestibular, visual) and highly adaptable triggered reactions, rather than stereotyped short-latency reflexive responses arising primarily from a single source of afferent drive (e.g., vestibulo-spinal reflex, myogenic stretch reflex). The triggering afferent signals depend on the nature of the perturbation, e.g., whether the

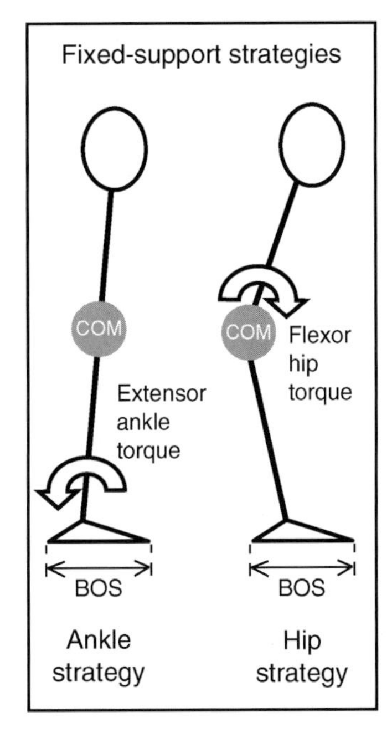

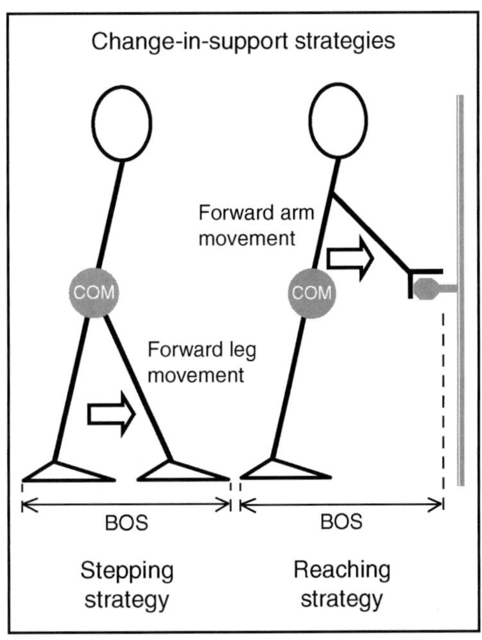

Postural Strategies. Figure 1 Postural strategies. Static postural equilibrium requires the body center of mass (*COM*) to be positioned over the base of support (*BOS*). The *stick figures* illustrate potential responses evoked by a perturbation that induces forward falling motion of the COM. Note the large increase in BOS associated with the change-in-support strategies.

perturbation involves the movement of the support surface or a force applied to the upper body. Responses to support surface perturbations may involve ankle muscle spindles (Ia afferents); however, the role of the vestibular system and/or somatosensory inputs from other joints cannot be ruled out [4] and it appears that even visual inputs (which are generally thought to require much longer processing times) can modulate the earliest postural muscle activation in some situations. The earliest muscle activation associated with the postural reaction typically occurs at a latency of about 80 to 140 ms after perturbation onset.

The control of these reactions is mediated via polysynaptic spinal and supraspinal neural pathways [5]. Furthermore, although balance reactions are often considered to be "automatic," it appears that control of later phases of the response (e.g., 200 to 500 ms after perturbation onset) may involve transcortical pathways and high-level cognitive and attentional systems [6]. This is supported by evidence from dual-task studies, where later phases of the balance response can be affected by performance of a concurrent cognitive task. Measurements of perturbation-evoked cortical potentials also support cognitive involvement. Thus, it appears that reactions to perturbation are not as stereotyped as once believed, but are in fact highly adaptable to the functional demands of maintaining stability, i.e., regulating the relationship between COM and BOS. Although some researchers have speculated that critical afferent information for determining the state of the COM may arise from load receptors such as Golgi tendon organs, maintaining stability also requires information about the state of the BOS and the direction of a gravitational reference vector. In contrast to the traditional view that the vestibular otoliths provide the reference vector, some researchers have suggested a construct based on multiple sensory modalities. Determination of the BOS state is also probably dependent on multiple sensory inputs, although it appears that the plantar cutaneous mechanoreceptors may play a particularly critical role in providing information pertaining to BOS stability limits and the state of contact between foot and ground [1].

Fixed-Support (Feet-in-Place) Strategies

In ►fixed-support strategies, the COM motion that is induced by the postural perturbation is decelerated via generation of active muscle torques at the joints of the supporting leg or legs, as well as the trunk and upper limbs. These joint torques create antero-posterior "environment reaction" forces (i.e., shear force at the foot ground interface, as well as forces occurring at the hand if it is in contact with a stable object or surface) and it is these forces that act to decelerate and arrest the horizontal COM motion (Fig. 2). Fixed-support reactions occur very rapidly (e.g., onset latency of 80 to 140 ms in ankle muscles) and are essentially the first line of defense against postural perturbation.

Biomechanically, the upright human body has redundant degrees of freedom. This means that there are potentially many different combinations of muscle

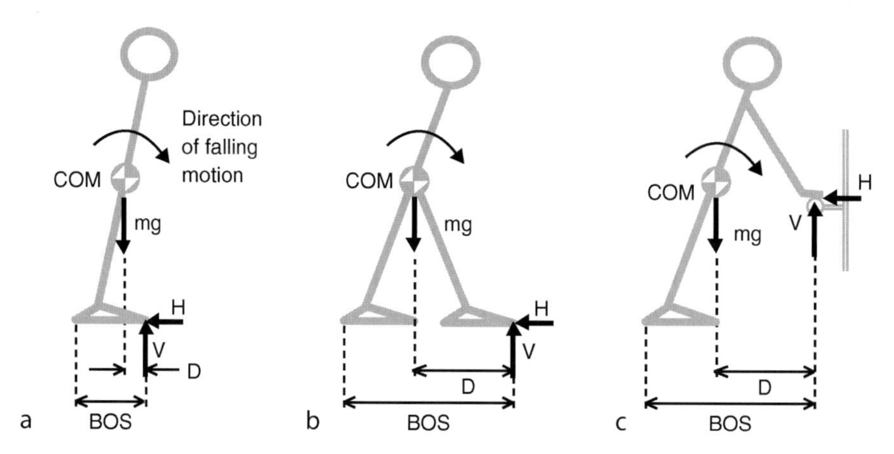

Postural Strategies. Figure 2 Biomechanical advantages of change-in-support strategies (*panels* b and c) in comparison to fixed-support strategies (*panel* a). The postural reaction must generate a horizontal ground-reaction shear force (*H*) in order to decelerate the horizontal motion of the center of mass (*COM*). Note that the stepping (b) and grasping (c) reactions can greatly amplify the moment arm (D) between the COM and the contact force (*V*), which allows greater shear force to be generated (H is approximately proportional to D). In addition, the increase in base of support (*BOS*) allows a larger range of COM motion to be accommodated without loss of stability. Adapted from [1].

torques at the various joints that could be used to re-establish postural equilibrium, in responding to a given postural perturbation. It has been proposed that the CNS deals with this redundancy, and simplifies the control problem, by restricting the response to a finite number of specific response patterns (often referred to as synergies as well as strategies) or weighted combinations of these patterns.

Early research identified two major strategies for responding to antero-posterior perturbations: (i) the ►ankle strategy and (ii) the ►hip strategy [7]. Although not included in the original definitions of these strategies, activation of knee muscles can also play an important role in both ankle and hip strategies, particular for perturbations that induce a backward falling motion and hence tend to cause the knee to "collapse" in flexion [4].

In the ankle strategy, the predominant stabilizing action involves active generation of ankle torque. This strategy is characterized by activation of ventral muscles (i.e., ankle dorsiflexors, hip flexors, abdominals) in response to backward falling motion and activation of dorsal muscles (i.e., ankle plantarflexors, hip extensors, paraspinals) in response to forward falling motion. In essence, the body is controlled to behave predominantly in the manner of a single-link inverted pendulum, with joint rotation primarily occurring at the ankle. For responses to support-surface perturbations, the involved muscles fire in a distal to proximal sequence.

In the hip strategy, the predominant stabilizing action involves active generation of hip torque. This strategy is characterized by activation of muscles on the opposite side of the thigh and trunk in comparison to the ankle strategy, i.e., dorsal muscles in response to backward falling motion and ventral muscles in response to forward falling motion. There is relatively little activation at the ankle. The main effect biomechanically is to generate stabilizing shear force (at the foot ground interface) that is larger than the shear force that can be generated using the ankle strategy.

The ankle strategy predominates at small levels of perturbation, whereas increasing levels of hip activation are added as the postural challenge increases [8]. These latter "mixed strategy" responses were originally thought to involve a weighted combination of the ankle and hip strategies; however, the "pure" hip strategy is seldom if ever observed during natural behavior (in experiments, it was learned over the course of repeated trials that involved standing on a shortened support surface, which limited ability to generate stabilizing ankle torque) [3]. This suggests that there is actually a continuum of postural responses formed by the addition of hip torque to the ankle strategy, rather than two distinct strategies.

Feet-in-place strategies for responding to medio-lateral perturbation primarily involve the hip abductors and adductors, due to the limited capacity at the ankle and knee for motion and torque generation in the frontal plane. Although one might expect responses to perturbations in "off axis" or "diagonal" directions to involve a weighted combination of the medio-lateral hip strategy and the antero-posterior ankle strategy, it appears that this construct is inadequate to explain the complex muscle responses that are evoked by multi-directional perturbations. Rather, there appears to be a continuum of strategies that are modifiable in a task-dependent manner [9]. Thus, for example, there may be co-contraction of agonist and antagonist ankle muscles (which serves to stiffen the ankle joint) and latencies for some muscles may differ according to the perturbation direction.

Although the feet in place strategies described above primarily involve activation of the lower limb and axial (trunk and neck) musculature, postural reactions involving the upper limbs often occur in parallel. In fact, activation of the arm muscles can occur as rapidly as the earliest activation at the ankle (80 ms after perturbation onset). The functional role of these arm-reaction strategies appears to vary, depending on task conditions. In situations where the hand is in contact with a supporting object or surface at time of perturbation onset, the arm reaction can rapidly generate stabilizing "environment reaction" forces at the hand. In situations where the arms are free to move, the arm movements may serve to augment the stabilization achieved by the lower limb reactions. For example, raising the arms can help to stabilize the body by acting as a counterweight, by inducing inertial joint torques at the shoulder (due to the acceleration and/or deceleration of the arm segments) or by increasing the rotational inertia of the body. In some situations, it appears that the raising of the arms may also serve a protective function, to absorb energy and reduce the risk of injury in the event that a fall does occur. It is also possible that the arm activation is related to an aborted ►reach-to-grasp ►change-in-support strategy (see below).

Change-in-Support Strategies

In change-in-support reactions, the BOS is altered via rapid stepping or reaching movements of the limbs. Increase in the BOS allows: (i) a larger range of perturbation-induced COM motion to be tolerated without loss of equilibrium, (ii) more time for this COM motion to be decelerated (via "environment reaction" forces generated at the foot and/or hand) and (iii) larger decelerating "environment reaction" forces to be generated (Fig. 2). Reach to grasp reactions provide a further biomechanical advantage in that they allow the body to be anchored with respect to the grasped object, provided that a sufficiently strong grip can be maintained. Change-in-support reactions can provide a much larger degree of stabilization, in comparison to fixed support reactions and are the only recourse in responding to large perturbations [1,3].

The neural control of change-in-support (compensatory) and volitional limb movements appear to differ in some fundamental ways. First, the compensatory postural movements are much more rapid. A compensatory stepping movement is typically completed in about 500 ms, approximately half the time required to step as fast as possible in reaction to a visual cue. Similarly, compensatory reaching reactions are more rapid than volitional arm movements. For example, compensatory arm activation typically begins at a latency of 80 to 140 ms after perturbation onset, whereas the most rapid latency for volitional arm activation (single-choice reaction-time task) is about 150–200 ms (and is slower when the task involves multiple choices).

Another fundamental distinction between compensatory stepping and volitional stepping pertains to the presence or absence of an "anticipatory postural adjustment" (APA), prior to the lifting of the swing leg. Volitional movements that involve stepping or raising a leg are invariably preceded by an APA which acts to propel the COM toward the stance limb and thereby serves to reduce the tendency of the COM to fall laterally toward the unsupported side during the subsequent foot movement. The APA is typically either absent or severely truncated during compensatory stepping reactions. This allows more rapid step execution; however, the consequence is that the lateral COM motion arising during the swing phase must be arrested after the swing foot lands. A large APA does occur when task demands require a prolonged swing phase (e.g., when the compensatory step must clear an obstacle). Presumably, the APA is required in such situations because the COM has much greater opportunity to fall laterally.

Whereas antero-posterior perturbations typically (in healthy young adults) evoke a single step that is predominantly directed forward or backward, perturbations in other directions evoke a wider variety of stepping patterns. If the perturbation-includes a medio-lateral component, then the perturbation induced COM motion will cause an increase in loading of one leg and a decrease in loading of the contralateral leg. Typically, when the perturbation is unpredictable, the unloaded leg is used to execute the stepping reaction. This has the advantage of allowing a more rapid foot lift, as less time is required to complete the unloading of the swing foot; however, the consequence is that a more complex stepping movement is required. This may involve a crossover step, in which the swing foot is moved across the body (either in front of or behind the stance leg). Alternatively, the unloaded leg may be used to execute a small initial medial step, which is then followed by a large lateral step with the contralateral leg (a "side-step sequence"). See Fig. 3.

The direction of perturbation-evoked reach-to-grasp reactions is highly dependent on both the direction of the perturbation and the location of potential handholds that can be grasped or touched for support. Remarkably, even the earliest portion of the arm trajectory (at a latency as early as 80 ms) is directed toward the nearest available handhold. Such a rapid response does not permit visual scanning of the environment; hence, it appears that the initial arm trajectory must either involve peripheral vision or "remembered" visuospatial information. In order to use "remembered" information,

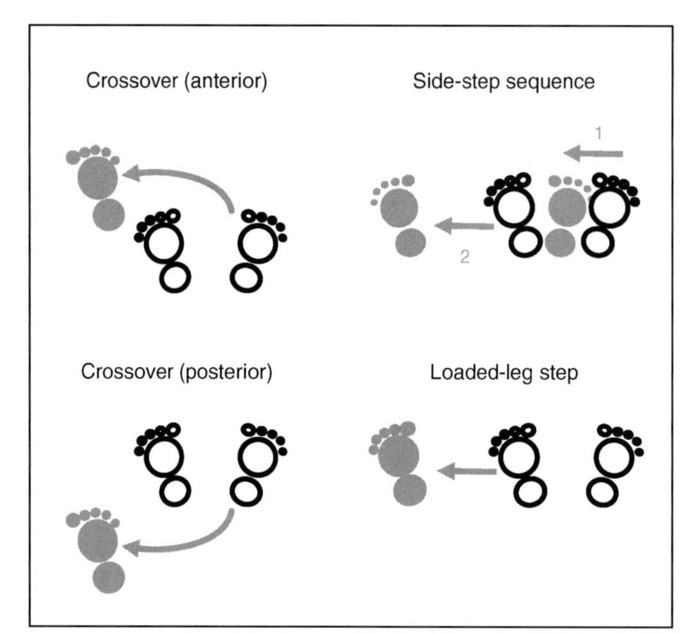

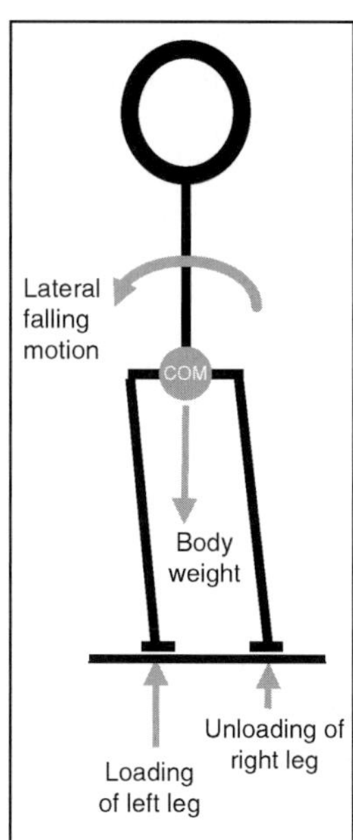

Postural Strategies. Figure 3 ►Stepping strategies evoked when the postural perturbation includes a lateral component. The footprint drawings illustrate responses evoked by a perturbation that induces leftward falling motion of the center of mass (*COM*). The *unshaded* footprints indicate the starting position at time of perturbation onset, and the *shaded* footprints indicate the landing position of each step. The *stick figure* illustrates how the initial COM motion induced by this perturbation creates increased loading of the left leg. As a consequence of this loading, the stepping responses are most commonly initiated with the unloaded (right) leg, and involve either a crossover movement or a side step sequence (a small medial step followed by a large lateral step). Reactions where the initial step is taken with the loaded foot have been observed to occur very rarely in some studies, but more commonly in others. The "loaded leg" steps seem more likely to occur when the individual can preplan to step with a specific leg (e.g., if the direction of the perturbation is known in advance).

the CNS would need to maintain and automatically update an egocentric spatial map of the immediate surroundings as the person moves about in daily life. This would then allow the hand to be directed very rapidly toward the nearest available handhold, if and when sudden unexpected loss of balance occurs. A similar control mechanism would allow rapid perturbation-evoked stepping reactions to be directed appropriately, so as to avoid obstacles and accommodate other constraints on foot movement.

Although the neural pathways involved in the control of change-in-support reactions are not well established, it seems likely that the planning of the limb trajectory makes use of the same neural pathways as those thought to be involved in planning volitional limb movements, i.e., spinal pattern generators in the case of stepping reactions and cortical pathways in the case of reaching reactions. The very rapid initiation and scaling of the trajectory could then be triggered by subcortical pathways similar to those thought to be involved in triggering the early fixed-support postural reactions.

Strategy Selection

A traditional view has been that change-in-support reactions are only used as a last resort when earlier fixed-support reactions fail to keep the COM within the stability limits of the BOS. The basic idea is that the ankle strategy is used to respond to antero-posterior perturbation when the induced COM motion is small in relation to the BOS stability limits, the hip strategy comes into play when the COM motion is larger, and stepping and reaching emerge only when the COM motion exceeds the stability limits of the BOS. This may be true when individuals are instructed to try not to step or move the arms; however, this is clearly not the case when the person is allowed to respond naturally. It is now clear that compensatory stepping and reaching are commonly initiated very early, with the COM well within the stability limits and in fact often seem to be the preferred response. For example, individuals will almost always step and begin to move the arms when the perturbation is unexpected or novel (e.g., the very first trial in an experiment), even if the perturbation is relatively small.

It appears that initiation of reach-to-grasp reactions occurs in parallel with the earliest fixed-support reaction, as evidenced by the similarity in timing of the arm and ankle activation (latency of ~100 ms). Typically, the initiation of compensatory stepping is not quite as rapid; however, the onset of the stepping reaction (i.e., active changes in leg loading) can occur as early as 130 ms after perturbation onset. Interestingly, the early fixed-support reaction is not eliminated, even when a rapid reach-to-grasp or stepping reaction is initiated. Presumably, the fixed-support reaction persists as an important safeguard against instability. Early initiation of the change-in-support response, regardless of perturbation magnitude, may also reflect the high priority given by the CNS to the task of safeguarding stability.

Ultimately, strategy selection can be influenced by a number of factors, including (i) the features of the perturbation (timing, direction, magnitude, predictability), (ii) the "►central set" of the individual (affect, arousal, attention, expectations, prior experience), (iii) ongoing activity (cognitive or motor) and (iv) environmental constraints (e.g., slippery surfaces that limit reaction-force generation, obstacles that constrain limb movement) [1,7,10].

Modulation of Strategies

Both fixed-support and change-in-support reactions are highly modifiable and adaptations to meet task demands can be learned rapidly (e.g., over the course of one to five experimental trials). The same factors that affect strategy selection listed above can also modulate many of the features of the reactions *per se*.

One of the features that is not highly modifiable is the timing of the early activation of the ankle muscles in the ankle strategy. This early activation typically persists, at a similar latency, even when the individual preplans to use a stepping reaction to recover balance, despite the fact that the ankle activation may interfere with the execution of the step. In keeping with the persistent and apparently automatic nature of this response, the early ankle activation is commonly referred to as the early "automatic postural response" (APR). Nonetheless, the APR is not a stereotyped response. The magnitude of the activation is scaled according to the direction and magnitude of the perturbation and is influenced by the predictability of the perturbation, the "central set" of the individual and the instructions given (e.g., whether or not to step). The functional task demands can also have a profound effect. For example, backward horizontal support surface movements evoke activation in the ankle plantarflexors, which serves to stabilize the body. When similar ankle rotation is evoked by toes-up tilt of the support-surface, the evoked plantarflexor activation acts to amplify the destabilization, but this non-functional response quickly habituates over the course of repeated trials in healthy individuals [10].

Change-in-support reactions exhibit an even greater degree of modifiability, in both magnitude and timing [1,3]. For example, initiation of stepping reactions can be delayed substantially by instructions to try not to step. Conversely, stepping and reaching reactions can be initiated more quickly when the person is given prior instruction to step or reach in response to the perturbation. Furthermore, the limb movements are heavily dependent on environmental constraints (i.e., location of obstacles and potential handholds). There also appears to be capacity for online modulation. For example, the limb trajectory can be altered (to at least some degree) to deal with additional perturbations or changes in environmental constraints arising after initiation of the reaction and reactions that are initiated can be aborted prior to completion.

References

1. Maki BE, McIlroy WE (2003) Effects of aging on control of stability. In: Luxon L et al. (eds) A textbook of audiological medicine: clinical aspects of hearing and balance. Martin Dunitz, London, pp 671–690
2. Pai YC, Patton J (1997) Center of mass velocity-position predictions for balance control. J Biomech 30:347–354
3. Maki BE, McIlroy WE (1997) The role of limb movements in maintaining upright stance: the "change-in-support" strategy. Phys Ther 77:488–507
4. Allum J, Carpenter MG, Honegger F (2003) Directional aspects of balance control in man. IEEE Eng Med Biol Mag 22:37–47
5. Dietz V (1992) Human neuronal control of automatic functional movements: Interaction between central programs and afferent input. Physiol Rev 72:33–69
6. Maki BE, McIlory WE (2007) Cognitive demands and cortical control of human balance-recovery reactions. Journal of Neural Transmission 114:1279–1296
7. Nashner LM, McCollum G (1985) The organization of human postural movements: a formal basis and experimental synthesis. Behav Brain Sci 8:135–172
8. Runge CF, Shupert CL, Horak FB, Zajac FE (1999) Ankle and hip postural strategies defined by joint torques. Gait Posture 10:161–170
9. Henry SM, Fung J, Horak FB (1998) EMG responses to maintain stance during multidirectional surface translations. J Neurophysiol 80:1939–1950
10. Horak FB (1995) Adaptation of automatic postural responses. In: Bloedel J et al. (eds) Acquisition of motor behavior in vertebrates. MIT, Cambridge MA

Postural Sway

Definition

In stance, a process of continuous, small corrections of the upright body position takes place to oppose the destabilizing effect of gravity. This creates a pattern known as spontaneous sway, or postural sway.

►Stabilometry

P

Postural Synergies

LENA H. TING
Laboratory for Neuroengineering, The W. H. Coulter Department of Biomedical Engineering, Emory University and Georgia Institute of Technology, Atlanta, GA, USA

Synonyms

Muscle synergies; functional muscle synergies; motor primitives; M-modes

Definition

A ►postural synergy is a preferred pattern of muscle co-activation that is used by the nervous system to maintain standing balance. Each postural synergy specifies a pattern of muscle activation across many muscles. Through flexible combinations of postural synergies, a repertoire of postural behaviors is produced. By eliminating the need to control each muscle independently, postural synergies are thought to simplify the neural control task of selecting and coordinating multiple muscles across the body. A postural strategy defines the overall goals involved in the maintenance of balance; these can vary depending on the particular postural task, the context in which the task is performed and the postural configuration. Postural synergies define the muscle activation patterns that are used by the nervous system to implement various postural strategies.

Description of the Theory

Introduction

The theory of postural synergies address the basic question of whether the nervous system activates each muscle independently when it performs a task or whether the multiple muscles are activated together, thus reducing the total number of neural command signals necessary. Currently, postural synergies are thought to represent neural "building blocks" for generating a wide range of postural behaviors. Each postural synergy specifies a pattern of muscle activation across many muscles and is purportedly controlled by one neural command signal. By combining muscle synergies in various proportions, a continuum of muscle activation patterns for postural control can be generated using just a few neural command signals.

The long-standing debate within the general motor control field over the concept of muscle synergies is exemplified by the specific debate over the existence of postural synergies. Nashner first described "fixed" postural synergies in subjects standing on a moving perturbation platform [1]. Distinct patterns of muscle activation across the ankle, knee, hip and trunk were reliably observed when the platform was moved either forwards or backwards (Fig. 1a) and were thought to represent two different postural synergies [2]. Originally, it was thought that postural synergies were activated in a mutually exclusive fashion and that each muscle was activated by only one postural synergy (Fig. 2a).

These conclusions were challenged by later studies that showed flexibility in patterns of muscle activation in response to a backward perturbation. Depending on the perturbation amplitude and prior experience of the subject, two types of responses were observed, the "ankle strategy" and the "hip strategy," so named for the major joint motions involved. Each strategy elicits a very different pattern of muscle activation, demonstrating that postural synergies to a particular direction of perturbation are not fixed. Moreover, when perturbations were given in many directions in the horizontal plane, even more complex patterns of muscle activation emerged in both humans and cats [2]. Each different perturbation direction elicited a unique pattern of muscle activation (Fig. 1b), suggesting that muscles must be controlled independently to perform multidirectional balance control. It was for this reason that the notion of muscle synergies was then rejected as being too constraining and inflexible for the production of natural movements [3].

Recently, new computational techniques have helped to demonstrate that a motor control architecture based on muscle synergies can both simplify neural control as well as provide flexibility in motor output. In the new framework, more than one muscle synergy can be activated during a postural response and each muscle can also be activated by more than one synergy. By varying the magnitude of the neural command signals to just a few muscle synergies, many different muscle activation patterns can be generated (Fig. 2b), including the responses to multidirectional postural responses describe above [4]. The neural substrates of muscle synergies for postural control remain unknown. Where muscle synergies are encoded within the neural control hierarchy is a complex topic and may also be task dependent. It is hypothesized that postural synergies are formed in the brainstem, based on observations of postural control following neural impairment.

Degrees of Freedom Problem

To maintain standing balance, the nervous system must confront the classic "degrees of freedom" problem posed by Nikolai Bernstein [5], where many different solutions are available due to the large number of elements or degrees of freedom involved. In postural control, a large number of muscles and joints across the limbs, trunk and neck must be coordinated to maintain the body's ►center of mass (CoM) over the base of support, typically formed by the feet. The large

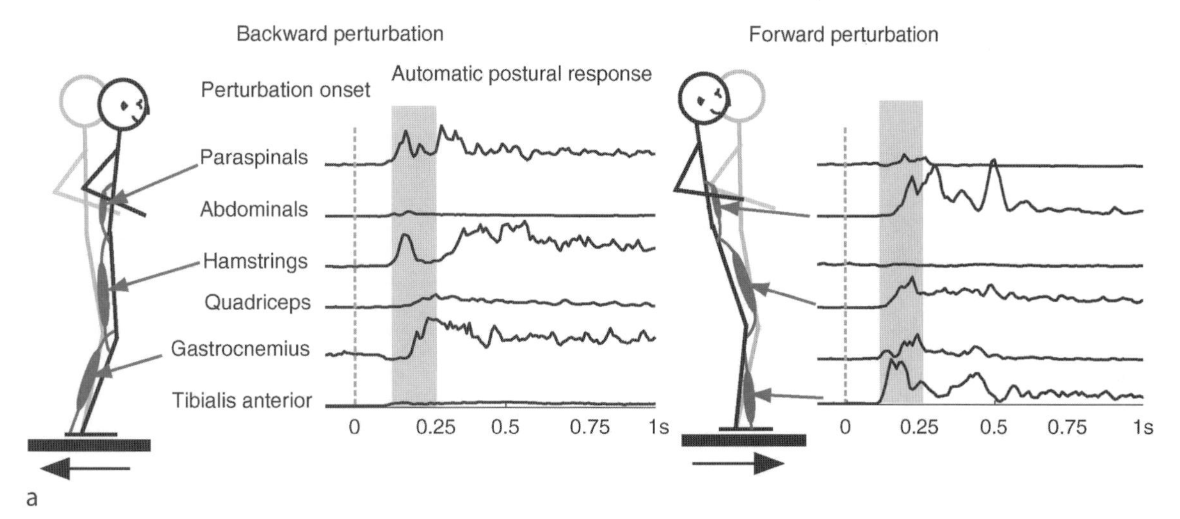

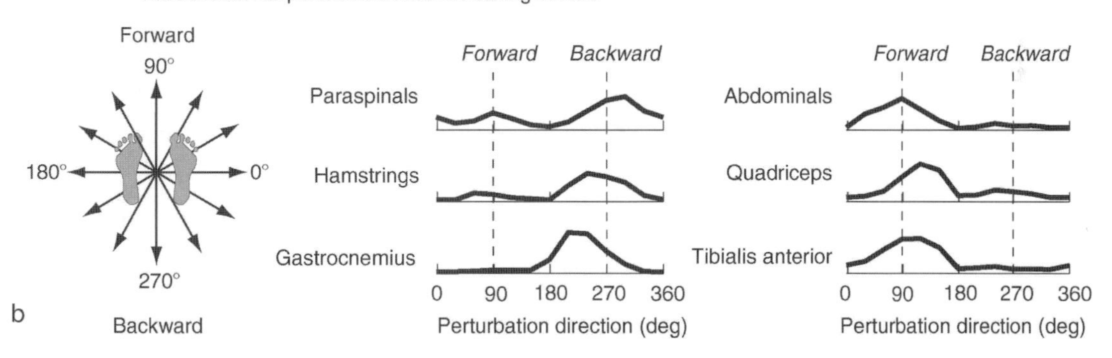

Postural Synergies. Figure 1 Muscle activity evoked following perturbations to the support-surface. (a) Backward perturbations elicit activity in muscles on the posterior side of the body. (b) Forward perturbations elicit activity in muscles on the anterior side of the body. The gray area represents the initial muscular response to perturbation, called the automatic postural response (APR). (c) The magnitude of the response during the APR varies as a function of direction and can be plotted as a tuning curve. Each muscle has a unique tuning curve, suggesting that each muscle is activated by a separate neural command signal.

number of degrees of freedom afforded by the multiple joints and muscles in the body thus allow for many solutions that can accomplish the task goals equally well. This multiplicity or redundancy of solutions allows flexibility in performing the postural task; it also poses the problem that the nervous system must choose from a large set of possible solutions. In contrast, if the body were a simple rigid stick balanced on one end, then the angle of the stick in space would completely determine the location of the center of mass with respect to the base of support. Moreover, if only one muscle is available, there is no ambiguity as to how to activate the muscle in order to move the center of mass. Thus, the "degrees of freedom problem" occurs only when overall task requirements are not sufficient to specify multiple output variables controlled by the nervous system.

Bernstein proposed the existence of synergies as a neural strategy for simplifying the control of multiple degrees of freedom by coupling or grouping output variables [5]. This scheme was based on experimental observations that many joint angles appear to be controlled together rather than independently during motor tasks. For example, during locomotor tasks such as running, the hip, knee and ankle joints all flex and extend at the same time, suggesting that they are not controlled independently. However, such observations only identify correlations between the joint motions. A variety of muscle activation patterns can produce similar joint movements. Therefore, joint angle changes do not necessarily have a direct relationship to neural command signals activating muscles. Since muscle activation is directly caused by motoneuron firing, correlations between muscle activation patterns can be more plausibly derived from a single neural command that is distributed across the various motoneuron pools. Thus, muscle synergies may represent a mechanism by which the nervous system can achieve repeatable multijoint coordination.

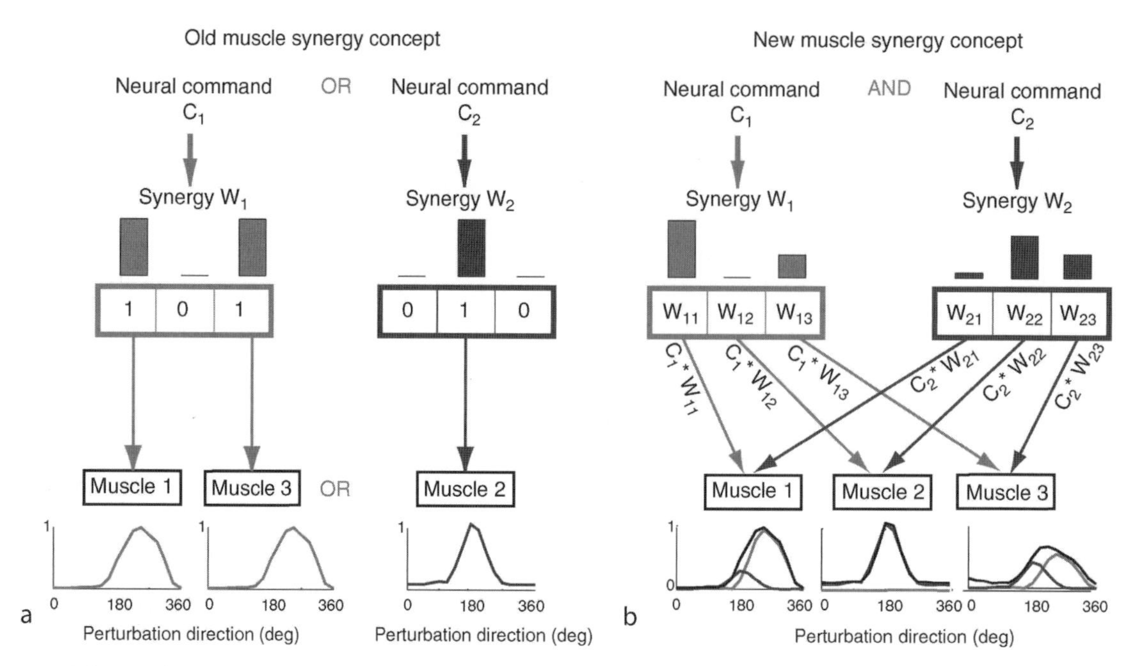

Postural Synergies. Figure 2 Illustrations of two different muscle synergy concepts. (a) In the original muscle synergy concept, only one muscle synergy was elicited at a time, and muscles could only be activated by one synergy. Therefore, all muscles activated by the same synergy would have the same directional tuning curve, determined by the neural command c that activated it. (b) In the new concept, more than one synergy can be activated at a time. Further, muscles can participate in multiple synergies, and have different weightings in each synergy. Therefore, each muscle's tuning curve is a weighted average of the two tuning curves of each muscle synergy.

Computational Methods for Identifying Postural Synergies

Recent computational techniques have redefined the working hypothesis of how muscle synergies can allow for flexible motor coordination while also simplifying the degrees of freedom problem. In this new formulation, a single synergy specifies a fixed muscle activation pattern that is modulated by a single neural command signal, but multiple muscle synergies can be activated at one time [4,6,7]. Mathematically, each muscle activation pattern is thus composed of a linear combination of a few (n) muscle synergies $\mathbf{W}_i$, each activated by one neural command c_i. The net muscle activation pattern vector **M** is therefore hypothesized to take the form:

$$\mathbf{M} = c_1\mathbf{W_1} + c_2\mathbf{W_2} + \ldots. + c_n\mathbf{W}_n$$

M is a vector where each element is the resulting level of activation in each muscle (Fig. 3a). $\mathbf{W}_i$ is a vector that specifies the pattern of muscle activity defined by that muscle synergy. Each element of $\mathbf{W}_i$ takes a value between 0 and 1, representing the relative contribution of each muscle to that muscle synergy. Each muscle synergy is then activated by a single, scalar neural command signal c_i, which determines the relative contribution of the muscle synergy $\mathbf{W_i}$ to the overall muscle activation pattern, **M**.

The above formulation allows for flexible "mixing" of a set of muscle synergies to produce the final output muscle activation pattern. Therefore, if two muscle synergies are present, rather than defining just two output muscle activation patterns, as in previous definitions (Fig. 2a), an entire continuum of output muscle activation patterns can be generated by varying the commands c_1 and c_2. Within this continuum, individual muscle activations are not strictly correlated to each other because most muscles belong to more than one muscle synergy and are thus activated independently by two different neural commands (Fig. 2b).

Linear decomposition techniques can be used to identify muscle synergies from experimentally measured muscle activation patterns. Because the number of muscle synergies is smaller than the number of muscles for any given task, the spectrum of muscle activation patterns that can be generated using muscle synergies is more limited than the case where muscles are controlled independently. However over the entire behavioral repertoire the number of muscle synergies could exceed the number of muscles. This dimensional reduction, which simplifies the degrees of freedom problem, can be identified using several mathematical analysis techniques such as principal components analysis (PCA), independent components analysis (ICA) and factor analysis (FA) [6]. Another such technique, non-negative matrix factorization (NMF),

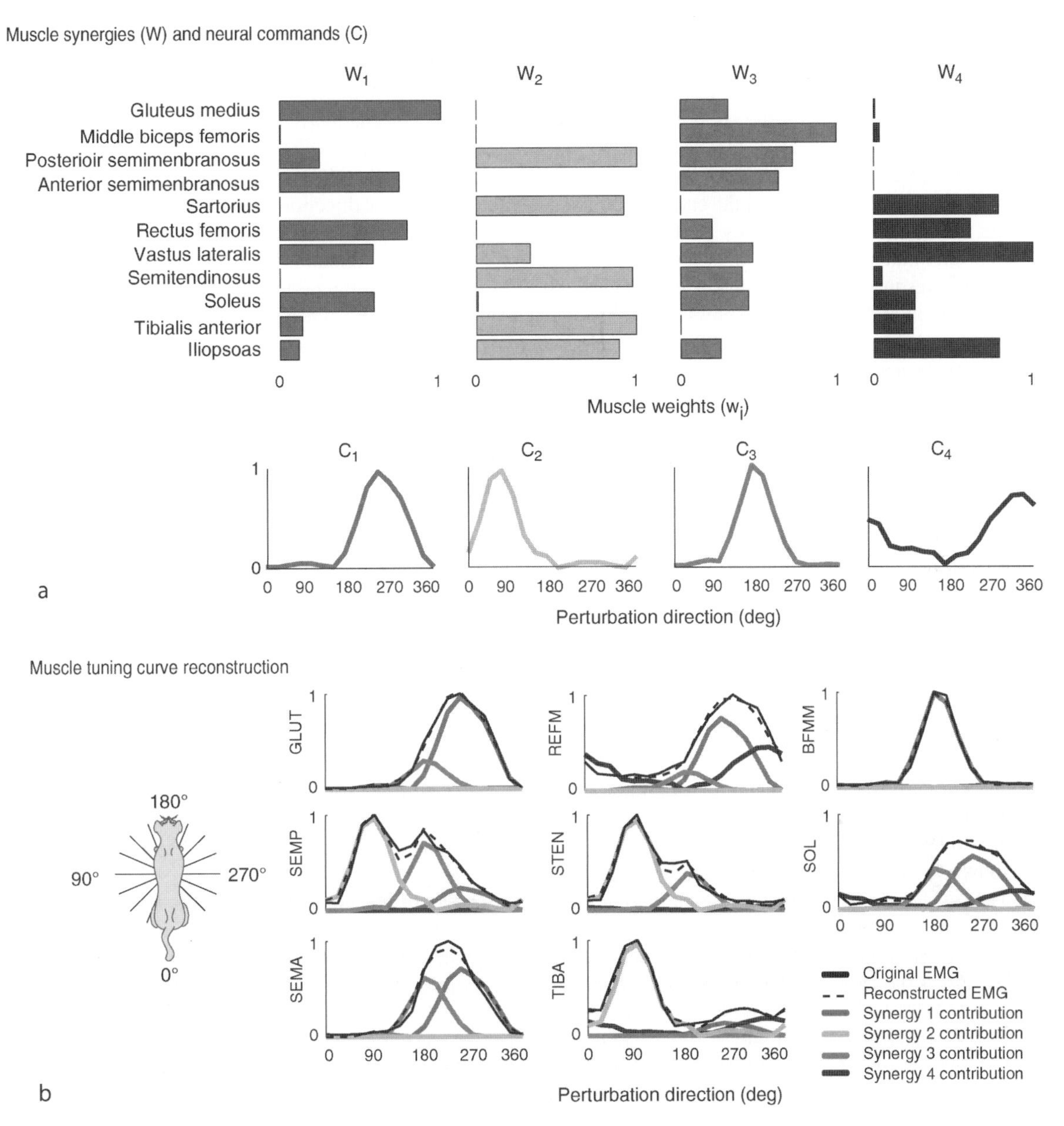

Postural Synergies. Figure 3 Muscle synergies and neural commands used to generate muscle tuning curves during postural responses in cats. (a) Each muscle can participate in each muscle synergy with a different weight, indicated by the bars. (b) Neural commands to each muscle synergy can also be illustrated as tuning curves. Each muscle synergy therefore has preferred direction of activation. (c) EMG tuning curves can be reconstructed using muscle synergies. Each muscle's tuning curve is found by summing the product of each tuning curve, c_i and the weighting of each muscle within the synergy W_i. All muscle tuning curves are thus constrained to be weighted averages of the synergy tuning curves. Therefore, the muscle tuning curves have more varied and complex shapes than the synergy tuning curves.

allows complex data sets to be more successfully partitioned into meaningful parts [4,6,8]. NMF is particularly useful for data that are inherently positive valued, such as neural spike trains or muscle activations. The extracted elements are based on the components forming the data set rather than on more holistic features. For example, when applied to images of faces, a non-negative extraction routine generates vectors representing noses, ears and eyes, whereas PCA generates components that all tend to look roughly like an entire face [8].

Muscle Synergies in Postural Control

During postural responses to perturbations in different directions, multiple muscles across the body are activated and for each different direction of the perturbation, a different pattern of muscle activation is elicited (Fig. 1b). In both humans and in cats, a stereotyped, directionally specific pattern of muscle activity called ►automatic postural response (APR) is evoked after perturbations to the support surface. The muscle activation occurs after the platform motion begins, but before the center of mass moves appreciably, with a latency of around 50 ms in the cat and 100 ms in humans. In both cases, this latency is about twice the ►stretch reflex latency for distal muscles and evokes a much larger response than the stretch reflex [2]. Each muscle's activation level can be expressed in terms of a muscle tuning curve, which shows the variation of the muscle activation with perturbation direction (Fig. 1b – human, Fig. 3b – cat). Thus, for some directions, a muscle may have high activation and for others it may not be active at all. These muscle tuning curves define the complex patterns of muscle activation evoked across many perturbation directions [2,4,9].

Although each direction of perturbation evokes a slightly different pattern of muscle activation over all muscles, these variations can be explained by a combination of just a few muscle synergies in the cat [4]. Over 95% of the variability in as many as fourteen muscle tuning curves can be explained by combining just four muscle synergies (Fig. 3b). Instead of activating each muscle independently for each perturbation direction, only four neural commands, each activating a synergy $\mathbf{W_i}$, need to be specified with amplitude $\mathbf{c_i}$ for any perturbation direction. The net muscle activation pattern is thus found by adding up the contributions of each muscle synergy to each muscle's activation level.

Muscle synergies may coordinate the limb to produce a specific biomechanical function for stabilizing the body. In the cat, it has been suggested that each muscle synergy allows the leg to produce a force in a particular direction in order to stabilize the leg (Fig. 4a). Variations in the components of active force generated by each leg are correlated to the variations in the neural commands of each muscle synergy (Fig. 4b). Each muscle synergy can generate a specific direction of force; the forces are distributed so that upward, downward, anterior, posterior and lateral force direction can be produced. Thus, muscle synergies may be organized to produce specific task-level biomechanical functions [4].

Even for postural perturbations of the same direction, multiple muscle synergies may exist. In backward perturbations of the support surface in humans, two types of responses can be elicited. One is called the "ankle strategy" where the body remains upright and most of the motion occurs around the ankle joint. The other is called the "hip strategy," where the trunk tilts forwards and the hip angle motion is most predominant. Each strategy can be defined by a specific pattern of joint torques. Because joint torques directly relate to the force generation of the musculature, this suggests that there are muscle synergies underlying these two strategies. While these two strategies were initially thought to be mutually exclusive, they in fact represent two different postural synergies that can be combined to produce a whole continuum of intermediate responses [2,11]. Therefore, rather than having a simple repertoire of just two response patterns, the flexible combination of these postural synergies allows the APR to be tuned and varied with perturbation amplitude, prior experience and anticipation.

Encoding of Muscle Synergies in the CNS

If muscle synergies reflect neural control mechanisms, then what are the neural substrates that generate muscle synergies? It is now understood that postural synergies

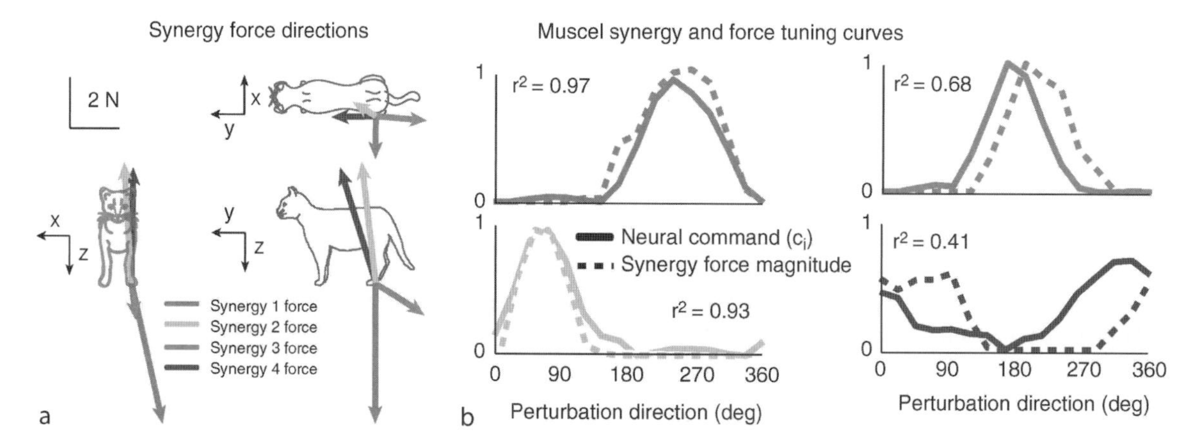

Postural Synergies. Figure 4 Forces produced during the automatic postural response correlate with muscle synergy activations. (a) Forces produced during postural responses can be decomposed into four force vectors. (b) During postural response, the magnitude of each force vector required to reproduce the total force varies as a function of direction and can be illustrated as a tuning curve. The tuning curves of force magnitude are highly correlated with the tuning curves of the neural commands c_i activated the muscle synergies. Thus, each force vector may represent the functional output of the muscle synergy.

cannot be explained just by reflexes acting in response to muscle stretch. In both humans and cats, it has been shown that perturbations that stretch the muscles differently can activate the same muscle synergies. For example, Nashner originally demonstrated that for a backward translation of the support surface, the calf muscle is stretched as the subject falls forward and that the same muscle is subsequently activated to maintain balance, consistent with a stretch reflex. In contrast, if a toes up rotation of the support surface is given, the calf muscle is stretched but the subject falls backwards, so that the antagonist muscle is activated to restore balance, in direct opposition to the stretch reflex [1]. This same principle has been demonstrated in multidirectional perturbations in both cats and humans [9]. Moreover, the loss of a single sensory modality, such as proprioceptive, vestibular or visual loss, does not appear to significantly affect muscle activation patterns, only their activation levels. Therefore muscle synergies are not a direct response to local sensory input, but appear to be related to more global variables, such as the direction of CoM displacement caused by the perturbation, that require multisensory integration [2,9].

How postural synergies are encoded in the nervous system is not known. For locomotor tasks, the encoding of muscle synergies appear to be located within the neural circuitry of the spinal cord [7], as animals can produce locomotor activity from a spinal cord that is isolated from the brain following spinal cord transection. These same animals can support their own weight while standing, but direction specific responses to postural perturbations are lost. This suggests that postural synergies are generated within the spinal cord [10]. It is known that the brainstem is essential to the maintenance of postural orientation and equilibrium and it is possible that neural mechanisms producing postural synergies reside there. Moreover, postural synergies appear intact in patients with postural impairments due to lesions in higher brain centers. For example, Parkinson's disease is characterized by pathology of the basal ganglia, which project to brainstem areas that are important for postural control. Individuals with Parkinson's disease have the aility to generate postural synergies that are similar to control subjects, but have difficulty changing the muscle synergy that is activated when perturbation conditions change. Similarly, in individuals with cerebellar dysfunction, postural synergies are similar to control subjects, but their activation levels do not decrease with repeated perturbations as in control subjects. Therefore, the muscle synergy structure appears intact, but the ability to correctly activate the neural commands to those muscle synergies is compromised, which impairs the postural stability in these individuals [2]. The theory of postural synergies therefore contributes to our understanding of the role of various nervous system structures in postural control and can guide experimental investigations that may further the validity of the theory.

►Postural Strategies

References

1. Nashner LM (1977) Fixed patterns of rapid postural responses among leg muscles during stance. Exp Brain Res 30(1):13–24
2. Horak FB, Macpherson JM (1996) Postural orientation and equilibrium. In: Handbook of physiology, Section 12. American Physiological Society, New York, pp 255–292
3. Macpherson JM (1991) How flexible are muscle synergies? In: Humphrey DR, Freund H-J (eds) Motor control: concepts and issues. Wiley, New York, pp 33–47
4. Ting LH, Macpherson JM (2005) A limited set of muscle synergies for force control during a postural task. J Neurophysiol 93(1):609–613
5. Bernstein N (1967) The coordination and regulation of movements. Pergamon, New York
6. Tresch MC, Cheung VC, d'Avella A (2006) Matrix factorization algorithms for the identification of muscle synergies: evaluation on simulated and experimental data sets. J Neurophysiol 95:2199–2212
7. Flash T, Hochner B (2005) Motor primitives in vertebrates and invertebrates. Curr Opin Neurobiol 15(6):660–666
8. Lee DD, Seung HS (1999) Learning the parts of objects by non-negative matrix factorization. Nature 401 (6755):788–791
9. Ting LH, Macpherson JM (2004) Ratio of shear to load ground-reaction force may underlie the directional tuning of the automatic postural response to rotation and translation. J Neurophysiol 92(2):808–823
10. Macpherson JM, Fung J (1999) Weight support and balance during perturbed stance in the chronic spinal cat. J Neurophysiol 82(6):3066–3081
11. Torres-Oviedo G, Ting L (2007) Muscle synergies characterizing Human postural responses. J Neurophysiol 98:2144–2156

Postural Tone

Definition

Background tension developed by the antigravity muscles. It represents a prerequisite for the maintenance of posture. The postural tone is regulated by intrinsic properties of spinal motoneurons, by the tonic activity of the corresponding muscle spindle afferents and by signals arising from brainstem systems projecting to the spinal cord, including the vestibular nuclei and the reticular formation.

►Postural Synergies
►Vestibulo-Spinal Reflexes

Postural Tremors

Definition

These tremors occur while trying to keep a body part in a constant position, such as an arm in outstretched posture.

Posture

Definition

A particular position assumed by the body.

Posture – Sensory Integration

ROBERT J. PETERKA
Neurological Sciences Institute, Oregon Health & Science University, Portland, OR, USA

Synonyms

Sensor fusion

Definition

Sensory integration, as it pertains to posture, refers to the process by which ►kinematic (orientation and motion) and ►kinetic (force related) information from multiple sensory sources is combined in the nervous system for the purpose of generating motor action to compensate for the destabilizing effect of gravity and to resist external perturbations.

Description of the Theory

The Task of Sensory Integration

Sensory information that is relevant to postural control is available from various sensory systems. These include the visual system, vestibular system, various aspects of somatosensation (proprioception signaling muscle stretch and joint angle, tendon force sensors, pressure sensors in the feet and other parts of the body signaling contact with the ground or the environment and tactile sensors in the skin around the joints) and the auditory system. Within each system, different subsystems encode physical variables related to different aspect of motion and orientation and related to forces applied to the body and within the body. Furthermore, within each subsystem, the individual sensory receptors typically have a variety of static and dynamic response characteristics. The monumental task of the sensory integration process is to somehow combine this information and make it available to the motor control system so that the organism generates coordinated motor actions that maintain stability, respond appropriately to external perturbations and permit the expression of voluntary actions.

Benefits of Sensory Integration

In many situations the orientation cues provided by different sensory systems are redundant. Consider the simplest possible case where the legs, trunk and head of a human subject move together as a single mechanical unit with body sway consisting of a rotation about the ankle joints. In this case, during stance on a level surface while viewing an earth fixed visual scene, body sway relative to earth vertical is accurately sensed by the visual system, which signals body motion relative to the visual scene, by the vestibular system, which signals body motion in space and by proprioceptors, which signal ankle joint angle. However, there is variability and therefore uncertainty associated with orientation estimates derived from each of these sensory systems. In this common situation with redundant sensory information, an orientation estimate with reduced overall variability can be obtained by appropriately combining the redundant sensory information.

What is the appropriate way to combine redundant sensory information? Previous theoretical and experimental work suggests that the nervous system may employ the principle of maximum likelihood estimation to combine sensory inputs [1]. For the case of two sensory sources, S_a and S_b, the combined maximum likelihood estimate, $\hat{S}$,is given by a weighted combination of the individual sensory estimates

$$\hat{S} = w_a \cdot S_a + w_b \cdot S_b$$

where the sensory weights w_a and w_b are equal to

$$w_a = \frac{\sigma_b^2}{\sigma_a^2 + \sigma_b^2}, \quad w_b = \frac{\sigma_a^2}{\sigma_a^2 + \sigma_b^2}$$

with σ_a^2 being the variance of S_a, and $\sigma_b^2$the variance of S_b. In words, the sensory source with the smaller variance will have the larger weight and will make a larger contribution to the overall sensory estimate $\hat{S}$. Although an intuitive method for combining sensory information might be to ignore information from the more variable source, the maximum likelihood principle shows that the best (lowest variance) estimate is obtained by a weighted combination, even though one source may be considerably less reliable (i.e., large noise or variance) than the other source.

It is not currently known if the maximum likelihood principle strictly applies to sensory integration for postural control. However, it is known that a sensory weighting mechanism can account for experimental results in humans where body sway was provoked by tilts of the support surface or visual surround [2].

Constraints on Sensory Integration

Because postural control involves motor action as well as sensory integration, there are additional constraints on the sensory integration process as well as opportunities for the sensory integration process to facilitate postural control. Consider a simplified representation (Fig. 1) of a ►postural control system where orientation information is provided by proprioceptors and ►graviceptors.

Graviceptors yield the sensory signal S_{bs} that encodes the physical variable, *BS* (i.e., body in space angular orientation). Proprioceptors yield the sensory signal S_{bf} that encodes the physical variable *BF* (i.e., body orientation relative to the feet). It can be hypothesized that a weighted combination of these sensory sources contributes to an overall estimate of orientation, $\hat{S}$,and this overall estimate is compared to an internal reference orientation, S_{ref}, that represents the desired body orientation. Without loss of generality, it can be assumed that $S_{ref} = 0$ to symbolize the desired goal of remaining in an upright orientation. The difference between the orientation estimate and the reference orientation gives a sensory "error" signal, *e*. This process is represented by the equation

$$e = \hat{S} - S_{ref} = w_p \cdot S_{bf} + w_g \cdot S_{bs} \text{ (for } S_{ref} = 0)$$

where w_p is the proprioceptive weighting factor and w_g the graviceptive weighting factor. Then, through a sensory to motor transformation, a corrective torque, T_c, is generated as a function of *e*, $T_c = f(e)$, and T_c is applied to the body to control body orientation in space.

The net result of the process of sensory encoding, sensory integration and sensory to motor transformation is that T_c is generated as a function of a weighted combination of the physical variables *BF* and *BS*

$$T_c = f(w_p \cdot BF + w_g \cdot BS)$$

During stance on a level surface, the physical variables *BF* and *BS* are equal and the torque generated in relation to both proprioceptive and graviceptive cues facilitates maintenance of a stable vertical body orientation. However, on a tilting surface, *BF* is equal to *BS-FS*, where *FS* is the orientation of the feet in space (and equal to the tilted surface orientation assuming that the feet remain in contact with the surface). Therefore the above equation can be rewritten as

$$T_c = f(w_p \cdot (-FS) + (w_p + w_g) \cdot BS)$$

This equation shows that there are two components to T_c and these components have opposite signs. One

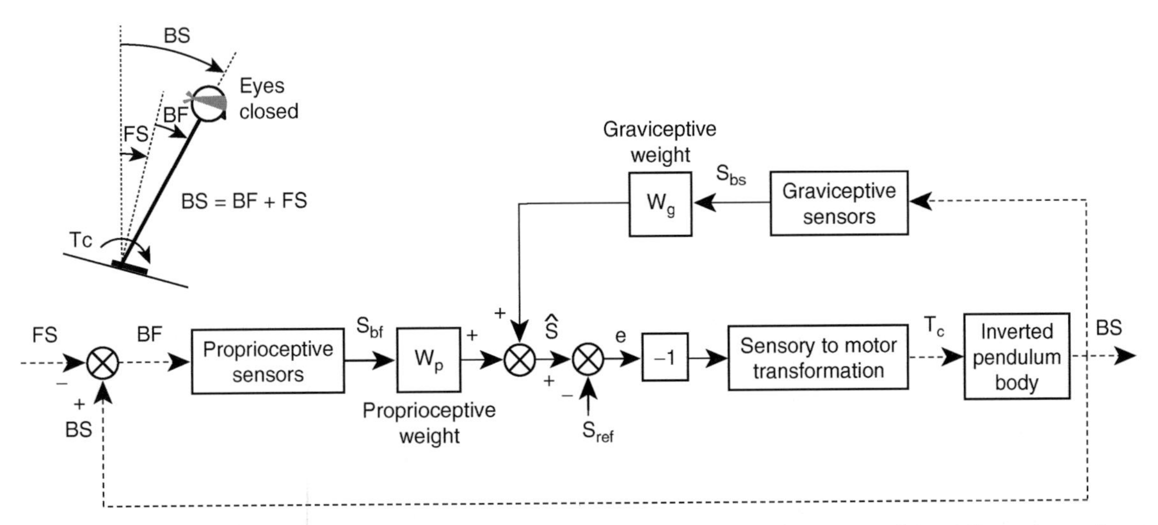

Posture – Sensory Integration. Figure 1 A block diagram representation of a sensory integration scheme for postural control based on a weighted combination of sensory orientation signals. In this example, proprioceptors are assumed to encode body orientation relative to the feet (*BF*) and graviceptors encode body in space orientation (*BS*). A weighted combination of these sensory signals is compared to an internal reference orientation (S_{ref}) and the resulting error, e, is used to generate a corrective torque, T_c, via a sensory to motor transformation process. The corrective torque acts on the body to change body orientation in space and relative to the feet. The block with –1 indicates a sign inversion such that a positive value of e produces a negative T_c that tends to drive the body back toward an upright orientation. The *dashed lines* connecting boxes indicate physical variables, and the *solid line* represents neural signals. The *inset stick figure* defines the positive direction of the physical variables.

component is related to *FS*. This component can be considered to be an undesirable disturbance torque that causes the body to align with the tilted surface and therefore would be destabilizing. The other component is related to *BS*. This component can be considered to be a desirable, stabilizing torque that causes the body to remain oriented with respect to earth vertical.

Now it is clear that an adjustment in the sensory weights can have an influence that goes beyond the consideration of optimal maximum likelihood estimation. Specifically, a reduction of w_p reduces the disturbance torque produced by surface tilt. A reduction in the disturbance torque would seem to be a desirable effect, except that a reduction in w_p also reduces the magnitude of the stabilizing torque related to *BS*. The magnitude of this stabilizing torque must be maintained above the level needed to counteract torque due to gravity. Furthermore, analysis of the postural control system in Fig. 1 indicates that the overall stabilizing torque level must be closely regulated in order to maintain stable, non-resonant behavior [3]. The dual task of reducing the destabilizing torque associated with surface tilt and maintaining adequate stabilizing torque can be accomplished by increasing w_g in equal proportion to the reduction in w_p [4]. This reciprocal adjustment of sensory weights is termed a ▶sensory re-weighting strategy and is also related to the concept of ▶sensory substitution. Therefore, a sensory integration mechanism that uses a weighted combination of sensory orientation sources can facilitate postural control by selecting weights to provide a low variance estimate of orientation in conditions where sensory systems provide redundant information and by adjusting weights to limit the effects of external disturbances while simultaneously maintaining stability.

The sensory integration mechanism shown in Fig. 1 is easily extended to include sensory information from vision [2] and other sensory systems by adding additional feedback loops, each with its own sensory weighting factor.

Combining Kinematic and Kinetic Information

The above discussion focused on combining information from kinematic sensors signaling body motion relative to the surface and relative to earth vertical. Integrating kinetic (force related) sensory information with kinematic information affords a further opportunity to enhance the capabilities of the postural control system. Fig. 2a shows a sensory integration scheme that includes a feedback path whereby a sensory signal encoding corrective torque contributes to the sensory error signal *e*.

Note that for the kinematic sensors, a forward (positive sign) body sway on a level surface produces a negative corrective torque that tends to restore body orientation to the upright position. That is, the kinematic sensors are organized within the postural control system to provide ▶negative feedback control. However, a negative corrective torque sensed by the kinetic sensors produces an even larger negative torque. That is, the kinetic sensors are organized to provide ▶positive feedback control.

The benefit of integrating a kinetic contribution with kinematic sensory sources is illustrated in Fig. 2b. With only kinematic control, the Fig. 1 model predicts that a surface tilt of 1° produces a large body tilt of about 2.4° if the sensory integration relies primarily on proprioception (w_p = 0.8, w_g = 0.2). Body tilt is reduced to about 0.8° if a sensory re-weighting occurs that shifts toward increased reliance on ▶graviception (w_p = 0.2, w_g = 0.8). When kinetic sensory information is integrated with the kinematic information using the positive feedback mechanism shown in Fig. 2a, the surface tilt induces only a transient body tilt. Note that the time course of responses to surface tilt depends on the sensory integration mechanism, the properties of the sensory to motor transformation and body mechanics. For the results shown in Fig. 2b, the sensory to motor transformation generated T_c in proportion to *e* and to the rate of change of *e*, and included a time delay representing the combined delays attributable to sensory transduction, neural transmission, central processing and muscle activation. The torque feedback loop included a low pass filter, which implies that torque feedback mainly influences tonic or low frequency behavior of the postural control system [4].

Alternative Mechanisms for Sensory Integration

Although sensory integration can be modeled as a sensory weighting process, there are alternative representations that may correspond more closely to actual central nervous system processes. One idea is that the nervous system uses sensory information to reconstruct an internal representation of the external physical reality. As an example, Fig. 3 shows a sensory integration mechanism whereby proprioceptive and graviceptive sensory signals are used to reconstruct an internal representation of foot in space orientation.

Even though there are no direct sensors of foot in space orientation, the nervous system now has access to a derived representation of this important physical variable that encodes the surface orientation (assuming the feet are in contact with the surface). This internal representation of surface orientation provides an internal base upon which the nervous system can apply a hierarchical set of transformations that can be used to encode the orientation of various body segments relative to the surface [6].

Figure 3 also shows a scheme for combining the various internal representations of body orientation for the purpose of generating an overall error signal, *e*, which is the basis for generating the corrective torque

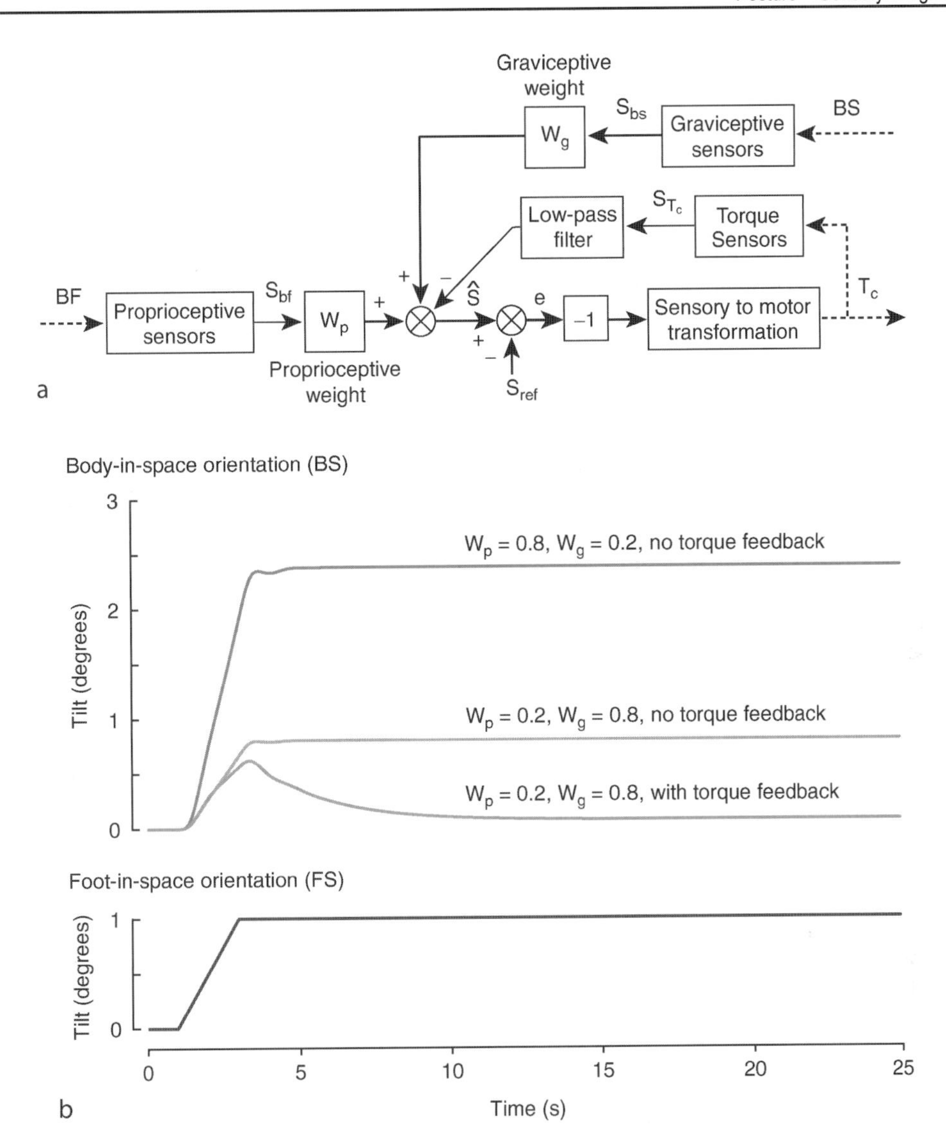

Posture – Sensory Integration. Figure 2 a Representation of a sensory integration scheme that includes a torque feedback contribution in addition to the weighted combination of proprioceptive and graviceptive cues shown in Fig. 1. The corrective torque, T_c, is assumed to be encoded, low pass filtered and then combined with other sensory signals via a summation with a sign opposite to those of the other sensory signals. Thus a positive value of T_c produces an even larger positive value of T_c, which facilitates the return of the body toward an upright position. b Examples of body sway responses evoked by a 1° tilt of the surface orientation (*lower blue trace*) for different sensory integration configurations. The predicted body sway is shown for different combinations of weighted sensory feedback both with and without torque feedback. See [4] for details regarding the postural control model and model parameters.

required for balance control as shown in Fig. 1. This sensory integration process includes gain factors and thresholds that effectively perform a sensory reweighting as a function of the amplitude of the internal sensory related signals [5].

A final sensory integration scheme that has been applied to postural control is based on the engineering concepts of optimal estimation and control [7,8,9] (Fig. 4). This scheme assumes that the nervous system possesses internal models of the body and sensor dynamics. An efference copy of the motor command generated by the postural control system is also applied to the internal model. The internal model is used to estimate body orientation and to predict the expected sensory signals associated with motor commands applied to the body. The predicted sensory signals are compared to the actual sensory signals and any sensory error is used to improve the estimate of body orientation. This improved orientation estimate is used to generate corrective motor responses via feedback control. Furthermore, this optimal estimation and control scheme is able to account for the noise

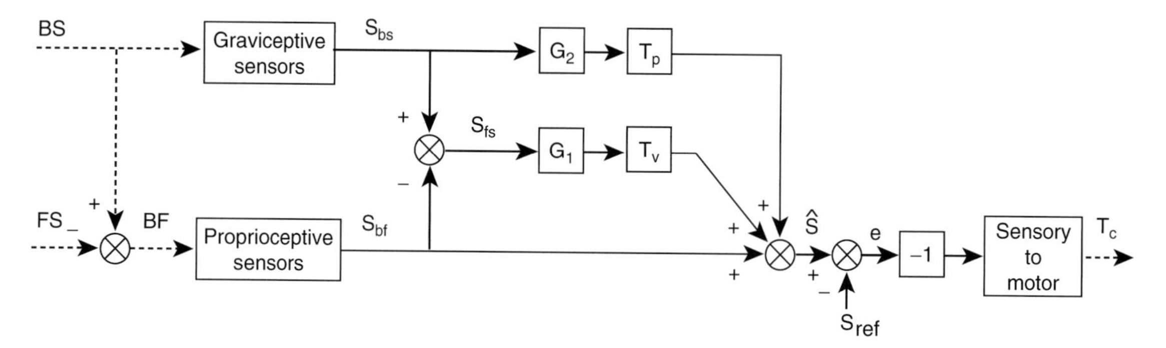

Posture – Sensory Integration. Figure 3 A sensory integration scheme based on a internal reconstruction of external physical variables. In this example, a neural representation of foot in space orientation, S_{fs}, is formed by combining graviceptive sensory information signaling body in space orientation, S_{bs}, and proprioceptive sensory information signaling body orientation relative to the feet, S_{bf}. These three sensory signals are combined to form an overall orientation signal, $\hat{S}$, which is used to generate corrective torque, T_c. The boxes labeled G_1 and G_2 are multiplying factors, T_p is a position related threshold, and T_v is a velocity related threshold. These multiplying factors and thresholds produce a change in $\hat{S}$ as a function of the amplitude and frequency of the sensory signals and effectively perform a re-weighting of these signals. See [5] for details.

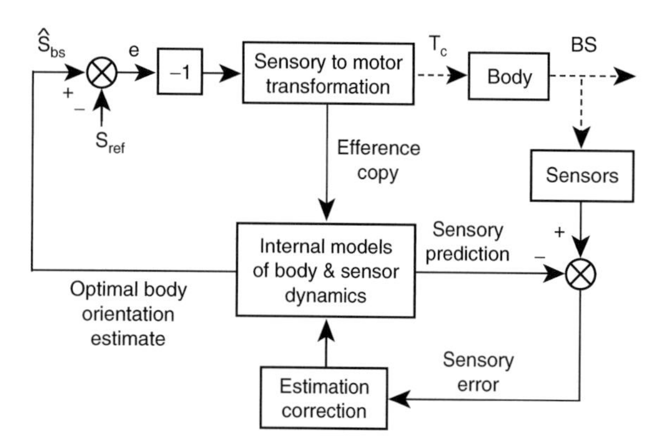

Posture – Sensory Integration. Figure 4 A block diagram representation of a sensory integration scheme for postural control based on optimal estimation of sensory orientation information. An optimal estimate of body orientation in space, $\hat{S}_{bs}$, is used to generate a corrective torque, T_c, via a sensory to motor transformation process. The block with −1 indicates a sign inversion such that a positive value of e produces a negative T_c, which tends to drive the body back toward an upright orientation. The optimal orientation estimate is derived via a process that accounts for the dynamic characteristics of the body and the various sensory systems that contribute information related to body in space orientation, *BS*. The *dashed lines* connecting boxes indicate physical variables and the *solid line* represents neural signals.

properties and dynamic characteristics of sensory and motor systems. Modifications of this scheme can also be used to generate internal estimates of external perturbations [9]. These optimal estimation and control models have been successful in predicting a variety of experimentally observed phenomena including the apparent sensory re-weighting that occurs in response to external perturbations of varying amplitude [9] and changes in the statistical properties of spontaneous sway caused by exposure to environments which limit access to accurate sensory orientation information [8].

Summary

The maximum likelihood principle provides an excellent foundation for understanding why the nervous system would benefit from using a weighted combination of sensory information when more than one sensory source is available. However, when sensory information is used for motor action, the physics of the body and its interaction with the environment place additional constraints on the sensory integration process. Sensory re-weighting and combined use of kinematic and kinetic sensory information provide a flexible mechanism for minimizing the effects of external disturbances while maintaining stability. Relatively simple models based on sensory reconstruction and re-weighting via threshold operations account for a wide variety of experimental data. Optimal estimation methods, developed for engineering applications and applied to postural control, also account for many experimentally observed features of sensory integration. The actual neural mechanisms for sensory integration remain to be determined.

References

1. Ernst MO, Banks MS (2002) Humans integrate visual and haptic information in a statistically optimal fashion. Nature 415:429–433
2. Peterka RJ (2002) Sensorimotor integration in human postural control. J Neurophysiol 88:1097–1118
3. Peterka RJ, Loughlin PJ (2004) Dynamic regulation of sensorimotor integration in human postural control. J Neurophysiol 91:410–423

4. Cenciarini M, Peterka RJ (2006) Stimulus-dependent changes in the vestibular contribution to human postural control. J Neurophysiol 95: 2733–2750
5. Maurer C, Mergner T, Peterka RJ (2006) Multisensory control of human upright stance. Exp Brain Res 171: 231–250
6. Mergner T (2002) The matryoshka dolls principle in human dynamic behavior in space: a theory of linked references for multisensory perception and control of action. Curr Psychol Cogn 21:129–212
7. Carver S, Kiemel T, van der Kooij H, Jeka JJ (2005) Comparing internal models of the dynamics of the visual environment. Biol Cybern 92:147–163
8. Kuo AD (2005) An optimal state estimation model of sensory integration in human postural balance. J Neural Eng 2:S235–S249
9. van der Kooij H, Jacobs R, Koopman B, van der Helm F (2001) An adaptive model of sensory integration in a dynamic environment applied to human stance control. Biol Cybern 84:103–115

Posture-Movement Problem

Definition

The problem of how the nervous system prevents the posture-stabilizing mechanisms from generating resistive forces when an active movement from an initial to a final posture is produced.

►Equilibrium Point Control

Posture Role of Cerebellum

Dagmar Timmann
Department of Neurology, University of Duisburg-Essen, Essen, Germany

Definition

The cerebellum is critical for motor coordination and motor learning. The cerebellum is involved both in voluntary movement control, for example upper limb coordination, and postural control.

►Ataxia of stance and gait are characteristic signs of cerebellar disease. Cerebellar disorders result in enhanced postural sway. As a compensatory response, the stance is overly wide based. If the subject attempts to stand on a narrow base, there is increase in postural sway and a tendency to fall (Fig. 1).

Many features of cerebellar gait are related to balance disorders and ways to compensate for them. For example, step length is decreased and step width is increased. Furthermore, the coordination between posture and rhythmic movements of locomotion is impaired. Likewise, postural adjustments are disordered prior to voluntary limb movements.

This chapter focuses on findings of disordered postural control during quiet stance and in response to balance disturbances in subjects with cerebellar lesions. Localizing signs of postural disturbances in cerebellar disease are reviewed first. Next, physiology and pathophysiology of cerebellar postural control is discussed.

Description of the Theory

The structure of the cerebellar cortex is the same all over the cerebellum. Various parts of the cerebellum differ in function because of differences in fiber connections. The cerebellar cortex receives afferent input from many parts of the peripheral and central nervous system. Proprioceptive and vestibular afferents are of particular importance in cerebellar control of posture. These sensory informations are relayed to differents parts of the cerebellum and probably related to different aspects of postural control.

The relative simplicity and quasicrystalline microstructure of the cerebellar cortex suggests a common computational function of this structure. As yet there is no unifying theory for cerebellar function. A number of theories and models have been proposed, including the coordination of movement across different joints, timing, an internal model for sensorimotor control or the cerebellum as a motor learning machine. These possibilities are not mutually exclusive. In the following, references to current theories of cerebellar function are made where applicable.

Functional Compartimentalization

Gross subdivision of the cerebellum into the lateral hemispheres and medial vermis gives a first idea of functional localisation within the cerebellum. The vermis is involved in the control of posture and equilibrium as well as eye movements. The hemispheres are involved in motor execution and planning of voluntary movements. Lesions of the vermis result in disturbances of stance, gait, and ocular movements, whereas lesions of the cerebellar hemispheres primarily affect limb movements.

On the basis of the efferent projections from the cerebellar cortex to the cerebellar nuclei the cerebellum has been subdivided into a medial zone (that is the vermis) projecting to the fastigial nuclei, an intermediate zone projecting to the interposed nuclei and a lateral zone projecting to the dentate nucleus. Animal lesion studies show that lesions of the fastigial nuclei are followed by impaired or prevented sitting,

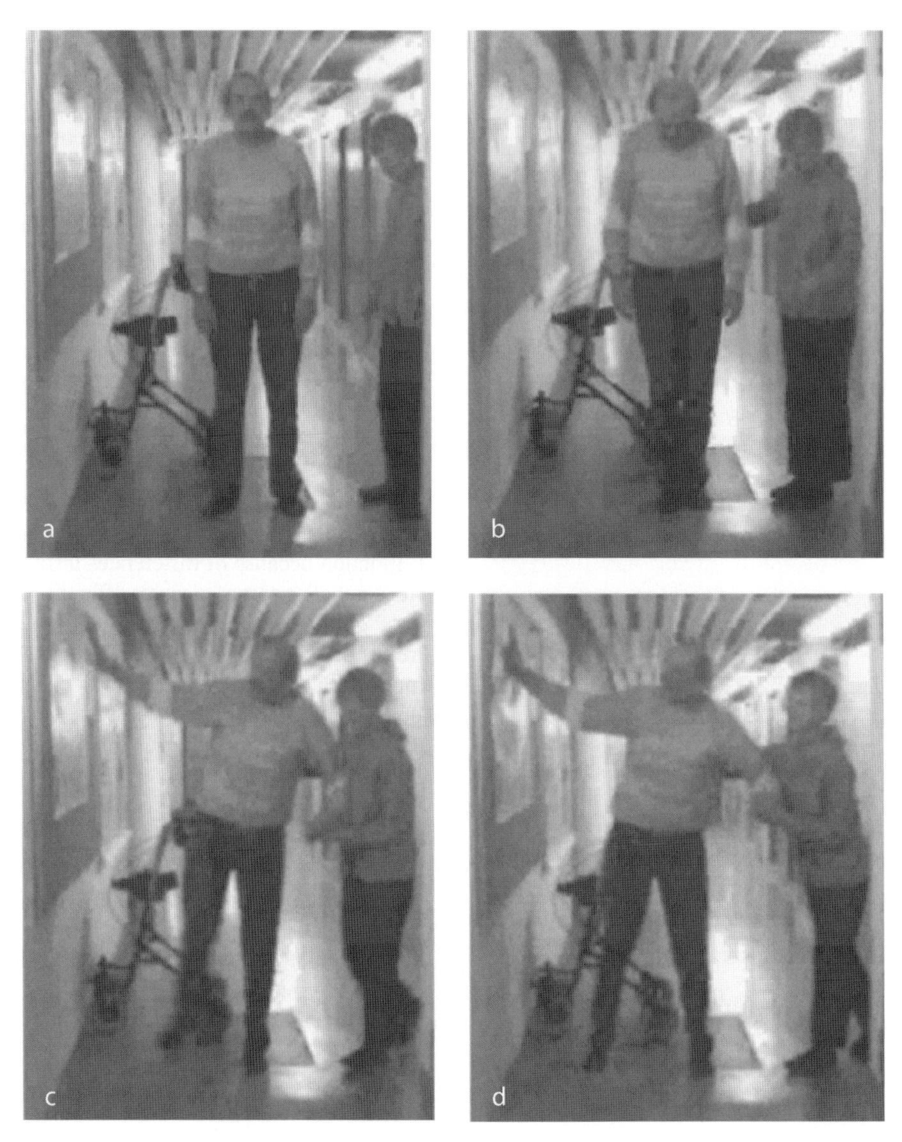

Posture Role of Cerebellum. Figure 1 Ataxia of stance in a cerebellar subject suffering from spinocerebellar type 6 (SCA6). a Stance is wide-based. b If the subject attempts to stand on a narrow base, balance is lost (c) and subject has to make use of the wall to prevent a fall (d).

standing and walking, because of falls to the side of the lesion. This was interpreted as a deficit in equilibrium [1]. Efferents from the fastigial nuclei project to the brain stem and modify vestibular and reticular influences on posture.

The flocculonodular lobe and adjoining parts of the caudal vermis have been named the ►vestibulocerebellum because of heavily projecting vestibular afferents. Lesions of the vestibulocerebellum cause postural ataxia of head and trunk during sitting, standing and walking. Patients frequently fall while sitting. The classic example is medulloblastoma, which occurs most often in the cerebellum in children between 5 and 10 years of age. In subjects with such lesions, visual stabilization of posture, as evaluated by comparing sway with eyes closed and sway with eyes open, is impaired (absence of ►Romberg's sign). Severe postural sway is present with eyes open and is essentially unchanged with eyes closed. Intersegmental movements are diminished.

The anterior and posterior parts of the vermis and paravermal parts of the cerebellar hemispheres are called the ►spinocerebellum because of their spinal afferents. Damage to the spinocerebellar parts of the ►anterior lobe is characterized by ataxia of stance and gait. The classic example is alcoholic cerebellar degeneration. Visual stabilization of posture is relatively preserved and the tremor is provoked by eye closure (presence of Romberg's sign). Patients rarely fall because the body tremor is opposite in phase in head, trunk, and legs, resulting in a minimal shift of the center of gravity.

Chronic damage to the lateral cerebellar hemispheres that is the cerebro- or pontocerebellum, does not result in significant postural disorders. The lateral hemispheres receive the main input from the cerebral cortex, synaptically interrupted in the pontine nuclei.

Diener and coworkers [2] measured body sway by means of a force-measuring platform in human subjects with lesions of the ponto-, vestibulo- or spinocerebellum (Fig. 2).

Postural sway was basically unaffected in subjects with lesions of the lateral cerebellar hemispheres. Lesions of the lower vermis caused omnidirectional postural sway with frequency components below 1 Hz. Lesions of the anterior lobe led to anterior-posterior body sway with a frequency of about 3 Hz. A more recent human lesion study questioned if 3 Hz body oscillations occur exclusively in lesions of the anterior lobe. Likewise, assessement of trunk sway in patients with spinocerebellar ataxias showed that postural instability was generally more pronounced in the pitch than in the roll plane, corresponding with predominant involvement of the spinocerebellum [3].

Posture Role of the Spinocerebellum

Postural muscle tone is a primary contributor to the maintenance of upright stance. Damage to the anterior lobe in experimental animals primarily produces a change of muscle tone. In decerebrate animals, the decerebrate rigidity increases, as do the postural reflexes. In humans it is doubtful whether lesions of the anterior lobe produce increased muscle tone. Postural sway following lesions of the anterior lobe has been explained by an increased gain of posturally stabilizing (long loop) reflexes [4].

Sway can be provoked by platform perturbations or electrical stimulation of the tibial nerve. Subjects with anterior lobe atrophy show hypermetric postural responses and overshooting of the initial posture, with larger than normal surface reactive torque responses and exaggerated and prolonged muscle activity (Fig. 3; [5]).

Latencies of postural responses provoked by platform perturbations are normal in patients with cerebellar disorders. Increased gain and prolonged duration of ►long loop reflexes result in an overcompensation of the postural tasks and are believed to evoke (exaggerated) postural responses of the corresponding antagonists. The postural tremor supposedly continues by the same mechanism.

A more recent model of the (spino-) cerebellum supports the notion that the cerebellum may contribute to balance by long loop feedback with scheduling of linear gains at the same joint and interjoint responses between ankle, knee, and hip [6]. Explicit internal dynamics models within the cerebellum, which have been hypothezised to control voluntary limb movements, do not necessarily contribute to spinocerebellar balance control.

Posture Role of the Vestibulocerebellum

Studies in primates have helped to understand the specific contributions of the vestibulocerebellum to postural control. Knowledge based on human studies is more limited.

The dominant afferent inputs to the vestibulocerebellum come from the semicircular canals, which signal changes in head position, and the otolith organs, which signal the orientation of the head with respect to gravity. Semicircular canal information is relayed to the flocculus and otolith information to the caudal cerebellar vermis (►nodulus and ►uvula).

Sensory input from the otoliths evoke vestibulocollic and vestibulospinal reflexes that maintain the head vertical with respect to gravity. Vestibulocollic and vestibulospinal reflexes are primarily static. The semicircular canals, however, have weaker influences on spinal

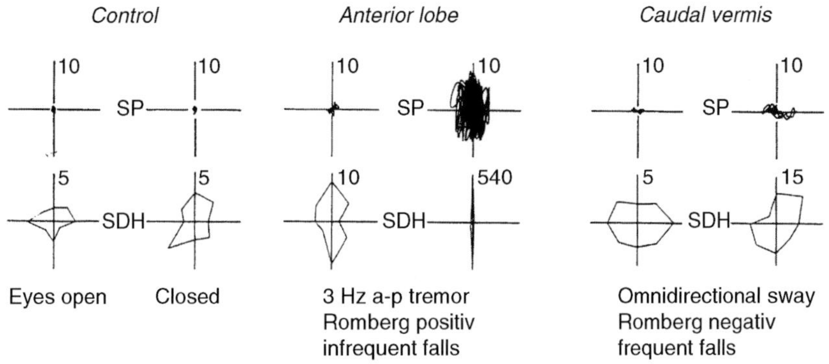

Posture Role of Cerebellum. Figure 2 Sway path (*SP*) and sway direction histograms (*SDH*) in a control, a patient with anterior lobe atrophy and a patient with a vestibulocerebellar lesion. Note the strong preference of anterior-posterior sway in the patient with anterior lobe atrophy. (Adapted from [2]; with permission).

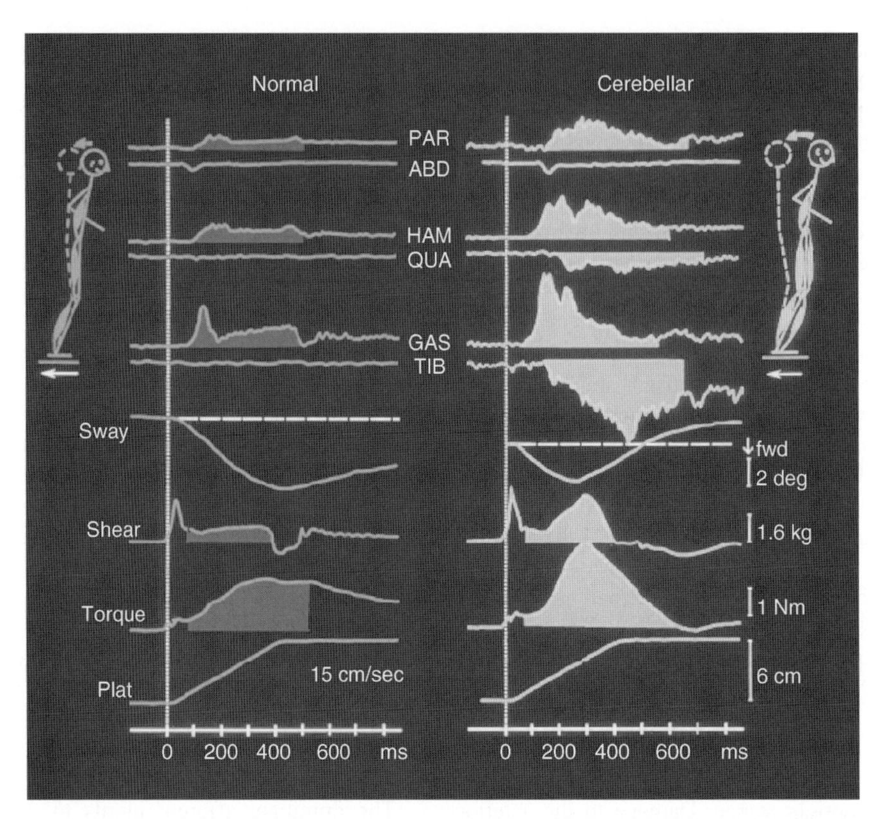

Posture Role of Cerebellum. Figure 3 Mean postural responses evoked by a backward translation of the supportive platform in a control group and a cerebellar group with anterior lobe atrophy. Cerebellar subjects show hypermetric postural responses, exaggerated and prolonged muscle activity and larger than normal surface reactive torque responses. Latencies of postural responses are normal. Traces show (*top to bottom*): electromyographic (*EMG*) recordings from paraspinal (*PAR*), rectus abdominis (*ABD*), biceps femoris (*HAM*), rectus femoris (*QUA*), gastrocnemius (*GAS*), and anterior tibial (*TIB*) muscles; sway, shear forces, surface torque and platform displacement. (Adapted from [5]; with permission).

circuits and serve predominantly to control extraocular muscles and coordinate head and eye movements.

It has been assumed that a lesion of the vestibulocerebellum leads to disturbed gravitational set values and therefore to a loss of spatial orientation vs. gravity. The set value of determining the upright is lost.

More recent single cell recording studies in the primate provide evidence that the rostral fastigial nucleus represents a main processing center of otolith driven information for inertial motion detection and spatial orientation. Angelaki and coworkers [7] showed that cerebellar and brainstem motion sensitive neurons encode dynamically processed otholith signals appropriate to construct an internal model of inertial motion detection.

A study in children and adolescents with chronic surgical cerebellar lesions underscores the importance of the fastigial nuclei in human postural control. High-resolution magnetic resonance imaging allowed detailed analysis of the lesion side. The ability to control upright stance based on vestibular information alone (that is without visual information and unreliable proprioceptive information) was only impaired in subjects with cerebellar lesions that included the fastigial nuclei (Fig. 4; [8]).

Findings further showed that the lesion site was critical for the motor recovery. Lesions affecting the cerebellar nuclei (but not the cerebellar cortex) were not compensated at any developmental age.

Interestingly, otolith dysfunction has been demonstrated in patients with spinocerebellar ataxia type 6 (SCA6). SCA6 is a hereditary disorder, which affects the vestibulocerebellum early in the disease.

Role of Cerebellum in Postural Adaptation

Many studies show that the cerebellum plays an important role in motor learning, in particular in adaptation and automatization of movement. Disordered adaptation probably contributes to ataxia of stance, but has been assessed by few studies only.

Most studies investigated adaptation of early automated postural responses to changes in surface perturbations. Because the contribution of the somatosensory system is much greater than that of the vestibular system when compensating for transient surface perturbations, contributions of the spinocerebellum are assessed. Initial findings of Nashner [9] showed that healthy subjects but not cerebellar patients adapt automated postural responses depending on context. Nashner compared postural responses to backward translations and upward rotations. Both lead to the same ankle rotation. Upright stance however, is maintained by contraction of the anterior tibial muscle in upward rotations, but of the gastrocnemius muscle in backward translations. In controls, but not cerebellar subjects, responses that stabilize posture were facilitated progressively with repeated trials, whereas responses that destabilize posture were diminished. These findings were, however, challenged in later studies. When the type of perturbation changed from translation to rotation and vice versa both controls and cerebellar subjects showed an immediate change in the response amplitudes of the gastrocnemius and anterior tibial muscles.

Horak and Diener [5] studied whether cerebellar subjects could learn to adjust for predictable postural perturbations during standing (Fig. 5).

When different displacement amplitudes were presented in a serial (predictable) format, healthy subjects were able to appropriately scale their initial postural responses. In contrast, cerebellar subjects were unable to learn to use predictive feedforward control (in other words central set from prior experience) to scale their early automated postural responses

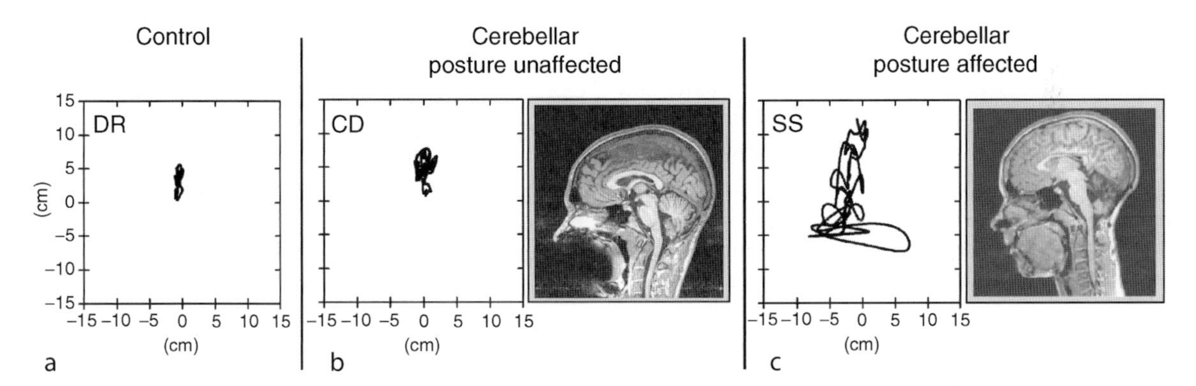

Posture Role of Cerebellum. Figure 4 Effects of fastigii lesions on postural sway. Postural sway is increased in patient *SS*, but not in patient *CD* compared to the control subject *DR*. In both patients an astrocytoma had been surgically removed. The sagittal MRI images reveal that area above the 4th ventricle, that is the location of the fastigial nuclei, was affected in SS but not in CD. Shown are the center of gravity sway paths over 20 s in a condition that is dependent on vestibular function (that is eyes closed and surface sway referenced). (Adapted from [8]; with permission).

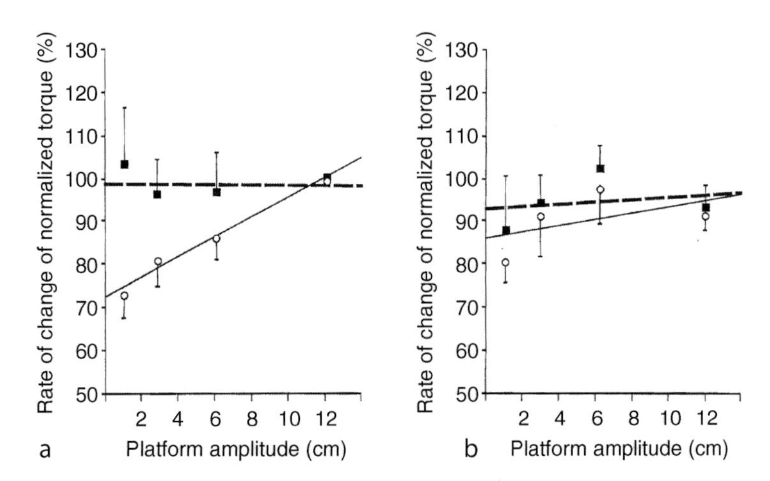

Posture Role of Cerebellum. Figure 5 Scaling of torque responses to platform displacement amplitude in controls (*continuous line*) and cerebellar subjects (*broken line*). The mean ±SE of normalized torque responses are indicated for serial (predictable) (a) and random (b) amplitude presentation. Control subjects scale to predictable but not random presentation of displacement amplitudes, whereas the cerebellar group scaled neither to predictable or random presentation. (Adapted from [10]; with permission).

to expected perturbation amplitudes. A subsequent study showed that cerebellar patients could predict perturbation amplitudes based on prior experience, but they could not use this prediction to modify precisely the gain of early automated postural responses [10]. The spinocerebellum may be important for accurate tuning of response gain based on prediction.

Cerebellar contribution to adaptation of vestibular reflexes is likely. Cerebellar contributions have been investigated in great detail for adaptation of the vestibulo-ocular reflex (VOR). The function of the VOR is to stabilize retinal images by generating smooth eye movements that are equal and opposite to each head movement. Learning occurs whenever image motion occurs persistently during head turns; as a result image stability is gradually restored. The cerebellar role in retention is disputed, but there is a consensus on the need of an intact cerebellum (that is flocculus) for acquisition.

The role of the caudal cerebellar vermis (nodulus and uvula) for adaptation of the "static" vestibulocollic and vestibulospinal reflexes has not been assessed in detail.

Summary

The medial cerebellum (that is vermis) is of particular importance in postural control. The contribution of the cerebellum is two-fold. Vermal parts of the anterior lobe (that is spinocerebellum) are involved in control of automated postural reflexes evoked by proprioceptive feedback. Hypermetric postural responses and disordered adaptation can be explained by inaccurate tuning of reflex gain. Disordered coordination between head, trunk and legs (asynergia) probably contributes but is less well understood. Caudal parts of the vermis (that is the vestibulocerebellum) play an important role in spatial orientation of the body against gravity and detection of inertial body motion evoked by otolith feedback. The vestibulocerebellum appears to be important in building an internal model of inertial body motion.

References

1. Thach WT, Bastian AJ (2004) Role of the cerebellum in the control and adaptation of gait in health and diesease. Prog Brain Res 143:353–366
2. Diener HC, Dichgans J, Bacher M, Gompf B (1984) Quantification of postural sway in normals and patients with cerebellar diseases. Electroencephalogr Clin Neurophysiol 57:134–142
3. Van de Warrenburg BP, Bakker M, Kremer BP, Bloem BR, Allum JH (2005) Trunk sway in patients with spinocerebellar ataxia. Mov Disord 20:1006–1013
4. Mauritz KH, Schmitt C, Dichgans J (1981) Delayed and enhanced long latency reflexes as the possible cause of postural tremor in late cerebellar atrophy. Brain 104:97–116
5. Horak FB, Diener HC (1994) Cerebellar control of postural scaling and central set in stance. J Neurophysiol 72:479–493
6. Jo S, Massaquoi SG (2004) A model of cerebellum stabilized and scheduled hybrid long-loop control of upright balance. Biol Cybern 91:188–202
7. Angelaki DE, Shaikh AG, Green AM, Dickman JD (2004) Neurons compute internal models of the physical laws of motion. Nature 430:560–564
8. Konczak J, Schoch B, Dimitrova A, Gizewski E, Timmann D (2005) Functional recovery of children and adolescents after cerebellar tumour resection. Brain 128:1428–1441
9. Nashner LM (1976) Adapting reflexes controlling the human posture. Exp Brain Res 26:59–72
10. Timmann D, Horak FB (1997) Prediction and set-dependent scaling of early postural responses in cerebellar patients. Brain 120:327–337

Posturography

Definition

Analysis of body posture, and postural control, typically using computerized image analysis of the position and movement of body segments over time.

Potential Energy

Definition

Certain systems of forces (called conservative systems) are such that the forces can be obtained as (minus) the spatial derivatives of a single scalar function of position. This function, if it exists, is called the potential energy of the system. It is of paramount importance in analytical mechanics.

►Mechanics

Potentiation

Definition

Potentiation refers to the combination of two stimuli resulting in a larger response than the sum of responses to each stimulus alone.

Potentiometer

Definition
A variable resistor used to control an electronic device.

► Hearing Aids

Power

Definition
Power is the amount of energy produced per unit time. The mechanical power a skeletal muscle produces is the product of the force in the muscle and the velocity of the muscle.

Power Density Spectrum

Definition
A power density spectrum is a plot of the power (watts) in a signal as a function of frequency; also referred to as the autospectrum. Fast Fourier transform is used most commonly to construct the power density spectrum.

► Signals and Systems

Power Stroke

Definition
The power stroke of the cross-bridge cycle is the phase of force production. In the current cross-bridge thinking, the power stroke is initiated by the release of the free phosphate from ATP hydrolysis.

► Sliding Filament Theory

PPARγ

Definition
Peroxisome proliferator-activated receptors act as ligand-activated transcription factors, similar to other nuclear hormone receptors. One of its isoforms, PPARγ, exerts anti-inflammatory activities in brain cells by reducing proinflammatory cytokines.

► Neuroinflammation: Chronic Neuroinflammation and Memory Impairments

p55-R

► Brain Inflammation: Tumor Necrosis Factor Receptors in Mouse Brain Inflammatory Responses

Praxis Navigation Strategy

Definition
Behavior relying on an egocentric reference frame and directed by a specific motor sequence to get a goal location.

► Spatial Memory

PRC

► Phase Response Curve

Pre-Bötzinger Complex

Definition
A physiologically defined region within the ventrolateral medulla of mammals that is critical for the

generation of inspiratory activity. The pre-Bötzinger complex can be functionally isolated in brainstem slice preparations, and is still capable of generating three specific rhythmic activities that have many characteristics of normal respiratory activity (eupnea), gasping and sighing.

►PreBötzinger Complex Inspiratory Neurons and Rhythm Generation

Pre-Bötzinger Complex Inspiratory Neurons and Rhythm Generation

Christopher A. Del Negro, John A. Hayes, Ryland W. Pace
The Department of Applied Science, The College of William and Mary, Williamsburg, VA, USA

Definition

The ensemble of neurons in the ventral medulla that plays an important role in generating breathing behavior in mammals by synchronously producing large-magnitude bursts that drive the inspiratory phase of the respiratory cycle.

Characteristics

The ►preBötzinger Complex (preBötC) of the ventral medulla plays an important role in generating breathing behavior in mammals. Anatomical studies have demarcated the borders of the preBötC and experiments in vivo have defined its functional purview. However an unsolved problem pertains to which neurons – and which intrinsic and synaptic properties – make up the rhythmogenic kernel? In vitro models of respiration, particularly slice preparations from neonatal rodents, make experimental tests possible. Transverse slices isolate the preBötC and provide unprecedented access to preBötC neurons for electrophysiology and imaging while maintaining spontaneous inspiratory motor activity, which can be monitored via the ►hypoglossal (XII) cranial nerve root. A diverse array of intrinsic and synaptic properties have been found, which subsequently motivated numerous attempts to subdivide preBötC neurons into various 'types' to assign roles in respiratory rhythm generation. Here we review and critique these classification schemes and proffer objective criteria to distinguish rhythmogenic neural properties.

Inspiratory Drive Latency and Peptide Receptors

Rekling and colleagues [1] proposed a classification scheme that considered sensitivity to ►neuropeptides and ►inspiratory drive latency, defined as the time interval consisting of a crescendo of EPSPs and spiking activity preceding XII motor output (Fig 1). They argued that inspiratory neurons with the earliest drive latency and highest levels of excitability, which also responded to inspiratory drive latency that modulate respiratory rhythm, were most likely to be rhythmogenic. Earliest to activate were type 1 neurons with ~400 ms drive latency and highly excitable membrane properties. Type 2 neurons also exhibited high excitability with ~170 ms drive latency. The least excitable were type 3 neurons with drive latency of ~100 ms. Type 1 neurons were proposed to be rhythmogenic and to activate type 2 neurons downstream, followed by type 3, which were postulated to have a motor or premotor function.

Transient K^+ current, i.e., A-current (I_A), was expressed exclusively in type 1 neurons whereas the hyperpolarization-activated mixed cation current (I_h) was associated only with type 2 neurons. These data suggested a genuine disparity that could distinguish a hierarchy of rhythmogenic subtypes. We measured I_A in ~60% of preBötC inspiratory neurons and found that its selective blocker, 4-aminopyridine (2 mM), caused profound disruptions in the respiratory rhythm (Fig. 2), consistent with the proposal that I_A expression is a hallmark of rhythmogenic neurons (i.e., type 1-like). I_h is present in ~15% of preBötC inspiratory neurons and blocking it with Cs^+ or organic agents speeds up the rhythm, which is consistent with I_h expression

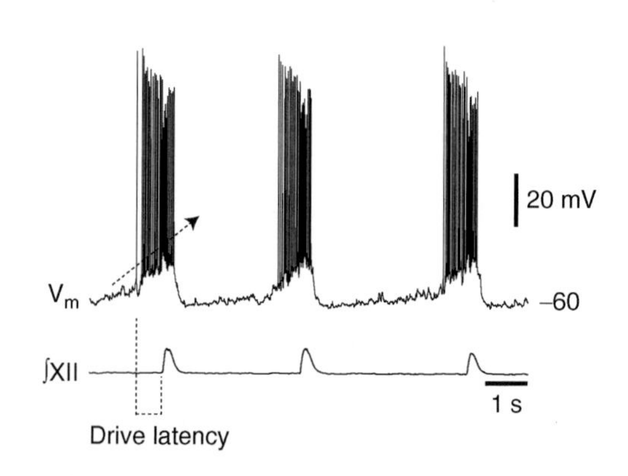

Pre-Bötzinger Complex Inspiratory Neurons and Rhythm Generation. Figure 1 A typical voltage trajectory for a preBötC neuron putatively involved in rhythm generation shown with respiratory-related XII motor output. Inspiratory drive latency is illustrated with *dotted lines* to mark the onset of inspiratory drive and the XII motor discharge. A *dotted-line arrow* emphasizes the incremental depolarization and spike discharge pattern that distinguish relatively small neurons with early drive latency.

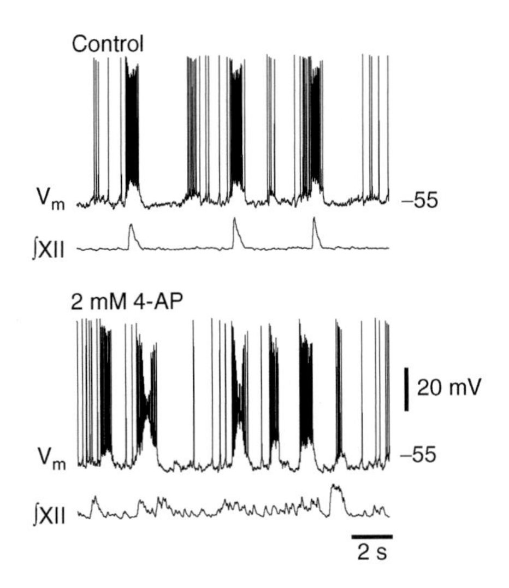

Pre-Bötzinger Complex Inspiratory Neurons and Rhythm Generation. Figure 2 Pharmacological blockade of I_A disrupts respiratory rhythm *in vitro*. The preBötC neuron in control showed early inspiratory drive latency and spike discharge as well as robust inspiratory bursts. After 2 mM 4-AP application the respiratory rhythm was erratic and noisy, and the inspiratory neuron generated bursts that were not necessarily associated with a collective inspiratory burst at the XII motor output level. Blockade of I_A furthermore increased inspiratory burst amplitude to such an extent that spiking activity inactivated transiently.

in neurons that may also be rhythmogenic, but not preeminent (i.e., type 2-like).

Whether I_A and I_h expression maps one-to-one with differences in drive latency, thus validating the type 1 versus 2 classification scheme, has not yet been resolved. Our measurements revealed a continuous drive latency distribution with a mean of ~300 ms (Fig. 3a), rather than bimodal with peaks at ~200 and ~400 ms, as originally suggested. Therefore, types 1 and 2 could reflect the same underlying population of rhythmogenic neurons, but neurons with I_A may be more important for rhythmogenesis simply because I_A plays a more important role than I_h in rhythmogenesis (Fig. 2).

Rekling's classification scheme recognized peptide sensitivity as a criterion to distinguish rhythmogenic neurons. In a watershed study, Gray *et al.* showed that ►neurokinin-1 receptor (NK1R) expression demarcated the borders of the preBötC and that ►substance P (SP), the endogenous ligand for NK1Rs, directly excited rhythmogenic neurons, i.e., with properties consistent with types 1 and/or 2, and also profoundly excited respiratory rhythm [1]. These results led to the hypothesis that NK1R-expressing ($NK1R^+$) neurons, distinct by anatomical and physiological criteria, comprised the rhythmogenic kernel in the preBötC. In support of this hypothesis, ablation of $NK1R^+$ neurons in awake intact adult rats abolishes normal breathing behavior [3].

Guyenet and colleagues used in vivo electrophysiology and neuroanatomy to show $NK1R^+$ neurons concentrated in a region coextensive with the preBötC, where there was an abundance of inspiratory neurons and few ►expiratory neurons [4]. Inspiratory neurons with early drive latency in vivo could be antidromically activated from the contralateral medulla 70% of the time, consistent with interneurons in a local circuit. 34% of these early inspiratory neurons were $NK1R^+$, whereas expiratory or phase-spanning neurons did not appear to have the NK1R (i.e., $NK1R^-$). These data are consistent with the idea that drive latency and peptide receptor expression might specify rhythmogenic neurons in the preBötC.

$NK1R^+$ neurons in the preBötC showed no immunoreactivity for tyrosine hydroxylase (TH) nor choline acetyl-transferase (ChAT) [4]. Therefore, these neurons were unlikely to be catecholaminergic or cholinergic (motoneurons). As a control against these negative findings in $NK1R^+$ neurons of the preBötC, co-labeled $NK1R^+/TH^+$ neurons were found in the noradrenergic A5 and C1 regions outside the preBötC and $NK1R^+/ChAT^+$ motoneurons were profusely labeled in the ►nucleus ambiguous [1,4]. The $NK1R^+$ neurons in the preBötC area rarely showed immunoreactivity to GAD67 or GlyT2, which would have distinguished them as GABAergic or glycinergic inhibitory neurons, and were later shown to contain vesicular glutamate transporter mRNA. In sum, $NK1R^+$ neurons concentrated in the preBötC are locally interconnected excitatory interneurons that discharge prior to the onset of the inspiratory phase, hallmark properties for a role in rhythmogenesis (Fig. 3b).

But the hypothesis that $NK1R^+$ neurons define the kernel needs to be reevaluated because the $NK1R^+$ population can be subdivided. The smallest $NK1R^+$ neurons express preprosomatostatin mRNA, do not project to the spinal cord, nor express preproenkephalin (PPE) mRNA, and are rostrally sited in the preBötC area [5]. 75% of these $NK1R^+$ neurons project contralaterally and may be co-extensive with the $NK1R^+$ cells whose destruction abolishes normal breathing [3]. However, much larger $NK1R^+$ neurons found in the caudal preBötC area express PPE mRNA and project to the spinal cord, consistent with a premotor role, not rhythmogenesis [5].

Combining whole-cell patch clamp with a fluorescent labeling technique that tags all SP-sensitive $NK1R^+$ neurons, we found rhythmogenic properties, e.g., early drive latency, small size, in $NK1R^-$ neurons (Fig. 3b). Early drive latency (~300 ms, type 1- or type 2-like) was associated with small preBötC inspiratory neurons (C_M ~ 45 pF) of which 36% were $NK1R^+$. In contrast, larger preBötC inspiratory neurons (C_M ~ 86 pF)

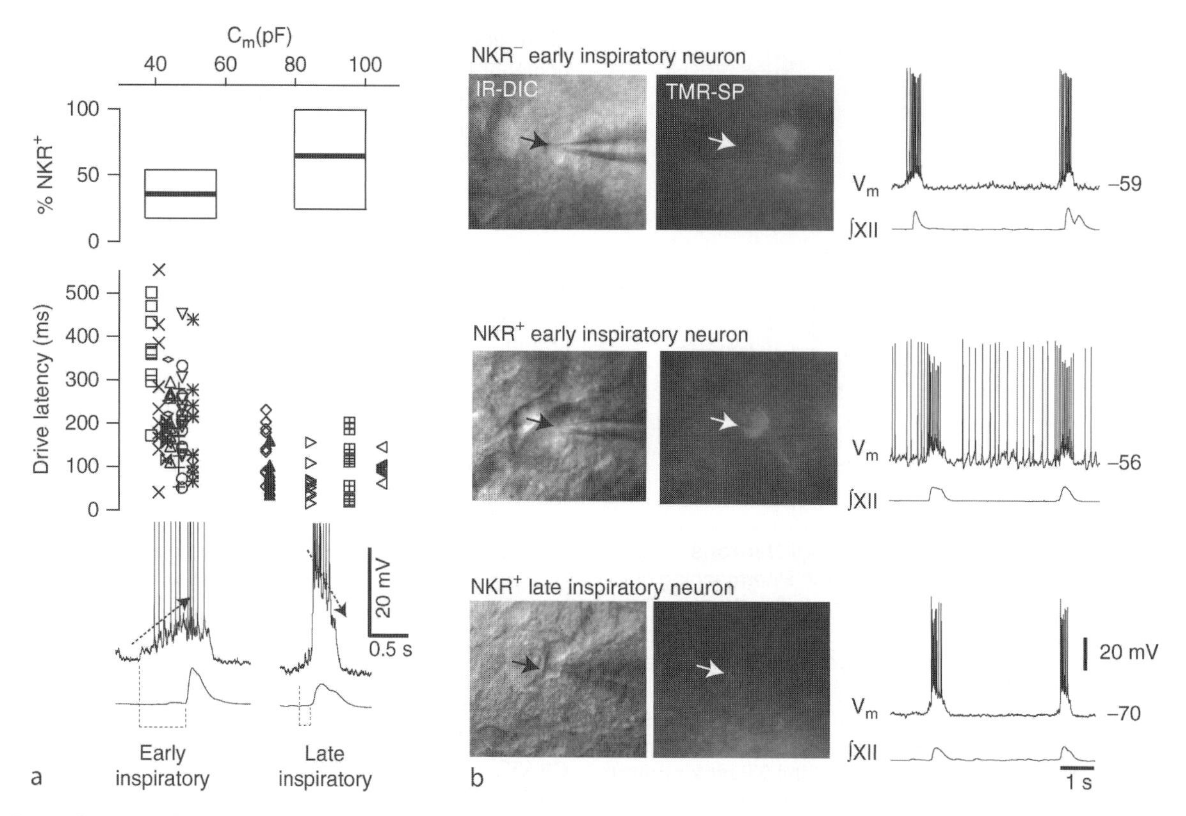

Pre-Bötzinger Complex Inspiratory Neurons and Rhythm Generation. Figure 3 Criteria to distinguish inspiratory preBötC neurons that are involved in rhythm generation. (a), Inspiratory drive latency (lower ordinate) and neurokinin receptor (NKR) expression (upper ordinate) are plotted as a function of whole-cell capacitance C_M. Box plots show mean and 95/5% credible interval range. Typical examples of early as well as late inspiratory phenotypes are shown beneath their characteristic C_M range. (b), NKR expression is shown for three typical inspiratory preBötC neurons. Two early inspiratory neurons, which are hypothesized to be rhythmogenic on the basis of membrane properties, are shown with NKR expression (NKR^+) and without (NKR^-). An NKR^+ late expiratory neuron, hypothesized to have motor or premotor function, is also illustrated. Modified from [2].

showed drive latency of ~100 ms and were NKR^+ 67% of the time (Fig. 3). The majority of the preBötC neurons we recorded were small with early inspiratory activity. The early latency, small size, and incremental discharge trajectory are characteristic of glutamatergic interneurons, and resemble types 1 and 2 [1] and therefore are probably glutamatergic, not GABA- or glycinergic neurons [4]. The fraction of $NK1R^+$ neurons with early drive latency in adult rats in vivo is also near 36%, so the fraction of NKR^+ neurons in the preBötC appears consistent in neonate and adult rodents.

Large preBötC inspiratory neurons that activate latest in the respiratory cycle (Fig. 3a) may be premotor neurons [5]. A large fraction of these cells were NKR^+ (Fig. 3b) thus sensitive to saporin lesions [3]. However, because the small neurons with early drive latency are more numerous saporin lesions will probably cause a greater total reduction in NKR^+ rhythmogenic-like neurons. However, the destruction of a large fraction of NKR^+ respiratory premotoneurons must be considered as a factor in explaining apneas resulting from saporin lesions [3].

Finally, we showed that peptide sensitivity and receptor expression were not necessarily synonymous. Even though only 36% of the putatively rhythmogenic neurons were NKR^+, a much larger fraction (87%) showed a SP-evoked inward current (I_{SP}) in voltage clamp, which suggests that ►gap junctions may provide a means for NKRs to evoke inward current in both NKR^+ as well as NKR^- preBötC inspiratory neurons [2]. This suggests that most of the putatively rhythmogenic neurons still satisfy the three objective criteria: small size, early inspiratory drive latency, and peptide sensitivity.

Pacemaker Properties: Not Specialized Phenotypes, Cannot Explain Rhythmogenesis

Evidence for a ►pacemaker cell-type accompanied the discovery of the preBötC [6]. In the absence of synaptic transmission, some neurons with a baseline membrane potential between −57 and −45 mV spontaneously depolarize and generate rhythmic bursts, dubbed ►conditional pacemaker properties. Voltage-dependent pacemaker properties were attractive from the standpoint of rhythmogenesis because conditional bursting in isolated cells had the same duty cycle as the network-intact XII rhythm. Pacemaker neurons were postulated to form a specialized phenotype that periodically excites so-called follower neurons and synchronizes both sets of neurons through excitatory synaptic interconnections.

PreBötC inspiratory neurons have been classified as either pacemakers or followers, yet this binary classification is more apparent than real. Voltage-dependent bursting depends on a requisite region of negative slope in the current-voltage relationship endowed by ►persistent Na^+ current (I_{NaP}). We now know that I_{NaP} is a generic property, ubiquitously expressed throughout the preBötC and ventral medulla. Since baseline membrane potential must be within a specific voltage window, ►leakage potassium current ($I_{K\text{-}Leak}$) also becomes important to maintain voltage-dependent bursting. I_{NaP} and $I_{K\text{-}Leak}$ are distributed continuously among preBötC inspiratory neurons, thus a small subset with conditional bursting properties arises naturally for neurons with the proper I_{NaP}/I_K ratio and is a byproduct of heterogeneity in membrane properties.

The real question is whether I_{NaP} is important for rhythmogenesis. To avoid caveats associated with bath-applications [7], we performed micropressure injections to apply riluzole – and low doses of TTX (20 nM) that preferentially block I_{NaP} – directly into the preBötC without affecting premotor or XII motoneurons (Fig. 4). Riluzole and 20 nM TTX microinjections did not stop the rhythm nor affect XII motor output, indicating that I_{NaP} is not obligatory for rhythmogenesis [8].

Voltage-dependent pacemaker properties do not constitute a specialized phenotype in the preBötC because I_{NaP} is commonplace and nonessential for rhythmic function. Nevertheless, I_{NaP} contributes to baseline membrane potential and facilitates high-frequency spiking, and thus helps maintain neural excitability [7,8].

Pacemaker properties unrelated to I_{NaP} were documented in 2001. Bursting was sensitive to blockade by Cd^{2+}, indicating dependence on intrinsic Ca^{2+} currents; later a ►Ca^{2+}-activated nonspecific cation current (I_{CAN}) was shown to be involved [17,20]. Cd^{2+}-sensitive bursting neurons were posited to form an additional rhythmogenic subpopulation that could drive rhythmogenesis in combination with, or in lieu of, I_{NaP} pacemaker neurons [9]. However, two key facts falsify this hypothesis: first, Cd^{2+}-sensitive pacemaker properties are sparse or nonexistent in early in post-natal development [9], yet riluzole does not stop the rhythm. Second, doses of flufenamic acid (FFA, 10–100 μM) that attenuate I_{CAN} to an extent that precludes Cd^{2+}-sensitive bursting – but do not completely block I_{CAN} – do not stop rhythmogenesis in the presence of riluzole or TTX at any post-natal age [8].

Unless a miniscule number of pacemaker neurons that are insensitive to I_{NaP} and I_{CAN} antagonists can drive the rhythm – or a heretofore undiscovered pacemaker phenotype exists – these observations invalidate the hypothesis that pacemaker neurons are the basis for rhythmogenesis.

I_{CAN} Activates Synaptically and Generates Rhythmic Inspiratory Bursts in Prebötc Neurons

►AMPA receptors (AMPARs) are necessary for rhythmogenesis. Moreover, group I ►metabotropic glutamate receptors (mGluRs) and ►NMDA receptors

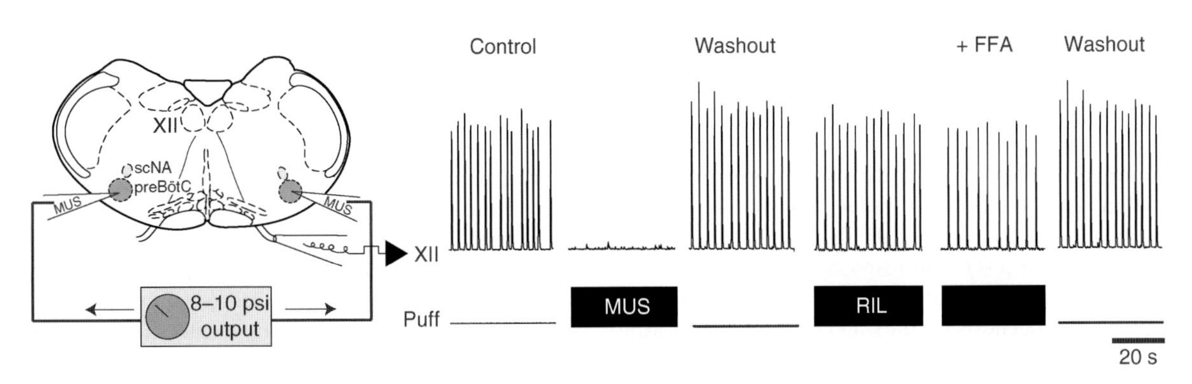

Pre-Bötzinger Complex Inspiratory Neurons and Rhythm Generation. Figure 4 Sequential drug application experiments using local microinjection of 10 μM riluzole (RIL). Top traces show XII motor output, lower traces labeled "puff" reflect TTL pulses at 3 Hz that gate micropressure drug-delivery injections. Bilateral injection of 15 μM muscimol (MUS) is used to verify the effective microinjection pipette locations. After recovering from MUS, RIL is microinjected for >20 min, followed by bath application of 100 μM flufenamic acid (FFA). The rhythm did not stop, nor became perturbed in any noticeable form, after >20 min of bath-applied FFA (cumulative RIL exposure >40 min). Recovery from all drugs occurred within 1–2 h and is shown as washout. Modified from [8].

(NMDARs) provide a substantial – yet heretofore unrecognized – contribution to generating inspiratory bursts and rhythm. Both ionotropic, and metabotropic glutamate receptors activate I_{CAN} via voltage-gated Ca^{2+} channels and inositol (1,4,5)-triphosphate (IP_3)-mediated intracellular Ca^{2+} release. Additionally, I_{CAN} serves to generate robust inspiratory drive potentials in all preBötC inspiratory neurons [10].

Group I mGluRs consist of subtypes mGluR1 and mGluR5. While mGluR5 triggers I_{CAN} activation, via IP_3-mediated intracellular Ca^{2+} release, mGluR1 appears to promote inspiratory drive potentials by transiently closing K^+ channels. In contrast, group II mGluRs modulate interburst period but do not contribute to inspiratory burst generation.

Ca^{2+} influx via NMDARs may also contribute to I_{CAN} activation. AMPAR-mediated depolarization recruits ►voltage-gated Ca^{2+} channels and may partially relieve the voltage-dependent Mg^{2+} block of NMDARs and thus indirectly activate I_{CAN} in the preBötC. This is the first example of a convergent activation of I_{CAN} involving NMDARs, Ca^{2+} channels and intracellular Ca^{2+} release to serve burst generation in a ►central pattern generator.

How important is I_{CAN}? We used two strategies to test its role. In the first, intracellular drug application was employed to test the role of I_{CAN} in drive potential generation in a single preBötC neuron without disrupting respiratory rhythm in the rest of the network. We recorded inspiratory drive in control using ►perforated patches (Fig. 5), which does not modify the intracellular milieu. Then we ruptured the patch and dialyzed the cytosol with patch solution containing 30 mM BAPTA, a high-affinity Ca^{2+} chelator. BAPTA abolishes the ability to activate I_{CAN} and reduced drive potentials by 70% after ≥20 min suggesting I_{CAN} that I_{CAN} is the major charge carrier underlying inspiratory drive potentials [10].

We examined whether I_{CAN} (Fig. 6) was crucial for rhythmogenesis in the network as a whole using bath application of the selective antagonist flufenamic acid (FFA). Dose is a critical issue: 100 μM FFA incompletely blocks I_{CAN} and does not stop the network rhythm, but nonetheless reduces inspiratory drive potentials by ~40% [10]. Higher concentrations of FFA (300–350 μM) stop respiratory rhythmogenesis. While an attractive conclusion is that FFA at ≥300 μM stops rhythmogenesis by fully and selectively blocking I_{CAN}, FFA doses exceeding 100 μM exert numerous side effects [10], and thus confound such a straightforward interpretation. Nevertheless, we can conclude that I_{CAN} contributes enormously to inspiratory drive on a cycle-to-cycle basis by transforming synaptic input into long-lasting membrane depolarization, and thus plays an important role in inspiratory burst generation.

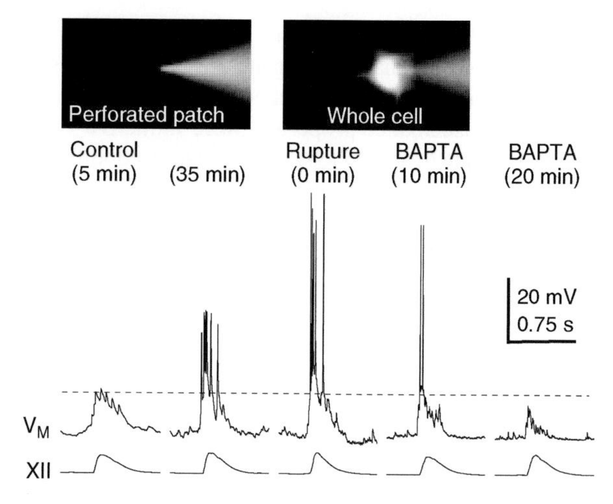

Pre-Bötzinger Complex Inspiratory Neurons and Rhythm Generation. Figure 5 Perforated-patch control recordings and subsequent intracellular dialysis with 30 mM BAPTA patch solution demonstrate the importance of Ca^{2+} transients and I_{CAN}. Control conditions were recorded in the perforated-patch configuration (5 and 35 min shown). BAPTA was introduced into the cytosol via patch rupture (0 min) and caused a progressive attenuation of the drive potential. Baseline membrane potential was −60 mV throughout the experiment. The perforated-patch configuration (PP; *left* picture) is confirmed by the failure of the Lucifer yellow in the patch-pipette solution to dialyze the cell. The whole-cell configuration (WC; *right* picture) allows Lucifer yellow to fill the cell. The V_M traces verify that the underlying inspiratory dive potential can be accurately measured in PP mode as compared to the first minute of WC mode. In WC, the spikes are less truncated because the access impedance decreases. Modified from [10].

The Group-Pacemaker Hypothesis of Respiratory Rhythm Generation

I_{CAN} activation depends on AMPA, NMDA, and metabotropic glutamate receptors during endogenous respiratory behavior, and thus is properly considered a network property. The group-pacemaker hypothesis (Fig. 7) postulates a mechanism for rhythmogenesis wherein recurrent synaptic excitation linked to postsynaptic intrinsic currents plays a special role. In the group pacemaker, a fraction of neurons in the preBötC are spontaneously firing; baseline voltage during the expiratory phase exceeds spike threshold. In the waning expiratory phase, active neurons excite silent neurons, which in turn excite additional silent neurons and also provide positive feedback to re-excite the neurons already spiking. I_{CAN} is normally latent and unavailable except when recruited by synaptic excitation. Glutamatergic input sufficient to evoke I_{CAN} is ultimately achieved a few hundred milliseconds prior to inspiratory burst discharge. The negative feedback process that causes burst termination remains unknown. After burst termination, neurons

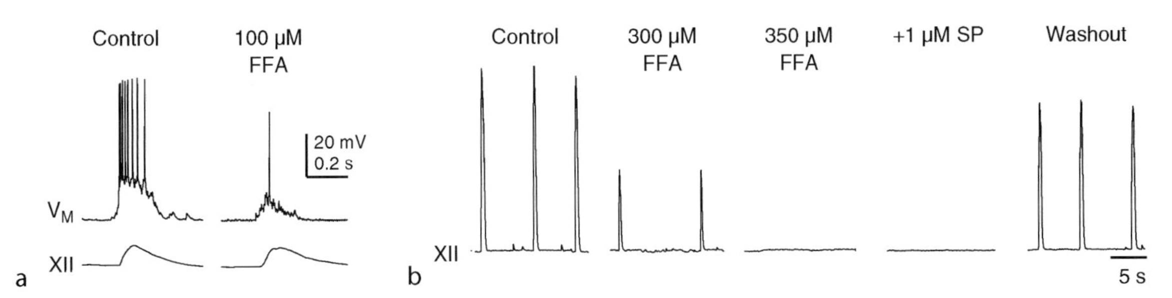

Pre-Bötzinger Complex Inspiratory Neurons and Rhythm Generation. Figure 6 Effects of I_{CAN} antagonist flufenamic acid (FFA) on inspiratory bursts and rhythmogenesis. (a), Typical inspiratory bursts recorded in whole-cell conditions for control and 100 μM FFA, respectively. (b), An experiment showing that 300 μM FFA severely perturbed the respiratory rhythm (monitored via XII discharge) and 350 μM FFA stopped it altogether. In the presence of 350 μM FFA, even the excitatory peptide substance P (which normally stimulates rhythmogenesis profoundly) fails to revive rhythm generation. The effects of FFA were reversible, as shown by the full recovery in washout conditions. Modified from [10].

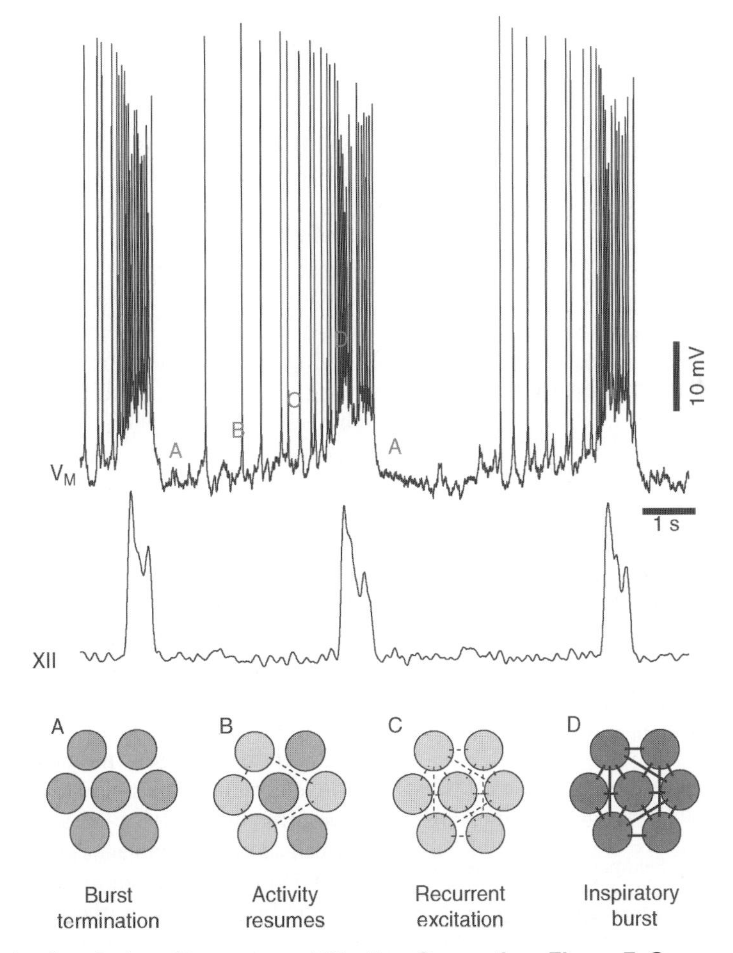

Pre-Bötzinger Complex Inspiratory Neurons and Rhythm Generation. Figure 7 Group-pacemaker hypothesis of rhythm generation. The membrane potential of an inspiratory preBötC neuron is shown (V_M) with XII motor output. Images at the bottom depict neuronal activity at different stages of the cycle. (a), the refractory state following the inspiratory burst. (b), some preBötC neurons recover their excitability and begin to spike low rates. (c), Spiking preBötC neurons begin to synaptically activate other silent preBötC neurons, leading to aggregation of network activity via recurrent synaptic excitation, which is positive feedback. (d), the inspiratory burst occurs when recurrent synaptic activity is sufficiently strong to recruit postsynaptic inward currents such as I_{CAN}. Inspiratory bursts terminate due to intrinsic properties that remain unknown. Modified from: Feldman and Del Negro, *Nat Rev Neurosci* 7: 232–242, 2006.

undergo a recovery phase in which they gradually approach a baseline membrane potential largely determined by $I_{K\text{-}Leak}$ and the by tonic inputs. The most excitable neurons spontaneously cross threshold and begin the next positive feedback cycle.

Given the diminishing evidence in support of the obligatory role of pacemaker properties in respiratory rhythm generation [8], a framework in which recurrent synaptic excitation evokes postsynaptic burst-generating membrane properties available to all preBötC inspiratory neurons, such as the group pacemaker hypothesis, is a viable mechanism that can explain key aspects of respiratory rhythmogenesis.

Acknowledgments

US National Science Foundation Integrative and Organismal Biology Award 0616099, The Suzann Wilson Matthews Faculty Research Award, and The Jeffress Memorial Trust.

References

1. Gray PA, Rekling JC, Bocchiaro CM, Feldman JL (1999) Modulation of respiratory frequency by peptidergic input to rhythmogenic neurons in the preBötzinger Complex. Science 286:1566–1568
2. Hayes JA, Del Negro CA (2007) Neurokinin receptor-expressing preBötzinger Complex neurons in neonatal mice studied in vitro. J Neurophysiol 97:4215–4224
3. Gray PA, Janczewski WA, Mellen N, McCrimmon DR, Feldman JL (2001) Normal breathing requires preBötzinger Complex neurokinin-1 receptor-expressing neurons. Nat Neurosci 4:927–930
4. Wang H, Stornetta RL, Rosin DL, Guyenet PG (2001) Neurokinin-1 receptor-immunoreactive neurons of the ventral respiratory group in the rat. J Comp Neurol 434:128–146
5. Guyenet PG, Sevigny CP, Weston MC, Stornetta RL (2002) Neurokinin-1 receptor-expressing cells of the ventral respiratory group are functionally heterogeneous and predominantly glutamatergic. J Neurosci 22:3806–3816
6. Smith JC, Ellenberger HH, Ballanyi K, Richter DW, Feldman JL (1991) Pre-Bötzinger Complex: A brainstem region that may generate respiratory rhythm in mammals. Science 254:726–729
7. Del Negro CA, Morgado-Valle C, Feldman JL (2002) Respiratory rhythm: an emergent network property? Neuron 34:821–830
8. Pace RW, Mackay DD, Feldman JL, Del Negro CA (2007) Role of persistent sodium current in mouse preBötzinger complex neurons and respiratory rhythm generation. J Physiol 580:485–496
9. Pena F, Parkis MA, Tryba AK, Ramirez JM (2004) Differential contribution of pacemaker properties to the generation of respiratory rhythms during normoxia and hypoxia. Neuron 43:105–117
10. Pace RW, Mackay DD, Feldman JL, Del Negro CA (2007) Inspiratory bursts in the preBötzinger complex depend on a calcium-activated nonspecific cationic current linked to glutamate receptors. J Physiol 582:113–125

Precentral Gyrus

Definition

The precentral gyrus is the cerebral gyrus immediately anterior and parallel to the central sulcus. It is part of the frontal lobe and is the primary motor cortex.

►Primary Motor Cortex
►Gyrus precentralls

Precerebellar Long-Lead Burst Neurons

Charles Scudder
Portland, OR, USA

Definition

Precerebellar neurons, in general, are neurons that have their somata in the brainstem or spinal cord and send their axons to the cerebellum. Precerebellar ►long-lead burst neurons (PCbLLBNs) (►burst cells – long lead (LLBNs)) are precerebellar neurons that have long-lead burst discharges during saccades. They comprise the major route for the transmission of saccadic commands to the cerebellum. All PCbLLBNs that have been identified to date have their somata in the pontine reticular formation or the pontine nuclei.

Characteristics

Higher Order Processes

PCbLLBNs receive information from higher order saccadic command centers and relay this information to the ►oculomotor vermis and the ►floccular lobe of the cerebellum (see ►Cerebellum – Role in Eye Movements). Projections to the oculomotor vermis comprise the first leg of a trans-cerebellar route by which saccadic commands are processed in the cerebellum and then conveyed to the saccadic burst generator. At the burst generator, these processed commands are combined with a raw command conveyed directly from the same higher command centers, and are thought to provide the detail needed for generating accurate saccadic eye movements (see cerebellum – role in eye movements). PCbLLBNs provide the major, but not the only, input to the oculomotor vermis, and provide a minor input to the floccular lobe.

Command centers that provide the major input to PCbLLBNs are the deep and intermediate layers of the ▶superior colliculus, the ▶frontal eye fields (FEF), the ▶supplementary eye fields (SEF), and the ▶lateral intraparietal area (LIP). The superior colliculus collects information from the above three cortical areas (FEF, SEF, LIP), the ▶basal ganglia, the superficial layers of the superior colliculus, and other smaller projections, and is considered the final common path for the saccadic command.

Parts of the LLBN Pathway to the Cerebellum

Groups of PCbLLBNs

Somata of PCbLLBNs reside in four principal areas within the pons, and each area receives somewhat different information. By far the largest population is in the caudal ▶nucleus reticularis tegmenti pontis (NRTP), both in the medial group of cells that extends dorsally from the main body of NRTP (Fig. 1), and to a lesser extent, in the medial part of the body of NRTP. The major input to this part of NRTP is from the superior colliculus, with a smaller input from the FEF and SEF. NRTP also receives a major feedback signal from the output of the midline cerebellum, namely, the fastigial nucleus.

A second group of PCbLLBNs resides in a diffuse collection of somata amid the fascicles of the medial longitudinal fasciculus (the intrafascicular nucleus of the MLF, or IFN) just rostral to the abducens nucleus (Fig. 1). The third group resides in raphe pontis, which is located below the MLF immediately rostral to the abducens nucleus (not illustrated). The known inputs to these areas based on anatomical data arise from the superior colliculus and from the ▶inhibitory burst neurons (IBNs) of the saccadic burst generator (see ▶brainstem burst generator), but there may be other sources of input.

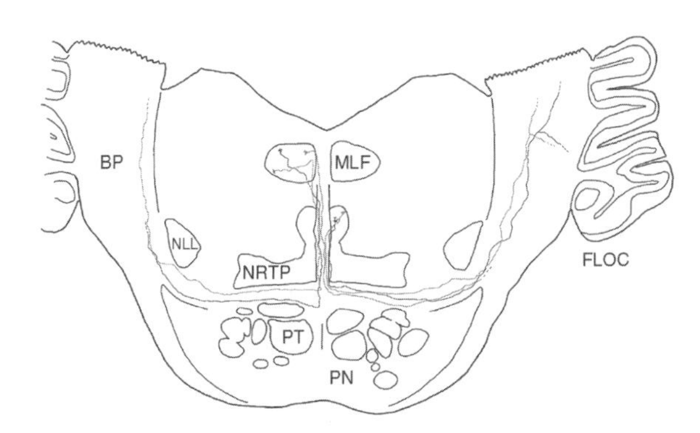

Precerebellar Long-Lead Burst Neurons. Figure 1 Transverse section of the monkey brainstem at a level just rostral to the abducens nucleus is illustrated along with the somata and axons of five precerebellar LLBNs. Neurons were visualized using intraaxonal staining and were reconstructed using a camera lucida. Three somata are located in a fasciculated portion of the medial longitudinal fasciculus (MLF), and two others are located in a dorsomedial extension of nucleus reticularis tegmenti pontis (NRTP). Axons course to the midline and then ventrally to a lateral fiber tract just ventral to NRTP, where some cross and some do not (follow axon colors). Axons course laterally and then back dorsally in brachium pontis (BP). Staining faded before reaching the cerebellum (dotted lines). The NRTP neurons sent collaterals toward the floccular lobe (FLOC). NLL, nucleus of the lateral lemniscus; PN, pontine nuclei; PT, pyramidal tract.

Locations of cell bodies and axonal projections for two of these three groups are illustrated in Fig. 1, which was derived from intraaxonally labeled neurons [1]. Axons of all NRTP, IFN, and raphe pontis PCbLLBNs course to the midline where they travel ventrally with other fiber bundles. Just above the pontine nuclei, they sharply turn either ipsilaterally or contralaterally and travel laterally in a well defined band of fibers between NRTP and the pontine nuclei. Subsequently, they enter the brachium pontis and travel dorsally to the cerebellum without ever branching in the brainstem. Several neurons originating in NRTP were observed to send collaterals to the floccular lobe, while none originating in the IFN or raphe pontis were so observed.

One group of investigators has found a fourth group of PCbLLBNs in the dorsolateral pontine nuclei [2]. The principal input to this group is from LIP, with a smaller input from the FEF. The dorsolateral pontine nuclei also receive heavy input from visual-motion sensitive cortical areas MST and MT, but PCbLLBNs presumably receive little of this input. This latter input goes mainly to neurons discharging during ▶smooth pursuit eye movements, which are intermixed with PCbLLBNs in this region.

Discharges of PCbLLBNs

Discharges of PCbLLBNs with somata in the reticular formation have been recorded extracellularly in medial NRTP, intra-axonally along their projection route, and extracellularly in the white matter of the oculomotor vermis and floccular lobe [1,3–5]. Like other LLBNs, they are mostly silent during fixations between saccades, but have a presaccadic firing that begins 21–300 ms before saccades in a preferred direction. Firing rate usually builds up towards a peak that usually also occurs before saccade onset. Firing ends near the time of saccade end.

PCbLLBNs in the reticular formation exhibit spatial properties that are intermediate between those of two common classes, namely ▶directional and ▶vectorial LLBNs. Like vectorial neurons, the range of directions for which a PCbLLBN may fire is less than a full hemifield, but it is usually larger than that of the superior-colliculus neurons that provide its major input. Unlike vectorial neurons, only a few PCbLLBNs have closed ►movement fields. That is, the discharge rate and number of spikes of typical PCbLLBNs first increase as saccade size increases but then might plateau with further increases in size. A few act like directional neurons, in that discharge parameters continually increase as saccade size increases. For neurons that do not have vertical preferred directions, all have an ipsilateral component to their preferred direction and none have a ►contraversive component. PCbLLBN responses have also been shown to depend on the torsional component of the saccade, making movement fields three dimensional [6]. These spatial characteristics are probably produced by convergent input from appropriately selected superior-colliculus efferents.

In the dorsolateral pontine nuclei, LLBNs are one population within a range of burst neurons [2]. Discharges of some burst neurons lag saccade onset, and some burst neurons also exhibit responses during smooth pursuit eye movements. All possible preferred directions for bursts are present in the population.

Lower Level Processes

The major targets of PCbLLBNs originating in the NRTP, IFN, and raphe pontis are the "oculomotor vermis" (lobules VIc and VII) [7] and the ►fastigial oculomotor region (FOR; see cerebellum – role in eye movements). The fastigial nucleus is presumably innervated by collaterals of ►mossy fibers projecting to the vermis, but retrograde tracing data suggests that not all PCbLLBNs issue these collaterals [8,9]. PCbLLBN signals are used in the cerebellum to produce a burst in FOR neurons, whose timing and amplitude affect the duration and amplitude of the discharges of burst neurons in the saccadic burst generator. The cerebellum thus provides fine control of the size and direction of saccades (see brainstem burst generator).

Some PCbLLBNs with somata in NRTP project to the ►flocculus and paraflocculus as collaterals of axons projecting to the oculomotor vermis [1]. It is uncertain whether all LLBN input to the floccular lobe arises in this fashion. IFN and raphe pontis neurons also project to the floccular lobe, but these may be the burst-tonic neurons that are also found in these nuclei. There is no firm data about how PCbLLBN signals are used in the floccular lobe, but one possibility is that they are used to help remove saccade-related signals from the many inputs with mixed saccade and smooth eye-movement signals, leaving mainly the latter in the output pathway.

Pathology

There have only been preliminary studies using experimental lesions of NRTP at the location of PCbLLBNs. Temporary inactivation of caudal NRTP with lidocaine or muscimol in monkeys has produced hypometric saccades, deficits in convergence and accommodation to a near target after a saccade, and aberrant torsional eye-position after a saccade [6,10]. Experimental lesions of the dorsolateral pontine nuclei produced deficits in smooth pursuit with no observable deficits in saccades. Specific lesions of raphe pontis have not been performed.

References

1. Scudder CA, Moschovakis AK, Karabelas AB, Highstein SM (1996) Anatomy and physiology of saccadic long-lead burst neurons recorded in the alert squirrel monkey. II. Pontine neurons. J Neurophysiol 76:353–370
2. Dicke PW, Barash S, Ilg UJ, Thier P (2004) Single-neuron evidence for a contribution of the dorsal pontine nuclei to both types of target-directed eye movements, saccades and smooth-pursuit. Eur J Neurosci 19:609–624
3. Crandall WF, Keller EL (1985) Visual and oculomotor signals in nucleus reticularis tegmenti pontis in alert monkey. J Neurophysiol 54:1326–1345
4. Kase M, Miller DC, Noda H (1980) Discharges of Purkinje cells and mossy fibers in the cerebellar vermis of the monkey during saccadic eye movements and fixations. J Physiol (London) 300:539–555
5. Ohtsuka K, Noda H (1992) Burst discharges of mossy fibers in the oculomotor vermis of macaque monkeys during saccadic eye movements. Neurosci Res 15:102–114
6. van Opstal AJ, Hepp K, Suzuki Y, Henn V (1996) Role of monkey nucleus reticularis tegmenti pontis in the stabilization of Listing's plane. J Neurosci 16:7284–7296
7. Thielert CD, Thier P (1993) Patterns of projections from the pontine nuclei and the nucleus reticularis tegmenti pontis to the posterior vermis in the rhesus monkey: A study using retrograde tracers. J Comp Neurol 337:113–126
8. Noda H, Sugita S, Ikeda Y (1990) Afferent and efferent connections of the oculomotor region of the fastigial nucleus in the Macaque monkey. J Comp Neurol 302:330–348
9. Yamada J, Noda H (1987) Afferent and efferent connections of the oculomotor cerebellar vermis in the macaque monkey. J Comp Neurol 265:224–241
10. Kaneko CRS, Fuchs AF (2004) Muscimol inactivation of the nucleus reticularis tegmenti pontis (nrtp) eliminates ipsilesional saccades. Soc Neurosci Abs 880.1

Precocial

Definition

Young born with hair or feathers, eyes open, the ability to move about immediately after birth and capable of leaving the nest within a few days.

►Neural Correlates of Imprinting

Precocious Puberty

Definition

The appearance of any sign of secondary sexual maturation, such as pubic hair, before the age of 8 years (or menarche before the age of 9 years) in girls and 9 years in boys.

▶Neuroendocrinology of Tumors

Precuneus

Definition

Part of the parietal lobe visible in a median section. Has a virtually square shape (hence also called quadrate lobe). The precuneus appears to be implicated in complex, sensory evaluation processes, language processing as well as spatial and temporal orientation.

▶Telencephalon

Precursor Cells

Definition

(Neural) precursor cells occur in both fetal and adult brains and are partially specialized; they undergo cell division and give rise to differentiated cells in a site-specific manner. In their normal states, adult precursor cells do not generate a wide variety of neurons. In the injured brain, adult precursor cells can partially replace neurons that are damaged or dead.

▶Adult Neurogenesis

Predicate (Attribute)

Definition

A predicate is what can be said of something, truly or falsely. The linguistic predicate "is red" can truly be applied to red things. The relational predicate "is larger than" can be used to state a relation between two things.

An attribute is a feature, like being red, for which a linguistic predicate stands. Realists claim, while nominalists deny, that an attribute is a special entity (universal) that can be exemplified by several particular things.

▶Argument
▶Logic

Prediction Error

Definition

A discrepancy between the expected and the experienced outcome of a behavioral situation. The prediction error is a relevant variable determining recruitment of dopaminergic action in the brain and formation of associations between events (stimuli and or behaviors).

▶Dopamine
▶Neuroethological Aspects of Learning

Predictive Eye Movements

Definition

Predictive eye movements occur whenever the motion of a target exhibits regular temporal features. Particularly with periodic movements, they compensate for the lag of the eye on the target implied by the reaction time, achieving either a nearly perfect synchronisation with the target (smooth pursuit of sinusoids at < 0.5 Hz) or even a lead (saccades tracking a target stepping back and forth at regular pace).

▶Oculomotor Control
▶Saccade, Saccadic Eye Movement

Preferential Motor Reinnervaton (PMR)

Definition

The bias and modest selectivity that motor axons exhibit in choosing motor endoneurial pathways and end-organs over sensory pathways and end-organs during reinnervation.

►Peripheral Nerve Regeneration and Nerve Repair

Preferred Direction (of a Neuron)

Definition

Neurons that respond to motion of either discrete visual stimulus or of a structured background modulate the strength of their response as function of the direction of motion. Plots of the strength of the response, e.g., mean firing rate, against the angle of stimulus trajectory are called tuning curves. The angle corresponding to the peak of the curve defines the preferred direction of the neuron. Visuomotor neurons, such as tectoreticulospinal neurons (TRSNs) have a directional tuning for visual stimulus, as well as to the direction of orienting movement associated to motor components of their bursts. Preferred movement direction of a given neuron is determined from the tuning curves, in the same way as for visual or, more generally, sensory preferred directions.

►Reaching Movements
►SC-Tectoreticulospinal neurons (TRSNs)

Prefrontal Cortex

Susan R. Sesack
Departments of Neuroscience and Psychiatry,
University of Pittsburgh, Pittsburgh, PA, USA

Synonyms

Frontal cortex; Dorsolateral cortex; Orbitofrontal cortex

Definition

The prefrontal cortex was so named because it was discovered in electrical stimulation studies to be a "silent" region in front of the motor areas of the frontal lobe. It appears to have increased in size and complexity in the course of evolution, culminating in the human brain as approximately 30% of the cortical mantle and at least 11 different cytoarchitectonically defined areas according to Brodmann's nomenclature. It is involved in many of the cognitive and ►executive control functions necessary for goal directed behavior, including ►working memory, shifting of ►selective attention, decision-making and ►response inhibition. Disturbances in function are thought to contribute to the pathophysiology of several mental conditions, including ►schizophrenia, major ►depression, ►post-traumatic stress disorder (PTSD), ►attention deficit hyperactivity disorder (ADHD) and susceptibility to ►addiction.

Characteristics

Anatomy

In common with all cortical areas, the PFC is a layered structure, with six layers in this case, and has the same major cellular constituents [1–5]. These include spiny ►pyramidal neurons with an excitatory glutamate phenotype whose axons ramify locally in addition to entering the white matter and projecting to other cortical and subcortical regions. The second major cell class consists of relatively aspiny, non-pyramidal ►interneurons whose inhibitory phenotype is GABAergic and whose axonal connections are strictly local. Several major subclasses of these local circuit neurons are distinguished by their content of calcium binding proteins or ►neuroactive peptides and by differences in their axonal targets, providing the necessary circuitry for feedback and feedforward inhibition within the cortex. Subsets of interneurons are also thought to entrain pyramidal cells to fire in ►oscillations, at frequencies deemed essential for specific forms of information processing [4].

In the past, cortical territories were defined by their inputs from specific nuclei of the ►thalamus, and this still constitutes a reasonable starting point for identifying cortical regions across species [3]. For the PFC, the principal thalamic division is the mediodorsal nucleus (MDTN). The MDTN is itself divided into three main subregions, each of which maintains reciprocal connections with portions of the PFC. (i) The paralamellar division is the most lateral part of the MDTN and innervates the cortical territory around the arcuate sulcus corresponding to Brodmann's area 8, also known as the ►frontal eye fields. This serves as the motor command region for voluntary (►saccadic) eye movements. (ii) Medial to the paralamellar division is the parvocellular region of the MDTN, which connects to the cortex along the dorsolateral convexity (DLPFC), including Brodmann's areas 9, 10 and 46, the latter lying within and along the banks of the principal sulcus. (iii) The most medial portion of the MDTN is the magnocellular division, which is interconnected to orbital regions of the cortex (named for their position above the eye socket) and to other ventral and medial surfaces of the frontal lobe, Brodmann's areas 11, 13, 14, 24, 25, 32 and 47/12. Collectively, these are termed the orbitomedial PFC (OMPFC). Areas 24, 25 and 32 lie within the ►cingulate gyrus and so are also considered parts of the ►anterior cingulate cortex.

In addition to specific inputs from the MDTN, the PFC is also innervated by other thalamic nuclei within the anterior, ventral, medial and midline divisions as well as the pulvinar nucleus. As an association cortex, the PFC receives extensive inputs from other cortical regions organized in a hierarchical fashion [3]. Major afferents arise from other association areas, including those in the posterior parietal and inferior temporal regions. The sensory streams innervating the PFC appear to involve mainly visual, auditory and somatosensory projections to the DLPFC and primarily olfactory, gustatory and viscerosensory projections to the OMPFC. The OMPFC is also innervated by important structures within the ►limbic system, the ►hypothalamus, ►hippocampus and ►amygdala [1,3,5]. The amygdala is a quasi-cortical structure that regulates behavior by conditioned associations and its reciprocal connections with the PFC are important for facilitating appropriate emotional behaviors. It has been argued that within the OMPFC, orbital divisions are the main sites of sensory termination, whereas medial regions are the main origin of descending projections to autonomic regions [5]. Hence, these two subdivisions may function as viscerosensory and visceromotor regions, respectively.

The PFC, like most cortical structures, is extensively innervated by ►ascending neuromodulatory projection systems (see Essay of same name) arising in the brainstem and basal forebrain, including pathways conveying ►acetylcholine, ►dopamine, ►norepinephrine and ►serotonin inputs [3,6,7]. These projections provide essential modulation of cortical firing patterns and many studies demonstrate that either decreases or increases in critical levels of monoamines are sufficient to disrupt PFC function [6,7]. Conversely, pharmacological therapies for mental disorders (e.g., antipsychotic, antidepressant and anxiolytic medications) are often designed to alter monoamine transmission in an effort to restore more balanced activity in this region.

Many of the efferent projections of the PFC reciprocate the afferent inputs [1,3], including those to association cortices and to thalamic nuclei. Another major output of the PFC, as with many cortical regions, is to the ►basal ganglia. Cortical projections to the ►striatopallidum (see Essay of same name) are relayed back to the cortex via the thalamus, forming functional "loops" for the selection of appropriate actions and suppression of maladaptive responses. The DLPFC targets mainly the head of the ►caudate nucleus, forming a major associative loop that supports cognitive and executive functions, whereas the OMPFC projects to more ventral parts of the striatal complex, including the ►nucleus accumbens and participates in limbic circuitry for motivated behaviors. The brainstem projections of the PFC include the ►superior colliculus for saccadic eye movement control and the pontine nucleus, which relays executive commands to the ►cerebellum for controlling the timing and coordination of movement and cognition. Interestingly, the OMPFC is the source of extensive projections to diencephalic and brainstem structures that are important regulators of autonomic output, including the amygdala, hypothalamus, ►periaqueductal gray, ►parabrachial nucleus and ►nucleus of the solitary tract [3,5] Moreover, portions of the OMPFC are among the only cortical regions that project directly to brainstem monoamine neurons [3], placing this division of the PFC in a position to regulate the level of modulatory drive to the cortex generally. The latter connections are no doubt important for understanding the pathophysiology of mood disorders, in which brainstem monoamines are implicated in both cause and treatment.

On the basis of these differential inputs and outputs, the DLPFC is considered to be the main region within which spatial and object working memory and other executive functions are carried out [1–4,8]. The OMPFC has been deemed a viscerosensory and visceromotor network that guides behavior based on emotional experience [1,3,5,9]. Of course, the major subdivisions of the PFC are interconnected, so that goal directed behavior is accomplished by consideration of both external and internal perceptions [3].

Function

The PFC performs many essential cognitive functions whose complexity increases in the course of evolutionary development. The PFC guides behavior particularly when situations are novel or complex and is probably not involved in functions that are routine or well learned [1]. There is considerable consensus that the PFC is a major contributor to the processes underlying ►working memory, a short form of memory in which information is held temporarily in mind until it can be used to guide immediate behavior [1,2,4]. Such information is also available for mental manipulation and the PFC has sometimes been described as the brain's "scratch pad." The PFC provides a mechanism for bridging time between the near past and the actions that proceed from keeping this information "in mind." In this way, the PFC is essential for preparing the motor systems for action and hence for future planning and logical sequences of action and thought [1].

Behavioral tasks in which a delay is interposed between cue and choice (►delayed–response tasks) are important tools for evaluating working memory performance in experimental animals and humans. Such tasks assess the ability to maintain a representation of spatial location, object identity or stimulus associations in the absence of the original cue and performance of these tasks is highly sensitive to PFC damage [1,2]. The oculomotor variant of this test, the ►delayed saccade task was created to minimize movement in behaving animals and so to facilitate electrophysiological recording of neuronal activity during working memory. The resultant studies have depicted cells that respond to the presentation of the cue or to the go signal for the motor response. More importantly, many of these studies have described neurons that increase their activity during the delay period between the cue and the go signal. These "delay" cells appear to represent the maintenance of critical information obtained from the cue in order to guide future responding and so are widely considered as a neural correlate of working memory. Many studies report that delay period cells are spatially tuned, responding best to cues that dictate subsequent motor responses into specific regions of space [2]. The exact cellular mechanisms that underlie delay period activity are not yet known. Collateral synapses between pyramidal cells may form local reverberatory connections that could sustain such firing. Alternatively or in addition, delay period activity probably reflects long-range interconnections between the PFC and posterior association areas where similar firing patterns have been described [1,2,4,8]. Other electrophysiological recording studies, particularly those in the OMPFC, have described neurons that appear to encode the salience or reward value of stimuli, which is likely to subsequently influence motivation in motor responses [1].

Many of these observations from animal physiology have been verified to the extent possible in humans using variants of working memory or decision making tasks and functional imaging methods [1,8,9]. Functional magnetic resonance imaging (►fMRI) and positron emission tomography (►PET) monitor signals that reflect altered blood flow as indirect measures of activation in brain regions. Performance of tasks for which working memory is an essential component produce signals consistent with an increase in blood flow to the PFC. Moreover, many studies using this methodology have produced findings that extend theories regarding the cognitive functions of this region to include stimulus ►encoding, sustained attention, decision making and motor preparation.

Disorders

Although lesions caused by stroke, tumor or accidental damage are rarely circumscribed within subdivisions of the PFC, some relatively distinct syndromes have been characterized [1]. Lesions that are centered in orbital regions of the OMPFC tend to cause loss of ►response inhibition [9], with changes in personality that include impulsiveness and recklessness. Patients have problems with selective attention and exhibit inappropriate decision making to the point of self-destructive and sociopathic behavior. Attentional problems are also seen with damage to medial portions of the OMPFC that include the anterior cingulate cortex, in this case accompanied by affective blunting, apathy and withdrawal [1]. Lesions of the DLPFC produce a "dysexecutive syndrome" characterized by deficits in working memory and related cognitive functions, poor planning capability, difficulty in the execution of logically ordered sequences of behavior or language and attention deficits [1]. The consistent observation of attentional problems with damage to the PFC suggests that this cortical territory is important for regulating the attentional processes necessary for goal directed behavior [1]. Functional imaging studies of the anterior cingulate cortex in particular support a major role for this region in attentional mechanisms. Partly as a result of these observations, the PFC is considered a principal site of pathophysiology in ADHD [6].

There is ample evidence from functional imaging, postmortem and neuropsychological analyses for a significant malfunction of the PFC in ►schizophrenia [2,4]. Schizophrenic patients show deficits in working memory and other cognitive skills, consistent with reduced blood flow to the PFC during performance of tasks designed to test these functions. These and other "negative" symptoms form the core features of the illness; they are the most predictive of long-term prognosis and are the least susceptible to pharmacological intervention. Postmortem studies show structural alterations in both pyramidal and non-pyramidal neurons and reductions in markers of synaptic communication. Some of these deficits are specific to the PFC, while others are observed throughout the cortex. Most appear to be associated with the illness and not with treatment with antipsychotic medications [4].

Functional imaging and postmortem studies also report changes in the PFC in major ►depressive disorders, including ►bipolar depression [5]. Altered metabolism/blood flow is particularly observed in the medial portions of the OMPFC and such changes are consistent with the symptom complex that accompanies lesions to this region. The earliest studies reported increased blood flow in the OMPFC with major depression, although later investigations have also shown decreased blood flow in this region. Some postmortem analyses have also reported reduced tissue volume in portions of the midline frontal cortex, primarily due to loss of glia and neuropil as opposed to a loss of neurons.

The PFC is highly susceptible to the impact of acute and chronic ►stress and it is well known that stress can exacerbate the symptoms of mental disorders [6,7,10]. Behavioral studies indicate that working memory performance is degraded in animals subjected to stress and several investigations report structural alterations in pyramidal neurons following exposure to chronically stressful conditions. Theories regarding the likely pathophysiology of PTSD often focus on the amygdala as a probable site of dysfunction, but these models also suggest loss of inhibitory regulation of the amygdala by the OMPFC in this condition [10]. Stress is also a known contributor to recidivism in addiction disorders. Loss of inhibitory control of behavior by a weakened PFC system has been implicated in susceptibility to addiction and the impact of acutely stressful events on PFC activity may further destabilize individuals toward relapse [7].

References

1. Fuster JM (2001) The prefrontal cortex – an update: time is of the essence. Neuron 30:319–333
2. Goldman-Rakic PS (1999) The physiological approach: functional architecture of working memory and disordered cognition in schizophrenia. Biol Psychiatry 46:650–661
3. Groenewegen HJ, Uylings HBM (2000) The prefrontal cortex and the integration of sensory, limbic and autonomic information. Prog Brain Res 126:3–28
4. Lewis DA, Hashimoto T, Volk DW (2005) Cortical inhibitory neurons and schizophrenia. Nat Rev Neurosci 6:312–324
5. Ongur D, Price JL (2000) The organization of networks within the orbital and medial prefrontal cortex of rats, monkeys and humans. Cereb Cortex 10:206–219
6. Arnsten AF, Li BM (2005) Neurobiology of executive functions: catecholamine influences on prefrontal cortical functions. Biol Psychiatry 57:1377–1384
7. Moghaddam B (2002) Stress activation of glutamate neurotransmission in the prefrontal cortex: implications for dopamine-associated psychiatric disorders. Biol Psychiatry 51:775–787
8. Postle BR (2006) Working memory as an emergent property of the mind and brain. Neuroscience 139:23–38
9. Bechara A, Damasio H, Damasio AR (2000) Emotion, decision making and the orbitofrontal cortex. Cereb Cortex 10:295–307
10. Rauch SL, Shin LM, Phelps EA (2006) Neurocircuitry models of posttraumatic stress disorder and extinction: human neuroimaging research – past, present, and future. Biol Psychiatry 60:376–382

Preganglionic Neuron

Definition

Within the autonomic nervous system – which is the set of central and peripheral nerve structures that control the activity of viscera (such as bladder, heart, intestines, trachea, glands) and of blood vessels – there are large arrays of neurons located in the spinal cord and the brain stem, arranged in columns and nuclei, that project their axons outside the central nervous system (CNS) and are part of the efferent (motor) pathways leading from brain centers to peripheral organs. They are known as preganglionic neurons because their axons gather into small nerves that reach autonomic ganglia (sympathetic and parasympathetic) and terminate with innumerable nerve endings synapsing on ganglion neurons.

In the brain stem the main bilateral groups of preganglionic neurons are the accessory oculomotor nuclei of Edinger-Westphal projecting to the ciliary ganglia, the superior salivatory nuclei projecting to the pterygo-palatine and submandibular ganglia, the inferior salivatory nuclei projecting to the otic ganglia, the dorsal vagal nuclei projecting to ganglia in the trachea, oesophagus and gastro-intestinal tract, and part of the nucleus ambiguus projecting to cardiac ganglia.

In the spinal cord, autonomic neurons are grouped into bilateral columns in the ventral horn; the largest ones are part of the sympathetic pathways extending from the last cervical level (C8) to the second lumbar level (L2). Other columns, which are part of the parasympathetic pathways, occupy sacral levels (between S2 and S4).The preganglionic neurons, which are all cholinergic, receive two main inputs: from higher centers, for example the pontine micturition centre, the cardiovascular centers, several autonomic nuclei in the hypothalamus, and from the periphery via afferent (sensory) autonomic neurons, directly or through an interneuron.

►Ageing of Autonomic/Enteric Function
►Hypothalamus
►Parasympathetic Nervous System
►Parasympathetic Pathways
►Sympathetic Nervous System
►Sympathetic Pathways

Preganglionic Neurotransmitters

Definition

Neurotransmitters released from preganglionic neurons to influence the activity of postganglionic neurons within autonomic ganglia. Acetylcholine is the primary neurotransmitter used by probably all preganglionic neurons. Co-transmitters in preganglionic neurons modulate the excitability of postganglionic neurons, or have presynaptic actions to affect further release of preganglionic neurotransmitters.

►Acetylcholine
►Autonomic Ganglia
►Preganglionic Neuron

Pregeniculate Nucleus (Primates)

►Intergeniculate Leaflet

Prehension

Definition

The act of taking hold, typically with the hand as in grasping.

►Coordination
►Motor Cortex – Hand Movements and Plasticity

Prejunctional

Definition

Prejunctional refers to the axon varicosity proximal to a neuroeffector junction, releasing neurotransmitters.

►Postganglionic Neurotransmitter

Prelude Neurons

►SC – Buildup Neurons

Premotor Areas

Definition

A term used to refer to secondary motor areas of cerebral cortex, particularly the dorsal and ventral areas on the lateral surface of the hemisphere. This term may also be used to refer to brain areas containing neurons that have synaptic linkages to motoneurons.

►Motor Cortex: Output Properties and Organization
►Visual Space Representation for Reaching

Premotor Cortex (Area 6)

Definition

The premotor cortex (area 6) appears to play a role in regulating grasping actions. In the caudal segments fibers arise from the pyramidal tract. Lesions frequently result in impaired grasping actions. Strength regulation of the grasping hand is likewise impaired.

►Telencephalon

Premotor Cortex (Area 8)

Definition

Frontal eye field. Plays an important role in voluntary control of eye movement. Lesions in this area cause loss of voluntary control of eye movement.

►Telencephalon

Premotor Interneurons

►Excitatory CPG Interneurons

Premotor Neurons

Definition

Neurons that have monosynaptic linkages to motoneurons.

Premotor Processing

Definition

Processing of information prior to generating movement that includes planning, anticipation of outcomes, evaluation of rewards and decision making as to the best pattern of motor outputs to achieve particular goals.

►Visual Space Representation for Action
►Visual Space Representation for Reaching

Pre-mRNA

Definition

Transcribed RNA prior to the splicing process, containing both introns and exons.

►Alternative Splicing and Glial Maturation

Prenatal Brain Injury by Chronic Endotoxin Exposure

Sandra Rees
Department of Anatomy and Cell Biology,
University of Melbourne, Melbourne, Australia

Synonyms

Fetal brain injury; Brain inflammation

Definition

Many factors contribute to perinatal brain injury, with the major causes likely to be hypoxia-ischemia/reperfusion [1] and infection of the fetus and/or mother [2]. Clinical studies have indicated that there are significant associations between maternal infection, preterm birth, neonatal brain damage and increased levels of ►proinflammatory cytokines in the amniotic fluid and umbilical cord; the strength of specific correlations are still being deduced. Fetal inflammatory responses can occur in the absence of overt signs of infection or fetal compromise and it is possible that subclinical inflammation underlies otherwise unexplained brain injury. Preterm birth occurs in 7–10% of pregnancies; advances in perinatal care have led to a significant improvement in the survival of very premature (less than 30 weeks) and very low birth weight (less than 1,500 g) infants. However up to 10% of these infants develop spastic motor deficits and 20–50% suffer developmental and behavioral disabilities. The most common cerebral neuropathology observed in premature infants is white matter injury, referred to as periventricular leukomalacia. This includes diffuse gliosis (►microgliosis and astrogliosis) extending throughout the white matter and also focal cystic infarction adjacent to the lateral ventricles in about 5% of cases. It is now increasingly recognized that the cerebral cortex and deep grey matter are also likely to be affected.

Invading microorganisms are thought to gain access to the amniotic cavity and fetus, most commonly, by ascending through the vagina and cervix [3]. This induces an innate immune response with inflammation of the chorioamniotic membranes and the production of proinflammatory cytokines. The cytokines and/or other inflammatory mediators then gain access to the fetus via swallowed amniotic fluid or fetal lungs, eyes or nasal membranes. It has been suggested that these agents increase the permeability of the blood–brain barrier with enhanced leucocyte infiltration of the brain mediated by brain chemokines; brain microglia and astrocytes will be stimulated to upregulate the production of cytokines and alterations to brain structures and/or overt injury will ensue depending on the severity of the inflammatory response.

Characteristics

Description of the Process

Lipopolysaccharide and the Inflammatory Response

In order to more fully understand the mechanisms involved in inflammatory-induced brain damage animal models are required with one of the specific goals being the development of therapeutic strategies. The bacterial endotoxin, lipopolysaccharide (LPS) is most commonly used to model the systemic inflammatory response induced by infection, although live bacteria have also been used. LPS is a major component of the outer membrane of gram negative bacteria. Since fetal inflammation associated with maternal infection is likely to be chronic in nature [3] models involving repeated exposure to LPS are most likely to mimic the human situation [4,5] although other models stand to contribute to an understanding of the inflammatory process. LPS has been administered to fetal rodents and sheep and neonatal rats, kittens and monkeys at a developmental age which equates to 25–30 weeks in the human fetus, a period of high vulnerability for white matter damage.

Cytokines

To contend with invading organisms, as typified by lipopolysaccharide, the innate immune response includes the generation of reactive oxygen species, phagocytosis and the production and release of cytokines and chemokines. LPS binds initially to LPS-binding protein (LBP), an acute phase protein released into the bloodstream from the liver [6]. LBP then appears to aid LPS in docking with the LPS receptor complex by forming a ternary complex with CD14 thus enabling LPS to be

P

presented to the LPS receptor complex consisting of Toll-like receptor-4 (TLR4) and MD-2 an extracellular adaptor protein, on the surface of myeloid and other cells. Binding to the complex activates transmembrane signaling pathways (Fig. 1) which result in the activation of the transcription factor nuclear factor (NF)-κB, containing subunits p50 and p65 as homo-or heterodimers. NF-κB binding activity can be induced by changes in the intracellular redox status and LPS is known to be one of the most potent activators of this pathway. NF-κB is normally present in the cell cytoplasm forming an inactive complex with an inhibitor I-κBα. Following cellular activation, I-κBα is phosphorylated by I-κB kinase; its subsequent degradation by cytoplasmic proteases releases NF-κB which is translocated to the nucleus where it regulates transcription of several genes including cell adhesion molecules, metalloproteinases and the pro-inflammatory cytokines, IL-1β, IL-6 and TNF-α. For example TNF-α is elevated within 2 h of LPS exposure in the preterm ovine fetus (Fig. 2).

Blood–Brain Barrier

Systemic cytokines (and LPS) can then bind to receptors on cerebral endothelial cells or periventricular cells initiating downstream signaling events which include an increase in prostaglandin synthesis and altered cGMP and nitric oxide levels. This results in increased blood–brain barrier (BBB) permeability for at least 24 h after LPS exposure particularly in white matter tracts; the exact molecular site within the blood vessels at which the leak to proteins occurs is not yet clear. Cytokines and proinflammatory substances can then pass into the brain activating microglia and

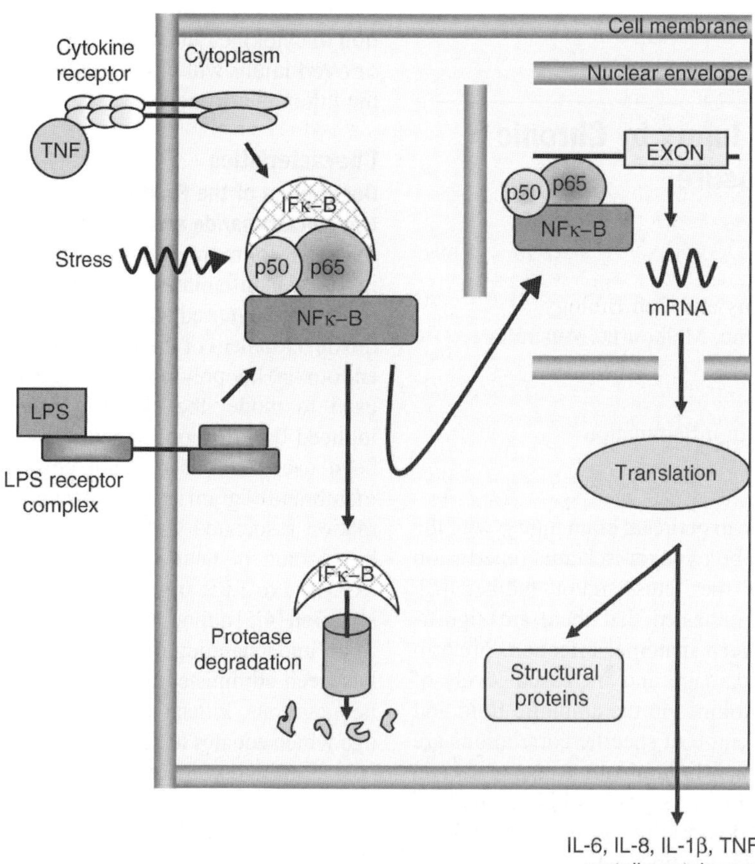

Prenatal Brain Injury by Chronic Endotoxin Exposure. Figure 1 In response to proinflammatory cytokines, endotoxins or environmental stress, the nuclear factor (NF) -kB pathway is activated. NF-kB (containing subunits p50 and p 65) is normally present in the cytoplasm forming an inactive complex with inhibitory factor (IF-kB). Following cellular activation, IF-kB is phosphorylated and subsequently degraded by cytoplasmic proteases. This releases NF-kB which is translocated to the nucleus where it regulates the transcription of several genes including proinflammatory cytokines.

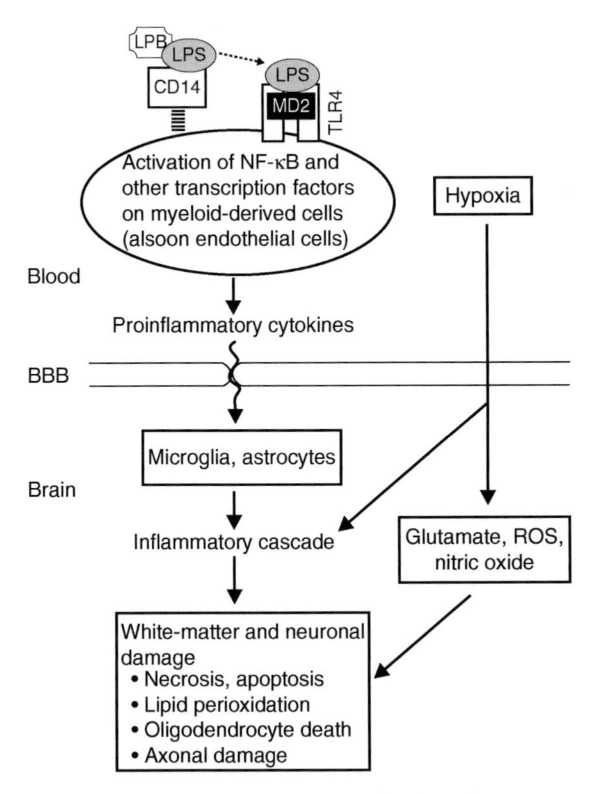

Prenatal Brain Injury by Chronic Endotoxin Exposure. Figure 2 Possible activation of the inflammatory cascade by endotoxin (lipopolysaccharide, LPS). LPS binds to LPS binding protein (LBP) which then forms a ternary complex with CD14, enabling LPS to be presented to the LPS receptor complex consisting of Toll-like receptor 4 (TLR4) and MD-2. This activates NF-κB and other signaling pathways. Proinflammatory and other proteins are transcribed; they affect the permeability of the blood–brain barrier, perhaps via the up-regulation of prostaglandins, and allow the entry of proinflammatory substances and circulating leukocytes into the brain. Here there is activation of microglia and astrocytes to up-regulate cytokine production; this ultimately leads to white-matter and neuronal damage. If hypoxia is also involved, excess extracellular glutamate will result in further nitrosative and oxidative stress, exacerbating neuronal and axonal damage.

astrocytes to release cytokines, oxygen free radicals and other factors which will have multiple effects on the brain ranging from altered neuronal development to overt injury (Fig. 3). A recent microarray study in the LPS-exposed neonatal rat demonstrated altered gene expression of hundreds of inflammatory molecules and also cell-death associated molecules in the brain. In the adult brain, TLR4 receptors are found in regions devoid of the BBB namely the circumventricular organs (CVOs) and also the choroid plexus, meninges and scattered cells within the brain parenchyma. It has been postulated that LPS binds to TLR4 in the CVOs inducing an inflammatory reaction which is then disseminated throughout the brain. In the neonatal brain mRNA for CD14 receptors has been identified although their cellular distribution is not known.

Oxidative and Nitrosative Stress

Reactive oxygen species (ROS) are toxic oxygen metabolites produced in small quantities during the normal cellular metabolic processes in mitochondria via univalent reduction of molecular oxygen. In response to stimuli, including an imbalance in the cellular redox status, the formation of ROS increases dramatically. ROS interact with the lipid components of cell membranes, initiating lipid peroxidation resulting in the breakdown of lipid constituents into highly reactive by-products including lipid aldehydes, for example 4-hydroxynonenal (4-HNE). These reactive aldehydes then bind to and modify protein creating protein adducts. In addition to oxidative stress, nitrosative stress results from nitric oxide (NO) released from reactive microglia reacting with superoxide anions to form peroxynitrite which targets tyrosine residues of proteins to form nitrotyrosine residues. Both of these processes are highly damaging to cell membranes. Lipid peroxidation of membranes can occur in the cerebral hemispheres within 6 h of LPS exposure in the ovine fetus [7]. The major antioxidant defence system that prevents the intracellular accumulation of ROS is the glutathione system. Reduced glutathione (GSH) acts as both a free radical scavenger and as a substrate in the GSH redox cycle. During late

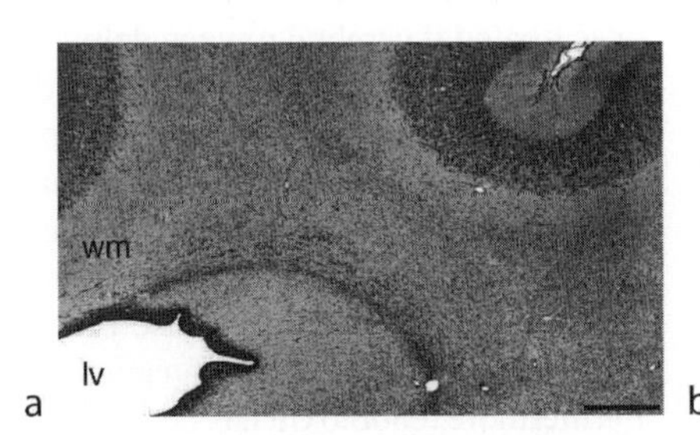

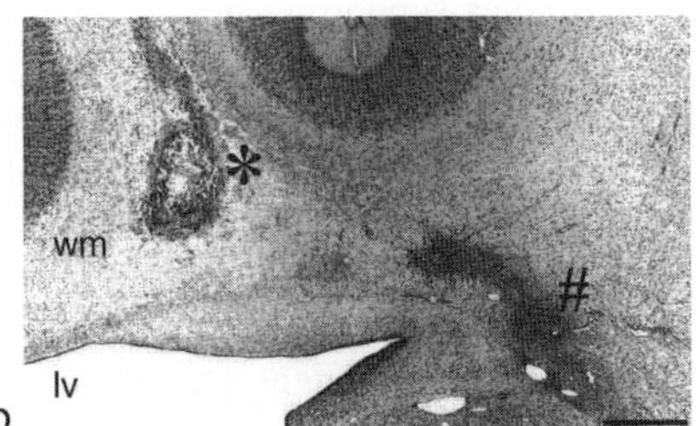

Prenatal Brain Injury by Chronic Endotoxin Exposure. Figure 3 Section of the cerebral hemispheres from control (a) and LPS-exposed (b) fetal sheep at 70% of gestation (term ~147 days). Repeat bolus doses of LPS resulted in both focal (*) and diffuse (#) white matter (wm) injury; lv – lateral ventricle. Image courtesy of Dr Jhodie Duncan. Scale = 528 μm.

gestation, there are marked increases in the antioxidant protective capacity. The combination of an immature defence system, LPS challenge and localized oxidative and nitrosative stress exposes the fetal brain to a high risk of ROS-mediated tissue injury.

Glial Response

Within the preterm ovine and rat brain activation of microglia is the most prominent glial response after LPS exposure; there is a significant correlation between the intensity of microglial/macrophage invasion and the extent of white matter injury indicating a substantial role for microglial activation in the manifestation of and/or response to injury [4,5]. Currently it is not certain whether these cells are mainly microglia already resident in the brain or macrophages invading from the circulation. Of interest is the recent suggestion that the high density of activated microglia in the white matter during normal fetal human brain development might potentially prime this area for diverse brain insults characterized by the activation of microglia; this also appears to be the case in the ovine and rodent fetus.

Axonal Injury

This could occur as a result of increased glutamate levels in the extracellular space induced by cytokine activation of glutamate–containing astrocytes and the subsequent activation of glutamate receptors on oligodendrocytes or the myelin sheath itself causing a toxic influx of Ca^{2+}. Excess Ca^{2+} will likely cause disruption to mitochondrial function and damage to the structural integrity of the axon. During the peak gestational age for white matter damage, developing oligodendrocytes (preoligodendrocytes) predominate [8]. The basis for this maturational susceptibility could be related to preferential vulnerability to free radicals, cytokines, glutamate toxicity or a mismatch of anti-oxidant enzymes and a subsequent imbalance of oxidative metabolism.

NF-kB Activation in the Brain

Within the brain the role of NF-κB in neuronal survival is controversial with studies supporting both a neuro-destructive or neuroprotective role, the latter possibly by upregulation of anti-apoptotic genes [9]. Perhaps this dichotomy is due to differences in stages of neuronal development, different monomeric composition, degrees of NF-κB activation or different combinations of cellular stimuli. It has been suggested that activation of NF-κB in neurons might protect them against degeneration whereas activation in microglia promotes neuronal degeneration via cytokine production [9]. Chronic systemic administration of LPS in the ovine fetus leads to increased binding activity of NF-kB subunits in specific regions of the fetal brain and placenta within 6 h of exposure [7]. There was no clear-cut relationship however between alterations in levels of NF-kB and the vulnerability to endotoxic damage. In the cerebral hemispheres where damage occurs NF-κB was not upregulated whereas in the hippocampus where damage does not occur NF-kB was elevated. Clearly the activation of NF-κB can result in several down-stream signaling pathways and an upregulation does not necessarily signal that brain damage will ensue. Therefore blocking NF-κB transcription might not be a means of protecting against brain injury.

Higher Level Processes

Cerebral white matter injury resulting from ►chronic endotoxin exposure in the prenatal brain (►prenatal brain injury (PBI)) is likely to have consequences on the overlying cortex; the cell bodies associated with damaged cortical efferent axons are likely to die. Cortical neurons in layers 5 and 6 demonstrate an upregulation of beta amyloid precursor protein (a marker of cellular stress) after chronic LPS exposure not only in regions of overt white matter damage as might be expected, but also in adjacent gyri. This suggests that chronic endotoxin exposure also induces damage to cortical neurons either directly or secondarily to sensory deafferentation. Progressive post-injury reorganization of the undamaged cerebral cortex will play a role in the underlying mechanisms of ensuing neurological sequelae. It is becoming increasingly recognized in human MRI studies of premature infants that subtle alterations in cerebral structure are common, particularly in the extremely immature infant. Effects of LPS on other aspects of development such as neurogenesis remains to be elucidated.

Interaction Between Inflammation and Hypoxemia

Elucidating the mechanisms involved in inflammatory–induced brain injury is made more complex by the observation that inflammation can interrupt hemodynamic stability and that activation of inflammatory pathways is involved in the neural responses to hypoxia/ischemia. In animal models the inflammatory cascade and associated brain damage appears to be exacerbated if cerebral oxygen delivery is also reduced. LPS administered in repeat bolus doses at 70% of gestation results in fetal hypoxemia, hypotension, acidemia and associated brain damage in the ovine fetus [4]. A chronic infusion of the same or higher total dose of LPS does not result in significant alterations in fetal physiology but does cause brain injury, which is not as severe as when accompanied by hypoxemia [5]. Furthermore, endotoxin has been shown to sensitize the immature brain to hypoxic-ischemic injury [10]. In human pregnancies where intrauterine inflammation is accompanied by fetal asphyxia, there appears to be a significant increase in the risk of cerebral palsy.

Thus there are likely to be synergistic pathways between hypoxia and infection/inflammation which potentiate the evolution of brain damage although it is evident that inflammation alone can indeed cause fetal brain injury. It is possible that infection could impair oxygen delivery to the fetal brain via effects on gas exchange in the placenta. Histologic chorioamnionitis when compounded with a placental perfusion defect has been shown to enhance the risk of abnormal neurologic outcome in extremely low birth weight infants.

Function

It is likely that altered brain structure and white matter lesions resulting from ►brain inflammation will have long term consequences for brain development and function, including cognition, behavioral and motor abilities. Inflammation might also have long term effects on autonomic and neuroendocrine responses, including the manifestation of fever. It has recently been shown that neonatal exposure to LPS attenuates the febrile and CNS neurochemical responses to an LPS challenge in later life; this might equate to an altered ability to combat diseases after birth.

References

1. Volpe JJ (2001) Perinatal brain injury: from pathogenesis to neuroprotection. Ment Retard Dev Disabil Res Rev 7(1):56–64
2. Dammann O, Leviton A (1997) Maternal intrauterine infection, cytokines and brain damage in the preterm newborn. Pediatrics Res 42:1–8
3. Romero R, Chaiworapongsa T et al. (2003) Micronutrients and intrauterine infection, preterm birth and the fetal inflammatory response syndrome. J Nutr 133 (5 Suppl 2):1668S–1673S
4. Duncan JR, Cock ML et al. (2002) White matter injury after repeated endotoxin exposure in the preterm ovine fetus. Pediatric Res 52:1–9
5. Duncan JR, Cock ML et al. (2006) Chronic endotoxin exposure causes brain injury in the ovine fetus in the absence of hypoxemia. J Soc Gynecol Invest 2006 (13):87–96
6. Palsson-McDermott EM, O'Neill LA (2004) Signal transduction by the lipopolysaccharide receptor, Toll-like receptor-4. Immunology 113:153–162
7. Briscoe TA, Duncan JR et al. (2006) Activation of NF-κB transcription factor in the preterm ovine brain and placenta after acute LPS exposure. J Neurosci Res 83:567–574
8. Back SA, Han BH et al. (2002) Selective vulnerability of late oligodendrocyte progenitors to hypoxia-ischemia. J Neurosci 22(2):455–463
9. Mattson MP, Camandola S (2001) NF-κB in neuronal plasticity and neurodegenerative disorders. J Clin Invest 107(3):247–254
10. Eklind S, Mallard C et al. (2001) Bacterial endotoxin sensitizes the immature brain to hypoxic-ischaemic injury. Eur J Neurosci 13(6):1101–1106

Prenatal Brain Injury (PBI)

Definition

It involves fetal brain injury and inflammation.

►Prenatal Brain Injury by Chronic Endotoxin Exposure

Preoptic Area

Synonyms

►Area preoptica

Definition

Situated at the lateral wall of the third ventricle, close to the optic recess. The region comprises three nuclear areas: periventricular nucleus, medial preoptic nucleus and lateral preoptic nucleus.

The area is also in the direct vicinity of the organum vasculosum of the lamina terminalis and plays a role in thermoregulation, hypovolemic thirst, male sexual behavior, brood care, gonadotropin secretion and locomotion.

►Diencephalon
►Hypothalamus
►Nocturnal/Diurnal

Preparatory Postural Adjustments

►Anticipatory Postural Responses

Prepositus Hypoglossal Nucleus

Synonyms

►Nucl. prepositus hypoglossi

Definition

The rod-shaped nucleus rostral to the hypoglossal nerve in the oculomotor control center which plays an important role in tracking moving objects – with the eyes, coordination of fast eye movements and fixation on objects. The nucleus has afferents from the cerebellum, vestibular „ nuclei as well as the interstitial nucleus (Caj al). Efferents to the nuclei of the eye muscles.

►Myelencephalon

Prepulse Inhibition

Definition

A decrease of the response to a startle inducing stimulus, when the stimulus is preceded by a weak stimulus that does not induce a startle response.

►Startle Response

Prepyriform Cortex

►Olfactory Cortex

Pressure Ejection

Definition

►Microiontophoresis and Micropressure Ejection

Pressure Micro-ejection

►Microiontophoresis and Micropressure Ejection

Pressure Receptors

►Cutaneous Mechanoreceptors, Anatomical Characteristics

Presynaptic Inhibition

JORGE N. QUEVEDO
Department of Physiology, Biophysics and Neuroscience, Centro de Investigación y de Estudios Avanzados del I.P.N, Mexico City, Mexico

Synonyms

Decrease of synaptic effectiveness

Definition

Presynaptic inhibition (PSI) refers to a decrease of transmitter release at central synapses. Five decades ago, it was reported that activation of afferent fibers originating in flexors led to depression of monosynaptic group Ia ►excitatory postsynaptic potentials (EPSPs) evoked on extensor motoneurones in the cat spinal cord [1]. This depression occurred with no detectable changes in the time course of monosynaptic EPSPs, membrane potential or motoneurone excitability [1,2]. It is now known that PSI occurs broadly within the central nervous system of both vertebrates and invertebrates, and that synaptic efficacy at axon terminals from sensory afferents, descending systems or interneurones [3] can be subject to an inhibitory control by a number of neurotransmitters and presynaptic receptors [4].

Characteristics

PSI Associated with Primary Afferent Depolarization (PAD)

PAD as a Measure of PSI in the Spinal Cord

Stimulation of sensory nerves produces a slow negative potential recorded in dorsal roots via depolarization of intraspinal terminals of afferent fibers. This depolarization, termed ►primary afferent depolarization (PAD), is electrotonically propagated to the dorsal roots and can be recorded as a ►dorsal root potential (DRP). It can also be recorded intra-axonally in the dorsal horn. When the intra-axonal membrane potential produced by PAD reaches firing threshold, back-propagating action potentials are seen, and are called ►dorsal root reflexes (DRRs). Owing to the parallel time course of PSI and PAD, it has been assumed that PAD is responsible for PSI and that both phenomena are mediated by the same mechanisms [2,3].

Mechanisms Involved in the Generation of PAD

Pharmacological and electrophysiological studies have lead to the conclusion that PAD is primarily mediated by last-order GABAergic interneurones, and generated by the activation of presynaptic ionotropic $GABA_A$ receptors [2,3]. $GABA_A$ receptors are Cl^--permeable, and their activation drives the membrane potential towards the Cl^- reversal potential. Because a Na^+, K^+, 2-Cl^- cotransporter (NKCC1) maintains the Cl^- reversal potential less negative than the membrane potential, $GABA_A$ receptor activation leads to an efflux of Cl^- ions and consequent depolarization of afferent terminals, i.e. PAD (Fig. 1) [4]. The increase in Cl^- conductance reduces transmitter release by inactivation of voltage-gated Na^+ and Ca^{2+} channels, or by shunting the membrane current to prevent spike invasion into the axon terminals [3]. The depolarization produced by activation of $GABA_A$ receptors can also enhance spontaneous transmitter release in the dorsal horn neurons by activating voltage-gated Ca^{2+} channels [5].

Interneurones Mediating PAD

There is clear anatomical evidence for the existence of GABAergic axo-axonic synapses on ▶muscle spindle group Ia and II, muscle ▶Golgi tendon organ group Ib, and large diameter cutaneous afferents. However, a direct identification of interneurones mediating PAD is still lacking [3]. In addition to activation of conventional interneuronal circuits, PAD on myelinated and unmyelinated cutaneous afferents may be also accounted for by non-spiking microcircuits that involve dendroaxonic GABAergic contacts or GABA spill-over [6].

The shortest pathway mediating PAD of group I, group II and large diameter cutaneous afferents fibers is thought to be trisynaptic (Fig. 2), including first order excitatory and last-order GABAergic interneurones interposed [2,3]. The long duration of PAD (Fig. 1), suggests that last-order interneurones mediating PAD produce a prolonged burst of spikes following a brief afferent input. However, since a single interneurone spike is able to generate a similarly long-lasting DRP,

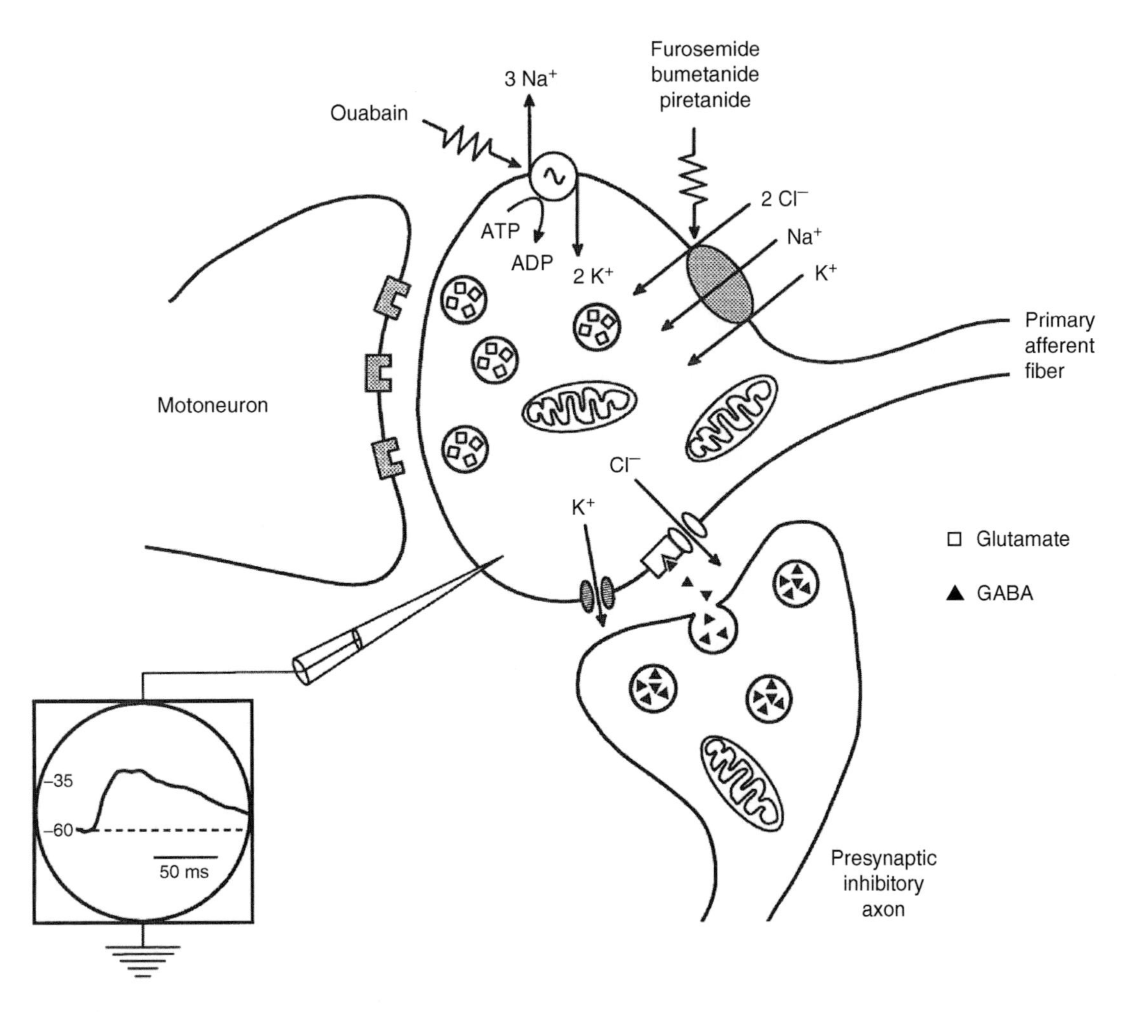

Presynaptic Inhibition. Figure 1 Schematic representation of a presynaptic inhibitory axon from a GABAergic interneurone contacting a primary afferent fiber. The activation of $GABA_A$ receptors present on afferent terminals drives the membrane potential towards the Cl^- reversal potential. Because a Na^+, K^+, 2-Cl^- cotransporter maintains the Cl^- reversal potential less negative than the membrane potential, $GABA_A$ receptor activation leads to an efflux of Cl^- ions and consequent depolarization of afferent terminals (i.e. PAD). This depolarization could be recorded theoretically by a microelectrode inserted in the terminal of the primary afferent. Na^+ concentration inside the terminal is regulated by the Na^+-K^+-ATPase pump. Several drugs that can inhibit the cotransporter are indicated. (From [4]).

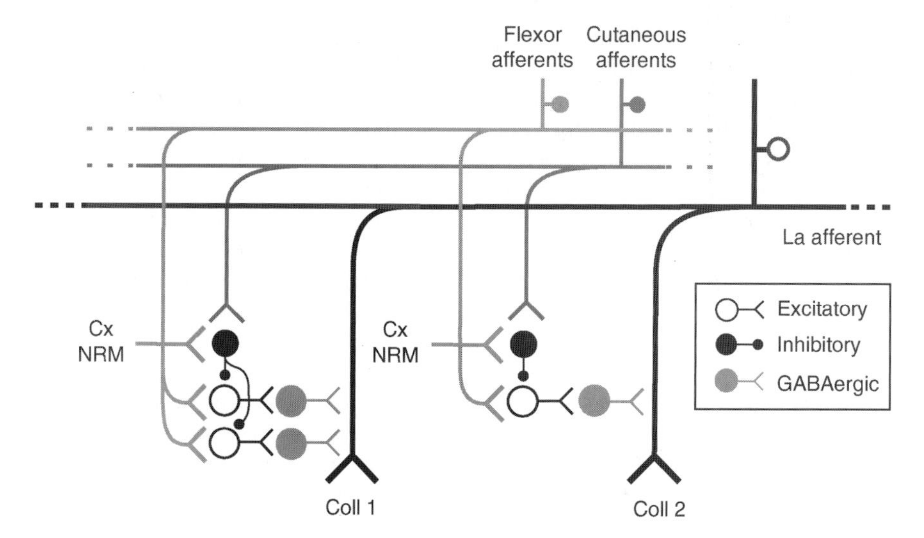

Presynaptic Inhibition. Figure 2 Schematic diagram illustrating the local character of primary afferent depolarization (PAD) in collaterals of a single muscle spindle (Ia) afferent (*black lines*) ending at two different spinal levels. PAD produced by stimulation of flexor afferents (*blue lines*) is mediated by at least two interneurones interposed. First-order and last-order interneurones are excitatory (*open circles*) and GABAergic (*gray circles*), respectively. PAD evoked in collateral 1 (Coll 1) and collateral 2 (Coll 2) is mediated by separate sets and number of interneurones. PAD evoked by stimulation of flexor afferents in each collateral is reduced by stimulation of cutaneous (*red lines*), and the corticospinal (Cx) and raphespinal (NRM) (*gray lines*) systems via different sets of inhibitory interneurones (*black circles*). From [7].

PAD may be due to slow kinetics of GABA release, action, or uptake [3].

Interneurones mediating PAD are spatially segregated in the spinal cord, generally located in regions corresponding to the termination sites of the various afferent modalities. PAD of cutaneous and group I muscle afferent fibers is mediated by interneurones located in the middle and lateral parts of Rexed laminae III-IV, and medial Rexed laminae V-VI, respectively. PAD of group II afferents is produced by separated sets of interneurones into dorsal horn and intermediate zone, and the PAD of group II afferents seen in mid-lumbar and sacral spinal segments is also mediated by distinct populations of interneurones [3,8].

Organization of Pathways Mediating PAD

The patterns of PAD evoked by stimulation of sensory and supraspinal inputs differ for each afferent modality, suggesting that interneuronal pathways mediating PAD are modulated in a rather selective and complex manner. However, a primary mechanism of recruitment of PAD appears associated with a negative feedback control of their activated afferents.

Muscle Afferents

PAD of Ia afferents is most effectively produced by stimulation of group Ia and group Ib flexor afferents, and vestibulospinal inputs; it is also inhibited by stimulation of various supraspinal (corticospinal, rubrospinal, reticulospinal and raphespinal serotonergic systems) and peripheral systems (large cutaneous and joint afferents) (Fig. 2). PAD of group Ib afferents occurs predominantly following activation of group Ib afferents from flexors and extensors, joint, and large cutaneous afferents (but may also be inhibited by these cutaneous afferents). In contrast to PAD of group Ia afferents, PAD of group Ib afferents is evoked by corticospinal, reticulospinal and raphespinal serotonergic systems, and also by the noradrenergic nucleus locus coeruleus, cerebellum and the vestibulospinal tract [3].

The different patterns of PAD of group Ia and group Ib afferent fibers produced by segmental and supraspinal inputs suggest that PAD is mediated by different sets of last-order GABAergic interneurones. Yet PAD of group I muscle afferents can be modulated differentially in afferent fibers ending at two separate spinal levels, or even in proximal collaterals of the same afferent fiber (Fig. 3). This local character of PAD is likely due to the activation of different sets of interneurons mediating PAD [3,7].

PAD of group II afferents is generally evoked by stimulation of group II, cutaneous, joint and pudendal afferents, as well as by the noradrenergic locus coeruleus and serotonergic raphespinal systems. It is most effectively evoked by group II afferents ending at the same spinal sacral or midlumbar segment. This implies that separate populations of interneurones mediate PAD of group II affetents projecting to these segments. PAD of group II afferents is generally stronger than PAD of group I afferents, indicating that PAD of group I and group II afferents is mediated by different sets of interneurons [3,8].

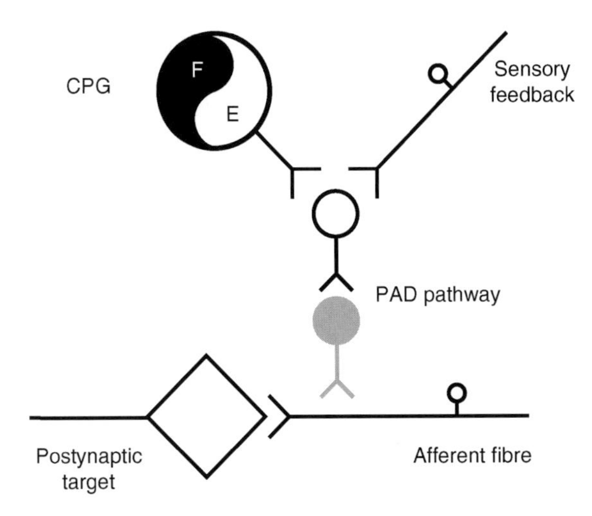

Presynaptic Inhibition. Figure 3 Diagram of systems modulating presynaptic inhibitory pathways during locomotion. Synaptic transmission from afferent fibers to postsynaptic targets (open diamond) is modulated by PAD-related PSI mechanisms. Inputs from the locomotor pattern generator (CPG) circuitry and the sensory feedback from limbs in movement modulate PAD pathways, which are mediated by first-order excitatory (open circles) and last-order GABAergic (gray circles) interneurones. PAD-related PSI mechanisms participate in shaping motor pattern activity to adapt movement limbs to the environment. Adapted from [9].

Cutaneous Afferents

PAD of low threshold cutaneous afferents is produced by stimulation of low threshold cutaneous, group Ib, group II, and high threshold muscle afferents, as well as by stimulation of corticospinal, reticulospinal, rubrospinal and the raphespinal serotonergic system. PAD of low threshold cutaneous afferents is larger in response to activation of their specific afferent modality (e.g. rapidly-adapting vs. slowly adapting). Therefore, PAD of low threshold cutaneous afferents plays an important role in sensory discrimination by enhancing contrast and eliminating surplus excitation. High threshold cutaneous afferents (Aδ and C) are more effectively depolarized by noxious stimulation, as well as by stimulation of reticulospinal and raphespinal serotonergic systems [3]. However, because of the scarcity of axo-axonic GABAergic contacts on Aδ and C fibers, spillover mechanisms may be involved in the generation of PAD in these fibers [3,6].

PAD Evoked by Non-GABAergic Mechanisms

The accumulation of K^+ ions in the extracellular space following activation of spinal interneurones was proposed as a mechanism to explain PAD. However, changes in extracellular K^+ cannot account for all of the characteristics of PAD associated with stimulation of peripheral nerves or supraspinal pathways [3].

In addition to $GABA_A$ receptors, primary afferents contain a diversity of receptors whose activation can evoke depolarization of terminals and in consequence PSI. Stimulation of high threshold afferent fibers evokes DRPs including a large component evoked by activation of NMDA and AMPA/kainate ionotropic glutamate receptors [6]. Activation of presynaptic AMPA and kainate receptors depresses excitatory transmission in dorsal horn neurons, but increases spontaneous transmitter release, both kinds of effects being apparently produced by PAD [5]. Capsaicin depresses excitatory synaptic transmission evoked by stimulation of C afferent fibers and this depression is also related to PAD [5].

The spinal cord receives diffuse projections from several descending monoaminergic nuclei that modulate PAD pathways. For instance, stimulation of the serotonergic nucleus raphe magnus produces PAD on group Ib afferent fibers [3]. However, because iontophoretic application of serotonin and noradrenaline produces no change in the intraspinal threshold of group I afferents, it is unlikely that PAD occurs by a direct effect of these monoamines on afferent fibers [3].

Stimulation of the noradrenergic locus coeruleus and the serotonergic raphe nucleus produces PAD on group II afferent fibers associated with a concomitant depression of synaptic transmission of the same afferents, suggesting participation of a presynaptic inhibitory mechanism. Nevertheless, the decrease in transmitter release appears not to be mediated by direct actions of monoamines on group II afferents, but indirectly via modulation of the interneuronal pathways mediating PAD [3,8].

Localization of serotonin and noradrenaline receptor subtypes is apparently restricted to high threshold cutaneous afferents [3,8]. Indeed, Aδ and C fibers are depolarized by direct application of serotonin, yet there is no clear association between PAD and decrease in synaptic effectiveness of these afferents [3].

PSI not Associated with PAD

While PAD is generated predominantly by the activation of $GABA_A$ receptors, it is well known that activation of presynaptic G-protein coupled receptors, such as $GABA_B$, can lead to PSI. This type of PSI is linked to intracellular signaling pathways that inhibit voltage-gated Ca^{2+} channels, with no associated presynaptic depolarization. Moreover, in contrast to the activation of $GABA_A$ receptors, activation of $GABA_B$ receptors produces a longer-lasting component of inhibition of the monosynaptic EPSPs [3].

Activation of metabotropic adenosine and cannabinoid receptors also inhibits transmitter release by inhibition of voltage-gated Ca^{2+} channels coupled to G-protein receptors [5]. The inhibitory effects of monoamines on the synaptic efficacy of high threshold cutaneous afferents may be also produced by activation of metabotropic receptor subtypes coupled to signal transduction pathways [3,8].

A long-lasting inhibition of the monosynaptic reflex has been associated with a ►homosynaptic depression of the synapse between group Ia afferents and motoneurones. This depression also occurs in the absence of changes in motoneurone excitability and results from postactivation transmitter depletion [3].

PSI and Function

PSI pathways are modulated during real and fictive motor behaviors. During ►fictive locomotion (i.e. in the absence of sensory information), there is a tonic PAD accompanied by an overall synaptic depression of group I, group II and cutaneous afferents. Depression of group I afferent transmission is associated with the reductions in gain of stretch reflexes and the H-reflex observed at the onset of locomotion in humans. Superimposed on the tonic PAD, there are phase-dependent fluctuations of the membrane potential in group I, group II and cutaneous afferents, with more depolarization during the flexion phase. Synaptic transmission from group I, group II and cutaneous afferents also displays a phase-dependent modulation during fictive locomotion. However, unlike the synaptic depression observed during tonic PAD, phase-dependent PAD is not correlated with a depression of synaptic transition from Ia afferents onto motoneurones. Both tonic and phase-dependent oscillations of PAD are mediated by as yet unknown mechanisms [9].

PAD evoked by stimulation of sensory afferents is also rhythmically modulated during fictive locomotion, suggesting that the locomotors central pattern generator (CPG) circuitry has access to the pathways mediating PAD (Fig. 3). The pattern of modulation of sensory evoked-PAD is complex and appears to depend more on the specific postsynaptic targets than on the peripheral origin of the afferents. During real as opposed to fictive locomotion, the patterns of PAD depend on interactions between supraspinal, sensory feedback and CPG inputs on pathways mediating PAD. PAD-related PSI mechanisms play an important role during patterning generation and contribute in shaping motor pattern activity to adapt limb movements to the environment [9].

During voluntary muscle contractions in humans there is a decrease of PSI of afferent terminals contacting the active motoneurone pools and a concomitant enhancement of PSI of afferent terminals ending on inactive motoneurone pools. These observations indicate that corticospinal control of PSI selectively "opens" transmission in group Ia afferents to voluntarily activated motoneurones, while "closing" transmission to motoneurones of relaxed muscles [10]. Findings in humans are supported by studies in the cat showing that different collaterals of the same afferent fiber are selectively controlled by separate sets of last-order interneurones mediating PAD (Fig. 2) [8]. The differential PAD-related PSI of individual collaterals of muscle afferents allows a selective control of sensory information according to the neuronal targets and to the motor task to be performed [8,10].

References

1. Frank K, Fortes MGF (1957) Presynaptic and postsynaptic inhibition of monosynaptic reflexes. Fed Proc 16:39–40
2. Eccles JC (1964) Presynaptic inhibition in the spinal cord. Prog Brain Res 12:65–91
3. Rudiment P, Schmidt RF (1999) Presynaptic inhibition in the vertebrate spinal cord revisited. Exp Brain Res 129:1–37
4. Alvarez-Legman's FJ, Nana A, Marquez S (1998) Chloride transport, osmotic balance, and presynaptic inhibition. In: Rudiment P, Room R, Mendel L (eds) Presynaptic inhibition and neural control. Oxford University Press, New York, pp 50–79
5. Engelmann HS, McDermott AB (2004) Presynaptic ionotropic receptors and control of transmitter release. Nat Rev Neurosci 5:135–145
6. Russo RE, Delgado-Lucama R, Hungary J (2000) Dorsal root potential produced by a TTX-insensitive microcircuitry in the turtle spinal cord. J Physiol 528:115–122
7. Lomelí J, Quevedo J, Linares P, Rudomin P (1998) Local control of information flow in segmental and ascending collaterals of single afferents. Nature 395:600–604
8. Jankowska E (2001) Spinal interneuronal systems: identification, multifunctional character and reconfigurations in mammals. J Physiol 533:31–40
9. Rossignol S, Dubuc R, Gossard JP (2006) Dynamic sensorimotor interactions in locomotion. Physiol Rev 86:89–154
10. Pierrot-Deseilligny E, Burke D (2005) Presynaptic inhibition of Ia terminals. In: Pierrot-Deseilligny E, Burke D (eds) The circuitry of spinal cord. Cambridge University Press, England, pp 337–383

Presynaptic Proteins

Michael Seagar
INSERM/Université de la Méditerranée, UMR641, Faculté de Médecine Nord, Marseille, France

Definition

Proteins that participate in neurotransmission by ensuring functions specific to the presynaptic compartment (i.e., nerve terminals). Typically, these proteins control the progression of synaptic vesicles through the steps (docking, priming, vesicle fusion) that lead up to the release of neurotransmitters into the synaptic cleft.

Characteristics

Nerve terminals concentrate neurotransmitters in synaptic vesicles. Neurotransmitter release into the synaptic cleft then occurs when Ca^{2+} influx triggers the fusion of these vesicles with the presynaptic plasma membrane. However, only a small fraction of the total vesicle population within a nerve terminal is available to respond to the Ca^{2+} signal [1]. The presynaptic proteins described in this section, control transmitter release by holding vesicles in reserve or alternatively preparing and then accomplishing fusion.

Synaptic SNARE Proteins and SNARE Complexes

Vesicle fusion involves assembly of *trans* ►SNARE complexes at the interface between a docked synaptic vesicle and the plasma membrane, a process that pulls the opposing membranes together. These *trans* complexes contain three proteins: one vesicle protein, the vesicular- or v-SNARE VAMP 2 (also known as ►synaptobrevin) and two plasma membrane proteins, the target- or t-SNAREs: ►syntaxin 1 and ►SNAP-25 [1,2]. VAMP and syntaxin have a similar topology with membrane anchors at their C-terminal extremities and the N-termini orientated towards the cytoplasm. SNAP-25 is attached to the membrane via palmitoylation of amino acids located in the middle of the sequence, thus both its N- and C-termini are cytoplasmic. SNARE proteins carry a characteristic 70 amino acid sequence, the SNARE motif. This sequence mediates the interactions with partner SNAREs to form complexes [3]. SNAP-25 has two SNARE motifs while VAMP and syntaxin have only one each.

The N-terminal portion of syntaxin constitutes an auto-inhibitory domain, which folds back to interact with and mask the membrane-proximal helical (H3) domain containing the SNARE motif. In this "closed" conformation, syntaxin cannot bind to SNAP-25 and VAMP. When syntaxin "opens," four SNARE motifs (1 from VAMP, 1 from syntaxin, 2 from SNAP-25) can assemble into a four-helical bundle forming the SNARE complex. In this bundle, the alpha helices are in a parallel orientation with all the N-termini at one end and all the C-termini at the other. Zipping-up of the complex, starting at the N-termini and progressing towards the C-termini pulls the C-terminal membrane anchors of v- and t-SNAREs, and consequently the vesicular and plasma membranes, towards each other. This leads to mixing of the outer lipid leaflet of the vesicle with the inner leaflet of the plasma membrane, and can then progress to full fusion upon Ca^{2+} influx. However, the SNARE proteins themselves do not directly bind Ca^{2+}. The fusion machine probably comprises of a radial array of SNARE complexes at the synaptic vesicle/plasma membrane interface, associated with Ca^{2+}-sensor proteins such as the synaptotagmins that confer Ca^{2+}-dependency on the release process.

The synaptic SNARE proteins are the targets of botulinum neurotoxins (of which there are seven distinct types: BoNT/A to G) and tetanus neurotoxin (TeNT) [4]. These extremely potent toxins inhibit transmitter release by introducing their light chains into the nerve terminal cytoplasm. The light chains are metalloproteases that specifically cleave synaptic SNAREs at defined peptide bonds (e.g., BoNT/B & TeNT - VAMP2, BoNT/A & E - SNAP-25, BoNT/C - syntaxin 1). BoNTs and TeNT do not affect vesicle docking but block synaptic vesicle fusion by a selective proteolytic attack on the fusion machine.

In addition to driving membrane fusion, *trans* v-SNARE/t-SNARE pairing may also contribute to proofreading, ensuring that vesicles can only fuse with the appropriate target membrane. Families of SNAREs have been identified in many other subcellular compartments in a variety of eukaryotic cells. Vesicular transport thus uses the same basic fusion machinery, conserved through evolution from yeast to human nerve terminals.

Presynaptic proteins include several molecules that bind to SNARE proteins and regulate the assembly of SNARE complexes.

Munc-18

►Munc-18 (also called nsec1 or rbsec1 in mammals) is a 67 kDa hydrophilic protein, which associates with nerve terminal membranes via it's interaction with syntaxin 1. Munc-18 genes are the mouse members of the SM (Sec1/Munc-18) gene family, which includes yeast Sec1 and the nematode and drosophila orthologues Unc-18 and Rop [5]. The precise mode of action of Munc18 is unknown. Munc-18 binds tightly to the "closed" conformation of syntaxin 1 preventing SNARE complex assembly in vitro, but it is not simply a negative regulator of vesicle fusion. Genetic manipulations that decrease SM protein levels generally diminish secretion, consistent with the notion that these proteins are required for fusion. Yeast *SEC1* was initially identified as a gene required for exocytosis. However, in contrast to Munc-18, sec1p does not bind to monomeric yeast syntaxins but only to assembled SNARE complexes. SM proteins may act as chaperones to favor SNARE complex formation. Thus, although Munc-18 binds to "closed" syntaxin, it might, in concert with additional factors, be involved in opening it and thus participate in priming. Munc-18 binds to additional presynaptic proteins including the Munc interacting proteins (Mint 1 and 2). Mints link the vesicle fusion apparatus to adhesive proteins in the presynaptic plasma membrane that are involved in establishing synaptic connections.

Munc-13

►Munc-13 proteins constitute a family of three high molecular weight (200 kDa) molecules. Munc13–1 and

Munc13–3 are brain specific, while Munc13–2 is ubiquitous. At the N-terminal extremity, Munc-13 contains a conserved Ca^{2+}-dependent calmodulin binding site followed by a C1 motif, which binds the lipid messenger diacylglycerol, and two C2 domains [6,7]. Munc-13 (Unc-13 in the nematode) is involved in synaptic vesicle priming. Gene deletions in both mice and nematodes principally affect the readily releasable pool of synaptic vesicles. Munc-13 probably acts by binding to the auto-inhibitory N-terminal domain of syntaxin 1. In the nematode Unc13 displaces Unc-18, which is bound to the closed state of syntaxin. Furthermore, expression of mutant syntaxin, frozen in a permanently open conformation, rescues worms with an Unc-13 mutation. Thus, munc-13 appears to act downstream of munc-18 by driving syntaxin from a closed to an open conformation. Regulation of Munc-13 action via Ca^{2+}/calmodulin or diacylglycerol binding has been implicated in presynaptic plasticity.

Synaptophysin

►Synaptophysin was the first synaptic vesicle membrane protein to be isolated and cloned [8]. It is a major component of synaptic vesicles, thus anti-synaptophysin antibodies are widely used in immunocytochemistry to identify nerve terminals, and to evaluate diagnostically the neuroendocrine phenotype of a variety of tumors. Synaptophysin is an N-glycosylated 38 kDa protein. Like its homologue physins, synaptoporin, pantophysin and mitsugumin 29, it contains four membrane-spanning segments with N- and C-termini orientated towards the cytoplasm. It is the major cholesterol-binding protein in synaptic vesicles, and may contribute to the induction of vesicle curvature during vesicle biogenesis. Synaptophysin forms a complex in the vesicle membrane with the v-SNARE VAMP and subunits of the vacuolar proton pump (V-ATPase). The synaptophysin – VAMP2 complex prevents VAMP from entering into SNARE complex assembly. It thus constitutes an additional molecular mechanism for regulating transmitter release by determining v-SNARE availability. The amount of VAMP sequestered by synaptophysin is modulated by neuronal activity, indicating that it is functionally implicated in plasticity. However, deletion of the synaptophysin gene in mice does not result in a significant phenotype, which may reflect compensation by other members of the physin family. Mice that lack both synaptophysin and the structurally related tetraspan vesicle protein synaptogyrin do display defects in synaptic plasticity.

Additional presynaptic proteins mediate interactions of synaptic vesicles with the cytoskeletal elements. In this way, they seem to retain vesicles in a reserve pool that does not participate in regular vesicle exo-endocytotic recycling, unless mobilized by intense stimulation.

Synapsins

►Synapsins are neuron-specific phosphoproteins that are among the most abundant synaptic vesicle proteins. Three synapsin genes (I-III) have been identified in mammals; each undergoes differential splicing to yield at least nine isoforms [1]. The N-terminal domains of synapsins are highly conserved, containing sites for phosphorylation by cAMP- and Ca^{2+}/calmodulin-dependent protein kinase while the C-terminal portions are variable. The central domain, which accounts for more than half the sequence, forms multimers and binds ATP. This region has structural similarities to some ATPases, suggesting that synapsin may have enzymatic activity requiring ATP.

Neither deletion of the single synapsin gene in the Drosophila genome, nor the deletion of individual synapsin genes in the mouse genome, has strong effects on synaptic transmission. Deletion of all three synapsin mouse genes is not lethal but does affect behavior. It induces changes in synaptic transmission and plasticity that differs in excitatory versus inhibitory synapses. The functions of the synapsins are unclear, but may include recruiting synaptic vesicles to a reserve pool. Synapsins associate with the surface of vesicles, and bind to both vesicular phospholipids and proteins. They also interact with cytoskeletal proteins including actin, spectrin and tubulin. These binding reactions are regulated by phosphorylation. Synapsins thus seem to be involved in tethering vesicles to the cytoskeleton and defining a reserve pool that is not immediately available for docking and fusion. Synaptic activity leads to phosphorylation of synapsins, which dissociate from vesicles allowing their mobilization and migration to the plasma membrane to prepare for fusion. However, synapsins may have additional functions. Mice with deleted synapsin I and II genes show a global decrease in the number of synaptic vesicles, suggesting that synapsins play a role in stabilizing vesicles. Finally, synapsins may also act at a step closer to fusion by regulating priming or fusion competence.

Synaptic Vesicle Protein 2 (SV2)

►SV2 is a synaptic vesicle glycoprotein of about 90 kDa, containing twelve potential transmembrane segments and N- and C-termini oriented towards the cytoplasm [1]. Sequences linking transmembrane regions are fairly short, although one highly glycosylated intraluminal loop is considerably larger. The sugar chains in this loop may contribute to the intravesicular matrix that is thought to bind neurotransmitters. Three SV2 genes in vertebrates encode homologous SV2A, SV2B and SV2C proteins that display distinct expression patterns in the brain, although SV2A is present in most neurons. The transmembrane and cytoplasmic linker sequences are highly homologous between different SV2s, whereas the N-terminal domain and the intraluminal linkers diverge.

SV2s have significant homology to the Major Facilitator Superfamily of transporters for organic anions and cations, phosphates and sugars, and were initially thought to be transporters of neurotransmitters. However, their ubiquitous distribution rules out this possibility and indicates that they fulfill a more general function common to all synaptic vesicles. This function is still unknown, although a reasonable hypothesis is that they transport an unidentified substrate from the cytoplasm into the vesicle lumen.

Mice with deletions of the SV2A gene and the double SV2A/SV2B knockout display severe seizure activity and die postnatally. Cultured neurons from knock-out mice display increases in Ca^{2+}-dependent synaptic transmission that can be blocked by buffering cytoplasmic Ca^{2+}. Lack of SV2 thus seems to lead to abnormally high cytoplasmic Ca^{2+} levels. SV2 might therefore be involved in Ca^{2+} transport into synaptic vesicles to counterbalance the effects of burst firing in the nerve terminal. Furthermore, the N-terminal domain of SV2 interacts with, and may regulate the function of, the synaptic vesicle calcium sensor protein synaptotagmin. SV2 also constitutes the binding site for the anti-epileptic drug levetiracetam (KEPPRA), which perhaps enhances the ability of SV2 to limit electrical hyperexcitability [9].

Pathology

Botulism and Tetanus

These two potentially fatal diseases result from the inhibition of neurotransmitter release by BoNT and TeNT, which proteolyze presynaptic SNARE proteins. BoNT and TeNT are produced by the soil-borne bacteria *Clostridium botulinum* and *Clostridium tetani* [4]. Human botulism, mainly due to BoNT serotypes A, B and E, is typically caused by either eating contaminated food, wound infection, or in infants by intestinal proliferation of ingested clostridial spores. Botulism is a flaccid paralysis with classic symptoms: double/blurred vision, drooping eyelids, slurred speech, difficulty in swallowing, dry mouth, and muscle weakness. In severe cases, death can result from respiratory failure. Tetanus, which typically results from contamination of a deep puncture wound with *C. tetani*, involves muscular hypercontraction with symptoms including headache, muscular stiffness in the jaw (lockjaw) and neck, difficulty in swallowing, rigidity of abdominal muscles, spasms, sweating, fever and convulsions.

The differences between botulism (flaccid paralysis) and tetanus (muscle contractions and spasms) are due to differences in the neuronal populations affected. BoNT inhibits motoneurone terminals, blocking acetylcholine release and subsequent muscle contraction. In contrast, TeNT acts specifically on inhibitory neurones in the spinal cord, diminishing tonic inhibition that indirectly activates acetylcholine release from motoneurones inducing muscle contractions.

Therapy

BoNTs (usually BoNT/A or B) are used as therapeutic agents that can be injected into muscles to reduce hypercontractions in dystonia, blepharospasm, strabism and a multitude of other indications. They are also widely used cosmetically (e.g., Botox) to reduce the muscle activity that underlies facial wrinkles.

References

1. Sudhof TC (2004) The synaptic vesicle cycle.Annu Rev Neurosci 27:509–547
2. Sollner T, Whiteheart SW, Brunner M, Erdjument-Bromage H, Geromanos S, Tempst P, Rothman JE (1993) SNAP receptors implicated in vesicle targeting and fusion. Nature 362:318–324
3. Sutton RB, Fasshauer D, Jahn R, Brunger AT (1998) Crystal structure of a SNARE complex involved in synaptic exocytosis at 2.4 A resolution. Nature 395:347–353
4. Turton K, Chaddock JA, Acharya KR (2002) Botulinum and tetanus neurotoxins: structure, function and therapeutic utility. Trends Biochem Sci 27:552–558
5. Toonen RFG, Verhage M (2003) Vesicle trafficking: pleasure and pain from SM genes. Trends Cell Biol 13:177–186
6. Gerst JE (2003) SNARE regulators: matchmakers and matchbreakers. Biochim Biophys Acta 1641:99–110
7. Junge HJ, Rhee JS, Jahn O, Varoqueaux F, Spiess J, Waxham MN, Rosenmund C, Brose N (2004) Calmodulin and Munc13 form a Ca^{2+} sensor/effector complex that controls short-term synaptic plasticity. Cell 118:389–401
8. Valtorta F, Pennuto M, Bonanomi D, Benfenati F (2004) Synaptophysin: leading actor or walk-on role in synaptic vesicle exocytosis? Bioessays 26:445–453
9. Lynch BA, Lambeng N, Nocka K, Kensel-Hammes P, Bajjalieh SM, Matagne A, Fuks B (2004) The synaptic vesicle protein SV2A is the binding site for the antiepileptic drug levetiracetam. Proc Natl Acad Sci USA 101:9861–9866

Pretectal Area

Synonyms

►Area pretectalis

Definition

Situated immediately behind the superior colliculus, this nucleus plays a vital role in pupillary reflex and adaptation. Afferents come from the retina and

P

occipital cortical fields. Efferents go to the ipsi- and contralateral accessory oculomotor nucleus and superior colliculus.

The pretectal area includes the following nuclei: pretectal olivar nucleus, medial, anterior and posterior pretectal nuclei and optic tract nucleus.

►Mesencephalon

Pretectal Nuclei

Definition

Part of the subcortical visual shell. The pretectum consists of 7 nuclei in the visual midbrain, the n. of the optic tract, posterior limitans n., the olivary pretectal n., the anterior pretectal n., the posterior pretectal n., the medial pretectal n. and the commissural pretectal n. All connect substantially with the intergeniculate leaflet (IGL) and with each other.

They serve a variety of different functions. Perhaps the most well known is control of the pupillary light reflex by the olivary pretectal nucleus. This reflex is mediated by intrinsically photoreceptive retinal ganglion cells, as well as the classical rod/cone photoreceptors.

►Intergeniculate Leaflet
►Neural Regulation of the Pupil
►Photoreceptors
►Pupillary Light Reflexes
►Retinal Ganglion Cells

Pretectum

Definition

A midbrain region just rostral (forward) of the superior colliculus. It consists of a superficial nucleus (the nucleus lentiformis in frogs) containing migrated, large-celled neurons and a deep subnucleus. Cells of the nucleus lentiformis mesencephali are involved in horizontal optokinetic nystagm. Pretectal neurons respond more selectively to slowly moving vertical patterns, although horizontally sensitive neurons likewise been reported.

Directional information is encoded in a large population of motion-sensitive units, which includes both narrowly and broadly tuned individual response profiles.

►Evolution of the Visual System: Amphibians
►Optokinetic Nystagmus
►Superior Colliculus

Prevertebral Ganglia

Definition

Prevertebral ganglia are autonomic ganglia that are found in the abdomen, around the origins from the aorta of the major vessels – the coeliac, superior mesenteric and inferior mesenteric arteries – and in the pelvis, close to the pelvic organs. The ganglia have various names applied to them the most common being: coeliac ganglia (also semilunar ganglia, solar plexus), superior mesenteric ganglia (inter-renal ganglia), inferior mesenteric ganglia and anterior (hypogastric) and posterior pelvic ganglia.

►Autonomic Ganglia

Prey-catching Behavior

Definition

Hunting animals show this type of behavior when trying to catch prey. This involves searching behavior, localization of prey and striking as well as killing of prey.

PRG

Definition

The Pontine Respiratory Group (PRG) is the region of respiratory neuron groups situated in the nucleus parabrachialis medialis (NPBM) and Kölliker-Fuse (KF) nuclei. The PRG is also termed as the pneumotaxic center.

►Respiratory Neurotransmitters and Neuromodulators

Primary Acoustic Cortex

Definition

Cerebral cortex areas in which the auditory tract terminates and which are involved in the first cortical processing steps for auditory signals. These include especially Brodmann areas 41 and 42 on the temporal plane.

► Telencephalon

Primary Afferent Depolarization (PAD)

Definition

A prolonged decrease in membrane potential usually produced by the activation of ionotropic $GABA_A$ receptors at the presynaptic terminals of afferent fibers.

This depolarization is propagated antidromically in an electrotonic manner and can be detected either (i) directly by intra-axonal recordings of afferent fibers, (ii) as a dorsal root potential (DRP) from the central stump of a cut dorsal root filament, or (iii) as a positive potential (P wave) from the cord dorsum. Given that intra-axonal recordings are normally obtained in the dorsal horn, PAD recorded intra-axonally reflects the summed membrane potential changes occurring in all the collaterals of an individual afferent fiber.

PAD is accompanied by an enhanced excitability of afferent fiber terminals. Changes in excitability of intrapinal terminals can be estimated from threshold changes in response to intraspinal current pulses. The activation of pathways mediating PAD produces a threshold decrease and a consequent increase in amplitude of the antidromic compound action potential (Wall's technique).

► Action Potential
► GABA
► Presynaptic Inhibition

Primary Afferents

Definition

Primary afferents are tracts ascending without interneurons.

One example is the posterior spinocerebellar tract which conducts impulses without interneurons from Clarke's nucleus of the thoracic cord to the cerebellar hemispheres.

► General CNS

Primary Cultures

Definition

Primary cultures are the first stage of cell culture in which cells removed from tissue are cultured but before cells are removed from the primary culture to start the next culture.

Primary Hyperalgesia

Definition

Hyperalgesia at a site of injury or inflammation.

► Hyperalgesia and Allodynia
► Joint Pain

Primary Lateral Sclerosis (PLS)

Definition

PLS designates a syndrome of progressive upper motor neuron dysfunction and shows consistent differences from ► amyotrophic lateral sclerosis (ALS), which is characterized by progressive degeneration of both upper and lower ► motoneurons. Unlike ALS, which is familial in 5-10% of the cases, PLS appears to be sporadic in adults. Initially, stiffness is more prevalent in PLS than in ALS patients, but limb wasting, pronounced in ALS patients, is rare in PLS patients. ► Spasticity is the most prominent sign in PLS. Cortical atrophy is most pronounced in the ► precentral gyrus and expands into the ► parietal-occipital region. The course of PLS is very slowly progressive.

► Amyotrophic Lateral Sclerosis (ALS)

Primary Motor Cortex (M1)

Definition

A part of the cerebral cortex that is most directly involved in controlling the activity of motoneurons. In primates, primary motor cortex is located in the precentral gyrus of the frontal lobe just anterior to the central sulcus. An orderly representation of movements by body part exists within primary motor cortex. From medial to lateral the representation is lower extremity, trunk, upper extremity, face and tongue. Many corticospinal neurons in primary motor cortex make monosynaptic linkages with motoneurons. A distinctive feature of primary motor cortex is the presence of large pyramidal cells (Betz cells) in cortical layer V.

►Motor Cortex: Output Properties and Organization

Primary Progressive Aphasia

Definition

Presenile progressive degenerative disorder characterized by initially isolated loss of language abilities for at least two years, but mostly transcending into muteness and ►dementia (often ►frontotemporal dementia). It is due to a degeneration of cerebro-cortical regions around the ►sulcus lateralis (Sylvii) of the left hemisphere. The most frequent variants are: (i) progressive non-fluent aphasia (characterized by labored speech, agrammatism); (ii) semantic aphasia (characterized by loss of word and object meaning and dyslexia); (iii) logopenic progressive aphasia (characterized by word-finding problems and impaired sentence comprehension).

Primary Reinforcer

Definition

Sensory stimulus with intrinsic rewarding properties such as food or water.

►Operant Conditioning

Primary Sjögren's Syndrome

Definition

A chronic, multisystem autoimmune disorder characterized by dryness of mouth and other mucous membranes that occurs in the absence of an associated rheumatic disease (secondary Sjögren's syndrome) and may involve extraglandular manifestations such as arthralgias, Raynaud's syndrome, pulmonary involvement, renal tubular acidosis, peripheral and central nervous system (CNS) disease.

►Central Nervous System Disease in Primary Sjögren's Syndrome

Primary Somatosensory Cortex (S1)

Definition

Primary somatosensory cortex comprises four cytoarchitectonic regions, areas 3a, 3b, 1 and 2 (going from rostral to caudal), located in the anterior portion of the parietal lobe. Each region has a complete representation of the body. Areas 3a and 2 receive inputs mainly from deep receptors (the hand representation is characterized by receiving inputs from both skin and deep receptors), while neurones in areas 3b and 1 largely respond to skin stimulation.

►Somatosensory Cortex I (SI)
►Somatosensory Cortex, Plasticity

Primary Visual Cortex

Definition

Cerebral cortex in the occipital lobe of the mammalian brain.

The primary visual cortex (also known as Brodmann area 17, V1 or striate cortex) is the principal site at which visual information enters the cortex. It lies in the calcarine sulcus of the occipital lobe and receives visual signals from the retina relayed via the lateral geniculate nucleus (LGN) of the thalamus. Efferent projections terminate in various visual areas in the cortex. Primary visual cortex has a complete map of visual space.

▶Geniculo-Striate Pathway
▶Striate Cortex Functions
▶Evolution of the Visual System: Mammals-Color Vision and the Function of Parallel Visual Pathways in Primates

Priming

Definition

Priming refers to the effect where a prior exposure to a stimulus exerts influences on a subsequently given stimulus. Typically, priming produces a faster and/or a more accurate response to a stimulus associated with a previously presented stimulus called prime.

▶Latent Learning
▶Long-Term Memory

Primordium Hippocampi

▶Evolution of the Pallium: in Amphibians

Principle of Neurological Minimization

Definition

Underlies responses of the neuromuscular system to changes in control variables and/or external forces; a tendency of system's elements to minimize, in the limits defined by neural, biomechanical and environmental constraints, the imposed activity by returning the system to the same or bringing it to a new steady state, depending on conditions; a solution to the redundancy problem.

▶Equilibrium Point Control

Principle of Virtual Work

Marcelo Epstein
Schulich School of Engineering, University of Calgary, Calgary, AB, Canada

Definition

From the physical point of view, the ▶principle of virtual work is an attempt to characterize unequivocally an equilibrium ▶configuration of a mechanical system (as defined in statics (q.v.)) by observing how it reacts to a small kinematical perturbation, called a *virtual displacement*.

Description of the Theory

In ▶classical mechanics (q.v.), a mechanical system is defined from the kinematical point of view by means of its *configuration space*, whose dimension is an expression of the number of *degrees of freedom* of the system. A (local) system of coordinates in the configuration space is known as a set of *generalized coordinates*. It is important to stress that the generalized coordinates $q^i (i = 1, \ldots, n)$ are mutually independent, since they already incorporate any geometrical constraints that the system may have. In the case of the double planar pendulum discussed in the article on classical mechanics (q.v.), for example, the number of degrees of freedom of the system is exactly two and so is the number of its generalized coordinates, whether the angular deviations from the vertical or any other set of appropriate geometrical parameters that are mutually independent and sufficient to define a state of the system are chosen. Another way to state this independence is to say that any description of a possible configuration of the system by means of coordinates, $x^I (I = 1, \ldots, n' > n)$ say, can be eventually boiled down to n' (smooth) functions of n generalized coordinates:

$$x^I = x^I(q^1, \ldots, q^n), \ (I = 1, \ldots, n'). \qquad (1)$$

In the case of the planar double pendulum, a system of Cartesian coordinates may be chosen and an arbitrary configuration of the system expressed by means of the two pairs of coordinates, each corresponding to the position of one of the two masses. Thus, in this simple case, $n' = 4$. Nevertheless, it is clear that (given the lengths of the two links) all four Cartesian coordinates can be expressed in terms of the two angles, q^1 and q^2, formed by the links and the vertical direction. The original four coordinates are interdependent, since they must satisfy the two constraints imposed by the assumed rigidity of the links. The generalized coordinates, on the other hand, by their very nature, already take these constraints into consideration and can, therefore, be varied independently without violating these constraints. A *virtual displacement* of a mechanical system consists precisely in a set of small (infinitesimal) perturbations or *variations* $\delta q^i (i = 1, ..., n)$ of the generalized coordinates away from a given configuration $q^i (i = 1, ..., n)$. The forces acting on the system, whether internal or external, will in general perform work on any given virtual displacement, a scalar quantity appropriately designated as *virtual work*. In many cases, however, the forces necessary to maintain the constraints will not perform any virtual work. For example, in the case of the pendulum it is clear that the constancy of the length of the links is physically attained by intermolecular forces, which result in an internal state of tension. The work of these tensile forces would be equal to the magnitude of the force multiplied by the change of length of the corresponding link. Since, by their very definition, virtual displacements *respect the geometrical constraints* (the constancy of the length of the links) the virtual work of these internal *forces of constraint* will vanish. For a different example, consider a body that is constrained to slide freely on a surface. As long as there is no friction, the force necessary to maintain the contact between the body and the surface will be perpendicular to the surface and, therefore, will perform no virtual work on any virtual displacement. If friction is considered, however, this will no longer be the case, and it is a philosophical point of view whether or not the frictional force should be called a force of constraint. Be that as it may, it is clear that by purposely defining a virtual displacement as a possible displacement of the system, namely one that respects its geometrical constraints, the virtual work of some forces will automatically cancel out. If therefore, an equilibrium configuration is characterized by means of some property of the virtual work, it will follow that the forces of constraint will automatically play no role in the determination of equilibrium, a feature that would be almost unthinkable in an approach based on the concept of a free-body diagram of statics (q.v.). The principle of virtual work, in fact, provides such a characterization.

According to the principle of virtual work, a mechanical system in classical mechanics (q.v.) is in an equilibrium configuration if, and only if, the virtual work of all the forces acting on the system vanishes *identically* for all possible virtual displacements that can be impressed away from this configuration. This principle, therefore, characterizes an equilibrium configuration as one for which the system is work-wise indifferent to small perturbations compatible with its constraints. It is important to notice that the principle of virtual work is not an equation but an identity. Only so can it be understood that a single scalar statement is equivalent to a number of vector equations. For consistency, the principle of virtual work, when applied to a rigid plate in two dimensions, will be shown to deliver the classical equations of equilibrium. The plate is assumed to be free to move in the *x*-*y* plane (no other constraints) and to be acted upon by N concentrated forces $\mathbf{F}^{\alpha} (\alpha = 1, ..., N)$ acting at points with coordinates $x_{\alpha}, y_{\alpha} (\alpha = 1, ..., N)$. In this case, the two Cartesian coordinates, x_p and y_p, of a point P of the plate and the (counter-clockwise, say) angular deviation θ of a material line drawn in the plate from, say, the x-axis can be adopted as generalized coordinates. A virtual displacement consists of any arbitrary combination of variations $\delta x_p, \delta y_p, \delta\theta$. The virtual displacement of the point of application of the force $\mathbf{F}^{\alpha}$ is the vector with components:

$$(\delta x_p - (y_{\alpha} - y_p)\delta\theta,\ \delta y_p + (x_{\alpha} - x_p)\delta\theta), \quad (2)$$

so that the principle of virtual work can be stated as:

$$VW = \sum_{\alpha=1}^{N} F_x^{\alpha} (\delta x_p - (y_{\alpha} - y_p)\delta\theta) + F_y^{\alpha}(\delta y_p + (x_{\alpha} - x_p)\delta\theta) \equiv 0, \quad (3)$$

with an obvious notation. Since this is an identity, any combination of $\delta x_p, \delta y_p, \delta\theta$ may be chosen. Choosing first $\delta x_p \neq 0, \delta y_p = \delta\theta = 0$:

$$\sum_{\alpha=1}^{N} F_x^{\alpha} = 0. \quad (4)$$

The choice $\delta x_p = 0, \delta y_p \neq 0, \delta\theta = 0$ yields:

$$\sum_{\alpha=1}^{N} F_y^{\alpha} = 0. \quad (5)$$

Finally, the choice $\delta x_p = \delta y_p = 0, \delta\theta \neq 0$ yields:

$$\sum_{\alpha=1}^{N} -F_x^{\alpha} (y_{\alpha} - y_p) + F_y^{\alpha}(x_{\alpha} - x_p) = 0. \quad (6)$$

Equations 4, 5 and 6 are immediately recognized as the standard equations of equilibrium of a rigid body in two dimensions. In particular, Eq 6 is the equation of moment equilibrium around point P. Considering now the case in which the plate is hinged at point P by means of a frictionless hinge, the configuration space of the system becomes one-dimensional and the rotational coordinate θ is a valid generalized coordinate. Obviously, in this case Eqs. 4 and 5 are no longer valid, and the equilibrium of the system is governed by Eq. 6 alone. Notice the essential difference with the conventional free-body diagram approach. In a free-body diagram the reactive forces at the hinge form an essential part of the package. The reactions thereat will intervene in the equations of equilibrium. In the virtual work approach, on the other hand, it is imperative not to disengage the plate from the hinge. Quite to the contrary, the presence of the hinge, by reducing the number of degrees of freedom of the system, reduces the number of equilibrium equations. Naturally, this simple example alone would not justify the use of the principle of virtual work, but in more complicated situations its use, by eliminating the participation of reactive forces, results in a considerable simplification of the equations of equilibrium.

The virtual work expression can always be brought to the form:

$$VW = \sum_{i=1}^{N} Q_i \, \delta q^i, \tag{7}$$

by simply collecting all the terms that affect the variation of each generalized coordinate. The multipliers Q_i are called *generalized forces* and, in statics, can be at most functions of the generalized coordinates. The forces acting on a mechanical system are said to be *conservative* if there exists a scalar function V of the generalized coordinates such that:

$$Q_i = -\frac{\partial V}{\partial q^i}, i = 1, ..., n. \tag{8}$$

This function, if it exists, is called a ▶*potential energy* of the forces, and the forces are said to *derive from this potential* (which is determined uniquely up to an additive constant). It is important to realize that the property of the forces being conservative is independent of the particular choice of generalized coordinates, as can be verified easily by using the chain rule of differentiation. Combining Eqs. 7 and 8, it may be concluded that in a conservative system the virtual work of the forces can be calculated as the exact differential of the negative of the potential, namely:

$$VW = -\delta V. \tag{9}$$

The principle of virtual work for conservative systems establishes, therefore, that the potential energy is *stationary* at a position of equilibrium. It can be shown that the equilibrium is *stable* if the potential energy attains a strict *minimum* at the equilibrium configuration. A simple example is that of a cherry at the bottom of a wine glass (stable equilibrium, minimum gravitational potential energy), as opposed to a ball at a mountaintop (unstable equilibrium, maximum gravitational potential energy). The idea of virtual displacements as small perturbations of a putative equilibrium configuration is particularly clear in such simple examples.

In the case of continuum mechanics (q.v.), the principle of virtual work is also applicable provided the so-called *internal virtual work* is carefully accounted for, that is the work of the stress tensor upon the small virtual variations $\delta\mathbf{F}$ of the ▶deformation gradient. To see how this thought can be formalized, the Lagrangian equation of equilibrium is obtained from the Lagrangian equation of balance of ▶linear momentum (Eq. 15 in ▶balance laws (q.v.), whose notation is adopted) by eliminating the ▶acceleration terms as:

$$T^I_{i,\,I} + B^i = 0. \tag{8}$$

Multiplying this equation by a *virtual displacement field*, that is, a field of the form $\delta x^i(X^1, X^2, X^3)$ and integrating over the referential volume of the body:

$$\int_{\Omega} (T^I_{i,\,I} + B_i)\delta x^i d\Omega = 0, \tag{9}$$

where the summation convention is used. This equation is obviously valid for any virtual displacement field. Considering the first term, the divergence operator is eliminated by invoking the divergence theorem. The price to pay is the appearance of a term at the boundary. Indeed:

$$\begin{aligned}\int_{\Omega} T^I_{i\,,I}\,\delta x^i d\Omega &= \int_{\Omega} (T^I_i\,\delta x^i)_{,I}\,d\Omega - \int_{\Omega} T^I_i\,\delta x^i{}_{,I} d\Omega \\ &= \int_{\partial\Omega} T^I_i\,\delta x^i N_I\,dA - \int_{\Omega} T^I_i\,\delta F^i_I\,d\Omega \\ &= 0.\end{aligned} \tag{10}$$

In the boundary integral the ▶surface traction is immediately recognized, namely, $S_i = T^I_i N_I$. Putting together the results of Eqs. 9 and 10 the identity:

$$\int_{\Omega} B_i\,\delta x^i\,d\Omega + \int_{\partial\Omega} S_i\,\delta x^i\,dA \equiv \int T^I_i\,\delta F^i_I\,d\Omega. \tag{11}$$

is finally obtained.

The left-hand side represents the *external virtual work (EVW)* of the ▶body forces and the surface tractions, while the right-hand side can be identified with the *internal virtual work (IVW)* of the (first Piola-Kirchhoff) stress. It is not difficult to reverse the steps of this derivation and conclude that the identical satisfaction of the principle of virtual work in the form:

$$EVW \equiv IVW, \tag{12}$$

is equivalent to the equations of equilibrium of a deformable continuum. No restrictions have been specified upon the virtual displacement fields. Assume, however, that the ▶deformation of part of the boundary is prescribed (for example, part of the boundary is fully supported). Then, the virtual displacement fields must vanish at that part of the boundary. One of the useful features of the principle of virtual work in continuum mechanics is that it delivers not only the interior equations of equilibrium, but also the appropriate boundary conditions of the problem. The Lagrangian formulation has been presented, but a similar treatment delivers also the Eulerian form of the principle. In this case, the internal virtual work is given by $IVW = \int_{\omega} t^{ij} \delta D_{ij} \, d\omega$.

In the case of a deformable continuum, the concept of potential energy can be extended to the stresses themselves. If the constitutive equation of the material is such that the ►first Piola-Kirchhoff stress tensor is derived form a scalar potential function of the deformation gradient, the material is said to be *hyperelastic* (►*hyperelasticity*). Configurations of equilibrium can be associated with a stationary value of the total potential energy of the system.

References

1. Truesdell C, Toupin R (1960) The classical field theories. In: Flügge S (ed) Handbuch der Physik, vol III/1. Springer, Berlin, pp 226–793
2. Lanczos C (1970) The variational principles of mechanics, 4th edn. Toronto University Press, Toronto
3. Epstein M, Herzog W (1998) Theoretical models of skeletal muscle. Wiley, Chichester

Principle of Virtual Work

Definition

In analytical mechanics, the statement that the equilibrium of a system is equivalent to the identical vanishing of the work done by the external forces on all possible virtual displacements of the system.

►Mechanics

Prion (Proteinaceous Infectious Particle)

Definition

Prion (proteinaceous infectious particle) is a proteinaceous infectious agent that cause bovine spongiform encephalopathy (BSE) and human variant Creutzfeldt Jakob disease in humans.

►Creutzfeldt-Jakob Disease

Prism Adaptation

Definition

When a person wears laterally displacing prism glasses, he or she initially experiences difficulty in reaching an object. However, he or she soon adapts to the prism so that the object can be reached. Such a phenomenon is called prism adaptation, and involvement of the cerebellum in the adaptation is known.

►Sensory Motor Learning/Memory and Cerebellum

Probabilistic Inference

►Bayesian Statistics (with Particular Focus on the Motor System)

Procedural Learning

Definition

Training a sensori-motor task, such as playing the piano.

►Sensory Plasticity and Perceptual Learning

Procedural Memory

Definition

Memory of skill or movement improved by practice or experience is called the procedural memory. It has been regarded distinct from the declarative memory about episodes or notions etc. These two types of memories are supported with independent neuronal mechanisms.

Patients impaired in the declarative memory retain the unimpaired procedural memory and vice versa.

►Long-Term Memory
►Sensory Motor Learning/Memory and Cerebellum

Process S

►Sleep Homeostasis

Processing of Tactile Stimuli

HEATHER E. WHEAT, ANTONY W. GOODWIN
Department of Anatomy and Cell Biology, University of Melbourne, Parkville, VIC, Australia

Definition

Tactile here refers to that which is concerned with the sense of touch or the perception of touch. When we touch or manipulate objects, four classes of low-threshold cutaneous mechanoreceptor sensors are activated by the stresses and strains arising from the interaction between skin and object. These four receptor types have different response characteristics because of differences in their structure and in their location in the skin. During manipulation, the receptor responses are determined by many different features of the manipulated object and of the hand movements. Such features include: contact or grip force, the presence of shear forces, the area and shape of the contact region, the surface texture, the shape and compliance of the object, and the presence of slip between the object and the skin. This information is conveyed to the central nervous system by mechanoreceptive afferents (peripheral nerve fibers) which innervate these mechanoreceptors. A population of mechanoreceptive afferents is able to simultaneously encode multiple stimulus parameters because each parameter has a different representation in the spatial and temporal patterns of activity across the afferent population.

Characteristics

Tactile Sensors of the Glabrous Skin

The glabrous (ridged) skin of the fingers consists of two main layers, an outer epidermis and an inner dermis, which are arranged in an interlocking pattern of grooves and ridges (Fig. 1).

Within the skin are thermoreceptors, nociceptors and low-threshold mechanoreceptors; these vary in structure from free nerve endings to mechanically complex end organs. In this chapter the focus is on the four low-threshold mechanoreceptors: Merkel cell-neurite complexes at the epidermal-dermal junction, Meissner corpuscles in the outer papillary layer of the dermis, and Pacinian and Ruffini corpuscles in the deeper reticular layers of the dermis. These receptors are innervated by large-diameter myelinated nerve fibers, classified as Aβ fibers, which conduct action potentials rapidly. A single fiber may end in a single mechanoreceptor, but more commonly will innervate many mechanoreceptors. Conversely, a single mechanoreceptor may be innervated by a single fiber or by multiple fibers [1].

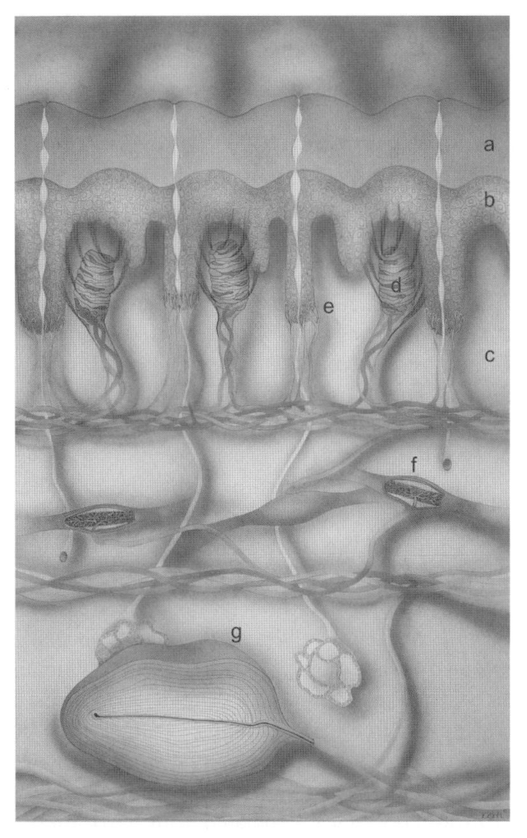

Processing of Tactile Stimuli. Figure 1 A cross-section through glabrous skin. The upper layer, or epidermis (b), is covered in thick keratin (a). Both the epidermis and the underlying layer, the dermis (c), have a wave-like structure. Peripheral nerve fibers, shown in blue, terminate on four types of mechanoreceptors, the Meissner corpuscles (d), the Merkel endings (e), the Ruffini organs (f) and the Pacinian corpuscles (g). Reproduced with permission from Ian Darian-Smith.

In humans, the fibers (also termed afferents) fall into four distinct groups classified by their response properties. Slowly-adapting type I (SAI) and slowly-adapting type II (SAII) afferents are associated with Merkel endings and Ruffini corpuscles respectively and respond to both the static and dynamic components of the stimulus. Fast-adapting type I and II (FAI and FAII) afferents are associated with Meissner corpuscles and Pacinian corpuscles respectively and respond only to dynamic components of the stimulus. Monkey glabrous skin lacks SAII afferents, but the remaining three afferent types, common to humans and monkeys, have similar response properties in both humans and monkeys. Fast adapting afferents are also referred to as rapidly adapting afferents. The density of SAI and FAI afferents innervating the hand decreases proximally from a maximum at the fingertip of 0.7 mm^{-2} for SAI afferents and 1.4 mm^{-2} for FAI afferents. The proximo-distal gradient is less pronounced for SAII and FAII afferents which in the human fingertip have lower densities than the type I afferents (0.1 and 0.21 mm^{-2} respectively). The receptive fields of type I mechanoreceptive afferents (with more superficial receptors) are small and well demarcated whereas those of the type II afferents (with deeper receptors) are large and diffuse with some extending across the entire hand.

Tactile Sensors of the Hairy Skin

The mechanoreceptors of the hairy skin are either associated with the hairs on the skin or are situated between them. There are three hair types – specialized sinus or vibrissae (which are not present in humans), vellus or pelage, and guard. The hair follicles are innervated by: free nerve endings, Merkel nerve endings associated with SAI afferents, lanceolate nerve endings associated with rapidly-adapting afferents, and Ruffini corpuscles associated with SAII afferents. Most sinus hairs are also innervated by lamellated corpuscle of Pacinian type which are associated with rapidly-adapting afferents. Between the hair follicles are dome-like elevations, the so called "touch domes" or *Haarscheiben* of Pinkus which are innervated by slowly-adapting Merkel endings and free nerve endings [2].

Central Nervous System Pathways

The peripheral afferent nerve fibers described above are the peripheral processes of bipolar cells with cell bodies in the dorsal root ganglia, which lie in close proximity to the spinal cord. The central processes of these bipolar cells enter the spinal cord via the dorsal roots and it is at this point that axons conducting information from the low-threshold mechanoreceptive afferents become segregated from those carrying nociceptive or thermal information. The predominant spinal pathway for the low-threshold mechanoreceptive afferents is via the dorsal columns (cuneate and gracile fascicles); most axons enter the spinal cord and ascend to the brainstem without synapse.

The dorsal column axons project principally to, and synapse on, cells within the ipsilateral cuneate and gracile nuclei (the dorsal column nuclei) in the brainstem. Axons of cells in these nuclei cross to the other side of the brainstem and ascend as the medial lemniscus, synapsing primarily in the ventroposterior nuclear complex of the thalamus. The axons of these thalamic cells project predominantly to the primary somatosensory cortices in the postcentral gyrus – Brodmann areas 3a, 3b, 1 and 2. Throughout, the ascending pathways are organized largely on somatotopic principles. Furthermore, neurons to this point retain many of the functional properties of their primary afferent input in terms of basic response functions and receptive field characteristics.

Overview of Tactile Processing

Tactile processing falls into two separate, but overlapping, broad categories. In the first category, touch informs pattern and form perception which allows us to identify objects. Such information includes the object's shape, size, softness and its surface texture. Commonly this is a conscious process. In the second category, touch forms a vital component of the sensorimotor integration that is essential for precision grip and effective manipulation. Object characteristics such as compliance and surface microgeometry, as well as task related characteristics such as position of contact on the skin, linear and torsional loads, and contact or grip forces are relayed by touch to the central nervous system, often without deliberate or conscious tactile perception. To assess what tactile information is signaled to and used by the brain, psychophysical measurements of the human capacity to detect, discriminate, scale and identify specific features of a stimulus have been conducted. Such behavioral studies provide an indication of the minimum information about a particular stimulus parameter that must be received by the brain for conscious perception. Many of these studies have been accompanied by neurophysiological experiments, employing the same stimuli, in order to determine how the essential stimulus parameters are represented in the responses of different neural populations.

The responses of single mechanoreceptive afferents innervating the skin have been recorded by inserting micro-electrodes through the skin into human peripheral nerves and by micro-dissection (fiber splitting) of monkey peripheral nerves. The response characteristics of the different afferent types will be discussed within the context of a range of functionally relevant parameters. Important to note is that all four low-threshold cutaneous mechanoreceptive afferent types are potentially activated by contact with an object but the extent to which each class contributes to the coding of essential task-related information varies depending on the stimulus and task parameters. Furthermore, for all but the simplest stimuli the parameters of the stimulus and task are only represented or encoded unambiguously in the responses of a population of fibers, and not in the responses of isolated single fibers. The studies described in the following sections are based on single fiber responses but for the most part, because of the manner in which the stimuli were presented, they can be extrapolated to represent population responses.

Neural Representation of Object and Task Parameters

Simple Stimuli

The principal value of simple stimuli, such as punctate probes indenting the skin or vibrating in and out of the skin, has been a clear-cut classification of the afferent types (Fig. 2). The temporal characteristics of the responses divide the afferents into slowly and fast adapting groups and the spatial characteristics (mainly size) of the receptive fields define the type I and type II groups. Thresholds of the responses correspond to human detection thresholds measured psychophysically.

Pattern and Form

The earliest measure of spatial resolution in the tactile system was the two-point limen, a simple but unreliable measure that underestimates the capacity of the system. More recently, tactile coding of spatial characteristics or fine form has been investigated using a range of relatively simple stimuli such as edges, bars and gratings as well as more complex patterns such as embossed letters and Braille-type characters [3]. Human psychophysics experiments are matched with single fiber recordings, commonly in monkeys and more rarely in humans. Such studies have found that SAI afferents resolve the spatial details of the stimuli with greater clarity than either FAI or FAII afferents in monkeys or FAI, FAII or SAII afferents in humans. Recognition and discrimination of patterned stimuli are enhanced when they are scanned across the finger, rather than pressed into the finger. This can be explained in part by the increased resolution in SAI responses when such stimuli are moved tangentially across the skin compared to stationary presentation. Variations in scanning speed and contact force have little effect on the spatial resolution of fiber responses or on human perception. Although there is strong evidence for the role of SAI afferents in encoding fine spatial detail, the other afferents do play a role, particularly the FAI afferents when there is lateral movement between the stimulus and the skin.

Texture

The most common perceptual descriptor of surface texture is the sensation of roughness. In early psychophysics experiments common materials, such as sandpaper, were used. More recent stimuli consist of precisely manufactured three-dimensional features, which can be defined mathematically. These allow quantitative neural studies, usually in monkeys, which can be compared with human psychophysics. Stimuli commonly used are either gratings of alternating grooves and ridges, or embossed dot arrays.

Perceived roughness increases monotonically with the spatial period of the texture up to spacings of around 3 mm for both gratings and dot arrays. The most critical feature underlying perceived roughness is the groove width for gratings and the dot spacing for dot arrays. Tangential movement between the stimulus and the skin affects roughness perception. The difference threshold for the spatial period of gratings is degraded from a threshold of 5% when there is movement to a threshold of 10% when there is no movement. Roughness perception is independent of whether the tangential movement is active or passive and is little affected by variations in the rate of movement, but perceived roughness increases when contact force increases [4].

Textures affect both the spatial and temporal patterns in peripheral neural population responses as well as intensive parameters like total neural response. How these population features are used by the central nervous system appears to depend on the nature of the stimulus and the task.

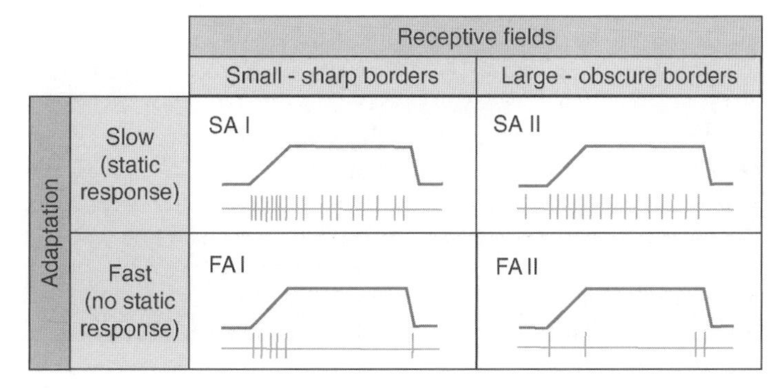

Processing of Tactile Stimuli. Figure 2 Classification of the four types of low-threshold mechanoreceptive afferent fibers innervating the glabrous skin of the human fingerpad. Upper trace (*green*) in each panel shows the indentation amplitude (*vertical axis*) as a function of time (*horizontal axis*) for a probe indented into the skin for 1 s. Traces below (*orange*) show neural responses; each vertical tick represents an action potential. Adapted with permission from Johansson RS and Vallbo AB (1983) Trends in Neuroscience 6:27–32.

Shape

The identification of an object's shape is essential for dexterous manipulation and often involves active exploration of the object. This brings into play a spectrum of sensory mechanisms which relay information about the position and movement of the joints in the hand and arm, as well as cutaneous tactile information. Objects range in shape from a simple sphere to the complex shapes of eating utensils and tools we routinely use. All shapes can be described in terms of their local curvatures. Shapes such as spheres, cylinders, and toroids can be quantified easily; this facilitates analysis of the neural representation of their shape and subsequent correlations with human performance. When simple shapes such as spheres are applied to the skin so that only cutaneous information can be utilized, humans are able to discriminate a difference in curvature (reciprocal of the radius) of the order of 10%. When spheres are indented into a monkey's fingerpad, a clear monotonic relationship exists between single fiber responses and stimulus curvature; both the SAI and FAI afferent responses are modulated but the effect is more pronounced and more reliable for the SAI afferents [5]. Comparable responses in human afferents to spheres and cylinders applied to the fingerpad have been reported; curvature significantly modulates responses in SAI, SAII, and FAI afferents.

For any shape more complex than a sphere, more than one parameter is needed to define it. For example, a cylinder is defined by the orientation of its long axis and the radius or curvature in the orthogonal direction. A toroid is defined by three parameters – two orthogonal curvatures and an orientation. SAI afferent responses are monotonically related to the curvature of these shapes and are modulated by stimulus orientation [6]. When stimuli such as toroids, cylinders and more complex arrays of alternating convex and concave cylindrical shapes of differing curvature are scanned across the monkey's fingerpad, both SAI and FAI afferent responses are modulated. FAI afferents however require higher stroke velocities to elicit responses. The major geometrical features of three-dimensional objects scanned across the skin such as spheres, cylinders and toroids varying in shape, orientation and stroke trajectory are reproduced in the spatiotemporal responses of SAI afferents and to a lesser degree FAI afferents.

When a complex shape is explored or handled, all the parameters of the stimulus and the manipulation affect the responses of individual primary afferent fibers. Nevertheless, independent information about each of the parameters is relayed to the central nervous system because each parameter is represented or encoded uniquely in the primary afferent population response (Fig. 3).

Manipulation

When we manipulate objects, the sensorimotor system optimizes the forces applied by the fingers. The motor control system must adopt forces that are sufficiently large to prevent slips but are not excessive in order to avoid fatigue and to ensure that delicate objects are not damaged. In addition, force magnitudes and directions must suit the shape of the object and the load conditions, and must take account of parameters such as friction. Also, the control system must rapidly and automatically adjust the grip forces to meet the demands imposed by any unexpected changes in loads during a task. Our ability to manipulate objects with dexterity is itself

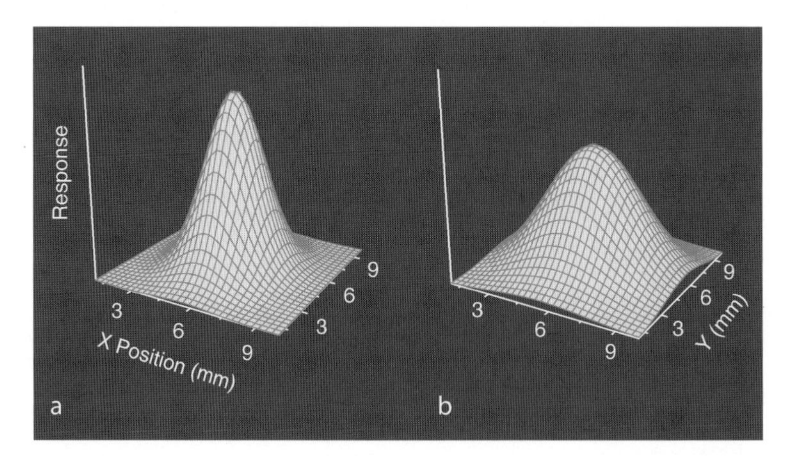

Processing of Tactile Stimuli. Figure 3 Representation of object shape in an ideal SAI afferent population response. The stimuli are spherical with radii *A*, 1.44 mm (curvature 694 m^{-1}) and *B*, 3.9 mm (curvature 256 m^{-1}). The differences in object shape are reflected in the population response profiles. The smaller, more curved sphere (a) elicits a response profile which is narrower and more peaked than that elicited by the larger, less curved sphere (b). These response profiles will be distorted by the characteristics of real peripheral neural populations including the pattern of innervation and the responsiveness of the individual fibers.

testimony to the accuracy of the underlying sensorimotor control system. This is only possible because of sensory feedback from the hand, a large component of which arises from the cutaneous mechanoreceptive fibers. Studies which mimic the forces occurring in everyday manipulations, such as lifting a container of liquid and tilting it, show that humans are able to scale and discriminate, independently, forces acting normal to the skin surface (grip force), forces acting tangential to the skin (load force) and rotational forces (torques) [7]. The magnitudes and directions of these forces and torques are encoded in the responses of whole populations of peripheral afferents along with precise information about the shape of the object being manipulated and the positions of contact on the skin [8]. Such studies demonstrate that the mechanoreceptors in the glabrous skin of the digits provide a rich and accurate source of information during complex manipulations. This information underlies both feedforward and feedback motor control.

Representation in the Central Nervous System

Many of the object and task parameters discussed above have been used in studies of the response properties of cells in the somatosensory cortices of non-human primates [9]. Studies of single cells in somatosensory cortex SI have shown that responses reproduce essential object features presented to the skin and show response patterns similar in many cases to the primary afferent input signals. For example, the configuration of embossed dot ensembles is clearly evident in the response patterns across arrays of single cortical units. Some single cortical units have characteristics which appear to combine response characteristics of more than one afferent type, e.g., SAI and FAI afferents, providing an additional layer of information. The SII somatosensory cortex appears to have some higher level functions than SI, such as extracting the orientation of a stimulus independent of the position of contact on the finger [10]. How input from tactile sources is combined with other sources of sensory input such as afferents from the joints, muscles and hairy skin is not known but information from all of these sources are potentially important in controlling precise hand movements.

References

1. Darian-Smith I (1984) The sense of touch: performance and peripheral neural processes. In: Handbook of physiology. The nervous system III. American Physiology Society, Bethesda, MD, pp 739–788
2. Halata Z (1993) Sensory innervation of the hairy skin light- and electronmicroscopic study. J Invest Dermatol 101:75S–81S
3. Johnson KO, Hsiao SS (1992) Neural mechanisms of tactile form and texture perception. Annu Rev Neurosci 15:227–250
4. Sathian K (1989) Tactile sensing of surface features. Trends Neurosci 12:513–519
5. Goodwin AW, Wheat HE (2004) Sensory signals in neural populations underlying tactile perception and manipulation. Annu Rev Neurosci 27:53–77
6. Khalsa PS, Friedman RM, Srinivasan MA, LaMotte RH (1998) Encoding of shape and orientation of objects indented into the monkey fingerpad by populations of slowly and rapidly adapting mechanoreceptors. J Neurophysiol 79:3238–3251
7. Wheat HE, Salo LM, Goodwin AW (2004) Human ability to scale and discriminate forces typical of those occurring during grasp and manipulation. J Neurosci 24:3394–3401
8. Birznieks I, Jenmalm P, Goodwin AW, Johansson RS (2001) Encoding of direction of fingertip forces by human tactile afferents. J Neurosci 21:8222–8237
9. Tremblay F, Ageranioti-Belanger SA, Chapman CE (1996) Cortical mechanisms underlying tactile discrimination in the monkey. 1. Role of primary somatosensory cortex in passive texture discrimination. J Neurophysiol 76(5):3382–3403
10. Thakur PH, Fitzgerald PJ, Lane JW, Hsiao SS (2006) Receptive field properties of the macaque second somatosensory cortex: nonlinear mechanisms underlying the representation of orientation within a finger pad. J Neurosci 26:13567–13575

Progenesis

Definition

Paedomorphosis (retention of formerly juvenile characters by adult descendants during evolution) produced by precocious sexual maturation of an organism still in a morphologically juvenile stage.

►Evolution and Phylogeny: Chordates

Progenitor Cell

Definition

A progenitor cell is a cell maintaining its capacity for self-renewal and differentiation. Although the distinction between progenitor cells and stem cells is often ambiguous, the term "progenitor cell" includes undifferentiated cells with more limited plasticity and in some cases is used for cells in which multipotency is difficult to demonstrate.

►Adult Neurogenesis

Programmed Cell Death

HIROYUKI YAGINUMA
Department of Anatomy, School of Medicine,
Fukushima Medical University, Fukushima, Japan

Synonyms

Naturally occurring cell death; Physiological cell death; Developmental cell death; PCD

Definition

►Programmed cell death (PCD) in the developing nervous system is defined as spatially and temporally reproducible and species–specific loss of large numbers of individual cells (neurons and glia) during development [1].

Characteristics

Types and Extent of PCD

Development of the neuronal cells can be divided into three phases, (i) proliferating neuronal precursors or founder cells, (ii) postmitotic young neurons before contacting their targets and (iii) maturing neurons after establishing synaptic contact with their targets. In early developmental stages, the neural tube consists of a population of proliferating cells organized into a pseudostratified columnar epithelium, known as the ventricular zone (VZ). Postmitotic young neurons derived from the VZ aggregate and mature between the VZ and the pial surface to form the intermediate zone (IZ). In some areas in the forebrain, such as the dorsal thalamus, a second proliferative zone, the subventricular zone (SVZ) is formed between the VZ and the IZ. In the cerebral cortex, young neurons derived from both the VZ and SVZ migrate through the IZ to form a cortical plate. Programmed cell death can be observed in all three phases of neuronal development [1,2].

After the generation of neuronal cells, glial cells differentiate from glial precursor cells that are derived from the VZ and SVZ. A considerable number of differentiated glial cells are also known to undergo PCD.

PCD of Neuronal Precursors

In the mouse embryo, PCD of proliferating precursor or founder cells can be observed as early as embryonic day (E) 8.5, when the closure of the neural tube is not yet completed. In early developmental stages, massive PCD occurs in specific regions of the developing neural tube. These include ventral and dorsal regions of the spinal cord, floor plate, neural crest, the lamina terminals, ventral region of the forebrain, dorsal region of the hindbrain and the optic vesicle [3,4]. Sporadic small amounts of PCD continue to occur in the VZ and SVZ from early to later developmental stages. But because cell proliferation occurs at the same time, it is difficult to elucidate the quantity of PCD. Estimated amounts of PCD vary from 0.3% to more than 50% depending on the regions analyzed and the methods that were used to detect dying cells. However, since it is reported that inhibition of PCD of neuronal precursors resulted in malformations of the nervous system, such as exencephaly and spina bifida, cell death of proliferating neural precursors is significant for normal development.

PCD of Postmitotic Young Neurons

Although this type of PCD is less common, there are some examples. Approximately 25% of postmitotic motoneurons in the non-limb innervating cervical spinal cord die before they establish synaptic contact with their target muscles [5]. In the dorsal root ganglion and the retina, a significant percentage of postmitotic neurons die before their axons reach their targets [4].

PCD After Establishing Synaptic Contacts with the Targets

Neurons that have survived earlier phases of PCD begin to establish synaptic connections with other neurons and target cells. During this period, postmitotic differentiating neurons undergo PCD. This type of PCD has been found to occur virtually everywhere that it has been looked for and includes motoneurons, sensory neurons, autonomic neurons, retinal neurons, optic tectum, isthmo-optic nucleus, basal ganglia, cerebellum and cerebral cortex. Despite this widespread occurrence of PCD, it is also known that PCD does not occur in spinal interneurons and neurons in the medial and lateral pontine nuclei of the chick embryo. Quantitative analyses performed in the neuronal groups that are easily defined revealed that ~20–80% of postmitotic neurons undergo cell death. This type of neuronal death has been well studied and it is known that inadequate neurotrophic support derived from their targets, afferent inputs and other sources regulate this type of cell death [1].

PCD of Glial Cells

Both oligodendrocytes and astrocytes are known to undergo PCD. About 50% of oligodendrocytes normally die in the developing rat optic nerve and significant numbers of dying cells in the neonatal rat cerebellum are astrocytes [2].

In most cases, cells die by ►apoptosis. They shrink in size, the nuclear chromatin becomes pyknotic and condenses against the nuclear membrane (Fig. 1 a–c) and cytoplasmic organelles remain intact.

Eventually, the cytoplasm and nucleus break up into apoptotic bodies that are phagocytized and digested by macrophages or by adjacent healthy cells. In contrast to ►necrosis (necrotic cell death), which is caused by

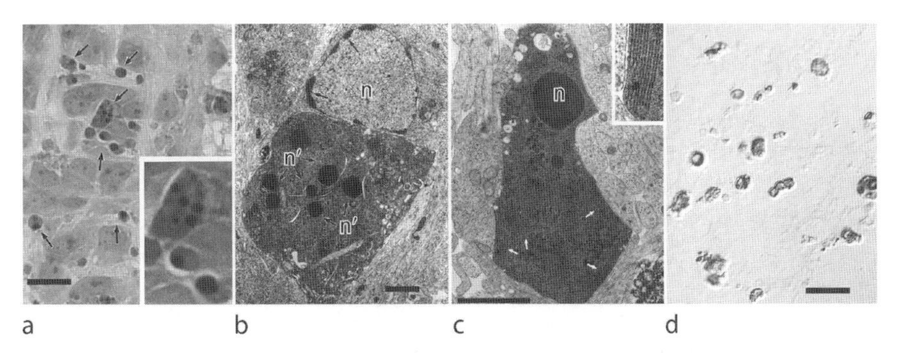

Programmed Cell Death. Figure 1 Light and electron micrographs of the ventral horn of the cervical spinal cord of the E4.5 chick embryo showing dying motoneurons. (a) Hematoxylin and eosin staining showing pyknotic dying neurons (*arrows*). *Inset* shows higher magnification of typical pyknotic neurons. (b, c) Electron micrographs showing typical examples of apoptotic degeneration. In the nucleus (*n*) of the upper cell in b, chromatin begins to condense against the nuclear membrane (*arrows*), suggesting that this cell is in the earliest stage of apoptosis. In the lower cell in b, the nucleus has been fragmented (*n'*) and the cytoplasm has become electron dense. The condensed nucleus and cytoplasm are conspicuous in the cell in (**c**). Aggregated ribosomes are often observed (*arrows and insets* in (c)). (d) TUNEL staining. Bar in (a) and (d), 10 µm. Bars in b and c, 2 µm.

acute cellular injury, apoptosis occurs as an ordered process without evoking inflammation. Another feature of apoptosis is nucleosomal DNA fragmentation and degradation. This can be observed in tissue sections by terminal deoxynucleotidyl transferase-mediated dUTP nick end labeling (TUNEL) (Fig. 1d). The estimated time for the whole process of apoptosis is from a few hours to at most one half day.

Lower Level Processes

Intra-Cellular Mechanisms of PCD

Cells in the developing nervous system share the same basic apoptosis mechanisms with all other cell types (Fig. 2).

The characteristic morphology of apoptosis is the result of activation of executioner ►caspases, such as caspases-3, -6 and -7. They activate a DNase (CAD) and acinus, which in turn cause DNA fragmentation and chromatin condensation (Fig. 2). Cytoplasmic and nuclear skeletal proteins are also cleaved by the executioner caspases, leading to cellular and nuclear shrinkage. The executioner caspases are activated by initiator caspases including caspase-8 and caspase-9. Caspase-8 and caspase-9 are activated through two different pathways, the death receptor pathway and the mitochondrial pathway, respectively. Association of a death receptor ligand (e.g. FasL) with its receptor (Fas) leads to recruitment of a death domain containing protein, FADD. This, in turn, recruits procaspase-8. Procaspase-8 undergoes auto-proteolysis to release active caspase-8 (Fig. 2). The key step in the mitochondrial pathway is the release of ►cytochrome *c* through pores formed in the mitochondrial outer membrane. Cytochrome *c*, ►Apaf1, procaspase-9 and ATP form the apoptosome, which is a multimeric active holoenzyme that activates executioner caspases (Fig. 2).

►Bcl-2 family proteins play fundamental roles in the process of mitochondrial pore formation. The Bcl-2 family consists of three subgroups, anti-apoptotic, pro-apoptotic and BH3-only proteins. When pro-apoptotic Bax or Bak is activated, they homo-oligomerize within the mitochondrial outer membrane to form large enough pores for cytochrome *c* release. Alternatively, Bax can change large channel proteins that reside in the outer mitochondrial membrane to allow cytochrome *c* to escape. Anti-apoptotic Bcl-2 family members, including Bcl-2 and Bcl-xL, can inhibit the effects of pro-apoptotic Bax or Bak through combining with them to form dimers. BH3-only proteins can bind to Bcl-2 and Bcl-xL to release pro-apoptotic, Bax or Bad, resulting in apoptosis. Therefore, BH3-only proteins serve as a key in the mitochondrial pathway (Fig. 2). It is also known that one BH3-only protein, Bid, can be cleaved (activated) by caspase-8 to form tBid, which in turn activates Bax and Bak. This allows cross talk between the two pathways (Fig. 2).

Besides cytochrome *c*, several other cell death-inducing molecules are known to escape through the pores formed in the outer mitochondrial membrane. These include ►AIF (apoptosis inducing factor), EndoG (Endonuclease G), Smac/DIABLO and Omi/HtrA2. Reactive oxygen species (ROS) are also released from dysfunctional mitochondria. These molecules activate caspase-dependent or caspase-independent cell death pathways serving as initiators of collateral cell death pathways (Fig. 2). It is also known that there are intrinsic molecules that antagonize caspase activities to prevent apoptosis. These include ►NAIP (Neuronal Apoptosis Inhibitory Protein) and XIAP.

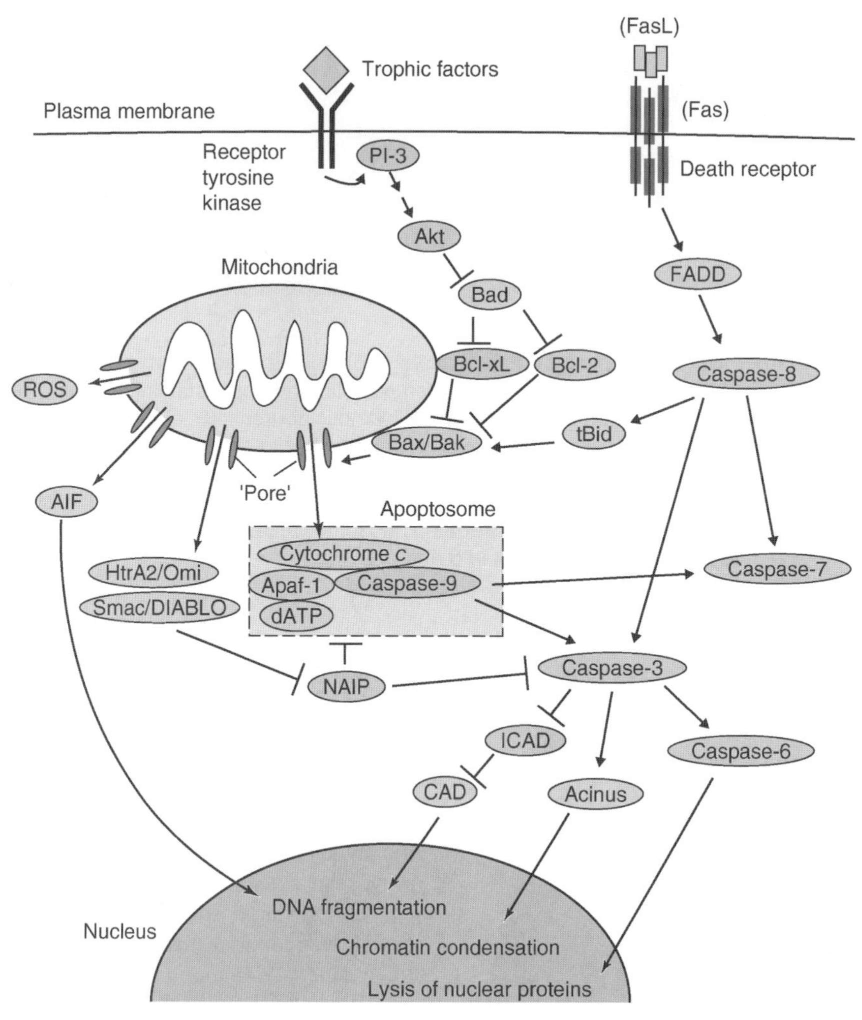

Programmed Cell Death. Figure 2 A simplified scheme of cell death pathways in cells in the developing nervous system. ►Caspases are indicated in *pink*, pro-apoptotic members in *yellow*, anti-apoptotic members in *green*. For details, see text.

Process Regulation

Inter-Cellular Regulation of PCD

Regulation by Trophic Factors (the Neurotrophic Theory)

The favored explanation as to why neuronal and glial cells die in the developing nervous system is "the neurotrophic hypothesis." This proposes that developing neurons and glial cells require trophic support for survival and compete for a limited supply of trophic factors. Cells that cannot obtain enough trophic support undergo cell death. This hypothesis has been supported by many experimental studies in which the quantitative relations between neuronal groups and their synaptic targets or afferent inputs were altered. For example, removal of a limb bud on E2 in the chick embryo resulted in total disappearance of limb innervating motoneurons by E10 because of excessive motoneuron death. On the other hand, addition of a supernumerary limb bud resulted in the survival of more motoneurons. Similarly, it has been proved that alteration of afferent inputs also affects the number of surviving neurons. However, it is not yet clear whether neurons compete for limited amount of trophic factors or limited sites where trophic factors are transferred to neurons. An important source of trophic support for developing neurons is their targets. In addition, signals derived from afferent inputs as well as from nonneuronal cells, such as central and peripheral glia, are recognized as possible sources of trophic regulation of cell death and survival.

Trophic factors that are required for survival of developing neurons vary depending on the neuronal population. For example, neurons in the sympathetic ganglia require target-derived NGF as a survival factor, whereas sensory neurons in peripheral ganglia require one or more of the ►neurotrophins, NGF, BDNF, NT-3 or NT-4/5. Several candidates for muscle-derived trophic factors for motoneurons have been identified and include BDNF, NT-4/5, IGF, HGF, CT-1 and GDNF [1].

The binding of trophic factors to their specific receptors induces rapid protein phosphorylation and the activation of complex cascades of intracellular signals. Among them, the phosphatidylinositol 3-kinase (PI3-K)-Akt pathway is known to be directly involved in inhibition of cell death. Activated Akt phosphorylates the BH3-only protein, Bad. Phosphorylation of Bad dissociates Bad and the anti-apoptotic Bcl-2 family, Bcl-xL, allowing Bcl-xL to prevent cell death by blocking pore formation in the mitochondrial outer membrane (Fig. 2).

The neurotrophic hypothesis also proposes that the intrinsic default fate of developing neurons is to undergo PCD. Although the mechanism of the intrinsic cell death pathways is not known for most kinds of neurons, it has been suggested that the death receptor pathway that is activated by FasL and Fas may play a role in the PCD of motoneurons [6].

Regulation of Target-independent PCD

Little is known about the mechanisms that regulate PCD of proliferative precursor cells or young postmitotic neurons. Since massive cell death of proliferative cells often occurs in specific regions of the neural tube, cell death may be the result of determination of regional specificity. In fact, following perturbation of sonic hedgehog signaling, which determines the regional specificity of the ventral half of the spinal cord, distribution of dying cells was altered [7]. Moreover, death-factor signaling may also induce early neuronal death. For example, in early retinal development, postmitotic young neurons are known to undergo cell death by NGF signaling through ►$p75^{LNTR}$ and by TGFβ, which are provided by macrophages and local surrounding cells respectively [8,9].

Function

Role(s) of PCD

Since perturbation of normal PCD results in various kinds of defects in the nervous system, cell death during development plays a significant adaptive role. Possible functions of cell death in the developing nervous system are:

1. Pattern formation and morphogenesis
2. System or size matching
3. Removal of cells of an appropriate phenotype or cells that have no function
4. Error correction
5. Removal of harmful cells that have defective DNA

Pathology

One genetic disease in humans in which cell death has been implicated is infantile ►spinal muscular atrophy (SMA). A clear link between SMA and failed inhibition of cell death has been proposed. In severe SMA, the neuronal specific inhibitor of apoptosis (IAP) family member known as NAIP is often dysfunctional due to missense and truncation mutations. IAPs such as NAIP potently block the enzymatic activity of executioner caspases (3 and 7) suggesting that NAIP mutations may permit unopposed developmental apoptosis to occur in sensory and motor systems resulting in lethal muscular atrophy [1].

Since neurons need their targets for their survival, loss of one neuronal group or target tissue often results in a secondary loss of other neuronal groups. For example, the "cerebelless" mutant mouse lacks the entire cerebellar cortex. The primary defect is a specific inhibition of GABAergic neurons including Purkinje cells, resulting in secondary and complete loss of external germinal layer, pontine and olivary nuclei during development [10].

References

1. Oppenheim RW, Johnson JE (2003) Programmed cell death and neurotrophic factors. In: Squire LR, Bloom FE, McConnell SK, Roberts JL, Spitzer NC, Zigmond MJ (eds) Fundamental neuroscience. Academic Press, San Diego, SA, pp 499–532
2. Roth KA (2005) Programmed cell death. In: Rao MS, Jacobson M (eds) Developmental neurobiology. Kluwer Academic/Plenum, New York, pp 317–328
3. Kuan CY, Roth KA, Flavell RA, Rakic P (2000) Mechanisms of programmed cell death in the developing brain. Trends Neurosci 23:291–297
4. de la rosa EJ, de Pablo F (2000) Cell death in early neural development: beyond the neurotrophic theory. Trends Neurosci 23:454–458
5. Yaginuma H, Tomita M, Takashita N, McKay SE, Cardwell C, Yin QW, Oppenheim RW (1996) A novel type of programmed neuronal death in the cervical spinal cord of the chick embryo. J Neurosci 16:3685–3703
6. Raoul C, Pettmann B, Henderson CE (2000) Active killing of neurons during development and following stress: a role for p75(NTR) and Fas? Curr Opin Neurobiol 10:111–117
7. Oppenheim RW, Homma S, Marti E, Prevette D, Wang S, Yaginuma H, McMahon AP (1999) Modulation of early but not later stages of programmed cell death in embryonic avian spinal cord by sonic hedgehog. Mol Cell Neurosci 13:348–361
8. Dechant G, Barde YA (2002) The neurotrophin receptor p75(NTR): novel functions and implications for diseases of the nervous system. Nat Neurosci 5:1131–1136
9. Duenker N (2005) Transforming growth factor-beta (TGF-beta) and programmed cell death in the vertebrate retina. Int Rev Cytol 245:17–43
10. Hoshino M, Nakamura S, Mori K, Kawauchi T, Terao M, Nishimura YV, Fukuda A, Fuse T, Matsuo N, Sone M, Watanabe M, Bito H, Terashima T, Wright CV, Kawaguchi Y, Nakao K, Nabeshima Y (2005) Ptf1a, a bHLH transcriptional gene, defines GABAergic neuronal fates in cerebellum. Neuron 47:201–213

Programmed Cell Death

Definition
Cell death by design, in which the cell uses specialized cellular machinery to kill itself. Programmed cell death is a process essential to cell termination, homeostasis, and development. This process allows metazoans to control cell number and eliminate surplus or erroneous cells.

Progressive Bulbar Palsy

Definition
►Motoneuron disease of the ►brainstem (*bulb* stands for ►medulla oblongata and *palsy* for weakness) with dysarthria (difficulty articulating) and dysphagia (difficulty swallowing).

Progressive Multifocal Leukoencephalopathy

Definition
Infrequent disorder of the nervous system that primarily affects individuals with suppressed immune systems (including, allograft recipients such as kidney transplant patients; patients with cancers such as leukemia or lymphoma; and nearly 10% of patients with ►acquired immune deficiency syndrome (AIDS). The disorder, which is caused by a common human polyomavirus, JC virus, is characterized by ►demyelination or destruction of the ►myelin sheath that covers nerve fibers.

►Acquired Immune Deficiency Syndrome (AIDS)

Progressive Supranuclear Palsy

Definition
Progressive degenerative disease belonging to the family of tauophaties with widespread pathology involving cortical and subcortical structures. In Progressive supranuclear palsy, oculomotor disturbance, early postural instability with falls, and frontal dementia predominate. There is symmetric onset of parkinsonism, early postural instability, severe axial rigidity, absence of tremor, and poor response to dopaminergic therapy. supranuclear gaze palsy, especially of downgaze, is the defining characteristic. Blepharospasm and eyelid opening apraxia are also common.

►Parkinsonism

Pro-inflammatory Cytokines

Definition
Pro-inflammatory cytokines such as interleukin (IL)-1β, IL-6, and tumor necrosis factor (TNF)-α, are overexpressed at the lesion site for several hours to days after central nervous system (CNS) injury. The cells of origin of these cytokines are neurons, astrocytes, microglial cells, infiltrated macrophages, and neutrophils. They are involved in the secondary tissue damage that is produced through a series of autodestructive events (e.g., apoptosis) initiated by the primary trauma. Low concentrations of pro-inflammatory cytokines can be beneficial; however, high concentrations mediate cell death and widespread tissue disruption. Thus, manipulation of this inflammatory response is one of the major therapeutic approaches for CNS injury.

►Transplantation of Neural Stem Cells for Spinal Cord Regeneration
►Tumor Necrosis Factor- α (TNF-α)

Pro-inflammatory Mediators

►Central Nervous System Disease – Natural Neuroprotective Agents as Therapeutics

Projection Neurons

Definition
Neurons that send ("project") their main axon outside a morphologically defined area or nucleus in which their cell bodies are located. The length of the axon depends on the distance between the structures they connect.

Thus, in the cerebral cortex, association axons linking adjacent areas are short compared to corticospinal axons. In the superior colliculus, projection axons to the pretectum are short compared to tectoreticulospinal axons. By opposition, neurons participating exclusively in the intrinsic connections of a given structure are called "local neurons" or "interneurons."

►Modulatory Projection Neurons

Projections

Definition

Axonal extensions, ranging from short to very long, that possess chemical synapses allowing for the electrochemical communication of neurons with other neurons over distance. A neuron with an axon possessing a chemical synapse in relation to another neuron at some distance is said to project to that neuron.

Axons carrying impulses away from a structure comprise efferent projections. Axons carrying impulses into a structure comprise afferent projections.

Prokineticin 2

Definition

A 102 amino acid polypeptide that may function as an output molecule from the suprachiasmatic nucleus (SCN) in transmitting timing behavioral circadian rhythms. It may also function locally within the SCN to synchronize cellular clocks. Receptors for PK2 (PKR2) are abundantly expressed in major target nuclei of the SCN output pathway. Intracerebroventricular infusion of PK2 at night, when endogenous PK2 mRNA levels are low, markedly reduces the nocturnal increase in locomotion.

Mice with a disruption of the PK2 gene display significantly reduced rhythmicity for a variety of circadian physiological and behavioral parameters including sleep-wake cycle, locomotor activity, body temperature, and circulating glucocorticoid as well as glucose levels.

►Circadian Rhythm
►Clock Coupling Factors
►Locomotion
►Sleep-wake Cycle
►Suprachiasmatic Nucleus

Proliferation

Definition

Production of new daughter cells by cell division.

►Neural Development

Promoter

Julia Kim, David H. Farb, Shelley J. Russek
Laboratory of Molecular Neurobiology, Department of Pharmacology and Experimental Therapeutics, Boston, MA, USA

Definition

A promoter contains all the gene regulatory information that is necessary for the expression of its protein product in vivo. The DNA sequence of the promoter can be contiguous lying upstream of the transcriptional start site or it can be separated by large distances dependent upon the presence or absence of key regulatory elements, such as enhancers or silencers, in introns and exons.

The binding of transcription factors to the sequence(s) of the promoter are believed to alter DNA conformation in such a manner that stabilizes the binding of RNA polymerase to enable regulated transcription. A promoter can also be regulated at the epigenetic level through the recruitment of chromatin modulators such as HAT, HDAC, mSin3A, and Swi/snf, that control the accessibility of transcription factors to DNA elements. Cell-and developmental-specific expression often relies on the presence of promoter elements, such as enhancers or silencers, that are located at a distance from the start site in exons, introns, or intergenic regions.

Characteristics

Structure

Core: Core promoters are the minimal elements needed for RNA polymerase II to initiate transcription at basal levels using a particular ►transcriptional start site (TSS). The most well-known core promoter region contains a TATA box, around 35 bp upstream of the TSS, and an Initiator element (Inr) that contains the TSS. Not all core promoter regions contain TATA boxes. In fact, many of the regulated genes in the human genome are TATA-less. TATA-less core regions often contain other elements such as the GC box or a binding site for a strong activator protein such as Specificity Protein 1 (SP1). Diversity in core promoter regions is just beginning to be identified suggesting that this region of the gene is highly specialized and plays an important role in the regulated nature of gene transcription.

Proximal Region: Proximal promoter region contains the Core and around 500 bp of sequence upstream of the ►TSS. They contain upstream binding sites for activators that are necessary to increase transcription above basal levels.

Distal Region: Distal promoter region is upstream of the Proximal and in most cases contains the recognition sites for activators and repressors responsible for full transcriptional activation, consistent with the levels of gene expression *in vivo*. The Distal region usually contains a few kilobase pairs of DNA lying upstream of the Proximal promoter region that contains the core.

Cis-acting elements and trans-acting factors: A particular set of regulatory sequences that are found within the promoter of a gene are referred to as cis-acting elements. The proteins that bind to these elements, as well as to related elements in multiple genes, are referred to as trans-acting factors, also known as transcription factors.

Gene clusters: A set of two or more genes that are derived from a common ancestor and are functionally related, or encode related gene products, are often found in gene clusters on a particular chromosome. For example, subunit genes coding for the major inhibitory neurotransmitter receptor, type A γ aminobutyric acid (GABA, $GABA_AR$) are organized as β–α–α–γ or β–α–γ on four human chromosomes [1]. This unique genomic structure suggests that there may be regulatory elements shared by the promoter regions or a single locus of control (as seen for the β-globin genes) that has been preserved throughout evolution.

Alternative Promoters: Many genes use more than one promoter to control either development or cell specific expression. The transcripts produced from these promoter regions increase the diversity of protein products or the stability of their mRNAs. The alternative promoters can be located in a downstream intron or in a distant region upstream from the dominant ►TSS.

Examples of Promoters, Their Elements and Transcription Factors Studied in the Brain

- Neural Specific Genes

GABRB1 [2] – A gene that codes for the beta 1 subunit of the $GABA_AR$, the major inhibitory receptor in the nervous system that contains an integral chloride ion channel gated by GABA. GABRB1 is organized in a head-to-head orientation with the α4 subunit gene (GABRA4). The human GABRB1 promoter lacks a ►TATA-box and contains an Inr element that by itself can determine the cell specific expression of the gene and its autologous regulation by GABA.

GLUR2 [3] – GLUR2 has multiple ►TSSs that are development and cell-specific. None of the TSSs identified contain TATA boxes and their transcription is modulated by Sp1, within a methylated CpG island, and a GLUR2 silencer that shares 71% similarity to the restrictive silencing factor REST/NRSF (see below).

GABBR1 [4] – A gene that codes for the R1 subunit of the $GABA_BR$, the G-protein coupled receptor that is activated by GABA. Use of alternative promoters, rather than alternative splicing as previously speculated, contributes to the generation of isoforms for the $GABA_BR1$ subunit. $GABA_BR1a$ expression is marked in the fetal brain and may regulate the formation of presynaptic $GABA_B$ receptors while it is modest in the adult brain. GABABR1 expression is low during development and marked in the adult where it may regulate the formation of postsynaptic $GABA_B$ receptors. The promoter regions of both isoforms are TATA-less and contain functional DNA sequence elements for regulation by the cAMP regulatory binding protein (►CREB).

Neuropeptide Y (NPY) [5] – NPY is the most abundant neuropeptide in the brain and is expressed in a wide variety of cell populations of the nervous system. Its promoter region contains the partial consensus sequence for transcription factors such as SP1, Activator Protein 1 (AP1) and CAAT box. However, CT-rich, instead of GC-rich, sequences are used by Sp1 to promote transcription.

- A signal-induced transcription factor

Early growth response gene (Egr) and Egr response element (ERE) [6,7]: While the most well known signal activated transcription factor in the nervous system is the ubiquitously expressed CREB, there is a family of immediate early gene (IEG) products known as Egrs that are increasingly being identified in the dynamics of brain function. Egrs bind to the consensus motif (depicted below) in response to signals elicited by neurotransmitters, altering the expression level of certain genes that contain a GC-rich ERE in their promoter region. Egrs link a short-term change in neurotransmission to a functional change via the synthesis of new proteins.

ERE consensus sequence

$$5' - \mathrm{GCG(G/T)GGGCG} - 3'$$

Egrs are implicated in the transcriptional control of multiple genes via their zinc-finger motif located in the C-terminal region. So far, four family members have been identified (Egr1–4). All of them recognize the same ERE sequence and are homologous, although different spatial and temporal expression is observed. Of the four members, Egr1 is the most well defined and its expression is dependant on activation of the *N*-methyl-D-aspartate receptor (NMDAR) and mitogen activated protein kinase (MAPK) signaling. However, the target or the specific mechanisms controlling transcriptional regulation of the other Egr family members is not completely known. Egr3 has recently been identified as a seizure-induced protein controlled by brain derived neurotrophic factor (BDNF, see below) and it plays an important role in learning and memory. Multiple forms of Egr3 that differ in their N-terminus have been described and are being investigated.

- A major target gene for Egr 1 and 3

Activity-regulated cytoskeleton-associated (Arc) protein [8] – A putative target of Egr1 and Egr3 transcription factors, Arc expression is upregulated by robust synaptic activity, such as observed in kainic-acid induced seizures or exploration of a novel environment by rodents. Egrs bind to the ERE site in the Arc promoter. Similar to Egrs, Arc expression is dependant on NMDAR activation and MAPK signaling, highlighting its association with long-term potentiation (LTP) and learning and memory.

- Repressor factor involved in neural specific gene expression

Neuron-Restrictive Silencer Element (NRSE) and Neuron-Restrictive Silencer Factor/RE1 Silencing Transcription factor (NRSF/ REST) [9] – Neuronal specificity can be conferred via unique transcriptional activators or via repressors that silence transcription in non-neuronal cells. Identification of the silencer element NRSE is one of the most important accomplishments in neural specific gene regulation today and its importance to brain function as well as disease is accumulating in the literature. NRSE is found primarily in non-coding sequence and is evolutionarily conserved. NRSEs are found in neuronal genes coding for proteins such as the Na^+ channel, Synapsin I, BDNF, NMDAR, nAchR, GABAAR, and Neurofilament M, and in non-neuronal genes coding for proteins such as Keratin, human/bovine P450–11β, and skeletal actin. NRSF binds to a 21-nucleotide sequence called the neuron-restrictive silencer element (NRSE/RE1).

NRSE consensus sequence

$$5' - \text{ttCAGCACCaaGGAcAGcgcC} - 3'$$

Uppercase letters: conserved Lowercase letters: less conserved

Although it was initially believed that NRSE containing genes are silenced only in non-neuronal cells via the binding of NRSF, more recent studies have revealed that they also bind to genes within neurons. NRSF contains a zinc-finger DNA-binding domain and interacts with the co-repressors CoREST and mSin3a via two repressor domains found at each end. Once bound to NRSF, the co-repressors recruit histone modifying proteins such as Histone Deacetylase (HDAC) and Methyl CpG binding protein 2 (MeCP2), altering conformation of the ►chromatin to a transcriptionally silent form (heterochromatin).

NRSF expression in neuronal cells decreases as they become more differentiated and its gene targets are involved in neuronal function, such as ion channels, neurotransmitter receptors, neurotransmitter-synthesizing enzymes, neuronal cytoskeleton, neuropeptides and neurotrophic factors. A truncated splice variant of NRSF, REST 4, has been shown to bind to NRSE rather weakly in neurons and is also found at high levels in biopsies of Small Cell Lung Carcinoma patients.

- A ubiquitous transcription factor with important brain function

Sp1 [10] – Sp1 is a transcription factor that regulates the expression of genes throughout the body. Particular to the brain, it regulates expression of the acetylcholine receptor (AchR) and Huntingtin (Htt). Sp1 is known to bind specifically to GC-rich DNA elements via its zinc-finger motif. With its strong glutamine-rich activation domain, Sp1 recruits basal transcription factor TFIID to DNA and induces marked transcription. TFIID contains TATA box binding protein (TBP) and TBP-associated factors (TAFs) of variable sizes. Of these TAFs, human TAFII130 is of particular interest due to its specific interaction with Sp1. Both Sp1 and TAFII130 bind to the mutant form of Htt, an association that is implicated in Huntington's Disease.

- Activity dependent neural specific signaling molecule

BDNF [11] – There are five exons in the rat BDNF gene and exon I through IV contain promoters that are upstream of one another (designated promoters I-IV). Use of alternative promoters ensures specificity and diversity of gene regulation. For example, BDNF transcription via promoter I is activated after Ca^{2+} influx through L-type voltage-dependent Ca^{2+} channels (L-VDCC), whereas promoter III is activated when Ca^{2+} influx occurs through the NMDAR. The activation of both promoters occurs via binding of phosphorylated ►CREB.

Techniques Used to Study Promoters

Electrophoretic Mobility Shift Assay (EMSA, also referred to as gel shift assay) – EMSA is an in vitro assay used to determine whether a particular transcription factor binds to a known sequence of DNA by identifying if there is specific binding activity in a given nuclear extract when exposed to a particular sequence of DNA or RNA. Extracts of nuclear proteins are incubated with a radiolabeled DNA probe that contains a sequence of interest and the resulting DNA/protein complex is resolved using non-denaturing acrylamide gel electrophoresis. In the absence of extract the probe migrates rapidly towards the bottom, however, in the presence of specific binding protein, the probe's migration is slowed in the gel. Competing cold oligonucleotides of various sequences are used to demonstrate sequence specificity for the binding interaction. In addition, to determine the identify of nuclear proteins, a specific antibody is added to the binding reaction and presence of a "supershift" (because of the increased size of the complex with the antibody attached, the probe will migrate even more slowly) occurs if the epitope of the binding protein is accessible.

Luciferase Reporter Assay – The luciferase reporter assay is especially useful in studying mammalian transcription because the natural expression of the luciferase gene product is restricted to fireflies. In molecular neurobiology, a vector containing the promoter sequence of interest is placed upstream of the luciferase reporter and promoter activity is studied in transfected primary neuronal cultures derived from different embryonic brain regions by assaying for the amount of light production from cleavage of the luciferase substrate luciferin.

Chromatin Immunoprecipitation (ChIP) – ChIP is utilized to examine the *in vivo* binding of a given protein to a specific promoter segment. After DNA is crosslinked to the DNA-binding proteins, the genomic DNA is sheared into small fragments of 300 bp or less and immunoprecipitated with a specific antibody that recognizes the protein of interest. Upon reverse-crosslinking, detection of the resulting precipitated DNA fragments are done using standard polymerase chain reaction (PCR) or quantitated via real-time PCR with taqman probe and primers. This assay is useful for identifying potential endogenous genes regulated by a particular transcription factor.

Disease

Single Nucleotide Polymorphism (SNP) within the promoter region. Single Nucleotide Polymorphism is characterized as a genetic deviation or change in DNA of more than 1% of a population. Generally, SNPs do not result in any phenotypic changes. However, when the SNPs occur within a coding region or a promoter of a gene, consequences such as a differential response to a drug or an increased predisposition to certain diseases have been observed.

Brain-derived Neurotrophic factor (BDNF). In particular, a genetic variance in the BDNF promoter I was recently identified. This novel variation has a cytosine replaced by adenine at 281 bp upstream of the TSS and causes reduced DNA binding by factors yet to be identified. In addition, this "A" allele decreases BDNF promoter activity in rat hippocampal neurons and an association of the allele with a decrease in psychopathology is reported in a phenotype-genotype analysis using human samples.

Neuropeptide Y (NPY). Lowered activity of NPY has been implicated in the pathophysiology of Schizophrenia. A novel polymorphism within the Japanese population at −485T>C in the promoter region of the NPY gene has been reported in patients suffering from Schizophrenia. The −485 nucleotide is contained within the Sp1 consensus site and abolishes potential binding site detection.

References

1. Russek SJ (1999) Evolution of GABA(A) receptor diversity in the human genome. Gene 227(2):213–222
2. Steiger JL, Russek SJ (2004) GABA-A receptors: building the bridge between subunit mRNAs, their promoters, and cognate transcription factors. Pharmacol Ther 101(3):259–281
3. Myers SJ, Peters J, Huang Y, Comer MB, Barthel F, Dingledine R (1998) Transcriptional regulation of the GluR2 gene: neural-specific expression, multiple promoters, and regulatory elements. J Neurosci 18(17):6723–6739
4. Steiger JL, Bandyopadhyay S, Farb DH, Russek SJ (2004) cAMP response element-binding protein, activating transcription factor-4, and upstream stimulatory factor differentially control hippocampal GABABR1a and GABABR1b subunit gene expression through alternative promoters. J Neurosci 24(27):6115–6126
5. Minth CD, Dixon JE (1990) Expression of the human neuropeptide Y gene. J Biol Chem 265(22):12933–12939
6. Beckmann A, Wilce P (1997) Egr transcription factors in the nervous system. Neurochem Int 31(4):477–510
7. Li L, Carter J, Gao X, Whitehead J, Tourtellotte WG (2005) The neuroplasticity-associated Arc gene is a direct transcriptional target of early growth response (Egr) transcription factors. Mol Cell Biol 25(23):10286–10300
8. Ooi L, Wood IC (2007) Chromatin crosstalk in development and disease: lessons from REST. Nat Rev Genet 8 (7):544–554 (Review)
9. Dunah AW, Jeong H, Griffin A, Kim YM, Standaert DG, Hersch SM, Mouradian MM, Young AB, Tanese N, Krainc D (2002) Sp1 and TAFII130 transcriptional activity disrupted in early Huntington's disease. Science 296(5576):2238–2243
10. Tabuchi A, Nakaoka R, Amano K, Yukimine M, Andoh T, Kuraishi Y, Tsuda M (2000) Differential activation of brain-derived neurotrophic factor gene promoters I and III by Ca^{2+} signals evoked via L-type voltage-dependent and *N*-methyl-D-aspartate receptor Ca^{2+} channels. J Biol Chem 275(23):17269–17275

Property

VOLKER GADENNE
Department of Philosophy and Theory of Science, Johannes-Kepler-University, Linz, Germany

Definition

Properties are ways things are; they are features, attributes, traits, characteristics or aspects.

Description of the Theory

Property and Object

Suppose that Jack is bald. Jack belongs to the category of *things* (or *objects*, including persons). Baldness, by contrast, is a *property*. Jack has this property and many other individuals possess it too. Things are bearers of properties and different objects may share the very same property. Things are also said to *exemplify* or *instantiate* properties.

But things are not the only entities that possess properties. Properties can themselves have properties and so can events. "Bald" is exemplified by individual persons and has itself the property of being physical. "Bold" and "shy" are properties of persons too, they are however mental. An event, like having a toothache, may have the property of being short or of occurring after breakfast.

When philosophers speak of objects and properties, they use these concepts in a wide sense. The category of objects may include tables, trees and persons, but also electrical fields or points in space-time. As to properties, many of them are denoted by adjectives, like "green" or "round". But there are also properties denoted by rather complex predicates. For example, a neuron instantiates (at certain times) a special property denoted by the term "resting potential." A person may be ascribed the property of being in pain, of believing that snow is white or of wanting to study philosophy. In such cases a person can also be said to be in the *state* of being in pain or in the *state* of believing that snow is white [1].

Ontological Status of Properties

The ontological status of properties has been discussed since Plato [2,3]. *Realists* (in a specific sense of the term) believe that words like "white" or "horse" name the ►universals "whiteness" and "horseness." Unlike particulars, universals are abstract entities (►abstract entity). They belong to the real (►Realism (as an ontological position)) world though they are not located in space and time. Some realists (Platonists) think a universal can exist even if it has no instances. Others (following Aristotle) assume that for a universal to exist it has to be exemplified by at least one individual thing.

According to ►nominalism, abstract entities do not exist at all. There are only particulars (and perhaps classes of particulars), like individual white things or individual horses. Different things are said to have the same property or belong to the same sort or kind, if the same predicate applies to, or *is true of*, these different things. (Exchange "predicate" with "concept" to get concept nominalism.) Realists object that for a predicate to be true of a particular (or a particular to fall under a concept), that particular must be connected with some real ►entity to which the predicate (or concept) refers. Otherwise the ascription of predicates to particulars would be an arbitrary matter. The nominalists' standard reply is that predicates do not function like names (see for the controversy between realists and nominalists [4,5]).

Some philosophers believe there are *essential* and *accidental* properties. A property F is said to be essential to an entity a, if a could not exist without exemplifying F. For example, the number two has essentially the property of being even. It is hard to see how it could lack that property and still be the number two. With respect to concrete objects, it is more difficult to decide whether a property is essential or accidental. It seems that Jack is essentially a human being. In contrast, his property of being a philosopher is accidental, since he could instead have become a postman. In ►possible world semantics, the idea of an essential property is expressed as follows; a has the property F essentially if and only if a has F in every ►possible world [6]. For a critique of ►essentialism, see [7].

Property and Predicate

If properties are regarded as parts of the real world, they must be distinguished from predicates. Properties are designated by predicates. In the sentence "Jack is bald," the word "Jack" denotes an individual and the predicate "is bald" refers to a property. Unlike names and predicates, properties are not linguistic entities but real features of the world (an assumption not shared by austere nominalists). It is less clear whether properties are different from concepts (meanings of words). They are different if properties are taken as real and concepts as something in the minds of persons (or as something existing in dependence on minds). However, properties may come close to concepts, if concepts too are conceived as mind independent entities or if both properties and concepts are conceived as mind dependent.

Intrinsic and Extrinsic (Relational) Properties

Objects possess some of their properties in their own right. For example, an object may be round (spherical) and have a mass of two kilograms and it has these features independently of how other things are. Such features are called *intrinsic* properties or *qualities*. Other properties are *extrinsic* or *relational* [1]. Socrates is a teacher of Plato and he is married to Xanthippe. These are relational properties of Socrates, since he has them not independently of other things. (According to some philosophers there are no relational properties; they regard properties and relations as ontologically different, and categorize both as attributes.) Though intrinsic properties are not themselves relational, their specification or measurement usually involves relations between objects. Consider for example how mass (an intrinsic property) is specified. To say that this rod has a mass of two kilograms is to say that it would balance, on an equal arm balance, two objects each of which balances the standard kilogram.

Dispositional Properties, Causal Powers, and Functional Properties

Special kinds of relational properties are *dispositional properties* and *causal powers*. This billiard ball has the qualities (intrinsic properties) of being spherical and solid. In virtue of these qualities, it has the disposition to roll when placed on an inclined surface. Other dispositional properties are solubility in water or fragility. Having a special disposition is to produce certain behavior under certain conditions. It seems that dispositions are always grounded in non-dispositional properties. For example, being fragile is grounded in a particular molecular structure. An object is fragile in virtue of having that molecular structure. If an object has the dispositional property G in virtue of the non-dispositional (intrinsic, qualitative) property F, F is called a *first-order property* and G a *second-order property*.

Another special kind of relational properties are *functional properties*. Many things are defined by functional descriptions, for example, a knife, a clock or an eye. An object has the property of being a clock, not because of the material of which it is made, but because it satisfies a certain job description – it keeps time. Similarly, an eye can be made of different materials and take different forms (compare our eye to that of the horse or the honeybee). What makes it an eye is a special functional property or functional role – it extracts information from light radiation and makes that information available to the system it subserves. Functional roles can best be described with the help of causal relations. Therefore, philosophers often speak of "causal roles" instead of functional roles.

Mental Properties

A major problem in the philosophy of mind is the status of mental properties. Physicalists have tried to demonstrate that the mental is nothing over and beyond the physical organism and its physical/biological properties. This view includes the theory that all mental properties are realized as physical/biological properties; according to advocates of reductionist physicalism, mental properties are even reducible to physical/biological properties. Similarly, functionalists have held that mental properties can be understood as functional properties (which are themselves realized as physical properties). Consider for example the property of being in pain. According to functionalism, roughly speaking a mental state is a pain if it is caused by specific stimuli (e.g. tissue damage), gives rise to specific thoughts and wants (e.g. about how to end the unpleasant state) and causes specific behavior (e.g. withdrawal).

It has been objected against physicalism and functionalism that they could not adequately account for *phenomenal qualities* or ►qualia (singular: quale). Some mental states, especially, sensations and feelings, are characterized by specific phenomenal qualities. Think of the painful, hurting character of a sharp headache. Such phenomenal qualities seem to be intrinsic properties not extrinsic (relational) ones. It is hard to see how phenomenal qualities could by identified with or reduced to physical properties, causal powers or functional roles. But if they cannot be reduced to the physical domain, it seems difficult to understand how they can have any effect on the physical organism and its behavior. The status of mental properties, especially phenomenal ones, is still controversially discussed.

References

1. Kim J (1996) Philosophy of mind. Westview, Boulder
2. Plato (1961) Parmenides. In: Hamilton E, Cairns H Plato: the collected works. Pantheon Books, New York
3. Mellor DH, Oliver A (eds) (1997) Properties. Oxford University Press, Oxford
4. Armstrong DM (1989) Universals. Westview, Boulder
5. Loux MJ (1998) Metaphysics. Routledge, London
6. Kripke SA (1980) Naming and necessity. Harvard University Press, Cambridge
7. Grossmann R (1983) The categorial structure of the world. Indiana University Press, Bloomington

Proposition, Propositional Attitudes

Definition

A proposition is what is asserted as the result of uttering a sentence; propositions are considered to be the meanings of closed sentences (as opposed to open sentences or predicates), and they are considered to be the bearers of truth and falsity. Propositions are expressed by that-clauses (the sentence "Hannah laughs," for instance, expresses the proposition (means) that Hannah laughs) and can be thought of as pairs consisting of objects and properties (like <Hannah; laughs>) or relations (like <<Hannah, Fred>; is taller than>).

A subject S can have different mental postures towards a proposition P, for example believing, remembering, desiring, intending, fearing, hoping etc.

In that case, S has a propositional attitude. A propositional attitude is an intentional relation R between a subject S and a proposition P, such that S bears R to P. Propositional attitude ascriptions are made up of a name for some thing, like "Fred," followed by a name for an attitude, like "believes," followed by an expression for a proposition, like "that Hannah laughs."

▶Epiphenomenalism
▶Possible World

Propositional Knowledge

Definition

Propositional knowledge is the kind of knowledge one can communicate using some proposition, i.e. in English using some phrase of the form "that p" where "p" can be replaced by some complete declarative sentence.

▶Knowledge

Proprioception

SIMON GANDEVIA
Prince of Wales Medical Research Institute, Sydney, NSW, Australia

Introduction

The production of ▶voluntary movements usually occurs effortlessly with few errors. We rarely drop objects held in the hand or fall while walking. This requires coordinated activity involving cognitive, sensory and motor areas of the cerebral cortex. Ultimately the correct output is generated by ▶motoneurons in the ▶spinal cord and ▶brainstem which then produce the correct changes in force and length of muscles. For muscles to exert continuously the proper forces on the skeleton requires the brain to plan, initiate and control movements. A set of sensory processes underlies this ability. It is variously termed "proprioception" or "▶kinaesthesia." These terms are now often used interchangeably, although proprioception is broader in its scope.

Proprioception and kinaesthesia refer to a class of sensations or sensory-motor processes which include the ability to detect movement and position at different joints and the ability to judge forces exerted by muscles and the heaviness of objects they lift. Knowledge about the timing of muscle contractions and knowledge about the overall body image are now also covered by these umbrella terms. Deficits that impair proprioception impair the control of voluntary movement and posture, whereas diseases that impair movement and posture usually have deficits in some aspects of proprioception. This synopsis focuses on the sensations of joint position and movement although some other areas are briefly mentioned.

The neural mechanisms underlying proprioception have been hotly debated for more than a century [1]. Much of this controversy has arisen because experimentalists and clinicians have taken a narrow view of what proprioception encompasses. In contrast, theoreticians have proposed complex models of how proprioception might contribute to movement control, but these have been difficult to validate.

On the "input" side, groups of specialized sensory receptors in the skin, joint and muscles respond to mechanical strains and signal the state of the limb to the spinal cord and brain. Evidence of the contribution of each of these receptor groups is summarized below. For these signals to be transformed into useful sensations which can be reported verbally requires central processing so that the different input signals are calibrated to the Newtonian state of the body, and combined or amalgamated to provide a body map (or "representation") which can help guide movement (see also ▶Phantom limb sensation and pain).

Because a particular proprioceptive state (e.g., the angle of a joint) can arise under passive conditions with muscles relaxed, or active conditions with muscles contracting, most proprioceptive models also include an input from motor command signals. Peripheral signals provide feedback that is evaluated against what movement has been commanded and what proprioceptive state might be expected. Various terms for these command signals have been coined including an ▶efference copy or ▶corollary discharge. A requirement for such signals is not unique to evaluation of proprioceptive signals, but is equally necessary in other modalities such as cutaneous, visual and ▶vestibular sensation, in which changes in the peripheral signals can be produced by self-generated forces or externally-generated events. An interesting example occurs when a tickling skin sensation is felt when generated externally but not when the same stimulus is self-generated [2].

Peripheral Proprioceptive Signals

There are two steps in the establishment of a proprioceptive role for a particular class of sensory receptors. First, the receptors must encode variables such as the forces exerted by muscles or changes in joint angles in their discharge. Second, the discharge must be capable of producing a change in perception of that proprioceptive variable. The first step is achieved by recording the discharge of the receptors. The second requires that changes in the discharge evoke changes in sensation. This can be achieved by finding a proprioceptive illusion when the receptors are activated selectively. Alternatively, performance in a proprioceptive test may diminish when the receptors are eliminated, such as when a diseased or damaged joint is replaced by an artificial one.

Of the peripheral proprioceptive signals, those arising in specialized muscle receptors are considered the most important [1,3,4] (see ►Proprioception: Roles of Muscle Receptors, and ►Movement Sense). ►Muscle spindle endings can be activated by local length changes and their firing can also be modified by a specialized ►fusimotor system that can modulate the background firing rate and gain of the endings. The ►primary muscle spindle endings are exquisitely responsive to muscle length changes including vibration. When a muscle (or its tendon) is vibrated at ~100 Hz, subjects experience an illusion consistent with the muscle lengthening, but also occupying a more lengthened position [5]. When the vibration is adjusted so that secondary spindle endings are more effectively activated, the illusion is more one of position than of movement. The central nervous system interprets these unexpected proprioceptive signals according to the current body map (see ►Phantom limb sensations and pain), and this leads to errors in voluntary movements and posture [6]. Muscles also contain specialized receptors (►Golgi tendon organs) which encode the active forces generated by voluntary contraction. These signals probably contribute to sensation of muscle force.

Until recently, specialized receptors in the skin were not considered to have a role in proprioceptive sensations, although it had long been realized that such receptors discharge with movement of nearby joints and that their activity contributed to the reflex control of movement in activities such as walking and grasping. Some populations of ►cutaneous receptor encode the pattern of local skin strain and its change with changes in joint position [7], while others encode the timing of voluntary movement and any disturbances to it (see ►Proprioception: Role of cutaneous receptors). Stretch receptors in the skin (probably innervating ►Ruffini endings) discharge when the skin around a joint is stretched by moving it via threads attached to tape stuck to the skin. Subjects then commonly report that the underlying joints are moving in a direction consistent with the pattern of artificial skin strain. These illusory sensations of movement are amplified when combined with muscle spindle signals evoked in appropriate muscle spindle ending populations by tendon vibration. Illusions evoked by skin stretch can be evoked at joints in the hand as well as at large proximal joints such as the elbow and knee [8]. Insight into the predictive capacity of the cortical proprioceptive decoders comes from observing that when skin and muscle receptors are activated artificially to produce proprioceptive illusions, subjects may report completely impossible anatomical positions. To the subject of the illusion, this extraordinary disruption of reality seems completely natural. Supportive evidence for the proprioceptive role of non-muscle, presumably skin receptors, derived from studies of the hand when nerves innervating the fingers (but not their muscles) have been anaesthetized [9].

Throughout most of the last century, joint receptors were considered so influential in proprioception that the term "►joint position sense" was adopted by clinicians in the belief that such receptors were crucial contributors. Evidence for the belief was much weaker than admitted. Specialized slowly adapting receptors in joint capsules and ligaments usually discharge only at one (or more) extreme of the usual physiological range of joint movement, or they discharge when abnormal stresses are put upon a joint [10]. Relatively few receptors discharge progressively across the movement range in a way that allows them to signal the angular motion (see ►Proprioception: Role of joint receptors). Electrical stimulation of digital nerves or single joint receptors can evoke illusory sensations of joint distortions and movement [11,12]. The difficulty of selectively activating a natural population of these receptors precludes a quantitative assessment of their role. Intra-articular anesthesia or joint replacement does not produce major deficits in proprioceptive sensation, which may simply indicate the redundancy in afferent sources of information (see ►Proprioception and orthopedics).

Motor Commands and Central Processing

In the planning and execution of movement, signals related to the voluntary command are generated. These signals have a number of roles to play in the control of movement and posture [13], but exactly how this is achieved for limb movement is uncertain. Lessons from robotics and engineering, as well as studies in insects and fish, reveal that access to command signals for movement control through predictive modeling of their action on the body, and through monitoring of their outcome via feedback is likely to be important in control of human movement and posture [14].

The oculomotor system is one example where central command signals help ensure the perceptual stability of our visual world. A second example is found in the judgment of exerted force and the heaviness of objects lifted by muscle contraction. Here, signals related to the size of the outgoing command bias the judgments. As a result, we are familiar with the increase in apparent heaviness of objects we lift when muscles become fatigued [15]. This type of perceptual illusion has been rigorously investigated under many circumstances affecting the relationship between the motor command required and the actual muscle force achieved, and the maxim holds that objects appear heavy wherever the efficiency of the neuromuscular apparatus is impaired [3]. However, the perceived force is not a simple readout of the motor command delivered (nor of the motor cortex output), because this readout must be interpreted in the light of ongoing afferent signals.

Experimental studies have usually focused on the role of command-related signals in highly volitional tasks requiring, or triggered by, an external cue. In activities such as walking in which the contractions occur more automatically, the access and processing of command signals may differ. Thus, the voluntary effort required to generate the force that can lift you onto the ball of one foot exceeds that during walking when similar forces are required. An additional complication that has not been resolved is how the muscle spindle signals from voluntarily contracting muscles are interpreted, because their signal is a resultant of the local strain on the receptor induced by the environment and that induced by the fusimotor system. This independent motor supply to muscles can alter the gain and set point of the spindle's response [16]. One way to resolve this complication is the reliance on several afferent sources of input (from the skin, joints and non-contracting muscles) rather than a signal arising solely in the active muscles.

It had long been thought that motor commands did not play a major role in sensing the position and movement of joints. However, recent work with normal subjects in whom a phantom hand has been induced by deafferentation and paralysis challenges this view. The phantom hand moves in the direction in which it is willed to move, and the size of the movement increases with the level of motor command and effort [17]. Hence, the isolated command to move, in the absence of any input from proprioceptors, generates an illusion. If this applies, as seems likely, when there are normal sensory signals, it will be clear that this group of proprioceptive sensations, like force-related sensations, is also biased by the command signals.

Presumably, the brain devises its best estimate of the proprioceptive state based on an amalgam of inputs, some command-related, some originating in different classes of sensory receptors, and some derived from internal predictive models. Visual signals and vestibular signals provide additional frames of reference against which the proprioceptive state of the limbs in extra personal space can be gauged. Vestibular signals extend the proprioceptive system by providing information about the head's position and its movement relative to gravity. Visual signals extend the proprioceptive system by calibrating it against external objects and in some circumstances visual inputs override proprioceptive ones.

This proprioceptive amalgam can adapt to rapidly changing conditions brought about, for example, by muscle fatigue or by changes in a limb's orientation to gravity occurring acutely with natural movements and posture, and chronically in the microgravity of an orbiting spacecraft (see ►Proprioception: Effect of gravity). Here, some receptors may act as "load" sensors. Proprioception must also adapt to other changes brought about by musculoskeletal growth, damage and disease, as well as the impairments accompanying ageing (see ►Proprioception: Effect of ageing). In the elderly, reduced proprioception contributes to postural instability and falling.

Studies using non-invasive imaging of the human brain have revealed that proprioceptive inputs from muscle receptors have specific projections to major cortical areas, particularly the primary motor cortex and its adjacent area (area 3a in the so-called "somatosensory" cortex). These areas have a direct role in movement production [18] as they contain cells projecting to the spinal cord and even to motoneurons. In addition, some other cortical areas are activated in a less specific way, for example, part of the supplementary motor area is active when either the left or right hand is vibrated to produce illusory movements and part of the cingulate motor area is active when the hand or foot is vibrated on the contralateral side. Other parietal and frontal cortical areas are active in producing the proprioceptive "amalgam" that is integrated with our "body image," with visual circuits, and with cognitive centers.

Conclusions

Proprioception is a broad term encompassing several sensations which are needed for normal movement. It is likely that the brain optimizes the use of both peripheral signals from muscle, skin and joint receptors together with signals about the timing, destination and strength of centrally generated command signals. The proprioceptive system must continually determine what changes in the proprioceptive state are self-induced and what represent changes brought about by external forces. It must also adapt to changes occurring to the musculoskeletal system.

While many of the afferent mechanisms are now well established, the central mechanisms are less well understood. Current studies are using a range of non-invasive methods to extract the location, strength and timing of the responsible neural activity. Further studies will focus on the precise deficits that occur when movement control is disrupted by a range of diseases, from schizophrenia to Parkinson's disease. Although it is clear that central processing in traditional somatosensory, motor and associative frontal and parietal areas is involved, the way in which cerebellar and basal ganglia circuits contribute is poorly understood.

References

1. McCloskey DI (1978) Kinesthetic sensibility. Physiol Rev 58:763–820
2. Blakemore SJ, Wolpert DM, Frith CD (1998) Central cancellation of self-produced tickle sensation. Nat Neurosci 1:635–640
3. Gandevia SC (1996) Kinesthesia: roles for afferent signals and motor commands. In: Rowell LB, Shepherd JT (eds) Handbook on integration of motor, circulatory, respiratory and metabolic control during exercise, American Physiological Society, Bethesda, pp 128–172
4. Proske U (2006) Kinesthesia: the role of muscle receptors. Muscle Nerve 34:545–558
5. Goodwin GM, McCloskey DI, Matthews PB (1972) The contribution of muscle afferents to kinaesthesia shown by vibration induced illusions of movement and by the effects of paralysing joint afferents. Brain 95:705–748
6. DiZio P, Lathan CE, Lackner JR (1993) The role of brachial muscle spindle signals in assignment of visual direction. J Neurophysiol 70:1578–1584
7. Edin BB, Abbs JH (1991) Finger movement responses of cutaneous mechanoreceptors in the dorsal skin of the human hand. J Neurophysiol 65:657–670
8. Collins DF, Refshauge KM, Todd G, Gandevia SC (2005) Cutaneous receptors contribute to kinesthesia at the index finger, elbow, and knee. J Neurophysiol 94:1699–1706
9. Gandevia SC, McCloskey DI (1976) Joint sense, muscle sense, and their combination as position sense, measured at the distal interphalangeal joint of the middle finger. J Physiol 260:387–407
10. Burke D, Gandevia SC, Macefield G (1988) Responses to passive movement of receptors in joint, skin and muscle of the human hand. J Physiol 402:347–361
11. Gandevia SC (1985) Illusory movements produced by electrical stimulation of low-threshold muscle afferents from the hand. Brain 108:965–981
12. Macefield G, Gandevia SC, Burke D (1990) Perceptual responses to microstimulation of single afferents innervating joints, muscles and skin of the human hand. J Phsiol 429:113–129
13. Gandevia SC (1987) Roles for perceived motor commands in motor control. Trends Neurosci 10:81–85
14. Frith CD, Blakemore SJ, Wolpert DM (2000) Abnormalities in the awareness and control of action. Philos Trans R Soc Lond B Biol Sci 355:1771–1788
15. Gandevia SC, McCloskey DI (1977) Sensations of heaviness. Brain 100:345–354
16. Matthews PBC (1981) Evolving views on the internal operation and functional role of the muscle spindle. J Physiol 320:1–30
17. Gandevia SC, Smith JL, Crawford M, Proske U, Taylor JL (2006) Motor commands contribute to human position sense. J Physiol 571:703–710
18. Naito E, Nakashima T, Kito T, Aramaki Y, Okada T, Sadato N (2007) Human limb-specific and non-limb-specific brain representations during kinesthetic illusory movements of the upper and lower extremities. Eur J Neurosci 25:3476–3487

Proprioception and Orthopedics

Kathryn M. Refshauge
Faculty of Health Sciences, University of Sydney, Sydney, NSW, Australia

Synonyms

Kinesthesia; Movement detection; Position sense; Movement discrimination

Definition

Proprioception refers to the group of sensations related to limb position and movement, muscle force, timing of muscle contractions and posture and size of a body "schema" of one or more joints. Peripheral proprioceptive signals arise from discharge of receptors located in muscle, joints and cutaneous tissue, although there is continuing debate concerning their relative contribution to conscious proprioceptive sensations and the nature of their contribution. Proprioceptive sensations are also derived from centrally generated motor commands and the interaction between the afferent and efferent signals.

Characteristics

Quantitative Description

Each of the proprioceptive sensations requires a different method of testing, with consequent different levels of acuity according to the test method. Acuity varies among joints in a single individual, even when tested with the same method. The only proprioceptive sensation to be systematically measured for human joints in the same individual is the sensation of joint movement, measured using detection levels for perception of passive movement. In both the leg and the arm, proximal joints have better acuity than distal joints when expressed as angular displacement. For example, in the leg, for passive movements imposed at 0.5°/s, movements of ~0.2° can be detected at the hip, ~0.6° at the knee, ~0.5° at the ankle and ~21° at the interphalangeal joint of the big toe. Similarly in the arm, for passive movements imposed at 1°/s, movements of ~0.2° can be detected at the shoulder, ~0.6° at the elbow and ~5° at the distal interphalangeal joint of the finger.

Test conditions can be manipulated to increase or decrease proprioceptive acuity. Small contractions of the muscles around the joint can improve acuity tenfold whereas fatigue increases the error in position matching tasks. Muscle history can also alter acuity. Finally, proprioceptive acuity improves with increasing velocity and decreases with increasing age.

Proprioceptive performance cannot be generalized across the wide variety of tests in common use because there is no relationship between the different tests. Therefore a single test is not adequate to make judgments about general proprioceptive status in health or after injury. This lack of relationship is likely to account, at least in part, for the conflicting results found for the effect of injury on proprioception.

Higher Level Components

Cortical projections must exist for a role to be assigned to the peripheral receptor discharge measured during movements. Such projections have been established for all proprioceptors.

Lower Level Components

Three classes of afferent contribute to proprioceptive sensibility, including cutaneous, muscle and tendon and joint capsule and ligament afferents. The adequate stimulus for all three is stretch of the tissue in which they are located, although cutaneous and joint receptors also respond to other stimuli. The major debate over the last century has concerned the contribution of each to proprioceptive sensations. It is now clear that all classes contribute to proprioceptive sensations and the debate is now directed at the nature of the contribution. Although it was hypothesized that cutaneous afferents may facilitate muscle afferents, more recent evidence suggests that such facilitation is unlikely. Based on both psychophysical and microneurographic evidence, the contemporary view is that cutaneous input is integrated with that from muscle spindles to provide accurate perception of joint position and movement.

Function

The ability to perceive limb movement and position is essential to normal movement and posture. Without this ability, function is severely disturbed because attempts to move must be guided and monitored using vision. Proprioception underlies all normal movement and its control.

Orthopedic Pathology

Interest in the effect of orthopedic pathology on proprioception has most often been directed at osteoarthritis, joint replacement and ligament injuries and rarely at non-specific conditions such as low back and neck pain. The effect of orthopedic pathology on proprioception varies with the pathology but a proprioceptive deficit has been found consistently in very few conditions. In most orthopedic pathologies, a deficit is not evident, the literature is inconsistent or a deficit has been found in some but not all of the different proprioceptive tests. The more important question concerns the relevance of a deficit and the magnitude of deficit required for it to impact on function.

Osteoarthritis (OA)

Proprioception has been extensively investigated in knees with OA. All studies have consistently found impairments in both joint position sense and detection of movement compared with healthy controls [1]. The magnitude of the deficit in movement detection was ~1.3° (normal ~2.3°) and in joint position sense ~2.9° (normal ~2.6°). Such a deficit has also been found in the contralateral unaffected knee, with proprioception in both knees being impaired compared with healthy controls [1], leading to speculation that this could indicate central changes. However, no correlation has been found between proprioceptive impairment and severity of disease [1], function or pain. Given the lack of correlation between proprioceptive deficit and function and the small magnitude of the deficit, the consequence of this deficit is unlikely to be functionally relevant.

Total Joint Replacement

It is of particular interest that total joint replacement for knee OA does not necessarily impair proprioception, despite removal of the articular surface. Different studies have found total joint replacement to variously impair, not change or even improve proprioception, although probably not to normal acuity. The magnitude of reported deficits was small, being between 0.3° and 0.72° for threshold to movement detection. However, even when proprioception was found to be impaired after surgery, balance measurements remained normal and clinical outcome was reported as excellent for replaced knees [2], indicating that small proprioceptive impairments do not manifest as functional deficits.

The most common *ligament* injuries involve the anterior cruciate ligament (ACL) at the knee and the lateral ligament complex at the ankle. Proprioception has been extensively investigated in both injuries, motivated by the belief that proprioceptive deficits are associated with persisting disability.

Anterior Cruciate Ligament Rupture

It has been suggested that the ACL has an important proprioceptive role via the fusimotor system and that its loss should therefore cause a significant proprioceptive deficit. However, the evidence does not uniformly support this proposal. A deficit in threshold to movement detection of 1.5° has been found compared with the contralateral uninjured knee and of 0.5° compared with healthy controls [3], but others found no impairment. Furthermore, no impairment was found in joint position sense in ACL deficient knees compared with healthy controls by most authors, although a deficit of up to 4.5° was found by others [4] and the deficit was significantly worse in the mid-range of movement [4]. When both threshold to movement detection and joint position sense were measured in the same subjects, there was no correlation in performance [5], suggesting that a deficit may be specific and that performance on a single proprioceptive test cannot be generalized. However, no study investigated proprioception in rotation movements, the direction of instability in ACL-deficient knees.

The effect of reconstructive surgery after ACL rupture on proprioception is also unclear. After surgical reconstruction, some authors found that movement detection was restored to normal, others found improved acuity but not to normal levels and unexpectedly, Co et al. [6] found that reconstructed knees were even more accurate than healthy controls by 0.4°. The findings for joint position sense were also inconsistent; reconstruction was variously found to confer no benefit [6] or to improve acuity, but not to normal [3]. These conflicting results potentially arise because not all ACL-deficient knees have a proprioceptive deficit to address. It is tempting to suggest that any improvement is due to the improved stability of the joint, but proprioceptive acuity correlated poorly with mechanical stability [4,3]. Alternatively, improved joint kinematics may normalize input from the muscles, thereby enhancing proprioceptive signals.

Even when proprioceptive deficits were found, they were generally small. Nevertheless the impact on function is uncertain. Although some authors found a high correlation (range, r = 0.6–0.9) between proprioception and function [4], others failed to find a relationship [3]. It is difficult to explain these conflicting results. It is possible that the different findings relate to different comorbid pathologies included in the studies, such as associated meniscal damage, although again the findings are inconclusive; concomitant meniscal or chondral damage was found to relate to proprioceptive deficits by some but not others [3].

Lateral Ligament Sprain at the Ankle

Persisting symptoms of pain and instability are common after ankle inversion sprain and are frequently attributed to a deficit in proprioception. Impaired movement detection has been found by most authors in the inversion-eversion plane after recurrent sprain compared with healthy controls, the deficit in movement detection being <1° and in some cases <0.5°. However, in the plantarflexion-dorsiflexion plane most authors found no impairment [7]. Findings for joint position sense in the inversion-eversion plane are less consistent; a significant impairment was found by some authors but not by others. Joint position sense in the plantarflexion-dorsiflexion plane has yet to be investigated. Consistent with other ligament injuries, there was no correlation between proprioceptive performance and mechanical stability.

These findings suggest that any proprioceptive deficit after ankle inversion sprain is specific to the inversion-eversion plane of movement and to particular proprioceptive tests. The deficit is potentially related to the anatomical structures involved in the sprain, particularly those that resist inversion rather than plantarflexion forces. Damage to these structures may therefore cause impairment only in proprioceptive tests specific to the inversion-eversion plane of movement.

Therapy

Proprioceptive training is considered an integral part of rehabilitation after many orthopedic injuries to improve assumed deficits and to prevent further injury, despite lack of knowledge about: the magnitude of deficit that is clinically meaningful; whether proprioception can be trained; and, if a deficit does exist and proprioception can be trained, whether improved proprioception is associated with improved function and reduced risk of re-injury.

The magnitude of proprioceptive deficit, when it exists in orthopedic pathology, is small in most test paradigms, generally being within one standard deviation of the average performance of control subjects. Although this difference is statistically significant in some cases, it is unlikely to be functionally relevant, because a deficit has not consistently been associated with pain or loss of function. It is possible, however, that a small deficit may be critical in situations where small errors could have considerable impact on highly skilled performance.

The most common forms of "proprioceptive training" are performance of tasks on an ankle disc or wobble board and agility training. It is indisputable that such training improves motor control; however, whether proprioception is improved is largely unknown. In the two studies that examined the effect of training on joint position sense at the ankle, one found selective improvement of ≥0.5° in error reduction at some, but not all, test angles [8], while the other found no change with training. There is currently insufficient evidence to conclude that either joint position sense or movement detection can be improved.

The major purpose of "proprioceptive" training, however, is to prevent re-injury. There is no direct evidence of a role for proprioception in prevention of injuries because the relationship between proprioception and injury risk has not been established. The few studies that have investigated the efficacy of "proprioceptive" training for injury prevention, have found the risk to be reduced, although the relationship between proprioception and injury was not investigated. Risk of ACL injury was reduced by 88% by agility training compared with control intervention. The pooled results of Wedderkop et al. [9] and Verhagen et al. [10] show that ankle disc training programs significantly reduced incidence of ankle injury, with an odds ratio for the pooled data of 0.41 (95% CI 0.268–0.629). This reduction in risk of injury has not been correlated with changes in proprioception and proprioception has not been used as an outcome measure.

Conclusions

Taken together, the findings suggest that a proprioceptive deficit has been consistently found for degenerative joint disease, but not for any other pathology and these deficits may be improved, perhaps to normal, by surgical intervention. Selective proprioceptive deficits occur in unstable joints, but the deficit is generally small. The lack of consistent findings may be attributed to the lack of relationship in proprioceptive acuity among different proprioceptive tests and therefore identification of a deficit is likely to depend on the specific proprioceptive test measured.

Given the poor relationship between proprioception and function, the relevance of small proprioceptive deficits is unknown. Potentially these small deficits could impair highly skilled performance, but are unlikely to manifest as functional problems in normal daily activities and therefore may not require remedial therapy.

References

1. Koralewicz LM, Engh GA (2000) Comparison of proprioception in arthritic and age-matched normal knees. J Bone Joint Surg 82-A:1582–1588
2. Pap G, Meyer M, Weiler HT, Machner A, Awiszus F (2000) Proprioception after total knee arthroplasty: a comparison with clinical outcome. Acta Orthop Scand 71:153–159
3. Reider B, Arcand MA, Lee H et al. (2003) Proprioception of the knee before and after anterior cruciate ligament reconstruction. Arthroscopy 19:2–12
4. Fremery R, Lobenhoffer P, Zeichen J, Skutek M, Bosch U, Tscherne H (2000) Proprioception after rehabilitation and reconstruction in knees with deficiency of the anterior cruciate ligament: a prospective, longitudinal study. J Bone Joint Surg 82:801–806
5. Friden T, Roberts D, Zatterstrom R, Lindstrand A, Moritz U (1997) Proprioception after an acute knee ligament injury: a longitudinal study on 16 consecutive patients. J Orthop Res 15:37–644
6. Co FH, Skinner HB, Cannon WD (1993) Effect of reconstruction of the anterior cruciate ligament on proprioception of the knee and the heel strike transient. J Orthop Res 11:696–704
7. Refshauge KM, Kilbreath SL, Raymond J (2003) The effect of recurrent ankle sprain on proprioceptive acuity for inversion-eversion movements. J Orthop Sports Phys Ther 33:166–173
8. Eils E, Rosenbaum D (2001) A multi-station proprioceptive exercise program in patients with ankle instability. Med Sci Sports Exerc 33:1991–1998
9. Wedderkopp N, Kaltoft M, Lundgaard B et al. (1999) Prevention of injuries in young female players in European team handball: a prospective intervention study. Scand J Med Sci Sports 9:41–47
10. Verhagen E, van der Beek A, Twisk J, Bouter L, Bahr R, van Mechelen W (2004) The effect of a proprioceptive balance board training program for the prevention of ankle sprains. Am J Sports Med 32:1385–1393

Proprioception: Effect of Aging

Stephen R. Lord
Prince of Wales Medical Research Institute and University of New South Wales, Sydney, NSW, Australia

Definition

Proprioception is the discrimination of the positions and movements of body parts based on information other than visual, auditory or verbal. The immediate stimuli come from changes in length and from tension, compression and shear forces arising from the effects of gravity, from the relative movements of body parts and from muscular contraction [1]. Studies assessing proprioceptive age changes have used the terms proprioception, kinaesthesis and joint position sense interchangeably and both passive and active movement paradigms have been used to assess proprioceptive judgments.

Ageing is an orderly or regular transformation with time of representative organisms living in representative environments.

Characteristics

Function

In 1937, Laidlaw and Hamilton [2] were the first researchers to demonstrate an age related decline in proprioception. They measured the ability of 60 subjects to detect movement in the shoulder, elbow, wrist, metacarpophalangeal joints in the upper limb and hip, knee, ankle and first metatarsophalangeal joints in the lower limb. They found that subjects aged 17–35 years had lower thresholds and superior ability to detect direction of joint movements than subjects aged 50–85 years. The hip was the most sensitive joint and the metatarsophalangeal joint was the least sensitive. They also found a wide range in perceived movement threshold and that this variability was most marked in the older age group.

More recent studies and referenced in Lord et al. [3] have, in general, confirmed Laidlaw and Hamilton's findings. Kokmen et al. [3] measured the threshold at which subjects could detect metacarpophalangeal and metatarsophalangeal joint motion. These joints were passively moved in an increasing sinusoidal flexion extension mode (a nonspecific test of joint movement only). They found that there was no difference in metacarpophalangeal joint motion detection between young subjects (aged 19–34 years) and older subjects (aged 61–84 years) at all frequencies between 0.5 and 8 Hz. They did find, however, a significant difference in metatarsophalangeal joint motion detection between young and old subjects at the lowest frequency of 0.5 Hz. At higher frequencies, the older age group showed higher but insignificant increases in metatarsophalangeal joint motion thresholds. Variability in threshold perception performance increased with age and joint motion thresholds for the metacarpophalangeal joint were significantly smaller than for the metatarsophalangeal joint in the older group at all frequencies except for 8 Hz. No such trend was evident in the young age group, which suggests that joint motion sense may decline more with age in the lower limb compared with the upper limb.

In a recent study, Ferrell et al. [3] used a position-matching task to assess the ability of subjects to detect displacements at the proximal interphalangeal joint of the index finger. These displacements were imposed at an angular velocity of 2°/min, which is below the threshold for movement detection. An older group of subjects (mean age = 57 years) showed significantly poorer performance in detecting the position of the index finger than a group of younger subjects (mean age = 24 years). The size of the matching error had correlation of 0.47 with age.

Two studies have examined the effect of age on the joint position sense of the knee. Skinner et al. [3] measured joint position sense of the knee in 29 subjects with normal knee joints ranging in age from 20 to 82 years. Joint position sense was determined by the threshold of detection of an experimenter-induced movement of the knee joint and by the ability to reproduce passive knee positioning. They found that joint position sense as measured by both tests deteriorated significantly with age.

Kaplan et al. [3] used a clinical goniometer to measure the ability of 29 women to match the position of each knee with the other knee and like Skinner et al. to reproduce a knee joint position after a period of rest. They found that an older age group (aged 60–80 years) performed significantly worse than a young group (aged 22–27 years) in both tests. In the experimental paradigm where subjects had to match the position of a knee with the other knee, they found no difference between the number of trials undershot and overshot in the young group, but a tendency to underestimate knee joint angle in the older group. The reported failure of the older group to reproduce the knee joint position may however have resulted from motor deficits as well as proprioceptive deficits.

In a study of 550 women aged 20–99 years Lord et al., there was a decline in proprioception with age as measured with a seated lower limb-matching task [4]. The average error in aligning the great toes either side of a vertical protractor placed between the legs increased from 0.7 degrees (SD = 0.3) in subjects in their twenties to 1.9 degrees (SD = 1.7) in subjects aged 65 plus years. Overall, there was a weak but significant association between error size and age across all age groups, $r = 0.20$.

Three clinical studies have investigated whether there is a decline in proprioception beyond 65 years of age. Howell [3] examined the ability of 200 patients aged 65 years and over to detect joint position sense of the great toes and to touch their noses with their eyes closed. He found that most patients could determine experimenter-induced movements of their toes but that approximately 44% of patients in all age groups above 65 showed abnormalities in the nose touching test. There were no significant changes with age in either test.

MacLennan et al. [5] also examined the ability of 308 elderly persons to identify the position of their great toes after they were manually moved by the experimenter. They found that the prevalence of inability to appreciate changes in position of the toe was significantly greater in the age group 75 years and over than in the age group aged 65–74 years for both males and females. Similarly, Brocklehurst et al. [6] measured the ability of 151 persons aged 65 years and over to detect experimenter-induced movements of the toe and ankle. They found no association between age and proprioception, but acknowledged that this may be due to the imprecision of their test.

It has been suggested that caution should be used in assessing proprioception in the lower limb when assessments are made while subjects are in the seated position [7], as is the case in the above studies. This is because thresholds in the ankle and knee are much lower when measured in the standing, weight bearing position where engagement of the leg muscles is greatly increased [1,3].

In a recent study, Bullock-Saxton et al. [7] assessed the effects of age on the accuracy of a knee joint repositioning test in both full and partial weight-bearing conditions in 60 healthy, pain-free subjects from three age groups (young: 20–35 years old, middle-aged: 40–55 years and older: 60–75 years). They found that subjects in all three groups performed better when full weight bearing than when partial weight bearing, and significant age related increases were found only in the partial weight bearing condition. However, other studies assessing position sense of the ankle joint when weight bearing have reported increased thresholds with age. Thelen et al. [3] compared the ability of young and older women to detect dorsiflexion and plantarflexion movements of the foot when weight bearing on a moveable platform and reported that the threshold for movement detection was 3–4 times larger in the older group. High detection thresholds for inversion and eversion movements of the ankle when standing either unipedally or bipedally on a rotating platform have also been found in older people [8]. Finally, Blaszczyk et al. [3] have reported that old subjects are significantly worse than young subjects in reproducing ankle joint positions when standing on a rotating platform.

Pathology

Reduced proprioception is associated with instability and falls in older people. Hurley et al. found that reduced proprioceptive acuity at the knee was significantly associated with an increased aggregate time to perform a range of function tasks comprising a timed walk, the get up and go test and stair ascent and descent in a combined group of young, middle aged and older people. In large prospective studies Lord et al., impaired lower limb proprioception was also associated with multiple falls in older people living independently in the community [9] and in those living in residential care [10]. Brocklehurst et al. [6] also reported a significant association between impaired proprioception in the ankle and/or great toe and falls in people aged 75–84 years.

Conclusion

There is a significant decline and an increase in variability in proprioception with age, particularly in the lower limbs. Proprioceptive thresholds for the ankle and knee are much lower when measured in the standing, weight bearing position. One study has reported no significant age related change in knee proprioception in this condition, but other studies have reported age related declines in the ankle joint when weight bearing. Reduced peripheral sensation is associated with falls in older people, when the measures of peripheral sensation are accurately and quantitatively ascertained.

References

1. Howard IP, Templeton WB (1966) Human spatial orientation. Wiley, London
2. Laidlaw RW, Hamilton NA (1937) A study of thresholds in apperception of passive movement among normal control subjects. Bull Neurol Inst 6:268–273
3. Lord SR, Sherrington C, Menz H (2001) Falls in older people: risk factors and strategies for prevention. Cambridge University Press, Cambridge
4. Lord SR, Ward JA (1994) Age-associated changes in sensori-motor changes and balance in community dwelling women. Age Ageing 23:452–460
5. MacLennan WJ, Timothy JI, Hall MPH (1980) Vibration sense, proprioception and ankle reflexes in old age. J Clin Exp Gerontol 2:159–171
6. Brocklehurst JC, Robertson D, James-Groom P (1982) Clinical correlates of sway in old age – sensory modalities. Age Ageing 11:1–10
7. Bullock-Saxon JE, Wong WJ, Hogan N (2001) The influence of age on weight-bearing joint reposition sense of the knee. Exp Brain Res 136:400–406
8. Gilsing MG, Vanden Bosch CG, Lee SG et al. (1995) Association of age with the threshold for detecting ankle inversion and eversion in upright stance. Age Ageing 24:58–66
9. Lord SR, Ward JA, Williams P, Anstey K (1994) Physiological factors associated with falls in older community-dwelling women. J Am Geriatr Soc 42:1110–1117
10. Lord SR, Clark RD, Webster IW (1991) Physiological factors associated with falls in an elderly population. J Am Geriatr Soc 39:1194–1200

Proprioception: Effect of Gravity

James R. Lackner, Paul DiZio
Ashton Graybiel Spatial Orientation Laboratory, Brandeis University, Waltham, MA, USA

Synonyms

Kinesthesia

Definition

Proprioception is the sensory registration of the ongoing spatial configuration of the body. Sherrington invented the term to denote responsivity to the internal, self-generated state of the body based on mechanoreceptors embedded within body tissues, subserving both conscious awareness and automatic control of posture and movement [1]. The term "kinesthesia" refers more specifically to the awareness of the position and movement of body parts. The perception of force, effort or heaviness relies on some of the same sensory and central mechanisms as proprioception and kinesthesia.

On earth, posture and locomotion are always carried out against the omnipresent force of gravity that accelerates objects downwards toward the earth's surface. The musculature of the body has to exert forces across the joints to keep the body straight so that it does not buckle and fall to the ground. In this context, proprioception is highly dependent on interrelating signals from the muscle spindle receptors within intrafusal muscle fibers to patterns of alpha and gamma motoneuron activation innervating the muscles that support the whole body or an individual limb against the acceleration of gravity [2]. Golgi tendon organs and Ruffini endings of the joints may participate as well, though their contribution is less well understood. Slowly adapting cutaneous stretch receptors are stimulated by distension of skin overlying moving joints and contribute to the sense of position and movement. The fingertip is especially rich in cutaneous mechanoreceptors and dragging or holding the finger lightly over a stable surface augments the perception and control of arm movements and of locomotor activity. During standing posture, the support surface exerts a contact force on the soles of the feet that counteracts the downward acceleration of the body's center of mass, and the distribution of the contact forces on the soles of the feet provides information about the orientation of the body to gravity. The vestibular system is an extension of the proprioceptive system because it senses head movement relative to gravity, and various sensor types in the viscera and vasculature [3] also contribute to estimates of body orientation. Although vision and audition are not typically considered part of the proprioceptive system, there are strong interactions among seen, heard and felt body configurations and the localization of external objects [4]. A rough estimation of body configuration and motion could perhaps even be acquired through the sense of smell. These examples illustrate the domain and sources of proprioception.

Characteristics

Higher Level Processes

Dynamic Sensory-Motor Calibrations to 1g

Until relatively recently in history, humans were not exposed to force levels other than the $1g$ ($g = 9.8$ m/s^2) background acceleration of Earth gravity, except momentarily, for example during jumping or running or horse riding. With the advent of powered vehicles including aircraft and space craft and orbiting space stations, exposure to weightlessness and to high force fields has become commonplace. On Earth, if an arm is raised, it is subjected to a gravity torque that must be opposed. This torque is not constant but depends on arm angle, see Fig. 1.

Nevertheless, it seems to require roughly constant effort – except in cases of fatigue – to position the arm relative to gravity. Under normal conditions, one does not sense the different force demands associated with changes in body configuration or body movements relative to gravity. The effects of gravity on movement control seem relatively transparent – as if they were not there. This transparency actually represents a form of

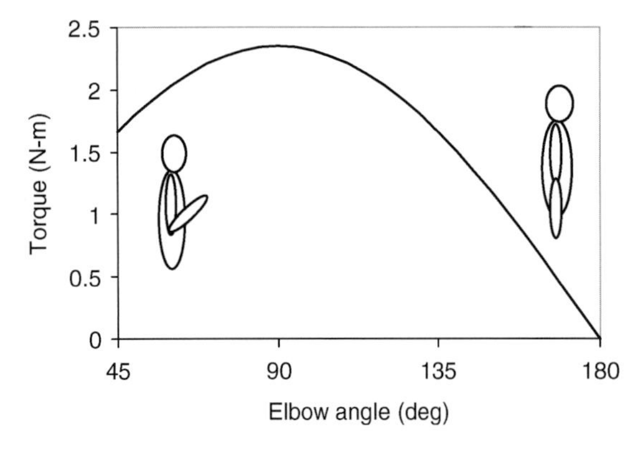

Proprioception: Effect of Gravity. Figure 1 Joint torque.

sensory-motor calibration to earth gravity – based on an internal model of the consequences of gravity on body actions, so that equal extents of movement seem equivalent regardless of the actual muscle force demands to achieve them [5].

Muscle spindle gain is affected by background force level. As a consequence, tonic vibration reflexes are attenuated in weightless conditions and augmented in greater than $1g$ background force levels. These reflexes occur because mechanical vibration of a muscle can excite muscle spindle primary and secondary endings, which leads to a reflexive contraction of the vibrated muscle. The g-dependent modulation of the tonic vibration reflex probably reflects a vestibulo-spinal modulation of the gain of spindle receptors of postural muscles, but could include altered patterns of alpha-gamma coactivation as the load demands on the muscles vary with force level. Studies of arm movement control in non-earth gravity force levels indicate patterns of change consistent with variations in spindle discharge levels [6]. When individuals practice in normal conditions to make arm movements of particular velocities and amplitudes and then attempt the same arm movements in parabolic flight in $0g$ and $1.8g$ force backgrounds, they exhibit systematic changes. Movements made at a natural speed in $0g$ are smaller in amplitude and have more dynamic overshoots of final end positions than movements of the same attempted velocity and amplitude in $1.8g$. This pattern is consistent with decreased spindle gain in $0g$. With long-term exposure to weightlessness, astronauts display an adaptive modification of their internal model of the effects of gravity on movement, which also may involve a recalibration of proprioception.

Self/Environment Discrimination

Proprioception figures prominently not just in the awareness of ongoing body configuration but also in the maintenance of a stable perceptual distinction between changes in afferent input due to self-motion versus motion of or within the environment. This is reflected in the adaptive resetting of internal models of the environment. The function of these tuning mechanisms becomes apparent when the body is exposed to a background force level different in magnitude from earth gravity. Then, until adaptation occurs, misperceptions of body motion and of the stability of the support surface result. Changes in background force level can be achieved in an aircraft flown in a parabolic trajectory to generate alternating periods of $1.8g$ (nearly twice Earth gravity) and of weightlessness ($0g$). In this circumstance, the body alternately becomes much heavier and much lighter than usual respectively. The consequences of this sudden weight change become apparent when an individual attempts to make deep knee bends in the $1.8g$ background force level [7]. Lowering the body involves eccentric contraction (controlled lengthening) of the body's antigravity muscles (e.g. the quadriceps and gastrocnemius muscles of the legs). More activation of these muscles will be required during the course of the movement than is the case on earth, otherwise the body would descend too rapidly because of its increased weight. In this situation, the individual will experience him or herself as moving downward too rapidly and the support surface, the deck of the aircraft, as simultaneously rising upward against the feet. To return to the upright, much greater muscular force than normal is required, the body seems to rise too slowly and the aircraft deck seems to move downward. In 0.5*g* background force levels, the opposite pattern is experienced during deep knee bends. Body weight is less than normal and less muscle force is required throughout the movements. As the body is lowered, it seems to move downward too slowly and the deck simultaneously seems to move downward under the feet. On rising back to the upright, the body feels as if it has moved too rapidly and too far and that the deck floor has simultaneously moved upward under the feet, causing the whole body to displace farther than intended.

If an individual makes repeated deep knee bends during exposure to a steady state non-1g background force level, the movements soon begin to feel more normal and natural and the illusory displacement of the stance surface is correspondingly attenuated. After 50–100 repetitions, the movements will again feel normal and the support surface stable [8]. But then on re-exposure to a normal 1g-force background, deep knee bends will again feel abnormal and the support surface will seem to displace. Fig. 2 shows the patterns of motion and effort experienced on (a) initial exposure to 1.8*g* force levels in parabolic flight, (b) after the execution of many deep knee bends and (c) on return to normal 1*g* levels. As illustrated, full adaptation is achieved and the aftereffect experienced is opposite in sign to the effects experienced during initial exposure to the abnormally high force level.

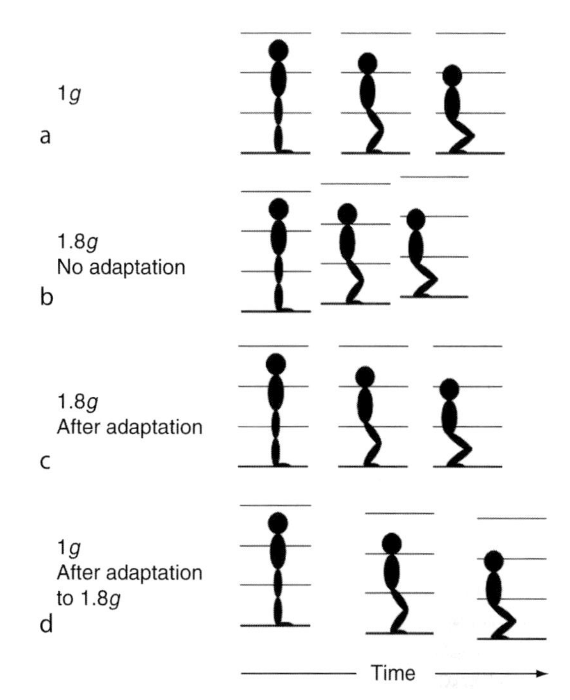

Proprioception: Effect of Gravity. Figure 2
(a) Schematic illustration of the sequence of actual and perceived events as one desends in a deep knee bend. (b) During the first deep knee bend made in 1.8*g*, one's body seems to descend too fast while the floor (*thick lines*) and visual surround (*thin lines*) rise. (c) After adaptation to 1.8*g*, self motion is perceived normally and the environment feels stable. (d) Returning to 1.0*g* elicits after effects opposite to the original illusion in 1.8*g*, until re-adaptation is complete.

This pattern means that motor control and position sense are recalibrated during exposure to the 1.8*g* force level and that this recalibration carries over on return to 1*g* where it is inappropriate. Animals evolved in the context of a 1*g*-background force level without any experience with steady state non-1*g* force levels. Nevertheless, significant adaptation to an unusual background force occurs within seconds or a few minutes.

The factors underlying adaptation to an increased force background can be understood as follows. As the body lowers in a deep knee bend in a 1.8*g* force background, the antigravity muscles that undergo controlled lengthening will be longer than normally would be the case for the levels of alpha and gamma motoneuron activation present, because body weight has nearly doubled. Consequently, these muscles need to be innervated at higher than normal levels. Signals from the spindle receptors within the muscles will be higher than normal also, because the spindles are under a greater degree of loading than normal. Position sense is generated in part by the CNS comparing patterns of spindle feedback with patterns of alpha-gamma activity. If the pattern of spindle feedback is aberrant in relation to the alpha-gamma activation patterns, a distortion of position sense occurs. Abnormally high levels of spindle activity in skeletal muscles evoke misperceptions of the joint positions controlled by the muscles, with the muscles being interpreted as being longer than they actually are. For example, an abnormally high level of quadriceps activity is associated with the knee joint being perceived as more flexed than it actually is.

When deep knee bends are initially made during exposure to 1.8*g*, the abnormally high spindle activity levels present in relation to the alpha motoneuron activity are interpreted as the legs being more flexed during the course of the lowering movement than they actually are. The CNS attributes this to external motion, the aircraft deck rising under the feet causing the knees to be more flexed than intended. With repetition the deep knee bends feel progressively more normal as the relationship between body movement, motor commands and expected spindle feedback is remapped in the internal model of movement control to be the "normal" state of affairs. On return to normal $1g$ background force levels, the relationship between motor commands and spindle feedback will again be abnormal. As the body is lowered, less innervation of the antigravity muscles will be necessary than in $1.8g$ to control the lowering of the body. Consequently, the spindle feedback generated will be less than expected; this is interpreted as the legs being less flexed than intended and attributed to the deck moving downward under the feet, preventing the knees from bending appropriately.

Proprioception and the Body Map

That individuals born without a limb can still experience a phantom limb (see Phantom limb sensations and pain) attests to the fact that a map of body segments and their connectedness exists in the brain, which is complementary to proprioception. We normally experience this map as a sense of body topology, the connectedness of a set of body segments. This map both governs proprioception and can be altered transiently or be semi-permanently remapped by proprioceptive inputs, as shown by vibratory myesthetic illusions. For example, illusory forearm extension is perceived if 120 Hz vibration is applied over the biceps of the restrained arm. The perceived finger trajectory is constrained by the internal map, which represents the forearm with a specific length and the biceps muscle as acting across the elbow joint. However, if the index finger of the vibrated arm is touching the nose then illusory elongation of the nose is perceived simultaneously with arm extension, violating the internal mapping of the nose as a non-jointed semi-rigid body part [9]. Longer-term vibrotactile skin stimulation alters the cortical somatosensory map.

Multisensory Body Maps

Multiple, interdependent representations of hand and arm position exist which are influenced by visual and muscle spindle inputs. This interdependence has been demonstrated in experiments involving vibratory myesthetic illusions of the forearm in a dark room with the test subject's finger made visible by phosphorescent paint [4]. When subjects feel their forearm move, they also see their finger move as well but through a smaller distance. The magnitude of felt motion of the forearm is greater in darkness than with the finger visible. Thus, vision affects the proprioceptive representation because the optically stationary finger attenuates felt motion, and proprioception affects the visual representation because the vibration-induced muscle spindle afferent signal makes the optically stationary finger appear to move.

If a visual-proprioceptive discrepancy is introduced for a longer exposure period, then semi-permanent adaptive changes in proprioception can occur. The nature of these changes depends on the exposure conditions. For example, a seated subject wearing laterally displacing prism spectacles will initially miss when pointing to a visual target. However, if the subject continues to reach and is allowed visual feedback of the reaches, accuracy will soon be regained. The adaptive change involves an arm-torso recalibration in terms of an internal shift in the felt position of the exposed arm but not of the other arm [10]. The adapted subject still wearing the prisms will reach accurately to the real target but see and feel the arm in its optically displaced location. The felt arm position then does not correspond to the real arm position but it does correspond to the visual arm position. By contrast, if subjects are allowed to walk around wearing the prism spectacles they are clumsy at first but ultimately learn to navigate. They also gradually adopt a deviated head posture but feel their head to be straight on their torso. This internal proprioceptive remapping of head to torso allows accurate body-relative localization and normal control of locomotion.

Pathology

With the advent of manned space flight, humans have been exposed to weightless conditions for prolonged periods. On return to earth, they exhibit a variety of derangements of posture and gait. These disturbances are related to adaptive changes in neuromuscular structure and control occurring in weightless conditions that are inappropriate for terrestrial conditions. Post-flight, when astronauts walk, the ground feels unstable under their feet and their movements require greater than normal effort. The simplest movement, e.g. raising the arm, can feel as if it requires immense effort and force. If the astronaut does a deep knee bend, then it feels as if he or she has moved downward too rapidly and that the ground has simultaneously moved upward. Precisely the same patterns are experienced as those by individuals, as described above, who are adapted to earth gravity and make deep knee bends in the high force phases of parabolic flight. These changes reflect the decreased muscle spindle gain associated with weightless conditions and the lack of appropriate contact cues to signal body orientation and configuration.

Proprioceptive loss is characteristic of a variety of diseases and injuries of the nervous system. The most severe cases involve loss in adulthood of large somatic sensory fibers, including muscle spindle, joint, tendon and cutaneous mechanoreceptor afferents. An affected individual with eyes closed has no ability to localize his or her limbs even during vigorous active or passive movement. Standing and walking in darkness are not possible. Affected individuals can reacquire functional movement using visual guidance. Even partial loss of proprioception has severe consequences for balance and locomotion.

In summary, the human body has a dynamic sensory-motor spatial calibration to Earth gravity. This calibration involves an internal model of the force background and its consequences for voluntary movement control. The calibration is continuously updated based on experience within the environment. Somatosensory, proprioceptive and visual signals contingent on self-produced movements are key elements contributing to updating. The continuous ongoing nature of the calibration process is probably why movement control adjusts so rapidly when the body is exposed to force backgrounds greater or less than 1*g* in magnitude and why patients can adapt to proprioceptive loss.

References

1. Sherrington CS (1948) The integrative action of the nervous system. Cambridge University Press, London
2. Matthews PBC (1988) Proprioceptors and their contribution to somatosensory mapping: complex messages require complex processing. Can J Physiol Pharmacol 66:403–438
3. Magnus R (1924) Körperstellung. Springer, Berlin Heidelberg New York
4. Lackner JR, Taublieb A (1984) Influence of vision on vibration-induced illusions of limb movement. Exp Neurol 85:97–106
5. Lackner JR, DiZio P (2000) Aspects of body self-calibration. Trends Cogn Sci 4:279–288
6. Fisk J, Lackner JR, DiZio P (1993) Gravitoinertial force level influences arm movement control. J Neurophysiol 69:504–511
7. Lackner JR, Graybiel A (1981) Illusions of postural, visual, and substrate motion elicited by deep knee bends in the increased gravitoinertial force phase of parabolic flight. Exp Brain Res 44:312–316
8. Lackner JR, DiZio P (1993) Spatial stability, voluntary action and causal attribution during self-locomotion. J Vestib Res 3:15–23
9. Lackner JR (1988) Some proprioceptive influences on the perceptual representation of body shape and orientation. Brain 111:281–297
10. Harris CS (1963) Adaptation to displaced vision: visual, motor, or proprioceptive change? Science 140:812–813

Proprioception: Effect of Neurological Disease

Jonathan Cole
COPMRE, University of Bournemouth and Poole Hospital, Dorset, UK

Synonyms

Large-fiber sensory neuronopathy and neuropathy; Sensory ataxia; Dorsal column ataxia; Sensory stroke

Definition

Loss of movement and position sense due to disease of the peripheral or central nervous system.

Characteristics

The perceptions of movement and position sense, as well as that of touch, are dependent on afferent information relayed by large myelinated sensory nerve fibers, with cell bodies in the dorsal root ganglia and which then ascend in the ipsilateral dorsal columns (DCs). They synapse in the dorsal column nuclei at the head of the spinal cord. The pathway then projects through the medulla in the medial lemniscus to synapse in the ventral-posterior thalamus and hence, via the internal capsule, to the sensory cortex. Within the brain information is relayed to coordinate movement as well as give consciousness of the position and movement states of the body and limbs. Loss of ►proprioceptive information can lead to in-coordination due to dysfunction in the motor cortex and the cerebellum. In-coordination due to peripheral ►deafferentation and cerebellar lesions can sometimes be difficult to distinguish completely.

Disease anywhere along the above pathways can result in ►proprioceptive loss, usually with cutaneous sensory loss. While selective losses are possible, in particular in the peripheral nerve cell and to an extent in the dorsal columns, more central lesions may also affect other areas, making the resultant physiological deficit more complex. Often, for instance, cortical lesions lead to reduced stereognosis (inability to manipulate and recognize objects placed in the hand without vision) and graphesthesia (ability to recognize figures touched over the hand without vision), as well as in reductions in cutaneous touch and ►proprioception.

►Proprioception is such a deep sense that in folk psychology terms it is almost unknown. People may imagine being blind or deaf, but not being without proprioception or touch. Diderot considered that of all the senses touch (and one presumes the then undiscovered ►proprioception) the most profound and philosophical. To understand what ►proprioception does therefore it is important to consider those rare individuals who have to live without it – without their witness the experience, and its effects on movement, are almost impossible to know.

Total loss of ►proprioception is rare but has been described in the ►acute sensory neuronopathy syndrome. Its effects initially are a complete inability to control or coordinate movement. When movements are made they are inappropriate in size and direction with poor coordination between both limbs and joints. With time, however, and with considerable conscious attention towards, and visual observation of, movement, people without movement or position sense may be able to move in such a fashion as to be almost able to pass unnoticed, in uncluttered, well lit areas (see Section on ►Large-Fiber Sensory Neuropathy).

P

Though these complete losses of ►proprioception are rare their importance is that they reveal both the consequences of the loss and the ability of subjects, at least partially, to mitigate its effects. In addition, there are less severe yet still important losses of ►proprioception associated with various neurological problems, from median nerve compression at the wrist to central strokes in which motor weakness predominates which may have clinical consequences and yet are often poorly recognized.

Pathology

Peripheral Neuropathy/Neuronopathy

The most pure form of ►proprioceptive loss is disease of the large myelinated sensory nerve fibres, either in their axons or their cell bodies (see Section on ►Large-Fiber Sensory Neuropathy). Lesser, but still important, losses of ►proprioception can occur with less selective sensory peripheral neuropathy, especially affecting the legs, leading to ataxia and a feeling of not knowing where the feet are unless they are looked at. Patients may volunteer that they are much worse in the dark and that they find it difficult to drive a car (because of the need to control foot pedals).

The division between cutaneous light touch and ►proprioception can also be somewhat artificial. In the hands and the face cutaneous information about stretch and deformation of the skin gives information about position. Then, loss of cutaneous sensation has effects on movement. This is sometimes seen in carpal tunnel syndrome. Patients with sensory loss in the thumb and first and second fingers may say they drop things, unless they look at them and think about it, without knowing why. Even small losses of cutaneous touch in the fingers may have affects on the coordination of grip strength and active touch.

Dorsal Column Disease

►Proprioceptive (and cutaneous) afferents ascend with the spinal cord in the dorsal columns; disease of these structures is a well-known cause of loss of movement and position sense. One well-known cause of ►ataxia due to DC loss is vitamin B_{12} deficiency.

Sub-acute combined degeneration of the spinal cord involves the dorsal and lateral columns of the spinal cord and presents often, in middle age, with ►parasthesiae of the feet, with tingling and even burning [1]. Difficulties in walking follow, if untreated, due to sensory ►ataxia and motor weakness, reflecting lateral (motor) column involvement. Loss of postural sense can be more severe than the cutaneous sensory deficit. These may be accompanied by optic atrophy and cognitive impairment. The pathology involves both demyelination and axonal loss and treatment is usually effective, especially if given early [2].

Historically, syphilis was a major pathogen affecting the dorsal columns (Wilson, 1940 for a clinical description of neurosyphilis) [3]. Tabes dorsalis usually emerges 10–25 years after the initial infection and can present with progressive ►ataxia and a high stepping gait, and with severe, intermittent, "lightning" pains in the legs and abdomen. More rarely, it can also been seen far earlier in congenital syphilis. The pathology may be demyelination and destruction of the dorsal roots. In this form of syphilis it is not solely the DCs that are affected; failing vision, impotence, deafness and arthropathy and ulceration, reflecting small fiber loss too, are often associated with it.

Others causes of DC loss are multiple sclerosis with a plaque involving the dorsal cord, tumor and, most rarely, penetrating wounds in the neck with DC dissection. It has also been associated with a rare form of retinitis pigmentosa, which offers the possibility to understand the genetic nature of DC delineation [4].

Intracerebral Lesions

Pure sensory ►ataxias after sub-cortical stroke, or plaque or tumor are rare, since these diseases do not affect single pathways or classes of neuron. For instance, Russmann et al. [5] analyzed the Lausanne Stroke Registry between 1986 and 1998 for those with an ischemic episode affecting the area of the lenticular nucleus. Of 820 consecutive patients, 13 had pure lenticular infarction with four having an ataxic sensorimotor hemisyndrome, presumably due to a lesion of the internal capsule. Though examples of pure sensory stroke after small lacunar infarcts have been described after infarcts in the ventral posterior thalamus, brain stem, internal capsule and cerebral cortex, more often central lesions lead to more complex and subtle deficits.

It can, for instance, be difficult to distinguish, at the two extremes, between a pure motor weakness from an ►ataxia from in-coordination due to sensory loss. Kim [6] analyzed a group of patients with thalamic stroke with motor and sensory problems. ►Ataxia and involuntary movements were seen most after sensory loss, with ►dystonia, and in-coordination of movements associated with proprioceptive loss. Central sensory loss due to stroke can also be associated with dysesthesia and pain.

Though the commonest pathology associated with pure sensory stroke is ischemia, this syndrome occurs after hemorrhagic stroke too [7]. Pontine strokes involving the medial lemniscus, but sparing the spinothalamic tracts, lead to selective loss of vibration sensitivity and proprioception without an effect on pinprick and temperature sensation. More rostral strokes do not always lead to pure loss of proprioception as the dorsal column/medial lemniscal and the spinothalamic tracts become closer within the brain. Thus some thalamic strokes may lead to loss of all modalities of sensation over the opposite side of the body, with reduced temperature sensation and pin prick sensation as well as dysesthesia.

In central pure stroke syndromes Fisher [8] suggested that though objective sensory disturbances can be mild, larger losses are seen in fine motor control and coordination of the hand and fingers. This may reflect on the one hand the relative insensitivity of clinical tests of sensation, and the possible use of other cortical areas for these discriminations, and on the other the crucial importance of parietal cortex in the coordination of finger movements. Occasionally this role of the cortex in movement leads to confusion as to whether weakness is peripheral, say due to an ulnar nerve lesion, or cortical [9]. A relative lack of sensory symptoms and preserved power with clumsiness and reduced coordination may point towards a cortical deficit.

In addition it is important to realize that mixed sensory-motor strokes can prove difficult to rehabilitate due to underlying ▶proprioceptive loss over and above the more obvious problems with motor weakness. Loss of information about movement and position superimposed on loss of power can add a significant additional problem, especially if not recognized. Correspondingly, there are some reports that exercises designed to improve coordination can add to any spontaneous recovery that occurs [10].

Though syndromes involving exclusively loss of movement and position sense are very rare in neurology, an awareness of the effects of proprioceptive loss may allow greater understanding of mixed sensory and motor syndromes and lead to different rehabilitation strategies.

References

1. Hemmer B, Glocker FX, Schumacher M, Deuschl G, Lucking CH (1998) Subacute combined degeneration: clinical, electrophysiological, and magnetic resonance imaging findings. J Neurol Neurosurg Psychiatry 65:822–827
2. Thomas PK (1998) Subacute combined degeneration. J Neurol Neurosurg Psychiatry 65(6):607
3. Wilson SAK (1940) Neurosyphilis. In: Bruce AN (ed) Neurology, vol 1. Edward Arnold, London, pp 455–469
4. Higgins JJ, Morton DH, Loveless JM (1999) Posterior column ataxia with retinitis pigmentosa (AXPC1) maps to chromosome 1q31-q32. Neurology 1:46–50, 52
5. Russmann H, Vingerhoets F, Ghika J, Maeder P, Bogousslavsky J (2003) Acute infarction limited to the lenticular nucleus: clinical, etiologic, and topographic features. Arch Neurol 60:351–355
6. Kim JS (2001) Delayed onset mixed involuntary movements after thalamic stroke: clinical, radiological and pathophysiological findings. Brain 124:299–309
7. Shintani S, Tsuruoka S, Siigai T (2000) Pure sensory stroke caused by a cerebral hemorrhage: clinical-radiologic correlations in seven patients. Am J Neuroradiol 21:515–520
8. Fisher CM (1965) Lacunes: small deep cortical infarcts. Neurology 15:774–784
9. Timsit S, Logak M, Manai R, Rancurel G (1997) Evolving isolated hand palsy: a parietal lobe syndrome associated with carotid artery disease. Brain 120:2251–2257
10. Smania N, Montagnana B, Faccioli S, Fiaschi A, Aglioti SM (2003) Rehabilitation of somatic sensation and related deficit of motor control in patients with pure sensory stroke. Arch Phys Med Rehabil 84:1692–1702

Proprioception: Role of Cutaneous Receptors

David F. Collins
Faculty of Physical Education and Recreation, Centre for Neuroscience, University of Alberta, Edmonton, AB, Canada

Definition

Proprioception and Kinesthesia

The term proprioception (proprius = Latin for "one's own") describes abilities related to the perception of different aspects of one's own movement. These include the sense of the position and movement of the body segments (kinesthesia) as well as sensations related to tension in muscles and tendons, balance and voluntary effort [1]. This essay reviews the evidence that cutaneous receptors contribute to kinesthesia, the ability to detect the position and movement of the body segments without using vision. Cutaneous receptors are sensory receptors located in the skin. During movement, the skin is stretched and compressed, activating cutaneous receptors in large areas of skin around the moving joint (Fig. 1) [2].

The pattern of skin strain is different for movements of different speeds and amplitudes and this information is encoded in signals from large numbers of receptors that discharge during movement. These signals travel along peripheral nerves to the spinal cord and pass to the brain where, along with feedback from sensory receptors in muscles and joints, they provide the information responsible for kinesthetic awareness and proprioceptive judgments (see ▶Proprioception: Role of Muscle Receptors and Proprioception: Role of Joint Receptors). Kinesthetic information is used for the planning and execution of movement.

Characteristics

Sensory Receptors and Kinesthesia

During movement, sensory receptors in muscles, joints and the skin send movement-related information to the brain where it is used for kinesthesia. For more than 30 years, it was thought that the muscle spindle, a specific type of muscle receptor, was the most important source of kinesthetic information [1,3]. Muscle spindles respond to muscle stretch and thus signal the changes in muscle length that occur during joint movement. Many experiments have shown that this information is important for kinesthesia, such as those investigating the strong illusions of movement that are produced when vibration is applied over a muscle or its tendon. Vibration causes spindles to discharge rapidly and results in powerful illusions of movement consistent with lengthening of the vibrated muscle. Recently, experiments have shown that movement illusions can also be generated by activating cutaneous receptors by stretching the skin around a stationary joint to mimic skin stretch patterns that occur during normal movement [4,5]. An example of one experiment is shown in Fig. 2 where skin stretch applied around the right elbow resulted in the illusion of elbow flexion (left panel).

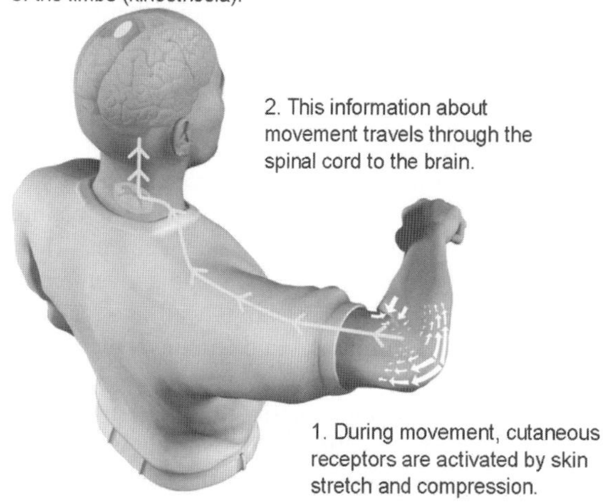

Proprioception: Role of Cutaneous Receptors. Figure 1 Schematic representation of the role of cutaneous receptors for kinesthesia at the elbow.

The subject matched these illusory movements of the right elbow with voluntary movements of the left arm. These movement illusions initiated by skin stretch show that cutaneous receptors are also important for kinesthesia. Movement illusions also occur when skin stretch is applied around the fingers [4,5] (see Fig. 3) and knee [5]; thus cutaneous feedback is important for kinesthesia at joints throughout the body [5].

When skin stretch and vibration are applied simultaneously, movement illusions are larger than when either stimulus is applied alone (see Figs. 2 and 3). Hence, the brain uses a combination of information from muscle spindles and cutaneous receptors for kinesthesia. The role played by joint receptors for kinesthesia is unclear, but the decrease in the ability to detect movement when joint receptor feedback was removed by anesthesia suggests they also contribute [3].

Cortical Structures and Kinesthesia

Kinesthetic information from sensory receptors enters the spinal cord through the dorsal horn and ascends to the brain along the dorsal columns. Before entering the brain the pathway crosses over to the opposite side of the body and projects to the main receiving area for kinesthetic information, the primary somatosensory cortex (shown in red in Fig. 1). Here, information from different parts of the body is received in different regions, resulting in a representative map of the body called a homunculus (Latin for "little man"). In the homunculus, regions of the body with the greatest

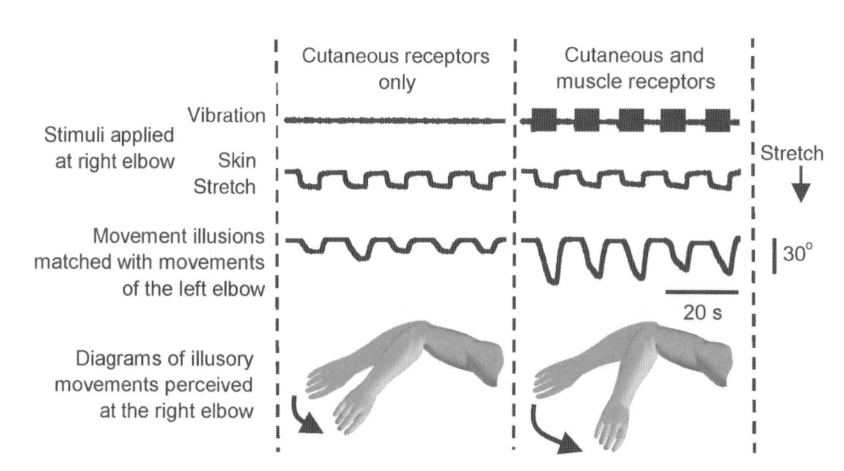

Proprioception: Role of Cutaneous Receptors. Figure 2 Movement illusions of the elbow produced by stimulating cutaneous receptors alone (skin stretch) or with simultaneous stimulation of receptors in muscle (vibration). Stimuli were applied to the right elbow and subjects matched perceived movements with voluntary movements of the left elbow. Adapted from Collins et al. 2005 [5].

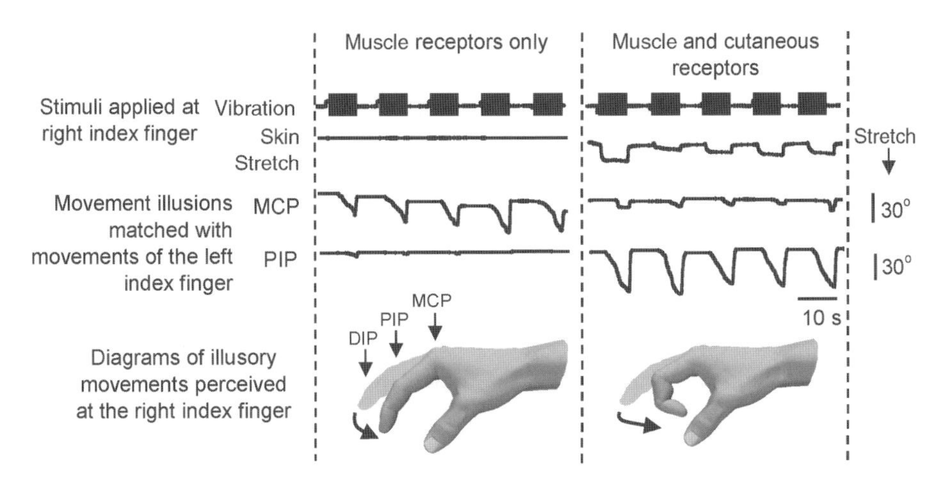

Proprioception: Role of Cutaneous Receptors. Figure 3 Movement illusions produced at the index finger by stimulating muscle receptors alone (vibration) or with simultaneous stimulation of cutaneous receptors (skin stretch). The skin stretch was applied around the two distal joints (PIP, DIP) and the vibration was applied over a tendon on the back of the hand near the MCP joint. Subjects matched perceived movements with voluntary movements of the left index finger. Adapted from Collins et al. 2005 [5]. (*MCP* metacarpophalangeal joint; *PIP* proximal interphalangeal joint; *DIP* distal interphalangeal joint)

numbers of receptors, such as the hands, have the largest representation. The perception of movement, however, is not localized to this one part of the brain but involves activity in many brain regions [6]. Thus, kinesthesia is a distributed process involving the somatosensory cortex as well as the motor cortex, cerebellum and frontal parts of the cortex with a particular importance of the right hemisphere [6].

Lower Level Components

Cutaneous Receptors

Much of our knowledge about cutaneous receptors comes from work on the hand where four morphologically distinct types of receptor have been identified [7]. Merkel discs and Meissner corpuscles lie in the superficial layers of the skin and have relatively small receptive fields (≤ 10 mm^2). Densities for these receptors are highest in the distal skin, with the digit tips containing 70 and 140 receptors/cm^2 for Merkel discs and Meissner corpuscles, respectively [7]. Pacinian corpuscles and Ruffini endings reside in the deeper layers and have larger receptive fields (≤ 25 mm^2) and densities that are lower and more uniform [7]. Meissner and Pacinian corpuscles adapt rapidly to sustained stimuli such as constant velocity movements and discharge primarily at the beginning and end of movement. In contrast, Merkel discs and Ruffini endings adapt slowly to sustained stimuli and continue to discharge throughout movement with a discharge frequency that is often closely related to the amplitude and velocity of the movement. Receptors similar to those in the hand have also been found in the skin of the arm and leg, although there are regional differences in receptor densities.

Higher Level Processes

Sensory Feedback and Movement Control

The control of human movement involves a complex interaction between motor output from the brain and sensory feedback from peripheral sensory receptors. Sensory feedback from the limbs provides information about movement execution, sending an error signal to the brain when movements do not proceed as planned. Sensory feedback also contributes to movement control at the level of the spinal cord. Spinal reflexes provide automatic responses to external disturbances that help to prevent for example slips of hand-held objects [7] or trips during walking. Sensory feedback also influences the timing and amplitude of muscle contractions during rhythmic movements by modifying the activity of central pattern generators in the spinal cord that are responsible for generating the basic patterns of muscle activity.

Lower Level Processes

Communication in the Nervous System

Information travels through the nervous system in the form of action potentials, tiny electrical impulses generated by the transient passage of ions across the cell membrane. During movement, action potentials are initiated in a cutaneous receptor during movement of the skin within its receptive field. These signals travel along myelinated axons of the nerve cells (neurons) at 35–80 m/s to synaptic junctions with other neurons. At the synapse, the site for communication between neurons, the arrival of the action potential triggers the release of neurotransmitter, a chemical signal that activates receptors on the membrane of the post-synaptic neuron. Receptor activation initiates ion movement across the post-synaptic membrane where, with a sufficiently large stimulus, another action potential is generated. Thus, communication between neurons (synaptic transmission) involves the conversion of an electrical signal (action potential) into a chemical signal at the synapse and back to an electrical signal in the post-synaptic cell.

Process Regulation

Gating of Cutaneous Feedback During Movement

Transmission along pathways from cutaneous receptors to the brain is suppressed during movement [8]. This suppression occurs in the spinal cord and brain and is thought to prevent these structures from becoming saturated by the massive amount of sensory traffic that is generated during movement. Control of information flow through the nervous system is selective; irrelevant signals are suppressed more than signals from receptors that provide information that is important for the successful performance of the movement [8].

Function

The Cutaneous Contribution to Kinesthesia

For many years it had been thought that cutaneous receptors play only a minor role in kinesthesia, perhaps acting to facilitate movement-related feedback from muscle spindles. This "facilitation" hypothesis was tested by measuring the ability to detect small finger movements when cutaneous feedback from adjacent digits was removed (by anesthesia) or enhanced (by skin stimulation) [9]. Movement detection was not decreased by cutaneous anesthesia nor was it enhanced by increasing cutaneous receptor discharge. Thus, the role for cutaneous feedback for kinesthesia is not one of general facilitation [9]. Instead, evidence has been mounting over the last 10 years that cutaneous receptors provide specific information about the movement itself [2] and this is used for kinesthesia [4,5]. Some idea of the relative roles of feedback from cutaneous receptors and muscle spindles has come from experiments investigating illusory movements of the fingers using vibration to activate muscle spindles and skin stretch to activate cutaneous receptors around specific finger joints (Fig. 3) [4,5]. When vibration was applied by itself, subjects perceived movements primarily at the metacarpophalangeal (MCP) joint (Fig. 3, left panel). When the skin around the proximal and distal interphalangeal joints was stretched and compressed during vibration, to mimic patterns that occur during movements of those joints, the perception of movement decreased at the MCP joint and increased at the joints where the skin stretch was applied. This supports the idea that one role for cutaneous feedback may be to help to distinguish which finger joint is moving [5]. Most muscles that control movements of the fingers are located in the forearm and cross more than one joint. Hence, the changes in muscle length detected by muscle spindles could arise from movement at one or more joints. The proximity of cutaneous receptors around individual joints enables them to provide specific information about which joint is moving. The receptors most likely to provide this information are the rapidly adapting Pacinian and Meissner corpuscles [2]. Information from the slowly adapting receptors, particularly the Ruffini endings, is more likely to be responsible for providing ongoing information about the position and velocity of the moving joint [2]. Cutaneous feedback is also important for kinesthesia at the elbow and knee and it is likely that feedback from muscle spindles and cutaneous receptors is used for kinesthesia for joints throughout the body [5].

Pathology

Clinical Implications

The idea that cutaneous receptors contribute significantly to kinesthesia is relatively recent and the clinical implications have yet to be fully explored. It had been suggested that improvements in joint stability associated with bandaging the knee might be due to enhanced cutaneous proprioception, but this was not supported by experiments assessing movement detection with and without a bandaged knee [10]. However, the cutaneous contribution to kinesthesia should be considered for tests of proprioception that make up part of standard neurological examinations. Clinicians should be careful to avoid generating conflicting signals from cutaneous receptors when applying passive joint movements. Finally, one would predict that persons who have suffered burns to large regions of skin would have reduced kinesthetic ability due to decreased cutaneous feedback and this may influence movement performance. An increased understanding of the importance of cutaneous receptors for kinesthesia may lead to new ideas for the treatment and assessment of movement disorders.

References

1. Proske U (2005) What is the role of muscle receptors in proprioception? Muscle Nerve 31:780–787
2. Edin BB, Abbs JH (1991) Finger movement responses of cutaneous mechanoreceptors in the dorsal skin of the human hand. J Neurophysiol 65:657–670
3. Gandevia SC (1996) Kinesthesia: roles for afferent signals and motor commands. In: Rowell LB, Shepherd JT (eds) Handbook of physiology. Oxford University Press, New York, pp 128–172
4. Collins DF, Prochazka A (1996) Movement illusions evoked by ensemble cutaneous input from the dorsum of the human hand. J Physiol 496:857–871
5. Collins DF, Refshauge KM, Todd G, Gandevia SC (2005) Cutaneous receptors contribute to kinesthesia at the index finger, elbow, and knee. J Neurophysiol 94:1699–1706

6. Naito E, Roland PE, Grefkes C, Choi HJ, Eickhoff S, Geyer S, Zilles K, Ehrsson HH (2005) Dominance of the right hemisphere and role of area 2 in human kinesthesia. J Neurophysiol 93:1020–1034
7. Johansson RS (1996) Sensory and memory information in the control of dexterous manipulation. In: Lacquaniti F, Viviani P (eds) Neural bases of motor behaviour. Kluwer, Netherlands, pp 205–260
8. Brooke JD (2004) Somatosensory paths proceeding to spinal cord and brain – centripetal and centrifugal control for human movement. Canadian J Physiol Pharmacol 82:723–731
9. Refshauge KM, Collins DF, Gandevia SC (2003) The detection of human finger movement is not facilitated by input from receptors in adjacent digits. J Physiol 551:371–377
10. Hewitt BA, Refshauge KM, Kilbreath SL (2002) Kinesthesia at the knee: the effect of osteoarthritis and bandage application. Arthritis Rheum 47:479–483

Proprioception: Role of Joint Receptors

VAUGHAN G. MACEFIELD
Professor of Integrative Physiology, School of Medicine, University of Western Sydney, Sydney, NSW, Australia
Conjoint Principal Research Fellow, Prince of Wales Medical Research Institute, Sydney, NSW, Australia

Synonyms

Kinesthesia; Joint sense; Position sense; Movement sense

Definition

Proprioception, literally the "sense of self" (from Latin "*proprius*" = "own"), is the group of sensory modalities that allow us to know the positions of our limbs in space and to detect and assess the magnitudes of movements and forces without vision. Classically, it had been assumed that this sense was subserved by joint receptors, specialized mechanoreceptors located in the capsular tissues of a joint. Indeed, the term "joint sense" – the capacity to detect movements or changes in position about a joint – is still used in clinical practice, though it is generally now accepted that joint receptors themselves do not bear primary responsibility for our proprioceptive acuity. Muscle spindles – exquisitely sensitive intramuscular stretch receptors – are generally believed to be the sensory endings primarily responsible. Perceptions of joint movement evoked by small amplitude vibration applied to muscles or tendons, either intact or surgically exposed, were the first convincing demonstrations that these receptors contribute importantly to proprioception (see ►Proprioception, Role of Muscle Receptors). The role of stretch receptors in the skin has come to increased prominence of late and there are many convincing studies implicating significant roles for cutaneous afferents in proprioception (see ►Proprioception, Role of Cutaneous Receptors).

Characteristics

Quantitative Description

The consensus from the animal literature is that joint afferents mostly respond at the extremes of joint rotation [1]. Microelectrode recordings from the median and ulnar nerves of conscious human subjects have shown that mechanoreceptors associated with the interphalangeal joints and metacarpophalangeal joints do not respond to forces applied to bone when there is no movement of the joint and have very high mechanical thresholds to indentation applied over the joint capsule [2,3]. While they do respond to joint movements, they respond primarily at the limits of angular excursion. An example of a recording from a human joint receptor is shown in Fig. 1. This unit, located within the metatarsophalangeal joint of the fourth toe was spontaneously active at rest and responded to imposed extreme plantarflexion and dorsiflexion of the joint and to (unphysiological) internal and external rotation of the joint. Because joint afferents often respond in both directions (e.g. flexion and extension) and in more than one axis of rotation (e.g. abduction/adduction and extorsion/intorsion), as a group, joint afferents have a very limited capacity to encode changes in joint position. Nevertheless, a small proportion of interphalangeal joint afferents do respond across the physiological range [2,3]. Moreover, there is evidence that afferents associated with the dorsal aspects of human metacarpophalangeal joints consistently respond throughout the physiological range of joint rotation [4].

Lower Level Components

Much of the early work on joint receptors, carried out mostly in the knee joint of the cat, supported the idea that joint receptors could encode changes in joint position [5]. However, it turned out that many of these were actually muscle spindle afferents coursing through the nerve supplying the joint. Nevertheless, a few receptors – unequivocally articular in origin – have been found that respond throughout the physiological range of angular excursion [6]. Histological and physiological studies have shown that mechanoreceptors in the posterior capsule of the cat knee joint, identified as Ruffini organs (i.e. similar to the stretch receptive SAII endings in skin) respond in a slowly adapting manner to strains applied in the plane of the tissue but have very high thresholds to compressive stresses applied perpendicularly – the adequate stimulus is the increase in tensile strain in the immediate environment of the receptor [7]. Ruffini endings have also been identified

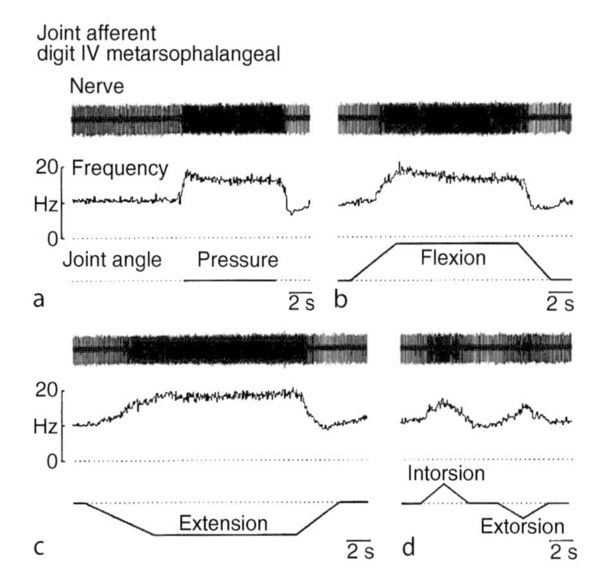

Proprioception: Role of Joint Receptors.
Figure 1 Microelectrode recording from a joint afferent associated with the metatarsophalangeal joint of the fourth toe in conscious human subject. The mechanoreceptor responded to pressure over its receptive field (a) and to passive flexion (b), extension (c) and longitudinal rotation of the joint (d). Changes in joint angle are represented schematically.

in the human finger and knee joints and the discharge behavior of human joint afferents associated with the interphalangeal and metacarpophalangeal joints fits with their being Ruffini endings; they have very high mechanical thresholds to indentation applied over the joint capsule and respond to joint movements that cause an increase in tensile strain within the joint capsule [2].

Higher Level Processes

Proprioceptive acuity is optimal when inputs from muscle, skin and joints are intact [8]. Anesthesia of the joint capsule attenuates and intra-articular fluid expansion augments proprioceptive acuity at the interphalangeal joint, arguing for a role of joint afferents in proprioception [9]. Intraneural microstimulation of a single joint afferent in conscious human subjects can be perceived as pressure over the joint or as a small movement, so joint afferents do have a strong synaptic coupling to higher order sensory neurons [3]. This means that, should a joint receptor be exposed to an adequate tensile strain within the joint capsule or extracapsular ligament, it could provide useful information. But given that these tensile strains are only reached during extreme joint rotations or direct pressure, their role in normal proprioception can be considered to be limited.

Function

As noted above, joint receptors are considered to primarily encode changes in joint angle at the extremes of angular excursion. However, given that some joint afferents can encode joint rotation throughout the physiological range [2,6] they may play a role in proprioception when inputs from muscle and skin cannot contribute [9].

Pathology

Loss of joint afferent input is not important; proprioceptive acuity is not greatly affected when people are fitted with prosthetic joints [10].

References

1. Burgess PR, Clark FJ (1969) Characteristics of knee joint receptors in the cat. J Physiol 203:317–333
2. Burke D, Gandevia SC, Macefield G (1988) Responses to passive movement of receptors in joint, skin and muscle of the human hand. J Physiol 402:347–361
3. Macefield G, Gandevia SC, Burke D (1990) Perceptual responses to microstimulation of single afferents innervating joints, muscles and skin of the human hand. J Physiol 429:113–129
4. Edin BB (1990) Finger joint movement sensitivity of non-cutaneous mechanoreceptor afferents in the human radial nerve. Exp Brain Res 82:417–422
5. Eklund G, Skogland S (1960) On the specificity of the Ruffini like joint receptors. Acta Physiol Scand 49:184–191
6. Ferrell WR (1980) The adequacy of stretch receptors in the cat knee joint for signalling joint angle throughout a full range of movement. J Physiol 199:85–99
7. Grigg P, Hoffman AH (1996) Stretch-sensitive afferent neurones in cat knee joint capsule – sensitivity to axial and compression stresses and strains. J Neurophysiol 75:1871–1877
8. Gandevia SC, McCloskey DI (1976) Joint sense, muscle sense, and their combination as position sense, measured at the distal interphalangeal joint of the middle finger. J Physiol 260:387–407
9. Ferrell WR, Gandevia SC, McCloskey DI (1987) The role of joint receptors in human kinaesthesia when intramuscular receptors cannot contribute. J Physiol 386:63–71
10. Cross MJ, McCloskey DI (1973) Position sense following surgical removal of joints in man. Brain Res 55:443–445

Proprioception: Roles of Muscle Receptors

UWE PROSKE
Department of Physiology, Monash University, Melbourne, VIC, Australia

Synonyms

Kinesthesia; Proprioception

Definition

When limbs are moved, under circumstances where vision cannot be used to monitor the movements, there is a quite accurate sense of where the limbs are in space and whether they are moving or not. This is the ►kinesthetic or the ►proprioceptive sense [1][9].

The neural basis of ►kinesthesia and ►proprioception has been the subject of debate for many years. In the mid nineteenth century it was believed that sensations arising from movements produced by contracting muscles were associated with the motor commands that produced the movements – a "sensation of innervation." This view was not shared by Sherrington [2] who believed that the kinesthetic sense arose from afferent signals generated in the muscles themselves. Interestingly, the debate about the central and peripheral origin of the kinesthetic sense has not entirely subsided to the present day.

During most of the twentieth century it was believed that ►kinesthetic sensibility was provided predominantly by peripheral signals, although some uncertainty persisted over what kinds of receptors were involved. At one stage it was thought that joint receptors were the main source of input. This idea has not been laid to rest and in recent years, for some joints, contributions from joint receptors [3] as well as skin receptors [4] have been emphasized.

The experiments of Goodwin et al. [5] provided the first direct evidence for a role of a peripheral signal of muscle origin in ►proprioception when they described illusions of both forearm position and movement during 100 Hz vibration of elbow flexor muscles. They concluded that muscle spindles were the principal ►kinesthetic receptor.

Characteristics

Higher Level Structures

A difficulty with muscle receptors as the ►kinesthetic sensors was that until the second half of the twentieth century it remained uncertain whether muscle afferents had access to the cerebral cortex. It is generally accepted that a cortical projection is a necessary prerequisite for access by sensory receptors to consciousness. In the event, it was shown that both spindle Group I afferents (primary muscle spindle endings) and Group II afferents (secondary muscle spindle endings) projected to areas 3a and 4 of somatosensory cortex [6].

Lower Level Structures

The illusions reported by Goodwin et al. [5] were of both position and movement, although vibration at 100 Hz produced principally a movement illusion. When vibration frequency was reduced to 20–40 Hz, the illusion faded from one of movement to one of limb position [7].

Animal studies had demonstrated that the primary endings of spindles had both a dynamic length sensitivity and a static length sensitivity [8]. Furthermore, primary endings were very sensitive to high frequency muscle vibration. It was therefore inferred that the movement illusions experienced during high frequency vibration were the result of signals generated by primary endings.

Secondary endings had no dynamic sensitivity but they responded with maintained changes in discharge during changes in muscle length. They responded only to low frequency vibration. It was concluded that spindle primary endings signaled both movement and position information, while spindle secondary endings contributed only positional information.

Lower Level Processes

Since muscle spindles are stretch receptors, whenever a muscle was stretched during movement of a limb, signals would be generated in both primary and secondary endings. A proportional relationship could be described between the size and rate of stretch and the dynamic and static components of spindle discharge [8]. All of this related to responses of spindles during length changes in the passive muscle. It was known that during a graded voluntary contraction, the ►fusimotor neurons to muscle spindles were ►co-activated [10]. So impulses can be generated in spindles by two fundamentally different processes, stretch and intrafusal contraction.

Higher Level Processes

According to the ►hypothesis that kinesthesia represents two distinct senses, the sense of position and the sense of movement, there is some evidence that central projection sites for dynamic and static spindle signals are separate [6]. Further processing of spindle information must take place to distinguish between ►fusimotor evoked spindle activity (reafference) and muscle stretch evoked activity (exafference). A central subtraction process was postulated to take place [11]. According to this scheme, the motor command goes to both alpha and gamma motoneurons as part of the ►co-activation strategy. Gamma-evoked impulses are subtracted from the total spindle signal to extract movement and position related components. It means that spindles could provide this information at all levels of voluntary activation [7].

Process Regulation

At the present time, we know little about the central processes and their regulation in the generation of ►kinesthetic sensations. Signals from area 3a (dynamic) and area 4 (static) must combine with exafferent information to produce the felt sensation. Where this takes places remains to be shown.

Function

The ►kinesthetic sense, making reference to a central map, allows us to locate the body and body segments in space, in the absence of vision. It also provides us with information about movement of the body. Since in everyday life many of our movements are carried out without visual guidance, the ►kinesthetic sense is important for normal motor activities.

Pathology

An indication of the importance of ►kinesthesia and ►proprioception for motor control is provided by subjects with a large fiber sensory neuropathy. These subjects carry out all activities under close visual control. In the absence of vision they exhibit large errors in ►kinesthetic tests and they are unaware of obstacles encountered during movements, unless they can see them [1].

►Large-Fiber Sensory Neuropathy: Effect on Proprioception

References

1. McCloskey DI (1978) Kinesthetic sensibility. Physiol Rev 58:763–820
2. Sherrington CS (1906) On the proprioceptive system, specially in its reflex aspects. Brain 29:467–482
3. Ferrell WR, Gandevia SC, McCloskey DI (1987) The role of joint receptors in human kinaesthesia when intramuscular receptors cannot contribute. J Physiol 386:63–71
4. Collins DF, Refshange KM, Todd G, Gandevia SC (2005) Cutaneous receptors contribute to kinesthesia at the index finger, elbow and knee. J Neurophysiol 94:1699–1706
5. Goodwin GM, McCloskey DI, Matthews PBC (1972) The contribution of muscle afferents to kinaesthesia shown by vibration induced illusions of movement and by the effects of paralysing joint afferents. Brain 95:705–748
6. Hore J, Preston JB, Cheney PD (1976) Responses of cortical neurons (areas 3a and 4) to ramp stretch of hindlimb muscles in the baboon. J Neurophysiol 39:484–500
7. McCloskey DI (1973) Differences between the senses of movement and position shown by the effects of loading and vibration of muscles in man. Brain Res 61:119–131
8. Matthews PBC (1972) Mammalian muscle receptors and their central actions. Arnold, London
9. Proske U (2006) Kinesthesia: the role of muscle receptors. Muscle Nerve 34:545–558
10. Vallbo AB (1974) Human muscle spindle discharge during isometric voluntary contractions. Amplitude relations between spindle frequency and torque. Acta Physiol Scand 90:319–336
11. McCloskey DI, Gandevia SC, Potter EK, Colebatch JG (1983) Muscle sense and effort: motor commands and judgements about muscular contractions. In: Desmedt JE (ed) Motor control mechanisms in health and disease. Raven, New York

Proprioceptor

Definition

Sensory receptor providing information about the mechanical state of the body.

►Proprioception
►Sensory Systems

Pro-Saccades

Definition

►Oculomotor Control

►Saccade, Saccadic Eye Movement

Prosencephalon

Definition

Forebrain. Anterior part of the brain. Composed of telencephalon and diencephalon.

►Evolution and Embryological Development of the Forebrain
►General CNS

Prosomere

Definition

Transverse, segmental divisions of the forebrain.

►Evolution and Embryological Development of Forebrain

Prosopagnosia

Definition

►Visual Neuropsychology

Prospective Monitoring

Definition

Prospective monitoring is defined as metacognitive experiences that on the basis of evaluation of the current states of memory and learning, one adjust one's behaviors (e.g., by adaptively modifying and/or changing currently inadequate learning strategies and figuring out new strategies), so as to prepare for achieving learning goals.

►Metacognition

Prospero: Protein Localization in Neuroblasts

XIAOHANG YANG
Institute of Molecular and Cell Biology,
Agency for Science, Technology and Research,
Proteos, Singapore

Definition

Prospero belongs to a unique group of proteins that are enriched in either the apical or basal cell cortex, just under the plasma membrane, of dividing neuroblasts. The asymmetric cortical localization of Prospero protein is controlled by a sophisticated ►asymmetric cell division mechanism that secures the exclusive segregation of Prospero to the future ►ganglion mother cell during mitosis.

Characteristics

The *Drosophila* embryonic central nervous system (CNS) is derived from an array of approximately 30 unique types of neuroblast. The highly diversified neurons and supporting cells in the CNS are the direct progenies of ganglion mother cells produced by neuroblast asymmetric divisions. The *Drosophlia* gene *prospero* (*pros*) is required for the generation of the ganglion mother cell.

pros [1] was identified in the early 1990s and named after the protagonist in *The Tempest*, a play by William Shakespeare. *pros* is located on the right arm of the third chromosome (86E2–86E4) and encodes three slightly varied large polypeptides (1535, 1403 and 1374 aa) due to alternative spicing. The Pros protein is a divergent homeodomain transcription factor that acts as a switch between self-renewal or terminal differentiation of a neural precursor cell. *pros* is evolutionally conserved and vertebrate homologs of *pros*, Prox1, have been identified in human, mouse, zebrafish and frog.

During early *Drosophila* embryonic neurogenesis, neuroblasts delaminate from the neuroectoderm and divide asymmetrically along the apicobasal axis to produce one large and one small daughter cell. The size difference between the two daughters is so prominent that it appears as if the small daughter buds off from the basal cortex of the telophase neuroblast (Fig. 1c).

The large apical daughter cell remains as the neuroblast and continues to divide asymmetrically. The small basal cell adopts the ganglion mother ►cell fate and divides terminally to make either two neurons or glial cells.

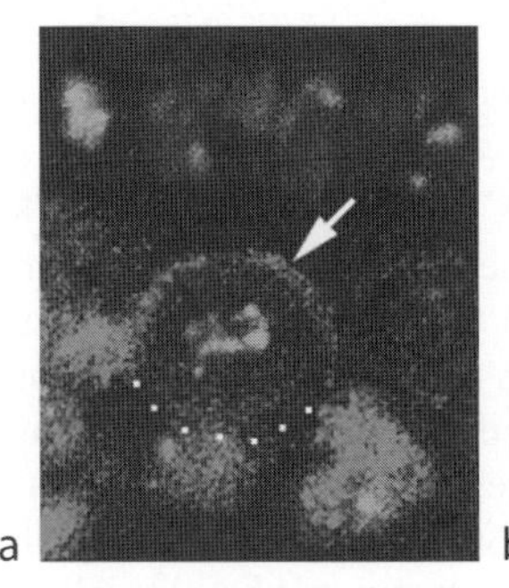

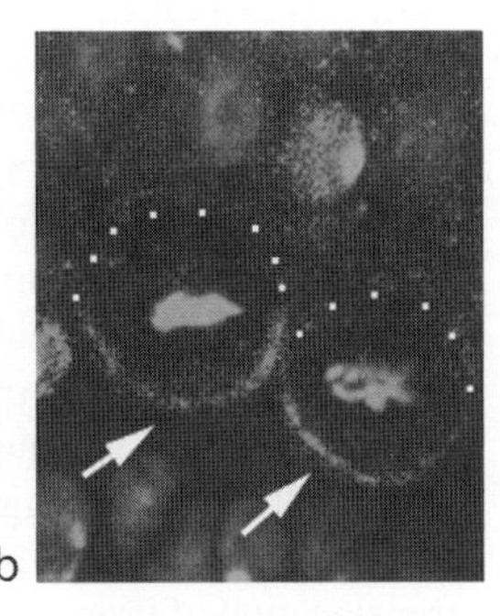

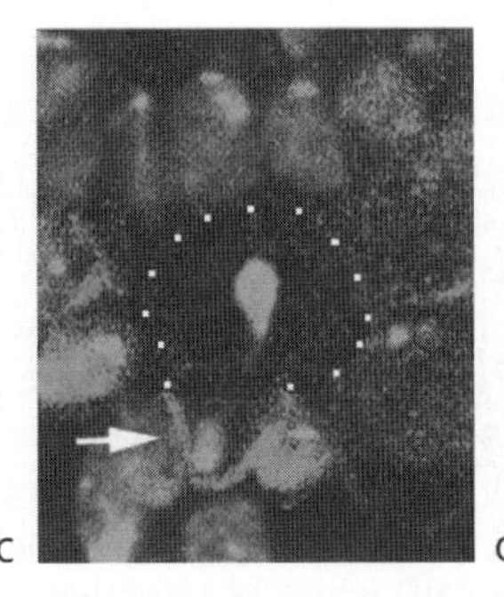

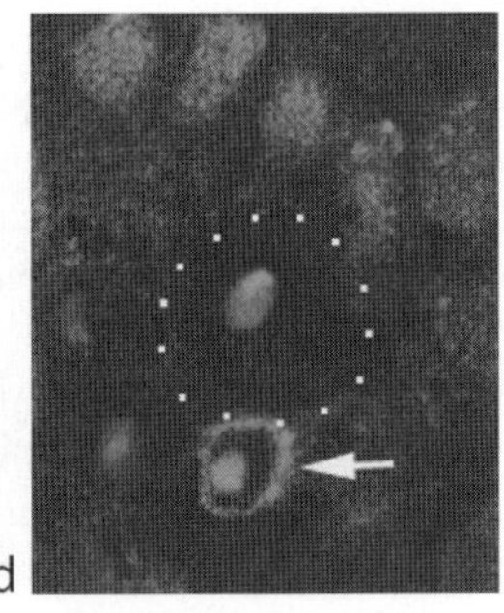

Prospero: Protein Localization in Neuroblasts. Figure 1 Confocal images of cell cycle-dependent asymmetric distribution of Pros protein in mitotic neuroblasts. Embryos at developmental stage 10 were stained with anti-Pros (*green*) and ToPro3 (chromosomes, *cyan*). At late interphase or early prophase, Pros protein is transiently enriched in the apical cortex and forms an apical crescent (Panel a, *arrow*). Starting from late prophase, Pros is relocated to the basal cortex and remains basal (Panel b, *arrows*; late prophase and metaphase) throughout the rest of mitosis. At telophase, Pros protein is sequestered into the future ganglion mother cell (Panel c, *arrow*). After cytokinesis, Pros protein is exclusively inherited by the ganglion mother cell (Panel d, *arrow*) and translocated into the nucleus later. Apical is up. Neuroblast cell bodies are outlined with *white dots*.

Pros protein is made in neuroblasts and its cellular distribution is tightly regulated by the cell cycle [1]. Anti-Pros immunofluorescence staining shows that Pros protein transiently enriches in the apical cortex (membrane-associated or "apical crescent") (Fig. 1a) in late interphase and early prophase neuroblasts. Starting from late prophase, Pros is relocated to the basal cortex of mitotic neuroblasts and remains basal ("basal crescent," Fig.1b) throughout the rest of mitosis. In telophase, Pros protein is sequestered into the future ganglion mother cell (Fig. 1c). After cytokinesis, Pros protein is exclusively inherited by the ganglion mother cell (Fig. 1d). In newly formed ganglion mother cells, Pros is cortical (membrane-associated, Fig. 1d) and is translocated into the nucleus later, which is essential for its function. The membrane-associated form of Pros is highly phosphorylated and its translocation to the nucleus requires dephosphorylation [2]. Interestingly, *pros* mRNA exhibits similar asymmetric cellular localization patterns in mitotic neuroblasts and gets sequestered into ganglion mother cells [3].

Pros protein appears to act as a master player (reminiscent of Prospero who has the magic power in *The Tempest*) that controls the ganglion mother cell fate. In the absence of Pros, or failure to translocate Pros to the nucleus, the smaller daughter cell fails to adopt the ganglion mother cell fate and does not produce proper neurons. Thus, Pros is also called a "cell fate determinant." Another well studied cell fate determinant in neuroblast asymmetric division is Numb [4], which exhibits similar asymmetric basal localization and exclusive segregation into future ganglion mother cells as Pros. Numb antagonizes the Notch signaling pathway.

A number of protein complexes are involved in regulating Pros asymmetric localization in mitotic neuroblasts [5]. Proteins such as Bazooka (Baz), DaPKC, Par6, Inscuteable (Insc), Partner of Insc and the α-subunit of trimeric G-protein, as well as Locomotion defects, co-localize to the apical cortex of mitotic neuroblasts from late interphase and form a functional complex (apical complex). This apically localized complex exhibits three major functions: (i) regulating the basal localization of cell fate ►determinants such as Pros and Numb; (ii) reorientating the mitotic spindle along the apicobasal axis by metaphase; and (iii) generating an asymmetric spindle (the apical half of the spindle arm is much longer than that of the basal half) which is responsible for the distinct cell size difference between the two daughters. Loss of function of any single protein (for example, in *baz* or *insc* mutants) from the apical complex usually causes mislocalization (randomized Pros crescent position) of Pros protein in early (prior to anaphase) mitotic neuroblasts. However, starting in anaphase, Pros protein in the majority of these mutant neuroblasts is redistributed as a crescent to the cell cortex where the future ►ganglion mother cell buds off. This redeployment of Pros protein late in mitosis in mutant neuroblasts has been referred to as "telophase rescue."

The asymmetric localization of Pros in mitotic neuroblasts also depends on a protein called Miranda (Mira) [6], which was named after Prospero's daughter who went into exile with him in *The Tempest*. In mitotic neuroblasts, Mira binds to Pros and functions as an adaptor protein for Pros asymmetric cortical distribution. In the absence of Mira, Pros loses its asymmetric localization and becomes cytoplasmic.

The cell-cycle dependent translocation of Pros protein from apical to basal cortex is regulated by the Snail family proteins of Zn-finger transcription factors Snail (Sna), Escargot (Esg) and Worniu (Wor) [5]. The members of the Snail family share homologous sequences and exhibit redundant functions. In the absence of all three Snail family proteins, two major players of the apical complex in the embryonic CNS, Baz and Insc, are down-regulated in the segmented CNS but not the brain. In these *snail* triple-mutant (*esg wor sna*) embryos, Pros protein remains tethered to its apical position and does not relocate late in mitosis to the basal cortex from where the future ganglion mother cells bud off. Since these small daughter cells do not inherit Pros, they do not adopt the ganglion mother cell fate. These observations indicate that in addition to the absence of Baz and Insc, the compensatory telophase rescue mechanism is also defective in *snail* triple-mutant (*esg wor sna*) neuroblasts. The detailed mechanism of telophase rescue remains largely unknown. A recent study of telophase rescue has indicated that two *Drosophila* homologs of mammalian TNF/TNFR molecules, Eiger and DTRAF1, are involved in Pros protein telophase rescue [7]. Interestingly, Numb telophase rescue requires neither Eiger nor DTRAF1, which suggests that a different pathway is involved.

Two tumor-suppressor genes, *lethal giant larvae* (*lgl*) and *discs large* (*dlg*), are also involved in the proper basal localization of Pros in dividing neuroblasts [8,9]. In contrast to its asymmetric localization to either the apical or basal cortex in wild type neuroblasts, in the absence of Lgl/Dlg Pros protein is evenly distributed in the cortex and is heavily associated with the mitotic spindle. It has been suggested that Lgl/Dlg may act in a secretory pathway involved in the basal intracellular translocation of Pros. The asymmetric basal localization of Pros protein also depends on the actin cytoskeleton, but not the microtubule structures inside cells. Disruption of the actin cytoskeleton results in loss of Pros from the membrane.

It appears that Pros asymmetric localization is only observed at embryonic stages. During late nervous system development, the asymmetric localization and segregation mechanism do not seem to play a major role in Pros function and the conventional transcription/ translation regulation mechanism prevails. For example, in the 3rd instar larval brain, Pros protein is expressed only in ganglion mother cells and not in neuroblasts, although Mira remains asymmetrically localized in mitotic larval neuroblasts and is exclusively partitioned into ganglion mother cells at telophase. Another example is the development of the adult external sensory organ. Pros protein is cytoplasmic in dividing sensory organ precursor cells and distributes equally to two daughter cells after cytokinesis [10]. Later the Pros protein level is down-regulated in one of the daughter cell (IIa), which divides to produce a hair and a socket. The daughter cell maintaining Pros protein (IIb) produces a neuron and a glial cell.

In summary, the cell fate determinant Pros protein is made in the mother cell (neuroblast) and only functions as a transcription factor in the nucleus of one of the two daughter cells (ganglion mother cell). The cell cycle-dependent asymmetric distribution of Pros protein during mitosis is a critical mechanism employed by dividing neuroblasts to secure the exclusive segregation of this protein to the ganglion mother cells. In vertebrates, it is not known whether Prox1 is involved in asymmetric divisions. Prox1 knock-out mice show embryonic lethality with impaired development of the lens, lymphatic system, liver and pancreas.

References

1. Fuerstenberg S, Broadus J, Doe CQ (1998) Asymmetry and cell fate in the Drosophila embryonic CNS. Int J Dev Biol 42:379–383
2. Srinivasan S, Peng CY, Nair S, Skeath J, Spana E, Doe CQ (1998) Biochemical analysis of Prospero protein during asymmetric cell division: cortical Prospero is highly phosphorylated relative to nuclear Prospero Dev Biol 204:478–487
3. Li P, Yang X, Wasser M, Cai Y, Chia W (1997) Inscuteable and Staufen are required for the asymmetric localisation and segregation of Prospero RNA during Drosophila neuroblast cell divisions. Cell 90:437–447
4. Knoblich JA, Jan LY, Jan YN (1995) Asymmetric segregation of numb and prospero during cell division. Nature 377:624–627
5. Chia W, Yang X (2002) Asymmetric division of Drosophila neural progenitors. Curr Opin Genet and Dev 12:459–464
6. Jan YN, Jan LY (1998) Asymmetric cell division. Nature 392:775–778
7. Wang H, Cai Y, Chia W, Yang (2006) Drosophila homologs of mammalian TNF/TNFR-related molecules regulate segregation of Miranda/Prospero in neuroblasts. EMBO J 25:5783–5793
8. Peng CY, Manning L, Albertson R, Doe CQ (2000) The tumour-suppressor genes lgl and dlg regulate basal protein targeting in Drosophila neuroblasts. Nature 408:596–600
9. Ohshiro T, Yagami T, Zhang C, Matsuzaki F (2000) Role of cortical tumour-suppressor proteins in asymmetric division of Drosophila neuroblast. Nature 408:593–596
10. Manning L, Doe CQ (1999) Prospero distinguishes sibling cell fate without asymmetric localization in the Drosophila adult external sense organ lineage. Development 126:2063–2071

Prostaglandins

Definition

A group of hormone-like substance derived from arachidonic acid, which participate in a wide range of neuronal and organismal functions.

Prosthesis

Definition

An artificial body part.

►Joints
►Measurement Techniques (Pressure)

Proteases

Definition

Proteases are enzymes that break down other proteins in a cell. Normally, proteases are carefully regulated. If unregulated, severe cell damage and death can occur.

Proteasome

Definition

Proteasome in eukaryotes, proteasomes are large nuclear and cytoplasmic protein complexes that proteolytically degrade proteins.

Protective Autoimmunity

MICHAL SCHWARTZ, JONATHAN KIPNIS
The Weizmann Institute of Science, Rehovot, Israel

Synonyms

Immune based-self maintenance, repair and renewal

Definition

Immune response recognizing auto-antigens that contribute to maintenance, protection, repair and renewal of the relevant tissue; so far proven with respect to the central nervous system.

Characteristics

Description

As a result of tissue injury, physiological compounds flood the body in toxic amounts. ►T cells (T lymphocytes) directed to self-antigens serve as the fighting force against these self-derived threatening factors ("the 'enemy within'") [1,2], much as the classical immune response fights invading enemies, such as microbes.

T-cell specificity for self-antigens (►autoimmunity) is required for homing of the T cells to the autoantigen-populated site of damage and their activation there [5]. Following activation by their relevant antigens, the T-cell effect is independent of both antigen specificity and the existing toxicity. These "►autoimmune T cells," via secreted factors such as cytokines, activate the local ►microglia (brain-resident immune cells) to support neuronal survival and repair [2].

Background

Damage to the central nervous system (CNS) results in irreversible functional loss that is often more extensive than would be expected from the severity of the primary insult [1]. This heightened impairment results from the inevitable degeneration of neurons that sustained the primary injury, as well as from failure of the tissue to withstand the self-perpetuating process of damage that spreads to neighboring neurons. The latter is due to a pathological disruption in homeostasis caused by self-compounds in concentrations that are beyond the buffering capacity of the tissue. In addition, the ability of CNS tissue to tolerate a defense mechanism mediated by its resident immune cells (microglia) is poor; thus, unless tightly controlled, microglial activation is itself likely to exacerbate the damaging conditions rather than help the tissue to withstand them. Poor recovery of the CNS is therefore a reflection of (i) the limited ability to regrow new fibers from cell bodies with damaged axons (regeneration), (ii) failure of the CNS to tolerate adverse conditions, including those caused by ineffectual attempts at repair, and (iii) the limited capacity of the adult CNS for spontaneous ►neurogenesis (cell renewal). T cell-based protective autoimmunity can be viewed as a means of controlling the brain's microglia-based system of self-defense in a way that the brain can tolerate.

That the effect of the ►immune system on CNS repair processes might be beneficial was not considered a possibility until the studies done by the Michal Schwartz group at the Weizmann Institute of Science in 1998 [3]. Up to that time, the consensus was that the healthy CNS is refractory to immune cells, and that the successful infiltration of immune cells into the damaged CNS would result in CNS malfunction and should therefore be minimized. Schwartz and her group discovered that – contrary to the traditional dogma – immune cells are pivotal for CNS repair [1,3]. T cells control the microglial response by producing growth factors and cytokines that empower the microglia infiltrating blood-brone macrophages to buffer the injury-induced toxicity [2], thereby avoiding overwhelming damage to neighboring cells. Thus, by acting as a well-controlled link between the damaged CNS and the healing immune system, suitably activated microglia/macrophages can carry out defensive tasks even in a tissue as fragile as the CNS [4].

Contrary to common wisdom, Schwartz's group suggested that protection, repair, and recovery after a CNS insult necessitate an immune response, but that the response produced is often too weak (and requires boosting) or inappropriate (and in need of modulation and rigorous control); in both cases the resulting modification must be one that the CNS can safely tolerate [4]. In rats or mice devoid of the ability to manifest an adaptive (T cell-mediated) response to CNS injury, the post-injury loss of neural tissue is significantly increased. Moreover, immunization of injured animals with self-antigens, a procedure that boosts accumulation of self-reactive (autoimmune) T cells at the site of injury, decreases neuronal cell death [5]. This phenomenon of ►neuroprotection mediated by autoimmune T cells residing at the site of injury, or "protective autoimmunity," was found to be physiological response to CNS injury that spontaneously occurs but is apparently insufficient for the repair [4].

Quantitative Aspects

Injury to the CNS induces activation of microglia at the damaged site, beginning at the time of injury and reaching a peak 7–10 days later. Astrocytes, however, disappear from the site of injury, as demonstrated by immunohistochemical analysis of glial fibrillary acid protein, an astrocyte marker. Besides activated microglia, the injury site is also populated by T cells and infiltrating blood-borne monocytes. Their accumulation starts immediately after the injury, reaches a peak 7–10 days later, and resolves itself by 2–3 weeks following the injury. A well-synchronized response in terms of amount, time and location is pivotal for the repair. Lately protective autoimmunity has been extented to include brain plasticity, not only in disease but also in health. Neurogenesis and spatial learning/memory capabilities were found to be dependant on the availability of T cells recognizing CNS antigens [6].

Higher-Level Structures

T cells that participate in the healing process within the injured CNS are part of the adaptive arm of the immune system. The site of the insult determines the specificity of the T cells needed for protection and repair. In the case of injuries to white matter, the protective T cells must be reactive to antigens associated with CNS myelin whereas after gray matter injuries, the protective T cells are specific to other proteins that are abundantly expressed at the injury site.

Lower-Level Components

Upon reactivation by encountering their specific antigens at the site of injury, the T cells can produce a variety of soluble proteins that are required for normal neuronal function. These include nerve growth factor, brain-derived neurotrophic factor, glial cell-derived neurotrophic factor, and others [7].

Anatomy and Physiology of Regulation

According to the concept of protective autoimmunity, autoimmune T cells in healthy individuals exist in a state that represents a compromise between the need for autoimmune protection and the risk of autoimmune disease. Schwartz's group discovered that a subpopulation of T cells, identified as the naturally occurring regulatory CD4 + CD25 + T cells (Treg cells), constitutively suppressed the activity of autoimmune T cells, and that this population is itself regulated (i.e. the suppression can be abolished or at least weakened) by brain-derived compounds such as the stress-related neurotransmitter dopamine [5].

The Regulatory Process

Naturally occurring Treg cells exert tight control over circulating autoimmune T cells, so that the latter normally exist in a state often described by immunologists as "nonresponsiveness." Treg cells are produced in the thymus and are released to the periphery, where they suppress the activity of autoimmune T cells. Depletion of Treg cells predisposes animals to the spontaneous development of autoimmune diseases. Animals depleted of Treg cells also demonstrate more efficient rejection of tumors, because their autoimmune Lately protective autoimmunity has been evtended to include brain plashcity, not only indisease but also in health. Neurogenesis and spatial learning/memory capabilities were found to be dependant on the availability of T cells recognizing CNS antigens [insert ref ziv et al] [6] T cells can be more easily activated [15]. In line with findings on cancer rejection, depletion of the Treg cell subpopulation enhances the animals' ability to withstand conditions of CNS neurodegeneration [5]. Thus, endogenous compounds such as dopamine, because of their ability to attenuate the activity of the Treg cells, might be useful as potential components of neuroprotective therapies.

Pathology

After an injury, the damaged CNS tissue becomes accessible to immune cells. Although the post-injury inflammatory process that is mediated by self-reactive T cells is a protective physiological response, these autoimmune T cells appear to include the very cells that can also induce experimental autoimmune encephalomyelitis (EAE), an animal model for the autoimmune disease, multiple sclerosis [8]. Therefore, proper regulation is critical if the autoimmune response is to protect the organism from injury-associated damage without the concomitant induction of another kind of pathology, namely autoimmune disease [9].

Therapy

The following immune-based therapies for CNS repair are currently under investigation:

- *T cell-based therapeutic vaccination*: This experimental treatment involves the boosting of T cells directed against weak agonists of self-antigens that cross-react with antigens residing at a site of stress or disease. The risk of autoimmune disease is avoidable by using weak agonists that exhibit only partial cross-reactivity with their relevant self-antigen. One such antigen is glatiramer acetate, also known as copolymer 1 (Cop-1), a synthetic copolymer composed of four amino acids (glutamate, lysine, alanine, and tyrosine). Immunization with Cop-1 was found to be neuroprotective in several models of CNS injury and neurodegenerative disorders [7]. Another possible way to boost autoimmune T cells is through the use of altered peptide ligands. An altered dominant peptide of myelin basic protein (MBP) that does not induce EAE upon immunization was found to be significantly neuroprotective in a rodent model of spinal cord injury [10].
- *Weakening of the activity of ►regulatory T cells*: Neuroprotection is enhanced in mice depleted of Treg cells [5]. In human subjects such depletion is

impossible, as most of these cells are located in lymph nodes, and there is no known way to selectively remove them. Therefore, compounds that can selectively weaken the suppressive activity of Treg cells are potential candidates for therapy. One such compound is the neurotransmitter dopamine, which alleviates the suppressive activity of naturally occurring Treg cells. Injection of dopamine or dopamine-receptor type-1 agonists after CNS injury was found to be neuroprotective [5]. An additional compound found to be effective in modulating T reg is a copolymer of glutamate tyrosine, poly YE, show to be effective in stroke [5].
- *Autologous macrophages:* These cells constitute the basis of an immune-based cell therapy for spinal cord repair. Depending on their environmental conditions, cells of the macrophage lineage can acquire either the type of activation appropriate for fighting off invading microorganisms, or – activation leading to tissue maintenance and repair. Schwartz's group discovered that macrophages with the characteristic features of antigen-presenting cells can facilitate CNS repair [3]. Once activated by autologous tissue such as skin, these macrophages produce cytokines and neurotrophic factors, but not tumor necrosis factor-α or other cytotoxic agents. In a therapy currently under development, macrophages are prepared from the patient's own blood, activated on the patient's own skin in such a way that they adopt a repair-promoting phenotype, and reintroduced into the margin of the lesion site at the specific time and dosing, facilitates repair. Such macrophages are reminiscent of the ones recruited by immunization with CNS self-antigen following spinal cord injury [11].
- *Dendritic cells:* Bone marrow-derived dendritic cells are professional antigen-presenting cells, which can be loaded with antigens and used to treat the acutely injured CNS. Dendritic cells prepared from autologous blood and loaded with relevant CNS antigens, can (like macrophage therapy) be applied locally to the severed spinal cord, or administered systemically as a vaccination [12].

Concluding Remarks

The immune system is the body's primary source of defense against danger and remedial assistance in the event of injury. The CNS, however, was long thought to have relinquished its claim on the immune system for such help, possibly as an evolutionary trade-off that protected its extraordinarily complex neural network from harmful immune (inflammatory) intervention. Similarly, immune cells directed against self-antigens, traditionally viewed as autoaggressors causing autoimmune disease, acquired an unfavorable reputation as an outcome of immune system malfunction. The studies described here, as well as studies from many other laboratories, provide persuasive evidence that the negative reputation of the immune system vis-à-vis the CNS is undeserved. Nevertheless, spontaneous activation of the CNS-resident microglia is often not sufficiently effective. Thus, additional assistance is needed from cells that constitute the adaptive arm of peripheral immunity, namely T cells specific to CNS self-antigens. Based on this concept, several approaches can be translated into workable therapies. The treatment of choice will ultimately be based on safety, feasibility, and efficacy, and the specific indication.

References

1. Moalem G et al. (1999) Autoimmune T cells protect neurons from secondary degeneration after central nervous system axotomy. Nat Med 5:49–55
2. Schwartz M, Shaked I, Fisher J, Mizrahi T, Schori H (2003) Protective autoimmunity against the enemy within: fighting glutamate toxicity. Trends Neurosci 26:297–302
3. Rapalino O et al. (1998) Implantation of stimulated homologous macrophages results in partial recovery of paraplegic rats. Nat Med 4:814–821
4. Yoles E et al. (2001) Protective autoimmunity is a physiological response to CNS trauma. J Neurosci 21:3740–3748
5. Kipnis J et al. (2004) Dopamine, through the extracellular signal-regulated kinase pathway, downregulates CD4 + CD25 + regulatory T-cell activity: implications for neurodegeneration. J Neurosci 24:6133–6143
6. Ziv Y et al. (2006) Immune cells contribute to the maintenance of neurogenesis and spatial learining abilities in adulthood. Nat Neurosci 9(2):268–275
7. Kipnis J et al. (2000) T cell immunity to copolymer 1 confers neuroprotection on the damaged optic nerve: possible therapy for optic neuropathies. Proc Natl Acad Sci USA 97:7446–7451
8. Sakaguchi S et al. (2001) Immunologic tolerance maintained by CD25 + CD4 + regulatory T cells: their common role in controlling autoimmunity, tumor immunity, and transplantation tolerance. Immunol Rev 182:18–32
9. Schwartz M, Kipnis J (2002) Autoimmunity on alert: naturally occurring regulatory CD4(+)CD25(+) T cells as part of the evolutionary compromise between a 'need' and a 'risk'. Trends Immunol 23:530–534
10. Hauben E et al. (2001) Postraunmatic therapcutie vaccination wih modified myelin self-antigen prevents Complete paralysis while avoiding autoimmune disesea. J Clin Invest 108:591–599
11. Rossignol S, Schwob M, Schwartz M, Fehlings MG (2007) Spinal cord injury: Time to move? J Neurosci 27(44):11782–1192
12. Hauben E et al. (2003) Vaccination with dendritic cells pulsed with peptides of myelin basic protein promotes functional recovery from spinal cord injury. J Neurosci 23:8808–8819

Protein Kinase A (PKA)

Definition

Also known as cAMP-dependent protein kinase. A family of protein kinases whose activity are dependent on the level of cAMP in the cell.

Protein Kinase C (PKC)

Definition

An enzyme that exists in many different forms, which are central to many signal transduction mechanisms in brain cells.

Protein Phosphatase 1, PP-1

►Protein Serine/Threonine Phosphatases in the Nervous System

Protein Phosphatase 2A, PP-2A

►Protein Serine/Threonine Phosphatases in the Nervous System

Protein Phosphatase 2B, PP-2B

►Protein Serine/Threonine Phosphatases in the Nervous System

Protein Serine/Threonine Phosphatases in the Nervous System

Qin Yan[1,2], Ying-Wei Mao[3], David W. Li[1,2,4]
[1]Department of Biochemistry and Molecular Biology, College of Medicine, University of Nebraska Medical Center, Omaha, NE, USA
[2]Key Laboratory of Protein Chemistry and Developmental Biology of National Education Ministry of China, College of Life Sciences, Hunan Normal University, Changsha, Hunan, China
[3]Howard Hughes Medical Institute, Massachusetts Institute of Technology, Boston, MA, USA
[4]Department of Ophthalmology and Visual Sciences, College of Medicine, University of Nebraska Medical Center, Omaha, NE, USA

Synonyms

Protein Phosphatase 1, PP-1; Protein Phosphatase 2A, PP-2A; Protein Phosphatase 2B, PP-2B

Definition

Protein Serine/threonine phosphatases are a family of protein enzymes, which specifically remove the phosphate group from the serine or threonine residues of the substrate proteins. These phosphatases termed PPPs include ►PP-1, ►PP-2A, ►PP-2B, ►PP-2C, PP-4, PP-5, PP-6 and PP-7.

Characteristics

The reversible ►phosphorylation and ►dephosphorylation of proteins at the serine, threonine and tyrosine residues play fundamental roles in mediating different signaling transduction pathways and thus act as a major mechanism to regulate eukaryotic cellular events such as gene expression, cell proliferation, differentiation, homeostasis and apoptosis. The two processes executed by ►protein kinases and ►protein phosphatases can trigger conformational changes in regulated proteins that finally alter their biological properties and functions. In this regard, phosphorylation may endow a protein with a favorable conformation for binding by its relative substrates or partners, which in the end brings about some cellular and morphological changes; on the contrary, its dephosphorylated form may possess these features. And at any instant, a protein's specific phosphorylation status is a balanced result between its regulatory kinases and phosphatases.

Genome research has identified 147 genes coding for the catalytic subunits of protein phosphatases. According to differences in substrates, the protein phosphatases can be divided into three groups, including ►protein serine/threonine phosphatases (PPPs), protein ►tyrosine phosphatases (PTPs), and dual-specific phosphatases (DSPs). On the basis of the selective inhibition of type-1 phosphatases (PP-1) by endogenous inhibitor-1 (►I-1) and inhibitor-2 (►I-2), the PPP family can be further divided into type-1 (PP-1) and type-2 (►PP-2) subgroups and a few minor phosphatases that include PP-4, -5, -6 and -7, although these phosphatases have a broad and overlapping substrate specificity. The PP-1 enzyme is composed of a catalytic subunit and one or more regulatory subunits, and it is hard to distinguish the regulatory subunit from its inhibitor or substrate. The PP-2 subgroup consists of three enzyme members, PP-2A, PP-2B and PP-2C, which can be distinguished by their substrate specificity, subunit number and the specific cations needed for function. For instance, PP-2A can dephosphorylate phosphorylase-a, while PP-2B and PP-2C do not. In addition, calcium is necessary for the function of PP-2B, which is also known as calcineurin. While PP-2A is composed of three subunits and PP-2B two subunits, PP-2C is a monomeric enzyme. In eukaryotes, 98% of dephosphorylation occurs at the serine/threonine residues, and more than 90% of protein ►serine/threonine phosphatase activity is contributed by PP-1 and PP-2A [1].

Protein Phosphorylation in the Nervous System

Based on morphology and physiology, the adult nervous tissue is composed of two main cell types: neurons and glial cells. Neurons transmit high speed nerve signals through depolarization potentials, while glial cells are in direct contact with neurons to provide a basic support. Existing in variable sizes and shapes, neurons are the functional units of the nervous system. All neurons consist of three parts: dendrite, cell body and axon. Dendrites receive information from another cell and transmit information to the cell body. The cell body contains the nucleus, mitochondria and other organelles typical of eukaryotic cells, and processes the neuronal signals for further transduction. The axon sends information away from the cell body. Under normal conditions, glial cells surround neurons to modulate neuronal homeostasis and detoxification, and produce neuronal survival factors. The functions of crucial proteins in the nervous system are modulated by kinases and phosphatases. This essay focuses on the functions of three major protein phosphatases in the nervous system, PP-1, PP-2A and PP-2B.

Protein Phosphatase-1 (PP-1)

Early studies showed that PP-1 accounts for 25–40% of the phosphorylase phosphatase activity in brain crude tissue extracts. In isolated synaptic junctions from rat forebrain, PP-1 accounts for 80% of the membrane-associated phosphatase activity. PP-1 can be inactivated by the inhibitor I-1 when phosphorylated by CAMP-dependent protein kinases, or by the inhibitor I-2 in the dephosphorylated form. In addition to the ubiquitous I-1 and I-2 inhibitors, brain tissue also contains a unique PP-1 inhibitor, ►DARPP-32, present predominantly in neural tissues. The localization of DARPP-32 is unique, for it is abundant in specific areas of the brain, including the basal ganglia which contains as much as 130 pmol of DARPP-32 per mg protein, but absent from most other tissues. Compared with the basal ganglia, the thalamus and cerebellum contain much less DARPP-32. In brain areas rich in DARPP-32, discrete populations of neurons show particularly high level of expression, such as the D1-receptor dopaminoceptive neurons, and the Purkinje cells of the cerebellum. The high phosphatase activity of PP-1 in the brain and its co-localization with the cognate inhibitors suggest that PP-1 might serve as an important regulatory enzyme in various neuronal events.

Neurons produce and transduce signals through alternative shifts in membrane potentials, which depend on different ion channels distributed in the plasma membranes. Ion channel activity can be regulated by closely associated kinases or phosphatases, which serves as a key mechanism for orchestrating neuromodulation. PP-1 is responsible for the dephosphorylation and modulation of some neuronal ion channels. In many instances, dephosphorylation is required for channel activity. An exemplary case is the peptide neurotransmitter Phe-Met-Arg-PheNH2 (FMRFamide) related serotonin-sensitive K + (S-K+) channel in Aplysia sensory neurons. Two PP-1 isoforms, with apparent MW of 170,000 and 38,000, were extracted from crude membrane preparations from the Aplysia nervous system. Biochemical and physiological studies suggest that of the two major protein phosphatases, PP-1 and PP-2A, PP-1 is associated preferentially with neuronal membranes and its activity is required for the induction of outward K + currents in Aplysia sensory neurons by FMRFamide [2]. Similarly, in rat brain and cerebellar Purkinje neurons, activation of the calcium-dependent potassium channels (BK channels) by neurotrophin-3 depends on their prior dephosphorylation by PP-1 or PP-2A. In addition, certain protein phosphatases are involved in the function of the nonselective cation channel that drives the after-discharge in Aplysia bag cell neurons. It has been shown that the reversal of the nonselective cation channel after addition of ATP was prevented by the PP inhibitor, ►microcystin-LR.

Various receptors in the nervous system play important roles in mediating signal transduction. Protein phosphatases can modulate the functions of these receptors through dephosphorylation. An example is the regulation of AMPA receptor by PP-1. The filamentous actin binding protein neurabin I (Nrb I) targets PP-1 to specific postsynaptic microdomains, where it can exert critical control over AMPA receptor-mediated synaptic transmission. PP-1 dephosphorylates

AMPA subtype glutamate receptor (GluR) 2 at Ser 880- to stabilize the basal transmission; while with long-term depression (LTD), PP-1 dephosphorylates GluR1 at both serine 845 and serine 831. These results suggest that in response to distinct synaptic activities post-synaptic targeted PP-1 could regulate the synaptic trafficking of specific AMPA receptor subunits [3]. In the western painted turtle, PP-1 or PP-2A dephosphorylates N-methyl-d-aspartate (NMDA) receptors and decreases their activity, which attenuates calcium influx and thus, excitotoxic cell death (ECD), a characteristic of mammalian brain following anoxia.

Physiological modulation of other target molecules by protein serine/threonine phosphatases also plays a role in the nervous system. Neural cell adhesion molecule (NCAM) has important functions during development and maintenance of the nervous system. NCAM 140 and NCAM 180 are transmembrane glycoproteins with large cytoplasmic domains of different lengths. PP-1 and PP-2A bind to both NCAM 140 and NCAM 180, suggesting possible functions of these two protein phosphatases during the development and maintenance of the nervous system. Further, doublecortin (DCX) is a highly phosphorylated protein in migrating neurons, whose mutation causes X-linked lissencephaly ("smooth brain") and double cortex syndrome in humans. It was reported that PP-1, by recruitment of Neurabin II, dephosphorylates the specific sites in DCX phosphorylated by JNK, which plays an important role in normal neuronal migration [4]. Finally, PP-1 may play an important in regulating brain and eye development. We recently demonstrated that PP-1 acts as a major phosphatase dephosphorylating the transcription factor Pax-6, which is highly expressed in the developing nervous system, and attenuates its transcriptional activity.

Protein Phosphatase-2A (PP-2A)

PP-2A, also known as a polycation-stimutated phosphatase, is very abundant in brain where it is mainly cytosolic and represents 60–75% of the phosphorylase phosphatase activity. The PP-2A core enzyme is composed of a catalytic subunit (C) and a tightly complexed scaffolding subunit A. This core enzyme associates with a regulatory subunit B, whose function is to target the whole enzyme to specific intracellular locations and substrates. There are four subgroups of the B family subunits including B/PR55, B'/PR56/PR61, B"/PR72 and B'"/PR93/PR110. The major function of the ubiquitously expressed PP-2A seems to be related with a wide range of events under both physiological and pathological conditions in the nervous system.

PP-2A is implicated in Alzheimer's disease (AD), a progressive brain disorder that gradually destroys a person's memory and ability to learn, reason and communicate. The major brain abnormalities in patients with AD include extracellular deposits of beta-amyloid, intraneuronal neurofibrillary lesions, and the massive loss of specific subsets of telencephalic neurons. Dysregulation of the regulatory subunit of ►PP2A, B/PR55, has been implicated in AD. Since one of the major functions of the B subunit in PP-2A is to target the whole enzyme to its substrates, its dysregulation might lead to a failure of dephosphorylation of relevant brain proteins and finally AD. Further, researchers have found that the abnormal phosphorylation of paired helical filaments (PHF) tau resulting from the failure of PP-2A and PP-2B may disrupt the microtubule network, impair axonal transport, and compromise the viability of neurons, thereby also contributing to the onset and progression of AD [5]. In the non-diseased nervous system, research at the Drosophila neuromuscular junction suggests that the B' regulatory subunit of ►protein phosphatase 2A may regulate normal cytoskeletal organization, synaptic growth, and synaptic function.

The physiological function of PP-2A is also being revealed. It was reported that PP-2A is required for the differentiation of pluripotent neural crest (NC) cells into the sympathoadrenal lineage. A moderate activation of cAMP signaling induces both the transcription and activity of the proneural transcription factor Phox2a in NCs, whereas treatment with the ►organic phosphatase inhibitor, ►okadaic acid, at a concentration (1–10 nM) effective to inhibit PP-2A, suppresses sympathoadrenal lineage development. PP-2A also functions in the mature nervous system. During spinal cord ►central sensitization, PP-2A may play an important role in determining the excitability of ►nociceptive neurons in the spinal cord by modulating the phosphorylation state of certain critical proteins. For instance, infusion of the phosphatase inhibitor okadaic acid or ►fostriecin, a specific PP-2A inhibitor, into the ►subarachnoid space enhances secondary mechanical ►hyperalgesia and ►allodynia [6]. In addition, it was observed that blockade of protein phosphatase activity potentiates central sensitization of nociceptive transmission in the spinal cord following capsaicin injection, indicating that PP-2A may be involved in determining the duration of capsaicin-induced central sensitization. Finally, since cyclin G1 and the B' subunits of PP-2A are co-localized in neurons, the function of PP-2A in the nervous system may depend partly on its regulation of cell cycle in neurons [7].

Protein Phosphatase-2B (Calcineurin)

PP-2B belongs to the family of Ca^{2+} /calmodulin-dependent protein phosphatases and it is the only protein phosphatase regulated by a second messenger, Ca^{2+}. Furthermore, PP-2B is highly localized in the central nervous system, especially in those neurons vulnerable to ischemic and traumatic insults. For these reasons, PP-2B is considered to play important roles in neuron-specific functions.

PP-2B has been extensively studied in nervous tissue. It was first purified as a major calmodulin-binding protein in brain (accounting for up to 1% of the total protein), and later identified as a protein phosphatase. The whole enzyme consists of two subunits, the catalytic and calmodulin-binding subunit calcineurin A, and the calcium binding regulatory subunit calcineurin B. PP-2B is essentially a neuronal protein due to its abundance in brain, where levels are 10–20 times higher than in other tissues. Nevertheless, it is distributed amongst a wide set of tissues in many species from yeast to mammals.

Some ion channels and receptors are regulated by PP-2B. In sympathetic neurons, the M current regulates neuronal excitability. Intracellular application of a preactivated form of PP-2B inhibited the macroscopic M current, while its application to excised inside-out patches reduced high open probability M channel activity. Thus, it is suggested that PP-2B selectively regulates M channel modal gating. PP-2B may also regulate the coupling between G protein and calcium channels in rat sympathetic neurons, as the alpha2 noradrenergic and somatostatin receptor-induced inhibition of N-type Ca^{2+} channels was greatly reduced by inhibition of PP-2B. Interestingly, PP-2B in some cases may regulate both the function and the expression of a protein. For instance, PP-2B directly regulates type 1 inositol 1, 4, 5-triphosphate receptor (IP3R) function by dephosphorylation on a short-term time scale and IP3R expression over more extended periods.

The physiological functions of PP-2B in the nervous system seem to be quite diverse. On one hand, PP-2B regulates certain signaling transduction pathways in neurons by directly dephosphorylating target protein substrates. An example is its regulation of the NF-κB pathway in astrocytes of the injured brain. After brain injury a reactive phenotype of astrocytes is triggered which produces inflammatory cytokines and neurotoxic free radicals and leads to brain trauma. Under such circumstances, insulin-like growth factor-I (IGF-I) regulates the NF-κB pathway by activating PP-2B to dephosphorylate IκBα, which leads to a stabilization of IκBα and retention of NF-κB in the cytosol and hence inhibition of the inflammatory reaction [8].

In other systems, the signal transduction pathways regulated by PP-2B result in the activation of different transcription factor cascades. For instance, in the mammalian nervous system, PP-2B appears to regulate the number of excitatory synapses via myocyte enhancer factor 2 (MEF2). After dephosphorylation and activation by PP-2B, MEF2 promotes the transcription of a set of genes that restricts synapse number. Alternatively, regulation of other signaling transduction pathways by PP-2B involves the downstream transcription factor NF-AT. In calcium-regulated pathways related to learning and memory process, NF-ATc4/NF-AT3 in hippocampal neurons can be translocated rapidly from the cytoplasm to the nucleus to activate NF-AT-dependent gene transcription in response to electrical activity or potassium depolarization. The PP-2B-mediated translocation is critically dependent on calcium entry through L-type voltage-gated calcium channels. This indicates that PP-2B/NF-AT-mediated gene expression may be involved in the induction of hippocampal synaptic plasticity and memory formation. In addition, the PP-2B/NF-AT pathway is involved in neuronal axon outgrowth, the first step in the formation of neuronal connections. After stimulation by growth factors, PP-2B caused nuclear localization of NF-ATc4 and the activation of NF-AT-mediated gene transcription in cultured primary neurons. Blockade of this pathway prevented axon outgrowth, indicating that PP-2B/NF-AT signaling is required during nervous system development [9]. PP-2B also enhances NGF-induced neurite outgrowth.

The list of roles for PP-2B is ever growing in that PP-2B activity regulates synaptic function. For instance, PP-2B activity is required for three forms of synaptic plasticity in the hippocampus and associative learning in *Caenorhabditis elegans* and also reward-related learning in mice. The importance of PP-2B in synaptic function is illustrated by the fact that conditional calcineurin knockout mice exhibit multiple abnormal behaviors related to schizophrenia [10], and that PP-2B is also implicated in epileptogenesis.

Conclusions

In summary, the protein serine/threonine phosphatases seem to be involved extensively in the regulation of neuronal development, differentiation, physiology and pathogenesis. The unscrambling of the signaling transduction pathways in nervous system would be a feasible solution for the ultimate conquering of various brain diseases such as Alzheimer's disease, epileptogenesis, and schizophrenia. Elucidation of the functional mechanisms of the protein phosphatases in the nervous system will provide a better understanding of neuronal function and brain disease.

Acknowledgments

This work is supported in part by the NIH/NEI grant 1R01EY015765, the startup funds from the University of Nebraska Medical Center, the Changjiang Scholar Team Project Funds from the National Education Ministry of China and the Lotus Scholar Program Funds from Hunan Province Government and Hunan Normal University.

References

1. Cohen P (1989) The structure and regulation of protein phosphatases. Annu Rev Biochem 58:453–508
2. Endo S, Critz SD, Byrne JH, Shenolikar S (1995) Protein phosphatase-1 regulates outward K + currents in sensory neurons of Aplysia californica. J Neurochem 64(4):1833–1840
3. Hu XD, Huang Q, Yang X, Xia H (2007) Differential regulation of AMPA receptor trafficking by neurabin-targeted synaptic protein phosphatase-1 in synaptic transmission and long-term depression in hippocampus. J Neurosci 27(17):4674–4686
4. Shmueli A, Gdalyahu A, Sapoznik S, Sapir T, Tsukada M, Reiner O (2006) Site-specific dephosphorylation of doublecortin (DCX) by protein phosphatase 1 (PP1). Mol Cell Neurosci 32(1–2):15–26
5. Trojanowski JQ, Lee VM (1995) Phosphorylation of paired helical filament tau in Alzheimer's disease neurofibrillary lesions: focusing on phosphatases. FASEB J 9(15):1570–1576
6. Zhang X, Wu J, Fang L, Willis WD (2003) The effects of protein phosphatase inhibitors on nociceptive behavioral responses of rats following intradermal injection of capsaicin. Pain 106(3):443–451
7. van Lookeren Campagne M, Okamoto K, Prives C, Gill R (1999) Developmental expression and co-localization of cyclin G1 and the B' subunits of protein phosphatase 2a in neurons. Brain Res Mol Brain Res 64(1):1–10
8. Sebastian Pons and Ignacio Torres-Aleman (2000) Insulin-like growth factor-I stimulates dephosphorylation of IγB through the serine phosphatase calcineurin (protein phosphatase 2B). J Biol Chem 275(49):38620–38625
9. Graef IA, Wang F, Charron F, Chen L, Neilson J, Tessier-Lavigne M, Crabtree GR (2003) Neurotrophins and netrins require calcineurin/NFAT signaling to stimulate outgrowth of embryonic axons. Cell 113(5):657–670
10. Miyakawa T, Leiter LM, Gerber DJ, Gainetdinov RR, Sotnikova TD, Zeng H, Caron MG, Tonegawa S (2003) Conditional calcineurin knockout mice exhibit multiple abnormal behaviors related to schizophrenia. Proc Natl Acad Sci USA 100(15):8987–8992

Protein Synthetic Machinery

Definition

The protein synthetic machinery of a eukaryotic cell consists of ribosomes, endoplasmic reticulum (ER) and Golgi apparatus. Proteins destined to stay within the cytoplasm (cytosolic proteins) are synthesized on ribosomes and the product is released into the cytoplasm. Luminal proteins (proteins packaged in vesicles) and integral membrane proteins are synthesized on ribosomes that are docked on the endoplasmic reticulum, and the polypeptide chain is secreted in the lumen of the ER (luminal proteins), or is inserted into the membrane of the ER (integral membrane proteins).

After trafficking through the Golgi apparatus the proteins are either released (secretory proteins), inserted into the cytoplasmic membrane (integral membrane proteins), or stored within vesicles in the cytoplasm (lysosomes).

►Extrasomal Protein Synthesis in Neurons

Protein Tyrosine Kinase

Definition

A protein that contains a domain that acts as a kinase, which transfers a phosphate group from ATP to a tyrosine of a protein.

Protein Zero (P0)

Definition

P0 was identified as the major protein component of Schwann cell myelin. The P0 gene is expressed throughout the Schwann cell lineage, from precursors in embryonic development to myelinating Schwann cells in the adult animal. During development, neurons do not express P0 gene, thus P0 gene expression can be used as an early marker of glial specification in neural crest development. In normal adult nerves, expression of P0 gene and protein is restricted to myelin-forming Schwann cells.

►Myelin
►Schwann Cell
►Schwann Cells in Nerve Regeneration

Proteoglycan

Definition

Proteoglycans are molecules that possess a protein core to which are attached one or more glycosaminoglycan chains at serine residues through a four sugar linker. In chondroitin sulfate proteoglycans the chains are made of repeating glucuronic acid and n-acetyl galactosamine, while in heparan sulfate proteoglycans the two sugar repeat is glucuronic acid (which may be epimerized to iduronic acid) and n-acetyl glucosamine.

The glycosaminoglycan chains are modified by sulfation at various points on the two sugars. Particularly in heparan sulfates sulfation occurs in patches. The pattern of sulfation defines the charge pattern of the glycan and therefore its binding properties. Chondroitin sulfate proteoglycans often have barrier functions, while heparan sulfate proteoglycans are often found as part of a ternary complex presenting growth factors to their receptors.

Proteomics and the Study of the Nervous System

JENS R. COORSSEN[1,2]
[1]Hotchkiss Brain Institute, Faculty of Medicine, University of Calgary, Calgary, Alberta, Canada
[2]Chair of Molecular Physiology, School of Medicine, University of Western Sydney, Campbelltown, NSW, Australia

Definition

Proteomics encompasses a number of techniques and approaches. At its simplest, it could be described as the large-scale analysis of the proteins encoded by a given genome, or of those proteins present, at the time of sampling, in a given biological milieu (e.g., tissue/cell extract, bodily fluid, etc). The term, coined by Marc Wilkins and colleagues in 1994, describes a new discipline that has moved forward from genomics to enable a more mechanistic understanding of biological processes and disease states. While the genome provides information regarding the proteins (and mutations) potentially expressed by a specific organism, it does not identify which proteins are locally expressed at a specific time, under specific conditions, nor does it provide information concerning the myriad of potentially critical post-translational modifications that can affect specific aspects of protein localization and function. Furthermore, although assessment of mRNA levels (►transcriptomics) is sometimes used as a surrogate for more detailed protein analyses, this often does not accurately reflect the actual protein complement (e.g., functional players) at the time of sampling. Thus, the definition of proteomics has grown to include not only the presence of an array of proteins in a sample, but also the localization, post-translational modifications, activities, and interactions of these proteins with each other and with other molecules. In many ways, the original approach has spawned a variety of sub-disciplines that might be broadly defined as follows:

Functional proteomics – the tight coupling of quantitative functional assays with detailed protein analyses as a direct route to dissecting molecular mechanisms underlying physiological processes.

Structural proteomics – perhaps most akin to structural biology, with the goal of defining the three dimensional or atomic structure of a protein as a route to understanding functional interactions/mechanism(s) of action within the cellular environment.

Discovery or Clinical proteomics – scanning of ►proteomes to identify alterations that potentially underlie a disease pathway or that could serve as biomarkers for specific clinical conditions. The resulting "catalogues" generally serve as important databases for other ongoing work as well.

Characteristics

What then separates proteomics from more traditional protein biochemistry/physical chemistry and cell physiology? This is generally seen to be a matter of scale, throughput, and automation. Rather than asking about the potential role of a single protein in a specific process, proteomics takes a global or systems approach to the integrated roles of proteins in physiological processes. Thus, as a discipline, proteomics has really arisen from the necessary interactions of protein (bio)chemists, cell physiologists, computer scientists, and engineers. The goal of this inter-disciplinary approach is to enhance the ►resolution and throughput of large-scale protein analysis, while simultaneously reducing any potential user bias in the assessment process. With the introduction of computer-driven automation, robotics, and data analysis, the result has been enhanced reproducibility, sensitivity, higher throughput, more objective analyses of the resulting protein data (particularly via subtractive analysis to identify critical differences between data sets), and the establishment of a myriad of databases, making information sharing one of the more important and successful priorities in proteomics.

Techniques

Until relatively recently, proteomics was almost synonymous with two-dimensional polyacrylamide gel electrophoresis (2DE). Building from earlier work, this technique was introduced in the mid-1970s and immediately revolutionized protein analysis. Proteins are separated by ►isoelectric focusing in the first dimension, according to their ►pI (Isoelectric Point), and in the second dimension according (approximately) to their molecular weight (MW), using sodium dodecyl sulphate polyacrylamide gel electrophoresis ►SDS-PAGE (Sodium Dodecyl Sulfate Polyacrylamide Gel Electrophoresis) [1]. In principle, any physical attribute of a protein can be used to establish a separation/purification strategy; hydrophobicity is another commonly used characteristic to aid in protein isolation. For several years, identification of specific protein spots could only be done by Western blotting (immunodetection) or Edman degradation. The real revolution in terms of proteomics developing as a discipline was the integration of ►mass spectrometry for protein identification. With an appropriately "clean" protein sample (see below), it is now routine to excise a spot of interest from a 2DE gel, digest it with a protease (usually trypsin) or another selective agent (such as cyanogen bromide), and analyze the resulting peptides using a variety of mass spectrometric methods. The simplest analysis is to derive a peptide fingerprint and attempt to identify the protein using computerized searches of available databases; essentially, archived protein sequences are digested *in silico*, and the best matches to the unknown protein identified. This approach has fallen out of favor due to a high rate of false-positives, although with a well-defined genome to search, this can be a useful scanning tool. More effort is now placed in using mass spectrometry for *de novo* protein sequencing and thus definitive protein identifications, as well as for the analysis of post-translational modifications [2].

Within the last several years, non-gel based approaches to large-scale protein resolution have been developed based on a combination of liquid chromatography and mass spectrometry (or other techniques "hyphenated" with mass spectrometry). Basically, a gross protein extract is proteolytically digested to yield a complex peptide mixture and this is then fractionated, generally by sequential cation-exchange and reverse-phase chromatography, and random samples of the separated peptides are then sequenced in a mass spectrometer [3]. Possible protein identifications are then made based on the presence of peptides "unique" to specific proteins. This so-called "shotgun" proteomics approach has proven to be an effective scanning tool. Most recently, additional proteomic applications for mass spectrometry have included analysis directly from tissue samples ("imaging" mass spectrometry); although not of high resolution, this technique does provide direct information on the localization of specific proteins [4].

Pros and Cons

It is probably safe to say that 2DE remains the workhorse of proteomics and the current gold-standard in terms of protein resolution. It certainly remains the only method currently available with the capacity to resolve thousands of proteins in a single run. The major complaint that seems to be leveled against it is that it is "old," yet it is precisely this maturity that yields analytical power: problems have been identified and effectively addressed; in the case of newer approaches, most problems have likely not even yet been effectively identified. High throughput always comes at a cost; in the case of shotgun methods, information concerning the ►pI, MW, and post-translational modifications of proteins is lost, and the approach requires substantial computing infrastructure. Nonetheless, it is likely an integration of these scanning and higher resolution approaches that will define future proteomic efforts, particularly considering the current pace of developments in mass spectrometry.

Regardless of the analytical approach used, the biggest difficulty in proteomics is that of protein detection. While it is likely that we are currently capable of resolving most proteins, those present at lower abundance (e.g., a few to tens of copies per cell) are not detectable using currently available total protein stains [5]; if a protein of interest is known, it can however be detected by a quantitative, high sensitivity Western blotting protocol [6]. There is currently no simple solution to this general detection problem except to fractionate the sample in some way and thus increase the total protein concentration of a specific fraction prior to analysis. Such prefractionation techniques are currently under intense development, particularly for samples such as plasma and cerebrospinal fluid that are dominated by a few high abundance species that obviate the effective analysis of the rest of the proteome. Such prefractionation can be as simple as lysing a tissue sample or cell pellet and using ultracentrifugation to isolate total soluble and membrane protein fractions that can then be separately resolved [7,8]. A myriad of alternate techniques, including sequential extractions with different detergents and/or solvents, selective precipitation, or immuno-isolation of select components, each resulting in the concentration of certain proteins, are all being used. Notably, care must be taken to avoid potential non-specific alterations to the proteome using these more aggressive approaches. In this regard, it has also been found that 2DE gels can be post-fractionated to further enhance the resolution of this technique [8].

Process

The primary caveat when embarking on any proteomic analysis is cleanliness. In the past, if all that was to be done with a 2DE gel for instance was staining or Western blotting, minor contaminants (most commonly, human skin keratin!) were of little or no consequence. Now however, the goal is to identify specific resolved proteins using mass spectrometry; more often than not, the first painful lesson is to find keratin everywhere, obviating any possible protein identification. All buffers and equipment must be scrupulously clean and maintained as such. The Brain Proteome Project initiative of the Human Proteome Organization is helping to establish criteria for sample handling and analysis (www.hbpp.org).

The goal of proteomics is thus an analysis that accurately represents the underlying biological complexity of the sample at the instant of sampling. Thus, regardless of the analytical technique eventually used to resolve the proteins, there must be a focus on optimal sample handling. All tissue dissections, isolations, and extractions are done in the presence of broad-spectrum protease, kinase, and phosphatase inhibitors [7,8]. In the case of tissue samples, including brain and spinal cord, we have found that the only way to ensure integrity of the protein complement is to snap freeze the sample at the instant of sampling; subsequent automated frozen disruption effectively powders the sample, ensuring optimal protein extraction [7]. In this regard, optimal solubilization of membrane proteins has always been a major concern, spurring development of multiple different detergents to supplement/supplant the most widely used, CHAPS (3 [(3-cholamidopropyl) dimethylammonio]-1-propanesulfonate); we have found that simply supplementing the CHAPS-based buffer with lysophosphatidylcholine substantially enhances the extraction and subsequent analysis of membrane proteins, particularly from neural tissue [9].

With an acceptable isolate in hand, the proteome can, as discussed above, be resolved by different techniques. In the case of 2DE, after the proteins are resolved on the second dimension gel, they must be detected. Currently, the favored total protein stain is Sypro Ruby due to its sensitivity, low variation in staining across protein species, and compatibility with subsequent mass spectrometric analysis. However, developments in this area are a major focus, and it is likely that another stain with enhanced sensitivity will be identified and widely used. Following staining, the gel is imaged, and it is this image that is then analyzed using specific software packages; these image analysis programs can warp and match images within sets of gels and then essentially overlay these, and carry out subtractive or differential analysis to identify statistically significant alterations in the proteome. In an effort to get as much information as possible from a single gel, "multiplexing" is gaining in popularity; here at least one other stain specific for a class of proteins (e.g., phosphorylated or glycosylated) is first used to identify select changes in the proteome, and then the total protein pattern is analyzed for more global changes. Proteins of interest are then excised, digested, and analyzed by mass spectrometry, as described above.

Applications in Neuroscience

With respect to Basic research, the use of proteomics has focused largely on the analysis of synaptic vesicles and the post-synaptic density. Here the effort has been to define the protein components and, in some cases, more specifically the phosphoprotein sub-proteome due to the importance of this ▶post-translation modification in neural function.

To date, the more widespread application of proteomics has been in the area of Neurological and related disorders [10]. This has included studies on Alzheimer's disease, spinal cord injury, muscular dystrophy, Parkinson's disease, and epilepsy, but this is far from an exhaustive list. The focus has largely been on identifying (phospho)protein alterations that may underlie the susceptibility, progression, or fundamental mechanism of a given disorder. Both in Neuroscience and other Clinical disciplines, the initial goal is most often to identify biomarkers of the condition in question so as to ensure more objective and accurate diagnoses and prognoses. The limitation is clearly accessibility, particularly in terms of acquiring samples from age-matched, "normal" controls; simply put, it is all but impossible to obtain samples from unaffected individuals. Thus, researchers must develop experimental designs that take into account the fact that human neural tissue is an unlikely sample source except in some specific instances of surgery. Although often used, post-mortem sampling must be recognized as insufficient in many regards due principally to the fact that tissue resection generally occurs at variable time intervals after death and that the expression levels of different proteins undergo significant alterations (both increasing and decreasing) during the time that precedes tissue sampling. In most cases though, a more serious and often overlooked caveat is that of disease progression; in most cases, sampling is only possible after clinical diagnosis, and thus after the damage has already occurred. What is to be reasonably learned at this stage? It would thus seem that, without appropriate animal models, much of the application of proteomics to questions in Clinical Neuroscience/Neurology will focus on identifying alterations that occur "after the fact." Although this is important from the standpoint of understanding the molecular mechanisms of disease progression or recovery from injury, it will unlikely be informative with regard to alterations underlying disease susceptibility and inception. Critical study design and focused application of techniques is mandatory to ensure successful experimental outcomes that will include the identification of effective diagnostic/prognostic biomarkers as well as components underlying fundamental disease mechanisms. This will largely hinge on the analysis of accessible diagnostic biofluids (e.g., cerebrospinal fluid, urine, plasma, and saliva). Thus, coupled with transcriptomics, ▶metabolomics, and other, developing, molecular analytical tools, proteomics promises to drive substantial advances in Basic, Clinical, and Translational Neuroscience research.

References

1. Carrette O, Burkhard PR, Sanchez JC, Hochstrasser DF (2006) State-of-the-art two-dimensional gel electrophoresis: a key tool of proteomics research. Nat Protoc 1:812–823

2. Olson MT, Epstein JA, Yergey AL (2006) De novo peptide sequencing using exhaustive enumeration of peptide composition. J Am Soc Mass Spectrom 17:1041–1049
3. Lohaus C, Nolte A, Blüggel M, Scheer C, Klose J, Gobom J, Schüler A, Wiebringhaus T, Meyer HE, Marcus K (2007) Multidimensional chromatography: a powerful tool for the analysis of membrane proteins in mouse brain. J Proteome Res 6:105–113
4. Altelaar AF, Luxembourg SL, McDonnell LA, Liersma SR, Heeren RM (2007) Imaging mass spectrometry at cellular length scales. Nat Protoc 2:1185–1196
5. Harris LR, Churchward MA, Butt RH, Coorssen JR (2007) Assessing detection methods for gel-based proteomic analyses. J Proteome Res 6:1418–1425
6. Coorssen, JR, Blank, PS, Albertorio F, Bezrukov L, Kolosova I, Backlund PS Jr., Zimmerberg J (2002) Quantitative femto- to attomole immunodetection of regulated secretory vesicle proteins critical to exocytosis. Anal Biochem 307:54–62
7. Butt RH, Coorssen JR (2006) Pre-extraction sample handling by automated frozen disruption significantly improves subsequent proteomic analyses. J Proteome Res 5:437–448
8. Butt RH, Coorssen JR (2005) Postfractionation for enhanced proteomic analyses: routine electrophoretic methods increase the resolution of standard 2D-PAGE. J Proteome Res 4:982–991
9. Churchward MA, Butt RH, Lang JC, Hsu KK, Coorssen JR (2005) Enhanced detergent extraction for analysis of membrane proteomes by two-dimensional gel electrophoresis. Proteome Sci 3:5
10. Drabik A, Bierczynska-Krzysik A, Bodzon-Kulakowska A, Suder P, Kotlinska J, Silberring J (2007) Proteomics in neurosciences. Mass Spectrom Rev 26:432–450

Protostome/Deuterostome

Definition

The Deuterostome clade in the animal kingdom including chordates, Hemichordates and Echinoderms.

In Deuterostome animals the Blastopore develops into the anus of the animal. The protostome clade contains most invertebrate phyla. In Protostome animals the Blastopore develops into the mouth of the animal. The Protostome clade is further subdivided into Lophotrochozoans and Ecdysozoans. The Lophotrochozoan clade most animals pass a trochophora-larva stage or use a Lophophor (tentacle like structure). In the Ecdysozoan clade most animal go through a molting process by Ecdysis.

►Evolution of the Brain: Urbilateria

Proust Effect

Gordon M. Shepherd[1], Kirsten Shepherd-Barr[2]

[1]Department of Neurobiology, Yale University School of Medicine, New Haven, CT, USA

[2]Faculty of English, Oxford University, Oxford, UK

Definition

"But when from a long-distant past nothing subsists, after the people are dead, after the things are broken and scattered, taste and smell alone, more fragile but more enduring, more unsubstantial, more persistent, more faithful, remain poised a long time, like souls, remembering, waiting, hoping, amid the ruins of all the rest; and bear unflinchingly, in the tiny and almost impalpable drop of their essence, the vast structure of recollection." (*Swann's Way*, trans. Moncrieff, 50–51).

The best known example of the power of smell to evoke memories and emotions is the "Proust effect," as described in the quotation above. This is based on a sensory experience described by Marcel Proust in *Remembrance of Time Past* [1]. The episode involves tasting a small cake dipped in tea, and being suddenly flooded with a childhood memory. The suddenness of the memory, and the strong emotions attached to it, have become symbolic, in both the realm of literary criticism and in sensory physiology, for the power of human taste and smell. Here we review the episode, and show that there is more than meets the eye (and nose).

Characteristics

The Madeleine Incident

Proust's character Marcel is in a depressed mood when he visits his mother. To soothe him she serves him a cup of tea and a madeleine cake. He dips the cake in the tea, tastes it, and is immediately seized by an overwhelming emotion: "an exquisite pleasure had invaded my senses, something isolated, detached, with no suggestion of its origin." He is aware that behind this is an as yet unidentified memory. After several more tastes and an intense period of searching his memory, he is suddenly transported back to his childhood summers in a seaside resort, Combray, which are associated with happy times. With repeated tasting the power of the memory fades.

The Theory of Pure Memory

In Proust's view, he had uncovered an essential truth about himself, elicited by taste and smell from a "pure memory" unadulterated by being remembered during the intervening years. From this he erected a theory of involuntary (unconscious) versus voluntary (conscious) memory. Samuel Beckett [2] followed with the view that voluntary (i.e., adulterated) memory cannot recall the exact (i.e., pure) sensation as it was originally experienced. Many literary critics have taken up this view and built on it, regarding it as showing how memory can spring suddenly and purely into mind, triggered after a long period of being forgotten. As a byproduct, they have maintained that conscious autobiographical memory is therefore artificial, a created fiction, compared with true, involuntary memory of the type evoked from the unconscious by the tea-soaked madeleine.

Re-Examining the Theory

We have re-examined Proust's account to test this view [3]. A close textual analysis has revealed that the memory was not sudden, but was the result of considerable effort, as described by the author. Indeed, a page and a half of text is required to describe the repeated efforts of the author to recall the memory. Thus, it appears that as soon as involuntary memory is brought into the arena of consciousness, it too, like voluntary recall, is subjected to similar processes of selection, editing and revision. We have further argued that an understanding of those processes requires knowledge of the neural pathways involved.

What does Neuroscience Tell Us?

Given its iconic status, the madeleine incident thus serves as a useful model for asking: What was happening in Marcel's brain, from the initial sensing of the taste of the tea-soaked madeleine cake to the emotions it called forth, and the memory that finally appeared? Recent research on the sensory mechanisms involved provides several insights that are generally unappreciated even among most neuroscientists.

First, we realize that the smell of an ingested item, cake or otherwise, is due to stimulation of the smell receptors in the nose not by sniffing in, but by breathing out. The volatiles released from the ingested food and liquid at the back of the mouth are carried by exhaled air by the retronasal route through the nasopharynx to reach the receptors (Sun and Halpern, 2005). This olfactory stimulation carries the major part of what appears to be the "taste," but it goes largely unrecognized as smell because it is referred to the mouth where the food and liquid are located. To avoid the confusion, the term "flavor" can be used to refer to the combined sense of taste and smell, as mentioned in the initial quotation from Proust. In fact, flavor also includes all the other senses associated with food intake: touch, pressure, pain, temperature, sound, and even the sight of the food before ingestion.

Second, the olfactory receptor cells connect to the olfactory bulb, where their responses are converted into spatial activity patterns, called variously odor images or odor maps [4]. Smell perception, by itself or as part of flavor, thus involves discrimination of different spatial images, which also vary with time. We can therefore hypothesize that Proust's effort to identify the memory of the smell of the madeleine involved matching the spatial pattern aroused by that smell with the similar spatial pattern contained in his memory. It may be similar to recognizing complex visual patterns such as faces.

Third, after processing by microcircuits in the olfactory bulb, the odor image is sent to the olfactory cortex, where it is converted by microcircuits into a distributed form called a content-addressable memory [5]. The output of this memory representation has two main destinations. One is to the prefrontal cortex, where it is processed by the highest cognitive centers of the brain [6]. The other is to centers in the limbic system, where emotions are generated. The olfactory pathway has privileged direct access to both these sites; hence the combination of intense emotion and intense cognitive effort experienced by Proust in recollecting the memory of Combray.

Finally, the olfactory pathway contains a series of mechanisms to reduce the stimulation with repeated odor exposure, called adaptation or desensitization, which appear to be reflected in Proust's account of the fading of the sensations with repeated tasting of his madeleine. Rather than intensifying over time, the memory/emotion diminishes with repeated tasting of the triggering stimulus.

The Neural Basis of the Proust Effect

How is the entire memory of Combray recalled from a single sip of tea? This should not be surprising; our cognitive mechanisms have a gestalt quality, in which we perceive and recall things as integrated wholes. According to Nadel and Moscovitch [7], "recovery of remote memories always depends on re-activation of the hippocampal–neocortical complex that constitutes the memory trace." One may hypothesize that this trace, representing the memory of Combray, is distributed in the olfactory cortex and the hippocampal–neocortical complex in a content-addressable form so that a small part of it – the flavor of the madeleine – activates it in its entirety. The brain's ability to fill in the rest of the memory image is turned by Proust into metonym.

Conclusion

We have here summarized only briefly the Proust effect, the literary superstructures built on it, and the recent neuroscience research that gives insight into the relevant brain mechanisms. Further information may be gained from the cited references. The present account may indicate the increasing attraction of this experience as a model for the brain circuits underlying the power of human senses in general, and the chemical senses in particular. Indeed, the Proust effect may be one of the best instances in which literary criticism can benefit from brain research, and neuroscientists can benefit from literary studies that extend the significance of their experiments into the domain of public discourse.

References

1. Proust M (1987) A la Recherche du Temps Perdu. Du Cote de Chez Swann. Flammarion, Paris
2. Beckett S (1931) Proust. Grove, New York
3. Shepherd-Barr K, Shepherd GM (1999) Madeleines and neuromodernism: reassessing mechanisms of autobiographical memory in Proust. Autobiography Stud 13:40–59
4. Xu FQ, Greer CA, Shepherd GM (2000) Odor maps in the olfactory bulb. J Comp Neurol 422:489–495
5. Neville KR, Haberly LB (2004) Olfactory cortex. In: Shepherd GM (ed) The synaptic organization of the brain. Oxford University Press, New York, 415–454
6. Rolls ET (2006) Brain mechanisms underlying flavour and appetite. Philos Trans R Soc Lond B Biol Sci 361:1123–1136
7. Nadel L, Moscovitch M (1997) Memory consolidation, retrograde amnesia and the hippocampal complex. Curr Opin Neurobiol 7:217–227
8. Sun BC, Halpern BP (2005) Identification of air phase retronasal and orthonasal odorant pairs. Chem Senses 30:693–706

Pseudo-bulbar Paralysis

Definition

Disease characterized by ▶dysarthria, ▶dysphonia, ▶dysphagia, bifacial ▶paralysis, forced crying and laughing, resulting from bilateral lesions of the ▶corticospinal tract.

▶Corticospinal Tract

Pseudogene

Definition

Pseudogene is a defunct version of a known gene that has lost its ability to be transcribed.

Pseudohypertrophy

Definition

Increase in limb size not attributed to hypertrophy (increased size) of muscle fibers but due to adipose and connective tissue deposition; occurs in Duchenne Muscular Dystrophy.

▶Duchenne Muscular Dystrophy

Pseudorandom

Definition

A process that appears random but is not. Pseudorandom sequences typically exhibit statistical randomness while being generated by an entirely deterministic computational process.

Pseudotumor Cerebri

Definition

Idiopathic intracranial hypertension (opening pressure >200–250 mmH_2O on CSF examination), with or without papilledema, and/or transient visual obscurations.

Headaches may have a mild to moderate intensity occurring on any part of the head or globally.

▶Headache

Psychedelics

▶Hallucinogens

Psychic Gaze Paresis

Definition

Staring gaze for minutes or sometimes hours, mostly upwards. It cannot by suppressed arbitrarily and is due to unconscious intrapsychical processes.

▶Psychic Gaze Spasm
▶Personality Disorder

Psychoacoustics

THOMAS N. BUELL[1], CONSTANTINE TRAHIOTIS[1,2], LESLIE R. BERNSTEIN[1,2]
[1]Department of Neuroscience, University of Connecticut Health Center, Farmington, CT, USA
[2]Department of Surgery (Otolaryngology), University of Connecticut Health Center, Farmington, CT, USA

Synonyms

Auditory psychophysics; Psychological acoustics

Definition

Psychoacoustics is the area of auditory research in which behavioral methods are used to discern and describe how, and how well, listeners perceive sounds. Combining information obtained from such experiments with corresponding information obtained from physiological and anatomical experiments has resulted in several important descriptions, explanations, and quantitative models of auditory function. Many of the relevant data and phenomena are discussed elsewhere in this volume. Therefore, this discussion is limited to two examples chosen to help the reader understand the kinds of questions addressed by psychoacousticians. In order to provide an intuitive understanding of the field, several practical applications that incorporate knowledge derived from psychophysical experiments will also be highlighted. For further information, the reader is referred to reviews of many pertinent topic areas [1] and two excellent introductory textbooks, one at a more advanced level [2] than the other [3].

Characteristics

Types of Measures

Some experiments in psychoacoustics concern thresholds of detection. They measure the smallest *amounts* of physical stimulation required to produce a reliable behavioral response indicating that the signal is present. Other experiments concern thresholds of discrimination. Their goal is to measure the smallest *changes* in physical stimulation required for listeners to indicate reliably that two signals are different. Still other, "scaling", methods use listeners as meters and measure one of a number of perceptual attributes of sound, such as ▶loudness and pitch.

Ties to Physiology

It should be stressed that a comprehensive understanding of hearing also necessarily requires an appreciation of the whole auditory system. Consequently, psychoacousticians have traditionally attempted to understand behavioral data in light of knowledge concerning the anatomy and physiology of the auditory system. We will illustrate such attempts in the context of two important areas of auditory processing; intensity perception and frequency discrimination/▶masking.

Stimulus Intensity

In terms of sensitivity to small changes in intensity, there are two general outcomes to be stated and understood. First, people are quite sensitive to changes in intensity, and can typically discriminate between stimuli differing by about one dB or so. This "just noticeable difference" of 1 dB corresponds to a difference of 25% in acoustic power. In turn, this degree of sensitivity is maintained over about 10–12 orders of magnitude of intensity (i.e. over a range of about 100–120 dB). One goal of psychoacousticians has been to explain this pair of findings in physiological terms. This has proved challenging, because individual neural elements that compose the peripheral auditory nerve have been shown to respond differentially only over the limited and small "dynamic range" of 20–25 dB. To overcome this difficulty, models of intensity perception have focused on ways in which information from a number of individual nerve fibers might be "pooled". The psychoacoustic and physiological findings are fairly well reconciled, using mathematical models that combine neural responses across small sets of neural elements. Portions of the set of neural elements are characterized as operating over different, albeit limited, ranges of intensity.

In order to understand changes in the perception of loudness, much larger changes of intensity must be considered. For example, it is universally found that one must increase the intensity of a signal by a factor of ten (or 10 dB) in order for it to be judged as "twice as loud". That is, perceptual increases along a loudness dimension are not "one-to-one" with increases in the physical stimulus, but rather increase at a much slower rate. This empirical outcome, which indicates a high degree of "compression" of the stimulus, has recently become to be understood at a physiological level. Compression, much like that suggested by the psychoacoustic data has been found in the mechanical-to-neural transduction that occurs in the cochlea. It is of further interest that such peripherally-based effects of compression must also be considered in order to account for a variety of auditory phenomena (see the recent reviews in 4).

Frequency Discrimination and Masking

In a sense, listeners are even more sensitive to changes in frequency than they are to changes in intensity. While the just noticeable increase of 1 dB discussed above corresponds to an increment of 25% in acoustic power, the just noticeable difference for listeners to discriminate between two different frequencies is typically much less. Over the range of frequency extending from about 200 to 8,000 Hz, the just noticeable differences in frequency are found to be less that 0.5%. That is, normal listeners can discriminate between tones of 1,000 and 1,005 Hz, 2,000 and 2,010 Hz, and so on.

A complementary question is how well a listener can analyze or resolve among the set of frequencies composing complex sounds (including noise, speech, music, etc). In effect, psychoacousticians determine how well a listener can "pick out" the presence of the particular frequency or set of frequencies that define the "signal". A classic type of experiment asks how intense a signal must be in order to be detected when it is presented along with a potentially obscuring background sound, the "masker". The number and variety of psychoacoustic experiments investigating such questions is remarkably large and constitutes the bulk of the published work in the field. One time-honored finding is that only a small fraction of frequency components composing the background masker actually affects detection of a tonal signal. For example, the detectability of a 1,000-Hz tone is determined by only the 160 Hz-wide portions of the masker that immediately surround the signal. That is, one can add or subtract large amounts of acoustic energy outside that region and not affect the signal's detectability. This means that other attributes of the masker, including its loudness and pitch, do not, by themselves, determine the degree of masking. It is handy to know that the bandwidth of the frequencies that are "critical" for masking tonal signals (nowadays termed the auditory filter) is approximately 16% of the frequency of the signal to be detected.

The links to physiology in these realms are both strong and pervasive. First and foremost, a "place" mapping of frequency exists within the cochlea, in that it is most sensitive to higher frequencies in its more basal portions, and most sensitive to lower frequencies in its more apical portions. This place mapping is preserved and reiterated throughout the auditory nervous system, including the auditory cortex. In addition, the frequency-tuning characteristics, or bandwidths, of individual nerve fibers are commensurate with the widths of psychoacoustically-determined auditory filters as a function of frequency. A second encoding of frequency is provided in the timing of neural discharges. That is, the time between neural discharges becomes smaller and smaller as the frequency of the signal is increased.

Many auditory scientists have debated the relative salience of place vs. time codes, while modeling a number of important auditory phenomena including frequency discrimination, masking, and the perception of pitch. Over time, one or the other view has been found to prevail. In our judgment, neither the acceptance nor the rejection of one of the two candidate codes may be logically possible, because changes in "place" and changes in "neural timing" both occur concomitantly as frequency is changed.

Applications

Knowledge and techniques derived from psychoacoustic research are relevant and useful in a variety of practical endeavors. While many applications may be familiar to us all (e.g., the design of telephones for the efficient transmission and understanding of speech), others are much less apparent. Beginning with the familiar, many of the basic findings from psychoacoustic research have been used to define and depict "normal hearing." Such findings, in turn, have been used to create and refine tests of hearing that are used in many ways to aid in the diagnosis and treatment of various kinds of hearing impairment. Otolaryngologists and audiologists use such information in order to differentiate among various hearing deficits, and in the selection and fitting of patient-appropriate hearing aids. For more than a decade, psychoacousticians have played an integral part in the development, design, understanding and evaluation of complex issues surrounding the use of cochlear implants. Knowledge from psychoacoustics has also proved beneficial in all phases of the discovery and application of specialized, non-invasive screening tests of infant hearing. Similar types of tests and guidelines have been usefully applied in industrial and governmental settings, both as part of hearing conservation programs and to develop standards concerning acceptable occupational and community noise levels.

Knowledge obtained from psychoacoustics has always been useful in designing and evaluating high-fidelity sound systems. Today, because of advances in computing power and miniaturization, psychoacoustic data have become highly relevant, if not ubiquitous. For example, one can purchase inexpensive devices that digitally store incredibly large amounts of music. This is possible because psychoacousticians and electrical engineers found ways of using knowledge concerning auditory sensitivity in order to greatly reduce the amounts of information required for satisfying reproduction. Specifically, only the necessary information, on a moment-by-moment basis, is preserved. Success of this technique is evidenced by the fact that listeners are typically unaware that the original sound has been modified. The same types of algorithms have also proved useful in the transmission of music and speech. These include the schemes used to send audio information to cellular telephones, satellite and digital television and radio devices and over the internet.

Another important arena for application of psychoacoustics is the concert hall. Here the thrust has been first to understand what physical variables differentiate successful concert halls from less successful ones. Such knowledge has been used to establish psychoacoustic-related dimensions that guide the design and determine the perceived quality of the halls. Finally, computer-based systems now exist that enable the designer to simulate candidate designs and to "preview" them. Similar schemes and technology are being applied to other aspects of architectural acoustics including improvements in the design of highly reverberant spaces such as churches and classrooms.

References

1. Crocker MJ (1997) Encyclopedia of acoustics, vol 3. Wiley, New York
2. Moore BCJ (2003) An introduction to the psychology of hearing, 5th edn, Academic Press, San Diego, CA
3. Yost WA (2000) Fundamentals of hearing: an introduction, 4th edn. Academic Press, San Diego, CA
4. Bacon SP, Fay RR, Popper AN (eds) (2004) Compression: from cochlea to cochlear implants, Springer handbook of auditory research, vol 17. Springer Verlag, New York

Psychogenic

Definition

Psychogenic refers to phenomena arising from thought processes.

Psychological Acoustics

►Psychoacoustics

Psychometric Curve/Psychometric Function

Definition

Relationship between physical properties of a stimulus and a measured behavior. Typically these take the form of a sigmoid function in which low levels of stimulus produce a small psychophysical result and at some point increases produce a faster rise in response, which saturates at high stimulus levels.

Psychomotor Seizures (Temporal-lobe Seizures)

Definition

►Complex Partial Seizures (Temporal-lobe or Psychomotor Seizures)

Psychopathic Personality

►Personality Disorder

Psychopathy

Persons who often exhibit aggressive behavior and try to achieve their personal aims by illegal actions and criminal lifestyle.

►Forensic Neuropsychiatry

Psychopharmacology

►Behavioral Neuropharmacology

Psychophysics

Walter H. Ehrenstein
Leibniz Research Center for Working Environment and Human Factors, University of Dortmund, Dortmund, Germany

Synonyms

Phenophysics

Definition

Psychophysics, as first established by Gustav Theodor Fechner in 1860, concerns the science of the relations between body and mind, or, to put it more precisely, between *physical* and *phenomenal* worlds [1–6]. Objectively we have physical events as reflected by brain processes, subjectively these processes appear as processes of our mind. Fechner already distinguished between *inner psychophysics* or the relation of sensory experience to its corresponding neural activity, and *outer psychophysics*, which deals with the relation between percepts and variations of the ►stimulus itself (see Fig. 1).

For much of the century following Fechner's publication of *Psychophysik* in 1860, inner psychophysics remained just a theoretical option until objective methods for the study of brain functions became available. Thus, outer psychophysics has shaped not only the development of experimental psychology and phenomenology [1,6], but also of sensory physiology, affording its pioneers (Aubert, Exner, Helmholtz, Hering, von Kries, Mach, Purkinje, Weber) to gain basic insights into sensory mechanisms [2–4]. Today, sensory and brain processes can be investigated by objective methods such as electrophysiology (single-unit recordings, EEG), functional magnetic resonance imaging (fMRI), magnetoencephalography (MEG), and positron emission tomography (PET). The relative ease of use and non-invasiveness of most of these techniques has made a new interplay between classic psychophysics and modern neuroscience possible. The complementary research approach that concerns itself with subjective and objective correlates of sensory and neural processes has come to be called ►correlational research [2,7], see Fig. 1.

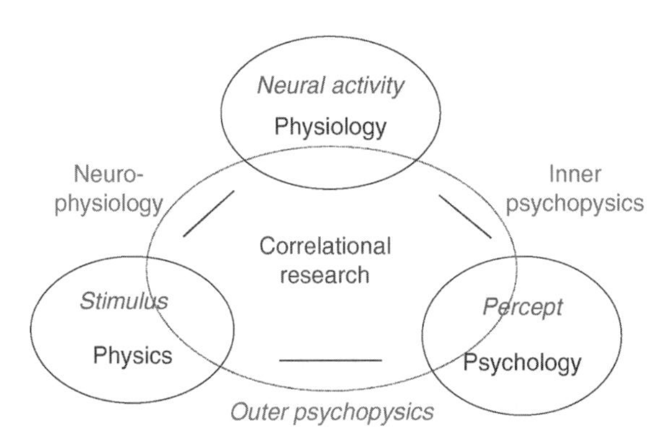

Psychophysics. Figure 1 *Interdisciplinary setting of psychophysics:* Percepts are related to corresponding physical stimulus properties (*outer psychophysics*) or to corresponding neural activity *(inner psychophysics)*. For a long time inner psychophysics remained a merely theoretical concept, whereas outer psychophysics provided the basis for methods to study sensory and brain processes. Now various objective methods afford a direct study of sensory and brain processes. This has made possible a new interplay between classic psychophysics and modern neuroscience that has come to be called ►correlational research [2,7].

Description of the Theory

Psychophysics accounts for the problem of measuring sensory experience by closely linking percepts to physical stimuli. No apparatus is necessary to obtain percepts; they are immediately present and available to each of us. The problem is thus not how to obtain sensory experience, but how to describe and investigate individual percepts so that they can be reliably communicated. An investigation might begin with a simple question such as "can you hear the tone?". This task requires *detection*. If we want to further determine which stimulus characteristics an observer can sense (e.g., the quality or spatial location of a sound), we arrive at the problem of *identification*. Detection and identification are solved quickly and almost simultaneously when the stimuli are strong and clear. However, under conditions of weak and noisy signals we often experience a stage at which we first detect only that something is there, but fail to identify what or where it is, exactly. In such a situation we try to filter out the consistent signal attributes, for instance, the sound of an approaching car, from inconsistent background noise. This task requires *discrimination* of the stimulus, or signal, from a noisy background, and the task is performed under uncertainty. As the car approaches and its sound becomes stronger, the probability of correct discrimination between signal and noise is enhanced. Even if we clearly detect and identify an object, we may still be faced with a further problem of ►perception, such as: "Is this car dangerously close?" or "Is the rattle under the hood louder than normal?" Questions such as these, concerning "How much x is there?", are part of another fundamental perceptual issue, that of *scaling*, i.e. to locate the magnitude of the stimulus on a psychophysical scale.

►Characteristics

Quantitative Laws of Psychophysics

Psychophysics emerged out of the contributions of the sensory physiologist E. H. Weber (1795–1878) who established perception as an experimental rather than observational discipline. Working with the discrimination of lifted weights, he demonstrated that the smallest difference between two weights *(ΔI)* which we can distinguish by way of feeling changes in muscle tension was a constant fraction *(k)* of the reference weight (I):

$$\Delta I/I = k \text{ (Weber's law)}$$

For example, if an increase of 1 g is needed in order to just experience that a weight is heavier than its reference of 40 g, an increase of 10 g would already be required for a weight to just appear heavier if its reference is 400 g. Accepting the validity of Weber's law, Fechner assumed that equal stimulus differences corresponded to equal perceptual differences or units. Sensory magnitudes could be assigned values according to the number of just noticeable differences. Inspired by Leibniz's concept of subliminal units of experience (*petites perceptions* or minute percepts as differential units of experience), Fechner postulated a relation between infinitesimal increase of stimulus intensity *(dI)* and corresponding subliminal increase in perceptual intensity *(dS)*, by deriving his fundamental formula: $dS = dI/I$. By integration we obtain a logarithmic relation between stimulus and sensation:

$$S = c \log I \text{ (Fechner's law)}$$

where S is the subjective and I the objective intensity and c refers to a constant depending on the respective sensory modality.

Fechner's law roughly predicts that a doubling of perceived intensity requires a 10 times increase of physical intensity. It holds pretty well for stimulus intensities in a middle range of the stimulus dimension. With very low or very high stimulus intensities, however, observed and predicted values deviate markedly. As an alternative to threshold measurements supra-threshold psychophysical scaling procedures were already developed by Tobias Mayer in 1754 [4] and more systematically later by J. A. F. Plateau (1872) and Delboeuf (1873, 1875) who studied the lightness of different gray levels. S. S. Stevens [10] popularized scaling procedures, especially the procedure of magnitude estimation, in which the experimenter assigns a value to a standard stimulus, e.g., the number 100, and the observers then rate the magnitude of other stimuli in proportion to the standard. So, if a stimulus appears half or twice as intense as the standard, it would be given the numbers 50 or 200, respectively. The results obtained with these and other direct scaling methods, such as ►cross-modality matching, are best described by a power function:

$$S = c\, I^n \text{ (Stevens's law)}$$

where S is scaled sensory intensity, c is a constant, I is physical intensity, and n is an exponent that varies for different sensory continua.

Psychophysical Methods

The most basic function of any sensory system is to detect changes of energy which can consist of chemical (taste, smell), *electromagnetic* (vision), *mechanical* (audition, proprioception, touch) or *thermal* events. In order to be noticed, the stimulus has to reach a minimal amount of stimulus intensity that, according to Fechner [3], "lifts its sensation over the threshold of consciousness." This is the *absolute threshold*, which is the intensity an observer can just barely detect, whereas the *difference threshold* is based on stimulus intensities above the absolute threshold and refers to the minimum intensity by which a variable or *comparison* stimulus must deviate from a constant or *standard* stimulus to produce a just noticeable difference in perception.

Method of Adjustment

The easiest way to determine thresholds is to let a subject adjust the stimulus intensity until it is just noticed or until it just becomes unnoticeable (absolute threshold); or until it appears to be just noticeably different from, or to just match some other standard stimulus (difference threshold). The observer is typically provided with a control of some sort that can be used to adjust the intensity, say of a sound, until it just becomes audible (or louder than a standard sound), and then the stimulus intensity is recorded to provide one estimate of the observer's threshold. Alternatively, the observer can adjust the sound from clearly audible to just barely inaudible (or to match the standard sound), providing another estimate of the threshold. Typically, the two series of measurement, one in which the signal strength is increased (ascending series) and one of decreasing signal strengths (descending series) are alternated several times and the results are averaged to obtain the threshold estimate. For example, if a tone is first heard at 5.0 or 5.5 dB on two ascending trials and first not heard at 4 or 4.5 dB on two descending trials, the resulting threshold estimate is 4.75 dB.

Method of Limits

In the ►method of limits the stimulus is varied by the experimenter (or a computer program) and the observer's response is recorded to each stimulus presentation. As in the previous method, the stimulus should initially be too weak to be detected and increased until it is perceived (ascending trials). Conversely in a descending trial, stimulus intensity, say a light, is reduced from clearly seen until it becomes invisible.

Method of Constant Stimuli

In the ►method of constant stimuli the experimenter selects a number of stimulus values (usually from five to nine) which, on the basis of previous exploration (e.g., using the ►Method of Adjustment), are likely to encompass the threshold value. This fixed set of stimuli is presented multiple times in a quasi-random order that ensures each will occur equally often. After each stimulus presentation, the observer reports whether or not the stimulus was detected (for the absolute threshold) or whether its intensity was stronger or weaker than that of a standard (for computing a difference threshold). Once each stimulus intensity has been presented multiple times (usually not less than 20), the proportion of "detected" and "not detected" (or, "stronger" and "weaker") responses is calculated for each stimulus level. The data are then plotted with stimulus intensity along the abscissa and percentage of perceived stimuli along the ordinate (see Fig. 2).

If there was a fixed threshold for detection, the psychophysical function should show an abrupt transition from "not perceived" (0%) to "perceived" (100%). Psychometric functions, however, seldom conform to this all-or-none rule. We usually obtain a sigmoid (S-shaped) curve that reflects that lower stimulus intensities are detected occasionally and higher values more often, with intensities in the intermediate region being detected on some trials but not on others. There are various reasons for obtaining an S-shaped rather than a sharp step function. A major source of variability is the continual fluctuations in sensitivity that occur in any biological sensory system (due to spontaneous activity or internal noise). Those inherent fluctuations mean that activity elicited by external stimulation is to be detected against a background level of activity. The threshold sensation that occurs with a certain *probability* can be defined statistically. By convention, the absolute threshold measured with the method of constant stimuli is defined as the intensity value that elicits "perceived" responses on 50% of the trials. Notice that in the example shown in Fig. 2, no stimulus level was detected on exactly 50% of the trials. However, level 4 was detected in 40% and level 5 in 74% of the trials. Consequently, the threshold value of 50% lies between these two points. If we assume that the percentage of trials in which the stimulus is detected increases linearly between these intensities (which is not unreasonable given that sigmoid functions are approximately linear in the middle range), we can determine the threshold intensity by linear interpolation as follows:

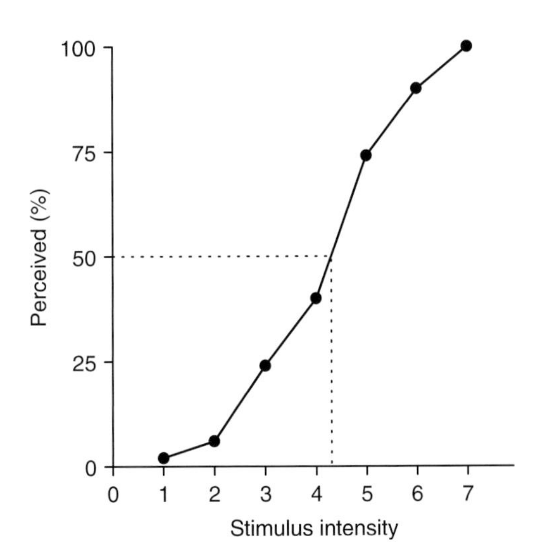

Psychophysics. Figure 2 Psychophysical function which shows the relationship between the percentage of times that a stimulus is perceived and the corresponding stimulus intensity. The threshold is defined as the intensity at which the stimulus is detected 50% of the time [2].

$$T = a + (b - a) \cdot \frac{50 - p_a}{p_b - p_a},$$

where *T* is the threshold, *a* and *b* are the intensity levels of the stimuli that bracket 50% detection (with *a* being the lower intensity stimulus), and p_a and p_b the respective percentages of detection. For the present case we obtain the following result:

$$T = 4 + (5 - 4) \cdot \frac{50 - 40}{74 - 40} = 4 + \frac{10}{34} = 4.29.$$

Although the method of constant stimuli is assumed to provide the most reliable threshold estimates, its major drawback is that it is rather time consuming and requires a patient, attentive observer because of the many trials required.

Adaptive Testing

Adaptive testing procedures are used to keep the test stimuli close to the threshold by adapting the sequence of stimulus presentations according to the observer's response. Since a smaller range of stimuli needs to be presented, adaptive methods are relatively efficient. Adaptive procedures were first introduced by Georg von Békésy in 1947, who applied his ►staircase method to audiometry [2]. The staircase method is a modification of the Method of Limits. As long as the observer says "yes" (I perceive) stimulus intensity is decreased by one step, until the stimulus becomes too weak to be detected. At this point we do not, as in the method of limits, end the series, but rather reverse its direction by increasing the stimulus intensity by one step. This continues with increasing the intensity if the observer's response is "no" and decreasing the intensity if it is "yes." In this way, the stimulus intensity flips back and forth around the threshold value. Usually six to nine such reversals in intensity are taken to estimate the threshold, which is defined as the average of all the stimulus intensities at which the observer's responses changed.

Research Fields and Applications

Psychophysics allows investigation of how living creatures perceive sensory stimuli in relation to their environmental setting as well as to compare the phenomenal worlds of humans with closely related species (monkey, cat) or even fairly distant species (birds, insects). Comparative relational psychophysical research across different species and across different stages of development provides the intriguing option of an integrated framework of behavioral and brain research [9]. For instance, the combined psychophysical and electrophysiological study of sensory performance has established the concept of the ►perceptive field as psychophysical counterpart of receptive field organization with the conjecture that perceptual organization is largely determined by single-cell activity. Meanwhile, various perceptual properties, including contrast of brightness, color, orientation, and motion as well as Gestalt phenomena, such as illusory contours or border-ownership, have been shown to be linked to the highly specialized functions of receptive fields at various levels of the visual system [7]. Furthermore, computer-assisted adaptive psychophysical methods are increasingly used in clinical routine diagnostics to assess impairment of sensory and neuropsychological function [1,2]. Psychophysics can also assist in elementary decisions of daily life. For example, scaling of auditory intensity can provide firm evidence for the effectiveness of noise protection measures. Industrial and ergonomic norms, such as of lighting on streets or at workplaces, or individual norms, e.g., accounting for age-dependent changes of sensory performance, are or are to be based on psychophysical evaluation. Consequently, psychophysics is of utmost importance for the approach of Human Factors [10], a joint profession of engineers and psychologists to design and evaluate simple and complex systems in order to optimize working and life conditions.

References

1. Ehrenstein WH (2003) Psychophysik. In: Neuser-von Oettingen k (ed) Psychologie von A–Z: Die wichtigsten 60 Disziplinen, Spektrum, Heidelberg, pp 156–159
2. Ehrenstein WH, Ehrenstein A (1999) Psychophysical methods. In: Windhorst U, Johansson H (eds) Modern techniques in neuroscience research. Springer, Berlin Heidelberg New York, pp 1211–1241
3. Fechner GT (1860/1966) Elemente der Psychophysik. Leipzig: Breitkopf & Härtel (reprinted in 1964 by Bonset, Amsterdam); English translation by HE Adler (1966): Elements of psychophysics. Holt, Rinehart & Winston, New York
4. Grüsser OJ (1993) The discovery of the psychophysical power law by Tobias Mayer in 1754 and the psychophysical hyperbolic law by Ewald Hering in 1874. Behav Brain Sci 16:142–144
5. Hensel H (1974) Thermoreceptors. Ann Rev Physiol 36:234–249
6. Horst S (2005) Phenomenology and psychophysics. Phenomenol Cogn Sci 4:1–21
7. Spillmann L (2006) From perceptive fields to Gestalt. Prog Brain Res 155:67–92
8. Stevens SS (1975) Psychophysics: introduction to its perceptual, neural, and social prospects. Wiley, New York
9. Sarris V (2006) Relational psychophysics in humans and animals. A comparative-developmental approach. Psychology Press, Hove & New York
10. Proctor RW, Van Zandt T (2008) Human factors in simple and complex systems, 2nd edn, CRC Press, Boca Raton, FL

Psychostimulants

►Stimulants

Psychotic Features

Definition

These include delusions (irrational thoughts and fears) and hallucinations (seeing or hearing things that are not really there). Often psychotically depressed people become paranoid or come to believe that their thoughts are not their own (thought insertion) or that others can “hear” their thoughts (thought broadcasting). People with psychotic depression are usually aware that these thoughts are not true.

►Major Depressive Disorder

Psychotomimetics

►Hallucinogens

Pterosaurs

Definition

Extinct flying reptiles (e.g. Pterodactylus, Pteranodon, Quetzalcoatlus, etc.), not ancestors of birds.

►Evolution, of the Brain: At the Reptile-Bird Transition

Ptosis

Definition

Drooping of the upper eyelid from paralysis of the third nerve.

►Endocrine Disorders of Development and Growth

Pudendal Nerve

Definition

A somatic nerve with cell bodies of the motoneurons located in Onuf's nucleus in the S2-S4 spinal segments; contains afferent and efferent pathways to the urethral and anal sphincters and afferent innervation to the perineum, urethra and the sex organs (penis and clitoris).

►Micturition
►Neurophysiology of Sexual Spinal Reflexes

Pulleys

Definition

When the eyes move from the primary position, the eye muscles do not slide freely within the orbital tissue.

Instead their paths are restricted, possibly by rings of connective tissue and smooth muscle that have been termed orbital pulleys.

►Eye Orbital Mechanics

Pulmonary Reflexes

►Respiratory Reflexes

Pulvinar

Definition

A component of the thalamus, relatively enlarged in carnivores and primates. It deals with salience (visual salience) and is involved in selective attention. It is reciprocally connected with much of the cerebral cortex. Its inferior and lateral parts also receive input from the superior colliculus. The homologous structure in lower animals is the Lateral Posterior thalamic group (LP-Pulvinar).

►Visual Attention
►Visual Role of the Pulvinar

Punisher

Definition

A punisher is any stimulus an animal will work to avoid, such as somatosensory pain, foot shock, timeout, or an unpleasant taste. In humans, monetary loss may also act as a punisher.

►Value-based Learning

Punishment

Definition

Punishment is a procedure whereby an aversive event (usually pain) following a response causes the reduction of that response. Punishment is considered by some behavioral psychologists to be a "primary process" – a completely independent phenomenon of learning, distinct from reinforcement.

►Aversive Learning

Pupillary Light Reflexes

Definition

Constriction of the pupils in response to increased light falling onto the retina. When light is shone on one retina, the pupil of the same eye constricts (direct response) as well as the pupil of the other eye (consensual response).

►Consensual Light Reflex
►Pretectal Nuclei
►Neural Regulation of the Pupil

Purinergic Receptors

Definition

Purinergic receptors are receptors that respond to adenosine triphosphate (ATP) and related agents.

Metabotropic (G-protein coupled receptors, P2Y) and ionotropic receptors (ion channels, P2X) have been identified. Excitatory P2X receptors are present in urinary bladder smooth muscle (P2X1) and in bladder afferent nerves (P2X2/3).

►G-protein Coupled Receptors (GPCRs): Key Players in the Detection of Sensory and Cell-Cell Communication Messages

Purinergic Receptors in Urinary Bladder

Definition

Excitatory P2X receptors are present in urinary bladder smooth muscle (P2X1) and in bladder afferent nerves (P2X2/3).

►Micturition
►Purinergic Receptors

Purkinje Cell, Neuron

Definition

Large, pear-shaped, GABAergic neuron located in the cerebellar cortex characterized by an intricate dendritic arbour and many dendritic spines.

The Purkinje cell sends the sole output from the cerebellar cortex. It receives excitatory synaptic inputs from parallel and climbing fibers and inhibitory synaptic inputs from basket, stellate, and other Purkinje neurons, and sends inhibitory outputs to the cerebellar or vestibular nucleus.

►Cerebellum
►Cerebellar Functions

Pursuit Eye Movement

Definition

An eye movement that matches the speed and direction of the eye to that of a moving target.

►Pursuit-Saccade Coordination

Pursuit-Saccade Coordination

ROBERT H. WURTZ
Laboratory of Sensorimotor Research, National Eye Institute, National Institutes of Health, Bethesda, MD, USA

Definition

Rapid or saccadic eye movements (►saccade, ►saccadic eye movement) allow us to shift our gaze from one part of a visual scene to another. This allows us to position objects in the scene of greatest interest on the region of the retina with the highest resolution, the central or foveal region. For example, we would make these saccadic eye movements several times per second as we looked out of a window onto a busy street scene. Smooth ►pursuit eye movements allow us to keep the fovea on objects that are moving so that we can have the sharpest vision of these objects in spite of their motion. As we look out the window, these pursuit eye movements would allow us to keep a person's face centered on the fovea even though they were walking along as we watched them. We can also make a saccade to the walking person and then immediately use the pursuit eye movements to keep that person's image on our fovea. As we continually use both of these eye movements in our everyday vision, these eye movements must be coordinated and the brain mechanisms underlying this coordination are beginning to be understood.

Characteristics

Upstream Event/Conditions

Control of saccadic eye movements depends upon a circuit within the brain that extends from the highest levels of visual processing in the cerebral cortex, to the neurons connecting to the eye muscles that lie in the midbrain and pons in the brainstem [1] The cortical areas devoted to this processing are centered in the lateral intraparietal area of parietal cortex and the frontal eye field area of frontal cortex. Both areas project to the brainstem, particularly to the structure on the roof of the brainstem, the ►superior colliculus (SC). Another pathway from the cortex passes through the basal ganglia to the superior colliculus. Pursuit eye movements also depend on activity in cerebral cortex for a critical feature of their function: the determination of the speed and direction of the object that is about to be followed by the eye. Areas providing this information are also distributed in cortex, but are concentrated in the visual region of cortex referred to as the middle temporal area and the medial superior temporal area, and in the frontal eye field in a region distinct from that related to saccadic eye movements. The processing related to saccades and to pursuit appears to be kept separate in the cerebral cortex.

Downstream Events/Condition

The generation of saccades requires connections to nuclei in the midbrain and pontine reticular formations, which in turn connect to the motor neurons that innervate the eye muscles [1]. The connections necessary for pursuit eye movements are conveyed to visuomotor nuclei in the pons, primarily the dorsolateral pontine nuclei. The pathway reaches the oculomotor neurons via projections through the cerebellum to the vestibular nucleus.

Involved Structures

The superior colliculus has a representation of the visual field spread out in each of its major layers [2]. In the superficial layers the neurons respond to visual stimulation, and in the intermediate layers the neurons are active before eye movements to that part of the visual field where the visual stimulation is located. The collicular neurons are organized into a map of the contralateral visual field, so that each neuron is active in relation to visual stimulation or eye movement in just one region of the visual field. It is in the intermediate layers that there is evidence for the coordination between saccades and pursuit. The intermediate layer neurons not only increase their discharge just before the saccade (or more generally an eye and head movement), but they show a buildup of activity that precedes that burst [3,4]. This buildup or prelude activity occurs whether or not the saccade related to the target actually occurs. If the saccade does occur, the buildup activity blends into the burst of activity preceding the saccade. If the saccade is not made, the activity decreases. Thus, the buildup activity can be largely independent of the actual generation of the movement, and it seems to be related to what must necessarily precede the movement including target selection.

One possibility is that the buildup activity simply indicates that there is an error between where the eye is looking now and where the target for the next eye movement is located – an error signal, not a saccadic movement signal. That is, each neuron on this map might indicate the error between where the eye is and where the target is, with large differences in the caudal colliculus and smaller ones in the rostral colliculus [5]. If the difference between where the eye is looking is very small (a fraction of a degree visual angle), the monkey frequently does not make any saccade. This activity has been referred to as fixation activity, but the most parsimonious interpretation of the activity is that it is the same buildup activity seen throughout the colliculus. The difference in eye and target position is just small, the error is small, and no movement need be made. With slightly larger errors, the buildup activity precedes a saccade to the target. If the target is moving, the monkey might make a not a saccade but a pursuit movement to the target, and this pursuit movement is also preceded by the increased buildup neuron activity [6]. Increased activity for both pursuit and saccades occurs only for movements in the visual field contralateral to the colliculus in which the neuron is found. Thus, the buildup neurons can be regarded as a shared error signal available to both the pursuit and the saccadic system, and this type of shared signal has been proposed as part of the mechanisms that coordinate these two movements. Further evidence supporting this view comes from experiments which showed that electrical stimulation or chemical inactivation altered pursuit eye movements [7]. Activity of the neurons can also predict the timing of pursuit as well as saccadic eye movements [8]. Thus, the same neurons that are involved in the preparation to make saccades might also contribute to the preparation to initiate pursuit. The test of this view would be to verify this pursuit – saccade interaction at points in the pathway after the signals for movement leave the superior colliculus, but this remains to be done.

Methods to Measure This Event/Condition

All of the observations described are derived from single neuron recording in awake behaving monkeys who select the targets and move their eyes to them [9]. The monkeys are trained on tasks that allow the comparison of neuronal activity to the same behavior on a series of similar movements, and this reveals the consistency of the relationship of the neuronal activity and the eye movement behavior. The eye movements are recorded using a precise method for recording eye position, velocity and acceleration. The monkey's behavior is controlled by online computer systems that also collect and store all the data from the experiments.

References

1. Krauzlis RJ (2004) Recasting the smooth pursuit eye movement system. J Neurophysiol 91:591–603
2. Sparks DL, Hartwich-Young R (1989) The neurobiology of saccadic eye movements. The deep layers of the superior colliculus. Rev Oculomot Res 3:213–256

3. Glimcher PW, Sparks DL (1992) Movement selection in advance of action in the superior colliculus. Nature 355:542–545
4. Munoz DP, Wurtz RH (1995) Saccade-related activity in monkey superior colliculus. I. Characteristics of burst and buildup cells. J Neurophysiol 73:2313–2333
5. Wurtz RH, Basso MA, Paré M, Sommer MA (1998) Contributions of the superior colliculus to the cognitive control of movement. In: Gazzaniga MS (ed) The cognitive neurosciences, MIT Press, Cambridge, pp 573–587
6. Krauzlis RJ, Basso MA, Wurtz RH (2000) Discharge properties of neurons in the rostral superior colliculus of the monkey during smooth-pursuit eye movements. J Neurophysiol 84:876–891
7. Basso MA, Krauzlis RJ, Wurtz RH (2000) Activation and inactivation of rostral superior colliculus neurons during smooth-pursuit eye movements in monkeys. J Neurophysiol 84:892–908
8. Krauzlis R, Dill N (2002) Neural correlates of target choice for pursuit and saccades in the primate superior colliculus. Neuron 35:355–363
9. Wurtz RH, Goldberg ME (1972) Activity of superior colliculus in behaving monkey. III. Cells discharging before eye movements J Neurophysiol 35:575–586

Putamen

Definition
The caudate nucleus and putamen together form the corpus striatum. Both are derived ontogenetically from the same anlagen, but are separated by incoming fibers from the internal capsule.

The corpus striatum is an important inhibitory component of motor movement programs and has manifold connections with the globus pallidus, substantia nigra and the motor cortex.

►Basal Ganglia
►Striatopallidum
►Telencephalon

Pyramidal

Definition
Refers to the triangular shape of an object; in neuroscience either to the shape of neocortical cell bodies in various laminae, or to the shape of the tract in the medulla oblongata of humans that is made up of axons derived from cortical upper motor neurons and destined for the spinal cord as the cortico-spinal path.

►Evolution of the Wulst

Pyramidal Decussation

Synonyms
►Decussatio pyramidum; ►Decussation of pyramids

Definition
In the pyramidal decussation, 70–90% of the pyramidal tract decussate to the contralateral side. The decussation lies directly below the pyramid (of the myelencephalon), in the middle of the myelencephalon. The decussation joins the lateral pyramidal tract.

►Myelencephalon

Pyramidal System Organization

►Motor Cortex: Output Properties and Organization

Pyramidal Tract

Synonyms
►Tractus pyramidalis

Definition
The largest descending motor tract is formed by the axons of the pyramidal cells of the motor cortex and is thus called the pyramidal tract. It courses nonstop from the cortex to the corresponding segments in the spinal cord, explaining why it is also called the corticospinal tract. At the upper margin of the myelencephalon, this tract rises to the surface as the pyramid (of myelencephalon).

70–90% of the fibers then cross in the pyramidal decussation to the contralateral side, and continue to run in the spinal cord as the lateral pyramidal tract. The remaining fibers descend in the medial pyramidal tract. The fibers project directly or indirectly to the alpha motoneurons, especially of the distal extremities (hand/forearm), hence the pyramidal tract plays a vital role in fine motor control.

►Corticospinal Neurons
►Motor Cortex: Output Properties and Organization

Pyramidotomy

Definition

The act of sectioning the pyramidal tract.

►Motor Cortex: Output Properties and Organization
►Pyramidal Tract

Pyrexia, Hyperpyrexia

►Endotoxic Fever

Pyridoxine Intoxication

Definition

Overdose intake of pyridoxine may lead to chronic ►sensory polyneuropathy, with symptoms of numbness, tingling and pain in the extremities, and sensory ataxia.

Pyridoxine (Vitamin B6) Deficiency

Definition

Not a common disorder anymore, except in the elderly populations, where it may be associated with impaired cognitive function, ►Alzheimer's disease, cardiovascular disease, and some types of cancer.

►Alzheimer's Disease

Pyriform Cortex

►Olfactory Cortex

Pythons

Definition

A group of giant snakes (Boidae) in Africa and South-Asia.

►Evolution, of the Brain: At the Reptile-Bird Transition

Q_{10}

Definition

Quantification of temperature dependence of the rate of a process across a limited (10°C) temperature range. Passive rate changes of biological processes typically have a Q10 between 2 and 3. Deviating values for biological processes indicate active intervention by regulatory processes.

Q of a Filter

Definition

A relative estimate of the width of the filter's passband. Q often equals the center frequency of the passband divided by an estimate of the width of the filter's passband.

►Acoustics

Quadrantanopsia/Quadrantanopia

Definition

Quadrantic visual field defect resulting from lesion of the ►optic radiation. For example, lesion of the optic radiation fibers looping into the ►temporal lobe causes visual loss in the upper quadrant of the contralateral half of the visual field of both eyes.

►Optic Radiation
►Visual Field

Quadrigeminal Plate

Synonyms

Lamina tecti; Tectalplate

Definition

Also called tectum or quadrigeminal plate. Composed of two pairs of hills: superior colliculus: the two upper hills belong to the visual system (control of eye movements). inferior colliculus: the two lower hills belong to the auditory system and are an integral part of information exchange from inner ear to auditory cortex.

►Mesencephalon

Quadriparesis

Definition

Mild form of ►quadriplegia.

►Quadriplegia

Quadriplegia

Definition

Bilateral paralysis involving all four limbs, trunk etc., with the site of lesion being at least as high as cervical level. One common cause is the ►Guillain-Barré syndrome.

►Guillain-Barré Syndrome

Qualia

Definition

Phenomenal, qualitative, experiential features of mental states are called "qualia" in philosophy of mind. These features show something of what it is like to undergo experiences, that experiences have their particular "feel."

►Cognitive Elements in Animal Behavior
►Emergence
►Property
►Reductionism (Anti-Reductionism, ►Reductive Explanation)
►The Knowledge Argument

Qualitative Sex Differences

Definition

Sex differences in the fundamental mechanisms underlying pain or analgesic responses (to be contrasted with quantitative sex differences). For example, that stress-induced analgesia can be reversed by opioid and/ or NMDA-receptor antagonists in males, but not females indicates the presence of a qualitative sex difference.

►Gender/sex Differences in Pain

Quantal

Definition

Quantal refers to the mechanism by which transmitters are secreted from nerve terminals in packets released by the exocytosis of the contents of a single synaptic vesicle (a quantum). Individual quanta are released spontaneously when the Ca^{2+} concentration briefly becomes high enough at a release site in the nerve terminal. When an action potential reaches the terminal, Ca^{2+} entry is sufficient to release a quantum from many release sites, leading to a larger multiquantal response. These events are studied postsynaptically as currents or potentials.

Quantal Release and Excitatory/ Inhibitory Miniature Potential

Definition

Quantal release refers to the mechanism by which transmitters are secreted from nerve terminals in packets released by the exocytosis of the contents of a single synaptic vesicle (a quantum). Individual quanta are released spontaneously when the Ca^{2+} concentration briefly becomes high enough at a release site in the nerve terminal. When an action potential reaches the terminal, Ca^{2+} entry is sufficient to release a quantum from many release sites, leading to a larger multi-quantal response.

These events are studied postsynaptically as currents or potentials. These potentials are spontaneous activities of membrane potentials either more positive (excitatory) or more negative (inhibitory) than the resting membrane potential. Miniature potentials are due to irreducible units of transmitter release, namely quantal release.

►Membrane Potential – Basics
►Quantal Transmission
►Synaptic Transmission: Model Systems

Quantal Transmission

Hiromu Yawo
Department of Developmental Biologiy and Neuroscience, Tohoku University Graduate School of Life Sciences, Sendai, Japan

Definition

Bernard Katz and his colleague found in their series of experiments in the 1950s that synaptic transmission at the frog neuromuscular junction consists of a unit, which they referred to as a quantum [1–3]. An endplate potential (EPP) was measured from the frog muscle fiber (postsynaptic cell) by a glass microelectrode, and evoked by the electrical stimulation applied onto the innervating motor nerve (presynaptic axon). The amplitude of EPP was dependent on the Ca^{2+} concentration in the extracellular Ringer solution and was attenuated by the reduction of Ca^{2+} and the increase of Mg^{2+}. The EPP attenuation was not continuous, but rather stepwise and fluctuated between steps (Fig. 1a). Statistically, the EPP amplitude is distributed with several peaks of even gaps (Fig. 1b).

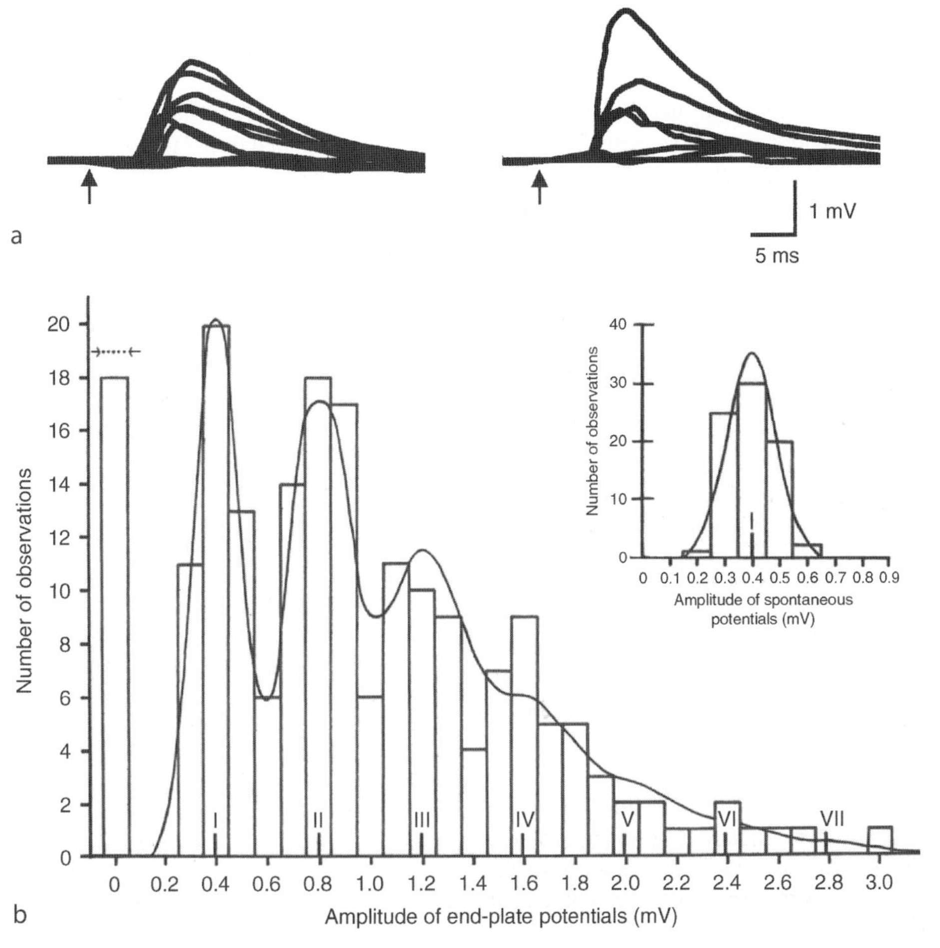

Quantal Transmission. Figure 1 Fluctuations in the end-plate potential (EPP). (a) Superimposed sample records of EPP. The amplitude of EPP fluctuates in a stepwise manner and sometimes resulted in null response (synaptic failure) to the nerve stimulation (applied at time point indicated by arrows). Modified after Fatt and Katz (1952). (b) Amplitude histograms of EPP (blue lines) and miniature EPP (mEPP, inset). The peak of the histogram occurs at 0 (failure), one (I), two (II), ≈ times of mEPP mean. The solid line represents the expected distribution according to Poisson statistics with mEPP mean and variance as units. On the experimentally obtained mean EPP amplitude (0.933 mV), the quantal content, m = 2.33 is predicted. Using the number of trials (198), the expected number of "failures" may be calculated from the Poisson equation as 19 (arrows), being almost identical to the experimental value. Modified after Martin (1966).

The gap distribution was almost equal to that of miniature EPPs (mEPPs), small changes of postsynaptic membrane potential spontaneously occurs under the blockade of presynaptic activity by e.g., tetrodotoxin. They put forward the quantal hypothesis that: (i) EPP consists of a unit (quantum), (ii) the number of quanta is stochastic around a mean, and (iii) mEPPs are spontaneous quanta. It is a generally accepted idea that each quantum is the synaptic transmission induced by the exocytotic release of a single presynaptic vesicle containing a certain amount of neurotransmitters (vesicular hypothesis) [2]. Similar stochastic transmission is found in neuro-neuronal synapses in both the peripheral and central nervous system [2,3]. The fundamental nature of this chemical synaptic transmission has provided a remarkable idea that, like the flip of coin, how a neural network performs is unreliable.

According to the quantal hypothesis, a synaptic event is mathematically described. This description enables one to estimate both presynaptic and postsynaptic parameters (►quantal analysis), thus, enabling one to presume the presynaptic and postsynaptic contributions quantitatively, which otherwise are very difficult to estimate. The quantal analysis is applied on the peripheral and invertebrate nervous systems and provides an important quantitative description of synapses. When the synaptic transmission was changed by some neural activities or some chemicals as neuromodulators and drugs, the primary cause could be attributed to either presynaptic, postsynaptic or both elements (pre/post problem). The pre/post problems have been revealed by quantal analysis in many synapses of peripheral and even central nervous systems. However, the quantal analysis has recently been challenged by "►silent synapse" phenomena, and its application should be careful for the CNS synapses.

Characteristics

Quantitative Description

According to the Katz model, neurotransmitter is prepackaged in discrete quantities of fixed size, called quanta, and each quantum is released probabilistically and independently of the others in response to the activation of presynaptic terminal. Hence, the size of the postsynaptic response, such as the EPP and the postsynaptic current, vary at random because it is assumed to be proportional to the number of quanta simultaneously released. The random process is quantitatively described by the application of probability theory. The probability that k quanta are simultaneously released, $P(k)$, is thus assumed to follow a binomial distribution:

$$P(k) = {}_nC_k p^k (1-p)^{n-k},$$

where p is the probability of a quantum to turn on release and n is the total number of quanta. In other words, p is the probability of a vesicle to be release and n is either the total number of docked vesicles or the number of release sites if one vesicle is predocked in one release site. When p/n is very small, as in the case of Katz' experiment with extracellular low Ca^{2+} and high Mg^{2+}, this distribution is approximated to the Poisson distribution. That is;

$$P(k) = (m^k/k!)e^{-k},$$

and

$$m = np.$$

The parameter m corresponds to the mean number of quanta simultaneously released (called "►quantal content" in the original works of Katz and collaborators). A Poisson process is one in which some event (like exocytosis) is unlikely to occur in any brief time window. The most familiar example of a Poisson process is the number of telephone calls made per day in your office. There are a large number of telephones connected to your office (n), and the probability of any one of them being active (p) is very low. On some days your office is quiet with no calls ("failure" response), on other days, one or perhaps two calls. Over a year, the number of days in which the number of telephone calls (k) is 0, 1, 2, 3 or more is distributed according to the above Poisson equation, using only the mean number of telephone calls a day (m) to determine the theoretically expected distribution without knowing n or p. The number of "failures," for example, should be given by:

$$n_0 = \mathrm{Ne}^{-\mathrm{m}},$$

where N is the total days of observation (e.g., 365). The number of single-call days by

$$n_1 = Ne^{-m} \cdot m$$

and so on.

The applicability of Poisson distribution to the ►quantal release of transmitter has been tested as illustrated in Fig. 1b. The quantal hypothesis predicts that the mean ►quantal size, q, a unit amplitude of quantum is equivalent to the mean amplitude of mEPP. The average size of postsynaptic response, E is thus expressed as,

$$E = qm.$$

Since mEPPs appear to be distributed normally around a mean of q with variance σ^2, the predicted n_1 also follows this distribution. Similarly the predicted k unit responses (n_k) are distributed normally around a mean amplitude of k q with variance k σ^2. The convoluted predicted distribution produces a smooth curve, as shown in Fig. 1b, which may be compared with the experimental histogram.

Since the value m consists of pure presynaptic factors, n and p, it has been widely used as a parameter sizing presynaptic releasing ability. The value m can be independently calculated either by the following equations.

If one can measure q on the ►miniature synaptic responses:

$$m = E/q.$$

If one can count the number of "failures," n_0:

$$m = ln(N/n_0).$$

Since the variance of Poisson distribution is equal to m, m is related to the coefficient of variation (CV = standard deviation/mean) as;

$$m = (CV)^{-2}.$$

If these values are all identical, the experimental distribution of synaptic responses almost follows the Poisson process. If not, there should be some deviations.

Higher Level Structures

The structure of synapse has been described in the accompanying essay.

Lower Level Components

The synaptic vesicles and the molecular mechanisms of transmitter release have been described in the accompanying essay.

Higher Level Processes

Hebb proposed in 1949 that, at the level of the synapse, learning follows a fundamental rule: "*When an axon of cell A is near enough to excite a cell B and repeatedly and persistently takes part in firing it, some growth or metabolic change takes place in one or both cells such that A's efficiency, as one of the cells firing B, is increased.*" This theory led to the discovery of a cellular process, long-term potentiation (LTP), as one of the neuronal mechanisms underlying learning and memory. Since LTP requires coincidence of presynaptic activity with depolarization of the postsynaptic cell, it is indicated that a process very akin to that proposed by Hebb is involved. LTP is classically induced by the Ca^{2+} influx through postsynaptic NMDA (*N*-methyl-D-aspartate) receptor, which works as a coincidence detector of synaptic transmission and postsynaptic depolarization.

Although the postsynaptic elevation of Ca^{2+} triggers LTP induction, there have been debates on the cellular mechanisms of potentiation; either (i) enhancement of the AMPA (α-amino-3-hydroxy-5-methyl-4-isoxazole propionic acid) receptor responses at the postsynaptic membrane (postsynaptic mechanism), or (ii) increase in the number of quanta of glutamate detected by the postsynaptic neuron in response to a presynaptic action potential (presynaptic mechanism). The latter mechanism is supported by the findings that LTP is associated with a decrease in the *CV* and in the failure rate, evidence for a presynaptic expression mechanism. Some retrograde signals may induce changes in presynaptic release mechanisms on the postsynaptic elevation of intracellular Ca^{2+}. However, LTP was accompanied by a relatively selective increase in the signal mediated by AMPA receptors, with little change in the NMDA receptor-mediated components. This finding is most easily explained by an enhancement in the AMPA receptor responses at the postsynaptic membrane.

To explain this apparent paradox, the idea of "silent synapse" is proposed [4]. The synapses are non-uniform in both morphology and physiology. Some synapses have NMDA receptor responses with no AMPA receptor responses (silent synapses) whereas others have both responses. After LTP induction, silent synapses may be replaced by dual component (AMPA and NMDA receptor-mediated) signals through one or a combination of the following mechanisms. (i) AMPA receptors are preserved in the intracellular vesicles and incorporated in the postsynaptic membrane by the LTP induction. (ii) LTP induction makes postsynaptic AMPA receptors sensitive to glutamate by accumulating them in a site just opposite the presynaptic releasing apparatus. (iii) Functionally inactive AMPA receptors are made active by biochemical reactions activated by LTP induction. (iv) Glutamate released by incomplete exocytosis, through the fusion

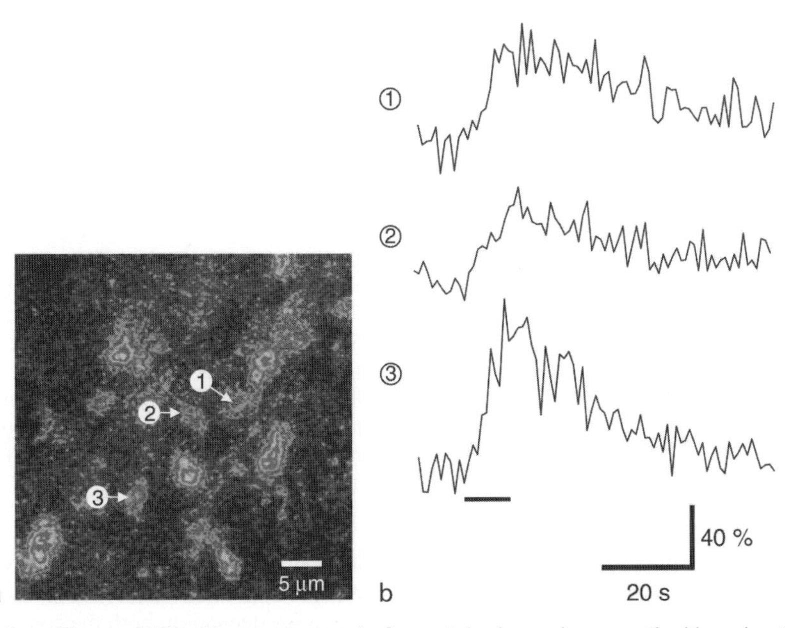

Quantal Transmission. Figure 2 Direct measurement of quantal release by an optical imaging technique. (a) Mossy fiber presynaptic terminals were fluorometrically identified under confocal microscopy in the hippocampal slice obtained from a mouse genetically encoding synapto-pHluorin [Araki et al. 2005]. (b) The fluorescent intensity of the presynaptic terminal increased by repetitive stimulation of mossy fibers (10 Hz for 10 s, horizontal bar), indicating the relative magnitude of quantal release at the individual presynaptic terminal (numbers correspond to those in a).

pores between synaptic vesicles and presynaptic plasma membrane, ("►kiss and run" mechanism) elevates the glutamate concentration to a low level that is enough to activate NMDA receptors without activating AMPA receptors. (v) The weak elevation of glutamate concentration is induced by the activation of neighboring synapses and activates NMDA receptors without activating AMPA receptors. (vi) The glutamate concentration at the postsynaptic AMPA receptors is increased by either the change of synaptic cleft geometry or the reduction of glutamate uptake by glia.

Therefore, the cellular mechanisms of LTP expression are not simple and involve both presynaptic (released amount of glutamate, mode of exocytosis) and postsynaptic mechanisms (number, distribution, functional modification). To test this, direct measurements of quantal release are promising by the use of presynaptic imaging techniques [5]: ►FM 1–43 and its analogue, and pH-sensitive form of green fluorescent protein (GFP) attached to the lumenal portion of a synaptic vesicle protein (e.g., synapto-pHluorin ►(synapto-pHluorin method)). The latter method has several advantages; (i) the optical signal changes instantaneously with the formation of a fusion pore between the vesicle membrane and the presynaptic plasma membrane, (ii) since recycled vesicles are rapidly re-acidified, both exocytosis and endocytosis of quanta are repeatedly measured at an individual synapse, and (iii) the reporter protein gene can be genetically introduced in experimental animal lines (Fig. 2).

These optical methods enable one to measure quantal release directly at individual synapses and to solve the pre/post problem of LTP.

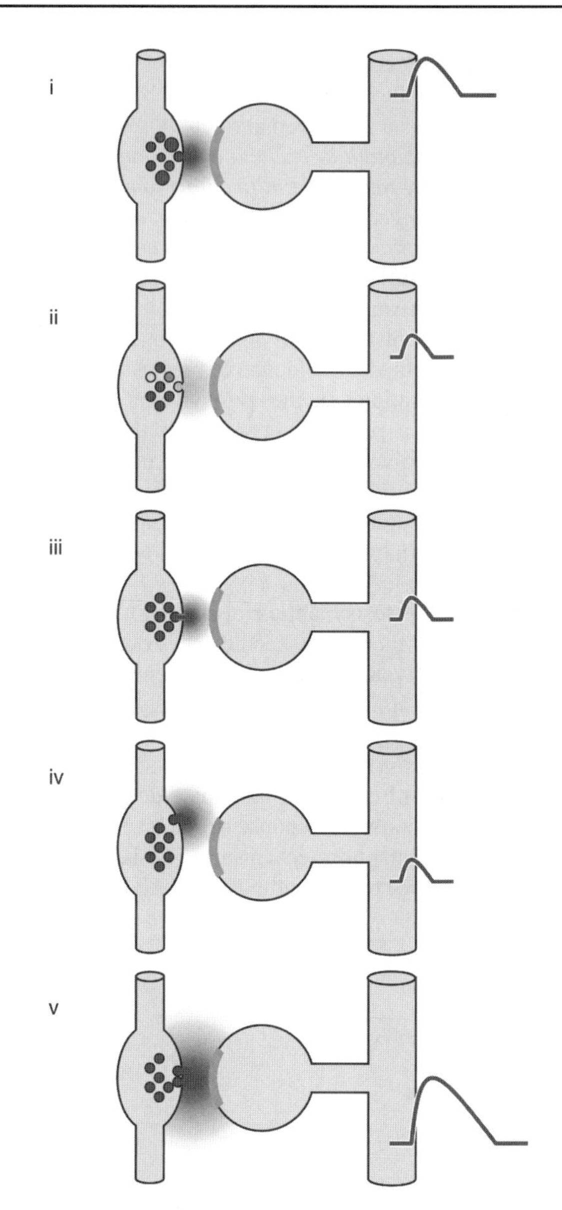

Quantal Transmission. Figure 3 Possible mechanisms underlying variable quantal size: (i) non-uniform vesicle volume, (ii) non-uniform vesicle filling, (iii) incomplete exocytosis of transmitter via a fusion pore with variable diameter, expansion rate, or lifetime, (iv) release site variation relative to postsynaptic receptors (green), and (v) synchronous release of multiple vesicles. Modified after Kullmann (1999).

Lower Level Processes

The quantal hypothesis is based on the prediction that there is a unit in the synaptic transmission. This is supported by the finding that mEPP is normally distributed around the mean. However, the miniatures are variable in amplitude and distributed in a highly skewed form in CNS synapses. Since focal application of glutamate at individual postsynaptic sites evoked currents less variable than the quantal EPSCs, the concentration of glutamate is assumed to be variable in magnitude or in kinetics in the synaptic cleft [6]. Several mechanisms are proposed to explain variable quantal size (Fig. 3) (i) non-uniform vesicle volume, (ii) non-uniform vesicle filling of transmitters, (iii) incomplete exocytosis of transmitters via a fusion pore with variable diameter, expansion rate, or lifetime, (iv) variable site of release relative to postsynaptic receptors, and (v) synchronous release of multiple vesicles or release of pre-fused vesicles.

Since NMDA receptors have a 100-fold higher affinity for glutamate, slower unbinding rate of glutamate and slower rate of desensitization than do AMPA receptors, the occupancy and opening probability of receptors depend on both the concentration and the time course of the glutamate transient [4]. If the glutamate transient is large and extremely brief, both NMDA and AMPA receptors are activated. On the other hand, if the glutamate wave is small and slow, NMDA receptors might be selectively activated. This would be one of the possible underlying mechanisms of "silent synapse" phenomenon [4].

Process Regulation

Quantal transmitter release from the presynaptic terminal is regulated by various neuromodulators, which have their specific receptors on the presynaptic terminal. The inhibition of presynaptic voltage-dependent Ca^{2+} channels plays a key role in presynaptic neuromodulations [7]. Up- and down-regulation of the exocytotic machinery are also involved in neuromodulations.

Use-dependent modifications such as ►paired-pulse facilitation, ►post-tetanic potentiation and LTP are also accompanied with changes in quantal release. While induction of LTP generally requires postsynaptic activation of NMDA receptors in the brain, LTP at the hippocampal mossy fiber (MF) pathway, the synapse between the dentate granule cells and the CA3 pyramidal neurons, is NMDA-independent [8]. The synaptic transmission can be measured from the postsynaptic CA3 pyramidal cell as an EPSC. Similar to EPPs at the neuromuscular junction, the EPSC amplitude was stochastically fluctuated from trial to trial and sometimes resultant in synaptic failure. When tetanic stimulation was applied on the mossy fiber pathway, the mean amplitude of EPSCs was increased with the reduction of synaptic failures for tens of minutes. The increase in the EPSC amplitude was accompanied by a linear increase in the CV^2 value. These results are consistent with a presynaptic mechanism for LTP expression in this type of synapse.

Function

The quantal release of neurotransmitter provides unreliable processes in the neural network of brain. Although at first glance such unreliability seems detrimental to brain functions, it is hypothesized to be utilized by the brain for the purposes of learning, in analogy to the way in which unreliable genetic replication is used for evolution [9].

Pathology

►Lambert-Eaton myasthenic syndrome (LEMS) is a disorder of neuromuscular transmission that is characterized by muscle weakness and autonomic dysfunction. The etiology of LEMS is unknown, but is often associated with small cell lung carcinoma, and is highly prevalent with autoantibodies, particularly to thyroid and stomach constituents. Neuromuscular transmission can be investigated by microelectrode methods in biopsied patients' skeletal muscles. The quantal analysis revealed that the quantal content (m) of the nerve-muscle transmission (the number of acetylcholine-containing quantal release per nerve impulse) was markedly reduced, so that *in vivo* the transmission-evoked depolarization would fail to exceed the threshold required for generating an action potential in the muscle membrane. Passive transfer of LEMS IgG to mice by daily intraperitoneal injection induced the reduction of quantal content of nerve-muscle transmission (Fig. 4).

Both quantal content and the human IgG level in the serum recovered in correlation after injections were terminated. These and other experiments support the hypothesis that the quantal content of nerve-muscle transmission is decreased by the reduction of Ca^{2+} entry during presynaptic activation, probably through the increased degradation of voltage-dependent Ca^{2+} channels in the presynaptic terminals [10].

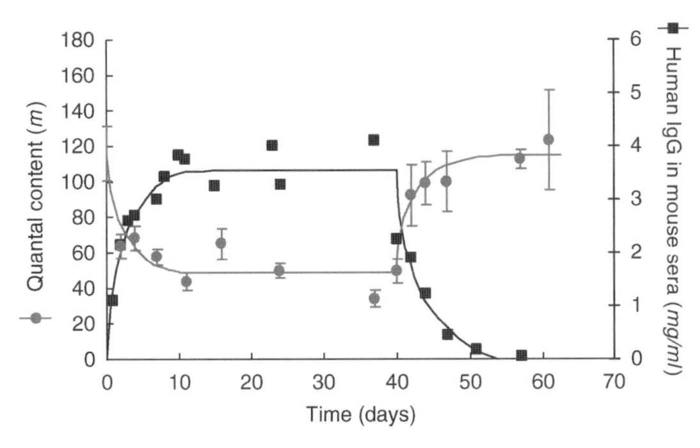

Quantal Transmission. Figure 4 Time course of the effect on quantal content of passively transferred Lambert-Eaton myasthenic syndrome (LEMS) IgG in mice. Daily injection of LEMS IgG increased the serum level of human IgG in the mice (black squares). IgG injection was stopped after 40 days administration. The quantal content was reduced with the increase of LEMS IgG and recovered with the reduction (red circles, mean and SEM). Modified after Prior et al. (1985).

References

1. Katz B (1966) Nerve, muscle and synapse. McGraw-Hill, New York
2. Kuno M (1995) The synapse: function, plasticity, and neurotrophism. Oxford University Press, New York
3. Stevens CF (1993) Quantal release of neurotransmitter and long-term potentiation. Cell 72 Neuron 10 (suppl):55–63
4. Kullmann DM (2003) Silent synapses: what are they telling us about long-term potentiation? Phil Trans R Soc Lond B 358:727–733
5. Ryan TA (2001) Presynaptic imaging techniques. Curr Opin Neurobiol 11:544–549
6. Liu G, Choi S, Tsien RW (1999) Variability of neurotransmitter concentration and nonsaturation of post-synaptic AMPA receptors at synapses in hippocampal cultures and slices. Neuron 22:395–409
7. Wu L-G, Saggau P (1997) Presynaptic inhibition of elicited neurotransmitter release. Trends Neurosci 20:204–212
8. Zalutsky RA, Nicoll RA (1990) Comparison of two forms of long-term potentiation in single hippocampal neurons. Science 248:1619–1624
9. Seung HS (2003) Learning in spiking neural networks by reinforcement of stochastic synaptic transmission. Neuron 40:1063–1073
10. Vincent A, Lang B, Newsom-Davis J (1989) Autoimmunity to the voltage-gated calcium channel underlies the Lambert-Eaton myasthenic syndrome, a paraneoplastic disorder. Trends Neurosci 12:496–502

Quantitative Sensory Testing (QST)

Definition

A sophisticated neurophysiological technique to test systems is quantitative sensory testing (QST). It uses a battery of standardized mechanical and thermal stimuli (graded v. Frey hairs, several pinprick stimuli, pressure algometers, quantitative thermotesting etc.). A standardized protocol for QST was recently proposed by the nationwide German Network on Neuropathic Pain including 13 parameters of sensory testing procedures for the analysis of the somatosensory phenotype. To judge plus or minus symptoms in patients an age- and gender-matched data-base for absolute and relative QST reference data in healthy human subjects was established. The QST data are then used to create zscore sensory profiles to judge altered somatosensation in patients.

►Neuropathic Pain
►Pain Psychophysics

Quantitative Sex Differences in Pain

Definition

Sex differences in the magnitude of pain or analgesic responses (to be contrasted with qualitative sex differences). For example, higher prevalence of pain or lower pain threshold in females represent quantitative sex differences.

►Gender/Sex Differences in Pain

Quick Phase of Nystagmus

Definition

One of the two types of eye movement that comprise nystagmus resembling a saccade. A fast resetting eye movement that returns the globe toward center position.

►Nystagmus
►Saccade
►Saccadic Eye Movement

RA Afferents

Definition

Rapidly adapting (RA) mechanoreceptive afferents (also called fast adapting type I afferents) found in the skin. They are thought to be associated with Meissner corpuscles in glabrous skin and hair follicle and field receptors in hairy skin. Their receptive fields are generally small, and they have a low threshold to mechanical stimulation, particularly low frequency sinusoids (flutter, <60 Hz).

►Active Touch
►Cutaneous Mechanoreceptors
►Functional Behavior
►Processing of Tatcile Stimuli

Rabies

Definition

Rabies is an acute, usually fatal encephalomyelitis caused by Rhabdoviridae. Highly endemic in parts of Africa, Asia, and Central and South America, rabies is almost always transmitted by an infected animal bites.

Infected people first develop fever, headache and skin sensation abnormalities (paresthesias) followed by paralysis ("dumb" form), hydrophobia, delirium or psychosis ("furious" form), then coma and death.

Confirmatory diagnosis is made by PCR assay of skin or saliva, but a negative result does not exclude the diagnosis. Pre-exposure vaccination is recommended for people who work with wild animals, travelers who anticipate prolonged stays in rural areas with high levels of endemic rabies as well as for cave explorers (spelunkers).

►Encephalomyelitis
►Polymerase Chain Reaction (PCR)

Radial-Arm Maze

Definition

An elevated maze with a central platform and, typically, eight radially-arranged alleys. The goal of a rat or mouse is to retrieve food hidden at the end of each alley without repeating an alley choice.

►Spatial Learning/Memory

Radial Glia

Definition

The radial glia is morphologically defined as a type of cell that possesses an elongated fiber spanning the developing cerebral cortex from the ventricular surface to the pial surface with an ovoid cell body located within the ventricular zone. The radial glia retains a neurogenic capacity and also its processes serve as a scaffold for migrating neurons.

►Cortical Development
►Cortical Development and its Disorders
►Neural Development

Radial Histogenetic Division

Definition

A radially arranged region or territory of the brain, whose neurons primarily derive from a specific morphogenetic field (i.e. from a restricted ventricular sector of the neural plate/tube). The radial feature of brain histogenetic divisions is based on the predominant glial fiber-guided migration of immature neurons in their way from the ventricular (proliferative) zone to the mantle during development. Nevertheless, radial histogenetic divisions can contain immigrant cells coming from other fields by tangential migration.

►Evolution and Embryological Development of the Forebrain

Radial Migration

Definition

Projection neurons are produced locally in the telencephalic wall and migrate to the overlying cortical plate perpendicular to the pial surface.

Radiation Term

Definition

The volumetric source or sink of non-mechanical power in the balance of energy.

►Mechanics

Radiculopathy

Definition

Radiculopathy refers to disease of the spinal nerve roots (from the Latin radix for root). Damage to the spinal nerve roots can lead e.g. to pain, numbness, weakness, and paresthesia (abnormal sensations in the absence of stimuli) in the limbs or trunk. Pain may be felt in a region corresponding to a dermatome, an area of skin innervated by the sensory fibers of a given spinal nerve.

►Neuropathic Pain

Radioisotope

Definition

A radioactive isotope of an element.

Radioligand

Definition

A radiolabeled molecular probe for the visualization of a particular receptor sub-type; see Positron Emission Tomography (PET).

►Positron Emission Tomography

Radiopharmaceutical (Radiotracer)

Definition

A specific pharmaceutical, labeled with radioactive isotope.

Radiotracer Imaging

Definition

Radiotracer imaging techniques involve intravenously injecting various short-lived radiolabelled molecules and then using positron emission tomography (PET) or single photon emission computed tomography (SPECT) to measure one or more biological functions of dopaminergic neurons in a resting state.

►Dopamine

Raf

Definition

A protein kinase and member of the MAPKK Kinase family. As a result of neurotrophic factor binding, MAPKKK is activated and phosphorylates MAPKK on its serine and threonine residues. The MAPKK then activates a MAPK through phosphorylation on its serine and tyrosine residues.

►Mitogen Activated Protein Kinase (MAPK)
►Neurotrophic Factors in Nerve Regeneration

RAGs (Regeneration-Associated Genes)

Definition

A series of changes in gene expression that occur in cell bodies (perikarya) of neurons with axon damage.

►Axon Degeneration and Regeneration of Peripheral Neurons

Random Process

Definition

The term "random process" denotes a series of uncorrelated events that are distributed either exponentially or in a Gaussian fashion.

►Circadian Rhythm

Raphé Interpositus

Definition

A collection of neurons lined up on either side of the midline ventral to the abducens nucleus. The neurons in raphé interpositus are the saccade-related omnipause neurons.

►Omnipause Neurons
►Saccade, Saccadic Eye Movement

Raphé Nuclei

Definition

The raphé nuclei are traditionally considered to be the medial portion of the reticular formation, and they appear as a ridge of cells in the center and most medial portion of the brain stem. The raphé nuclei have a vast impact upon the central nervous system. The raphé nuclei can be of particular interest to neurologists and psychologists since many of the neurons in the nuclei (but not the majority) are serotonergic, i.e. contain serotonin – a type of monoamine neurotransmitter.

Serotonin, also called 5-HT, seems to be the culprit in many of our modern psycho-pharmaceutical problems, such as anorexia, depression, and sleep disorders. It is not the sole culprit in the aforementioned disorders, but it is the area that the pharmacologists know how to affect in the best manner. It is important to note that pharmacology traditionally affects global serotonin levels, while the actions of the raphe nuclei are dependent on the complex interplay between nuclei.

►Serotonin

Raphé Nuclei and Circadian Rhythm

Definition

The midbrain dorsal and median raphé nuclei known for their widespread, extensively overlapping, ascending serotonergic projections. The projections of each nucleus, serotonergic or not, contribute to a great many different brain functions. In the context of the circadian rhythm system, the innervation by the dorsal and median raphé is somewhat unique because the raphé efferent projections of those two nuclei do not overlap in the two primary components of the system, the suprachiasmatic nucleus (SCN) and the intergeniculate leaflet (IGL). The SCN is very heavily innervated by neurons with cell bodies in the median raphé nucleus.

The majority of these contain the neurotransmitter, serotonin, but many median raphé neurons projecting to the SCN contain a different, currently unknown, neurotransmitter. Neurons of the median raphé do not project to the IGL. In contrast, both serotonergic and non-serotonergic neurons in the dorsal raphé nucleus project to IGL, but not to the SCN. In addition, the median and dorsal raphé nuclei reciprocally connect to one another via serotonergic and non-serotonergic connections. The direct serotonergic median raphé-SCN projection has been implicated as an inhibitor of retinohypothalamic tract transmission of photic input to the SCN, while the dorsal raphé serotonergic projection to the IGL has been implicated in the non-photic regulation of circadian rhythm phase.

►Circadian Rhythm
►Intergeniculate Leaflet
►Serotonin
►Suprachiasmatic Nucleus

Raphespinal Tract

Synonyms
Tractus raphespinalis

Definition
Projections of the magnocellular raphe nuclei (median zone of the reticular formation) to the gray matter of the spinal cord.

►Pathways

Rapid Eye Movement (REM) Sleep

Definition
REM sleep (also called paradoxical sleep (PS) and activated sleep) is a distinctive sleep stage in mammals. Normally this stage of sleep appears after a period of non-REM (NREM) sleep and then alternates with episodes of NREM sleep throughout the sleep period. REM sleep is characterized by a constellation of events including the following: (i) low-amplitude synchronization of fast oscillations in the cortical electroencephalogram (EEG) (also called activated EEG); (ii) very low or absent muscle tone (atonia) in the electromyogram (EMG). The atonia is observed to be particularly strong on antigravity muscles, whereas the diaphragm and extra-ocular muscles retain substantial tone; (iii) singlets and clusters of rapid eye movements (REMs) in the electrooculogram (EOG); (iv) theta rhythm in the hippocampal EEG; and (v) spiky field potentials in the pons (P-wave), lateral geniculate nucleus, and occipital cortex (called ponto-geniculo-occipital (PGO) spikes). Supplemental to these polysomnographic signs, other REM sleep-specific physiological signs are: myoclonic twitches, most apparent in the facial and distal limb musculature; pronounced fluctuations in cardio-respiratory rhythms and core body temperature; penile erection in males and clitoral engorgement in females (tumescence). In humans, awakening from REM sleep typically yields detailed reports of hallucinoid dreaming, even in subjects who rarely or never recall dreams spontaneously.

REM sleep is critical for memory processing and improvement of learning. REM sleep is not identifiable in the fish, amphibian, or reptile classes. In birds REM sleep is seen only for brief periods of time, especially following hatching. Generally, REM sleep is considered to be a highly evolved behavioral stage of terrestrial mammals.

►Atonia
►EEG in Sleep States
►Electroencephalography
►Electromyogram
►Electrooculogram (EOG)
►Non-REM Sleep
►Sleep States

Rapid Eye Movement (REM) Sleep Disorder

Definition
►REM Sleep Behavior Disorder

Rapidly Adapting Pulmonary Receptors

►Respiratory Reflexes

Rapidly Adapting Type I Mechanoreceptors

Definition
A mechanically sensitive sensory ending in the skin that adapts rapidly to a sustained indentation and therefore is sensitive to dynamic events such as vibration. It has small, well-defined receptive fields and the sensory terminal is believed to innervate the Meissner corpuscle.

Also known as FAI (fast-adapting type I) afferents in humans, RA (rapidly-adapting) receptors in the cat and QA (quickly-adapting) receptors in the primate.

►Cutaneous Mechanoreceptors
►Functional Behavior
►Processing of Tactile Stimuli
►Electric Fish

Rapidly Adapting Type II Mechanoreceptors

Definition

A mechanically sensitive sensory ending in the skin that adapts rapidly to a sustained indentation and therefore is sensitive to dynamic events such as vibration. It has large, poorly-defined receptive fields and the sensory terminal is believed to innervate the Pacinian corpuscle.

Also known as FAII (fast-adapting type II) afferents in humans and PC (Pacinian Corpuscle) receptors in the cat and primate.

►Cutaneous Mechanoreceptors
►Functional Behavior
►Pacinian Corpuscle
►Processing of Tactile Stimuli
►Vibration Sense
►Electric Fish

Rapsyn

Definition

Rapsyn (Receptor associated protein of the synapse) is important for initiating postsynaptic differentiation (pre-patterning) and is tightly associated with acetylcholine receptors suggesting that this complex becomes aggregated and stabilized at postsynaptic membranes.

►Synapse Formation: Neuromuscular Junction Versus
►Central Nervous System

Rarefaction

Definition

Areas of a propagating sound pressure wave of maximal decreased pressure (decrease below the static pressure).

►Acoustics

Ras GTPases

Definition

A family of molecules that include RhoA, Rac and CDC42, signals within growth cones.

►Axon Degeneration and Regeneration of Peripheral Neurons

Rate Coding in Motor Units

Definition

Control of force output from an individual motor unit by regulation of motoneuron firing frequency.

►Motor Units

Rate of Cross-Bridge Detachment

Definition

In the cross-bridge theory, cross-bridge attachment and detachment to the actin filament are quantitatively described by position-dependent rate functions. The detachment rate describes the first order kinetics of cross-bridge detachment from actin, while the attachment rate describes the first order kinetics of cross-bridge attachment to actin. In order for force production and contraction to always be in the same direction (i.e. a muscle always tends to shorten upon contraction and to produce tensile forces), these rate functions have to be asymmetric relative to the equilibrium point of the cross-bridge.

►Actin
►Force Depression/Enhancement in Skeletal Muscles

Rathke's Pouch

Definition

The pituitary anlage from which a craniopharyngioma may arise.

►Neuroendocrinology of Tumors

Rating Task

Definition

A psychophysical task in which a subject is asked to state the magnitude of a stimulus either in absolute terms or relative to a reference.

Ratiometric Dye

Definition

Some dyes respond to a metabolic change with both increase and decrease of fluorescence, depending on how they are measured. For example, the fluorescence of the calcium sensitive dye fura increases with increasing calcium when excited at 340 nm, and decreases when excited at 380 nm. FRET-dyes (FRET means Fluorescence Resonance Energy Transfer) shift their emission spectrum, with the result that fluorescence decrease in one band, and increases in another.

These dyes can be evaluated by creating the ratio (hence the name ratiometric dye) of the two signals, creating a number that is independent of the absolute fluorescence strength.

►Functional Imaging

Ray-finned Fishes

Definition

Also known as actinopterygian fishes. So named because of the flexible rays that provide the structural support of their fins. They make up approximately 95% of all living fishes and about half of all living vertebrate species.

►Evolution of the Spinal Cord

RC Circuit

Definition

Electrical circuit consisting of a resistor and a capacitor.

►Cable Theory

RCS Rat

Definition

Royal College of Surgeons rat model of Retinitis Pigmentosa has a mutation affecting retinal pigment epithelium. The mutation leads to an inability to phagocytose the photoreceptor outer segment. The same gene mutation is found in human patients with Retinitis Pigmentosa.

►Inherited Retinal Degenerations
►Retinitis Pigmentosa

rd/rd or rd1 Mouse

Definition

A mouse model of Retinitis Pigmentosa with a naturally occurring mutation of the beta-subunit of phosphodiesterase (an enzyme important in the visual transduction cascade). The same gene mutation is found in human patients with Retinitis Pigmentosa (see Inherited Retinal Degenerations).

►Inherited Retinal Degenerations
►Retinitis Pigmentosa

Reach to Grasp Postural Strategy

Definition

A change in support reaction to postural perturbation in which a rapid reaching movement of the arm permits a stable object to be touched or grasped for support, in order to restore equilibrium.

►Postural Strategies
►Reaching Movements

Reaching Behavior

Definition

Goal-directed behavior of humans and animals that requires visual information for movement of arms and hands in reach for objects.

Reaching Movements

Andrea d'Avella
Department of Neuromotor Physiology,
Santa Lucia Foundation, Rome, Italy

Definition

The act of reaching, bringing a part of the body in contact with an object, is a crucial component of many animal behaviors. Several vertebrate species use the distal portions of their forelimbs to explore and feed. Reaching movements are particularly important for primates, whose hands are capable of grasping and manipulating objects, and consequently these movements have been extensively investigated in humans and monkeys.

Characteristics

To reach for an object with the hand, the central nervous system (CNS) must map sensory input, which provides information about the object and hand locations in space, into motor output, comprising activations of shoulder and arm muscles that move the hand towards the target. Considering visually guided reaching, the location of a visual target is specified in retinal coordinates, proprioception gives information on the initial hand location in terms of arm muscle lengths, and muscle activations generate forces between arm segments. Thus, the CNS must transform sensory information into motor commands that are encoded in different ►frames of reference. It is usually assumed that the CNS performs these sensorimotor transformations in two stages. First, sensory information is used to define a kinematic plan. Target location and hand location are mapped into a common reference frame and a difference vector or motor error is computed. Second, the movement is executed by mapping the plan into muscle activations. This transformation may be performed using sensory signals for correcting the motor commands while they are generated (►Feedback control) or by pre-computing the appropriate commands (►Feedforward control). Since the delays involved in the conduction and processing of sensory signals may create instabilities in a feedback controller, the control of fast reaching movements requires feedforward control. Knowledge of the dynamical behavior of the musculoskeletal system necessary for pre-computing the appropriate motor commands is thought to be incorporated into the controller either explicitly as an ►internal model of the motor apparatus or implicitly as a collection of motor programs. The kinematic and dynamic characteristics as well as the muscle activation patterns observed during reaching movements have provided the experimental bases for the elaboration of these and other models of the computations involved in controlling reaching movements (►Motor control models).

Kinematics

The motion of the arm during reaching, arm kinematics, is fully specified by the rotational motion of all the joints in the arm. Considering the wrist as the arm end-point to be positioned in space, three rotations at the shoulder (flexion-extension, adduction-abduction, internal-external rotation), and one at the elbow (flexion-extension) are required to characterize reaching kinematics. Since there are four joint angles, or degrees-of-freedom, for three spatial coordinates of the wrist, the system is redundant, i.e. the same spatial location can be reached with the arm in many different configurations. For example, it is possible to raise the elbow without moving the wrist. Moreover, there are infinite paths along which the wrist can be moved to reach a target location from a given start location. Thus, to plan a reaching movement the CNS has to select one out of infinite possible kinematics.

Simple invariant features have been observed in the kinematics of reaching movements and they have provided an indication of the strategy used by the CNS for the planning stage. When performing a point-to-point reaching movement between two points on a horizontal plane, the wrist paths are straight and the wrist tangential velocity has a "bell-shaped" profile with a single peak [1]. For unrestrained movements in a vertical plane, the hand path is not always straight but it is independent of the speed of the movement [2] and of the hand-held load [3] (Fig. 1b–c).

Moreover, the tangential velocity profiles for movement at different speeds have the same shape when normalized for speed (Fig. 1d). The existence of invariant kinematic features has been interpreted as evidence for kinematic planning of reaching movements. Moreover, the straightness of the wrist path has been interpreted as evidence for planning end-point trajectories or displacements. However, since movements are executed by changing joint angles, end-point planning also requires mapping desired end-point positions into joint angles (inverse kinematics).

Dynamics

Arm movements are generated by the forces applied on the arm segments by the contraction of the muscles interconnecting them as well as by gravity. Since the arm is a chain of articulated segments, the motion of one joint depends not only on the forces directly applied to it but also on the motion of the other joints and the forces applied to them. For example, during a sagittal-plane reach to a target at shoulder level, from a starting posture with the forearm at waist level and the upper arm vertical along the trunk, the shoulder flexes and the

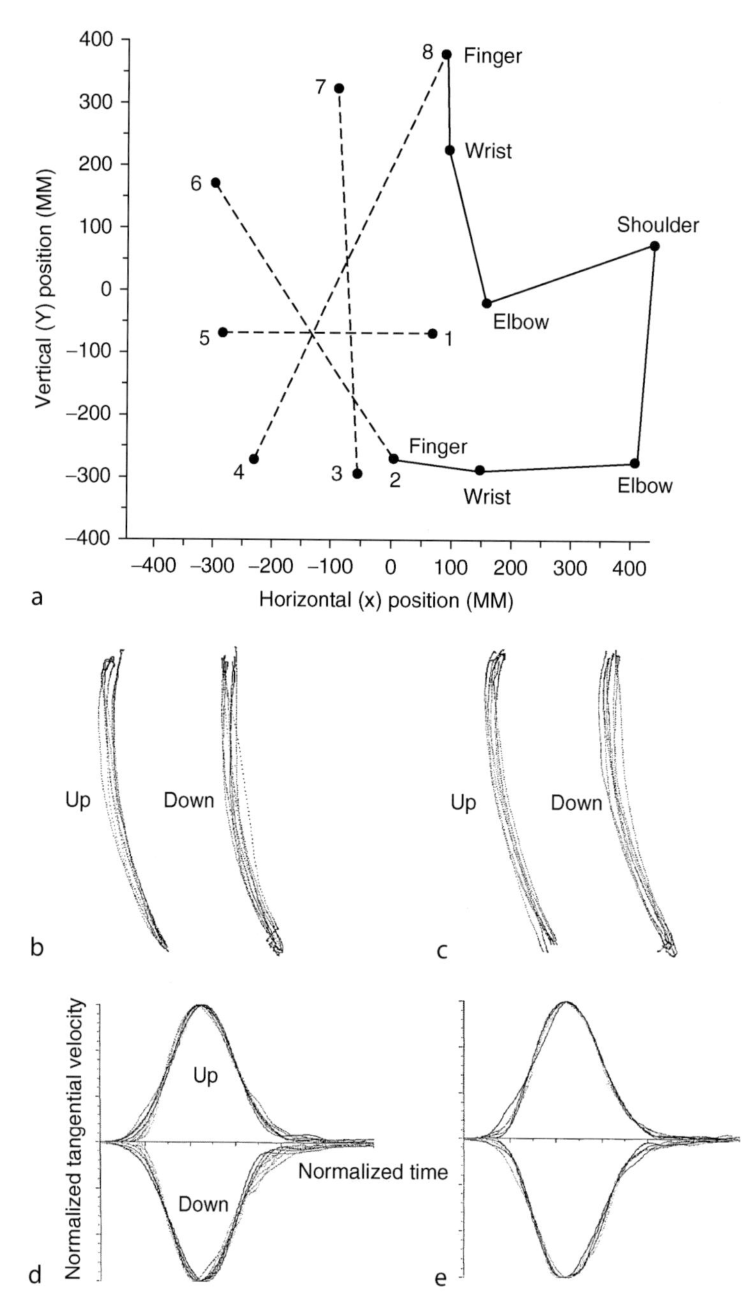

Reaching Movements. Figure 1 Invariant wrist path and tangential velocity for point-to-point movements across speeds and loads. (a) The position in space of markers placed on the arm of subjects performing unrestrained reaching movements between two points in the sagittal plane is recorded. (b–c) The path, on the sagittal plane, of the wrist for upward and downward movements (between points 3 and 7) does not change with the speed (b, where 6 slow, 6 medium, and 6 fast movement paths are overlapped) and the hand-held load (c, where 6 unloaded, 6 with 2 lb load, and 6 with 4 lb load movement paths are overlapped). (d–e) Similarly, the tangential velocity profile for upward and downward (with inverted ordinates) movements between the same two points, once normalized for speed, does not change with speed (d) and load (e). Adapted from [3] copyright © 1985 by the Society of Neuroscience, with permission.

elbow extends. However, because of the intersegmental dynamics, the muscles generate a flexor torque at *both* shoulder and elbow joints. Thus, the transformation between kinematics and dynamics (inverse dynamics) is not trivial and how the CNS implements this transformation is still an open question.

The characteristics of the torque profiles generated by the muscle contractions suggest that the CNS uses simple rules to find approximate yet adequate solutions to the inverse dynamic problem. The net torque generated at each joint by all muscles acting on it can be estimated from the arm kinematics using a simplified dynamic model of the arm based on the Newtonian equations of motion. For point-to-point movements in the sagittal plane, from one central location to several peripheral locations arranged on a circle, the dynamic muscle torque (expressed as the net muscle torque minus the torque required to counteract gravity) at the shoulder and at the elbow are related almost linearly to each other [4] (Fig. 2).

Both shoulder and elbow dynamic torque profiles have similar biphasic and synchronous shapes. Moreover, the relative amplitude of the two torque profiles changes with ▶movement direction, with the same biphasic torque profile scaled at each joint by a coefficient that varies as a linear function of the angular displacement at both joints. Simple torque scaling rules have also been proposed as a mechanism to generate movements with invariant paths and tangential velocity with different speeds and loads [3]. These rules derive from the observation that scaling in time the anti-gravity torque profiles and both in amplitude and in time the dynamic torque profiles generates invariant kinematics.

Muscle Patterns

The patterns of muscle activation observed during reaching movements have a complex dependence on the movement direction and speed. For reaching in vertical planes, the electromyographic (EMG) waveforms are constructed by combining components related to both dynamic and gravitational torques [5]. The waveform components responsible for the dynamic torques (phasic activations) have an intensity and a timing that changes with the movement direction in a complex manner [6]. Each muscle has a distinct spatial and temporal pattern, with a recruitment intensity maximal in multiple directions and a recruitment timing changing gradually across directions. Moreover, the phasic activations scale in time with movement speed differently for different muscles.

Despite their complex dependence on the movement parameters, the muscle patterns for reaching are generated according to relatively simple rules. The changes in the muscle patterns for fast reaching movements in different directions on vertical planes are well captured by the combinations of a few time-varying ▶muscle synergies [7] (Fig. 3).

A muscle synergy represents the coordinated activation of a group of muscle with specific activation profiles. Each synergy is modulated in intensity and

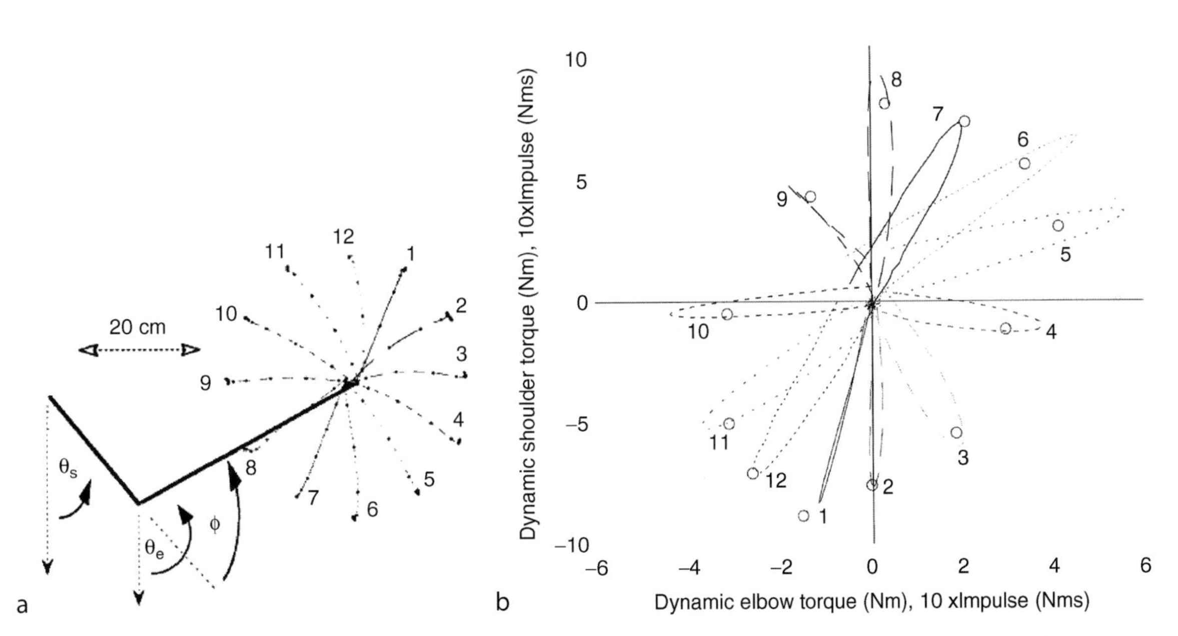

Reaching Movements. Figure 2 Scaling of dynamic muscle torques as a function of movement direction. (a) The elbow and shoulder muscle torques necessary for performing center-out reaching movements to 12 targets in the sagittal plane are estimated from the movement kinematics. (b) The average dynamic torque at the elbow and at the shoulder, obtained removing the torque required for resisting gravity from the total muscle torque at each joint, are plotted against each other, during the initial accelerating phase, for the 12 different directions (*solid and dashed lines*; open symbols represent the integrated torque, or impulse, at elbow and shoulder). The dynamic elbow and shoulder torque are approximately linearly related for all movements with a slope depending on the movement direction. Adapted from [4] copyright © 1997 by the American Physiological Society, with permission.

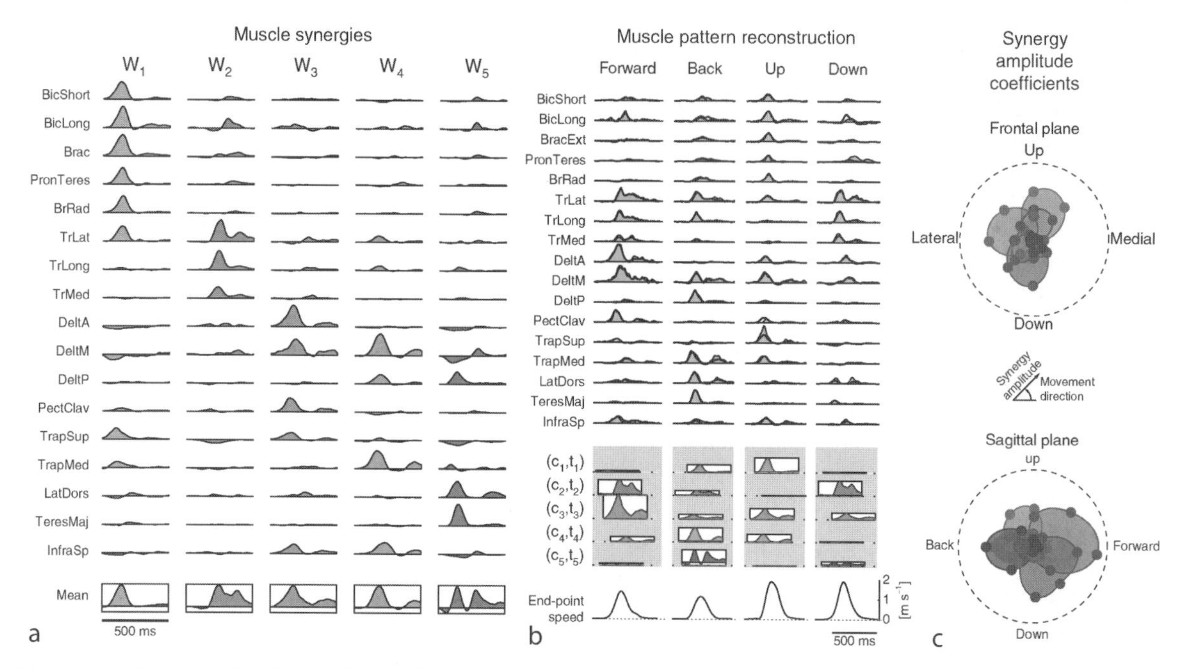

Reaching Movements. Figure 3 Muscle synergies for reaching. (a) A set of five time-varying synergies, identified from the muscle patterns recorded during point-to-point movements between one central location and eight peripheral locations in the frontal and sagittal planes. (b) The activation waveforms of 17 shoulder and arm muscles are reconstructed (*top*, where the gray area represents the averaged EMG activity and the solid black line the synergy reconstruction) by scaling in amplitude and shifting in time (*bottom*, where the amplitude scaling coefficient is represented by the height of a rectangle and the onset latency by its horizontal position) and combining, muscle by muscle, each one of the five synergies. Different movements are reconstructed with different synergy combination coefficients. (c) The amplitude scaling coefficients are directionally tuned (▶Directional tuning), with a tuning in most cases well captured by a cosine function. Adapted from [7] copyright © 2006 by the Society of Neuroscience, with permission.

delayed in time differently across movement directions and multiple synergies are combined to generate the observed muscle patterns. Such a combination mechanism may simplify the sensorimotor transformations for reaching by allowing a direct, low-dimensional mapping between kinematic plans and muscle patterns, and, thus, an implicit implementation of approximate inverse kinematics and inverse dynamic computations.

Neural Control

A distributed network of cortical areas in the parietal and frontal cortex and subcortical structures (spinal cord, cerebellum, basal ganglia) is involved in the neural control of reaching movements. This network functions in an integrated manner and it has not been possible to associate specific stages of the sensorimotor transformations to specific areas or neuronal populations. However, each area has a different degree of involvement into the different aspects of the control process. Spatial representation of limb position, target locations, and potential motor actions are highly expressed in the parietal cortex which is thought to be mainly involved in the early sensorimotor transformations. Selection and execution of motor actions are strongly expressed in the motor areas of the frontal cortex, from which most of the descending axons to the brain stem and the spinal cord originate, and which are believed to play a major role in transforming kinematic plans into descending commands closely related to the muscle patterns.

To understand the neural mechanisms underlying the sensorimotor transformations involved in reaching, the characteristics of the activity of individual neurons in many of the cortical areas involved have been investigated in monkeys. Recordings in the motor areas of the frontal cortex, composed by the primary motor cortex and six distinct premotor areas, have shown that the activity of most neurons is broadly tuned to the direction of movement [8]. The activity of each cell depends on the movement direction approximately as a cosine function, with a maximum in a ▶preferred direction that varies from cell to cell. Thus, each cell is active for a broad range of movement directions. Conversely, each movement direction is associated by a pattern of graded activation of the entire neural population. In fact, the direction of movement, either during movement preparation or movement execution, can be approximately estimated using a "►population vector," the sum of the preferred direction vector of each recorded cell weighted by its firing rate change from baseline. These observations have been interpreted as an indication that the motor cortex in mainly involved in high-level movement representation in terms of spatial location of the hand. However, the activity of most of the cells in the motor cortex is also modulated by the posture of the arm [9] and by the movement dynamics [10]. Thus, the representation of both kinematic and dynamic features are likely to coexist in the motor cortex, as expected in a neural network implementing a coordinate transformation from a kinematic motor plan to dynamic motor commands.

References

1. Morasso P (1981) Spatial control of arm movements. Exp Brain Res 42(2):223–227
2. Soechting JF, Lacquaniti F (1981) Invariant characteristics of a pointing movement in man. J Neurosci 1(7):710–720
3. Atkeson CG, Hollerbach JM (1985) Kinematic features of unrestrained vertical arm movements. J Neurosci 5(9):2318–2330
4. Gottlieb GL, Song Q, Almeida GL, Hong DA, Corcos D (1997) Directional control of planar human arm movement. J Neurophysiol 78(6):2985–2998
5. Flanders M, Herrmann U (1992) Two components of muscle activation: scaling with the speed of arm movement. J Neurophysiol 67:931–943
6. Flanders M, Pellegrini JJ, Geisler SD (1996) Basic features of phasic activation for reaching in vertical planes. Exp Brain Res 110(1):67–79
7. d'Avella A, Portone A, Fernandez L, Lacquaniti F (2006) Control of fast-reaching movements by muscle synergy combinations. J Neurosci 26(30):7791–7810
8. Georgopoulos AP, Kalaska JF, Caminiti R, Massey JT (1982) On the relations between the direction of two-dimensional arm movements and cell discharge in primate motor cortex. J Neurosci 2(11):1527–1537
9. Scott SH, Kalaska JF (1997) Reaching movements with similar hand paths but different arm orientations. I. Activity of individual cells in motor cortex. J Neurophysiol 77(2):826–852
10. Kalaska JF, Cohen DA, Hyde ML, Prud'homme M (1989) A comparison of movement direction-related versus load direction-related activity in primate motor cortex, using a two-dimensional reaching task. J Neurosci 9(6):2080–2102

Reaction

►Feedback Control of Movement

Reaction Time

Definition

The time from the presentation of a stimulus to the onset of the movement. Movement onset is usually defined either as the time a threshold in speed is exceeded or as the beginning of a burst of electromyographic activity, the latter criterion yielding smaller values.

►Eye-Hand Coordination

Reaction Time Task

Definition

A class of experimental paradigms in which a response (a movement) occurs reflexively in response to the appearance of a sensory stimulus. Movement onset is usually defined either as the time a threshold in speed is exceeded or as the beginning of a burst of electromyographic activity, the latter criterion yielding smaller values. The reaction time is shorter in contrast to voluntary tasks in which the response requires the selection of a response goal that is dependent on other cognitive factors.

Reactive Astrocyte

Definition

When the cebtral nervous system (CNS) is damaged, inflamed or infected the astrocytes undergo a characteristic set of changes known as reactive gliosis. The cells may proliferate.

Morphologically they hypertrophy and generally put out more and longer processes. There are characteristic changes in the cytoskeleton with upregulation of GFAP, vimentin and nestin. The cells may secrete a range of cytokines and may express class II major histocompatibility complex (MHC) receptors.

After injury the cells may be neuroprotective, play a part in controlling inflammation and in resealing the blood-brain barrier.

►Astrocytes
►Cytokines
►Cytoskeleton
►Major Histocompatibility Complex
►Glial Scar

Reactive Gliosis

►Glial Scar

Reactive Oxygen Species: Superoxide Anions

►Neuroinflammation: Modulating Pesticide-Induced Neurodegeneration

Readily Releasable Secretory Vesicles

►Neurotransmitter Release: Priming at Presynaptic Active Zones

Reafference

Definition

Sensory input resulting from an animal's own motor output.

►Reafferent Control in Electric Communication

Reafferent Control in Electric Communication

Bruce A. Carlson
Department of Biology, Washington University in St. Louis, St. Louis, MO, USA

Synonyms

Electrocommunication; Electrical communication

Definition

Every motor act that an animal produces will elicit sensory input from its own receptors [1]. Termed ►reafference, this self-generated sensory input can be quite useful. For example, bats listen to the echoes of their own ultrasonic calls to navigate through the night, and sensory feedback from skeletal muscles can be used to improve motor control. On the other hand, reafferent input is often not informative, and it can even interfere with the detection of external sensory input. A major problem faced by all animals is distinguishing reafferent sensory input from external sensory input. This issue is particularly relevant to the subject of animal communication. A communicating animal must produce its own signal as well as detect the signals produced by other individuals. A central question in the neurobiology of communication behavior is how sensory systems are able to discriminate self-generated from externally produced signals.

Consider the problem of reafference for visual perception. Any movement of the eyes, either directly or indirectly, due to movements of the head or body, causes the visual input to the retina to shift dramatically. How does the visual system compensate for this shift and maintain sensitivity to external visual stimuli? Early experiments suggested that every time a motor command that induces eye movement is issued, a copy of that command is also sent to the visual system, which generates a negative image of the visual input expected to result from that movement [1,2]. Combining this negative image with actual visual input eliminates any self-induced changes. As a result, the perceived visual world maintains its stability and only externally generated visual inputs are detected.

This basic mechanism relies on two distinct features. First, the timing of motor output must be relayed to the sensory system through what is referred to as a ►corollary discharge [2]. Second, the corollary discharge must activate a negative image of the reafferent input, a so-called ►efference copy [1]. Research on weakly electric fish has provided insight into the neuronal implementation of these two features [3,4].

Characteristics

Quantitative Description

African mormyrid fish possess an electromotor system that generates weak electric signals from a specialized ►electric organ, as well as an electrosensory system for detecting these signals (Fig. 1a). This unique sensorimotor system serves two functions. Through ►active electrolocation, mormyrids are able to detect distortions in their own electric field caused by nearby objects and thereby locate and identify various features of those objects, as well as navigate through their environment. By sensing the electric signals generated by other individuals, mormyrids are also able to communicate within the electric modality.

Electric signals in mormyrids consist of a fixed ▶electric organ discharge (EOD) separated by a variable ▶sequence of pulse intervals (SPI) (Fig. 1b). The EOD waveform conveys several aspects of the sender's identity, such as its species, sex, dominance, and possibly even its individual identity [5]. The total duration of the EOD is a particularly salient variable across species, ranging from as little as 100 μs to over 10 ms, and it may also exhibit sex- and status-related differences, with dominant males having a two- to three-fold longer EOD than females. By contrast, the SPI is involved in communicating contextual information about the sender's behavioral state and motivation. A variety of different patterns in the SPI have been linked with behaviors such as courtship and aggression [5].

In order for mormyrids to utilize the information available to them in these electric signals, however, they must first be able to distinguish their own EODs from those of other individuals. This distinction is made possible by a corollary discharge pathway that relays the timing of EOD production to central electrosensory regions (Figs. 1a and 2). By comparing incoming electrosensory information with an internal copy of their electromotor commands, they are able to distinguish their own electric signals from those of other nearby fish [4].

Higher Level Structures

Electromotor Pathway

Each EOD is initiated by a group of neurons in the ventral hindbrain that together constitute the electric organ ▶command nucleus (CN) [5]. The neurons in the CN project both directly and indirectly to an adjacent group of neurons that make up the medial relay nucleus (MRN). The neurons in the MRN receive the command from the CN and relay it down the spinal cord to electromotor neurons that drive the electric organ (Fig. 2). The activity in the CN, and therefore the SPI, is determined by a number of descending inputs, foremost of which is a precommand nuclear complex (PCN) consisting of two adjacent, but physiologically and anatomically distinct neuronal populations [5].

Electrosensory Pathway

The electrosensory system of mormyrids consists of three distinct pathways, one of which is relevant for electric communication (Fig. 2). The primary sensory afferents in this pathway receive input from so-called ▶knollenorgan electroreceptors, and project to a region of the dorsal hindbrain termed the nucleus of the electrosensory lateral line lobe (nELL) [6]. The neurons in the nELL relay this electrosensory input to a large midbrain structure termed the ▶torus semicircularis, a sensory processing region considered homologous to the inferior colliculus of mammals.

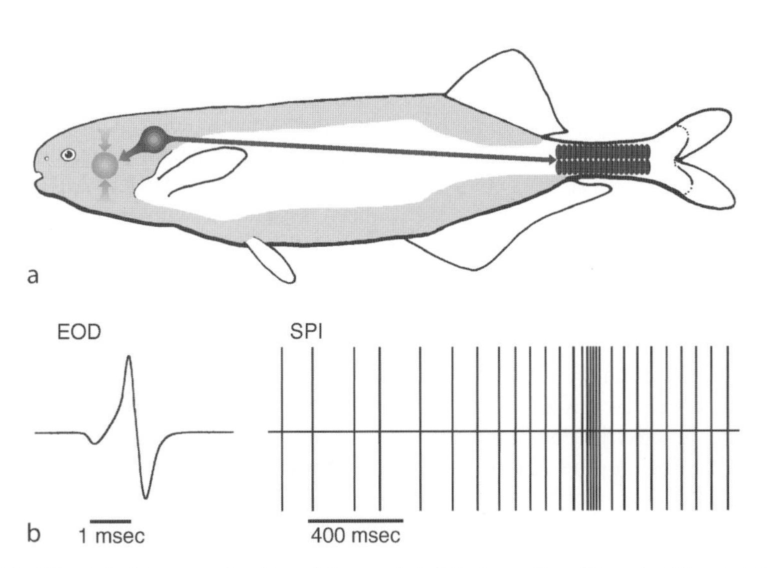

Reafferent Control in Electric Communication. Figure 1 (a) Schematic of the electric communication system in the mormyrid *Brienomyrus brachyistius*. The electric organ, shown in blue, is controlled by a command center in the hindbrain. Each descending command drives the production of a single electric organ discharge (EOD). External electric fields are detected by electroreceptors, whose distribution on the body surface is indicated by turquoise shading. Input from the electroreceptors converges onto an electrosensory region in the hindbrain, which also receives input from the electric organ command center. (b) Structure of electric signals in mormyrids. Head positive voltage is plotted upward. The electric organ discharge (EOD) has a fixed, characteristic waveform, while the pattern of EOD production, indicated by the sequence of pulse intervals (SPI), is variable.

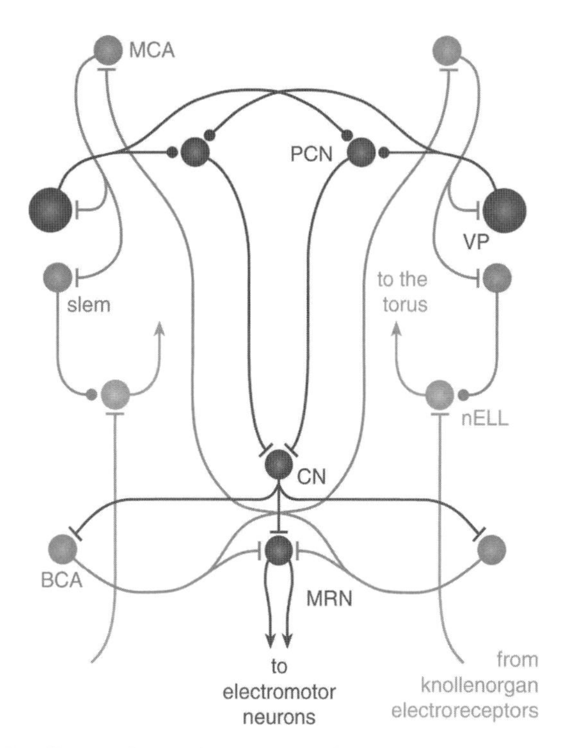

Reafferent Control in Electric Communication. Figure 2 Electric communication pathways in mormyrids. The electromotor pathway is shown in blue, the electrosensory pathway in green, and the corollary discharge pathway in red. Excitatory connections are indicated by flat lines, inhibitory connections by solid circles. Abbreviations: *BCA*, bulbar command-associated nucleus; *CN*, command nucleus; *MCA*, mesencephalic command-associated nucleus; *nELL*, nucleus of the electrosensory lateral line lobe; *PCN*, precommand nuclear complex; *MRN*, medial relay nucleus; *slem*, sublemniscal nucleus; *VP*, ventroposterior nucleus.

Electric Organ Corollary Discharge Pathway

The EOD command issued by the CN is relayed not just down the spinal cord to the electric organ, but also to higher brain centers that provide a precise timing reference of EOD production (Fig. 2) [3,5]. This electric organ corollary discharge (EOCD) pathway plays an important role in electric communication. For electrosensory processing in the knollenorgan pathway, it gives rise to an inhibitory input to the nELL that serves to block responses to reafferent electrosensory input (Fig. 2) [4]. In addition, the EOCD pathway helps regulate EOD production, as it projects to an electromotor region that provides inhibitory input to the PCN (Fig. 2). As a result, the region that drives the CN to fire is inhibited each time an EOD is generated. This negative feedback, referred to as recurrent inhibition, plays a critical role in controlling the SPI [5].

Lower Level Components

Electric Organ

The electric organ of mormyrids is located at the base of the tail and consists of a homogenous population of disc-shaped, modified muscle cells called ►electrocytes (Fig. 1a) [7]. When they are activated in synchrony by input from spinal electromotor neurons, their individual electrical potentials summate and give rise to the EOD, the amplitude of which is typically a few volts. Differences in the EOD waveform across species and between the sexes are directly related to variations in electrocyte morphology [7].

Electroreceptors

The knollenorgans involved in electric communication typically contain a few receptor cells that are housed together within a single large capsule [8]. Knollenorgan receptors are broadly tuned to the spectrum of the species-specific EOD and are extremely sensitive, with thresholds as low as 0.1 mV. In response to outside positive-going voltage steps, they fire a single spike at a short fixed latency. This phase-locked activity is relayed by primary sensory afferents to the nELL.

Specialized Features of Time-Coding Circuitry

The electromotor and electrosensory pathways of mormyrids are characterized by several unique anatomical specializations. Both pathways contain high levels of calcium-binding protein and consist of large, spherical, adendritic cell bodies that give rise to thick, heavily myelinated axons. Synapses in both pathways are typically mixed chemical-electrical, and often form large terminals that envelope a significant portion of the postsynaptic soma. Unlike most brainstem nuclei that occur in bilateral pairs, the CN and MRN form unpaired, midline nuclei. All of these features have been associated with neural circuits in which spike timing precision is of the utmost importance [9]. For the electromotor system, this precision is critical for activating the electrocytes in synchrony and thereby maintaining a constant EOD waveform. For the electrosensory system, it is involved in accurate temporal coding of the EOD waveform.

Higher Level Processes

Distinguishing Self-Generated EODs from External EODs

Knollenorgan receptors respond equally to any EOD that is above threshold, whether it is generated by the fish's own electric organ or that of another fish. In both cases, primary knollenorgan afferents generate a single spike that gives rise to an excitatory input to nELL [4]. However, the neurons in nELL also receive inhibitory input from the EOCD pathway [4], which causes the nELL neurons to respond quite differently to self-generated and external EODs (Fig. 3).

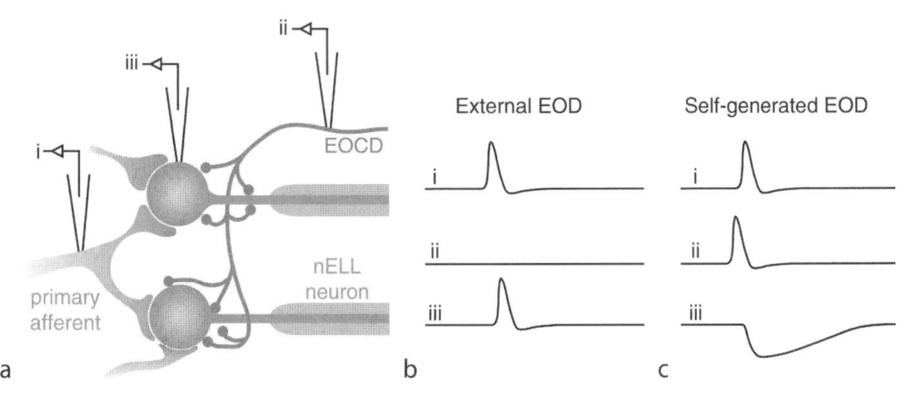

Reafferent Control in Electric Communication. Figure 3 Corollary discharge-mediated inhibition of reafferent electrosensory input in the nucleus of the electrosensory lateral line lobe (nELL). (a) Primary knollenorgan afferents form large, excitatory, mixed chemical-electrical synapses onto the soma of large, adendritic spherical nELL neurons. The electric organ corollary discharge (EOCD) pathway also provides inhibitory input onto the soma and initial segment of nELL neurons. (b) Patterns of activity recorded from the electrode locations shown in (a) in response to an external EOD. (c) Patterns of activity recorded from the electrode locations shown in (a) in response to a self-generated EOD.

When knollenorgan afferents respond to an external EOD, the EOCD pathway is not active. As a result, the nELL neurons only receive the excitatory afferent input, which they relay to the midbrain (Fig. 3b). By contrast, when the fish generates its own EOD, the EOCD pathway also becomes active, providing inhibitory input to nELL neurons. This inhibition blocks the response of nELL neurons to afferent electrosensory input (Fig. 3c), and the signal therefore does not get relayed to the midbrain [4]. As the reafferent input for this system is simply a brief excitation, the corollary discharge-driven efference copy is simply a brief inhibition.

Temporal Coding of the EOD Waveform

The EOD of a neighboring fish will cause current to flow into one half of the body surface and out the other, meaning that knollenorgans on these two surfaces will be exposed to opposite stimulus polarities. As knollenorgans only respond to positive-going voltage steps, those located where current is entering the skin respond to the rising edge of the stimulus, while those located where current is exiting the skin respond to the falling edge. Thus, by comparing spike times from opposite sides of the body, a mormyrid can, in principle, determine the duration of the EOD waveform [6].

A primary projection site of nELL axons is the anterior exterolateral nucleus (ELa) in the torus semicircularis (Fig. 4a). Within the ELa, there are two distinct types of neurons, large cells and small cells, both of which receive excitatory input from nELL axons. Upon entering the ELa, the nELL axons immediately terminate onto 1 or 2 large cells, and then wind their way throughout the nucleus over distances of 3 to 4 mm before branching and terminating onto a large number of small cells [6]. The large cells project exclusively within the ELa, terminating on small cells with large inhibitory synapses [6]. Thus, the small cells receive phase-locked input from two different sources: excitatory input from nELL axons and inhibitory input from ELa large cells (Fig. 4b). However, the excitatory input is significantly delayed by the time it takes an action potential to propagate down the long, winding path of the nELL axon, suggesting an "anti-coincidence detection" model for comparing spike times from knollenorgans on opposite sides of the body [6].

As an example, the small cell shown in Fig. 4b receives delayed excitatory input in response to stimulus onset and inhibitory input in response to stimulus offset. For short duration stimuli, this delayed excitatory input will arrive during the inhibition, and the small cell will not fire (Fig. 4c). As stimulus duration increases, however, there will be a greater delay before the inhibitory input reaches the small cell. If the duration is long enough such that the delayed excitatory input arrives before the inhibitory input, then the small cell will fire (Fig. 4c). Thus, a given small cell will only respond to EODs that are longer than some threshold duration. Assuming that different small cells receive input from nELL axons of varying delays, each small cell will have a different threshold value, and EOD duration will be reflected in the total number of active small cells [6].

Function

The Reafference Principle

Dealing with reafferent sensory input is a problem faced by all animals [1]. In the communication system of mormyrid electric fish, this problem is solved by a very simple, yet effective solution: incoming sensory input is blocked by inhibition every time the fish produces a

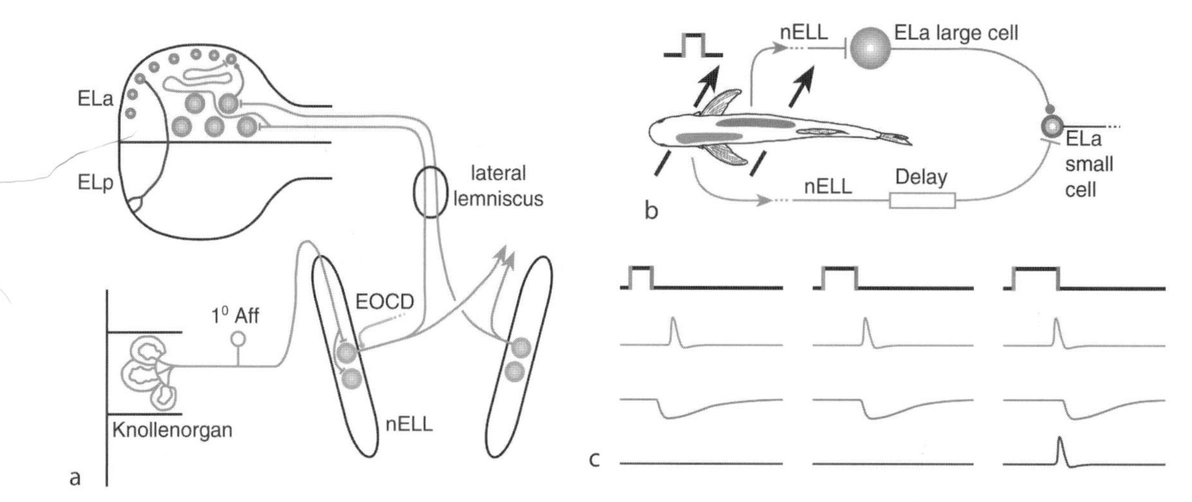

Reafferent Control in Electric Communication. Figure 4 Model of EOD waveform discrimination in mormyrids. (a) Neuroanatomy of the knollenorgan pathway. Excitatory connections are indicated by flat lines, inhibitory connections by solid circles. Primary afferents from knollenorgans project ipsilaterally onto the nucleus of the electrosensory lateral line lobe (nELL), which also receives inhibitory input from the electric organ corollary discharge pathway (EOCD). Axons from nELL ascend through the lateral lemniscus to project bilaterally to the anterior exterolateral nucleus (ELa) of the torus semicircularis, first onto large cells, then after winding throughout the nucleus, onto small cells. The large cells provide inhibitory input to the small cells. The small cells project ipsilaterally to the posterior exterolateral nucleus (ELp). (b) Schematic diagram showing the inputs to the small cell shown in (a) in response to a transverse square pulse. The ipsilateral side responds to the pulse onset, providing delayed excitatory input to the small cell, while the contralateral side responds to the pulse offset, providing inhibitory input to the small cell. c, Responses of the small cell shown in (b) to square pulses of varying duration. The green traces show the excitation provided by the nELL axon, while the red traces show the inhibition provided by the large cell. The blue traces show the resulting output of the small cell.

signal. Thus, the fish only senses the electric signals produced by other individuals. Recent studies have shown that this same strategy is used by singing crickets to block auditory responses to their own song [10]. Thus, corollary discharge-driven inhibition may be a widespread solution to dealing with the problem of reafference.

However, reafferent stimuli may often be much more complex, and the temporary blanking of responses afforded by simple inhibition may not be an effective solution. The earlier description of the effects of eye movement on visual processing is an illustrative example. Rather than brief excitation, the reafferent input in this case is a complex pattern of excitation and inhibition across many neurons over time, which is dependent on the specific eye movement undertaken. It is not sufficient to simply block incoming visual input during any movement, because this would result in complete blindness. In this case, rather than simple inhibition, the corollary discharge activates a spatiotemporally complex efference copy that cancels out the sensory input arriving from each portion of the visual field in response to the movement [1].

For active electrolocation in mormyrids, the fish's own EOD is the signal of interest, while those of other individuals constitute noise. Not surprisingly, then, the EOCD pathway provides excitatory, rather than inhibitory, input to the electrosensory pathway involved in active electrolocation and thereby facilitates reafferent sensory input [3]. However, much of this input is not informative, as it signals the presence of unchanging, or predictable, environmental features. In contrast to the hard-wired inhibition provided to the knollenorgan pathway, this corollary discharge-driven excitatory input can be altered through experience so that expected sensory input is nullified and only novel, informative input gets through [3]. This system provides an example of a modifiable efference copy, one that may be adjusted to compensate for changes in the sensory consequences of motor production.

Temporal Coding

Early research on electric communication in mormyrids focused on the SPI, because it was assumed that the EOD acted simply as a carrier signal for information encoded in a temporal pattern. The reasoning behind this was that EODs must be too brief to transmit any information. However, field recordings from mormyrids in the field revealed incredible species-specific diversity in the EOD waveform, as well as sex differences in many species [5]. Playback experiments in the field later demonstrated that these differences were behaviorally significant. In particular, EOD duration, or the relative timing of positive and negative voltage deflections were especially important [6]. These experiments therefore demonstrated that EOD recognition was mediated by a temporal code. In this chapter, we have seen a remarkable, yet simple, example of how the information contained within such a temporal code may be extracted through dedicated neuronal circuitry.

References

1. von Holst E, Mittelstaedt H (1950) Das reafferenzprinzip. Naturwissenschaften 37:464–476
2. Sperry R (1950) Neural basis of spontaneous optokinetic response produced by visual inversion. J Comp Physiol Psychol 43:482–489
3. Bell C, Bodznick D, Montgomery J, Bastian J (1997) The generation and subtraction of sensory expectations within cerebellum-like structures. Brain Behav Evol 50:17–31
4. Bell CC, Grant K (1989) Corollary discharge inhibition and preservation of temporal information in a sensory nucleus of mormyrid electric fish. J Neurosci 9:1029–1044
5. Carlson BA (2002) Electric signaling behavior and the mechanisms of electric organ discharge production in mormyrid fish. J Physiol (Paris) 96:403–417
6. Xu-Friedman MA, Hopkins CD (1999) Central mechanisms of temporal analysis in the knollenorgan pathway of mormyrid electric fish. J Exp Biol 202:1311–1318
7. Bass AH (1986) Electric organs revisited: Evolution of a vertebrate communication and orientation organ. In: Bullock TH, Heiligenberg W (eds) Electroreception. Wiley, New York, pp 13–70
8. Zakon HH (1986) The electroreceptive periphery. In: Bullock TH, Heiligenberg W (eds) Electroreception. Wiley, New York, pp 103–156
9. Carr CE, Friedman MA (1999) Evolution of time coding systems. Neural Comput 11:1–20
10. Poulet JFA, Hedwig B (2002) A corollary discharge maintains auditory sensitivity during sound production. Nature 418:872–876

Realism (Metaphysical, Internal, Common Sense, Naïve, Scientific)

Alexander Staudacher
Otto-von-Guericke-Universitaet Magdeburg, Institut für Philosophie, Magdeburg, Germany

Definition

Realism is a metaphysical position concerning the status of objects, facts and properties which can be of the most different kinds. One may be a realist concerning objects in space and time like trees, rocks, and molecules, concerning abstract objects like numbers or values, properties like being red or facts like the fact that the earth is round. What does realism with respect to one or more of these types of items amount to? Unfortunately there is no shared view among the experts in the field as to how realism is best defined. The question is especially disputed among adherents of the various brands of realism and their critics, the so-called anti-realists. According to the definition shared by most (but not all) philosophers considering themselves realists, realism with respect to a certain item implies the following two claims: First, *the existence claim (EC):* The items in question exist. Secondly, *the independence claim (IC):* The items in question are neither themselves something mental (mere ►ideas or representations) nor is their existence in any way dependent on whether we represent them (that is, perceive them or think of them) in a particular way or not. If you believe, for example, that the earth exists independently of whether there is a being with mental states able to represent it then you are a realist about the earth. Realism is often restricted to certain types of items: one may be a realist concerning physical objects in space and time without being a realist concerning moral values. According to the two defining claims one might dispute realism concerning a certain item in two ways: by denying either (EC) or (IC). For example, realism about moral values can be denied either by denying that there are any such values in the first place or by admitting their existence but taking it to be completely dependent on our ability to devise such values.

According to the alternative definition put forward by anti-realists realism is not so much a theory about the nature of objects, facts or properties but a doctrine concerning the question of how the truth of sentences is best understood. The relevant conception of truth implies that truth is verification-transcendent, that is, a sentence might be true although we don't have the slightest possibility to find out that it is true. Anti-realists use this definition to criticize realism, because they take the verification-transcendent conception of truth to be at odds with their preferred accounts concerning the question of what is implied when a speaker understands a proposition [1]. Realists have objected to this characterization of their position that they see no need to commit themselves to any substantial notion of truth whatsoever by endorsing (EC) and (IC) [2]. This essay will therefore follow the first definition.

Description of the Theory

Realism cannot only be held with respect to different items, it can also be formulated with varying strength. These variations in strength are mainly due to the fact that (IC) can be interpreted in various ways. According to the strongest reading, (IC1), the items in question exist independently whether *any* mind (not only human minds but also more powerful minds) has even the *ability* to represent them. It is then not only possible that there are items with nobody represented at a certain time but with could have been represented in principle, as was probably the case with the earth one billion years ago, but that there might even be items which lie completely outside of any representational power. A somewhat weaker reading, (IC2), would restrict this claim to human minds. Still, the world might contain many items we will not even have the possibility to form a conception of just as a chimpanzee is unable to form a conception of an electron [3]. In this sense our conception and a fortiori our knowledge of the world might always be limited and partial even if we lack the slightest evidence to suppose that they are limited and partial in that way. A considerably weaker reading, (IC3), would allow that there are many items we have never represented and we will never be able to discover but could at least form a conception of, so that we could at least speculate about their existence. A still weaker reading, (IC4), would allow that there are many items we have never represented but could have represented and would have been able to discover. The weakest reading, (IC5), only allows that the items in question can exist independently of whether someone actually represents them, but not independently of whether we can discover them or not.

The last three readings all make items in the world dependent in a certain way on our abiliy to represent them. Therefore, one might argue that they are too weak to convey the idea of independence which is inherent in realism. Realism is generally contrasted with ►**idealism** which holds that everything is in some sense dependent on our minds. True enough, there are forms of idealism which even contradict the weakest reading as it is the case with the idealism of Bishop Berkeley (1685–1753), who identified the existence of things with their actually being perceived. But there are many less radical forms of idealism (laying their emphasis on different kinds of *dependence* of the world of our mind or our representational capacities) which are compatible with these two readings. Note also, that the last two readings atleast don't allow for a verification-transcendent notion of truth, because they imply that truths about the world have to be discoverable by us. Therefore, they would also fail to count as reconstructions of realism according to the second, anti-realist definition of realism. This explains why the term "realism" is generally associated with the stronger readings, but as will become clear below, Hilary Putnam's ►**internal realism** forms a notable exception.

The first two readings allow for insurmountable ignorance about parts of the world and the first three allow for certain kinds of radical error concerning parts of the world we have a conception of. It is not only possible that we err simply in mistaking something green for something blue or something spherical for a flat disc, we might even err in ascribing whole classes of properties to things that don't possess. In this case, the concepts we make use of in our characterizations of the world (our "conceptual schemes") don't correspond to the internal structure of the world: we take the world to be coloured in the way it appears to our eyes, but it might be that in fact nothing is coloured in that sense. In fact, science tells us that the surfaces of tomatoes aren't red in the way they appear red to our eyes, but that this appearance is largely due to the structure of our perceptual apparatus [3]. If science is right about these matters that we can say that it gives a more adequate picture of the world as it is than our everyday view. Considerations like these lead to interesting consequences concerning the question of how to deal with competing conceptions of reality which are not compatible with each other: According to realism there is a fact of the matter, how the world is. Therefore, either one of them will get closer to the true story about the world or both will fail in this attempt. Consequently, realism is opposed to various forms of relativism according to which truth and knowledge have to be relativized to culture, historical epoch, conceptual schemes and the like. Competing claims concerning the shape of the earth might then be correct relative to their specific cultural, historical and conceptual context and there might be no fact of the matter beyond these contexts allowing us to ask whether a claim, a theory or a conceptual scheme is correct or not. In contrast with these claims realism allows us to hold that the replacement of one theory or conceptual scheme by another scheme may be interpreted as progress in our endeavour to gain a picture of the world as it is independently of any of our representations of it.

Furthermore, there might be possibilities of large-scale error which open the door for certain notorious sceptical scenarios: the stronger versions of (IC) seem to allow that we could even be wrong about reality as a whole. Accordingly, it might be the case that we are always dreaming or, to cite another famous example, we might be all brains in vat filled with a nutritient and supplied not with real information about the world but only with hallucinations induced by a super-computer connected to us by nefarious neuroscientists [2,3] And considering that our revisions of our former world views also have to be couched in our conceptions of the world we can ask again whether these tensions will tell us the true story about the world as it really is [3]. In this sense realism leads to the consequence that all our epistemic accomplishments are in different respects fallible. Therefore a sceptical position which puts into doubt whether we will ever be able to gain knowledge about the world could possibly be true. A strong enough realism seems even to be one of the central presuppositions needed in order to make these kinds of sceptical hypotheses intelligible in the first place. Most philosophers supporting such a strong kind of realism don't embrace scepticism, however. The fact that we have to admit the possible truth of scepticism should not be confounded with the fact that we have to take it seriously [2]. To the contrary, realists typically hold that they have the best explanation of how knowledge and scientific progress are possible inthe first place.

"▶Metaphysical realism" is often used as a name for the kind of realism based on stronger readings of (IC) like (IC1) and (IC2). The term was originally coined by Hilary Putnam who refuses this kind of realism, because he takes the idea that a conception or theory of the world might be wrong, although it fulfilling all our predictions and following all our methodological constraints (coherence, elegance, simplicity etc.) to be incoherent. Additionally he has argued that metaphysical realism has to give up a commitment not only metaphysical realists would like to subscribe to: the claim that our representations of items in the world are connected with these items in a way which gives them a definite reference (e.g., that the concept "cat" refers to cats and not to rats) [4].

Internal realism is Putnam's alternative to metaphysical realism and can be characterized roughly by following two claims: (IR1) A description of the world is true if it can be justified under epistemically ideal conditions. A description is justified if it is internally coherent and can be in principle verified, so that it is at least in principle possible for us to detect its truth. This implies that its truth does not consist of a kind of correspondence to facts in the world which are completely independent of our way of conceiving them and which might be completely inaccessible to us. (IR2) We have to acknowledge a certain kind of conceptual relativity according to which questions as to what kinds of objects there are or how many there are can't be answered independently of the choice of a certain conceptual framework. If someone asks for example "How many objects are in this room?" the right answer depends on certain decisions concerning our concept of "object." If we admit as objects only things which are not attached to other things my nose or a lampshade will not count as objects, if we do without this restriction, they will. In this sense there is no fact of the matter of how many objects are in the room which is independent of our concept of an object [4,5].

Conceptual relativity puts internal realism close to relativism. Putnam has emphasized, however, that internal realism is to be distinguished from relativism which he takes as holding a wrong conception of truth and considers even to be self-refuting. In his eyes, relativists typically their truth to mere rational acceptability. Therefore, according to relativism, the claim of the ancients that the earth was a flat disc was true at their time (although false today) because it was rationally acceptable in light of the available methods of investigation and evaluation at that time (but not in the light of the methods available today). However, the claim was not *ideally* rationally acceptable even at that time, because the conditions of verification were not ideal. A claim may lose its rational acceptability over time, but it can not lose its ideal rational acceptability. Relativism is self-refuting because in claiming its own absolute truth it exempts itself from the claim that all truths have to be relativized to certain historical conditions, conceptual schemes and so on [4].

Internal realism obviously only allows for weak readings of (IC) such as (IC4) and (IC5) because it takes the existence of the relevant items to be dependent on our conceptual resources and decisions and our ability in principle to verify what is the case. It can allow the existence of a certain rock in the desert even if it isn't represented by anybody at any time. But the existence of rocks remains relative to the fact that we have the concept of a rock. It can also admit that there might be facts (e.g., in the past) we are not able to verify. But it can't allow the possibility that reality might be a certain way if we can't verify this under ideal conditions. Therefore, one might ask whether internal realism should be seen as a form of realism at all. It is no wonder that many have seen internal realism as a form of anti-realism [2].

Critics of internal realism have questioned among other things (i) whether it can be successfully distinguished from relativism [6], (ii) whether the specific examples Putnam gives of conceptual relativity cannot be accommodated within metaphysical realism, so that they don't conflict with the claim that there are facts which are completely independent of our conceptual schemes [6], and (iii) whether ideal rational acceptability makes truth really accessible to us Putnam himself admitting that we can never tell whether we have reached ideal conditions and comparing this kind of idealization in question with unattainable idealizations such as frictionless surface. More recently Putnam himself has given up the claim that truth can be explained as idealized rational acceptability [7].

It is often assumed that realism with respect to spatio-temporal objects like rocks, chairs, etc., is a view dictated by common sense and held independently of any sophisticated knowledge about philosophical matters by "the plain man or woman on the street." Realism of this kind is therefore often called "▶common sense realism." Since common sense isn't a developed philosophical doctrine it is not easy to decide to which reading of (IC) common sense realism is committed to. Arguably, common sense is not sophisticated enough to make the necessary distinctions required for any decision on these matters. Note, however, that philosophers with wildly diverging views also use this label for their own account of realism [2,7].

►Naïve realism is often taken to be a position quite similar to common sense realism. In philosophical debates on perception Naïve realism is often taken to be a view according to which perception presents us the world by and large as it really is. For example, things not only appear to us as coloured (because of the specific nature of our perceptual apparatus) they really *are* coloured.

►Scientific realism is a theory concerning the correct understanding of theoretical terms in scientific theories. Scientific theories make intensive use of theoretical terms like “molecule” “atom,” “electron” and the like which don't refer to observable phenomena but play an indispensable role in the scientific explanation of such phenomena. We may say that with the help of these terms respective theoretical entities have been introduced into the scientific theory in question: molecules, atoms, electrons and so on. The behaviour of the observable phenomena is explained with the help of certain claims about the behaviour or state of these theoretical entities. The fact that water begins to boil at sea-level at 100°C is for example explained with the help of claims concerning the properties and the behaviour of H_2O-Molecules. Because theoretical entities like molecules or atoms are not among the things which can be observed, the question arises as to whether we ever have any good reason to believe in their existence and to accept the respective claims about their properties and their behaviour as true. Scientific realism gives an affirmative answer to these questions. According to one of its classical formulations [8] we have to interpret theoretical terms as putatively referring expressions and we often have enough reason to accept claims containing such terms as at least approximately true. Furthermore, we can see scientific progress as a steady approximation toward the truth of the observable and the unobservable. The reality described by scientific theories is largely independent of our thoughts and theoretical commitments. Therefore, we can say that we not only introduce theoretical terms in order to facilitate empirical predictions and the organization of our observation-knowledge, we also discover that there are molecules and electrons etc. In this sense scientific realism clearly endorses (EC) and a strong version of (IC), although the precise strength is often left open because the discussion concentrates more on whether theoretical entities are claimed to exist at all. One of the main arguments put forward in favour of this position is based on the claim that we can only plausibly explain why scientific theories have the predictive success they have if we suppose that the theoretical claims referring to theoretical entities are approximately true [8,2]. A classical objection to this claim is the historical observation that theories can be predictively successful although they are largely wrong [9]. A further general objection to scientific realism is that it cannot deal with the fact that two successful theories with commitments to different theoretical entities might lead to the same empirical predictions. It is argued that in such cases there is no evidential basis allowing a decision between these theories. If theoretical statements can be literally true, however, as Scientific Realism would have it, such a decision must be possible in principle. Against this, Scientific Realists have argued that we should allow for a conception of evidential support that is not restricted to positive outcomes concerning prediction [8]. Sometimes scientific realism is and to imply a further claim which puts it into strong opposition to naïve realism or common sense realism. If there is, e.g., an irreconcilable collision between the common sense view of physical objects as continuous solids and the scientific view that they are swarms of molecules, the commitment to the existence of the theoretical entities of science demands that we give up our naïve and common sense views concerning the nature of reality [10].

References

1. Dummett M (1993) The seas of language. Oxford University Press, Oxford
2. Devitt M (1997) Realism and truth, 2nd edn with an new afterword by the author. Princeton University Press, Princeton
3. Nagel T (1986) The view from nowhere. Oxford University Press, Oxford
4. Putnam H (1981) Reason, truth, and history. Cambridge University Press, Cambridge
5. Putnam H (1987) The many faces of realism. Open Court, La Salle, Ill
6. Field H (1982) Realism and relativism. J Philos 79:553–567
7. Putnam H (1999) The threefold cord mind, body, and world. Columbia University Press, New York
8. Boyd R (1984) The current status of scientific realism. In: Leplin J (ed) Scientific realism. University of California Press, Berkeley and Los Angeles, pp 41–82
9. Laudan L (1984) A confutation of convergent realism. In: Leplin J (ed) Scientific realism. University of California Press, Berkeley and Los Angeles, pp 218–249
10. Churchland PM (1979) Scientific realism and the plasticity of mind. Cambridge University Press, Cambridge

Reality Monitoring

Definition

Reality monitoring is defined as the ability of distinguishing between external memories (e.g., those of events directly perceived or actions actually performed) and internal memories (e.g., those of events imagined or actions planned or intended to perform).

►Metacognition

Realization

Definition

Mental properties, although not identical to physical properties, are still said to be physical properties in a broad sense in virtue of being realized by physical properties, just as a machine table, for instance, is implemented by but not identical to the states of its physical implementation. A central idea is that if property F realizes property G, then G is not something distinct from or something over and above F. Unlike identity, realization is asymmetric: F realizes G only if the instantiation of F in o necessitates or determines the instantiation of G in o but not vice versa, where the necessity in question is at least nomological necessity.

►Epiphenomenalism

Reasoning

MARKUS KNAUFF
Center for Cognitive Science, University of Freiburg, Freiburg, Germany

Definition

Reasoning is a process of drawing inferences from information that is taken for granted. Formal reasoning is within the scope of mathematics and philosophy. It is the study of inferences whose validity only derives from its formal structure. Mental reasoning is a function of the human brain. It comes into play whenever people go beyond what is explicitly given. It is the cognitive activity to infer that something must be true, or is likely to be true, given that the known information is true. The problem information is given by a number of statements which are called ►premises, and the task is to find a ►conclusion that follows from these premises. The following inference is a typical reasoning problem:

If a patient's left hemisphere is damaged, then he has impaired reasoning abilities.
Alan's left hemisphere is damaged.

- -

Therefore, Alan has impaired reasoning abilities.

Although the premises (above the line) do not say anything about Alan's reasoning abilities, most people immediately agree with what is stated in the conclusion (below the line). The conclusion necessarily – logically – follows from the premises. Another inference is given in the following example.

Mammals have a nervous system.
Birds have a nervous system.
Fishes have a nervous system.

- -

All animals have a nervous system.

Although a reasoner might form the belief that the conclusion could be true, the premises do not warrant the truth of the conclusion. The reasoner is generating the ►hypothesis that the conclusion is true. The former inference is an example of ►deductive reasoning, while the latter is an instance of ►inductive reasoning.

Characteristics

Deductive and Inductive Reasoning

Mental deductive reasoning is strongly related to formal logic. The latter serves as the normative model for the former (a critical assessment of this account from a neuroscience perspective can be found in [1]). To explore deductive reasoning in the psychological laboratory, people are typically asked to draw conclusions from given premises and later their responses are evaluated for logical validity. This evaluation is based on logical correctness only and does not account for the content of the statements (the deductive inference above is logically valid, although the content concerning the role of the left hemisphere is probably wrong; see below). In ►conditional reasoning, the premises of the problem consist of an "if A then B" construct that posits B to be true if A is true. The two logically valid inferences are the Modus Ponens (if p then q; p; q, MP) and the Modus Tollens (if p then q; not-q; not-p, MT). Humans are pretty good in making inferences of the form MP, but they make many mistakes in the form MT [2], In ►syllogistic reasoning, the premises of the problem consist of quantified statements such as "All A are B," "Some A are B," "No A are B," and "Some A are not B." People often make many mistakes in syllogistic reasoning, in part because of the existence of a variety of biases [2]. The most frequently used sort of inferences in daily life (and in the psychological lab) are based on relations. In ►relational reasoning, at least two relational terms

A r_1 B and B r_2 C are given as premises and the goal is to find a conclusion A r_3 C that is consistent with the premises. The relations represent spatial (e.g., left of), temporal (e.g., earlier than), or more abstract information (e.g., is akin to). People are pretty good in making such inferences, but the difficulty depends on the number of premises, the order of terms and premises, the content, and the ease to envisage the content of the problem [3,4]. Moreover, in cases where a reasoning problem has multiple solutions, reasoners consistently prefer the same subset of possible answers – and often just a single solution [5].

Inductive reasoning has not as much to do with logic because the conclusion goes beyond the information given in the premises. The premises only provide good reasons for accepting the conclusion. Thus, inductive reasoning is not truth-preserving but it is the most important basis of our ability to create new knowledge. This new knowledge is often based on a limited number of observations from which we formulate a law recurring to a set of phenomenal experiences. Cognitive theories of induction typically describe it as a process in which hypotheses are generated, selected, and evaluated [6,7]. Although there is no generally accepted definition of the term "induction," the majority of psychologists adopt the very broad definition that mental inductions are "all inferential processes that expand knowledge in the face of uncertainty" [6, p. 1]. Given that almost nothing is known about the neural basis of inductive reasoning this review is restricted to deductive reasoning. An easily accessible summary of behavioral findings on inductive reasoning can be found in Manktelow [8]. The main problems of research on inductive reasoning are summarized in Sloman and Lagnado [9].

Cognitive Theories of Reasoning

There are two main theories of deductive reasoning. They differ in the postulated underlying mental representations and the computational process that work on these representations. In one theory, it is believed that people think deductively by applying mental ►rules which are similar to rules in computer programs. In the other theory, deductive reasoning is conceived as a process in which the reasoner constructs, inspects, and manipulates ►mental models. The ►rule-based theory is a syntactic theory of reasoning, as it is based on the form of the argument only, whereas, the ►mental models theory is a semantic theory, because it is based on the meaning (the interpretation) of the premises.

The ►rule-based theories are primarily represented by the work of Rips [10] and Braine and O'Brian [11]. These theories claim that reasoners rely on formal rules of inference akin to those of formal logic, and that inference is a process of proof in which the rules are applied to mental sentences (but cf. Stenning and Oberlander [12]). The formal rules govern sentential connectives such as "if" and quantifiers such as "any," and they can account for relational inferences when they are supplemented with axioms governing transitivity, such as: For any a, b, and c, if a is taller than b and b is taller than c, then a is taller than c. The rules are represented in the human brain and the sequence of applied rules results in a mental proof, or derivation, which is seen as analogous to the proofs of formal logic [10].

The ►theory of mental models has been developed by Johnson-Laird and colleagues [13–15]. According to the model theory, human reasoning relies on the construction of integrated mental representations of the information that is given in the reasoning problem's premises. These integrated representations are models in the strict logical sense. They capture what is common to all the different ways in which the premises can be interpreted. They represent in "small scale" how "reality" could be – according to what is stated in the premises of a reasoning problem. The model theory distinguishes between three different mental operations. In the construction phase, reasoners construct the mental model that reflects the information from the premises. In the inspection phase, this model is inspected to find new information that is not explicitly given in the premises. In the variation phase, reasoners try to construct alternative models from the premises that refute the putative conclusion. If no such model is found, the putative conclusion is considered true [14].

Reasoning and the Brain

The two reasoning theories are related to different brain areas. The rule theory implies that reasoning is a linguistic and syntactic process, and so reasoning should depend on regions located in the left hemisphere. The model theory, in contrast, postulates that a major component of reasoning is not verbal, and so the theory predicts that the right cerebral hemisphere should play a significant role in reasoning [16]. More detailed predictions are related to specific brain areas. Here the rule theory assumes that the neural computations during reasoning are implemented in the language processing regions and here specifically in the temporal cortex, while the model theory predicts that the parietal and occipital cortical areas involved in spatial working memory, perception, and movement control are evoked by reasoning [17]. The lateralization of the reasoning process has been primarily investigated in patient studies, while brain imaging techniques allow for a more detailed localization of reasoning processes.

Patient Studies

Early studies of patients with brain-damages seemed to support the rule-based theories of reasoning. Conditional reasoning has been studied by Golding [18]. The author used the Wason-Selection-Task, which is probably the most important paradigm in behavioral research on human reasoning [19]. In the task, four

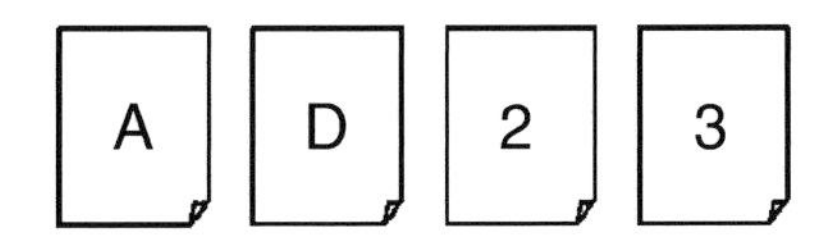

Reasoning. Figure 1 The Wason selection task.

cards are presented to the participants (see Fig. 1) and they are instructed to verify the rule "If there is a vowel on one side of the card, then there is an even number on the other side." The participants are allowed to turn over the cards in order to verify the rule. The visible letters and numbers on the card correspond to the four possible propositions p, not-p, q, and not-q. According to the propositional calculus of formal logic the only correct choices are p (according to the MP a q must be on the other side) and not-q (according to the MT a not-p must be on the other side). However, only one of the left-hemisphere-damaged patients but half of the right-hemisphere-damaged patients selected the two correct cards. Deglin and Kinsbourne [20] studied syllogistic reasoning with psychiatric patients while recovering from transitory ictal suppression of one hemisphere by electroconvulsive therapy (ECT; that simulates a short-term lesion). The premises were familiar or unfamiliar and true or false. When the right hemisphere was suppressed, the participants tended to perform deductive inferences even when the factual answer was obviously false. While their left hemisphere was suppressed, the same participants used their prior knowledge and if the content was unfamiliar they completely refused to answer. Patient studies on relational reasoning have been reported by Caramazza et al. [21] and Read [22]. Caramazza et al. [21] presented relational premises such as "Mike is taller than George" to brain-damaged patients. After reading the statements they had to answer either a congruent ("Who is taller?") or incongruent ("Who is shorter?") question. The left-hemispheric patients showed impaired performance in all problems no matter they were congruent or incongruent. Right-hemispheric patients, in contrast, showed impaired performance only in the incongruent problems. Read [22] used two relational premises and asked patients who suffered from temporal-lobectomy to generate a conclusion from these statements. Overall, the left-hemispheric patients again performed weaker than the patients with right-temporal-lobectomy, but the right-hemispheric patients were more impaired with the incongruent conclusions.

The reported findings have been frequently used by neuroscientists to support the idea that reasoning is mainly a linguistic and syntactic process, but this interpretation seems awkward to many cognitive oriented reasoning researchers. Although lesions to the left hemisphere might result in a deficit in the processing of the linguistic elements of the problem and, thus, impair overall performance, it does not necessarily follow that the damage will also affect the reasoning process. It is likely that left-hemisphere lesions lead to an inability to process the linguistic aspects of a reasoning problem, but that for the pure reasoning process the right hemisphere is important. This interpretation would also explain most of the findings. For instance, in the studies by Caramazza et al. [21] and Read [22] the patients had problems in logically deducing the converse of relations. Moreover, Whitaker et al. [23] examined conditional reasoning in patients that had undergone a unilateral anterior temporal lobectomy, one group to the right hemisphere and the other group to the left hemisphere. The content of the problems was related to the participants' prior knowledge of the world. Given the premises

If it rained, the streets will be dry.
It rained.

The right-hemisphere-damaged patients had a strong tendency to conclude "The streets will be wet" while the left-hemisphere-damaged patients concluded "The street will be dry." In other words, these right-hemispheric patients were unable to perform the deduction in isolation from their prior knowledge, while the left-hemisphere patients relied on the linguistic content of the problem.

Brain Imaging Studies

Brain imaging studies have been conducted on all the main types of deductive inferences. As with the patient studies the early findings have been frequently interpreted in favor of the rule-based theories of reasoning, as they have shown that reasoning activates a fronto-temporal neural network often just in the left hemisphere [24,25]. However, more sophisticated experimental paradigms suggest this might be due to the confounding of linguistic processing and deductive reasoning. Knauff et al. [17] studied conditional reasoning problems by presenting premises such as "If the teacher is in love, then he likes pizza" to the participants. In half of the problems the second premise was "The teacher is in love" and the participants had to conclude (by MP) "The teacher likes pizza." In the other half of problems the second premises was "The teacher does not like pizza" and the participants had to conclude (by MT) "The teacher is not in love." Both types of problems activated a bilateral occipito-parietal-frontal network, including parts of the prefrontal cortex and the cingulate gyrus, the superior and inferior parietal cortex, the precuneus, and the visual association cortex. These finding are difficult to explain based on purely linguistic processes, as the activated brain areas are implicated in the processing of visual and spatial information and visuo-spatial working memory (→) (cf. [26–28]). Similar findings have been reported from a study on syllogistic reasoning. Goel et al. [29] used problems with semantic content (e.g., "All apples are red; all red fruit are sweet; therefore all apples are sweet") and without semantic content (e.g., "All A are B; all B are C; therefore all A are C"). They found evidence for the engagement of both linguistic and spatial systems. The role of linguistic and spatial systems has been largely investigated by means of relational reasoning problems. In the study by Knauff et al. [30] such problems activated similar brain areas as the conditional problems did. However, the activity in visual association areas was even higher than during conditional reasoning. Goel and Dolan [31] addressed the question by using sentences with a spatial content. They again were either concrete (e.g., "The apples are in the barrel; the barrel is in the barn; therefore the apples are in the barn") or abstract (e.g., "A are in B; B is in C; therefore A is in C"). They reported that all problems activated a similar bilateral occipito-parietal network no matter if they were concert or abstract.

Reasoning and Visual Mental Imagery

Many of the reported experiments seem to support the model theory of reasoning. However, it is essential not to confuse mental models with visual images (→) [32,33]. Visual images are structurally similar to real visual perceptions, and can represent objects, their colors and shapes, and the metrical distances between them. They have a limited resolution, but they can be scanned and mentally manipulated [34]. They are often accompanied by neural activity in visual association areas (→) and under certain conditions also activate the primary visual cortex (→) (e.g., [30,35,36]). In contrast, mental models are likely to exclude visual detail, to represent only the information relevant to inference and to take the form of multi-dimensional arrays that maintain ordinal and topological properties [33]. Visual images represent information in a modality-specific format, whereas spatial models are abstract and not restricted to a specific modality. To clarify the role of visual images in reasoning Knauff, et al. [37] conducted a combined behavioral and brain imaging study with four sorts of relations: (i) visuo-spatial relations that are easy to envisage visually and spatially, (ii) visual relations that are easy to envisage visually but hard to envisage spatially, (iii) spatial relations that are hard to envisage visually but easy to envisage spatially, and (iv) control relations that are hard to envisage either visually or spatially. This study highlighted two important findings: First, reasoners were significantly slower with the visual relations than with the other sorts of relations. This is called the visual-impedance effect [38]. And second: On the brain level, all types of reasoning problems evoked activity in the parietal cortices and this activity seems to be a "default mode" of brain functioning during reasoning. However, only the problems based on visual relations also activated areas of the visual cortices. Obviously, in the case of visual relations, reasoners cannot suppress a spontaneous visual image but its construction calls for additional activity in visual cortices and retards the construction of a mental model that is essential for the inferential process. Interestingly, congenitally totally blind people are immune to the visual-impedance effect, since they do not tend to construct disrupting visual images from the premises [39]. For a more detailed explanation on how visual images and mental models interact in reasoning the interested reader is directed to Knauff [4].

Content Effects and Belief Biases

How easy it is to visualize is only one aspect of the content of a reasoning problem. Another aspect is how well the content agrees with the reasoners previous experiences and prior knowledge. Many behavioral studies have shown that prior knowledge can significantly influence how efficiently a reasoning problem is solved. Technically speaking, the abstract (logical) truth value of an inference can be the same as the truth value of our prior knowledge – in this case the inference is supported. Or, the formal truth value conflicts with the truth value of the prior knowledge – then the inference is more difficult, which means it results in more errors or takes significantly longer. If an inference generated by a person is biased towards the truth value of the prior knowledge or even overwritten by it, this is called belief bias [40]. Some patient studies, as described, have therefore explored the effects of brain injuries on reasoning with concrete and abstract materials. Their findings agree with the brain imaging study by Goel et al. [29] in which evidence for the engagement of both linguistic and spatial systems have been found. Reasoning with a semantic content activated a left-hemispheric temporal system, whereas problems without semantic content activated an occipito-parietal network distributed over both hemispheres. Goel and Dolan [41] brought logic and belief into conflict and found evidence for the engagement of a left temporal lobe system during belief-based reasoning and a bilateral parietal lobe system during belief-neutral reasoning. Activation of right prefrontal cortex was found when the participants inhibited a response associated with belief-bias and correctly completed a logical task. When logical reasoning, in contrast, was overwritten by a belief-bias, there was engagement of ventral medial prefrontal cortex, a region implicated in affective processing. In the dual-processing theory, Goel, et al. therefore suggests that deductive reasoning is implemented in two separate systems whose engagement is primarily a function of the presence or absence of semantic content. Content-free reasoning seems to be stronger related to visuo-spatial cortical areas in the right hemisphere, whereas content-based reasoning recruits language-related areas in the left temporal cortex. If the content of the reasoning problem results in a conflict between belief and logic, this conflict recruits additional areas in the right prefrontal cortex.

Evaluation of Reasoning Theories

For a long time the psychology of reasoning was strongly committed to the assumption that reasoning should be studied in term of computational processes. How these computations are biologically implemented in the human brain has been conceived to be not sufficient, because each computational function can be computed on each hardware (and, thus, also the brain) that is equivalent to a Turing machine (e.g., [42]). However, reasoning research is a good example of where the assumption of implementation-independency fails. As there are many mappings possible between cortical regions and cognitive functions, neuroscientific data alone are certainly too weak to test cognitive theories. But, if such data are consistent with behavioral findings this can provide strong support for a cognitive theory of human reasoning. An outstanding example is the field of relational reasoning, where hardly any researchers defend an approach based on inference rules (e.g., [3]). The behavioral and neuroscientific evidence showing that people use their visuospatial system to preserve the structural properties of the world are too overwhelming. In other fields of reasoning the situation is more complicated (cf. [43]). Many researchers will agree that mental models play a key role if humans perform inferences based on conditionals and quantifiers [8]. On the other hand, there is also evidence that verbal, linguistic, and syntactic processes are also involved. The most reasonable corollary from the field of research is that human think deductively by applying different mental algorithms and that these algorithms are implemented in different brain areas. Content-free inferences are "real logical" inferences and they seem to rely on neural computations in the right parietal cortices the precuneus, and the extrastriate and (sometimes) striate cortex. They are accompanied by executive functions and control processes in the prefrontal cortex. When the logical problem is embedded into a semantic content or related to the reasoners' beliefs additional linguistic and semantic processes in the left temporal cortex come into play. Another corollary from the neuro-cognitive research is that reasoning is a multi-component process and that the diverse components strongly overlap with the components of other cognitive functions. There is no single "cheater detection module" as proposed for reasoning about social contracts [44,45] much as there are no "pragmatic schemas" [46] that completely spare human to reason.

References

1. Cosmides L, Tooby J (2005) Neurocognitive adaptations designed for social exchange. In: Buss DM (ed) Evolutionary psychology handbook. Wiley, New York, pp 584–627
2. Evans JStBT, Newstead SE, Byrne RMJ (1993) Human reasoning. The psychology of deduction. Lawrence Erlbaum Associates, Hove
3. Goodwin GP, Johnson-Laird PN (2005) Reasoning about relations. Psychol Rev 112:468–493
4. Knauff M (2006) A neuro-cognitive theory of relational reasoning with visual images and mental models. In: Held C, Knauff M, Vosgerau G (eds) Mental models and the mind: a concept at the intersection of cognitive psychology, neuroscience, and philosophy of mind. Elsevier, North-Holland
5. Rauh R, Hagen C, Knauff M, Kuß T, Schlieder C, Strube G (2005) From preferred To alternative mental models in spatial reasoning. Spatial Cognition and Computation 5:239–269
6. Holland JH, Holyoak KJ, Nisbett RE, Thagard PR (1986) Induction: processes of inference, learning, and discovery. MIT, Cambridge, MA
7. Holyoak KJ, Nisbett RE (1988) Induction. In: Sternberg RJ, Smith EE (eds) The psychology of human thought. Cambridge University Press, New York, pp 50–91
8. Manktelow KI (1999) Reasoning and Thinking. Psychology, Hove
9. Sloman SA, Lagnado D (2005) The problem of induction. In: Morrison R, Holyoak K (eds) Cambridge handbook of thinking and reasoning. Cambridge University Press, New York, pp 95–116
10. Rips LJ (1994) The psychology of proof. MIT, Cambridge, MA
11. Braine MDS, O'Brian DP (eds) (1998) Mental logic. Lawrence Erlbaum, Mahwah, NJ
12. Stenning K Oberländer K (1995) A cognitive theory of graphical and linguistic reasoning: logic and implementation. Cogn Sci 19:97–140
13. Johnson-Laird PN (1983) Mental models. Cambridge University Press, Cambridge
14. Johnson-Laird PN, Byrne RMJ (1991) Deduction. Erlbaum, Hove
15. Johnson-Laird PN (2001) Mental models and deduction. Trend Cogn Sci 5:434–442
16. Johnson-Laird PN (1994) Mental models, deductive reasoning, and the brain. In: Gazzaniga S (ed) The cognitive neurosciences. MIT, Cambridge, MA, pp 999–1008
17. Knauff M, Mulack T, Kassubek J, Salih HR, Greenlee MW (2002) Spatial imagery in deductive reasoning: a functional MRI study. Cognitive Brain Research, 13:203–212
18. Golding E (1981) The effect of unilateral brain lesion on reasoning. Cortex 17:31–40
19. Wason PC (1966) Reasoning. In: Foss B (ed) New horizons in psychology. Penguin, Harmondsworth, UK, pp 135–151
20. Deglin VL, Kinsbourne M (1996) Divergent thinking styles of the hemispheres: how syllogisms are solved during transitory hemisphere suppression. Brain Cogn 31:285–307
21. Caramazza A, Gordon J, Zurif EB, DeLuca D (1976) Right-hemispheric damage and verbal problem-solving behavior. Brain Lang 3:41–46
22. Read DE (1981) Solving deductive-reasoning problems after unilateral temporal lobectomy. Brain Lang 12:116–127

23. Whitaker HA, Markovits H, Savary F, Grou C, Braun C (1991) Inference deficits after brain damage. J Clin Exp Neuropsychol 13:38
24. Goel V, Gold B, Kapur S, Houle S (1997) The seats of reason: a localization study of deductive and inductive reasoning using PET (O15) blood flow technique. Neuroreport 8:1305–1310
25. Goel V, Gold, B, Kapur S, Houle S (1998) Neuroanatomical correlates of human reasoning. J Cogn Neurosci 10:293–302
26. Andersen RA (1997) Multimodal integration for the representation of space in the posterior parietal cortex. Philos Trans R Soc London B Biol Sci 352:1421–1428
27. Smith EE, Jonides J (1997) Working memory: a view from neuroimaging. Cogn Psychol 33:5–42
28. Jonides J, Smith EE, Koeppe RA, Awh E, Minoshima S (1993) Spatial working memory in humans as revealed by PET. Nature 363:623–625
29. Goel V, Büchel C, Frith C, Dolan RJ (2000) Dissociation of mechanisms underlying syllogistic reasoning. Neuroimage 12:504–514
30. Knauff M, Kassubek J, Mulack T, Greenlee MW (2000) Cortical activation evoked by visual mental imagery as measured by functional MRI. Neuroreport 11:3957–3962
31. Goel V, Dolan RJ (2001) Functional neuroanatomy of three-term relational reasoning. Neuropsychologia 39:901–909
32. Johnson-Laird PN (1998) Imagery, visualization, and thinking. In: Hochberg J (ed) Perception and cognition at century's end. Academic, San Diego, CA, pp 441–467
33. Knauff M, Schlieder C (2004) Spatial inference: no difference between mental images and models. Behav Brain Sci 27:589–590
34. Kosslyn SM (1994) Image and brain: the resolution of the imagery debate. MIT, Cambridge, MA
35. Kosslyn SM, Ganis G, Thompson WL (2001) Neural foundations of imagery. Nat Rev Neurosci 2:635–642
36. Kosslyn SM, Thompson WL (2003) When is early visual cortex activated during visual mental imagery? Psychol Bull 129:723–746
37. Knauff M, Fangmeier T, Ruff CC, Johnson-Laird PN (2003) Reasoning, models, and images: behavioral measures and cortical activity. J Cogn Neurosci 4:559–573
38. Knauff M, Johnson-Laird PN (2002) Visual imagery can impede reasoning. Mem Cogn 30:363–371
39. Knauff M, May E (2006) Mental imagery, reasoning, and blindness. Quart J Exp Psychol 59:161–177
40. Evans, JStBT (1989) Bias in human reasoning. Lawrence Erlbaum, Hove
41. Goel V, Dolan RJ (2003) Explaining modulation of reasoning by belief. Cognition 87(1):B11–B22
42. Pylyshyn ZW (1984) Computation and cognition: toward a foundation for cognitive science. MIT, Cambridge, MA
43. Held C, Knauff M, Vosgerau G (eds) (2006) Mental models and the mind. A conception in the intersection of cognitive psychology, neuroscience, and philosophy of mind. Elsevier, North-Holland
44. Cosmides L (1989) The logic of social exchange: has natural selection shaped how humans reason? Studies with the Wason selection task. Cognition 31:187–276
45. Stone V, Cosmides L, Tooby J, Kroll N, Knight R (2002) Selective impairment of reasoning about social exchange in a patient with bilateral limbic system damage. Proc Natl Acad Sci 99(17):11531–11536
46. Cheng PW, Holyoak KJ (1985) Pragmatic reasoning schemas. Cogn Psychol 17:391–416
47. Fangmeier T, Knauff M, Ruff CC, Sloutsky V (2006) The neural correlates of logical thinking: an event-related fMRI study. J Cogn Neurosci (in press)

Rebound Bursting

Definition

Discharge of a burst of action potentials after the end of a hyperpolarizing influence, such as an inhibitory postsynaptic potential.

►Action Potential

Recall

Definition

Recall is the ability to not only recognize something as having been experienced in the past, but also to retrieve, on demand, spatiotemporal details of the context in which the stimulus or event was originally encountered.

►Recognition Memory

Receiver

Definition

In general an instrument that is able to register a signal. In communication theory, the receiver registers a signal, decodes it and reacts accordingly.

Recency

Definition

With respect to recognition, recency refers to the capacity to remember more accurately information which has just been experienced, as compared to events or items encountered further in time from retrieval.

►Recognition Memory

Recent and Remote Memory

►Long-Term Memory

Receptive Field

Definition

The aspect (for example, a location or a temporal frequency) of the outer world that is represented by a given neuron in the brain is referred to as its receptive field.

Receptive Field, Visual

Definition

The receptive field of a "visual" neuron is the region of the visual field in which the presentation of a stimulus exerts a response of the neuron.

►Visual Cortical and Subcortical Receptive Fields

Receptive Field of Retinal Ganglion Cell

Definition

In physiological studies the visual field area over which a cell responds to light. The receptive field area corresponds roughly to the dendritic field area.

►Retinal Ganglion Cells

Receptive Field Selectivity

►Contrast Enhancement

Receptor

Definition

The term receptor is an ambiguous term because, on the one hand, it is used as shorthand for sensory receptor cell. A sensory receptor (in physiology) is any structure which, on receiving environmental stimuli, produces an informative nerve impulse. The receptor recognizes a stimulus in the external or internal environment, initiates a transduction process by producing graded potentials (receptor potentials), from which all-or-none action potentials are elicited, that are conducted along afferent fibers originating in the same or adjacent cells. On the other hand, a membrane receptor, neurotransmitter receptor, etc. (in biochemistry/pharmacology) is a transmembrane glycoprotein, which is activated by ligands. Receptors to neurotransmitters are located at the plasma membrane. Upon binding by the specific transmitter, receptors can allow the passage of ions or activate enzymes, which ultimately modify the membrane potential. Ionotropic receptors are fast neurotransmitter-gated receptors formed by homomeric or heteromeric subunits outlining a channel, which allows influx or outflux of monovalent or divalent ions. Instead, metabotropic receptors are slow neurotransmitter-gated receptors, which are coupled to G proteins activating diverse effector mechanisms. Excitotoxicity is caused by ionotropic glutamate receptors of the AMPA, kainate and NMDA classes.

►Action Potential
►Glutamate Receptor Channels
►Sensory Systems

Receptor Agonist

Definition

A chemical substance that binds to a cell membrane receptor and mimics the regulatory effects of endogenous signaling compounds such as neurotransmitters, neuromodulators and hormones.

►Membrane Components

Receptor Cell

Definition

►Sensory Receptor

►Sensory Systems

Receptor Channel

►Ionotropic Receptor

Receptor Current

Definition

Receptor current denotes the transmembrane current evoked, at the receptor membrane of a sensory receptor cell, by an impinging sensory stimulus through opening or closing of specific ion channels.

►Sensory Systems

Receptor Desensitization

Definition

Receptor desensitization is a reduced response to a neurotransmitter or agonist drug due to a decrease in number of receptors available, or decreased activity of intracellular signaling pathways and ion channels, after prolonged exposure to the neurotransmitter or drug.

Desensitization also results from receptor internalization, the removal of receptors from a plasma membrane by endocytosis. Agonist-binding receptors are pinched off and drawn into the cytoplasm with membrane vesicles and either recycled to the cell surface or degraded.

►Ionotropic Receptor

Receptor Membrane

Definition

Receptor membrane denotes that region of a sensory receptor cell, where the transformed physico-chemical stimulus is converted, by a specific process called sensory transduction, into receptor current and receptor potential.

►Receptor Current
►Receptor Potential
►Sensory Receptor
►Sensory Systems

Receptor Potential

Definition

Receptor potential denotes the membrane potential change evoked, at the receptor membrane of a sensory receptor cell, by an impinging sensory stimulus through opening or closing of specific ion channels.

►Sensory Receptor
►Sensory Systems

Receptor Regulation, Editing

Definition

A novel channel regulation of the ionotropic glutamate receptors. It occurs at a specific CAG codon for glutamine which changes to a CGG codon for arginine in the pre-mRNAs for a specific subtype of glutamate receptor subunits. The edited codon determines the Ca^{2+} permeability of the receptors containing the edited subunit.

►Ionotropic Receptor

Receptor Regulation, Phosphorylation

Definition

Certain types of neurotransmitter receptors, such as G protein-coupled receptors and ionotropic receptors, may be regulated by phosphorylation. These regulations are controlled by a combination of kinase and phosphatase, both of which are receptor-selective.

▶Ionotropic Receptor

Receptor Regulation, Splicing

Definition

Alternative exon selection changes the receptor structure, modifying the property of the receptor. For example, the "flip/flop structures" of splice isoformes in the AMPA receptors determine the desensitization rate, and the C-terminal splice isoformes of many glutamate receptor subunits control their distribution in the subsynaptic membrane.

▶Ionotropic Receptor

Receptor Regulation, Subunit Change

Definition

Almost all ionotropic receptors consist of different kinds of subunits. Their composition determines the functional properties and diversity of each receptor.

▶Ionotropic Receptor

Receptor Trafficking

Fiona K. Bedford
Department of Anatomy and Cell Biology, McGill University, Montreal, QC, Canada

Synonyms

Post-synaptic receptor trafficking; Neurotransmitter receptor trafficking

Definition

▶Receptor trafficking is a term used to describe the movement of receptors within a neuron. It is used broadly to describe several distinct stages of receptor movement in neurons; the movement of newly synthesized receptors through the secretory pathway (Fig. 1), the movement of receptors into ▶axons or ▶dendrites and their targeting to the pre and postsynaptic domains of ▶synapses respectively. In addition it is also used to describe the internalization of receptors from the plasma membrane as well as their subsequent intracellular trafficking. In all these incidents it should be noted receptor trafficking literally means the movement of receptors between distinct membrane and vesicular compartments of a neuron (Fig. 1).

Characteristics

Description of the Process

In neurons, receptor trafficking is a process by which numerous neural functions, such as neuronal migration and ▶synaptic transmission can be regulated. ▶Receptor trafficking controls these functions by setting the capacity of a neuron to respond to an external cue. At the ▶post-synaptic density for example, receptor trafficking can regulate the number of receptors available at any one time to respond to the ▶presynaptic release of ▶neurotransmitter molecules. As the amount of released neurotransmitter molecules often outweighs the number of available ▶post-synaptic receptors, the process of receptor trafficking can control the efficiency and amplitude of the post-synaptic response. Over the past decade our understanding of the cellular mechanisms utilized by neurons to control postsynaptic receptor trafficking has increased significantly. We have discovered amongst other things that the basic mechanisms that control postsynaptic receptor trafficking are largely conserved with those controlling receptor trafficking in general. Therefore we have chosen to use examples of postsynaptic receptor trafficking below to illustrate the process and regulation of general ▶receptor trafficking in neurons.

As ▶receptor trafficking involves the movement of receptors between distinct membrane and vesicular compartments, it is intuitive that receptors must be recognized by components of these distinct compartments or by the trafficking machinery at each step. Simply put, receptor trafficking involves the recognition of discrete motifs within receptors by components of the different compartments. The motifs within receptors are defined as ▶trafficking motifs and the components of the distinct compartments as ▶trafficking adaptors. Hence if a receptor contains a specific trafficking motif, it has the potential to be recognized by a specific adaptor molecule of the compartment/ trafficking pathway defined by that motif and to be moved there. The

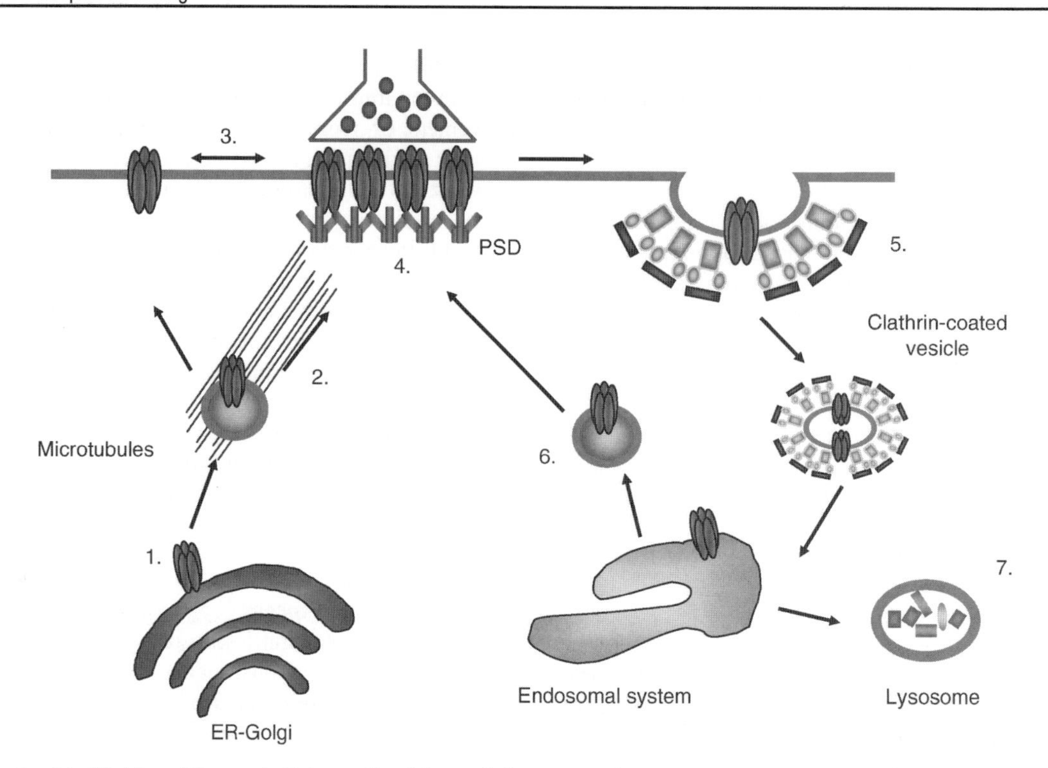

Receptor Trafficking. Figure 1 Schematic of the cellular steps of post-synaptic receptor trafficking in neurons. (1) Post-synaptic receptors are generally synthesized and processed within the endoplasmic reticulum (ER) and Golgi of the neuronal cell body. (2) Receptors are then inserted locally into the plasma membrane or trafficked along dendritic microtubules to distal synapses. (3) In both cases receptors are either inserted directly at synapses or in the extra-synaptic membrane followed by later diffusion to the synapse. (4) Anchoring proteins within the post-synaptic density (PSD) then limit trafficking from the synapse. (5) Following release from the PSD anchoring proteins, receptors are internalized via clathrin-mediated endocytosis directly at the synapse or by lateral diffusion to designated endocytic zones. Internalized receptors then traffic within the endosome system and are either recycled back to the plasma membrane (6) or targeted for degradation within lysosomes (7). At each of these steps a specific protein-protein interaction between a trafficking adaptor(s) and a trafficking motif(s) within post-synaptic receptors dictate their trafficking itineraries.

physical movement in most cases is mediated by an indirect interaction with the cytoskeleton.

Postsynaptic receptors are multi subunit receptors frequently composed of different sub-classes of subunits. Because of this heteromeric nature, it is common that different subunits within a single receptor may contain distinct trafficking motifs dictating specific movement to discrete compartments of the neuron. In the sections below I will describe examples that illustrate our present understanding of the motifs and adaptor molecules controling postsynaptic receptor trafficking.

Newly Synthesized Receptor Trafficking

In the synthesis of heteromeric receptors, active retention in the ►endoplasmic reticulum (ER) is a commonly used process, whereby individual subunits are retained within the ER until they are correctly assembled into the mature receptor. This process acts primarily as a quality control measure ensuring the release of only functional molecules from the ER. The precise motifs and ER-proteins directing this retention are largely unknown for the vast majority of receptors. However it is thought and indeed the case in certain proteins that highly charged residues, which may be exposed in individual subunits but masked in the assembled receptor, play an important role in this retention process. In the case of the postsynaptic ►NMDA receptors, an additional layer of regulation in ER receptor trafficking exists [1].

The predominant NMDA-type glutamate receptor in the brain is composed of 2 NR1 and 2 NR2 type subunits and the trafficking of this NMDA receptor from the ER into the secretory pathway is controlled by differential splicing of the NR1 subunit. The NR1 subunit gene contains the splicing cassettes C1, C2, C2′ that produce different cytoplasmic C-termini of the NR1 subunit. The NR1–1 splice variant, which is the main splice variant in the brain, encodes a C2 cassette and a C1 cassette, which contains a triple arginine (RRR) ER retention motif. If this NR1–1 splice variant assembles with NR2 subunits in the ER, mature assembled receptors will not exit the ER efficiently, due to the presence of this ER retention motif. However if the C1 cassette is expressed in combination with a C2′ cassette (NR1–3 splice variant) the assembled receptor successfully exits the ER. This is due to the presence of a specific trafficking motif within the C2′ cassette (Serine-Threonine-Valine-Valine, STVV) that overrides the RRR retention motif of the C1 cassette. This STVV trafficking motif is able to override the RRR retention motif, as it belongs to the family of ►PDZ domain interacting motifs and specifically binds to the PDZ domain containing protein Sec23. Sec23 is a component of the ►COPII coat complex that promotes ER exit via COPII vesicle formation. Differential expression of these different splice variants is in addition controlled via ►neuronal activity. For example, increased neuronal activity promotes the insertion of the C2 cassette. This favors the C1 retention mechanism; reducing NMDA receptor exit from the ER and thus lowering overall surface expression of NMDA receptors and neuronal activity. In contrast blocking neuronal activity promotes C2′ insertion over C2, which promotes ER exit, the secretory trafficking of NMDA receptors and increases neuronal activity. In essence, the above example neatly illustrates how a commonly used mechanism (ER retention) can be used to control receptor trafficking from the ER and how an additional process can be employed to override this retention in order to respond to the needs of the system.

Trafficking Along Dendrites and Targeting to Synapses

Most postsynaptic receptors are synthesized in the ►soma and once they have passed through the ER and Golgi are either inserted locally within the plasma membrane or travel large distances within ►dendrites to reach the distal dendritic synapses. The latter is achieved predominantly via vesicular transport along ►microtubules within the dendrite. To traffic receptors in this manner, specific linker proteins that bind to trafficking motifs within the cytoplasmic domain of receptors and to ►microtubule motor proteins move the vesicle along the microtubule. The microtubule motor proteins controlling this distal direction of vesicle movement (anterograde trafficking) belong to the ►kinesin superfamily proteins. It is estimated that a large anterograde transport vesicle may be associated on average with between 100–200 kinesin molecules. Once the target of this vesicle is reached (the dendritic synapse) exocytosis of these receptors occurs and the kinesin molecules are either degraded or recycled back to the soma for future use.

With respect to the linker proteins involved, an obvious pre-requisite is their ability to interact with kinesins. Of the linker proteins identified thus far, many have the ability to bind only a single kinesin protein but to broadly bind a variety of receptor types. This occurs as these linker proteins generally contain multiple interaction domains that can bind distinct trafficking motifs within different proteins. Furthermore some receptors, such as the ►AMPA receptors, can bind multiple kinesins via different linker proteins. AMPA-receptors can bind convention kinesin via the linker protein the glutamate-receptor-interacting protein (GRIP1) and the kinesin family member, KIF1 via liprin-α [2]. Both interactions are mediated via carboxyl-terminal PDZ motifs in the cytoplasmic domains of the GluR2 and GluR3 AMPA receptor subunits. Thus a single postsynaptic receptor may encode multiple dendritic vesicular trafficking motifs, which can bind different linker proteins, which all participate in the anterograde movement of the receptor.

Under certain circumstances, postsynaptic receptors can also be trafficked in a retrograde fashion back to the soma. This involves a different set of linker proteins binding alternate trafficking motifs that have an affinity for retrograde microtubule motor proteins, such as the ►dynein superfamily of proteins. Finally exocytosis and insertion of receptors into the neuronal surface can occur at the synapse or at extra-synaptic sites followed by lateral diffusion of the receptors within the neuronal plasma membrane to the synaptic site. Once at the synapse, the binding of postsynaptic anchoring proteins subsequently retard further post-synaptic receptor trafficking and movement.

Endocytosis and Intracellular Trafficking

The next and penultimate step in the journey of neurotransmitter receptor trafficking is ►endocytosis, where the receptor is targeted for internalization from the plasma membrane. The vast majority of postsynaptic receptors are internalized via ►clathrin-mediated endocytosis (CME). Endocytosis of postsynaptic receptors can occur at the synapse, but preferentially occurs at designated internalization sites lateral to the postsynaptic density. In which case, postsynaptic receptors need to disengage from their postsynaptic anchoring proteins prior to internalization. This occurs constitutively or in a signal-regulated manner, involving a post-translational modification that alters the association of the receptor with the postsynaptic density anchoring protein.

In general CME is mediated by endocytic trafficking motifs located in the cytoplasmic domains of postsynaptic receptors, which are recognized by the tetrameric clathrin-binding endocytic adaptor protein 2 complex (AP-2). Two classical endocytic motifs exist, a tyrosine (Y) based motif, YxxΦ, where –x- represents any amino acid and Φ is a hydrophobic amino acid and an acidic di-leucine motif, D/ExxxLL, (D = aspartic acid,

E = glutamic acid and L = Leucine). These motifs, which have been found in the cytosolic domains of many neurotransmitter receptors, are recognized by different subunits of the cytosolic AP-2 complex. AP-2 binding in both cases however promotes membrane invagination and the recruitment of the clathrin lattice, leading to the formation of endocytic vesicles and receptor internalization. Some postsynaptic receptors encode a single endocytic motif, which is recognized by the AP-2 complex whilst others contain multiple. The inhibitory ►$GABA_A$ receptor heteropentamer for example, is composed of subunits that contain both tyrosine and di-leucine endocytic trafficking motifs, which all may be relevant in mediating $GABA_A$ receptor ►endocytosis [3].

Once internalized these postsynaptic receptor containing endocytic vesicles quickly mature, losing their clathrin lattice to first form early endosomes, which then subsequently mature into late/sorting endosomes. It is in late/sorting endosomes that internalized receptors are targeted either to be recycled back to the plasma membrane or alternatively for degradation in lysosomes. Again, the interaction between trafficking motifs within the receptor and compartment specific adaptors plays an important role. In the vast majority of cases the recycling of receptors is the default pathway, while trafficking to the lysosomal pathway involves an additional sorting step. The di-leucine motif mentioned above has been implicated in this step, through the binding of a related lysosomal specific adaptor complex to AP-2, the AP-3 complex. In addition to these peptide specific endocytic sorting motifs, the modification of receptor subunits by the addition of a 7 kilodalton(kDa) protein called ►ubiquitin has been identified as a targeting signal for endosomal-lysosomal sorting, which is discussed in more detail below.

Regulation of the Process

►Post-translation modifications, the addition of a molecule to a protein after it has been synthesized, provides an additional layer of regulation in receptor trafficking. Receptors can be modified by a number of different post-translational additions, such as lipids, inorganic ions or by the covalent attachment of small proteins. Listed below are some examples of how these different modifications can alter receptor trafficking itineraries.

Phosphorylation

►Phosphorylation, the addition, or de-phosphorylation, the removal of an inorganic phosphate, are mediated by kinase or phosphatase enzymes respectively. This type of post-translational modification is used widely in neurons to regulate receptor trafficking, by altering the binding specificity of a trafficking adaptor to its linear trafficking motif. For example, the rate of $GABA_AR$ endocytosis is regulated by phosphorylation. All $GABA_AR$ heteropentamers contain a beta-type subunit, which encodes an atypical endocytic AP-2 binding motif that encompasses an established regulatory phosphorylation site [4]. Phosphorylation at this site results in the reduced affinity of the AP-2 complex for the endocytic motif, which interferes with the rate of $GABA_AR$ endocytosis. This example neatly demonstrates a phospho-dependent regulation of endocytosis via regulation of the endocytic adaptor protein AP-2 binding affinity.

Palmitoylation

►Palmitoylation is the reversible post-translational attachment of a saturated fatty acid, palmitic acid, to cysteine residues in a membrane protein via a thiol-ester bond. Palmitoyl acyl transferase enzymes mediate this reaction and in neurons the action of one such enzyme, the Golgi-specific DHHC zinc finger protein GODZ directs the palmitoylation of two classes of postsynaptic neurotransmitter receptors. These are the AMPA-type glutamate receptor and the $GABA_AR$ [5,6]. Palmitoylation of these receptors can modulate their membrane trafficking, by enhancing their hydrophobicity, which leads to an enhanced rate of surface expression. Palmitoylation of post-synaptic receptors therefore facilitates enhanced secretion of newly synthesized receptors.

Ubiquitination and Sumoylation

►Ubiquitin (Ub) is a highly conserved 76 amino acid polypeptide that is covalently conjugated to lysine residues on target proteins or to itself by a reaction involving three classes of enzymes: an E1 activating enzyme, an E2 conjugating enzyme and an E3 ligase that also determines substrate specificity. Protein ►ubiquitination is reversible and this is controlled by the action of de-ubiquitinating enzymes that cleave the Ub-protein bond. Modification of proteins by Ub chains ($Ub^n > 4$) primarily targets them for degradation by the multi-subunit proteolytic complex called the ►proteasome. In contrast, a single ubiquitin (mono-ubiquitination) as well as multi-site mono-ubiquitination, functions as a signal in the endocytic pathway controlling proteins internalization and lysosomal degradation. Mono-ubiquitination functions as an efficient endocytic-lysosomal trafficking signal, as several endocytic adaptor proteins encode ubiquitin-binding domains, which specifically recognize, bind to and traffic ubiquitinated membrane proteins.

There is now compelling evidence demonstrating a role for ubiquitination in regulating the abundance of glutamate receptors and synapse-associated proteins. Ubiquitin dependent proteasomal degradation of PSD95 for example, a glutamate receptor synaptic anchoring protein, enhances the endocytosis rate of AMPA receptors.

It does so by releasing AMPA receptors from the post-synaptic density and thereby enabling their interaction with the endocytic machinery [7]. This leads to a reduction in excitatory synaptic responses. With respect to the direct ubiquitination of glutamate receptors, a strong role for an E3 ligase complex of proteins called the Cullin E3 ligase (cul) complex has been discovered. The Cul3 adaptor protein actinfilin has been found to bind and mediate a Cul3 dependent ubiquitination of the GluR6 subunit of ►kainate receptors, controlling GluR6 levels and receptor accumulation at excitatory synapses [8]. The NR1 subunit of the NMDA-type of glutamate receptor is also targeted for ubiquitination by a cullin complex protein, the F-box protein Fbx2 [9]. Ubiquitination of NR1 by Fbx2 alternatively controls its retro translocation from the ER and ubiquitin-proteasome mediated degradation in an activity dependent manner, leading to a reduction in NMDA-dependent currents. Of the third class of mammalian glutamate receptors, the AMPA receptor, subunit ubiquitination has yet to be demonstrated, but is highly probable as the signal sequences targeting ubiquitination of the related *C. elegans* GLR-1 subunit, are conserved in all mammalian AMPA type glutamate receptors.

►Sumo (small ubiquitin-related modifier), also named "sentrin," is a 101-amino acid protein that can also be covalently attached to cytosolic lysine residues, in a process that is analogous to ubiquitination. ►Sumoylation was originally associated only with the functions of nuclear proteins, however more recently sumoylation of neurotransmitter receptors has been demonstrated and this modification on these proteins, like ubiquitination can regulate receptor trafficking. Sumoylation of the GluR6 kainate-type of glutamate receptor for example, regulates the rate of kainate receptor endocytosis and modifies the efficiency of synaptic transmission [10]. How precisely sumoylation facilitates this process is at present unknown, but it is likely it either release kainate receptors from an anchoring protein or alter the binding affinity of the receptor to endocytic adaptors. Sumoylation of the GluR6 subunit in neurons is rapidly enhanced in response to kainate treatment, which implies this modification is employed in an autoregulatory type of response to agonist application.

Closing Comments

In this essay, I described the main cellular steps of neurotransmitter receptor trafficking that occur in a neuron (Fig. 1). Using specific examples of receptor trafficking for certain post-synaptic receptors, I have described how the different processes of receptor trafficking function at each step. In addition, I have outlined several post-translational modifications, which can regulate these processes of receptor trafficking and have explained how this regulation occurs and controls receptor trafficking.

References

1. Mu Y, Otsuka T, Horton A-C, Scott D-B, Ehlers M-D (2003) Activity-dependent mRNA splicing controls ER export and synaptic delivery of NMDA receptors. Neuron 240(3):581–594
2. Shin H, Wyszynski M, Huh K-H, Valtschanoff J-G, Lee J-R, Ko J, Streuli M, Weinberg R-J, Sheng M, Kim E (2003) Association of the kinesin motor KIF1A with the multimodular protein liprin-alpha. J Biol Chem 278:11393–11401
3. Kittler J-T, Delmas P, Jovanovic J-N, Brown D-A, Smart T-G, Moss S-J (2000) Constitutive endocytosis of $GABA_A$ receptors by an association with the adaptin AP2 complex modulates inhibitory synaptic currents in hippocampal neurons. J Neurosci 20:7972–7977
4. Kittler J-T, Chen G, Honing S, Bogdanov Y, McAinsh K, Arancibia-Carcamo I-L, Jovanovic J-N, Pangalos M-N, Haucke V, Yan Z, Moss SJ (2005) Phospho-dependent binding of the clathrin AP2 adaptor complex to $GABA_A$ receptors regulates the efficacy of inhibitory synaptic transmission. Proc Natl Acad Sci USA 102:14871–14876
5. Hayashi T, Rumbaugh G, Huganir R-L (2005) Differential regulation of AMPA receptor subunit trafficking by palmitoylation of two distinct sites. Neuron 47:709–723
6. Keller C-A, Yuan X, Panzanelli P, Martin M-L, Alldred M, Sassoè-Pognetto M, Lüscher B (2004) The gamma2 subunit of GABA(A) receptors is a substrate for palmitoylation by GODZ. J Neurosci 24:5881–5891
7. Colledge M, Snyder E-M, Crozier R-A, Soderling J-A, Jin Y, Langeberg L-K, Lu H, Bear M-F, Scott J-D (2003) Ubiquitination regulates PSD-95 degradation and AMPA receptor surface expression. Neuron 40:595–607
8. Salinas G-D, Blair L-A, Needleman L-A, Gonzales J-D, Chen Y, Li M, Singer J-D, Marshall J (2006) Actinfilin is a Cul3 substrate adaptor, linking GluR6 kainate receptor subunits to the ubiquitin-proteasome pathway. J Biol Chem 281:40164–40173
9. Kato A, Rouach N, Nicoll R-A, Bredt D-S (2005) Activity-dependent NMDA receptor degradation mediated by retrotranslocation and ubiquitination. Proc Natl Acad Sci USA 102:5600–5605
10. Martin S, Nishimune A, Mellor J-R, Henley J-M (2007) Sumoylation regulates kainate-receptor-mediated synaptic transmission. Nature 447:321–325

R

Receptor Tyrosine Kinase

Definition

Receptor tyrosine kinase is a transmembrane receptor whose intracellular domain contains a kinase that is capable of transferring a phosphate group from ATP to a tyrosine of a protein. Many growth factors and extrinsic signaling molecules act through receptor tyrosine kinases.

►Growth Factor

Reciprocal Activation

Definition

Simultaneous activation of muscles with a mechanical action on a joint (agonist) and inhibition of muscles with the opposite mechanical action (antagonists).

Reciprocal activation may be of both central and peripheral origin and it is mediated by excitation of agonist motoneurons and reciprocal inhibition of antagonist motoneurons, via inhibitory spinal interneurons, by descending fibers (in voluntary movements) and by sensory afferents (in reflexive responses).

►Reaching Movements

Reciprocal Dendrodendritic Synapse

Definition

Principal synapse of the olfactory bulb. This is a synapse between mitral cell lateral dendrites and granule cell dendrites. Depolarization of a mitral cell releases glutamate, which excites the postsynaptic granule cell. This cell releases GABA at the same synapse and inhibits the mitral cell in a reciprocal fashion.

►Olfactory Bulb

Reciprocal Inhibition

Definition

A pattern of synaptic connection between neurons or groups of neurons where they each make an inhibitory synapse on the other neuron or group of neurons in a mutual or reciprocal fashion. This pattern of connectivity assures reciprocity/alternation in the activity of the two neurons/groups. The term also refers to inhibition of antagonist motoneurones when the agonist motoneurones are activated either as part of a stretch reflex or as part of a voluntary movement. Ia inhibitory interneurones play an important role in ensuring reciprocal inhibition, but other mechanisms also contribute.

►Integration of Spinal Reflexes
►Intersegmental Coordination
►Reciprocal Activation

Recognition

Definition

Recognition is the act during which a specific item or event is identified as having been experienced or encountered on a previous occasion. Variants of items or events that are not exactly like those previously experience can also be recognized by generalization or inference.

►Recognition Memory

Recognition Memory

Luke Woloszyn, David Sheinberg
Department of Neuroscience, Brown University, Providence, RI, USA

Synonyms

Familiarity; Recollection; Recency; Declarative memory; Explicit memory; Old/new recognition

Definition

►Recognition memory refers to the ability of a system to classify a specific item or event as having been experienced or encountered on a previous occasion. Behavioral, neuropsychological, electrophysiological, and neuroimaging evidence indicate that recognition memory may be dissociable into the distinct processes of ►familiarity and recollection.

Characteristics

The ability to form, retain and manipulate memories is necessary for an organism to adapt to its environment. The broad concept of memory is generally divided into procedural and declarative branches. Whereas procedural, or implicit, memory underlies the unconscious learning of motor, perceptual and habitual tasks, declarative, or explicit, memory refers to the conscious memory for facts and events. Declarative memory has been regarded as critical for providing temporal contiguity to conscious experience. Subsumed under the umbrella of declarative memory is recognition memory – the psychological ability to judge a particular item or event as having been previously encountered in the past. A hallmark of recognition memory is that it can both lead to a distinct feeling of oldness, or familiarity, as well as evoke vivid spatiotemporal details associated with the prior experience of the item or event. For example, imagine having a conversation with a stranger on a subway and then seeing that same stranger in a different setting a month later. Not only will you most likely have a strong sense of recognizing this person as familiar, but you also might be able to ►recall the specific train your were on during your previous encounter. Most current theoretical models maintain that there is qualitative distinction between recognition memory that is or is not complemented by contextual recall [1].

Such dual-process models label recognition unembellished with episodic content as "familiarity" and recognition with accompanying contextual particulars as "recollection." This view can be contrasted with the less common single-process models which posit that the difference between familiarity and recollection is purely quantitative [2]. In this latter framework, familiarity and recollection vary along a continuum that captures how much background information is recalled at the time of recognition, with recollection obviously summoning more details than familiarity. Dual-process models also posit that memory traces can vary in strength, especially within the domain of familiarity, but there is a qualitative, as opposed to a quantitative, leap when one transitions from familiarity to recollection. Although dual-process models dominate the literature, and as we will show are supported by a wealth of converging evidence, the parsimony of single-process models makes them a viable alternative.

Several different experimental paradigms have been employed to independently assess the contribution of familiarity and recollection to recognition memory [3]. The remember/know procedure asks subjects to first study a list of words or pictures and then identify, within a list including both old and new items, those items which have been previously studied. If an item is judged as old, the subject further indicates whether he or she simply "knows" (K) this or can specifically "remember" (R) studying it (Fig. 1). Although the K judgment presumably reflects familiarity and the R judgment is analogous to recollection, it is difficult to tease these components apart because it is often the case that R judgments are preceded by implicit K judgments. Adding to the confound is the observation that the proportion of K and R judgments can be biased by manipulating the subjects' decision criteria. A similar procedure aimed at differentiating familiarity from recollection asks subjects to rate the confidence of their old judgments, with higher confidence ratings indicating recollection. Another common paradigm used to explore familiarity and recollection is a source judgment task. In this setup, subjects again encode items that the experimenter has specifically embedded in a specific context (e.g., a particular quadrant of the screen). In the test phase, subjects may be required to not only signal whether an item is new, but to also report some aspect of the context in which the item originally appeared. In this paradigm, the proportion of trials with incorrect or correct source judgments can provide indices of familiarity or recollection, respectively.

Through the application of the above-mentioned and other experimental manipulations, an emerging consensus is that familiarity and recollection are, to a certain degree, functionally distinct. So, for example, it has been shown that recollection benefits more than does familiarity from undivided attention at the time of encoding. Similarily recollection improves when subjects are given the opportunity to encode the initial list of items in more depth, which happens, for example, when subjects generate words instead of passively reading them. Conversely, familiarity judgements are

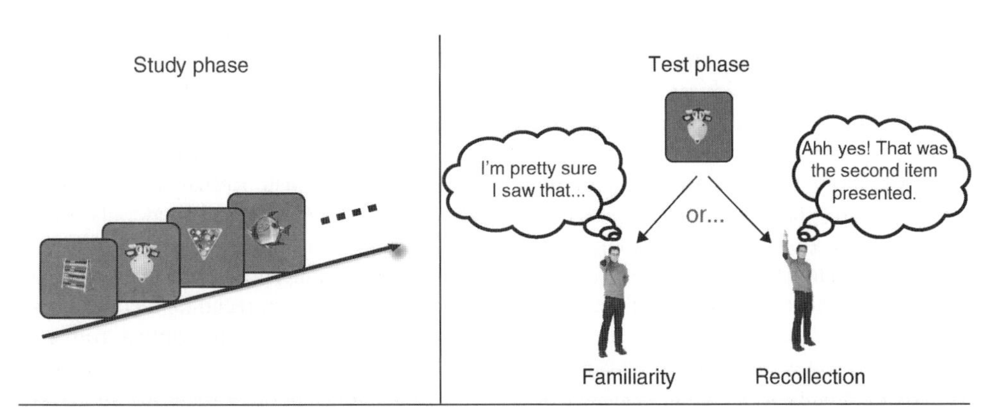

Recognition Memory. Figure 1 A subject first encodes a set of words or pictures. Following a delay interval that is usually on the order of several minutes, subjects are presented with a list of words or pictures that include the initially studied items as well as new items. The subject's task is to indicate which items are old and which items are new. If the subject responds old, he or she is asked to elaborate if the old judgment is simply due to some intuitive sense of "knowing" or if it is due to "remembering" specific details associated with the original encoding of the item in question. A "know" response presumably reflects a familiarity-based judgment whereas a "remember" response reveals a recollection-based judgment.

R

more common when duration of encoding is limited. In addition, priming has been shown to have an effect on familiarity but not on recollection. While a full discourse on the relevant findings dissociating the two types of recognition memory is beyond the scope of this article, it is generally assumed that familiarity operates automatically and quickly whereas recollection requires voluntary control and time.

Given that these distinct cognitive phenomena must have neural correlates, a logical question is whether differences between familiarity and recollection are evident at the electrophysiological level. Event-related potentials (ERP) are one means to non-invasively measure the stimulus-locked neural activity of human subjects. Indeed, early ERP studies found that recognized items, at test, generated more positive ERPs than to unrecognized items. A series of studies then demonstrated a double dissociation between familiarity- and recollection-driven ERPs [1,4,5]. Specifically, recognition judgments based on familiarity elicit ERP effects that are more localized to frontal electrode sites and have a relatively short latency, starting around 300 ms and continuing for another 100–200 ms. Not surprisingly, frontally localized ERPs corresponding to familiar items have more positive amplitudes than the ERPs of correctly rejected new items. The hypothesized electrophysiological correlates of recollection, on the other hand, appears more localized to parietal sites. It, like the early frontal effect, is more positive for correctly recognized items than for correctly rejected items but is slower to develop, usually beginning at 400 ms post-stimulus onset and persisting for another 300–400 ms. The differential latencies of the ERPs parallel the behavioral observations that familiarity is more instinctive and thus quicker than the more effortful and slower recollection. Compellingly, old items receiving incorrect "new" classifications do not elicit the early frontal effect whereas false alarms do. This suggests that familiarity allows as to perceive oldness.

ERPs have relatively poor spatial resolution, which does not allow for a finer localization of the neural substrates of recognition. To overcome this particular limitation, investigators have turned to functional magnetic resonance imaging (fMRI). Making use of behavioral paradigms similar to the ones described above, several groups have started to delineate circumscribed areas of the brain responsible for familiarity and recollection [3,6]. These results have revealed distinctions between familiarity and recollection that parallel those obtained with ERPs. Familiarity has been shown to cause activity in a network of areas, including various frontal/prefrontal regions, medial temporal lobe (MTL) structures excluding the hippocampus, and the superior parietal lobule. The most convincing evidence centers on the MTL structures, which include the perirhinal, parahippocampal and entorhinal cortices. These structures not only exhibit reduced responses to old items, but this reduction correlates with the strength of the familiarity judgment, suggesting a neural substrate for the continuously scaled familiarity signal. Recollection-based judgments also tend also to activate several brain regions, most notably the hippocampus and a left lateral parietal cortex. These observations are consistent with the widely held belief that the hippocampus is critical for the recall of associations, which presumably constitutes a large component of the episodic retrieval inherent to recollection. Crucially for dual-process models, activity in the hippocampus does not increase for familiarity judgments. Intriguingly, the activity in the left lateral parietal region has been shown to increase even when items are mistakenly classified as old, again lending support to the existence of a brain network responsible for the subjective experience of oldness.

Another localization approach, complementing the fMRI studies, has been to examine patient groups with focal lesions. Because the hippocampus is particularly susceptible to hypoxia, patients suffering from lesions of the hippocampus are relatively easy to find. Investigations of these patients have revealed mixed results in that whereas some reveal significant impairment in tasks requiring recollection, others show equivalent deficits for both familiarity and recollection [3,7]. This has led some to postulate that while there may be important distinctions between the two kinds of recognition memory, all the structures within the MTL act as one integrated unit, which in turn contributes to both familiarity and recollection.

Notwithstanding the inconclusiveness of the human results, more controlled lesion studies in nonhuman primates have revealed that some division of labor is present within the classically defined MTL structures. Because we cannot ask monkeys to introspect about their recognition processes, it is difficult to tap into their different types of recognition memory; nevertheless, ►delayed match-to-sample (DMS) and ►delayed non-match-to-sample (DNMS) paradigms have been extensively used to explore familiarity judgments. The basic format for this task requires the animal to encode an object (generally, but not always, by looking at it), to retain its identity throughout a delay interval, which can range from seconds to minutes or longer, and to then recall its identity by selecting the matching (or non-matching) object from a choice array containing two or more possible alternatives.

A meta-analysis of primate lesion studies, with the lesions restricted either to the hippocampus or to the surrounding perirhinal cortex, showed ablation of the perirhinal cortex impairs DNMS performance more than lesions of the hippocampus. Additionally, a correlation analysis showed that the magnitude of the DNMS performance deficit scaled positively with the extent of the perirhinal damage. While hippocampal damage also had a detrimental effect on the DNMS task, the size of the hippocampal lesion was negatively correlated with performance deficit, with the deficit becoming essentially negligible for monkeys totally lacking the hippocampus. This supports the view that the surrounding MTL structures play a more vital role in familiarity judgments than does the hippocampus itself.

The lesion studies described above are consistent with single cell recordings performed in the perirhinal cortex of nonhuman primates [8]. The monkeys in these experiments performed a serial recognition task in which the goal was to press one button if the stimulus was familiar (having been seen repeatedly on a daily basis or just recently presented) and to press another button if the stimulus was novel. Three distinct classes of neurons have been reported. ►Recency neurons decrease their response when a displayed stimulus is one that was seen a few trials back, regardless of whether it is highly familiar or novel for that session. Familiarity neurons exhibit a reduced response to stimuli which are familiar to the monkey. Finally, novelty neurons show a marked enhanced response to the first presentation of a novel stimulus, with subsequent presentations of it or other familiar stimuli evoking much lower or shorter responses. Remarkably, some of these neurons appear to have memory spans of up to 24 h, providing persuasive evidence that they are involved in the extraordinary capacity of primates to remember stimuli for long periods of time, even following single exposures. Interestingly, the general tendency of these neurons to decrease their response to known stimuli mirrors results of fMRI studies in which the strength of the familiarity signal determines the amount of activity decrease observed in MTL structures. Additional studies have shown that the response of PFC neurons to familiar items is less affected by noise manipulations, suggesting that these items are more accurately and efficiently represented.

Recently, neurophysiologists have begun describing in more detail the properties of the neural activity underlying familiarity in primates. ERP studies have shown results similar to those obtained in humans, namely that the magnitude of the ERP differentiates between novel and familiar stimuli, with familiar stimuli eliciting a more positive ERP. These differences are present even in simple fixation conditions, which speaks to the fundamental contribution of familiarity to everyday behavior. Furthermore, as the monkey becomes more familiar with the novel set, the ERPs to the two sets of images become systematically more similar. Familiar and novel items also produce differences in the temporal dynamics of single cell responses. Specifically, both elicit a similar initial peak in the firing rate (~100 ms), but whereas the familiar response then quickly declines, the response to the novel images persists for an extended period of time. This suggests, again, that familiar items may be encoded more efficiently than novel ones, and that the prolonged response observed for novel images might contribute to the creation of new memories.

Although recollection is difficult to study in nonverbal animals, preliminary evidence does indicate that the hippocampus is essential [9,10]. Numerous studies have shown that removing the hippocampus in rats impairs their ability to use relational knowledge, particularly with regard to spatial relationships, which has been interpreted as reflecting deficits in associative recollection. Single cell recordings in monkeys have also demonstrated that the firing pattern of hippocampal neurons correlates with the simultaneous acquisition and immediate expression of arbitrary stimuli associations. It should be pointed out, however, that the long-term storage of associations is thought to rely on the surrounding MTL structures, particularly the perirhinal cortex. Hence, a precise and accurate account of the specific neural substrates of recollection will require more time and research.

In sum, there is strong evidence suggesting a dichotomy between familiarity- and recollection-based recognition memory. The precise nature of these differences is difficult to characterize given that the two processes interact extensively at the cognitive level; however, neurophysiological data obtained from both humans and primates has begun to elucidate possible neural substrates and processes that contribute to this distinction.

References

1. Rugg MD, Curran T (2007) Event-related potentials and recognition memory. Trends Cogn Sci 11(6):251–257
2. Donaldson W (1996) The role of decision processes in remembering and knowing. Mem Cognit 24(4):523–533
3. Rugg MD, Yonelinas AP (2003) Human recognition memory: a cognitive neuroscience perspective. Trends Cogn Sci 7(7):313–319
4. Jager T, Mecklinger A, Kipp KH (2006) Intra- and inter-item associations doubly dissociate the electrophysiological correlates of familiarity and recollection. Neuron 52(3):535–545
5. Wheeler ME, Buckner RL (2003) Functional dissociation among components of remembering: control, perceived oldness, and content. J Neurosci 23(9):3869–3880
6. Gonsalves BD et al. (2005) Memory strength and repetition suppression: multimodal imaging of medial temporal cortical contributions to recognition. Neuron 47 (5):751–761
7. Squire LR, Wixted JT, Clark RE (2007) Recognition memory and the medial temporal lobe: a new perspective. Nat Rev Neurosci 8(11):872–883
8. Brown MW, Aggleton JP (2001) Recognition memory: what are the roles of the perirhinal cortex and hippocampus? Nat Rev Neurosci 2(1):51–61
9. Wirth S et al. (2003) Single neurons in the monkey hippocampus and learning of new associations. Science 300(5625):1578–1581
10. Eichenbaum H (2000) A cortical-hippocampal system for declarative memory. Nat Rev Neurosci 1(1):41–50

Recognition Neurons

Definition

Neurons that are able to signal by their response that a specific, generally complex stimulus has been recognized. Face neurons in the inferior temporal cortex are an example for recognition neurons (see also Grandmother Neuron).

Recollection

►Recognition Memory

Recruitment

Definition

Abnormal increase in perceived loudness.

►Hearing Aids

Recruitment in Acoustics

Definition

Abnormal increase in perceived loudness.

►Hearing Aids

Recruitment of Motor Units

Definition

Central nervous system control of force output from a muscle by regulating the numbers and identities of motor units that are activated or de-activated (de-recruitment) during a movement.

►Motor Units

Rectifying Gap Junctions

Definition

Rectifying gap junctions conduct ionic current better in one direction than in the other. In contrast to electrical synapses comprised of non-rectifying gap junctions, electrical synapses comprised of rectifying gap junctions transmit electrical signals in a unidirectional fashion.

►Electrical Synapses

Recurrent Brief Depression

Definition

Depressive episodes lasting at least 2 days but less than 2 weeks occurring at least once a month for 12 consecutive months. They often occur unpredictably but frequently. An average of two attacks a month is typical. Although the episodes are brief, symptoms are severe. Intense suicidal ideation is common.

►Major Depressive Disorder

Recurrent Facilitation

Definition

►Recurrent Inhibition

Recurrent Hypersomnia

Definition

Also known as Kleine-Levin syndrome, this disorder is characterized by periods of excessive sleepiness and long sleep times lasting days to weeks. It is most commonly found in adolescent males. During symptomatic periods, there is often increased food intake as well as cognitive and emotional dysfunction.

►Sleep – Developmental Changes

Recurrent Inhibition

Definition

Recurrent inhibition is a basic type of neuronal circuit throughout the central nervous system. In the spinal cord, motoneurons give off axon collaterals to excite GABAergic/glycinergic Renshaw cells, which mediate recurrent inhibition of motoneurons, Ia inhibitory interneurons, Renshaw cells and some cells of origin of the ventral spinocerebellar tract. Since the Renshaw cells also inhibit Ia inhibitory interneurons, which inhibit antagonistic motoneurons, activation of the Renshaw cells leads to removal of the inhibition of the antagonist motoneurons, and this phenomenon is termed recurrent facilitation.

►Ia Inhibitory Interneuron
►Integration of Spinal Reflexes
►Renshaw Cell

Recurrent Network

Definition

A neural network architecture in which both feedforward and feed-back connections between neurons are present. In a fully recurrent architecture, the neurons are fully interconnected.

►Neural Networks

Recurrent Processing

Definition

Information is processed in a set of stages in which activation can propagate in feedback loops.

Red Nucleus

Definition

The red nucleus consists of two functionally entire separate divisions, the caudal magnocellular red nucleus and the more rostral parvocellular red nucleus. The magnocellular red nucleus is a large spherical nucleus in the midbrain containing very large neurons. It is the level of the oculomotor nucleus and the superior colliculus. It receives a very large input from the cerebellum by way of the superior cerebellar peduncle and also a small descending input from motor areas of the cerebral cortex. It projects to the contralateral facial and trigeminal nuclei in the brainstem and to the contralateral spinal cord, primarily to the cervical and lumbar enlargements that control the limbs. In fresh sections the magnocellular division of the red nucleus appears red or pinkish-yellow because of its marked vascularization, hence its name.

The parvocellular red nucleus, just rostral to the magnocellular division, consists of smaller neurons. In lower animals it is smaller than the magnocellular division but in primates, it is much larger. The parvocellular red nucleus receives a large input from the motor regions of the cerebral cortex and projects to the principle division of the ipsilateral inferior olive.

In summary, the magnocellular red nucleus is a premotor nucleus for muscles in the contralateral head and limb while the parvocellular red nucleus is a relay between the cerebral cortex and the inferior olive.

Reductionism (Anti-Reductionism, Reductive Explanation)

Robert C. Richardson[1], Achim Stephan[2]
[1]Department of Philosophy, University of Cincinnati, Cincinnati, OH, USA
[2]Institute of Cognitive Science, University of Osnabrück, Lower Saxony, Germany

Definition

In philosophy of science and in the sciences, reductionism is sometimes a methodological stance; sometimes it is a substantive position. As a methodological stance, it is committed to understanding a system's behavior analytically, i.e., in terms of the system parts and their interactions. As a substantive position, it anticipates the success of the reductionist methodology. Reduction sometimes is thought of in ontological terms; that is, as a commitment to the idea that, e.g., living systems consist of nothing other than physical constituents. This minimal commitment may be coupled with a claim that reduction holds *in principle*, even if not in practice, but does not require even this much. Reduction sometimes is an explanatory relation, either between theories or between domains. The former is *theory reduction* while the latter is *reductive or mechanistic explanation*.

Description of the Theory

Reductionism has been a persistent attractor for scientific thought. In physics, it led to some of the most striking successes of classical physics, such as the particle theory of light in the eighteenth century or ►statistical mechanics in the nineteenth century. In the biological sciences, the rapid growth of molecular genetics is a reductionistic triumph. In the ►cognitive sciences, reductionism has been no less attractive, as is evidenced by the rise of what is called "cognitive neuroscience" in the twentieth century, or the dramatic successes of the neurosciences more generally. Opposition to reductionist approaches also has been prominent, especially in cognitive psychology. The successes of cognitive psychology and of ►artificial intelligence in the middle decades of the twentieth century encouraged the idea that reductionist methods were not necessary in order to explain human psychological capacities. In opposition to rather brute reductionistic claims such as that there is nothing but elementary particles, or that mental states are nothing but brain processes, champions of the special sciences feared they would lose their particular subject matter. Vitalistic or dualistic alternatives are not fashionable any more, but there are a number of antireductionist positions that have been recently defended.

Varieties of Reduction: Theory reduction

Theory reduction claims to deduce or "explain" one theory (the secondary or reduced theory) from another (the primary or reducing theory), perhaps as a limiting case. Theory reduction may involve theories at the same level of organization or at different levels of organization. A classical example for same level reduction, as discussed by Ernest Nagel [1], is the reduction of Newton's mechanics to Einstein's relativity theory; for reduction of theories at different levels Nagel refers to the relation of classical ►thermodynamics to statistical mechanics. The reducing theory is typically thought to be more general or more exact, or both. The reduced theory correspondingly is thought of as more restricted in its domain of application, or as an approximation, by comparison with the reducing theory. Thus, Newton's mechanics applies only to velocities far from the velocity of light. The classical gas laws in thermodynamics apply only to gases at intermediate temperatures and pressures. When reducing and reduced theories are on the same level, or cover the same domains, we have "homogeneous" reductions. If reduction is possible at all, it can be achieved relatively easily, because the theories at least appear to have the same concepts. For example, both Newton and Einstein appeal to *mass* and *velocity*, although they in fact may not be the same concepts (insofar as the theory defines its concepts). When reducing and reduced theories refer to different levels, or cover different domains, we have "heterogeneous" reductions. If reduction is possible at all, it is more difficult in these cases, since the theories do not share the same vocabulary. For example, statistical mechanics has neither the notion of temperature nor that of pressure, though these are the key concepts of classical thermodynamics. So, in these cases, connections between the two domains or levels have to be forged, in the form of "►bridge laws." These connections are usually thought to be identifications between levels; e.g., mean kinetic energy is identified with temperature, at least within the domain of classical thermodynamics.

These identifications often lead to so-called "nothing buttery," i.e., to the claim that the phenomena the upper level concepts originally referred to in fact are *nothing but* the entities the lower level concepts refer to. So, it is often said, temperature is *nothing but* mean molecular motion, and genes are *nothing but* nucleotide sequences. This sometimes leads to eliminativist claims concerning the reduced theory – that is, to denying the existence of things reduced [2]. In case there are suitable identifications available, reduction guarantees that the explanatory work done by the higher level theory can, at least in principle, be done by the lower level theory. The higher level theory can then, in principle, be eliminated without explanatory loss. In case the concepts of upper level theory are vaguer or more inexact, they cannot be identified with the more exact concepts of a lower level theory. Once again the original theory can be eliminated though in this case because it could *not* be reduced [3] ("phlogistaded").

To take an example closer to the neurosciences, the trichromatic theory was an important case historically. Newton's experiments with prisms revealed a visible spectrum ordered by wavelength. It was subsequently shown that any specific spectral hue could be matched by combining three different primary colors in different intensities. This led Thomas Young to propose, in 1801, that the retina contained exactly three different color-sensitive elements. This was subsequently confirmed and consolidated by the great physiologist, Hermann Helmholtz, with three classes of cones differing in their central sensitivities, though with substantial overlap of responses. This looked like a reduction of color theory to a trichromatic theory, which had broad acceptance into the middle of the last century. Unfortunately, not all the colors perceived by human subjects are represented in the physical spectrum. *Brown* is the most salient example of a color that is outside the physical spectrum. Hering eventually posited an "opponent process" theory which was aimed at more exactly describing the subjective phenomena of color perception. The existence of opponent processes at the neural level was eventually confirmed in the macaque. We still lack a decisive direct test, but this sort of case also is suggestive of theory reduction.

To consider a more contemporary example, John Bickle [2] argues that contemporary neuroscience captures the phenomena on offer from psychology concerning memory. In particular, he observes that Eric Kandel's landmark work on ►aplysia has forged a link between the mechanisms of long term potentiation (►LTP) and ►memory consolidation. Memory consolidation is particularly important as the link between short-term and long-term memory. The key psychological phenomena include the importance of stimulus repetition, the time course relevant to fixation, and ►retrograde interference. The mechanisms, typically molecular, which characterize LTP capture these phenomena, at least to a first approximation. Bickle concludes that the "intended empirical applications of the two theories are virtually identical," even if the theories differ in substantial ways.

Varieties of Reduction: Reductive or Mechanical Explanation

Reductive or mechanical explanations tell us *why* a certain entity instantiates a certain property, typically a property that is only attributed to the system as a whole [4,5]. So we might want to explain, e.g., how humans recognize faces, or how we acquire language. In trivial cases it suffices to simply add the corresponding properties of the system's components. So we get the weight of brains from the weight of its parts. However, the behavior of the brain in vision or perception – both definitely more interesting properties – cannot be deduced in an equally simple manner. Here we need to know the arrangement of the components, particularly those of the visual areas, what properties they have, and how they interact among each other and with the visual stimuli. The capacity to see would be reductively explained if it followed from the above-mentioned factors and the natural laws that hold generally. Of course, at the moment, we do not yet have such a reduction in place. That need not trouble reductionists, insofar as they are committed to what we will or (perhaps) could achieve.

So, the aim of each reductive explanation is to explain (or predict) a system's dispositions and properties solely by reference to its components, their properties, arrangement, and interactions. To be successful as a reductive or mechanistic explanation, several conditions must be met:

i. There must be a systemic property to be explained
ii. The property to be reduced must be functionally construable or reconstruable
iii. The specified functional role must be filled by the system's parts and their mutual interactions
iv. The behavior of the system's parts must follow from the behavior they show in isolation or in simpler systems than the system in question

These are demanding constraints, but without them no reductive explanation will be complete. What we are looking for, then, are functional characterizations of the properties to be reduced, and explanations of those functional properties in more fundamental terms. Usually we refer to these properties by concepts that classify properties at the system level, where specific patterns such as, e.g., learning a new person's face brings the instantiation of a capacity to our notice. The time course of memory consolidation is one such functional property. The functional analysis is a precondition for constructing appropriate explanatory connections between the components and systemic behavior. If, however, this conceptual "priming procedure" fails, as it may in the case of ►qualia, the corresponding reductive explanation fails, too [6].

Reductive or mechanistic explanations can be forged in two opposite directions. Given that we already know that some system S has a systemic property P one task is to provide a reductive explanation for P. For that we refer to the microstructure MS(S) of S, to the behavior of the components, and to the interaction among the components C_i of S. With these resources, we try to show that S must have P *given the analysis of its structure*. So, in the case described above, it was important to the trichromatic theory that humans had exactly three color sensitive cones, and that these had appropriate spectral sensitivities. We also knew that humans could, typically, discriminate among spectral hues. So there was a systemic property (human discriminatory abilities) and there were microstructural features (three types of cones) that were appropriate. All this depended on adequate conceptual preparations. That is, we needed to know the range of discriminable colors. When it turned out that the systemic properties were different from those predicted by the trichromatic theory, that required a change in the understanding of the physiology.

In converse cases, we might at first only know the microstructure MS(S) of a system S and be uncertain concerning its exact capacities. The task, then, is to verify theoretically, or to forecast whether or not S has (a desired) systemic property P. To do so, we again must make use of adequate conceptual preparations and refer to the microstructure MS(S) of S, to the capacities and interactions of the components C_i of S. This is a difficult procedure, since the combinatorial possibilities are daunting. Experimental procedures that follow from imposing deficits, whether experimental or accidental, fall into this category. Ablation studies in nineteenth century physiology follow this simple approach, though there are more sophisticated methods (e.g., knock-out

R

genes) available to recent physiologists. So when Broca discovered a patient (Tan) that lacked the ability to produce coherent speech but who had normal comprehension, and who had damage to the left temporal lobe, he concluded that the temporal lobe was the location of articulate speech. Many fMRI experiments still follow this research strategy. They do not provide reductive explanations.

Of course, a system can be looked at from a "top down" perspective, a "bottom up" perspective, or from both simultaneously. In that case, we can stitch together the perspectives we garner from the bottom and the top.

Anti-Reductionistic Positions

If it turns out that some purported phenomenon is not real, the property to be reduced is not accepted. In that case, condition (i) fails, and reductive explanation fails right at the outset. If telepathy is not real, there is nothing to explain or reduce. At the extreme, this becomes ▶eliminative materialism, in which all psychological phenomena are rejected as real. Paul and Patricia Churchland [3], followed by John Bickle [2], defend largely eliminative positions, supported by an emphasis on theory reduction. If it turns out that one cannot functionalize a psychological property accepted for reductive explanation, even in principle, then condition (i) is satisfied, while (ii) is rejected. In this case, the systemic property is real but *irreducible*, and thereby it is *synchronically* (i.e., in a strong sense) *emergent*. Strong emergentism (▶Emergence) is at least a type of property dualism [7]. The case in point is qualia. Commonly, it is held that, e.g., our sensation of red has an intrinsic character that cannot be captured in terms of function; and if this is so, then (ii) fails in this case [6]. If it turns out that the system components with their particular arrangement and interactions are not sufficient to account for the system's behavior, then there is a failure of condition (iii). Again, a mechanistic explanation fails. If this is a consequence of limitations on our knowledge, then this is not a principled limitation. If it is a principled limitation, as the great neurophysiologists Sherrington and Eccles [8] maintained, then that is a failure of the reductionist/mechanist program. We would be driven to substantive dualism, ▶vitalism or Cartesianism (▶Cartesian dualism). If the behavior of components, embedded in the appropriate context, is sufficient to explain the system behavior, but the behavior of the components in simpler constellations cannot account for their interactions within the more complex system, then condition (iv) fails, In that case, we are forced to a kind of ▶holism; which establishes one type of emergentism. Irreducibility can coexist with mechanistic explanations [9].

Theory reduction also has its detractors, who usually contend that there will be a failure to provide appropriate bridge laws to connect the theories in heterogeneous reductions. Functionalists such as Jerry Fodor [10] and hold that psychological states are functional states (▶Functionalism), which can be realized in multiple ways, and conclude that therefore there will not be appropriate identities available. They conclude that this supports a kind of autonomy for psychological explanations relative to physiological explanations. A more radical failure would be the absence of sufficient physical conditions to explain psychological capacities. In this case, we would be driven again to either dualism or eliminativism.

References

1. Nagel E (1961) The structure of science: problems in the logic of scientific explanation. Routledge & Kegan Paul, London
2. Bickle J (2003) Philosophy and neuroscience: a ruthlessly reductive account. Kluwer, Dordrecht
3. Churchland PS (1986) Neurophilosophy: towards a unified understanding of the mind-brain. MIT, Cambridge
4. Bechtel W, Richardson RC (1993) Discovering complexity: decomposition and localization as strategies in scientific research. Princeton University Press, Princeton
5. Craver C, Darden L (2001) Discovering mechanisms in neurobiology: the case of spatial memory. In: Machamer PK, Grush R, McLaughlin P (eds) Theory and method in neuroscience. University of Pittsburgh Press, Pittsburgh
6. Stephan A (2001) Phänomenaler Pessimismus. In: Pauen M, Stephan A (eds) Phänomenales Bewusstsein – Rückkehr zur Identitätstheorie? Paderborn, mentis, pp 342–363
7. Stephan A (1999) Emergenz. Von der Unvorhersagbarkeit zur Selbstorganisation. 2nd printing, mentis, Paderborn 2005
8. Popper K, Eccles J (1977) The self and its brain: an argument for interactionism. Springer, Berlin Heidelberg New York
9. Boogerd FC, Bruggeman FJ, Richardson RC, Stephan A, Westerhoff HV (2005) Emergence and its place in nature: a case study of biochemical networks. Synthese 145:131–164
10. Fodor J (1975) The language of thought. Thomas Crowell, New York

Redundancy of Degrees of Freedom

Definition

An excess of elemental variables (degrees –of freedom) within a system as compared to a number of constraints imposed on the system in typical tasks; this term assumes that redundant elemental variables need to be eliminated to make the system controllable.

▶Coordination
▶Degrees of Freedom

Redundant Set, Redundancy Problem

Definition

Also called Bernstein's problem, named after the scientist who has considered it as a major problem in motor control research; ambiguity in transforming a set of variables into a set of more numerous variables so that a unique solution of many possible solutions of the motor task must be chosen, for example, when it is necessary to distribute total torque acting on a joint into individual torques of muscles spanning this joint, or to choose one of many possible ways of combining different degrees of freedom of the body to reach the motor goal. See also the principle of neurological minimization.

►Equilibrium Point Control

Reference

Definition

Reference is the relation between an expression and what it refers to. A typical unambiguous concrete singular term like "the highest mountain" or "Gottlob Frege" refers to a certain concrete particular like a thing or a person. A [→] predicate like "is red" distributively refers to all those particulars to which it applies, i.e. to the red things. Sometimes the relation between a predicate and the set of things it applies to is also called "reference."

►Argument
►Logic

Reference Frame

Definition

►Sensory Systems

Reference Model

Definition

The Reference Model is used for a certain type of control design. It is a mathematical model that represents the desired behavior of the controlled physical system.

►Adaptive Control

Referent Configuration

Definition

A position of the body or its segments at which muscles are silent in the absence of co-activation or, otherwise, produce net zero joint torques but generate activity and resistive forces in response to deviations from it; modified by control levels to produce motor actions.

►Equilibrium Point Control

Referent Trajectory of an Effector

Definition

Comprised of the positions of the effector's associated with threshold configurations of the body at each instant of movement (see Threshold control).

►Equilibrium Point Control

Referential

Definition

Pointing to the meaning of an utterance.

►Cognitive Elements in Animal Behavior

Referred Pain

Definition

Referred pain is the phenomenon wherein nociceptive stimulation in one location results in the perception of pain in another location. In clinical practice, this phenomenon is most often thought of as involving the projection of pain from a visceral structure to the body surface. However, nociceptive stimulation of muscle, and possibly other somatic tissues, can also lead to referred pain. A number of mechanisms have been proposed to explain referred pain. These include, most importantly, the convergence of afferent neurons from the site of insult and the site of perceived pain. This may occur through the dichotomization of afferent fibers such that one branch of an axon terminates on, for example, a visceral structure, while another branch of the same axon terminates in the skin. Alternatively, two distinct peripheral sensory neurons may both terminate on the same dorsal horn neuron. In either instance, it is proposed that the brain would have difficulty in determining the true source of nociceptive input, and would preferentially attribute pain to the more familiar source of sensory input – hence the body surface.

►Ascending Nociceptive Pathways
►Somatosensory Projections to the Central Nervous System

Reflex

Definition

Involuntary modification of activity in motoneurons in response to activation of sensory receptors.

►Motor Control
►Feedback Control of Movement

Reflex Adaptation

Definition

One of the simplest forms of motor learning. Inborn reflexes, such as an eyeblink reflex, are adaptable to new environmental conditions. For example, the force required for eyelid closure self-adjusts when the eyelid movement is impeded by an external load.

►Motor Learning

Reflex Chain

Definition

A sequence of reflexes where the action of the first reflex activates a set of receptors (e.g. proprioreceptors) that trigger a second reflex, and so on. The peripheral control hypothesis proposed that complex behaviours consist of simple reflexes that are linked together in a reflex chain. This hypothesis has been superseded by the central control hypothesis.

►Central Pattern Generator

Reflex Sympathetic Dystrophy (RSD)

Definition

►Complex Regional Pain Syndromes (CRPS)

Reflexes

Arthur Prochazka
Centre for Neuroscience, University of Alberta, Edmonton, AB, Canada

Definition

(Taken from Dr. Wilfrid Jänig's essay on ►Autonomic reflexes) Reflexes are functionally defined by an efferent (motor) output system that generates a distinct effector response when activated and by the population of afferent neurons stimulated. Reflexes are fragments of more complex somatomotor behaviors and are used in the laboratory as tools to study experimentally the central organization of neural regulation of movement.

Background

There is no universally accepted definition of reflexes that distinguishes them from voluntary responses to stimuli [1]. The Roman poet Ovid used the word *reflex* to describe "turning or bringing back." Substitute "feed" for "bring" and we have the word "feedback." In engineering, feedback refers to information about a process monitored by sensors and supplied to the controller of that process (see ►Feedback control of movement).

In biology, the first attempt at a definition of reflex is attributable to Georgiy Prochaska [2]: a behavior in response to an excitation, mediated by separate motor and sensory nerves. Prochaska saw the function of reflexes to be *conservation of the individual,* later called *homeostasis* by Claude Bernard [3]. The psychologist Herbert Spencer posited that reflexes were the atoms of the psyche; that the psyche was an assemblage of reflexes, and that instincts were reflex assemblies consolidated by repetition and transmitted in an hereditary manner [4]. The Russian clinical physiologist Ivan Sechenov went one step further, proposing that all motor acts in humans as well as animals were simply chains of elemental reflexes [5]. He argued that the appearance of spontaneity and volition was illusory and that all movements were in principle predicted by the history of prior events, sensory inputs and associated thoughts. His conception of reflexes included complex responses that involved choice, as well as learned responses that his successor Pavlov would later call conditioned reflexes. The ideas of Spencer and Sechenov were taken to their literal conclusion in the behaviorist theories of Watson and Skinner. These theories rejected all non-measurable explanations of behavior and replaced voluntary movement with operants: conscious arbitrary acts that have become associated with arbitrary stimuli through learning and arbitrary reinforcement.

Hughlings Jackson argued from clinical observations that movements ranged in a continuum from the most automatic or evolutionarily primitive to the least automatic or most evolutionarily advanced [6]. Primitive reflexes in humans were unmasked or released when the higher centers were damaged. The Jacksonian continuum from automatic to voluntary probably best encapsulates the current view of most neuroscientists. In this view, reflexes are brief, automatic and invariant responses to stimuli. But even this definition is problematic: Goldstein (1939) reviewed various responses called reflexes and found them all to be variable, state-dependent and mutable [1].

Jonathan Wolpaw recently argued that a single comprehensive hypothesis related to this question of distinguishing reflexes from voluntary actions developed in the nineteenth century, namely that the whole function of the nervous system is to convert sensory input into appropriate motor output [1]. Neuroscientists who say they are studying reflex behaviors are studying behaviors in which the connections from stimulus to response, from experience to behavior, are known to be, or at least believed to be, short and simple enough to be accessible to description with presently available methods, and they are excluding by one means or another voluntary behaviors, or behaviors involving connections so long and complex as to defy present-day analysis. Implicit in these definitions is the expectation that, as methodology and understanding advance, the class of reflex behaviors will grow larger and larger and the class of voluntary behaviors smaller and smaller.

When reading the essays on specific reflexes in this *Encyclopedia* it is useful to bear in mind the following comments of Francois Clarac [1]:

1. In normal behavior, reflexes are simple, fast reactions to the environment. The term should be confined to the simplest input-output reactions mediated by monosynaptic (or oligosynaptic) pathways at the lowest level, i.e. at the motoneuronal level. Reflexes should be viewed as elements of feedback control that each species possesses to react automatically to the environment.
2. The experimenter can induce a reflex artificially.... Reflexes then reduce to informative tests of CNS state. A reflex might be seen as a physiological "scalpel" permitting entrance into simple workings of the CNS, while not being a distinct and separable element when normal movements are considered. Thus although the understanding of motor behavior has benefited from reflex experiments, the normal functioning of the CNS, in which many afferent messages are integrated, should never be viewed as reflexive behavior even in the case of the "automatic" movements of invertebrates and lower vertebrates.

Overview of the Essays Grouped within the Topic "Reflexes"

In the following synopsis, key points are extracted or paraphrased from each of the essays in this volume related to reflexes.

►*Autonomic Reflexes*. Reflexes related to autonomic regulation of pelvic organs, cardiovascular system, functions of skin, gastrointestinal tract, airways, eye and pineal gland are mediated by spinal cord, brain stem or hypothalamus and are functionally defined by their afferent input and efferent output. They are di- or polysynaptic, organized at the segmental propriospinal or propriobulbar level and form the building blocks of autonomic regulations. Interneurons are important for the integration and coordination of different autonomic and somatomotor systems. Command signals from higher centers act primarily via these interneurons rather than directly with the final autonomic pathways.

►*Conditioned Reflexes*. The fact that reflexes are affected by activity-dependent plasticity throughout life (and even in utero) implies that the traditional distinction between unconditioned and conditioned reflexes is merely an artificial distinction imposed by an experimenter. In reality, most and probably all reflexes are conditioned in the sense that they have been shaped by activity. Those traditionally designated as "unconditioned," such as the normal flexion withdrawal

R

reflex that withdraws a limb from a painful stimulus, are reflexes that have undergone standard conditioning in the course of earlier life, and thus are similar in most normal individuals. In essence, "unconditioned reflexes" are simply reflexes that were conditioned before the experimenter began to observe them.

►*Feedback control of movement*. The word feedback is used extensively in engineering. In neurophysiology, it is used to describe the sensory signals used by the CNS to control a large number of bodily functions to maintain constancy of the internal environment (homeostasis). Signals from mechanoreceptors in muscles, joints and skin are involved in the control of movement, as well those from the eyes, ears and vestibular apparatus. All levels of the CNS from the spinal cord to the cerebellum and cerebral cortex receive feedback from mechanoreceptors and all these levels are involved in controlling even the simplest movements.

►*Integration of reflexes*. Despite more than a hundred years of research, the integration of spinal reflex circuitries and descending motor commands remains a challenge. For example, there is still no consensus about the mechanism by which the much-studied monosynaptic stretch reflex contributes to the activation of muscles during walking, if it makes a meaningful contribution at all. An understanding of spinal reflex networks is a requirement for developing useful therapeutic strategies in the rehabilitation of neurological disorders.

►*Locomotor reflexes*. Locomotor reflexes play an essential role in the patterning of motor activity for walking. These reflexes have three major functions: (i) to regulate the timing of motor commands according to the mechanical state of the limbs and body, (ii) to control the magnitude of ongoing muscle activity, and (iii) to initiate corrective responses when the limbs or body are unexpectedly perturbed by events in the environment.

►*Long loop reflexes*. transcortical reflex. By definition, long loop reflexes occur at latencies longer than the simplest reflexes mediated by *segmental* circuits within the spinal cord yet the latencies are too short to be mediated volitionally. For muscles in the hand, the fastest (spinal) reflexes to muscle stretch occur at latencies ~35 ms; long loop reflexes occur at latencies ~60 ms, whereas volitional responses occur at ~140 ms. However, automatic motor responses of comparable latencies can also be generated by tactile (cutaneous) stimuli that do not involve muscle stretch, so the term "long loop reflex" should not be restricted to those generated by muscle stretch.

►*Presynaptic inhibition*. Presynaptic inhibition (PSI) refers to a decrease of transmitter release at central synapses. For example, activation of afferent fibers originating in flexor muscles attenuates monosynaptic ►EPSPs in extensor motoneurons without detectable changes in the time course of the EPSPs, membrane potential or motoneurone excitability. PSI occurs widely within the CNS of both vertebrates and invertebrates. Synaptic efficacy at axon terminals from sensory afferents, descending systems or interneurons can be subject to PSI control by a number of neurotransmitters and presynaptic receptors.

►*Respiratory reflexes*. Generating an optimal breathing pattern for O_2 and CO_2 homeostasis requires the integration of sensory information from a variety of receptors including central and peripheral chemoreceptors for adjusting the magnitude of alveolar ventilation and stretch receptors for regulating the depth and rate of breathing. Sensory input is also important in the coordination of breathing with other systems, such those required for speaking, eating, walking, running, and vomiting. Finally, sensory information is necessary for protection of the airways and lungs. Receptors in the nose, pharynx, larynx and lower airways elicit a variety of reflexes including coughs, sneezes and apnea that protect the airways from inhalation of noxious substances and increase mucous secretion that aids in their removal.

►*Sexual reflexes*. Spinal reflexes consisting of afferent and efferent components instigate the genital vasocongestion and neuromuscular tension responsible for sexual arousal (erection in men and vaginal lubrication and elongation in women), the triggering of ejaculation in men, and possibly orgasm in both sexes. While the spinal cord contains all the neural circuitry involved in the generation of genital arousal, many other body senses, emotions and cognitive processes mediating social awareness determine whether an individual person will orient towards sexual activity or lose interest.

►*Concluding thoughts*. The difficulty in classifying motor acts as either voluntary, involuntary or reflexive is the inevitable consequence of the overlap in the attributes that describe these categories as well as the brain mechanisms that control them. The various essays summarized above all demonstrate this overlap in one way or another. The Jacksonian continuum, "most automatic" to "most voluntary," should always be borne in mind when considering reflexive control of bodily functions.

References

1. Prochazka A, Clarac F, Loeb GE, Rothwell JC, Wolpaw JR (2000) What do reflex and voluntary mean? Modern views on an ancient debate. Exp Brain Res 130:417–432
2. Prochaska G (1784) The principles of physiology, Prochaska on the nervous system. (De functionibus systematis nervosi. Commentatio. Wolfgang Gerle, Prague (English translation: (1851) A Dissertation on the Functions of the Nervous System). The Sydenham Society, London, p 463

3. Bernard C (1878) Leçons sur les phénomènes de la vie communes *aux animaux et aux des végéraux*. Baillière, Paris, pp 114–121
4. Spencer H (1855) Principles of psychology, 2nd edn. Published in 1873. Appleton, New York, p 648
5. Sechenov IM (1863) Reflexes of the brain (Refleksy Golovnogo Mozga). In: Subkov AA (ed) Im sechenov, selected works. State Publishing House, Moscow
6. Hughlings Jackson J (1884) On the evolution and dissolution of the nervous system. Croonian lectures 3, 4 and 5 to the Royal Society of London. Lancet 1:555–739

Reflexive Saccades

Definition

These are also called reactive saccades, pro-saccades, or visual grasp reflex. A saccade elicited by visual, auditory and even tactile events, and direct gaze at the perceived location of these events. In freely behaving subjects they tend to be accompanied by head rotation when the eccentricity of this location exceeds 20° or so.

Depending on the modality and intensity of the stimuli, they occur at latencies of 150 to –350 ms. Considerable attentional effort is required if reflexive saccades are to be voluntarily suppressed.

▶Oculomotor Control
▶Saccade, Saccadic Eye Movement

Refractory Period

Definition

When, during an action potential, the membrane has undergone a full-blown depolarization (up to several tens of millivolts positive), the Na^+ system is subsequently in a state of reduced responsiveness, from which it recovers slowly over several milliseconds. This period is called refractory period. There is often an initial short period during which the Na^+ system cannot be activated at all, however strong the depolarization.

This is called the absolute refractory period. During the subsequent relative refractory period, the Na^+ system responds in part.

▶Action Potential
▶Sodium Channels

Regeneration

Chizuka Ide[1,2], Mari Dezawa[1],
Naoya Matsumoto[1,3], Yutaka Itokazu[1]
[1]Department of Anatomy and Neurobiology, Kyoto University Graduate School of Medicine, Yoshidakonoe-cho, Sakyo-ku, Kyoto, Japan
[2]Department of Occupational Therapy, Aino University, Ibaraki, Osaka, Japan
[3]Department of Traumatology and Acute Critical Medicine, Osaka University Graduate School of Medicine, Osaka, Japan

Regenerating axons grow well in the peripheral nervous system (PNS), but, in contrast, effective nerve regeneration rarely occurs in the central nervous system (CNS). Following ▶axotomy, the distal segment of the injured axon degenerates (▶Wallerian degeneration), whereas the proximal segment usually remains intact. The most prominent change of the neural cell body following axotomy is the disintegration of ▶Nissl bodies [1]. This phenomenon is called ▶chromatolysis (▶Chromatolysis). Chromatolysis that can occur in neural cell bodies in both the PNS and CNS involves not degenerating, but regenerating reactions of neurons to axotomy. Though morphological changes of neurons in relation to chromatolysis have been extensively studied, the molecular basis behind this phenomenon has not yet been clarified. Chromatolysis is a sign for neurons, when their axons are injured, to shift from the normal condition to the regenerating phase, leading to axonal growth. In the normal condition, neurons are mainly involved in the synthesis and release of ▶transmitters. However, following injury, neurons should change their machinery to produce molecules that contribute to regeneration. Molecular changes in association with axonal degeneration and regeneration have been studied, and gene expression involving in axonal elongation is crucial in understanding the molecular mechanism of nerve regeneration (▶Neuronal changes in axonal degeneration and regeneration). The molecular changes of neurons responding to axonal injury have been studied [2].

In the PNS, ▶Schwann cells and their basal laminae act as efficient scaffolds and sources of ▶neurotrophic factors required for the growth of regenerating axons, and adhesion molecules present on the surface of Schwann cells contribute to nerve regeneration. On the other hand, glial cells including ▶astrocytes and ▶oligodendrocytes in the CNS play no supporting role in the growth of regenerating axons. In addition, there exist no extracellular matrices such as basal laminae in the CNS. Although axons in the CNS

R

have the ability to regenerate after injury, the microenvironment appropriate for the growth of regenerating axons is not provided in the CNS. Cell transplantation using Schwann cells, ►olfactory ensheathing cells, ►neural stem cells, ►choroid plexuses, and ►macrophages has been extensively studied to overcome this difficulty, as it could provide an efficient environment to enable regenerating axons to grow. Other studies have also been conducted on how to facilitate nerve regeneration via suppressing inhibitory factors using specific ►antibodies or via supplying trophic factors to the lesion using genetically altered cells to produce specific trophic factors.

Nerve Regeneration in the PNS

Axons are ensheathed by Schwann cells in the PNS, where each cell forms a ►myelin sheath segment around the axon with ►nodes of Ranvier intervening between the neighboring myelin sheath segments. Schwann cells of myelinated and unmyelinated fibers possess basal laminae on the surface facing the connective tissue, which are continuous even at the nodes of Ranvier. Therefore, the axon and associated Schwann cells reside within a basal lamina tube along its entire length.

Peripheral nerve fibers are located in the connective tissue compartment, an essential difference from the central nerve fibers, which are tightly packed within the brain and spinal cord without any structural extracellular component present between the nerve fibers. During Wallerian degeneration following axonal injury, remaining Schwann cells temporally proliferate and form cell strands called "►Schwann cell columns" within the basal lamina tube. Axonal sprouts that emerge at the nodes of Ranvier adjacent to lesions of the proximal stump extend as regenerating axons through the connective tissue compartment to the distal stump, in which they further elongate along Schwann cell columns.

The tip of the regenerating axon is specialized as a ►growth cone [3]. Growth cones are formed at the growing tip of axons during development and regeneration. Growth cones emit filopodia and lamellipodia on the surface, which actively move in various directions to survey the surrounding environment. The structure of living growth cones and their impressive movement were first observed in the culture of single-dissociated neurons under cine-microscopy in Nakai's pioneering work [4]. He proposed that filopodia and lamellipodia might represent sensors searching for appropriate targets during their extension [5]. Numerous studies have been performed regarding the structure and function of growth cones ►Growth cone.

The growth of axonal sprouts is facilitated or suppressed depending on the conditions of myelin sheath degradation that occurs during Wallerian degeneration in the myelin sheath of the distal stump [6]. In the distal stump, regenerating axons come into contact with Schwann cell columns. Schwann cells play a critical role in nerve regeneration in the PNS; they provide cellular scaffolds for regenerating axons to grow through, and express trophic factors as well as adhesion molecules for promoting the extension and maintenance of regenerating axons. In Wallerian degeneration, Schwann cells disrupt myelin sheaths that have lost contact with axons into fragments called myelin balls that are subsequently transferred to and phagocytosed by macrophages. Thus, macrophages contribute to successful nerve regeneration. Schwann cells cooperate with macrophages not only in myelin sheath removal, but also in trophic factor production for promoting axon growth during nerve regeneration (►Schwann cells in nerve regeneration). Schwann cells that remain "quiescent" in Schwann cell columns during Wallerian degeneration become "active" by coming into contact with regenerating axons, following which they gradually ensheath axons and finally form myelin sheaths in myelinated fibers.

Schwann cells are primary sources of trophic factors for nerve regeneration. A large number of neurotrophic factors have been identified. A well-known neurotrophic factor is ►nerve growth factor (NGF), a member of the neurotrophins that include ►brain-derived nerve growth factor (BDNF), ►neurotrophin 3, and ►neurotrophin 4/5. ►Glial cell line-derived neurotrophic factors (GDNF) belong to another family that promotes the survival and neurite extension of neurons. ►Cytokines also have neurotrophic activity, which include ►ciliary neurotrophic factor (CNTF) and ►interleukin 6 (IL-6). Neurotrophic molecules have been studied in relation to their corresponding receptors, effects on axonal cytoskeletons, and gene expression to explore the regeneration mechanisms of injured axons in the PNS and CNS (►Neurotrophic factors in nerve regeneration).

On the other hand, in one experiment in which Schwann cells were killed by freeze-treatment, the damaged cells were removed by macrophages. Regenerating axons were then observed to vigorously grow through such basal lamina tubes in contact with the inner surface of the basal lamina [7]. This result indicates that peripheral nerve axons can grow through the acellular matrices (►Role of basal lamina in nerve regeneration). This is the theoretical basis for the use of artificial materials for nerve regeneration in the PNS. Basal laminae serve not only as the scaffold for growing axons, but also as a supply of trophic/nutritional factors that are adsorbed by heparan sulfate present on the outer surface of the basal laminae [8]. Thus, peripheral nerves are provided with dual structural insurance, Schwann cells and basal laminae, for successful nerve regeneration.

In a regular nerve suture, regenerating axons randomly enter distal Schwann cell columns. Therefore, it is not definite that regenerating axons can reach their original targets. The clinical estimation of functional recovery is important [9]. Several different methods of treatment have been developed and used clinically. In the case of crush injury, in which basal lamina tubes (endoneurial tubes) remain undisrupted, axonal sprouts can extend through the original Schwann cell tubes to the original targets. Therefore, nerve regeneration occurs readily, and the high correspondence in axon-target reinnervation is secured for accurate functional recovery. On the other hand, in the case of transection, the proximal and distal stump should be sutured by apposing directly, or using grafts including autologous nerve grafts and artificial tubes. In stump suturing, fascicular repair to connect the corresponding nerve fasciculi has been recommended to ensure that regenerating axons can reinnervate the original targets as accurately as possible (►Regeneration: clinical aspects).

Neural connections in the ►somatosensory area of the ►cerebral cortex can be reorganized depending on the input from the peripheral nervous system. The cortical neural organization including ►sensory cortex and ►motor cortex is not fixed, but can be modified subject to the functional demand. When the sensory inputs are lost due to peripheral nerve damage including the amputation of limbs or fingers, the corresponding areas of the somatosensory cortex are reorganized to receive inputs from the neighboring skin areas including the stump. A similar reorganization occurs in the motor cortex. Thus, motor and sensory representations of the cerebral cortex become remodeled following nerve injury and regeneration in the PNS. This means that patients should relearn how to appropriately perform an action through the remodeled cerebral cortex in rehabilitation. Reorganization of the cerebral cortex is a basis for the rehabilitation of limb activity (►Somatosensory reorganization; ►Regeneration: clinical aspects).

Artificial materials have been studied as guides for the growth of regenerating axons in the PNS. Collagen gel has been most commonly used. Other biodegradable polymers that have been utilized for nerve regeneration are polyglycolic acid, polylactic acid, poly-ε-caprolactone, alginate, and chitosan. Alginate is derived from brown seaweed, and chitosan is from the crustacian exoskeleton. These materials have been used as substrates for nerve regeneration in the PNS and CNS (►Transplantation of artificial materials for nerve regeneration). A polyglycolic acid-collagen tube has been developed with good results [10].

Blood supply is a critical point for successful nerve regeneration. Blood capillaries that readily regenerate in the connective tissue compartment greatly contribute to peripheral nerve regeneration. Blood vessels, once damaged, rarely regenerate in the CNS. This is probably because, unlike in the PNS, there is no extracellular matrix which can act as a scaffold for developing vessels in the CNS. The loss of blood supply results in severe ischemia, which in turn leads to the suppression of tissue repair and the promotion of cavity formation in the CNS.

Motor Nerves: The Neuromuscular Junction

Motor nerves terminate at ►neuromuscular junctions, which consist of presynaptic axon terminals and postsynaptic folds of the muscle fiber plasma membrane. In addition, terminal Schwann cells cover the presynaptic axon terminals, and the basal lamina, a continuation of the ordinary basal lamina of muscle fibers, is present on the folds of the muscle fibers. Following Wallerian degeneration, the presynaptic terminal disappears, and the postsynaptic folds at the endplate become gradually less distinct, but remain as a remnant of small folds. At the same time, terminal Schwann cells persist in the preterminal region. Regenerating axons enter the empty endplate, and become presynaptic terminals, thus forming new neuromuscular junctions. The presynaptic terminal can be formed in the absence of terminal Schwann cells, such as in the acellular scaffold [11]. In addition, regenerating axons can develop a presynaptic structure when they come into contact with the basal lamina at the site of the original postsynaptic folds. This indicates that the basal lamina at the endplate contains molecules that induce the regeneration of axons and cause postsynaptic specialization.

When muscle is partially denervated, ►axonal sprouting occurs from the intact neuromuscular junction. Following motor neuron injury, axon terminals are lost from the endplates of muscle fibers belonging to injured motor neurons. Responding to the denervation of endplates, axonal sprouts emanate from axon terminals of neighboring intact endplates, reinnervating denervated endplates. Thus, the ►motor unit is enlarged, and the muscle activity can be kept almost unchanged. Terminal Schwann cells that have lost contact with axon terminals by denervation extend cell processes toward the neighboring intact endplates. Such Schwann cell processes act as guide tubes for axonal sprouts to elongate from the intact endplate to the denervated one (►Role of sprouting in sustaining neuromuscular function in health and disease).

Sensory Nerves: Sensory Corpuscles in the Skin

Sensory nerve terminals occur as various types of organized corpuscles present in the skin. Representative sensory terminals include ►Pacinian corpuscles and ►Meissner corpuscles, which are composed of axon terminals and modified Schwann cells called lamellar cells. Following Wallerian degeneration, lamellar cells in these corpuscles become atrophic owing to the loss of contact with axons, but continue to exist for a long period of time. Upon the arrival of regenerating axons, lamellar cells begin to proliferate and take on the same structures as those found in the original corpuscles. For the acellular corpuscle, corpuscular basal laminae deprived of lamellar cells can also serve as a scaffold that promotes the regeneration of the original corpuscle following reinnervation. However, no new corpuscle regeneration in regions other than at the original corpuscle in the skin occurs (►Meissner corpuscle Regeneration; ►Pacinian corpuscle Regeneration).

►Merkel cell-neurite complexes are different from Pacinian and Meissner corpuscles in that their axons make direct contact with Merkel cells. Merkel cells deprived of axon terminals tend to disappear, probably due to degeneration and/or movement to the surface of the epidermis. Upon reinnervation, Merkel cells reappear, partly due to the differentiation of keratinocytes, and then form Merkel-neurite complexes similar to the original ones present at the base of the epidermis (►Merkel cell-neurite complex Regeneration).

Nerve Regeneration in the CNS

Nerve regeneration in the CNS, especially in the spinal cord, has been extensively studied for more than 100 years. No effective nerve regeneration or tissue repair occurs in lesions of the spinal cord, which usually results in cavity formation without distinct tissue repair. Following injury to the spinal cord, strong regenerative responses occur including the formation of growth cones and associated glial cell migration. However, such neural reactions do not develop as efficiently as they do in the PNS, and result in cavity formation without axonal extension into the lesion. As described above, the essential difference between the PNS and CNS is the presence of an extracellular matrix in the PNS, which is composed of basal laminae and collagen fibers. This means that there is no effective scaffold available for the growth of regenerating axons in the CNS. The same can be said for blood vessel regeneration, which requires an extracellular matrix scaffold to regenerate.

The ►glial scar is usually produced around the cystic cavity at a chronic stage after injury. If the ►pia mater is cut open, the fibroblast-like cells of the meningeal layer invade the lesion, contributing to the formation of dense glial scar of connective tissues composed of extracellular matrices including collagen fibers. In such cases, astrocytes form a barrier between the connective tissue and adjacent CNS tissue. Basal laminae are formed on the surface of astrocyte processes facing the connective tissues, and thus, the CNS tissue tends to segregate itself from the invading connective tissues using astrocytic scar tissue [12]. This kind of glial scar is regarded as the main obstacle preventing the growth of regenerating axons in the CNS. Connective tissue invasion followed by glial scarring is therefore an undesirable phenomenon for nerve regeneration in the CNS. On the other hand, in lesions in which the pia mater is not cut open, the tissue reaction is different. When the spinal cord is crush-injured, the pia mater is usually not damaged, with the pial basal lamina kept intact. Cavities of various sizes are usually formed in the lesion at chronic stages. In such cases, there is no cell invasion from the outside. Astrocytic proliferation is found along the margin of the cavity, and oligodendrocytes line the inner surface of the cavity margin.

Usually, astrocytes, oligodendrocytes, macrophages, and microglia all contribute to the formation of glial scars. The mechanisms for scar formation, roles of contributing cells, and expression of specific molecules including proteoglycans are complicated, and yet to be understood (►Glial scar). Glial scars are thought to be the main impediments to the growth of regenerating axons. Regenerating axons from the proximal stump have to surpass the glial scar at the distal stump to invade the host spinal cord tissue. The digestion of proteoglycans has been proposed to promote axonal growth through glial scar in the spinal cord.

Cavities resulting from the degeneration of impaired tissues hamper axonal extension through the lesion. The margin of the cavity is not as thick as the astrocytic scar tissue as previously thought in the crush-injured spinal cord. Using appropriate techniques including cell transplantation, it may be possible to induce regenerating axons to grow through a region with cavities.

Functional assessment is important for nerve regeneration in the CNS, for which BBB scoring and other types of estimation of behavioral recovery have been employed [13].

At present, cell transplantation is being extensively studied to facilitate nerve regeneration in the injured spinal cord. Several varieties of cells have been used for transplantation, among which the major cell types include: Schwann cells, bone marrow stromal cells, olfactory ensheathing cells, choroid plexus ependymal cells, neural stem cells, macrophages, and those found in embryonic spinal cord tissue. Other studies have also been carried out which focused on the suppression of inhibitory factors such as ►Nogo A by the specific antibody.

Schwann Cells

Aguayo and colleagues showed that neurons within the CNS can induce the elongation of regenerating axons into peripheral nerves which were inserted at one end into the spinal cord and at the other end into the medulla oblongata [14]. This shows that regenerating axons from neurons in the CNS can extend if they are provided with an appropriate environment. Since this report, cell implantation aimed at CNS nerve regeneration has been extensively studied.

Schwann cells play a role in axonal elongation in grafted peripheral nerves, where regenerating axons come into contact with Schwann cells, which support the growth and maturation of regenerating axons, as in the case of the PNS. The utility of peripheral nerve grafts has prompted the use of cultured Schwann cells for transplantation in the CNS. Here, cultured Schwann cells mingle in matrigel, and are then placed into an artificial tube, after which the tube is subsequently grafted into the lesion of the spinal cord. Many regenerating axons extend through the tube, and some enter the distal stump. A few axons then form synapses with neurons present in the distal segment of the spinal cord, and behavioral improvement subsequently takes place [15].

Although Schwann cells also serve as effective conduits for regenerating axons in the CNS, extracellular matrices including basal laminae are inefficient as scaffolds for the growth of regenerating axons in the CNS, unlike in the PNS. Since Schwann cells possess basal laminae and the ability to produce collagen matrices, extracellular matrices can be brought into the lesion following Schwann cell transplantation. Therefore, astrocytes at the border of the lesion are apt to form barriers composed of cell processes with basal laminae on the surface facing the Schwann cell compartment, and regenerating axons cannot penetrate such astrocyte scar tissue. Studies have been concentrated to overcome this difficulty in Schwann cell transplantation (►Transplantation of Schwann cells).

The ►optic nerve is frequently used as an experimental model of nerve regeneration in the CNS. Since it is an isolated bundle composed of central nerve fibers, nerve regeneration can be more precisely evaluated than by using the spinal cord. Transplantation of peripheral nerves and Schwann cells into the optic nerve has been well studied [16]. A long peripheral nerve transplanted into the optic nerve can serve as a conduit for regenerating axons traveling from the optic nerve to the ►superior colliculus, in which some synaptic connections are formed. Some functional recovery of light sensation has also been reported using this technique.

Many ►retinal ganglion cells undergo retrograde degeneration after optic nerve injury, and this poses another major problem for optic nerve regeneration. The administration of trophic factors into the optic cup has been studied with the aim of promoting the survival of ganglion cells (►Regeneration of optic nerve).

Bone Marrow Stromal Cells

Although they do not belong to the CNS, bone marrow stromal cells (BMSCs) have been used for transplantation into the spinal cord [17]. BMSCs are grafted by directly injecting them into the lesion or by infusing them through the cerebrospinal fluid (CSF) [18,19]. Some BMSCs then gather in the lesion and survive there for 2–3 weeks after grafting. BMSCs do not differentiate into neural cells after grafting into the spinal cord. In the rat, they tend to disappear from the spinal cord more than 4 weeks after grafting. Although BMSCs do not become integrated into lesions of the spinal cord, tissue repair including the suppression of cavity formation is facilitated. In addition, behavioral improvement is obvious in the spinal cord-injured rat. These findings imply that BMSCs can be used in transplants to treat ►spinal cord injury.

Since BMSCs can be obtained from the patients themselves and are not stem cells but ordinary functioning cells present in the bone marrow, they show promise for the clinical treatment of spinal cord injuries. Transplantation by infusing BMSCs into the CSF is more effective than direct injection into the lesion. Also, since BMSCs disappear several weeks after transplantation, they might release trophic factors that reverse the degeneration of damaged neural tissue. The clinical application of BMSCs by injecting them into the cerebrospinal fluid via lumbar puncture has progressed (►Transplantation of bone marrow stromal cells for spinal cord regeneration).

Bone marrow stromal cells can be trans-differentiated into Schwann cells and neurons, and the transplantation of trans-differentiated Schwann cells and neurons may be a promising technique for CNS as well as PNS regeneration [20].

Neural Stem Cells

Neural stem cells (NSCs) have been regarded as appropriate for cell transplantation to treat spinal cord injuries. After transplantation, these cells survive, migrate, and become integrated into the host tissue. They also have the ability to differentiate into neurons, astrocytes, and oligodendrocytes after transplantation.

There is hope that NSCs can be used after differentiation into neurons, astrocytes, and oligodendrocytes in vitro. Although NSCs are attractive for use in clinical therapy, the source of these cells is the most critical problem. There are strict limitations regarding the use of human embryos as sources of NSCs. It is also difficult to obtain NSCs from adult tissues. However, it is possible that NSCs can be acquired from the brains of deceased human bodies within a short period after death [21].

Another difficulty in using NSCs is the potential for unlimited cell proliferation, and, thus, the formation of cancer. Even in cases for which NSCs are used after differentiation in vitro, it is possible that a few undifferentiated cells may remain, and cause undesirable cell proliferation. Neural stem cells have been transplanted for spinal cord regeneration (►Transplantation of neural stem cells for spinal cord regeneration).

The Choroid Plexus

The choroid plexus is the main region where the cerebrospinal fluid (CSF) is produced. It consists of epithelial cells and associated connective tissue containing plenty of sinusoidal blood vessels, and few fibroblasts and macrophages in a scanty collagen matrix. Grafting of the choroid plexus into the injured spinal cord promotes tissue repair including the growth and regeneration of axons in lesions [22]. Other in vitro and in vivo studies have indicated that choroid plexus epithelial cells (modified ependymal cells) might have the ability to promote nerve regeneration by rescuing neural tissues from degeneration and facilitating axonal growth. Considering that the CSF plays an important role in maintaining normal brain function, it should contain a variety of factors that promote the survival and proper activity of neural cells including neurons and glial cells.

Olfactory Ensheathing Cells

Recently, olfactory ensheathing cells (OECs) have been extensively studied for use in transplantation to promote nerve regeneration in the spinal cord. These cells possess the properties of Schwann cells and astrocytes, and grafted OECs become Schwann cells associated with axons and perineurial cells surrounding nerve fibers, as seen in the PNS [23]. Furthermore, functional recovery can be observed in accordance with histological improvement. It is obvious that OECs provide the guide for the extension of regenerating axons. OECs have had a great impact on the study of spinal cord regeneration; however, their effects have varied among such studies. In spite of the many studies performed so far, how OECs exert their effects in spinal cord regeneration has not yet been fully elucidated. Identification of OECs in in vivo experiments might be a crucial requisite for accurately understanding their roles in spinal cord regeneration. Genetically labeled OECs might be useful for long-term observation after transplantation (►Transplantation of olfactory ensheathing cells).

Embryonic Spinal Cord Tissue

Satisfactory regeneration can be induced in the spinal cord in which a segment of the spinal cord obtained from an embryo has been transplanted in rats during the early postnatal period [24]. It has been proposed that embryonic tissue can overcome glial scaring of the cystic cavity in chronic injury of the spinal cord. Unfortunately, the use of embryonic tissues is greatly limited due to ethical problems.

Genetically Modified Cells

Experiments in which cells that were genetically transformed to secrete trophic factors have been transplanted into the injured CNS tissue have been conducted. Fibroblasts and other kinds of cells including OECs have been used for such experiments [25]. However, at present, genetically modified cells involve problems of ethics and safety that should be overcome before clinical application.

Immune System

Following injury to the CNS, microglia and macrophages invade the lesion. Microglia, macrophages, and T-cells have been demonstrated to play important roles in CNS protection and repair. Regulatory and autoimmune T-cells are contradictory regarding CNS protection. Autoimmune T-cells activate the microglia to secrete trophic factors for CNS repair (►Protective autoimmunity). Macrophages preincubated with sciatic nerve fragments or autologous skin promoted repair of the optic nerve and spinal cord. Microglia and bone marrow-derived macrophages with the property of antigen-presenting cells secrete trophic factors that contribute to neuroprotection in the CNS (►Autologous macrophages for central nervous system repair). Studies on the functions of immune cells such as microglia, macrophages, and T-cells provide insights into the CNS mechanisms of protection against injury, and contribute to achieve CNS regeneration.

Inhibitory Molecules

It is believed that CNS regeneration cannot occur partly due to the presence of inhibitory molecules in the CNS. The main inhibitory molecules are associated with myelin sheaths and oligodendrocytes, including myelin-associated glycoprotein (MAG), oligodendrocyte-myelin glycoprotein (OMgp), and Nogo-A. Growth cones collapse when they encounter these inhibitory molecules. The administration of anti-Nogo A antibody promotes the growth of regenerating axons in the injured spinal cord and improves behavioral function in rats [26].

Another major inhibitor is chondroitin sulfate proteoglycan (CSPG) produced in the glial scar. CSPGs serve as barriers to growing axons. Regenerating axons are thought to be blocked at the boundary of the glial scar in the spinal cord injury (►Inhibitory molecules in regeneration).

References

1. Lieberman AR (1971) The axon reaction: a review of the principal features of perikaryal responses to axon injury. Int Rev Neurobiol 14:49–124
2. Kiryu-Seo S, Gamo K, Tachibana T, Tanaka K, Kiyama H (2006) Unique anti-apoptotic activity of EAAC1 in injured motor neurons. EMBO J 25:3411–3421
3. Dontchev VD, Letourneau PC (2003) Growth cones integrate signaling from multiple guidance cues. J Histochem Cytochem 51:435–444
4. Nakai J (1956) Dissociated dorsal root ganglia in tissue culture. Am J Anat 99:81–129
5. Nakai J, Kawasaki Y (1959) Studies on the mechanism determining the course of nerve fibers in tissue culture. I. The reaction of the growth cone to various obstructions. Z Zellforsch Mikrosk Anat 51:108–122

6. Torigoe K, Lundborg G (1998) Selective inhibition of early axonal regeneration by myelin-associated glycoprotein. Exp Neurol 150:254–262
7. Ide C, Tohyama K, Yokota R, Nitatori T, Onodera S (1983) Schwann cell basal lamina and nerve regeneration. Brain Res 288:61–75
8. Fujimoto E, Mizoguchi A, Hanada K, Ide C (1997) Basic fibroblast growth factor promotes extension of regenerating axons of peripheral nerve. In vitro experiment using a Schwann cell basal lamina tube model. J Neurocytol 26:511–528
9. Lundborg G, Rosen B, Dahlin L, Holmberg J, Rosen I (2004) Tubular repair of the median or ulnar nerve in the human forearm: a 5-year follow-up. J Hand Surg [Br] 29:100–107
10. Nakamura T, Inada Y, Fukuda S, Yoshitani M, Nakada A, Itoi S, Kanemaru S, Endo K, Shimizu Y (2004) Experimental study on the regeneration of peripheral nerve gaps through a polyglycolic acid-collagen (PGA-collagen) tube. Brain Res 1027:18–29
11. Sanes JR (2003) The basement membrane/basal lamina of skeletal muscle. J Biol Chem 278:12599–12600
12. Silver J, Miller JH (2004) Regeneration beyond the glial scar. Nat Rev Neurosci 5:146–156
13. Coumans JV, Lin TT, Dai HN, MacArthur L, McAtee M, Nash C, Bregman BS (2001) Axonal regeneration and functional recovery after complete spinal cord transection in rats by delayed treatment with transplants and neurotrophins. J Neurosci 21:9334–9344
14. David S, Aguayo AJ (1981) Axonal elongation into peripheral nervous system "bridges" after central nervous system injury in adult rats. Science 214:931–933
15. Xu XM, Zhang SX, Li H, Aebischer P, Bunge MB (1999) Regrowth of axons into the distal spinal cord through a Schwann-cell-seeded mini-channel implanted into hemisected adult rat spinal cord. Eur J Neurosci 11:1723–1740
16. Inoue T, Hosokawa M, Morigiwa K, Ohashi Y, Fukuda Y (2002) Bcl-2 overexpression does not enhance in vivo axonal regeneration of retinal ganglion cells after peripheral nerve transplantation in adult mice. J Neurosci 22:4468–4477
17. Chopp M, Zhang XH, Li Y, Wang L, Chen J, Lu D, Lu M, Rosenblum M (2000) Spinal cord injury in rat: treatment with bone marrow stromal cell transplantation. Neuroreport 11:3001–3005
18. Wu S, Suzuki Y, Noda T, Bai H, Kitada M, Kataoka K, Chou H, Ide C (2003) Bone marrow stromal cells enhance differentiation of co-cultured neurosphere cells and promote regeneration of injured spinal cord. J Neurosci Res 72:343–351
19. Ohta M, Suzuki Y, Noda T, Ejiri Y, Dezawa M, Kataoka K, Chou H, Ishikawa N, Matsumoto N, Iwashita Y, Mizuta, Kuno S, Ide C (2004) Bone marrow stromal cells infused into the cerebrospinal fluid promote functional recovery of the injured rat spinal cord with reduced cavity formation. Exp.Neurol 187:266–278
20. Dezawa M, Kanno H, Hoshino M, Cho H, Matsumoto N, Itokazu Y, Tajima N, Yamada H, Sawada H, Ishikawa H, Mimura T, Kitada M, Suzuki Y, Ide C (2004) Specific induction of neuronal cells from bone marrow stromal cells and application for autologous transplantation. J Clin Invest 113:1701–1710
21. Xu Y, Kimura K, Matsumoto N, Ide C (2003) Isolation of neural stem cells from the forebrain of deceased early postnatal and adult rats with protracted post-mortem intervals. J Neurosci Res 74:533–540
22. Ide C, Kitada M, Chakrabortty S, Taketomi M, Matumoto N, Kikukawa S, Mizoguchi A, Kawaguchi S, Endo K, Suzuki Y (2001) Grafting of choroid plexus ependymal cells promotes the growth of regenerating axons in the dorsal funiculus of rat spinal cord – A preliminary report. Exp Neurol 167:242–251
23. Li Y, Decherchi P, Raisman G (2003) Transplantation of olfactory ensheathing cells into spinal cord lesions restores breathing and climbing. J Neurosci 23:727–731
24. Kawaguchi S, Iseda T, Nishio T (2004) Effects of an embryonic repair graft on recovery from spinal cord injury. Prog Brain Res 143:155–162
25. Blesch A, Tuszynski MH (2003) Cellular GDNF delivery promotes growth of motor and dorsal column sensory axons after partial and complete spinal cord transections and induces remyelination. J Comp Neurol 467:403–417
26. Schwab ME (2004) Nogo and axon regeneration. Curr Opin Neurobiol 14:118–124

Regeneration Associated Genes (RAGs)

Definition

Genes that are upregulated within the neuron following axotomy. The protein product of these genes such as tublin, GAP43 and others are anterogradely transported to and are critical to the elongation of the growth cone and regenerating axon.

►Neurotrophic Factors in Nerve Regeneration
►Peripheral Nerve Regeneration and Nerve Repair

Regeneration: Clinical Aspects

Göran Lundborg
Hand Surgery/Department of Clinical Sciences, Malmö University Hospital, Lund University, Lund, Sweden

Definition

Outgrowth of ►axons following clinical nerve injury and repair, resulting in functional restoration in denervated body parts.

Characteristics

Background

Injuries to peripheral nerve trunks constitute a major clinical problem [1,2]. Such injuries are most frequently seen in the upper extremity. The consequences are severe and the result is often permanent disturbances in sensory and motor functions of the hand. Normally, there is an interaction between the hand and the brain so that the hand is very well represented in the somatosensory cortex as well as the motor cortex [1,3,4]. A nerve injury implies a sudden de-afferentiation with arrest in inflow of sensory impulses to the brain. This results in a rapid cortical remodelling process where the "vacant" cortical area, previously representing the innervated area of the hand, is invaded by expanding adjacent cortical areas [1,3,4]. An analogous phenomenon occurs in the motor cortex.

If the nerve injury is surgically repaired, there is a regeneration of axons downstream of the distal nerve segment aiming at reinnervation of the denervated body part. To regain normal function, axons have to reinnervate their "correct" peripheral targets. However, there is always, in spite of meticulous surgical techniques, a large extent of misorientation of regenerating axons at the repair site and consequently an incorrect peripheral reinnervation [1]. With the reinnervation process, the cortical hand representation is again restored, however in a new and distorted pattern due to the peripheral mal-orientation. A relearning process is required that can be easily managed by the child's brain but usually not by the adult brain [5]. Therefore, fine tactile discriminative functions are seldom or never fully restored in an adult patient. The process of clinical regeneration is influenced by a number of intrinsic and extrinsic factors, some of them reviewed below.

The Nerve Trunk and the Regeneration Process

The nerve trunk represents a composite tissue structure constructed to maintain continuity, nutrition and protection of its basic elements – the axons (Fig. 1). An axon is a long tubular process of the nerve cell body, which may be situated in a dorsal root ganglion (sensory axons), or the anterior horn of the spinal cord (motor axons). The nerve cell and its processes is called a neuron. The axons are ensheathed by Schwann cells that may produce a myelin sheath. The Schwann cell basal lamina contributes to constitute an "endoneurial tube." The axons are closely packed within the endoneurial connective tissue inside ►fascicles [1]. Each fascicle is surrounded by a perineurium, which is a multicellular laminated sheath of considerable mechanical strength, providing a diffusion barrier. The fascicles are embedded within an ►epineurium, which is a supporting and protective connective tissue sheath carrying a longitudinal network of epineurial blood vessels.

Nerve injuries may be of several types and magnitudes. A severe compression or a crush lesion

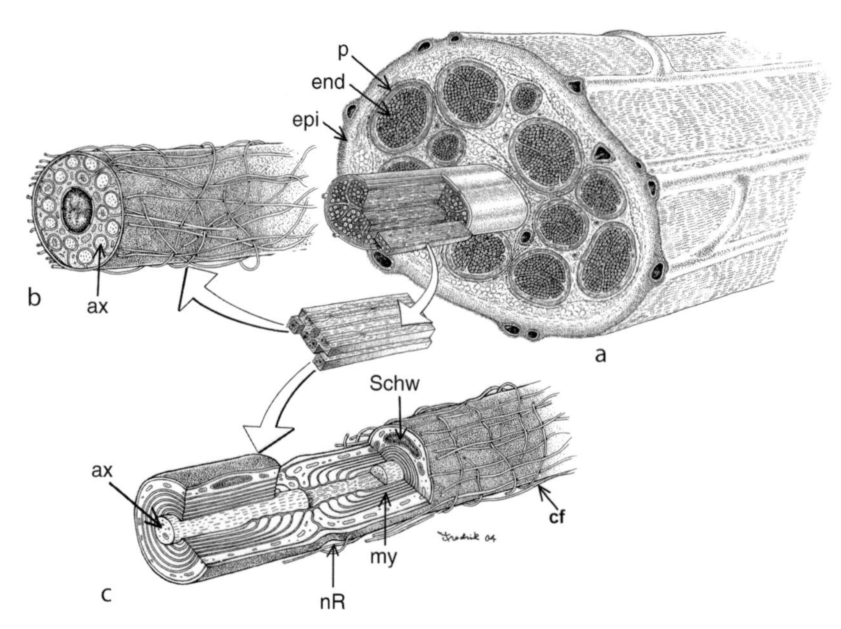

Regeneration: Clinical Aspects. Figure 1 Microanatomy of peripheral nerve trunk and its component. (a) Fascicles surrounded by a multi-laminated perineurium (p) are embedded in loose connective tissue, the epineurium (epi). The outer layers of the epineurium are condensed into a sheath. (b and c) The appearance of unmyelinated and myelinated fibres, respectively, is shown. Schw Schwann cell; my myelin sheath; ax axons; nR node of Ranvier; end endoneurium. Reproduced with permission from Lundborg 2004.

may disrupt axons while the ensheathing chain of Schwann cells and their basal lamina may be preserved. Disruption of axons results in degeneration of their distal segments, implying disintegration of the axonal elements and the myelin sheath. A regeneration process is then required where axons grow distally, following their original pathways, hereby reinnervating correct peripheral targets. The normal cortical representation of the body part is hereby re-established [3,4]. With transection of a nerve trunk, however, the situation is quite different: the "sprouts" that are formed by the transected proximal part of the axons may orient themselves into incorrect distal "Schwann cell tubes" that result in reinnervation of incorrect peripheral targets. Before the regeneration process is initiated, an "initial delay" may last for days or weeks. As a result of the injury a large number of nerve cell bodies in dorsal root ganglia may die, which excludes possibilities for regeneration of their corresponding axons [1]. Several physical, biochemical and other factors influence the course and functional outcome of the regeneration process [1].

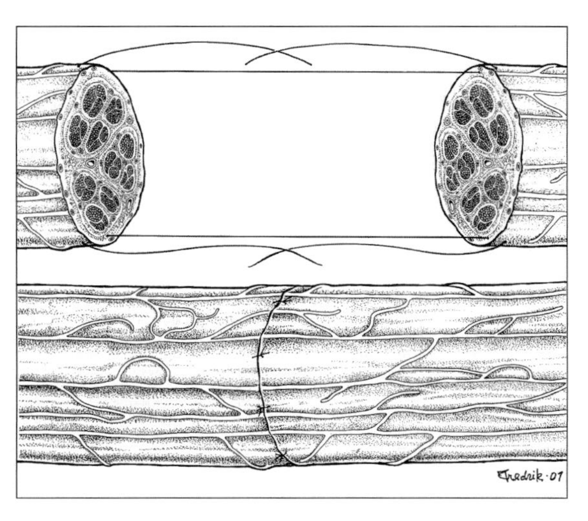

Regeneration: Clinical Aspects. Figure 2 Epineurial suture. Adaptation of the nerve ends is achieved by single stitches in the superficial part of the epineurium along the circumference of the nerve. Reproduced with permission from Lundborg 2004.

Clinical Nerve Repair and Reconstruction

Repair and reconstruction of injured peripheral nerves span over a variety of techniques such as direct repair, ►nerve grafting, use of conduits, ►nerve transfer and ►end-to-side (ETS) anastomosis [1,6,7,8].

With ►epineurial repair, the nerve stumps are approximated by suturing the epineurial sheath, using external landmarks like longitudinal blood vessels to ensure a correct rotational adaptation of the nerve ends. Although the external aspect of the repair site may look perfect, this technique does not, however, ensure an absolutely correct matching of the fascicular structures inside the nerve trunk. A more correct mechanical adaptation may be achieved by fascicular repair or ►group fascicular repair, which requires an internal dissection of the nerve so that separate bundles of fascicles are adapted towards each other. This technique may be justified in cases where sensory and motor fibres are running in separate, well-defined fascicles or fascicular groups, but otherwise the technique has no advantages over the epineurial technique. The various suture techniques may be combined and supplemented by fibrin glue. In ►tubular repair, a small distance is left between the nerve ends that are enclosed in a tubular structure of biological or synthetic type. Such tubes, which may be biodegradable or non-degradable, may give equally good results as direct sutures of the nerve ends (Figs. 2, 3).

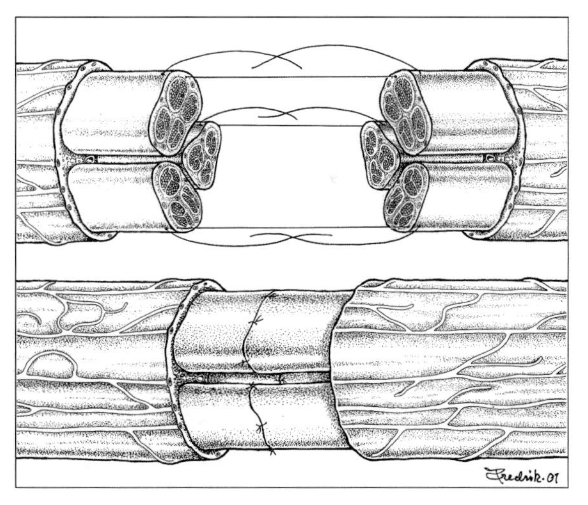

Regeneration: Clinical Aspects. Figure 3 Group fascicular suture. After resection of the epineurial tissue fascicular groups are approximated with single sutures in the connective tissue between separate fascicles or in the outer layer of the perineurium. Reproduced with permission from Lundborg 2004.

With more severe injury, there may be a defect in continuity of the nerve trunk. Such a situation may be seen in severe lacerations in the extremities or as a result of severe traction in the brachial plexus. Well-known examples are traction injuries occurring in difficult obstetrical situations or in adults involved in motorcycle or other types of traffic accidents. In such cases, the defect has to be bridged with a conduit to allow overgrowth of nerve fibres from the proximal to the distal nerve end. The most commonly used technique is to insert a nerve graft, usually harvested from the lower limb. Several cables of thin nerve grafts such as these are inserted between both of the nerve ends, using microsurgical techniques [7].

Nerve Transfers

In severe nerve injuries, a proximal nerve segment may not always be available. An alternate "donor nerve" may then be required to provide the distal segment of the injured nerve with axonal input from a proximal nerve segment, a so-called nerve transfer. The situation requires sacrifice of the donor nerve, which can then be transferred to the distal segment of the injured nerve [6]. Such nerve transfers are widely used in brachial plexus surgery for restoring function in paralysed muscle by using adjacent intact nerves as donors, but can also be applied to more distal nerve injuries, for instance, to achieve motor or sensory functions in the hand by transferring an intact, nearby non-injured nerve.

End-to Side (ETS) Nerve Repair

For more than 10 years it has been known that a distal nerve segment, when sutured in an end-to-side fashion to an adjacent intact nerve, can be reinnervated by sprouts from axons in the healthy donor nerve [8]. It was assumed that the intact axons in such cases may send out lateral sprouts that may reinnervate the sutured distal nerve segment. It was soon realised that this might be a new and promising possibility in clinical cases when routine nerve-grafting procedures where not possible, such as in root avulsions in brachial plexus injuries. The clinical results from these operations, as reported in the literature, are very variable.

Functional Remodelling of Brain Cortex

A nerve transection represents an acute deafferentiation with immediate and longstanding influence on the cortical representation of the innervated body part. For instance, deafferentiation due to median nerve transection results in rapid expansion of adjacent cortical areas, which then occupy the former median nerve cortical territories. If no regeneration occurs, as after an amputation, the extensive cortical reorganisation persists so that the cortical area, previously receiving input from the median nerve, remains occupied by expanding adjacent cortical areas. In amputation, severe cortical reorganisations in such cases may result in persistent ►*phantom sensations* and *phantom* ►*pain*. After a ►*crush injury,* regenerating axons are guided by their original Schwann cell tubes so that they reach their original skin locations, and the corresponding cortical hand representation is normalised. However, after ►*transection and repair,* this scenario is quite different due to peripheral axonal misorientation. The previously well-organised cortical representation is changed to a mosaic-like pattern [1,3,4] and the nerve does not recapture all of its original cortical territory.

Sensory Relearning and Sensory Re-Education

The outcome from nerve repair in adults is far from satisfactory and often disappointing, especially with respect to recovery of tactile discrimination [5]. One major factor is the new and distorted cortical hand representation – "the hand speaks a new language to the brain." A relearning process is required, and it can be a difficult task for adults to require their lost functional sensibility. In hand rehabilitation, ►*sensory relearning* is based on the use of ►sensory re-educational protocols [9,10]. According to these strategies, the brain is reprogrammed based on a relearning process. First, the perception of different touch modalities and the capacity to localise touch is trained, followed by touching and exploration of items, presenting shapes and textures of varying and increasing difficulty to the patient with eyes open or closed. In this way, an alternate sense (vision) trains and improves the deficient sense "sensation."

Factors Influencing the Outcome from Nerve Repair

The functional outcome of nerve repair may vary considerably between patients although identical techniques may be used. There are several factors that are known to influence the outcome of nerve repair.

Age

Although the functional recovery in adults is disappointing, especially with reference to recovering sensory functions, the situation is quite different among children who consistently show superior functional results after nerve repair [5]. This has usually been attributed to superior plasticity of the brain in children, with a better ability in central adaptation to the new pattern of afferent impulses presented by misdirected axons. A critical age period for recovery of functional sensibility in hands after nerve repair can be defined, the best results being seen in those younger than 10 years, followed by a rapid decline levelling out after late adolescence [5].

Cognitive Brain Capacities

In adults, specific cognitive capacities of the brain such as verbal learning capacity and visuo-spatial logic capacity may help to explain variations in the recovery of functional sensibility after nerve repair [1].

Timing of Repair

Nerve injuries should be repaired as soon as possible – if the condition allows. The posttraumatic nerve cell death, which usually occurs following nerve injury, can be reduced in this way. With early repair, the surgery is easier to perform since tissues may not yet be swollen, and the natural landmarks such as blood vessels can still be used to ensure a correct matching of the nerve ends. With increasing preoperative delay, there is a fibrosis of the distal nerve segment, atrophy of Schwann cells and there may be a progressive loss of neurons. After nerve transection the corresponding muscle atrophy rapidly, and after two years the muscle fibres may fragment and disintegrate.

Type of Nerve

The type of nerve that is injured considerably influences the functional outcome. If a pure motor nerve is injured, there is no risk of mismatch between motor and sensory cutaneous nerve fibres, thus optimising the accuracy in reinnervation. For pure sensory nerves, the situation is analogous. With mixed nerves, however, the situation is quite different with obvious risks for motor/sensory mismatch.

Level of Injury

After nerve transection, there is an initial delay of days or weeks followed by sprouting and axonal outgrowth. The regeneration in humans has been reported to be non-linear, with a gradually decreased regeneration rate in distal parts. In humans the average outgrowth rate is at most 1–2 mm/day. When digital nerves in fingers are injured, there is only a short distance separating the regenerating axons from their distal target, while more proximal lesions may have a very substantial distance to grow. Lesions to the median nerve at wrist level may require 3–4 months before the first signs of reinnervation of the hand occur. In brachial plexus lesions reinnervation of the hand seldom or never occurs because of the long regeneration distance.

References

1. Lundborg G (2004) Nerve injury and repair. Regeneration, reconstruction and cortical re-modelling. Elsevier, Philadelphia
2. McAllister RM et al. (1996) The epidemiology and management of upper limb peripheral nerve injuries in modern practice. J Hand Surg Br 21:4–13
3. Kaas JH, Florence SL (1997) Mechanisms of reorganization in sensory systems of primates after peripheral nerve injury. Adv Neurol 73:147–158
4. Merzenich MM, Jenkins WM (1993) Reorganization of cortical representations of the hand following alterations of skin inputs induced by nerve injury, skin island transfers, and experience. J Hand Ther 6:89–104
5. Lundborg G, Rosen B (2001) Sensory relearning after nerve repair. Lancet 358:809–810
6. Dvali L, Mackinnon S (2003) Nerve repair, grafting, and nerve transfers. Clin Plast Surg 30:203–221
7. Millesi H (1984) Nerve grafting. Clin Plast Surg 11:105–113
8. Viterbo F (1993) A new method for treatment of facial palsy: the cross-face nerve transplantation with end-to-side neurorraphy. Rev Soc Bras Cir Plast Estet Reconstr 8:29
9. Dellon AL (1981) Evaluation of sensibility and re-education of sensation in the hand. Williams and Wilkins, Baltimore
10. Rosén B, Balkenius C, Lundborg G (2003) Sensory re-education today and tomorrow. Review of evolving concepts. Br J Hand Ther 8:48–56

Regeneration of Optic Nerve

MARI DEZAWA[1], KWOK-FAI SO[2]
[1]Department of Anatomy and Neurobiology, Kyoto University Graduate School of Medicine, Kyoto, Japan
[2]Department of Anatomy, The University of Hong Kong, Hong Kong, People's Republic of China

Definition

Regeneration of retinal ganglion cell axons.

Characteristics

Higher Level Structure of Optic Nerve

The optic nerve is part of the central nervous system (CNS) and has a structure similar to other CNS tracts. The axons that form the optic nerve originate in the ►ganglion cell layer of the retina and extend through the ►optic tract. As a tissue, the optic nerve has the same organization as the white matter of the brain in regard to its glia. There are three types of glial cells: oligodendrocytes, astrocytes and microglia.

Structural Regulation

Little structural and functional regeneration of the CNS takes place spontaneously following injury in adult mammals. In contrast, the ability of the mammalian peripheral nervous system (PNS) to regenerate axons after injury is well documented. A number of factors are involved in the lack of CNS regeneration, including: (i) the response of neuronal cell bodies against the damage, (ii) myelin-mediated inhibition by oligodendrocytes, (iii) glial scarring, by astrocytes, (iv) macrophage infiltration, and (v) insufficient trophic factor support.

Higher Level Process

The fundamental difference in the regenerative capacity between CNS and PNS neuronal cell bodies has been the subject of intensive research. In the CNS, the target normally conveys a retrograde trophic signal to the cell body. CNS neurons die because of trophic deprivation. Damage to the optic nerve disconnects the neuronal cell body from its target-derived trophic peptides, leading to the death of retinal ganglion cells (RGCs). Furthermore, the axotomized neurons become less responsive to the peptide trophic signals they do receive. The survival of certain types of CNS neurons depends on physiological activation of electrical activity or elevation of intracellular ►cyclic AMP (cAMP). On the other hand, adult PNS neurons are intrinsically responsive to neurotrophic factors and do not lose trophic responsiveness after axotomy [1].

Oligodendrocytes, which represent the myelinating glia in the CNS, carry on their surface axon growth-inhibiting molecules [2]. The hypothesis states that neurons in the CNS begin to lose their axonal regenerative capacity at roughly the period with the onset of myelination. Specific components of the myelin produced by the oligodendrocytes, such as ►Nogo A and ►myelin associated glycoprotein (MAG), have been shown to inhibit axonal growth, and antibodies against these proteins resulted in axonal regrowth in the CNS.

The glial scar at the injury site is a biochemical and physical barrier to successful regeneration. It contains large numbers of reactive astrocytes, oligodendrocyte precursor cells, and CNS meningeal cells. A recent study suggests that injury-upregulated ►bone morphogenetic protein 7 (BMP7) synthesized within the CNS induces differentiation of astrocytes from neural progenitors, which may also contribute to glial scar formation after CNS injury [3]. The expression of repulsive molecules such as ►semaphorin-3A, ►tenascin, ►NG2, ►neurocan, ►phosphacan, ►chondroitin and keratan sulfate proteoglycans are related to the repulsive nature of glial scars [4]. The reactive glial extracellular matrix is directly associated with the failure of axonal regeneration, whereas the myelinated white matter beyond the glial scar is rather permissive for regeneration. Nevertheless, Moon and Fawcett [5] have shown that despite the reduction of scar formation by treatment with antibodies to ►transforming growth factors (TGFs), sufficient enhancement of spontaneous CNS regeneration was not obtained. There is no doubt that glial scars have a negative impact on CNS regeneration, although their precise contribution to the inhibitory nature of the CNS environment needs to be ascertained.

The injury is very slowly and poorly infiltrated by macrophages. The importance of macrophage infiltration is illustrated by the observation that it stimulates regenerative responses in the transected rat optic nerve axons [6]. However, microglial activation is considered to be a double-edged response. The first stage of activation includes a non-phagocytic state, where microglia become hypertrophic and produce molecules that are cytotoxic to neuronal cells, such as ►tumor necrosis factor (TNF)-alpha. However, microglia also release cytokines to promote regeneration, for example ►TGF-beta, to promote tissue repair by reducing astrocytic scar formation. In addition, trophic factors including ►BDNF and ►GDNF, secreted by microglia, may also support regeneration.

Process Regulation

The ability of the mammalian PNS to regenerate axons after injury is well documented. Studies in the past decade have shown that the Schwann cell, one of the most important myelin components of the peripheral glia, plays a key role in regeneration. The proliferation and activation of Schwann cells leads to the production of various kinds of factors and other related molecules, to enhance the axons of the proximal nerve stump to grow through the distal stump. Activated Schwann cells express a variety of cell adhesion molecules including ►neural cell adhesion molecules (NCAM), ►L1 and their close homologues ►CHL1, ►N-cadherin and integrins, represented by ►alpha1-beta1 and alpha6-beta1-integrin ($\alpha1\beta1$ and $\alpha6\beta1$-integrin), which mediate interactions between Schwann cells and axons, including growth cones. Besides these trophic factors and cell adhesion molecules, the Schwann cell supplies molecules to the extracellular matrix, such as ►fibronectin, ►laminin, ►J1/tenascin and ►merosin (laminin-2), to the injured axons, which then extend their processes. Among these extracellular molecules, ►laminin-alpha2 is known to play an important role in establishing remyelination, since its absence in mice led to reduced compactness and delay of myelination [7].

Therapy

One strategy to elicit optic nerve regeneration is to provide a favorable environment by supplying neurotrophic factors and the transplantation of cells known to support axonal regeneration. Schwann cell is a strong candidate for transplantation, because optic nerve axons are known to regenerate, when the usual glial milieu is experimentally replaced by Schwann cells and/or peripheral nerve segments. Indeed, several experiments, involving CNS, have shown that exogenous supply of Schwann cells can improve axonal growth across the injured site [7].

Some cells such as gene-transfected astrocytes, ►olfactory ensheathing cells, ►ependymal cells, differentiated embryonic stem (ES) cells, and neuronal stem cells, can induce elongation of CNS nerve fibers, however, it has not been established that the elongated nerve fibers are remyelinated. Many CNS axons are myelinated by oligodendrocytes. The optic nerve tract is a typical example. Myelinating cells, either of Schwann cell or oligodendrocyte origin, mediate the spacing of sodium channel clusters at the nodes of Ranvier to enable saltatory conduction, which is a prerequisite for normal neuronal activity and function. Therefore, even if the CNS can elongate its axons, remyelination of regenerated axons is indispensable for the re-establishment of CNS function.

Schwann cells myelinate in peripheral axons, they also remyelinate CNS axons when transplanted. They are "cells with a purpose" and amongst the best candidates for implantation to support CNS regeneration. Thus, it is expected that transplantation of Schwann cells could become a feasible clinical treatment in the future if the technical and surgical issues can be overcome.

In addition to Schwann cell implantation, various other approaches have been attempted, as mentioned above, but a single approach alone does not appear to provide an optimal condition for optic nerve regeneration. Instead, recent studies using combined approaches, for example, ►CNTF with ►Nogo-neutralizing antibody IN-1 [8], and CNTF with cAMP [9] have shown a synergistic effect on RGC axon regeneration. It is therefore suggested that combining various experimental approaches including neutralizing inhibitory molecules (e.g. ►IN-1 or ►Nogo receptor blocker), blocking inhibitory signaling pathways (e.g. ►Rho pathway inhibitor), supplementing appropriate neurotrophic factors (e.g. BDNF, ►NT-4/5 or CNTF), providing a favorable environment for axon regeneration (e.g. peripheral nerve graft or Schwann cells/ olfactory ensheathing glia transplantation), preventing scar tissue formation (e.g. ►Chondroitinase ABC), and elevating intrinsic regrowth capability (e.g. cAMP elevation), will help to provide the most favorable condition for optic nerve regeneration.

References

1. Goldberg JL, Barres BA (2000) The relationship between neuronal survival and regeneration (Review). Annu Rev Neurosci 23:579–612
2. Fournier AE, Strittmatter SM (2001) Repulsive factors and axon regeneration in the CNS. Curr Opin Neurobiol 11:89–94
3. Setoguchi T, Yone K, Matsuoka E, Takenouchi H, Nakashima K, Sakou T, Komiya S, Izumo S (2001) Traumatic injury-induced BMP7 expression in the adult rat spinal cord. Brain Res 921:219–225
4. Pasterkamp RJ, Giger RJ, Ruitenberg MJ, Holtmaat AJ, De Wit J, De Winter F, Verhaagen J (1999) Expression of the gene encoding the chemorepellent semaphorin III is induced in the fibroblast component of neural scar tissue formed following injuries of adult but not neonatal CNS. Mol Cell Neurosci 13:143–166
5. Moon LD, Fawcett JW (2001) Reduction in CNS scar formation without concomitant increase in axon regeneration following treatment of adult rat brain with a combination of antibodies to TGFbeta1 and beta2. Eur J Neurosci 14:1667–1677
6. Schwartz M, Moalem G, Leibowitz-Amit R, Cohen IR (1999) Innate and adaptive immune responses can be beneficial for CNS repair. Trends Neurosci 22:295–299
7. Dezawa M, Adachi-Usami E (2000) Role of Schwann cells in retinal ganglion cell axon regeneration. Prog Retin Eye Res 19:171–204
8. Cui Q, Cho K-S, So K-F, Yip H (2004) Synergistic effect of Nogo-neutralizing antibody IN-1 and ciliary neurotrophic factor on axonal regeneration in adult rodent visual systems. J Neurotrauma 21:617–625
9. Cui Q, Yip HK, Zhao RC et al. (2003) Intraocular elevation of cyclic AMP potentiates ciliary neurotrophic factor-induced regeneration of adult rat retinal ganglion cell axons. Mol Cell Neurosci 22:49–61

Regeneration of the Central Nervous System

Definition

Regeneration in general represents the replacement of lost body parts. Regeneration of the central nervous system (CNS) classically referred mainly to the regrowth of damaged neuronal axons. However, it has been realized that the replenishment of lost neural cells, and furthermore, the recovery of lost neural function, can be included in the concept of CNS regeneration. In fact, the attempt to recapitulate normal neural development has become a vital strategy for CNS regeneration.

►Regeneration

Regionalization of the Vertebrate Central Nervous System

HARUKAZU NAKAMURA
Department of Molecular Neurobiology, Graduate School of Life Sciences and Institute of Development, Aging; Cancer, Tohoku University, Aoba-ku, Sendai, Japan

Definition

The vertebrate central nervous system first arises as a simple neural plate, which then forms a neural tube. The neural tube is divided into functionally and morphologically distinct regions. The first sign of regionalization in the central nervous system is the appearance of primary brain vesicles such as the prosencephalon, mesencephalon and rhombencephalon (Fig. 1).

As a result of the subdivision of the prosencephalon into the telencephalon and diencephalon and the rhombencephalon into the metencephalon and myelencephalon, five secondary brain vesicles are formed, which are the fundamental brain plan. The telencephalon differentiates into the cerebral cortex and nuclei. The diencephalon differentiates into the thalamus and hypothalamus. The retina, neurohypophysis and pineal body are also derivatives of the diencephalon. The mesencephalon differentiates into the optic tectum and tegmentum. The metencephalon differentiates into the cerebellum and the pons. The myelencephalon differentiates into the medulla oblongata.

R

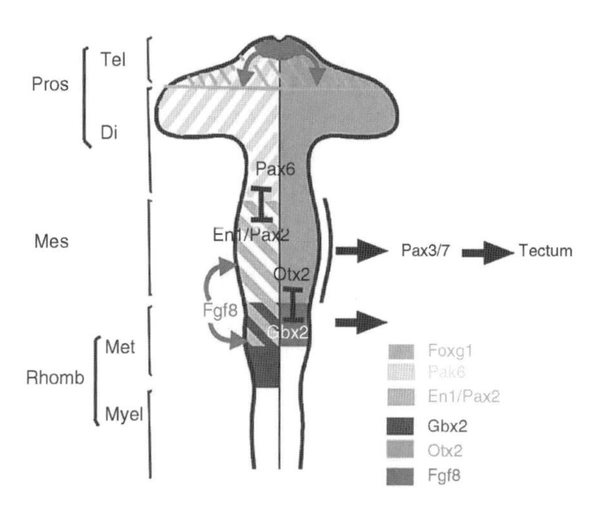

Regionalization of the Vertebrate Central Nervous System. Figure 1 Brain vesicles The fundamental brain plan is in the brain vesicles. Primary brain vesicles (prosencephalon, mesencephalon and rhombencephalon) are transformed into secondary brain vesicles. The fate of the brain vesicles is determined by a combination of expression of transcription factors. The anterior neural ridge and mes-metencephalic boundary function as signaling centers. Otx2 is expressed down to the mes-metencephalic boundary. Foxg1 is expressed in the prospective telencephalon. The di-mesencephalic boundary is determined by repressive interaction between Pax6 and En1/Pax2 and the mes-metencephalic boundary is determined by repressive interaction between Otx2 and Gbx2. The region where Otx2, En1 and Pax2 are expressed is the mesencephalon. Additional expression of Pax3/7 in the mesencephalic alar plate confers differentiation into the optic tectum. *di* diencephalons; *mes* mesencephalon; *met* metencephalon; *pros* prosencephalon; *rhomb* rhombencephalon; *tel* telencephalon.

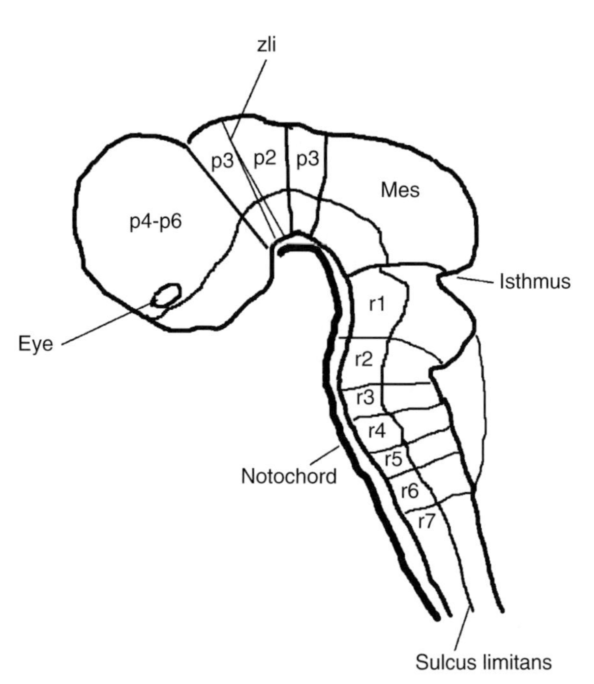

Regionalization of the Vertebrate Central Nervous System. Figure 2 Neuromeres In the rhombencephalon, rhombomeres are formed, which are characterized by bulges and constrictions. Rhombomeres are compartments whose boundaries cells do not cross. In the prosencephalon, prosomere models are proposed. Zli is formed between p2 and p3 and functions as a signaling center. *p1-p6* prosomere 1–6; *r1-r6* rhombomere 1–6; *zli* zona limitans interthalamica.

The sulcus limitans divides the neural tube into the dorsal alar plate and the ventral basal plate (Fig. 2).

Characteristics

Description of the Process

The fate of the region is determined by a combination of transcription factors expressed in the region. For the antero-posterior (AP) axis, boundaries function as organizing centers. Signals from the organizing center regulate expression of the transcription factors, thus regulating the fate of the adjacent region. For the dorso-ventral axis (DV), signaling centers are in the outside of the neural tube, the notochord and the dorsal midline ectoderm. Transcription factors that are homologous to those of *Drosophila melanogaster* are expressed in the vertebrate brain anlage and define the fate of the brain region.

Regionalization of the Prosencephalon

Otx1 and *Otx2*, homologs of *orthodenticle* (*otd*) of *Drosophila*, are expressed in the prosencephalon and mesencephalon. Expression of these genes differs with time; *Otx2* is expressed from a very early stage whereas the *Otx1* expression window is later. *Otx2* plays a more important role in defining the region; *Otx2* null mutant mice lack prosencephalon, mesencephalon and anterior rhombencephalon, although *Otx1* null mutant mice show abnormality in the dorsal telencephalic cortex. Since *Otx1* could be replaced by *Otx2*, it was suggested that the difference in the phenotypes of *Otx1* and *Otx2* null mutant mice stems from differences in expression patterns [1–3].

Emx1 and *Emx2*, homologues of *empty spiracle* (*ems*) are expressed in the telencephalon. These molecules may play a crucial role in arearization in the telencephalon, rather than defining the telencephalic region. *Emx2* is expressed in a gradient, posterior high and anterior low. Fgf8 signal from the anterior and Wnt signal from the posterior (cortical hem) determine the pattern of *Emx2* [1–3]. *Wnt* genes are homologs of *Drosophila wingless* (*wg*).

Foxg1 (*BF1*) is expressed in the telencephalon and defines the telencephalic region (Fig. 1). *Six3* is expressed anterior to the zona limitans interthalamica (zli) and confers competence to express *Foxg1* in response to Fgf8. *Irx3* is expressed posterior to the zli and confers competence to express *En1* and *En2* in response to Fgf8. *Six3*, *Irx* and *En* are homologs of *sine oculis, iroqois* and *engrailed* respectively [4].

Puelles and Rubenstein had proposed that the prosencephalon consisted of six prosomeres (p1-p6), but then reduced it to four prosomeres [5] (Fig. 2).

P1 corresponds to the synencephalon, which is a prospective pretectum. P2 and P3 correspond to the parencephalon and are prospectively thalamus and prethalamus respectively. P4-P6 are the secondary prosencephalon, which gives rise to the telencephalon and hypothalamus. The zli is formed between p2 and p3 and functions as a signaling center [4–6]. P1-P3 are the epichordal part and P4-P6 are the prechordal part.

Regionalization of the Mesencephalon

The mesencephalon is characterized by a combinatorial expression of *Otx2*, *En1* and *Pax2* [7]. *Otx2* is expressed down to the mes-metencephalic boundary (Fig. 1). The posterior limit of the mesencephalon corresponds to that of the *Otx2* expression domain. Misexpression of *Otx2* in the metencephalon changes the fate of the metencephalon to that of mesencephalon, i.e. the metencephalon differentiates into the tectum instead of the cerebellum after misexpression of *Otx2*. *Otx2* knockout mice lack prosencephalon and mesencephalon. Misexpression of *Gbx2,* which is expressed in the metencephalon, causes an anterior shift in the posterior limit of the tectum. Fgf8, Pax2/5, En1/2 are in a positive feedback loop for their expression, so that misexpression of one of these molecules in the diencephalon activates the loop. Since *Otx2* is intrinsically expressed in the diencephalon, misexpression of one of these genes changes the fate of the diencephalon to that of the mesencephalon [3,7,8]. *Gbx2* is a vertebrate homolog of *Drosophila unplugged* and *Pax* genes contain a paired box, which was originally identified in the *Drosophila paired* gene.

Regionalization of the Rhombencephalon

The rhombencephalon is characterized by seven or eight swellings called rhombomeres (r) (Fig. 2). It was shown that the ►neuromeres in the hindbrain are ►compartments [2,6]. The spinal cord also shows a metameric pattern, which is characterized by motor nerves and dorsal root ganglia. ►Metamerism in the spinal cord is not however intrinsically formed, but is a reflection of the ►segmentation of the somite [6]. It was shown that rhombomeres are true segments and form compartments whose boundaries are cell lineage restricting ones. Eph receptor tyrosine kinases and their ligands may be involved in lineage restriction. Receptors (EphA4, EphB2, EphB3) are expressed in odd-numbered rhombomeres (r3, r5) and their ephrin B ligands are expressed in the even numbered rhombomeres (r2, r4 and r6). The ephrin-Eph system is shown to produce repulsion, since the cells in the odd-numbered rhombomeres and those in the even-numbered rhombomeres do not intermingle [2]. Each rhombomere is characterized by a set of motor neurons. Orthologues of *Drosophila Hox* genes are expressed in an ordered and nested manner [2,6].

The identity of the rhombomere is determined by the combination of the expression of *Hox* genes [6]. Regulation of rhombomere identity by *Hox* genes has been shown by gain- and loss-of-function studies. *Hoxb1* is uniquely expressed in r4. Some of the facial motor neurons that are produced in r4 migrate to r6 and vestibuloacoustic neurons migrate to the contralateral side in wild type mice. In *Hoxb1*-knock out mice, neurons produced in r4 do not migrate either to r4 or to the contralateral side, which suggests that the r4 is transformed to r2 in the mutant mice. On the other hand, misexpression of *Hoxb1* in r2 changed its fate to that of r4.

Regulation of the Process by Signaling Centers

The fate of the brain vesicles is determined by a combination of transcription factors. Signals from the boundary regulate expression of the transcriptionm factors. The mes-metencephalic boundary (isthmus) was first recognized to function as a secondary organizer for the tectum and cerebellum [2,7,8]. This was first shown by ectopic transplantation of the brain vesicles. The alar plate of the diencephalon changed its fate and differentiated into the tectum when it was transplanted to the posterior part of the mesencephalon. The fate change did not occur in the anterior part of the mesencephalon. Transplantation of the isthmus to the diencephalon induced the tectum around the transplant, showing that the isthmus functions as the organizer.

Implantation of an Fgf8-soaked bead into the diencephalon mimicked transplantation of the isthmus, i.e. tectum was induced ectopically in the diencephalon by Fgf8 [7,8]. Furthermore, *Fgf8* mutants in zebra fish and mice showed a disruption of the mes and r1. Later work all supported the idea that Fgf8 is a major organizing molecule in the isthmus (Fig. 1). Another secreted factor *Wnt1* is first expressed widely in the mesencephalon and restricted to the posterior margin of the mesencephalon. Wnt1 null mutant mice show a severe deficit in midbrain and hindbrain. But later studies indicated that *Wnt1* functions as a growth-accelerating factor. Fgf17 and Fgf18 have also been shown to function as growth promoting factors. *Wnt1* is a homolog of *Drosophila* wingless (wg).

Among eight splicing isoforms of *Fgf8, Fgf8a* and *Fgf8b* are expressed in the isthmus. Misexpression by *in ovo* electroporation in chick embryos showed that Fgf8a changed the fate of the diencephalon to

R

that of the mesencephalon and that Fgf8b changed the fate of the mesencephalon to that of the metencephalon. Since electroporation with a 1/100 dilution of Fgf8b expression vector exerted Fgf8a type effects, the difference in the effects of Fgf8a and Fgf8b may be due to difference in the intensity of the signal. A strong Fgf8 signal may activate the genes for cerebellar differentiation [7].

Signaling via FGF receptors, tyrosine kinase receptors (RTK), can activate the mitogen-activated protein kinase (MAPK) and the phosphatidylinositol 3-kinase (PI3K). Blocking of the Ras-ERK (MAPK) signaling pathway by the dominant negative form of Ras changed the fate of the metencephalon to that of the mesencephalon, i.e. the tectum developed instead of the cerebellum in the metencephalic region after misexpression of the dominant negative form of Ras. These results indicate that the strong Fgf8 signal activates the Ras-ERK pathway to cause differentiation into cerebellum [7]. Ras-ERK signaling is so strong that it may need negative regulators. Sprouty2, Sef (similar expression of Fgf8) and Mkp3 are induced by Fgf8, but regulate the pathway negatively. Sprouty2 is expressed overlappingly with Fgf8 and can be induced very rapidly by Fgf8 [7]. Repression of the Ras-ERK signaling pathway by misexpression of Sprouty2 changes the fate of the alar metencephalon to become the tectum. On the contrary, excess Ras-ERK signaling by application of dominant negative form of Sprouty2 results in an anterior shift in the mid-hindbrain boundary. Application of a specific inhibitor of the PI3K pathway indicated that this pathway is also activated by Fgf8 to induce Mkp3 and En2.

The anterior neural ridge also expresses Fgf8 and functions as a secondary organizer for the telencephalon. Fgf8 induces *Foxg1* in the telencephalon (anterior to the zli), but induces En in the region posterior to the zli. It was shown that the difference in competence is dependent on the transcription factors expressed in the region. *Six3* confers ability to express *Foxg1* in response to Fgf8, whereas *Irx3* confers ability to express *En2* in response to Fgf8 [4]. *Six3* is a homolog of *Drosophila sine oculis*.

The zli is another signaling center. There Shh is expressed and regulates the differentiation of the thalamic nuclei. *Sox14* and *Gbx2* are expressed in the young neurons of specific nuclei in the dorsal thalamus (*Sox14*: interstitial nucleus of the optic tract, perirotundic area; *Gbx2*: nucleus rotundus, posterior nucleus). High doses of Shh induce GliI, which in turn mediates expression of *Sox14*. On the other hand, low doses of Shh induce GliII, which in turn mediates expression of *Gbx2* [4]. *Shh* (*sonic hedgehog*) is one of vertebrate homologs of *Drosophila hedgehog,* and *Gli* is the homolog of *Drosophila Cubitus interuptus*.

Regionalization Along Dorsoventral (DV) Axis

The floor plate and roof plate, which are situated at the ventral and dorsal midline respectively, segregate the bilateral halves of the neural tube. On each side, motor neurons differentiate in the ventral third, relay neurons in the middle third and smaller interneurons in the dorsal third. *Pax3/7* and *Pax6* are expressed in the dorsal and middle thirds respectively and *Nkx2.2* is expressed in the most ventral part. Class II homeodomain proteins such as Nkx2.2, Nkx6.1, Nkx6.2 and bHLH transcription factor Olig2 are expressed in the most ventral part of the neural tube and Class I homeodomain (HD) proteins such as Pax6, Dbx2, Irx3 and Dbx1 are expressed dorsal to the Class II HD protein. Combination of these transcription factors defines the cell types along the DV axis.

Notochord was shown to have ventralizing activity. Implantation of the notochord lateral to the neural tube could induce floor plate and motor neurons near the implant. On the other hand, removal of the notochord results in extension of the dorsal markers to the ventral and motor neurons and the floor plate disappear. When notochord formation is genetically perturbed in mouse and zebra fish, ventral cell types are absent. For the ventralizing signal, Shh signaling was shown to play a crucial role. Shh is first expressed in the notochord, then the floor plate expresses Shh. Shh could elicit floor plate and motor neuron development ectopically. Centrally, Shh null mutant mice lack floor plate and motor neurons. Shh induces expression of ventral markers and the motor neuron marker islet1, but represses the ventral markers.

BMP4 and BMP 7 emanate from the roof plate and the dorsal ectoderm and antagonize the Shh signal. Thus, cell fate along the DV axis is determined by these signals.

For the DV axis in the mesencephalon, Pax3/7 are expressed in the alar plate of the mesencephalon and force it to differentiate as a tectum [7]. Shh is expressed in the floor plate of the mesencephalon. Misexpression of Shh in the mesencephalon represses Pax3/7 expression and changes the fate of the alar plate of the mesencephalon to the tegmentum (Fig. 1). After misexpression of Shh, motor neurons and dopaminergic neurons differentiate in the dorsal part of the mesencephalon.

References

1. Boncinelli E (1999) Otx and Emx homeobox genes in brain development. Neuroscientist 5:164–172
2. Brown M, Keynes R, Lumsden A (2001) The developing brain. Oxford University Press, New York
3. Simeone A (2000) Positioning the isthmic organizer where *Otx2* and Gbx2 meet. Trends Genet 16:237–240

4. Kobayashi D, Kobayashi M, Matsumoto K, Ogura T, Nakafuku M, Shimamura K (2002) Early subdivisions in the neural plate define distinct competence for inductive signals. Development 129:83–93
5. Puelles L, Rubenstein JL (2003) Forebrain gene expression domains and the evolving prosomeric model. Trends Neurosci 26:469–476
6. Kiecker C, Lumsden A (2005) Compartments and their boundaries in vertebrate brain development. Nat Rev Neurosci 6:553–564
7. Nakamura H, Katahira T, Matsunaga E, Sato T (2005) Isthmus organizer for midbrain and hindbrain development. Brain Res Brain Res Rev 49:120–126
8. Alvarado-Mallart RM (2005) The chick/quail transplantation model: discovery of the isthmic organizer center. Brain Res Brain Res Rev 49:109–113

Regulation of Neurotransmitter Release by Protein Phosphorylation

Robert D. Burgoyne, Alan Morgan
The Physiological Laboratory, School of Biomedical Sciences, University of Liverpool, Liverpool, UK

Definition

The release of neurotransmitters at ►synapses is brought about by the process of regulated ►exocytosis, whereby a rise in the concentration of cytosolic Ca^{2+} triggers the fusion of a ►synaptic vesicle with the plasma membrane and the release of the vesicle's neurotransmitter content into the extracellular space. The proteins responsible for the sensing of the Ca^{2+} signal (synaptotagmin) and for vesicle docking and fusion (the SNARE proteins, SNAP-25, syntaxin and VAMP), have been identified and well characterized. In addition, several other proteins characterized through genetic approaches in flies, worms or mice are known to be either essential for neurotransmitter release (such as Munc13, Munc18-1 and NSF) or have important regulatory roles (such as cysteine string protein (CSP), Rabs and Rab effector proteins). In many neuronal cell types and various other kinds of secretory cells, exocytosis can be modulated through signalling pathways that result in phosphorylation of one or more of the key proteins involved in the exocytotic machinery [1]. The extent or kinetics of neurotransmitter release has been found to be modified by the action of several different ►protein kinases including PKA, PKC, Cdk5 and calmodulin-dependent protein kinase II (CaMKII). It has also been shown that ►protein phosphorylation by these, and other kinases, is required for various forms of ►synaptic plasticity. The changes in ►neurotransmission that underlie synaptic modification are in part post- and in part pre-synaptic. Regulation of protein phosphorylation is important for presynaptic changes in neurotransmitter release which is, therefore, likely to be involved in learning and memory formation. Neurotransmitter release could be modified via phosphorylation of channels or receptors, but it is clear that components of the release machinery are direct targets for protein phosphorylation and these will be the focus of this review. The protein targets for the various protein kinases, and the particular amino acids within these substrates that are phosphorylated, are increasingly being identified, and this is allowing the physiological roles of specific phosphorylation events to be established. The strategy that is being used and is most informative is the expression of mutated forms of the proteins, in which the identified phosphorylated amino acid is rendered non-phosphorylatable (e.g. by changing serine to alanine) or phosphomimetic, by mutation of serine to the acidic amino acid glutamate or aspartate. We have concentrated here on proteins known to be important, based on genetic manipulation, as part of the exocytotic machinery for neurotransmitter release, for which the mutation strategy has defined a significant functional role for protein phosphorylation on a defined amino acid. Other presynaptic proteins that are substrates for protein phosphorylation are known [1], but their importance for neurotransmission has not yet been validated genetically.

Characteristics

Many aspects of neurotransmitter release can be modified following activation of protein kinases. These include an increase in release probability of vesicles, an increase in the size of the ►ready-releasable pool of synaptic vesicles, changes in Ca^{2+}-sensitivity of the release mechanism or changes in the kinetics of individual fusion and release events [1]. This has led to the search for the protein substrates involved. Several key exocytotic proteins have been shown to be substrates for protein kinases *in vitro*, and some of these have been confirmed to be phosphorylated in intact cells in response to physiological stimuli. In even fewer cases has the phosphorylation of a specific protein been convincingly linked to one of the known effects on neurotransmitter release of activation of a specific kinase. Nevertheless, a number of examples of well defined regulation by protein phosphorylation are now known. The key presynaptic proteins involved in neurotransmitter release, which have been shown to be protein kinase substrates, are shown in Fig. 1, and the identified phosphorylation sites that are known are listed in Table 1.

We will concentrate on those proteins that have been confirmed to be important for neurotransmission through genetic approaches, and whose phosphorylation has been shown to be physiologically significant for exocytosis. Phosphorylation of several proteins has been

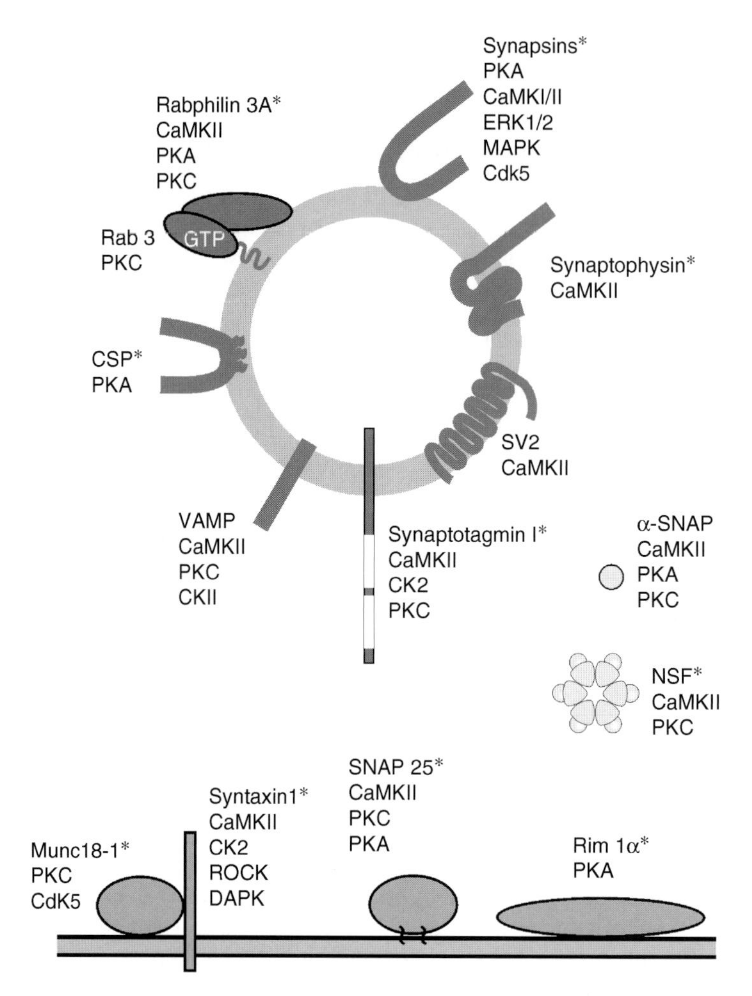

Regulation of Neurotransmitter Release by Protein Phosphorylation. Figure 1 Protein kinase substrates with established roles in the machinery for neurotransmitter release. Key synaptic proteins present on synaptic vesicles, the presynaptic membrane or the cytosol are shown, along with the protein kinases known to phosphorylate them *in vitro*. Only those proteins that have been confirmed through genetic approaches to be required for, or to regulate neurotransmitter release, are included. Proteins that are known to be phosphorylated in intact cells are indicated by asterisks.

linked to modification at various stages in the exocytotic process including vesicle mobilization (synapsins), vesicle recruitment into a releasable pool (RIM1), the maintenance of the ready-releasable pool size (SNAP-25) and late events during membrane fusion (CSP and Munc18-1).

The first presynaptic proteins whose phosphorylation was found to regulate neurotransmitter release were the synapsins, which have been extensively characterized both biochemically and functionally [1]. These proteins cross-link the reserve pool of synaptic vesicles to each other and to the cytoskeleton. Their phosphorylation by CaMKII following nerve terminal depolarization allows the release of the vesicles and, thereby, increases their availability for exocytosis. The functional significance of synapsins in the control of vesicle availability and recycling has been well established through the study of synapses from synapsin I and synapsin II knock-out mice.

A study using neurons from knock-out mice has shown that the Rab effector Rim1, which is localized on the presynaptic plasma membrane, is required to maintain the normal level of release probability in synapses and for ►long term potentiation (LTP) at parallel fibre/ Purkinje cell synapses of the cerebellum [2]. LTP at these synapses is dependent on presynaptic PKA. RIM1 is phosphorylated both *in vitro* and *in vivo* by PKA on Ser-413. The ability of this residue to be phosphorylated is necessary for the recovery of the wild-type phenotype when expressed in neurons from null mutant mice, as expression of non-phosphorylatable mutants was ineffective. In contrast, mutation of another putative PKA phosphorylation site, Ser- 1548 was without effect on the recovery in knock-out mice. This study suggests that phosphorylation of Ser-413 of RIM1 is a significant mechanism for the PKA-dependent plasticity that

Regulation of Neurotransmitter Release by Protein Phosphorylation. Table 1 Identified protein kinase substrates involved in exocytosis and the kinases that phosphorylate them

Protein	*In vitro* phosphorylation sites	*In vivo* phosphorylation sites	Functional significance tested?
CSP	PKA: S10	S10	Yes
Munc18-1	PKC: S306, S313 Cdk5: T574	S313	Yes
Rabphilin 3A	PKA and PKC: S234, S274	S234, S274	No
Rim 1	PKA: S413, S1548 CaMKII: S241, S287 (indirect)	S413	Yes
SNAP-25	PKA: T138 PKC: S187	T138 S187	Yes
Synapsin	PKA and CaMKI: S9	As for *in vitro*	Yes
	CaMKII: S566, S603		
	ERK1: S62		
	ERK2: S67		
	MAPK and Cdk5: S549		
	Cdk5: S551		
Synaptotagmin I	PKC and CaMKII: T112	T112	No
Syntaxin 1A	CK2 and ROCK: S14 DAPK: S188	S14	Yes

Key synaptic proteins are listed that have been shown to be phosphorylated *in vitro* and whose phosphorylation sites have been identified. Only those proteins that have been confirmed through genetic approaches to be required for, or to regulate neurotransmitter release, are included.

exists in certain types of synapses and that involves changes in neurotransmitter release. The mechanistic basis for the effect of RIM1 phosphorylation is, however, unknown.

SNAP-25 is one of the key SNARE proteins that associates with syntaxin and VAMP and mediates vesicle docking/fusion at the plasma membrane. Phosphorylation of SNAP-25 has been suggested to regulate the size of the ready-releasable pool of vesicles, based on data from studies on ►adrenal chromaffin cells [3,4]. SNAP-25 [5] can be phosphorylated both *in vitro* and *in vivo* by PKA and PKC on identified sites (Table 1), and this has been implicated in the functional effects of PKA and PKC activation on exocytosis. Activation of PKC has multiple effects on exocytosis, one of which is an increase in the rate of refilling of the ready releasable pool of vesicles. SNAP-25 is phosphorylated by PKC on Ser-187 and this reduces its association with other SNARE proteins. Expression of SNAP-25 in adrenal chromaffin cells with mutations in this residue either increased (phosphomimetic mutant) or impaired (non-phosphorylatable mutant) the rate of refilling of the ►vesicle pools [3]. This suggests that this effect of PKC activation could be through phosphorylation of Ser-187 of SNAP-25. In contrast, a study on hippocampal pyramidal neurons did not find any effect of mutating Ser-187 of SNAP-25 on neurotransmitter release, suggesting the existence of other functionally important PKC substrates that increase neurotransmitter release in hippocampal synapses. Neurotransmitter release can also be increased by activation of PKA. In addition, the tonic activity of PKA is linked to the maintenance of the pool of ready releasable vesicles in adrenal chromaffin cells, and this was revealed by the use of PKA inhibitors [4]. Expression of SNAP-25 mutated at Thr138, the identified PKA phosphorylation site, to produce a non-phosphorylatable mutant reduced the size of the initial fast burst of exocytosis in chromaffin cells, suggesting that this might be the target for PKA's action on the ready releasable pool size. Phosphorylation of Thr138 has no effect on SNAP-25 binding to other SNAREs, although the effect on other protein interactions made by SNAP-25 is unknown.

CSP is a synaptic and secretory vesicle protein that has a chaperone role in the synapse. The phosphorylation status of CSP has been shown to affect late events in exocytosis that lead to changes in vesicle release kinetics and quantal size [6]. Overexpression of CSP in adrenal chromaffin cells was found to reduce the number of exocytotic events and also slowed vesicle release kinetics. CSP is phosphorylated *in vitro* on Ser-10 and this site was found to be phosphorylated *in vivo*. CSP phosphorylated on Ser-10 shows a reduced affinity for the syntaxin 1A and synaptotagmin I [6]. Expression of CSP with a mutation in Ser-10 (a non-phosphorylatable mutant) still reduced exocytosis but no longer modified the release kinetics. This suggests that Ser-10 is a target for protein phosphorylation, and that its phosphorylation can regulate neurotransmitter release through an effect on the time course of release from individual vesicles. It is currently unclear, however, whether the regulation of CSP is a consequence of PKA-mediated phosphorylation or phosphorylation by some other kinase that recognizes the motif at Ser-10.

Munc18-1 is a member of the Munc18/Sec1 family of proteins that function in essentially all intracellular membrane fusion events. It is essential for neurotransmission in mice and knock-out animals are paralysed and die *in utero*. Munc18-1 and its orthologues in other species appears to have both negative and positive functions exerted in part through its interaction with syntaxin. Munc18-1 is phosphorylated *in vitro* by PKC on Ser-306 and Ser-313, and by Cdk5 on Thr-574. Only phosphorylation on Ser-313 has so far been confirmed to occur *in vivo* [7]. Phosphorylation of Munc18-1 by PKC or mutation of Ser-306 and Ser-313 to glutamates reduces the affinity of Munc18-1 for binding syntaxin 1A. Significantly, expression of Munc18-1 with the phosphomimetic mutations in Ser-306 and Ser-313 mimics the effects of PKC in increasing the speed of single vesicle release events in chromaffin cells [8]. Another effect of PKC, to increase the number of exocytotic events was not observed as a consequence of expressing phosphomimetic Munc18-1, suggesting that another PKC substrate (SNAP-25?) must be involved in this effect of PKC. In contrast, expression of a phosphomimetic mutation of Munc18-1 at Thr574 only partially reproduced the effect of Cdk5 activation. The evidence suggests that Munc18-1 may be a physiological substrate for PKC, but the significance of the putative phosphorylation by Cdk5 is still unclear.

Phosphorylation of syntaxin 1A *in vivo* has been demonstrated and recently implicated in the exocytotic events that are involved in the insertion of new membrane in growing neuronal processes [9]. Syntaxin 1A was found to be the substrate for both casein kinase II [10] and the Rho-associated serine/threonine kinase (ROCK) [9], and was phosphorylated on Ser-14 by both kinases. This phosphorylation site was demonstrated to be functionally important as its phosphorylation increased the affinity of binding of tomosyn to syntaxin, and thereby inhibited its ability to form productive SNARE complexes. It was suggested that this would provide a mechanism for the spatial regulation of exocytosis. Other work has shown, however, that phosphorylation of Ser-14 of syntaxin increases its binding to synaptotagmin and its recovery in complexes with SNAP-25, which would be more consistent with a stimulatory effect on exocytosis. It is not known whether phosphorylation of syntaxin 1 does regulate neurotransmitter release.

The changes in neurotransmitter release that occur following phosphorylation of synapsin, RIM1, SNAP-25, CSP and Munc18-1 are also believed to be due to modifications in specific protein-protein interactions between these proteins and others in the exocytotic machinery. The molecular basis for the effects of phosphorylation, are in general still to be resolved. In particular, the effect of PKA phosphorylation of RIM1 and SNAP-25 on their protein-protein interactions is unknown. As noted above, more information is available on the molecular effects of other phosphorylation events. It is also possible that other substrate proteins could be crucial for mediating the protein kinase effects. We still have only limited knowledge of the significance of the phosphorylation of other exocytotic proteins, and the physiological conditions under which specific protein phosphorylation events occur and become relevant for changes in neurotransmission and synaptic plasticity. It is clear, however, that these are important mechanisms that contribute to learning and memory.

References

1. Leenders AGM, Sheng Z-H (2005) Modulation of neurotransmitter release by the second messenger-activated protein kinases: implications for presynaptic plasticity. Pharmacol Ther 105:69–84
2. Lonart G, Schoch S, Kaeser PS, Larkin CJ, Sudhof TC, Linden DJ (2003) Phosphorylation of RIM1α by PKA triggers presynaptic long-term potentiation at cerebellar parallel fiber synapses. Cell 115:49–60
3. Nagy G, Matti U, Nehring RB, Binz T, Rettig J, Neher E, Sorensen JB (2002) Protein kinase C-dependent phosphorylation of synaptosome-associated protein of 25 kDa at Ser^{187} potentiates vesicle recruitment. J Neurosci 22:9278–9286
4. Nagy G, Reim K, Matti U, Brose N, Binz T, Rettig J, Neher E, Sorensen JB (2004) Regulation of releasable vesicle pool sizes by protein kinase A-dependent phosphorylation of SNAP-25. Neuron 41:351–365
5. Shimazaki Y, Nishiki T-I, Omori A, Sekiguchi M, Kamata Y, Kozaki S, Takahashi M (1996) Phosphorylation of 25-kDa synaptosome-associated protein. Possible involvement in protein kinase C-mediated regulation of neurotransmitter release. J Biol Chem 271:14548–14553
6. Evans GJO, Wilkinson MC, Graham ME, Turner KM, Chamberlain LH, Burgoyne RD, Morgan A (2001) Phosphorylation of cysteine string protein by PKA: implications for the modulation of exocytosis. J Biol Chem 276:47877–47885
7. Craig TJ, Evans GJO, Morgan A (2003) Physiological regulation of Munc18/nSec1 phosphorylation on serine-313. J Neurochem 86:1450–1457
8. Barclay JW, Craig TJ, Fisher RJ, Ciufo LF, Evans GJO, Morgan A, Burgoyne RD (2003) Phosphorylation of Munc18 by protein kinase C regulates the kinetics of exocytosis. J Biol Chem 278:10538–10545
9. Sakisawa T, Baba T, Tanaka S, Izumi G, Yasumi M, Takai Y (2004) Regulation of SNARES by tomosyn and ROCK: implication in extension and retraction of neurites. J Cell Biol 166:17–25
10. Foletti DL, Lin R, Finley MA, Scheller RH (2000) Phosphorylated syntaxin 1 is localized to discrete domains along a subset of axons. J Neurosci 20:4535–4544

Regulatory Region

Definition

Regulatory region is a promoter, enhancer or other DNA sequence of a gene that is bound by transactivating factors that control gene expression. The best-studied regulatory regions are in DNA, but they also exist in RNA where they can be bound for example by micro RNA.

Regulatory Route

Definition

Regulatory route refers to the pathway whereby after synthesis in the rough endoplasmic reticulum, proteins are transported via the Golgi apparatus to be stored in granules, and exocytosed from granular storage.

Regulatory T Cells

Definition

A sub-population of T cells, which in their resting state bear markers, e.g. CD25, characteristic of activated T lymphocytes. These cells, which originate in the thymus, ensure peripheral tolerance of autoimmune T cells by a mechanism known to be characterized by cell-cell contact and cytokine secretion, but are not yet fully understood.

►Protective Autoimmunity

Reinforcement

Definition

This term was first used in classical conditioning by Pavlov to describe the process by which a conditioned stimulus (CS) came to substitute for the unconditioned stimulus (US) and thus elicit a conditioned response. Its current use in the classical conditioning literature is rather casual and denotes trials on which the CS is followed by the US in contrast to the term non-reinforcement which denotes trials where the CS is presented without the US.

Reinforcement is a core concept in operant conditioning. It has traditionally been used to refer to the process of strengthening a response or behavior. Positive reinforcement is often used to describe situations in which the occurrence of an instrumental response leads to a desirable outcome (or reward). Its counterpart, negative reinforcement, is used to refer to situations in which the occurrence of an instrumental response leads to either the removal or the postponement of an undesirable event, although, depending on the particular theoretical orientation, the terms escape and avoidance learning, respectively, are more likely to be employed.

►Classical Conditioning (Pavlovian Conditioning)
►Operant Conditioning
►Theory on Classical Conditioning

Reinforcement Learning in Neural Networks

Definition

An approach for training noisy networks based on increasing the likelihood of outputs that yield greater reward on average.

►Neural Networks for Control

Reinforcement Learning in Animals

Definition

Reinforcement learning is a learning rule to search optimal value based on a reward signal, signifying to the organism which conditions are desirable and which are the undesirable ones.

►Reinforcement

Reinforcer

Definition

In associative conditioning theory, a reinforcer is an event that modifies the frequency of the behavior that preceded it. The term refers to operant learning (also called Skinnerian or instrumental learning), a form of associative learning in which animals learn to associate a behavioral action (for instance pressing a lever) to an outcome, either positive (a food reward) or negative (a punishment with an electric shock). Typically, the probability of the bar pressing response would increase if it is associated to the reward, but would decrease if it is associated to the punishment, a principle termed "the Law of Effect" by Edward Thorndike at the end of the 19th century. Note however, that the removal of a punishment can also act as a reinforcer: for instance, if bar pressing induces the end of a very loud noise, this behavior can increase because the end of the loud noise acts as a positive reinforcer.

►Reinforcement

Reinnervation

Definition

Return of lost nerve fibers (innervation) to a cell, tissue, organ.

►Neuronal Changes in Axonal Degeneration and Regeneration
►Regeneration

Reinnervation of Muscle

Definition

The nerve supply to denervated muscle fibers can be restored by reinnervation; injured nerve fibers regrow their axons to reach and resupply or reinnervate the denervated muscle fibers.

►Axonal Sprouting in Health and Disease

Reinstatement

Definition

The return of a conditioned response following re-exposure to the unconditioned stimulus after extinction training.

►Learning and Extinction

Relation

Definition

In the basic binary case, a relation R is a rule specifying when an object a is related by R to b. In an abstract mathematical sense, a binary relation is simply the set of ordered pairs (a,b) upon which it holds. The domain of the relation is the set of a which are related to some b.

The range of the relation is the set of b which are related to by some a. The relation is reflexive if every object a is related to itself. The relation is symmetric if whenever a is related to b, then b is also related to a. The relation is transitive if whenever a is related to b, and b is related to c, then a is related to c. An equivalence relation is a relation that is reflexive, symmetric and transitive.

A relation f is a function if to each a in its domain, there is exactly one b to which a is related, and this b is said to be the value of the function at a, written b = f(a).

A function is one-to-one if whenever a1 and a2 are distinct, then so also are f(a1) and f(a2). The function is onto B if every object in B is in the range of the function.

A one-to-one correspondence between A and B is a oneto- one onto function with domain A and range B. That is, a one-to-one correspondence provides a way of matching objects in A to objects in B in such a way that every object in A corresponds to a unique object in B and every object in B is corresponded to by a unique object in A.

The concept of binary relation can be generalized to trinary relations, which holds of triples (a,b,c), and so on to any dimension, even to infinite dimensions.

►Physicalism

Relational or Configural Navigation Strategy

Definition

Behavior relying on an allocentric reference frame and oriented by an internal spatial representation.

►Spatial Memory

Relative Pain Unpleasantness

Definition

The amount of pain unpleasantness associated with a specific intensity of a pain sensation. It is a measure of how much a specific pain sensation bothers an individual. Equivalent intensities of pain may vary in unpleasantness, such as laboratory pain versus the pain of childbirth, or the pain of childbirth versus late stage cancer pain.

►Emotional/Affective Aspects of Pain

Releasing Values

Definition

Each stimulus has a certain attractiveness and may elicit a certain behavior. The attractiveness is measured by the releasing value. The natural, adequate stimulus has a releasing value of 100. Many stimulus have a lower releasing value, but some may have higher or supernormal releasing values.

Releasing-Hormone and Release-Inhibiting Hormone

Definition

These chemicals (mostly peptides) are produced by specific cells (neurosecretory cells) situated mainly in the hypothalamus and transported to the anterior pituitary gland. There they stimulate or inhibit a release of various anterior pituitary hormones. They include thyrotropin (TSH)-releasing hormone (TRH), which also acts as prolactin-releasing factor (PRF); adrenocorticotropin (ACTH) – releasing hormone (CRH); growth hormone (GH) – releasing hormone, (GHRH); growth hormone release-inhibiting hormone (somatostatin); gonadotropin (GnH) – releasing hormone (GnRH), sometimes called luteinizing-hormone (LH)- releasing hormone (LHRH), and prolactin release-inhibiting factor, now considered to be dopamine.

►Homeostasis
►Hypothalamo-pituitary-adrenal Axis
►Stress and Depression
►Hypothalamo-pituitary-thyroid Axis
►Hypothalamus
►Pituitary gland

Reliabilism

Definition

Reliabilists claim that knowledge is true belief arrived at in a reliable manner, i.e. in a manner that makes it likely that the resulting belief is true.

►Knowledge

REM

Definition

Rapid Eye Movement Sleep.

►EEG in Sleep States

REM-off Cells

Definition

Extracellular single-unit-recording studies show that many neurons discharge at their highest rates (<5 spikes/sec) during waking, diminish their activity during non-REM (NREM) sleep, and become silent during REM sleep. Since these cells stop firing during REM sleep, they are called REM-off cells (also called PS-off cells). The majority of these REM-off cells are located in the locus coeruleus (LC) and raphé nuclei (RN). REM-off cells located in the LC contain the neurotransmitter noradrenaline and REM-off cells located in the RN contain the neurotransmitter serotonin. Although few in number, this type of cell is also present in the caudal part of the pedunculopontine tegmentum (PPT) and laterodorsal tegmentum (LDT).

►Locus Coeruleus
►Non-REM Sleep
►Noradrenaline
►Raphé nuclei
►Rapid Eye Movement (REM) Sleep
►Serotonin

REM-on Cells

Definition

Neurons that exhibit increases in extracellularly recorded discharge rate during ►rapid eye movement (REM) sleep, rather than during waking and non-REM (NREM) sleep. This population of neurons is characterized by a progressively increasing mean tonic discharge rate as the animal moves from wake to NREM sleep and finally to REM sleep. This type of cell is mostly located in the pontine reticular formation, pedunculopontine tegmentum (PPT) and laterodorsal tegmentum (LDT). REM-on cells located within the pontine reticular formation contain the neurotransmitter glutamate and REM-on cells located in the PPT and LDT contain the neurotransmitter acetylcholine.

►Acetylcholine
►Glutamate
►Non-REM Sleep
►Rapid Eye Movement (REM) Sleep

REM Sleep

Definition

►Rapid Eye Movement (REM) Sleep

REM Sleep Behavior Disorder (RBD)

Definition

A parasomnia (a disorder involving abnormal behavior during sleep) characterized by the acting out of vivid, sometimes violent, confrontational or belligerent dreams during REM sleep. This happens because the REM-related atonia of voluntary muscles is lacking, allowing the muscles to move during dreaming. REM behavior disorder causes sleep disruption and potential injury to self or to others (e.g., a bed partner).

►Alertness Level

Remapping in Hippocampus

Definition

Remapping is the process of replacing one map representation with another. A representation is "remapped" when the map elements are scrambled. In the hippocampus, changing environments, or contexts, is associated with remapping.

►Spatial Learning/Memory

REMO

Definition

Episodic retrieval mode; a component process of episodic retrieval that is required for remembering earlier experiences.

►Hemispheric Asymmetry of Memory

Remote Memory

Definition

The long-term representation of information that was learned months to years earlier.

►Memory and Dementia

Remyelination

Definition

Myelin sheaths are formed by Schwann cells in the peripheral nervous system, and by oligodendrocytes in the central nervous system. If myelin sheaths are degraded due to injury or pathological changes, new myelination develops on the surviving axons. In the central nervous system, remyelination is usually incomplete with a reduced number and irregular configuration of myelin lamellae over a long period of time.

►Myelin
►Oigodendrocyte
►Regeneration
►Schwann cell
►The Role of Basal Lamina in Nerve Regeneration
►Autoimmune Demyelinating Disorders: Stem Cell Therapy

Renshaw Cell

Definition

Renshaw cells are inhibitory interneurons (using glycine and GABA as transmitters), which are located in the ventral horn of the spinal cord, receive their main excitatory inputs from collaterals of motoneurons and mediate recurrent inhibition of motoneurons, Ia inhibitory interneurons, Renshaw cells and cells of origin of the ventral spinocerebellar tract. Other inputs to Renshaw cells arise from sensory afferent fibers and tracts descending from supraspinal structures.

►Ia Inhibitory Interneuron
►Recurrent Inhibition

Repetition Maximum (RM)

Definition

Repetition Maximum represents the load used in resistance training. Performing sets of ten repetition maximum (RM) loads or less are typically used for resistance training, with one RM being the maximum weight an individual can lift once, and ten RM being the weight an individual can lift exactly ten times. These values represent 100% and ~70% of maximum capability for one RM and ten RM, respectively.

►Muscle – Age-Related Changes

Replacement Neuromast

Definition

A superficial neuromast (hair cell of the lateral line system) having phylogenetic continuity with a neuromast found inside a canal in other taxa. This superficial configuration is most likely due to retarded development of canals in that taxon.

►Evolution of Mechanosensory and Electrosensory Lateral Line Systems
►Neuromast

Repolarization

Definition

Repolarization is the return of membrane potential to its resting value. The term refers mostly to repolarization of the action potential, although more generally it also means the return to a more negative value after (forced) depolarization. Repolarization of the action potential is often carried by the outward flux of potassium ions mainly through delayed rectifying, voltage-gated potassium channels.

►Action Potential
►Neuronal Potassium Channels

Report

►Feedback Control of Movement

Representation (Mental)

MICHAEL SCHÜTTE
Institut für Philosophie, Universität Magdeburg, Magdeburg, Germany

Definition

The term "mental representation" is sometimes used to cover any mental item which is semantically evaluable,

R

i.e. which has content, ►truth-value, refers to something, possesses ►truth-conditions, or is about something. Under this broad construal its extension includes beliefs, thoughts, memories, desires, perceptions and all other mental phenomena exhibiting the feature of intentionality.

But there also is a narrower construal of "mental representation" closely associated with the agenda of cognitive science. Under this narrower construal, mental representations are certain theoretical entities, i.e., semantically evaluable particulars which are postulated by classical or other types of cognitive architectures in order to explain processes and states which count as mental representations only in the broad sense.

Description of the Theory

Mental representations as theoretical entities postulated by cognitive scientists come in very different shapes. Think of a cognitive scientist working in the paradigm of classical architectures. Insofar as she tries to understand the mind as a complex system that receives, transforms and stores information, that is, as a complex symbol-manipulating system, her approach is based on the idea that mental phenomena should be explained by postulating mental representations (symbols). Or think of a cognitive scientist working in the paradigm of ►connectionist architectures. Insofar as he tries to understand human behavior and mentality as based on the activity of neural networks, his approach, too, is based on the idea that mental phenomena should be explained by postulating mental representations, although these are notably different from those of classical architectures.

Mental representations as theoretical entities postulated by cognitive scientists come in great variety including, e.g., activity vectors in connectionist networks, Marr's 2½ -D sketches in his theory of vision, Kosslyn's mental images, the mental models of Johnson-Laird, or Fodor's "sentences in the language of thought" [1]. Ironically, most mental representations are misleadingly labeled "mental" representations, for they are explicitly understood as certain neuronal or other physical structures. But certainly, they all qualify as mental in the weak sense that they are postulated in order to explain mental features. The mental features to be explained range from pattern recognition to intentional behavior, but in the following we will concentrate on the so called propositional attitudes like believing or desiring that something is the case. Generally speaking, propositional attitudes are mental states which can be ascribed with the help of "that"-clauses.

Representation

Representations are, of course, ubiquitous: There are words, photographs, paintings, maps, traffic signs, diagrams, graphs, music notes, X-ray photographs, digital images, and much more. They are not bound to a specific medium or syntax, and there are virtually no limits as to which things can represent which. A representation, as a self-representation, can represent itself, and two different things can represent each other. Representation should be distinguished from mere information, at least if the latter is taken to include things like the universe containing information about the big bang or the smoky sky containing the information that there is a fire. The idea behind distinguishing representation from information is that information cannot be false (otherwise it would not be information at all), whereas misrepresentation is possible. If this is correct, smoke does not represent fire, although it "indicates" or "means" fire [2]. It is notoriously hard to spell out precisely the necessary and together sufficient conditions which make something a representation, but there is a kind of consensus that every representation purports to stand for, denote, refer to, or be about something. Another important aim of philosophical thinking about representation is to build a useful classification of the many different forms of representation (see [3] and [4] for two very influential classificatory schemes).

Representational Theory of Mind

The best-known representational theory of propositional attitudes is developed by Fodor [1]. It is a paradigm instance of a ►classical architecture in cognitive science, and is best seen as an attempt to explain how propositional attitudes and reasoning processes can be physically realized. Strictly speaking, this is a two-step enterprise: The first step is concerned with the question how propositional attitudes and cognitive processes can be realized by computational relations and processes in which symbols are manipulated. The second step consists in explaining how these computational relations and processes can in principle be physically implemented. Fodor thinks that the second step is already established by the theoretical work of Turing and others and, of course, by the actual development of computers. Therefore, he sees his main task in making intelligible how propositional attitudes can be realized by computational relations. The central features of propositional attitudes which are to be explained include the following: They are (i) semantically evaluable, (ii) causally efficacious, and (iii) opaque.

In order to account for these features, Fodor does several things. First, he assumes that there are mental representations. These are held to be sentence-like physical structures in a "language of thought." This means four things: Like sentences mental representations have propositional content; like sentences they are structured entities which have parts that themselves possess meaning; like sentences they have a compositional semantics, i.e., their meaning is a function of the meanings of their parts and the order of these parts; and these parts are "transportable" which means that the same parts can appear in many mental formulas (Fodor [1]: 137). Fodor calls the conjunction of these claims the Language of Thought Hypothesis. Second, Fodor then uses these views to formulate the representational thesis according to which propositional attitudes are relations between organisms and mental representations. For example, to believe that grass is green means, according to the representational thesis, to stand in a certain relation (the belief-relation) to a mental representation which means that grass is green. More generally, for any organism O, and any attitude A toward the proposition P, there is a computational relation R and a mental representation M such that (i) M means that P, and (ii) O has A if and only if O bears R to M (Fodor [1]: 17). That an organism bears a certain computational relation to a mental representation M is spelled out in the following way: The representation M occupies a certain causal or functional role in the organism; i.e. it is tokened in a special functionally defined area (e.g. the "belief-box" or the "desire-box") and will be manipulated in a specific way. Third, Fodor claims that mental processes are causal sequences of tokenings of mental representations. This is best conceived as the view that causal relations between propositional states rely on computational processes that are sensitive to the structure of the involved mental representations.

This theory neatly explains the central features of propositional attitudes mentioned above as follows. (i) That propositional attitudes are semantically evaluable is accounted for by the fact that they are realized with the help of mental representations which are semantically evaluable. (ii) The deeper point in connection with second feature (causal efficacy) is this: Often causal relations between propositional attitudes contrive to respect their relations of content. For example, my thoughts that p and that (if p, then q) often cause me to think that q. This is explained by the fact that cognitive processes are computational processes which are structure-sensitive. This sensitivity to the syntax of mental representations is enough to explain the possibility of the parallel structure of logical and causal relations, because, as is well-known from logic, logical relations can be characterized syntactically. (iii) The belief that p and the belief that p* can be different beliefs, such that it is possible to believe that p without at the same time believing that p* (and vice versa), even when p and p* are both true or both false (or even possess the same truth-conditions). This feature of ►opacity can be explained by the representational theory under the assumption that the mental representations of p and p* are syntactically different. Because syntactically different representations are typically manipulated in different ways, it is no mystery how at a certain time, there can be the mental representation p, but not the mental representation p* in someone's belief-box.

Main objections to Fodor's representational theory concern two issues. The first is the issue whether there is empirical evidence against its implication that causally efficacious attitudes require actual tokenings of mental representations. The second is the issue whether the assumption that there are physical structures which have semantic content can be made plausible at all. This is the topic of the next section.

Physical Structures as Mental Representations

The representational theory as outlined above simply assumes that mental representations have a semantic content. Therefore, it remains another task for its proponents to explain how these representations being neural or physical entities can be semantically evaluable at all: How can physical or neuronal structures actually represent some state of affairs? This matter is not only of interest to the proponents of the representational theory, but is also of crucial interest for anyone else who takes a realistic stance on mental representations. Over the last two and a half decades, philosophers have developed several approaches to answer this question [1,5,6].

1. Dretske's information-theoretic approach is rendered in terms of information, and analyzes the property of having semantic content as a form of carrying information: A certain structure S (e.g. a representation) has the semantic content that p if and only if S carries the information that p and the information that p is the most specific piece of information which S carries, i.e. S carries no other piece of information in which the information that p is nested (see Dretske [5]: 177). That S carries information about something X at all basically means that there is a certain causal correlation between S and X. The attractiveness of this approach lies in the fact that in principle it is no mystery how physical structures (rocks, clouds, or brains) can carry information. In order to explain how misrepresentation is possible, Dretske appeals to the learning period in which a representation is acquired. In a nutshell, the explanation is that only causal correlations during the learning period determine what S represents, whereas after the learning period S can be caused by different things which, then, are misrepresented by S. This information-theoretic account faces two major problems: (i) it only works for representations which are acquired through an individual learning history, and (ii) it cannot deal with the ►disjunction problem (Fodor [1]: 101f). The latter problem is a fundamental problem for every broadly causal account of meaning or content: If the contents of A-representations are determined by their being caused by

C-states, but sometimes A's are brought about by some other cause C*, then how can the causal theory account for the difference between the case in which A's represent C, but sometimes are caused by C*-states, and the case in which A's represent the disjunction C-or-C*? Dretske's appeal to a learning period would only be of help here if it were guaranteed that during this period all and only those factors cause A's which should enter into the semantic content of A's. But this seems to be false for empirical reasons.
2. Teleological accounts try to solve the disjunction problem by appeal to the biological function or purpose of representational states [6]. According to them, the mental representation R, although sometimes caused by dogs, represents wolfs and not wolfs-or-dogs, because it is the biological function of R-type representations to indicate wolfs and not wolfs-or-dogs. A major difficulty for this type of accounts is to explicate the notion of biological function in a non-semantic way. This is typically tried to be accomplished by an appeal to the (evolutionary) history of the organism. But it has turned out over the years that this is by far no easy task. Another problem lies in the fact that teleological accounts which appeal to evolution have counterintuitive consequences. For example, most people have the intuition that a perfect physical and behavioral duplicate of a human being would also have beliefs and desires. But if this duplicate has the wrong kind of history (or literally no history at all), it is thereby precluded from having mental representations. This arguably leads to an epistemological difficulty as well: If it is the evolutionary history of an organism which determines whether and which things are represented by its inner states, we can only know whether someone believes something if we know enough about its evolution.
3. Fodor [1] developed a third type of account which is based on the notion of asymmetric dependence. It runs along the following lines: A-states (of an organism O) represent C if and only if (i) under optimal conditions all C's cause an A-token, and (ii) all A-tokens which are caused by a state C* are asymmetrically dependent on the causation of A-tokens by C's. The idea behind Fodor's notion of dependence can be caught by a question: Would the C*-state also have caused A-tokens if C-states did not cause A-tokens? If not, the causal relation between C*-states and A-tokens is dependent on the relation between C-states and A-tokens. This dependence is asymmetrical if it is not the other way round, i.e. if it is not true that the relation between C-states and A-tokens is dependent on the relation between C*-states and A-tokens. Although this is an ingenious proposal, as some philosophers have pointed out, it might be entirely misguided. Let us assume that tokens of A in an organism O mean "bird," but sometimes are caused by big insects. According to Fodor's proposal, this is the case because big insects would not cause A's if birds did not cause A's in O. But, now, what is it that precludes big insects to cause A's in O even if there were no birds around? Why, in other words, should there be an asymmetric dependence relation at all? Certainly, if A were to mean "bird" in the first place, it would be quite plausible that some big insects cause A's in O only because normally birds cause A's in O. But that A means "bird" is exactly what Fodor's theory tries to explain and, therefore, cannot be assumed by it.

Intentional Realism Versus Eliminativism

The representational theory of propositional attitudes and the project of naturalizing mental representations as sketched in the last section are committed to ►Intentional Realism. Intentional Realism is the view that humans have intentional states which (i) more or less obey the laws of folk psychology and which (ii) have a semantic content that is (iii) causally efficacious. But these assumptions can, of course, be denied. Most prominently some philosophers favor an eliminative stance towards propositional attitudes. Churchland [7] argues that folk psychology is a rather unsuccessful theory which in the long run will be substituted by a much better explanation of human behavior developed by scientific psychology or neuroscience which is incompatible with the folk assumptions. But because intentional states like beliefs and desires are only theoretical entities postulated by folk psychology, Churchland argues, they will be eliminated when folk psychology is abandoned.

Intentional Realism and Eliminative Materialism are opposing views on the ends of a spectrum and there are positions in-between. A very prominent one is Dennett's [8]. On the one hand, Dennett agrees with the Eliminative Materialists that there are in principle neuroscientific explanations of human behavior which are superior and incompatible with ►intentional explanations. On the other hand, he stresses that we cannot abandon intentional explanations altogether, because there are certain patterns in human behavior which can only be discovered from the intentional stance. Whether this or other "in-between" positions (as [9,10]) can be coherently defended is difficult to evaluate. What they all try to show is that the semantic contents of intentional states are real enough to underpin the autonomy of intentional explanations, but at the same time are not real enough to require their physical implementation. This can be put it in another way which perhaps is an exaggeration: These approaches aim to preserve mental representations in the broad sense (intentional states) without committing themselves to the existence of mental representations in the narrow sense. Faced with the empirical and conceptual problems of Intentional Realism and the smell of absurdity of eliminativism, this may be an attractive direction of inquiry for further theories of mental representation.

References

1. Fodor J (1987) Psychosemantics. MIT Press, Cambridge, MA
2. Dretske F (1995) Naturalizing the mind. MIT Press, Cambridge, MA
3. Goodman N (1976) Languages of art. Hackett, Indianapolis
4. Peirce CS (1935) Collected papers, vol. 2. Harvard University Press, Cambridge, MA
5. Dretske F (1981) Knowledge and the flow of information. MIT Press, Cambridge, MA
6. Millikan R (1984) Language, thought, and other biological categories. MIT Press, Cambridge, MA
7. Churchland P (1981) Eliminative materialism and the propositional attitudes. J Philos 78:67–90
8. Dennett D (1987) The intentional stance. MIT Press, Cambridge, MA
9. Beckermann A (1996) Is there a problem about intentionality? Erkenntnis 45:1–23
10. Cummins R (1989) Meaning and mental representation. MIT Press, Cambridge, MA

Repressed Memories

Definition

Memories for traumatic experiences that the mind supposedly banishes from conscious awareness due to their threatening nature. According to this theory, once repressed, these memories may return to consciousness.

The resulting memories are thought to be accurate in detail, and involve processes that are different from ordinary forgetting and remembering. Credible scientific support is lacking for these notions.

►Memory Distortion

Repressor

Definition

A transcription factor that negatively regulates the expression of a gene.

Reproduction

Definition

Production of offspring.

Reproductive Organs

►Visceral Afferents

Reptilia

Definition

The amniote clade incorporating the last common ancestor of turtles, lizards, crocodiles and birds, and all descendents of that common ancestor.

►The Phylogeny and Evolution of Amniotes

Repulsive Guidance Molecule

Definition

The molecule by which growth cone movement is repelled.

►Axon Pathfinding

RER

Definition

►Rough Endoplasmic Reticulum

R

Res Cogitans

Definition

Latin phrase meaning "thinking thing," introduced by Descartes to refer to the mind as opposed to the body (the res extensa or "extended thing").

► Emergence

Rescorla-Wagner Model

Definition

This model of classical conditioning attributes variations in the effectiveness of conditioned stimulus-unconditioned stimulus (CS-US) pairings to variations in US processing. The model asserts that an US must be surprising for learning to occur. An US is defined as surprising if the discrepancy term ($\lambda - V_T$) is different from zero. This discrepancy reflects the difference between the maximum conditioning the US can support (λ) and the associative strength of all the stimuli on the trial (V_T). The equation for calculating a change in the associative strength of a CS is:

$$\Delta V_{CS} = \alpha_{CS}\beta_{US}(\lambda - V_T)$$

where V_T represents the total or sum of the individual associative strengths of all CSs present on that trial; α and β are fixed rate parameters (values from 0 to1) determined by the salience (physical properties) of the CS and US, respectively; λ is the maximum conditioning that the US can support.

► Theory on Classical Conditioning

Resetting

Definition

Alteration of a circadian rhythm such that it occurs earlier (advance) or later (delay) than predicted in subsequent cycles.

► Circadian Cycle
► Circadian Rhythm
► Clock

Residual Brain Cells

Definition

Astrocytes, microglia and neurons are the residual brain cells. Along with cells of the immune system which migrate towards the brain in CNS disorders, these also play an important role in the etiology of these disorders.

► Central Nervous System Disease – Natural Neuroprotective Agents as Therapeutics

Residual Hearing

Definition

The amount of hearing left after hearing loss.

► Hearing Aids

Residual Schizophrenia

Definition

Constellation of symptoms which often occur after many years of a chronic course. Beside psychotic symptoms, patients suffer from symptoms of general cognitive impairment and affective flattening.

► Schizophrenia

Resistance (Electrical)

Definition

Resistance (electrical) is the reciprocal of conductance and a measure of the resistance of an object to electric current flow.

► Ohm's Law

Resistance Training

Definition
Resistance training or strength training can be defined as progressively overloading the neuromuscular system using near maximal muscle contractions against high resistance. Its purpose is to increase the ability to perform maximal contractions and increase muscle size.

►Muscle – Age-Related Changes

Resonance

Definition
The frequency at which maximum output occurs.

►Hearing Aids

Respiration – Neural Control

PETER M. LALLEY
Department of Physiology, Medical Sciences Center, University of Wisconsin School of Medicine and Public Health, University Avenue, Madison, WI, USA

Synonyms
Respiratory system neurophysiology; Neuroanatomy and neurotransmitter/neuromodulator control

Definition
The neural control of respiration refers to functional interactions between networks of neurons that regulate movements of the lungs, airways and chest wall and abdomen, in order to accomplish (i) effective organismal uptake of oxygen and expulsion of carbon dioxide, airway liquids and irritants, (ii) regulation of blood pH.

Introduction
The neural control of respiration is still not completely understood, although remarkable progress has been made as instrumentation, technology and analytical procedures continue to improve at an accelerated pace. (Many excellent reviews are available that have followed progress in the field, and the interested reader is encouraged to consult them for particular areas of interest [1–14,16–18,20,21,23–26,28–31,33].)

It is axiomatic that biological cells are dependent on respiration for survival, proper function and ►homeostasis. They require an efficient transport system that provides oxygen (O_2) for aerobic metabolism and energy production and for extrusion of its end products, carbon dioxide (CO_2) and water.

In mammals respiration takes on an additional, organismal meaning and significance, synonymous with ventilation; i.e., the act of breathing ambient air in and out to deliver O_2 from the mouth and nasal passages to the lungs, and to transport CO_2 from lungs to atmosphere.

The Respiratory Apparatus in Mammals
Effective organismal exchange of O_2 and CO_2 in mammals requires finely coordinated interactions between the organs of breathing and the ►cardiovascular system. The latter consists of the heart acting as the pump for blood in which O_2 and CO_2 are dissolved and the vascular network, from capillaries to major arteries and veins, which serves as the transport line between the lungs and other O_2-consuming, CO_2-producing organs. A system of valves in the heart and in the sphincters surrounding arterioles regulates the flow of blood.

The organs of breathing (Fig. 1) can also be subdivided into pump and flow components. The pump machinery consists of the diaphragm, chest wall and abdominal muscles, while the transport system involves the mouth, nasal passages, bronchi, bronchioles and lungs. In the lungs, the alveoli, a vast network of air-filled sacs, are in intimate contact with blood capillaries where exchange of O_2 and CO_2 takes place. Airway resistance and airflow in and out of the lungs is affected by altering the tone of bronchiole smooth muscle, pharyngeal skeletal muscle, nasal musculature, as well as tongue position and the tone of the laryngeal (vocal fold) abductor and adductor muscles.

Performance of the Respiratory Apparatus During Inspiration and Expiration
During inspiration, airflow into the lungs and alveoli is produced as the diaphragm contracts during periodic discharges of the phrenic and inspiratory intercostal nerves (Fig. 2); the discharges are activated by excitatory synaptic drive within the ►brainstem-spinal cord respiratory network. Contraction of the diaphragm changes its configuration from dome shaped to relatively flat. This increases the intrathoracic volume, resulting in an increase of negative intrapleural pressure that promotes lung inflation and inward airflow from the mouth to the lungs. Discharges in intercostal nerves contract the inspiratory muscles of the rib cage,

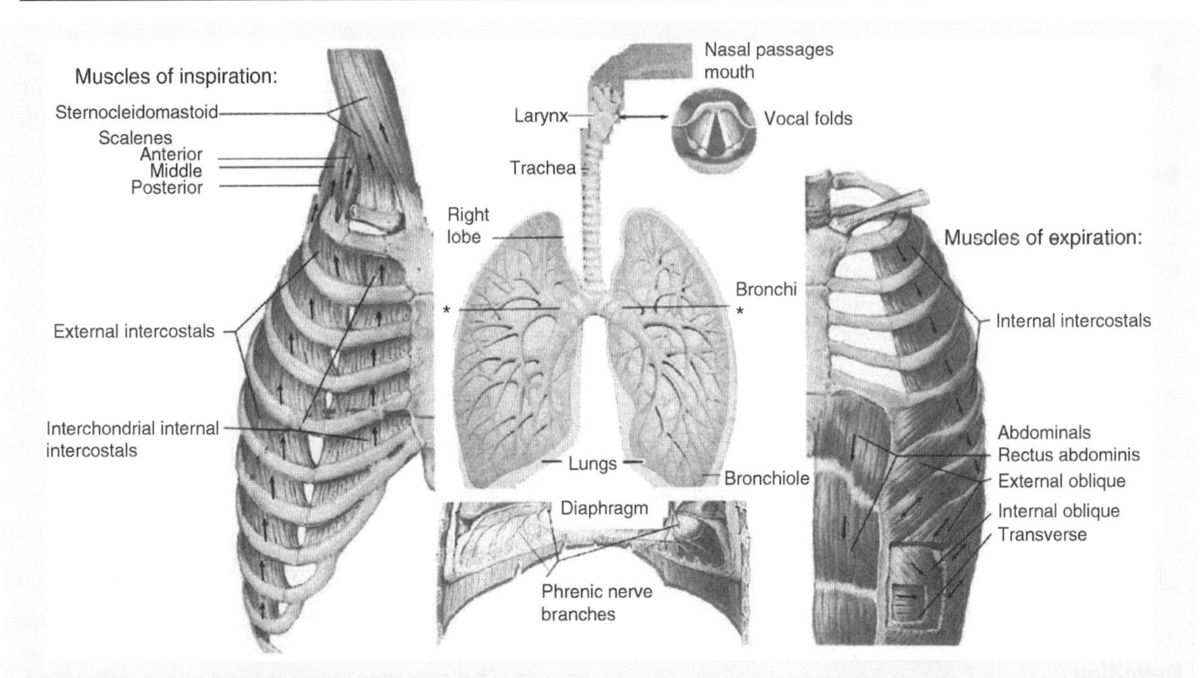

Respiration – Neural Control. Figure 1 The respiratory apparatus. Left side, muscles that expand the chest for lung inflation during inspiration are illustrated. Arrows show the upward direction of rib movement. The middle upper segment illustrates the airways, from nasal passages to lungs. A cross section through the larynx shows the laryngeal folds in an open (abducted) state during inspiration. The middle lower segment shows the diaphragm, which contracts downward to inflate the lungs during inspiration. Phrenic nerve branches that innervate the diaphragm and cause the musculature to contract are also seen. Right, muscles of the chest wall and abdomen that contract to aid lung deflation during active expiration. Arrows show the direction of rib and abdominal muscle movements. Modified from [19].

moving the ribs upward and outward to further increase intrathoracic volume and inward airflow. (►Spinal respiratory neurons) Movement of air into the lungs is further supported by cranial ►motoneuron discharges (►Action potential) that reduce upper airway resistance by contracting the muscles of the nasal passages and pharynx, move the tongue forward in the mouth (►Respiratory control of hypoglossal motoneurons during sleep and wakefulness) and dilate the vocal folds by contracting abductor laryngeal muscles.

Until the end of the inspiratory phase, phrenic and inspiratory intercostal nerve discharges progressively increase, causing a gradual flattening of the diaphragm and expansion of the chest wall.

At the end of the inspiratory phase, phrenic, inspiratory intercostal and abductor laryngeal motoneurons stop discharging. A very brief silent period is followed by resumption of discharges that is less intense and declining as it progresses. During this transitional stage, referred to as either the postinspiratory or early expiratory phase, adductor motoneurons of the superior laryngeal nerve also discharge with declining intensity. The decrementing phrenic and laryngeal nerve discharge patterns result in a more gradual relaxation of the diaphragm, a reduced rate of outward airflow and thus a slowing in the rate of lung deflation. Alveolar collapse is opposed and ►functional residual capacity (FRC) is maintained, which has beneficial pulmonary consequences. Normally, only about 15% of air in the lungs is replaced by new air during each normal inspiration, and about the same amount of old air is expired. The slow replacement of air prevents sudden changes in blood gases that would destabilize respiratory control. Excessive increases and decreases of blood pO_2, pCO_2 and pH are also prevented when respiration is temporarily interrupted, for example during swallowing or phonation. Partial inflation during a normal FRC also maintains surfactant release and thus lung compliance, because the principal stimulus for liberation of surfactant appears to be direct mechanical distortion of type II alveolar cells [35].

Expiration during quiet breathing is mainly passive. Discharges of inspiratory cranial, phrenic and inspiratory intercostal nerves are silenced by synaptic inhibition in the brainstem and spinal cord. The chest wall and diaphragm return to their resting configurations and airway patency is maintained to allow passive outward airflow from the lungs to the atmosphere.

During active expiration, for example during exercise or coughing, discharges of the internal (expiratory) intercostal nerves move the lower ribs downward and inward. In addition, lumbar motoneuron discharges

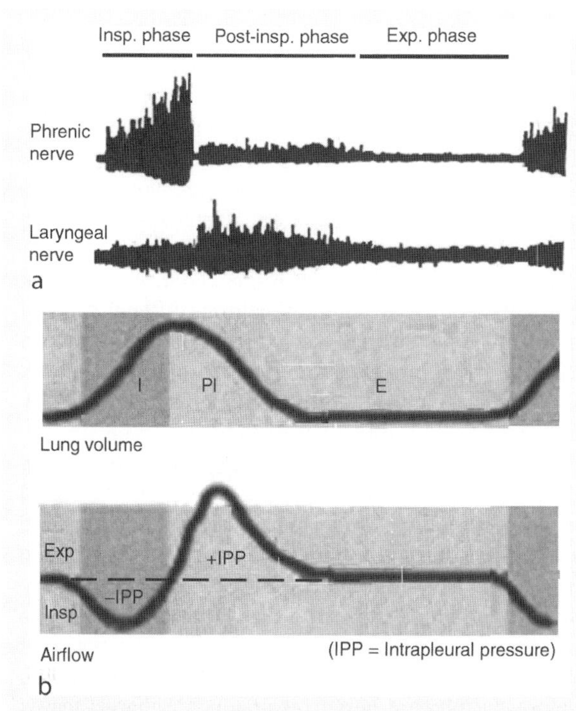

Respiration – Neural Control. Figure 2 Discharges of the phrenic and laryngeal nerves (A) and changes in lung volume and airflow (B) during one respiratory cycle. In part B, I = inspiration, PI = post-inspiration, E = expiration. Part A adapted from Bianchi et al. 1995 [2]. Part B adapted from [24].

contract the abdominal muscles and compress the abdominal contents, pushing up the diaphragm and actively expelling air from the lungs. Upper airway patency is maintained by discharges of laryngeal abductor nerves and pharyngeal constrictor nerves.

The cycle of inspiration, postinspiration and expiration in the adult human is repeated, on average, about 15 times during quiet breathing.

Respiratory Muscles Contract in an Ordered Sequence that Optimizes Mechanical Advantage

The inspiratory pump muscles in both humans and quadrupeds discharge with a set temporal order that optimizes the mechanical advantage, or leverage of the muscles, and reduces the work of breathing [8]. Electromyographic studies in humans show that, relative to the onset of airflow into the airways, the diaphragm and the third dorsal external intercostal muscles are the first to contract, followed by the second parasternal intercostal and scalene muscles and lastly by the fifth dorsal external intercostal muscles. The order of recruitment is consistent with the relative inspiratory mechanical advantage that each of the muscles has, and the degree of inspiratory opening pressure exerted on the airways by each. The intensity and duration of unit discharges are greater for the diaphragm and third dorsal external intercostal ►motor units than for the other pump muscles. The pattern of recruitment of intercostal muscles not only expands the rib cage, but also applies stretch to the diaphragm to increase contractile strength.

Several factors have been posited for the recruitment and discharge patterns of the different pump muscles, including: (i) recruitment order of bulbospinal and spinal interneurons, (ii) different degrees of persistent and rhythmic inward currents and (iii) their spatial distribution over the soma and dendrites of motoneurons and (iv) graded distribution of inhibitory central respiratory drive potentials.

Central Nervous Control of the Respiratory Apparatus

Aggregates of neurons that discharge periodically during inspiration, post-inspiration or expiration are distributed bilaterally in the bulbar brainstem, from the rostral ►pons (►Pontine control of respiration) to the caudal border of the ►medulla (►Anatomy and function in the respiratory network). Synaptic interactions among the neurons establish the network respiratory rhythm, and their connections with cranial and spinal motoneurons and interneurons set up the timing and patterns of contraction in the muscles of respiration. Three regions of the medulla in particular have been studied for their roles in respiratory rhythmogenesis: the Pre►Bötzinger Complex (►PreBötzinger Complex Inspiratory Neurons and Rhythm Generation) and the para-facial and retrotrapezoid regions (►Respiratory network analysis and isolated respiratory centre functions; [12]). Their functional integrity is essential for a normal respiratory rhythm, and in the PreBötzinger and para-facial areas neurons with autorhythmic pacemaker properties have been identified (►Respiratory network analysis, isolated respiratory centre functions; ►Pacemaker neurons and respiratory control).

Respiratory neurons of the brainstem receive modulatory synaptic input from non-respiratory regions such as the ►motor cortex, pontine and medullary ►reticular formation, ►cerebellum, ►hypothalamus, other ►limbic and cardiovascular regions of the brainstem as well as from extrapyramidal motor areas (►Anatomy and function in the respiratory network). These non-respiratory modulatory inputs adapt breathing rhythm and pattern to accommodate activities such as phonation, swallowing, coughing, physical exertion, defecation and postural change.

Rhythm Formation in Bulbar Respiratory Neurons

The ►membrane potential of medullary respiratory neurons normally oscillates between cycles of depolarization and hyperpolarization. The pattern of depolarization or hyperpolarization may be augmenting (increasing in intensity from onset to termination), decrementing (declining in intensity) or plateau (constant from onset to termination). In association with the patterns of depolarization, periodic discharges can be augmenting, decrementing or of constant intensity [9,25].

The rhythm and pattern of discharge in bulbar respiratory neurons result from a combination of intrinsic membrane ion ►conductances, synaptic interactions among the neurons, and input from other CNS neurons and peripheral sensory afferents.

Intrinsic membrane ion conductances initiate membrane depolarization that triggers action potential discharge, control the rate of depolarization and hyperpolarization, and terminate action potential discharge [25,26] (►PreBötzinger Complex Inspiratory Neurons and Rhythm Generation). Tonic excitatory drive comes from at least two sources. One is from CO_2-sensitive neurons in the medulla (►Central nervous chemoreceptors and respiratory drive; ►Medullary raphe nuclei and respiratory control). A second is from non-respiratory reticular activating neurons. These excitatory inputs can be suppressed or reinforced by feedback synaptic input from medullary and pontine respiratory neurons. ►Chemoreceptor and ►mechanoreceptor afferents from the ►carotid bodies (►Carotid body chemoreceptors and respiratory drive), heart, lungs, chest wall and upper airways also influence discharge properties of bulbar respiratory neurons.

All afferent inputs and synaptic interactions among the respiratory and non-respiratory neurons are regulated chemically by ►neurotransmitters and neuromodulators, including excitatory and inhibitory amino acids, acetylcholine, peptides, monoamines and adenosine (►Respiratory neurotransmitters and neuromodulators).

Rhythmic Properties, Connections and Functions of Respiratory Neurons in the Pons and Medulla

The roles that various types of bulbar neurons play in respiratory control have been investigated by: (i) measuring membrane potential and discharge properties during various phases of the respiratory cycle, (ii) identifying synaptic connections among them and (iii) observing their responses to changes in respiratory rhythm and ventilation. From such studies, theories of how the neurons interact as a network to generate rhythm have been proposed [2,16,25] (►Anatomy and function in the respiratory network). Elegant computer modeling studies based on the experimental findings have tested and support many of the proposed connections and predict additional ones (►Computational modeling of the respiratory network).

Neurons in the respiratory-related regions of the rostral pons discharge with waves of excitatory and inhibitory postsynaptic potentials and intermittent bursts of action potentials and that are coincident with one of the three phases of the respiratory cycle, and some discharge more intensely at the transitions between phases (►Pontine control of respiration). In unanesthetized cats prepared for chronic studies, respiratory rhythm is not obvious during sleep [2], and rhythmicity is diminished when nervous input from the lungs is intact. The pontine network of neurons modulates amplitude and timing of the respiratory muscles, and seems to promote inspiratory termination if pulmonary afferent feedback is impaired.

In respiratory regions of the medulla, six different types of medullary respiratory neurons differentiated by membrane potential and discharge properties have been identified: (i) Early-Inspiratory, (ii) Ramp-Inspiratory, (iii) Late-Inspiratory, (iv) Post-Inspiratory, (v) Late-Expiratory, (vi) Pre-Inspiratory (Fig. 3).

Early-Inspiratory neurons are propriobulbar, that is, their cell bodies, dendrites and axons are restricted to bulbar regions. They depolarize suddenly to threshold shortly before phrenic nerve discharge begins at the onset of inspiration. One type of Early-Inspiratory neuron exhibits a discharge that is intense but decrementing, in association with a declining pattern of depolarization, until it terminates late in inspiration, before the phrenic nerve inspiratory discharge ceases [25]. The other type exhibits a constant, or plateau pattern of depolarization and discharge that begins and ends with the phrenic nerve inspiratory discharge [9]. In either case, two processes seem to be responsible for discharge termination. One is weak synaptic inhibition, possibly from Late-Inspiratory neurons, but a more prominent source is the development of a ►Ca^{2+}-activated K^+ conductance that builds up as Ca^{2+} enters during cell discharge through high voltage-regulated channels. Thereafter, membrane potential is hyperpolarized by synaptic inhibition produced by the discharge of Post-Inspiratory and Late-Expiratory neurons. The proposed role of Early-Inspiratory neurons of the decrementing type is to impose synaptic inhibition on Late-Inspiratory and Post-Inspiratory neurons, which prevents premature termination of phrenic motoneuron discharges [25]. As for the plateau type, they seem to augment inspiratory phase excitatory synaptic drive, because input to Ramp-Inspiratory neurons has been demonstrated electrophysiologically [9].

Ramp-Inspiratory neurons are either propriobulbar and/or bulbospinal and provide excitatory synaptic drive to phrenic and intercostal inspiratory motoneurons and interneurons. They depolarize and discharge with an augmenting pattern at the beginning of inspiration, in parallel with phrenic nerve discharges. The pattern of depolarization and discharge in Ramp-I neurons is attributed to a combination of mutual recurrent excitation among the neurons and declining inhibitory synaptic input from Early-I neurons. Discharge is terminated by inhibitory synaptic input from Late-Inspiratory and Post-Inspiratory neurons.

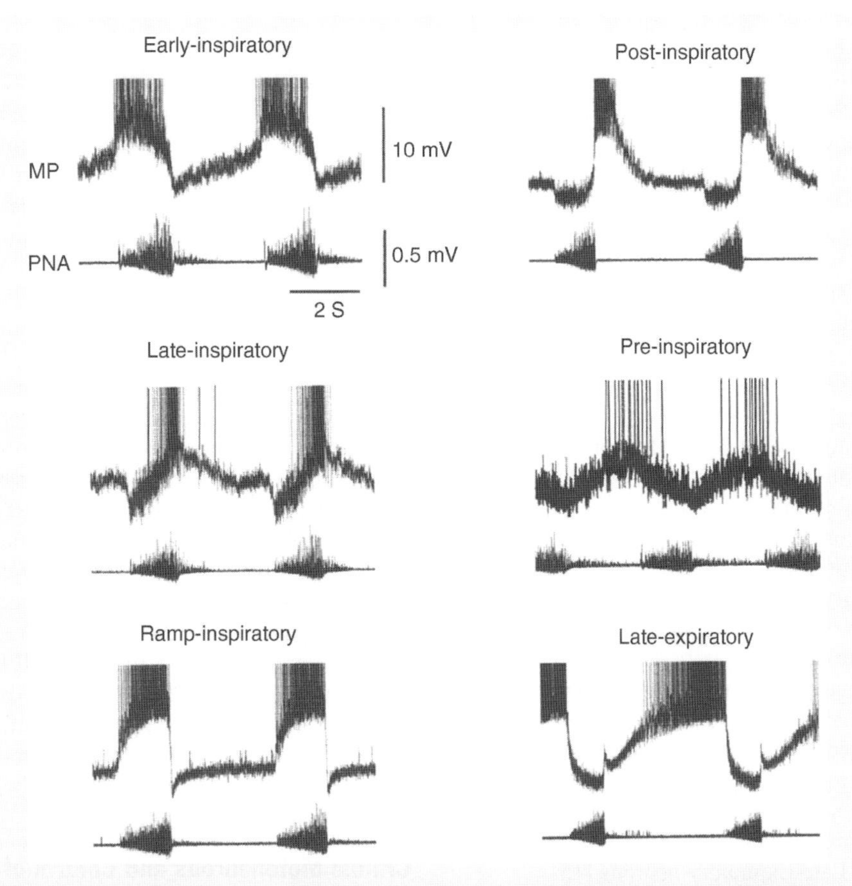

Respiration – Neural Control. Figure 3 Medullary respiratory neurons thought to be involved in respiratory rhythmogenesis. The six types of neurons were recorded intracellulary in adult cats *in vivo* in studies performed by A. Haji and coworkers. Each pair of tracings shows neuron membrane potential (MP) and phrenic nerve activity (PNA). Figure courtesy of Prof. Dr. Akira Haji, Laboratory of Neuropharmacology, School of Pharmacy, Aichi Gakuin University, Nagoya.

Late-Inspiratory neurons are propriobulbar. During most of the inspiratory phase, their discharges are prevented by declining synaptic inhibition, probably set up by Early-Inspiratory neurons. They discharge coincident with the termination of Ramp-Inspiratory and phrenic nerve discharges and cease firing due to synaptic inhibition that seems to derive from discharges of Post-Inspiratory and Late-Expiratory neurons. Their proposed function is to initiate inspiratory phase termination [2,25] and mediate reflex inspiratory inhibition by slowly adapting ►lung stretch afferents (Respiratory Neurotransmitters and Neuromodulators).

Post-Inspiratory neurons are of two types. Some are cranial motoneurons that contract the pharyngeal constrictor and laryngeal adductor muscles. Others are propriobulbar and are thought to contribute to rhythm formation by securing termination of the inspiratory phase, and imposing a delay to the onset of expiratory neuron discharges. According to the network theory of rhythmogenesis, reciprocal inhibitory interactions between propriobulbar Post-I and Early-I neurons constitute the primary rhythm generator, and contribute to shaping the discharge patterns of inspiratory and expiratory neurons [25]. Post-Inspiratory neurons depolarize abruptly and discharge action potentials coincident with the termination of firing in Ramp-I neurons and with the arrest of phrenic nerve inspiratory phase discharges. Membrane depolarization and discharges in Post-I neurons exhibit declining patterns, which seem to be activity-dependent and mediated by a Ca^{2+}-activated K^+ conductance. Throughout the inspiratory phase, the neurons receive a declining wave of inhibitory postsynaptic potentials, evidently set up by Early-Inspiratory neurons. The rapid onset of depolarization and discharge in Post-I neurons is attributed to release from Early-I neuron inhibition and reactivation of low ►voltage-dependent Ca^{2+} currents and ►non-selective cationic currents that bring the membrane to threshold for action potential discharge. Throughout the expiratory phase, Post-I neurons receive synaptic inhibition set up by the discharge of Late-Expiratory neurons.

Late-Expiratory neurons are of two types, according to location and function. One group is located in the Bötzinger region of the rostral ventrolateral respiratory column (VRC) of the medulla, the other in the caudal region of the VRC (►Alheid and McCrimmon: anatomy and function in the respiratory network). Late-E neurons in both regions have similar membrane potential and discharge properties but play different roles in respiratory control. Late-E neurons located in the caudal VRC are bulbospinal and provide excitatory synaptic drive to expiratory intercostal and lumbar motoneurons and interneurons. During inspiration and postinspiration, synaptic inhibition set up by Early-Inspiratory and Post-Inspiratory neurons prevents Late-E neuron discharge. The expiratory neurons depolarize gradually to threshold during the post-inspiratory phase and then discharge steadily during the expiratory phase. The Late-E neurons in the Bötzinger region of the VRC are both propriobulbar and bulbospinal. Their discharges result in inhibition of medullary, phrenic and intercostal inspiratory neurons during the expiratory phase.

Pre-Inspiratory neurons are propriobulbar. They discharge in short bursts at the end of expiration and sometimes at the end of inspiration. They are subject to augmenting synaptic inhibition during the inspiratory phase and declining inhibition during the postinspiratory phase. Based on these time-intensity profiles, they are thought to receive inhibitory synaptic input from Ramp-Inspiratory and Post-Inspiratory neurons. One function they might have is to initiate termination of discharge in Late-Expiratory neurons [15].

Network Hypothesis: How Bulbar Respiratory Neurons Express a Three-phase Rhythm that Controls Respiration in the Mature Respiratory Network [2,9,25] (Computational Modeling of the Respiratory Network)

According to the network model, bulbar respiratory neurons receive tonic excitatory drive from pH/CO_2-sensitive chemoreceptor neurons and from neurons of the brainstem reticular activating system. Inspiration begins when Ramp-I neurons are released from inhibition and low-voltage activated (LVA) Ca^{2+} currents and non-selective cationic currents are activated. Ramp-like inspiratory discharges begin, driven by mutual recurrent excitation and by declining inhibition from Early-I neurons. Early-I discharges are terminated by Ca^{2+}-activated K^+ conductances, allowing disinhibition of Late-I and Post-I neurons that terminate discharge of Ramp-I neurons and arrest inspiration. Buildup of Ca^{2+}-activated K^+ conductances ends Post-I neuron inhibitory discharges, allowing discharge of Late-E neurons. Arrest of Late-E neuron discharge is initiated by Pre-I neurons and sustained by Early-I and Post-I neurons.

A key element in the cyclic regeneration of the inspiratory and expiratory phases is the termination of Early-I and Post-I discharges by activity-dependent Ca^{2+}-activated K^+ conductances.

Post-Inspiratory Discharges in Phrenic and Intercostal Inspiratory Motoneurons

The origin of postinspiratory discharge activity in inspiratory spinal motoneurons is not firmly established, but it does not arise from intrinsic membrane currents. Hypoxia increases its duration and hypercapnea shortens it. ►Pulmonary irritant receptors and stretch receptors inhibit it, whereas withholding of lung inflation prolongs it.

Bulbospinal neurons have been identified in the medulla of the cat that discharge with two bursts; one beginning simultaneously with the onset of phrenic nerve inspiratory discharges and, a second that begins after a very brief pause and declines in synchrony with the postinspiratory discharge of phrenic nerve activity The discharge of these *Inspiratory-Post-I* (*IPI*) *neurons* and that of the phrenic nerve respond identically to activation of tracheal and pulmonary afferents. Thus, the postinspiratory discharge component in spinal inspiratory motoneurons may be linked to excitatory synaptic input from the medullary IPI neurons. The slow decline of the postinspiratory discharge in IPI neurons is attributed to integration of excitatory and inhibitory synaptic inputs coming from two populations of postinspiratory neurons within the medulla, one excitatory and the other inhibitory [27].

Cranial Motoneurons and Control of Flow in the Upper Airways

Cranial motoneurons with respiratory periodicity in the ventrolateral medulla have axons in the trigeminal (5th), facial (7th), glossopharnygeal (9th), vagal (10th) and hypoglossal (12th) cranial nerves. They innervate the muscles of the nostrils, pharynx, tongue and larynx, coordinating their positions and movements with those of the diaphragm, chest wall and abdominal muscles during breathing. Trigeminal motoneuron discharges open the mouth during breathing, whereas facial motoneurons flare the nostrils. Glossopharyngeal motoneurons discharge with an augmenting pattern during inspiration and contract the muscles of the pharynx and palate. Laryngeal motoneurons with an augmenting inspiratory discharge pattern contract the abductor muscles, those with a decrementing postinspiratory discharge contract and narrow the opening of the glottis and slow outward airflow as the lungs gradually deflate. Vagal motoneurons with an augmenting expiratory discharge contract the pharyngeal constrictor muscles during expiration to lower upper airway resistance. Hypoglossal motoneurons (►Respiratory control of hypoglossal neurons during sleep and wakefulness) also regulate airway resistance and flow patterns by controlling tongue position.

Respiratory rhythm and pattern is derived from periodic excitatory and inhibitory synaptic input from the propriobulbar respiratory neurons described above.

For laryngeal and hypoglossal motoneurons, at least, important sources of synaptic excitation and inhibition are respiratory neurons of the preBötzinger Complex and surrounding ventrolateral medulla [20,32].

The motoneurons are assigned other duties related to sneezing, coughing, movements of the mouth, swallowing, vomiting, etc., during which their discharge properties are appropriate for the movements they promote.

Spinal Motoneurons and Contraction of the Pump Muscles

The location, synaptic connections and electrophysiological properties of spinal respiratory neurons are described in detail elsewhere, with special emphasis on α-motoneurons that innervate the extrafusal muscles of respiration (▶Spinal respiratory neurons and respiratory control). Interested readers can also consult an earlier review [17].

Phrenic motoneurons innervating the diaphragm are located in the ventral horns of the lower cervical spinal segments, and receive bilateral excitatory synaptic drive from medullary bulbospinal Ramp-I neurons and inhibitory synaptic input from bulbospinal Late-E neurons of the Bötzinger Complex. The neurons depolarize and fire with an augmenting discharge pattern during inspiration and hyperpolarize with an augmenting pattern during expiration. Not all phrenic motoneurons reach threshold simultaneously, rather, there is a scatter in the discharge latencies (recruitment times) with respect to the onset of the population discharge in the phrenic nerve. It appears that the order of recruitment derives, at least in part, from similar temporal properties of the bulbospinal neurons that provide excitatory synaptic input. Some of the excitatory drive also comes from interneurons located in the phrenic motoneuron pool and at higher cervical levels, which are driven by excitatory input from bulbospinal inspiratory neurons [34].

Intercostal motoneurons that contract the scalene, sternocleidomastoid and intercostal muscles are located in the ventral horn at all levels of the thoracic spinal cord. Medullary bulbospinal neurons that control phrenic motoneurons are also responsible for the periodic depolarization, discharge and hyperpolarization of intercostal motoneurons. Some of the synaptic connections are direct and others are made through interneurons located in the same segment as the motoneuron pool and in other cervical and thoracic segments.

Lumbar spinal motoneurons innervate the rectus abdominis, external and internal oblique and transverse abdominal muscles of the abdomen, and produce contraction during active expiration.

Recurrent Inhibitory Interneurons with a very high frequency of action potential firing are found near to phrenic and intercostal motoneurons. They are activated to discharge by motor axon collaterals and in turn suppress motoneuron discharges [17]. The high-frequency discharge and the resulting feedback suppression of motoneuron discharges are reminiscent of the ▶Renshaw inhibition that modulates discharge properties of limb motoneurons.

The Effects of Hypoxia on the Respiratory Apparatus: A Multiphase Reaction Involving the Carotid Bodies and the CNS Respiratory Network

Hypoxia produces dramatic disturbances of respiration. The initial response to acute hypoxia is increased breathing in an attempt to replenish O_2. The immediate source of respiratory augmentation is stimulation of type 1 (glomus) cells of the carotid bodies (CB) located bilaterally in the bifurcation of the common carotid arteries, which leads to activation of CB afferents and reflex stimulation of the CNS respiratory network. (▶Carotid Body Chemoreceptors and Respiratory Drive).

If hypoxia is maintained, disturbances of synaptic function within the CNS convert breathing rhythm to gasping and apnea. (▶Neural respiratory control during acute hypoxia). The CNS-derived hypoxic ventilatory response is a 5-component process, consisting of augmentation, apneusis or breath holding, protective apnea, gasping and terminal apnea. For each component, there are related changes in endogenous neurotransmitter and neuromodulator release and alterations in ion channel permeabilities.

Here, metabolic, enzymatic, ion channel and chemical neuromodulatory mechanisms that control the carotid body oxygen sensor and trigger the respiratory response to acute hypoxia are presented. The role of the CB in CO_2/pH sensing and its importance in health and disease are also discussed (Carotid Body Chemoreceptors and Respiratory Drive). Sites and mechanisms within the CNS respiratory network that engender ventilatory disturbances and ultimate apnea are described. In one essay (▶Respiratory network responses to hypoxia) the hypoxic response is defined in terms of two phases and a comprehensive description of neural pathways, neurotransmitters and neuromodulators that mediate respiratory depression during the late hypoxic ventilatory response (HVR) is presented. In another essay (▶Neural respiratory control during acute hypoxia), the energy cost of hypoxia to cells and its effects on ionic homeostasis are discussed. In addition, the HVR is presented as a 5-component process: augmentation, apneusis or breath holding, protective apnea, gasping and terminal apnea. For each component, accompanying changes in endogenous neurotransmitter and neuromodulator release and ion channel changes are described.

Concluding Comments

This overview has focused on respiratory control mechanisms in the adult mammal. Other contributions will show how the respiratory network develops before and shortly after birth (►Development of the respiratory network) and how autorhythmic pacemaker neurons in the rodent medulla regulate respiration in the postnatal period (►Respiratory pacemakers). Disturbances of respiratory control that are gene-related are also reviewed elsewhere (►Gene-related respiratory control disturbance).

Other important aspects of respiratory control not considered in this overview include: (i) neuroplasticity in the respiratory network, which allows readjustments of network responsiveness to injury and other environmental challenges (►Respiratory neuroplasticity), (ii) ►laryngeal chemoreflexes responding to liquid and chemical stimulation of laryngeal receptors, and (iii) respiratory control during sleep (►obstructive sleep apnea), (►Respiratory control of hypoglossal neurons during sleep and wakefulness).

Hopefully, the reader will appreciate the valuable insights of the contributing authors into how respiration is controlled, and the innovative methods they utilize, including imaging techniques, (►Respiratory network analysis, functional imaging) computer modeling (►Computational modeling of the respiratory network) and the use of novel preparations to study integrated cardio-respiratory regulation (►Central integration of cardiovascular and respiratory activity studied in situ).

References

1. Ballanyi K, Onimaru H, Homma I (1999) Respiratory network function in the isolated brainstem-spinal cord of newborn rats. Prog Neurobiol 59:583–634
2. Bianchi AL, Denavit-Saubie M, Champagnat J (1995) Central control of breathing in mammals: neuronal circuitry, membrane properties and neurotransmitters. Physiol Rev 75:1–45
3. Bonham AC (1995) Neurotransmitters in the CNS control of breathing. Respir Physiol 101:219–230
4. Butera RJ Jr, Rinzel J, Smith JC (1999) Models of respiratory rhythm generation in the pre-Bötzinger complex. I. Bursting pacemaker neurons. J Neurophysiol 82:382–397
5. Butera RJ Jr, Rinzel J, Smith JC (1999b) Models of respiratory rhythm generation in the pre-Botzinger complex. II. Populations Of coupled pacemaker neurons. J Neurophysiol 82:398–415
6. Cohen MI (1979) Neurogenesis of respiratory rhythm in the mammal. Physiol Rev 59:1105–1173
7. Del Negro CA, Johnson SM, Butera RJ, Smith JC (2001) Models of respiratory rhythm generation in the pre-Botzinger complex. III. Experimental tests of model predictions. J Neurophysiol 86:59–74
8. De Troyer A, Kirkwood PA, Wilson TA (2004) Respiratory action of the intercostals muscles. Physiol Rev 86:717–756
9. Ezure K (1990) Synaptic connections between medullary respiratory neurons and considerations on the genesis of the respiratory rhythm. Prog Neurobiol 35:429–450
10. Feldman JL (1986) Neurophysiology of breathing in mammals. In: Handbook of physiology. The nervous system. intrinsic regulatory system in the brain. Am Physiol Soc, Sect. 1, vol. IV, Chap. 9. Washington, DC, pp 463–524
11. Feldman JL, McCrimmon DR (2003) Neural control of breathing. In: Squire LR, Bloom FE, McConnell SK, Roberts JL, Spitzer NC, Zigmond MJ (eds) Fundamental neuroscience, 2nd edn, Chap. 37. Academic, Amsterdam
12. Feldman JL, Mitchell GG, Nattie EE (2003) Breathing: rhythmicity, plasticity, chemosensitivity. Annu Rev Neurosci 26:239–266
13. Greer JJ, Funk GD, Ballanyi K (2005) Preparing for the first breath; prenatal maturation of respiratory neural control. J Physiol 570:437–444
14. Haji A, Takeda R, Okazaki M (2000) Neuropharmacology of control of respiratory rhythm and pattern in mature mammals. Pharmacol Ther 86:277–304
15. Klages S, Bellingham MC, Richter DW (1993) Late expiratory inhibition of stage 2 expiratory neurons in the cat – a correlate of expiratory termination. J Neurophysiol 70:1307–1315
16. Long S, Duffin J (1986) The neuronal determinants of respiratory rhythm. Prog Neurobiol 27:101–182
17. Monteau, R, Hilaire G (1991) Spinal respiratory motoneurons. Prog Neurobiol 37:144–191
18. Nattie EE (2001) Central chemosensitivity, sleep and wakefulness. Respiration Physiology and Neurobiology 129:257–268
19. Netter FH (1997) Atlas of Human Anatomy, 2nd Edition. In: Dalley AE (ed) Novartis, Hanover, NJ. p. 523 pp. 967–990
20. Ono K, Shiba K, Nakazawa K, Shimoyama I (2006) Synaptic origin of the respiratory-modulated activity of laryngeal motoneurons. Neuroscience 140:1079–1088
21. Pierrefiche O, Shevtsova NA, St-John WM, Paton JF, Rybak IA (2004) Ionic currents and endogenous rhythm generation in the pre-Botzinger complex: modelling and in vitro studies. Advances in Experimental Medicine and Biology 551:121–6
22. Purvis LK, Smith JC, Koizumi H, Butera RJ (2007) Intrinsic bursters increase the robustness of rhythm generation in an excitatory network. Journal of Neurophysiology 97:1515–26
23. Ramirez JM, Viemari JC (2005) Determinants of inspiratory activity. Respiration Physiology and Neurobiology 147:145–157
24. Richerson GB, Boron WF (2005) Control of ventilation. In: *Medical Physiology*. WF Boron and EL Boulpaep (eds) Philadelphia, Elsevier Saunders, chapter 31, pp. 712–734
25. Ritchter DW (1996) Neural regulation of respiration: Rhythmogenesis and afferent control. In: *Comprehensive Human Physiology*. R. Gregor and U. Windhorst (eds) Berlin, Springer, vol. 2, chapt. 105, pp. 2079–2095
26. Richter DW, Ballanyi K, Schwarzacher S (1992) Mechanisms of respiratory rhythm generation. Current Opinions in Neurobiology 26:788–93
27. Richter DW, Ballantyne D (1988) On the significance of post-inspiration. Funtonanalyse biologischer Systeme 18:149–156
28. Richter DW, Pierrefiche O, Lalley PM, Polder HR (1996) Voltage-clamp analysis of neurons within deep layers of the brain. Journal of Neuroscience Methods 67:121–131
29. Richter DW, Spyer KM (2001) Studying rhythmogenesis of breathing: comparison of *in vivo* and *in vitro* models. Trends in Neuroscience 24:464–472

30. Rybak IA, Shevtsova NA, Paton JF, Dick TE, St.-John WM, Morschel M., Dutschmann M (2004a) Modeling the ponto-medullary respiratory network. Resp. Physiol. Neurobiol 143:307–319
31. Rybak IA, Shevtsova NA, Paton JF, Pierrefiche O, St-John WM, Haji A (2004b) Modelling respiratory rhythmogenesis: focus on phase switching mechanisms. Advances in Experimental Medicine and Biology 551:189–94
32. Shao XM, Feldman JL (2005) Cholinergic neurotransmission in PreBötzinger Complex modulates excitability of respiratory neurons and regulates respiratory rhythm. Neuroscience 130:1069–1081
33. Smith JC, Butera RJ, Koshiya N, Del Negro C, Wilson CG, Johnson SM (2000) Respiratory rhythm generation in neonatal and adult mammals: the hybrid pacemaker-network model. Respiration Physiology 122:131–47
34. Tian GF, Duffin J (1996) Connections from upper cervical inspiratory neurons to phrenic and intercostal motoneurons studies with cross-correlation in the decerebrate rat. Experimental Brain Research 110:196–204
35. Wood PG, Daniels CB, Orgeig S (1995) Functional significance and control of release of pulmonary surfactant in the lizard lung. American Journal of Physiology, Regulatory, Integrative and Comparative Physiology 269:R838–R847

Respiratory Bursting Neurons

►Respiratory Pacemakers

Respiratory Central Pattern Generator

►Computational Modeling of the Respiratory Network

Respiratory Control of Hypoglossal Neurons During Sleep and Wakefulness

John H. Peever[1], James Duffin[2]
[1]Department of Cell and Systems Biology, University of Toronto, Toronto, ON, Canada
[2]Department of Anaesthesia and Physiology, University of Toronto, Medical Sciences Building, Toronto, ON, Canada

Synonyms

Respiratory control of the tongue and airway

Definition

Hypoglossal ►motoneurons control the tongue muscles; genioglossus muscle (tongue protrudor), hyoglossus and styloglossus (tongue retractors) and intrinsic muscles. Because the genioglossus as well as other tongue muscles participate in a range a motor acts (e.g., drinking, licking, swallowing and breathing), the hypoglossal motoneurons that innervate them receive numerous inputs from a variety of relevant brain areas. Since the tongue position affects airway patency and resistance, respiratory control is essential to coordinate respiratory muscle effort and airway resistance, minimizing resistance during inspiration and using the control of resistance during expiration to modify expiratory flow patterns. The hypoglossal control of the tongue assumes a greater importance during sleep when relaxation of the tongue can occlude the airway; a condition of ►obstructive sleep apnoea (OSA).

Characteristics

Hypoglossal Motoneurons Anatomical Location

Hypoglossal ►motoneurons are ►somatotopically organized bilaterally in compact columns extending along the midline of the brainstem above and below the obex. For example a recent study in dogs [1] showed that the nuclei extend from 0.75 mm caudal to 3.45 mm rostral to the obex, with cells 0.37–2.12 mm below the dorsal surface, and symmetrically centered between 0.66 and 1.33 mm from the midline. There are a number of detailed locations studies including those in rats, cats, rabbits, frogs and monkeys as well as dogs. The hypoglossal motoneurons have varied and extensive dendritic arborizations that provide the potential for a wide range of afferent contacts. (Fig. 1) shows a schematic of the medullary respiratory neurons illustrating their general location and pattern of respiratory activity.

Projections of Hypoglossal Motoneurons

The axons of these motoneurons course ventrally and slightly laterally to emerge from the medulla in the preolivary sulcus separating the olive and the pyramid, and form the twelfth cranial nerve (XII). The hypoglossal nerve innervates all the muscles of the tongue except for the palatoglossus muscle which is innervated by the vagus nerve (X) and the accessory nerve (XI). The tongue is a complex muscle, and the hypoglossal nerve bifurcates to form medial and lateral branches, with the medial branch innervating the protrusor tongue muscles, and the lateral branch innervating the retractor tongue muscles [2].

The ventrolateral and ventromedial subnuclei contain motoneurons that innervate the geniohyoid and genioglossus muscles respectively, and the dorsal subnucleus motoneurons innervate the hyoglossus and styloglossus muscles. The genioglossus muscle is of particular importance clinically because it is considered the main protruder and depressor muscle of the tongue.

R

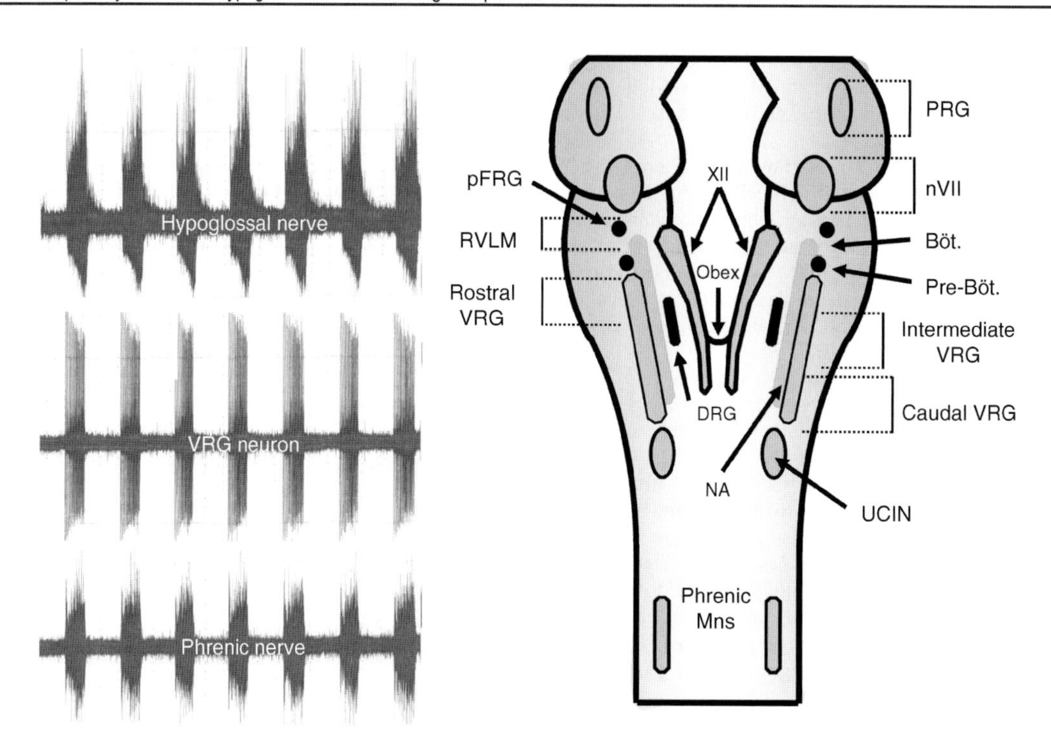

Respiratory Control of Hypoglossal Neurons During Sleep and Wakefulness. Figure 1 A schematic showing the location of hypoglossal motoneurons relative to medullary respiratory neuron groups with representative recordings from the hypoglossal nerve, a ventral respiratory group inspiratory neuron and the phrenic nerve in an adult decerebrate rat. Abbreviations: PRG = pontine respiratory group, nVII = Facial nucleus, pFRG = parafacial respiratory group, XII = hypoglossal motor nucleus, RVLM = rostro ventrolateral medulla, Böt = Bötzinger complex, Pre-Böt = preBötzinger complex, VRG = ventral respiratory group, DRG = dorsal respiratory group, NA = nucleus ambiguus, UCIN = upper cervical inspiratory neurons.

It is co-activated during inspiration with tongue retractor muscles [3] so that airway compliance is reduced, and upper airway patency increased during inspiration.

Respiratory Inputs to Hypoglossal Motoneurons

Hypoglossal motoneurons participate in rhythmic oro-facial motor acts such as mastication, licking, swallowing and breathing; receiving neural inputs from the brainstem rhythm-generating networks that generate these behaviors. They play a major role in respiratory airway control and the following focuses on the neural mechanisms by which respiratory drive is communicated to them.

A major function of the genioglossus muscle is to stiffen the pharyngeal airspace during inspiration so that it does not collapse during diaphragmatic contraction. Therefore, hypoglossal motoneurons are synaptically excited during inspiration (i.e., discharge action potential during inhalation); however, they are activated 200–300 ms before the ►phrenic motoneurons that innervate the ►diaphragm [3]. This pre-activation of hypoglossal motoneurons ensures that the airway is stiffened before the diaphragm contracts.

Although both the hypoglossal and phrenic motoneurons receive inspiratory signals from the medullary network that generates breathing, they receive their respective inspiratory commands from separate premotor populations. Inspiratory drive is transmitted to phrenic motoneurons by premotor neurons located in the ventral respiratory group (parallel to the nucleus ambiguus), while hypoglossal motoneurons receive respiratory drive from premotor neurons located in the medullary lateral tegmental field, lateral to the hypoglossal motor nucleus (Fig. 2) [4].

Although separate premotor populations relay inspiratory drive to phrenic and hypoglossal motoneurons, both release glutamate to activate post-synaptic non-NMDA receptors to induce inspiratory activation [5]. Unlike phrenic motoneurons, which are silenced during expiration by GABAergic and glycinergic inhibition, hypoglossal motoneurons receive no such inhibitory drive; instead, they are passively disfacilitated (withdrawal of excitation) during expiration [4]. Because the genioglossus is not only involved in the control of breathing, but is also in the control of speech, the lack of inhibition during expiration enables more effective modulation of their activity and thereby tongue muscles because motoneurons are more easily excited by other non-respiratory inputs.

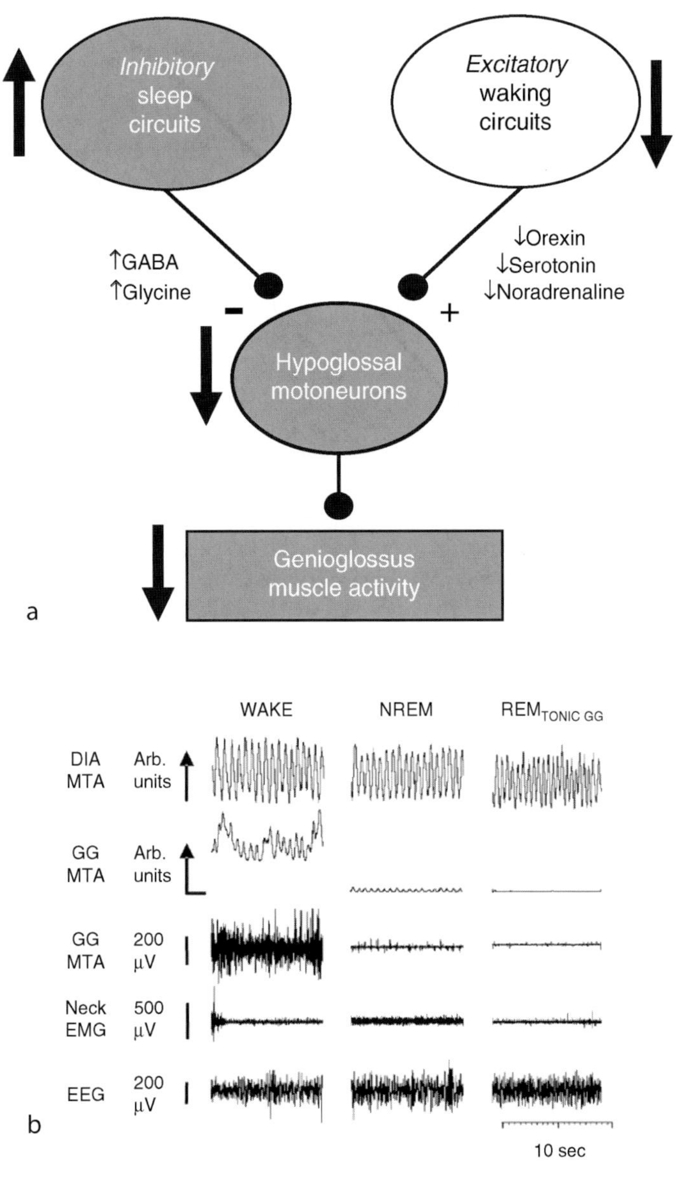

Respiratory Control of Hypoglossal Neurons During Sleep and Wakefulness. Figure 2 A schematic representation of the neural mechanisms responsible for suppression of genioglossus muscle activity in sleep. (a) It is hypothesized that active inhibition and passive disfacilitation reduce hypoglossal (airway) motoneuron activity and hence genioglossus muscle tone in sleep. Several lines of evidence indicate that inhibitory processes play a predominant role in controlling airway motoneuron and muscle activity in sleep, particularly in REM sleep. There is also evidence indicating that withdrawal of excitatory sleep-related inputs (e.g., serotonin, orexin, noradrenaline) in sleep may reduce airway motoneuron excitability. (b) A typical example from a naturally behaving rat demonstrating that genioglossus (and neck) muscle activity is depressed in sleep, and particularly REM sleep. This figure was modified (with permission) from Morrison et al. Journal of Physiology, 2003, 552.3, pp. 975–991.

Hypoglossal motoneurons are not only controlled by premotor neurons in the lateral tegmental field, they also receive respiratory signals from a population of interneurons located directly within the hypoglossal motor nuclei [4]. Hypoglossal interneurons are significantly smaller than motoneurons, they are active only during inspiration and they make inhibitory (e.g., GABA) connections with hypoglossal motoneurons. The precise role that interneurons play in transmitting inspiratory drive to hypoglossal motoneurons is unclear; however, it is hypothesized that they gate or filter presynaptic inputs.

Impact of Sleep on Hypoglossal Motoneuron Activity

►Sleep suppresses the excitability of hypoglossal motoneurons. Although this review focuses on hypoglossal motoneurons, it should be made clear that the excitability of all somatic motoneurons including other airway motoneurons (e.g., trigeminal and facial) are also affected by sleep [6]. Understanding how hypoglossal motoneuron activity is regulated in sleep is clinically important because sleep-related reductions in their activity lead to reduced airway motor tone, airway narrowing and collapse in predisposed individuals (e.g., small airway). The suppression of airway motoneuron activity in sleep, and particularly during rapid-eye-movement (►REM) sleep, is the primary cause of OSA.

The root cause of hypoglossal motoneuron suppression in sleep is unknown; however, recent work from our laboratory and others has begun to shed light on potential mechanisms. It is hypothesized that the neurocircuitry generating sleep (e.g., REM and ►non-REM) also innervates and regulates hypoglossal motoneuron activity. Although multiple neural circuits are involved in sleep generation, they can be subdivided into two categories – those that are excitatory and promote wakefulness and those that are inhibitory and promote sleep. Excitatory circuits are active during waking and project to both the cortex and motoneurons to promote behavioural arousal and high levels of motor tone [7]. Inhibitory circuits are active in sleep and project to and switch-off both the wake-promoting circuits as well as inhibiting airway motoneurons. Therefore, it is hypothesized that hypoglossal motoneurons are both actively inhibited and passively disfacilitated (withdrawal of excitatory inputs) during sleep (Fig. 3).

The primary wake-promoting circuits consist of excitatory neurons located in the noradrenergic locus coeruleus, the ►serotonergic dorsal ►raphe, the ►orexinergic (also called hypocretin) lateral hypothalamus, the histaminergic tuberomammillary nucleus and the ►dopaminergic ►periaquaductal grey and ventral tegmental area. The activity of these neural populations is highest in waking and minimal or silent in sleep. Because they project to motoneurons, it is hypothesized that withdrawal of noradrenergic, orexinergic, and serotonergic inputs may be a contributing factor to the reduction of motoneuron excitatory and hence muscle activity in sleep [6]. The major sleep-promoting

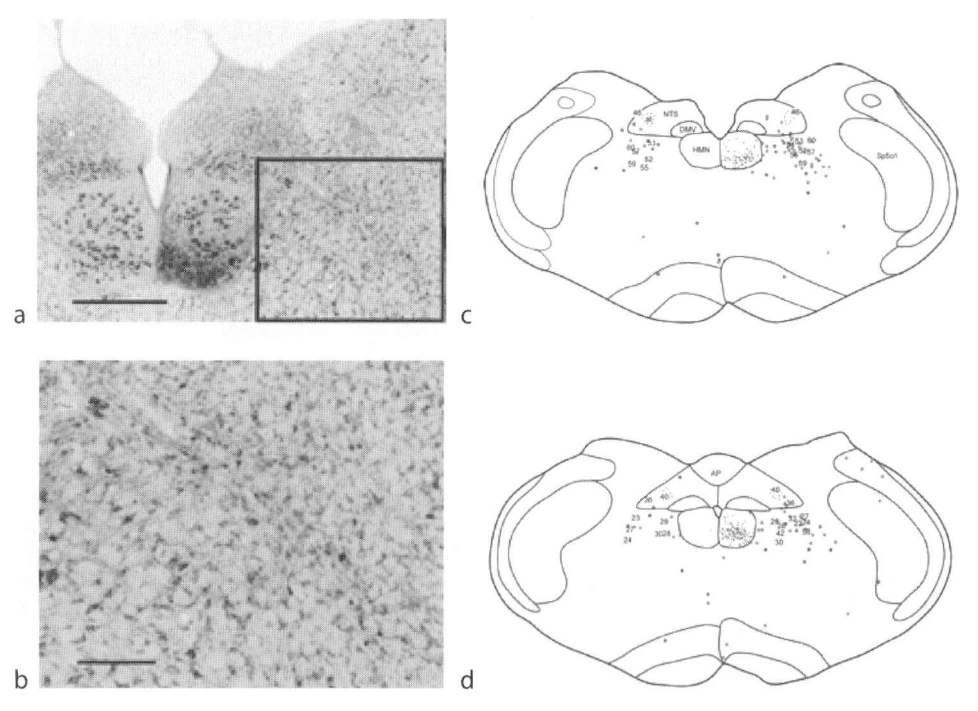

Respiratory Control of Hypoglossal Neurons During Sleep and Wakefulness. Figure 3 Location of the premotor neurons that relay inspiratory drive to hypoglossal motoneurons. (a) Photograph of the premotor neurons in lateral tegmental field that are hypothesized to relay inspiratory drive to hypoglossal motoneurons. Premotor neurons were identified by visualizing the location of pseudorabies virus that was retrogradely transported from the genioglossus muscle where it had been injected. (b) Higher magnification of the black box in (a); brown cells represent hypoglossal premotor neurons. C and D, represent the anatomical distribution of hypoglossal premotor neurons (*small dots*) and hypoglossal motoneurons (*small dots*) plotted on schematic cross-sections of rat brainstem at 0–500 μm rostral to obex (c) and −100 to −600 μm caudal to obex (d). This figure was modified (with permission) from Chamberlin et al., J. Physiology, 579.2, 2007, pp 515–526.

system consists of inhibitory neurons located in the GABAergic ventrolateral and median preoptic areas of the anterior hypothalamus. The activity of these neurons is lowest in waking and highest in sleep (particularly non-REM sleep). Because they project to motoneurons, it is hypothesized that they actively inhibit (via GABA) hypoglossal motoneuron activity and thereby suppress airway motor tone in non-REM sleep.

Another source of motoneuron inhibition comes from the medial medullary reticular formation. Unlike neurons in the ventrolateral and median preoptic areas, these neurons contain both GABA and glycine, and are maximally active in REM sleep when genioglossus muscle tone is minimal or absent [7]. It therefore appears that together GABA and glycine play a role in regulating hypoglossal motoneuron activity in both non-REM and REM sleep. Although GABAergic and glycinergic neurons in the medial medullary reticular formation project to hypoglossal motoneurons and are active in REM sleep, they are not responsible for the ►muscle atonia that typifies this state. Rather GABAergic and glycinergic inhibition are responsible for suppressing the phasic muscle twitches that characterize REM sleep [8]. The muscle atonia of REM sleep can not be explained by disfacilitation of excitatory because application of glutamate or glutamatergic receptor agonists (e.g., AMPA or NMDA) directly into airway (e.g., trigeminal) motoneurons can not reverse REM sleep atonia. Therefore, the cause of airway motoneuron inactivity and hence airway muscle atonia in REM sleep has yet to be identified. Identification of the neurochemical responsible for REM atonia requires attention because OSA is most common and severe during REM sleep.

Plasticity of Hypoglossal Motoneurons

Somatic motoneurons are generally considered to be passive neural relays that monotonically respond to pre-synaptic inputs; however, they are able to undergo remarkable degrees of plasticity. One type of motoneuron plasticity (►neural plasticity) that is particularly relevant to hypoglossal control is respiratory ►long-term facilitation (LTF). LTF is characterized by a progressive increase in the inspiratory amplitude of the hypoglossal nerve (or genioglossus muscle) activity following exposure to ►intermittent hypoxia (an example is shown in Fig. 4) [9].

LTF can only be evoked by intermittent hypoxia; exposure to continuous hypoxia does not evoke LTF. The central serotonergic system is required for LTF because blocking either serotonin release or serotonin receptors nullifies LTF. It is hypothesized that LTF is induced because intermittent hypoxia activates the serotonergic medullary raphe nuclei to release serotonin onto hypoglossal motoneurons. Intermittent serotonin receptor activation subsequently causes plastic changes in the excitability of hypoglossal motoneurons perhaps via group-I ►metabotropic glutamate receptors [10].

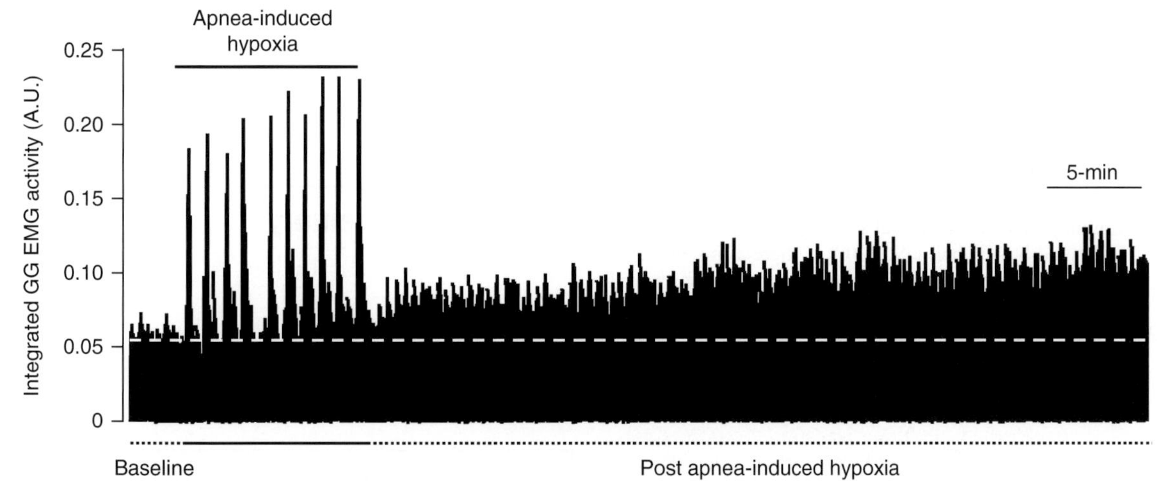

Respiratory Control of Hypoglossal Neurons During Sleep and Wakefulness. Figure 4 A typical example of apnea-induced long-term facilitation (LTF) of genioglossus motor outflow in an anaesthetized rat. Basal levels of inspiratory genioglossus muscle activity were recorded 5-min before (*baseline*) and for 60-min after obstructive apneas (ten 15-s apneas each separated by 45 s). Each obstructive apnea caused a reflexive increase in the inspiratory activity of the genioglossus muscle (see under apnea-induced hypoxia). The cluster of ten apneas induced a persistent and progressive increase in peak inspiratory genioglossus muscle activity that lasted for over 60-min (i.e., LTF). The *dotted line* represents the average magnitude of genioglossus inspiratory efforts before apnea-induced LTF; note that genioglossus activity returned to *baseline* levels after apneas and then progressively increased to reach maximal levels at 60-min post-apnea.

Since intermittent hypoxia causes persistent increases in hypoglossal motoneuron excitability and hence genioglossus motor output, and since hypoglossal motoneuron activity is lost in sleep and contributes to airway obstructions, then inducing LTF of during sleep may be an effective method for minimizing or reversing the root cause of obstructive sleep apnea. Therefore understanding the cellular mechanisms of LTF may provide the basis for rationale development of therapeutics for treating this prevalent (it affects about 5% of adults) sleep disorder.

References

1. Brandes IF, Zuperku EJ, Dean C, Hopp FA, Jakovcevic D, Stuth EA (2007) Retrograde labeling reveals extensive distribution of genioglossal motoneurons possessing 5-HT2A receptors throughout the hypoglossal nucleus of adult dogs. Brain Res 1132:110–119
2. Dobbins EG, Feldman JL (1995) Differential innervation of protruder and retractor muscles of the tongue in rat. J Comp Neurol 357:376–394
3. Peever JH, Mateika JH, Duffin J (2001) Respiratory control of hypoglossal motoneurones in the rat. Pflugers Archiv 442:78–86
4. Peever JH, Shen L, Duffin J (2002) Respiratory premotor control of hypoglossal motoneurons in the rat. Neuroscience 110:711–722
5. Steenland HW, Liu H, Sood S, Liu X, Horner RL (2006) Respiratory activation of the genioglossus muscle involves both non-NMDA and NMDA glutamate receptors at the hypoglossal motor nucleus in vivo. Neuroscience 138:1407–1424
6. Peever JH, McGinty D (2007) Why do we sleep? In: Lavigne G, Choinière M, Sessle Band oja P (eds) Sleep and Pain. International Association for the Study of Pain (IASP) Press, pp. 10–15
7. Siegel JM (2005) REM sleep. In: Kryger MH, Dement WC (eds) Principles and practice of sleep medicine. W.B. Saunders, Philadelphia, pp 120–135
8. Brooks P, Peever JH (in press) Role of chloride-mediated inhibition in the neurochemical control of airway motoneurons during natural sleep. Adv Exp Biol Med
9. Mahamed S, Mitchell GS (2007) Is there a link between intermittent hypoxia-induced respiratory plasticity and obstructive sleep apnoea? Exp Physiol 92:27–37
10. Bocchiaro CM, Feldman JL (2004) Synaptic activity-independent persistent plasticity in endogenously active mammalian motoneurons. Proc Natl Acad Sci USA 101:4292–4295

Respiratory Control of the Tongue and Airway

►Respiratory Control of Hypoglossal Neurons During Sleep and Wakefulness

Respiratory CPG

►Computational Modeling of the Respiratory Network

Respiratory Cycle (Phase)

Definition

The neuronal cycle of respiration consists of three phases: inspiration in which inspiratory muscles contract, post-inspiration or passive expiration (stage 1 expiration) in which inspiratory muscles cease progressively to contract while activity of the adductor muscles of the upper airway reduces exhalation, and active expiration (stage 2 expiration) in which expiratory muscles contract.

Respiratory Kernel

Definition

An aggregate of neurons that is essential for a respiratory function.

►Development of the Respiratory Network

Respiratory Memory

►Respiratory Neuroplasticity

Respiratory Network

Definition

The central respiratory network consists of the respiratory neurons and generates respiratory rhythm and pattern.

►Anatomy and Function in the Respiratory Network
►Computational Modeling of the Respiratory Network

Respiratory Network Analysis, Functional Imaging

KLAUS BALLANYI[1], HIROSHI ONIMARU[2], BERNHARD KELLER[3]
[1]Department of Physiology, University of Alberta, Edmonton, AB, Canada
[2]Department of Physiology, Showa University School of Medicine, Tokyo, Japan
[3]Department of Neurophysiology, University of Göttingen, Göttingen, Germany

Synonyms

Visualization of respiratory centers; Microscopy of structure-function relationships in mammalian respiratory networks

Definition

Functional imaging of ►respiratory networks means optical analysis, with (►Confocal microscopy), ►multiphoton microscopy (two-photon microscopy) or cameras (e.g. CCD- or CMOS-type) of structure-function relationships in rhythmically active neuronal brainstem networks involved in the control of breathing in mammals. Such imaging has so far only been carried out *in vitro*, specifically in (*i*) respiratory active slices, (*ii*) *en bloc* brainstem preparations from perinatal rodents and (*iii*) arterially-perfused "working-heart-brainstem" preparations of both newborn and adult rodents. ►Voltage-sensitive dye imaging of the activity of neuronal populations is feasible in the latter three *in situ* models. In contrast, simultaneous ►Ca^{2+} imaging of both the activity and basic morphology of single respiratory neurons, or clusters of such cells, has not been reported so far in the perfused preparation, and only in one case for *en bloc* medullas. In rhythmic slices, ►Ca^{2+} imaging is primarily done in inspiratory active interneurons and hypoglossal (XII) motoneurons. The latter neuron populations are also being used for optical analyses of subcellular processes such as activity- or metabolism-related changes of mitochondrial membrane potential, Ca^{2+} or redox state. Finally, expression of fluorescent proteins in transgenic mice and ►fluorescence labeling of neurotransmitter receptors are currently used for targeted electrophysiological recording from respiratory neurons in the slices. As outlined below, these approaches have provided important insights into both the structural organization and functional properties of respiratory centers. The following (yet mostly hypothetical scenario) would likely answer, with imaging techniques, most relevant questions regarding the neural control of breathing: Subpopulations of respiratory interneurons or output cells will be identified via a specific pattern of intrinsically-expressed fluorescent proteins and acutely fluorescence-labeled ion channels, receptors and/or transporters. Populations of these identified cells will then be loaded with cell-permeant fluorescent dyes for, e.g., simultaneous imaging of dynamic changes of signaling factors such as Ca^{2+} or pH in the cytosol and cellular organelles. At the same time, one or few of these cells will be whole-cell recorded (with further functional or morphological dyes in the patch electrode solution) for a correlation of dynamic changes of (sub) cellular activities with respiratory-related membrane potential oscillations or underlying membrane currents. Further improvement of computerized data processing and of the spatiotemporal resolution of fluorescent microscopy will enable simultaneous 4D-imaging of all these events for a thorough structure-function relationship of interactions between respiratory centers and (pre) motor circuits.

Characteristics

Medullary Respiratory Networks

Three major bilaterally-organized respiratory groups have been identified in the lower brainstem of mammals [1]. The dorsal respiratory group is primarily involved in the transmission of (chemosensory and mechanosensory) inputs to the medullary respiratory control system. The pontine respiratory group plays a major role in orchestrating the highly complex (pre/post)inspiratory-expiratory synaptic neuronal activity pattern for the innervation of diverse groups of respiratory muscles that are active during one or several of these phases. The ventral respiratory column contains various aggregations of respiratory neurons, among which some are capable of generating primary ►respiratory rhythms. Specifically, the ►pre-Bötzinger complex (preBötC) appears to be pivotal for the generation of inspiratory-related interneuronal and motor activities. Conversely, the ►parafacial respiratory group (pFRG), located between the pons and the more caudal preBötC, generates preinspiratory (and postinspiratory) activities that drive, e.g., expiratory abdominal muscles [1,2]. Both, the preBötC and pFRG remain active in distinct transverse brainstem slices from newborn rodents (see ►isolated respiratory centers). In more intact preparations, the pFRG and preBötC constitute presumably a dual respiratory center that may interact with the pontine and dorsal respiratory groups for generation of the full spectrum of respiratory activities [1]. As described below, voltage-sensitive dye imaging has been used for characterization of all three respiratory groups, whereas other imaging approaches were so far primarily used for studying the ventral respiratory column, in fact mostly presumptive or histologically-identified preBötC interneurons and inspiratory XII motoneurons.

Voltage-Sensitive Dye Imaging of Spatiotemporal Respiratory Patterns

Voltage-sensitive dye imaging is principally suitable to measure the activity of single cells. Though, this technique has been applied yet only to monitor, at low optical magnification, spatiotemporal activity patterns in large populations of respiratory neurons. In most studies, fluorescence signals were collected from the intact ventral brainstem surface. Two dyes are preferentially used for this approach. The agent Di-4-ANEPPS stains primarily superficial tissue layers, whereas the less lipophilic Di-2-ANEPEQ stains deeper brainstem structures in addition. Bath-application of these agents for time periods of 0.5 h to sometimes >1 h is necessary for effective staining of respiratory active regions in newborn rodent brainstems [2,3]. Decreases and increases in fluorescence intensity of the above voltage-sensitive dyes correlate with enhanced and attenuated neuronal activity, respectively, due to the fact that the fluorescence of Di-2-ANEPEQ and Di-4-ANEPPS is proportional to (neuronal) membrane potential [2]. Accordingly, in an area corresponding to the *locus coeruleus*, which is located between the pontine respiratory group and the pFRG, initial inspiratory-related voltage-sensitive dye imaged activity is followed by a pronounced period of decreased activity [3]. The time course of these optical signals is similar to that of the inspiratory depolarization and the subsequent postinspiratory hyperpolarization of single neurons in this area. In contrast, optical activity in the main area of the pFRG is primarily preinspiratory-related, with less pronounced postinspiratory activity compared to that observed with whole-cell recording from single pFRG neurons [2]. In the latter study, inspiratory-related activity was in particular pronounced in the region of cervical spinal motoneurons, and in an area located ~0.4–0.6 mm caudal to the caudal end of the facial motor nucleus [2]. The latter region corresponds well with the presumptive rostrocaudal extension of the preBötC [4]. Most areas of the ventral respiratory column, including the pFRG and preBötC, contain different classes of respiratory neurons. It is not clear at present, why no prominent respiratory-related optical activity is revealed in regions of the ventral respiratory column other than the pFRG and preBötC.

Despite these caveats, respiratory voltage-sensitive dye imaging provided important information regarding the structural organization and function of respiratory networks. In *en bloc* medullas from newborn rodents, voltage-sensitive dye imaging revealed that both the dorsal and pontine respiratory groups are active in regions similar to those previously identified in adult mammals with combined electrophysiological and histological techniques [3]. The seminal imaging study on the discovery of the pFRG [2] suggested that the area of the location of these rhythmogenic preinspiratory neurons may extend further rostrally than previously assumed. Regarding respiratory functions, voltage-sensitive dye imaging showed that anoxia-induced slowing of ►respiratory rhythm in newborn rat brainstems is accompanied by depressed preinspiratory and augmented postinspiratory medullary optical activities that coincide with the occurrence of inspiratory-related cervical nerve burst doublets (Fig. 1).

The latter findings suggest that anoxia synchronizes the dual respiratory rhythm generators. This may result in enhanced postinspiratory medullary activity that triggers repetitive inspiratory motor bursts. All this shows that voltage-sensitive dye imaging is a potent tool for the analysis of normal and pathologically disturbed spatiotemporal patterns of respiratory activities, in particular when this technique is combined with electrophysiological recording of cellular respiratory bursting.

Imaging of Respiratory Ca^{2+} Oscillations and (Sub) Cellular Signaling Factors

In excitable cells, a notable rise of the free cytosolic [Ca^{2+}] is associated with the Ca^{2+} influx caused by activity-related depolarization. Accordingly, Ca^{2+} imaging is a widely used tool for monitoring neuronal activity (see ►neuron-glia imaging). The vast majority of Ca^{2+} imaging studies of the respiratory system has focused on preBötC neurons or preBötC-driven interneurons and XII motoneurons in rhythmic slices from perinatal rodents. In most of these reports, clusters of preBötC neurons were loaded with the membrane-permeant acetoxy-methyl (AM) form of ►$Ca^{2\pm}$-sensitive dyes, in particular Fluo-4-AM, Calcium-Green-1-AM and Fura-2-AM [4–8].

In the first respiratory Ca^{2+} imaging study, cytosolic Ca^{2+} oscillations in preBötC neurons occurred synchronously with inspiratory-related XII activity [7]. Following pharmacological blockade of glutamatergic synaptic transmission, asynchronous Ca^{2+} oscillations persisted in a subpopulation of these cells, in line with the hypothesis that preBötC neurons have intrinsic bursting ("pacemaker") properties [1,7]. For this CCD camera imaging study, Calcium-Green-1-AM was microinjected near the midline contralateral to the preBötC side to be imaged. This ensured that only cells were imaged that project their axons to the contralateral preBötC. As a caveat, such loading of cells via diffusion of dye from their axon to the cell bodies required "overnight," i.e. >10 h, incubation times. This substantial delay between the generation of the acute slice and the start of recording may affect functional properties of rhythmogenic preBötC networks although basic inspiratory activity was preserved, at least in solution of artificially-elevated (7–8 mM) [K^{+}] [7]. Alternatively, inspiratory Ca^{2+} oscillations can be monitored within <20 min after the injection of AM Ca^{2+} dye into the "online histologically-identified" preBötC [4] (Fig. 1).

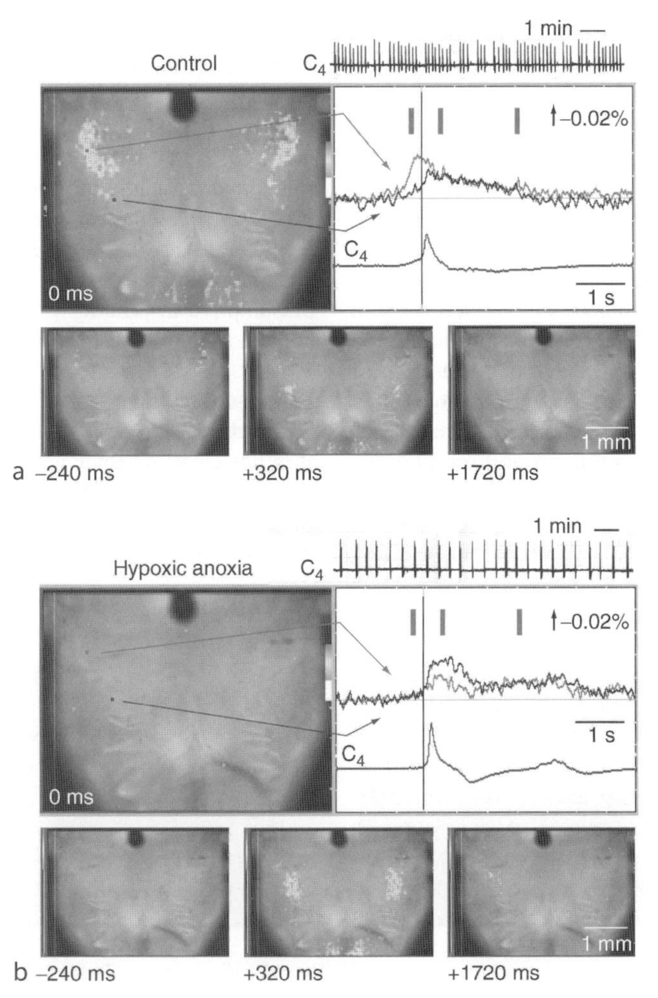

Respiratory Network Analysis, Functional Imaging. Figure 1 Anoxic respiratory pattern shifts in newborn rat brainstem-spinal cords. (a) traces in box (50 averaged optical signals) show that pFRG activity (red dot) preceded preBötC activity (blue dot) in control. Images below box show fluorescence signals of the voltage-sensitive dye Di-2-ANEPEQ during time periods indicated by vertical red bars. (b) Hypoxic anoxia due to N_2-gassed solution decreased inspiratory-related cervical nerve (C_4) burst rate and induced double bursts (compare integrated C_4 activities in (a) and (b)). Anoxia also suppressed preinspiratory optical pFRG activity (50 averages) and elicited a second optical peak in the pFRG, preBötC and intermittent regions. The latter activity appeared in the postinspiratory phase after the augmented C_4 peak and coincided with secondary C_4 activity. In original C_4 traces, the amplitude of the second C_4 peak was similar to that of initial activity, but was attenuated by averaging due to variation in the time of burst onset. Scale bar indicates percentage change in fluorescence. (From K. Ballanyi & H. Onimaru, unpublished).

As a major advantage of this approach, preBötC rhythms can be studied for several hours in physiological (3 mM) K^+ which ensures a higher sensitivity of preBötC rhythms to clinically-relevant agents such as opioids [4]. Although multiphoton microscopy was used in that report, neither the activity nor the gross morphology of preBötC neurons could be imaged at high spatial resolution at depths >80 μm into the tissue for yet unknown reasons. Similarly, CCD camera imaging and confocal laser scan microscopy are restricted to recording depths <70 μm [4–8]. Focal injection of Ca^{2+} dye limits the monitored area of active respiratory neurons to a circular spot with a diameter of 150–300 μm [4]. This limitation is overcome by loading cells unspecifically in the entire slice via bath-application of the AM Ca^{2+} dye which, however, penetrates <50 μm into the tissue. As further caveats, Ca^{2+} imaging may induce phototoxic effects (►Phototoxicity) on (respiratory) neurons and is subjected to bleaching of dye (►Photobleaching). The extent of these effects increases at both higher sampling rates and enhanced optical magnification for visualizations of (sub)cellular structures. Though, at scan rates of 1.5–3 scans/s most of the peak of respiratory-related Ca^{2+} oscillations is captured and continuous multiphoton or confocal imaging is possible for time periods >30 min

[4] (Fig. 2), similar to CCD camera imaging (Fig. 3) (B.U. Keller, unpublished observations).

The correlation between inspiratory-related membrane potential oscillations (or underlying membrane currents) and intracellular Ca^{2+} can be analyzed during whole-cell recording of respiratory neurons that are loaded with Ca^{2+} dye via the recording patch-electrode (Fig. 3). The first study in that regard showed with photomultiplier-based "point" imaging in presumptive preBötC neurons that somatic Ca^{2+} rises with a magnitude of maximally 300 nM occur during the inspiratory drive potential, and that a major portion of this signal is due to Ca^{2+} influx via high voltage-gated Ca^{2+} channels [5]. The relation of membrane excitability with cytosolic Ca^{2+} transients and their buffering, or with other cellular signaling factors, has been studied thoroughly in inspiratory active XII cells [8] (Fig. 3). These cells are not only a potent model to study the inspiratory drive from the preBötC to motor networks, but also for analysis of the selective vulnerability of particular motoneurons in amyotrophic lateral sclerosis [8]. Disturbances of (sub)cellular signaling processes such as (respiratory-related oscillations of) mitochondrial Ca^{2+} and membrane potential or redox processes are studied in XII motoneurons with regard to the latter disease [8]. Rhythmic changes of (sub)cellular signaling factors have also been optically monitored in presumptive inspiratory preBötC neurons, e.g., in the context of anoxic respiratory depression [9].

Targeted Recording from Fluorescence-Tagged Respiratory Neurons

It is not clear, whether yet unidentified rhythmogenic respiratory neurons, in particular of the preBötC and pFRG, have specific morphological features such as a particular size and shape of the soma or the number and

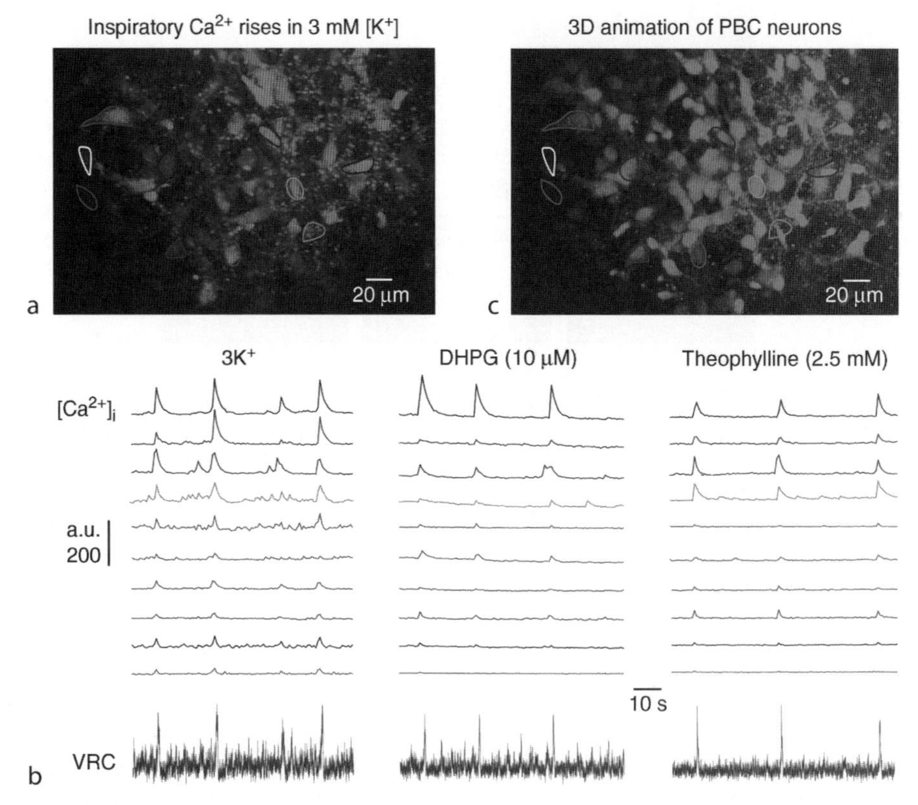

Respiratory Network Analysis, Functional Imaging. Figure 2 Multiphoton/confocal imaging of the activity and morphology of inspiratory preBötC neurons. (a) cells located 0.59 mm caudal to the caudal end of facial nucleus in the preBötC of a 600 µm thick newborn rat brainstem slice were stained by pressure-injection (0.7–1.0 psi, 10 min) with Ca^{2+} sensitive dye, Fluo-4-AM. Movie (*supplemental material*) shows 90 s recording in 3 mM $[K^+]$ of Ca^{2+} oscillations in these neurons that were in phase with inspiratory population activity recorded from the contralateral PBC; bottom left trace in (b). Fluo-4-AM fluorescence intensity is plotted in arbitrary units (a.u.) against time. After washout of rhythm in 3 mM $[K^+]$, preBötC bursting and Ca^{2+} oscillations were restored by low concentrations of the metabotropic glutamate receptor agonist dihydroxyphenylglycine (DHPG) and the clinically used respiratory stimulant theophylline. (c) 3D animation (*supplemental material*) showing gross morphology of preBötC neurons and neighboring non-rhythmic cells obtained from z-stack (0.5 µm single step) image series encompassing areas starting 7.5 µm above to 7.5 µm below image plane of (a) (reproduced from [4] with permission).

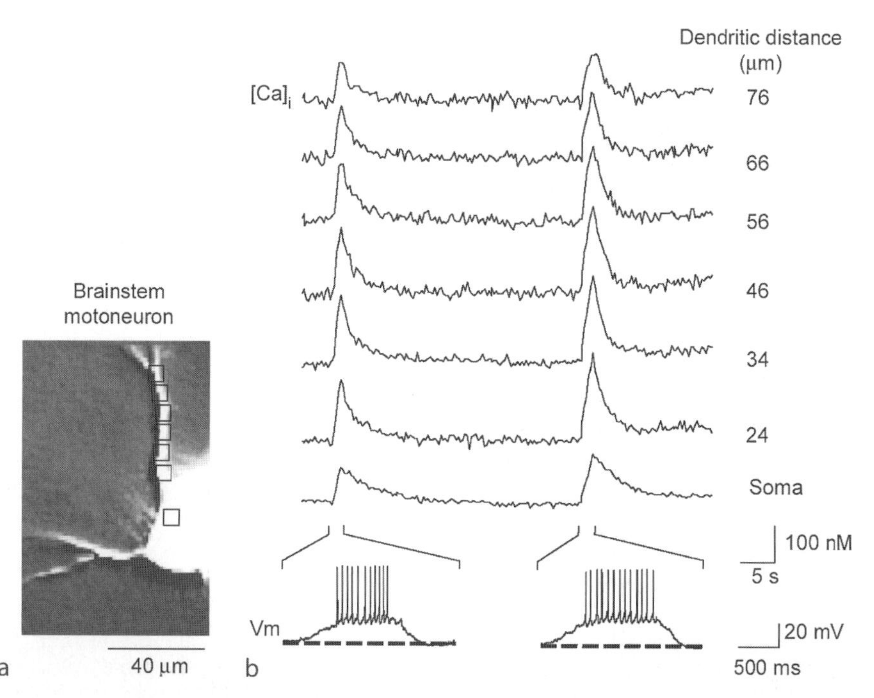

Respiratory Network Analysis, Functional Imaging. Figure 3 Whole-cell patch-clamp recording and simultaneous Ca^{2+} imaging of a rhythmically active hypoglossal motoneuron. (a) Fluorescence image of the soma and proximal dendrite of a patch-clamped hypoglossal motoneuron in a 700 μm thick mouse brainstem slice. (b) Whole-cell recording in current-clamp mode and simultaneous ratiometric CCD camera imaging of cytosolic Ca^{2+} concentrations. Rhythmic electrical discharges are paralleled by notable Ca^{2+} transients in the soma and six dendritic compartments, selected for analysis at distances of 24–76 μm from the soma (compartment positions are represented by boxes in (a)). Spontaneous bursts of action potentials, shown in the bottom trace as changes in membrane voltage (Vm) are accompanied by cytosolic Ca^{2+} rises up to 200 nM (reproduced from [8] with permission).

array of (primary) dendrites. If this were the case, such cells could be selectively targeted after fluorescence labeling for intracellular electrophysiological recording in the rhythmic slices. The above mentioned Ca^{2+}-sensitive fluorescent dyes, and also other functional dyes such as the marker for mitochondrial potential Rhodamine-123, can principally be used as morphological markers (Fig. 2–4) (see neuron-glia imaging).

Regarding Ca^{2+} dyes, Fura-2 provides a robust fluorescence signal at low (resting) cytosolic $[Ca^{2+}]$, whereas both Fluo-4 and Calcium-Green-1 fluoresce during rises of cytosolic Ca^{2+} (Fig. 2–4) [4–7]. Although recording from inspiratory (preBötC) interneurons in the rhythmic slices is routinely done under visual control, these superficial cells are surprisingly not routinely labeled during recording with high resolution morphological dyes, e.g., of the "Alexa" family.

Alternatively, subpopulations of respiratory neurons in acute medullary slices can be labeled with fluorescent markers for proteins in the plasma membrane or cytosol. In that context, it was assumed that rhythmogenic preBötC neurons are characterized by postsynaptic neurokinin-1 receptors that are normally activated by substance P [1,6]. Accordingly, fluorescence-tagging of substance P uptake via these receptors revealed that preBötC neurons can indeed be targeted [6] (Fig. 4). However, as shown in the latter study, the labeling was not specific and included various other types of (respiratory) brainstem neurons. In addition, respiratory neurons can be targeted in acute slices from transgenic mice that express a fluorescent protein coupled to a promoter such as glutamic acid decarboxylase [10]. Also this approach is yet not specific enough for identifying one particular population of candidate rhythmogenic respiratory neurons. Currently, fluorescence-tagged transcription factors that may be specific for rhythmogenic respiratory neurons, are being constructed in transgenic mice. However, it may turn out that such cells are not characterized by a single characteristic feature that can be visualized in the *in vitro* models, but rather by a unique pattern of transcription factors, (pacemaker) ion channels plus neurotransmitter receptors and/or transporters.

Acknowledgments

Work contributing to this study was supported by the Canadian Institutes of Health Research (CIHR), the

R

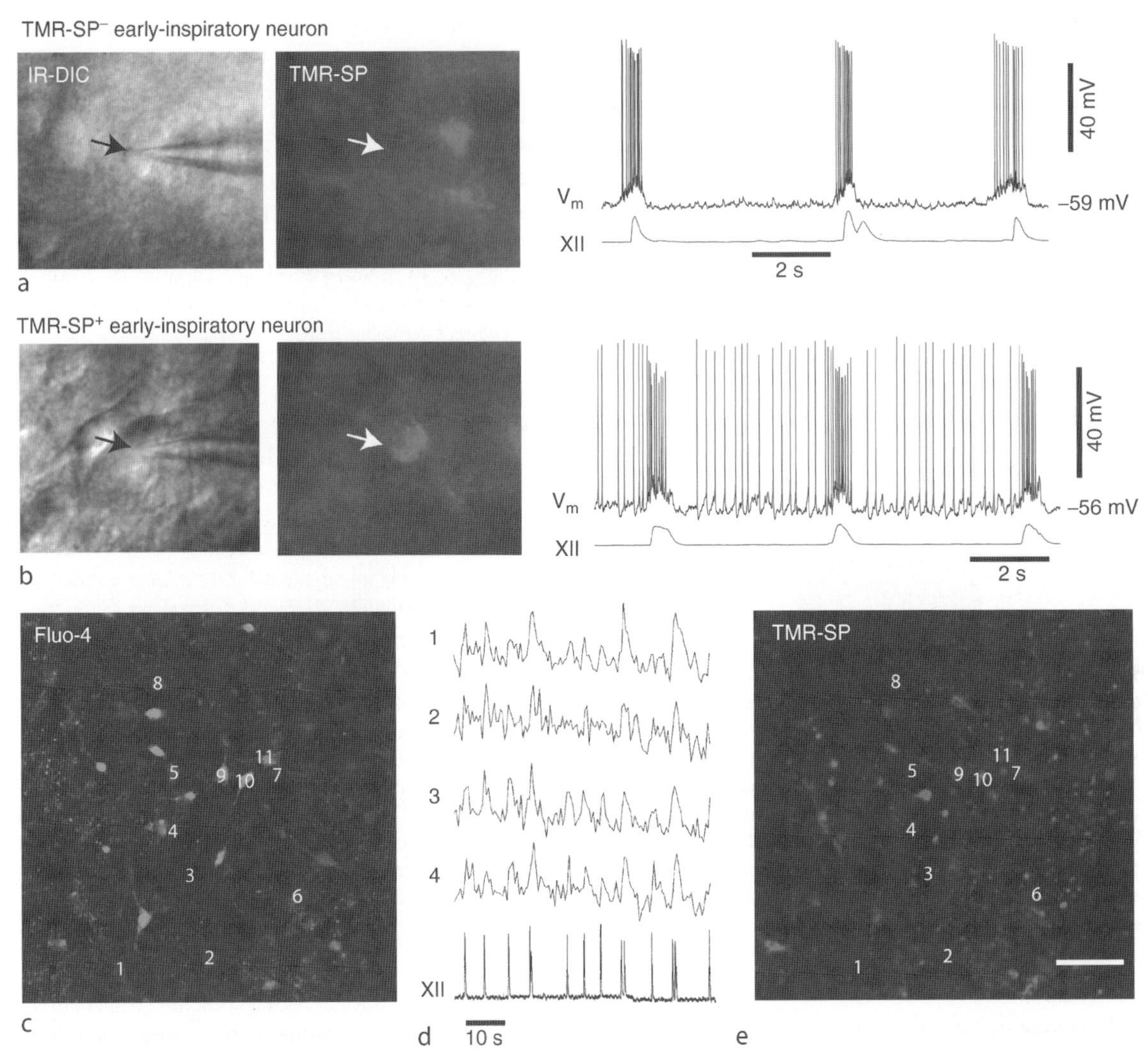

Respiratory Network Analysis, Functional Imaging. Figure 4 Targeted recording from acutely fluorescence-tagged preBötC neurons. (a, b) tetramethylrhodamine conjugated to substance P (TMR-SP) labeling in preBötC neurons with different phenotypic properties. Infrared differential interference contrast (IR-DIC) and epifluorescence images are shown in left columns, with corresponding intracellular traces to the right. (a) TMR-SP$^-$ early inspiratory neuron with silent interburst intervals. (b) TMR-SP$^+$ early inspiratory neuron with tonic low-frequency spiking properties. Scale bar (25 μm) applies to all images in (a, b). (c, e) simultaneous measurements of inspiratory activity and TMR-SP labeling in preBötC neurons. (a) fluo-4 image shows a peak acquisition of Ca^{2+} labeling; TMR-SP image shows TMR-SP$^+$ cells in the same region. Scale bar: 50 μm. (d) changes in fluorescence intensity from regions of interest (ROIs) indicated by numerals in (c), plotted with synchronized XII activity (reproduced from [6] with permission).

Alberta Heritage Foundation for Medical Research (AHFMR), the Canada Foundation for Innovation (CFI-ASRIP), the Bundesministerium für Bildung und Forschung (BMBF) and the Berstein Center for Computational Neuroscience (BCCN).

References

1. Ballanyi K (2004) Neuromodulation of the perinatal respiratory network. Curr Neuropharmacol 2:221–243
2. Onimaru H, Homma I (2003) A novel functional neuron group for respiratory rhythm generation in the ventral medulla. J Neurosci 23:1478–1486
3. Onimaru H, Homma I (2005) Optical imaging of respiratory neuron activity from the dorsal view of the lower brainstem. Clin Exp Pharmacol Physiol 32:297–301
4. Ruangkittisakul A, Schwarzacher SW, Ma Y, Poon B, Secchia L, Funk GD, Ballanyi K (2006) High sensitivity to neuromodulator-activated signaling pathways at physiological [K$^+$] of confocally-imaged respiratory center neurons in online-calibrated newborn rat brainstem slices. J Neurosci 26:11870–11880

5. Frermann D, Keller BU, Richter DW (1999) Calcium oscillations in rhythmically active respiratory neurones in the brainstem of the mouse. J Physiol 515:119–131
6. Hayes JA, Del Negro CA (2007) Neurokinin receptor-expressing pre-botzinger complex neurons in neonatal mice studied in vitro. J Neurophysiol 97:4215–4224
7. Koshiya N, Smith JC (1999) Neuronal pacemaker for breathing visualized in vitro. Nature 400:360–363
8. von Lewinski F, Keller BU (2005) Ca^{2+}, mitochondria and selective motoneuron vulnerability: implications for ALS. Trends Neurosci 28:494–500
9. Mironov SL, Richter DW (2001) Oscillations and hypoxic changes of mitochondrial variables in neurons of the brainstem respiratory centre of mice. J Physiol 533:227–236
10. Kuwana S, Tsunekawa N, Yanagawa Y, Okada Y, Kuribayashi J, Obata K (2006) Electrophysiological and morphological characteristics of GABAergic respiratory neurons in the mouse pre-Botzinger complex. Eur J Neurosci 23:667–674

Respiratory Network Analysis, Isolated Respiratory Center Functions

KLAUS BALLANYI[1], HIROSHI ONIMARU[2]
[1]Department of Physiology, University of Alberta, Edmonton, AB, Canada
[2]Department of Physiology, Showa University School of Medicine, Tokyo, Japan

Synonyms

Dual respiratory center organization; Mammalian respiratory rhythm generators

Definition

Analysis of isolated respiratory centers means a reductionistic approach for studying the neural control of breathing using *in vitro* brainstem preparations (mostly from ►perinatal rodents) in which rhythmogenic respiratory networks remain active. The ►pre-Bötzinger complex (preBötC) has been identified as a brainstem region that generates respiratory rhythm in mammals and remains active in a transverse medullary slice preparation. In a more rostral transverse medullary slice without the preBötC, the pre/postinspiratory active ►parafacial respiratory group (pFRG) continues to drive facial (VII) motoneurons rhythmically. Although both groups of rhythmogenic neurons operate autonomously in the distinct brainstem slices, they appear to constitute a ►dual respiratory center, at least in less reduced " ►*en bloc*" brainstem-spinal cord preparations from perinatal rodents and juvenile rats *in vivo*. This hypothesis is based on a differential action of opioids on functionally inspiratory (preBötC-driven) or expiratory (pFRG-driven) motor activities *in vivo* and *in vitro* suggesting that the pFRG provides a pivotal excitatory drive to the preBötC. Conversely, anoxia appears to synchronize and enhance the activities of both rhythm generators, resulting in pronounced postinspiratory medullary activities and lumbar/facial motor bursting that are accompanied by inspiratory-related nerve burst doublets. The latter findings suggest that the preBötC and pFRG are capable of adjusting their interactions to cope in particular with pathological disturbances of breathing.

Characteristics

Respiratory Network Organization *in vitro* and *in vivo*

Three major anatomically defined respiratory groups have been identified in the lower brainstem of mature mammals *in vivo* by microelectrode analysis of respiratory-related extra- or intracellular activities and subsequent histological identification of the recording sites [1] (Fig. 1). The ►dorsal respiratory group is a relay site of peripheral mechano- and chemoreceptor inputs to primary respiratory networks, whereas the pontine respiratory group is important for the generation of the multiphase neuronal activity pattern, which is projected to the respiratory muscles [1]. The ►ventral respiratory group (or rather column) contains arrays of interneurons, which are involved in the generation of the basic rhythm [1,3,4] (Fig. 1). In 1984, Suzue reported that respiratory activity in mammals is retained *in vitro*, specifically in isolated brainstem-spinal cord preparations from newborn rats [3] (Fig. 1). Extra- and intracellular electrophysiological recording of rhythmic drive potentials and/or action potential discharge in histologically-identified brainstem sites established that different classes of respiratory neurons are active in this *en bloc* model in areas corresponding to those in adult mammals *in vivo* [1,3,4] (Fig. 1). Neonatal rat ventral respiratory column neurons have been classified according to the phase relation of their cellular bursting with inspiratory-related cervical nerve bursting. This revealed that such neurons are active during one or several phases of the *in vitro* respiratory rhythm, which is comprised of a preinspiratory, inspiratory, postinspiratory and an active expiratory ("E-2") component in brainstem-spinal cord preparations [3,4–7] (Fig. 1).

Isolation of Respiratory Centers

The findings from a large number of studies using the *en bloc* brainstem model greatly advanced the understanding of cellular mechanisms involved in the neural control of breathing [3,4]. In that regard, findings from one seminal study [8] supported in 1991 the long-standing hypothesis that breathing movements originate from a limited area, a "*noeud vitale,*" in the "upper neck" [1]. Specifically, microsection of the newborn

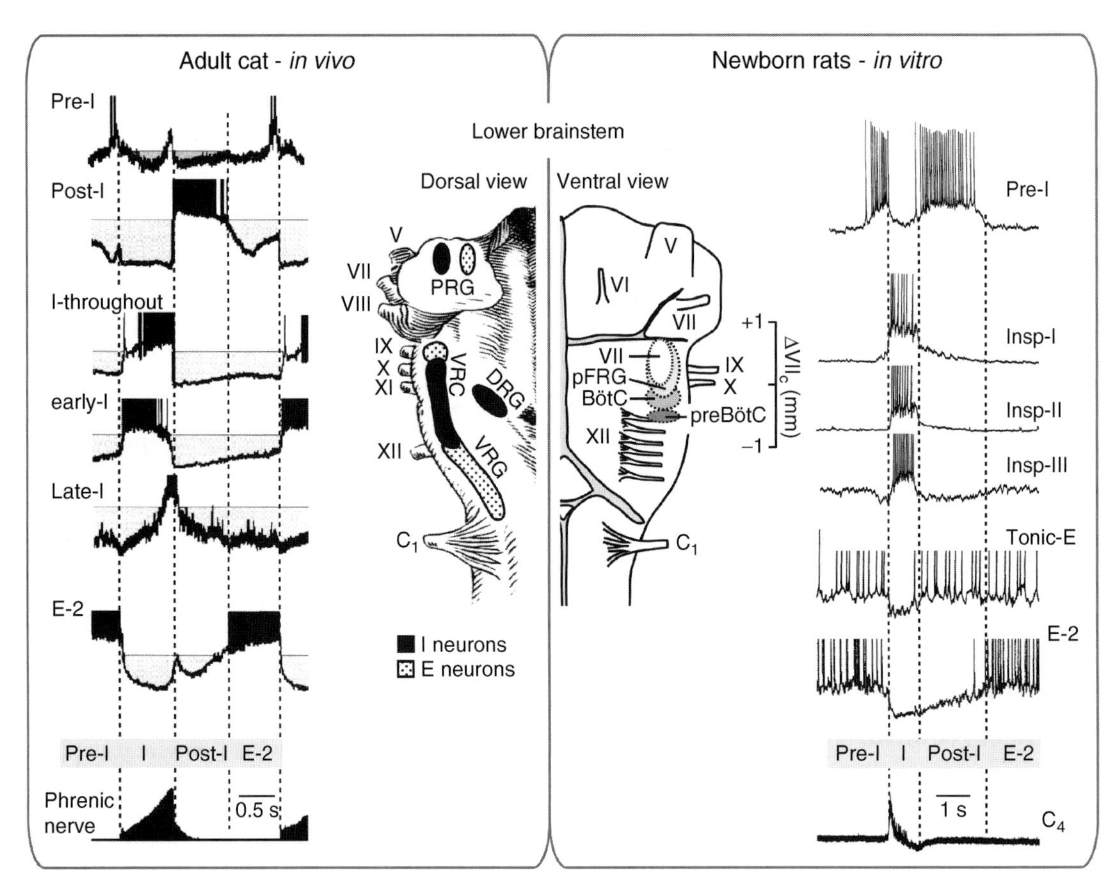

Respiratory Network Analysis, Isolated Respiratory Center Functions. Figure 1 Respiratory groups, centers and neuron classes in mammals. Sharp microelectrode membrane potential recordings in the left part of the figure revealed rhythmic depolarizations in six classes of adult cat medullary respiratory neurons. These cells discharge action potentials (black bars) in a specific phase relation with the inspiratory (I) plus postinspiratory (post-I) activities of the phrenic nerve which activates the diaphragm, i.e. the main inspiratory muscle. Grey areas indicate periods of inhibition via $GABA_A$ and glycine receptors. Modified with permission from D.W. Richter (in Comprehensive Human Physiology, eds. R. Greger, U. Windhorst; Springer-Verlag, Berlin Heidelberg, 1996). The dorsal schematic view on an adult cat brainstem shows the simplified distribution of I neurons (black areas) and expiratory (E, dotted areas) neurons in the pontine, dorsal and ventral respiratory groups (PRG, DRG, VRG). Note that the rostral portion of the VRG is named ventral respiratory column (VRC). Modified with permission from [1]. The attached ventral view on a newborn rat brainstem shows the locations and rostrocaudal extensions of the parafacial respiratory group (pFRG) and pre-Bötzinger Complex (preBötC) rhythmogenic respiratory centers with reference to the caudal end of the facial (VII) motonucleus, VII_c. The constancy of the rostrocaudal extensions of respiratory marker nuclei such as the VII nucleus and the inferior olive allowed the generation of "calibrated" newborn rat brainstem-spinal cord ("*en bloc*") preparations with a defined content of (respiratory) brainstem tissue [2]. The ventral brainstem view also shows the location of cranial nerves and blood vessels, which are used as landmarks for insertion of "patch-clamp" electrodes in the *en bloc* model for "whole-cell" recordings of membrane potentials from different types of newborn rat VRC neurons. In the right part of the figure, the activity patterns of such neurons are aligned with reference to inspiratory-related activity of ventral cervical nerve roots ($C_{3–6}$) forming the phrenic nerve. Specifically, these neurons are I neurons (sublabeled Insp-I, Insp-II, Insp-III after [3]), two types of E neurons, and preinspiratory (plus postinspiratory) active "Pre-I"-type pFRG neurons. Modified with permission from H. Onimaru, A. Arata, I. Homma (Respiration & Circulation 46, 773–782, 1998). Abbreviations: *E-2*, active expiratory phase; *BötC*, Bötzinger Complex; *V-XII*, cranial nerves, specifically *V*, trigeminus; *VI*, abducens; *VIII*, vestibulocochleat; *IX*, glossopharyngeus; *X*, vagus; *XI*, accessorius; *XII*, hypoglossus; C_1, 1st ventral cervical nerve.

rat brainstem-spinal cord preparation was combined with suction electrode recording of cranial and spinal nerve activities to consolidate first the conclusion from previous findings on that model [3] that neither the pontine nor the dorsal respiratory group are necessary for fictive inspiratory-related rhythm [8]. Instead, inspiratory rhythm was irreversibly blocked when microsection affected a medullary area, named the preBötC, which extended ~200 μm in rostrocaudal direction (Fig. 2). Finally, this study demonstrated that the preBötC remains inspiratory active after isolation in a transverse brainstem slice (Fig. 2).

The view that the preBötC constitutes an essential respiratory center is supported since its discovery by numerous *in vivo* and *in vitro* studies [3,4]. However, the preBötC is not the only autonomous respiratory

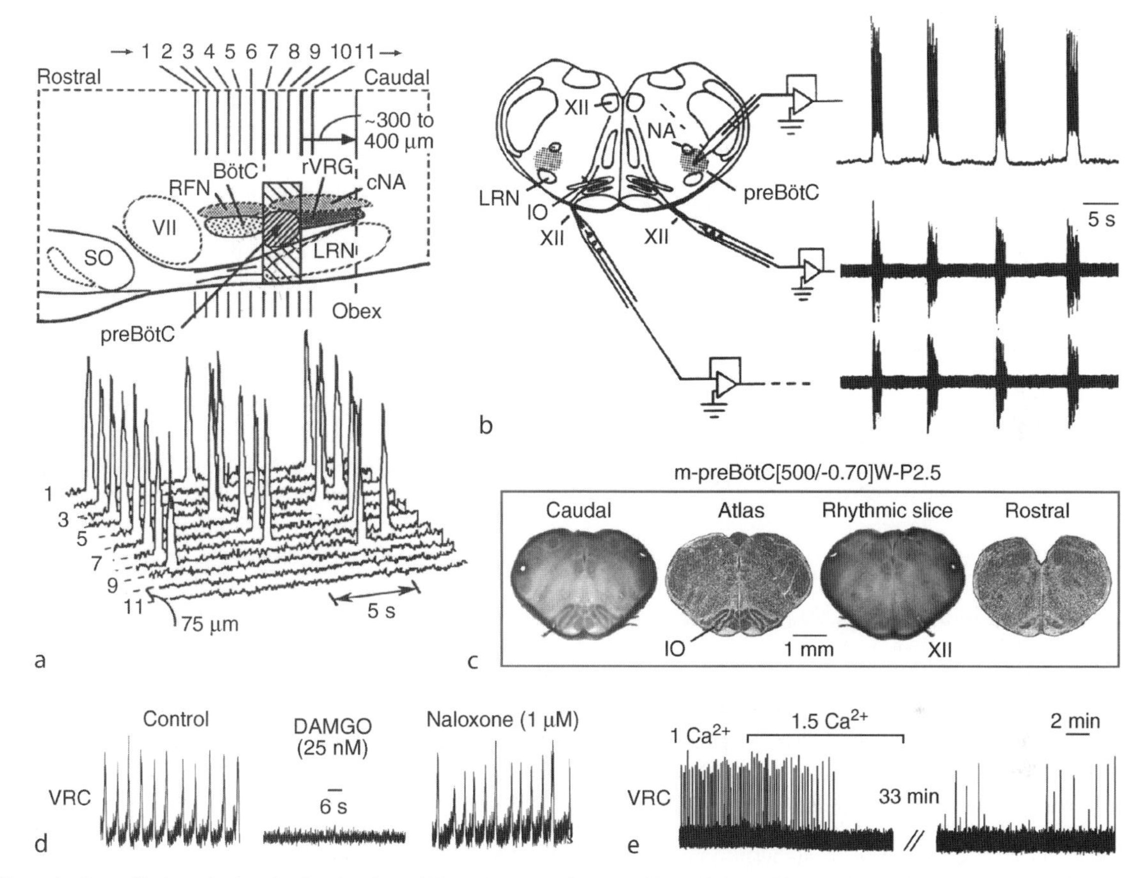

Respiratory Network Analysis, Isolated Respiratory Center Functions. Figure 2 Isolation of the inspiratory center. (a) the upper panel shows a schematic lateral sagittal section through the ventral aspect of the newborn rat brainstem. The numbers correspond to consecutive 75 μm transverse sections through the brainstem-spinal cord model (Fig. 1). The sections were carried out in rostral to caudal direction, while recording with suction electrodes inspiratory-related cervical nerve bursts which are displayed, after integration, in the lower panel. Such experiments revealed an irreversible block of rhythmic discharge following section 8. These findings were complemented by results from corresponding recordings from inspiratory active cranial nerves during sectioning in caudorostral direction. This identified the preBötC as a brainstem region with a rostrocaudal extension by ~200 μm (see box) which is important for generation of respiratory rhythm. Abbreviations others than those in Fig. 1: *LRN*, lateral reticular nucleus; *RFN*, retrofacial nucleus; *SO*, superior olive; *rVRG*, rostral ventral respiratory group; *cNA*, caudal nucleus ambiguus. (b) The preBötC remains inspiratory active in a newborn rat brainstem slice with a thickness >175 μm. Rhythmically active neurons in the area of the ventrolateral medulla rhythmically drive XII motoneurons which innervate via the XII nerve genioglossal tongue muscles for patency of the upper airways during inspiration. (a and b modified with permission from [8].) (c) The constancy of respiratory marker nuclei such as the inferior olive enables the generation of preBötC slices with rostrocaudal boundaries which are "calibrated" by comparison with a newborn rat brainstem atlas [9]. The latter study introduced a terminology for such slices. In this example the "m-preBötC[500/-0.70]W-P2.5" slice contains the preBötC in the middle ("m-preBötC"), is 500 μm thick with the caudal boundary 0.70 mm caudal to VII_c, and was produced from a 2.5 days-old (P2.5) Wistar (W) rat. (d) calibrated preBötC slices generate robust inspiratory-related rhythm in the area of the VRC in superfusate with physiological (3 mM) instead of routinely used 7–11 mM [K^+]. This rhythm is very sensitive to low concentrations of the μ-opioid receptor agonist DAMGO. Note that the 3 mM K^+ rhythm is effectively restored within few minutes of application of the opioid receptor antagonist naloxone (1 μM). c and d modified with permission from [9]. (e) in a different preBötC slice, 3 mM K^+ rhythm was abolished shortly after raising superfusate Ca^{2+} from 1 mM (the lower limit of the proposed physiological range) to 1.5 mM (the upper limit). Note the incomplete recovery of inspiratory rhythm following washout of raised Ca^{2+}. Modified with permission from [2].

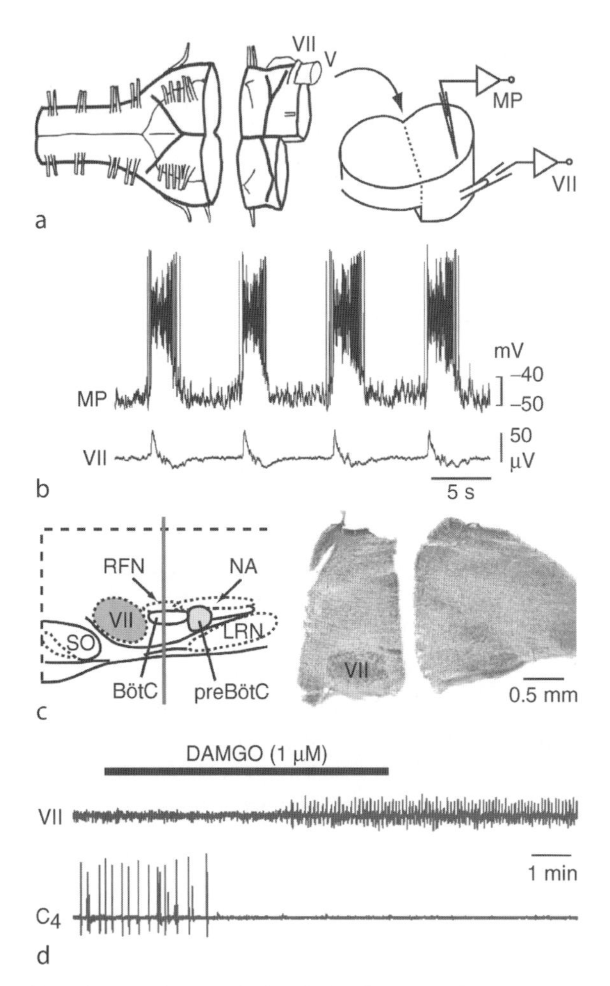

Respiratory Network Analysis, Isolated Respiratory Center Functions. Figure 3 Rhythmic pFRG activity in a transverse slice of brainstem tissue rostral to the preBötC. (a) the newborn rat *en bloc* model was transected slightly rostral to the most rostral XII root to obtain a transverse slice for simultaneous suction electrode recording from the VII nerve and whole-cell recording of membrane potential (MP) from neurons within the ventrolateral medulla. (b) membrane potential oscillations of putative Pre-I neuron in the rostral block were synchronous with VII nerve activity in "Suzue-type" solution with 6.2 mM K^+ and 2.4 mM Ca^{2+}. Shortly before the recording, the preparation was treated for 10–15 min with DAMGO (1 μM) for enhancement of such bursting. (c) histological reconstruction revealed that the transection was between VII_c and the rostral boundary of the preBötC, thus in the region corresponding to the BötC in adult mammals (compare Fig. 1). (d) in a preparation transected at a level similar to that in the experiment of a–c, but without removing the rostral block, DAMGO in 6.2 mM K^+ and 2.4 mM Ca^{2+} abolished inspiratory-related C_4 activity in the caudal aspect of the transected preparation, but restored VII nerve rhythm, which was transiently depressed due to the transection procedure. These findings suggest that pFRG neurons in rostral medullary slices, not including the preBötC, produce rhythmic bursting which is facilitated by opioids, in contrast to a strong depressing action of such drugs on more caudal preBötC-driven rhythms. Modified with permission from [7,8].

center in mammals. Already more than 20 years ago, the hypothesis has been proposed that pre/postinspiratory active "Pre-I" neurons are important for maintenance of the rhythmic activity of inspiratory medullary networks in the newborn rat *en bloc* preparation [3]. More recent findings from voltage-sensitive dye imaging of spatiotemporal respiratory patterns and concomitant electrophysiological recording of membrane potential indicated that Pre-I neurons form the pFRG, a functionally and anatomically defined respiratory group [6] (Fig. 1). The pFRG remains rhythmically active and drives VII motoneurons in a transverse slice of brainstem tissue that rostrally neighbors the preBötC [7] (Fig. 3). Most pFRG neurons are active during both the preinspiratory and postinspiratory phase, but are subject to pronounced inhibition via hyperpolarizing $GABA_A$ and glycine receptor-mediated inhibitory postsynaptic potentials during the inspiratory phase [3,5–7] (Fig. 1). In contrast, pFRG neurons are continuously active for a time period of several seconds in the slices with rhythmic VII nerve activity [7] (Fig. 3). This supports earlier assumptions that the preBötC is responsible for inspiratory inhibition of Pre-I cells in the *en bloc* medullas [3]. Although "reference" inspiratory motor activity is missing in the rhythmic pFRG slices, it is likely that the sustained neuronal and VII nerve activities (Fig. 3) span the preinspiratory, inspiratory plus postinspiratory phases.

Bursting of pFRG neurons during these phases is in accordance with the finding that branches of the VII nerve innervate muscles of the *alae nasi* that decrease the nasal airway resistance in cats and dogs before, during and after inspiration [7]. Conversely, the finding that interneurons in the ventrolateral aspect of preBötC slices induce rhythmic activity of XII motoneurons (Fig. 2) strongly suggests that the rhythm in that model is inspiratory-related [8], because subgroups of XII motoneurons innervate the tongue during inspiration for patency of the upper airways [1,4].

Determinants of Isolated Respiratory Center Activities

The rhythmic activities of the isolated respiratory centers are not identical with those in intact animals or less reduced *in vitro* preparations such as the "working heart brainstem preparation" of rodents [see corresponding chapters]. Though, activities in the rhythmic slices share several features with respiratory behaviors *in vivo*. For example, in juvenile rats preBötC-driven inspiratory activity is blocked by opioids, whereas rhythmic contractions of pFRG-driven expiratory abdominal muscles are not inhibited [4,5]. Similar to these *in vivo* findings, opioids depress preBötC-driven (motor) rhythms *in vitro*, whereas pFRG-driven cellular and nerve activities are not inhibited [4,5,7]. The effects of various neuromodulators on the *in vitro* respiratory-related rhythms are influenced by the experimental conditions, which differ notably between laboratories. In particular the superfusate concentrations of K^+ and Ca^{2+} vary between 3–11 mM and 0.8–2.4 mM, respectively, despite the notion that these cations strongly modulate neuronal excitability [2]. Regarding the action of opioids, preBötC slice rhythms in physiological K^+ (3 mM) and 1 mM Ca^{2+} are blocked by low nanomolar concentrations of opioids (Fig. 2), whereas close to micromolar concentrations are needed to depress rhythms in preBötC slices or *en bloc* medullas in superfusate with elevated K^+ (Fig. 3) [7,9]. Furthermore, preBötC slices generate long-term and robust rhythm in 3 mM K^+ and 1 mM Ca^{2+} (corresponding to the lower range of the physiological spectrum), whereas rhythm is depressed by elevation of Ca^{2+} to 1.5 mM (the proposed upper limit of the physiological range) [2,9] (Fig. 2). In 1.5 mM Ca^{2+}, preBötC slice rhythm is reactivated by raised K^+ leading to the view that isolated inspiratory center activity is determined by an extracellular "Ca^{2+}/K^+ antagonism" [2].

Respiratory center rhythms depend also critically on the physical dimensions of the *in vitro* models. For example, findings in the newborn rat *en bloc* model indicated that the pFRG drives pre/postinspiratory bursting of lumbar motoneurons in spinal L_{1-2} segments via premotoneurons which are located caudal to the preBötC [5]. This view was substantiated by the observation in juvenile rats *in vivo* that brainstem transection at the caudal end of the VII nucleus, which partially overlaps with the pFRG [6] (Figs. 1 and 3), abolished pre/postinspiratory bursting of expiratory abdominal muscles innervated by L_{1-2} lumbar motoneurons [4]. Similar transection experiments in the newborn rat *en bloc* model revealed that the transection level critical for blocking pre/postinspiratory lumbar bursting is quite close to the rostral instead of the caudal end of the VII nucleus [2]. The absence of respiratory lumbar bursting in such transected preparations suggests that pFRG neurons responsible for this motor behavior are located in the most rostral aspect of the pFRG [2] (Fig. 1). However, it is for example also possible that axons from more caudal pFRG neurons inducing lumbar respiratory bursting project first rostrally and may thus have been transected [2,10]. This indicates that results from transection experiments need to be considered with caution. Furthermore, Pre-I neurons are not only found in the main area of the pFRG, but also within the preBötC, and even in regions caudal to the preBötC. In addition to Pre-I and expiratory cells, the preBötC in both perinatal rodent *en bloc* medullas and *in vivo* contains various subclasses of inspiratory neurons [3] (Fig. 1). Conversely, inspiratory (and expiratory) neurons are also active in the main area of the pFRG. While the rostral portion of the pFRG co-locates with the VII nucleus, the caudal part of the pFRG partially covers an area corresponding to the Bötzinger Complex in mature mammals. Finally, the pFRG also more or less overlaps medially with the Retrotrapezoid nucleus, which is one presumptive site of central respiratory chemosensitivity [4].

Due to the overlap of primary (rhythmogenic) areas and secondary (chemosensitive) respiratory drive regions, in concert with a rostrocaudally dispersed distribution of distinct classes of respiratory neurons, the respiratory centers can not be isolated without portions of functionally and/or anatomically different structures that may interact with these centers. Despite these caveats, the reductionistic approach has already, and will further, provide important information on the neural control of breathing. For example, ►multiphoton/confocal Ca^{2+} imaging has been adapted to study the activity and gross morphological features such as soma size or shape of preBötC neurons, located in a histologically defined rostrocaudal area of "calibrated" preBötC slices that operate in physiological cation solution [9] (Fig. 2) [see ►Respiratory network analysis, Functional imaging]. A structure-function relationship of the isolated respiratory centers may be feasible by using the calibrated *in vitro* models in combination with nerve and intracellular electrophysiological recording plus Ca^{2+} and voltage-sensitive dye imaging, which was crucial for identification of the pFRG [6] [see ►Respiratory network analysis, Functional imaging].

A Dual pFRG-preBötC Respiratory Center

Findings in the *en bloc* brainstem model and juvenile rats *in vivo* suggest that the pFRG and the preBötC constitute a dual respiratory center. This hypothesis has been first proposed according to the above described distinct effects of opioids on inspiratory and pre/postinspiratory motor behaviors. These findings led to the conclusion that opioids inhibit breathing, at least partly, due to depression of synaptic excitatory transmission between the pFRG and the preBötC [4,5]. This view that excitatory drive from the pFRG to the preBötC is necessary for robust activity of the preBötC has already been proposed much earlier based on the finding in the *en bloc* model that focal lesion of the area including the pFRG impairs inspiratory cervical motor output [3]. However, as stated above the preBötC slices are capable of generating robust rhythm in physiological ion solution for several hours in the absence of structures corresponding to the main location of the pFRG [2,9]. The finding that rhythmic VII nerve activity in the pFRG slices is not depressed, but rather stimulated, by μ-opioid receptor agonists [7] (Fig. 3) supports the view that this respiratory center may be in particular important for breathing during the perinatal period, when the brainstem is presumably subject to a surge by endogenous opioids [4].

Anoxia represents a further approach for analyzing the cooperativity of the pFRG and the preBötC. In the *en bloc* model, both hypoxic and chemical anoxia synchronize and enhance the activities of the pFRG and preBötC rhythm generators. This results in pronounced and persistent (>20 min) postinspiratory medullary activities and lumbar/facial motor nerve bursting during anoxia which is accompanied by inspiratory-related nerve burst doublets [10]. A causal relation between the latter anoxia-related phenomena is suggested by the finding that control pre/postinspiratory lumbar bursting is absent (similar to anoxia-induced enhancement of postinspiratory lumbar bursting and inspiratory-related nerve burst doublets) upon transection of the newborn rat *en bloc* preparation between the preBötC and the caudal end of the VII nucleus [10]. These results suggest that the anoxic postinspiratory augmentation of medullary interneuronal and lumbar/facial nerve bursting as well as inspiratory motor burst doublets require an interaction between the preBötC and the pFRG [10].

In summary, the rhythmogenic preBötC and pFRG appear to constitute a dual respiratory center, which adjusts its activity to cope with pathological disturbances of breathing. Under the influence of opioids, boosted pFRG activity may partly compensate for the depressed intrinsic preBötC interneuronal activity and provide enhanced drive for breathing efforts. During oxygen depletion, a functional reorganization of the preBötC and pFRG for synchronized and augmented bursting may optimize uptake of oxygen by enhancing single breaths. However, the extent of cooperativity between these respiratory centers during normal breathing is not clear yet. In particular, it remains to be shown that the pFRG has a similarly important role for breathing in mature mammals compared to newborns. That the pFRG is active *in vivo* after birth is indicated by the above finding of pre/postinspiratory abdominal muscle activity in juvenile rats [4,5]. It may be important to study whether the pFRG in mature mammals is closely related to the chemosensitive ►Retrotrapezoid nucleus [4]. That this may be the case is suggested by the anatomical overlap of these brainstem regions and by the findings that neurons in both regions are excited by raised levels of CO_2 and H^+ and project to other respiratory areas including the preBötC [3,4].

References

1. Feldman JL (1986) Neurophysiology of breathing in mammals. In: Handbook of physiology. The nervous system. Intrinsic regulatory systems in the brain, vol 4, chapter 9. Am Physiol Soc,Washington, DC, Sect 1, pp 463–524
2. Ruangkittisakul A, Secchia L, Bornes TD, Palathinkal DM, Ballanyi K (2007) Dependence on extracellular Ca^{2+}/K^+antagonism of inspiratory centre rhythms in slices and en bloc preparations of newborn rat brainstem. J Physiol 584:489–508
3. Ballanyi K, Onimaru H, Homma I (1999) Respiratory network function in the isolated brainstem-spinal cord of newborn rats. Prog Neurobiol 59:583–634
4. Feldman JL, Del Negro CA (2006) Looking for inspiration: new perspectives on respiratory rhythm. Nat Rev Neurosci 7:232–242
5. Janczewski WA, Onimaru H, Homma I, Feldman JL (2002) Opioid-resistant respiratory pathway from the preinspiratory neurones to abdominal muscles: in vivo and in vitro study in the newborn rat. J Physiol 545:1017–1026
6. Onimaru H, Homma I (2003) A novel functional neuron group for respiratory rhythm generation in the ventral medulla. J Neurosci 23:1478–1486
7. Onimaru H, Kumagawa Y, Homma I (2006) Respiration-related rhythmic activity in the rostral medulla of newborn rats. J Neurophysiol 96:55–61
8. Smith JC, Ellenberger HH, Ballanyi K, Richter DW, Feldman JL (1991) Pre-Bötzinger complex: a brainstem region that may generate respiratory rhythm in mammals. Science 254:726–729
9. Ruangkittisakul A, Schwarzacher SW, Secchia L, Poon BY, Ma Y, Funk GD, Ballanyi K (2006) High sensitivity to neuromodulator-activated signaling pathways at physiological [K^+] of confocally imaged respiratory center neurons in on-line-calibrated newborn rat brainstem slices. J Neurosci 26:11870–11880
10. Taccola G, Secchia L, Ballanyi K (2007) Anoxic persistence of lumbar respiratory bursts and block of lumbar locomotion in newborn rat brainstem-spinal cords. J Physiol 585:507–524

Respiratory Network Responses to Hypoxia

Narong Simakajornboon[1], David Gozal[2]
[1]Cincinnati Children's Hospital Medical Center, Cincinnati, OH, USA
[2]Kosair Children's Hospital Research Institute, Department of Pediatrics, University of Louisville, Louisville, KY, USA

Definition

Respiratory network responses to ►hypoxia refer to the complex interactions between groups of neurons located mainly in the medulla, pons and midbrain that are responsible for control of ventilation during hypoxia. The physiologic mechanisms underlying these processes involve intricate interplay of ►neuromodulators released from respiratory neurons and glial cells.

The ventilatory response to hypoxia is precisely controlled and distinctive patterns emerge during postnatal development. The functional role of this process is not only to adapt the respiration during hypoxic conditions, but also to ensure cell survival, especially during the vulnerable period of development.

Characteristics

The mammalian ventilatory response to hypoxia is biphasic. It consists of an initial increase in ►minute ventilation, followed by a later decline in ventilation which is termed hypoxic ventilatory depression (Fig. 1).

In developing mammals, the magnitude of hypoxic ventilatory depression is particularly prominent such that minute ventilation decreases below normoxic level [1]. The early component of hypoxic ventilatory response is mediated mainly through the ►peripheral chemoreceptor in the ►carotid body. The type I glomus cells in the carotid body are the primary site of oxygen sensing signal. The afferent signal is then transmitted to sensory terminals of the carotid sinus nerve and is subsequently projected to the nucleus of solitary tract (nTS). The nTS is located in the brainstem region and provides the first central synaptic relay to peripheral chemoreceptor afferent inputs. Other nuclei at this level of brainstem that play a role in respiratory control and the HVR include the nucleus ambiguus, the area postrema, the dorsal motor nucleus of vagus, the hypoglossal nucleus and the pre-Botzinger complex. A variety of neuromodulators in these areas play a crucial role in the central mediated hypoxic ventilatory response. Several studies have shown that the early response to hypoxia is mediated through platelet activating factor receptor pre-synaptically [2,3] and post-synaptically by N-methyl-D-aspartate (NMDA) glutamate receptors [4,5], which then activate downstream signaling pathways such as protein kinase C, tyrosine kinases, and calcium calmodulin kinase [6]. The hypoxic ventilatory depression is mediated through several complex mechanisms. In addition to hypoxia-induced reductions in metabolism, several neuromodulators have been thus far identified as playing a role in the hypoxic ventilatory depression, namely adenosine, γ-aminobutyric acid (GABA), serotonin (5-HT), opioids, and platelet derived growth factor (PDGF-β) receptors.

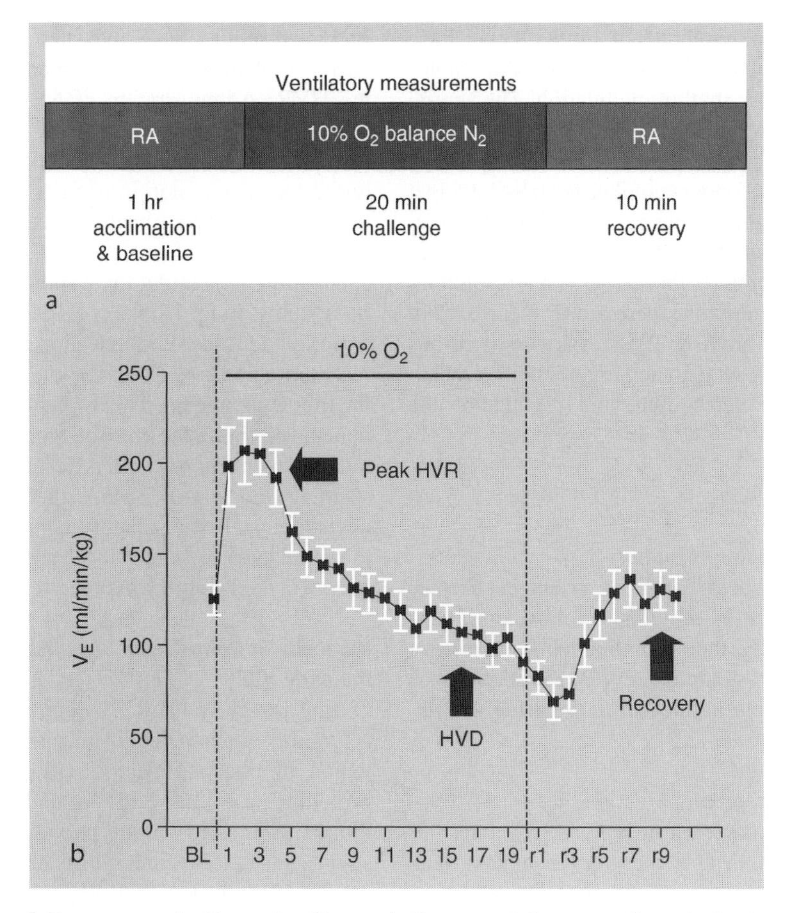

Respiratory Network Responses to Hypoxia. Figure 1 Representative recording of minute ventilation during a 20-min hypoxic challenge followed by 10 min recovery in normoxia in a 14-day old rat pup. Please note initial increase in minute ventilation (HVR) followed by progressive time-dependent reduction in ventilatory output (HVD).

In this section, we will discuss the respiratory neuronal network with particular emphasis in the caudal brainstem, and will delineate specific neuromodulators mediating each component of hypoxic ventilatory response from a developmental perspective.

Respiratory Neuronal Network

The peripheral chemoreceptors are located in carotid bodies and aortic bodies. These areas contain glomus cells of 2 types. The type I glomus cells in the carotid bodies is oxygen sensing cells of the peripheral chemoreceptor. The hypoxic stimulus is then transmitted to the sensory terminal of the carotid sinus nerve (CSN). From CSN, the signal projects to the several regions of the nucleus of solitary tract (nTS), the first synapses for primary afferents originating from peripheral chemoreceptors. Retrograde tracer studies reveal that the medial, dorsomedial, lateral and commissural regions of the nTS receive dense innervations from peripheral chemoreceptor afferent fibers. The nTS, the main neuonal nucelus of the ►dorsal respiratory group, has interconnections to other respiratory neurons including the pontine respiratory and the ►ventral respiratory group. The pontine respiratory group is composed of the lateral and medial parabrachial and Kollicker-Fuse nucleus, which play a role in diaphragmatic motor control and respiratory rhythm modulation. The ventral respiratory group is divided into the rostral and the caudal group. The rostral part of ventral respiratory group includes the Botzinger complex, the pre-Botzinger complex, and the parafacial respiratory group. The pre-Botzinger complex and the parafacial respiratory group are believed to encompass the kernel for ►respiratory rhythmogenesis. In addition, there are influences from many rostral brain areas to the respiratory neurons including the suprapontine nuclei, midbrain, diencephalons, hypothalamus, cerebellum, and regions of the cerebral cortex.

Neuromodulators

Early HVR

Platelet Activating Factor Receptors

Platelet activating factor and its cognate receptor (PAFR) are proposed to modulate glutamatergic signaling presynaptically, thereby influencing the release of glutamate into the synaptic cleft. PAFR activity has now been conclusively implicated in the acute ventilatory response to hypoxia [2,3].

NMDA Glutamate Receptors

In the cardiorespiratory control regions, N-methyl-D-aspartate glutamate (NMDA) receptors mediate critical components of the respiratory pattern generation, cardiovascular regulation and HVR. The early response to hypoxia is mediated through NMDA glutamate receptors [4,5]. NMDA receptors are widely expressed throughout the brain, including the respiratory control areas such as nTS. Previous studies have demonstrated that systemic and targeted brainstem administration of NMDA glutamate receptor antagonists is associated with attenuation of HVR in adult and developing animals. In addition, hypoxia induces an increase in glutamate concentration within the nTS of conscious rats, which is associated with an increase in minute ventilation. However, the hypercapneic (►hypercapnia) ventilatory response is not affected by NMDA glutamate receptors [5]. The structure of NMDA glutamate receptors consists of heterodimeric, mandatory subunits that include one or more of the splice variants of the NMDA NR1 subunit and additional NR2 and NR3 subunits. Activation of NMDA receptors in the caudal brainstem of conscious rats involves tyrosine phosphorylation of both NR1 and NR2A/B subunits. In addition, the role of NMDA receptor in HVR is developmentally regulated such that an increasing dependency on NMDA glutamate receptor emerges over time and transition from an immature to a more mature hypoxic response requires NMDA receptor-bearing neurons within the nTS.

Non-NMDA Receptors

Previous studies have indicated the potential role of AMPA glutamate receptors in the ventilatory control and the HVR. Administration of NBQX (a selective non-NMDA receptor antagonist) did not affect ventilatory output in adult conscious mice and cats, but led to marked respiratory depression in neonatal animals. Microinjection of the AMPA glutamate receptor blocker NBQX within the nTS of anesthetized adult rats resulted in attenuation of ventilatory responses following carotid body stimulation. However, NBQX failed to modulate hypoxia-induced c-Fos activation in the adult rat. AMPA receptors appear to influence the respiratory pattern in the immature animal. The respiratory rhythm generation in neurons within the pre-Botzinger complex of neonatal rats is dependent on AMPA receptor activity. Notwithstanding, the role of AMPA glutamate receptors in the developmental of respiratory control may be limited to the regulation of timing mechanisms during normoxia, but not in mediating the hypoxic ventilatory responses [7].

Intracellular Downstream Signaling Pathways Underlying the early HVR

During the early HVR, NMDA receptors (NMDA-R) activation elicits calcium influx, and subsequent activation of phospholipase C (PLC), mitogen-activated protein kinase kinase (SEK) and calcium calmodulin kinase 2 (CaCmII). Our previously proposed model suggests that activation of PLC leads to translocation of protein kinase C, and phosphorylation of serine/threonine residues in the intracellular domain of NMDA receptors. SEK phosphorylates stress activated protein kinase 2 (JNK-2) leading to activation of the activator

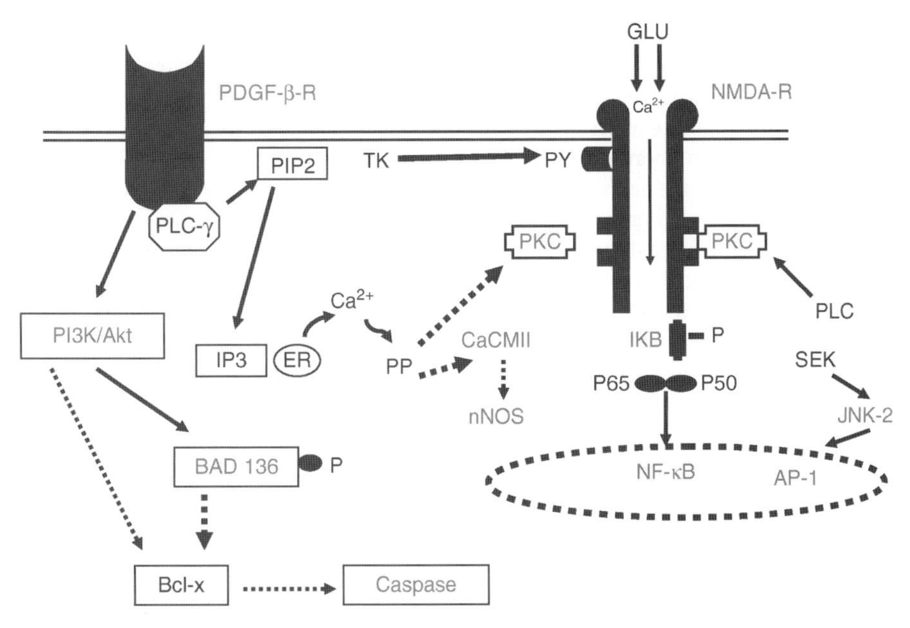

Respiratory Network Responses to Hypoxia. Figure 2 Schematic diagram of signaling pathways that are operational in respiratory neurons within the nucleus of the solitary tract during hypoxia. Signal transduction proteins for which there is definitive evidence are shown in red. (See text for more details).

protein-1 complex (AP-1). CaCmII activates neuronal nitric oxide (NO) synthase resulting in NO formation. NMDA receptor activation will also lead to phosphorylation of IκB with subsequent activation of nuclear kappa B (NF-κB) and activation of tyrosine kinase (TK) by tyrosine phosphorylation (PY) (Fig. 2).

Of all the downstream signaling pathways, the functional role of PKC on respiratory control neurons has been studied extensively. PKC activation within the respiratory neurons of the ventral medullary group is associated with increased respiratory drive potentials. Endogenous PKC activity modulates tonic activity and excitability of the expiratory neurons in the cat. PKC within the caudal brainstem underlies critical components of both tonic respiratory drive and the HVR [6]. Most of the known PKC isoforms are expressed with in the dorsocaudal brainstem, and activation of both Ca^{2+}-dependent and Ca^{2+}- independent PKC isoforms occurs in the nTS during hypoxia [6]. PKC exerts a significant influence on respiratory timing during normoxia in the early postnatal period, and the effect decreases with advancing age. In contrast, hypoxia-induced PKC activation is absent in the immature animal and emerges concomitantly with the appearance of NMDA dependency. Nitric oxide (NO) is another important neuromodulator with a dual role in hypoxic chemotransduction. While NO derived from endothelial nitric oxide synthase (eNOS) exerts an inhibitory effect at the carotid body level, NO derived from neuronal nitric oxide synthase (nNOS) in the caudal brainstem plays a significant role in sustaining ventilation during the second phase of the HVR. Activation of NMDA receptors will lead to opening of a voltage-dependent calcium channel, calcium calmodulin binding and subsequent nNOS activation. The intracellular NO will in turn modulate glutamate release, either through activation of cGMP-dependent protein phosphorylation cascades, or by retrograde activation of the pre-synaptic neuron. Therefore, nNOS acts an excitatory neurotransmitter during the HVR and may prevent the early onset of hypoxia-induced ventilatory depression. In addition, we have recently identified a mechanism whereby deoxyhemoglobin elicited by the presence of environmental hypoxia activates the formation of S-nitrosothiols through a very tightly regulated process, and that these compounds lead to excitation of respiratory-related neurons within the nTS, and thus contribute to the early phase of HVR [8].

Late HVR

As the duration of hypoxia is extended, some degree of ventilatory depression will develop. This component of HVR is extremely prominent in developing animals. Several neuromodulators including γ-amino-butyric acid (GABA), serotonin (5-HT), adenosine, opioid receptors, and platelet-derived growth factor (PDGF)-β receptors have all been shown to play contributory roles to the emergence of the hypoxic ventilatory depression associated with prolonged hypoxia. We will briefly delineate the role of each neuromodulator in the late phase of HVR. GABA acts through two GABA receptor subtypes, GABA-A and GABA-B. GABA-A receptors modulate tidal volume, whereas GABA-B receptors modulate respiratory frequency and pattern of breathing. It is postulated that the hypoxic ventilatory depression is the result of imbalance between the excitatory glutamate and the inhibitory GABA [4]. In addition, hypoxic ventilatory depression of developing animals is partly mediated through the neuro-depressant effect of GABA. Another neuromodulator, adenosine plays an important role in hypoxic ventilatory depression during the early postnatal period, and the effect decreases with maturation. Among the major 4 adenosine receptors (A_1, A_{2A}, A_{2B} and A_3), adenosine A_1 and A_{2A} receptors are postulated to play a role in the hypoxic ventilatory depression. The inhibitory effect of adenosine A_1 receptors may involve postsynaptic hyperpolarization, presynaptic depression of synaptic transmission, modulation of cAMP mediated pathway and activation of potassium channels. While adenosine A_1 receptors are involved in cardiorespiratory control during normoxia, adenosine A_{2A} receptors play a critical role in the hypoxic ventilatory depression. Serotonin (5-HT) has been shown to play a role in hypoxic ventilatory depression in both adult and developing animals. This neurotransmitter exerts multiple effects on respiratory control, and modulates both the respiratory rhythm generator and the respiratory motoneurons. Among the myriad of 5-HT receptor subtypes, 5-HT_{1A} and 5-HT_2 receptors have been shown to play a role in respiratory control. While 5-HT exerts an excitatory effect on the central respiratory rhythm generator within the rostral medulla area, it inhibits the hypoglossal inspiratory output in developing animals, possibly through activation of 5-HT_2 receptors. The release of endogenous 5-HT may signal the termination of the early hypoxic augmentation, possibly through activation of 5-HT_{1A} receptors. 5-HT_{1A} receptors are present in the hypoglossal nucleus of developing rats. Their density is high in the newborn and decreases with increasing postnatal age. Since the use of morphine led to occasional onset of respiratory depression, it became apparent that opioids are involved in respiratory functions within the CNS. Interestingly, endogenous opioids have been shown to play a role in the late phase of hypoxic ventilatory response, whereby ventilatory depression may be partly mediated through opioid-mediated neuronal inhibition. Opioids modulate the respiratory frequency and tidal volume through activation of μ- and δ-opioid receptors respectively. The caudal brainstem, especially the nTS and the nucleus ambiguus, seem to be important sites for opioid-induced inhibition of respiration. Opioid receptors display a distinct maturation pattern during the early postnatal period. The μ-opioid receptor binding sites are present during the mid-fetal period and are located in the cardiorespiratory-related brainstem nuclei, whereas the δ-opioid receptors primarily appear during the postnatal period. Both μ(1) and μ(2) opioid receptors are involved in opioid-induced respiratory depression in early postnatal period.

Finally, we have shown that hypoxia specifically triggers the release of the PDGF polypeptide isoform called PDGF-BB from glial cells, which in turn leads to subsequent activation of PDGF-β receptor in the nTS, where it reduces ventilatory output [9]. Both PDGF-B chains and PDGF-β receptors are abundantly expressed in nTS neurons of adult rats [9], and activation of the receptors leads to down-regulation of ligand-gated ion channels, such as NMDA glutamate receptors. PDGF-β receptor activation is an important contributor to the hypoxic ventilatory depression at all postnatal ages, but is more critical in the immature animals. The increased expression of PDGF-β receptors in the caudal brainstem of immature animals may provide additional protection against hypoxia-induced ►apoptosis. In fact, PDGF-β receptors exert their role in promoting neuronal cells survival via two major signaling pathways, namely the phosphoinositide 3 kinase (PI3K)/Akt and the MEK/MAPK pathways. Activation of PDGF-β receptors leads to tyrosine phosphorylation of sites that will activate Ras kinase. Ras can activate PI3K, which in turn may phosphorylate PI3K, which in turn may phosphorylate the serine-threonine protein kinase called Akt, the latter phosphorylating BAD at serine 136. Phosphorylated BAD binds to cytosolic 14-3-3 protein, whereas dephosphorylated BAD binds elements of the Bcl-2 complex such as Bcl-x to promote apoptosis (Fig. 2). In fact, hypoxia-induced phosphorylation of PDGF-β receptors in the caudal brainstem of adult rats is temporally associated with activation of an anti-apoptotic mechanism via the PI3 kinase-dependent phosphorylation of both Akt and BAD pathways [10]. This mechanism may prevent induction of apoptosis in the respiratory neurons during hypoxia, and may contribute to the well known increased hypoxic tolerance of the brainstem neurons.

References

1. Eden GJ, Hanson MA (1987) Maturation of the respiratory response to acute hypoxia in the newborn rat. J Physiol 392:1–9
2. Simakajornboon N, Graff GR, Torres JE, Gozal D (1998) Modulation of hypoxic ventilatory response by systemic platelet-activating factor receptor antagonist in the rat. Respir Physiol 114(3):213–225
3. Gozal D, Holt GA, Graff GR, Torres JE (1998) Platelet-activating factor modulates cardiorespiratory responses in the conscious rat. Am J Physiol 275(2):R604–R611
4. Kazemi H, Hoop B (1991) Glutamic acid and gamma-aminobutyric acid neurotransmitters in central control of breathing. J Appl Physiol 70:1–7
5. Ohtake PJ, Torres JE, Gozal YM, Graff GR, Gozal D (1998) NMDA receptors mediate peripheral chemoreceptor afferent input in the conscious rat. J Appl Physiol 84:853–861

6. Gozal D, Graff GR, Gozal E, Torres JE (1998) Modulation of the hypoxic ventilatory response by Ca^{2+}-dependent and Ca^{2+}-independent protein kinase C in the dorsocaudal brainstem of conscious rats. Respir Physiol 112(3):283–290
7. Whitney GM, Ohtake PJ, Simakajornboon N, Xue YD, Gozal D (2000) AMPA glutamate receptors and respiratory control in the developing rat: anatomic and pharmacological aspects. Am J Physiol Regul Integr Comp Physiol 278:R520–R528
8. Lipton AJ, Johnson MA, Macdonald T, Lieberman MW, Gozal D, Gaston B (2001) S-nitrosothiols signal the ventilatory response to hypoxia. Nature 413(6852):171–174
9. Gozal D, Simakajornboon N, Czapla MA, Xue YD, Gozal E, Vlasic V, Lasky JA, Liu JY (2000) Brainstem activation of platelet-derived growth factor-beta receptor modulates the late phase of the hypoxic ventilatory response. J Neurochem 74:310–319
10. Simakajornboon N, Szerlip NJ, Gozal E, Anonetapipat JW, Gozal D (2001) In vivo PDGF beta receptor activation in the dorsocaudal brainstem of the rat prevents hypoxia-induced apoptosis via activation of Akt and BAD. Brain Res 895:111–118

Respiratory Neuroplasticity

Kendall F Morris[1], Donald C. Bolser[2]
[1]Department of Molecular Pharmacology and Physiology, Neuroscience, University of South Florida College of Medicine, Tampa, FL, USA
[2]Department of Physiological Sciences, University of Florida College of Veterinary Medicine, Gainesville, FL, USA

Synonyms

Respiratory memory; Respiratory recovery from injury; Lasting alteration of breathing reflexes

Definition

Respiratory Neuroplasticity has been defined as "a persistent change in the neural control system (morphology and/or function) based on prior experience" [1]. Plasticity exists in various forms in all of the segments of the respiratory network; the afferent, central control and efferent segments. The plasticity may be classed as recovery from injury, respiratory memory, and lasting alterations of protective reflexes such as cough.

Characteristics

The basic respiratory rhythm and pattern is generated by neural networks in what is called the ►ventral respiratory column (VRC) of the medulla. The column spans several groupings of cells, called nuclei, in the ventral lateral medulla from nearly the beginning of the cervical spinal cord almost to the pons. However, breathing is modulated by many other regions of the medulla, such as the midline raphe, the main source of the neurotransmitter serotonin. Neural networks in the pons, cerebellum and cerebrum can also affect breathing and play roles in breathing plasticity.

The central respiratory neural networks use information, or afferent input, from various sensors. Important examples include the carotid and ►aortic bodies that sense carbon dioxide concentrations, as well as acidity (pH) but are the predominant sensors of oxygen in the blood. There are also sensors in the lung that transmit information about lung distension, irritation and excess fluid. Sensors within the medulla itself feedback pH of the brain tissue. Since carbon dioxide and pH are in chemical balance in the body, and brain tissue only functions well in a narrow range of pH, it is important that the respiratory control system tightly control elimination of carbon dioxide produced by metabolism as well as provide oxygen.

Neurons that connect with muscles are called motor neurons. Two important groups of such neurons that reside in each side of the spinal cord are known as phrenic motor neurons. Phrenic motor neurons receive drive, or efferent input, from the VRC directly and in many species indirectly from another group of medullary cells called the dorsal respiratory group (Fig. 1a) Phrenic motor neurons send axons through the phrenic nerves to provide drive to the diaphragm, the major muscle that expands the lungs and produces inspiration of air.

Recovery from Injury

In response to loss of partial function, such as removal of important afferent input, the respiratory neural networks can recover significant normal function. This plasticity varies among species and with age. After surgical removal of ►carotid bodies, normal response to low oxygen, hypoxia, is temporarily lost, but can be recovered in some species, such as cats or rats. Human asthmatics who have had carotid bodies removed show little or no recovery of function. Dogs, ponies and goats recover much less of this function while one-day-old goats that have carotid bodies removed completely recover normal responses to hypoxia. Mechanisms of recovery include up-regulation of alternative input, in this case that from the aortic bodies and perhaps other tissues as well as up-regulation of the efferent limb so that phrenic motor neurons become more responsive to drive.

In response to spinal cord injury, pathways that have little or no activity normally can become active. If one half of the spinal cord is cut, the diaphragm on that side becomes paralyzed because the efferent output of the VRC for phrenic motor neurons is interrupted.

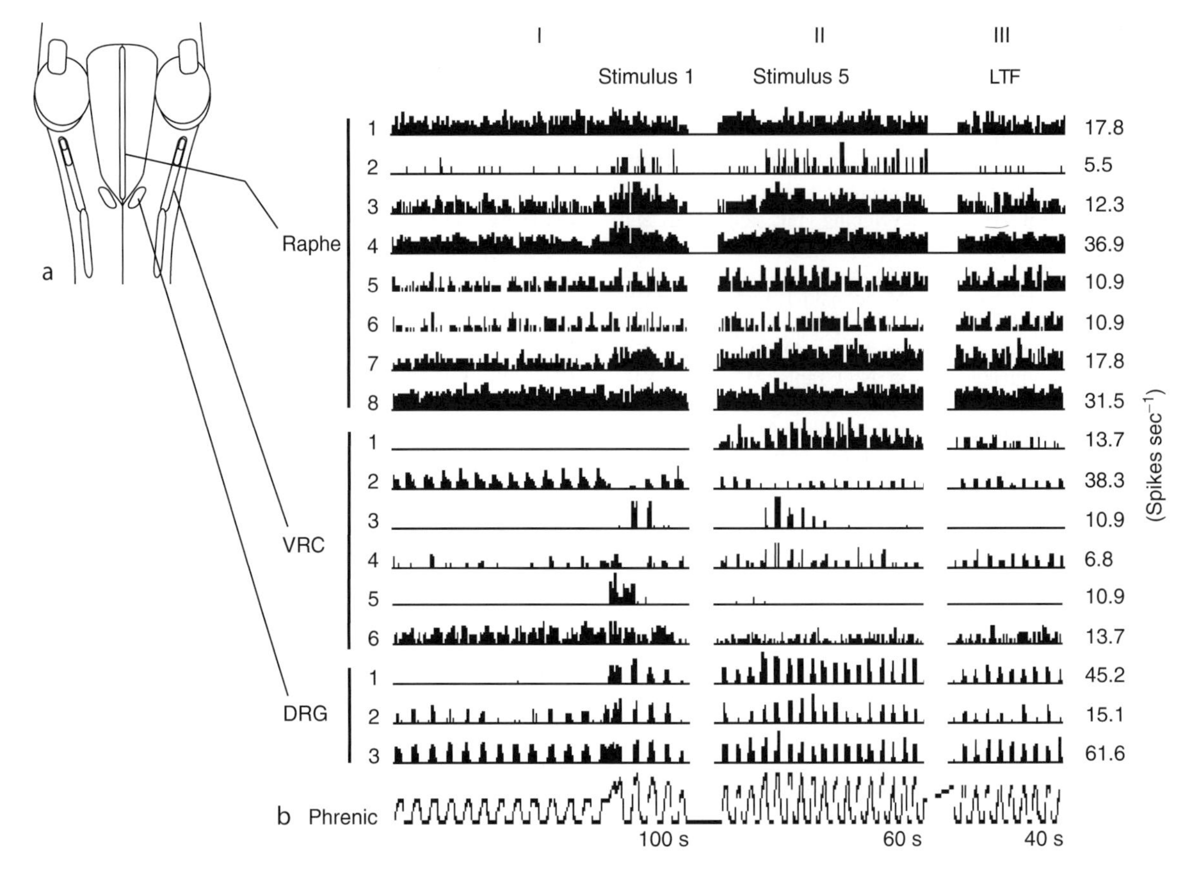

Respiratory Neuroplasticity. Figure 1 Firing rate histograms from multi-site recordings during induction of long-term facilitation (LTF). a, schematic dorsal view of the cat brainstem showing midline raphe, Ventral Respiratory Column (VRC), and the Dorsal Respiratory Group (DRG) b, Data segments show firing rate histograms of 17 neurones and integrated phrenic nerve activity, recorded simultaneously at the indicated sites during the first (I) and fifth (II) period of carotid chemoreceptor stimulation and 6 min following the fifth and final stimulus and induction of LTF (III). Numbers on the right are the firing rates that correspond to the highest "bin." This demonstrates that phrenic amplitude has nearly doubled, while the rate of cycling has increase. Note that some cells have persistent greater peak activity with concomitant shorter durations of activity corresponding to the shorter respiratory phases. Other cells that may be inhibitory to inspiratory activity have persistent decreases in activity (adapted from Fig. 2 of [4] with permission).

However, under increased respiratory drive, alternate, usually inactive pathways from the uninjured side can be activated resulting in partial recovery of function. This is known as the "crossed phrenic phenomenon" [1,2].

Memory

There is a transient plasticity associated with the offset of a respiratory response to a stimulus; e.g., hypoxia, hypercapnia, or many other stimuli with an excitatory effect on breathing. This transient effect is termed short-term potentiation (STP) a slow decay of breathing back to baseline after stimulation. The exact mechanism that produces STP, although neural, remains unknown.

In many rodent strains, there is a short-term decline in respiratory cycle frequency (STFD) following an episode of hypoxia, but not other stimuli, that coincides with STP. STFD results entirely from prolongation of time in expiration. Both post-hypoxic STFD and STP have a similar time course. These changes in pattern have been referred to as "activity-dependent" plasticity [1].

Damage or chemical blockade of regions of the pons removes the prolongation of time in expiration and the short-term decline in respiratory frequency after hypoxia, with no effect on the response to hypoxia. STFD may be mediated by changes in network connections between the medullary and pontine respiratory networks [2].

Long-term facilitation (LTF, Fig. 1b), an increase in respiratory motor output that persists more than 1 h, is another type of plasticity and a robust type of memory. Induction and expression of this memory can be blocked by serotonin and brain derived neurotrophic factor antagonists [3]. LTF is induced by repeated brief, intermittent, but not extended, hypoxia, chemical stimulation of carotid chemoreceptors, or electrical stimulation of the carotid sinus nerve or brain stem midline but not by hypercapnia. In some experiments with cats, rats, dogs and goats, LTF can increase measures of phrenic nerve activity to approximately twice that of baseline.

Altered activity and connectivity among neurons in the VRC and raphe neurons have been identified in spike train data sets in which; (i) the constituent neurons had respiratory-modulated firing patterns, (ii) significant changes in firing rate during carotid chemoreceptor stimulation were correlated with altered respiratory efferent activity, (iii) persistent firing rate changes were expressed during LTF, (iv) there was evidence of effective connectivity between the recorded neurons appropriate to contribute to LTF, and (v) changes in measures of effective connectivity between these neurons after induction of LTF were greater than those during different control periods [2,4].

LTF has been demonstrated in rats in both the activity of the phrenic nerve and in sympathetic nerve activity that is involved in control of blood pressure [9]. Human beings who have sleep apnea are exposed to brief periods of hypoxia each night, similar to many protocols that produce LTF in animal experiments. These people have increased sympathetic nerve activity during the day when they are not hypoxic, and that activity has increased modulation with their breathing. The increased sympathetic nerve activity probably contributes to their increased incidence of high blood pressure, cardiovascular disease and stroke. Normal awake human subjects experimentally exposed to brief, intermittent hypoxia show persistent changes in breathing pattern, they breathe more shallowly and faster as well as with less variability [5]. However, they do not have an over-all increase in breathing similar to LTF in some animals. In contrast, subjects with sleep apnea have a persistent increase in breathing in response experimental intermittent hypoxia during sleep [3].

The plasticity of the neural network expressed as LTF can therefore be both adaptive and maladaptive. It may act to stabilize upper airways and prevent further hypoxia. However, if the intermittent hypoxia persists it may contribute to hypertension and attendant illness.

Sudden infant death syndrome (SIDS) is the most common cause of death in infants between 2 weeks and 1 year of age. Some SIDS cases appear to result from fetal neural damage that later compromises responses to breathing or blood pressure challenges during sleep. A major risk factor is pre- or post-natal tobacco smoke exposure. The deficits appear to involve alterations in neural network function within regions involved in oxygen-sensing and cardiovascular control. A developmental abnormality in serotonergic neurons in the caudal raphe, i.e., a major part of the network implicated in LTF, may result in a failure of protective responses to life-threatening stressors during sleep [6].

Finally, the respiratory networks demonstrate a "metaplasticity" in that early exposure to hypoxia or hyperoxia can produce life long changes in respiratory behaviors, responses and plasticity [1]. Neonatal hyperoxia produces plastic changes that lead to blunted responses to hypoxia in later life, whereas hypoxia in infancy produces greater adult hypoxic ventilatory response and increased expression of LTF [1].

Airway Defensive Reflexes

Cough is an essential component of pulmonary defense and is the most common manifestation of pulmonary disease. Cough is the single most common reason why sick patients visit physicians in the United States. The function of cough is to remove fluids, mucus, and/or foreign bodies from the respiratory tract by the generation of high velocity airflows. These airflows during cough are generated by a complex motor pattern involving three phases: inspiration, compression, and expulsion. The inspiratory phase of cough is generated by a large burst of activity in inspiratory muscles, such as the diaphragm. The compressive phase of cough is produced by laryngeal closure caused by a burst in expiratory laryngeal muscles during rapidly increasing expiratory thoracic and abdominal muscle activity. The resulting large increase in lung air pressure produces very high airflows (up to 12 L s^{-1} in humans) when the larynx opens and the expulsive phase begins. The expulsive phase is characterized by extremely large bursts of activity in expiratory thoracic and abdominal muscles.

In the lower airways, slowly adapting receptors (SARs), rapidly adapting receptors (RARs), and pulmonary C-fibers all can influence the production of cough. There is little doubt that RARs can elicit cough. SARs have a permissive role in the production of cough. The exact role of C-fibers in the production of cough is more controversial, with some groups supporting an excitatory role and others supporting an inhibitory role. Sensory information is processed in the ►brainstem, where the basic elements responsible for the production of cough are located. Pulmonary afferent information is processed by second order interneurons located near to and in various subnuclei of the nucleus of the tractus solitarius.

It was once thought that neural networks separate from those controlling normal breathing, eupnea, controlled other reflexes that defend the upper airways and lungs, such as cough. Recent research has revealed that the brainstem neural networks that produce eupnea

R

also are involved in coughing and sneezing as well as other less well known reflexes. The process by which the brainstem neural network for breathing can be involved in the production of other behaviors is known as *reconfiguration*. That is, the breathing network changes its "circuit diagram" to allow for the generation of a non-breathing behavior [2]. This process involves alteration of the discharge patterns and effective connectivity of neurons in the respiratory network. The reconfiguration process may also involve "conscription" of neurons that have little to do with breathing but have activity patterns that are selective for certain behaviors, such as cough. This conscription may include recruitment of previously silent neurons and significant modification of the activities of neurons during cough that were not modulated during breathing. There is good evidence that these processes take place and that, in addition to the network that controls breathing, coughing is also controlled by another brainstem control mechanism known as a gate. In essence, the system can be functionally subdivided into a controller (the gate) and an effector (the brainstem respiratory network). The controller regulates the excitability of the behavior and the effector is responsible for the coordination of motor drive to respiratory muscles for cough. During breathing the gate, or controller, is functionally quiescent and the respiratory network is primarily involved in the production of breathing. When RARs are stimulated, the gate becomes active and the brainstem respiratory network reconfigures to produce coughing [7].

Plasticity of the Cough Reflex

There is considerable evidence that cough can undergo significant plasticity in both humans and animals, especially during induced or naturally occurring airway disease. This plasticity is usually manifest in the form of an increased number of coughs in response to a given stimulus and/or increased sensitivity to inhaled irritants. The relative role of central and peripheral mechanisms in this plasticity is less well understood.

In humans, chronic spontaneous cough lasting for years is well documented and can occur in a variety of conditions, such as smoking, asthma, chronic obstructive pulmonary disease, upper airway disorders, and gastro-esophageal reflux. In many of these conditions, the sensitivity of humans to inhaled irritants is elevated but this enhanced cough sensitivity resolves with successful treatment of the underlying disorder. However, tobacco smoke exposure during childhood is associated with cough in adulthood, suggesting that there is a permanent alteration of some important part of the cough reflex [8]. The increased sensitivity of the cough reflex in these patients is consistent with plasticity. It is presumed that hyperexcitable peripheral afferents are responsible for the enhanced coughing in these patients, but the potential contribution of central mechanisms has been difficult to address in humans.

Very similar observations have been made in animal models of airway disease and it is well established that airway sensory afferents responsible for cough undergo significant plasticity in many of these conditions. It also has been shown in an animal model that inflammation of one region of the airway will elicit an enhanced cough response to stimulation of a non-inflamed region of the airway. Presumably the sensory afferent responsiveness in the non-inflamed region of the airway was normal. This suggests that plasticity can occur in the central cough neural networks.

Cough can undergo hypoexcitability during neurological diseases. Stroke, Parkinson's Disease and Multiple Sclerosis are all associated with cough weakness or an inability to cough at all. This cough impairment can contribute to an increased susceptibility to aspiration in these patient groups. In stroke, cough impairment can occur even if the lesion does not include the brainstem. This fact, in combination with the knowledge that cough can be produced voluntarily suggest that suprapontine mechanisms can be important in the regulation of cough excitability in awake humans. The extent to which these mechanisms can be subject to plasticity is unknown. More research must be performed to gain a greater understanding of these mechanisms.

References

1. Mitchell GS, Johnson SM (2003) Neuroplasticity in respiratory motor control. J Appl Physiol 94(1):358–374
2. Morris KF, Baekey DM, Nuding SC, Dick TE, Shannon R, Lindsey BG (2003) Invited review: Neural network plasticity in respiratory control. J Appl Physiol 94 (3):1242–1252
3. Mahamed S, Mitchell GS (2007) Is there a link between intermittent hypoxia-induced respiratory plasticity and obstructive sleep apnoea? Exp Physiol 92(1):27–37
4. Morris KF, Shannon R, Lindsey BG (2001) Changes in cat medullary neurone firing rates and synchrony following induction of respiratory long-term facilitation. J Physiol (Lond) 532(Pt 2):483–497
5. Morris KF, Gozal D (2004) Persistent respiratory changes following intermittent hypoxic stimulation in cats and human beings. Respir Physiol Neurobiol 140(1):1–8
6. Sahni R, Fifer WP, Myers MM (2007) Identifying infants at risk for sudden infant death syndrome. Curr Opin Pediatr 19(2):145–149
7. Bolser DC, Poliacek I, Jakus J, Fuller DD, Davenport PW (2006) Neurogenesis of cough, other airway defensive behaviors and breathing: A holarchical system? Respir Physiol Neurobiol 152(3):255–265
8. Joad JP, Sekizawa S, Chen CY, Bonham AC (2007) Air pollutants and cough. Pulm Pharmacol Ther 20 (4):347–354
9. Dick TE, Hsieh YH, Wang N, Prabhakar N (2007) Acute intermittent hypoxia increases both phrenic and sympathetic nerve activities in the rat. Exp Physiol 92(1):87–97

Respiratory Neurotransmitters and Neuromodulators

Akira Haji
Laboratory of Neuropharmacology, School of Pharmacy, Aichi Gakuin University, Chikusa, Nagoya, Japan

Synonyms

Neurochemicals; Endogenous receptor agonists

Definition

Chemicals synthesized by neurons involved in generating rhythm and pattern in the ▶respiratory network, and in transmitting input signals from central and peripheral chemoreceptors and mechanoreceptors to respiratory interneurons and motoneurons (Fig. 1).

Active Phase

The period of the ▶respiratory cycle (phase) in neurons characterized by membrane depolarization to threshold for the generation of action potentials [2,9].

Silent Phase

The period of the respiratory cycle characterized by membrane hyperpolarization that prevents action potential generation [2,9].

Characteristics

Neurotransmitter Functions in the Respiratory Network

Rhythmic fluctuations of membrane potential in respiratory neurons evolve from neurotransmitter- and membrane conductance-dependent, periodic barrages of ▶IPSPs and ▶EPSPs that occur with precise timing during the respiratory cycle. Neurotransmitter-dependent excitatory synaptic connections between synchronously active neurons evoke bursts of action potential discharge, while discharges of reciprocally activated neurons periodically release inhibitory ▶neurotransmitters that hyperpolarize membrane potential away from firing threshold. Tonic neurotransmitter release provides a continual excitatory bias on what appears to be all types of respiratory neurons, whereas tonic release of inhibitory neurotransmitter can have a stabilizing effect on membrane potential.

Inhibitory Amino Acids (GABA and glycine)

There are three types of phasic inhibition in respiratory neuron activities; reciprocal inhibition, recurrent inhibition and phase-transition inhibition, as well as tonic inhibition. All four types are characterized by membrane hyperpolarization and lowered input resistance [1,3,4,7].

Phasic inhibition during the inactive phase

GABA initiates IPSPs during the inactive phase by binding to a $GABA_A$-type of receptor in respiratory neurons. During the inactive phase, temporal summation of IPSPs hyperpolarizes membrane potential near to the equilibrium potential for chloride ions (Cl^-).

Inhibition during Phase Transitions

The $GABA_A$ receptor-mediated mechanism plays an essential role during transition from one respiratory phase to another. During transition from the inspiratory to the expiratory phase, postinspiratory (early expiratory) IPSPs occur in augmenting inspiratory (aug -I) neurons. During transition from late expiration to inspiration, inspiratory IPSPs are observed in augmenting expiratory (aug-E) neurons. In the latter case, activation of $GABA_B$ receptors is partially involved, leading to increased potassium (K^+) conductances.

Inhibition during the Active Phase

Glycine mediates IPSPs during the later part of stage 2 expiration in aug-E neurons, and IPSPs during late inspiration in aug-I neurons. Aug-I neurons also show IPSPs during early part of inspiration, but whether GABA, glycine or both are involved is unclear.

Tonic Inhibition

$GABA_A$, $GABA_B$ and glycine receptor-mediated postsynaptic inhibitions are active in respiratory neurons to help stabilize membrane potential level.

Inhibition of Spinal Motoneurons

GABA mediates the inhibition of phrenic motoneurons during expiration through $GABA_A$ receptors [6,8]. This inhibition comes from aug-E neurons of the Bötzinger complex. $GABA_A$ mechanisms are also involved in the raphe-stimulated inhibition of the respiratory neuronal activity. The $GABA_B$ mechanism decreases transmitter release by acting at presynaptic site and hyperpolarizing phrenic motoneurons through activation of K^+ conductances at postsynaptic sites [6].

Excitatory Amino Acid (Glutamate)

All types of respiratory neurons exhibit glutamate-activated EPSPs, in association with lowered input resistance and induction of action potential discharge. They summate temporally to bring membrane potential to discharge threshold [1,3,4,7].

Phasic Excitation during the Active Phase

Glutamate initiates EPSPs during the active phase in most types of respiratory neurons by activating both AMPA- and NMDA-type glutamate receptors. The sequential activation of the two types of postsynaptic receptor is required for production of respiratory-related

R

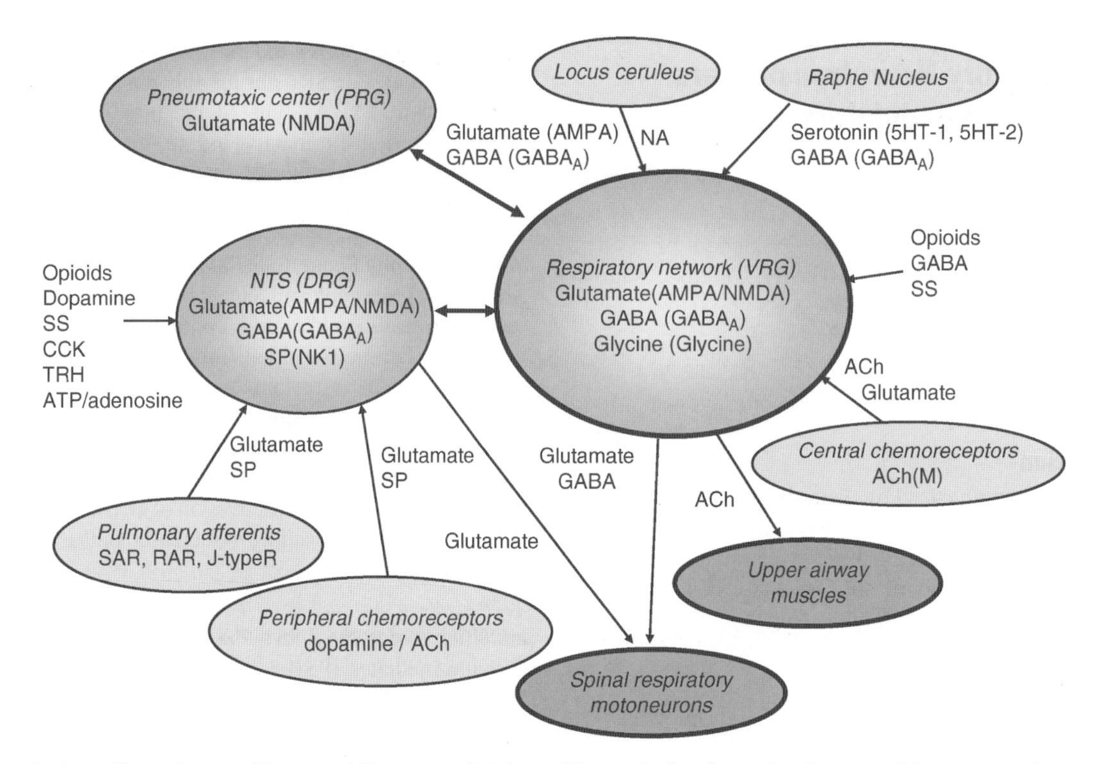

Respiratory Neurotransmitters and Neuromodulators. Figure 1 A schematic diagram of the neuronal interactions between the respiratory center and other modulatory structures, and of putative neurotransmitters and neuromodulators involved in the neuronal control of respiratory rhythm and pattern generation. The main neurotransmitters of respiratory neurons are glutamate, GABA and glycine. Glutamate mediates excitatory transmissions through AMPA and NMDA receptors and GABA mediates inhibitory transmissions through GABA$_A$ receptors. They are also utilized in the phase-switching process which is regulated by the pneumotaxic descending inputs and pulmonary SAR afferents. Transduction of hypoxic signals within the peripheral chemoreceptors depends primarily on O_2-sensitive K^+ channels associated with neuroactive substances such as ACh and dopamine. These signals are transmitted to relay neurons in the NTS by glutamate through AMPA receptors. Hypercapnia stimulates preferentially the central chemosensitive area near the ventral surface of the medulla, where ACh acts as a mediator of such signals. Many neuroactive substances other than amino acids have been implicated in modulating the respiratory rhythm. Serotonergic, noradrenergic and dopaminergic inputs modulate the respiratory neuronal discharge and respiratory rhythm. Muscarinic and nicotinic cholinergic modulations are also apparent. Several neuroactive peptides including SP, SS, CCK and opioids affect respiration. They act either presynaptically or postsynaptically to modulate synaptic transmission within the primary neuronal network as well as at the input and output relay nuclei. These substances are not involved in generation of a basic eupnea. Interactions among various neuroactive substances are essential for precise control of the normal functioning and adaptive processes in the central organization of respiratory rhythm and pattern.

bursts of discharge. In aug-E neurons, depolarization and discharge are due to AMPA receptor activation, as is phasic recurrent excitation in aug-I neurons. Metabotropic glutamate receptor-mediated mechanisms have no significant effect on membrane potential of ►bulbar respiratory neurons.

Inspiratory Off-Switch (►IOS) Mechanism

The NMDA mechanism plays an important role in IOS of pontine as well as medullary respiratory neurons. IOS is accomplished by a sequential activation of late inspiratory (late-I) and postinspiratory (post-I) neurons to produce barrages of IPSPs in aug-I neurons. During ►apneusis caused by NMDA blockade, active phase depolarization of late-I and post-I neurons and their firing activity are decreased. The discharge activity of aug-I neurons is also decreased during apneusis.

Tonic Excitation

Respiratory neurons receive glutamatergic tonic inputs that activate both NMDA and AMPA receptors.

Pneumotaxic Descending Inputs

Termination of the IOS can be produced by afferents originating from the ►PRG. Glutamate through the NMDA mechanism responsible for IOS is present in the pontine structure. However, the pontine descending inputs generating fast EPSPs in bulbar respiratory neurons are not mediated by NMDA receptors, but by AMPA receptors.

Pulmonary Mechanoreceptor Afferent Inputs

Glutamate mediates the primary afferent excitation of the NTS neurons in the ►Hering-Breuer reflex pathways, primarily through the activation of AMPA receptors. The Hering-Breuer inspiratory promotion reflex induced by application of lung deflation during expiration is mediated by NMDA receptors.

Excitatory Drive to Bulbospinal Motoneurons

Glutamate mediates the bulbospinal transmission of respiratory drive acting on both AMPA and NMDA receptors [6,8]. Contribution of the former is greater than that of the latter to motor outputs. AP4 receptors are located at the presynaptic sites of the inspiratory bulbospinal terminals. Short term potentiation is mediated by NMDA receptors, which augment EPSPs and prolong depolarization of phrenic motoneurons. Activation of metabotropic glutamate receptors affects the inspiratory-modulated activity of phrenic motoneurons via distinct mechanisms at pre- and postsynaptic sites.

Excitatory Drive to Spinal Motoneurons

The major glutamatergic excitatory drives to phrenic, intercostal and abdominal motoneurons come from bulbospinal neurons that activate AMPA/Kainate- and NMDA-types of receptors

Neurotransmitters, Neuromodulators and Responsiveness to Hypercapnea and Hypoxia

Carbon dioxide and its acid byproduct, H^+ ion, constitute the primary respiratory stimulus within the central nervous system. Sensitivity to CO_2/pH is up-regulated or down-regulated by several different neurotransmitters and ►neuromodulators [5,7]. Acetylcholine (ACh) activation of muscarinic receptors on neurons close to the ventrolateral surface of the medulla increases respiratory responsiveness to CO_2/pH^+. Adrenergic cell groups have also been reported to increase respiratory responsiveness to hypercapnea/acidosis in the rostral ventrolateral medulla. Glutamate is also a neurotransmitter candidate for tonic excitation of respiratory neurons mediated by central CO_2/pH^+ sensitive neurons. A GABAergic mechanism in the caudal hypothalamus dampens respiratory responsiveness to hypercapnea/acidosis.

Respiratory responsiveness to hypoxia is modulated peripherally by neuromodulators in the carotid bodies and within the central respiratory network. Generally, excitatory transmission is cholinergic, whereas dopamine (DA) plays an inhibitory role in carotid body chemoreceptors. Hypoxic release of ACh activates nicotinic receptors, leading to augmentation of hypoxia-induced depolarization and further release of ACh and other neurotransmitters. DA release, on the other hand, suppresses carotid body discharges by activating D_2-type receptors. Glomus cells contain other compounds that are released during hypoxia, including serotonin (5-HT), enkephalins, prostaglandins, ATP, adenosine, substance P (SP), cholecystokinins (CCK), nitric oxide and atrial natriuretic peptide.

SP localized within vagal afferent fibers appears to act as a neurotransmitter or modulator of chemo- and baroreceptor fibers. Central dopaminergic mechanisms are also involved in modulating the chemoreflex respiratory control.

In the central respiratory network, glutamate release from the afferent glossopharyngeal terminals in the nucleus of the solitary tract (NTS) activates AMPA receptors on relay neurons in response to hypoxia. Release and local accumulation of GABA and/or neuromodulators, including catecholamines and opioids, are important factors leading to late hypoxic depression. GABA mediates inhibition in the NTS neurons that respond to stimulation of the carotid sinus nerve. Accumulation of metabolic byproducts such as adenosine may increase K_{ATP} channel currents in postsynaptic neurons. Hypoxia also increases endogenous 5-HT levels and increases K^+ currents via $5\text{-}HT_{1A}$ receptors in respiratory neurons, resulting in depression of respiratory neurons.

Respiratory Neuromodulation by Monoamines and Peptides

It has been more difficult to assess the neuromodulatory roles of serotonin, catecholamines and peptides in the central respiratory network. Their actions are generally slow in onset, discreet, state-variable, and dependent on a vast array of receptor subtypes. Assessment of function is often assumed from the effects of exogenous receptor agonists and antagonists, not all of which are suitably selective [1,4,7].

Serotonergic Agents

5-HT has diverse effects on respiratory neurons. Respiratory neuron excitability is increased postsynaptically by $5\text{-}HT_2$, $5\text{-}HT_{1C}$ and $5HT_4$ receptor agonists, and decreased postsynaptically via $5\text{-}HT_{1A}$ and $5HT_7$ agonists. Activation of $5\text{-}HT_{1A}$ and $5HT_7$ receptors depresses the cAMP-protein kinase A pathway, $5HT_4$ receptors activate it, and $5\text{-}HT_2$ receptor activation stimulates the PLC/PLA- protein kinase C pathway.

Catecholaminergic Inputs

The effects of catecholamines (noradrenaline, adrenaline, DA) are also diverse and dependent on which subtypes of receptor are affected. Noradrenaline and adrenaline have a predominantly depressant effect on bulbar respiratory neurons. Dopaminergic mechanisms exert a tonic inhibitory influence in the central pathways involving in the hypoxic ventilatory responses through

D_2 receptors, and increase central respiratory responsiveness to CO_2 via D_1 receptors.

Peptides and Hormones

SP and thyrotropin-releasing hormone (TRH) have excitatory, and somatostatin (SS) and ►opioid peptides have depressant effects on respiration. CCK produces either excitatory or inhibitory effects, depending on its receptor types activated. Individual neuropeptides often coexist and interact with classical neurotransmitters in respiratory neurons. They play some roles in the central control of respiratory activity including the chemoreflex [1,4,7].

Substance P

SP mediates excitatory neurotransmission and integration of the peripheral chemoreflex in the NTS through NK_1 receptors. SP reverses the respiratory neuronal depression induced by SS or opioids. Hypoxia induces desensitization of NK_1 receptors to SP in the NTS neurons, leading to a decline of hyperventilation during hypoxia. SP is a transmitter of the pulmonary C fiber-mediated reflex which induces a rapid shallow breathing and/or apnea.

Thyrotropin-Releasing Hormone

TRH had postsynaptic excitatory effects on neurons in the NTS, nucleus ambiguus and pre-Bötzinger complex. TRH enhances glutamatergic transmissions and counteracts the inhibitory effects of opioids. TRH also seems to be involved in central (CO_2/pH^+) chemoreception (►central chemoreception).

Somatostatin

SS has an inhibitory effect on respiratory neurons. Anesthesia or sleep enhances the effects of SS. SS metabolites potentiate the voltage-dependent, non-inactivating outward K^+ currents (I_M current) in the NTS neurons.

Opioid Peptides

Opioid peptides may be the most important peptides endogenously involved in respiratory modulation in the brainstem. Opioid peptides depress respiratory neuron activity by increasing K^+ conductances through μ receptors. Endogenous opioids negatively interact with the CO_2-sensitive cholinergic transmission, and with the glutamatergic transmission of respiratory neuronal activity.

Cholecystokinin

CCK octapeptide (CCK_8) causes various effects on respiration; It stimulates respiration by stimulating either the forebrain or the medullary region and depresses ventilation by stimulating vagal afferents. The activation of CCK_A receptors causes inhibition of respiratory neurons due to an increase of K^+ conductances, while that of CCK_B receptors produces excitation due to a decrease in K^+ conductances. Further, activation of CCK_B receptors by endogenous CCK reduces the GABA-mediated fast inhibitory responses in the NTS neurons.

Progesterone

Endogenous progesterone is a respiratory stimulant [10]. It increases ventilatory responsiveness to CO_2. In pregnancy and during the luteal phase of the menstrual cycle, it accounts for hyperventilation and low CO_2. Endogenous progesterone also has a beneficial effect on the upper airways. It increases tonic and phasic geneoglossus activities.

References

1. Ballanyi K, Onimaru H, Homma I (1999) Respiratory network function in the isolated brainstem-spinal cord of newborn rats. Prog Neurobiol 59:583–634
2. Bianchi AL, Denavit-Saubié M, Champagnat J (1995) Central control of breathing in mammals: neuronal circuitry, membrane properties, and neurotransmitters. Physiol Rev 75:1–45
3. Bonham AC (1995) Neurotransmitters in the CNS control of breathing. Respir Physiol 101:219–230
4. Denavit-Saubié M, Foutz AS (1997) Neuropharmacology of respiration. In: Miller AD, Bianchi AL, Bishop BP (eds) Neural control of the respiratory muscles. CRC press, New York, pp 143–157
5. Eyzaguirre C (2005) Chemical and electric transmission in the carotid body chemoreceptor complex. Biol Res 38:341–345
6. Feldman JL, McCrimmon DR (1999) Neuronal control of breathing. In: Zigmond MJ, Bloom FE, Landis SC, Roberts JL, Squire LR (eds) Fundamental Neuroscience, Academic press, New York, pp 1063–1090
7. Haji A, Takeda R, Okazaki M (2000) Neuropharmacology of control of respiratory rhythm and pattern in mature mammals. Pharmacol Ther 86:277–403
8. Hilaire G, Montau R (1997) Brainstem and spinal control of respiratory muscles during breathing. In: Miller AD, Bianchi AL, Bishop BP (eds) Neural control of the respiratory muscles. CRC press, New York, pp. 91–105
9. Richter DW (1996) Neural regulation of respiration: rhythmogenesis and afferent control. In: Greger R, Windhorst W (eds) Comprehensive Human Physiology, Vol. 2. Springer, Berlin, pp 2079–2095
10. Saaresranta T, Polo O (2002) Hormones and breathing. Chest 122:2165–2182

Respiratory Pacemaker Neuron

Definition

Neuron with an intrinsic ability to generate rhythmic bursts.

►Respiratory Pacemakers

Respiratory Pacemakers

Jan-Marino Ramirez
Department of Organismal Biology and Anatomy, Committees on Neurobiology, Computational Neuroscience and Molecular Medicine, The University of Chicago, Chicago, IL, USA

Synonyms

Respiratory bursting neurons; Respiratory pacemaker neurons

Definition

A respiratory pacemaker is a neuron with the intrinsic ability to generate rhythmic ►bursts that emerge through voltage-, time- and calcium- dependent ion fluxes [1]. These ion fluxes give rise to rhythmic membrane fluctuations that are defined as "►drive potentials" or "pacemaker potentials." The ion fluxes leading to drive potentials are carried by sodium, calcium, and/or non-specific cations, but many ionic conductances contribute to shape and frequency of these potentials. The drive potentials can, but do not always give rise to a series of action potentials. A drive potential that gives rise to action potentials is called a "burst". Bursts that are generated by ionic conductances intrinsic to the neuron are often referred to as "►intrinsic bursting." Two types of pacemaker neurons have been described in the respiratory network: (i) Pacemakers that depend on CAN current and are blocked by cadmium are referred to as "►Cadmium-sensitive pacemakers" [2]. (ii) Pacemakers dependent on the persistent sodium current burst even in the presence of cadmium. These neurons are referred to as "►Cadmium-insensitive pacemakers" [2].

A neuron that generates pacemaker activity only in the presence of a neuromodulator is generally called a "►conditional pacemaker." As described in the next paragraph, the dynamic regulation of pacemaker neurons is the rule, and not the exception. Hence, it is conceivable that all respiratory pacemakers are "conditional".

Characteristics

Quantitative Description

The identity of a respiratory pacemaker is not fixed, but dynamically regulated. Non-pacemakers can be turned into respiratory pacemakers, and conversely respiratory pacemakers can become non-pacemakers [1]. This dynamic process is not all-or-none. A non-pacemaker can turn into a weakly bursting or strongly bursting pacemaker, and weak pacemakers can turn into strong pacemakers. Weak pacemakers are characterized by small amplitude, intrinsically generated ►drive potentials that give rise to one or two action potentials. Bursts can occur irregularly in some, while regularly in other neurons. Thus, the discharge properties of respiratory neurons cover a wide range from non-bursting, to weak-irregular bursting to strong-regular bursting [1,3]. Transformation of pacemakers and non-pacemakers and their dynamic regulation are not unique for the respiratory system. For example the discharge patterns of neocortical and thalamic neurons change dramatically during the transition from wake to sleep [1,4,5] and there are numerous other examples in which neurons can loose or attain pacemaker properties.

Mechanistically, this is not surprising as the pacemaker property emerges through a complex, and modifiable ratio of different ionic currents in which inward currents (typically carried by Na^+ or Ca^{++}) are larger than outward currents (typically carried by K^+ or Cl^- currents).

In the functional network the ratio of these ion channels is continuously modulated by endogenously released neuromodulators, such as amines and peptides. In the respiratory system, induction of pacemaker properties has been demonstrated for serotonin, acetylcholine, norepinephrine, TRH (►TRH: thyrotropin releasing hormone) and substance P [1,3,6]. It is likely that many still unexamined neuromodulators can induce and suppress pacemaker properties in respiratory neurons.

In the functional network intrinsically generated drive potentials are also dynamically regulated by synaptic transmission: Intrinsically generated bursts can be activated by excitatory synaptic inputs, and thus function as a mechanism to boost or amplify synaptic inputs. But the boosting mechanism must also be considered dynamically, since concurrently occurring synaptic inhibitory mechanisms can also suppress these intrinsically generated drive potentials. Thus, in the functional network concurrent inhibitory synaptic mechanisms can regulate the bursting mechanism to the extent that excitatory synaptic drive is necessary to activate bursting in a pacemaker neuron. Synaptic mechanisms play also critical role in timing the onset of a burst, even in pacemakers that have strong intrinsic bursting properties. Thus, the bursting property must be considered as a dynamic property that is highly influenced by fine balance of synaptic as well as neuromodulatory mechanisms.

Due to the tight interaction between synaptic and intrinsic membrane properties, demonstrating pacemaker properties is challenging. It must be shown that the rhythmicity recorded in a neuron is generated intrinsically and is not the result of rhythmic synaptic input that emerges through network interactions. In the respiratory network pharmacological approaches are typically used to isolate pacemaker neurons [6,7]. Exogenously applied neurotransmitter antagonists can block inhibitory and excitatory neurotransmission which eliminates rhythmic synaptic population inputs. The pharmacological approaches are usually combined with electrophysiological approaches that take advantage of the voltage-dependency of ion channels [7]. Brief de- or hyperpolarizing current injections can reset ongoing pacemaker activity by advancing or delaying the generation of a pacemaker burst. Long-lasting de- or hyperpolarizing current injections can accelerate or slow the frequency of pacemaker activity. Brief depolarizing current pulses can prematurely trigger, while hyperpolarizing current pulses can prematurely terminate ongoing pacemaker bursts. It is important to be aware that pharmacological approaches can be misleading. For example low concentrations of extracellular calcium can block synaptic transmission, but at the same time, this manipulation can induce pacemaker properties in non-pacemakers by blocking ►calcium-dependent potassium currents [2]. Conversely, low calcium concentrations could block the activation of the CAN current, which plays a major role in evoking bursts in some respiratory neurons [2]. Bicuculline, is a substance that blocks ►GABAergic synaptic transmission, but at higher concentration it can also block potassium channels which could induce pacemaker properties.

An even greater challenge is the interpretation of lesion experiments in a functional network. Pacemaker, synaptic and modulatory properties are highly integrated elements of the functional network and provide the respiratory network with the necessary adaptability and flexibility for survival. The removal of any of these elements will change the overall network property, and whether its removal abolishes rhythmicity is neither an indicator for its importance nor its specific role in respiratory rhythm generation [1].

Higher Level Structures

The majority of neuronal networks in the brain generate rhythmic activity, and pacemakers are found in the majority of these networks including networks within the spinal cord, medulla, ►neocortex, ► basal ganglia, thalamus, ►locus coeruleus, ventral tegmentum area (VTA), ►hippocampus and ►amygdala [1,8]. In many cases it is unclear how the rhythmicity in general and how pacemakers in particular contribute to these network functions, and the respiratory network is no exception. Pacemaker neurons have been identified in various areas that belong to the respiratory network, including the NTS [6], the pre-Bötzinger complex [1] and the ►parafacial nucleus [9]. Pacemakers are also found in areas that are driven by the respiratory network including the locus coeruleus [10]. While most pacemakers were identified in vitro slices, it is likely that recordings in more intact networks reveal that pacemaker neurons are more abundant than generally expected. Intact networks have more active modulatory systems that can promote pacemaker properties, such as norepinephrine and serotonin.

Lower Level Components

In general terms, the ionic mechanisms that give rise to pacemaker activity are very heterogeneous. These mechanisms typically involve a complex interaction between voltage-dependent and voltage-independent components of ion channels within their intra- and extracellular environment [1]. In general, a neuron depolarizes and ultimately bursts either in response to the activation of inward currents that are carried by sodium and/or calcium ions, or in response to the cessation of outward currents that are carried by potassium ions. The inward currents include the hyperpolarization-activated current (Ih current), the persistent sodium current, various low- and high-voltage activated calcium currents and the calcium-activated non-specific cation (CAN) current [1]. The ongoing burst is commonly terminated by either of two principal ionic mechanisms. (i) The channels responsible for the inward current inactivate. Such properties may play a major role in determining bursting in neurons dependent on persistent sodium current. (ii) The calcium or sodium influx during the ongoing burst can activate calcium- or sodium-dependent potassium currents that hyperpolarize the membrane and thereby terminate the burst. Possible mechanisms that cause a repolarization can include voltage-independent intracellular signals, and slow activation or inactivation properties of inward or outward currents.

Structural Regulation

There is currently no characteristic anatomical structure that defines a ►respiratory pacemaker neuron. Similarly there are many different discharge patterns that characterize a respiratory pacemaker neuron. "Irregular-" and "regular-bursting" neurons are differentiated by the regularity of the burst periodicity. Given that rhythmic drive potentials can arise through a variety of ionic mechanisms, it is not surprising that the same anatomical region may contain different types of pacemaker neurons. The fact that the same anatomical region contains more than one type of pacemaker neuron is not the exception, but presumably the rule [1]. It is assumed that different types of pacemaker neurons play different roles in the generation of network activity, an issue of much ongoing research [1,2,3,8,9]. This complexity is not unique to the nervous system: cardiac pacemakers for example are also very diverse.

Higher Level Processes

Pacemaker neurons are embedded in complex neuronal networks. Hence there are many synaptic and modulatory processes that govern the activity of a pacemaker neuron [1,3]. Many principle insights into the interactions between pacemaker neurons, synaptic transmission and ►neuromodulators were gained from studying small neuronal networks of invertebrates. It can be expected that medullary pacemakers are modulated by various inputs from networks outside the ►medulla, and vice versa that pacemakers influence networks outside the medulla. Unfortunately, recordings from pacemakers in more intact networks are sparse. Thus very little is known about these potential network interactions.

Lower Level Processes

Neuromodulators play a critical role in modulating the cellular events that govern the discharge pattern of a pacemaker neuron. Endogenously released neuromodulators can phosphorylate voltage-dependent ion channels, or alter second messenger pathways and intracellular calcium thereby changing ion channel properties. This complex interplay between neuromodulators, the intracellular milieu and voltage-dependent ion fluxes will significantly alter pacemaker activity. In doing so, neuromodulators can determine the burst frequency, the amplitude and shape of the drive potential [3]. Neuromodulators are also responsible for the fact that the pacemaker property itself is not a fixed property as described above.

Function

As described above, pacemakers are embedded in synaptically organized networks and therefore pacemaker activity itself is influenced by synaptic inputs [1]. Thus, in general pacemaker activity can not be regarded as a "driver" of network activity. Thus assigning a specific function to a pacemaker neuron becomes difficult if not impossible, since this property can not be separated from the other properties that determine its discharge. Tonic excitatory or inhibitory synaptic inputs can determine the frequency of pacemaker activity. Excitatory synaptic input can prematurely trigger pacemaker activity, which means that synaptic inputs can determine the timing of pacemaker activity. A pacemaker burst can act as a non-linear amplifier of synaptic excitatory inputs, while synaptic inhibitory inputs can act as leak currents that will greatly suppress pacemaker activity.

Various functions have been ascribed to pacemaker neurons. It is thought that pacemakers can influence regularity, burst amplitude and frequency of respiratory activity. Due to differences in their voltage-dependence cadmium-sensitive and insensitive pacemakers may assume different roles in regulating frequency versus amplitude of respiratory bursts [3].

Process Regulation

The number, the types of pacemakers, and the degree of their bursting properties in a functional neuronal network will be continuously regulated by neuromodulators and synaptic interactions. Consequently, the contribution of pacemaker properties to the overall network output will not be fixed [1,2]. By altering for example the number of active pacemaker neurons a network can assume different configurations that can lead to different network outputs. These complex modulatory interactions imbue neuronal networks with a high degree of plasticity. This is an essential prerequisite for generating a rhythmic behavior that has to continuously adapt to changes in behavioral, environmental and metabolic conditions.

Pathology

It has been hypothesized that the suppression of pacemaker properties may lead to the failure of gasping and possibly Sudden Infant Death Syndrome [1,2].

Therapy

In vitro studies suggest that pacemaker neurons are important in regulating regularity of respiratory rhythmic activity. Hence, a better understanding of the ionic basis of pacemaker activity and their modulation may be an important step towards developing rational therapies or strategies to prevent neurological disorders associated with erratic breathing.

References

1. Ramirez JM, Tryba AK, Peña FP (2004) Pacemaker neurons: an integrative view. Curr Opin Neurobiol 14(6):665–674
2. Peña F, Parkis MA, Tryba AK, Ramirez JM (2004) Differential contribution of pacemaker properties to the generation of respiratory rhythms during normoxia and hypoxia. Neuron 43:105–117
3. Viemari JC, Ramirez JM (2006) Norepinephrine differentially modulates different types of respiratory pacemaker and non-pacemaker neurons. J Neurophysiol 95(4):2070–2082
4. Pape HC, McCormick DA (1989) Noradrenaline and serotonin selectively modulate thalamic burst firing by enhancing a hyperpolarization-activated cation current. Nature 340:715–718
5. Steriade M (2004) Neocortical cell classes are flexible entities. Nat Rev Neurosci 5:121–134
6. Dekin MS, Richerson GB, Getting PA (1985) Thyrotropin-releasing hormone induces rhythmic bursting in neurons of the nucleus tractus solitarius. Science 229(4708):67–69
7. Tryba AK, Pena F, Ramirez JM (2003) Stabilization of bursting in respiratory pacemaker neurons. J Neurosci 23(8):3538–3546
8. Arshavsky YI (2003) Cellular and network properties in the functioning of the nervous system: from central pattern generators to cognition. Brain Res Brain Res Rev 41:229–267

9. Onimaru H, Arata A, Homma I (1989) Firing properties of respiratory rhythm generating neurons in the absence of synaptic transmission in rat medulla in vitro. Exp Brain Res 76(3):530–536
10. Ballantyne D, Andrzejewski M, Mückenhoff K, Scheid P (2004) Rhythms, synchrony and electrical coupling in the Locus coeruleus. Respir Physiol Neurobiol 143(2–3):199–214

Respiratory Plasticity

►Respiratory Reflexes

Respiratory Recovery from Injury

►Respiratory Neuroplasticity

Respiratory Reflexes

Donald R. McCrimmon[1], George F. Alheid[2]
[1]Department of Physiology, Feinberg School of Medicine, Northwestern University, Chicago, IL, USA
[2]Department of Physiology and Northwestern University Feinberg School of Medicine, Chicago, IL, USA

Synonyms

Pulmonary reflexes; Breuer-Hering reflexes; Deflation reflex; Central chemoreceptors; Peripheral chemoreceptors; Carotid chemoreflex; Slowly adapting pulmonary stretch receptors; Rapidly adapting pulmonary receptors; Irritant receptors; Cough; Respiratory plasticity

Definition

Respiratory reflexes encompass a significant repertoire of responses to a variety of sensory receptors regulating the depth and frequency of individual breaths and participating in the protection of airways from potentially damaging inhaled substances. Specifically, receptors in the airways and lungs sense the relative inflation or deflation of the lungs as well as the presence of inhaled irritants, and elicit appropriate responses via brainstem respiratory circuits to maintain the integrity and efficient function of the lungs and airways. Central (brain) chemoreceptors and peripheral chemoreceptors in contact with arterial blood, evoke changes in breathing to maintain appropriate levels of oxygen and carbon dioxide as well as pH.

Characteristics

Continuous, rhythmic breathing movements are essential for the homeostatic regulation of arterial blood gases, acid-base balance and, ultimately, for the maintenance of life itself. Neurons responsible for generating respiratory rhythm and shaping it into the detailed pattern of activity evident on respiratory motor output are located predominantly in two brainstem regions (see Alheid & McCrimmon this volume). One group forms a long column of cells in the ventrolateral medulla in close proximity to the nucleus ambiguus. Termed the ventral respiratory column (VRC), this group extends rostrally from the spinal-medullary junction to a region ventral to facial nucleus. A second group, termed the dorsal respiratory group, is localized in the dorsomedial medulla, mainly within the ventrolateral nucleus of the solitary tract (NTS). These neurons fire in bursts phase locked to the breathing rhythm. Most fire either during inspiration (inspiratory neurons) or expiration (expiratory neurons) although some fire in bursts that span phase transitions between inspiration and expiration.

The magnitude and rate of respiratory efforts generated by brainstem respiratory neurons are regulated to maintain brain and arterial tensions of oxygen (O_2), carbon dioxide (CO_2), and pH within narrow limits despite large variations in metabolic requirements. A variety of chemical and mechanical receptors located in the airways, lungs, and chest-wall provide the sensory feedback essential to optimization of the breathing pattern. Sensory feedback from arterial and central (brain) chemoreceptors as well as lung mechanoreceptors modulates the breathing pattern, e.g., tidal volume and breathing frequency.

Receptors distributed throughout the airways also help defend the respiratory system. Afferent-evoked protective reflexes include apnea (a transient cessation of breathing), shallow rapid breathing, coughing, sneezing, mucus secretion, and airway constriction. These reflexes protect the airways from irritants and facilitate the removal of inhaled substances potentially harmful to the lungs and airways.

Chemoreceptors

Regulating the level of the metabolic product CO_2 and maintenance of tissue oxygenation are principal roles of the respiratory system. In performing these tasks, the central circuitry generating respiratory pattern receives sensory input from chemoreceptors located in the brain (central chemoreceptors) and the arterial system (peripheral chemoreceptors). Through the regulation of CO_2, the respiratory system also adjusts pH and hence contributes importantly to acid-base balance.

Central chemoreceptors have a relatively greater role than peripheral receptors in regulating CO_2 and pH and they have been identified in several regions of the brain [1]. Most are in the medulla and pons including: (i) regions at the ventral surface of the medulla, particularly in the retrotrapezoid nucleus, (ii) midline raphe serotonergic neurons, (iii) the NTS, (iv) the preBötzinger complex [a subregion of the VRC] and (v) noradrenergic neurons in the locus coeruleus. Additional chemosensitive sites have been identified in: (vi) the fastigial nucleus of the cerebellum, and (vii) the posterior hypothalamus. The relative importance of several sites including the NTS, retrotrapezoid nucleus and caudal raphe may vary with physiological conditions such as the sleep–wake state. Central O_2 chemoreceptors appear to exist but little ventilatory response to hypoxia is observed after peripheral deafferentation when CO_2 levels are held constant.

Peripheral chemoreceptors have a dominant role in eliciting the ventilatory increases in response to hypoxia [2]. Peripheral chemoreceptors are located in both the carotid and aortic bodies but the carotid bodies are quantitatively much more important in regulating breathing. The aortic bodies contribute relatively more to cardiovascular adjustments. Afferent fibers emanating from the carotid bodies have a low discharge rate at normal resting levels of arterial O_2 and CO_2 but increase their discharge in response to a decrease in the partial pressure of arterial O_2 (PO_2) or to an increase in the partial pressure of arterial CO_2 (PCO_2) or to decreases in pH.

Overall, chemoreceptors are remarkably sensitive to PCO_2. Increasing arterial PCO_2 by 1–3 mm Hg from a normal resting value of about 40 mm Hg can cause a doubling of ventilation. About 60% of this response is contributed by central chemoreceptors. In contrast, there is little ventilatory response to hypoxia until arterial PO_2 decreases below 60 mm Hg from a normal resting value of 80–100 mm Hg.

Chemoafferent fibers from the carotid body are contained in the carotid sinus nerve (CSN), a branch of the glossopharyngeal nerve, with the cell bodies of (first order) CSN chemoafferent neurons found mainly within the petrosal ganglia. CSN chemoreceptor afferent fibers are a mixture of unmyelinated C-fiber axons and myelinated A-fibers. The principal fast transmitter in these afferents appears to be glutamate, however, dopamine appears to be present in about 40% of the C-fibers. Other potential chemoafferent transmitters include substance P and ATP.

Carotid body afferents target 2nd-order caudal NTS neurons [3], specifically within its commissural division (SolC; Fig. 1).

Within the NTS, intrinsic polysynaptic pathways provide recurrent excitatory feedback that can initially increase the excitability of 2nd- and higher-order NTS chemoafferent interneurons. Inhibitory GABAergic neurons in SolC are also activated during hypoxic stimulation of breathing and the initial activation of NTS neurons is followed by local inhibition that ultimately limits the excitatory responses of the NTS 2nd and higher order neurons.

Multiple pathways emanating from the NTS are involved in processing carotid chemoreceptor afferent activity. Both second and higher order neurons in SolC are the source of extrinsic projections to the brainstem and forebrain. Among brainstem targets are VRC respiratory neurons. Additionally, SolC neurons relay chemoafferent input to the rostral dorsolateral pons in the region of the parabrachial and Kölliker-Fuse nuclei. These nuclei contain respiratory neurons and are collectively referred to as the pontine respiratory group (PRG). Some neurons in the PRG are likely relays for chemoafferent (and other visceral) inputs to higher brain structures (in addition to the direct forebrain projections from the NTS). PRG neurons also provide descending inputs that coordinate respiratory activity with other systems such as cardiovascular control as well as with nociceptive afferent input.

Plasticity in Chemoreceptor Breathing Responses

Short and long-term modifications (plasticity) occur in the breathing response to chemoreceptor activation. Respiratory plasticity accommodates the changing demands of development as well as environmental demands such as changing PO_2 levels at varying altitudes, and physiological demands created by pathological changes in the efficiency of the airways and lungs.

Plasticity of the hypoxia reflex is evident in changes in respiratory pattern occurring in multiple stages. The acute response to hypoxia is characterized by increases in both respiratory frequency and the volume of each breath (tidal volume). There is a progressive increase in tidal volume (termed short-termed potentiation) over a period of seconds to minutes. Upon returning to normal oxygen levels there is a slow return to the normal tidal volume. There is also a post-hypoxic decline in breathing frequency lasting several minutes in which respiratory frequency declines below pre-hypoxic levels. While the mechanisms underlying these changes are not well understood, short term potentiation may involve recurrent excitation within the chemoreceptor pathway in the NTS as well as pre-synaptic changes (e.g. calcium accumulation) in NTS or downstream neurons. Post hypoxic frequency decline, on the other hand, may require participation of neurons in the ventrolateral pons.

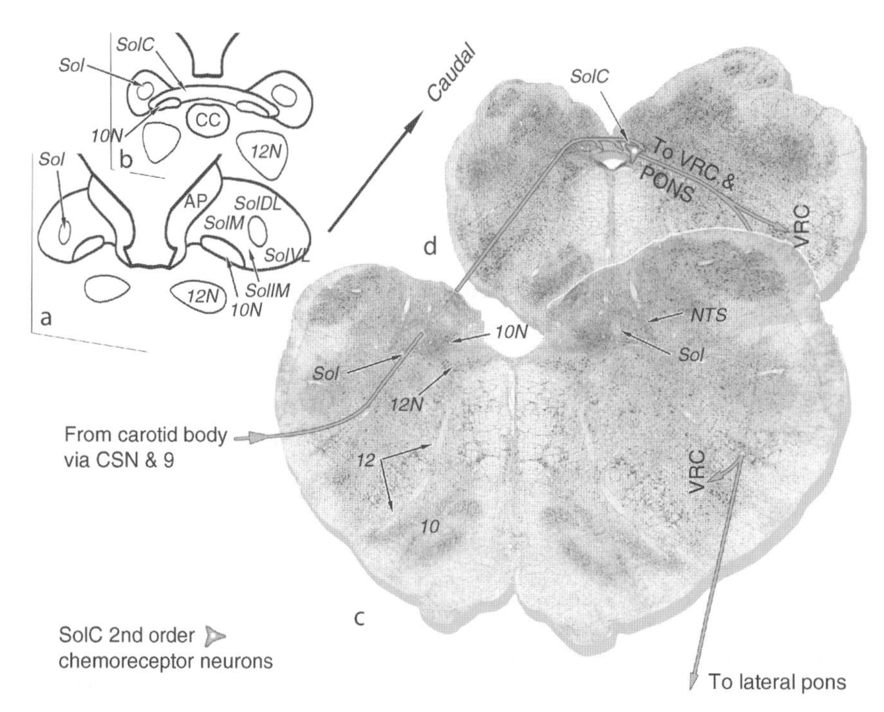

Respiratory Reflexes. Figure 1 Carotid body chemoreceptor afferent terminations within the nucleus of the solitary tract (NTS). (a & b) NTS subnuclei at two rostrocaudal levels. (c & d) Central pathways of carotid chemoreceptors projected onto Nissl-stained coronal sections of the medulla at the levels approximating those diagramed in A & B. The axons of these sensory fibers are carried by the carotid sinus nerve (CSN), a branch of the glossopharyngeal nerve [9]. They enter the medulla near the level of the facial nucleus and run caudally within the solitary tract (sol) terminating predominantly within the commissural subregion (SolC) in the caudal aspect of the NTS. NTS 2nd-order neurons relay this afferent input directly (or indirectly via NTS higher-order neurons) to respiratory regions in the ventrolateral medulla and pons. Compare with the distribution of lung mechanoreceptor afferents in Fig. 3. Abbreviations: *10N* dorsal motor nucleus of the vagus; *12* hypoglossal nerve; *12N* hypoglossal nucleus; *AP* area postrema; *CC* central canal; *IO* inferior olive; *sol* solitary tract; *SolDL* dorsolateral subnucleus; *SolIM* intermediate subnucleus; *SolM* medial subnucleus; *SolVL* ventrolateral subnucleus; *VRC* ventral respiratory column of the ventrolateral medulla.

As illustrated in Fig. 2, repeated episodes of hypoxia lasting from seconds to minutes result in an additional form of plasticity consisting of a long-term facilitation (LTF) of respiratory motor output that can persist for several hours [1].

LTF is observed at motoneurons innervating respiratory pump muscles (e.g. phrenic and external intercostal motoneurons in the cervical and thoracic spinal cord) and upper airway muscles (e.g. hypoglossal motoneurons) and has been related to brainstem serotonergic afferents to these cells (via 5HT-2A receptors) as well as to up-regulation of the peptide, brain derived neurotrophic factor (BDNF), and the molecular signalling proteins activated by BDNF receptors (e.g. TrkB).

In humans, an etiologic role for central chemoreceptors in the medullary arcuate nucleus (located at the medial ventral surface of the brain) has been postulated in sudden infant death syndrome (SIDS) and this has been supported by observations of abnormal development of arcuate serotonin receptors [4]. Disruption of normal chemoreceptor function is also suggested in congenital central hypoventilation syndrome (CCHS). CCHS patients with a polyalanine expansion mutation in the *PHOX2B* gene have negligible sensitivity to elevated PCO_2 or hypoxemia [5].

Airway Receptors

Respiratory reflexes arising from the airways serve both in protecting the airways as well as in regulating the depth and frequency of breathing. Protective reflexes include apnea, cough, sneeze, mucus secretion, and airway constriction that both protect the airways from irritants and facilitate the removal of potentially harmful substances. Sensory feedback also helps coordinate breathing with other behaviours such as locomotion or vocalization.

Receptors in the Nasal Passages and Pharynx

Receptors in the mucous membranes of the nasal cavities are sensitive to cold and pressure changes associated with breathing as well as to inhaled irritants such as cigarette smoke and ammonia. Branches of the trigeminal nerve, including the anterior ethmoidal and

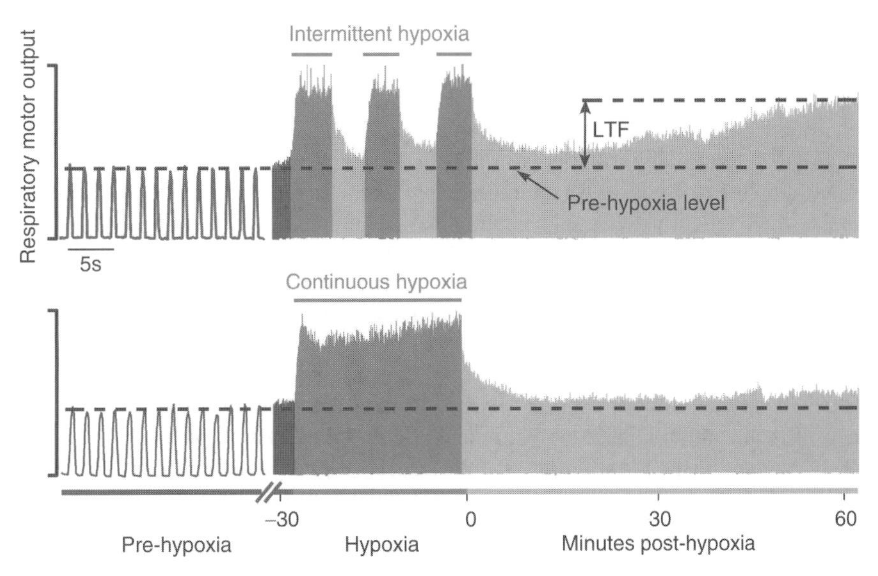

Respiratory Reflexes. Figure 2 Respiration-related neural activity in an anesthetized and artificially ventilated rat demonstrates a form of respiratory plasticity known as respiratory long-term facilitation (LTF). On the left in each trace, electrical activity in the phrenic nerve (the nerve innervating the diaphragm, the principal inspiratory muscle) is integrated such that each peak represents a "fictive breath" that the artificially ventilated rat had intended to make. Under conditions of normal arterial oxygen and carbon dioxide (pre-hypoxia baseline; blue traces), the phrenic nerve bursts are rhythmic and relatively constant in frequency and amplitude. Ventilating the rats with lowered oxygen levels (red traces) causes differential effects on phrenic nerve activity depending upon the pattern of hypoxia exposure (intermittent, upper trace; continuous, lower trace). In both conditions, phrenic bursts increase in amplitude, primarily reflecting the increased activation of carotid body (peripheral) chemoreceptors. Following either pattern of hypoxia exposure, phrenic nerve activity also asymptotes toward baseline levels over several minutes following the return to normoxia (green traces). However, after intermittent, but not after continuous hypoxia, phrenic nerve burst amplitude again increases slowly and progressively over the next hour, even though arterial oxygen and carbon dioxide are at pre-hypoxia levels. This slow increase reflects LTF that is elicited in response to the repetitive exposure to hypoxia. This facilitation requires release of serotonin in the region of the motor neurons (Data provided by T.L. Baker-Hermann and G.S. Mitchell, reproduced with permission from Fundamental Neuroscience, 2nd Edition, edited by LR Squire, FE Bloom, SK McConnell, JL Roberts, NC Spitzer, MJ Zigmond, Academic Press, San Diego, 2003).

maxillary nerves convey the sensory information to the central nervous system. Respiratory motor responses to activation of these receptors include sneezing and apnea. Additional reflex components occur secondarily to alterations in the activity of the autonomic nervous system and include mucus secretion, bradycardia, and increased blood pressure.

The diving reflex is also elicited by receptors with afferent fibers in the trigeminal nerve. This reflex is elicited by water on the face or in nasal passages, and consists of apnea and peripheral vasoconstriction, along with marked increases in arterial pressure and bradycardia. Stimulation of the anterior ethmoidal nerve, which innervates the nasal passages, mimics the diving reflex. Its central terminations are found mainly in ventral aspects of the spinal trigeminal nucleus at levels caudal to the facial nucleus with additional terminations in the NTS and paratrigeminal nucleus.

Laryngeal Receptors

The larynx is richly innervated by several subgroups of sensory receptors [6,7]. Their afferent fibers are mainly in the recurrent and superior laryngeal branches of the vagus nerve with terminations in the NTS. The receptors are located at the entrance to the trachea and lower airways and elicit strong protective reflexes including cough and apnea. The pronounced apneas elicited from laryngeal receptors has also suggested that abnormal development of their reflex pathways could contribute to SIDS. In contrast, a subset of laryngeal receptors promotes airway patency via activation of airway dilating muscles such as the genioglossus and posterior cricoarytenoid.

Receptors in the Lower Airways

Receptors within the lungs and lower airways, *i.e.*, those below the larynx, are classified into two main types based on whether the sensory afferent fibers are myelinated or unmyelinated [6,8,9]. The afferent axons arising from both groups travel in the vagus nerves and terminate in middle and caudal aspects of the NTS. Receptors with myelinated axons constitute airway mechanoreceptors and are activated by distension of the airway during lung inflation or by a reduction in airway dimensions during lung deflation, especially deflations below the normal end-expiratory volume. Two groups of receptors, slowly (SAR) and rapidly (RAR) adapting receptors, are identified based on their sensitivity to distension of the airways during lung inflation and the rate of accommodation in their response. An additional group of receptors, termed deflation activated receptors (DARs), are more prominent in small animals (e.g. rats). DARs share several common properties with RARs, which are more readily observed in larger animals and activation of either RARs or DARs elicits augmented inspiratory efforts.

SARs are located in airway smooth muscle. Their activation by lung inflation inhibits inspiratory motor activity, thereby shortening inspiratory duration and reducing tidal volume (termed the *Breuer-Hering inspiration-inhibiting reflex*). Maintaining inflation into the expiratory period prolongs expiratory duration (*Breuer-Hering expiratory facilitatory reflex*). The Breuer-Hering reflexes are activated during normal resting breathing in most mammals while in humans they appear to only significantly influence breathing pattern when tidal volumes increase to 2–3 times their resting values as may occur during muscular exercise. Activation of SARs has several additional effects, including reductions in airway smooth muscle tone resulting in bronchodilation, and reductions in heart rate and systemic vascular resistance.

RARs are located in airway epithelium and submucosa. While they are activated by lung inflation they are less responsive than SARs and tend to have little activity under normal breathing conditions. Their discharge adapts rapidly to lung inflation and they typically fire irregularly, giving rise to one or a few action potentials during lung inflation. They respond with a significantly more sustained discharge to inhaled irritants. RARs are implicated in a number of potent airway protective reflexes, including augmented breaths (sighs), airway constriction, mucus secretion and laryngeal closure. While they are generally believed to trigger coughing, this function has recently been related to a specific subset of polymodal A∂-fibers (termed cough receptors; [10]).

DARs. Some researchers group DARs with RARs and the degree to which these represent the same or separate populations requires further examination. Nevertheless, lung deflation triggers reflexes that shorten expiratory duration and increase inspiratory effort. There is also a reflex narrowing (adduction) of the glottis and stimulation of low intensity diaphragm activity during expiration that slows expiration. Together these responses help prevent alveolar collapse, and are particularly important in human infants as well as in other small mammals that have highly compliant chest walls.

C-Fibers constitute the largest class of pulmonary vagal afferent fibers (~75%). They are polymodal and nociceptive, responding to a variety of inflammatory mediators and inhaled irritants. Reflex changes in breathing in response to C-fiber activation involve rapid shallow breathing interspersed with apneas; additional reflex effects include bronchoconstriction, mucus secretion, hypotension, bradycardia and airway mucosal vasodilation. Beyond identification of the regions of C-fiber termination within the NTS, little is known concerning central pathways mediating these responses.

Second and Higher-Order Neurons in Reflexes From Airway Mechanoreceptors

Central Pathways of SARs

SAR primary afferent fibers terminate within mid to caudal portions of the NTS (Fig. 3) [6]. Only two functional classes of NTS neurons receive monosynaptic input from SARs (Fig. 3).

One type, termed Iβ neurons, exhibits inspiratory discharge patterns, and activation of SARs increases the discharge rate of these neurons. Iβ neurons are bulbospinal premotor neurons that monosynaptically excite phrenic motoneurons. These motoneurons innervate the diaphragm and this reflex accordingly provides positive feedback excitation of diaphragm inspiratory activity.

A second group of NTS neurons, termed pump-neurons, mediate Breuer-Hering reflex changes in respiratory rhythm. Pump neurons receive monosynaptic SAR input but are readily distinguished from Iβ neurons by the general absence of an inspiratory discharge when SAR input is removed. Consistent with the broad effects they elicit on respiratory pattern, their axons arborize extensively within pontomedullary regions (ventrolateral medulla and rostral dorsolateral pons) containing neurons responsible for generating the respiratory pattern. Many pump neurons are inhibitory, containing GABA with only a small percentage also containing glycine. Experimental evidence also suggests that there may be excitatory pump neurons, however, these cells have not yet been directly identified.

Central Pathways of RARs

RAR primary afferent fibers terminate in caudal aspects of the NTS where they monosynaptically excite neurons termed RAR interneurons (Fig. 3) [6]. Like pump neurons, these interneurons provide extensive axonal arborizations to pontomedullary regions involved in respiratory pattern generation. RAR interneurons are believed to be excitatory and presumably facilitate the discharge of inspiratory neurons. RAR activation accordingly elicits augmented inspirations such as sighs, and the large inspiration preceding a cough. RARs are also stimulated by decreases in lung compliance that result from alveolar collapse. The resulting RAR-triggered large inspiration stretches the lung, reopening collapsed alveoli.

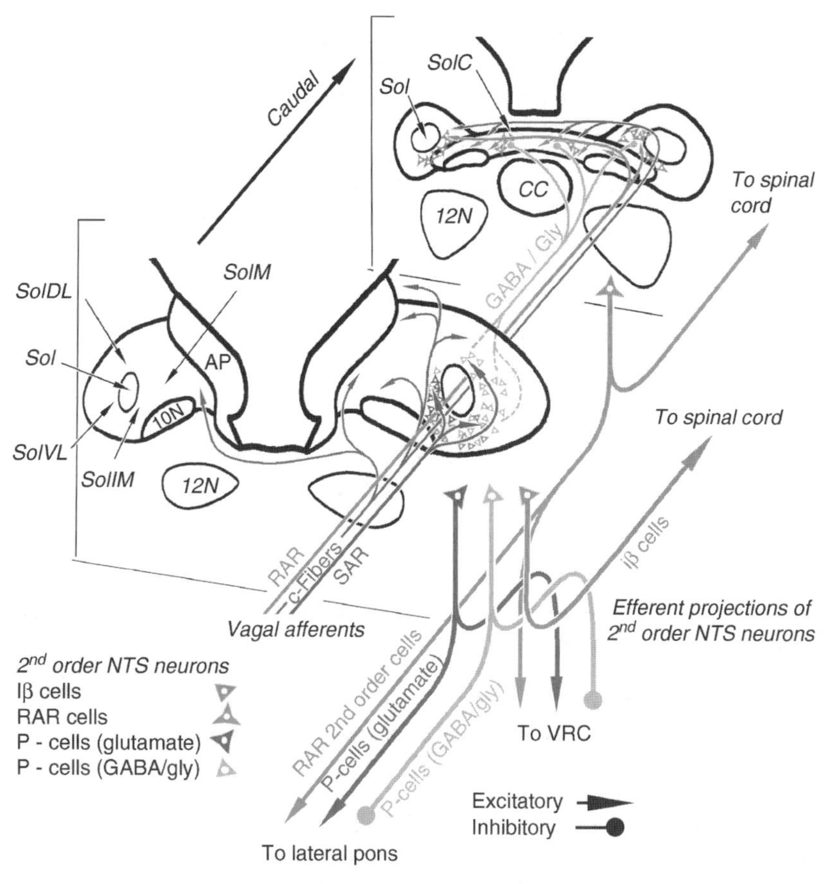

Respiratory Reflexes. Figure 3 The topographical distribution within the NTS of three classes of pulmonary afferents. The terminal distribution slowly and rapidly adapting stretch receptors (SARs and RARs, respectively) and bronchopulmonary C-fibers is shown at two rostrocaudal levels of the NTS. The three afferent systems in general project to topographically separate NTS targets. The principal projections of their 2nd order neurons are also indicated. Among the known projections is an inhibitory projection of pump (P-) cells to RAR relay neurons in the NTS commissural subnucleus. See Fig. 1 for abbreviations (used with permission from [6]).

Summary

Although seemingly effortless in the healthy individual, generating an optimal breathing pattern for O_2 and CO_2 homeostasis requires the integration of sensory information arising from a variety of receptors. These include multiple central and peripheral chemoreceptors for adjusting the magnitude of alveolar ventilation as well as multiple mechanoreceptors that respond to stretch of the airways and regulate the relative depth and rate of breathing to reduce energy expenditure. Sensory input is also important in the coordination of breathing with other systems, such those required for speaking, eating, walking, running, vomiting. Finally, sensory information is necessary for protection of the airways and lungs. Receptors in the nose, pharynx, larynx and lower airways elicit a variety of reflexes including coughs, sneezes and apnea that protect the airways from inhalation of noxious substances and increase mucous secretion that aids in their removal. Failure of any of these systems can be catastrophic, severely compromising the quality of life for an individual or ultimately leading to their death.

References

1. Lee L-Y, Pisarri TE (2001) Afferent properties and reflex functions of bronchopulmonary C-fibers. Respir Physiol 125:47–65
2. Yu J (2005) Airway mechanosensors. Respir Physiol Neurobiol 148:217–243
3. Berry-Kravis EM, Zhou L, Rand CM, Weese-Mayer DE (2006) Congenital central hypoventilation syndrome: PHOX2B mutations and phenotype. Am J Respir Crit Care Med 174(10):1139–1144
4. Paterson DS, Trachtenberg FL, Thompson EG, Belliveau RA, Beggs AH, Darnall R, Chadwick AE, Krous HF, Kinney HC (2006) Multiple serotonergic brainstem abnormalities in sudden infant death syndrome. JAMA 296(17):2124–2132
5. Powell FL, Milsom WK, Mitchell GS (1998) Time domains of the hypoxic ventilatory response. Respir Physiol 112(2):123–134

R

6. Kubin L, Alheid GF, Zuperku EJ, McCrimmon DR (2006) Central pathways of pulmonary and lower airway vagal afferents. J Appl Physiol 101:618–627
7. Feldman JL, Mitchell GS, Nattie EE (2003) Breathing: rhythmicity, plasticity, chemosensitivity. Annu Rev Neurosci 26:239–266
8. Canning BJ, Mazzone SB, Meeker SN, Mori N, Reynolds SM, Undem BJ (2004) Identification of the tracheal and laryngeal afferent neurones mediating cough in anaesthetized guinea-pigs. J Physiol (Lond) 557:543–558
9. Paton JFR, Deuchars J, Li Y-W, Kasparov S (2001) Properties of solitary tract neurones responding to peripheral arterial chemoreceptors. Neuroscience 105 (1):231–248
10. Sant'Ambrogio G, Tsubone H, Sant'Ambrogio FB (1995) Sensory information from the upper airway: role in the control of breathing. Respir Physiol 102:1–16

Respiratory Sinus Arrhythmia

►Central Integration of Cardiovascular and Respiratory Activity Studied In?Situ

Responding Conditioning

►Classical Conditioning (Pavlovian Conditioning)

Response, Instrumental

Definition

A voluntary, conditioned response to a cue performed for reinforcement.

►Reinforcement

Response Acquisition

►Learning and Motivation

Response Extinction

Definition

The result of extinction, observed as a decrease in conditioned responses to a conditioned stimulus following non-reinforced presentations of the conditioned stimulus.

►Learning and Extinction

Response Inhibition

Definition

Inhibitory control is the ability to suppress behaviors that are inappropriate under the circumstances. Neuropsychological studies of the prefrontal cortex indicate that this function arises from the orbitomedial divisions, most probably via descending projections to structures such as the amygdala, basal ganglia and hypothalamus.

►Prefrontal Cortex

Rest-Activity Cycle

Definition

The fundamental alternation between extended periods of activity and rest that define a complete circadian cycle. Can also define ultradian (much less than 24 h) cycles.

►Circadian Cycle
►Internal Desynchrony
►Sleep-wake Cycle

Resting Membrane Potential

Definition

The resting potential is a stable membrane potential in non-excitable cells or, in excitable cells, the most stable membrane potential between action potentials without excitatory or inhibitory inputs. In some excitable tissues, a resting potential cannot be defined because of continuous changes in membrane potential.

►Membrane Potential: Basics
►Action Potential

Resting Tremor

Definition

Approximately 70% of patients notice tremor as the first symptom of Parkinson disease. Onset of tremor is usually in one hand and it may later involve the contralateral upper limb or ipsilateral lower limb.

Typically, the tremor is 3–5 Hz rhythmic "pill-rolling" movements of the thumb and index finger while the hand is at rest. There may be abduction and adduction of the thumb, or flexion and extension of the wrist, or of the metacarpophalyngeal joints. The tremor may also extend to the forearm with pronation–supination or even to the elbow and upper arm. During early disease, tremor is often intermittent and is evident only under stress. Tremor is worsened by anxiety, fatigue, and sleep deprivation. It diminishes with voluntary activity but may reappear with static posture (e.g. outstretched hands) and is absent during sleep. Resting tremor is enhanced by mental task performance, such as serial seven subtractions, and by motor task performance in a different body part. The hand tremor may also be enhanced during ambulation. Compared to essential tremor, the resting tremor of Parkinson disease is generally less prone to exacerbation by caffeine or improvement with alcohol.

►Parkinson Disease

Restless Legs Syndrome

Definition

Restless limbs syndrome is a common disorder with a prevalence of 5–15% in western countries. It is characterized by a distressing desire to move the legs, motor restlessness brought on by rest, worsening symptoms in the evenings and at night, and periodic limb movements during sleep. Although it can be seen with peripheral neuropathy, most cases of restless legs are not accompanied by neuropathy. There is an association between restless limbs and brain dopamine and iron deficiency. Therefore, checking iron and ferritin levels is part of the workup for restless legs syndrome. If iron deficiency is detected, it should be evaluated (anemia workup, etc.) and treated with iron supplementation. If a sleep study reveals sleep apnea together with periodic limb movements during sleep, the apnea component should be treated. Symptomatic treatment of restless legs and limb movements during sleep usually begins with a dopamine agonist (pramipexole or ropinirole). Dopamine agonists have longer durations of action compared to levodopa and one or two evening/bedtime dose(s) may suffice. If dopamine agonists are not well tolerated, a controlled release formulation of carbidopa/levodopa should be tried next, and the dose titrated as tolerated. Other adjunct medications include gabapentin, benzodiazepines, and low potency opioids as a last resort.

►Sleep – Developmental Changes

Ret

Definition

The signaling component of the glial cell line-derived neurotrophic factor (GDNF) family receptor complex. Ligand-binding to GPI-linked GFRα receptors (1–4) and subsequent induction of Ret dimerization results in Ret phosphorylation and activation of several signaling cascades, including those involving MAPK, PI3K and PLCγ.

►Glial Cell Line-derived Neurotrophic Factor (GDNF)
►Neurotrophic Factors in Nerve Regeneration

Retention

Definition

Retention is the ability to maintain in mind information about a stimulus when that stimulus is no longer physically present. Retention span can vary from on the order of seconds or minutes to months or even years.

►Recognition Memory

R

Reticular Core

►Reticular Formation

Reticular Formation

Stefanie Geisler
Behavioral Neuroscience Branch, National Institute on Drug Abuse, Intramural Research Program, Baltimore, MD, USA

Synonyms

Formatio reticularis; Substantia reticularis; Isodendritic core; Reticular core

Definition

The reticular formation is a netlike structure of cells and fibers that extends throughout the core of the brainstem. It maintains vegetative functions, plays an essential role in coordinated motor behaviors and exerts control on cortical and thalamocortical activation. Extensive damage to the reticular formation is incompatible with survival.

Characteristics

Anatomical Organization and Concepts

The term reticular formation was coined by anatomists in the last century to describe a region in the core of the ►brainstem characterized by scattered neurons of various sizes and shapes with long, sparsely branching dendrites lying in and among multiple, differently oriented fiber systems. The dendrites of neighboring neurons overlap extensively with each other, giving the structure a netlike ("reticular") appearance. The reticular formation extends continuously from the ►caudal medulla oblongata to the ►rostral mesencephalon (and possibly beyond, see below).

Based on differences in cytoarchitecture, cytochemistry and connections, the reticular formation has been divided into three longitudinal zones, a median zone, which contains the ►serotoninergic raphe nuclei (Greek *raphe* = seam), a medial zone, characterized by big cells intermingling with many small ones and a lateral zone, which consists predominantly of (Fig. 1) small neurons and has fewer fibers than the medial zone (Fig. 1a–c).

More recently, an intermediate zone has been delineated between the medial and lateral ones (Fig. 1a, [1]). The intermediate zone comprises large, medium and small cells and exhibits slightly stronger ►acetylcholine esterase staining as compared to the adjacent medial and lateral zones. ►Cranial nerve nuclei (e.g. facial and cochlear) and relay nuclei (e.g. cuneate and red nuclei), which largely consist of densely packed neurons of similar appearance (Fig. 1d) are not included in the reticular formation.

Nuclei of the Recticular Formation

By analyzing Nissl stained (►Nissl Stain) sections of the human and rabbit brainstem Olszewski and colleagues noted cytoarchitectural heterogeneities, which led them (and subsequently others in other species) to suggest that the reticular formation consists of different nuclei [2,10]. An overview of reticular formation nuclei is given in Table 1.

At the medullary level of the medial reticular formation, the ►gigantocellular nucleus has been delineated, which is composed of prominent multipolar giant cells, as well as large, medium sized and some small neurons (Fig. 1a, b). It is surrounded by the gigantocellular nucleus pars alpha and the lateral and dorsal paragigantocellular nuclei, all of which also contain large neurons. The gigantocellular nucleus extends from the obex to the rostral medulla oblongata and is continuous caudally with the ventral reticular nucleus and rostrally with the ►caudal pontine reticular nucleus. The caudal pontine reticular nucleus extends to the rostral pons where it merges with the oral pontine reticular nucleus, which in turn extends rostralward to the level of the decussation of the superior cerebellar peduncle. Dorsal and ventral to the caudal pontine reticular nucleus are the dorsomedial tegmental and ventral pontine reticular nuclei respectively. The lateral zone consists, in order from ►caudal to ►rostral, of the dorsal reticular (caudal medulla), parvocellular (medulla; Fig. 1c), subceruleus (pons) and cuneiform nucleus (pons, ►mesencephalon). The mesencephalic reticular formation is dominated by the deep mesencephalic nucleus, which some authors regard as part of the medial and others as part of the lateral zone of the reticular formation. Some of the above mentioned nuclei have been further subdivided (see [2,3]).

In addition to the above-mentioned nuclei are some that are regarded by some, but not all, authors to be part of the reticular formation, e.g. the lateral reticular nucleus, ►parabrachial nucleus, locus ceruleus, pedunculopontine and laterodorsal tegmental nuclei, retrorubral field, and ventral tegmental area. It sometimes seems to be a matter of personal taste whether a nucleus is included in the reticular formation or not.

The Veticular Formation and the Ascending Reticular Activity System

In 1949, Moruzzi and Magoun published a seminal paper describing studies in which they demonstrated that stimulation of the reticular formation in lightly

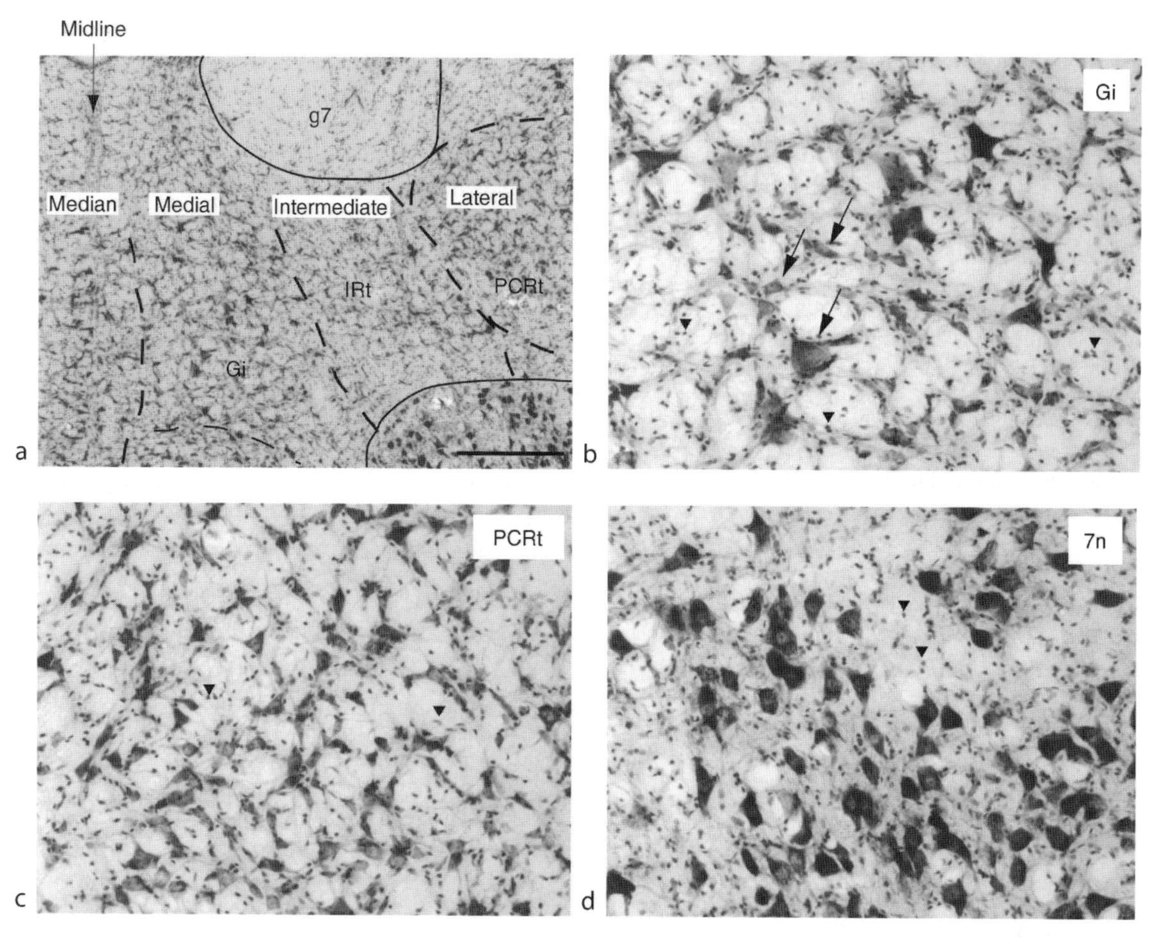

Reticular Formation. Figure 1 (a) Photomicrographs showing some reticular formation nuclei in the median, medial, intermediate and lateral zones. Prominent, big neurons of the gigantocellular nucleus can be easily recognized. The gigantocellular nucleus is enlarged in (b), showing large and small neurons (*arrows*) side-by-side, one of the characteristics of the reticular formation. The very small cells (*arrowheads*) are not neurons, but glial cells. (c) shows an enlargement of the parvocellular reticular nucleus, which has neurons that are obviously smaller than those in the gigantocellular nucleus. Note also the different sizes and shapes of neurons. The facial nerve nucleus, which is not included in the reticular formation, is shown in (d). Its neurons are all of similar size and appearance and are more densely packed than those in the gigantocellular and parvocellular reticular nuclei (compare b, c, d). Abbreviations: *7n* facial nucleus, *g7* genu of facial nerve; *Gi* gigantocellular nucleus; *IRt* intermediate reticular nucleus; *PCRt* parvocellular reticular nucleus. *Arrows* point to neurons; *arrowheads* point to glial cells. The scale bar in (a) represents 0.5 mm in (a), 0.125 mm in (b–d).

anaesthetized cats evokes a desynchronization of the cortical ►EEG, closely resembling the changes observed in the human EEG upon transition from sleep to wakefulness or relaxation and drowsiness to alertness and attention [4]. Such EEG changes could be elicited by stimulation of the medial medullary, pontine and mesencephalic reticular formation and dorsal hypothalamus and subthalamus. They proposed a series of relays in the reticular formation ascending to the basal diencephalon and exerting influence on widespread areas of the cortex via a "diffuse thalamic projection system." These and other observations led to the concept of an ascending reticular activating system (often referred to in the literature as ARAS), which spurred extensive research and had an enormous influence on subsequent views concerning the neural basis of consciousness. Because potentials could be recorded from large cortical territories following stimulation of afferents from a particular source and impulses from several sources were found to reach the same region, it was assumed that the system would be entirely diffusely organized. Unfortunately, the term "ascending reticular activating system" describing a physiological concept was soon frequently used synonymously with the morphological term "reticular formation," which led Olszewski to comment [5]: "There are presently two reticular formations – an anatomic one and a physiologic one – and they do not correspond to each other."

The Concept of the Isodendritic Core

Due to the confusion surrounding the term reticular formation and the lack of a generally accepted definition, some scientists suggested that use of the construct be discontinued altogether (e.g. [5,6]), whereas others attempted to define it more precisely [1 and refs. therein]. Based on analyses of Golgi stained (▶Golgi Stain) brain sections of different species Ramon-Moliner and Nauta described not one defining characteristic of reticular formation, but rather an aggregation of several morphological features that they concluded are found together only in the reticular formation: i) cytological polymorphism, large and small neurons are side by side (Fig. 1b); ii) generalized dendrites, long, radiating, relatively rectilinear and sparsely branching processes, iii) considerable dendritic overlap and iv) free intermingling of dendrites and passing myelinated and unmyelinated fiber bundles. Ramon-Moliner and Nauta called neurons (Fig. 2) with the generalized dendritic patterns found in the reticular formation "isodendritic" (from the Greek isos: similar, uniform, Fig. 2a) and distinguished them from allodendritic (allos: different) and idiodendritic (idios: peculiar) neurons (Fig. 2b).

Idio- and allo-dendritic neurons, also sometimes referred to as "hodophob" (hodos: path, pathway; phob: fearing), are characterized by short dendrites that ramify

Reticular Formation. Table 1 Nuclei of the brainstem reticular formation

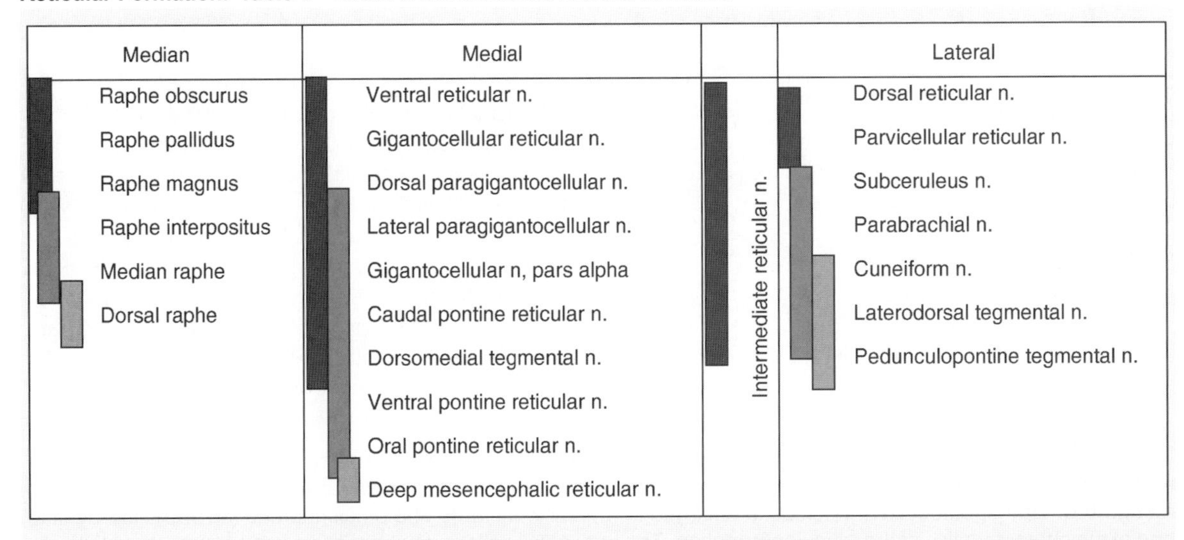

Median	Medial		Lateral
Raphe obscurus	Ventral reticular n.	Intermediate reticular n.	Dorsal reticular n.
Raphe pallidus	Gigantocellular reticular n.		Parvicellular reticular n.
Raphe magnus	Dorsal paragigantocellular n.		Subceruleus n.
Raphe interpositus	Lateral paragigantocellular n.		Parabrachial n.
Median raphe	Gigantocellular n, pars alpha		Cuneiform n.
Dorsal raphe	Caudal pontine reticular n.		Laterodorsal tegmental n.
	Dorsomedial tegmental n.		Pedunculopontine tegmental n.
	Ventral pontine reticular n.		
	Oral pontine reticular n.		
	Deep mesencephalic reticular n.		

Nuclei are ordered from caudal to rostral, colored boxes indicate where the nuclei are situated in the brainstem: ▶blue – Medulla oblongata; ▶orange – Pons; ▶green – Mesencephalon. At medullary levels an intermediate zone has been delineated. Whereas the zones are easily recognizable at medullary levels (caudal), it is more difficult to distinguish them at rostral pontine and mesencephalic levels. Thus, at mesencephalic levels the allocation of nuclei to distinct zones is fairly tentative. Abbreviation: *n* nucleus.

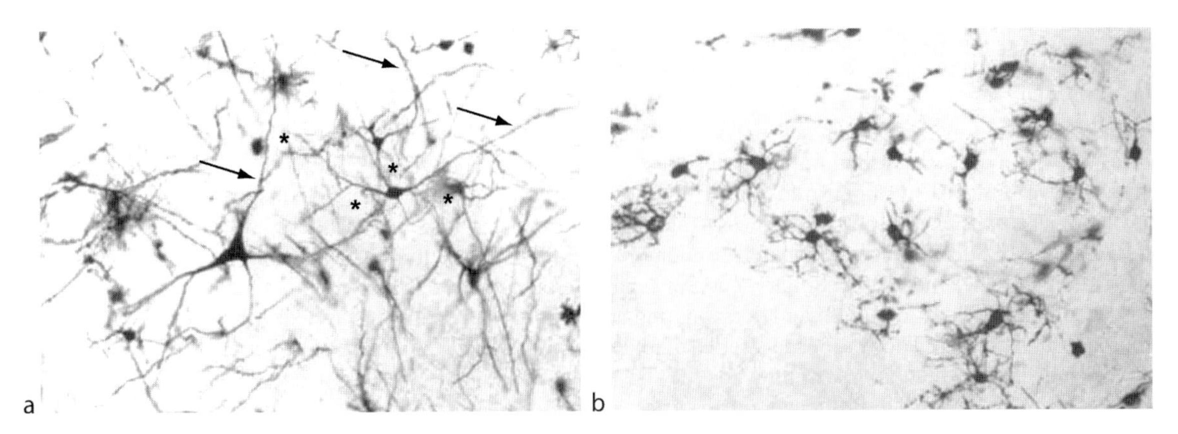

Reticular Formation. Figure 2 (a) Isodendritic neurons have long, sparsely branching dendrites (*arrows*) and overlapping dendritic fields (*asterisks*). (b) Allodendritic neurons have short dendrites, which ramify close to the cell body. Dendrites of neighbouring neurons largely do not overlap with each other. These Golgi preparations are adapted from figs. 6 and 7 of [1], with permission from Wiley Interscience.

close to the cell body and do not extend into passing fiber bundles (Fig. 2b). These neuron types are found, e.g. in sensory nuclei, like the cuneate nucleus. In contrast, the "reticular" isodendritic neurons are regarded as "hodophil" (phil: friendly).

Ramon-Moliner and Nauta's concept of reticular formation does not necessarily preclude the existence of regional differences and is not incompatible with parcellations suggested by Olszewski and others. Thus, although the extensively overlapping dendritic fields of reticular neurons make it difficult, if not impossible, to draw definite lines around nuclei, the division of the reticular formation into nuclei is nevertheless helpful and necessary (and widely employed today) to describe the locations of nerve cells, electrode placements and lesions within the reticular formation. Applying the criteria of Ramon-Moliner and Nauta, the lateral reticular nucleus is not part of the reticular formation because it consists of allodendritic neurons and has a restricted set of connections with the ►spinal cord and ►cerebellum. In contrast, e.g. the ►parabrachial nucleus, locus ceruleus, pedunculopontine and laterodorsal tegmental nuclei, retrorubral field and ventral tegmental area have sufficient reticular characteristics to be included. Furthermore, if these criteria are applied, the reticular formation is not confined to the brainstem, but also includes structures in the forebrain, such as the ►lateral hypothalamic and preoptic areas and the magnocellular ►basal forebrain [7]. Even if these areas are conservatively excluded from the reticular formation, their neuroanatomical organization suggests that they process information in a manner similar to that which occurs in the reticular formation of the brainstem.

Connections and Functions of the Reticular Formation of the Brainstem

The reticular formation maintains pivotal vegetative functions (e.g. respiration, heart beat, blood pressure) and plays an essential role in coordinated motor behaviors (e.g. swallowing, chewing and ►locomotion). In addition, it functions as an intermediary through which amygdala, septum, and basal ganglia gain access to the autonomic and motor systems [8]. Via its ascending projections, the reticular formation exerts control on cortical and thalamocortical activation and functions.

Medial Reticular Formation

Reticular neurons give rise to long ascending (e.g. to the cerebral ►cortex) and descending (e.g. to the sacral level of the spinal cord) axons that give off several collaterals along their course, thereby interconnecting different parts of the reticular formation [3]. For example the oral and caudal pontine reticular nuclei receive half of their afferents from other parts of the brainstem reticular formation. Other main afferents to the medial reticular formation arise in the ►prefrontal and ►sensory cortices, zona incerta, fields of Forel, lateral hypothalamus, preoptic area, substantia nigra pars reticulata, superior colliculus, central gray, cerebellum and spinal cord. The medial reticular formation projects mainly to the spinal cord and motor cranial nerve nuclei and to the laterodorsal and pedunculopontine tegmental nuclei, intralaminar thalamic nuclei, fields of Forel, parafascicular thalamic nucleus, zona incerta and lateral hypothalamus. The projections of the individual nuclei of the medial reticular formation differ in degree rather than kind. Thus, all parts of the medial reticular formation project to the telencephalon, but the largest number of neurons projecting there is located at the mesencephalic level, whereas only a few are situated at caudal pontine and medullary levels. Similarly, whereas many neurons in the gigantocellular nucleus project to the spinal cord, only a few in the mesencephalic reticular formation do so [3].

Neurons in the dorsal two thirds of the caudal pontine and medullary medial reticular formation project to the intermediate and ventral horn of the spinal cord, where the premotor interneurons and motorneurons for the axial and proximal musculature (i.e. trunk, hip, back, shoulder and neck) are situated. Thus, they play an important role in the control of posture, integration of the movements of body and limbs and the orientation of body and head. Because orienting movements of the head and eye are tightly linked, it might not be surprising that parts of the medial reticular formation that project to the spinal cord also possess neurons that innervate eye muscle motor neurons (oculomotor and ►abducens nuclei), which are necessary for horizontal ►gaze control. Neurons controlling the phylogenetically later developed vertical gaze control (trochlear and oculomotor nuclei) are situated further rostrally in the medial reticular formation.

Neurons in the medial reticular formation also influence wide areas of the cerebral cortex via their strong ascending projections to the laterodorsal and pedunculopontine tegmental nuclei and, to a lesser extent, to intralaminar thalamic nuclei (see Essay on Mesopontine Tegmentum).

Lateral Reticular Formation

The lateral reticular formation receives afferents from the primary motor and ►somatosensory cortex, ►insular cortex, ►central nucleus of amygdala, bed nucleus of stria terminalis, central gray, trigeminal nuclei, rostral part of the ►nucleus of the solitary tract, red nucleus, other parts of the reticular formation and cerebellum. Neurons in the lateral reticular formation project predominantly to motor cranial nuclei (e.g. trigeminus, facial and ►hypoglossal) and the medial reticular formation.

Sensory information from jaw muscle spindle afferents and oral cavity (via trigeminal nuclei), taste (via the rostral part of the nucleus of the solitary tract), visceral information (via the insular cortex) and oral motor information (from the motor cortex) are relayed to and integrated within the lateral reticular formation, which in turn projects to motor cranial nuclei containing neurons for muscles involved in swallowing, chewing and salivation. In addition, inputs from structures commonly regarded as being involved in emotional processing, such as ►amygdala, bed nucleus of stria terminalis and ►periaqueductal gray are integrated in the lateral reticular formation and relayed to, e.g. facial motor neurons and autonomic centers.

The connections conceptualized

These anatomical data provide a general idea about the connections, but are insufficient to explain the precise circuits that underlie the functions of the reticular formation. Against the common perception that the reticular formation is organized in a diffuse way, the complex behaviors it is involved in require very specific and precisely tuned connections. Breathing, for example, requires the coordinated movements of jaws, lips, tongue, pharynx and larynx, can be controlled voluntarily, but usually happens automatically and has to adapt for eating, speaking, fighting or fleeing. Thus, a high level of specificity must exist in the reticular formation, but it is the very structure of the reticular formation that makes such a specificity very difficult to detect.

So, what might such an apparently disorganized structure be good for? To appreciate the functional anatomical organization of the reticular formation, it might be useful to compare it to a system that is quite differently organized, such as the primary motor or sensory system. The so-called lemniscal system carries sensory information via two synapses to the primary sensory cortex. Such a sensory pathway of minimal interruption rigorously maintains the topography of the sensory periphery and thus permits the exact localization of the source of the sensory stimulus. Hence, this system is very well suited for discriminative functions. The reticular formation, in contrast, consisting of neurons with long dendrites receiving multiple heterogeneous inputs and emitting axons that collateralize a lot, is especially well suited for integrative functions. As Nauta and Feirtag [9] state "Life depends on the innervation of the viscera: in a way all the rest is biological luxury. And vital systems ought to be organized on the principle that no single excitation should greatly affect their workings." It remains an exciting challenge to untangle the precise underlying neuronal networks.

References

1. Ramon-Moliner E, Nauta WJH (1966) The isodendritic core of the brain stem. J Comp Neurol 126:311–336
2. Paxinos G, Watson C (1986) The rat brain in stereotaxic coordinates. Academic, Sydney
3. Jones BE (1995) Reticular formation: cytoarchitecture, transmitters, and projections. In: Paxinos (ed) The rat nervous system, 2nd edn. Academic, Sydney, pp 155–171
4. Moruzzi G, Magoun HW (1949) Brain stem reticular formation and activation of the EEG. Clin Neurophysiol 1:455–473
5. Olszewski (1957) Reticular formation of the brain. In: Jasper HH, Proctor LD, Knighton RS, Noshay WC, Costello RT (eds) Little, Brown, Boston, MA, p 56
6. Blessing WW (1997) The lower brainstem and bodily homeostasis. Oxford University Press, Oxford, NY
7. Leontovich TA, Zhukova GP (1963) The specificity of the neuronal structure and cography of the reticular formation in the brain and spinal cord of carnivore. J Comp Neurol 121:347–379
8. Zahm DS (2006) The evolving theory of basal forebrain functional-anatomical 'macrosystems'. Neurosci Biobehav Rev 30:148–72
9. Nauta WJH, Feirtag M (1979) The organization of the brain. Sci Am 241:88–111
10. Olszewski J, Baxter D (1954) Cytoarchitecture of the human brain stem. Karger, Basel

Reticulospinal Cells (Neurons)

Definition

Neurons located within the reticular formation with an axonal projection to the spinal cord. Reticulospinal cells project to different levels of the spinal cord, some activating exclusively the most rostral segments whereas others activate the caudal ones.

Reticulospinal Long-Lead Burst Neurons

Charles Scudder
Portland, OR, USA

Definition

Reticulospinal neurons, in general, are neurons that have their somata in the mesencephalic, pontine, or medullary reticular formation and have axons that project at least as far as the upper-most spinal cord. Reticulospinal ►long-lead burst neurons (RS-LLBNs) (►burst cells – long lead (LLBNs)) are reticulospinal neurons that discharge before and during ►gaze saccades in a preferred direction that depends on the particular cell. Like other reticulospinal neurons, RS-LLBNs integrate input from multiple sources and project to multiple targets. In the case of RS-LLBNs, the purpose is to generate coordinated eye, head, and sometimes trunk movements.

Characteristics

Higher-Order Processes

RS-LLBNs are part of a system of descending pathways that mediate orienting toward areas or objects of interest. The stereotypical ►orienting response depends on the species. Macaque monkeys move only their eyes if the target of interest is near the current direction of gaze, but will make a combined eye and head ►gaze movement for targets further away. Cats preferentially move the eyes and head together, and typically also redirect their pinnae and maybe their trunk. Rodents do all of the above with greater movement of the trunk. Human orienting is more flexible, but usually includes combined eye and head movements. The descending system that mediates this behavior includes the reticulospinal pathway, the tectospinal pathway, the corticospinal pathway, as well as less direct pathways.

The command to execute an orienting movement originates in cerebral cortex and involves many of the same structures as saccade generation. The ►lateral intraparietal cortex (LIP), the ►supplementary eye fields (SEF), and the ►frontal eye fields (FEF), which are themselves interconnected, project to the deep and intermediate layers of the ►superior colliculus as well as to other brainstem locations. Signals for these cortical areas, as well as those from subcortical areas, are integrated in the superior colliculus, and the colliculus issues the principal command to move the eye, head, etc. This command is conveyed to the brainstem, including the ►saccadic burst generator and reticulospinal neurons, and directly to the spinal cord. In cats, this "tectospinal" pathway is strong and consists of well studied ►tecto-reticulo-spinal neurons (TRSNs). In primates, the tectospinal pathway is weak and probably includes neurons that differ from the cat TRSNs. This means that primates rely heavily on reticulospinal pathways to orient.

The superior collicular projection to the brainstem provides the principal input to the RS-LLBNs and a smaller input to other reticulospinal neurons. Efferents from cerebral cortex provide additional input to RS-LLBNs, but not as much as to other reticulospinal neurons [1]. Cortical inputs arise from motor cortex and possibly premotor cortex. Another input to reticulospinal neurons, including RS-LLBNs, is from the ►fastigial nucleus of the midline cerebellum. The fastigial nucleus, in turn, receives indirect input from the superior colliculus, the FEF, the SEF, and motor cortex, and represents another pathway by which these cortical structures can influence RS-LLBNs. In fact, orienting movements in cats and monkeys are severely disrupted by chemical inactivation of the fastigial nucleus. Many reticulospinal neurons that receive input from the superior colliculus also receive input from the vestibular system [2], although the relevance for RS-LLBNs is controversial. Caudal reticulospinal neurons receive input from collaterals of more rostral reticulospinal neurons [1,3–5].

Parts of the RS-LLBN System

RS-LLBNs as well as other reticulospinal neurons are found in the medullary ►nucleus reticularis gigantocellularis (NRG), pontomedullary ►nucleus reticularis pontis caudalis (NRPc), and in the mesencephalic reticular formation in and near the ►interstitial nucleus of Cajal (INC) and the H-fields of Forel (called the ►riMLF in the monkey). Neurons have different properties in different areas, and in different species according to current incomplete data. In general, neurons in NRPc and dorsal NRG have horizontal preferred directions and appear to innervate horizontal eye and head movers, while those in ventral NRG and in the midbrain have vertical preferred directions and appear to innervate vertical eye and head movers. Behaviorally identified RS-LLBNs have mainly been studied using intra-axonal recording and subsequent injection with horseradish peroxidase.

Cats

The best studied RS-LLBNs are "eye-neck neurons," which have their somata in the NRPc rostral and/or ventral to the abducens nucleus [3]. Collaterals of the descending axon arborize in the abducens nucleus, the prepositus nucleus, the medial vestibular nucleus (all related to horizontal eye movements), the facial nucleus (mediating pinna movement), the dorsal NRG and the nucleus reticularis ventralis (containing reticulospinal neurons), and paramedian cell groups projecting to the cerebellum (Fig. 1) [3].

During attempted gaze shifts in head-fixed cats having electromyographic (EMG) electrodes implanted in the neck muscles, a typical eye-neck neuron exhibits a burst of spikes that begins 66–132 ms before the saccade (hence the long-lead designation), peaks during the saccade, and slowly decays to an end after saccade end but roughly coinciding with the end of the phasic component of neck EMG activity [6]. Eye-neck neurons also have a moderate firing rate during sustained EMG activity when the head is held eccentrically. Phasic and sustained activity increases with increasing eye movement towards the ►ipsiversive side and with increasing EMG activity in the ipsilateral neck muscles, but is not perfectly correlated with either alone. These discharges and the axonal arborization of these neurons both suggest a role in the coordinated activation of eye and neck muscles.

A second group of more caudally located RS-LLBNs has similar discharges but with very little sustained activity when the head is stationary [4]. Somata are

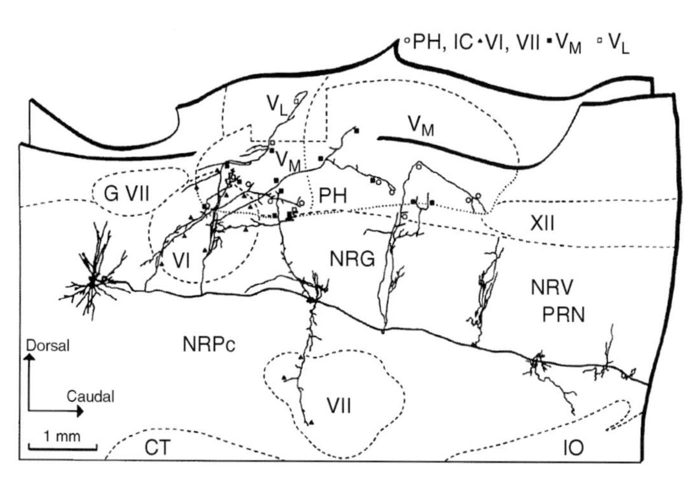

Reticulospinal Long-Lead Burst Neurons. Figure 1 Camera lucida drawing of part of a cat reticulospinal neuron drawn in a parasagittal plane. The large soma is located in the PPRF just rostral to the abducens nucleus (VI), and the descending axon gives off branches that terminate in the abducens and facial motor nuclei (VI and VII), nucleus prepositus hypoglossi (PH), nucleus intercalatus (IC), the medial and lateral vestibular nuclei (V_M and V_L), the nucleus reticularis gigantocellularis (NRG), and the nucleus reticularis ventralis (NRV). Symbols show where the branches left the plane of section, and are coded according to where the branches terminated (see Key). Other abbreviations; G VII = genu of the VIIth nerve, NRPc = nucleus reticularis pontis caudalis, PRN = paramedian reticular nucleus, XII = hypoglossal nucleus, IO = inferior olive, CT = corticospinal tract. Adapted from Grantyn et al. [3].

mostly located caudal to the abducens nucleus, but one was located rostral to the abducens. The axons do not collateralize to innervate any cranial motor nuclei, but rather innervate the paramedian reticular nucleus, interstitial cell groups in the MLF (which both project to the cerebellum), and nucleus reticularis ventralis. Preferred directions could have large vertical components, and one neuron had a contralateral preferred direction and an axon that descended in and innervated the contralateral medulla and spinal cord. The lack of projections to oculomotor structures, the projection to the spinal cord, and the phasic burst all imply these neurons function as head burst-neurons.

A third group of RS-LLBNs have their somata in the H-fields of Forel – the cat counterpart of the monkey ►rostral interstitial nucleus of the MLF. Called "augmenting neurons" by some, these neurons have a long build up beginning about 200 ms before upward gaze movements. The burst lasts about as long as the head movement in head-free alert cats and has a peak rate that is proportional to head velocity [7]. They also have a low-rate spontaneous discharge that is independent of eye position, and thus are only borderline LLBNs. Projections of augmenting neurons were studied electrophysiologically. Besides projecting to the spinal cord, augmenting neurons project to the midbrain cuneiform nucleus (containing saccade-related neurons) [8] and strongly to reticulospinal neurons both rostral and caudal to the abducens nucleus (NRPc and NRG) [5]. There are more typical LLBNs located at the caudal border of the H fields, but they do not have projections to the spinal cord.

Monkey

Very little research has been devoted to reticulospinal neurons in monkeys. LLBNs have been recorded in NRPc that fire throughout the duration of the head movement in gaze saccades, but their potential spinal projections were not explored. Two groups of confirmed RS-LLBNs have been found in head-stabilized monkeys, but there are surely more than this.

Two RS-LLBNs with their somata in NRPc anterior and ventral to the abducens nucleus had preferred directions that were ipsiversive for one and down for the other [9]. Burst leads averaged 17 and 42 ms, and the number of spikes in the burst increased weakly with saccade size. As the monkeys had their heads stabilized, burst parameters could not be analyzed in relation to head movements. Axons traveled outside the paramedian tracts on their way to the spinal cord, and issued collaterals that innervated NRG, the middle of the prepositus nucleus, the paramedian reticular nuclei (both of which project to the cerebellum), and to the nucleus reticularis ventralis (Fig. 2). With the exception of the soma locations, these RS-LLBNs are similar to the head burst-neurons in the cat (group 2, above).

Three RS-LLBNs had their somata just lateral to the interstitial nucleus of Cajal [10]. All discharged preferentially for upward saccades beginning 25–105 before saccade onset, and the number of spikes increased with increasing saccade amplitude. The axons descended in the MLF and issued no collaterals until the caudal pons. They subsequently innervated raphe pontis, raphe obscurus, and the paramedian reticular nucleus, all of which project to the cerebellum, including oculomotor-related areas

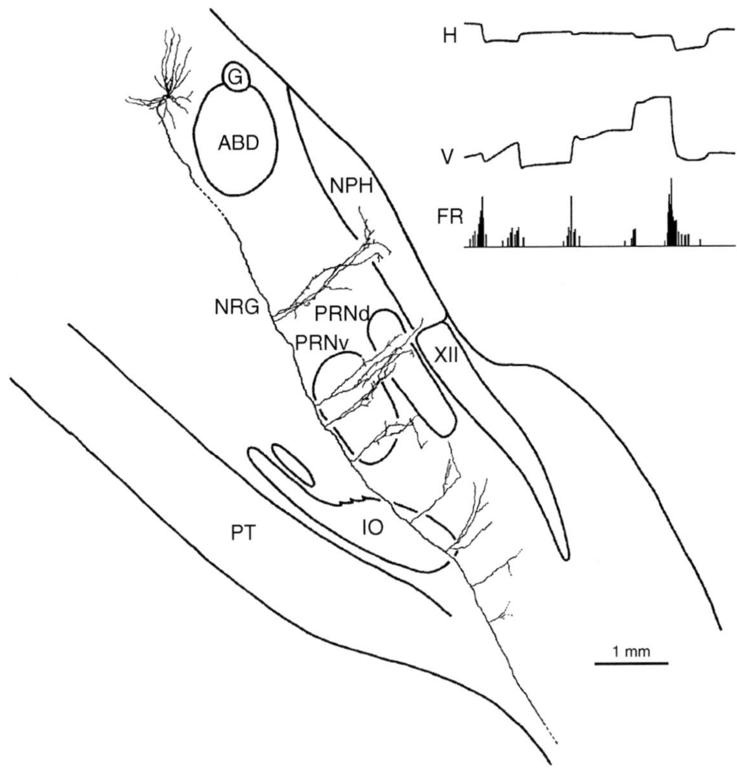

Reticulospinal Long-Lead Burst Neurons. Figure 2 Camera lucida drawing of a monkey reticulospinal neuron drawn in a parasagittal plane. The large soma is located in the PPRF just rostral to the abducens nucleus (ABD), and the descending axon gives off dorsally-coursing branches that terminate in the medullary reticular formation, the nucleus prepositus hypoglossi (NPH), and the dorsal and ventral paramedian reticular nucleus (PRNd and PRNv). Staining faded as the main axon entered the spinal cord (dotted line). Inset shows the firing rate (FR) of the neuron during saccades, shown as a horizontal component (H) and vertical component (V). Other abbreviations; G = genu of the VIIth nerve, NRG = nucleus reticularis gigantocellularis, XII = hypoglossal nucleus, IO = inferior olive, PT = pyramidal tract. Adapted from Scudder et al. [9].

in the first two cases. Axons additionally innervated ►raphe interpositus and the dorsal NRG, both containing neurons that are part of the saccadic burst generator, and more ventral parts of NRG containing reticulospinal neurons.

Both types of RS-LLBNs have heavy projections to the cerebellum via relay nuclei. These projections are perhaps especially appropriate in the monkey relative to the cat because primates are credited with more flexible eye-head strategies, and could rely on the cerebellum to insure the different combinations of eye and head movements end with gaze directed at the target.

Lower level Processes

The innervation of the spinal cord has been studied in detail for cat reticulospinal neurons, but not for RS-LLBNs specifically. This is because staining of the intra-axonally filled RS-LLBNs described above always faded before reaching their terminations in the cord. Doubtlessly, RS-LLBNs share some features of the more general innervation. Reticulospinal neurons with somata in NRPc and NRG and receiving input from the superior colliculus project to upper cervical segments, including interneuron and motoneuron pools for the neck (Rexed laminae VII, VIII, IX). Motoneurons that are monosynaptically contacted by these reticulospinal neurons include those innervating the multisegmental dorsal muscles, splenius (a horizontal neck rotator) and biventer cervicis and complexus (a vertical neck rotator). Individual neurons in NRG innervate either splenius or biventer/complexus, but poorly innervate the other. Some NRPc and NRG reticulospinal neurons project to interneurons in the lower cervical segments (innervating forelimbs) and possibly lumbar segments (innervating hindlimbs). Activity in fore- and hind-limb muscles has been observed during orienting movements, but these could be a postural reaction as much as a direct product of orienting behavior. Reticulospinal neurons with somata in the H fields and receiving input from the superior colliculus make monosynaptic connections with biventer cervicis and complexus motoneurons, but few with splenius, confirming the specificity of the H fields for vertical movements [1,5].

Pathology

Excitotoxic destruction of the somata of neurons in the cat NRPc and NRG on one side produced severe to moderate deficits in producing ipsiversive gaze saccades, depending on the extent of the lesion. Vertical gaze saccades were also impaired when the head was oriented to the ipsilesional side. The loss of eye-saccades is presumably due to the destruction of the cells of the ►brainstem burst generator, and the loss of head-saccades is presumably due to destruction of the RS-LLBNs. ►Excitotoxic lesions of the H-fields in cats produced debilitating asymmetries in neck muscle tone when done unilaterally, and major deficits in vertical head saccades when done bilaterally. Lesions in humans produced by tumors or infarcts would likely result in more severe deficits due to the destruction of the many fibers of passage as well as destruction of reticulospinal neurons.

References

1. Isa T, Sasaki S (2002) Brainstem control of head movements during orienting; organization of the premotor circuits. Prog Neurobiol 66:205–241
2. Peterson BW, Fukushima K (1982) The reticulospinal mechanism and its role in generating vestibular and visuomotor reflexes. In: Sjolund B, Bjorklund A (eds) Brain stem control of spinal mechanisms. Elsevier, New York, pp 225–251
3. Grantyn AA, Ong-Meang Jacques V, Berthoz A (1987) Reticulo-spinal neurons participating in the control of synergic eye and head movements during orienting in the cat II. Morphological properties as revealed by introaxonal injections of horseradish peroxidase. Exp Brain Res 66:355–377
4. Grantyn AA, Hardy O, Olivier E, Gourdon A (1992) Relationships between task-related discharge patterns and axon morphology of brainstem projection neurons involved in orienting eye and head movements. In: Shimazu H, Shinoda Y (eds) Vestibular and brain stem control of eye, head, and body movements. Karger, New York, pp 255–273
5. Isa T, Sasaki S (1992) Mono- and disynaptic pathways from Forel's field H to dorsal neck motoneurones in the cat. Exp Brain Res 88:580–593
6. Grantyn AA, Berthoz A (1987) Reticulo-spinal neurons participating in the control of synergic eye and head movements during orienting in the cat I. Behavioral properties. Exp Brain Res 66:339–354
7. Isa T, Naito K (1994) Activity of neurons in Forel's field H during orienting head movements in alert head-free cats. Exp Brain Res 100:187–199
8. Isa T, Itouji T (1992) Axonal trajectories of single Forel's field H neurones in the mesencephalon, pons and medulla oblongata in the cat. Exp Brain Res 89:484–495
9. Scudder CA, Moschovakis AK, Karabelas AB, Highstein SM (1996) Anatomy and physiology of saccadic long-lead burst neurons recorded in the alert squirrel monkey. II. Pontine neurons. J Neurophysiol 76:353–370
10. Scudder CA, Moschovakis AK, Karabelas AB, Highstein SM (1996) Anatomy and physiology of saccadic long-lead burst neurons recorded in the alert squirrel monkey. I. Descending projections from the mesencephalon. J Neurophysiol 76:332–352

Reticulospinal Neurons in Eye Movements

Definition

Neurons of the brainstem reticular formation that project to the spinal cord. Anatomical studies have demonstrated that, on their way to the spinal premotor centers, axons of resticulospinal neurons (RSNs) emit numerous collaterals to various oculomotor and vestibular centers.

Together with some superior colliculus (SC) output neurons showing a similar pattern of projections, those RSNs that receive input from the SC are thus ideally positioned to decompose the collicular "desired gaze displacement" signal into motor commands for the eye and head platforms.

►Eye-Head Coordination
►Superior Colliculus

Reticulospinal Tract

Synonyms

Tractus reticulospinal ant; Anterior reticulospinal tract

Definition

The medial reticulospinal tract begins in the caudal pontine reticular nucleus and in the caudal portion of the oral pontine reticular nucleus. It descends in the medial longitudinal fasciculus in the spinal cord. Its fibers terminate mostly in lamina VII and VIII of the spinal gray matter; but they also run in lamina IX in which the motoneurons for the trunk musculature lie.

►Pathways

Reticulotectal Long-Lead Burst Neurons

ADONIS MOSCHOVAKIS
Institute of Applied and Computational Mathematics, and Department of Basic Sciences, University of Crete, Heraklion, Crete, Greece

Synonyms

RTLLBs

Definition

Saccade related long-lead burst neurons of the primate mesencephalic reticular formation (►cMRF) that project to the SC.

Characteristics

Higher Order Structure

Although they belong to the reticular formation, due to their location and projections RTLLBs can be thought to belong to a satellite system of the SC.

Parts of This Structure

Figure 1 illustrates the salient morphological and physiological features of RTLLBs following the intracellular study of their discharge in alert squirrel monkeys, their subsequent injection with a tracer and the camera lucida reconstruction of their axons in the frontal plane [1]. RTLLB somata (Fig. 1b, arrow) are located in an area receiving strong input from saccade related neurons of the SC (see TLLBs). The territory occupied by RTLLB somata probably receive input from additional classes of SC neurons, as indicated by axonal reconstructions of single X neurons of the cat [2,3] and the monkey [4,1]. The axonal terminations of RTLLBs are contained within the intermediate and deeper layers of the SC, in both sides of the brain (Fig. 1b).

Functions of the Structure

RTLLBs emit bursts of discharge which precede contraversive saccades, whether upward, downward or horizontal (Fig. 1a) by about 20 ms on average. They often do not discharge for saccades in the opposite direction (Fig. 1a, right-most example). Although the on-direction of many RTLLBs is roughly horizontal, the existence of RTLLBs with vertical and oblique on-directions has been documented [1]. The activity of RTLLBs provides a good estimate of the metrics of impending saccades; the correlation coefficient between the number of spikes in the burst, and the amplitude of saccades in their on-direction can be as high as 0.85.

Higher Order Function

RTLLBs could belong to a highly conserved satellite system of the SC as cells with quite similar morphology have been encountered in lower phyla. Neurons with bilateral projections largely confined to the optic tectum have been found in the nucleus lateralis profundus mesencephali of the snake [5] and the dorsolateral tegmental nucleus in fish [6]. Also, saccade related signals were discovered in the intertectal commissure of fish [7] before the existence of saccade related signals in the superior colliculus of mammals became known. Given their discharge pattern and projections, RTLLBs are eminently qualified to supply the SC with an efference copy signal

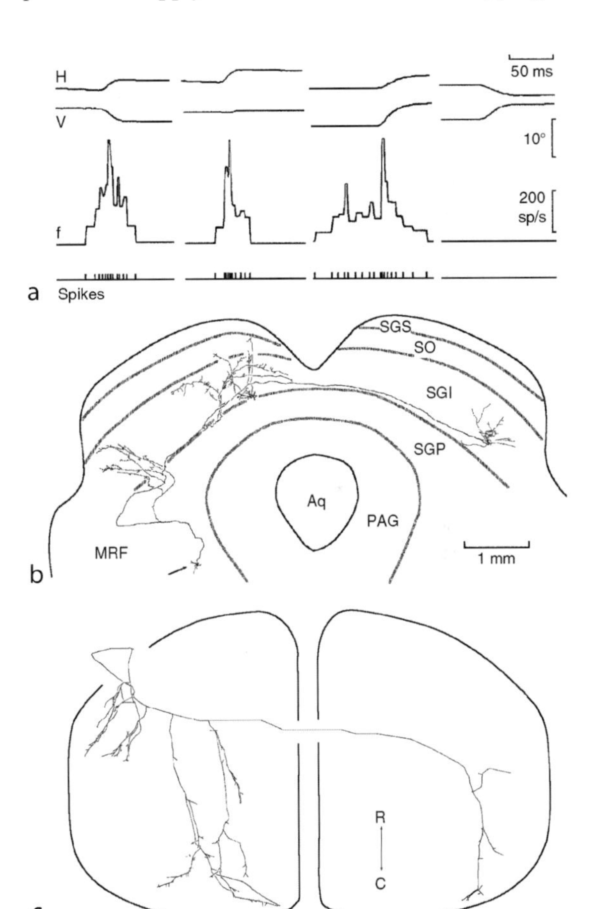

Reticulotectal Long-Lead Burst Neurons.
Figure 1 Salient morphological and physiological features of RTLLBs (reprinted from [1], with permission). (a) Saccade related discharge pattern for one RTLLB. (b) Frontal reconstruction of the axonal system of the same neuron from serial sections. (c) The axonal system of the same neuron as it appears when looking down upon the surface of the SC. Scale in (b) applies to both (b) and (c). Abbreviations: *Aq*, aqueduct; *C*, caudal; *MRF*, mesencephalic reticular formation; *PAG*, periaqueductal grey; *R*, rostral; *SGI*, stratum griseum intermediale; *SGP*, stratum griseum profundum; *SGS*, stratum griseum superficiale; *SO*, stratum opticum.

R

indicative of the metrics of ongoing saccades. Together with the visual input carried by L neurons (see ►SC – interlayer neurons) RTLLBs endow the SC with the machinery needed to implement the Vector Subtraction hypothesis (see ►Foveation hypothesis).

Quantitative Measure for This Structure

Besides detailed quantitative descriptions of the pattern of their discharge, the 3D spatial distribution of the terminals deployed in the SC by single functionally identified RTLLBs has been described in squirrel monkeys [8]. When overlaid on a horizontal map of the SC, they can occupy a considerable portion of the rostrocaudal and mediolateral extent of its deeper layers (Fig. 1c).

References

1. Moschovakis AK, Karabelas AB, Highstein SM (1988) Structure-function relationships in the primate superior colliculus. II. Morphological identity of presaccadic neurons. J Neurophysiol 60:263
2. Grantyn A, Grantyn R (1982) Axonal patterns and sites of termination of cat superior colliculus neurons projecting to the tecto-bulbo-spinal tract. Exp Brain Res 46:243
3. Moschovakis AK, Karabelas AB (1985) Observations on the somatodendritic morphology and axonal trajectory of intracellularly HRP-labeled efferent neurons located in the deeper layers of the superior colliculus of the cat. J Comp Neurol 239:276
4. Moschovakis AK, Karabelas AB, Highstein SM (1988) Structure-function relationships in the primate superior colliculus. I. Morphological classification of efferent neurons. J Neurophysiol 60:232
5. Dacey DM, Ulinski PS (1986) Optic tectum of the eastern garter snake, *Thamnophis sirtalis*. V. Morphology of brainstem afferents and general discussion. J Comp Neurol 245:423
6. Niida A, Ohono T (1984) An extensive projection of fish dorsolateral tegmental cells to the optic tectum revealed by intra-axonal dye marking. Neurosci Lett 48:261
7. Johnstone JR, Mark RF (1969) Evidence for efference copy for eye movements in fish. Comp Biochem Physiol 30:931
8. Bozis A, Moschovakis AK (1998) Neural network simulations of the primate oculomotor system. III. A one-dimensional one-directional model of the superior colliculus. Biol Cybern 79:215

Retina

Definition

The light-sensitive layered neural tissue that lines the posterior hemisphere of the eye and contains the photoreceptors (rods and cones) and the initial processing machinery for the primary visual pathways. It is a highly organized structure whose function is to capture, process, and transmit visual images to the brain. The signals generated by photoreceptors are then processed by other neurons in the retina before being transmitted to the brain as trains of action potentials by the axons of the retinal ganglion cells. Additionally, the retina serves to detect changes in ambient levels of light to regulate a multitude of non-visual photoresponses such as the pupillary light reflex.

►Pupillary Light Reflex
►Photoreceptors
►Retinal Ganglion Cells
►Vision

Retinal Bipolar Cells

Ulrike Grünert[1,2]
[1]The National Vision Research Institute of Australia
[2]Department of Optometry & Vision Sciences, The University of Melbourne, Australia

Synonyms

Retinal bipolar neurons

Definition

Bipolar cells are interneurons in the ►retina (►Vision), which transfer visual information from photoreceptors (rods and cones; ►Photoreceptors) to amacrine (►Retinal direction selectivity: Role of starburst amacrine cells) and ganglion cells (►Retinal ganglion cells). Bipolar cells consist of multiple (9–12) subtypes that differ in their morphology, synaptic connectivity, and response properties. Different types of bipolar cells process different visual modalities in parallel pathways. The following article describes the structure, distribution, synaptic connectivity, and function of bipolar cells in the mammalian retina.

Characteristics

Quantitative Description

Morphology of Bipolar Cells

The name "bipolar cell" is derived from its morphology. Bipolar cells have a cell body in the ►inner nuclear layer from which a primary dendrite extends into the ►outer plexiform layer and an axon extends into the ►inner plexiform layer (►Vision). Morphologically two major types, cone bipolar and rod bipolar cells, can be distinguished with respect to their connections with photoreceptors (►Photoreceptors). The dendrites of cone bipolar cells contact cone photoreceptors almost

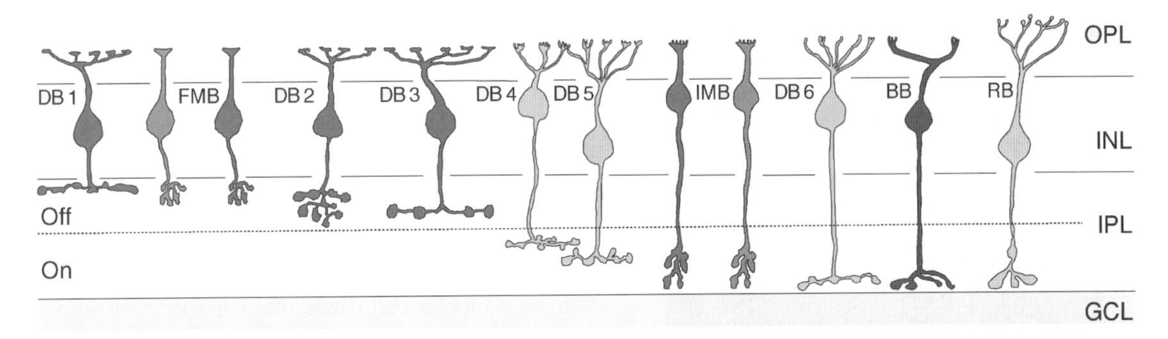

Retinal Bipolar Cells. Figure 1 Bipolar cell types in primate retina as analysed from Golgi-impregnated macaque retina. The axon terminals stratify at different levels of the inner plexiform layer (IPL). Diffuse bipolar cells (DB1 – DB6) non-selectively contact multiple cones in the outer plexiform layer (OPL). There are two types of midget bipolars, flat midget bipolar (FMB or OFF midget) and invaginating bipolar cells (IMB or ON midget). Both types contact single M- or L-cones and carry a chromatic signal. Blue cone bipolar (BB) cells contact S-cones selectively and carry an S-cone ON signal. Rod bipolar cells contact rod spherules and transfer scotopic signals.

exclusively and are thus involved in high-acuity daytime vision and colour vision (►Color processing). The dendrites of rod bipolar cells contact rod photoreceptors and are involved in night or ►scotopic vision.

Bipolar cells make up about 40% of all retinal neurons (apart from photoreceptors) and are the most numerous interneurons in the retina [1]. Each bipolar type is found across the retina and forms a regular mosaic.

Diffuse Bipolar Cells

Cone bipolar cells are subdivided into OFF and ON bipolar cells [2]. The OFF bipolar cells hyperpolarise in response to light, whereas the ON bipolar cells depolarise. The axons of OFF bipolar cells stratify in the *outer* half of the inner plexiform layer. The axons of ON bipolar cells stratify in the *inner* half of the inner plexiform layer.

The OFF and ON cone bipolar types are further subdivided into at least nine subtypes with respect to their stratification in the inner plexiform layer and their connections with cones (Fig. 1) [1–6].

Most cone bipolar types contact five to ten cones, and have thus been named *diffuse bipolar* (DB) cells [3]. Some of these subtypes can be selectively labelled with immunohistochemical methods. An example of an immunohistochemically labelled ON bipolar type (DB6) from macaque retina is shown in Fig. 2.

Figure 3 shows the mosaic formed by DB6 cells in whole mount view. The fine dendrites form a dense meshwork across the outer plexiform layer. The somata are located in the outer half of the inner nuclear layer, and the axons tile the retina in a regular mosaic.

In recent years, a number of studies have estimated the density of bipolar cell types using a variety of methods including electron microscopy, Golgi-impregnation, immunohistochemistry, intracellular injection and photofilling [3–5]. The major conclusions from these quantitative studies are that all types have comparable densities

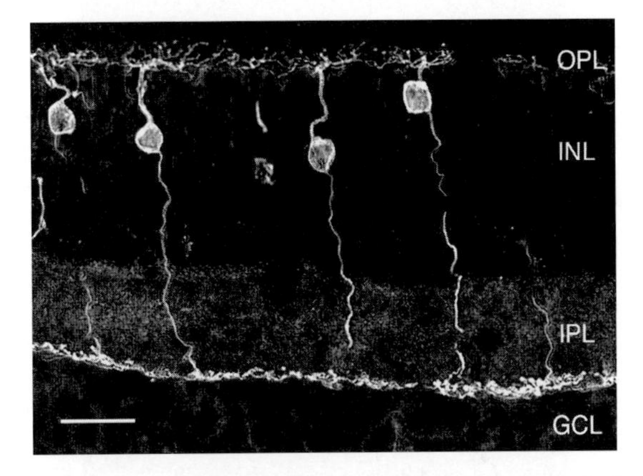

Retinal Bipolar Cells. Figure 2 Diffuse bipolar cells in primate retina. DB6 cells (Fig. 1) are immunohistochemically labelled. Scale bar: 25 μm.

ranging between 1,000 and 3,000 cells/mm^2, and thus no type of bipolar cell usually predominates. The only exception to this rule is the midget bipolar cell in primate fovea (see below). The density of cone bipolar cells is usually lower than the cone density. However, since most bipolar cell types contact multiple cones, it is assumed that all cones provide input to all types of cone bipolar cells [1,3,6].

Bipolar Cell Types Involved with Colour Vision

Most mammals possess two types of cones, one that is maximally sensitive to short wavelengths of the visible spectrum (S or blue cone), and one that is maximally sensitive to long wavelengths [7]. Mammals with two cone types have dichromatic colour vision. Trichromatic colour vision is based on the presence of a third cone type that is maximally sensitive to medium wavelengths. Among placental mammals, trichromatic colour vision has only been described in primates (►Color processing) [7].

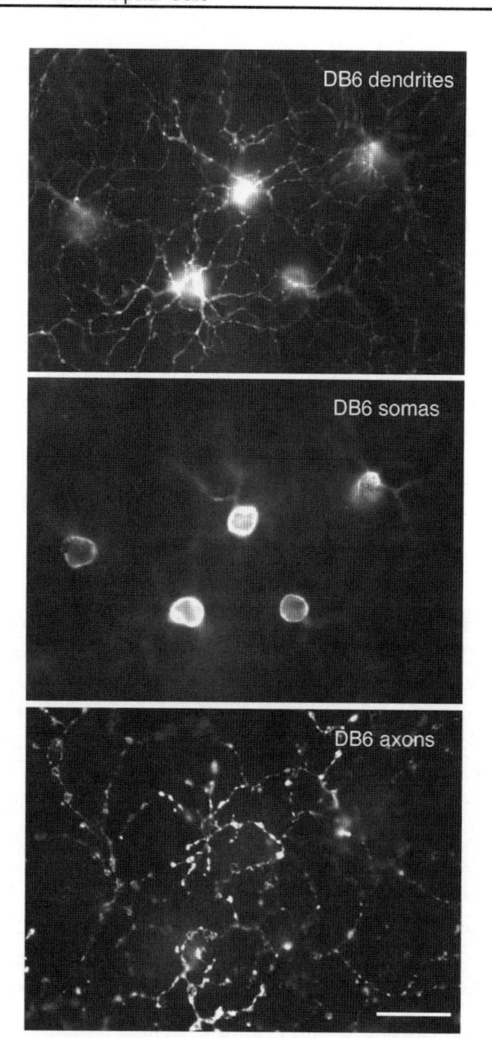

Retinal Bipolar Cells. Figure 3 Diffuse bipolar cells in primate retina. Whole mount view of immunohistochemically labelled DB6 cells. The cells are shown at different focal levels. Scale bar: 20 μm.

Three types of cone bipolar cells are involved with colour vision: blue cone bipolar cells, and two types of midget bipolar cells. The blue cone bipolar cells (BB, Fig. 1) are ON cells and receive input exclusively from S-cones. The axons of blue cone bipolar cells stratify close to the ganglion cell layer where they provide output to a special type of ganglion cell, the blue ON/yellow OFF ganglion cell [1,3,4,6,8].

Each blue cone bipolar cell receives input from between one and five S-cones (convergence), and each S cone contacts between one and five blue cone bipolar cells (divergence). The maximal density of blue cone bipolar cells is approximately 800 cells/mm^2, thus they are among the least numerous bipolar types in the retina.

In mammals, midget bipolar cells [1,3,4,6,9] are probably unique to primate retina where they are thought to play a role in trichromatic colour vision. Midget bipolar cells can be subdivided into ON and OFF types. The ON midget bipolar dendrites make ►invaginating synapses with cones, and have thus been named invaginating midget bipolar (IMB, Figure 1) cells. The OFF midget bipolar dendrites make flat synapses with cones, and are thus called flat midget bipolar (FMB, Fig. 1) cells. In the central retina, each midget bipolar cell receives input from a single cone and in turn contacts one ON or one OFF midget (parvocellular projecting) ganglion cell. Thus, each midget bipolar cell carries the chromatic signal of the cone type it contacts to the inner plexiform layer.

In central retina, midget bipolar cells are the most numerous cone bipolar type. The density of FMB cells was estimated in macaque retina. In central retina, their density follows the cone density (> 10,000 cells/mm^2). In peripheral retina, FMB cells contact more than one cone, and thus their density (< 5000 cells/mm^2) drops below the cone density [9].

Rod Bipolar Cells

In all mammalian retinae only one type of rod bipolar cell studied to date has been described (RB, Fig. 1). Each rod bipolar cell contacts between ~20 and 100 rod ►spherules (►Retinal ribbon synapses) in the outer plexiform layer. Rod bipolar cells are depolarised by light, and thus are ON cells [2]. The axon terminals of rod bipolar cells are located at the border of the inner plexiform layer with the ganglion cell layer. The axon terminals of rod bipolar cells tile the retina in a non-overlapping mosaic (Fig. 4).

Rod bipolar axons provide output to AII amacrine cells, which then feed the rod signal into cone pathways [2,5,6].

The density of rod bipolar cells varies depending on the retinal location and between species. For example, in cat retina the density ranges between ~20,000 and 46,000 cells/mm^2, whereas in rabbit the density ranges between ~2,000 and 5,000 cells/mm^2. In macaque retina, rod bipolar cells are absent from the fovea, and the maximal density is ~15,000 cells/mm^2 at about 1–3 mm distance from the fovea. Thus, the density of rod bipolar cells is higher than that of individual cone bipolar cell types, but cone bipolar cells outnumber rod bipolar cells in total [1].

The ratio between rods and rod bipolar cells (numerical convergence) varies between species. It is 10:1 in central macaque retina, 15:1 in the *area centralis* of cat retina, and 50:1 in rabbit retina. However, the divergence between rods and rod bipolar cells is relatively constant. For all species studied, and at all eccentricities, between one and four rod bipolar cells are postsynaptic to each rod.

Description of the Structure

Synaptic Connectivity in the Outer Plexiform Layer

The synaptic terminal of a rod photoreceptor (rod spherule) is a relatively simple structure. It contains one or two synaptic ribbons (►Retinal ribbon synapses), which are presynaptic to usually two rod bipolar and two horizontal cell processes at a single synaptic invagination. The rod bipolar cell dendrites form the central elements at this synapse [5].

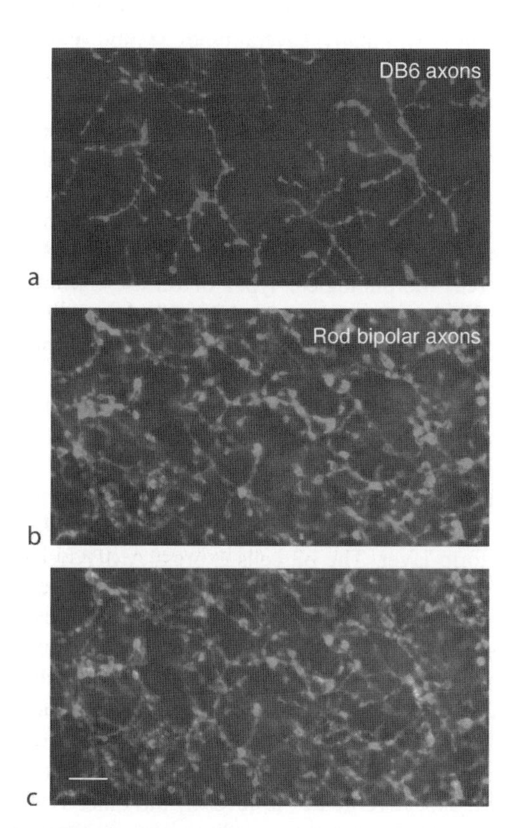

Retinal Bipolar Cells. Figure 4 Whole mount view of the axonal terminals of two types of ON bipolar cells in primate retina (courtesy of Patricia Jusuf, NVRI, Melbourne). The two types of cell (rod bipolar and DB6) form separate mosaics but might share some postsynaptic cell types. Scale bar: 10 μm.

The synaptic terminal of a cone photoreceptor (cone ►pedicle; ►Retinal ribbon synapses) is a much more complex synapse (Fig. 5a).

It comprises of between 20 and 50 invaginating and several hundred flat contacts. The invaginating contacts consist of a presynaptic ribbon, one or two central elements deriving from ON cone bipolar cells, and two lateral processes deriving from horizontal cells (triad). The ►flat contacts are located at the base of the cone pedicle and derive mostly from OFF bipolar cells. In total, each cone pedicle makes about 500 contacts. The number of postsynaptic cells is lower as some cells receive multiple contacts. About 10–15 individual cone bipolar cells, comprising of a variety of different types, are postsynaptic to a cone pedicle [3]. Thus, at the cone pedicle, the first synapse in the retina, the light signal is distributed into multiple pathways [1,6,10].

Until recently, all invaginating bipolar processes were thought to belong to ON bipolar cells, whereas all flat contacts were thought to belong to OFF bipolar cells. However, in primate this rule applies strictly only to ON bipolar and OFF midget bipolar cells. Most OFF bipolar types make a mixture of both types ofcontact [3]. It is now known that it is the type of ►glutamate receptor expressed at bipolar dendrites that determines whether they are ON or OFF types.

In addition to the layers of invaginating and flat contacts, a third postsynaptic layer has been detected at a distance of about 1.5 μm from the cone pedicle base. This layer consists of junctions that have the appearance of desmosomes ("desmosome-like junctions") but do not contain any known protein normally present at junctions. Instead, these structures are postsynaptic densities that are located on horizontal cell processes and express ionotropic glutamate receptor subunits [10].

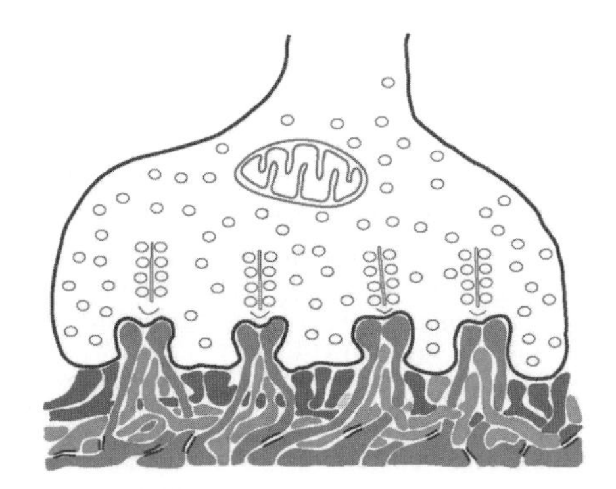

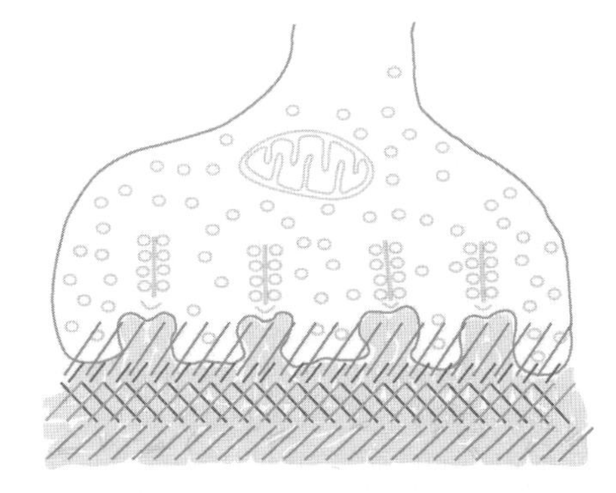

Retinal Bipolar Cells. Figure 5 Schematic drawing of a cone pedicle in macaque monkey retina. (a) Four presynaptic ribbons and four triads are shown. Invaginating dendrites of horizontal cells (red) form the lateral elements, invaginating dendrites of ON bipolar cells (green) form the central elements of the triads. Flat contacts (blue) are mainly made by OFF bipolar cells. Desmosome-like junctions (black bars) are located at a distance of about 1.5 μm underneath the pedicle. (b) The same pedicle with the laminated expression of postsynaptic glutamate (red and blue) and GABA receptors (blue and green) is shown.

Glutamate and GABA Receptors in the Outer Plexiform Layer

The neurons postsynaptic to photoreceptors express different types of glutamate receptors [6]. The ON bipolar cells express sign inverting metabotropic glutamate receptors (mGluR6) that act via G-proteins. Horizontal cells and OFF bipolar cells express sign conserving ionotropic (►AMPA, α-amino-3-hydroxy-5-methyl-4-isoxazole propionic acid and ►kainate) glutamate receptors.

The ►neurotransmitter of horizontal cells is ►GABA (gamma amino butyric acid). $GABA_A$ and $GABA_c$ receptors are consistently found on the dendrites of bipolar cells. The presence of different types of postsynaptic receptors on different processes at the cone pedicle base results in a laminated arrangement (Fig. 5b) [10].

Connectivity of Bipolar Cells in the Inner Plexiform Layer

In the inner plexiform layer, bipolar axon terminals form ribbon synapses (►Retinal ribbon synapses) onto two postsynaptic processes (►dyads). Cone bipolar axons usually contact one amacrine and one ganglion cell process, whereas rod bipolar axons do not contact ganglion cells but form dyads onto two amacrine processes, one of which belongs to the ►AII amacrine cell [2]. Different types of diffuse bipolar cells form different numbers of output synapses [5], but relatively little is known about the identity of their postsynaptic targets.

Bipolar axons make excitatory (sign-conserving) glutamatergic synapses. The two postsynaptic processes at a bipolar dyad usually express two different types of ionotropic glutamate receptors (AMPA, kainate and ►NMDA) [6].

The synaptic input to bipolar cells in the inner plexiform layer derives from amacrine cells containing the inhibitory neurotransmitters GABA or ►glycine. The axon terminals of different bipolar types vary with respect to the expression, subunit composition and frequency of $GABA_A$, $GABA_C$ and glycine receptor clusters [1,6].

Higher Level Structures

OFF bipolar cells transfer their signals to OFF ganglion cells, whereas ON bipolar cells contact ON ganglion cells (►Retinal ganglion cells). The ganglion cells then transfer the bipolar signals to distinct regions in the brain. As bipolar axons stratify in distinct strata of the inner plexiform layer, they can only contact ganglion cells stratifying in the same stratum [1]. Some bipolar types provide input to only one type of ganglion cell (e.g., midget bipolar to midget ganglion cells in primate; CD15-OFF bipolar cells to ON-OFF direction-selective ganglion cells in rabbit (►Retinal ganglion cells; ►Retinal direction selectivity: Role of starburst amacrine cells) [4]. Some ganglion cell types receive input from only one bipolar type (e.g., midget ganglion cells). Other bipolar cell types contact more than one type of ganglion cell, e.g., DB3 cells in primate retina presumably contact OFF parasol (►magnocellular pathway) and the outer (OFF) tier of small bistratified cells (►koniocellular pathway) [3]. Finally, several types of bipolar cells can provide input to the same type of ganglion cell, e.g., cat alpha and beta cells, primate parasol cells and direction-selective cells in rabbit retina (►Retinal ganglion cells; ►Retinal direction selectivity: Role of starburst amacrine cells) [1].

Function

Different morphological types of diffuse bipolar cells play different functional roles [6]. Physiological differences could be based on the presence of different types of glutamate receptors on bipolar dendrites, as well as distinct patterns of inhibitory input from horizontal and amacrine cells. For example, the type b2 and b3 OFF bipolar cells in the ground squirrel stratify at different levels of the inner plexiform layer and express distinct types of ionotropic glutamate receptors in the outer plexiform layer. The b2 cells express AMPA receptors and are involved in the transmission of transient responses to light. The b3 cells express kainate receptors and are involved with sustained light response. Studies in tiger salamander showed that different types of bipolar cells, with axon terminals at different levels of the inner plexiform layer, differ with respect to their light response characteristics. The idea that parallel streams in the visual system first diverge at the level of the outer flexiform layer is thus supported in all retinas studied so far (►Visual processing streams in primates). The question why the brain requires multiple afferent signals streams, and how these distinct streams are processed in the brain, remains a major challenge for visual neuroscience.

References

1. Masland RH (2001) The fundamental plan of the retina. Nat Neurosci 4:877–886
2. Nelson R, Kolb H (2003) ON and OFF pathways in the vertebrate retina and visual system. In: Chalupa LM, Werner JS (eds) The Visual Neurosciences. The MIT Press, Cambridge, pp 260–278
3. Boycott B, Wässle H (1999) Parallel processing in the mammalian retina. The Proctor lecture. Investig Ophthalmol Vis Sci 40:1313–1327
4. Masland RH (2001) Neuronal diversity in the retina. Curr Opin Neurobiol 11:431–436
5. Sterling P (2003) How retinal circuits optimize the transfer of visual information. In: Chalupa LM, Werner JS (eds) The Visual Neurosciences. M.I.T. Press, Cambridge, pp 243–268
6. Wässle H (2004) Parallel processing in the mammalian retina. Nat Rev Neurosci 5:747–757
7. Jacobs GH (1993) The distribution and nature of colour vision among the mammals. Biol Rev 68:413–471

8. Dacey DM, Lee BB (1994) The ‘blue-on’ opponent pathway in primate retina originates from a distinct bistratified ganglion cell type. Nature 367:731–735
9. Wässle H, Grünert U, Martin PR, Boycott BB (1994) Immunocytochemical characterization and spatial distribution of midget bipolar cells in the macaque monkey retina. Vis Res 34:561–579
10. Haverkamp S, Grünert U, Wässle H (2000) The cone pedicle, a complex synapse in the retina. Neuron 27:85–95

Retinal Color Vision in Primates

Paul R. Martin
National Vision Research Institute of Australia, Carlton, VIC, Australia

Synonyms

Color vision; Color processing

Definition

Color Vision in the Retina

The eyes of nearly all animals contain multiple classes of cone photoreceptor (►Photoreceptors), and the ability to discriminate objects by their ►spectral reflectance (►Color processing) is an almost universal feature of animal visual systems studied so far. The neural processes that give rise to color sensations begin in the retina, where certain neurons show selectivity for distinct regions of the visible spectrum. Color is not a property of objects or of retinal processes, but is a result of the brain's ability to interpret these neural signals. Humans and other primates normally show tri-variant (►trichromacy) color vision whereas most other mammals show more rudimentary (►dichromacy or "red-green color blind") color vision.

Characteristics

Description of the Process

Encoding Spectral Signals

Most humans show trichromatic color vision. This means that the eye contains three distinct classes of cone photoreceptors that are sensitive to different wavelength ranges in the visible spectrum, and that differential activation of these photoreceptors can be analyzed by the brain to yield color sensations. The cone photoreceptors are active under ►photopic (daylight) conditions whereas ►scotopic (night) vision is served by a single class of rod photoreceptor and thus is color blind (►color blindness).

Figure 1 gives an overview of spectral signal processing in the primate retina. Objects in the environment (in this example, a fruit-bearing tree) reflect more or less strongly the incident ►photons in the visible spectrum (the visible range is approximately 400–700 nm in the electromagnetic spectrum). For example (Fig. 1), a ripe red fruit on the tree reflects more long-wavelength photons than a green leaf on the tree, but the leaf will reflect more medium-wavelength photons than the fruit does. The shapes of these reflectance curves are determined by the physical-

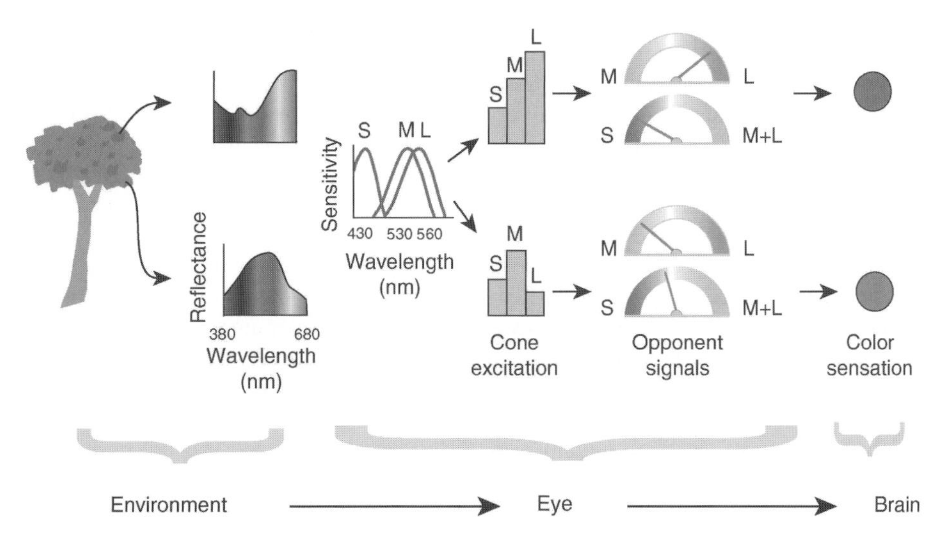

Retinal Color Vision in Primates. Figure 1 First stages of color vision. From *left* to *right*: objects in the environment show different reflectance spectra (in this example, the *upper row* shows fruit and the *lower row* shows foliage). These spectra are integrated into three separate wavelength bands by the short (S), medium (M) and long-wavelength (L) sensitive cones. The cone excitations are transformed into two cone-opponent signals (M − L) and (S − [M + L]) for transmission to the brain. The brain interprets these signals to yield color sensations.

chemical properties of the objects, and the spectrum and intensity of the light that illuminates the objects. Thus, color is not a property of the objects *per se* but is a result of the brain's ability to interpret the spectral reflectance of an object, relative to the reflectance of other objects in the ►visual field. This is the most important fact to learn about color vision and is an essential prerequisite for understanding spectral processing in the retina.

The first stage of vision is the transduction of light by photoreceptors (►Phototransduction). Each cone photoreceptor expresses one of three types of proteins called cone ►opsins, which together with the vitamin A derivative 11-cis-retinal form the only light-dependent stage of vision. The amino acid sequence of the opsin "tunes" its spectral sensitivity towards the short (peak ~430 nm), medium (~530 nm) or long (~560 nm) wavelength regions of the spectrum [1]. In other words, the probability of photon absorption is maximal at one of three constant positions in the spectrum, yet each receptor absorbs photons across a wide band of the visible spectrum (Fig. 1). Because the absorption spectra are broad, wavelength is confounded with intensity in each receptor's response. For example, a bright light at 500 nm or a dim light at 430 nm could produce identical photon absorptions by the short-wavelength sensitive (S) cones. It is only by comparing the output of different receptor classes that specific information about wavelength can be recovered. The photoreceptor responses or "cone excitations" are thus shown as the relative heights of the bars in Fig. 1, because each receptor effectively ignores the wavelength of absorbed photons across its sensitivity range. In summary, the ripe fruit on the tree yields a red sensation because it activates long wavelength sensitive (L) cones more than it activates the medium– (M) and short wavelength-sensitive (S) cones, whereas a leaf on the tree appears green because it activates the M cones more than the L or S cones. The reflectance spectra of these objects are collapsed into this trivariant, or trichromatic, signal at the first stage of color vision.

Neural Signals Underlying Color Sensations

How are the signals that enable color sensations transmitted to the brain? The retinal structures described in the following sections yield two main signals called "►cone opponent" signals (Fig. 1). This term conveys the idea that activity of one cone type is subtracted from or "opposes" the activity of another cone type. One opponent channel pits the activity of M against L cones, the other pits the activity of S cones against a combination of M and L cones. Distinct anatomical pathways in the retina form the substrate of these opponent channels. For simplicity these are often referred to as "red-green" and "blue-yellow" pathways in the retina, although the perceptual color axes do not correspond exactly to the cone-opponent axes. It is also important to note that most retinal neurons respond to changes in either brightness (intensity) or color (relative spectral reflectance) of a stimulus, so do not form exclusive color-detecting channels [2] (►Color processing).

Red-Green Pathways in the Retina

There is general agreement that the so-called "midget-parvocellular projecting" retinal pathway carries the M-L opponent signals which serve the red-green axis of color vision. This pathway forms the dominant output of the primate retina (it comprises ~80% of all ►optic nerve fibers), and carries signals serving high-acuity spatial vision in addition to signals for red-green color vision. The genes encoding M and L cone opsins diverged relatively recently in the evolutionary history of primates, yielding trichromatic color vision from the primordial dichromatic system thought to be common to most mammals [1,3]. The evolution of high-acuity spatial vision in primates may have enabled the more recent evolution of red-green color vision [4,5]. The idea is illustrated in Fig. 2. Figure 2a shows a schematic drawing of the primate eye, with the eye's output neurons (►Retinal ganglion cells) concentrated near the point of highest visual acuity (the ►fovea). The M and L cones are very tightly packed near the fovea to enable this high acuity, and each cone makes contact with two bipolar cells (►Retinal bipolar cells) of the midget-parvocellular pathway in addition to contacting other bipolar cells [6]. One of the midget bipolar cells responds to brightness decrements (off-type response) and the other responds to brightness increments (on-type response). These connections show one-to-one specificity (Fig. 2b), so the spatial acuity of the cone array can be preserved at subsequent processing stages. The spectral (M or L) signature of each cone in the array will likewise be preserved by this chain of excitatory synaptic connections, and yields four response signatures in midget-parvocellular cells: red-on, red-off, green-on and green-off. The red-green color signals thus are "piggybacked" on the system for high-acuity spatial signals.

The anatomical organization of the midget-parvocellular pathway in the peripheral retina is shown in Fig. 2c. Here, many of the midget bipolar cells make contact with two or more cones and will receive a mixed spectral signal if they contact cones of different type. The bipolar cells likewise make convergent connections with ganglion cells (Retinal ganglion cells) [2,7]. The spatial acuity and spectral purity of ganglion cell signals should thus decrease in the peripheral retina, and both spatial acuity and red-green color vision in the peripheral visual field are correspondingly poor [8].

How do the cone opponent properties of midget-parvocellular ganglion cells arise? Inhibitory interneurones (►horizontal cells and ►amacrine cells) are the most

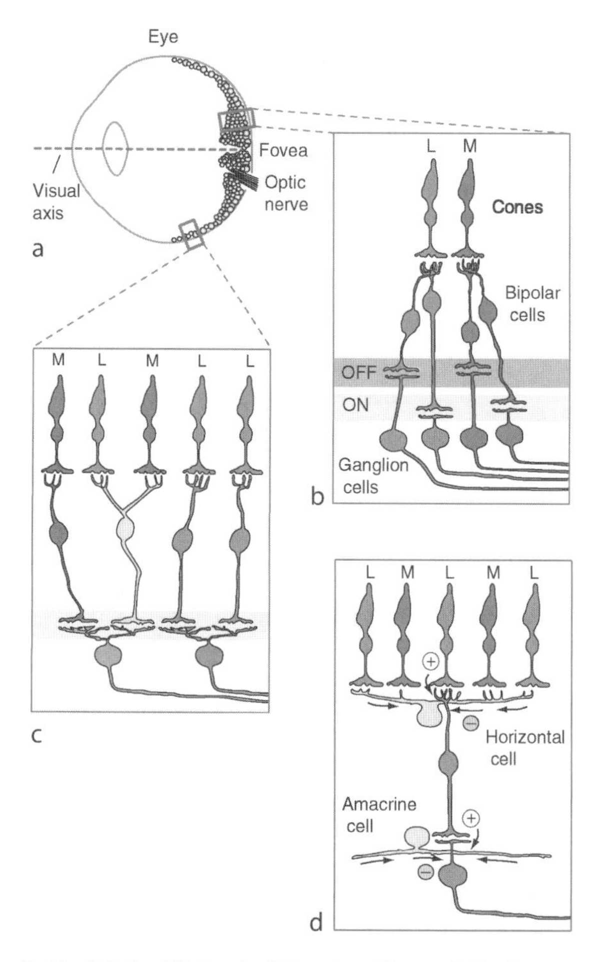

Retinal Color Vision in Primates. Figure 2 Red-green chromatic pathways in the eye. (a) schematic cross section of the eye showing concentration of ganglion cells near the fovea (F) on the visual axis (dashed red line). Ganglion cell axons form the optic nerve. (b) connections of midget-parvocellular pathway neurons near the fovea. Each cone is contacted in a one-to-one fashion by both on-type and off-type midget bipolar cells. The bipolar cells in turn contact midget ganglion cells. (c) connections of midget-parvocellular pathway neurons in mid-peripheral retina. For simplicity, only on-type connections are shown. Some midget bipolar cells receive convergent input from multiple cones, and most ganglion cells receive convergent input from multiple bipolar cells. Both these convergent steps will degrade the spectral purity of the ganglion cell response. (d) organization of inhibitory inputs to midget-parvocellular pathway ganglion cells. Horizontal cells and amacrine cells make widespread connections and feed mixed spectral signals to ganglion cells.

likely source of the opponent cone signals (Fig. 2d). One subtype of horizontal cell in primate retina (the H1 subtype) collects signals from M and L cones and provides feedback inhibition to these cones as well as to bipolar cells [2,7]. Anatomical connections of amacrine cells with midget-parvocellular ganglion cells are less well-understood, but as is the case with horizontal cells, amacrine cells should pool spatial signals originating in large numbers of M and L cones. Both these cell types thus provide a spectrally mixed inhibitory input to antagonize or "oppose" the spectrally pure (in the fovea) or biased (in the peripheral retina) excitatory inputs from midget bipolar cells.

There is functional (physiological) evidence that both excitatory and inhibitory inputs to parvocellular ganglion cells show greater spectral selectivity than predicted by the pure "random wiring" scheme outlined above, but to date there is no direct anatomical evidence for cone-selective connections in the midget-parvocellular pathway [2,7]. This functional selectivity likely arises from subtle changes in the strength of synaptic connections rather than specific wiring between specialized cell classes, of the type described below for the blue-yellow retinal pathway.

Blue-Yellow Pathways in the Retina

The short wavelength sensitive (S) cones form less than 10% of all cones in the primate retina (Fig. 3a). Molecular studies show that a distinct S cone pigment emerged before the mammalian radiation, and thus could form the basis for a primordial color vision system in mammals [1,5]. The main synaptic output of S cones in primates is to a single class of bipolar cell called the *blue cone bipolar cell.* These bipolar cells form a specific network contacting S cones (Fig. 3b) and transmit on-type signals (corresponding to increases in the ►photon catch of S cones) from the S cone array. The blue cone bipolar cells contact a specific class of ganglion cell called the small bistratified (blue-on) cell. The small bistratified cells get excitatory input from S cones and inhibitory input from M and L cones, to yield a "blue-on, yellow off" cone opponent response [9]. There are two sources of inhibition from M and L cones. First, the small bistratified cell receives synaptic input from off-type diffuse bipolar cells, which contact predominantly M and L cones. Second, the S cones and the blue-cone bipolar cells receive inhibitory input from the H2 subtype of horizontal cell, which gets synaptic input from all cone types. This spectrally mixed inhibitory input opposes the spectrally pure S cone input to blue cone bipolar cells [7]. The selective connections of S cones with blue cone bipolar cells and H2 horizontal cells are preserved across the primate retina, so the deterioration in blue-yellow chromatic sensitivity with increasing visual field eccentricity is not as marked as the deterioration in red-green sensitivity [8].

The question how off-type signals from S cones are transmitted to the brain has not been fully resolved. At least two types of sparsely-branched or "wide field"

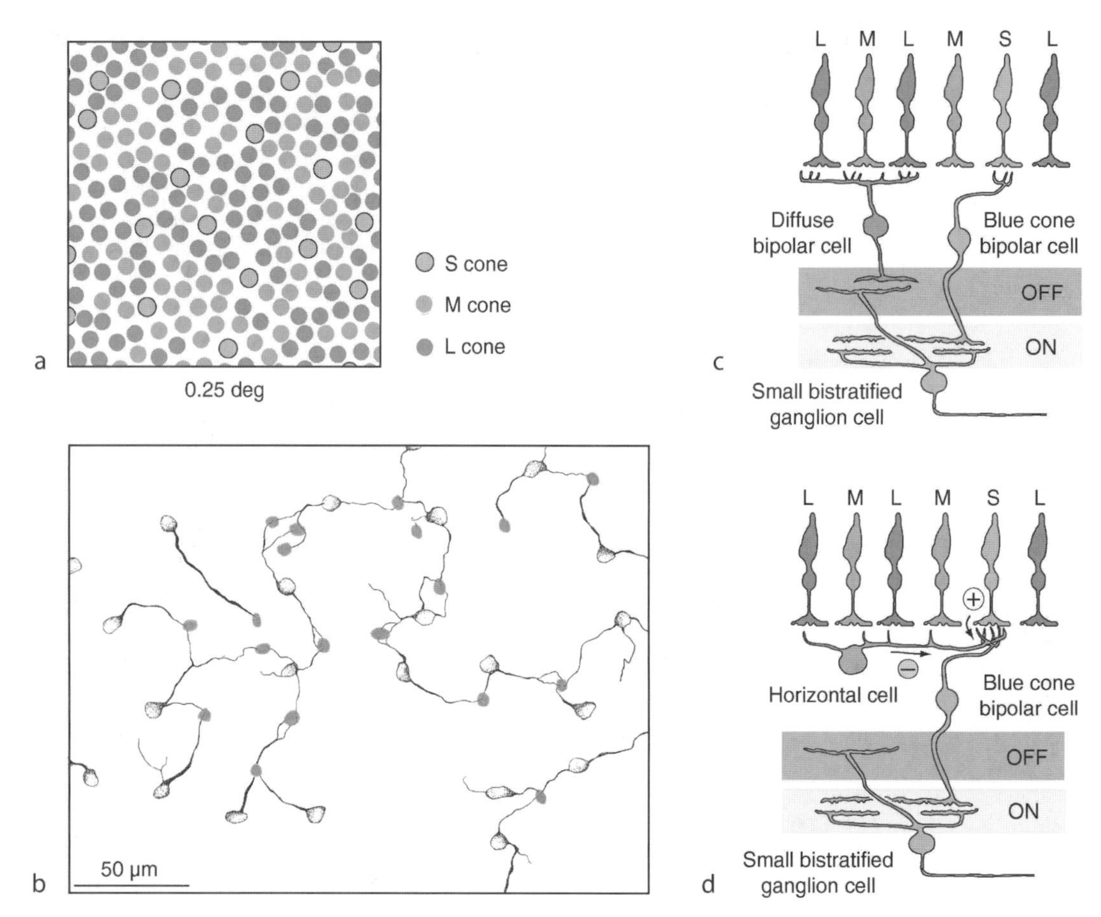

Retinal Color Vision in Primates. Figure 3 Short-wavelength sensitive (S) cone pathway. (a) fragment of the cone photoreceptor mosaic in macaque retina [modified from ref. 15]. (b) connections of blue-cone bipolar cells (*grey*) with S cones (*blue plaques*) in marmoset retina [modified from ref. 16]. The dendrites of blue cone bipolar cells make dominant contact with S cones. (c) schematic vertical section through primate retina showing connections of a small bistratified (blue-on) ganglion cell with blue cone bipolar and diffuse bipolar cells. (d) a second source of cone opponent inputs to blue-on ganglion cells arises from inhibitory feedback by the H2 subclass of horizontal cell.

ganglion cells show blue-off type responses, and presumed targets of these cells in the brain show large receptive fields [7,10]. One of these cell types also shows intrinsic, melanopsin-based photosensitivity and may contribute to ►circadian entraining as well as to color vision [7,11]. Connections from S cones to off-type midget bipolar cells have been reported in macaque fovea, but S cones in marmoset retina make negligible connections to midget cells and only sparse connections with diffuse, off-type bipolar cells [6,12]. Whether these differences reflect true species differences or methodological differences is not yet clear.

In summary, there is clear evidence for a selective network transmitting on-type signals from S cones to the brain to yield a "primordial" dichromatic color vision channel. A specific network of S cone connections to bipolar cells has also been shown in mouse retina [13], and the question whether this or other elements of S-cone circuits are common to other diurnal mammals has become an interesting topic in comparative neurology.

Higher Level Processes

Central Targets for Chromatic Signals

The main target for all ganglion cell axons in primates is the dorsal ►lateral geniculate nucleus (LGN) of the ►thalamus. The dominant input to the parvocellular layers of the LGN is from midget ganglion cells, and accordingly in trichromatic primates most parvocellular cells show red-green chromatic opponent properties. Geniculocortical relay cells (►Geniculostriate connections) in the parvocellular layers project to granular layer 4Cβ in the ►primary visual cortex. By contrast, blue-on and blue-off type responses are segregated to the intercalated or koniocellular division of the LGN [10]. Koniocellular relay cells show relatively diffuse cortical projections including supragranular layers 3 and 4A in the primary visual cortex: consistently, blue-on and blue-off type responses of presumed geniculate afferents can be recorded from these layers [14]. The two chromatic streams thus remain segregated at least to the early stages of cortical processing, and the mechanism by which these channels combine to enable color perception remains an outstanding and fascinating topic in neuroscience.

References

1. Nathans J (1999) The evolution and physiology of human color vision: insights from molecular genetic studies of visual pigments. Neuron 24:299–312
2. Solomon SG, Lennie P (2007) The machinery of colour vision. Nat Rev Neurosci 8:276–286
3. Jacobs GH (1993) The distribution and nature of colour vision among the mammals. Biol Rev 68:413–471
4. Wässle H (2004) Parallel processing in the mammalian retina. Nat Rev Neurosci 5:747–757
5. Mollon JD (1989) "Tho' she kneel'd in the place where they grew." The uses and origins of primate colour vision. J Exp Biol 146:21–38
6. Klug K, Herr S, Ngo IT, Sterling P, Schein S (2003) Macaque retina contains an S-cone OFF midget pathway. J Neurosci 23:9881–9887
7. Dacey DM, Packer OS (2003) Colour coding in the primate retina: diverse cell types and cone-specific circuitry. Curr Opin Neurobiol 13:421–427
8. Sakurai M, Mullen KT (2006) Cone weights for the two cone-opponent systems in peripheral vision and asymmetries of cone contrast sensitivity. Vision Res 46:4346–4354
9. Dacey DM, Lee BB (1994) The 'blue-on' opponent pathway in primate retina originates from a distinct bistratified ganglion cell type. Nature 367:731–735
10. Szmajda BA, Buzás P, FitzGibbon T, Martin PR (2006) Geniculocortical relay of blue-off signals in the primate visual system. Proc Natl Acad Sci USA 103:19512–19517
11. Dacey DM, Liao HW, Peterson BB, Robinson FR, Smith VC, Pokorny J, Yau KW, Gamlin PD (2005) Melanopsin-expressing ganglion cells in primate retina signal colour and irradiance and project to the LGN. Nature 433:749–754
12. Lee SCS, Grünert U (2007) Connections of diffuse bipolar cells in primate retina are biased against S-cones. J Comp Neurol 502:126–140
13. Haverkamp S, Wässle H, Duebel J, Kuner T, Augustine GJ, Feng G, Euler T (2005) The primordial, blue-cone color system of the mouse retina. J Neurosci 25:5438–5445
14. Chatterjee S, Callaway EM (2003) Parallel colour-opponent pathways to primary visual cortex. Nature 426:668–671
15. Roorda A, Metha AB, Lennie P, Williams DR (2001) Packing arrangement of the three cone classes in primate retina. Vision Res 41:1291–1306
16. Luo X, Ghosh KK, Martin PR, Grünert U (1999) Analysis of two types of cone bipolar cells in the retina of a New World monkey, the marmoset, Callithrix jacchus. Visual Neurosci 16:707–719

Retinal Direction Selectivity: Role of Starburst Amacrine Cells

Susanne Hausselt, Thomas Euler
Department of Biomedical Optics, Max-Planck Institute for Medical Research, Heidelberg, Germany

Definition

Direction-Selectivity in the Retina

Computing the direction of image motion is an essential task for the visual system. The retina's ability to detect the direction of image motion (►direction selectivity or DS) was first described more than 40 years ago [1]. ►Direction-selective ganglion cells (DSGCs) (►Retinal ganglion cells) fire vigorously when a stimulus is moving in a certain direction ("preferred"), while remaining silent when the same stimulus moves in the opposite ("null") direction (Fig. 1). ►Starburst amacrine cells (SACs) [2,3] are retinal interneurons closely intertwined with the direction-selective circuitry in the retina. While there is common consensus that SACs are crucial for the computation of motion direction, the exact nature of their involvement is still controversial. For review and further reading see [4,5].

Characteristics

(Quantitative) Description

Direction-Selective Ganglion Cells (DSGCs)

Direction-selective ganglion cells (►Retinal ganglion cells) have been primarily studied in rabbit retina, where they account for 10% of the ganglion cells. They have functionally and morphologically equivalent counterparts in other mammals and in non-mammalian vertebrates. Most research has focused on the ON/OFF DSGC (Fig. 1a), which has a bistratified dendritic tree with one arborization in the outer half (the OFF sublamina) of the ►inner plexiform layer (IPL), and another arborization in the inner half (the ON sublamina) of the IPL.

This arrangement allows responses to the direction of image motion of dark objects on a light background – mediated by the OFF arbor – as well as to objects brighter than the background – mediated by the ON arbor. The cell comes in four functional subtypes, each preferring one particular direction of motion (as an example see Fig. 1b). A second type of retinal direction-selective cell is the monostratified ON DS ganglion cell, which prefers one of three particular directions (see *Function*). Each DSGC subtype tiles the retina with little dendritic overlap making directional information for any of the preferred directions available at every retinal location.

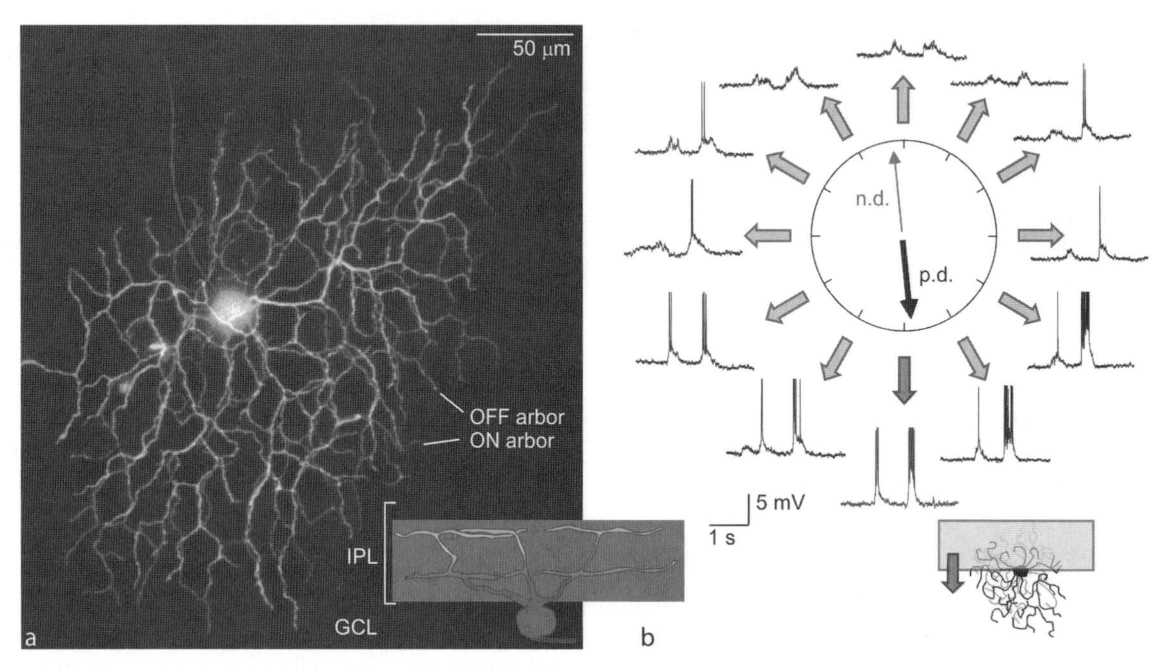

Retinal Direction Selectivity: Role of Starburst Amacrine Cells. Figure 1 (a) Fluorescent dye-injected ON/OFF direction-selective ganglion cell in a flat mounted rabbit retina (pseudo-color codes retinal depth). The two dendritic arbors stratify in the ON (red) and OFF (green) sublaminae of the IPL. (b) Electrical responses of a DSGC to a bar moving in 12 directions across its receptive field (see inset): both the leading (ON) and the trailing edge (OFF) of the bar elicit responses (p.d.: preferred, n.d.: null direction).

Starburst Amacrine Cells (SACs)

The dendrites of SACs closely co-fasciculate with the dendrites of DSGCs, and therefore have long been implicated in the computation of direction selectivity. (Strictly speaking, SAC neurites are not dendrites as they not only receive input, but also make output synapses – this is the case for most amacrine cells.) Unlike most other retinal neurons, SACs display a tremendous dendritic overlap (30–70 fold coverage) and, hence, offer plenty of "substrate" to provide the different DSGC subtypes with adequate neural circuitry. When SACs are removed from the circuitry, e.g., by gene-targeted cell ablation [6], direction-selective responses in DSGCs are abolished, confirming that SACs play a crucial role for direction-selectivity.

Starburst cells have been found in non-mammalian and mammalian species including primates. SAC morphology is well conserved among species: several primary dendrites radiate symmetrically from the soma before dividing into smaller branches (Fig. 2). The distal third of the branches is decorated with bead-like swellings (varicosities) [3]. SACs contain two transmitters, ►γ-aminobutyric acid (GABA) and ►acetylcholine (ACh). Due to the presence of ACh, they are also called "cholinergic amacrine cells." Two subtypes of SACs exist: OFF SACs co-stratify with the OFF dendritic arbor of the ON/OFF DSGCs, whereas ON SACs co-stratify with the ON dendritic arbor of the ON/OFF DSGC and the ON DSGC dendritic arbor. Despite the opposite polarity of their light responses, ON- and OFF-SACs are considered functionally equivalent.

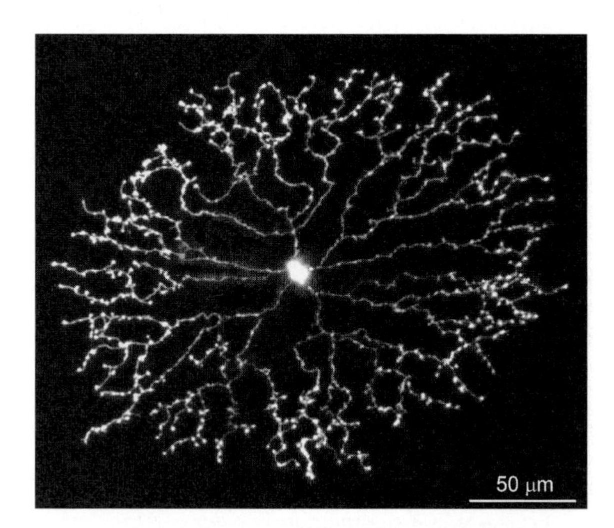

Retinal Direction Selectivity: Role of Starburst Amacrine Cells. Figure 2 Starburst amacrine cell in a flat mounted rabbit retina injected with a fluorescent dye. The cell's morphology is reminiscent of a "starburst" fireworks display.

Description of the Process

As recognized in the first studies on DSGCs in rabbit retina [1], the generation of direction-selectivity could be attributed to spatially offset inhibition biased towards the "null" direction. This proposal assumes asymmetrical wiring, such that the DSGC receives inhibition preferentially from interneurons displaced to one side of its ▶receptive field (Fig. 3a). It was proposed that motion in the "null" direction triggers inhibition (via the spatially displaced interneurons) before the stimulus reaches and excites the DSGC directly. If the inhibition is sufficiently delayed or, alternatively, long-lasting, it will coincide with the direct excitation and prevent the DSGC from responding. Motion in the opposite ("preferred") direction, on the other hand, will also trigger inhibition, but too late to prevent the DSGC from responding.

Pre- or Postsynaptic

It was originally proposed that direction-selectivity is computed by such a delay-based "veto"-mechanism from non-directional inputs in the DSGC itself. Detailed electrical recordings from DSGCs (reviewed in [5]), however, revealed that the inputs that DSGCs receive are already directionally tuned, with inhibitory input being larger for "null" direction motion, and excitatory input being larger for "preferred" direction motion. While is consistent with spatially offset inhibition in general, this finding suggests that the DSGC's response is substantially determined by the ratio of excitation and inhibition, and less by their temporal sequence in the DSGC, as originally proposed. More importantly, this indicates that direction-selectivity is already computed in interneurons presynaptic to the DSGC. Postsynaptic processing in the dendrites of DSGCs essentially supplements the

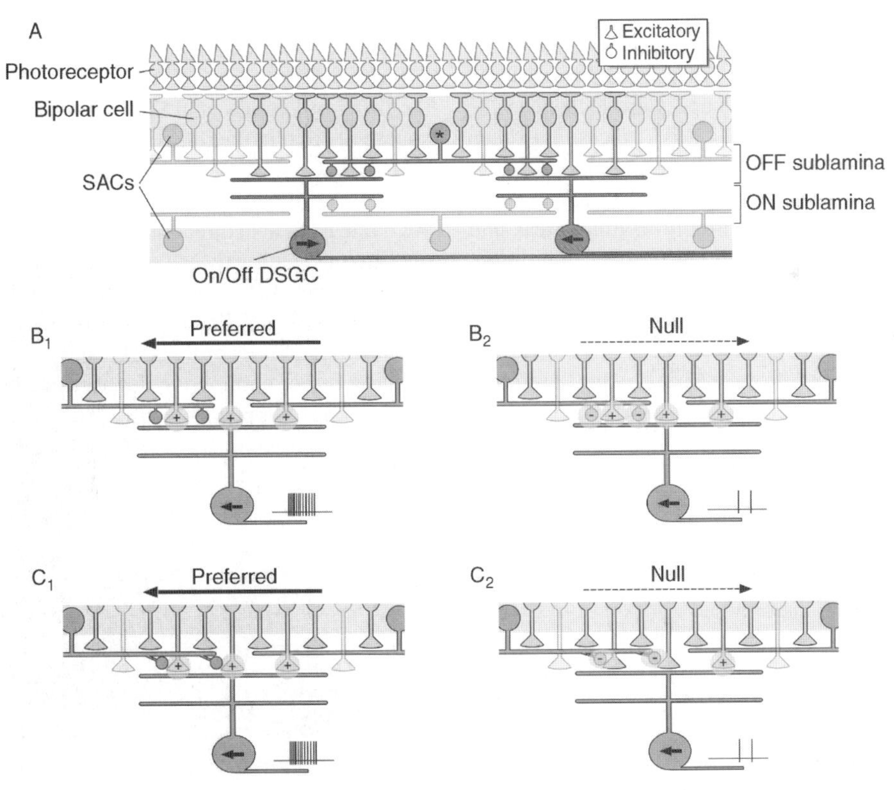

Retinal Direction Selectivity: Role of Starburst Amacrine Cells. Figure 3 (a) Schematic retinal cross section illustrating the direction-selectivity circuitry. The central SAC (*) serves (here: inhibits) two DSGCs with opposite preferred directions (indicated by arrow). For simplicity, cholinergic input, other amacrine cells involved and connections in the ON sublamina of the IPL are omitted. Proposed direction-selectivity mechanisms: (b) *Direction-selective inhibition*: preferred motion (i) elicits a response in the DSGC, because the left SAC's dendrite, which is connected to the DSGC, is not activated by this motion direction, while the right SAC's dendrite is activated but not connected. Null direction motion (ii) elicits no response in the DSGC, because the left SAC's dendrite is activated and inhibits the DSGC. (c) *Direction-selective excitation*: the left SAC (or, alternatively, another amacrine) inhibits the excitatory input from bipolar cells presynaptically to the DSGC for null direction motion [2], but not for preferred direction motion [1]. A similar mechanism could also be implemented via excitatory (cholinergic) connections from SACs.

presynaptic direction-selective mechanisms by sharpening the directional tuning.

Multiple Mechanisms

Retinal direction-selectivity is largely independent of contrast, mean brightness and velocity of the stimulus. This robustness alone suggests that different mechanisms at multiple levels participating in the generation and amplification of direction-selectivity [reviewed in 5]. This is also reflected by the fact that both inhibitory and excitatory inputs to DSGCs are by themselves direction-selective.

Direction-Selective Inhibition

Pharmacologically blocking ►$GABA_A$ receptors has consistently been shown to abolish direction-selectivity in DSGCs, indicating that the direction-selective inhibition is mediated by GABAergic input [8]. This GABAergic input is at least partially provided by SACs (Fig. 3b). More importantly, light-evoked Ca^{2+} signals optically recorded in the distal SAC dendrites are direction-selective (Fig. 3a) [9]. Thus, it is highly likely that GABA release from SACs is also direction-selective, because the SACs' output synapses are located in the distal dendrites [3], and their GABA release has been shown to be Ca^{2+}-dependent [10]. Furthermore, SACs appear to make direct GABAergic synapses with DSGCs, which seem to be highly asymmetrical and fit the requirements for spatially offset inhibition. Paired recordings have shown that DSGCs receive inhibitory synaptic input from SACs located on the "null side" of the DSGC's receptive field, but do not receive synaptic input from SACs on the "preferred side" [11]. Alternatively, (additional) directionally tuned inhibitory input may come from other, yet to be identified amacrine cells; however, direct evidence for this is lacking so far.

Direction-Selective Excitation

The excitatory input to DSGCs is ►glutamatergic as well as cholinergic. The glutamatergic input comes from bipolar cells (►Retinal bipolar cells) and appears to be directionally tuned, but it is not yet clear by which synaptic circuitry. Such tuning could result from suppression of bipolar cell activity for "null" direction motion (Fig. 4c); however, it is not known which amacrine cell could supply the required spatially offset inhibition to the bipolar cell terminals. The SAC is a potential candidate, and ultrastructural evidence for SAC output onto bipolar cell terminals does exist, but such contacts seem to be too sparse [3]. Alternatively, bipolar cell activity could be enhanced by motion in the "preferred" direction (facilitation), but again an appropriate pathway has not yet been unequivocally identified.

It is likely that DSGCs receive excitatory input from SACs via cholinergic synapses, because DSGCs express ►ACh receptors. ACh receptor blockers reduce the firing of DSGCs, and SACs are considered the only source of ACh in the retina [2]. Since ACh release from SACs is Ca^{2+} dependent and SAC dendritic Ca^{2+} signals are direction-selective, one would expect that ACh release is also direction-selective. Nonetheless, the cholinergic pathway is enigmatic in several ways (reviewed in [4]): Laser ablation of SACs on the preferred side of a DSGC reduced its excitatory input. However, paired recordings of SACs and DSGCs failed to show direct cholinergic connections [11]. Blocking ACh receptors appears to reduce direction-selectivity in

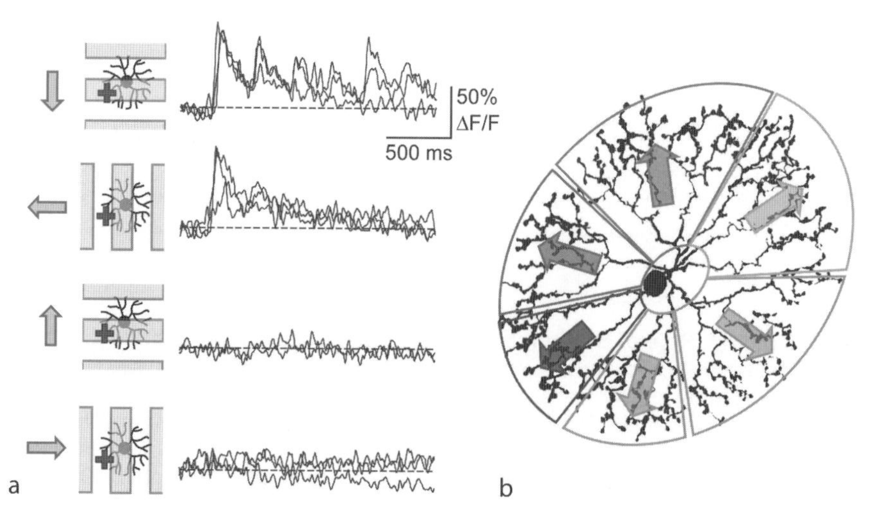

Retinal Direction Selectivity: Role of Starburst Amacrine Cells. Figure 4 (a) Dendritic Ca^{2+} responses (as relative change in fluorescence ΔF/F) optically recorded from a distal SAC dendrite to a bar grating moving in four different directions. Only motion from the soma roughly towards the tip of the imaged dendrite (indicated by a blue cross) elicits a response (from [9], Fig. 4a, modified). (b) SAC dendritic branches signal centrifugal motion.

DSGCs for certain stimuli (like bar gratings). On the other hand, in the presence of GABA receptor blockers, ACh antagonists reduce DSGC responses independent of motion direction, suggesting that cholinergic excitation is symmetrical [12].

Higher Level Processes

Direction Selectivity from Network Interactions

Various models of retinal direction-selectivity are based on network interactions to explain the generation of the direction-selective signals observed in the DSGCs (reviewed in [4,5,13]. The models differ in the complexity of interactions, and by the number and types of neurons recruited, but the basic principles are similar: spatially offset inhibition and asymmetrical wiring. Depending on the model, the spatially offset inhibition is provided by SACs or by not yet identified amacrine cells. In either case, most, if not all, network models assign some central role to SACs, ranging from simply relaying signals to DSGCs to providing essential direction-selective output. As a general mechanism for rendering signals direction-selective, both facilitation and suppression have been suggested.

SAC Networks

Starburst cells form a network with attractive properties. Reciprocal cholinergic excitation among SACs is prominent in the developing retina, but seems to be strongly reduced during retinal maturation. GABAergic interactions between neighboring SACs are prominent in the adult retina, and have long been suspected to be crucial for the generation of direction-selectivity (e.g., [13]). An excited SAC inhibits its neighbors, which in turn reduces their inhibition onto the first SAC, and effectively amplifies excitation. Thus, SAC dendrites pointing in the same direction could stabilize each other's responses. Such network interaction may well serve to enhance DS [10]. Nonetheless, GABAergic inhibition appears not to be pivotal to render SAC output direction-selective, because blocking GABA receptors does neither abolish direction selectivity in the SACs' dendritic Ca^{2+} not in their somatic volatage response [14].

Lower Level Processes

Direction Selectivity from Intrinsic SAC Properties

While in the network models direction selectivity arises primarily from neuronal interactions, a second group of models suggests that direction selectivity is initially generated in SAC dendrites as a result of intrinsic properties. Note that network models and intrinsic models are not necessarily mutually exclusive, but rather complementary.

At first glance SACs appear to be very symmetrical neurons (Fig. 2), which seems to disagree with the role of a detector of motion direction. In fact, SACs are better viewed as a collection of "wedge-shaped" direction detectors represented by the primary dendrites (Fig. 4b). The dendritic branches are largely electrically isolated and respond independently to local light stimulation [8]. Thus, they can be considered as largely "autonomous" computational units [14]. In contrast to the whole cell, the dendrites are indeed highly polarized structures. Synaptic inputs and outputs are differentially distributed along the dendrites: input synapses are located along the whole length, whereas output synapses are associated with the varicosities on the distal third of the branches [3]. Each principal branch responds more strongly to centrifugal motion (towards the dendritic tips) than to centripetal motion (towards the soma) [8], thus displaying dendritic direction selectivity. The mechanism underlying dendritic direction selectivity in SACs is not yet fully understood. Several "cell-autonomous" models of SAC dendritic direction selectivity have been proposed, each of them employing different (but not necessarily mutually exclusive) intrinsic mechanisms and properties.

Morphology and Amplification. Computational studies have suggested that the starburst dendrite morphology, with its steady increase of input synapses towards the dendritic tips, would generate a weak direction-selective dendritic signal by itself (e.g., [13]). This small direction-dependent difference in membrane potential (►Membrane potential – basics) could be amplified by other mechanisms, such as differential activation of voltage-gated channels. Starburst cells express several types of voltage-gated Ca^{2+} channels (►Calcium channels – an overview) that would be suitable. Support for this comes from the fact that blocking Ca^{2+} channels that are predominant on SACs abolishes direction selectivity in DSGCs, while leaving the DSGC's responsiveness to light intact. It is also possible that voltage-gated Na^+ channels (►Sodium channels) play a role; however, it is uncertain whether SACs generate spikes carried by Na^+.

Chloride Gradient along the Dendritic Branches. Immunocytochemical and physiological experiments suggest that a differential expression of the chloride transporters NKCC1 and KCC2 (►Chloride channels and transporters) leads to a chloride concentration ($[Cl^-]$) gradient within the SAC dendrite (with high $[Cl^-]$ in and near the soma) [15]. Such a gradient would render the proximal GABAergic inputs excitatory, whereas the distal GABAergic inputs would remain inhibitory. In this model, the temporal sequence in which proximal and distal GABAergic inputs occur (centrifugal vs. centripetal) lead to the generation of direction-selective transmitter release from the SAC distal dendrites.

Intracellular "Calcium Wave". An important question is how spatially offset inhibition can be delayed long enough so that it coincides with excitation in the DSGC. A similar question arises for the signal propagation within the SAC dendrite: If constructive and/or destructive interactions of inputs create dendritic direction selectivity in SACs, what causes the required delay? As electrical propagation seems too fast, it has been proposed that an intracellular Ca^{2+} "wave" supported by Ca^{2+}-induced Ca^{2+} release (within the SAC) may provide suitable delays.

Other, Yet Unidentified Interneurons in the DS Circuitry

A general problem with proposing that interneurons other than the SAC perform essential direction-selectivity computations is lack of neuronal "substrate." Retinal neurons usually tile the retina; such that their density appears too low to build local circuitries dedicated to each of the seven DSGC subtypes at every retinal location (see *Function*). Local dendritic processing – as implemented in SACs – may be a solution. In the case of bipolar cells, as has been proposed, this would be demanding. To tune the output of single branches of a bipolar cell axon to different preferred directions would require a tremendous locality of processing.

Involved Structures

Ultrastructural Basis of Direction Selectivity

While it is commonly agreed that retinal direction selectivity requires spatial asymmetries in the wiring of the circuitry, a directly corresponding anatomical correlate has not been unequivocally identified. It has so far not been possible to predict a DSGC's preferred direction from its anatomy. Although at the ultrastructural level complex arrangements of amacrine cell synapses onto DSGC dendrites have been described [16], systematic asymmetries in the wiring of the direction-selectivity circuitry have not been found. The only available evidence for asymmetrical wiring comes from paired recordings [10]. Another puzzle is the location of GABA release sites on SACs. Synapses with ACh-containing vesicles could be located in the varicosities, whereas, due to the lack of conventional ultrastructural features, GABA release sites have not yet been localized on SAC dendrites.

Function

For visually oriented animals, it is a matter of survival to swiftly detect moving objects (►local image motion) and reliably discriminate their direction of motion. In addition, motion of the whole visual field (►global image motion) provides important information about head/body movements (►Visual motion processing; ►Optic flow). Hence, coding the direction of image motion is an important task for the visual system. The three preferred motion directions of the ON DSGCs correspond to rotation around the axes defined by the three ►semicircular canals in the inner ear (reviewed in [4]). The cells respond preferentially to global motion and project to the ►accessory optic system (AOS), thus, ON DSGCs are thought to provide a correctional signal for ►eye movement and gaze-stabilization. This is supported by the finding that ablating SACs leads to a loss of the ►optokinetic reflex [6]. The four preferred directions of the ON/OFF DSGCs are roughly aligned with the extraocular rectus muscles and, therefore, with the cardinal ocular rotation directions. ON/OFF DSGCs seem to contribute some input to the optokinetic system, however, they send projections to the ►superior colliculus and the ►lateral geniculate nucleus (LGN), indicating that their signals also serve other functions, possibly including control of spatial attention (►Visual attention). The responses of ON/OFF DSGCs to moving objects are attenuated by synchronous motion of the background, indicating that they preferentially signal local motion. Nonetheless, there is no functional evidence so far that signals from retinal DSGCs take part in higher visual processing of motion direction.

Development

Up to now there is no satisfactory explanation of how the direction selectivity circuitry is wired during retinal development. That direction selectivity is established at the time of eye-opening suggests that the wiring process does not require visual stimulation.

References

1. Barlow HB, Hill RM, Levick WR (1964) Rabbit retinal ganglion cells responding selectively to direction and speed of image motion in the rabbit. J Physiol 173:377–407
2. Masland RH, Mills JW (1979) Autoradiographic identification of acetylcholine in the rabbit retina. J Cell Biol 83:159–178
3. Famiglietti EV Jr (1983) "Starburst" amacrine cells and cholinergic neurons: mirror-symmetric on and off amacrine cells of rabbit retina. Brain Res 261:138–144
4. Vaney DI, He S, Taylor WR, Levick WR (2001) Direction-selective ganglion cells in the retina. In: Zanker JM, Zeil J (eds) Motion Vision – Computational, Neural, and Ecological Constraints. Springer, Berlin, pp. 13–56
5. Taylor WR, Vaney DI (2003) New directions in retinal research. Trends Neurosci 26:379–385
6. Yoshida K, Watanabe D, Ishikane H, Tachibana M, Pastan I, Nakanishi S (2001) A key role of starburst amacrine cells in originating retinal directional selectivity and optokinetic eye movement. Neuron 30:771–780
7. Ariel M, Daw NW (1982) Pharmacological analysis of directionally sensitive rabbit retinal ganglion cells. J Physiol 324:161–185
8. Euler T, Detwiler PB, Denk W (2002) Directionally selective calcium signals in dendrites of starburst amacrine cells. Nature 418:845–852

9. Zheng JJ, Lee S, Zhou ZJ (2004) A developmental switch in the excitability and function of the starburst network in the mammalian retina. Neuron 44:851–864
10. Fried SI, Münch TA, Werblin FS (2002) Mechanisms and circuitry underlying directional selectivity in the retina. Nature 420:411–414
11. He S, Masland RH (1997) Retinal direction selectivity after targeted laser ablation of starburst amacrine cells. Nature 389:378–382
12. Chiao CC, Masland RH (2002) Starburst cells nondirectionally facilitate the responses of direction-selective retinal ganglion cells. J Neurosci 22:10509–10513
13. Borg-Graham LJ, Grzywacz NM (1992) A model of the directional selectivity circuit in retina: transformations by neurons singly and in concert. In: McKenna T, Zornetzer SF, Davis JL (eds) Single Neuron Computation. Academic Press, San Diego, CA, pp 347–375
14. Miller RF, Bloomfield SA (1983) Electroanatomy of a unique amacrine cell in the rabbit retina. Proc Natl Acad Sci USA 80:3069–3073
15. Gavrikov KE, Dmitriev AV, Keyser KT, Mangel SC (2003) Cation–chloride cotransporters mediate neural computation in the retina. Proc Natl Acad Sci USA 100:16047–16052
16. Dacheux RF, Chimento MF, Amthor FR (2003) Synaptic input to the ON-OFF directionally selective ganglion cell in the rabbit retina. J Comp Neurol 456:267–278

Retinal Flow

►Optic Flow

Retinal Ganglion Cells

LEO PEICHL
Max Planck Institute for Brain Research, Frankfurt am Main, Germany

Definition

The retinal ganglion cells (RGCs) are the output stage of retinal information processing. They are the only cells in the retina with axons that leave the eye. The ganglion cell axons form the ►optic nerve and transmit retinal information – in the form of spike trains – to the visual target areas in the brain. The name "ganglion cell" derives from the anatomical notion that these cells constitute the "*ganglion nervi optici*," i.e. the cluster of somata that give rise to the fibers of the optic nerve. The ganglion cell somata are located in the ►ganglion cell layer (GCL), the innermost layer of the retina. The dendrites of the ganglion cells ramify in the ►inner plexiform layer (IPL) where they are postsynaptic to bipolar cell axons and amacrine cell processes (►Retinal bipolar cells). There are more than a dozen different types of ganglion cell in all mammalian retinae studied so far. They differ in dendritic field size and dendritic branching pattern, and they receive input from different bipolar and amacrine cell types, and hence have different functional properties. The various types are specialized to encode different aspects of a visual scene, e.g. fine spatial detail (visual resolution), brightness, color, or movement. These ganglion cell types are the basis of distinct parallel visual pathways relaying a decomposed representation of the visual scene to distinct target areas in the brain [1–5].

Characteristics

Description

Among mammals, ganglion cells have been morphologically and functionally most thoroughly studied in the cat, the rabbit, and some primates [1,2,4,6,7]. More recently, detailed morphological classifications of mouse and rat ganglion cells have been added [8]. These comparative studies suggest that there is a basically similar set of about 15–20 ganglion cell types in all mammals, but formal proofs of homology are still lacking (Figs. 1–3). The accounting of types in any one species differs between authors depending on the classification criteria applied. For historical reasons, the nomenclature of types differs between species. The present article focuses on the most thoroughly studied ganglion cell types in the cat and primates, which continue to serve as benchmarks for ganglion cell classifications.

Alpha Ganglion Cells/Primate Parasol Cells

Alpha ganglion cells have been identified morphologically in all mammalian species studied to date [6]. At every retinal location, they are the type with the largest soma, the largest-caliber, fastest-conducting axon, and a large dendritic field. The dendritic tree is circular-to-oval with stout radial, relatively densely branched dendrites that rarely overlap, and it is monostratified in the IPL (Figs. 1, 2, and 4).

Alpha cells have been identified as the brisk-transient (Y) cells of physiology [1,6,9]. They comprise two functional subtypes, ON alpha cells and OFF alpha cells, as defined by their response to light (see *dendritic stratification* below). The ►receptive fields of alpha/Y cells are large and have a concentric organization with an excitatory centre and a larger antagonistic, inhibitory surround [9,10]. Alpha/Y cells show a vigorous transient (phasic) response whenever there is a stimulus change; their response to stationary, constant stimuli decays within the first few tenths of a second following stimulus onset. Alpha/Y cells respond best to rapid

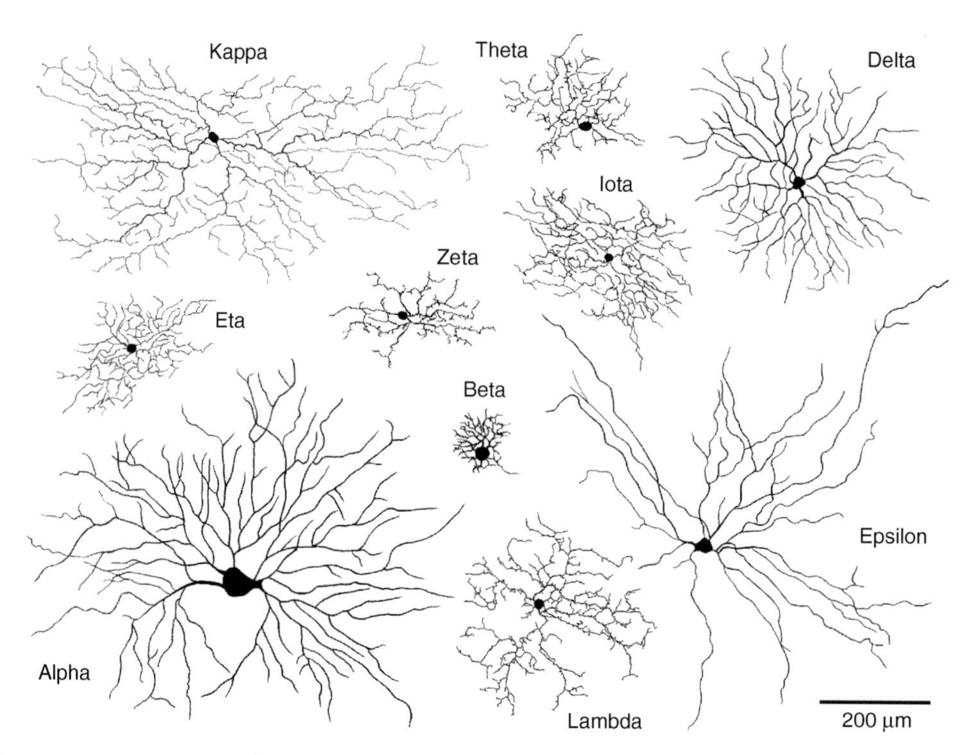

Retinal Ganglion Cells. Figure 1 A selection of ganglion cell types in the cat retina. The cells have been dye-injected and are seen in flat view, all drawn at the same scale. Cat ganglion cell types are named by Greek letters. Each type is characterized by a specific morphology, e.g. alpha cells have a large soma and a large dendritic tree formed by stout dendrites, whereas kappa cells have a small soma and a large dendritic tree formed by fine dendrites. The axons exiting from the soma have been omitted. Cell drawings kindly provided by David M. Berson.

spatial or temporal changes of coarse patterns, hence they can be considered as "novelty detectors." They are very sensitive to low luminance contrasts, but not to chromatic stimuli.

The corresponding cell type in primate retinae is thought to be the ►parasol (= umbrella-shaped) cell [4,5]. Due to its axonal projection to the ►magnocellular (M) layers (of the lateral geniculate nucleus), it is also termed ►M cell. The parasol cell is morphologically very similar to the alpha cells in other species (Fig. 3).

Like the alpha cells of non-primates, the parasol cells comprise two functional subtypes, ON and OFF, and their relatively large receptive fields have an ►antagonistic centre-surround organization [10]. The response to visual stimuli is transient. Parasol cells respond well to achromatic "luminance" stimuli and poorly to chromatic stimuli. There are also, however, functional differences between primate parasol and cat alpha cells. For example, alpha cells show non-linear spatial summation whereas parasol cells show linear summation. Thus, the question of correspondence between parasol and alpha cells remains contentious [6].

Beta Ganglion Cells/Primate Midget Cells

The beta ganglion cells have medium-sized somata and medium-caliber axons. Their dendritic trees are very small and circular-to-oval, with radial, relatively densely branched dendrites that rarely overlap (Figs. 1, 2, and 4). Like those of alpha cells, they are monostratified in the IPL. In fact, beta cells look like miniature versions of alpha cells. Accordingly, they collect input from only a few bipolar cells and are the high-resolution (visual acuity) system of the retina. Beta cells have been identified as the brisk-sustained (X) cells of physiology [1,9]. Like the alpha/Y cells, the beta/X cells comprise two functional subtypes, ON and OFF, and their small receptive fields show an antagonistic centre-surround organization [9,10]. However, their response to visual stimuli is sustained (tonic). Beta/X cells respond well to small, high-contrast, stationary stimuli.

In primate retinae, the ganglion cell type with the smallest dendritic and thus receptive field is termed midget ganglion cell [4,5]. Due to its axonal projection to the ►parvocellular (P) layers (of the lateral geniculate nucleus), it is also termed P cell. The midget cells may be homologous to the beta cells of cat and other mammals. Midget ganglion cells have medium-sized somata and very small dendritic trees that are relatively densely branched and monostratified in the IPL (Fig. 2). They comprise two functional subtypes, ON and OFF, with concentric antagonistic receptive fields [10]. The dendritic trees of the central-most

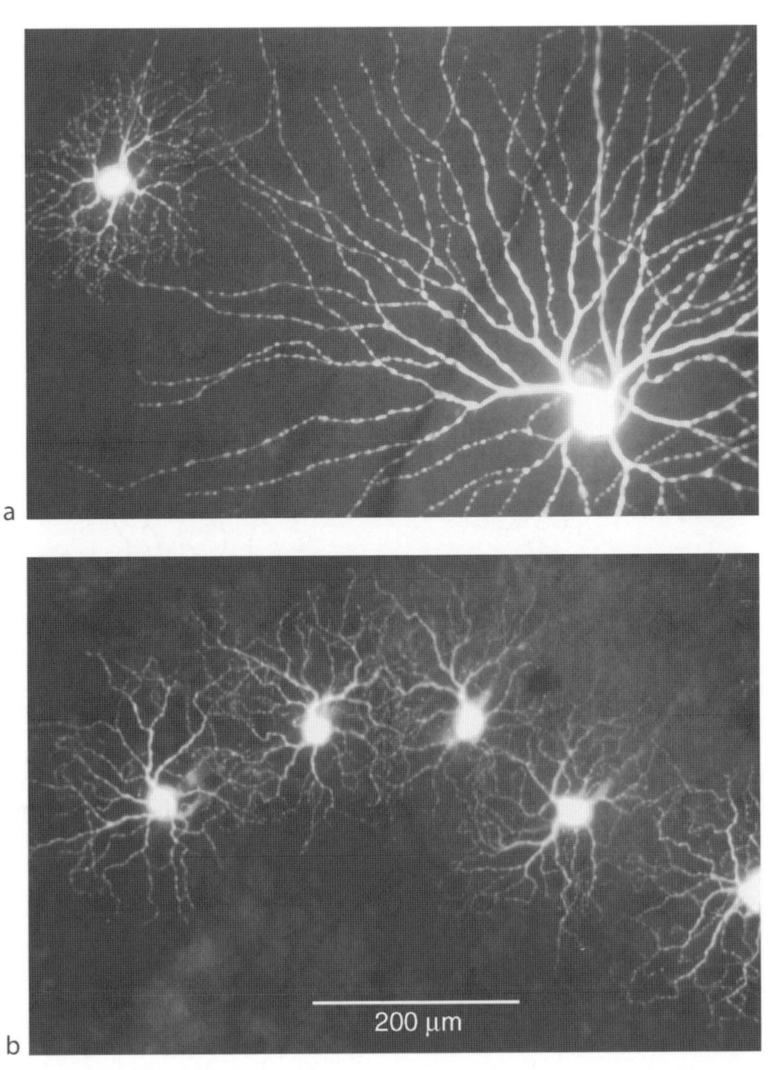

Retinal Ganglion Cells. Figure 2 Examples of ganglion cells that have been individually injected with the fluorescent dye lucifer yellow. (a) Flat view of a neighboring alpha (*right*) and beta (*left*) ganglion cell in peripheral dog retina. (b) Flat view of a group of neighboring alpha cells in central rat retina. The scale bar applies to A & B. A, modified from Peichl (1992) J. Comp. Neurol. 324:590–602; B, modified from Peichl (1989) J. Comp. Neurol. 286:120–139.

R

midget cells near the ►fovea are so small that each contacts just one midget bipolar cell (►Retinal bipolar cells), which likewise contacts just one cone. This foveal 1:1:1 connectivity, termed the midget system, is the anatomical basis for the high visual acuity of diurnal primates and man. As a corollary, the single cone input conveys to a midget cell the spectral tuning of that cone. The midget cells are widely considered to be the retinal keystone of the red-green chromatic channel of trichromatic primates, doing "double duty" in spatial resolution and color vision. The blue-yellow chromatic channel is implemented by other ganglion cell types [4] (►Color processing).

Interestingly, rabbit, mouse and rat do not possess a ganglion cell type that easily fits the morphological characteristics of beta cells [6–8]. It is possible that the beta cell is not as ubiquitous as the alpha cell, and that in some species another type of ganglion cell subserves spatial resolution.

Intrinsically Photosensitive, Melanopsin-Containing Ganglion Cells

A recently discovered ganglion cell type that appears to subserve "non-image-forming" functions is the "intrinsically photosensitive retinal ganglion cell" (ipRGC) [4,11]. The most intriguing feature of ipRGCs is that they contain the putative photopigment melanopsin and are directly sensitive to light. In addition, they receive conventional photoreceptor input via bipolar cells. The ipRGCs seem to play a major role in

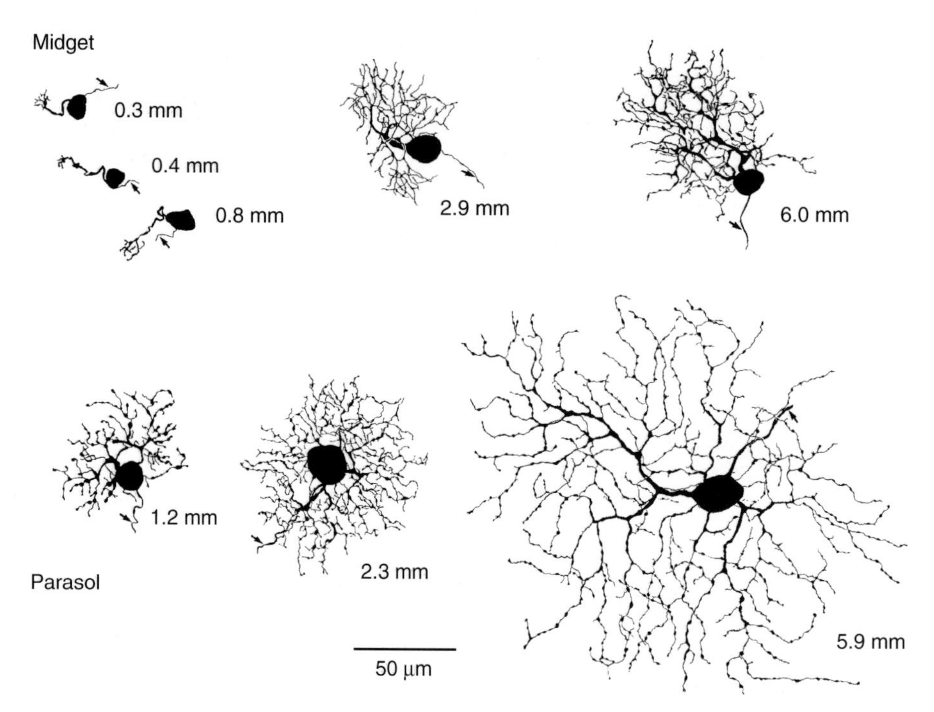

Retinal Ganglion Cells. Figure 3 Midget (*upper row*) and parasol (*lower row*) ganglion cells in the retina of the marmoset, a diurnal primate. The cells have been dye-injected and are seen in flat view, all drawn at the same scale. Both cell types increase in size with increasing distance from the fovea (distance given for each cell), but at each location the midget cells are smaller than the parasol cells. Axons are marked by arrows. Note that these marmoset cells are smaller than their presumed counterparts in cat, the beta and alpha cells (Fig. 4); 6 mm is peripheral in the smaller marmoset retina. Modified from [5], courtesy Paul R. Martin.

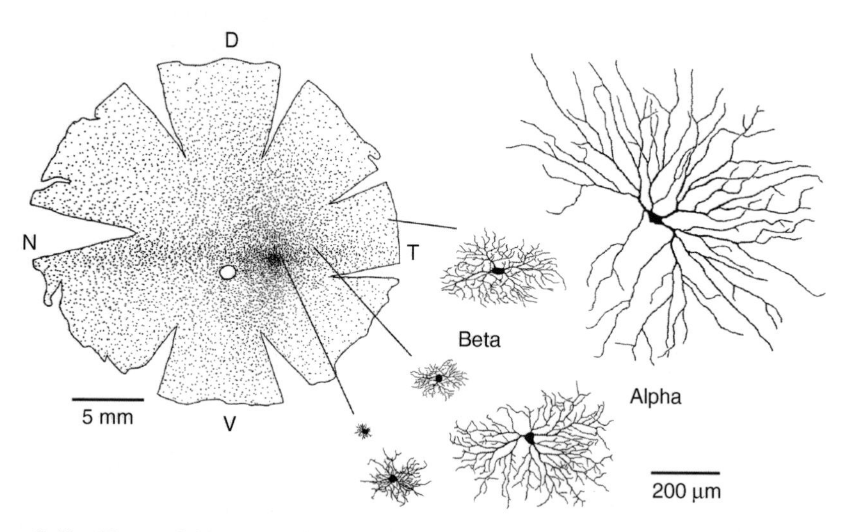

Retinal Ganglion Cells. Figure 4 Variation of ganglion cell population density and ganglion cell size across the cat retina. *Left*: Schematic map of the ganglion cell density in a whole retina; radial cuts were made to flatten the retina. From the region of highest ganglion cell density, the *area centralis* in temporal retina, cell density decreases monotonically towards the retinal periphery. The open circle signifies the optic nerve head. *Right*: Dye-injected alpha and beta ganglion cells at three distances from the *area centralis*, showing that dendritic field size increases as cell density decreases, while the size difference between alpha and beta cells is maintained. D, dorsal; V, ventral; N, nasal; T, temporal. The left scale bar applies to the map, the right scale bar to the individual cells. Modified from Peichl (1990) Optometrie 3/1990:3–12.

the entrainment of ►circadian rhythms, as evidenced by their projection to the circadian pacemaker, the ►suprachiasmatic nucleus (SCN), and in regulating ►pupil constriction, as evidenced by their projection to the ►pretectum. The ipRGCs constitute a small fraction of the ganglion cells (1–3% in rodent retinae, ~0.2% in primate retina). Their dendritic trees in the IPL are relatively large but sparsely branched. The cells show a sluggish, tonic ON response that is monotonically increasing with the light intensity.

Other Ganglion Cell Types

Many of the non-alpha and non-beta ganglion cell types occur at low densities, each constituting a small fraction of the total ganglion cell population. In primates, these other types mostly have larger dendritic fields than the parasol cells; some are sparsely and some rather densely branched [4]. In the cat, there is a wider range from rather small to very large types (Fig. 1). Some of these types have a concentric antagonistic receptive field organization, others do not, and for some morphologically recognized types, functional data are lacking.

Examples of further well-studied ganglion cell types are two kinds of direction-selective (DS) ganglion cells, the monostratified ON-DS cell and the bistratified ON/OFF-DS cell (►Retinal direction selectivity: Role of starburst amacrine cells). They are specialized to detect retinal image movement. Another example is the primate "small bistratified" cell, which shows a blue-yellow antagonism and serves in color processing [4] (►Color processing).

Dendritic Stratification and Light Response

ON ganglion cells have a receptive field centre that is activated by a light increase, OFF ganglion cells are activated by a light decrease in the receptive field center. These response characteristics are determined by their input neurons. The inner plexiform layer (IPL) is morphologically and functionally clearly stratified to keep these different processing circuits segregated (Fig. 5).

Cells stratifying in the *outer* part of the IPL are OFF cells, whilst those stratifying in the *inner* part are ON cells. The stratification level of the ganglion cell's dendritic field ensures appropriate (ON or OFF) bipolar cell input. The actual thickness of the two strata differs from species to species, in some they are nearly equal, in others the ON stratum may be twice as thick as the OFF stratum (because it also contains the axon terminals of the rod bipolar cells, which are ON cells; ►Retinal bipolar cells).

Both the alpha/parasol ganglion cells and the beta/midget ganglion cells are dichotomous, comprising equally numerous subpopulations of ON and OFF cells. Their monostratified dendritic trees ramify in the inner sublamina and the outer sublamina of the IPL, respectively (Fig. 5). The ON-DS ganglion cells also monostratify in the inner sublamina, while the ON/OFF-DS ganglion cells have bistratified dendritic trees.

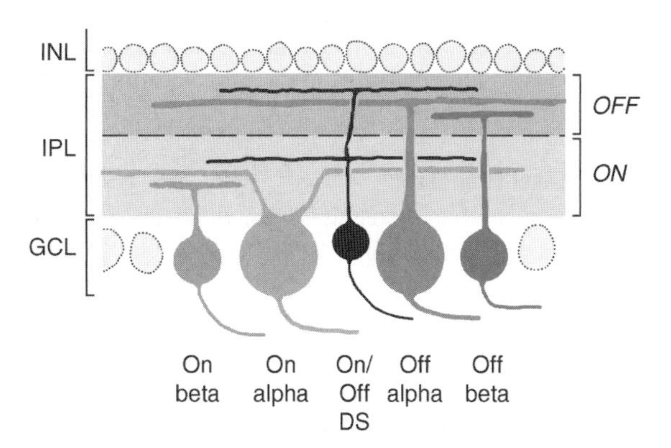

Retinal Ganglion Cells. Figure 5 Schematic drawing of the functional stratification of the inner plexiform layer (IPL) in the mammalian retina as seen in a transverse section. The dendritic trees of OFF ganglion cells stratify in the *outer* (distal) part of the IPL, those of ON ganglion cells in the *inner* (proximal) part. The ON/OFF direction selective (DS) ganglion cell has a bistratified dendritic tree with dendrites in the ON and the OFF stratum. Ganglion cell types are rendered in different colors. *Abbreviations*: INL, inner nuclear layer; GCL, ganglion cell layer.

Retinal Topography and Population Properties

A basic requirement of retinal organization is that the complete computing machinery has to be present at each retinal location in order to process a visual stimulus wherever it hits the retina. To achieve this, ganglion cells (and most other retinal neurons) are distributed economically across the retina. The members of each functional type (e.g. ON alpha cells, OFF alpha cells, ON beta cells, and OFF beta cells) tile the retina, such that their dendritic trees cover the retinal surface without gaps and without too much overlap [1]. The grain of this tessellation is finest in the central retina (the ►*area centralis* of most mammals, and the fovea of primates), where highest visual acuity is achieved. It becomes coarser toward the retinal periphery, where complete coverage is obtained by a less dense spacing and larger individual dendritic trees for each ganglion cell type (Figs. 3 and 4). Relative size differences between the ganglion cell types are preserved across the retina.

The alpha/parasol ganglion cells with their large dendritic trees require a low packing density. Depending on species, they only constitute between 1 and 10% of the total ganglion cell population [6]. Nevertheless, they exhibit uniform coverage and contribute to processing at each retinal point. The beta/midget ganglion cells have much smaller dendritic trees and correspondingly higher packing densities. In the cat retina, about 50% of the ganglion cells are beta cells, in the primate retina, about 80% of the ganglion cells are midget cells.

Lower Level Processes

The response characteristics of the different ganglion cell types are thought to be set mainly by the specific mix of synaptic inputs from different bipolar and amacrine cell types [1–3,5]. The sustained (tonic) beta/midget cells are probably driven by tonic bipolar cells, and the transient (phasic) alpha/parasol cells by phasic bipolar cells (Retinal bipolar cells). A ganglion cell's specific dendritic geometry, and hence electrotonic properties (►Electrotonic spread), also contribute to its response characteristics. The antagonistic surround of a ganglion cell's receptive field is mediated by lateral inhibitory input, partly directly from amacrine cells to the ganglion cell, and partly from horizontal cells to the bipolar cells (►Lateral interactions in the retina). The ganglion cells operate at high (►photopic) as well as low (►scotopic) light levels, i.e. the cone pathway and the rod pathway converge onto the same ganglion cells. The luminance-dependent functional switch from the cone pathway to the rod pathway is regulated by dopaminergic amacrine cells [2].

Higher Level Processes

There is homologous electrical coupling by ►gap junctions between the dendrites of neighboring alpha/parasol cells of the same centre sign (ON with ON, OFF with OFF), and heterologous ►gap junctional coupling between alpha/parasol cells and certain amacrine cell types. ON/OFF-DS cells also show homologous gap junctional coupling, likely to be restricted to partners with the same direction tuning. Modulated electrical coupling allows changing associations between neighboring ganglion cells. It probably is the basis for the high incidence of synchronous firing among homotypic neighbors. Synchronous firing, i.e. concerted activity among multiple ganglion cells, which has been observed in retinal multineuron recordings, may represent a "population coding" or "multiplexing" that is more powerful in encoding visual stimuli than the independent activity of individual cells [12]. Beta/midget cells do not show such electrical coupling, probably because signal spread across neighbors would decrease spatial resolution.

Each ganglion cell type projects to specific thalamic and/or midbrain target structures and specific subdivisions within these target structures [4,5]. This indicates segregation of the various retinal processing channels also at higher processing levels. The ►lateral geniculate nucleus (LGN) and ►superior colliculus (SC) are targets for several ganglion cell types. Other nuclei are more selectively targeted, and some ganglion cell types innervate several targets by axon collaterals.

Function

Visual acuity, dominantly mediated by the beta/midget system, is of more importance to some species than to others. In the diurnal primates, the midget system provides the anatomically possible maximum of acuity by fully exploiting the tight foveal cone packing. In the crepuscular-to-nocturnal cat, there is a considerable convergence from cones to beta cells even in the *area centralis*, "giving away" some of the acuity theoretically possible with the cone packing density. Here perhaps, evolutionary pressure was less on high acuity and more on a good signal-to-noise ratio at dim light. The latter requires signal summation over several photoreceptors, in addition to the presence of rods in the *area centralis*.

Alpha/parasol cells, with their large receptive fields, are thought to contribute little to acuity or to color vision. Their particular responsiveness to changing stimuli (novelty detectors), and their fast conduction velocity, would make them an "alarm" or "warning" system to direct visual attention (►Visual attention) to objects entering the visual field. Alpha/parasol cells also play an important role in global form perception (►Form perception) and depth perception (►Binocular vision). Both alpha/parasol and beta/midget cells project to the lateral geniculate nucleus and hence provide major inputs for cortical, "higher" visual processing.

The ON direction-selective ganglion cells project to the ►accessory optic system, the ON/OFF direction-selective ganglion cells project to the ►optokinetic system, the superior colliculus and the lateral geniculate nucleus. Both types are thought to contribute to the discrimination between self-movement and object movement (Retinal direction selectivity: Role of amacrine cells). The intrinsically photosensitive ganglion cells are specialized to encode ambient light intensities, they are involved in synchronizing circadian rhythms with the solar day and in regulating the pupillary light reflex. These three ganglion cell types are examples for retinal channels feeding into subcortical, "lower" visual processing systems.

For many of the less well-characterized ganglion cell types, we lack a clear idea of their role in image analysis. On the one hand, it is clear that there are basic parallel processing channels feeding specific parts of visual information into ganglion cell types with different response properties. On the other hand there is increasing awareness that retinal image processing works by a finely tuned and stimulus-dependent interplay of these many different channels, rather than by a strict dedication of "one type for this task, one type for that task." For example, primate parasol cells, despite their large receptive fields, play a significant role in hyperacuity (►Visual acuity, hyperacuity). Actually, there seem to be more types of ganglion cell (and of other retinal neurons) than we need to account for the functions we currently attribute to the retina [2].

We still have to learn a lot about this intriguing piece of neural tissue.

►Retinal Bipolar Cells
►Lateral Interactions
►Color Processing
►Direction Selectivity
►Gap Junctional Coupling

References

1. Wässle H, Boycott BB (1991) Functional architecture of the mammalian retina. Physiol Rev 71:447–480
2. Masland RH (2001) The fundamental plan of the retina. Nat Neurosci 4:877–886
3. Wässle H (2004) Parallel processing in the mammalian retina. Nat Rev Neurosci 5:747–757
4. Dacey D (2004) Origins of perception: retinal ganglion cell diversity and the creation of parallel pathways. In: Gazzaniga MS (ed) The cognitive neurosciences III. MIT Press, Cambridge Mass, pp 281–301
5. Martin PR, Grünert U (2004) Ganglion cells in mammalian Retinae. In: Chalupa LM, Werner JS (eds) The visual neurosciences. MIT Press, Cambridge Mass, pp 410–421
6. Peichl L (1991) Alpha ganglion cells in mammalian retinae: common properties, species differences, and some comments on other ganglion cells. Vis Neurosci 7:155–169
7. Rockhill RL, Daly FJ, MacNeil MA, Brown SP, Masland RH (2002) The diversity of ganglion cells in a mammalian retina. J Neurosci 22:3831–3843
8. Sun W, Li N, He S (2002) Large-scale morphological survey of mouse retinal ganglion cells. J Comp Neurol 451:115–126
9. Levick WR (1996) Receptive fields of cat retinal ganglion cells with special reference to the alpha cells. Progr Ret Eye Res 15:457–500
10. Rodieck RW (1998) The first steps in seeing. Sinauer Associates, Sunderland Mass
11. Berson DM (2003) Strange vision: ganglion cells as circadian photoreceptors. Trends Neurosci 26:314–320
12. Meister M, Berry MJ (1999) The neural code of the retina. Neuron 22:435–450

Retinal Implant

Definition

Device intended to restore useful vision in pathologies that selectively affect the photo-detectors of the retina while leaving relatively intact the other retina neurons and the fibers of the optic nerve (such as retinitis pigmentosa and macular degeneration). These devices share with cochlear implants the basic design principles and requirements, but are at a much earlier stage of development than cochlear implants (there is at least a 20–30 years gap), because of the greater information density (thousands of hair cells on the basilar membrane vs. millions of photodetectors on the retina) and the more complex structure of the sensor.

►Computer-Neural Hybrids

Retinal Lateral Interactions

Martin Wilson
Department of Neurobiology, Physiology and Behavior, College of Biological Sciences, University of California Davis, Davis, CA, USA

Synonyms

Lateral interactions in the retina

Definition

The influence of signals generated by retinal neurons on the activity of other neurons laterally distant in the retina.

Many processes are covered by this term, some acting over short distances between immediately adjacent neurons, and others acting over virtually the entire extent of the retina [1]. Lateral interactions occur at all levels in the retina, from the ►photoreceptors, the input neurons of the retina, through to ganglion cells (►Retinal ganglion cells), the output neurons of the retina. Lateral interactions may be positive or negative. Light falling on a patch of retina may augment the signal generated in neurons in an adjacent patch or it may diminish this signal. Lateral interactions are often time-dependent and may also depend on special features of the stimulus such as stimulus velocity or the wavelength of the illuminating light.

There are several different mechanisms known to mediate lateral interactions. Some forms of lateral interaction, particularly those occurring in the inner retina, are known only from indirect and incomplete evidence and are not well understood.

Characteristics

The best known form of lateral interaction in the retina is lateral inhibition. Lateral inhibition was inferred by Ernst Mach in the 1860s on the basis of psychophysical experiments. Much later it was shown that lateral inhibition was one of the first neural operations performed in the compound eye of the horseshoe crab, *Limulus*, where eccentric cells, roughly the equivalent of ganglion cells in the vertebrate retina, are thought to make inhibitory ►synapses with their neighbors (Fig. 1). Hartline and his colleagues quantified the effect of lateral inhibition on the steady-state firing rate of eccentric cells by writing simultaneous equations [2]. The firing of every eccentric cell is influenced by the firing of every other eccentric cell, so that, considering only a pair of illuminated eccentric cells, A and B, the responses of A and B, r_A, and r_B are given by:

$$r_A = e_A - K_{A,B}\left(r_B - r_B^0\right)$$

$$r_B = e_B - K_{B,A}\left(r_A - r_A^0\right),$$

where e_A is the response of A in the absence of stimulation to B, $K_{A,B}$ is a term representing the strength of inhibitory connection from B to A, r_B is the response of B, and r_B^0 is a threshold firing rate for B below which it exerts no inhibition on A. Terms for the response of B have analogous meanings. These simple piecewise linear equations give a good approximation of the responses of eccentric cells for different patterns of light falling on the *Limulus* eye and capture the essential features of lateral inhibition.

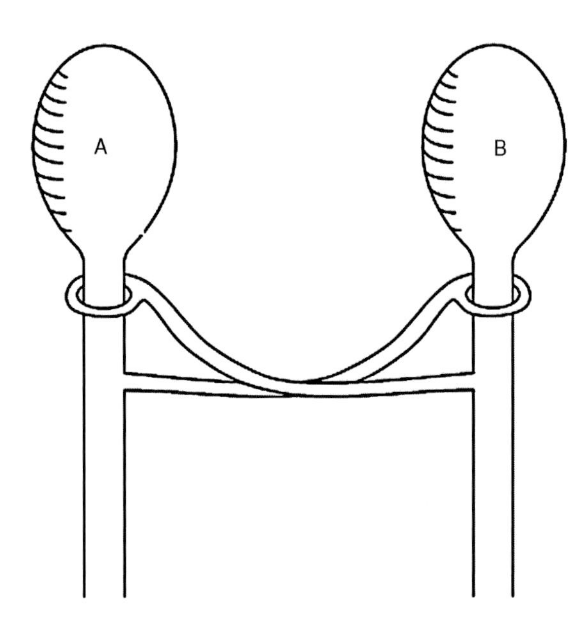

Retinal Lateral Interactions. Figure 1 The idea underling the quantitative description of inhibitory interactions in the *Limulus* retina [2]. A pair of adjacent eccentric cells, A and B, each capable of independent excitation by light, are shown as cell bodies with axons sending information to the brain and collaterals that make local inhibitory connections with each other. The excitation of one cell tends to reduce the response of its neighbor. A subtle but important feature of the inhibition, reflected in this diagram and in the descriptive equations, is that it is recurrent. By this is meant that for any cell, the inhibitory effects of neighbors are exerted upstream of any inhibitory output from that cell.

The organization of the vertebrate retina is radically different from that of *Limulus* and yet lateral inhibition is similarly one of its first processing steps. To understand how lateral inhibition operates in the outer vertebrate retina it is necessary to consider all three general classes of neuron found there, their lateral interactions, and their ►receptive fields.

Photoreceptors have very small receptive fields, though in many instances these are at least slightly broader than expected from the anatomical dimensions of the photoreceptor. The explanation for this enlargement is that photoreceptors are weakly coupled together via ►gap junctions (Photoreceptors) so that signals leak from one photoreceptor to its neighbors. Quantitative models of signal spread through the photoreceptor network have been based on square or hexagonal networks in which each photoreceptor is represented by a resistor and a capacitor, coupled to its neighbors through resistors representing gap junctions (Fig. 2). This kind of model gives a reasonable approximation for the spread of small signals through the network but misses some of the network's time-dependent properties. More sophisticated models, incorporating the voltage-dependent conductances of photoreceptors, show that, at least for the amphibian and turtle retina, the network has strongly time-dependent properties [4,5].

Photoreceptors pass signals to both horizontal cells and bipolar cells (►Retinal bipolar cells) (Fig. 3). Horizontal cells are strongly coupled homotypically via gap junctions, with the result that signals spread widely throughout the network of horizontal cells, giving these neurons very broad receptive fields. Because the coupling between horizontal cells is so strong, individual cells can be ignored and the network approximated

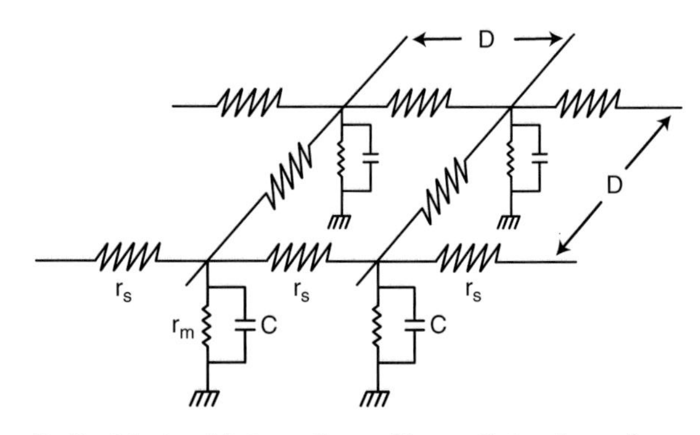

Retinal Lateral Interactions. Figure 2 A schematic representation of part of a network of coupled photoreceptors based on [3]. In this model, photoreceptors are arranged in a square grid separated by a distance, D. The resistors, r_s, represent the coupling resistance of the gap junctions between adjacent photoreceptors while the photoreceptors themselves, positioned at every node of the grid, are represented by a capacitance and a membrane resistance.

to a continuous and homogeneous sheet. The distribution of voltage in this model is given by a second-order partial differential equation for which useful analytical solutions have been found [6].

Bipolar cells, in contrast to horizontal cells, are not strongly coupled and also differ from horizontal cells in having a receptive field comprising two distinct regions. A relatively narrow central region receives input from overlying photoreceptors but outside this lies an annulus of inhibitory, or antagonistic surround. If a spot of light falling in the central region depolarizes the bipolar cell, a spot falling in the surround would hyperpolarize the cell, and a large patch of light covering both center and surround would produce no sustained response (although it would produce transient responses when the light comes on and when it goes off). Loosely speaking, the bipolar cell receptive field is generated from the sum of the inputs from overlying photoreceptors and a sign-reversed input from horizontal cells of much larger spatial extent, as summarized in Fig. 3. How horizontal cell inputs "sum" is not entirely clear and there is evidence supporting both inhibitory synaptic feedback to photoreceptor terminals and inhibitory synaptic feedforward to bipolar cells as well. Neither of these proposed mechanisms is entirely satisfactory and an alternative idea based on extracellular current flow has recently been revived [7].

The lateral spread of signal between neurons coupled together via gap junctions can be inferred from intracellular recording from a pair of cells. Current injected into one cell produces a voltage in the other, and by examining pairs of neurons at known separations, it is possible to measure signal spread directly and estimate the electrical parameters of the network. This direct approach has been applied to photoreceptors and horizontal cells but a less direct method has proved useful between other cell types. The injection of small tracer molecules, often fluorescent molecules, into a single cell, implies the presence of coupling gap junctions if fluorescence appears in nearby cells. This method is not easily amenable to quantitative analysis, moreover gap junctions differ in the size of tracer molecule they allow to pass, nevertheless it has the huge advantages that intracellular injection into only a single neuron is required, and even very sparse connections can be revealed this way. The tracer molecules Biocytin and Neurobiotin have revealed a previously unsuspected wealth of coupling, both homo- and heterotypic, between neurons in the inner retina. Although this coupling implies the lateral spread of signal, electrical coupling between neurons is known to serve other

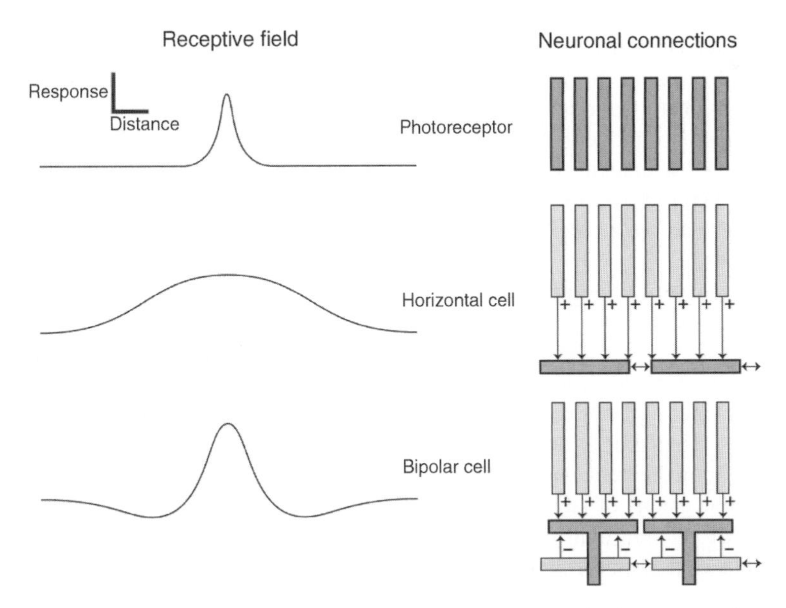

Retinal Lateral Interactions. Figure 3 The receptive fields of the neurons in the outer retina and an explanatory schematic showing how these receptive fields are generated. For simplicity only one spatial dimension is shown, equivalent to mapping receptive fields using a long thin bar of light moved orthogonal to its long axis. Photoreceptors, shown in green, have narrow receptive fields, only slightly wider than the width of an individual receptor. Horizontal cells (blue), in contrast, have wide receptive fields because they collect signals from all their overlying photoreceptors and are strongly coupled to each other, thereby allowing signals to spread laterally. Bipolar cells (orange) collect signals from a relatively small group of overlying photoreceptors but also receive signals of opposite sign from nearby horizontal cells. This arrangement gives rise to a receptive field in which light falling in the center of the field, dominated by direct input from overlying photoreceptors, produces a response whose sign is opposite to that produced by more lateral stimuli in which horizontal cell input dominates.

functions, such as the synchronization of spiking, and it may transpire that the spatial aspects of this coupling are inconsequential.

Higher Level Processes

Perceptual illusions (►Visual illusions), such as Mach bands and the Hermann grid illusion, originate from lateral interactions in the retina [8].

Lower Level Components

Lateral interactions are mediated through gap junctions and synapses. In some instances it is possible to say which specific gap junction subunits are present in which retinal neurons but in many instances this is not yet known. The synapses and their ►transmitters mediating lateral interactions are known in outline but, as always, the details of the inner retina are much less clear than those of the outer retina.

Process Regulation

A large body of evidence shows that many, but not all, of the gap junctions that mediate lateral signal spread in the retina can be functionally closed by transmitters and ►neuromodulators. Best known is the example of coupling between horizontal cells in the retinas of fish and turtle [9,10]. ►Dopamine, which is released by neurons in the inner retina, and plays a role in switching the retina from ►scotopic to ►photopic conditions, clearly uncouples horizontal cells by elevating intracellular ►cAMP, thereby decreasing the receptive field size of these neurons and greatly restricting the ability of injected dye molecules to move between cells. A similar effect is seen in AII ►amacrine cells [11] that, in low-light conditions when rod photoreceptors dominate, are coupled together and are a crucial link in the transmission of rod signals.

In addition to dopamine, ►nitric oxide also closes gap junctions, in this case acting through ►cGMP, and probably also plays a role in light/dark adaptation, though how these two agents, dopamine and nitric oxide and perhaps others, act together and what their different functional roles might be, is not known. In general though, there are good theoretical arguments for modulating the spatial dimensions of lateral inhibition to optimize it for different ambient light intensities. In flies, the measured spatial extent of lateral inhibition is reasonably well matched to theory [12] but in vertebrate retina no similar comparison has been attempted.

Function

Lateral inhibition clearly enhances edges, which, very likely, is a necessary step in the task of parsing the visual world into separate, identifiable objects. Unfortunately, this idea does not lead easily to a quantitative theory.

Another way of looking at the function of lateral inhibition, initiated by H. B. Barlow and subsequently built on by others, has its roots in ►information theory and is known as *predictive coding* (►Sensory systems). The starting point for this approach is the postulate that the visual system is interested in all spatial information, i.e. edges are accorded no special *a priori* value. Key ideas are (1) that the retina has been optimized for efficiency of information transfer to the brain. This is a plausible conjecture, if only because there are obvious anatomical constraints on the thickness of the ►optic nerve and therefore the number of output channels from the retina. A more subtle constraint is exercised by intrinsic noise within a neuron that limits its information carrying capacity. (2) There exist statistical regularities in the spatial and temporal structure of the images falling on the retina. For example, the intensity of light falling on any point in the retina is likely to be similar to the intensities found adjacent to that point, first, because of blurring caused by the eye's optical limitations, and second, because the visual world is not a random pattern of dots. (3) From point 2, we can say that some information is redundant and, on the basis of point 1, ought not to be transmitted to the brain.

Lateral inhibition performs this task of redundancy removal and permits the efficient use of the limited bandwidth provided by retinal neurons. Simply put, this view of lateral inhibition is that it is a way of computing a best guess for the signal expected at any point in the retina, based on a weighted sum of the signals from points around it. This best guess is then subtracted from the actual signal, thereby allowing any deviations from the guess to fill the bandwidth of that neuron.

On the basis of these arguments we would predict that the strength and extent of lateral inhibition should be matched to the statistics of different visual environments. An important contributor to these statistics is the variance in the number of photons arriving, which is a function of light intensity. At low light intensities a statistically reliable best guess requires that a large number of points be included, in other words lateral inhibition should become broader [12]. At higher light levels, where signal-to-noise ratios are higher, the best guess would include only those points in the immediate neighborhood.

The spatial characteristics of different scenes, a beach versus a forest for example, ought, from the principles of predictive coding, to elicit different lateral interactions within the retina. Some evidence from ganglion cell recordings suggests that this may be occurring in the inner retina, probably mediated by amacrine cells, though the mechanism is presently unknown [13].

While lateral inhibition has provided a path into some of the deepest questions concerning the design of the nervous system, positive lateral interactions have also stimulated a careful consideration of function.

The function of coupling between horizontal cells is fairly apparent, but what about photoreceptor coupling? At first sight this is puzzling, not only because photoreceptor coupling is apparently the opposite of lateral inhibition, for which there is a good understanding, but also because it must degrade spatial resolution.

A crucial concept in understanding photoreceptor coupling is that photoreceptors are very high-gain detectors for which noise is an inescapable problem. Some of this noise is produced by the thermal activation of the molecules involved in transduction (►Phototransduction) but a large contribution comes from chance fluctuations in the number of photons caught. Even for cones operating in the photopic range, few enough photons are caught per integration time that these fluctuations compromise the reliability of the signal. Coupling between photoreceptors is a form of signal averaging that improves the signal-to-noise ratio by canceling out some of the uncorrelated noise found in every cell. This engineering trick necessarily involves trading a gain in reliability (improved signal-to-noise ratio) for spatial resolution. But the trade is definitely worthwhile. In the case of mammalian cones, coupling increases the signal-to-noise ratio by about 70% by allowing some signal to leak into neighboring photoreceptors. The spatial blurring this causes actually turns out to be less than the blurring caused by the imperfect optics of the eye, so in fact no resolution is sacrificed by coupling [14].

While mammalian cones have coupling that is, in effect, strong enough to spread signals only to immediate neighbors but no further, poikilothermic vertebrates have somewhat stronger coupling between their photoreceptors. Undoubtedly the same arguments about signal-to-noise ratio apply, but as first noticed by Hodgkin and his collaborators, signal spreads a long way through the rod network shortly after a light comes on, but then contracts down to a much smaller area. A plausible, though unproven idea stemming from this finding is that the retina, at the level of bipolar cells, might employ two sampling strategies to read the rod network: one with good temporal resolution but poor spatial resolution, and the other with the converse properties [5].

References

1. Passaglia CL, Enroth-Cugell C, Troy JB (2001) Effects of remote stimulation on the mean firing rate of cat retinal ganglion cells. J Neurosci 21:5794–5803
2. Hartline HK, Ratliff F (1957) Inhibitory interaction of receptor units in the eye of Limulus. J Gen Physiol 40:357–376
3. Lamb TD, Simon EJ (1976) The relation between intercellular coupling and electrical noise in turtle photoreceptors. J Physiol 263:257–286
4. Attwell D, Wilson M (1980) Behaviour of the rod network in the tiger salamander retina mediated by membrane properties of individual rods. J Physiol 309:287–315
5. Detwiler PB, Hodgkin AL, McNaughton PA (1980) Temporal and spatial characteristics of the voltage response of rods in the retina of the snapping turtle. J Physiol 300:213–250
6. Lamb TD (1976) Spatial properties of horizontal cell responses in the turtle retina. J Physiol 263:239–255
7. Kamermans M, Fahrenfort I, Schultz K, Janssen-Bienhold U, Sjoerdsma T, Weiler R (2001) Hemichannel-mediated inhibition in the outer retina. Science 292:1178–1180
8. Spillmann L (1994) The Hermann grid illusion: a tool for studying human perspective field organization. Perception 23:691–708
9. Teranishi T, Negishi K, Kato S (1983) Dopamine modulates S-potential amplitude and dye-coupling between external horizontal cells in carp retina. Nature 301:243–246
10. Dowling JE (1992) The Charles F. Prentice Medal Award Lecture 1991: dopamine; a retinal neuromodulator. Optom Vis Sci 69:507–514
11. Xia XB, Mills SL (2004) Gap junctional regulatory mechanisms in the AII amacrine cell of the rabbit retina. Vis Neurosci 21:791–805
12. Srinivasan MV, Laughlin SB, Dubs A (1982) Predictive coding: a fresh view of inhibition in the retina. Proc R Soc Lond B Biol Sci 216:427–459
13. Hosoya T, Baccus SA, Meister M (2005) Dynamic predictive coding by the retina. Nature 436:71–77
14. DeVries SH, Qi X, Smith R, Makous W, Sterling P (2002) Electrical coupling between mammalian cones. Curr Biol 12:1900–1907

Retinal Photoreceptors

Nathan S. Hart
School of Biomedical Sciences, University of Queensland, St. Lucia, Brisbane, QLD, Australia

Synonyms

Visual cells; Rods; Cones

Definition

Photoreceptor cells are light-sensitive neurons that respond with a graded change in the release of ►neurotransmitter (►Glutamate) from their ►synaptic terminal. Photoreceptors are found in most classes of metazoan organism and vary widely in structure and function. Photoreceptors are predominantly located in the ►retina of both invertebrate and vertebrate eyes, where they are involved in vision, but they may also be located extraocularly on the integument or in the brain.

Characteristics

Quantitative Description

Photoreceptors are neurons that respond with a graded change in transmembrane potential (►Membrane potential – basics), to the absorption of photons by light-sensitive proteins (►Photopigments) embedded in specialized regions of their plasma membrane. The size,

R

shape, ultrastructure, biochemistry, electrical properties and developmental origins of photoreceptors vary considerably across the animal kingdom. For the sake of brevity, the following descriptions are related to vertebrate visual photoreceptors (rather than non-visual retinal or extraocular photoreceptors), and in most cases, draw upon mammalian examples.

Photoreceptor Types

The vertebrate retina is inverted by virtue of its developmental origins and, consequently, the photoreceptors are situated close to the ►sclera at the back of the eye. They point away from the incident light, which must traverse the other retinal layers before reaching the photoreceptors [1]. There are two distinct types of retinal photoreceptor: ►rods that operate under low light (►Scotopic) levels of illumination and ►cones that operate under bright light (►Photopic conditions). The rod-cone nomenclature reflects differences in aspects of photoreceptor morphology in the mammalian retina: the outer segments of rods are rod-like whereas cone outer segments are typically conical (Fig. 1).

However, it should be remembered that this is not always the case – even within the same retina – and it can be misleading to classify photoreceptors using morphology alone; rods and cones also differ in their biochemistry, physiology and function.

Retinal Photoreceptors. Figure 1 Schematic representations of vertebrate rod and single cone retinal photoreceptors. Oil droplets – found in the photoreceptor ellipsoid in some species – are not shown. See text for details.

The human retina contains a single type of rod photoreceptor but three spectrally distinct cone types, which each contain a different cone photopigment maximally sensitive to blue, green and red light (see ►Photopigments). The cones of humans and other placental mammals, lungfish and elasmobranchs are all of the "single" type. Amphibians, marsupials, monotremes, birds, reptiles and teleost fish possess an additional photoreceptor type that consists of a pair of closely opposed cone cells. They are usually referred to as double cones when the two members are unequal in size – with a larger "principal" and a smaller "accessory" member – and twin cones when the two members are of similar size and shape. The closely opposed outer segments of the two members are separated from those of other photoreceptors – but not each other – by the processes of ►retinal pigmented epithelium (RPE) cells; the two members are thought to be both optically and electrically coupled.

Double cones are more widely distributed among the vertebrate classes, and both members usually contain the same spectral type of photopigment. Often, but not always, the principal member (and occasionally the accessory member) contains an oil droplet (see below). Twin cones occur predominantly in teleost fishes, and may have identical or distinct photopigments in each member, even within the same retina.

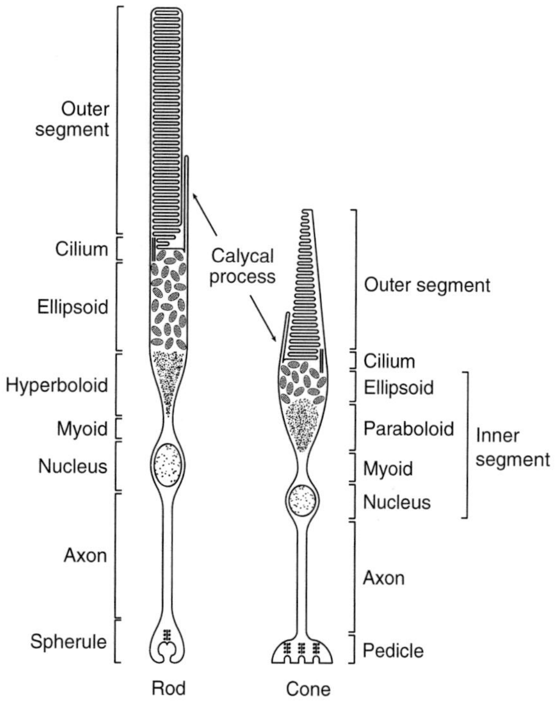

Retinal Topography

The absolute density and relative proportion of rods and cones varies as a function of visual ecology. Strongly diurnal species (e.g. most birds, turtles) have a lower rod:cone ratio than nocturnal species (e.g. many rodents) and, in some cases, lack rods altogether (e.g. lizards). The average human retina contains about 4.6 million cones and 92 million rods, giving a rod to cone ratio of about 20:1 (excluding the ►fovea). In contrast, nocturnal rodents have rod to cone ratios of around 100:1 [2].

Moreover, the topographic distribution of photoreceptors across the retina is usually non-uniform. Most animals have a retina with one or more areas of increased photoreceptor density, which permit increased spatial sampling of the retinal image and provide high acuity vision over a defined region of visual space. The areas are often circular and usually occur in the central (►Area centralis) or temporal (►Area temporalis) retina, but may also form horizontal streaks (►Area horizontalis).

In some cases, the vitreal surface of the retina is indented above the area of highest photoreceptor density. This type of area is called a fovea, and the foveal indentation or pit is a result of the lateral displacement of secondary retinal neurons from the optical path, to provide the incident light with unimpeded access to the photoreceptor layer. The shape of the foveal pit varies in curvature, and may serve to magnify the image projected onto the retina by virtue of the difference in ►refractive index between the vitreous and retina. The human fovea contains approximately 200,000 cones, with reported peak densities ranging from 100,000 to 320,000 cells mm^{-2}. Cone density falls rapidly with increasing eccentricity: at 10° eccentricity – or 3 mm from the centre of the fovea – it averages 7,000 cones mm^{-2}. Rods are excluded from the human fovea but attain their highest density of around 160,000 cells mm^{-2} in a perifoveal ring located at 18° eccentricity or about 5 mm from the centre of the fovea [3].

Description of the Structure

Vertebrate photoreceptors are comprised of three morphologically and functionally distinct regions: the outer segment, inner segment and synaptic terminal.

Outer Segment

Human rod outer segments are 2 μm in diameter and vary in length with retinal eccentricity from 24 μm at the periphery to 40 μm in the parafoveal region. Human foveal cone outer segments are 30–40 μm in length, are less tapered/conical than cones located in the periphery and have a diameter of about 0.8 μm [1]. The outer segments of cones in the peripheral retina are more conical but only about half as long. Photoreceptor outer segment dimensions also vary widely with habitat and life history. For example, nocturnal species, or those inhabiting light-limited environments such as the deep sea, tend to have longer and wider outer segments than diurnal species in order to capture more of the available light.

Rod outer segments consist of stacks of isolated membranous discs or "saccules" bounded by the plasma membrane. Each saccule is around 19 nm in thickness. The periodicity of saccule spacing is 28 nm and, in humans, there are around 1,000 saccules in each rod outer segment. In contrast, cone outer segments are formed by multiple infolding of the plasma membrane. Both the saccule and infolded plasma membranes are packed with photopigment molecules that absorb the incident light. The high lipid content of the outer segment endows it with a high refractive index relative to the surrounding extracellular space. Consequently, the outer segment acts as a waveguide, and light entering at the base is contained and propagated along its length by total internal reflection.

The base of the outer segment is connected to the inner segment by a modified non-motile ►cilium, which projects through a narrow cytoplasmic bridge ("ciliary stalk") that is 1 μm in length. In some species, such as rodents, a second contiguous cytoplasmic bridge is observed. A number of fine processes originating from the inner segment, called calycal processes, extend distally along the outer segment for approximately one third of its length and probably provide mechanical support. In rod outer segments, the calycal processes are contiguous, with indentations in the plasma membrane that overlie – but do not invaginate – scalloped incisures in the radial edge of the saccule membrane.

Inner Segment

The inner segment contains all the structures necessary for cellular metabolism and protein synthesis. It may also contain organelles that help to capture and/or spectrally filter the incident light and to focus it onto the outer segment. The most important structures are described below:

Ellipsoid

Photoreceptors contain numerous ►mitochondria in the distal portion of their inner segment. These are packed together in a highly refractile body called the ellipsoid, and may be oriented with their long axis parallel to that of the photoreceptor, as in mammals, or clumped randomly. Photoreceptors are energetically demanding cells and the mitochondria must generate sufficient adenosine triphosphate (ATP) to support cellular function, in particular the maintenance of ►sodium/potassium (Na^+/K^+) pumps (Ion transport) in the plasma membrane, the production of photopigment molecules and the turnover of guanosine 3′,5′ cyclic monophosphate (cGMP; see ►Phototransduction).

In the mammalian retina, cones contain many more mitochondria than rods, but it is not readily apparent why they should do so on a metabolic basis [4]. It is possible that they also have an optical function by increasing the refractive index of the inner segment. Both the inner and outer segments have a higher refractive index than the surrounding medium and, at least in cones, the inner segment is significantly wider than the outer segment. Light striking the inner segment is funnelled by total internal reflection into the outer segment, a physical phenomenon known as waveguiding. Consequently, the cross-sectional area of the inner segment, rather than that of the outer segment, defines the photon capture area of the photoreceptor. Human foveal cones are 2.3 μm in diameter whereas those in the peripheral retina are 7.9 μm in diameter, a difference that confers a 12-fold increase in photon capture area and, therefore, optical sensitivity [1].

The ellipsoids of some species show further specializations. The proximal region of the cone ellipsoid in some lizards contains aggregations of extended endoplasmic reticula, known as refractile bodies, which may well have a light-gathering function in addition to any putative metabolic storage role. Similarly, tree shrews (*Tupaia* sp.) have large distended mitochondria ("megamitochondria") with visible but irregular cristae, which may confer an enhanced light-gathering ability to the ellipsoid. This structure could replace the colorless oil droplets present in the cones of their ancestors and retained by marsupials but lost by other placental mammals [5]. Other modified mitochondria (ellipsosomes) are present in the ellipsoids of teleost fish and contain filtering pigments that resemble reduced cytochrome C.

Oil Droplets

The photoreceptors of a number of species contain an inclusion located either within or just distal to the ellipsoid, known as an oil droplet. As their name suggests, oil droplets are composed predominantly of lipids, but they may also contain light-absorbing ►carotenoid pigments. Pigmented oil droplets are found in the cones of birds, turtles, lizards and lungfish; colorless oil droplets occur in some geckos, anuran amphibians, chondrostean fishes, marsupials and some monotremes. Oil droplets are absent from lampreys, teleosts, elasmobranchs, snakes, crocodilians and placental mammals [6].

Pigmented oil droplets act as filters that tune the spectral sensitivity of the photoreceptor, in most cases narrowing the spectral sensitivity function and shifting the wavelength of peak sensitivity to a wavelength longer than the wavelength of maximum absorbance (λ_{max}) of the photopigment in the outer segment. Non-pigmented oil droplets probably serve a similar function to ellipsosomes in capturing and focusing light into the outer segment.

Paraboloid and Hyperboloid

The rod and cone inner segments of some holosteans, amphibians, birds and reptiles contain a granular structure proximal to the ellipsoid known as the paraboloid (cones) or hyperboloid (rods). It is thought to act as a store of glycogen, presumably supporting the high metabolic activity of the cell.

Myoid

The myoid region lies proximal to the ellipsoid (and paraboloid/hyperboloid if present) but distal to the nucleus. It contains free ribosomes, rough endoplasmic reticulum (RER) and the Golgi apparatus, and it is the site of protein synthesis in the photoreceptor. Photopigment ►messenger RNAs from the nucleus migrate to the RER and are translated into opsin proteins (see ►Photopigments) that are eventually packaged into small vesicles by the Golgi apparatus. These vesicles migrate to the ciliary stalk, pass through the cytoplasmic bridge into the outer segment and become incorporated in the saccule and plasma membranes. In some lamprey and lizard species, the entire myoid contains a diffuse yellow-orange pigment that, like the pigmented oil droplets, spectrally filters the incident light before it reaches the outer segment.

In some "lower" vertebrates (e.g. fish, amphibians) the myoid is motile. Cone myoids contract in the light and elongate in the dark; the opposite is true for the rod myoid. These so called "retinomotor movements" are substantial (50–70 μm) and represent a form of light/dark adaptation, the function of which is to shield the rods in the retinal pigmented epithelium during the day but fully expose them at night. Myoid contraction and elongation is controlled by both light-dependent and endogenous (circadian) mechanisms [7].

Synaptic Terminal

Beneath the myoid lies the nucleus and, proximal to this, a thin fiber (axon) that connects the inner segment to the synaptic terminal. The synaptic terminal of a photoreceptor cell ramifies in the outer plexiform layer of the retina and is the site of communication with other retinal neurons. The graded changes in outer segment transmembrane potential that occur as a result of the phototransduction process propagate electrotonically (►Electrotonic spread) to the synaptic terminal and are communicated to other retinal neurons in two ways. Firstly, the electrical potential may be transferred passively to adjacent photoreceptors via low-resistance intercellular junctions, called ►gap junctions. Secondly, changes in membrane potential at the synaptic terminal alter the internal Ca^{2+}concentration and modulate the rate at which ►synaptic vesicles fuse with the plasma membrane and release neurotransmitter (glutamate) into the synapse.

The morphology of rod and cone synaptic terminals differs markedly. The rod synaptic terminal ("►spherule") is roughly spherical and contains a single invagination ("synaptic cleft") within which lie the processes of two to five rod bipolar cells and two horizontal cells (►Retinal bipolar cells; ►Inherited retinal degenerations; ►Vision). The cone synaptic terminal ("►pedicle") is much larger and almost pyramidal in shape. In the mammalian retina, peripheral cones have much larger pedicles than foveal cones and display as many as 50 synaptic clefts. Like the rod spherule, the synaptic clefts of cone pedicles contain the processes of two or more cone bipolar cells and two horizontal cells [8].

In both rod and cone synaptic terminals, the active zone immediately above the synaptic clefts contain one (cones) or two (rods) synaptic ribbons that modulate the release of neurotransmitter (see ►Ribbon synapses). Rod and cone bipolar cell processes entering the synaptic cleft are termed "central" or "invaginating" processes. Other bipolar cell processes contact the cone pedicle on either side of the invaginating processes ("semi-invaginating" processes) or diffusely across the base of the pedicle ("flat" processes). Consequently, each cone pedicle may make several hundred synapses with 10 or more postsynaptic neurons.

Higher Level Structures

In the mammalian retina, rods contact only one type of bipolar cell, the rod ON bipolar (►Retinal bipolar cells). Cone photoreceptors synapse with up to 11 different types of ON and OFF cone bipolar cells. In other vertebrates, both rods and cones may contact the same bipolar cell, which may be either ON or OFF and either rod- or cone-dominated. Photoreceptors are presynaptic to horizontal cells in a sign-conserving manner, i.e. ►hyperpolarization of the photoreceptor transmembrane potential results in a hyperpolarization of the horizontal cell. However, horizontal cells feed back onto photoreceptors in a sign-inverting manner that antagonizes the effects of transmembrane hyperpolarization, reduces glutamate release and allows the outer retina to adapt to steady illumination [1].

Regulation of the Structure

In the dark, a balanced flow of cations into the outer segment and out of the inner segment (known as the dark current) maintains a moderate depolarization of the transmembrane potential. The magnitude of the dark current in the rods (−34 pA) and cones (−30 pA) is similar in the macaque monkey. Stimulation of photoreceptors with light causes a reduction in the dark current due to phototransduction. This change in dark current is called the photocurrent and, although both rods and cones are capable of signaling the absorption of a single photon of light, the photocurrent produced by a single photoisomerization event (see ►Phototransduction) is much smaller in cones (33 fA) than it is in rods (700 fA) [1]. Consequently, more photons per unit time are required by cones than rods to provide a large enough change in glutamate release at the synaptic terminal to be reliably detected by the bipolar cells and, therefore, rods are better for vision under low levels of illumination.

There are also qualitative differences in the time course of the photocurrent between rods and cones. The magnitude of the photocurrent peaks about 50 ms after the onset of light in primate cones but takes up to four times longer in the rods. Moreover, the rod photocurrent decays more slowly, taking up to one second to return to zero, whereas cones recover up to five times faster. These differences in response and recovery kinetics enable cone pathways to respond to higher temporal frequencies than the rods, although at the expense of absolute sensitivity.

Photoreceptor outer segments are renewed continuously. Rod saccules are shed from the distal tip of the outer segment in packets of 8–30, where they are taken up in phagocytotic vesicles ("phagosomes") by RPE cells. Shed saccules are replaced by the synthesis of additional membrane at the base of the outer segment, where new photopigment molecules generated in the inner segment are incorporated. Cones shed and regenerate membrane in a similar fashion to rods, although newly synthesized photopigment molecules can diffuse to any location in the membrane, as the outer segment is not compartmentalized. Membrane shedding follows a circadian rhythm; cone membranes are shed after dusk when the visual system begins to rely on rods, and rod saccules are shed at dawn when cones become dominant [9].

Function

The retina is a two-dimensional sensor array that extracts salient features from the image of a three-dimensional world projected onto the back of the eye by the lens and cornea (►Vision). The size, packing density, spectral sensitivity and electrical response characteristics of the photoreceptors limit the spatial, temporal and chromatic information that can be extracted from that image.

Rods are more sensitive than cones and are used in dim light (scotopic) conditions. Most vertebrates, including humans, have only one spectral class of rod and, consequently, are essentially color blind at night. Cones, on the other hand, function at higher (photopic) light levels. The possession of both scotopic and photopic photoreceptor types – otherwise known as a ►duplex retina – has the primary function of extending the range of light intensities over which the visual system is operational. For most animals, this range varies over 10 log units from starlight to bright sunlight [1].

Where multiple spectral types of cone are present, their outputs can be compared by secondary neurons to extract chromatic information from the retinal image. The ability to distinguish objects based on their color (spectral reflectance) independently of brightness is called color vision (►Color processing). Most mammals have only two spectral types of cone and have a dichromatic color vision system. Humans and some primates have three spectral cone types and are trichromats. Birds have a tetrachromatic color vision system based on four spectral types of single cone photoreceptor [10]. The function of double/twin cones is unclear, despite being the most numerous photoreceptor types in many species. Limited evidence suggests that, at least in the avian retina, double cones mediate purely achromatic (brightness discrimination) tasks, including motion detection, and are not involved in color vision.

R

References

1. Rodieck RW (1998) The first steps in seeing. Sinauer Associates, Sunderland, MA
2. Ahnelt PK, Kolb H (2000) The mammalian photoreceptor mosaic – adaptive design. Prog Retin Eye Res 19:711–777
3. Curcio CA, Sloan KR, Kalina RE, Hendrickson AE (1990) Human photoreceptor topography. J Comp Neurol 292:497–523

4. Hoang QV, Linsenmeier RA, Chung CK, Curcio CA (2002) Photoreceptor inner segments in monkey and human retina: mitochondrial density, optics, and regional variation. Vis Neurosci 19:395–407
5. Knabe W, Skatchkov S, Kuhn HJ (1997) "Lens mitochondria" in the retinal cones of the tree-shrew *Tupaia belangeri*. Vis Res 37:267–271
6. Walls GL (1942) The vertebrate eye and its adaptive radiation. Hafner, New York
7. Burnside B, Wang E, Pagh-Roehl K, Rey H (1993) Retinomotor movements in isolated teleost retinal cone inner-outer segment preparations (CIS-COS): effects of light, dark and dopamine. Exp Eye Res 57:709–722
8. Haverkamp S, Grünert U, Wassle H (2000) The cone pedicle, a complex synapse in the retina. Neuron 27:85–95
9. Nguyen-Legros J, Hicks D (2000) Renewal of photoreceptor outer segments and their phagocytosis by the retinal pigment epithelium. Int Rev Cytol 196:245–313
10. Hart NS (2001) The visual ecology of avian photoreceptors. Prog Retin Eye Res 20:675–703

Retinal Pigment Epithelium

Definition

These cells form the outer part of the vertebrate retina, forming the outer blood-retinal barrier and providing key roles in chromophore recycling and transport of metabolites and metabolic byproducts.

►Inherited Retinal Degenerations
►Photoreceptors

Retinal Ribbon Synapses

Catherine W. Morgans, Philippa R. Bayley
Neurological Sciences Institute, Oregon Health and Science University, Beaverton, OR, USA

Definition

Ribbon synapses are specialized synapses of the primary sensory neurons of the eye (retinal ►photoreceptors) and ear (►cochlear hair cells and ►vestibular hair cells). They are also formed by ►retinal bipolar cells, ►vestibular receptor cells, and fish ►electroreceptors. Morphologically, these synapses are characterized by a presynaptic electron-dense bar, the ►synaptic ribbon, at the site of ►neurotransmitter release (Fig. 1). Ribbon synapses support continuous release of the neurotransmitter ►glutamate, and modulate the rate of release in response to graded changes in membrane potential (►Membrane potential – basics).

Characteristics

Description of the Structure

Anatomically, ribbon synapses are characterized by the presence of a structural specialization of the ►active zone, the synaptic ribbon. The synaptic ribbon is a planar structure in retinal photoreceptors and bipolar cells, and a spheroid structure in inner ear hair cells where it is also referred to as a "synaptic body" (Fig. 1). The ribbons are attached to the plasma membrane and typically extend perpendicularly into the cytoplasm up to several hundred nanometers. The ribbon is surrounded by a monolayer of ►synaptic vesicles that are tethered to the ribbon by fine filaments. Vesicles at the base of the ribbon are docked on the plasma membrane and represent the readily releasable pool of synaptic vesicles. Those tethered further up the ribbon provide a reserve pool of synaptic vesicles. Synaptic ribbons vary in size and shape depending on the cell, but for a particular cell type, such as the mammalian rod photoreceptor, the total ribbon area and geometry vary comparatively little, even between species.

Photoreceptor ribbon synapses are the most structurally complex of ribbon synapses. Photoreceptor ribbon synapses are formed by an invagination into the photoreceptor terminal, over the apex of which lays the synaptic ribbon, and into which postsynaptic processes from ►horizontal cells (►Lateral interactions in the retina, ►inherited retinal degenerations, ►vision) and bipolar cells protrude (Fig. 1). Rod photoreceptor terminals (referred to as ►spherules) contain a single invagination that gives rise to one or two ribbon synapses [1], whereas cone photoreceptor terminals (referred to as ►pedicles) contain 10–30 invaginations, depending on whether they are in the central or peripheral retina, each invagination corresponding to a single ribbon synapse (Fig. 1). The large synaptic ribbons of rod photoreceptors curve around the invagination in a characteristic horseshoe shape, which can be clearly discerned at the light microscopic level by immunofluorescent staining of ribbon proteins (such as bassoon or RIBEYE, see Fig. 2).

The photoreceptor synaptic ribbon is attached to the plasma membrane via a linear, trough-shaped structure, the ►arciform density. The arciform density defines the site of neurotransmitter release, or active zone, of the photoreceptor ribbon synapse (Fig. 2b). The plasma membrane is pinched where it contacts the arciform density to form a ridge along the underside of the invagination. Synaptic vesicles at the base of the ribbon are docked on the plasma membrane, fusing on either side of the arciform density. Two postsynaptic horizontal cell processes, known as "lateral elements," extend

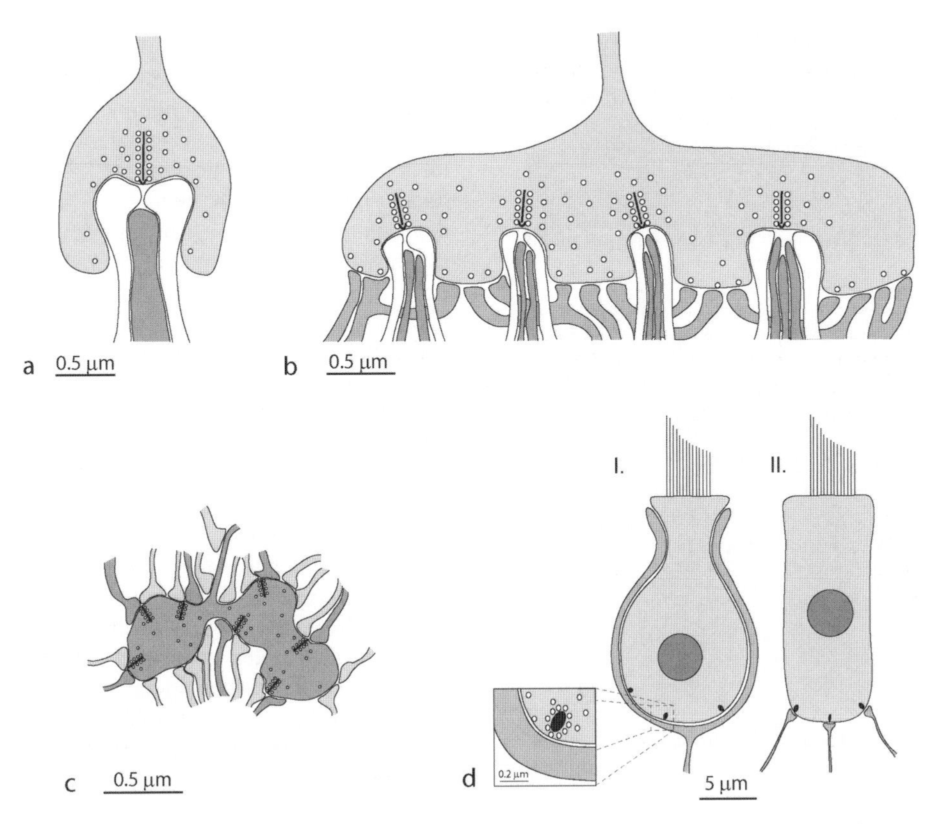

Retinal Ribbon Synapses. Figure 1 Schematic diagrams of ribbon synapses in the retina and inner ear, drawn approximately to scale (as indicated). (a). Rod photoreceptor terminal (spherule) containing a vesicle-covered synaptic ribbon with an arciform density at its base. Postsynaptic horizontal cell (white) and rod bipolar cell (orange) processes protrude into the invagination, opposed to the site of transmitter release. (b) Cone photoreceptor terminal (pedicle) containing four separate ribbons with their associated arciform densities, and postsynaptic invaginations of horizontal cell (white) and ON-cone bipolar cell (orange) processes. OFF-cone bipolar cells (purple) make synapses at flat contacts, outside the invaginations. (c) Cone bipolar cell terminal (orange) containing multiple small ribbons. Postsynaptic amacrine cell (yellow) and ganglion cell (green) processes are opposed to the sites of neurotransmitter release. (d) Type I (I.) and type II (II.) inner ear hair cells contain spherical synaptic bodies, covered with tethered vesicles. Type I hair cells form a single calyceal terminal, whereas type II hair cells make synapses with individual afferent boutons, each bouton receiving input from a single ribbon.

their tips close to the synaptic ribbon. One or more bipolar cell processes, known as "central elements", occupy the central region of the invagination, with their tips located at a further distance from the ribbon. An electron micrograph of a photoreceptor ribbon synapse will often exhibit a "triad" arrangement, with two horizontal cell processes flanking a single bipolar cell process within the invagination postsynaptic to the ribbon. Rods make synaptic contacts with only one type of bipolar cell, the rod bipolar cell. Cones contact both ON-bipolar cells (depolarize in light) and OFF-bipolar cells (hyperpolarize in light) (►Retinal bipolar cells). At cone ribbon synapses it is only the ON-bipolar cells that extend their dendrites into the invagination; OFF-bipolar cells contact cone pedicles outside the invaginations at basal contacts.

Bipolar cell and hair cell ribbon synapses are simpler in structure (Fig. 1). They both lack invaginations and arciform densities. It is unclear how the ribbons are anchored to the plasma membrane in these cells. As in photoreceptors, bipolar cell synaptic ribbons are planar structures sitting perpendicular to the plasma membrane above a linear active zone, although they are smaller than photoreceptor ribbons. Two postsynaptic processes from amacrine cells (►Retinal direction selectivity: role of starburst amacrine cells) and ganglion cells (►Retinal ganglion cells) abut the bipolar cell membrane, one on either side of the active zone. Each bipolar cell terminal forms multiple ribbon synapses.

Hair cells also contain multiple ribbons, averaging ~20/cell, each covered with a layer of one hundred to several hundred tethered synaptic vesicles. Each hair cell ribbon synapse releases glutamate onto a single postsynaptic ending/contact. Type I vestibular hair cells are enveloped by a single calyceal terminal, which receives the output from all of the hair cell's ribbon synapses;

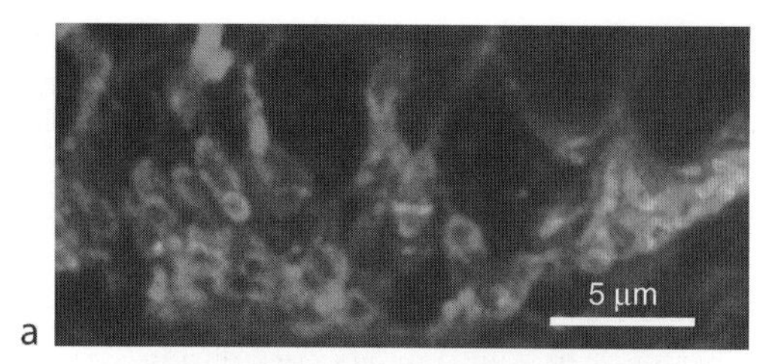

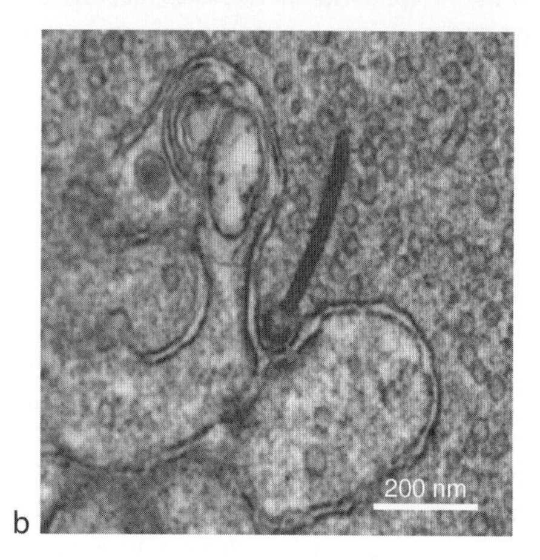

Retinal Ribbon Synapses. Figure 2 (a) Individual arc-shaped synaptic ribbons, labeled with an antibody to RIBEYE (red), are seen within rod and cone photoreceptor terminals (green) in a confocal micrograph through the ►outer plexiform layer of the rat retina. (b) Rod ribbon synapse ultrastructure is observed in more detail by transmission electron microscopy. The synaptic ribbon, in transverse section, is perpendicular to the cell membrane, and overlies an arciform density (red arrowhead). Vesicles of neurotransmitter are evident throughout the terminal, with a subset of these tethered to the ribbon. Postsynaptic horizontal and rod bipolar cell processes invade the terminal, opposed to the site of neurotransmitter release.

whereas each ribbon synapse of a type II vestibular hair cell has a separate post-synaptic dendrite [2]. Hair cells in the cochlea also have only one post-synaptic afferent/dendrite per ribbon [3]. Hair cell synaptic ribbons are often spherical, but elongated and planar ribbons are also found, especially in type II vestibular hair cells.

Quantitative Description of Ribbon Synapses in the Retina

Reconstruction of rod spherules in the cat retina reveal a planar synaptic ribbon ~2 μm in length (or two half-sized ribbons), extending from the plasma membrane into the cytoplasm by ~0.4 μm. Each face of the ribbon binds approximately 385 synaptic vesicles, 65 of which lie along the bottom of the ribbon docked at the plasma membrane on each side of the arciform density. Cone synaptic ribbons are smaller but more numerous than in rods. Cone ribbons are ~1 μm long and extend from the plasma membrane into the cytoplasm by ~0.2 μm. Although different ribbons within a cone terminal vary in length, the total ribbon length and number of tethered vesicles per cone is quite uniform within a retinal locus. The 20 or so ribbons in a primate foveal cone tether a total of ~3,600 synaptic vesicles, ~720 of which are docked at the plasma membrane [4]. Bipolar cells, depending on their type, contain from 30–100 synaptic ribbons, but they are small compared to photoreceptor ribbons, tethering several dozen vesicles each. The total number of ribbons and associated docked vesicles per ►synaptic terminal may be related to the information content that has to be transmitted by the neuron [4]. For example, rod photoreceptors transmit the detection of single photons in very dim light, like starlight, whereas cones must transmit finely graded changes over orders of magnitude of light intensity (►Photoreceptors, ►Phototransduction).

The most detailed and quantitative description of a ribbon synapse has been obtained for the mammalian rod photoreceptor. Digital reconstruction of primate and cat rod spherules from serial electron micrographs has revealed the following invariant features of rod ribbon synapses [1]: Each spherule contains a single invagination approximately 1μm in diameter, and two synaptic units. Post-synaptically, each unit is made up of two lateral elements (horizontal cell dendrites) and at least one central element (rod bipolar cell dendrite). Although there are always two sets of postsynaptic processes, presynaptically a spherule may contain either one or two synaptic ribbons. However, even in cases where a spherule contains a single ribbon, the ribbon contacts two discrete arciform densities, each ~1.0 μm in length. Typically, the two arciform densities are in different planes, so that a single ribbon will twist to contact both. The volume of extracellular space within the invagination is small, roughly 0.1μm^3, and the most distant bipolar cell glutamate receptors from the active zone (as defined by the arciform density) are within 1.5 μm. Thus, diffusion of glutamate to receptors on all postsynaptic processes should be rapid.

Higher Level Structures

The principal component of synaptic ribbons is the ribbon-specific protein, RIBEYE, composed of a unique N-terminal domain possessing a self-aggregating activity that may be important for the polymerization of the ribbons, and a C-terminal domain that is identical (minus the first 20 amino acids) to the transcriptional co-repressor, C-terminal binding protein 2 (CtBP2) [5]. Other proteins associated with the ribbons are components of the presynaptic matrix at conventional synapses including RIM (rab3-interacting protein), and the large (>400 kD) scaffolding proteins bassoon and piccolo, suggesting that at least some aspects of ribbon function are equivalent to that served by the presynaptic matrix at the active zone of conventional synapses. Bassoon has been localized by post-embedding immuno-electron microscopy to the base of photoreceptor ribbons close to the plasma membrane, whereas piccolo and RIM are localized towards the distal portion of the ribbon [6]. In mice deficient in bassoon, photoreceptor and hair cell synaptic ribbons are either absent or free-floating in the cytoplasm [7,8] suggesting that bassoon is required for anchoring the ribbons to the plasma membrane.

The presynaptic Ca^{2+} channels at ribbon synapses are localized to the plasma membrane at the base of the ribbons [6]. Ribbon synapse Ca^{2+} channels belong to the L-type Ca^{2+} channel family (▶Calcium channels – an overview), and are of the subtypes Cav1.3 (α_{1D}) in cone photoreceptors and hair cells, and Cav1.4 (α_{1F}) in rod and cone photoreceptors and some bipolar cells. A distinctive characteristic of the ribbon synapse Ca^{2+} channels is that they are non-inactivating, an essential property for maintaining transmitter release during prolonged depolarization, as occurs in darkness.

The majority of synaptic vesicle-associated proteins are identical between ribbon and conventional synapses; however, there are a few significant differences [9]. For example, ribbon synapses are the only synapses known not to contain ▶synapsins, peripheral membrane proteins of synaptic vesicles at conventional synapses. Synapsins immobilize synaptic vesicles in the absence of action potentials by linking them to the actin-based cytoskeleton, and then release them upon Ca^{2+} influx to replenish the readily releasable pool. At ribbon synapses, which are tonically active, synaptic vesicles are in constant flux and synapsins are not needed. Retinal ribbon synapses also differ in which ▶syntaxin gene they express. At conventional synapses, syntaxin 1 is one of the key proteins catalyzing synaptic vesicle fusion. Retinal ribbon synapses are the only synapses known not to contain syntaxin 1 but to contain syntaxin 3 instead. This substitution may reflect differences in the regulation of synaptic vesicle fusion between ribbon and conventional synapses.

Function

Ribbon synapses are formed by sensory neurons of the visual, auditory, and vestibular systems. These neurons are electrotonically compact (▶Electrotonic spread), and track changes in external stimuli with graded changes in their membrane potential (in contrast to axon-bearing neurons that fire action potentials). The graded changes in membrane potential in turn modulate the rate of tonic release of the neurotransmitter, glutamate. Synaptic ribbons are not found in neurons that undergo action potential-driven transmitter release. As at other chemical synapses, neurotransmitter is released at ribbon synapses by the Ca^{2+}-dependent fusion of neurotransmitter-filled synaptic vesicles with the plasma membrane. Calcium enters the nerve terminal through voltage-sensitive Ca^{2+} channels that open in response to membrane depolarization and close upon hyperpolarization. Photoreceptors are depolarized in darkness and undergo a continuous stream of synaptic vesicle fusion in this state. Absorption of photons hyperpolarizes the photoreceptor, closing Ca^{2+} channels and reducing glutamate release (▶Phototransduction). The synaptic ribbons most likely serve to maintain a pool of synaptic vesicles in close proximity to the active zone, and ensure the continual replenishment of synaptic vesicles during prolonged depolarizations. They have also been proposed to facilitate compound fusion of vesicles, or simultaneous fusion of multiple vesicles [10].

The critical role of the synaptic ribbons in ▶synaptic transmission at ribbon synapses is illustrated by the phenotype of the bassoon knockout mouse. Elimination of bassoon prevents synaptic ribbons from anchoring at the plasma membrane. The physiological consequence is a drastic impairment of synaptic transmission between photoreceptors and depolarizing bipolar cells [7]. Likewise, in hair cells of the bassoon knockout mouse, patch-clamp recordings (▶Intracellular recording) indicate a 50% reduction in fast ▶exocytosis. The hair cell recordings also reveal substantially smaller Ca^{2+} currents in the bassoon knockout mouse compared to wild-type, suggesting that recruitment and stabilization of Ca^{2+} channels at hair cell active zones may be dependent on association with synaptic ribbons [8].

Retinal Essays: (photoreceptor outer segments), (▶retinal bipolar cells) and (lateral interactions in the retina).

References

1. Migdale K, Herr S, Klug K, Ahmad K, Linberg K, Sterling P, Schein S (2003) Two ribbon synaptic units in rod photoreceptors of macaque, human, and cat. J Comp Neurol 455:100–112
2. Lysakowski A, Goldberg J (1997) A regional ultrastructural analysis of the cellular and synaptic architecture in the chinchilla cristae ampullares. J Comp Neurol 389:419–433
3. Fuchs PA, Glowatzki E, Moser T (2003) The afferent synapse of cochlear hair cells. Curr Opin Neurobiol 13:452–458
4. Sterling P, Matthews G (2005) Structure and function of ribbon synapses. Trends Neurosci 28:20–29
5. Schmitz F, Konigstorfer A, Sudhof T (2000) RIBEYE, a component of synaptic ribbons: a protein's journey through evolution provides insight into synaptic ribbon function. Neuron 28:857–872
6. tom Dieck S, Altrock WD, Kessels MM, Qualmann B, Regus H, Brauner D, Fejtova A, Bracko O, Gundelfinger ED, Brandstatter JH (2005) Molecular dissection of the photoreceptor ribbon synapse: physical interaction of Bassoon and RIBEYE is essential for the assembly of the ribbon complex. J Cell Biol 168:825–836
7. Dick O, tom Dieck S, Altrock W, Ammermuller J, Weiler R, Garner C, Gundelfinger E, Brandstatter J (2003) The presynaptic active zone protein bassoon is essential for photoreceptor ribbon synapse formation in the retina. Neuron 37:775–786

8. Khimich D, Nouvian R, Pujol R, tom Dieck S, Egner A, Gundelfinger ED, Moser T (2005) Hair cell synaptic ribbons are essential for synchronous auditory signalling. Nature 434:889–894
9. Morgans CW (2000) Presynaptic proteins of ribbon synapses in the retina. Microsc Res Tech 50:141–150
10. Prescott ED, Zenisek D (2005) Recent progress towards understanding the synaptic ribbon. Curr Opin Neurobiol 15:431–436

Retinal Slip

Definition

Motion of the visual image on the surface of the retina. Slip of the visual image across large portions of the retina is the stimulus that stimulates optokinetic eye movements, and also the stimulus that produces the adaptation (improvement) of the optokinetic system.

►Optokinetic Response Adaptation

Retinitis Pigmentosa

Definition

A group of hereditary retinal degenerations characterized by loss of peripheral vision (constricted visual fields) and night vision (nyctalopia). Although a vast variety of genetic mutations have been identified, patients with Retinitis Pigmentosa display similar symptoms and ocular fundus appearance in the end stage of the disease: bone spicule pigment deposits, pale atrophic optic nerve head and attenuated blood vessels.

►Inherited Retinal Degenerations

Retino-geniculo-cortical Pathway

Definition

A large fraction (at least 90%) of retinal ganglion cells project to visual cortex through the lateral geniculate nucleus. This pathway is thought to be most important for visual perception. Non-geniculo-cortical pathways, such as the pathways through the suprachiasmatic nucleus or pre-tectum involve smaller numbers of ganglion cells and in some cases have specialized functions (e.g. suprachiasmatic nucleus is involved in diurnal rythyms).

►Retinal Ganglion Cells
►Suprachiasmatic Nucleus
►Vision
►Visual Processing Streams in Primates

Retinohypothalamic Tract

Ignacio Provencio
Department of Biology, University of Virginia, Charlottesville, VA, USA

Synonyms

RHT

Definition

A monosynaptic neural projection that extends from the ►retina to the hypothalamus.

Characteristics

Origin and Projections

The retinohypothalamic tract (RHT) originates from a subset of ►retinal ganglion cells (RGCs). In rodents, this small cohort of cells represents only about 1–2% of all RGCs of which 80% or more express the ►photopigment melanopsin, thereby making them intrinsically photosensitive. However, compared to ►rod and cone photoreceptors, these photoreceptive cells are relatively insensitive to light. The sparse, varicose dendritic arbors of murine ►intrinsically photosensitive RGCs (ipRGCs) arise from 2 or 3 primary dendrites emanating from perikarya about 20 μm in diameter. ipRGC dendritic fields have a mean diameter of 450 μm and tile the retina with substantial overlap. The arbors themselves contain melanopsin and are capable of ►phototransduction, giving each of these cells a capture radius of approximately 15° of visual space. These overlapping dendrites coupled with the relative photic insensitivity of ipRGCs are consistent with the sensory characteristics observed in many non-visual responses to light such as the synchronization (►entrainment) of circadian rhythms to the astronomical day. Relative to the visual system, the circadian axis requires higher levels of light to elicit a response and can integrate photic stimuli over longer temporal intervals and broader spatial domains. In essence, while the visual system functions like a camera, the "photoreceptive net" arising from the widely distributed ipRGCs functions like a photographer's light meter [1,2].

In most animals examined to date, the ipRGCs are distributed evenly throughout the retina. In rats and primates, however, there is a shallow density cline peaking in the superiotemporal and parafoveal retinal domains, respectively. The principal target of the RHT is the ►suprachiasmatic nucleus (SCN), the site of a primary ►circadian pacemaker. The RHT is the anatomical route by which information about environmental light levels is conveyed from the eye to the SCN, where it is processed and used to entrain a multitude of circadian rhythms to the prevailing ►light:dark cycle. Among these are circadian rhythms of activity, daily variations in core body temperature, and 24-h rhythms in levels of hormones such as ►melatonin and cortisol [2,3].

The topography of the retinal innervation to the SCN is highly variable across mammalian species and even varies significantly among the rodents. In general, the ventrolateral aspect of the SCN receives the most dense innervation from the retina and is coextensive with the distribution of vasoactive intestinal peptide (VIP)-positive cells. However, it should be noted that the entire SCN receives some retinal afferents [3,4].

The degree to which the retina projects to each SCN also varies significantly among mammals. Contralaterality of retinal projections to central visual structures is highly correlated with the lateral placement of the eyes on the head. For example, mammals such as rodents with very limited binocular vision due to laterally positioned eyes show a high degree of contralaterality of the retinal projections to the dorsal lateral geniculate nucleus (LGN). By contrast, the retinal projections of primates and cats that have frontally positioned eyes are not contralaterally dominant, but rather extend an equal number of contralateral and ipsilateral retinal projections to the geniculate body.

The degree of bilaterality of the RHT demonstrates no such correlation to the lateralization or frontalization of the eyes. Perhaps the most striking example is that of the scaly anteater whose laterally placed eyes results in a total binocular field of only 15°, a feature that is reflected in the optic chiasm where greater than 99% of retinal fibers cross the midline and project to the contralateral lateral geniculate. However, both eyes project an equal number of axons to each SCN. As previously mentioned, primates, due to very frontalized eyes, have an extensive binocular visual field and accordingly, have equally weighted bilateral projections to central visual sites. This is in stark contrast to the projections to the SCN which are heavily ipsilaterally weighted. For example, in the gibbon, almost 90% of fibers emerging from one retina project to the ipsilateral SCN, whereas visual projections are balanced [5].

In the hamster, about 5% of the RGCs of the RHT bifurcate and send axonal collaterals to both SCN. In addition, some RGCs that project to the SCN via the RHT also send axonal collaterals to other non-visual retinorecipient sites in the brain such as the ►intergeniculate leaflet and the ►olivary pretectal nucleus. Whether this is a common feature across all mammals remains to be determined [4]. It should be noted that while the SCN is the primary and best-studied target of the RHT, other hypothalamic sites receive RHT innervation. These include the retrochiasmatic area, the subparaventricular zone, the perisupraoptic nucleus and the lateral hypothalamus [2,3]. Understanding how these regions mediate responses to light is a subject of ongoing investigation.

Neurotransmitters

There is abundant evidence that ►glutamate is the primary neurotransmitter of the RHT [6,7]. Among this evidence is the presence of anti-glutamate immunoreactivity in presynaptic terminals within the SCN of rat and mouse. Furthermore, glutamate release can be induced in the SCN by electrically stimulating the optic nerve and glutamate application mimics that effect of light on the SCN. However, some investigators have observed that application of glutamate onto the SCN does not accurately mimic the circadian phase shifting effect of light, although administration of the glutamate agonist *N*-methyl-*D*-aspartate (NMDA) does indeed mimic light and these effects can be blocked by NMDA antagonists. Intraperitoneal injection of MK801, a competitive NMDA receptor antagonist, blocks the phase-shifting effect of light on mouse circadian locomotor rhythms [4].

A hallmark of the lateral geniculate, a well-characterized target of glutamatergic neurotransmission, is an abundance of the ►glutamate vesicular transporters VGluT1 and VGluT2. The SCN, however, shows low expression of both of these proteins. The presence of glutamate receptors in the SCN, however, reinforces a prominent role for glutamate as the primary neurotransmitter of the RHT. In addition to NMDA receptors, ionotropic receptors sensitive for the glutamate agonist α-amino-3-hydroxy-5-methylisoxazole-4-propionic acid (AMPA) are found throughout the SCN. In particular, receptors composed of the GluR2 subunits are localized to the ventrolateral aspect of the SCN where most axonal terminals of RGCs are found. It must be considered that glutamate receptors are frequently expressed in astrocytes, thus raising the possibility that glutamate functions indirectly as a neuromodulator. Alternatively, astrocytes themselves may serve as a source of glutamate, released in response to some other signal originating from the RHT. Finally, derivatives of glutamate, such as *N*-acetylaspartylglutamate (NAAG) may play a role in RHT neurotransmission. NAAG is released from RGC terminals in a calcium-dependent manner and can be converted to its constituent amino acids within the synaptic cleft. The prospect of NAAG as a primary RHT neurotransmitter is diminished by the fact that it is only found in a fraction of the retinorecipient aspect of the SCN [4].

Nitric oxide (NO) also plays a critical role in transferring information about environmental light levels from the retina to the SCN. The phase shifting effect of light (or glutamate) on SCN slice preparations can be faithfully mimicked by NO generators such as sodium nitroprusside. Moreover, these effects are blocked by the application of the competitive nitric oxide synthase (NOS) inhibitor N^G-nitro-*L*-arginine methyl ester (L-NAME). This inhibition, however, can be reversed by increasing the availability of *L*-arginine, the natural substrate of NOS. The emerging evidence suggests that glutamate released from the presynaptic terminals of retinal afferents binds to NMDA receptors in postsynaptic SCN cells. This, in turn, results in a transient increase in intracellular calcium concentrations, thereby leading to activation of nNOS and the subsequent production of NO [4,8]. Because of the inherent instability of NO, such a mechanism could provide very fine spatial and temporal resolution to signaling occurring at the synapses of retinal afferents.

►Pituitary adenylyl cyclase activating peptide (PACAP) is one of a handful of peptide transmitters that may modulate the glutamatergic-based signaling of the RHT. PACAP is colocalized with glutamate and it has been suggested that all melanopsin-containing RGCs express PACAP [9]. Evidence from other systems where a small-molecule and a peptide neurotransmitter coexist in presynaptic terminals has shown that the small molecule transmitter is released upon weak or transient presynaptic stimulation. Stronger tonic stimulation results in release of both classes of transmitter, thereby encoding a broad dynamic range of stimulus strength. Similarly, weak retinal illumination may result in the release of glutamate from RHT terminals, while higher levels of light induce the release of PACAP.

Several labs have conducted experiments where PACAP has been applied *in vitro* to SCN slices or infused *in vivo* to hamsters. These experiments have not produced a consensus regarding PACAP's role in RHT neurotransmission. Additionally, mice null for PACAP show modest circadian effects. They show no re-entrainment deficits when exposed to a shifted light:-dark cycle, no diminution in the lengthening of circadian period in response to constant light, and no loss of ►masking behavior, i.e., acute light-induced suppression of nocturnal locomotor activity. However, these animals did exhibit attenuated circadian responses to ►phase-advancing or ►phase-delaying pulses of light and they displayed a modestly shortened ►free-running circadian ►period. The PACAP-specific receptor PAC1 has been knocked out in mice. The behavioral phenotype of these mice is rather subtle; they exhibit light-induced phase delays of circadian activity rhythms that are about 30% longer than wild-type mice [4]. By contrast, mice null for the VPAC2 receptor, which binds PACAP and VIP, are ►arrhythmic. This dramatic phenotype, however, is difficult to attribute to PACAP signaling because of the non-specificity of the receptor [10].

Finally, the tachykinin, substance P (SP), has been implicated as a neuromodulator of the RHT. This implication has been challenged because localization of SP to any RGCs, much less RGCs that comprise the RHT, has been equivocal. It has been reported that SP does not colocalize with PACAP in the retina and that bilateral enucleation does not abolish SP immunoreactivity. SP's proposed neuromodulatory role in RHT transmission remains to be shown conclusively [4].

Concluding Remarks

The identification and characterization of ipRGCs has provided new insight into a major constituent of the RHT. The specifics of how information is transferred from the retinal afferents to SCN neurons, and how this information is subsequently processed by these neurons remains a critical gap in our body of knowledge. However, it also represents a fertile field for future investigations.

References

1. Berson DM (2007) Phototransduction in ganglion-cell photoreceptors. Pflugers Arch 454:849–855
2. Provencio I (2007) Melanopsin Cells. 1.20:423–431 In: Basbaum AI, Bushnell M, Smith D, Beauchamp G, Firestein S, Dallos P, Oertel D, Masland R, Albright T, Kaas J, Gardner E (eds) *The senses: a comprehensive reference*, Elsevier, Amsterdam, 4640 p
3. Moore RY, Leak RK (2001) Suprachiasmatic Nucleus. In: TakahashiJS, Turek FW, Moore RY (eds) Circadian clocks: handbook of behavioral neurobiology, vol 12. Kluwer Publishers, New York, p xxiii, 770 p
4. Morin LP, Allen CN (2006) The circadian visual system, 2005. Brain Res Brain Res Rev 51:1–60
5. Magnin M, Cooper HM, Mick G (1989) Retinohypothalamic pathway: a breach in the law of Newton-Muller-Gudden? Brain Res 488:390–397
6. Ebling FJ (1996) The role of glutamate in the photic regulation of the suprachiasmatic nucleus. Prog Neurobiol 50:109–132
7. Hannibal J (2002) Neurotransmitters of the retinohypothalamic tract. Cell Tissue Res 309:73–88
8. Gillette MU, Mitchell JW (2002) Signaling in the suprachiasmatic nucleus: selectively responsive and integrative. Cell Tissue Res 309:99–107
9. Hannibal J (2006) Roles of PACAP-containing retinal ganglion cells in circadian timing. Int Rev Cytol 251:1–39
10. Harmar AJ (2003) An essential role for peptidergic signalling in the control of circadian rhythms in the suprachiasmatic nuclei. J Neuroendocrinol 15:335–338

Retinotopic

Definition
Topographic arrangement of visual pathways and visual centers that reflects the spatial organization of the neurons responding to visual stimuli in the retina.

►Evolution of the Optic Tectum: In Amniotes

Retinotopic Frame of Reference

Definition
Also, "Oculocentric frame of reference." A reference frame specifying the location of a visual target with respect to the position of the eyes in space.

Retinotopy

Definition
The visual field is represented in orderly maps in occipital cortex in retinal centered coordinates. The left half of the visual field is represented in the right hemisphere and vice versa, whereas the upper visual field is mapped to the inferior occipital cortex (below the calcarine sulcus) and vice versa. Therefore, when subjects hold their eye position constant (e.g. when they maintain central fixation), the region of occipital cortex that responds to a particular stimulus in the visual field can be determined within these maps. This is known as the retinotopic representation of the visual stimulus.

►Striate Cortex Functions
►Vision
►Visual Field

Retroambiguus Nucleus

Synonyms
►Nucl. Retroambiguus; ►Retro-ambiguus nucleus

Definition
Nuclear region of the myelencephalon continuing to the upper cervical cord and integrated in cardiorespiratory functions.

►Myelencephalon
Prosencephalon

Retrograde Amnesia

Definition
There is memory loss for events prior to the incident (e.g., trauma), but memories from the distant past and the period following the incident are intact.

►Amnesia
►Memory Improvement

Retrograde Degeneration

►Chromatolysis

Retrograde Interference

Definition
The disruption of transfer from short- to long-term memory by distractions introduced after the initial items are acquired.

Retrograde Messenger

Takako Ohno-Shosaku
Department of Cellular Neurophysiology, Graduate School of Medical Science, Kanazawa University, Kanazawa, Japan

Definition
Retrograde messenger is a chemical substance that is released from postsynaptic neurons and acts on

presynaptic neurons. In the nervous system, information coded by action potentials is transferred from neuron to neuron at a specialized site called "synapse" (Fig. 1). The transmission at ►chemical synapses is generally one-directional. Neurotransmitters are released from presynaptic terminals on the arrival of action potentials, and transmit a signal to postsynaptic neurons by activating the corresponding receptors (Fig. 1a). In contrast to this fundamental anterograde information transfer, the signaling from postsynaptic to presynaptic neurons is called retrograde signaling (Fig. 1b) [1,2]. The retrograde signaling can be mediated by either a diffusible factor that is called "retrograde messenger," or a direct interaction of presynaptic and postsynaptic membrane-bound elements. In typical cases, a retrograde messenger is released from the postsynaptic site lacking morphologically specialized structures for release (e.g. ►active zone), activates the receptors located on presynaptic terminals, and influences the function of presynaptic terminals (i.e. transmitter release) (Fig. 1b).

Characteristics

Quantitative Description

The substances so far proposed as retrograde messengers are ►endogenous cannabinoids (endocannabinoids) [3], ►nitric oxide (NO), ►carbon monoxide (CO), ►arachidonic acid, platelet-activating factor, neurotrophic factors, and some classical neurotransmitters or neuropeptides [2]. Among them, endocannabinoids are the most widely accepted substances as retrograde messengers in the brain. Major endocannabinoids are arachidonoylethanolamide (anandamide) and 2-arachidonoylglycerol (2-AG) (Fig. 2).

Anandamide is the amide of arachidonic acid with ethanolamine, and the molecular weight is 347.54. 2-AG is the glycerol derivative in which the second hydroxyl group is linked to arachidonic acid residue by an ester bond, and the molecular weight is 378.55. The structural features of endocannabinoids are quite different from classical neurotransmitters, and shared by lipid messengers such as eicosanoids, which mediate signals of inflammation and pain.

Lower Level Components

Arachidonoylethanolamide

Arachidonoylethanolamide, known as "anandamide," was the first endocannabinoid to be identified. The name of anandamide is based on the Sanskrit word for bliss and tranquility, *ananda*. Anandamide binds to both CB1 and CB2 ►cannabinoid receptors, but displays lower affinity for CB2 compared to CB1 receptors.

2-Arachidonylglycerol

2-Arachidonoylglycerol (2-AG) is another major endocannabinoid, and binds to both CB1 and CB2 receptors. 2-AG is widely distributed in the brain and periphery. The level of 2-AG is reported to be much higher (ca. 170 times) than anandamide in brain tissues. This molecule is the most likely candidate for the endocannabinoid that mediates retrograde synaptic modulation at hippocampal and cerebellar synapses.

Other Putative Endocannabinoids

Other putative endocannabinoids include noladin ether and virodhamine. Noladin ether is an ether-linked analogue of 2-AG. Virodhamine is the ester of arachidonic acid with ethanolamine. It is not determined whether these molecules actually mediate retrograde signals.

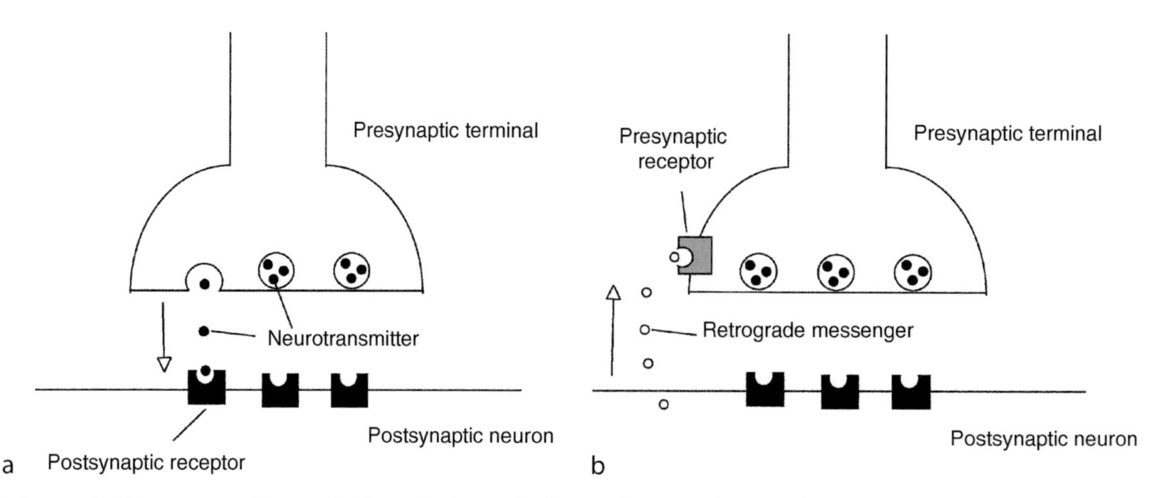

Retrograde Messenger. Figure 1 Synaptic transmission and retrograde signaling. In synaptic transmission (a), neurotransmitters are released from presynaptic terminals, and bind to postsynaptic receptors. In retrograde signaling (b), retrograde messengers are released from postsynaptic neurons, and activate presynaptic receptors.

Higher Level Processes

Interneuronal Communication

In the anterograde signaling, classical neurotransmitters, neuropeptides, neurotrophic factors, and some other substances are released from presynaptic terminals, and produce rapid changes in membrane potential (i.e. generation of postsynaptic potentials) as well as long-term structural and metabolic changes in the postsynaptic cells. In the retrograde signaling, retrograde messengers including some of the substances used for the anterograde signaling are released from postsynaptic cells, and influence the function or morphology of presynaptic neurons. In addition to these diffusible factors, direct interactions of presynaptic and postsynaptic membrane-bound elements (e.g. cadherins) are also involved in interneuronal communication between presynaptic and postsynaptic neurons.

Lower Level Processes

Synthesis of Endocannabinoids (Anandamide and 2-AG)

Anandamide formation in neurons is a two-step process. The first step is the transfer of an arachidonic acid group from the *sn*-1 position of phosphatidylcholine to the head group of phosphatidylethanolamine (PE) by the enzyme *N*-acyltransferase, resulting in the formation of *N*-arachidonyl-PE. The second step is the cleavage of *N*-arachidonyl-PE by phospholipase D (PLD), which produces anandamide and phosphatidic acid (Fig. 3a) 2-AG is formed through two distinct pathways. The first step of the main pathway is the cleavage of phosphatidylinositol (PI) by the enzyme PI-specific phospholipase C (PI-PLC), producing 1,2-diacylglycerol (DAG). The second step is the further cleavage of DAG by DAG lipase, yielding 2-AG (Fig. 3b). The alternative pathway involves phospholipase A_1 and lyso-PLC. The PLD and DAG lipases responsible for endocannabinoid formation have been identified. Among several types of PI-PLC, β type PLC (PLCβ) is the most important for 2-AG formation.

Anandamide

2-AG

Retrograde Messenger. Figure 2 Chemical structures of two major endocannabinoids.

Breakdown of Endocannabinoids (Anandamide and 2-AG)

Anandamide is broken down into arachidonic acid and ethanolamine by the enzyme fatty acid amide hydrolase (FAAH) (Fig. 3a). 2-AG is broken down into arachidonic acid and glycerol by the enzyme monoacylglycerol lipase (MGL) (Fig. 3b). These enzymes have been identified. They exhibit unique distributions in the brain. In general, FAAH is a postsynaptic enzyme, whereas MGL is presynaptic.

CB1 Receptor Signaling at Presynaptic Terminals

The CB1 receptor is densely distributed on presynaptic axons and terminals in various regions of the brain. They include excitatory synapses on cerebellar Purkinje cells (both climbing fiber and parallel fiber synapses), inhibitory synapses on cerebellar Purkinje cells, part of hippocampal inhibitory synapses including CCK-positive basket cell to pyramidal cell synapses, and inhibitory synapses from the striatum to globus pallidus. Activation of presynaptic CB1 receptors suppresses the release of the

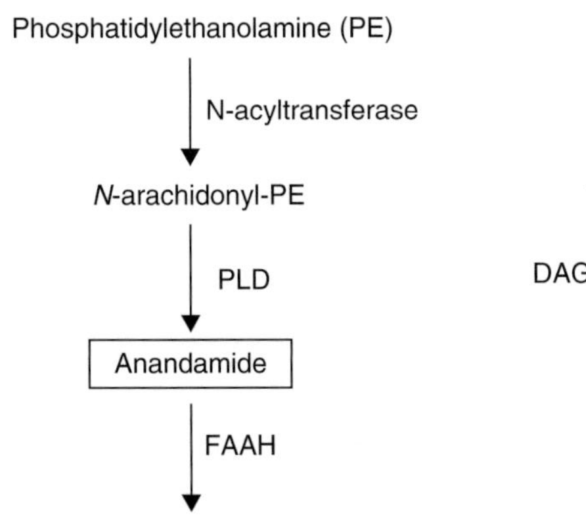

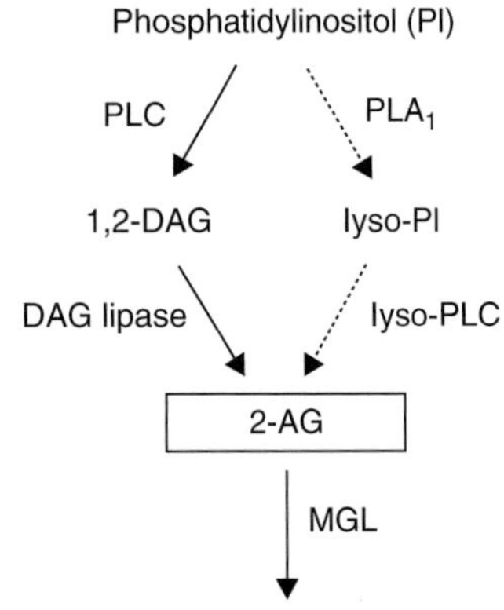

Retrograde Messenger. Figure 3 Pathways of formation and degradation of two major endocannabinoids. *PLD* phospholipase D; *FAAH* fatty acid amide hydrolase; *PLC* phospholipase C; *PLA_1* Phospholipase A_1; *DAG* diacylglycerol; *MGL* monoacylglycerol lipase.

transmitters (glutamate or GABA). This CB1-mediated suppression can be caused by inhibition of voltage-gated Ca^{2+} channels, activation of K^+ channels, direct effect on release machinery, or some other unknown mechanisms.

Process Regulation

The release of retrograde messengers from postsynaptic neurons is generally controlled by neural activity. For example, endocannabinoids are produced on demand in response to depolarization-induced elevation of intracellular Ca^{2+} concentration [4,5], or activation of Gq-coupled receptors such as group I metabotropic glutamate receptors [6] and M_1/M_3 muscarinic receptors [7] (Fig. 4). As endocannabinoids are membrane-permeable, they are considered to diffuse across the plasma membrane to the extracellular space immediately after production (Fig. 4).

Endocannabinoids are produced much more effectively when Ca^{2+} elevation and receptor activation coincide [8]. The released endocannabinoids then work as retrograde messengers (Fig. 4), and are removed from the extracellular space through uptake and enzymatic degradation. Other membrane-permeable factors such as NO and CO can be produced in response to certain conditions, and released to the extracellular space. In contrast, membrane-impermeable factors including classical neurotransmitters and neuropeptides are supposed to be stored in vesicles, and secreted through exocytotic mechanisms in response to certain triggering stimuli such as an elevation of intracellular Ca^{2+} concentration.

Function

Retrograde messengers play important roles in formation, maturation, and plasticity of synaptic connections. Among them, most attention has been focused on their crucial roles in activity-dependent modulation of synaptic transmission, including both short-term and long-term forms of synaptic plasticity.

Endocannabinoids mediate retrograde signals involved in several forms of short-term synaptic plasticity including ►DSI, ►DSE, and ►receptor-driven retrograde suppression. The CB1 receptor is widely distributed in the brain, and located densely on many, but not all, types of presynaptic terminals. At CB1-expressing synapses, endocannabinoids released from the postsynaptic neurons activate presynaptic CB1 receptors, and thereby suppress the transmitter release. Endocannabinoid release is triggered by elevation of intracellular Ca^{2+} concentration, activation of Gq-coupled receptors, or combination of the two [8] (Fig. 4). Under physiological conditions, endocannabinoid-mediated retrograde suppression is triggered by synaptic activity that can produce postsynaptic Ca^{2+} elevation and Gq-coupled receptor activation. Thus, the endocannabinoid-mediated retrograde suppression provides a feedback mechanism, by which the postsynaptic neurons receiving synaptic inputs can retrogradely influence the function of CB1-expressing presynaptic terminals. This endocannabinoid-mediated retrograde suppression is reversible and thus classified as a form of short-term synaptic plasticity. However, endocannabinoids are also involved in long-term synaptic plasticity, especially long-term depression (LTD), in some brain regions including the striatum and amygdala [9,10].

It has been reported that dopamine can be released from dendrites of dopaminergic neurons. Some other classical neurotransmitters or neuropeptides are also proposed to be released from dendrites [2]. They

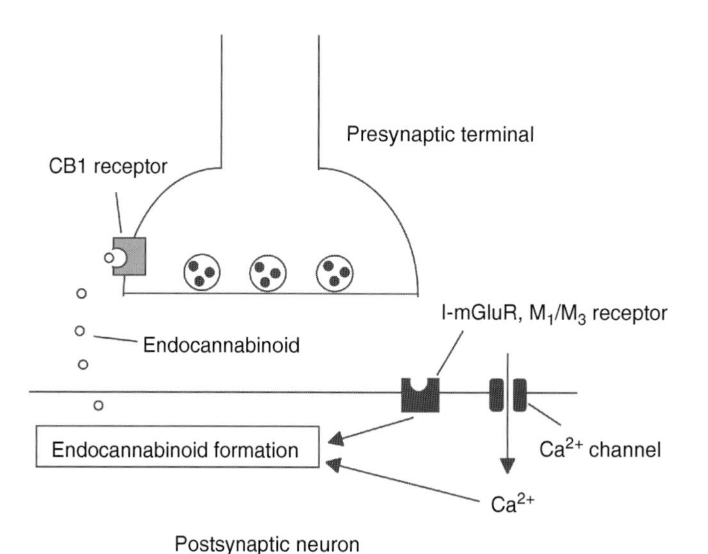

Retrograde Messenger. Figure 4 Endocannabinoid-mediated retrograde signaling. Endocannabinoid production is induced by postsynaptic depolarization-triggered Ca^{2+} influx, or activation of Gq-coupled receptors such as group I metabotropic glutamate receptors (I-mGluRs) and M_1/M_3 muscarinic receptors. The endocannabinoids that are produced are then released from postsynaptic neurons, and activate presynaptic CB1 receptors to suppress the transmitter release.

include glutamate, GABA, and dynorphin. Although the actual release mechanisms of these substances are not fully elucidated, it is likely that these substances also contribute to short-term plasticity.

Synaptic activity can induce long-term potentiation (LTP) or depression (LTD) depending on the pattern of activity. Membrane-permeable factors such as NO, CO, arachidonic acid and platelet-activating factor have been proposed to contribute to long-term synaptic plasticity, especially LTP. However, there are many controversial results, and general consensus has not yet been reached as to the possible roles of these factors as retrograde messengers in long-term synaptic plasticity.

Retrograde messengers also play important roles in formation, maturation and refinement of synaptic connections, especially at early developmental stages. These processes require information exchange between presynaptic and postsynaptic cells through anterograde and retrograde signals. Although molecular mechanisms of these signals are not clearly elucidated, it has been proposed that neurotrophic factors, including nerve growth factor (NGF) and brain-derived neurotrophic factor (BDNF), could play important roles in these processes.

Therapy

Cannabinoids for Therapies

Marijuana, which contains the natural cannabinoid Δ^9-tetrahydrocannabinol, as well as several commercially-available synthetic cannabinoids (nabilone and dronabinol) are clinically effective for several disorders such as nausea from cancer chemotherapy, chronic pain, and exhaustion in AIDS patients. However, these drugs also have psychoactive side effects such as dizziness and thinking abnormalities. They are inherent problems, because the CB1 receptor is widely distributed in the brain. To avoid these problems, it has been attempted to develop drugs that enhance the endocannabinoid signaling in a target-specific manner.

References

1. Tao HW, Poo M (2001) Retrograde signaling at central synapses. Proc Natl Acad Sci USA 98:11009–11015
2. Alger BE (2002) Retrograde signaling in the regulation of synaptic transmission: focus on endocannabinoids. Prog Neurobiol 68:247–286
2. Piomelli D, Giuffrida A, Calignano A, Rodriguez de Fonseca F (2000) The endocannabinoid system as a target for therapeutic drugs. Trends Pharmacol Sci 21:218–224
4. Wilson RI, Nicoll RA (2002) Endocannabinoid signaling in the brain. Science 296:678–682
5. Maejima T, Ohno-Shosaku T, Kano M (2001) Endogenous cannabinoid as a retrograde messenger from depolarized postsynaptic neurons to presynaptic terminals. Neurosci Res 40:205–210
6. Maejima T, Hashimoto K, Yoshida T, Aiba A, Kano M (2001) Presynaptic inhibition caused by retrograde signal from metabotropic glutamate to cannabinoid receptors. Neuron 31:463–475
7. Fukudome Y et al. (2004) Two distinct classes of muscarinic action on hippocampal inhibitory synapses: M2-mediated direct suppression and M1/M3-mediated indirect suppression through endocannabinoid signaling. Eur J Neurosci 19:2682—2692
8. Hashimotodani Y et al. (2005) Phospholipase Cbeta serves as a coincidence detector through its $Ca2^+$ dependency for triggering retrograde endocannabinoid signal. Neuron 45:257–268
9. Marsicano G et al. (2002) The endogenous cannabinoid system controls extinction of aversive memories. Nature 418:530–534
10. Gerdeman GL, Ronesi J, Lovinger DM (2002) Potsynaptic endocannabinoid release is critical to long-term depression in the striatum. Nat Neurosci 5:446–451

Retrograde Neuron Reaction

►Neuronal Changes in Axonal Degeneration and Regeneration

Retrograde Tracing

Definition

A method for visualizing the neurons of origin of an axonal pathway. A dye is injected into the region of the nervous system where axons are thought to terminate. The dye is absorbed by the axons and transported back to the cell bodies allowing them to be visualized.

Retrograde Tracing Techniques

Definition

Neuron tracing techniques take advantage of the fact that axoplasmic transport of materials goes in both directions within the axon. Retrograde flow of materials is toward the cell body. Early studies of retrograde transport used the enzyme horseradish peroxidase (HRP), which after injection into nerve tissue is picked up at axon terminals by micropinocytosis. The HRP is transported back to the cell body of the neuron where it can be made visible with a variety of chemical reactions. More recent studies use retrograde immunocytochemical and fluorescent labeling techniques to visualize neuron morphology.

Retrograde Transport

Definition

Retrograde transport is a process, mediated by the microtubule motor dynein, by which chemical messages are sent from the axon back to the cell. Fast retrograde transport can cover over 100 mm/day.

Retronasal/Orthonasal Olfaction

Definition

Orthonasal smell perception occurs when volatile molecules are pumped in through the external nares of the nose and activate the sensory cells in the olfactory epithelium. This is the route used to sense odors in the environment.

Retronasal stimulation occurs during food ingestion, when volatile molecules released from the food in the mouth are pumped, by movements of the mouth, from the back of the oral cavity up through the nasopharynx to the olfactory epithelium. It is activated only when breathing out through the nose between mastications or swallowings. This is the route used to sense aromas of food.

►Flavor
►Olfactory Epithelium

Retrospective Monitoring

Definition

Retrospective monitoring refers to metamemory experiences when one searches for and retrieves the origin and content of information stored in long-term memory.

It has been studied by examining the experiences of "tip-of-the-tongue" (TOT), which one has when the information one tries to remember feels like right on the edge of the tongue), and of "feeling-of-knowing" (FOK).

►Metacognition

Retroviral Vectors

Definition

These are retroviruses used to insert novel genes into neurons or other cells by infecting them.

Rev-erbα

Definition

Member of the nuclear hormone receptors with hemin as a potential ligand. Identified as the main circadian repressor of the *Clock* and *Bmal1* genes. Nuclear orphan receptor that binds to the consensus sequence ([A/T]A[A/T]NT[A/G]GGTCAtermedRORE, in the promoter of target genes. The gene is transcribed from the opposite strand of the erba gene, which is a cellular homolog of the viral oncogene v-erbA.

►Clock-Controlled Genes
►Clock Genes

Reverberation

Definition

The persistence of sound in an enclosed space, as a result of multiple reflections, after the sound from the source has stopped.

►Acoustics

Reverberation Time

Definition

The time between the offset of the originating sound and when the reverberant sound remaining in the enclosed space is 60 dB (a factor of 1,000 in pressure) less than the level of the originating sound.

►Acoustics

Reversal Learning

Definition

In reversal learning, a particular discrimination task is first learned and then the reinforcement contingencies are reversed. In other words, once the subject has learned to discriminate a reinforced from a non-reinforced stimulus, it has to learn to reverse its response to such stimuli. Such reversals tend to be difficult since there are negative transfer effects; e.g. the individual tends to persist in responding to the stimulus that was originally reinforced. Eventually, however, this tendency becomes weaker, and the response to the alternative stimulus becomes more frequent until it is consistently evoked.

►Reinforcement

Reversal Potential

Definition

Reversal potential (also called Nernst potential) is the membrane voltage at which there is no net flow of a particular ion from one side of the membrane to the other.

►Membrane Potential: Basics
►Synaptic Transmission

Reverse Real-Time quantitative PCR (RT-qPCR)

Definition

Complementary DNA (cDNA) is first made from an RNA template, using a reverse transcriptase enzyme. A specific sequence of cDNA is then amplified and the amount of product produced at the end of each PCR cycle is evaluated by measuring signal strength of fluorescent markers. Since PCR generates products at an exponential rate, the relative abundance of template can be compared between samples. Alternatively the absolute abundance of template can be determined if reference dilutions of template are used. RT-qPCR can also refer to Reverse Transcriptase quantitative PCR, Real-time quantitative PCR (PCR).

►Serial Analysis of Gene Expression

Reverse Signaling: Nervous System Development

Li Yu, Yong Rao
Centre for Research in Neuroscience, Department of Neurology and Neurosurgery, McGill University Health Centre, Montreal, QC, Canada

Definition

Reverse signaling refers to the signaling mechanism by which a known membrane-bound ligand also functions as a receptor to trigger intracellular signaling events in the ligand-bearing cell, thereby modifying its behavior. Such dual function of a membrane protein as both ligand and receptor allows the ligand-receptor system to mediate bi-directional signal exchange between two neighboring cells, thus greatly increasing the plasticity of intercellular communications. Two major reverse signaling pathways mediated by ephrin and semaphorin have been implicated in regulating the development of the nervous system.

Characteristics

Description of Reverse Signaling Mechanism

Ephrin Reverse Signaling

Ephrins were originally identified as ligands for the Eph-family receptor tyrosine kinases. Ephrins are divided into A and B subclasses. A subclass ephrins are tethered to the cell membrane by a glycosylphosphatidylinositol (GPI) anchor, while each B subclass ephrin contains a hydrophobic transmembrane segment spanning the plasma membrane followed by a short cytoplasmic domain [1–3]. Based on sequence homology and their preference for binding ephrins, Eph receptor tyrosine kinases are also classified into two subgroups, including ephrin-A-binding Eph-As and ephrin-B-binding EphBs [1–3].

Early studies of the ephrin-Eph interaction focused on understanding the signaling mechanism by which activation of Eph by ephrin modulates downstream signaling proteins to regulate cellular behaviors. This type of Eph-mediated signaling in response to ephrin binding is called ►forward signaling (Fig. 1). Each Eph family member has a highly conserved domain architecture in the cytoplasmic region that is comprised of a

R

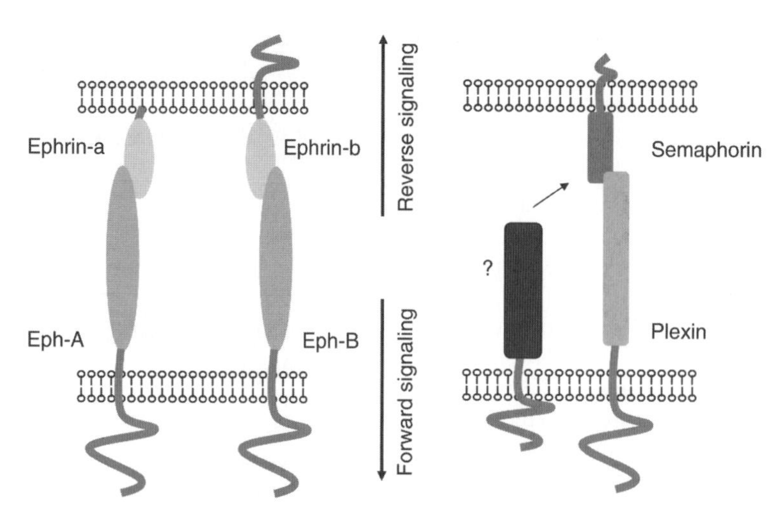

Reverse Signaling: Nervous System Development. Figure 1 Bi-directional signaling mediated by two ligand-receptor systems. Binding of ephrin and semaphorin activates Eph and Plexin forward signaling, respectively. Binding of Eph and Plexin can also activate ephrin and transmembrane semaphorin to trigger reverse signaling, respectively. The identity of cell surface protein (*colored in blue*) that activates semaphorin reverse signaling during neural development is unknown.

juxtamembrane region, a kinase domain, a SAM (sterile α domain) domain and a PDZ-domain binding motif [1–3]. Binding of ephrin to the extracellular region of Eph induces conformational changes, allowing the Eph cytoplasmic domain to modulate the activity and/or subcellular localization of downstream signaling proteins, which then modulate cytoskeletal reorganization leading to an attractive or repulsive response.

Later studies show that the binding between ephrin and Eph is capable of triggering downstream signaling events in both ephrin-expressing- (i.e., reverse signaling) and Eph-expressing (i.e., forward signaling) cells (Fig. 1), and thus mediates ►bi-directional signaling [1–3]. The action of ephrins as receptors to mediate reverse signaling relies on their ability to modulate the activity and/or subcellular localization of downstream signaling proteins in response to Ephs. EphrinBs can utilize both phospho-tyrosine residues and the PDZ domain-binding motif in its cytoplasmic domain to recruit downstream signaling proteins. For instance, Eph-induced ephrin-B1 phosphorylation of the conserved tyrosine residues provides a docking site for the ►SH2 domain (Src Homology 2 Domain) of the adaptor protein Grb4 [4]. Grb4 in turn links ephrin-B1 to multiple signaling pathways through the binding of its multiple ►SH3 domain (Src Homology 3 Domain) to proline-rich proteins (e.g., Abl-interaction protein-1 (Abi-1), c-Cbl-associated protein (CAP) and scaffolding protein axin), thus modulating ►focal adhesion and cytoskeletal reorganization [1,3]. Ephrin-Bs can also recruit downstream signaling proteins (e.g., glutamate-receptor-interacting-protein-1 (GRIP1), GRIP2, syntenin, protein kinase C-interacting protein-1 (PICK1), tyrosine phophatase PTP-BL, and PDZ-RGS3) in a phosphorylation-independent way via the PDZ domain-binding motif in their C-terminus [1,3]. Ephrin-B-mediated reverse signaling can also directly stimulate the enzymatic activity of Fak (focal adhesion kinase) to regulate focal adhesion [5].

Although ephrin-As do not have a cytoplasmic domain, they are also capable of mediating reverse signaling in response to Eph binding [1–3]. Ephrin-As are localized to ►lipid rafts, specific plasma membrane microdomains consisting of glycosphingo-lipids and cholesterol, where ephrin-As presumably associate with other membrane protein complexes. The activation of ephrin-As can lead to the recruitment of intracellular signaling proteins such as the ►Src family kinase Fyn to lipid rafts, which in turn regulates downstream proteins to modulate cytoskeletal reorganization and cell adhesion [6].

Semaphorin Reverse Signaling

Semaphorins are a large family of secreted and transmembrane proteins that share a conserved ~500 amino-acid Semaphorin (Sema) domain at the amino terminus [7]. Semaphorins can also mediate bi-directional signaling. Semaphorins mediate forward signaling by functioning as ligands to bind and activate their receptors ►Plexin and ►neuropilin, which in turn initiate a cascade of signaling events in Plexin and/or neuropilin-expressing cells to regulate cytoskeletal changes for directed axonal growth and cell movement. Like ephrin-Bs, some transmembrane semaphorins can also mediate reverse signaling by utilizing their cytoplasmic domains to recruit intracellular signaling proteins (Fig. 1). For instance, binding of Plexin-A1 to the extracellular region of Semaphorin-6D increases the association of its cytoplasmic domain with the Abl tyrosine kinase, but decreases its association with Mena, a member of the enabled (Ena)/vasodilator-stimulated phosphoprotein (Vasp) family proteins [7]. In Drosophila, it also appears likely that direct association of the transmembrane semaphorin-1a via its cytoplasmic domain with the fruitfly Enabled protein mediates semaphorin-1a-dependent reverse signaling [8]. The changes in the activity and/or localization of these semaphorin-associating intracellular signaling proteins contribute to the cytoskeletal reorganization necessary for cell migration and axonal projections.

Function of Reverse Signaling in Neural Development

The function of reverse signaling in a ligand-receptor system mediating bi-directional signaling in neural development can be specifically assessed in several ways [2]. For instance, wild-type ligand or receptor in mice would be replaced with a mutant version incapable of interacting with intracellular signaling proteins by gene targeting, thus selectively inactivating reverse- or forward signaling, respectively. The contribution of reverse or forward signaling would then be assessed by comparing the phenotype displayed in the above mutants to that caused by the loss of both forward- and reverse signaling that occurs in receptor or ligand null mutants. Reverse signaling can also be selectively inactivated in zebrafish and Xenopus by expressing a dominant-negative version of a transmembrane ligand in which the cytoplasmic domain mediating reverse signaling is deleted. The dominant-negative mutant is still able to mediate forward signaling through binding to its receptor, but interferes with reverse signaling through competing with wild-type counterpart for receptor binding. In Drosophila, the contribution of reverse signaling can be determined by assessing whether null ligand mutants are rescued by expression of a mutant ligand that is capable of activating its receptor but defective in reverse signaling. The phenotype that is not rescued by the reverse-signaling-defective mutant ligand likely reflects the function of reverse signaling mediated by this ligand.

Ephrin Reverse Signaling in the Vertebrate Neural Development

The Formation of Anterior Commissure Tract

The projection of both ►acP axon tract and ►acA axon tract of ►anterior commissure in mice requires ephrin reverse signaling [1,2]. Both ephrin-B2 and Eph-B2 are required for the guidance of acP axons. However, ephrin-B2 is expressed in acP axons and functions in a cell-autonomous manner, whereas Eph-B2 is expressed in cells underlying acP axons and is required non-cell-autonomously for the projection of acP axons. That the guidance of acP axons requires an intact cytoplasmic domain of ephrin-B but not the kinase domain of Eph-B2 supports the involvement of ephrin-B2 reverse signaling in the guidance of acP axons. The activation of ephrin-B2 reverse signaling by Eph-B2 appears to initiate a repulsive response that guides acP axons toward the midline. The guidance of both acP and acA axons also requires the activation of ephrin-A reverse signaling by Eph-A4, which functions non-cell-autonomously as a ligand to attract acP and acA axons. The identity of the ephrin activated by Eph-B4 in acP and acA axons remains unclear.

Retinotectal Mapping Along Dorsal-ventral Axis

In the vertebrate visual system, retinal ganglion cells in the eye project axons into the optic tectum in a topographic fashion along both anterior-posterior and dorsal-ventral axes. While ►topographic projections of retinal ganglion axons along the anterior-posterior axis are directed by the ephrin-As-Eph-As forward signaling [1,3], dorsoventral topographic projections appear to require the ephrin-B-mediated reverse signaling [2]. In Xenopus, retinal ganglion cell axons display a decreasing dorsal-to-ventral expression gradient of ephrin-B2 and B3 in the retina, while Eph-B1 shows a complementary expression pattern (i.e., decreasing ventral-to-dorsal gradient) on cells in the optic tectum. In vivo and in vitro studies suggest strongly that Eph-B1 in the ventral tectum activates ephrin-B2 and B3 expressed on dorsal retinal ganglion axons to initiate reverse signaling leading to an attractive response, which targets dorsal ganglion axons toward the ventral tectum.

Establishment of Vomeronasal Map

The vomeronasal organ (VNO) is involved in detecting pheromones. VNO axons projected from the vomeronasal epithelium are targeted topographically to specific glomeruli comprised of sensory projections in the accessory olfactory bulb (AOB) during embryonic development. Topographic projections of VNO axons appear to involve an attractive response mediated by ephrin-A5 reverse signaling when activated by Eph-A6 [2]. Ephrin-A5 is expressed differentially in VNO axons and required for their topographic projections. The difference in the expression level of ephrin-A5 appears to dictate onto which regions of the accessory olfactory bulb VNO axons are targeted: i.e., axons with higher level of ephrin-A5 are targeted to regions with higher level of Eph-A6 and vice versa.

Semaphorin Reverse Signaling in Neural Development in Drosophila

Several recent studies, including our own, suggest strongly that transmembrane semaphorin-1a-mediated reverse signaling plays an important role in regulating neural development in Drosophila [8–10]. While it has been shown that reverse signaling mediated by transmembrane semaphorin-6D is required for the guidance of myocardial patterning in vertebrates [7], future studies are needed to determine if semaphorin reverse signaling is also involved in regulating the vertebrate neural development.

The Formation of Giant-fiber-motor Neuron Synapse in Drosophila

The ►giant fiber system of Drosophila is involved in controlling the jump-and-flight response. During development, a giant interneuron in the giant fiber system of the brain projects an axon into the second thoracic segment where the axon forms synapses with a motor neuron, which in turn controls the activity of the jump muscle. Semaphorin-1a is required both pre- and postsynaptically for the formation of giant-fiber-motor neuron synapses, suggesting a role for semaphorin to mediate bi-directional signaling between pre- and postsynaptic partners [8]. Overexpression of wild-type semaphorin-1a, but not a truncated semaphorin-1a mutant protein lacking the cytoplasmic domain, causes a gain-of-function phenotype. These data suggest that the participation of semaphorin-1a in synapse formation involves the action of semaphorin-1a reverse signaling. It remains to be determined whether Plexin or other semaphorin-interacting proteins function as a ligand to activate semaphorin-1a in synapse formation.

Photoreceptor Axon Guidance in Drosophila

In the Drosophila adult visual system, R1-R6 photoreceptors project their axons from the retina to the superficial layer of the optic lobe, the lamina. During development, R1-R6 axons temporally stop at their intermediate target region in between two layers of glial cells prior to establishing synaptic connections with lamina neurons. We found recently that the transmembrane semaphorin-1a functions cell-autonomously in photoreceptor axons for the proper arrangement at the intermediate target region [9]. The function of semaphorin-1a in photoreceptor axons requires its cytoplasmic domain, consistent with a role for semaphorin-1a as a receptor to mediate reverse signaling. The identity of cell surface proteins that activate semaphorin-1a reverse signaling in photoreceptor axons remains to be determined.

Dendritic Targeting of Olfactory Projection Neurons in Drosophila

In the development of the Drosophila olfactory system, projection neurons project their dendrites onto discrete units called glomeruli in the ►antennal lobe, the first olfactory information relay center equivalent to olfactory bulb in vertebrates. Different types of odorant receptor axons form one-to-one precise connections to dendrites projected from different types of projection neurons at glomeruli. Semaphorin-1a displays a differential expression pattern on dendrites of projection neurons. Genetic analysis showed that semaphorin-1a is required differentially in projection neurons for the targeting of their dendrites onto discrete glomeruli in a cell-autonomous manner, indicating a role for semaphorin-1a as a receptor for dendritic targeting [10]. Consistently, the cytoplasmic domain of semaphorin-1a is shown to be indispensable for its function, which likely reflects its role in recruiting downstream signaling proteins within the dendrites of projection neurons. The guidance cue that activates semaphorin-1a reverse signaling in dendrites is unknown.

References

1. Kullander K, Klein R (2002) Mechanisms and functions of Eph and ephrin signalling. Nat Rev Mol Cell Biol 3:475–486
2. Davy A, Soriano P (2005) Ephrin signaling in vivo: look both ways. Dev Dyn 232:1–10
3. Murai KK, Pasquale EB (2003) "Eph" ective signaling: forward, reverse and crosstalk. J Cell Sci 116:2823–2832
4. Cowan CA, Henkemeyer M (2001) The SH2/SH3 adaptor Grb4 transduces B-ephrin reverse signals. Nature 413:174–179
5. Murai KK, Nguyen LN, Irie F, Yamaguchi Y, Pasquale EB (2003) Control of hippocampal dendritic spine morphology through ephrin-A3/EphA4 signaling. Nat Neurosci 6:153–160
6. Davy A, Gale NW, Murray EW, Klinghoffer RA, Soriano P, Feuerstein C, Robbins SM (1999) Compartmentalized signaling by GPI-anchored ephrin-A5 requires the Fyn tyrosine kinase to regulate cellular adhesion. Genes Dev 13:3125–3135
7. Tran TS, Kolodkin AL, Bharadwaj R (2007) Semaphorin regulation of cellular morphology. Annu Rev Cell Dev Biol 23:263–292
8. Godenschwege TA, Hu H, Shan-Crofts X, Goodman CS, Murphey RK (2002) Bi-directional signaling by Semaphorin-1a during central synapse formation in Drosophila. Nat Neurosci 5:1294–1301
9. Cafferty P, Yu L, Long H, Rao Y (2006) Semaphorin-1a functions as a guidance receptor in the Drosophila visual system. J Neurosci 26:3999–4003
10. Komiyama T, Sweeney LB, Schuldiner O, Garcia KC, Luo L (2007) Graded expression of semaphorin-1a cell-autonomously directs dendritic targeting of olfactory projection neurons. Cell 128:399–410

Reward

Definition

A reward or positive reinforcer is any stimulus an animal will work to obtain. Often these stimuli have biological significance to the animal such as food, shelter or sex, a class of stimuli sometimes referred to as primary or unlearned reinforcers. Other types of reward are initially affectively neutral but acquire value through being associated with a primary reinforcer.

An example of such a stimulus for humans is money which acquires value by virtue of its capacity to be exchanged for other kinds of primary reinforcers such as food or shelter.

►Reinforcer
►Value-based Learning

Reward Signal in Neural Networks

Definition

A scalar performance measure used for reinforcement learning of networks.

►Neural Networks for Control

RFLP

Definition

Restriction fragment length polymorphism. These are polymorphisms that change restriction sites. RFLPs with known chromosomal locations were used in linkage analysis, with Southern blotting, to map disease genes until the advent of microsatellite markers.

►Bioinformatics

Rheobase

Definition

Strength of a rectangular depolarizing direct current (DC) current necessary to elicit an action potential.

►Action Potential

Rheological Models

Marcelo Epstein
Schulich School of Engineering, University of Calgary, Calgary, AB, Canada

Definition

A rheological model consists of an assembly of one-dimensional *rheological elements*, each of which can be seen as a black box with two protruding terminals.

Description of the Theory

Each rheological element is characterized by a deterministic relationship between the (history of the) relative displacement (or *elongation*) $u(t)$ between the terminals and the (history of the) applied force $f(t)$. In other words, knowing the displacement function $u(t)$ for all past times t up to the present, the present value of the force can be uniquely determined by some mathematical operation or vice versa. The merit of this one-dimensional oversimplification of the complexity of a true continuum constitutive theory (q.v.) lies in the fact that, by combining a small number of rheological elements in series and in parallel, a wealth of mathematically tractable, surprisingly varied and suggestive force-elongation responses is obtained. Two rheological elements are said to be *connected in series* if one of the terminals of the first is connected to one of the terminals of the second, in such a way that the remaining two unconnected terminals are considered as the terminals of the new combined-element black box. Let $[f_1(t), u_1(t)]$ and $[f_2(t), u_2(t)]$ denote, respectively, the force-elongation pairs of the first and second elements connected in series. The main feature of a series combination is that both elements experience the same force, while the elongation of the combined black box is the sum of the elongations of the original elements. Denoting by $[f(t), u(t)]$ the force-elongation pair of the combined series black box:

$$u(t) = u_1(t) + u_2(t) \qquad f(t) = f_1(t) = f_2(t). \quad (1)$$

Two elements are said to be *connected in parallel* if each terminal of one element is connected to a counterpart in the other element. The two common terminals thus obtained are considered as the terminals of the combined black box. As a result, both elements experience necessarily the same elongation, while the forces are added to produce the response of the combined element. Using the same notation as before, for the ►parallel arrangement:

$$u(t) = u_1(t) = u_2(t) \qquad f(t) = f_1(t) + f_2(t). \quad (2)$$

Some of the most common elements in use are: the ►*linear spring*, the ►*linear damper* (or *linear dashpot*) and the ►*contractile element*. The linear spring is characterized by two material constants: the *rest length* L_0 and the *stiffness constant* k. By convention, in the linear spring the elongation u is measured with respect to the rest length (in other words, when the elongation vanishes, the distance between the terminals is equal to the rest length). The force at time t is then proportional to the elongation at that time, namely:

$$f(t) = k\,u(t). \tag{3}$$

Thus, the linear spring provides a purely elastic response, whereby the past history of the ►deformation plays no role, except for the fact that the material always remembers its "original" rest length. The linear damper, on the other hand, is completely characterized by a single material constant c called the *viscous constant*. There is no rest length. The force-elongation relation is given by the equation:

$$f(t) = c\,\dot{u}(t). \tag{4}$$

In a linear damper the only fact that counts in determining the force between the terminals at a given time is the speed of elongation $\dot{u}(t)$ at that time. In other words, the past history plays a role, albeit limited to the very immediate past. More sophisticated history elements can be defined. Finally, the contractile element is a useful device with important applications to muscle ►mechanics. It can be thought of as a frictionless slider that produces no force, whatever the value of the elongation may be. In muscle mechanics applications, however, it is usually assumed that this behavior is characteristic of the *inactive state* only and that the contractile element may be *activated*, so that in the *active state* the force-elongation response abides by an ad-hoc law (for example, a so-called *force-length relation*).

To illustrate the variety of material responses that can be obtained by means of rheological models, the *Maxwell model*, obtained by placing a linear spring and a linear dashpot in series is considered. It is not difficult to show that the response of the Maxwell model is completely contained in the following first-order ordinary differential equation:

$$\dot{u} = \frac{\dot{f}}{k} + \frac{f}{c}. \tag{5}$$

If a force f_0 is suddenly applied, an instantaneous elongation of value $u_\infty = \frac{f_0}{k}$ develops completely at the expense of the spring, while the dashpot does not have time to react. As time goes on, however, if the force is kept at a constant value, the elongation will keep growing steadily at the expense of the deformation of the damper. If the force is suddenly removed, the spring goes instantaneously back to its original length, while the damper immediately stops deforming. At the end of the process, therefore, the Maxwell model retains a residual deformation. The response of a system to a suddenly applied load that remains constant in time is known as ►*creep*.

If a linear spring and a damper are combined in parallel, the result is the *Voigt model*. It is governed by the differential equation:

$$f = ku + c\dot{u}. \tag{6}$$

A sudden application of a force is met with no instantaneous response, since the damper cannot react immediately. As time goes on, however, an exponential growth of the elongation is observed which approaches asymptotically the value $u_\infty = \frac{f}{k}$. If the force is suddenly removed, the spring will slowly bring back the system to its original rest length. This is the type of behavior observed when sitting on a feather- or down-filled cushion. This effect is sometimes described as "delayed ►elasticity," although the response is anything but elastic.

A more realistic description of the behavior of many materials is obtained by combining in parallel a Maxwell model with a linear spring. The result is known as the *Kelvin model* or the *standard linear solid*. Its creep response is similar to that of a Voigt model, except that, just like a Maxwell model, it also has an instantaneous elastic response, governed by the spring added in parallel.

If instead of subjecting the various models to a sudden force they are subjected to a sudden elongation, the response obtained in terms of the decay of the resulting force as time goes on is known as ►*relaxation*.

References

1. Pipkin AC (1972) Lectures on viscoelasticity theory. In: Applied mathematics series, vol 7. Springer, Berlin
2. Fung YC (1981) Biomechanics. Spinger, New York
3. Epstein M, Herzog W (1998) Theoretical models of skeletal muscle. Wiley, Chichester

Rheostasis

Definition

Regulation of the internal environment of an animal to a stable condition with a changing reference setpoint.

►Hibernation

Rheumatoid Arthritis (RA)

Definition

RA is traditionally considered a chronic, inflammatory autoimmune disorder that causes the immune system to attack the joints. It is a disabling and painful inflammatory condition, which can lead to substantial loss of mobility due to pain and joint destruction. RA is a systemic disease, often affecting extra-articular tissues throughout the body including the skin, blood vessels, heart, lungs, and muscles.

Rhinencephalon

The olfactory bulb and those structures that receive afferents form the olfactory bulb are classified as being part of the rhinencephalon. They include primarily the olfactory tract and the basal olfactory area, parts of the amygdaloid body, septum verum and prepiriform cortex.

►General CNS

Rhizotomy

Definition

A surgical procedure in which spinal nerve roots are cut.

Rho

Definition

The Greek letter ρ, used to denote the rest phase of the circadian rest-activity cycle. Occurs at night in diurnal animals, and in the day in nocturnal animals.

►Alpha (Activity Phase) in Circadian Cycle
►Circadian Cycle
►Rho GTPases

Rho Family of Small Guanosine Triphosphatases (Rho GTPases)

Definition

Rho family of small guanosine triphosphatases (Rho GTPases) are important intracellular signaling proteins involved in various aspects of neuronal morphogenesis including migration, polarity, axon growth and guidance, dendrite arborization, spine plasticity, and synapse formation. Acting as intramolecular switches, the Rho GTPases transduce signals from various extracellular ligands to the cytoskeleton. They exist in two states: a GTP-bound active state, and a GDP-bound inactive state. Guanine nucleotide exchange factors turn on Rho GTPases by facilitating the exchange of GDP for GTP, and GTPase activating proteins increase their GTPase activity, helping to turn them off.

►Cytoskeleton
►Growth Cones
►Neural Development

Rho GTPases

Ryan Petrie, Nathalie Lamarche-Vane
Department of Anatomy and Cell Biology, McGill University, Montreal, Quebec, Canada

Synonyms

Rho Family of Small GTP-Binding Proteins

Definition

The Rho family of small GTP-binding proteins consists of 22 mammalian proteins related to each other based on the similarity of their amino acid sequence to the first family member to be identified, RhoA. These proteins are relatively small (less than 25 kDa) and all possess an intrinsic GTPase activity, which hydrolyzes the guanosine triphosphate (GTP) into guanosine diphosphate (GDP). The bound nucleotide regulates the activity of the GTPase, rendering it inactive or active in the case of GDP or GTP, respectively.

Characteristics

The ►Rho GTPases are expressed in all cells from fertilization through adulthood and their activity is critical to many aspects of cell biology that are required for normal functioning of the organism [1]. When active, the Rho GTPases can activate a diverse array of intracellular signaling pathways. RhoA, Rac1, and Cdc42, the best characterized members of the Rho family of small GTPases, regulate the assembly of the ►actin and ►microtubule cytoskeleton, a filamentous network of proteins within a cell that control cell shape, cell adhesion, cell polarity and cell migration. Specificity of Rho GTPase signaling is achieved by the coordinated regulation of the nucleotide bound state of the GTPase and the region of the cell to which the active form of the protein is targeted.

The Cycling of GTP and GDP Regulates Rho GTPase Activity

Each of these monomeric GTPases acts as a molecular switch to control the downstream signaling pathways. The cycling of GDP-GTP binding to Rho GTPases is tightly regulated by three families of proteins. The GTPase is unable to exchange the bound GDP for free GTP in the cytosol without the help of proteins termed ►guanine nucleotide exchange factors (GEFs). In response to external stimuli, GEFs are activated and bind the GDP-bound conformation of Rho GTPases leading to the release of GDP. Then, the Rho GTPases in a free-nucleotide transition state are able to bind GTP in the cytosol and activate downstream effectors in their active GTP-bound conformation. The activity is terminated by ►GTPase-activating proteins (GAPs), which enhance the intrinsic GTPase activity, leading to the inactive state of the GTPase. Additionally, ►guanine nucleotide dissociation inhibitors (GDIs) can sequester Rho GTPases in their GDP-bound, inactive state (Fig. 1).

Subcellular Localization of Rho GTPases

While inactive GTPases are usually restricted to the cytoplasm, the active forms are localized to the surface of membranes within the cell, such as plasma, golgi, or endosomal membranes. The correct intracellular localization of active GTPases is critical for their regulation and coupling to downstream signaling pathways. The targeting information is contained in the ►CAAX box present at the C-terminal end of the GTPase amino acid sequence [2]. This sequence is modified by the addition of an isoprenyl group on the side chain of the cysteine, which is then able to insert into membrane lipid bilayers. Additionally, there is a short stretch of basic residues upstream of the CAAX box that confers specificity to each Rho GTPase and influences into which membrane the isoprenyl chain inserts. The specificity in action is also achieved by the restricted subcellular distribution of GAPs and GEFs, which results in the inactivation/activation of the Rho GTPases being tightly coupled to specific regions of the cell.

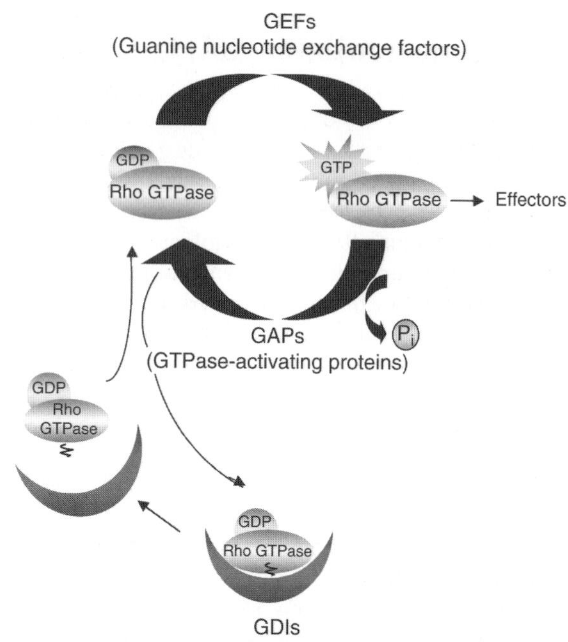

Rho GTPases. Figure 1 *Cycling of Rho GTPases.* Rho GTPases exist in either an inactive, GDP-bound state or an active, GTP-bound state. Three families of proteins tightly regulate this GDP/GTP cycle. GEFs stimulate the exchange of GDP for GTP, thereby activating the GTPases. GAPs enhance the intrinsic GTPase activity, leading to the inactive GDP-bound form. GDIs bind to the GDP-bound GTPase and sequester the protein in the cytosol.

Functions of Rho GTPases During Development of the Nervous System

Neuron Polarization and Axon Specification

During the initial stages of the development of the nervous system when neural precursors are generated by cell division, the cells lack features such as dendrites or axons and are non-polarized [3]. To begin the process of forming axons and dendrites the cells must specify which region of the plasma membrane will begin to extend away from the cell body in order to form elongated structures known as ►neurites. These structures will ultimately become the single axon and several dendrites of the nascent ►neuron, a process known as neuronal polarization. In vitro studies have revealed key roles for the Rho GTPases Cdc42 and Rac1 in the initial polarization of a neuron [4]. Cdc42 is targeted to a sub-region of the plasma membrane and recruits the polarity complex of Par3, Par6, and the atypical protein kinase C, a conserved multiprotein complex used throughout evolution to polarize different cell types. The polarity complex in turn recruits the GEFs Tiam1 and STEF, which catalyze the exchange of GDP for GTP on Rac1. The active Rac1 mediates the formation of filamentous actin and neurite formation ultimately polarizing the cell.

Axon Guidance

As the nervous system develops, newborn neurons extend axons towards their cognate targets. The neuronal growth cone, located at the tip of the growing axon, is a highly motile structure acting as a sophisticated signal transduction device, capable of recognizing extracellular guidance cues and translating them into directed neurite outgrowth [5]. Over the past 15 years, a combination of cellular and genetic studies has led to the identification of highly conserved families of guidance molecules that can be either membrane-bound factors or secreted molecules, acting over short or long distances, respectively, to guide the growth of axons. They include the classical molecular cues: netrins, slits, ephrins, and semaphorins [6].

Cytoskeletal rearrangements are crucial during growth cone guidance. The growth cone is enriched in the cytoskeletal elements F-actin and microtubules that are rapidly remodeled in response to environmental cues and direct the migration of the growth cone [5]. There is now compelling evidence demonstrating a role for RhoA, Rac1, and Cdc42 as important signaling elements downstream of most, if not all, guidance cue receptors [7]. Indeed, Rho GTPases mediate a cascade of responses from receptors to actin remodeling within the neuronal growth cone. For instance, Rac1 interacts directly with the semaphorin receptor plexin-B1, suggesting that Rac1 plays a role in mediating the repulsive activities of semaphorins. On the other hand, the Rho-specific GEF, PDZ-RhoGEF LARG, interacts with plexin B to activate RhoA signaling, provoking growth cone repulsion and collapse. Ephexin, a GEF for RhoA and Cdc42, binds to Eph receptors to modulate Ephrin-induced growth cone collapse, whereas the slit receptors Robo mediate axon repulsion by interacting with srGAP that inhibits Cdc42. Finally, Rac1 and Cdc42 are important mediators of the signaling response of axons to the netrin-1 receptor DCC. Overall, it is clear that Rho GTPases are important regulators in axon pathfinding and guidance, and it is the correct balance of localized Rho GTPase activities through GEFs and GAPs that will determine the appropriate attractive and repulsive response of an axon to extracellular cues.

Dendritogenesis and Structural Plasticity of the Dendritic Arbor

During the maturation of connectivity within the central nervous system (CNS), dendrites are highly dynamic. The branched structure of the dendrites is highly enriched in the cytoskeletal elements and it is the controlled remodeling of these structures that is responsible for the shape and complexity of the dendritic arbor. In many cases, dendritic spines are the location of the synaptic connections and their size and number correlate with their ability to transmit the correct information from axon to dendrite. The ability of Rho GTPases to convert the upstream signals into cytoskeletal changes leads to remodeling of the dendrites [8]. For example, N-methyl D-aspartate (NMDA) receptors transmit excitatory transmissions mediated by L-glutamate. Activation of NMDA receptors increases the activity of Cdc42 and Rac1 while decreasing the activity of RhoA, leading to the stabilization of the dendritic spines [7]. The regulation of the growth and elaboration of the whole dendritic architecture represent an important mechanism of plasticity in the central nervous system.

Axon Regeneration in the CNS

The activity of the Rho GTPase family member RhoA is known to induce acto-myosin contractility within the cell and to produce mechanical forces that can retract actin-dependent structures such as neurites [9]. Inhibition of RhoA in neuronal cells leads to neurite extension over substrates that would not normally be permissive for neurite outgrowth. Many reports are now suggesting that RhoA is highly activated following lesions in the CNS and mediates many of the signals associated with the growth suppressive environment of the CNS following injury. For example, myelin associated glycoprotein (MAG) is released from the damaged myelin sheath and binds to the NOGO receptor to activate RhoA, preventing regrowth of the damaged axons. This is accomplished because RhoA activates in turn Rho-kinase (ROCK), which is able to induce intracellular acto-myosin contractility that prevents axon outgrowth following injury. Therefore, RhoA is now a promising pharmacological target for therapy in an aim to promote ▶axon regeneration in the CNS following injury.

Implication of Rho GTPase Dysfunction as Being Causative for Mental Retardation

The importance of Rho GTPases in the nervous system development is further highlighted by research studies linking dysfunction in Rho GTPase signaling pathways and mental retardation in the adult. In this situation mutations in specific genes affect the developmental program such that the organism develops abnormally [7]. Malformed dendrites and dendritic spines are common in these conditions and are hypothesized to be important indicators of the mechanisms of impaired cognition in mental retardation. For example, in non-syndromic x-linked mental retardation, three genes out of a total of 13 have been pinpointed to be mutated and are encoding either regulators of Rho GTPases or downstream mediators of the pathways activated by Rho GTPases. In particular, the *OPHN-1* gene encodes the protein oligophrenin-1, which contains a RhoGAP domain shown to negatively regulate RhoA, Rac1, and Cdc42. It is specifically expressed at high levels in both axons and dendrites throughout the brain and a mutation causing decreased levels of the oligophrenin-1 mRNA is associated with mental retardation. In addition, the *ARHGEF6* gene encodes a GEF that activates Rac1 and Cdc42 and a truncation mutant protein missing the first 28 amino acids is associated with mental retardation. Finally, the serine/threonine kinase activity of PAK3, a member of the p21-activated protein kinase family (PAK) family of proteins, acting downstream of Rac1 and Cdc42, has been found to be compromised in individuals with this neurological disorder.

Concluding Remarks

RhoA, Rac1, and Cdc42 regulate a wide variety of intracellular signaling pathways to mediate many aspects of the development of the nervous system. The Rho family of GTPases includes 19 other members, the roles of which are relatively unknown. Future research will undoubtedly reveal many novel functions of this diverse family of proteins in the development of the nervous system.

References

1. Jaffe AB, Hall A (2005) Rho GTPases: biochemistry and biology. Annu Rev Cell Dev Biol 21:247–269
2. Michaelson D, Silletti J, Murphy G, D'Eustachio P, Rush M, Philips MR (2001) Differential localization of Rho GTPases in live cells: regulation by hypervariable regions and RhoGDI binding. J Cell Biol 152:111–126
3. Arimura N, Kaibuchi K (2005) Key regulators in neuronal polarity. Neuron 48:881–884
4. Watabe-Uchida M, Govek EE, Van Aelst L (2006) Regulators of Rho GTPases in neuronal development. J Neurosci 26:10633–10635
5. Dent EW, Gertler FB (2003) Cytoskeletal dynamics and transport in growth cone motility and axon guidance. Neuron 40:209–227
6. Dickson BJ (2002) Molecular mechanisms of axon guidance. Science 298:1959–1964
7. Govek EE, Newey SE, Van Aelst L (2005) The role of Rho GTPases in neuronal development. Genes Dev 19:1–49
8. Schubert V, Dotti CG (2007) Transmitting on actin: synaptic control of dendritic architecture. J Cell Sci 120:205–212
9. Ellezam B, Dubreuil C, Winton M, Loy L, Dergham P, Selles-Navarro I, McKerracher L (2002) Inactivation of intracellular Rho to stimulate axon growth and regeneration. Prog Brain Res 137:371–380

Rhodopsins

►Photopigments

Rhombencephalon

Definition

The rhombencephalon (Greek for rhombus-shaped brain) is the caudal part of the developing neural tube, the hindbrain, which is composed of the metencephalon (pons and cerebellum) and myelencephalon (medulla).

Rhombomere

►Evolution of the Vestibular System

RHT

►Retinohypothalamic Tract

Rhynchocephalia

Definition

Sister taxon to the Squamata (lizards, snakes, amphisbaenians) and incorporating the living Tuatara of New Zealand, Sphenodon. The term Sphenodontia refers to a subset of Rhynchocephalia that excludes the most basal forms.

►The Phylogeny and Evolution of Amniotes

Rhythm

Definition

Periodic change of an entity. In behavior, several rhythms are distinguished by their periods: circadian (about 24 h), circalunar (about 28 days), circannual (about 1 year).

►Hippocampus: Organization, Maturation, and Operation in Cognition and Pathological Conditions

Rhythmic Jaw Movements

►Mastication

Rhythmic Movements

OLE KIEHN
Mammalian Locomotor Laboratory, Department of Neuroscience, Karolinska, Stockholm, Sweden

The Behavior

Rhythmic movements are motor acts that are characterized by the activation of groups of muscles in a recurring or cyclic pattern. Rhythmic movements are found in all animals ranging from invertebrates to man and include various behaviors that are continuously ongoing, like ►respiration, are episodic, like ►swimming, ►mastication and ►walking, or brief like ►scratching and the ►startle response. The rhythmic movements are generated by localized neuronal networks, called ►central pattern generators, or CPGs. Activity in the CPGs directly controls the timing (rhythm) and phasing (pattern) of ►motoneurons, whose activity in turn activates the ►muscles needed to generate the rhythmic movements; e.g., the limb muscles acting on the leg during walking or intercostals muscles and the diaphragm acting on the lungs during respiration. Thus, the term CPG alludes to the fact that these neuronal networks are restricted to specific regions of the central nervous system and, when appropriately activated, are capable of generating both the timing and phasing of rhythmic movements without receiving patterned ►sensory information. The CPG for respiration or mastication in vertebrates is, for example, localized in the ►brainstem while the CPGs for swimming, scratching and walking are localized in the ►spinal cord. The CPGs controlling rhythmic movements in invertebrates are typically localized to ganglionic structures, like to the ►stomatogastric ganglion that controls the foregut movements in Crustaceans or to the chain of midbody ganglia that control ►leech swimming. Traditionally, rhythmic movements were studied in intact or semi-intact animals. Because of the distinct network localization, numerous preparations have been developed where the part of the central nervous system that contains the CPG network can be studied in isolation in vitro.

External Control

Although CPGs are able to intrinsically generate the timing and phasing of rhythmic movements, they do not function in isolation. Rather, in most cases their activity is often turned on and off by an external command signal. For example, when a cat starts to walk or a fish starts to swim, the spinal CPGs are activated by activity in descending fiber tracts originating in the brainstem [1,2]. Initiating signals originating in the forebrain are funneled through the ►basal ganglia and conveyed to nuclei in the diencephalon (►diencephalic locomotor region) and mesencephalon (►mesencephalic locomotor region) and then to excitatory ►reticulospinal neurons in brainstem ►locomotor regions in the midbrain and diencephalon. The reticulospinal neurons then project to the spinal locomotor network and provide the external ►excitatory drive needed to initiate and maintain the rhythmic activity. Similarly, descending inputs from "head ganglia" to CPGs in the stomatogastric ganglion or the swimming CPG in leech can turn specific rhythmic motor behaviors on and off [3]. Sensory inputs, like loud sound or sudden changes in light, might be the direct trigger for the descending signal leading to escape or startle responses [4], which is followed by more persistent rhythmic movements. Sensory inputs are also triggers for the scratching movements where the motor behavior is initiated by tactile stimulation applied to the skin [5].

Sensory Information

Although rhythmic motor outputs can be generated in the absence of sensory information CPGs receive sensory feedback. Some of these sensory signals cause corrections of the rhythmic movements, as when a person is stumbling over an object. In this case ►cutaneous sensory receptors mediate the corrective signal to the CPG circuit [6]. Other sensory signals are involved in phase transitions and amplitude modulation of the rhythmic movements and are caused by ►proprioceptive feedback from the moving appendages. Examples of sensory inputs that are involved in phase transitions are feedback from ►stretch receptors in the lungs that regulate the transition from inspiration to expiration, joint receptors in the hip in mammals that regulate the transition from stance to swing and sensory receptors in the wing of the locust that influence the transition from wing depression to wing elevation [6,7]. Load receptors, like ►tendon organs, or stretch receptors in muscles, like ►muscle spindles, also provide proprioceptive cues for phase transitions and are actively involved in modulating the amplitude variation of rhythmic movements [7–9].

Basic Network Features

The intrinsic function of rhythmic motor networks is defined by the synaptic interconnections between the

R

CPG neurons in the network and the membrane properties of the neurons. A minimal characterization of the network function therefore requires that CPG component neurons are identified, that the connectivity between individual neurons is established, and that the salient membrane properties are described. An analysis at this level of detail has only been obtained in a limited number of rhythmic motor systems both in invertebrates and in vertebrates.

Notable examples in invertebrates of CPGs that have been characterized in detail are the swim CPGs in the mollusks *Clione* [10], and *Tritonia* [11], the heart-beat network in leech [3,12], and the CPG circuits in the stomatogastric ganglion controlling foregut movements in Crustacea [13]. Because of the small number of cells (less than 30 neurons), the complete network connectivity has been worked out and the cellular properties of individual CPG neurons have been determined in great detail. From the analysis of these small CPG networks, it is clear that each CPG network has its specific characteristics and that none of them are alike. However, several basic network and cellular building blocks can be extracted from the analysis [11]. Network elements include extensive ►reciprocal inhibition in a ►half-center fashion, delayed ►feed-forward excitation, ►electrical coupling and ►graded synaptic release. These network elements alone do not determine the timing and phasing that the CPG network produces but they interact with cellular properties that actively interpret the synaptic activity and contribute to timing and phasing. Such cellular elements are ►bursting pagemaker properties that can provide sustained rhythmic drive, post-inhibitory rebound firing that helps escape inhibition and is generated ►by h-channels (►HCN) and ►T-type calcium channels, ►plateau potentials generated by ►persistent calcium or sodium channels, ►calcium-activated calcium channels (►CAN channels) that amplify and prolong synaptic inputs, delayed activation generated by activation of potassium channels with slow kinetics (►A-Type channels), and spike-frequency adaptation generated by ►calcium-activated potassium channels [14].

In vertebrates, the CPG organizations for ►swimming in lamprey and in ►*Xenopus* tadpole have been revealed in great detail [15,16]. The core of these networks consists of ►excitatory ►CPG interneurons and ►inhibitory glycinergic ►interneurons. The glutamatergic interneurons project ipsilaterally and provide the excitatory drive to other CPG interneurons and motoneurons necessary to produce sustained rhythmic drive on one side of the cord. The glycinergic interneurons are ►commissural interneurons projecting to the contralateral side where they connect to all CPG neurons and motor neurons and mediate reciprocal inhibition segmentally so that when one side is active the other side is inactive. This half center organization is the basis for the side-to-side undulatory swimming. The rhythm itself is not dependent on inhibitory connections, but can be generated in a network of mutually coupled excitatory neurons. Each one of the about 100 spinal segments that makes up the lamprey spinal cord appear to contain such a basic CPG unit [17]. These units are coupled both in the ascending and descending directions. These connections provide the basis for the ►intersegmental coordination of muscular activity along the length of the body that is required for the animal to swim. Similarly to invertebrate CPGs, a large number of intrinsic membrane properties influence the rhythmogenic capability of the swim CPG neurons and participate in patterning of the motor output [18,19].

The large number of neurons controlling any given behavior in mammals has made it difficult to reveal the detailed network organization of, for example, the CPGs controlling mammalian ►walking [20], ►mastication [21] or ►respiration [22–24]. Knowledge about the functionally of these CPGs is, however, advancing rapidly. For the walking CPG, the key network functions are the rhythm generation, ipsilateral coordination of flexors and extensors across the same or different joints in a limb, and ►left-right coordination [20]. The rhythm is generated by glutamatergic ipsilateral projecting interneurons [25]. The exact identity of these neurons has not been determined. The circuits underlying coordination of flexors and extensors segmentally and intersegmentally include inhibitory ►Ia interneurons and ►Renshaw cells, as well as a number of unidentified interneurons. Functional analysis of left-right circuitries in the mammals suggests that ►intersegmental coordination provided by ►commissural interneurons is involved in binding motor synergies along the cord, while inhibitory ►intrasegmental commissural connections control segmental alternation and excitatory commissural connections control synchronous activity [20]. It thus appears that some basic characteristics of swimming CPG and walking CPG network structure are preserved. However, the commissural circuitries seem more complex in the walking CPG than what has been described for the swimming CPG. Additionally, while network elements in the swimming CPG appear to be composed of homogenous populations of neurons, similar network elements, such as the rhythm generation network in the walking CPG, appear to be composed of more heterogeneous populations of neurons. Thus, additional network layers are added when moving from a non-limbed to a limbed CPG. Similar to what is seen in invertebrate and lower vertebrate neurons, mammalian CPG neurons express to a variable degree rhythmogenic/pacemaker-like membrane properties or phase-regulating membrane properties [14,22–24,26].

A new addition to the CPG network analysis is methods for genetically dissecting the neuronal circuits.

Such methods include genetic silencing or activation of molecularly defined populations of neurons and have been applied to both invertebrate and vertebrate CPGs. In networks with many neurons, such manipulations can more directly link a population of CPG neurons to a specific network function than traditional electrophysiological methods are able to do [27].

Neuromodulation of Rhythmic Movements

A lesson that has been learned from studies of CPGs in both invertebrates and vertebrates is that the overall network function is flexible and can be changed by ►neuromodulation that acts on individual CPG neurons and connectivity [28,29]. The neuromodulation may be the result of neurotransmitters and hormones released from sources outside the CPG circuits (►extrinsic neuromodulation) or the neuromodulation may be the result of neurotransmitters and signals released from active CPG neurons (►intrinsic neuromodulation). Examples of extrinsic ►neuromodulatory systems are the numerous ►amine and ►peptide containing systems that project to stomatogastric ganglion [28,30] and the descending 5-HT or dopamine systems in vertebrates [31]. Examples of intrinsic ►neuromodulatory systems are neurotransmitters released from swim CPG neurons in *Tritonia* [32], adenosine released from CPG neurons tadpole [18] and endocannabinoids from CPG neurons in the lamprey [33]. In all cases, the targets for neuromodulators are ligand-gated ion channels and/or chemical or electrical synaptic transmission. Because neuromodualtors have these ubiquitous network targets, they can change timing, phasing or amplitude of the rhythmic movements separately or all of these parameters at ones. Thus in some cases, the neuromodulation is a fine tuning of the rhythmic motor behavior, while in others the neuromodulation causes dramatic switching in the motor coordination.

Disorders of Rhythmic Movements

Primary disorders of rhythmic movements are not common, although a number of respiratory dysfunctions involve defects in network function and/or its modulation [22]. Secondary disorders of rhythmic movements are due to injury of the external control systems, for example, damage to the spinal cord that leads to loss of ambulatory ability below the lesion. The ultimate way of restoring the rhythmic motor behavior after spinal cord injury is to promote re-growth or ►regeneration of the severed fibers across the injury. An alternative approach is ►neuro-rehabilitation. Experiments in animals with ►spinal cord injury have shown that sustained ►locomotor training on a treadmill in combination with drugs that activate the spinal locomotor CPG can lead to substantial recovery of the lost locomotor capability [34,35]. Clinical trials have shown that humans with partial spinal cord injury also can benefit from such ►locomotor rehabilitation therapy [34,35].

References

1. Brocard F, Dubuc R (2003) Differential contribution of reticulospinal cells to the control of locomotion induced by the mesencephalic locomotor region. J Neurophysiol 90:1714–27
2. Jordan LM (1998) Initiation of locomotion in mammals. Ann N Y Acad Sci 860:83–93
3. Marder E, Calabrese RL (1996) Principles of rhythmic motor pattern generation. Physiol Rev 76:687–717
4. Fetcho JR (1991) Spinal network of the Mauthner cell. Brain Behav Evol 37:298–316
5. Stein PS (2005) Neuronal control of turtle hindlimb motor rhythms. J Comp Physiol A Neuroethol Sens Neural Behav Physiol 191:213–29
6. Grillner S (1981) Control of locomotion in bipeds, tetrapods, and fish. In Brooks V (ed) Handbook of Physiology, Bethesda, pp 1176–236
7. Pearson KG (1995) Proprioceptive regulation of locomotion. Curr Opin Neurobiol 5:786–91
8. Hultborn H (2001) State-dependent modulation of sensory feedback. J Physiol 533:5–13
9. McCrea DA (2001) Spinal circuitry of sensorimotor control of locomotion. J Physiol 533:41–50
10. Orlovsky GN, Deliagina TG, Grillner S (1999) Neuronal control of locomotion. From mollusc to man. Oxford University Press, Oxford
11. Getting P (1988) Comparative analysis of invertebrate spinal pattern generators. In Cohen A, Rossignol S, Grillner S (ed) Neural control of rhythmic movements in vertebrates, Wiley, New York, pp 101–28
12. Calabrese RL, Peterson E (1983) Neral control of heartbeat in the leech, Hirido medicinalis. In Roberts A, Roberts B (ed) Neural origin of rhythmic movements, Cambridge University Press, Cambridge, pp 195–221
13. Selverston A, Moulin M (1987) The crustacean stomatogastric system. Springer Berlin
14. Harris-Warrick RM (2002) Voltage-sensitive ion channels in rhythmic motor systems. Curr Opin Neurobiol 12:646–51
15. Grillner S, Wallen P, Brodin L, Lansner A (1991) Neuronal network generating locomotor behavior in lamprey: circuitry, transmitters, membrane properties, and simulation. Annu Rev Neurosci 14:169–99
16. Roberts A, Soffe SR, Wolf ES, Yoshida M, Zhao FY (1998) Central circuits controlling locomotion in young frog tadpoles. Ann N Y Acad Sci 860:19–34
17. Grillner S (2003) The motor infrastructure: from ion channels to neuronal networks. Nat Rev Neurosci 4:573–86
18. Dale N, Kuenzi FM (1997) Ion channels and the control of swimming in the Xenopus embryo. Prog Neurobiol 53:729–56
19. Grillner S (1997) Ion channels and locomotion. Science 278:1087–8
20. Kiehn O (2006) Locomotor circuits in the mammalian spinal cord. Annu Rev Neurosci 29:279–306
21. Lund J, Kolta A (2006) Generation of the central masticatory pattern and its modification by sensory feedback. Dysphagia 21(3):167–74
22. Feldman JL, Mitchell GS, Nattie EE (2003) Breathing: rhythmicity, plasticity, chemosensitivity. Annu Rev Neurosci 26:239–66
23. Ramirez JM, Tryba AK, Pena F (2004) Pacemaker neurons and neuronal networks: an integrative view. Curr Opin Neurobiol 14:665–74

24. Richter DW, Spyer KM (2001) Studying rhythmogenesis of breathing: comparison of in vivo and in vitro models. Trends Neurosci 24:464–72
25. Kiehn O, Quinlan KA, Restrepo CE, Lundfald L, Borgius L, et al. (2008) Excitatory components of the walking mammalian locomotor CPG. Brain Res Rev 57(1):56–63
26. Kiehn O, Kjaerulff O, Tresch MC, Harris-Warrick RM (2000) Contributions of intrinsic motor neuron properties to the production of rhythmic motor output in the mammalian spinal cord. Brain Res Bull 53:649–59
27. Kiehn O, Kullander K (2004) Central pattern generators deciphered by molecular genetics. Neuron 41:317–21
28. Harris-Warrick RM, Marder E (1991) Modulation of neural networks for behavior. Annu Rev Neurosci 14:39–57
29. Kiehn O, Katz P (1999) Making circuits dance: modulation of motor pattern generation. In Katz P (ed) Beyond neurotransmission. neuromodulation and its importance for information processing. Oxford University Press, Oxford, pp 275–317
30. Nusbaum M, Beenhakker M (2002) A small-systems approach to motor pattern generation. Nature 417 (6886):343–250
31. Schmidt BJ, Jordan LM (2000) The role of serotonin in reflex modulation and locomotor rhythm production in the mammalian spinal cord. Brain Res Bull 53:689–710
32. Katz P, Getting P, Frost W (1994) Dynamic neuromodulation of synaptic strength intrinsic to a central pattern generator circuit. Nature 367:729–31
33. El Manira A, Kyriakatos A, Nanou E, Mahmood R (2008) Endocannabinoid signaling in the spinal locomotor circuitry. Brain Res Rev 57(1):29–36
34. Edgerton VR, Kim SJ, Ichiyama RM, Gerasimenko YP, Roy RR (2006) Rehabilitative therapies after spinal cord injury. J Neurotrauma 23:560–70
35. Rossignol S, Drew T, Brustein E, Jiang W (1999) Locomotor performance and adaptation after partial or complete spinal cord lesions in the cat. Prog Brain Res 123:349–65

Ribonuclease (RNase)

Definition

An enzyme that catalyses the hydrolysis of an RNA resulting in either cleavage to smaller RNA units or by degradation to constituent nucleotides.

Ribosome

Definition

A ribosome is a non-membranous organelle that translates of a mRNA molecule into a polypeptide chain. It consists of 65% ribosomal RNA and 35% ribosomal proteins.

Riddoch Phenomenon or Syndrome

Definition

When visual cerebro-cortical area V5 is disconnected from area V1 (with which it is reciprocally connected and from which it normally receives its visual input) but has a secondary visual input that reaches it without passing through area V1, the subject can still experience visual motion consciously though crudely.

►Blindsight
►Visual Perception

Rigidity

Definition

An increased resistance to passive stretch that is nearly equal in both agonist and antagonist muscles and generally uniform throughout the range of motion of the joint being tested. It may be sustained (plastic or lead pipe) or intermittent and rachetty (cogwheel). Although cogwheel rigidity is usually thought to be Parkinsonian rigidity complicated by Parkinsonian tremor, it may occur in the absence of tremor and the frequency felt by the examiner tends to be higher than that of the visible resting tremor.

►Parkinson Disease
►Resting Tremor

Rigor Configuration

Definition

The rigor configuration in the cross-bridge cycle is associated with the end-state of the power stroke with the nucleotide products (ADP and P) having been released. In order to advance from the rigor configuration, ATP is required to release the cross-bridge from actin.

►Molecular and Cellular Biomechanics
►Power Stroke
►Sliding Filament Theory

RNA Interference

Definition

RNA interference – this procedure is abbreviated RNAi. It consists of the down-regulation of gene expression by using specific double-stranded ribonucleic acids. The specific or chosen RNA base pairs with its complementary strand of mRNA resulting in the degradation of the latter.

►GAL4/UAS

RNA Localization

►mRNA Targeting: Growth Cone Guidance

RNA Synthesis

►DNA Transcription

RNA Translation

Andrew Simmonds
Department of Cell Biology, Faculty of Medicine and Dentistry, University of Alberta, Edmonton, AB, Canada

Synonyms

Protein synthesis; Polypeptide synthesis

Definition

RNA translation is the process whereby the genetic information encoded in messenger RNA (►mRNA) is translated by specialized cytoplasmic complexes (ribosomes) into a polypeptide. This process is essential for the function of all cell types, and each step of translation represents a highly regulated event.

Characteristics

Eukaryotic mRNAs are initially transcribed from genetic information encoded in the nuclear DNA. After processing by various nuclear proteins (e.g., splicing factors), mRNAs are exported to, and subsequently translated within, the cytoplasm. The basic process translates the sequence of bases organized in three letter codons within the mRNA into a specific polypeptide string of linked amino acids. As more is learned about mRNA translation, it has become apparent that this process is a highly regulated event with several levels of control including storage of mRNAs in a non-translating state and restriction of translation to specific sub-domains within the cytoplasm. Many of these regulatory events are mediated by sequences encoded within the mRNA itself. Each mRNA can be functionally divided into several regions including the 5′ cap, the 5′ untranslated region (UTR), the ►open reading frame (ORF), the 3′ UTR and a 3′ domain consisting of repeated adenosine bases that are added post-transcriptionally (Fig. 1).

In a generalized model, the ORF is the most important region, with the information within encoding a specific protein, while the UTRs are thought to play roles in regulating localization, degradation and restrictions in the translation of the mRNA.

The core component of the cellular complex that reads the genetic information in the ORF and translates it into a specific protein is the ribosome. Several ribosomes can bind to a single mRNA to form a larger translation complex called a polysome. The genetic information in the ORF of the mRNA is read in three base segments by a collection of ►transfer RNA (tRNA) molecules that contain a complementary RNA code (anti-codon) that recognizes and binds a particular nucleotide triplet (codon) in the mRNA. A parallel cellular process specifically links each tRNA to the amino acid that corresponds to the codon (Fig. 2). The ribosomal complex processes along the mRNA and through the progressive recruitment of specific amino-acid linked tRNAs. The amino acids delivered by the tRNAs to the ribosome are covalently joined to produce the polypeptide encoded by the mRNA. This process is regulated at several levels. For example, proteins that will remain inside the cell are translated on free ribosomes in the cytoplasm while the ribosomes-mRNA-nascent polypeptide complexes translating exported or membrane bound proteins are trafficked to the endoplasmic reticulum. However, there are several other regulatory events during mRNA translation including: editing of the mRNA after it is transcribed (►RNA editing), sub-cellular localization of mRNAs, sub-cellular restriction of mRNA translation, sequestering multiple non-translating mRNAs in protein complexes and selective mRNA degradation.

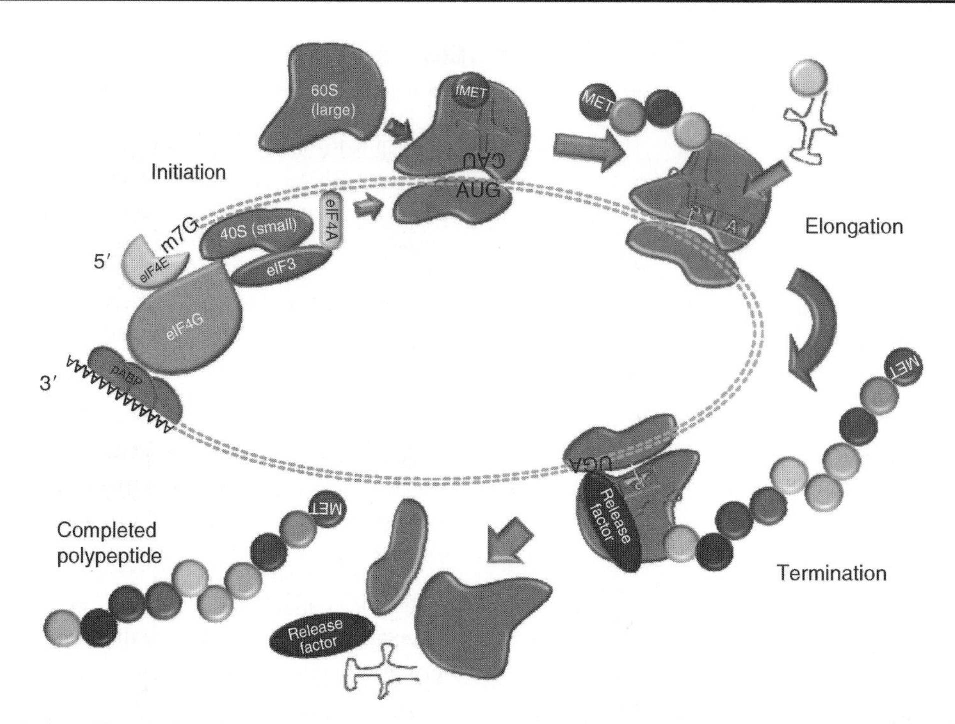

RNA Translation. Figure 1 RNA translation: Initiation – RNA binding proteins that specifically interact with the 5′ m7G cap and 3′ UTR regions of mRNAs form the translation initiation complex. A complex of poly A binding protein (pABP, *red*) interacts with the 3′ poly A tail, while eukaryotic Initiation Factors (eIFs) form a complex at the 5′ end of the mRNA. The major proteins in this complex include the cap-binding protein eIF4E (*yellow*), and the eIF4G (*orange*), eIF3 (*green*) and eIF4A (*pale blue*). They serve to position the 40S (small) ribosomal subunit on the mRNA. In most cases of mRNA translation, the 40S subunit is thought to scan along the mRNA in a 5′-3′ direction until it reaches an AUG codon where a specialized tRNA (fMET) brings in the first amino acid and the 60S (large) ribosomal subunit joins the complex to form a complete ribosome.

Details of mRNA Translation

The translation process is often divided into three stages (Fig. 1): Initiation –where a complex of proteins (►eukaryotic initiation factors, eIFs) recruits ribosome(s) to the mRNA where they move to the beginning of the ORF, Elongation – which encompasses the processing of the ribosome along the coding region of the mRNA linking amino acids brought by specific transfer RNAs (tRNAs) into a growing polypeptide chain and Termination – which occurs at the end of the ORF characterized by the disassembly of the ribosomal complex and release of the full-length polypeptide.

Initiation

Eukaryotic ribosomes contain a large and small subunit, each formed from a specific collection of ribosomal proteins and non-coding ribosomal RNAs (rRNAs). When an mRNA is to be translated, the small subunit of the ribosome first binds to a site "upstream" (on the 5′ side) of the ORF. This activity is mediated by an initiation-complex consisting of ►eukaryotic initiation factor (eIF) proteins that bind to the 7 methylguanosine cap (m7G) [►RNA binding proteins Structures/ Processes/Conditions] in the 5′ and UTR. This also requires interaction with proteins that bind the polyA+ sequence of the 3′ UTR (Fig. 1). The small ribosome subunit then proceeds along the 5′ UTR region of mRNA in a 5′ -> 3′ direction until it encounters an AUG-►start codon (Fig. 1). At this point it joins with a large ribosomal subunit to form a functional ribosome. Ribosomes contain two sites that contain tRNA molecules, the P-site which usually contains a peptidyl-tRNA molecule (i.e., a tRNA with the growing peptide attached) and an A-site which normally recruits aminoacytled –tRNAs which bring in additional amino acids for incorporation into the polypeptide (Fig. 1). Translation is initiated by an "initiator" tRNA, the only tRNA that can bind directly to an empty P-site of the ribosome. Most often in eukaryotes, this initiator tRNA encodes a methionine (Met) amino acid. At this point the ribosome is ready to recruit additional aminoacytled –tRNAs and proceed to synthesise a full length polypeptide. In special cases, other mechanisms can also initiate mRNA translation. This includes termination re-initiation, where ribosomes are re-directed to translate the same mRNA multiple times. Also, initiation can occur at an AUG codon other than the one nearest the m7G cap (leaky scanning), and ribosome shunting. Finally, there is an alternative to m7G cap-dependent translation initiation where internal ribosome entry sites (IRES) along the mRNA direct translation.

1st position		2nd position U	2nd position C	2nd position A	2nd position G
1st position	U	UUU Phenylalanine Phe F	UCU Serine Ser S	UAU Tyrosine Tyr Y	UGU Cysteine Cys C
	U	UUC Phenylalanine Phe F	UCC Serine Ser S	UAC Tyrosine Tyr Y	UGC Cysteine Cys C
	U	UUA Leucine Leu L	UCA Serine Ser S	UAA *STOP*	UGA *STOP*
	U	UUG Leucine Leu L	UCG Serine Ser S	UAG *STOP*	UGG Tryptophan Trp W
	C	CUU Leucine Leu L	CCU Proline Pro P	CAU Histidine His H	CGU Arginine Arg R
	C	CUC Leucine Leu L	CCC Proline Pro P	CAC Histidine His H	CGC Arginine Arg R
	C	CUA Leucine Leu L	CCA Proline Pro P	CAA Glutamine Gln Q	CGA Arginine Arg R
	C	CUG Leucine Leu L	CCG Proline Pro P	CAG Glutamine Gln Q	CGG Arginine Arg R
	A	AUU Isoleucine Ile I	ACU Threonine Thr T	AAU Asparagine Asn N	AGU Serine Ser S
	A	AUC Isoleucine Ile I	ACC Threonine Thr T	AAC Asparagine Asn N	AGC Serine Ser S
	A	AUA Isoleucine Ile I	ACA Threonine Thr T	AAA Lysine Lys K	AGA Arginine Arg R
	A	AUG Methionine START Met M	ACG Threonine Thr T	AAG Lysine Lys K	AGG Arginine Arg R
	G	GUU Valine Val V	GCU Alanine Ala A	GAU Aspartic acid Asp D	GGU Glycine Gly G
	G	GUC Valine Val V	GCC Alanine Ala A	GAC Aspartic acid Asp D	GGC Glycine Gly G
	G	GUA Valine Val V	GCA Alanine Ala A	GAA Glutamic acid Glu E	GGA Glycine Gly G
	G	GUG Valine Val V	GCG Alanine Ala A	GAG Glutamic acid Glu E	GGG Glycine Gly G

RNA Translation. Figure 2 Codon usage table: Codons are grouped by the first (*left*) and second (*top*) position. Note that in many cases the third position is redundant and that codons with a common first and second position base base often encode encode the same amino acid.

Elongation

In eukaryotic cells, there are 20 amino acids commonly used for protein synthesis. Each tRNA contains a region with three unpaired nucleotides (anti-codon) which binds the corresponding codon in the mRNA. The use of three bases to encode each amino acid means that there are actually (4^3) different possibilities that can be used to uniquely encode them providing 64 unique identities. Thus, there are more codons than amino acids and therefore considerable redundancy in the code, with some amino acids encoded by four or more tRNAs with different anticodons. A separate process couples each specific amino acid to their representative encoding tRNA (s) via an activating enzyme (aminoacyl-tRNA synthetase). Also, some tRNAs can recognize more than one codon. This modified base pairing at the third nucleotide of a codon is called "wobble-pairing." For example, the phenylalanine tRNA with the anticodon 3′ AAG 5′ recognizes UUC and UUU. Also, some codons are reserved for specialized functions. AUG signals for the beginning of each ORF and for the amino acid methionine (Fig. 2). As a result, there is usually a methionine at the amino terminal of the polypeptide synthesized from the mRNA. Any AUG codons after this point are interpreted for the insertion of an internal methionine.

Elongation of a nascent ribosome associated polypeptide is an iterative process in which a series of specific aminoacytled –tRNAs, are recruited to their respective three-base codons within the A-site adjacent to the P-site (Fig. 1). Once this occurs, elongation factors hydrolyze GTP (an energy source) to covalently link to the incoming amino acid to the existing polypeptide (Fig. 1). At this point the tRNA at the P site is released and the ribosome moves one codon (3 bases) downstream. The newly-arrived tRNA at the A-site, while still attached to the nascent polypeptide, shifts to the P site and opens the A site for recruitment of an aminoacyl-tRNA that carries an amino acid corresponding to the next codon. This occurs via another protein elongation factor and requires the energy of hydrolysis of another molecule of GTP. Once the ribosome complex clears the recruitment site, it is possible to recruit additional ribosome complexes to a single mRNA. Often a single mRNA molecule is translated simultaneously by many ribosomes. This multi-ribosome complex is called a polysome.

Termination

Translation of a protein is normally finished when the ribosome reaches one or more ►STOP codons (UAA, UAG, UGA) (Fig. 2). Normally, there are no corresponding tRNA molecules with anti-codons for STOP codons. Instead, specific proteins (release factors) recognize these codons when they arrive at the A site of the ribosome and release the completed polypeptide (Fig. 1). During this process the ribosome complex is dissociated back into its corresponding subunits, which are then available for translation of additional mRNAs (Fig. 1).

R

Regulation of mRNA Translation

Regulation of mRNA translation appears to be a much more common mechanism for regulating protein expression that was previously thought. There are several mechanisms that limit mRNA translation to specific cellular sub-domains. In addition, cytoplasmic mRNAs have a finite life and like mRNA transcription, degradation of mRNAs is also a regulated process. This includes the surveillance and destruction of aberrant mRNAs as the proteins translated from these mRNAs would potentially be detrimental to cell function. Also, there is an entire ►RNA interference [Structures/Processes/Conditions] regulatory system that employs small ►non-coding RNA molecules that also regulate translation and/or mRNA degradation (i.e., ►RNAi). These events are regulated through specific RNA binding proteins [Structures/Processes/Conditions] that interact with the RNA either before or after it is exported into the cytoplasm for translation. It is now thought that there is a dynamic balance between translation by ribosomes and cytoplasmic RNA regulatory complexes for access to mRNAs.

RNA Transport Particles

In many cell types, mRNAs are transported to specific sub-domains to regulate protein expression. This process is important for establishing cellular asymmetry during cell division or directed cell migration. This process also appears critical for aspects of neuronal function, including axon guidance and nerve regeneration [1]. The RNA particles that contain these transported mRNAs also contain several of the proteins involved in translation including ribosomal subunits (Fig. 1). It is thought that translation is suppressed during mRNA transport.

Processing-bodies

Processing bodies (P-bodies) are thought to be primarily sites of mRNA degradation although there are cases where they also sequester mRNAs away from the translational protein complexes to be released later. Degradation-specific P-bodies contain a complex that break down the mRNA by removing the 7mG cap and a 5′-3′ exonuclease that subsequently breaks down the mRNA [2]. These structures also appear to be a destination for mRNAs that are being regulated by the RNAi pathway.

Stress Granules

In mammalian cells, stress granules appear during events that seem to require a rapid shift in the translational events within a cell [3]. They are proposed to be sites where pools of translationally arrested mRNAs are stored as they contain several proteins also involved in mRNA translation such as the small ribosomal subunit. Functionally, these structures seem to have a role in sorting, remodeling and exporting mRNAs either for subsequent translation or storage in complexes such as P-bodies.

Preventing Translation of Mutated mRNAs

Eukaryotic cells possess several surveillance mechanisms to ensure the fidelity and appropriate translation of cytoplasmic mRNAs. Translation of aberrant mRNAs would obviously present a disadvantage to a cell. Defects in mRNA molecules can arise via mutations in the DNA of the encoding gene, or by errors that occur during transcription. Some of these errors would have no effect on translation, especially if they occur in the third (wobble) position of a codon due to codon redundancy (Fig. 2). However, mRNAs with incorrect amino acid codons or premature STOP codons would produce truncated proteins that would be ineffective or even harmful.

Nonsense Mediated mRNA Decay

One process that removes non-sense mutations (premature stop codons) is the nonsense mediated decay (NMD) pathway [4]. Briefly, the complex of the RNA binding proteins that mediate joining of the last exons during nuclear RNA splicing remain associated with the mRNA as it is exported to the cytoplasm. During the first round of mRNA translation, these complexes are removed. If the ribosome complex dissociates from an mRNA via a premature termination codon, one or more of these complexes are not removed, and their retention marks the aberrant mRNA for selective destruction.

Nonstop mRNA Decay and the Exosome

Nonstop transcripts contain no functional STOP codon. These mutations often occur during mRNA processing either by abnormal splicing or premature addition of the polyA+ tail. During translation of these mRNAs, termination does not occur and the ribosome complex reaches the end of the poly(A) tail. If this occurs, the Ski7 protein binds to the empty A site overhanging the 3′ end of the mRNA and recruits it to the exosome [5]. This exosome complex is the primary mediator of 3′-5′ mRNA degradation that targets old mRNAs for degradation due to shortening of their 3′ poly A+ tail.

References

1. Willis D, Li KW, Zheng JQ, Chang JH, Smit A, Kelly T, Merianda TT, Sylvester J, van Minnen J, Twiss JL (2005) Differential transport and local translation of cytoskeletal, injury-response, and neurodegeneration protein mRNAs in axons. J Neurosci 25:778–791
2. Sheth U, Parker R (2003) Decapping and decay of messenger RNA occur in cytoplasmic processing bodies. Science 300:805–808
3. Cheng J, Maquat LE (1993) Nonsense codons can reduce the abundance of nuclear mRNA without affecting the abundance of pre-mRNA or the half-life of cytoplasmic mRNA. Mol Cell Biol 13:1892–1902

4. Kedersha NL, Gupta M, Li W, Miller I, Anderson P (1999) RNA-binding proteins TIA-1 and TIAR link the phosphorylation of eIF-2 alpha to the assembly of mammalian stress granules. J Cell Biol 147:1431–1442
5. Frischmeyer PA, van Hoof A, O'Donnell K, Guerrerio AL, Parker R, Dietz HC (2002) An mRNA surveillance mechanism that eliminates transcripts lacking termination codons. 95:2258–2261

RNase H

Definition

Ribonuclease is a general term for enzymes which catalyse the hydrolysis of RNA into smaller fragments. Ribonucleases are grouped into several sub-classes within Enzyme Class 2.7. (phosphorolytic enzymes) and Enzyme Class 3.1 (hydrolytic enzymes). RNase H cleaves the 3′-O-P-bond of RNA in a DNA/RNA duplex to produce 3′-hydroxyl and 5′-phosphate terminated products. Since RNase H hydrolyses RNA in a DNA:RNA duplex without degrading the DNA, it is commonly used to remove the RNA template after first strand cDNA synthesis.

►Serial Analysis of Gene Expression

Rod and Frame, Rod and Disc

Definition

Protocols to assess the impact of a static frame or a rotating disk on the visual perception of verticality (svv).

►Verticality Perception

Rod Photoreceptor

Definition

Rod-shaped photoreceptors in the vertebrate retina that mediate vision at low intensity light levels. Rods are far more numerous than cones in humans and are of only one type with peak spectral sensitivity ~500 nm.

►Photoreceptors

Rod Spherule

Definition

Synaptic terminal of a rod photoreceptor that provides output to rod bipolar and horizontal cells.

►Horizontal Cells
►Photoreceptors
►Retinal Bipolar Cells

Roll Off

Definition

The amount of attenuation of sound magnitude provided by a filter for waves outside of the filter's passband, usually expressed in dB/octave.

►Acoustics

Romberg Sign

Definition

Simply put, while standing and with the feet together, a positive Romberg sign is when the subject sways or steps out with the eyes closed but not with the eyes open. The amount of sway should make the examiner worried that the subject will fall since most people will sway a little with this test.

The interpretation of an abnormal Romberg test is not simple. A great deal of the nervous system is involved when performing the Romberg test and multiple small deficits in many systems could be present. In other words, the nervous system functions together. For instance, adequate strength in the legs is required. Proximal or distal weakness will prevent the subject from making corrections in stance. Significant cerebellar dysfunction will cause the subject to be unsteady with feet together and eyes open, while mild cerebellar dysfunction might only reveal itself with the eyes closed.

With intact cerebellar and motor function, one maintains balance using the visual, vestibular and proprioceptive systems. Since the Romberg test is performed with the eyes closed, abnormalities in either of the other two systems can cause an abnormal Romberg test. Classically, the Romberg test assumes that the vestibular system is intact. That may not be a good assumption if dizziness is the primary complaint. In the classic situation, if the subject closes their eyes and significantly sways or steps out, the implication is that proprioception is impaired. If proprioceptive and vestibular functions were both intact, closing the eyes would not be a problem.

Proprioceptive function is transmitted through large diameter myelinated peripheral nerves and the dorsal columns of the spinal cord to nuclei in the brain stem. Peripheral neuropathy, vitamin B12 deficiency, multiple sclerosis and neurosyphilis are examples of diseases to consider when the Romberg sign is positive.

►Vestibular Tests: Romberg Test

Root Neurons

Definition

Special neurons unique to the rodent cochlear nucleus located in the auditory nerve root.

►Cochlear Nucleus

Ror

Definition

Nuclear orphan receptor related to the retinoic acid receptors. Binds to the consensus sequence ([A/T]A[A/T]NT[A/G]GGTCA termed RORE, in the promoter of target genes.

►Clock Genes

Rostral Interstitial Nucleus of the MLF (riMLF)

Definition

Located at the mesodiencephalic junction, this is the most rostral of the interstitial nuclei of the MLF.

►Eye Movements Field

Rostral Ventrolateral Medulla (RVLM)

Definition

The RVLM is part of the ventrolateral medulla and located just caudal to the facial nucleus. It is a vasomotor nucleus and contains (in addition to interneurons) sympathetic premotor neurons related to the sympathetic cardiomotor, cutaneous vasoconstrictor, muscle vasoconstrictor, renal vasoconstrictor, visceral (non-renal) vasoconstrictor pathways and to the adrenal medulla (probably cells secreting noradrenaline). These populations of sympathetic premotor neurons are viscerotopically organized.

►Autonomic Reflexes
►Blood Volume Regulation

Rotation Vector

Definition

Three dimensional eye positions can be expressed in rotation vector form, where the direction of the vector specifies the axis of rotation from primary position and its length specifies the rotation angle.

►Vestibulo-ocular Reflexes

Rough Endoplasmic Reticulum (RER)

Definition

The endoplasmic reticulum (ER) is an extensive membrane network of tubes and sac-like structures held together by the cytoskeleton. Some parts of the ER are covered with ribosomes on the surface, which give them a rough appearance at the level of the electron microscope. Ribosomes assemble amino acids into proteins based on instructions from the nucleus. and insert the freshly produced proteins directly into the ER, which processes them and then passes them on to the Golgi apparatus.

Route Navigation (or Taxon Navigation)

Definition
In route navigation the traveler is guided by a memorized set of turns and straight paths to reach a goal. Rule sets such as "at the big rock turn left; next, at the fallen tree bear right and proceed…" are examples of route learning. In animals route learning is based on operant conditioning. In humans it is usually accomplished by "following directions" generated by another (cultural transmission). Route learning can be effective but is inflexible.

►Operant Conditioning
►Spatial Learning/Memory

Route Navigation Strategy

Definition
Behavior relying on an egocentric reference frame and directed by chaining sequences of taxon and praxis strategies.

►Spatial Memory

RTLLBs

►Reticulotectal Long-Lead Burst Neurons

Rubrobulbar Tract

Synonyms
Tractus rubrobulbaris

Definition
Descending fibers of the rubrobulbar tract and rubrospinal tract terminate on the interneurons in the lateral reticular formation and the dorsolateral intermediate zone of the spinal cord and directly on motoneurons of the nucleus of the facial nerve. The rubrobulbar tract and rubrospinal tract have a somatotopic arrangement.

►Mesencephalon

Rubrospinal Tract

Synonyms
Tractus rubrospinalis

Definition
Somatotopically arranged fiber bundles between the red nucleus and spinal cord. Runs in the lateral column of the spinal cord, originating in the magnocellular portion of the red nucleus, going to the spinal cord segments as far as the thoracic cord. Regulates the tone of important flexors.

►Mesencephalon

RVOR (Rotational VOR)

►Vestibulo-Ocular Reflex